Neuromuscular Disorders of Infancy, Childhood, and Adolescence

A Clinician's Approach

Neuromuscular Disorders of Infancy, Childhood, and Adolescence

A Clinician's Approach

Second Edition

Edited by

Basil T. Darras, MD

Joseph J. Volpe Professor of Neurology, Harvard Medical School, Associate Neurologist-in-Chief, Chief, Division of Clinical Neurology, Director, Neuromuscular Program, Boston Children's Hospital, Boston, Massachusetts, USA

H. Royden Jones, Jr., MD

Jaime Ortiz-Patino Chair in Neurology, Lahey Clinic, Burlington, Massachusetts Clinical Professor of Neurology, Harvard Medical School, Director Emeritus, Electromyography Laboratory, Boston Children's Hospital, Boston, Massachusetts, USA, d. 2013

Monique M. Ryan, M Med BS, FRACP

Associate Professor, Paediatric Neurologist, Children's Neurosciences Centre, Research Fellow, Murdoch Childrens Research Institute, Clinical Associate Professor, University of Melbourne, Royal Children's Hospital, Victoria, Australia

Darryl C. De Vivo, MD

Sidney Carter Professor of Neurology Professor of Pediatrics Associate Chairman (Neurology) for Pediatric Neurosciences, Director, Pediatric Neurology, Emeritus Director, Colleen Giblin Research Laboratories, Director, Pediatric Neuromuscular Disease Center, Co-Director, Center for Motor Neuron Biology and Disease, Columbia University Medical Center, The Neurological Institute, New York, New York, USA

ELSEVIER

AMSTERDAM • BOSTON • HEIDELBERG • LONDON
NEW YORK • OXFORD • PARIS • SAN DIEGO
SAN FRANCISCO • SINGAPORE • SYDNEY • TOKYO
Academic Press is an imprint of Elsevier

Academic Press is an imprint of Elsevier
32 Jamestown Road, London NW1 7BY, UK
525 B Street, Suite 1800, San Diego, CA 92101-4495, USA
225 Wyman Street, Waltham, MA 02451, USA
The Boulevard, Langford Lane, Kidlington, Oxford OX5 1GB, UK

First edition 2003
Second edition 2015

Notices
Knowledge and best practice in this field are constantly changing. As new research and experience broaden our understanding, changes in research methods, professional practices, or medical treatment may become necessary.

Practitioners and researchers must always rely on their own experience and knowledge in evaluating and using any information, methods, compounds, or experiments described herein. In using such information or methods they should be mindful of their own safety and the safety of others, including parties for whom they have a professional responsibility.

To the fullest extent of the law, neither the Publisher nor the authors, contributors, or editors, assume any liability for any injury and/or damage to persons or property as a matter of products liability, negligence or otherwise, or from any use or operation of any methods, products, instructions, or ideas contained in the material herein.

ISBN: 978-0-12-417044-5

British Library Cataloguing-in-Publication Data
A catalogue record for this book is available from the British Library

Library of Congress Cataloging-in-Publication Data
A catalog record for this book is available from the Library of Congress

For information on all Academic Press publications
visit our website at store.elsevier.com

Typeset by MPS Limited, Chennai, India
www.adi-mps.com

Printed and bound in China

15 16 17 18 10 9 8 7 6 5 4 3 2 1

Dedication

The editors dedicate this book to the man who was the architect of the first edition and the inspiration for the second. Roy Jones was an excellent physician, a caring, generous and inspiring teacher to a generation of neurologists, and respected by professionals around the world. Most of all, Roy was cherished as a dear friend, a valued colleague, and a loving husband and parent. His courage, endless optimism and infectious enthusiasm will always be remembered by those of us fortunate enough to have known and worked with him.

The editors wish to acknowledge the late Julianna Wittrédi of Geneva, Switzerland, whose elegant drawings of the Gowers maneuver appear on the book's front and back covers. Mrs. Wittrédi, the mother of contributing author Dr. Caroline Menache-Starobinski, passed away during the preparation of this edition, not knowing that her artwork would grace our cover.

To my late parents, Theodore and Anastasia, for instilling in me intellectual curiosity and ambition; and to my daughters, Natasha and Irene-Olympia, for your genuine love.

– **Basil T. Darras**

To my husband Peter, with gratitude for your generosity, support and love, for Annabel and Campbell who I love dearly, and for Patrick, who will one day read this book and wonder how you could write such a long book without putting in any dragons at all.

– **Monique M. Ryan**

To Kristin, Jessica, Cynthia and Alicia for your continuing love, support and understanding; and to the memory of our daughter Alicita, who left us on Resurrection Sunday for a better place.

–**Darryl C. De Vivo**

ABOUT THE COVER IMAGES

Front cover: Top: Gowers figure by Julianna Wittrédi of Geneva, Switzerland. Bottom four figures: Muscle biopsies have been immunostained with a combination of antidystrophin (DYS2) and ankyrin 1 (ANK1) antibodies; the latter stain the sarcomere green. Figures courtesy of Matthew Alexander, PhD, Division of Genetics, Boston Children's Hospital.

Back cover: Gowers figure by Julianna Wittrédi of Geneva, Switzerland.

Contents

Part IV
Peripheral Nerve Disorders

16. Congenital and Early Infantile Neuropathies

Eppie M. Yiu and Jonathan Baets

17. Hereditary Neuropathies in Late Childhood and Adolescence

Kathryn M. Brennan and Michael E. Shy

21. Chronic Inflammatory Demyelinating Polyradiculoneuropathy

Hugh J. McMillan and Peter B. Kang

22. Neuropathies Secondary to Systemic Disorders

Jo M. Wilmshurst and Robert A. Ouvrier

23. Toxic Neuropathies

David Pleasure

24. Hansen's Disease (Leprosy): Leprous Neuropathy

Wilson Marques Jr.

Part V
Neuromuscular Junction Disorders

25. Acquired Presynaptic Neuromuscular Junction Disorders

Ai Sakonju and Thomas O. Crawford

Foreword

This foreword to the second edition should begin with a heartfelt testimonial to the founder of this book, the late Dr. Roy Jones. Roy was a personal friend and colleague for more than 20 years, and his larger-than-life personality and dedication to everything that is good in medicine was a source of inspiration and admiration for all of us.

The symptoms and signs of neuromuscular disease are extraordinarily common in infancy, childhood, and adolescence. Hypotonia in the infant, weakness in the child, and cramps in the adolescent are only a few of the complaints frequently encountered. Indeed, in general pediatric practice, apparent disturbances of neuromuscular function are among the most frequent reasons for consultation, and the pediatrician is often faced with vexing questions concerning the appropriate detail of the evaluation. As a consequence, referral to neurological specialists for evaluation of such symptoms of neuromuscular disease is very common indeed. Therein lies the need for a comprehensive text, with detail and sophistication for the specialist as well as practical clinical insights for the primary physician. Such a single source became available with the first edition of this book. Several excellent monographs on anterior horn cell disease, peripheral neuropathies, and related neuromuscular disorders in children have been published, and comprehensive textbooks on neuromuscular disorders in adults are also available. However, this marvelous book, the current edition co-edited by Drs. Darras, Ryan and De Vivo with important initial contributions by Roy Jones before his untimely passing, fills the need for a comprehensive, state-of-the-art text on "Neuromuscular Disorders of Infancy, Childhood, and Adolescence" in an extraordinary manner.

The co-editors are extremely well-suited for the task of preparing this comprehensive text, and, indeed, their skills complement each other very effectively. Dr. Darras brings the strengths not only of a superb clinical child neurologist with a focus on neuromuscular disease, but, additionally, the skills of a fully trained and certified geneticist. His understanding of the molecular genetics of disorders of the neuromuscular system is broad and deep and ensures that the explosion of molecular genetics into this field is an intricate part of this book. His extensive experience with children, gained at Boston Children's Hospital in Massachusetts is unparalleled anywhere. The late Dr. Jones brought his extraordinary skills in clinical assessment and in electrophysiological studies of muscle, nerve, neuromuscular junction, and anterior horn cell. His indelible stamp on the book remains. Dr. Ryan, who benefited greatly from the mentorship of Dr Jones and Dr. Darras, brings her considerable clinical strengths, especially in the many facets of peripheral nerve disorders. Dr. De Vivo is a world leader in child neurology and, in particular, in metabolic disorders and their impact on the neuromuscular system. His exceptional knowledge of the complex interactions in the biochemical pathways of glucose, fatty acid, and mitochondrial metabolism in neuromuscular diseases has provided this book with unique strengths in this crucial aspect of the field.

The organization, topics, and authors chosen for the second edition of *Neuromuscular Disorders of Infancy, Childhood, and Adolescence* are superb. Thus, rather than plodding in an encyclopedic fashion through every disease, the book begins with five excellent chapters on the clinical approach, modern genetics, EMG, muscle biopsy, and nerve biopsy. After overviews of the evaluation of the "floppy infant" and the infant with arthrogryposis, the book then is organized logically in major sections based on the anatomical level of the disorder. Within these sections, the specific categories of disease are described in depth. The book concludes with a superb section on therapy in neuromuscular disorders.

The authors who have contributed to this book are extraordinary. The list reads like a "who's who" in modern neuromuscular disease. The co-editors are to be congratulated for enlisting the best. The individual chapters are superbly written and edited, such that the approach is uniform and up-to-date. Appropriate emphasis is placed on the impact of molecular genetics on understanding of pathogenesis, diagnosis, and therapy.

This book will have a very broad audience. Primary physicians, neurologists, neuromuscular experts, residents, and fellows will learn very much from this volume. The great success of the first edition and the amplifications, elaborations and state-of-the-art updating of the current edition should firmly establish this book as the bible of the field.

Joseph J. Volpe, MD
Bronson Crothers Distinguished Professor of Neurology, Harvard Medical School,
Neurologist-in-Chief Emeritus
Boston Children's Hospital, Boston, Massachusetts, USA

Preface to the Second Edition

More than a decade has passed since the first edition of this textbook was published. Since 2003, much has happened in medicine and science, and in publishing. Simply stated, these professions have been transformed: medicine and science by the explosion of molecular genetics, and publishing by the explosion of the Internet and social media. These transformations have been equivalent to forces of nature; and understandably, these forces have been imposed upon the architects of the second edition of this textbook, the editors, the contributors, and the publisher.

As we said in the Preface to the First Edition, "We [the editors and contributors] have learned other lessons while assembling this textbook." We can restate this comment today as we complete the second edition, but this time punctuated with an exclamation point! Physicians and academicians are pulled in more directions today than ever before, even though each present day has the same number of minutes as past days. It is increasingly difficult to find the time to review and reflect on the burgeoning literature, both old and new, because of the greater demands imposed upon us. The days seem longer, the nights shorter, and the information overload simply exhausting. And the reason for these feelings is the exploding universe of information. We seem to be "drowning" in this sea of information, desperately in need of management strategies that will allow us to capitalize on the flood of new ideas and knowledge and thereby care for our patients in more efficient and effective ways. This reality has been the main challenge presented during the preparation of the second edition of this textbook, and we believe the contributors have risen to this challenge. They have achieved the editors' primary goal of the second edition: to maintain focus on the patient, emphasizing the clinical approach through, for example, one of the signatures of the first and second editions, the clinical vignettes that "humanize" the text.

We also recognize the importance of the extraordinary scientific advances of the past decade and melding these advances with the rich clinical experience and history surrounding the field of neuromuscular diseases. The practice of medicine today is truly translational, a dynamic interplay between the bench and the bedside. The genotype, as defined by next generation sequencing, is an equal partner with the clinical phenotype, as defined traditionally by careful clinical evaluation. This dynamic interplay is equivalent to the nature-nurture story in neuroscience; it is not either-or, but both. As a result, the professional experience is much more rewarding and demanding, and our patients will be the ultimate beneficiaries. Neurologists now have the opportunity to truncate the "diagnostic odyssey" and in fact diagnose an individual before the onset of clinical symptoms. These advances do not make the practice of neurology easier; they simply change the paradigm and introduce new challenges. They do mean that we can do our job for the patient even more effectively, more efficiently, and hopefully more cost-effectively. The contributors have done a superb job in capturing and incorporating these scientific advances into their revised or new chapters, of which there are 54 in this second edition.

Neuromuscular disorders are a diverse group of conditions that affect one in 2000 children worldwide. We now know the genetic basis of more than 40 muscular dystrophies, 38 congenital myopathies, 25 congenital myasthenic syndromes, 26 metabolic myopathies, and 75 hereditary motor and sensory neuropathies, to name just a few of the disease categories seen by the child neurologist. Clinicians now routinely access novel next-generation sequencing techniques, resulting in rapid identification of new syndromes and, in many cases, expansion of our understanding of the molecular basis of known conditions. This improved understanding has led to the development of specific therapies for some diseases. Enzyme replacement therapy, immunosuppression, and cofactor or vitamin supplementation are now reversing or significantly ameliorating potentially devastating neuromuscular disorders of childhood. Gene therapy and other molecular-based therapeutic strategies are now entering our clinics as treatments for spinal muscular atrophy, the dystrophinopathies, and other genetically determined neuromuscular diseases. For many other diseases, specific therapeutics remain on the horizon but are likely to become realities before the third edition of this textbook is published. These are truly exciting times.

The identification of specific syndromes and development of new therapies for pediatric neuromuscular disorders has, in the last decade, been facilitated by expanded collaborations among medical centers throughout the world, as well as collaborations among patient and parent support groups, academia, industry and government. These collaborations by the diverse stake-holders in the US, Europe and other continents have standardized clinical care, advanced outcomes research, improved clinical trial design, and led to the establishment of disease registries. These synergistic initiatives have accelerated the search for effective therapies and harmonized the standard of care across the world in a way that could not have been anticipated when the first edition of this book was published.

The value in these modern advances depends on astute clinicians who are the first to examine the symptomatic child, and who are conversant with all of the diagnostic and therapeutic tools at their disposal. This textbook is written for such physicians and therapists involved in the day-to-day care of children with neuromuscular disorders, who share the ultimate goal of improving their multidisciplinary care and rescuing patients therapeutically from their diseases whenever possible.

Each chapter has been prepared by experts in the area, and we are very grateful to our many colleagues and friends for their generosity in sharing their knowledge and expertise for this textbook. Every first edition chapter in this book has required significant revisions or extensive rewriting, and every new chapter similarly reflects the extent of scientific progress in the last 11 years, often shifting the focus from clinical diagnosis, to molecular mechanisms, to exciting new therapies.

Some of these changes since the first edition have been truly breathtaking: witness just one example, Duchenne muscular dystrophy. Our improved understanding of this disorder has forced us to expand the relevant sections in this book, devoting specific chapters to diagnosis, clinical management, and clinical trials. The current emphasis on clinical trials and novel therapeutics has prompted us to devote a new chapter to translational research into neuromuscular disorders. We hope that these changes in focus are of value in providing a detailed background to a rapidly changing field.

Textbooks such as this take much work. The editors have dedicated hours to this book, competing with time otherwise spent caring for patients, teaching students, or enjoying quality time with our families. We acknowledge with gratitude the support of our colleagues, fellows, spouses, and children. Without their selfless understanding, this work could not have been finished. It truly has been a labor of love, and a measure of our commitment to the patients we serve. Shaye Moore and Elizabeth Jarvis, from the Career Development and Medical Writing Team of the Department of Neurology at Boston Children's Hospital, have provided hundreds of hours of outstanding editorial support, and have been patient, efficient, dedicated and exacting in their invaluable contributions to the editing process of each of the 54 chapters in this book. Dr. Darras is grateful for the assistance of Jenan D'Isola at Boston Children's Hospital, and Dr. De Vivo for the assistance of Rosangel Cruz of Columbia University Medical Center. From Elsevier, we acknowledge with thanks the assistance provided by Mica Haley, Publisher, Neuroscience, and Kristi Anderson, Editorial Project Manager. Significant funding for this effort was provided to Dr. Darras from the Joseph J. Volpe Chair fund at Boston Children's Hospital.

Sadly, some authors from the first edition of this textbook were not able to contribute to this edition. We note with sadness the passing of our respected colleagues Guillermo A. Suarez, Asa Wilbourn, John W. Griffin, and Hugo Moser, and we acknowledge with gratitude their great wisdom and their outstanding, long-lasting contributions to the field.

Finally, during the preparation of this textbook H. Royden Jones, Jr., editor-in-chief of the first edition and the driving force behind this new edition, died of pancreatic cancer. Any important accomplishment must begin with an original idea, and the idea of a textbook dedicated to pediatric neuromuscular diseases emerged from Roy's creative mind. Roy was an outstanding clinician who defined the field of neurophysiologic evaluation of neuromuscular function in children. He wrote or co-wrote most of the seminal works in his area, and was a teacher, friend, and mentor to many of the experts who have contributed to this book. The impact of his contribution is reflected in the number of references to his papers in these chapters. He was a dedicated clinician and a wonderful neurologist. Although not a pediatrician by training, he loved working with children. One might say that he was "a natural," comfortably interacting with children and their parents and leaving them with a feeling of hope. He had a wonderful sense of humor and great personal warmth, and will always be missed by those lucky enough to have known him.

Basil T. Darras, Editor-in-Chief
Monique M. Ryan
Darryl C. De Vivo
Boston, Melbourne, and New York
September 2014

Preface to the First Edition

The hypotonic infant and the weak child have captured the attention of physicians for centuries. In many ways, the seminal reports of Duchenne, Werdnig, and Hoffmann, and many others have served as the historical underpinnings of the modern discipline of childhood neuromuscular disorders as we currently appreciate them. A detailed knowledge of molecular neurogenetics and protein chemistry is now essential for our comprehension of the numerous disorders resulting from alterations of the dystrophin complex or the survival motor neuron (SMN) complex. And it does not end there—witness the extraordinary advances of the past decade in our understanding of the many disease mechanisms underlying the inherited and acquired neuropathies, disorders of neuromuscular transmission, and other myopathies. The rapidly expanding knowledge base related to the limb-girdle myopathies and the hereditary motor and sensory neuropathies has been nothing short of spectacular.

Early on, this clinical discipline was captured partly in the writings of several scholars within the field. And what we know today is the product of a collective effort by many physicians and scientists who have labored tirelessly on behalf of the many children afflicted with neuromuscular disorders. Although there are a few well-respected individual textbooks dedicated to specific portions of the child's peripheral motor unit, these concentrate on just one area, such as motor neuronopathies, peripheral neuropathies, or myopathies. Each of the editors of this text, early on and subsequently in our careers, appreciated the lack of a comprehensive and substantial textbook of neuromuscular disorders dedicated to the pediatric perspective. Therefore, child and adult neurologists, pediatricians, orthopedic surgeons, and rehabilitation medicine specialists have often needed to seek specific seminal individual papers from the pediatric literature. Unfortunately, in this setting the practitioner is often left uncertain, both to the science and the precise clinical approach, despite having a well-known expert's opinion in this ever-expanding field of pediatric neuromuscular disorders.

Therefore, we set out to achieve our goal—the production of a textbook that concentrates on all aspects of neuromuscular disorders presenting in infancy, childhood, and adolescence. Prior to and concomitantly at an early stage of the evolution of this book, we recognized the enormity of the collective input of our colleagues over the past several decades. We wanted this volume to be comprehensive, covering the child's neuromuscular system, and particularly clinically complete, focusing on treatment issues, rehabilitation, and bioethics. To be sure, a textbook that focuses primarily on the pediatric domain would still inevitably overlap with some existing textbooks that cover similar clinical material. But we were determined to have our emphasis differ.

The clinical perspective is captured in the numerous case examples that appear throughout most of the 65 chapters. Each of these vignettes allows the student of this medical discipline to appreciate the clinical presentation in the classical sense, or in certain instances the difficulty initially making a correct diagnosis. Many of these case examples represent clinical descriptions taken from the first published reports in an effort to be faithful to the original observations. Hopefully, as pediatric physicians, we will always start with the child and end with the child, a cycle that takes us from the bedside to the bench and back to the bedside. Additionally, very important ethical issues are now receiving their long overdue emphasis in many medical formats. These are especially germane to the care of children with many neuromuscular disorders. The multiple new diagnostic genetic tests, as well as the splendid advances in intensive care management, add to the complex bioethical issues confronting the physician and the parents of the child with many forms of pediatric neuromuscular disorders. These sometimes challenge our societal mores, as discussed in the concluding chapter of this monograph. Overall congenital issues and developmental nuances are given primary attention in this monograph, whereas adult disorders, such as amyotrophic lateral sclerosis, are discussed only when there is an instructive lesson that can be applied to its very rare pediatric presentation.

Many of our colleagues have been called upon to help us develop this text. We are greatly indebted to each of them for their very enthusiastic support. These world-renowned clinicians and neuromuscular researchers have provided a very broad perspective for the reader. Their chapters provide the child's physician with an in-depth review including genetic, pathophysiologic, electrodiagnostic, and histologic correlations, differential diagnosis, case examples, and a comprehensive analysis of general therapeutic principles. The latter encompasses not only pharmacologic but also orthopedic and rehabilitative protocols. Future therapies will certainly include biogenetic formats; an introduction to these modalities is also provided.

We sincerely believe we have achieved our goal in this first edition, and we know that a second edition will be even better. We have tried to keep the focus of attention on the patient, emphasizing the clinical approach, even though the scientific breakthroughs are so overwhelmingly dominant. Clearly, it is easier to discuss the new insights into the molecular genetics underlying these human diseases. But we simultaneously run the risk of losing the clinical touch. Our understanding of pathogenesis and phenotype-genotype correlation remains limited, and an uplifting discussion of effective treatments remains to be written for a future edition—hopefully sooner rather than later.

The editors and contributors also have learned other lessons while assembling this textbook. It is a demanding effort, and the time that is necessary to write is taken from time that should be devoted to other equally important responsibilities. We have neglected our families, our students, and our colleagues during this period. The editors have prevailed on the many contributors to work even harder to meet yet another deadline. And everyone has responded magnificently. We acknowledge that despite our diligence occasional mistakes will be identified scattered throughout these pages. Recognizing that such errors do occasionally occur in endeavors of this magnitude, we, the editors, assume full responsibility for these imperfections as we had the "last crack" at the galley proofs. Hopefully the readers will identify such and bring them to our attention so that we can address them in a subsequent edition. Future knowledge will also be incorporated to reflect the anticipated major advances that most definitely will occur in these many fields.

The commitment of our spouses as well as our own children, who compassionately and patiently understand our need to pursue more effective care for our patients as well as our dedication to this venture, deserves our most sincere appreciation. Without their understanding and encouragement we would not have had the opportunity to bring some of this work to its fruition. One also has to recognize the very important early encouragement of our parents as well as our dedicated and very skillful teachers, too many to specifically name. We also give special thanks to those individuals who provided funds to support our endowed chairs, thus providing DCD and HRJ further time to carry this project to completion.

Others have played vital roles as well but generally receive little or no formal recognition. It is impossible to identify everyone who has participated in this project. Our secretaries, assistants, colleagues, residents, and fellows have been invaluable in obvious and subtle ways. The professional staff at Butterworth-Heinemann, a division of Elsevier Science, ultimately facilitated the completion of this work. Susan Pioli has been with us from the beginning as the Executive Publisher. We have developed heightened admiration for her managerial skills and her recognition of the importance of deadlines. It is possible that the work would still be incomplete without her firm guidance and determination. Similar plaudits are due Jennifer Shreiner, Developmental Editor, and Robin Davis, Senior Project Manager. Both worked closely with Susan Pioli, the editors, and the many contributors to ensure the success of this endeavor. In summary, we are indebted to all who made this publication a reality, and are particularly appreciative of our families.

In conclusion, and most importantly, we wish to give thanks to our many patients and their families who have provided us with these various challenges during our careers. We hope that what we have learned will serve as a foundation for future care of similar children. These children have taught us much of what we know and have volunteered their clinical stories and their photographs to make these many pages more accurate and true to life.

H. Royden Jones, Jr., MD
Darryl C. De Vivo, MD
Basil T. Darras, MD
Boston, Massachusetts,
and New York, New York
July 23, 2002

List of Contributors

Hasan O. Akman, PhD, Department of Neurology, Columbia University Medical Center, New York, New York, USA

Felicia B. Axelrod, MD, FAAP, Division of Autonomic Disorders, Department of Neurology, New York University, Langone Medical Center, New York, New York, USA

Jonathan Baets, MD, PhD, Neurogenetics Group, VIB, Department of Molecular Genetics, Laboratory of Neurogenetics, Institute Born-Bunge, University of Antwerp; Department of Neurology, Antwerp University Hospital, Antwerpen, Belgium

Alan H. Beggs, PhD, Division of Genetics and Genomics, The Manton Center for Orphan Disease Research, Boston Children's Hospital; Harvard Medical School, Boston, Massachussetts, USA

Carsten G. Bönnemann, MD, Neuromuscular and Neurogenetic Disorders of Childhood Section, National Institute of Neurological Disorders and Stroke/ National Institutes of Health, Porter Neuroscience Research Center, Bethesda, Maryland, USA

Kathryn M. Brennan, MBChB, PhD, Department of Neurology, Carver College of Medicine, University of Iowa, Iowa City, Iowa, USA

Robert H. Brown, Jr., DPhil, MD, Neurology, University of Massachusetts Medical School, Worcester, Massachusetts, USA

Kate Bushby, MD, Institute of Genetic Medicine, International Centre for Life, Newcastle University, Newcastle upon Tyne, United Kingdom

Stirling Carpenter, MD, Department of Anatomic Pathology, Centro Hospitalar São João, Porto, Portugal

Wendy Chung, MD, PhD, Institute of Human Nutrition, Columbia University; New York-Presbyterian Hospital, New York, New York, USA

Emma Ciafaloni, MD, Department of Neurology, Neuromuscular Disease Center, University of Rochester Medical Center, Rochester, New York, USA

Pedro D.S.C. Ciarlini, MD, Neuropathology, Department of Pathology, Brigham and Women's Hospital, Boston, Massachusetts, USA

Thomas O. Crawford, MD, The Johns Hopkins Children's Center, Baltimore, Maryland, USA

Basil T. Darras, MD, Department of Neurology, Harvard Medical School; Department of Neurology, Boston Children's Hospital, Boston, Massachusetts, USA

William S. David, MD, PhD, Harvard Medical School; Neuromuscular Diagnostic Center and EMG Laboratory, Massachusetts General Hospital, Boston, Massachusetts, USA

Umberto De Girolami, MD, Division of Neuropathology, Brigham and Women's Hospital; Harvard Medical School, Boston, Massachusetts, USA

Darryl C. De Vivo, MD, Departments of Neurology and Pediatrics, The Neurological Institute, Columbia University Medical Center, New York, New York, USA

Feza Deymeer, MD, MS (Epid.), Department of Neurology, Istanbul Faculty of Medicine, Istanbul University, Istanbul, Turkey

Salvatore DiMauro, MD, H. Houston Merritt Clinical Research Center, Department of Neurology, Columbia University Medical Center, New York, New York, USA

James J. Dowling, MD, PhD, Division of Neurology, Genetics and Genome Biology Program, Hospital for Sick Children, Departments of Paediatrics and Molecular Genetics, University of Toronto, Toronto, Ontario, Canada

Hacer Durmus, MD, Department of Neurology, Istanbul Faculty of Medicine, Istanbul University, Istanbul, Turkey

Andrew G. Engel, MD, Mayo Clinic College of Medicine; Department of Neurology, Mayo Clinic, Rochester, Minnesota, USA

Elizabeth C. Engle, MD, Neurology and Ophthalmology, Boston Children's Hospital; Harvard Medical School, Boston, Massachusetts, USA; Howard Hughes Medical Institute, Bethesda, Maryland, USA

Hans H. Goebel, Department of Neuropathology, Charité-Universitätsmedizin Berlin, Campus Charité Mitte, Berlin, Germany

Padraic J. Grattan-Smith, MB, BS, MRCP, FRACP Southern Neurology, Sydney, Australia

Michela Guglieri, MD, Institute of Genetic Medicine, International Centre for Life, Newcastle University, Newcastle upon Tyne, United Kingdom

Debra Guntrum, MS, FNP, Department of Neurology, Neuromuscular Disease Center, University of Rochester Medical Center, Rochester, New York, USA

Judith G. Hall, MD, Departments of Pediatrics and Medical Genetics, University of British Columbia, BC Children's Hospital, Vancouver, British Columbia, Canada

Veronica Hinton, PhD, G. H. Sergievsky Center and Department of Neurology, Columbia University, New York, New York, USA

Susan T. Iannaccone, MD, University of Texas Southwestern Medical Center, Children's Medical Center, Dallas, Texas, USA

H. Royden Jones, Jr., MD, Department of Neurology, Lahey Clinic, Burlington, Massachusetts; Department of Neurology, Harvard Medical School, Boston Children's Hospital, Boston, Massachusetts, USA (d. 2013)

Karin Jurkat-Rott, MD, PhD, Division of Neurophysiology, University of Ulm, Ulm, Germany

Peter B. Kang, MD, Division of Pediatric Neurology, University of Florida College of Medicine, Gainesville, Florida, USA

Horacio Kaufmann, MD, FAAN, Department of Neurology and Medicine, New York University, Langone Medical Center, New York, New York, USA

Petra Kaufmann, MD, MSc, National Institutes of Health/NINDS, Bethesda, Maryland, USA

Werner Klingler, MD, PhD, Department of Neuroanaesthesia, Ulm University, Günzburg, Germany

Louis M. Kunkel, PhD, Division of Genetics and Genomics, The Manton Center for Orphan Disease Research, Boston Children's Hospital; Harvard Medical School, Boston, Massachussetts, USA

Frank Lehmann-Horn, MD, PhD, Division of Neurophysiology, University of Ulm, Ulm, Germany

Kerry H. Levin, M.D., Neuromuscular Center, Department of Neurology, Cleveland Clinic, Cleveland, Ohio, USA

Wendy K.M. Liew, MBChB, MRCPCH, KK Women's and Children's Hospital, Singapore; Boston Children's Hospital, Boston, Massachusetts, USA

Deirdre Logan, PhD, Department of Psychiatry, Harvard Medical School; Psychological Services for Pain Medicine, Boston Children's Hospital, Boston, Massachusetts, USA

Stephen M. Maricich, MD, PhD, Richard King Mellon Institute for Pediatric Research, Department of Pediatrics, University of Pittsburgh School of Medicine, Division of Child Neurology, Children's Hospital of Pittsburgh of UPMC, Pittsburgh, Pennsylvania, USA

Jennifer A. Markowitz, MD, Department of Neurology, Boston Children's Hospital, Boston, Massachusetts, USA

Wilson Marques, Jr., Professor of Neurology, Department of Neurosciences and Behavior Sciences, School of Medicine of Ribeirão Preto, University of São Paulo, São Paulo, Brazil

Thornton B.A. Mason, II, MD, PhD, MSCE, Perelman School of Medicine at the University of Pennsylvania, Children's Hospital of Philadelphia, Philadelphia, Pennsylvania, USA

Andrew McKeon, MD, Departments of Neurology and Laboratory Medicine and Pathology, College of Medicine, Neuroimmunology Laboratory, Mayo Clinic, Rochester, Minnesota, USA

Hugh J. McMillan, MD, MSc, University of Ottawa, Division of Pediatric Neurology, Children's Hospital of Eastern Ontario, Ottawa, Ontario, Canada

Caroline C. Menache-Starobinski, MD, Clinique des Grangettes, Geneva, Switzerland

Jerry R. Mendell, MD, The Research Institute at Nationwide Children's Hospital, Columbus, Ohio, USA

Eugenio Mercuri, MD, PhD, Catholic University, U.O.C. Neuropsichiatria Infantile, Policlinico Universitario "A. Gemelli", Rome, Italy

Payam Mohassel, MD, Neuromuscular and Neurogenetic Disorders of Childhood Section, National Institute of Neurological Disorders and Stroke/National Institutes of Health, Bethesda, Maryland, USA

Umrao R. Monani, Ph.D, Department of Pathology and Cell Biology, Department of Neurology, Columbia University Medical Center, New York, New York, USA

Jacqueline Montes, PT, EdD, NCS, Columbia University, Department of Neurology, New York, New York, USA

Manikum Moodley, MD, Center for Pediatric Neurology, Neurological Institute, Cleveland Clinic, Cleveland, Ohio, USA

Richard T. Moxley, III, MD, University of Rochester School of Medicine and Dentistry, Department of Neurology, Neuromuscular Disease Center, University of Rochester Medical Center, USA

Francesco Muntoni, MD, Dubowitz Neuromuscular Centre, UCL Institute of Child Health & Great Ormond Street Hospital for Children, London, United Kingdom

Kathryn N. North, Murdoch Childrens Research Institute, The Royal Children's Hospital, Victoria, Australia

Anders Oldfors, MD, Department of Pathology, Sahlgrenska University Hospital, Gotheburg, Sweden

Maryam Oskoui, MD, McGill University Health Centre, Montreal, Quebec, Canada

Robert A. Ouvrier, The Children's Hospital at Westmead, The University of Sydney, Sydney, Australia

Lauren M. Pachman, MD, Division of Pediatric Rheumatology, Stanley Manne Children's Research Institute, affiliated with Ann and Robert H Lurie Children's Hospital of Chicago, Feinberg School of Medicine, Northwestern University, Chicago, Illinois, USA

Massimo Pandolfo, MD, Université Libre de Bruxelles, Chief of Neurology, Erasme Hospital, Brussels, Belgium

Carmen Paradas, MD, Unidad de Enfermedades Neuromusculares, Servicio de Neurologia, Hospital Universitario Virgen del Rocío/Instituto de Biomedicina de Sevilla, Universidad de Sevilla, Seville, Spain

Marc C. Patterson, MD, Division of Child and Adolescent Neurology, Mayo Clinic College of Medicine, Rochester, Minnesota, USA

Alan K. Percy, MD, Department of Pediatrics, Neurology, Neurobiology, and Genetics, University of Alabama at Birmingham, School of Medicine, Birmingham, Alabama, USA

Matthew Pitt, MD, FRCP, Department of Clinical Neurophysiology, Great Ormond Street Hospital for Children National Health Service Foundation Trust, London, United Kingdom

David Pleasure, MD, Shriner's Hospital for Children; Institute for Pediatric Regenerative Medicine, University of California Davis, Neurology Research, Shriner's Hospital, Sacramento, California, USA

Susana Quijano-Roy, MD, Hôpital Raymond Poincaré (Garches)-Hôpitaux universitaires Paris Ile-de-France Ouest, Université de Versailles Saint-Quentin-en-Yvelines, Hôpital Necker Enfants Malades, Paris, France

Francis Renault, MD, Unité de Neurophysiologie Clinique de l'Enfant, Hôpital Armand-Trousseau, Paris, France

Louise R. Rodino-Klapac, PhD, The Research Institute at Nationwide Children's Hospital, Columbus, Ohio, USA

Manuel Roig-Quilis, MD, PhD, Pediatric Neurology Service and Pediatric Neurology Research Laboratory, Vall d'Hebron University Hospital; Quiron-Dexeus University Hospital, Barcelona, Spain

David P. Roye, MD, Columbia University; New York-Presbyterian Hospital; Pediatric Orthopaedic Surgery, Morgan Stanley Children's Hospital of New York; Columbia Cerebral Palsy Center, New York, New York, USA

Reinhardt Rüdel, PhD, Division of Neurophysiology, University of Ulm, Ulm, Germany

Barry S. Russman, MD, Department of Pediatrics and Neurology, Oregon Health Science and University, Portland, Oregon

Monique M. Ryan, MMed, BS, FRACP, Children's Neurosciences Centre, Murdoch Childrens Research Institute, University of Melbourne, Royal Children's Hospital, Victoria, Australia

Ai Sakonju, MD, University of Utah School of Medicine, Salt Lake City, Utah

Harvey B. Sarnat, MSc, MD, FRCPC, Department of Paediatrics, Pathology (Neuropathology), and Clinical Neurosciences, Division of Paediatric Neurology, Alberta Children's Hospital, Calgary, Alberta, Canada

William A. Scott, MD, Division of Cardiology, University of Texas, Southwestern Medical Center; Heart Center, Children's Medical Center, Dallas, Texas, USA

Alan R. Seay, MD, Department of Pediatrics, University of Colorado Denver School of Medicine, Children's Hospital Colorado, Aurora, Colorado, USA

Piraye Serdaroglu-Oflazer, MD, Department of Neurology, Istanbul Faculty of Medicine, Istanbul University, Istanbul, Turkey

Navil F. Sethna, MD, FAAP, Mayo Family Pediatric Pain Rehabilitation Center, Boston Children's Hospital, Boston, Massachusetts, USA

Evan D. Sheha, Columbia University, New York, New York, USA

Michael E. Shy, MD, Division of Neuromuscular Medicine, Neurology, Carver College of Medicine, University of Iowa, Iowa City, Iowa, USA

Jemeen Sreedharan, PhD, MRCP, The Babraham Institute, Cambridge, United Kingdom; Neurology, Neurobiology, UMass Medical School, Worcester, Massachusetts, USA

Nancy E. Strauss, MD, Columbia University, Rehabilitation and Regenerative Medicine, New York, New York, USA

Rabi Tawil, MD, University of Rochester Medical Center, Rochester, New York, USA

Ingrid Tein, BSc, MD, FRCP(C), Neurometabolic Clinic and Research Laboratory, Division of Neurology, Genetics and Genome Biology Program, The Research Institute; The Hospital for Sick Children, University of Toronto, Toronto, Ontario, Canada

Jennifer A. Tracy, MD, Department of Neurology, College of Medicine, Mayo Clinic, Rochester, Minnesota, USA

Bjarne Udd, MD, PhD, Neuromuscular Research Center University and University Hospital of Tampere, Folkhalsan; Institute of Genetics, Biomedicum, University of Helsinki, Helsinki; Department of Neurology, Vasa Central Hospital, Vasa, Finland

Silvère M. van der Maarel, PhD, Leiden University Medical Center, Leiden, The Netherlands

Angela Vincent, MBBS, FRCPath FMedSci FRS, The Nuffield Department of Clinical Neurosciences, University of Oxford; John Radcliffe Hospital, Oxford, United Kingdom

Joseph J. Volpe, MD, Harvard Medical School; Department of Neurology, Boston Children's Hospital, Boston, Massachusetts, USA

Jo M. Wilmshurst, MBBS, MD, MRCP(London), FCPaed(SA), Department of Paediatric Neurology and Neurophysiology, Red Cross Children's Hospital; School of Child and Adolescent Health, University of Cape Town, Cape Town, South Africa

Nanfang Xu, Columbia University, New York, New York, USA

Eppie M. Yiu, MB BS, PhD, FRACP, Children's Neurosciences Centre Royal Children's Hospital, Murdoch Childrens Research Institute and University of Melbourne, Victoria, Australia

Huda Y. Zoghbi, MD, Departments of Pediatrics, Neurology, Molecular and Human Genetics, and Neuroscience, Baylor College of Medicine; Howard Hughes Medical Institute; and the Jan and Dan Duncan Neurological Research Institute at Texas Children's Hospital, Houston, Texas, USA

Part I

Clinical and Laboratory Approach to the Infant and Child with a Neuromuscular Problem

Chapter 1

Introduction: Historical Perspectives

Darryl C. De Vivo, Basil T. Darras, Monique M. Ryan, and H. Royden Jones, Jr.

INTRODUCTION

The role of the clinician in the diagnosis and treatment of a weak child is as important today as it was in the 19th century, when pediatric neuromuscular diseases were first being recognized by such luminaries as Meryon (1852),[1] Duchenne (1861),[2] Werdnig (1891),[3] and Hoffmann (1893),[4] and later by Batten (1903).[5] The clinician needs to react to the concerns of the patient and their parents at the first encounter. A detailed medical history and carefully performed physical examination remain the fundamental tools and starting point for assessing various symptoms and signs. This initial clinical assessment has three possible immediate outcomes: (1) the concerns of the patient and family appear to be unfounded, and the clinician can be reassuring. A follow-up visit is advised to substantiate this outcome and to give the parents peace of mind that all, indeed, is well; (2) the clinician shares the concerns of the patient and family but decides to use time as the first test to determine the natural history of the process; or (3) the clinician recognizes the importance of the clinical symptoms and realizes the need to perform diagnostic procedures as soon as possible.

This clinical approach places great demands on the physician at the first and subsequent encounters. Here, the difference between the 19th-century clinician and the 21st-century clinician is enormous. This difference is a measure of the scholarly advances of the past century, particularly over the past three decades, since the advent of molecular neurogenetics.[6–11] Examples of such advances include major discoveries in the genetic analysis of the muscular dystrophies, congenital myasthenic syndromes, spinal muscular atrophies, and hereditary neuropathies. These advances in molecular neurogenetics have led to new diagnostic tests, which may confirm the clinical diagnosis accurately and noninvasively. Traditional diagnostic tests such as nerve conduction studies, electromyography (EMG), and open muscle biopsy are less often required in the evaluation of neuromuscular disorders in infancy, childhood, and adolescence. Nevertheless, these traditional tests remain useful in certain situations: to evaluate patients in whom molecular genetic analyses fail to confirm the initial clinical impression, to facilitate the rapid electrophysiologic diagnosis of a neuropathy or spinal muscular atrophy, or to focus more precisely on molecular mechanisms before ordering relatively expensive molecular genetic tests. The dramatic advances in DNA diagnostics have added to the complexity of a challenging clinical field and have left physicians occasionally uncertain about the relative indications for traditional tests such as EMG and muscle biopsy. Clearly, all these diagnostic tests remain useful, and it is up to the modern clinician to make informed decisions, after the initial clinical evaluation, to facilitate an accurate biomolecular diagnosis as quickly and as economically as possible.

TRADITIONAL DIAGNOSTIC TESTS

The technique of muscle biopsy was first introduced by Duchenne in 1868, using a harpoon-like device to sample the affected tissue during life.[12] This crude device was the harbinger of the biopsy needle. Earlier and later investigators used muscle tissues obtained after the patient's death. These techniques have continued to be refined, and analysis of involved tissue has become increasingly elegant and precise, as demonstrated by advances in electron microscopy, enzyme histochemistry, immunocytochemistry, and single myofiber analysis. Today, a biopsied specimen can be processed for several studies, including light microscopy, electron microscopy, enzyme histochemistry, immunohistochemistry, biochemistry, tissue culture, and molecular studies,[13] and protein expression patterns can be assessed in single 8-micron sections taken from muscle biopsies.[14] Advances in electrophysiology have been equally dramatic. The first application of these physiologic principles to neuromuscular diseases dates to the latter part of the 19th century.[2] However, the practical use

B.T. Darras, H. Royden Jones, Jr., M.M. Ryan & D.C. De Vivo (Eds): Neuromuscular Disorders of Infancy, Childhood and Adolescence, Second edition.
DOI: http://dx.doi.org/10.1016/B978-0-12-417044-5.00001-9

of nerve conduction studies and needle EMG did not emerge until the mid-20th century. At that time, observations by Lambert at the Mayo Clinic in Rochester, Minnesota, Gilliatt at Queens Square in London, and Buchthal at the Institute of Neurophysiology in Copenhagen defined the electrical parameters of the normal peripheral motor and sensory unit as well as the potential for widespread application of these findings to the evaluation of neuromuscular disorders. These three clinical neurophysiologists, among others, also recognized the importance of carefully defined normal neurophysiologic values, which are still used today.[15–20] These studies defined the significant maturational changes during early development that remain important to the performance and interpretation of pediatric EMG, as detailed in Chapter 3. The more universal application of EMG to pediatric motor unit disorders had attracted increasing interest during the two decades that preceded the mapping and discovery of the SMN gene mutated in classic spinal muscular atrophy.[20,21] Byers' and Banker's classification of the spinal muscular atrophies at Boston Children's Hospital was an important early contribution[22] later expanded by Dubowitz.[23]

Other seminal reports continued to appear. Dyck and Lambert separated Charcot-Marie-Tooth disease into demyelinating and axonal forms, presaging a similar DNA classification that followed a quarter century later.[24] Gutrecht and Dyck performed important anatomic studies defining the maturation of peripheral nerve myelination.[25] These findings mirrored the normal increase in the speed of peripheral nerve conduction from infancy to age 5 years.[16] Appreciation of these normal ranges for motor nerve conduction velocities also led to the definition of a subacute acquired polyneuropathy: chronic inflammatory demyelinating polyneuropathy (CIDP). These observations were important both because of CIDP's predominant proximal presentation, often mimicking a myopathy, but particularly because CIDP is a treatable condition. These maturational distinctions also emphasize the importance of traditional testing modalities such as nerve conduction studies to enable diagnosis in complex cases with dual pathologies, such as the rare instances of CIDP superimposed upon genetic neuropathies.[26,27]

Engel and Lambert applied microelectrode methodology *in vitro*, extending our understanding of the various uncommon congenital neuromuscular transmission defects (NMTDs).[28] These observations helped define the normal physiology of neuromuscular transmission, as well as enhancing our ability to differentiate congenital, inherited NMTDs from acquired, immunologically mediated NMTDs. However, it remains difficult to make a precise clinical neurophysiologic diagnosis of some of these congenital NMTDs. For example, some electrophysiologic parameters are not well defined for the immature neuromuscular junction. Additionally, precise microelectrode analysis of a biopsied neuromuscular junction is still necessary for diagnosis of some conditions.

Infant motor unit potentials (MUPs) are typically quite small, mimicking the size of abnormal adult "myopathic" motor units. This maturational distinction sometimes challenges the clinical neurophysiologist who is attempting to differentiate a myopathy in a newborn from a normal response. The rapid acquisition of a full recruitment pattern encompassing very small MUPs may provide the initial clue to the presence of a myopathy. On some occasions, it is equally important for the clinical neurophysiologist to advise the referring physician that even though the MUPs may appear to be normal in a floppy infant, such a finding does not exclude a myopathy.

Similarly, the finding of fibrillation potentials may not define a neuromuscular disorder as either neurogenic or myopathic. Generally, these potentials signify a denervating process, but similar abnormalities are sometimes seen in myopathies.[29] In reality, the fibrillation potential results from the disconnection of all or part of a myofiber from the nerve, so the principle remains intact; that is, fibrillation potentials signify denervation of the myofibers. It is important to emphasize that the MUP is the key to making the neurophysiologic distinction between a primary anterior horn cell process and one that relates to a dysfunction of the muscle fiber.[29] Differentiating a normal immature MUP from a congenital myopathy or early dystrophy is a bigger challenge for the clinical neurophysiologist.

DNA analysis now frequently resolves the confusion in these clinical settings. The benefit of DNA analysis is apparent in the evaluation of Duchenne muscular dystrophy (DMD), where EMG and muscle biopsy were state-of-the-art diagnostic studies three decades ago. Today, the clinical diagnosis of children with suspected DMD or spinal muscular atrophy can be quickly confirmed by specific DNA testing. Similar changes are occurring in a number of the other hereditary neuromuscular disorders, as reviewed elsewhere in this book.

Clinical chemistry methods emerging after World War II provided another important means of distinguishing between neuropathies and myopathies. The serum transaminases and aldolase measures were introduced first, but measurement of serum creatine kinase (CK) activity has proved to be more tissue specific, CK expression being limited to muscle and brain. Elevated values—in excess of 1000 IU—favor a primary myopathic process.[30] Other serum enzyme values usually parallel the serum CK values, but add little to our understanding of the precise disease process. As is true of every diagnostic procedure, the exceptions continue to accumulate. Thus, many "myopathies" often have normal serum enzyme values—witness congenital myopathies, metabolic myopathies, the periodic paralyses, and the various NMTDs. But the generalization is still helpful diagnostically, and a serum CK value is useful when first evaluating a patient with weakness, hypotonia, fatigue, or pain. A high serum CK value suggests a myopathy and mandates further investigation, but a normal value does not exclude primary pathology of the myofiber (e.g. congenital myopathies).

Other blood and urine tests have become available and provide valuable information as screening procedures. The serum and CSF lactate and pyruvate may be elevated in mitochondrial diseases, and carnitine and its acylated profile may be altered in defects of fatty acid oxidation. Immunologic tests may be informative, such as acetylcholine receptor and anti-MuSK antibodies in myasthenia gravis. As with all tests, abnormal results may be informative, but normal results may be consistent with the initial clinical diagnosis.

MODERN DIAGNOSTIC TESTING

The clinical landscape of modern medicine will continue to change as a result of the breathtaking advances in molecular genetics. These advances can be traced back to several seminal contributions over the past decades since the elucidation of the structure of DNA. Newborn screening was introduced in the 1960s by Robert Guthrie[31]; and now, with the increasing application of molecular diagnostics, expanded newborn screening has the potential of eliminating the often expensive "diagnostic odyssey" that begins with the onset of clinical symptoms postnatally. Even more exciting is the possibility of treating the genotype, rather than the clinical phenotype, as clearly witnessed by the phenylketonuria (PKU) model. This proactive approach obviates the need to "rescue" the phenotype since the treatment intervention will precede the onset of symptoms. Undoubtedly, there must exist a window of therapeutic opportunity for genetically determined neuromuscular diseases, and this window probably continues to close as the postnatal timeline is extended. For example, we know that the disease process is progressing for a period of time before clinical symptoms become manifest. Children with Duchenne muscular dystrophy have very high serum CK values in infancy, long before they develop symptoms in early childhood. Again, the PKU experience reinforces these basic principles. The longer the patient is symptomatic, the more difficult it will be to react to the symptoms, rescue the clinical phenotype and restore the patient to good health.

In 1969, Tay-Sachs disease was shown to be caused by hexosaminidase deficiency[32]; and the next year, the concept of preconception testing and counseling emerged. As a long-term result, Tay-Sachs disease has largely been eliminated in the at-risk Ashkenazi Jewish population.[33] Now, over 100 recessive diseases, most untreatable and some affecting the neuromuscular system, can be prevented using this same model. Applying next-generation sequencing to preconception screening has the potential of eliminating many recessively transmitted neuromuscular diseases that are discussed throughout the several chapters of this book, analogous to the elimination of smallpox, poliomyelitis, and other infectious diseases by effective vaccine programs.

Next-generation molecular testing also will allow for a more complete interrogation of the human genome in puzzling clinical conditions like myoadenylate deaminase deficiency. This condition is present in 1% to 2% of the population, but most of the carriers are clinically asymptomatic, allowing us to speculate as to whether the condition is truly disease-causing, or whether it is a genetic susceptibility factor that remains silent in the absence of another genetic modifier. In some clinical settings, patients with myoadenylate deaminase deficiency are clearly symptomatic with aches, pain, fatigue, and weakness, and the forearm ischemic exercise test is abnormal with unchanging venous ammonia values. It is possible that these patients have another genetic factor that, when present (or absent), produces a clinical phenotype. The complex molecular mechanisms recently uncovered to explain "reversible cytochrome oxidase deficiency" are another example of multiple molecular factors acting in concert to produce clinical symptomatology.[34]

When a specific genetic condition currently is being considered, DNA studies are increasingly the first laboratory test performed after the clinical evaluation and measurement of serum CK activity. The DNA studies may be targeted to the sequencing of the suspected gene when the clinical phenotype is essentially diagnostic, or broadened to include whole exome or next generation sequencing when the clinical phenotype is less specific. The traditional diagnostic tests mentioned earlier are increasingly reserved for the evaluation of acquired disorders (e.g. toxic, immune-mediated, and/or inflammatory) or genetic disorders that have failed initial DNA screening. Rarely, however, when DNA analysis in a boy with a DMD phenotype fails to reveal a dystrophin gene mutation and the family history is negative, one proceeds to a muscle biopsy. Immunostaining determines whether dystrophin is present or absent, and Western blot analysis allows the size and abundance of the protein to be determined.

Molecular diagnostics are increasingly valuable in the reclassification of clinical disease groups. One excellent example is limb-girdle muscular dystrophies (LGMDs). This category includes autosomal recessive (LGMD2) and autosomal dominant (LGMD1) forms.[35] The autosomal recessive forms usually have an earlier onset, more progression, and higher serum CK activity, with a phenotype that overlaps with the dystrophinopathies. Cognitive involvement, when present, favors a dystrophinopathy, and a serum CK value in excess of 1000 IU/L usually favors a myopathic process rather than a neuropathic process such as spinal muscular atrophy type III.

DNA testing for LGMD1 and LGMD2 subtypes emerged from several research laboratories over the past decade and is now clinically available.[36–38] Immunohistochemistry of biopsied skeletal muscle tissue remains useful in demonstrating abnormalities of the α-, β-, γ-, and δ-sarcoglycans; dystroglycans; dysferlin; and other proteins. The gene localization, mutated protein, and pattern of inheritance of the various LGMDs are discussed in Chapter 34.[29]

Patients with the autosomal dominant forms of LGMD (type 1) are usually older, with a slower clinical progression and less elevated serum CK values, with the possible exceptions of LGMDs types 1B and 1C.

These few examples emphasize the clinician's ability to use modern molecular and traditional diagnostics to confirm a clinical diagnosis rapidly and economically. The neurologist has always been an expert in recognizing the clinical phenotypes, but the advances in molecular diagnosis now demand a sophisticated diagnostic approach to the causative genotypes. Although the phenotype-genotype correlation has sometimes remained elusive, probably because the biological rules still remain incompletely understood, the genotypic approach to diagnosis complements the phenotypic approach. While reliable phenotyping of patients will remain the gold standard in the field of neuromuscular medicine, DNA studies will continue to pave the way for a molecular classification of neuromuscular diseases.

CLINICAL CLASSIFICATION

The classic phenotypes represent the cornerstone of clinical diagnosis. Adherence to the classic phenotype is mandatory if one needs a clinically pure sample to identify a candidate gene. Misdiagnosis will affect the resulting logarithm of the odds (LOD) score in linkage analysis studies (discussed in Chapter 2). An experienced clinician can diagnose a child with DMD by inspection shortly after he enters the room, but modern neurogenetics has taught us that the phenotypic range of the dystrophinopathies is very broad, ranging from the classic Duchenne and Becker phenotypes to patients with myalgias, cramps, hyperCKemia, and possibly even isolated cognitive deficits.[39] The expanded clinical spectrum of genetic conditions can challenge even the experienced clinician, and demand an appreciation of phenotypic and genotypic homogeneity and heterogeneity. Several different gene mutations may cause the same phenotype (e.g. emerin and lamin A/C gene mutations causing Emery-Dreifuss muscular dystrophy), and several different phenotypes may result from the same genotype (e.g. LGMD2B, Miyoshi distal myopathy, and distal myopathy with anterior tibial onset caused by dysferlin gene mutations; and autosomal dominant Emery-Dreifuss muscular dystrophy, LGMD1B, cardiomyopathy with conduction system disease, and partial lipodystrophy caused by mutations within the lamin A/C gene). As a result, phenotype-genotype correlations often remain a puzzle. The lack of correlation between gene defect or residual tissue enzyme activity and clinical condition implies that there are other genetic and environmental factors modifying the expression of the primary mutation.

Neuromuscular disorders are conveniently classified according to the anatomic structure of the motor unit. Diseases of the anterior horn cell are referred to as neuronopathies; of the peripheral nerve as neuropathies; of the neuromuscular junction as myasthenic syndromes or, more commonly today, neuromuscular transmission disorders; and of the myofiber as myopathies. Classically, each of these subgroups presents with distinctive clinical features that orient the clinician during the initial patient evaluation.

Neuronopathies and neuropathies represent a continuum of denervating diseases. Neuronopathies classically involve the cell body, and neuropathies classically affect their extensions and the investing myelin sheath. The dominant neuronopathies in the pediatric age group are the genetically determined spinal muscular atrophies. Their clinical picture varies to some degree, depending on the age at presentation. Infants with spinal muscular atrophy are typically weak and areflexic. The alert infant lying quietly on the examining table with a wide-eyed expression and tongue fasciculations, and predominantly distal movements of the limbs is easily recognized. The older child with the juvenile presentation has more obvious proximal weakness of the shoulder and pelvic girdle muscles and hyporeflexia, simulating the clinical presentation of a dystrophinopathy.[40] However, joint contractures are less common in children with juvenile spinal muscular atrophy, and serum CK values tend to be normal or only slightly elevated. These distinctions allow one to quickly arrive at an initial clinical impression of a neuronopathy versus an active, progressive myopathy. Classically, the extensor digitorum brevis muscle is atrophied in juvenile spinal muscular atrophy and hypertrophied in DMD, another subtle finding that quickly leads the experienced clinician to a presumptive diagnosis. Neurogenic disease causes more wasting than myopathic disease does, and the loss of muscle bulk is more distal. However, these generalizations can be misleading in certain clinical entities. As mentioned previously, in the juvenile phenotype of spinal muscular atrophy, there may be more proximal weakness.

In contrast, certain myopathies are associated with predominantly distal weakness, as is seen, for example, with myotonic muscular dystrophy, desmin myopathy, Miyoshi myopathy, and the myopathy associated with nephropathic cystinosis. Fasciculations of the tongue are prominent in anterior horn cell diseases and may be seen occasionally in neuropathies. Certain metabolic diseases such as Pompe disease also involve the anterior horn cell and may produce fasciculations, but the EMG is distinctive, revealing myotonic discharges that are not seen with infantile spinal muscular atrophy or congenital neuropathies. Combined upper and lower motor neuron signs, the hallmark of amyotrophic lateral sclerosis, are seen infrequently in pediatric patients, but rare examples of juvenile motor neuron disease and neuronal intranuclear hyaline inclusion disease may be encountered, with upper and lower motor neuron signs, bulbar weakness, and fasciculations.[41,42]

Associated sensory loss implicates the peripheral nerves and argues against motor neuron diseases, NMTDs, and myopathies. Loss of muscle stretch reflexes is also the hallmark of a peripheral neuropathy. Areflexia is the rule when sensory involvement is present. However, muscle stretch reflexes are often reduced or absent in patients with congenital nonprogressive myopathies such as central core disease and nemaline myopathy.

Cramps are the hallmark of denervating diseases and need to be distinguished from muscle contractures. Typically, cramps are associated with intense muscle pain and may cause a palpable mass in the muscle. These symptoms typically occur with the muscle at rest and are brief in duration and sudden in onset. Passive muscle stretching often leads to relief. EMG of a cramping muscle reveals high-frequency motor unit discharges similar to those seen during maximal muscle contraction. Cramps may occur in the absence of definable disease and are generally described as benign, often occurring at night. Otherwise, cramps usually indicate disease of the anterior horn cell, nerve roots, or peripheral nerve elements. Alternatively, cramping may signify the presence of a metabolic derangement altering the neuronal microenvironment, as is seen with renal failure, hypothyroidism, hepatic failure, adrenal insufficiency, or disturbances of electrolyte balance. Cramps and pain, however, are not limited to neurogenic diseases; myalgias and cramps may be seen as the minimal clinical expression of a dystrophinopathy,[39] and painful cramps may accompany caveolinopathy (LGMD1C) or glycogen storage disease, as discussed in Chapters 30, 34, and 39. Pain and cramping have also been described in mitochondrial diseases and in inflammatory diseases such as dermatomyositis, polymyositis, and Guillain-Barré syndrome. Inflammation of the nerve roots may produce intense pain with the slightest movement, making examination of the child impossible. This discomfort may be so pronounced at times that the child becomes irritable and uncooperative, leading to an initial clinical impression of an acute encephalopathy rather than Guillain-Barré syndrome.[43]

Contractures differ from cramps clinically and electrically. The contracture is electrically silent and may cause muscle pain and localized swelling of the muscle that persists for hours. Unlike cramps, contractures generally occur with exercise and suggest an underlying metabolic myopathy such as phosphorylase deficiency or other glycolytic enzyme defects. Contractures also may occur in patients with hypothyroidism, rippling muscle syndrome, Brody's disease, and paramyotonia congenita.

Disorders of the neuromuscular junction characteristically present with intermittent symptoms, including weakness and fatigue. In contrast, disorders of the anterior horn cell, peripheral nerve, and muscle generally present with fixed symptoms that are often progressive over time. Fatigue has been underappreciated as a symptom of denervating diseases, particularly spinal muscular atrophy.[44] Recent research has highlighted the early targeting of the synaptic region in both conditions, which may underpin their common symptomatology.[45]

Inflammatory diseases of nerve and muscle may evolve, plateau, and then regress, whereas genetically determined diseases of the motor unit emerge and steadily progress over time. In the pediatric population, disorders of the neuromuscular junction include genetically determined NMTDs, acquired disorders such as infant botulism, and immunologically mediated disorders such as transient neonatal myasthenia gravis, fetal acetylcholine receptor inactivation syndrome[46] and immune-mediated juvenile myasthenia gravis. Each of these disorders is distinctive and often recognizable clinically by age at presentation and symptoms. EMG studies of the motor unit, particularly neuromuscular junction testing, as mentioned earlier, may be useful as an initial diagnostic study in these disorders.

Congenital disorders of neurotransmission are described in detail in Chapter 26. These disorders produce varying degrees of weakness and fatigability, often beginning during infancy. Typical symptoms include hypotonia, ptosis, ocular motility disturbances and intermittent apnea. To some extent, these disorders overlap symptomatically with disorders of central neurotransmission, such as aromatic L-amino acid decarboxylase deficiency and tyrosine hydroxylase deficiency (see Chapter 6). Transient neonatal myasthenia gravis and infant botulism are acquired disorders of peripheral neurotransmission. The first follows the transplacental transfer of maternal antibodies in the setting of maternal myasthenia gravis; the second results from the ingestion of *Clostridium botulinum* spores that germinate in the intestinal tract and elaborate the botulinum toxin. Again, the clinical picture is distinctive in each situation. The diagnosis of transient neonatal myasthenia gravis is determined primarily by a family history and examination of the mother. Infant botulism is diagnosed by the clinical symptoms, which include dilated, poorly reactive pupils; constipation; decreased bowel sounds; limpness; apnea, often while feeding at the breast; and weakness, with diminished muscle stretch reflexes. Both conditions improve with time, and no specific treatment may be necessary beyond supportive care.

Acquired, immunologically mediated myasthenia gravis is more frequently encountered in later childhood or adolescence, although we have seen patients as young as 15 months with antibody-positive myasthenia gravis. The intermittent nature of the symptoms is informative, and repetitive motor nerve stimulation is essentially diagnostic, with a characteristic decrement in evoked compound muscle action potential responses. In the morning and at rest, patients are often less symptomatic or asymptomatic. Fatigue associated with repetitive stimulation or with the passage of time during the day is an important clinical characteristic suggesting a defect of neuromuscular transmission.

Intermittent symptoms also raise the diagnostic possibility of a periodic paralysis. The channelopathies are often associated with episodic weakness and myotonia. The myotonias, as a group of diseases, are subdivided into dystrophic and nondystrophic disorders. The dystrophic disorders include myotonic dystrophy and proximal myotonic myopathy. The nondystrophic myotonias and the periodic paralyses, now commonly referred to as channelopathies, result from genetic mutations of various ion channels in muscle. The channelopathies are subdivided according to the ion channel involved in the molecular defect. Sodium channelopathies include the hyperkalemic periodic paralyses and paramyotonia congenita, both transmitted as autosomal dominant conditions. The potassium-aggravated myotonias (myotonia fluctuans, myotonia permanens, and acetazolamide-responsive myotonia) are also transmitted as autosomal dominant conditions and are associated with sodium channel mutations.

Chloride channelopathies include myotonia congenita. This disorder is further subdivided into the autosomal dominant form, known as Thomsen's disease, and the autosomal recessive form, known as Becker's disease. Hypokalemic periodic paralysis is the best-known calcium channelopathy. Other channelopathies include Schwartz-Jampel syndrome, rippling muscle disease, Andersen Tawil syndrome, Brody's disease, and malignant hyperthermia. Andersen Tawil syndrome is associated with periodic paralysis, cardiac arrhythmias, and dysmorphic facial features; Brody's disease is associated with delayed relaxation and no myotonia; and malignant hyperthermia is an anesthetic-induced delayed relaxation of muscle, one form of which is transmitted as an autosomal dominant trait resulting from a mutation of the ryanodine receptor on chromosome 19.

Evaluation of patients with periodic paralysis is facilitated by an awareness of the phenotype. For example, patients with Andersen Tawil syndrome have characteristic dysmorphic features, including hypertelorism, short stature, low-set ears, and clinodactyly.[47] These dysmorphic features, in the setting of prolonged Q-T interval and life-threatening ventricular arrhythmias, permit an accurate diagnosis in the office. Similarly, patients with Schwartz-Jampel syndrome are phenotypically distinctive, with short stature, bone and joint deformities, chondrodystrophy, hypertrichosis, blepharophimosis, and muscle stiffness.[48] EMG shows nonvariable, continuous high-frequency electrical discharges with delayed muscle relaxation.

Myopathies also are characterized by loss of strength, but the degree of weakness is disproportionate to the degree of muscular atrophy, particularly early in the clinical course. As mentioned previously, the extent of muscular atrophy appears disproportionate in neurogenic diseases. Patients with myopathies appear to be unduly weak, without significant loss of muscle bulk. DMD stands out as a striking example. The muscular-appearing child with DMD appears remarkably weak, struggling to rise from the floor or walk up and down stairs. The large proximal muscles are differentially affected, with relative preservation of the distal muscles. Children with myotonia congenita often appear quite muscular, but struggle to keep up with their peers in sporting activities. Again, there are numerous exceptions to these generalizations, such as the many myopathies that affect distal muscles, including Welander's and Miyoshi distal myopathies and telethonin deficiency (LGMD2G). Myotonic dystrophy also differentially affects the distal muscles. Gowers' sign is a manifestation of pelvic girdle muscle weakness, most commonly seen in the setting of DMD (Figure 1.1), but it can also be seen in other neuromuscular disorders, such as juvenile spinal muscular atrophy (type III), CIDP, and mitochondrial diseases.

Gowers' description of this maneuver occurred in his writings on DMD, and little can be added to the original description.[49] Patients "first put the hands on the ground, then stretch out the legs behind them far apart, with the chief weight of the trunk resting on the hands, by keeping the toes on the ground and pushing backwards, they manage to get the knees extended so that the trunk is supported by the hands and feet, all placed as widely apart as possible. Next the hands are moved alternately along the ground backwards so as to bring a larger portion of the weight of the trunk over the legs. Then, one hand is placed on the knee, and a push with this and with the other hand on the ground is sufficient to enable the extensors of the hip to bring the trunk into the upright position." Gowers thought that this maneuver was pathognomonic for pseudohypertrophic muscular paralysis.[50] Since the original description, however, clinicians have come to understand that this sign is present whenever there is significant weakness of the hip

FIGURE 1.1 Duchenne muscular dystrophy and Gowers' sign. This series of photographs shows the components of a "one-handed" Gowers' maneuver. The patient uses hand support on the floor, initially bilateral then unilateral (A), and hand support on the thighs, either unilateral (B, C) or bilateral, to attain the standing position (D, E).

and knee extensors, regardless of whether the underlying disease process affects primarily nerve or muscle.

Muscle stretch reflexes tend to be relatively preserved with myopathic diseases and are roughly proportionate to the degree of atrophy. When lost, the proximal reflexes are more affected than the distal reflexes. However, patients with congenital myopathies often have diminished reflexes or areflexia. Clinically, if the patient is notably weak with preserved muscle bulk and loss of muscle stretch reflexes, the condition is most likely a myopathy.

CLINICAL APPROACH

The clinical approach to the evaluation of a weak child demands a thorough understanding of the many rules that describe diseases of the motor unit and the several exceptions and overlapping features of the symptoms, as discussed previously. Nothing is more important than a careful history and physical examination. Asking patients or their parents to describe the chronology of the clinical syndrome is of inestimable value. The temporal profile (the onset, duration, and evolution of the symptoms and signs) usually suggests one or more diagnostic possibilities. For example, knowing that patients with DMD typically present with weakness at age 3 years is important. Other conditions presenting in early childhood include Becker's dystrophy; Emery-Dreifuss dystrophy; facioscapulohumeral dystrophy; limb-girdle dystrophy; myotonic dystrophy; inflammatory myopathies; various metabolic diseases, including lipid storage myopathies; mitochondrial diseases; and various endocrine and metabolic disorders.

The presentation and the pattern of disease over time allow one to categorize the possible clinical conditions. As a rule, the genetically determined neuronopathies and the muscular dystrophies are inexorably progressive from the time of onset. However, in infants and young children, disease progression is often mitigated by normal childhood development. As a result, at certain points in early development, the parents may report that the child has stabilized or actually improved functionally. Similarly, some patients may have seasonal improvement owing to increased outdoor activities, such as swimming. This seasonal effect is particularly evident in patients with juvenile spinal muscular atrophy. Water activity and outdoor play generally have a beneficial effect on all patients with neuromuscular disorders. In contrast, other children have episodic or saltatory patterns to their clinical symptoms, which generally suggest an underlying ion channel disturbance or metabolic disease. In addition, inflammatory diseases of the neuromuscular system may wax or wane symptomatically. Children with dermatomyositis frequently present with this type of history. When the serum CK is markedly elevated, saltatory progression of the illness favors an inflammatory disease of muscle rather than a muscular dystrophy.

Weakness evident during the newborn period raises other possibilities, such as spinal muscular atrophy, congenital muscular dystrophy, myotonic dystrophy, the several congenital myopathies defined by distinctive histochemical abnormalities, and certain metabolic diseases such as acid maltase deficiency, phosphorylase deficiency, and carnitine palmitoyltransferase type II deficiency. These disorders need to be considered, along with the congenital myasthenic syndromes and genetic peripheral neuropathies.

CIDP may mimic many of these disorders. Where an older child has weakness coupled with prominent pain and misery, dermatomyositis is likely, particularly if there are cutaneous abnormalities, including a violaceous discoloration of the upper eyelids and punctate ulcerations of the extensor surfaces of the limbs. Electrical studies of the motor unit, EMG, muscle biopsy, and occasionally nerve biopsy may be valuable in diagnosing these treatable conditions.

Family history may provide valuable insight into the patient's condition. Most disorders of the motor unit are genetically determined autosomal dominant, autosomal recessive, or X-linked disorders. Others are transmitted as maternal, non-Mendelian traits, pathognomonic for mitochondrial DNA mutations. Nothing may be more informative than examining the mother of a weak newborn infant to determine whether she has evidence of myotonic dystrophy, myasthenia gravis, inflammatory bowel disease, or another immune-mediated condition.[51]

Similarly, identifying precipitating factors that may trigger the onset of symptoms is informative. A history of pain, weakness, or myoglobinuria provoked by exercise quickly leads to the consideration of a metabolic disease, including the several glycolytic enzyme defects and mitochondrial and lipid storage myopathies. Weakness associated with fever or fasting leads to suspicion of a defect of fatty acid oxidation. Dietary factors, such as the ingestion of a high-carbohydrate meal, lead to the consideration of periodic paralysis. Patients with paramyotonia congenita may report that cold exposure precipitates their symptoms of muscle stiffness.

Involvement of other organs may lead to diagnostic possibilities. Cardiac disease often accompanies DMD, Becker's muscular dystrophy, myotonic dystrophy, Emery-Dreifuss dystrophy, LGMD1B, LGMD1D, Andersen Tawil syndrome, and various metabolic disorders, including mitochondrial diseases, acid maltase deficiency, and carnitine deficiency. Inflammatory diseases of muscle also may affect cardiac muscle. In contrast, diseases affecting anterior horn cell, peripheral nerve, and neuromuscular junction spare the heart. Multisystemic involvement is common in mitochondrial diseases; strokes or stroke-like episodes, migraine headaches, short stature, pigmentary retinopathy, sensorineural hearing loss, proximal limb weakness, and lactic acidosis are common findings in children with the MELAS (mitochondrial encephalopathy and lactic acidosis with stroke-like episodes) phenotype. Muscle biopsy is distinctive in MELAS, classically showing ragged red fibers (Figure 1.2).

Similarly, respiratory failure leads to the consideration of various diseases that affect the muscle fiber, including the spinal muscular atrophies, muscular dystrophies, metabolic myopathies such as acid maltase deficiency and carnitine deficiency, mitochondrial diseases, congenital myopathies such as nemaline and centronuclear myopathy, and inflammatory myopathies such as polymyositis and dermatomyositis.

Liver involvement may be seen with mitochondrial DNA depletion syndrome, acid maltase deficiency, debranching enzyme deficiency, and carnitine deficiency. Ocular involvement may be expected with myotonic dystrophy, congenital muscular dystrophies, and mitochondrial diseases. Dysmorphic features may be seen with the congenital myopathies, Andersen Tawil syndrome, and Schwartz-Jampel syndrome. Fixed musculoskeletal contractures are characteristic of certain long-standing myopathies, such as DMD, Emery-Dreifuss dystrophy, and Bethlem myopathy.

Scoliosis is uncommon in ambulatory patients but is characteristic of Friedreich's ataxia. Spinal curvature may develop and progress alarmingly fast once the child becomes wheelchair-dependent. A multidisciplinary approach to these patients is ideal, and complications can be presented or managed early in the clinical course (see Chapters 52 and 53). Children with neuromuscular disorders benefit from input

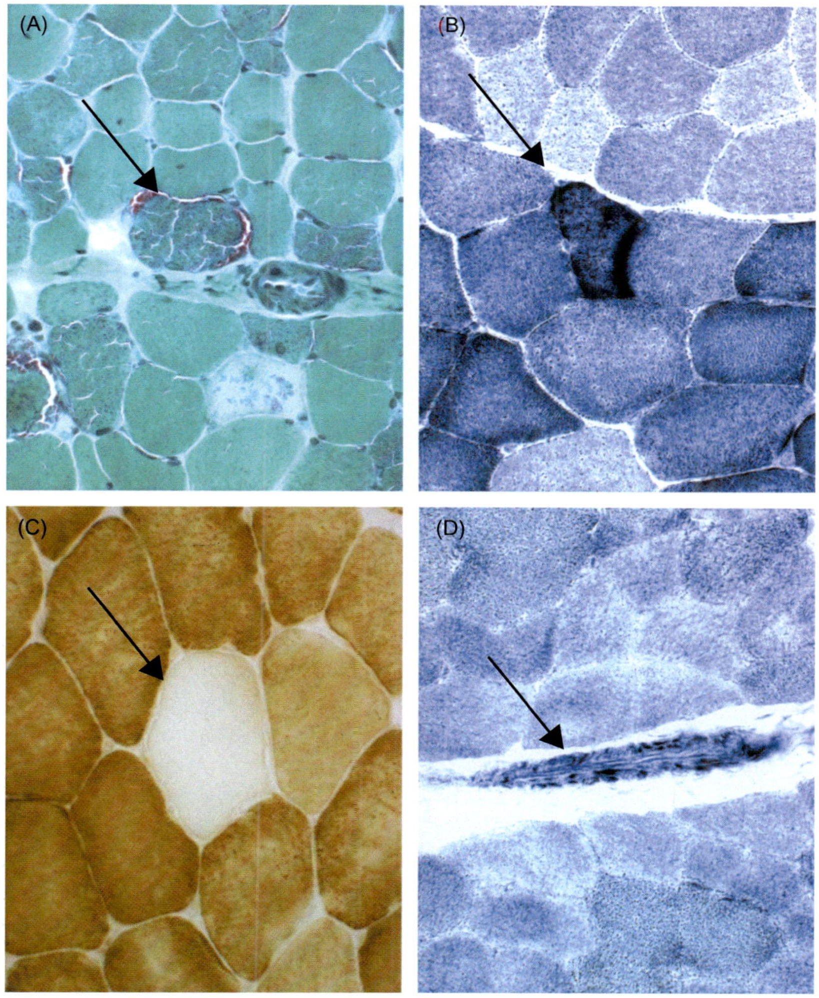

FIGURE 1.2 (**A**) Muscle biopsy specimen from a child with myopathy and the A3243G mtDNA mutation commonly associated with the MELAS phenotype shows a ragged red myofiber (arrow). The reddish granular material in the subsarcolemmal zone reflects proliferation of mitochondria (modified Gomori trichrome). (**B**) A similar ragged red fiber exhibits intense histochemical staining of succinate dehydrogenase (complex II of the electron transport chain) (arrow). Complex II is entirely encoded by nuclear DNA. (**C**) The histochemical reaction for cytochrome-c oxidase (COX or complex IV) of another fiber is unstained (arrow). Three subunits of complex IV are encoded by mitochondrial DNA and are adversely affected by the point mutation (3243) in MELAS to produce this COX-deficient fiber. (**D**) The smooth muscle cells of a small blood vessel (arrow) show prominent punctate staining of succinate dehydrogenase. This finding indicates that abnormal blood vessels are part of the pathology of MELAS syndrome. *(Histopathology courtesy of Dr. Arthur P. Hays.)*

from several subspecialties, including general pediatrics, neurology, psychiatry, orthopedics, physical medicine, rehabilitation, cardiology, pulmonary medicine, and genetics. Social services, physical therapy, occupational therapy, and speech therapy are important interventions, assisting in the management of daily living activities. A multidisciplinary clinic is an ideal treatment setting for these patients. Unfortunately, some of these rehabilitative services are economically challenging in today's health care climate.

Gastrointestinal disturbances may be life-threatening in mitochondrial diseases. Oromotor dysfunction, constipation, diarrhea, malabsorption, and intestinal pseudo-obstruction are well-recognized complications. Less specific complications such as gastroesophageal reflux, with resulting erosive esophagitis, and functional constipation are common to many neuromuscular disorders, particularly when symptoms occur in infancy.

PRESENTING COMPLAINTS

Most children with neuromuscular disorders present with hypotonia, weakness, fatigue, pain, or an elevated serum CK value. Fatigue and pain are symptoms, and the others are signs. The age of the patient influences the presentation. Infants and young children usually present with signs, whereas older children and adolescents may have symptoms that dominate the clinical picture. If symptoms are disproportionate or exist in the absence of signs, psychogenic issues must be considered, particularly depression.

The floppy infant is hypotonic and also may be weak. The skill of the examiner is often tested in this setting. Sorting out weakness from hypotonia can be challenging, and occasionally hypotonia and weakness exist without any primary pathology of the peripheral motor unit. Many genetic syndromes are dominated by congenital hypotonia, so-called cerebral or central hypotonia. Examples include Prader-Willi syndrome, Down syndrome, Smith-Lemli-Opitz syndrome, Zellweger syndrome, and Coffin-Siris syndrome, to name just a few. These conditions need to be considered as alternatives to primary neuromuscular disorders such as spinal muscular atrophy, congenital myasthenic syndromes, and congenital muscular dystrophies. Increased muscle stretch reflexes immediately direct attention to the more common central nervous system mechanisms for hypotonia. However, the relationship between tendon reflex activity and limb tone is generally not fixed. Dysmorphic features may help distinguish a newborn with Prader-Willi syndrome from one with Werdnig-Hoffmann syndrome and allow immediate molecular confirmation. Electrophysiologic and morphologic studies of muscle are no longer necessary in most cases.

Determining whether hypotonia is present can be challenging. Several signs aid the clinician, such as the classic scarf sign, in which the hand is drawn across the chest to the opposite ear. Other maneuvers and measures can be used, but tone is qualitative and subjective. Tone itself is nebulous; it is the subliminal muscle contraction that opposes gravity and permits a person to maintain posture. Thus, careful observation of an infant's posture provides information about resting tone. Obviously, weakness and fatigue contribute to hypotonia, and time of day and relationship to sleep also influence the degree of tone. We have all experienced relative hypotonia at the conclusion of a long and fatiguing day.

Tone is developmentally determined. A 28-week gestation premature infant is normally hypotonic, with minimal resistance to passive manipulation in all limbs. Flexor tone emerges during the remaining period of gestation, and at birth, a full-term infant has strong flexor tone that is evident on passive manipulation of the limbs.

Infantile postures and spontaneous limb movements are best observed before intruding on the patient. A full-term infant demonstrates a flexed limb posture at rest and, with advancing postnatal age, shows more spontaneous movements of the limbs and trunk. By 6 months, a normal infant should be strong enough to sit and maintain an appropriate posture. By 12 months, most healthy infants are crawling, pulling to stand, and taking early steps. These "motor milestones" vary from infant to infant, but weakness is easier to detect with advancing age.

One cannot ascertain with certainty the presence of sensory deficits at this young age. As a result, a sensorimotor neuropathy may be difficult to distinguish from a neuronopathy because the sensory loss may be difficult to define. The behavioral response is most valuable. A noxious stimulus elicits a prompt withdrawal of the limbs. If this reflex response is not accompanied by a grimace or cry, one should suspect a sensory disturbance. Deep sensory disturbances affecting proprioception may disturb the early motor milestones. These infants progress normally to the crawling and cruising stages (9–12 months) but then fail to walk independently and continually seek external support to maintain an erect posture. A young girl with a congenital sensory neuropathy is shown in Figure 1.3 as she constantly places her hand on the wall to achieve better balance.

Eliciting tendon reflexes in such young patients also requires experience. Tapping on one's own fingers held over the appropriate tendon is useful and limits discomfort. Ankle jerks are particularly useful to elicit, as these responses are often absent in infantile-onset neuromuscular disorders. However, tendon reflexes are difficult to elicit in other disorders as well, such as Prader-Willi syndrome—a relatively common cause of neonatal cerebral hypotonia.

Signs and symptoms of neuromuscular disease are more obvious in older infants and children, and symptoms can be elicited more readily after age 2 years, when most children are beginning to speak. Again, the examiner should take advantage of observation before intruding on the child. Observing the child's behavior and motor activities while taking a history from the parents often provides significant information that leads to a clinical diagnosis. An appreciation of the expected motor milestones during late infancy and early childhood is of paramount importance. For example, one expects most children to be walking around 1 year of age. By age 18 months, children are walking independently, and some are starting to run and climb stairs without assistance. By age 2 years, the child is able to run quite well, kick a ball, and travel up and down stairs without hesitation. Standing on one leg and attempting to jump off a step is often accomplished by age 3 years, and hopping on one foot is attempted by age 4 years. By age 5 years, the child is able to hop well on either leg.

Many important observations regarding movement in the supine, sitting, and standing positions can be made while the child is fully clothed. The stance and gait can be observed, and one can determine whether the child is rising up onto his or her toes or walking on flat feet. Engaging a young child in play with a ball or other object of interest can allow additional observations, such as the child arising from the floor, reaching over the head, or

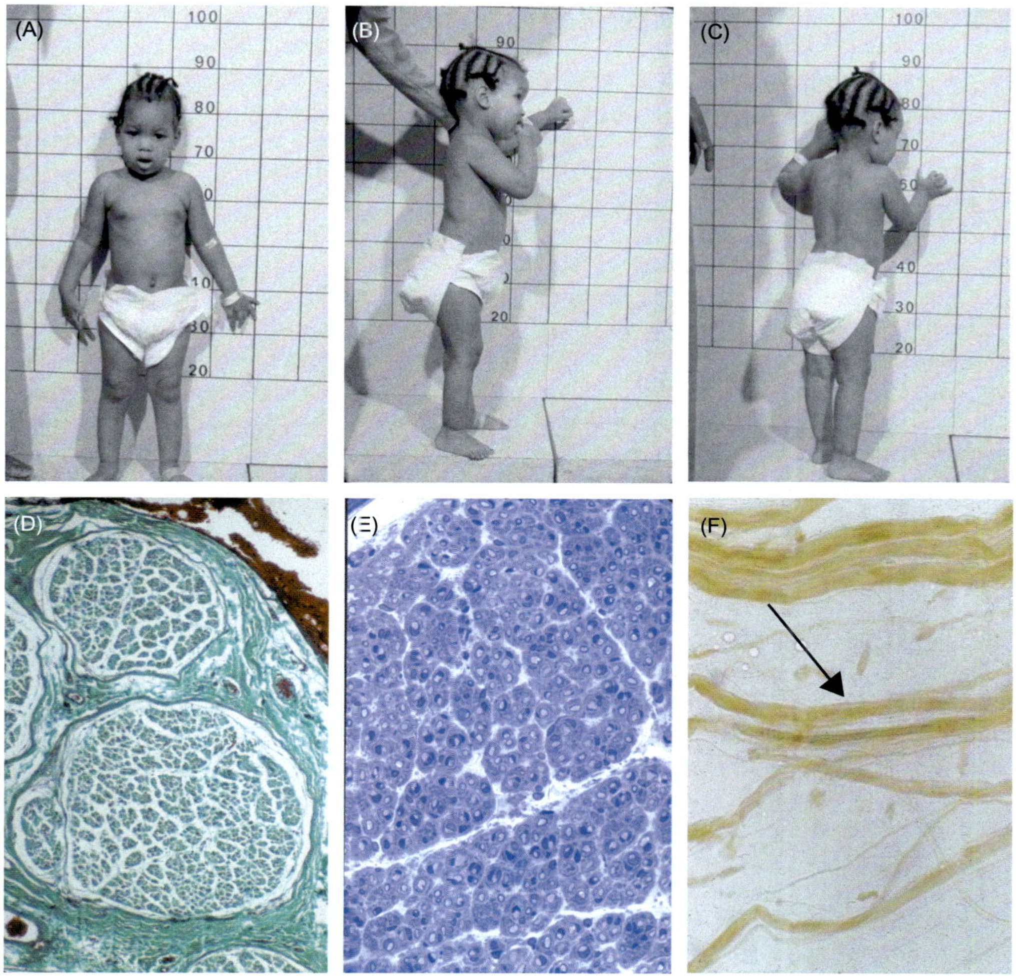

FIGURE 1.3 Congenital hypomyelination neuropathy and sensory ataxia. **(A–C)** Clinical features in this child include poor balance and pes planus and valgus deformities of the feet. She has areflexia and slowed nerve conduction velocities. **(D)** A transverse section of the nerve biopsy shows no discernible myelinated fibers by routine histology (trichrome). **(E)** A transverse thin section (1 μm thick) of epoxy resin-embedded tissue has greater resolution than the paraffin section (6 μm thick) and demonstrates a barely visible thin, dark myelin sheath around each large, pale axon (toluidine blue). **(F)** A teased myelinated nerve fiber (arrow) shows a very thin myelin sheath as a double-contoured structure resembling a railroad track. Myelin sheaths of other fibers are too thin to identify clearly (osmium tetroxide). *(Panels D, E, and F, courtesy of Dr. Arthur P. Hays.)*

pulling an object from the examiner's hand. One can also note eye movements, the position of the upper eyelids, and the facial expression under these conditions, particularly if the child can be encouraged to smile or laugh, or if the child becomes distressed and demonstrates facial grimacing. In fact, most of the important observations regarding the neuromuscular system can be made under these circumstances, and little may be added by the formal examination.

Muscle testing can be accomplished in increasing detail with advancing age, although the functional measures of strength are often the most informative at any age. Determining the child's strength is central to the neuromuscular evaluation. We never analyze all of the 434 muscles in the human body. Rather, we select certain muscle groups to evaluate routinely, knowing that most diseases of the neuromuscular system are relatively symmetrical and involve limb and axial muscle groups to a greater or lesser degree. Clearly, there are exceptions to this statement. Some diseases are distinguished by the fact that they are quite asymmetrical, such as facioscapulohumeral muscular dystrophy, in which one might find prominent involvement of one side of the body or the congenital absence of a pectoral muscle. Nevertheless, a quick survey of major muscles is often informative and sufficient. Testing of the neck flexors is particularly useful, because these muscle groups are preferentially affected in many myopathies. Weakness of muscle groups in the shoulder and pelvic girdles is a useful finding, as is weakness in the biceps, triceps, iliopsoas, quadriceps, hamstrings, and distal muscles of the hands and feet. Strength in these muscles can be surveyed rather quickly. More formal assessment of all accessible muscle groups is done using the grading system originally developed by the Medical Research Council in 1943.[52] This system has withstood the test of time and is still valuable in recording degree of weakness at presentation and over time.

Physicians have a tendency to be unnecessarily precise regarding clinical observations. The Medical Research Council system has five grades: 0 for no movement of the muscle, 1 for a flicker or trace of movement, 2 for active movement with gravity eliminated, 3 for active movement against gravity, 4 for active movement against gravity and some applied resistance, and 5 for normal power. Strictly speaking, only the 0 grade is unequivocal. Even a grade of 5 can be debated, because each examiner has his or her own idea of normal power. However, to further subdivide these categories by adding a plus or minus sign accomplishes little.

Disrobing the child after initial observations have been made frequently provides important clues. Children are innately modest, and a compassionate clinician takes the time to reassure the patient while performing a careful physical examination. Appreciating the presence of dysmorphic features may be essentially diagnostic, for the reasons mentioned earlier. Patterns of weakness and wasting, and the presence of fasciculations or other spontaneous movements of muscle are important. Eye movements, eyelid posture, facial expression, wasting of the temporalis muscle, failure to close the eyes completely, inability to purse the lips or whistle during expiration, nasality of voice, wasting of the sternocleidomastoid and trapezius muscles, wasting of the tongue, presence of a deep crease running from the axilla obliquely toward the neck, a step-like appearance where the base of the neck and the clavicles meet, winging of the scapulae at rest or when the patient attempts to raise the arms in front of the body, atrophy of the intrinsic muscles of the hand or a semiflexed posture of the weakened fingers with some extension at the metacarpal phalangeal joints, exaggerated lumbar lordosis or curvature of the spine, protuberance of the abdomen, wasting of the quadriceps or the anterior compartment muscles of the legs, tapering of the legs distally, tightness of the heel cords, abnormalities of the foot such as pes cavus and pes planus deformities, and presence of foot-drop are informative and often quickly bring one or more diagnostic possibilities to mind.

The physical examination should end with a search for other diagnostic clues. Retinopathy, deafness, cardiac dysfunction, respiratory insufficiency with paradoxical breathing pattern, visceral enlargement, or cutaneous abnormalities should be noted. A careful evaluation of the sensory system is important, searching for evidence of superficial or deep sensory loss. Limb tone and tendon reflex activity should be assessed, but again, one need not determine a precise grade. The patient has hypotonia, hypertonia, or normal tone; the tendon reflexes are absent, diminished, normal, or hyperactive, with or without clonus. The presence of Babinski's signs clearly indicates an upper motor neuron disease. A small number of patients with DMD have initial extension of the great toe after stimulation of the plantar surface of the foot. Whether this represents evidence of upper motor neuron disease, or differential weakness within the foot that limits the response of the great toe to one of extension, can be debated. Assessing the response to plantar stimulation in an infant can be challenging and is not critical in the overall assessment. Applying the stimulus laterally on the foot (Chaddock's reflex) avoids some of the other competing reflexes seen in this area during infancy.

The clinician needs to consider all the diagnostic clues provided by the medical history and clinical examination and then decide whether additional testing is necessary. Often blood studies, including a serum CK measurement, are sufficient. Occasionally, electrophysiologic studies of the motor unit are indicated, particularly if the clinical evaluation points toward a neuronopathy, neuropathy, or neuromuscular transmission disorder. These studies can be performed quickly and relatively noninvasively by nerve conduction studies and EMG. DNA testing may quickly confirm the clinical impression of most muscular dystrophies, including the dystrophinopathies; channelopathies, including many of the periodic paralyses and myotonias; and spinal muscular atrophies. Other testing may be valuable in selected instances. For example, brain magnetic resonance imaging scans may be informative in evaluating congenital muscular dystrophies such as Fukuyama muscular dystrophy, merosin-deficient congenital muscular dystrophy, muscle-eye-brain syndrome, and Walker-Warburg syndrome. Magnetic resonance imaging and magnetic resonance spectroscopy may be informative in mitochondrial diseases, in which selective involvement of the basal ganglia is classic, and signal elevations of brain and ventricular lactate may be seen.

Finally, the modern-day clinician may be overwhelmed by the explosion of new information and can be assisted by several valuable websites: Online Mendelian Inheritance of Man (http://www.ncbi.nlm.nih.gov/Omim/searchomim.html), National Library of Medicine: PubMed (http://www.ncbi.nlm.nih.gov/PubMed/), Gene Clinics (http://www:geneclinics.org), Emery-Dreifuss Muscular Dystrophy Mutation Database (http://www.path.cam.ac.uk/emd/), Leiden Muscular Dystrophy (http://www.dmd.nl/), the Neuromuscular Disease Center at Washington University School of Medicine, St. Louis (http://www.neuro.wustl.edu/neuromuscular), and Muscular Dystrophy Association, USA (http://www.mdausa.org).

CONCLUSION

The field of pediatric neuromuscular disorders has continued to expand scientifically since the era of molecular neurogenetics began in the mid-1980s. The rapid changes in the field may be overwhelming to busy, practicing clinicians. Older children and their families are increasingly aware of these extraordinary advances through their own access to the Internet, and they challenge us to remain informed and updated. They wait impatiently for us to translate these scientific achievements into clinical research that will lead to more meaningful treatments and ultimately to cures. The chapters that follow represent an effort to capture this dynamic process at one point in time. The frustration of the editors and the authors is similar to that of clinicians and their patients. On the one hand, much is happening, and the knowledge base is expanding at a breathtaking pace. On the other hand, our daily management of patients is closer to that of our professional predecessors who initially described many of the disorders discussed in this text. Advances in molecular genetics have been breathtaking, and these advances will likely transform the clinical approach from reactive (to presenting symptoms) to proactive (to genotypic lesions) in the near future. Patients will be identified before the onset of clinical complaints, and treatment will be anticipatory and preventive, emulating the successes of newborn screening since the 1960s.

We can, currently, cite several treatment successes, such as intravenous immunoglobulin in autoimmune myasthenia gravis, CIDP, Guillain-Barré syndrome, and inflammatory myopathies; advances in intensive care; and the triumph of immunizations. Infantile poliomyelitis is no longer a significant concern in the industrialized world, but it continues to challenge us in other parts of the world, as do other infectious neuromuscular disorders such as Hansen's disease, tetanus, and rabies. It is our hope that these treatable, preventable illnesses will eventually become a worldwide footnote in the history of pediatric neuromuscular disorders. For the present, however, they are still an important challenge for our generation. We can cite fewer therapeutic successes in genetically determined diseases of the neuromuscular system, but many promising attempts are underway. Molecular therapies currently are being explored in clinical trials for spinal muscular atrophy,[53] DMD[54] and other neuromuscular diseases. Undoubtedly, future editions of this text will need to devote more pages to the molecular classification of neuromuscular diseases and to a description of their specific treatments and cures.

REFERENCES

1. Meryon E. On granular and fatty degeneration of the voluntary muscles. *Med Chir Trans* 1852;**35**(73–84):1.
2. Duchenne GBA. *De l'electrisation localisee et son application a la pathologie et a la therapeutique*. 2nd ed. Paris: Bailliere; 1861.
3. Werdnig G. Zwei frühinfantile hereditäre Fälle von progressiver Muskelatrophie unter dem Bilde der Dystrophie aber auf neurotischer Grundlage. *Arch Psych Nervenkrankh* 1891;**22**:437–80.
4. Hoffmann J. Überchronische spinale Muskelatrophie im Kindesalter, auf familiärer Basis. *Dtsch Zeitschr Nervenheilk* 1893 **3**:427–70.
5. Batten FE. Three cases of myopathy, infantile type. *Brain* 1903;**27**:147–8.
6. Botstein D, White RL, Skolnick M, Davis RW. Construction of a genetic linkage map in man using restriction fragment length polymorphisms. *Am J Hum Genet* 1980;**32**:314–31.
7. Murray JM, Davies KE, Harper PS, Meredith L, Mueller CR, Williamson R. Linkage relationship of a cloned DNA sequence on the short arm of the X chromosome to Duchenne muscular dystrophy. *Nature* 1982;**300**:69–71.
8. Davies KE, Pearson PL, Harper PS, Murray JM, O'Brien T, Sarfarazi M, et al. Linkage analysis of two cloned DNA sequences flanking the Duchenne muscular dystrophy locus on the short arm of the human X chromosome. *Nucleic Acids Res* 1983 **11**:2303–12.
9. Francke U, Ochs HD, de Martinville B, Giacalone J, Lindgren V, Disteche C, et al. Minor Xp21 chromosome deletion in a male associated with expression of Duchenne muscular dystrophy, chronic granulomatous disease, retinitis pigmentosa, and McLeod syndrome. *Am J Hum Genet* 1985;**37**:250–67.
10. Monaco AP, Bertelson CJ, Middlesworth W, Colletti CA, Aldridge J, Fischbeck KH, et al. Detection of deletions spanning the Duchenne muscular dystrophy locus using a tightly linked DNA segment. *Nature* 1985;**316**:842–5.
11. Ray PN, Belfall B, Duff C, Colletti CA, Aldridge J, Fischbeck KH, et al. Cloning of the breakpoint of an X;21 translocation associated with Duchenne muscular dystrophy. *Nature* 1985;**318**:672–5.
12. Duchenne GBA. Recherches sur la paralysie musculaire pseudohypertrophique ou paralysie myo-sclerosique. *Arch Gen Med* 1868;**11** 5–25, 179–209, 305–21, 429–43, 552–88.
13. Carpenter S, Karpati G. *Pathology of Skeletal Muscle*. 2nd ed. New York: Oxford University Press; 2001.
14. Cooper ST, Lo HP, North KN. Single section Western blot: improving the molecular diagnosis of the muscular dystrophies. *Neurology* 2003;**61**:93–7.
15. Lambert FH, Eaton LM, Rooke ED. Defect of neuromuscular conduction associated with malignant neoplasm. *Am J Physiol* 1956;**187**:612–13.
16. Thomas JE, Lambert EH. Ulnar nerve conduction velocity and H-reflex in infants and children. *J Appl Physiol* 1960;**15**:1–9.
17. Gilliatt RW, Sears TA. Sensory nerve action potentials in patients with peripheral nerve lesions. *J Neurol Neurosurg Psychiatry* 1958;**21**:109–18.
18. Sacco G, Buchthal F, Rosenfalck P. Motor unit potentials at different ages. *Arch Neurol* 1962;**6**:366–73.
19. Wagner AL, Buchthal F. Motor and sensory conduction in infancy and childhood: reappraisal. *Dev Med Child Neurol* 1972;**14**: 189–216.
20. Buchthal F, Olsen PZ. Electromyography and muscle biopsy in infantile spinal muscular atrophy. *Brain* 1970;**93**:15–30.
21. Hausmanowa-Petrusewicz I, Karwanska A. Electromyographic findings in different forms of infantile and juvenile proximal spinal muscular atrophy. *Muscle Nerve* 1986;**9**:37–46.
22. Byers RK, Banker BQ. Infantile muscular atrophy. *Arch Neurol* 1961;**5**:140–64.
23. Dubowitz V. Infantile muscular atrophy—a broad spectrum. *Clin Proc Child Hosp Dist Columbia* 1967;**23**:223–39.

24. Dyck PJ, Lambert EH. Lower motor and primary sensory neuron diseases with peroneal muscular atrophy. I. Neurologic, genetic, and electrophysiologic findings in hereditary polyneuropathies. *Arch Neurol* 1968;**18**:603–18.
25. Gutrecht JA, Dyck PJ. Quantitative teased-fiber and histologic studies of human sural nerve during postnatal development. *J Comp Neurol* 1970;**138**:117–29.
26. Sladky JT, Brown MJ, Berman PH. Chronic inflammatory demyelinating polyneuropathy of infancy: a corticosteroid-responsive disorder. *Ann Neurol* 1986;**20**:76–81.
27. DeVivo DC, Engel WK. Remarkable recovery of a steroid-responsive recurrent polyneuropathy. *J Neurol Neurosurg Psychiatry* 1970; **33**:62–9.
28. Engel AG, Lambert EH. Structure-function correlations in myasthenia gravis and a new myasthenic syndrome. *Electroencephalogr Clin Neurophysiol Suppl* 1978;469–77.
29. Darras BT, Jones HR. Diagnosis of pediatric neuromuscular disorders in the era of DNA analysis. *Pediatr Neurol* 2000;**23**:289–300.
30. Brooke MH. *A Clinician's View of Neuromuscular Diseases.* Baltimore: Williams & Wilkins; 1976.
31. Guthrie R, Susi A. A simple phenylalanine method for detecting phenylketonuria in large populations of newborn infants. *Pediatrics* 1963;**32**:338–43.
32. Okada S, O'Brien JS. Tay-Sachs disease: generalized absence of a beta-D-N-acetylhexosaminidase component. *Science* 1969;**165**: 698–700.
33. Kaback MM. Population-based genetic screening for reproductive counseling: the Tay-Sachs disease model. *Eur J Pediatr* 2000;**159** (Suppl. 3):S192–5.
34. Boczonadi V, Smith PM, Pyle A, Gomez-Duran A, Schara U, Tulinius M, et al. Altered 2-thiouridylation impairs mitochondrial translation in reversible infantile respiratory chain deficiency. *Hum Mol Genet* 2013;**22**:4602–15.
35. Bonnemann CG, McNally EM, Kunkel LM. Beyond dystrophin: current progress in the muscular dystrophies. *Curr Opin Pediatr* 1996;**8**:569–82.
36. Moreira ES, Wiltshire TJ, Faulkner G, Nilforoushan A, Vainzof M, Suzuki OT, et al. Limb-girdle muscular dystrophy type 2G is caused by mutations in the gene encoding the sarcomeric protein telethonin. *Nat Genet* 2000;**24**:163–6.
37. Minetti C, Sotgia F, Bruno C, Scartezzini P, Broda P, Bado M, et al. Mutations in the caveolin-3 gene cause autosomal dominant limb-girdle muscular dystrophy. *Nat Genet* 1998;**18**:365–8.
38. McNally EM, de Sa Moreira E, Duggan DJ, Bonnemann CG, Lisanti MP, Lidov HG, et al. Caveolin-3 in muscular dystrophy. *Hum Mol Genet* 1998;**7**:871–7.
39. Gospe Jr. SM, Lazaro RP, Lava NS, Grootscholten PM, Scott MO, Fischbeck KH. Familial X-linked myalgia and cramps: a nonprogressive myopathy associated with a deletion in the dystrophin gene. *Neurology* 1989;**39**:1277–80.
40. Kugelberg E, Welander L. Heredofamilial juvenile muscular atrophy simulating muscular dystrophy. *AMA Arch Neurol Psychiatry* 1956;**75**:500–9.
41. Ben Hamida M, Hentati F, Ben Hamida C. Hereditary motor system diseases (chronic juvenile amyotrophic lateral sclerosis). conditions combining a bilateral pyramidal syndrome with limb and bulbar amyotrophy. *Brain* 1990;**113**(Pt 2):347–63.
42. Goutieres F, Mikol J, Aicardi J. Neuronal intranuclear inclusion disease in a child: diagnosis by rectal biopsy. *Ann Neurol* 1990;**27**:103–6.
43. Bradshaw DY, Jones HR. Pseudomeningoencephalitic presentation of pediatric Guillain-Barré syndrome. *J Child Neurol* 2001;**16**: 505–8.
44. Montes J, Blumenschine M, Dunaway S, Alter AS, Engelstad K, Rao AK, et al. Weakness and fatigue in diverse neuromuscular diseases. *J Child Neurol* 2013;**28**:1277–83.
45. Kariya S, Obis T, Garone C, Akay T, Sera F, Iwata S, et al. Requirement of enhanced Survival Motoneuron protein imposed during neuromuscular junction maturation. *J Clin Invest* 2014; **124**:785–800.
46. Oskoui M, Jacobson L, Chung WK, Haddad J, Vincent A, Kaufmann P, et al. Fetal acetylcholine receptor inactivation syndrome and maternal myasthenia gravis. *Neurology* 2008;**71**: 2010–2.
47. Tawil R, Ptacek LJ, Pavlakis SG, DeVivo DC, Penn AS, Ozdemir C, et al. Andersen's syndrome: potassium-sensitive periodic paralysis, ventricular ectopy, and dysmorphic features. *Ann Neurol* 1994;**35**:326–30.
48. Schwartz O, Jampel RS. Congenital blepharophimosis associated with a unique generalized myopathy. *Arch Ophthalmol* 1962;**68**:52–7.
49. Gowers WR. *Pseudo-hypertrophic muscular Paralysis—A clinical lecture.* London: J & A Churchill; 1879.
50. Gowers WR. Clinical lecture on pseudo-hypertrophic muscular paralysis. *Lancet* 1879;**2** 1–2, 37–9, 73–5, 113–16.
51. Bamford NS, Trojaborg W, Sherbany AA, De Vivo DC. Congenital Guillain-Barré syndrome associated with maternal inflammatory bowel disease is responsive to intravenous immunoglobulin. *Eur J Paediatr Neurol* 2002;**6**:115–19.
52. Medical Research Council. *Aids to the Investigation of Peripheral Nerve Injuries.* London: Her Majesty's Stationery Office; 1943.
53. Singh P, Liew WK, Darras BT. Current advances in drug development in spinal muscular atrophy. *Curr Opin Pediatr* 2013; **25**:682–8.
54. Mercuri E, Muntoni F. Muscular dystrophy: new challenges and review of the current clinical trials. *Curr Opin Pediatr* 2013;**25**:701–7.

Chapter 2

Genetics of Neuromuscular Disorders

Wendy Chung

INTRODUCTION

Many neuromuscular disorders in children are single gene disorders, and therefore genetics plays a primary role in the diagnosis of many neuromuscular conditions. Establishing a definitive molecular genetic diagnosis provides clarity about prognosis; suggests possible therapeutic strategies; allows patients, families, and providers to access appropriate clinical trials, research studies, centers of excellence, and family support organizations; and provides parents and other family members with critical information to accurately determine risk of recurrence and reproductive options to increase the probability of having a healthy child in the future. Increasingly, the exact type of mutation may also suggest possible therapeutic strategies, and it will become increasingly important to identify the exact mutation(s) in clinically affected individuals even when the clinical diagnosis is not in question. Molecular genetic diagnostics have been revolutionized by the decreased cost of sequencing and elimination of patents on genes, and improvements in genetic diagnostics should allow for increasingly efficient, expedient, and cost-effective methods to establish a diagnosis so patients can move beyond diagnosis and on to treatment.

This chapter will briefly review the basic principles of genetics and the principles of molecular genetics, and then focus largely on the clinical application of genetics to provide new approaches to diagnosis, counseling, and treatment. Given the uneven access to clinical genetics services outside major academic medical centers, the burden of the increased demand of genetic evaluation will fall largely on the neurologist who diagnoses and manages the neuromuscular patient. Thus, neurologists need to be conversant with the technologies and principles of genetic evaluation, testing, and counseling.

PATTERNS OF GENETIC TRANSMISSION

The human genome consists of approximately 3 billion base pairs of DNA packaged into 23 pairs of chromosomes. The 23 pairs of nuclear chromosomes consist of 22 non-sex chromosomes, called autosomes, and two sex chromosomes, the X and the Y. In addition, there are approximately 16,500 base pairs of DNA located on circular double-stranded molecules within mitochondria, referred to as the mitochondrial genome. Recessively inherited diseases require mutations in both copies of the gene whereas dominantly inherited conditions require mutations in just one of the two copies of the gene. X-linked conditions such as Duchenne muscular dystrophy more commonly manifest in males since males have only a single X chromosome and lack the second copy of most of the genes on the X chromosome as a backup if there is a mutation in an X-linked gene. Mitochondria are exclusively maternally inherited, so mutations of the mitochondrial genome are maternally inherited.

PEDIGREE ANALYSIS AND RISK ASSESSMENT

Taking a three-generation family history should be a standard part of any initial clinical evaluation. This is most efficiently performed by having patients or their parents complete a brief tabular questionnaire about the family history at home or in the waiting room so the provider can review the pertinent positives while taking the medical history. Some practices have a patient portal to facilitate collection of family history data. A sample family history screening tool is shown in Figure 2.1, and an online tool is also available.[1] It is important to remember to ask about deceased relatives, since some people forget to mention them when a family history is taken.

B.T. Darras, H. Royden Jones, Jr., M.M. Ryan & D.C. De Vivo (Eds): Neuromuscular Disorders of Infancy, Childhood and Adolescence, Second edition.
DOI: http://dx.doi.org/10.1016/B978-0-12-417044-5.00002-0

	Child's Family			Mom's Family					Dad's Family				
	Mom	Dad	Brothers/ sisters	Mom	Dad	Brothers	Sisters	Nieces/ Nephews	Mom	Dad	Brothers	Sisters	Nieces/ Nephews
Birth defects (heart, spine, missing fingers, clubfeet etc)													
Born deaf or blind													
Carrier of a genetic condition													
Clotting/ bleeding/ blood disorder													
Autism, learning disability													
Early death/ sudden death <50													
Heart attack <50 years males <65 years Females													
Known genetic condition													
Special education services													
Seizures													

FIGURE 2.1 Family history screening tool patients can complete prior to the clinical encounter.

Individuals who have had multiple partners should be noted, and full siblings should be distinguished from half siblings as this may provide important clues to a dominantly inherited disorder such as myotonic dystrophy. Consanguinity can be screened by asking if the child's parents are related by blood, which is common in some cultures. The ethnic background of both sides of the family should be noted since hereditary diseases vary in prevalence by ethnicity.

Assessment of recurrence risk is extremely important for many families who are considering having more children and is often difficult in small families. Genetic testing can be helpful to confirm the diagnosis, establish a precise risk of recurrence, and provide the family with options for future family planning. For an autosomal recessive disorder, parents are usually both asymptomatic carriers for the condition, and neither parental side may report a prior family history of the condition. If the affected child has no siblings, or is a member of a small family with no affected siblings, parents will often dismiss the possibility of a genetic etiology because they have never observed this condition within their family. However, parents who are asymptomatic carriers of an autosomal recessive condition have a 25% risk of recurrence with each pregnancy.

Increasingly, couples considering having children are offered carrier screening for a variety of autosomal recessive conditions, and some obstetricians include carrier screening for spinal muscular atrophy (SMA) in that panel. However, it is important to remember that screening tests are usually not comprehensive and are not 100% sensitive; therefore, even if parents have had normal carrier screening for a condition within the child's differential, the child could still have the disease due to a mutation not included within the carrier screening panel or due to technical limitations of the carrier screening test.

Autosomal recessive conditions are more frequent when there is a family history of consanguinity, but consanguinity does not guarantee that the condition will be recessive. When the child has two copies of the same mutation, the child is homozygous for the mutation. When the child has two different mutations, the child is a compound heterozygote. Occasionally, a child will be found to carry more than two genetic variants in a gene, and the inheritance of the variants can be most readily determined by testing the parents to determine which variants are in cis (contained on the same copy of the gene and inherited together from one parent) and which are in trans (contained on separate copies of the gene and inherited from each parent separately).

Individuals with an autosomal dominant disorder have a 50% chance of passing the mutation on to any offspring, as is the case for conditions such as myotonic dystrophy or Charcot-Marie-Tooth type IA. In children, a significant percentage of severe and lethal diseases in childhood that are dominantly inherited are due to *de novo* or new mutations in the child. Although genetic testing from blood in the parents is normal, there is a small chance that one of the parents is a gonadal mosaic and has multiple germ cells carrying the mutation even though the blood cells do not carry the mutation. A 1% risk of recurrence is given to such parents who have a child with a *de novo* mutation. It is important to note that if a child carries a *de novo* mutation for a dominant condition, the mutation becomes a germline mutation in the child and the risk of recurrence for that child's children will be 50%.

For X-linked disorders, the recurrence risk depends solely upon the mother's genetic status and does not rely upon the father's genetic status. The lack of male-to-male transmission of the genetic disease is the hallmark of X linkage. If the mother is a carrier for the genetic mutation, there is a 50% risk that her sons (regardless of her partner) will inherit the condition and will be symptomatic while 50% of her daughters will be carriers and unlikely to have classical symptoms for most disorders due to the random process of lyonization or X inactivation in the relevant tissue. However, milder manifestations in female carriers are more common. As an example, some women who are carriers for Duchenne muscular dystrophy will have a mild elevation of CPK and a small percentage will develop signs and symptoms of cardiomyopathy, although most will not develop clinically significant muscular dystrophy. For some X-linked disorders, again using Duchenne muscular dystrophy as an example, there is a high frequency of spontaneous mutations. Women who have sons with spontaneous mutations have a 1% risk of recurrence in future pregnancies due to the small chance of gonadal mosaicism as described previously.

Risk assessment in mitochondrial disorders is more difficult because the genes responsible for mitochondrial function can be encoded in the nuclear or mitochondrial genes, so that inheritance can be either autosomal, rarely X linked, or maternal. The majority of proteins involved in oxidative phosphorylation in the mitochondria are encoded in the nucleus, but there are 13 proteins involved in oxidative phosphorylation, as well as a full set of transfer and ribosomal RNAs encoded in the mitochondrial genome (see Chapter 41). Because the brain and skeletal muscle have high energetic demands, clinical manifestations of mitochondrial disorders are often neuromuscular. Nuclear encoded genes lead to straightforward risk of recurrence, but mitochondrial encoded genes are difficult to predict in the next generation due to heteroplasmy; that is, the mitochondria are, in fact, a population of organelles with only a subset of the organelles carrying the mutation (Figure 2.2). Only the maternal contribution is relevant for disorders encoded within the mitochondria since all mitochondria in the embryo and fetus are derived from the egg. From one egg to the next, the percentage of mitochondria with the mutation varies and is not predictable, and within the developing fetus as cells divide by mitosis the distribution of mutant mitochondria between tissues can vary and give rise to significant heterogeneity in severity and affected organs. Thus, prenatal diagnosis for mutations encoded within the mitochondria is imprecise and can be inaccurate.

For all modes of inheritance, the genetic information for a gene is referred to as the genotype; the manifestations

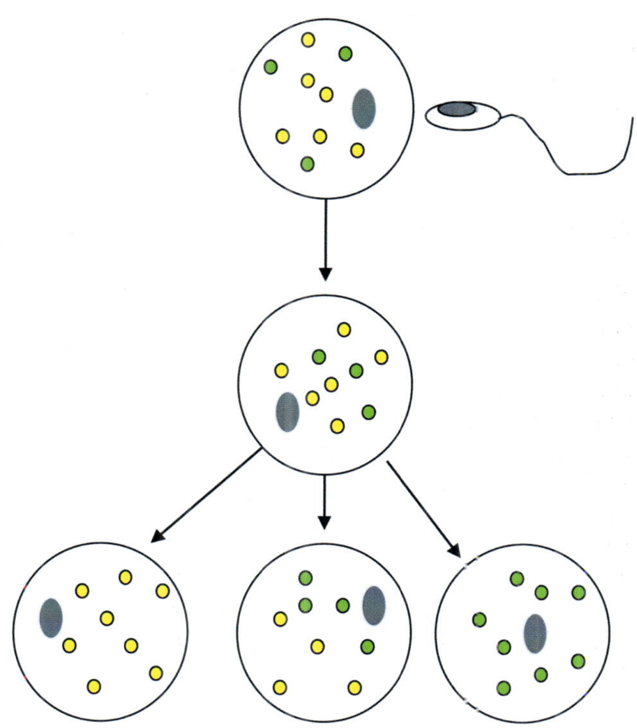

FIGURE 2.2 Heteroplasmy of mitochondrial encoded mutations.

of that gene are referred to as the phenotype, and in medicine this largely refers to the clinically observable signs and symptoms of disease. For most pediatric neuromuscular conditions, the penetrance or the probability of expressing a manifestation of a genetic mutation is usually high and often 100%. However, in many cases, manifestations may not be present at birth, so penetrance can be age dependent. In genetic conditions, the phenotypic manifestations may differ from person to person, a phenomenon referred to as *variable expressivity*.

CHROMOSOMES AND CHROMOSOMAL ABNORMALITIES

Aside from mutations in single genes, genetic abnormalities may result from contiguous gene defects of entire chromosomes or parts of chromosomes. Children can be born with extra or missing chromosomes (called aneuploidy), often associated with neonatal hypotonia. With the exception of the X and Y chromosomes, having a missing chromosome tends to be lethal during embryonic or fetal development. Trisomy is lethal for most autosomes during the prenatal period but is compatible with life for chromosomes 13, 18, and 21, although each of these is associated with moderate to severe intellectual disability. Chromosome rearrangements may occur within a chromosome, such as deletion, duplication, or inversion, or may involve exchange of genetic material between a pair of chromosomes, referred to as translocation. Translocations are balanced when no genetic material is gained or lost overall. Balanced translocations are less likely to result in disease compared to unbalanced translocations, but disruption of a gene at the translocation breakpoint can lead to clinical manifestations that are often quite specific and well defined. Deletions are generally more severe than duplications, and in general the larger the deletion or duplication and the greater the number of genes involved, the more severe are the clinical manifestations. New, extremely sensitive technologies called chromosome microarrays are now available to detect segmental aneuploidies or copy number variants associated with microdeletions or microduplications.

GENETIC IMPRINTING

For a small number of genes, the parent of origin is paramount because only one copy of the gene is expressed. Whether the maternal or paternal copy of the gene is expressed is consistent for each imprinted gene, but some genes along a chromosome segment will be maternally expressed and paternally imprinted (silenced) while other genes in close proximity on the same chromosome will be paternally expressed and maternally imprinted (silenced). An example of imprinting is the Prader-Willi syndrome/Angelman syndrome locus on chromosome 15. Prader-Willi syndrome often presents as neonatal hypotonia and is due to defects in the paternal copy of the SNRPN locus on 15p, including deletions of the paternal locus, two maternal copies of chromosome 15 (uniparental disomy for chromosome 15), or rarely mutations in the imprinting locus. Angelman syndrome is due to defects in the maternal copy of the nearby *UBE3A* gene on chromosome 15p, including deletions of the maternal gene, two paternal copies of chromosome 15, mutations in *UBE3A*, or rarely mutations in the imprinting locus. The imprint is erased and reset at each generation. Imprinting disorders, when inherited, can thus have a peculiar pattern of inheritance because of appearing to skip generations. However, most imprinted disorders in pediatric neuromuscular conditions arise as *de novo* mutations, and there is rarely a family history of the disorder.

MOLECULAR GENETICS

Understanding the basic anatomy of a gene is necessary to appreciate the nature and consequences of genetic mutations and to understand the limitations of genetic tests. A prototypical gene is shown in Figure 2.3. Transcription begins with the binding of RNA polymerase to the promoter, and the full length of the gene is then copied into RNA. There may be enhancers of gene expression located at some distance from the promoter located upstream or downstream of the gene, or embedded in the middle of the gene in an intron. The gene is composed of exons and introns. After transcription is complete, the introns are removed by splicing to produce the messenger RNA (mRNA), which is exported from the nucleus and translated into a protein in the cytoplasm. The sequences at the junction of the intron and exon are critical for correct splicing, and consensus sequences mark the splice donor and splice acceptor sites. Splice acceptors always end in an AT sequence, preceded by a "consensus sequence," and donors begin with GC, followed by a conserved consensus sequence. Mutations in these sequences at the splice junctions can result in aberrant splicing, leading to an abnormal mRNA and abnormal protein product. Mutations can also occur deep within introns or as synonymous changes (which do not change the amino acid encoded by the codon) within the exons leading to alternative splice sites that can also lead to aberrant splicing.

Splice mutations at the splice junction are readily detected by most molecular genetic methods used in genetic testing. However, with the exception of the small number of genetic tests that are preformed directly on the mRNA, mutations leading to aberrant splicing outside the splice junction can easily be missed because they may not be included within the target for sequencing. Mutations in

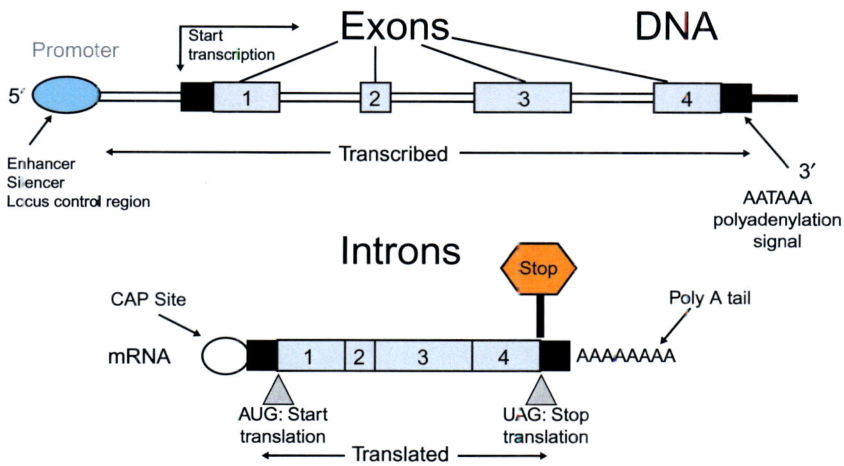

FIGURE 2.3 Structure of a gene, demonstrating exons and introns making up a gene and processing of the RNA transcript to splice out the introns and make the mRNA.

enhancers are also frequently missed by most genetic tests because they have not been well defined and are not included in the sequence analyzed in most genetic tests.

GENE MUTATIONS AND MODES OF INHERITANCE

The mechanism by which a sequence change affects the protein and its function in part determines the mode of inheritance. Most recessive disorders are due to loss of function mutations. In general, the more severe the loss of function, the more severe the clinical manifestations, or the earlier the age of onset for recessive diseases. Certain mutations lead to complete gene disruptions and include whole gene deletions, multiple exon deletions/duplications, most premature termination mutations except those at the end of the protein, and most frameshift mutations except those at the end of the protein. Exon deletions/duplications are more deleterious when they also result in a change in the reading frame of the protein distal to the deletion. Splice mutations disrupt genes to a variable degree depending on how leaky the mutation is, and how much of the normal splice product is made. Promoter and enhancer mutations have variable severity. Missense mutations that result from the substitution of one amino acid for another can have a wide range of effects from severe to minimal change in function of the protein depending on how critical that amino acid is in the protein and how significant the difference is in the properties of the amino acid substitution (e.g. size, charge, ability to make disulfide bonds). Loss of function mutations are inherited in either a dominant or recessive manner depending on how much of the protein product is necessary for normal function. For many enzymes as little as 10–20% of the enzymatic activity may be compatible with a normal phenotype. Disorders encoded by such enzymes are usually inherited as autosomal recessive conditions in which carriers are asymptomatic. In general, the lower the level of enzymatic activity, the earlier the age of onset and the more severe the clinical features. For other genes, the gene product is much more tightly regulated, and loss of 50% of the normal gene product can produce significant problems. Disorders encoded by these tightly regulated genes are usually inherited as dominant conditions and include many developmental disorders in which gene expression is finely tuned to support normal organogenesis.

Missense mutations can rarely result in gain of a novel function to the protein or may result in an increase in activity or expression of the protein. Such mutations are usually inherited in a dominant manner. In addition, some loss of function mutations are within protein subunits that make up multimeric proteins. In such cases, because only one aberrant subunit can significantly impair function of the entire complex, a mutation in one allele may be amplified to affect more than 50% of the complexes and can be inherited in a dominant manner.

Triplet repeat expansions represent a unique type of mutation. Triplet repeats consist of three nucleotides such as CGG, CTG, or CAG, which are repeated multiple times. These repeats can be located in various parts of the gene, including a CGG repeat in the promoter in fragile X syndrome that interferes with the promoter of FMR,[2] a CTG repeat in the 3′ untranslated region in myotonic dystrophy that interferes with RNA processing of many proteins,[3,4] a GAA repeat within an intron in Friedreich's ataxia which interferes with RNA splicing,[5] or a CAG repeat within an exon encoding a polyglutamine tract in Huntington's disease[6] or spinocerebellar ataxias[7] leading to accumulation of mutant protein that is toxic to the cells.[8] All of these triplet repeats are inherited as autosomal dominant disorders except for Friedreich's ataxia. There can be normal variation in size in the number of repeats for each of these triplet repeats, but when the repeat sizes go beyond a specific threshold that is unique for each gene, pathology ensues: the larger the repeat, the earlier the age of onset and the more severe the clinical phenotype. For some repeat disorders, there is anticipation of the disease and younger age of symptom onset with expansion of the repeat from one generation to the next (typically associated with transmission by a specific gender parent that is specific for each gene). As an example, myotonic dystrophy is associated with anticipation with maternal transmission, and a mildly affected mother can have a child with a severe form of congenital myotonic dystrophy.

DEVELOPMENTS IN MOLECULAR DIAGNOSTICS

In the past several years, genetic diagnosis has become an important tool in confirming clinical diagnoses, providing improved prognostic information, predicting risk of disease for unaffected family members, and offering reproductive options for families to avoid transmission of heritable diseases to the next generation. The number of available clinical genetic tests has increased significantly in the last decade, and they are increasingly becoming much more comprehensive, even including genomic analyses in some cases. Previously, the heretofore relatively high cost of dideoxy sequencing had limited the potential impact of medical sequencing to diseases of small genes and diseases caused by one or two genes. The availability of next-generation sequencing methods and the elimination of patents on genes by the Supreme Court have dramatically reduced the cost of sequencing, which should continue to significantly reduce the cost of clinical genetic tests, increase the clinical utility of currently offered tests by expanding genetic coverage of currently tested genes or adding additional genes, and enable sequencing tests for diseases with extreme clinical heterogeneity such as myopathies and neuropathies.

A genetic test must meet several requirements before it is offered to the medical community. The primary requirement is that the test must have clinical utility. This means that the result should be useful to the physician in one or more of the following ways: (1) by establishing or confirming the clinical diagnosis or predicting an increased future risk of disease, (2) providing the physician with more accurate information that allows development of a plan for surveillance or response to therapy that improves clinical outcome, (3) providing prognostic information, and (4) providing the patient and his/her family with information that can be used for family planning purposes (e.g. prenatal diagnosis, preimplantation genetic diagnosis). As molecularly based intervention for specific mutation types becomes available using molecular chaperones for missense mutations or read through of premature terminations, it will be critical to understand the genetic basis for all genetics conditions to facilitate molecular therapeutics.

A genetic test should also have high analytic and clinical sensitivity. Analytic sensitivity is a measure of the proportion of tested samples that correctly give a positive result by the test method and is a measure of how well the test is technically performing in the laboratory. Clinical sensitivity measures the proportion of individuals who demonstrate a positive result using the test method as a fraction of those who have the disease. In addition, the genetic test should have high analytic and clinical specificity. Analytic specificity is a measure of the proportion of tested samples that are truly negative of all negative results by the test method. Clinical specificity measures the proportion of all unaffected individuals identified by the test method as test negative.[9]

The development of a clinical test becomes especially complex in the case of a disorder in which a mutation in any one of several genes may underlie the disease (genetic heterogeneity). In many cases, mutations in any one of several genes can lead to similar clinical manifestations (phenocopies) that are clinically indistinguishable. Genetic testing for these diseases has proven difficult because the number of genes to be tested is often large, and may require complete analysis of as many as 10 or more genes. In addition, because mutations are "private" or unique to each family in many diseases, including those that are genetically heterogeneous, complete sequence analysis is required to ensure high clinical sensitivity. Examples of disorders for which any one of multiple genes with private mutations could underlie the phenotype in a given family include intellectual disabilities, hypotonia, Charcot-Marie-Tooth, congenital muscular dystrophies, or muscle eye brain disease.

Usually, genes that account for a large proportion of a clinical phenotype are the first to be identified and reported in the scientific literature. Heretofore, most laboratories have provided testing of only those most common genetic causes, in an effort to minimize the sequencing burden and to provide a diagnosis for the largest fraction of patients at the lowest cost for the test. However, as new genes are identified, the number of disease-causing genes for which testing is appropriate also increases. As a result, a cascade of tiered (or "reflex") genetic tests has been offered for neuropathies such as Charcot-Marie-Tooth, starting with genes for which mutations are more common and then proceeding to the genes for which mutations account for a smaller proportion of cases. However, with the advent of next-generation sequencing, it is now possible to quickly and cost effectively perform testing for as many as 50–100 genes in a panel for a specific clinical indication. All genes are analyzed simultaneously in parallel rather than sequentially. With the advent of targeted genetic therapy based upon molecular etiology and mutation type, such as antisense oligonucleotide therapy for Duchenne muscular dystrophy[10] to induce skipping of the targeted exon by modulating pre-mRNA splicing, restore the open reading frame of Dystrophin, and produce internally deleted, yet partially functional dystrophin protein isoforms, it has become increasingly important to be able to accurately define the specific gene and mutation causing a patient's disorder. Thus, there is a need to offer extensive testing of as many genes as possible so that therapeutic options, as well as accurate information about recurrence risk and effective reproductive options, are available to individuals with the disease and their family members.

METHODS OF MOLECULAR GENETIC DIAGNOSTICS

Current methods of molecular genetic diagnostics can be roughly divided into the "macro" and "micro" categories. "Macro" genetic diagnostics is mostly represented by whole-genome screening for large mutations with the use of microarrays. "Micro" genetic diagnostics targets specific genes with small mutations down to the level of the nucleotide. Macro whole-genome screening, offered in various formats, usually surveys the entire genome for copy number variations. Bacterial artificial chromosome (BAC) arrays were once the most common clinical assay in this category, but the focus has shifted to oligonucleotide microarrays with the ability to genotype single nucleotide polymorphisms (SNPs). The resolution of the genome-wide microarrays can be quite high; however, this type of analysis only identifies abnormalities of copy number of chromosomal regions, genes, or individual exons (deletions, duplications) and does not address the issue of errors at the level of the primary DNA sequence due to alterations in one or a small number of nucleotides.

Whole-genome screening to survey the genome for copy number variations (deletions and duplications) provides a molecular karyotype and is equivalent to performing a genome-wide fluorescence *in situ* hybridization (FISH) study (Table 2.1). Oligonucleotide microarrays are now commonly available through a number of clinical laboratories and are routinely covered by insurance. The genome-wide microarrays range in probe density from 40,000 to over 2,500,000 oligonucleotide probes to generate a high-resolution molecular karyotype. These arrays have replaced the karyotype as a first line test because of the improved resolution of a chromosome microarray (routinely at least <0.5 Mb and often as high as 100 kb) relative to a karyotype (>5 Mb).

Chromosome microarrays are recommended as the first tier test in patients with unexplained developmental delay/intellectual disability, autism spectrum disorders, or multiple congenital anomalies.[11] Although generally less informative, a karyotype is complementary to a chromosome microarray since only a karyotype can provide structural information about the chromosome (i.e. location of duplicated material, ring chromosomes) and identify balanced rearrangements such as balanced translocations. However, it is infrequent that a karyotype will identify a clinically relevant anomaly that cannot also be identified on a chromosome microarray.

Laboratories are currently utilizing two types of oligonucleotide chromosome microarrays: (1) one that includes single nucleotide polymorphisms (SNPs) or (2) oligonucleotides designed only to quantify copy number. The advantage of the single nucleotide polymorphism oligonucleotide microarray analysis (SOMA) is the ability to distinguish parental genotypes and hence detect uniparental disomy, which is necessary for the diagnosis of Angelman and Prader-Willi syndrome. Availability of the SNP genotypes is also useful to detect stretches of homozygosity (two copies of the same genotype) that may suggest a recessive condition with a homozygous mutation from a common ancient ancestor. Oligonucleotide microarray analysis has approximately a 10–15% yield in children with intellectual disability and a normal karyotype,[12] with

TABLE 2.1 Comparison of Genetic Testing Platform and Limits of Resolution

Test	Limit of Resolution	Types of Mutations Detected
Karyotype	5 Mb	Differences in chromosome number (aneuploidy), large deletions, translocations, ring chromosomes
Fluorescence *in situ* hybridization	100 kb	Deletions, duplications, translocations, ring chromosomes
Chromosome microarray analysis	100 kb	Deletions and duplications and uniparental disomy for SNP arrays
Exon array	~500–1000 bp	Deletions and duplications
Sequencing	1 bp	Changes in single or small numbers of nucleotides

increasing yield as the severity of intellectual disability increases. Other clinical features that increase the yield are dysmorphic features, intrauterine growth restriction/failure to thrive, birth defects, seizures, or a history of recurrent miscarriages in the parents (usually associated with one parent who is a balanced translocation carrier). With the recent advances in chromosome microarray analysis, we are beginning to define many novel genetic syndromes associated with recurrent deletions and duplications, many of which are associated with hypotonia and developmental delay in infancy.

It should be noted that we are still defining the normal genomic architecture, including the location and frequency of normal, benign polymorphic deletions and duplications that are present throughout the genome. Therefore, it is not uncommon to receive a test report from a chromosome microarray analysis that identifies a deletion or duplication "variant of unknown clinical significance." In such cases, it has not been clearly established whether this copy number variant (CNV) is a normal benign polymorphism or a pathogenic, disease-associated mutation.[13] To distinguish between these two possibilities, it is recommended to test the parents for the copy number variant. If both parents are normal, the copy number variant is interpreted to be disease-associated because the copy number variant has risen *de novo* from unaffected parents. If one of the parents is affected, the copy number variant may be pathogenic if it is inherited from an affected parent. Note that many copy number variants often are associated with different neuropsychiatric phenotypes, so the offspring and parent may both be "affected" but manifesting different symptoms. With additional population characterization, the number of test reports with variants of unknown clinical significance should decrease significantly.

Genome wide microarrays are not generally designed to detect small genomic alterations of less than ~100 kb. However, targeted "exon arrays" and multiplex ligation-dependent probe amplification (MLPA) are now being developed to detect small deletions or duplications involving only portions of single genes associated with specific symptoms.[14] These types of mutations are generally on the order of 500–100,000 base pairs and are routinely "invisible" to molecular methods that rely on polymerase chain reaction (PCR) amplification or genomic capture followed by sequence analysis. Increasingly, exon arrays and MLPA are being designed to complement sequence-based tests[14] to ensure that all types of mutations are detected. This test is particularly useful for recessive disorders for which only one mutation is identified by sequence analysis, suggesting that there is another mutation not detected by sequencing methods.

Genetic testing for a limited number of specific genes is more commonly performed if a neuropathy or myopathy is suspected. In contrast to the "macro" methods above, this approach targets "micro" changes in the DNA sequence, down to individual nucleotides. There are two distinct approaches to micro testing, depending upon whether one or a small number of specific mutations are known for which the patient is at risk. In the case of common mutations, analysis can be targeted to detect only those known mutation(s). Targeted mutation testing is widespread for the triplet repeat disorders and for genes with specific founder mutations (e.g. the c.1167insA mutation in the *FKTN* gene causing Walker-Warburg syndrome in Ashkenazi Jewish individuals)[15] but is limited in application to those conditions for which specific mutations account for the majority of patients with a clinical diagnosis. When the majority of patients are likely to have a "private" unique mutation, testing must examine the entire gene, nucleotide by nucleotide, and this represents a greater challenge.

A typical clinical genetic resequencing test involves polymerase chain reaction (PCR) amplification or capture of each exon of a given gene, along with its splice junctions, and sometimes the promoter region of the gene, followed by some method of sequencing. The sequence obtained is compared to the published reference sequence, and any nucleotide changes are then interpreted and compared with databases of known normal genetic variants (polymorphisms) and of disease-causing mutations.

The traditional method for generating sequence data until 2010 was PCR amplification and dideoxy sequencing, and the cost of a test was linearly related to the number of exons and size of the gene. The so-called next-generation sequencing platforms are based on highly parallel data generation, and such instruments produce up to 600 Gb per sequencing run, requiring up to a week to complete a run. The processes of data acquisition, data extraction, processing, and analysis are qualitatively and quantitatively different from the conventional capillary approach[16] and are clearly more complex due to the large amounts of data generated. Turnaround times are often 8–10 weeks from the time a sample is submitted until reports are issued, due to the complexity of the data generation and test interpretation. As a result, next-generation sequencing has opened new horizons in clinical genetic diagnostics.

The format of next-generation platforms differs from the format of conventional sequencing in many aspects. In particular, next-generation sequencing instruments produce a large number of shorter sequencing reads (50–100 bases), thus changing the process of sequence alignment and data analysis (Figure 2.4). This difference in data generation and analysis requiring alignment of short sequencing reads leads to technical difficulties with detecting specific types of mutations. A challenge in data analysis for next-generation sequencing relates to the

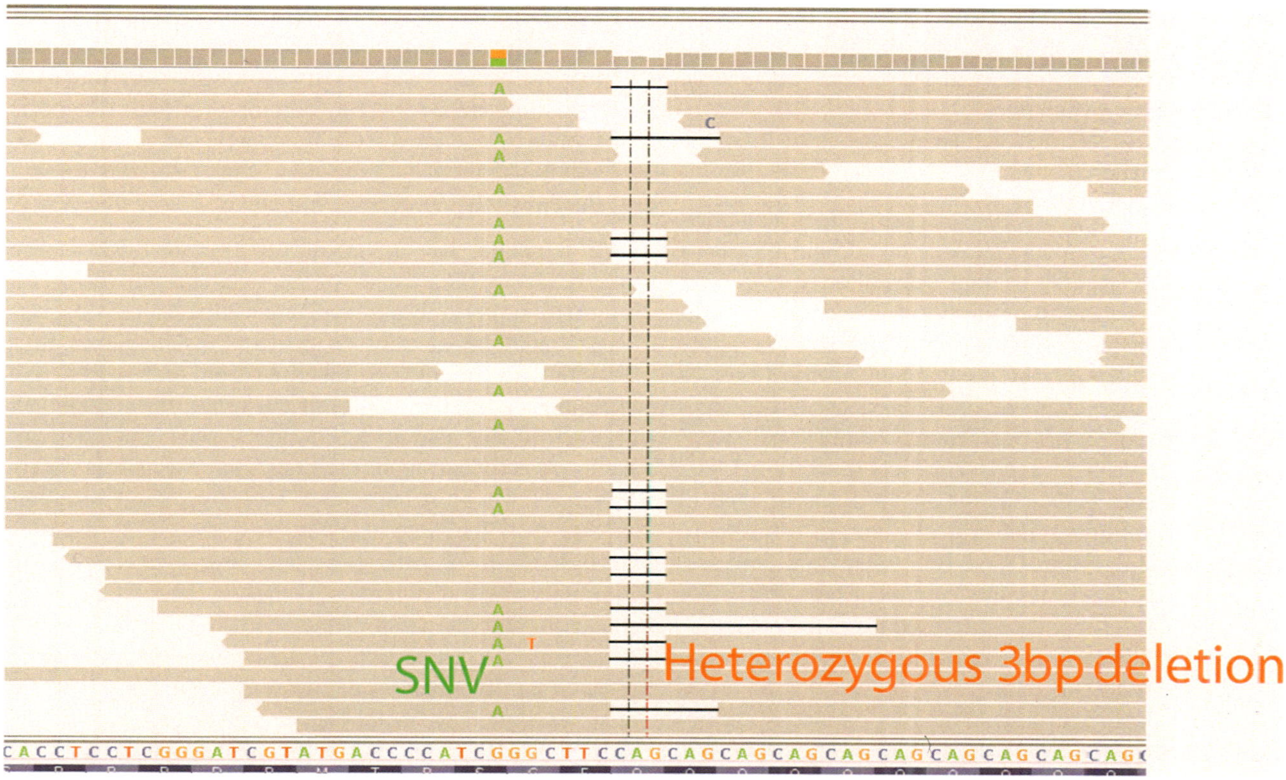

FIGURE 2.4 Alignment of short sequencing reads in next-generation sequencing.

algorithmic nucleotide mismatch penalties vs. correctly mapped nucleotides. If the patient's DNA contains a mutation, the alignment must allow for an ambiguity, so that the sequencing read can correctly align to the reference sequence. At the same time, if too many mismatches are allowed, then poorly sequenced data, or data that belong to another homologous sequence, may align incorrectly. The balance between the length of the sequencing reads (e.g. ~100 bases) and the alignment penalties results in alignment algorithms allowing for only 2–3 mismatches. Typically, up to 5 consecutive mismatches are allowed between a sequence read and the reference sequence, and a sequence read containing more mismatches is eliminated from the alignment. This alignment model inevitably fails when a series of sequence changes are located close to each other (less than ~20 bp apart), or in the case of an insertion or deletion greater than a few base pairs. Large insertions/deletions can be addressed by other means such as monitoring the level of sequence coverage or using other platforms such as chromosome microarrays. In addition, since portions of the human genome consist of repetitive sequences, it is impossible to correctly map some short sequencing reads back to a unique location in the genome. Due to sequence similarities among many related genes, it may still be impossible to correctly assign the sequencing read. Sequence similarities between the genes can produce situations in which a sequence variation could map to either gene, thus making it impossible to unambiguously determine the gene involved. Such genes are better interrogated with classical Sanger sequencing.

Clinical testing is most useful when there is clear interpretation of the data, including novel variants. Interpretation of variations depends critically on identifying the same variant in reported mutations databases and databases of normal individuals to identify normal genetic variants called polymorphisms and differentiate them from mutations. The polymorphism databases have become increasingly robust with the completion of the 1000 Genomes Project[17] and increasing availability of more and more sequencing data in the public domain.

The results of genetic testing are reported as one of five categories:

1. *Mutation*: Previously reported in the literature and clearly pathologic.
2. *Variant, favor mutation*: Novel variant but extremely likely to be pathogenic based upon the type and location of the mutation.
3. *Variant of uncertain significance*: Unclear whether this is benign or disease-causing.

4. *Variant, favor polymorphism*: Very likely to be a normal variant based upon the type of variant and location but is not present in the normal population, in >1% of the population.
5. *Polymorphism*: Normal variant present in at least 1% of the population.

Classification in many cases is based upon computational predictions of pathogenicity of variants. Variants common in the general healthy population (>1%) are generally assumed not to be disease-causing and are usually classified as polymorphisms. Variants that are highly conserved across species by prediction algorithms such as PhyloP[18] are more likely to be deleterious. Other algorithms including SIFT,[19] Pmut,[20] MutationTaster,[21] PhastCons,[22] GERP,[23] and Likelihood Ratio Test[24] attempt to predict pathogenicity on the basis of type of amino acid substitution and critical domains of proteins. Although these algorithms have utility, they are only predictions and ultimately require functional data and larger genetic data sets of replication to establish pathogenicity.

The implications of the test results depend on the clinical situation. When the patient is symptomatic, the test is considered a diagnostic test. When the patient is asymptomatic but has a family history of the disease, the test is considered predictive. Predictive testing is most informative when a mutation for the disease has been identified in the family. Interpretation of test results depends on whether testing is diagnostic or predictive and whether a familial mutation has been identified.

Positive Result

A positive result indicates that a disease-causing mutation was identified in that individual. Knowledge of a positive result provides valuable information to the patient, physician, and family members. Knowledge of a patient's genotype can assist in making management and treatment decisions. Furthermore, testing of the patient's family members can allow for accurate predictions of disease risks or reproductive risks.

Negative Result

A negative result means that a disease-causing mutation was not identified in the individual tested. A negative result can have different interpretations based on the following scenarios:

1. *True Negative Result*: An individual who tests negative for a known familial mutation and is not a carrier of a disease-predisposing mutation that has been identified in another family member. The risk for disease in this individual is generally not expected to be higher than in the general population.
2. *Uninformative Negative Result*: If testing was performed on an individual with symptoms, this means that we do not have an explanation for why this individual developed these symptoms. A genetic diagnosis cannot be excluded since no genetic test is 100% sensitive and there may be other genes for this disease that have not yet been identified.

If a patient has a variant of uncertain significance (VUS), testing the parents for the VUS can provide useful information. For a recessive disorder, disease-causing variants should be in trans to confirm the diagnosis in a symptomatic patient (one inherited from each parent and therefore on different copies of the gene). For a dominant disorder, if one parent carries the variant and does not share the clinical manifestations, the variant is likely benign. However, if the variant is *de novo* (and both parents are truly the biological parents of the child), the variant is likely to be pathogenic and disease-causing. Good laboratories will test parents for free in these situations to clarify the interpretation of the VUS. Beyond the parents, if there is still a question of pathogenicity, testing other affected family members can be useful. Similarly affected relatives should share the same genetic variant(s) if the variant is pathogenic. As more data about variants are gathered across laboratories, variants should be reclassified and new reports should then be issued to the referring physician.

Next-generation sequencing technology can also be used to analyze an entire genome or, more commonly, the exome that is the collection of all the exons or all the coding portions of genes in the genome.[25] During the process of exome sequencing, the ~2% of the genome that encodes the exons is selectively captured and sequenced. Ideally, the sequence of the symptomatic child is analyzed concurrently with the parents and/or other affected family members to allow for efficient filtering of the thousands of genetic variants that are identified from any one individual exome. Normal genetic variants or polymorphisms are filtered out. Previously identified disease-causing mutations are annotated by filtering the variants through mutation databases. *De novo* variants present in the child but not in either parent are readily identified by comparing the sequence of the child to both parents. Other rare or novel genetic variants are identified and computational prediction algorithms are used to predict pathogenic variants. Genes with two rare genetic variants predicted to be pathogenic, one inherited from each parent, are also flagged as possible recessive disorders.

Exome sequencing is an extremely efficient way to screen the entire genome when there is clearly a familial condition based upon multiple affected individuals, and especially if a gene has not yet been identified for the condition or if the condition is so genetically heterogeneous that the number of possible genes is large (e.g. intellectual disability or autism). However, the clinical sensitivity of exome sequencing for familial apparently monogenic disorders is still likely <50% due to limitations in our understanding of disease association for only about 3000 of our 20,000 genes, our limited ability to recognize pathogenic variants among all the novel variants, limitations in our sensitivity to identify all insertions and deletions, limitations with sequence alignments for genes with homology with other genes, and the fact that not all mutations lie within the captured exons. Indications for exome sequencing include: a clearly familial condition for which a gene has not yet been identified, a condition for which genetic testing is not otherwise available, and a condition with genetic heterogeneity for which exome sequencing is more cost effective than other testing alternatives.

COMPARISON BETWEEN PANELS AND WHOLE EXOME SEQUENCING

In some cases, the clinician may have the choice between ordering a genetic test for a defined panel of genes for conditions, such as congenital muscular dystrophy or peripheral sensory neuropathy, or an exome test. If the clinical sensitivity of the targeted gene panel is high, it is preferable to use a targeted gene panel to minimize expense and increase the probability of identifying the mutation. Targeted panels are designed to cover all the coding regions of the targeted genes, whereas an exome test will often only cover 90% to 95% of the average gene. However, if the clinical sensitivity of the targeted panel is <50%, it may be worthwhile to order an exome test with the initial analysis focused on known disease genes (referred to as an exome slice by some laboratories) with the option to analyze the remainder of the exome if a mutation is not identified in the first round of analysis.

SECONDARY FINDINGS

The American College of Medical Genetics (ACMG) has recommended mandatory return of secondary findings for 56 genes that are medically actionable (largely cancer and cardiac conditions) and potentially life threatening to patients of all ages, including children.[26] It is important that patients and families understand this procedure prior to ordering exome testing, and the ACMG guidelines recommend genetic counseling prior to exome testing so that patients and families can understand the secondary findings that may be reported. Some clinical laboratories are offering the option to opt out of return of these secondary findings for patients who prefer not to have these results returned.

CLINICAL INDICATIONS FOR GENETIC TESTING

Genetic testing in a symptomatic patient with a muscular dystrophy or neuropathy is described as "diagnostic," while testing in a patient who is asymptomatic but at increased future risk of disease due to a family history is termed "predictive." Most genetic testing for neurological conditions in children will be diagnostic unless there is an older affected sibling or other family member.

The clinical indications for genetic testing are to establish a diagnosis, clarify the prognosis, assist with treatment and management, and clarify risk of disease for family members and future children for the parents and the patient. Identifying a specific genetic etiology may save the patient from further diagnostic evaluation and potentially invasive testing (including muscle and nerve biopsy) and should shorten the diagnostic odyssey that many families endure. As molecular genetic testing improves, it is likely to increasingly become a first line test and perhaps even replace muscle and nerve biopsies. In the majority of patients with myopathies and neuropathies, the results of genetic testing will not radically change the medical management, but they do often provide families and providers with greater certainty that helps with planning.

APPROACH TO GENETIC EVALUATION

A genetic evaluation consists of review of the patient's history and medical records and documentation of at least a three-generation pedigree to identify patterns of inheritance. The physical examination should include assessment of growth, birth defects, dermatologic findings, and dysmorphic features in addition to the neurologic exam. Electromyogram and nerve conduction studies, CPK, and other laboratory tests will frequently be the next step. Genetic testing may consist of examination of metabolites in the urine, blood, and cerebrospinal fluid; enzymatic activity in samples of blood, fibroblasts, or muscle; molecular genetic testing; and/or cytogenetics including chromosome microarray. In general, most molecular genetic tests can be performed noninvasively on blood.

An updated list of genetic tests available and laboratories performing testing is available online from GeneTests[27] (Table 2.2), which provides an excellent resource for clinicians. For the patient facing a genetic diagnosis, the United States National Library of Medicine provides a comprehensive home reference to understanding genetic conditions, inheritance patterns, and related resources.[28] In the United States, only Clinical Laboratory Improvement Amendments (CLIA) certified clinical laboratories should be used for clinical diagnosis of patients.

Although uncommon with symptomatic children, some families have concerns about the ethical, legal, and social

TABLE 2.2 Major Internet Sites of Value in Medical Genetics

Title	Description	URL
Online Mendelian Inheritance in Man (OMIM)	The standard and authoritative source of information on the current state of knowledge of single gene disorders. Provides links to relevant literature, gene mapping and structure information, and a clinical summary.	http://www.ncbi.nlm.nih.gov/omim/
National Organization for Rare Disorders	A database of information for patients and families with genetic disorders. Provides access to support groups and lay information.	http://www.rarediseases.org/
Genetic Test Registry	The Genetic Testing Registry (GTR) provides a central location for voluntary submission of genetic test information by providers. The scope includes the test's purpose, methodology, validity, evidence of the test's usefulness, and laboratory contacts and credentials.	http://www.ncbi.nlm.nih.gov/gtr/
GeneTests	Database of laboratories that perform molecular diagnosis for specific disorders. Requires password for access, which is freely available to professionals.	www.genetests.org
GeneReviews	Online genetics textbook, providing guidelines for diagnosis and management of genetic conditions.	http://www.ncbi.nlm.nih.gov/books/NBK1116/
Human Mutation Database	A compendium of databases of mutations in various genes responsible for human genetic disorders.	http://www.hgmd.org/
Genetic Alliance	Organization of support groups for genetic disorders, a good source of information for families.	www.geneticalliance.org

implications of genetic testing, including the potential for stigma, distress, adverse labeling, and loss of confidentiality. Genetic testing should always be performed with patient consent, or parental consent in the case of a minor. A team approach including a genetic counselor is helpful whenever possible, and increasingly genetic counselors are embedded within child neurology programs at large medical centers. A list of available genetic counselors is available at the National Society of Genetic Counselors website.[29]

POTENTIAL BARRIERS TO TESTING

There are multiple potential and actual barriers to diagnostic genetic testing in child neurology. A major barrier to genetic testing is often financial. The cost of many of these tests is on the order of hundreds to several thousands of dollars per test, some portion of which may be covered by insurance companies. Increasingly, multiple laboratories are available that perform the genetic testing, many of which accept insurance. Therefore, careful comparison between the laboratories will often identify a laboratory that will accept insurance for part or all of the cost of testing. Genetic information may lead to psychological and social consequences, including blame or guilt, stigmatization, and feelings of isolation.[30] As with other heritable conditions, patients may be concerned about possible discrimination for themselves and other family members. The federal Genetic Information Non-Discrimination Act (GINA) was passed by the United States Congress in May 2008, and began to take effect in May 2009. GINA protects Americans from having their health insurance rates raised, being denied health insurance coverage, or being denied a job or job promotion on the basis of a genetic predisposition (i.e. predictive genetic testing). This protection does not extend to life insurance, disability insurance, or long-term care insurance. Some states also have additional legislation that affords more extensive protection for patients. GINA should provide additional reassurance to families seeking genetic testing.

ATTITUDES TOWARDS GENETIC TESTING

Families concerned about reproductive risk of recurrence for themselves or their family members may be more likely to pursue genetic testing. Families of children with more severe neurological problems, and families with more than one affected family member, may be especially motivated to purse genetic testing because of the risks and burden to have another affected child. Some families may be interested specifically in genetic testing that alters the treatment course but may be ambivalent to genetic testing that will not change the treatment. Many families seek a diagnosis to clarify prognosis and find an appropriate support group of other families with similarly affected children.

RECURRENCE RISK IN FAMILIES AND REPRODUCTIVE OPTIONS

The issue of risk of recurrence for the parents, siblings, and other family members as well as the patient him/herself is often of concern to the family. Usually, the immediate concern is the risk of recurrence to the parents if children are young. Many severe pediatric neurogenetic conditions are the result of *de novo* or spontaneous mutations. If this is proven after testing the parents' blood, then the risk of recurrence to the parents is $<1\%$, and any recurrence is due to undetectable gonadal mosaicism. In such families, prenatal testing with chorionic villus sampling or biopsy (CVS or CVB) or amniocentesis may be an appropriate route and provides welcome reassurance to parents.

There is a significant risk of recurrence if a parent is (i) a balanced chromosomal translocation carrier; (ii) a mother who is a carrier for an X-linked condition; (iii) a mother who is a carrier of a mitochondrial mutation; (iv) both parents who are carriers for an autosomal recessive condition; or (v) a parent who carries an autosomal dominant disorder (Table 2.3). Parents desiring additional unaffected children have multiple options including (i) adoption; (ii) use of donor egg or donor sperm to remove the genetic contribution of the parent transmitting the mutation; (iii) prenatal testing (either chorionic villus sampling at 10 to 12 weeks of gestation or amniocentesis after 16 weeks of gestation); or (iv) preimplantation genetic diagnosis (PGD) in which only embryos without the familial mutation are transferred into the uterus.

Prenatal diagnosis without consideration of termination of pregnancy may be helpful in certain metabolic disorders for which immediate neonatal management is effective and can avoid neurological damage. Donor sperm is practically easier than donor egg due to greater availability of sperm in large sperm banks. Donors of sperm or eggs should not be genetically related to the parents unless they are found not to carry the familial mutation. Preimplantation genetic diagnosis is an appealing option for couples in that it avoids the difficulty of terminating an at-risk fetus. However, it requires *in vitro* fertilization, has only a 25% success rate/cycle in couples without fertility issues,[31,32] and may not be covered in part or whole by insurance. Preimplantation genetic diagnosis should be viewed as significantly risk-reducing but is associated with rare errors in ~2% of cases.[30]

TABLE 2.3 Risk of Recurrence and Effective Reproductive Options Available to Avoid Recurrence

Mode of Inheritance	Risk of Recurrence	Reproductive Options
Balanced translocation carrier	Variable depending on gender of parent and exact translocation	Donor sperm Donor egg Prenatal testing PGD
X linked	50% of males Rarely in females	Donor egg Prenatal testing PGD
Mitochondrial	Variable and unpredictable	Donor egg
Autosomal recessive	25%	Donor egg Donor sperm Prenatal testing PGD
Autosomal dominant	50%	Donor egg Donor sperm Prenatal testing PCD
Complex	Empirical	None available

Thus, although preimplantation genetic diagnosis usually avoids the necessity to terminate an affected pregnancy, the process should be carefully considered by couples prior to initiation. Preimplantation genetic diagnosis is available in the USA, but is not available in all countries.

Attitudes toward these family planning options vary by culture, in part based upon religious beliefs and acceptability of pregnancy termination.[33] For extended family members, the risk should first be defined by genetic testing for the familial mutation(s).[34] For those found to carry a familial mutation, similar reproductive options are available. Notably, although a genetic condition may have been *de novo* in the affected patient and associated with a benign family history, there may be a significant risk of recurrence to the children of an affected *de novo* mutation carrier, as high as 50%.

Mutations encoded within the mitochondrial genome pose a unique problem in terms of risk of recurrence due to heteroplasmy (a mixture of mitochondrial genotypes in the cell). Risk cannot be accurately predicted for a woman carrying a mitochondrial mutation nor is prenatal testing accurate since there may be differential mitochondrial mutation load by tissue that is not accurately reflected in the placenta at chorionic villus sampling or in amniocytes. For women in such families with mitochondrial epilepsies such as Myoclonic Epilepsy, Lactic Acidosis, and Stroke-like episodes (MELAS), the only way to guarantee avoidance of recurrence is to use a donor egg.[35]

TREATMENT OF GENETIC DISORDERS

Historically, only a relatively small number of genetic disorders have been amenable to definitive treatment. Most notable in this group are many inborn errors of metabolism that can be treated by manipulation of diet, administration of coenzymes, removal of toxins, bone marrow transplant, or liver transplant. For most other neuromuscular genetic disorders, treatment has been symptomatic. Medical management focuses on early detection of treatable complications and anticipatory guidance of the family. Major advances in understanding the molecular basis of genetic disorders, however, have brought new hope that improved means of treatment will be developed. Identification of the gene responsible for a disorder leads to elucidation of the dysfunctional cellular and physiological pathways, and may lead to the discovery of new targets for drug therapy. There is also hope that new technologies will permit, in some cases, replacement or repair of defective genes as with the early clinical trials in using antisense mediated exon skipping in DMD[36,37] and correction of ineffective splicing of *SMN2* in SMA.[38–40]

The molecular diagnostic genetic testing platforms reviewed in this chapter will be increasingly relevant in the diagnostic evaluation for children with neuromuscular conditions. Novel genetic syndromes are being defined through advances in chromosomal microarray analysis and exome sequencing. Oligonucleotide microarrays now complement structural information from karyotype and provide much higher resolution than karyotypes to identify deletions and duplications. Targeted exon arrays allow detection of small genomic alterations invisible to genome wide microarrays and complement sequence-based tests for analysis of single genes. Sequence-based testing for specific genes is widely available and increasingly will be multiplexed by clinical indication to improve clinical diagnosis. Exome sequencing is helping to elucidate novel causes of disease and more efficiently and expeditiously diagnose complex neuromuscular conditions. Molecular genetic diagnostics are rapidly evolving in step with the discovery of new genetic syndromes and should reduce the time to diagnosis so efforts can be focused on treatment rather than diagnosis. Ability to utilize these tests effectively in the clinical setting and to interpret these tests in a cost-efficient manner is essential for the practicing child neurologist in the 21st century.

REFERENCES

1. My Family Health Portrait [Internet]. Office of the Surgeon General and the National Human Genome Research Institute; [cited March 8, 2014]. Available from: <https://familyhistory.hhs.gov/>; 2014.
2. Rousseau F, Heitz D, Tarleton J, MacPherson J, Malmgren H, Dahl N, et al. A multicenter study on genotype-phenotype correlations in the fragile X syndrome, using direct diagnosis with probe StB12.3: the first 2,253 cases. *Am J Hum Genet* 1994;**55**:225–37.
3. Brook JD, McCurrach ME, Harley HG, Buckler AJ, Church D, Aburatani H, et al. Molecular basis of myotonic dystrophy: expansion of a trinucleotide (CTG) repeat at the 3′ end of a transcript encoding a protein kinase family member. *Cell* 1992;**68**:799–808.
4. Conne B, Stutz A, Vassalli JD. The 3′ untranslated region of messenger RNA: A molecular 'hotspot' for pathology? *Nat Med* 2000;**6**:637–41.
5. Campuzano V, Montermini L, Molto MD, Pianese L, Cossee M, Cavalcanti F, et al. Friedreich's ataxia: autosomal recessive disease caused by an intronic GAA triplet repeat expansion. *Science* 1996;**271**:1423–7.
6. Gusella JF, MacDonald ME, Ambrose CM, Duyao MP. Molecular genetics of Huntington's disease. *Arch Neurol* 1993;**50**:1157–63.
7. Banfi S, Servadio A, Chung MY, Kwiatkowski Jr. TJ, McCall AE, Duvick LA, et al. Identification and characterization of the gene causing type 1 spinocerebellar ataxia. *Nat Genet* 1994;**7**:513–20.
8. Zoghbi HY, Orr HT. Glutamine repeats and neurodegeneration. *Annu Rev Neurosci* 2000;**23**:217–47.
9. Maddalena A, Bale S, Das S, Grody W, Richards S, ALQA Committee. Technical standards and guidelines: molecular genetic testing for ultra-rare disorders. *Genet Med* 2005;**7**:571–83.
10. Koo T, Wood MJ. Clinical trials using antisense oligonucleotides in Duchenne muscular dystrophy. *Hum Gene Ther* 2013;**24**:479–88.

11. Miller DT, Adam MP, Aradhya S, Biesecker LG, Brothman AR, Carter NP, et al. Consensus statement: chromosomal microarray is a first-tier clinical diagnostic test for individuals with developmental disabilities or congenital anomalies. *Am J Hum Genet* 2010;**86**:749–64.
12. Edelmann L, Hirschhorn K. Clinical utility of array CGH for the detection of chromosomal imbalances associated with mental retardation and multiple congenital anomalies. *Ann NY Acad Sci* 2009;**1151**:157–66.
13. Mefford HC, Muhle H, Ostertag P, von Spiczak S, Buysse K, Baker C, et al. Genome-wide copy number variation in epilepsy: novel susceptibility loci in idiopathic generalized and focal epilepsies. *PLoS Genet* 2010;**6**:e1000962.
14. Tayeh MK, Chin EL, Miller VR, Bean LJ, Coffee B, Hegde M. Targeted comparative genomic hybridization array for the detection of single- and multiexon gene deletions and duplications. *Genet Med* 2009;**11**:232–40.
15. Chang W, Winder TL, LeDuc CA, Simpson LL, Millar WS, Dungan J, et al. Founder Fukutin mutation causes Walker-Warburg syndrome in four Ashkenazi Jewish families. *Prenat Diagn* 2009;**29**:560–9.
16. Holt RA, Jones SJ. The new paradigm of flow cell sequencing. *Genome Res* 2008;**18**:839–46.
17. 1000 Genomes Project [cited 3/10/14]; Available from: <http://www.1000genomes.org/>.
18. phyloP. [cited 9/30/2013]; Available from: <http://compgen.bscb.cornell.edu/phast/help-pages/phyloP.txt/>.
19. Craig Venter J, Institute [Internet]. SIFT. [cited 9/30/2013]; Available from: <http://sift.jcvi.org/>.
20. Molecular Modeling and Bioinformatics Group [Internet]. PMut. [cited 9/30/2013]; Available from: <http://mmb.pcb.ub.es/PMut/>.
21. MutationTaster [Internet]. [cited 9/30/2013]; Available from: <http://www.mutationtaster.org/>.
22. Siepel, A. PhastCons HOWTO [Internet]. [cited 9/30/2013]; Available from: <http://compgen.bscb.cornell.edu/phast/phastCons-HOWTO.html/>; 2005.
23. Genomic Evolutionary Rate Profiling: GERP [Internet]. Sidow Lab [cited 9/30/2013]; Available from: <http://mendel.stanford.edu/SidowLab/downloads/gerp/>.
24. Knudsen B, Miyamoto MM. A likelihood ratio test for evolutionary rate shifts and functional divergence among proteins. *Proc Natl Acad Sci USA* 2001;**98**:14512–17.
25. Bamshad MJ, Ng SB, Bigham AW, Tabor HK, Emond MJ, Nickerson DA, et al. Exome sequencing as a tool for Mendelian disease gene discovery. *Nat Rev Genet* 2011;**12**:745–55.
26. Green RC, Berg JS, Grody WW, Kalia SS, Korf BR, Martin CL, et al. ACMG recommendations for reporting of incidental findings in clinical exome and genome sequencing. *Genet Med* 2013;**15**:565–74.
27. GeneTests [Internet]. Elmwood Park, NJ. GeneTests.org [cited 3/10/2014]; Available from: <www.genetests.org/>.
28. United States National Library of Medicine [Internet]. Genetics Home Reference. [cited 3/4/2014]; Available from: <http://ghr.nlm.nih.gov/>.
29. National Society of Genetic Counselors [Internet]. [cited 3/10/2014]; Available from: <http://www.nsgc.org/>.
30. Shostak S, Ottman R. Ethical, legal, and social dimensions of epilepsy genetics. *Epilepsia* 2006;**47**:1595–602.
31. Harper JC, Boelaert K, Geraedts J, Harton G, Kearns WG, Moutou C, et al. ESHRE PGD Consortium data collection V: cycles from January to December 2002 with pregnancy follow-up to October 2003. *Hum Reprod* 2006;21:3–21.
32. Feyereisen E, Steffann J, Romana S, Lelorc'h M, Ray P, Kerbrat V, et al. Five years' experience of preimplantation genetic diagnosis in the Parisian Center: outcome of the first 441 started cycles. *Fertil Steril* 2007;**87**:60–73.
33. Dehlendorf C, Rodriguez MI, Levy K, Borrero S, Steinauer J. Disparities in family planning. *Am J Obstet Gynecol* 2010;**202**:214–20.
34. Pal DK, Pong AW, Chung WK. Genetic evaluation and counseling for epilepsy. *Nat Rev Neurol* 2010;**6**:445–53.
35. Marchington D, Malik S, Banerjee A, Turner K, Samuels D, Macaulay V, et al. Information for genetic management of mtDNA disease: sampling pathogenic mtDNA mutants in the human germline and in placenta. *J Med Genet* 2010;**47**:257–61.
36. Goemans NM, Tulinius M, van den Akker JT, Burm BE, Ekhart PF, Heuvelmans N, et al. Systemic administration of PRO051 in Duchenne's muscular dystrophy. *N Engl J Med* 2011;**364**:1513–22.
37. Goyenvalle A, Seto JT, Davies KE, Chamberlain J. Therapeutic approaches to muscular dystrophy. *Hum Mol Genet* 2011;**20**:R69–78.
38. Hua Y, Sahashi K, Hung G, Rigo F, Passini MA, Bennett CF, et al. Antisense correction of SMN2 splicing in the CNS rescues necrosis in a type III SMA mouse model. *Genes Dev* 2010;**24**:1634–44.
39. Passini MA, Bu J, Richards AM, Kinnecom C, Sardi SP, Stanek LM, et al. Antisense oligonucleotides delivered to the mouse CNS ameliorate symptoms of severe spinal muscular atrophy. *Sci Transl Med* 2011;**3**:72ra18.
40. Hua Y, Sahashi K, Rigo F, Hung G, Horev G, Bennett CF, et al. Peripheral SMN restoration is essential for long-term rescue of a severe spinal muscular atrophy mouse model. *Nature* 2011;**478**:123–6.

Chapter 3

Electromyography in Pediatrics

Matthew Pitt and Peter B. Kang

INTRODUCTION

This chapter discusses the use and techniques of pediatric electromyography (EMG). In a comprehensive text on neuromuscular disease such as this, targeted towards clinicians in neuromuscular medicine, the intention of this chapter is not to make readers experts in the art of the EMG, but rather to present a brief summary of the use of EMG in children to foster awareness of what is possible. In that way, readers will be able to anticipate what they are going to expose their patients to.

DEALING WITH MISCONCEPTIONS

It may seem strange to begin a chapter on a diagnostic investigation by discussing the misconceptions that exist regarding its use. This kind of discussion is unlikely to take place when discussing muscle biopsy, muscle MRI, or genetic testing, but it is a significant problem for EMG. Some bodies of opinion in the neuromuscular world hold that EMG has no role in the future, that it has perhaps gone the way of the myelogram, which has been superseded by MRI. This is disappointing, because EMG provides, very quickly and with, in reality, a minimum amount of discomfort, information that can have a major influence on the management of children with neuromuscular disorders. The misconceptions held fall into one of three categories: first that it is too painful, secondly that it is too difficult, and finally that it is redundant in the era of molecular genetics.

It Is too Painful

Most pediatric electromyographers have experienced the referring practitioner telling parents their child was about to have "the most painful test they will have ever experienced." This is an extraordinary thing for any colleague to communicate to a parent, but paradoxically the fact that they consider sending the child for such an appalling investigation suggests that they, perhaps subconsciously, realize that the test has an important role in the investigation of the child. It is not hard to imagine the kind of anxiety this produces in the parents and child and the effort needed to persuade them of the benign nature of the investigation.

In considering where this misconception arose, one has to acknowledge that the first pioneers of pediatric EMG, in the early 1970s, were working in much less favorable conditions. The initial needles used were far larger in diameter than those used currently and were reused multiple times. Those needles became blunt after repeated use and, despite admonishments to sharpen them, it was often difficult to do so. To add further difficulty, inserting the needles repeatedly often meant that they would bend and become permanently bent with the possible risk of breakage. This was particularly a problem if they were used in children. Today, disposable needles are universally used in the Western world. However, larger-bore needles are still in widespread use in adult EMG laboratories, perpetuating misconceptions about how the testing is performed on children in the hands of a skilled and thoughtful examiner.

The way we regard children has changed a great deal over the years. In England and the United States, the traditional view was that children were to be "seen but not heard," but enlightenment regarding the rights of children has changed that attitude considerably in recent decades. Children are very much allowed to be heard, and if an investigation is uncomfortable for them we should recognize that and make strenuous efforts to diminish the discomfort. There is also now a realization, rather reluctant, that some of the discomfort produced in children resulted from inadequate knowledge among the people performing the pediatric EMG. There was a lack of appreciation of how children react and of the diseases they suffer. As an example, the correct investigational strategy to be used when discovering anterior horn cell disease in children depends on a differential diagnosis between whether the

B.T. Darras, H. Royden Jones, Jr., M.M. Ryan & D.C. De Vivo (Eds): Neuromuscular Disorders of Infancy, Childhood and Adolescence, Second edition.
DOI: http://dx.doi.org/10.1016/B978-0-12-417044-5.00003-2

condition is segmental or generalized, which is quite different from encountering these findings in adults where the diagnostic question often involves motor neuron disease versus cervical and lumbar radiculopathy. In children, this question can be answered straightforwardly by sampling as few as two muscles.

So what is the reality with regard to the pain of pediatric EMG? Some studies have tried to answer this question.[1,2] At Great Ormond Street Hospital in London, an informal analysis asked participants to rate the pain on a sliding scale after the examination had been completed, comparing it with blood tests. We also included venous cannulation as a further comparison, but the numbers for that comparison were less. To our satisfaction, the results of this unpublished analysis showed that the discomfort rests somewhere just above a blood test and a long way short of the cannulation. As a further observation, it was interesting to talk to parents about the stress of other maneuvers, some of which would not immediately be associated with discomfort. Among that list is the insertion of a nasogastric tube, which causes huge discomfort to both parents and children alike and is one of the major factors that upset children when lying down for investigations such as bulbar EMG and single fiber EMG. Since the first of these tests is often requested when a child is having feeding difficulty, it is important to realize the association and to be able to reassure parents.

What can one do about the pain of EMG? A variety of strategies is available to the practitioner. The first perhaps is the use of local anesthetic. This is now commonplace for nearly all children having blood tests and some very effective local anesthetics exist, some based on amethocaine and others using xylocaine as the active agent. The problem with their application in EMG is that in many ways the needle EMG, however it might be regarded by the parents or child, is not really the most painful experience. Motor nerve stimulation can be more painful. Perhaps because needle EMG takes longer than the motor stimulation and the child often becomes more upset, it is hard not to feel that local anesthetic would help. Of course, it only helps the passage of the needle through the skin, which if done skillfully only takes a matter of milliseconds.

Next there is a tendency to use sedation, sometimes termed conscious sedation. The problem of sedation, which is well recognized, is that a child who is asleep will be awakened by any painful experience and this situation is no different for a child who is sedated and having EMG. To achieve a degree of analgesia with sedation is only possible when you approach an anesthetic application. Clearly this will have important implications with regard to respiratory depression and inhalation and other serious complications. It is quite possible for children to be nearly anesthetized and then left to recover from their sedation with no awareness of how severely depressed they are. So, if sedation is used it is mandatory that care is taken to ensure there is adequate training for staff and adequate monitoring. Resuscitation equipment also needs to be readily available.

The next stage in the escalation of methods of reducing the discomfort is the use of inhalation anesthetics such as nitrous oxide. This in many ways is a logical step, as it is an anesthetic with significant analgesic effects, which are perhaps more than its sedative effects. It is perhaps salutary to remember that there are requirements for trained personnel to deliver this and significant dangers if the equipment is not maintained. During the administration of nitrous oxide to mothers in labor, they are required to clasp the mask across their face; when the anesthetic reaches a certain level the sedative effects cause the mother to then be unable to hold the mask against her face. This is a built-in safety measure, which prevents overdose. Unfortunately, children will not cooperate in this technique as well as adults and therefore the mask has to be held on the face by an anesthetic assistant. There are dangers if attendants do not realize how deep the children are.

At Boston Children's Hospital, propofol general anesthesia has been used for EMG for over a decade. This medication is generally safe in the proper circumstances, which typically include monitoring by trained anesthesiologists and/or nurse anesthetists with cardiorespiratory telemetry in a properly equipped day surgery procedure room. The only significant effect on the quality of nerve conduction studies is that F responses are typically obliterated by propofol.

The downside of sedation and general anesthesia, if the previous discussions are not sufficient to curb enthusiasm, is that while they help the quality of nerve conduction studies, they have a deleterious effect on the needle EMG examination, as children will not respond to commands and will not contract as much as desired under these circumstances.

None of the above methods is entirely safe, even the use of local anesthetic, as a child may develop an allergy to the anesthetic and, at worst, an anaphylactic reaction, although fortunately this is very rare.[3] This is more common with amethocaine as opposed to xylocaine.

At this juncture, it is perhaps opportune to make a case for not using sedation at all. In order for this to work, one must engage very actively with the children themselves, if they are old enough, and, if not, with their parents. In truth, regardless of the child's age, it is imperative that the parents be brought on board with the procedure. Clearly it is essential that they not be excluded from the examination room and, indeed, any person accompanying the child should be encouraged to join in. The first step towards engagement is the requirement of complete honesty. There is no point in telling children or their parents that the procedure to follow will produce no feelings of discomfort. Certainly it does produce discomfort and it is important to be candid about this. Reassurance that it is perhaps no worse than a blood test is a good starting point. It is also very important to put all parties at their ease, and the social interactions which would normally be employed when talking to children or their parents outside a medical setting are perfectly acceptable methods. Some might regard this as unprofessional, but it can be very effective. Children will respond after a certain age to questions about what they love to do or their favorite subjects at school, and an anxious child can quite suddenly change and sometimes smile while discussing enjoyable experiences. It is preferable to talk to a child about their dreams of becoming a disco dancer rather than try to calm them by repeated reassurances that what they are experiencing is not uncomfortable, when they clearly disagree. Getting children to talk about favorite subjects can be immensely effective. Many instances of needle-phobic children passing through the EMG without any anxiety have occurred using this approach.

Another invaluable distraction technique is the use of portable electronic devices. Almost every parent will have games or music or even videos on their mobile phone, and many children will come with electronic tablets or similar devices. Whatever amuses them must be employed. Our experience has been that battery-powered devices (i.e. those powered by direct current) generate little to no electrical artifact, whereas nearly any device that is plugged into an electrical outlet (i.e. those powered by alternating current) tends to generate 60/50 Hz interference during the study. Although toys seem to be rather old-fashioned nowadays, a good supply of these is also important.

Another point that is important to emphasize, particularly in the younger patients, is that not all reactions are pain related. A lot of children become angry when they are investigated and asked to do things they do not want to do, even such things as simple as lying down on the couch. Parents recognize this kind of upset and can easily distinguish it from that caused by pain. If they appreciate that the practitioner working with them also understands this, it will firmly establish yet another measure of trust. Many of us have examples of instances when this has occurred. One of the authors (MCP) will always remember a child about two years old, who could only be described as "creating" during the EMG. The noise and anger generated were truly remarkable. Her grandmother, who was of Italian descent, was circling around the doctor, who feared the worst, anticipating a tirade of fury against him for being so unkind to her granddaughter. Instead, all she said was, "The Latin temperament!"

It Is too Difficult

The next misconception is that EMG is difficult. In part this leads from the first misconception, and certainly if the EMG is regarded as a painful experience it is more likely to feel like a struggle. In particular, the EMG practitioner may become frightened of pursuing the examination with adequate determination if they feel they are hurting the child. Whatever one's beliefs about the discomfort of the test, there is a measure of general acceptance that recording reliable results in children is a more skillful exercise than in adults. There are many ways in which skill can be enhanced, not least of which is being slick but also by determining which tests are essential and not performing unnecessary tests that perhaps only corroborate others that have already been done. Modifications to the equipment are important but one of the major problems in pediatric EMG is training. The number of pediatric EMGs performed in any center is small by comparison with the number of adult EMGs. Even in large referral centers with dedicated pediatric electromyographers, the volumes do not compare to those found in many adult EMG laboratories. Thus, the opportunities for a debutant EMG practitioner to learn from the examination of children, whether compliant or not, are limited for a variety of reasons, the most important of which is that children should not have to undergo repeated investigations because the first one was not performed well. Therefore pediatric electromyographers, unlike in adult practice, cannot leave trainees loosely supervised, no matter how valuable it might be for improving confidence and gaining experience. This means that EMG training in pediatrics is unique among all forms of physician training, as a period of observation of the techniques and strategies is absolutely crucial. In this way perhaps pediatric EMG is more like a surgical speciality than a medical one. At Great Ormond Street the period of observation varies according to the aptitude of the debutant. They are gradually allowed to handle cases on their own, but with a senior member of staff either observing and helping them, or near enough to be able to step in and complete the examination seamlessly if the situation becomes untenable.

Another important point with regard to acquiring skills in pediatric EMG is interpretation of the results. While the recording and analysis of the nerve conduction studies is a skill that can be attained over a reasonable period, that of analysis of the EMG is particularly difficult. A specific difficulty is that the muscle fibers in children are smaller than in adults and increase their diameters over the period of maturation. Thus, for example in infants, a normal EMG pattern can appear myopathic compared to those of older children and adults. The measurement of the motor unit duration is absolutely crucial in these circumstances.

To help trainees acquire the experience needed to distinguish pathological abnormalities from normal findings, one of the best methods is to learn from an archive of previously studied cases. Many EMG systems now allow storage of all the data collected including the EMG. At Great Ormond Street Hospital data from selected studies are archived prior to analysis. These data are graded according to difficulty into either *easy*, *moderate*, or *difficult*. In the meantime the study is reported in the usual way and stored in a patient results database. The trainees select what level they feel they want to study and choose a patient from that category. They are then able to go to the patient results database and bring up the data supplied at the time of the referral. Next they go to the EMG database, assess the abnormality, and then return to the database to see if their conclusions concur with those of the consultant or attending neurophysiologist. If they do concur, then the trainee can move on to another case and simply repeat the process, noting instances where there is disagreement, which can then form the basis of in-depth discussions of key topics. This is an invaluable resource, as even trainees who are assigned to a pediatric EMG laboratory for only short periods can still acquire a great deal of knowledge. Regardless of how busy the clinical service is, it is certainly possible for a keen trainee to review the data from upwards of 20 cases in a morning. The benefits to their skill and confidence are hard to overestimate.

EMG Has no Future in the Molecular Age

The impact of molecular genetics on all diagnostic tests has been considerable. One of the reasons that it has not been used perhaps to its fullest extent is the cost of the sequencing and the intensity of the data storage and analysis required. As with many technological advances, this cost will come down with time. Next-generation sequencing technologies are already being used in clinical practice, especially targeted sequence capture panels, which are less expensive and more efficient than traditional Sanger sequencing panels. There are some concerns about the consistency of the depth of coverage for exome sequencing to be used routinely in clinical practice, but exome sequencing will soon become a widespread clinical resource as well, followed by whole genome sequencing. However, these powerful new techniques do not eliminate the dreaded "variant of unknown significance" seen in genetic test reports. Such ambiguities will, if anything, proliferate as more genetic data are generated for each patient. Accurate phenotype information, including EMG, will thus continue to be crucial to help interpret genetic test data.

One common example seen in pediatric EMG laboratories is the child with pes cavus in whom the question of Charcot-Marie-Tooth disease has been raised. There are clinical findings that can help predict whether the child has a polyneuropathy[4] but in many instances an EMG is crucial to determine definitively whether this is the case and, if so, whether the neuropathy is demyelinating or axonal. Unfortunately, because genetic tests are promoted as being less invasive and painful than studies such as EMG, children with pes cavus or toe-walking sometimes undergo an expensive genetic test that reveals a variant of unknown significance in an obscure Charcot-Marie-Tooth disease gene, only to be found later by a pediatric neuromuscular specialist and/or pediatric electromyographer not to have a polyneuropathy at all. Similar situations occur when congenital myasthenic syndrome and various channelopathies are considered in the differential diagnosis, as valuable data regarding the presence and likely subtypes of these disorders can still be obtained more readily from EMG testing than from other sources. It is salutary to note that in a highly specialized congenital myasthenia clinic run at one of our hospitals, some children whose condition has defied genetic classification, when eventually referred for assessment of the neuromuscular junction, are found to have normal jitter. It is therefore highly likely that, while the role of EMGs will continue to evolve, it will still have an important function in defining the phenotype along with clinical examination. Another way of looking at this might be that its role will change from "phenotype down" to "genotype up" analysis, as available genetic data become more widespread among patient populations.

EMG will also remain central to the diagnostic evaluation of children and adolescents with acquired neuromuscular disorders, including inflammatory/autoimmune, toxic, and traumatic insults. These include Guillain-Barré syndrome,[5–7] chronic inflammatory demyelinating polyradiculoneuropathy,[8,9] myasthenia gravis,[10,11] toxic neuropathies[12–14] and neonatal brachial plexopathies.[15–18]

Lastly, more clinical research is being conducted on diseases such as Charcot-Marie-Tooth disease and spinal muscular atrophy, and EMG will likely be an important outcome measure. A specialized nerve conduction study technique, motor unit number estimation, is being used routinely now in clinical research on spinal muscular atrophy.[19,20]

NORMATIVE DATA

It is a fundamental problem, and it does not matter what parameter is measured, whether in clinical neurophysiology or in any other quantitative speciality, that the dataset of normative values available is never truly comprehensive. Factors that inhibit the collection of rigorous normative data for pediatric EMG include the cost, ethical issues (it is hard to imagine a modern institutional review board approving such a study in children and hard to imagine that many young children and their families would volunteer), and the many variables that would have to be accounted for. If we look at, for example, nerve conduction it is recognized that such varying factors as handedness, height, age, and sex might all influence the measurement.[21–30] Temperature is well recognized as being something that affects velocity and yet the truth is that over the range of temperatures seen in most laboratories, its influence most probably is minimal, particularly as the relationship is nonlinear with little if any change over the normal temperatures encountered in the laboratory.[31] Yet this is something we all correct for, perhaps just because we can. However, when considering some of the factors that we do not correct for, again returning to conduction velocity, it is almost certain from our experience in a central London hospital with a wide range of ethnic subgroups that the Caucasian population has the slowest conduction velocity of any group. The velocities experienced by one of the authors (MCP) when visiting Bangladesh were truly astounding when compared with the normal population in the United Kingdom. Trying to cover every single parameter with every single factor that might affect it could, if one is not careful, become a life's work.

Some medical specialities have abandoned the slavish collection of normative data. For example, when a new sequence is devised in MRI it is uncommon for the findings in normal populations to be studied. Instead, what is more likely to happen is that the reporting clinicians make a judgment based on previous data and also the incidence of disease being studied as to what is likely to be abnormal and what is likely to be normal. Experience in these circumstances is the absolute key.

A further problem, perhaps peculiar to children, is that collecting any data on normal children by techniques perceived as being uncomfortable (even if not as painful as some would have it) is highly unlikely to be granted ethical permission. Fortunately, we are lucky to have some historical data, which is well validated, including the studies on nerve conduction velocity changes with age. There are many different studies showing the increase in conduction velocity over the first years of life and, although there are many different sources, they all roughly show the same change and are a useful basis for determining normal velocity.[32–41] Another group on which exhaustive studies have taken place is the motor unit duration values that were first produced by Buchthal and his team in Copenhagen.[42] This was truly a gargantuan piece of work, with individual muscles studied in innumerable normal subjects from very early in life to old age. Unfortunately, this data had every danger of being redundant as it was performed using the standard EMG needle, as the facial needle was either not available or not used in the study. Pediatric electromyography was the particular beneficiary when work by Sanjev Nandedkar and colleagues discovered that the facial needle when used to measure the duration of the motor unit produces a value which is very close to that produced with either a monopolar needle or standard concentric needle[43] but there are significant differences with regard to the amplitudes recorded. This means this precious resource is available for pediatric electromyography to use, hugely improving the accuracy of assessment of conditions such as myopathy and neurogenic change.

However, not such a rosy view can be taken of other normative data. One of the particular problems is amplitude of both the compound muscle action potential and also the sensory nerve action potentials. It is one of the mysteries to all collected wisdom in EMG that, while the EEG world made as its top priority, at the inception of the speciality, standardization not only of the position of the electrodes but also the dimensions of the electrodes themselves, the EMG world did no such thing. EEG is a specialty where amplitude, which is affected by the size of the recording surface, is not as crucial as it is in the nerve conduction studies and yet standardization across the whole specialty in EMG and nerve conduction studies has not yet taken place, even if some laboratories have their own internal standards. Measurements therefore exist but if the recording system used is not similar to your own, the results are essentially unusable. To make the problem worse, in publications detailing the recordings of the amplitude measurements, the exact dimensions of the recording electrodes may not be given.

If there are not enough problems already in the measurement of amplitude, the range of amplitudes encountered between subjects may not be appreciated. To take one example, in certain ethnic groups seen in London, median sensory amplitudes reaching as high as 160 μV have been recorded regularly using an orthodromic palm-to-wrist technique. However, in other ethnic groups without any evidence of peripheral nerve damage values of 30 to 35 μV are more common. This means that a sensory nerve action potential that reaches a modest 40 μV might be passed as normal in one of the former group, with the tester completely unaware that this is perhaps one quarter of what it should be. In fact the child may have lost 75% of their sensory fibers.

From what has been said previously, it is obvious that the interpretation of results is very difficult indeed. It would be wonderful to be able to look at the single criterion and say this is abnormal, but this is an extremely dangerous technique. All of us in pediatric EMG, perhaps no more than our adult colleagues, recognize that only if there is a concordance between our results are we able to accurately assess individual measurements, but only in that context. Needle EMG has been described as being a technique combining auditory pattern recognition with

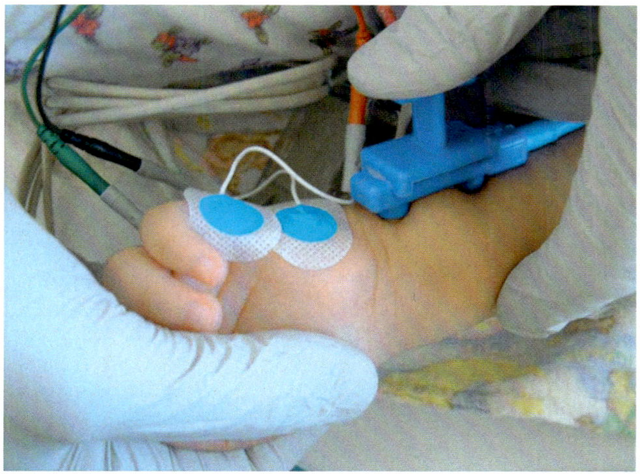

FIGURE 3.1 Ulnar nerve motor studies recording from abductor digiti minimi. Stimulation at the wrist.

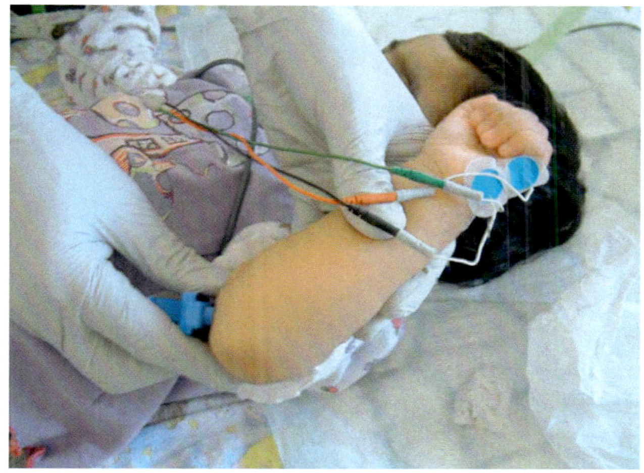

FIGURE 3.2 Stimulation above the elbow.

semiquantitation,[44] but there are elements of this approach that apply to the whole electrodiagnostic examination. It is important to remember this. Unfortunately—and this is one of the major difficulties with the specialty and not the first time that it has been said—experience is key.

TECHNIQUES

Technical advances have improved the quality of data obtained from pediatric EMG studies and have improved the comfort of patients who undergo this testing. First and foremost has been the huge improvement in the recording electrodes, which are placed on the surface. The days when every laboratory had its own homemade recording electrodes are rapidly disappearing. Now there are single-use self-adhesive pre-gelled electrodes, which are high quality at a relatively low price. The recording area is of standard size; it is possible to cut down the whole electrode for use in infants without affecting this, but it is often not necessary even in the smallest limbs (Figures 3.1 and 3.2). Stimulating electrodes also are now better than those formerly used. Unfortunately, there is a tendency to produce what can be best described as "cattle prod" electrodes, large devices often with an intensity control within their handles, which are placed on the limb. While these are probably acceptable in an adult environment, where the patient can be relied upon to remain still, in a child environment they are not always practical. One of the authors (MCP) has a personal preference for using somewhat older electrodes, which allow the limb to be encircled at the same time as delivering the stimulus (Figure 3.3). Even if the child should move, the relationship between stimulus and the nerve is maintained and the test can be performed successfully.

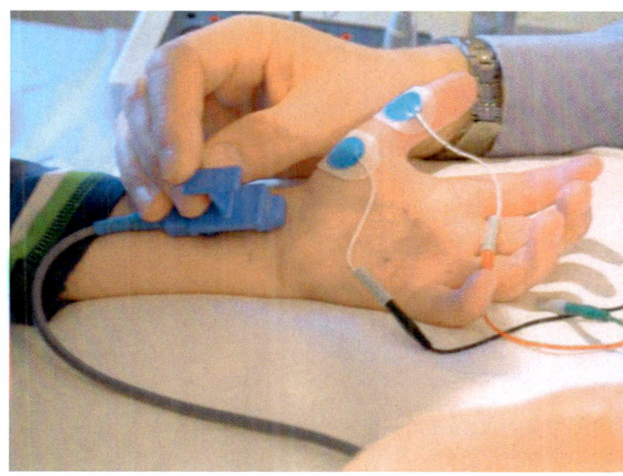

FIGURE 3.3 Stimulation of the median nerve at the wrist recording from abductor pollicis brevis for motor studies demonstrating the use of the stimulating electrode.

The needles have been touched on already, but they are now very high quality, so sharp that just the weight of the electrode cable will allow their passage through normal muscle. For one of the authors (MCP), a key factor in success when performing stimulated single fiber EMG (StimSFEMG) has been that it is now possible to obtain the stimulating electrode at 15 mm lengths rather than the standard 30 mm, which means that the electrode will remain in position without having to be held (Figure 3.4). This leaves the practitioner with both hands free to manipulate the recording electrode. For some time now, this procedure has been performed with facial concentric needle electrodes using various manipulations of the low frequency bandpass to produce something that approaches, but is not exactly, single fiber

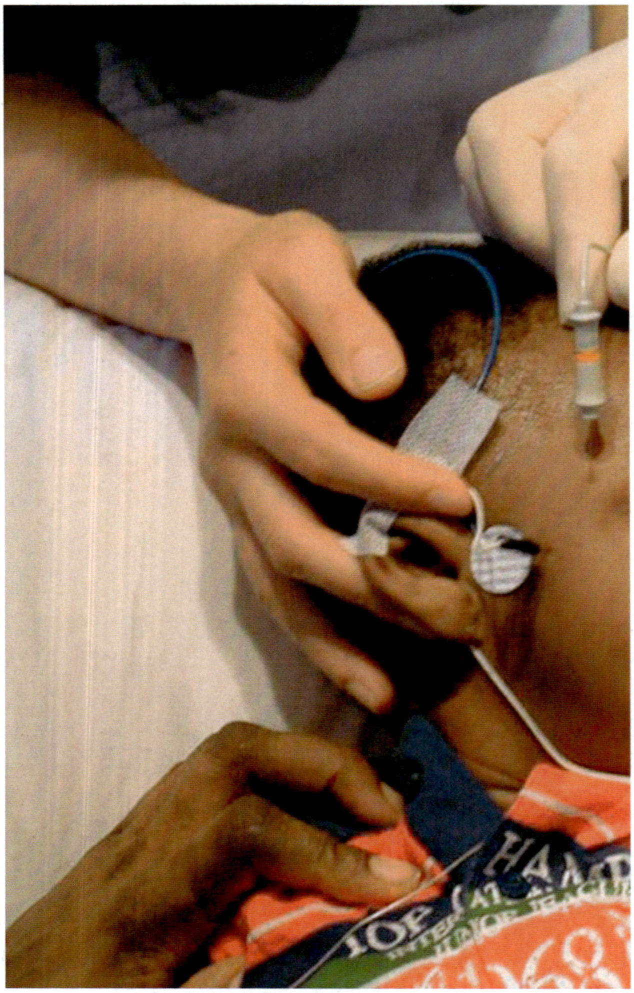

FIGURE 3.4 Electrode positioning for stimulating single fiber electromyography.

electromyography.[45–54] The single fiber electrodes are now being produced as single-use devices, with the price coming down as the technology improves. This development would be a huge advantage for the continued use of StimSFEMG, but the lack of availability of these needles has not been the only reason facial needles are preferred. The problem particularly relevant in children is that, because the recording area is set back from the needle in a child or particularly a baby with a very thin muscle, it is very possible to be in the muscle itself or even through it, but not have the recording surface in the muscle. With the facial needle, where the recording surface is at the leading edge, this is not a problem. Earlier, when using reusable needles it was feasible to sand down the needle to shorten the distance between its tip and the recording area, but clearly this is not possible when using single-use electrodes, which are not meant to be resterilized.

Finally, EMG machines have become much smaller, and they are portable in the true sense of the word. They have successfully made the transition from analog oscilloscopes to digital recordings, and this has led to a very important feature being incorporated in most machines, which is that the EMG is stored. This is hugely important for pediatric EMG as it allows the practitioner to make the analysis of the EMG after the child has left. Some store the whole EMG, others significant samples.

Choice of Nerves

Nowadays, using the stick-on surface electrodes that are available, it is entirely feasible to perform almost every test that one performs in adults on even on the smallest of babies. However, certain nerves are preferentially chosen when studying children. For example, the sural nerve may be difficult to find and stimulate effectively, particularly in small babies. In these, there is often a small pad of fat overlying the lateral malleolus, making this nerve difficult to record in infants. In contrast, the medial plantar, uncomfortable and difficult to stimulate in adults, is by contrast very easy to stimulate in babies and, because the distances involved are greater than when attempting the sural, less likely to be disturbed by artifact. The superficial peroneal is also useful not only because the distance can be varied to reduce the stimulus artifact but also because in combination with the medial plantar it is possible to obtain information on the two branches of the sciatic nerve separately. The sural is well recognized as being a composite nerve with varying contribution from the tibial and peroneal branches of the sciatic nerve.[55,56] Its use in determining the difference between a peroneal and a sciatic nerve palsy should be discouraged.

With motor nerve stimulation in the leg, while the peroneal nerve is useful, being easy to stimulate at the fibula head, often the amplitude of the compound muscle action potential obtained from EDB is quite low at around 1.5 to 2.0 mV, certainly in those under 5 years, which makes any subtle changes difficult to determine. The tibial nerve stimulated at the ankle and recorded from abductor hallucis is perhaps better to use, particularly in the youngest children as the proximal stimulation; something that is quite uncomfortable in the adult is easy and well tolerated in babies.

When considering a motor nerve in the arm, the median is the mainstay of many investigations but the ulnar is the better nerve in children because the median proximal stimulation can be uncomfortable. The ulnar by contrast, with the nerve passing subcutaneously around the elbow, is easy to stimulate using less current.

The choice of nerve chosen will vary according to the clinical question but in most instances in children a generalized abnormality is looked for rather than specific mononeuropathies. Recording from the muscles described above therefore will often suffice, but if a specific nerve is under suspicion the examiner will need to do a standard examination for that particular mononeuropathy.

When looking at muscles, of course one can sample any muscle that can be reached, but a major problem in children, compounded if they are significantly sedated, is activation. The muscle that reigns supreme with regard to effective activation is tibialis anterior. In even the youngest child, once the needle is placed in the muscle it is easy to get very effective activation by tickling the sole of the foot. It is also easy for more aware children to respond to instructions to move the foot upwards and backwards at the ankle joint on command. Although what is often asked for is a more widespread examination of muscles and in particular the examination of proximal muscles, over the years experience of vastus medialis or any other of the quadriceps group has been generally disappointing. It is difficult enough to activate in a cooperative adult and these problems are compounded several-fold in children. In babies it is virtually impossible. A colleague attempted to get around these problems by placing the needle in the quadriceps and then asking the helpers to make the baby take their weight on their legs. Instead of extending the knee, children placed in this position would tend to demonstrate a stepping reflex, flexing both knee and hip joints. Iliopsoas is another proximal muscle, which is perhaps easier but not without its difficulties in sampling. It requires the baby or young infant to be undressed and it is perfectly possible in very small babies to cause damage to the neurovascular bundle; for these reasons it is not used routinely. In the upper limb biceps can be easily sampled and can provide useful information. As an observation though, babies who seem to be moving the arms freely often will not move the arm at all when biceps is sampled, much to the frustration of the electromyographer, likely due to the discomfort that seems to be quite prominent for the muscle.

It is one of the author's opinions (MCP) that it is much more important to sample one muscle in great detail than to use brief samplings of several muscles. The discomfort of the EMG is mainly related to the passage of the needle through the skin, although it must be acknowledged that there is some discomfort within the muscle. This technique of studying one muscle in detail does reduce the trauma of the examination. We have found that even in a situation where the clinical picture is suggestive of proximal abnormality, if the tibialis anterior is sampled adequately, which can mean up to 5 minutes recording, it is possible to detect abnormalities in most neuromuscular conditions, even if showing a proximal prominence of the abnormality such as in spinal muscular atrophy.

The technique used in sampling is as follows and is based loosely on the techniques used by Buchthal in Copenhagen. After initial insertion of the needle, it is passed progressively deeper into the muscle with the recording being made throughout this period. It is then withdrawn to almost the skin surface and angled and passed in another direction, again sampling all the time. This can be done in four quadrants, and the information obtained if the recording is made of several different depths is extensive and far more than what is recorded with a needle muscle biopsy. The final recording is made with maximum contraction and this allows the automatic interference pattern analysis programs to assess the quality of the interference pattern.[57–59] This technique will provide adequate data to employ the motor duration program, which is now available on all commercial EMG machines. A much shorter period of examination is entirely feasible in adults who are able to cooperate fully and give the exact amount of activation needed, but in children one must be patient and wait for the periods when the contraction is not so strong as to make isolation of individual motor units impossible.

INVESTIGATION STRATEGIES

When presented with the investigation of generalized weakness as the clinical problem, there is a simple strategy which involves assessment of one sensory nerve and one motor nerve in the legs followed by EMG of tibialis anterior. If a neuropathy is found, it is important to make an assessment of the arm nerves as well. It is important, when presented with generalized weakness and if one is convinced by your colleagues that this is a neuromuscular disorder, not to stop until an abnormality is found. This may mean that after having examined several nerves and EMG of tibialis anterior without finding abnormality, you may go on to assess neuromuscular junction. In Great Ormond Street Hospital the preference is to do this with StimSFEMG of orbicularis oculi.[10] There are occasions, however, when generalized weakness is thought most likely to be due to a central nervous system problem, and the request for EMG is in part to reassure the examiner that this is indeed a case. In these instances, the simple strategy suffices.

Neuropathies can obviously be discovered when looking for the cause of generalized weakness, but sometimes one is given a very strong clue from the clinical picture. Whatever the setting, the major function in an EMG examination is to determine whether this is an axonal abnormality or demyelination. If the axonal form is suggested by perhaps reduction or even absence of the sensory nerves' responses, it is absolutely crucial, irrespective of the findings on the motor nerve compound muscle action potential assessment, that EMG is performed. It is perfectly feasible for the compound muscle action potential not to show significant reduction because of compensatory reinnervation, yet the denervation is obvious when doing EMG. It is very important not to allow a diagnosis of a sensory neuropathy to stand unless EMG is normal.

The discovery of a demyelinating polyneuropathy may be one setting where more extensive studies are needed. First there is a requirement to demonstrate conduction block, which may not be seen in all nerves and has to be pursued, and if found does give some neurophysiological support to a diagnosis of CIDP, particularly Lewis Sumner syndrome[60] rather than hereditary neuropathy. This distinction, however, is far from foolproof, with some HMSN genotypes[61,62] and other inherited neuropathies demonstrating conduction block; all of us have had experience of seeing it in proven HMSN type 1, likely artifactual from nerve thickening, making effective stimulation difficult. Furthermore, more extensive studies may be needed to demonstrate preservation of the velocity gradient, with stimulation of proximal nerve segments showing increased velocity compared with more distal segments and finally demonstrating that the same segments have similar velocities. These latter two findings are more commonly associated with hereditary neuropathy rather than an acquired etiology.

F waves can add to the accuracy of nerve conduction studies but their absence may be a crucial finding, the only demonstrable neurophysiological change in the case of Guillain-Barré syndrome. Sometimes the children are sufficiently weak to be in intensive care and yet this is the only abnormality. This study needs to be pursued if clinical suspicion of Guillain-Barré syndrome is high.

A suspicion of motor neuronopathy itself is seldom the reason for EMG. The commonest form of motor neuronopathy in children, SMA, in the ideal situation with genetic screening available within a few days, is usually a clinical diagnosis backed by support from the genetic abnormality, and in this circumstance EMG has little or no place. Undoubtedly, particularly where the child is already demonstrating respiratory compromise and the gene screening may take a few weeks, cases may be referred for EMG to confirm the clinical suspicions. These EMG investigations are an invaluable aid to management decisions should questions about ventilation arise before the gene result has returned. As an incidental finding motor neuronopathy is usually discovered, and increasingly so in the context of generalized weakness with obvious changes, usually on tibialis anterior. What is done next is quite different from what is done in adults, as alluded to above. When finding a motor neuronopathy in adults, the differential diagnosis typically lies between motor neuron disease and concurrent cervical and lumbar radiculopathy. This differential does not occur in children; instead, the differential diagnosis is between a generalized motor neuronopathy and a segmental myelopathy in either the cervical or lumbar regions. It is not necessary to follow the most recent recommendations for the EMG diagnosis of motor neuron disease.[63–65] Having demonstrated motor neuronopathy in tibialis anterior if one of the bulbar muscles is sampled, particularly the tongue, which is well tolerated from a submental approach (Figure 3.5), and found to be abnormal as shown, it is possible to give clear indication that this is generalized. Finding that the tongue is denervated means that further examination of other muscles is redundant. If the tongue is normal, we then need to determine whether this is a lumbar myelopathy or a myelopathy restricted to the spinal cord only. EMG examination of the arm follows, and if the arm is normal, examination of the opposite leg. Even in the situation where a segmental myelopathy is present, the number of needle insertions is a great deal less than is typically performed in an adult.

Myopathies are often diagnosed clinically, supported by genetics and other important markers such as raised CK, particularly relevant for diagnoses such as muscular dystrophy, and backed up with the muscle biopsy. Therefore, in many instances the cases that are referred for EMG are those in whom the diagnosis of myopathy

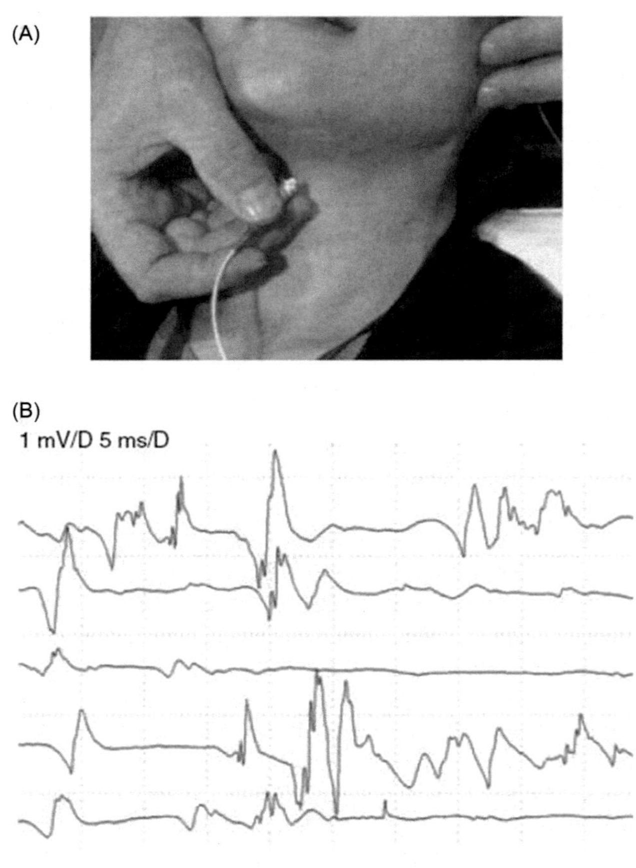

FIGURE 3.5 EMG of the tongue approached by the submental route (A), with a neurogenic pattern seen on EMG (B).

is more difficult or perhaps more doubtful. You are therefore seeing the difficult cases. In these instances a very detailed EMG as described above with up to 5 minutes of sampling is the norm. There are times when the findings will be at variance with the muscle biopsy for the reasons of sampling but also, more interestingly, at variance with the MRI. It is important not to be overawed by this lack of concordance and perhaps not trust the EMG findings. This information is still of value to clinicians who, reassured by the absence of abnormality on the muscle biopsy and MRI, are still able to give some assurances to the parents that they are indeed correct in suspecting that their children's difficulties do have a pathological basis, even if the abnormality is mild. For some parents this is all they want to know. Some conditions which do not seriously affect the mobility of the child, such as ligamentous laxity and collagen disorders, may have EMG abnormalities.

Disorders of the neuromuscular junction are particularly important to determine because, uncommon in most neuromuscular disease, they do raise the possibility of treatment. It is therefore entirely justifiable to be rigorous in pursuit of the diagnosis. Single fiber EMG is the most sensitive technique, and this should not really be the subject of further debate.[66–69] The choice of the different forms of single fiber EMG, either voluntary or stimulation, in children does not really exist. Voluntary single fiber EMG is only practical on a small proportion of children. The Great Ormond Street Hospital in London has had a long experience in investigating disorders of neuromuscular junction and thus abandoned using voluntary single fiber EMG in children regardless of the age, turning instead to StimSFEMG of orbicularis oculi as the preferred technique.[10,70] Although perfecting the technique has taken some years with the application of Ametopp (local anaesthetic preparation of amethocaine) over the points of stimulation and recording, it has been their experience that it is possible to do this in nearly every child without any other form of sedation or recourse to general anesthetic, which is the fallback plan. It is well tolerated and the results appear to be reproducible and reliable.

It is important to realize, though, that in children the finding of an abnormal single fiber EMG does not equate to a diagnosis of a neuromuscular junction disorder, and therefore myasthenia. In adults the differential diagnosis between myasthenia and other conditions that are similar is much narrower, and once motor neuron disease has been excluded, an abnormal single fiber result essentially equates to the diagnosis of myasthenia, particularly in the elderly. In children it is very important to consider the discovery of an abnormality on stimulation single fiber EMG as the beginning of the investigative strategy. In the strategy followed in Great Ormond Street Hospital, the examination of orbicularis oculi is followed by an analysis of the EMG of that muscle, and if any doubt remains as to its normality, EMG of the genioglossus. Certainly in children under one year of age, bulbar palsies are more likely to cause the abnormality of jitter than myasthenia. This examination is then followed by a general EMG examination covering sensory and motor nerves and a peripheral muscle EMG. Minor abnormalities of jitter are more commonly associated with myopathies[71,72] than myasthenia, although the changes in Dok7 myasthenia may be less than other forms.[73,74] In their strategy repetitive nerve stimulation may be used, but only after all the other tests have been completed, as undue stress of the children at this point will not jeopardize the results of the StimSFEMG, and if abnormal gives further support to a possible myasthenic disorder, but if normal does not exclude it. The sensitivity of StimSFEMG is high with this strategy but specificity, something that is a concern when using the results of StimSFEMG on their own, is greatly enhanced by incorporating of StimSFEMG into this comprehensive investigative strategy.

It is relevant at this point to further discuss repetitive nerve stimulation. It is undoubtedly a useful technique with high specificity but if normal, needs to be followed up by single fiber EMG, in order to completely rule out a disorder of the neuromuscular junction.[75–77] Unfortunately, the test itself is uncomfortable and if used as one of the first interventions with the child, an analysis using StimSFEMG will at least be more difficult and possibly not possible to do at all. It was for this reason that Great Ormond Street hospital developed a strategy that began with StimSFEMG. If the repetitive nerve stimulation follows this, it allows distinction between those cases which are due to a myopathy with an associated neuromuscular junction abnormality, in whom the repetitive nerve stimulation will be normal, and those cases due to myasthenia where it is more likely but not inevitably to be abnormal.

Sometimes there are requests for investigation of other presentations. One such example is a facial palsy, which can be seen as a result of trauma or congenital conditions such as Möbius syndrome. The EMG is required to document the abnormality. While the blink reflexes and EMG of the affected muscles are useful examinations, recording the amplitude of the evoked compound muscle action potential and comparing with the unaffected side gives a further indication of the degree of axonal loss.

Diaphragmatic EMG is sometimes specifically requested. It can safely be performed in any child and is most common, because of the likely presentation, to be done in a critical care unit. Approached via the transcostal route, as recommended by Bolton,[78] information about the intercostals, which are traversed as the needle progresses towards the diaphragm, can also be useful (Figure 3.6).

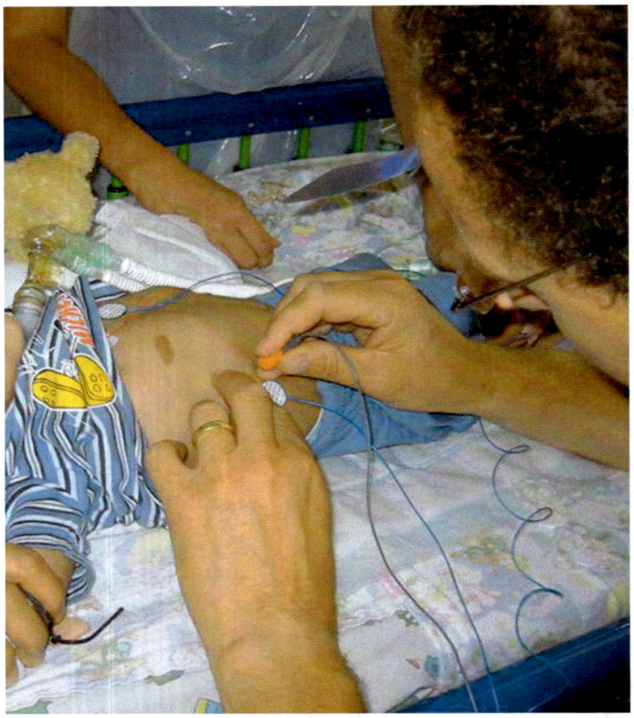

FIGURE 3.6 Positioning for trans-costal examination of the diaphragm.

Despite the ease of performing the test, its contributions to diagnosis are relatively few. A particular instance is the diagnosis of SMARD[79–83] where demonstration of diaphragmatic denervation is crucial to separate that condition from the more common SMA in which the diaphragm should not be involved. Sometimes the electromyographer is asked to look at the diaphragm because of failure to wean, but perhaps the most common reason for diaphragmatic palsy in a pediatric hospital is a postoperative palsy following cardiac operations. Strangely, this is the one area where it is not used. The reason for this is that for the EMG of the diaphragm to be discriminatory between neurotmesis and neurapraxia requires a few days to have passed. EMG on its own is not the usual way to determine nerve damage because of the delay before fibrillation potentials appear. However, one is forced to rely solely on this as phrenic nerve stimulation is only possible proximally, and without distal stimulation an absent response cannot be distinguished between neurapraxia, for which an expectant policy could be followed, and complete separation of the nerve for which surgical intervention is needed. In adults and adolescents, the appearance of fibrillation potentials may take as long as two or three weeks and in a baby is almost certain to be at least for five days. Unfortunately, the pressure to wean children off the ventilators in cardiac intensive care means that, for the very best of reasons, cardiac surgeons will, on finding an absence of a phrenic nerve response, rapidly refer a child for a plication procedure.

Obstetric brachial plexus palsy, also known as neonatal brachial plexus palsy, is a difficult condition to treat and to justify, within its management, the role of EMG.[17,18] There is agreement that obstetric brachial plexus palsy is not disappearing and this is most probably due in part to the increase in birthweight.[84,85] Also, some children will do better with operation than not and the majority will, while showing varying degrees of improvement, all show some residual defect.[86] Disagreement remains about whether this is always the obstetrician's fault[87,88] and evidence suggests that some cases of neonatal brachial plexus injury are caused by intrauterine biomechanical forces.[89] The role of EMG in the evaluation and monitoring of neonatal brachial plexus injury has also been questioned.[90] Work has been done over the years using the results of needle EMG.[91] A further issue is when EMG should be done. Traditionally it has always been suggested that examination at three months of age,[15,92] a time when clinical assessment is thought to be most predictive of later outcome,[93,94] produces the most useful information. A recent paper has challenged this, suggesting that the information at one month is as good if not better than that at 3 months.[16]

One of the major concerns is that the parents often don't have accurate information as to the likely prognosis early on, particularly if the prognosis for recovery is poor, making the management difficult. Even if documenting a severe abnormality early on does not lead to any change in treatment, as operating on these babies may be dangerous when they are very young because of the trauma they have experienced, it is still is something that we should strive towards.

Finally, examination in critical care units is a particular specialty within pediatric EMG. It is very much the investigation of the very young.[95] In a study at Great Ormond Street Hospital, the majority were under one year of age and the largest proportion under three months of age. EMG and nerve conduction studies being performed in children so young means interpretation of results can be difficult, particularly for the EMG because the individual muscle fibers in infants are very small, making normal muscle take on the appearance of a myopathy. There are also challenges presented by electromagnetic interference in critical care settings, which are often unavoidable as these patients usually cannot travel to the EMG laboratory and the interference is often caused by intravenous or enteral nutrition pumps that cannot easily be turned off. Incessant 60 Hz artifact is the unfortunate result in many cases. Despite these problems, invaluable information can be discovered, such as abnormalities of the neuromuscular junction, spinal muscular atrophy, or congenital myopathy when demonstrating a demyelinating neuropathy.

FINAL WORDS

This chapter maintains that EMG, which includes nerve conduction study, is an investigation that will continue to be valuable in the management of neuromuscular disorders. It is likely to continue to deliver important information. This information will be less commonly diagnostic on its own but has an important role in directing further investigations and often delivers this information very quickly with a minimum of effort. The problems of its acceptability may persist, and neurophysiologists need to continue to emphasize the positive aspects of its use. Unquestionably, like many ancillary investigations, it will evolve in its role with the increasing impact of molecular genetics, particularly as next-generation tools become more affordable and accessible. EMG has particular strengths that should not be overlooked. One such strength is in the recognition of motor neuronopathy. Clinicians may not be aware of this because for a time early in these diseases effective reinnervation will prevent wasting. Its identification has acquired particular importance with the recent discovery that some cases of Brown-Vialetto-Van Laere syndrome have a riboflavin transporter defect, which if corrected can lead to remarkable improvement.[96–99] The motor neuronopathy, which appears often in the bulbar muscles first, may be identifiable before other manifestations of the disease, such as deafness, are noted.

Clearly EMG continues to have important roles not only in classification of neuropathies but also in screening for myopathy. In the latter, EMG is likely to be used for those myopathies not clearly identifiable clinically. The continual problems of trying to diagnose myasthenia in very young children will remain, in which EMG and, in particular, StimSFEMG will have a pivotal role.

REFERENCES

1. Hays RM, Hackworth SR, Speltz ML, Weinstein P. Physicians' practice patterns in pediatric electrodiagnosis. *Arch Phys Med Rehabil* 1993;**74**:494–6.
2. Hays RM, Hackworth SR, Speltz ML, Weinstein P. Exploration of variables related to children's behavioral distress during electrodiagnosis. *Arch Phys Med Rehabil* 1992;**73**:1160–2.
3. Russell SC, Doyle E. A risk-benefit assessment of topical percutaneous local anaesthetics in children. *Drug Saf* 1997;**16**:279–87.
4. Karakis I, Gregas M, Darras BT, Kang PB, Jones HR. Clinical correlates of Charcot-Marie-Tooth disease in patients with pes cavus deformities. *Muscle Nerve* 2013;**47**:488–92.
5. Devos D, Magot A, Perrier-Boeswillwald J, Fayet G, Leclair-Visonneau L, Ollivier Y, et al. Guillain-Barre syndrome during childhood: particular clinical and electrophysiological features. *Muscle Nerve* 2013;**48**:247–51.
6. Ryan MM. Guillain-Barre syndrome in childhood. *J Paediatr Child Health* 2005;**41**:237–41.
7. Sladky JT. Guillain-Barre syndrome in children. *J Child Neurol* 2004;**19**:191–200.
8. McMillan HJ, Kang PB, Jones HR, Darras BT. Childhood chronic inflammatory demyelinating polyradiculoneuropathy: combined analysis of a large cohort and eleven published series. *Neuromuscul Disord* 2013;**23**:103–11.
9. Bromberg MB. Review of the evolution of electrodiagnostic criteria for chronic inflammatory demyelinating polyradiculoneuropathy. *Muscle Nerve* 2011;**43**:780–94.
10. Pitt M. Neurophysiological strategies for the diagnosis of disorders of the neuromuscular junction in children. *Dev Med Child Neurol* 2008;**50**:328–33.
11. Liew WK, Kang PB. Update on juvenile myasthenia gravis. *Curr Opin Pediatr* 2013;**25**:694–700.
12. Fleming FJ, Vytopil M, Chaitow J, Jones Jr. HR, Darras BT, Ryan MM. Thalidomide neuropathy in childhood. *Neuromuscul Disord* 2005;**15**:172–6.
13. Priolo T, Lamba LD, Giribaldi G, De NE, Grosso P, De GE, et al. Childhood thalidomide neuropathy: a clinical and neurophysiologic study. *Pediatr Neurol* 2008;**38**:196–9.
14. Ho DT, Rodig NM, Kim HB, Lidov HG, Shapiro FD, Raju GP, et al. Rapid reversal of uremic neuropathy following renal transplantation in an adolescent. *Pediatr Transplant* 2012;**16**:E296–300.
15. Pitt M. Why wait 3 months before doing electromyography in obstetric brachial plexus lesions? Challenging the norm. *Dev Med Child Neurol* 2012;**54**:682.
16. van Dijk JG, Pondaag W, Buitenhuis SM, Van Zwet EW, Malessy MJ. Needle electromyography at 1 month predicts paralysis of elbow flexion at 3 months in obstetric brachial plexus lesions. *Dev Med Child Neurol* 2012;**54**:753–8.
17. Pitt M, Vredeveld JW. The role of electromyography in the management of the brachial plexus palsy of the newborn. *Clin Neurophysiol* 2005;**116**:1756–61.
18. Pitt M, Vredeveld JW. The role of electromyography in the management of obstetric brachial plexus palsies. *Suppl Clin Neurophysiol* 2004;**57**:272–9.
19. Swoboda KJ, Prior TW, Scott CB, McNaught TP, Wride MC, Reyna SP, et al. Natural history of denervation in SMA: relation to age, SMN2 copy number, and function. *Ann Neurol* 2005;**57**:704–12.
20. Kang PB, Gooch CL, McDermott MP, Darras BT, Finkel RS, Yang ML, et al. The motor neuron response to SMN1 deficiency in spinal muscular atrophy. *Muscle Nerve* 2014;**49**:636–44.
21. Gupta N, Sanyal S, Babbar R. Sensory nerve conduction velocity is greater in left handed persons. *Indian J Physiol Pharmacol* 2008;**52**:189–92.
22. Kommalage M, Gunawardena S. Influence of age, gender, and sidedness on ulnar nerve conduction. *J Clin Neurophysiol* 2013;**30**:98–101.
23. Rivner MH, Swift TR, Malik K. Influence of age and height on nerve conduction. *Muscle Nerve* 2001;**24**:1134–41.
24. Robinson LR, Stolov WC, Rubner DE, Wahl PW, Leonetti DL, Fujimoto WY. Height is an independent risk factor for neuropathy in diabetic men. *Diabetes Res Clin Pract* 1992;**16**:97–102.
25. Bodofsky E, Tomaio A, Campellone J. The mathematical relationship between height and nerve conduction velocity. *Electromyogr Clin Neurophysiol* 2009;**49**:155–60.
26. Rivner MH, Swift TR, Malik K. Influence of age and height on nerve conduction. *Muscle Nerve* 2001;**24**:1134–41.

27. Hyllienmark L, Ludvigsson J, Brismar T. Normal values of nerve conduction in children and adolescents. *Electroencephalogr Clin Neurophysiol* 1995;**97**:208–14.
28. Stetson DS, Albers JW, Silverstein BA, Wolfe RA. Effects of age, sex, and anthropometric factors on nerve conduction measures. *Muscle Nerve* 1992;**15**:1095–104.
29. Bhorania S, Ichaporia RB. Effect of limb dominance on motor nerve conduction. *Indian J Physiol Pharmacol* 2009;**53**:279–82.
30. Campbell Jr. WW, Ward LC, Swift TR. Nerve conduction velocity varies inversely with height. *Muscle Nerve* 1981;**4**:520–3.
31. Todnem K, Knudsen G, Riise T, Nyland H, Aarli JA. The non-linear relationship between nerve conduction velocity and skin temperature. *J Neurol Neurosurg Psychiatry* 1989;**52**:497–501.
32. Garcia A, Calleja J, Antolin FM, Berciano J. Peripheral motor and sensory nerve conduction studies in normal infants and children. *Clin Neurophysiol* 2000;**111**:513–20.
33. Smit BJ, Kok JH, De Vries LS, Dekker FW, Ongerboer de Visser BW. Motor nerve conduction velocity in very preterm infants. *Muscle Nerve* 1999;**22**:372–7.
34. Hyllienmark L, Ludvigsson J, Brismar T. Normal values of nerve conduction in children and adolescents. *Electroencephalogr Clin Neurophysiol* 1995;**97**:208–14.
35. Lang HA, Puusa A, Hynninen P, Kuusela V, Jantti V, Sillanpaa M. Evolution of nerve conduction velocity in later childhood and adolescence. *Muscle Nerve* 1985;**8**:38–43.
36. Vecchierini-Blineau MF, Guiheneuc P. Maturation of nerve conduction velocity from birth to 5 years of age. *Arch Fr Pediatr* 1979;**36**:563–72.
37. Martinez AC, Ferrer MT, Conde MC, Bernacer M. Motor conduction velocity and H-reflex in infancy and childhood. II. Intra and extrauterine maturation of the nerve fibres. Development of the peripheral nerve from 1 month to 11 years of age. *Electromyogr Clin Neurophysiol* 1978;**18**:11–27.
38. Moosa A. Further studies of motor nerve conduction velocity in newborn infants. *Arch Dis Child* 1969;**44**:782–3.
39. Dunn HG, Buckler J, Morrison GC, Emery AW. Conduction velocity of motor nerves in infants and children. *Pediatrics* 1964;**34**:708–27.
40. Monacelli G, Faienza C, Squarcia U. Determination of conduction velocity of the motor fibers of the ulnar nerve and median nerve in the healthy child. *Lattante* 1964;**35**:651–66.
41. Gamstorp I. Conduction velocity of peripheral nerves in children. *Trans Am Neurol Assoc* 1964;**89**:198–9.
42. Ludin H-P. *Normal Values, Electromyography in Practice*. New York: Georg Thiem Verlag Thieme-Stratton Inc; 2013. p. 124
43. Barkhaus PE, Roberts MM, Nandedkar SD. "Facial" and standard concentric needle electrodes are not interchangeable. *Electromyogr Clin Neurophysiol* 2006;**46**:259–61.
44. Daube JR, Rubin DI. Needle electromyography. *Muscle Nerve* 2009;**39**:244–70.
45. Kouyoumdjian JA, Stalberg EV. Concentric needle jitter on stimulated Orbicularis Oculi in 50 healthy subjects. *Clin Neurophysiol* 2011;**122**:617–22.
46. Orhan EK, Deymeer F, Oflazer P, Parman Y, Baslo MB. Jitter analysis with concentric needle electrode in the masseter muscle for the diagnosis of generalised myasthenia gravis. *Clin Neurophysiol* 2013;**124**:2277–82.
47. Stalberg E. Jitter analysis with concentric needle electrodes. *Ann N Y Acad Sci* 2012;**1274**:77–85.
48. Sanders DB. Measuring jitter with concentric needle electrodes. *Muscle Nerve* 2013;**47**:317–18.
49. Kouyoumdjian JA, Stalberg EV. Concentric needle jitter on stimulated Orbicularis Oculi in 50 healthy subjects. *Clin Neurophysiol* 2011;**122**:617–22.
50. Stalberg EV, Sanders DB. Jitter recordings with concentric needle electrodes. *Muscle Nerve* 2009;**40**:331–9.
51. Farrugia ME, Weir AI, Cleary M, Cooper S, Metcalfe R, Mallik A. Concentric and single fiber needle electrodes yield comparable jitter results in myasthenia gravis. *Muscle Nerve* 2009;**39**:579–85.
52. Cattaneo L, Cucurachi L, Pavesi G. Concentric needle recording of neuromuscular jitter in the temporalis muscle. *Neurophysiol Clin* 2007;**37**:50–1.
53. Sarrigiannis PG, Kennett RP, Read S, Farrugia ME, Single-fiber EMG. with a concentric needle electrode: validation in myasthenia gravis. *Muscle Nerve* 2006;**3**:61–5.
54. Ertas M, Baslo MB, Yildiz N, Yazici J, Oge AE. Concentric needle electrode for neuromuscular jitter analysis. *Muscle Nerve* 2000;**23**:715–19.
55. Huelke DF. The origin of the peroneal communicating nerve in adult man. *Anat Rec* 1958;**132**:81–92.
56. Huelke DF. A study of the formation of the sural nerve in adult man. *Am J Phys Anthropol* 1957;**15**:137–47.
57. Nandedkar SD, Sanders DB, Stalberg EV. Automatic analysis of the electromyographic interference pattern. Part II: Findings in control subjects and in some neuromuscular diseases. *Muscle Nerve* 1986;**9**:491–500.
58. Nandedkar SD, Sanders DB, Stalberg EV. Simulation and analysis of the electromyographic interference pattern in normal muscle. Part II: Activity, upper centile amplitude, and number of small segments. *Muscle Nerve* 1986;**9**:486–90.
59. Nandedkar SD, Sanders DB, Stalberg EV. Automatic analysis of the electromyographic interference pattern. Part I: Development of quantitative features. *Muscle Nerve* 1986;**9**:431–9.
60. Lewis RA. Neuropathies associated with conduction block. *Curr Opin Neurol* 2007;**20**:525–30.
61. Yger M, Stojkovic T, Tardieu S, Maisonobe T, Brice A, Echaniz-Laguna A, et al. Characteristics of clinical and electrophysiological pattern of Charcot-Marie-Tooth 4C. *J Peripher Nerv Syst* 2012;**17**:112–22.
62. Cottenie E, Menezes MP, Rossor AM, Morrow JM, Yousry TA, Dick DJ, et al. Rapidly progressive asymmetrical weakness in Charcot-Marie-Tooth disease type 4J resembles chronic inflammatory demyelinating polyneuropathy. *Neuromuscul Disord* 2013;**23**:399–403.
63. Carvalho MD, Swash M. Awaji diagnostic algorithm increases sensitivity of El Escorial criteria for ALS diagnosis. *Amyotroph Lateral Scler* 2009;**10**:53–7.
64. Costa J, Swash M, de CM. Awaji criteria for the diagnosis of amyotrophic lateral sclerosis: a systematic review. *Arch Neurol* 2012;**69**:1410–16.
65. Dengler R. El Escorial or Awaji Criteria in ALS diagnosis, what should we take? *Clin Neurophysiol* 2012;**123**:217–18.
66. Benatar M. A systematic review of diagnostic studies in myasthenia gravis. *Neuromuscul Disord* 2006;**16**:459–67.

67. Zivkovic SA, Shipe C. Use of repetitive nerve stimulation in the evaluation of neuromuscular junction disorders. *Am J Electroneurodiagnostic Technol* 2005;**45**:248–61.
68. AAEM Quality Assurance Committee. American Association of Electrodiagnostic Medicine. Literature review of the usefulness of repetitive nerve stimulation and single fiber EMG in the electrodiagnostic evaluation of patients with suspected myasthenia gravis or Lambert-Eaton myasthenic syndrome. *Muscle Nerve* 2001;**24**:1239–47.
69. Witoonpanich R, Dejthevaporn C, Sriphrapradang A, Pulkes T. Electrophysiological and immunological study in myasthenia gravis: Diagnostic sensitivity and correlation. *Clin Neurophysiol* 2011;**122**:1873–7.
70. Pitt M. Workshop on the use of stimulation single fibre electromyography for the diagnosis of myasthenic syndromes in children held in the Institute of Child Health and Great Ormond Street Hospital for Children in London on April 24th, 2009. *Neuromuscul Disord* 2009;**19**:730–2.
71. Forrest KM, Al-Sarraj S, Sewry C, Buk S, Tan SV, Pitt M, et al. Infantile onset myofibrillar myopathy due to recessive CRYAB mutations. *Neuromuscul Disord* 2011;**21**:37–40.
72. Munot P, Lashley D, Jungbluth H, Feng L, Pitt M, Robb SA, et al. Congenital fibre type disproportion associated with mutations in the tropomyosin 3 (TPM3) gene mimicking congenital myasthenia. *Neuromuscul Disord* 2010;**20**:796–800.
73. Jephson CG, Mills NA, Pitt MC, Beeson D, Aloysius A, Muntoni F, et al. Congenital stridor with feeding difficulty as a presenting symptom of Dok7 congenital myasthenic syndrome. *Int J Pediatr Otorhinolaryngol* 2010;**74**:991–4.
74. Klein A, Pitt MC, McHugh JC, Niks EH, Sewry CA, Phadke R, et al. DOK7 congenital myasthenic syndrome in childhood early diagnostic clues in 23 children. *Neuromuscul Disord* 2013;**23**:883–91.
75. Srivastava A, Kalita J, Misra UK. A comparative study of single fiber electromyography and repetitive nerve stimulation in consecutive patients with myasthenia gravis. *Electromyogr Clin Neurophysiol* 2007;**47**:93–6.
76. Sonoo M, Uesugi H, Mochizuki A, Hatanaka Y, Shimizu T. Single fiber EMG and repetitive nerve stimulation of the same extensor digitorum communis muscle in myasthenia gravis. *Clin Neurophysiol* 2001;**112**:300–3.
77. Padua L, Stalberg E, LoMonaco M, Evoli A, Batocchi A, Tonali P. SFEMG in ocular myasthenia gravis diagnosis. *Clin Neurophysiol* 2000;**111**:1203–7.
78. Bolton CF, Grand'Maison F, Parkes A, Shkrum M. Needle electromyography of the diaphragm. *Muscle Nerve* 1992;**15**:678–81.
79. Giannini A, Pinto AM, Rossetti G, Prandi E, Tiziano D, Brahe C, et al. Respiratory failure in infants due to spinal muscular atrophy with respiratory distress type 1. *Intensive Care Med* 2006;**32**:1851–5.
80. Pitt M, Houlden H, Jacobs J, Mok Q, Harding B, Reilly M, et al. Severe infantile neuropathy with diaphragmatic weakness and its relationship to SMARD1. *Brain* 2003;**126**:2682–92.
81. Appleton RE, Hubner C, Grohmann K, Varon R. Congenital peripheral neuropathy presenting as apnoea and respiratory insufficiency: spinal muscular atrophy with respiratory distress type 1 (SMARD1). *Dev Med Child Neurol* 2004;**46**:576.
82. Rudnik-Schoneborn S, Stolz P, Varon R, Grohmann K, Schachtele M, Ketelsen UP, et al. Long-term observations of patients with infantile spinal muscular atrophy with respiratory distress type 1 (SMARD1). *Neuropediatrics* 2004;**35**:174–82.
83. Grohmann K, Varon R, Stolz P, Schuelke M, Janetzki C, Bertini E, et al. Infantile spinal muscular atrophy with respiratory distress type 1 (SMARD1). *Ann Neurol* 2003;**54**:719–24.
84. Pondaag W, Allen RH, Malessy MJ. Correlating birthweight with neurological severity of obstetric brachial plexus lesions. *BJOG* 2011;**118**:1098–103.
85. Ecker JL, Greenberg JA, Norwitz ER, Nadel AS, Repke JT. Birth weight as a predictor of brachial plexus injury. *Obstet Gynecol* 1997;**89**:643–7.
86. Eng GD, Binder H, Getson P, O'Donnell R. Obstetrical brachial plexus palsy (OBPP) outcome with conservative management. *Muscle Nerve* 1996;**19**:884–91.
87. Doumouchtsis SK, Arulkumaran S. Are all brachial plexus injuries caused by shoulder dystocia? *Obstet Gynecol Surv* 2009;**64**:615–23.
88. Jennett RJ, Tarby TJ, Kreinick CJ. Brachial plexus palsy: an old problem revisited. *Am J Obstet Gynecol* 1992;**166**:1673–6.
89. Gherman RB, Ouzounian JG, Goodwin TM. Brachial plexus palsy: an in utero injury? *Am J Obstet Gynecol* 1999;**180**:1303–7.
90. van Dijk JG, Malessy MJ, Stegeman DF. Why is the electromyogram in obstetric brachial plexus lesions overly optimistic? *Muscle Nerve* 1998;**21**:260–1.
91. Smith SJ. The role of neurophysiological investigation in traumatic brachial plexus lesions in adults and children. *J Hand Surg [Br]* 1996;**21**:145–7.
92. van Dijk JG, Pondaag W, Malessy MJ. Obstetric lesions of the brachial plexus. *Muscle Nerve* 2001;**24**:1451–61.
93. Gilbert A, Khouri N, Carlioz H. [Birth palsy of the brachial plexus—surgical exploration and attempted repair in twenty one cases (author's transl)]. *Rev Chir Orthop Reparatrice Appar Mot* 1980;**66**:33–42.
94. Gilbert A, Tassin JL. [Surgical repair of the brachial plexus in obstetric paralysis]. *Chirurgie* 1984;**110**:70–5.
95. Pitt MC. Nerve conduction studies and needle EMG in very small children. *Eur J Paediatr Neurol* 2012;**16**:285–91.
96. Spagnoli C, de Sousa C. Brown-Vialetto-Van Laere syndrome and Fazio-Londe disease – treatable motor neuron diseases of childhood. *Dev Med Child Neurol* 2012;**54**:292–3.
97. Ciccolella M, Corti S, Catteruccia M, Petrini S, Tozzi G, Rizza T, et al. Riboflavin transporter 3 involvement in infantile Brown-Vialetto-Van Laere disease: two novel mutations. *J Med Genet* 2013;**50**:104–7.
98. Bosch AM, Stroek K, Abeling NG, Waterham HR, Ijlst L, Wanders RJ. The Brown-Vialetto-Van Laere and Fazio Londe syndrome revisited: natural history, genetics, treatment and future perspectives. *Orphanet J Rare Dis* 2012;**7**:83
99. Ciccolella M, Catteruccia M, Benedetti S, Moroni I, Uziel G Pantaleoni C, et al. Brown-Vialetto-van Laere and Fazio-Londe overlap syndromes: a clinical, biochemical and genetic study. *Neuromuscul Disord* 2012;**22**:1075–82.

Chapter 4

Muscle Biopsy for Diagnosis of Neuromuscular and Metabolic Diseases

Harvey B. Sarnat and Stirling Carpenter

INTRODUCTION

In the 1960s and early 1970s, histochemical stains were for the pathological examination of the routine muscle biopsy to define myofiber types and certain metabolic disturbances, and they continue to be the standard of diagnostic examination of the muscle biopsy.[1–7] Transmission electron microscopy (EM) of the muscle biopsy began in earnest a decade earlier, but the findings and their interpretation were greatly expanded during the same period of histochemistry.[6] The further introduction of immunocytochemistry to specify the expression of specific sarcolemmal regional proteins and distinguish types of inflammatory cells in the late 1980s and 1990s served to further enhance the value of the muscle biopsy as the primary diagnostic technique in neuromuscular diseases. Only a few specific disorders, notably disorders of neuromuscular transmission and a few myopathies such as myotonia congenita, remain better defined by electrophysiological testing than by muscle biopsy. Though muscle biopsy remains a primary diagnostic tool for myopathies and neuropathies, genetic testing for etiological mutations, deletions, and expansions in specific genes, performed from leukocytes in blood samples, has superseded muscle biopsy in recent years in the diagnosis of spinal muscular atrophy (SMA), many muscular dystrophies, and other myopathies because genetic testing is less invasive and more specific at the molecular genetic level. For many disorders in which the genetic basis remains unknown or testing is inaccessible because of cost, and also in disorders not of genetic origin, muscle biopsy remains a valuable and primary diagnostic method.

This chapter summarizes the most frequently employed neuropathological techniques in examining the muscle biopsy and the principal findings in the various disorders in which it contributes to a definitive diagnosis.

TISSUE SELECTION AND PREPARATION

Muscle selection for biopsy should avoid both end-stage muscles that are likely to show only nonspecific fibrosis and loss of myofibers, and clinically uninvolved muscles that may not yet be pathologically diagnostic. The vastus lateralis of the quadriceps femoris muscle is most often selected in children because it is a large muscle from which adequate tissue can be taken without causing weakness, it is involved in nearly all myopathic disorders, it is easily accessible away from major nerves and blood vessels, it shows relatively equal numbers of types I and II myofibers, and it is the best-documented muscle in terms of normal controls. In young children, the risk of general anesthesia can sometimes be avoided by a skilled femoral nerve block and mild oral or intravenous sedation.

The specimen should be dealt with promptly after removal to minimize autolytic loss of important enzymatic activities and of ultrastructural details by delayed freezing or improper fixation. Four portions of the muscle biopsy tissue are usually taken:

(1) A small portion frozen in isopentane (2-methylbutane) cooled to the temperature of liquid nitrogen (−160°C) for cryostat sections oriented transversely for light microscopy and histochemistry.
(2) A larger portion frozen in a freezer at −80°C for biochemical assay. Preservation of histological detail is not an important consideration for biochemical studies; therefore, isopentane is not needed and would interfere with some assays.
(3) A small portion is immersed in 2% buffered glutaraldehyde for electron microscopy. Use of an isometric clamp has advantages: the muscle is relatively relaxed, so features in the I-band are clearly seen. Orientation is made easy. Dicing makes orientation difficult. Gluteraldehyde penetrates between the muscle fibers

B.T. Darras, H. Royden Jones, Jr., M.M. Ryan & D.C. De Vivo (Eds): Neuromuscular Disorders of Infancy, Childhood and Adolescence, Second edition.
DOI: http://dx.doi.org/10.1016/B978-0-12-417044-5.00004-4

sufficiently to allow good fixation as long as the piece of tissue removed is not too thick. After sufficient time in fixative, the specimen can remain refrigerated in buffer for several days if desired, to optimally schedule blocking and embedding. The semithin plastic sections should be carefully examined because they give the best light-microscopic morphology. The very best resolution comes from paraphenylenediamine-stained sections viewed with a phase microscope, but toluidine-blue stained sections are more widely used because phase microscopes are not readily available in all laboratories. Both transverse and longitudinal sections should be cut, and at least 2 transverse and 1 longitudinal examined. From the semithin sections, it can be decided whether or not electron microscopy is needed. The unused tissue can be kept in reserve in buffer for a few weeks or can be embedded in epoxy resin for future study, and preserved indefinitely.

(4) If enough tissue is available, a small portion can be fixed in 10% buffered formalin for paraffin embedding. Paraffin sections are not essential. The histology can be better studied from the frozen sections, which, when competently done, have fewer artifacts than paraffin sections, though nuclear detail is not as well preserved.

An open biopsy is generally preferable to needle biopsies of muscle because of the volume of tissue required for the various studies. Unlike needle biopsies of the liver or kidney, muscle needle biopsies still require a small skin incision and a large-bore needle, and they are not less traumatic than open biopsies. Needle biopsies are more likely to lead to the complication of focal hematoma formation after the biopsy.

Histopathological Aspects of Muscle Biopsy Preparation and Examination

Histological Examination

Hematoxylin and eosin (HE) stain has been the most universal and traditional method for examination of formalin-fixed, paraffin-embedded tissue sections of all tissues for more than a century, and its application to cryostat sections of the muscle biopsy is no exception. The amount of diagnostic information yielded by histological examination alone is limited, but it can detect the presence of inflammation; foreign tissues including *Trichinella* and other parasites; neoplasia, both primary and metastatic; myofiber necrosis; perimysial and endothelial connective tissue proliferation; and patterns of distribution of atrophic myofibers. Subcellular structural abnormalities, such as ragged-red fibers, are subtler and harder to see than with histochemical stains and some, such as nemaline rods, are almost impossible to detect because of lack of color contrast.

Histochemical Examination

The purpose of histochemistry is to distinguish myofiber types and subtypes, certain metabolic products such as glycogen and neutral lipids, and many subcellular abnormalities not well seen with HE. Histochemical studies are traditionally performed on freshly frozen sections only, though some enzymatic activities and fiber typing using myosin and actin antibodies can also now be demonstrated in paraffin sections by immunocytochemistry. The most useful and frequently applied histochemical stains are:

(1) Modified Gomori trichrome. The chromotrope-2 R component of this stain has a high affinity for phospholipids and is perceived as a red color; mitochondrial and endoplasmic reticulum membranes have high phospholipid content in contrast with most plasma membranes that have a higher ratio of glycolipids. In normal myofibers in transverse section, a thin subsarcolemmal rim of red staining may be seen because of mitochondrial concentration, particularly evident at motor endplates. The intermyofibrillar sarcoplasm appears as a red reticular network because of mitochondria and endoplasmic reticulum against the green staining of myofibrils from the fast green FCF component. The third component of this trichrome stain is hematoxylin, which stains nuclear chromatin (DNA) purple. This stain is ideal for demonstrating ragged-red fibers of some mitochondrial cytopathies. Chromotrope-2 R also stains tropomyosin, so nemaline rods arising from the Z-band also are well demonstrated. Myelin sheaths of normal intramuscular nerve twigs are also intensely red stained.

(2) Myofibrillar calcium-mediated adenosine triphosphatase (ATPase) stains. These stains are highly reliable for distinguishing fiber types and subtypes, depending on the pH of preincubation. At alkaline pH (usually 9.4 to 10.2), type II fibers are dark and type I fibers are light. At acid pH (usually 4.6–4.3), the reverse is true, and subtypes of II can be distinguished. Types of intrafusal myofibers of muscle spindles differ from those of extrafusal fibers. These stains are very helpful for those myopathies showing type-specific atrophies. In chronic neurogenic atrophy with many cycles of denervation and reinnervation, a phenomenon of histochemical type-grouping is seen, in which the normal intermixed checkerboard distribution of types I and II myofibers is replaced by small and large groups of myofibers, all of the same type, adjacent to other groups of the opposite type. Some groups may be so large as to occupy an entire fascicle.

(3) Respiratory chain complexes, or "oxidative mitochondrial enzymes." The three most relevant stains include nicotinamide adenine dinucleotide tetrazolium reductase (NADH-TR; respiratory complex I), succinate dehydrogenase (SDH; complex II) and cytochrome c oxidase (COX; complex IV). These activities are greatly increased in ragged-red zones. In many mitochondrial diseases, scattered myofibers that are otherwise well preserved and not degenerating show loss of COX activity, even if complex IV is not selectively impaired. One caution is that the intensity of histochemical activity of COX is generally lower than either NADH-TR or SDH, particularly in type II myofibers, hence normally low activity should not be misinterpreted as pathologically absent. Modification of the SDH stain by the addition of phenazine methosulfate is useful because this compound suppresses activity in normal mitochondria but does not affect SDH expression in abnormal mitochondria. Because complex II has only four subunits, all encoded by nDNA, it remains reactive in all mitochondrial diseases regardless of any mtDNA point mutations. The other four complexes include a combination of mtDNA and nDNA encoded subunits. These mitochondrial enzymes also are markers of fiber types, showing more intense activity in type I fibers than those of type II, but caution must be exercised in interpretation because severely atrophic myofibers of either type show intense activity due to loss of myofibrils and concentration of sarcoplasmic mitochondria. NADH-TR marks not only mitochondria but sarcoplasmic (i.e. endoplasmic) reticulum as well, unlike SDH and COX, which are pure mitochondrial stains.

(4) Acid phosphatase. This stain is useful for showing abnormal lysosomal activity in diseases such as infantile (Pompe) and adult acid maltase deficiencies. It appears as granular material within the myofiber. Nonspecific esterase can used as an alternative. Alkaline phosphatase is not a suitable alternative.

(5) Oil Red O and Sudan Black B. These stains are used to demonstrate neutral lipids within myofibers and in the connective tissue, but complex phospho- and glycolipids, such as sphingomyelin and galactocerebroside, are not stained.

(6) Periodic acid-Schiff (PAS) reaction. Increased intramuscular glycogen occurs in some glycogenoses, but may also represent a physiological condition of a patient who is receiving good nutrition, even parenterally for an extended time, and who has little physical activity to utilize the glycogen that accumulates in myofiber, as in children in intensive care units. PAS also is good for demonstrating peripheral nerve myelin in both cryostat and paraffin sections.

Immunocytochemical Reactivities

These techniques are usually applied to frozen cryostat sections, but also may be applied to paraffin-embedded sections, except for dystrophin. They can demonstrate features not shown by the routine histochemical battery of stains, including sarcolemmal-region proteins both of the sarcolemma itself and those that attach the sarcolemma to the basal lamina (basement membrane). They are highly specific for many muscular dystrophies, such as the dystrophinopathies, merosin deficiency, and dystroglycanopathies. Other immunoreactivities can substitute for histochemistry in distinguishing fiber types if only formalin-fixed, paraffin-embedded sections are available, for example heavy- and light-chain ("slow" and "fast") myosin immunoreactivities that provide similar information as myofibrillar ATPase stains. Immunocytochemistry can also distinguish types of inflammatory cells, such as T-lymphocytes (CD-43), B-lymphocytes (CD-20), and macrophages (CD-68). Endothelial cells of capillaries and other small intramuscular blood vessels are well demonstrated by vimentin and nestin immunoreactivities, or by the more specific CD-31 and CD-34 antibodies.

Silver Impregnations

This group is useful for showing axons in peripheral nerve, including intramuscular nerves, and includes variations of technique such as those of Bielschowsky, Bodian, Sevier-Munger, and others. Reticulin stains demonstrate reticulin fibers around intramuscular capillaries.

Transmission Electron Microscopy (EM)

Since the 1950s, EM has provided ultrastructural insight into the subcellular anatomy of myofibers (Figure 4.1) and of aberrations in many myopathies, to both confirm and verify the interpretation of histochemical stains and to offer additional detail at a resolution not possible by light microscopy. In recent years, it has been used more sparsely and selectively because of its labor intensity for both technologists and pathologists and because of cost, including the maintenance of the equipment. Special tissue preparation is required, with fixation in glutaraldehyde, post-fixation in osmium tetroxide, and staining with uranyl acetate and lead citrate. Ultrafine sections must be cut on special diamond knife blades. EM can, however, demonstrate many features of diagnostic importance if used for selective muscle biopsies. Nemaline rods and disarrays of myofilaments are well shown. One of its most important contemporary applications is to demonstrate ultrastructural changes in mitochondrial cytopathies. There are several mitochondrial alterations, including simplification of the cristae, stacked and whorled cristae, increased numbers of osmophilic spheroids within the mitochondrial matrix and paracrystalline structures.

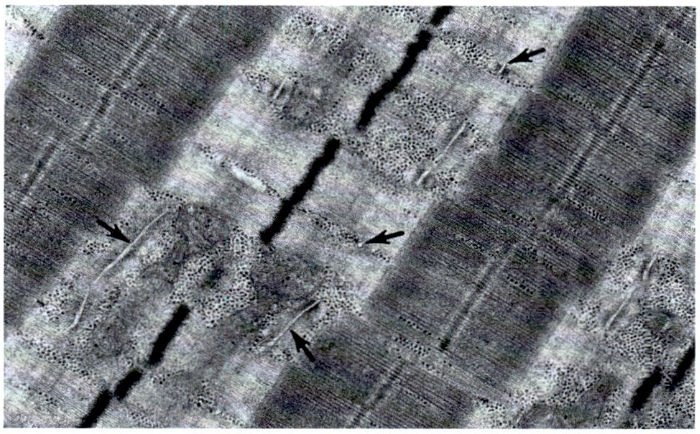

FIGURE 4.1 This electron micrograph (EM) shows a longitudinal view of part of a normal muscle fiber fixed at resting length. The moderately dark A-bands contain the myosin (thick) filaments. In the part of the A-bands that adjoin the I-bands, the actin (thin) filaments are interspersed with the thick filaments. The very dark Z-disc in the middle of the I-band marks the boundary of the sarcomere. The plane of section passes through the trajectory of the T-tubules, some of which are marked by arrows. The darkish zone next to the T-tubule walls marks the ryanodine receptor, which is part of the wall of each cistern of sarcoplasmic reticulum, two cisterns and the T-tubule between them forming the triads. The longitudinal part of the SR is not visible. The small round granules are glycogen. The section grazes two mitochondria where one I-band goes out of the plane of section. Reproduced with permission from Oxford University Press, USA.[6]

The latter are an array of geometrical crystal-like structures within the mitochondria and are a highly specific finding. Within myofibers, paracrystalline structures (sometimes called "inclusions," though this is incorrect because they are not foreign particles) often assume a brick shape with sharp corners and elongation. The ultrastructural periodicity of these structures is constant and predictable; these structures are in the space between the cristae and are made of creatine kinase. Often, fragments of preserved cristae can also be seen within the same mitochondrion that has a paracrystalline structure, confirming their location as intramitochondrial. Endothelial cells of intramuscular capillaries in young infants with mitochondrial disease also contain paracrystalline-like structures suggestive of intramitochondrial location but without preserved cristae.[8] Normal Weibel-Palade granules of endothelial cells that contain von Willebrand factor V consist of parallel microtubules that, on oblique or longitudinal sections, have a repetitive pattern reminiscent of a crystal.[9]

Biochemical Assay of Respiratory Chain Complexes

The measurement of specific activities of each of the five respiratory chain complexes, both individually and in combination (e.g. complexes 1 + III), may be performed from homogenates of frozen tissue and provides quantitative numbers. Citrate synthetase is usually examined as a control because if it is decreased, the other specific activities may be unreliable due to delayed freezing of the tissue. At times, one may find convincing histochemical and ultrastructural evidence of mitochondrial disease, associated with normal activities of the complexes, particularly in infants. This anomaly is due to the measurement of an average of the muscle tissue homogenate rather than an examination of individual myofibers. If only a few scattered myofibers are affected, they may be diluted by the normal activities of the majority of myofibers. Interpretation of these assays must be made cautiously, and clinicians must understand that normal quantitative assay results do not necessarily exclude mitochondrial disease. Similarly, decreased specific activity in one complex may be an artifactual result and, in the absence of supportive evidence in other aspects of the muscle biopsy, should not be taken as conclusive for mitochondrial cytopathy. Diminished activity in all or most complexes with normal citrate synthetase activity is a primary criterion of the "mitochondrial depletion syndrome" of infancy, a genetically heterogeneous group with many clinical phenotypical features in common.

Molecular Genetic Studies

Muscle tissue can be used for molecular genetic studies of many muscular dystrophies. It is also useful in showing both mitochondrial DNA (mtDNA) point mutations and nuclear DNA (nDNA) mutations or deletions in mitochondrial disorders. Ideally, the tissue should be freshly frozen for such studies, but DNA extraction can also be made from formalin-fixed tissue, though this is less reliable. Details of techniques and discussion of the many mutations that have been demonstrated in mitochondrial diseases are beyond the scope of this chapter.

ONTOGENESIS OF STRIATED MUSCLE

Muscle originates embryologically from two principal sources. The myoblasts themselves are derived from the myotomes, a portion of the mesodermal somite (somitomeres before segmentation occurs) that also gives origin to the dermatomes and the sclerotomes that form the vertebrae. Myoblasts destined to become intrafusal myofibers are of myotomic origin, similar to those that form the extrafusal fibers. The remainder of muscle architecture is derived from neural crest tissue: the collagenous and adipose connective tissue (including fibroblasts), Schwann cells and intramuscular nerve sheaths, and blood vessels within muscle. A third origin might be cited for axons of intramuscular nerves, which arise from motor neurons in the spinal cord or brainstem, though the cell bodies of afferent nerves from muscle spindles reside in the dorsal root ganglia and are neural crest derivatives.

As with other developmental processes, the ontogenesis of striated muscle is precisely time-linked by genetic programming and specific myogenic genes are identified.[10–12] Muscle development can therefore be divided into several broad periods:

(1) **The myoblastic phase** (3–8 weeks of gestation) involves the peripheral distribution of myoblasts along programmed migratory pathways in synchrony with neural crest migrations.
(2) **The myotubular phase** (8–15 weeks of gestation) is characterized by fusion of myoblasts to form early myotubes. Myoblastic fusion is largely mediated by the gene *myogenin* and also by *myomaker*, a late membrane activator of myoblast fusion.[13] Myotubes are a cylinder of striated muscle with a core free of myofilaments but containing a single column of nuclei from myoblast fusion as well as mitochondria in the core spaces between the nuclei; primitive endoplasmic reticulum and ribosomes also form in this internuclear sarcoplasm. Motor neuron innervation of striated muscle occurs at 9 to 11 weeks, in the middle of the myotubular stage. Initially, this innervation of myotubes is diffuse, with multiple innervations of the same myotube at various sites along its length and a diffuse distribution of sarcolemmal membrane receptors. Both collateral axons of the same motor neuron and axons of other motor neurons then retract to leave a single innervation in the middle of the length of the myofiber. The fascicular architecture of the muscle is already evident at this stage with the formation of perimysial septa. Muscle spindles become enclosed by collagenous capsules.
(3) **The myocytic stage** (15–20 weeks of gestation) involves the redistribution of organelles, including mitochondria, to the intermyofibrillar sarcoplasm and the peripheral movement of nuclei to the subsarcolemmal region. Intermediate filaments of vimentin are numerous during the myotubular stage and help to hold the nuclei in their central position, but vimentin regresses in the myocytic stage, enabling the nuclei to move centrifugally. Many sarcomeres are added with the rapid longitudinal growth of the myofibers and of the long bones. Additional axonal retractions from the transitory multiple innervations continue. Muscle fascicles become more defined with more definite perimysial septa.
(4) **The histochemical differentiation stage** (20–28 weeks of gestation) is the period when slow and fast myofibers and their histochemical counterparts as types I and II myofibers become distinguishable. Prior to this stage, all myofibers are stained uniformly as subtype IIc fibers, the least well differentiated type. During the initial period of histochemical differentiation, type I fibers increase in transverse diameter more than those of type II and at about 24 weeks' gestation are readily distinguishable by size in transverse sections: type I fibers represent about 10% of total myofibers, are 2 to 3 times larger, and react as type I with myofibrillar ATPase stains. These large type I fibers are called *Wohlfart b-fibers;* the larger proportion of smaller IIc fibers are *Wohlfart a-fibers* (Figure 4.2). *Wohlfart b-fibers* are uniformly distributed within the fascicles and never grouped, as are the giant type I fibers in infantile SMA. As this stage proceeds, the type I fibers become more numerous as more *Wohlfart a-fibers* are converted to b-fibers. The a-fibers themselves undergo subtype differentiation and grow in size so that by 28 weeks, size alone no longer distinguishes the fiber types and the ratio of types I and II myofibers becomes relatively equal, with regional predominance in some muscles that persists at maturity (e.g. type I predominance in the deltoid, type II predominance in the gastrocnemius). Scattered centronuclear fibers reminiscent of the myotubular stage may persist in the second trimester, but become less frequent with age.

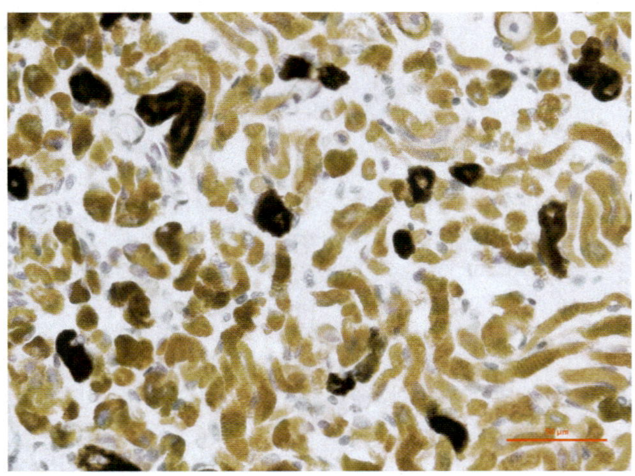

FIGURE 4.2 Quadriceps femoris muscle from the autopsy of a 22-week fetus. About 10% of total myofibers are larger than the rest, scattered and not grouped, and react as type I fibers (dark): these are designated "Wohlfart b-fibers." The majority of fibers are smaller, rounded without angularity, and react as type II (light); these are "Wohlfart a-fibers." With further maturation, more a-fibers will be converted to b-fibers and the a-fibers will grow in size with the addition of more myofibrils until distinction by size no longer is evident and the ratio of fiber types becomes relatively equal. Two fibers with central nuclei and a thin rim of myofibrils at the top of the figure are scattered persistent myotubular forms that are still found at this gestational age. The oblique orientation of some fibers is artifactual and unavoidable in autopsy fetal muscle because of the delicate endomysium. Paraffin section with antibody against slow myosin (corresponds to myofibrillar ATPase pre-incubated at acid pH in frozen sections).

(5) **The maturing myofiber stage** (28–40+ weeks of gestation) includes further refinement in histochemical differentiation of subtypes of II. Also, at about 30 weeks, an increase in desmin intermediate filaments is associated with registry between Z-discs of adjacent myofibrils to facilitate smoother coordination of contraction. The lack of registry before this stage and its presence thereafter is well demonstrated by EM. After 30 weeks, triads (the terminal cisternae of two sarcoplasmic reticula flanking a transverse tubule) rotate 90° from a parallel orientation to the long axis of the myofiber to perpendicular, usually near a Z-disk.

Myofibers continue to add sarcomeres to grow longitudinally throughout fetal life, infancy, and childhood, and also increase in transverse diameter. Tables are available to denote mean myofiber diameters at all ages for use in interpreting muscle biopsy. The mean diameter in the full-term neonate is 15 μm. The adult diameter of about 65 μm is achieved at about 13 years of age. The rate is not linear. There is only modest increase in fiber diameter in the first year and a half of infancy because most of the muscle growth is longitudinal.

CELLULAR DEATH AND REGENERATION

Death of cells can occur through necrosis or apoptosis. In mature muscle cells with their multiple nuclei, the role of apoptosis after fetal life is probably quite limited. This discussion will, therefore, focus upon necrosis and regeneration.

Necrosis of a myofiber means that it has lost its ability to maintain homeostasis and passed the point of no return, en route to becoming debris.[13] Necrosis of skeletal muscle fibers usually begins focally in one segment of the fiber. A stereotyped sequence of changes takes place once necrosis is initiated. Some of these changes can be detected by light microscopy and others only by EM. The fiber appears pale with general tissue stains such as HE. Dystrophin loses its affinity for the antibody. The Z-discs disappear, as seen in semithin sections by light microscopy or by EM. This phenomenon is attributed to the activation of calpain, normally localized within the Z-disc, by calcium flooding through the now permeable plasma membrane. This membrane itself shows breaks and discontinuities by EM. Mitochondria become more spherical, gain fluffy matrix densities, lose their regulated position near the Z-disk, and start to aggregate. Degradation of organelles progresses in segments undergoing necrosis independent of the arrival of phagocytes. Thin (actin) myofilaments disappear sooner than the thick (myosin) filaments. Some degree of spontaneous contraction may take place in the myofibrils before the Z-discs fragment.

Hypercontracted fibers, the unusually large, round, dark fibers that stain strongly with almost any histological or histochemical stain, are not a sign of necrosis but a biopsy artifact. The basal lamina persists over the necrotic segment for up to 5 weeks and provides a scaffold for regeneration. Necrosis not followed by regeneration can sometimes be visualized by finding a circular skein of basal lamina with no muscle fiber or regenerative myoblast within it. Examples may be seen in Duchenne muscular dystrophy and polymyositis.

Necrosis of a skeletal muscle fiber usually starts focally in one segment. This focality is seldom visualized, in part because most biopsies are only studied in transverse sections. The process of necrosis, once initiated focally, extends into neighboring, initially intact segments until a barrier forms.[14] In experimental animal muscle, this usually occurs by 6 hours after injury. The extension of necrosis is probably carried to a considerable extent by calcium ions. Necrosis extends along the fiber until a barrier is formed, first by matted myofilaments on which calcium is precipitated and then by the formation of a delimiting membrane.

Sodium freely enters the necrotic segment, just as calcium does, and diffuses more rapidly. In an experimental model of focal necrosis, the delimited surviving segments of myofibers contain many large clear vacuoles, which are huge dilatations of T-tubules.[15] Sodium, freely diffusing out of the necrotic segment, meets the Na+/K+ ATPase on the transverse tubule (T-tubule) membrane and is ejected into the T-tubule in exchange for potassium. Since 3 sodium ions go out for each 2 potassium ions that enter, the osmolarity in the T-tubule begins to rise and water is pulled in. The result is impressive T-tubule swelling. Similar T-tubular swelling occurs in surviving stumps in human muscle, exemplified in dermatomyositis. However, it is surprisingly not seen in Duchenne dystrophy. Any human muscle biopsy, if allowed to rest for a sufficient time in physiological saline, will develop T-tubule swelling as sodium enters the cut ends of fibers and diffuses within them.

Phagocytes normally invade necrotic segments by 10 hours. Recognition of necrosis is greatly facilitated when phagocytosis has begun. In a condition such as Duchenne[16] dystrophy, where necrosis is recurrent, phagocytes may already be present in the muscle, enabling phagocytosis to begin more promptly. Invasion of non-necrotic fibers by lymphocytes in polymyositis must be distinguished from phagocytosis of necrosis.

Regeneration occurs through the migration, proliferation, and fusion of satellite cells.[16] The process of regeneration recapitulates the process of fetal ontogenesis of muscle. In normal muscle, roughly 7% of the nuclei one sees on the periphery of myofibers are really nuclei of resting progenitor satellite cells. They migrate into the necrotic segment, where they appear next to the basal lamina. They divide, several times if necessary, to fill the old basal lamina. At the same time, phagocytes come and go. The satellite cells, which now would be called myoblasts, fuse with one another and with the surviving stumps and begin to form new myofibrils. The number of divisions a satellite cell can perform is limited.[17] This becomes evident in Duchenne dystrophy, in which regenerated muscle fibers, once they mature, are susceptible to further cycles of necrosis, and so the satellite cells in time exhaust their capacity to divide further, resulting in stunted or absent regeneration.

The *Pax7* and *WNT3a* genes that are essential for normal fetal myogenesis are also critical for normal function of satellite cells in older children and adults.[18,19] *WNT3a* upregulates *frizzled* (*Fzd7*) and induces symmetrical expansion of satellite stem cells.[12,18,19] Adult satellite stem cells enclosed by the sarcolemma of a myofiber remain quiescent throughout life, but their intrinsic regenerative function declines with aging so that muscle repair in elderly individuals is not able to respond to injury as easily as in younger individuals; the maintenance of quiescence in adult life depends upon active repression of senescence pathways.[20] This same feature of aging may play a role in some degenerative myopathies, particularly in muscular dystrophies, so that regeneration is limited.

DENERVATION, DISUSE, AND HYPOTROPHY

As with the muscular dystrophies, identification of the genetic defect in the SMN gene on chromosome 5 has supplanted the need for muscle biopsy in most cases suspected on clinical grounds. Some rarer motor neuron diseases of childhood are not due to the same genetic defect, and the muscle biopsy may provide diagnostic information. The same is true of the group of hereditary neuropathies, in which both muscle and nerve biopsy are performed at the same time.

The emblematic pathological pattern of denervation of muscle is the presence of angular atrophic fibers of both fiber types. The angular shape of the atrophic fibers is probably the result of pressure from the nearby normally contracting fibers. When the number of atrophic fibers is so high that they predominate, the angular appearance is lost. In denervating disease of early infancy (essentially spinal muscular atrophy [SMA] type I), atrophic fibers are typically rounded, but this may be related more to a lack of growth than to shrinkage, since the disease process probably begins before birth. In SMA type I, the very small fibers are usually type I, as are the less frequent hypertrophic fibers, whereas normal sized fibers are usually type II.

A second useful sign of denervation is the presence of histochemical "type grouping." When denervation occurs, nearby intact nerve twigs coming from other neurons are often stimulated to sprout and reinnervate the denervated muscle fibers. This results in blocks of fibers of type I or type II that replace the normal checkerboard pattern seen with the ATPases, a phenomenon called "type grouping." It must be distinguished from the situation when there is simply numerical predominance of one or the other fiber type. Normally the motor unit is widely dispersed in the muscle so that two adjacent fibers are seldom innervated by the same neuron. Type grouping does not indicate how recently denervation occurred; it can be an old lesion. For type grouping to take place, various cycles of denervation and reinnervation may be required. Large groups might denote many cycles of denervation and reinnervation, and small groups might denote fewer cycles.

Denervation also causes a number of more subtle changes in myofibers. Mitochondria become rearranged in denervated fibers, which may produce tiny foci of mitochondrial absence that are highlighted in oxidative enzyme reactions. Fibers with this abnormality are called "moth-eaten." Mitochondrial relocation can also lead to larger, rounded foci of mitochondrial absence known as "targets," best demonstrated with oxidative enzymatic reactions. Within an area of mitochondrial absence, the myofibrils tend to show a change known as Z-disc streaming, where the material that gives the Z-disc its density streams over other components of the myofibrils. In itself, Z-disc streaming is fairly nonspecific, but the occurrence of well-formed targets is a fairly reliable sign of denervation. They must, however, be distinguished from the cores of central core disease and various diseases with minicores.

Electron microscopy offers an additional sign of denervation, useful when other signs are sparse (Figure 4.3). The basal lamina of muscle fibers normally clings tightly to the plasma membrane. Shrunken denervated fibers usually display at least two or three places on their periphery where empty sleeves of superfluous basal lamina extend out from the fiber surface. Myofibrils in reinnervated fibers also have a tendency to get out of line, leading to individual myofibrils oriented perpendicular to one another.

Intramuscular nerves are sometimes encountered in muscle biopsies. It is worthwhile to look at them carefully in semithin plastic sections, since it is sometimes possible to find ovoids of Wallerian degeneration in the terminal perineurial sheaths.

When angular atrophic fibers are found in a muscle biopsy, the main distinction to make is between denervation and type II fiber atrophy. The latter can be caused by disuse, cachexia, immobilization (as with plaster casting for a long bone fracture), and chronic steroid administration.

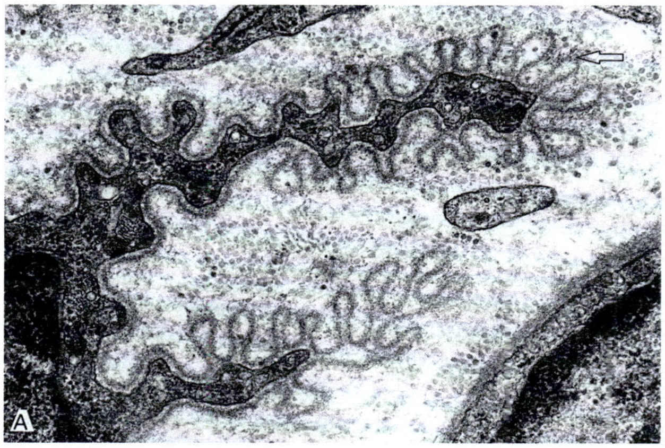

FIGURE 4.3 This EM shows part of an angular atrophic denervated fiber. Note the empty sleeve and loops of basal lamina (arrow), typical of denervated fibers. Reproduced with permission from Oxford University Press, USA.[6]

The atrophy is usually not as severe as in denervation, but the atrophic fibers are angular and could be confused with denervation in histological sections stained with HE. Myofibers do not lose their glycogen, the oxidative enzyme profile does not darken, targets do not occur, and the basal lamina does not form empty sleeves. In animal experiments, it has been found that denervated type II fibers tend to show signs of atrophy faster than the type I fibers. This may explain the rare biopsies in which denervation atrophy seems to affect only type II fibers but, in general, neurogenic atrophy is not type-selective.

Muscle fibers that are smaller than normal but rounded or polygonal generally indicate lack of growth. This may result from some poorly understood reaction to an altered gene, as in the type I fibers of various congenital myopathies. It can also result from deficient regeneration, as occurs ultimately in Duchenne dystrophy.

MUSCLE BIOPSY IN SPECIFIC MYOPATHIES

Muscular Dystrophies

Muscle biopsy is no longer performed as the definitive diagnostic test in all patients with a suspected muscular dystrophy because of the development of less invasive and more specific genetic tests utilizing blood samples,[21] but it is still of high diagnostic value in ambiguous cases or if genetic testing fails to provide a diagnosis. The muscular dystrophies form a prominent group with many members among the myopathies. While the term "dystrophy" is not precisely defined, it implies a process of repeated fiber necrosis with gradual exhaustion of regeneration, leading to loss of myofibers accompanied by fibrosis and adipose replacement, without other distinctive features. A "dystrophic picture" by microscopy thus generally means patchy myofiber loss with necrosis, phagocytosis, some regeneration, and a mixture of fibers of various sizes. Specific diagnosis (specific as to the protein genetically altered, which gives its name to the disease) is usually not possible solely from the morphology and histochemistry. Antibodies are now available to apply to tissue sections to identify some of the dystrophies. At present, there are antibodies to individually demonstrate each of the three domains of the dystrophin molecule (carboxy- or C-terminus; rod domain; amino or N-terminus), and the four sarcoglycans: merosin, caveolin, dysferlin, and calpain. Stainable dystrophin is lost rapidly when a fiber becomes necrotic. This should not be interpreted as a primary deficiency. In addition, one must be aware that secondary losses of reactivity can occur. For example, with loss of dystrophin, sarcoglycan and dystroglycan staining is impaired, although laminin is preserved. With deficiency of any one sarcoglycan, all the other sarcoglycans tend to be impaired. In caveolin or calpain deficiency, staining of dysferlin may be reduced. Loss of α-dystroglycan from the surface of muscle fibers occurs in deficiency of fukutin and fukutin-related protein, in muscle-brain-eye disease, and in Walker-Warburg syndrome; this can be visualized with the anti-α-dystroglycan antibody.

As a classic example of what is meant by dystrophy, consider a biopsy from a patient with Duchenne dystrophy, from a still reasonably strong muscle, from a 5-year-old boy. It will show some mixture of fiber sizes with hypertrophied, normal-sized, and small fibers (see Chapter 30). The normal fiber type mixture is preserved, though the proportion of type IIB fibers tends to diminish, and type IIC fibers (resulting from regeneration) may be present. The small fibers are rounded, and they have not resulted from atrophy but from imperfect regeneration. Reconstruction of these small fibers from serial semithin sections has shown how they may branch or end abruptly, and fail to achieve a normal girth.[22] Some necrotic fibers are usually found, often in groups. Grouped fibers in various stages of regeneration are also seen. By electron microscopy, it is striking that fibers that are not necrotic and not regenerating have normal internal ultrastructure (Figure 4.4). The only abnormality they show is separation of segments of basal lamina. This is also seen in various other dystrophies. Collagen is laid down focally inside fascicles parallel to the muscle fibers. As the disease progresses, regeneration is less and less successful while necrosis continues, so that more and more fiber loss follows, with fibrosis and breakdown of the fascicular architecture.

An autosomal recessive limb-girdle muscular dystrophy phenotype may be due to novel mutations in the desmin gene without the histopathological features of myofibrillar or other desmin storage myopathies.[23]

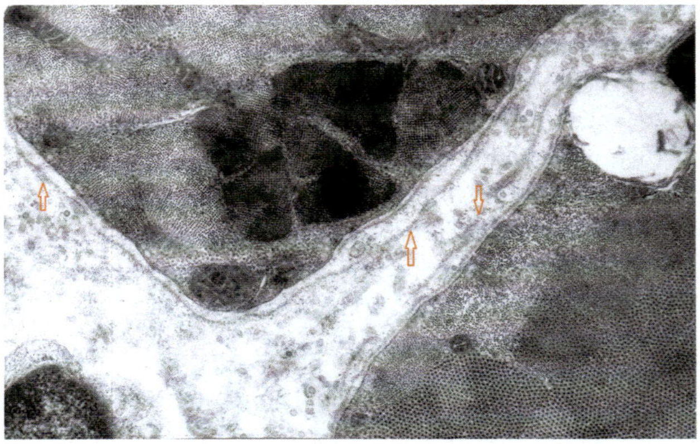

FIGURE 4.4 This EM shows a transverse view of two muscle fibers from a male child with Duchenne muscular dystrophy. The myofibrillar component is completely intact, but note the stretches of lifted and reduplicated basal lamina, which are characteristic of Duchenne and some other dystrophies. Reproduced with permission from Oxford University Press, USA.[6]

Congenital Muscular Dystrophies

The congenital muscular dystrophies (CMDs) account for many of the autosomal recessive muscular diseases present at birth or in the first year of life. Many of the congenital dystrophies have allelic forms with later onset and milder symptoms. This is true of deficiencies of laminin, fukutin-related protein (FKRP), and collagen VI. The spectrum of congenital muscular dystrophies is expanding as new genetic data correlated with metabolic defects become available.[24] An important feature of the congenital muscular dystrophies is that they are often associated with brain malformations, and it is difficult at times to separate the clinical expressions between myopathy and cerebral dysfunction. Muscle biopsies are sometimes requested for this reason, even if the predominant symptoms and signs are more suggestive of cerebral origin (see Chapter 29).

A number of the congenital dystrophies have a disturbance in the glycosylation of α-dystroglycan. Alpha-dystroglycan binds to β-dystroglycan, which is embedded in the muscle plasma membrane, and projects from the fiber surface to link with laminin-2 in the basal lamina. Other proteins link the basal lamina to the interstitial connective tissue. In a group of congenital dystrophies, the glycosylation of α-dystroglycan is perturbed, resulting in partial or complete absence of α-dystroglycan staining by immunocytochemistry. These dystrophies include muscle-eye-brain disease and deficiency of fukutin and FKRP, as well as varieties of Walker-Warburg syndrome. Cobblestone lissencephaly (related to abnormalities in the pial membrane) and other problems of brain development are often related. In FKRP deficiency, loss of α-dystroglycan staining may be only partial.

The basal lamina of muscle fibers has two parts: a pale zone next to the plasma membrane and a dense zone, called the reticular lamina, which is filled with networks of collagen IV and laminin-2. Laminin-2 (also called merosin) binds to α-dystroglycan and to integrin-7 in the plasma membrane, thus forming links in the chain that connects the actin cytoskeleton of the fiber to the extracellular matrix. Mutations in the gene for laminin-2 are responsible for nearly half the cases of congenital muscular dystrophy. The anti-laminin-2 antibody may show complete or partial loss. Laminin-5, normally absent, may become visible. The muscle shows some necrosis, fiber loss, and fibrosis. Peripheral neuropathy may be present, and asymptomatic abnormalities may be seen in imaging studies of the brain. Secondary loss of laminin-2 with a normal gene can be seen in the glycosylation disorders. Mutations in the integrin-7 gene to which laminin-2 connects also cause a type of dystrophy with hypotonia and proximal weakness, and later with respiratory involvement.

Collagen VI forms microfibrils, which are present in the interstitium between muscle fibers. In the skin, they are known as *oxytalan fibers*. For light microscopy, they can be shown by antibodies, and electron microscopy shows them as 10–12 nm nonrigid microfibrils often arranged in bundles. They should not be mistaken for amyloid. Interaction with the basal lamina of the muscle cells has been shown. Collagen VI is made of three proteins. Mutations in all three genes (*COL6A1* and *COL6A2* on chromosome 21q22 or *COL6A3* on chromosome 2q37) have been reported. Patients with autosomal dominant disease, which is relatively mild, are said to have Bethlem myopathy; patients with severe disease, autosomal recessive until proven otherwise, are said to have Ullrich myopathy; however, it has been shown recently that both disorders can be caused by either autosomal dominant or recessive mutations. Although there is phenotypic overlap between the collagen VI CMDs, the Bethlem patients typically have contractures, which can appear in the first or second decade. Muscle biopsy usually shows mild nonspecific features, such as variation in fiber size, and possibly an isolated necrotic fiber. The antibody to collagen VI gives normal staining. The Ullrich patients typically have severe early muscle weakness, proximal joint contractures, and hyperelasticity of distal joints. Biopsies give variable results ranging from nonspecific features to a dystrophic picture. The antibody usually gives absent or markedly reduced staining.

Myotonic Muscular Dystrophies

The myotonic dystrophies have special features clinically, genetically, and pathologically (see Chapter 37). Myotonic dystrophy 1 results when repeats of CGT in a noncoding part of a gene called myotonic dystrophy protein kinase (DMPK) exceed 50. The number of repeats often goes into the thousands. Myotonic dystrophy 2 (also called proximal myotonic myopathy) results from an analogous excess of quadrinucleotide repeats in an intron (thus noncoding) of a gene called *zinc finger protein 9* (*ZNF9*).

The syndrome of infantile myotonic dystrophy arises when a mother with myotonic dystrophy 1 passes on to her offspring an exceptionally large number of repeats. The mother is not always aware of her disease. This may delay recognition of the diagnosis in the child. Transmission of such a large number of repeats from the father almost never occurs. The result is neonatal weakness, hypotonia, and difficulty in swallowing and feeding. Myotonia is almost invariably absent in infancy. Biopsy in the first few months of life is liable to show signs of delayed maturation of muscle, retarded fiber type differentiation (sometimes a single fiber type), increased central nuclei, occasionally ring fibers, and small subsarcolemmal areas without ATPase activity.[25] A majority of fibers will show multiple small acid-phosphatase-positive sites. This abnormal lysosomal activity is slowly lost over the next few years, but may appear focally in some fibers when the patient eventually and inevitably develops signs of classical myotonic dystrophy. Electron microscopy in the infants may show sarcoplasmic masses, containing disorganized normal structures, beneath a nucleus lightly separated from the sarcolemma. Tubules with dense cores correspond to the acid phosphatase activity. Nonprogressive mental retardation accompanies infantile myotonic dystrophy. Infantile myotonic dystrophy can be and has been misdiagnosed as congenital fiber type disproportion and centronuclear myopathy.

Classical myotonic dystrophy is, in general, a disease of adults, but some manifestations, particularly myotonia, can occur at the end of the first decade. Muscle biopsy is often not clearly diagnostic. Type I muscle fibers are usually small. Nuclear clump fibers are found. Internal nuclei are increased. Ring fibers may be present, though they are not specific. Necrotic fibers are rare. Fiber loss tends to result in adipose replacement rather than in fibrosis. Electron microscopy shows reduplication of the basal lamina of many fibers, as in many other dystrophies.

Proximal myotonic myopathy (DM2), with its tetranucleotide repeats, does not have an infantile counterpart, but myotonia can begin at the end of the first decade. Muscle pain is a frequent symptom. Biopsy shows atrophy of the type II fibers, some nuclear clumps, increased internal nuclei and, rarely, ring fibers.

Congenital Myopathies

This heterogeneous category encompasses several nondystrophic myopathies that are *defined by the muscle biopsy* either as an alteration in ratio and size of the two major types of myofibers or as subcellular structural abnormalities (see Chapter 28). Though they are nonprogressive, these diseases nevertheless can be severe even at birth and even life-threatening because of dysphagia and respiratory insufficiency, as in severe neonatal nemaline myopathy. Some congenital myopathies may be viewed as developmental disorders of muscle ontogenesis. The recent elucidation of the genetic mutations in the congenital myopathies demonstrates that many of the phenotypically distinct clinical and pathological entities are actually genetically related. For example, *TPM3* mutations may cause isolated congenital muscle fiber-type disproportion, nemaline myopathy, cap myopathy, and central core disease.

Congenital Muscle Fiber-Type Disproportion

The first of two obligatory muscle biopsy criteria[26,27] is disproportion in size of myofibers, with type I fibers being uniformly smaller than those of type II, which often show compensatory hypertrophy, especially of subtype IIb; the small type I fibers retain their polygonal contour in transverse section and are not angular as in denervation or in the atrophic type II fibers of disuse atrophy. The second histochemical criterion is disproportion in the ratio of fiber types, so that there is an overall predominance of type I fibers of up to 80%; over years, there may be no further type II fibers demonstrated, so that there is a progression in this aspect. Congenital muscle fiber-type disproportion (CMFTD) may be isolated, though uniform in all striated muscles, or may be associated with other myopathies, such as nemaline myopathy and the severe neonatal form of myotonic muscular dystrophy. Mutations of some genes involved in nemaline myopathy, such as the *TPM3* gene that encodes α-tropomyosin, also occur in isolated CMFTD without nemaline rods,[28] though the typical rods may appear later over time and be demonstrated by repeat muscle biopsy after the neonatal period. Mutations in *tropomyosin-3* have been reported with disproportion in size of type I fibers, but without predominance. Mutations of *ACTA-1* and of *SELN* also have been reported. CMFTD is also seen in genetic/metabolic encephalopathies with peripheral neuropathy, such as Krabbe disease, metachromatic leukodystrophy, and multiple sulfatase deficiency.

CMFTD may occur secondary to brain malformations, particularly those of the posterior fossa such as cerebellar hypoplasia, presumably because of abnormal suprasegmental innervation of spinal motor neurons from the bulbospinal tracts during fetal life.[29] By contrast, spastic diplegia due to corticospinal tract damage at any level during fetal development or from birth injury results in selective type II myofiber atrophy but not CMFTD. CMFTD is therefore a syndrome, not a specific single myopathy.

Nemaline Myopathy

This congenital myopathy is characterized by the presence of nemaline rods (*nema* = thread, Greek) within the myofibrils. The incidence of this myopathy is 1:50,000 and the majority of affected patients are already symptomatic from the neonatal period. Nine genes are now known to be responsible, though the clinical and pathological phenotype does not predict the genotype. There is also poor correlation between the "rod burden" within myofibers and the severity of clinical expression.[30] Clinical and pathological aspects are reviewed by Wallgren-Pettersson.[31]

Microscopically, nemaline rods are eosinophilic and thus difficult to detect among the normal eosinophilic myofibrils with H&E stain. With modified Gomori trichrome in frozen sections, they stain red. Atrophy of the rod-bearing fibers may occur, rendering the rods harder to see. They have no enzymatic activity. They are most easily seen on semithin sections, where their osmophilia makes them stand out darkly. They tend to accumulate in masses under the sarcolemma, though they first form in continuity with myofibrils. CMFTD is almost a uniformly accompanying feature and the type I predominance may be so great that type II fibers are essentially absent.

Ultrastructurally, nemaline rods are a proliferation of the lattice of the Z-disc. EM on longitudinal views shows both cross- and longitudinal striations. Rods preferentially occur in type I myofibers, but about half of patients have some rods in type II fibers as well. Antibodies to α-actinin (normally present in the Z-disc) and α-actin stain the rods. Intranuclear rods occur rarely and probably only in neonates with severe weakness; they usually indicate *ACTA1* mutations. Cytoplasmic rods tend to coexist with intranuclear rods, but may be absent. Nemaline myopathy due to *ACTA1* mutation is one of a spectrum of "actinopathies."[32]

Nemaline myopathy can be caused by mutations in at least nine genes, including α-actin (*ACTA-1*), α-tropomyosin 3 (*TPM3*), β-tropomyosin (*TPM2*), troponin-T, nebulin, cofilin-2, *KBTBD13,* and Kelch-like family members (*KLHL40, KLHL41*). The latter causes severe neonatal autosomal recessive nemaline myopathy. The proteins encoded by all these genes are related to the thin filaments of myofibrils. Dominant nemaline myopathy is most often caused by mutations in *ACTA-1* or α-tropomyosin; recessive mutations usually result from nebulin mutations. A number of these genes can, with different mutations, give rise to different diseases. For example, some mutations in the α-actin gene give rise to a severe neonatal myopathy with masses of exclusively thin filaments, with or without rods. Mutations in *TPM3* can cause a congenital myopathy with nonspecific findings, as well as **cap myopathy**, in which accumulations of distorted myofilaments are focally present on the periphery of fibers.[33] Somatic mosaicism is demonstrated in *TPM2*-related nemaline myopathy with cap structures. Clinical and myopathological effects of *TPM2* and *TPM3* are very similar.

Nemaline bodies are relatively specific for nemaline myopathy, though rods occasionally occur in other conditions. Rods along with cores and internalized nuclei are seen with some mutations of the ryanodine receptor, and cores were reported accompanied by rods in a boy with recessive mutations in the nebulin gene. Rods are described in myofibrillar myopathies, especially with cofilin mutations. There is also a syndrome with myosin loss and rods that occurs in HIV-infected adults. Experimentally, some rods can be produced in rats by tenotomy, which leads to focal myofibrillar loss starting with loss of thick filaments.

Central Core and Minicore/Multicore Myopathy

Central cores are sharply demarcated myofibril-bearing areas in muscle fibers. They lack mitochondria, and the degree of myofibrillar contraction within them is independent of that in the rest of the fiber. By EM, the myofibrils in cores are less distinct from one another, and the Z-disc tends to be irregular. Cores must be distinguished from targets, which are associated with denervation and which also lack mitochondria. Any myofibrillar area that lacks mitochondria is prone to show Z-disc streaming, which is always present in targets and can occur in cores. Cores extend from one end of the myofiber to the other; targets do not. Cores are generally much more widespread than targets, more uniform, and larger. Despite their name, cores are not always located in the center of fibers, and there can be more than one core in a single fiber.

Cores show up well as negative areas on oxidative enzyme reactions. They also lack phosphorylase and are negative with PAS. They contain less desmin than the normal areas of the fiber. A circle of increased oxidative activity may surround them, as well as a ring of lipid droplets. They occur in type I fibers, and there is usually marked type I predominance or universality.

Central core disease most often manifests itself in children with slow motor development who are hypotonic and have mild weakness, especially in the hip girdle. Congenital dislocation of the hip and progressive scoliosis may be associated.

Central core disease is caused by mutations in the ryanodine receptor,[34] the large tetrameric protein that lies at the junction of the t-tubule with the cistern of the sarcoplasmic reticulum. It contains the channel by which calcium is released among the myofilaments. Ryanodine itself is a pharmacological agent used to identify the channel. Central core disease is almost always an autosomal dominant trait. Thus, both mutated and nonmutated proteins are present. It is possible that the mutated proteins are sequestered in the cores. Mutations in this protein also cause the malignant hyperpyrexia syndrome (MHS). Patients with central core disease are therefore at risk for this catastrophic complication, but not all patients susceptible to MHS have central cores. In addition to cores, nemaline bodies have been found in a few patients with mutations in the ryanodine receptor gene.

Minicores or multicores are small areas, usually multiple within muscle fibers that lack mitochondria and show Z-disc streaming. They can be caused by mutations in the gene *SEPN,* which encodes selenoprotein N (a protein localized to triads), especially if they are associated with a rigid spine syndrome, but autosomal recessive mutations are found in the ryanodine receptor in some families. Levels of the protein are markedly reduced. These patients have ophthalmoplegia, ptosis, and axial myopathy or distal weakness and wasting. In one family with autosomal recessive disease, there were many internal nuclei as well as large poorly delimited areas of myofibrillar disorganization instead of cores.

Centronuclear Myopathies

In considering the centronuclear myopathies, one must distinguish between isolated nuclei (or occasionally a group of tightly packed nuclei) in the geometrical center of fibers versus internalized nuclei without any consistent location. Five genes have been found to be responsible for most cases of centronuclear myopathy.[35] Cases remain in which no mutations have been found. Aside from the central nuclei, almost all cases show clear predominance of histochemical type I fibers. On longitudinal section, the central nuclei tend to form rows. On transverse section, ATPase stains may show central parting of the myofibrils to accommodate a nucleus that is out of the plane of section. Similarly, oxidative enzymatic activities and PAS often show strong focal positivity in the center of the fibers because of mitochondria and glycogen trailing the central nuclei.

Mutations in the *myotubularin* gene cause an X-linked centronuclear myopathy that most commonly manifests in the neonatal period. Onset at that age is a sign of poor prognosis. This form has been called *myotubular myopathy*, since it resembles fetal muscle at 8 to 15 weeks, when myofibers have central nuclei surrounded by myofibrils; however, there is no general arrest in muscle development. The central nuclei tend to be very numerous and are often larger and more vesicular than normal mature nuclei. Type I fibers may be universal or exhibit a massive predominance. The average size of myofibers is smaller than in controls. Oxidative activity of mitochondria associated with the central nuclei contrasts with a relative pallor in the surrounding ring of cytoplasm. The fibers tend to express vimentin and stain somewhat more strongly than normal for desmin. EM shows disorganized triads and focal loss of myofilaments. Patients with much milder symptoms or a much later age of onset with mutations in the same gene are now known. Some of these are manifesting carriers. An interesting histological abnormality may be present, "necklace fibers," which have a basophilic ring where nuclei accumulate about 3 μm below the sarcolemma. EM shows small obliquely oriented myofibrils in this ring along with mitochondria and sarcoplasmic reticulum.

Mutations in the gene for *dynamin-2* give autosomal dominant disease and account for about half of all patients with centronuclear myopathy.[36] Some of the severe neonatal cases have *dynamin-2* mutations. Significant numbers of centronuclear fibers are usually present, but the proportion varies with age and the specific muscle, and even with the side of the body. Central nuclei occur in both fiber types. In some fibers, oxidative reactions may show a "spoked wheel" pattern because of a radial arrangement of mitochondria and sarcoplasmic reticulum. This feature has been described as "radiating sarcoplasmic strands." The myofibrils in the periphery of fibers tend to be larger than those in the center. Small foci of fibrosis may be present, though not associated with signs of necrosis. Below 5 years, the characteristic changes may be incomplete or lacking in some muscles, and this may lead to a suspicion of dystrophy or a diagnosis of congenital muscle fiber-type disproportion.

Mutations in the *BIN-1* gene interfere with the binding of its protein, called amphiphysin, to *dynamin-2*. Inheritance is autosomal recessive. The type I fibers consistently predominate or are the sole fiber type. Most have central nuclei. Occasionally packs of central nuclei are found. Radiating strands are rarely seen. Fibro-adipose replacement may be found without visible necrosis.

Myotubularin (*MTM1*), *dynamin-2*, and amphiphysin are all localized by antibody studies to the T-tubule wall in triads. This crucial region is where the action potential releases a signal to the ryanodine receptor to release calcium. Myotubularin is an inositol phosphatase that regulates the levels of a muscle-specific inositol phosphate lipid with a role in membrane recycling. *Dynamin-2* and amphiphysin regulate membrane trafficking in endosomes. In all three types of centronuclear myopathy, normal myotubularin is lost, resulting in alterations in triad organization and of T-tubules.[37] Large duplications of *MTM1* also can cause centronuclear myopathy.[38] Sarcotubular disorganization impairs excitation coupling needed for contraction of myofibrils. How this could lead to the characteristic centronucleation is less clear. It has also been found, however, that myotubularin normally binds to desmin, which has a role in stabilizing the position of organelles.

Recently, compound heterozygous mutations in the giant *titin* gene have been shown to cause a myopathy in children with very extensive internalization of nuclei and enough central nuclei to suggest a diagnosis of centronuclear myopathy.[39] The normal titin protein extends from the Z-discs to the M-line in the middle of the A-band, and appears to be important in myofibril formation and maintenance. Type 1 smallness and predominance are present and many fibers exhibit areas of myofibrillar disorganization with Z-disc streaming.

There are other patients with prominent internalization of nuclei in combination with some centralization and large areas of myofibrillar disorganization and variable oxidative activity. Ophthalmoparesis and both axial and proximal weakness are found. Many of these patients have mutations in the ryanodine receptor, better known in connection with central core disease and malignant hyperpyrexia.

Genetic and neuropathological studies of X-linked centronuclear ("myotubular") myopathy have led to effective gene therapy in mice[40] and in dogs, so that even after a single dose the animals are more ambulatory and have much improved weakness. Gene replacement therapy in this disease is now being studied in humans.

Segmental Amyoplasia

Congenital absence of individual muscles is common and is often asymmetric. A common aplasia is the palmaris longus muscle of the ventral forearm, which is in 30% of normal subjects and is fully compensated by other flexors of the wrist. Unilateral absence of a sternocleidomastoid muscle is one cause of congenital torticollis. Absence of one pectoralis major muscle is part of the Poland anomaly.

When innervation does not develop, as in the lower limbs in severe cases of *myelomeningocele,* muscles can fail to develop. In *sacral agenesis,* the abnormal somites that fail to form bony vertebrae can also fail to form muscles from the same defective mesodermal plate, a disorder of induction resulting in segmental amyoplasia. Skeletal muscles of the extremities fail to differentiate from embryonic myomeres if the long bones do not form. Absence of one long bone, such as the radius, is associated with variable aplasia or hypoplasia of associated muscles, such as the flexor carpi radialis. End-stage neurogenic atrophy of muscle is sometimes called *amyoplasia,* but this use is semantically incorrect. In segmental amyoplasia, the connective tissue structure of the muscle, blood vessels and nerves are present, but the myofibers themselves are replaced by adipose tissue. This pathological picture suggests a defect in the myomere with preservation of the neural-crest derived components of muscle.

Generalized amyoplasia usually results in fetal death, and liveborn neonates rarely survive. A mutation in one of the myogenic genes is the suspected etiology because of genetic knockout studies in mice, but it has not been proven in humans.

Arthrogryposis multiplex congenita is a syndrome of multiple joint contractures that can result from lack of fetal movement or intrauterine immobilization, as with oligohydramnios, sometimes called *fetal akinesia syndrome*[41] (see Chapter 7). It can result from fetal neuronopathy, neuropathy, or severe myopathy, such as type 0 or type I spinal muscular atrophy and in some cases of congenital myotonic dystrophy and of congenital myasthenia gravis.

Muscular Dysgenesis (Proteus Syndrome Myopathy)

Proteus syndrome is a disturbance of cellular growth involving ectodermal and mesodermal tissues. The cause is a mosaic activating mutation in the *AKT1* gene[42] and it is often associated with hemimegalencephaly, an *AKT3* mutation. It manifests as asymmetric overgrowth of the extremities, verrucous cutaneous lesions, angiomas of various types, thickening of bones, and excessive growth of muscles without weakness. It is regarded as one of the epidermal nevus syndromes. Histologically, the muscle demonstrates a unique *muscular dysgenesis.*[43] Abnormal zones are adjacent to zones of normal muscle formation and do not follow anatomical boundaries. The muscle biopsy may resemble a muscular dystrophy at first glance, but it lacks the histopathological features of a true dystrophy.

MYOFIBRILLAR MYOPATHIES

Myofibrillar myopathies are a group of progressive myopathies with distinctive histopathological alterations in the myofibrils consisting of dissolution of contractile filaments and disintegration of the Z-disk, accumulation of granular-filamentous material next to myofibrils and proteinaceous deposits.[44,45] The diagnosis is established by muscle biopsy including special immunoreactivities and ultrastructure.

Most myofibrillar myopathies are not symptomatic in childhood, but occasionally older children and adolescents show early symptoms of nonspecific proximal and distal weakness. Infantile autosomal recessive forms also occur and can cause mild neonatal hypotonia and weakness with disproportionately severe dysphagia and respiratory insufficiency, at times leading to early death. Some patients show improvement in later infancy and early childhood, acquiring the ability to swallow by 3 years of age. Cardiomyopathy is a complication.

Early changes in myofibrillar myopathy are foci of Z-disc streaming. All myofibrillar myopathy mutations studied to date have appeared in Z-disk-associated proteins.[44,45] By EM, the streaming is associated with the accumulation of granular-filamentous material that is as dense as the Z-disc but is found between myofibrils. More advanced lesions show myofibrillar disruption and accumulation of proteinaceous material that displaces normal organelles and that characteristically is immunoreactive for several normal proteins, including desmin, the small heat-shock protein α-B-crystallin, dystrophin, N-CAM, myotilin, ZASP, filamin-C and beta-amyloid precursor protein. In histochemical preparations, these masses interrupt the distribution of normally staining structures, leading to the descriptive term of "rubbed out" areas. Rounded cytoplasmic bodies, formed of disoriented thin filaments, are often associated. Occasionally, there are rimmed vacuoles in cryostat sections and membranous whorls by EM.

Mutations in several genes have been found responsible for cases of myofibrillar myopathy and more genes clearly wait to be discovered. Almost all the respective proteins are normally localized to thin filaments of the Z-disc. Desmin, which is one of the culpable proteins, surrounds the Z-disc. Mutations in some proteins seem to cause disease only in adults. Those that may cause disease in children are desmin,[46] α-B-crystallin, *BAG-3*, cofilin, and FHL1. The rare autosomal recessive mutations in desmin cause severe disease in infants. Autosomal dominant mutations can occur in older children or in adults. Cardiomyopathy is present in about 60% of patients. Alpha-B-crystallin is a chaperone molecule aside from its structural role in the crystallin lens. An autosomal recessive myopathy in Canadian Cree aboriginal infants is characterized by severe generalized muscular hypertonia that is not relieved by neuromuscular blockade.[47] The proteinaceous deposits of myofibrillar myopathy usually are strongly positive for α-B-crystallin. In these children, they were negative with an antibody to the whole protein, giving a clue to the homozygous mutation of α-B-crystallin. Cofilin mutations can cause autosomal recessive disease in young children. *Bag3* mutations give rise to dominant disease in children. These same mutations also can cause giant axonal neuropathy.[48] The axons are packed with neurofilaments. Rigid spine and cardiopathy occur. In addition to the myofibrillar lesions in the biopsies, dense, presumably apoptotic, nuclei have been described. Mutations of the FHL1 gene can cause various forms of X-linked dominant myopathy that include myofibrillar myopathy. Biopsy usually shows reducing bodies, which are eosinophilic deposits that are positive with the menadione-nitro blue-tetrazolium reaction.[49] By electron microscopy they are seen to be made up of packed strongly osmiophilic tubular filaments. They often surround nuclei and rarely are found within nuclei. An Emery-Dreifuss-like syndrome with joint contractures, mild weakness, and later-onset cardiac symptoms is also caused by *FHL1* mutations, but reducing bodies are absent in this disease.

GLYCOGENOSES

The most important entity here is *α-glucosidase* or *acid maltase deficiency*, also known by the eponym *Pompe disease*, originally applied to the infantile form, but now often extended to the entire age range (see Chapter 39). This disease is the only lysosomal type of glycogen storage. In the infantile form, storage of glycogen in muscle is massive, appearing both in vacuoles and as cytosolic lakes, probably resulting from rupture of lysosomes. Acid phosphatase marks the lysosomes. The wall of some vacuoles is positive for dystrophin, indicating that they have become connected to the extracellular space. Storage in endothelium can be seen in semithin sections. EM shows masses of glycogen particles surrounded by a single membrane. Some of the lysosomal vacuoles may appear empty because the glycogen can be leached out during tissue processing. Heart, liver, and eccrine glands of the skin show storage. In juvenile patients, storage is far more discrete. The heart is not involved. Acid phosphatase reaction is useful in finding the muscle fibers with storage. PAS is also useful, combined with PAS after amylase digestion, to prove the identity of glycogen, since mucopolysaccharides within the intramuscular connective tissue also are PAS-positive but are not digested by diastase. EM will show relatively small lysosomes filled with glycogen particles. The presence of some excess of cytosolic glycogen in the vicinity of lysosomes by EM is common and not of prognostic significance.

The only disease whose pathology could be confused with that of acid maltase deficiency is *Danon disease*, deficiency of LAMP-2, a protein of the lysosomal wall. Enzymatically incompetent lysosomes proliferate in various muscle fibers. The contents are variable, but glycogen predominates. Cardiac involvement is present at all ages. Inheritance is X-linked dominant.

Among the nonlysosomal glycogen storage diseases, deficiency of *branching enzyme* stands out because the storage is of a fibrillary polysaccharide ultrastructurally distinct from ordinary glycogen. The abnormality is in the synthesis of glycogen. The deposits are PAS-positive, sharply demarcated, relatively resistant to α-amylase but digestible by pectinase. Isoforms of the enzyme exist, one apparently limited to the nervous system, and there are a number of clinical syndromes.

Deficiency of the debranching enzyme leads to the development of PAS-positive cytosolic lakes of glycogen that can replace myofibrils. This must be distinguished from a nonspecific mild excess of glycogen that is encountered sometimes without explanation. Occasionally also in muscle that has been allowed to hypercontract before processing, normal glycogen particles can be pushed into subsarcolemmal aggregates, which can be confused with storage. Debranching enzyme deficiency leads to excess glycogen in eccrine glands of skin. This can serve as a check.

McArdle disease, or myophosphorylase deficiency, can be diagnosed by a specific histochemical reaction. The product fades on exposure to light, so that slides must be examined promptly after being reacted, otherwise they might show a false lack of activity. Storage of glycogen is usually not conspicuous.

A histochemical method exists for phosphofructokinase, but its reliability in demonstrating lack of the enzyme is not well established. Two histochemical substrates must be used: fructose-6-phosphate and fructose-1, 6-diphosphate. For this and for the other, rarer, glycogen storage diseases, diagnosis depends on clinical, biochemical, and genetic data.

LIPID STORAGE MYOPATHIES

Muscle biopsy is useful to demonstrate the presence of a lipid storage myopathy (see Chapter 40), though the etiology and mechanisms involved must be revealed by biochemical and molecular genetic methods. Globules of triglyceride, displayed with oil red O, Sudan 4, or Sudan black, are most numerous in type I fibers. The 2A fibers have slightly less and 2Bs the least. Some type I fibers may be atrophic. Necrotic fibers with massive amounts of lipid have been reported in a riboflavin-responsive lipid storage myopathy. In general, however, the light microscopic picture is similar whether there is deficiency of carnitine or problems with β-oxidation or a neutral lipid storage myopathy. Electron microscopy in β-oxidation problems or carnitine deficiency often shows some mitochondrial abnormalities, such as clumping of mitochondria between myofibrils, abnormal cristal patterns, or crystals of creatine kinase.

One must also remember that lipid droplets in muscle reflect to some degree the level of plasma fatty acids. With fasting, increased triglyceride breakdown in adipose cells leads to increased plasma fatty acids and increased lipid globules in skeletal muscle. Occasional muscle biopsies show a moderate excess of lipid droplets that is never explained. A focal increase in lipid droplets is sometimes found in ischemic situations, for instance, in dermatomyositis.

Mitochondrial Cytopathies

Mitochondrial disorders are increasingly identified as a more frequent cause of encephalomyopathies than was previously recognized (see Chapter 41). For decades, the muscle biopsy has been indispensable for diagnosis of this diverse group of genetic diseases. It often provides clues to the type of mitochondrial disease, but does not define the specific genetic mutation. Muscle biopsy is also useful in some patients who exhibit no weakness or other clinical signs of myopathy, but in whom the CNS, heart, or kidney may be the predominantly involved organ in a mitochondrial cytopathy. The muscle biopsy also documents secondary altered mitochondrial structure and function as secondary metabolic defects in some genetic diseases not primarily of mitochondrial origin, such as in the lysosomal diseases of infantile (Pompe) and adult-onset acid maltase deficiencies. Toxins and some drugs also can induce mitochondrial alterations.[50] Mutations in the mitochondrial genes generally result in proliferation of muscle mitochondria, producing "ragged-red" fibers. Mutations in the nuclear genome do not, unless the proteins produced are essential for the maintenance or translation of the mtDNA.[51] In children, less time has been present since birth for mitochondria to proliferate. For this reason, even with a mutation in the mitochondrial genome, ragged red fibers may be difficult to identify in the biopsy. Also, ragged-red fibers tend to be associated with combination respiratory complexes I and IV defects, the most frequent in adult mitochondrial cytopathies, but much less frequent in infants and young children in whom complexes III and V are usually more involved.[52–54]

Mitochondria continuously change shape through the actions of fission and fusion, and the equilibrium may be altered in mitochondrial diseases. Mitochondrial fusion proteins are regulated by proteolysis and ubiquitination; dynamins are inner mitochondrial membrane fusion factors and include the Zn-protease Oma1, which may cause changes in the architecture of the cristae.[55]

Advances in molecular genetics now enable the definitive diagnosis of some mitochondrial diseases from lymphocytes in blood samples, which are less invasive than muscle biopsy. Most mitochondrial diseases in which genetic diagnosis is now possible are in the category of the "ragged-red fiber" diseases, involving a combination of defects in respiratory chain complexes I and IV, but these are mostly adult diseases that are much less frequent in infants and children than the non-ragged-red fiber diseases that mostly involve complexes III and V, as previously mentioned. For these diseases, the molecular genetic defect often remains unknown or is a novel mutation not previously reported. For infants and children, therefore, the muscle biopsy remains the diagnostic procedure of choice. Other clinical parameters, such as elevated serum and CSF lactate and changes in the basal ganglia and periventricular regions demonstrated by magnetic resonance imaging (MRI) of the brain, are useful adjuncts to diagnosis but are too nonspecific to be definitively diagnostic, though they are helpful in selecting patients for muscle biopsy and in justifying this procedure. Muscle biopsy is of particular value in investigating infants with suspected Leigh disease and other mitochondrial encephalopathies presenting with seizures, developmental regression, and motor disturbances.

HE-stained sections yield limited information for the diagnosis of mitochondrial disease and are never definitive, but are useful for excluding other neuromuscular diseases. Ragged red fibers may appear as amorphous basophilic finely granular cytoplasm in the subsarcolemmal regions, corresponding to the red-staining regions with modified Gomori trichrome stain. Inflammation and zones of necrosis are not features. The application of the modified Gomori trichrome stain in frozen sections is a sensitive means of demonstrating ragged-red fibers, which appear as red-staining rims with irregular margins and discontinuities in subsarcolemmal regions of myofibers and sometimes in the intermyofibrillar sarcoplasm. Masson trichrome stain, an older and more generally used trichrome for other tissues, is not specific for ragged-red fibers because myofibrils also are stained red. In mitochondrial myopathies at any age, more type I myofibers are often selectively atrophic than those of type II, a difference from disuse atrophy and from neurogenic atrophy that is nonselective for fiber type. Indeed, in some chronic mitochondrial diseases, there may be small areas of type-grouping, indicating denervation with reinnervation, because of mitochondrial involvement of Schwann cells or axons in peripheral nerve. The "oxidative enzyme" histochemical stains are, together with Gomori trichrome, the most important histochemical stains for the diagnosis of mitochondrial cytopathy. This panel of findings of mitochondrial enzymatic activity together confirms mitochondrial cytopathy, but cannot denote the specific genetic point mutations of mtDNA or of nDNA encoding subunits of the respiratory chain complexes.

The muscle biopsy is valuable for the investigation of suspected mitochondrial encephalopathies.[56,57] Vascular endothelial cells of the brain, as well as in muscle, often are severely involved and contribute to encephalopathy, particularly in infantile mitochondrial diseases.[8,58–60] The same histochemical stains used in muscle biopsies for mitochondrial enzymes also can be applied to freshly frozen brain tissue, for example in surgical resections of epileptic foci, to demonstrate hypermetabolic neurons in focal cortical dysplasias, hemimegalencephaly, tuberous sclerosis, and other encephalopathies including mitochondrial.[61] Peripheral nerve also can be affected in mitochondrial cytopathies, the Schwann cells in particular.[62]

A condition termed the *mitochondrial depletion syndrome* is usually expressed in the fetus, neonate, or infant, but rarely may be delayed until later childhood or even early adult life. Abnormally low levels of respiratory chain enzymatic activities are demonstrated in most or all of the five complexes, but histochemical changes in frozen sections of muscle are milder than anticipated from the biochemical assays; ultrastructural changes in the mitochondria further confirm the diagnosis. This syndrome occurs in several forms: myopathic; encephalomyopathic; hepatoencephalopathic; and intestinal encephalopathic.[63,64] *Alper disease* is one of the most frequent encephalomyopathic forms[65] and *Leigh encephalopathy* accounts for some other infantile cases. Several genetic mutations have been demonstrated, particularly in the later-onset forms (e.g. *POLG1, TK2, TYMP, RRM2B, PEO1/twinkle*), hence the mitochondrial depletion syndrome is not a single genetic disease.

In summary, the majority of mitochondrial diseases of infants and children cannot presently be diagnosed by genetic testing; hence, the muscle biopsy remains the procedure of choice for definitive diagnosis. Alterations in muscle can usually be demonstrated even if clinical signs of myopathy are absent and the principal manifestations are encephalopathic, cardiac, hepatic, or referable to other organ systems. Mitochondrial structural and functional alterations also can be diagnosed as secondary phenomena in other genetic diseases and in nongenetic acquired toxic or drug-induced encephalomyopathies.

INFLAMMATORY MYOPATHIES

Inflammatory myopathies are important, being serious diseases with treatment possibilities[66] (see Chapter 42). Before addressing them directly, the reader should be aware that certain genetic diseases, especially facioscapulohumeral dystrophy and dysferlinopathy, may show perimysial and perivascular lymphocytic infiltrates in the muscle biopsy to the extent that misdiagnosis is possible. The antibody to the major histocompatibility antigen class 1 (MHC-1) is quite useful. It should be done with both positive and negative controls. In normal muscle, the level of expression on muscle fibers is too low to be visible. Regenerating fibers, however, are positive. In dermatomyositis fibers in damaged areas on the periphery of fascicles are positive and sometimes all fibers are positive with perifascicular enhancement. In polymyositis, all fibers are equally positive, whether they show signs of damage or not. In muscular dystrophies with inflammation, MHC-1 antigen is not turned on. A number of muscle biopsies will show some small inflammatory infiltrates with no characteristic location, no capillary pathology, and absent or very focal activation of MHC-1. Myositis-specific antibodies are liable to be absent in such cases. Myositis-associated antibodies are too high in the general population to be significant. In some cases, a viral cause is found; in others, a specific myopathy; in most, no specific diagnosis emerges.

Dermatomyositis

The pathological picture of dermatomyositis is usually quite characteristic and specific. Most biopsies in children will show atrophy of fibers on the periphery of fascicles, while the fibers in the center remain normal or show less advanced stages of damage.[6,67] Rarely, all the fibers in a fascicle seem equally affected. This is seen more often in adults. Dermatomyositis is basically a vascular disease with poorly understood autoimmune roots. Capillaries are destroyed in the muscle, usually starting at the periphery of fascicles, which is the area furthest from the source; the process can also extend to involve arteries, in which case infarcts may occur in the muscle. Vascular damage in the intestine is an occasional complication. Immunoreactivity for the lytic component of complement marks the capillaries, even before the muscle fibers show changes.[68] The capillaries also tend to be abnormally positive for acid phosphatase, a marker for lysosomes. Loss of capillaries can be visualized using an antibody to endothelial cells, such as factor VIII, or by the lectin UEAI (Ulex Europaeus agglutinin-I). The capillary loss in dermatomyositis can also detected by careful study of semithin sections (Figure 4.5). Evaluation of capillary loss necessitates familiarity with the normal pattern. Until almost 9 years of age, the number of capillaries per muscle fiber and per transverse myofiber area is normally less than in adults, and in small infants the ratio per fiber is much less.

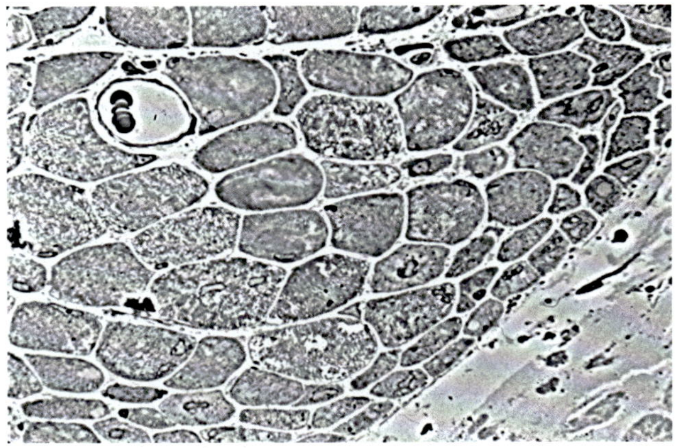

FIGURE 4.5 This is a semithin section stained with paraphenylene diamine and viewed through phase optics. The biopsy is from an 18-year-old girl who had suffered from dermatomyositis for several months without effective treatment. Note the perifascicular atrophy and the almost complete depletion of capillaries.

Electron microscopy shows tubuloreticular structures in the smooth endoplasmic reticulum (or in the equivalent perinuclear cistern) of endothelium (Figure 4.6) and occasionally in some lymphocytes or macrophages.[6] These structures are seen in other tissues in various viral diseases and in lupus erythematosus. They can be produced *in vitro* in lymphocytes by interferon. In dermatomyositis, they can also be demonstrated in endothelial cells in skin, whether or not macroscopic lesions are present. By EM, the endothelial cells in the muscle show various reactive changes: increased cytoplasmic volume and increased complexity of processes. Capillaries in necrosis may be found, as well as the empty basal lamina of defunct capillaries.

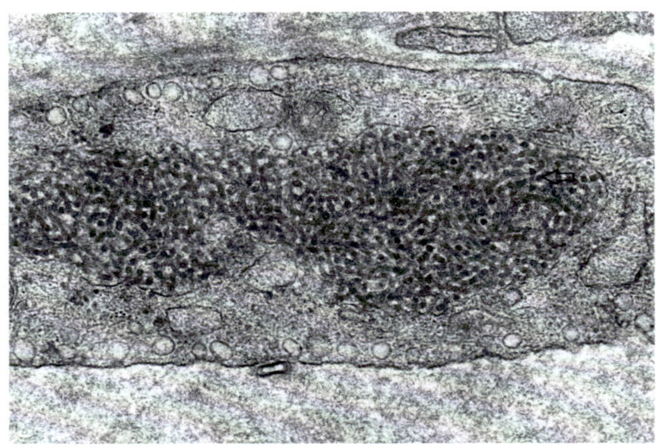

FIGURE 4.6 This EM shows a large collection of tubuloreticular structures in the smooth endoplasmic reticulum of an endothelial cell in the muscle of a child with dermatomyositis. Reproduced with permission from Oxford University Press, USA.[6]

The atrophic muscle fibers are often angular and flattened parallel to the surface of the fascicle. With H&E or the modified trichrome, they show circumscribed areas of pallor among the myofibrils and smudgy areas of dark staining that correspond to Z-disc streaming. They react strongly with oxidative enzyme reactions, probably because the atrophy pushes the mitochondria together. On semithin sections, the muscle fibers show punched-out myofibrillar loss and multifocal Z-disc streaming. Nuclei become centralized. EM shows that the areas where myofibrils have been lost are filled with glycogen particles along with a few abnormally shaped mitochondria. Necrotic fibers are not usually found, but they do occur in two situations. Rarely, there can be scattered necrotic fibers in a perifascicular distribution, or there can be a mass of necrotic fibers reaching into the center of a fascicle, a picture highly suggestive of infarction. Some viable fibers close to such necrotic groups may show an excess of neutral lipid globules, as if the lipid could not be utilized in that location. A special feature of necrotic fibers in dermatomyositis is a delay in macrophage invasion and phagocytosis. These processes depend on the vascular supply. Satellite cell activation and proliferation, on the other hand, are not particularly delayed, and so an anomalous picture of regeneration preceding phagocytosis may occur.

Distinction should be made between regeneration and repair. Regeneration is the genesis of a new muscle fiber after necrosis. Repair is what happens when a cell suffers nonlethal damage, and is subsequently able to reconstruct damaged substructures. Perifascicular necrosis can occur, but is rare. If a healing phase of the disease occurs, the atrophic cells on the periphery of the fascicle will start to produce new myofibrils. One biopsy from a patient with inactive disease showed small fibers filled with myofibrils in this location, and the distribution of capillaries was normal in relation to the cross-sectional fiber area.

Inflammatory cells, lymphocytes, and monocytes are usually present in moderate numbers in the interfascicular space. The lymphocytes consist of both B- and T-cells (usually with more helper cells than suppressor/killer cells). Recent work has revealed that there are many plasmacytoid dendritic cells in the muscle. These cells, which have a plasmacytoid morphology, are identified by a combination of antibodies. The plasmacytoid dendritic cells produce type 1 interferon as well as IL-2. Proteins whose production is stimulated by interferon are increased in the tissue, as well as in blood. The tubuloreticular inclusions in endothelium can be experimentally produced *in vitro* in lymphocytes by interferon.

Type 1 major histocompatibility antigens, normally undetectable, become strongly expressed on the surface of muscle cells, either in a perifascicular distribution or on all fibers with enhancement in the perifascicular areas. In polymyositis, by contrast, all fibers in the biopsy are equally affected.

Many patients, particularly adults, have various myositis-associated or myositis-specific autoantibodies (MSAs). The individual MSAs in particular tend to be associated with various clinical characteristics. No autoantibodies to endothelium have yet been demonstrated. Perhaps because of this, in most immunological studies of dermatomyositis, there is no reference to the pathological data, or if there is, the flimsiest rather than the most solid work is cited. One study in adults successfully combines the two approaches.[69] A very few patients without cutaneous signs have pathological evidence of dermatomyositis.

Recent serological studies suggest that a high proportion of cases of inflammatory myopathy are overlap myositis. Such cases appear to include cases with features of systemic sclerosis and others with lupus. Our experience is that when a patient with lupus has myositis, the pathological picture in the muscle is indistinguishable from that of dermatomyositis. Lupus, of course, shares with dermatomyositis the ultrastructural presence of tubuloreticular structures in endothelium. Patients with antisynthetase antibodies, in particular anti-Jo1, tend to have interstitial lung disease, which might lead them to be classified as having overlap myositis.

Histological damage in dermatomyositis is multifocal. It is possible to have a normal area and a highly abnormal area within one biopsy. This means that the severity of changes in a biopsy is not a guide to prognosis.

Polymyositis

Polymyositis, as has been recently recognized, is less common than dermatomyositis. It does occur in children but not as often as in adults. Pathologically, it is characterized by extensive inflammatory infiltrates inside fascicles and invasion of nonnecrotic muscle fibers by lymphocytes, especially CD8 suppressor/cytotoxic lymphocytes and monocytes. This can occur with isolated fibers or those at the borders of the inflammatory infiltrates. A few isolated necrotic fibers are usually seen. Fibrosis, with no particular orientation of collagen, may be present. The capillary bed is customarily normal. Anti-MHC-1 antibodies mark the surface of all fibers in the biopsy. This is particularly useful because the nonnecrotic invaded fibers are not numerous and may be absent in any particular biopsy. They are also not completely specific, since they also occur in inclusion body myositis, a disease of adults. EM often shows reduplications of the basal lamina. Small, rounded osmiophilic granular structures encircled by basal lamina are associated. They may be composed of pinched-off superficial cytoplasm. They tend to be numerous where a fiber has died without regeneration occurring.

Autoimmune Necrotizing Myopathy

Autoimmune necrotizing myopathy is a recently described entity that occurs mainly in adults and rarely in children. Biopsy shows extensive necrosis with very little in the way of lymphocytic infiltrates. The lytic component of complement has been demonstrated in the wall of small vessels, and immune complexes have been seen. Nearly half of the cases have an autoantibody to the signal recognition peptide (it recognizes the signal sequence indicating that a protein should enter the rough endoplasmic reticulum). Other causes of rhabdomyolysis should be ruled out, for example toxins and carnitine palmitoyltransferase (CPT) deficiency.

REFERENCES

1. Dubowitz V, Sewry CA. *Muscle Biopsy: A Practical Approach.* 3rd ed. London: Elsevier Saunders; 2007.
2. Adams RD. *Diseases of Muscle. A Study in Pathology.* 3rd ed. Hagerstown, MD: Harper & Row; 1975.
3. Mastaglia FL, Walton JN. *Skeletal Muscle Pathology.* 2nd ed. Edinburgh: Churchill Livingstone; 1992.

4. Sarnat HB. *Muscle Pathology and Histochemistry*. Chicago: American Society of Clinical Pathologists Press; 1983.
5. Heffner Jr. RR. *Muscle Pathology (Contemporary Issues in Surgical Pathology Ser*, vol. 3. Edinburgh, NY: Churchill Livingstone; 1984.
6. Carpenter S, Karpati G. *Pathology of Skeletal Muscle*. 2nd ed. Oxford, NY: Oxford University Press; 2001.
7. Goebel HH, Sewry CA, Weller RO. *Muscle Disease: Pathology and Genetics*. 2nd ed. London: Wiley-Blackwell; 2013.
8. Sarnat HB, Flores-Sarnat L, Casey R, Scott P, Khan A. Endothelial ultrastructural alterations of intramuscular capillaries in infantile mitochondrial cytopathies: "mitochondrial angiopathy". *Neuropathology* 2012;**32**:617–27.
9. Valentijn KM, Sadler JE, Valentijn JA, Voorberg J, Eikenboom J. Functional architecture of Weibel-Palade bodies. *Blood* 2011;**117**: 5033–43.
10. Sarnat HB. Ontogenesis of striated muscle. In: Polin RA, Fox WW, Abman SH, editors. *Fetal and Neonatal Physiology*. 4th ed. Philadelphia: Elsevier; 2011. pp. 1924–47.
11. Romero NB, Mezmezian M, Finziańska A. Main steps of skeletal muscle development in the human: morphological analysis and ultrastructural characteristics of developing human muscle. In: Dulac O, Lassonde M, Sarnat HB, editors. *Handbook of Clinical Neurology,* vol. 113: *Pediatric Neurology*. Edinburgh: Elsevier B.V.; 2012. pp. 1299–310.
12. Bentzinger CF, Wang YX, Rudnicki MA. Building muscle: molecular regulation of myogenesis. *Cold Spring Harb Perspect Biol* 2012;**4**(2): pii: a008342.
13. Millay DP, O'Rourke JR, Sutherland LB, Bezprozvannaya S, Shelton JM, Bassel-Duby R, et al. Myomaker is a membrane activator of myoblast fusion and muscle formation. *Nature* 2013;**499**:301–5.
14. Carpenter S, Karpati G. Segmental necrosis and its demarcation in experimental micropuncture injury of skeletal muscle fibers. *J Neuropathol Exp Neurol* 1989;**48**:154–70.
15. Casademont J, Carpenter S, Karpati G. Vacuolation of muscle fibers near sarcolemmal breaks represents T-tubule dilatation secondary to enhanced sodium pump activity. *J Neuropathol Exp Neurol* 1988;**47**:618–28.
16. Yin H, Price F, Rudnicki MA. Satellite cells and the muscle stem cell niche. *Physiol Rev* 2013;**93**:23–67.
17. Webster C, Blau HM. Accelerated age-related decline in replicative life-span of Duchenne muscular dystrophy myoblasts: implications for cell and gene therapy. *Somat Cell Mol Genet* 1990;**16**:557–65.
18. von Maltzahn J, Jones AE, Parks RJ, Rudnicki MA. Pax7 is critical for the normal function of satellite cells in adult skeletal muscle. *Proc Natl Acad Sci USA* 2013;**110**:16474–9.
19. von Maltzahn J, Chang NC, Bentzinger CF, Rudnicki MA. Wnt signaling in myogenesis. *Trends Cell Biol* 2012;**22**:602–9.
20. Sousa-Victor P, Gutarra S, Garcia-Prat L, Rodriguez-Ubreva J, Ortet L, Ruiz-Bonilla V, et al. Geriatric muscle stem cells switch reversible quiescence into senescence. *Nature* 2014;**506**:316–21.
21. Barresi R. From proteins to genes: immunoanalysis in the diagnosis of muscular dystrophies. *Skelet Muscle* 2011;**1**:24.
22. Schmalbruch H. Regenerated muscle fibers in Duchenne muscular dystrophy: a serial section study. *Neurology* 1984;**34**:60–5.
23. Cetin N, Balci-Hayta B, Gundesli H, Korkusuz P, Purali N, Talim B, et al. A novel desmin mutation leading to autosomal recessive limb-girdle muscular dystrophy: distinct histopathological outcomes compared with desminopathies. *J Med Genet* 2013;**50**:437–43.
24. Mercuri E, Muntoni F. The ever-expanding spectrum of congenital muscular dystrophies. *Ann Neurol* 2012;**72**:9–17.
25. Sarnat HB, Silbert SW. Maturational arrest of fetal muscle in neonatal myotonic dystrophy. A pathologic study of four cases. *Arch Neurol* 1976;**33**:466–74.
26. Sarnat HB. Congenital muscle fiber-type disproportion. San Diego: MedLink Corporation; [updated 27 Mar 2013; cited 13 Apr 2014]; Available from: <http://www.medlink.com/medlinkcontent.asp>; 1994.
27. Clarke NF. Congenital fiber-type disproportion. *Semin Pediatr Neurol* 2011;**18**:264–71.
28. Lawlor MW, Dechene ET, Roumm E, Geggel AS, Moghadaszadeh B, Beggs AH. Mutations of tropomyosin 3 (TPM3) are common and associated with type 1 myofiber hypotrophy in congenital fiber type disproportion. *Hum Mutat* 2010;**31**:176–83.
29. Sarnat HB. Cerebral dysgeneses and their influence on fetal muscle development. *Brain Dev* 1986;**8**:495–9.
30. Romero NB, Sandaradura SA, Clarke NF. Recent advances in nemaline myopathy. *Curr Opin Neurol* 2013;**26**:519–26.
31. Wallgren-Pettersson C, Sewry CA, Nowak KJ, Laing NG. Nemaline myopathies. *Semin Pediatr Neurol* 2011;**18**:230–8.
32. Nowak KJ, Ravenscroft G, Laing NG. Skeletal muscle alpha-actin diseases (actinopathies): pathology and mechanisms. *Acta Neuropathol* 2013;**125**:19–32.
33. Malfatti E, Schaeffer U, Chapon F, Yang Y, Eymard B, Xu R, et al. Combined cap disease and nemaline myopathy in the same patient caused by an autosomal dominant mutation in the TPM3 gene. *Neuromuscul Disord* 2013;**23**:992–7.
34. Zhou H, Jungbluth H, Sewry CA, Feng L, Bertini E, Bushby K, et al. Molecular mechanisms and phenotypic variation in RYR1-related congenital myopathies. *Brain* 2007;**130**:2024–36.
35. Romero NB. Centronuclear myopathies: a widening concept. *Neuromuscul Disord* 2010;**20**:223–8.
36. Catteruccia M, Fattori F, Codemo V, Ruggiero L, Maggi L, Tasca G, et al. Centronuclear myopathy related to dynamin 2 mutations: clinical, morphological, muscle imaging and genetic features of an Italian cohort. *Neuromuscul Disord* 2013;**23**:229–38.
37. Dowling JJ, Vreede AP, Low SE, Gibbs EM, Kuwada JY, Bonnemann CG, et al. Loss of myotubularin function results in T-tubule disorganization in zebrafish and human myotubular myopathy. *PLoS Genet* 2009;**5**:e1000372.
38. Amburgey K, Lawlor MW, Del Gaudio D, Cheng YW, Fitzpatrick C, Minor A, et al. Large duplication in MTM1 associated with myotubular myopathy. *Neuromuscul Disord* 2013;**23**:214–18.
39. Ceyhan-Birsoy O, Agrawal PB, Hidalgo C, Schmitz-Abe K, DeChene ET, Swanson LC, et al. Recessive truncating titin gene, TTN, mutations presenting as centronuclear myopathy. *Neurology* 2013;**81**:1205–14.
40. Lawlor MW, Armstrong D, Viola MG, Widrick JJ, Meng H, Grange RW, et al. Enzyme replacement therapy rescues weakness and improves muscle pathology in mice with X-linked myotubular myopathy. *Hum Mol Genet* 2013;**22**:1525–38.
41. Haliloglu G, Topaloglu H. Arthrogryposis and fetal hypomobility syndrome. In: Dulac O, Lassonde M, Sarnat HB, editors. *Handbook of Clinical Neurology, vol. 113: Pediatric Neurology*. Edinburgh: Elsevier B.V.; 2013. pp. 1311–19.
42. Lindhurst MJ, Sapp JC, Teer JK, Johnston JJ, Finn EM, Peters K, et al. A mosaic activating mutation in AKT1 associated with the Proteus syndrome. *N Engl J Med* 2011;**365**:611–19.

43. Sarnat HB, Diadori P, Trevenen CL. Myopathy of the Proteus syndrome: hypothesis of muscular dysgenesis. *Neuromuscul Disord* 1993;**3**:293–301.
44. Selcen D. Myofibrillar myopathies. *Neuromuscul Disord* 2011;**21**: 161–71.
45. Selcen D, Engel AG. Myofibrillar myopathies. *Handb Clin Neurol* 2011;**101**:143–54.
46. Clemen CS, Herrmann H, Strelkov SV, Schroder R. Desminopathies: pathology and mechanisms. *Acta Neuropathol* 2013;**125**:47–75.
47. Del Bigio MR, Chudley AE, Sarnat HB, Campbell C, Goobie S, Chodirker BN, et al. Infantile muscular dystrophy in Canadian aboriginals is an alphaB-crystallinopathy. *Ann Neurol* 2011;**69**: 866–71.
48. Jaffer F, Murphy SM, Scoto M, Healy E, Rossor AM, Brandner S, et al. BAG3 mutations: another cause of giant axonal neuropathy. *J Peripher Nerv Syst* 2012;**17**:210–16.
49. Selcen D, Bromberg MB, Chin SS, Engel AG. Reducing bodies and myofibrillar myopathy features in FHL1 muscular dystrophy. *Neurology* 2011;**77**:1951–9.
50. Dykens JA, Will Y. *Drug-induced Mitochondrial Dysfunction.* Hoboken, New Jersey: Wiley; 2008.
51. DiMauro S. Mitochondrial encephalomyopathies—fifty years on: the Robert Wartenberg Lecture. *Neurology* 2013;**81**:281–91.
52. Marin-Garcia J, Goldenthal MJ, Sarnat HB. Probing striated muscle mitochondrial phenotype in neuromuscular disorders. *Pediatr Neurol* 2003;**29**:26–33.
53. Sarnat HB, Marin-Garcia J. Pathology of mitochondrial encephalomyopathies. *Can J Neurol Sci* 2005;**32**:152–66.
54. Delonay P, Rotig A, Sarnat HB. Respiratory chain deficiencies. In: Dulac O, Lassonde M, Sarnat HB, editors. *Handbook of Clinical Neurology,* vol. 113: *Pediatric Neurology*. Edinburgh: Elsevier B.V.; 2013. pp. 1651–66.
55. van der Bliek AM, Shen Q, Kawajiri S. Mechanisms of mitochondrial fission and fusion. *Cold Spring Harb Perspect Biol* 2013;**5**(6): pii:a011072.
56. DiMauro S, Schon EA. Mitochondrial disorders in the nervous system. *Annu Rev Neurosci* 2008;**31**:91–123.
57. Miles L, Miles MV, Horn PS, Degrauw TJ, Wong BL, Bove KE. Importance of muscle light microscopic mitochondrial subsarcolemmal aggregates in the diagnosis of respiratory chain deficiency. *Hum Pathol* 2012;**43**:1249–57.
58. Ohama E, Ohara S, Ikuta F, Tanaka K, Nishizawa M, Miyatake T. Mitochondrial angiopathy in cerebral blood vessels of mitochondrial encephalomyopathy. *Acta Neuropathol* 1987;**74**:226–33.
59. Mizukami K, Sasaki M, Suzuki T, Shiraishi H, Koizumi J, Ohkoshi N, et al. Central nervous system changes in mitochondrial encephalomyopathy: light and electron microscopic study. *Acta Neuropathol* 1992;**83**:449–52.
60. Marin-Garcia J, Ananthakrishnan R, Goldenthal MJ, Filiano JJ, Sarnat HB. Skeletal muscle mitochondrial defects in nonspecific neurologic disorders. *Pediatr Neurol* 1999;**21** 538–42.
61. Sarnat HB, Flores-Sarnat L, Hader W Bello-Espinosa L. Mitochondrial "hypermetabolic" neurons in paediatric epileptic foci. *Can J Neurol Sci* 2011;**38**:909–17.
62. Stickler DE, Valenstein E, Neiberger RE, Perkins LA, Carney PR, Shuster JJ, et al. Peripheral neuropathy in genetic mitochondrial diseases. *Pediatr Neurol* 2006;**34**:127–31.
63. Tesarova M, Mayr JA, Wenchich L, Hansikova H, Elleder M, Blahova K, et al. Mitochondrial DNA depletion in Alpers syndrome. *Neuropediatrics* 2004;**35**:217–23.
64. Finsterer J, Ahting U. Mitochondrial depletion syndromes in children and adults. *Can J Neurol Sci* 2013;**40**:635–44.
65. Khan A, Trevenen C, Wei XC, Sarnat HB Payne E, Kirton A. Alpers syndrome: the natural history of a case highlighting neuroimaging, neuropathology, and fat metabolism. *J Child Neurol* 2012;**27**:636–40.
66. Hohlfeld R, Dornmair K. Revisiting the immunopathogenesis of the inflammatory myopathies. *Neurology* 2007;**69**:1966–7.
67. Sarnat HB. Juvenile dermatomyositis. In: Mastaglia FL, editor. *Inflammatory Diseases of Muscle*. Oxford: Blackwell Scientific Publications; 1988. pp. 71–86.
68. De Visser M, Emslie-Smith AM, Engel AG. Early ultrastructural alterations in adult dermatomyositis. Capillary abnormalities precede other structural changes in muscle. *J Neurol Sci* 1989;**94**:181–92.
69. Fernandez C, Bardin N, De Paula AM, Salort-Campana E, Benyamine A, Franques J, et al. Correlation of clinicoserologic and pathologic classifications of inflammatory myopathies: study of 178 cases and guidelines for diagnosis. *Medicine (Baltimore)* 2013;**92**:15–24.

Chapter 5

Nerve Biopsy

Jo M. Wilmshurst, Pedro D.S.C. Ciarlini, and Umberto De Girolami

INTRODUCTION

Although the rapidly expanding area of molecular genetics has reduced the necessity of invasive investigations, peripheral nerve biopsy remains a valuable diagnostic aid when used appropriately.[1,2] There are now specific DNA tests for a number of the demyelinating hereditary sensorimotor neuropathies.[3] On the other hand, many axonopathies remain poorly understood and difficult to categorize. Nerve biopsy is particularly useful in the evaluation of a child with an undetermined peripheral neuropathy when clinical assessment, laboratory investigation, and electrophysiological studies are not diagnostic.[4] If a decision is made to proceed to nerve biopsy, the indications and expectations must be established in advance of the procedure. This is particularly critical when looking after a chronically ill and debilitated patient, considering first whether other data would be sufficient to guide therapeutic management.[5] Argov and colleagues indicated that a diagnosis could be established on clinical grounds and laboratory studies, other than a nerve biopsy, in 48% of patients aged 12 to 78 years with peripheral neuropathy. On the other hand, the information derived from nerve biopsy contributed greatly to the diagnosis in 38% of patients in that series. In half of these cases, clinical management was directly affected by the demonstration of specific histopathologic findings. The importance of careful patient selection was emphasized.[6]

There have been only a few large studies of nerve biopsies in childhood. At the Nerve Research Laboratory of Sydney University, 362 biopsies were performed in children aged 1 month to 17 years over a 33-year period.[7] These biopsies confirmed and defined 260 peripheral neuropathies. In the remaining 102 children, all of whom had clinical features or nerve conduction studies suggestive of a neuropathy (such as distal spinal muscular atrophy), the nerve biopsy was normal.

Although electrophysiologic studies are useful in confirming the presence of a peripheral neuropathy, they are not always consistent or diagnostic. For example, conventional nerve conduction studies are often normal in children with small-fiber sensory neuropathy. Similarly, patients with axonopathies sometimes have peripheral electrophysiologic features suggestive of a demyelinating condition.[8] Nerve biopsy is sometimes the definitive test elucidating the exact nature of neuropathies. Nerve biopsy also has an important role in the diagnosis of a number of rare progressive diseases which also have CNS manifestations, such as infantile neuroaxonal dystrophy and giant axonal neuropathy.

Nerve biopsies need to be analyzed at a center with both the necessary technical laboratory resources and professional staff skilled in the interpretation of all aspects of the morphologic study of the biopsy,[9] including both light and electron microscopy, immunocytochemistry, nerve teasing, and immunofluorescence capability. In addition, the laboratory must have the ability to perform quantitative analysis of light microscopic and ultrastructural images, when these are needed.

TECHNIQUES

The peripheral nerve most often chosen for biopsy is the distal sural nerve.[10] This primarily sensory nerve supplies sensation to the lateral aspect of the foot. It contains both myelinated and unmyelinated nerve fibers having both sensory and autonomic functions. Severing a segment of the distal sural nerve at the time of biopsy typically results in a localized, variably perceptible functional deficit, depending on the degree of preceding sensory loss in its distribution.

In children, nerve biopsy is usually performed under general anesthesia. To reduce the number of anesthetic

B.T. Darras, H. Royden Jones, Jr., M.M. Ryan & D.C. De Vivo (Eds): Neuromuscular Disorders of Infancy, Childhood and Adolescence, Second edition.
DOI: http://dx.doi.org/10.1016/B978-0-12-417044-5.00005-6

procedures a child undergoes, if a muscle biopsy is also indicated, it is best performed concomitantly with the nerve biopsy. This investigation is often useful in its own right and can complement nerve biopsy findings, such as by confirming the presence of a primary neurogenic atrophy secondary to chronic denervation. It is sometimes opportune to perform the nerve/muscle biopsy when the patient is undergoing some other surgical procedure under general anesthesia.

Special care must be taken in processing and handling the nerve, avoiding damage to the fascicles by overzealous manipulation at the time of surgery or subsequently. A whole nerve biopsy is usually obtained, particularly when the sural nerve is being sampled, but fascicular biopsy can also be performed, depending on the center's experience, the degree of neurologic dysfunction, and the nerve selected. When screening specifically for certain pathologic conditions, such as chronic inflammatory demyelinating polyradiculoneuropathy, sarcoidosis, and amyloidosis, it is advisable to obtain whole nerve sections.[11] When sampling a mixed nerve, however, a fascicular biopsy limits the resulting motor deficit and maximizes the potential for re-innervation of a muscle denervated as a result of the biopsy.

At the time of the surgical procedure, it is important to maintain the nerve such that undue stretching does not occur, as this can result in substantial artifact. Some centers use a disposable muscle biopsy clamp, applied to the nerve before it is cut. Others gently place the severed nerve over an appropriate surface and pin the nerve at both ends. Special care must be taken in transportation of the specimen from the operating room to the laboratory; it is sometimes expedient to immediately transport the biopsy to the laboratory in saline-moistened gauze. Also, the peripheral nerve specimen must be maintained moist (though not submerged) in saline, to prevent drying during the dissection and apportioning of the biopsy. To avoid artifacts the nerve biopsy must be processed within 15 minutes of removal. This urgency also applies to autopsy samples, because of the propensity for spontaneous autolysis. Unless the samples are taken soon after death, there is a significant possibility that the findings will be impossible to interpret, although data extrapolated from rats indicate that myelinated fibers can be reliably assessed up to 24 hours if the cadaver is refrigerated.[12]

Detailed descriptions of the various techniques used to assess nerve histology are available.[9,11] The following technique is used at our laboratories and many other specialized centers.[13]

1. Paraffin and Light Microscopy

   The nerve sample is divided into five portions, each 0.5 to 1 cm in length, depending on the processing procedure. One piece of each nerve is fixed in 10% buffered formalin, embedded in paraffin, and cut transversely and longitudinally into sections 5 μm thick. Sections are stained with hematoxylin-eosin as the initial survey stain. The paraffin-embedded specimen can then be used for specific histological and immunocytochemical preparations.

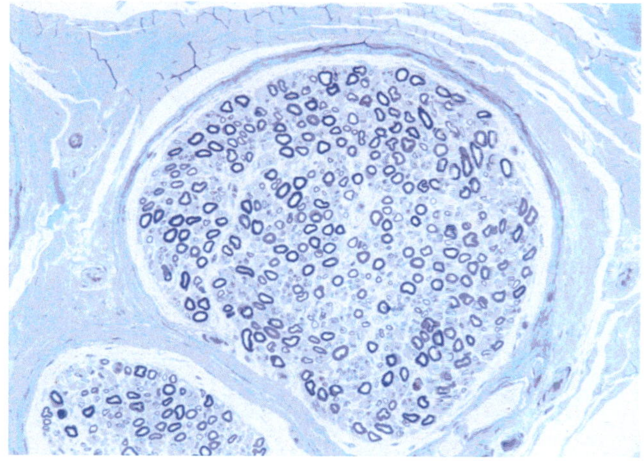

FIGURE 5.1 Semi-thin section of peripheral nerve showing epineurium, perineurium, and endoneurium. The nerve has a normal complement of large and smaller myelinated fibers (toluidine blue).

2. Nerve Teasing

   A second piece is cut longitudinally into strips of about 1 mm diameter, which are fixed in 10% buffered formalin or 2–3% buffered glutaraldehyde for at least 24 hours, stained overnight (for at least 5 hours) in 1% aqueous osmic acid, transferred to a 2:1 mixture of glycerol and water for 12 hours, and individually teased out under a dissecting microscope.
3. Plastic and Light/Electron Microscopy

   A third piece is fixed in 2–3% buffered glutaraldehyde overnight at 4°C, followed by 2% osmium tetroxide for 90 minutes. The tissue is dehydrated in graded concentrations of ethanol, passed through acetone, and embedded in plastic. Semi-thin sections measuring 0.25–1.00 μm, are cut transversely and stained with toluidine blue (Figure 5.1). When electron microscopic study is needed, selected areas can be identified in the semi-thin sections, and ultrathin sections can then be processed for ultrastructural analysis (Figure 5.2).
4. Frozen Sections

   A small portion of the nerve is snap-frozen in liquid nitrogen, and cryostat sections are cut for immunohistochemical staining and visualization of abnormal deposits by immunofluorescence or other methods.
5. The remainder of the specimen is kept frozen for possible biochemical studies.

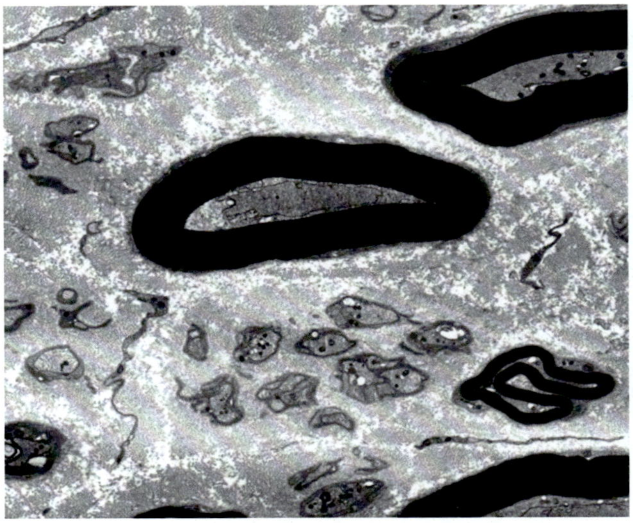

FIGURE 5.2 Electron micrograph showing myelinated and unmyelinated fibers. (Courtesy of Douglas C. Anthony, MD, PhD.)

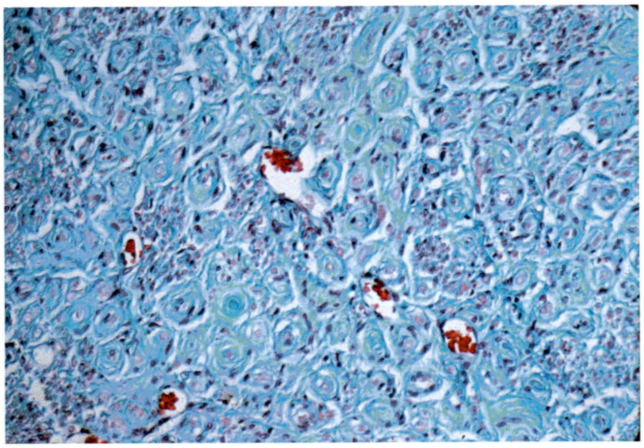

FIGURE 5.4 Cross section of peripheral nerve illustrating onion bulb formation (Masson's trichrome).

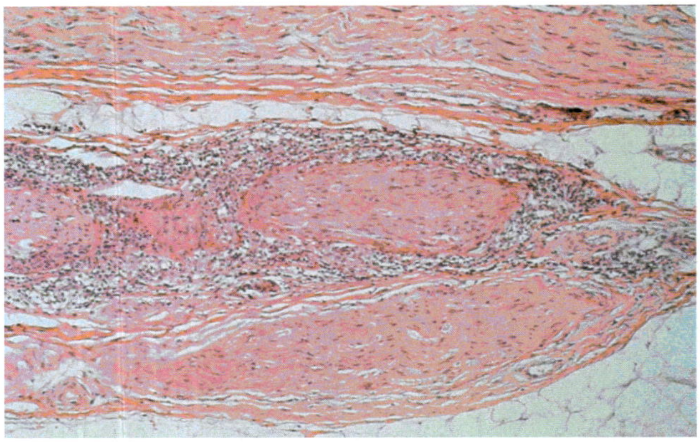

FIGURE 5.3 Peripheral nerve involved by chronic inflammatory infiltrate, in a case of Lyme disease (H&E).

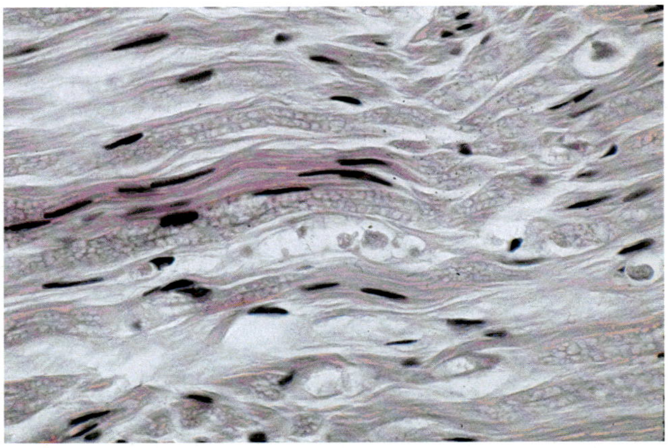

FIGURE 5.5 Longitudinal section of a nerve showing Wallerian degeneration with myelin ovoids.

Light Microscopy Assessment

Certain stains are preferred to identify specific pathologic processes.

Transverse Sections

Hematoxylin and eosin (H&E) sections may identify perineurial edema and signs of inflammation (as occurs in chronic inflammatory demyelinating polyradiculoneuropathy, vasculitis, or other infectious and inflammatory conditions; Figure 5.3). Cuffing with lymphocytes around endoneurial vessels is a significant marker of inflammation, which is well demonstrated with hematoxylin and eosin sections. Congo red stains are used to screen for amyloid infiltration, although this is more relevant for adults. Toluidine blue stains provide data on both demyelinating and axonal disorders. Demyelinating disorders are typically associated with reduced myelinated fiber density, increased Schwann cell and inflammatory cell numbers, reduced myelin thickness, evidence of myelin debris, and onion bulb formations (Figure 5.4). Axonopathies are typified by reduced myelinated fiber density, fibers undergoing active axonal degeneration (Figures 5.5–5.7), and cluster formation when regeneration is occurring.

Teased Fibers

Teased fiber studies, though time-consuming to prepare, provide invaluable nerve pathology information. Ideally, 30 to 100 fibers need to be separated out to give a

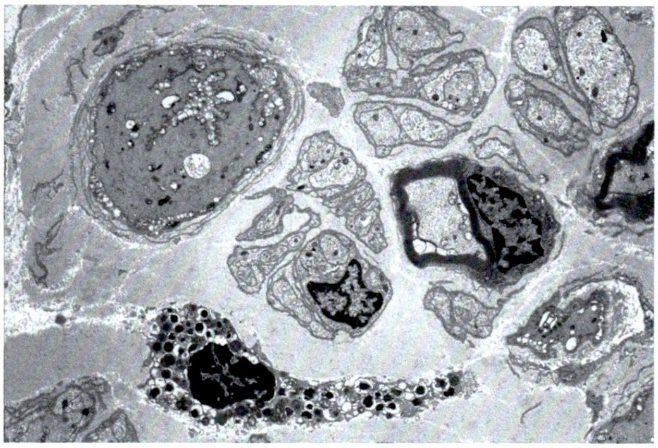

FIGURE 5.6 Electron micrograph showing Wallerian degeneration with degenerating axon. Also note the presence of a macrophage with multiple vacuoles.

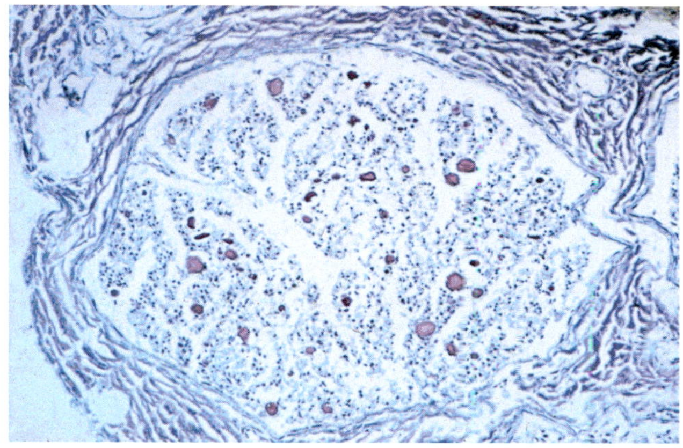

FIGURE 5.7 Enlarged axons in a biopsy of a patient with giant axonal neuropathy (Bodian stain).

reasonable selection of nerve characteristics. This technique allows the examination of myelinated fiber changes in a longitudinal axis, which transverse sections may not identify. It is possible to look for segmental demyelination, widening of the nodal gap with paranodal demyelination, segmental remyelination (often appearing as two or more short segments between segments of normal length), and regenerated fibers in which all internodal segments are short (0.2–0.6 mm) and have disproportionately thick diameters. The finding of myelin ovoids suggests active axonal degeneration.

Teased fiber sections are particularly helpful in differentiating primary segmental demyelination from a secondary demyelinating process due to axonal disease. The finding of "clustering" of areas of segmental demyelination along a single diseased axon, where many of the remaining axons appear well myelinated, suggests axonal degeneration as the primary process, whereas apparently random regions of segmental demyelination throughout many fibers is more compatible with a primary demyelinating process.[14] Preparation of teased fiber preparations is technically taxing and labor-intensive, and should only be undertaken after careful consideration.

Electron Microscopy

Electron microscopy (EM) allows the detailed ultrastructural assessment of myelinated and unmyelinated fibers (Figure 5.2). The technique is particularly important in evaluating unmyelinated fiber density (which is much too difficult by light microscopy alone). EM is also essential for the evaluation of structural abnormalities, organelle number and structure, myelin thickness, and abnormal lamellation.

EM provides detailed pictures of the intrinsic nature of the fascicle and its basement membrane configuration and continuity. Further, it may be the only way to detect evidence of Büngner's bands (early regenerating large myelinated fibers) and Schwann cell subunits (in which a single Schwann cell profile, or a group of profiles, is enclosed by a continuous basal lamina that may or may not include axons). Subtle changes in the myelin folds may suggest early demyelination or failure of compaction. In contrast, early axoplasmic shrinkage, associated with degradation products, suggests axonal degeneration. Artifact is a common problem with even the most carefully prepared nerves. In such circumstances, electron microscopy frequently enables the nerve pathologist to differentiate between genuine disease pathology and artifactual changes.

Quantitative Analysis

Myelinated fiber morphometric studies are performed using photomicrographic enlargements of toluidine blue-stained sections. Electron micrographs also allow assessment of unmyelinated nerve fibers. Various computer-assisted methods are available for the rapid assessment of large numbers of fibers, but even the most sensitive programs require observer verification to ensure that the fiber counts include only axons.

An alternative method uses visual counts on electron micrographs or magnified light microscopic fascicles. At least 10 electron microscopic plates, at a magnification of around 3500 magnitude, are taken from random sections

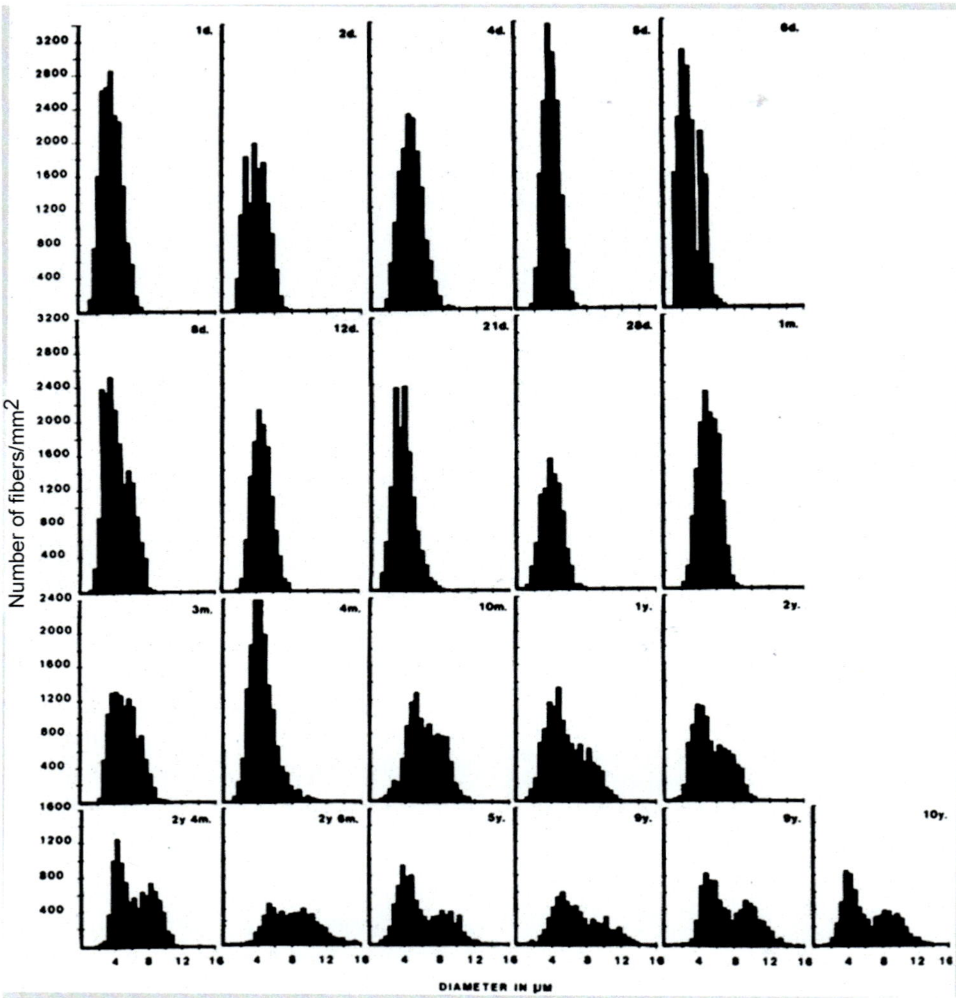

FIGURE 5.8 Normal changes in myelinated fiber diameter with age (0–10 years). In early life, the myelinated fibers are of small diameter (2–5 μm), and the histograms are unimodal. By 1 year of age, a bimodal trend appears as larger-diameter fibers form (8–12 μm).[20]

of the same fascicle. Myelinated fibers are counted, and the density per square millimeter is calculated from this figure and the area. Similarly, the fiber diameters are assessed through the average of the maximum and minimum diameter, or by conversion from the circumference of the fiber (diameter = circumference/π). From these values, the equivalent number per square millimeter can be extrapolated, and a histogram of fiber diameter distribution produced. Counting the total number per plate, until a value of more than 80 is reached, generally provides a reasonably accurate estimate of unmyelinated fiber density. The diameters are measured by the same method as for myelinated fibers. Other useful measurements from electron microscopy include the number of Schwann cell nuclei, denervated Schwann cell subunits, and collagen pockets, the last two being indicators of unmyelinated fiber loss.

The ratio of axon diameter (d), excluding the myelin, to total myelinated fiber diameter (D) indicates the thickness of myelin around a specific fiber. This confirms the presence of excessively thin myelin (high d/D ratio), as occurs with large numbers of regenerated or hypomyelinated fibers, or excessively thick myelin (low d/D ratio), as is seen in the tomaculous neuropathies (particularly hereditary neuropathy with liability to pressure palsies).

COMPLICATIONS

About 40% of adult patients complain of pain or paresthesia persisting 1 year after sural nerve biopsy. The pain is moderate to severe in a quarter of these individuals, and is commonly described as being like a pinprick or a tight sensation during standing or walking.[15–17] The outcome is not influenced by whether the biopsy was whole nerve or fascicular.[18] Prolonged follow-up after 2 years suggests that there is a gradual improvement in symptoms over time.[19] Most of these studies are derived from adults; there are few comparative data from children. However, complaints following biopsy were usually not significant among cases seen in the peripheral neuropathy clinics at the Children's Hospital at Westmead in Sydney, Australia. Only one case stands out, that of a boy who underwent biopsy of the sural nerve and muscle at the mid-calf level. He subsequently suffered discomfort in that region, which was believed to be secondary to adhesions around the nerve.

AGE-RELATED DIFFERENCES IN SURAL NERVE MORPHOLOGY

The structure of peripheral nerve in children differs from that of adults.[20–27] Between birth and age 5 years, important peripheral nerve maturational changes occur in fiber density and myelinated fiber size, including thickness. There is a gradual evolution in these parameters from early infancy until the mature adult pattern is approached in the preschool years.[20–25] Similarly, nerve conduction parameters do not reach adult values until 3 to 5 years of age.[26,27]

Myelination of peripheral nerve commences from 18 weeks' gestation. By term, most of the fibers that are destined to become myelinated have started the process.[11] The total number of myelinated fibers per sural nerve increases from 4000 at birth to 7000 to 12,000 by adulthood. However, the density of myelinated fibers per square millimeter follows the opposite trend. This is because the endoneurial area is small at birth (0.2–0.5 mm^2), whereas the adult size increases to 0.6–1.2 mm^2. Densities at birth are as high as 12,000 to 20,000 myelinated fibers per square millimeter and rapidly decrease to near 10,000 by 1 year of age. Adult levels are attained around 10 years of age.

There is relative hypomyelination at birth, with d/D ratios above 0.8 being common.[28] By 5 years of age, the axons reach their adult size, whereas final myelin thickness is not evident until 10 years of age. The myelinated fiber diameter histogram is unimodal at birth and evolves into a bimodal appearance toward the end of the first year, as large myelinated fibers become established. Internode lengths are 200 to 300 μm at birth and increase to 200 to 1800 μm by adulthood (Figure 5.8).[20]

BIOPSY FINDINGS IN RELEVANT DISEASES

Tables 5.1 to 5.5 summarize the predominant nerve biopsy findings in the various childhood peripheral neuropathies.[3] Molecular DNA genetic testing confirms diagnosis in many of the hereditary demyelinating neuropathies. However, a substantial number of the axonal neurodegenerative disorders still require a nerve biopsy for diagnosis. The leukocyte (white cell) enzyme screen may confirm a leukodystrophy, but in metachromatic leukodystrophy borderline arylsulfatase A levels can be misleading, making biopsy the only diagnostic test available other than specific genetic testing.[29] Vasculitis may occur without systemic features, in which case a confirmed diagnosis—leading to treatment with immunosuppressive therapy—is contingent on performance of a nerve biopsy.[30]

CONCLUSION

Peripheral nerve biopsy is a valuable investigative modality when appropriately used and interpreted in a reliable, experienced laboratory. It is an invasive test and must have well-defined indications; it should not be performed without good reason. The data derived can provide diagnostic information as well as prognostic advice and patient management information. In the setting of a hereditary peripheral neuropathy, a sural nerve biopsy from a single affected family member can provide very helpful data for the entire family.

TABLE 5.1 Demyelinating Forms of CMT

Disorder (OMIM)	Age of Onset	Key Clinical/Differentiating Features	NCS	Biopsy Findings	Causative Gene
CMT1 (Demyelinating)					
CMT 1 A-F (HMSN type I) AD Summary	1st–4th decade	Predominantly distal weakness, loss of deep tendon reflexes, mild sensory loss. Hypertrophy of nerves. Most common form of CMT	Slowing: motor NCV <38 m/s	Segmental demyelination and remyelination with onion bulb formations and axonal loss	
1A (118220)	All ages	About 5% of CMT1 group. Clinically more severe than CMT1A	NCV < 38 m/s	As above	*PMP22*
1B (118200)	1st–2nd decade		NCV < 38 m/s	As above. Occasional myelin outfolding	*MPZ*
1C (601098)	Childhood onset	Abnormal gait. Occasional nerve hypertrophy. Rarely early-onset deafness	NCV < 38 m/s	As above. Hypertrophic nerve changes	*LITAF*
1D (607678)	1st–2nd decade	Possible cranial nerve involvement	NCV < 38 m/s		*EGR2*
1E (118300)	Childhood onset	Associated with deafness	NCV < 38 m/s		*PMP22*
1F (607734)	Onset 1–13 years	Rare cases with recessive inheritance. Occasional deafness	NCV < 38 m/s Rare cases with severe slowing	As above. Reduced MF density, loss of large diameter myelinated fibers. Clusters of axonal degeneration	*NEFL*
CMT3 (Demyelinating)					
CMT 3 (Déjerine-Sottas syndrome (145900) AR	First 2 years	Severe disability. Hypotonia, proximal weakness, motor delay, ataxia. By second decade proximal weakness, hand and feet deformities, scoliosis. Nerve hypertrophy. Moderate to severe sensory loss. Nystagmus. Deafness. Mild bifacial weakness. Raised CSF protein	Motor NCV <10 m/s SNAPs absent EMG chronic denervation	Decreased MF density. Thin MF. Multiple OB formations. Increased transfascicular diameter due to OB formations (Figure 5.4), and collagen and Schwann cell proliferation	*PMP22*, *MPZ* *PRX* *EGR2* *FIG4*
CMT4 (Demyelinating)					
CMT 4 (A-J) AR		Similar to or slightly more severe than CMT1. Ataxia. Areflexia. Scoliosis. Nerve hypertrophy rare	Moderate slowing of NCS	Marked reduction in MF density. Some well myelinated fibers. Segmental demyelination OB few to many	
4A (214400) allelic to CMT2K	Onset <2 years	Tunisian, Libyan, Algerian, and Moroccan families. Severe progressive. Less severe European phenotypes	25–35 m/s	Demyelinating or axonal. Hypomyelination and OB	*GDAP1*
4B1 (601382)	Mean 34 months	Ophthalmoplegia. Vocal cord paralysis. Facial/bulbar weakness. Scoliosis. Proximal/distal weakness. Abnormal auditory evoked potentials. Death in 4th–5th decade	9–20 m/s	Myelin outfolding. Loss of MF	*MTM2* (*MPZ*)

4B2 (604563)	4–13 years	Glaucoma (with *SBF2* mutations). Kyphoscoliosis. Raised CSF protein	15–30 m/s	Myelin outfolding, loss of MF	*SBF2*, *MTM13*
4C (601596)	Early-onset. 1st and 2nd decade	Scoliosis, deafness, abnormal pupillary responses. Tongue fasciculations	4–37 m/s	Basal membrane OB	*SH3TC2* (*KIAA1985*)
4D (601455) HMSN-Lom form	Onset <10 years	Deafness (by 2nd–3rd decade)	10–20 m/s	Hypomyelination. OB formations. Myelin decompaction. Axonal inclusions with curvilinear profiles. Loss of MF	*NDRG1*
4E (605253) AR (rare AD forms)	Birth	Congenital hypotonia, respiratory failure. Arthrogryposis multiplex congenita	5–20 m/s	Severe loss of myelinated and unmyelinated fibers, OB formations and focally folded myelin sheaths	*ERG2/ KROX 20*; *MPZ*
4F (145900)	Birth	Slowly progressive, prominent sensory involvement	<5 m/s	Severe loss MF. Congenital hypomyelination. Virtually absent OBs	*PRX*
4G (605285) type Russe	8–16 years	Prominent sensory loss	30–35 m/s	Hypomyelination Regenerating clusters. Loss of larger MF	*HK1*
4H (609311)	<2 years	Delayed motor milestones. Scoliosis, severe course	<10 m/s or absent	Hypomyelination. Small OB Sometimes myelin outfoldings	*FDG4*
4J (611228)	By 5 years	Severe disorder, similarities to motor neuron disease	2–7 m/s	Profound axonal loss, thinly myelinated fibers, de- and remyelination	*FIG4*
CCFDN (604168)	Birth	Congenital cataract, microcornea, dysmorphism. Mental retardation. Distal motor peripheral neuropathy	19–33 m/s	Diffuse hypomyelination	*CTDP1*

DTR: deep tendon reflexes; NCV: nerve conduction velocity; MF: myelinated fibers; OB: onion bulb; CCFDN: Congenital cataracts, facial dysmorphism, and neuropathy. Reprinted with permission.[3]

TABLE 5.2 Axonal Forms of CMT

Disorder (OMIM)	Age of Onset	Key Clinical / Differentiating Features	NCS	Biopsy Findings	Causative Gene
CMT2 (Axonal)					
CMT 2 A-N (HMSN type II) Summary AD: A,B,D,E,F,G,I, M,N AR: BI, B2, H,K	All ages (more in adults)	Clinically similar to CMT1, but later onset. No peripheral nerve enlargement	Normal NCVs (>38 m/s)	Neuronal atrophy, degeneration of peripheral motor and sensory neurons	
2A1 (118210) AD	20–50 years	Slowly progressive	Normal / mildly dec. NCVs (>38 m/s)	Axonal degeneration. Small OBs. Reduced MF density	*KIF1B* one family
2A2 (609260) A2 (=HMSN VI)AD (AR form allelic to EOHMSN)	10–50 years	Proximal weakness, severe distal weakness, optic atrophy	Normal / mildly dec. NCVs (>38 m/s)	As for CMT2A1	*MFN2*
HMSN II with onset in early childhood (EOHMSNA) Severe early-onset axonal neuropathy (SEOAN) AR(allelic CMT2A2 – dominant form)	<5 years	Rapid progression. Very severe distal weakness, moderate sensory changes. Occasional optic atrophy	Severe axonal neuropathy. Absent SNAPs and lower extremity motor responses	Marked reduction in MF density, especially of large MF. No demyelination or active axonal degeneration, occasional cluster formations. Increased Schwann cell nuclei. Unusual mitochondrial features. OB formations	*MFN2* *GDAP1*
2B (600882) AD	Second decade	Severe sensory loss. Foot ulcers, mutilating arthropathy	Chronic sensory axonal neuropathy	Large and small sensory fiber loss	*RAB7*
2B1 (605588) AR	2nd decade	Kyphosis. Severe	NCVs >38 m/s	Small OBs. Decreased MF density	*LMNA*
2B2 (605589) AR	3rd decade		NCVs > 38 m/s	–	19q13.3 *MED25*
2C (606071) AD	Birth to 60 years	Pediatric cases more severe. Vocal cord, diaphragm, respiratory involvement	NCV >38 m/s EMG denervation	Axonal polyneuropathy	*TRPV4*
2D (601472) AD allelic dSMA/dHMN5	2nd decade	Upper limb predominance	Normal NCSEMG denervation	–	*GARS* *BSCL2*
2E (607684)AD (allelic CMT 1F)	1st–3rd decade	High stepping gait. Clawing of hands. Hyperkeratosis	Intermediate/slow NCS	Giant fibers, neurofilament accumulation	*NEFL*
2F (606595) AD allelic dHMN2B	Adult onset	Mild motor involvement	Axonal pattern	Chronic axonal neuropathy	*HSPB1* (*HSP27*)

2G (608591) AD	Variable	Severe subform with early onset	NCVs >38 m/s	Axonal atrophy, axonal degeneration / regeneration. Reduced MF density	12q12–q13
2H (607731) AR (allelic CMT4A)	1st decade	Pyramidal and vocal cord involvement	Normal NCS	Reduced MF density. Thin myelin. Rare axonal degeneration	*GDAP1*
2I (607677) AD	Late onset >40 years	Prominent sensory disease	NCVs 39–40 m/s	Axonal degeneration. Reduced MF density	*MPZ*
2J (607736) AD	Late onset (youngest 13 years)	Pupillary anomalies. Marked sensory involvement. Pain. Hearing loss. Dysphagia	NCVs >38 m/s	Cluster formations, axonal degeneration	*MPZ*
2K (607831) AR (AD rarely)	Severe early onset form	Severe form. Vocal cord paralysis. Kyphoscoliosis. Milder dominant form	NCVs >38 m/s	Reduced MF density. Axonal degeneration. Pseudo OB formations	*GDAP1*
2L (608673) AD allelic to dHMN2A	Onset 15–33 years	Occasional proximal weakness. Scoliosis	NCVs >38 m/s	Axonal neuropathy, loss of large MF and thin myelin axons	*HSPB8*; (*HSP22*) 12q24
2M (606482) AD allelic DI-CMTB CMTDI 1	Variable	Occasional neutropenia, cataracts	NCS 26–54 m/s	Diffuse loss of MF density. Axonal regeneration. Focal myelin thickness	*DNM2*
2N (613287) AD	Variable	Asymmetrical motor involvement	Mean MNCVs 41 m/s	–	*AARS*

MF: myelinated fibers; OB: onion bulb. Reprinted with permission.[3]

TABLE 5.3 Inflammatory and Infectious Disorders

Disorder	Age of Onset	Key Clinical / Differentiating Features	NCS	Biopsy Findings	Comment
AIDP[31]	Childhood to adulthood	Rapidly progressive weakness, sensory loss and hyporeflexia	Demyelinating	Active demyelination, variable lymphocytic infiltration and complement activation products	Biopsy not routinely recommended for diagnosis
AMAN[32]	Childhood to adulthood	Similar to AIDP, but no sensory deficits	Axonal	Wallerian degeneration in motor fibers without inflammation, sensory nerves normal	Biopsy not routinely recommended for diagnosis
AMSAN[33]	Childhood to adulthood	Similar to AIDP but often more severe	Axonal	Wallerian degeneration (both sensory and motor) without inflammation	Biopsy not routinely recommended for diagnosis
CIDP[34]	Infancy to late-adulthood	Progressive muscle weakness over at least 4 weeks, hyporeflexia	Demyelinating	Inflammatory infiltrates (often perivascular in epineurium). Subperineurial edema; macrophages; onion bulbs; variable axonal loss	Biopsy not routinely recommended; more useful in atypical cases
Vasculitis[35]	Childhood–adulthood	Mononeuritis multiplex. Most common peripheral manifestation	Axonal	Vasculitis in endoneurium and perineurium; acute axonal degeneration	Rare in children—consider SLE, Sjögren's disease, JIA
Leprosy[36]	Childhood to adulthood	Asymmetric nerve enlargement, anesthetic macules or plaques in tuberculoid leprosy; multiple nonanesthetic macules or papules and bilateral symmetrical distal neuropathies in lepromatous leprosy	Mixed demyelinating and axonal	Segmental demyelination, axonopathy; granulomas with rare bacilli in tuberculoid leprosy; macrophages and Schwann cells filled with organisms (Fite stain, Figure 5.9) and debris (foamy cells); much of architecture preserved in lepromatous leprosy	PCR is a very important tool especially in pure neural leprosy
Diphtheria[37]	Childhood and early adulthood	Polyneuropathy can mimic GBS; biphasic course with craniofacial followed by diffuse involvement	Demyelinating	Minor focal segmental demyelination	Very rare, due to vaccination
Lyme disease[38]	Children (5–14 years) and adults (30–50 years)	Multisystem (skin, joints, heart, eye, nervous system) Early disseminated: meningitis, cranial neuropathy Late: cognitive dysfunction, peripheral neuropathy	Axonal	Lymphocyte and plasma cell infiltration (Figure 5.3); thick pericapillary cuffs without necrosis of the vessel wall	Acute peripheral facial palsy and aseptic meningitis most frequent manifestations in pediatric population
HIV[39]	Infancy to late adulthood	Prevalence estimated at 5% in children. Distal symmetric axonal polyneuropathy most common. –Therapy-related (drug toxicity) –Opportunistic infections (CMV, VZV). –Other forms (AIDP, CIDP, lumbosacral polyradiculopathy)	Axonal; often NCSs are abnormal even in young children with HIV	Loss of large myelinated fibers; mild epineurial and endoneurial perivascular inflammation. CMV-related: axonal degeneration, segmental demyelination; necrotizing endoneurial and epineurial vasculitis with prominent neutrophil infiltrate; on EM, CMV virions in endoneurial cells	Although neuropathy in HIV is rarely reported, it should be considered, especially in older children; biopsy data, however, are scarce

AIDP = acute inflammatory demyelinating polyradiculoneuropathy; AMAN = acute motor axonal neuropathy; AMSAN = acute motor and sensory axonal neuropathy; CIDP = chronic inflammatory demyelinating polyradiculoneuropathy; CMAP = compound muscle action potential; CMV = cytomegalovirus; CV = conduction velocity; DL = distal latency; EM = electron microscopy; JIA = juvenile idiopathic arthritis; GBS = Guillain-Barré syndrome; HIV = human immunodeficiency virus; HMSN, hereditary motor sensory neuropathy; MF = myelinated fiber; ML = minimal latency; NCS = nerve conduction study; SLE = systemic lupus erythematosus; SNAP = sensory nerve action potential.

TABLE 5.4 Neurodegenerative, Metabolic, and Other Genetic Disorders

Disorder (OMIM)	Age of Onset	Key Clinical Features	NCS	Biopsy Findings	Causative Gene
Friedreich ataxia[40] (229300) AR	10–40 years	Ataxia, dysmetria, dysarthria; proprioceptive and superficial sensory loss; weakness and atrophy of the extremities; depressed reflexes; Babinski responses	Axonal	Decreased density; selective loss of large MFs; clusters rare; decreased number of Schwann cells	*FXN*
Mitochondrial DNA depletion syndrome 7 [IOSCA] (271245)[41] AR	9–18 months	Ataxia, hypotonia, deafness, ophthalmoplegia, optic atrophy	Axonal	Progressive loss of large MFs; mild axonal degeneration; rare OBs and clusters	*C10orf2*
NBIA2A[INAD] (256600)[42] AR	<3 years	Psychomotor deterioration, long tract signs, truncal hypotonia	Axonal	Globular, ovoid, or fusiform swellings within axons, axonal degeneration; typical neurofilaments on EM sections (Figure 5.10). Can also be diagnosed from other biopsy sites (e.g. skin, conjunctiva)	*PLA2G6*
Fabry disease[43] (301500) X-linked	10–25 years	Acroparesthesia and pain crises, stroke, angiokeratomas, cornea verticillata	Axonal	Selective loss of small and unMFs; glycolipid granules in perineurial and endoneurial cells	*GLA*
Niemann-Pick disease Infantile form type A[44] (257200) AR	Infancy	Failure to thrive, hypotonia, hepatosplenomegaly, neurodegeneration, cherry red spots	Demyelinating	Loss of MFs; thin myelin sheaths; within Schwann cells, numerous round or fusiform bodies	*SMPD1*
Farber disease[45] (228000) AR	Infancy	Growth failure, hoarse voice, subcutaneous and periarticular nodules, progressive arthropathy, psychomotor deterioration	Axonal	Loss of axons; round, membrane-bound, clear inclusions in Schwann cells and histiocytes	*ASAH1*
Leigh syndrome[46] AD/X-linked	Infancy	Hypotonia, developmental regression or arrest, ophthalmoplegia, respiratory and bulbar dysfunction, ataxia, sensorimotor polyneuropathy	Demyelinating	Hypomyelination and segmental demyelination	Various genes involved[87]; CNS imaging helpful
Krabbe disease[47] (245200) AR	Infancy, rarely later-onset forms	Severe irritability, progressive spasticity, developmental regression, seizures, widespread sensorimotor peripheral neuropathy	Demyelinating	Reduced MF density; inclusions in Schwann cells; diagnostic picture on biopsy; macrophages	*GALC*
Metachromatic leukodystrophy[48] (250100) AR	Late-infantile, juvenile, adult-onset	Lower extremity weakness, difficulty walking, absence of deep tendon reflexes, regression of motor and cognitive abilities, bulbar paralysis, seizures	Demyelinating	Reduced MF density; OBs; on acid cresyl violet stain, metachromatic material; on EM, macrophages, Schwann cells, and occasional axons containing prismatic inclusions, tuffstones, zebra bodies, and vacuoles	*ARSA*
Cockayne syndrome[49] (133540, 216400) AR	Infancy–early childhood	Growth failure, premature aging, leukodystrophy, cutaneous photosensitivity, sensorineural hearing loss, dental caries	Demyelinating	Demyelinating fibers; Schwann cells containing granular inclusions	*ERCC8, ERCC6*
Zellweger Syndrome Spectrum (NALD/IRD)[50] AR	Infancy–childhood	Hypotonia, developmental delay, seizures, leukodystrophy, sensorineural hearing loss, retinitis pigmentosa, liver dysfunction, peripheral neuropathy	Demyelinating	Demyelinating fibers with no Schwann cell inclusions	*PEX* gene family

(Continued)

TABLE 5.4 (Continued)

Disorder (OMIM)	Age of Onset	Key Clinical Features	NCS	Biopsy Findings	Causative Gene
Abetalipo-proteinemia (200100)[51] AR	Childhood–young adults	Malabsorption of fat and fat-soluble vitamins, retinal and spinocerebellar degeneration, peripheral neuropathy	Axonal	Large MF density; clusters; paranodal demyelination	*MTP*
Tangier disease[52] (205400) AR	Childhood–young adults	Enlarged yellow tonsils, hepatosplenomegaly	Demyelinating	Demyelination, lipid vacuoles, especially concentrated in Remak's cells	*ABCA1*
Chédiak-Higashi syndrome[53] (214500) AR	Infancy–early childhood	Partial oculocutaneous albinism, photophobia, increased susceptibility to infection and lymphomas, seizures	Axonal	Giant lysosomal inclusions in Schwann cells of peripheral nerves (and throughout the nervous system)	*LYST*
Xeroderma pigmentosum[54] AR	Infancy–early childhood	Hypersensitivity to sunlight, cutaneous malignancies, photophobia, sensorineural deafness	Axonal and demyelinating	MFs and unMFs; denervated Schwann cell subunits and increase collagen pockets	*XPA, XPC, ERCC2-5, POLH, DDB2*
Cerebrotendinous xanthomatosis[55] (213700) AR	Infancy–adulthood	Tendon xanthomata, ataxia, cataracts, diarrhea, xanthelasmata	Axonal and/or demyelinating	Large MFs; hypertrophic demyelinating neuropathy; cholesterol deposits; occasional lipid accumulation in the nerves	*CYP27A1*
Porphyrias[56] AD (most), AR, X-linked	Usually after puberty[1,20]	Neurovisceral crises, abdominal pain	Axonal	Loss of large MFs with active axonal degeneration	*HMBS (AIP), CPOX (HCP), PPOX (VP), ALAD (ADP)*
Tyrosinemia type I[57] (276700) AR	Infancy–early childhood	Hepatic and renal dysfunction; neurological crises	Axonal	Active axonal degeneration with secondary demyelination	*FAH*
Neurofibromatosis type 2[58] (101000) AD	Adolescence–adulthood	Vestibular schwannomas, meningiomas, ependymomas	Axonal/demyelinating	Schwann cell clusters, increased collagen, endoneurial edema; decreased number of myelinated and unmyelinated fibers	*NF2*
Achalasia-addisonianism-alacrima syndrome (Allgrove)[59] (231550) AR	Early childhood–early adulthood	Palmoplantar and punctate hyperkeratosis, short stature, osteoporosis, xerostomia, angular cheilitis, glossitis, microcephaly	Axonal	Axonal degeneration; minor loss of MFs	*AAAS*
Multiple endocrine neoplasia type IIB[60] (162300) AD	Childhood–adulthood	Marfanoid body habitus, mucocutaneous neuromas, medullary carcinoma of thyroid, pheochromocytomas, parathyroid hyperplasia	Axonal	Moderate loss of MFs; Characteristic adventitious plaque of tissue composed of hyperplastic interlacing bands of Schwann cells and MFs overlying the posterior columns of the spinal cord	*RET*

AD = autosomal dominant; ADP = ALA-dehydratase-deficient porphyria; AIP = acute intermittent porphyria; AR = autosomal recessive; EM = electron microscopy; FAP = familial amyloidotic polyneuropathy; HCP = hereditary coproporphyria; HMSN = hereditary motor sensory neuropathy; INAD = infantile neuroaxonal dystrophy; IOSCA = infantile-onset spinocerebellar ataxia with sensory neuropathy; IRD = infantile Refsum disease; MF = myelinated fiber; MLD = metachromatic leukodystrophy; NALD = neonatal adrenoleukodystrophy; NBIA2A = neurodegeneration with brain iron accumulation 2A; OB = onion bulb formation; unMF = unmyelinated fiber; VP = variegate porphyria.

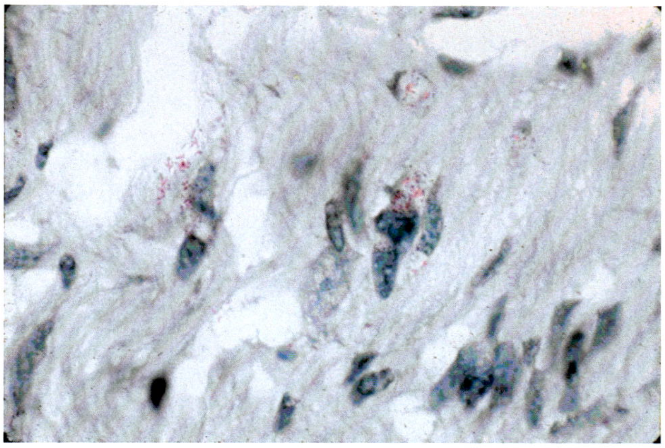

FIGURE 5.9 Sections showing presence of multiple acid-fast bacilli in a patient with leprosy (Fite stain).

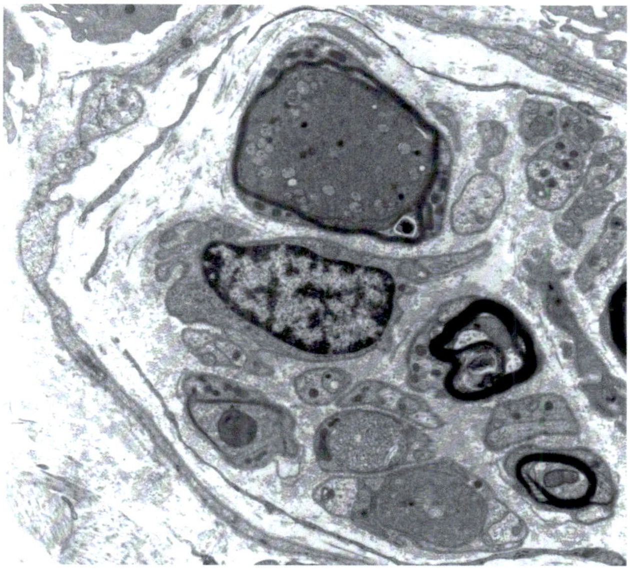

FIGURE 5.10 Electron micrograph showing axons with increased neurofilaments, a spheroid body, and abnormal mitochondria and vacuoles in a patient with neuroaxonal dystrophy.

TABLE 5.5 Pathologic Findings in Toxic, Vitamin-Deficient, and Other Neuropathies

Disorder	Key Clinical / Differentiating Features	NCS	Biopsy Findings
Glue-sniffing neuropathy[61]	Progressive predominantly motor neuropathy associated with chronic inhalation of organic solvents (especially n-hexane)	Axonal	Swelling of axons and thinning of myelin; on EM, axonal swellings, densely packed whorled masses of neurofilaments
Lead[62]	Classic form: motor neuropathy with weakness of wrist and finger extensors; in children, almost always associated with encephalopathy. Bone marrow suppression, renal failure, hypertension, gout	Axonal	Chronic and active axonal degeneration
Chemotherapy (vincristine)[63,64]	Primarily sensory, distal predominant neuropathy, frequent autonomic involvement. Can lead to GBS-like picture especially in patients with underlying hereditary neuropathy (CMT)	Axonal	Axonal degeneration
Chemotherapy (cisplatin)[63,64]	Primarily sensory, distal predominant neuropathy	Axonal	Axon loss with atrophy of the DRG
Ciguatera[65]	Acute symptoms within 12 h of ingestion of ciguatoxic fish: acute gastroenteritis, circumoral/ limb paresthesias, paradoxical dysesthesia, itch, myalgia, arthralgia, weakness, bradycardia, hypotension, shock	Axonal	Edema of the adaxonal Schwann cell cytoplasm, demyelinating internodes
Vitamin B1 deficiency[66]	Predominantly distal sensorimotor polyneuropathy, occasional vagal neuropathy resulting in dysphonia	Axonal	Axonal degeneration with central distal ("dying back") axonopathy
Vitamin B12 deficiency[67]	Can present with predominantly distal sensorimotor peripheral neuropathy	Axonal	Axonal degeneration
Vitamin E deficiency[68]	Polyneuropathy, spinocerebellar degeneration, dysautonomia, myopathy, bulbar palsy	Axonal/ Demyelinating	Loss of large MFs, axonal degeneration; vacuoles in myelin sheath; focal demyelination

CMT = Charcot-Marie-Tooth disease; DRG = dorsal root ganglion; EM = electron microscopy; GBS = Guillain-Barré syndrome; HMSN = hereditary motor sensory neuropathy; MF = myelinated fiber.

REFERENCES

1. Lacomis D. Clinical utility of peripheral nerve biopsy. *Curr Neurol Neurosci Rep* 2005;**5**:41–7.
2. Oh SJ. Diagnostic usefulness and limitations of the sural nerve biopsy. *Yonsei Med J* 1990;**31**:1–26.
3. Wilmshurst JM, Ouvrier R. Hereditary peripheral neuropathies of childhood: an overview for clinicians. *Neuromuscul Disord* 2011;**21**:763–75.
4. Schweikert K, Fuhr P, Probst A, Tolnay M, Renaud S, Steck AJ. Contribution of nerve biopsy to unclassified neuropathy. *Eur Neurol* 2007;**57**:86–90.
5. McLeod JG. Investigation of peripheral neuropathy. *J Neurol Neurosurg Psychiatry* 1995;**58**:274–83.
6. Argov Z, Steiner I, Soffer D. The yield of sural nerve biopsy in the evaluation of peripheral neuropathies. *Acta Neurol Scand* 1989;**79**:243–5.
7. Wilmshurst JM. *Peripheral neuropathies in children [thesis]*. Cape Town, South Africa: University of Cape Town; 2009.
8. Behse F. Buchthal. Sensory action potentials and biopsy of the sural nerve in neuropathy. *Brain* 1978;**101**:473–93.
9. Weis J, Brandner S, Lammens M, Sommer C, Vallat JM. Processing of nerve biopsies: a practical guide for neuropathologists. *Clin Neuropathol* 2012;**31**:7–23.
10. Dyck PJ, Lofgren EP. Nerve biopsy. Choice of nerve, method, symptoms, and usefulness. *Med Clin North Am* 1968;**52**:885–93.
11. Dyck PJ, Dyck PJB, Engelstad J. Pathologic alterations of nerves. In: Dyck PJ, Thomas PK, editors. *Peripheral Neuropathy*. Philadelphia: W B Saunders; 2005. pp. 733–829.
12. Ouvrier RA. *Chronic peripheral neuropathy in childhood [thesis]*. Sydney, Australia: University of Sydney; 1985.
13. Ouvrier RA, McLeod JG, Pollard JD. *Peripheral Neuropathy in Childhood*. 2nd ed. London: MacKeith Press; 1999.
14. Dyck PJ, Lais AC, Karnes JL, Sparks M, Hunder H, Low PA, et al. Permanent axotomy, a model of axonal atrophy and secondary segmental demyelination and remyelination. *Ann Neurol* 1981;**9**:575–83.
15. Solders G. Discomfort after fascicular sural nerve biopsy. *Acta Neurol Scand* 1988;**77**:503–4.
16. Rappaport WD, Valente J, Hunter GC, Rance NE, Lick S, Lewis T, et al. Clinical utilization and complications of sural nerve biopsy. *Am J Surg* 1993;**166**:252–6.
17. Hilton DA, Jacob J, Househam L, Tengah C. Complications following sural and peroneal nerve biopsies. *J Neurol Neurosurg Psychiatry* 2007;**78**:1271–2.
18. Pollock M, Nukada H, Taylor P, Donaldson I, Carroll G. Comparison between fascicular and whole sural nerve biopsy. *Ann Neurol* 1983;**13**:65–8.
19. Flachenecker P, Janka M, Goldbrunner R, Toyka KV. Clinical outcome of sural nerve biopsy: a retrospective study. *J Neurol* 1999;**246**:93–6.
20. Ouvrier RA, McLeod JG, Conchin T. Morphometric studies of sural nerve in childhood. *Muscle Nerve* 1987;**10**:47–53.
21. Jacobs JM, Love S. Qualitative and quantitative morphology of human sural nerve at different ages. *Brain* 1985;**108**(Pt 4):897–924.
22. O'Sullivan DJ, Swallow M. The fibre size and content of the radial and sural nerves. *J Neurol Neurosurg Psychiatry* 1968;**31**:464–70.
23. Gibbels E. Morphometry of unmyelinated nerve fibers. *Clin Neuropathol* 1989;**8**:179–87.
24. Behse F. Morphometric studies on the human sural nerve. *Acta Neurol Scand Suppl* 1990;**132**:1–38.
25. Gutrecht JA, Dyck PJ. Quantitative teased-fiber and histologic studies of human sural nerve during postnatal development. *J Comp Neurol* 1970;**138**:117–29.
26. Thomas JE, Lambert EH. Ulnar nerve conduction velocity and H-reflex in infants and children. *J Appl Physiol* 1960;**15**:1–9.
27. Miller RG, Kuntz NL. Nerve conduction studies in infants and children. *J Child Neurol* 1986;**1**:19–26.
28. Ferriere G, Denef JF, Rodriguez J, Guzzetta F. Morphometric studies of normal sural nerves in children. *Muscle Nerve* 1985;**8**:697–704.
29. Vos AJ, Joosten EM, Gabreels-Festen AA, Gabreels FJ, Notermans SL, Lamers KJ. The diagnostic value of sural nerve biopsy in metachromatic leucodystrophy and other conditions with low leucocyte arylsulphatase A activities. *Neuropediatrics* 1982;**13**:42–6.
30. Kararizou E, Davaki P, Karandreas N, Davou R, Vassilopoulos D. Nonsystemic vasculitic neuropathy: a clinicopathological study of 22 cases. *J Rheumatol* 2005;**32**:853–8.
31. Hicks CW, Kay B, Worley SE, Moodley M. A clinical picture of Guillain-Barré syndrome in children in the United States. *J Child Neurol* 2010;**25**:1504–10.
32. Chowdhury D, Arora A. Axonal Guillain-Barré syndrome: a critical review. *Acta Neurol Scand* 2001;**103**:267–77.
33. Feasby TE, Gilbert JJ, Brown WF, Bolton CF, Hahn AF, Koopman WF, et al. An acute axonal form of Guillain-Barré polyneuropathy. *Brain* 1986;**109**(Pt 6):1115–26.
34. McMillan HJ, Kang PB, Jones HR, Darras BT. Childhood chronic inflammatory demyelinating polyradiculoneuropathy: combined analysis of a large cohort and eleven published series. *Neuromuscul Disord* 2013;**23**(2):103–11.
35. Andrade R, Moya-Machado A, Gómez-Conde S, García-Espinosa A, Molina-Díaz M, Machado-Rojas A, et al. Neuropatías por vasculitis en la infancia. *Rev Neurol* 2004;**38**(7):619–24.
36. Ooi WW, Srinivasan J. Leprosy and the peripheral nervous system: basic and clinical aspects. *Muscle Nerve* 2004;**30**:393–409.
37. Fisher CM, Adams RD. Diphtheritic polyneuritis—a pathological study. *J Neuropathol Exp Neurol* 1956;**15**:243–68.
38. López-Alberola RF. Neuroborreliosis and the pediatric population: a review. *Rev Neurol* 2006;**42**(3):S91–6.
39. Govender R, Eley B, Walker K, Petersen R, Wilmshurst JM. Neurologic and neurobehavioral sequelae in children with human immunodeficiency virus (HIV-1) infection. *J Child Neurol* 2011;**26**:1355–64.
40. Koeppen AH, Mazurkiewicz JE. Friedreich ataxia: neuropathology revised. *Neuropathol Exp Neurol* 2013;**72**(2):78–90.
41. Suomalainen A, Isohanni P, Mitochondrial DNA. Depletion syndromes—many genes, common mechanisms. *Neuromuscul Disord* 2010;**20**:429–37.
42. Nardocci N, Zorzi G, Farina L, Binelli S, Scaioli W, Ciano C, et al. Infantile neuroaxonal dystrophy: Clinical spectrum and diagnostic criteria. *Neurology* 1999;**52**(7):1472–8.
43. Altarescu GM, Goldfarb LG, Park KY, Kaneski C, Jeffries N, Litvak S, et al. Identification of fifteen novel mutations and genotype-phenotype relationship in Fabry disease. *Clin Genet* 2001;**60**:46–51.
44. Gumbinas M, Larsen M, Liu HM. Peripheral neuropathy in classic Niemann-Pick disease: Ultrastructure of nerves and skeletal muscles. *Neurology* 1975;**25**:107–13.

45. Pellissier JF, Berad-Bradier M, Pinsard N. Farber's disease in two siblings: Sural nerve and subcutaneous biopsies studied by light and electron microscopy. *Acta Neuropathol (Berl)* 1986;**72**:178–88.
46. Finsterer J. Leigh and Leigh-like syndrome in children and adults. *Pediatr Neurol* 2008;**39**:223–35.
47. Siddiqi ZA, Sanders DB, Massey JM. Peripheral neuropathy in Krabbe disease: Electrodiagnostic findings. *Neurology* 2006;**67**:263–7.
48. Haberlandt E, Scholl-Bürgi S, Neuberger J, Felber S, Gotwald T, Sauter R, et al. Peripheral neuropathy as the sole initial finding in three children with infantile metachromatic leukodystrophy. *Eur J Paediatr Neurol.* 2009;**13**(3):257–60.
49. Weidenheim KM, Dickson DW, Rapin I. Neuropathology of Cockayne syndrome: Evidence for impaired development, premature aging, and neurodegeneration. *Mech Ageing Dev* 2009;**130**:619–36.
50. Waterham HR, Ebberink MS. Genetics and molecular basis of human peroxisome biogenesis disorders. *Biochim Biophys Acta* 2012;**1822**:1430–41.
51. Rampoldi L, Danek A, Monaco AP. Clinical features and molecular bases of neuroacanthocytosis. *J Mol Med* 2002;**80**:475–91.
52. Brooks-Wilson A, Marcil M, Clee SM, Zhang LH, Roomp K, van Dam M, et al. Mutations in ABC1 in Tangier disease and familial high-density lipoprotein deficiency. *Nature Genetics* 1999;**22**:336–45.
53. Roy A, Kar R, Basu D, Srivani S, Badhe BA. Clinico-hematological profile of Chediak-Higashi syndrome: experience from a tertiary care center in south India. *Indian J Pathol Microbiol.* 2011;**54**(3):547–51.
54. Lehmann AR, McGibbon D, Stefanini M. Xeroderma pigmentosum. *Orphanet J Rare Dis* 2011;**6**:70.
55. Frederico A, Dotti MT. Cerebrotendinous xanthomatosis: Clinical manifestations, diagnostic criteria, pathogenesis, and therapy. *J Child Neurol* 2003;**18**:633–8.
56. Balwani M, Desnick RJ. The porphyrias: advances in diagnosis and treatment. *Blood* 2012;**120**:4496–504.
57. Gibbs TC, Payan J, Brett EM, Lindstedt S, Holme E, Clayton PT. Peripheral neuropathy as the presenting feature of tyrosinaemia type I and effectively treated with an inhibitor of 4-hydroxyphenylpyruvate dioxygenase. *J Neurol Neurosurg Psychiatry* 1993;**56**:1129–32.
58. Schulz A, Baader SL, Niwa-Kawakita M, Jung MJ, Bauer R, Garcia C, et al. Merlin isoform 2 in neurofibromatosis type 2-associated polyneuropathy. *Nat Neurosci* 2013;**16**(4):426–33.
59. Vallet AE, Verschueren A, Petiot P, Vandenberghe N, Nicolino M, Roman S, et al. Neurological features in adult Triple-A (Allgrove) syndrome. *J Neurol* 2012;**259**(1):39–46.
60. Ramos-Leví AM, Díaz-Pérez A, Sobrido MJ, Piñeiro-Hermida S, Blanco-Arias P, Cabezas-Agrícola JM, et al. Axonal neuropathy, long limbs and bumpy tongue: think of MEN2B. *Muscle Nerve* 2012;**46**(6):961–4.
61. Korobkin R, Asbury AK, Sumner AJ, Nielsen SL. Glue-sniffing neuropathy. *Arch Neurol* 1975;**32**:158–62.
62. Thomson RM, Parry GJ. Neuropathies associated with excessive exposure to lead. *Muscle Nerve* 2006;**33**(6):732–41.
63. Quasthoff S, Hartung HP. Chemotherapy-induced peripheral neuropathy. *J Neurol* 2002;**249**:9–17.
64. Gilchrist LS, Tanner L. The pediatric-modified total neuropathy score: a reliable and valid measure of chemotherapy-induced peripheral neuropathy in children with non-CNS cancers. *Support Care Cancer* 2013;**21**:847–56.
65. Achaibar KC, Moore S, Bain PG. Ciguatera poisoning. *Pract Neurol* 2007;**7**:316–22.
66. Kril JJ. Neuropathology of thiamine deficiency disorders. *Metab Brain Dis* 1996;**11**(1):9–17.
67. Saperstein DS, Barohn RJ. Peripheral neuropathy due to cobalamin deficiency. *Curr Treat Options Neurol* 2002;**4**(3):197–201.
68. Puri V, Chaudhry N, Tatke M, Prakash V. Isolated vitamin E deficiency with demyelinating neuropathy. *Muscle Nerve* 2005;**32**:230–5.

Part II

Infantile Hypotonia and Arthrogryposis

Chapter 6

Neonatal Hypotonia

Joseph J. Volpe

INTRODUCTION

Neonatal hypotonia, manifested by the clinical appearance of a "floppy infant" (Figure 6.1) and by diminished resistance to passive movement, is the principal presenting feature of most neuromuscular disorders of the newborn.[1] Thus, the disorders responsible for the hypotonia are discussed throughout this book. This chapter reviews the major features of the examination of a newborn with neonatal hypotonia and a suspected neuromuscular disorder, and then discusses the distinguishing features of the major categories of neuromuscular disorders. Such important laboratory studies as assessment of cerebrospinal fluid and serum muscle enzyme levels; electromyography; determination of nerve conduction velocity; muscle biopsy and genetic studies; and imaging and other assessments of the central nervous system, are discussed elsewhere in this book in relation to the specific entities.

HISTORY AND GENERAL EXAMINATION

As with other assessments of an infant with a neurological disorder, the evaluation begins with a careful history and general examination. The neonatal neuromuscular examination, the focus of this chapter, then follows.

History

The importance of acquiring a careful history frequently is overlooked in the evaluation of an infant with a motor disorder. The pertinent historical features will become apparent in subsequent discussions of the various disorders elsewhere in the book. However, it should be emphasized here that certain findings that initially might not be considered relevant to a motor abnormality, such as polyhydramnios, may prove to be a valuable clue in diagnosis (e.g. myotonic dystrophy). Moreover, the family history should be supplemented by examination of the infant's parents. The myotonia and facial weakness of myotonic dystrophy or the pes cavus and leg weakness of familial polyneuropathy are easily overlooked in many affected adults.

The peripartum and early neonatal history may provide evidence for an asphyxial insult, e.g. fetal heart rate abnormalities, intrapartum "sentinel event" (e.g. cord prolapse, uterine rupture, placental abruption), depressed Apgar scores, fetal acidosis, signs of encephalopathy. Hypoxic–ischemic encephalopathy is a very common cause of neonatal hypotonia.[1] The early neonatal history may provide a clue for a serious metabolic disorder (e.g. metabolic acidosis, hyperammonemia), in which hypotonia is a common feature. The gestational age of the infant is important because prematurity is associated with a wide variety of cerebral and cerebellar disorders that result in hypotonia.[1] Evidence for intrauterine growth retardation may suggest a scenario that includes hypoglycemia, a common cause of neonatal hypotonia. Finally, detection of seizures, which often are subtle in the newborn, will suggest structural or metabolic involvement of the central nervous system, a frequent anatomical substrate for neonatal hypotonia.

Physical Examination

The physical examination must be performed carefully and completely. As will become apparent later, dysmorphic features, cardiac abnormalities, respiratory insufficiency, hepatomegaly, and the like may be features of certain disorders of the motor system. Congenital hip dislocation and other joint contractures are particularly common features in neonatal motor disorders; these and related joint abnormalities are discussed in more detail later. (see "Arthrogryposis Multiplex Congenita").

The neurological evaluation of the motor system is discussed in detail next. It need only be emphasized here, as explained later, that the anatomical site of the disorder of the motor system is determined best by a careful determination of muscle bulk, power, tone, tendon reflexes, primary neonatal reflexes, and the presence or absence of myotonia, myasthenia, and fasciculations. Other neurological features,

B.T. Darras, H. Royden Jones, Jr., M.M. Ryan & D.C. De Vivo (Eds): Neuromuscular Disorders of Infancy, Childhood and Adolescence, Second edition.
DOI: http://dx.doi.org/10.1016/B978-0-12-417044-5.00006-8

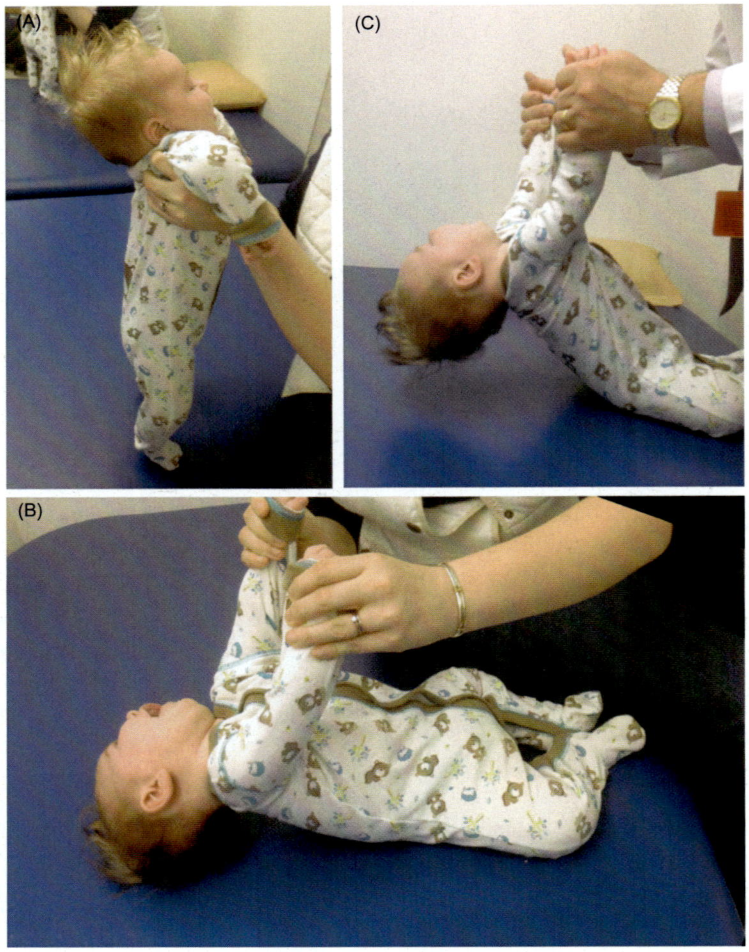

FIGURE 6.1 Major features of neonatal hypotonia ("floppy infant"). With vertical suspension (**A**), note the dangling lower limbs with lack of hip flexion, tendency of upper limbs to slip through the examiner's hands, and lack of neck flexion with resulting head lag. When subject is supine, note the lack of traction response (**B**) and the lag of head (**C**) with attempts by examiner to pull infant to sitting position. With horizontal suspension (not shown), the child drapes over the examiner's hand.

such as abnormalities of cranial nerve function, sensory discrimination, or the occurrence of seizures, provide important supplementary information in selected instances.

NEONATAL NEUROMUSCULAR EXAMINATION

Examination of a floppy infant of course should emphasize the motor examination, the evaluation of the primary neonatal reflexes, and the sensory examination. Careful attention to central nervous system signs (seizures, focal motor deficits, impaired level of alertness, spinal cord signs, etc.) is also important and is discussed in detail elsewhere.[1]

Motor Examination

The major features of the motor examination in the neonatal period are muscle tone and posture of limbs, motility and muscle power, muscle stretch reflexes, and the plantar response. The infant's postnatal age and level of alertness have an important bearing on essentially all these features. Most of the observations described are applicable to infants more than 24 hours old with an optimal level of alertness, unless otherwise indicated.

Tone and Posture

Muscle tone is assessed best by passive manipulation of the limbs with the head placed in the midline. Because the tone of various muscles partly determines the posture of the limbs at rest, careful observation of posture is valuable for the proper evaluation of tone. Some investigators have devised various maneuvers for the passive manipulation of limbs (e.g. approximation of heel to ear, hand to opposite ear [scarf sign], measurement of such joint angles as the popliteal angle) to quantitate tone.[2,3] I have not found these maneuvers particularly useful and do not discuss them in detail here.

Developmental Aspects. Saint-Anne Dargassies described an approximate caudal-rostral progression in the development of tone, particularly flexor tone, with maturation.[4] Thus, at 28 weeks' gestational age, there is minimal resistance to passive manipulation in all limbs, and indeed, premature infants of ≤28 weeks' gestational age appear clinically like "floppy infants." By 32 weeks, distinct flexor tone becomes apparent in the lower extremities. By 36 weeks, flexor tone is prominent in the lower extremities and palpable in the upper extremities. By term, passive manipulation affords an appreciation of strong flexor tone in all extremities.

The posture of an infant in repose reflects these changes in tone, to some extent. In my experience, these postures are apparent principally in the slightly drowsy state. An alert infant at these various gestational ages is more active and motile, and fixed postures or so-called preference postures are difficult to define. This fact has been documented by Prechtl and coworkers and by others.[5–9] Nevertheless, a very quiet infant at 28 weeks' gestational age often lies with minimally flexed limbs, whereas by 32 weeks there is distinct flexion of the lower extremities at the knees and hips. By 36 weeks, flexor tone in the lower extremities results in a popliteal angle of 90 degrees, and there is consistent and frequent flexion at the elbows. By term, the infant assumes a flexed posture of all limbs.[4] The evolution of hip (and knee) flexor tone with maturation is reflected in the developmental increase in pelvic elevation when the infant is in the prone position.[10]

Preference of Head Position. A consistent and interesting aspect of posture in newborn infants is a preference for position of the head toward the right.[5,11–13] Prechtl and coworkers demonstrated head position toward the right 79% of the time, versus 19% toward the left, and only 2% toward the midline.[5] The head orientation preference may be less prominent in the first 24 hours of life.[14] This preference is not attributable to differences in lighting, nursing practices, or other factors; it appears to reflect a normal asymmetry of cerebral function at this age.

Motility and Power

The quantity, quality, and symmetry of motility and muscle power are the parameters of interest. Hadders-Algra and Prechtl combined videotape and electrophysiologic methods to describe the postnatal development of motor activity in term infants.[15] In the first 4 weeks, movements with a writhing quality predominate; in the period from 4 to 12 weeks, "fidgety" movements are prominent; and after 8 to 12 weeks, rapid large-amplitude "swipes" and "swats" are the predominant movements. In general, preterm infants exhibited similar patterns of motor development when they attained comparable postmenstrual ages.[9,16] Prechtl and others have emphasized that the *quality of spontaneous* movements in preterm and term infants is of major importance in evaluating the status of the central nervous system.[17–25]

Saint-Anne Dargassies, using less sophisticated techniques, also noted a writhing quality of the initial movements of preterm infants.[4] By 32 weeks of gestation, movements are predominantly flexor, especially at the hips and knees, often occurring in unison.[4] Although head turning is present, neck flexor and extensor power is negligible, as judged by complete head lag on pull-to-sit maneuvers or when the infant is held in the sitting position. By 36 weeks, the active flexor movements of the lower extremities are stronger and often occur in an alternating rather than symmetrical fashion. Flexor movements of the upper extremities are prominent. For the first time, definite neck extensor power can be observed. When the infant is supported in the sitting position, the head is lifted off the chest and remains upright for several seconds. By term, the awake infant is particularly active if stimulated with a gentle shake. Limbs move in an alternating manner, and neck extensor power is better. Neck flexor power also becomes apparent; when the infant is pulled to a sitting position with a firm grasp of the proximal upper limbs, the head is held in the same plane as the rest of the body for several seconds.[4]

The importance of a fixed program in motor development is suggested by the similarities among (at the same postmenstrual age) the fetus, the premature infant, and the term infant.[4,5,9,21,25–27] These similarities markedly outweigh the rather small differences when one compares similar age infants.

Muscle Stretch Reflexes

The muscle stretch reflexes readily elicited in term newborns are the pectoralis, biceps, brachioradialis, knee, and ankle jerks. I have considerable difficulty obtaining triceps jerks in term infants. Most of these reflexes can be elicited but are less active in preterm infants. The knee jerk is often accompanied by crossed adductor responses, which should be considered a normal finding in the first several months of life (<10% of normal infants demonstrate crossed adductor responses after 8 months of age).[28]

Ankle clonus of 5 to 10 beats should also be accepted as a normal finding in newborn infants if no other abnormal neurologic signs are present and the clonus is not distinctly asymmetrical. Ankle clonus usually disappears rapidly, and more than a few beats beyond 3 months of age is abnormal.

Plantar Response

The plantar response is usually stated to be extensor in newborn infants,[29] but this is clearly related to the manner in which the response is elicited. Using drag of thumbnail along the lateral aspect of the sole, the stimulus I prefer, Hogan and Milligan observed flexion in 93 of 100 newborn infants examined.[30] We observed a similar result in 116 of 124 infants (94%).[31] In contrast, Ross and associates, using the nociceptive stimuli of drag of a pin or pinstick, observed a predominance of extensor responses, with flexion in only about 5% of patients.[32]

When evaluating the neonatal plantar response, it is necessary to consider at least four competing reflexes leading to movements of the toes. Two reflexes that result in extension are nociceptive withdrawal (often accompanied by triple flexion at hip, knee, and ankle) and contact avoidance (elicited best by stroking the dorsum of the foot, which often occurs inadvertently when holding the foot to elicit the plantar response). Two responses that lead to flexion are plantar grasp and positive supporting reaction (both elicited by pressure on the plantar aspect of the foot). Because of these competing reflexes and the relative inconsistency of responses, I consider the plantar response to be of limited value when evaluating newborn infants to determine the presence of an upper motor neuron lesion.

Primary Neonatal Reflexes

Many primary neonatal reflexes have been described in the classic writings on the neonatal examination. The ones that I find moderately useful are the Moro reflex, the palmar grasp, and the tonic neck response. In general, these reflexes are more valuable in assessing disorders of the lower motor neuron, nerve, and muscle than of the upper motor neuron.

Moro Reflex

The Moro reflex, elicited by the sudden dropping of the baby's head in relation to the trunk (the falling head should be caught by the examiner), consists of opening of the hands and extension and abduction of the upper extremities, followed by anterior flexion ("embracing") of the upper extremities and an audible cry. Hand opening is present by 28 weeks of gestation, extension and abduction by 32 weeks, and anterior flexion by 37 weeks.[4] Audible cry appears at 32 weeks. The Moro reflex disappears by 6 months of age in normal infants.[28,29]

Palmar Grasp

Palmar grasp is clearly present at 28 weeks of gestation and strong at 32 weeks. At 37 weeks, it is strong enough and associated with enough extension of upper extremity muscles to allow the infant to be lifted from the bed.[4] The palmar grasp becomes less consistent after about 2 months of age, when voluntary grasping begins to develop.

Tonic Neck Response

The tonic neck response, elicited by rotating the head, consists of extension of the upper extremity on the side to which the face is rotated and flexion of the upper extremity on the side of the occiput (the lower extremities respond similarly but often not as strikingly). The term *fencing posture* is an apt description. The response appears by 35 weeks of gestation[4] but is most prominent about 1 month postterm; it disappears by approximately 6 to 7 months of age,[28,29,33] although the changes in tone may be palpable for several additional months.[34]

Sensory Examination

Careful evaluation of sensory function is an important part of the neonatal neurological examination in the context of hypotonia. Unfortunately, the imprecise term *withdrawal* is often used to describe the infant's response to stimulation. It is noteworthy that a premature infant of just 28 weeks' gestational age discriminates touch and pain; the former results in alerting and slight motor activity, and the latter in withdrawal and cry.[35] The rooting reflex, elicited by tactile stimulation of the perioral region, is well established by 32 weeks of gestation. By approximately 36 weeks, there is rapid turning of the head away from pinprick over the side of the face.

I routinely assess the infant's responses to multiple (five) pinpricks over the medial aspect of the extremities. Responses to be observed are latency, limb movement, facial movement (i.e. grimace), vocalization (i.e. cry), and habituation. A lower-level response is extremely rapid, consisting of a stereotyped response (e.g. triple flexion at hip, knee, and ankle), and is not accompanied by grimace or cry. There is no clear response decrement with repeated trials (i.e. no habituation). A normal, higher-level response has a recognizable latency and usually consists of an apparently purposeful avoidance maneuver, lateral withdrawal, and grimace or cry. The response "dampens" with repeated trials; this characteristic of habituation is an important feature of the normal neonatal response.[36] In a systematic study of 130 healthy newborn infants (124 full term), we observed the higher-level motor response in 94%.[37]

In an infant with hypotonia and possible peripheral neuropathy, careful evaluation of the distal extremities, especially the lower extremities, is important. The responses to be observed are as just discussed.

DISTINGUISHING FEATURES OF MOTOR SYSTEM DISORDERS

The most critical task for the physician evaluating an infant with hypotonia is defining the anatomic level of the underlying abnormality and the likely pathology (see Table 6.1).

TABLE 6.1 Disorders Associated with Neonatal Hypotonia

Levels above the lower motor neuron	Level of the lower motor neuron
Congenital (Nonprogressive) Encephalopathies	Spinal muscular atrophy type I (Werdnig-Hoffman disease; also type 0)
Hypoxia-ischemia	Spinal muscular atrophy variants (not linked to chromosome 5q) (SMARD 1, X-linked recessive SMA, A-dominant lower limb SMA, A-dominant proximal SMA, pontocerebellar hypoplasia, type 1)
Intracranial hemorrhage	Glycogen storage disease type II (Pompe's disease)
Intracranial infection	Hypoxic-ischemic injury
Metabolic	Neonatal poliomyelitis (other enteroviruses?)
Multiple (mitochondrial, organic, or amino acid disorders, etc.)	Neurogenic arthrogryposis multiplex congenita
Endocrine	**Level of the peripheral nerve**
Hypothyroid	***Chronic Motor-Sensory Polyneuropathy***
Trauma	Myelin
Developmental disturbance	Hypomyelination
Cerebral (e.g. Prader-Willi syndrome, neuronal migration disorders)	Autosomal recessive, sporadic
Cerebellar	Hypomyelination + demyelination-remyelination (onion bulbs)
Degenerative (Progressive) Encephalopathies	Autosomal recessive, Autosomal dominant, sporadic
Gray matter and white matter degenerations	Other (associated with axonal disease or focal hypermyelination)
Peroxisomal disorder	Chronic inflammatory demyelinating polyneuropathy
Spinal Cord Disorders	Synaptic abnormalities (acetylcholinesterase deficiency)
Trauma	Postsynaptic abnormalities (acetylcholine receptor defects, slow-channel and fast-channel syndromes)
Developmental	***Toxic-Metabolic***
Neuronal-axonal (rare)	Hypermagnesemia
Autosomal recessive, Autosomal dominant, sporadic	Antibiotics, aminoglycosides
Subcellular	***Infantile Botulism***
Mitochondrial	**Level of the muscle**
Cytoskeletal structures: intermediate filaments (e.g. giant axonal neuropathy), neurofilaments (e.g. infantile neuroaxonal dystrophy)	***Histology Not Diagnostic***
Lysosomal-Krabbe's disease	Congenital myotonic dystrophy
Congenital Sensory Neuropathies	Congenital muscular dystrophies
Familial dysautonomia, congenital sensory neuropathy ± anhidrosis and mental retardation	Merosin-deficient (primary [LAMA 2] and secondary
Acute Polyneuropathy	Merosin-positive (classic [pure], Ullrich, rigid spine syndrome)
Guillain-Barré syndrome	With overt CNS abnormalities (Fukuyama, Walker-Warburg, muscle-eye-brain; LARGE-related)
Level of the neuromuscular junction	Facioscapulohumeral dystrophy
Myasthenia	Polymyositis
Neonatal transient	Minimal change myopathy
Congenital (hereditary) myasthenic syndromes	
Presynaptic abnormalities (defects in synaptic vesicles)	
Histology Diagnostic	
Central core disease	
Nemaline (rod body) myopathy	
Myotubular (centronuclear) myopathy	
Congenital fiber-type disproportion	
Other specific congenital myopathies (e.g. minicore-multicore, fingerprint body, spheroid body, myofibrillar inclusion, sarcoplasmic body, zebra body, reducing body, Mallory body-like)	
Mitochondrial myopathies	
Cytochrome-c oxidase deficiency	
Mitochondrial DNA depletion	
Metabolic myopathies	
Glycogen disorders (acid maltase, phosphorylase, phosphofructokinase deficiencies)	
Lipid disorders (fatty acid oxidation defects)	

The major disorders leading to neonatal hypotonia, classified according to the level of the abnormality, are listed in the table and are discussed elsewhere in this book, as well as in other sources.[1] Several recent informative reviews of selective categories of the disorders shown in the table are also available.[38–42]

The anatomic level of the abnormality and the likely pathology in neonatal disorders of the motor system can

TABLE 6.2 Distinguishing Features of Disorders of the Motor System

	Weakness*							
Locus of Lesion	Face	Arms	Legs	Proximal-Distal	Muscle Stretch Reflexes	EMG / NCS	Muscle Biopsy†	Other
Central	0	+	+	> or =	Normal or ↑	Normal	Normal	Seizures, hemiparesis, delayed development
Anterior horn cell	Late	++++	++++	< or =	0	Markedly decreased numbers of MUPs, fibrillations, normal SNAPs	Denervation pattern	Fasciculations (tongue)
Peripheral nerve	0	+++	+++	<	↓	Mildly decreased numbers of MUPs, fibrillations, abnormal / absent SNAPs	Denervation pattern	Sensory deficit, elevated CSF protein, depressed nerve conduction velocity, abnormal nerve biopsy
Neuromuscular junction	+++	+++	+++	=	Normal	Decremental response (myasthenia); incremental response and myopathic MUPs (botulism)	Normal	Response to neostigmine or edrophonium (myasthenia); constipation and fixed pupils (botulism)
Muscle	Variable (+ to ++++)	++	+	>	↓	Short-duration, low-amplitude polyphasic MUPs	Myopathic pattern†	Elevated muscle enzyme levels (variable)

**+ to++++, indicating increasing degrees of severity.*
†May also show unique features, such as in central core disease, nemaline myopathy, myotubular myopathy, and congenital fiber-type disproportion.
CSF = cerebrospinal fluid; EMG/NCS = electromyography/nerve conduction studies; MUPs = motor unit potentials; SNAPs = sensory nerve action potentials.

be determined by keeping in mind just a few distinguishing features (Table 6.2). Thus, the major levels to be defined include those above the lower motor neuron (i.e. "central"), the anterior horn cell, the peripheral nerve, the neuromuscular junction, and the muscle. Distinguishing clinical features include the pattern of weakness, activity of muscle stretch reflexes, presence of fasciculations or sensory loss, cerebrospinal fluid (CSF) findings, level of serum muscle enzymes, electromyogram (EMG) pattern, nerve conduction studies, and histologic appearance of the muscle or, less commonly, nerve biopsy.

Central Disorders

Disorders related to central causes (e.g. hypoxic-ischemic encephalopathy) are characterized by weakness of the limbs (the degree of weakness is usually less striking than the degree of hypotonia), relatively preserved muscle stretch reflexes (which become hyperactive after the first few months of life), and absence of fasciculations (Tables 6.1, 6.2; see Case Example 6.1). In our experience this group of disorders, especially hypoxic-ischemic disease, constitutes the most common causes of neonatal hypotonia *per se.* There are no abnormalities in CSF (with a few exceptions), serum enzymes, EMG, nerve conduction parameters, and muscle biopsy. Other indicators of central disease, such as seizures, hemiparesis, and diffuse developmental delay, are important to identify.

Anterior Horn Cell Disorders

Diseases of the anterior horn cell (e.g. type I spinal muscular atrophy (SMA) or Werdnig-Hoffman disease) are characterized by generalized weakness, absent reflexes, and fasciculations (especially of the tongue), but normal sensory function and CSF examination; serum muscle enzyme levels are usually normal or slightly elevated (Tables 6.1, 6.2; see Case Example 6.2). EMG demonstrates motor

CASE EXAMPLE 6.1 Central Hypotonia: Aromatic L-Amino Acid Decarboxylase Deficiency

The patient was first evaluated at age 6½ months for profound hypotonia and severe delays in early motor development. Her birth weight was 6-pound, 15-ounce after a 37½-week gestation, born to a 25-year-old mother. The pregnancy was complicated by preterm contractions at 27 weeks, treated with bed rest and terbutaline; the gestation was notable for decreased fetal movements. At 37½ weeks, the terbutaline was discontinued, and labor began spontaneously 1 week later. At birth, a nuchal cord was present, and there was evidence of meconium aspiration. The infant was noted to feed poorly, but otherwise she appeared healthy and was discharged at age 2 days. She was readmitted to the hospital at age 5 days with a slow heart rate, lowered blood pressure, and hypothermia. She had also developed jaundice and was treated with antibiotics for presumed sepsis. After 3 days of treatment in the hospital, the infant was discharged at age 8 days. Since that time, and up to the first neurologic evaluation at age 6½ months, the infant made essentially no developmental progress. Early on, the mother had noticed marked feeding difficulty, weak suck, and poor coordination of swallowing movements. Feeding was a struggle, and the infant gagged or choked when food was presented to her. The mother reported that the infant spent up to an hour for a 3-ounce feeding. Gross and fine motor milestones had been significantly delayed, and no progress had been noted since the neonatal period. The infant had decreased spontaneous activity, as evidenced by difficulty in eye opening and elevation of her eyelids.

At age 2 months, she developed possible seizures, characterized by crossing of her eyes and tonic posturing of the limbs; she would also giggle or laugh during the episodes. She was treated with phenobarbital and later topiramate, which proved to be effective in controlling her "seizures." Subsequently, the infant remained somnolent, sleeping up to 20 hours each day. Various blood and urine studies failed to yield any significant diagnostic clues. A brain MRI scan was normal, and an electroencephalogram showed some normal background slowing without any epileptiform discharges. Nerve conduction studies and EMG were normal. In addition, repetitive motor nerve stimulation studies showed no evidence of a decremental response.

A genetic-metabolic screen had been negative for many conditions, and fluorescence *in situ* hybridization studies for Prader-Willi and Angelman's syndromes were negative. The ocular fundi and brain stem auditory evoked potentials were normal. At age 6½ months she was passive, hypotonic, and somnolent. She lay quietly in the supine position and had difficulty opening her eyelids. When the eyelids were open, she would look at her parents and smile. Vision and hearing were present. Eye movements were full in all directions but could be appreciated only after the eyelids were held open. Axial and appendicular hypotonia was evident. She had no head control and demonstrated minimal spontaneous movements. Her cry was weak and not sustained for more than a few seconds. Muscle stretch reflexes were present inconsistently at 2+ in the arms and legs. A repeat EMG was normal, and repetitive motor nerve stimulation failed to demonstrate any decrement of the CMAP. Assay of CSF biogenic amines produced an abnormal pattern consistent with aromatic L-amino acid decarboxylase deficiency. Specifically, the CSF levels of 5-hydroxyindoleacetic acid and homovanillic acid were decreased, and 3-O-methyldopa was elevated; the tetrahydrobiopterin-neopterin profile was normal. The activity of aromatic L-amino acid decarboxylase in plasma was just barely detectable, confirming that this child had a deficiency of the enzyme. Blood obtained from her parents confirmed that they were heterozygous for the enzyme deficiency.

On subsequent follow-up at 9 months of age, she was started on high-dose vitamin B6 and bromocriptine, which led to better eye opening and improved facial expression. At 10 months, the parents reported more facial expression, better eye opening, improved muscle tone, wakefulness for longer periods (usually 8 hours per day), fewer oculogyric crises, and less limb stiffening. The dose of bromocriptine was increased with further improvement.

Comment

Genetic defects of central neurotransmission are emerging as a new category of diseases associated with profound infantile hypotonia. When ptosis and eye movement abnormalities are present, the phenotypic overlap between defects of peripheral neurotransmission (e.g. congenital myasthenia gravis) and defects of central neurotransmission can be appreciated. Aromatic L-amino acid decarboxylase deficiency, for example, may be confused with congenital myasthenia gravis. Electrodiagnostic studies of neurotransmission and CSF analysis for biogenic amines can easily distinguish one from the other.

units that are significantly diminished in number but increased in size, that fire rapidly, and that are associated with widespread fibrillation potentials. Muscle biopsy demonstrates marked panfascicular atrophy (only uncommonly the type grouping or grouped atrophy observed in later-onset cases).

Many "non-5q" spinal muscular atrophies also have been recognized in recent years.[43] These entities are rarer than the chromosome 5q-linked SMA noted above, although some can present in the neonatal period (Table 6.1).

Several other disorders may lead to neonatal hypotonia and weakness mediated at the level of the anterior horn cell (Table 6.1). These entities are discussed elsewhere in the book.

Peripheral Nerve Disorders

Diseases of peripheral nerve are typified by generalized hypotonia similar to that seen with lesions at any other level of the peripheral motor unit. These infants also have

CASE EXAMPLE 6.2 Anterior Horn Cell Disease: Spinal Muscular Atrophy Type I

This 7-pound infant was the product of a normal pregnancy. She was first noted to be floppy at age 1 month. She had diminished leg movement and poor control of her head. At 2 months of age, she underwent needle EMG at another laboratory; the results were borderline but "negative for motor neuron disease." At 3 months of age, she was unable to roll over and could not lift her head. She was referred for a second opinion and was found to be an alert infant with generalized hypotonia, weak facial muscles, increased head lag, a tendency to slip through the examiner's hands with vertical suspension, tongue fasciculations, lack of traction response, and areflexia (Figure 6.1). The creatine kinase level of 291 IU/L was slightly elevated.

Nerve conduction studies demonstrated low amplitude of CMAPs of 0.4 and 0.5 mV for the peroneal and median nerves, normal-for-age motor conduction velocities (MCVs) of 29 and 23 m/s, and normal distal latencies of 2.4 and 1.6 ms, respectively. Median SNAP was normal, with a 16-μV amplitude and a 1.4-ms latency. In multiple muscles in three extremities, needle EMG demonstrated a severe deficiency in the number of MUPs recruited. Some were of longer duration and of higher amplitude (2–3 mV). In addition, a similar distribution of well-defined active denervation was characterized by the presence of fibrillation potentials and positive waves.

Muscle biopsy demonstrated grouped atrophy and hypertrophy mostly of type I fibers, consistent with spinal muscular atrophy (SMA) type I, or Werdnig-Hoffman disease. Genetic testing was not yet available. Her course was characterized by the need for tube feedings, frequent suctioning, recurrent aspiration, and pneumonia. This infant died 6 months later.

Comment

This case illustrates several important clinical, electrodiagnostic, and pathologic features of SMA. Hypotonia, tongue fasciculations, and areflexia are typical clinical findings in SMA and other diseases of the anterior horn cell or peripheral nerve. However, in the current era, EMG is generally not necessary in the initial evaluation of infants with possible anterior horn cell disease. If the clinical presentation is consistent with SMA, the diagnosis can be confirmed by genetic testing of the survival motor neuron gene on chromosome 5q (deletion of exons 7 and 8) using peripheral blood-derived DNA. Also, a muscle biopsy is not needed in cases confirmed by genetic testing.

depressed muscle stretch reflexes, abnormal sensory examination, and elevated CSF protein concentration (Tables 6.1, 6.2; see Case Example 6.3). The abnormalities of sensation require care in the examination for detection but are generally consistent and diagnostically useful. EMG demonstrates slowing of motor nerve conduction or low compound muscle action potential (CMAP), amplitudes, with low or absent sensory nerve action potentials (SNAPs). Muscle biopsy demonstrates signs of denervation, and nerve biopsy usually demonstrates predominant hypomyelination or amyelination, most commonly without onion bulb formation. Axonal neuropathies also cause neonatal hypotonia but are less frequent.

Neuromuscular Junction Disorders

Diseases of the neuromuscular junction (e.g. myasthenia gravis) are characterized by weakness that involves the face, eyelids, and extraocular muscles, but there are often normal muscle stretch reflexes (Tables 6.1, 6.2; see Case Example 6.4). Sensory function, CSF, and serum enzyme levels are normal (Table 6.2). EMG reveals the characteristic decremental response to repetitive motor nerve stimulation. An infant with botulism usually has a relatively acute onset of generalized hypotonia, commonly preceded by a few days of difficulty feeding, often associated with a poor suck and constipation. On examination, in addition to generalized weakness, the pupillary light reactions are often absent and ptosis is frequently present. EMG typically demonstrates a characteristic incremental response to rapid repetitive motor nerve stimulation. Myopathic motor unit potentials (MUPs) are often an associated finding with needle examination.

Muscle Disorders

Diseases of muscle are typified by weakness that is often greater in the proximal rather than distal limbs (Table 6.2; see Case Examples 6.5 and 6.6). Involvement of the face occurs in several such disorders, including myotonic dystrophy, congenital muscular dystrophy, facioscapulohumeral dystrophy, central core disease, nemaline myopathy, myotubular myopathy, congenital fiber-type disproportion, and mitochondrial myopathies (especially cytochrome-c oxidase deficiency) (Table 6.1). Muscle stretch reflexes are depressed in proportion to the degree of weakness; there are no fasciculations or sensory or CSF abnormalities. Serum muscle enzyme concentrations may be elevated, although in the slowly progressive disorders, this change is often not detectable. EMG exhibits myopathic changes and perhaps more distinctive abnormalities (e.g. myotonia). Muscle biopsy demonstrates myopathic changes and, in certain disorders, abnormalities that are diagnostic-central cores, rod bodies, myotubular fibers, disproportionate distribution and size of muscle fiber types, abnormal mitochondria, or accumulation of glycogen or lipid.

CASE EXAMPLE 6.3 Peripheral Nerve Disease: Congenital Hypomyelinating Neuropathy

A 16-day-old infant with a "paucity of spontaneous movements" since birth had a weak cry but was able to breast-feed. The mother thought she had perceived normal fetal movements *in utero*. On examination, this infant was alert but severely hypotonic, with a frog-leg posture, no spontaneous movement of the legs, wrist drop, weak palmar grasp, and arthrogryposis multiplex congenita. When held prone, he had an inverted U posture. He did not hold the pacifier tightly, and the gag reflex was weak. Tongue fasciculations were present. He was areflexic, and the plantar responses were absent. He responded to noxious stimuli applied to all extremities, but more careful assessment of sensation was not done. The clinicians' differential diagnosis included SMA type I, acid maltase deficiency, and an unusual neonatal poliomyelitis; however, disease at the peripheral nerve, neuromuscular junction, or muscle cell could not be excluded. The CSF protein was elevated at 190 mg/dL.

Nerve conduction studies failed to demonstrate a median nerve SNAP. The median and peroneal motor CMAPs were widely dispersed and of low amplitude (0.05 and 0.02 mV), had prolonged distal latencies of 3.6 and 8.2 ms, and had slow MCVs of 6 and 8.2 m/second. Needle EMG demonstrated mild diminution in the number of MUPs, which were of slightly longer duration and normal amplitude, polyphasic, and fired at increased frequency. The most distal intrinsic hand and foot muscles were the only areas with occasional fibrillation potentials and positive waves. These EMG findings were compatible with a diffuse hypomyelinating peripheral neuropathy.

Sural nerve biopsy demonstrated hypomyelination and onion bulb formation. Because it was difficult to distinguish between an inborn error of myelination and an unusual acquired demyelinating peripheral neuropathy possibly starting *in utero*, a trial of prednisone was attempted but was not helpful. Today, one would attempt intravenous immunoglobulin therapy. Progressive respiratory failure led to death at 2 months of age. Autopsy was declined.

Comment

Congenital hypomyelinating neuropathology can easily be confused with SMA type I because their clinical presentations are similar; tongue fasciculations and areflexia are present in both conditions and support an anterior horn cell or peripheral nerve origin for the hypotonia. Nonetheless, arthrogryposis multiplex congenita is not common in SMA type I (although it may occur rarely), and the CSF protein is normal. Also, in SMA, MCVs are normal and distal latencies are not prolonged. Further, SNAPs are easily detected and are always normal in SMA, because the sensory system is preserved in pure anterior horn cell disorders.

CASE EXAMPLE 6.4 Neuromuscular Junction Defect: Infantile Botulism

A 5-month-old infant was admitted with a 2-day history of poor feeding and irritability. He was no longer able to roll over, lift his head, or sit up. There was no maternal history of neuromuscular disease. The paternal history was unknown because the infant was a product of artificial insemination. Development before the present illness was normal. The only unusual point in his history was chronic obstipation with stool once a week.

On admission, the infant was afebrile and had no sign of systemic illness. Neurologic evaluation demonstrated an alert infant with expressionless facies, mild ptosis, fixed pupils nonreactive to light, weak cry and suck, and diffuse hypotonia and hypoactive muscle stretch reflexes. Results of EMG (including repetitive stimulation), multiple blood studies, CSF analysis, MRI scan of the head, radiographic analyses, and rectal biopsy for ganglion cells were normal (with the exception of multiple distended bowel loops without intussusception on radiographs).

Initially, some clinical improvement was observed. A brief period of ptosis and esotropia then occurred. Edrophonium (Tensilon) did not affect the generalized hypotonia, cry, or suck. Twelve days later, he had slight improvement with feeding and no further ptosis, ophthalmoplegia, or other signs of cranial nerve involvement. However, he was still profoundly weak, hypotonic, and areflexic.

This infant was referred for further neurophysiologic evaluation 18 days after the onset of neurologic symptoms. Left median SNAP was normal. Motor nerve conduction studies of the left median nerve demonstrated a CMAP of 1.6 mV (normal, 3.5–6 mV), with a right peroneal CMAP of 1.7 mV (normal, 1.6–8 mV) and normal MCVs and distal latencies. Repetitive stimulation of the left median nerve at 2 Hz demonstrated no decrement; however, a 30% facilitation was documented with 200-Hz stimulation. Needle EMG demonstrated an increased recruitment of short-duration, low-amplitude, at times polyphasic MUPs, with some increased insertional activity in each of the four muscles examined and a few fibrillation potentials limited to the medial gastrocnemius.

Stool samples obtained at the time of initial evaluation and sent to the Centers for Disease Control and Prevention in Atlanta tested positive for botulinum toxin.

Comment

Hypotonia, hyporeflexia, fixed pupils or other cranial nerve involvement, poor feeding, and constipation in a previously healthy infant are characteristic clinical features of infantile botulism. The diagnosis is supported by the finding of facilitation of the CMAP upon repetitive motor nerve stimulation and confirmed by direct *in vivo* assay of botulinum toxin in stool.

CASE EXAMPLE 6.5 Muscle Disease: Congenital Myotonic Dystrophy

This infant with arthrogryposis multiplex congenita developed respiratory distress soon after birth, requiring intubation for the first 5 weeks of life. She was referred to the EMG laboratory when she was almost 3 months old for evaluation of "neurogenic arthrogryposis." Motor nerve conduction studies demonstrated low-amplitude CMAPs but normal distal latencies, MCVs, and median SNAPs. Profound and prolonged decrescendo myotonic discharges were noted on EMG, associated with scattered fibrillation potentials and normal MUPs. The mother had no myotonia, but she recalled that her father had difficulty releasing his grip when shoveling snow.

Comment

Congenital myotonic dystrophy is one of the most common dystrophies that presents with hypotonia in infancy. Infants with congenital myotonic dystrophy may have severe hypotonia and a classic facies, with tenting of the mouth. A family history may not be recognized, as the disease is often milder in the mother, as occurred here. Maternal inheritance is by far more common than paternal inheritance.

EMG-definable myotonia may be present in floppy infants as early as age 5 days.[44,45] However, the myotonia of infants may be somewhat atypical, with a higher pitch and tending to be less sustained than in adults.[45] Some floppy infants have no myotonia on EMG and demonstrate only nonspecific changes. In these circumstances, a maternal EMG can define previously unsuspected myotonia.[44,45] If this fails, a subsequent EMG may demonstrate later-onset myotonia in some infants. An EMG is not needed, however, if the diagnosis has been confirmed in the mother clinically or by genetic testing. In that case, DNA testing in the infant can easily and noninvasively detect the genetic mutation (large expansion of a CTG repeat in the myotonin-protein kinase gene on chromosome 19q13) and unequivocally confirm the diagnosis.

CASE EXAMPLE 6.6 Muscle Disease: Phosphofructokinase Deficiency

A 39-week-gestation newborn with a history of diminished fetal movements had immediate respiratory distress at birth. He had marked distal arthrogryposis with severe generalized and bifacial weakness, a high arched palate, and areflexia. At 2 days of age, nerve conduction studies demonstrated low-amplitude to absent CMAPs with otherwise normal motor and sensory conduction parameters. EMG demonstrated low-amplitude, short-duration MUPs primarily in proximal muscles. Muscle biopsy demonstrated an absence of phosphofructokinase.

Swoboda[46] suggested that a ketogenic diet might be beneficial in this disorder, and the latter was begun at age 4 months. Repeat EMG at age 9 months demonstrated improved CMAP amplitudes. The infant gained some motor improvement through age 2 years, but he died of other complications later that year.

Comment

Enzymatic myopathies (e.g. glycogen storage disorders) only rarely present in the newborn period as a floppy infant. A fatal case of phosphorylase deficiency presenting at 4 weeks of age with feeding problems, generalized weakness, areflexia, hypotonia, and respiratory distress had a rapidly downhill course.[47]

CONCLUSION

With these distinguishing features relative to the motor and sensory examination, electrophysiological assessment, and CSF evaluation in mind, the clinician should be able to define the level of the lesion in the neuromuscular apparatus that accounts for the infant's hypotonia and weakness. This critical initial accomplishment usually leads to the correct pathologic and etiologic diagnosis, and where relevant, genetic characterization. Appropriate prognostic counseling and therapeutic intervention, specific or nonspecific (discussed elsewhere in this book), then becomes a relatively straightforward next step.

REFERENCES

1. Volpe JJ. *Neurology of the Newborn*. 5th ed. Philadelphia: Elsevier; 2008.
2. Amiel-Tison C. Neurological evaluation of the maturity of newborn infants. *Arch Dis Child* 1968;**43**:89–93.
3. Amiel-Tison C, Maillard F, Lebrun F, Breart G, Papiernik E. Neurological and physical maturation in normal growth singletons from 37 to 41 weeks' gestation. *Early Hum Dev* 1999;**54**:145–56.
4. Saint-Anne Dargassies S. *Neurological Development in the Full-term and Premature Neonate*. New York: Excerpta Medica; 1977.
5. Prechtl HF, Fargel JW, Weinmann HM, Bakker HH. Postures, motility and respiration of low-risk pre-term infants. *Dev Med Child Neurol* 1979;**21**:3–27.
6. Vles JS, Kingma H, Caberg H, Daniels H, Casaer P. Posture of low-risk preterm infants between 32 and 36 weeks postmenstrual age. *Dev Med Child Neurol* 1989;**31**:191–5.
7. Cioni G, Ferrari F, Prechtl HF. Posture and spontaneous motility in fullterm infants. *Early Hum Dev* 1989;**18**:247–62.
8. Vles JS, van Oostenbrugge R, Kingma H, Caberg H, Casaer P. Posture during head turning in pre-term infants: a longitudinal study of 15 low-risk infants of 32–36 weeks of conceptional age. *Neuropediatrics* 1989;**20**:25–9.

9. Cioni G, Prechtl HF. Preterm and early postterm motor behaviour in low-risk premature infants. *Early Hum Dev* 1990;**23**:159–91.
10. Lacey JL, Henderson-Smart DJ, Edwards DA. A longitudinal study of early leg postures of preterm infants. *Dev Med Child Neurol* 1990;**32**:151–63.
11. Turkewitz G, Gordon EW, Birch HG. Head turning in the human neonate: spontaneous patterns. *J Genet Psychol* 1965;**107**:143–58.
12. Gardner J, Lewkowicz D, Turkewitz G. Development of postural asymmetry in premature human infants. *Dev Psychobiol* 1977;**10**:471–80.
13. Konishi Y, Mikawa H, Suzuki J. Asymmetrical head-turning of preterm infants: some effects on later postural and functional lateralities. *Dev Med Child Neurol* 1986;**28**:450–7.
14. Vles J, van Zutphen S, Hasaart T, Dassen W, Lodder J. Supine and prone head orientation preference in term infants. *Brain Dev* 1991;**13**:87–90.
15. Hadders-Algra M, Prechtl HFR. Developmental course of general movements in early infancy. 1. Descriptive analysis of change in form. *Early Hum Dev* 1992;**28**:201–13.
16. Konishi Y. Prechtl HFR. Finger movements and fingers postures in pre-term infants are not a good indicator of brain development. *Early Hum Dev* 1994;**36**:89–100.
17. Hayes MJ, Plante L, Kumar SP, Delivoria-Papadopoulos M. Spontaneous motility in premature infants: features of behavioral activity and rhythmic organization. *Dev Psychobiol* 1993;**26**:279–91.
18. Hayes MJ, Plante LS, Fielding BA, Delivoria-Papadopoulos M. Functional analysis of spontaneous movements in preterm infants. *Dev Psychobiol* 1994;**27**:271–87.
19. Van Der Meer AL. Keeping the arm in the limelight: Advanced visual control of arm movements in neonates. *Eur J Paediatr Neurol* 1997;**4**:103–8.
20. Prechtl HFR, Einspieler C, Cioni G, Bos AF, Ferrari F, Sontheimer D. An early marker for neurological deficits after perinatal brain lesions. *Lancet* 1997;**349**:1361–3.
21. Prechtl HFR. State of the art of a new functional assessment of the young nervous system. An early predictor of cerebral palsy. *Early Hum Dev* 1997;**50**:1–11.
22. Cioni G, Ferrari F, Einspieler C, Paolicelli PB, Barbani MT. Prechtl HFR. Comparison between observation of spontaneous movements and neurologic examination in preterm infants. *J Pediatr* 1997;**130**:704–11.
23. Cioni G, Prechtl HFR, Ferrari F, Paolicelli B, Einspieler C, Roversi MF. Which better predicts later outcome in fullterm infants: quality of general movement or neurological examination? *Early Hum Dev* 1997;**50**:71–85.
24. Bos AF, Martijn A, Okken A. Prechtl HFR. Quality of general movements in preterm infants with transient periventricular echodensities. *Acta Paediatr* 1998;**87**:328–35.
25. Hadders-Algra M, Groothuis AM. Quality of general movements in infancy is related to neurological dysfunction, ADHD, and aggressive behaviour. *Dev Med Child Neurol* 1999;**41**:381–91.
26. Parmelee AH, Schultz HR, Disbrow MA. Sleep patterns of the newborn. *J Pediatr* 1961;**58**:241.
27. Walters CE. Reliability and comparison of four types of fetal activity and of total activity. *Child Dev* 1964;**35**:1249.
28. Paine RS, Brazelton TB, Donovan DE. Evolution of postural reflexes in normal infants and in the presence of chronic brain syndromes. *Neurology* 1964;**14**:1036.
29. Gingold MK, Jaynes ME, Bodensteiner JB, Romano JT, Hammond MT. The rise and fall of the plantar response in infancy. *J Pediatr* 1998;**135**:568–70.
30. Hogan GR, Milligan JE. The plantar reflex of the newborn. *N Engl J Med* 1971;**285**:502–3.
31. Rich E, Marshall R, Volpe J. Letter: Plantar reflex flexor in normal neonates. *N Engl J Med* 1973;**289**:1043.
32. Ross ED, Velez-Borras J, Rosman NP. The significance of the Babinski sign in the newborn—a reappraisal. *Pediatrics* 1976;**57**:13–15.
33. Futagi Y, Tagawa T, Otani K. Primitive reflex profiles in infants: differences based on categories of neurological abnormality. *Brain Dev* 1992;**14**:294–8.
34. Capute AJ, Palmer FB, Shapiro BK, Wachtel RC, Ross A, Accardo PJ. Primitive reflex profile: A quantitation of primitive reflexes in infancy. *Dev Med Child Neurol* 1984;**26**:375–83.
35. Saint-Anne Dargassies S. Neurological maturation of the premature infant of 28–41 weeks' gestational age. In: Falkner F, editor. *Human Development*. Philadelphia: WB Saunders; 1966. pp 306–25.
36. Moreau T, Birch HG, Turkewitz G. Ease of habituation to repeated auditory and somesthetic stimulation in the human newborn. *J Exp Child Psychol* 1970;**9**:193–207.
37. Rich EC, Marshall RE, Volpe JJ. The normal neonatal response to pin-prick. *Dev Med Child Neurol* 1974;**16**:432–4.
38. Markowitz JA, Singh P, Darras BT. Spinal muscular atrophy: a clinical and research update. *Pediatr Neurol* 2012;**46**:1–12.
39. Farrar MA, Vucic S, Johnston HM, du Sart D, Kiernan MC. Pathophysiological insights derived by natural history and motor function of spinal muscular atrophy. *J Pediatr* 2013;**162**: 155–9.
40. Domingo RM, Haller JS, Gruenthal M. Infant botulism: two recent cases and literature review. *J Child Neurol* 2008;**23**:1336–46.
41. Engel AG. Current status of the congenital myasthenic syndromes. *Neuromuscul Disord* 2012;**22**:99–111.
42. Mercuri E, Muntoni F. The ever-expanding spectrum of congenital muscular dystrophies. *Ann Neurol* 2012;**72**:9–17.
43. Darras BT. Non-5q spinal muscular atrophies: The alphanumeric soup thickens. *Neurology* 2011;**77**:312–14.
44. Swift TR, Ignacio OJ, Dyken PR. Neonatal dystrophia myotonica. *Am J Dis Child* 1975;**129**:734.
45. Kuntz NL, Daube JR. Electrophysiology of Congenital Myotonic Dystrophy. *AAEE Course E*. Rochester, Minn: American Association of Electromyography and Electrodiagnosis; 1984. p. 23.
46. Swoboda KJ, Specht L, Jones HR, Shapiro F, DiMauro S, Korson M. Infantile phosphofructokinase deficiency with arthrogryposis: Clinical benefit of a ketogenic diet. *J Pediatr* 1997;**131**:932–4.
47. DiMauro S, Hartlage PL. Fatal infantile form of muscle phosphorylase deficiency. *Neurology* 1978;**28**:1124–9.

Chapter 7

Arthrogryposis

Judith G. Hall and Angela Vincent

INTRODUCTION

Arthrogryposis is a term used to describe the presence of multiple contractures of joints in more than one area that are present at birth. The extended term, arthrogryposis multiplex congenita (AMC), has also been used historically to describe multiple congenital contractures (arthro = joint; gryp = curved; multiplex = multiple; congenita = present at birth). Such contractures are usually nonprogressive and may respond to physical therapy and orthopedic procedures. Arthrogryposis has been recognized for centuries in folklore, art, and medical reports. In the past, the term arthrogryposis was used as a diagnosis for any child born with multiple congenital contractures; however, over the years it has become apparent that there are many different types of conditions with multiple congenital contractures. Thus, the term has come to be used as a clinical sign or as a general category of disorders.[1,2] For the purposes of this chapter, the term arthrogryposis is used as a generic term that encompasses many different conditions with varied causes. The common pathway to congenital contractures appears to be decreased fetal movement, which may occur for many different reasons. The challenge is to determine the specific cause of decreased fetal movement and then to be able to predict natural history, recurrence risk, and best therapies for a specific case. This chapter attempts to present a functional approach.

HISTORICAL BACKGROUND

The first medical report of arthrogryposis was by Adolph Wilhelm Otto in 1841.[3] In the 1930s, 1940s, and 1950s, orthopedists gathered together cases to report the effects of surgery. In the 1950s and 1960s, specific recognizable disorders began to emerge.[4] In 1983, Moessinger proposed the unifying theory, based on animal studies, that lack of fetal movement leads to a whole set of secondary deformations including congenital contractures.[5] Also in the 1980s, Hall and coworkers developed a clinical approach to congenital contractures.[6–8] Many specific types of arthrogryposis began to be enumerated. In the 1990s, specific genes responsible for inherited forms of arthrogryposis began to be mapped and defined; in addition, multiple mechanisms leading to decreased fetal movement began to be elucidated.[2]

EPIDEMIOLOGY

Approximately 1 in 100 children is born with some type of congenital contracture (clubfoot: 1/300; dislocated hip: 1/200; camptodactyly: 1/200; and so forth), but arthrogryposis implies more generalized involvement. The incidence of newborns with multiple congenital contractures from epidemiological studies is about 1 in 3000 live births.[9] The most common specific condition is Amyoplasia, or "classical arthrogryposis," which occurs in about 1 in 10,000 births and, as noted later, has a relatively good prognosis.[10]

Some data suggest multiple congenital contractures are common among miscarriages and stillbirths;[11] thus, the *in utero* frequency may be higher than at birth. Evaluation of fetal movement is not routinely a part of fetal ultrasound.[12] However, affected babies often come to attention when clubfoot or an unusual positioning of a limb is noted during routine prenatal ultrasonography and further real-time investigations indicate a more generalized limitation of a joint movement *in utero*.

Difficult births occur in many cases of arthrogryposis. Abnormal positioning, such as breech and transverse lies, often requires cesarean section. Fracture of long bones during vaginal deliveries of infants with arthrogryposis is common both because of the abnormal, inflexible positioning of the limbs and the decreased ossification often seen related to decreased *in utero* fetal movement. Environmental events, including trauma to mother, maternal viral illnesses, bleeding, and medications taken by mother, have now been implicated as causative agents for

B.T. Darras, H. Royden Jones, Jr., M.M. Ryan & D.C. De Vivo (Eds): Neuromuscular Disorders of Infancy, Childhood and Adolescence, Second edition.
DOI: http://dx.doi.org/10.1016/B978-0-12-417044-5.00007-X

some cases of arthrogryposis.[2,9,10,13,14] Uterine abnormalities and oligohydramnios are rare findings, and even more rarely causative.

Arthrogryposis can be divided into three groups as defined by Hall[8] in a large study of affected individuals: (1) primarily involving limbs; (2) involving limbs plus other body areas; and (3) involving central nervous system (CNS) abnormalities or leading to early death, or both. Clinically, this approach helps lead to the differential diagnosis and recognition of specific conditions.[2]

PATHOGENESIS

It has become increasingly apparent, both from animal and human studies, that anything that leads to decreased fetal movement *in utero* may also lead to congenital contractures or fixation of the fetal joints. With decreased fetal movement, a series of abnormalities are seen to a greater or lesser extent: (1) contractures of joints; (2) polyhydramnios, apparently secondary to lack of fetal swallowing, which leads also to immature intestinal tract development with secondary feeding problems after birth; (3) pulmonary hypoplasia secondary to lack of fetal breathing with failure of maturation of alveolae and surfactin; (4) short umbilical cord; (5) intrauterine growth restriction; (6) osteoporosis of bones; and (7) craniofacial abnormalities including micrognathia, cleft palate (both hard and soft palate), underdeveloped maxilla, prominent bridge of the nose, and depressed tip of the nose. In its most severe form, this combination is called the fetal akinesia deformation sequence.[5,15] Because it was first described phenotypically by Pena and Shokeir, when defining a specific familial syndrome, it is also sometimes called the Pena-Shokeir phenotype.[6,15] However, the combination of morphological and pathological features can be seen with generalized decreased movement from many causes and is present in many other specific conditions.[16]

It appears that any *in utero* process that leads to decreased movement can secondarily lead to fetal contractures. Swinyard and Bleck[17] described a connective tissue response to the lack of joint movement with thickening of the joint capsule and fiber deposition. The earlier such a process starts in fetal development, the more severe and immobilizing the contractures. Thus, the timing during fetal development probably plays a critical role with regard to the position of the joint (i.e. flexion versus extension) and the severity of contractures at birth. Also, when the muscles of the limbs are not used *in utero* they atrophy and fail to mature or may be replaced by fat and fibrous tissue. Secondary changes can occur to other organ systems as described earlier in the fetal akinesia sequence. It would appear that lack of normal joint movement, or even an unusual degree of hypotonia that is present for more than a month during the third trimester, can produce contractures. It may take an even shorter period of time during the first and second trimesters.

The potential causes of limitation of movement *in utero* are numerous. At least seven categories exist: (1) muscle abnormalities, including myopathies and abnormal muscle structure or function; (2) neuropathic processes, including abnormalities in nerve structure and function, either central or peripheral, including the failure of the CNS to myelinate and/or mature; (3) abnormalities of connective tissue including skin, bone, cartilage, and joint and tendon attachments; (4) limitation of space leading to restriction of fetal movement such as a structural abnormality of the uterus, amniotic bands, chronic leakage of amniotic fluid, multiple fetuses, or uterine masses such as fibroids; (5) those related to maternal illness such as myasthenia gravis, maternal infection, compromise of uterine blood supply, prolonged maternal fever, and other maternal illnesses; (6) those related to medications, drugs, exposures, and other environmental agents or deficiencies to which mother is exposed; and (7) compromise of maternal or fetal vascular supply, which leads to hypoxia or hypotension in fetal tissues.[18] Multiple congenital contractures can occur as a manifestation of a large number of syndromes including chromosomal abnormalities, structural abnormalities of the CNS, and metabolic disorders; the exact cause is often hard to determine.

Whether a particular type of arthrogryposis has a recurrence risk depends, of course, upon whether it has a genetic basis or an environmental basis or has been described as sporadic. The most common form of arthrogryposis, Amyoplasia, appears to be sporadic, although it is seen with increased frequency in one of monozygotic twins.[10] In Amyoplasia, contractures are thought to occur on the basis of placental vascular compromise leading to hypotension and oxygen deprivation at critical periods in development of nerve and muscle. This sequence could theoretically recur in a second pregnancy, although this has not been observed as yet. Autosomal dominant, autosomal recessive, X-linked recessive single gene, and even mitochondrial inheritance have all been observed, including new mutations in the affected individual.[2] Evidence of a genetic contribution in some cases is the higher frequency than expected of clubfoot or dislocated hip in the first- and second-degree relatives of affected individuals. Environmental factors such as an abnormal uterine structure, maternal use of misoprostol, or maternal illnesses such as the development of maternal antibodies to fetal neurotransmitters (see later) can all lead to recurrence in further children even though a genetic basis is not present.

Arthrogryposis is usually not due to an abnormality in formation of structures during the embryonic stage because bones and joints almost always have formed in a normal way (the exception is some types of synostosis and chondrodysplasias). Thus, arthrogryposis is not usually a problem with embryogenesis *per se,* but rather a failure to develop normal movement owing to abnormal mechanical forces or dysplasias (failure of a specific tissue to mature or function), which then lead to secondary changes in the bone and joints, as well as possibly secondary changes in other organ systems.

CLINICAL FEATURES

The workup of a child with multiple congenital contractures is complex because there can be so many possible causes. Box 7.1 outlines some of the important differential features. The history (including family, prenatal, pregnancy, and delivery histories) may give clues to the particular type of arthrogryposis. The newborn examination is extremely helpful in defining whether there are other problems in addition to the contractures, whether the contractures

BOX 7.1 Clinical Evaluation

History

Pregnancy (anything decreasing in utero movement leads to congenital contractures)

- Illness in mother, chronic or acute (diabetes, myasthenia gravis, myotonic dystrophy, etc.)
- Infections (rubella, rubeola, coxsackie, enterovirus, Akabane, etc.)
- Fever (>39°C, determine timing in gestation)
- Nausea (viral encephalitis, position of baby, etc.)
- Drugs (curare, robaxin, alcohol, dilantin, addictive drugs, misoprostol, etc.)
- Fetal movement (polyhydramnios, fetal kicking in one place, "rolling," decreased)
- Oligohydramnios, chronic leakage of amniotic fluid
- Polyhydramnios, hydrops
- Trauma during pregnancy (blow to the abdomen, attempted termination, car accident, etc.)
- Other complications during pregnancy such as bleeding, abnormal lie, threatened abortion, etc.
- Prenatal diagnosis (early amniocentesis, ultrasound studies, etc.)

Delivery History

- Presentation (breech, transverse, etc.)
- Length of gestation
- Traumatic delivery (limb position, CNS, fracture, etc.)
- Intrauterine mass (twin, fibroid, etc.)
- Abnormal uterine structure or shape
- Abnormal placenta, membranes, or cord length or position
- Time of year, geographic location

Family History

- Marked variability within family
- Change with time – degenerate vs. improve
- Increased incidence of congenital contractures in second- and third-degree relatives
- Hyperextensibility or hypotonia present in family member
- R/O myotonic dystrophy, myasthenia gravis in parents (particularly mother)
- Consanguinity
- Advanced parental (mother or father) age
- Increased stillbirths or miscarriages
- If more than one consecutively affected child, consider maternal antibodies to fetal neurotransmitter

Newborn Evaluation

Description of Contractures

- Which limbs and joints
- Proximal vs. distal
- Flexion vs. extension
- Amount of limitation (fixed vs. passive vs. active movement)
- Characteristic position at rest
- Severity (firm vs. some give)
- Complete fusion or ankylosis vs. soft tissue contracture

Other Anomalies (contractures are most obvious, look carefully for other anomalies)

- Deformities
  - Genitalia (cryptorchid, lack of labia, microphallus, etc.)
  - Limbs (pterygium, shortening, webs, cord wrapping, absent patella, dislocated radial heads, dimples, etc.)
  - Jaw (micrognathia, trismus, etc.)
  - Facies (asymmetry, flat bridge of nose, hemangioma, movement, etc.)
  - Scoliosis and kyphosis (fixed or flexible)
  - Dimples (over specific joints or bones)
  - Skin (hemangioma, defects, hirsutism)
  - Dermatoglyphics (absent, distorted, crease abnormalities, etc.)
  - Hernias, inguinal and umbilical, abdominal wall defect
- Other features of fetal akinesia sequence:
  - Intrauterine growth restriction
  - Pulmonary hypoplasia
  - Craniofacial anomalies (hypertelorism, cleft palate, depressed tip of nose, high bridge of nose)
  - Functional short gut with feeding problem
  - Short umbilical cord
- Malformations
  - Eyes (small, corneal opacities, malformed, ptosis, strabismus, etc.)
  - CNS (structural malformation, seizures, MR, etc.)
  - Palate (high, cleft, submucous, etc.)
  - Limb (deletion anomalies, radioulnar synostosis, etc.)
  - GU (structural anomalies of kidneys, ureters, and bladder)
  - Skull (craniosynostosis, asymmetry, microcephaly, etc.)
  - Heart (congenital structural anomalies vs. cardiomyopathy)

(Continued)

BOX 7.1 (Continued)

- Lungs (hypoplasia vs. weak muscles or hypoplastic diaphragm)
- Tracheal and laryngeal clefts and stenosis
- Changes in vasculature (hemangiomas, cutis marmorata, blue cold distal limbs, etc.)
- Other visceral anomalies

Other Features

Neurologic examination (detailed)

- Vigorous vs. lethargic
- Deep tendon reflexes (present vs. absent, slow vs. fast)
- Sensory intact or not

Muscle

- Mass (normal vs. decreased)
- Texture (soft vs. firm)
- Fibrous bands
- Normal tendon attachments or not
- Changes with time

Connective tissue

- Skin (soft, doughy, thick, extensible)
- Subcutaneous (decreased fat, increased fat)
- Hernias (inguinal, umbilical, diaphragmatic or eccentric)
- Joints (thickness, symphalangism, etc.)
- Tendon attachment and length

Course

Changes with Time

- Developmental landmarks (motor vs. social and language)
- Growth of affected limbs
- Progression of contractures
- Lethal vs. CNS damage vs. stable vs. improvement
- Asymmetry (decreases or progresses)
- Trunk vs. limb changes
- Intellectual abilities
- Socialization
- Feeding problems

Response to Therapy

- Spontaneous improvement
- Response to physical therapy
- Response to casting
- Which surgery at which time
- Development of motor strength proportionate to limb size
- Abnormal reaction to drugs

Laboratory Evaluation

Tests

Documentation of range of motion and position with photographs

Radiographs if:

- Bony anomalies (gracile, fusions, extra or missing carpals and tarsals, etc.)
- Disproportionate
- Scoliosis
- Ankylosis
- Dislocation (hips, radial head, patella, etc.)

CT scan, MRI, or functional MRI to evaluate CNS or muscle mass obscured by contractures

Ultrasonic evaluation of CNS, other anomalies, or to establish potential muscle tissue

Chromosome studies or CGH if:

- Multiple system involvement
- CNS abnormality (eye, microcephaly, MR, lethargic degenerative)
- Streaky or segmental involvement
- Consider fibroblasts if lymphocytes were normal and patient has MR with no diagnosis
- Gene testing if fits known disorder in which gene testing available

Video of movement including facial, range of movement, strength; repeat at regular intervals

Viral culture ± specific antibodies or IgM levels in newborn

Muscle biopsy in normal and affected areas at time of surgery to distinguish myopathic from neuropathic (do special histopathology and electron micrographic studies; if CPK or unusual muscle response, do muscle biopsy earlier, examine mitochondria)

EEG if seizures or episodes

EMG in normal and affected area

Nerve conduction in normal and affected area

CPK if:

- Generalized weakness
- Doughy or decreased muscle mass
- Progressively worse

Eye examination (opacities, retinal degeneration, etc.)

Maternal antibodies to neurotransmitters, if myasthenia gravis or recurrent affected pregnancies without diagnosis

Spinal muscular atrophy (SMN) DNA testing if accompanying hypotonia

Mitochondrial DNA if other suggestions of mitochondropathy

Metabolic screening

Autopsy

Visceral anomalies

CNS—brain neuropathology

Spinal cord (number and size of anterior horn cells, presence or absence of tracts at various levels)

Ganglion, peripheral nerve

Eye (neuropathology)

Muscle tissue from different muscle groups (EM & special stains, R/O ragged red fibers)

Diaphragm for thickness or hernia

Fibrous bands replacing muscle

Cartilagenous or bony fusion

Tendon attachments

Other malformations, deformations, or disruptions

CGH array if multiple congenital anomalies

Save DNA for molecular testing

Abbreviations: CGH, comparative genomic hybridization; CNS, central nervous system; CPK, creatine phosphokinase; CT, computed tomography; EM, electron microscopy; EMG, electromyography; GU, genitourinary; IgM, immunoglobulin M; MR, mental retardation; MRI, magnetic resonance imaging; R/O, rule out.
Adapted from Hall.[2]

involve all four limbs, whether they are distal or proximal, and whether the positioning is characteristic (as in the case of Amyoplasia and the distal arthrogryposes).

It is very useful to separate patients into three groups once the initial evaluation has been done in order to develop a differential diagnosis: (1) those with primarily limb involvement; (2) those with limb plus other body areas; and (3) those with severe CNS structural abnormalities or dysfunction. Box 7.2 gives a differential diagnosis of specific entities within each of these categories.[2]

Case Examples 7.1, 7.2, and 7.3 and Figures 7.1, 7.2, and 7.3 illustrate the variety of problems that present as arthrogryposis. The three case examples are used to illustrate categories I, II, and III. Table 7.1 provides differential diagnoses for each of these categories. It should be noted that a number of chondrodystrophies may present with congenital contractures, but the bony abnormalities usually make themselves obvious on radiography or because of the disproportion of the limbs. A number of maternal illnesses are associated with the birth

BOX 7.2 Various Specific Types of Arthrogryposis with Their Known Genes

Primarily Limb Involvement

Absence of finger prints [4q22.3 (*SMARCADI*)]

Bruck syndrome [17p12 (TLH) (PLOD2), 17q21.2 (FKBP10) and 3q23-24 linkage]

Clasped thumbs [1q25-q31(PRG4)]

Continuous contractures [12p13.3 (*KCNA1*)]

Distal arthrogryposis types 1 and 2B [9p11.3 (TPM2), 11p15.5 (TNNI2),11p15.5 (TNNT3), 17p13MYH3 12q23.2 (MYBPC1)]

Lower limb DA—Fleury type [12q23-24 (*TRPV4*)]

Mesomelic dysplasia [Xq24-32]

Patella aplasia-hypoplasia [17q21-22 (*PTLAH*)]

Symphalangism/synostosis, coalitions [17q21-22 (NOG), 20q (GDF5)]

X-linked arthrogryposis, lower limbs only [Xq23-27]

Musculoskeletal Involvement Plus Other System Anomalies

Camptodactyly-arthropathy–coxa vara–pericarditis syndrome [(*ARC*) 1q24-q25 (*PRG4*)]

Congenital fiber disproportion [1q42-13 (*ACTA1*)]

Conradi-Hunermann [(Xp11.23) (EBP)]

Contracture arachnodactyly [5q23-31 (*FBN2*) 12q13]

Craniosynostosis syndrome [4p16.3 (FGFR3), 8p11.23 (FGFR1)]

Deafness, camptodactyly (DA 6) [11q25 (*CATSAL, CATSHL*)]

Diastrophic dysplasia [5q32 (SLC26A2)]

Digital arthrogryposis I [9q21.2 (*TNNT3*), 11p15.5 (*TNN12, TNNT3*), and 17p13 (*MYH3*)]

Distal arthrogryposis type 2B (Sheldon-Hall) [9q21.2 (TNNT3), 11p15.5 (TNN12, TNNT3), and 17p13 (MYH3)]

Dundar-Sonoda syndrome TARP [15q15.1 (*CHST14*)]

Ehlers-Danlos VI [2q32.2 (COL3A1)]

Ehlers-Danlos VIII [12p13]

Ehlers-Danlos VIB-2 [15q15 (*CHST14*)]

Epidermolysis bullosa dystrophica [3p21.31 (COL7A1)]

Freeman-Sheldon syndrome (DA 2A) [17p13 (*MYH3*)]

Genitopatallar syndrome [10q22.1 (KAT6B)]

Gordon syndrome (DA IIA, DA3) [17p13.1 (*MYH3*)]

Hand-foot-uterus syndrome [7p15-p14.2 (HOXA13)]

Holt-Oram syndrome [12q24.1 (*TBX5*)]

Inclusion-body myopathy [17p13.1 (MYH2, MYH3)]

Kniest dysplasia [12q13.1 (*COLA21*)]

King-Denborough syndrome/multiple pterygium and malignant hyperthermia [Genes for malignant hyperthermia in 19q13.1 (RYR1) and 17q (MHS3)]

Larsen syndrome [3p14.3 (*FLNB*), 3p21.31 (*COL7A1*), AR 8q12.1 (*IMPAD*), 17q25.3 (*CANT*), 5q32 (*DTDST*)]

Marfan syndrome (severe neonatal) [15q21.1 (FBN1), 3p24.2-p25?]

MASP1 (3M, Carnevale) [3q27.3 (*MASP1*), 2p25.3 (*COLEC11*)]

Metaphyseal dysplasia (Jansen) [3p21.31 (PTH1R)]

Metatrophic dysplasia [12q24.1 (*TRPV4*)]

Mobius syndrome [13q12.2-q13]

Moore-Weaver [11p15 (*TNN12*)]

Multiple pterygium (Escobar type) [2q31.1 (CHRNA1), 2q37.1 (CHRNG, CHRND)]

Multiple synostosis, symphalangism [17q21-22 (*NOG*), 20q11.22 (*GDF5*), 13q11 (*FGF9*), 4p16.3 (*DOK7*)]

Myelinopathies [1q23.3 (MPZ), 10q21.1 (ERG2), 17p12 (PMP22), 19q13.2 (PRX)]

Myopathies [22q12.3 (*LARGE*), 6q25 (SYNE1), many others]

Nail-patella syndrome [9q34.1 (LMX1B)]

Nemaline myopathy [2q22 (*NEB*), 1q42.13 (*ACTA1*)]

Neurofibromatosis [17q11-q12 (NF1)]

Neuropathic Israeli-Arab (autosomal recessive) [5qter]

Nevo syndrome [1p36.22 (*PLOD1*)]

Oculo-dental-digital syndrome [6q22.31 (GJA1)]

Otospondylomegaepiphyseal dysplasia [6p21.32 (*COL11A2*)]

Parastremmatic dwarfism [12q24.1 (TRPV40)]

Pfeiffer cardiocranial syndrome [10q26 (*FGFR2*), 8p11.23 (*FGFR1*)]

Popliteal pterygium syndrome [1q32-q41 (IRF6)]

Proteus syndrome [14q32.32 (*AKT1*)]

Puretic-Murray syndrome (infantile hyalinosis) [4q21 (ANTXR2)]

Rigid spine muscular dystrophy [1q36.11 (*SEPN1*)]

Robino syndrome [3p14.3 (WNT5A)]

Sacral agenesis [7q36 (*MNX1*, Currarino syndrome)]

Schwartz-Jampel syndrome [1p36.1 (HSPG2)]

Spondyloepiphyseal dysplasia congénita [12q13.1 (*COLA21*)]

(*Continued*)

BOX 7.2 (Continued)

Stiff man syndrome [5q33.1 (GLRA1), 4q32.1 (GLRAB)]
Trismus-pseudocamptodactyly [17p12-p13.1 (*MYH8*)]
Tuberous sclerosis [9q (TSC1) and 16q (TSC2)]
Ullrich congenital muscular dystrophy [2q37 (*COL6A3*), 21q22.3 (*COL6A1*), 21q22.3 (*COL6A2*)]
VATER Association [2q31.1 (HOXD13)]
Van den Ende-Gupta syndrome [22q11.21 (*SCARF2*)]
Waardenburg-Klein [2q25-q37.12 (PAX3)]
Weill-Marchesani syndrome [19p13.3-13.2 (*ADAMS10*)]
Winchester syndrome [16q13 (MMP2)]
X-linked moderately severe [Xp11.3-q11.2]

Muscoloskeletal Involvement Plus Central Nervous System Dysfunction and/or Mental Retardation

Acrocallosal syndrome [7p13 (*GLI3*), 15q26.1 (*K1F7*)]
Adenyl succinate lyase deficiency [22q13.1-13.2 (ADSL)]
Aicardi-Goutierres syndrome [3p21 (*TREX1*), 11q13.2 (*RNASEH2C*), 13q14.3 (*RNASEH2A*), 19p13.13 (*RNASEH2A*), 20Q11.23 (*SAMHD1*)]
Al-Awadi syndrome [3p25 (WNT7A)]
Antley-Bixler syndrome [7q11.2 (*POR*), 10q26 (*FGFR2*)]
ARC (Nezelof syndrome) [15q26.1 (VPS33B), 14q24.3 (VIPAR)]
Bartsocas-Papas syndrome [20Q11.21 (*ASXL1*)]
Bohring-Opitz syndrome [21q22.3 (RIPK4)]
Bowen-Conradi syndrome [12p13.3 (*EMG1*)]
Camptomelic dysplasia [17q24.3-q25.1 (SOX9)]
Carbohydrate deficient glycoprotein syndrome [16p13 (*PMM2*), 15q22-qter (*PM11*)]
Central core disease [19q13.1 (RYR1)]
Cerebro-oculo-facial-skeletal (COFS) syndrome [19q13.2 (*ERCC1*), 13q33.1 (*ERCC5*), 10q11.23 (*ERCC6*), 19q13.32 (*ERCC2*)]
Chondrodysplasia punctata (rhizomelic) [6q22-q24 (PEX7)]
Clasped thumb & mental retardation [Associated with X-linked hydrocephaly]
Cocoon syndrome [10q24.31 (*CHUK*)]
Congenital muscular dystrophy (merosin deficient) [6q22-23 (LAMA2)]
Contractural arachnodactyly [5q23-31 (*FEN2*)]
Crisponi syndrome [19p13.11 (CRLF1)]
Dandy-Walker, MR, basal ganglion seizures [Xq25-27]
Dyggve-Melchior-Clausen syndrome [18q12-q21.1 (*DYM*)]
Episodic ataxia/myokymia syndrome [12p13.32 (KCNA1)]
FG syndrome [Xq13 (*MED12*), Xq28 (*FLNA*)]
Fowler type hydranencephaly [14q24.3 (FLVCR2)]
Fukuyama congenital muscular dystrophy [9q31.2 (*FKTN*); 14q24.3 (*POMPT2*), 9q31-33 (*FCMD*), 19q13.2 (*FKRP*)]
Gaucher disease—perinatal lethal [1q22 (GBA)]
Geleophysic dysplasia [9q342 (*ADAMTSG2*), 15q21.1 (*FBN1*)]
Glycogen storage IV [9q34 (GLE1)]
Leprechaunism [19p13.2 (*INSR*)]
Lethal arthrogryposis with AHC disease [9q34.11 (GLE1)]
Lethal congenital contractures syndrome (Finnish) I [9q34.11 (*GLE1*)]
Lethal congenital contractures syndrome (Finnish) II [12q13 (ERBB3)]
Lethal congenital contractures syndrome (Finnish) III [19p13.3 (*PIP5K1C*)]
Lissencephaly with FADS [7q22.1 (RELN), 17p13.3 (PAFAH1B1)]
Martsoff syndrome [1q41 (*RAB3GAP2*), 2q21.3 (*RAB3GAP1*), 10p12.2 (*RAB18*)]
MASA syndrome [Xq28 (L1CAM)]
Miller-Dieker syndrome [17q13.2 (*L1S1*)]
Multiple pterygium syndrome—lethal [2q31.1 (CHRNA1), 2q37.1 (CHRND; CHRNG), 22q12.3 (LARGE)]
Muscular dystrophy, congenital, rigid spine [1p35-36 (*SEPN1*)]
Myasthenia, congenital [11p11.2-p11.1 (RAPSN), 17p12-p11 (CHRNB1), 17p13-p12 (CHRNE)]
Myelinopathies [1q23.3 (*MP2*), 10q21.1 (*ERG2*), 17p12 (*PMP22*), 19q13.2 (*PRX*)]
Myopathy centronuclear [2q14.2 (BIN1)]
Myotonic dystrophy [19q13.3 (*DMPK*)]
Neuromuscular disease of larynx [5q31]
Nevo syndrome
Oculodentodigital syndrome [6q21-23.2 (*GJA1*)]
Osteogenesis imperfecta [7q21.3-q22.1 (COL1A2), 17q21.23 (COL1A1)]
Otopalatodigital syndrome type II [Xq28 (*FLNA*)]
Pfeiffer syndrome [10q26 (FGFR2), 8p11.23 (FGFR1)]
Phosphofructokinase deficiency infantile (glycogen storage VII) [12p13.3 (*PFKM*)]
Potter syndrome [11q11.2 (RET), 22q13.31 (UPK3A)]
Raine syndrome [7q22.3 (*FAM20C*)]
Renal adysplasia [10q11.21 (RET), 22q13.31 (UPK3A)]
Restrictive dermopathy [4p16.3 (*DOK7*), 11p11.2 (*RAPSN*), 1p34.2 (*2MPSTE24*), 1q22 (*LMNA*)]
Robert syndrome [8p21.1 (ESCO2)]
Schinzel-Gideon syndrome [18q21.1 (*SETBP1*)]
Schwartz-Jampel syndrome [1p36.12 (HSPG2)]
Smith-Lemli-Opitz syndrome [11q12-q13 (*DHCR7*)]
Spinal muscular atrophy [5q13.2 (SMN)]
Spastic paraplegia (Goldblatt syndrome) [Xq21-q22]
Spinal muscular atrophy [5q13.2 (*SMN*)]
Spastic paraplegia (Goldblatt syndrome) [Xq21-q22]
Spinal muscular atrophy [5q13.2 (SMN)]
Trigonocephaly (C syndrome) [3q13.3-q13.2 (*CD96*)]
Waardenburg Shah syndrome [22q13.1 (SOX10)]
Walker-Warburg syndrome [3p22.1 (*GTDC2*)]
Weaver syndrome [5q35 (NSD1)]
X-linked lethal arthrogryposis 1 [Xp11.3-q11.2 (*UBE1*)]
Zellweger syndrome [Multiple PEX genes]
5q deletions (spinal muscular atrophy) [5q13.3 (SMN) microdeletions]

Abbreviations: DA, distal arthrogryposis; MR, mental retardation
Adapted from Hall.[2]

CASE EXAMPLE 7.1

This was the third pregnancy for a 28-year-old mother and 29-year-old father with no family history of any congenital abnormalities or consanguinity. In the two previous pregnancies, the resulting children were perfectly normal. The mother noted somewhat unusual fetal movement during the third trimester in that the baby seemed to kick mainly in the left lower part of her abdomen. At 32 weeks of pregnancy, the baby was noted to be in breech position and an ultrasound was done. A question of clubfeet was raised. A second, more extensive ultrasound study demonstrated, in addition to clubfeet, that the knees remained flexed and the elbows remained extended. No other structural abnormalities of viscera were noted and the pregnancy progressed normally aside from unusual fetal movement. At term, cesarean section was required when the attempts at converting to a vertex delivery were unsuccessful.

At birth, multiple anomalies were noted (Figure 7.1). However, the baby cried vigorously, had borderline low weight and normal head size (the length was difficult to measure because of contractures), and nursed well immediately. The abnormalities of limbs included internal rotation of the shoulders, fixed extension contractures of the elbows, flexion contractures of the wrists, and bilaterally symmetrical camptodactyly of the fingers. Both hips were dislocated and the hips were held in flexion and adducted. The knees were in 90 degrees fixed flexion contractures and both feet were fixed in severe equinus position. The baby was noted to have a large flat vascular birthmark over the forehead and midface as well as on the back of the neck. Physical therapy and stretching were begun in the nursery with splinting of the wrists. There was very little muscle mass. The baby went home at 3 days feeding well. At 3 months of age, a muscle biopsy of the quadriceps revealed fatty, fibrous change and no muscle tissue. Casting of feet resulted in some improvement in position and physical therapy increased the range of motion of elbows and knees by 20 degrees. The diagnosis of Amyoplasia was made because of the unique positioning of the elbows, the findings on muscle biopsy,[10,19,20] and the alert, vigorous responses. The child had surprisingly good response to physical therapy, but required multiple surgical procedures in order to be ambulatory. The family considered her to be brighter than the other children in the family. It is anticipated that she will be able to live an independent life. The family was counseled that this was a specific type of arthrogryposis known as Amyoplasia. It is considered a sporadic condition without a significant recurrence risk to parents or the affected child.

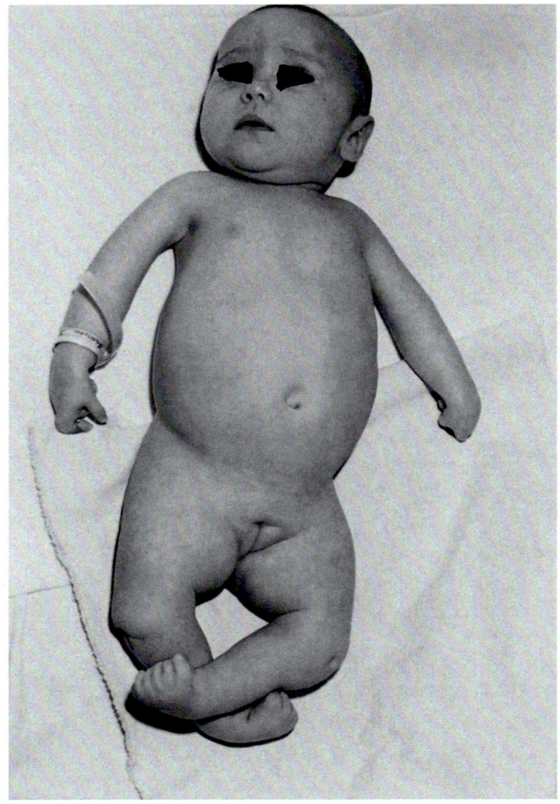

FIGURE 7.1 Baby with Amyoplasia; note fixed extension deformity of elbow and severe equinovarus deformity of feet.

of children with arthrogryposis.[14] Specific body areas may be involved (e.g. skin, eye, or heart abnormalities) that can give clues for particular types of arthrogryposis. Specific physical findings such as trismus, pterygia, or particular positioning of the contracture in the newborn period may also help to arrive at a specific diagnosis. If multiple organ systems are involved, comparative genomic hybridization (CGH) array studies should be done, as chromosomal abnormalities can be associated with congenital contractures. Many cases of chromosomal mosaicism have been observed in arthrogryposis with developmental delay. If blood chromosome studies do not reveal mosaicism for a chromosome abnormality, it may be worthwhile to do fibroblast chromosomal studies.[26]

GENOTYPE/PHENOTYPE CORRELATION

With the Human Genome Project completed, it may be possible to find the specific gene responsible for or predisposing to familial types of arthrogryposis. Many specific genes associated with multiple congenital contractures have been mapped and sequenced.[2,27] However, the actual pathogenetic mechanisms are just beginning to be elucidated (e.g. distal arthrogryposis and sarcolemmic

CASE EXAMPLE 7.2

This was the third child born to a 32-year-old mother and 33-year-old father. Their first child had been born with multiple congenital contractures as well as a cleft palate and fusion of some cervical vertebrae (Figure 7.2A). No consanguinity was known. The family had been told there might be some increased risk for recurrence and consequently had had careful, real-time prenatal ultrasound studies done starting at 16 weeks of pregnancy. The initial ultrasound suggested flexion of elbows and extension of knees, but otherwise normal structures and no cleft palate. On ultrasound at 22 weeks, a suggestion of bilateral vertical talus was seen, as were the contractures of the elbows and knees. However, some movement was seen in both areas. The mother noted that the baby kicked less than during her second pregnancy, which had produced a normal girl. At 38 weeks' gestation, the mother ruptured membranes and went into labor. The baby was born head first and was noted at birth to be mildly hirsute with flexion contractures of hips and elbows, extension contractures of knees, and clenched hands and vertical talus with rocker bottom feet (see Figure 7.2B). There was no cleft of the palate, and her contractures were somewhat milder than her brother's. She was a normal weight and length with normal occipitofrontal circumference. Radiography showed two areas of vertebral fusion, but no scoliosis. She was vigorous, fed well, and went home at 3 days of age. Physical therapy improved her range of motion and she walked at 14 months. Over time, she developed increasing webbing around the neck, axilla, and groin area. Both she and her brother were noted to have a somewhat myopathic facies with mild micrognathia and antimongoloid slant to the eyes and mild ptosis. The child had good response to physical therapy. In this case, the diagnosis of multiple pterygium syndrome of the Escobar type was considered; molecular testing was done because of recurrence.[21,22] It had not been available earlier. Two different mutations were found in *CHRND*, which encodes the fetal subunit of the neuromuscular junction acetylcholine receptor, confirming the clinical impression.[23–25] The finding of vertebral fusion was helpful in suggesting the diagnosis, as was the presence of two affected children with normal parents. This type of pterygium syndrome progresses with webs or pterygia more evident over time, particularly those at the neck and axilla. The hands are usually very functional without surgery; however, they usually continue to have some camptodactyly. Kyphosis and scoliosis may be progressive at adolescence.

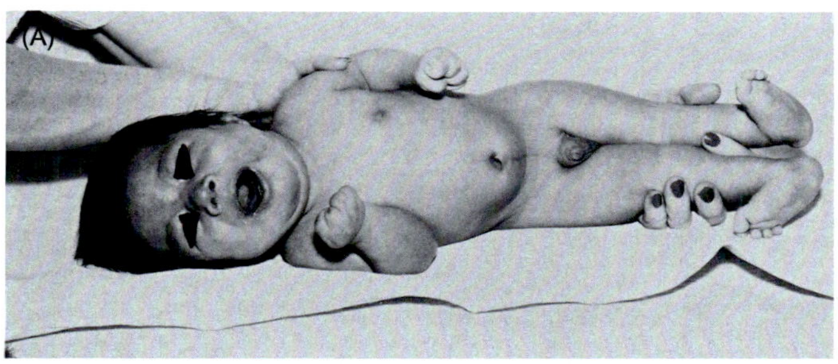

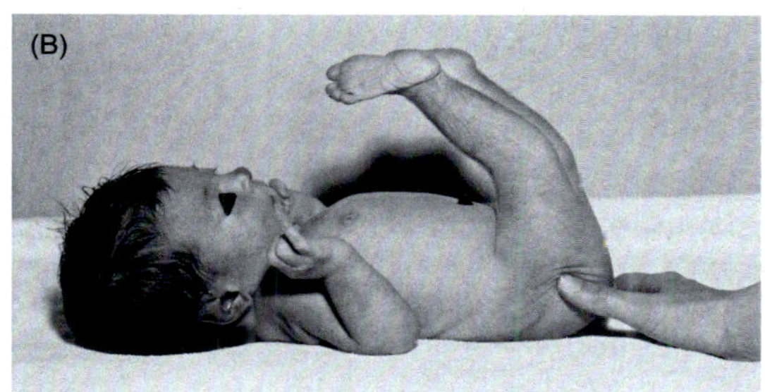

FIGURE 7.2 (**A** and **B**) Two siblings with multiple pterygium syndrome of the Escobar type. Note multiple contractures and lack of severe webbing in the newborn period. Both do have mild ptosis and mild myopathic facies.

proteins; multiple pterygium syndromes and embryonic receptor pathway). Nevertheless, we can anticipate that multiple pathways will be defined over the next decade. Approximately 5% of cases of arthrogryposis are secondary to myopathies.[28] Obviously, in those cases, muscle biopsy is extremely useful. Abnormalities of muscle texture, myopathic facies, and weakness in spite of normal muscle mass all suggest a myopathic workup with electromyography and muscle biopsy, keeping in mind mitochondrial disease.

Many cases of arthrogryposis will appear to have a neuropathic basis in that there are nonspecific muscle changes suggesting anterior horn cell loss as well as neuropathic nerve conduction studies. In general, histopathology has not been very useful. Some mild fiber disproportion is often seen, but only rarely is a specific finding such as central cores, nemaline rods, or ragged red fibers found. However, if a child is going to surgery, performing a muscle biopsy with appropriate stains and electron microscopic studies is certainly worthwhile for documentation of normal and abnormal muscle.

As mentioned earlier, if anomalies are seen in multiple systems and a patchy distribution of changes is observed, chromosomal abnormalities using CGH array should be considered. If blood chromosome studies are normal, fibroblast chromosome studies maybe considered.[26] Gene testing is appropriate for many distal arthrogryposis disorders, those with pterygium (see stars in Box 7.2), and for all lethal conditions. Over 130 specific genes have been identified in different disorders with multiple congenital contractures.[2]

CASE EXAMPLE 7.3

This was the second child born to a 24-year-old mother and 24-year-old father. The first child was perfectly normal. There is no past history of problems, family history of any abnormalities, or consanguinity. The mother had some spotting at 12 weeks. A 16-week ultrasound was said to be normal, but by 20 weeks' gestation, the mother's abdomen was increasing in size and polyhydramnios was noted. Mild intrauterine growth restriction was also present, but no structural abnormalities were described. By 26 weeks' gestation, the head size had fallen to the 10th percentile and fetal movement seemed decreased. Although there was some movement of the limbs, both arms and legs seemed to be held in flexed position and the feet were clubbed. By 28 weeks' gestation, the polyhydramnios had increased.

Amniocentesis revealed that the lungs were immature and chromosomes were normal. At 32 weeks' gestation, the mother went into labor. She was given tocolytics and steroids. Two days later, her labor became even more active and her membranes ruptured. She vaginally delivered a boy who had multiple flexion contractures and clubbed feet as well as intrauterine growth restriction, mild hirsutism, and an Apgar score of only 4. Resuscitation was difficult because the lungs were very stiff. The baby's height and weight were at the third percentile and occipitofrontal circumference was below the third percentile (Figure 7.3). An ultrasound showed no specific abnormalities, but the brain was thought to be small. The baby could not breathe independently and on the third day had a cardiac arrest and could not be resuscitated. At autopsy, the child was found to have pulmonary hypoplasia, short gut, and a small brain with no other specific structural abnormality. The family was told that this was a nonspecific, neuropathic type of arthrogryposis with features of fetal akinesia deformation sequence (FADS). With CNS abnormalities or dysfunction and prenatal demise, the empiric recurrence risk would be 15% for subsequent pregnancies even though a specific diagnosis could not be achieved.[2,8]

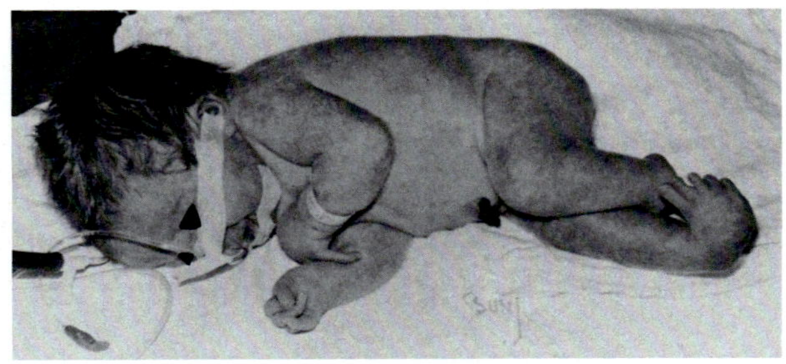

FIGURE 7.3 Note multiple flexion contractures, clubbed feet, and mild hirsutism. By measurements at birth, the baby has intrauterine growth restriction and microcephaly. The small chest reflects pulmonary hypoplasia.

ARTHROGRYPOSIS ASSOCIATED WITH MATERNAL ILLNESSES

A number of maternal illnesses and exposures are associated with children who have multiple congenital contractures (Box 7.3). For this reason, attempting to make a diagnosis in a mother is appropriate if she has any symptoms of disease. The disorders may be associated with substances or physical conditions that are detrimental to the fetus (such as maternal acidosis), or be evidence of hereditary conditions (such as myotonic dystrophy). In some cases, such as maternal multiple sclerosis, there are only a few case reports and the connection is not altogether clear. Perhaps the most important of the known maternally related abnormalities is the presence of maternal antibodies to fetal acetylcholine receptors. This is a particularly important consideration because of the possibility of treatment, and it is discussed later.

THERAPY

The type of therapy provided for the child with multiple congenital contractures depends very much on the specific diagnosis. For most types, physical therapy mobilizing contractures in the newborn period is extremely important. It appears that during the first 3 to 4 months, an excellent response to physical therapy often occurs. Thereafter, surgical intervention is often needed. However, there are certain types of congenital contractures, such as those seen in diastrophic dysplasia, that actually get worse with physical therapy. The cartilage in diastrophic dysplasia seems to be particularly sensitive to damage and actually calcifies and fuses with vigorous physical therapy. Thus, a specific diagnosis should be made if at all possible when considering treatment. On the other hand, the best response to physical therapy is during the first few months, so time should not be lost. In the past, individuals with arthrogryposis were generally casted and operated on in the first few years of their lives, and as a consequence, whatever muscle they did have often atrophied. A cautious combination of stretching, casting, and splinting is encouraged.[1] A number of new techniques for orthopedic surgery seem to have remarkably improved the outcome of these children,[29] and excellent reviews of therapy are available in the orthopedic literature.[10,29] The importance of splinting after surgery cannot be overemphasized. The contractures of most individuals will return to their previous position without conscientious night splinting. Approaches to physical therapy, occupational therapy, and schooling are well covered in the book *Arthrogryposis: A Text Atlas*.[1]

One consideration in terms of therapy is the potential for *in utero* therapy. In familial cases of arthrogryposis,

TABLE 7.1 Distal Arthrogryposes Differential Diagnosis with Additional Types of Distal Arthrogryposis (not including camptodactyly syndromes)

Hall Classification	Description	Bamshad Classification	Genes
I	1. Distal (AD, simplex) some families mainly involve hands, includes digitotarsal dysmorphism	1A	*TPM2, MYBPC1, TNN12, MYH3*
IIA	2. Gordon syndrome (AD, short stature ± CP)	3	*PIEZO2*
IIB	3. AD or AR with eye involvement such as ophthalmoplegia and/or ptosis – may not be congenital; one family included retinal pigment changes and MR; may have Dandy-Walker in family	5	*ECEL1, PIEZO2*
IIC	4. Clefting, AD, Trismus, may be Gordon syndrome		
IID	5. Scoliosis (may include Goodman syndrome)	4	
IIE	6. Trismus and unusual hand ~ Amyoplasia spectrum (See Figure 7.5)		
	7. Freeman-Sheldon syndrome	2A	*MYH3*
	8. Sheldon-Hall (includes Moore-Weaver syndrome)	2B	*TNNT3, TNN12, MYH3, TPM2*
	9. Sheldon-Hall look alike	2C	
	10. Deafness, Camptodactyly syndrome	6	*FGFR3*
	11. Trismus Pseudocamptodactyly (Beal's syndrome, Kentucky Dutch syndrome)	7	*MYH8*
	12. AD, Multiple pterygium	8	
	13. Contractural arachnodactyly	9	*Fibrillin 2*
	14. Distal with plantar flexion contractures	10	
	15. Dundar Sonoda syndrome (MR, unusual face)		
	16. Absent teeth and Abnormal facies (may be Sheldon-Hall)		
	17. Chitayat syndrome (AR, MR, hypopituitary)		
	18. X-linked distal arthrogryposis		
	19. Shalev (MacMillan) syndrome (AR, mainly upper limbs, ptosis)		

Abbreviations: AD, autosomal dominant; AR, autosomal recessive; CP, cleft palate; MR, mental retardation.
Adapted from Hall.[2]

there can be remarkable differences between affected siblings regarding how severely they are involved. This may reflect modifying genes, but one new approach taken is to try to do "physical therapy" *in utero*. This means getting the baby to move, stretch, and extend joint movement *in utero*. It turns out that there are three or four things that increase fetal movement. The first is deep breathing by the mother, which leads to increased movement in the fetus. The second is exercise by the mother, which leads to increased movement in the fetus. The third is caffeine, which leads to increased movement of the fetus, but should not be considered in the first trimester, and thereafter, with careful discussion with mother's obstetrician weighing the pros and cons. A drink containing caffeine could be taken by mothers who are at risk for having a child affected with arthrogryposis. These mothers may consider doing some type of moderately strenuous exercise each day such as vigorous walking; stop 5 to 6 times a day and take 10 to 20 deep breaths; and, although we would not recommend large amounts of caffeine, having a drink containing caffeine (e.g. coffee, tea, soft drink, etc.) two or three times a day after the first trimester as long as it is tolerated.

Interestingly, babies affected by multiple congenital contractures who are delivered early (if lungs are mature) may have a better outcome because physical therapy can be started early. The contractures seem to get worse the longer the baby remains *in utero*. However, early delivery must obviously be weighed against maturation of lung and the maturity and involvement of the infant in other ways.

PROGNOSIS

Prognosis in arthrogryposis clearly depends on the specific type of arthrogryposis. Many forms have an excellent outcome. In Amyoplasia,[10] the most common type of arthrogryposis, intelligence is usually normal or superior and the ability to live independently is very much dependent on whether the affected individual can function in an independent way in terms of mobility and self-care. Many individuals with pterygium syndromes, Freeman-Sheldon syndrome, and distal arthrogryposis respond extremely well to rehabilitation measures and function completely independently. Other forms of arthrogryposis that involve the CNS with structural or functional abnormalities may leave the affected individual quite dependent and often involve early death. Thus, making a specific diagnosis and determining the natural history of that form are extremely important in terms of determining the prognosis and possible therapies. It is often difficult in the newborn period to tell whether an individual seems to be responsive and has the necessary motor power to be independent. Usually by 2 years of age, it becomes clear whether an affected individual has the motivation and the musculoskeletal involvement that will allow independence. Usually by 4 years of age, quite a reasonable overview of where things are headed can be established.

To many families that have affected individuals, the question of recurrence risk is important. If a specific diagnosis can be made, the recurrence risk can usually be determined. However, if no specific diagnosis can be made, empirically, it has been observed that there is a 3% to 5% recurrence risk both for the parents of the affected child and for the affected individual himself or herself to have a child affected with the same type of arthrogryposis.[8] Prenatal diagnosis in subsequent pregnancies should be offered to all families who have had a child with arthrogryposis. It is essential to use real-time ultrasound evaluation of movement, and this should be offered every 2 to 4 weeks starting at 14 weeks' gestation. Often, the diagnosis can be made already at 14 weeks' gestation from lack of normal movement. However, sometimes lack of movement does not actually occur until 24 or 26 weeks' gestation. If congenital contractures are observed *in utero*, as noted earlier, certain kinds of activities may improve the outcome.

BOX 7.3 Maternal Problems Associated with Children Who Have Multiple Congenital Contractures

- Abnormal uterus shape or structure
- Attempted termination of pregnancy and/or trauma
- Maternal acidosis
- Maternal antibodies against fetal neurotransmitter receptor subunits
- Maternal fluconazole ingestion
- Maternal hyperthermia
- Maternal hypothermia
- Maternal misoprostol ingestion or vaginal suppositories
- Maternal myasthenia gravis
- Maternal multiple sclerosis
- Maternal muscle relaxant
- Myotonic dystrophy
- Prenatal diagnosis—early amniocentesis or chorionic villi sampling

RELATIVELY COMMON SPECIFIC CONDITIONS ASSOCIATED WITH MULTIPLE CONGENITAL CONTRACTURES

Mainly Limb Involvement

Amyoplasia

Amyoplasia (a = no; myo = muscle; and plasia = growth) or "classical arthrogryposis" is the most common condition with multiple congenital contractures.[10,19,30] It occurs once in approximately 10,000 live births and makes up about one-third of all affected liveborns. Most affected individuals have all four limbs symmetrically involved with the trunk spared. On biopsy, the affected muscles are seen to be hypoplastic and replaced by fatty fibrous tissue, interspersed with normal muscle, whereas unaffected muscles are normal. Upper limb involvement has a unique position with internally rotated and adducted shoulders, fixed extended elbows, pronated forearms, and flexed wrists and fingers (in a "waiter's tip hand" or "policeman's hand" position) (Figure 7.1). Lower limb involvement always involves severe talipes equinovarus deformity while the knees may be flexed or extended and the hips flexed or extended, abducted or adducted, and often dislocated. About 25% of affected individuals have only upper limbs or lower limbs involved, and rarely three limbs may be involved. Amyoplasia has a relatively good response to early physical therapy.[20] Eighty percent of individuals have no other system involvement, but many other anomalies may occur with Amyoplasia.[10,14] For example, 10% may also have an abdominal wall defect (gastroschisis or lateral wall hernias) or bowel atresia. It is also common for affected individuals to have a hemangioma-type birthmark over the midface (as do many children with limb defects) and for the face to be somewhat round and flat with mild micrognathia. In addition, approximately 14% may have amniotic bands or terminal digit reduction anomalies on one or more limbs. Scoliosis does occur rarely in severely involved individuals or at adolescence. Intelligence is usually normal, unless there has been trauma or anoxia, or both, at birth. With the advent of computers, most affected individuals can be gainfully employed. Most people with Amyoplasia can achieve walking if they have reasonable hip muscles and motivation.[20] However, many choose to get around outside of the home by using a mobile wheelchair. Amyoplasia is increased among monozygotic twins, with one unaffected twin and one affected twin.[10,30] Amyoplasia seems to be a totally sporadic disorder that has not been observed to recur in siblings or offspring (although obviously it could occur simply on a frequency basis). Amyoplasia is likely to occur secondary to some type of vascular compromise leading to fetal hypotension and hypoxia of anterior horn cells at a critical time in the first trimester.

Bony Fusions Leading to Apparent Congenital Contractures

Disorders in which the bones are fused together have often been confused with arthrogryposis. The only way to make the distinction between congenital contractures and fused bones is through radiographs.[2] Three bony fusions often confused with arthrogryposis are symphalangism (fusion of phalanges),[31] coalition (fusion of the carpals and tarsal bones), and synostosis (fusion of other bones). Symphalangism can also be associated with more generalized contractures and should be considered whenever deafness is present because the ossicles may also fuse.[2] Radioulnar and radiohumeral synostosis may be confused with congenital contractures of the elbow. Some bony fusions run in families, whereas others occur sporadically.

Contractural Arachnodactyly

Contractural arachnodactyly, also known as Beals' syndrome,[31–33] is an autosomal dominantly inherited condition associated with abnormalities of the fibrillin gene located on chromosome 5q23–q31. Affected individuals have congenital contractures, are unusually long and thin for their age, and have crumpling or overfolding of the ear. Kyphoscoliosis is often seen. In addition, affected individuals often have chest deformities, with pectus excavatum or carinatum. Although 90% of those affected improve with age, 10% seem to be severely affected with progressive scoliosis. Contractural arachnodactyly resembles Marfan syndrome, but lacks the typical cardiovascular and ocular abnormalities of Marfan syndrome and involves a different fibrillin gene. However, really severe congenital Marfan syndrome may have contractures at birth.

The Distal Arthrogryposes

The distal arthrogryposes are a heterogeneous group of disorders primarily involving the distal limbs with clubbed hands and feet. Great progress has been made in distinguishing subtypes and identifying their genes (see Table 7.1).[34,35] See Figures 7.4 and 7.5.

Distal Arthrogryposis Type I

Distal arthrogryposis type I is inherited as an autosomal dominant trait with several responsible genes (*TNN12, TNNT3, MYBPC1,* and *TPM2*),[27] and has been found to be due to mutations in sarcolemma proteins. Common features include a unique position of the hands (medially overlapping fingers, clenched fists, ulnar deviation of the fingers when they open out, and contractures of the fingers [camptodactyly]) and foot contractures (Figure 7.4A and B). These hand and foot abnormalities seem to be due to misplaced tendons. The knees and hips may also be mildly involved, and intelligence is usually normal. Patients usually have a good response to physical therapy and surgery if necessary.

Gordon Syndrome (Distal Type IIA)

Mutations in *MYH3* and *PIEZO2* are responsible for Gordon syndrome, an autosomal dominant disorder.[27] Affected individuals often have distal contractures (hand and foot involvement), short stature, and may have cleft palate.[7] They tend to have short stature (approximately 30% to 50%) compared with their relatives. The presence of cleft palate and short stature seems to be proportionate to the severity of distal contractures. Mutations in *PIEZO2* are responsible for some cases.

Distal Arthrogryposis Type 5 (IIB)

Distal arthrogryposis type 5 is heterogeneous with both autosomal dominant and recessive inheritance.[27] It consists of distal contractures, muscles that feel hard and woody, ptosis, ophthalmoplegia, thickened skin, and absence of some palmar creases. Several families have been described with male-to-male transmission. The ophthalmoplegia may develop over time, and it appears to be a progressive disorder with increasing restrictive lung involvement.

Limbs Plus Other Body Areas

Freeman-Sheldon Syndrome (Whistling Face Syndrome) and Sheldon-Hall Syndrome[36]

Freeman-Sheldon syndrome, also known as whistling face syndrome or craniocarpotarsal dystrophy, was first described by Freeman and Sheldon in 1938.[37] Normally, it occurs as an autosomal dominantly inherited trait, although it can also be seen in families consistent with autosomal recessive inheritance, in which case it tends to be more severe and at times lethal.[38] Affected individuals tend to have a masklike face with a small puckered mouth or whistling mouth appearance; high arched palate; small tongue; an H-shaped cutaneous dimpling on the chin; mid-face hypoplasia; deep-set eyes; epicanthic folds;

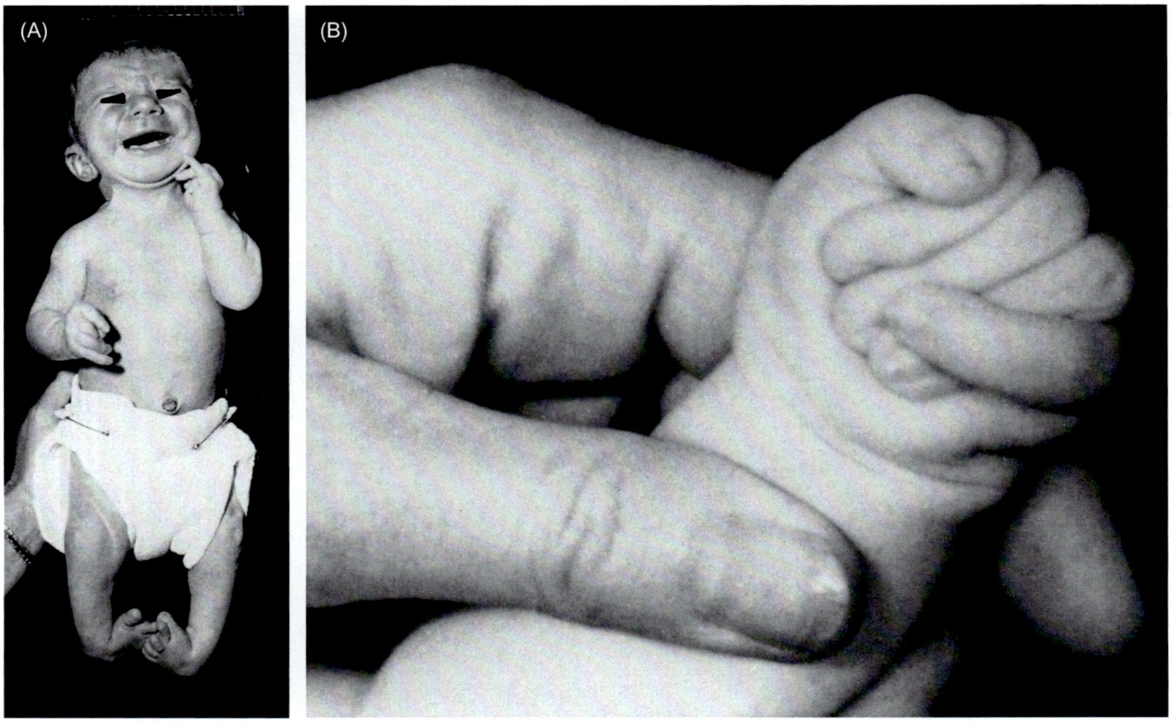

FIGURE 7.4 **(A)** An infant with distal arthrogryposis type I. Note primarily hand and foot involvement. The hand has a characteristic appearance when clasped. Here the hands are open, with ulnar drift to the fingers. **(B)** Characteristic positioning of the clasped hand in newborn period with overlapping fingers—the little finger over third and fourth. Often the second finger and thumb overlap the third finger as well. As the hand opens with use and therapy, the fingers develop an ulnar drift. The fingers often lack well-developed flexion creases. These positional changes seem to be related to misplaced tendons.

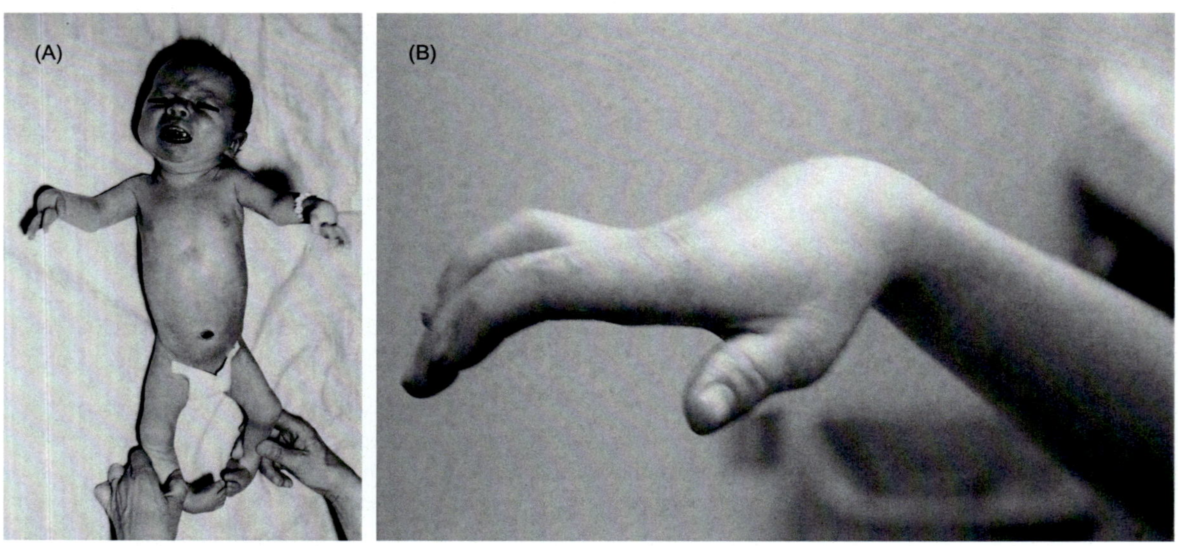

FIGURE 7.5 **(A)** An infant with the distal IIE type of arthrogryposis with limited jaw opening (trismus), clubfeet, flexion contractures of the knees, and the typical unusual positioning of the hands. The characteristic position of distal IIE is flexion of the wrist with flexed hyperextension of the metacarpophalangeal joint. **(B)** An older child with distal IIE. Note the typical positioning persists, although the hand is quite usable.

strabismus; ptosis; antimongoloid slant of the eyes; lateral coloboma of the alae nasae; small nose with a broad nasal bridge; overlapping, flexion, and ulnar deviation of the fingers; equinovarus feet with contracted toes; kyphosis; and scoliosis (Figure 7.6). Mutations in *MYH3* have been described. In addition, there may be postnatal growth restriction, inguinal hernias, and incomplete descended testes. Milder forms of distal arthrogryposis with mild facial involvement and distal contractures have been called Sheldon-Hall syndrome. Both involve mutations of sarcolemma fast twitch proteins.

Multiple Pterygium Syndromes

Many conditions with arthrogryposis have pterygium (see Box 7.4).

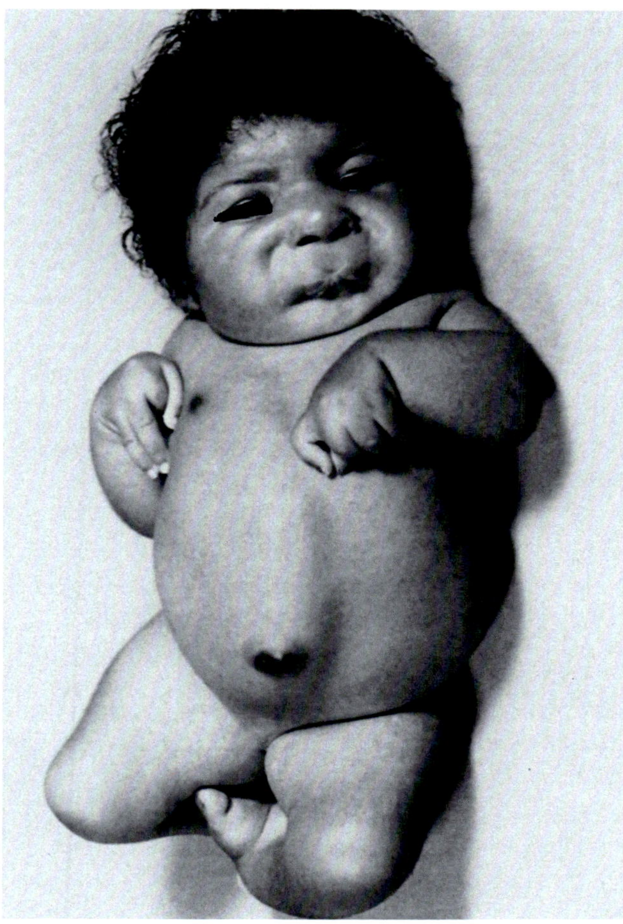

FIGURE 7.6 An infant with Freeman-Sheldon syndrome. Note ulnar drift of fingers, puckered appearance of face, pursed lips, small mouth (which will just barely open), equinovarus deformity of feet, scoliosis, and diastasis rectus. The neck and trunk appear shortened. The face is quite immobile.

Escobar Type

The majority of multiple pterygium syndrome (Escobar type) cases are sporadic in nature and inherited as an autosomal recessive trait.[2,21–25,39] However, there have been families reported with marked variability and possible dominant inheritance.[40,41] Common characteristics include flexion contractures, cleft palate, deafness, scoliosis, segmentation defect of the vertebrae, and short stature. Webs of skin (triangular membrane that forms across a body joint) may or may not be present at birth, but with age develop at the neck, elbows, knees, and intercrural areas. Interestingly, this syndrome may be a progressive condition because about 20% of affected individuals seem to get worse as they enter puberty, with decreasing pulmonary capacity and increasing thoracic lordosis. Cleft palate and vertebral fusions are occasionally seen and help to confirm the diagnosis. Specific genes have been identified, particularly those related to the fetal ACh receptor (see myasthenia syndromes).

Popliteal pterygium syndrome or Gorlin type is an autosomal dominant condition that has marked variability in its expression.[42,43] Affected individuals usually have

BOX 7.4 Multiple Pterygium Syndromes

1. Myasthenia spectrum
   a. Myasthenia gravis congenital
   b. Antibodies against fetal neurotransmitters
   c. Escobar type of multiple pterygium syndrome (CHRNG, CHRNA1, CHRNB1, CHRND, RAPSYN, DOK-7)
   d. Lethal multiple pterygium syndromes
2. Popliteal pterygium syndromes
   a. Popliteal pterygium (Gorlin) syndrome (IRF6)
   b. Bartsocas-Papas syndrome (RIPK4)
3. Syndromes not to miss
   a. Pterygia with malignant hyperthermia
   b. X-linked and autosomal dominant pterygium syndrome
4. Severe Amyoplasia
5. Frequent disorders which occasionally have pterygia
   a. Sacral agenesis/neural tube defects
   b. Fetal akinesia sequence disorders
   c. Pterygia and chromosomal disorders
   d. Boney disorders that lead to lack of joint movement
   e. Synostosis/chondrodystrophies and vertebral fusions
   f. Myopathies
6. Other less frequent disorders known to develop pterygia
   a. Cutis laxa (LOX, FBLN2, FBLN4, FBLN5)
   b. Kuskokwim syndrome
   c. Pterygium colli medium
   d. Shalev syndrome
   e. Haspeslagh syndrome
7. Interesting sporadic reports

dense webs in the popliteal area (the hollow part of the leg behind the knee joint), cleft lip and palate, syndactyly, nail anomalies, reduction deformities of digits, and other contractures. Expression is often quite variable within a family. The responsible gene is *IRF6*.

Limbs Plus Central Nervous System Dysfunction

Lethal Multiple Pterygium Syndrome

Many lethal multiple pterygium syndromes have been reported as autosomal recessive disorders that can be diagnosed prenatally.[44,45] However, there is also evidence of an X-linked recessive form.[46] The condition consists of intrauterine growth restriction, severe flexion contractures, webs *in utero*, cystic hygromas, hydrops, and hypoplastic lungs. In addition, polyhydramnios, ocular hypertelorism, cryptorchidism, cardiac hypoplasia, and ambiguous genitalia may be present. Other subtypes related to bony fusions and when the severity is recognized in gestation also exist.[45] Many specific genes have been identified.[15]

Neu-Laxova Syndrome

Neu-Laxova is a lethal, autosomal, recessively inherited trait that can be diagnosed during the prenatal period.[47] Common features include intrauterine growth restriction with short limbs, microcephaly, open eyes, tight ichthyotic skin, and severe CNS anomalies.[48] Additionally, kyphosis, syndactyly, and hydrops are often seen. The appearance is quite bizarre because of the open eyes. Several other types of constricting dermopathies are now recognized.[49]

Pena-Shokeir Phenotype

The Pena-Shokeir phenotype is a prenatally diagnosable condition that was first described by Pena and Shokeir in 1970. Also known as fetal akinesia (a = no; kinesis = movement), it is an autosomal recessive disorder in which the fetus does not move, or stops moving, during early pregnancy. Pena-Shokeir is lethal because no movement *in utero* prevents proper lung development (the lungs are hypoplastic) and leads to respiratory failure after the infant is born. This lack of fetal movement *in utero* also leads to congenital contractures, and the degree of congenital contractures is determined by the time of onset of the Pena-Shokeir phenotype during the pregnancy (Figure 7.7). In addition, intrauterine growth restriction, craniofacial abnormalities, micrognathia, small mouth, and a distinctive nose with a very high bridge and depressed tip are often seen.[15] Further, there may also be cleft palate and polyhydramnios because the fetus does not swallow.[2] Not swallowing amniotic fluid prevents proper development and function of intestines, leading to failure to thrive and malabsorption in the newborn.[2] Pena-Shokeir phenotype is also called fetal akinesia sequence, is extremely heterogenous, and relatively common.[15]

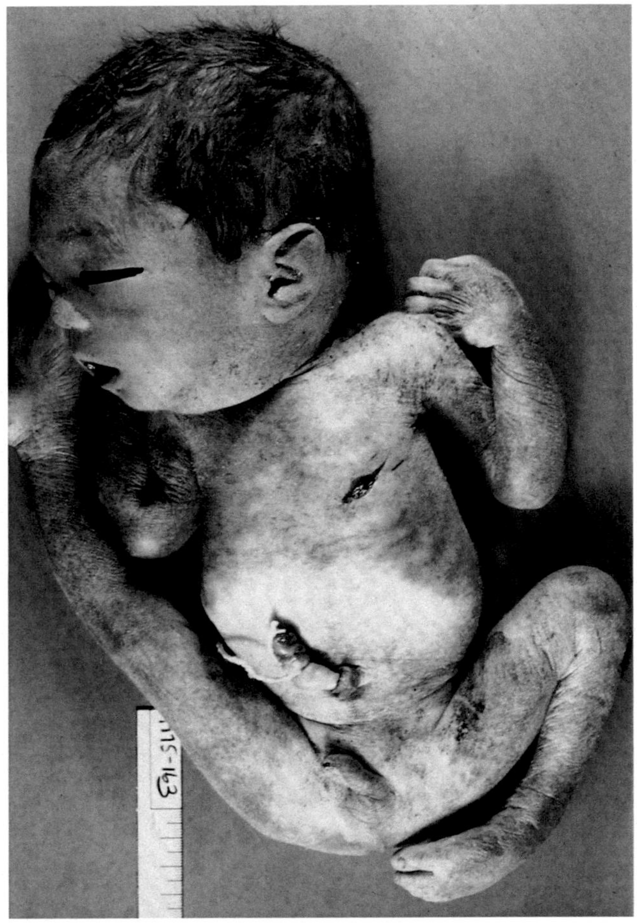

FIGURE 7.7 An infant with the Pena-Shokeir/fetal akinesia phenotype. The infant died from pulmonary insufficiency. Note one knee is held in flexion, and one is in extension. Both elbows are held in flexion. Hands and feet are stiff. The jaw is small, the palate high, and the mouth small. The bridge of the nose is prominent and the tip of the nose underdeveloped. The infant was small for dates and in breech position.

CHROMOSOMAL ANOMALIES

Trisomy 18 almost always has contractures at birth including clubfeet and camptodactyly with overlapping fingers medially, intrauterine growth restriction, and cardiac anomalies, and it usually leads to death in the first year.

Trisomy 8 mosaicism is a chromosomal anomaly that often presents with congenital contractures.[2,8] Affected individuals often have a long face with a high prominent forehead, a thick everted lower lip, and large ears. They also tend to have moderate intellectual disability, skeletal malformations, an absent patella, and deep furrows in the palms or soles with camptodactyly of fingers and toes.

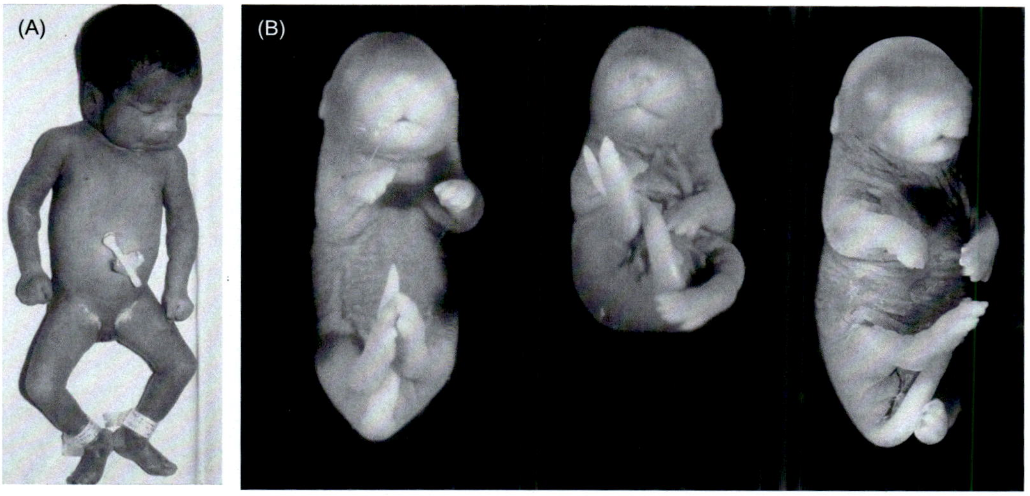

FIGURE 7.8 **(A)** A newborn baby with antibody-mediated arthrogryposis, showing knees fixed in flexion, clenched fists, and micrognathia. The baby died 7 hours after birth from respiratory insufficiency. (Photograph courtesy of Dr. Susan Huson.) (From Barnes *et al.*[52]) **(B)** Mice with passively transferred arthrogryposis. The pregnant dams were injected for 8 days with 0.5 mL of plasma, and the fetuses removed under terminal anesthesia on the day before the expected delivery. Left, a control mouse from a mouse dam treated with plasma from a mother of healthy offspring, negative for acetylcholine receptor antibodies. Note the normal posture with symmetrical positions of the fore and hind limbs. Center and right, mice from a mouse dam treated with plasma from two acetylcholine receptor antibody-positive mothers of offspring with arthrogryposis. Note the abnormal posture particularly evident in the position of the hind limbs.

Many chromosomal anomalies (both deletions and duplications) require fibroblast studies (chromosomes or CGH array) to identify the defect. These should be undertaken when true intellectual disability and multiple congenital contractures are present and the specific diagnosis cannot be made.

MATERNAL ANTIBODIES WITH AND WITHOUT MYASTHENIA GRAVIS

Since 1971, there have been a number of reports of arthrogryposis associated with myasthenia gravis (MG).[50] MG is an autoimmune disease that occurs quite commonly in young women (incidence around 2 per million in women of childbearing age). It is caused by serum autoantibodies to the acetylcholine receptors at the neuromuscular junction. The antibodies lead to loss of the acetylcholine receptors, failure of neuromuscular transmission, and muscle weakness. About 10% of women with MG transmit a transient "neonatal" form of MG to their babies; the babies become weak at birth or shortly after, but usually recover within a few weeks (see Chapter 27). Much more rarely, arthrogryposis and its accompanying features are found, and in these cases survival is uncommon, with death *in utero* or after a few hours from respiratory deficiency. In these women, a high proportion of the circulating autoantibodies is directed specifically against the fetal isoform of the acetylcholine receptor and blocks its function, thus paralyzing the baby. It is thought that the level of fetal subunit-specific antibodies determines the severity of the phenotype, which does not relate to the severity of the maternal symptoms. In some cases, the antibodies may activate complement deposition and endplate destruction, which will not necessarily be dependent on the presence of the fetal AChR.[51]

Fetal movements may be normal until immunoglobulin transport across the placenta increases substantially around 16 to 18 weeks' gestation. Because maternal antibodies disappear with time after birth, and in humans the fetal isoform of the AChR is replaced by the adult isoform during the later stages of pregnancy, the outcome is generally good if the baby survives.[50]

Severe antibody-mediated arthrogryposis usually conforms to the Pena-Shokeir type with contractures, hypoplasia of the lung, and micrognathia (Figure 7.8A), but other abnormalities are frequently also present. Among the reported cases, the features have included hypotonia and ptosis; polyhydramnios or oligohydramnios; cerebral and cerebellar atrophy; palatal deformities, hypertelorism, scoliosis, and pectus excavatum; abnormal genitalia; and atrial septal defects.[50] Curiously, the two surviving children from two women whose subsequent pregnancies had fatal outcomes had sensorineural deafness.[52,53]

An important feature of maternal antibody-mediated arthrogryposis is that it recurs in subsequent pregnancies (unless the mother is treated to reduce the autoimmunity), and that it is possible for the mother herself to be asymptomatic even though the baby is severely affected.[53]

For this reason, maternal acetylcholine receptor antibodies should be suspected if two or more consecutive fetuses have reduced fetal movements or arthrogryposis, even if a diagnosis of myasthenia in the mother has been excluded. Most immunology laboratories use a mixture of fetal and adult receptors for screening for antibodies to the acetylcholine receptor for the diagnosis of myasthenia gravis, so the test is widely available and suitable for use in the investigation of these mothers; although it will not specifically demonstrate the presence of fetal-specific antibodies, this is usually unnecessary as most MG patients have antibodies to both adult and fetal isoforms. It is not usually necessary to test the blood of the live newborn or infant, unless they are showing signs of lack of response to treatment or deterioration.

The finding of antibodies causing AMC raised the possibility that this mechanism might be a common cause. However, in a study of 179 mothers of AMC babies, only three sera (1.5%) from asymptomatic AMC mothers had AChR antibodies. Nevertheless, reactivity with muscle or neuronal antigens was found in 33% of the maternal sera, and was more common when the children had CNS involvement.[54] Other cases survive often with minimal evidence of joint involvement, only to subsequently fail to thrive because of persistent facial and bulbar weakness. These infants require careful management during the neonatal period.[55]

Proof that a disease is caused by autoantibodies requires the presence of an autoantibody against a defined target, the ability to cause disease in an experimental animal by injection of the autoantibodies, and prevention of disease by immunosuppressive treatments. The antibodies to fetal acetylcholine receptors directly inhibit the function of the acetylcholine receptor in *in vitro* assays.[56] Injection of maternal serum into pregnant mice resulted in transfer of disease to many of the fetuses; these were born paralyzed and with deformities of the joints (see Figure 7.8B).[57] In addition, two women with four and two babies with severe arthrogryposis, respectively, gave birth to healthy babies following treatment for their MG with plasma exchange or intravenous immunoglobulin therapy, or both, during the second trimester of pregnancy[58]; later treatments may not be so helpful.[59] Why do some women develop these particularly harmful antibodies, whereas other women with myasthenia do not? There is some preliminary evidence that antibodies specific for fetal acetylcholine receptor are particularly common in the relatively few young women who develop myasthenia during or after their first pregnancies.[60] If confirmed, this would suggest that, in these patients, myasthenia may sometimes be secondary to a primary immunization against the fetus and has implications for other fetal and neonatal diseases.

ACKNOWLEDGMENTS

Dr. Judith G. Hall is supported by the British Columbia Children's Hospital, the Children and Family Research Institute, and the University of British Columbia's Departments of Pediatrics and Medical Genetics. Work in Dr. Angela Vincent's laboratory on arthrogryposis was supported by the Medical Research Council of Great Britain and Action Research. Further studies are supported by the NIHR Oxford Biomedical Research Centre. Thanks to Kimi Tanaka for her technical assistance.

REFERENCES

1. Staheli LT, Hall JG, Jaffe KM, Paholke DO. *Arthrogryposis: A Text Atlas*. Cambridge, UK: Cambridge University Press; 1998. Available from: <http://www.global-help.org/publications/books/book_arthrogryposis.html>.
2. Hall JG. Arthrogryposes (Multiple congenital contractures). In: Rimoin DL, Connor JM, Pyeritz RE, Kork BR, editors. *Emery and Rimoin's Principles and Practice of Medical Genetics*. 6th ed. New York: Churchill Livingstone; 2013, pp. 1–101.
3. Otto AW. *Monstrorum sexcentorum descriptio anatomica in Vlatislaviae Museum*. Anatomico-Patholgicum Vratislaviaevise; 1841.
4. Drachman DB, Coulombre AJ. Experimental clubfoot and arthrogryposis multiplex congenita. *Lancet* 1962;**2**:523–6.
5. Moessinger AC. Fetal akinesia deformation sequence: an animal model. *Pediatrics* 1983;**72**:857–63.
6. Hall JG. Analysis of Pena Shokeir phenotype. *Am J Med Genet* 1986;**25**:99–117.
7. Hall JG. Overview of arthrogryposis. In: Staheli LT, Hall JG, Jaffe KM, Paholke DO, editors. *Arthrogryposis: A Text Atlas*. Cambridge, UK: Cambridge University Press; 2008, pp. 1–26.
8. Hall JG. Arthrogryposis multiplex congenita: etiology, genetics, classification, diagnostic approach, and general aspects. *J Pediatr Orthop B* 1997;**6**:159–66.
9. Lowry RB, Sibbald B, Bedard T, Hall JG. Prevalence of multiple congenital contractures including arthrogryposis multiplex congenita in Alberta, Canada, and a strategy for classification and coding. *Birth Defects Res A Clin Mol Teratol* 2010;**88**:1057–61.
10. Hall JG, Aldinger KA, Tanaka KI. Amyoplasia revisited. *Am J Med Genet A* 2014;**164A**:700–30.
11. Christianson C, Huff D, McPherson E. Limb deformations in oligohydramnios sequence: effects of gestational age and duration of oligohydramnios. *Am J Med Genet* 1999;**86**:430–3.
12. Filges I, Hall JG. Failure to identify antenatal multiple congenital contractures and fetal akinesia—proposal of guidelines to improve diagnosis. *Prenat Diagn* 2013;**33**:61–74.
13. Hall JG, Reed SD. Teratogens associated with congenital contractures in humans and in animals. *Teratology* 1982;**25**:173–91.
14. Fahy MJ, Hall JG. A retrospective study of pregnancy complications among 828 cases of arthrogryposis. *Genet Couns* 1990;**1**:3–11.
15. Hall JG. Pena-Shokeir phenotype (fetal akinesia deformation sequence) revisited. *Birth Defects Res A Clin Mol Teratol* 2009;**85**:677–94.
16. Rodriguez JI, Palacios J. Pathogenetic mechanisms of fetal akinesia deformation sequence and oligohydramnios sequence. *Am J Med Genet* 1991;**40**:284–9.

17. Swinyard CA, Bleck EE. The etiology of arthrogryposis (multiple congenital contracture). *Clin Orthop Relat Res* 1985;15–29.
18. Robertson WL, Glinski LP, Kirkpatrick SJ, Pauli RM. Further evidence that arthrogryposis multiplex congenita in the human sometimes is caused by an intrauterine vascular accident. *Teratology* 1992;**45**:345–51.
19. Reid CO, Hall JG, Anderson C, Bocian M, Carey J, Costa T, et al. Association of amyoplasia with gastroschisis, bowel atresia, and defects of the muscular layer of the trunk. *Am J Med Genet* 1986;**24**:701–10.
20. Sells JM, Jaffe KM, Amyoplasia Hall JG. The most common type of arthrogryposis: the potential for good outcome. *Pediatrics* 1996;**97**:225–31.
21. Escobar V, Bixler D, Gleiser S, Weaver DD, Gibbs T. Multiple pterygium syndrome. *Am J Dis Child* 1978;**132**:609–11.
22. Hall JG, Reed SD, Rosenbaum KN, Gershanik J, Chen H, Wilson KM. Limb pterygium syndromes: a review and report of eleven patients. *Am J Med Genet* 1982;**12**:377–409.
23. Vogt J, Harrison BJ, Spearman H, Cossins J, Vermeer S, ten Cate LN, et al. Mutation analysis of CHRNA1, CHRNB1, CHRND, and RAPSN genes in multiple pterygium syndrome/fetal akinesia patients. *Am J Hum Genet* 2008;**82**:222–7.
24. Hoffmann K, Muller JS, Stricker S, Megarbane A, Rajab A, Lindner TH, et al. Escobar syndrome is a prenatal myasthenia caused by disruption of the acetylcholine receptor fetal gamma subunit. *Am J Hum Genet* 2006;**79**:303–12.
25. Michalk A, Stricker S, Becker J, Rupps R, Pantzar T, Miertus J, et al. Acetylcholine receptor pathway mutations explain various fetal akinesia deformation sequence disorders. *Am J Hum Genet* 2008;**82**:464–76.
26. Reed SD, Hall JG, Riccardi VM, Aylsworth A, Timmons C. Chromosomal abnormalities associated with congenital contractures (arthrogryposis). *Clin Genet* 1985;**27**:353–72.
27. Bamshad M, Watkins WS, Zenger RK, Bohnsack JF, Carey JC, Otterud B, et al. A gene for distal arthrogryposis type I maps to the pericentromeric region of chromosome 9. *Am J Hum Genet* 1994;**55**:1153–8.
28. Banker BQ. Arthrogryposis multiplex congenita: spectrum of pathologic changes. *Hum Pathol* 1986;**17**:656–72.
29. Staheli LT, Chew DE, Elliott JS, Mosca VS. Management of hip dislocations in children with arthrogryposis. *J Pediatr Orthop* 1987;**7**:681–5.
30. Hall JG, Reed SD, McGillivray BC, Herrmann J, Partington MW, Schinzel A, et al. Part II. Amyoplasia: twinning in amyoplasia—a specific type of arthrogryposis with an apparent excess of discordantly affected identical twins. *Am J Med Genet* 1983;**15**: 591–9.
31. Matthews S, Farnish S, Young ID. Distal symphalangism with involvement of the thumbs and great toes. *Clin Genet* 1987;**32**:375–8.
32. Beals RK, Hecht F. Congenital contractural arachnodactyly. A heritable disorder of connective tissue. *J Bone Joint Surg Am* 1971;**53**:987–93.
33. Viljoen D, Ramesar R, Behari D. Beals syndrome: clinical and molecular investigations in a kindred of Indian descent. *Clin Genet* 1991;**39**:181–8.
34. Hall JG, Reed SD, Greene G. The distal arthrogryposes: delineation of new entities—review and nosologic discussion. *Am J Med Genet* 1982;**11**:185–239.
35. Bamshad M, Jorde LB, Carey JC. A revised and extended classification of the distal arthrogryposes. *Am J Med Genet* 1996;**65**: 277–81.
36. Sung SS, Brassington AM, Grannatt K, Rutherford A, Whitby FG, Krakowiak PA, et al. Mutations in genes encoding fast-twitch contractile proteins cause distal arthrogryposis syndromes. *Am J Hum Genet* 2003;**72**:681–90.
37. Freeman EA, Sheldon JH. Cranio-carpo-tarsal dystrophy. *Arch Dis Child* 1938;**13**:277–83.
38. Stevenson DA, Carey JC, Palumbos J, Rutherford A, Dolcourt J, Bamshad MJ. Clinical characteristics and natural history of Freeman-Sheldon syndrome. *Pediatrics* 2006;**117**:754–62.
39. Thompson EM, Donnai D, Baraitser M, Hall CM, Pembrey ME, Fixsen J. Multiple pterygium syndrome: evolution of the phenotype. *J Med Genet* 1987;**24**:733–49.
40. Carnevale A, Lopez Hernandez A, De los Cobos L. [Familial pterygium syndrome with probably dominant transmission linked to the X chromosome]. *Rev Invest Clin* 1973;**25**:237–44.
41. Frias JL, Holahan JR, Rosenbloom AL, Felman AH. *An autosomal dominant syndrome of multiple pterygium, ptosis and skeletal abnormalities. Fourth International Conference on Birth Defects*, vol. 19. Vienna: Excerpta Medica; 1973, 41
42. Gorlin RJ, Sedano HO, Cervenka J. Popliteal pterygium syndrome. A syndrome comprising cleft lip-palate, popliteal and intercrural pterygia, digital and genital anomalies. *Pediatrics* 1968;**41**: 503–9.
43. Escobar V, Weaver D. Popliteal pterygium syndrome: a phenotypic and genetic analysis. *J Med Genet* 1978;**15**:35–42.
44. Van Regemorter N, Wilkin P, Englert Y, El Khazen N, Alexander S, Rodesch F, et al. Lethal multiple pterygium syndrome. *Am J Med Genet* 1984;**17**:827–34.
45. Herva R, Leisti J, Kirkinen P, Seppanen U. A lethal autosomal recessive syndrome of multiple congenital contractures. *Am J Med Genet* 1985;**20**:431–9.
46. Tolmie JL, Patrick A, Yates JR. A lethal multiple pterygium syndrome with apparent X-linked recessive inheritance. *Am J Med Genet* 1987;**27**:913–19.
47. Muller LM, de Jong G, Mouton SC, Greeff MJ, Kirby P, Hewlett R, et al. A case of the Neu-Laxova syndrome: prenatal ultrasonographic monitoring in the third trimester and the histopathological findings. *Am J Med Genet* 1987;**26**:421–9
48. Curry CJ. Further comments on the Neu-Laxova syndrome. *Am J Med Genet* 1982;**13**:441–4.
49. Holbrook KA, Dale BA, Witt DR, Hayden MR, Toriello HV. Arrested epidermal morphogenesis in three newborn infants with a fatal genetic disorder (restrictive dermopathy). *J Invest Dermatol* 1987;**88**:330–9.
50. Polizzi A, Huson SM, Vincent A. Teratogen update: maternal myasthenia gravis as a cause of congenital arthrogryposis *Teratology* 2000;**62**:332–41.
51. Reimann J, Jacobson L, Vincent A, Kornblum C. Endplate destruction due to maternal antibodies in arthrogryposis multiplex congenita. *Neurology* 2009;**73**:1806–8.
52. Barnes PR, Kanabar DJ, Brueton L, Newsom-Davis J, Huson SM, Mann NP, et al. Recurrent congenital arthrogryposis leading to a diagnosis of myasthenia gravis in an initially asymptomatic mother. *Neuromuscul Disord* 1995;**5**:59–65.
53. Brueton LA, Huson SM, Cox PM, Shirley I, Thompson EM, Barnes PR, et al. Asymptomatic maternal myasthenia as a cause of the Pena-Shokeir phenotype. *Am J Med Genet* 2000;**92**:1–6.

54. Dalton P, Clover L, Wallerstein R, Stewart H, Genzel-Boroviczeny O, Dean A, et al. Fetal arthrogryposis and maternal serum antibodies. *Neuromuscul Disord* 2006;**16**:481–91.
55. D'Amico A, Bertini E, Bianco F, Papacci P, Jacobson L, Vincent A, et al. Fetal acetylcholine receptor inactivation syndrome and maternal myasthenia gravis: a case report. *Neuromuscul Disord* 2012;**22**:546–8.
56. Riemersma S, Vincent A, Beeson D, Newland C, Hawke S, Vernet-der Garabedian B, et al. Association of arthrogryposis multiplex congenita with maternal antibodies inhibiting fetal acetylcholine receptor function. *J Clin Invest* 1996;**98**: 2358–63.
57. Jacobson L, Polizzi A, Morriss-Kay G, Vincent A. Plasma from human mothers of fetuses with severe arthrogryposis multiplex congenita causes deformities in mice. *J Clin Invest* 1999;**103**:1031–8.
58. Huson S, Newsom-Davis J, Vincent A. Unpublished observations. 1998.
59. Chieza JT, Fleming I, Parry N, Skelton VA. Maternal myasthenia gravis complicated by fetal arthrogryposis multiplex congenita. *Int J Obstet Anesth* 2011;**20**:79–82.
60. Matthews I, Sims G, Ledwidge S, Stott D, Beeson D, Willcox N, et al. Antibodies to acetylcholine receptor in parous women with myasthenia: evidence for immunization by fetal antigen. *Lab Invest* 2002;**82**:1407–17.

Part III

Motor Neuron Disorders

Chapter 8

Spinal Muscular Atrophies

Basil T. Darras, Jennifer A. Markowitz, Umrao R. Monani, and Darryl C. De Vivo

INTRODUCTION

Spinal muscular atrophies (SMAs) are genetic disorders that are clinically characterized by progressive muscle weakness and atrophy, associated with degeneration of spinal and, in the most severely affected patients, lower bulbar motor neurons. Classic proximal SMA is the most common form of SMA and the leading cause of infant mortality; it seems to be found in practically all populations but is diagnosed more frequently in infants and children, rather than adults. In the early 1890s, the Austrian clinician Guido Werdnig and the German physician Johann Hoffman were the first to describe the severe form of SMA, at the University of Graz, Austria, and in Heidelberg, Germany, respectively.[1,2] Their reports described the neuromuscular phenotype of the disease and the associated loss of anterior horn cells in the spinal cord.

The disease results from homozygous deletions or mutations involving the "survival of motor neuron" (*SMN*) gene at locus 5q13. The *SMN* gene is present in 2 copies on each chromosome 5, designated *SMN1* and *SMN2*, forming an inverted duplication (Figure 8.1). Five nucleotide changes which do not change amino acids differentiate *SMN2* from *SMN1*. A single nucleotide change that creates an exonic splicing suppressor in exon 7 of *SMN2* is the critical one and leads to exclusion of exon 7 in most transcripts[3]; thus the duplicated (*SMN2*) gene produces less functional SMN protein. Most patients with 5q proximal recessive SMA harbor homozygous deletions involving exon 7 of the *SMN1* gene, but maintain at least 1 copy of *SMN2*. There is a rough correlation between *SMN2* gene copy number, which varies normally in the population, the level of SMN protein, and severity of disease. The role of the SMN protein is still under active investigation. Modifying genes appear to be present as well, serving other roles in motor neuron function. SMA may involve dysfunction of more than lower motor neurons, with abnormalities of the neuromuscular junction noted in animal models, and abnormal muscle development in the most severely affected patients. There are also many other types of SMA, known as non-5q SMAs, related to mutations in various genes that are expressed in a wide range of tissues, including the nervous system.

EPIDEMIOLOGY

The incidence of SMA has been estimated at 1 in 6000–11,000 live births[4–7] or about 7.8–10 per 100,000 live births[8–10] and at 4.1 per 100,000 live births for SMA type I.[8] A 2005 study in Cuba found a reduced incidence of type I SMA among the Cuban population overall (3.53 per 100,000 live births) and especially among those of African ancestry (0.89 to 0.93 per 100,000 live births).[11] The estimated pan-ethnic disease frequency is approximately 1 in 11,000.[7] The carrier frequency for mutations in the *SMN1* gene has been estimated at from 1:38 to 1:70. A 2009 epidemiologic study sought to determine the carrier frequency in different ethnic groups in North America. The carrier frequency was highest in Caucasians (1 in 37, or 2.7%) and was lowest in Hispanics (1 in 125, 0.8%). Ashkenazi Jews (1 in 46, 2.2%) and African Americans (1 in 56, 1.8%) were of intermediate frequency.[12] It should be noted that, despite the high carrier frequency, the incidence of SMA is lower than expected. It has been postulated that this may reflect that some fetuses have a 0/0 *SMN1/SMN2* genotype (i.e. no SMN protein is present at all), which is known in other species to be embryonic lethal.[13]

CLINICAL CHARACTERISTICS

Though most patients with spinal muscular atrophy have deletions or mutations involving the *SMN1* gene, a range of phenotypic severity permits division into four broad clinical subtypes. It is recognized that the subtypes represent a phenotypic continuum extending from the very severe, with onset *in utero*, to the very mild, with onset

B.T. Darras, H. Royden Jones, Jr., M.M. Ryan & D.C. De Vivo (Eds): Neuromuscular Disorders of Infancy, Childhood and Adolescence, Second edition.
DOI: http://dx.doi.org/10.1016/B978-0-12-417044-5.00008-1

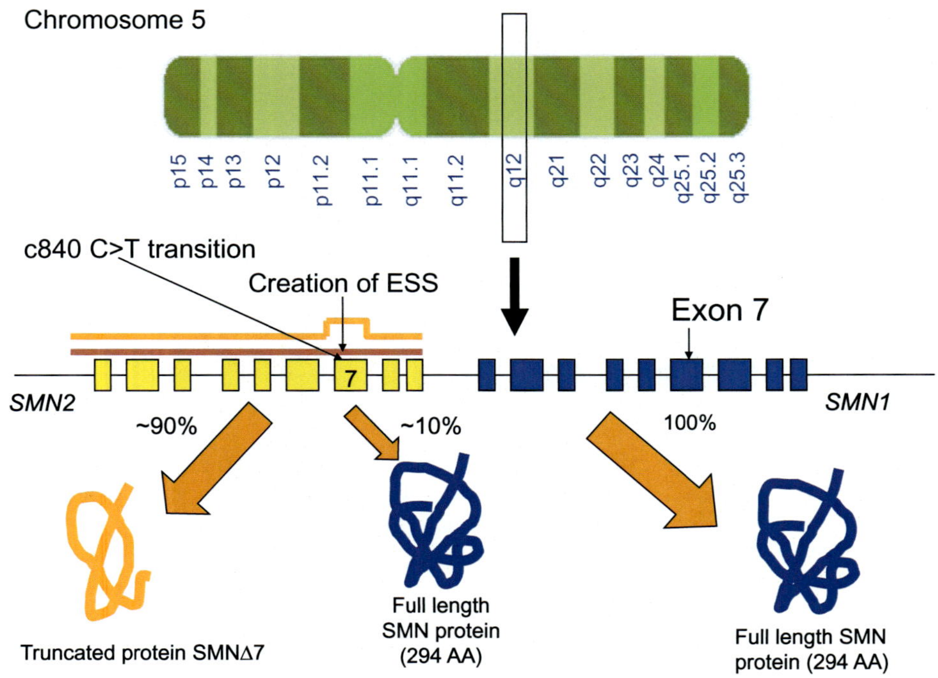

FIGURE 8.1 Schematic diagram of human *SMN1* and *SMN2* genes on chromosome 5. Patients with spinal muscular atrophy have deletions or mutations in both copies of *SMN1*. A C-to-T transition at position 6 of *SMN2* creates an exonic splicing suppressor (ESS) that leads to skipping of exon 7 during transcription and production of truncated, nonfunctional SMN protein. However, a small amount (~10%) of full-length messenger RNA is produced from the SMN2 gene, resulting in functional, full-length SMN protein. AA = amino acids.

during adulthood; there is also a spectrum of severity within each of these groups (Table 8.1).[14,15] For the purposes of clinical classification, or of guidelines developed for standards of care, the "maximal functional status achieved" approach, which classifies type I patients as "non-sitters," type II patients as "sitters," and type III patients as "walkers," has been used.[16,17] Patients with a mild phenotype and onset during middle or late age are classified as type IV. The age at onset is also considered in the classification but because of potential overlap between subtypes and the difficulty in accurately determining the onset of symptoms, it has not been considered as the sole determinant of disease subtype.

Type I SMA

Following the initial description of infantile SMA by Werdnig and Hoffman in the early 1890s and further descriptions made by Sylvestre in 1899 and Beevor in 1903, infantile or type I SMA was described again in detail, both clinically and pathologically, by Randolph Byers and Betty Banker at Boston Children's Hospital in 1961.[18] Patients with type I SMA, also known as Werdnig-Hoffman disease, present between birth and 6 months of age. SMA type I has been further subdivided into three groups: type IA (or type 0 in certain reports[19]) with onset *in utero* and presentation at birth, type IB with onset of symptoms prior to 3 months of age, and type IC with onset between 3 and 6 months of age. Infants with SMA type I have progressive proximal weakness that affects the legs more than the arms. They have poor head control, hypotonia that causes them to assume a "frog-leg" posture when lying and to "slip through" on vertical suspension, and areflexia. They are never able to sit ("non-sitters") (Table 8.1). There is also weakness of the intercostal muscles with relative sparing of the diaphragm, producing a bell-shaped chest and a pattern of paradoxical or "belly"-breathing. Infants with type I SMA also classically exhibit tongue fasciculations and eventually have difficulty swallowing, with risk for aspiration and failure to thrive. Other cranial nerves are not as affected, although facial weakness does occur at later stages of the disease. Cognition is normal, and they are often noted at diagnosis to have a bright, alert expression that contrasts with their generalized weakness. Infants with type I SMA usually develop respiratory failure by age 2 years, or much earlier, and in the past most did not survive past 2 years; however, there has been some increase in survival in recent years with the use of assisted ventilation (to be described in more detail in later sections) and other interventions.[14,20–22] Prior natural history studies have reported shortened life span, with 68% mortality within 2 years and 82% by 4 years of age.[23,24] The application of nutritional and respiratory interventions has reduced the mortality to approximately 30% at 2 years, but about half of the survivors are fully dependent on noninvasive ventilation.[25]

It was previously believed that SMA is a purely motor neuron disorder. However, recent studies have shown that severe type I SMA can result in various organ manifestations, apart from spinal cord motor neurons such as brain,

TABLE 8.1 Clinical Classification of Spinal Muscular Atrophy

SMA Type	Other Names	Age at Onset	Life Span	Highest Motor Milestone Achieved	Other Features	Proportion of Total SMA (%)
Type IA	Prenatal, Congenital SMA, Werdnig-Hoffman disease	Prenatal	<6 months	Mostly unable to achieve motor milestones	• Severe weakness at birth • Profound hypotonia • Facial diplegia • Areflexia • Early respiratory failure • Joint contractures	60[a]
Type IB, Type IC	Werdnig-Hoffman disease, Severe SMA ("non-sitters")	Type IB (0–3 months) Type IC (3–6 months)	<2 years without respiratory support	Never sits supported	• Weakness • "Frog leg" posture, hypotonia • Tongue fasciculations • Hyporeflexia, areflexia • Suck and swallow difficulties • Respiratory failure	60[a]
Type II	Intermediate SMA ("sitters"), Dubowitz disease	6–18 months	>2 years ~70% alive at 25 years of age	Sits independently. Never stands or walks	• Proximal weakness, hypotonia • Postural hand tremor • Hyporeflexia • Average or above average intellectual skills by adolescence • Scoliosis	27
Type III	Kugelberg-Welander disease, Mild SMA ("walkers")	>18 months Type IIIA (prior to 3 years) Type IIIB (after 3 years)	Almost normal	Stands and walks	• May have hand tremor • Resembles muscular dystrophy	12
Type IV	Adult SMA	>21 years	Normal	Normal		1

[a]*SMA types I, IA, IB, and IC all have a 60% proportion of total SMA.*
Modified with permission from Markowitz et al.[155]

cardiac, vascular, and even sensory nerve involvement. Recent autopsy studies have shown increasing evidence of congenital heart disorders in severe SMA, with the most common association being hypoplastic left heart syndrome.[26] In a study, 3 out of 4 (75%) patients with 1 *SMN2* copy had congenital SMA (SMA type 0) associated with hemodynamically significant atrial or ventricular septal defects, which underscores the relevance of the SMN protein for normal cardiac development[27]; despite the reported cases, however, a chance association has not been firmly excluded. Severe SMA mouse models have shown autonomic disturbances including bradyarrhythmia and progressive heart block, as well as dilated cardiomyopathy and decreased contractility[28–30]; in humans with molecularly proven SMA there has been no evidence of dilative or congestive cardiomyopathy, nor have there been reports of AV conduction anomalies or arrhythmias besides bradyarrhythmias. Studies done on various mouse models have also found that severe SMN protein deficiency might present as vasculopathy. This was also noted in the case reports of two unrelated patients with severe SMA type I. Both of these infants developed ulcerations and necroses of the fingers and toes. Sural nerve biopsy studies on them showed thrombotic occlusion of small vessels without inflammation resulting in perfusion abnormality and subsequent tissue necrosis in the absence of a significant peripheral neuropathy.[31] Autonomic dysfunction is thought to be the primary cause of this vasculopathy.[32] While some of the reported patients had only one *SMN2* copy, two cases of infants with SMA type I in whom a distal necrosis developed had two *SMN2* copies.[32] In a different biopsy study done by Rudnik-Schöneborn et al.[33] on 19 patients with infantile SMA, significant sensory nerve pathology was found in severe SMA type I patients, whereas no sensory involvement was found in SMA type II and type III patients. Out of these, axonal degeneration was noted in seven and abnormal sensory conduction in five SMA type I patients.

The abnormal nerve conduction study patients were followed with biopsy, which showed marked fiber density reduction in four patients, thus indicating that severe SMA type I patients can show both clinical and morphological sensory nerve involvement.

Type II SMA

This intermediate form of SMA was first reported in 1893 at the University of Edinburgh, United Kingdom, and described again by Byers and Banker in 1961[18] and in detail by Dubowitz in 1964.[34,35] Patients with type II SMA, also known as intermediate SMA or Dubowitz disease, are able to sit unsupported at some point ("sitters"), but are never able to stand alone or walk (Table 8.1). The onset of symptoms is between 6 and 18 months of age. They have progressive proximal weakness affecting legs more than arms, hypotonia, and areflexia. They also develop progressive scoliosis which, in combination with intercostal muscle weakness, results in significant restrictive lung disease as they grow older. They develop joint contractures and can have ankylosis of the mandible. They exhibit tremor, or polyminimyoclonus, of the hands.[14,22] Although their body mass index may be low (at the 3rd percentile or less when compared with normal children), the high-functioning, nonambulatory patients have a higher relative fat mass index and are at risk of becoming overweight.[36] Cognition is normal and verbal intelligence may be above average.[37] In a study done on 240 type II patients, survival rates were found to be 98.5% at 5 years and 68.5% at 25 years.[38] Patients may live into the third decade, but life expectancy is shortened due to the risk of respiratory compromise.[14,22]

Type III SMA

In 1956, Kugelberg and Welander described a much milder form of SMA characterized by prolonged ambulation.[39] Patients with type III SMA, also known as Kugelberg-Welander disease, are able to stand alone and walk at some point ("walkers"). The onset of symptoms occurs after the age of 18 months; it has further been subdivided into type IIIA (onset between 18 months and 3 years) and type IIIB (onset after 3 years). They have progressive proximal weakness affecting legs more than arms and may ultimately need to use a wheelchair, but they generally develop little to no respiratory muscle weakness or severe scoliosis. Loss of ambulation increases the risk of these complications. They may have tremor or polyminimyoclonus of the hands. Sometimes, the calves of these patients can be prominent and hence type III SMA can be confused with Becker muscular atrophy. Life expectancy is not significantly different compared with the normal population (Table 8.1).[14,22,38] Case Example 8.1 describes the disease progression of a particular type III patient.

There has been much debate about the appropriate classification of patients into these three types of SMA because, as mentioned, there are patients within these categories who exhibit phenotypes of differing severities. A classification system based on a continuous rather than discrete variable (e.g. SMA "type 1.8" in the case of a less severely affected type I patient) has been proposed to better capture the clinical spectrum of these patients.[15]

Outliers

There are also patients who are outliers on either end of the phenotypic spectrum. As discussed above, an SMA "type IA" (formerly known as type 0) has been used to describe neonates who present with severe weakness and profound hypotonia, probably of prenatal onset, as well as with a history of decreased fetal movements. The majority do not attain any motor milestones. Other findings include areflexia, facial diplegia, atrial septal defects, and joint contractures. In SMA type IA, respiratory failure forms an important cause of morbidity and mortality, requiring noninvasive ventilation and endotracheal intubation at birth. Life expectancy is reduced and most of them are unable to survive beyond 6 months of age (Table 8.1).[19,40] Furthermore, arthrogryposis multiplex congenita (congenital joint contractures involving at least two regions of the body) has been noted in SMA patients with *SMN1* gene deletions,[41] and congenital axonal neuropathy involving motor and sensory nerves in conjunction with facial weakness, joint contractures, ophthalmoplegia, and respiratory failure at birth has been reported in three newborn siblings with deletions in the SMA chromosomal region.[42]

A milder adult-onset SMA, or type IV SMA, has also been described with onset of symptoms after age 21 years and essentially normal life span. Most patients with the SMA type IA and IV phenotypes have homozygous deletions of exon 7 in *SMN1*, but as it will be discussed in the Genetics section, the *SMN2* copy number is usually only 1 in SMA type IA and 4 to 5 in SMA type IV.[14,43]

Other "Spinal Muscular Atrophies"

The non-5q13 associated spinal muscular atrophies are a heterogeneous group of motor neuron diseases associated with mutations in a variety of different genes (e.g. X-linked and autosomal dominant or recessive spinal muscular atrophies), distal or segmental spinal muscular atrophies, or distal hereditary motor neuropathies.[14,43,44] Patients with these disorders generally have some clinical characteristics that can help differentiate them from those with 5q13-associated or classic SMA.[45] The differential diagnosis of SMA is detailed in the respective section.

CASE EXAMPLE 8.1 SMA, TYPE III

An 11-year-old girl was evaluated at age 19 months for delayed motor development. At age 4 months, hypotonia was noted, but her early gross motor development had been normal. She sat at 8 months, crawled at 9 months, and pulled up to stand and cruised at 12 months. By age 25 months, she was taking a few steps without assistance and initiating walking. At 19 months, after the clinical diagnosis of SMA was confirmed by molecular testing, she was enrolled in early intervention and aqua therapy programs, and physical therapy was continued through preschool. An intensive PT and aqua therapy program was initiated after she entered elementary school.

At age 19 months, she was evaluated by pediatric neurology because of delayed motor milestones, hypotonia, and weakness. The bilateral weakness involved the axial and proximal limb muscles to a greater degree than the distal muscles, and the weakness was asymmetric with greater involvement of the right arm and left leg. The muscle bulk was reduced. Eye, facial and tongue muscles were normal. Tendon reflexes and Babinski signs were absent. The spinal contour was normal but the lumbar lordosis was accentuated. She was remarkably intelligent. Electromyography showed ongoing denervation with evidence of reinnervation in multiple muscles of legs and arms; her sensory nerve conduction studies were normal whereas the motor responses were of low amplitude but with normal conduction velocities; these findings suggested motor neuron disease. The diagnosis of spinal muscular atrophy was confirmed by DNA testing which showed a homozygous deletion of *SMN1* and three copies of *SMN2* gene.

At the age of 5 years, she was enrolled in a prospective multicenter natural history study at the Columbia University Spinal Muscular Atrophy Clinical Research Center; she was evaluated at baseline (age 6 years) and subsequently at months 2, 4, 6, 9, and 12. She continued to be followed every 6 months for a total of 84 months (age 13 years). The outcome measures included motor function, pulmonary function, muscle strength, electrophysiological measures (CMAP and MUNE), and patient-reported outcomes. From 5 to 7.5 years, her motor function improved but from 7.5 to 11 years, her motor function declined with continued body growth. Manual muscle testing showed only minimal decrease in strength. This patient walked with assistance until age 9 years, longer than expected, probably due to the intensive rehabilitation she continued to receive. In fact, her motor function improved during the first 18 months of observation, but it was then followed by a slow functional decline. These observations are consistent with the results of published natural history studies of spinal muscular atrophy types II and III, which show stability over the initial 12-month period (with younger patients often showing improvement), followed by gradual functional decline over years. After infancy and with continued weight bearing, the patient developed a progressive thoracolumbar scoliosis, pelvic obliquity, and recurrent painful hip subluxation ultimately requiring multiple orthopedic procedures to correct the advancing musculoskeletal deformities and persistent joint pain. The orthopedic deformities accelerated after ambulation was lost and the patient spent her waking hours in a motorized wheelchair. Respiratory complaints also supervened with intercurrent pulmonary infections demanding aggressive proactive management to protect the airway and to ensure adequate ventilation.

Comment

This case illustrates several points. Dating the onset of symptoms is often in the "eye of the beholder." This patient has SMA type IIIA with onset before age 3 years. These children often regress functionally and lose the ability to stand and walk after years. In fact, this patient had neuromuscular symptoms dating back to early infancy with hypotonia. She was cruising by 12 months, but she did not take steps until age 19 months, suggesting underlying weakness, sensory disturbance, or incoordination. There were no cerebellar signs and the sensory nerve studies were normal. Her remarkable motor accomplishments, despite the early signs of hypotonia and weakness, underscore possible mitigating factors that can alter the natural history of SMA. Early intensive rehabilitation clearly benefited this patient's natural course. It also highlights the fact that motor function can worsen over long periods of time despite relatively preserved strength and optimal rehabilitative care. Growth during development also adds additional physical burden to a compromised neuromuscular system, and contributes to the subtle functional decline during childhood and adolescence. Whether the disease process worsens in parallel with the functional decline remains a subject of debate, and the answer is important particularly as it relates to the anticipated response to therapeutic intervention. Respiratory insufficiency and musculoskeletal deformities, often with considerable associated pain, are inevitable complications of SMA and should be addressed proactively to forestall and minimize the resulting acute and chronic morbidity that otherwise erodes the patient's quality of life.

GENETICS

The SMN Gene

Linkage analysis studies showed that all three forms of SMA map to chromosome 5q11.1-13.3. In 1995, Lefebvre et al. identified the *SMN* gene within this region, which was absent or interrupted in 98.6% percent of the patients in their group.[46] The structure of this region is complex, with a large inverted duplication of a 500 kb element. This contains the *SMN1* gene, which is deleted or interrupted in patients with SMA and is evolutionarily older, in the telomeric portion of the region; and the *SMN2* gene, a duplication of *SMN1* that differs from it by only 5 nucleotides, in the centromeric portion (Figure 8.1).[46] The critical difference between *SMN1* and *SMN2* is a C to T transition that creates an exon splicing suppressor in exon

TABLE 8.2 Genetic Diagnostic Testing in Spinal Muscular Atrophy

Type of Mutation	Test applied	Mutation Detection Rate
Homozygous deletion of exon 7[a]	*SMN1* Targeted mutation analysis PCR/restriction enzyme analysis or multiplex ligation probe amplification (MLPA) methodologies	~95–98%
Compound heterozygosity (Deletion of *SMN1* exon 7 [allele 1] and an intragenic mutation of *SMN1*[b] [allele 2])	Targeted mutation analysis combined with *SMN1* gene sequence analysis[c]	2–5%
SMN2 copy number[d]	Quantitative PCR analysis and other methodologies[e]	N/A

[a]*Testing for exon 8 deletion is not necessary.*
[b]*Small intragenic deletions/insertions and nonsense, missense, and splice site mutations.*
[c]*Whole-gene deletions/duplications are not detected.*
[d]SMN2 *copy number ranges from 0 to 5.*
[e]*MLPA, long-range PCR, chromosomal microarray (CMA) that includes the* SMN1, SMN2 *chromosomal segment.*
Modified with permission from Markowitz et al.[155]

7 of *SMN2*. While this splice modulator change is translationally silent (i.e. it does not change the amino acid sequence), it affects the alternative splicing of the gene, so that exon 7 is spliced out of or excluded from most *SMN2* mRNA transcripts. This altered mRNA results in production of a truncated version of the SMN protein, which does not oligomerize efficiently and is degraded. Since exon 7 is not always spliced out of all *SMN2* mRNA, a small amount of full-length transcript and hence functional protein is produced by *SMN2*, but it yields only on average about 10% as much as that produced by *SMN1*.[47] In patients with SMA both copies of the *SMN1* gene are deleted or disrupted, so the individual is left with only the small amount of SMN protein produced by the remaining copies of *SMN2*. The amount of SMN protein is inversely correlated with the severity of disease.[48]

About 95% to 98% of patients with SMA harbor deletions of the telomeric *SMN1* gene (Table 8.2). The remainder have small intragenic mutations or have undergone gene conversions from *SMN1* to *SMN2*. *De novo* mutations occur at a rate of about 2%, which is relatively high, and explained by the fact that this region of chromosome 5 is unstable, containing not only the inverted repeat of *SMN1* and *SMN2*, but other surrounding low copy number repeats. This leads to a high rate of unequal crossover and *de novo* mutations (especially paternally derived), which could explain the relatively high carrier frequency despite the mortality rate for the most severe forms of the disease.[49] More mildly affected patients seem more likely to undergo gene conversions of *SMN1* to *SMN2* rather than deletion of *SMN1*.[50–52]

The number of copies of *SMN2* per chromosome 5 varies among normal individuals, and 10–15% of the population possess no copies of *SMN2*.[6,53] Among patients with SMA, a clear correlation has been established between *SMN2* copy number and phenotypic severity. Feldkotter et al. in 2002 found that in their series 80% of patients with type I SMA had one or two copies of *SMN2*, 82% of patients with type II had three copies of *SMN2*, and 96% of patients with type III had three or four *SMN2* copies (Figure 8.2).[53] Studies by Mailman et al. in 2002 and Arkblad et al. in 2009 found similar results (95–100% of type I patients had one or two copies of *SMN2*, and all type III patients had at least three copies of *SMN2*).[10,54] However, this correlation is not so perfect as to permit the absolute prediction of clinical severity based on *SMN2* copy number, especially in the intermediate forms of the disease where there is some overlap (patients with three copies of *SMN2* have been described with all three phenotypes).[55,56] One reason for the overlap is that not all copies of the *SMN2* gene are equal; in terms of full-length SMN protein production, some copies probably produce less and some more than 10% functional protein. In general, though, it can be said that a patient with one or two copies of *SMN2* is highly likely to present with SMA type IA, IB, or IC.[10,53] Interestingly, unaffected family members with homozygous *SMN1* deletions and five copies of *SMN2* have been described, which suggests that *SMN2* copy number alone cannot be the sole modifying factor in disease severity, since there are patients with type III SMA who also have five copies of *SMN2*.[10,57] In *Smn*(−/−) knockout mice, the introduction of eight copies of human *SMN2* can rescue the phenotype and result in essentially normal mice. The presence of only two copies of *SMN2* in mice results in a severe SMA-like phenotype and death at 5 days. The addition of an *SMN* transgene lacking the exon 7 sequence (*SMNΔ7*) along with two copies of *SMN2* prolongs the lifespan of the transgenic mouse to about 14 days.[58–60]

A positive modifier has been identified in the *SMN2* gene.[61] A single base substitution in exon 7 in three

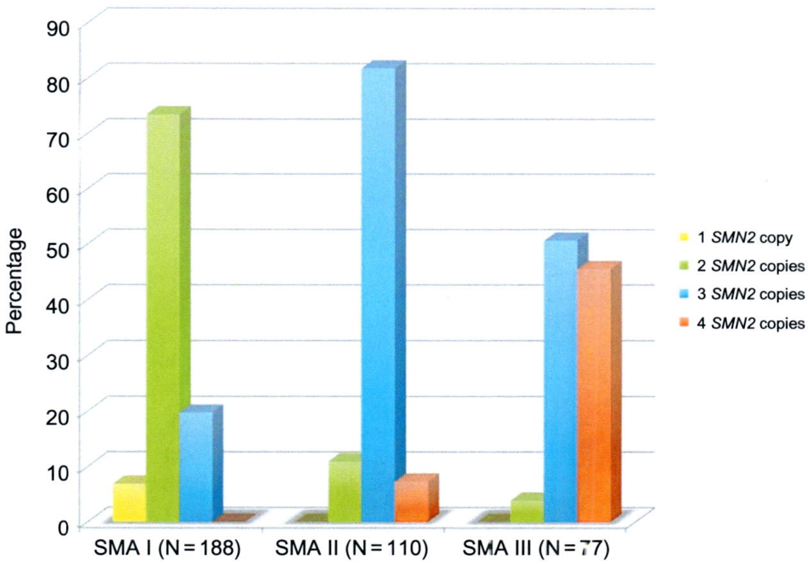

FIGURE 8.2 Frequency of patients with SMA types I, II, and III and *SMN2* copy numbers. In SMA type I, 80% of patients had 1 or 2 copies of *SMN2*, 82% of patients with type II had 3 copies of *SMN2*, and 96% of patients with type III carried 3 or 4 *SMN2* copies. *Modified with permission. This image was published in* American Journal of Human Genetics, *copyright Elsevier (2002).*[53]

unrelated patients' DNA (c.859G > C) creates a new exonic splicing enhancer that increases exon 7 inclusion and thus the amount of full-length protein by about 20%. The phenotype in these patients was less severe and did not correlate with their *SMN2* copy numbers, supporting the positive modifying effect of this sequence change and the notion that not all *SMN2* copies are equal. Of note, this variant has been identified in two copies in type IIIB patients with a mild phenotype, in one copy in type II patients, and not at all in severe type I patients.[61–63] It seems that a 20% increase in full-length protein in patients with two copies of *SMN2* will probably result in mild type IIIB SMA; therefore, it is fair to speculate that a 25% increase in full-length SMN protein in the same patients may result in total rescue of the phenotype.[61,63] There are also modifiers that do not involve the *SMN1* or *SMN2* genes. It is well known that haploidentical siblings with the same *SMN2* copy number can have SMA of different severities.[64–66] Higher expression of plastin 3 (*PLS3*) was also identified as a gender-specific protective modifier of SMA in asymptomatic *SMN1*-deleted females, carrying the same number of *SMN2* copies as their affected sibling.[67] A subsequent study, however, showed that *PLS3* may be an age- and/or puberty-specific and sex-specific modifier as the expression of the gene was greatest in postpubertal female patients with SMA type III, intermediate in SMA type II, and lowest in SMA type I patients, underscoring an association with disease severity as well.[68] Further, high plastin levels have also been detected in female siblings with more severe SMA presentation.[69] It is possible that the plastin 3 modifier is not an important modifier of SMA phenotype or it may be female-specific and incompletely penetrant.[70] No changes in the plastin 3 gene itself have been reported; thus, the modifier role of plastin 3 in SMA remains uncertain. Sequence variants within splicing regulators of the *SMN1* and *SMN2* genes, either polymorphisms or mutations, may explain divergent phenotypes in haploidentical siblings.[70,71]

Regarding the level of SMN protein itself, although there is in general an inverse correlation between the level of SMN protein and severity of disease, this correlation does not seem to be quite as close as that between *SMN2* copy number and phenotype.[72]

Genetic Diagnosis

Genetic testing for SMA can be performed with PCR-based targeted mutation analysis using a restriction enzyme that digests exon 7 (Table 8.2). If *SMN1* exon 7 is homozygously deleted (as it is in about 95–98% of patients) then the restriction fragments will be absent, the DNA will not be amplified, and a band will not be seen on the gel. If an individual is a carrier (heterozygous deletion) then a lighter band will be present (Figure 8.2). However, multiplex ligation probe amplification (MLPA) methodology is currently applied in most DNA diagnostic laboratories for deletion analysis of exon 7 of the *SMN1* gene in potential probands and carriers. This type of targeted mutation testing in conjunction with sequence analysis can also detect individuals who are compound heterozygotes[53] with a deletion of exon 7 in one *SMN1* allele and an intragenic point mutation in the other allele. In such a case (approximately 2–5% of individuals with clinical diagnosis of SMA), sequence analysis of the *SMN* gene will detect the mutation; this sequence testing, however, will not detect exonic deletions or duplications[73] and will not determine whether the point mutation is in the *SMN1* gene or *SMN2* gene (if one of these genes is not deleted). Fortunately, certain point mutations have been described in more than one SMA patient already and thus

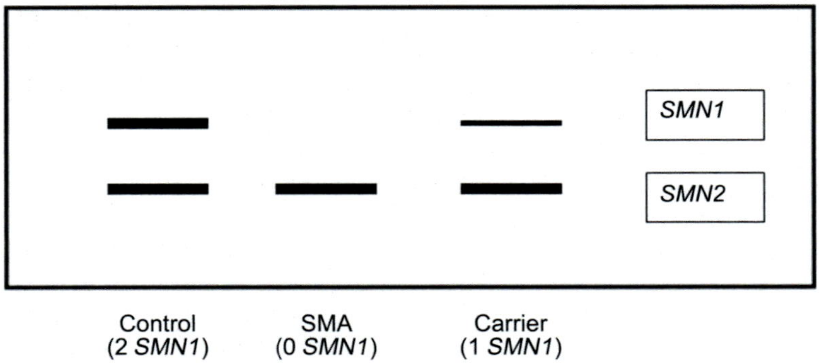

FIGURE 8.3 PCR diagnosis of *SMN1* deletion. PCR-based targeted mutation analysis using a restriction enzyme that digests exon 7. In a patient with a homozygous deletion of exon 7, the restriction fragment is absent (0 copies of *SMN1*). In a carrier (heterozygous deletion), a lighter *SMN1* band will be present. Similar targeted mutation analysis can also be performed with multiplex ligation probe amplification methodology. *Reprinted with permission from Markowitz et al.*[155]

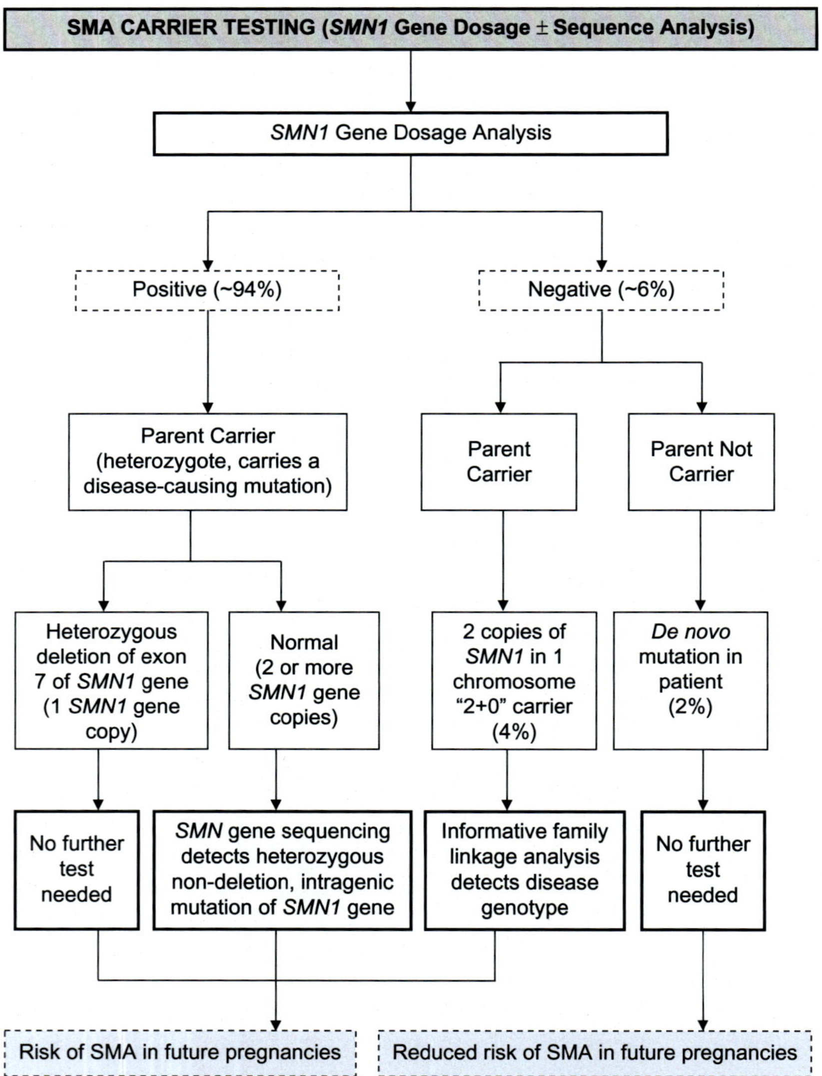

FIGURE 8.4 SMA carrier testing. This algorithm shows the diagnostic investigation of carrier status in potential carrier parents with *SMN1* gene dosage analysis plus/minus sequencing of the *SMN1* gene. Rarely, family linkage analysis may need to be employed in informative families, particularly if a deletion mutation has been transmitted more than once from a parent with two copies of *SMN1* on gene dosage testing. *Reprinted with permission from Markowitz et al.*[155]

the detection of a previously reported mutation supports its pathogenicity and its location in the *SMN1* gene. The use of long-range PCR or subcloning can also allow specific analysis of *SMN1* and confirm that the mutation is present in *SMN1*.[73] See Figure 8.3.

Carrier testing is feasible and accurate in the parents of patients with homozygous deletions of exon 7 or compound heterozygosity using a PCR-based dosage assay, known as "SMN gene dosage analysis" (Figure 8.4). It can also be performed using other techniques such as long-range PCR,

MLPA, and chromosomal microarray (CMA) that includes this gene segment. Sequencing of the *SMN* gene will detect point mutations in nondeletion carriers. Rarely, carriers may have two copies of *SMN1* on one chromosome (a so-called "2 + 0 carrier"); the incidence of this genotype is about 4% of the general population. In the "2 − 0 carriers", the SMA dosage carrier test will be falsely normal and thus one may need to pursue other methods such as family linkage analysis to identify the disease-associated genotype in families in which a deletion mutation has been transmitted more than once from a parent with two copies of the *SMN1* gene on gene dosage testing.[49,74–76] Due to the occurrence of *de novo* mutations in 2% of patients with SMA, one of the parents may not be a carrier.[73] So, in approximately 6% of parents of a child with SMA secondary to a homozygous *SMN1* deletion, *SMN1* gene dosage testing will be normal.

As mentioned, a general correlation has been found between *SMN2* copy number and disease severity, and it is relatively straightforward to determine *SMN2* copy number in individual patients using various methodologies. However, this correlation is not so strict that the severity or type of disease can be reliably predicted based on copy number, and it is hence not advisable to offer families prognostic information based on *SMN2* copy number assays.

Newborn Screening

The potential for newborn screening for SMA has been of great interest because the ideal time to initiate therapy would be prior to the initial degeneration of motor neurons, and newborn screening may help to identify presymptomatic individuals.[13,49] Swoboda et al. performed a prospective study in prenatally diagnosed infants with type I SMA and found that, associated with the initial onset of symptoms or decline in function in these young infants, there was electrophysiologic evidence of precipitous denervation (assessed with serial measurements of compound motor action potential amplitude and motor unit number estimation), suggesting that motor neuron loss occurs very early.[77] This provides further support for the potential usefulness of newborn screening to identify patients prior to the period of greatest motor neuron loss, should a therapeutic intervention become available.

Regarding the issue of routine preconception and prenatal screening for SMA, the American College of Obstetricians and Gynecologists Committee on Genetics published an opinion recommending against preconception and prenatal screening for SMA in the general population, citing insufficient evidence as to its cost-effectiveness and concerns regarding the education of the public and physicians about the complex issues raised by prenatal diagnosis of this disorder.[78] However, various other professional medical societies have stressed the importance of screening. To address this issue and create a consensus, a meeting was organized by National Institutes of Health in late 2009. The conclusion of the meeting was that a carrier screening is technically feasible but its implementation will require addressing broader issues of screening in general.[13,79]

OTHER DIAGNOSTIC TESTS

In patients with spinal muscular atrophy the serum creatine kinase may be 2- to 4-fold elevated, but not more than 10 times normal.[22] Nerve conduction studies demonstrate normal sensory potentials, but may show diminished compound motor action potential amplitudes.[77] Needle EMG in type II/III patients demonstrates a neurogenic pattern with high amplitude, long duration motor unit potentials with reduced recruitment pattern. Needle EMG in type I patients shows denervation changes but may not show evidence of reinnervation, as there may not have been enough SMN protein and/or time for this to occur yet. As will be discussed in more detail in the Pathology section, muscle biopsy in all types of SMA demonstrates a neurogenic pattern, with grouped atrophy; the small atrophic fibers are of both types (type I and type II) whereas the large ones are always type I fibers. Although a neurogenic pattern with large group atrophy is common, notably, a fetal appearance has been noted in muscle biopsy specimens from patients with type I SMA, leading some to question whether type I SMA may actually result from arrested development of the motor unit rather than degeneration of the motor neuron. Indeed, a morphometric analysis of fetuses with type I SMA at 12–15 weeks showed delayed maturation of myotubes when compared with controls.[80] While EMG continues to be used in the diagnosis of SMA in selected atypical cases, the use of muscle biopsy has become essentially obsolete.

MOLECULAR FUNCTION OF SMN

SMN is a 38 kDa protein found in all cells, located in both the cytoplasm and in the nucleus, where it localizes to structures known as Gemini of Cajal bodies, or simply "gems"[81]; therefore, its expression is ubiquitous. In patient fibroblasts, the number of gems correlates with SMA severity. SMN is also localized in motor neuron axons.[82] The SMN protein in conjunction with several Gemin proteins forms an SMN complex whose chaperone function facilitates the assembly of spliceosomal small nuclear ribonucleoprotein particles (snRNPs),[83–86] essential components of the spliceosome complex, and hence plays a critical role in pre-mRNA splicing. The SMN protein may also be essential in assisting arginine methylation of some splicing-related proteins,[87] transporting axonal mRNAs in motor neurons[88] and perhaps in other processes in muscle and neuromuscular junction. The role of SMN protein in axonal mRNA trafficking and mRNA splicing may explain the selective vulnerability of spinal cord motor neurons to decreased SMN protein; albeit not the only ones, these neurons have long axons and many targets, particularly in large muscles, and thus may be very dependent on axonal mRNA transport. Further, they may be extremely vulnerable to the mRNA splicing defect. Nevertheless, it continues to be unclear whether a splicing defect due to SMN protein deficiency, disruption of an additional axonal SMN function, an unknown function, or a combination of these is responsible for the pathogenesis of SMA.[70]

The complete absence of the SMN protein in cells is embryonic lethal in mice and other organisms.[58,85] The reason that motor neurons are specifically vulnerable to partial SMN deficiency is unclear. Mouse and zebrafish models have been used to explore this question, but because the *SMN2* gene is present only in humans, animal models have been created via knockout of *SMN* with insertion of a human *SMN2* transgene.[58,59,85] In addition to its role in spliceosome assembly, which is essential in all cells, SMN seems to associate with actin, which is involved in motor axon pathfinding and outgrowth during development. Two groups have demonstrated aberrant motor axon outgrowth and pathfinding in zebrafish models of SMA.[89,90] It has been shown using laser capture microdissection that the amount of full-length SMN from SMN2 is lower in motor neurons of normal transgenic mice than in other neuronal cell types; this may explain the selective vulnerability of motor neurons.[91] In both *Drosophila* and mice, SMN reduction leads to disrupted splicing of a minor intron in the *Stasimon* gene[92,93] and hence reduction of a spliced isoform in motor neurons and proprioceptive neurons. Expression of Stasimon in both sensory and cholinergic neurons of the SMN-deficient fly corrects some of the larval neuromuscular junction and muscle defects[92] and suggests that Stasimon has a role similar to that of SMN in the motor circuit function. Further, SMN seems to be required for the correct splicing of *Stasimon* and probably of other genes. Furthermore, SMN is required for normal sensory-motor circuit function in *Drosophila*; this circuit includes not only proprioceptive sensory neurons but also interneurons. However, whether this circuit is operative in the pathogenesis of SMA in humans remains unknown.

Patients with SMA also appear to have structural and physiological abnormalities of the neuromuscular junction.[94–98] Kong et al. showed that in severely affected SMA mice, prior to anterior horn cell death and/or axonal degeneration there is synaptic dysfunction at the neuromuscular junction in the form of decreased synaptic vesicle density at motor terminals, reduced quantal content, and slowed maturation of the acetylcholine receptor with prolonged retention of fetal characteristics.[96] Kariya et al. also demonstrated structural and functional abnormalities at the level of the neuromuscular junction that precede overt symptoms in mice, as well as structural abnormalities in the neuromuscular junctions of humans with SMA, and hence proposed that SMA may be a "synaptopathy."[99] Based on this concept and the electrophysiologic data suggesting dysfunction of the neuromuscular junction in patients with SMA types II and III, treatment approaches directed towards enhancing neuromuscular transmission may also be beneficial to patients with SMA.[100]

DIFFERENTIAL DIAGNOSIS

The differential diagnosis of 5q SMA is listed in Box 8.1 and is described in detail in this section.

Non-5q Spinal Muscular Atrophies

Non-5q SMAs are genetically heterogeneous, clinically diverse, and rare compared with 5q SMA.[45,101] Various published classifications of non-5q SMAs have proved either inadequate, or encyclopedic and confusing. Classification by distribution of weakness (distal, proximal, bulbar) and mode of inheritance is currently employed by most experts in this field.[102,103] The classification scheme outlined in Table 8.3, although based on these premises, has its own limitations. It does not include all non-5q motor axonopathies or neuronopathies or SMA plus syndromes, and especially excludes conditions with uncertain nosology or those with no gene/locus information. Table 8.3 is therefore offered as an aid to understanding non-5q SMAs by illustrating the major categories of a classification scheme.[45]

Distal SMAs (DSMA) present with *predominantly distal weakness* and exhibit significant phenotypic overlap with distal hereditary motor neuropathies or neuronopathies (dHMN). In spinal muscular atrophy with respiratory distress type I (SMARD1), most patients have low birth weight and present within the first 3 to 6 months of life with diaphragmatic paralysis, hypotonia, and distal more than proximal weakness, as well as with sensory and autonomic nerve involvement; these infants become ventilator dependent. In a series of 141 *SMN1* mutation-negative patients presenting with respiratory distress, and diaphragmatic and intercostal weakness, three distinct phenotypes were noted[104]: (1) congenital contractures with associated respiratory failure at birth; (2) respiratory distress with onset between 3 and 6 months, and hip flexion with minimal or no movements of the distal muscle groups; and (3) respiratory distress after age 6 months with congenital contractures. In a series of 10 patients with SMARD1 from the Netherlands, significant phenotypic variability was noted with no clear phenotype-genotype correlations.[105] CPK is usually normal and there is no cardiac involvement. In children with SMARD1, life expectancy is limited but rarely patients can have only mild sleep hypoventilation.[106] SMARD1 is caused by mutations

BOX 8.1 Differential Diagnosis of 5q Spinal Muscular Atrophy

Spinal Cord Disorders

- Neoplasms (SMA types I, II, III)
- Other myelopathies (SMA types I, II, III)

Other Motor Neuron Disorders

- SMARD1 (SMA type I)
- Juvenile muscular atrophy of distal upper extremity (Hirayama disease)
- Fazio-Londe disease, Brown-Vialetto-van Laere syndrome
- Other non-5q SMAs (SMA types I, II, III)
- Juvenile ALS (SMA types I, II, III)

Neuropathies

- Congenital hypomyelinating or axonal neuropathies (SMA types I, II)
- Hereditary motor and sensory neuropathies (SMA types I, II, III)
- CIDP (SMA types II, III)

Neuromuscular Junction Disorders

- Botulism (SMA type I)
- Congenital myasthenic syndromes (SMA types I, II, III)
- Lambert-Eaton myasthenic syndrome (SMA type III)
- Autoimmune myasthenia gravis (SMA types II, III)

Myopathies

- Congenital myopathies (SMA types I, II, III)
- Congenital myotonic dystrophy (SMA type I)
- Congenital muscular dystrophies (SMA types I, II)
- Muscular dystrophies (DMD/BMD, LGMD) (SMA type III)
- Mitochondrial myopathies (SMA types I, II, III)
- Acid maltase/Pompe disease (SMA types I, II, III)
- Other metabolic myopathies (SMA types I, II, III)
- Inflammatory myopathies (SMA type III)
- Channelopathies (SMA type III)

Other Disorders

- Chromosomal abnormalities (SMA types I, II, III)
- Prader-Willi syndrome (SMA type III)
- Central nervous system abnormalities (SMA types I, II, III)
- Hexosaminidase A deficiency (SMA types III, IV)

Abbreviations: ALS, amyotrophic lateral sclerosis; SMARD1, spinal muscular atrophy with respiratory distress 1; CIDP, chronic inflammatory polyneuropathy; DMD, Duchenne muscular dystrophy; BMD, Becker muscular dystrophy; LGMD, limb-girdle muscular dystrophy.
Modified with permission from Markowitz et al.[155]

in the gene *IGHMBP2* encoding the immunoglobulin μ-binding protein.[106]

HMN5A and Charcot-Marie-Tooth (CMT), type 2D are allelic conditions due to *GARS* gene mutations, and are characterized initially by upper-limb predominance of the weakness, with selective atrophy of the thenar eminence and first dorsal interosseous muscles; by contrast, the hypothenar eminence is spared until later in the course of the disease. Mutations in the *BSCL2* gene result in allelic phenotypes including spastic paraplegia with amyotrophy of hands and feet (Silver syndrome/SPG17), congenital generalized lipodystrophy, type 2, and distal SMA with early hand involvement (HMN5B).

Non-5q SMAs with proximal or diffuse weakness include adult-onset Finkel type SMA, associated also with cramps, generalized fasciculations, and sometimes with typical severe amyotrophic lateral sclerosis presentation (ALS8). TRPV4-related disorders are another example of diverse phenotypes, ranging from congenital SMA with contractures to CMT2C associated with vocal cord and phrenic nerve paralysis and to SMA with scapuloperoneal and laryngeal distribution of weakness. There are also non-5q spinal and bulbar muscular atrophies and SMA plus types (Table 8.3). Some of the arthrogryposis multiplex congenita syndromes need to be considered in the differential diagnosis of SMA type IA due to *SMN1* mutations but also other non-5q SMAs presenting at birth with arthrogryposis, like the lethal congenital contracture syndrome related to *GLE1* gene mutations. Brown-Vialetto-van Laere syndrome (BVVLS) and Fazio-Londe disease are clinically overlapping motor neuron diseases involving primarily the lower cranial nerves and presenting with ponto-bulbar palsy, sensorineural deafness (BVVLS), and respiratory failure. BVVLS and Fazio-Londe disease have been linked to mutations in the *RFT2* (*C20ORF54*) gene and defective riboflavin transport.[107] These syndromes are discussed in detail in Chapter 9. Juvenile ALS may also present with lower motor neuron signs early on.

The autosomal recessive pontocerebellar hypoplasias (PCH) feature postnatal progressive microcephaly combined with brainstem and cerebellar hemispheres that are hypoplastic at birth, with the cerebellar vermis relatively spared. To date, seven PCH clinical syndromes have been described (PCH1–7).[108]

When PCH is combined with anterior horn cell degeneration, the designation PCH1 is applied; PCH1 has been linked to mutations in the *VRK1* gene.[109] This is an unusual phenotype, in that a characteristic brain malformation is combined with abnormalities of motor neurons in the spinal cord. Recently, Rudnik-Schöneborn and colleagues sequenced *EXOCS3* in a cohort of 27 families with PCH1, and found mutations in 37%.[110] A common c.395A > C, p.D132A mutation was present in about half of these families. While there was some variation in clinical features, enough phenotype-genotype correlation was present to underscore the continued usefulness of PCH1 as a clinical category.

Other non-5q spinal and bulbar SMAs include X-linked spinal and bulbar muscular atrophy (SBMA/

TABLE 8.3 Simplified Classification of Non-5q Spinal Muscular Atrophies

Gene/Locus	Disease/Phenotype, Selected Distinguishing Features	Selected OMIM Designations
Distal spinal muscular atrophy/distal hereditary motor neuropathy or neuronopathy (DSMA/distal HMN)		
Autosomal Recessive		
IGHMBP2	SMA with respiratory distress or diaphragmatic SMA	SMARD1/HMN6 or DSMA1
9p21.1-pl2	Distal HMN	DSMA2/HMNJ
11q13	Distal SMA	DSMA3/HMN3,4
PLEKHG5	Lower motor neuron syndrome with childhood onset	DSMA4
Autosomal Dominant		
7q34-q36	Distal HMN/SMA, juvenile	HMN1
HSPB8	Distal adult HMN, type IIA	HMN2A
HSPB1	Distal HMN, type IIB	HMN2B
HSPB3	Distal HMN, type IIC	HMN2C
GARS	Distal SMA with upper-limb predominance, type VA Charcot-Marie-Tooth disease 2D	HMN5A* CMT2D
BSCL2	Distal SMA with upper-limb predominance, type VB Silver syndrome/SPG17	HMN5B*
SLCA7	Distal HMN with vocal cord paralysis	HMN7A
Dynactin1	Distal HMN with vocal cord paralysis	HMN7B
Proximal spinal muscular atrophy (± distal involvement)		
Autosomal Dominant		
VAPB	SMA with late-onset, Finkel type/ALS8	
TRPV4	Congenital SMA with contractures/SMA, congenital, nonprogressive, with lower limb predominance Scapuloperoneal SMA Charcot-Marie-Tooth, type 2C	SPSMA HMSN2C
DYNC1H1	SMA with lower extremity predominance (early onset)	SMALED
TFG	HMSN, proximal (Okinawa type)	HMSNP
Other non-5q spinal and bulbar muscular atrophies, SMA plus types		
Autosomal recessive		
GLE1	Lethal arthrogryposis with anterior horn cell disease or lethal congenital contracture syndrome	LAAHD
VRK1, EXOCS3	Pontocerebellar hypoplasia with SMA	SMA-PCH1
RFT2 (C20ORF54)	Brown-Vialetto-Van Laere syndrome Fazio-Londe disease, bulbar palsy	BVVLS
X-linked recessive		
Androgen receptor	Bulbo-SMA, Kennedy disease	SBMA/SMAX1
UBA1	Infantile SMA with arthrogryposis	SMAX2
ATP7A	Distal SMA, X-linked	SMAX3

**Listed in OMIM as HMN5.*

SMAX1), also known as Kennedy disease. SBMA has an onset between 20 and 50 years; adult men with SBMA often present with proximal muscle weakness and atrophy, bulbar symptoms, fasciculations, gynecomastia, testicular atrophy, and reduced fertility. It is related to an expansion of a CAG trinucleotide repeat in the androgen receptor gene.[111] Further, infantile SMA with arthrogryposis ± bone fractures, dysmorphic features, myopathic facies, hypotonia, areflexia, and digital contractures may represent a lethal X-linked form of SMA (SMAX2) and be confused with SMA type IA; mutations in the *UBA1* gene have been detected in this group of patients.[112] A clinical and pathological study of a *UBA1* gene mutation-positive SMAX2 case showed white matter abnormalities on MRI brain imaging, prominent motor and sensory system, as well as cerebellar involvement and widespread inflammatory changes on muscle biopsy.[113]

Within the complex field of non-5q SMAs, the classification soup of letters and numbers poses an added challenge to clinicians who need to make a specific diagnosis and determine which genetic tests to consider. Fortunately, several excellent websites interconnect and provide clinicians with a full range of helpful information. One may, for example, search clinical details at the Washington University site[114] and be directed from there to Online Mendelian Inheritance in Man (OMIM)[115] for genetic information, and from OMIM to GeneTests[116] for available reviews and even to laboratories that offer clinical testing. In the future, widespread availability of genetic testing will facilitate accurate diagnosis, with important implications for genetic counseling, determining prognosis, and development of future clinical trials that may lead to treatments for these rare conditions.

Congenital Neuropathies

As discussed earlier, Korinthenberg et al.[42] reported three newborn siblings with motor and sensory axonal neuropathy who had *SMN1* mutations. Because the presentation can be similar, congenital hypomyelinating and axonal neuropathies, as well as other hereditary and acquired neuropathies, should be considered in the differential diagnosis of SMA (Box 8.1).

Hexosaminidase A Deficiency

Patients with late-onset hexosaminidase A deficiency (GM2 gangliosidosis) may show progressive muscle wasting, weakness, fasciculations, and dysarthria, sometimes associated with upper motor neuron signs, abnormal saccades, and nonspecific cerebellar atrophy.[117] The presentation may be indistinguishable for SMA type III or type IV, or early-onset amyotrophic lateral sclerosis.[118,119]

Juvenile Muscular Atrophy of Distal Upper Extremity (Hirayama Disease)

Juvenile muscular atrophy of distal upper extremity (JMADUE) or Hirayama disease was first described in Japan in 1959.[120] The clinical criteria for the diagnosis of Hirayama disease include (1) gradual onset of predominantly distal weakness and atrophy of the forearm and hand muscles; (2) onset of symptoms between the ages of 10 years to the early 20s; (3) unilaterally dominant symptoms and signs; (4) gradual progression for the first several years followed by stabilization of the clinical course; (5) lack of any significant sensory changes, abnormal muscle stretch reflexes, or Babinski signs; (6) lack of symptoms in the lower extremities; and (7) exclusion of other disorders including spinal cord tumors, juvenile ALS, syringomyelia, multifocal motor neuropathy, and polymyelitis.[121,122] The intrinsic hand muscles (interossei, thenar, and hypothenar muscle groups) as well as the ulnar side of the forearm are predominantly involved. There is no atrophy or weakness of the brachioradialis or biceps brachii muscles. This creates a clearly defined margin between the atrophic distal muscles and the well-preserved brachioradialis, and thus the term "oblique amyotrophy." In about 15% of the cases there might be slight triceps brachii atrophy as well, but in most cases atrophy and weakness are not present in the upper arms.[122] Although this condition has a predominantly unilateral presentation (approximately 70%), in about 30% of the patients bilateral upper extremity involvement may be identified. Muscle stretch reflexes are usually preserved in all extremities, and there are no pyramidal tract signs in the lower extremities. Of note, there appears to be exacerbation of the hand muscle weakness in a cold environment. There is also a fine finger tremor noted in the outstretched position of the arms and hands. Fasciculations may be observed not only in the primarily affected muscles but also in nonatrophic ones in the upper extremities. Although the condition may progress fairly rapidly, in most cases there is slow progression after the onset with stabilization of the symptoms and signs after about 5 years. Slight improvement in muscle strength may be noted after the cessation of progression.[122]

Nerve conduction studies may show low amplitude CMAPs of the median and ulnar nerves, but the sensory nerve action potentials are preserved. Needle EMG shows evidence of fibrillation potentials, positive waves, and signs of reinnervation in the distribution of C7, C8, and T1 myotomes, which can also be present in more proximal, nonatrophic muscles such as the deltoid and brachioradialis but also in clinically unaffected contralateral hand muscles.[123–125]

Flexion myelopathy has been proposed as a probable pathogenetic mechanism, with MRI showing anterior shift of the posterior dural sac and compression of the lower cervical cord consistent with "tight dural canal in flexion."[122,126–128] The lower cervical cord moves forward during neck flexion and contacts the posterior surface of the vertebra, and thus it becomes flattened at the contact segment. A pathological study of an autopsy case showed anterior-posterior flattening of the spinal cord focused at the C7 to C8 level with a decreased number of the C7–C8 anterior horn cells predominantly on the left side in a patient who had left-hand muscle weakness for 23 years before his death from lung cancer.[129]

Dr. Hirayama has used the term "focal cervical ischemic poliomyelopathy."[130] A three-generation Greek family with four members affected by asymmetric upper limb amyotrophy has been reported.[131] DNA testing for DSMA-V was negative. Assuming the members of this family had a typical phenotype, this familial occurrence of Hirayama disease suggests the possible involvement of genetic factors in its pathogenesis.

The management consists of restriction of excessive neck flexion, use of low pillows, and avoidance of sports that require frequent neck flexion. Anterior fusion of vertebrae and use of a cervical collar have also been reported, but the indications for and the efficacy of these interventions remain controversial[122,132,133] or unproven.[134,135]

Other Disorders

Other conditions such as juvenile amyotrophic lateral sclerosis, which involves both upper and lower motor neurons, are discussed elsewhere in this text and are included in Box 8.1.

THE PATHOLOGY OF SMA

SMA is characterized by spinal motor neuron loss and skeletal muscle atrophy. However, a precise understanding of how the cellular pathology of the disease evolves over its course continues to be handicapped by a paucity of suitable patient material.[136] Our current appreciation of SMA pathophysiology stems largely from analyses of autopsy material, reflecting pathology at the end of the disease.[137] Analyses of early disease pathology have relied mostly on murine models. Still, from the current studies, certain key findings emerge.[137] These, almost invariably, include a profound loss of the large anterior horn cells of the spinal cord (Figure 8.5). More infrequently, motor neurons are found mispositioned.[138] Anterior horn motor neurons that remain often appear swollen or chromatolytic, containing phosphorylated neurofilaments, ribosomes, and vesicles. In fewer instances, such pathology has also been observed in dorsal root neurons, cells comprising Clark's column, and the thalamus of the brain. Intramuscular nerves in SMA patients sometimes appear beaded or are reported to undergo Wallerian degeneration.

Considering the degeneration of the spinal motor neurons, an expected pathological characteristic is neurogenic atrophy of the skeletal muscles.[139] Muscle from severely (type I) affected patients typically consists of a sea of small, rounded, atrophic fibers within which are sometimes seen islands of one or more large hypertrophic cells (Figure 8.5). In some instances, the atrophic fibers also appear immature with centrally rather than peripherally located nuclei, giving the appearance of a myopathic condition.[140] Owing to the severity of SMA type I and thus the inability of motor neurons in this disease to re-innervate fibers, muscle transverse sections often continue to appear checkerboard with equal numbers of type I and type II fibers. In contrast to muscle from severely affected SMA patients, muscle from more mildly affected individuals exhibits evidence of repeated cycles of denervation and innervation. Accordingly, atrophic and hypertrophic fibers appear grouped by type when examined in transverse sections. Muscle from mildly affected SMA patients has also been observed to exhibit myopathic features such as the presence of central nuclei, fiber splitting, basophilia, and an increase in endomysial connective tissue.[139] These characteristics are accompanied by somewhat elevated levels of serum creatine kinase.

While SMA has historically been described as a disease characterized by motor neuron cell body loss, more recent data suggest the earliest pathological effects of reduced SMN are observed distally, at the neuromuscular junctions.[99] This early pathology that affects the pre- as well as post-synapse was first observed in mouse models of the disease and has now been confirmed in human samples as well (Figure 8.6).[99,141]

ANIMAL MODELS

Animal models are important tools to understand human disease. Accordingly, a number of animal models of SMA have been developed in the last two decades.[142] SMA mice, in particular, have served as the basis of preclinical development of therapies for the human disease.[136] In conjunction with invertebrate models of the disease, they have also provided the community with a wealth of information about the cellular basis of human SMA.

Worm, Fly, and Fish Models of SMA

An accurate genetic model of SMA requires the presence of constant, ubiquitous, low levels of the SMN protein. In human patients, these low levels of SMN protein derive from the copy gene, *SMN2*. However, the genomes of model organisms harbor only one *SMN* gene, an *SMN1* equivalent. Ablating this gene in the worm, *C. elegans*; the fly, *D. melanogaster*; or zebrafish, *D. rerio* is therefore embryonic lethal, once the maternal contribution of

(A)

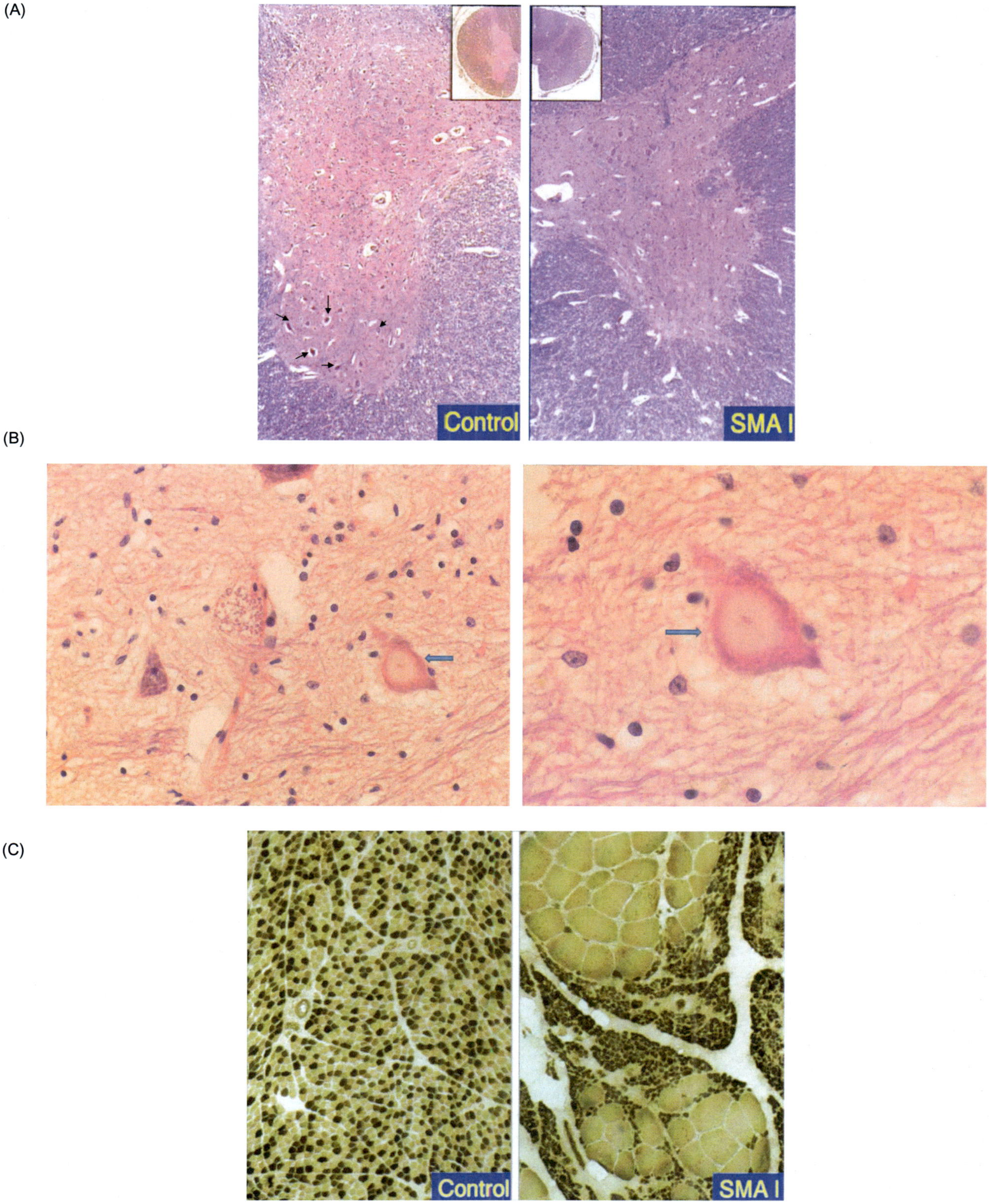

FIGURE 8.5 The pathology of SMA. (**A**) Spinal cord. Loss of large motor neurons (arrows) in the anterior horns of the spinal cord in tissue obtained from autopsy material from a patient with SMA type I (stained with cresyl echt violet). (**B**) Motor neuron undergoing chromatolysis appears swollen at lower and higher magnification (arrows). (**C**) Skeletal muscle biopsy from a patient with SMA type I showing large group atrophy. The small fibers have staining characteristics of both types (I and II), whereas the large hypertrophic fibers all stain as type I (ATPase stain, pH 9.4; dark fibers are type II fibers). A control muscle biopsy is shown on the left for comparison.

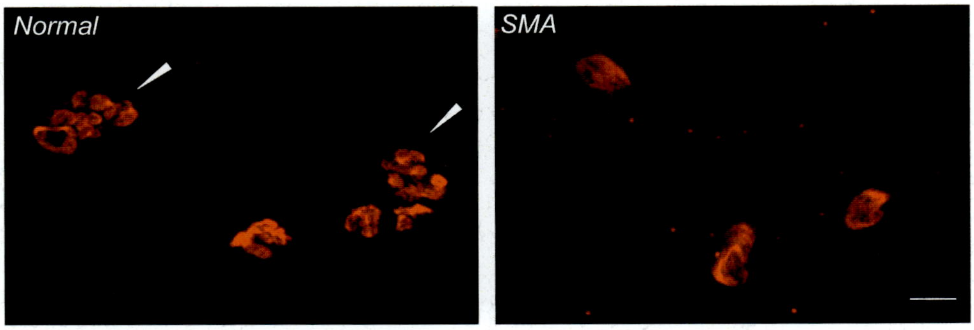

FIGURE 8.6 The neuromuscular junction defects in SMA. The postsynaptic motor endplates are smaller and less complex compared to those of a control normal individual (arrowheads). Tissue was obtained from autopsy and stained as a whole mount preparation with fluorescently labeled α-bungarotoxin. Scale bar: 30 μM. SMA, spinal muscular atrophy. *Reprinted with permission from Monani and De Vivo.*[136]

SMN protein to the embryo is depleted.[142] The absence of the *SMN2* gene in many of the reported worm, fly, and fish models is therefore a handicap, and findings that have emerged from a study of these models must be interpreted with caution. Nevertheless, a careful calibration of SMN levels in these organisms by newer technologies such as RNA interference has enabled scientists to make considerable progress in addressing important questions about human SMA.[142]

Murine Models of SMA—Additional Insights into Disease Pathology

The most relevant current models of human SMA are, arguably, those involving transgenic mice. These were generated by introducing varying copies of the human *SMN2* gene onto the null murine (Smn−/−) genetic background. Two copies of the *SMN2* gene rescued the Smn−/− embryonic lethality and resulted in a model of severe SMA.[58,59] Such mice were found to express greatly reduced SMN protein, mimicking findings in SMA type I patients. In contrast, mice harboring eight copies of the transgene expressed near-normal levels of the protein and were found to be no different from wild-type littermates.[58] The distinct phenotypes of the two sets of mice provided clear evidence of the ability of *SMN2* to modify disease outcome, justifying the use of the copy gene as a molecular target in treating SMA.

In the period since the first *SMN2*-transgenic mice were generated, a number of additional transgenes harboring mild as well as severe mutations have been employed to model the spectrum of phenotypes observed in humans and to study disease pathology. These studies have shed substantial light on human SMA. For instance, an inability to reconcile early muscle weakness with the continued presence of motor neuron cell bodies within the spinal cord led to the discovery of distal defects of the motor unit. These experiments revealed that low SMN has its earliest effects at the neuromuscular synapse, preventing proper nerve terminal arborization and precluding the transition of the post-synapse to its normal, mature state.[99] Importantly, the findings were subsequently confirmed in tissue from human patients.[141] In a series of follow-up mouse studies, it was determined that central synapses onto the motor neurons were also affected by reduced SMN, implying that eventual motor neuron loss in SMA may be a consequence of events initiated at the synapse.[143–145] Subsequent detailed analyses of severely affected SMA mice revealed defects that extended beyond motor neurons to the muscle, brain, and even heart.[28,146,147] Considering the housekeeping role of SMN in the splicing cascade, perhaps such findings are not unexpected, although the multisystemic nature of the disease reported in these mice is unlikely to be characteristic of all but the most severely affected SMA patients.

Murine Models of SMA—Revealing the Cellular and Temporal Requirements of SMN

Precisely defining the cellular sites of action of a protein and determining its temporal requirements is critical to effectively treating any associated disease. The SMA mice described above have been an excellent resource in allowing scientists to address these issues. Considering the neuromuscular nature of the human disease, initial studies to define the cellular sites of action of the SMN protein focused on muscle and nerve. These studies revealed that while depleting SMN in motor neurons of healthy mice caused disease, restoring SMN pan-neuronally or to motor neurons alone of SMA mice failed to fully rescue the disease phenotype.[143,148–150] Selectively restoring SMN to myofibers also failed to benefit the mice.[151] However, more recent experiments to restore SMN to muscle stem cells were found to mitigate disease.[148] These studies suggest that motor neurons and muscle independently contribute to the disease phenotype. Restoring the protein to both

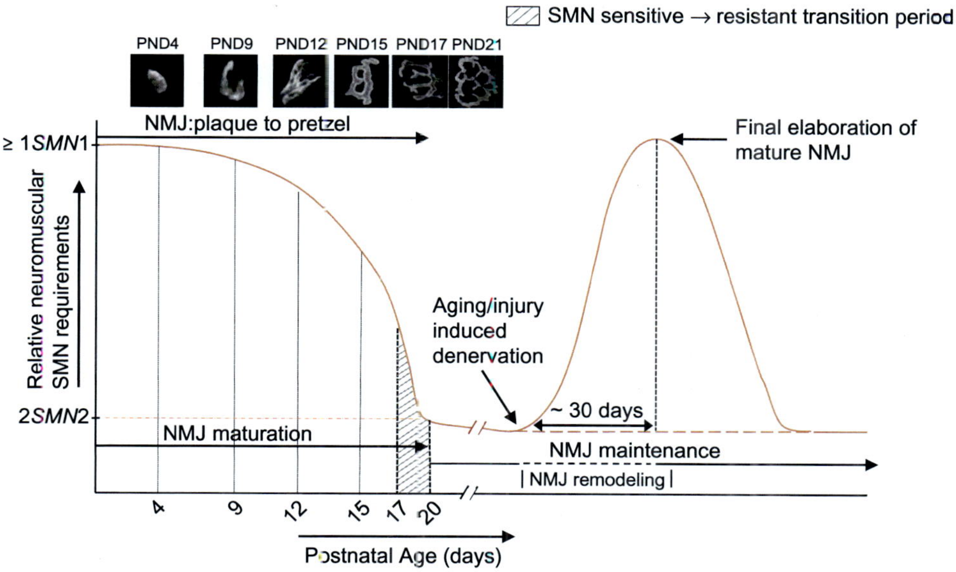

FIGURE 8.7 A model depicting the postnatal requirements for the SMN protein. The need for SMN is satisfied by one or more *SMN1* copies, or *SMN2* copies that express equivalent levels of the protein to the *SMN1* gene(s). The requirement for SMN is highest during neonatal life, when the neuromuscular junctions are maturing. The requirement declines by ~PND20 when the neuromuscular junctions are fully mature, to levels that can be satisfied by two copies of the *SMN2* gene. An approximately 3-day window defines the transition from an SMN-sensitive to SMN-resistant state. Disruption of the NMJ during the normal process of aging or as a consequence of injury triggers a surge in demand for SMN in the motor unit, which is not met by two *SMN2* copies. The enhanced requirement for SMN reaches its zenith as the neuromuscular junction matures and precludes, in SMN-depleted mutants, an ability to repair the synapse. Thus, SMN is postulated to play dual roles at the neuromuscular junction: its initial maturation and its continued maintenance. *Reprinted with permission from Kariya et al.*[154]

tissue types may be sufficient to fully rescue the SMA phenotype, but will likely depend on precise levels of the protein in the patient being treated. This prediction stems from the housekeeping role of SMN and thus the expectation that in the most severely affected mice/patients, selectively restoring protein to cells may not be fully protective against the inability to efficiently assemble snRNP particles in nontargeted tissue. By exploiting *Smn* alleles that can be induced in a temporal manner, investigators have also shed light on when the protein might have to be reinstated in gene or protein replacement-type therapies.[152–154] These studies have arrived at two important conclusions. First, restoring SMN early during the course of the disease is more effective than restoring protein late.[153] Nevertheless, remarkable therapeutic benefit may be derived from augmenting SMN even after the onset of a severe clinical phenotype. Second, the requirements for the protein change profoundly with age (Figure 8.7). Indeed, studies have demonstrated that once the neuromuscular system matures, levels of protein may be allowed to fall dramatically without causing overt disease.[154] From a clinical standpoint, this finding suggests that maintaining high levels of the SMN protein using small molecule-type drugs may not be essential once adulthood is reached, precluding long-term, off-target effects characteristic of many chemical entities. Collectively, the data gathered exemplify the utility of the SMA models in developing rational therapeutic strategies.

TREATMENT

Currently, there is no cure for spinal muscular atrophy. However, despite the presence of homozygous deletions of *SMN1* in the majority of patients with SMA,[155] the unique structure of the 5q11.1-13.3 inverted duplication provides potential therapeutic targets. There has been great interest in identifying agents that can increase the amount of full-length SMN protein by upregulating expression of the *SMN2* gene or promoting inclusion of exon 7. Researchers are also actively exploring several other approaches to treatment.

Clinical Trials in SMA—Outcome Measures

Several challenges face those designing and implementing clinical trials in SMA. The natural history of the disease, especially in patients with the intermediate forms of the disease, is that of early decline and then a prolonged plateau period followed by modest functional decline noted over a number of years. In SMA types II and III, a 48-month natural history study initiated by the Pediatric Neuromuscular Clinical Research (PNCR) network in 2005 showed no measurable disease progression over 12 months and slow functional decline beyond 12 months of observation.[156,157] Hence, it may be extremely difficult to detect an effect of a therapy on the rate of progression

TABLE 8.4 Outcome Measures Used in Spinal Muscular Atrophy Clinical Trials

Outcome Measures	Measure
• Myometry	• Assessment of strength
• Motor function measure (MFM)	• Functional scale
• GMFM	• Functional scale
• CHOP INTEND	• Functional scale
• TIMP	• Functional scale
• HFMS	• Functional scale
• HFMSE	• Functional scale
• Six-min walk test	• Measurement of function and endurance
• Pulmonary function tests	• Functional scale
• MUNE	• Measurement of functional motor units
• Quantitative muscle ultrasound	• Correlation with strength
• Serologic biomarkers (blood SMN messenger RNA, protein levels)	• Correlation with disease severity
• PedsQL, neuromuscular module	• Quality of life questionnaire

Abbreviations: CHOP INTEND = Children's Hospital of Philadelphia infant test of neuromuscular disorders; GMFM = gross motor function measure; HFMS = Hammersmith functional motor scale; MFM = motor function measure; HFMSE = Expanded HFMS; MUNE = motor unit number estimation; TIMP = test of infant motor performance.
Adapted with permission from Lippincott Williams and Wilkins/Wolters Kluwer Health.[169]

over a 12- or 18-month period. An intervention will have to cause improvement in function rather than slowing/arrest of decline in order for a drug effect to be detected.

Quantitative strength measures may be insensitive because the symptoms tend to stabilize over time, and may also be unsuitable for very young children. Functional status scales may be difficult to interpret as development progresses and children learn to compensate for their weakness. Contractures may also complicate the picture.[16,158] However, several new outcome measures have shown encouraging results in different subsets of SMA patients (Table 8.4). For ambulatory SMA patients the Six-Minute Walk Test has shown good correlation with other established outcome measures such as the Hammersmith Functional Motor Scale Expanded ($r = 0.83$, $p < 0.0001$), 10 minute walk/run ($r = -0.87$, $p < 0.0001$), and knee flexor strength ($r = 0.62$, $p = 0.01$). The Six-Minute Walk Test is an objective and easy-to-administer test, and it also detects motor fatigue in SMA.[159] Studies of therapies for patients with type I SMA have been partially limited by the fact that rigorous motor skills assessment in infants has been difficult. However, the Children's Hospital of Philadelphia Infant Test of Neuromuscular Disorders (CHOP INTEND) was developed for this purpose and could be a reliable outcome measure in infants.[157] Several other outcome measures for clinical trials have been established to be reliable in SMA populations including the Children's Hospital of Philadelphia Test of Strength in Spinal Muscular Atrophy, the Gross Motor Function Measure™ (GMFM),[161] Hammersmith Functional Motor Scale (HFMS),[162] and Modified[163] or Expanded HFMS.[20] Montes et al. reviewed these, concluding that a consensus needs to be reached as to the most reliable outcome measures for clinical trials (Chapter 54).[161,164]

Electrophysiologic outcome measures have included motor unit number estimation (MUNE) and maximum compound motor action potential (CMAP) amplitude. An increase in CMAP amplitude captures both an increase in the number of motor units, or the presence of collateral sprouting of axons from adjacent surviving motor neurons. In a study of these two parameters in SMA patients, Swoboda et al. found a statistically significant correlation between compound motor action potential amplitude at initial assessment and the patients' functional outcome. Patients with type II SMA showed a decline in maximum compound motor action potential amplitude over time, and type I and II patients showed an age-dependent decline in motor unit number estimation and maximum compound motor action potential amplitude. As described above, prospective studies using motor unit number estimation and compound motor action potential amplitude in infants diagnosed prenatally with type I SMA found precipitous denervation associated with the onset of symptoms or decline in function, suggesting that these are useful parameters to follow in individual patients over time.[77] Of interest, the recent PNCR natural history longitudinal study showed mild increase in MUNE and stable CMAP amplitude in a 62-patient cohort of type II and III SMA patients observed prospectively for up to 42 months.[165] At baseline, as shown before, these measures detected significant motor neuron loss and compensatory collateral reinnervation.

Lastly, serologic biomarkers in the form of SMN mRNA and protein levels can be measured reliably from peripheral blood leukocytes in patients, but they do not necessarily correlate with disease severity. One group found that patients with type II and III SMA had relatively normal peripheral blood SMN protein levels.[72] Another group

TABLE 8.5 Status of Therapeutics Development in Spinal Muscular Atrophy

Therapeutic Targets	Approaches	Clinical Trials
Increase SMN transcript	• Histone deacetylase inhibitors • Nonhistone deacetylase inhibitors • Quinazoline • Prolactin	• Valproic acid, sodium 4-phenylbutyrate • Hydroxyurea • Repligen RG3039
SMN2 exon 7 inclusion	• Antisense oligonucleotides	• ISIS SMNRx
Stablization of SMN protein	• Aminoglycoside • Proteasome inhibitors • Indoprofen	
Neuroprotection	• Neurotrophic factors	• Riluzole • Gabapentin • Olesoxime (TRO19622)
Cell therapy	• Stem cells	
Replacement of *SMN1*	• Gene therapy	

Reprinted with permission from Lippincott Williams and Wilkins/Wolters Kluwer Health.[169]

found a high correlation between mouse leukocyte and spinal cord *SMN* expression, but only a moderate correlation between leukocyte *SMN* expression and age of disease onset in patients with SMA. Thus, it has been proposed that peripheral blood SMN mRNA levels and/or protein levels may be useful to follow in individual patients as a response to therapy, but they may not be useful to compare mRNA and protein levels between patients.[72,166,167]

Clinical Trials in SMA—Therapeutics

SMA is a unique "translational" disease because of the presence of the *SMN2* gene, which is a well-validated target for therapeutic interventions. The status of therapeutics development is shown in Table 8.5. The effort in therapeutics is aimed at finding pharmaceutical compounds that can upregulate *SMN2* expression or affect other modifying genes to produce more functional SMN protein.

Agents That Upregulate *SMN2* Gene Expression and Promote Exon 7 Inclusion

Small Molecules

A class of drugs known as histone deacetylase inhibitors (HDACIs) has been investigated extensively as potential therapeutic agents in SMA.[168,169] Histones, which are core proteins in chromatin, play a role in epigenetic regulation of gene expression via their acetylation status. Several compounds that are HDACIs have been shown to increase full-length *SMN2* transcript levels in cell lines from patients,[72,170,171] usually by activating the human *SMN2* promoter, enhancing transcription, and correcting the splicing pattern.

Phenylbutyrate is an HDACI that has been shown to increase the amount of full-length SMN mRNA, SMN protein levels, and the number of nuclear gems in fibroblast cultures from type I, II, and III patients *in vitro*.[172] Other pilot studies have also shown increased SMN gene expression in peripheral leukocytes[170] and improvement in HFMS scores[174] from treatment with oral phenylbutyrate. However, a randomized placebo-controlled trial found no significant improvement on the HFMS with a phenylbutyrate regimen.[175]

Valproic acid, which is approved for the treatment of epilepsy, is another HDACI that has been shown to increase the amount of SMN protein in fibroblasts derived from type I SMA patients.[176,177] A phase II open label study was conducted in a group of type I, II, and III SMA patients treated with doses of valproic acid sufficient to achieve overnight trough levels of 50–100 mg/dL for 12 months. These patients showed no evident hepatotoxicity but a significant proportion of the first patients enrolled developed low free or total plasma carnitine levels, associated with increased weakness in two patients, which prompted a recommendation for carnitine supplementation in all patients. Fat mass increased by 38% between baseline and 12 months for the type II patients, as opposed to a 14% increase in the type III patients. Weight was associated with a decline in motor function, measured by modified HFMS, in several patients. Significantly increased mean modified HFMS score was seen in 27 type II patients, most of whom were under 5 years of age; however, the possibility of developmental gain in motor function cannot be excluded. Bone mineral density and maximum ulnar compound motor action potential scores also increased significantly.[178] A phase II randomized double blind placebo controlled CARNIVAL TRIAL has shown no difference in modified HFMS after 6 months of treatment with valproate compared with placebo group in a nonambulatory cohort of type II SMA.[179]

A proprietary small molecule developed by PTC Therapeutics, Inc. (NJ, USA), Roche (Basel, Switzerland), and the SMA Foundation (New York, NY, USA) extends the life expectancy of the SMN Δ7 severe mouse model from 14 days to more than 6 months. Along with the quinazoline derivatives, this compound is further developed for upcoming clinical trials in humans. The distinct advantage of these small molecules is that they can cross the blood-brain barrier (BBB).

The FDA-approved drug hydroxyurea, a non-HDACI, was identified in the course of drug screens using cell lines from SMA patients to increase the amount of full-length SMN transcript and protein *in vitro*.[180–182] However, a small pilot study of hydroxyurea at three different doses for 8 weeks in 33 type II and III patients showed no statistically significant benefit.[181] A randomized, double-blind, placebo-controlled trial failed to show any improvement over an 18-month period.[183]

Albuterol, a beta-adrenergic agonist, was evaluated in a pilot study of 13 SMA type II and III patients due to its reported positive effect on muscle strength in healthy individuals. At 6 months a significant improvement was noted in myometry, FVC, and DEXA scores, but there was no significant change in Medical Research Council strength score.[184] Another pilot study of 23 SMA type II patients treated with salbutamol (a form of albuterol) for 12 months showed improved functional scores on the HFMS after 6 and 12 months, but this was not a placebo-controlled study and must be interpreted with caution.[185] On an *in vitro* level, salbutamol has been shown to increase full-length SMN mRNA, SMN protein, and gem numbers by promoting inclusion of exon 7.[186] The response was directly proportional to *SMN2* gene copy number.[187] This finding prompts further interest in exploring the effects of beta agonists on SMA with randomized controlled trials.

Quinazoline compounds increase expression of *SMN2* transcripts with one study demonstrating reduction of motor neuron loss and improvement in mobility and lifespan of severe SMA mouse models following administration of quinazoline derivatives.[188] A recent study of Repligen RG3039 (a leading quinazoline derivative) in mice was found to increase *SMN2* transcript levels, SMN protein, and lifespan, with no significant adverse effects. A single-dose phase I clinical trial of RG3039 in healthy volunteers found the drug to be well tolerated at all the tested doses.[189] A multidose safety study is currently underway.

Another HDACI, LBH589 (hydroxamic acid), a C5-quinazoline derivative, already widely used in cancer clinical trials, could be a promising candidate for SMA therapy. It targets the DcpS enzyme and induces a 10-fold increase in SMN levels, 2- to 3-fold increase in full length SMN2, and marked increase in number of gems in SMA fibroblasts that were unresponsive to valproate.[190] Clinical trials have not been conducted yet.

Prolactin has recently been tested as a potential therapeutic agent for the treatment of SMA.[191,192] Systemic administration of prolactin in severe SMA mouse models increased SMN levels in the brain and spinal cord, leading to enhanced motor function and survival.[192]

N-methyl-D-aspartate receptor activation was found to increase *SMN2* expression in a dose-dependent manner.[193] In an SMA mouse model it was shown to increase motor unit postnatal maturation, decrease apoptosis in the spinal cord, induce a marked increase of *SMN* expression, and extend life span. Thus, the agents discussed here hold promise as future therapeutic tools.

Neuroprotective, SMN Protein Stabilization Agents

Other Small Molecules

Aminoglycosides appear to promote read-through of the stop codon in exon 8 and thereby stabilize the SMN protein, providing another way to increase SMN protein levels in patient fibroblasts.[194] Heier treated SMA mice with the aminoglycoside geneticin (G418), resulting in increased SMN protein level and improved motor function, but unfortunately this agent caused significant toxicity in mice.[195]

Riluzole and Gabapentin, "the neuroprotective agents," were studied to assess the effect on motor performance in SMA. Unfortunately, the results were not found encouraging or the studies were not adequate to show efficacy.[196–199] A phase II multicenter, randomized, double-blind study is currently being conducted to assess the efficacy and safety of riluzole in young adults with SMA types II and III.

Olesoxime (TRO19622), a novel neuroprotective compound, has been tested in a randomized, multicenter, parallel group, double-blind, placebo-controlled trial conducted in Europe. Over 150 nonambulant type II and III SMA patients aged from 3 to 25 years were enrolled in this efficacy and safety trial with the primary endpoint being a change in the Motor Function Measure scale.[200]

Other Approaches

Antisense Oligonucleotides

Antisense oligonucleotides (ASOs) have been developed that block an intronic splicing suppressor element, which in turn prevents skipping of exon 7 (Figure 8.8).[201]

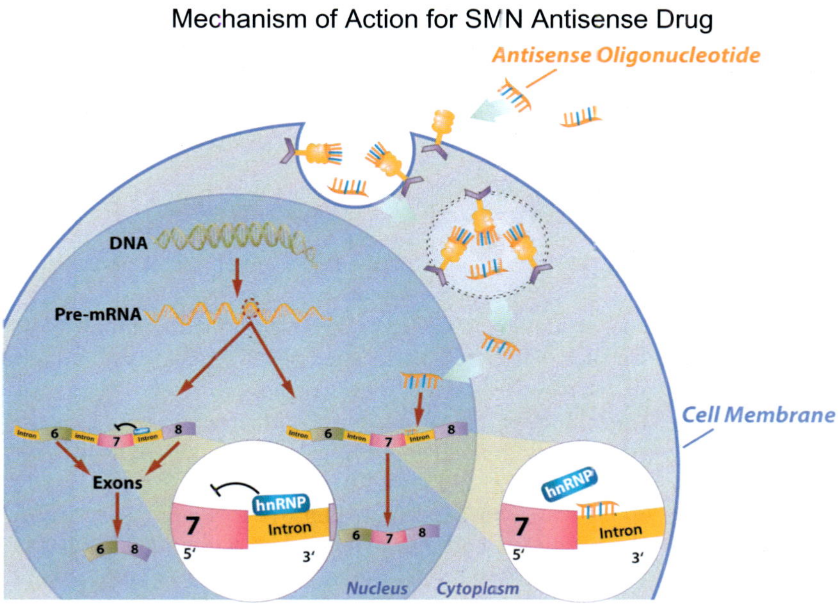

FIGURE 8.8 Mechanism of action for SMN antisense oligonucleotide. Single-stranded antisense oligonucleotides (ASOs) interact with proteins on the surface of cells and enter the cytoplasm by endocytosis. Subsequently, they escape the endosome and enter the nucleus, where they bind to the *SMN2* pre-mRNA and block an intronic splicing suppressor element located in intron 7. The binding of the ASO to the RNA displaces protein hnRNP, which normally suppresses splicing of exon 7; thus, it enhances exon 7 inclusion during splicing events and results in the production of full-length SMN protein. *Courtesy of Frank Bennett, PhD, ISIS Pharmaceuticals, Carlsbad, CA, USA.*

Hua et al. in their work on transgenic mice found these intronic splicing suppressors to be located in intron 7 as tandem motifs, namely hnRNP A1/A2.[202] Blocking of these motifs by ASOs enhanced exon 7 inclusion in the SMA mouse model. Similar work by Singh et al. demonstrated enhanced production of full-length SMN mRNA in fibroblasts from patients treated by antisense oligonucleotides.[203] Periodic intracerebroventricular deliveries of ASOs in SMA mice models have been found to improve the motor phenotype.[204,205] Based on the promise of preclinical studies, a multicenter phase I trial developed by ISIS Pharmaceuticals was recently conducted to assess the safety, tolerability, and dose-range finding of multiple doses of ASO (ISIS-SMNRx), which was delivered intrathecally into the subarachnoid space of SMA patients.

Stem Cells

Interest in the use of stem cells also continues, both as potential treatment for SMA and for use in constructing model systems for therapeutics development. During the last few years, pluripotent stem cells with the capacity to differentiate into motor neurons that lacked *SMN1* expression were induced from a type 1 SMA patient and his mother; this could serve as an important model system for testing of new compounds and eventually for stem cell approaches to the treatment of SMA.[206,207]

Gene Therapy

Among other therapeutic targets, gene therapy has shown potential in animal models.[208,209] Foust et al. demonstrated that self-complementary adeno-associated virus 9 (scAAV9) could cross the BBB and infect approximately 60% of motor neurons when injected intravenously into neonatal mice. It was more beneficial when administered on postnatal day 1 when compared with postnatal day 5 and 10.[208] Similar results were obtained in nonhuman primates. With the discovery of novel serotypes of adeno-associated virus (AAV) such as AAV9, which not only infects numerous cell types but also appears to traverse the BBB, there has been renewed interest in applying gene replacement protocols to treat SMA. Neonatal SMA model mice injected systemically with scAAV9 engineered to carry the wild-type SMN gene expressed high levels of protein in multiple tissues and derived remarkable therapeutic benefit, surviving in some instances to 12 months or more with no evidence of muscle weakness.[210–212] Considering proof-of-concept studies demonstrating the ability of scAAV9 to penetrate the mature BBB and infect adult motor neurons,[213] an SMN replacement strategy to treat SMA is a promising alternative to pharmacologic approaches. A drawback, however, based on the design of the vector to be maintained episomally, is the dilution and eventual loss of the therapeutic gene in dividing cells.

Those involved in the care of patients with SMA and SMA families who seek further information regarding ongoing clinical trials should visit the ClinicalTrials.gov website and search for "spinal muscular atrophy."

CARE OF THE PATIENT WITH SMA

Patients with SMA and their families benefit greatly from a multidisciplinary approach to care. This approach involves members from neurology/neuromuscular medicine, orthopedics, physical and occupational therapy, pulmonology, nutrition, and gastroenterology. For severely affected patients with type I SMA, early involvement of the pediatric advanced care or palliative care team can provide parents with support and assistance in making decisions that are consonant with their values and help to maximize their child's quality of life. In 2007 a Consensus Statement for Standard of Care in Spinal Muscular Atrophy was released by a multidisciplinary team regarding the current best recommendations for management of patients with SMA.[17]

Pulmonary

Respiratory failure is the major cause of mortality in patients at the more severe end of the disease spectrum, namely types I and II SMA. Infants with type I SMA have weak intercostal muscles with relatively preserved diaphragm strength, resulting in a bell-shaped chest, pectus excavatum, and in some cases underdevelopment of the lungs. Patients with type II SMA have weak intercostal muscles with scoliosis contributing to progressive restrictive lung disease.[214,215] The restrictive lung disease results in insidious onset of sleep hypoventilation. In general, sleep disordered breathing is often the first sign of respiratory muscle weakness in neuromuscular disease.[216] Mellies found evidence of sleep hypoventilation on overnight polysomnography in 7/12 SMA patients who had vital capacities below 60%; these patients had improvement in their symptoms (of fatigue, headaches, disturbed sleep) with initiation of noninvasive ventilation (BiPAP).[216] Regarding the use of noninvasive ventilation in patients with type I SMA, Oskoui reviewed 143 type I patients enrolled in the International SMA Registry and found significantly increased survival for those born between 1995 and 2006 versus those born between 1980 and 1994, with a 70% reduced risk of death over mean follow-up of 49.9 months, that was significantly affected by the increased use of ventilation (noninvasive [BiPAP] and invasive), mechanical cough assist devices, and gastrostomy feeding.[20,25] Proper use of BiPAP, with correct pressure adjustments and mask placement, has no significant side effects on patient hemodynamics.[217] Patients with SMA and respiratory muscle weakness also have a weak cough, which impairs their ability to clear the airway of secretions. This places them at risk for hypoxemia from mucus plugging, especially during times of acute illness, and for recurrent infections. It is recommended that patients at risk for mucus plugging be monitored with overnight oximetry during acute illnesses, and that assisted airway clearance methods be employed. These may be in the form of a mechanical in-/ex-sufflator device, manual assisted airway clearance, and postural drainage. There should also be a low threshold for the use of antibiotics during acute illnesses in these patients, due to the risk of pneumonia.[17,214,215] Patients should be followed regularly by a pulmonologist experienced in caring for patients with neuromuscular diseases, with home visits by a home ventilation team if available.

Gastrointestinal

Patients with type I SMA are extremely weak and hence tire during feedings, which can lead to failure to thrive and aspiration with recurrent respiratory infections.[214] It has been proposed that patients with SMA have a high incidence of silent gastroesophageal reflux as well, which can contribute to aspiration.[218] In a small retrospective study Durkin et al. found that early laparoscopic Nissen fundoplication and gastrostomy in patients with type I SMA was associated with improved nutritional status in these patients, and also perhaps with a trend toward fewer long-term aspiration events.[218] Another group (Yuan et al.) also had positive outcomes with a series of type I and one severe type II patient who underwent laparoscopic Nissen fundoplication and gastrostomy tube placement followed by postoperative noninvasive ventilation, in that there were no adverse events and there were fewer pneumonias following the procedure in some patients.[219] Lastly, patients with SMA are at risk of constipation, which can, if severe (especially in young type I patients) worsen reflux or even respiratory symptoms.[214]

Nutrition

Failure to thrive or growth failure are common in infants with type I SMA and in some severely affected type II patients. However, though many type II patients plot as having a "normal" BMI (often as low as 3rd percentile for a healthy child of their age), they may actually have excessive fat mass relative to their muscle mass. Clinically high functioning nonambulatory SMA patients (Hammersmith score ≥ 12) were more prone to adiposity compared with both low functioning nonambulatory (Hammersmith score <12) and ambulatory patients (likely due to excess calories provided despite low resting energy expenditure) and are at risk of becoming overweight.[36,220] Hence close attention must be paid to nutritional status in patients with all types of SMA, and consultation with a dietician who is aware of these special concerns is vital.

Orthopedic

Patients with SMA at the severe end of the disease spectrum require close orthopedic follow-up for the development of scoliosis and contractures. Surgical intervention for scoliosis is often required, and careful coordination of perioperative respiratory and nutritional support can help to minimize complications.[17] Fractures and hip subluxation are commonly seen in milder type II and type III SMA patients. Distal femur is the commonest fracture site, followed by lower leg, ankle, and upper arm. Most of the fractures can be treated conservatively.[221]

Fatigue

Physiological fatigue is a common complaint in milder SMA patients and can be measured by the decrement in distance walked from the first to sixth minute of the Six-Minute Walk Test.[159,222,223] The mechanism(s) underlying fatigue in SMA remains to be elucidated, but it may be related at least in part to the neuromuscular junction defects described earlier. One medication already approved for the treatment of fatigue in multiple sclerosis, 4-aminopyridine, is currently being tested in ambulatory adult SMA patients. Anecdotal evidence suggests that albuterol PO in usual pediatric doses is an effective agent in the treatment of fatigue in SMA patients and thus it is frequently used; however, fatigue has not been studied directly in the conducted albuterol pilot studies that showed improvement in motor function.[184,185]

CONCLUSION

Spinal muscular atrophy is a chronic, inherited motor neuron disease for which there is no established treatment. Yet there is cause for optimism, as it is an area of active research and knowledge about the molecular genetics and pathogenesis of SMA is ever increasing. Several groups are actively exploring pharmacologic treatments, whether through the use of approved drugs, identification of new agents via high throughput screens, or development of novel pharmaceutical compounds. Consortia of clinicians and researchers are working together to organize multicenter trials and identify the best outcome measures. The SMA Patient Registry has helped to facilitate inclusion of patients in these studies. Standards of care have also been developed to optimize the long-term multidisciplinary management of patients with SMA. Patient support and advocacy groups (SMA Foundation, Families of SMA, Fight SMA, Project Cure, and others) have played a vital role in supporting research efforts and providing a community for children and families affected by SMA. Although it may seem at times that a treatment for SMA is far in the future, the advances made since the gene was identified in 1995 permit a modicum of hope to patients, their families, and those with the privilege to care for SMA patients.

ACKNOWLEDGMENTS

This chapter was funded in part by NINDS R01 NS057482 and by the SMA Foundation (New York, NY).

REFERENCES

1. Werdnig G. Zwei fruihnfantile hereditare Falle von progressive Muskelatrophie unter dem Bilde der Dystrophie, aber auf neurotischer Grundlage. *Arch Psychiatr Neurol* 1891;**22**:437–81.
2. Hoffmann J. Ueber chronische spinale Muskelatrophie im Kindesalter, auf familiärer Basis. *Deutsch Z Nervenheilk* 1893;**3**:427–70.
3. Monani UR, Lorson CL, Parsons DW, Prior TW, Androphy EJ, Burghes AH, et al. A single nucleotide difference that alters splicing patterns distinguishes the SMA gene SMN1 from the copy gene SMN2. *Hum Mol Genet* 1999;**8**:1177–83.
4. Pearn J. Incidence, prevalence, and gene frequency studies of chronic childhood spinal muscular atrophy. *J Med Genet* 1978;**15**:409–13.
5. Ogino S, Leonard DG, Rennert H, Ewens WJ, Wilson RB. Genetic risk assessment in carrier testing for spinal muscular atrophy. *Am J Med Genet* 2002;**110**:301–7.
6. Ogino S, Wilson RB, Gold B. New insights on the evolution of the SMN1 and SMN2 region: simulation and meta-analysis for allele and haplotype frequency calculations. *Eur J Hum Genet* 2004;**12**:1015–23.
7. Sugarman EA, Nagan N, Zhu H, Akmaev VR, Zhou Z, Rohlfs EM, et al. Pan-ethnic carrier screening and prenatal diagnosis for spinal muscular atrophy: clinical laboratory analysis of >72,400 specimens. *Eur J Hum Genet* 2012;**20**:27–32.
8. Mostacciuolo ML, Danieli GA, Trevisan C, Muller E, Angelini C. Epidemiology of spinal muscular atrophies in a sample of the Italian population. *Neuroepidemiology* 1992;**11**:34–8.
9. Thieme A, Mitulla B, Schulze F, Spiegler AW. Epidemiological data on Werdnig-Hoffmann disease in Germany (West-Thuringen). *Hum Genet* 1993;**91**:295–7.
10. Mailman MD, Heinz JW, Papp AC, Snyder PJ, Sedra MS, Wirth B, et al. Molecular analysis of spinal muscular atrophy and modification of the phenotype by SMN2. *Genet Med* 2002;**4**:20–6.
11. Zaldivar T, Montejo Y, Acevedo AM, Guerra R, Vargas J, Garofalo N, et al. Evidence of reduced frequency of spinal muscular atrophy type I in the Cuban population. *Neurology* 2005;**65**:636–8.
12. Hendrickson BC, Donohoe C, Akmaev VR, Sugarman EA, Labrousse P, Boguslavskiy L, et al. Differences in SMN1 allele frequencies among ethnic groups within North America. *J Med Genet* 2009;**46**:641–4.
13. Prior TW, Snyder PJ, Rink BD, Pearl DK, Pyatt RE, Mihal DC, et al. Newborn and carrier screening for spinal muscular atrophy. *Am J Med Genet A* 2010;**152A**:1608–16.
14. Zerres K, Davies KE. 59th ENMC International Workshop: Spinal Muscular Atrophies: recent progress and revised diagnostic criteria. 17–19 April 1998, Soestduinen, The Netherlands. *Neuromuscul Disord* 1999;**9**:272–8.
15. Dubowitz V. Chaos in the classification of SMA: a possible resolution. *Neuromuscul Disord* 1995;**5**:3–5.
16. Iannaccone ST, Russman BS, Browne RH, Buncher CR, White M, Samaha FJ. Prospective analysis of strength in spinal muscular atrophy. DCN/Spinal Muscular Atrophy Group. *J Child Neurol* 2000;**15**:97–101.
17. Wang CH, Finkel RS, Bertini ES, Schroth M, Simonds A, Wong B, et al. Consensus statement for standard of care in spinal muscular atrophy. *J Child Neurol* 2007;**22**:1027–49.

18. Byers RK, Banker BQ. Infantile muscular atrophy. *Arch Neurol* 1961;**5**:140–64.
19. Dubowitz V. Very severe spinal muscular atrophy (SMA type 0): an expanding clinical phenotype. *Eur J Paediatr Neurol* 1999;**3**:49–51.
20. O'Hagen JM, Glanzman AM, McDermott MP, Ryan PA, Flickinger J, Quigley J, et al. An expanded version of the Hammersmith Functional Motor Scale for SMA II and III patients. *Neuromuscul Disord* 2007;**17**:693–7.
21. Thomas NH, Dubowitz V. The natural history of type I (severe) spinal muscular atrophy. *Neuromuscul Disord* 1994;**4**:497–502.
22. Crawford TO. Spinal muscular atrophies. In: Jones Jr. HR, De Vivo DC, Darras BT, editors. *Neuromuscular Disorders of Infancy, Childhood, and Adolescence: A Clinician's Approach*. Philadelphia: Butterworth Heinemann; 2003. pp. 145–66.
23. Zerres K, Rudnik-Schöneborn S. Natural history in proximal spinal muscular atrophy. Clinical analysis of 445 patients and suggestions for a modification of existing classifications. *Arch Neurol* 1995;**52**:518–23.
24. Munsat T, Davies K. Spinal muscular atrophy. 32nd ENMC International Workshop. Naarden, the Netherlands, 10–12 March 1995. *Neuromuscul Disord* 1996;**6**:125–7.
25. Oskoui M, Levy G, Garland CJ, Gray JM, O'Hagen J, De Vivo DC, et al. The changing natural history of spinal muscular atrophy type 1. *Neurology* 2007;**69**:1931–6.
26. Menke LA, Poll-The BT, Clur SA, Bilardo CM, van der Wal AC, Lemmink HH, et al. Congenital heart defects in spinal muscular atrophy type I: a clinical report of two siblings and a review of the literature. *Am J Med Genet A* 2008;**146A**:740–4.
27. Rudnik-Schöneborn S, Heller R, Berg C, Betzler C, Grimm T, Eggermann T, et al. Congenital heart disease is a feature of severe infantile spinal muscular atrophy. *J Med Genet* 2008;**45**:635–8.
28. Heier CR, Satta R, Lutz C, DiDonato CJ. Arrhythmia and cardiac defects are a feature of spinal muscular atrophy model mice. *Hum Mol Genet* 2010;**19**:3906–18.
29. Bevan AK, Hutchinson KR, Foust KD, Braun L, McGovern VL, Schmelzer L, et al. Early heart failure in the SMNDelta7 model of spinal muscular atrophy and correction by postnatal scAAV9-SMN delivery. *Hum Mol Genet* 2010;**19**:3895–905.
30. Shababi M, Habibi J, Yang HT, Vale SM, Sewell WA, Lorson CL. Cardiac defects contribute to the pathology of spinal muscular atrophy models. *Hum Mol Genet* 2010;**19**:4059–71.
31. Rudnik-Schöneborn S, Vogelgesang S, Armbrust S, et al. Digital necroses and vascular thrombosis in severe spinal muscular atrophy. *Muscle Nerve* 2010;**42**:144–7.
32. Araujo Ade Q, Araujo M, Swoboda KJ. Vascular perfusion abnormalities in infants with spinal muscular atrophy. *J Pediatr* 2009;**155**:292–4.
33. Rudnik-Schöneborn S, Goebel HH, Schlote W, Graul-Neumann L, Fusch C, Zerres K. Classical infantile spinal muscular atrophy with SMN deficiency causes sensory neuronopathy. *Neurology* 2003;**60**:983–7.
34. Thompson J, Bruce A. Progressive muscular atrophy in a child with a spinal lesion. *Edinb Hosp Rep* 1893;**1**:372.
35. Dubowitz V. Infantile muscular atrophy: A prospective study with particular reference to a slowly progressive variety. *Brain* 1964;**87**:707–18.
36. Sproule DM, Montes J, Montgomery M, Battista V, Koenigsberger D, Shen W, et al. Increased fat mass and high incidence of overweight despite low body mass index in patients with spinal muscular atrophy. *Neuromuscul Disord* 2009;**19**:391–6.
37. von Gontard A, Zerres K, Backes M, Laufersweiler-Plass C, Wendland C, Melchers P, et al. Intelligence and cognitive function in children and adolescents with spinal muscular atrophy. *Neuromuscul Disord* 2002;**12**:130–6.
38. Zerres K, Rudnik-Schöneborn S, Forrest E, Lusakowska A, Borkowska J, Hausmanowa-Petrusewicz I. A collaborative study on the natural history of childhood and juvenile onset proximal spinal muscular atrophy (type II and III SMA): 569 patients. *J Neurol Sci* 1997;**146**:67–72.
39. Kugelberg E, Welander L. Heredofamilial juvenile muscular atrophy simulating muscular dystrophy. *AMA Arch Neurol Psychiatry* 1956;**75**:500–9.
40. MacLeod MJ, Taylor JE, Lunt PW, Mathew CG, Robb SA. Prenatal onset spinal muscular atrophy. *Eur J Paediatr Neurol* 1999;**3**:65–72.
41. Bingham PM, Shen N, Rennert H, Rorke LB, Black AW, Marin-Padilla MM, et al. Arthrogryposis due to infantile neuronal degeneration associated with deletion of the SMNT gene. *Neurology* 1997;**49**:848–51.
42. Korinthenberg R, Sauer M, Ketelsen UP, Hanemann CO, Stoll G, Graf M, et al. Congenital axonal neuropathy caused by deletions in the spinal muscular atrophy region. *Ann Neurol* 1997;**42**:364–8.
43. Russman BS. Spinal muscular atrophy: clinical classification and disease heterogeneity. *J Child Neurol* 2007;**22**:946–51.
44. Guillot N, Cuisset JM, Cuvellier JC, Hurtevent JF, Joriot S, Vallee L. Unusual clinical features in infantile Spinal Muscular Atrophies. *Brain Dev* 2008;**30**:169–78.
45. Darras BT. Non-5q spinal muscular atrophies: The alphanumeric soup thickens. *Neurology* 2011;**77**:312–14.
46. Lefebvre S, Burglen L, Reboullet S, Clermont O, Burlet P, Viollet L, et al. Identification and characterization of a spinal muscular atrophy-determining gene. *Cell* 1995;**80**:155–65.
47. Lorson CL, Androphy EJ. An exonic enhancer is required for inclusion of an essential exon in the SMA-determining gene SMN. *Hum Mol Genet* 2000;**9**:259–65.
48. Lefebvre S, Burlet P, Liu Q, Bertrandy S, Clermont O, Munnich A, et al. Correlation between severity and SMN protein level in spinal muscular atrophy. *Nat Genet* 1997;**16**:265–9.
49. Prior TW. Spinal muscular atrophy diagnostics. *J Child Neurol* 2007;**22**:952–6.
50. DiDonato CJ, Ingraham SE, Mendell JR, Prior TW, Lenard S, Moxley 3rd RT, et al. Deletions and conversion in spinal muscular atrophy patients: is there a relationship to severity? *Ann Neurol* 1997;**41**:230–7.
51. van der Steege G, Grootscholten PM, Cobben JM, Zappata S, Scheffer H, den Dunnen JT, et al. Apparent gene conversions involving the SMN gene in the region of the spinal muscular atrophy locus on chromosome 5. *Am J Hum Genet* 1996;**59**:834–8.
52. Campbell L, Potter A, Ignatius J, Dubowitz V, Davies K. Genomic variation and gene conversion in spinal muscular atrophy: implications for disease process and clinical phenotype. *Am J Hum Genet* 1997;**61**:40–50.
53. Feldkotter M, Schwarzer V, Wirth R, Wienker TF, Wirth B. Quantitative analyses of SMN1 and SMN2 based on real-time lightCycler PCR: fast and highly reliable carrier testing and prediction of severity of spinal muscular atrophy. *Am J Hum Genet* 2002;**70**:358–68.

54. Arkblad E, Tulinius M, Kroksmark AK, Henricsson M, Darin N. A population-based study of genotypic and phenotypic variability in children with spinal muscular atrophy. *Acta Paediatr* 2009;**98**:865–72.
55. Saito M, Chen Y, Mizuguchi M, Igarashi T. [Quantitative analysis of SMN2 based on real-time PCR: correlation of clinical severity and SMN2 gene dosage]. *No To Hattatsu* 2005;**37**:407–12.
56. Rudnik-Schöneborn S, Berg C, Zerres K, Betzler C, Grimm T, Eggermann T, et al. Genotype-phenotype studies in infantile spinal muscular atrophy (SMA) type I in Germany: implications for clinical trials and genetic counselling. *Clin Genet* 2009;**76**:168–78.
57. Prior TW, Swoboda KJ, Scott HD, Hejmanowski AQ. Homozygous SMN1 deletions in unaffected family members and modification of the phenotype by SMN2. *Am J Med Genet A* 2004;**130A**:307–10.
58. Monani UR, Sendtner M, Coovert DD, Parsons DW, Andreassi C, Le TT, et al. The human centromeric survival motor neuron gene (SMN2) rescues embryonic lethality in Smn(-/-) mice and results in a mouse with spinal muscular atrophy. *Hum Mol Genet* 2000;**9**:333–9.
59. Hsieh-Li HM, Chang JG, Jong YJ, Wu MH, Wang NM, Tsai CH, et al. A mouse model for spinal muscular atrophy. *Nat Genet* 2000;**24**:66–70.
60. Le TT, Pham LT, Butchbach ME, Zhang HL, Monani UR, Coovert DD, et al. SMNDelta7, the major product of the centromeric survival motor neuron (SMN2) gene, extends survival in mice with spinal muscular atrophy and associates with full-length SMN. *Hum Mol Genet* 2005;**14**:845–57.
61. Prior TW, Krainer AR, Hua Y, Swoboda KJ, Snyder PC, Bridgeman SJ, et al. A positive modifier of spinal muscular atrophy in the SMN2 gene. *Am J Hum Genet* 2009;**85**:408–13.
62. Vezain M, Saugier-Veber P, Goina E, Touraine R, Manel V, Toutain A, et al. A rare SMN2 variant in a previously unrecognized composite splicing regulatory element induces exon 7 inclusion and reduces the clinical severity of spinal muscular atrophy. *Hum Mutat* 2010;**31**:E1110–25.
63. Bernal S, Alias L, Barcelo MJ, Also-Rallo E, Martinez-Hernandez R, Gamez J, et al. The c.859G > C variant in the SMN2 gene is associated with types II and III SMA and originates from a common ancestor. *J Med Genet* 2010;**47**:640–2.
64. Cobben JM, van der Steege G, Grootscholten P, de Visser M, Scheffer H, Buys CH. Deletions of the survival motor neuron gene in unaffected siblings of patients with spinal muscular atrophy. *Am J Hum Genet* 1995;**57**:805–8.
65. Hahnen E, Forkert R, Marke C, Rudnik-Schoneborn S, Schonling J, Zerres K, et al. Molecular analysis of candidate genes on chromosome 5q13 in autosomal recessive spinal muscular atrophy: evidence of homozygous deletions of the SMN gene in unaffected individuals. *Hum Mol Genet* 1995;**4**:1927–33.
66. Burghes AH, Ingraham SE, Kote-Jarai Z, Rosenfeld S, Herta N, Nadkarni N, et al. Linkage mapping of the spinal muscular atrophy gene. *Hum Genet* 1994;**93**:305–12.
67. Oprea GE, Krober S, McWhorter ML, Rossoll W, Muller S, Krawczak M, et al. Plastin 3 is a protective modifier of autosomal recessive spinal muscular atrophy. *Science* 2008;**320**:524–7.
68. Stratigopoulos G, Lanzano P, Deng L, Guo J, Kaufmann P, Darras B, et al. Association of plastin 3 expression with disease severity in spinal muscular atrophy only in postpubertal females. *Arch Neurol* 2010;**67**:1252–6.
69. Bernal S, Also-Rallo E, Martinez-Hernandez R, Alias L, Rodriguez-Alvarez FJ, Millan JM, et al. Plastin 3 expression in discordant spinal muscular atrophy (SMA) siblings. *Neuromuscul Disord* 2011;**21**:413–19.
70. Arnold WD, Burghes AH. Spinal muscular atrophy: development and implementation of potential treatments *Ann Neurol* 2013;**74**:348–62.
71. Bebee TW, Gladman JT, Chandler DS. Splicing regulation of the survival motor neuron genes and implications for treatment of spinal muscular atrophy. *Front Biosci (Landmark Ed)* 2010;**15**:1191–204.
72. Sumner CJ, Kolb SJ, Harmison GG, Jeffries NO, Schadt K, Finkel RS, et al. SMN mRNA and protein levels in peripheral blood: biomarkers for SMA clinical trials. *Neurology* 2006;**66**:1067–73.
73. Prior TW, Russman BS, Spinal muscular atrophy. In: Pagon RA, Bird TC, Dolan CR, Stephens K, editors. *GeneReviews [Internet]*. Seattle, WA: University of Washington, Seattle, 1993–2000. Feb 24 [updated 2013 Nov 14].
74. McAndrew PE, Parsons DW, Simard LR, Rochette C, Ray PN, Mendell JR, et al. Identification of proximal spinal muscular atrophy carriers and patients by analysis of SMNT and SMNC gene copy number. *Am J Hum Genet* 1997;**60**:1411–22.
75. Eggermann T, Zerres K, Anhuf D, Kotzot D, Fauth C, Rudnik-Schöneborn S. Somatic mosaicism for a heterozygous deletion of the survival motor neuron (SMN1) gene. *Eur J Hum Genet* 2005;**13**:309–13.
76. Smith M, Calabro V, Chong B, Gardiner N, Cowie S, du Sart D. Population screening and cascade testing for carriers of SMA. *Eur J Hum Genet* 2007;**15**:759–66.
77. Swoboda KJ, Prior TW, Scott CB, McNaught TP, Wride MC, Reyna SP, et al. Natural history of denervation in SMA: relation to age, SMN2 copy number, and function. *Ann Neurol* 2005;**57**:704–12.
78. ACOG Committee opinion No. 432. Spinal muscular atrophy. *Obstet Gynecol* 2009;**113**:1194–6.
79. Gitlin JM, Fischbeck K, Crawford TO, Cwik V, Fleischman A, Gonye K, et al. Carrier testing for spinal muscular atrophy. *Genet Med* 2010;**12**:621–2.
80. Martínez-Hernández R, Soler-Botija C, Also E, Alias L, Caselles L, Gich I, et al. The developmental pattern of myotubes in spinal muscular atrophy indicates prenatal delay of muscle maturation. *J Neuropathol Exp Neurol* 2009;**68**:474–81.
81. Liu Q, Dreyfuss G. A novel nuclear structure containing the survival of motor neurons protein. *EMBO J* 1996;**15**:3555–65.
82. Zhang H, Xing L, Rossoll W, Wichterle H, Singer RH, Bassell GJ. Multiprotein complexes of the survival of motor neuron protein SMN with Gemins traffic to neuronal processes and growth cones of motor neurons. *J Neurosci* 2006;**26**:8622–32.
83. Meister G, Hannus S, Plottner O, Baars T, Hartmann E, Fakan S, et al. SMNrp is an essential pre-mRNA splicing factor required for the formation of the mature spliceosome. *EMBO J* 2001;**20**:2304–14.
84. Pellizzoni L, Yong J, Dreyfuss G. Essential role for the SMN complex in the specificity of snRNP assembly. *Science* 2002;**298**:1775–9.
85. Kolb SJ, Battle DJ, Dreyfuss G. Molecular functions of the SMN complex. *J Child Neurol* 2007;**22**:990–4.
86. Pellizzoni L. Chaperoning ribonucleoprotein biogenesis in health and disease. *EMBO Rep* 2007;**8**:340–5.
87. Cheng D, Cote J, Shaaban S, Bedford MT. The arginine methyltransferase CARM1 regulates the coupling of transcription and mRNA processing. *Mol Cell* 2007;**25**:71–83.

88. Rossoll W, Jablonka S, Andreassi C, Kroning AK, Karle K, Monani UR, et al. Smn, the spinal muscular atrophy-determining gene product, modulates axon growth and localization of beta-actin mRNA in growth cones of motoneurons. *J Cell Biol* 2003;**163**:801–12.
89. Beattie CE, Carrel TL, McWhorter ML. Fishing for a mechanism: using zebrafish to understand spinal muscular atrophy. *J Child Neurol* 2007;**22**:995–1003.
90. McWhorter ML, Monani UR, Burghes AH, Beattie CE. Knockdown of the survival motor neuron (Smn) protein in zebrafish causes defects in motor axon outgrowth and pathfinding. *J Cell Biol* 2003;**162**:919–31.
91. Ruggiu M, McGovern VL, Lotti F, Saieva L, Li DK, Kariya S, et al. A role for SMN exon 7 splicing in the selective vulnerability of motor neurons in spinal muscular atrophy. *Mol Cell Biol* 2012;**32**:126–38.
92. Lotti F, Imlach WL, Saieva L, Beck ES, Hao le T, Li DK, et al. An SMN-dependent U12 splicing event essential for motor circuit function. *Cell* 2012;**151**:440–54.
93. Imlach WL, Beck ES, Choi BJ, Lotti F, Pellizzoni L, McCabe BD. SMN is required for sensory-motor circuit function in *Drosophila*. *Cell* 2012;**151**:427–39.
94. Voigt T, Meyer K, Baum O, Schumperli D. Ultrastructural changes in diaphragm neuromuscular junctions in a severe mouse model for Spinal Muscular Atrophy and their prevention by bifunctional U7 snRNA correcting SMN2 splicing. *Neuromuscul Disord* 2010;**20**:744–52.
95. Ruiz R, Casanas JJ, Torres-Benito L, Cano R, Tabares L. Altered intracellular Ca^{2+} homeostasis in nerve terminals of severe spinal muscular atrophy mice. *J Neurosci* 2010;**30**:849–57.
96. Kong L, Wang X, Choe DW, Polley M, Burnett BG, Bosch-Marce M, et al. Impaired synaptic vesicle release and immaturity of neuromuscular junctions in spinal muscular atrophy mice. *J Neurosci* 2009;**29**:842–51.
97. Boon KL, Xiao S, McWhorter ML, Donn T, Wolf-Saxon E, Bohnsack MT, et al. Zebrafish survival motor neuron mutants exhibit presynaptic neuromuscular junction defects. *Hum Mol Genet* 2009;**18**:3615–25.
98. Rossoll W, Bassell GJ. Spinal muscular atrophy and a model for survival of motor neuron protein function in axonal ribonucleoprotein complexes. *Results Probl Cell Differ* 2009;**48**:289–326.
99. Kariya S, Park GH, Maeno-Hikichi Y, Leykekhman O, Lutz C, Arkovitz MS, et al. Reduced SMN protein impairs maturation of the neuromuscular junctions in mouse models of spinal muscular atrophy. *Hum Mol Genet* 2008;**17**:2552–69.
100. Wadman RI, Vrancken AF, van den Berg LH, van der Pol WL. Dysfunction of the neuromuscular junction in spinal muscular atrophy types 2 and 3. *Neurology* 2012;**79**:2050–5.
101. Zerres K, Rudnik-Schöneborn S. 93rd ENMC International Workshop: non-5q-spinal muscular atrophies (SMA)—clinical picture (6–8 April 2001, Naarden, the Netherlands). *Neuromuscul Disord* 2003;**13**:179–83.
102. Pestronk A. *Hereditary motor syndromes*. St. Louis, MO: Washington University; 1996 [updated 21 Dec 2013; cited 27 Dec 2013]; Available from: <http://neuromuscular.wustl.edu/synmot.html>.
103. Harms MB, Allred P, Gardner Jr. R, Fernandes Filho JA, Florence J, Pestronk A, et al. Dominant spinal muscular atrophy with lower extremity predominance: linkage to 14q32. *Neurology* 2010;**75**:539–46.
104. Guenther UP, Varon R, Schlicke M, Dutrannoy V, Volk A, Hubner C, et al. Clinical and mutational profile in spinal muscular atrophy with respiratory distress (SMARD): defining novel phenotypes through hierarchical cluster analysis. *Hum Mutat* 2007;**28**:808–15.
105. Stalpers XL, Verrips A, Poll-The BT, Cobben JM, Snoeck IN, de Coo IF, et al. Clinical and mutational characteristics of spinal muscular atrophy with respiratory distress type 1 in the Netherlands. *Neuromuscul Disord* 2013;**23**:461–8.
106. Joseph S, Robb SA, Mohammed S, Lillis S, Simonds A, Manzur AY, et al. Interfamilial phenotypic heterogeneity in SMARD1. *Neuromuscul Disord* 2009;**19**:193–5.
107. Ciccolella M, Catteruccia M, Benedetti S, Moroni I, Uziel G, Pantaleoni C, et al. Brown-Vialetto-van Laere and Fazio-Londe overlap syndromes: a clinical, biochemical and genetic study. *Neuromuscul Disord* 2012;**22**:1075–82.
108. Namavar Y, Barth PG, Poll-The BT, Baas F. Classification, diagnosis and potential mechanisms in pontocerebellar hypoplasia. *Orphanet J Rare Dis* 2011;**6**:50.
109. Renbaum P, Kellerman E, Jaron R, Geiger D, Segel R, Lee M, et al. Spinal muscular atrophy with pontocerebellar hypoplasia is caused by a mutation in the VRK1 gene. *Am J Hum Genet* 2009;**85**:281–9.
110. Rudnik-Schöneborn S, Senderek J, Jen JC, Houge G, Seeman P, Puchmajerova A, et al. Pontocerebellar hypoplasia type 1: clinical spectrum and relevance of EXOSC3 mutations. *Neurology* 2013;**80**:438–46.
111. La Spada AR, Wilson EM, Lubahn DB, Harding AE, Fischbeck KH. Androgen receptor gene mutations in X-linked spinal and bulbar muscular atrophy. *Nature* 1991;**352**:77–9.
112. Ramser J, Ahearn ME, Lenski C, Yariz KO, Hellebrand H, von Rhein M, et al. Rare missense and synonymous variants in UBE1 are associated with X-linked infantile spinal muscular atrophy. *Am J Hum Genet* 2008;**82**:188–93.
113. Dlamini N, Josifova DJ, Paine SM, Wraige E, Pitt M, Murphy AJ, et al. Clinical and neuropathological features of X-linked spinal muscular atrophy (SMAX2) associated with a novel mutation in the UBA1 gene. *Neuromuscul Disord* 2013;**23**:391–8.
114. Washington University. Neuromuscular Disease Center [website]. 2014. Available from: <http://neuromuscular.wustl.edu>.
115. Johns Hopkins University School of Medicine. McKusick-Nathans Institute of Genetic Medicine. Online Mendelian Inheritance in Man (OMIM). [website]. 2014. Available from: <http://www.ncbi.nlm.nih.gov/omim>.
116. GeneTests [website]. 2014. Available from <http://www.genetests.org>.
117. Neudorfer O, Pastores GM, Zeng BJ, Gianutsos J, Zaroff CM, Kolodny EH. Late-onset Tay-Sachs disease: phenotypic characterization and genotypic correlations in 21 affected patients. *Genet Med* 2005;**7**:119–23.
118. Johnson WG. Motor neuron diseases resulting from hexosaminidase deficiency. *Semin Neurol* 1993;**13**:369–74.
119. Jamrozik Z, Lugowska A, Golebiowski M, Krolicki L, Maczewska J, Kuzma-Kozakiewicz M. Late onset GM2 gangliosidosis mimicking spinal muscular atrophy. *Gene* 2013;**527**:679–82.
120. Hirayama K, Toyokura Y, Tsubaki T. Juvenile muscular atrophy of unilateral upper extremity: a new clinical entity. *Psychiatr Neurol Jpn* 1959;**61**:2190–7.

121. Hirayama K. Juvenile non-progressive muscular atrophy localized in hand and forearm: observation in 38 cases. *Clin Neurol (Tokyo)* 1972;**12**:313–24.

122. Kikuchi S, Tashiro K. Juvenile muscular atrophy of distal upper extremity (Hirayama's disease). In: Jones HRJ, De Vivo DC, Darras BT, editors. *Neuromuscular Disorders of Infancy, Childhood, and Adolescence: A Clinician's Approach.* 1st ed. Philadelphia: Butterworth-Heinemann; 2003. p. 167–81.

123. Hirayama K. Non-progressive juvenile spinal muscular atrophy of the distal upper limb (Hirayama's disease). In: Vinken PJ, Bruyn GW, Klawans HL, editors. *Diseases of the Motor System; Handbook of Clinical Neurology*. Amsterdam: Elsevier; 1991. pp. 107–20.

124. Guo XM, Qin XY, Huang C. Neuroelectrophysiological characteristics of Hirayama disease: report of 14 cases. *Chin Med J (Engl)* 2012;**125**:2440–3.

125. Ghosh PS, Moodley M, Friedman NR, Rothner AD, Ghosh D. Hirayama disease in children from North America. *J Child Neurol* 2011;**26**:1542–7.

126. Kikuchi S, Tashiro K, Kitagawa M, Iwasaki Y, Abe H. [A mechanism of juvenile muscular atrophy localized in the hand and forearm (Hirayama's disease)—flexion myelopathy with tight dural canal in flexion]. *Rinsho Shinkeigaku* 1987;**27**:412–19.

127. Iwasaki Y, Tashiro K, Kikuchi S, Kitagawa M, Isu T, Abe H. Cervical flexion myelopathy: a "tight dural canal mechanism." Case report. *J Neurosurg* 1987;**66**:935–7.

128. Huang YL, Chen CJ. Hirayama disease. *Neuroimaging Clin N Am* 2011;**21**:939–50, ix–x.

129. Hirayama K, Tomonaga M, Kitano K, Yamada T, Kojima S, Arai K. Focal cervical poliopathy causing juvenile muscular atrophy of distal upper extremity: a pathological study. *J Neurol Neurosurg Psychiatry* 1987;**50**:285–90.

130. Hirayama K. Juvenile muscular atrophy of distal upper extremity (Hirayama disease): focal cervical ischemic poliomyelopathy. *Neuropathology* 2000;**20**(Suppl.):S91–4.

131. Andreadou E, Christodoulou K, Manta P, Karandreas N, Loukaidis P, Sfagos C, et al. Familial asymmetric distal upper limb amyotrophy (Hirayama disease): report of a Greek family. *Neurologist* 2009;**15**:156–60.

132. Paredes I, Esteban J, Ramos A, Gonzalez P, Rivas JJ. A severe case of Hirayama disease successfully treated by anterior cervical fusion. *J Neurosurg Spine* 2013;**20**:191–5.

133. Agundez M, Rouco I, Barcena J, Mateos B, Barredo J, Zarranz JJ. Hirayama disease: Is surgery an option? *Neurologia* 2013. Available from: http://dx.doi.org/10.1016/j.nrl.2013.05.005.

134. Hosokawa T, Fujieda M, Wakiguchi H, Oosaki Y. Pediatric Hirayama disease. *Pediatr Neurol* 2010;**43**:151–3.

135. Verma R, Lalla R, Patil TB, Gupta A. Hirayama disease: a frequently undiagnosed condition with simple inexpensive treatment. *BMJ Case Rep* 2012;2012 ID: 23220833.

136. Monani UR, De Vivo DC. Neurodegeneration in spinal muscular atrophy: From disease phenotype and animal models to therapeutic strategies and beyond. *Future Neurol* 2014;**9**:49–65.

137. Crawford TO, Pardo CA. The neurobiology of childhood spinal muscular atrophy. *Neurobiol Dis* 1996;**3**:97–110.

138. Simic G, Mladinov M, Seso Simic D, Jovanov Milosevic N, Islam A, Pajtak A, et al. Abnormal motoneuron migration, differentiation, and axon outgrowth in spinal muscular atrophy. *Acta Neuropathol* 2008;**115**:313–26.

139. Dubowitz V. *Muscle Disorders in Childhood.* 2nd ed. Philadelphia: Saunders; 1995.

140. Nadeau A, D'Anjou G, Debray G, Robitaille Y, Simard LR, Vanasse M. A newborn with spinal muscular atrophy type 0 presenting with a clinicopathological picture suggestive of myotubular myopathy. *J Child Neurol* 2007;**22**:1301–4.

141. Martínez-Hernández R, Bernal S, Also-Rallo E, Alías L, Barcelo MJ, Hereu M, et al. Synaptic defects in type I spinal muscular atrophy in human development. *J Pathol* 2013;**229**:49–61.

142. Schmid A, DiDonato CJ. Animal models of spinal muscular atrophy. *J Child Neurol* 2007;**22**:1004–12.

143. Park GH, Maeno-Hikichi Y, Awano T, Landmesser LT, Monani UR. Reduced survival of motor neuron (SMN) protein in motor neuronal progenitors functions cell autonomously to cause spinal muscular atrophy in model mice expressing the human centromeric (SMN2) gene. *J Neurosci* 2010;**30**:12005–19.

144. Ling KK, Lin MY, Zingg B, Feng Z, Ko CP. Synaptic defects in the spinal and neuromuscular circuitry in a mouse model of spinal muscular atrophy. *PLOS ONE* 2010;**5**:e15457.

145. Mentis GZ, Blivis D, Liu W, Drobac E, Crowder ME, Kong L. Early functional impairment of sensory-motor connectivity in a mouse model of spinal muscular atrophy. *Neuron* 2011;**69**:453–67.

146. Wishart TM, Huang JP, Murray LM, Lamont DJ, Mutsaers CA, Ross J, et al. SMN deficiency disrupts brain development in a mouse model of severe spinal muscular atrophy. *Hum Mol Genet* 2010;**19**:4216–28.

147. Mutsaers CA, Wishart TM, Lamont DJ, Riessland M, Schreml J, Comley LH, et al. Reversible molecular pathology of skeletal muscle in spinal muscular atrophy. *Hum Mol Genet* 2011;**20**:4334–44.

148. Martinez TL, Kong L, Wang X, Osborne MA, Crowder ME, Van Meerbeke JP, et al. Survival motor neuron protein in motor neurons determines synaptic integrity in spinal muscular atrophy. *J Neurosci* 2012;**32**:8703–15.

149. Gogliotti RG, Quinlan KA, Barlow CB, Heier CR, Heckman CJ, Didonato CJ. Motor neuron rescue in spinal muscular atrophy mice demonstrates that sensory-motor defects are a consequence, not a cause, of motor neuron dysfunction. *J Neurosci* 2012;**32**:3818–29.

150. Lee AJ, Awano T, Park GH, Monani UR. Limited phenotypic effects of selectively augmenting the SMN protein in the neurons of a mouse model of severe spinal muscular atrophy. *PLOS ONE* 2012;**7**:e46353.

151. Gavrilina TO, McGovern VL, Workman E, Crawford TO, Gogliotti RG, DiDonato CJ, et al. Neuronal SMN expression corrects spinal muscular atrophy in severe SMA mice while muscle-specific SMN expression has no phenotypic effect. *Hum Mol Genet* 2008;**17**:1063–75.

152. Le TT, McGovern VL, Alwine IE, Wang X, Massoni-Laporte A, Rich MM, et al. Temporal requirement for high SMN expression in SMA mice. *Hum Mol Genet* 2011;**20**:3578–91.

153. Lutz CM, Kariya S, Patruni S, Osborne MA, Liu D, Henderson CE, et al. Postsymptomatic restoration of SMN rescues the disease phenotype in a mouse model of severe spinal muscular atrophy. *J Clin Invest* 2011;**121**:3029–41.

154. Kariya S, Obis T, Garone C, Akay T, Sera F, Iwata S, et al. Requirement of enhanced survival motor neuron protein imposed during neuromuscular junction maturation. *J Clin Invest* 2014;**124**:785–800.

155. Markowitz JA, Singh P, Darras BT. Spinal muscular atrophy: a clinical and research update. *Pediatr Neurol* 2012;**46**:1–12.
156. Kaufmann P, McDermott MP, Darras BT, Finkel R, Kang P, Oskoui M, et al. Observational study of spinal muscular atrophy type 2 and 3: functional outcomes over 1 year. *Arch Neurol* 2011;**68**:779–86.
157. Kaufmann P, McDermott MP, Darras BT, Finkel RS, Sproule DM, Kang PB, et al. Prospective cohort study of spinal muscular atrophy types 2 and 3. *Neurology* 2012;**79**:1889–97.
158. Crawford TO. Concerns about the design of clinical trials for spinal muscular atrophy. *Neuromuscul Disord* 2004;**14**:456–60.
159. Montes J, McDermott MP, Martens WB, Dunaway S, Glanzman AM, Riley S, et al. Six-Minute Walk Test demonstrates motor fatigue in spinal muscular atrophy. *Neurology* 2010;**74**:833–8.
160. Glanzman AM, Mazzone E, Main M, Pelliccioni M, Wood J, Swoboda KJ, et al. The Children's Hospital of Philadelphia Infant Test of Neuromuscular Disorders (CHOP INTEND): test development and reliability. *Neuromuscul Disord* 2010;**20**:155–61.
161. Nelson L, Owens H, Hynan LS, Iannaccone ST. The gross motor function measure is a valid and sensitive outcome measure for spinal muscular atrophy. *Neuromuscul Disord* 2006;**16**:374–80.
162. Main M, Kairon H, Mercuri E, Muntoni F. The Hammersmith functional motor scale for children with spinal muscular atrophy: a scale to test ability and monitor progress in children with limited ambulation. *Eur J Paediatr Neurol* 2003;**7**:155–9.
163. Krosschell KJ, Maczulski JA, Crawford TO, Scott C, Swoboda KJ. A modified Hammersmith functional motor scale for use in multi-center research on spinal muscular atrophy. *Neuromuscul Disord* 2006;**16**:417–26.
164. Montes J, Gordon AM, Pandya S, De Vivo DC, Kaufmann P. Clinical outcome measures in spinal muscular atrophy. *J Child Neurol* 2009;**24**:968–78.
165. Kang PB, Gooch CL, McDermott MP, Darras BT, Finkel RS, Yang ML, et al. The motor neuron response to SMN1 deficiency in spinal muscular atrophy. *Muscle Nerve* 2014;**49**:636–44.
166. Tsai LK, Yang CC, Ting CH, Su YN, Hwu WL, Li H. Correlation of survival motor neuron expression in leukocytes and spinal cord in spinal muscular atrophy. *J Pediatr* 2009;**154**:303–5.
167. Tiziano FD, Pinto AM, Fiori S, Lomastro R, Messina S, Bruno C, et al. SMN transcript levels in leukocytes of SMA patients determined by absolute real-time PCR. *Eur J Hum Genet* 2010;**18**:52–8.
168. Darras BT, Kang PB. Clinical trials in spinal muscular atrophy. *Curr Opin Pediatr* 2007;**19**:675–9.
169. Singh P, Liew WK, Darras BT. Current advances in drug development in spinal muscular atrophy. *Curr Opin Pediatr* 2013;**25**:682–8.
170. Lunke S, El-Osta A. The emerging role of epigenetic modifications and chromatin remodeling in spinal muscular atrophy. *J Neurochem* 2009;**109**:1557–69.
171. Chang JG, Hsieh-Li HM, Jong YJ, Wang NM, Tsai CH, Li H. Treatment of spinal muscular atrophy by sodium butyrate. *Proc Natl Acad Sci USA* 2001;**98**:9808–13.
172. Andreassi C, Angelozzi C, Tiziano FD, Vitali T, De Vincenzi E, Boninsegna A, et al. Phenylbutyrate increases SMN expression in vitro: relevance for treatment of spinal muscular atrophy. *Eur J Hum Genet* 2004;**12**:59–65.
173. Brahe C, Vitali T, Tiziano FD, Angelozzi C, Pinto AM, Borgo F, et al. Phenylbutyrate increases SMN gene expression in spinal muscular atrophy patients. *Eur J Hum Genet* 2005;**13**:256–9.
174. Mercuri E, Bertini E, Messina S, Pelliccioni M, D'Amico A, Colitto F, et al. Pilot trial of phenylbutyrate in spinal muscular atrophy. *Neuromuscul Disord* 2004;**14**:130–5.
175. Mercuri E, Bertini E, Messina S, Solari A, D'Amico A, Angelozzi C, et al. Randomized, double-blind, placebo-controlled trial of phenylbutyrate in spinal muscular atrophy. *Neurology* 2007;**68**:51–5.
176. Sumner CJ, Huynh TN, Markowitz JA, Perhac JS, Hill B, Coovert DD, et al. Valproic acid increases SMN levels in spinal muscular atrophy patient cells. *Ann Neurol* 2003;**54**:647–54.
177. Brichta L, Hofmann Y, Hahnen E, Siebzehnrubl FA, Raschke H, Blumcke I, et al. Valproic acid increases the SMN2 protein level: a well-known drug as a potential therapy for spinal muscular atrophy. *Hum Mol Genet* 2003;**12**:2481–9.
178. Swoboda KJ, Scott CB, Reyna SP, Prior TW, LaSalle B, Sorenson SL, et al. Phase II open label study of valproic acid in spinal muscular atrophy. *PLOS ONE* 2009;**4**:e5268.
179. Swoboda KJ, Scott CB, Crawford TO, Simard LR, Reyna SP, Krosschell KJ, et al. SMA CARNI-VAL trial part I: double-blind, randomized, placebo-controlled trial of L-carnitine and valproic acid in spinal muscular atrophy. *PLOS ONE* 2010;**5**:e12140.
180. Grzeschik SM, Ganta M, Prior TW, Heavlin WD, Wang CH. Hydroxyurea enhances SMN2 gene expression in spinal muscular atrophy cells. *Ann Neurol* 2005;**58**:194–202.
181. Liang WC, Yuo CY, Chang JG, Chen YC, Chang YF, Wang HY, et al. The effect of hydroxyurea in spinal muscular atrophy cells and patients. *J Neurol Sci* 2008;**268**:87–94.
182. Xu C, Chen X, Grzeschik SM, Ganta M, Wang CH. Hydroxyurea enhances SMN2 gene expression through nitric oxide release. *Neurogenetics* 2011;**12**:19–24.
183. Chen TH, Chang JG, Yang YH, Mai HH, Liang WC, Wu YC, et al. Randomized, double-blind, placebo-controlled trial of hydroxyurea in spinal muscular atrophy. *Neurology* 2010;**75**:2190–7.
184. Kinali M, Mercuri E, Main M, De Biasia F, Karatza A, Higgins R, et al. Pilot trial of albuterol in spinal muscular atrophy. *Neurology* 2002;**59**:609–10.
185. Pane M, Staccioli S, Messina S, D'Amico A, Pelliccioni M, Mazzone ES, et al. Daily salbutamol in young patients with SMA type II. *Neuromuscul Disord* 2008;**18**:536–40.
186. Angelozzi C, Borgo F, Tiziano FD, Martella A, Neri G, Brahe C. Salbutamol increases SMN mRNA and protein levels in spinal muscular atrophy cells. *J Med Genet* 2008;**45**:29–31.
187. Tiziano FD, Lomastro R, Pinto AM, Messina S, D'Amico A, Fiori S, et al. Salbutamol increases survival motor neuron (SMN) transcript levels in leucocytes of spinal muscular atrophy (SMA) patients: relevance for clinical trial design. *J Med Genet* 2010;**47**:856–8.
188. Butchbach ME, Singh J, Thorsteinsdottir M, Saieva L, Slominski E, Thurmond J, et al. Effects of 2,4-diaminoquinazoline derivatives on SMN expression and phenotype in a mouse model for spinal muscular atrophy. *Hum Mol Genet* 2010;**19**:454–67.
189. Van Meerbeke J, Gibbs R, Plasterer H, Feng Z, Lin M, Wee C, et al. The therapeutic effects of RG3039 in severe spinal muscular atrophy mice and normal human volunteers [abstract]. *Neurology* 2012;**78**(S25):003.
190. Garbes L, Riessland M, Holker I, Heller R, Hauke J, Trankle C, et al. LBH589 induces up to 10-fold SMN protein levels by several independent mechanisms and is effective even in cells from SMA patients non-responsive to valproate. *Hum Mol Genet* 2009;**18**:3645–58.

191. Ting CH, Lin CW, Wen SL, Hsieh-Li HM, Li H. Stat5 constitutive activation rescues defects in spinal muscular atrophy. *Hum Mol Genet* 2007;**16**:499–514.
192. Farooq F, Molina FA, Hadwen J, MacKenzie D, Witherspoon L, Osmond M, et al. Prolactin increases SMN expression and survival in a mouse model of severe spinal muscular atrophy via the STAT5 pathway. *J Clin Invest* 2011;**121**:3042–50.
193. Biondi O, Branchu J, Sanchez G, Lancelin C, Deforges S, Lopes P, et al. In vivo NMDA receptor activation accelerates motor unit maturation, protects spinal motor neurons, and enhances SMN2 gene expression in severe spinal muscular atrophy mice. *J Neurosci* 2010;**30**:11288–99.
194. Wolstencroft EC, Mattis V, Bajer AA, Young PJ, Lorson CL. A non-sequence-specific requirement for SMN protein activity: the role of aminoglycosides in inducing elevated SMN protein levels. *Hum Mol Genet* 2005;**14**:1199–210.
195. Heier CR, DiDonato CJ. Translational readthrough by the aminoglycoside geneticin (G418) modulates SMN stability in vitro and improves motor function in SMA mice in vivo. *Hum Mol Genet* 2009;**18**:1310–22.
196. Haddad H, Cifuentes-Diaz C, Miroglio A, Roblot N, Joshi V, Melki J. Riluzole attenuates spinal muscular atrophy disease progression in a mouse model. *Muscle Nerve* 2003;**28**:432–7.
197. Russman BS, Iannaccone ST, Samaha FJ. A phase 1 trial of riluzole in spinal muscular atrophy. *Arch Neurol* 2003;**60**:1601–3.
198. Miller RG, Moore DH, Dronsky V, Bradley W, Barohn R, Bryan W, et al. A placebo-controlled trial of gabapentin in spinal muscular atrophy. *J Neurol Sci* 2001;**191**:127–31.
199. Merlini L, Solari A, Vita G, Bertini E, Minetti C, Mongini T, et al. Role of gabapentin in spinal muscular atrophy: results of a multicenter, randomized Italian study. *J Child Neurol* 2003;**18**:537–41.
200. Lloyd A, Hunter N. Trophos completes patient enrollment in pivotal efficacy study of olesoxime in spinal muscular atrophy. Trophos; 2011 [updated 8 Sep 2011; cited Jan 4 2014]; Available from: <http://www.trophos.com/news/pr20110908.htm>.
201. Burghes AH, McGovern VL. Antisense oligonucleotides and spinal muscular atrophy: skipping along. *Genes Dev* 2010;**24**:1574–9.
202. Hua Y, Vickers TA, Okunola HL, Bennett CF, Krainer AR. Antisense masking of an hnRNP A1/A2 intronic splicing silencer corrects SMN2 splicing in transgenic mice. *Am J Hum Genet* 2008;**82**:834–48.
203. Singh NN, Shishimorova M, Cao LC, Gangwani L, Singh RN. A short antisense oligonucleotide masking a unique intronic motif prevents skipping of a critical exon in spinal muscular atrophy. *RNA Biol* 2009;**6**:341–50.
204. Hua Y, Sahashi K, Hung G, Rigo F, Passini MA, Bennett CF, et al. Antisense correction of SMN2 splicing in the CNS rescues necrosis in a type III SMA mouse model. *Genes Dev* 2010;**24**:1634–44.
205. Williams JH, Schray RC, Patterson CA, Ayitey SO, Tallent MK, Lutz GJ. Oligonucleotide-mediated survival of motor neuron protein expression in CNS improves phenotype in a mouse model of spinal muscular atrophy. *J Neurosci* 2009;**29**:7633–8.
206. Corti S, Nizzardo M, Nardini M, Donadoni C, Salani S, Ronchi D, et al. Embryonic stem cell-derived neural stem cells improve spinal muscular atrophy phenotype in mice. *Brain* 2010;**133**:465–81.
207. Ebert AD, Yu J, Rose Jr. FF, Mattis VB, Lorson CL, Thomson JA, et al. Induced pluripotent stem cells from a spinal muscular atrophy patient. *Nature* 2009;**457**:277–80.
208. Foust KD, Wang X, McGovern VL, Braun L, Bevan AK, Haidet AM, et al. Rescue of the spinal muscular atrophy phenotype in a mouse model by early postnatal delivery of SMN. *Nat Biotechnol* 2010;**28**:271–4.
209. Passini MA, Bu J, Roskelley EM, Richards AM, Sardi SP, O'Riordan CR, et al. CNS-targeted gene therapy improves survival and motor function in a mouse model of spinal muscular atrophy. *J Clin Invest* 2010;**120**:1253–64.
210. Valori CF, Ning K, Wyles M, Mead RJ, Grierson AJ, Shaw PJ et al. Systemic delivery of scAAV9 expressing SMN prolongs survival in a model of spinal muscular atrophy. *Sci Transl Med* 2010;**2**:35ra42.
211. Dominguez E, Marais T, Chatauret N, Benkhelifa-Ziyyat S, Duque S, Ravassard P, et al. Intravenous scAAV9 delivery of a codon-optimized SMN1 sequence rescues SMA mice. *Hum Mol Genet* 2011;**20**:681–93.
212. Duque S, Joussemet B, Riviere C, Marais T, Dubreil L, Douar AM, et al. Intravenous administration of self-complementary AAV9 enables transgene delivery to adult motor neurons. *Mol Ther* 2009;**17**:1187–96.
213. Gowing G, Svendsen CN. Stem cell transplantation for motor neuron disease: current approaches and future perspectives. *Neurotherapeutics* 2011;**8**:591–606.
214. Iannaccone ST. Modern management of spinal muscular atrophy. *J Child Neurol* 2007;**22**:974–8.
215. Schroth MK. Special considerations in the respiratory management of spinal muscular atrophy. *Pediatrics* 2009;**123**(Suppl. 4): S245–9.
216. Mellies U, Dohna-Schwake C, Stehling F, Voit T. Sleep disordered breathing in spinal muscular atrophy. *Neuromuscul Disord* 2004;**14**:797–803.
217. Markstrom A, Cohen G, Katz-Salamon M. The effect of long term ventilatory support on hemodynamics in children with spinal muscle atrophy (SMA) type II. *Sleep Med* 2010;**11**:201–4.
218. Durkin ET, Schroth MK, Helin M, Shaaban AF. Early laparoscopic fundoplication and gastrostomy in infants with spinal muscular atrophy type I. *J Pediatr Surg* 2008;**43**:2031–7.
219. Yuan N, Wang CH, Trela A, Albanese CT. Laparoscopic Nissen fundoplication during gastrostomy tube placement and noninvasive ventilation may improve survival in type I and severe type II spinal muscular atrophy. *J Child Neurol* 2007;**22**:727–31.
220. Sproule DM, Montes J, Dunaway S, Montgomery M, Battista V, Koenigsberger D, et al. Adiposity is increased among high-functioning, non-ambulatory patients with spinal muscular atrophy. *Neuromuscul Disord* 2010;**20**:448–52.
221. Fujak A, Kopschina C, Forst R, Gras F, Mueller LA, Forst J. Fractures in proximal spinal muscular atrophy. *Arch Orthop Trauma Surg* 2010;**130**:775–80.
222. Montes J, Dunaway S, Montgomery MJ, Sproule D, Kaufmann P, De Vivo DC, et al. Fatigue leads to gait changes in spinal muscular atrophy. *Muscle Nerve* 2011;**43**:485–8.
223. Montes J, Blumenschine M, Dunaway S, Alter AS, Engelstad K, Rao AK, et al. Weakness and fatigue in diverse neuromuscular diseases. *J Child Neurol* 2013;**28**:1277–83.

Chapter 9

Juvenile Amyotrophic Lateral Sclerosis

Jemeen Sreedharan and Robert H. Brown, Jr.

INTRODUCTION

The motor neuron diseases (MNDs) constitute a debilitating range of disorders that cause variable disability and affect all ages from neonates to the elderly. The best known, most aggressive, and most common MND in adults is amyotrophic lateral sclerosis (ALS, or Lou Gehrig's disease). In rare cases it can affect children and young (<25 years) adults, in which case it may be designated juvenile ALS (jALS). Juvenile ALS overlaps with classical ALS but has significant differences. By comparing and contrasting the two we may be able to understand the biology of motor neuron degeneration and the role of perturbations in developmental events in the genesis of motor neuron disease. It is likely that the juvenile instances of ALS will offer intriguing genetic insights into motor neuron degeneration with considerable relevance to ALS in general.

Critically, both jALS and typical ALS involve the same general populations of motor neurons, including primarily those that reside in the spinal cord and brainstem to innervate muscle ("lower motor neurons") and those in the frontal regions that descend to influence the lower motor neurons. The latter project to the spinal motor neurons via both mono- and polysynaptic pathways, predominantly in the cortico-bulbar and cortico-spinal tracts. That this selective set of neurons is comparably affected in both jALS and ALS strongly suggests that these different diseases share some common molecular mechanisms. An understanding of even the rarest of motor neuron diseases may well yield insight into more common forms. In this review we will draw attention to the expanding field of jALS with an emphasis on genetic highlights potentially of broad relevance to all types of ALS.

NOSOLOGY OF MOTOR NEURON DISEASES

A number of different terms are commonly used to describe the progressive, degenerative motor neuron diseases. Classification under these headings is largely dependent on the predominant and earliest noted clinical feature. Lower motor neuron (LMN) syndromic terms are progressive muscular atrophy (PMA), spinal muscular atrophy (SMA), and distal hereditary motor neuropathy (dHMN) or distal SMA. Upper motor neuron (UMN) syndromes include primary lateral sclerosis (PLS) and hereditary spastic paraparesis (HSP). However, there is often a great deal of phenotypic overlap between the diseases that these terms are used to describe, rendering them confusing if not frankly redundant. In many cases mutations in a single gene can give rise to one or another of the canonical syndromes.

The clinical manifestations of LMN and UMN failure are quite distinct. In LMN degeneration, there is weakness, flaccid dysarthria (laxity of tongue and orolabial muscles), and flaccid atrophy of the limbs, usually with fasciculations, and often accompanied by premonitory cramps. In UMN degeneration, there is weakness, limb stiffness, spastic dysarthria (tightness of tongue and lip movement, like eating a hot potato), and emotional lability (pseudobulbar affect). PMA has a clinical and temporal course that mimics ALS but without UMN features. Although the course is generally slowly progressive, there are numerous important exceptions to this rule, and some patients may go on to develop UMN features, fulfilling the criteria for a diagnosis of ALS. In dHMN, by contrast with PMA, weakness and atrophy are more obviously peripheral; these features may overlap with axonal forms of Charcot-Marie-Tooth disease (CMT). In dHMN, juvenile onset is common. In addition, atypical features are often present, such as sensory and (infrequently) UMN features. For this reason, it is reasonable to consider some forms of dHMN as atypical ALS syndromes rather than pure dHMN.

As reviewed in detail elsewhere in this volume, SMA is a degenerative disorder selectively affecting LMNs. There is a striking disparity in the clinical severity in these cases. In some, the onset is apparently intrauterine;

B.T. Darras, H. Royden Jones, Jr., M.M. Ryan & D.C. De Vivo (Eds): Neuromuscular Disorders of Infancy, Childhood and Adolescence, Second edition.
DOI: http://dx.doi.org/10.1016/B978-0-12-417044-5.00009-3

there are joint contractures at birth (arthrogryposis) attesting to restriction of limb movement before birth. In others, the disorder declares itself within the first few postnatal days or weeks, heralded by severe newborn weakness, hypotonia, respiratory insufficiency and difficulty crying and sucking. In these aggressive newborn cases (Werdnig-Hoffman disease) tongue fasciculations are common, and there is tremulousness of the limbs, presumably also reflecting abnormal firing of motor neurons. In still other cases, the clinical manifestations are not well-defined until late teen or early adult years. SMA is the commonest genetic cause of neonatal mortality and is caused by loss of function of the *survival motor neuron 1* gene (*SMN1*). The clinical severity is dictated by the number of copies of the *SMN2* gene (see Chapter 8). Moreover, it is now appreciated that there are multiple genes whose mutations cause the phenotype of SMA.

Juvenile ponto-bulbar palsies include Brown-Vialetto-Van Laere (BVVL) syndrome and Madras Motor Neuron Disease (MMND). The former is often a disorder of early childhood, while the latter may not show onset until the early teens. The central element in both disorders is bulbar LMN degeneration affecting speech and swallowing. However, MMND can entail nonmotor aspects, such as deafness and visual dysfunction. Indeed, there are cases of MMND in which deafness is the first manifestation. Because these disorders are very rare, it is difficult to find large sets of pedigrees from which one can confidently garner insight into the mode of transmission of the affected genes. Both probably have recessively inherited forms, and there are some pedigrees of MMND that suggest inheritance may be dominant.

PLS is essentially ALS without LMN features. The course of PLS varies. While in some cases the disease evolves slowly, in others it proves to be quite disabling with bulbar involvement and a time course resembling that of ALS. It is not uncommon for cases that begin insidiously with a PLS phenotype to develop LMN features.

The HSPs constitute a huge and growing group of diseases (reviewed by Fink[1]). Defects in multiple genes have been defined as causing HSP; in many instances, the genetic locus but not the specific defect has been identified. In some classifications, these are designed as SPG (***S**pastic **P**araple**G**ia*) and numbered in order of discovery. Multiple inheritance patterns are seen. HSP may be autosomal dominant, recessive, X-linked or, in at least one case, mitochondrially inherited. Clinically, patients develop mainly lower limb spasticity and weakness as well as dorsal column impairment causing reduced vibratory sensation. Weakness is usually symmetric. Urinary urgency is common.

HSP phenotypes vary considerably from essentially "uncomplicated" forms with spastic paraparesis alone to "complicated" variants associated with involvement of other central and/or peripheral parts of the nervous system and, rarely, systemic manifestations. Clinical deterioration in HSP is variable but usually very slow, though often nonuniform. In many patients the upper limbs remain uninvolved and lifespan is not affected. Disease may reach a plateau phase after 5–10 years of progression. Pathological data, though limited, have shown distal axonopathy in the corticospinal tract and the gracile fascicle (worst in the cervical cord).[2–5]

Molecular Classification of jALS—Syndromes of Juvenile ALS

With these general points as background, we have compiled a list of disorders that in our view arise as differential considerations in the category of jALS (Table 9.1). This classification is intended to be inclusive, spanning some disorders that, as described previously, are not always considered as jALS. We discuss the major entities on this list in the following sections. We define jALS as any progressive motor neuron disease affecting both upper and lower motor neurons with onset at age 25 years or less. There are many syndromes fitting this description, all of which are extremely rare. Unlike typical adult-onset ALS, jALS is often very slowly progressive, nonfatal, and may affect nonmotor systems. Inheritance is often autosomal recessive rather than dominant, the more typical pattern in adult-onset familial ALS (fALS). Disease is often manifested in closed populations in which consanguineous marriages are the norm. In recent years, detailed family trees and genetic analyses of blood DNA extraction have yielded many new genes linked to juvenile forms of motor neuron disease. Because of the diversity of forms of jALS, the organization of any review is necessarily somewhat arbitrary. We have elected to consider four categories of disorders: (1) ALS overlaps, (2) lower motor neuron predominant; (3) upper motor neuron predominant; and (4) miscellaneous forms of jALS.

DISCUSSION OF TABLE 9.1

ALS Overlap Disorders

ALS2 (Alsin)

ALS2 is a rare, recessively inherited, juvenile-onset form of ALS. It is predominantly a UMN syndrome characterized by slowly progressive spasticity beginning in the lower limbs, with gradual extension upwards to involve the upper limbs and bulbar musculature.[5] Truncation mutations in the *ALS2* gene (coding for alsin) were found in kindreds from Tunisia and Kuwait, and also in a Saudi Arabian family with juvenile PLS.[7,8] Mutations in *alsin*

TABLE 9.1 Molecular Genetics of Selected Types of Juvenile ALS

Disease	Gene	Locus	Protein	Inheri-Tance	Non-jALS Aspect	Gene/Protein Function	References
ALS Overlap Disorders							
ALS2	ALS2	2q33.1	alsin	AR		Guanine nucleotide exchange, endosomal trafficking	6–20
ALS4	SETX	9q34.13	senataxin	AD		DNA/RNA helicase	21–32
ALS5	SPG11	15q15.1	spatacsin	AR	HSP	Transmembrane protein; role in axon maintenance	33–35
ALS6	FUS	16p11.2	fused in sarcoma	AD	ALS	DNA/RNA binding protein	36–38
ALS16	SIGMAR1	9p13.3	opioid sigmoid receptor	AR		Non-opioid receptor, ER chaperone	39–45
SPG18	ERLIN2	8p11.22	ER lipid raft protein 2	AR	kyphoscoliosis, saccadic eye movements	Functions in protein degradation in the endoplasmic reticulum	46
jALS		6p25, 21q22	unknown	AR	ptosis, gynecomastia		47
BVVL1*	C2orf54]	20p13	SLC52A3 riboflavin transporter	AR	deafness	Transports riboflavin, the central component of cofactors FMN and FAD	48–50
BVVL2	GPR172A	8q24	SLC52A2 riboflavin transporter	AR	deafness, optic atrophy	Transports riboflavin, the central component of cofactors FMN and FAD	51
BVVL	GPR172B	17p13.2	SLC52A1 riboflavin transporter	AD		Transports riboflavin, the central component of cofactors FMN and FAD	52
BVVL	UBQLN1	9q21.32	ubiquilin 1	AD	deafness	Targeting ubiquitinated proteins for degradation	53
Madras MND			unknown	AR, S	deafness, optic atrophy		54–58

Lower Motor Neuron Predominant							
dHMN	BSCL2**	11q13	seipin	AD	jALS – upper limb	ER*** glycoprotein modulates unfolded protein response	61–66
dHMN	DCTN1	2p13	dynactin 1	AD		Retrograde axonal transport	67–69
dHMN	HSPB1	7q11.23	HSP27	AD/AR			70
dHMN		7q34-36	unknown	AD			70
dHMN		4q34-35	unknown	AD	neuropathy		71
dHMN (Jerash)		9p21,1-12	unknown	AR			72
Upper Motor Neuron Predominant							
SPG3A	ATL1	14q11-q21	atlastin	AD	HSP-like	Enhances ER membrane fusion, interacts with spastin, REEP1, NIPA1	73–82
SPG4	SPAST	2p22.3	spastin	AD	HSP-like	Microtubule severing	83–84
SPG20	Spartin	13q	spartin	AR	HSP-like	Endosomal trafficking, microtubule binding dynamics	85–88
SPG39	PNPLA6	19p13.2	neuropathy target esterase	AR	HSP-like	Deacetylates phosphatidylcholine, may reduce organophosphate toxicity	89–95
SPOAN****		11q13	gene unknown	AR	HSP-like		96
SPG57	TFG	3q12	TRK-fused gene	AR	HSP/HSAN*****	Oncogene, with role in RNA sensing and NF-kappaB pathway	97–99
SPG31	REEP1	2p11.2	Receptor expression-enhancing protein 1 (REEP1)	AD	HSP-like	Functions in ER and in vesicle transport	99

**Brown-Vialeto-van Laere*
***Berardinelli-Seip Congenital Lipodystrophy*
****endoplasmic reticulum*
*****spastic paraplegia, optic atrophy, neuropathy*
******hereditary sensory-motor autonomic neuropathy*
AR: autosomal recessive
AD: autosomal dominant
S: sporadic

have also been found in several families with a syndrome described as infantile-onset ascending hereditary spastic paraplegia (IAHSP).[9–11] Most mutations generate truncated *alsin* products, which are unstable leading to loss of function. Homozygous missense mutations have also been described, in association with alsin protein mislocalization, leading to loss of function.[12] There are no published reports of the neuropathology of ALS2 patients.

The *alsin* gene is ubiquitously expressed in the CNS (especially cerebellum) and periphery (especially kidney). Alsin appears to protect cultured neurons from mutant SOD1-mediated toxicity[13,14]; its overexpression promotes hippocampal neurite outgrowth through Rac1 activation.[15] Alsin may also have a role in preventing glutamate mediated excitotoxicity by appropriate expression of the Ca^{2+} impermeable Glur2 subunit of AMPA receptors.[16]

The best characterized properties of alsin relate to its functions in endosomal dynamics. Present in the cytoplasm, alsin is relocalized to endosomes by Rac1. Alsin then acts as a guanine-nucleotide exchange factor (GEF) to activate the GTPase Rab5. Rab5 is important for endocytosis, endosome fusion and endosomal trafficking.[17,18] Alsin knockout mice have relatively subtle motor phenotypes and pathological changes are limited to mild corticospinal tract degeneration, suggesting that ALS2 may be a distal axonopathy.[19] However, *in vitro* studies of alsin null neurons and fibroblasts confirm reduced endosomal trafficking and altered endosomal fusion. Furthermore, the rare missense mutations of alsin appear to act through loss of function due to impairment of Rac1-mediated relocalization to endosomes.[20]

ALS4 (Senataxin)

ALS4 is a rare, nonfatal, autosomal dominant, juvenile-onset dHMN characterized by limb weakness, severe muscle wasting and pyramidal tract signs including brisk reflexes and extensor plantar responses. Bulbar and respiratory muscles are spared and lifespan is unaffected. The mean age of onset is 17 years. Symptoms begin with difficulty walking followed by weakness and wasting of intrinsic hand and foot muscles. By the fourth to fifth decade, patients display significant proximal weakness and are often wheelchair-bound. By the sixth decade, many lose hand function entirely. Sensation remains normal both clinically and electrophysiologically, although pathological studies have shown dorsal column degeneration in addition to corticospinal tract degeneration and anterior horn cell loss.[21]

ALS4 was first characterized in four kindreds from America and Europe and linked to a locus on Chr 9q.[22] Dominant missense mutations in the *SETX* gene encoding senataxin were subsequently found.[23] In all, as of 2013 only seven mutations in senataxin affecting 64 individuals were listed on the ALS Online Genetics Database (ALSoD). These affect the N-terminus and helicase domain. A dominant *SETX* mutation has also been found in a family with dominantly transmitted proximal SMA (ADSMA).[24] Interestingly, mutations of *SETX* cause ataxia-oculomotor apraxia 2 (AOA2),[25] although in these cases the mutations are recessive and mostly truncations. A toxic gain of senataxin function may be responsible for ALS4, while loss of function may lead to AOA2.

The full range of functions of senataxin remains to be determined. It seems clear that it has a putative DNA/RNA helicase domain and demonstrates punctate expression in the nucleus and diffuse expression in the cytoplasm.[26] Senataxin is important in preventing oxidative damage to DNA,[27] interacts with a variety of RNA processing proteins (including RNA polymerase II and SMN) and influences gene transcription and mRNA splicing.[28] Senataxin also shows homology to another putative DNA/RNA helicase, immunoglobulin μ-binding protein 2 (IGHMBP2). Notably, recessive (probably loss of function) mutations of *IGHMBP2* cause a childhood onset LMN disease, spinal muscular atrophy with respiratory distress type 1 (SMARD1).[29]

Senataxin knockout mice have no motor phenotype but the observation that male animals are sterile and females are subfertile led to detailed studies demonstrating critical roles for senataxin in crossing over in meiosis, meiotic sex chromosome inactivation, and DNA damage response.[30]

In spite of this research on loss of senataxin function, which is clearly relevant to AOA2, the mechanisms by which dominant mutations cause ALS4 remain relatively unexplored. *In vitro* studies demonstrate a role for senataxin in neuritogenesis through fibroblast growth factor 8 and suggest that the dominant mutants lack this ability. Thus, haploinsufficiency is a possible mechanism of disease in ALS4.[31]

Finally, caution should be taken in interpreting SETX sequencing results. Several variants described in the literature as ALS4-causing mutations may actually be benign polymorphisms, although it remains possible that they may contribute to disease risk or phenotype without being completely penetrant.[32]

ALS5/SPG11 (Spatacsin)

ALS5, on chromosome 15q, was originally described in several kindreds from Tunisia and Germany with autosomal recessive jALS. Families were linked to chromosome 15q15.1-21.1.[33] Patients develop progressive gait disturbance, presenting with mixed UMN and LMN features between 8 and 18 years of age. Dysarthria begins after 3 to 4 years and survival is generally 10–25 years from symptom onset. In parallel studies, a form of autosomal recessive hereditary spastic paraplegia (ARHSP) with thin corpus callosum (TCC), designated SPG11, was reported to be caused by mutations in a novel gene, labelled *spatacsin*.[34] This is a transmembrane protein that is phosphorylated after DNA damage; its function is not well understood. Of interest, Orlacchio et al.[35] collected a series of 25 families (from Italy, Japan, Turkey, Canada) with recessively inherited jALS; 10 of the 25 kindreds were found to have mutations in *spatacsin*. While we do not know if the gene defect in the original ALS5 families has been confirmed to be in *spatacsin*, the general premise now accepted is that the ALS5 gene is indeed *spatacsin*.

ALS6 (Fused in Sarcoma/Translocated in Liposarcoma)

This form of ALS most commonly arises in adults with a typical, if somewhat aggressive, phenotype.[36,37] However, it is now well documented to arise in individuals in the second decade, usually with a devastatingly rapid course. These cases are all caused by mutations in the gene encoding FUS (fused in sarcoma), a multifunctional RNA- and DNA-binding protein of 525 amino acids. FUS is normally located almost exclusively within the nucleus but is observed to exit to the cytosol after exposure to some cellular stresses (e.g. hyperosmolar shock). Mutant FUS is observed to mislocalize to the cytosol where it sometimes participates in forming aggregates. Rapidly progressive jALS has been associated with several different *FUS* gene mutations; among the most aggressive is a particular dominant missense variant that substitutes lysine for proline at codon 525, thereby disrupting the nuclear localization signal.[38] At autopsy these cases are associated with distinctive basophilic intranuclear aggregates.

ALS16 (Sigma-1 Intracellular Receptor) and SPG18 (ERLIN2)

A homozygous missense mutation (E102Q) in *SIGMAR1* (encoding sigma nonopioid intracellular receptor 1, Sig-1 R) was comparatively recently reported in a consanguineous Saudi family with autosomal recessively inherited jALS.[39] The phenotype was slowly progressive, with predominantly UMN involvement.

Sig-1 R regulates K^+ channels and is involved in Ca^{2+} signaling through IP3R. It is a receptor for a variety of ligands including steroids, psychostimulants, and haloperidol. It also has chaperone activities in the ER, implicating the unfolded protein response in neurodegeneration. *In vitro* studies showed that the E102Q mutation abrogates the ability of SigR1 to suppress apoptosis induced by ER stress.[39] The E102Q mutation occurs in a predicted transmembrane domain and subcellular fractionation suggests that the mutation causes a shift to lower density membrane fractions where mutant protein forms detergent-resistant complexes.

A role for SigR1 in sporadic ALS has also been suggested by pathological studies showing altered distribution of SigR1 at C-boutons (excitatory postsynaptic densities at spinal cord motor neurons) and cytoplasmic accumulation of SigR1.[40]

The homozygous E102Q mutation is the only clearly pathogenic mutation yet described. Heterozygous 3′ untranslated region (UTR) variants described in familial ALS-FTLD have been identified but it remains unclear whether these are actually benign polymorphisms or whether they do contribute to disease risk.[41,42] No animal studies of these mutations have yet been conducted, but a recent study in mice showed that the native Sig-1 R protein is located exclusively in motor neurons, and that knockout causes motility problems.[43] Furthermore, the Sig-1 R agonist PRE-084 has recently been shown to be protective in both SOD1-G93A and wobbler models of ALS, supporting a loss of function as the mechanism by which Sig-1 R mutation may cause disease.[44,45]

In the context of ER stress it is worth noting the recent discovery of recessive mutations in ERLIN2, an endoplasmic reticulum protein that is associated with lipid rafts, in an inbred family with juvenile PLS.[46] Patients were normal at birth but developed impaired crawling at 8 months of age, toe walking at two years with progressively impaired mobility such that they were bedridden by just 15 years of age. In addition to severe generalized spasticity by end stage, patients also developed severe dysarthria and pseudobulbar palsy, skeletal deformities including kyphoscoliosis, saccadic eye movements, and moderate distal limb wasting. Brain PET scan showed hypometabolism in the temporal and parietal lobes. Erlin2 is similar to Sig-1 R in that it is a component of ER lipid rafts. Splice junction mutation of the *ERLIN2* gene results in nonsense-mediated decay of aberrantly spliced transcripts causing a loss of function. It will be interesting to determine whether Erlin2 and Sig-1 R interact and how they contribute to distinct clinical phenotypes.

jALS Linked to Chromosome 6p25 and 21q22

A family from Utah has been reported in which four of six children from a consanguineous first cousin marriage had jALS. The phenotype was characterized by prominent upper motor neuron signs as well as amyotrophy of limb and bulbar musculature. Atypical features included mild ptosis, gynecomastia in one patient, and mild distal impairment of vibration sensation. Onset was between 4–10 years of age with slow progression to loss of ambulation after 10 years. Homozygosity mapping identified linkage to two loci (6p25 and 21q22) but no disease-causing mutation has yet been identified.[47]

Brown-Vialetto-Van Laere Syndrome

The Brown-Vialetto-Van Laere (BVVL) and Fazio Londe (FL) syndromes are pontobulbar palsies usually of juvenile onset. The conditions appear identical, except that BVVL syndrome cases also have hearing loss. The occurrence of both phenotypes within the same kindreds demonstrate that BVVL and FL syndromes are one disease (hereafter referred to as BVVLS).[48] The condition is extremely rare—a 2012 literature review found only 101 reported cases.[49] It is both fascinating and clinically quite significant that some inherited cases of BVVLS are eminently curable, emphasizing the importance of a molecular diagnosis (see below).

The clinical presentation of BVVLS is usually with bulbar palsy (92%), sensorineural hearing loss (81%), facial weakness (77%), and respiratory compromise (64%). The age of onset varies widely from infancy to adulthood, with around 80% of cases having onset before 18 years of age with a mean age of onset of 8 years. Early motor milestones are reported to be normal.[50] Two thirds of patients are female. Around half of teenage-onset patients have muscle weakness and UMN signs. Around a quarter of patients may have oculomotor palsies (cranial nerves III, IV, and VI). Less commonly involved are the trigeminal and optic nerves and the diaphragm. Around 20% may have limb LMN signs. Ataxia and tremor are seen in 10% of cases, while behavioral changes are seen in 4%. Disease progression can vary markedly between and within families. Survival varies from a few months to 32 years (mean survival 5 years) from presentation and is worst in those with early onset. Death is usually from respiratory insufficiency.

Clues to the genetic underpinnings of some cases of BVVLS came from the apparent autosomal recessive pattern of inheritance seen in consanguineous families of Arabic and Pakistani origin.[50] Using linkage analysis (autozygosity mapping) and candidate gene sequencing, homozygous and compound heterozygous mutations in *C20orf54* (*SLC52A3*) were identified in these cases. *SLC52A3* encodes an intestinal riboflavin transporter. These disease-linked mutations impair riboflavin absorption, as evidenced by low serum flavin and acylcarnitine. Amazingly, striking clinical improvement was seen in these patients following oral or intravenous riboflavin supplementation, with improvement seen from a few days to several months after initiation of therapy. Some patients were able to come off mechanical ventilation. Of note, mutations in related riboflavin transporter genes *SLC52A2* and *SLC52A1* have now also been detected in other BVVLS cases, with similar responses to riboflavin.[51,52] Thus, vitamin B2 deficiency seems to be the common pathogenic mechanism in these BVVLS cases. Riboflavin (vitamin B2) is an essential component of the nucleic acid cofactors FAD (flavin adenine dinucleotide) and FMN (flavin mononucleotide) and accordingly is important in numerous cellular functions including electron transport and a wide range of reactions including synthesis of niacin and folate.

Ubiquilin1

A missense mutation of *UBQLN1* has also been described in one individual with the BVVLS phenotype whose motor neuron symptoms began before the age of 10 years.[53] This mutation may impair the ability of ubiquilin1 to carry out its normal function of shuttling ubiquitinated proteins for degradation through the proteasome or autophagosomes.[53]

Madras Motor Neuron Disease

This rare form of jALS seen predominantly in South India was first described in 1970 in the city of Madras (now known by its precolonial name of Chennai).[54] Madras motor neuron disease (MMND) has also rarely been reported in Thailand, Korea, and Turkey.[55–57] There is phenotypic overlap with BVVLS though a recent study has shown no mutations in the riboflavin transporter genes linked with BVVLS.[58] A genetic cause has yet to be identified but is a likely contributor given that around a quarter of patients have a family history of MMND (familial MMND [FMMND]).[59] Incidence in males and females appears equal in a case series of 116 patients from Southern India seen over 36 years (1971–2007).[59] Patients with MMND typically present at a mean age of 16 years (range 1–39 years); deafness often heralds the onset.

The features that set MMND apart from typical ALS are a high incidence of hearing loss (seen in 90% of patients, and the initial presenting feature in 50%) and optic atrophy (25% of MMND patients, 100% of MMND variant [MMNDV] patients). The overall picture is of predominantly LMN disease with striking wasting, particularly of proximal and distal upper limbs, and lower cranial nerve involvement causing profound facial weakness and bulbar palsy. A minority of patients (6%) were observed to have schizophrenia.[59] UMN signs are seen in most patients, more commonly in the lower limbs, with an extensor plantar reflex in 62% of patients. Survival is of the order of 30 years from onset, though females notably have a shorter survival of around 20 years.

EMG findings are typical of ALS with active and chronic denervation and reinnervation findings in most cases. Muscle biopsy demonstrates neurogenic atrophy with no evidence of mitochondrial myopathy. Brain imaging is normal. Visual evoked potentials in those with MMNDV are consistent with optic atrophy.

No mutations in any known ALS genes have so far been found in MMND, including the riboflavin transporter genes implicated in BVVLS.[58] Although consanguinity is not clearly apparent in the majority of familial cases, most of these families demonstrate an autosomal recessive pattern of inheritance and the parents of these families tend to come from the same regional location in south India, suggesting that a closed gene pool influenced by unknown environmental factors could underpin the disease.

Lower Motor Neuron Predominant Disorders

There are multiple inherited disorders that cause degeneration predominantly of LMNs but with subtle dysfunction and pathology in corticospinal motor neurons. Many are distal in onset and so are designated distal hereditary motor neuropathy (dHMN). Because of the pyramidal features, several of these dHMNs could be considered forms of jALS. Salient examples of mutant genes whose phenotypes share this overlap feature are *SETX* (described previously), *BSCL2*, *DCTN1*, and *HSPB1*, which are described below.

Berardinellie-Seip Congenital Lipodystrophy Type 2 Gene

Recessive loss of function mutations in the Berardinellie-Seip Congenital Lipodystrophy type 2 (*BSCL2*) gene were originally found to cause congenital generalized lipodystrophy type 2.[60] Dominant mutations in *BSCL2* were subsequently identified in several Austrian kindreds demonstrating a variety of motor neuron phenotypes, including dHMN with pyramidal features (dHMN5), Silver syndrome (late childhood onset of hand weakness and limb spasticity, in association with pes cavus, also designated SPG17) and HSP.[61,62] *BSCL2* mutations are among the commonest known causes of the dHMN phenotype. The mean age of onset is typically 15 years with a range of 2–40 years. LMN features tend to start in the hands with wasting of the intrinsic palmar muscles (e.g. thenar eminence). Most patients eventually develop lower limb features including foot deformities. Reflexes may be brisk. Some sensory involvement is seen in the form of reduced vibration sense in the lower limbs. Progression is very slow over decades. Nerve conduction studies essentially show a motor axonal neuropathy.

BSCL2 encodes seipin, an integral ER membrane protein. While the originally defined cases of *BSCL2* mutations were recessively inherited, it is now appreciated that some dHMN-associated mutations are dominantly transmitted. These appear to enhance glycosylation and are more prone to aggregation.[61] Activation of the unfolded protein response has also been seen in cells expressing mutant seipin *in vitro*.[63] Transgenic mice expressing mutant seipin show enhanced activation of the autophagy pathway.[64] These observations implicate impaired protein homeostasis as a key component of the pathogenesis of dHMN5. As an interesting point, it has also recently been shown through exome sequencing that one of the original CMTX3 families linked to chromosome X harbors a mutation in *BSCL2*.[65] Determining how *BSCL2* mutations can result in such diverse phenotypes (from lipodystrophy to dHMN to CMT) will likely yield important therapeutic targets.

DCTN1

The G59S dominant missense mutation in the *DCTN1* (Dynactin) gene is associated with a rare form of dHMN with vocal cord paresis characterized by progressive facial and arm weakness, with later leg weakness.[66] Onset may be in early adulthood, but UMN features are generally not seen, meaning that this is not a true jALS. However, other missense mutations in *DCTN1* are actually associated with a typical ALS phenotype, with disease duration of 4 to over 9 years, and also a frontotemporal dementia (FTD) phenotype.[67,68] However, the onset in these typical ALS cases is from age 48–64 years, indicating that this is not a juvenile disease. *DCTN1* encodes the p150 glued subunit of dynactin, which is involved in retrograde axonal transport.

HSPB1

Dominant or recessive missense mutations in Heat Shock Protein 27 (*HSPB1*) (encoding HSP27, a chaperone) are associated with dHMN in families of European ancestry.[69] Onset is between 21 and 54 years. Distal lower leg involvement is seen early with arms involved 5–10 years after onset. Brisk reflexes are often noted.

Other dHMN Loci

Autosomal dominant dHMN with UMN features has also been linked to three other genetic loci. Firstly, an Australian family has been linked to Chr 7q34–36.[70] Age of onset is 10 years (range 3–40 years) with initial onset in the lower limbs causing impaired gait. As well as wasting, there is increased lower limb tone and extensor plantar responses in around half of patients. There may be slight sensory symptoms. Progression is slow. Axonal loss is seen on biopsy.

Secondly, an Italian family with dominantly inherited dHMN with pyramidal features has been linked to chromosome 4q34.3-q35.2.[71] Onset is from age 25 to 40 years with spastic gait, brisk reflexes, distal leg wasting, and weakness. Again, progression is slow. Nerve conduction studies demonstrate axonal neuropathy.

Thirdly, autosomal recessive Jerash type dHMN (seen in Jordanians) has been linked to chromosome 9p21.1-p12.[72] Onset is age 6–10 years. Weakness is distal at onset and seen initially in the legs and then in the upper limbs. There is wasting in the hands and feet. Reflexes tend to be brisk at the knees and absent at the ankles with extensor plantars.

Upper Motor Neuron Predominant Disorders

Juvenile-onset HSP with Lower Motor Neuron Involvement

Some early onset HSP variants demonstrate distal amyotrophy or lower motor neuropathy and may therefore be described as jALS. The HSP variants associated with peripheral neuropathy are HSP types SPG2, 3 A, 5, 6, 7, 10, 25, 27, 30, 31, 55, and 56; other related genes in which a usual HSP phenotype is coupled with LMN dysfunction are SPOAN, Cct5 (epsilon subunit mutation), and mitochondrial ATP6 gene mutation. Variants associated with distal amyotrophy are HSP types 5, 10, 14, 15, 17, 20, 26, 30, 38, 39, 41, 43, 55, and 57. Distal amyotrophy is also rarely seen in HSP types 3A and 4. A full description of all of these entities is beyond the scope of this review. However, we offer here a description of the better-understood variants.

SPG3A (Atlastin)

SPG3A (atlastin) is essentially an uncomplicated HSP though patients can sometimes develop an axonal motor neuropathy and/or distal muscle wasting. Sensory neuropathy has also been described and interestingly SPG3A is allelic with hereditary sensory neuropathy I.[73] Symptoms usually begin in childhood (and may not progress), with a mean age of onset of four years (range from one year to the seventh decade). Adolescent and adult onset cases tend to progress insidiously. Nonetheless, wheelchair dependency is rare. SPG3A is caused by dominant mutations in *ATL1*, which encodes atlastin. Over 40 *ATL1* mutations have been described, mostly missense. Atlastin is most highly expressed in the CNS, especially layer V of the cortex, which is where the primary motor neurons reside. Atlastin is a dynamin-like GTPase that mediates endoplasmic reticulum (ER) membrane fusion and interacts with a number of other proteins mutated in HSP, including spastin, REEP1, and NIPA1 (all of which may be associated with LMN degeneration).[74–76,77,78] Atlastin has also been shown to be required for Golgi apparatus morphogenesis.[79]

Zebrafish atlastin has been shown to control motility and spinal motor neuron architecture through inhibition of bone morphogenetic (BMP) signaling.[80] A study in *Drosophila* has recently shown that atlastin interacts with valosin containing protein (VCP), a multifunctional AAA-ATPase with roles in the ubiquitin proteasome system, membrane fusion, and DNA repair.[81] This is particularly interesting given that VCP mutations are a rare cause of ALS.[82]

SPG4 (Spastin)

SPG4 (spastin) is the most common autosomal dominant form of HSP, accounting for around 40% of cases.[83] The *SPG4* gene encodes the protein spastin, which is important for microtubule severing and thus critical for axonal organelle transport. Mutant spastin lacks the normal microtubule severing capacity, and its binding to wild-type spastin may inhibit the normal allele, thus having a dominant negative influence on wild type spastin.[84]

SPG20 (Spartin)

SPG20 is an autosomal recessive, complicated HSP caused by mutations in spartin. The phenotype is called Troyer syndrome.[85] Spartin has some similarity at the N-terminus to spastin and is homologous to proteins involved in the morphology and trafficking of endosomes. It localizes to mitochondria and may have roles in endocytosis, vesicle trafficking, or mitogenic activity.[86] Spartin also inhibits BMP signaling and interacts with a number of proteins involved in ER stress, ubiquitination, and the nucleolar protein nucleolin.[87,88]

SPG39 (Patatin-like Phospholipase Domain-containing Protein 6, Also Known as Neuropathy Target Esterase)

Rare autosomal recessive families with a Troyer-like phenotype but without developmental, cognitive, skeletal, extrapyramidal or cerebellar features have also been described. Screening for spartin mutations was negative in these kindreds, but linkage and candidate gene studies identified homozygous and compound heterozygous mutations in the gene encoding neuropathy target esterase (NTE).[89] Onset is in childhood with progressive lower limb spasticity and weakness, and distal upper and lower limb wasting and weakness. Electrophysiology demonstrates motor axonal neuropathy and MRI scans identify spinal cord atrophy.

NTE is a phospholipase B localized to ER membranes, which may reduce cytotoxic phosphatidylcholine. Its activity may be important in preventing toxic organophosphorus compound induced delayed neurodegeneration (OPIDN). It may also function to regulate cyclic AMP-dependent protein kinase.[90–92] The offending SPG39 mutations in NTE appear to result in a loss of function, either by affecting the catalytic domain or by causing a frameshift and protein truncation. It can be argued that the NTE mutations provide an intriguing link to the organophosphate poisoning hypothesis of ALS in Gulf

War veterans,[93] consistent with the view that rare, ALS-specific paraoxonase variants (PON1, 2, and 3) contribute to ALS risk.[94,95]

Spastic Paraplegia, Optic Atrophy, and Neuropathy

Spastic Paraplegia, Optic Atrophy, and Neuropathy (SPOAN) is a complicated autosomal recessive HSP with juvenile onset described in a large inbred Brazilian family.[96] The earliest noted abnormality is congenital optic atrophy, followed in infancy by the onset of progressive spastic paraplegia. Motor and sensory axonal neuropathy develops in late childhood/early adolescence causing significant weakness in grip strength and reduced sensation. Dysarthria starts in the third decade of life. Other features are an exaggerated startle response, progressive joint contractures, and spinal deformities. Patients are wheelchair dependent by age 15. No mental dysfunction was described and brain MRI is normal. Although the family was convincingly linked to a genetic locus at chromosome 11q13, no mutation has yet been identified.

SPG57 (Tyrosine-kinase Fused Gene)

SPG57 was recently described in two Indian brothers in their teens. These boys had identical clinical histories with normal early milestones in the first year, but difficulty walking due to spasticity, visual impairment due to optic atrophy, and peripheral muscle wasting in all limbs, indicating lower motor neuron degeneration. Though there were no sensory symptoms or signs, nerve conduction studies demonstrated sensory axonal neuropathy. SPG57 is caused by a homozygous missense mutation, c316C > T substituting cysteine for arginine at codon 106 in the Trk-fused gene (TFG).[97]

TFG is a highly conserved regulator of protein secretion in the endoplasmic reticulum. It normally functions as an oligomer, but the disease-linked mutation is in a coiled-coil domain and disrupts the ability of TFG to oligomerize. Interestingly, a heterozygous TFG mutation (substituting leucine for proline at codon 285) causes autosomal dominant hereditary motor and sensory neuropathy (HMSN) with proximal dominant involvement.[98] The P285L mutation is in the C-terminal P/Q rich region. The implication is that this mutation may differ from that associated with SPG57 by virtue of an undetermined toxic gain of function effect. Intriguingly, histological studies demonstrate TDP-43 positive skein-like inclusions similar to those found in ALS. TFG positive inclusions are also seen, though these do not always coaggregate with TDP-43 and TFG inclusions were not seen in sporadic ALS cases.[98]

SPG31 (Receptor Expression-enhancing Protein 1)

Mutations in the REEP1 gene (encoding (receptor expression-enhancing protein 1)) cause diverse phenotypes including HSP at one extreme or predominant LMN axonopathy on the other.[99] This phenotypic diversity is reminiscent of BSCL2/dHMN. REEP1 is involved in ER dynamics, functioning in vesicle transport, ER formation, and interaction with the cytoskeleton.[78] It remains unclear why REEP1 mutants so selectively influence viability and function of motor neurons, although it is striking that other ER proteins (above) and proteins that influence vesicle transport are also implicated in motor neuron disease.

Miscellaneous Disorders

Other childhood-onset disorders may potentially mimic aspects of ALS, with involvement of both the corticospinal tracts and the spinal motor neurons. In general, these are quite rare and not likely to be mistaken for true jALS. Nonetheless, they bear brief mention because some of these variants are potentially treatable. For example, HTLV I and II are also not typically confused with jALS, because their neurological course is typically dominated by the corticospinal component, potentially with dementia as well. That said, it is striking that there are cases of HTLV I that resemble ALS but with a very beneficial response to antiretroviral therapy. Poliomyelitis can afflict both upper and lower motor neurons. However, it will usually be distinguished by the initial febrile meningitis, sometimes preceded by gastrointestinal symptoms, with subsequent, nonprogressive motor weakness that is focal or multifocal. Another infectious entity that frequently enters this differential diagnostic discussion is Lyme disease. In our experience, Lyme disease duplicating jALS (or adult ALS for that matter) is exceedingly uncommon. Nonetheless, because of the therapeutic implications, we would propose screening new-onset cases of jALS for Lyme infection. While vitamin deficiency states are rare at any age, it is important to recall that insufficiency of vitamin B12 can confer both spinal cord (including corticospinal) and peripheral nerve (usually motor and sensory) pathology. Importantly, serum B12 levels may not be obviously deficient, in which case elevated levels of methylmalonic acid may be diagnostic. Inherited deficiency of the vitamin E transporter produces slowly progressive neurodegeneration, generally evident as cerebellar dysfunction reminiscent of Friedreich's ataxia; like B12 deficiency, vitamin E deficiency may be remarkably responsive to repletion of the missing vitamin. Heavy metal intoxication is usually also enumerated in this set of differential considerations, with lead as the main suspect. This is easy to screen with 24-hour urinary heavy metals testing, but the yield has been vanishingly low.

Finally, we note that there are rare or even single jALS families with multisystem disorders that can mimic jALS. For example, triple A syndrome, originally identified in association with achalasia, alacrima, and adrenal insufficiency, as well as autonomic dysfunction, can infrequently present as jALS (e.g. in a 22-year-old female patient with a two-year history of distal muscular atrophy and weakness in all four limbs with brisk reflexes[100]). This syndrome is a consequence of mutations in a gene that encodes aladin, a nuclear membrane protein.

MANAGEMENT

There are currently no disease-modifying treatments for any of the preceding disorders, with the exception of the vitamin deficiencies and perhaps autoimmune neuropathies. Riluzole, the only FDA-approved drug for ALS, has not been documented to benefit any of these disorders. Accordingly, management is essentially supportive and multidisciplinary, encompassing occupational therapy and physical therapy to stretch and strengthen muscles and prevent limb contractures. Medical management of spasticity includes oral baclofen or tizanidine, intramuscular botulinum toxin, or intrathecal baclofen infusion pumps. Care should be taken not to overtreat, as the patient may depend on some of their spasticity to provide support when mobilizing and transferring. Distal ankle dorsiflexion weakness can be managed with ankle foot orthoses. Anticholinergic agents may be useful for urinary urgency.

GENETIC COUNSELING

The rarity of many of the mutations causing jALS variants makes genetic counseling a complex process. Where there are clear patterns of inheritance and a mutation has been identified there is still the issue of penetrance, which can be variable and incomplete. Preimplantation diagnosis can still be of value, of course, for couples with a background of a known mutation (or even when no mutation is known, as exome sequencing is becoming increasingly available).

In cases of HSP with LMN involvement genetic counseling can be very complicated, in part due to the age-dependency and variable penetrance. Many HSP variants are of course very rare; private mutations and limited understanding of these diseases can make counseling an uncertain business. Counseling in individuals with paraplegin mutations (SPG7) is further complicated by the fact that HSP may manifest as both a dominantly and recessively inherited disorder. As yet it is not clear which specific mutations are associated with these modes of inheritance. It remains to be determined if other forms of jALS demonstrate similar bimodal inheritance.[1]

CONCLUSIONS

Juvenile ALS is a broad diagnostic term that encompasses a core entity of upper and lower motor neuron disease with, in some cases, a wide realm of additional clinical features. And, as the foregoing attempts illustrate, the last 10–15 years have witnessed an unprecedented proliferation of information about the molecular genetics of disorders that can be described as jALS; in many respects, the enormous genetic diversity mirrors and anticipates the clinical heterogeneity. At this point, it is difficult to find unifying molecular themes in the diverse set of jALS genes, a quandary that reflects in part our relatively poor understanding of the genesis of this type of neurodegenerative disorder. One anticipates that the exploration of the molecular biology of jALS will ultimately also provide insight into the nature of adult-onset ALS. That the two can converge is clear in the example of jALS caused by FUS mutations. In the long term, one also anticipates that new insights into the molecular pathology of jALS will provide targets for therapy. Moreover, independently of the full delineation of the molecular pathways, one also anticipates that strategies such as gene silencing and gene replacement may prove beneficial for cases arising respectively from gain-of-function (dominant) and loss-of-function (recessive) mutations. Recent early successes in gene replacement (e.g. lipoprotein lipase deficiency mediated by adeno-associated virus) and exon skipping for muscular dystrophy (e.g. Sarepta Inc. eteplirsen trial in Duchenne muscular dystrophy) permit cautious optimism that meaningful therapies will be possible for inherited forms of jALS.

REFERENCES

1. Fink JK. Hereditary spastic paraplegia: clinico-pathologic features and emerging molecular mechanisms. *Acta Neuropathol* 2013;**126** (3):307–28.
2. Behan WMH, Maia M. Strumpell's familial spastic paraplegia: genetics and neuropathology. *J Neurol Neurosurg Psychiatry* 1974;**37**:8–20.
3. Deluca GC, Ebers GC, Esiri MM. The extent of axonal loss in the long tracts in hereditary spastic paraplegia. *Neuropathol Appl Neurobiol* 2004;**30**(6):576–84.
4. Harding AE. Hereditary spastic paraplegias. *Semin Neurol* 1993;**13** (4):333–6.
5. White KD, Ince PG, Lusher M, Lindsey J, Cookson M, Bashir R, et al. Clinical and pathologic findings in hereditary spastic paraparesis with spastin mutation. *Neurology* 2000;**55**(1):89–94.

6. Hentati A, Bejaoui K, Pericak-Vance MA, Hentati F, Speer MC, Hung W, et al. Linkage of recessive familial amyotrophic lateral sclerosis to chromosome 2q33-q35. *Nat Genet* 1994;**7**(3):425–8.
7. Hadano S, Yanagisawa Y, Skaug J, Fichter K, Nasir J, Martindale D, et al. Cloning and characterization of three novel genes, ALS2CR1, ALS2CR2, and ALS2CR3, in the juvenile amyotrophic lateral sclerosis (ALS2) critical region at chromosome 2q33–q34: candidate genes for ALS2. *Genomics* 2001;**71**(2):200–13.
8. Yang Y, Hentati A, Deng HX, Dabbagh O, Sasaki T, Hirano M, et al. The gene encoding alsin, a protein with three guanine-nucleotide exchange factor domains, is mutated in a form of recessive amyotrophic lateral sclerosis. *Nat Genet* 2001;**29**(2):160–5.
9. Devon RS, Helm JR, Rouleau GA, Leitner Y, Lerman-Sagie T, Lev D, et al. The first nonsense mutation in alsin results in a homogeneous phenotype of infantile-onset ascending spastic paralysis with bulbar involvement in two siblings. *Clin Genet* 2003;**64**(3):210–15.
10. Eymard-Pierre E, Lesca G, Dollet S, Santorelli FM, di Capua M, Bertini E, et al. Infantile-onset ascending hereditary spastic paralysis is associated with mutations in the alsin gene. *Am J Hum Genet* 2002;**71**(3):518–27.
11. Gros-Louis F, Meijer IA, Hand CK, Dube MP, MacGregor DL, Seni MH, et al. An ALS2 gene mutation causes hereditary spastic paraplegia in a Pakistani kindred. *Ann Neurol* 2003;**53**(1):144–5.
12. Panzeri C, De Palma C, Martinuzzi A, Daga A, De Polo G, Bresolin N, et al. The first ALS2 missense mutation associated with JPLS reveals new aspects of alsin biological function. *Brain* 2006;**129**(Pt 7):1710–19.
13. Kanekura K, Hashimoto Y, Niikura T, Aiso S, Matsuoka M, Nishimoto I, et al. Alsin, the product of ALS2 gene, suppresses SOD1 mutant neurotoxicity through RhoGEF domain by interacting with SOD1 mutants. *J Biol Chem* 2004;**279**(18):19247–55.
14. Jacquier A, Buhler E, Schafer MK, Bohl D, Blanchard S, Beclin C, et al. Alsin/Rac1 signaling controls survival and growth of spinal motoneurons. *Ann Neurol* 2006;**60**(1):105–17.
15. Tudor EL, Perkinton MS, Schmidt A, Ackerley S, Brownlees J, Jacobsen NJ, et al. ALS2/Alsin regulates Rac-PAK signaling and neurite outgrowth. *J Biol Chem* 2005;**280**(41):34735–40.
16. Lai C, Xie C, McCormack SG, Chiang HC, Michalak MK, Lin X, et al. Amyotrophic lateral sclerosis 2-deficiency leads to neuronal degeneration in amyotrophic lateral sclerosis through altered AMPA receptor trafficking. *J Neurosci* 2006;**26**(45):11798–806.
17. Otomo A, Hadano S, Okada T, Mizumura H, Kunita R, Nishijima H, et al. ALS2, a novel guanine nucleotide exchange factor for the small GTPase Rab5, is implicated in endosomal dynamics. *Hum Mol Genet* 2003;**12**(14):1671–87.
18. Topp JD, Gray NW, Gerard RD, Horazdovsky BF. Alsin is a Rab5 and Rac1 guanine nucleotide exchange factor. *J Biol Chem* 2004;**279**(23):24612–23.
19. Cai H, Shim H, Lai C, Xie C, Lin X, Yang WJ, et al. ALS2/alsin knockout mice and motor neuron diseases. *Neurodegener Dis* 2008;**5**(6):359–66.
20. Otomo A, Kunita R, Suzuki-Utsunomiya K, Ikeda JE, Hadano S. Defective relocalization of ALS2/alsin missense mutants to Rac1-induced macropinosomes accounts for loss of their cellular function and leads to disturbed amphisome formation. *FEBS Lett* 2011;**585**(5):730–6.
21. Rabin BA, Griffin JW, Crain BJ, Scavina M, Chance PF, Cornblath DR. Autosomal dominant juvenile amyotrophic lateral sclerosis. *Brain* 1999;**122**(Pt 8):1539–50.
22. Chance PF, Rabin BA, Ryan SG, Ding Y, Scavina M, Crain B, et al. Linkage of the gene for an autosomal dominant form of juvenile amyotrophic lateral sclerosis to chromosome 9q34. *Am J Hum Genet* 1998;**62**(3):633–40.
23. Chen YZ, Bennett CL, Huynh HM, Blair IP, Puls I, Irobi J, et al. DNA/RNA helicase gene mutations in a form of juvenile amyotrophic lateral sclerosis (ALS4). *Am J Hum Genet* 2004;**74**(6):1128–35.
24. Rudnik-Schöneborn S, Arning L, Epplen JT, Zerres K. SETX gene mutation in a family diagnosed autosomal dominant proximal spinal muscular atrophy. *Neuromuscul Disord* 2012;**22**(3):258–62.
25. Moreira MC, Klur S, Watanabe M, Nemeth AH, Le Ber I, Moniz JC, et al. Senataxin, the ortholog of a yeast RNA helicase, is mutant in ataxia-ocular apraxia 2. *Nat Genet* 2004;**36**(3):225–7
26. Chen YZ, Hashemi SH, Anderson SK, Huang Y, Moreira MC, Lynch DR, et al. Senataxin, the yeast Sen1p orthologue: characterization of a unique protein in which recessive mutations cause ataxia and dominant mutations cause motor neuron disease. *Neurobiol Dis* 2006;**23**(1):97–108.
27. Suraweera A, Becherel OJ, Chen P, Rundle N, Woods R, Nakamura J, et al. Senataxin, defective in ataxia oculomotor apraxia type 2, is involved in the defense against oxidative DNA damage. *J Cell Biol* 2007;**177**(6):969–79.
28. Suraweera A, Lim Y, Woods R, Birrell GW, Nasim T, Becherel OJ, et al. Functional role for senataxin, defective in ataxia oculomotor apraxia type 2, in transcriptional regulation. *Hum Mol Genet* 2009;**18**(18):3384–96.
29. Grohmann K, Schuelke M, Diers A, Hoffmann K, Lucke B, Adams C, et al. Mutations in the gene encoding immunoglobulin mu-binding protein 2 cause spinal muscular atrophy with respiratory distress type 1. *Nat Genet* 2001;**29**(1):75–7.
30. Becherel OJ, Yeo AJ, Stellati A, Heng EY, Luff J, Suraweera AM, et al. Senataxin plays an essential role with DNA damage response proteins in meiotic recombination and gene silencing. *PLoS Genetics* 2013;**9**(4):e1003435.
31. Vantaggiato C, Bondioni S, Airoldi G, Bozzato A, Borsani G, Rugarli EI, et al. Senataxin modulates neurite growth through fibroblast growth factor 8 signalling. *Brain* 2011;**134**(Pt 6):1808–28.
32. Arning L, Epplen JT, Rahikkala E, Hendrich C, Ludolph AC, Sperfeld AD. The SETX missense variation spectrum as evaluated in patients with ALS4-like motor neuron diseases. *Neurogenetics* 2013;**14**(1):53–61.
33. Hentati A, Ouahchi K, Pericak-Vance M, Nijhawan D, Ahmad A, Yang Y, et al. Linkage of a commoner form of recessive amyotrophic lateral sclerosis to chromosome 15q15-q22 markers. *Neurogenetics* 1998;**2**:55–60.
34. Stevanin G, Santorelli FM, Azzedine H, Coutinho P, Chomilier J, Denora PS, et al. Mutations in SPG11, encoding spatacsin, are a major cause of spastic paraplegia with thin corpus callosum. *Nat Genet* 2007;**39**(3):366–72.
35. Orlacchio A, Babalini C, Borreca A, Patrono C, Massa R, Basaran S, et al. SPATACSIN mutations cause autosomal recessive juvenile amyotrophic lateral sclerosis. *Brain* 2010;**133**(Pt 2):591–8.

36. Vance C, Rogelj B, Hortobagyi T, De Vos KJ, Nishimura AL, Sreedharan J, et al. Mutations in FUS, an RNA processing protein, cause familial amyotrophic lateral sclerosis type 6. *Science* 2009;**323**(5918):1208–11.
37. Kwiatkowski Jr. TJ, Bosco DA, Leclerc AL, Tamrazian E, Vanderburg CR, Russ C, et al. Mutations in the FUS/TLS gene on chromosome 16 cause familial amyotrophic lateral sclerosis. *Science* 2009;**323**(5918):1205–8.
38. Conte A, Lattante S, Zollino M, Marangi G, Luigetti M, Del Grande A, et al. P525L FUS mutation is consistently associated with a severe form of juvenile amyotrophic lateral sclerosis. *Neuromuscul Disord* 2012;**22**(1):73–5.
39. Al-Saif A, Al-Mohanna F, Bohlega S. A mutation in sigma-1 receptor causes juvenile amyotrophic lateral sclerosis. *Ann Neurol* 2011;**70**(6):913–19.
40. Prause J, Goswami A, Katona I, Roos A, Schnizler M, Bushuven E, et al. Altered localization, abnormal modification and loss of function of Sigma receptor-1 in amyotrophic lateral sclerosis. *Hum Mol Genet* 2013;**22**(8):1581–600.
41. Luty AA, Kwok JB, Dobson-Stone C, Loy CT, Coupland KG, Karlstrom H, et al. Sigma nonopioid intracellular receptor 1 mutations cause frontotemporal lobar degeneration-motor neuron disease. *Ann Neurol* 2010;**68**(5):639–49.
42. Belzil VV, Daoud H, Camu W, Strong MJ, Dion PA, Rouleau GA. Genetic analysis of SIGMAR1 as a cause of familial ALS with dementia. *Eur J Hum Genet* 2013;**21**(2):237–9.
43. Mavlyutov TA, Epstein ML, Andersen KA, Ziskind-Conhaim L, Ruoho AE. The sigma-1 receptor is enriched in postsynaptic sites of C-terminals in mouse motoneurons. An anatomical and behavioral study. *Neuroscience* 2010;**167**(2):247–55.
44. Mancuso R, Olivan S, Rando A, Casas C, Osta R, Navarró X. Sigma-1 R agonist improves motor function and motoneuron survival in ALS mice. *Neurotherapeutics* 2012;**9**(4):814–26.
45. Peviani M, Salvaneschi E, Bontempi L, Petese A, Manzo A, Rossi D, et al. Neuroprotective effects of the Sigma-1 receptor (S1R) agonist PRE-084, in a mouse model of motor neuron disease not linked to SOD1 mutation. *Neurobiol Dis* 2014;**62**:218–32.
46. Al-Saif A, Bohlega S, Al-Mohanna F. Loss of ERLIN2 function leads to juvenile primary lateral sclerosis. *Ann Neurol* 2012;**72**(4):510–16.
47. Butterfield RJ, Ramachandran D, Hasstedt SJ, Otterud BE, Leppert MF, Swoboda KJ, et al. A novel form of juvenile recessive ALS maps to loci on 6p25 and 21q22. *Neuromuscul Disord* 2009;**19**(4):279–87.
48. Dipti S, Childs AM, Livingston JH, Aggarwal AK, Miller M, Williams C, et al. Brown-Vialetto-Van Laere syndrome; variability in age at onset and disease progression highlighting the phenotypic overlap with Fazio-Londe disease. *Brain Dev* 2005;**27**(6):443–6.
49. Bosch AM, Stroek K, Abeling NG, Waterham HR, Ijlst L, Wanders RJ. The Brown-Vialetto-Van Laere and Fazio Londe syndrome revisited: natural history, genetics, treatment and future perspectives. *Orphanet J Rare Dis* 2012;**7**:83.
50. Green P, Wiseman M, Crow YJ, Houlden H, Riphagen S, Lin JP, et al. Brown-Vialetto-Van Laere syndrome, a ponto-bulbar palsy with deafness, is caused by mutations in c20orf54. *Am J Hum Genet* 2010;**86**(3):485–9.
51. Johnson JO, Gibbs JR, Megarbane A, Urtizberea JA, Hernandez DG, Foley AR, et al. Exome sequencing reveals riboflavin transporter mutations as a cause of motor neuron disease. *Brain* 2012;**135**(Pt 9):2875–82.
52. Ho G, Yonezawa A, Masuda S, Inui K, Sim KG, Carpenter K, et al. Maternal riboflavin deficiency, resulting in transient neonatal-onset glutaric aciduria Type 2, is caused by a microdeletion in the riboflavin transporter gene GPR172B. *Hum Mutat* 2011;**32**(1):E1976–84.
53. Gonzalez-Perez P, Lu Y, Chian RJ, Sapp PC, Tanzi RE, Bertram L, et al. Association of UBQLN1 mutation with Brown-Vialetto-Van Laere syndrome but not typical ALS. *Neurobiol Dis* 2012;**48**(3):391–8.
54. Meenakshisundaram E, Jagannathan K, Ramamurthi B. Clinical pattern of motor neuron disease seen in younger age groups in Madras. *Neurol India* 1970;**18**(Suppl. 1):109+.
55. Phanthumchinda K, Supcharoen O, Mitrabukdi E. Madras pattern of motor neuron disease: case report from Thailand. *J Med Assoc Thai* 1996;**79**(6):399–402.
56. Im SH, Kim NH. First case of Madras motor neuron disease from Korea. *Muscle Nerve* 2008;**38**(1):941–2.
57. Isak B, Uluc K, Tanridag T, Ozsahin S, Dengler R, Us O, et al. Madras motor neuron disease in Turkey. *Amyotroph Lateral Scler* 2009;**10**(5–6):347–9.
58. Nalini A, Pandraud A, Mok K, Houlden H. Madras motor neuron disease (MMND) is distinct from the riboflavin transporter genetic defects that cause Brown-Vialetto-Van Laere syndrome. *J Neurol Sci* 2013;**334**(1–2):119–22.
59. Nalini A, Thennarasu K, Yamini BK, Shivashankar D, Krishna N. Madras motor neuron disease (MMND): clinical description and survival pattern of 116 patients from Southern India seen over 36 years (1971–2007). *J Neurol Sci* 2008;**269**(1–2):65–73.
60. Magre J, Delepine M, Khallouf E, Gedde-Dahl Jr. T, Van Maldergem L, Sobel E, et al. Identification of the gene altered in Berardinelli-Seip congenital lipodystrophy on chromosome 11q13. *Nat Genet* 2001;**28**(4):365–70.
61. Windpassinger C, Auer-Grumbach M, Irobi J, Patel H, Petek E, Horl G, et al. Heterozygous missense mutations in BSCL2 are associated with distal hereditary motor neuropathy and Silver syndrome. *Nat Genet* 2004;**36**(3):271–6.
62. Auer-Grumbach M, Schlotter-Weigel B, Lochmuller H, Strobl-Wildemann G, Auer-Grumbach P, Fischer, et al. Phenotypes of the N88S Berardinelli-Seip congenital lipodystrophy 2 mutation. *Ann Neurol* 2005;**57**(3):415–24.
63. Ito D, Suzuki N. Seipinopathy: a novel endoplasmic reticulum stress-associated disease. *Brain* 2009;**132**(Pt 1):8–15.
64. Guo J, Qiu W, Soh SL, Wei S, Radda GK, Ong WY, et al. Motor neuron degeneration in a mouse model of seipinopathy. *Cell Death & Disease* 2013;**4**:e535.
65. Chaudhry R, Kidambi A, Brewer MH, Antonellis A, Mathews K, Nicholson G, et al. Re-analysis of an original CMTX3 family using exome sequencing identifies a known BSCL2 mutation. *Muscle Nerve* 2013;**47**(6):922–4.
66. Puls I, Jonnakuty C, LaMonte B, Holzbaur E, Tokito M, Mann E, et al. Mutant dynactin in motor neuron disease. *Nat Genet* 2003;**33**(4):455–6.
67. Munch C, Sedlmeier R, Meyer T, Homberg V, Sperfeld AD, Kurt A, et al. Point mutations of the p150 subunit of dynactin (DCTN1) gene in ALS. *Neurology* 2004;**63**(4):724–6.
68. Munch C, Rosenbohm A, Sperfeld AD, Uttner I, Reske S, Krause BJ, et al. Heterozygous R1101K mutation of the DCTN1 gene in a family with ALS and FTD. *Ann Neurol* 2005;**58**(5):777–80.

69. Evgrafov OV, Mersiyanova I, Irobi J, Van Den Bosch L, Dierick I, Leung CL, et al. Mutant small heat-shock protein 27 causes axonal Charcot-Marie-Tooth disease and distal hereditary motor neuropathy. *Nat Genet* 2004;**36**(6):602–6.
70. Gopinath S, Blair IP, Kennerson ML, Durnall JC, Nicholson GA. A novel locus for distal motor neuron degeneration maps to chromosome 7q34-q36. *Hum Genet* 2007;**121**(5):559–64.
71. Muglia M, Magariello A, Citrigno L, Passamonti L, Sprovieri T, Conforti FL, et al. A novel locus for dHMN with pyramidal features maps to chromosome 4q34.3-q35.2. *Clin Genet* 2008;**73**(5):486–91.
72. Christodoulou K, Zamba E, Tsingis M, Mubaidin A, Horani K, Abu-Sheik S, et al. A novel form of distal hereditary motor neuronopathy maps to chromosome 9p21.1-p12. *Ann Neurol* 2000;**48**(6):877–84.
73. Guelly C, Zhu PP, Leonardis L, Papic L, Zidar J, Schabhuttl M, et al. Targeted high-throughput sequencing identifies mutations in atlastin-1 as a cause of hereditary sensory neuropathy type I. *Am J Hum Genet* 2011;**88**(1):99–105.
74. Barlowe C. Atlasin GTPases shape up ER networks. *Dev Cell* 2009;**17**(2):157–8.
75. Botzolakis EJ, Zhao J, Gurba KN, Macdonald RL, Hedera P. The effect of HSP-causing mutations in SPG3A and NIPA1 on the assembly, trafficking, and interaction between atlastin-1 and NIPA1. *Mol Cell Neurosci* 2011;**46**(1):122–35.
76. Connell JW, Lindon C, Luzio JP, Reid E. Spastin couples microtubule severing to membrane traffic in completion of cytokinesis and secretion. *Traffic* 2009;**10**(1):42–56.
77. Evans K, Keller C, Pavur K, Glasgow K, Conn B, Lauring B. Interaction of two hereditary spastic paraplegia gene products, spastin and atlastin, suggests a common pathway for axonal maintenance. *Proc Natl Acad Sci USA* 2006;**103**(28):10666–71.
78. Park SH, Zhu PP, Parker RL, Blackstone C. Hereditary spastic paraplegia proteins REEP1, spastin, and atlastin-1 coordinate microtubule interactions with the tubular ER network. *J Clin Invest* 2010;**120**(4):1097–110.
79. Rismanchi N, Soderblom C, Stadler J, Zhu PP, Blackstone C. Atlastin GTPases are required for Golgi apparatus and ER morphogenesis. *Hum Mol Genet* 2008;**17**(11):1591–604.
80. Fassier C, Hutt JA, Scholpp S, Lumsden A, Giros B, Nothias F, et al. Zebrafish atlastin controls motility and spinal motor axon architecture via inhibition of the BMP pathway. *Nat Neurosci* 2010;**13**(11):1380–7.
81. O'Sullivan NC, Drager N, O'Kane CJ. Characterization of the *Drosophila* atlastin interactome reveals VCP as a functionally related interactor. *J Genet Genomics* 2013;**40**(6):297–306.
82. Johnson JO, Mandrioli J, Benatar M, Abramzon Y, Van Deerlin VM, Trojanowski JQ, et al. Exome sequencing reveals VCP mutations as a cause of familial ALS. *Neuron* 2010;**68**(5):857–64.
83. Hazan J, Fonknechten N, Mavel D, Paternotte C, Samson D, Artiguenave F, et al. Spastin, a new AAA protein, is altered in the most frequent form of autosomal dominant spastic paraplegia. *Nat Genet* 1999;**23**(3):296–303.
84. Eckert T, Link S, Le DT, Sobczak JP, Gieseke A, Richter K, et al. Subunit Interactions and cooperativity in the microtubule-severing AAA ATPase spastin. *J Biol Chem* 2012;**287**(31):26278–90.
85. Patel H, Cross H, Proukakis C, Hershberger R, Bork P, Ciccarelli FD, et al. SPG20 is mutated in Troyer syndrome, an hereditary spastic paraplegia. *Nat Genet* 2002;**31**(4):347–8.
86. Renvoise B, Parker RL, Yang D, Bakowska JC, Hurley JH, Blackstone C. SPG20 protein spartin is recruited to midbodies by ESCRT-III protein Ist1 and participates in cytokinesis. *Mol Biol Cell* 2010;**21**(19):3293–303.
87. Edwards TL, Clowes VE, Tsang HT, Connell JW, Sanderson CM, Luzio JP, et al. Endogenous spartin (SPG20) is recruited to endosomes and lipid droplets and interacts with the ubiquitin E3 ligases AIP4 and AIP5. *Biochem J* 2009;**423**(1):31–9.
88. Milewska M, McRedmond J, Byrne PC. Identification of novel spartin-interactors shows spartin is a multifunctional protein. *J Neurochem* 2009;**111**(4):1022–30.
89. Rainier S, Bui M, Mark E, Thomas D, Tokarz D, Ming L, et al. Neuropathy target esterase gene mutations cause motor neuron disease. *Am J Hum Genet* 2008;**82**(3):780–5.
90. Atkins J, Glynn P. Membrane association of and critical residues in the catalytic domain of human neuropathy target esterase. *J Biol Chem* 2000;**275**(32):24477–83.
91. Rainier S, Bui M, Mark E, Thomas D, Tokarz D, Ming L. Cellular function of neuropathy target esterase in lysophosphatidylcholine action. *Toxicol Appl Pharmacol* 2008;**232**(3):376–83.
92. Winrow CJ, Hemming ML, Allen DM, Quistad GB, Casida JE, Barlow C. Loss of neuropathy target esterase in mice links organophosphate exposure to hyperactivity. *Nat Genet* 2003;**33**(4):477–85.
93. Horner RD, Kamins KG, Feussner J, Horner RD, Kamins KG, Feussner JR, et al. Occurrence of amyotrophic lateral sclerosis among Gulf War veterans. *Neurology* 2003;**61**(6):742–9.
94. Ticozzi N, LeClerc AL, Keagle PJ, Glass JD, Wills AM, van Blitterswijk M, et al. Paraoxonase gene mutations in amyotrophic lateral sclerosis. *Ann Neurol* 2010;**68**(1):102–7.
95. Saeed M, Siddique N, Hung WY, Usacheva E, Liu E, Sufit RL, et al. Paraoxonase cluster polymorphisms are associated with sporadic ALS. *Neurology* 2006;**67**(5):771–6.
96. Macedo-Souza LI, Kok F, Santos S, Amorim SC, Starling A, Nishimura A, et al. Spastic paraplegia, optic atrophy, and neuropathy is linked to chromosome 11q13. *Ann Neurol* 2005;**57**(5):730–7.
97. Beetz C, Johnson A, Schuh AL, Thakur S, Varga RE, Fothergill T, et al. Inhibition of TFG function causes hereditary axon degeneration by impairing endoplasmic reticulum structure. *Proc Natl Acad Sci USA* 2013;**110**(13):5091–6.
98. Ishiura H, Sako W, Yoshida M, Kawarai T, Tanabe O, Goto J, et al. The TRK-fused gene is mutated in hereditary motor and sensory neuropathy with proximal dominant involvement. *Am J Hum Genet* 2012;**91**(2):320–9.
99. Zuchner S, Wang G, Tran-Viet KN, Nance MA, Gaskell PC, Vance JM, et al. Mutations in the novel mitochondrial protein REEP1 cause hereditary spastic paraplegia type 31. *Am J Hum Genet* 2006;**79**(2):365–9.
100. Strauss M, Koehler K, Krumbholz M, Huebner A, Zierz S, Deschauer M, et al. Triple A syndrome mimicking ALS. *Amyotroph Lateral Scler Other Motor Neuron Disord* 2008;**9**(5):315–17.

Chapter 10

Infectious or Acquired Motor Neuron Diseases

Manikum Moodley, William S. David, and Alan R. Seay

INFANTILE POLIOMYELITIS

Historical Background

To many physicians in the developed world, poliomyelitis serves as a dramatic example of how basic virologic methods can lead to vaccine development and almost complete eradication of a human viral infection and crippling neuromuscular disease.[1–3] Today, with a few geographic exceptions, the formidable anxiety and fear that polio once evoked, especially during the late 19th and early 20th centuries, have been relegated to a closed chapter in the history of medicine.[4] Caution, however, should be exercised, because this human viral infection has not yet been eliminated completely, and lack of vigorous surveillance and sustained efforts to immunize susceptible populations around the world could allow the reemergence of devastating epidemic poliomyelitis.

Although endemic, sporadic paralytic poliovirus infections had occurred for centuries, widespread epidemics did not begin to appear until the 19th and early 20th centuries.[5] The shift from sporadic to epidemic infections, somewhat paradoxically, paralleled improvements in public water supplies and sanitation. As a result of these improvements, exposure to poliovirus was delayed from early childhood to late childhood and adolescence. Delayed exposure eventually led to increasingly larger populations of susceptible older children, adolescents, and adults and eventually to the evolution of seasonal epidemics. It was apparent by the late 19th century that clinical disease caused by poliovirus was more severe and more likely to be paralytic and fatal in older people than it was in young children. At the turn of the century, 80% of paralytic cases occurred in children younger than 5 years old whereas, by the 1950s, the majority of paralytic cases were in children 5 to 9 years old.[6] By some estimates, the risk of paralytic disease and death was almost 10 times greater in adolescents and adults than in young children.

By the first half of the 20th century, this trend for older people to become infected, perhaps with a shift in poliovirus neurovirulence, resulted in epidemics of massive proportions, with thousands of cases of paralytic disease occurring annually. Between 1950 and 1955, for example, 25,000 to 50,000 cases of paralytic polio were reported each year in the United States, with about one-third of paralytic disease and more than half the deaths occurring in people older than 15 years.[7,8]

Throughout the 19th century, the cause of poliomyelitis was not known. However, by 1951 the three immunologically distinct serotypes of poliovirus had been identified, and all three had been shown to cause paralytic disease. Soon thereafter, Salk's killed-virus vaccine (1954), Sabin's live-virus oral vaccine (1961), and trivalent oral vaccine (1963) were developed and became widely distributed.[9]

Remarkably, within only 20 years after the introduction of these vaccines, wild-type poliovirus infections were eliminated in the United States, and major reductions, but not total elimination, had been achieved throughout the world. Today, endemic wild poliovirus infections continue to be detected primarily in southern Asia, central and western Africa, and regions of the Middle East.[10–16] Despite the continued occurrence of natural paralytic poliovirus infections in these geographic regions, global eradication of all human poliovirus infections appears achievable, though continued outbreaks in undeveloped countries with poor vaccine penetrance continues to pose a problem.[17] Continuous surveillance and early detection programs will be necessary in the future to prevent the reemergence and spread of wild poliovirus infections.[18]

B.T. Darras, H. Royden Jones, Jr., M.M. Ryan & D.C. De Vivo (Eds): Neuromuscular Disorders of Infancy, Childhood and Adolescence, Second edition.
DOI: http://dx.doi.org/10.1016/B978-0-12-417044-5.00010-X

Epidemiology

In the past, poliovirus infections, including paralytic poliomyelitis, had worldwide distribution, with a higher prevalence in temperate climates. Endemic infections occurred sporadically throughout the year. Epidemic outbreaks, however, occurred most frequently in the late summer and early fall in temperate climates but could occur at any time of the year in tropical regions.

As with other human enteroviral infections, the principal mechanism of spread is oral ingestion of virus, usually in fecally contaminated food or water. Less commonly, virus is transmitted by oropharyngeal secretions or respiratory droplets. Close contact, overcrowding, poor sanitation, contaminated water supplies, and inadequate hygiene increase the risk of transmission and promote epidemic outbreaks.

Formerly, poliovirus infection was one of the most common human neurologic viral infections. Before 1956, poliovirus caused 25,000 to 50,000 cases of paralytic disease annually in the United States.[5,7,8] By 1957, however, only 2 years after the killed-virus vaccine was licensed in the United States, the incidence of poliovirus infections dropped precipitously from 20 to 40 cases per 100,000 population to fewer than 5 cases per 100,000. After the introduction of oral live-virus vaccine in 1961 and trivalent vaccine in 1963, the incidence was reduced further. By 1970, only 32 cases of paralytic disease were reported in the United States. One of the last cases of poliomyelitis caused by wild-type virus in the Western Hemisphere was reported in 1991 from Peru. During the first half of 2000, 678 cases of paralytic polio were identified worldwide, with only 154 being wild-type, natural infections. In the few regions where natural poliovirus infections still occur, campaigns to eliminate the disease are underway but are hampered by geographic isolation, reduced access to vaccines, limited resources, lax containment standards, and sociopolitical turmoil and unrest. In the past 15 to 20 years, with the gradual elimination of natural paralytic disease, live-virus vaccine-induced neurologic disorders and paralysis have emerged as major risks and concerns.[19–21] Because of this shift in risk, the use of killed-virus vaccines and revised immunization schedules to achieve maximal protection of populations has been advocated.[21]

Recently, on May 5, 2014, the World Health Organization (WHO) declared a global public health emergency because of the spread of polio to at least 10 countries in Africa, Asia, and the Middle East. In Pakistan, because of a reduced immunization rate, the number of cases increased by 60% in 2013. Polio outbreaks have also occurred recently in unstable conflict zones, such as Syria. WHO has recommended immunization for people travelling to or from 10 countries; further, regardless of age, all residents of Pakistan, Syria, and Cameroon should be vaccinated before travelling abroad. For more information, see the related *New York Times* article.[22]

Virology, Pathogenesis, and Genetics

Poliovirus, echovirus, coxsackievirus, and other related viruses that are now classified numerically make up the genus Enterovirus, which belongs to the family Picornaviridae.[8,23–25] Other human pathogens in the Picornaviridae family include rhinovirus and hepatovirus (hepatitis A). The picornaviruses all share structural and compositional features. They are small (27–30 nanometer diameter), nonenveloped, single-stranded RNA virions that have an icosahedral structure. The virion's RNA core is enclosed by an outer shell (capsid) composed of four major structural proteins, referred to as VP1, VP2, VP3, and VP4. A fifth precursor protein, VP0, is present in trace amounts and undergoes cleavage to form VP2 and VP4 during virus assembly. Neutralization antigenic sites and attachment sites for host cell receptors are created by these proteins and their geometric conformation on the surface of the virion. After site-specific attachment to the host cell occurs, the virion, by mechanisms incompletely understood, penetrates the host cell membrane, uncoats, and releases viral RNA into the cytoplasm. Cytoplasmic viral RNA and viral protein synthesis take over the host cell's synthetic machinery and cause inhibition of host cell protein and nucleic acid synthesis. This takeover ultimately leads to progressive failure of vital cellular metabolic and regulatory processes, possible activation of apoptotic processes, and cell death.

Poliovirus and other enteroviruses are typically transmitted by ingestion of virus in fecally contaminated food and water.[26–28] The stability of enteroviruses at low pH environments and their ability to readily replicate at 37°C allow them to survive passage through the human gastrointestinal tract without becoming inactivated or destroyed and to replicate efficiently. Once ingested, poliovirus attaches to specific receptor sites on specific cells on the mucosal surfaces of the oropharynx and intestine.[28,29] After attachment, the virus passes through the mucosal surface by pinocytotic and endocytotic mechanisms and is transported to local lymphoid tissue, where local replication occurs. After an incubation period, usually 2 to 3 days, virus is disseminated hematogenously throughout the body, with subsequent secondary replication within susceptible nonneural tissues. Four to five days later, a secondary viremia ensues and results in additional spread of the virus throughout the body, including into the central nervous system (CNS).

The precise mechanisms by which poliovirus passes from the bloodstream across the blood-brain barrier into the CNS are not understood completely. Virus may gain direct access to the CNS in regions normally lacking a blood-brain barrier, such as the area postrema. Circulating cytokines and cerebral endothelial injury may also play a role in allowing virus to reach the CNS. In addition to hematogenous spread, experimental and clinical evidence suggests that virus may gain access to the CNS through axonal transport along nerves that innervate the gastrointestinal tract. Regardless of the route of access, once virus is within the CNS, axonal transport is probably the principal mechanism by which virus is disseminated throughout the CNS.

Within the CNS, discrete cell populations that are specifically susceptible to poliovirus infection include anterior horn cells, medullary (bulbar) motor nuclei, brain stem reticular formation neurons, and a few cell populations rostral to the brain stem in the dentate nucleus, thalamus, hypothalamus, and precentral motor cortex. The specific tropism of poliovirus for certain host cells is determined not only by surface proteins on the virus but also by the presence of specific poliovirus receptors expressed on the surface of host cells. These poliovirus receptor proteins are genetically controlled host cell surface proteins belonging to the immunoglobulin superfamily.[29,30,173] Once infected, specifically vulnerable neurons undergo progressive cytolytic changes and ultimately cell death.

Host responses to poliovirus infection play an important part in determining the outcome of infection. After virus enters the body and primary replication begins, serotype-specific immunoglobulin M-neutralizing antibodies become detectable in blood within 1 to 3 days, increase to peak titers by 2 to 3 weeks, and then decline to undetectable levels by 2 to 3 months. Serotype-specific immunoglobulin G-neutralizing antibodies are first detectable 7 to 10 days after infection, increase to peak titers by 2 to 3 months, and then persist for years. Immunoglobulin A antibodies are detectable on mucosal surfaces and in blood 2 to 4 weeks after infection. Humoral antibody responses are serotype specific. Antibodies generated against one strain of poliovirus do not provide cross-immunity against the other two strains. Serotype-specific antibodies are responsible for limiting primary and secondary viremias and preventing hematogenous spread of virus to susceptible tissues. In addition, circulating neutralizing immunoglobulin G and mucosal immunoglobulin A antibodies provide prolonged serotype-specific protection against reinfection.

The outcome of poliovirus infections and the diseases they produce are determined not only by viral cytolytic effects and the host's humoral and cell-mediated immune responses but also by other important host factors. Inherited and acquired immunodeficiency diseases or drug-induced immunosuppression increase the risk of paralytic disease caused by wild-type or vaccine-associated virus. Age is also an important factor, with increasing age being associated with increasing risk of severe disease, including paralysis and death. For reasons that are not well understood, tonsillectomy and pregnancy also increase the risk that poliovirus infection will cause paralytic disease.

Clinical Features

Infection with the poliovirus typically results in one of four major types of clinical conditions (Box 10.1). First and most common is an asymptomatic, subclinical infection. Second is a mildly symptomatic, monophasic, nonspecific illness. These patients experience only low-grade fever, malaise, headache, diarrhea, or other mild constitutional, nonneurologic symptoms that correlate with the occurrence of the primary viremia. Symptoms resolve in a few days, and the patients recover completely. Asymptomatic and mildly symptomatic illness accounts for more than 95% of all cases of poliovirus infections.

The third clinical illness is aseptic meningitis, sometimes referred to as abortive nonparalytic infection. This accounts for about 1 to 2% of all poliovirus infections. This brief illness is characterized by modest fever, malaise, headache, neck and back aches, muscle stiffness, and vomiting. Patients do not develop focal deficits or paralysis, and they recover quickly and completely, usually within 2 to 4 days.

The fourth condition and the one of greatest clinical concern is paralytic poliomyelitis, which fortunately accounts for only 1% or less of all poliovirus infections. Although anterior horn cells throughout the spinal cord and brain stem are at risk, the sites of highest predilection are the lumbar, cervical, and medullary regions. One to two weeks after exposure, the illness begins as a minor, nonspecific illness characterized by fever (101–103°F), malaise, sore throat, and vomiting (Box 10.2). This initial phase of the illness accompanies the primary viremia. The patient may begin to improve for 1 to 3 days, only to relapse acutely into the second phase of the illness, which is characterized by high fever, malaise, headache, nuchal rigidity, back and limb pain, and painful muscle spasms in the back and extremities. There

BOX 10.1 Poliovirus-related Neurologic Disorders

Major syndromes

- Asymptomatic infection
- Mild symptomatic transient illness
- Aseptic meningitis (abortive nonparalytic infection)
- Paralytic poliomyelitis

Uncommon syndromes

- Encephalitis
- Acute cerebellar ataxia
- Acute transverse myelitis
- Guillain-Barré syndrome

BOX 10.2 Typical Clinical Features of Paralytic Poliomyelitis

- Fever, meningeal signs, muscle pain
- Muscle spasms precede paralysis
- Generalized weakness (fever present) followed by asymmetrical flaccid paralysis
- Normal sensory functions
- Spinal fluid abnormalities
  - Increased white cell count (20–200 cells)
  - Increased protein (50–100 mg/dL)
  - Normal glucose
  - Cultures usually negative
- Throat swab and stool cultures positive
- Electromyographic and nerve conduction velocity studies
  - Early: diminished recruitment pattern
  - Subacute: fibrillations, positive waves
  - Late: motor units very high amplitude, long duration, diminished members, increased firing rate
  - Motor conduction-decreased compound action potential amplitudes: normal sensory conduction parameters

is rapid evolution of generalized flaccid weakness that reaches maximal severity in 1 to 2 days. After a plateau phase of several days, the patient's generalized symptoms improve, and the patient is left with asymmetrical flaccid paresis or paralysis of the extremities or, less commonly, the bulbar muscles. In epidemics that occurred in the United States during the 1940s and early 1950s, bulbar paralysis accounted for 10% to 15% of cases, and a combined bulbar and spinal distribution accounted for another 15%.[6,31,32]

Osler provides detailed accounts of the clinical features of poliomyelitis.[33] He notes, for example, that the most severely affected muscle groups in the lower extremities are the anterior tibialis, leg extensors, and thigh flexors. Hamstrings and gluteal muscles are often relatively spared. Facial muscles are infrequently involved, and extraocular muscles are usually spared. On examination, muscle stretch reflexes are lost in affected limbs, whereas cerebellar and sensory functions remain intact. Patients with bulbospinal involvement of cranial nerves IX and X or involvement of diaphragmatic and thoracic muscles are at risk of apnea, aspiration, and airway obstruction. Autonomic disturbances constitute less common but clinically important and potentially dangerous manifestations of poliomyelitis. Infections involving brain stem reticular formation neurons may cause severe, sometimes fatal, vasomotor instability, irregular respiratory drive with apnea and cyanosis, hypertension, and cardiac arrhythmias. Patients with involvement of lower brain stem nuclei and reticular formation neurons may require prolonged mechanical ventilation. In contrast to the 5% to 10% mortality in spinal forms of the disease, the mortality rate in bulbar polio reaches 25% to 35%.

During the 1940s, comparisons of poliomyelitis in infants younger than 1 year, children, and adults disclosed important clinical and epidemiologic differences.[34] In all age groups, the annual incidence was higher for males than females, except for 25- to 29-year-olds, where females outnumbered males. Poliomyelitis was infrequent in infants younger than 1 year of age, compared with older children and adults. The incidence was less than 5 per 100,000 in infants younger than 6 months old, was approximately 30 per 100,000 by 1 year, peaked in 5- to 9-year-olds at 103 per 100,000, and then decreased to less than 10 per 100,000 by 25 years. When poliovirus infection did occur in children younger than 1 year old, paralytic disease occurred in 100% of those younger than 2 months old and in 85% to 95% of those 3 to 11 months. In the group of patients younger than 1 year old, almost all cases were of the spinal type; bulbar involvement occurred in less than 10%. In contrast, infections in older children resulted in paralytic disease in approximately 50% to 55% of cases, with 15% to 20% of children and up to 36% of adults having bulbar-type paralysis (see Case Example 10.1). Early manifestations of poliomyelitis in young infants included fever, nuchal rigidity, and spasms of the back muscles. The case-fatality ratio in infants younger than 1 year with paralytic disease was 8% to 10%, similar to that seen in other age groups. In patients with bulbar-type paralysis, the case fatality was 30% to 35%.

In general, adults with poliomyelitis were noted to have more widespread and severe weakness and paralysis than children had. Adults tended to develop quadriplegia frequently, whereas children usually had paralysis confined to one or two extremities. Bulbar and brain stem involvement, including cranial nerve nuclei III, IV, and VI and reticular formation neurons, was more frequent in adults. Paralysis of respiratory muscles occurred in 40% of adults, in contrast to 10% of children. In some comparison studies, people between the ages of 16 and 50 years had a death rate nine times higher than that observed in children younger than 15 years old. This difference in mortality was accounted for in part by the fact that more adults than children developed bulbar involvement, and most deaths occurred in patients with bulbar-type paralysis.

Postpolio Syndrome

An unusual late manifestation of polio has been observed in adults who had paralytic poliomyelitis 15 to 20 years or more after the acute illness.[175–179] It is characterized by a

CASE EXAMPLE 10.1

A 7-year-old boy was admitted to the hospital with complaints of fever, headache, and vomiting for the previous week and stiffness and spasms of the neck and back muscles for the previous 2 days. Examination confirmed nuchal rigidity and tenderness and spasms of muscles of the back and hips. On admission, Brudzinski's and Kernig's signs were present, and muscle stretch reflexes at the knees and ankles were hyperactive. As a result of weakness of abdominal and truncal muscles, the child could not go from a supine position to sitting without assistance and could not sit erect without support. His speech, swallowing, and other bulbar functions were normal. Spinal fluid protein concentration was 20 mg/dL, and glucose concentration was 59 mg/dL. In the cerebrospinal fluid, there were 78 white cells/mm^3, predominantly lymphocytes. The clinical diagnosis of poliomyelitis was made, and the patient was treated with supportive measures. During the next 3 months, the patient's strength improved, and there were no residual functional deficits.

Approximately 1 year later, this boy again developed fever, headache, vomiting, and neck pain. Within the first 2 to 3 days of illness, he also developed inability to swallow without nasal regurgitation, nasal speech, inability to cough, and double vision. Examination revealed tenderness and spasms of the muscles of the neck and upper back and prominent nuchal rigidity. His tongue protruded to the left, and there was paralysis of soft palate elevation. Cough reflex was absent. Spinal fluid contained 32 lymphocytes/mm^3. Protein and glucose concentrations were normal. Bulbospinal poliomyelitis was diagnosed, and the patient was treated with supportive and rehabilitative measures. Gradually, this boy improved with prolonged convalescent care.

This child, originally seen in 1946 and reported in 1948, illustrates many common features, as well as some rare clinical features, of acute poliomyelitis during epidemics in the 1940s.[35] Each acute illness experienced by this patient was initially characterized by the common triad of fever, nuchal rigidity, and muscle spasms of the back and neck. In the initial illness, muscular weakness was most severe in the lower back and proximal lower extremity muscles. Although this boy initially displayed hyperreflexia, most patients eventually lose muscle stretch reflexes in the paralytic extremities. The second episode illustrates the development and progression of bulbar polio with prominent early deficits in chewing, swallowing, and coughing. Extraocular muscle weakness was uncommon in children but was not unusual in older adolescents and adults. Patients with bulbar involvement were at risk of developing vocal cord and diaphragmatic weakness, autonomic instability, and respiratory failure. Intubation and tracheostomies were required to provide mechanical ventilatory support.

This youngster was particularly unusual in developing recurrent paralytic disease. By 1948, almost 40 cases of recurrent polio were reported. Statistically, the risk of recurrent polio was estimated to be about 1 recurrence per 1000 cases of paralytic disease.[180] The immunity that developed during illness with one strain did not seem to provide protection against disease caused by other strains. A few years later, laboratory techniques were developed demonstrating the absence of cross-immunity among the three strains of poliovirus and confirming these earlier clinical observations.

Comment

Today, cases of wild-type and vaccine-induced poliomyelitis continue to occur in a few areas of the world. Occasionally such cases are seen in the United States in patients who have traveled from those regions. The clinical symptoms, signs, and course followed by contemporary patients are the same as those experienced by patients during the first half of the 20th century and illustrated by the case example. Until eradication of all human poliovirus infections has been achieved, clinicians worldwide need to continue to include viral poliomyelitis in the differential diagnosis of acute-onset flaccid weakness accompanied by fever, headache, nuchal rigidity, and muscle spasms of the back. Today, patients suspected of having poliomyelitis should undergo a thorough evaluation using virologic and immunologic diagnostic techniques capable of isolating and typing the infective agent and quantifying the patient's immunologic responses. Unfortunately, our ability to treat patients has progressed little during the past 50 years. General measures to assist breathing, detect and correct autonomic dysfunction, protect the airway, relieve pain, reduce muscle spasms, and provide rehabilitation continue to be the main treatment modalities available to patients with viral poliomyelitis.

gradual or, in rare cases, a sudden onset of progressive and persistent new muscle weakness, decreased muscle endurance, fatigue, pain, and atrophy in the same muscle groups that appeared to be involved in the acute infection or in muscle groups not clinically involved originally. This slowly progressive anterior horn cell degenerative disorder may continue for long periods and in a few patients may lead to new, severe functional impairments and incapacitation. Since there are no specific diagnostic tests for postpolio syndrome (PPS), diagnosis is based on exclusion of other possible causes for the new symptoms. Several diagnostic criteria for PPS have been proposed but that recommended at the March of Dimes International Conference on PPS in 2000 is the most plausible[34,36] (Box 10.3).

The total number of polio survivors is estimated to be about 20 million by the World Health Organization. The prevalence of PPS has been reported to range from 15% to 80% of all people with previous paralytic poliomyelitis depending on the criteria applied and population studied.[37] Given the large numbers of survivors of the 1940 and early 1950s poliomyelitis epidemics when vaccination programs were launched, this frequency of new weakness makes PPS the most prevalent motor neuron disease in North America.[38]

BOX 10.3 March of Dimes Criteria for Post-poliomyelitis Syndrome

1. Prior paralytic poliomyelitis with evidence of motor neuron loss, as confirmed by history of the acute paralytic illness, signs of residual weakness and atrophy of muscles on neurologic examination, and signs of denervation on electromyography (EMG).
2. A period of partial or complete functional recovery after acute paralytic poliomyelitis, followed by an interval (usually 15 years or more) of stable neurologic function.
3. Gradual or sudden onset of progressive and persistent new muscle weakness or abnormal muscle fatigability (decreased endurance), with or without generalized fatigue, muscle atrophy, or muscle and joint pain. (Sudden onset may follow a period of inactivity, trauma, or surgery.) Less commonly, symptoms attributed to PPS include new problems with breathing or swallowing.
4. Symptoms persist for at least 1 year.
5. Exclusion of other neurologic, medical, and orthopedic problems as causes of symptoms.

Source: Reference 36.

BOX 10.4 Differential Diagnosis of Acute Flaccid Paralysis

Neuronopathies/Neuropathies
- Viral poliomyelitis syndrome (see Box 10.5)
- Parainfectious disorders
  - Guillain-Barré syndrome
  - Acute motor axonal neuropathy
  - Plexopathies
  - Isolated mononeuropathies

Myoneural junction disorders
- Myasthenia gravis
- Botulism
- Biologic toxins and venoms
- Chemical neurotoxins

Myopathy
- Polymyositis
- Glycogen storage

Spinal cord or brain stem lesions
- Tumors
- Abscesses
- Vascular lesions
- Infarcts
- Hemorrhages
- Arteriovenous malformations

Trauma
Asthma-associated poliomyelitis syndrome
Postpolio syndrome

The pathogenesis of PPS is still unclear. It may be related in part to the loss of terminal axonal sprouts in pathologically large motor units. Over time previously damaged anterior horn cells may be unable to metabolically support and maintain the increased number of axon sprouts that developed during recovery after the initial acute infection. Degeneration and loss of these sprouts lead secondarily to muscle necrosis and progressive weakness. Although results of previous virologic and immunologic studies failed to show evidence for active viral replication or viral reactivation in this syndrome, results of recent investigations using polymerase chain reaction techniques have suggested that the development of postpolio syndrome may be related to persistence of mutated poliovirus in the CNS.[39]

In addition to paralytic disease and the other three major clinical outcomes of poliovirus infection, other less common neurologic disorders can result. Poliovirus encephalitis has been described and is similar clinically to other viral encephalitides in causing lethargy, confusion, disorientation, focal and lateralized neurologic deficits, and occasionally seizures. Acute cerebellar ataxia, acute transverse myelitis, and Guillain-Barré syndrome have all been reported with or following natural poliovirus infections and after use of oral live-virus vaccines.[8,23,24,35]

Differential Diagnosis

The differential diagnosis of acute flaccid weakness is outlined in Box 10.4. The diagnosis of poliomyelitis is usually based on information derived from a detailed history and thorough neurologic examination (see Box 10.2). At the onset of acute flaccid weakness, Guillain-Barré syndrome may be considered, but this can usually be differentiated by its ascending pattern of weakness, sensory involvement, acellular spinal fluid with high protein content, and impaired motor nerve conduction velocities. Additionally, paralytic poliomyelitis is more typically asymmetric in presentation. Acute motor axonal neuropathy is a frequent cause of acute flaccid paralysis in China and other Asian countries, but it differs clinically from poliomyelitis in that the weakness is bilaterally symmetrical and involves bulbar muscles in 40% to 50% of cases.[40,41] Botulism may produce acute-onset flaccid weakness of the extremities or bulbar musculature, but the associated pupillary abnormalities, extraocular muscle involvement, gastrointestinal symptoms, normal spinal fluid, and abnormal electromyography results and repetitive nerve stimulation studies distinguish it from polio.

Acute transverse myelitis and acute disseminated encephalomyelitis can resemble viral poliomyelitis initially, with acute symmetrical or asymmetrical flaccid paralysis, fever, and back pain. Occasionally, immunizations with poliovirus vaccine have preceded the onset of acute transverse myelitis.[27,42] The detection of a sensory level and the emergence of sensory deficits indicative of spinothalamic tract and dorsal column involvement separate this disorder from polio on clinical grounds. Myasthenia gravis, neurotoxins, and spinal cord lesions, such as tumors, abscesses, hemorrhages, and vascular malformations, can cause acute flaccid weakness but can be distinguished from poliomyelitis by their associated clinical features, spinal fluid findings, electrophysiologic abnormalities, and neuroimaging abnormalities.

Currently, patients with acute flaccid weakness who have a history of recent travel through regions known to harbor wild poliovirus need to be evaluated for poliomyelitis. Similarly, a history of recent live-virus vaccination or close contact with a recently vaccinated person raises the possibility of vaccine-induced paralytic polio. This is especially important in patients who may have an underlying inherited or acquired immunodeficiency state.[43–46] Viruses other than poliovirus rarely cause the poliomyelitis syndrome including other enteroviruses, adenoviruses, mumps virus, members of the herpesvirus family, and several alpha- and flavivirus togaviruses (Box 10.5).[47–58] Virologic and serologic techniques are necessary to isolate, identify, and differentiate these viruses.

BOX 10.5 Viral Causes of Poliomyelitis Syndrome

Enteroviruses
- Poliovirus serotypes 1, 2, 3
- Coxsackievirus
  - Group A (A7, A10)
  - Group B (B3, B5)
- Echovirus types 2, 6, 9
- Other human enterovirus types 70, 71

Adenovirus
Mumps virus
Herpesvirus
- Herpes simplex virus type 1
- Varicella-zoster virus
- Epstein-Barr virus

Togaviruses
- Alphaviruses
  - Eastern equine encephalitis virus
  - Western equine encephalitis virus
  - Venezuelan equine viruses

Laboratory Evaluation

Blood, Cerebrospinal Fluid, and Urine Studies

During the acute, nonspecific early stages of infection, no specific or diagnostically helpful changes occur in the blood chemistries or urinalysis. In paralytic cases, the spinal fluid pressure may be normal or mildly elevated. Pleocytosis, normal or modestly increased protein concentration, and normal glucose concentration characterize the spinal fluid formula. The spinal fluid cell count usually ranges from 10 to 200 leukocytes/mm^3 but occasionally reaches or exceeds 500 cells. In the first few days of illness, neutrophils predominate in the cerebrospinal fluid, but as the disease evolves, lymphocytes become predominant. The protein concentration typically ranges from 50 to 100 mg/dL, though normal or very high levels are occasionally encountered.

Poliovirus is cultured from the cerebrospinal fluid in less than 1% of cases. In contrast, this virus is frequently recovered from throat swabs and fecal specimens. Once isolated in culture, viruses are specifically identified and typed using type-specific neutralizing antibodies. Monoclonal antibodies, polymerase chain reaction techniques, and restriction fragment length polymorphism analysis are also used to detect the presence of poliovirus and determine whether the virus is vaccine derived or wild type.[54,55] Of the three poliovirus serotypes, types 1 and 3 are most frequently associated with epidemic poliomyelitis, but all three serotypes can cause paralytic disease.

Other diagnostically useful laboratory procedures include detection and titration of virus-specific neutralizing and complement-fixing antibodies in the patient's blood. These antibodies are serotype specific and do not provide cross-immunity among the three serotypes of poliovirus. Evidence of poliovirus infection includes detection of antipoliovirus serotype-specific immunoglobulin M and immunoglobulin A neutralizing antibodies and the demonstration of a fourfold or greater rise in immunoglobulin G serotype-specific neutralizing antibodies over a 2- to 4-week interval. The use of polymerase chain reaction techniques capable of detecting small amounts of viral nucleic acid in human tissues and fluids provides a rapid, specific, and sensitive method for the early diagnosis of poliovirus and other enteroviruses.[59,60]

Electromyography

In the early stage of paralytic disease, electromyography and nerve conduction velocity studies demonstrate increased numbers of polyphasic potentials and a generalized reduction in the motor unit recruitment pattern.[61,62] In later stages, fibrillation and fasciculation potentials develop in affected muscle groups. With recovery, reinnervation produces motor unit potentials that are prolonged and eventually of profoundly increased amplitude (10–20 mV). Nerve conduction velocities are normal at the beginning of the illness, but compound muscle action potentials may be quite small and polyphasic. Throughout the course of the illness, sensory nerve action potentials usually remain normal.

Pathology

During the early stages of acute disease, there is extensive hyperemia, vessel engorgement, and perivascular lymphocytic infiltration in the meninges of the spinal cord. These changes are most intense and extensive in the meninges along the anterolateral aspect of the spinal cord. In the spinal cord parenchyma, most notably in the lumbar region, perivascular lymphocytic infiltration may be associated with multifocal sites of hemorrhagic necrosis. In very early stages, the inflammatory infiltrate may contain many polymorphonuclear leukocytes, whereas in later stages, the infiltrates contain a mixture of lymphocytes, plasma cells, and macrophages. As the lesions advance, microglia proliferate, and regions of hemorrhagic necrosis evolve into small cystic lesions. Chromatolytic and other intracellular degenerative changes affect anterior horn neurons. Rarely, Cowdry type B intranuclear inclusion bodies may be seen. Neuronophagia, neuronal cell loss, and gliosis lead to atrophy of the anterior horn and corresponding anterior spinal root in the late, chronic phases of the disease. The involvement, however, is asymmetrical between the two sides of the cord and is quite variable from level to level rostrocaudally. Pathologic changes are usually most prominent in the lumbar region of the spinal cord, though identical features may be present in variable degrees in the cervical cord, thoracic cord, or brain stem motor nuclei and reticular formation. The sacral spinal cord is usually minimally involved or spared.

In addition to the spinal cord, other regions of the CNS may contain foci of perivascular and parenchymal inflammation and neuronal degeneration. These sites include the brain stem reticular formation, dentate nucleus, globus pallidus, thalamus, hypothalamus, and precentral motor cortex. Other cortical regions and the meninges over the cortical hemispheres show little or no involvement.

Pathologic changes in affected muscle groups are secondary and nonspecific, typical of a disease affecting primarily motor neurons. Early in the course, there may be isolated angulated fibers, myofiber degeneration, and necrosis. Subsequently, type grouping and grouped atrophy develop. In paralytic muscle, fatty, fibrous connective tissue may completely replace muscle tissue. Severe motor neuron loss and secondary muscle atrophy markedly impair growth of the affected extremities.

Therapy

Specific antiviral pharmacotherapy is not available for poliomyelitis. Treatment of acute poliomyelitis, as well as postpolio syndrome, is directed at managing fever, relieving pain, reducing muscle spasms, and providing supportive treatment.[63,64] Patients with bulbar polio or generalized weakness may need assistance to protect the airway, support breathing, and maintain nutrition. Once the acute illness subsides, physical and occupational therapy help maintain and promote limb and joint mobility and prevent functionally limiting contractures and joint deformities. Later, depending on how much strength and voluntary movement return, various mechanical and prosthetic devices, along with ongoing physical and occupational therapy programs, assist the patient in achieving maximal functional recovery.

Experience during the past 50 years has demonstrated how a well-organized worldwide vaccination program can dramatically and effectively control and almost completely eliminate a serious human viral disease. In the recent past, most poliovirus vaccination programs have relied on live-virus vaccines to effectively provide immunity for large populations. However, with the near elimination of wild-virus infections worldwide, the shift to the use of killed-virus or inactivated vaccines, as advocated by the Centers for Disease Control and Prevention and enforced in the United States since 2000, has eliminated the risk of vaccine-acquired poliomyelitis but continues to provide effective immunity to children and other susceptible populations. Of note, live-virus polio vaccine is still used in many parts of the world.

Prognosis

The long-term outcome of acute poliomyelitis is dependent on the severity and extent of neuronal damage and loss. Some patients, though quite ill during the acute illness, have minimal neuronal loss and show good functional recovery during the ensuing months. Other patients, especially those with bulbar polio or those with involvement of the diaphragm, may experience minimal recovery and require continued respiratory and physical rehabilitation and intensive supportive care. For most patients with paralytic polio involving lower or upper extremities, a modest amount of steady improvement may continue for 6 to 18 months after the acute illness. These patients, however, commonly have permanent functional impairments that affect their gait, posture, extremity use, and growth.

Asthma-related Poliomyelitis Syndrome

In 1974, Hopkins reported on a series of 10 children he had observed from 1968 through 1974 in Melbourne, Australia. Each child had developed a poliomyelitis syndrome in association with acute asthma attacks.[65,66] Similar cases were reported from around the world during the next few years.[67–76] The majority of patients were 1 to 12 years old, although patients as old as 73 years have been described.[74] There was no definite sex predilection. Typically, a flaccid paralysis, without any associated sensory disturbance, began 4 to 7 days after the onset of a moderate to severe attack of asthma. Spinal fluid from most patients was abnormal, with mild to moderate lymphocytic pleocytosis, mild protein elevation, or both. Cerebrospinal fluid glucose concentrations were normal.

Detailed electrophysiologic studies carried out in a few of these patients yielded results consistent with motor neuronopathy and denervation in affected muscles.[67] Virologic, serologic, and epidemiologic studies did not uncover any evidence to implicate recent or active, persistent poliovirus infection in these patients. In a few cases, however, these studies provided evidence that infections with adenovirus, other enteroviruses, or members of the herpesvirus family were causally related to the paralytic illness. Whether an underlying, subtle immunodeficiency state related to asthma or asthma treatment rendered these patients susceptible to the development of paralytic disease is unknown.

There is no specific treatment for this paralytic syndrome. Some reports suggest that if herpesviruses are isolated or suspected, acyclovir or other antiviral agents might modify the course of the disease. The prognosis for asthma-related poliomyelitis is poor. Most patients have a moderate to severe permanent flaccid weakness in one or two extremities.

Poliomyelitis in Patients with Immunodeficiency

The clinical illness in these patients differs significantly from that which develops in otherwise normal, immunocompetent individuals. In immunodeficient patients, vaccine-associated poliovirus infection causes a chronic, slowly progressive neurologic disease with asymmetrical flaccid paralysis. The distribution of paralysis is similar to that seen in typical cases of poliomyelitis but evolves slowly over several weeks or months. Occasionally, tremors, myoclonus, and seizures are observed in some patients, indicating more widespread encephalitic disease. Evidence suggests that virus may persist in both the gastrointestinal tract and the CNS and be excreted in feces for months or years. During this period of persistent infection and chronic excretion, virus may mutate or revert to more neurovirulent forms.[46] The pathologic lesions in the CNS of immunodeficient patients are similar in type and distribution to those found in typical cases but are more numerous, extensive, and severe.[44]

Because of the risk of vaccine-associated poliomyelitis, immunodeficient patients should not receive live-virus vaccines and should be separated for at least 3 weeks from people who have recently been immunized with live-virus vaccines. Killed-virus vaccines eliminate the risk of vaccine-induced poliomyelitis and are preferred for immunodeficient patients who have not received poliomyelitis immunization.

VACCINE-ASSOCIATED POLIOMYELITIS

Historical Background

Development of Poliovirus Vaccines

The first clinical description of poliomyelitis was provided in the 1840s,[77,181] and poliovirus was discovered to be the etiologic agent of poliomyelitis in 1908.[78]

In the late 1930s, the first human trials of poliovirus-infected central nervous system tissue vaccines were attempted, although these studies did not incorporate measures of safety and efficacy.[79] Additionally, some cases of paralytic poliomyelitis resulted from use of the vaccine.[79] Several advancements in the late 1940s and early 1950s facilitated the development of the modern vaccines. In 1949, Enders demonstrated that poliovirus could be grown in cultures of human embryonic tissues (including brain).[80] That same year, Bodian reported the existence of three distinct antigenic types of poliovirus.[81]

In the 1950s, Salk took pathogenic, wild-type poliovirus strains grown in primary monkey kidney cell cultures and rendered them noninfectious by treatment with formaldehyde.[82] The inactivated viruses retained their immunogenicity. Large-scale national field trials commenced in 1954, with more than 1.8 million children involved.[83] Overall, protection rates were high, as measured by effective coverage for the various viral serotypes. Unfortunately, during the vaccine campaign, 260 cases of paralytic poliomyelitis developed in vaccine recipients.[84] Analysis identified defective vaccine lots originating from a single manufacturer, Cutter Laboratories.[79,84] In these preparations, there was incomplete formaldehyde inactivation. New safety requirements and another filtration step during vaccine preparation were added, and no further cases of vaccine-associated paralytic poliomyelitis (VAPP) have occurred with inactivated poliovirus vaccine (IPV).[79] IPV was licensed in the United States in 1955, and until 1961, it was the only poliovirus vaccine used in the USA.

In the late 1950s, Sabin developed an oral poliovirus vaccine (OPV) consisting of live attenuated strains selected by numerous passages of wild-type polioviruses in monkey tissues, both *in vivo* and *in vitro*. Repeated passages produced populations of viruses with diminished neurovirulence in monkeys, and eventually purified clones of all three serotypes were obtained.[85,86] Oral administration produced a natural gut infection, leading to resistance to reinfection. Once a child has completed a primary series (three doses of OPV), greater than 95% of recipients have long-lasting immunity to all three viral serotypes.[87] Between 1959 and 1962, several large-scale studies ensued, particularly in the Soviet Union, Mexico, and the United States, with demonstration of efficacy and high levels of coverage.[79] In 1959 to 1960, over 100 million people in Eastern Europe received OPV, with almost complete elimination of poliomyelitis (a drop from 10.6 to 0.43 cases per 100,000 by 1963).[88] In 1960, Sabin demonstrated the safety and high seroconversion rate of a single type of each OPV given in sequence, and in 1961, 700,000 children in the United States received OPV safely, with reduced epidemics as a result.[79] OPV was licensed in 1961. Starting in 1965, trivalent OPV became the preferred vaccine (over IPV) in the United States, because it was easy to administer, was low in cost, conferred long-lasting immunity, and produced bowel immunity, which interrupted the cycle of community transmission.

Development of a third vaccine, an enhanced immunogenic IPV (eIPV), began in the 1970s.[79] In 1978, van Wezel described a technique for making a more immunogenic vaccine, and subsequent use demonstrated a high seroconversion rate to all three serotypes[79]; after three doses of eIPV, 99 to 100% of recipients developed immunity to all three serotypes at 6 months.[89] Numerous clinical trials in several countries, including Sweden, Finland, Africa, the United States, and Canada, demonstrated high seroconversion rates.[79] The eIPV was licensed in 1987. More recent trials in the United States, using a sequential eIPV-OPV schedule (1 to 3 doses of eIPV followed by 1 to 2 doses of OPV), have demonstrated high levels of immunity[89,90]; safety and efficacy data have been provided.[91]

Impact of Poliovirus Vaccines

IPV-only strategies are employed successfully in numerous countries, including Sweden, Finland, Iceland, and the Netherlands.[79,91] Sweden eliminated wild-type poliovirus infection by 1962, with the exception of a single outbreak in members of an unvaccinated religious group.[79,91] Similarly, Finland eliminated wild-type poliovirus infection with the exception of an isolated outbreak in 1984, attributed to a single poorly immunogenic lot of vaccine.[79] The Netherlands also achieved successful control with an IPV-only protocol, apart from two outbreaks, one in 1978 (110 cases) and a second in 1992 (71 cases), both outbreaks were confined to members of a religious sect that refused vaccination.[92]

In the United States, mass community vaccination programs and mandatory school immunization requirements greatly affected the incidence of poliomyelitis.[93] From 1950 to 1955, there was an average of 16,000 new cases of poliomyelitis reported annually in the United States,[94] with the peak occurring in 1952, when there were over 20,000 cases. By 1957, the annual figures had declined to 2499, by the mid-1960s to less than 100 cases, and from 1980 to 1991 only 8 to 9 cases had been reported yearly[95,96]; the latter cases were all VAPP, as are discussed later (Figure 10.1).

The last outbreaks occurring in the United States appeared in 1972 and 1979.[97] In 1972, 11 of 128 students in a Connecticut Christian Scientist school contracted poliomyelitis; 10 of the 11 had never received poliomyelitis

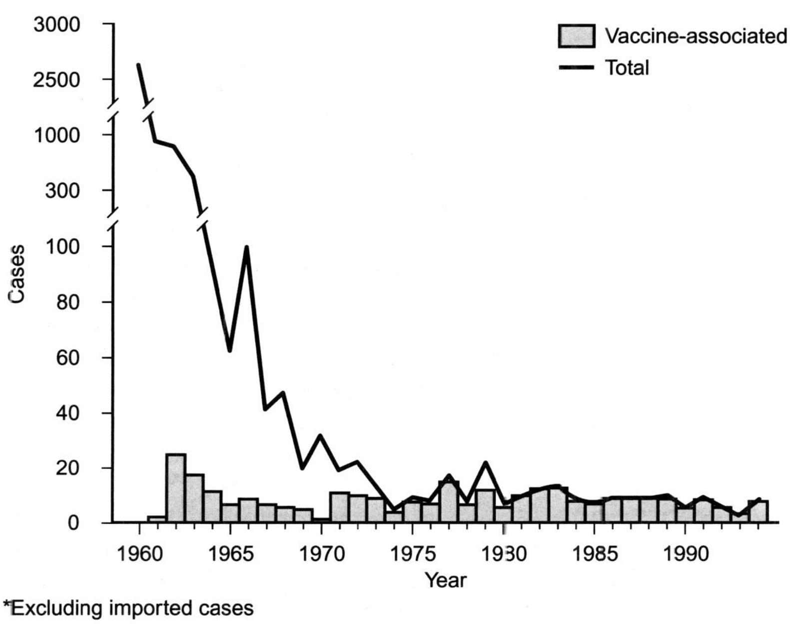

FIGURE 10.1 New cases of poliomyelitis reported annually in the United States. *(From Paralytic poliomyelitis—United States, 1980–1994. MMWR Morb Mortal Wkly Rep 46:79–83, 1997.)*

immunization. In 1979, 13 Amish developed poliomyelitis; again, the victims had never received poliovirus vaccine. The last case of indigenously acquired wild-type poliomyelitis infection in the United States occurred in 1979[96]; the last to appear in the Western Hemisphere was recorded in Peru in August 1991.[98] In 1994, an international commission certified that wild-type poliovirus transmission was now interrupted in the Western Hemisphere.[89] Concomitantly, the global eradication initiative of the World Health Organization (WHO) has resulted in an 80% reduction in the incidence of reported poliomyelitis cases worldwide since 1988.[99] By 1996, 81% of the infants in the world had been appropriately vaccinated.[100] In 1997, there were only 5160 cases reported to WHO, and in 1998, all countries with endemic poliomyelitis staged "immunization day" campaigns to provide additional OPV to children.[101] But the existence of occasional outbreaks in the world, as witnessed in the Netherlands,[92] Africa,[102] and the Middle East[103] as well as the persistence of some unvaccinated or inadequately vaccinated groups, will require that vaccination programs continue world wide in order to prevent outbreaks induced by accidental reintroduction of a wild-type poliovirus strain. The recent outbreaks in Pakistan, Syria, Cameroon, and other countries were discussed in the previous section.

Epidemiology of Vaccine-Associated Paralytic Poliomyelitis

Because OPV contains live attenuated viruses, spontaneous reversion to a more virulent state can cause a vaccinated child, or a close contact of a vaccinated child, to contract paralytic poliomyelitis. The temporal association of a case of paralytic poliomyelitis with administration of the OPV vaccine, or the recovery of a "vaccine-related" poliovirus from the victim's stool, has typically identified these cases. National surveillance is largely coordinated by the Centers for Disease Control and Prevention (CDC) in the United States.[89,94,98,104,105] However, a National Vaccine Injury Compensation Program (NVICP) was established by Congress in 1986, to help assist individuals suffering adverse reactions to vaccines.[95] Some cases (average 1.4 cases per year[106]) have been reported only to the NVICP, and not to the CDC. In general, the CDC risk estimates of VAPP are not altered by inclusion of previously unidentified VAPP cases (reported to the NVICP only[95]).

The earliest vaccine-associated cases occurred with the poliovirus-infected central nervous system tissue vaccines in the 1930s[79] and the Cutter Laboratories IPV incident in the 1950s.[79,84] The earliest OPV-associated cases appeared in 1962. At a meeting of the OPV Advisory Committee in Washington in 1962, 11 cases of paralytic poliomyelitis following administration of monovalent OPV vaccine were reviewed: it was felt that sufficient evidence existed to conclude that these cases were vaccine-related.[107] Through 1964, 123 cases of VAPP had been reported,[107] with a calculated risk of 0.4 per 1 million doses of monovalent type 3 OPV, and 0.17 per 1 million doses of monovalent type 1 OPV.

Several large epidemiologic studies reviewed the US experience with VAPP between 1961 and 1984.[94,96,105,107] There were 229 cases of VAPP reported between 1961 and 1984. From 1961 to 1964, most VAPP cases resulted from monovalent type 1 or type 3 OPV, while from 1965 to 1972 more than half the cases were associated with trivalent OPV. When an interim analysis in 1964 disclosed that many VAPP recipients were older than 14 years of age, vaccination guidelines were revised to exclude the routine vaccination of persons older than 18. After 1965, only 2 of 16 recipient cases were older than 14.[105]

From 1973 to 1984, 138 cases of paralytic poliomyelitis were reported in the United States, and of these, 105 (76%) were cases of VAPP.[93] Of the 105, 85 occurred in immunocompetent individuals: 35 in recipients of OPV, and 50 in contacts of OPV recipients. Fourteen cases occurred in immunodeficient individuals, and six victims were not recipients and had no close contacts (presumed "community" contact). Ninety-four percent of recipient cases, 82% of contact cases, and 36% of immunodeficiency cases were associated with the first dose of OPV. The interval between vaccination and illness onset for the three groups was 4 to 34 days (median 20 days) for recipients, 11 to 58 days (median 35 days) for contacts, and 12 days to 8 months (median 41 days) in immunodeficient victims. The calculated risk at that time was 1 per 2.6 million doses overall, or 1 per 520,000 first doses and 1 per 12.3 million subsequent doses.[94]

From 1980 to 1994, 133 cases of paralytic poliomyelitis were reported to the CDC, with 125 representing cases of VAPP; six cases were imported, and two were declared indeterminate.[104] Forty-nine (39%) occurred in healthy recipients (91% within the first year of life), 40 (32%) in healthy close contacts of OPV recipients, 30 (24%) in immunodeficient recipients (23) and contacts (7), and 6 in "community" contacts (nonrecipients with no close contacts, but in whom vaccine virus was recovered from stool). Ninety-seven cases (78%) were associated with the first dose of OPV. The overall risk was calculated at 1 per 750,000 children receiving the first OPV dose, with a 1 per 2.4 million dose risk overall.[103] This estimated risk was similar to earlier estimates,[94–96,105] indicating that the overall risk of VAPP in the United States was relatively unchanged from 1965 to 1994. In general, 8 to 9 cases of VAPP are reported annually in the United States (see Figure 10.1),[104] though possibly some underreporting exists.[95]

Immunodeficiency significantly increases the risk of VAPP in infants 3200- to 6800-fold.[108] Most cases occur either among individuals with disorders of humoral immunity, or those with B-lymphocyte disorders that inhibit immune globulin synthesis, such as agammaglobulinemia, hypogammaglobulinemia, and combined immune deficiency.[109–112] Other cases are described in association with thymic abnormalities,[112] human immunodeficiency virus (HIV) infection,[102] prednisone treatment,[113,182] and cancer treated with chlorambucil/prednisone.[114]

Epidemiologic studies are also available from other countries. In Latin America, from 1989 to 1991, the overall risk of VAPP was 1 per 1.5 to 2.2 million doses of OPV, compared with 1 per 1.4 million doses in England and Wales (1985 to 1991).[115]

As vaccine policy in the United States has evolved in recent years, there has been increased usage of IPV over OPV, with a resultant decline in the number of VAPP cases. From January 1997 to December 1998, only four cases of VAPP were reported, all following the first or second dose of an all-OPV regimen.[116] Three cases occurred in recipients, one in a contact.

Laboratory Techniques

Since 1979, the CDC has characterized viral isolates (as wild-type or vaccine-derived) by molecular methods such as oligonucleotide fingerprinting and RNA sequencing.[89,104,117] Not only do these techniques allow for the differentiation of wild-type and vaccine strains but they can also demonstrate genetic homologies between polioviruses in different geographic areas, permitting a surveillance team to trace an outbreak.

In the evaluation of a patient with suspected poliomyelitis, the CDC recommends obtaining at least two stool and two throat swab specimens, 24 hours apart, as early in the course of the illness as possible (ideally within the first 14 days).[89,104] The stool has the greatest yield, followed by the throat; virus is rarely detected in the cerebrospinal fluid (CSF). The specimens should be cultured for enteroviruses, followed by serotyping. In addition, an acute-phase serologic specimen should be secured as soon as possible, followed by a convalescent specimen obtained 3 weeks later; each should be tested for neutralizing antibodies to each of the three serotypes of poliovirus, with a fourfold rise in titer indicating infection.[89,104]

Viral Attenuation and Reversion to Virulence

A number of mutations in the Sabin OPV strains attenuate the viruses during vaccine preparation, rendering them nonpathogenic. Mutations at numerous locations in the viral genome impede the ability of Sabin strains to replicate and spread in the nervous system. As Sabin OPV strains replicate in the gut, some strains revert (mutate) back toward wild type, and changes in the nucleotide sequences can lead to increased neurovirulence of the poliovirus vaccine strains.[91,118–120] Such neurovirulent-mutated progeny of vaccine virus are capable of producing paralytic disease in recipients and contacts, and neurovirulent vaccine strains have been found in the stool of vaccinees (and the central nervous system of patients with VAPP).

The various poliovirus vaccine serotypes are unstable to varying degrees. Type 1 is more likely to cause paralytic poliomyelitis in nonimmunized individuals, and types 2 and 3 are more often isolated in immunized groups; poliovirus types 2 and 3 are the most frequent serotypes isolated from VAPP cases (88% of cases).[79,96,105,118,121] Furthermore, VAPP appears to be primarily associated with type 3 vaccine virus in immunocompetent individuals (see Case Example 10.2), and type 2 vaccine virus is more commonly associated with contact cases and immunodeficient individuals.[96] In actuality, VAPP is probably caused by a variety of alterations of the attenuated viral genomes, including reverse mutations at key locations, genetic recombination (vaccine/vaccine and vaccine/nonvaccine), and host factors.[79,118,119,121,122,125–129]

Electrodiagnostic Abnormalities in Vaccine-Associated Poliomyelitis

Few detailed reports on the electrophysiologic abnormalities in VAPP exist.[77,130] In general, the electrodiagnostic findings in VAPP are identical to those reported for wild-type paralytic poliomyelitis described earlier.

Differential Diagnosis

Acute generalized weakness in a previously healthy infant can be caused by a variety of pathophysiologic processes, both central and peripheral. Fever, irritability and an abnormal CSF profile suggests a meningoencephalitis, while a rapidly evolving quadriplegia suggests a cervical spine epidural abscess. MRI studies are crucial to this diagnosis. The neuromuscular or peripheral causes of acute flaccid paralysis are outlined in Box 10.4 (poliomyelitis).

Recommended Vaccination Strategies

Since poliovirus vaccines became available in the 1950s, there has been a considerable evolution in the vaccination schedule recommendations. Prior to 1961, only IPV was available for use. After the licensing of OPV in the early 1960s, an all-OPV schedule became the standard primary vaccination sequence in the United States. Both an OPV-only and IPV-only schedule, as well as a sequential IPV-OPV regimen, confer high coverage rates and long-lasting immunity.[89,99]

Both OPV and IPV can effectively reduce viral replication in the pharynx, and thus impede oral–oral

CASE EXAMPLE 10.2

An 11-week-old infant was admitted with a 2-day history of fever, vomiting, and irritability.[76] Initial neurologic examination was notable only for irritability and head lag. A computed tomography (CT) scan of the head revealed mild ventriculomegaly. Lumbar puncture (LP) disclosed 580 white blood cells (WBC) and 8 red blood cells (RBC) per cu mm, a protein concentration of 143 mg/dL, and a glucose level of 56 mg/dL. The next day a repeat LP revealed 278 WBC per cu mm, protein 98 mg/dL, and glucose 52 mg/dL; cultures for acid-fast bacillus (AFB), bacteria, and fungi were normal, as were serologic tests for syphilis and cryptococcus. Ampicillin and cefotaxime were initially administered; streptomycin, isoniazid, and rifampin were added because of a positive purified protein derivative (PPD) test in the patient's mother (the child was anergic).

Two days following admission, the child developed a weak cry and suck, hypotonia, and generalized areflexia. There was minimal movement of the limbs with relative sparing of the left leg. The infant blinked sparsely and did not smile. There was no ptosis. Extraocular movements, pupillary size, and reactivity remained normal. Apneic periods prompted intubation. Magnetic resonance imaging (MRI) of the brain revealed findings similar to those of CT. Creatine kinase and serum immunoglobulin measurements were normal. Stool cultures were negative for *Campylobacter*. Human immunodeficiency virus (HIV) serologies were negative. Subsequently, urine culture grew out vaccine-related poliovirus type III; routine immunizations had been administered 3 weeks prior to admission. Twenty-four months later the child remained respirator dependent.

Comment

This typical case of VAPP is particularly common after the type III live virus immunization early in life. These infants present as an acute flaccid paralysis or floppy baby syndrome. Often there are specific clinical clues to help differentiate this from other causes for an acute floppy infant. Three cases of a variant of acute flaccid paralysis occurred within a 12- to 18-month period in the mid-1990s.[76,122,123]

All three babies were 3 to 4 months old, each having received type 3 poliovirus immunization 3 to 4 weeks earlier. They had all presented with an acute febrile illness that was typified by progressive asymmetric weakness, head lag, irritability, and lethargy. The CSF demonstrated 100 to 580 WBC, with protein values ranging between 82 and 143 mg/dL and glucose of 49 mg/dL in one. Most instances of VAPP occurred subsequent to the initial live poliomyelitis vaccine in babies less than age 1 year and within 2 weeks of immunization.

Recent changes in the CDC guidelines have stipulated that the initial immunization protocol now requires that the Salk killed vaccine become the primary immunization method for all children.[124,183] Hopefully, as with poliomyelitis in older children, this new infantile immunization protocol will also make vaccination-related poliomyelitis of historical interest. However, that change will be less likely in the economically underdeveloped countries where the Sabin vaccine is still very popular because of its ease of use.

The clinical features of VAPP are identical to those of wild-type poliomyelitis, apart from the obvious vaccine link.

transmission, but OPV is considered to be more effective at producing mucosal immunity in the intestines,[91,99,103,131,132] thus preventing a carrier state.[89] It is important to recognize that recipients of OPV can shed live vaccine from the feces. This vaccine virus could then spread to unvaccinated close contacts, imparting an immunologic boost to these individuals, thus promoting herd immunity. For these reasons, OPV has until recently been considered the vaccine of choice for eradicating wild-type disease in endemic areas.

The disadvantages of IPV are the necessity for additional injections, the theoretical possibility of poor compliance, added expense, and possibly less than optimal intestinal immunity.[99] IPV has not been associated with severe adverse side effects. The major disadvantage for OPV-only schedules is the possibility of the occurrence of VAPP, an entity not observed in IPV-only recipients.

In the 1980s, several developments shifted the risk/benefit balance of the various vaccines, and ultimately led to a change in the US vaccine schedule recommendations. When the last case of wild-type poliomyelitis was reported in 1979, it became clear that the circulation of wild-type poliovirus in the United States had been interrupted. There was a high immunization rate among school children in the United States. Global eradication efforts were proving quite successful, lessening the chances for reintroduction of wild-type poliovirus back into the United States (and, consequently, lessened the need for OPV). Concomitantly, there was the growing concern about VAPP. This form of immunization was then accounting for 8 to 9 cases of paralytic poliomyelitis annually in the United States.[79,99] In 1987, a sequential IPV-OPV regimen, consisting of two doses of IPV (at 2 and 4 months) followed by two doses of OPV (at 12 to 18 months and again at 4 to 6 years) was proposed for primary vaccination in order to decrease the risk of VAPP.[89,99]

In January 1997, the Advisory Committee on Immunization Practices (ACIP) of the CDC and the American Academy of Pediatrics (AAP) formally recommended adopting the sequential schedule for poliomyelitis immunization.[87,89,99,104,116] Three options, an OPV-only, eIPV-only, and sequential eIPV-OPV schedule, were offered; each is felt to be equally acceptable, with specific indications, although the sequential regimen is preferred. The rationale for the sequential regimen is simple: the initial two doses of eIPV promote sufficient humoral immunity to prevent the occurrence of VAPP (in the vaccine recipient) from the subsequent OPV doses, while the final OPV doses themselves will produce effective intestinal immunity and subsequently prevent vaccine recipients from becoming carriers of wild-type poliovirus, thus promoting herd immunity.[87,89,91,93,99,116]

Between 1997 and 1999, there was continued progress in global eradication efforts, as well as an increased use of IPV in the United States without a loss of compliance.[101,116] Given these developments, and the desire to further reduce the risk of VAPP, the AICP recommended in June 1999 that the United States switch to an eIPV-only schedule as of January 2000. Ultimately, when global eradication efforts are successful at eliminating wild-type poliovirus from circulation, no vaccinations will be required.[89,99,101] However, as recent polio outbreaks in at least 10 countries have shown, the eradication efforts have not been successful for various reasons. In the near future, VAPP will most likely be of historical interest only in many Western countries. Until OPV is no longer utilized in some countries, VAPP will still need consideration in the differential diagnosis of an acute febrile motor syndrome, possibly mimicking GBS, or even a chronic asymmetrical or monomelic motor syndrome with a relatively acute onset in prior months or even years.

TETANUS: A COMBINED PERSPECTIVE FROM SOUTH AFRICA AND NORTH AMERICA

Tetanus, including neonatal tetanus, is an ancient disease and is common in developing countries, although virtually extinct in the industrialized populations. In endemic areas, mortality rates from neonatal tetanus range from less than 5 to more than 60 per 1000 live births, representing half a million newborn children annually.[133] The mortality rate is often high, varying between 14.5% in a South African study[134] to 34% in a Nigerian study, and up to 60% in an Ethiopian study. In rural India, the mortality rate is 67 per 1000 live births, whereas in Pakistan and Bangladesh, the rate decreases to between 5 and 20 per 1000 live births.[135]

Neonatal tetanus is the second leading cause of vaccine preventable deaths in children after measles (Box 10.6). In the United States, the incidence of both adult and neonatal tetanus has declined steadily since the 1940s, with the most rapid decline occurring in the 1960s. With one exception, neonatal tetanus was not reported in the United States in the 1980s and 1990s.[136] This one child was born in 1995 to parents who had immigrated from Mexico.[137] This infant's mother had only received one tetanus immunization at age 12 years Even so, she had had two uneventful prior pregnancies and had received prenatal care. They lived near a cattle pasture.

BOX 10.6 Types of Tetanus

- Localized
- Generalized
  - Neonatal
  - Non-neonatal
- Cephalic
- Otogenic

This paucity of neonatal tetanus cases reflects the efficacy of the aggressive immunization program in the United States. Conversely, the World Health Organization reported more than 600,000 cases of neonatal tetanus in 1980 and 1981 in developing countries[138] but the estimated global number of neonatal deaths showed a decline from 510,000 in 1990 to 355,000 in 1997.[139]

Etiology

Tetanus is caused by a potent neurotoxin (tetanospasmin) released by the Gram-positive spore-forming obligate anaerobe *Clostridium tetani* from a focus of infection. In the newborn, the umbilicus is the common site of infection, whereas in older individuals this bacterial infection can be introduced at any site: contaminated wounds, burns, multiple ear piercing, tattooing, circumcision and retained foreign bodies are particularly dangerous.

Epidemiology

C. tetani has a worldwide distribution, leading to disease that is particularly common in developing countries, including those in Africa as well as in Asia, in particular India and Pakistan. It is not surprising given the underdeveloped health services and concomitant low hygienic standards created by poverty and cultural beliefs. In developed countries, neonatal tetanus was actually eradicated before the development of passive immunization against tetanus. For example, in Finland, the last case of neonatal tetanus was reported in 1915. This is attributable most likely to good birth practices because the systematic immunization of infants against tetanus did not commence there until relatively recently—in 1957.[133] In North America, most cases of tetanus occur in adults older than 60 years. However, rare cases in neonates and older children do occasionally occur.[136,137] It is estimated that every year 500,000 children and 50,000 adults die from tetanus worldwide.[140]

There are also a few case reports of patients who have contracted tetanus despite adequate immunization and a protective level of antibody. Usually in such instances the disease is mild, short lasting, or modified in its clinical picture. Most non-neonatal pediatric cases of *C. tetani* infection are associated with a traumatic injury, subsequent to a penetrating wound inflicted by a dirty object such as a nail splinter. These lead to a contaminated wound or retained foreign bodies.

Pathogenesis

Tetanus is caused by an exotoxin (tetanospasmin) released from a focus of infection by the anaerobe *C. tetani*. In the newborn the main route of infection is through the umbilical cord, either through contaminated cord cutting practices or through unhygienic cord care practices such as the application of animal dung dressing on the umbilical cord. The toxin is released into the bloodstream, and after binding to the neuromuscular junction, it is picked up by the motor nerve endings of the alpha motor neuron. Subsequently, it travels from the peripheral nerve perineural spaces to eventually reach the anterior horn cells of the spinal cord. Thereafter, it enters the adjacent spinal inhibitory interneurons where it prevents neurotransmitter release. This results in the classic muscular hypertonia and intermittent muscle spasms.

The relatively short cranial nerves, particularly the primary motor seventh (facial) as well as the motor part of the fifth (trigeminal), respectively innervating the powerful facial and masseter muscles, mediate both risus sardonicus (a spasmodic tetanic involuntary smile) and trismus (lockjaw), which are early and constant clinical signs of *C. tetani* infection. Muscles that are close to the site of infection appear to be more severely affected early on and are the last to lose their hypertonicity.[141]

Clinical Features

Tetanus may be localized or generalized, and its clinical presentation varies from a relatively mild presentation to very severe. Generalized tetanus is the most common and severe form of the disease. By convention, tetanus is best described as either neonatal or non-neonatal because their presentations are different. Almost all newborns have severe disease; however, sympathetic overactivity is not described. On the other hand, in older infants and children, the severity and prognosis of the disease vary depending on the incubation period and onset period (i.e. the time from the first symptom to the first generalized spasm). Furthermore, the concomitant presence of sympathetic overactivity invariably results in a less favorable outcome.

The typical incubation period is usually between 2 and 14 days, but it can be delayed as long as a few weeks to many months after an injury. A severe clinical picture is more likely to occur when the incubation period is shorter than 8 days and the onset period is less than 48 hours (Box 10.6).

Neonatal Tetanus

In developing countries, neonatal tetanus remains a major and familiar health problem (see Case Example 10.3). Although this is a distinct rarity in developed countries, neonatal tetanus occasionally still occurs. In one more comparatively recent instance in the United States, the child's parents were immigrant farm workers.[136,137] Therefore, pediatricians, pediatric neurologists, and intensivists still need to keep this uncommon disorder in their differential diagnosis. The focus of tetanus infection in neonates is always the umbilicus.

Neonatal tetanus is typically a generalized form of this disorder that is manifested by irritability, crying, and refusal to feed within 3 to 12 days of birth. These symptoms are secondary to the combination of trismus followed by tetanic spasms that are often mistaken for convulsions. Marked rigidity of the abdominal muscles, associated with opisthotonus and risus sardonicus are the primary hallmarks of this form of the disease. A high index of

CASE EXAMPLE 10.3

Thandi, a 6-day-old native African girl, was brought to the emergency department of the University Hospital in Durban with a 2-day history of lethargy, decreased activity, and refusal to feed for the past 24 hours. She also cried and stiffened her body at regular intervals, experiencing frequent apneic episodes during the preceding 8 hours.

Her mother, who lived in a rural village, had a full-term and uncomplicated pregnancy with vaginal delivery. She had received no antenatal care. The village midwife performed the delivery of this initially healthy infant girl. There was no problem during the delivery or in the immediate neonatal period. The cord was cut with an old razor and smeared with cow dung, the usual practice in this village. Initially, this infant was breast-fed, sleeping well until 4 days of age.

At the hospital, this infant was difficult to examine. She had marked intermittent facial grimacing with body stiffening, crying, and drooling of saliva when her body was disturbed. Her weight was 3.7 kg. She was alert and well perfused, and she had no jaundice. Her fontanelle was normotensive. Tone appeared generally increased and her muscle stretch reflexes were brisk. The remainder of her neurologic and systemic examinations was normal.

Comment

This infant represents a typical example of neonatal tetanus. Symptoms commenced within her first week of life, manifesting as generalized muscular hypertonicity and frequent episodic apnea.

suspicion for this condition, in certain demographic settings, may be rewarded with an early diagnosis leading to appropriate treatment with potential for complete recovery. Laboratory tests may be required to exclude other medical conditions that may mimic or complicate tetanus. The differential diagnosis for tetanus includes stiff person syndrome, meningitis, neuroleptic malignant syndrome, and rabies. The progression as well as symptoms helps to distinguish tetanus from other diseases.

Non-neonatal Tetanus

In children with partial or complete immunization, tetanus may manifest in a relatively mild form with generalized hypertonia, extension of the neck and spine (opisthotonus), and the characteristic smile (i.e. risus sardonicus) (Box 10.7). This results from severe spasm of the facial and buccal muscles. In nonimmunized children, the illness is often much more severe.

Trismus (masseter muscle spasm, or lockjaw) is the presenting symptom in approximately 50% of cases of generalized tetanus. Subsequently, shocklike painful spasms of all muscles are observed. These spasms are usually provoked by the slightest disturbance, including sight, sound, or touch. Occasionally, the tetanic spasms also occur spontaneously. Between the intermittent severe spasms, there is often continuous muscle rigidity characterized by clenched jaw, risus sardonicus, and a stiff back, neck, abdominal wall, and limbs. Occasionally, the spasms may be so severe that muscles may be detached from their insertions and sometimes bones, most commonly vertebrae, may fracture during the severe spasms. Pharyngeal spasms result in salivary secretion pooling with dribbling from the mouth. The major problems faced by the clinician caring for a child with tetanus are laryngeal and respiratory muscle spasms.

BOX 10.7 Clinical Features of Tetanus

- Trismus
- Risus sardonicus
- Dysphagia
- Laryngospasm
- Muscle rigidity
  - Spinal muscles
  - Abdominal muscles
- Muscle spasms
- Sympathetic overactivity
  - Tachycardia
  - Labile hypertension
  - Arrhythmias
  - Hyperpyrexia
  - Diaphoresis
  - Irritability

These often lead to airway obstruction causing apnea, cyanosis, diaphoresis, and even death from asphyxiation.

Because tetanus toxin does not affect cortical function or the sensory nerves, these patients are fully conscious, experiencing severe pain with every muscle contraction. These tetanic spasms become progressively severe in the first week after onset, stabilize in the second week, and gradually improve over the next 1 to 4 weeks.

Sympathetic overactivity (SOA) characterized by tachycardia, labile hypertension, and arrhythmias commonly occurs in the older child with tetanus. In contradistinction, SOA is not typically seen in neonates. The youngest patient with SOA in a South African series was 13 months old.[134] The occurrence of SOA in tetanus is a poor prognostic feature (Box 10.7).

Focal Tetanus

Localized or focal tetanus is an unusual manifestation of tetanus (Box 10.6). By definition, this form is limited to the muscles in the region of the wound site *per se*. This is thought to occur when circulating antitoxin neutralizes the toxin. However, the antitoxin does not prevent the spread of the tetanus toxin along the regional neurons. Focal wound tetanus is characterized by painful spasms of the muscles adjacent to the wound site. It may precede generalized tetanus. Usually these spasms last a few weeks; if generalization does not occur, eventually there is complete recovery.

Cephalic Tetanus

Cephalic tetanus is another more specific but rare form of focal wound tetanus (Box 10.6). This typically involves the cranial nerves when the initial wound site for this infection involves the face or its contiguous structures. For example, in developing countries, an otitis media is occasionally the locus for the *C. tetani* infection. This otogenic tetanus is typically characterized by a facial paralysis on the side of the infection. This results in a unilateral risus sardonicus on the contralateral (opposite) side of the face as a result of the ipsilateral facial palsy. The mortality rate is usually high in this form of tetanus.[141]

At the University College Hospital in Ibadan, Nigeria, 9 of 73 children with postneonatal tetanus had “atypical portals of entry” for the *C. tetani* organism.[142] Seven of the nine children had focal tetanus. This affected a single leg in six children. Five of these children’s primary leg involvement with *C. tetani* was initially induced through intramuscular injections into the buttock. Another child had an infected insect bite on the buttock. The seventh child had a lesion in the cephalic region. The other two children had an impacted foreign body in the ear. There was no concomitant otitis media.[142]

Unusual complications developed in five of these children during the course of their treatment for tetanus.[142] Arthritis of large joints was encountered in three children, and two had episodes of severe bronchospasm that responded to bronchodilator treatment.

BOX 10.8 Differential Diagnosis of Tetanus

- Intracranial hemorrhage
- Encephalitis, meningitis, encephalomyelitis, rabies
- Dental abscess, peritonsillar infection
- Hypocalcemia
- Epileptic seizures
- Phenothiazine toxicity
- Strychnine poisoning; narcotic/drug withdrawal
- Peritonitis
- Stiff Person Syndrome
- Neuroleptic Malignant Syndrome

Differential Diagnosis

The classic picture of tetanus, especially the abdominal rigidity and risus sardonicus, is highly characteristic and supports the clinical diagnosis of tetanus. However, a number of other conditions may occasionally mimic tetanus. These include dental abscesses, peritonsillar infections, and submaxillary lymphadenitis. Trismus, neck stiffness, and generalized spasms may be mistaken for encephalitis, encephalomyelitis, meningitis, or intracranial hemorrhage.

The absence of an altered level of consciousness as well as the normal cerebrospinal fluid differentiates tetanus from primary central nervous system infections.

Dysphagia may occasionally simulate rabies. However, the hyperexcitability, hydrophobia, aerophobia, and maniacal behavior are much more typical for rabies. Here there is also an associated cerebrospinal fluid pleocytosis that is particularly helpful in distinguishing rabies from tetanus. Spasm of the abdominal muscles in tetanus occasionally simulates peritonitis. Spinal muscle spasms may initially lead to the consideration of a spinal injury. Hypocalcemic tetany is usually distinguished by carpopedal spasms and positive Chvostek's and Trousseau's signs.

Phenothiazine toxicity sometimes results in dystonia and opisthotonus mimicking tetanus. However, the concomitant presence of an occasional associated oculogyric crisis and prompt response to intravenous diphenhydramine or Akineton differentiates this idiosyncratic medication reaction from tetanus.

Strychnine poisoning is rare and may mimic tetanus.[143] Facial stiffness develops first, sometimes even producing a risus sardonicus. Subsequently, one sees 30-second to 2-minute periods of severe generalized muscular rigidity, sometimes leading to opisthotonus in response to the most minimal stimuli. The patient is always fully alert, which helps to differentiate this toxicity from an epileptic event. Furthermore, the skeletal muscles completely relax between spasms. Epileptic seizures and drug withdrawal reactions also deserve consideration in the differential diagnosis of tetanus (Box 10.8).[140–143]

Laboratory

Although the clinical picture of tetanus is most characteristic, the *C. tetani* organisms are only isolated from the wound in one-third of the affected patients. There are no specific confirmatory blood studies or cerebrospinal fluid analyses for tetanus.

However, the diagnosis can be strongly supported by performing an electromyogram (EMG).[144,145] An EMG evaluation of 13 patients with suspected tetanus, albeit with somewhat atypical features, defined seven individuals with the two characteristic and very useful findings "typical for tetanus."[144]

1. Spontaneous motor unit activity could not be suppressed by voluntary activity.
2. There was a concomitant shortening of the silent period after a muscle stretch reflex or after electrical stimulation of the nerve.

In two other patients, one of these criteria was met. In each of these nine patients, full-blown clinical tetanus later developed, while the other four individuals remained symptom-free.

Prevention

1. Tetanus is an entirely preventable disease as is clearly shown in developed countries where an aggressive primary immunization program is in place. Triple vaccine (DTP) in infancy with a booster 1 year later and tetanus toxoid every 10 years thereafter affords good protective antibody levels against tetanus.
2. Prenatal immunization of women with tetanus toxoid vaccine (as a single booster dose to those who received two doses in the preceding 3 years and as two doses at a 1-month interval to those who had no previous history of tetanus toxoid immunization or who were previously only partially immunized) was the single most effective intervention in reducing the incidence of neonatal tetanus in an Indian study.[146] The World Health Organization is currently engaged in a global campaign to eliminate neonatal tetanus through prenatal immunization with at least two doses of tetanus toxoid.[147]

Furthermore, immunization of pregnant women with toxoid protects their infants against tetanus in the first 4 to 6 months after birth.

Wound Management

Patients who have been fully immunized need no tetanus prophylaxis for clean minor wounds. Tetanus toxoid is given to children who are fully immunized but with a previous booster more than 5 years previously. If the patient is not immunized or the status is unknown, the primary immunization series consists of three doses of tetanus toxoid given at six weekly intervals.

If the wound is penetrating, contains a foreign body or soil, or is more than 6 hours old, appropriate wound toilet is followed by human tetanus immunoglobulin 250 to 500 IU given intramuscularly in the opposite limb to the toxoid injection. An antibiotic (penicillin or erythromycin) is given for 5 to 7 days to eradicate *C. tetani* that may be present in the wound.

In developing countries, birth attendants should be taught aseptic techniques in cutting the cord. Furthermore, birth practices such as cutting the cord with glass or dirty instruments and applying mud or cow dung to the umbilical stump need to be vigorously discouraged.

Management

The cornerstones in the management of tetanus are as follows:

1. Total eradication of the infecting organism and its exotoxin with appropriate antibiotics (penicillin or erythromycin) and tetanus immunoglobulin, respectively. If tetanus immunoglobulin is not available, normal immunoglobulin for intramuscular or intravenous use can be administered as an alternative because of the universal presence of antibodies in these preparations.
2. Local wound care—cleansing and surgical débridement if necessary. All foreign bodies need to be removed.
3. Control of spasms with appropriate muscle relaxants/sedatives. Diazepam offers significant benefits in the treatment of tetanus.
4. Ventilatory support for severe cases with intermittent positive pressure ventilation (IPPV) and appropriate neuromuscular blocking agents such as vecuronium and pancuronium.
5. If sympathetic overactivity is present, appropriate pharmacologic therapy must be instituted.
6. Meticulous nursing care in a quiet room where all facilities for immediate resuscitation, if necessary, are readily available.
7. Maintenance of adequate nutrition and hydration.
8. Prevention of recurrences with tetanus toxoid as determined by immunization status.

Treatment

Ventilation

Most neonates have severe tetanus and invariably require IPPV. They also require total muscle paralysis using neuromuscular blocking agents such as vecuronium and pancuronium. Most neonates are weaned off the ventilator in 3 to 4 weeks.

Immunoglobulins

Human antitetanus immunoglobulin 500 IU is given as a single intramuscular injection. Intrathecal human antitetanus immunoglobulin given to neutralize tetanus toxin in the spinal cord is not effective as shown in a recent meta-analytic study.[147]

Muscle Relaxation/Sedation

Diazepam, midazolam, and chlorpromazine are the drugs of choice for neonatal sedation. If facilities for mechanical ventilation are not available and heavy sedation is a necessity, then a tracheostomy should be performed to facilitate suction and to prevent laryngospasm. In neonates, however, tracheostomy without IPPV is of no value.[141] In developing countries, phenobarbitone is an important additional drug for control of spasms and should be used if muscle relaxation and sedation is not achieved with the benzodiazepines and phenothiazines. Uncontrolled studies suggest that magnesium sulphate controls spasms in patients with established tetanus. However, a recent meta-analysis of controlled trials that compared magnesium sulfate with placebo or diazepam for the treatment of patients with tetanus showed no reduced mortality.[148]

When intensive care facilities including mechanical ventilation are scarce, intrathecal baclofen allows effective control of muscle rigidity and spasms and reduces the need for tracheal intubation or deep anesthesia.[149] Intrathecal baclofen has been shown to be most effective in older children, especially when they are evaluated before the onset of reflex spasms.[149]

Antibiotics

Penicillin G is the drug of choice for the effective elimination of *C. tetani*. The usual dose is 100,000 to 200,000 IU/kg/day in four divided doses given intravenously for 10 days. For penicillin-allergic patients, tetracycline in patients older than 8 years and erythromycin and cephalosprins are alternative agents.

Sympathetic Overactivity

A major problem in severely affected older children relates to the presence of sympathetic overactivity. This syndrome occurs in 20% to 70% of patients.[133] The mortality in tetanus associated with autonomic dysfunction is as high as 50%.

Unexpected cardiac arrest is the most common cause of death in patients admitted to the ICU. No single pharmacologic regimen is uniformly effective, although there have been attempts to use various therapeutic modalities. In children, alpha- and beta-blocking agents, including propranolol, bethanidine, and labetalol, given for a period of 2 to 3 weeks, is useful, but many deaths still occur. High-dose morphine therapy for sympathetic overactivity has been used in many special care units in the past 10 years, with improvement in the prognosis.[134,149] Recently, epidural blockade with a local anesthetic agent (bupivacaine) and a synthetic opioid (sufentanil) has been shown to be effective in controlling sympathetic overactivity.[150,184]

Fluids

All patients with significant spasms require intravenous fluids. In neonatal tetanus when spasms are well-controlled, expressed breast milk is given via a nasogastric tube. In developed countries, total parenteral nutrition is the preferred route to maintain adequate nutrition.

Active Immunization

It is important to emphasize that one episode of tetanus does not result in lifelong immunity. Therefore, active immunization with tetanus toxoid at the time of discharge is mandatory to prevent further attacks.

Prognosis

The most important predictors of prognosis in tetanus are the incubation period and the onset period. If the incubation period is less than 8 days, the onset period is less than 48 hours, and if reflex spasms have been present more than 12 per 24 hours, the prognosis is generally poor. However, with modern methods of treatment, the overall mortality rate from tetanus is substantially reduced throughout the world. In South Africa, the mortality rate is reduced to 14.5% from a previous figure of about 50% for older children and more than 90% for neonates.[134,142]

Conclusion

Neonatal tetanus is still an important cause of death of infants in developing countries. In Bangladesh, about 10% of all fatalities in young children age 1 to 4 years result from either tetanus or measles.[150] In the United States, neonatal tetanus is virtually eliminated with only one case being reported in the last 10 years.[136] Immunization coverage for tetanus, as for most other childhood viral illnesses, has increased in record high numbers during the past few years, even in the United States.[151] Even so, in the 3-year period from 1995 to 1997, there were 124 patients seen in the United States, six of whom were younger than 20 years of age,[136] including one with neonatal tetanus.[136,137] Therefore, tetanus is still a clinical consideration in the differential diagnosis of acute generalized or focal muscle rigidity. This especially becomes a potential issue in the occasional child whose parents do not believe in the use of immunizations.[137] Although it is difficult to culture the *C. tetani* organism, the appreciation of EMG techniques, such as silent period studies[144,152] and magnetic brain stimulation,[153] provides a means to offer the clinician a reasonable diagnostic modality. Although tetanus is an exceedingly rare condition, the differential diagnostic consideration of tetanus has presented to us (HRJ) intermittently at Children's Hospital of Boston; therefore, we have presented this in some detail to provide a base for the clinician who is occasionally concerned about this possibility. Lastly, one must always consider other possible mechanisms with this clinical presentation, especially ingestion of substances (e.g. strychnine) in the patient with possible tetanus.[143]

RABIES

Human rabies, though largely controlled in the developed world, remains a significant public health problem in developing countries. Despite rapid increases in our understanding of its pathogenesis it still remains an inevitably fatal disease. Public health campaigns with emphasis on rabies immunization of pets and development of effective pre- and postexposure immunization regimens, have reduced acquired human rabies fatalities in the USA to fewer than 3 per year.[154]

Etiology

Rabies virus is a neurotropic RNA virus in the virus family Rhabdoviridae and genus *Lyssavirus* that infects warm-blooded animals throughout the world. Human rabies is a viral infection of the central nervous system usually transmitted by the bite of an infected animal, but infection by inhalation of the virus has occurred in caves inhabited by rabid bats. Transmission also occurs by infected animal saliva coming into contact with abrasions on mucous membranes.[155] Rarely, human-to-human transmission occurs via organ transplantation, in particular, corneal transplantation.[156]

Epidemiology

The preferential animal vectors vary in different countries. In Asia and Africa, stray dogs are the main transmitters of the disease, followed by meerkats and fruit bats. In the United States, the relatively rare human rabies virus occurs mainly in bats, dogs, and skunks, although raccoons were recently heavily infested, and foxes may also be at risk. In central and western Europe, it is the red fox, whereas in South and Central America, the dog and the vampire bat are the major endemic hosts.[157]

Unfortunately, rabies is still a common disease, with an estimated 55,000 deaths from rabies in Africa and Asia where infection is endemic in dogs.[158] This is definitely an underestimate of its incidence worldwide, due to rudimentary reporting. In India alone, 15,000 people die of rabies and 3 million receive postexposure treatment each year.[159]

In the United States and its territories, human rabies has decreased markedly from an average of 40 cases during the 1940s to only 2 cases during the 1960s and 1970s.[160] However, a more recent discussion mentions 32 cases of rabies between 1980 and 1998 and only 19 cases between 2000 and 2006 in the United States.[158,161] Approximately 50% of these cases were acquired by people traveling in countries where rabies is much more heavily endemic, such as Nepal. The increased incidence of rabies in the United States has been attributed to the movement of rabid animals, such as the raccoon, to new geographic areas.[161] Through stringent and successful vaccination of dogs and programs for controlling stray animals, some developed countries such as the United Kingdom, Australia, and New Zealand have kept their countries free of rabies.[160]

Pathogenesis

The key sequential steps in rabies pathogenesis are shown in Figure 10.2. After penetration of the skin and inoculation with saliva from a rabid animal, the virus multiplies in striated muscle, where antibody, interferon, and other host factors may retard nerve invasion.[162] The virus then travels through the peripheral nerves and spinal cord to the brain, where it causes neuronal destruction, especially in the brain stem.

The virus next passes centrifugally to a wide variety of richly innervated sites, the most common being the mucus-secreting and salivary glands, where the viral load is variable, which may explain why only 50% of bites by proven rabid dogs resulted in rabies.[163] Characteristic viral inclusions (Negri bodies) are found in the neurons, especially in the pyramidal cells of the hippocampus, and in Purkinje's cells of the cerebellum. It is important to note that the absence of Negri bodies does not exclude rabies.

Clinical Features

Incubation Period

The incubation period of human rabies can be 2 weeks to 1 year, but it is usually 1 to 3 months. Within the peripheral nerves, intra-axonal transport occurs at a rate of 8 to 20 mm/day. In one adult case, the incubation period was 65 days, suggesting a rate of 10 mm/day.[161] Therefore, in children, the time from exposure to onset of clinical symptoms is relatively rapid. The incubation period is particularly shorter in children when the bite is on the face and is longest when the leg is bitten. This is well illustrated in Case Example 10.4, in which clinical signs of rabies appeared in less than 3 weeks.

Prodrome Phase

The early symptoms are fever, malaise, headache, anorexia, vomiting, and paresthesia at the site of the wound. These symptoms are likely secondary to infection and inflammatory changes in local sensory ganglia (dorsal root or cranial ganglia). These nonspecific symptoms usually last 2 to 10 days, followed by severe central nervous system disease.

Acute Neurologic Phase

This phase can have two clinical presentations. If excitement is predominant, the case is classified as "furious" rabies. When paralysis dominates the clinical picture, the case is classified as "dumb" or paralytic rabies.

Furious Rabies

Furious rabies is the most common form (80% of cases) and is characterized by excitement, fear, terror, bizarre aggressive behavior, disorientation, and hallucinations, all alternating with periods of lucidity. Signs of brain stem encephalitis include dysarthria, dysphagia, and hoarseness, resulting from pharyngeal, palatal, and laryngeal paralysis. Hydrophobia, aerophobia, or both occur in about 50% of cases; these are unique features that are considered to be clinically pathognomonic for rabies. In hydrophobia, the mention or appearance of water elicits terror and a choking sensation secondary to spasm of the pharyngeal and respiratory muscles.

Paralytic Rabies

The clinical presentation of paralytic rabies lacks the typical excitement stage of furious rabies. Paralytic rabies

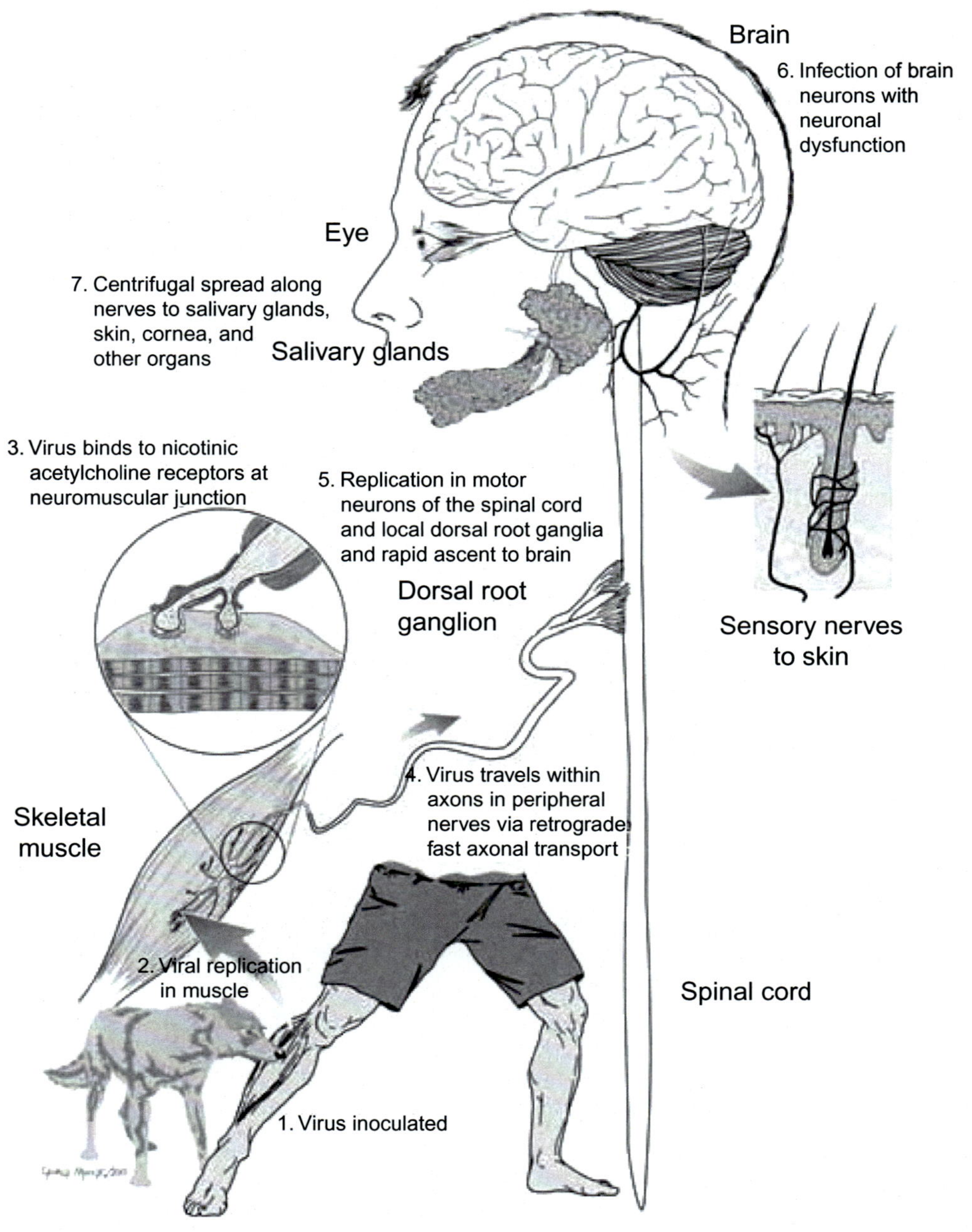

FIGURE 10.2 Schematic diagram showing the sequential steps in the pathogenesis of rabies after an animal bite/peripheral inoculation of rabies virus. *From Jack,*[174] © *Elsevier, reproduced with permission.*

accounts for less than 20% of cases of human rabies. There is an ascending, symmetrical paralysis resembling Guillain-Barré syndrome. Occasionally the sphincters are involved, and the condition may resemble acute transverse myelitis. The cerebrospinal fluid may show pleocytosis and an elevated protein level. On other occasions, the onset of symptoms may be more focal or consist of a combination of motor and encephalopathic features.[164] Over time, many if not all the cranial nerves become involved, the face becomes expressionless, paralysis increases, and coma ensues (see Case Example 10.5). Respiratory failure or cardiac arrhythmias are the usual cause of death.

Differential Diagnosis

The history of an animal bite combined with the classic clinical picture of hyperexcitability, hydrophobia, aerophobia, and bizarre aggressive behavior alternating with periods of lucidity, strongly supports the diagnosis of rabies. A number of conditions may mimic rabies, however.

CASE EXAMPLE 10.4

Alfred, a 15-year-old boy from a rural village in Natal, South Africa, was bitten on the leg by a stray dog that died 2 weeks later. Several days after the bite, Alfred developed redness, numbness, and tingling at the wound site. He presented to a regional hospital 1 week later, where the wound was cleaned and dressed. While there, he received a dose of penicillin. Rabies immunization with human diploid cell vaccine was also commenced. Unfortunately, he did not return to complete his course of antirabies vaccine.

Five weeks later, this boy presented with agitation, aggressive behavior, and hydrophobia to the emergency department of the university hospital. A clinical diagnosis of rabies was made. A high dose of antirabies immunoglobulin and appropriate sedation were commenced. Despite these therapies, his condition progressively deteriorated, culminating in death just 7 days later. Autopsy demonstrated Negri bodies typical of rabies in the pyramidal cells of the hippocampus and Purkinje cells of the cerebellum.

CASE EXAMPLE 10.5

A 10-year-old boy was bitten on the hand by a skunk while asleep in his tent during a camping trip in Minnesota.[163] He initially had pain at the bite site. The skunk was never found. He received penicillin and 14 daily injections of duck-embryo antirabies vaccine. A stiff neck, fever, and malaise occurred 19 days after the bite, leading to his hospitalization at the Mayo Clinic. He soon developed weakness and numbness in the hand that had been bitten. Two days later he still had weakness, stiffness of the neck, fever, and malaise. He also appeared "jumpy," had an unsteady gait and nuchal rigidity, and vomited several times. Soon he became confused and developed nasal speech, difficulty swallowing, cerebellar ataxia of the left arm, and flaccid weakness of the right shoulder and arm. His body temperature reached 40.3°C (104.5°F), and he became lethargic and exhibited perseveration. Bilateral, asymmetrical peripheral facial and abducens nerve palsies developed. He died of respiratory paralysis 26 days after symptom onset. Although Negri bodies were not detected in this child's brain, homogenates of brain and spinal cord tissue were injected into mouse brains. Negri bodies were found in the mice that received hippocampal and cerebellar tissue from the child's brain. No antibody studies were available at the time of this incident.[163]

These include viral encephalitis, vaccine encephalitis, cerebral malaria, and tetanus. Paralytic rabies may be misdiagnosed as Guillain-Barré syndrome, poliomyelitis, or postvaccine encephalomyelitis.[162] Hysteria in a bitten patient who knows the danger of rabies may represent a diagnostic challenge, but the absence of true hydrophobia or aerophobia usually helps exclude rabies.

Laboratory Analysis

Basic, routine investigations including cerebrospinal fluid analysis and electroencephalogram studies do not offer specific diagnostic findings in rabies. Magnetic resonance imaging, specifically T2 hyperintense signals in the brainstem, hippocampus, deep subcortical and cortical white and gray matter as well as gadolinium enhancement, can signify a rabies diagnosis.[158] Cerebrospinal fluid results may be rather nonspecific, but a significant pleocytosis, primarily a lymphocytosis (>100 lymphocytes), may occur. Sometimes there is hypoglycorrhachia, with cerebrospinal fluid glucose values as low as 24 mg/dL but normal protein.[164]

Antemortem laboratory diagnosis of rabies is now possible. Fluorescent monoclonal rabies antibody tests are used to detect viral antigen in tissues such as the cornea and skin from the nape of the neck, just above the hairline. There are ample cutaneous nerves surrounding hair follicles. Serum and cerebrospinal fluid antirabies antibodies are detectable in high titers after the first 8 days of illness. But even in vaccinated individuals, the cerebrospinal fluid titers rarely rise to any significant level. Post mortem fluorescent antibody testing of brain tissue and virus isolation in cell culture confirm a diagnosis.[158]

Prevention

Preexposure

Because human rabies in still incurable, measures to control the disease in endemic areas are the best means of dealing with this deadly illness. The most effective measures are vaccinating all domestic animals, decreasing the growth of the vector population, and avoiding contact with potentially infected animals. Persons at higher risk for rabies, such as veterinarians and laboratory workers, need to be immunized with rabies human diploid cell vaccine. The World Health Organization recently recommended pre-exposure rabies vaccination using intradermal injections of one-fifth the standard intramuscular dose of the current vaccine as a cost-saving alternative for developing countries where rabies is endemic.[165]

TABLE 10.1 Postexposure Antirabies Treatment Guide

Animal Species	Condition of Animal at Time of Attack	Treatment of Human
Wild: skunk, raccoon, bat, others	Regard as rabid	HRIG[1] and V[2]
Domestic: dog, cat	Healthy	None[3]
	Unknown (escaped)	HRIG[1] and V[2]
	Rabid or suspected rabid	HRIG[1] and V[2]

[1]*Administer 20 IU/kg body weight. HRIG should not be administered if previously vaccinated with any type of rabies vaccine and a documented history of antibody response to the prior vaccination.*
[2]*Human diploid cell vaccine (HDCV) or purified chick embryo vaccine (PCECV) 1.0 mL, I.M.1 (one) each on days 0, 3, 7, and 14 (4 doses) in not previously vaccinated and 1 each on days 0 and 3 in previously vaccinated patients. For details, see source below.*
[3]*Isolate for 10 days if bite was unprovoked.*
Abbreviations: HRIG, human rabies immunoglobulin; V, vaccine.
Adapted from Rupprecht CE et al. Use of a reduced (4-dose) vaccine schedule for postexposure prophylaxis to prevent human rabies. Recommendations of the Advisory Committee on Immunization Practices. MMWR Morbidity and Mortality Weekly Report 2010 March 19;59(RR-2):1–12; and from Report of the Committee on Infectious Diseases, ed. Committee on Infectious Diseases. Elk Grove Village IL: American Academy of Pediatrics, 1991, with permission from John Wiley and Sons.

Postexposure

Initially, one needs to determine whether there is a definite risk or just an understandable fear of rabies on the part of the parent or child. An animal bite carries a 50 to 100 times greater risk of transmitting rabies than does a simple scratch.[161] Consultation with local public health officials may be useful to ascertain which animals are particularly of risk in that locale.

Following an exposure, immediate and thorough washing of all bite wounds and scratches with soap and water, followed by application of a virucidal agent such as povidone-iodine, markedly reduces the likelihood of contracting rabies.[165,185] Postexposure antirabies treatment must proceed according to the protocol outlined in Table 10.1.

An increasingly important issue in North America is the fact that 17 of 32 cases of rabies reported in the United States between 1980 and 1996 (in patients aged 4 to 82 years) were related to bat-associated strains.[166] In only one instance was there a history of a bat bite. Eight of the other 15 cases included known exposure to a bat. If one finds a bat in a room with a sleeping child, post-exposure prophylaxis deserves serious consideration unless the bat is either dead or captured and pathologic evaluation fails to confirm the presence of rabies.[161] However, caution is urged, as "there is no evidence that transmission by aerosol is epidemiologically significant. Particular concern must be directed to those situations in which contact was possible but where a trivial bite may have gone unnoticed."[167,168] Because bats have extremely small, sharp teeth, exposure may involve an undetected bite.[167]

To ensure passive and active immunity, both human rabies immunoglobulin and rabies human diploid cell vaccine must be administered as soon as possible after any suspicious bite. Human rabies immunoglobulin is administered in a dose of 20 IU/kg, half of which is infiltrated at the site of the wound; the remainder is given intramuscularly. Human diploid cell vaccine is administered intramuscularly in 1-mL doses at the same time as the human rabies immunoglobulin and repeated on days 3, 7, and 14. If after 10 days of observation by a veterinarian the suspect domestic animal is found to be free of rabies, this regimen can be discontinued.[169]

For fully vaccinated individuals who have been exposed to rabies, two doses of vaccine given 3 days apart will confer adequate immunity against rabies.[162] Preventing human-to-human transmission of rabies is achieved by not transplanting corneas from donors who died of encephalitis and by using face masks, gloves, and gowns when in direct contact with a patient suspected of having rabies.[170]

Management

Once rabies develops, there is no specific treatment to improve survival. Death usually occurs within 2 to 7 days if no life-support measures are administered.[161] Intensive supportive care only prolongs the clinical course that inevitably ends in death. With the exception of survival in a single patient discussed in the following paragraph, there is no well-documented evidence of anyone surviving after clinical disease developed, in the absence of pre- or immediate postexposure prophylaxis.[161] In developing countries, improvement in public health measures to prevent the disease in dogs and the provision of better facilities for postexposure prophylaxis are better uses of resources than are attempts at intensive care.

An exception has been reported: a case of a 15-year-old girl who developed clinical rabies one month after a bat bite[171]; she was treated with induction of coma, but rabies vaccine was not administered. She survived, and at 5 months after her hospitalization, she was alert and able to communicate, but had neurological deficits (ataxia, dysarthria, and choreoathetosis).

Prognosis

Rabies is a uniformly fatal disease in humans. Only four symptomatic persons are known to have survived rabies infection, and three received rabies vaccination before exposure.[171,172] The prevalence of rabies in North American wildlife requires a concerted effort by caregivers to elevate public awareness of the risk of rabies associated with contacting wildlife, in particular bats.

REFERENCES

1. Drinker P, McKhann CF. The use of a new apparatus for the prolonged administration of artificial respiration: A fatal case of poliomyelitis. *JAMA* 1929;**92**:1658–60.
2. Eggers HJ. Milestones in early poliomyelitis research (1840 to 1949). *J Virol* 1999;**73**:4533–5.
3. Ogra PL. Poliomyelitis as a paradigm for investment in and success of vaccination programs. *Pediatr Infect Dis J* 1999;**18**:10–15.
4. Wilson DJ. A crippling fear: Experiencing polio in the era of FDR. *Bull Hist Med* 1998;**72**:464–95.
5. Paul JR. *A History of Poliomyelitis*. New Haven, Conn: Yale University Press; 1971.
6. Weinstein L, Shelokov A, Seltser R, Winchell GD. A comparison of the clinical features of poliomyelitis in adults and in children. *N Engl J Med* 1952;**246**:296–302.
7. Bell W, McCormick WF. *Neurologic Infections in Children*. 2nd ed. Philadelphia: Saunders; 1981, pp 308–21.
8. Johnson RT. *Viral Infections of the Nervous System*. 2nd ed. Philadelphia: Lippincott-Raven; 1998. pp. 87–132.
9. Lambert SM, Markel H. Making history: Thomas Francis, Jr, MD, and the 1954 Salk poliomyelitis vaccine field trial. *Arch Pediatr Adolesc Med* 2000;**154**:512–17.
10. John TJ. The final stages of the global eradication of polio. *N Engl J Med* 2000;**343**:806–7.
11. John TJ, Walker DH. Enterovirus infections, including poliomyelitis. In: Guerrant RL, Walker DH, Weller PF, editors. *Tropical Infectious Diseases: Principles, Pathogens, and Practice*. Philadelphia: Churchill Livingstone; 1999. pp. 1123–32.
12. Kuroiwa C, Vongphrachanh P, Chosa T, Murakami H, Hashizume M, Wakai S, et al. Risk of poliomyelitis importation and re-emergence in Laos. *Lancet* 2000;**356**:1487–8.
13. Sutter RW, Cochi SL. Enteric viral infections. In: Strickland GT, editor. *Hunter's Tropical Medicine and Emerging Infectious Diseases*. 8th ed. Philadelphia: Saunders; 2000. pp. 214–20.
14. Yoshida H, Horie M, Otani H, Takano M, Tsuji K, Kubota T, et al. Characterization of vaccine-derived polioviruses isolated from sewage and river water in Japan. *Lancet* 2000;**356**:1461–3.
15. Hennessey KA, Marx A, Hafiz R, Ashgar H, Hadler SC, Jafari H, et al. Widespread paralytic poliomyelitis in Pakistan: A case-control study to determine risk factors and implications for poliomyelitis eradication. *J Infect Dis* 2000;**182**:6–11.
16. Valente F, Otten M, Balbina F, van de Weerdt R, Chezzi C, Eriki P, et al. Massive outbreak of poliomyelitis caused by type-3 wild poliovirus in Angola in 1999. *Bull World Health Org* 2000;**78**:339–46.
17. Sutter RW, Prevots DR, Cochi SL. Poliovirus vaccines: Progress toward global poliomyelitis eradication and changing routine immunization recommendations in the United States. *Pediatr Clin North Am* 2000;**47**:287–308.
18. D'Souza RM, Kennett M, Antony J, Herceg A, Harvey B, Longbottom H, et al. Surveillance of acute flaccid paralysis in Australia, 1995–97: Australian paediatric surveillance unit. *J Paediatr Child Health* 1999;**35**:536–40.
19. Nkowane BM, Wassilak SGF, Orenstein WA, Bart KJ, Schonberger LB, Hinman AR, et al. Vaccine-associated paralytic poliomyelitis. United States: 1973–1984. *JAMA* 1987;**257**:1335–40.
20. Ridgeway D. The logic of causation and the risk of paralytic poliomyelitis for an American child. *Epidemiol Infect* 2000;**124**:113–20.
21. Committee on Infectious Diseases: Prevention of poliomyelitis: Recommendations for use of only inactivated poliovirus vaccine for routine immunization. *Pediatrics* 1999;**104**:1404–6.
22. McNeil, DG, Jr., Polio's return after near eradication prompts a global health warning. The New York Times. May 5, 2014. Available from: <http://www.nytimes.com/2014/05/06/health/world-health-organization-polio-health-emergency.html?hp&_r=0>.
23. Lipton HL, Jubelt B. Enterovirus infections of the central nervous system. In: Tyler KL, Martin JB, editors. *Infectious Diseases of the Central Nervous System*. Philadelphia: Davis 1993. pp. 103–30.
24. Jubelt B. Other enteroviral diseases. In: McKendall RR, Stroop WG, editors. *Handbook of Neurovirology*. New York: Marcel Dekker; 1994. pp. 503–19.
25. Haller AA, Semler BL. Enteroviruses. In: McKendall RR, Stroop WG, editors. *Handbook of Neurovirology*. New York: Marcel Dekker; 1994. pp. 481–91.
26. Faber HK. *The Pathogenesis of Poliomyelitis*. Springfield, Ill: Thomas; 1955.
27. Nordli DR, Bello JA, De Vivo DC. Myelitis. In: Schlossberg D, editor. *Infections of the Nervous System*. New York: Springer-Verlag; 1990. pp. 179–91.
28. Melinck JL. Poliomyelitis. In: McKendall RR, Stroop WG, editors. *Handbook of Neurovirology*. New York: Marcel Dekker; 1994. pp. 493–501.
29. Solecki D, Gromeier M, Harber J, Bernhardt G, Wimmer E. Poliovirus and its cellular receptor: A molecular genetic dissection of a virus/recptor affinity interaction. *J Mol Recognit* 1998;**11**:2–9.
30. Sicinski P, Rowinski J, Warchol JB, Jarzabek Z, Gut W, Szczygiel B, et al. Poliovirus type 1 enters the human host through intestinal M cells. *Gastroenterology* 1990;**98**:56–8.
31. Baker AB. *Bulbar poliomyelitis: Rocky Mountain Conference on Infantile Paralysis*. Denver: University of Colorado School of Medicine and National Foundation for Infantile Paralysis; 1946 pp. 110–13.
32. Grulee CG, Panos TC. Epidemic poliomyelitis in children. *Am J Dis Child* 1948;**75**:24–39.
33. Osler W. *Modern Medicine: Its Theory and Practice*. Philadelphia: Lea & Febiger; 1910. pp. 258–68.
34. Abramson H, Greenberg M. Acute poliomyelitis in infants under one year of age: Epidemiological and clinical features. *Pediatrics* 1955;**16**:478–87.

35. Friedrich F. Neurologic complications associated with oral poliovirus vaccine and genomic variability of the vaccine strains after multiplication in humans. *Acta Virol* 1998;**42**:187–94.
36. Gonzalez H, Olsson T, Borg K. Management of Post Polio Syndrome. *Lancet Neurol* 2010;**9**:634–42.
37. Koopman FS, Uegaki K, Gilhus NE, Beelen A, de Visser M, Nollet F. *Treatment for Post Polio Syndrome (Review). The Cochrane Collaboration.* John Wiley and Sons; 2011, Issue 2.
38. Trojan DA, Cashman NR. Post polio syndrome. *Muscle Nerve* 2005;**31**:6–19.
39. Julien J, Leparc-Goffart I, Lina B, Fuchs F, Foray S, Janatova I, et al. Postpolio syndrome: Poliovirus persistence is involved in the pathogenesis. *J Neurol* 1999;**246**:472–6.
40. McKhann GM, Cornblath DO, Griffith J, Ho TW, Li CY, Jiang Z, et al. Acute motor axonal neuropathy: A frequent cause of acute flaccid paralysis in China. *Ann Neurol* 1993;**33**:333–42.
41. Phadke MA, Gambhir PS, Deshpande AS, Kurlekar SU, Godbole KG. Asian paralysis syndrome. *Ann Trop Paediatr* 1999;**19**: 317–20.
42. Ozawa H, Noma S, Yoshida Y, Sekine H, Hashimoto T. Acute disseminated encephalomyelitis associated with poliomyelitis vaccine. *Pediatr Neurol* 2000;**23**:177–9.
43. Feigin RD, Guggenheim MA, Johnsen SD. Vaccine-related paralytic poliomyelitis in an immunodeficient child. *J Pediatr* 1971;**79**:642–7.
44. Davis LE, Bodian D, Price D, Butler IJ, Vickers JH. Chronic progressive poliomyelitis secondary to vaccination of an immunodeficient child. *N Engl J Med* 1977;**297**:241–5.
45. Chitsike I, van Furth R. Paralytic poliomyelitis associated with live oral poliomyelitis vaccine in child with HIV infection in Zimbabwe: Case report. *BMJ* 1999;**318**:841–3.
46. Bellmunt A, May G, Zell R, Pring-Akerbloma P, Verhagen W, Heim A. Evolution of poliovirus type I during 5.5 years of prolonged enteral replication in an immunodeficient patient. *Virology* 1999;**265**:178–84.
47. Jarcho LW, Fred HL, Castle CH. Encephalitis and poliomyelitis in the adult due to coxsackie virus group B, type 5. *N Engl J Med* 1963;**268**:235–8.
48. Kopel FB, Shore B, Hodes HL. Non-fatal bulbospinal paralysis due to ECHO 4 virus. *J Pediatr* 1965;**67**:588–94.
49. Foley JF, Chin TDY, Gravelle CR. Paralytic disease due to infection of ECHO virus type 9. *N Engl J Med* 1959;**260**:924–6.
50. Katiyar B, Misra S, Singh RB, Singh A, Gupta S, Gulati AK. Adult polio-like syndrome following enterovirus 70 conjunctivitis (natural history of the disease). *Acta Neurol Scand* 1983;**67**: 263–74.
51. Samuda GM, Chang W-K, Yeung C-Y, Tang PS. Monoplegia caused by enterovirus 71: An outbreak in Hong Kong. *Pediatr Infect Dis J* 1987;**6**:206–8.
52. Hayward JC, Gillespie SM, Kaplan KM, Packer R, Pallansch M, Plotkin S, et al. Outbreak of poliomyelitis-like paralysis associated with enterovirus 71. *Pediatr Infect Dis J* 1989;**8**:611–16.
53. Huang C-C, Liu C-C, Chang Y-C, Chen CY, Wang ST, Yeh TF. Neurologic complications in children with enterovirus 71 infections. *N Engl J Med* 1999;**341**:936–42.
54. Kyllerman MG, Herner S, Bergström TB, Ekholm SE. PCR diagnosis of primary herpesvirus type I in poliomyelitis-like paralysis and respiratory tract disease. *Pediatr Neurol* 1993;**9**:227–9.
55. Wong M, Connolly AM, Noetzel MJ. Poliomyelitis-like syndrome associated with Epstein-Barr infection. *Pediatr Neurol* 1999; **20**:235–7.
56. Seay AR. Alpha and flavivirus diseases. In: McKendall RR, Stroop WG, editors. *Handbook of Neurovirology*. New York: Marcel Dekker; 1994. pp. 391–411.
57. Likar M, Dane DS. An illness resembling acute poliomyelitis caused by a virus of the Russian spring-summer encephalitis louping ill group in Northern Ireland. *Lancet* 1958;**1**:456–8.
58. Schellinger PD, Schmutzhard E, Fiebach JB, Pfausler B, Maier H, Schwab S. Poliomyelitic-like illness in central European encephalitis. *Neurology* 2000;**55**:299–302.
59. Chezzi C. Rapid diagnosis of poliovirus infection by PCR amplification. *J Clin Microbiol* 1996;**34**:1722–5.
60. Georgopoulou A, Markoulatos P, Spyrou N, Vamvakopoulos NC. Improved genotyping vaccine and wild-type poliovirus strains by restriction fragment length polymorphism analysis: Clinical diagnostic implications. *J Clin Microbiol* 2000;**38**:4337–42.
61. Goodgold J, Eberstein A. *Electrodiagnosis of Neuromuscular Diseases*. 3rd ed. Baltimore: Williams & Wilkins; 1972. pp. 232–3.
62. Kimura J. *Electrodiagnosis in Diseases of Nerve and Muscle. Principles and Practice*. 2nd ed. Philadelphia: Davis; 1989. pp. 440–1.
63. Bower AG. *Diagnosis and Treatment of the Acute Phase of Poliomyelitis and Its Complications*. Baltimore: Williams & Wilkins; 1954.
64. Jubelt B, Agre JC. Characteristics and management of postpolio syndrome. *JAMA* 2000;**284**:412–14.
65. Hopkins IJ, Shield LK. Poliomyelitis-like illness associated with asthma in childhood. *Lancet* 1974;**1**:760.
66. Hopkins IJ. A new syndrome: Poliomyelitis-like illness associated with acute asthma in childhood. *Aust Paediatr J* 1974;**10**:274–6.
67. Danta G. Electrophysiological study of amyotrophy associated with acute asthma (asthmatic amyotrophy). *J Neurol Neurosurg Psychiatry* 1975;**38**:1016–21.
68. Ilett SJ, Pugh RJ, Smithells RW. Poliomyelitis-like illness after acute asthma. *Arch Dis Child* 1977;**52**:738–40.
69. Shapiro GG, Chapman JT, Pierson WE, Bierman CW. Poliomyelitis-like illness after acute asthma. *J Pediatr* 1979;**94**: 767–8.
70. Beede HE, Newcomb RW. Lower motor neuron paralysis in association with asthma. *Johns Hopkins Med J* 1980;**147**:186–7.
71. Blomquist HK, Bjorksten B. Poliomyelitis-like illness associated with asthma. *Arch Dis Child* 1980;**55**:61–2.
72. Liebeschuetz HJ. Poliomyelitis-type illness associated with severe asthma in a child. *J Roy Soc Med* 1981;**74**:71–2.
73. Nihei K, Naitoh H, Ikeda K. Poliomyelitis-like syndrome following asthmatic attack (Hopkins syndrome). *Pediatr Neurol* 1987;**3**:166–8.
74. Shahar EM, Hwang PA, Niesen CE, Murphy EG. Poliomyelitis-like paralysis during recovery from acute bronchial asthma: Possible etiology and risk factors. *Pediatrics* 1991;**88**:276–9.
75. Horiuchi I, Yamasaki K, Osoegawa M, Ohyagi Y, Okayama A, Kurokawa T, et al. Acute myelitis after asthma attacks with onset after puberty. *J Neurol Neurosurg Psychiatry* 2000;**68**:665–8.
76. Cohen HA, Ashkenasi A, Ring H, Weiss R, Wolach B, Paret G, et al. Poliomyelitis-like syndrome following asthmatic attack (Hopkins' syndrome): Recovery associated with IV gamma globulin treatment. *Infection* 1998;**26**:247–9.

77. David W, Doyle J. Acute infantile weakness: A case of vaccine-associated poliomyelitis. *Muscle Nerve* 1997;**20**:747–9.
78. Rowland L, editor. *Merritt's Textbook of Neurology*. 9th ed. Baltimore: Williams & Wilkins; 1995.
79. Soto N, Lutwick L. Poliovirus immunizations. What goes around, comes around. *Infect Dis Clin Am* 1999;**13**(1):265–78.
80. Enders J, Weller T, Robbins F. Cultivation of Lansing strain of poliomyelitis virus in cultures of various human embryonic tissues. *Science* 1949;**109**:85–7.
81. Bodian D, Morgan I, Howe H. Differentiation of types of poliomyelitis viruses. III. The grouping of 14 strains into three basic immunological types. *Am J Hyg* 1949;**49**:234–45.
82. Salk J. Consideration in the preparation and use of poliomyelitis virus vaccine. *JAMA* 1955;**158**(14):1239–48.
83. Francis T. Evaluation of the 1954 poliomyelitis vaccine field trial. *JAMA* 1955;**158**(14):1266–70.
84. LaForce F. Poliomyelitis vaccines. Success and controversy. *Infect Dis Clin North Am* 1990;**4**(1):75–83.
85. Sabin A. Prevention of poliomyelitis by vaccination. *Ann Pediatr* 1958;**10**:197.
86. Sabin A, Boulger L. History of Sabin attenuated poliovirus oral live vaccine strains. *J Biol Stand* 1973;**1**:115–18.
87. Prevention CfDCa. Recommended childhood immunization schedule—United States, 1997. *MMWR* 1997;**46**(2):35–41.
88. Grachev V. Long-term use of poliovirus vaccine from Sabin strains in the Soviet Union. *Rev Infect Dis* 1984;**6**(S2):S321–2.
89. Prevention CfDCa. Poliomyelitis prevention in the United States: Introduction of a sequential vaccination schedule of inactivated poliovirus vaccine followed by oral poliovirus vaccine. Recommendations of the Advisory Committee on Immunization Practices (ACIP). *MMWR* 1997;**46**:1–25.
90. Faden H. Poliovirus vaccination: A trilogy. *J Infect Dis* 1993; **168**:25–8.
91. Murdin A, Barreto L, Plotkin S. Inactivated poliovirus vaccine: Past and present experience. *Vaccine* 1996;**14**(8):735–46.
92. Oostvogel P, van Wijngaarden J, van der Avoort H, Mulders MN, Conyn-Van Spaendonck M, Rumke H, et al. Poliomyelitis outbreak in an unvaccinated community in the Netherlands, 1992–1993. *Lancet* 1993;**344**:665–70.
93. Sutter R, Onorato I, Patriarca P. Current poliomyelitis immunization policy in the United States. *Pediatr Ann* 1990;**19**: 702–6.
94. Nkowane B, Wassilak S, Orenstein W, Bart K, Schonberger L, Hinman A, et al. Vaccine-associated poliomyelitis in the United States: 1973 through 1984. *JAMA* 1987;**257**:1335–40.
95. Prevots DR, Sutter RW, Strebel PM, Weibel RE, Cochi SL. Completeness of reporting for paralytic poliomyelitis, United States, 1980 through 1991. *Arch Pediatr Adolesc Med* 1994;**148**: 479–85.
96. Strebel P, Sutter R, Cochi S, Brink E, Kew O, Pallansc M, et al. Epidemiology of poliomyelitis in the United States one decade after the last reported case of indigenous wild virus-associated disease. *Clin Infect Dis* 1992;**14**:568–79.
97. Schonberger L, Kaplan J, Kim-Farley R, Moore M, Eddins D, Hatch M. Control of paralytic poliomyelitis in the United States. *Rev Infect Dis* 1984;**6**:S424–6.
98. Prevention CfDCa. Update: Eradication of paralytic poliomyelitis in the Americas. *MMWR* 1992;**41**(36):681–3.
99. American Academy of Pediatrics CoID: Poliomyelitis prevention: Recommendations for use of inactivated poliovirus vaccine and live oral poliovirus vaccine. *Pediatrics* 1997;**99**(2):300–5.
100. Zimmerman R, Spann S. Poliovirus vaccine options. *Am Fam Physician* 1999;**59**(1):113–18.
101. American Academy of Pediatrics CoID: Poliomyelitis prevention: Revised recommendations for use of inactivated and live oral poliovirus vaccines. *Pediatrics* 1999;**103**(1):171–2.
102. Chitsike I, van Furth R. Paralytic poliomyelitis associated with live oral poliomyelitis vaccine in a child with HIV infection in Zimbabwe: Case report. *Br Med J* 1999;**318**:841–3.
103. Tulchinsky T, Abed Y, Handsher R, Toubassi N, Acker C, Melnick J. Successful control of poliomyelitis by a combined OPV/IPV polio vaccine program in the West Bank and Gaza, 1978–93. *Isr J Med Sci* 1994;**30**:489–94.
104. Centers for Disease Control and Prevention (CDC). Paralytic poliomyelitis—United States, 1980–1994. *MMWR* 1997;**46** (4):79–83.
105. Schonberger L, McGowan JJ, Gregg M. Vaccine-associated poliomyelitis in the United States, 1961–1972. *Am J Epidemiol* 1976;**104**(2):202–11.
106. Weibel R, Benor D. Reporting vaccine-associated paralytic poliomyelitis: Concordance between the CDC and the National Vaccine Injury Compensation Program. *Am J Publ Health* 1996;**86**:734–7.
107. Henderson D, Witte J, Morris L, Langmuir A. Paralytic disease associated with oral polio vaccines. *JAMA* 1964;**190**(1):153–60.
108. Sutter R, Prevots D. Vaccine-associated paralytic poliomyelitis among immunodeficient persons. *Infect Med* 1994;**426**:429–38.
109. Wyatt H. Poliomyelitis in hypogammaglobulinemics. *J Infect Dis* 1973;**128**:802–6.
110. Davis L, Bodian D, Price D, Butler I, Vickers J. Chronic progressive poliomyelitis secondary to vaccination of an immunodeficient child. *N Engl J Med* 1977;**297**:241–5.
111. Sakano T, Kittaka E, Tanaka Y, Yamaoka H, Kobayashi Y, Usui T. Vaccine-associated poliomyelitis in an infant with agammaglobulinemia. *Acta Pediatr Scand* 1980;**69** 549–51.
112. Wright PF, Hatch MH, Kasselberg AG, Lowry SP, Wadlington WB, Karzon DT. Vaccine-associated poliomyelitis in a child with sex-linked agammaglobulinemia. *Pediatrics* 1977;**91**(3):408–12.
113. Riker JB, Brandt CD, Chandra R, Arrobio JO, Nakano JH. Vaccine-associated poliomyelitis in a child with thymic abnormality. *Pediatrics* 1971;**48**(6):923–9.
114. Mathias R, Routley J. Paralysis in an immunocompromised adult following oral polio vaccination. *Can Med Assoc J* 1985; **132**:738–9.
115. Andrus J, Strebel P, de Quadros C, Olive J. Risk of vaccine-associated paralytic poliomyelitis in Latin America, 1989–91. *WHO Bulletin OMS* 1995;**73**:33–9.
116. Centers for Disease Control and Prevention (CDC). Impact of sequential IPV/OPV schedule on vaccination coverage levels—United States, 1997. *MMWR* 1998;**47**(47) 1017–19.
117. Kew O, Nottay B. Molecular epidemiology of polioviruses *Rev Infect Dis* 1984;**6**:S499–504.
118. Friedrich F. Neurologic complications associated with oral poliovirus vaccine and genomic variability of the vaccine strains after multiplication in humans. *Acta Virol* 1998;**42**:187–95.
119. Evans D, Dunn G, Minor P, Schild GC, Cann AJ, Stanway G, et al. Increased neurovirulence associated with a single

nucleotide change in a noncoding region of the Sabin type 3 poliovaccine genome. *Nature* 1985;**314**:548–50.
120. Kew OM, Nottay BK, Hatch MH, Nakano JH, Obijeski JF. Multiple genetic changes can occur in the oral poliovaccines upon replication in humans. *J Gen Virol* 1981;**56**:337–47.
121. Georgescu MM, Balanant J, Macadam A, Otelea D, Combiescu M, Combiescu AA, et al. Evolution of the Sabin Type 1 poliovirus in humans: Characterization of strains isolated from patients with vaccine-associated paralytic poliomyelitis. *J Virol* 1997;**71** (10):7758–68.
122. Li J, Zhang L, Yoneyama T, Yoshida H, Shimizu H, Yoshii K, et al. Genetic basis of the neurovirulence of type 1 polioviruses isolated from vaccine-associated paralytic patients. *Arch Virol* 1996;**141**:1047–54.
123. Beausoleil JL, Nordgren RE, Modlin JF. Vaccine associated paralytic poliomyelitis. *J Child Neurol* 1994;**9**:334–5.
124. Goldstein J. *Infantile poliomyelitis in a recently immunized baby seen at Yale, personal communication*, 1995.
125. Friedrich F, Da Silva E, Schatzmayr H. Type 2 poliovirus recombinants isolated from vaccine-associated cases and from healthy contacts in Brazil. *Acta Virol* 1996;**40**:27–33.
126. Furione M, Guillot S, Otelea D, Balanant J, Candrea A, Crainic R. Polioviruses with natural recombinant genomes isolated from vaccine associated paralytic poliomyelitis. *Virology* 1993;**196**: 199–208.
127. Georgescu MM, Delpeyroux F, Tardy-Panit M, Balanant J, Combiescu M, Combiescu AA, et al. High diversity of poliovirus strains isolated from the central nervous system from patients with vaccine-associated paralytic poliomyelitis. *J Virol* 1994;**68** (12):8089–101.
128. Lipskaya G, Muzychenko A, Kutitova O, Maslova S, Equestre M, Drozdov S, et al. Frequent isolation of intertypic poliovirus recombinants with serotype 2 specificity from vaccine-associated polio cases. *J Med Virol* 1991;**35**:290–6.
129. Macadam A, Arnold C, Howlett J, John A, Marsden S, Taffs F, et al. Reversion of the attenuated and temperature-sensitive phenotypes of the Sabin type 3 strain of poliovirus in vaccinees. *Virology* 1989;**172**:408–14.
130. Tate C, Johnson G. Case report: Acute vaccine-associated paralytic poliomyelitis. *Muscle Nerve* 1997;**20**:253–4.
131. Ogra P. Comparative evaluation of immunization with live attenuated and inactivated poliovirus vaccines. *Ann NY Acad Sci* 1995;**12**:97–107.
132. Willis E, Sherrod J. Childhood immunizations: Position on the enhanced inactivated poliovirus vaccine and live attenuated oral poliovirus vaccine dilemma. *J Natl Med Assoc* 1997;**89**:785–9.
133. Luisto M, Livanainen M. Tetanus of immunized children. *Dev Med Child Neurol* 1993;**35**:346–58.
134. Wesley AG, Pather M. Tetanus in children: An 11 year review. *Ann Trop Paediatr* 1987;**7**:32–7.
135. Cook R, Galazka A. Eliminating neonatal tetanus—An attainable goal. *Arch Dis Child* 1985;**60**:401–2.
136. Craig AS, Reed GW, Mohon RT, Quick ML, Swarner OW, Moore WL, et al. Neonatal tetanus in the United States: A sentinel event in the foreign born. *Pediatr Infect Dis J* 1997;**16**:955–9.
137. Bardenheimer B, Prevots DR, Khetsurani N, Wharton M. Tetanus surveillance—United States 1995–1997. *MMWR CDC Surv Summ* 1998;**47**:1–13.
138. Dunkle LM. Anaerobic infections. In: Oski FA, editor. *Principles and Practice of Pediatrics*. 2nd ed. Philadelphia: JB Lippincott; 1994. pp. 1157–60.
139. Demicheli V, Barale A, Rivetti A. *Vaccines for Women to Prevent Neonate Tetanus (Review). The Cochrane Collaboration.* Published by John Wiley and Sons, Ltd. Copyright; 2012.
140. Okoromah CAN, Lesi AFE. *Diazepam for Treating Tetanus (Review). The Cochrane Collaboration.* Published by John Wiley and Sons, Ltd. Copyright; 2012.
141. Wesley AG. Infectious diseases. In: Coovadia HM, Loening WE, editors. *Paediatrics and Child Health*. 3rd ed. Cape Town: Oxford University Press; 1992. pp. 177–9.
142. Fatunde OJ, Familusi JB. Atypical presentations in children with post-neonatal tetanus in Ibadan, Nigeria. *Ann Trop Paediatr* 2001;**21**:72–6.
143. Case records of the Massachusetts General Hospital. Weekly clinicopathological exercises. Case 12-2001. A 16-year-old boy with an altered mental status and muscle rigidity. *N Engl J Med* 2001;**344**:1232–9.
144. Steinegger T, Wiederker M, Ludin H, Roth F. Electromyography as a diagnostic aid in tetanus. *Schweiz Med Wochenschr* 1996;**126**:379–85.
145. Fernandez JM, Ferrandiz M, Larrea L, Ramio R, Boada M. Cephalic tetanus studied with single fiber EMG. *J Neurol Neurosurg Psychiatr* 1983;**46**:862–6.
146. Gupta SD, Keyl PM. Effectiveness of prenatal tetanus toxoid immunization against neonatal tetanus in a rural area in India. *Pediatr Infect Dis J* 1998;**17**:316–21.
147. Arnon SS. Tetanus. In: Behrman RE, Kliegman RM, Jenson HB, editors. *Nelson Textbook of Pediatrics*. 16th ed. Philadelphia: Saunders; 2000. pp. 878–81.
148. Rodrigo C, Samaakoon L, Fernando SD, Rajappakse S. A meta-analysis of magnesium for tetanus. *Anesthesia* 2012;**67**:1370–4.
149. Engrand N, Guerot E, Rouamba A, Vilain G. The efficacy of intrathecal baclofen in severe tetanus. *Anaesthesiology* 1999;**90**:1773–6.
150. Bhagwanjee S, Bosenberg AT, Muckart JJ. Management of sympathetic overactivity in tetanus with epidural bupivacaine and sufentanil. Experience with 11 patients. *Crit Care Med* 1999;**27**:1721–5.
151. Barker LE, Luman ET. Changes in vaccination coverage estimates among children aged 19–35 months in the United States, 1996–1999. *Am J Prev Med* 2001;**20**:28–31.
152. Leis AA. Silent period studies and long latency reflexes. In: Levin KH, Luders HO, editors. *Comprehensive Clinical Neurophysiology*. Philadelphia: WB Saunders; 2000. pp. 295–306.
153. Warren JD, Kimber TE, Thompson PD. The silent period after magnetic brain stimulation in generalized tetanus. *Muscle Nerve* 1999;**22**:1590–2.
154. WHO Expert Consultation on Rabies—First report, Geneva, World Health Organization; 2004. WHO Technical Report Series No. 931.
155. Grattan-Smith PJ, O'Regan WJ, Ellis PS, O'Flaherty SJ, McIntyre PB, Barnes CJ. Rabies: A second Australian case, with a long incubation period. *Med J Aust* 1992;**156**:651–4.
156. Jackson AC. Rabies. *Neurol Clin* 2008;**26**:717–26.
157. Duckett S. *Rabies in Paediatric Neuropathology*. 1st ed. Baltimore: Williams & Wilkins; 1995. pp. 411–2.

158. Wyatt J. Rabies—update on a Global Disease. *Pediatr Infect Dis J* 2007;**26**:351–2.
159. Lakhanpal U, Sharma RC. An epidemiological study of 177 cases of human rabies. *Int J Epidemiol* 1985;**14**:614–17.
160. Anderson LJ, Nicholson KG, Tauxe RV, Winkler WG. Human rabies in the United States, 1960 to 1979: Epidemiology, diagnosis and prevention. *Ann Intern Med* 1984;**100**:728–35.
161. Basgoz N. Case records of the Massachusetts General Hospital. *N Engl J Med* 1998;**339**:105–12.
162. Adams WG. Rabies. In: Behrman RE, Kliegman R, Jenson HB, editors. *Nelson Textbook of Pediatrics*. 16th ed. Philadelphia: Saunders; 2000. pp. 1014–17.
163. Wesley AG. Infectious diseases. In: Coovadia HM, Loening WEK, editors. *Paediatrics and Child Health*. 3rd ed. Cape Town, South Africa: Oxford University Press; 1992. pp. 192–3.
164. Gomez M, Siekert RG, Herrmann EC. A case of human-skunk rabies. *JAMA* 1965;**194**:101–3.
165. Lang J, Hoa DQ, Gioi NV, Vien NC, Nguyen CV, Rouyrre N, et al. Immunogenicity and safety of low dose intradermal rabies vaccination given during an expanded programme on immunization session in Vietnam: Results of a comparative randomized trial. *Trans R Soc Trop Med Hyg* 1999;**93**:208–13.
166. Noah DL, Drenzek CL, Smith JS, Krebs JW, Orciari L, Shaddock J, et al. Epidemiology of human rabies in the United States, 1980 to 1996. *Ann Intern Med* 1998;**128**:922–30.
167. Debbie JG, Trimarchi CV. Prophylaxis for suspected exposure to bat rabies. *Lancet* 1997;**350**:1790–1.
168. Ledwith M, Baltimore RS. Management of rabies exposure in pediatric practice. *Curr Opin Pediatr* 1997;**9**:478–82.
169. Rupprecht CE, Briggs D, Brown CM, Franka R, Katz SL, Kerr HD, et al. Use of a reduced (4-dose) vaccine schedule for postexposure prophylaxis to prevent human rabies. Recommendations of the Advisory Committee on Immunization Practices. *MMWR Morbidity and Mortality Weekly Report* 2010;**59**(RR-2):1–12.
170. Helmick CG, Tauxe RV, Vernon AA. Is there a risk to contact with patients with rabies? *Rev Infect Dis* 1987;**9**:511.
171. Willoughby Jr RE, Tieves KS, Hoffman GM, Ghanayem NS, Amlie-Lefond CM, Schwabe MJ, et al. Survival after treatment of rabies with induction of coma. *N Engl J Med* 2005;**352**: 2508–14.
172. Hattwick MA, Weis TT, Stechschulte CJ, Baer GM, Gregg MB. Recovery from rabies: A case report. *Ann Intern Med* 1972;**76**:931–42.
173. Gromeier M, Solecki D, Patel DD, Wimmer E. Expression of the human poliovirus receptor/CD155 gene during development of the central nervous system: Implications for the pathogenesis of poliomyelitis. *Virology* 2000;**273**:248–57.
174. Jack AC. Pathogenesis. In: Jackson AC, Wunner WH, editors. *Rabies*. San Diego: Academic Press; 2002. pp. 345–83.
175. March of Dimes. Post-polio syndrome: identifying best practices in diagnosis and care. White Plains, NY: March of Dimes Birth Defects Foundation; 2001. Available from: <www.modimes.org.>
176. Dalakas MC, Elder G, Hallett M. A long-term follow-up study of patients with post-poliomyelitis neuromuscular symptoms. *N Engl J Med* 1986;**314**:959–63.
177. Cashman NR, Maselli R, Wollmann RL, Roos R, Simon R, Antel JP. Late denervation in patients with antecedent paralytic poliomyelitis. *N Engl J Med* 1987;**317**:7–12.
178. Dalakas MC, Bartfeld H, Kurland LT. The post-polio syndrome: Advances in pathogenesis and treatment. *Ann NY Acad Sci* 1995;**753**:1–409.
179. Aurlien D, Strandjord RE, Hegland O. The postpolio syndrome—a critical comment to the diagnosis. *Acta Neurol Scand* 1999; **100**:76–80.
180. Fox MJ, Madden WJ, Kohn SE. Recurrent poliomyelitis. *Am J Dis Child* 1948;**75**:395–400.
181. Blondel B, Duncan G, Couderc T, Delpeyroux F, Pavio N, Colbère-Garapin F. Molecular aspects of poliovirus biology with a special focus on the interactions with nerve cells. *J Neurovirol* 1998;**4**:1–26.
182. Gross TP, Khurana RK, Higgins T, Nkowane BS, Hirsch RL. Vaccine-associated poliomyelitis in a household contact with Netherton's syndrome receiving long-term steroid therapy. *Am J Med* 1987;**83**:797–800.
183. Centers for Disease Control (CDC). Recommendations of the advisory committee on immunization practices: Revised recommendations for routine poliomyelitis vaccination. *MMWR* 1999;**48**:590.
184. Baqui AH, Sabir AA, Begum N, Arifeen SE, Mitra SN, Black RE. Causes of childhood deaths in Bangladesh: An update. *Acta Paediatr* 2001;**90**:682–90.
185. Fishbein DB, Robinson LE. Rabies. *N Engl J Med* 1993;**329**: 1632–8.

Chapter 11

The Stiff-man Syndrome in Children and Adolescents

Jennifer A. Tracy and Andrew McKeon

INTRODUCTION

The stiff-man syndrome (SMS, also known as stiff-person syndrome) is a rare central nervous system autoimmune disease, but is likely underrecognized. This disorder has only been rarely reported in children. Most of the available literature is in case report form, and a recent series from the Mayo Clinic reported 8 cases identified over a 25-year period.[1] These patients were frequently diagnosed initially with other disorders, including dystonia, musculoskeletal disorders, and functional disorders. In this chapter, we will review the clinical features, both neurological and immunological, as well as electrophysiological findings. We will also discuss differential diagnosis of SMS among children. There will be some discussion regarding pathophysiology. Finally, we will review current treatment options (symptomatic and immunotherapeutic) and prognosis.

HISTORICAL PERSPECTIVE

In 1956 Moersch and Woltman reported 14 patients (all adults) who presented to the Mayo Clinic with chronic fluctuating truncal and limb muscle rigidity and spasms; 8 were men.[2] They encountered their first case in an Iowan farmer in 1924, and acquired the other cases over the subsequent 32 years before publishing the report. The authors named the disorder "stiff-man syndrome,"[2] noting that muscle involvement was "fairly symmetrical," the lower extremities were affected more commonly than the upper extremities, and proximal limb segments were affected more severely. In 12 patients, back and abdominal muscles were involved ("hard," "firm," "board-like") but 2 lacked truncal symptoms. There was frequently significant disability from painful spasms and falls, and many patients had fixed spinal deformities from long-standing rigidity. Chest wall spasms caused respiratory impairment in some patients. In 1963, it was reported that diazepam brought about considerable improvement in stiffness and spasms of affected patients, a phenomenon that was incorporated into diagnostic criteria for SMS.[3–5] Distinctive neurophysiological findings were also reported and localized the disorder to hyperexcitability of brainstem and spinal motor neurons.[6–10] In 1988, De Camilli and colleagues recognized that SMS was an organ-specific autoimmune disorder, recognizing seropositivity of most patients for a neuroendocrine autoantibody specific for glutamic acid decarboxylase 65-isoform (GAD65).[11] This same antibody was subsequently recognized to also be a marker of a much more common autoimmune disease, type 1 diabetes mellitus.[12]

Since the original description, more limited forms of SMS (sometimes paraneoplastic) including "stiff-limb" and "stiff-trunk" have been reported (often GAD65 Ab seronegative).[13–26] Furthermore, some with high GAD65 Ab values have other neurological disorders coexisting with SMS, including epilepsy,[11,27] cerebellar ataxia, brainstem disorders, and myelopathies.[28] Another form has been described as rapidly progressive with diffuse CNS findings known as progressive encephalomyelitis with rigidity and myoclonus (PERM).[29] These patients frequently have whole body symptoms, and eye and bulbar signs in addition to more classic SMS symptoms. Other autoantibody markers that have been discovered and aid the diagnosis of the SMS spectrum include amphiphysin-IgG (only in adults), detected in patients with paraneoplastic neurological autoimmunity associated with breast adenocarcinoma or small-cell lung carcinoma,[24,23] and glycine receptor $\alpha 1$ subunit (GlyRα)-IgG, which was first encountered in patients with the PERM phenotype but is also encountered among 20% of patients with SMS, both GAD65-IgG positive and negative cases.[30,31]

The pediatric stiff-man literature is limited to 1 series of 8 patients and 12 other reported cases, and many of the data discussed herein also pertains to the adult literature.

B.T. Darras, H. Royden Jones, Jr., M.M. Ryan & D.C. De Vivo (Eds): Neuromuscular Disorders of Infancy, Childhood and Adolescence, Second edition.
DOI: http://dx.doi.org/10.1016/B978-0-12-417044-5.00011-1

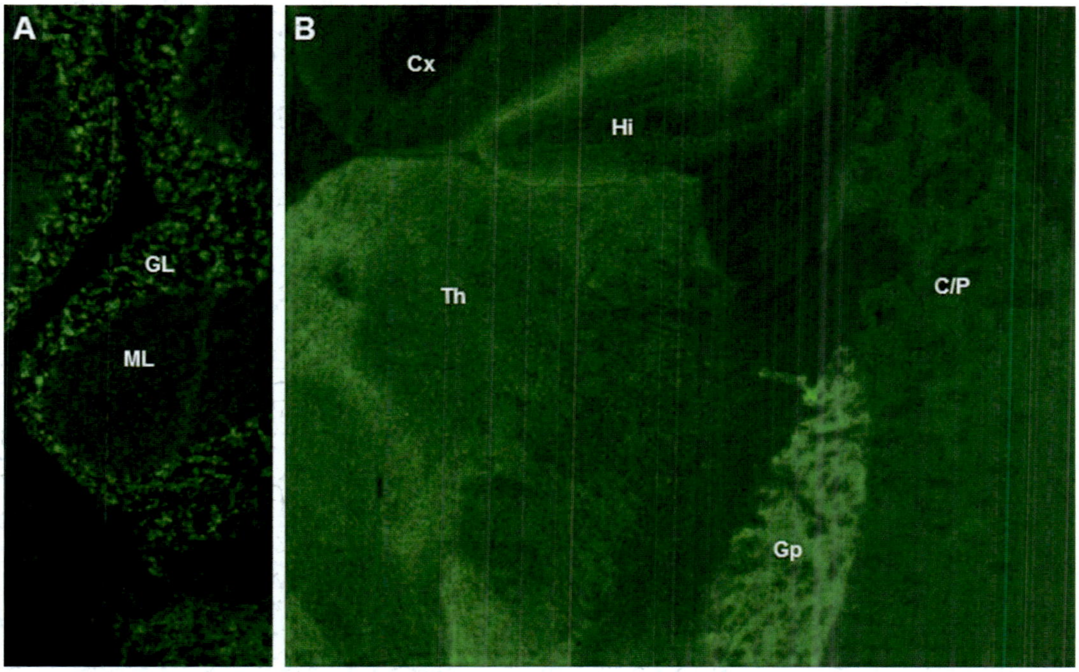

FIGURE 11.1 GAD65-IgG detected by indirect immunofluorescence on mouse cerebellum, **A,** and other brain regions (coronal section), **B.** Characteristic synaptic staining pattern is most prominent in the granular layer (GL) of cerebellum and globus pallidus (Gp). Less intensely stained regions include molecular layer of cerebellum (ML), thalamus (Th), caudate and putamen (C/P), hippocampus (Hi) and cerebral cortex (Cx). *Reproduced with permission from Lippincott, Williams and Wilkins.*[98]

EPIDEMIOLOGY

SMS is very rare, even in adults; in one review, the prevalence was estimated at approximately 1 in 1,250,000.[32] In one large series of stiff-man patients, 68% were female and 90% were white, with a median age of symptom onset at 40 years old. In that series, 67% had at least one additional autoimmune disease, 43% had type 1 diabetes mellitus and 35% had autoimmune thyroid disease.[33] The frequency of SMS in the pediatric population is unknown. Clardy et al. identified only 8 patients with pediatric-onset SMS in the Mayo Clinic from 1984 to 2012, representing 5% of the total SMS cases during that time period.[1] Of these patients, the median age of onset was 11 years, and half were female. Half of those children had type 1 diabetes mellitus. Only 12 additional individual cases of pediatric SMS have been reported in the literature.[34–45]

PATHOPHYSIOLOGY

Convergent clinical, serological, and electrophysiological evidence is indicative of SMS being an autoimmune disease of CNS inhibitory synaptic signaling. Patients have symptoms of exaggerated startle, and continuous muscle contractions, consistent with a loss of inhibitory control at a brainstem and spinal cord level. More often than not, SMS patients have a coexisting autoimmune disease, including type 1 diabetes. There is an association between SMS and the DQB1*0201 allele, which is also a susceptibility allele for type 1 diabetes mellitus.[46] Also, 80% of SMS patients are seropositive for GAD65-IgG,[33] an antibody that is also detected in 90% of patients with type 1 diabetes.[47] This serological finding implicates autoimmunity directed against GABA synthesis.[11,12] Other antigens pertinent to inhibitory synaptic transmission implicated in SMS pathophysiology include the α1 subunit of the glycine receptor, GABA receptor associated protein (GABARAP),[48] gephyrin (1 patient described),[49] and amphiphysin (relevant to paraneoplastic disorders in adults).[50]

GAD65-IgG (Figure 11.1) recognizes a cytoplasmic-facing synaptic vesicle protein, and thus is unlikely to be pathogenic.[11] Consistent with this, responses to antibody-depleting therapies in SMS patients seropositive for GAD65-IgG are variable, and post-therapy normalization is uncommon.[33,51] IgGs specific for cytoplasmic and nuclear autoantigens are considered surrogate markers of inflammatory organ-specific autoimmune disorders mediated by cytotoxic T cells, a hypothesis supported by autopsy findings in some SMS patients.[20] In contrast, the glycine receptor is a cell surface antigen, and thus is accessible to antibody. Although the pathogenicity of GlyRα1-IgG is yet unproven (Figure 11.2), detection of this may predict immunotherapy-responsiveness. In one series, immunotherapy responses were noted more frequently in GlyRα1-IgG-positive SMS

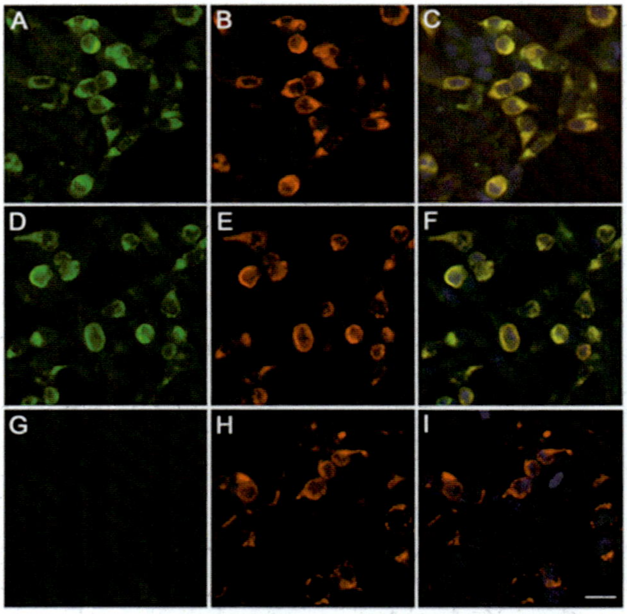

FIGURE 11.2 GlyRα1-IgG demonstrated in HEK293 cells expressing the human GlyRα1 subunit. The reactivity of serum from two representative patients with HEK293 cells expressing the human GlyRα1 subunit is shown in panels **A** and **D**. Serum from a control patient lacks reactivity with transfected cells (**G**). Panels **B**, **E** and **H** show the reactivity with a commercial GlyRα1 monoclonal antibody. Note that the immunostaining co-localizes with that of patient antibodies (**C** and **F**, but not **I**). Scale bar, 20 μm. *Reproduced with permission from the American Medical Association.*[31]

patients (6 of 7 improved) than in seronegative patients (7 of 25 improved).[31] Similar responses to immunotherapy have also been documented for other neurological disorders characterized serologically by IgGs that target neuronal cell surface proteins.[52] GABARAP antibody is detected primarily in patients who are also GAD65-IgG seropositive.[48]

Neurophysiology studies of SMS confirm a CNS hyperexcitable phenotype with impaired inhibition of H-reflexes[10] and excessive, poorly habituating motor activity in response to acoustic startle.[8]

NEUROLOGICAL PRESENTATION

Symptom onset is characteristically acute or subacute in all (see Case Examples 11.1–11.3). Lower extremity stiffness or spasms causing gait difficulties are the most common initial symptoms.[2] Other axial presenting symptoms include isolated severe lumbar paraspinal muscle spasms resulting in marked lumbar lordosis, intrascapular muscle spasms, and midthoracic back pain. Reported presentations also include exaggerated startle, limb muscle cramps, and breathing difficulties when exercising. Respiratory failure due to chest wall spasms is a rare

CASE EXAMPLE 11.1

A 7-year-old African American girl with a history of type 1 diabetes mellitus (diagnosed at age 3) developed right foot pain, followed by pain through the remainder of the right lower extremity, followed by the left lower extremity over a 3-week period. She had associated muscle stiffness and frequent muscle spasms, rendering her unable to walk. She was found to have GAD65 antibody positivity (285 nmol/L; normal value, ≤0.02 nmol/L). Diazepam provided no significant improvement. She had initial improvements with intravenous immune globulin (IVIg), which were unsustained. She was treated with baclofen, and then with plasma exchange, which gave significant benefit along with baclofen. Examination after plasma exchange and baclofen was normal aside from mild decreased distal pinprick sensation (EMG under conscious sedation was normal).

CASE EXAMPLE 11.2

A 17-year-old male patient, previously very active, developed severe aching pain and spasms in the musculature of his thoracic and lumbar back 3 years prior to presentation. The discomfort was constant, with superimposed exacerbations lasting up to 3 hours associated with increased physical activity. He had received multiple treatments with muscle relaxants, benzodiazepines, gabapentin, amitriptyline, and botulinum toxin, with incomplete response, though he found baclofen (at a total of 60 mg/day) and botulinum toxin to be most helpful. On examination, he had tenderness as well as tight upper and midthoracic muscles; on EMG, he was unable to fully relax thoracic paraspinal muscles. His GAD65 serum antibody level was found to be elevated at 0.06 nmol/L. He was diagnosed with SMS and was treated with IVIg, which was helpful, followed by prednisone and plasma exchange (in addition to symptomatic therapies) with marked improvement.

CASE EXAMPLE 11.3

A 14-year-old girl with a history of both type 1 diabetes mellitus and autoimmune thyroid disease had begun to walk late at 17 months, and since that time had been noted to walk abnormally in relation to her right leg. She then developed right knee pain 3 years prior to presentation. Six months prior to presentation, her right foot started to turn inward. She was found on examination to have hyperreflexia, worse on the right than on the left, and rigidity of the right lower extremity with slightly decreased muscle bulk. Trials of carbidopa/levodopa and tizanidine therapies did not help her symptoms. An EMG under conscious sedation was normal. GAD65 antibodies were significantly elevated, at 3847 nmol/L. She was subsequently treated with IVIg with significant improvement, though its use was limited by side effects. Ultimately, she had surgery for correction of her right foot deformity.

complication. Abdominal pain due to muscle spasms is also common. Emotional stress, startle and anxiety may precipitate or worsen symptoms, and can lead to falls. One patient has been reported to have exaggerated startle and limb symptoms brought on by being asked to speak in front of the class at school.[1] This frequently resulted in falling over "stiff as a board" without protective reflexes, resulting in facial injury.

The principal exam findings are of stiffness (increased muscle tone) in affected parts of the body, usually limbs and lower back. The classic exaggerated lumbar lordosis should not resolve completely with forward flexion or lying supine. Paraspinal muscles may be hypertrophied and tense to palpation. At initial evaluation, myelopathic findings (usually isolated brisk deep tendon reflexes) are documented in over half of patients. Rare manifestations include a focal jerking limb.[53] Coexisting neurological disorders (reported only in adults to date) include epilepsy, parkinsonism, and peripheral neuropathy (mainly in those with diabetes mellitus).[28,33]

Three main phenotypes are recognized. Classic SMS patients have stiffness and spasms largely confined to the low back and proximal lower extremities. Variants may include patients with stiffness and spasms confined to one lower extremity ("stiff-limb") or the axis ("stiff-trunk"), and patients with whole body spasms.[14,33] This latter group of patients includes those with progressive encephalomyelitis with rigidity and myoclonus (PERM), a severe disorder characterized by one or more of the following: encephalopathy, brainstem signs, and myelopathy, as well as axis and limbs stiffness and spasms.[29] All three phenotypes have been reported in children.[1]

COEXISTING AUTOIMMUNE DISEASES

Other autoimmune diseases are commonly encountered among patients with GAD65-IgG positive SMS, particularly type 1 diabetes and autoimmune thyroid disease. In one large series of adults and children, 67% had ≥1 coexisting autoimmune disease.[33] The onset of diabetes may predate or postdate the onset of SMS. Among the Mayo pediatric SMS series, 5 of 8 patients had a coexisting autoimmune disorder and all were GAD65-IgG positive. Four patients had type 1 diabetes mellitus, 2 patients had thyroid disease and 2 patients had vitiligo. Of the 4 pediatric SMS patients with diabetes mellitus, the diagnosis of diabetes preceded the onset of SMS symptoms in 2 patients, and in 2 patients diabetes was diagnosed after the first onset of SMS.[1] Other autoimmune diseases both organ specific (premature ovarian failure, Addison disease, Sjögren's syndrome, and celiac disease) and non-organ specific (systemic lupus erythematosus) have been reported to coexist with SMS.[33] In contrast, just 4 of 12 pediatric SMS patients reported individually in the literature had GAD65-IgG positivity and just one had a coexisting autoimmune disease (type 1 diabetes).

In the Mayo Clinic pediatric series, serological testing revealed GAD65-IgG in 7 patients (median value, 754 nmol/L; range, 0.06−3847 nmol/L; normal value 0.00−0.02 nmol/L). GlyRα1-IgG was detected in 3 of 8 patients (1/7 was positive in serum, 2/3 were positive in CSF). One or more other autoantibody specificities were detected in 3 patients: antinuclear antibody, 2; thyroid peroxidase antibodies, 2; insulin antibody, 1.[1]

ONCOLOGICAL FINDINGS

Cancer has not been reported among pediatric patients with SMS. Indeed cancer has only been rarely reported among adult patients with GAD65-IgG positive SMS. In the Mayo series, among 99 patients, both adults and children, just 3 had carcinoma (thyroid, renal cell, and colon).[33] Other rare single case reports among GAD65 antibody-positive SMS patients include carcinoma of kidney[54] and breast,[55] neuroendocrine carcinoma,[56] and thymoma.[26] Among GlyRα1-IgG-positive patients, only a minority (1 of 10 in 1 series) have cancer detected. Reported oncological associations include Hodgkin lymphoma,[31] thymoma,[30] and lung cancer.[19]

DIFFERENTIAL DIAGNOSIS

Disorders that increase muscle tone, whether producing spasticity or rigidity, can be mistaken for SMS. One of these disorders is hereditary hyperekplexia, a disease that is present at birth and is characterized by baseline increased muscle tone, which disappears during sleep. Hyperekplexia is also characterized by an overactive startle response, resulting in increased rigidity. In most cases, symptoms improve markedly after the first year of life, but residual milder symptoms, including rigidity in response to startle, can persist throughout life.[57] This is a genetic disorder; the most common gene mutation is in the *GLRA1* gene, which encodes the α1 subunit of the glycine receptor. These mutations uncouple receptor binding from the chloride channel function, disrupting normal neuronal inhibitory pathways.[58] The other identified genes are *SLC6A5, GLRB, GPHN,* and *ARHGEF9*.[59]

Dystonic disorders may have some similarities to SMS. Dopa-responsive dystonia (Segawa's disease) is due to a defect in the GTP-cyclohydrolase I gene. Early presentation includes an abnormal gait with posturing of a lower extremity, which can worsen later in the day and with exercise. Over time, more widespread clinical features with parkinsonism become more apparent, and can be helpful in distinguishing this entity. One large study of 66 patients reported the mean age of onset as 6.5 years.[60] Identification of this disorder is important, given the marked symptomatic response to levodopa in these patients. Spinal fluid evaluation, which shows low homovanillic acid, biopterin, and neopterin levels (dopamine metabolites) can be useful when this diagnosis is suspected. Early-onset primary dystonia (DYT1) can present with limb posturing and eventually progress to other limbs in a more generalized form.[61] This is an autosomal dominant disorder, caused by a mutation in the *TOR1A* gene, but has low penetrance (~30%), so a clear family history may not be present.[62]

Hereditary spastic paraparesis (HSP) is a familial condition which results in spasticity, predominantly involving the lower extremities. There are multiple genes identified, and the pattern of inheritance may be autosomal dominant, autosomal recessive, or X-linked. This can present in childhood, and an SPG3 mutation should be considered in autosomal dominant childhood onset HSP; this can also be associated with peripheral neuropathy, optic atrophy, and seizure activity.[63] Isaac's syndrome is characterized by muscle stiffness, often involving the trunk as well as limbs; this usually presents in adulthood but can also be seen in childhood, and can be associated with voltage-gated potassium channel complex antibodies.[64] Schwartz-Jampel syndrome is an autosomal recessive genetic disorder that can result in muscle stiffness, usually presenting within the first few years of life. It is typically associated with skeletal and cartilage deformities and a dysmorphic appearance, which can be an important clue to diagnosis. EMG can show continuous motor activity and myotonia. This is caused by a defect in the *HSPG2* (perlecan) gene.[65]

Tetanus is a rare disease in the United States because of widespread vaccination, but can occur after infection of even a minor wound by *Clostridium tetani*. This disease can result in spasms of muscles, including trismus, which can be stimulus sensitive. Patients can have opisthotonus, with profound spasms of paraspinal muscles. Infection can also be accompanied by fever and autonomic instability.[66] A very early case in the literature of SMS in a child was initially felt to represent recurrent tetanus.[34] Dystonic drug reactions should be considered if there is a time link to drug exposure. Strychnine poisoning (primarily from exposure to rodenticides) could be considered. Strychnine inhibits glycine receptors, leading to muscle spasms including opisthotonus, which can be stimulus sensitive.[67]

TESTING

Neurophysiological Findings

Electrophysiological studies performed in specialized neurophysiology laboratories usually reveal evidence of brainstem and spinal cord hyperexcitability. Exaggerated acoustic startle reflexes can be detected by using surface EMG to study the pattern and habituation of motor responses after applying binaural 105 dB intensity stimuli, 1 minute apart over 3–5 trials.[8] Compared to normals, SMS patients demonstrate excessive, poorly habituating motor activity predominately in axial and leg muscles. Similarly, exteroceptive (cutaneomuscular) responses are enhanced, habituate poorly, and are not found in normal subjects.[10] The response can be widespread and detected at anatomical sites remote from the site of stimulation. This reflex is elicited by stimulation of peripheral nerve (such as the medial plantar nerve), measuring the response with surface EMG in limb segments, paraspinal muscles, and cranial segments. Continuous motor unit activity is also characteristic of SMS, is predominantly found in axial muscles, and can be recorded through concentric needle EMG on clinically affected muscles.[6] Grouped irregular discharges of many motor units have been reported among some patients with stiff-limb syndrome.[13] Sensitivity of electrophysiological testing may be diminished by GABAergic treatments, including benzodiazepines and baclofen.

Serological Testing

In general, serum testing for GAD65-IgG should be undertaken; this will be positive in 80% of patients, and usually at a value of >20 nmol/L. For those patients who are GAD65-IgG seronegative, serum and CSF could be submitted for GlyRα1-IgG testing. At the time of writing, only two neuroimmunology laboratories, both academic, are offering this test: the University of Barcelona (Josep Dalmau) and Oxford University (Angela Vincent).

TREATMENT

Symptomatic Treatment

The data available are mainly in relation to adults with SMS, and this data will be discussed first, with specific information relating to the pediatric experience described at the end of this section.

One of the mainstays of symptomatic management in SMS is the use of benzodiazepines, particularly diazepam.[3,4,68] Many patients require high doses for symptomatic control. In the McKeon et al. study, a median dosage of 40 mg/day of diazepam was used for symptomatic management.[33]

Other symptomatic treatment options include muscle relaxants such as baclofen, to which good responses have been reported.[69,70] However, because of the side effect profile and poor CNS penetration, baclofen has also been administered intrathecally in some patients. A placebo-controlled trial of 3 patients with SMS, treating with either intrathecal baclofen or placebo on consecutive days, found that patients had an improvement in reflex EMG activity after intrathecal baclofen.[71] Stayer et al. reviewed the course of 8 patients with SMS for whom oral symptomatic medications were not sufficiently effective, either because of intolerable side effects or lack of clinical response, and who were given intrathecal baclofen pumps.[72] These patients were reported to have improvement of their rigidity and mobility, but no complete remissions. Stayer and co-workers also noted that a single patient died from pump failure (with dose reduction).

Tizanidine,[73] gabapentin,[74] pregabalin,[75] valproate,[76] levetiracetam,[77,78] and vigabatrin[79] have also been reported to provide varying degrees of symptomatic benefit. An SMS patient with severe leg spasms insufficiently responsive to multiple medications was treated with intravenous propofol with good response prior to intrathecal baclofen pump placement.[80] Botulinum toxin introduced into affected muscles has been shown to be beneficial in some patients.[81–84]

The pediatric literature regarding symptomatic responsiveness in SMS is limited by the small numbers of patients, but similar treatment strategies have been used in children. In the Clardy et al. study, all 6 patients treated with diazepam (median total dose 14 mg/day) and 3/3 treated with baclofen (median total dose 60 mg/day) received some symptomatic benefit.[1] An 8-year-old patient with SMS was reported to have some benefit from diazepam (up to 60 mg/day) and baclofen,[36] and a 16-year-old boy with SMS was reported as showing marked symptomatic improvement with diazepam (up to 100 mg/day).[37] A 13-year-old girl with SMS received some benefit from baclofen and intravenous diazepam.[38] A 14-month-old girl identified received marked improvement with diazepam and baclofen, and later had sustained improvement with valproate and steroids added to her regimen.[39] Another 6-year-old boy responded well to diazepam (up to 22 mg/day) and baclofen.[40] An 11-year-old boy improved with diazepam (2 mg three times a day).[41]

Immunotherapy

Most of the available data regarding immunotherapy for SMS derives from the adult literature. Again, the adult data will be discussed first, with specific reference to pediatric data at the end of this section Intravenous immunoglobulin treatment has been a mainstay of treatment for SMS. Dalakas et al. performed a randomized placebo-controlled trial of IVIg in 16 GAD65-antibody positive stiff-person syndrome patients; they received either IVIg (2 g/kg on two days per month for three months) or placebo followed by a washout period and then the alternate treatment.[51] There was a significant benefit from IVIg treatment in the distribution-of-stiffness score (a scale indicating the presence or absence of stiffness in multiple body regions) and 11 of the patients clinically improved after IVIg use. Plasmapheresis has been reported to be effective in treatment of SMS in several case reports.[85–89]

Rituximab is a monoclonal antibody that binds to the CD20 antigen of B-cells, resulting in B-cell depletion. There are some individual reports of SMS responsiveness to rituximab,[90–95] but these results were not confirmed in a controlled trial.[96] This randomized placebo-controlled trial of 24 patients with SMS did not show benefit from rituximab (2 doses of 1 gram, two weeks apart) in stiffness index, though 4 of the 12 patients who received rituximab were judged to have a clear improvement of function. It was postulated that a strong placebo effect may have compromised the ability to detect improvement in this study.[96] Venhoff et al. treated identical twins with SMS and autoimmune thyroiditis with rituximab and no improvement occurred.[97]

In the Clardy et al. pediatric SMS study, 3 out of 4 patients treated with IVIg, 3 out of 4 patients treated with plasmapheresis, and 1 patient treated with prednisone were treatment-responsive.[1] A 7-year-old girl described as having stiff-person syndrome, but with the atypical feature of striatal lesions on diffusion weighted MRI imaging and being GAD antibody negative, had good response to steroids and IVIg.[43] A 12-year-old boy with SMS was reported to have mild improvement with benzodiazepines and IVIg, no benefit from baclofen or gabapentin, but marked clinical improvement with rituximab (2 treatments of 500 mg/meter squared, 2 weeks apart).[44] A 14-month-old girl with PERM and glycine receptor antibodies received benefit from IVIg and intravenous steroids. After discontinuation, she relapsed and responded again to IV steroids, IVIg, and also IV levetiracetam, and has been reported to be doing well on maintenance oral prednisolone and IVIg treatments.[45]

PROGNOSIS

The prognosis of SMS is variable, but appears linked to the specific subtype of SMS. Regarding adult data, Barker et al. studied 23 adult patients with rigidity and spasms in association with continuous motor unit activity, and reported that the 2 patients with progressive encephalomyelitis with rigidity in their group had a rapid course with death within 6 and 16 weeks.[13] The classical SMS group (8 patients) all had good response to symptomatic medications and remained ambulatory (with the exception of a single patient who used a wheelchair when outdoors). The stiff-limb variant (13 patients) had a more frequent relapsing-remitting course and 6 of these became wheelchair-bound after a mean length of time of 3.5 years. Of the 8 patients described in the Clardy et al. study, all patients treated with diazepam or baclofen had improvement and most improved with immunotherapy.[1]

In that series, half of the patients had mild or no symptoms and half had moderate or severe symptoms at last follow-up.

CONCLUSION

SMS is a rare disorder among children and teens. Its distinction from the plethora of pediatric myelopathies, dystonias, and peripheral nervous system disorders is facilitated by recognition of the hyperexcitable clinical and electrophysiological phenotypes, supported by clinical and serological evidence of autoimmunity. Responses to GABAergic drugs and immunotherapies may also be diagnostic as well as therapeutic.

REFERENCES

1. Clardy SL, Lennon VA, Dalmau J, Pittock SJ, Jones Jr. HR, Renaud DL, et al. Childhood onset of stiff-man syndrome. *JAMA Neurol* 2013;**70**(12):1531–6.
2. Moersch FP, Woltman HW. Progressive fluctuating muscular rigidity and spasm ("stiff-man" syndrome): report of a case and some observations in 13 other cases. *Proc Staff Meet Mayo Clin* 1956;**31**:421–7.
3. Howard Jr. FM. A new and effective drug in the treatment of the stiff-man syndrome: preliminary report. *Proc Staff Meet Mayo Clin* 1963;**38**:203–12.
4. Olafson RA, Mulder DW, Howard FM. "Stiff-Man" Syndrome: a review of the literature, report of three additional cases and discussion of pathophysiology and therapy. *Mayo Clin Proc* 1964;**39**:131–44.
5. Lorish TR, Thorsteinsson G, Howard Jr. FM. Stiff-man syndrome updated. *Mayo Clin Proc* 1989;**64**:629–36.
6. Armon C, McEvoy KM, Westmoreland BF, McManis PG. Clinical neurophysiologic studies in stiff-man syndrome: use of simultaneous video-electroencephalographic-surface electromyographic recording. *Mayo Clin Proc* 1990;**65**:960–7.
7. Mamoli B, Heiss WD, Maida E, Podreka I. Electrophysiological studies on the "stiff-man" syndrome. *J Neurol* 1977;**217**:111–21.
8. Matsumoto JY, Caviness JN, McEvoy KM. The acoustic startle reflex in stiff-man syndrome. *Neurology* 1994;**44**:1952–5.
9. Molloy FM, Dalakas MC, Floeter MK. Increased brainstem excitability in stiff-person syndrome. *Neurology* 2002;**59**:449–51.
10. Floeter MK, Valls-Sole J, Toro C, Jacobowitz D, Hallett M. Physiologic studies of spinal inhibitory circuits in patients with stiff-person syndrome. *Neurology* 1998;**51**:85–93.
11. Solimena M, Folli F, Denis-Donini S, Comi GC, Pozza G, De Camilli P, et al. Autoantibodies to glutamic acid decarboxylase in a patient with stiff-man syndrome, epilepsy, and type I diabetes mellitus. *N Engl J Med* 1988;**318**:1012–20.
12. Solimena M, Folli F, Aparisi R, Pozza G, De Camilli P. Autoantibodies to GABA-ergic neurons and pancreatic beta cells in stiff-man syndrome. *N Engl J Med* 1990;**322**:1555–60.
13. Barker RA, Revesz T, Thom M, Marsden CD, Brown P. Review of 23 patients affected by the stiff man syndrome: clinical subdivision into stiff trunk (man) syndrome, stiff limb syndrome, and progressive encephalomyelitis with rigidity. *J Neurol Neurosurg Psychiatry* 1998;**65**:633–40.
14. Brown P, Marsden CD. The stiff man and stiff man plus syndromes. *J Neurol* 1999;**246**:648–52.
15. Brown P, Rothwell JC, Marsden CD. The stiff leg syndrome. *J Neurol Neurosurg Psychiatry* 1997;**62**:31–7.
16. Essalmi L, Meaux-Ruault N, Hafsaoui C, Gil H, Curlier E, Dupond JL. [Stiff person syndrome associated with thymoma. Efficacy of thymectomy]. *Rev Med Interne* 2007;**28**:627–30.
17. Gutmann B, Crivellaro C, Mitterer M, Zingerle H, Egarter-Vigl E, Wiedermann CJ. Paraneoplastic stiff-person syndrome, heterotopic soft tissue ossification and gonarthritis in a HLA B27-positive woman preceding the diagnosis of Hodgkin's lymphoma. *Haematologica* 2006;**91**:ECR59.
18. Iwata T, Inoue K, Mizuguchi S, Morita R, Tsukioka T, Suehiro S. Thymectomy for paraneoplastic stiff-person syndrome associated with invasive thymoma. *J Thorac Cardiovasc Surg* 2006;**132**:196–7.
19. Liu YL, Lo WC, Tseng CH, Tsai CH, Yang YW. Reversible stiff person syndrome presenting as an initial symptom in a patient with colon adenocarcinoma. *Acta Oncol* 2010;**49**:271–2.
20. McCabe DJ, Turner NC, Chao D, Leff A, Gregson NA, Womersley HJ, et al. Paraneoplastic "stiff person syndrome" with metastatic adenocarcinoma and anti-ri antibodies. *Neurology* 2004;**62**:1402–4.
21. McEvoy KM, Lennon VA. Stiffman syndrome: clinical aspects and anti-islet cell antibodies as a disease marker. In: Layzer RB, Sandrini G, Piccolo G, Martinelli P, editors. *Motor Unit Hyperactivity States*. New York: Raven Press; 1993. pp. 45–52.
22. Murinson BB, Guarnaccia JB. Stiff-person syndrome with amphiphysin antibodies: distinctive features of a rare disease. *Neurology* 2008;**71**:1955–8.
23. Pittock SJ, Lucchinetti CF, Parisi JE, Benarroch EE, Mokri B, Stephan CL, et al. Amphiphysin autoimmunity: paraneoplastic accompaniments. *Ann Neurol* 2005;**58**:96–107.
24. Rosin L, DeCamilli P, Butler M, Solimena M, Schmitt HP, Morgenthaler N, et al. Stiff-man syndrome in a woman with breast cancer: an uncommon central nervous system paraneoplastic syndrome. *Neurology* 1998;**50**:94–8.
25. Toothaker TB, Rubin M. Paraneoplastic neurological syndromes: a review. *Neurologist* 2009;**15**:21–33.
26. Vernino S, Lennon VA. Autoantibody profiles and neurological correlations of thymoma. *Clin Cancer Res* 2004;**10**:7270–5.
27. Quek AM, Britton JW, McKeon A, So E, Lennon VA, Shin C, et al. Autoimmune epilepsy: clinical characteristics and response to immunotherapy. *Arch Neurol* 2012;**69**:582–93.
28. Pittock SJ, Yoshikawa H, Ahlskog JE, Tisch SH, Benarroch EE, Kryzer TJ, et al. Glutamic acid decarboxylase autoimmunity with brainstem, extrapyramidal, and spinal cord dysfunction. *Mayo Clin Proc* 2006;**81**:1207–14.
29. Whiteley AM, Swash M, Urich H. Progressive encephalomyelitis with rigidity. *Brain* 1976;**99**:27–42.
30. Clerinx K, Breban T, Schrooten M, Leite MI, Vincent A, Verschakelen J, et al. Progressive encephalomyelitis with rigidity and myoclonus: resolution after thymectomy. *Neurology* 2011;**76**:303–4.
31. McKeon A, Martinez-Hernandez E, Lancaster E, Matsumoto JY, Harvey RJ, McEvoy KM, et al. Glycine receptor autoimmune spectrum with stiff-man syndrome phenotype. *JAMA Neurol* 2013;**70**:44–50.

32. Espay AJ, Chen R. Rigidity and spasms from autoimmune encephalomyelopathies: stiff-person syndrome. *Muscle Nerve* 2006;**34**:677–90.
33. McKeon A, Robinson MT, McEvoy KM, Matsumoto JY, Lennon VA, Ahlskog JE, et al. Stiff-man syndrome and variants: clinical course, treatments, and outcomes. *Arch Neurol* 2012;**69**:230–8.
34. Bowler D. The 'stiff man syndrome' in a boy. *Arch Dis Child* 1960;**35**:289–92.
35. Kugelmass IN. Stiff-man syndrome in a child. *NY State J Med* 1961;**61**:2483–7.
36. Isaacs H. Stiff man syndrome in a black girl. *J Neurol Neurosurg Psychiatry* 1979;**42**:988–94.
37. Daras M, Spiro AJ. 'Stiff-man syndrome' in an adolescent. *Pediatrics* 1981;**67**:725–6.
38. Maccario M, Baugh JR, Mena H. Sudden death in Moersch-Woltman. *Neurology* 1984;**34**:407.
39. Udani VP, Dharnidharka VR, Gajendragadkar AR, Udani SV. Sporadic stiffman syndrome in a young girl. *Pediatr Neurol* 1997;**17**:58–60.
40. Garzo C, Perez-Sotelo M, Traba A, Esteban A, Grandas F, Munoz-Blanco JL. Stiff-man syndrome in a child. *Mov Disord* 1998;**13**:365–8.
41. Markandeyulu V, Joseph TP, Solomon T, Jacob J, Kumar S, Gnanamuthu C. Stiff-man syndrome in childhood. *J R Soc Med* 2001;**94**:296–7.
42. Mikaeloff Y, Jambaque I, Mayer M, Ponsot G, Kalifa G, Carel JC. Benefit of intravenous immunoglobulin in autoimmune stiff-person syndrome in a child. *J Pediatr* 2001;**139**:340.
43. Sanefuji M, Torisu H, Kira R, Yamashita H, Ejima K, Shigeto H, et al. A case of childhood stiff-person syndrome with striatal lesions: a possible entity distinct from the classical adult form. *Brain Dev* 2013;**35**:575–8.
44. Fekete R, Jankovic J. Childhood stiff-person syndrome improved with rituximab. *Case Rep Neurol* 2012;**4**:92–6.
45. Damasio J, Leite MI, Coutinho E, Waters P, Woodhall M, Santos MA, et al. Progressive encephalomyelitis with rigidity and myoclonus: the first pediatric case with glycine receptor antibodies. *JAMA neurology* 2013;**70**:498–501.
46. Pugliese A, Solimena M, Awdeh ZL, Alper CA, Bugawan T, Erlich HA. Association of HLA-DQB1*0201 with stiff-man syndrome. *J Clin Endocrinol Metab* 1993;**77**:1550–3.
47. Bingley PJ. Clinical applications of diabetes antibody testing. *J Clin Endocrinol Metab* 2010;**95**:25–33.
48. Raju R, Rakocevic G, Chen Z, Hoehn G, Semino-Mora C, Shi W, et al. Autoimmunity to GABAA-receptor-associated protein in stiff-person syndrome. *Brain* 2006;**129**:3270–6.
49. Butler MH, Hayashi A, Ohkoshi N, Villmann C, Becker CM, Feng G, et al. Autoimmunity to gephyrin in Stiff-Man syndrome. *Neuron* 2000;**26**:307–12.
50. De Camilli P, Thomas A, Cofiell R, Folli F, Lichte B, Piccolo G, et al. The synaptic vesicle-associated protein amphiphysin is the 128-kD autoantigen of stiff-man syndrome with breast cancer. *J Exp Med* 1993;**178**:2219–23.
51. Dalakas MC, Fujii M, Li M, Lutfi B, Kyhos J, McElroy B. High-dose intravenous immune globulin for stiff-person syndrome. *N Engl J Med* 2001;**345**:1870–6.
52. Dalmau J, Tuzun E, Wu HY, Masjuan J, Rossi JE, Voloschin A, et al. Paraneoplastic anti-N-methyl-D-aspartate receptor encephalitis associated with ovarian teratoma. *Ann Neurol* 2007;**61**:25–36.
53. Leigh PN, Rothwell JC, Traub M, Marsden CD. A patient with reflex myoclonus and muscle rigidity: "jerking stiff-man syndrome". *J Neurol Neurosurg Psychiatry* 1980;**43**:1125–31.
54. McHugh JC, Murray B, Renganathan R, Connolly S, Lynch T GAD. Antibody positive paraneoplastic stiff person syndrome in a patient with renal cell carcinoma. *Mov Disord* 2007;**22**:1343–6.
55. Agarwal PA, Ichaporia NR. Glutamic acid decarboxylase antibody-positive paraneoplastic stiff limb syndrome associated with carcinoma of the breast. *Neurol India* 2010;**58**:449–51.
56. Saiz A, Blanco Y, Sabater L, Gonzalez F, Bataller L, Casamitjana R, et al. Spectrum of neurological syndromes associated with glutamic acid decarboxylase antibodies: diagnostic clues for this association. *Brain* 2008;**131**:2553–63.
57. Bakker MJ, van Dijk JG, van den Maagdenberg AM, Tijssen MA. Startle syndromes. *Lancet Neurology* 2006;**5**:513–24.
58. Zhou L, Chillag KL, Nigro MA. Hyperekplexia: a treatable neurogenetic disease. *Brain Dev* 2002;**24**:669–74.
59. Tijssen M, Rees MI. Hyperekplexia. *GeneReviews [Internet]* 2007.
60. Nygaard TG, Marsden CD, Fahn S. Dopa-responsive dystonia: long-term treatment response and prognosis. *Neurology* 1991;**41**:174–81.
61. Albanese A, Lalli S. Update on dystonia. *Curr Opin Neurol* 2012;**25**:483–90.
62. Pauls DL, Korczyn AD. Complex segregation analysis of dystonia pedigrees suggests autosomal dominant inheritance. *Neurology* 1990;**40**:1107–10.
63. Schule R, Schols L. Genetics of hereditary spastic paraplegias. *Semin Neurol* 2011;**31**:484–93.
64. Klein CJ, Lennon VA, Aston PA, McKeon A, O'Toole O, Quek A, et al. Insights from LGI1 and CASPR2 potassium channel complex autoantibody subtyping. *JAMA Neurol* 2013;**70**:229–34.
65. Ho NC, Sandusky S, Madike V, Francomano CA, Dalakas MC. Clinico-pathogenetic findings and management of chondrodystrophic myotonia (Schwartz-Jampel syndrome): a case report. *BMC Neurology* 2003;**3**:3.
66. Brook I. Tetanus in children. *Pediatr Emerg Care* 2004;**20**:48–51.
67. Smith BA. Strychnine poisoning. *J Emerg Med* 1990;**8**:321–5.
68. Cohen L. Stiff-man syndrome. Two patients treated with diazepam. *JAMA* 1966;**195**:222–4.
69. Whelan JL. Baclofen in treatment of the 'stiff-man' syndrome. *Arch Neurol* 1980;**37**:600–1.
70. Miller F, Korsvik H. Baclofen in the treatment of stiff-man syndrome. *Ann Neurol* 1981;**9**:511–12.
71. Silbert PL, Matsumoto JY, McManis PG, Stolp-Smith KA, Elliott BA, McEvoy KM. Intrathecal baclofen therapy in stiff-man syndrome: a double-blind, placebo-controlled trial. *Neurology* 1995;**45**:1893–7.
72. Stayer C, Tronnier V, Dressnandt J, Mauch E, Marquardt G, Rieke K, et al. Intrathecal baclofen therapy for stiff-man syndrome and progressive encephalomyelopathy with rigidity and myoclonus. *Neurology* 1997;**49**:1591–7.
73. Meinck HM, Conrad B. Neuropharmacological investigations in the stiff-man syndrome. *J Neurol* 1986;**233**:340–7.
74. Holmoy T. Long-term effect of gabapentin in stiff limb syndrome: a case report. *Eur Neurol* 2007;**58**:251–2.

75. Squintani G, Bovi T, Ferigo L, Musso AM, Ottaviani S, Moretto G, et al. Efficacy of pregabalin in a case of stiff-person syndrome: clinical and neurophysiological evidence. *J Neurol Sci* 2012;**314**:166–8.
76. Spehlmann R, Norcross K, Rasmus SC, Schlageter NL. Improvement of stiff-man syndrome with sodium valproate. *Neurology* 1981;**31**:1162–3.
77. Ruegg SJ, Steck AJ, Fuhr P. Levetiracetam improves paroxysmal symptoms in a patient with stiff-person syndrome. *Neurology* 2004;**62**:338.
78. Sechi G, Barrocu M, Piluzza MG, Cocco GA, Deiana GA, Sau GF. Levetiracetam in stiff-person syndrome. *J Neurol* 2008;**255**:1721–5.
79. Vermeij FH, van Doorn PA, Busch HF. Improvement of stiff-man syndrome with vigabatrin. *Lancet* 1996;**348**:612.
80. Hattan E, Angle MR, Chalk C. Unexpected benefit of propofol in stiff-person syndrome. *Neurology* 2008;**70**:1641–2.
81. Davis D, Jabbari B. Significant improvement of stiff-person syndrome after paraspinal injection of botulinum toxin A. *Mov Disord* 1993;**8**:371–3.
82. Liguori R, Cordivari C, Lugaresi E, Montagna P. Botulinum toxin A improves muscle spasms and rigidity in stiff-person syndrome. *Mov Disord* 1997;**12**:1060–3.
83. Szczepanska-Szerej A, Kulka M, Wojczal J, Stelmasiak Z. [Botulinum toxin A in the treatment of stiff man syndrome]. *Neurol Neurochir Pol* 2003;**37**(Suppl. 5):175–81.
84. Anagnostou E, Zambelis T. Botulinum toxin A in anti-GAD-positive stiff-limb syndrome. *Muscle Nerve* 2012;**46**:457–8.
85. Nakamagoe K, Ohkoshi N, Hayashi A, Hisahara S, Shoji S. [Marked clinical improvement by plasmapheresis in a patient with stiff-man syndrome: a case with a negative anti-GAD antibody]. *Rinsho Shinkeigaku* 1995;**35**:897–900.
86. Brashear HR, Phillips 2nd LH. Autoantibodies to GABAergic neurons and response to plasmapheresis in stiff-man syndrome. *Neurology* 1991;**41**:1588–92.
87. Hao W, Davis C, Hirsch IB, Eng LJ, Daniels T, Walsh D, et al. Plasmapheresis and immunosuppression in stiff-man syndrome with type 1 diabetes: a 2-year study. *J Neurol* 1999;**246**:731–5.
88. Fogan L. Progressive encephalomyelitis with rigidity responsive to plasmapheresis and immunosuppression. *Ann Neurol* 1996;**40**:451–3.
89. Vicari AM, Folli F, Pozza G, Comi GC, Comola M, Canal N, et al. Plasmapheresis in the treatment of stiff-man syndrome. *N Engl J Med* 1989;**320**:1499.
90. Lobo ME, Araujo ML, Tomaz CA, Allam N. Stiff-person syndrome treated with rituximab. *BMJ case reports 2010* 2010. PMID: 22802263.
91. Sevy A, Franques J, Chiche L, Pouget J, Attarian S. [Successful treatment with rituximab in a refractory stiff-person syndrome]. *Rev Neurol (Paris)* 2012;**168**:375–8.
92. Baker MR, Das M, Isaacs J, Fawcett PR, Bates D. Treatment of stiff person syndrome with rituximab. *J Neurol Neurosurg Psychiatry* 2005;**76**:999–1001.
93. Katoh N, Matsuda M, Ishii W, Morita H, Ikeda S. Successful treatment with rituximab in a patient with stiff-person syndrome complicated by dysthyroid ophthalmopathy. *Intern Med* 2010;**49**:237–41.
94. Dupond JL, Essalmi L, Gil H, Meaux-Ruault N, Hafsaoui C. Rituximab treatment of stiff-person syndrome in a patient with thymoma, diabetes mellitus and autoimmune thyroiditis. *J Clin Neurosci* 2010;**17**:389–91.
95. Bacorro EA, Tehrani R. Stiff-person syndrome: persistent elevation of glutamic acid decarboxylase antibodies despite successful treatment with rituximab. *J Clin Rheumatol* 2010;**16**:237–9.
96. Dalakas M, Rakocevic G, Dambrosia J, McElroy B. A double-blind, placebo-controlled study of rituximab in patients with stiff-person syndrome (SPS). *Ann Neurol* 2009;**66**:S21.
97. Venhoff N, Rizzi M, Salzer U, Bossaller L, Thoden J, Eibel H, et al. Monozygotic twins with stiff person syndrome and autoimmune thyroiditis: rituximab inefficacy in a double-blind, randomised, placebo controlled crossover study. *Ann Rheum Dis* 2009;**68**:1506–8.
98. O'Toole O, Lennon VA, Ahlskog JE, Matsumoto JY, Pittock SJ, Bower J, et al. Autoimmune chorea in adults. *Neurology* 2013;**80**:1133–44.

Part IV

Peripheral Nerve Disorders

Chapter 12

Radiculopathies and Plexopathies

H. Royden Jones, Jr., Monique M. Ryan, and Kerry H. Levin

ANATOMY

In total, there are 31 spinal nerve root pairs: 8 in relation to the 7 cervical vertebrae, 12 thoracic, 5 lumbar, 5 sacral, and 1 coccygeal. In general, one nerve root is adjacent to each vertebra, but the cervical spine has an extra set of nerve roots accompanying the seven cervical vertebrae. At the cervical level, the first seven roots exit the spinal column above their respective vertebrae. The C8 root exits below the seventh cervical vertebra, and all nerve roots caudal to this level leave the spinal canal below their respective vertebrae.

At each spinal level there is a ventral (anterior) and dorsal (posterior) root exit or entry zone, from or to the respective portions of the spinal cord. As shown in Figure 12.1, the anterior horn of the homologous spinal cord level contains the cells of origin for the anterior motor root, while the primary sensory neuron cell bodies are within the dorsal root ganglion. The anterior or ventral (somatic motor) and posterior or dorsal (somatic sensory) roots join in the intervertebral foramen to form a combined motor and sensory spinal nerve, which then bifurcates into anterior and posterior primary rami. The larger anterior primary rami are the most important, particularly those in the cervical and lumbosacral regions, which innervate the muscles of the extremities (myotomes) and are the source for the sensory innervation of the skin (dermatomes). The posterior primary rami innervate the paraspinal musculature and the skin of the neck and posterior trunk.

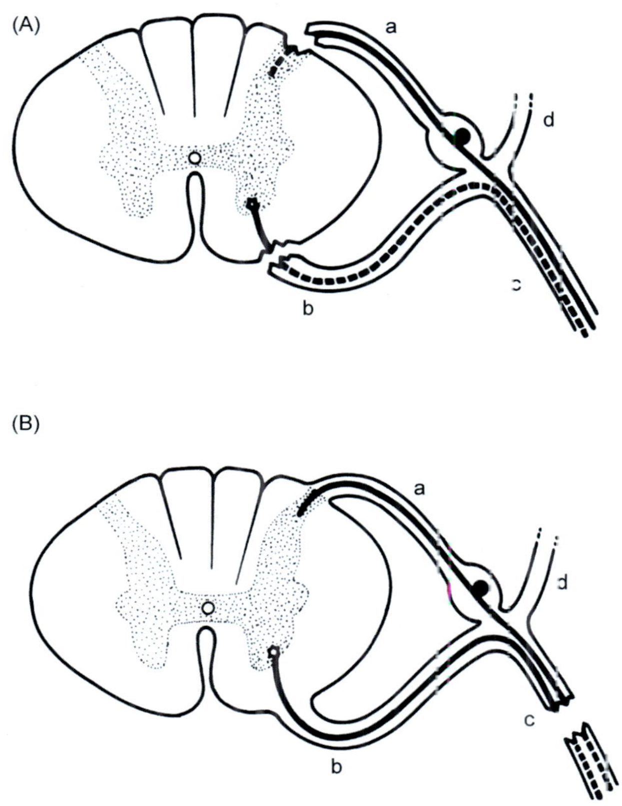

FIGURE 12.1 Configuration of vertebrae and nerve root. **(A)** Preganglionic. **(B)** Postganglionic. a, dorsal root; b, ventral root; c, anterior primary ramus; d, posterior primary ramus.

DIAGNOSTIC PROCEDURES

Magnetic Resonance Imaging

Magnetic resonance imaging (MRI) is the procedure of choice to evaluate and characterize lesions affecting spinal nerve roots. With disc protrusions, MRI enables visualization of effacement of the nerve root by the herniated nucleus pulposus. In children, MRI imaging is useful to identify and characterize intra- and extramedullary lesions causing nerve root compression and radiculitis.

Electromyography

Nerve conduction studies are generally normal in radiculopathies. Intraspinal lesions, such as those seen with a

B.T. Darras, H. Royden Jones, Jr., M.M. Ryan & D.C. De Vivo (Eds): Neuromuscular Disorders of Infancy, Childhood and Adolescence, Second edition.
DOI: http://dx.doi.org/10.1016/B978-0-12-417044-5.00012-3

herniated nucleus pulposus, occur prior to the bifurcation of the spinal nerve, affecting both its anterior and posterior primary rami. With compression of a single nerve root, motor nerve conduction studies tend to be normal because typically only a fraction of nerve fibers in the root undergo axon loss, and because of the overlapping segmental innervation of many muscles. At the same level the sensory nerve action potentials (SNAPs) are also normal, because the dorsal root ganglion is distal to the site of the damage, residing within the intervertebral foramen (Figures 12.2 and 12.3). This is true for all cervical and lumbosacral root levels except L5, where up to 30% of dorsal root ganglia are intraspinal and therefore vulnerable to compression from disc extrusion, resulting in an abnormal superficial peroneal SNAP amplitude.[1] Electromyography (EMG) in radiculopathies is more sensitive to changes affecting a single nerve root, showing signs of denervation (spontaneous discharges such as fibrillation potentials), in the paraspinal muscles as well as other muscles of the myotome corresponding to that spinal level.

In contrast, when the lesion is distal to the intervertebral foramen, the posterior ramus is spared and EMG of the paraspinal muscles is normal, while EMG of limb muscles shows the characteristic changes of denervation. One important caveat to the EMG assessment of radiculopathies in children is that, when they are anxious or upset, it can be difficult to obtain enough paraspinal muscle relaxation to adequately evaluate for abnormalities.

Children with peripheral nerve trauma may have both nerve root avulsion and plexus injuries. In the pediatric age group, therefore, EMG findings compatible with a primary plexus lesion (i.e. absent SNAPs) do not exclude additional damage at the nerve root level (Figure 12.4).[2] A combined process should always be considered in patients with possible proximal nerve lesions.

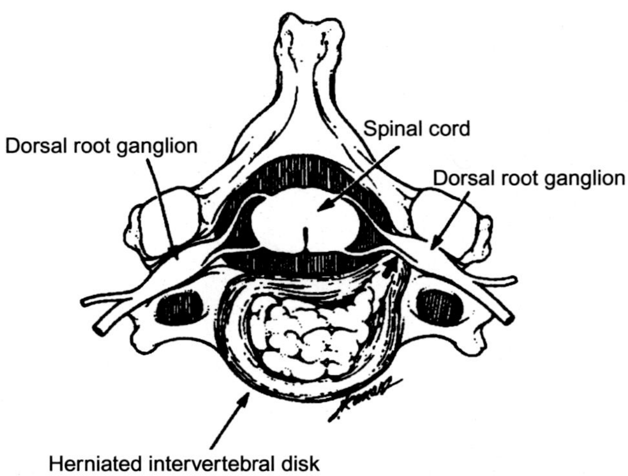

FIGURE 12.2 Radiculopathy with sparing of the dorsal root ganglion. Herniated invertebral discs are a common cause of cervical and lumbosacral radiculopathy. Disc herniation is most often lateral and posterior, with the dorsal root ganglion located distal to the herniation. This anatomic relationship results in injury to the roots but sparing of the dorsal root ganglion and the peripheral sensory nerves. This figure was published in Brown and Bolton (1993), © *Butterworth*.[239]

CERVICAL RADICULOPATHIES

Cervical nerve root lesions in children are rare, but tend to mimic the signs and symptoms of cervical radiculopathy in adults: local or radiating pain and focal weakness, with or without focal loss of the deep tendon reflexes. In adults C7 root lesions are most common, followed by C6 root lesions, but in children there are too few such lesions to judge their relative frequency.

Etiology

Most cervical radiculopathies in childhood follow trauma or traction injuries (see Case Example 12.1), but tumors,

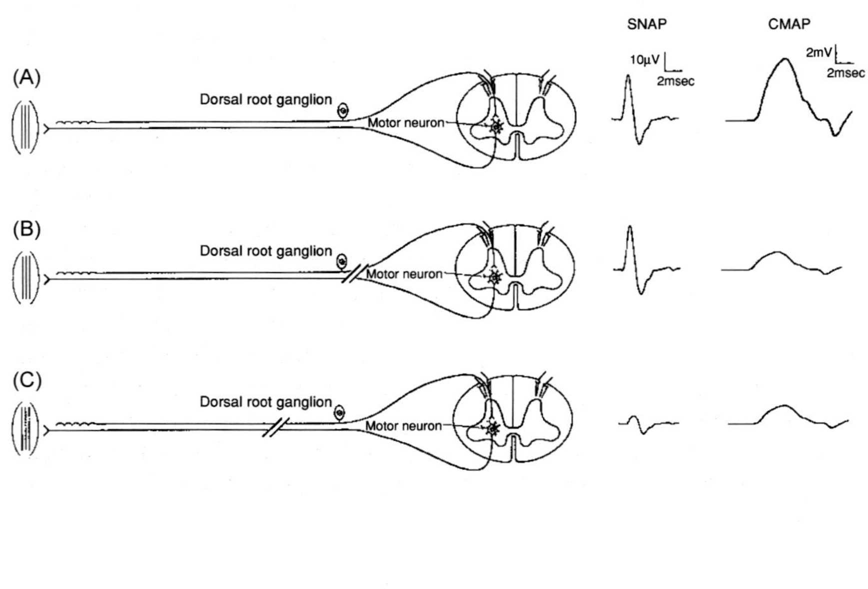

FIGURE 12.3 Sensory and motor potentials in axonal lesions distal and proximal to the dorsal root ganglion. **(A)** Normal. **(B)** Lesion proximal to the dorsal root ganglion. **(C)** Lesion distal to the dorsal root ganglion. This figure was published in Preston and Shapiro BE (1998), © *Butterworth*.[240]

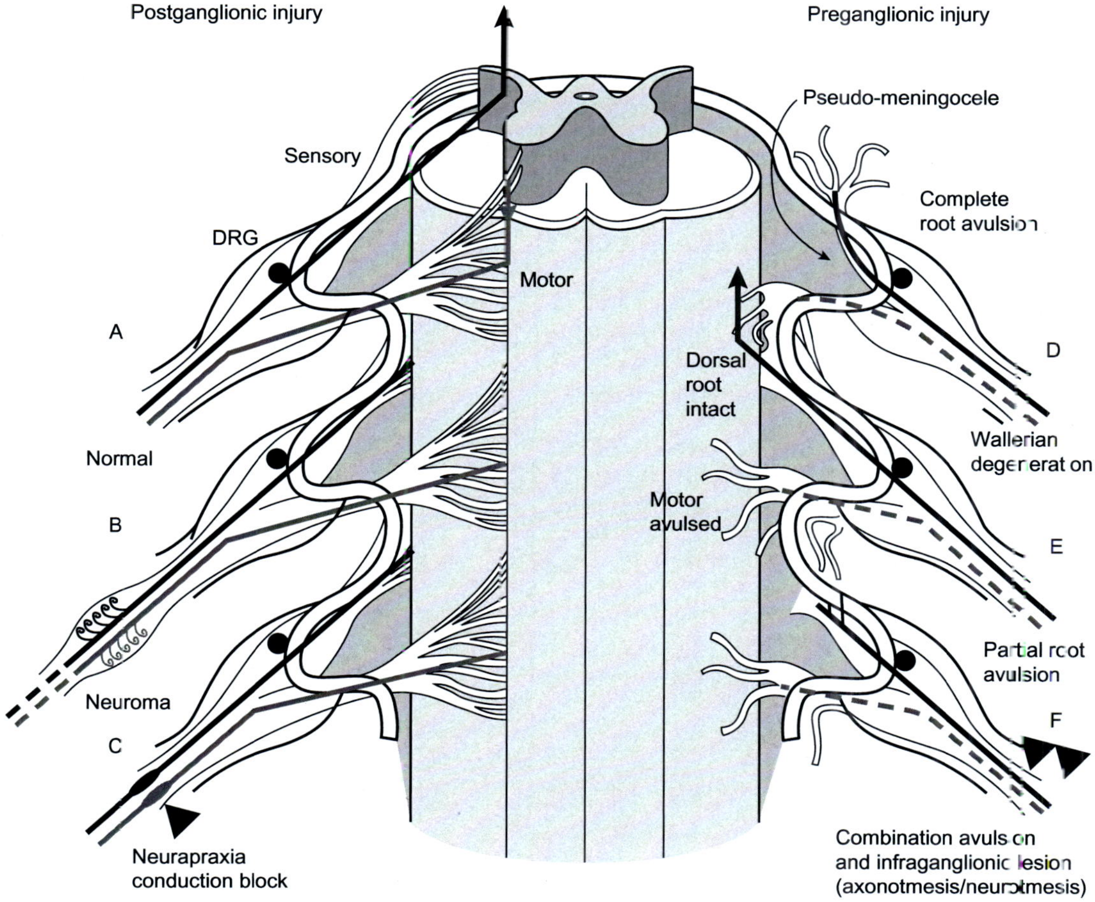

FIGURE 12.4 Schematic representation of the normal anatomy of and injuries to the roots of plexus, demonstrating typical preganglionic and postganglionic abnormalities. **(A)** Normal ventral (motor) rootlets and dorsal (sensory) rootlets. **(B)** Postganglionic injury (axonotmesis or neurotmesis) with neuroma formation and partial or complete disconnection of neural elements. Continuity can be determined only by intraoperative neurophysiologic studies across damaged segments. The compound muscle action potentials (CMAPs) and sensory nerve action potentials (SNAPs) from the corresponding territory will be abnormal (amplitude decreased owing to axonal loss). **(C)** A neurapraxic lesion resulting in abnormal sensation and weakness but normal SNAPs and CMAPs distal to the area of focal conduction block. Conduction across the segment may be blocked intraoperatively. **(D)** Preganglionic injury with complete avulsion and traumatic pseudomeningocele. The cell body of the sensory neuron is intact in the dorsal root ganglion; therefore, a normal SNAP is recorded in an asensate territory. CMAPs will be nonrecordable. **(E)** Partial root avulsion with dorsal root intact and ventral (motor) root avulsed. This produces weakness and normal sensation and a normal SNAP with absent or abnormal CMAPs in respective segmental distribution. **(F)** Combination of a preganglionic and postganglionic lesion, which often can be determined only intraoperatively and correlated with imaging. The roots (motor and sensory) have been avulsed, but the infraganglionic component is also abnormal (axonotmesis–neurotmesis), similar to **B**. Therefore, the SNAPs and CMAPs are absent or abnormal. This can mislead the electromyographer into thinking that the lesion is purely infraganglionic, and he or she might miss the preganglionic injury. *From Brown, Bolton, and Aminoff (2002), © Saunders*, p. 1604.[241]

inflammatory lesions[3] and other extramedullary lesions can also cause nerve root compression or injury (Box 12.1).[4]

Clinical Presentation

It is important to be familiar with the clinical characteristics of radiculopathies at each spinal level, because these lesions may mimic much more common clinical entities, such as mononeuropathies or plexopathies. Most individuals with cervical radiculopathy have cervical or periscapular pain, which often radiates into the arm. With traumatic radiculopathies, this pain may precede the onset of numbness by 4 to 8 weeks, can spontaneously improve, and can then be followed by paresthesia and weakness a few weeks later. C7 or C6 nerve root lesions typically present with numbness affecting digits two or three, in a similar distribution to a median neuropathy. In contrast, C8 radiculopathies cause sensory symptoms involving the fifth digit, mimicking the distribution of the ipsilateral ulnar nerve or medial cord of the brachial plexus. When there is weakness, the triceps muscle is particularly affected in C7 lesions and the pronator teres and flexor carpi radialis in C6 radiculopathies; and the intrinsic hand muscles, flexor pollicis longus, extensor pollicis longus,

CASE EXAMPLE 12.1

A 15-year-old boy, who participated on his high school wrestling team, developed intermittent numbness and tingling in his right shoulder, radiating down his upper lateral arm, and lasting 15 minutes at a time. Three months earlier he had had an episode of "whole arm" numbness, and he had a history of multiple right shoulder dislocations requiring surgery some years earlier. No specific neurologic abnormalities were identified on examination.

He was referred to the EMG laboratory. The right median, ulnar, radial, medial, and lateral antebrachial cutaneous sensory amplitudes and distal latencies were normal. Similarly his right median and ulnar motor conduction studies and associated F-latencies were normal. Concentric needle EMG demonstrated mildly reduced recruitment in several C6–7 innervated muscles including biceps, deltoid, triceps, and brachioradialis. There was no abnormal spontaneous activity in these muscles or in the right C5–6 paraspinals. These findings were suggestive of a mild patchy chronic right C6–7 radiculopathy with no significant signs of active or chronic denervation. He was restricted from wrestling, as this was thought to be the pathophysiologic mechanism for his injury, and his symptoms resolved.

Comment

Cervical radiculopathies are very rare in the pediatric population but may be triggered by athletic activities.[5] These injuries have to be differentiated from traumatic lesions at the level of the brachial plexus, including the so-called stinger injury.[6] The two major proposed etiologies for these athletic injuries are either cervical nerve root compression and traction or a mild stretch injury to the upper trunk of the brachial plexus. If weakness persists, and either MRI or EMG evidence of a nerve root lesion is identified, one may need to consider cervical laminectomy.

BOX 12.1 Reported Causes of Pediatric Cervical Radiculopathy

- Trauma: athletics injuries.[5]
- Inflammatory lesions: pseudoneoplasm.[3]
- Extramedullary lesions: hematoma.[4]

and extensor indicis proprius are typically weakened by C8 nerve root lesions.

Differential Diagnosis

Refer to Box 12.2 for a list of differential diagnoses, discussed in the following subsections.

BOX 12.2 Differential Diagnosis of Pediatric Cervical Radiculopathy

- Congenital cervical stenosis
- Hirayama disease[7,8]
- Motor neuron disease[10]
- Neck–tongue syndrome[14,15]
- Grisel syndrome[11,12]
- Neurogenic thoracic outlet syndrome.

Congenital Cervical Stenosis

Although the authors' experience has not included any typical cervical radiculopathies secondary to either spinal disk or osteophyte-induced foraminal stenosis,[5] we have seen occasional cases of pain in neck, shoulders or arms, or of painless progressive upper extremity weakness, with changes of chronic denervation on EMG and imaging evidence of congenital spinal stenosis, often in association with segmentation anomalies affecting the cervical spine.

Hirayama Disease

This cervical myelopathy causes progressive painless wasting and weakness of the distal upper extremities in adolescence.[7] Most affected children are teenage males, often of western Pacific Rim heritage. Weakness progresses for a few years then arrests spontaneously, for reasons which are poorly understood. Neuropathologic and radiologic findings in this condition suggest forward displacement of the posterior cervical dural sac during neck flexion, causing compression of the cervical cord and ischemic changes in the anterior horn.[8] Because Hirayama disease usually initially affects hand function, it can be confused with ulnar neuropathy, a medial cord brachial plexus lesion, or C8 radiculopathy.

Motor Neuron Disease

The authors have seen only one case of juvenile motor neuron disease, in a teenager with a one-year history of progressive painless weakness of one arm.[9] In this adolescent the median and ulnar compound muscle action potentials were absent, the SNAPs normal, and needle EMG demonstrated diffuse fibrillation potentials with almost no definable motor unit potentials (MUPs). Eighteen months later, when she died, an autopsy demonstrated classic ALS. In a major review of 123 ALS patients there was only one similar case of symptoms of MND commencing before age 20 years.[10]

Grisel Syndrome

Grisel syndrome is an atraumatic atlanto-axial subluxation which may follow inflammatory or infection conditions in the head and neck.[11,12] Affected children typically present with unrelenting neck or throat pain, followed by torticollis, usually with a preceding history of a flu-like illness, tonsillitis, or pharyngitis. Other neurologic manifestations may include greater occipital nerve paresthesias and development of long-tract signs due to cervical myelopathy.[12,13] Occasional patients develop cervical radiculopathy. The physical findings include tenderness to spinal percussion and restriction of neck movement leading to a torticollis. Plain X-rays and MRI imaging demonstrate atlanto-axial subluxation. Treatment with antibiotics, anti-inflammatories, immobilization and/or traction is important in order to prevent cord injury.

A similar condition, neck-tongue syndrome, presents with paroxysmal attacks of sharp unilateral neck pain associated with ipsilateral tongue numbness, which may be precipitated by rotation of the neck.[14] Some cases are associated with torticollis, tongue deviation, and dysarthria.[15] Symptoms are presumed related to intermittent compression of the C2 nerve root, and respond in most cases to conservative treatment.[14]

Neurogenic Thoracic Outlet Syndrome

This disorder is caused by a congenital aberrant fibrous band extending from an elongated C7 transverse process to the first thoracic rib, and can come to clinical attention in late adolescence or early adulthood. Patients describe atrophy and weakness of hand muscles, aching in the medial forearm, and occasionally loss of sensation along the ulnar aspect of the hand.[16,17] The initial consideration is of ulnar neuropathy or C8 radiculopathy. However, close inspection shows predominant weakness and atrophy of the thenar eminence. The pathology stems from progressive extraspinal stretching of the T1 anterior primary ramus over the fibrous band. Nerve conduction studies (NCS) show the classic pattern of severe reduction of the median CMAP to the thenar eminence (abductor pollicis brevis muscle), less reduction of the ulnar compound muscle action potential (CMAP) to the hypothenar eminence (abductor digiti minimi muscle), severe reduction of medial antebrachial cutaneous SNAP and less reduction of the ulnar SNAP to digit 5. Based on this electrical pattern, imaging should be performed to identify the elongated transverse process and band, and surgery is indicated to section the band and release the T1 nerve trunk.[18,19]

LUMBOSACRAL RADICULOPATHIES

Etiology

On the rare occasions that intervertebral disk herniation occurs in children, it usually occurs in the lumbosacral area.[20–26] See Box 12.3 for reported causes of pediatric lumbar radiculopathies.

Most cases are in adolescent boys. Trauma, related to sporting activities, lifting, falls, or occupational injuries, accounts for at least one-third to one-half of pediatric lumbosacral radiculopathies.[21–23] The youngest such case was of a 27-month-old boy who fell from his crib.[27] A significant proportion (13–57%) of adolescents with lumbosacral disc disease has a first-degree relative with a similar history, suggesting a genetic component to this condition.[26]

Scoliosis surgery also predisposes to development of pediatric lumbosacral radiculopathies. One series attributed a 29% incidence of these lesions postoperatively to either surgical trauma or traction in the lumbosacral area.[28] Intraoperative spinal monitoring may aid prevention of this surgical complication.

Developmental anomalies involving intramedullary and extramedullary structures can impinge on the spinal roots, causing pain, weakness and sensory changes.[29–31]

Nerve root tumors occasionally present with radicular pain radiating into the lower extremities, with or without sensory changes and focal weakness. Reported causes of this syndrome include schwannoma, leukemia, synovial sarcoma, osteoblastoma, osteosarcoma, and aneurysmal bone cysts.[32–37] Treatment of childhood malignancies with chemotherapy, radiotherapy, and bone marrow transplantation may also be followed by radiculitis.[38–40]

BOX 12.3 Reported Causes of Pediatric Lumbar Radiculopathy

- Trauma: sport-related, falls, heavy lifting.[21–23,27,235]
- Postoperative: direct trauma, or traction[28]
- Developmental anomalies: vertebral anomalies, synovial cysts,[236] diskal cysts,[30] arachnoid cysts.[29,31]
- Tumors: schwannoma,[32] osteblastoma,[33,34] osteosarcoma,[37] leukemia,[35] synovial sarcoma,[36] giant cell tumor,[237] aneurysmal bone cyst.[238]
- Immunosuppression: chemotherapy, bone marrow transplantation,[38,39] radiotherapy.[40]

Clinical Presentation

Most children present with back pain and sciatica.[21–23,41] Pain is not invariable; occasional children develop a painless scoliosis.[23,25,42] Children aged less than 10 years present with irritability, paravertebral muscle spasm, and leg weakness.[26,27]

Neurologic deficits are usually relatively subtle and apparent at only one spinal level.[41] A mild degree of focal weakness is seen in 32–60% of adolescents,[21–23,25] most often involving the L5 musculature (tibialis anterior, extensor hallucis, and/or extensor digitorum). This is not generally severe enough to cause a foot drop.[22] Well-defined sensory loss may or may not be present. In addition, paravertebral muscle spasm with concomitant limitation of spinal motion, cough-sneeze exacerbation (76%), and/or a positive Lasègue sign are often apparent.[23] Most adolescents have pain on straight-leg raise or back motion.[25,26] Disturbances of bladder and bowel function, however, are very rare and should prompt consideration of intrinsic cord dysfunction.

Evaluation

Routine spinal radiographs may demonstrate congenital bony and spinal segmentation anomalies, but the significance of these is often uncertain.[23,25] A combined diagnostic approach including lumbosacral X-rays, lumbosacral spine MRI, CT imaging in certain cases, and—when these studies are nondiagnostic—EMG, has been recommended for diagnosis of lumbar nerve root disease in teenage children.[24,42]

Differential Diagnosis

Infections, malignancies, and spinal injuries or developmental disorders such as spondylolisthesis should be considered in all children with spinal pain, lumbar nerve root symptoms, and an abnormal gait. As in adults, acute back pain is usually a self-limited musculoskeletal disorder that does not require investigation, and which responds to mobilization exercises and nonsteroidal anti-inflammatory medications. However, there are red flags that should trigger more aggressive evaluation. These include fever; pain that does not improve at night, at rest, or when supine; weight loss; known malignancy; or when symptoms extend beyond several weeks. See Box 12.4 for a list of differential diagnoses.

BOX 12.4 Differential Diagnosis of Pediatric Lumbar Radiculopathy

- Infections: diskitis.[43,44]
- Inflammatory conditions: ankylosing spondylitis, sacroiliac misalignment.[45]
- Developmental anomalies: Chiari malformation,[47] tethered cord, cord lipoma.
- Mass lesions: angiolipoma, arteriovenous hemangioma, cavernous hemangioma.[46]
- Sacral lesions: Ewing's sarcoma, osteoblastoma, hemangiopericytoma, chordoma.[48]
- Pelvic lesions: hematocolpos.[49]

Orthopedic Conditions

Diskitis, or intervertebral disk inflammation, is an uncommon and poorly understood disorder that generally presents in childhood with back pain, limp, and refusal to walk. It can be associated with decreased muscle strength and/or hyporeflexia. Although diskitis is believed to be infectious, positive bacterial cultures are rarely obtained. Imaging studies show narrowing of the intervertebral disk space and/or increased uptake in the area on bone scan, and most authors advocate antibiotic therapy. Disk protrusion into the spinal canal in diskitis may underpin radicular pain and focal weakness in a small proportion of children with this condition.[43,44]

In older children, especially adolescent boys, ankylosing spondylitis is an important consideration in the differential diagnosis of sciatica. There may be a history of early morning stiffness and "jelling," associated with diminished chest excursion on examination. Sacroiliac joint misalignment can also cause back pain.[45]

Intraspinal Lesions

Angiolipomas, arteriovenous hemangiomas and cavernous hemangiomas are benign extradural tumors which are occasionally seen around the spine.[46] (See Case Example 12.2.) Occasionally congenital anomalies, such as Chiari malformations and cord tethering, may present with increasing weakness and foot drop mimicking radiculopathy.[47]

Sacral and Pelvic Lesions

Primary sacral bone tumors or pelvic lesions may present with radicular pain, paresthesiae, and progressive weakness mimicking a lumbosacral radiculopathy.[48,49] Standard X-rays may be normal; MRI and CT are the best diagnostic studies. (See Case Example 12.3.)

Treatment

Conservative management of lumbosacral radiculopathy—bed rest, analgesia and antiinflammatories, and physiotherapy—may relieve pain and weakness, although these symptoms may later recur.[23,42] Surgery is indicated in children with severe pain refractory to conservative treatment, or with progressive neurological deficits. Surgical approaches include percutaneous endoscopic diskectomy or open diskectomy, with or without laminotomy or laminectomy. Outcomes of surgery are generally very good,[21–23,25,26] but reoperation is needed in about 10 to 25 percent of cases.[21,23]

CASE EXAMPLE 12.2

A 10-month-old girl was evaluated for failure to crawl and progressive inability to use her legs since age 6 months. Neurological examination showed hypotonia, weakness, and hyporeflexia in the legs, with relative preservation of motor function in the arms. She was referred for electrodiagnostic studies to exclude spinal muscular atrophy. NCS of the leg and arm were within normal limits, including sensory responses. The needle EMG showed fibrillation potentials in muscles below the knees bilaterally, associated with reduced recruitment of motor unit action potentials with neurogenic morphological changes. Because of sparing of L3–4 muscles and arm muscles, the pattern suggested an active cauda equina syndrome affecting the L5 and S1 roots bilaterally. Subsequently, spine MRI showed an enhancing paraspinal mass extending into the spinal canal from T12 to S1, with severe effacement of the thecal sac. Biopsy showed a poorly differentiated neuroblastoma.

CASE EXAMPLE 12.3

This 13-year-old boy developed a slowly progressive painless foot drop over a 2-year period, with development of pes cavus and mild calf atrophy. Neurologic examination at the Lahey Clinic demonstrated 3/5 weakness, primarily affecting his left tibialis anterior muscle. He was referred to the EMG laboratory for evaluation of a suspected peroneal neuropathy.

Peroneal and tibial NCS and sural SNAPs were normal. Needle EMG demonstrated findings of both S1 and L5 nerve root involvement including abnormal insertional activity in the lumbar paraspinal muscles. Spinal MRI demonstrated a tethered cord and congenital tumor affecting the L5 and S1 nerve roots.

Comment

The neurophysiologic examination in this case was vital, excluding a (much more common) peroneal neuropathy and indicating a proximal polyradiculopathy, which guided imaging and diagnostic surgical exploration.

PLEXOPATHIES

Brachial Plexopathies

Pediatric brachial plexus lesions include birth-related lesions and other acquired disorders sustained from infancy through late adolescence. Neonatal brachial plexus palsy (NBP) complicates a very small proportion of births, but continues to be a challenge in the setting of complicated labor and childbirth. Most brachial plexopathies in older children result from trauma.

Brachial Plexus Anatomy

There are five major components of the brachial plexus; the roots, trunks, divisions, cords, and terminal nerves (Figure 12.5). Cervical roots C5–8 and the first thoracic root provide the primary source for the three trunks (upper, middle, and lower); two divisions (consisting of six elements in all); three cords (lateral, posterior, and medial); and the various terminal branches. In the arm and hand, these include the axillary, musculocutaneous, radial, median, and ulnar nerves.

The long thoracic and dorsal scapular nerves are early (supraclavicular) derivatives of the brachial plexus, which form from the anterior primary rami proximal to the brachial plexus. The dorsal scapular nerve arises from the C5 anterior primary ramus near the intervertebral foramen, and innervates the rhomboid muscle. Denervation of the rhomboids is generally a poor prognostic sign in a child with a neonatal brachial plexopathy, as it suggests a very proximal lesion, possibly including C5 root avulsion. The long thoracic nerve originates from the C5, C6, and C7 anterior primary rami, and innervates serratus anterior.

Most peripheral nerves supplying the upper extremity originate from the infraclavicular portion of the brachial plexus. The three trunks each divide into an anterior and posterior division at the level of the clavicle. The three posterior divisions form the posterior cord, the anterior divisions of the upper and middle trunks form the lateral cord, and the anterior division of the lower trunk continues distally as the medial cord. The three major cords each provide the primary source for the individual nerves of the arm, with the exception of the conjoint origin of the median nerve from both the lateral and medial cords.

The lateral cord (C5–6) gives off the lateral pectoral nerve to the pectoralis major muscle, and then gives rise to the musculocutaneous nerve, innervating the biceps brachii and the brachialis muscles, and contributes to the median nerve, innervating pronator teres and the forearm flexor muscles.

Nerves formed from the posterior cord (C6, 7, 8) include the subscapular nerve (to subscapularis), thoracodorsal (to latissimus dorsi), lower subscapular (to teres major), axillary (to deltoid and teres minor), and the radial nerve (supplying the triceps, anconeus, and all wrist and finger extensors).

The medial cord (C8–T1) gives rise to the medial pectoral, medial brachial, and medial antebrachial cutaneous

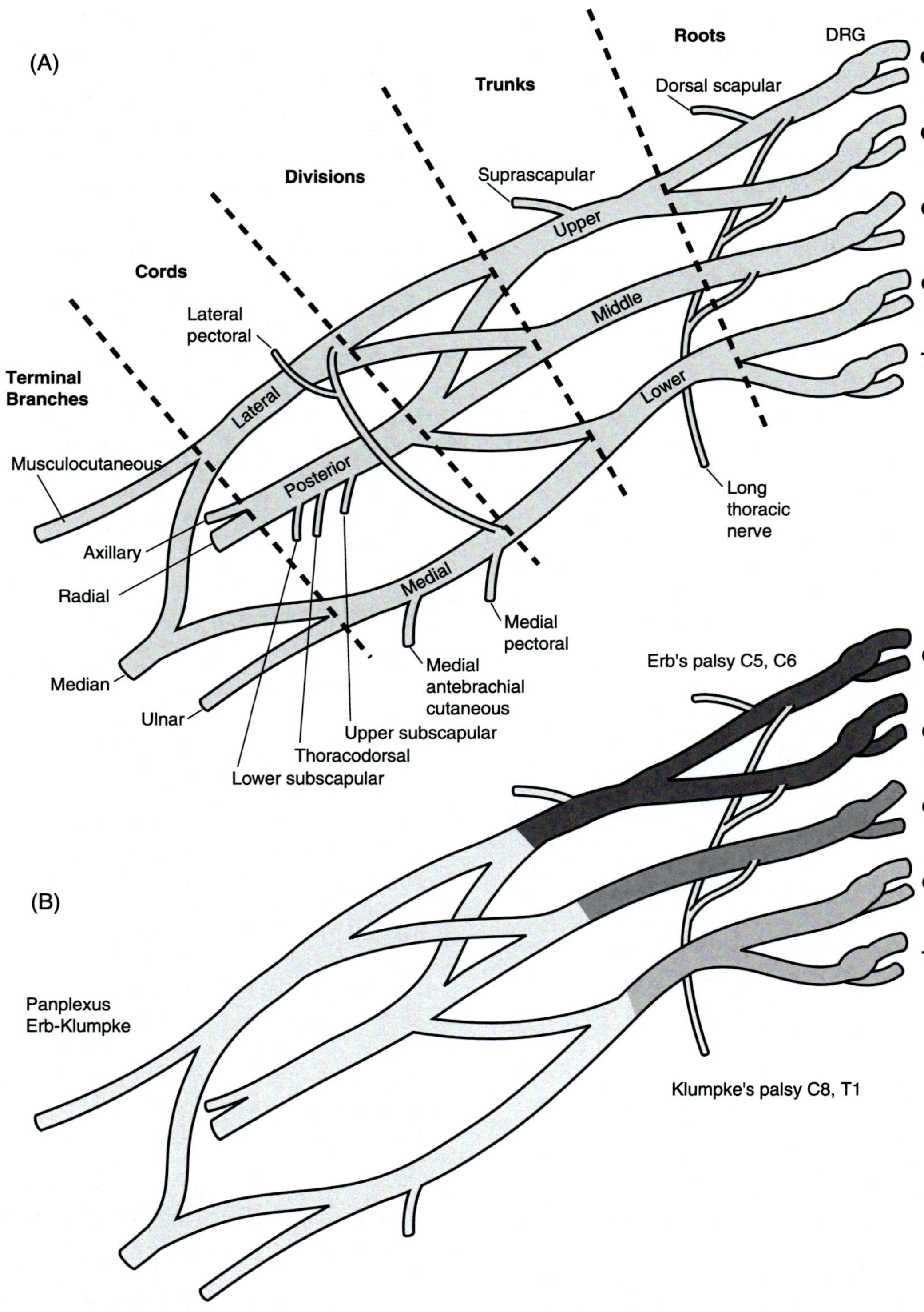

FIGURE 12.5 **(A)** Schematic representation of the brachial plexus with its terminal branches. **(B)** The most common pediatric plexus injury is the C5–6 palsy also known as Erb's palsy; occasionally, C7 is also affected. The second most common palsy is a panplexus injury (Erb-Klumpke). A rare isolated palsy of the C8–T1 segment is commonly referred to as Klumpke's palsy. *From Brown, Bolton, and Aminoff (2002),* © *Saunders.*[241]

nerves, and contributes to both the median nerve (innervating flexor pollicis longus and most thenar intrinsic hand muscles) and the ulnar nerve (which innervates the hypothenar hand intrinsics and adductor pollicis).

At or very near its origin, the anterior primary ramus of each nerve root receives a contribution from the corresponding sympathetic trunk ganglion, the gray ramus communicans. In addition, the first thoracic anterior primary ramus contributes preganglionic sympathetic fibers to the inferior cervical (stellate) ganglion via a white ramus communicans (Figure 12.6A). A lesion affecting these fibers produces a Horner syndrome (unilateral miosis, ptosis, enophthalmos, and facial anhidrosis) (Figure 12.6C).

Diagnostic Procedures

Imaging

Magnetic resonance imaging (MRI) is one of two primary diagnostic modalities for evaluation of the child with a suspected brachial plexus lesion. This evaluation must differentiate between (preganglionic) nerve root lesions, and (postganglionic) brachial plexus lesions. Avulsion of ventral or dorsal nerve rootlets from the spinal cord results in preganglionic nerve root injury with no potential for spontaneous nerve regeneration (Figure 12.6). The diagnosis of nerve root avulsion is typified by the finding of a pseudo-meningocele, documented by demonstration of deformity of the subarachnoid space, most often at the C7 level.[2,50] MRI is noninvasive, gives excellent depiction of pseudo-meningoceles, and also assesses the spinal cord and soft tissue surrounding the brachial plexus (Figure 12.7). Although there is a strong association between nerve root avulsion and development of pseudo-meningoceles, this association is not invariable: a significant percentage of avulsions are not associated with pseudo-meningoceles, and vice versa.[51] Imaging studies have variable sensitivity in the diagnosis of root avulsion. CT myelography provides higher spatial resolution and may be necessary in some situations to confirm the presence of a pseudo-meningocele. CT myelography can be especially useful in demonstrating absent rootlets within the pseudomeningocele, a better predictor of nerve root avulsion and absent intraoperative evoked responses.[51–53] In older children, combined spinal and brachial plexus MRI is often useful in differentiating brachial plexus and nerve root lesions.

With the continued advances in instrumentation and improvement in resolution, MR imaging has supplanted CT myelography for investigation and preoperative evaluation of NBP.[50,54–61]

Electrodiagnosis

When evaluating the infant or child with a possible brachial plexus lesion there are three specific questions to address. These include: (1) establishment of the level and extent of plexus involvement; (2) accurate identification of the presence of a preganglionic lesion and/or root avulsion; and (3) defining the nature of the lesion, in terms of its underlying pathophysiological response to injury as described by Seddon and Sunderland (i.e. neurapraxia versus axonotmesis/neurotmesis). These results provide information important in identifying patients who might benefit from surgery.

The electrodiagnostic evaluation of brachial plexopathies typically includes median and ulnar sensory and motor nerve conduction studies (NCS), radial sensory NCS, and musculocutaneous motor NCS. The lateral anterior brachial cutaneous nerve sensory response is informative but technically difficult to obtain in infants. With all these studies, comparison with the response on the opposite side may provide useful information. CMAP and SNAP amplitudes provide some indication of the degree of axonal damage. CMAP and SNAP amplitudes drop progressively after axonotmesis. In adults, loss of CMAP amplitude is apparent by day 3 and maximal at day 5–8. SNAP amplitude loss begins at day 5 and is maximal by day 9–11.[62] Data for infants and young children have not been published.

Needle electromyographic examination (NEE) provides additional information. The NEE can assess the presence and severity of nerve injury in muscles from which CMAPs cannot be recorded, and can identify more subtle degrees of motor axon loss than is apparent from determination of CMAP amplitude. The precise timing of the onset of positive waves and fibrillation potentials after axonal injury in the older child is similar to that of the adult (10–14 days).[63,64] There are very few studies specifically defining the temporal profile of EMG signs of axonal damage in the neonate. Because of the very short distances from the site of injury to the muscle, as well as the immature, poorly myelinated peripheral nervous system, the timing of these changes is likely to be very different from that in adults. Several case reports have suggested that EMG findings suggestive of acute motor axonal injury in neonatal brachial plexopathy can be seen as early as four days after delivery.[2,65,66] In other cases nerve injury prior to delivery has also been associated with very early EMG abnormalities.[67,68]

Paraspinal muscle denervation is the primary criterion for differentiation between a nerve root and plexus lesion. Because of the baby's inherent inability to relax, however, needle EMG of the paraspinal muscles is usually not technically satisfactory in infants. Preservation of SNAPs in the absence of clinical sensory function is a classic finding suggesting avulsion of the corresponding preganglionic nerve root(s), but in infants and toddlers identification of clinical sensory loss can be very difficult. Additionally, it is not uncommon for neonatal plexus lesions to have combined preganglionic and postganglionic components. An avulsion, with its inherently poor prognosis, may be present despite the absence of SNAPs in the corresponding segment (Figure 12.6). These factors illustrate the technical limitations of electrodiagnosis.[50] In addition, precise

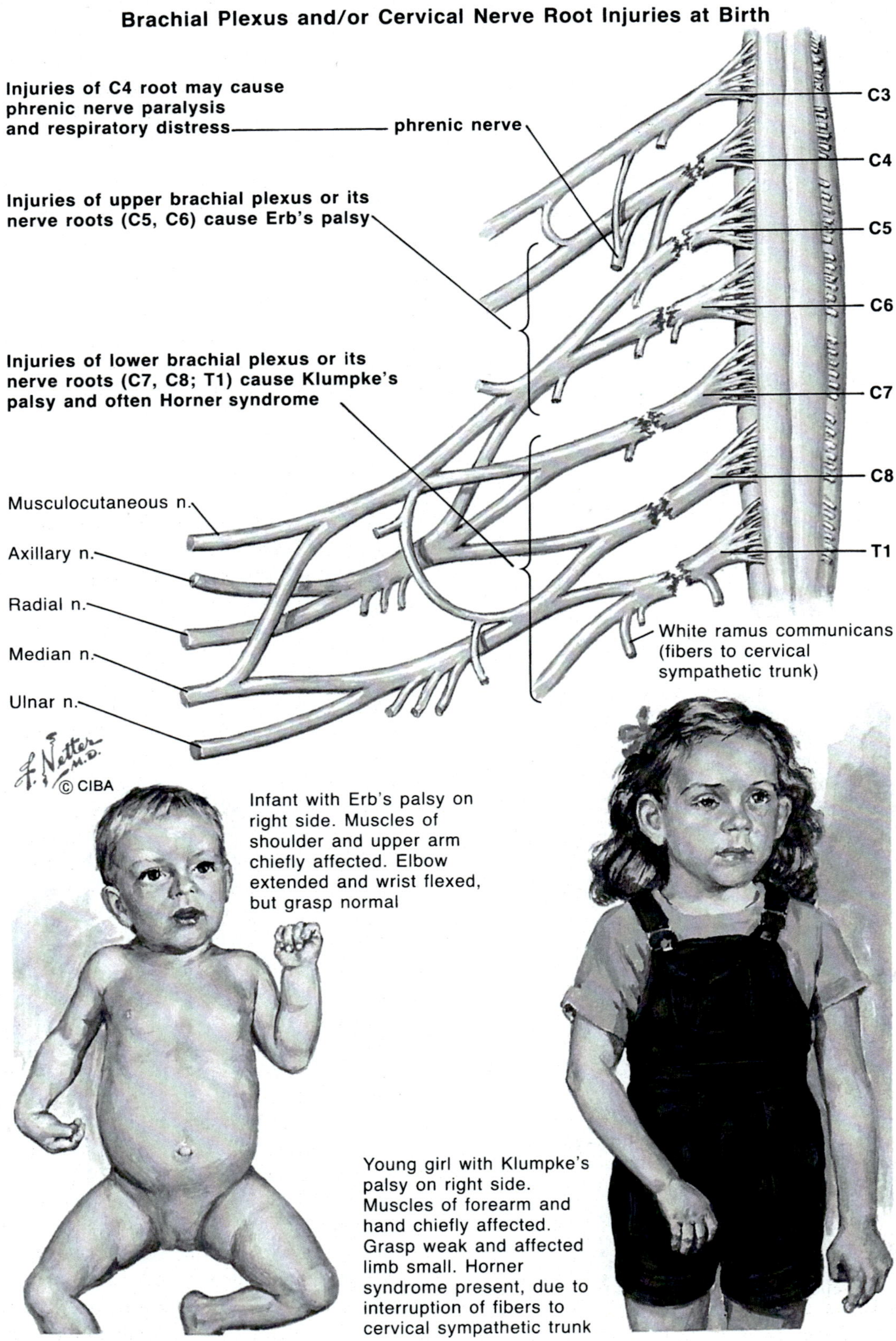

FIGURE 12.6 **(A)** Brachial plexus anatomy after a severe stretch injury, with major roots being avulsed from the spinal cord. Note that the anterior primary ramus of each nerve root receives a contribution from the corresponding sympathetic trunk ganglion, the gray ramus communicans. In addition, the first thoracic anterior primary ramus contributes preganglionic sympathetic fibers to the inferior cervical (stellate) ganglion via a white ramus communicans. **(B)** Classic "waiter's tip" posture of an infant with neonatal brachial plexopathy. **(C)** An older child with associated Horner syndrome. *© 1997, Icon Learning Systems, LLC. A subsidiary of MediMedia, USA, Inc. All rights reserved.*

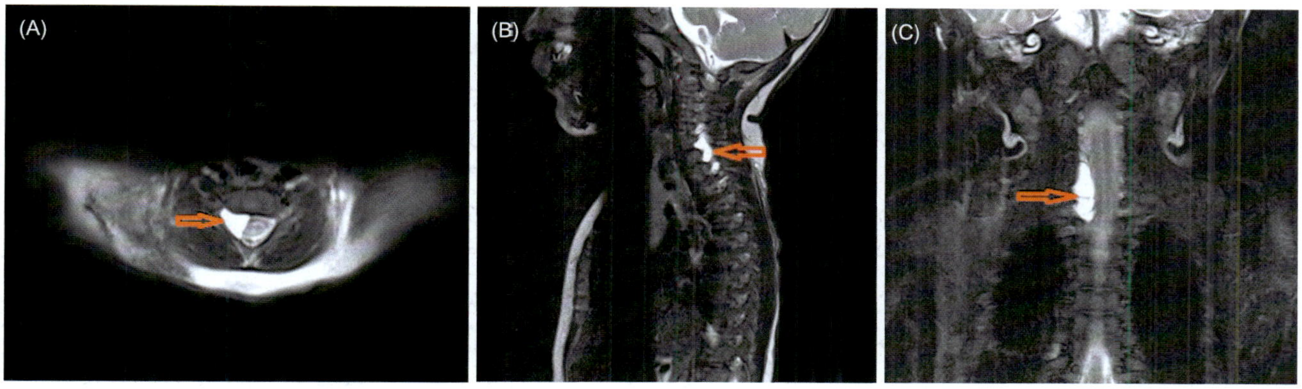

FIGURE 12.7 **(A)** Axial T2-weighted magnetic resonance imaging (MRI) scan shows a pseudomeningocele from nerve root avulsion (arrow). **(B)** Sagittal T2 spine MRI image showing multiple pseudomeningoceles in the cervical spine from C5 to T1 (arrow). **(C)** Coronal T2 fat-suppression MRI scan indicative of multiple pseudomeningoceles secondary to nerve root avulsions on the right side (arrow). *Courtesy Janet S. Soul, MD, Department of Neurology, Boston Children's Hospital.*

localization of the root(s) affected is limited by the overlapping segmental innervation of many muscles in the arm. To achieve accurate localization complementary testing, such as magnetic resonance imaging, is required.

Although somatosensory evoked potentials (SSEPs) were initially thought to be valuable in the evaluation of neonatal brachial plexopathy (NBP), particularly for the assessment of nerve root continuity, they have not proven to add more to diagnosis than EMG. SSEPs may be helpful in the intraoperative assessment of BP lesions, but their clinical utility is limited by the fact that they assess only dorsal (sensory) root integrity.[69,70] Intraoperative transcranial electrical motor evoked potentials have been used to evaluate the functional status of anterior spinal roots and spinal nerves during brachial plexus repair in adults, but their value in pediatric practice is unclear.[71,72]

Neonatal Brachial Plexus Lesions

Historical Aspects

Robotti et al.[73] provided an excellent historical review of this subject. They attribute the first citation of NBP to Smellie in 1768. Duchenne de Boulogne described the typical unilateral NBP in his 1872 treatise on localized electrical stimulation of four neonates, and attributed the injury to traction on the arm at the time of birth. In 1874, Erb reported electrical stimulation of the brachial plexus in NBP, and acknowledged Duchenne's prior descriptions. Thus, the classic upper brachial plexus palsy, described first by Duchenne, bears the name of "Erb's palsy" although some still refer to this entity as Erb-Duchenne palsy. Klumpke, in 1885, described the paralysis of the lower portion of the BP with associated Horner sign secondary to concomitant sympathetic fiber involvement. When the entire plexus is involved, this is referred to as an Erb-Klumpke paralysis (Figure 12.5).

Incidence and Risk Factors

Sever's seminal 1925 study of 1100 newborns with brachial plexopathies provided the first detailed description of neonatal brachial plexus palsy.[74] A decreasing incidence of NBP was reported between 1938 (1.56 per 1000 live births) and 1963 (0.38 per 1000 live births) in New York, possibly reflecting improved obstetric care.[75] The incidence of NBP incidence in industrialized countries now varies from 0.5 to 2 per 1000 live births.[76–81]

The greatest risk factor for NBP is macrosomia. Brachial plexopathy is almost three times more common in newborns weighing >4500 g.[79] Other major risk factors include gestational diabetes, shoulder dystocia, and delivery by vacuum or forceps extraction (Box 12.5). (See Case Example 12.4.) Children with a combination of these risk factors have up to a 52–76-fold increase in risk of NBP in comparison to the baseline population rate.[82,83] Less significant risk factors include preeclampsia, hypertension, multiparity, duration of labor, breech presentation, and previous pregnancy with NBP.[51,79,82,84–88] Parents may not be able to provide a full account of the delivery details, but other relevant postnatal information, including respiratory difficulties, evidence of fracture (particularly clavicular), Horner syndrome, and the extent of the infant's paralysis during the first days of life, contribute to an objective assessment of risk and severity of NBP.

Prenatal diagnosis of macrosomia, and prediction of the risk of shoulder dystocia and NBP, is difficult and imprecise.[79,89,90] Bryant et al. evaluated a policy of elective caesarean delivery for macrosomic infants. In mothers without diabetes, they found that, at an estimated fetal weight of 4500 grams, 155–588 caesarean deliveries were required to prevent a single case of permanent injury.[90] Although McFarland et al. demonstrated a 52-fold increased risk of NBP in macrosomic newborns of diabetic mothers, 92% of these babies were delivered vaginally without complications.[91] Mathematic modeling of forces

BOX 12.5 Risk Factors for Neonatal Brachial Plexopathy

Major

- Macrosomia (birth weight >4500 gm)[79,83]
- Gestational diabetes[82]
- Instrumented vaginal delivery[83]
- Shoulder dystocia.[83]

Minor

- Preeclampsia
- Hypertension
- Multiparity
- Duration of labor
- Breech presentation
- Previous pregnancy with NBP.

TABLE 12.1 Classification of Neonatal Brachial Plexopathies

	Level	Clinical Deficit
Group I	C5–6	Paralysis of shoulder and biceps
Group II	C5–7	Paralysis of shoulder, biceps and forearm extensors
Group III	C5–T1	Complete paralysis of limb
Group IV	C5–T1	Complete paralysis of limb with Horner syndrome

CASE EXAMPLE 12.4

A 4300-gram baby girl was born to a gravida V, para III mother via a vaginal delivery complicated by shoulder dystocia. The mother's previous babies weighed over 4000 grams at birth. One had a transient Erb's palsy and the other cerebral palsy. At birth, the infant had a flaccid left arm, Horner syndrome, and absent muscle stretch reflexes, and was diagnosed with Erb-Klumpke palsy (Group IV).

An EMG was performed at 5 days of age. No CMAPs or SNAPs could be elicited from the left median and ulnar nerves. NEE demonstrated active denervation with fibrillation potentials or positive waves in the deltoid, biceps, triceps, extensor carpi radialis, and flexor carpi radialis. There were no voluntary MUPs in any of these muscles.

Four years later the child had significant shoulder girdle improvement but impaired (4/5) finger grip and finger extension (2/5).

Comment

There were typical predisposing factors for NBP, including birth weight over 4000 grams, shoulder dystocia, and a sibling with a similar history. Her course was one of gradual improvement with significant residual weakness. The question of a possible intrauterine onset of nerve injury remains unresolved, as does nerve root avulsion as an explanation for the early-onset fibrillation potentials, as no MRI was performed to confirm that diagnosis.

associated with shoulder dystocia demonstrates that the pressures generated from endogenous forces (uterine and maternal expulsive efforts) are 4 to 9 times greater than exogenous forces (applied traction to the fetal head).[92]

NBP rarely occurs in the setting of caesarean section. When it does, consideration of other mechanisms for limb paralysis is appropriate.[93] Intrauterine causes of NBP include fetal positioning and amniotic bands.[94,95]

Classification

Neonatal brachial plexopathies are best classified into four anatomic groups on the basis of the neurological examination at age 2 to 3 weeks (Table 12.1).[96,97] Paralysis of the upper roots (group I, Erb's palsy) is by far (81–98%) the most common form of neonatal brachial plexopathy.[81,84,98,99] In contrast, Klumpke's palsy (pure C8–T1 with Horner syndrome) is seen in only 0.6% of all cases of NBP.[99]

An infant with an Erb's palsy (proximal upper trunk and lateral cord lesion) typically lies with his/her affected arm limp at the side with the shoulder adducted and internally rotated, elbow extended, forearm pronated, and fingers and wrists flexed—a posture commonly referred to as "waiter's tip" (Figure 12.6B). The shoulder adduction results from paralysis of the deltoid and supraspinatus muscles. There may also be scapular winging due to serratus anterior muscle weakness from concomitant long thoracic nerve injury, but in newborns this is often difficult to recognize. Internal rotation of the shoulder occurs due to the unopposed contraction of the pectoralis and latissimus dorsi and the unopposed pull of subscapularis, in the setting of teres minor and infraspinatus weakness. Elbow extension is due to both the effects of gravity and weakness of the elbow flexors [biceps, brachialis, and brachioradialis (C5/6)]. The pronated forearm is due to paralysis of the supinator and biceps muscles (C5/6). The flexed wrist and fingers result from weakness of the wrist extensors (C7).

Horner syndrome (ptosis, miosis, and decreased facial sweating) can accompany involvement of the T1 (and sometimes C8) nerve roots, accompanied by severe intrinsic hand weakness typical of Klumpke-type NBP (Figure 12.6C). Involvement of the long thoracic, dorsal scapular, and C8–T1 sympathetic nerves suggests a very proximal injury, such as a lesion at the nerve root entry zone from complete avulsion of the preganglionic spinal nerves. Children with this constellation of findings have a poor prognosis for recovery, in contrast to those with postganglionic lesions, in which nerve regeneration is more likely.

Management of Neonatal Brachial Plexopathies

Nonsurgical

Affected infants should be followed at a clinic with expertise in peripheral nerve injuries, and should be assessed regularly over a period of months in order to assess the extent of recovery, if any. A reliable and reproducible method of quantifying motor function is probably the most challenging aspect of assessing the newborn with a brachial plexopathy. The MRC muscle grading system is insufficient for evaluation of infants. Several groups have established their own grading scales. One scale ranges from M-0 (no contraction) to M-3 (complete movement against the weight of the corresponding segment of the extremity).[100] Mallett proposed an 8-item scale designed to capture subtle but significant changes in movement of the arm (Figure 12.8). This system focuses on joint movement, in contrast to individual muscle testing, and is helpful in both the pre- and postoperative settings, providing a practical means of following infants longitudinally.[101,102] Alternatively the standardized clinical examination defined by the Hospital for Sick Children, Toronto,

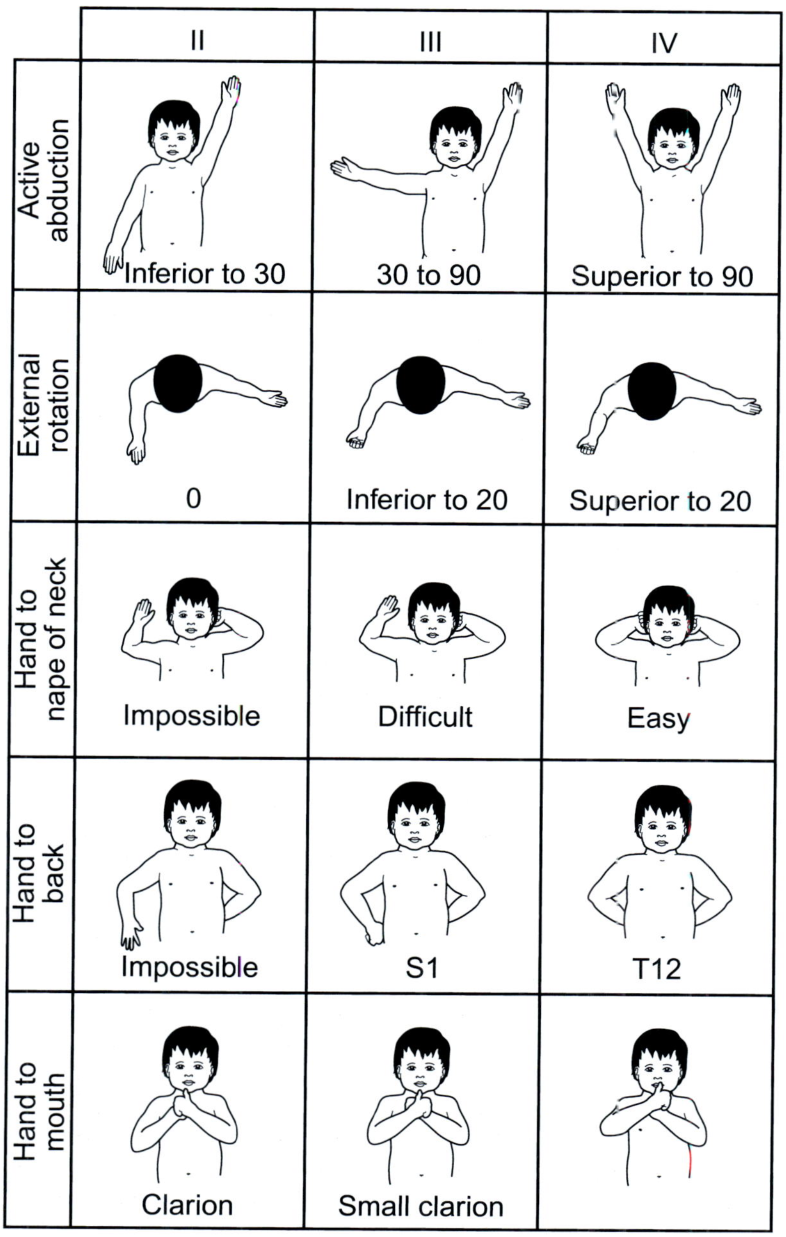

FIGURE 12.8 Grading system for evaluation of infantile brachial plexus lesions, as proposed by Mallet. *Modified from Gilbert, 1993, p. 579.*[242]

TABLE 12.2 Muscle Grading System for Neonatal Brachial Plexopathies: Hospital for Sick Children Toronto (HSCT)

Motion	Score	Muscle Grade Score
Gravity eliminated		
No contraction	0	0
Contraction (no motion)	0.3	1
Motion < {1/2} range	0.3	2
Motion > {1/2} range	0.6	3
Full motion	0.6	4
Against gravity		
Motion < {1/2} range	0.6	5
Motion > {1/2} range	1.3	6
Full motion	2	7

The scores for 5 separate joint motions are assessed at 3 months age as above. These include elbow flexion, together with elbow, wrist, thumb, and finger extension with a maximum score of 2 for each joint. The results are then added together with a maximum score of 10. If the score is <3.5 out of 10 at 3 months, a poor recovery is expected.[101,102]

evaluates extension of the elbow, wrist, fingers and thumb, and flexion of the elbow.[102] A positive score on this clinical exam is associated with a 95% confidence for prognosis of recovery by one year of age (Table 12.2).

Physiotherapy is a critical weapon in the war on joint contractures and is focused on the elbow and forearm, where flexion, extension, and full rotation must be maintained.[103,104] The most common acquired shoulder contracture with NBP is internal rotation due to shortening of the external rotators (teres minor and infraspinatus). Without continued physiotherapy, late consequences of NBP include internal rotation contracture, hypoplasia of the arm, altered sensibility, flexion contracture of the elbow, dislocation of the radial head, and their unrecognized and often disabling psychological and social sequelae.[105] Mothers must be taught to perform stretches with each diaper change. When secondary reconstructive surgery is planned, results are enhanced by preservation of a free and mobile glenohumeral joint.[51,78,100,103,106–108]

Surgical

In the last two decades, advances in microsurgical techniques enabling primary plexus repair have offered potentially improved outcomes in infants with NBPs, but it remains a challenge to identify the best surgical candidates. A detailed review of the surgical approach to NBP has been recently published.[109]

Keeping in mind that some neonatal brachial plexopathies—such as those at the upper trunk / lateral cord level—have a relatively good prognosis, a standard of practice has been to observe clinically for six months.[78,98,107,110,111] More recently, however, several groups have defined groups who may benefit from earlier surgical intervention.[100,105,106,112,113] Gilbert and Razaboni proposed three indications for proceeding with surgical repair of NBP: (1) complete brachial plexopathy with a flail arm and Horner syndrome, (2) C5/6 plexopathy without clinical muscle contraction by 3 months, with EMG evidence of an axon loss lesion, and (3) C5/6 palsy with no recovery in biceps at 3 months (biceps being chosen because examination of the deltoid isolated from the pectoralis major is difficult at this age).[100,105,106] These authors reported a much better outcome (80% good to near-normal) in children with severe C5–6 level injuries treated surgically, as opposed to 0% spontaneous recovery. For more diffuse plexopathies encompassing C5, 6, and 7, similar results were achieved in only 45% of those operated versus 0% for the nonoperated infants. Other authors have proposed alternative indications for earlier surgery, such as global paralysis, persisting abnormalities of biceps or deltoid function at 3 months,[114] or failure of recovery in two of biceps, triceps, and deltoid muscles by 4–6 months.[115]

Gilbert and Razaboni, who operated on 178 NBP infants between 1977 and 1986, reported good or near normal shoulder results in 80%. However, useful hand function returned in only 30% of cases.[100,106,116] Narakas' NBP classification (according to the number of roots affected) provides a means of dividing patients into groups likely to benefit from surgery.[96] Group I shows evidence of only C5 and C6 damage, whereas Group IV had loss of function in all roots (C5–T1) and Horner syndrome. All in Groups I and II (C5–7 lesions) made some recovery, but in a few cases, ultimate function was poor. In general, their Group IV lesions had a very poor prognosis, if untreated.

Functional restoration in cases of preganglionic root avulsion is best attempted by neurotization, the transfer of a "donor" nerve to the distal portion of the plexus or nerve. These procedures are best done soon after the injury, in contrast to palliative approaches such as tendon or muscle transfer surgery. Neurotization procedures may include use of the hypoglossal nerve (XII), spinal accessory (XI), phrenic nerve (C3, C4, C5), intercostal nerves, the long thoracic nerve, and the ipsilateral C7 and contralateral C7 root.[96,107,114,117,118]

Nerve transfers to assist in shoulder abduction include multiple or double neurotization to the axillary and/or suprascapular nerve. Nerve transfers for elbow flexion often include neurotization of the intercostal nerves to the musculocutaneous nerve. More recent work has suggested that, to best assist elbow flexion and extension, the surgeon may try an upper trunk transfer

TABLE 12.3 Prognostic Factors for Neonatal Brachial Plexopathy

Poor Prognostic Factors	Good Prognostic Factors	Unknown/Indeterminate Factors
Completely flail upper limb with Horner syndrome (Narakas Group IV)	Recovery of elbow flexion by 2 m	Motor unit recovery: volitional MUAPs in biceps at age 4 m without antigravity elbow flexion
Horner syndrome	Antigravity elbow flexion by age 3m	Clavicular fracture
Phrenic nerve paralysis	No EMG evidence of axonal loss	Narakas Group II (C5, 6, 7)
No recovery of biceps function (elbow flexion) by 3–4 m	Score of >3.5 at 3 m (HSCT scale)	Early recovery of elbow flexion to less than antigravity by age 4 m with subsequent plateau
Score of <3.5 at age 3 m (HSCT scale)	No root avulsion	
Root avulsion	No pseudomeningoceles on MR imaging	
Pseudomeningocele (nerve rootlets absent)		

Abbreviations: EMG, electromyography; HSCT, Hospital for Sick Children, Toronto; MRI, magnetic resonance imaging; MUAP, motor unit action potential.

or contralateral C7 nerve transfer, with or without functioning free muscle transplantation.[107,114,118] Isolated section of the C7 root for neurotization procedures does not result in significant loss of any specific muscle function. This was initially described in the Chinese literature and has recently been reported in North America.[114,117–119] It may be more applicable in children due to their nervous system plasticity and greater potential for adaptation.

The phrenic nerve has been used in combination with intercostal nerve transfer for neurotization in adult brachial plexopathy. Because the adult diaphragm is fixed to the vertebra, sacrificing the phrenic nerve results in only slight elevation of the diaphragm, with mild dyspnea being the only respiratory complication. However, in infants the diaphragm is not yet fixed to the vertebral bodies. Sacrificing the phrenic nerve in infants causes significant diaphragmatic elevation and severe respiratory distress. This procedure is therefore not recommended under the age of 2 years.[118]

Carlstedt and colleagues recently reported their experience with intraspinal implants for root avulsion in infants and adults. This procedure involves placing a graft into a slit in the spinal cord and connecting it more distally to the avulsed root. Some nerve fiber growth has been demonstrated, but functional improvement has been limited.[120–124]

The electromyographer can provide important information to the surgeon by assessing function of cranial nerve XI (trapezius), the phrenic nerve (diaphragm), and the contralateral C7 innervated musculature. If EMG analysis of those muscles demonstrates abnormalities consistent with axon loss, those nerves are less useful for neurotization. Intraoperative EMG techniques can complement this assessment.

Late plexus repair and rehabilitation

The deficits caused by neonatal brachial plexopathy vary in their extent and severity. A persisting loss of function at C5/6 results in poor external rotation and elevation of the shoulder, and decreased elbow flexion and forearm supination. Patients may develop internal rotation and adduction contractures at the shoulder, which can be addressed by transfers of teres major and latissimus dorsi to the posterior aspect of the greater tuberosity at the insertion of infraspinatus.[96,97,125–127]

Persisting shoulder deformity in Erb's palsy can be overcome by surgical release of the shoulder with reconstruction of the glenohumeral capsule. Simultaneous neurolysis of the upper plexus and bypass grafting from C5 or C6—using end-to-side repair—to the suprascapular nerve may augment external rotation. Postoperative rehabilitation strategies include arm protection in full external rotation, and the use of chemodenervation (by botulinum toxin) to the internal rotators, pectoralis major and minor.[107,125,128]

Prognosis

In neonates with brachial plexopathies, lesions confined to the upper trunk and lateral cord have the best prognosis, although outcome is not uniformly good. A poor outcome is predicted by lower plexus lesions with associated Horner syndrome, total plexus involvement with Horner sign in the absence of improvement within the first week, and failure to improve by six months of age (Table 12.3).[96] Associated phrenic nerve palsy is also a very poor prognostic sign.[78,96,129,130] Although a clavicular fracture in a newborn supports a traumatic process, its presence when associated with NBP does not significantly alter prospects for spontaneous recovery.[93]

Malessy and colleagues studied clinical predictors of outcome of NBP. The presence of elbow extension, elbow flexion, and voluntary motor unit activation in the biceps at age 1 month correctly predicted mild vs. severe lesions in >90% of an initial group, and in 88% and 85% respectively in two additional patient cohorts. Results were based on surgical findings, histopathological review, and outcomes over two years.[131]

Overall, at least 70% of infants with NBP have a good to excellent outcome; 22% have moderate residual damage and 8% severe residual damage,[98,111] although as many as 50% of adults feel that their deficits are significant.[76,132] A Swedish study found that a complete recovery occurred in only 49% of children, with severe impairment in 22%.[76] Other series found that 8–15% of children continue to have a significant permanent disability.[78,102,111,133] Gilbert and Razaboni, who operated on 178 NBP infants between 1977 and 1986, reported good or near normal shoulder results in 80%. However, useful hand function returned in only 30% of cases.[100,106,116]

A systematic review of the literature on prognosis produced fewer than 10 studies meeting at least two inclusion criteria including prospective design, demographic population, 3-year follow up, and end-stage reproducible grading system. In the two studies that came closest to meeting these criteria, 20–30% of patients demonstrated residual deficits.[134]

EMG and Prognosis

The value of neurophysiologic studies in defining prognosis of neonatal brachial plexopathies is debated. The size of the median mixed nerve action potential (NAP) may be a useful indicator for surgical repair. An absent or low amplitude median NAP may reflect significant axonotmesis and a poor prognosis for spontaneous recovery.[135]

In general, EMG evidence of reinnervation usually precedes the clinical evidence of return of function by at least 1 month.[50,98,110,114,136] EMG is also valuable for determination of severity of NBP.[51,72,114,135,137] However, there are concerns that EMG sometimes offers an overly optimistic prognosis in NBP.[50,138]

EMG results sometimes show more voluntary activation of motor unit potentials than would be expected for the degree of weakness and disability apparent in infants with NBP. There are several theories regarding this discrepancy. Firstly, infant muscle fibers are less than one-third the diameter of adult fibers. Because they have more motor units within the catchment area of the EMG needle electrode, more muscle fiber action potentials contribute to each motor unit compared with adults.[139] A second line of reasoning suggests that decreased volitional movement in NBP may result from aberrant reinnervation (misrouting) of regenerating nerve fibers. Some nerve fibers meant for the biceps but instead innervating the deltoid and/or the triceps muscles would give rise to co-contraction of agonists and antagonists, resulting in reduced mobility. A third postulate is that a significant degree of the disability from NBP may relate to aberrant central nervous system coordination of movement—a developmental dyspraxia.[140] In the absence in early life of normal movement and accompanying patterns of proprioceptive and cutaneous sensory input, the brain and spinal cord may not undergo the activity-dependent maturation needed for skilled movements.[141,142] Some patients with NBP with agonist/antagonist–bicep/triceps co-contraction have persisting benefit from botulinum toxin injections, a phenomenon more consistent with the theory that the co-contractions are due to a central abnormality rather than to the effect of peripheral regeneration.[143]

Van Dijk and colleagues undertook serial EMGs in 48 infants with NBP, to identify the best predictors of prognosis.[144] The best prediction of absence of elbow flexion at age 3 months was provided by EMG performed at age 1 month. The absence of biceps motor units at one month had 95% sensitivity, and absence of deltoid motor units had 100% sensitivity, for elbow flexion paralysis at 3 months. EMG at 3 months was misleading, showing voluntary motor unit activation in 19/20 infants with peristing paralysis of elbow flexion.

In one study, 16% of infants with NBP who lacked any functional return at age 4 months were found to have conduction block on nerve stimulation at Erb's point.[135] With a conservative therapeutic approach, a full functional recovery occurred in each baby by 1 year of age. Without EMG, these infants may have been referred for surgery.

Prognosis and Indications for Surgery

Surgery is generally reserved for those with a poor prognosis for spontaneous recovery.[80] This makes it difficult to compare outcome in series of children treated conservatively, and those undergoing surgical brachial plexus repair. Optimal timing of assessment is debated but assessments including a larger number of key muscles and reliant on movement, rather than power, appear most valuable.[101] One report analyzed the clinical course of 39 babies examined at 3-month intervals from birth over a 3-year period.[102] In this study, absent elbow flexion at 3 months wrongly predicted recovery quality in 13% of children. However, when five different functions (see Table 12.2) were scored and summed to a total of <3.5 at 3 months, a poor recovery was invariable. While most babies lacking biceps contraction by age 3 to 4 months required surgery, a small percentage did not achieve this milestone until 4 to 6 months but still eventually recovered completely without surgery.[78,102,107,111] The authors therefore recommended observation over 9 months to

1 year before deciding on surgery. Similar findings were reported in another recent review.[145]

A Swedish study described 247 children who received nonoperative supportive therapy for neonatal brachial plexopathies and underwent extensive functional assessments at 5 years of age.[80] This study excluded babies who had some movement in the deltoid and biceps muscle at age 3 months of age, as all achieved full recovery by age 6 months. The authors concluded that significantly better shoulder movement was apparent in operated C5–6 palsies, but that there were no other significant differences between the operated and the nonoperated groups. Their results did not support operating on children with no biceps and deltoid activity at age 3 months. Other findings in the study included: (1) all children with complete recovery by 5 years had some recovery by 2 months, (2) infants operated on before 6 months of age fared no better than those who had later surgery, and (3) surgery provided no advantage for Erb-Klumpke type lesions.

Other Infantile Brachial Plexopathies

Acquired infantile brachial plexopathies are much less common than NBP. The differential diagnosis of early-onset brachial plexus palsy in a previously healthy infant includes amyoplasia congenita, anterior horn cell lesions, and peripheral nerve lesions.[68,146–152]

A number of reports have described infantile brachial plexopathy secondary to adjacent osteomyelitis.[152–156] This uncommon etiology must be differentiated from a previously unrecognized NBP, or "pseudopalsy" due to a humeral fracture. Findings suggesting a septic plexopathy include fever and an elevated ESR or CRP. The initial bone scan may be normal or equivocal.[153,156,157]

Other rare causes of infantile brachial plexopathy include pyogenic cervical osteomyelitis and paraspinal abscess,[158] neonatal hemangiomatosis,[156,159] exostosis of the first rib,[157] and therapy for childhood acute lymphoblastic leukemia.[160–162] Neoplasm is a rare cause of infantile brachial plexopathy which can be distinguished from typical NBP by (1) the onset of weakness after delivery, (2) progressive course, (3) history of normal delivery and birth weight, (4) the appearance of a mass in the supraclavicular region without radiographic evidence of clavicle fracture, and (5) scratch marks on a weak arm.[157,162]

Brachial Plexopathies after the Neonatal Period

Familial causes of pediatric brachial plexopathy include hereditary neuralgic amyotrophy and hereditary neuropathy with tendency to pressure palsies. Trauma—via compression, traction, ischemia, or laceration—is the primary mechanism for acquired nonhereditary pediatric brachial plexopathy. This is particularly so because of the superficial anatomic location of the brachial plexus, close to bony and vascular structures in the shoulder and neck. Motorcycle and other motor vehicle accidents, sporting activities (burners or stingers from football and other contact sports injuries), gunshot and knife wounds, and, rarely, falls, can result in brachial plexus injury. Child abuse must always be considered in any patient with a presumed idiopathic lesion, and in patients with suggestive neurophysiologic findings neurogenic thoracic outlet syndrome (see above) should also be considered.

Hereditary Neuralgic Amyotrophy

Hereditary neuralgic amyotrophy (HNA), also termed hereditary brachial plexus neuropathy, is a genetically heterogeneous condition most commonly caused by point mutations or duplications in the *SEPT9* gene on chromosme 17q25.[163] HNA presents with repeated episodes of pain in the shoulder and arm, followed by development of patchy weakness and atrophy. Initial episodes of HNA can occur as early as 3–4 years of age,[164] although this condition most often presents in adolescence.[165] Attacks may be triggered by viral infections or physical exertion, and are thought to be autoimmune in nature.[164] The pain can be significant, increases over the first week of an attack, and is often refractory to treatment.[163] Weakness develops within a few days to a week, most often affecting the upper brachial plexus distribution, often in association with scapular winging. Bilateral symptoms are seen in one-third of cases. Muscle wasting is prominent. Recovery is slow and often incomplete. Occasionally, patients have atypical findings such as phrenic or cranial nerve involvement or concomitant lumbar plexitis.

Nerve conduction studies in HNA demonstrate low-amplitude median and radial sensory responses.[164] While signs of motor axon loss were evident in most children with HNA, in some without signs of denervation on needle EMG a proximal demyelinating lesion was proposed as a cause of weakness.

The clinical picture of hereditary neuralgic amytrophy is almost undistinguishable from isolated idiopathic neuralgic amyotrophic (INA): the two cannot be separated on the basis of clinical features, course or recurrence.[163] In the absence of a family history, the diagnosis of HNA is suggested by the typical dysmorphic features (slender facies, ocular hypotelorism, long nasal bridge, cleft [bifid] uvula or cleft palate, long philtrum, facial asymmetry, partial syndactyly of the fingers or toes, skin folds or creases on the neck or forearms, and short stature) associated with the common c262C > T (p.Arg88Trp) mutation in the *SEPT9* gene (Figure 12.9).[165,166] The pain seen in HNA distinguishes it from hereditary neuropathy with

FIGURE 12.9 The facial features of patients with hereditary neuralgic amyotrophy resemble portraits painted by artist Amedeo Modigliani, as in this portrait. Similar features that can be discerned are those of hypotelorism, long nasal bridge, mild facial asymmetry, long philtrum, and a slender facies. Girl with Braids (detail), Amedeo Modigliani. *Courtesy of www.modigliani-foundation.org.*

liability to pressure palsies (HNPP), which is painless.[165,167,168] The *SEPT9* gene gives rise to multiple messenger RNA transcripts though alternative splicing. These variant forms of septin-9 are thought to participate in cell division, cell polarity, and cytokinesis, but their exact function is unknown.[169] Other genes are likely to be implicated in HNA.[163]

Hereditary Neuropathy with Tendency to Pressure Palsies

Hereditary neuropathy with tendency to pressure palsies (HNPP) is caused by interstitial deletions causing complete loss of one allele of the peripheral myelin protein 22 (*PMP22*) gene, and usually presents with painless mononeuropathies.[165,167,168] In 10% of childhood cases, however, presentation is with painless weakness of the arm, often with acquired scapular winging, or with mononeuritis multiplex.[170] The inciting trauma is often minimal or seemingly coincidental.[171] Most cases recover completely.

Traumatic Brachial Plexopathies

Traumatic causes of brachial plexoapthy include contact sports injuries ("stingers and burners"), bike and motorcycling accidents, perioperative events, and gunshot wounds.[6,63,172–189] There are two proposed etiologies of stingers and burners: compression and traction of the C5/6 cervical nerve roots, and mild stretch injury to the upper trunk of the brachial plexus. The clinical presentation is consistent with changes at either of these anatomic sites.[174] An electrodiagnostic examination can be helpful in children with persisting pain and/or weakness.[6,182–191] With transient symptoms the EMG is likely to be normal. With persistent motor dysfunction, in the absence of EMG signs of conduction block or denervation, the differential diagnosis should include musculoskeletal causes such as a rotator cuff tear or fracture.

Pressure Palsies

Compressive brachial plexopathies have been reported in two contexts: automobile safety seats[66] and prolonged wearing of back packs.[172,192] The lack of recent cases is likely the result of better padding and ergonomic design.

Child Abuse

The possibility of child abuse as a cause of brachial plexopathy must be considered when no other plausible etiology is apparent.

Brachial Plexus Neoplasms

Tumors affecting the brachial plexus are extremely uncommon in childhood, but include a variety of pathologies, including neurofibromas, lymphomas, sarcoma/rhabdosarcomas, neuroectodermal tumors, neuroblastomas, and metastatic lesions. Mass lesions affecting the brachial plexus have been reported in children as young as 7 days, and can present with a supraclavicular mass, arm wasting or weakness, or sensory loss.[147,157] See Case Example 12.5.

Idiopathic Neuralgic Amyotrophy

Idiopathic neuralgic amyotrophy (INA), most likely a primary autoimmune acute brachial plexopathy, has multiple synonyms, including the Parsonage-Turner syndrome, neuralgic amyotrophy, idiopathic brachial plexus neuropathy and idiopathic brachial neuritis. Pain is an early and universal symptom of neuralgic amyotrophy, followed by weakness in the shoulder girdle and arm.[193–196] Weakness most commonly affects the shoulder girdle but can be patchy. Distal involvement and scapular winging are common. About a third of patients with INA have bilateral but asymmetric involvement.[63,163,195] INA is much more commonly recognized in adults, but is occasionally seen in children of all ages, and occasional infantile cases are described.[163,197]

INA is a diagnosis of exclusion, so other causes of weakness should be sought. Hereditary neuralgic amotophy and HNPP are important differential diagnoses, as are Hopkins' syndrome and other acute polio-like illnesses.

CASE EXAMPLE 12.5

A previously healthy 3-year-old boy developed increasing pain and weakness of the right arm over a three-month period. An orthopedic surgeon diagnosed a hairline right clavicular fracture. He was referred to a neurologist who noted flaccid paralysis with the exception of trace finger flexion. All muscle stretch reflexes in the arm were absent. Sensation appeared decreased to pinprick and light touch in the right arm and hand.

Nerve conduction studies demonstrated low amplitude median and absent radial SNAPs. Motor NCS of the right median nerve demonstrated mild slowing, with impersistent and prolonged F waves. Ulnar motor studies were normal. Needle EMG demonstrated patchy involvement of muscles in the distribution of the posterior and lateral cords more than the medial cord.

MR imaging of the right brachial plexus and spinal cord revealed a large enhancing mass, originating around the brachial plexus and extending into the spinal canal through the neural foramina, with compression of the spinal cord. On spinal cord decompression and biopsy of the tumor an undifferentiated sarcoma was identified.

Comment

The differential diagnosis of brachial plexus tumors in childhood in order of frequency includes neuroblastoma, sarcoma/rhabdosarcoma or neuroectodermal tumor, lymphoma, neurofibroma, and metastasis. During the 10 days subsequent to the biopsy, this tumor grew more than 100%, becoming superficially palpable. Chemotherapy was begun but was complicated by a respiratory arrest and subsequent severe cognitive disability. [Unpublished communication, D Escolar]

Lumbosacral Plexopathies

Lesions of the lumbosacral plexus are among the most uncommon peripheral nerve problems in children. Most are the result of trauma, such as from motor vehicle accidents, pelvic fractures, and gunshot wounds.[175,198–208] Immune-related post viral plexitis has been reported.[200,209,210] Tumors are another unusual cause of lumbosacral plexopathy in children. Iatrogenic lesions have been described.[211–215] Finally, patients with HNPP and Ehlers-Danlos syndrome have developed lumbosacral plexopathy in the setting of minor trauma or nerve compression.[216,217]

Anatomy

Lesions may affect the lumbar plexus, lumbosacral trunk and/or sacral plexus (Figure 12.10). There is significant overlap between these structures in children.

The femoral nerve is the major derivative of the lumbar plexus. It predominantly innervates the iliopsoas and quadriceps femoris muscles. Its primary sensory distribution includes the superficial sensory branches to the anterior and lateral thigh, and the medial aspect of the foreleg below the knee in the form of the saphenous nerve. The obturator nerve is also derived from the lumbar plexus, innervating the adductor longus muscle and the proximal 80% of the adductor magnus, the distal extent of which receives innervation from the sciatic nerve. This mixed innervation of the adductor can lead to mislocalization of a diffuse plexus lesion in patients with sciatic neuropathy.

The sacral plexus innervates the remainder of the lower extremity, including the posterior thigh and buttock muscles, and all muscles of the leg below the knee. The superior and inferior gluteal nerves are the most proximal nerves originating from the sacral plexus, innervating the gluteus medius and gluteus maximus muscles, respectively. Initial branches of the sciatic nerve lead to the medial and lateral hamstring muscles. The sciatic nerve subsequently bifurcates into the peroneal and tibial nerves, providing all motor and most of the sensory innervation below the knee. The peroneal (fibular) nerve gives off its first branch to the short head of the biceps femoris in the mid-posterior thigh, a useful muscle for EMG differentiation of peroneal neuropathy above the knee from that at the fibular head.

Below the knee the peroneal nerve bifurcates into the superficial and the deep peroneal (fibular) nerves. The deep peroneal nerve supplies all of the anterior compartment muscles of the lower leg; the superficial peroneal nerve supplies the lateral compartment. The calf muscles, posterior tibial, flexor digitorum longus, and all intrinsic foot muscles (except the extensor digitorum brevis) are all supplied by the tibial nerve. Aside from the superficial peroneal sensory nerve, the sural, medial, and lateral plantar nerves are innervated by the tibial nerve.

Histologically, the lumbosacral plexus is similar to that of other peripheral nerves.[218,219] The epineurium occupies approximately 45–65% of the total cross-sectional area of the lumbosacral plexus nerve roots. This is more condensed toward the periphery.[220] There is a relatively higher connective tissue content in the L5 and S1 roots, potentially contributing to their strength and providing a cushion against compressive forces from the sacrum and sacroiliac joint.[220]

Evaluation of Pediatric Lumbosacral Plexopathies

Depending on the clinical situation, an imaging study may be the logical first step in assessment of suspected

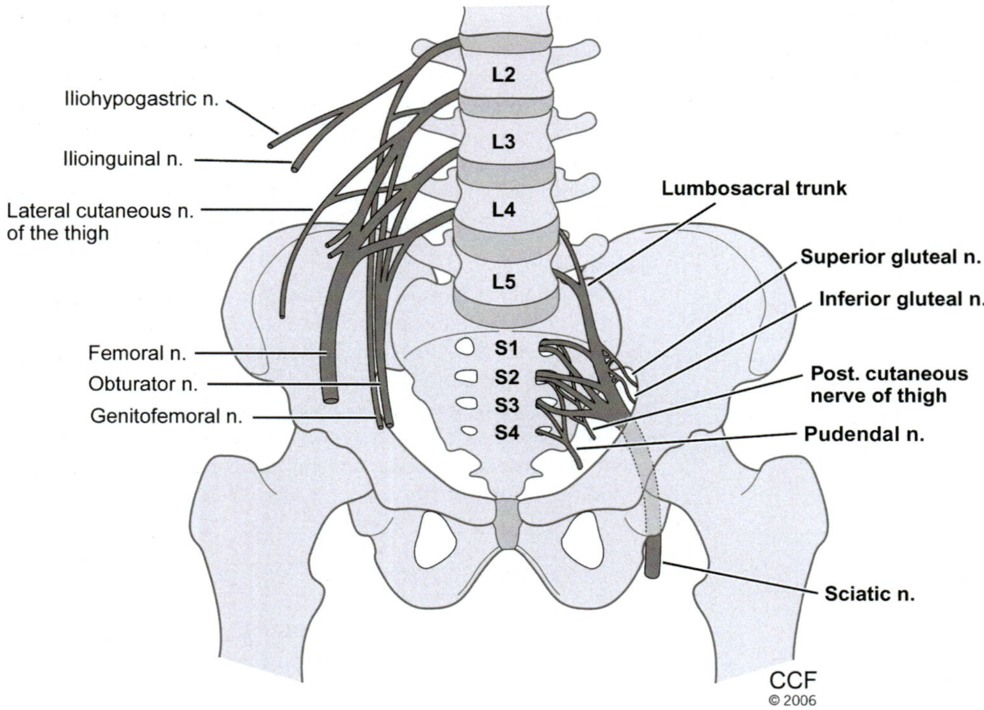

FIGURE 12.10 The lumbosacral plexus. *From Brown, Bolton, and Aminoff (2002), © Saunders, p. 1623.*[241]

lumbosacral plexopathy in a child. When localization of the neurological deficits is unclear and when imaging studies are nondiagnostic, electrodiagnostic studies are indicated.

The femoral motor and saphenous sensory conduction studies are the only NCS available to directly assess function of the lumbar plexus. Both are technically challenging in the awake child. Needle EMG of anterior thigh muscles is a more expedient way to get similar information. In contrast, the sural, plantar, peroneal, and tibial NCS provide an easy means of assessing function of the sacral plexus.[221] In sacral plexopathies, needle EMG of the glutei and tensor fascia lata is normal, distinguishing these from sciatic nerve lesions. Sural, plantar, and superficial peroneal sensory NCS are normal in L5 and S1 root lesions, but typically abnormal in sciatic and sacral plexus lesions. One report has pointed out that the L5-derived superficial peroneal sensory response can be affected in L5 root lesions when the L5 dorsal root ganglion is intraspinal (Figure 12.4).

Neonatal Lumbosacral Plexopathies

In contrast to the high frequency of brachial plexus damage in newborns, lumbosacral plexopathies are rare.[174,222,223] Rare lumbosacral "traction" injuries after complicated breech delivery are generally followed by incomplete recovery.[136] There have been several other reports of presumed lumbosacral plexopathy in newborns.[223,224] The differential diagnosis includes nerve root avulsion, an asymmetric myelomeningocele, parenteral injection injury, or a central nervous system (CNS) lesion.

CASE EXAMPLE 12.6

A 4-month-old boy was dragged by his leg by an older sibling. Later that day, his mother noted that the leg was limp and cool to the touch. Neurologic examination demonstrated a flaccid leg with no elicitable muscle stretch reflexes. Within ten days, modest return of proximal motor function was noted, without improvement distally. The knee muscle stretch reflex remained absent. NCS demonstrated a low amplitude tibial CMAP with a normal peroneal response. Needle EMG demonstrated fibrillation potentials and reduced numbers of motor units in the tibialis anterior, medial gastrocnemius, and vastus lateralis. Examination at 7 months showed normal proximal strength but persistent peroneal distribution weakness.

Lumbosacral Plexopathies after the Neonatal Period

In infants, trauma is the leading cause of lumbosacral plexopathy. Posterior dislocation of the hip and sacral fractures may be associated with stretch injuries to the lumbosacral plexus.[225–227] In infants and toddlers a stretch injury to the lumbosacral plexus is more common than nerve root avulsion.[199,202,205,228] (See Case Example 12.6.) The sciatic portion of the lumbosacral lexus is affected most often, tending to follow pelvic and sacral fracture, and hip dislocation.[205,229,230] Other reported mechanisms of injury include knife injuries, gunshot wounds, accidents caused by motor vehicles, and abdominal crush injuries.[175,201,203,205]

Acute Lumbosacral Plexitis

Acute lumbosacral plexitis (acute lumbosacral plexus neuropathy, analogous to idiopathic neuralgic amyotrophy (INA, Parsonage-Turner syndrome) is manifested by rapid onset of pain, leg weakness, and atrophy.[200,210] This very uncommon condition has only occasionally been described in children. Most experience a viral illness 3 to 10 days before onset of symptoms. Weakness is primarily proximal. Most recover completely or with mild residual weakness.[209,231,232] The differential diagnosis of sudden leg weakness and pain in children includes poliomyelitis.[224,233]

Neoplasms of the Lumbosacral Plexus

Primary benign and malignant and secondary metastatic malignancies rarely affect the lumbosacral plexus in children.[160,205,234] Symptoms suggesting this condition include insidious onset of pain, paresthesia, and leg weakness. Diagnostic modalities include MRI, CT, and ultrasound.

REFERENCES

1. Levin KH. L5 radiculopathy with reduced superficial peroneal sensory responses: intraspinal and extraspinal causes. *Muscle Nerve* 1998;**21**(1):3–7.
2. Mancias P, Slopis JM, Yeakley JW, Vriesendorp FJ. Combined brachial plexus injury and root avulsion after complicated delivery. *Muscle Nerve* 1994;**17**(10):1237–8.
3. Bartanusz V, Ziu M, Jimenez DF, Henry JM. Calcifying pseudoneoplasm of the atlantoaxial joint in a child. *J Neurosurg Spine* 2013;**18**(4):367–71.
4. Ravid S, Schneider S, Maytal J. Spontaneous spinal epidural hematoma: an uncommon presentation of a rare disease. *Childs Nerv Syst* 2002;**18**(6–7):345–7.
5. Radhakrishnan K, Litchy WJ, O'Fallon WM, Kurland LT. Epidemiology of cervical radiculopathy. A population-based study from Rochester, Minnesota, 1976 through 1990. *Brain* 1994;**117**(Pt 2):325–35.
6. Poindexter DP, Johnson EW. Football shoulder and neck injury: a study of the "stinger." *Arch Phys Med Rehabil* 1984;**65**(10):601–2.
7. Hirayama K, Tsubaki T, Toyokura Y, Okinaka S. Juvenile muscular atrophy of unilateral upper extremity. *Neurology* 1963;**13**:373–80.
8. Huang YL, Chen CJ. Hirayama disease. *Neuroimag Clin N Am* 2011;**21**(4):939–50, ix–x.
9. Jones HR. *Pediatric Case Studies. American Association of Electrodiagnostic Medicine (AAEM) Plenary Session. New Developments in Pediatric Neuromuscular Disease*. Rochester MN: AAEM; 1993. pp. 51–60.
10. Dantes M, McComas A. The extent and time course of motoneuron involvement in amyotrophic lateral sclerosis. *Muscle Nerve* 1991;**14**(5):416–21.
11. Harma A, Firat Y. Grisel syndrome: nontraumatic atlantoaxial rotatory subluxation. *J Craniofac Surg* 2008;**19**(4):1119–21.
12. Nozaki F, Kusunoki T, Tomoda Y, Hiejima I, Hayashi A, Kumada T, et al. Grisel syndrome as a complication of Kawasaki disease: a case report and review of the literature. *Eur J Pediatr* 2013;**172**(1):119–21.
13. Mathern GW, Batzdorf U. Grisel's syndrome. Cervical spine clinical, pathologic, and neurologic manifestations. *CORR* 1989;**244**:131–46.
14. Chedrawi AK, Fishman MA, Miller G. Neck-tongue syndrome. *Pediatric Neurol* 2000;**22**(5):397–9.
15. O'Meara M, Wise G. Painful torticollis with tongue atrophy—a different neck-tongue syndrome. *Neuropediatrics* 1995;**26**(5):276–80.
16. Chang K, Graf E, Davis K, Demos J, Roethle T, Freischlag JA. Spectrum of thoracic outlet syndrome presentation in adolescents. *Arch Surg* 2011;**146**(12):1383–7.
17. Chan KH, Gitomer SA, Perkins JN, Liang C, Strain JD. Clinical presentation of cervical ribs in the pediatric population. *J Pediatr* 2013;**162**(3):635–6.
18. Levin KH, Wilbourn AJ, Maggiano HJ. Cervical rib and median sternotomy-related brachial plexopathies: a reassessment. *Neurology* 1998;**50**(5):1407–13.
19. Teddy PJ, Johnson RD, Cai RR, Wallace D. Surgery for paediatric thoracic outlet syndrome. *J Clin Neurosci* 2012;**19**(2):235–40.
20. King AB. Surgical removal of a ruptured intervertebral disc in early childhood. *J Pediatr* 1959;**55**(1):57–62.
21. Kurihara A, Kataoka O. Lumbar disc herniation in children and adolescents. A review of 70 operated cases and their minimum 5-year follow-up studies. *Spine* 1980;**5**(5):443–51.
22. Fisher RG, Saunders RL. Lumbar disc protrusion in children. *J Neurosurg* 1981;**54**(4):480–3.
23. DeOrio JK, Bianco Jr. AJ. Lumbar disc excision in children and adolescents. *J Bone Joint Surg Am* 1982;**64**(7):991–6.
24. Epstein JA, Epstein NE, Marc J, Rosenthal AD, Lavine LS. Lumbar intervertebral disk herniation in teenage children: recognition and management of associated anomalies. *Spine* 1984;**9**(4):427–32.
25. Shillito Jr. J. Pediatric lumbar disc surgery: 20 patients under 15 years of age. *Surg Neurol* 1996;**46**(1):14–18.
26. Dang L, Liu Z. A review of current treatment for lumbar disc herniation in children and adolescents. *Eur Spine J* 2010;**19**(2):205–14.
27. Revuelta R, De Juambelz PP, Fernandez B, Flores JA. Lumbar disc herniation in a 27-month-old child. Case report. *J Neurosurg* 2000;**92**(1 Suppl.):98–100.
28. Dunne JW, Silbert PL, Wren M. A prospective study of acute radiculopathy after scoliosis surgery. *Clin Exp Neurol* 1991;**28**:180–90.
29. Lee HJ, Cho DY. Symptomatic spinal intradural arachnoid cysts in the pediatric age group: description of three new cases and review of the literature. *Pediatr Neurosurg* 2001;**35**(4):181–7.
30. Khalatbari MR, Moharamzad Y. Discal cyst in pediatric patients: case report and review of the literature. *Neuropediatrics* 2012;**43**(5):289–92.
31. Bond AE, Zada G, Bowen I, McComb JG, Krieger MD. Spinal arachnoid cysts in the pediatric population: report of 31 cases and a review of the literature. *J Neurosurg Pediatr* 2012;**9**(4):432–41.
32. Lahat E, Rothman AS, Aron AM. Schwannoma presenting as lumbar disc disease in an adolescent boy. *Ann Neurol* 1986;**20**(5):643–4.
33. Rothschild EJ, Savitz MH, Chang T, Worcester D, Peck HM. Primary vertebral tumor in an adolescent girl. *Spine* 1984;**9**(7):695–701.

34. Patel AJ, Fox BD, Fahim DK, Fulkerson DH, Whitehead WE, Curry DJ, et al. A clinicopathologic correlation in osteoblastoma of the spine in a child. *J Clin Neurosci* 2011;**18**(12):1728–30.
35. Anderson SC, Baquis GD, Jackson A, Monteleone P, Kirkwood JR. Ventral polyradiculopathy with pediatric acute lymphocytic leukemia. *Muscle Nerve* 2002;**25**(1):106–10.
36. Greene S, Hawkins DS, Rutledge JC, Tsuchiya KD, Douglas J, Ellenbogen RG, et al. Pediatric intradural extramedullary synovial sarcoma: case report. *Neurosurgery* 2006;**59**(6):E1339, discussion E.
37. Yamamoto T, Fujita I, Kurosaka M, Mizuno K. Sacral radiculopathy secondary to multicentric osteosarcoma. *Spine* 2001;**26**(15):1729–32.
38. Vazquez E, Lucaya J, Castellote A, Piqueras J, Sainz P, Olive T, et al. Neuroimaging in pediatric leukemia and lymphoma: differential diagnosis. *Radiographics* 2002;**22**(6):1411–28.
39. Colombo AA, Giorgiani G, Rognoni V, Villani P, Furione M, Bonora MR, et al. Differential outcome of neurological HCMV infection in two hematopoietic stem cell transplant recipients. *BMC Infect Dis* 2012;**12**:238.
40. Antunes NL, Wolden S, Souweidane MM, Lis E, Rosenblum M, Steinherz PG. Radiation myelitis in a 5-year-old girl. *J Child Neurol* 2002;**17**(3):217–19.
41. Kumar R, Kumar V, Das NK, Behari S, Mahapatra AK. Adolescent lumbar disc disease: findings and outcome. *Childs Nerv Syst* 2007;**23**(11):1295–9.
42. Ghabrial YA, Tarrant MJ. Adolescent lumbar disc prolapse. *Acta orthop Scand* 1989;**60**(2):174–6.
43. Crawford AH, Kucharzyk DW, Ruda R, Smitherman Jr. HC. Diskitis in children. *Clin Orthop Relat Res* 1991;**266**:70–9.
44. Nussinovitch M, Sokolover N, Volovitz B, Amir J. Neurologic abnormalities in children presenting with diskitis. *Arch Pediatr Adolesc Med* 2002;**156**(10):1052–4.
45. Stoev I, Powers AK, Puglisi JA, Munro R, Leonard JR. Sacroiliac joint pain in the pediatric population. *J Neurol Pediatr* 2012;**9**(6):602–7.
46. Rovira A, Capellades J, Zauner M, Bella R, Rovira M. Lumbar extradural hemangiomas: report of three cases. *AJNR Am J Neuroradiol* 1999;**20**(1):27–31.
47. Laufer I, Engel M, Feldstein N, Souweidane MM. Chiari malformation presenting as a focal motor deficit. *Report of two cases. J Neurol Pediatr* 2008;**1**(5):392–5.
48. Kozlowski K, Barylak A, Campbell J, Hoeffel JD, Beluffi G, Masel J, et al. Primary sacral bone tumours in children (report of 16 cases with a short literature review). *Australas Radiol* 1990;**34**(2):142–9.
49. Deathe AB. Hematometra as a cause of lumbar radiculopathy. A case report. *Spine* 1993;**18**(13):1920–1.
50. Yilmaz K, Caliskan M, Oge E, Aydinli N, Tunaci M, Ozmen M. Clinical assessment, MRI, and EMG in congenital brachial plexus palsy. *Pediatr Neurol* 1999;**21**(4):705–10.
51. Laurent JP, Lee RT. Birth-related upper brachial plexus injuries in infants: operative and nonoperative approaches. *J Child Neurol* 1994;**9**(2):111–17, discussion 8.
52. Chow BC, Blaser S, Clarke HM. Predictive value of computed tomographic myelography in obstetrical brachial plexus palsy. *Plast Reconstr Surg* 2000;**106**(5):971–7, discussion 8–9.
53. Hashimoto T, Mitomo M, Hirabuki N, Miura T, Kawai R, Nakamura H, et al. Nerve root avulsion of birth palsy: comparison of myelography with CT myelography and somatosensory evoked potential. *Radiology* 1991;**178**(3):841–5.
54. Azouz EM, Oudjhane K. Disorders of the upper extremity in children. *Magn Reson Imaging Clin N Am* 1998;**6**(3):677–95.
55. Birchansky S, Altman N. Imaging the brachial plexus and peripheral nerves in infants and children. *Semin Pediatr Neurol* 2000;**7**(1):15–25.
56. Francel PC, Koby M, Park TS, Lee BC, Noetzel MJ, Mackinnon SE, et al. Fast spin-echo magnetic resonance imaging for radiological assessment of neonatal brachial plexus injury. *J Neurosurg* 1995;**83**(3):461–6.
57. Gomez-Anson B. MR imaging of the brachial plexus. *J Neurol Neurosurg Psychiatry* 2000;**68**(6):801B.
58. Hems TE, Birch R, Carlstedt T. The role of magnetic resonance imaging in the management of traction injuries to the adult brachial plexus. *J Hand Surg* 1999;**24**(5):550–5.
59. Miller SF, Glasier CM, Griebel ML, Boop FA. Brachial plexopathy in infants after traumatic delivery: evaluation with MR imaging. *Radiology* 1993;**189**(2):481–4.
60. Nakamura T, Yabe Y, Horiuchi Y, Takayama S. Magnetic resonance myelography in brachial plexus injury. *J Bone Joint Surg Br* 1997;**79**(5):764–9.
61. Uetani M, Hayashi K, Hashmi R, Nakahara N, Aso N, Ito N. Traction injuries of the brachial plexus: signal intensity changes of the posterior cervical paraspinal muscles on MRI. *J Comput Assist Tomogr* 1997;**21**(5):790–5.
62. Chaudhry V, Cornblath DR. Wallerian degeneration in human nerves: serial electrophysiological studies. *Muscle Nerve* 1992;**15**(6):687–93.
63. Wilbourn AJ. Brachial plexus disorders. In: Dyck PJ, Thomas PK, Griffin JW, Low PA, Paduslo JF, editors. *Peripheral Neuropathy*. 3rd ed. Philadelphia: WB Saunders; 1993. pp. 903–50.
64. Luco JV, Eyzaguirre C. Fibrillation and hypersensitivity to ACh in denervated muscle: effect of length of degenerating nerve fibers. *J Neurophysiol* 1955;**18**(1):65–73.
65. Koenigsberger MR. Brachial plexus palsy at birth: Intrauterine or due to delivery trauma? [Abstract]. *Ann Neurol* 1980;**8**:228.
66. Peterson CR, Peterson CM. Brachial-plexus injury in an infant from a car safety seat. *N Engl J Med* 1991;**325**(22):1587–8.
67. Jones Jr. HR, Herbison GJ, Jacobs SR, Kollros PR, Macones GA. Intrauterine onset of a mononeuropathy: peroneal neuropathy in a newborn with electromyographic findings at age one day compatible with prenatal onset. *Muscle Nerve* 1996;**19**(1):88–91.
68. Ross D, Jones Jr. R, Fisher J, Konkol RJ. Isolated radial nerve lesion in the newborn. *Neurology* 1983;**33**(10):1354–6.
69. Sugioka H, Tsuyama N, Hara T, Nagano A, Tachibana S, Ochiai N. Investigation of brachial plexus injuries by intraoperative cortical somatosensory evoked potentials. *Arch Orthop Trauma Surg* 1982;**99**(3):143–51.
70. Landi A, Copeland SA, Parry CB, Jones SJ. The role of somatosensory evoked potentials and nerve conduction studies in the surgical management of brachial plexus injuries. *J Bone Joint Surg Br* 1980;**62-B**(4):492–6.
71. Turkof E, Millesi H, Turkof R, Pfundner P, Mayr N. Intraoperative electroneurodiagnostics (transcranial electrical motor evoked potentials) to evaluate the functional status of anterior spinal roots and spinal nerves during brachial plexus surgery. *Plast Reconstr Surg* 1997;**99**(6):1632–41.

72. Papazian O, Alfonso I, Yaylali I, Velez I, Jayakar P. Neurophysiological evaluation of children with traumatic radiculopathy, plexopathy, and peripheral neuropathy. *Semin Pediatr Neurol* 2000;**7**(1):26–35.
73. Robotti E, Longhi P, Verna G, Bocchiotti G. Brachial plexus surgery. An historical perspective. *Hand Clin* 1995;**11**(4):517–33.
74. Sever JW. Obstetric paralysis: Report of eleven hundred cases. *JAMA* 1925;**85**:1862–5.
75. Adler JB, Patterson Jr. RL. Erb's palsy. Long-term results of treatment in eighty-eight cases. *J Bone Joint Surg Am* 1967;**49**(6):1052–64.
76. Bager B. Perinatally acquired brachial plexus palsy—a persisting challenge. *Acta Paediatr* 1997;**86**(11):1214–19.
77. Brown KL. Review of obstetrical palsies. Nonoperative treatment. *Clin Plast Surg* 1984;**11**(1):181–7.
78. Kay SP. Obstetrical brachial palsy. *Br J Plast Surg* 1998;**51**(1):43–50.
79. Rouse DJ, Owen J, Goldenberg RL, Cliver SP. The effectiveness and costs of elective cesarean delivery for fetal macrosomia diagnosed by ultrasound. *JAMA* 1996;**276**(18):1480–6.
80. Strombeck C, Krumlinde-Sundholm L, Forssberg H. Functional outcome at 5 years in children with obstetrical brachial plexus palsy with and without microsurgical reconstruction. *Dev Med Child Neurol* 2000;**42**(3):148–57.
81. Vassalos E, Prevedourakis C, Paraschopoulou-Prevedouraki P. Brachial plexus paralysis in the newborn: An analysis of 169 cases. *Am J Obstet Gynecol* 1968;**101**:554–6.
82. Gilbert WM, Nesbitt TS, Danielsen B. Associated factors in 1611 cases of brachial plexus injury. *Obstet Gynecol* 1999;**93**(4):536–40.
83. McFarland LV, Raskin M, Daling JR, Benedetti TJ. Erb/Duchenne's palsy: a consequence of fetal macrosomia and method of delivery. *Obstet Gynecol* 1986;**68**(6):784–8.
84. Gordon M, Rich H, Deutschberger J, Green M. The immediate and long-term outcome of obstetric birth trauma. I. Brachial plexus paralysis. *Am J Obstet Gynecol* 1973;**117**(1):51–6.
85. Ballock RT, Song KM. The prevalence of nonmuscular causes of torticollis in children. *J Pediatr Orthop* 1996;**16**(4):500–4.
86. Benjamin B, Khan MR. Pattern of external birth trauma in southwestern Saudi Arabia. *J Trauma* 1993;**35**(5):737–41.
87. Dodds SD, Wolfe SW. Perinatal brachial plexus palsy. *Curr Opin Pediatr* 2000;**12**(1):40–7.
88. Ubachs JM, Slooff AC, Peeters LL. Obstetric antecedents of surgically treated obstetric brachial plexus injuries. *Br J Obstet Gynaecol* 1995;**102**(10):813–17.
89. Bryant DR, Leonardi MR, Landwehr JB, Bottoms SF. Limited usefulness of fetal weight in predicting neonatal brachial plexus injury. *Am J Obstet Gynecol* 1998;**179**(3 Pt 1):686–9.
90. Ecker JL, Greenberg JA, Norwitz ER, Nadel AS, Repke JT. Birth weight as a predictor of brachial plexus injury. *Obstet Gynecol* 1997;**89**(5 Pt 1):643–7.
91. McFarland MB, Langer O, Piper JM, Berkus MD. Perinatal outcome and the type and number of maneuvers in shoulder dystocia. *Int J Gynaecol Obstet* 1996;**55**(3):219–24.
92. Gonik B, Walker A, Grimm M. Mathematic modeling of forces associated with shoulder dystocia: a comparison of endogenous and exogenous sources. *Am J Obstet Gynecol* 2000;**182**(3):689–91.
93. Al-Qattan MM, Clarke HM, Curtis CG. The prognostic value of concurrent clavicular fractures in newborns with obstetric brachial plexus palsy. *J Hand Surg* 1994;**19**(6):729–30
94. Iffy L, McArdle JJ. The role of medico-legal reviews in medical research. *Med Law* 1996;**15**(3):399–406.
95. Pollack RN, Buchman AS, Yaffe H, Divon MY. Obstetrical brachial palsy: pathogenesis, risk factors, and prevention. *Clin Obstet Gynecol* 2000;**43**(2):236–46.
96. Narakas AO. Obstetrical brachial plexus injuries. In: Lamb DW editor. *The Paralyzed Hand.* Edinburgh: Churchill Livingstone, 1987. pp. 116–35.
97. Egloff DV, Raffoul W, Bonnard C, Stalder J. Palliative surgical procedures to restore shoulder function in obstetric brachial palsy. Critical analysis of Narakas' series. *Hand Clin* 1995;**11**(4):597–606.
98. Eng GD, Koch B, Smokvina MD. Brachial plexus palsy in neonates and children. *Arch Phys Med Rehabil* 1978;**59**(10):458–64.
99. Al-Qattan MM, Clarke HM, Curtis CG. Klumpke's birth palsy. Does it really exist? *J Hand Surg* 1995;**20**(1):19–23.
100. Gilbert A. Long-term evaluation of brachial plexus surgery in obstetrical palsy. *Hand Clin* 1995;**11**(4):583–94, discussion 94–5.
101. Clarke HM, Curtis CG. An approach to obstetrical brachial plexus injuries. *Hand Clin* 1995;**11**(4):563–80, discussion 80–1.
102. Michelow BJ, Clarke HM, Curtis CG, Zuker RM, Seifu Y, Andrews DF. The natural history of obstetrical brachial plexus palsy. *Plast Reconstr Surg* 1994;**93**(4):675–80, discussion 81.
103. Hoeksma AF, Wolf H, Oei SL. Obstetrical brachial plexus injuries: incidence, natural course and shoulder contracture. *Clin Rehabil* 2000;**14**(5):523–6.
104. Ramos LE, Zell JP. Rehabilitation program for children with brachial plexus and peripheral nerve injury. *Semin Pediatr Neurol* 2000;**7**(1):52–7.
105. Gilbert A, Tassin JL. Obstetrical palsy: A clinical, pathologic, and surgical review. In: Terzis JK, editor. *Microreconstruction of Nerve Injuries.* Philadelphia: WB Saunders Co; 1987. pp. 529–53.
106. Gilbert A, Razaboni R, Amar-Khodja S. Indications and results of brachial plexus surgery in obstetrical palsy. *Orthop Clin North Am* 1988;**19**(1):91–105.
107. Grossman JA. Early operative intervention for birth injuries to the brachial plexus. *Semin Pediatr Neurol* 2000;**7**(1):36–43.
108. Hoffer MM. The shoulder in neonatal brachial palsy. *Clin Orthop Relat Res* 1999;**368**:101–4.
109. Malessy MJ, Pondaag W. Obstetric brachial plexus injuries. *Neurosurg Clin N Am* 2009;**20**(1):1–14, v.
110. Eng GD. Brachial plexus palsy in newborn infants. *Pediatrics* 1971;**48**(1):18–28.
111. Eng GD, Binder H, Getson P, O'Donnell R. Obstetrical brachial plexus palsy (OBPP) outcome with conservative management. *Muscle Nerve* 1996;**19**(7):884–91.
112. Boome RS, Kaye JC. Obstetric traction injuries of the brachial plexus. Natural history, indications for surgical repair and results. *J Bone Joint Surg Br* 1988;**70**(4):571–6.
113. Kay SP, Yaszemski MJ, Rockwood Jr. CA. Acute tear of the rotator cuff masked by simultaneous palsy of the brachial plexus. A case report. *J Bone Joint Surg Am* 1988;**70**(4):611–12.
114. Terzis JK, Papakonstantinou KC. Management of obstetric brachial plexus palsy. *Hand Clin* 1999;**15**(4):717–36.

115. Laurent JP, Lee R, Shenaq S, Parke JT, Solis IS, Kowalik L. Neurosurgical correction of upper brachial plexus birth injuries. *J Neurosurg* 1993;**79**(2):197–203.
116. Gilbert A, Brockman R, Carlioz H. Surgical treatment of brachial plexus birth palsy. *Clin Orthop Relat Res* 1991;**264**:39–47.
117. Chuang DC, Wei FC, Noordhoff MS. Cross-chest C7 nerve grafting followed by free muscle transplantations for the treatment of total avulsed brachial plexus injuries: a preliminary report. *Plast Reconstr Surg* 1993;**92**(4):717–25, discussion 26–7
118. Chuang DC. Neurotization procedures for brachial plexus injuries. *Hand Clin* 1995;**11**(4):633–45.
119. Holland NR, Belzberg AJ. Intraoperative electrodiagnostic testing during cross-chest C7 nerve root transfer. *Muscle Nerve* 1997; **20**(7):903–5.
120. Carlstedt T, Grane P, Hallin RG, Noren G. Return of function after spinal cord implantation of avulsed spinal nerve roots. *Lancet* 1995;**346**(8986):1323–5.
121. Carlstedt T, Anand P, Hallin R, Misra PV, Noren G, Seferlis T. Spinal nerve root repair and reimplantation of avulsed ventral roots into the spinal cord after brachial plexus injury. *J Neurosurg* 2000;**93**(2 Suppl.):237–47.
122. Glasby MA, Hems TE. Repairing spinal roots after brachial plexus injuries. *Paraplegia* 1995;**33**(6):359–61.
123. Kline DG. Spinal nerve root repair after brachial plexus injury. *J Neurosurg* 2000;**93**(2 Suppl.):336–8.
124. Carlstedt T. Nerve root replantation. *Neurosurg Clin N Am* 2009;**20**(1):39–50, vi.
125. Birch R. Brachial plexus injuries. *J Bone Joint Surg Br* 1996; **78**(6):986–92.
126. Birch R, Achan P. Peripheral nerve repairs and their results in children. *Hand Clin* 2000;**16**(4):579–95.
127. Liggio FJ, Tham S, Price A, Ramos LE, Mulloy E, Grossman JA. Outcome of surgical treatment for forearm pronation deformities in children with obstetric brachial plexus injuries. *J Hand Surg* 1999;**24**(1):43–5.
128. Birch R. Surgery for brachial plexus injuries. *J Bone Joint Surg Br* 1993;**75**(3):346–8.
129. Jeffery AR, Ellis FJ, Repka MX, Buncic JR. Pediatric Horner syndrome. *J Aapos* 1998;**2**(3):159–67.
130. Al-Qattan MM, Clarke HM, Curtis CG. The prognostic value of concurrent Horner's syndrome in total obstetric brachial plexus injury. *J Hand Surg* 2000;**25**(2):166–7.
131. Malessy MJ, Pondaag W, Yang LJ, Hofstede-Buitenhuis SM, le Cessie S, van Dijk JG. Severe obstetric brachial plexus palsies can be identified at one month of age. *PLOS ONE* 2011;**6**(10): e26193.
132. Gjorup L. Obstetrical lesion of the brachial plexus. *Acta Neurol Scand* 1966;**42**(Suppl. 18):1–80.
133. Rust RS. Congenital brachial plexus palsy: where have we been and where are we now? *Semin Pediatr Neurol* 2000;**7**(1):58–63.
134. Pondaag W, Malessy MJ, van Dijk JG, Thomeer RT. Natural history of obstetric brachial plexus palsy: a systematic review. *Dev Med Child Neurol* 2004;**46**(2):138–44.
135. Smith SJ. The role of neurophysiological investigation in traumatic brachial plexus lesions in adults and children. *J Hand Surg* 1996;**21**(2):145–7.
136. Eng GD. Neuromuscular disease. In: Avery GB, editor. *Neonatology: Pathophysiology and Management of the Newborn.* Philadelphia: WB Saunders; 1981. pp. 989–92.
137. Jahnke Jr. AH, Bovill DF, McCarroll Jr. HR, James P, Ashley RK. Persistent brachial plexus birth palsies. *J Pediatr Orthop* 1991;**11**(4):533–7.
138. van Dijk JG, Malessy MJ, Stegeman DF. Why is the electromyogram in obstetric brachial plexus lesions overly optimistic? *Muscle Nerve* 1998;**21**(2):260–1.
139. van Dijk JG, Pondaag W, Malessy MJ. Obstetric lesions of the brachial plexus. *Muscle Nerve* 2001;**24**(11):1451–61.
140. Johnson EW, Alexander MA, Koenig WC. Infantile Erb's palsy (Smellie's palsy). *Arch Phys Med Rehabil* 1977;**58**(4):175–8.
141. Brown T, Cupido C, Scarfone H, Pape K, Galea V, McComas A. Developmental apraxia arising from neonatal brachial plexus palsy. *Neurology* 2000;**55**(1):24–30.
142. Rapalino OA, Levine DN. Developmental apraxia arising from neonatal brachial plexus palsy. *Neurology* 2000;**55**(11):1761.
143. Rollnik JD, Hierner R, Schubert M, Shen ZL, Johannes S, Troger M, et al. Botulinum toxin treatment of cocontractions after birth-related brachial plexus lesions. *Neurology* 2000;**55** (1):112–14.
144. Van Dijk JG, Pondaag W, Buitenhuis SM, Van Zwet EW, Malessy MJ. Needle electromyography at 1 month predicts paralysis of elbow flexion at 3 months in obstetric brachial plexus lesions. *Dev Med Child Neurol* 2012;**54**(8):753–8.
145. Al-Qattan MM. The outcome of Erb's palsy when the decision to operate is made at 4 months of age. *Plast Reconstr Surg* 2000;**106**(7): 1461–5.
146. Agboatwalla M, Kirmani SR, Sonawalla A, Akram DS. Nerve conduction studies and its importance in diagnosis of acute poliomyelitis. *Indian J Pediatr* 1993;**60**(2):265–8.
147. Alfonso I, Papazian O, Prieto G, Alfonso DT, Melnick SJ. Neoplasm as a cause of brachial plexus palsy in neonates. *Pediatric neurology* 2000;**22**(4):309–11.
148. Beausoleil JL, Nordgren RE, Modlin JF. Vaccine-associated paralytic poliomyelitis. *J Child Neurol* 1994;**9**(3):334–5.
149. Bell WE. Acquired immobility of the right arm. *Semin Pediatr Neurol* 1996;**3**(3):173–6.
150. Escolar DM, Jones Jr. HR. Pediatric radial mononeuropathies: a clinical and electromyographic study of sixteen children with review of the literature. *Muscle Nerve* 1996;**19**(7):876–83.
151. Felice KJ, Royden Jones Jr. H. Pediatric ulnar mononeuropathy: report of 21 electromyography-documented cases and review of the literature. *J Child Neurol* 1996;**11**(2):116–20.
152. Gabriel SR, Thometz JG, Jaradeh S. Septic arthritis associated with brachial plexus neuropathy. A case report. *J Bone Joint Surg Am* 1996;**78**(1):103–5.
153. Clay SA. Osteomyelitis as a cause of brachial plexus neuropathy. *Am J Dis Child* 1982;**136**(12):1054–6.
154. Gainor BJ, Olson S. Combined entrapment of the median and anterior interosseous nerves in a pediatric both-bone forearm fracture. *J Orthop Trauma* 1990;**4**(2):197–9.
155. Jones HR. Infantile plexopathies and neuropathies with intrauterine pathophysiologic mechanisms. Xth *International Congress of EMG and Clinical Neurophysiology*. Amsterdam: Elsevier Science; 1996. pp. 492–7.
156. Sadleir LG, Connolly MB. Acquired brachial-plexus neuropathy in the neonate: a rare presentation of late-onset group-B streptococcal osteomyelitis. *Dev Med Child Neurol* 1998;**40**(7):496–9.
157. Alfonso I, Alfonso DT, Papazian O. Focal upper extremity neuropathy in neonates. *Semin Pediatr Neurol* 2000;**7**(1):4–14.

158. Sharma RR, Sethu AU, Mahapatra AK, Pawar SJ, Nath A. Neonatal cervical osteomyelitis with paraspinal abscess and Erb's palsy. A case report and brief review of the literature. *Pediatric neurosurgery* 2000;**32**(5):230–3.

159. Lucas JW, Holden KR, Purohit DM, Cure JK. Neonatal hemangiomatosis associated with brachial plexus palsy. *J Child Neurol* 1995;**10**(5):411–13.

160. Harila-Saari AH, Vainionpaa LK, Kovala TT, Tolonen EU, Lanning BM. Nerve lesions after therapy for childhood acute lymphoblastic leukemia. *Cancer* 1998;**82**(1):200–7.

161. Alessandri AJ, Pritchard SL, Massing BG, Sargent MA, Theissen P, Schultz KR. Misleading leads: Bone pain caused by isolated paraspinal extramedullary relapse of childhood acute lymphoblastic leukemia. *Med Pediatr Oncol* 1999;**33**(2):113–15.

162. Inoue M, Kawano T, Matsumura H, Mori K, Yoshida T. Solitary benign schwannoma of the brachial plexus. *Surgical neurology* 1983;**20**(2):103–8.

163. Van Alfen N. Clinical and pathophysiological concepts of neuralgic amyotrophy. *Nat Rev Neurol* 2011;**7**(6):315–22.

164. Dunn HG, Daube JR, Gomez MR. Heredofamilial branchial plexus neuropathy (hereditary neuralgic amyotrophy with branchial predilection) in childhood. *Dev Med Child Neurol* 1978;**20**(1):28–46.

165. Chance PF, Lensch MW, Lipe H, Brown Sr. RH, Brown Jr. RH, Bird TD. Hereditary neuralgic amyotrophy and hereditary neuropathy with liability to pressure palsies: two distinct genetic disorders. *Neurology* 1994;**44**(12):2253–7.

166. Ueda M, Kawamura N, Tateishi T, Sakae N, Motomura K, Ohyagi Y, et al. Phenotypic spectrum of hereditary neuralgic amyotrophy caused by the SEPT9 R88W mutation. *J Neurol Neurosurg Psychiatry* 2010;**81**(1):94–6.

167. Gouider R, LeGuern E, Emile J, Tardieu S, Cabon F, Samid M, et al. Hereditary neuralgic amyotrophy and hereditary neuropathy with liability to pressure palsies: two distinct clinical, electrophysiologic, and genetic entities. *Neurology* 1994;**44**(12):2250–2.

168. Stogbauer F, Young P, Kuhlenbaumer G, De Jonghe P, Timmerman V. Hereditary recurrent focal neuropathies: clinical and molecular features. *Neurology* 2000;**54**(3):546–51.

169. Hall PA, Russell SE. The pathobiology of the septin gene family. *J Pathol* 2004;**204**(4):489–505.

170. Verhagen WI, Gabreels-Festen AA, van Wensen PJ, Joosten EM, Vingerhoets HM, Gabreels FJ, et al. Hereditary neuropathy with liability to pressure palsies: a clinical, electroneurophysiological and morphological study. *J Neurol Sci* 1993;**116**(2):176–84.

171. Behse F, Buchthal F, Carlsen F, Knappeis GG. Hereditary neuropathy with liability to pressure palsies. Electrophysiological and histopathological aspects. *Brain* 1972;**95**(4):777–94.

172. Hirasawa Y, Sakakida K. Sports and peripheral nerve injury. *Am J Sports Med* 1983;**11**(6):420–6.

173. Barisic N, Perovic D, Mitrovic Z, Jurenic D, Zagar M. Assessment of war and accidental nerve injuries in children. *Pediatr Neurol* 1999;**21**(1):450–5.

174. Dumitru D. Brachial plexopathies and proximal mononeuropathies. In: Dumitru D, editor. *Electrodiagnostic Medicine*. Philadelphia: Hanley & Belfus Inc; 1995. pp. 585–642.

175. Kline DG. Operative management of major nerve lesions of the lower extremity. *Surg Clin North Am* 1972;**52**(5):1247–65.

176. Kline DG, Judice DJ. Operative management of selected brachial plexus lesions. *J Neurosurg* 1983;**58**(5):631–49.

177. Kline DG. Civilian gunshot wounds to the brachial plexus. *J Neurosurg* 1989;**70**(2):166–74.

178. Kline DG, Hudson AR. Diagnosis of root avulsions. *J Neurosurg* 1997;**87**(3):483–4.

179. Mehlman CT, Scott KA, Koch BL, Garcia VF. Orthopaedic injuries in children secondary to airbag deployment. *J Bone Joint Surg Am* 2000;**82**(6):895–8.

180. Menegaux F, Keeffe EB, Andrews BT, Egawa H, Monge H, Concepcion W, et al. Neurological complications of liver transplantation in adult versus pediatric patients. *Transplantation* 1994;**58**(4):447–50.

181. Schwartz DM, Drummond DS, Hahn M, Ecker ML, Dormans JP. Prevention of positional brachial plexopathy during surgical correction of scoliosis. *J Spinal Disord* 2000;**13**(2):178–82.

182. Chrisman OD, Snook GA, Stanitis JM, Keedy VA. lateral-flexion neck injuries in athletic competition. *JAMA* 1965;**192**:613–15.

183. Clancy WG. Brachial plexus and upper extremity peripheral nerve injuries. In: Torg JS, editor. *Athletic Injuries to the Head, Neck, and Face*. Philadelphia: Lea & Febiger; 1982. pp. 215–20.

184. Di Benedetto M, Markey K. Electrodiagnostic localization of traumatic upper trunk brachial plexopathy. *Arch Phys Med Rehabil* 1984;**65**(1):15–17.

185. Robertson Jr. WC, Eichman PL, Clancy WG. Upper trunk brachial plexopathy in football players. *JAMA* 1979; **241**(14):1480–2.

186. Rockett FX. Observations on the "burner": traumatic cervical radiculopathy. *Clin Orthop Relat Res* 1982 **164**:18–19.

187. Watkins RG. Neck injuries in football players. *Clin Sports Med* 1986;**5**(2):215–46.

188. Wilbourn AJ. Electrodiagnostic testing of neurologic injuries in athletes. *Clin Sports Med* 1990;**9**(2):229–45.

189. Wroble RR, Albright JP. Neck and low back injuries in wrestling. *Clin Sports Med* 1986;**5**(2):295–325.

190. Clancy Jr. WG, Brand RL, Bergfield JA. Upper trunk brachial plexus injuries in contact sports. *Am J Sports Med* 1977;**5**(5):209–16.

191. Pellegrino JE, George RA, Biegel J, Farlow MR, Gardner K, Caress J, et al. Hereditary neuralgic amyotrophy: evidence for genetic homogeneity and mapping to chromosome 17q25. *Hum Genet* 1997;**101**(3):277–83.

192. Rothner AD, Wilbourn A, Mercer RD. Rucksack palsy. *Pediatrics* 1975;**56**(5):822–4.

193. Parsonage MJ, Turner JW. Neuralgic amyotrophy; the shoulder-girdle syndrome. *Lancet* 1948;**1**(6513):973–8.

194. To WC, Traquina DN. Neuralgic amyotrophy presenting with bilateral vocal cord paralysis in a child: a case report. *Int J Pediatr Otorhinolaryngol* 1999;**48**(3):251–4.

195. Tsairis P, Dyck PJ, Mulder DW. Natural history of brachial plexus neuropathy. Report on 99 patients. *Arch Neurol* 1972;**27**(2):109–17.

196. Turner JW, Parsonage MJ. Neuralgic amyotrophy (paralytic brachial neuritis); with special reference to prognosis. *Lancet* 1957;**273**(6988):209–12.

197. Bale Jr. JF, Thompson JA, Petajan JH, Ziter FA. Childhood brachial plexus neuropathy. *J Pediatr* 1979;**95**(5 Pt 1):741–2.

198. Stoehr M. Traumatic and postoperative lesions of the lumbosacral plexus. *Arch Neurol* 1978;**35**(11):757–60.

199. Barquet A. Traumatic anterior dislocation of the hip in childhood. *Injury* 1982;**13**(5):435–40.

200. Chad DA, Bradley WG. Lumbosacral plexopathy. *Semin Neurol* 1987;**7**(1):97–107.
201. Chiou-Tan FY, Kemp Jr. K, Elfenbaum M, Chan KT, Song J. Lumbosacral plexopathy in gunshot wounds and motor vehicle accidents: comparison of electrophysiologic findings. *Am J Phys Med Rehabil* 2001;**80**(4):280–5, quiz 6–8.
202. Christie J, Jamieson EW. Traction lesion of the lumbosacral plexus. *J R Coll Surg Edinb* 1974;**19**(6):384–5.
203. Egel RT, Cueva JP, Adair RL. Posttraumatic childhood lumbosacral plexus neuropathy. *Pediatr Neurol* 1995;**12**(1):62–4.
204. Jellis JE, Helal B. Childhood sciatic palsies: congenital and traumatic. *Proc R Soc Med* 1970;**63**(7):655–6.
205. Kline DG, Kim D, Midha R, Harsh C, Tiel R. Management and results of sciatic nerve injuries: a 24-year experience. *J Neurosurg* 1998;**89**(1):13–23.
206. Marra TA. Recurrent lumbosacral and brachial plexopathy associated with schistosomiasis. *Arch Neurol* 1983;**40**(9):586–7.
207. Rai SK, Far RF, Ghovanlou B. Neurologic deficits associated with sacral wing fractures. *Orthopedics* 1990;**13**(12):1363–6.
208. Switzer JA, Nork SE, Routt Jr. ML. Comminuted fractures of the iliac wing. *J Orthop Trauma* 2000;**14**(4):270–6.
209. Evans BA, Stevens JC, Dyck PJ. Lumbosacral plexus neuropathy. *Neurology* 1981;**31**(10):1327–30.
210. Thomson AJ. Idiopathic lumbosacral plexus neuropathy in two children. *Dev Med Child Neurol* 1993;**35**(3):258–61.
211. Brown JA, Braun MA, Namey TC. Pyriformis syndrome in a 10-year-old boy as a complication of operation with the patient in the sitting position. *Neurosurgery* 1988;**23**(1):117–19.
212. MacDonald NE, Marcuse EK. Neurologic injury after vaccination: buttocks as injection site. *CMAJ* 1994;**150**(3):326.
213. MacDonald NE. Does immunization in the buttocks cause sciatic nerve injury? [Abstract]. *Pediatrics* 1994;**93**:351.
214. Marin R, Bryant PR, Eng GD. Lumbosacral plexopathy temporally related to vaccination. *Clin Pediatr (Phila)* 1994;**33**(3):175–7.
215. Villarejo FJ, Pascual AM. Injection injury of the sciatic nerve (370 cases). *Childs Nerv Syst* 1993;**9**(4):229–32.
216. Gabreels-Festen AA, Gabreels FJ, Joosten EM, Vingerhoets HM, Renier WO. Hereditary neuropathy with liability to pressure palsies in childhood. *Neuropediatrics* 1992;**23**(3):138–43.
217. Galan E, Kousseff BG. Peripheral neuropathy in Ehlers-Danlos syndrome. *Pediatr Neurol* 1995;**12**(3):242–5.
218. Sunderland S. The connective tissues of peripheral nerves. *Brain* 1965;**88**(4):841–54.
219. Sunderland S. Anatomical features of nerve trunks in relation to nerve injury and nerve repair. *Clin Neurosurg* 1970;**17**:38–62.
220. Ebraheim NA, Lu J, Yang H, Huntoon M, Yeasting RA. Lumbosacral plexus: a histological study. *Acta Anat (Basel)* 1997;**158**(4):274–8.
221. Bye A, Fagan E. Nerve conduction studies of the sural nerve in childhood. *J Child Neurol* 1988;**3**(2):94–9.
222. Dumitru D. Lumbosacral plexopathies and proximal mononeuropathies. In: Dumitru D, editor. *Electrodiagnostic Medicine*. Philadelphia: Hanley & Belfus Inc; 1995. pp. 643–88.
223. Hope EE, Bodensteiner JB, Thong N. Neonatal lumbar plexus injury. *Arch Neurol* 1985;**42**(1):94–5.
224. Volpe JJ. Injuries of extracranial, cranial, intracranial, spinal cord, and peripheral nervous system structures. In: Volpe JJ, editor. *Neurology of the Newborn*. Philadelphia: JB Lippincott; 1987. pp. 638–58.
225. Barnett HG, Connolly ES. Lumbosacral nerve root avulsion: report of a case and review of the literature. *J Trauma* 1975;**15**(6):532–5.
226. Kolawole TM, Hawass ND, Shaheen MA, Badr AH, Rahman NU. Lumbosacral plexus avulsion injury: clinical, myelographic and computerized tomographic features. *J Trauma* 1988;**28**(6):861–5.
227. Verstraete KL, Martens F, Smeets P, Vandekerckhove T, Meire D, Parizel PM, et al. Traumatic lumbosacral nerve root meningoceles. The value of myelography, CT and MRI in the assessment of nerve root continuity. *Neuroradiology* 1989;**31**(5):425–9.
228. Young NL, Davis RJ, Bell DF, Redmond DM. Electromyographic and nerve conduction changes after tibial lengthening by the Ilizarov method. *J Pediatr Orthop* 1993;**13**(4):473–7.
229. Jones Jr. HR, Gianturco LE, Gross PT, Buchhalter J. Sciatic neuropathies in childhood: a report of ten cases and review of the literature. *J Child Neurol* 1988;**3**(3):193–9.
230. Yuen EC, Olney RK, So YT. Sciatic neuropathy: clinical and prognostic features in 73 patients. *Neurology* 1994;**44**(9):1669–74.
231. Awerbuch GI, Nigro MA, Dabrowski E, Levin JR. Childhood lumbosacral plexus neuropathy. *Pediatr Neurol* 1989;**5**(5):314–16.
232. Awerbuch GI, Levin GR, Dabrowski E. Lumbosacral plexus neuropathy of children [Abstract]. *Ann Neurol* 1989;**26**:452.
233. Chin CH, Chew KC. Lumbosacral nerve root avulsion. *Injury* 1997;**28**(9–10):674–8.
234. Pasternak JF, Volpe JJ. Lumbosacral lipoma with acute deterioration during infancy. *Pediatrics* 1980;**66**(1):125–8.
235. Shaw BA, Holman M. Traumatic lumbosacral nerve root avulsions in a pediatric patient. *Orthopedics* 2003;**26**(1):89–90.
236. Gelabert-Gonzalez M, Prieto-Gonzalez A, Maria Santin-Amo J, Serramito-Garcia R, Garcia-Allut A. Lumbar synovial cyst in an adolescent: case report. *Childs Nerv Syst* 2009;**25**(6):719–21.
237. Metkar U, Wardak Z, Katz DA, Lavelle WF. Giant cell tumor of a lumbar vertebra in a 7-year-old child: a case report. *J Pediatr Orthop* 2012;**32**(8):e76–80.
238. Novais EN, Rose PS, Yaszemski MJ, Sim FH. Aneurysmal bone cyst of the cervical spine in children. *J Bone Joint Surg Am* 2011;**93**(16):1534–43.
239. Wilbourn AJ, Aminoff MJ. Radiculopathies. In: Brown WF, Bolton CF, editors. *Clinical Electromyography*. 2nd ed. Boston: Butterworth-Heinemann; 1993. pp. 175–205.
240. Preston DC, Shapiro BE. *Electromyography and Neuromuscular Disorders: Clinical-Electrophysiologic Correlations*. Boston: Butterworth-Heinemann; 1998.
241. Brown WF, Bolton CF, Aminoff MJ, editors. *Neuromuscular Function and Disease*. 2nd ed. Philadelphia: Saunders; 2002.
242. Gilbert A. Obstetrical brachial plexus palsy. In: Tubiana R, editor. *The Hand*, Vol. 4. Philadelphia: Saunders; 1993.

Chapter 13

Congenital and Acquired Facial Palsies

Francis Renault and Susana Quijano-Roy

INTRODUCTION

Facial palsy is one of the most frequent mononeuropathies in the pediatric age group. The diagnosis of idiopathic benign Bell's palsy is based on exclusion of other acquired central or peripheral nervous system conditions, some of which have more ominous implications. Facial paralysis in a newborn may be due to prenatal or obstetric stress to the nerve. Congenital facial weakness or asymmetry may also reveal a recognizable malformation syndrome or congenital neuromuscular disorder. Any child who has either a congenital or acquired facial weakness requires a detailed general and neurologic examination complemented with neurophysiologic, neuroimaging, auditory, and other laboratory studies. Because of significant functional and esthetic consequences, the evaluation of a child with facial paralysis requires that major emphasis be placed on early etiologic definition, therapeutic decisions, and a reliable prognostic assessment.

EVALUATION OF THE FACIAL NERVE IN CHILDREN

Anatomy

The facial nerve (Figure 13.1) emerges from multiple functionally specialized ventrolateral brainstem nuclei. The motor nucleus of cranial nerve VII originates in the caudal pons. Five groups of cells are defined, corresponding to a topographic organization of motor neurons innervating different muscles. The ventral groups of cells supply the periorbital muscles; the dorsal groups supply the perioral muscles.[1,2] Taste fibers for the anterior two-thirds of the tongue, and sensory fibers from the external acoustic canal, project to the upper part of the nucleus of the tractus solitarius in the vicinity of the motor nucleus of cranial nerve VII. The lacrimal, sublingual, and submandibular preganglionic parasympathetic fibers of the facial nerve originate in the nearby superior salivary nucleus. The twin sensory and motor roots of the facial nerve emerge from the brainstem at the level of the bulbopontine sulcus, between the sixth and eighth cranial nerves, at the cerebellopontine angle, then enter the pars petrosa ossis temporalis via the internal acoustic meatus and follow a common path in the first part of the facial canal. The sensory components enter the geniculate ganglion at this level. Motor fibers of the facial nerve traverse peripherally through the internal auditory canal in the temporal bone, in company with the nervus intermedius, consisting of the sensory fibers of cranial nerve VII as well as with the VIIIth cranial nerve. The greater petrosal nerve arises from the labyrinthine segment of the facial nerve and carries parasympathetic fibers to the pterygopalatine ganglion, where it synapses and sends fibers to the lacrimal gland. More distally, the facial nerve's mastoid segment travels inferiorly and laterally to the jugular fossa. Here, the stapedius nerve arises to innervate the stapedius muscle. The facial nerve gives off the chorda tympani in the distal mastoid segment. The chorda tympani then passes through the middle ear to supply afferent taste fibers to the anterior two-thirds of the tongue, as well as parasympathetic fibers to the submandibular and sublingual glands. The primary motor bundle of the facial nerve traverses the second and third parts of the facial canal alone. The facial motor nerve emerges from the skull at the stylomastoid foramen, gives off the posterior auricular branch, and then divides into two main terminal branches in the parotid gland. The inferior cervicofacial branch passes down along the mandible to supply the muscles in the lower part of the face, giving off a buccal branch to the risorius, the buccinator, and the orbicularis oris muscles; a mental branch to the depressor anguli oris, the depressor labii inferioris, and the mentalis muscles; and a cervical branch to platysma. The superior temporofacial branch runs horizontally forward, giving off frontal branches to the frontalis and orbicularis oculi muscles; suborbital branches to the levator labii superioris, zygomaticus, levator anguli oris, and dilatator naris muscles; and buccal branches to the buccinator and

B.T. Darras, H. Royden Jones, Jr., M.M. Ryan & D.C. De Vivo (Eds): Neuromuscular Disorders of Infancy, Childhood and Adolescence, Second edition.
DOI: http://dx.doi.org/10.1016/B978-0-12-417044-5.00013-5

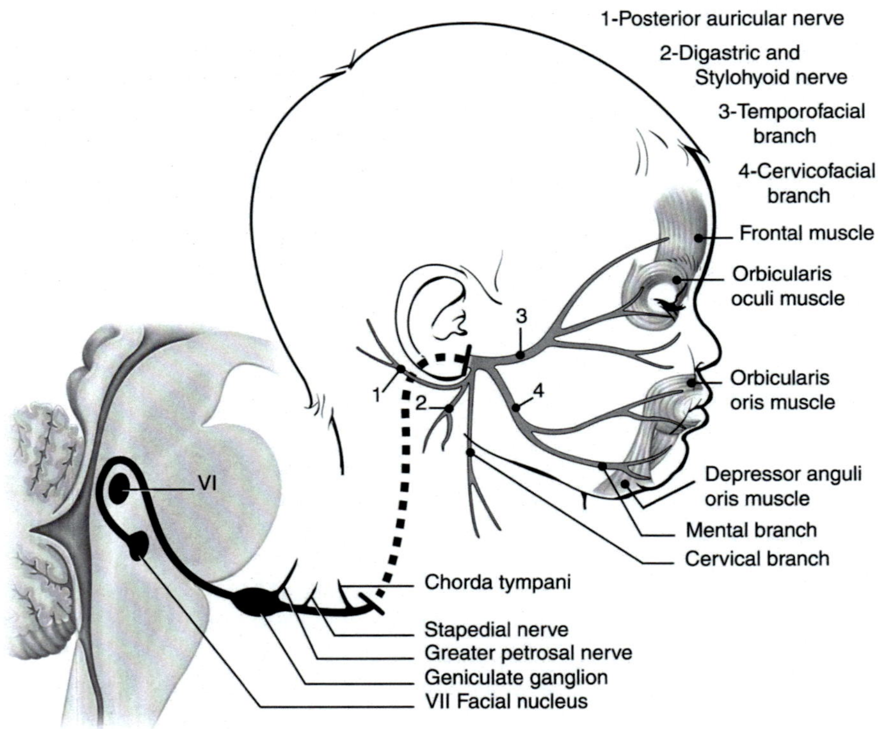

FIGURE 13.1 Functional anatomy of the facial nerve.

orbicularis oris muscles. Although there is considerable diversity in its trajectory and divisions, cranial nerve VII innervates all muscles of facial expression except the levator palpebrae superioris. The close anatomic proximity of cranial nerves V, VI, VII, and VIII in the brainstem and cerebellopontine angle explains the involvement of multiple nerves in malformations, and in ischemic or compressive disorders. Interneuronal and synaptic V–VII connections in the brainstem provide the basis for the neurophysiologic studies exploring the blink reflex.[3,4] The geniculate ganglion, located within the internal auditory canal, contains the soma of sensory facial fibers. Reactivation of viral particles latent in the geniculate ganglion is believed to be involved in the pathogenesis of idiopathic and Ramsay Hunt facial palsies. Lesions of the facial nerve trunk localized distal to the ganglion may involve different facial nerve branches (i. e. greater superficial petrosal nerve, nerve to stapedius muscle, and chorda tympani), causing variable dysfunction of lacrimation, salivation, taste, and/or hearing. Nerve entrapment is most likely to occur at the narrowest intraosseus segment, the meatal foramen, where it is not protected by epineurium and perineurium.[5,6]

Clinical Examination

Newborn and Young Infants

In a newborn patient with unilateral facial palsy, the face may appear symmetrical at rest. Examination on crying, however, reveals mouth deviation, failure to completely close the eye, and a wider palpebral fissure on the nonaffected side. Absence of the fine dilatory movements of the nostril synchronous with breathing, an asymmetrical searching reflex, asymmetrical closing of the lips over the pacifier, and absence of frontal wrinkling on the nonaffected side may be also noted. Clinical features suggesting a possible traumatic origin include mastoid, pre-auricular, or temporal hematoma; otorrhagia; or hemotympanum. If there is fever and sensitivity to pressure over the preauricular area, or a context of maternal-fetal infection, neonatal otitis media must be excluded. Associated dysfunction of other cranial nerves, mandibulofacial or ocular malformations, or signs of any systemic dysfunction may provide information relevant to both etiology and prognosis. Also, an ophthalmologic examination may detect ocular malformations, subtle oculomotor dysfunction, or a retinal defect that may characterize a specific congenital malformation syndrome or embryofetopathy.

Older Children

Facial palsies in older children are typically characterized by inability to close the eye, disappearance of the nasolabial fold, and deviation of the mouth. Associated features such as decreased tearing, hyperacusis, and loss of taste sensation may help to localize the seventh nerve lesion. The topography and severity of a facial palsy is assessed by observing the response to commands activating the different branches of the facial motor nerve: closing the eyes, elevating the eyebrows, frowning, showing the teeth, puckering the lips, and tensing the soft tissues of the neck. A standardized assessment scale, such as the House-Brackmann grading system, may be useful to define severity and record progress.[7] A full neurologic examination aims to rule out involvement of other cranial nerves or a more widespread neurologic process. The general physical examination includes evaluation of mastoid and parotid areas, visualization of the external auditory canal, and inspection of the tympanic membrane. Mass lesions, inflammation, or infection may lead to facial nerve injury; vesicles or scabby skin may implicate zoster virus infection. The ophthalmologic examination searches for concomitant conjunctival or corneal complications, or signs of increased intracranial pressure.[8]

Clinical-Anatomic Relationships

When partial or complete facial paralysis has developed, associated features may help in distinguishing central from peripheral palsies, and determining the specific site of the lesion in the cranial nerve VII pathways or nucleus. Lesions in the pons may induce hyperacusis due to dysfunction of motor fibers to the stapedius muscle, and may progress to other features of brainstem involvement. A lesion between pons and geniculate ganglion may produce hyperacusis and impairment of lacrimation, salivation, and taste of the anterior two-thirds of the tongue. Between the geniculate ganglion and stapedius nerve, hyperacusis and impairment of salivation and taste are expected, with preservation of lacrimation. Between the stapedius nerve and chorda tympani, salivation and taste will be impaired but hearing is normal. Beyond the exit of chorda tympani, the only symptom will be facial weakness.

Central facial palsies (suprabulbar) show more widespread clinical features and a progressive course. Upward and outward eye movement when blinking (Bell's sign) is usually absent. Typically, spontaneous facial expression, such as smiling as an emotional reaction, is preserved, while no voluntary movement is obtained in response to a command, such as asking the patient to smile.

Complementary Investigations

Congenital facial palsies (CFPs) have a more standardized diagnostic algorithm than acquired facial palsies (AFPs). Most CFPs are either related to a congenital malformation or are of traumatic origin. A combination of neurophysiologic tests and neuroimaging provides an opportunity for early diagnosis, management, and determination of prognosis,[9] but their use in AFPs is still controversial. Globally, electrodiagnostic studies are used to search for prognostic indicators. Magnetic resonance imaging (MRI) is mandatory in traumatic cases or whenever an intracranial lesion is suspected. Cerebrospinal fluid (CSF) examination is not a routine test but may be useful to rule out neoplastic, infectious, or inflammatory meningeal involvement. Polymerase chain reaction (PCR) of serum, CSF, or saliva may allow early detection of Lyme disease, varicella or other herpes viruses, taking into account geographical and seasonal differences.

Middle Ear and Auditory Investigations

Several noninvasive tests are useful in young children with no localizing symptoms and in older patients with associated hearing loss or nonspecific ear complaints. These studies provide a means to localize the facial nerve lesion, search for middle ear infection or trauma, and define concomitant eighth cranial nerve involvement. Otoscopy will determine the presence or absence of outer and middle ear disease. Impedance audiometry evaluates the middle ear system. The acoustic-stapedial reflex (AR) is easy to perform at any age and detects the contraction of the ipsilateral stapedius muscle in response to a high level stimulus (70 to 90 dB) at different frequencies. Results are shown as absent reflex, reflex requiring thresholds over 90 dB, and normal response. Absence of AR is expected if the lesion is proximal to the branching of the nerve to stapedius muscle. Abnormal AR during the first week after clinical onset has been associated with worse recovery of AFP.[10] Pure-tone and speech audiometry can detect hearing loss. In infants or uncooperative children, hearing thresholds may be evaluated by transiently evoked otoacoustic emissions. Brainstem auditory evoked responses (BAERs) can be performed at any age, and are useful for measuring auditory thresholds and studying retrocochlear auditory pathways. The Schirmer lacrimation test, the salivary flow test, and electrogustometry are not routinely undertaken in young children.

Imaging Studies

In CFPs, MRI is used mainly to rule out associated cerebral lesions or posterior fossa malformations, although it may reveal abnormal signal in the area of the seventh cranial nerve nucleus or other minor cerebral anatomic defects.[11] In AFPs, imaging is particularly valuable when there are otoscopic findings of a mass in the middle ear or a history of chronic otitis media, previous mastoid surgery, or trauma. High-resolution computed tomography (CT) is the best method to study the course of the facial nerve through the petrous bone, and detect a fracture line or posttraumatic hemorrhage. Pre- and postcontrast MRI is used to investigate the brainstem, cranial nerves, and parotid gland. MRI will be required in the presence of complicated otitis media, multiple cranial mononeuropathies, recurrent ipsilateral facial paralysis, progression beyond three weeks, absence of improvement after 6 months, development of facial hemispasm or other neurologic signs, or retrocochlear abnormalities in the BAERs.[12] Contrast enhancement of the geniculate ganglion within the labyrinthine segment of the facial nerve is a common finding in Bell's palsy, although some enhancement, especially of the first genu and proximal tympanic segment, may also be seen in normal subjects.[13,14] Herpes zoster may be suspected, even in the absence of vesicular eruption, if the enhancement is localized to the inner ear structures.[15]

Facial Nerve Conduction Studies and Electroneuronography

These techniques are used to determine the characteristics and severity of the lesion, and give clues regarding the timing of injury. Facial nerve conduction studies (FNCSs) assess function of the extracranial portion of the facial nerve. For FNCSs, surface electrodes are placed over facial muscles to record the response elicited by an electrical stimulus applied to the facial nerve. A brief square wave electric shock (0.2 ms) of supramaximal intensity (20–60 mA) is used for cervicofacial branch FNCS, applied first at a point anterior to the tragus, and then at a point along the horizontal portion of the mandible, recording responses from the orbicularis oris muscle. Normal values are available from birth to age 15 years.[16,17] The facial nerve conduction velocity (NCV) increases markedly during growth, particularly within the first year of life. Electroneuronography (ENOG) uses surface electrodes placed along the nasolabial fold over the nasalis and perioral muscles to record the response elicited by a brief supramaximal electrical stimulus applied to the facial nerve near the stylomastoid foramen. The peak-to-peak amplitude of the compound muscle action potential (CMAP) is recorded as a percentage of the amplitude of the contralateral normal side.[18] This percentage is presumed to correspond to the number of surviving motor neurons. Asymmetry greater than 30% is considered abnormal. FNCSs and ENOG have shown prognostic value in complete acute nontraumatic unilateral facial paralysis in childhood.[9,18] In idiopathic AFPs, amplitudes reach their nadir 7 to 14 days after the onset of weakness. Hence, the extent of Wallerian degeneration may be defined as early as 7 days after the onset of Bell's palsy.[10] In CFPs, serial FNCSs and ENOGs help distinguish fixed developmental defects from recovering traumatic insults.

Needle Electromyography of Facial Muscles

Conventional needle electromyography (EMG) confirms neurogenic dysfunction of the facial muscles and provides a means to evaluate outcome and assist surgical decisions. At birth, a prenatal denervation process should be suspected when fibrillation potentials are recorded before the fourth day of life.[19,20] In newborn and young infants, EMG can detect facial palsies that are not apparent clinically. At any age, EMG is particularly useful during the second and third weeks after onset of the facial palsy, once Wallerian degeneration has occurred. Later on in the course of recovery, changing EMG patterns demonstrate nerve regeneration and muscle reinnervation. Needle EMG examination is performed at rest, and with voluntary contraction or after stimulation. Stimulation-detection EMG is very useful in young children who do not cooperate, as well as in very severe palsies with no apparent voluntary recruitment of motor unit potentials (MUPs). The EMG activity is recorded by a concentric needle electrode in at least one muscle innervated by the temporofacial branch (orbicularis oculi, frontalis), and in another innervated by the cervicofacial branch (orbicularis oris, depressor anguli oris). Fasting the baby for up to 4 hours before the test may allow recording bursts of activity arising from the facial muscles while the baby is crying or sucking a pacifier. In older children, voluntary contraction can be recorded after asking the child to imitate the examiner when closing the eyes, smiling, and whistling.

Blink Responses

This technique explores both facial and trigeminal nerve functions, as well as the pathways between their respective cranial nuclei within the brainstem. Numerous studies indicate the value of blink responses (BRs) for detecting brainstem dysfunction and predicting clinical outcome in adult patients with facial palsies.[21,22] This study is particularly useful during the early stages of nerve regeneration, when the persistence or return of BRs suggests satisfactory recovery. The blink reflex results from the contraction of the orbicularis oculi muscle provoked by stimulation of the supraorbital branch of the trigeminal nerve.[23,24] A single electric shock is applied to the supraorbital foramen to elicit BRs recorded from the orbicularis oculi muscle on both sides. Stimulation of the ipsilateral cranial nerve V provokes a direct response with two components: R_1, which is immediate and brief, and R_2, which is delayed and longer lasting. Stimulation of the contralateral nerve V provokes a crossed R_2 response. The maturation from birth of the electrically elicited BRs has been studied.[25–27] The R_1 response is always present, whatever the age, and its latency achieves normal adult values at term birth. The ipsilateral R_2 response can be evoked in most newborns and infants, but the contralateral R_2 response is not always obtained before the age of 8 months. The R_1 component corresponds to an oligosynaptic reflex arc involving at least 2 and no more than 3 synapses in the pons between the main sensory nucleus of cranial nerve V and the motor nucleus of the ipsilateral cranial nerve VII. The R_2 component follows polysynaptic medullary pathways, which are more caudal and closer to the bulbar formations. The spinal trigeminal nucleus has projections to the adjacent paramedian reticular formation and the motor nuclei of both seventh nerves.[28] Injury in the trigeminal pathway leads to delayed or absent ipsilateral R_1 and R_2 responses, and contralateral R_2 response, while a lesion in the facial pathway shows delayed ipsilateral R_1 and R_2, but normal contralateral R_2.[29] Lesions at the pons show ipsilateral abnormal R_1 response, while contralateral stimulus has normal R_1.[24] Lateral medullary lesions show abnormal R_2 response on the affected side, stimulating either the nonaffected or the affected side.[30]

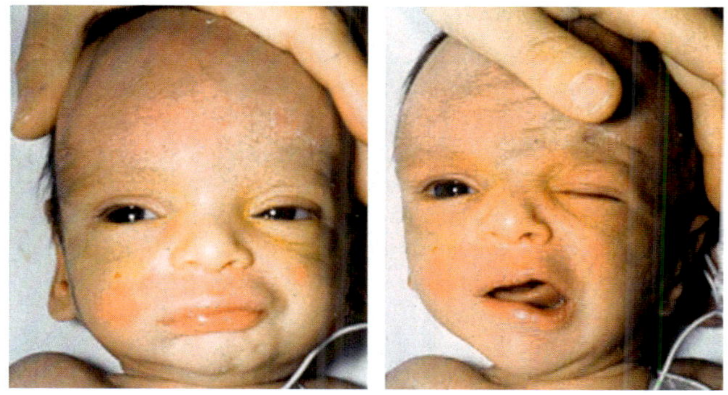

FIGURE 13.2 Congenital facial palsy, at rest and on crying, in a full-term boy at 3 days of age.

CONGENITAL FACIAL PALSIES

The incidence of CFPs ranges from 1.8 to 7.5 per thousand births.[31,32] The differential diagnosis includes intrauterine compression, obstetric trauma, and congenital malformation syndromes (Box 13.1). Most children with a facial palsy secondary to a congenital malformation present with other associated physical anomalies.[33] Congenital facial asymmetry may also result from hypoplasia of the depressor anguli oris muscle (DAOM).

BOX 13.1 Congenital Facial Weakness: Principal Causes

Birth Trauma

- Facial nerve injury: obstetric trauma or intrauterine stress
- Neonatal encephalopathy
- Posterior fossa hematoma

Developmental Defect

- First and second branchial arch malformation syndromes: facio-auriculo-vertebral spectrum, hemifacial microsomia, mandibulofacial dysostosis
- Asymmetric crying facies: hypoplasia of the depressor anguli oris muscle, cardiofacial syndrome, 22q11 deletion
- CHARGE association – Colobomata, Heart disease, Atresia of the choanae, Retarded growth or development, Genital hypoplasia, Ear anomalies
- Moebius syndrome

Neuromuscular Disease

- Congenital myotonic dystrophy
- Congenital muscular dystrophy
- Congenital myopathies: nemaline, myotubular, fiber-type disproportion
- Facioscapulohumeral muscular dystrophy: early-onset type
- Congenital myasthenic syndromes
- Neonatal transient myasthenia gravis

Perinatal Facial Nerve Injuries

Forceps delivery, prolonged labor, a birth weight above 3500 g, and primiparity are significant risk factors for CFP.[32] In a retrospective study of 172 children with congenital facial asymmetry, 67% had facial mononeuropathy with no associated malformation.[34] The face was often symmetrical at rest but cry revealed the unilateral absence of brow puckering, incomplete eyelid closure, and asymmetry of the mouth (Figure 13.2). More than the half of patients with nonsyndromal CFP were born by forceps delivery. A neonatal purulent otitis media was diagnosed in two neonates. In the remaining patients, no causal factors were found. In such cases CFP may result from pre- or perinatal compression of the extracranial part of the VIIth nerve against the myometrium and fetal shoulder, or the sacral prominence of the maternal pelvis. In fact, there is often concordance between the side of CFP and that of cephalic presentation.[35] MRI shows developmental abnormalities of the facial nucleus in a small proportion of children with isolated unilateral CFP.[11]

Alterations in facial NCV and BRs, as well as needle EMG of facial muscles, precisely define the extent and severity of the facial nerve lesion. Moderate forms show only partial denervation and either affect the territories of the two main branches equally, or predominantly involve the cervicofacial branch. Severe forms show absent facial CMAPs and BRs, and very reduced or single EMG recruitment patterns.

Clinical and EMG monitoring of CFPs demonstrates that facial nerve regeneration and muscle reinnervation usually proceed slowly, over some months. The percentage values of facial function over time provided by serial ENOGs correlate with prognosis.[9] When the palsy seems

CASE EXAMPLE 13.1

This baby boy was born at term with a breech presentation. A complete unilateral CFP was noted at birth. At age 3 months, slight mobility of the nasolabial fold and labial commissure appeared. The first EMG was performed at age 3 months 10 days. The orbicularis oculi muscle was inactive and did not respond to facial nerve stimulation. BRs were absent. In the orbicularis oris, a slight contraction was detected. Stimulation of cranial nerve VII elicited a low-amplitude normal-latency motor response. NCV of the cervicofacial branch was normal (25 m/s). The results of a second EMG, at age 6 months, were identical to the first. At 7 months, there was no clinical improvement. Exploratory surgery at 8 months identified lesions due to chronic inflammation of the mastoid antrum with compression of the second section of the seventh nerve by a bony spur secondary to a fracture line. There was no improvement postoperatively. Repeated EMGs at ages 10 and 21 months did not demonstrate reinnervation of either the orbicularis oculi or orbicularis oris. Clinical improvement was poor and very slow. Conjunctival infections ceased only at the age of 4 years. Facial asymmetry was still significant at age 10 years.

Comment

Given the findings on surgical exploration, it is likely that an earlier intervention would have improved the prognosis in this case.

CASE EXAMPLE 13.2

A baby girl, born at term by forceps delivery, showed a complete unilateral CFP. The first EMG at 23 days of age revealed complete inexcitability of the orbicularis oris muscle, with no response to electrical stimulation of cranial nerve VII. An extremely reduced interference pattern was noted in the orbicularis oculi. Stimulation of nerve VII at the pretragal point elicited a low-amplitude CMAP in orbicularis oculi, with a markedly increased latency (16.1 ms). No BRs were elicited. At day 45, there was no significant clinical improvement and a second EMG was performed. The orbicularis oris muscle was still inactive and unresponsive. In the orbicularis oculi muscle, the interference pattern was still reduced, but showed a combination of several types of MUPs, and the CMAP showed a greater amplitude and shorter latency (10.0 ms). BRs were still absent. At 5 months, further clinical signs of recovery of facial motor activity were noted: the eyelids closed spontaneously with sleep and with stimulation during blink reflex testing, and the labial commissure elevated slightly in reaction to skin stimulation. The third EMG revealed signs of reinnervation in orbicularis oris; stimulation of the facial nerve at the pretragal point elicited a low-amplitude CMAP with moderately increased latency (7.6 ms). The cervicofacial branch NCV was normal (30 m/s). In orbicularis oculi, the contraction showed an intermediate recruitment pattern; the CMAP had normal morphology, amplitude, and latency, and the R1 component of the BRs had a normal latency. Recovery was clinically complete at 15 months.

Comment

This initially complete CFP in a newborn delivered by forceps resolved spontaneously. In this instance, early surgical intervention would have been pointless or counterproductive. The EMG recording of voluntary activity and a CMAP from the orbicularis oculi muscle in an early examination, even if it showed very poor activation and a very low-amplitude CMAP, can be considered predictive of a good outcome.

clinically to remain complete, EMG may indicate a more favorable outcome by showing nascent MUPs with the reappearance of a low-amplitude increased-latency motor response.

In our series, severe or moderate palsies showed a more favorable outcome in the group delivered without forceps than in the group delivered with forceps. Improvement in facial motor activity and EMG parameters may continue as long as 2 years after nerve injury. Surgery must be avoided if it is so early that it might preclude spontaneous recovery or if there is no focal lesion for which repair is possible. In contrast, one must not fail to recognize a definitive facial nerve lesion related to a specific area of focal compression. Even if CT and MR imaging fail to define a focal VII nerve lesion, surgical consideration is warranted if there are no signs of clinical or EMG improvement. In our series, surgical exploration was performed in eight infants with severe CFP who had no clinical improvement. In each case, two or three successive EMGs had demonstrated either the persistence of a major lesion of the two major facial nerve branches without any sign of recovery, or partial recovery of the cervicofacial branch with persistence of complete paralysis of the temporofacial branch. In two instances, surgical exploration identified facial nerve compression against a fracture line in the petrous temporal bone, associated in one case with chronic inflammation of the mastoid antrum. In the six other infants, the facial nerve trunk was altered—pale, thin, flattened—without any obvious local cause. In a retrospective study of children with unilateral isolated CFP who had incomplete, often very limited recovery by 5 years, there was no association between this poor outcome and risk factors for birth trauma to the VIIth cranial nerve.[36] Case Example 13.1 and Case Example 13.2 illustrate the difficulties experienced when one considers surgical exploration in an infant with a severe CFP.

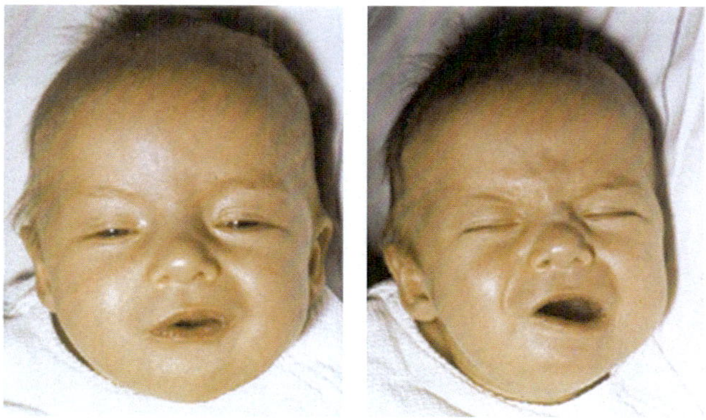

FIGURE 13.3 Asymmetric crying facies due to hypoplasia of the depressor anguli oris muscle in a full-term boy at 20 days of age. The face is symmetrical at rest; when the child is crying, the lower lip is asymmetrical, but the upper face and nasolabial folds remain symmetrical.

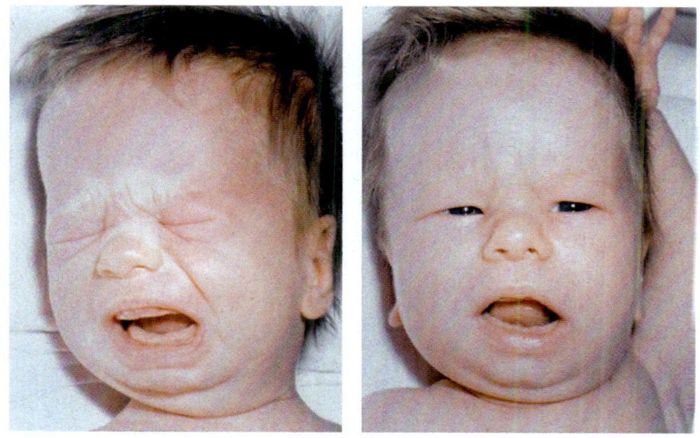

FIGURE 13.4 Pseudoparalysis: congenital facial asymmetry due to sustained abnormal intrauterine position. First day of life. Note the lack of parallelism of the gums.

Asymmetric Crying Facies

Congenital facial asymmetry obvious only when the child is crying, and involving only the lower lip, has several potential causes. Diagnosis requires EMG needle examination. Most of the time, asymmetric crying facies (ACF) is due to hypoplasia of the DAOM.[37,38] Typical findings are the absence of neurogenic EMG abnormalities, a low-amplitude interference pattern on crying, and normal facial NCV. Less often, EMG signs of partial denervation of the DAOM, depressor labii inferioris, and mentalis muscles are indicative of a distal partial facial palsy. The selective topography of these nerve lesions corresponds to the territory of the mental branch of the facial nerve. This branch, arising from the cervicofacial branch, has a course that in the fetus and newborn closely follows the horizontal part of the mandible and is thus easily compressed.[39,40] Regardless of the cause, this clinical presentation is characterized by a labial commissure that does not descend on the affected side, but the nasolabial folds and contraction of the frontalis muscles are symmetric (Figure 13.3). The lower part of the DAOM muscle inserts in a diffuse manner into the platysma muscle, along the base of the mandible, and enters as a narrow fascicle into the labial commissure, where its fibers merge with those of the orbicularis oris muscle. Its contraction draws the labial commissure and the external part of the lower lip downward and outward, producing an expression of displeasure.[41] Hypoplasia of the depressor labii inferioris muscle may also be present. The association of ACF with congenital heart malformation has been termed the cardiofacial syndrome.[42] Prospective studies of newborns with ACF have subsequently confirmed a higher risk of visceral malformation, particularly of the heart, great vessels, and genitourinary system.[43] A deletion at 22q11 has been reported both with ACF as a part of the cardiofacial syndrome, and with isolated ACF.[44,45] Thus, an EMG diagnosis of hypoplasia of the DAOM should lead to a search for associated visceral malformation(s). In a personal series of 51 children with ACF, EMG studies diagnosed a partial facial palsy in 10 patients, and revealed signs of hypoplasia of the DAOM in 41, associated with congenital malformations in 5 of the 41.

Finally, in certain cases of congenital facial asymmetry the EMG does not show any sign of paralysis or muscular hypoplasia. In these cases, asymmetry is noted both at rest and on crying, and no apparent malformations are observed. Such "pseudoparalysis" may be associated with asymmetric mandibular hypoplasia, but more often is the result of a restraint in intrauterine position or movement. A clinical feature characteristic of this situation is the loss of parallel symmetry of the gums (Figure 13.4). Pseudoparalysis was identified in 5 of 172 infants in our congenital facial asymmetry series.[34] EMG studies were normal in all, demonstrating the integrity of facial muscles and nerves. This functional asymmetry resolved rapidly in all 5 babies within the first 2 months.

Orofacial Malformations with Cranial Nerve Palsies

Hemifacial malformations involving the derivatives of the first and second branchial arches can result in lesions of cranial nerves V and VII (Figure 13.5). Facial paralysis is present in most cases of Goldenhar syndrome [OMIM#164210],[46] in 43% of infants with CHARGE syndrome (Colobomata, Heart disease, Atresia of the choanae, Retarded growth and development, Genital hypoplasia, and Ear anomalies) association due to *CHD7* mutations [OMIM#214800],[47] and in 22% of children with hemifacial microsomia.[48] The authors have reported a series of 33 infants with asymmetric facial motor activity

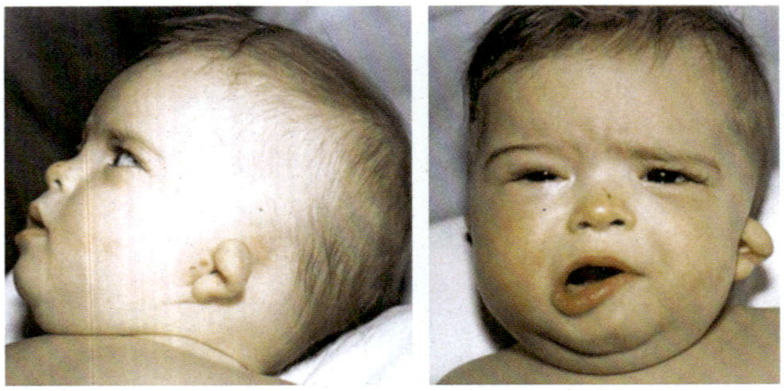

FIGURE 13.5 Congenital facial palsy in a three-month-old boy with facio-auriculo-vertebral syndrome.

associated with hemifacial hypoplasia and ear malformation.[34] In addition to Goldenhar syndrome and CHARGE association, clinical diagnoses in this series included the facio-auriculo-vertebral spectrum[49] and Treacher-Collins-Franceschetti syndrome [OMIM#154500].[50] Facial weakness was either diffuse or localized to one side of the face. EMG activity was variably normal, neurogenic, or low-amplitude. Among the 19 infants with totally or partially denervated muscles, MRI or CT scan showed a piliferous cyst of the cerebellum in one case, and cerebellar hypoplasia in another. In the 14 infants with no sign of neurogenic atrophy, no abnormalities of the central nervous system were found. In hemifacial malformations, involvement of the facial nerve is a predictive sign of an associated malformation of the central nervous system, in particular of the posterior fossa.[51,52]

Moebius Syndrome

Diagnosis of this syndrome is easy in cases of total or partial facial diplegia, whether symmetric or asymmetric, when it is associated with bilateral paralysis of abduction of the eye (Figure 13.6).[53,54] When the facial diplegia is partial, it usually predominantly involves the upper half of the face. A concomitant paralysis of the hypoglossal nerve with atrophy of the tongue is present in one-third of patients with Moebius syndrome. Most of these infants have congenital dysphagia, drooling, malocclusion, velopharyngeal incompetence, dysarthria, and delayed speech. Trigeminal nerve involvement, with temporal-masseter weakness or contracture, is less frequent. Club foot, malformations of the hands and fingers, and the Poland anomaly may be associated with the Moebius syndrome (Table 13.1). Children with Moebius syndrome later suffer from the psychological consequences of esthetic damage, but the disorder usually is not associated with mental retardation and is not progressive. Most cases are sporadic, but families with autosomal dominant inheritance have been reported.[55] Additionally, six patients with Moebius syndrome associated with an axonal peripheral

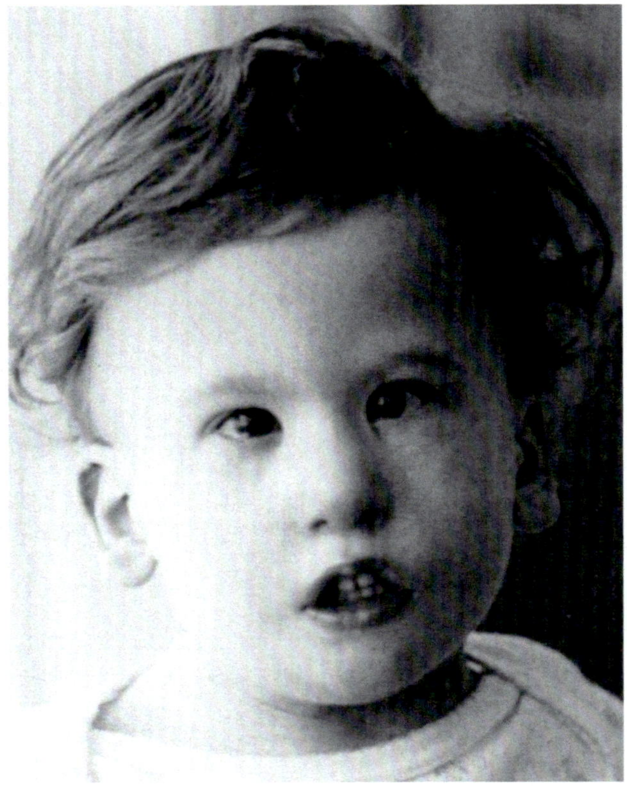

FIGURE 13.6 A child with Moebius syndrome at age 14 months.

neuropathy and hypogonadism have been described.[56] Facial EMG is an important diagnostic tool contributing to the differential diagnosis and elucidation of the multifaceted pathogenesis of Moebius syndrome. If facial EMG is difficult to interpret secondary to muscle atrophy, study of the lingual and palatal muscles can be useful in identifying a concomitant neurogenic process in the territories of the hypoglossal nerve and pharyngeal plexus. EMG of the limbs distinguishes Moebius syndrome from congenital myotonic dystrophy or other myopathies.

A spectrum of histologic findings has been described in Moebius syndrome[57]: hypoplasia or atrophy of cranial nerve nuclei, hypoplasia of the facial nerve, focal brainstem necrosis, brainstem hypoplasia, muscular dysplasia, and even dystrophic muscular changes. Most published EMG studies point towards a neurogenic origin, describing sparse motor units, increased latencies, slowing of facial nerve conduction, and sometimes altered BRs.[58] Such data support the hypothesis of a prenatal, possibly ischemic nuclear lesion affecting specific cranial nerves.[59] Our EMG results, in a large series of children with Moebius syndrome, were heterogeneous but showed signs of predominant axonal loss.[60,61] Most patients (55/58) showed reduced, even single, motor unit recruitment patterns in several muscles of the face, tongue, or soft palate, favoring the hypothesis of a developmental defect of

TABLE 13.1 Clinical Features in 75 Patients with Moebius Syndrome

Clinical Findings	Number of Patients (percent)
Cranial nerve palsies	
Facial diplegia	75/75 (100)
Symmetrical, 46 / asymmetrical, 29	
Complete, 46 / Upper face, 26 / Lower face, 3	
Abducens palsy	68/75 (91)
Bilateral, 65 / Unilateral, 3	
Ptosis	3/75 (4)
Temporal-masseter weakness or contracture	21/59 (35)
Sensory hearing loss	6/75 (8)
Laryngeal or pharyngeal weakness	28/72 (39)
Tongue atrophy	47/71 (62)
Associated malformations	
Epicanthus	73/75 (97)
Fingers	33/75 (44)
Club foot	30/75 (40)
Poland anomaly	12/75 (16)
Toes	3/75 (4)
Visceral	8/75 (11)
Cleft palate	2/75 (3)

Data from the French national referral center for rare facial and oral diseases. Renault F, Sergent B, Vazquez MP, 2013.

specific cranial nerve nuclei. In the three other children, motor unit patterns showed full interference and low amplitude; motor responses were of low amplitude with normal latencies, and facial NCVs were normal; although a myopathy was suspected, deltoid muscle biopsy and molecular studies were normal, and the clinical course was stable. Therefore, in patients with Moebius syndrome, myopathic EMG signs in facial muscles do not necessarily reveal a myopathic disease, but more probably indicate developmental muscular hypoplasia. See Case Example 13.3.

Neuromuscular Disorders with Congenital Facial Diplegia

Facial weakness is an important clinical sign in several neuromuscular disorders, including congenital myasthenic syndromes,[62] congenital myotonic dystrophy,[63] congenital myopathies,[64] and some metabolic myopathies. Correct diagnosis leading to genetic counseling in these hereditary conditions has important consequences for future pregnancies. Other conditions such as the Prader-Willi syndrome, and transient neonatal myasthenia gravis, should also be considered. Congenital myotonic dystrophy and nemaline myopathy typically cause facial diplegia with a characteristic elongated and expressionless face and tented upper lip, sometimes associated with arthrogryposis. Dysmorphic facial features likely result from prenatal muscle inactivity. Ptosis, bulbar weakness with dysphagia, generalized hypotonia, and various degrees of neck and limb weakness are usually present in patients with congenital myotonic dystrophy, but the extraocular muscles are spared. In contrast, centronuclear-myotubular myopathies, recessive central core disease, and certain myasthenic syndromes are often associated with ophthalmoplegia. Atypical cases of congenital myotonic dystrophy may present with facial weakness with transient or mild neonatal hypotonia. Electrodiagnostic studies show the normal or myopathic EMG in facial muscles, normal FNCSs, and normal BRs. EMG study of proximal and distal muscles of upper and lower limbs may show myopathic abnormalities, although myotonia is generally not apparent in infancy.

CASE EXAMPLE 13.3

A baby boy was born to a healthy 30-year-old mother. The pregnancy and delivery were uneventful. He showed a weak cry, poor suck, and decreased facia motility, with some asymmetry of the nasolabial folds when crying. Physical examination revealed asymmetric facia diplegia, esotropia, micrognathia, tongue atrophy, and a bifid uvula. Funduscopic examination was normal. Feeding remained slow but no aspiration or respiratory distress occurred. Complementary studies, including: X-rays of the skull, thorax, and limbs; electrocardiogram; EEG; brain and heart ultrasonography; visual evoked potentials; electroretinogram; and BAERs, were normal. A congenital malformation syndrome was hypothesized; the karyotype was normal and a 22q11 deletion was excluded. Serologic testing for toxoplasmosis, rubella, cytomegalovirus, and HIV infection was also negative. At 6 months the patient was referred to us for neurologic and EMG evaluation because his dysphagia persisted and he was unable to smile. Facial diplegia with predominance on the left side and bilateral abducens palsies were obvious at that time. Facial EMG showed the absence of frontal muscles and severe hypoplasia of the orbicularis oculi with no CMAP and no BRs. Neurogenic EMG responses were recorded from orbicularis oris, genioglossus, and levator veli palatine, with normal NCV of the cervicofacial branch of the VIIth nerve. A brain CT scan and MRI were normal. At 14 months, no significant abnormalities were found in the child's psychomotor development, and the patient showed the typical facial picture of Moebius syndrome (Figure 13.6).

CASE EXAMPLE 13.4

A full-term baby from a first pregnancy of a 21-year-old woman with no previous relevant history presented at birth with moderate respiratory distress and some cyanotic episodes. On examination, there was stridor, weak suck, and axial hypotonia. These symptoms progressively improved in a few days. Nonspecific facial dysmorphic features were noticed, and the motor development was slow. Walking was achieved at age 24 months. Later on, speech delay was noted and the intelligence quotient score was 65 at 30 months of age. The peculiar facies was considered a feature of a congenital malformation syndrome associated with encephalopathy. At age 34 months, limb muscle wasting and atrophy suggested a possible neuromuscular disorder. EMG examinations showed diffuse myopathic changes in the child and myotonic discharges in the mother. The diagnosis of myotonic dystrophy was confirmed by DNA testing.

Comment

In this case, the diagnosis was delayed several years because of the misinterpretation of facial weakness as a dysmorphic feature. Early diagnosis allows the opportunity for genetic counseling; the mother is at risk of having other children with congenital myotonic dystrophy. Finally, early diagnosis can avoid further costly and unnecessary tests.

CASE EXAMPLE 13.5

A full-term baby from a first pregnancy of a 35-year-old woman with no remarkable previous history presented with predominantly mild axial hypotonia and a symmetric facial diplegia. Reduced facial movements and slow sucking were observed. The facial weakness was more predominant in the upper part of the face, with no apparent function of the frontal muscles, and some limitation of the abduction of the eyes. No associated malformations were found. A clinical diagnosis of Moebius syndrome was made. The facial EMG at 3 weeks of age showed absence of denervation in the facial, lingual, and velopharyngeal muscles. BRs were normal. EMG of limb muscles demonstrated myopathic changes without myotonia. Physical examination of the mother showed typical myotonia. This mother had noted this symptom since adolescence and did not feel it was abnormal. The DNA testing confirmed the diagnosis of congenital myotonic dystrophy.

Facial diplegia, with or without dysphagia, associated with early development of marked joint contractures and increased creatine kinase levels, suggests a congenital muscular dystrophy, especially of the merosin-deficient variety.[65] An infantile form of facioscapulohumeral (FSH) muscular dystrophy with early facial diplegia and a severe clinical course has also been described.[66] The facial diplegia is obvious in the first year of life: the face is expressionless, patients have a peculiar horizontal smile, and they may be unable to close their eyes and mouth fully. In some babies with infantile FSH MD, a poor suck and lack of facial expression have been noted from birth. Sensorineural hearing loss and tortuosity of retinal arterioles have also been reported. Motor milestones may not be delayed, but cognitive involvement and later progression of weakness of limb and respiratory muscles often occur.[67]

Two case studies, Case Example 13.4 and Case Example 13.5, are included as a reminder that congenital myotonic dystrophy should be suspected in the presence of a congenital facial diplegia, even in the absence of severe neonatal hypotonia.

Congenital Facial Weakness Due to Neonatal Encephalopathy

Orofacial apraxia may be an early sign of cerebral palsy in newborn and young infants who suffer from hypoxic-ischemic encephalopathy. MRI may show thalamic or opercular lesions. EMGs of facial muscles, FNCSs, and BRs are normal, except when vascular brain injuries have involved brainstem nuclei and pathways. We have reported 12 infants with congenital dysphagia, aspiration-induced pulmonary disease, and failure to thrive.[68] Most had facial hypotonia (Figure 13.7). EMG showed neurogenic signs in the facial muscles, associated with altered BRs. Even though the brainstem and the spinal cord are less vulnerable to ischemia than the cerebral cortex, the diencephalon, the pons, and the medulla oblongata are structural targets of neuronal loss in cases

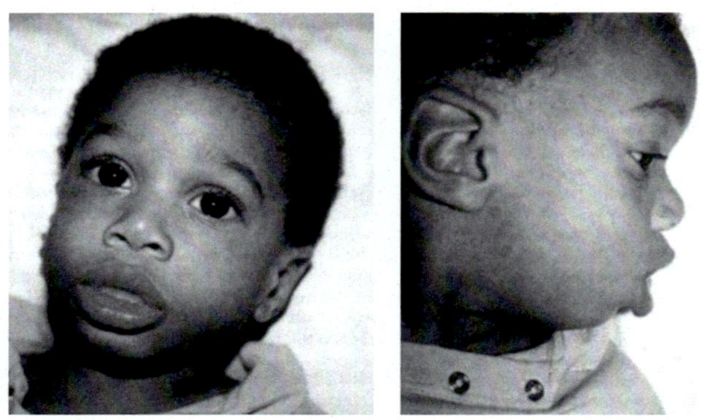

FIGURE 13.7 Bilateral facial weakness in a 14-month-old boy with neonatal brainstem encephalopathy.

BOX 13.2 Differential Diagnosis of Acquired Facial Weakness in Children

- Idiopathic Bell's palsy
- Infection and inflammation
  - Otitis media, mastoiditis, parotitis
  - Meningitis: viral, bacterial, tuberculosis, trichinosis, meningeal leukemia
  - Guillain-Barré syndrome, polyneuritis cranialis, Miller Fisher syndrome
  - Viral infections: Epstein-Barr, mycoplasma, herpes zoster/varicella, HIV, echovirus
  - Borrelia: Lyme disease
  - Kawasaki disease
  - Vasculitis: Henoch-Schönlein purpura
  - Neurosarcoidosis
- Recurrent facial palsies
  - Melkersson-Rosenthal syndrome
  - Familial alternating Bell's palsy
  - Hereditary neuropathy with liability to pressure palsies
- Trauma
  - Craniofacial trauma, penetrating injury
  - Iatrogenic surgery
- Mass lesions
  - Brainstem glioma and other posterior fossa tumors and vascular malformations
  - Cholesteatoma
  - Osteopetrosis
  - Lymphoma
  - Histiocytosis X
  - Facial tumor: vascular, osseous, parotid gland tumor; myofibroma; rhabdomyosarcoma
  - Schwannoma of the seventh cranial nerve
  - Benign intracranial hypertension—pseudotumor cerebri
- Malformation
  - Chiari malformation, syringobulbia
- Neuromuscular disorders
  - Muscular dystrophies: facioscapulohumeral, myotonic dystrophy
  - Progressive bulbar palsy
  - Myasthenia gravis
  - Botulism
  - Neoplastic, vascular, degenerative, metabolic, inflammatory
  - Central: suprabulbar palsy

HIV, human immunodeficiency virus; HSV, herpes simplex virus.

of severe birth asphyxia of the "cardiac arrest" type.[69] An autopsy of a child with facial diplegia, velopharyngeal incoordination, and ocular motor apraxia demonstrated neuronal depletion of cranial nerve nuclei and intact cerebral hemispheres.[70] When congenital facial diplegia is due to perinatal vascular damage involving both the brainstem and hemispheric structures, the clinical condition may progress to quadriparetic cerebral palsy.

ACQUIRED FACIAL PALSIES

About 90% of facial palsies in children are acquired.[71,72] The incidence of peripheral facial palsy in children varies from 5 to 21:100,000 per year,[73] with higher frequencies between 5 and 11 years of age, and no significant gender predominance. Various lesions may result in AFPs (Box 13.2), but in practice, Bell's palsy, local infection, and traumatic injuries account for more than 80%.[71,72] Because more severe conditions may initially resemble an isolated unilateral Bell's palsy, a close follow-up of the patient is required. For example, up to 20% of tumors involving the facial nerve have an acute onset mimicking Bell's palsy.[12] Gradual progression of paralysis beyond 3 weeks, no return of function after 6 months, ipsilateral recurrence, hemifacial spasm, associated cranial neuropathies and other neurologic signs, pain, and single branch involvement suggest the possibility of a neoplasm.[74] A history of trauma or ear infection, time interval since onset, rapidity of progression, associated symptoms, previous episodes, skin rashes or infectious antecedents, exposure to tick bites, triggering factors, systemic diseases such as diabetes mellitus or HIV risk factors, and family history should be noted. Close follow-up may reveal Ramsay Hunt syndrome (RHS), Lyme disease (LD), or a more widespread process involving multiple cranial nerves.

Bell's Palsy

Bell's palsy accounts for almost the half of all pediatric AFPs.[75–79] One-third of children have a history of previous upper airway infection or exposure to cold or a draft. Pain in the ear or surrounding area is followed within a few hours or days by a rapid evolution of facial weakness including the forehead, cheek, and perioral area. Taste sensation is lost in up to half of patients, but lacrimation is usually spared. A subset of patients also complains of postauricular pain and/or facial numbness, which may be accompanied by abnormal BAERs or BRs. Outcome in this group does not differ from the purely mononeuritic group.[8] Bell's idiopathic AFP may occur in the first months of life, but is most frequently seen after age 8 years. The annual incidence of Bell's palsy has been reported as 4.2 per 100,000 under the age of 10 years, and

15.3 from 10 to 19 years; this is much lower than in adults.[77] Significant differences in the nerve/canal diameter ratio are claimed as a reason for the reduced vulnerability of children to facial nerve entrapment.[6]

Bilateral Bell's palsy is very rare, accounting for less than 1% of all cases,[80] but up to 15% of patients with Bell's palsy present with recurrent episodes.[81] Familial cases have been variously estimated from 3% to 28%,[82] and may represent a distinct entity, as suggested by reports of dominant and recessive inheritance of Bell's palsies.[83] Anatomic abnormalities in the facial canal were found in some members of a family with multiple cases, and were associated with a worse outcome.[84] Herpes simplex virus type 1 (HSV-1) may be the causal agent of Bell's palsy in nonendemic areas of Lyme borreliosis.[85,86] Pathogenesis may relate to direct injury to the facial nerve by reactivated HSV-1, or a virus-associated immune-mediated mechanism. Serologic tests are not reliable, because about 50% of the general population is HSV-1 positive, low rates of seroconversion to HSV-1 are reported in patients with Bell's palsy, reactivation is not always accompanied with an antibody response, and there may be cross-reactivity between shared antigens of HSV-1 and varicella zoster virus (VZV).[87] PCR testing has identified HSV-1 in facial nerve, endoneurial fluid, and facial-innervated muscles of patients with Bell's palsy, but not in control subjects.[88–91] These results suggest axonal spread and multiplication of a reactivated neurotropic virus leading to inflammation and focal demyelination. Other viruses have been implicated less frequently in the pathogenesis of acute AFP: VZV, Epstein-Barr virus, echovirus, poliovirus, adenovirus, rubella, *Cytomegalovirus*, *Mycoplasma*, hepatitis virus, influenza A and B, and HIV.[92–95] Diagnosis may be confirmed by identification of a four-fold increase in specific antibodies between the acute and convalescent phases of the infection. There is increasing availability of highly sensitive microarray assays for DNA detection of these infectious agents in serum, saliva, or CSF samples. AFP has also been related to immunization against diphtheria-pertussis-tetanus and poliomyelitis.[96]

Complete spontaneous recovery is generally the rule in children with Bell's palsy, with percentages in different series around 90% (the younger the patient, the more favorable the prognosis) while in adults the percentages are closer to 80%.[97–99] These percentages may be lower if EMG is performed.[79] Normality of CMAP latency[100] and preserved BRs with the first weeks after onset[101] are predictive of complete recovery. Inexcitability of the facial nerve, or a marked decrease in CMAP amplitude or absent BR after several weeks, is associated with a poor outcome or delayed recovery.[100,102] ENOG performed from 8 to 12 days after onset may be used to predict recovery. A loss of CMAP of less than 90% compared to the contralateral normal side is usually followed by a complete recovery in 7 or 8 weeks. When the loss is between 90% and 98%, recovery will be slower but usually complete. However, a loss of CMAP of more than 98%, especially after the third week of onset, has a worse prognosis.[99,103] When the paralysis remains clinically complete after the sixth week, improvement in ENOG at successive examinations is a useful predictor of a favorable outcome.[104] Complications such as crocodile tears and synkinesis, seen in 10% of adults, are rarely observed in children.

In adults with Bell's palsy, the value of steroid therapy has been confirmed by Cochrane reviews[105] and a recent Korean study confirmed the additional benefit of treatment with acyclovir.[106] There is no current evidence of any clear benefit of these treatments in children.[107,108] Since the small percentage of children with poor outcome belongs to the group with initially complete paralysis, some authors recommend steroids only when the facial palsy is clinically very severe in the first 72 hours, especially in older patients. The dose is 1–2 mg/kg/day for 8 days, followed by a gradual taper. Symptomatic treatment consists in protection of an exposed cornea by lubrication and eye patching during sleep.

Ramsay Hunt Syndrome

Ramsay Hunt syndrome is caused by reactivation of latent varicella zoster virus in the geniculate ganglion, and consists of acute AFP with a vesicular eruption on the auricle, face, oral mucosa, or neck, frequently associated with vestibulocochlear dysfunction. If no eruption develops, VZV-related AFP is clinically indistinguishable from idiopathic Bell's palsy. The clinical diagnosis is often difficult because onset of vesicles may occur several days after AFP onset. Ramsay Hunt syndrome is a rare event, accounting for less than 15% of acute facial palsies in adults and even fewer in children. Prognosis seems to be worse than in idiopathic facial palsies, with less than 80% of patients showing complete recovery. Geographical differences have been observed; this condition is more common in Japan[109] and Northern Europe,[93] but very rare in Brazil[92]; therefore, the use of antiviral agents may depend on geographic issues. PCR allows early detection of VZV in saliva,[92] auricular skin exudates,[110] and CSF.[111,112]

Lyme Disease

Facial palsy may be the presenting sign of Lyme disease (LD), and is its most common neurologic complication in children (16–50%), while adults more frequently develop meningism.[113] LD is endemic in some areas of the United States and in northeastern countries of Europe, where it is the most common cause of pediatric AFP,[77,114,115] but in nonendemic countries, borreliosis does not seem to be a major cause of AFP.[116] Patients may not have other symptoms of LD, and the tick bite, one to three weeks before the first signs, may not be reported. Headache is the most common neurologic symptom in children, and facial palsy is frequently associated with occult meningitis, although neck stiffness may be minimal or absent. A lumbar puncture should therefore be included in the diagnostic workup of any AFP in an endemic region, especially in children under 8 years of age.[117] The finding of a peculiar skin rash extending centrifugally around the area of the tick bite (erythema chronicum migrans) characterizes the first stage of the infection in up to 90% of cases. Preceding the facial palsy, there may be associated auricular pain, hemifacial edema, or erythema of the face. The facial palsy is rarely complete and may occur on the same side as the erythema migrans or spread to the contralateral side, typically 10 to 15 days later. Associated symptoms such as general malaise, fever, myalgia, or sometimes limb paresthesias may be present but are seen less frequently than in adults. Meningoradiculitis and peripheral neuropathy syndromes are also rare. In contrast, a "pseudotumor cerebri" syndrome consisting of papilledema, mildly increased intracranial pressure, and slightly increased CSF protein content without pleocytosis is relatively common in the pediatric population in North America.[118,119] The diagnosis is based on the history, clinical manifestations, and blood and CSF workup, including the serologic demonstration of antibodies against *Borrelia burgdorferi* by ELISA, coupled with immunoblot if positive,[120] and the identification of the DNA of *Borrelia burgdorferi* in CSF by PCR. Early negative serologic testing does not rule out LD. The IgM response usually peaks during the third to sixth weeks, and IgG antibody typically arises around the fourth week.[121] In contrast to adults, most children with isolated LD-induced facial palsy have abnormal CSF.[122] The CSF may show pleocytosis, hyperproteinorachia, and increased IgG titers, which are more sensitive and specific than serum antibody titers alone.[123] Electrodiagnostic studies indicate demyelination in most cases; axonal loss is associated with a worse prognosis in adults.[124] EEG may show a diffuse or localized slowing. MRI may reveal patchy hyperintensities on T2-weighted images of the brain.[125] Isolated peripheral facial palsy in LD should be treated with doxycycline (100 mg daily in children over 8 years of age). Amoxycillin (50 to 60 mg/kg/day), penicillin (25 to 50 mg/kg/day), or cefuroxime (20 to 30 mg/kg/day) are recommended in younger children. Erythromycin (30 mg/kg/day) can be used in children with penicillin allergy. The duration of therapy is usually 4 weeks. In cases of neuroborreliosis, therapy with intravenous antibiotics is necessary for 2 to 4 weeks: ceftriaxone (75 to 100 mg/kg/day), cefotaxime (90 to 180 mg/kg/day), or penicillin G (300,000 U/kg/day, in six divided doses).[126] The facial palsy usually lasts less than 2 months and has a favorable outcome when treated. Prophylactic treatment is now possible. Two vaccines have been developed. Although both appear to be effective in 80% to 100% of cases, their long-term safety for children is not yet confirmed.

Otogenic Facial Palsy

Middle ear infection is a relatively common cause of AFPs in infants and young children.[127] A high level of suspicion should be maintained because they require active treatment and may have a worse prognosis than idiopathic and viral AFPs. Facial paralysis may develop quickly within the first 72 hours in association with acute otitis media (AOM). Inspection of the external meatus and tympanic membrane usually leads to the diagnosis. Acute mastoiditis is a relatively frequent complication of otitis media, and should be suspected when fever, retroauricular pain, and inflammation of the mastoid process develop. If the facial palsy develops weeks after onset of AOM, a coalescent subacute mastoiditis must be ruled out. Meningitis and temporal lobe abscess are rare complications.[12] A gradual onset of the facial palsy points towards a chronic otitis media, but other causes such as cholesteatoma or invasive processes are possible. Tuberculous mastoiditis must be kept in mind in refractory otitis media with AFP in older children from regions where tuberculosis is endemic, or in AIDS patients. Neuroimaging is useful to image the course of the facial nerve through the petrous portion of the temporal bone, and to rule out other processes such as middle ear rhabdomyosarcoma, intracranial abscess, or histiocytosis X. The prognosis of otogenic AFP is usually good if adequate treatment is instituted early. AOM with facial paralysis can be treated with intravenous antibiotics and paracentesis/surgical myringotomy,[127] with or without concomitant steroid therapy. If there is no improvement within a few days or if mastoiditis is diagnosed, mastoidectomy may be indicated. Mastoidectomy plus decompression of the third portion of the nerve may be required for severe seromucous otitis that does not respond to transtympanic tubes. Cholesteatomas have to be surgically removed and require antibiotic therapy based on cultures and sensitivity testing of specimens obtained from the middle ear or mastoid. Despite surgery, a small percentage of otogenic AFP patients may present incomplete recovery or develop synkinesias or facial hemispasm later in the course.[99]

Traumatic and Iatrogenic Facial Palsy

Paralysis of the VIIth cranial nerve may occur as a result of temporal bone fractures, penetrating head and neck wounds, or iatrogenic surgical injuries. The main iatrogenic cause is surgery of the posterior fossa, the middle ear, or the parotid area, in particular when there is an atypical nerve course, a tumor, or a vascular malformation. Traumatic facial nerve injuries occur mostly within the intratemporal course of the nerve. High-resolution CT scanning is helpful in determining the orientation of a fracture plane. Longitudinal fractures are the most common type of temporal bone fracture leading to facial nerve injuries, especially in the perigeniculate area, where the facial nerve is most susceptible to damage. Associated hearing loss is typically conductive or mixed. In transverse fractures, lesions are usually confined to the tympanic and geniculate segments of the facial nerve. Many cases resolve spontaneously, but surgical exploration is indicated when ENOG shows greater than 98% neural degeneration and no voluntary MUPs are recorded on needle EMG. Surgical exploration is reserved for cases that do not show recovery after 3 weeks. Postoperative paralysis requires immediate exploration not only for the psychological and forensic implications, but also in order to achieve the best outcome. Transection requires suturing of the nerve, or even placing inlay grafts or complete interpositional grafts when end-to-end suturing is not possible.[128] In traumatic facial nerve palsy, and also AFP of other origins and CFP, reconstructive surgical procedures may be performed to improve function or esthetics when deficits are fixed.[129,130]

Recurrent Facial Palsy

Fewer than 15% of children with Bell's palsy patients suffer recurrent ipsilateral or alternating facial paralysis.[131] Intracranial or extracranial tumors, meningeal leukemia, trauma, granulomatous and vascular diseases, malformations, and inflammatory and infectious causes should be excluded. Some inherited diseases may present with recurrent episodes of transient facial weakness. Most are transmitted in an autosomal dominant pattern or form part of a recognizable syndrome.[84] The Melkersson-Rosenthal syndrome (MRS) is a rare, noncaseating granulomatous disease characterized by the triad of persistent or recurrent orofacial edema, relapsing facial paralysis, and a fissured tongue. The etiology and mechanisms of MRS are still unknown, but genetic and allergic causes, infectious diseases, benign lymphogranulomatosis, and abnormal reactivity of the cranial neurovascular system have been implicated. Facial palsy is the initial sign in one-third of affected children; 75% have swollen lips during episodes and 50% have facial erythema and edema.[132] Patients are often of Mediterranean origin and may present in childhood.[133] Dominant inheritance with variable penetrance has been reported. Because residual facial weakness may increase with repeated episodes of paralysis, some authors have recommended prophylactic surgical decompression.[134] Another hereditary condition that may cause recurrent alternating acute facial palsy is osteopetrosis tarda, which presents in childhood with osseous fractures or neurologic complications from encroachment by bone growth on cranial nerve foramina. Skull X-rays show typical osseous hyperdensity. Finally, recurrent facial palsies have been reported in a family with hereditary neuropathy with liability to pressure palsies (HNPP) due to 17p11.2-12 deletion.[135,136] In patients with HNPP, electrodiagnostic studies including the upper and lower extremities may demonstrate widespread slowing of motor and sensory NCVs, more frequently across entrapment sites, with prolonged distal latencies. Surgical therapy of recurrent facial palsy is considered in patients with recurrent facial palsies who develop residual deficits, such as weakness, synkinesis, and contractures, and in patients in whom the interval between episodes of palsy is decreasing.[128]

Acquired Facial Diplegia

The more common causes of bilateral facial weakness in children are idiopathic Bell's palsy, LD, Guillain-Barré and Miller Fisher syndromes, multiple idiopathic cranial neuropathy, and brainstem encephalitis.[80] Trauma, tumors of the posterior fossa or meningeal leukemia and neuromuscular disorders such as FSH myopathy, structural myopathies, and congenital myasthenic syndromes are other possible causes of bifacial weakness. Very rarely in children, bilateral facial weakness is a manifestation of neurosarcoidosis, which associates swelling of the parotid gland and facial palsy with uveitis or, less commonly, with optic neuritis (Heerfordt syndrome). In adult series, other infections occasionally causing facial diplegia in endemic regions or in immunodepressed patients are poliomyelitis, tuberculous meningitis, leprosy, and cryptococcal meningitis with acquired immunodeficiency syndrome.[80] Electrodiagnostic studies of the cranial nerves and of the muscles and nerves of the limbs may contribute significantly to the etiologic diagnosis of bilateral facial palsy, by differentiating axonal loss from demyelination, and by distinguishing neurogenic from myopathic causes. Repetitive nerve stimulation enables diagnosis of myasthenia, but in children may be challenging due to movement artifacts. In the last decade, stimulated single fiber techniques have shown to be a very sensitive tool to detect abnormalities of neuromuscular transmission.[137]

Facial involvement may be the presenting signs of acute polyradiculoneuritis or cranial polyneuritis with sparing of the extremities. Electrodiagnostic studies help to investigate the process, showing prolonged latencies and temporal dispersion of CMAPs and BRs, and EMG signs of denervation of the facial muscles.[138] Miller Fisher syndrome is characterized by ophthalmoplegia, ataxia, and areflexia.[139] Electrodiagnosis may show reduced amplitude of sensory nerve action potentials in the extremities and low-amplitude, normal latency facial CMAPs. BRs may be either normal[140] or altered, in parallel with the decreased amplitude of the facial CMAPs and denervation of the facial muscles.[141] A clinical study of the general course of cranial nerve dysfunction during progression of the disease showed that there was significant overlap between polyneuritis cranialis, Miller Fisher syndrome, and classical ascending Guillain-Barré syndrome in adults as well as in children.[142]

REFERENCES

1. Courville J. The nucleus of the facial nerve; the relation between cellular groups and peripheral branches of the nerve. *Brain Res* 1966;**1**:338–54.
2. Chouard CH. *Anatomie, Pathologie et Chirurgie du Nerf Facial.* Paris: Masson et Cie; 1972.
3. Willer JC, Boulu P, Bratzlavsky M. Electrophysiological evidence for crossed oligosynaptic trigemino-facial connections in normal man. *J Neurol Neurosurg Psychiatry* 1984;**47**:87–90.
4. Soliven B, Meer J, Uncini A, Petajan J, Lovelace R. Physiologic and anatomic basis for contralateral R1 in blink reflex. *Muscle Nerve* 1988;**11**:848–51.
5. Fisch U, Esslen E. Total intratemporal exposure of the facial nerve Pathologic findings in Bell's palsy. *Arch Otolaryngol* 1972;**95**: 335–41.
6. Eicher SA, Coker NJ, Alford BR, Igarashi M, Smith RJ. A comparative study of the fallopian canal at the meatal foramen and labyrinthine segment in young children and adults. *Arch Otolaryngol Head Neck Surg* 1990;**116**:1030–5.
7. House JW, Brackmann DE. Facial nerve grading system. *Otolaryngol Head Neck Surg* 1985;**93**:146–7.
8. Ramos Lizana L, Seijas L, Carrasco Marina LI. Pseudotumor cerebri y paralisis facial. *An Esp Pediatr* 1995;**42**:377–8.
9. Shapiro NL, Cunningham MJ, Parikh SR, Eavey RD, Cheney ML. Congenital unilateral facial paralysis. *Pediatrics* 1996;**97**:261–4.
10. Qiu WW, Yin SS, Stucker FJ, Hoasjoe DK. Neurophysiological evaluation of acute facial paralysis in children. *Int J Pediatr Otorhinolaryngol* 1997;**39**:223–36.
11. Jemec B, Grobbelaar AO, Harrison DH. The abnormal nucleus as a cause of congenital facial palsy. *Arch Dis Child* 2000;**83**:256–8.
12. Grundfast KM, Guarisco JL, Thomsen JR, Koch B. Diverse etiologies of facial paralysis in children. *Int J Pediatr Otorhinolaryngol* 1990;**19**:223–39.
13. Jackson CG, von Doersten PG. The facial nerve. Current trends in diagnosis, treatment, and rehabilitation. *Med Clin North Am* 1999;**83**:179–95.
14. Swartz JD, Harnsberger HR, Mukherji SK. The temporal bone. Contemporary diagnostic dilemmas. *Radiol Clin North Am* 1998;**36**:819–53.
15. Kuo MJ, Drago PC, Proops DW, Chavda SV. Early diagnosis and treatment of Ramsay Hunt syndrome: the role of magnetic resonance imaging. *J Laryngol Otol* 1995;**109**:777–80.
16. Raimbault J, Les Conductions Nerveuses Chez. *L'enfant normal.* Paris: Expansion Scientifique Française; 1988.
17. Renault F, Raimbault J. [Facial, lingual and pharyngeal electromyography in children: a method to study sucking and swallowing disorders and their pathophysiology]. *Neurophysiol Clin* 1992;**22**:249–60.
18. Esslen E. Electrodiagnosis in facial palsy. In: Miehlke A, editor. *Surgery of the Facial Nerve.* Philadelphia: WB Saunders; 1973. pp. 45–51.
19. Mancias P, Slopis JM, Yeakley JW, Vriesendorp FJ. Combined brachial plexus injury and root avulsion after complicated delivery. *Muscle Nerve* 1994;**17**:1237–8.
20. Jones Jr. HR, Herbison GJ, Jacobs SR, Kollros PR, Macones GA. Intrauterine onset of a mononeuropathy: peroneal neuropathy in a newborn with electromyographic findings at age one day compatible with prenatal onset. *Muscle Nerve* 1996;**19**:88–91.
21. Dumitru D, Walsh NE, Porter LD. Electrophysiologic evaluation of the facial nerve in Bell's palsy. A review. *Am J Phys Med Rehabil* 1988;**67**:137–44.
22. Ghonim MR, Gavilan C. Blink reflex: prognostic value in acute peripheral facial palsy. *ORL J Otorhinolaryngol Relat Spec* 1990;**52**:75–9.
23. Kugelberg E. [Facial reflexes]. *Brain* 1952;**75**:385–96.
24. Kimura J. Alteration of the orbicularis oculi reflex by pontine lesions. Study in multiple sclerosis. *Arch Neurol* 1970;**22**:156–61.
25. Kimura J, Bodensteiner J, Yamada T. Electrically elicited blink reflex in normal neonates. *Arch Neurol* 1977;**34** 246–9.
26. Vecchierini-Blineau MF, Guiheneuc P. Maturation of the blink reflex in infants. *Eur Neurol* 1984;**23**:449–58.
27. Tomita Y, Shichida K, Takeshita K, Takashima S. Maturation of blink reflex in children. *Brain Dev* 1989;**11**:389–93.
28. Boulu P, Willer JC, Cambier J. [Electrophysiological study of blink reflex in man: segmental, intersegmental, auditory and visual interactions (author's transl)]. *Rev Neurol (Paris)* 1981;**137**:523–33.
29. Kiers L, Carroll WM. Blink reflexes and magnetic resonance imaging in focal unilateral central trigeminal pathway demyelination. *J Neurol Neurosurg Psychiatry* 1990;**53**:526–9.
30. Cruccu G, Deuschl G. The clinical use of brainstem reflexes and hand-muscle reflexes. *Clin Neurophysiol* 2000;**111**:371–87.
31. Levine MG, Holroyde J, Woods Jr. JR, Siddiqi TA, Scott M, Miodovnik M. Birth trauma: incidence and predisposing factors. *Obstet Gynecol* 1984;**63**:792–5.
32. Falco NA, Eriksson E. Facial nerve palsy in the newborn: incidence and outcome. *Plast Reconstr Surg* 1990;**85**:1–4.
33. Smith JD, Crumley RL, Harker LA. Facial paralysis in the newborn. *Otolaryngol Head Neck Surg* 1981;**89**:1021–4.
34. Renault F, Garabedian EN, Harpey JP. *Diagnostic des Paralysies et Asymétries Faciales Congénitales. Journées Parisiennes de Pédiatrie.* Paris: Flammarion Médecine-Sciences; 1993. pp. 155–161.
35. Hepner Jr. WR. Some observations on facial paresis in the newborn infant: etiology and incidence. *Pediatrics* 1951;**8**:494–7.
36. Laing JH, Harrison DH, Jones BM, Laing GJ. Is permanent congenital facial palsy caused by birth trauma? *Arch Dis Child* 1996;**74**:56–8.
37. McHugh HE, Sowden KA, Levitt MN. Facial paralysis and muscle agenesis in the newborn. *Arch Otolaryngol* 1969;**89**:131–43.
38. Nelson KB, Eng GD. Congenital hypoplasia of the depressor anguli oris muscle: differentiation from congenital facial palsy. *J Pediatr* 1972;**81**:16–20.
39. Hoefnagel D, Penry JK. Partical facial paralysis in young children. *N Engl J Med* 1960;**262**:1126–8.
40. Sammarco GJ, Ryan RF, Longenecker CG. Anatomy of the facial nerve in fetuses and stillborn infants. *Plast Reconstr Surg* 1966;**37**:566–74.

41. Duchenne de Boulogne G. *De L'électrisation Localisée et son Application à la Pathologie et à la Thérapeutique*. 2nd ed. Paris: JB Baillère et Fils; 1861.
42. Cayler GG, Blumenfeld CM, Anderson RL. Further studies of patients with the cardiofacial syndrome. *Chest* 1971;**60**:161–5.
43. Alexiou D, Manolidis C, Papaevangellou G, Nicolopoulos D, Papadatos C. Frequency of other malformations in congenital hypoplasia of depressor anguli oris muscle syndrome. *Arch Dis Child* 1976;**51**:891–3.
44. Giannotti A, Digilio MC, Marino B, Mingarelli R, Dallapiccola B. Cayler cardiofacial syndrome and del 22q11: part of the CATCH22 phenotype. *Am J Med Genet* 1994;**53**:303–4.
45. Stewart HS, Smith JC. Two patients with asymmetric crying facies, normal cardiovascular systems and deletion of chromosome 22q11. *Clin Dysmorphol* 1997;**6**:165–9.
46. Orobello P. Congenital and acquired facial nerve paralysis in children. *Otolaryngol Clin North Am* 1991;**24**:647–52.
47. Zentner GE, Layman WS, Martin DM, Scacheri PC. Molecular and phenotypic aspects of CHD7 mutation in CHARGE syndrome. *Am J Med Genet A* 2010;**152A**:674–86.
48. Carvalho GJ, Song CS, Vargervik K, Lalwani AK. Auditory and facial nerve dysfunction in patients with hemifacial microsomia. *Arch Otolaryngol Head Neck Surg* 1999;**125**:209–12.
49. Jones KL. First and second branchial arch syndrome. In: Jones K, editor. *Smith's Recognizable Patterns of Human Malformation*. Philadelphia: WB Saunders; 1997. pp. 642–5.
50. Jones KL. Treacher collins syndrome. In: Jones K, editor. *Smith's Recognizable Patterns of Human Malformation*. Philadelphia: WB Saunders; 1997. pp. 250–1.
51. Couly G, Aicardi J. [Associated morphological anomalies of the face and brain in infants]. *Arch Fr Pediatr* 1988;**45**:99–104.
52. Menenzes M, Coker SB. Charge and Joubert syndromes: are they a single disorder? *Pediatr Neurol* 1990;**6**:428–30.
53. Sudarshan A, Goldie WD. The spectrum of congenital facial diplegia (Moebius syndrome). *Pediatr Neurol* 1985;**1**:180–4.
54. Verzijl HT, Padberg GW, Zwarts MJ. The spectrum of Mobius syndrome: an electrophysiological study. *Brain* 2005;**128**:1728–36.
55. Verzijl HT, van den Helm B, Veldman B, Hamel BC, Kuyt LP, Padberg GW, et al. A second gene for autosomal dominant Mobius syndrome is localized to chromosome 10q, in a Dutch family. *Am J Hum Genet* 1999;**65**:752–6.
56. Baraitser M, Rudge P. Moebius syndrome, an axonal neuropathy and hypogonadism. *Clin Dysmorphol* 1996;**5**:351–5.
57. Towfighi J, Marks K, Palmer E, Vannucci R. Mobius syndrome. Neuropathologic observations. *Acta Neuropathol* 1979;**48**:11–17.
58. Jaradeh S, D'Cruz O, Howard Jr. JF, Haberkamp TJ, Konkol RJ. Mobius syndrome: electrophysiologic studies in seven cases. *Muscle Nerve* 1996;**19**:1148–53.
59. Govaert P, Vanhaesebrouck P, De Praeter C, Frankel U, Leroy J. Moebius sequence and prenatal brainstem ischemia. *Pediatrics* 1989;**84**:570–3.
60. Renault F. Cranial nerve studies in 22 children with Möbius syndrome. *Muscle Nerve* 2000;**23**:1631.
61. Nicot F, Renault F. Moebius syndrome, clinical and EMG data. In: Pitt M, editor. *Improving the Use of Electromyography in Paediatrics*. London: Institute of Child Health; 2013.
62. Eymard B, Hantai D, Estournet B. Congenital myasthenic syndromes. *Handb Clin Neurol* 2013;**113**:1469–80.
63. Aicardi J, Conti D, Goutieres F. [Neonatal forms of Steinert's mytonic dystrophy]. *J Neurol Sci* 1974;**22**:149–64.
64. Wallgren-Pettersson C, Sainio K, Salmi T. Electromyography in congenital nemaline myopathy. *Muscle Nerve* 1989;**12**:587–93.
65. Philpot J, Bagnall A, King C, Dubowitz V, Muntoni F. Feeding problems in merosin deficient congenital muscular dystrophy. *Arch Dis Child* 1999;**80**:542–7.
66. Yasukohchi S, Yagi Y, Akabane T, Terauchi A, Tamagawa K, Mizuno Y. Facioscapulohumeral dystrophy associated with sensorineural hearing loss, tortuosity of retinal arterioles, and an early onset and rapid progression of respiratory failure. *Brain Dev* 1988;**10**:319–24.
67. Brouwer OF, Padberg GW, Wijmenga C, Frants RR. Facioscapulohumeral muscular dystrophy in early childhood. *Arch Neurol* 1994;**51**:387–94.
68. Renault F, Couvreur J. [Congenital deglutition disorder revealing cerebral stem disorders. 12 cases with neurophysiological study]. *Arch Fr Pediatr* 1992;**49**:511–17.
69. Govaert P. *Cranial Haemorrhage in the Term Newborn Infant (Clinics in Developmental Medicine)*. London: Mac Keith Press; 1993.
70. Roig M, Gratacos M, Vazquez E, Del Toro M, Foguet A, Ferrer I, et al. Brainstem dysgenesis: report of five patients with congenital hypotonia, multiple cranial nerve involvement, and ocular motor apraxia. *Dev Med Child Neurol* 2003;**45**:489–93.
71. Desjardins R, Guerguerian AJ, Tabchy B. [Facial paralysis in children—study of 99 cases]. *J Otolaryngol* 1980;**9**:67–71.
72. Quijano-Roy S, Barcik U, Pascual SI, López LE, Blanco FM, Pascual CI. Paralisis facial en la infancia: una revision de 98 casos. *Rev Neurol* 1999;**28**:S172.
73. Tveitnes D, Oymar K, Natas O. Acute facial nerve palsy in children: how often is it Lyme borreliosis? *Scand J Infect Dis* 2007;**39**:425–31.
74. Jackson CG, Glasscock 3rd ME, Hughes G, Sismanis A. Facial paralysis of neoplastic origin: diagnosis and management. *Laryngoscope* 1980;**90**:1581–95.
75. Alberti PW, Biagioni E. Facial paralysis in children. A review of 150 cases. *Laryngoscope* 1972;**82**:1013–20.
76. Adour KK, Byl FM, Hilsinger Jr. RL, Kahn ZM, Sheldon MI. The true nature of Bell's palsy: analysis of 1,000 consecutive patients. *Laryngoscope* 1978;**88**:787–801.
77. Drack FD, Weissert M. Outcome of peripheral facial palsy in children—a catamnestic study. *Eur J Paediatr Neurol* 2013;**17**:185–91.
78. Adour KK, Ruboyianes JM, Von Doersten PG, Byl FM, Trent CS, Quesenberry Jr. CP, et al. Bell's palsy treatment with acyclovir and prednisone compared with prednisone alone: a double-blind, randomized, controlled trial. *Ann Otol Rhinol Laryngol* 1996;**105**:371–8.
79. Katusic SK, Beard CM, Wiederholt WC, Bergstralh EJ, Kurland LT. Incidence, clinical features, and prognosis in Bell's palsy, Rochester, Minnesota, 1968–1982. *Ann Neurol* 1986;**20**:622–7.
80. Keane JR. Bilateral seventh nerve palsy: analysis of 43 cases and review of the literature. *Neurology* 1994;**44**:1198–202.
81. English JB, Stommel EW, Bernat JL. Recurrent Bell's palsy. *Neurology* 1996;**47**:604–5.
82. Takahashi A, Fujiwara R. Familial Bell's palsy. *Clin Neurol Neurosurg* 1971;**11**:454–61.
83. Aldrich MS, Beck RW, Albers JW. Familial recurrent Bell's palsy with ocular motor palsies. *Neurology* 1987;**37**:1369–71.

84. Hageman G, Ippel PF, Jansen EN, Rozeboom AR. Familial, alternating Bell's palsy with dominant inheritance. *Eur Neurol* 1990;**30**:310–13.
85. Baringer JR. Herpes simplex virus and Bell palsy. *Ann Intern Med* 1996;**124**:63–5.
86. Schirm J, Mulkens PS. *Bell's Palsy and Herpes Simplex Virus.* APMIS; 1997;**105**:815–23.
87. Morgan M, Nathwani D. Facial palsy and infection: the unfolding story. *Clin Infect Dis* 1992;**14**:263–71.
88. Czornyj L, Freire C, Cisterna D, Fejerman N. Herpes virus simplex como probable factor etiologico de la paralisis facial periferica aguda idiopatica en la infancia. Estudio clinico y serologico en 169 niños. In: Fejerman N, editor. *Neurologia Pediatrica.* 2nd ed. Buenos Aires: Ed Med Panam; 1997. pp. 892–8.
89. Murakami S, Mizobuchi M, Nakashiro Y, Doi T, Hato N, Yanagihara N. Bell palsy and herpes simplex virus: identification of viral DNA in endoneurial fluid and muscle. *Ann Intern Med* 1996;**124**:27–30.
90. Sugita T, Murakami S, Yanagihara N, Fujiwara Y, Hirata Y, Kurata T. Facial nerve paralysis induced by herpes simplex virus in mice: an animal model of acute and transient facial paralysis. *Ann Otol Rhinol Laryngol* 1995;**104**:574–81.
91. Furuta Y, Fukuda S, Chida E, Takasu T, Ohtani F, Inuyama Y, et al. Reactivation of herpes simplex virus type 1 in patients with Bell's palsy. *J Med Virol* 1998;**54**:162–6.
92. Santos MA, Caiaffa Filho HH, Vianna MF, Almeida AG, Lazarini PR. Varicella zoster virus in Bell's palsy: a prospective study. *Braz J Otorhinolaryngol* 2010;**76**:370–3.
93. Kanerva M, Nissinen J, Moilanen K, Maki M, Lahdenne P, Pitkaranta A. Microbiologic findings in acute facial palsy in children. *Otol Neurotol* 2013;**34**:e82–7.
94. Mori H, Nakai Y. Bell's palsy and virus: ten years experience on serological examination. In: Portmann M, editor. *Proceedings of the Fifth International Symposium on Facial Nerve.* New York: Masson; 1984. pp. 230–3.
95. Murr AH, Benecke Jr. JE. Association of facial paralysis with HIV positivity. *Am J Otol* 1991;**12**:450–1.
96. Manning JJ, Adour KK. Facial paralysis in children. *Pediatrics* 1972;**49**:102–9.
97. Pettersen E. Natural history of Bell's palsy. In: Graham MD, House WF, editors. *Disorders of the Facial Nerve.* New York: Raven Press; 1982.
98. Devriese PP. Bell's palsy in children. *Acta Otorhinolaryngol Belg* 1984;**38**:261–7.
99. Truy E, Granade G, Bensoussan J, Kauffman I, Langue J, Morgon A. [Acquired peripheral facial palsy in children. Current data illustrated by 66 recent personal cases]. *Pediatrie* 1992;**47**:481–6.
100. Aminoff MJ. Facial neuropathies. In: Aminoff M, editor. *Electromyography in Clinical Practice: Clinical and Electrodiagnostic Aspects of Neuromuscular Disease.* New York: Churchill Livingstone; 1998. pp. 513–23.
101. Kimura J, Giron LT, Young SM. Electrophysiological study of Bell palsy: electrically elicited blink reflex in assessment of prognosis. *Arch Otolaryngol* 1976;**102**:140–3.
102. Jenkins HA, Herzog JA, Coker NJ. Bell's palsy in children. Cases of progressive facial nerve degeneration. *Ann Otol Rhinol Laryngol* 1985;**94**:331–6.
103. Esslen E. *The Acute Facial Palsies.* New York: Springer-Verlag; 1977.
104. Eavey RD, Herrmann BS, Joseph JM, Thornton AR. Clinical experience with electroneurography in the pediatric patient. *Arch Otolaryngol Head Neck Surg* 1989;**115**:600–7
105. Gronseth GS, Paduga R. American Academy of Neurology. Evidence-based guideline update: steroids and antivirals for Bell palsy: report of the Guideline Development Subcommittee of the American Academy of Neurology. *Neurology* 2012;**79**:2209–13.
106. Lee HY, Byun JY, Park MS, Yeo SG. Steroid-antiviral treatment improves the recovery rate in patients with severe Bell's palsy *Am J Med* 2013;**126**:336–41.
107. Prescott CA. Idiopathic facial nerve palsy in children and the effect of treatment with steroids. *Int J Pediatr Otorhinolaryngol* 1987;**13**:257–64.
108. Unuvar E, Oguz F, Sidal M, Kilic A. Corticosteroid treatment of childhood Bell's palsy. *Pediatr Neurol* 1999;**21**:814–16.
109. Hato N, Kisaki H, Honda N, Gyo K, Murakami S, Yanagihara N. Ramsay Hunt syndrome in children. *Ann Neurol* 2000;**48**:254–6.
110. Murakami S, Honda N, Mizobuchi M, Nakashiro Y, Hato N, Gyo K. Rapid diagnosis of varicella zoster virus infection in acute facial palsy. *Neurology* 1998;**51**:1202–5.
111. Becerra JC, Sieber R, Martinetti G, Costa ST, Meylan P, Bernasconi E. Infection of the central nervous system caused by varicella zoster virus reactivation: a retrospective case series study. *Int J Infect Dis* 2013;**17**:e529–34.
112. Furuta Y, Ohtani F, Kawabata H, Fukuda S, Bergstrom T. High prevalence of varicella-zoster virus reactivation in herpes simplex virus-seronegative patients with acute peripheral facial palsy. *Clin Infect Dis* 2000;**30**:529–33.
113. Pachner AR, Steere AC. The triad of neurologic manifestations of Lyme disease: meningitis, cranial neuritis, and radiculoneuritis. *Neurology* 1985;**35**:47–53.
114. Halperin J, Luft BJ, Volkman DJ, Dattwyler RJ. Lyme neuroborreliosis. Peripheral nervous system manifestations. *Brain* 1990;**113**:1207–21.
115. Christen HJ, Hanefeld F, Eiffert H, Thomssen R. Epidemiology and clinical manifestations of Lyme borreliosis in childhood. A prospective multicentre study with special regard to neuroborreliosis. *Acta Paediatr Suppl* 1993;**386**:1–75.
116. Ruel M, Arzouni JP, Raoult D, Postic D, Tailame G, Freyss G. Lack of association of Bell's palsy and *Borrelia burgdorferi* antibodies. *Arch Intern Med* 1993;**153**:1725–6.
117. Blin-Rochemaure N, Quinet B. [Should a lumbar puncture be performed in any child with acute peripheral facial palsy and clinical suspicion of Lyme borreliosis?]. *Arch Pediatr* 2012;**19**:1354–61.
118. Raucher HS, Kaufman DM, Goldfarb J, Jacobson RI, Roseman B, Wolff RR. Pseudotumor cerebri and Lyme disease: a new association. *J Pediatr* 1985;**107**:931–3.
119. Belman AL, Iyer M, Coyle PK, Dattwyler R. Neurologic manifestations in children with North American Lyme disease. *Neurology* 1993;**43**:2609–14.
120. Steere AC. Lyme disease. *N Engl J Med* 1989;**321**:586–96.
121. Craft JE, Grodzicki RL, Steere AC. Antibody response in Lyme disease: evaluation of diagnostic tests. *J Infect Dis* 1984;**149**: 789–95.
122. Albisetti M, Schaer G, Good M, Boltshauser E, Nadal D. Diagnostic value of cerebrospinal fluid examination in children

with peripheral facial palsy and suspected Lyme borreliosis. *Neurology* 1997;**49**:817–24.

123. Belman AL, Reynolds L, Preston T, Postels D, Grimson R, Coyle PK. Cerebrospinal fluid findings in children with Lyme disease-associated facial nerve palsy. *Arch Pediatr Adolesc Med* 1997;**151**:1224–8.
124. Angerer M, Pfadenhauer K, Stohr M. Prognosis of facial palsy in *Borrelia burgdorferi* meningopolyradiculoneuritis. *J Neurol* 1993;**240**:319–21.
125. Belman AL, Coyle PK, Roque C, Cantos E. MRI findings in children infected by *Borrelia burgdorferi*. *Pediatr Neurol* 1992;**8**:428–31.
126. Jenke AC, Stoek LM, Zilbauer M, Wirth S, Borusiak P. Facial palsy: etiology, outcome and management in children. *Eur J Paediatr Neurol* 2011;**15**:209–13.
127. Pollac R, Brown L. Facial paralysis in otitis media. In: Graham M, House W, editors. *Disorders of the Facial Nerve: Anatomy, Diagnosis, and Management*. New York: Raven Press; 1982. p. 221.
128. Bauer CA, Coker NJ. Update on facial nerve disorders. *Otolaryngol Clin North Am* 1996;**29**:445–54.
129. Labbe D, Huault M. Lengthening temporalis myoplasty and lip reanimation. *Plast Reconstr Surg* 2000;**105**:1289–97, discussion 98.
130. Zuker RM, Goldberg CS, Manktelow RT. Facial animation in children with Mobius syndrome after segmental gracilis muscle transplant. *Plast Reconstr Surg* 2000;**106**:1–8.
131. Pitts DB, Adour KK, Hilsinger Jr. RL. Recurrent Bell's palsy: analysis of 140 patients. *Laryngoscope* 1988;**98**:535–40.
132. Zimmer WM, Rogers 3rd RS, Reeve CM, Sheridan PJ. Orofacial manifestations of Melkersson-Rosenthal syndrome. A study of 42 patients and review of 220 cases from the literature. *Oral Surg Oral Med Oral Pathol* 1992;**74**:610–19.
133. Wadlington WB, Riley Jr. HD, Lowbeer L. The Melkersson-Rosenthal syndrome. *Pediatrics* 1984;**73**:502–6.
134. Graham MD, Kemink JL. Total facial nerve decompression in recurrent facial paralysis and the Melkersson-Rosenthal syndrome: a preliminary report. *Am J Otol* 1986;**7**:34–7.
135. Poloni TE, Merlo IM, Alfonsi E, Marinou-Aktipi K, Botti S, Arrigo A, et al. Facial nerve is liable to pressure palsy. *Neurology* 1998;**51**:320–2.
136. Pareyson D, Taroni F. Deletion of the PMP22 gene and hereditary neuropathy with liability to pressure palsies. *Curr Opin Neurol* 1996;**9**:348–54.
137. Tidswell T, Pitt MC. A new analytical method to diagnose congenital myasthenia with stimulated single-fiber electromyography. *Muscle Nerve* 2007;**35**:107–10.
138. Polo A, Manganotti P, Zanette G, De Grandis D. Polyneuritis cranialis: clinical and electrophysiological findings. *J Neurol Neurosurg Psychiatry* 1992;**55**:398–400.
139. Najim al-Din AS, Anderson M, Eeg-Olofsson O, Trontelj JV. Neuro–ophthalmic manifestations of the syndrome of ophthalmoplegia, ataxia and areflexia. Observations on 20 patients. *Acta Neurol Scand* 1994;**89**:87–94.
140. Sauron B, Bouche P, Cathala HP, Chain F, Castaigne P. Miller Fisher syndrome: clinical and electrophysiologic evidence of peripheral origin in 10 cases. *Neurology* 1984;**34**:953–6.
141. Fross RD, Daube JR. Neuropathy in the Miller Fisher syndrome: clinical and electrophysiologic findings. *Neurology* 1987;**37**:1493–8.
142. Ter Bruggen JP, van der Meche FG, de Jager AE, Polman CH. Ophthalmoplegic and lower cranial nerve variants merge into each other and into classical Guillain-Barré syndrome. *Muscle Nerve* 1998;**21**:239–42.

Chapter 14

Mononeuropathies

Monique M. Ryan and H. Royden Jones, Jr.

INTRODUCTION

Mononeuropathies are rare in children, accounting for fewer than 10% of pediatric referrals for electromyographic (EMG) testing.[1] This is in contrast to adults, in whom as many as 30% of EMG referrals are for investigation and characterization of discrete peripheral nerve lesions. Another major difference between pediatric and adult mononeuropathies is the distribution of nerve involvement (Figure 14.1). In children, neuropathies almost equally affect the median, ulnar, radial, peroneal, and sciatic nerves, whereas in adults, 65% of focal neuropathies affect the median nerve, mainly because of carpal tunnel syndrome (CTS). Trauma is the most common cause of focal nerve dysfunction in children, most often because of fractures and lacerations, many of which are related to sports injuries. Compressive lesions are the second most common mechanism for pediatric mononeuropathies, while nerve entrapment is relatively uncommon, in contrast to adults. Peripheral nerve entrapment and compression account for most adult mononeuropathies. This chapter will discuss the diagnosis and management of focal neuropathies in children. The reader may supplement this material by referring to larger texts dedicated to the topics of pediatric EMG,[1] pediatric neuropathies,[2] and focal neuropathies in children and adults.[3]

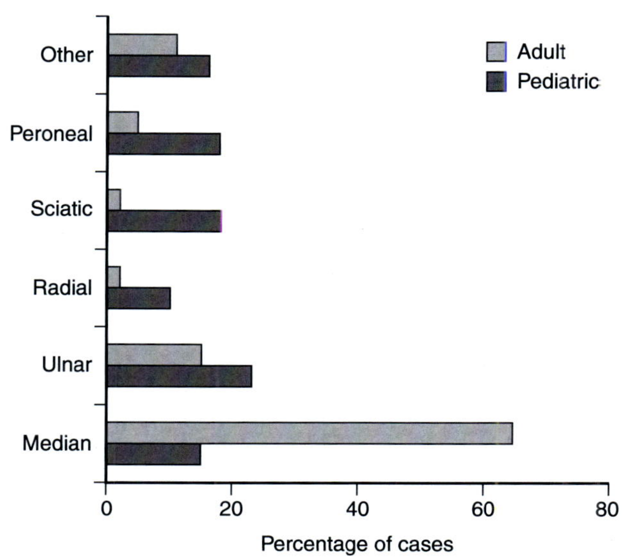

FIGURE 14.1 Distribution of peripheral mononeuropathies in pediatric and adult practice.

ASSESSMENT AND EVALUATION OF PERIPHERAL NERVE LESIONS IN CHILDHOOD

The clinical history is often very helpful in identifying the likely etiology of peripheral nerve lesions. An obvious cause is apparent for most acute pediatric mononeuropathies.[4–6] A targeted clinical examination confirms the clinical suspicion of a peripheral nerve lesion and localizes focal deficits. The neurophysiologic examination enables exclusion of generalized neuropathies, plexopathy, and radiculopathy. A detailed electromyographic examination is particularly useful in localizing focal lesions in idiopathic mononeuropathies. Neuroimaging, whether by ultrasound or magnetic resonance (MR) neurography, is an invaluable adjunct to the clinical and electrodiagnostic examinations.[7–11] Both modalities give excellent soft tissue resolution, are noninvasive, and do not involve ionizing radiation. Ultrasound may be even more sensitive than MRI in identifying focal nerve lesions when performed by a skilled technician in sonographically accessible regions.[8] Soft tissue changes such as high-signal changes in atrophic muscles on STIR- and T2-weighted MR images infer axonal injury rather than neuropraxia or conduction block,[12] the presence of which may affect management. Both ultrasound and MRI can be useful to guide fascicular biopsy in instances in which the diagnosis remains unclear.

B.T. Darras, H. Royden Jones, Jr., M.M. Ryan & D.C. De Vivo (Eds): Neuromuscular Disorders of Infancy, Childhood and Adolescence, Second edition.
DOI: http://dx.doi.org/10.1016/B978-0-12-417044-5.00014-7

UPPER EXTREMITY MONONEUROPATHIES

Median Nerve

Anatomy

The median nerve is formed in the axilla by branches of the lateral and medial cords of the brachial plexus (Figure 14.2).[3,13] In the forearm, the median nerve innervates the pronator teres (C6–C7), flexor carpi radialis (C6–C7), palmaris longus (C7–T1) and flexor digitorum superficialis (C7–C8). The anterior interosseous nerve, a primarily motor nerve, separates from the main trunk of the median nerve in the upper forearm and innervates the lateral head of the flexor digitorum profundus (C7–C8), flexor pollicis longus (C7–C8), and pronator quadratus (C7–C8). In the lower forearm, the median nerve gives off the palmar cutaneous branch. The main trunk then enters the wrist and travels through the carpal tunnel. Distal to the carpal tunnel, the median nerve divides into sensory and motor terminal branches. The motor branch supplies the first and second lumbricals (C8–T1) in the palm; in addition, a recurrent thenar motor branch supplies the abductor pollicis brevis (C8–T1), opponens pollicis (C8–T1), and superficial head of the flexor pollicis brevis (C8–T1). The terminal sensory branches supply sensation to the thumb, index, and middle fingers, and the lateral aspect of the ring finger.

Clinical Evaluation

In infants and young children, mononeuropathies may become apparent only when an observant parent or pediatrician notices focal weakness, atrophy, or an abnormal limb position. Children aged less than 5 years often have difficulty describing discomfort from median neuropathies. Many authors have reported a paucity of complaints of numbness, tingling, or nocturnal pain, and absence of Tinel's and Phalen's signs in children.[14–17] Diagnosis is especially difficult in children with cognitive impairment, in whom it becomes additionally important to recognize subtle symptoms such as decreased sweating, clumsiness and tremor, nocturnal waking, gnawing of fingers and apparent insensitivity to pain, withdrawal of hands from touch of others, alterations in grasp or playing pattern, and increasing difficulties with fine motor tasks, which may be associated with classic signs of pulp atrophy, wasting of the thenar eminence, and weakness of thumb abduction and opposition.[14–17]

Examination should include a careful search for atrophy and fasciculations. The hand may take on an abnormal appearance when there is atrophy of the thenar eminence (the "simian hand") and median-innervated finger flexors (benediction hand). To evaluate median motor function, the examiner should evaluate the patient's performance of the following maneuvers, without and against resistance: pronation of the forearm, flexion and abduction of the wrist, flexion of the fingers at the proximal and distal interphalangeal joints, extension of the index finger at the proximal interphalangeal joint, flexion of the distal phalanx of the thumb, abduction of the thumb at a right angle to the palm, and opposition of the thumb to the base of the small finger.

The sensory examination includes testing of light touch and pinprick sensation of the fingers and hand. The sensory abnormalities in proximal median neuropathies typically involve the lateral ring finger, the lateral aspect of the palm, and the palmar aspects of the thumb, index, and middle fingers. Sensory loss is generally incomplete. The sensory nerve examination is normal in lesions restricted to the anterior interosseous nerve. Median mononeuropathies at the wrist (e.g. carpal tunnel syndrome) spare palmar sensation, as the palmar cutaneous branch does not traverse the carpal tunnel. Occasionally, Phalen's maneuver (wrist flexion test) and Tinel's test (nerve tap test) provide additional help with localization, but these tests are neither specific nor sensitive.

In older children, the clinical evaluation of focal median nerve injuries is like that of adults. Symptoms of chronic mild or intermittent median nerve dysfunction are usually entirely sensory in nature and include paresthesias of the lateral hand, thumb, and index, middle, and ring fingers, which may radiate into the forearm, upper arm, and shoulder. With distal median mononeuropathies, symptoms often worsen with certain positions (e.g. wrist flexion) and activities (e.g. sleep, use of a computer keyboard).

The differential diagnosis of median mononeuropathies includes C8–T1 spinal segment or root lesions causing thenar muscle weakness and atrophy, C6–C7 root lesions for sensory symptoms, and brachial plexopathy (e.g. thoracic outlet syndrome). Thenar hypoplasia can also be present as a congenital defect in isolation (Cavanagh's syndrome) or in association with cardiac (Holt-Oram syndrome) or ocular lesions (Okihiro's syndrome).[18]

Proximal Median Neuropathies

Reported causes of pediatric proximal median mononeuropathies are shown in Box 14.1. Trauma is the most common cause of proximal median nerve injuries in children.[19] Fractures of the supracondylar humerus, midradius, and radioulnar joint can cause median nerve compression, entrapment, or laceration. Most of these injuries involve the main trunk of the median nerve, but the anterior interosseous nerve can be traumatized in isolation with supracondylar fractures.[20] Traumatic median nerve injuries may also result from elbow dislocations, lacerations, blunt soft tissue injury, and arterial or venous puncture,[1,21–23] or during nerve blocks for regional anesthesia.[24]

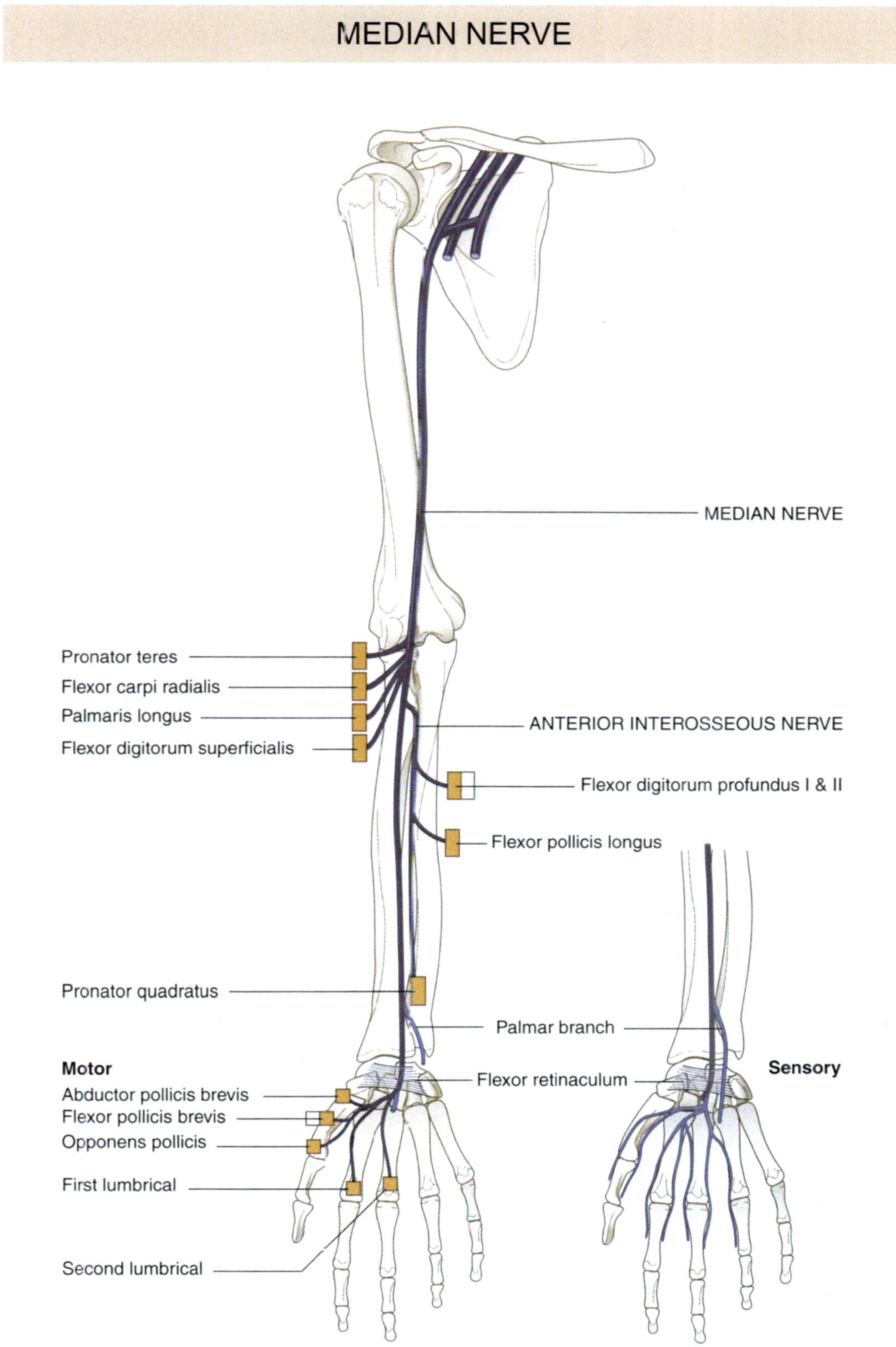

FIGURE 14.2 Diagram of the median nerve, its cutaneous branches and the muscles which it supplies. Note: the white rectangle signifies that the muscle indicated receives a part of its nerve supply from another peripheral nerve. Reproduced from *Aids to the Examination of the Peripheral Nervous System*, 4th edition, WB Saunders 2000, with permission from The Guarantors of Brain.

BOX 14.1 Reported Cases of Pediatric Proximal Median Neuropathy

- Trauma
- Fractures, blunt trauma, lacerations[23,197]
- Direct injury from arterial or venous puncture[1,21–23]
- Axillary nerve blocks[24]
- Entrapment
  - Ligament of Struthers[27]
  - Fibromuscular bands[198,199]
  - Pronator syndrome[28,29]
  - Bicipital aponeurosis[30]
- Soft tissue lesions/tumors
  - Lipofibromas[32]
  - Hamartomas[33]
  - Neurofibromas[34]
  - Hemangiomas[35]
  - Calcified flexor digitorum superficialis tendon[200]
  - Juvenile cutaneous mucinosis[1]
- Other causes
  - Osteoid osteoma[1]
  - Idiopathic
- Anterior interosseous neuropathy
  - Supracondylar humeral fracture[36]
  - Brachial neuritis[37]

BOX 14.2 Reported Causes of Pediatric Distal Median Neuropathy

Carpal tunnel syndrome (CTS)

- Idiopathic[45]
- Traumatic
  - Distal radial fractures[38]
- Overuse/abnormal posturing
  - Skiing, bicycling, golfing, computer games, work-related[1,19,201]
  - Cerebral palsy with dystonic hand movements[202]
  - Schwartz-Jampel syndrome[203]
- Inborn errors of metabolism
  - Mucopolysaccharidoses I, II, IV, VI[4,38,40]
  - Mucolipidosis II, III[38,40]
  - Fabry disease[204]
- Anatomic
  - Familial CTS[43,44,46–48]
  - Bone dysplasias: melorheostosis, Leri's pleonosteosis, acromicric dysplasia[205]
  - Weill-Marchesani syndrome,[206] Poland syndrome,[49] acrodactyly[38]
  - Scleroderma[19]
  - Hemangioma[207]
- Tumors
  - Lipofibromatous hamartoma[208,209]
  - Perineurioma[210]
  - Disseminated angiomatosis[44]
- Inflammatory
  - Scleroderma[1]
  - Trigger finger
  - Pyomyositis, tenosynovitis[211,212]
- Genetic
  - Hereditary neuropathy with liability to pressure palsies (HNPP)[38,54]
  - Dejerine-Sottas syndrome[213]
- Other distal median neuropathies
  - Congenital—constriction bands[25,26]
  - Traumatic—burns, compression[19,21,55]

Less common are median neuropathies caused by congenital soft tissue anomalies such as constriction bands,[25,26] or entrapment by the ligament of Struthers,[27] pronator teres,[28,29] or bicipital aponeurosis.[30] The ligament of Struthers, a fibrous band extending from a small supracondylar spur to the medial epicondyle of the humerus, forms the roof of a tunnel through which the median nerve and brachial artery pass. These spurs are relatively common but the ligament of Struthers is a rare cause of median nerve entrapment.[3] In pronator syndrome, the median nerve is compressed by the hypertrophied heads or thickened tendinous bands of the pronator teres muscle. Because this compression is distal to the origin of the motor branch innervating the pronator teres, this muscle is usually not involved, distinguishing this syndrome from more proximal median lesions (e.g. ligament of Struthers entrapment). Congenital or acquired soft tissue lesions can also cause compressive proximal or distal median neuropathies.[31–35]

Anterior interosseous neuropathy has been reported in several children, either spontaneously due to brachial neuritis[36,37] or after supracondylar fractures of the humerus.[20]

Distal Median Neuropathies

The causes of distal median mononeuropathy are shown in Box 14.2. In children, compressive median neuropathy at the wrist—carpal tunnel syndrome (CTS)—is rare, and generally associated with a predisposing disorder.[38] In a large epidemiologic study conducted over a 2-year period in Wisconsin, the diagnosis of probable or definite CTS was made in 309 patients.[39] Of these, only 7 (2.3%) were children under age 17 years, establishing an incidence rate of 0.26 per 1000 person-years.

Pediatric CTS is most often associated with lysosomal storage disorders (specifically, the mucopolysaccharidoses and mucolipidoses), in which it can be seen as early as

2 years of age[14,38,40] and may be associated with trigger finger.[14,41,42] Infantile CTS is also occasionally familial.[43,44] In distinction, idiopathic CTS is usually reported in older children and adolescents,[1] in whom this is a diagnosis of exclusion.[45]

Predisposing factors are common in pediatric CTS and include congenital canal stenosis (e.g. familial CTS)[43,46–48] or other anatomic anomalies,[49] wrist trauma or injuries,[19,50] thickening of the flexor retinaculum (e.g. mucopolysaccharidoses, mucolipidoses, trigger finger),[14–17,41,45,51–53] repetitive hand and wrist movements or abnormal posturing, inflammatory conditions, and hereditary neuropathy with liability to pressure palsies (HNPP).[54]

In addition to CTS, pediatric distal median mononeuropathies may result from congenital soft tissue anomalies,[25,26] or from focal compression from hematoma, casting, or burns.[19,21,55]

Evaluation

The EMG evaluation of the median nerve in infants includes measurement of a sensory nerve action potential (SNAP) from either the index or middle finger (the latter is preferable in babies because it allows longer interelectrode distance), the compound muscle action potential (CMAP) from the thenar muscles, median motor conduction velocity across the forearm segment, comparative ulnar sensory and motor conduction studies, and needle examination of the minimal number of muscles necessary.

In older children with mild symptoms, the examination should also include a more sensitive test for CTS, including either comparative transcarpal studies of either the median and ulnar mixed nerve action potential (MNAP) (palmar studies) peak latencies, or the median-to-second lumbrical and ulnar-to-interosseous distal latencies.[56] In mild CTS, one may find prolonged latencies of the median SNAP and median-to-second lumbrical CMAP. Prolongation of both the median CMAP distal latency and attenuation of the SNAP amplitude indicates moderate disease. Severe CTS is associated with an absent median SNAP and attenuated thenar CMAP, and needle examination abnormalities indicating both ongoing and chronic changes of denervation and reinnervation in the abductor pollicis brevis. The distal latencies are normal in proximal median neuropathies, but conduction velocities across the forearm segment may be reduced with demyelinating injuries, and needle examination abnormalities often extend into median-innervated forearm muscles with axonal or mixed injuries.

In congenital thenar hypoplasia (e.g. Cavanagh's syndrome), the median SNAP amplitude and peak latency are normal, the CMAP is of low amplitude or absent, and thenar motor unit action potentials (MUAPs) are reduced in number without associated changes of denervation.[18]

The clinical findings and EMG results may dictate the need for additional studies to evaluate median mononeuropathies. Plain X-rays of the hand may be indicated in suspected cases of congenital thenar hypotrophy, in order to assess for hypoplastic changes of hand bones.[18] X-rays of the distal humerus are recommended in suspected cases of ligament of Struthers entrapment, to identify the bony spur often seen in that disorder. Nerve ultrasound or magnetic resonance imaging (MRI) may offer more precise anatomic assessment of suspected soft tissue infiltrative or compressive lesions, especially with symptoms of slowly progressive nerve dysfunction or with focal tenderness or palpable swelling.[1,3] Metabolic studies for the mucopolysaccharidoses and mucolipidoses are indicated in cases of CTS associated with dysmorphic features, organomegaly, and other systemic findings. Unexplained multiple or repeated mononeuropathies should prompt a chromosomal microarray or genetic testing to exclude the chromosome 17p11 deletion seen in hereditary neuropathy with tendency to pressure palsies.[54]

Treatment and Prognosis

Traumatic median neuropathies require prompt attention. As with all peripheral nerve injuries, the main considerations in cases of acute median nerve trauma include nerve continuity, injury type (demyelinating, axonal, or mixed), and prognosis for meaningful recovery. Within 9 to 11 days following acute axonal injury, Wallerian degeneration is complete. EMG studies are recommended at this time in cases with severe weakness or when nerve function cannot be assessed clinically due to other factors related to the trauma (e.g. immobilization, casting).[57] In neurapraxia (focal demyelinating injury), the distal median SNAP and CMAP are preserved, but stimulation proximal to the site of injury evokes an attenuated or absent response (partial or complete conduction block). In mild or moderate axonotmesis (axonal injury), the distal SNAP and CMAP are attenuated or absent, and after 2 to 3 weeks electromyographic examination will show fibrillations and positive waves in median-innervated muscles distal to the site of injury. In severe axonotmesis or neurotmesis, the distal median SNAP and CMAP are absent. Because nerve continuity may be uncertain in these cases, early surgical exploration (and if necessary, repair) is usually recommended. Quite often, EMG findings indicate both demyelinating and axonal features (mixed nerve injury) in traumatic nerve injuries. In cases with preserved continuity and slow return of function, periodic EMG evaluation may be helpful in following nerve regeneration, collateral sprouting, and muscle reinnervation.

Pediatric median neuropathies generally have a good prognosis.[19] Treatment of mild idiopathic and activity-related CTS is focused on conservative measures such as

CASE EXAMPLE 14.1

A 14-year-old, right-handed boy described a 12-month history of intermittent numbness and pain in both hands, particularly affecting digits 2 and 3. He had a lumbar myelodysplasia with paraparesis but was able to self-propel his manual wheelchair. Prior to the onset of symptoms he had begun to play computer games, using a joystick and keyboard, for several hours per day. Neurologic examination demonstrated normal muscle bulk and strength (including the thenar muscles), symmetrically normal reflexes, and normal light touch sensation in his hands and arms.

Nerve conduction studies showed bilaterally prolonged median SNAP and CMAP peak latencies, and a prolonged distal latency difference between the median to second lumbrical and ulnar to interosseous CMAPs. Needle EMG examination of both abductor pollicis brevis and first dorsal interosseous muscles was normal.

Comment

The EMG findings were consistent with moderate bilateral median mononeuropathies localized to the carpal tunnel segments. These findings were due to focal demyelinating injuries that probably resulted from excessive use of a computer keyboard and joystick. His parents limited his computer games to 1 hour per day, and the symptoms resolved after several months.

avoidance of compromising hand positions and activities. (See Case Example 14.1.) Median nerve decompression is required in children with continuing or progressive symptoms refractory to conservative measures, or where the EMG study indicates axon loss. Surgical decompression is indicated in distal median neuropathies associated with inborn errors of metabolism and compression secondary to congenital or acquired soft tissue or bone lesions. This surgery includes division of the flexor retinaculum and release of the median nerve, with or without external neurolysis and/or synovectomy.[14,40] Surgical decompression usually affords some improvement in symptoms and function, with ultimate recovery related to the degree of preoperative axonal loss.

Ulnar Nerve

Anatomy

The ulnar nerve is derived from the medial cord of the brachial plexus (C8–T1) (Figure 14.3).[3,13] At the elbow, it travels first through the epicondylar groove, and then the cubital tunnel as it enters the forearm. Here the ulnar nerve innervates the flexor carpi ulnaris (C8–T1) and the medial head of flexor digitorum profundus (C8–T1). Prior to reaching the hand, the ulnar nerve gives off the palmar and dorsal cutaneous branches in the lower forearm. At the wrist, the nerve enters the Guyon canal and then bifurcates into the superficial and deep branches. The superficial branch innervates the palmaris brevis muscle (C8–T1), and then becomes the terminal sensory nerves, which supply sensation to the fifth digit and medial aspect of the ring finger. The deep branch supplies the hypothenar muscles (C8–T1), including the abductor digiti minimi, opponens digiti minimi, and flexor digiti minimi, and then curves along the palm providing motor branches to the third and fourth lumbricals (C8–T1), the 4 dorsal and 3 palmar interossei (C8–T1), adductor pollicis (C8–T1), and the deep head of the flexor pollicis brevis (C8–T1).

Clinical Evaluation

Mild or intermittent ulnar nerve dysfunction is often associated only with sensory symptoms, which involve the medial aspect of the hand and ring finger, and the entire small finger.[3,13] Sensory symptoms or signs involving the dorsomedial hand point to dysfunction proximal to the dorsal ulnar cutaneous branch, usually at the elbow or cubital tunnel. Specific arm positions or maneuvers may reproduce sensory symptoms due to proximal ulnar lesions. For example, symptoms produced by extreme elbow flexion point to dysfunction at the level of the elbow or cubital tunnel. In some patients with cubital tunnel entrapment, symptoms are reproduced by forceful flexion and medial deviation of the wrist. Tenderness or a positive Tinel's sign at the elbow or wrist may also help to localize the level of injury. Weakness of the flexor carpi ulnaris and flexor digitorum profundus of digit 5 also point to ulnar nerve involvement at the cubital tunnel or elbow; however, preserved function of these muscles is also seen with proximal lesions.

More severe ulnar lesions are associated with weakness and atrophy of ulnar-innervated muscles of the hand and, occasionally, the forearm, depending on localization. Weakness without atrophy is indicative of a demyelinating or early axon-loss injury. Chronic axonal injury and muscle atrophy may produce the so-called claw-hand deformity wherein the fourth and fifth fingers are hyperextended at the metacarpophalangeal joints, due to weakness of the interossei and ulnar lumbricals, and flexed at the interphalangeal joints due to the action of the unopposed flexor digitorum superficialis.

The differential diagnosis of ulnar mononeuropathies includes C8–T1 spinal segment or root lesions, lower trunk or medial cord brachial plexopathy, and mononeuropathy multiplex (e.g. HNPP).

Etiology

The many reported causes of pediatric ulnar mononeuropathy are listed in Box 14.3. Trauma is the most

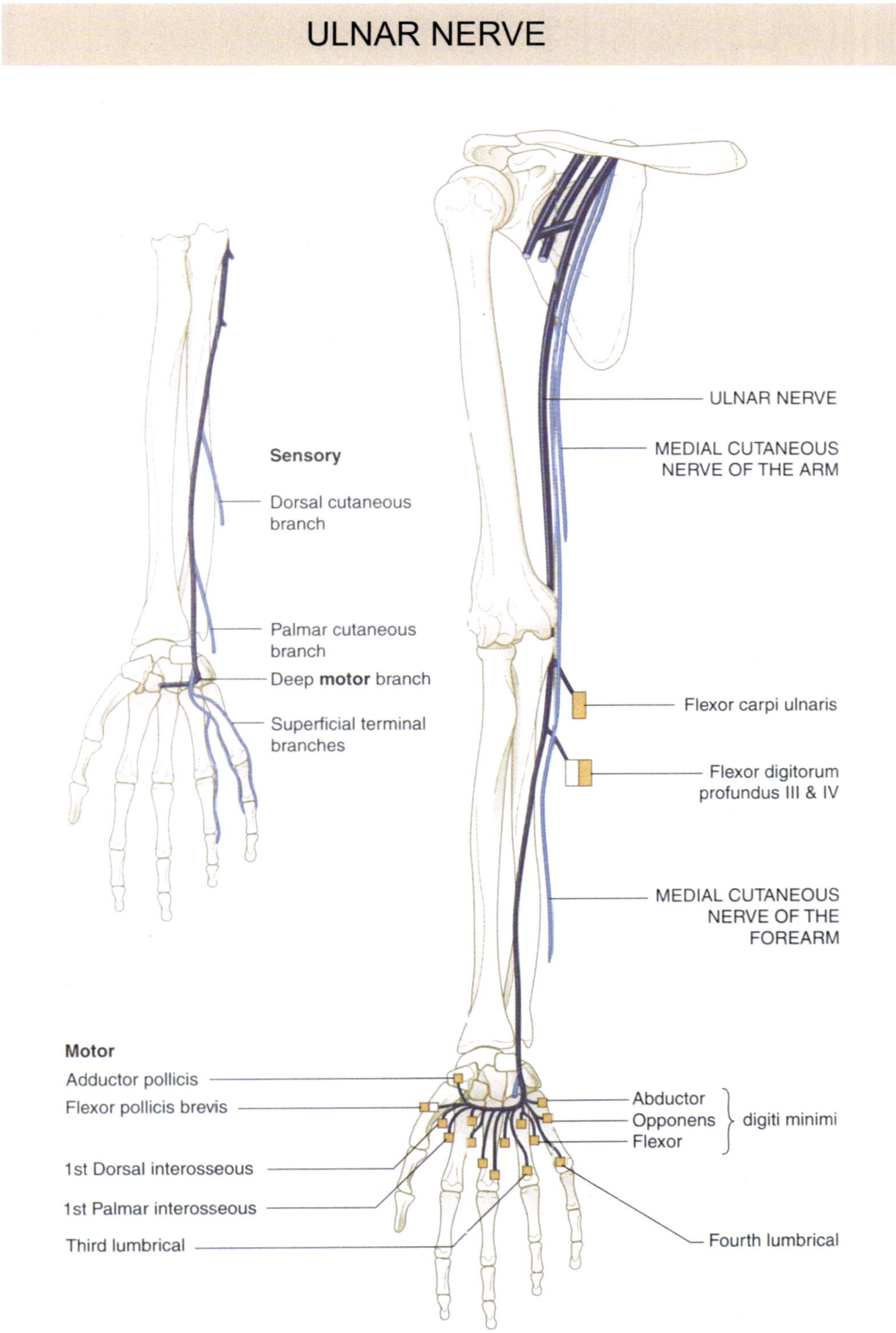

FIGURE 14.3 Diagram of the ulnar nerve, its cutaneous branches and the muscles which it supplies. Reproduced from *Aids to the Examination of the Peripheral Nervous System*, 4th edition, WB Saunders 2000, with permission from The Guarantors of Brain.

common cause of ulnar mononeuropathies in children.[1,58,59] Traumatic proximal ulnar neuropathies result from fractures (supracondylar, medial epicondylar or forearm).[60–62] (See Case Example 14.2.) Distal radial fractures can result in ulnar neuropathies via blunt injury, entrapment, compression, or laceration. The ulnar nerve may also be damaged as the result of surgery to repair an elbow or forearm fracture.[61] Finally, delayed (or "tardy") ulnar nerve palsy is occasionally described in children following elbow trauma, presumably due to posttraumatic

BOX 14.3 Reported Causes of Pediatric Ulnar Neuropathy

- Trauma
  - Fractures—supracondylar, medial epicondylar, forearm, distal radial[1,58,60–62]
  - Elbow dislocation[214]
  - Lacerations, puncture, and stab wounds
  - Axillary nerve blocks[24]
- Other injuries
  - Burns[215]
  - Repetitive throwing movements[216]
  - Nerve ischemia
- Entrapment
  - Cubital tunnel syndrome[7]
  - Persistent epitrochleoanconeus muscle
  - Congenital constriction bands
- Compression
  - Surgical compression
  - Wheelchair arm rests, bicycle hand rests
  - Weightlifting bars
  - Hematoma[217]
  - Infiltration of intravenous fluids[218]
  - Compartment syndrome
- Tumors
  - Hamartomas[31,219]
  - Neurofibromas[220]
  - Neurilemmoma
  - Perineurioma (localized idiopathic hypertrophy)[77,221]
- Other Causes
  - Idiopathic
  - Leprosy[222]
  - Recurrent dislocation
  - Hereditary neuropathy with liability to pressure palsies

CASE EXAMPLE 14.2

An 11-year-old girl developed left hand weakness and numbness 4 months after falling off her bicycle and fracturing her proximal radius and ulna. The upper extremity neurologic examination was remarkable for moderately severe weakness and atrophy of the ulnar hand muscles. The ulnar forearm and median hand muscles were normal. There was reduced light touch sensation along the medial hand (including the dorsal surface) and fingers 4 (with splitting) and 5.

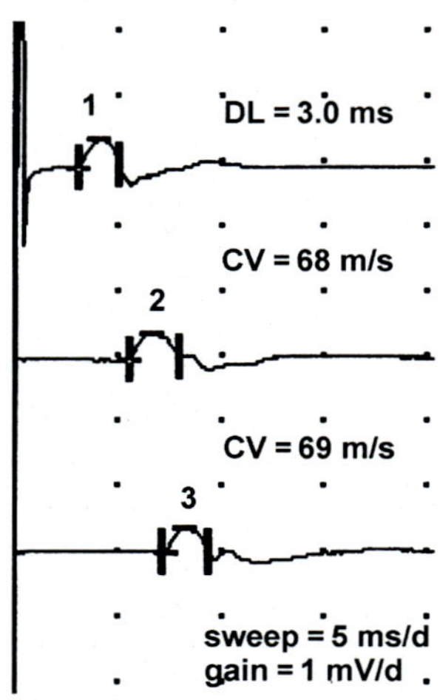

FIGURE 14.4 Nerve conduction studies for Case Example 14.2. The left ulnar nerve is stimulated at the wrist (waveform 1), below elbow (waveform 2), and above elbow (waveform 3) site with a recording electrode on the hypothenar muscles. Note the diffusely low-amplitude CMAPs with normal distal latency (DL) and proximal conduction velocities (CVs). CMAPs, compound muscle action potentials.

Nerve conduction studies demonstrated absent ulnar digital and dorsal cutaneous SNAPs on the affected side. The left ulnar/hypothenar and ulnar/first dorsal interosseous motor nerve conduction studies (Figure 14.4) revealed low-amplitude CMAPs with normal distal latencies and conduction velocities. Contralateral ulnar and ipsilateral median nerve studies were normal. The needle examination revealed fibrillations and positive waves, with reduced recruitment of MUAPs from the left first dorsal interosseous and abductor digiti minimi. The MUAPs were of increased amplitude and duration and fired at increased frequencies. Needle examination of the left flexor carpi ulnaris and abductor pollicis brevis was normal.

Comment

The EMG findings were consistent with a severe but incomplete axon-loss ulnar mononeuropathy, localized proximal to the dorsal cutaneous sensory branch (upper forearm or elbow), probably due to blunt nerve injury. One year following the injury, hand muscle strength had normalized and the only residual symptom was continuing numbness of the little finger.

bony changes and fibrosis.[3,63] The ulnar nerve can also be injured by intraneural injections during nerve blocks for regional anesthesia, elbow lacerations, and puncture wounds.[24]

Ulnar nerve entrapment within the cubital tunnel is occasionally reported in childhood, in some cases possibly related to overuse injury.[58] Other causes of ulnar nerve entrapment include a persistent epitrochleoanconeus muscle[64,65] and congenital constriction bands.[25]

Compressive injuries to the ulnar nerve may develop intraoperatively or as a result of pressure in sleep, from wheelchair rests, or from bicycle handlebars.[58,66] Less common causes of ulnar neuropathy are listed in Box 14.3.

Evaluation

The ulnar nerve EMG evaluation requires measurement of an ulnar SNAP from the little finger, ulnar CMAP from the hypothenar muscles, and motor conduction velocities across the forearm and elbow segments; comparative median sensory and motor nerve conduction studies; and needle examination of the first dorsal interosseous, abductor digiti minimi, and flexor carpi ulnaris muscles. Ulnar motor nerve conduction studies recorded from the first dorsal interosseous muscle are occasionally required as a more sensitive test for demyelinating injuries at the wrist or elbow segments.[67] Addition of the dorsal ulnar cutaneous SNAP may help in differentiating between ulnar lesions at the wrist and elbow. Study of the medial antebrachial cutaneous SNAP is helpful in identifying brachial plexus involvement in selected cases.

Proximal ulnar neuropathies at the elbow are associated with focal motor conduction block, or conduction slowing when there is a primary demyelinating process. If there is an axonal component to the nerve injury, the distal SNAP and CMAP responses are either attenuated or absent, with variable changes of denervation and reinnervation on needle examination of ulnar hand and forearm muscles. Mixed injuries are associated with a combination of demyelinating and axonal features.

Treatment and Prognosis

The management of pediatric ulnar focal neuropathies variously includes nerve decompression and repair of fractures following trauma, surgical nerve repair and grafting following nerve lacerations and severe traumatic injuries, resection of compressive masses and tumors, nerve decompression in cubital tunnel syndrome, and nerve transposition and decompression with progressive lesions localized at the elbow segment.[58,59] Prognosis for full recovery is better in those with atraumatic lesions.[58]

Radial Nerve

Anatomy

The radial nerve is derived from the posterior cord of the brachial plexus (C5–C8) (Figure 14.5).[3,13] The nerve descends in the upper arm between the long and medial heads of the triceps, posterior to the axillary artery. Proximal to the spiral groove, the radial nerve gives off the posterior cutaneous nerve of the arm, motor branches to the triceps (C6–C8) and anconeus (C6–C8), and the posterior cutaneous nerve of the forearm. At the spiral groove, the nerve travels from the medial to the posterolateral aspect of the lower arm. Distal to the groove, the radial nerve innervates the brachioradialis (C5–C6) and extensor carpi radialis longus (C5–C6), then bifurcates into superficial and deep branches. The superficial branch descends the forearm under the brachioradialis prior to emerging in the distal forearm as the superficial sensory branch, supplying sensation to the dorsomedial hand and first web space. The deep branch continues as the posterior interosseous nerve and enters the supinator muscle through an opening termed the arcade of Frohse. Within the extensor compartment of the forearm, the posterior interosseous nerve innervates the supinator (C6–C7), extensor carpi radialis brevis (C5–C7), extensor digitorum (C7–C8), extensor digiti minimi (C7–C8), extensor carpi ulnaris (C7–C8), abductor pollicis longus (C7–C8), extensor pollicis longus (C7–C8), extensor pollicis brevis (C7–C8), and extensor indicis (C7–C8).

Clinical Evaluation

Radial nerve lesions are typically localized to the axilla, spiral groove segment, or restricted to the terminal motor (posterior interosseous nerve) or superficial sensory branches.[3,13] Injuries in the axilla affect all sensory and motor branches of the radial nerve, causing weakness of arm extension, wrist extension (wrist drop), and finger/thumb extension; hypoactive triceps and brachioradialis reflexes; and sensory loss and paresthesias involving the posterolateral surface of the arm, forearm, hand, and fingers 1–4. Spiral groove segment lesions typically spare the motor branches supplying the triceps and anconeus muscles, and the posterior cutaneous nerve of the arm. Weakness resulting from radial nerve damage in the spiral groove usually affects the brachioradialis, wrist extensors, supinator, and finger and thumb extensors. Lesions restricted to the posterior interosseous nerve cause weakness without sensory loss, as this radial nerve branch is purely motor. Children present with finger and thumb drop without associated wrist drop, due to sparing of the extensor carpi radialis longus. Radial deviation of the wrist is noted, however, due to weakness of the extensor carpi ulnaris. Lesions of the superficial cutaneous sensory branch are associated with sensory loss or paresthesias affecting the dorsolateral hand (especially the first web space), and the dorsum of fingers 1 to 4.

The differential diagnosis of radial neuropathies includes C6–C8 spinal segment or root lesions, middle trunk or posterior cord brachial plexopathy, and mononeuropathy multiplex (e.g. HNPP and inflammatory nerve disease).

Etiology

Reported causes of pediatric radial neuropathies are shown in Box 14.4. Traumatic injuries are most common. Radial neuropathy may result from fractures (supracondylar, lateral epicondylar or Monteggia), lacerations, or direct injury.[68–70]

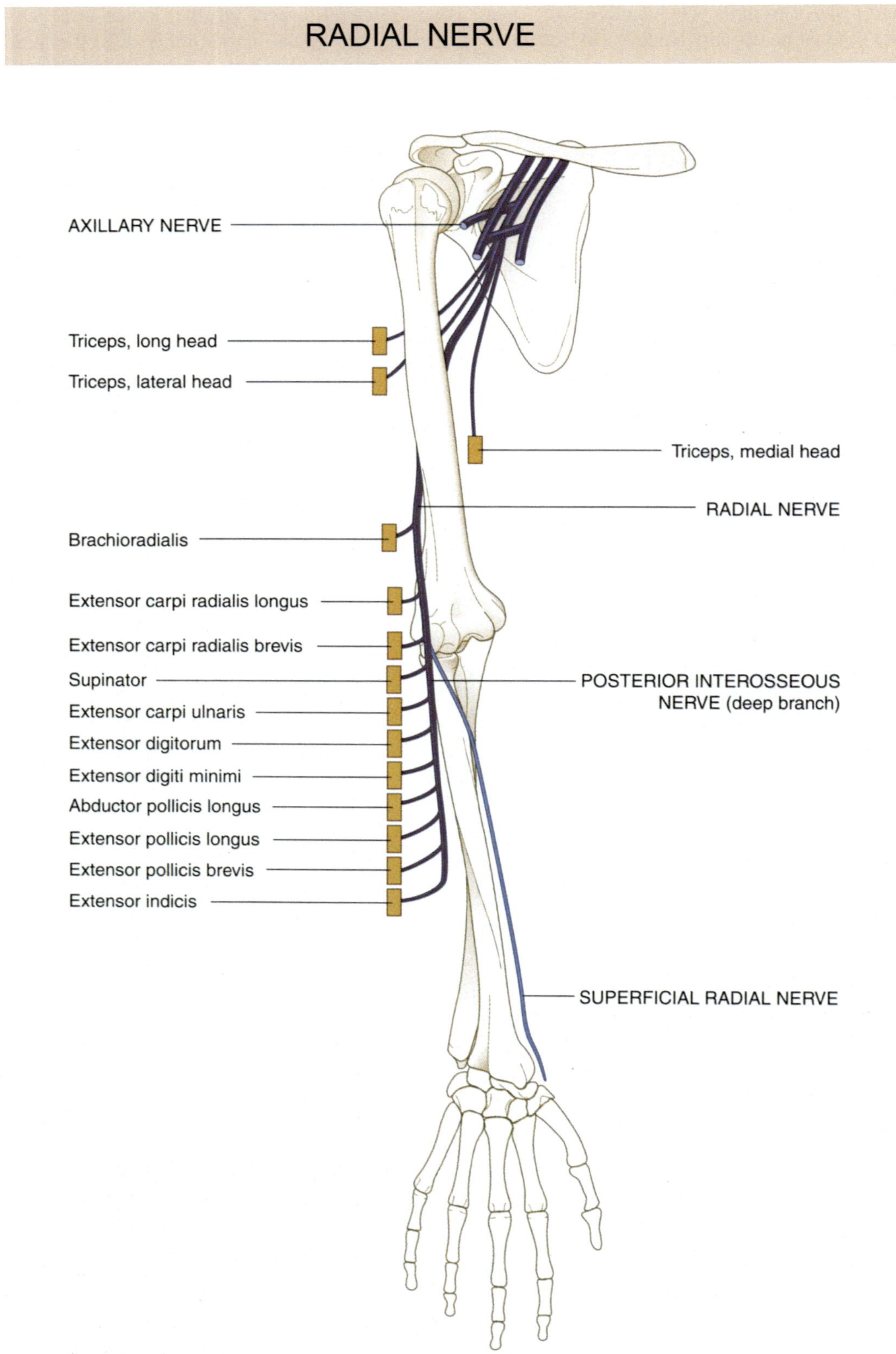

FIGURE 14.5 Diagram of the radial nerve, its major cutaneous branches, and the muscles which it supplies. Reproduced from *Aids to the Examination of the Peripheral Nervous System*, 4th edition, WB Saunders 2000, with permission from The Guarantors of Brain.

BOX 14.4 Reported Causes of Pediatric Radial Neuropathy

- Trauma
  - Fractures—supracondylar, lateral epicondylar, Monteggia[68]
  - Lacerations[68]
  - Injection injuries[70]
  - Arthroscopic elbow surgery[69]
  - Overuse (knitting)[223]
- Compression
  - Perioperative[68]
  - Hematoma[224]
  - Compartment syndrome[68]
  - Sleep palsy, crutch palsy[5,68]
  - Cortical hyperostosis, exostoses[224,225]
  - Soft tissue entrapment: post-sclerotherapy,[68] fibrous or muscular[78,226]
- Tumors
  - Lipomas
  - Gangliomas
  - Fibromas
  - Neuromas
  - Hemangiomas
  - Angioleiomyoma[86]
  - Perineurioma (localized idiopathic hypertrophy)[76,77]
  - Malignant nerve sheath tumor[74]
- Other Causes
  - Hereditary neuropathy with liability to pressure palsies[73]
  - Osteomyelitis[88]
- Congenital and Neonatal
  - Uterine contraction rings[79,80]
  - Subcutaneous abscess or hematoma[81]
  - Dystocia, presumed focal compression[82–85]
  - Humeral fractures[80]
  - Blood pressure monitoring[87]
  - Osteomyelitis[88]
  - Angioleiomyoma[86]

Posterior interosseous neuropathies are very rare in children but may develop after Monteggia fractures, as a result of instability of the head of the radius.[71,72]

As is the case in adults, in children the radial nerve in children is most susceptible to compression injuries in the spiral groove segment. In one large series compressive radial neuropathies developed in sleep, during surgery, from crutches and from infiltration of intravenous fluid.[68] Very rarely, radial neuropathies are the presenting feature of pediatric HNPP.[73]

Benign tumors and other compressive lesions including lipomas, ganglia, fibromas, neuromas, and hemangiomas may cause radial neuropathies.[1,74] Intraneural perineuriomas of the radial nerve are occasionally reported in childhood.[75–77] The radial nerve may also become entrapped within the triceps muscle.[1,78]

Neonatal radial neuropathies (Box 14.4) may result from intrauterine compression by uterine contraction rings,[79,80] subcutaneous abscess or hematoma.[81] Numerous reports have described infants with isolated neonatal radial nerve palsies, usually after a long labor, with clinical findings of a firm subcutaneous area of fat necrosis along the line of the radial nerve in the arm, with or without visible bruising to the area (Figure 14.6).[82–85] Newborn infants have developed radial neuropathies secondary to obstetric humeral fractures, hematomata and soft tissue lesions, blood pressure monitoring, and neonatal osteomyelitis.[80,86–88] Most neonatal radial neuropathies recover completely.[68,82–85]

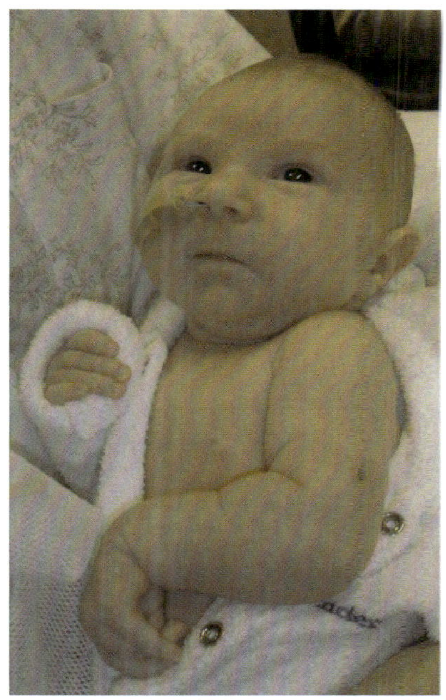

FIGURE 14.6 Neonatal radial neuropathy. There is weakness of finger and hand extension, with an area of subcutaneous fat necrosis due to compressive injury sustained during a prolonged labor. This child experienced complete spontaneous recovery within 3 months.

Evaluation

The EMG evaluation of the radial nerve includes measurement of a radial SNAP from the affected limb, comparative radial SNAP from the unaffected limb (or ipsilateral median SNAP if both limbs are affected), and needle EMG of proximal (e.g. triceps) and distal (e.g. brachioradialis) radial-innervated muscles and a muscle innervated by the posterior interosseous nerve (e.g. extensor indicis). Radial motor nerve conduction studies are technically difficult in neonates and small children. In older children, the radial motor study, recorded from the extensor indicis and stimulated at three sites (forearm, below and above spiral groove) is relatively easy to perform and reliably demonstrates conduction abnormalities across the spiral groove segment.

Pure demyelinating injuries in the spiral groove are associated with motor conduction block and/or slowing, normal distal radial SNAPs and CMAPs, and reduced recruitment of MUAPs without abnormal spontaneous activity in affected muscles. Axonal injuries are associated with absent or attenuated radial SNAPs and CMAPs as well as reduced recruitment of MUAPs, with fibrillations and positive waves in affected muscles. Posterior interosseous neuropathies are associated with a normal radial SNAP, a low-amplitude or absent CMAP from the extensor indicis, and needle examination abnormalities limited to radial-innervated muscles distal to the supinator.

Treatment and Prognosis

Early EMG studies are indicated with lacerations and fractures, in order to assess nerve continuity. An absent radial SNAP and CMAP distal to the site of injury 9–11 days after trauma indicates severe axonotmesis or neurotmesis, and strongly suggests the potential for either complete radial nerve laceration or severe disruption. Such findings warrant surgical exploration and, if necessary, nerve repair. Seven of eight children with traumatic radial neuropathies in one series improved or completely recovered within 7–17 months. Four had undergone surgical repair.[68]

Acute compression injuries in the spiral groove segment are usually primarily demyelinating or mixed, and have a more favorable prognosis for full recovery.[68] Slowly progressive radial or posterior interosseous nerve dysfunction may require MRI studies and surgical exploration to search for tumors, fascial bands, or other sources of nerve compression or entrapment.

Other Upper Extremity Nerves

Axillary Nerve

The axillary nerve derives from the posterior cord of the brachial plexus with the radial nerve, and lies in close proximity to the surgical neck of the humerus. The major branches of the axillary nerve include the lateral cutaneous nerve of the arm and motor branches to the deltoid and teres minor muscles (C5–C6). Axillary nerve injuries cause weakness of arm abduction, deltoid muscle atrophy with severe axonal injuries, and sensory loss along the upper lateral arm. In children, axillary neuropathies are reported after shoulder trauma and injury related to humeral exostosis or septic arthritis (Box 14.5).[89–91] The EMG evaluation of axillary mononeuropathies in children is usually limited to the needle examination of the deltoid muscle.

BOX 14.5 Reported Causes of Other Pediatric Upper Extremity Mononeuropathies

- Axillary nerve
  - Trauma[227]
  - Exostosis of the humerus[89]
  - Sports-related injuries[90]
  - Septic arthritis[91]
- Long thoracic nerve
  - Compressive[1]
  - Trauma[92]
  - Sports-related injury[93]
  - Idiopathic[92]
- Musculocutaneous nerve
  - Compressive[1]
  - Sports-related injury[94]
  - Hereditary neuropathy with liability to pressure palsy[54]
  - Idiopathic
- Suprascapular nerve
  - Entrapment at suprascapular notch[95]
  - Compression by ganglion[97]
  - Trauma[96]
- Thoracodorsal nerve
  - Chest wall empyema
- Spinal accessory nerve
  - Surgical injury
  - Hereditary neuropathy with liability to pressure palsy
  - Idiopathic

Long Thoracic Nerve

The long thoracic nerve originates from the C5–C7 roots and descends in the axilla, posterior to the brachial plexus, to innervate the serratus anterior muscle, which anchors the scapula to the chest wall. Injuries to the long thoracic nerve cause winging of the scapula, especially with the arm in anterior abduction. Reported causes of long thoracic neuropathies (see Box 14.5) in children include sports-related injuries (e.g. tennis, weightlifting), shoulder trauma, and compression from braces, backpacks, and school bags. Some lesions are idiopathic.[1,92,93] The EMG evaluation is usually limited to the needle examination of the serratus anterior at its digitations over the ribs in the midaxillary line.

Musculocutaneous Nerve

The musculocutaneous nerve is derived from the lateral cord of the brachial plexus and innervates the biceps brachii, brachialis, and coracobrachialis muscles (C5–C6), terminating as the lateral cutaneous nerve of the forearm.

Injuries to the musculocutaneous nerve are associated with weakness of arm flexion and sensory loss along the lateral forearm. In children, musculocutaneous neuropathies are rare and generally caused by compressive or overuse injuries, or are associated with HNPP.[54,94] The EMG evaluation of the musculocutaneous nerve may include the lateral antebrachial cutaneous SNAP, a biceps brachii CMAP, and needle examination of the biceps brachii and coracobrachialis.

Suprascapular Nerve

The suprascapular nerve originates from the upper trunk of the brachial plexus and innervates the supraspinatus and infraspinatus (C5–C6) muscles. Injuries to this nerve cause weakness of shoulder abduction and external rotation. Lesions to this nerve usually result from entrapment injuries or trauma (see Box 14.5). The suprascapular nerve may become entrapped by the transverse ligament in the suprascapular notch,[95] compressed by ganglion cysts, or injured in sporting activities.[96,97] The EMG evaluation includes needle EMG of the supraspinatus and infraspinatus.

Spinal Accessory Nerve

The spinal accessory nerve is derived from cranial fibers from the nucleus ambiguus, and spinal fibers from upper cervical motor neurons. The nerve exits the skull via the jugular foramen. The cranial fibers innervate the laryngeal muscles, while the spinal fibers innervate the sternocleidomastoid and trapezius muscles. Injury to the spinal accessory nerve is rare, and usually a complication of surgical procedures involving the posterior triangle of the neck (see Box 14.5).[1,3] Spinal accessory neuropathies can also develop in HNPP.[54] The EMG evaluation of spinal accessory mononeuropathies includes recording a CMAP from the trapezius, followed by needle examination of the trapezius and sternocleidomastoid muscles.

LOWER EXTREMITY MONONEUROPATHIES

Introduction

Mononeuropathies within the lower extremities may be seen at all ages, from the neonate to the adolescent.[1,98] Most are posttraumatic, but lower extremity mononeuropathies occasionally follow compression or entrapment. An important differential diagnosis is focal lower extremity weakness caused by spinal lesions, radiculopathies, or plexopathies.[99] When a mononeuropathy is diagnosed, efforts need to be made to define the exact location of the nerve lesion. This is usually done by means of neurophysiologic and imaging studies. Surgical exploration may be required, both to clarify the etiology of the lesion and sometimes to attempt nerve repair.[1]

Sciatic Nerve

Anatomy

The sciatic nerve is derived from the L4–5 and S1–2 segments of the spinal cord, the anterior rami of which form the lower portions of the lumbosacral plexus, uniting to form the sciatic nerve. This large nerve is about 2.0 cm in diameter. From the lumbosacral plexus, it inclines laterally beneath the gluteus maximus muscle, where it rests on the posterior surface of the ischium. On its medial side, it is accompanied by the posterior femoral cutaneous nerve, supplying cutaneous innervation to the posterior thigh. It emerges from the pelvis through the sciatic notch lying anterior to the piriformis muscle, where it is accompanied by the inferior gluteal artery, which provides the major blood supply to the nerve. On reaching a point about midway between the ischial tuberosity and the greater trochanter, the sciatic nerve turns downward over the gemelli, the obturator internus tendon, and quadratus femoris, and enters the thigh beneath the lower border of the gluteus maximus.

The sciatic nerve descends within the posterior thigh, where it innervates the semitendinosus (L4–S2), semimembranosus (L4–S2), biceps femoris (L4–S2), and distal part of the adductor magnus muscles. The sciatic nerve has two primary divisions: the laterally and more superficially placed peroneal (L4–S2), and the more medially placed tibial nerves (L4–S3) (Figure 14.7). The tibial division innervates all of the posterior thigh muscles except the short head of biceps femoris, which is supplied by the peroneal division. In approximately 90% of individuals, the two divisions share a common sheath from the pelvis to the popliteal fossa. In 10% of individuals, this division occurs at higher levels. Once the sciatic nerve reaches the superior popliteal fossa, it bifurcates into its two terminal branches, the common peroneal and tibial nerves.

The peroneal division of the sciatic nerve is more laterally placed and has fewer and larger fascicles, with less supportive tissue than the tibial nerve, which renders it more susceptible to stretch injury and compression.[3,98,100] Isolated involvement of the peroneal division of the sciatic nerve may mimic (the more common) peroneal lesions at the fibular head, particularly distal sciatic lesions, which may spare the knee flexors. The pediatric neurologist should therefore consider a primary sciatic nerve lesion in the differential diagnosis of children presenting with a foot drop.

Clinical Presentation

Sciatic neuropathies occur in all pediatric age groups, and represent about a quarter of all pediatric mononeuropathies.[101] In pediatric practice, sciatic neuropathies occur as

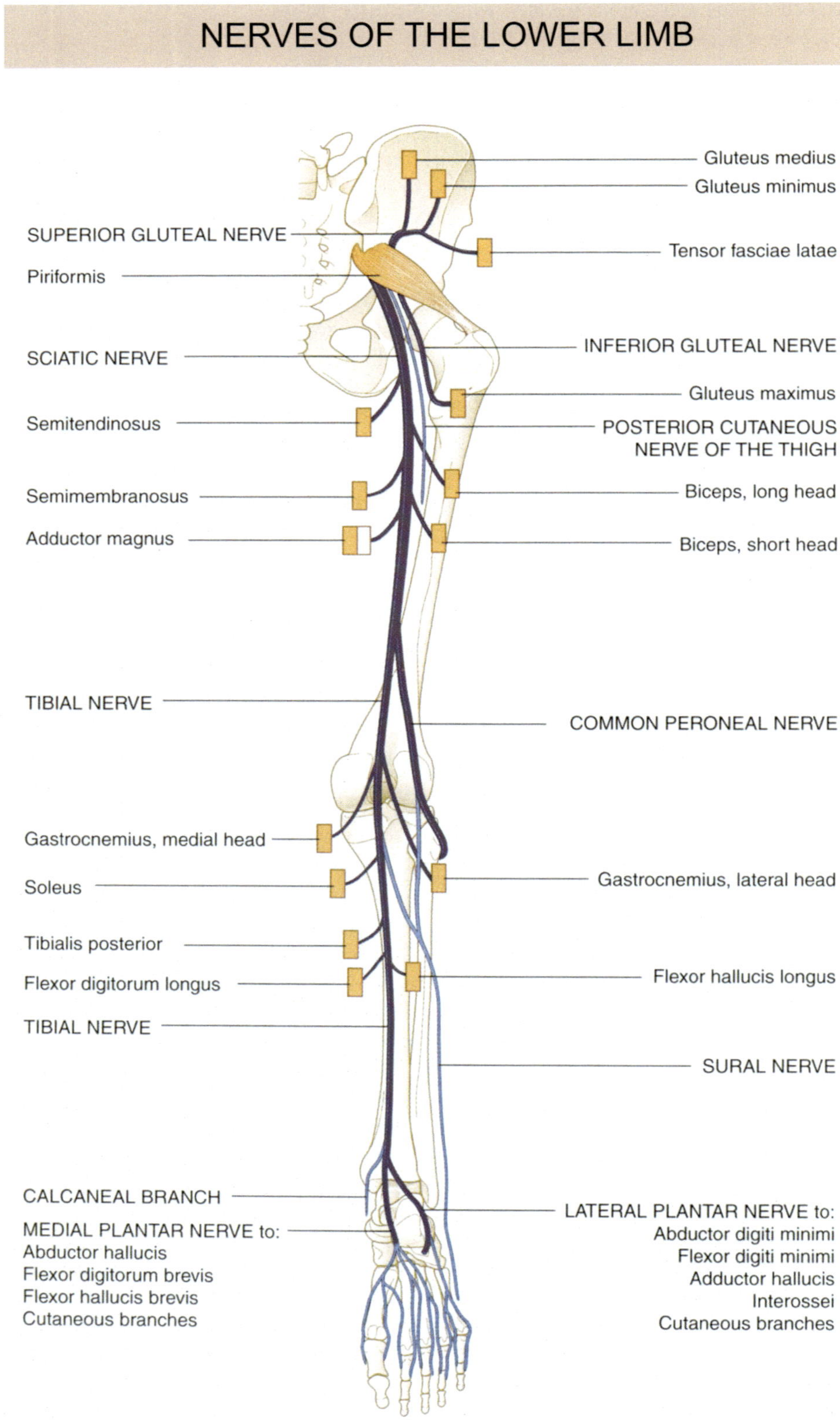

FIGURE 14.7 Diagram of the nerves of the posterior aspect of the lower limb, their cutaneous branches and the muscles which they supply. Reproduced from *Aids to the Examination of the Peripheral Nervous System*, 4th edition, WB Saunders 2000, with permission from The Guarantors of Brain.

BOX 14.6 Reported Causes of Pediatric Sciatic Neuropathy

- Trauma
  - Posterior hip dislocation; pelvic, femoral, and tibial fractures[4]
  - Traction (breech delivery, pelvic osteotomy, repair of hip dysplasia, leg lengthening procedures)
  - Penetrating injuries
  - Intramuscular injections[4,104,105,137]
  - Crush injuries[228]
  - Lacerations
- Compression
  - Prolonged immobilization: chemical paralysis, coma, overdose[4]
  - Lithotomy
  - Hereditary neuropathy with tendency to pressure palsies[229]
  - Prolonged sitting, severe weight loss, yoga
  - Orthopedic casts: long leg, body
  - Hematoma,[106] occult hematocolpos[230]
  - Iliac artery aneurysm
  - Uterine fibroids,[231] endometriosis[232]
  - Heterotopic calcification
- Ischemia
  - Umbilical artery catheterization[4,115]
  - Hypersensitivity vasculitis[4]
  - Meningococcemia[4,98]
  - Embolization of arteriovenous malformation[4]
  - Persistent congenital sciatic artery[146,233]
- Neoplasms
  - Neurofibroma, neurofibrosarcomas[234,235]
  - Rhabdomyosarcoma[4]
  - Neuroblastoma[116]
  - Perineurioma (localized idiopathic hypertrophy)[77]
  - Lymphoma, leukemia, chloroma[101,116,120,236]
  - Osteochondroma, osteosarcoma[4,116]
- Idiopathic[4]
- Post-viral
  - Neuralgic amyotrophy[128]
- Congenital and infantile
  - External compression due to obstetric factors[129,130]
  - Congenital iliac anomaly[141]
  - Myofascial bands
  - Obstetric traction
  - Intragluteal injections
  - Umbilical artery catheterization[4,115,237]

commonly as peroneal neuropathies, in contrast to adults, in whom the latter are more frequent. The diagnosis of sciatic neuropathy in a child depends on a methodical clinical examination, targeted to exclude more common pathologies and accurately localize the lesion. Sciatic nerve injury most commonly occurs within the pelvis, at the sciatic notch or in the thigh. The clinical course of sciatic neuropathies may be acute, subacute, or chronic.

The peroneal and tibial divisions of the sciatic nerve innervate the hamstrings, the distal adductor magnus, the anterior and posterior compartments of the leg, and the intrinsic foot musculature (Figure 14.7). Sensation to the skin of the foot and posterior lower leg is supplied through sensory branches of the tibial nerve (sural, medial, and lateral plantar, and calcaneal branches) and the superficial peroneal nerve, which supplies sensation to the foot and posterior lower leg.[3] Thus, sciatic neuropathies manifest with weakness of foot dorsiflexion (tibialis anterior), plantar flexion (medial and lateral gastrocnemii), inversion (tibialis posterior) and eversion (peroneus longus group), and of knee flexion (medial and lateral hamstrings), as well as decreased sensation or painful dysesthesia of the sole, dorsum of the foot, and posterolateral leg. More slowly progressive lesions may present with progressive foot drop or pes cavus.[102] Milder lesions may present only with weakness of tibialis anterior, with or without involvement of the gastrocnemii. The ankle muscle stretch reflex is usually absent or decreased. Back pain is more common with spinal or intraspinal lesions, but is occasionally seen in pure sciatic neuropathies.[103] Pain in the posterior aspect of the leg extending into the foot is more suggestive of a sciatic neuropathy than radiculopathy.[103]

Because of the fascicular anatomy of the sciatic nerve, proximal sciatic lesions can resemble more distal peroneal neuropathies. The presence of clinical or EMG evidence of hamstring weakness helps differentiate between sciatic and peroneal nerve lesions. Involvement of the gluteal muscles and buttock pain is indicative of very proximal lesions with combined sciatic and gluteal nerve involvement.

The most common cause of pediatric sciatic neuropathy is trauma, whether exogenous or iatrogenic.[4] Compressive sciatic neuropathies are also occasionally seen in childhood, while tumors and vascular lesions are rare but serious causes of this presentation (Box 14.6).

Traumatic Sciatic Neuropathies

The sciatic nerve is particularly vulnerable to trauma, as it is situated immediately behind the bony pelvis and adjacent to the hip joint, predisposing it to injury with pelvic or femoral fractures and/or posterior hip dislocations. Trauma—whether exogenous or iatrogenic—caused 26 of 53 pediatric sciatic neuropathies in the largest pediatric series reported to date.[4] Possible mechanisms of injury include laceration, stretching, and late entrapment in heterotopic calcification. Stretch injury of the nerve

CASE EXAMPLE 14.3

A 14-year-old boy underwent surgical repair of congenital hypospadias, during which he was placed in the lithotomy position. Immediately upon awakening he noted left leg weakness, which was initially presumed to be due to a peroneal nerve compression injury. When his weakness was not improved 8 months later, he was evaluated at the Boston Children's Hospital. Although a low-amplitude tibial CMAP was present, no left peroneal CMAPs or sural SNAPs were elicited. EMG demonstrated occasional fibrillation potentials and a moderate diminution in the number of MUAPs in muscles innervated by the sciatic nerve.

Comment

Postlithotomy sciatic neuropathies can be devastating, remain poorly understood, and may develop even where care is taken to avoid nerve compression at the sciatic notch. One postulate suggests that in some instances the blood supply to the sciatic nerve in the affected individuals represents an anatomic variant with a persistent fetal artery.[113] The lithotomy position may, in these individuals, somehow cause nerve infarction unrelated to pressure or stretch.

may also occur during repair of compound fractures.[4] Surgical exploration should be considered if no recovery is seen within three months, as the prognosis for subsequent recovery is poor.[104]

Sciatic nerve trauma can follow parenteral injections into the inferior medial quadrant of the buttock. Fortunately, this modality of administering intramuscular medications has been largely discontinued, but occasional cases are still reported.[104,105] This is an entirely preventable cause of sciatic neuropathy.

Compressive Sciatic Neuropathies

The sciatic nerve is susceptible to compressive injury in the pelvis and sciatic notch.[4] Within the pelvis, sciatic compressive neuropathy by hematoma is described in hemophiliacs.[106] Other pelvic masses can also cause focal sciatic compression (Box 14.6). Neuropathies at the sciatic notch may be seen in children who are comatose, pharmacologically paralyzed, or immobilized for protracted periods, for example in long leg and/or body casts.[4,98,107] Compressive sciatic neuropathies in childhood have also been attributed to sitting for prolonged periods on hard surfaces,[108] severe weight loss,[109] surgery[110,111] (see Case Example 14.3), and critical illness.[112]

Ischemic Sciatic Neuropathies

Ischemia was responsible for five of the 53 cases of pediatric sciatic neuropathy in a recent large series.[4] Sciatic neuropathies have been attributed to hypersensitivity or post-streptococcal vasculitides, and to purpura fulminans due to meningococcemia.[4] Nerve injuries developing in association with allergic vasculitides may be due to embolism to the vasa nervorum, or a toxic effect of eosinophilic granules on the peripheral nerve.[114] Sciatic and peroneal nerve injury can also follow umbilical artery catheterization; this occurs presumably because of thromboembolization of the inferior gluteal artery and resulting ischemic sciatic nerve injury.[4,115]

Ischemic sciatic nerve injury is associated with low-amplitude motor responses with relative preservation of sensory responses. This somewhat unusual pattern of neurophysiologic change may be a clue to etiology of the nerve injury.[98]

Sciatic Nerve Neoplasms

Primary sciatic nerve tumors are uncommon, but neoplasias can arise from a variety of neural and supporting tissues including bone and muscle. Overall, about 10% of pediatric sciatic neuropathies are associated with some form of neoplasm, which should be considered wherever there is no prodrome of trauma or surgery.[116] Leg pain and weakness—particularly foot drop—are the most common presenting symptoms of sciatic nerve tumors.[103,116] Sciatic neurofibromas can be the initial manifestation of neurofibromatosis,[117] and are palpable along the nerve course in up to two-thirds of cases.[103,117,118] Compression or invasion of the sciatic nerve by pelvic neuroblastoma,[119] chloroma,[120] and bone tumors (sessile osteochondromas, osteosarcoma)[121,122] are occasionally described in childhood. Diagnosis of such lesions may be delayed for as long as 3 to 5 years, even in those with malignant tumors.[103] This raises the question of the potential for malignant transformation of some tumors and emphasizes the necessity of accurate diagnosis in children presenting with chronic sciatic neuropathy, even in the absence of clinical progression. Most tumors can be diagnosed with MR imaging or ultrasonography.[7–10,123] Imaging studies are sometimes unrevealing, however, and surgical exploration may be required in order to exclude the possibility of a treatable lesion.

Special mention should be made of perineuriomas, which are an increasingly well-recognized cause of sciatic and other pediatric mononeuropathies. These benign

CASE EXAMPLE 14.4

A 14-year-old boy presented with a 5-month history of progressive left foot drop and paresthesia of the left side of the foot. Examination revealed wasting of the left calf muscles and weakness of foot dorsiflexion, plantar flexion, and eversion. A limited neurophysiologic assessment showed evidence of either a left sciatic or peroneal neuropathy. MRI of the left leg demonstrated abnormal thickening and enhancement of the sciatic nerve from its origin, extending 20 cm down the thigh. A fascicular biopsy revealed features consistent with perineurioma (Figure 14.8). Follow-up clinical examination and MR imaging 12 months later revealed no change in strength or further increase in sciatic nerve enlargement.

Comment

The sciatic nerve is the most common site for perineuriomas, which generally present with painless, slowly progressive weakness with much milder sensory deficits. Treatment is supportive once diagnosis is confirmed.

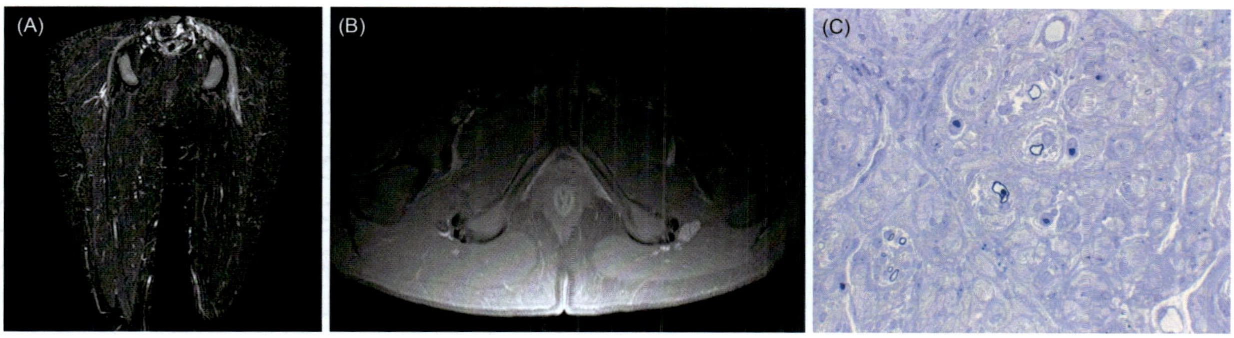

FIGURE 14.8 Sciatic nerve perineurioma in an adolescent. (**A**) Coronal T2WI turbo spin-echo MR imaging of the pelvis shows diffuse enlargement of the proximal sciatic nerve. (**B**) Marked enlargement of the sciatic nerve is seen on axial imaging through the pelvis. (**C**) Toluidine blue section of a fascicular biopsy of the sciatic nerve showed a marked reduction in the density of myelinated fibers with remaining myelinated axons demonstrating onion bulb-like formations, consistent with perineurioma.

hypertrophic peripheral nerve tumors can present at any age with progressive painless weakness with relatively mild or absent sensory deficits, and can affect single nerves or nerve plexuses.[124] Perineuriomas most often affect the sciatic nerve or lumbosacral plexus.[77] When they develop in the sciatic nerve, perineuriomas cause foot drop and pes cavus (see Case Example 14. 4).[103,116,125] Identification and localization of these tumors has been significantly aided by MRI imaging, which shows nerve enlargement, T1 isointensity with hyperintensity on T2-weighted images, and post-contrast enhancement. The diagnosis can be confirmed on fascicular biopsy.[77,116] These tumors tend to be static or only slowly progressive and remain localized, hence resection and reconstructive surgery are probably not warranted once a diagnosis is established.[77] Nerve grafts seem ineffective in restoring function in children with perineuriomas.[125]

Idiopathic Sciatic Neuropathies

Idiopathic sciatic neuropathies in children are very unusual,[126,127] but occasional cases defy pathologic explanation. Electrodiagnostic studies in such cases often show progressive axonal lesions without any cause identifiable on serial imaging or surgical exploration.[101,126]

Postviral Sciatic Neuropathies

Occasional pediatric sciatic neuropathies follow viral infections. This apparent association may be coincidental,[4] but reports of localized presentations mimicking neuralgic amyotrophy suggest that acute mononeuropathies may sometimes represent a focal postinfectious neuritis.[128]

Congenital and Infantile Sciatic Neuropathies

Prenatal injury to the sciatic nerve may result from external compression due to oligohydramnios, abnormal uterine contractions during prolonged labor, amniotic fluid bands, or uterine abnormalities including adhesions and synechiae.[129,130] Necrotic ischemic lesions (eschars) are occasionally apparent at the site of intrauterine compression.[101,129] In such cases it is important, if possible, to perform the EMG during the first days of life in order to demonstrate prenatal onset of the nerve injury.[131]

The sciatic nerve can also be injured during complicated breech deliveries, presumably as a result of traction injury.[132,133]

Postnatally, sciatic nerve injury can follow intragluteal injections to neonates,[134–138] or umbilical artery catheterization.[4,115,139,140]

Focal entrapment of the sciatic nerve is very rare in early childhood, but has occasionally been reported in patients with myofascial or fibrovascular bands or congenital iliac anomalies.[141–145] There is a single report of a persistent fetal sciatic artery causing entrapment of the sciatic nerve at the pelvic notch.[146]

Neonatal sciatic neuropathies often remain idiopathic.[147] In the largest series reported to date, an excellent recovery was seen in 16 of 21 cases. Interestingly, the babies with the worst outcome were all delivered by cesarean section.[147]

Evaluation

Electromyography is a fundamental step in the evaluation of a child with a sciatic neuropathy. Sparing of the paraspinal muscles and muscles innervated by the superior and inferior gluteal nerves distinguishes sciatic nerve lesions from lumbosacral plexopathies. This distinction may be more difficult in pediatric EMG because of the technical challenges of achieving paraspinal muscle relaxation in nervous or upset children. EMG also usually enables differentiation between lesions of the sciatic and peroneal nerves. The finding of neurogenic changes within the short head of the biceps femoris identifies lesions proximal to the origin of the common peroneal nerve, within the peroneal trunk of the sciatic nerve.

Most children with sciatic neuropathies have neurophysiologic evidence of involvement of the peroneal (83%) and/or tibial (67%) motor components of the nerve.[4] Sensory studies are usually also abnormal (sural 79%, and superficial peroneal 60%). These changes generally reflect axonal injury rather than demyelination. Where tested, the H-response is generally absent.[98] Needle EMG is abnormal in virtually all subjects within peroneal-innervated muscles, and slightly less frequently in the tibial-innervated muscles and hamstrings.[4]

It is important when testing any child with a presumed mononeuropathy that, where possible, contiguous unilateral nerves as well as the homologous nerve of the unaffected extremity are examined. The reported "normal" values for motor and sensory amplitudes in pediatrics are very conservative, and side–side differences in the amplitude of sensory and motor responses will be missed if not specifically tested for. As in the adult, needle examination should be used to localize the lesion. Involvement of tibial- and peroneal-innervated muscles distally in the leg, but not of the hamstrings, localizes the lesion to the distal sciatic nerve. Neurogenic changes limited to the peroneal-supplied distal muscles and the short head of the biceps localize the lesion to the peroneal trunk of the sciatic nerve.[100,118] Pure peroneal trunk changes in a pediatric sciatic neuropathy are uncommon but occasionally reported.[98,100] Involvement of the hamstring muscles, but not the glutei or vastus lateralis, localizes the lesion to the proximal sciatic nerve.[98] Involvement of the glutei is indicative either of additional involvement of the gluteal nerves or of a lumbosacral plexopathy.[98] Concomitant involvement of femoral or obturator supplied muscles suggests a lesion of the lumbosacral plexus. Abnormalities of the paraspinal muscles point to an intraspinal or nerve root lesion.

Once a sciatic nerve lesion is defined and localized, ultrasonography, MR neurography, or MRI with contrast can be very helpful in disclosing nerve enlargement or compression by mass lesions or tumors.[4,116,124] Owing to its size, the sciatic nerve is relatively easily visualized by these techniques.

Surgical exploration and nerve repair should be considered after traumatic sciatic neuropathies if no recovery is seen within 3 months.[104] Early exploration and neurolysis or direct nerve repair are effective in short-segment traumatic nerve injuries where early recovery is suboptimal, but nerve grafting over long sciatic segments is rarely effective.[104,125] Careful surgical exploration may also prove fruitful in cases that remain idiopathic despite targeted EMG studies and nerve imaging. [126,148]

Treatment and Prognosis

Prognosis is better in children with posttraumatic and iatrogenic sciatic neuropathies than in children with those caused by vascular lesions or neoplasia.[4,116] EMG findings do not necessarily correlate with eventual prognosis in pediatric sciatic mononeuropathy, but preservation of peroneal and tibial CMAPs appears predictive of partial or full recovery in most cases.[98,101] The presence or absence of denervation by needle examination is not always prognostically informative. A large series including both adult and pediatric patients suggested that patients with injury to the tibial division of the sciatic nerve have a better prognosis for complete recovery than those patients with lesions of the peroneal division, regardless of the level or mechanism of injury.[149] In children with poor recovery of sciatic neuropathies, long-term complications may include growth failure of the affected limb, pes cavus, lesions (callosities and ulceration) related to sensory loss in the feet, and tendon shortening, which frequently necessitates surgical lengthening or tendon transfer procedures.[150]

Femoral Nerve

Anatomy

The femoral nerve (Figure 14.9) is derived from the anterior primary rami of nerve roots L2, 3, and 4, and is the largest branch of the lumbar plexus. Within the pelvis, it innervates the psoas major and the iliacus (both L2, 3), and then exits the pelvis anteriorly, at the femoral triangle, descending beneath the inguinal ligament lateral to

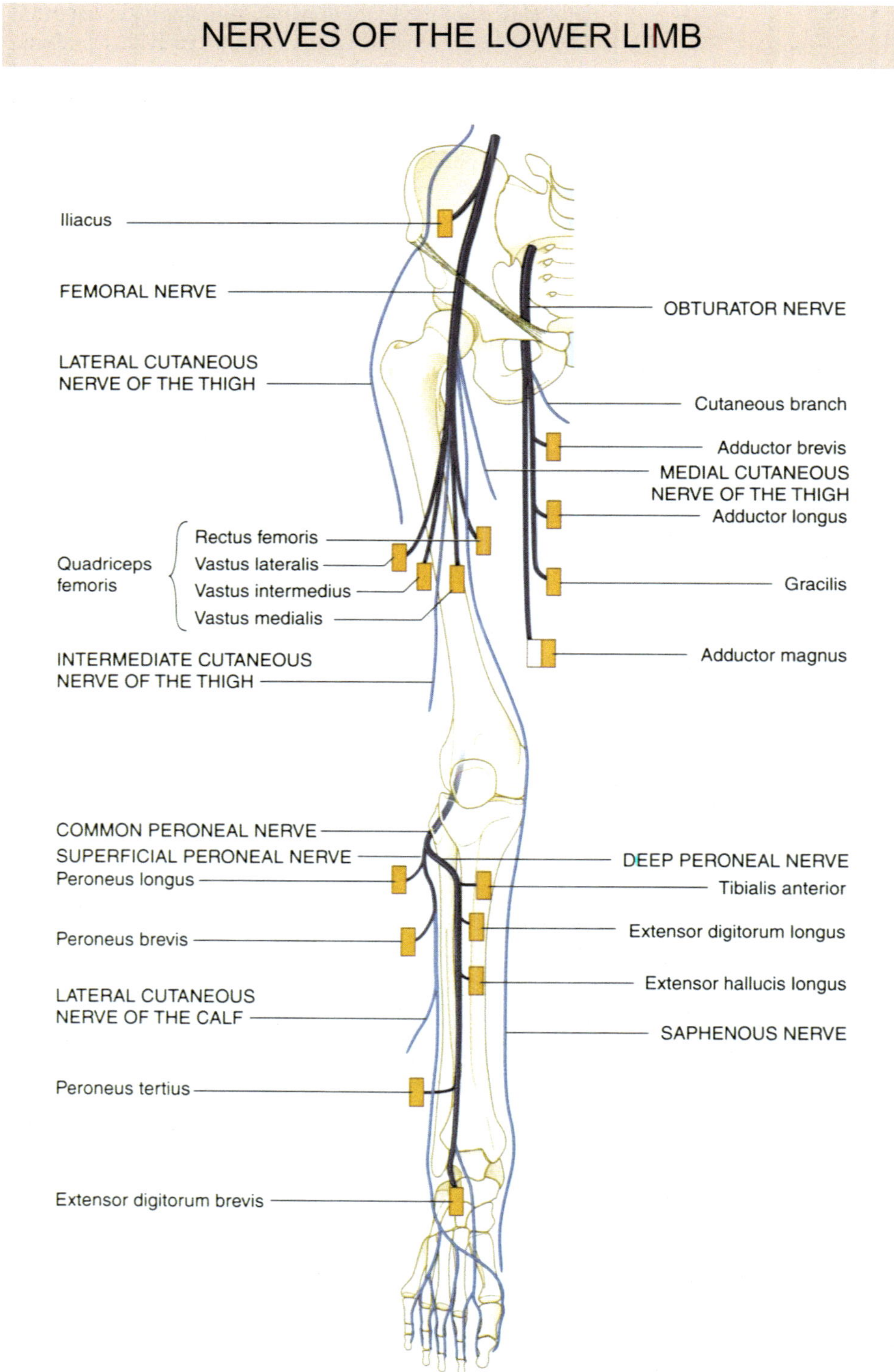

FIGURE 14.9 Diagram of the nerves of the anterior lower limb, their cutaneous branches and the muscles which they supply. Reproduced from *Aids to the Examination of the Peripheral Nervous System*, 4th edition. WB Saunders 2000 with permission from The Guarantors of Brain.

the femoral vascular sheath. Muscular branches supply the pectineus, sartorius, and quadriceps femoris muscles. The anterior femoral cutaneous nerves supply the skin and fascia of the front and medial sides of the thigh. The saphenous nerve communicates with the anterior femoral cutaneous and obturator nerves, and then gives off its infrapatellar branch, descends medial to the knee on to the medial foreleg close to the great saphenous vein, then dividing at the ankle to branches to the medial arch and dorsum of the foot.

BOX 14.7 Reported Causes of Pediatric Femoral Neuropathy

- Traumatic
  - Perioperative[238,239]
  - Femoral catheterization[240]
  - Reduction of hip dysplasia[152]
- Soft tissue lesions/tumors
  - Perineurioma (localized idiopathic hypertrophy)[241,242]
  - Neurofibroma[243,244]
  - Pelvic masses[245]
- Compression
  - Hematomas (intraneural, iliopsoas muscle)[106]
- Idiopathic[7,126,246]

Clinical

Femoral mononeuropathies are very uncommon in children. Acute femoral nerve deficits may occur as a result of compression from an expanding mass, such as hematoma in children with coagulopathies, or after surgical injury.[106] Rarely, infiltrating tumors cause slow compressive injury to the femoral nerve within the pelvis. Reported causes of femoral neuropathy in children are shown in Box 14.7.

Proximal femoral nerve lesions affecting the iliopsoas muscle present with weakness of hip flexion, usually with weakness of quadriceps femoris, which when severe leads to total loss of knee extension. This precludes walking. With partial lesions, patients first note difficulty going downstairs, as they are unable to lock their knee to support their weight. Sensory symptoms involve the anteromedial thigh and medial lower leg. The quadriceps femoris may be wasted, and the patellar muscle stretch reflex is diminished or absent. A very unusual pure motor syndrome with distal quadriceps weakness occurs with lesions distal to the origin of the saphenous nerve in the thigh.

Neonatal femoral neuropathies are extremely uncommon but have been reported as a complication of breech presentation and developmental hip dysplasia, subluxation, or dislocation. Occasional cases appear to develop atraumatically, as a result of positioning in utero,[151] but most have followed treatment with Pavlik-type harnesses.[152]

Treatment and Prognosis

Children with pain and weakness in the anterior thigh should be evaluated to rule out spinal and nerve root pathology. Lumbar polyradiculopathies may be more common as a cause of these complaints than femoral mononeuropathies. Electromyography is useful for precise localization of femoral nerve lesions, and for differentiation of femoral neuropathy from lesions at the level of the plexus or spine.

Normal CMAP amplitude and distal latency values are not determined for the femoral nerve in children, but motor studies of the femoral nerve, stimulating at the groin and recording from the vastus lateralis or rectus femoris, may still provide useful information if the values obtained are compared with those of the unaffected side. It is also possible to record a SNAP from the lateral femoral cutaneous nerve in children, as is done in adults. In contrast, needle examination is useful in identifying axonal pathology and in ruling out involvement of the obturator nerve, which would be suggestive of a lumbar plexopathy.

Treatment is based on the underlying cause. Most compressive femoral neuropathies respond well to removal of the causative lesion.

Lateral Femoral Cutaneous Nerve

The lateral femoral cutaneous nerve is a pure sensory nerve derived from the L2, L3 nerve roots, which emerges from the lateral psoas major, passing obliquely over the iliacus and then coursing toward the anterior superior iliac spine. It enters the thigh passing above or through the most lateral portion of the inguinal ligament, then crosses over or through the proximal part of the sartorius muscle and descends deep to the fascia lata. After a number of small branches are delivered to the overlying skin, the lateral femoral cutaneous nerve pierces the fascia about 10 cm below the inguinal ligament and innervates the anterolateral thigh.

Damage to this cutaneous nerve is common in adults but rare in children. Neuropathies of the lateral femoral cutaneous nerve have a classic presentation: the patient notes a vague numbness and annoying paresthesiae along the lateral thigh also known as *meralgia paresthetica*. On neurologic examination, the patient is able to easily outline the distribution of their abnormal sensation. This is typically a small, elliptical area of dysesthesia, which is easily reproduced on careful clinical examination. The neurologic examination is otherwise entirely normal.

Most lateral femoral cutaneous neuropathies develop insidiously, without an obvious cause.[153,154] Children may be predisposed by obesity.[155] More acute presentations can be triggered by acute compressive injuries (i.e. use of a tight harness or belt across the iliac crest) or by trauma (blunt trauma to the thigh in high energy sports or from repetitive impact). Transient meralgia paresthetica can also follow posterior spinal fusion.[156] There is usually no acute pain. Constant boring pain, leading to postural changes or difficulty sleeping, should prompt imaging to exclude the possibility of an underlying soft tissue or bone lesion.

The diagnosis of lateral femoral cutaneous neuropathy can usually be made clinically. Although sensory responses from the lateral femoral cutaneous nerve can be recorded in the thigh, this sort of testing is technically challenging and seldom necessary in childhood. Most cases resolve spontaneously.

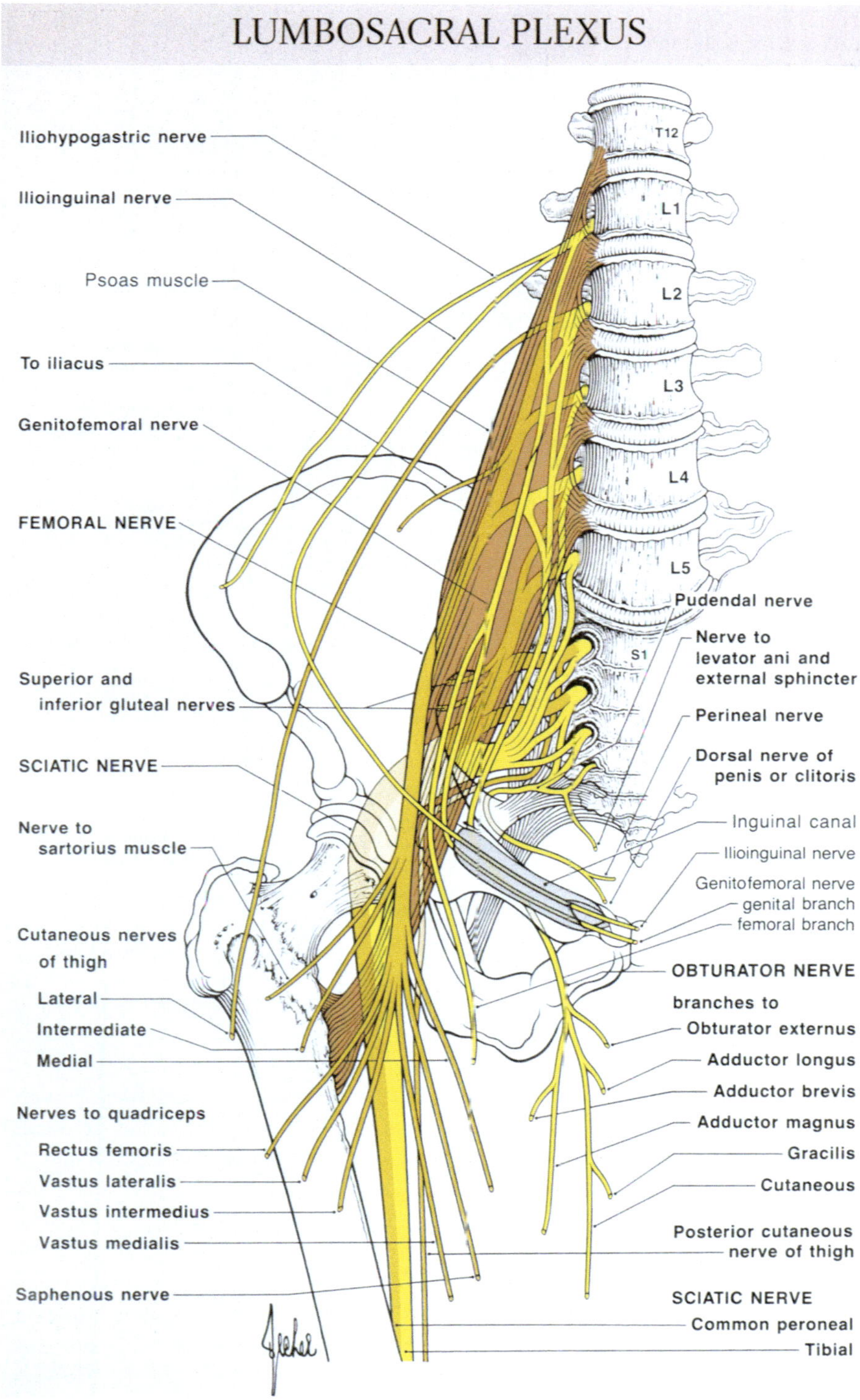

FIGURE 14.10 Diagram of the lumbar-sacral plexus, its branches and the muscles which they supply. Reproduced from *Aids to the Examination of the Peripheral Nervous System*, 4th edition, WB Saunders 2000, with permission from The Guarantors of Brain.

Iliohypogastric, Ilioinguinal, and Genitofemoral Nerves

The iliohypogastric, ilioinguinal, and genitofemoral nerves (Figure 14.10) have a primary sensory function, innervating the lower abdomen, inguinal region, the upper and medial anterior thigh, and part of the genitalia. Each has an L1 origin, with or without contributions from T12 and/or L2. These nerves are most likely to be affected by local trauma or surgery (i.e. for inguinal hernia repair or appendectomy).[157–159] These neuropathies are typified by iliac fossa and inguinal allodynia or hyperesthesia radiating to the genitalia. Pain is often exacerbated by walking and hip extension, and relieved by hip flexion. These neuropathies are difficult to prove; sometimes a local nerve block provides both diagnosis and therapy. Severe or persistent pain should prompt exclusion of retroperitoneal lymphoma or intrapelvic/inguinal endometriosis.

Obturator Nerve

This nerve originates in the lumbar plexus from the ventral divisions of the L2–4 nerve roots, forming in the posterior psoas muscle, emerging medially near the sacroiliac joint and then coursing along the pelvis lateral to the ureter and internal iliac vessels, bending anteroinferiorly and following the lateral pelvic wall anterior to the obturator vessels (Figure 14.10). It then enters the obturator canal via the obturator groove, descending into the medial thigh and supplying the obturator externus, pectineus, adductor longus, adductor brevis, and gracilis muscles, the hip and knee joints, and the medial thigh.

Obturator mononeuropathies are exceedingly rare in adults and children. Reported mechanisms include pelvic and hip fractures; malignancies; surgery involving the hip and pelvis, particularly when patients are placed in the lithotomy position; laparoscopic interventions; and obturator hernias. The differential diagnosis of obturator neuropathies includes L3–4 nerve root and lumbosacral plexus lesions.

Peroneal Nerve

Anatomy

The peroneal nerve fibers are primarily derived from the anterior primary rami of L4, L5, and to a lesser degree, the S1 nerve roots. The peroneal nerve is one of the two primary divisions of the sciatic nerve, which is formed in the pelvis from portions of the lumbosacral plexus. Despite their common nerve sheath, the fibers destined to become the peroneal and tibial nerves are anatomically segregated divisions throughout the length of the sciatic nerve. The peroneal division of the sciatic nerve is placed laterally and superficially where the sciatic nerve exits the pelvis, and hence is more prone to compressive and other injuries. Thus, a lesion at the sciatic notch can lead to a foot drop mimicking a primary peroneal lesion at the fibular head. Within the thigh, the only muscle supplied by the peroneal division is the short head of biceps femoris.

The peroneal and tibial nerves finally diverge immediately proximal to the popliteal fossa (Figure 14.11). The peroneal nerve then winds around the fibular head, splitting into its two main divisions; the deep portion supplies the tibialis anterior, the anterior compartment muscles including the extensor hallucis longus, and the extensor

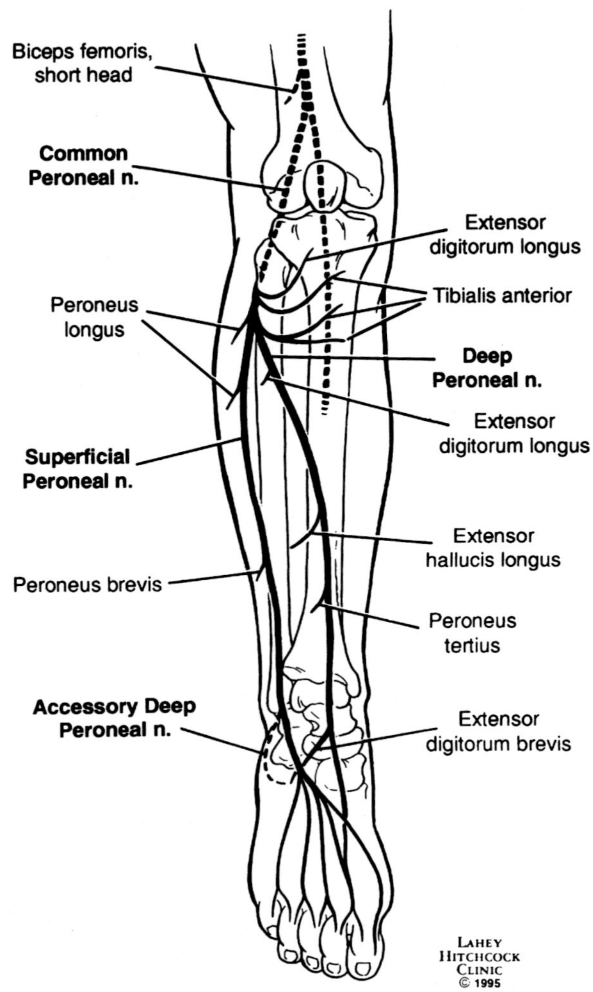

FIGURE 14.11 Peroneal nerve. *Reprinted with permission from Lahey Clinic, Burlington, Mass.*

digitorum brevis. The superficial peroneal nerve innervates peroneus longus and brevis in the lateral compartment of the leg, and supplies sensation to the dorsum of the foot.

Clinical Manifestations

Most children with peroneal mononeuropathies present with weakness and foot drop.[6] Presentation with pain and paresthesias over the dorsum of the foot, in the distribution of the superficial peroneal nerve, has also been reported.[6] In the largest pediatric series reported to date, of 17 children with peroneal mononeuropathies, the site of involvement was at the fibular head in 16 patients (94%) and at the ankle in just 1 case (6%).[6] The common peroneal nerve was primarily affected in 10 (59%) of the cases, deep peroneal nerve in 2 (12%), and superficial peroneal nerve in 1 (6%). Precise localization was limited

BOX 14.8 Reported Causes of Pediatric Peroneal Neuropathy

- Compression
  - Orthopedic appliances; traction apparatus, casts[247,248]
  - Anorexia nervosa[6,160–163]
  - Synovial cysts, intraneural ganglion cysts[249–253]
  - Prolonged leg crossing, strawberry picking, surfing and water skiing[164,165]
  - Compartment syndrome[254]
- Entrapment
  - Peroneus longus tendon, fascia of superficial peroneal nerve, myofascial bands[168–171]
- Neoplasms
  - Exostoses: fibular head, talotibial, osteochondromas[166,167]
  - Schwannoma[167]
  - Hemangioma[172]
  - Perineurioma (localized idiopathic hypertrophy)[255]
- Trauma
  - Lacerations[173]
- Systemic
  - Vasculitis: systemic lupus erythematosus, anaphylactic purpura[256]
  - Diabetes mellitus
  - Hereditary neuropathy with tendency to pressure palsies
  - Leprosy[179]
- Idiopathic
- Neonatal peroneal neuropathies
  - Umbilical artery catheterization[257]
  - Intrauterine compression, rings[180]
  - Breech presentation with stretch injury[181]
  - Compression against footboard[184]
  - Idiopathic[131]
  - Compartment syndrome, infiltration of intravenous fluids[183]

CASE EXAMPLE 14.5

A teenage football player dislocated his hip, requiring operative repair. His neurologic examination was normal immediately after the injury on the football field and in the recovery room. He was placed in traction with his leg elevated, and developed a foot drop on the second postoperative day. It was assumed that he had a peroneal neuropathy related to the traction, although he also complained of hip pain. On the fourth postoperative day, the hip area was examined and a large hematoma at the wound site was identified. When he was first seen at Boston Children's Hospital 3 months later, this adolescent had severe persistent distal leg weakness primarily affecting his peroneal musculature. EMG demonstrated a severe axonal sciatic neuropathy with peroneal predominance.

Comment

The peroneal nerve trunk is placed laterally and superficially within the sciatic nerve at its exit from the pelvis through the sciatic notch, and is susceptible to compression at this site. The possibility of a high sciatic lesion, rather than the more common lesions at the fibular head, should always be considered in children presenting with foot drop.

by technical factors in the remaining 4 (23%) children in this series.[6] (For reported causes of pediatric peroneal neuropathies, see Box 14.8.)

Etiology

Compressive Peroneal Neuropathies

Compressive lesions account for more than 50% of peroneal neuropathies in childhood. Many cases are iatrogenic, resulting from compression from casts or splints, peroneal nerve compression during prolonged immobilization, surgery, or traction.[6] (See Case Example 14.5.) Symptoms of weakness and foot drop may develop suddenly or subacutely. In most such cases, nerve conduction studies demonstrate conduction block at the fibular head, the degree of which usually correlates with that of clinical weakness.[6,160] Weight loss, such as that seen in anorexia nervosa, predisposes to peroneal neuropathy at the fibular head, a clinical syndrome known as "slimmer's palsy."[6,160–163] In such cases, children may be predisposed to neuropathy by nutritional deficiencies.[162]

Compressive peroneal mononeuropathies have also been reported in children with a variety of soft tissue and neural lesions, including synovial cysts, intraneural ganglion cysts, and ganglia within the anterior compartment of the leg. Chronic leg crossing, and prolonged compression during water skiing, surfing and skiing have also been identified as causative of peroneal mononeuropathies in childhood.[164,165]

Entrapment Peroneal Neuropathies

Common or deep peroneal nerve entrapment should be considered in children presenting with chronic progressive foot drop, repeated ankle sprains secondary to peroneal muscle weakness, or cavus foot deformities. Causes of entrapment reported in children include bony exostoses,[166,167] or tendinous and myofascial bands at the level

CASE EXAMPLE 14.6

An 8-year-old boy presented with a history of several months of gait changes that developed subsequent to a pneumonic illness, during which he received parenteral antibiotics in the left buttock. The local neurologist noted weakness of the left tibialis anterior and peroneal muscles, with normal muscle stretch reflexes. The peroneal motor response was normal, but EMG showed large MUAPs in the tibialis anterior.

At Boston Children's Hospital, examination confirmed a painless foot drop with weakness confined to muscles innervated by the peroneal nerve. Nerve conduction studies were again normal, but needle EMG showed changes consistent with denervation in the left tibialis anterior and peroneus longus. EMG of the short head of biceps femoris was normal. These findings suggested a peroneal neuropathy, but did not exclude the possibility of a focal peroneal division sciatic lesion. A CT scan of the pelvis, sciatic notch, and knee was unremarkable.

Eight months later, because of clinical progression, his nerve conduction studies were again repeated, and on this occasion showed slowing of peroneal nerve conduction on the left, with loss of amplitude and dispersion of the motor response. The left sural SNAP was normal. Needle EMG was unchanged. These findings were compatible with a left peroneal lesion distal to the innervation of the short head of biceps femoris.

On surgical exploration of the popliteal fossa, the common peroneal nerve was markedly attenuated and flattened by the tendon of origin of the peroneus longus and a number of fascial bands as the nerve tracked anteriorly for 2 to 3 cm around the fibular head (Figure 14.12). The entrapment was released.

Comment

This peroneal entrapment neuropathy evolved gradually. The initial symptoms developed after a parenteral injection into the buttock, causing initial suspicion of a very proximal injury. The value of repeat EMG studies is illustrated by this case, as the evolution of neurophysiologic abnormalities provided increasing evidence of localization, while surgical exploration enabled a specific diagnosis and treatment.

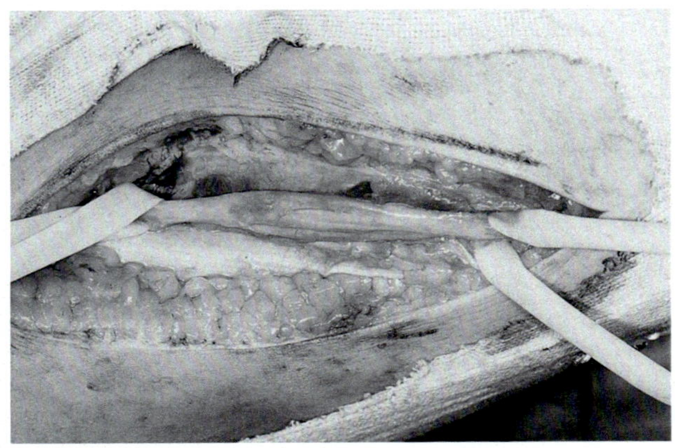

FIGURE 14.12 Peroneal nerve entrapment at the fibular head by the tendon of origin of the peroneus longus muscle (Case example 14.6). *Courtesy of John Hall, MD.*

origin of the peroneus longus.[6,168] Such lesions are sometimes seen on imaging, but in other cases are identifiable only on surgical exploration.[6,166,167,169] (See Case Example 14.6.) The superficial peroneal nerve can be entrapped by scarring due to repetitive ankle sprain, a situation amenable to surgical correction by limited fasciectomy.[170,171] Finally, as it leaves the lateral compartment distally, the superficial peroneal nerve can be entrapped by fascia.[171]

Other Causes of Peroneal Neuropathy

Occult nerve tumors sometimes present with protracted symptoms related to lesions affecting the common, superficial, or deep peroneal nerves.[172] Localized hypertrophic neuropathy (intraneural perineurioma) may also cause progressive peroneal neuropathy.[77] The advent of MR neurography has improved the pre-operative diagnosis of such uncommon lesions.

Trauma to the peroneal nerve is relatively uncommon, but it is sometimes injured with lacerations at the level of the fibular head.[6,173]

Pediatric peroneal neuropathies are rarely the presenting sign of such systemic disorders as vasculitis[174] and diabetes.[175–177] Peroneal nerve palsy can be the first manifestation of hereditary neuropathy with tendency to pressure palsy (HNPP).[178] Leprosy involving the peroneal nerve, common in other countries, was reported in an immigrant boy from Vietnam who presented with a 1-year history of unilateral foot drop.[179] In some children, peroneal lesions remain idiopathic despite extensive EMG and neuroimaging studies, surgical exploration, and pathologic analysis of the nerve.

Neonatal Peroneal Neuropathies

Peroneal neuropathy has been documented in a number of newborns, including one with a lesion that appeared most likely to be of antenatal onset.[131,180,181] The cause of these mononeuropathies is often unclear, but the rapid and complete recovery seen in most instances suggests a neuropraxic lesion, possibly related to stretching.[6,182] Neonatal peroneal palsies have also been related to infiltration of intravenous fluids[183] and compression from footboards.[184]

The role of EMG evaluation in newborns with mononeuropathies remains to be determined, but early perinatal electrophysiologic testing can clearly differentiate lesions of antenatal onset from those related to obstetric trauma, with its obvious medicolegal implications.

Evaluation

EMG examination is vital in the diagnosis of a pediatric peroneal neuropathy, and should include studies of the deep motor and superficial peroneal, posterior tibial, and sural nerves. Peroneal CMAPs are reduced in amplitude in about two-thirds of pediatric peroneal neuropathies. There may be conduction block at the fibular head.[6] Where the peroneal CMAP cannot be recorded distally over the extensor digitorum brevis, an attempt should be made to record a response from tibialis anterior or peroneus longus, as this may enable demonstration of conduction block or focal slowing at the fibular head.[185] Conduction block at the fibular head is less common in childhood than in adult peroneal neuropathies, but can be very helpful in localization of the lesion.

Sensory deficits vary in common peroneal neuropathies according to the degree of involvement of the different nerve fascicles.[185] The superficial peroneal response is sometimes preserved with common peroneal nerve injuries. Sural responses are normal in pediatric peroneal neuropathies.[6]

Needle EMG most commonly shows axonal changes in children with peroneal neuropathies.[6] EMG examination should include the short head of the biceps femoris, as well as the peroneal-innervated muscles below the knee, in order to exclude the possibility of sciatic nerve peroneal trunk lesions mimicking a peroneal neuropathy at the fibular head. This is especially important considering the similar incidence of sciatic and peroneal neuropathies in childhood.[4,6] L5 nerve root lesions may also cause foot drop, and can mimic peroneal neuropathy. EMG examination in suspected peroneal neuropathy should therefore also include muscles innervated by the tibial nerve with L5 input, such as tibialis posterior and flexor digitorum longus.

Imaging studies are increasingly helpful in the diagnosis and management of young patients with progressive mononeuropathies. MRI can assist in identifying the cause of nerve injuries and may guide management. Lesions potentially identifiable on MRI include ganglion cysts,[186] tumors,[187] neuromas, and perineuriomas.[77] Precise localization of nerve lesions enables more directed approaches with surgical exploration.[7,188,189] MR neurography may also distinguish intraneural from extraneural masses, and has the potential to detect discontinuity at the fascicular level in traumatic injuries, identifying an indication for immediate surgical repair.[190]

With recent ultrasound equipment improvements, ultrasonography also adds to the range of diagnostic imaging techniques available for characterization of peripheral nerve localization and pathology. There is a burgeoning literature on ultrasound characterization of normal nerve morphology as well as findings in nerve tumors,[191] traumatic injuries, nerve entrapment,[191] intraneural and mucous cysts,[192] and nerve inflammation.[193]

Treatment and Prognosis

Prognosis of peroneal mononeuropathies is variable. While most children have complete or significant improvement with time, a subset has incomplete recovery.[6] EMG findings are prognostically helpful in peroneal neuropathy. Absent or low-amplitude CMAPs are generally associated with an unfavorable outcome, whereas children with conduction block at the fibular head have the best prognosis.[6] Primary repair of childhood peripheral nerve lesions is often followed by significant recovery.[194] Early surgical repair is indicated in cases where significant improvement is not seen within 3 to 4 months of acute nerve injury. The most significant variable in determining results of surgery is delay in repair.[194] Where improvement in nerve function is not achieved despite surgical release or decompression, tendon transfer procedures may re-establish functional foot dorsiflexion, ameliorating foot drop.

Sural Nerve

Isolated sural mononeuropathies are extremely uncommon in children, as they are in adults. There is a single report of sural neuropathy secondary to compression from a tight ankle bracelet in a girl who experienced compete recovery within 1 month of removal of the bracelet.[195]

Tibial Nerve

Tibial mononeuropathies are rare, but the tibial nerve can be concomitantly affected with the peroneal nerve at the popliteal fossa. One report of "tarsal tunnel syndrome" in girls with foot pain was not confirmed by NCS or EMG.[196]

REFERENCES

1. Jones Jr. HR. Mononeuropathies. In: Jones Jr. HR, Bolton CF, Harper CM, editors. *Pediatric Clinical Electromyography*. Philadelphia: Lippincott-Raven; 1996. pp. 171–250.
2. Ouvrier RA, Shield L. Focal lesions of peripheral nerves. In: Ouvrier RA, McLeod JG, Pollard JD, editors. *Peripheral Neuropathy in Childhood*. 2nd ed. London: MacKeith Press; 1999. pp. 244–64.

3. Stewart JD. *Focal Peripheral Neuropathies*. Philadelphia: Lippincott; 2000.
4. Srinivasan J, Ryan MM, Escolar DM, Darras B, Jones HR. Pediatric sciatic neuropathies: a 30-year prospective study. *Neurology* 2011;**76**:976–80.
5. Jones Jr. HR. Compressive neuropathy in childhood: a report of 14 cases. *Muscle Nerve* 1986;**9**:720–3.
6. Jones Jr. HR, Felice KJ, Gross PT. Pediatric peroneal mononeuropathy: a clinical and electromyographic study. *Muscle Nerve* 1993;**16**:1167–73.
7. Weig SG, Waite RJ, McAvoy K. MRI in unexplained mononeuropathy. *Pediatr Neurol* 2000;**22**:314–17.
8. Zaidman CM, Seelig MJ, Baker JC, Mackinnon SE, Pestronk A. Detection of peripheral nerve pathology: comparison of ultrasound and MRI. *Neurology* 2013;**80**:1634–40.
9. De Smet K, De Maeseneer M, Talebian Yazdi A, Stadnik T, De Mey J. MRI in hypertrophic mono- and polyneuropathies. *Clin Radiol* 2013;**68**:317–22.
10. Cortes C, Ramos Y, Restrepo R, Restrepo JA, Grossman JA, Lee EY. Practical magnetic resonance imaging evaluation of peripheral nerves in children: magnetic resonance neurography. *Radiol Clin North Am* 2013;**51**:673–88.
11. Merlini L, Vargas MI, Anooshiravani M, Viallon M, Fluss J, Hanquinet S. Look for the nerves! MR neurography adds essential diagnostic value to routine MRI in pediatric practice: a pictorial overview. *J Neuroradiol* 2011;**38**:141–7.
12. West GA, Haynor DR, Goodkin R, Tsuruda JS, Bronstein AD, Kraft G, et al. Magnetic resonance imaging signal changes in denervated muscles after peripheral nerve injury. *Neurosurgery* 1994;**35**:1077–85, discussion 85–6.
13. Brazis PW, Masdeu JC, Biller J. *Localization in Clinical Neurology*. 3rd ed. Boston: Little, Brown; 1996.
14. Haddad FS, Jones DH, Vellodi A, Kane N, Pitt MC. Carpal tunnel syndrome in the mucopolysaccharidoses and mucolipidoses. *J Bone Joint Surg Br* 1997;**79**:576–82.
15. Yuen A, Dowling G, Johnstone B, Kornberg A, Coombs C. Carpal tunnel syndrome in children with mucopolysaccaridoses. *J Child Neurol* 2007;**22**:260–3.
16. Wraith JE, Alani SM. Carpal tunnel syndrome in the mucopolysaccharidoses and related disorders. *Arch Dis Child* 1990;**65**:962–3.
17. White K, Kim T, Neufeld JA. Clinical assessment and treatment of carpal tunnel syndrome in the mucopolysaccharidoses. *J Pediatr Rehabil Med* 2010;**3**:57–62.
18. Sonel B, Senbil N, Yavus Gurer YK, Evcik D. Cavanagh's syndrome (congenital thenar hypoplasia). *J Child Neurol* 2002;**17**:51–4.
19. Deymeer F, Jones Jr. HR. Pediatric median mononeuropathies: a clinical and electromyographic study. *Muscle Nerve* 1994;**17**:755–62.
20. Babal JC, Mehlman CT, Klein G. Nerve injuries associated with pediatric supracondylar humeral fractures: a meta-analysis. *J Pediatr Orthop* 2010;**30**:253–63.
21. Pape KE, Armstrong DL, Fitzhardinge PM. Peripheral median nerve damage secondary to brachial arterial blood gas sampling. *J Pediatr* 1978;**93**:852–6.
22. Floyd 3rd WE, Gebhardt MC, Emans JB. Intra-articular entrapment of the median nerve after elbow dislocation in children. *J Hand Surg Am* 1987;**12**:704–7.
23. Sumner JM, Khuri SM. Entrapment of the median nerve and flexor pollicis longus tendon in an epiphyseal fracture-dislocation of the distal radioulnar joint: a case report. *J Hand Surg Am* 1984;**9**:711–14.
24. Ben-David B, Barak M, Katz Y, Stahl S. A retrospective study of the incidence of neurological injury after axillary brachial plexus block. *Pain Pract* 2006;**6**:119–23.
25. Weeks PM. Radial, median, and ulnar nerve dysfunction associated with a congenital constricting band of the arm. *Plast Reconstr Surg* 1982;**69**:333–6.
26. Weinzweig N, Barr A. Radial, ulnar, and median nerve palsies caused by a congenital constriction band of the arm: single-stage correction. *Plast Reconstr Surg* 1994;**94**:872–6.
27. Bilge T, Yalaman O, Bilge S, Cokneseli B, Barut S. Entrapment neuropathy of the median nerve at the level of the ligament of Struthers. *Neurosurgery* 1990;**27**:787–9.
28. Danielsson LG. Iatrogenic pronator syndrome: case report. *Scand J Plast Reconstr Surg* 1980;**14**:201–3.
29. Hartz CR, Linscheid RL, Gramse RR, Daube JR. The pronator teres syndrome: compressive neuropathy of the median nerve. *J Bone Joint Surg Am* 1981;**63**:885–90.
30. Gessini L, Jandolo B, Pietrangeli A. Entrapment neuropathies of the median nerve at and above the elbow. *Surg Neurol* 1983;**19**:112–16.
31. Amadio PC, Reiman HM, Dobyns JH. Lipofibromatous hamartoma of nerve. *J Hand Surg Am* 1988;**13**:67–75.
32. Louis DS, Hankin FM, Greene TL, Dick HM. Lipofibromas of the median nerve: long-term follow-up of four cases. *J Hand Surg Am* 1985;**10**:403–8.
33. Callison JR, Thoms OJ, White WL. Fibro-fatty proliferation of the median nerve. *Plast Reconstr Surg* 1968;**42**:403–13.
34. Barfred T, Zachariae L. Neurofibroma in the median nerve treated with resection and free nerve transplantation. Case reports. *Scand J Plast Reconstr Surg* 1975;**9**:245–8.
35. Patel CB, Tsai TM, Kleinert HE. Hemangioma of the median nerve: a report of two cases. *J Hand Surg Am* 1986;**11**:76–9.
36. Fearn CB, Goodfellow JW. Anterior interosseous nerve palsy. *J Bone Joint Surg Br* 1965;**47**:91–3.
37. Nakano KK, Lundergran C, Okihiro MM. Anterior interosseous nerve syndromes. Diagnostic methods and alternative treatments. *Arch Neurol* 1977;**34**:477–80.
38. Van Meir N, De Smet L. Carpal tunnel syndrome in children. *J Pediatr Orthop B* 2005;**14**:42–5.
39. Nordstrom DL, DeStefano F, Vierkant RA, Layde PM. Incidence of diagnosed carpal tunnel syndrome in a general population. *Epidemiology* 1998;**9**:342–5.
40. Jadhav T, Peters H, Lee J, Ryan MM. Median neuropathy in pediatric mucopolysaccharidosis. *J ICNA* 2014.
41. MacDougal B, Weeks PM, Wray Jr. RC. Median nerve compression and trigger finger in the mucopolysaccharidoses and related diseases. *Plast Reconstr Surg* 1977;**59**:260–3.
42. Maurer K, Fenske A, Samii M. Carpal tunnel syndrome combined with trigger finger in early childhood. *J Neurol Neurosurg Psychiatry* 1980;**43**:1148.
43. Swoboda KJ, Engle EC, Scheindlin B, Anthony DC, Jones HR. Mutilating hand syndrome in an infant with familial carpal tunnel syndrome. *Muscle Nerve* 1998;**21**:104–11.
44. Poilvache P, Carlier A, Rombouts JJ, Partoune E, Lejeune G. Carpal tunnel syndrome in childhood: report of five new cases. *J Pediatr Orthop* 1989;**9**:687–90.

45. McArthur RG, Hayles AB, Gomez MR, Bianco Jr. AJ. Carpal tunnel syndrome and trigger finger in childhood. *Am J Dis Child* 1969;**117**:463–9.
46. Danta G. Familial carpal tunnel syndrome with onset in childhood. *J Neurol Neurosurg Psychiatry* 1975;**38**:350–5.
47. De Smet L, Fabry G. Carpal tunnel syndrome: familial occurrence presenting in childhood. *J Pediatr Orthop B* 1999;**8**:127–8.
48. McDonnell JM, Makley JT, Horwitz SJ. Familial carpal-tunnel syndrome presenting in childhood. Report of two cases. *J Bone Joint Surg Am* 1987;**69**:928–30.
49. Harpf C, Schwabegger A, Hussl H. Carpal median nerve entrapment in a child associated with Poland's syndrome. *Ann Plast Surg* 1999;**42**:458–9.
50. Koenigsberger MR, Moessinger AC. Iatrogenic carpal tunnel syndrome in the newborn infant. *J Pediatr* 1977;**91**:443–5.
51. Cruz Martinez A, Arpa J. Carpal tunnel syndrome in childhood: study of 6 cases. *Electroencephalogr Clin Neurophysiol* 1998;**109**:304–8.
52. Miner ME, Schimke RN. Carpal tunnel syndrome in pediatric mucopolysaccharidoses. Report of four cases. *J Neurosurg* 1975;**43**:102–3.
53. Starreveld E, Ashenhurst EM. Bilateral carpal tunnel syndrome in childhood. A report of two sisters with mucolipidosis III (pseudo-Hurler polydystrophy). *Neurology* 1975;**25**:234–8.
54. Felice KJ, Leicher CR, DiMario Jr. FJ. Hereditary neuropathy with liability to pressure palsies in children. *Pediatr Neurol* 1999;**21**:818–21.
55. Fissette J, Onkelinx A, Fandi N. Carpal and Guyon tunnel syndrome in burns at the wrist. *J Hand Surg Am* 1981;**6** 13–15.
56. Preston DC. Distal median neuropathies. *Neurol Clin* 1999;**17**:407–24.
57. Robinson LR. Traumatic injury to peripheral nerves. *Muscle Nerve* 2000;**23**:863–73.
58. Felice KJ, Royden Jones Jr. H. Pediatric ulnar mononeuropathy: report of 21 electromyography-documented cases and review of the literature. *J Child Neurol* 1996;**11**:116–20.
59. Baysefer A, Izci Y, Akay KM, Kayali H, Timurkaynak E. Surgical outcomes of ulnar nerve lesions in children. A retrospective clinical study. *Pediatr Neurosurg* 2004;**40**:107–11.
60. Uchida Y, Sugioka Y. Ulnar nerve palsy after supracondylar humerus fracture. *Acta Orthop Scand* 1990;**61**:118–19.
61. Slobogean BL, Jackman H, Tennant S, Slobogean GP, Mulpuri K. Iatrogenic ulnar nerve injury after the surgical treatment of displaced supracondylar fractures of the humerus: number needed to harm, a systematic review. *J Pediatr Orthop* 2010;**30**:430–6.
62. Garg S, Weller A, Larson AN, Fletcher ND, Kwon M, Schiller J, et al. Clinical characteristics of severe supracondylar humerus fractures in children. *J Pediatr Orthop* 2014;**34**:34–9.
63. Ouvrier R, McLeod J, Pollard J. *Peripheral Neuropathy in Childhood*. New York: Raven; 1990.
64. Gessini L, Jandolo B, Pietrangeli A, Occhipinti E. Ulnar nerve entrapment at the elbow by persistent epitrochleoanconeus muscle. Case report. *J Neurosurg* 1981;**55**:830–1.
65. Boero S, Senes FM, Catena N. Pediatric cubital tunnel syndrome by anconeus epitrochlearis: a case report. *J Shoulder Elbow Surg* 2009;**18**:e21–3.
66. Noth J, Dietz V, Mauritz KH. Cyclist's palsy: Neurological and EMG study in 4 cases with distal ulnar lesions. *J Neurol Sci* 1980;**47**:111–16.
67. Kothari MJ, Heistand M, Rutkove SB. Three ulnar nerve conduction studies in patients with ulnar neuropathy at the elbow. *Arch Phys Med Rehabil* 1998;**79**:87–9.
68. Escolar DM, Jones Jr. HR. Pediatric radial mononeuropathies: a clinical and electromyographic study of sixteen children with review of the literature. *Muscle Nerve* 1996;**19**:876–83.
69. Papilion JD, Neff RS, Shall LM. Compression neuropathy of the radial nerve as a complication of elbow arthroscopy: a case report and review of the literature. *Arthroscopy* 1988;**4**:284–6.
70. Gaur SC, Swarup A. Radial nerve palsy caused by injections. *J Hand Surg Br* 1996;**21**:338–40.
71. Holst-Nielsen F, Jensen V. Tardy posterior interosseous nerve palsy as a result of an unreduced radial head dislocation in Monteggia fractures: a report of two cases. *J Hand Surg Am* 1984;**9**:572–5.
72. Tubbs RS, Griessenauer C, Rizk E, Shoja MM, Pehler SF, Wellons 2nd JC, et al. Posterior interosseous nerve palsy in a child associated with recurrent dislocation of the head of the radius. *J Neurosurg Pediatr* 2013;**11**:389–91.
73. Meier C, Moll C. Hereditary neuropathy with liability to pressure palsies. Report of two families and review of the literature. *J Neurol* 1982;**228**:73–95.
74. Shamji MF, Purgina B, Flood T, Michaud J, Ventureyra EC. Pediatric malignant peripheral nerve sheath tumour of the radial nerve. *Can J Neurol Sci* 2010;**37**:524–7.
75. Isaac S, Athanasou NA, Pike M, Burge PD. Radial nerve palsy owing to localized hypertrophic neuropathy (intraneural perineurioma) in early childhood. *J Child Neurol* 2004;**19**:71–5.
76. Cortes W, Cheng J, Matloub HS. Intraneural perineurioma of the radial nerve in a child. *J Hand Surg Am* 2005;**30**:820–5.
77. Mauermann ML, Amrami KK, Kuntz NL, Spinner RJ, Dyck PJ, Bosch EP, et al. Longitudinal study of intraneural perineurioma—a benign, focal hypertrophic neuropathy of youth. *Brain* 2009;**132**:2265–76.
78. Manske PR. Compression of the radial nerve by the triceps muscle: a case report. *J Bone Joint Surg Am* 1977;**59**:835–6.
79. Feldman GV. Radial nerve palsies in the newborn. *Arch Dis Child* 1957;**32**:469–71.
80. Morgan L. Radial nerve paralysis in the newborn. *Arch Dis Child* 1948;**23**:137–9.
81. Ross D, Jones Jr. R, Fisher J, Konkol RJ. Isolated radial nerve lesion in the newborn. *Neurology* 1983;**33**:1354–6.
82. Hayman M, Roland EH, Hill A. Newborn radial nerve palsy: report of four cases and review of published reports. *Pediatr Neurol* 1999;**21**:648–51.
83. Deshmukh NV, Phillips GE. Isolated radial nerve palsy in a newborn: report of two cases. *Hand Surg* 2002;**7**:293–4.
84. Monica JT, Waters PM, Bae DS. Radial nerve palsy in the newborn: a report of four cases and literature review. *J Pediatr Orthop* 2008;**28**:460–2.
85. Alsubhi FS, Althunyan AM, Curtis CG, Clarke HM. Radial nerve palsy in the newborn: a case series. *CMAJ* 2011;**183**:1367–70.
86. Sunram F, Harms D, Mau G. [Neonatal paresis of the radial nerve due to benign angioleiomyoma (author's transl.)]. *Z Orthop Ihre Grenzgeb* 1977;**115**:124–8.
87. Tollner U, Bechinger D, Pohlandt F. Radial nerve palsy in a premature infant following long-term measurement of blood pressure. *J Pediatr* 1980;**96**:921–2.
88. Lejman T, Strong M, Michno P. Radial-nerve palsy associated with septic shoulder in neonates. *J Pediatr Orthop* 1995;**15**:169–71

89. Witthaut J, Steffens KJ, Koob E. Intermittent axillary nerve palsy caused by a humeral exostosis. *J Hand Surg Br* 1994;**19**:422–3.
90. Kline DG, Kim DH. Axillary nerve repair in 99 patients with 101 stretch injuries. *J Neurosurg* 2003;**99**:630–6.
91. Michelotti F, Camathias C, Gaston M, Rutz E. Severe septic arthritis of the shoulder with an axillary nerve lesion in a 4-year-old child. *J Orthop Sci* 2011;**16**:821–4.
92. Foo CL, Swann M. Isolated paralysis of the serratus anterior. A report of 20 cases. *J Bone Joint Surg Br* 1983;**65**:552–6.
93. Gregg JR, Labosky D, Harty M, Lotke P, Ecker M, DiStefano V, et al. Serratus anterior paralysis in the young athlete. *J Bone Joint Surg Am* 1979;**61**:825–32.
94. Henry D, Bonthius DJ. Isolated musculocutaneous neuropathy in an adolescent baseball pitcher. *J Child Neurol* 2011;**26**:1567–70.
95. Laulund T, Fedders O, Sogaard I, Kornum M. Suprascapular nerve compression syndrome. *Surg Neurol* 1984;**22**:308–12.
96. Montagna P, Colonna S. Suprascapular neuropathy restricted to the infraspinatus muscle in volleyball players. *Acta Neurol Scand* 1993;**87**:248–50.
97. Ogino T, Minami A, Kato H, Hara R, Suzuki K. Entrapment neuropathy of the suprascapular nerve by a ganglion. A report of three cases. *J Bone Joint Surg Am* 1991;**73**:141–7.
98. Escolar DM, Jones Jr. HR. Pediatric sciatic mononeuropathies: a clinical and electromyographic analysis [abstract]. *Muscle Nerve* 1994;**17**:108.
99. David WS, Jones Jr. HR. Soft tissue sarcoma presenting as atypical extremity pain. *Muscle Nerve* 1994;**17**:1071–2.
100. Katirji B, Wilbourn AJ. High sciatic lesion mimicking peroneal neuropathy at the fibular head. *J Neurol Sci* 1994;**121**:172–5.
101. Jones Jr. HR, Gianturco LE, Gross PT, Buchhalter J. Sciatic neuropathies in childhood: a report of ten cases and review of the literature. *J Child Neurol* 1988;**3**:193–9.
102. Bassett GS, Monforte-Munoz H, Mitchell WG, Rowland JM. Cavus deformity of the foot secondary to a neuromuscular choristoma (hamartoma) of the sciatic nerve. A case report. *J Bone Joint Surg Am* 1997;**79**:1398–401.
103. Thomas JE, Piepgras DG, Scheithauer B, Onofrio BM, Shives TC. Neurogenic tumors of the sciatic nerve. A clinicopathologic study of 35 cases. *Mayo Clin Proc* 1983;**58**:640–7.
104. Senes FM, Campus R, Becchetti F, Catena N. Lower limb nerve injuries in children. *Microsurgery* 2007;**27**:32–6.
105. Mayer M, Romain O. [Sciatic paralysis after a buttock intramuscular injection in children: an ongoing risk factor]. *Arch Pediatr* 2001;**8**:321–3.
106. Ehrmann L, Lechner K, Mamoli B, Novotny C, Kos K. Peripheral nerve lesions in haemophilia. *J Neurol* 1981;**225**:175–82.
107. Kubiak R, Wilcox DT, Spitz L, Kiely EM. Neurovascular morbidity from the lithotomy position. *J Pediatr Surg* 1998;**33**:1808–10.
108. Deverell WF, Ferguson JH. An unusual case of sciatic nerve paralysis. *JAMA* 1968;**205**:699–700.
109. Lee R, Fann AV, Sobus K. Bilateral sciatic nerve entrapment due to weight loss. *J Ark Med Soc* 1998;**95**:153–5.
110. Brown JA, Braun MA, Namey TC. Pyriformis syndrome in a 10-year-old boy as a complication of operation with the patient in the sitting position. *Neurosurgery* 1988;**23**:117–19.
111. Gozal Y, Pomeranz S. Sciatic nerve palsy as a complication after acoustic neurinoma resection in the sitting position. *J Neurosurg Anesthesiol* 1994;**6**:40–2.
112. Goh KJ, Tan CB, Tjia HT. Sciatic neuropathies—a retrospective review of electrodiagnostic features in 29 patients. *Ann Acad Med Singapore* 1996;**25**:566–9.
113. Romfh JH, Currier RD. Sciatic neuropathy induced by the lithotomy position. *Arch Neurol* 1983;**40**:127.
114. Dorfman LJ, Ransom BR, Forno LS, Kelts A. Neuropathy in the hypereosinophilic syndrome. *Muscle Nerve* 1983;**6**:291–8.
115. Fok TF, Ha MH, Leung KW, Wong W. Sciatic nerve palsy complicating umbilical arterial catheterization. *Eur J Pediatr* 1986;**145**:308–9.
116. McMillan HJ, Srinivasan J, Darras BT, Ryan MM, Davis J, Lidov HG, et al. Pediatric sciatic neuropathy associated with neoplasms. *Muscle Nerve* 2011;**43**:183–8.
117. Hruban RH, Shiu MH, Senie RT, Woodruff JM. Malignant peripheral nerve sheath tumors of the buttock and lower extremity. A study of 43 cases. *Cancer* 1990;**66**:1253–65.
118. Oberle J, Kahamba J, Richter HP. Peripheral nerve schwannomas—an analysis of 16 patients. *Acta Neurochir (Wien)* 1997;**139**:949–53.
119. Cruccetti A, Kiely EM, Spitz L, Drake DP, Pritchard J, Pierro A. Pelvic neuroblastoma: low mortality and high morbidity. *J Pediatr Surg* 2000;**35**:724–8.
120. Stillman MJ, Christensen W, Payne R, Foley KM. Leukemic relapse presenting as sciatic nerve involvement by chloroma (granulocytic sarcoma). *Cancer* 1988;**62**:2047–50.
121. Bickels J, Kahanovitz N, Rubert CK, Henshaw RM, Moss DP, Meller I, et al. Extraspinal bone and soft-tissue tumors as a cause of sciatica. Clinical diagnosis and recommendations: analysis of 32 cases. *Spine (Phila Pa 1976)* 1999;**24**:1611–16.
122. Favier T, Menei P, Rizk T, Alhayek G, Mercier P. [Osteosarcoma of the sacrum. Apropos of a case in a 14 year-old girl]. *Rev Rhum Ed Fr* 1993;**60**:365–6.
123. Simmons Z, Mahadeen ZI, Kothari MJ, Powers S, Wise S, Towfighi J. Localized hypertrophic neuropathy: magnetic resonance imaging findings and long-term follow-up. *Muscle Nerve* 1999;**22**:28–36.
124. Ostergaard JR, Smith T, Stausbol-Gron B. Intraneural perineurioma of the sciatic nerve in early childhood. *Pediatr Neurol* 2009;**41**:68–70.
125. Ferraresi S, Garozzo D, Bianchini E, Gasparotti R. Perineurioma of the sciatic nerve: a possible cause of idiopathic foot drop in children: report of 4 cases. *J Neurosurg Pediatr* 2010;**6**:506–10.
126. Engstrom JW, Layzer RB, Olney RK, Edwards MB. Idiopathic, progressive mononeuropathy in young people. *Arch Neurol* 1993;**50**:20–3.
127. Sawaya RA. Idiopathic sciatic mononeuropathy. *Clin Neurol Neurosurg* 1999;**101**:256–9.
128. Yamada K, Mano T, Toribe Y, Yanagihara K, Suzuki Y. MRI findings and steroid therapy for neuralgic amyotrophy in children. *Pediatr Neurol* 2011;**45**:200–2.
129. Sheth D, Gutmann L, Blumenthal DT, Mullett M, Bodensteiner JB. Compressive sciatic neuropathy due to uterine abnormality. *Muscle Nerve* 1994;**17**:1486–8.
130. Rombouts JJ, Debauche C, Verellen G, Lyon G. [Congenital paralysis due to compression. Apropos of 4 cases]. *Ann Chir Main Memb Super* 1993;**12**:39–44.
131. Jones Jr. HR, Herbison GJ, Jacobs SR, Kollros PR, Macones GA. Intrauterine onset of a mononeuropathy: peroneal neuropathy in a

newborn with electromyographic findings at age one day compatible with prenatal onset. *Muscle Nerve* 1996;**19**:88–91.

132. Sharrard WJ. Neonatal sciatic paralysis (two cases). *Proc R Soc Med* 1973;**66**:218–19.
133. Sriram K, Sakthivel A. Sciatic nerve palsy in the new born. *Ann Acad Med Singapore* 1981;**10**:472–5.
134. Curtiss Jr. PH, Tucker HJ. Sciatic palsy in premature infants. A report and follow–up study of ten cases. *JAMA* 1960;**174**:1586–8.
135. Pedrocca A, Nicosia U, De Guidi G. [Neonatal paralysis of the sciatic nerve caused by intragluteal injections]. *Minerva Pedictr* 1968;**20**:818–25.
136. Schneegans E, Amar G, Isch C, Schneegans D. [Sciatic paralysis of newborn and premature infants]. *Ann Pediatr (Paris)* 1968;**15**:657–63.
137. Merkulov VN. [Sciatic nerve injuries in children as a complication of intramuscular injections]. *Feldsher Akush* 1987;**52**:21–4
138. Villarejo FJ, Pascual AM. Injection injury of the sciatic nerve (370 cases). *Childs Nerv Syst* 1993;**9**:229–32.
139. Goven'ko FS, Babin AV. [Lesions of the sciatic nerve in newborn infants after administration of various drugs into the umbilical arteries]. *Akush Ginekol (Mosk)* 1990 May;(5):61–3.
140. de Sanctis N, Cardillo G, Nunziata Rega A. Gluteoperineal gangrene and sciatic nerve palsy after umbilical vessel injection. *Clin Orthop Relat Res* 1995;180–4.
141. Lester PD, McAlister WH. Congenital iliac anomaly with sciatic palsy. *Radiology* 1970;**96**:397–9.
142. Tani JC. Fibrous band compression of the tibial nerve branch of the sciatic nerve. *Am J Orthop (Belle Mead NJ)* 1995;**24**:910–12.
143. Tada K, Yonenobu K, Swanson AB. Congenital constriction band syndrome. *J Pediatr Orthop* 1984;**4**:726–30.
144. Søgaard I. Sciatic nerve entrapment. Case report. *J Neurosurg* 1983;**58**:275–6.
145. Sayson SC, Ducey JP, Maybrey JB, Wesley RL, Vermilion D. Sciatic entrapment neuropathy associated with an anomalous piriformis muscle. *Pain* 1994;**59**:149–52.
146. Gasecki AP, Ebers GC, Vellet AD, Buchan A. Sciatic neuropathy associated with persistent sciatic artery. *Arch Neurol* 1992;**49**:967–8.
147. Ramos-Fernandez JM, Oliete-Garcia FM, Roldan-Aparicio S, Kirchschlager E, Barrio-Nicolas A. [Neonatal sciatic palsy: etiology and outcome of 21 cases]. *Rev Neurol* 1998;**26**:752–5.
148. Venna N, Bielawski M, Spatz EM. Sciatic nerve entrapment in a child. Case report. *J Neurosurg* 1991;**75**:652–4.
149. Kline DG, Kim D, Midha R, Harsh C, Tiel R. Management and results of sciatic nerve injuries: a 24-year experience. *J Neurosurg* 1998;**89**:13–23.
150. Bigos SJ, Coleman SS. Foot deformities secondary to gluteal injection in infancy. *J Pediatr Orthop* 1984;**4**:560–3.
151. Szalay EA. Femoral nerve palsy and hip instability in infants with breech birth presentation: a review of the literature and report of 2 cases. *J Pediatr Orthop* 2010;**30**:739–41.
152. Murnaghan ML, Browne RH, Sucato DJ, Birch J. Femoral nerve palsy in Pavlik harness treatment for developmental dysplasia of the hip. *J Bone Joint Surg Am* 2011;**93**:493–9.
153. Richer LP, Shevell MI, Stewart J, Poulin C. Pediatric meralgia paresthetica. *Pediatr Neurol* 2002;**26**:321–3.
154. Edelson R, Stevens P. Meralgia paresthetica in children. *J Bone Joint Surg Am* 1994;**76**:993–9.
155. Fernandez-Mayoralas DM, Fernandez-Jaen A, Jareno NM, Perez BC, Fernandez PM, Sola AG. Meralgia paresthetica in the pediatric population: a propos of 2 cases. *J Child Neurol* 2010;**25**:110–13.
156. Tejwani SG, Scaduto AA, Bowen RE. Transient meralgia paresthetica after pediatric posterior spine fusion. *J Pediatr Orthop* 2006;**26**:530–3.
157. Seid AS, Amos E. Entrapment neuropathy in laparoscopic herniorrhaphy. *Surg Endosc* 1994;**8**:1050–3.
158. Suresh S, Patel A, Porfyris S, Ryee MY. Ultrasound-guided serial ilioinguinal nerve blocks for management of chronic groin pain secondary to ilioinguinal neuralgia in adolescents. *Paediatr Anaesth* 2008;**18**:775–8.
159. Soyer T, Tosun A, Keles I, Inal E, Cesur O, Cakmak M. Electrophysiologic evaluation of genitofemoral nerve in children with inguinal hernia repair. *J Pediatr Surg* 2008;**43**:1865–8.
160. Cruz Martinez A. Slimmer's paralysis: electrophysiological evidence of compressive lesion. *Eur Neurol* 1987;**26**:189–92.
161. Constanty A, Vodoff MV, Gilbert B, Dantoine F, Roche JF, Piguet C, et al. [Peroneal nerve palsy in anorexia nervosa: three cases]. *Arch Pediatr* 2000;**7**:316–17.
162. Sotaniemi KA. Slimmer's paralysis—peroneal neuropathy during weight reduction. *J Neurol Neurosurg Psychiatry* 1984;**47**:564–6.
163. Cruz-Martinez A, Arpa J, Palau F. Peroneal neuropathy after weight loss. *J Peripher Nerv Syst* 2000;**5**:101–5.
164. Vaccaro AR, Ludwig SC, Klein GR, McQuire M, Green D, Green NE. Bilateral peroneal nerve palsy secondary to a knee board: report of two cases. *Am J Orthop (Belle Mead NJ)* 1998;**27**:746–8.
165. Watemberg N, Amsel S, Sadeh M, Lerman-Sagie T. Common peroneal neuropathy due to surfing. *J Child Neurol* 2000;**15**:420–1.
166. Edlich HS, Fariss BL, Phillips VA, Chang DE, Smith JF, Hartigan C, et al. Talotibial exostoses with entrapment of the deep peroneal nerve. *J Emerg Med* 1987;**5**:109–13.
167. Levin KH, Wilbourn AJ, Jones Jr. HR. Childhood peroneal neuropathy from bone tumors. *Pediatr Neurol* 1991;**7**:308–9.
168. Sidey JD. Weak ankles. A study of common peroneal entrapment neuropathy. *Br Med J* 1969;**3**:623–6.
169. Daghino W, Pasquali M, Faletti C. Superficial peroneal nerve entrapment in a young athlete: the diagnostic contribution of magnetic resonance imaging. *J. Foot Ankle Surg* 1997;**36**:170–2.
170. Styf J, Morberg P. The superficial peroneal tunnel syndrome. Results of treatment by decompression. *J Bone Joint Surg Br* 1997;**79**:801–3.
171. McAuliffe TB, Fiddian NJ, Browett JP. Entrapment neuropathy of the superficial peroneal nerve. A bilateral case. *J Bone Joint Surg Br* 1985;**67**:62–3.
172. Bilge T, Kaya A, Alatli M, Bilge S, Alatli C. Hemangioma of the peroneal nerve: case report and review of the literature. *Neurosurgery* 1989;**25**:649–52.
173. Shevell MI, Stewart JD. Laceration of the common peroneal nerve by a skate blade. *CMAJ* 1988;**139**:311–12.
174. Jones Jr. HR, Darras BT. Acute care pediatric electromyography. *Muscle Nerve Suppl* 2000;**9**:S53–62.
175. el Bahri-Ben Mrad F, Gouider R, Fredj M, Ben Becher S, Mrad-Mazigh S, Mrabet A. Childhood diabetic neuropathy: a clinical and electrophysiological study. *Funct Neurol* 2000;**15**:35–40.

176. Barkai L, Kempler P, Vamosi I, Lukacs K, Marton A, Keresztes K. Peripheral sensory nerve dysfunction in children and adolescents with type 1 diabetes mellitus. *Diabet Med* 1998;**15**:228–33.

177. Lawrence DG, Locke S. Neuropathy in children with diabetes mellitus. *Br Med J* 1963;**1**:784–5.

178. Mouton P, Tardieu S, Gouider R, Birouk N, Maisonobe T, Dubourg O, et al. Spectrum of clinical and electrophysiologic features in HNPP patients with the 17p11.2 deletion. *Neurology* 1999;**52**:1440–6.

179. Choe W. Leprosy presenting as unilateral foot drop in an immigrant boy. *Postgrad Med J* 1994;**70**:111–12.

180. Craig WS, Clark JM. Of peripheral nerve palsies in the newly born. *J Obstet Gynaecol Br Emp* 1958;**65**:229–37.

181. Godley DR. Neonatal peroneal neurapraxia: a report of two cases and review of the literature. *Am J Orthop (Belle Mead NJ)* 1998;**27**:803–4.

182. Crumrine PK, Koenigsberger MR, Chutorian AM. Footdrop in the neonate with neurologic and electrophysioloic data. *J Pediatr* 1975;**86**:779–80.

183. Kreusser KL, Volpe JJ. Peroneal palsy produced by intravenous fluid infiltration in a newborn. *Dev Med Child Neurol* 1984;**26**:522–4.

184. Fischer AQ, Strasburger J. Footdrop in the neonate secondary to use of footboards. *J Pediatr* 1982;**101**:1003–4.

185. Sourkes M, Stewart JD. Common peroneal neuropathy: A study of selective motor and sensory involvement. *Neurology* 1991;**41**:1029–33.

186. Uetani M, Hashmi R, Hayashi K, Nagatani Y, Narabayashi Y, Imamura K. Peripheral nerve intraneural ganglion cyst: MR findings in three cases. *J Comput Assist Tomogr* 1998;**22**:629–32.

187. Loredo R, Hodler J, Pedowitz R, Yeh LR, Trudell D, Resnick D. MRI of the common peroneal nerve: normal anatomy and evaluation of masses associated with nerve entrapment. *J Comput Assist Tomogr* 1998;**22**:925–31.

188. Aagaard BD, Maravilla KR, Kliot M. MR neurography. MR imaging of peripheral nerves. *Magn Reson Imaging Clin N Am* 1998;**6**:179–94.

189. Kuntz C, Blake L, Britz G, Filler A, Hayes CE, Goodkin R, et al. Magnetic resonance neurography of peripheral nerve lesions in the lower extremity. *Neurosurgery* 1996;**39**:750–6, discussion 6–7.

190. Filler AG, Kliot M, Howe FA, Hayes CE, Saunders DE, Goodkin R, et al. Application of magnetic resonance neurography in the evaluation of patients with peripheral nerve pathology. *J Neurosurg* 1996;**85**:299–309.

191. Martinoli C, Bianchi S, Derchi LE. Ultrasonography of peripheral nerves. *Semin Ultrasound CT MR* 2000;**21**:205–13.

192. Marchiodi L, Mignani G, Stilli S. Mucous cysts of the external popliteal sciatic nerve during childhood: presentation of two cases and a review of the literature. *Chir Organi Mov* 1994;**79**:175–9.

193. Taniguchi N, Itoh K, Wang Y, Omoto K, Shigeta K, Fujii Y, et al. Sonographic detection of diffuse peripheral nerve hypertrophy in chronic inflammatory demyelinating polyradiculoneuropathy. *J Clin Ultrasound* 2000;**28**:488–91.

194. Birch R, Achan P. Peripheral nerve repairs and their results in children. *Hand Clin* 2000;**16**:579–95.

195. Reisin R, Pardal A, Ruggieri V, Gold L. Sural neuropathy due to external pressure: report of three cases. *Neurology* 1994;**44**:2408–9.

196. Albrektsson B, Rydholm A, Rydholm U. The tarsal tunnel syndrome in children. *J Bone Joint Surg Br* 1982;**64**:215–17.

197. Spinner M, Schreiber SN. Anterior interosseous-nerve paralysis as a complication of supracondylar fractures of the humerus in children. *J Bone Joint Surg Am* 1969;**51**:1584–90.

198. Marlow N, Jarratt J, Hosking G. Congenital ring constrictions with entrapment neuropathies. *J Neurol Neurosurg Psychiatry* 1981;**44**:247–9.

199. Uchida Y, Sugioka Y. Peripheral nerve palsy associated with congenital constriction band syndrome. *J Hand Surg Br* 1991;**16**:109–12.

200. Hotta T, Kanbara H, Sato S, Uchima S, Kobayashi K. [Case of carpal tunnel syndrome due to calcification of the flexor digitorum sublimis II in a child]. *Seikei Geka* 1970;**21**:948–50.

201. Senveli ME, Turker A, Arda MN, Altinors MN. Bilateral carpal tunnel syndrome in a young carpet weaver. *Clin Neurol Neurosurg* 1987;**89**:281–2.

202. Alvarez N, Larkin C, Roxborough J. Carpal tunnel syndrome in athetoid-dystonic cerebral palsy. *Arch Neurol* 1982;**39**:311–12.

203. Cruz Martinez A, Arpa J, Perez Conde MC, Ferrer MT. Bilateral carpal tunnel in childhood associated with Schwartz-Jampel syndrome. *Muscle Nerve* 1984;**7**:66–72.

204. Ghali J, Murugasu A, Day T, Nicholls K. Carpal tunnel syndrome in Fabry disease. *JIMD Rep* 2012;**2**:17–23.

205. Barfred T, Ipsen T. Congenital carpal tunnel syndrome. *J Hand Surg Am* 1985;**10**:246–8.

206. Dellon AL, Trojak JE, Rochman GM. Median nerve compression in Weill-Marchesani syndrome. *Plast Reconstr Surg* 1984;**74**:127–30.

207. Coessens B, De Mey A, Lacotte B, Vandenbroeck D. Carpal tunnel syndrome due to an haemangioma of the median nerve in a 12-year-old child. *Ann Chir Main Memb Super* 1991;**10**:255–7.

208. Ulrich D, Ulrich F, Schroeder M, Pallua N. Lipofibromatous hamartoma of the median nerve in patients with macrodactyly: diagnosis and treatment of a rare disease causing carpal tunnel syndrome. *Arch Orthop Trauma Surg* 2009;**129**:1219–24.

209. Cherqui A, Sulaiman WA, Kline DG. Resection and nerve grafting of a lipofibrohamartoma of the median nerve: case report. *Neurosurgery* 2009;**65**:A229–35.

210. Alfonso DT, Sotrel A, Grossman JA. Carpal tunnel syndrome due to an intraneural perineurioma in a 2-year-old child. *J Hand Surg Br* 2001;**26**:168–70.

211. Blennow G, Bekassy AN, Eriksson M, Rosendahl R. Transient carpal tunnel syndrome accompanying rubella infection. *Acta Paediatr Scand* 1982;**71**:1025–8.

212. Williams Jr. LF, Geer T. Acute carpal tunnel syndrome secondary to pyogenic infection of the forearm. *JAMA* 1963;**185**:409–10.

213. Al-Qattan MM, Thomson HG, Clarke HM. Carpal tunnel syndrome in children and adolescents with no history of trauma. *J Hand Surg Br* 1996;**21**:108–11.

214. Reed MW, Reed DN. Acute ulnar nerve entrapment after closed reduction of a posterior fracture dislocation of the elbow: a case report. *Pediatr Emerg Care* 2012;**28**:570–2.

215. Marquez S, Turley JJ, Peters WJ. Neuropathy in burn patients. *Brain* 1993;**116**(Pt 2):471–83.

216. Godshall RW, Hansen CA. Traumatic ulnar neuropathy in adolescent baseball pitchers. *J Bone Joint Surg Am* 1971;**53**:359–61.

217. Cordingley FT, Crawford GP. Ulnar nerve palsy in a haemophiliac due to intraneural haemorrhage. *Br Med J (Clin Res Ed)* 1984;**289**:18–19.

218. Dunn D, Wilensky M. Median and ulnar nerve palsies after infiltration of intravenous fluid. *South Med J* 1984;**77**:1345.
219. Drut R. Ossifying fibrolipomatous hamartoma of the ulnar nerve. *Pediatr Pathol* 1988;**8**:179–84.
220. Cavanagh NP, Pincott JR. Ulnar nerve tumours of the hand in childhood. *J Neurol Neurosurg Psychiatry* 1977;**40**:795–800.
221. Phillips II LH, Persing JA, Vandenberg SR. Electrophysiological findings in localized hypertrophic mononeuropathy. *Muscle Nerve* 1991;**14**:335–41.
222. Saxena U, Ramesh V, Misra RS, Mukherjee A. Giant nerve abscesses in leprosy. *Clin Exp Dermatol* 1990;**15**:349–51.
223. Yuksel S, Ozcakar L, Kayserili K, Yuksel G. Drop hand in a 12-year-old girl: radial nerve at peril due to knitting. *Eur J Paediatr Neurol* 2009;**13**:65–6.
224. Parratte S, Launay F, Jouve JL, Malikov S, Petit P, Bollini G. [A pediatric case of rupture of the brachial artery and radial nerve palsy secondary to proximal humeral exostosis]. *Rev Chir Orthop Reparatrice Appar Mot* 2007;**93**:186–9.
225. Finsterbush A, Husseini N. Infantile cortical hyperostosis with unusual clinical manifestations. *Clin Orthop Relat Res* 1979;276–9.
226. Lotem M, Fried A, Levy M, Solzi P, Najenson T, Nathan H. Radial palsy following muscular effort. A nerve compression syndrome possibly related to a fibrous arch of the lateral head of the triceps. *J Bone Joint Surg Br* 1971;**53**:500–6.
227. Tyagi A, Drake J, Midha R, Kestle J. Axillary nerve injuries in children. *Pediatr Neurosurg* 2000;**32**:226–9.
228. Yoshida T, Tada K, Uemura K, Yonenobu K. Peripheral nerve palsies in victims of the Hanshin-Awaji earthquake. *Clin Orthop Relat Res* 1999;208–17.
229. Ichikawa K, Nezu A. Hereditary neuropathy with liability to pressure palsies in childhood: report of a case and a brief review. *Brain Dev* 2005;**27**:152–4.
230. London NJ, Sefton GK. Hematocolpos. An unusual cause of sciatica in an adolescent girl. *Spine (Phila Pa 1976)* 1996;**21**:1381–2.
231. Bodack MP, Cole JC, Nagler W. Sciatic neuropathy secondary to a uterine fibroid: a case report. *Am J Phys Med Rehabil* 1999;**78**:157–9.
232. Dhote R, Tudoret L, Bachmeyer C, Legmann P, Christoforov B. Cyclic sciatica. A manifestation of compression of the sciatic nerve by endometriosis. A case report. *Spine (Phila Pa 1976)* 1996;**21**:2277–9.
233. Shah SK, Phan NB, Doshi S, Richardson CJ. Symptomatic persistent sciatic artery in a newborn. *J Pediatr Surg* 2008;**43**:1741–4.
234. Gulhan B, Kupeli S, Yalcin B, Akyuz C, Buyukpamukcu M. An unusual presentation of infantile fibrosarcoma in a male newborn. *Am J Perinatol* 2009;**26**:331–3.
235. Pascual-Castroviejo I, Pascual-Pascual SI, Viano J, Velazquez-Fragua R, Lopez-Gutierrez JC. Bilateral spinal neurofibromas in patients with neurofibromatosis 1. *Brain Dev* 2012;**34**:563–9.
236. Liu HC, Hung GY, Yen HJ, Hsieh MY, Chiou TJ. Acute sciatica: an unusual presentation of extramedullary relapse of acute lymphoblastic leukemia. *Int J Hematol* 2007;**86**:163–5.
237. Purohit DM, Levkoff AH, deVito PC. Gluteal necrosis with foot-drop. Complications associated with umbilical artery catheterization. *Am J Dis Child* 1978;**132**:897–9.
238. Van Hoff J, Shaywitz BA, Seashore JH, Collins WF. Femoral nerve injury following inguinal hernia repair. *Pediatr Neurol* 1985;**1**:195–6.
239. Crook TJ, Steinbrecher HA, Tekgul S, Malone PS. Femoral nerve neuropathy following the psoas hitch procedure. *J Pediatr Urol* 2007;**3**:145–7.
240. Kachko L, Ben Ami S, Liberman A, Birk E, Kronenberg S. Duloxetine contributing to a successful multimodal treatment program for peripheral femoral neuropathy and comorbid 'reactive depression' in an adolescent. *Pain Res Manag* 2011;**16**:457–9.
241. Takao M, Fukuuchi Y, Koto A, Tanaka K, Momoshima S, Kuramochi S, et al. Localized hypertrophic mononeuropathy involving the femoral nerve. *Neurology* 1999;**52**:389–92.
242. Emory TS, Scheithauer BW, Hirose T, Wood M, Onofrio BM, Jenkins RB. Intraneural perineurioma. A clonal neoplasm associated with abnormalities of chromosome 22. *Am J Clin Pathol* 1995;**103**:696–704.
243. Sharma SC, Ray RC. Femoral pain of solitary neurofibromatous origin. *Indian Pediatr* 1988;**25**:1221–3.
244. Shetty GM, Murari AS, Song HR, Lee SH, Yang JH. Neurofibromatous sensory neuropathy of the thigh in a 7-year-old boy. *Arch Orthop Trauma Surg* 2008;**128**:1093–7.
245. Kokavec M, Cingel V, Zilinek V, Bialik V. Rare peripheral femoral nerve neuropathy in children. *J Pediatr Orthop B* 2007;**16**:90–3.
246. Carter GT, McDonald CM, Chan TT, Margherita AJ. Isolated femoral mononeuropathy to the vastus lateralis: EMG and MRI findings. *Muscle Nerve* 1995;**18**:341–4.
247. Slawski DP, Schoenecker PL, Rich MM. Peroneal nerve injury as a complication of pediatric tibial osteotomies: a review of 255 osteotomies. *J Pediatr Orthop* 1994;**14**:166–72.
248. Ryan MM, Darras BT, Soul JS. Peroneal neuropathy from ankle-foot orthoses. *Pediatr Neurol* 2003;**29**:72–4.
249. Gayet LE, Morand F, Goujon JM, Pries P, Clarac JP. Compression of the peroneal nerve by a cyst in a seven-year-old child. *Eur J Pediatr Surg* 1998;**8**:61–3.
250. Nucci F, Artico M, Santoro A, Bardella L, Delfini R, Bosco S, et al. Intraneural synovial cyst of the peroneal nerve: report of two cases and review of the literature. *Neurosurgery* 1990;**26**:339–44.
251. Gurdjian ES, Larsen RD, Lindner DW. Intraneural cyst of the peroneal and ulnar nerves. Report of two cases. *J Neurosurg* 1965;**23**:76–8.
252. Nicholson TR, Cohen RC, Grattan-Smith PJ. Intraneural ganglion of the common peroneal nerve in a 4-year-old boy. *J Child Neurol* 1995;**10**:213–15.
253. Aulisa L, Tamburrelli F, Padua R, Lupparelli S, Tonali P, Padua L. Intraneural cyst of the peroneal nerve. *Childs Nerv Syst* 1998;**14**:222–5.
254. Sloane AE, Vajsar J, Laxer RM, Babyn PS, Murphy EG. Spontaneous non-traumatic anterior compartment syndrome with peroneal neuropathy and favorable outcome. *Neuropediatrics* 1994;**25**:268–70.
255. Johnson PC, Kline DG. Localized hypertrophic neuropathy: possible focal perineurial barrier defect. *Acta Neuropathol* 1989;**77**:514–18.
256. Ritter FJ, Seay AR, Lahey ME. Peripheral mononeuropathy complicating anaphylactoid purpura. *J Pediatr* 1983;**103**:77–8.
257. Giannakopoulou C, Korakaki E, Hatzidaki E, Manoura A, Aligizakis A, Velivasakis E. Peroneal nerve palsy: a complication of umbilical artery catheterization in the full-term newborn of a mother with diabetes. *Pediatrics* 2002;**109**:e66.

Chapter 15

Overview of Pediatric Peripheral Neuropathies

Hugh J. McMillan and Monique M. Ryan

INTRODUCTION

The term peripheral neuropathy refers to a disorder affecting one or more components of the peripheral nervous system, specifically axons, myelin, and/or Schwann cells. Peripheral neuropathy becomes increasingly prevalent with advancing age. In children, the prevalence of hereditary peripheral neuropathy (Charcot-Marie-Tooth disease) is approximately 0.2%,[1] with hereditary neuropathies accounting for 30 to 70% of all cases of pediatric peripheral nerve disease.[2,3] Among adults, the prevalence of peripheral neuropathy is 2.4%, increasing to as much as 8% in adults aged more than 55 years and/or populations with diabetes mellitus.[4,5] While peripheral neuropathy is less common in children, a wide range of disorders can be associated with this disease including hereditary, metabolic, inflammatory, and acquired causes. This chapter will: (1) review the normal maturation of peripheral nerves, (2) provide a practical approach to the child with suspected peripheral nerve disease, (3) discuss traumatic neuropathies of childhood, and (4) discuss the electrodiagnostic evaluation of the child with suspected neuropathy.

Peripheral Nerve Maturation

Peripheral nerve myelination begins as early as 15 weeks gestational age.[6] The diameter of axons and myelin sheaths gradually increases from birth, until adult values are attained at 4 to 5 years of age.[7] Normal reference ranges for electrodiagnostic tests also change with increasing age, reflecting the growth and progressive myelination of peripheral nerves that occurs in the first several years of life (Figure 15.1). Interpretation of pediatric nerve conduction studies should always include consideration of the child's age. Normal neonatal motor nerve conduction velocities are in the range of 20 to 35 m/s and should not be misinterpreted as demyelination. Adult nerve conduction velocities are reached in normal healthy children by 3 to 5 years of age (Figure 15.2).[8,9] Normal pediatric values for electrodiagnostic studies are provided later in this chapter.

APPROACH TO EVALUATING CHILDHOOD NEUROPATHIES

The evaluation of a child with suspected neuropathy requires an organized, step-wise approach, particularly in light of the broad spectrum of metabolic and genetic disorders associated with neuropathy in this age group. With a detailed history, physical examination, and the use of appropriate electrodiagnostic tests, the clinician can direct investigations appropriately. We recommend seven broad areas of inquiry when evaluating a child with suspected neuropathy (Box 15.1).

Confirm the Anatomical Site(s) Involved

Confirmation of which site(s) are involved along the neuroaxis is an important first step to narrowing the differential diagnosis in a patient with a suspected neuropathy. This is important to confirm the presence or absence of a peripheral nerve lesion and to evaluate for involvement of other sites along the neuraxis, which can help redirect investigations and/or provide important diagnostic clues.

Confirming the presence of peripheral nerve involvement is essential because some patients referred for evaluation of sensory symptoms, weakness, and/or ataxia will ultimately be found not to have any evidence of neuropathy. Patients must be evaluated to determine if their pattern of sensory loss is best explained by a lesion affecting one or more peripheral nerves, a plexus, or a dermatomal

B.T. Darras, H. Royden Jones, Jr., M.M. Ryan & D.C. De Vivo (Eds): Neuromuscular Disorders of Infancy, Childhood and Adolescence, Second edition.
DOI: http://dx.doi.org/10.1016/B978-0-12-417044-5.00015-9

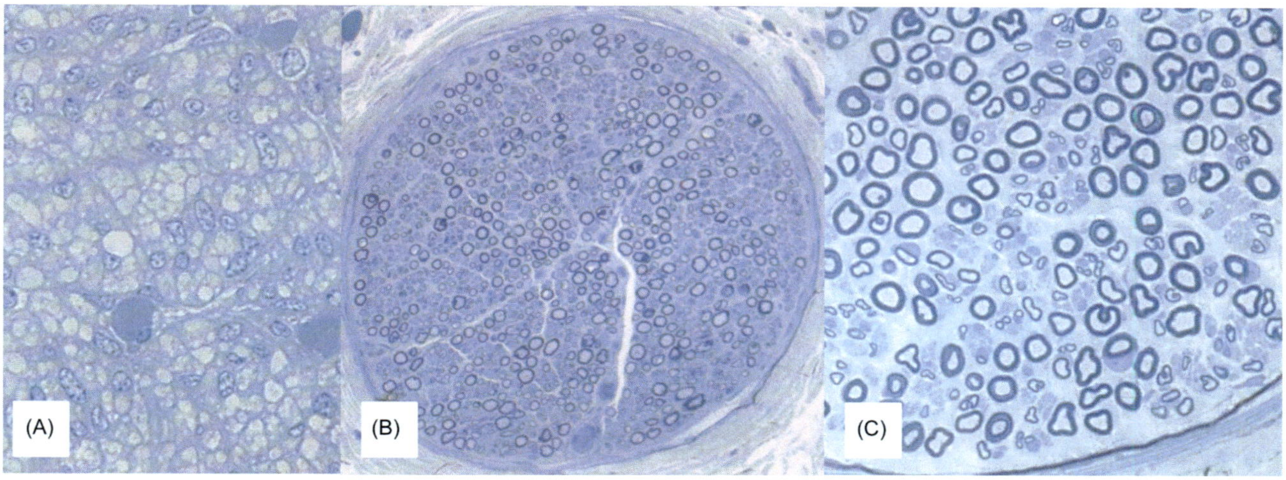

FIGURE 15.1 Nerve biopsies (toluidine blue staining, resin embedded) showing normal progression of peripheral nerve myelination with age. **(A)** 17-week-old fetus (normal lumbar plexus) shows essentially no myelination of large fiber axons; **(B)** 10-month-old infant (normal sural nerve) shows thin myelin; and **(C)** 10-year-old child (normal sural nerve) shows adult thickness of myelin. Biopsy **A** was performed at the time of postmortem study after spontaneous premature delivery. Biopsies **B** and **C** were surgical specimens in patients with a muscular condition or non-neurological systemic disease. *Photo credit: Dr. Jean Michaud, Department of Pathology, Children's Hospital of Eastern Ontario.*

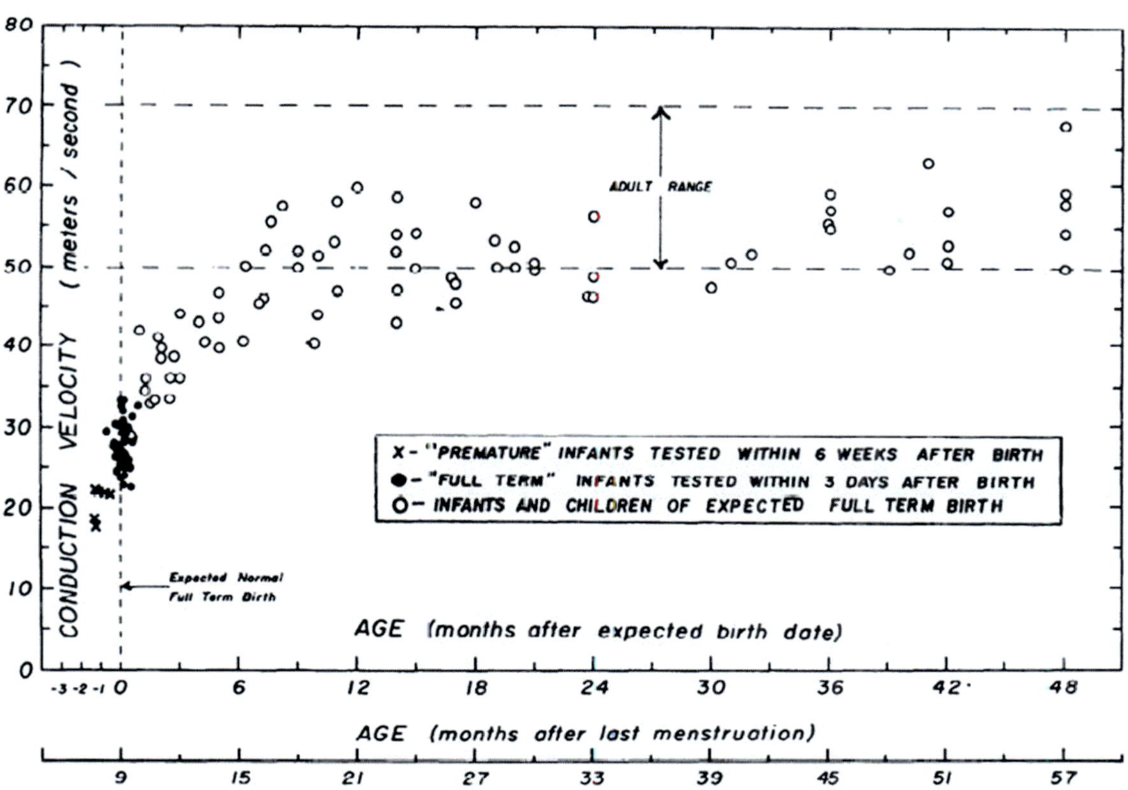

FIGURE 15.2 Normal change in ulnar motor nerve conduction velocity within the first few years of life, reflecting progressive growth and myelination.[9]

BOX 15.1 Approach to Evaluating Childhood Neuropathies

1. Confirm the anatomical site(s) involved:
   - Cranial nerve(s)
   - Plexus
   - Sensory ganglia
   - Nerve root(s)
   - Motor neuron
   - Peripheral nerve(s)
2. Confirm peripheral nerve type(s) and fiber size(s) involved:
   - Motor
   - Sensory, large fiber (vibration, proprioception, two-point discrimination)
   - Sensory, small fiber (pain, temperature, light touch)
   - Autonomic
3. Confirm pattern of peripheral nerve involvement
   - Polyneuropathy
   - Mononeuropathy
   - Mononeuritis multiplex
4. Confirm time course of disease (acute or chronic)
5. Confirm age of first symptom onset
6. Confirm electrodiagnostic features
   - Axonal versus demyelinating
   - Uniform versus nonuniform findings
7. Confirm associated clinical features
   - Non-neurological symptoms
   - Central nervous system involvement (developmental delay, seizures, psychosis)
   - Concomitant diseases (autoimmune disease, mitochondrial disease)
   - Toxin/medication exposures
   - Family history, consanguinity

pattern, the latter reflecting nerve root involvement. Clinical symptoms elicited by lesions within the central nervous system often involve much larger areas of the body. For example, in children with sensory loss due to a spinal cord syrinx or hydromyelia, sensory loss often corresponds to a "cape-like" distribution. In children with spinal cord lesions, a well demarcated sensory level, typically with associated bowel and bladder symptoms, will be described. Muscle weakness must similarly be carefully evaluated to see if the pattern of weakness points to involvement of one or more peripheral nerves, a plexus, or a myotome (reflecting nerve root injury). Deep tendon reflexes are typically depressed or absent in children with peripheral neuropathy.

Concomitant involvement of other sites along the neuroaxis (e.g. cranial nerves, nerve roots, and/or sensory ganglion) can provide important clues to the diagnosis in children with peripheral neuropathy.

Diphtheria, due to *Corynebacterium diphtheria* infection, is an example of a peripheral nerve disease which is frequently preceded by cranial neuropathies. Although diphtheria is rare in Western nations it remains an important cause of neuropathy world-wide, and occasional outbreaks are reported.[10] Initial symptoms include acute pharyngitis, often with an associated grayish-white pseudomembrane in the throat. Approximately 15% of patients develop neurological complications, with the first neurological symptom being bulbar dysfunction caused by lower cranial nerve paralysis.[10] Almost all patients who develop diphtheria-toxin mediated polyneuropathy (which resembles Guillain-Barré syndrome) have preceding cranial nerve involvement, making this an important diagnostic clue in this disorder.

Inflammatory neuropathies such as Guillain-Barré syndrome (GBS) may also be associated with cranial nerve and nerve root involvement. Children with the Miller Fisher variant of GBS typically present with the triad of ophthalmoplegia, ataxia, and areflexia.[11] About 50% of children with GBS complain of back, buttock, or leg pain, presumably reflecting nerve root or peripheral nerve inflammation.[12,13] In approximately 20% of cases of pediatric GBS, children actually present with back pain, which can be severe and may be associated with neck stiffness and positive Kernig and Brudzinski signs.[14]

Localizing a neuropathy to one or more sites of focal compression can be a further clue to diagnosis. Common sites of focal nerve compression include the carpal tunnel at the wrist for the median nerve, the cubital tunnel at the elbow for the ulnar nerve, and the fibular head at the lateral knee for the peroneal nerve. Genetic conditions such as hereditary neuropathy with liability to pressure palsy (HNPP) often present with recurrent episodes of compressive mononeuropathy.[15] Metabolic conditions such as the mucopolysaccharidoses and mucolipidoses may also present with pediatric carpal tunnel syndrome.[16,17]

Confirm Peripheral Nerve Type(s) and Fiber Size(s) Involved

Clinicians must ascertain if the patient's symptoms point to predominant dysfunction of sensory fibers, motor fibers, both sensory and motor fibers (sensorimotor), and/or autonomic fibers. Although most peripheral nerves are mixed and contain all fiber types, hereditary diseases can show a predilection for specific fiber types.

Sensory-predominant neuropathies can result from small-fiber or large-fiber nerve dysfunction. Small fiber sensory afferents include unmyelinated C-type and poorly myelinated Aδ-type fibers, which carry information about pain, temperature, and light touch. Patients with small fiber sensory dysfunction may complain of dysesthesia (spontaneous pain), hyperesthesia (a heightened perception of a stimulus), allodynia (a sensation of pain elicited by a stimulus that is not normally painful), paresthesia (pins and needles), or hypesthesia (decreased perception). Small fiber sensory disease must be identified on the basis of the clinical history and physical examination, since conventional electrodiagnostic tests (nerve conduction studies and electromyography) do not evaluate the integrity of small fibers. Standard electrodiagnostic testing selectively evaluates the large fiber sensory afferents, which are myelinated and conduct much more quickly than the small fiber sensory afferents.

Large fiber sensory afferents carry information about vibration, proprioception, and two point discrimination along large, myelinated Aα fibers. Patients with large-fiber sensory dysfunction typically complain of gait ataxia and clumsiness, and have difficulty with tasks requiring intact proprioception, such as moving about in the dark.

Hereditary sensory and autonomic neuropathies (HSAN) are genetic diseases characterized by small fiber sensory dysfunction, with varying degrees of autonomic fiber dysfunction depending upon the specific HSAN type. Due to their lack of pain perception, patients may present with painless self-mutilation, for example that caused by biting their own lips or fingers while teething. Painless finger and foot ulceration are also seen, resulting from pressure injury or unnoticed trauma. Patients may demonstrate decreased or absent pain from long bone fractures or skin laceration. Clinicians must have a high index of suspicion for these conditions, since electrodiagnostic testing will again yield normal sensory responses in all but one form of HSAN (HSAN type 2), due to the selective involvement of small fibers in sensory nerves.[18] Nerve biopsy may reveal selective loss of specific populations of small unmyelinated nerve fibers. Genetic testing is commercially available.

Fabry disease is an example of a late childhood to adult onset disease affecting unmyelinated and small myelinated sensory afferents. Fabry disease is an X-linked lysosomal disorder resulting from a deficiency in the enzyme alpha-galactosidase. Males with Fabry disease typically present in early childhood with episodic crises of severe hand and or foot pain, which can last from minutes to weeks.[19,20] Females often present in late adolescence or adulthood.[20] Once again, clinicians must maintain a high index of suspicion for this small fiber sensory neuropathy, since nerve conduction studies are typically normal.[21] Over time, patients will develop additional symptoms including abundant skin angiokeratomas or "cherry red spots" (particularly in the inguinal region), gastrointestinal symptoms, renal dysfunction, and cerebrovascular disease, including stroke.[22]

Pediatric diseases associated with predominantly large-fiber sensory loss are characterized by significant gait and limb ataxia and include diseases such as the autosomal recessive sensory ataxia of Charlevoix-Saguenay (ARSACS),[23] abetalipoproteinemia,[24] and Friedreich ataxia.[25]

Autonomic fibers innervate many visceral organs and can present with a range of clinical symptoms including orthostatic hypotension, syncope, sudomotor symptoms (impaired or excessive sweating), gastrointestinal symptoms (dysphagia, early satiety, nausea, vomiting, decreased appetite and/or weight loss due to impaired gastric emptying, constipation), and urinary symptoms (frequent micturition, urinary tract infections due to impaired bladder emptying). Erectile dysfunction may be seen in adolescent and adult males. Similarly to small-fiber sensory neuropathy, autonomic neuropathy is largely diagnosed by a careful history. Nerve conduction studies cannot be used to evaluate the integrity of autonomic nerves. Assessments of autonomic function, such as tilt table and quantitative sudomotor axon reflex testing, can be performed at some highly specialized pediatric centers. Some disorders give rise to a pure autonomic neuropathy, but in most there is overlap with sensory and/or motor neuropathies. HSAN type 3 (Riley-Day syndrome) is associated with significant autonomic dysfunction in addition to its sensory findings. Dysautonomia is also seen with childhood Guillain-Barré syndrome (in 20–40% of cases)[12,13] as well as drug exposure (e.g. vincristine, amiodarone).

Motor-predominant neuropathies are uncommon in childhood. Most children presenting with painless progressive muscle weakness will eventually be found to have a disease of motor neurons or anterior horn cells, such as spinal muscular atrophy (SMA). Inflammatory nerve disorders, such as the Guillain-Barré syndrome variant acute motor axonal neuropathy (AMAN), show selective involvement of motor nerves. AMAN is more common in Asian countries and has an association with preceding *Campylobacter jejuni* gastroenteritis.[26] Distal hereditary motor neuropathies (dHMNs) are an emerging group of hereditary disorders affecting motor nerves and showing some clinical and genetic overlap with Charcot-Marie-Tooth disease (CMT) type 2 as well as distal spinal muscular atrophies (dSMAs) and hereditary spastic paraplegia (HSP).[27] Motor-predominant neuropathy has also been linked to drug exposures including amiodarone, dapsone, and tacrolimus.[28] Lead poisoning causes a motor neuropathy in adults but in children is more likely to cause predominantly central nervous system symptoms (e.g. encephalopathy).

Sensorimotor polyneuropathy is the most common form of neuropathy in children. Not surprisingly, there remains a long differential diagnosis. By confirming the other steps (following), clinicians can direct appropriate testing for the many acquired, metabolic, and inherited causes of neuropathy in childhood.

Confirm Pattern of Peripheral Nerve Involvement

There are three common patterns of peripheral nerve injury. *Mononeuropathies* are characterized by sensory and/or motor complaints restricted to the distribution of a single peripheral nerve. Traumatic injury and/or extrinsic nerve compression are common causes of mononeuropathies and will be discussed in greater detail later in this chapter. *Mononeuritis multiplex* is defined by sensory and/or motor symptoms involving two or more peripheral nerves, and separated in space and/or time. Hereditary neuropathy with liability to pressure palsies (HNPP) is the most common cause of mononeuritis multiplex in childhood. Vasculitic neuropathies can present with one or more painful mononeuropathies or an asymmetric polyneuropathy, but are very rare in children.[29] Pain is the predominant presenting feature of vasculitic neuropathies, since vasculitis involves inflammation and destruction of the blood vessels supplying peripheral nerves (the vasa nervorum), resulting in painful ischemia and infarction of the peripheral nerves they supply. Vasculitic neuropathy has been reported in adolescents with Wegener granulomatosis,[30] Churg-Strauss syndrome,[31] and cutaneous polyarteritis,[32] systemic lupus erythematosus,[29,33] and Henoch-Schonlein purpura.[34] HNNP-related mononeuropathies differ from vasculitic cases in that they are typically painless.

Polyneuropathy is the third pattern of peripheral nerve disease. Polyneuropathies are sensorimotor and are typically symmetrical and length-dependent, affecting the longest nerves first in a "glove and stocking" distribution. Children typically present with foot drop and clumsiness, usually with minimal sensory deficits. In chronic polyneuropathies such as Charcot-Marie-Tooth disease, children often develop characteristic foot deformities as described in the next section. As symptoms progress, the hands become involved, resulting in weak grip and atrophy of intrinsic hand muscles. Autonomic nerve dysfunction can also be seen in some cases of chronic polyneuropathy affecting small-fiber sensory nerves (e.g. diabetic neuropathies), with patients complaining of cold hands and feet, shiny skin, and hypo- or hyperhydrosis.

Confirm Rapidity of Disease Progression (Acute or Chronic)

Polyneuropathies can be subdivided according to how rapidly a patient's symptoms progress (Table 15.1). Acute-onset neuropathies such as Guillain-Barré syndrome, vasculitic neuropathies, trauma, and some metabolic neuropathies (e.g. hereditary tyrosinemia and porphyria) can become apparent within days to weeks.

TABLE 15.1 Age and Temporal Pattern of Symptom Onset for Pediatric Peripheral Neuropathies

	Congenital to Early Childhood Onset	Late Childhood to Adolescent Onset
Acute onset	Hereditary tyrosinemia	Hereditary neuropathy with liability to pressure palsies (HNPP)
	Guillain-Barré syndrome	Guillain-Barré syndrome
	Mitochondrial (Leigh disease)	Porphyria
		Vasculitic neuropathy
Chronic onset	Charcot-Marie-Tooth disease type 4 (formerly Déjerine-Sottas syndrome) Hereditary sensory autonomic neuropathy (HSAN) types 2–5	Charcot-Marie-Tooth disease type 1 Charcot-Marie-Tooth disease type 2 Charcot-Marie-Tooth disease type X
	Krabbe disease	Hereditary sensory autonomic neuropathy (HSAN) type 1
	Metachromatic leukodystrophy	Friedreich ataxia
	Giant axonal neuropathy	Fabry disease
	Cockayne syndrome	Chronic inflammatory demyelinating polyradiculoneuropathy (CIDP)

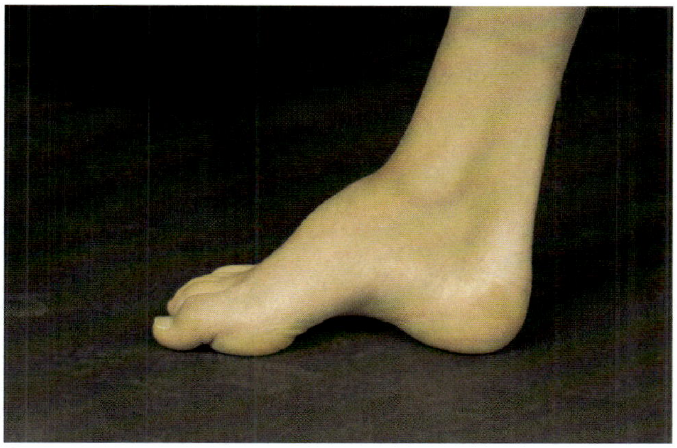

FIGURE 15.3 Pes cavus and hammer toes in a 12-year-old girl with Charcot-Marie-Tooth disease type 1A.

Chronic polyneuropathies include the various forms of Charcot-Marie-Tooth (CMT) disease, Friedreich ataxia, and chronic inflammatory demyelinating polyneuropathy (CIDP). Children with chronic peripheral nerve disease typically demonstrate slow onset and progression. Many patients become aware of problems only when these are pointed out by a relative or friend, and have difficulty specifying the precise time of symptom onset. Some patients may initially seek evaluation for orthopedic problems such as tight heel cords, hammer toe, or foot deformities (Figure 15.3), and callus or even ulcer formation,[35] before later being diagnosed with a peripheral nerve disorder. This is particularly true for adolescents with CMT, also known as hereditary motor and sensory neuropathy (HMSN). CMT is the most common cause of hereditary polyneuropathy, with a disease prevalence (all CMT types) of approximately 1 in 2500 people.[36] Many genes have been linked to CMT disorders showing autosomal dominant (types 1 and 2), autosomal recessive (type 4) and X-linked patterns of inheritance (Table 15.2). The autosomal dominant forms of CMT are differentiated on the basis of electrodiagnostic testing. CMT1 shows "demyelinating" features (e.g. severe slowing of nerve conduction), while CMT2 shows predominantly "axonal" features (low-amplitude sensory and motor responses, with normal conduction velocities). Subjects with CMT1 may complain of an upper extremity tremor additional to the symptoms listed previously. This tremor was initially thought to clinically represent a distinct disorder, the Roussy-Levy syndrome, but is now recognized as a symptom of many hereditary peripheral neuropathies.[37] CMT1 accounts for about 60% of all patients with CMT.[38] Six subtypes have been described, involving five genes (Table 15.2). CMT1A accounts for 70–80% of all CMT1 cases and results from a 1.5-Mb duplication at 17q11.2 involving the peripheral myelin protein 22 (*PMP22*) gene. CMT1B (5–10% of all CMT1 cases) results from point mutations in the myelin protein zero (*MPZ*) gene. The four remaining CMT1 subtypes are less common. Recommendations pertaining to CMT gene testing strategies have been published, to assist clinicians with sequential genetic testing for patients suspected to have CMT on clinical and electrodiagnostic grounds.[39] Nerve biopsies are rarely performed in patients suspected to have CMT, except for rare cases with unusual clinical features raising concern about possible vasculitic neuropathy or other treatable inflammatory nerve disorders.

CMT type 2, the predominantly axonal form of CMT disease, also exhibits autosomal dominant inheritance. The overall prevalence of CMT2 is about half that of CMT1. Its age of onset varies from childhood to late adulthood.[40] Fifteen clinical subtypes, involving at least 12 genes, are described (Table 15.2).[41] CMT2A accounts for 20–30% of all cases of CMT2, and is caused by a mutation in the mitofusin 2 (*MFN2*) gene, an important mediator of mitochondrial fusion. The remaining CMT2 genes are individually rare. CMTX accounts for 15% of CMT cases.[38] CMTX1 results from a mutation in the gap junction beta-1 (*GJB1*) gene. Affected males show a phenotype similar to, or more severe than, CMT1A. Female heterozygotes may appear clinically normal or demonstrate very mild to moderate clinical symptoms. Although all forms of CMT typically demonstrate a chronic disease course, disease onset can be more rapid in some cases, particularly in CMT patients exposed to neurotoxic agents such as chemotherapeutic drugs (e.g. vincristine), or with prolonged antibiotic therapy (e.g. metronidazole, nitrofurantoin).[42]

Guillain-Barré syndrome (GBS) is the most common form of acute flaccid paralysis in childhood.[2] GBS is divided into clinical subgroups: acute inflammatory demyelinating polyradiculoneuropathy (AIDP), Miller-Fisher syndrome (MFS), and acute motor axonal neuropathy (AMAN). GBS is an acute-onset, autoimmune disorder wherein T lymphocytes invade peripheral nerves and Schwann cells, activating antibody and complement deposition within these structures.[43] Since the majority of patients (65–75%) report a preceding viral illness in the weeks prior to symptom onset,[12] molecular mimicry has been proposed as a possible cause of this autoimmune nerve disorder. Children with GBS present with rapidly evolving, symmetrical muscle weakness associated with diminished or absent muscle tendon reflexes. Symptoms typically develop over several days. Children may complain of limb pain or paraesthesia. The degree of weakness can vary, from relatively mild gait difficulty to complete paralysis requiring intubation and mechanical ventilation. Respiratory failure occurs in 16–17% of pediatric GBS patients[12,13] compared to 25–30% of adult GBS patients.[44] Autonomic symptoms, including variability in heart rate and blood pressure, occur in 20–40% of children with GBS[12,13] compared to 65% of

TABLE 15.2 Classification of Charcot-Marie-Tooth Disease

Charcot-Marie-Tooth disease, type 1 Autosomal dominant; Demyelinating; Comprises >50% of all CMT cases		
CMT1A	*PMP22* duplication	60–70% of CMT1 cases
CMT1B[a]	*MPZ* mutation	5–10% of CMT1 cases
CMT1C	*LITAF* mutation	1–2% of CMT1 cases
CMT1D[a]	*EGR2* mutation	1–2% of CMT1 cases
CMT1E[a]	*PMP22* point mutation	<5% of CMT1 cases
CMT1F/2E	*NEFL* mutation	<5% of CMT1cases
Hereditary neuropathy with liability to pressure palsies (HNPP)		
HNPP	*PMP22* deletion	
Charcot-Marie-Tooth disease, type X X-linked; Demyelinating; Comprises 15–20% of all CMT cases		
CMTX1	*GJB1* (*Cx32*) mutation	90% of CMTX cases
CMTX5	*PRPS1* mutation	Rare
Charcot-Marie-Tooth disease, type 2 Autosomal dominant; Axonal; Comprises 20% of all CMT cases		
CMT2A2[a]	*MFN2* mutation	20–30% of CMT2 cases
CMT2A1	*KIF1B* mutation	Rare
CMT2B	*RAB7A* mutation	Rare
CMT2B1	*LMNA* mutation	Rare (allelic with LGMD1B)
CMT2B2	*MED25* mutation	Rare
CMT2C[a]	*TRPV4* mutation	Rare
CMT2D	*GARS* mutation	Rare
CMT2E/1F[a]	*NEFL* mutation	Rare (same disease as CMT1F)
CMT2F	*HSPB1* mutation	Rare
CMT2G	Not known	Rare
CMT2H/K	*GDAP1* mutation	Rare
CMT2I/J	*MPZ* mutation	Rare (allelic with CMT1B)
CMT2L	*HSPB8* mutation	Rare
Charcot-Marie-Tooth disease, type 4 Autosomal recessive; Comprises 5–10% of all CMT cases		
CMT4A[a]	*GDAP1* mutation	25% of all CMT4 cases
CMT4B1	*MTMR2* mutation	Rare
CMT4B2	*SBF2* mutation	Rare
CMT4C[a]	*SH3TC2* mutation	Rare
CMT4D	*NDRG1* mutation	Rare
CMT4E[a]	*EGR2* mutation	Rare
CMT4F	*PRX* mutation	Rare
CMT4H	*FGD4* mutation	Rare
CMT4J	*FIG4* mutation	Rare

[a]*Indicates that gene has been associated with more severe, infant-onset disease in some patients.*

adults with GBS.[45] Guillain-Barré syndrome must be differentiated from other causes of acute flaccid paralysis. Children presenting with rapidly progressive weakness (<24 hours), especially those with bowel or bladder symptoms, must be evaluated for other causes of these findings such as spinal lesions, drug or toxin exposure and metabolic causes. Chronic inflammatory demyelinating polyradiculoneuropathy (CIDP) is distinguished from GBS by its much more chronic nature. Children with CIDP demonstrate a progressive or relapsing sensorimotor polyneuropathy evolving over at least 1 to 2 months.[46] Although many children with CIDP exhibit a slow insidious onset of weakness and sensory loss, up to 20% have an initial acute onset mimicking GBS.[47,48]

Metabolic causes of acute weakness include hereditary tyrosinemia and porphyria. Hereditary tyrosinemia is a rare autosomal recessive disease. Patients present with acute and chronic liver disease, hepatocellular carcinoma, and Fanconi syndrome. Some subjects present with acute neurological crises that may include acute painful polyneuropathy crises.[49] Gastrointestinal dysmotility and muscle weakness can also occur. EMG and nerve biopsy demonstrate axonal degeneration with secondary demyelination. Acute intermittent porphyria (AIP) is the most common form of porphyria in childhood,[50] presenting with acute episodes of neuropathy and/or paralysis and respiratory failure, sometimes after exposure to particular medications.[51] AIP may also have a GBS-like presentation in some patients.[52]

Confirm Age at Symptom Onset

Age of first symptom onset can aid in differentiating childhood neuropathies (Table 15.1). Many hereditary or metabolic causes of polyneuropathy are symptomatic from birth, causing hypotonia, hypo- or areflexia, contractures of one or more joints (arthrogryposis multiplex congenita), feeding difficulties, and/or gross motor or global developmental delay. Early-onset peripheral neuropathies may be isolated (e.g. CMT) or can be associated with combined peripheral and central nervous system findings (e.g. Krabbe and metochromatic leukodystrophy, giant axonal neuropathy, and mitochondrial diseases). The severity of the disease phenotype can also vary significantly, particularly in metabolic disorders where symptoms relate to the amount of residual enzyme activity.

Charcot-Marie-Tooth disease may cause severe weakness which may be apparent from birth. Historically, infants with congenital hypomyelinating neuropathies were diagnosed with Déjerine-Sottas disease or CMT type 3 (CMT3), but with advances in molecular genetics it has become apparent that mutations in genes causative of CMT1 (autosomal dominant demyelinating type), CMT2 (autosomal dominant axonal type), and CMT4 (autosomal recessive type) (Table 15.2) can also cause severe congenital forms of the disease. Several genes have been implicated in infantile CMT, including autosomal dominant (*de novo*) mutations in *PMP22, MPZ, EGR2, GDAP, NEFL, MFN2*, and *TRPV4,* as well as autosomal recessive mutations in *SH3TC2*. Unlike the hereditary metabolic diseases discussed below, symptoms of CMT are generally limited to the peripheral nervous system.

Several rare hereditary disorders of metabolism present during infancy or early childhood with peripheral nerve involvement and varying degrees of involvement of the central nervous system. These conditions include Krabbe disease, metachromatic leukodystrophy, perioxisomal disorders, Cockayne syndrome, giant axonal neuropathy, and hereditary tyrosinemia.

Krabbe disease, or globoid cell leukodystrophy, is a lysosomal storage disorder which may present with an infant-onset polyneuropathy, although its severe central nervous system symptoms more often overshadow the concomitant peripheral neuropathy. Krabbe disease is inherited in an autosomal recessive manner, and results from deficiency of the enzyme galactocerebrosidase (GALC). Symptom onset is typically between 3 and 8 months of age. Most affected children have severe irritability, gross motor delay, hypertonia, and seizures.[53] A smaller subset of patients with higher (>5%) residual enzyme activity develop initial signs of Krabbe disease later in childhood or even adulthood. Stem cell or bone marrow transplantation may be beneficial if performed early in the course of this disease.[54,55] Untreated, infants with Krabbe disease typically die before 2 years of age.

Metachromatic leukodystrophy (MLD) is another autosomal recessive lysosomal storage disease, caused by a dysfunction in one of three enzymes: arylsulfatase A, sulfatase (multiple deficiency), or less commonly, saposin B. MLD typically presents in late infancy (1–3 years), but can present later in patients with higher residual enzyme activity. Affected children show signs of progressive central and peripheral demyelination, including progressive weakness, hypotonia, clumsiness, and dysarthria. Late signs include dementia, spasticity, seizures, and loss of vision and hearing. Bone marrow transplantation is offered for patients diagnosed early in disease course.[56]

Giant axonal neuropathy (GAN) is a rare disease caused by recessive mutations in the gigaxonin gene. GAN affects the central and peripheral nervous system. Symptoms of this autosomal recessive disease become apparent any time from shortly after birth to several years of age. Affected children typically present with developmental delay or regression and gait difficulty, subsequently developing progressive cognitive impairment, seizures, and dyscoordination (ataxia, dysarthria, dysmetria, nystagmus). Curly or kinky hair is frequently seen in children with GAN.[57] Nerve conduction studies reveal a sensorimotor polyneuropathy with axonal features.[57] Nerve biopsy shows axonal loss with some giant swollen axons filled with neurofilaments.[58]

Cockayne syndrome is a rare autosomal recessive disease associated with post-natal growth failure, acquired microcephaly and developmental delay, cutaneous photosensitivity, a progeriod appearance, sensorimotor polyneuropathy with demyelinating features, sensorineural hearing loss, and dental anomalies.[59] The disorder results from mutations in the *ERCC6* or *ERCC8* gene causing impaired DNA repair. The demyelinating polyneuropathy of Cockayne syndrome is usually not a presenting or prominent clinical feature.

Confirm Electrodiagnostic Features

Nerve conduction studies (NCS) and needle electromyography (EMG) have several important functions. Firstly,

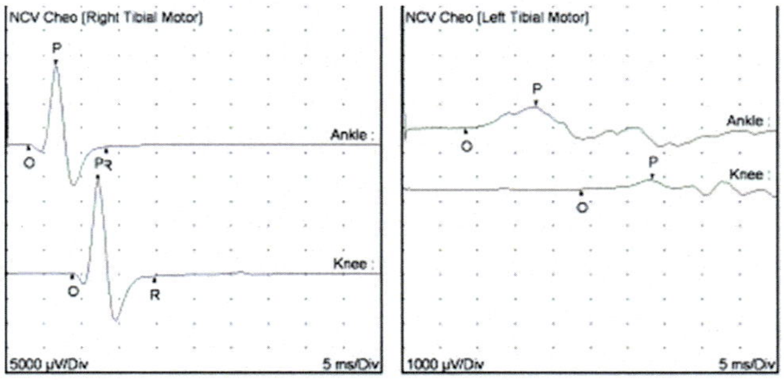

FIGURE 15.4 Tibial motor nerve responses in a normal child and a child with demyelinating polyneuropathy. Nerve conduction studies illustrate a normal tibial motor nerve study (at abductor hallucis) in a healthy 5-year-old child (left) (distal latency 2.9 msec, distal CMAP amplitude 16.0 mV, conduction velocity 52 m/s) compared to a 3-year-old child with chronic inflammatory demyelinating polyradiculoneuropathy, or CIDP (right) (distal latency 8.4 msec, distal CMAP amplitude 1.0 mV, conduction velocity 11 m/s). The tracing on the right demonstrates temporal dispersion, indicating nonuniform conduction slowing and conduction block (a >50% amplitude drop between distal and proximal stimulation sites) which is seen in acquired demyelinating polyneuropathies. The sensitivity is 5× greater on the right, illustrating the much lower CMAP amplitude. The sweep speed is identical between the two tracings.

these electrodiagnostic studies provide objective evidence of peripheral nerve abnormality and help to characterize the specific nerve fiber types involved (sensory, motor, or sensorimotor). Secondly they provide information regarding the type of nerve injury (demyelinating, axonal, or combined), as well as whether this abnormality is uniform versus nonuniform. Lastly, they provide information regarding the severity of a peripheral neuropathy.

Demyelinating features include prolongation of distal latencies, and moderate-to-severe slowing of nerve conduction, to less than 75% of normal values (upper limbs < 38 m/s and lower limbs < 30 m/s) (Figure 15.4). Conduction velocity slowing may be nonuniform, giving rise to temporal dispersion as different large myelinated fibers conduct at different rates depending upon the degree to which they have been affected. Nonuniform slowing is more commonly seen with acquired demyelinating disorders (such as GBS or CIDP), although it has been reported with some hereditary nerve disorders such as CMTX.[60] Uniform slowing implies that the myelin abnormalities are similar at all points along the length of the axon, and similar from fiber to fiber, so that the signal is conducted in an equally slow manner along all nerve fibers. Conduction block is an extreme example of nonuniform slowing, where the signal is partially or wholly blocked from transmitting. Conduction block is often a sign of an acquired (typically inflammatory) neuropathy, but can also be seen with hereditary neuropathy with liability to pressure palsies (HNPP) and toxin exposure such as gasoline sniffing[61] and diphtheria.[62]

Confirm Associated Clinical Features

Nonneurological symptoms can also provide clinicians with important clues to the diagnosis of many peripheral neuropathies. Skin changes accompanying peripheral neuropathies in childhood can help to greatly narrow the differential diagnosis. Angiokeratomas, or cherry red spots, raise suspicion of Fabry disease. Other rashes, pigmentary changes, photosensitivity, and nail bed findings may be important clues to specific disorders (Table 15.3). Involvement of organs such as the liver, spleen, heart, kidneys, tendons, and lymphatic tissues increases the likelihood of specific diseases in children with polyneuropathy. Signs of central nervous system involvement can include language delay or regression, cognitive difficulty, seizures, gait ataxia, spasticity, hyperreflexia, and/or extensor plantar responses, and should direct testing to investigate for cortical and/or white matter involvement. MR imaging of the brain is indicated in such cases, with additional metabolic testing to investigate for leukodystrophies (e.g. Krabbe, metachromatic), mitochondrial disorders (e.g. *POLG* mutations), peroxisomal disorders, and toxic exposures.

PEDIATRIC ELECTRODIAGNOSTIC TESTING

Nerve Conduction Studies

Nerve conduction studies (NCS) confirm the presence of peripheral nerve disease, characterize the fiber types involved (sensory and/or motor), and define the severity and nature of the key electrodiagnostic characteristics of pediatric neuropathies (demyelinating versus axonal; uniform versus nonuniform findings).

NCS are typically performed using a hand-held stimulator and surface electrodes to record peripheral nerve responses, in the same way as this testing is undertaken in adults. Recording electrodes should be selected to conform to small hand and foot size. Caution must be taken to ensure that limb temperature is monitored. Anxiety and/or small limb size increase the risk of limb cooling and temperature-related artifact. Skin temperature should be at least 32°C in the upper limb and 30°C in the lower limb.[63]

Nerve conduction testing typically begins with sensory nerve studies, as these require less stimulation and are better tolerated by children. Normal values are grouped by age, with conduction velocity and amplitudes increasing with advancing age, particularly over the first 2 years of life (see Chapter 3).

The selection of peripheral nerves amenable to NCS in infants is more limited, due to short limb and digit length, and a relative increase in subcutaneous fat

TABLE 15.3 Non-neurological Signs and Symptoms Providing Clues to Diagnosis of Pediatric Peripheral Neuropathies

Sign/Symptom		Possible Diagnoses
Eye	Optic atrophy	Mitochondrial disease (OPA1, Friedreich ataxia) CMT2A (*MFN2*), CMT4A (*GDAP1*)
	Retinitis pigmentosa	Peroxisomal disease (e.g. infantile Refsum disease), mitochondrial disease (NARP), ataxia with vitamin E deficiency (AVED)
	Ophthalmoplegia	Mitochondrial disease (progressive nuclear ophthalmoplegia), Miller Fisher variant (GBS)
	Lisch nodules	Neurofibromatosis
Ear	Sensorineural hearing loss	CMTX, Cockayne syndrome
Skin	Hypopigmentation	Leprosy
	Hyperpigmentation	Adrenoleukodystrophy (buccal), diabetes (acanthosis nigricans), neurofibromatosis (café au lait spots)
	Angiokeratomas	Fabry disease
	Purpura	Systemic vasculitis (Henoch-Schönlein purpura)
	Malar or discoid rash	Systemic lupus erythematosus (SLE)
	Photosensitivity	Xeroderma pigmentosa, Cockayne syndrome, SLE
	Desquamation	Arsenic exposure
	Nail changes	Arsenic and thallium (Mee's lines)
Hair	Alopecia	Connective tissue, thallium poisoning
	Tight kinky hair	Giant axonal neuropathy
Heart	Cardiomyopathy	Mitochondrial disease, Friedreich ataxia, ataxia with vitamin E deficiency (AVED)
	Conduction defect	Mitochondrial, glue/solvent abuse
Extremities	Arthritis	Lyme disease
	Tendon pigmentation	Cerebrotendinous xanthomatosis
Pharynx	Yellow-orange tonsils	Tangier disease
	Gray pseudomembrane	Diphtheria
GI	Abdominal pain	Porphyria, Fabry disease, mitochondrial disease (MNGIE), arsenic or lead toxicity
Lymph	Lymphadenopathy	Lymphoma
	Splenomegaly	Lymphoma
	Hepatomegaly	Toxins (e.g. amiodarone), mitochondrial disease (PEO), tyrosinemia, hemophagocytosis syndromes
Brain	Developmental regression or seizures	Krabbe disease, metachromatic leukodystrophy, mitochondrial disorders, lead toxicity, peroxisomal disorders

compared to older children. The sensory nerves typically studied in infants include the median, ulnar, and medial plantar nerves. Motor nerves generally studied in infants include median (at abductor pollicus brevis), ulnar (at abductor digiti minimi), peroneal (at extensor digitorum brevis), and tibial (at abductor hallucis). The common peroneal nerve response (recorded over extensor digitorum brevis) can be difficult to elicit in newborns due to the small size of the EDB muscle, but becomes technically easier to do after the first 12 months of age. The sural sensory response can be very difficult to record in small chubby infants.

Electromyography

Needle electromyography (EMG) is typically performed using a 30 gauge 25 mm disposable concentric needle. Centers will differ as to whether this study is performed with children awake or sedated, but our experience is that most children can tolerate an efficient and limited (2–3 muscle) evaluation while awake or with conscious sedation. Electromyography is important in defining whether motor units are typically large amplitude and rapidly firing (suggesting a possible neurogenic lesion), or smaller amplitude and polyphasic (suggesting a possible "myopathic" lesion). In children with a possible traumatic peripheral nerve injury, EMG is valuable for assessing the presence or absence of axonal continuity, which can help predict likelihood of spontaneous recovery.

Needle EMG must be interpreted with some caution when evaluating neonates and infants with traumatic neuropathies. In children with obstetric brachial plexus palsies (OBPP), EMG may provide an overly optimistic view of potential for recovery. There are several reasons for the discrepancy between EMG findings and clinical outcome in infants. One is that the average muscle fiber diameter of a 3-month-old infant is only 17 μm, which is more than three-fold smaller than the adult muscle fiber diameter. Due to this smaller cross-sectional area, the muscle of an infant contains 11 times more motor units compared to an adult. An EMG needle will therefore "listen" to many more motor units when recording the recruitment pattern of an infant's muscle, and may overestimate the normality of that pattern in small children.[64]

TRAUMA

Traumatic injury is an important cause of neuropathy in children. Specific injuries are commonly linked to certain mononeuropathies (e.g. axillary neuropathy after anterior shoulder dislocation and radial neuropathy after supracondylar fracture), reflecting the mechanism of injury and proximity of nerve to the site of trauma in each case. Nerve conduction studies and EMG can be particularly helpful in moderate and severe neuropathies, providing important information regarding whether or not axonal continuity exists, as well as the presence and degree of spontaneous reinnervation. Seddon[65] proposed a classification system to categorize the severity of nerve injury into one of three groups: neurapraxia, axonotmesis, and neurotmesis.

Neurapraxia is the least severe of these injuries and involves an area of segmental demyelination due to local traumatic and/or compression injury. The myelin is disrupted, and affected fibers do not convey afferent sensory or efferent motor signals. Axons remain intact. Clinically, this cannot be distinguished in the acute phase from more serious nerve injuries. Long-term outcome is typically favorable however, as the viable Schwann cells can repair the damaged segment (assuming the offending cause has been removed). Patients with certain acquired (e.g. GBS) or hereditary neuropathies (e.g. HNPP) may exhibit clinical and electrodiagnostic evidence of segmental demyelination. Drug abuse (specifically gasoline or glue sniffing, or "huffing") has also been reported to cause conduction block upon a superimposed sensorimotor polyneuropathy with demyelinating features.[61]

Axonotmesis is the next severe type of nerve injury. In axonotmesis, some or all axons within a nerve fiber are transected. When an axon is cut, the segment of the axon that has been separated from its cell body (which is in the dorsal root ganglion, in the case of sensory fibers, or in the anterior horn in the case of motor fibers) degenerates, through a process known as Wallerian degeneration. While axonotmesis is commonly seen with trauma, it can also be associated with atraumatic nerve injuries, including inflammatory neuropathies, vasculitides, and toxin exposures, among others. Although the patient's clinical symptoms are immediately apparent, it takes time for the changes of denervation to become apparent on EMG: anywhere from days to 3–4 weeks, depending upon the length of the nerve and the site of injury.[66] During the Wallerian degeneration phase, the nerve fibers distal to the site of transection die back. Approximately 4 to 5 days after the injury, the myelin begins to form linear arrays of ovoids and globules along the line of the degenerating axon (Figure 15.5). Axonal swelling becomes apparent. During the recovery phase, the axon of the proximal segment begins to regenerate, forming sprouts. These nerve twigs grow distally, growing into and through surviving distal neurilemmal tubes of Schwann cell basement membranes. The regenerating axons are then myelinated by the Schwann cells. With axonotmesis, some components of the nerve, including Schwann cells and potentially a percentage of the total nerve fibers, remain intact, subsequently acting as a conduit or path along which the regenerating nerve may grow. The fact that this path remains in place increases the likelihood that regenerating nerve fibers will locate and accurately reinnervate their target muscles distal to the site of nerve injury. Nerve regeneration proceeds at a slow pace, approximately 1 millimeter per day, or 1 inch per month.[67] Needle EMG can show some evidence of axonotmesis in situations where active denervation is noted (fibrillation potentials or positive sharp waves) accompanied by some evidence of ongoing axonal continuity (motor unit action potentials under volitional control). When no axonal continuity is noted, the electromyographer cannot distinguish axonotmesis from the most severe form of nerve injury, neurotmesis.

Neurotmesis is the most severe form of nerve injury and is characterized by complete transection of all nerve elements. In this situation the likelihood of spontaneous and meaningful reinnervation is low, as there is no path to guide the regenerating nerves.

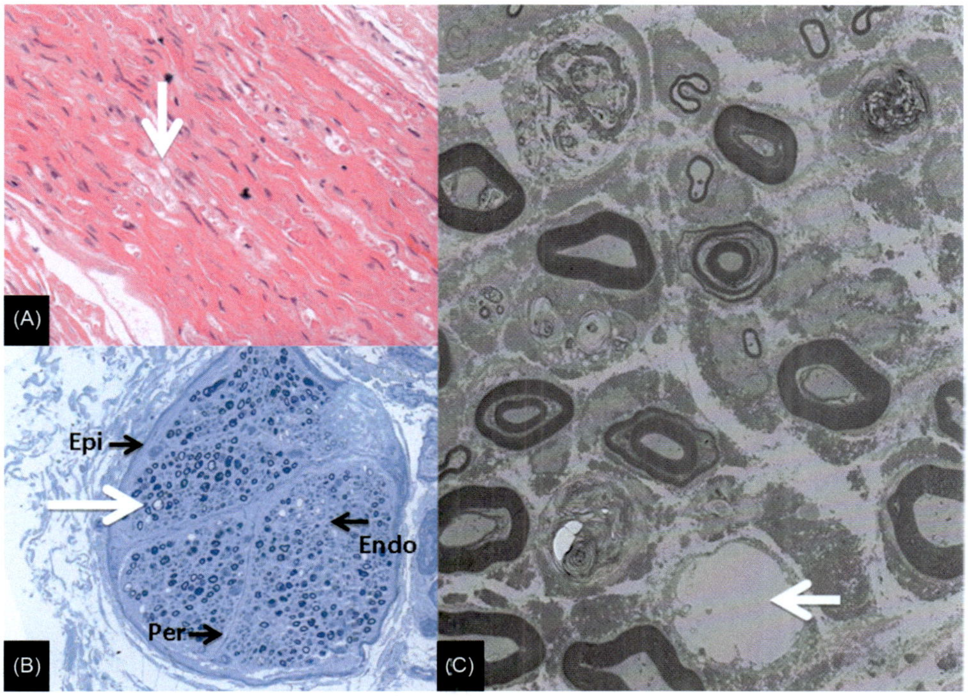

FIGURE 15.5 Peripheral nerve sections **(A)** longitudinal section, stained with hematoxylin and eosin, **(B)** cross-section, stained with toluidine blue, and **(C)** electron microscopy, showing Wallerian degeneration of an axon (large white arrow in each image). Epi = epineurium; Peri = perineurium, Endo = endoneurium. *Photo credit: Dr. Jean Michaud, Department of Pathology, Children's Hospital of Eastern Ontario.*

NERVE BIOPSY

Nerve biopsies are only rarely performed in children with peripheral nerve disease. With the recent advances in molecular testing, nerve biopsy is not required for the precise diagnosis of most pediatric polyneuropathies.

There are, however, still a few situations in which nerve biopsy remains an important and useful diagnostic test. The first is the patient suspected of having a vasculitic neuropathy. It is imperative to have a precise diagnosis, if possible, before initiating immunosuppressants, such as corticosteroids and cyclophosphamide, which carry a substantial risk of side effects. The MRI appearance of benign peripheral nerve tumors (perineuriomas, schwannomas) can appear very similar to vasculitic neuropathy. When the clinical history and ancillary biochemical and serological testing cannot confirm a definite diagnosis, it is important to consider a biopsy in these cases. The second situation is a patient with a chronic polyneuropathy due to either a hereditary (CMT) or inflammatory (CIDP) cause. The clinical history and electrodiagnostic testing cannot always differentiate the two, and again it can be helpful to obtain a definite neuropathological diagnosis before beginning long-term costly immunosuppressive therapies.

Given that nerve biopsies are only rarely performed, it is important that they are carried out at a center with sufficient expertise to adequately sample the correct tissue and carry out the correct immunologic and/or ultrastructural studies in addition to the routine staining. The yield of nerve biopsies is increased when an affected nerve or nerve segment is sampled. Electrodiagnostic testing and MR imaging can be very important in directing the site of biopsy. Nerve biopsies can be either whole nerve or fascicular. Whole nerve biopsies are reserved for distal sensory nerves (e.g. sural nerve) and involve transecting the nerve at two sites before removing a segment for neuropathological analysis. Because the nerve is transected, the patient is left with permanent sensory loss in the area formerly supplied by this nerve. Mixed or motor nerves cannot be studied in this manner. Fascicular biopsies involve sampling a portion of a mixed or motor nerve. Patients are left with varying degrees of sensory and motor deficit after fascicular biopsies although the potential for recovery is greater given that persisting axonal continuity in most of the nerve provides the sacrificed nerve fibers a path along which they may regrow.

NEUROIMAGING IN CHILDHOOD POLYNEUROPATHIES

Magnetic resonance imaging (MRI) and ultrasound can be useful ancillary tools in patients with select types of

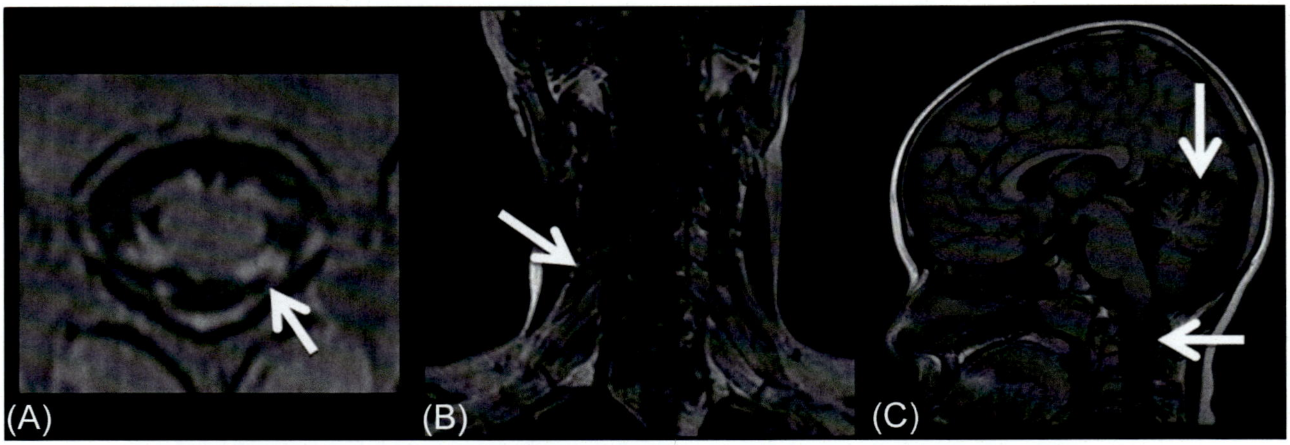

FIGURE 15.6 Nerve root thickening and enhancement in chronic inflammatory demyelinating polyneuropathy **(A)**. Similar enlargement of peripheral nerves can be seen in inherited neuropathies such as Charcot-Marie-Tooth disease **(B)**. Neuroimaging can also be useful in identifying central nervous system involvement in conditions such as Friedreich ataxia and autosomal recessive ataxia of Charlevoix-Saguenay (ARSACS). MR imaging in ARSACS may show thinning of the upper cervical spinal cord and progressive cerebellar atrophy, particularly affecting the superior cerebellar folia **(C)**.

peripheral neuropathy. This is particularly true for children suffering with idiopathic mononeuropathies, in whom nerve imaging rules out extrinsic nerve compression by adjacent tumors or soft tissue lesions, as well as intrinsic nerve lesions such as perineuriomas, schwannomas, and neurofibromata. On MRI, focal nerve thickening and enhancement may be seen after administration of contrast. Imaging can also be a useful accompaniment to electrodiagnostic testing, in guiding the specific site of nerve biopsy where required.

MRI is being increasingly used as a tool to aid in the diagnosis of some patients with suspected acute or chronic inflammatory polyradiculoneuropathies including Guillain-Barré syndrome and chronic inflammatory demyelinating polyradiculoneuropathy (CIDP). MRI shows thickening of the cauda equina and/or thickening and gadolinium enhancement of the spinal nerve roots in 90–95% of children and adults with GBS.[68–70] Nerve root thickening or enhancement (Figure 15.6A) is seen in approximately 40% of children[48] and 60% of adults[71] with CIDP. In some CIDP patients, particularly those with a protracted course, nerve root enlargement may be striking.[72] Adults with CIDP have undergone biopsy of their thickened cranial nerves out of concern for possible malignancy, or required laminectomy for decompression of the cauda equina.[73] While nerve root thickening can support a diagnosis of inflammatory neuropathy in the right clinical context, it is by no means specific to this disease, since nerve root thickening has also been reported with CMT[74] (Figure 15.6B) and neurofibromatosis[75] among other causes.

Characteristic cervical spine and posterior fossa abnormalities are noted in older children and adolescents with inherited ataxias including Friedreich ataxia and ARSACS (autosomal recessive ataxia of Charlevoix-Saguenay). MR imaging of these patients may be normal, or may show thinning of the upper cervical spinal cord and progressive cerebellar atrophy particularly affecting the superior cerebellar folia (Figure 15.6C).

REFERENCES

1. Hagberg B, Westerberg B. The nosology of genetic peripheral neuropathies in Swedish children. *Dev Med Child Neurol* 1983;**25** (1):3–18.
2. Evans OB. Polyneuropathy in childhood. *Pediatrics* 1979;**64**(1): 96–105.
3. Ouvrier RA, McLeod JG. Chronic peripheral neuropathy in childhood: an overview. *Aust Paediatr J* 1988;**24**(Suppl. 1):80–2.
4. Martyn CN, Hughes RA. Epidemiology of peripheral neuropathy. *J Neurol Neurosurg Psychiatry* 1997;**62**(4):310–18.
5. Baldereschi M, Inzitari M, Di CA, Farchi G, Scafato E, Inzitari D. Epidemiology of distal symmetrical neuropathies in the Italian elderly. *Neurology* 2007;**68**(18):1460–7.
6. Gamble HJ, Breathnach AS. An electron-microscope study of human foetal peripheral nerves. *J Anat* 1965;**99**(Pt 3):573–84.
7. Schroder JM. Introduction. In: Schroder JM, editor. *Pathology of Peripheral Nerves: An Atlas of Structural and Molecular Pathological Changes*. 1st ed. Berlin: Springer; 2001. pp. 1–14.
8. Wagner AL, Buchthal F. Motor and sensory conduction in infancy and childhood: reappraisal. *Dev Med Child Neurol* 1972;**14**(2): 189–216.
9. Thomas JE, Lambert EH. Ulnar nerve conduction velocity and H-reflex in infants and children. *J Appl Physiol* 1960;**15**:1–9.
10. Logina I, Donaghy M. Diphtheritic polyneuropathy: a clinical study and comparison with Guillain-Barré syndrome. *J Neurol Neurosurg Psychiatry* 1999;**67**(4):433–8.

11. Lin JJ, Hsia SH, Wang HS, Lyu RK, Chou ML, Hung PC, et al. Clinical variants of Guillain-Barré syndrome in children. *Pediatr Neurol* 2012;**47**(2):91–6.
12. Bradshaw DY, Jones Jr. HR. Guillain-Barré syndrome in children: clinical course, electrodiagnosis, and prognosis. *Muscle Nerve* 1992;**15**(4):500–6.
13. Vajsar J, Fehlings D, Stephens D. Long-term outcome in children with Guillain-Barré syndrome. *J Pediatr* 2003;**142**(3):305–9.
14. Bradshaw DY, Jones HR. Pseudomeningoencephalitic presentation of pediatric Guillain-Barré syndrome. *J Child Neurol* 2001;**16**(7):505–8.
15. Bayrak AO, Battaloglu E, Turker H, Baris I, Oztas G. Hereditary neuropathy with liability to pressure palsy (HNPP) in childhood: a case study emphasizing the relevance of detailed electrophysiological examination for suspected HNPP in the first decade. *Brain Dev* 2009;**31**(6):445–8.
16. Van Meir N, De Smet L. Carpal tunnel syndrome in children. *Acta Orthop Belg* 2003;**69**(5):387–95.
17. Kwon JY, Ko K, Sohn YB, Kim SJ, Park SW, Kim SH, et al. High prevalence of carpal tunnel syndrome in children with mucopolysaccharidosis type II (Hunter syndrome). *Am J Med Genet A* 2011;**155A**(6):1329–35.
18. Shorer Z, Moses SW, Hershkovitz E, Pinsk V, Levy J. Neurophysiologic studies in congenital insensitivity to pain with anhidrosis. *Pediatr Neurol* 2001;**25**(5):397–400.
19. Desnick RJ, Brady RO. Fabry disease in childhood. *J Pediatr* 2004;**144**(5 Suppl.):S20–6.
20. Mehta A, Ricci R, Widmer U, Dehout F, Garcia de LA, Kampmann C, et al. Fabry disease defined: baseline clinical manifestations of 366 patients in the Fabry Outcome Survey. *Eur J Clin Invest* 2004;**34**(3):236–42.
21. Sheth KJ, Swick HM. Peripheral nerve conduction in Fabry disease. *Ann Neurol* 1980;**7**(4):319–23.
22. Zarate YA, Hopkin RJ. Fabry's disease. *Lancet* 2008;**372**(9647):1427–35.
23. Peyronnard JM, Charron L, Barbeau A. The neuropathy of Charlevoix-Saguenay ataxia: an electrophysiological and pathological study. *Can J Neurol Sci* 1979;**6**(2):199–203.
24. Lowry NJ, Taylor MJ, Belknapp W, Logan WJ. Electrophysiological studies in five cases of abetalipoproteinemia. *Can J Neurol Sci* 1984;**11**(1):60–3.
25. Fogel BL, Perlman S. Clinical features and molecular genetics of autosomal recessive cerebellar ataxias. *Lancet Neurol* 2007;**6**(3):245–57.
26. Nishimoto Y, Susuki K, Yuki N. Serologic marker of acute motor axonal neuropathy in childhood. *Pediatr Neurol* 2008;**39**(1):67–70.
27. Rossor AM, Davidson GL, Blake J, Polke JM, Murphy SM, Houlden H, et al. A novel p.Glu175X premature stop mutation in the C-terminal end of HSP27 is a cause of CMT2. *J Peripher Nerv Syst* 2012;**17**(2):201–5.
28. Venance SL, Strong MJ. Toxic motor neuronopathies and neuropathies. In: Eisen A, editor. *Handbook of Clinical Neurophysiology.* Elsevier; 2004. pp. 437–50.
29. Ryan MM, Tilton A, De Girolami U, Darras BT, Jones Jr. HR. Paediatric mononeuritis multiplex: a report of three cases and review of the literature. *Neuromuscul Disord* 2003;**13**(9):751–6.
30. Orlowski JP, Clough JD, Dyment PG. Wegener's granulomatosis in the pediatric age group. *Pediatrics* 1978;**61**(1):83–90.
31. Wang SJ, Yang YH, Lin YT, Tsai MJ, Chiang BL. Childhood Churg-Strauss syndrome: report of a case. *J Microbiol Immunol Infect* 2000;**33**(4):263–6.
32. Draaisma JM, Fiselier TJ, Mullaart RA. Mononeuritis multiplex in a child with cutaneous polyarteritis. *Neuropediatrics* 1992;**23**(1):28–9.
33. Steinlin MI, Blaser SI, Gilday DL, Eddy AA, Logan WJ, Laxer RM, et al. Neurologic manifestations of pediatric systemic lupus erythematosus. *Pediatr Neurol* 1995;**13**(3):191–7.
34. Belman AL, Leicher CR, Moshe SL, Mezey AP. Neurologic manifestations of Schoenlein-Henoch purpura: report of three cases and review of the literature. *Pediatrics* 1985;**75**(4):687–92.
35. Shy M, Lupski JR, Chance PF, Klein CJ, Dyck PJ. Hereditary motor and sensory neuropathies: an overview of clinical, genetic, electrophysiological and pathological features. In: Dyck PJ, Thomas PK, editors. *Peripheral Neuropathy.* Philadelphia: Elsevier; 2005. pp. 1623–58.
36. Skre H. Genetic and clinical aspects of Charcot-Marie-Tooth's disease. *Clin Genet* 1974;**6**(2):98–118.
37. Auer-Grumbach M, Strasser-Fuchs S, Wagner K, Korner E, Fazekas F. Roussy-Levy syndrome is a phenotypic variant of Charcot-Marie-Tooth syndrome IA associated with a duplication on chromosome 17p11.2. *J Neurol Sci* 1998;**154**(1):72–5.
38. Ionasescu VV, Ionasescu R, Searby C. Screening of dominantly inherited Charcot-Marie-Tooth neuropathies. *Muscle Nerve* 1993;**16**(11):1232–8.
39. Saporta AS, Sottile SL, Miller LJ, Feely SM, Siskind CE, Shy ME. Charcot-Marie-Tooth disease subtypes and genetic testing strategies. *Ann Neurol* 2011;**69**(1):22–33.
40. Bienfait HM, Baas F, Koelman JH, de Haan RJ, van Engelen BG, Gabreels-Festen AA. Phenotype of Charcot-Marie-Tooth disease Type 2. *Neurology* 2007;**68**(20):1658–67.
41. Zuchner S, Vance JM. Molecular genetics of autosomal-dominant axonal Charcot-Marie-Tooth disease. *Neuromolecular Med* 2006;**8**(1–2):63–74.
42. Weimer LH, Podwall D. Medication-induced exacerbation of neuropathy in Charcot Marie Tooth disease. *J Neurol Sci* 2006;**242**(1–2):47–54.
43. Dalakas MC. Basic aspects of neuroimmunology as they relate to immunotherapeutic targets: present and future prospects. *Ann Neurol* 1995;**37**(Suppl. 1):S2–13.
44. Hahn AF. The challenge of respiratory dysfunction in Guillain-Barré syndrome. *Arch Neurol* 2001;**58**(6):871–2.
45. Zochodne DW. Autonomic involvement in Guillain-Barré syndrome: a review. *Muscle Nerve* 1994;**17**(10):1145–55.
46. Nevo Y, Topaloglu H. 88th ENMC International Workshop: Childhood chronic inflammatory demyelinating polyneuropathy (including revised diagnostic criteria), Naarden, the Netherlands, December 8–10, 2000. *Neuromuscul Disord* 2002;**12**(2):195–200.
47. McCombe PA, Pollard JD, McLeod JG. Chronic inflammatory demyelinating polyradiculoneuropathy. A clinical and electrophysiological study of 92 cases. *Brain* 1987;**110**(Pt 6):1617–30.
48. McMillan HJ, Kang PB, Jones HR, Darras BT. Childhood chronic inflammatory demyelinating polyradiculoneuropathy: combined analysis of a large cohort and eleven published series. *Neuromuscul Disord* 2013;**23**(2):103–11.
49. Mitchell G, Larochelle J, Lambert M, Michaud J, Grenier A, Ogier H, et al. Neurologic crises in hereditary tyrosinemia. *N Engl J Med* 1990;**322**(7):432–7.

50. Chisolm Jr. JJ. Pediatric aspects of the porphyrias. *J Pediatr* 1964;**64**:159–73.
51. Albers JW, Fink JK. Porphyric neuropathy. *Muscle Nerve* 2004;**30** (4):410–22.
52. McEneaney D, Hawkins S, Trimble E, Smye M. Porphyric neuropathy—a rare and often neglected differential diagnosis of Guillain-Barré syndrome. *J Neurol Sci* 1993;**114**(2):231–2.
53. Hagberg B. Krabbe's disease: clinical presentation of neurological variants. *Neuropediatrics* 1984;**15**(Suppl.):11–15.
54. Krivit W, Shapiro EG, Peters C, Wagner JE, Cornu G, Kurtzberg J, et al. Hematopoietic stem-cell transplantation in globoid-cell leukodystrophy. *N Engl J Med* 1998;**338**(16):1119–26.
55. Krivit W, Peters C, Shapiro EG. Bone marrow transplantation as effective treatment of central nervous system disease in globoid cell leukodystrophy, metachromatic leukodystrophy, adrenoleukodystrophy, mannosidosis, fucosidosis, aspartylglucosaminuria, Hurler, Maroteaux-Lamy, and Sly syndromes, and Gaucher disease type III. *Curr Opin Neurol* 1999;**12**(2):167–76.
56. Peters C, Steward CG. Hematopoietic cell transplantation for inherited metabolic diseases: an overview of outcomes and practice guidelines. *Bone Marrow Transplant* 2003;**31**(4):229–39.
57. Treiber-Held S, Budjarjo-Welim H, Reimann D, Richter J, Kretzschmar HA, Hanefeld F. Giant axonal neuropathy: a generalized disorder of intermediate filaments with longitudinal grooves in the hair. *Neuropediatrics* 1994;**25**(2):89–93.
58. Yiu EM, Ryan MM. Genetic axonal neuropathies and neuronopathies of pre-natal and infantile onset. *J Peripher Nerv Syst* 2012;**17** (3):285–300.
59. Nance MA, Berry SA. Cockayne syndrome: review of 140 cases. *Am J Med Genet* 1992;**42**(1):68–84.
60. Ryan MM, Jones Jr. HR. CMTX mimicking childhood chronic inflammatory demyelinating neuropathy with tremor. *Muscle Nerve* 2005;**31**(4):528–30.
61. Burns TM, Shneker BF, Juel VC. Gasoline sniffing multifocal neuropathy. *Pediatr Neurol* 2001;**25**(5):419–21.
62. Creange A, Meyrignac C, Roualdes B, Degos JD, Gherardi RK. Diphtheritic neuropathy. *Muscle Nerve* 1995;**18**(12):1460–3.
63. Kang PB. Pediatric nerve conduction studies and EMG. In: Blum AS, Rutkove SB, editors. *Clinical Neurophysiology Primer*. Totowa: Humana Press Inc; 2007. pp. 369–89.
64. Van Dijk JG, Malessy MJ, Stegeman DF. Why is the electromyogram in obstetric brachial plexus lesions overly optimistic? *Muscle Nerve* 1998;**21**(2):260–1.
65. Seddon HJ. A classification of nerve injuries. *Br Med J* 1942;**2** (4260):237–9.
66. Gooch CL, Weimer LH. The electrodiagnosis of neuropathy: basic principles and common pitfalls. *Neurol Clin* 2007;**25**(1):1–28.
67. Seddon HJ, Medawar PB, Smith H. Rate of regeneration of peripheral nerves in man. *J Physiol* 1943;**102**(2):191–215.
68. Gorson KC, Ropper AH, Muriello MA, Blair R. Prospective evaluation of MRI lumbosacral nerve root enhancement in acute Guillain-Barré syndrome. *Neurology* 1996;**47**(3):813–17.
69. Byun WM, Park WK, Park BH, Ahn SH, Hwang MS, Chang JC. Guillain-Barré syndrome: MR imaging findings of the spine in eight patients. *Radiology* 1998;**208**(1):137–41.
70. Yikilmaz A, Doganay S, Gumus H, Per H, Kumandas S, Coskun A. Magnetic resonance imaging of childhood Guillain-Barré syndrome. *Childs Nerv Syst* 2010;**26**(8):1103–8.
71. Duggins AJ, McLeod JG, Pollard JD, Davies L, Yang F, Thompson EO, et al. Spinal root and plexus hypertrophy in chronic inflammatory demyelinating polyneuropathy. *Brain* 1999;**122**(Pt 7): 1383–90.
72. McMillan HJ, Miller E. Cranial nerve hypertrophy in pediatric chronic inflammatory demyelinating polyradiculoneuropathy. *Pediatr Radiol* 2010;**40**(Suppl. 1):S176.
73. Schady W, Goulding PJ, Lecky BR, King RH, Smith CM. Massive nerve root enlargement in chronic inflammatory demyelinating polyneuropathy. *J Neurol Neurosurg Psychiatry* 1996;**61**(6):636–40.
74. Haubrich C, Krings T, Senderek J, Zuchner S, Schroder JM, Noth J, et al. Hypertrophic nerve roots in a case of Roussy-Levy syndrome. *Neuroradiology* 2002;**44**(11):933–7.
75. Lin J, Martel W. Cross-sectional imaging of peripheral nerve sheath tumors: characteristic signs on CT, MR imaging, and sonography. *AJR Am J Roentgenol* 2001;**176**(1):75–82.

Chapter 16

Congenital and Early Infantile Neuropathies

Eppie M. Yiu and Jonathan Baets

INTRODUCTION

The congenital and infantile neuropathies are an uncommon and complex group of conditions with broad phenotypic and genetic diversity. The overwhelming majority have a genetic basis, and will be the main focus of this chapter. Rarely, acquired inflammatory causes may occur. Although significant progress has been made in the genetic diagnosis of infantile neuropathies in the past two decades, more than half of children presenting in the first year of life remain without a specific diagnosis. With advances in molecular genetic techniques, the already rapidly expanding number of genes attributed to infantile neuropathies is set to further increase.

ETIOLOGY

In a large cohort of 260 children with biopsy-confirmed peripheral neuropathy over a 33-year period, 50 (19.2%) were symptomatic in infancy, of whom 48% had a demyelinating and 42% an axonal disorder.[1] While many of these cases presented prior to the era of genetic diagnosis, 20 children fell under the rubric of Charcot-Marie-Tooth disease (CMT)/ hereditary motor and sensory neuropathy (HMSN), 19 of whom had a demyelinating form of CMT/HMSN. Five children had spinal muscular atrophy (with axonal findings on nerve biopsy), three had severe infantile axonal neuropathy with respiratory failure, and four had hereditary sensory autonomic neuropathy. Other etiologies were either single cases (e.g. giant axonal neuropathy) or nonspecific diagnoses, such as peroxisomal or "neurodegenerative" disorders, and many of these infants had co-existent central nervous system involvement. More recently, in a series of 77 children with infantile onset (nonsyndromic) neuropathies (58% demyelinating, 20% axonal, and 22% undefined), 35 children (45%) received a molecular diagnosis involving mutations in one of eleven genes, most of which were associated with demyelinating forms of CMT (Figure 16.1).[2] This emphasizes the genetic heterogeneity of the infantile neuropathies, and that demyelinating phenotypes are more likely to receive a definitive genetic diagnosis, but that over half of children remain without one.

CLINICAL PRESENTATION AND APPROACH TO PERIPHERAL NEUROPATHIES IN EARLY INFANCY

Clinical Presentation

Infants with early onset neuropathies tend to display one of two main phenotypes.

The first group has clear prenatal/neonatal onset and presents at birth with hypotonia, arthrogryposis, and respiratory insufficiency. A history of decreased fetal movements, intrauterine growth retardation, and polyhydramnios may be obtained. Delivery of a hypotonic infant may result in obstetric complications and birth asphyxia, masking the underlying peripheral neuropathy. Ventilatory support is often required and early mortality common. *De novo* mutations in *MPZ*, *PMP22*, *EGR2*, and *NEFL* are typically found in this cohort[2] (Figure 16.2). Spinal muscular atrophy (SMA) and the non-5q early-onset SMA syndromes are important differential diagnoses for this type of presentation, and, as discussed below, are not always easily differentiated from a classic neuropathy.

The second (and more common) group presents later in infancy with delayed motor development and/or foot deformities, after an uneventful neonatal period.[2] In contrast to adults, in whom distal weakness and sensory loss are the predominant features of a neuropathy, delayed motor milestones (an indicator of infantile onset) and

B.T. Darras, H. Royden Jones, Jr., M.M. Ryan & D.C. De Vivo (Eds): Neuromuscular Disorders of Infancy, Childhood and Adolescence, Second edition.
DOI: http://dx.doi.org/10.1016/B978-0-12-417044-5.00016-0

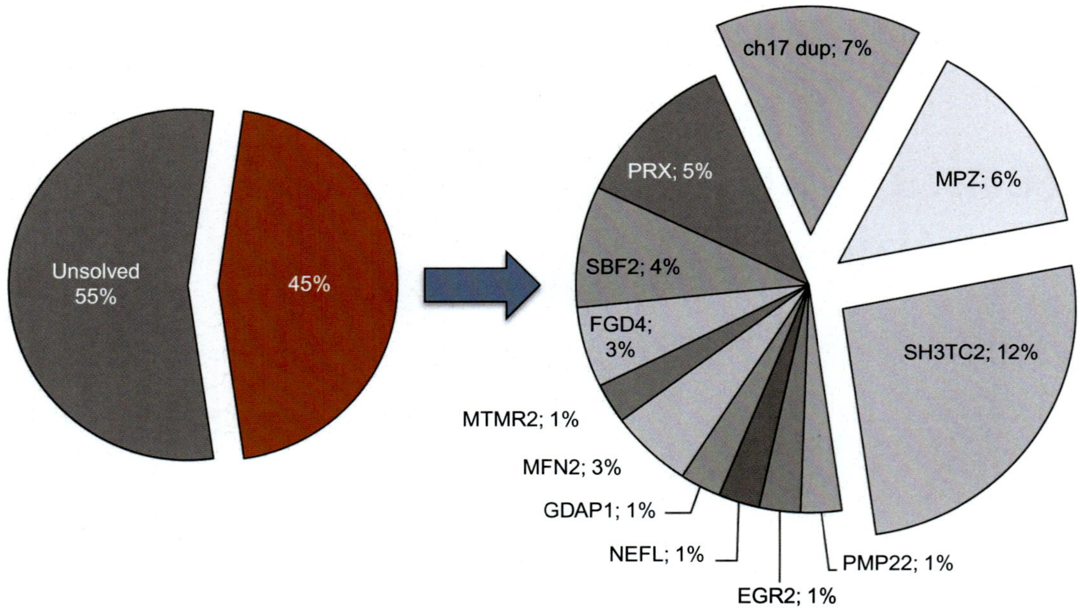

FIGURE 16.1 Gene distribution among 35 patients with pathogenic mutations, out of a cohort of 77 unrelated index patients with infantile-onset neuropathies. *Reprinted from reference 2 with permission from Oxford University Press.*

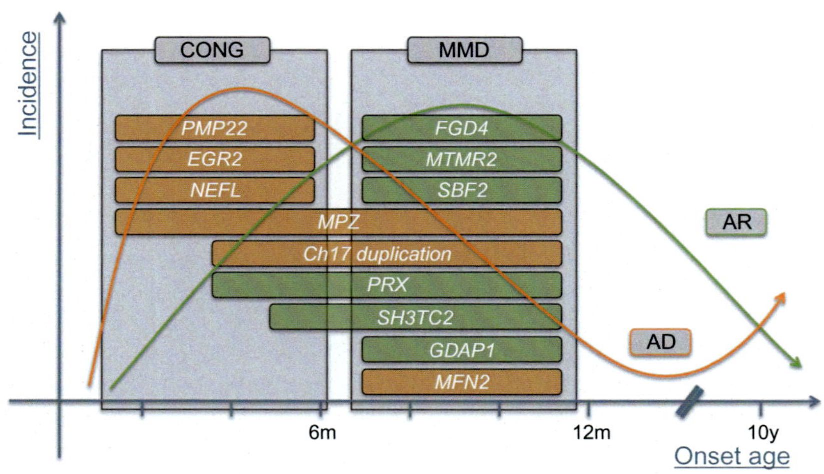

FIGURE 16.2 Gene distribution for patients with infantile-onset neuropathies with a known pathogenic mutation as function of disease onset, inheritance pattern, and phenotypic subgroup. The x-axis represents the age at onset in months and years, the y-axis displays the estimated incidence of patients with dominant (red line) and recessive subtypes (green line) of hereditary neuropathy. CONG, congenital onset phenotype characterized by neonatal hypotonia +/− breathing and feeding difficulties; MMD, motor milestone delay phenotype with or without early foot deformities and progressive delay in motor milestones within the first year of life; m, month; y; year. *Reprinted from reference 2 with permission from Oxford University Press.*

areflexia are more relevant clinical clues at this age. Proximal weakness, foot deformity, scoliosis and congenital hip dysplasia may be apparent, and the clinical features may mimic myopathy. Facial weakness and tongue fasciculations may be present, reflecting a severe neuropathic process. Recessive mutations in *FGD4*, *PRX*, *MTMR2*, *SBF2*, *GDAP1*, and *SH3TC2*, and dominant mutations in *MFN2*, are strongly represented in this subgroup[2] (Figure 16.2).

Diagnostic Approach

It is important to determine: (i) whether the neuropathy is primarily demyelinating/hypomyelinating or axonal; (ii) whether the infant has an isolated peripheral neuropathy or neuropathy in association with a central nervous system (CNS) disorder; (iii) the inheritance pattern; and (iv) if there are any distinguishing features on clinical, neurophysiologic, or neuropathologic examination.[3]

TABLE 16.1 Clinical Clues to the Diagnosis of Infantile Demyelinating Neuropathies

Clinical Feature	Clinical Phenotype	Gene
Pupillary abnormalities	CHN, DSD, CMT1B CMT4C	*MPZ* *SH3TC2*
Multiple cranial neuropathies	CHN, DSD, CMT4E	*EGR2*
Prominent foot and hand deformities	CMT4A, CMT2H/K CMT4C	*GDAP1* *SH3TC2*
Severe spinal deformities	CMT4C	*SH3TC2*
Asymmetric weakness	DSD, CMT4J GBS/CIDP	*FIG4* –
Post-traumatic rapid progression of weakness	DSD, CMT4J	*FIG4*
Vocal cord paresis	CMT4A, CMT2H/K CMT4B1	*GDAP1* *MTMR2*
Diaphragmatic involvement	CMT4A, CMT2H/K	*GDAP1*
Glaucoma	CMT4B2	*SBF2*
Prominent sensory involvement	DSD, CMT4F	*PRX*
Congenital cataracts	HCC CCFDN	*FAM126A* *CTDP1*
Raised creatine kinase	Merosin-deficient CMD	*LAMA2*

Abbreviations: CCFDN, congenital cataracts facial dysmorphism neuropathy syndrome; CHN, congenital hypomyelinating neuropathy; CMD, congenital muscular dystrophy; CMT, Charcot-Marie-Tooth disease; DSD, Dejerine-Sottas disease; GBS/CIDP, Guillain-Barré syndrome/chronic inflammatory demyelinating polyneuropathy; HCC, hypomyelination and congenital cataract.
Modified from Reference 3, with permission from Wiley.

The demyelinating and axonal infantile neuropathies will be presented separately in this chapter. Conditions with isolated peripheral nervous system involvement, versus those with associated CNS involvement, will be distinguished. Most of the isolated peripheral neuropathies in this age group have both motor and sensory involvement, and fall under the umbrella of CMT/HMSN, which is outlined in Chapters 15 and 17. The hereditary motor neuropathies overlap with both CMT/HMSN and the motor neuronopathies (such as SMA), and will be mentioned only briefly as they are covered in detail in Chapter 8. The hereditary sensory and autonomic neuropathies (HSAN), with preferential involvement of the sensory and autonomic systems, are discussed in Chapter 18. In general, the axonal neuropathies are less well characterized and more often associated with CNS or systemic manifestations, which can provide helpful diagnostic clues. Distinguishing clinical features of the congenital and infantile demyelinating and axonal neuropathies are presented in Tables 16.1 and 16.2.

Nerve conduction studies are central to determining whether a neuropathy is primarily demyelinating or axonal. As discussed in Chapter 3, nerve conduction velocities increase during the first 2 to 3 years of life, so normative values should be referenced.[4,5] As nerve conduction studies, particularly sensory studies, can be technically challenging in infants and neonates, absent sensory responses should be interpreted with caution. Infants with severe neuropathies may have unobtainable motor and sensory responses, making it difficult to discriminate between a primarily demyelinating or axonal process. Testing of more proximal nerves (such as the phrenic nerve) can be helpful in this instance, while needle electromyography (EMG) helps to distinguish myopathies from neuropathies. Conduction slowing may be so severe in some demyelinating neuropathies that the time base for nerve conduction testing needs to be adjusted to identify very delayed responses.

It is important to appreciate that the distinction between axonal neuropathies and neuronopathies is not always clear.[6] Involvement of both peripheral motor and sensory nerves is described in neuronopathies such as classic SMA and spinal muscular atrophy with respiratory distress type 1, while mutations in genes such as *TRPV4* and *DYNC1H1* can cause both an SMA phenotype and axonal CMT. Motor responses may be unobtainable, and reliance on absent sensory responses to localize pathology to the nerve rather than anterior horn cell (as may be done in adults or older children) is not always appropriate. Combined with the inability of electromyography to differentiate between axonal motor neuropathies and neuronopathies, this means that it can be difficult to distinguish

TABLE 16.2 Clinical Clues to the Diagnosis of Infantile Axonal Neuropathies

Clinical Feature	Clinical Phenotype	Gene	Inheritance
Optic atrophy	CMT2A *GDAP1*-related neuropathies IOSCA Infantile neuroaxonal dystrophy Riboflavin transporter deficiency	*MFN2* *GDAP1* *C10orf2* *PLA2G6* *SLC52A2, SLC52A3*	AD, AR AR AR AR
Ophthalmoparesis	Mitochondrial disorders	*SCO2* *C10orf2* *TK2*	AR AR AR
Skeletal abnormalities	CMT2C, SPSMA, congenital dSMA	*TRPV4*	AD
Congenital contractures/ arthrogryposis	Congenital dSMA SMARD1 X-linked SMA Pontocerebellar hypoplasia type I SMA with congenital bone fractures	*TRPV4* *IGHMBP2* *UBE1* *EXOSC3,* *VRK1,* *TSEN54,* *RARS2* Unknown	AD AR X-linked AR Unknown, presumed AR
Congenital fractures	X-linked SMA SMA with congenital fractures	*UBE1* Unknown	X-linked Unknown, presumed AR
Vocal cord paresis	CMT2A CMT2C , SPSMA, congenital dSMA *GDAP1*-related neuropathies Riboflavin transporter deficiency	*MFN2* *TRPV4* *GDAP1* *SLC52A2, SLC52A3*	AD, AR AD AR AR
Early infantile respiratory failure	SMA1 SMARD1 X-linked SMA Pontocerebellar hypoplasia type I SMA with congenital fractures Lethal neonatal AR axonal sensorimotor polyneuropathy Congenital axonal neuropathy with encephalopathy	*SMN1* *IGHMBP2* *UBE1* *EXOSC3,* *VRK1, TSEN54,* *RARS2* Unknown Unknown Unknown	AR AR AR AR Unknown, presumed AR AR Unknown, presumed AR
Predominant motor involvement	Congenital dSMA, SPSMA SMA1 CMT2O / SMA-LED X-linked SMA Pontocerebellar hypoplasia type I SMA with congenital fractures Mitochondrial disorders	*TRPV4* *SMN1* *DYNC1H1* *UBE1* *EXOSC3,* *VRK1, TSEN54,* *RARS2* Unknown *SCO2, TK2*	AD AR AD X-linked AR Unknown, presumed AR AR
Kinky hair	Giant axonal neuropathy	*GAN*	AR
Hepatopathy	Mitochondrial disorders MTP/LCHAD deficiency	*DGUOK* *C10orf2* *HADHA/HADHB*	AR AR AR

(*Continued*)

TABLE 16.2 (Continued)

Clinical Feature	Clinical Phenotype	Gene	Inheritance
Cardiomyopathy	Mitochondrial disorders	*SCO2* *TK2* *DGUOK*	AR AR AR
	MTP/LCHAD deficiency	*HADHA/ HADHB*	AR
CNS involvement	Pontocerebellar hypoplasia type I	*EXOSC3, VRK1, TSEN54, RARS2*	AR
	Giant axonal neuropathy	*GAN*	AR
	Infantile neuroaxonal dystrophy	*PLA2G6*	AR
	HMSN/ACC	*KCC3*	AR
	IOSCA	*C10orf2*	AR
	CMTX1	*GJB1*	X-linked
	CMT2O / SMA-LED	*DYNC1H1*	AD
	With encephalopathy		
	Congenital axonal neuropathy	Unknown	Unknown, presumed AR
	Mitochondrial disorders	*SCO2* *DGUOK* *TK2* *RMND1*	AR AR AR AR
	With cerebellar signs/involvement		
	Pontocerebellar hypoplasia type I	*EXOSC3, VRK1, TSEN54, RARS2*	AR
	Giant axonal neuropathy	*GAN*	AR
	Infantile neuroaxonal dystrophy	*PLA2G6*	AR
	IOSCA	*C10orf2*	AR

Abbreviations: AD, autosomal dominant; AR autosomal recessive; CMT, Charcot-Marie-Tooth disease; dSMA, distal spinal muscular atrophy; HMSN/ACC, hereditary motor and sensory neuropathy with agenesis of the corpus callosum; IOSCA, infantile-onset spinocerebellar ataxia; MTP/LCHAD, mitochondrial trifunctional protein/long-chain 3-hydroxyacyl-CoA dehydrogenase; SMA, spinal muscular atrophy; SMA-LED, spinal muscular atrophy with lower extremity predominance; SMARD, spinal muscular atrophy with respiratory distress type 1; SPSMA, scapuloperoneal spinal muscular atrophy.
Modified from reference 3 with permission from Wiley.

axonal sensorimotor neuropathies from motor neuronopathies in early childhood. Motor neuronopathies should therefore form part of the differential diagnosis of infants presenting with hypotonia, weakness, areflexia, and neurogenic electromyography findings (and vice versa).

Nerve biopsy can be very informative in determining the basis of prenatal and infantile neuropathies, particularly if neurophysiologic findings are unhelpful. Specific neuropathologic abnormalities may point to particular genetic syndromes. Distinguishing nerve biopsy features of the inherited demyelinating and axonal neuropathies are presented in Table 16.3. Acquired and potentially treatable causes of neuropathies such as infantile chronic inflammatory demyelinating polyneuropathy (CIDP), although rare, may mimic or occur superimposed on genetic neuropathies, and can be diagnosed on nerve biopsy.

Cerebral neuroimaging is not always required in the work-up of infants presenting with a purely peripheral process. However, magnetic resonance imaging (MRI) can sometimes provide important diagnostic clues, and is especially helpful in the investigation of syndromic disorders affecting the central nervous system.

Assessment and Management of Complications

Despite the genetic heterogeneity of infantile neuropathies, they tend to share typical management issues. Respiratory insufficiency may be related to diaphragmatic weakness (caused by involvement of the phrenic nerve), intercostal muscle weakness, or spinal deformity. Ventilatory support may be required from birth or later in the disease course, and assessment for respiratory insufficiency is warranted in most cases. Bulbar weakness may cause poor feeding and recurrent aspiration. Gastroesophageal reflux is probably more common than is recognized. Skeletal deformities including foot deformity, developmental hip dysplasia, and scoliosis are common and should be appropriately managed. Visual disturbance due to optic atrophy or cataracts occurs in some conditions, as does sensorineural hearing loss.

The long-term prognosis of individuals with early-onset demyelinating neuropathies is variable and not well documented. Outcomes are dependent not only on the underlying genetic mutation, but likely also environmental and

TABLE 16.3 Neuropathologic Clues to the Diagnosis of Inherited Infantile Neuropathies

Neuropathologic Feature	Clinical Phenotype	Gene
Demyelinating infantile neuropathies		
Focal myelin folding	*Frequently present* *MPZ*-related CHN/DSD CMT4B1 CMT4B2 CMT4B3 CMT4F CMT4H *Occasionally present* *EGR2*-related neuropathies *GDAP1*-related neuropathies CMT4C	*Frequently present* *MPZ* *MTMR2* *SBF2* *SBF1* *PRX* *FGD4* *Occasionally present* *EGR2* *GDAP1* *SH3TC2*
Uncompacted myelin	*MPZ*-related CHN/DSD	*MPZ*
Long Schwann cell cytoplasmic processes	CMT4C	*SH3TC2*
Node of Ranvier disorganization	CMT4C	*SH3TC2*
Paranodal abnormalities	CMT4F CMTX1	*PRX* *GJB1*
Axonal infantile neuropathies		
Giant axons	Giant axonal neuropathy CMT4C CMT2E / CMT1F Myofibrillar myopathy	*GAN* *SH3TC2* *NEFL* *BAG3*
Axonal spheroids	Infantile neuroaxonal dystrophy	*PLA2G6*
Mitochondrial abnormalities	CMT2A	*MFN2*

Abbreviations: CHN, congenital hypomyelinating neuropathy; CMT, Charcot-Marie-Tooth disease; DSD, Dejerine-Sottas disease.
Modified from Reference 3 with permission from Wiley.

epigenetic factors, given the inter- and intrafamilial variability often present. Individuals with significant respiratory involvement tend to have a poorer prognosis than those without respiratory compromise. Appropriate genetic counselling should always be provided.

GENETIC DEMYELINATING NEUROPATHIES WITH INFANTILE ONSET

Demyelinating Forms of CMT with Early Onset

Terminology

Congenital hypomyelinating neuropathy (CHN), Dejerine-Sottas disease (DSD, also referred to as CMT3 or HMSN III) and CMT4 (also referred to as autosomal recessive CMT1, ARCMT1) are all forms of demyelinating CMT with early onset. The nomenclature is very confusing, as the original descriptions and classification were made prior to the availability of genetic testing, and use of the terms (in particular, DSD and CHN) is inconsistent.[3] A single gene can cause multiple phenotypes of variable severity, and a single phenotype can be caused by mutations in more than one gene.

The features of DSD were defined by Dyck and Lambert[7] and then by Ouvrier et al.[8] as: recessive or sporadic inheritance, onset in infancy or the first two years of life, delayed motor development, extremely slow nerve conduction (motor nerve conduction velocity, MNCV $<$ 12 m/s), elevated cerebrospinal fluid (CSF) protein, and nerve biopsy features of a marked reduction in myelinated fiber density, thin myelin sheaths, and onion bulb formations. Most children are areflexic. Foot deformity is common.

Many children with DSD have *de novo* dominant mutations in *MPZ*,[9] *PMP22*,[10] or *EGR2*.[11] In addition, individuals with CMT4 (autosomal recessive CMT) sometimes have a DSD phenotype. Thus the term DSD is probably best used to describe the clinical syndrome of a severe demyelinating genetic neuropathy manifesting before two years of age with hypotonia or delayed motor development, with a median nerve MNCV of 12 m/s or less.[12,13]

CHN is rare and less well characterized than DSD. It is thought to reflect a defect in myelin synthesis rather than degeneration and/or demyelination of preexisting myelin. In its classic form, CHN is characterized pathologically by complete absence of peripheral nerve myelin without onion bulb formations, and the term *congenital amyelination* has also been used in these cases.[14] Some individuals with CHN have severe hypomyelination and/or unusual "basal lamina onion bulbs" consisting of concentric whorls of double-layered Schwann cell basement membrane.[15,16] In these "atypical" onion bulbs, Schwann cell processes are thought to have degenerated, leaving behind their basal laminae.[17] Evidence of myelin degeneration is absent. In comparison, nerve biopsies in DSD show evidence of both hypomyelination and demyelination-remyelination, classic onion bulb formation, and myelin breakdown products.[1,12]

Severe cases of CHN present at birth with profound weakness, hypotonia, arthrogryposis, and respiratory insufficiency.[15,18] Nerve conduction is very slow or responses are absent. Tongue fasciculations may be present.[19] Less severe cases present in infancy with hypotonia and motor delay.[16,20] Complete absence of peripheral nerve myelination does not always predict prenatal or neonatal onset, or early demise, and survival into adulthood is documented.[21,22] Hence the discrepancy between biopsy findings and clinical features, as well as overlap in genetic causation, makes the distinction between CHN and DSD of limited utility in the diagnosis and management of these patients, particularly in the era of genetic testing in which nerve biopsy may not be required. CHN and DSD are best thought of as a spectrum of early onset "myelinopathies." Variable mutations (sometimes even identical mutations) in *MPZ*,[23] *PMP22*,[10,24] and *EGR2*[11,25] can cause both phenotypes. It may be more appropriate to refer to these early onset disorders as infantile- or prenatal-onset neuropathies associated with a particular genetic cause if known. See Case Example 16.1.

A summary of the clinical and neurophysiologic phenotypes associated with the demyelinating CMT genes is presented in Table 16.4. Figure 16.3 illustrates the nerve conduction velocities and inheritance patterns associated with these genes.

Early Onset Neuropathies Due to *PMP22* Mutations

Heterozygous duplications involving *PMP22* cause CMT1A, the most frequent type of CMT,[26,27] while the reciprocal deletion causes hereditary neuropathy with pressure palsies (HNPP).[28] In the context of infantile neuropathies, heterozygous (often *de novo*) point mutations in *PMP22* are associated with DSD[10,29] and, less commonly, CHN.[24,30] *PMP22* duplications may also manifest in infancy.[2]

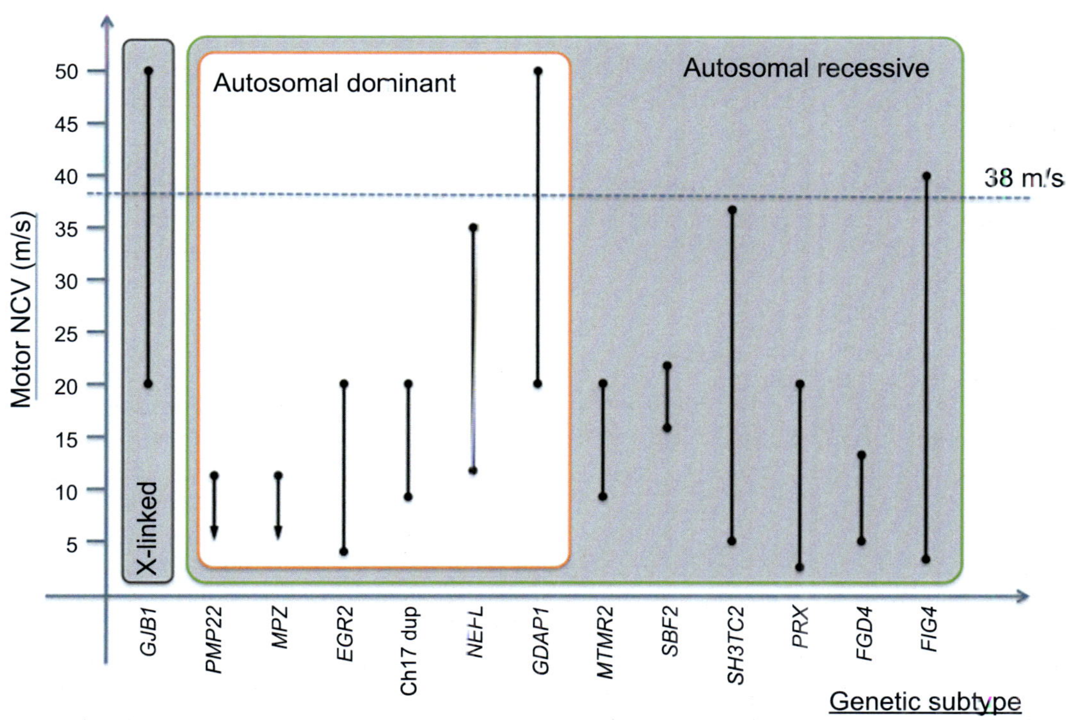

FIGURE 16.3 Motor nerve conduction velocities in demyelinating and intermediate forms of early-onset CMT. The x-axis represents the causal genes, and the y-axis displays the motor nerve conduction velocities in m/s. NCV, nerve conduction velocity; m/s, meters per second.

CASE EXAMPLE 16.1

This male infant was born at term via normal delivery after a pregnancy complicated by decreased fetal movements. He had early respiratory distress requiring ventilatory support. He was severely hypotonic, had a marked reduction in spontaneous movement, and was areflexic. Nerve conduction studies were consistent with a severe demyelinating neuropathy with a median nerve MNCV of 10 m/s. Sensory responses were unobtainable. Nerve biopsy showed a moderate reduction in myelinated fibers, and thinned/incomplete myelination of existing myelinated fibers. Electron microscopy showed generalized uniform thinning of the myelin sheaths in intermediate and large fibers (Figure 16.4). The patient died from respiratory failure at 6 weeks of age. Genetic testing for mutations in *MPZ*, *PMP22*, and *EGR2* was negative. While this infant had a severe congenital phenotype, myelin was still present, illustrating the discrepancy sometimes seen between clinical severity and degree of hypomyelination.

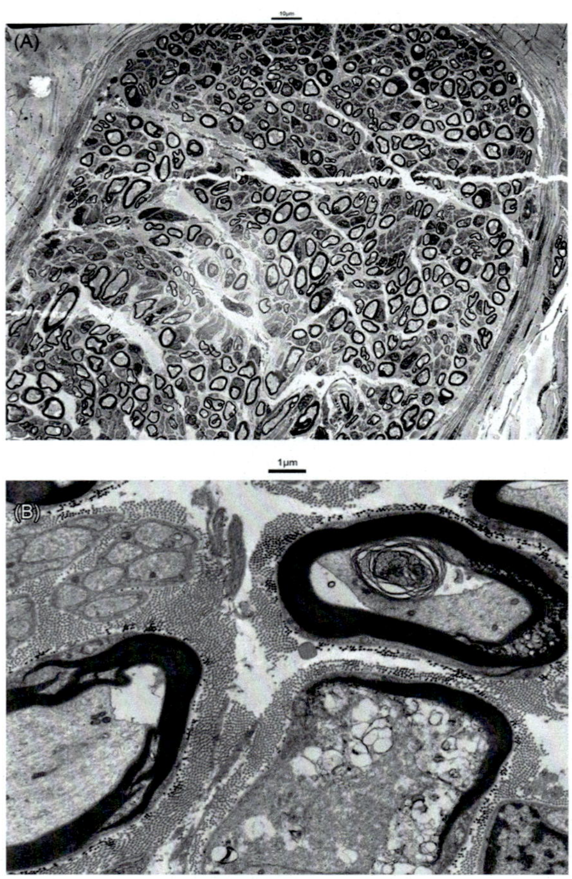

FIGURE 16.4 Sural nerve biopsy. Electron micrographs at low (**A**) and high (**B**) power show generalized uniform thinning of the myelin sheaths in intermediate and large fibers. Focal myelin breakdown is also seen in (**B**). Bar = 10 µm in (**A**) and bar = 1 µm in (**B**). *Reprinted from reference 3 with permission from Wiley.*

Almost all children with DSD/CHN due to *PMP22* point mutations have delayed development; many do not walk until 3–4 years of age, and some only with assistance.[31,32] A small number never achieve independent ambulation.[29,32] Increasing weakness may result in wheelchair dependency in later life.[12,33] Scoliosis and restrictive lung disease may occur[12]; rarely respiratory involvement leads to early death.[33] Other features reported with *PMP22* mutations include sensorineural hearing loss,[34,35] facial weakness,[31] and rarely nystagmus,[35] vestibular dysfunction,[36] ptosis, limited extraocular movements,[37] and bladder symptoms.[32]

On nerve conduction studies MNCV is almost always <10–12 m/s.[12,24] Sensory responses are often absent and compound muscle action potentials (CMAPs) are usually low amplitude, reflecting secondary axonal loss. Findings on nerve biopsy are usually typical of DSD with thin myelin sheaths, myelinated fiber loss, increased total transverse fascicular area (due to increased collagen fibers), and prominent onion bulb formations, composed of thin Schwann cell lamellae or basal membranes or both (Figure 16.5). In those cases designated as CHN, hypomyelination is more marked.[12,24]

Less commonly recessive *PMP22* mutations (homozygous point mutations[38] or compound heterozygous deletions[39,40]) cause classic DSD. Hence in children with a severe phenotype carrying a *PMP22* deletion, analysis for mutations in the other *PMP22* allele should be performed.

Heterozygous *PMP22* duplications (usually associated with a classic CMT1 phenotype) can also cause an early-onset phenotype, accounting for 7% of all infantile neuropathies in the series by Baets et al., compared to 1% for *PMP22* point mutations (Figure 16.1), although the MNCV may lie in the CMT1A rather than DSD range.[2,12] Congenital foot deformities in the presence of otherwise normal early development are common in these infants.[2] Nerve biopsies have a lower g-ratio (thicker myelin sheaths) compared to *PMP22* point mutations.[12] Homozygous duplications of *PMP22*, albeit rare, can also cause DSD,[26,41,42] although curiously, the homozygous state is not always associated with greater clinical severity or additional slowing of motor nerve conduction.[43] See Case Example 16.2.

PMP22, a 22-kDa glycoprotein, accounts for 2–5% of peripheral nerve compact myelin,[44,45] and may play a role in proliferation and differentiation of Schwann cells. Missense mutations within the transmembrane domains (where the majority of point mutations lie) are thought to result in toxic gain-of-function effects including impaired intracellular protein trafficking, accumulation of mutant PMP22 in the endoplasmic reticulum (ER), inhibition of post-endoplasmic reticulum protein degradation, and cell apoptosis.[46,47]

TABLE 16.4 Clinical Features of Demyelinating Forms of CMT with Early Onset

Gene	Clinical Phenotypes in Infancy	Inheritance	MNCV (m/s)	Other Features	Other Clinical Phenotypes
PMP22 (point mutation)	CHN, DSD	AD, AR	<12	SNHL, facial weakness, nystagmus, vestibular dysfunction	CMT1A, HNPP
PMP22 (duplication)	DSD	AD AR	9–20 10–17		CMT1A
PMP22 (deletion)	DSD	AR[a]	<10	SNHL	HNPP
MPZ	CHN, DSD	AD, AR	<12	Pupillary abnormalities, SNHL, facial weakness	CMT1B, Intermediate CMT, CMT2
EGR2	CHN, DSD, CMT4E	AD, AR	3–20	Cranial neuropathies frequent	CMT1D
NEFL	DSD, CMT1F, CMT2E	AD, AR	12–35	Intellectual disability, pyramidal signs, SNHL	CMT1F, CMT2E
GJB1	CMTX1	X-linked	20–50	SNHL, tremor, transient CNS disturbances	CMTX1
GDAP1	CMT4A, CMT2H, CMT2K (AR)	AR	20–60	Vocal cord paresis and diaphragmatic involvement. Severe distal weakness, prominent foot deformity and claw hands	CMT2K (AD)
MTMR2	CMT4B1	AR	9–20	Facial weakness	
SBF2/ MTMR13	CMT4B2	AR	16–21	Congenital or juvenile-onset glaucoma	
SH3TC2/ KIAA1985	CMT4C	AR	4–37	Early-onset spinal deformities. Foot deformities, SNHL, abnormal pupillary responses	
PRX	DSD, CMT4F	AR	2–20	Prominent sensory involvement	
FGD4	DSD, CMT4H	AR	5–13	Slow progression	
FIG4	DSD, CMT4J	AR	3–41	Asymmetric weakness, rapid progression following trauma	

[a]*Only compound heterozygous deletions associated with infantile neuropathies.*
Abbreviations: AD, autosomal dominant; AR, autosomal recessive; CHN, congenital hypomyelinating neuropathy; CMT, Charcot-Marie-Tooth disease; CNS, central nervous system; DSD, Dejerine-Sottas disease; HNPP, hereditary neuropathy with liability to pressure palsies; MNCV, motor nerve conduction velocity in meters/second; SNHL, sensorineural hearing loss.
Reproduced from Reference 3 with permission from Wiley.

Early Onset Neuropathies Due to *MPZ* Mutations

Clinical phenotypes associated with *MPZ* mutations include CHN/DSD, CMT1B, intermediate CMT, and a CMT2-like phenotype. Most subjects fall into one of two phenotypes—an early onset demyelinating neuropathy, or a later onset neuropathy with axonal features. Mutations are generally phenotype-specific.[48,49] *MPZ* mutations accounted for 6% of all infantile neuropathies in a large series (Figure 16.1).[2]

Children with early-onset demyelinating *MPZ* neuropathies usually have a DSD phenotype, presenting with hypotonia or delayed motor development. Most do not walk until three to four years of age, and some as late as ten years. Increasing weakness and scoliosis may develop and some become wheelchair-dependent.[48,49] A few cases designated as CHN have been reported[23,50], including two presenting at birth with hypotonia, arthrogryposis, respiratory failure, and almost no myelin on nerve biopsy.[22,51] Pupillary abnormalities may be seen with both early- and late-onset *MPZ* mutations, providing an important diagnostic clue.[12] Facial weakness, sensorineural hearing loss, bulbar problems, and, rarely, ptosis with limited extraocular movements are reported.[52] While data on long-term outcomes are limited, most subjects survive into adulthood, albeit requiring mobility assistance.[48,53]

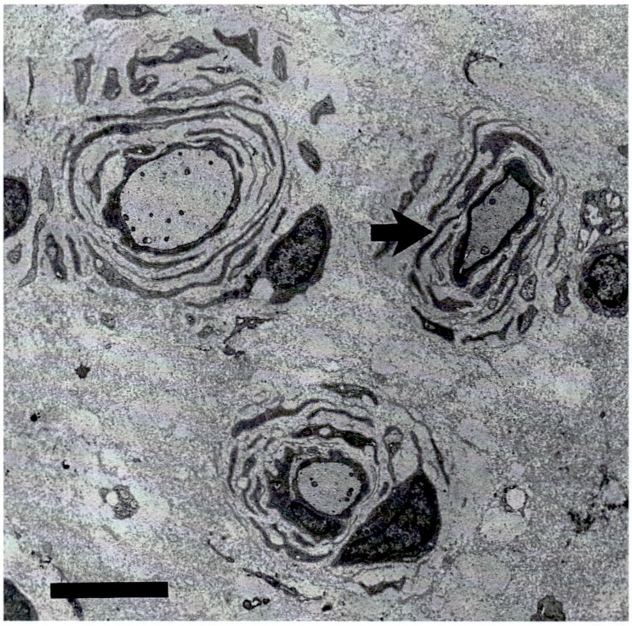

FIGURE 16.5 Sural nerve biopsy of 4½-year-old girl with a DSD phenotype due to a *PMP22* missense mutation. Large onion bulbs are seen around demyelinated and remyelinated (arrow) fibers. Marked increase in endoneurial collagen. Bar = 5 μm.

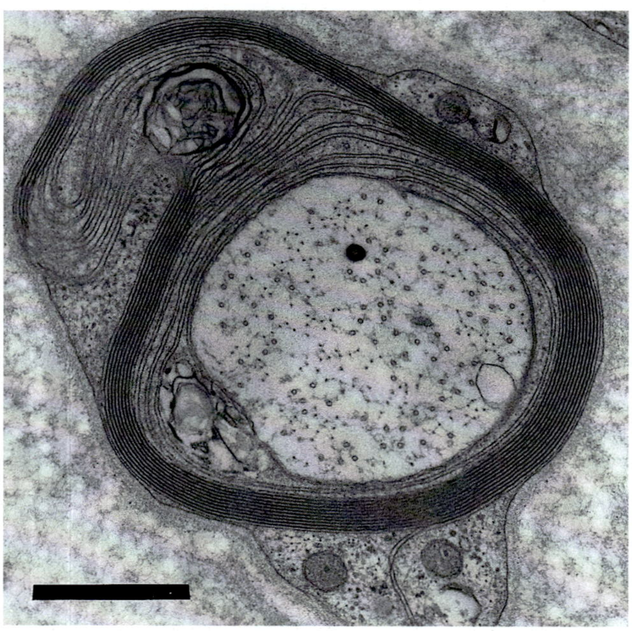

FIGURE 16.6 Uncompacted myelin lamellae in sural nerve biopsy of a 12-year-old girl with a missense mutation in the extracellular domain of *MPZ*. Bar = 1 μm.

Motor conduction velocities are usually less than 12 m/s. CMAP and SNAP (sensory nerve action potential) amplitudes are reduced (and sometimes unobtainable) in all patients.[48] Apart from the expected neuropathologic findings of myelinated fiber loss, myelin thinning (or absent myelin in some cases of CHN), and onion bulb formations, two distinct and mutually exclusive ultrastructural abnormalities have been described in some cases: focally folded myelin sheaths, or areas of uncompacted myelin, reflecting the role of MPZ in myelin adhesion and compaction[48,54] (Figures 16.6 and 16.7).

CASE EXAMPLE 16.2

A 2-year-old girl was born at term with bilateral foot deformities and mild generalized hypotonia. She had delayed motor development and walked at 19 months of age. At 2 years of age she had bilateral pes cavus, normal muscle bulk, hypotonia of the lower limbs, a steppage gait, and absent ankle reflexes. On nerve conduction studies her median nerve MNCV was 17 m/s. Chromosomal microarray testing revealed the typical 17p11.2 duplication seen in CMT1A. Testing of both parents indicated her father was also affected. He did not complain of any neurologic symptoms but on examination had bilateral pes cavus and absent ankle reflexes.

Comment

While most individuals with a *PMP22* duplication present during childhood or adolescence, some present during infancy with congenital/early onset foot deformities and/or motor delay. A tendency to toe-walk may be a feature. On nerve conduction studies the MNCV tends to lie in the CMT1A rather than DSD range. In infants, lower limb wasting may not be evident, or it may be obscured by subcutaneous fat. The CMT1A duplication should be considered in any child with an inherited demyelinating neuropathy, even in those presenting during infancy.

While the overwhelming majority of early onset neuropathies associated with MPZ are due to dominant heterozygous (often *de novo*) mutations, recessive homozygous or compound heterozygous mutations are occasionally identified. In these cases, heterozygous parents have a milder neuropathy, with MCNVs ranging from 29 to 54 m/s.[23,55]

MPZ is a member of the immunoglobulin gene superfamily. It functions as a homophilic adhesion molecule, playing an important role in myelin adhesion and compaction. Most mutations causing early-onset demyelinating neuropathies are in the extracellular domain and probably alter the MPZ tertiary structure and thus myelination.[49] A dominant-negative effect of mutant MPZ at the plasma membrane has been demonstrated, as has impaired mutant protein trafficking with ER accumulation, which can induce apoptosis and protein degradation.[56–58] See Case Example 16.3.

Early Onset Neuropathies Due to *EGR2* Mutations

Mutations in *EGR2* account for less than 1% of CMT cases overall[59] and 1% of infantile neuropathies.[2] Approximately

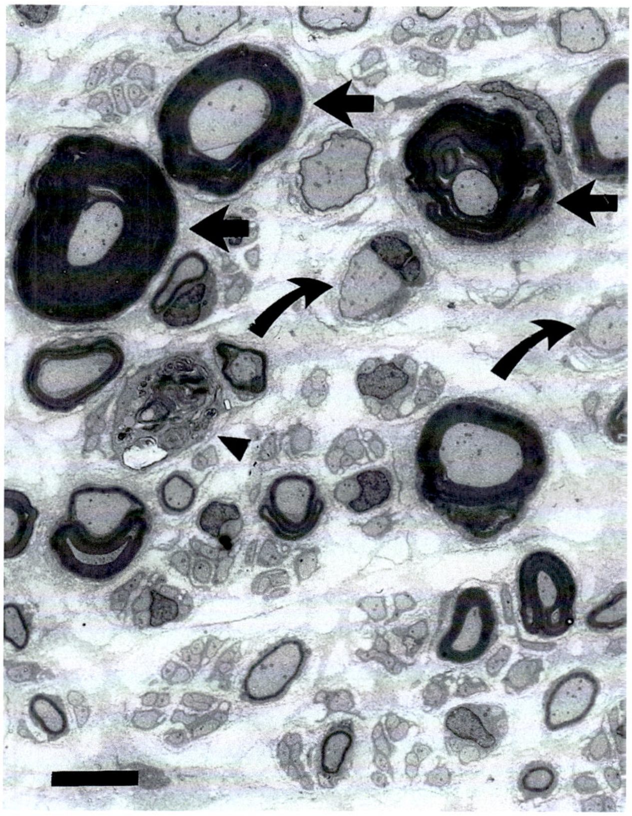

FIGURE 16.7 Sural nerve biopsy of a 3-year-old boy with a DSD phenotype due to a missense mutation in the extracellular domain *MPZ*. Excessive myelin foldings are present around several axons (arrows). Many demyelinated or early remyelinated fibers (curved arrows) are also present. One degenerating fiber (arrowhead). Bar = 5 μm.

11 cases of CHN/DSD have been reported, the majority due to *de novo* heterozygous missense mutations.[11,25,60–63] Recessive inheritance was identified in one family with three affected siblings (designated CMT4E),[11] and a clinical phenotype similar to the heterozygous cases. CMT1D is also associated with *EGR2* mutations.[11,64]

Most cases present in the first few months of life with hypotonia and delayed motor development.[11,25] Peculiar to *EGR2*-related neuropathies is a high frequency of cranial nerve involvement.[63,65] Ptosis,[11,63] eye movement abnormalities,[61,63] tongue fasciculations,[11] and facial weakness[25,63] are described. The course is slowly progressive in most cases. Motor conduction velocities are usually <12 m/s[11,60] but may be as high as 20 m/s.[63] Nerve biopsies show a range of hypomyelination (from almost complete myelin absence to thinly myelinated axons), as well as myelinated fiber loss, demyelination/remyelination, and onion bulb formations.[11,60] Focally folded myelin sheaths were described in one patient.[60]

EGR2 is a zinc finger transcription factor which binds regulatory domains of target genes, including those critical for myelin formation and maintenance such as *MPZ*, *PMP22*, *GJB1*, and *PRX*.[66–68] Most mutations lie in the zinc finger domains, likely affecting DNA binding and transcription of these genes.[69]

Early Onset Neuropathies Due to *NEFL* Mutations

Dominant mutations in neurofilament light chain polypeptide (*NEFL*) cause CMT1F and CMT2E.[70–72] While most

CASE EXAMPLE 16.3

An 11-month-old infant was referred with a 5-month history of generalized hypotonia and decreased movement of the left upper limb. Over the following 6 months he developed proximally predominant weakness of the other extremities. Language and social development were normal. He had generalized hypotonia but was able to sit without support when placed. He had no movement around the left shoulder girdle, reduced power in the lower limbs, and was areflexic. Nerve conduction studies indicated a sensorimotor demyelinating neuropathy, with MNCVs of 5–12 m/s, marked increase in distal motor latencies, and absent sensory responses. The CMAPs were of very low amplitude. The CSF protein was mildly elevated (0.48 g/L).

Sural nerve biopsy showed severe depletion of large myelinated fibers, with focal thickening of myelin sheaths. No onion bulbs or inflammatory cell infiltrates were seen. Numerous tomacula-like structures were identified on the teased fiber preparation. Focally folded myelin sheaths were noted on electron microscopy. Genetic testing revealed a *de novo* heterozygous Lys130Arg mutation in the *MPZ* gene. At 6 years of age he can ambulate short distances with a walker but otherwise requires a wheelchair for mobility. He has scoliosis and restrictive lung disease and has had a number of respiratory infections.

Comment

The initial presentation with an upper limb monoplegia in this case is unusual for an inherited neuropathy, and a diagnosis of infantile CIDP was considered in the first instance. A mildly elevated CSF protein is not specific for inflammatory neuropathies, as this finding may also be present in inherited neuropathies. The severe generalized slowing of nerve conduction and absence of inflammatory infiltrates on the nerve biopsy were more consistent with an inherited neuropathy, and the presence of focally folded myelin sheaths on electron microscopy provided an important diagnostic clue and directed sequencing of the *MPZ* gene. Excessive myelin outfoldings are seen in demyelinating neuropathies due to mutations in the *MPZ*, *SBF1*, *SBF2*, *MTMR2*, *PRX*, and *FGD4* genes[3] (Table 16.3). As the most common causes of infantile-onset demyelinating neuropathies are mutations in the *MPZ*, *PMP22*, and *SH3TC2* genes,[2] the *MPZ* gene was sequenced in the first instance.

individuals with *NEFL* mutations become symptomatic in childhood or adolescence, a number of severe early-onset cases have been caused by dominant, or occasionally recessive, mutations.[72–74] *NEFL* mutations accounted for 1% of infantile neuropathies in one series.[2] Affected children present before the age of two years with hypotonia, delayed motor development, or foot deformity. Distal weakness and wasting can be severe, and proximal weakness may be present. Additional features in early-onset patients include intellectual disability, pyramidal signs, ataxia, tremor, and hearing loss.[72–76]

Neurophysiologic findings are variable in *NEFL*-related neuropathies. Mixed axonal and demyelinating features may be present. In those with the early-onset phenotype, MNCVs are between 12 and 35 m/s. CMAPs are usually moderately to severely reduced.[72,74,75]

NEFL mutations probably do not cause a true "demyelinating" neuropathy, but rather an axonal neuropathy with secondary myelin loss. NEFL assembles into heteropolymers with heavy and medium neurofilaments, forming the major intermediate filament in axons. Neurofilaments determine axonal diameter and therefore conduction velocity.[77] *NEFL* mutations disrupt neurofilament assembly and axonal transport, resulting in decreased axonal diameter, contributing to slowed conduction.[78,79] On nerve biopsy the predominant finding is loss of large myelinated fibers with regenerating clusters.[72,73] Some individuals have giant axons containing accumulations of irregularly oriented neurofilaments.[75,80]

Early Onset Neuropathies Due to *GDAP1* Mutations

In infancy, recessive *GDAP1* mutations are associated with an early onset neuropathy with demyelinating and/or axonal features which has been designated CMT4A,[81] CMT2H/K,[82] and recessive "intermediate" CMT.[83] In reality these conditions probably lie on a pathologic spectrum. *GDAP1* mutations accounted for 1% of infantile neuropathy cases in Baets et al.[2]

The clinical phenotype associated with recessive *GDAP1* mutations is fairly homogeneous. Children present in the first year of life with prominent foot deformities, gait difficulties, and distal weakness.[84,85] Some are hypotonic at birth[86] but many have normal early motor development and walk by 2 years of age, although are often limited by foot deformity.[83,86,87] Progression is relatively rapid, leading to severe distal weakness and wasting of the feet and hands. Many patients develop "claw hands" within the first or second decade, and foot deformities are often severe.[82,83,86] Progressive proximal weakness renders many subjects wheelchair-dependent by the end of the second decade.[85,86] A more slowly progressive course is less common.[85] Vocal cord paresis, diaphragmatic dysfunction, and restrictive lung disease are common.[86] Facial weakness is described[88] and one patient had optic atrophy.[89]

Both axonal and demyelinating features are apparent. MNCVs range from 20 to 60 m/s, with MNCVs in the demyelinating,[84,90] axonal (most commonly),[82,86] and intermediate range in individual patients.[83] CMAPs are low amplitude or unrecordable.[86] On nerve biopsy some patients have a predominantly demyelinating process, with a decrease in the number of myelinated fibers, segmental demyelination, and frequent onion bulb formations.[81,90] Focal myelin folding is occasionally noted.[88] More commonly, a primarily axonal process is present, with axonal degeneration, regeneration, and pseudo-onion bulbs.[86,87] In some, the picture is mixed.[83]

The GDAP1 protein is expressed in axons, localized to the outer mitochondrial membrane,[91,92] and involved in mitochondrial fission and regulation of the mitochondrial network.[92,93]

Early Onset Neuropathies Due to *MTMR2* Mutations (CMT4B1)

Mutations in the *myotubularin-related protein 2* gene (*MTMR2*) are associated with 1% of infantile neuropathies,[2] and are an uncommon cause of recessive CMT (CMT4B1).[94] Focally folded myelin sheaths on nerve biopsy are characteristic. Redundant folds of myelin wrap irregularly around the axon and are seen as irregular focal thickenings on teased fiber analysis.[95,96] The thickenings are thought to be distinct from the tomacula described in HNPP, being smaller in diameter and more irregular in contour.[96,97] Focal myelin folding is not specific to CMT4B1; it has also been observed in neuropathies due to *SBF1*, *SBF2*, *MPZ*, *PRX*, and *FGD4* mutations (Table 16.3). Other biopsy features of CMT4B1 include a marked decrease in myelinated fiber density, evidence of demyelination-remyelination, and rare onion bulb formations.[95,96]

Onset is usually within the first 2 years of life.[95,98] Some patients have normal motor development during infancy.[95] Early complaints are of gait difficulties and proximal leg weakness. Progressive involvement of the upper and lower limbs occurs in the first and second decades, and foot deformities are sometimes severe. By adulthood many are wheelchair-dependent.[95,96] Bilateral facial weakness is common,[98] and vocal cord paresis and respiratory insufficiency may develop in adulthood.[95,96] MNCVs are in the demyelinating range (9–20 m/s) and CMAPs are usually low-amplitude and dispersed.[95,98]

The MTMR2 protein is a phospholipid phosphatase which dephosphorylates phosphoinositides (such as phosphatidylinositol 3-phosphate; PI(3)P), which are involved in membrane trafficking and homeostasis.[99] MTMR2 and MTMR13/SBF2 (a catalytically inactive pseudophosphatase associated with CMT4B2) form heterotetramers and function together[100,101]; loss of function probably leads to dysregulation of Schwann cell membrane homeostasis and abnormal myelin folding,[102,103] and explains the similar pathologic features of CMT4B1 and CMT4B2.

Early Onset Neuropathies Due to *SBF2* Mutations (CMT4B2)

An early-onset recessive demyelinating neuropathy with focally folded myelin can also be caused by mutations in *SET binding factor 2* (*SBF2*), also known as *myotubularin-related protein 13* (*MTMR13*).[104,105] Some kindreds also have early-onset glaucoma, which precedes onset of neuropathy in some individuals.[105–107] Symptom onset is in infancy or early childhood with delayed motor milestones. Anisocoria, facial weakness, and hypophonia have been described as additional features.[2] CMT4B2 accounted for 4% of infantile neuropathies in one series.[2] MNCVs range from 14 to 26 m/s.[2,105]

Neuropathies Due to *SBF1* Mutations (CMT4B3)

A single kindred with compound heterozygous mutations in the *SBF1* (*MTMR5*) gene was recently reported.[108] MTMR5 shares structural similarities to MTMR13/SBF2 and also interacts with MTMR2.[99,109] Not surprisingly, the clinical phenotype is similar to that of CMT4B1 and CMT4B2, with a demyelinating neuropathy and focally folded myelin sheaths. Onset in this family was in mid-childhood[108] and infantile cases have not yet been reported.

Early Onset Neuropathies Due to *SH3TC2* Mutations (CMT4C)

Mutations in *SH3TC2* (*KIAA1985*)[110] are a relatively frequent cause of CMT4, accounting for approximately 17% of all cases,[111] and 12% of infantile neuropathies (the highest frequency of all genes tested, Figure 16.1).[2] CMT4C is characterized by early-onset severe spinal deformities (affecting over 90% of individuals in some series), which are often the presenting symptom.[112,113] There is significant intra- and interfamilial variability. Symptoms usually begin in childhood or adolescence,[113] but some cases present with hypotonia and scoliosis in the first few months of life, or with delayed motor development.[2,110,114] Most children walk by 24–30 months of age,[2,110,114] although in rare instances ambulation is never achieved.[114,115] Foot deformity is common, often severe, and may precede onset of weakness.[112,113] Some have a normal neurologic examination at presentation despite prominent skeletal deformity (although nerve conduction tests are abnormal).[112] Hearing loss is the most common form of cranial nerve involvement,[113,116] but internal ophthalmoplegia,[117] facial weakness,[113] and tongue fasciculations[114,117] are also reported.

The clinical course is variable. In some, progression is slow and ambulation maintained past the fifth decade.[110,115] Others are wheelchair-dependent within the first decade.[116] The spinal deformities are also variably progressive, and surgery is required in some patients.[112,118] Some patients develop respiratory insufficiency.[111,112]

Nerve conduction studies confirm a demyelinating neuropathy, with highly variable MNCVs ranging from 4 to 37 m/s (mean 23 m/s).[110,113] Neuropathological features include loss of myelinated fibers, thinly myelinated fibers, basal lamina onion bulbs, and large cytoplasmic Schwann cell processes surrounding multiple unmyelinated axons; the latter are thought to be a distinguishing feature[110,118] (Figure 16.8). Abnormal organization of the

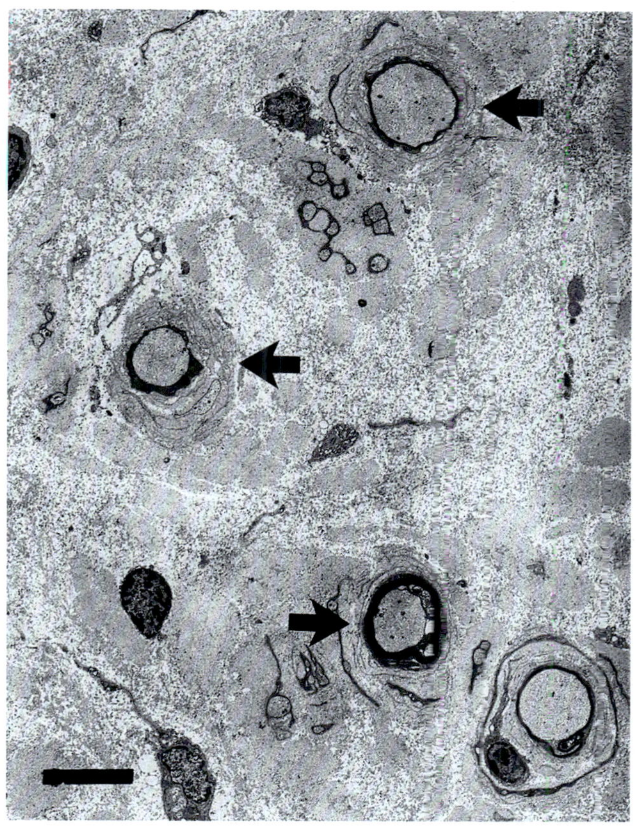

FIGURE 16.8 Sural nerve biopsy of an 8-year-old boy with an early onset demyelinating neuropathy with scoliosis, due to a recessive mutation in *SH3TC2*. Basal membranes are seen around some myelinated and nonmyelinated fibers (arrows). Bar = 5 μm.

CASE EXAMPLE 16.4

This child presented with delayed motor milestones. He walked at 2 years of age and had frequent falls and a steppage gait. On examination at 6 years of age he had bilateral pes cavus, moderate weakness and atrophy, walked with ankle-foot orthoses, and was areflexic. His median nerve MNCV was 25 m/s. Sural nerve biopsy revealed a hypertrophic demyelinating neuropathy with basal lamina membranes around myelinated and unmyelinated fibers. His 12-year-old brother had a severe progressive scoliosis. His neurologic and neurophysiologic examinations revealed a mild distal motor and sensory neuropathy. The younger brother subsequently also developed a rapidly progressive scoliosis at 10 years of age. Both required scoliosis surgery. Genetic testing later revealed a homozygous R954X nonsense mutation in the *SH3TC2* gene in both siblings.

Comment

The phenotype associated with recessively inherited *SH3TC2* mutations has a highly variable disease onset ranging from early infancy to adolescence. A striking and often pathognomonic feature is early and very pronounced scoliosis, often requiring surgery. Although scoliosis is common in severe early-onset neuropathies, in *SH3TC2*-related cases the scoliosis is often disproportionally severe and can manifest before other signs of neuropathy develop.

node of Ranvier,[119] and, rarely, giant distended axons[113] and focally folded myelinated fibers are seen.[117]

SH3TC, expressed exclusively in Schwann cells,[119] is thought to play a role in endosomal recycling and membrane trafficking processes required for normal myelin formation.[120,121] See Case Example 16.4.

Early Onset Neuropathies Due to *PRX* Mutations (CMT4F)

Periaxin (PRX)-related neuropathies are characterized by delayed motor development, prominent sensory involvement and extremely slow nerve conduction,[122–124] a phenotype consistent with DSD. The locus was first reported in a large inbred Lebanese family with a recessive demyelinating CMT designated CMT4F.[122] All cases demonstrate recessive inheritance.[123,125] In one cohort of 20 unrelated DSD patients, three had *PRX* mutations (four had *MPZ*, three *PMP22*, and two *EGR2* mutations),[123] and *PRX* mutations accounted for 5% of infantile neuropathies in another series.[2]

Most cases present in infancy with delayed motor development. Affected children may not walk until four to five years of age.[124,126] The neuropathy is slowly progressive and weakness generally limited to distal muscles. Ambulation is maintained into late adulthood.[122,124] Sensory involvement, including sensory ataxia, is common and often out of proportion to the degree of motor involvement.[123,124] Some patients complain of positive sensory symptoms.[122,123] Scoliosis is common but usually mild.[126,127] Hearing loss and tongue fasciculations are reported in some patients.[126]

Conduction velocities are extremely low (and often undetectable even in childhood) with MNCV usually less than 10 m/s (range 2–20 m/s).[123,124] Temporal dispersion is marked in some.[126] SNAPs are usually absent.[124,126] Typical neuropathologic features include moderate to severe loss of myelinated fibers, and classic and basal lamina onion bulbs. Focally folded myelin may be seen.[125,126] Paranodal abnormalities have been reported in some patients.[126,127]

PRX encodes periaxin, a cytoskeleton-associated protein expressed exclusively in myelinating Schwann cells.[128,129] Periaxin helps link the basal lamina to the Schwann cell cytoskeleton, and is important in elongation of Schwann cells and myelination during development.[129,130] Periaxin-deficient mice display peripheral demyelination and, interestingly, prominent sensory involvement.[131]

Early Onset Neuropathies Due to *FGD4* Mutations (CMT4H)

Mutations in *FGD4* or *frabin* (FGD1-related F-actin binding protein) cause a recessive demyelinating CMT designated CMT4H,[132,133] and account for 3% of infantile neuropathies.[2] Age at onset can range from infancy to childhood. Most affected children walk by two to three years of age and clinical progression is slow. Scoliosis and foot deformities are common.[132,134] Nerve conduction velocities are markedly reduced, with MNCV ranging from 5 to 13 m/s.[134,135] On nerve biopsy decreased myelinated fiber density, thin myelin sheaths, demyelination-remyelination, and classic onion bulbs are seen. Focal irregular myelin foldings are prominent, similar to those seen in CMT4B1 and 4B2.[132,136]

FGD4 is a Rho GDP/GTP nucleotide exchange factor expressed in many tissues including peripheral nerve.[133,137] FGD4 induces changes in Schwann cell shape.[134] Pathologic similarities between CMT4H and CMT4B, and the presence of phosphoinositide binding domains in frabin, suggest that altered phosphoinositide metabolism may be a shared mechanism for abnormal myelination in these forms of CMT.[136]

Early Onset Neuropathies Due to *FIG4* Mutations (CMT4J)

Mutations in FIG4 cause CMT4J, a rare recessive demyelinating neuropathy with variable severity and age of onset ranging from early childhood to mid- to late-adulthood. Some early-onset cases were diagnosed with DSD; these children had delayed motor milestones or early gait abnormalities and very slow nerve conduction (<12 m/s). CMAP amplitudes were markedly reduced or absent. One child had MNCVs of 29–41 m/s. Nerve biopsies in these early-onset cases reflect a primarily demyelinating process with extensive myelinated fiber loss, onion bulbs, and segmental demyelination on teased fiber analysis. Unusual features of CMT4J include prominent proximal weakness, asymmetric involvement, and sometimes rapid progression of muscle weakness precipitated by trauma. In those with early-onset disease, weakness is progressive and wheelchair dependency occurs by the third decade.[138–140]

FIG4 is a phosphoinositide phosphatase; CMT4J is thought to be a disorder of phosphoinositide signaling similar to CMT4B1, 4B2, 4B3, and 4H.[138,141]

Early Onset Neuropathies Associated with *GJB1* Mutations

Mutations in *GJB1* (*Gap junction protein beta-1*, or *connexin 32*) are associated with X-linked CMT (CMTX) type 1, which is classically associated with symptom onset in the first or second decade.[142] In one pediatric series however, delayed motor development in 38% of affected children reflected onset in infancy.[143] *GJB1* neuropathy is slowly progressive, and most patients remain ambulant.[144] Sensorineural hearing loss may occur.[143,144] While females generally have a milder course, onset in infancy or a severe phenotype can occur.[143,145] Nerve conduction velocities are in the "intermediate range," i.e. 25–45 m/s, and CMAP amplitudes are usually low.[143,146] Both demyelinating and axonal features are apparent on nerve biopsies.[143,147,148]

Other forms of CMTX occurring in infancy are associated with mental retardation (CMTX2),[149] hearing loss and cognitive deficits (CMTX4),[150] and distal weakness, foot deformities and significant conduction slowing.[143]

DEMYELINATING NEUROPATHIES OF INFANCY ASSOCIATED WITH CENTRAL NERVOUS SYSTEM INVOLVEMENT

These disorders can be divided into two broad groups: those with CNS hypomyelination (i.e. myelin-deficient), and those in which the white matter is abnormal (dysmyelinating).[151] In these disorders, CNS symptoms and signs may overshadow and mask the clinical signs of peripheral nerve involvement which, if detected, provide important clinical clues to the underlying diagnosis. Relevant clinical syndromes are presented in Table 16.5. As this chapter is focused on disorders in which neuropathy is the prominent feature, detailed descriptions of these conditions should be sought elsewhere. Disorders associated with CNS hypomyelination, and the congenital cataracts facial dysmorphism neuropathy syndrome (CCFDN), will be briefly discussed.

Metabolic conditions associated with peripheral nervous system involvement are discussed in Chapter 19. Some of these may present in infancy. Special mention should be made of metachromatic leukodystrophy (late-infantile form) and Krabbe disease, which are consistently associated with an early-onset demyelinating sensorimotor peripheral neuropathy.[152–154] Both can present in infancy with an apparently isolated neuropathy, prior to the development of other neurologic manifestations.[155–157] Conduction velocities are usually <30 m/s.[153,154] Mitochondrial disorders of infancy may also have peripheral nerve involvement which may be demyelinating or axonal (Tables 16.2 and 16.5).

Demyelinating Neuropathies of Infancy Associated with CNS Hypomyelination

Hypomyelination and Congenital Cataract

Hypomyelination and congenital cataract (HCC) is an autosomal recessive disorder characterized by congenital cataracts, demyelinating neuropathy, slowly progressive neurologic impairment (with spasticity and cerebellar signs), and mild to moderate intellectual disability.[158,159] Developmental delay is apparent in the first year. Most children walk with support by 24 months of age but lose ambulation by eight to nine years of age.[158–160] MNCVs range between 10 and 39 m/s.[159–161] Loss of myelinated fibers with thinly myelinated or amyelinated axons is seen on nerve biopsy; occasionally small onion bulb formations, uncompacted myelin, and irregularly folded myelin are also present.[159] Brain MRI shows diffuse supratentorial hypomyelination in all patients.[162]

HCC is caused by mutations in the *FAM126A* gene (previously called *DRCTNNB1A*) which encodes a neuronal membrane protein called hyccin, assumed to be essential for myelin initiation and/or maintenance.[158,163]

SOX10 Related Conditions

Heterozygous mutations in *SOX10* are associated with Waardenburg syndromes types 2 and 4 (hearing loss, pigmentary abnormalities, and in type 4, Hirschsprung disease),[164,165] as well as Kallmann syndrome with

TABLE 16.5 Infantile Demyelinating Neuropathies with CNS Involvement

Associated with CNS Hypomyelination	Gene	Other Features
Hypomyelination with congenital cataracts (HCC)	*FAM126A*	Congenital cataracts, pyramidal signs, cerebellar signs, intellectual disability
Peripheral demyelinating neuropathy, central dysmyelinating leukodystrophy, Waardenburg syndrome and Hirschsprung disease (PCWH)	*SOX10*	Waardenburg syndrome, Hirschsprung disease, spasticity, ataxia, dysautonomia, intellectual disability
Pelizaeus-Merzbacher disease	*PLP1*[a]	Early nystagmus and titubation, ataxia, spasticity, movement disorder, intellectual disability
Pelizaeus-Merzbacher-like disease	*GJA12*	Nystagmus, ataxia, delayed development
Cockayne syndrome	*ERCC6*, *ERCC8*	Growth failure, photosensitivity, retinopathy, progressive neurologic impairment
Associated with Abnormal CNS White Matter	**Gene**	**Other Features**
Metachromatic leukodystrophy	*ARSA*	Psychomotor regression, spasticity, seizures
Krabbe disease	*GALC*	Extreme irritability, spasticity, psychomotor regression
Niemann-Pick disease type C[b]	*NPC1*, *NPC2*	Hepatomegaly, vertical gaze palsy, progressive ataxia, dystonia, cataplexy
Merosin-deficient congenital muscular dystrophy	*LAMA2*	Proximal weakness, raised creatine kinase, muscular dystrophy
Navajo neurohepatopathy	*MPV17*	Liver disease, corneal scarring, recurrent metabolic acidosis, recurrent infections, failure to thrive
Associated with other CNS Involvement	**Gene**	**Other Features**
Congenital disorders of glycosylation	Multiple genes	Variable features
Congenital cataracts, facial dysmorphism and neuropathy (CCFDN)	*CTDP1*	Congenital cataracts, microretina, intellectual disability, facial dysmorphism, short stature, hypogonadism
POLG-related hepatocerebral mtDNA deletion syndromes	*POLG1*	Encephalopathy, refractory seizures, liver dysfunction
Leigh syndrome	Multiple genes	Psychomotor regression, brainstem and basal ganglia signs, raised lactate levels
Pyruvate dehydrogenase deficiency	*PDHA1*	Raised lactate levels, ophthalmoplegia, psychomotor regression

[a]*Peripheral neuropathy associated with PLP1 null mutations only.*
[b]*Peripheral neuropathy rarely seen in Niemann-Pick disease type C.*
Modified from Reference 3 with permission from Wiley.

deafness.[166] Some children with *SOX10* mutations have central and/or peripheral demyelination, a syndrome called "PCWH" (peripheral demyelinating neuropathy, central dysmyelinating leukodystrophy, Waardenburg syndrome, and Hirschsprung disease).[167,168] Severe cases present with hypotonia, weakness, arthrogryposis, and respiratory insufficiency in the first few months of life.[169,170] Milder cases have variable hypotonia, spasticity, ataxia, and growth retardation.[168,171]

Nerve conduction studies show MNCV ranging from 2 to 38 m/s.[168] Findings on nerve biopsy include absent peripheral myelin,[170] micropolyfasciculation with thin myelin sheaths,[171] and excessive myelin folding.[168] Cerebral MRI shows variable degrees of hypomyelination,[167,172] or dysmyelination.[168,172]

SOX10 is a transcription factor with a crucial role in neural crest development, glial cell development, and myelin formation and maintenance.[173,174]

Cockayne Syndrome

Cockayne syndrome (CS) is a recessively inherited multisystem disorder characterized by growth failure, cutaneous photosensitivity, progressive neurologic impairment, and CNS hypomyelination. Peripheral neuropathy is common. Other typical features include sensorineural hearing loss, pigmentary retinopathy, and a characteristic "cachectic dwarfism" appearance.[175,176] CS encompasses a spectrum of phenotypes. Types I (classic CS) and II (connatal CS) present in the first two years of life.[176] Overt signs of peripheral neuropathy are uncommon, but nerve conduction studies generally show a demyelinating neuropathy (MNCV 20–38 m/s).[175] Nerve biopsy findings include segmental demyelination and remyelination, with membrane-bound electron-dense inclusions in some Schwann cells.[177,178]

Congenital Cataracts Facial Dysmorphism Neuropathy Syndrome (CCFDN)

Congenital cataracts facial dysmorphism neuropathy syndrome (CCFDN) has, to date, been reported only in individuals of Balkan Gypsy ethnicity. CCFDN is caused by homozygous founder mutations in the *CTDP1* gene, which encodes the protein phosphatase FCP1 involved in the regulation of RNA polymerase II.[179]

This disorder is characterized by bilateral congenital cataracts, a progressive demyelinating/hypomyelinating neuropathy, other ophthalmologic abnormalities, mild intellectual disability, facial dysmorphism (evident from late childhood), short stature, and hypogonadotrophic hypogonadism.[180,181] Most children do not walk before two to three years. The neuropathy manifests early and is progressive, causing severe disability by the third decade. Skeletal deformities are common.[182,183]

Nerve conduction studies show conduction slowing (20–30 m/s) after 12 months of age. Sensory responses are often preserved.[182,183] Nerve biopsy shows thinly myelinated fibers with mildly reduced density but no onion bulb formations; in older patients superimposed demyelination/remyelination and axonal degeneration may be seen.[181,184]

GENETIC AXONAL NEUROPATHIES WITH INFANTILE ONSET

Axonal Forms of CMT with Early Onset

The CMT genes most often associated with early-onset axonal neuropathies are *MFN2*, *TRPV4*, *DYNC1H1*, *GDAP1*, and *NEFL*.

Severe Early Onset Axonal Neuropathy Due to *MFN2* Mutations

CMT2A, due to mutations in *mitofusin 2* (*MFN2*), includes two distinct phenotypes: an early-onset severe form with weakness beginning in early childhood and, less commonly, a later onset milder phenotype.[185–187] The severe early onset axonal phenotype was first described by Ouvrier et al.[188] and Gabreels-Festen et al.[189]; approximately 50% of cases in the original Ouvrier cohort were subsequently shown to carry *MFN2* mutations.[190] *MFN2* mutations accounted for 3% of infantile neuropathies in the series by Baets et al.[2]

> **CASE EXAMPLE 16.5**
>
> A 3-year-old boy presented with motor delay and pes planus. His early development was normal but he did not walk until 20 months of age, had never walked normally, and had frequent falls. On examination he had marked pes planus and absent reflexes. He had both a waddling and steppage gait, indicative of combined proximal and distal weakness. Nerve conduction studies showed a motor and sensory axonal neuropathy, with a median nerve MNCV of 47 m/s and absent sensory responses. His mother had had a severe motor and sensory axonal neuropathy since early childhood, with severe distal weakness and wasting. She had been wheelchair-bound since her late teens, and developed optic atrophy and vocal cord paresis in her early twenties. Testing of the *MFN2* gene revealed a heterozygous R364W mutation in both family members.

Early-onset *MFN2* neuropathy presents before 5 years of age, often in infancy, with distal weakness or gait abnormalities (see Case Example 16.5). While most children acquire the ability to walk, weakness is rapidly progressive, and wheelchair dependence is common by the end of the second decade. Severe weakness and atrophy below the knees and elbows, complete distal paralysis, and clawed hands are frequently found. Planus foot deformities are common. Early proximal weakness of the lower limbs is an important feature and may produce a characteristic waddling gait associated with foot drop in some children.[187,188,191]

Optic atrophy develops in 10–20% of individuals with *MFN2* mutations, and is particularly associated with the early-onset phenotype.[186,187,191] Visual deterioration is usually subacute, starting in late adolescence, but may occur in early childhood. Curiously, a number of patients experience some recovery of visual acuity.[192] In individuals with the early-onset phenotype, vocal cord paresis, respiratory insufficiency, and scoliosis may develop later in the disease course. Sensorineural hearing loss and tremor are also reported.[186,187] Neuroimaging abnormalities in individuals with *MFN2* mutations may include focal hyperintensities on T2-weighted sequences in the periventricular white matter, cerebellar peduncles, and thalami.[187,193]

Nerve conduction studies show an axonal motor and sensory neuropathy with MNCVs greater than 38 m/s. Motor and sensory responses are often unrecordable.[185–187] Characteristic features on nerve biopsy include severe loss of large myelinated fibers and

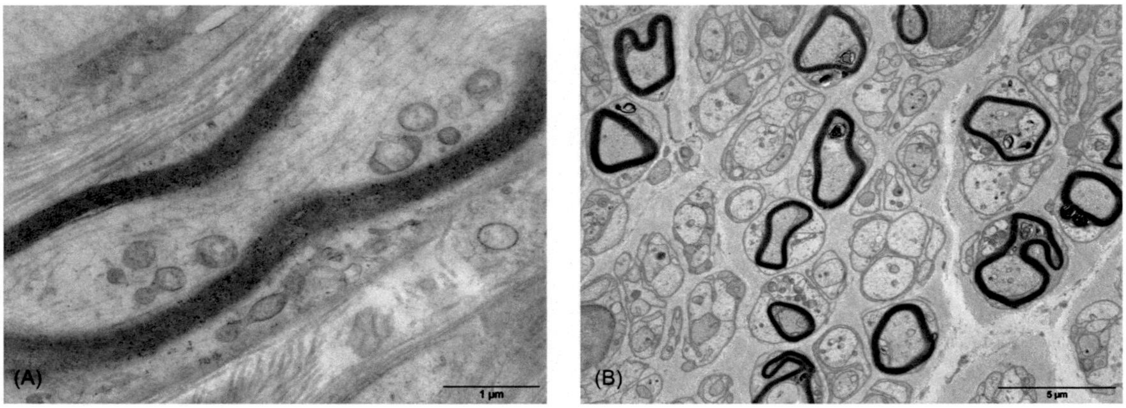

FIGURE 16.9 Sural nerve biopsy from a 2-year old boy with a heterozygous *MFN2* mutation. (**A**) On longitudinal sections of sural nerve, mitochondria are small, round, and abnormally aggregated. Bar = 1 μm. (**B**) Severe loss of myelinated fibers and moderate rarefaction of remaining myelinated fibers. Bar = 5 μm. *Reprinted from reference 3 with permission from Wiley.*

abnormal aggregation of small round mitochondria on electron microscopy[194] (Figure 16.9).

MFN2 is a transmembrane GTPase located in the outer mitochondrial membrane. It is important for mitochondrial fusion, mitochondrial DNA stability, axonal mitochondrial transport, and tethering of mitochondria to the endoplasmic reticulum.[195–198] Most *MFN2* mutations are dominant, with 30–40% being *de novo* mutations.[186,187] Compound heterozygous and homozygous mutations are also reported in individuals with the early-onset phenotype.[199] The phenotype tends to be concordant within families, and in kindreds with the same mutation, but marked intrafamilial variability has also been documented.[200]

Infantile Neuropathies Due to *TRPV4* Mutations

Dominant mutations in *TRPV4* (*transient receptor potential vanilloid 4*) are associated with a spectrum of neuromuscular disorders, including congenital distal spinal muscular atrophy (congenital dSMA),[201] scapuloperoneal spinal muscular atrophy (SPSMA),[201,202] and CMT2C.[202,203] Key features of these *TRPV4*-related disorders include a predominantly motor axonal neuropathy (with sensory involvement in CMT2C[204]), vocal cord paresis,[204,205] prominent contractures (particularly in congenital dSMA),[206] and scoliosis.[206,207] There is marked intrafamilial variability in age at onset, clinical phenotype, and severity.[201] *TRPV4* mutations are also associated with a number of skeletal dysplasias,[208–210] and individuals with both neuromuscular and skeletal abnormalities are reported.[205,206,211–213]

Infantile onset cases frequently have a congenital dSMA phenotype.[201,206] Affected infants may have congenital foot deformities, lower limb contractures, hip dysplasia, and scoliosis. Orthopaedic correction is often required in the first two years of life.[206,212,214] Weakness and wasting are more severe in the lower limbs, and upper limb sparing is described in some subjects.[206,212] Vocal cord paresis and respiratory insufficiency are less common in congenital dSMA than in CMT2C.[207] Reflexes are absent or diminished. Most patients can walk with crutches or assistance.[212] The weakness is strikingly nonprogressive, and survival into adulthood is usual.[206,214] Motor and sensory nerve conduction studies are normal, but EMG findings are neurogenic.[214,215] The underlying pathophysiology of congenital dSMA is poorly understood, and autopsy studies are lacking. Autopsy in an individual with *TRPV4*-related SPSMA showed a normal population of motor neurons in the spinal cord, suggesting that the primary pathologic process may be a motor neuropathy.[202]

SPSMA and CMT2C may also have infantile onset, often presenting with vocal cord paresis, which manifests as stridor or respiratory distress. Congenital contractures may be absent.[204,205,216] Diaphragmatic involvement may also contribute to respiratory insufficiency.[204,216,217] Nerve conduction studies show a motor and sensory axonal neuropathy. Findings on nerve biopsy include a moderate loss of myelinated fibers and clusters of regenerating axons.[204,216]

TRPV4 is a transmembrane Ca^{2+}-permeable nonselective cation channel responsive to a number of physical and chemical stimuli.[218,219] It is widely expressed in a number of tissues including cartilage, motor neurons, and dorsal root ganglia.[203,219]

Early Onset Axonal Neuropathies Due to *DYNC1H1* Mutations

Heterozygous mutations in *DYNC1H1* are associated with both an axonal CMT (designated CMT2O)[220] and dominant SMA with lower extremity predominance.[221,222] As with *TRPV4* mutations, both phenotypes may exist within the same family and there are overlapping features.[220] The phenotype is characterized by motor delay or childhood-onset gait difficulties, foot deformity (which may be congenital), lower limb predominance, and a slowly or minimally progressive course.[220–222] Other clinical features may vary. The initial CMT2O pedigree had length-dependent distal weakness with variable sensory involvement and hyporeflexia.[220] The SMA phenotype is characterized by proximally predominant leg weakness (with preferential involvement of the quadriceps), and absence of sensory involvement on neurophysiologic testing.[221,222] Interestingly, motor and sensory nerve conduction studies in the CMT2O pedigree were normal in many family members, with the EMG showing features of denervation. Sural nerve biopsy in one patient showed some axonal degeneration.[220]

Mutations in *DYNC1H1* also cause malformations of cortical development, and a number of individuals with features of both peripheral and central nervous system involvement have been reported.[223,224] *DYNC1H1* encodes cytoplasmic dynein heavy chain 1, a subunit of the main motor protein responsible for retrograde axonal transport in neurons.[225]

Early Onset Axonal Neuropathy Due to *TRIM2* Mutations

Compound heterozygous mutations in *TRIM2* were recently identified in a single individual with motor delay, hypotonia, progressive distal wasting and weakness, and poor weight gain. Nerve conduction studies indicated a primarily axonal motor and sensory neuropathy. Sural nerve biopsy showed reduced myelinated fiber density, axonal degeneration, and swollen large myelinated fibers, which on electron microscopy showed accumulation of neurofilaments.[226] Subsequent investigation of a large cohort of individuals with early-onset neuropathies for *TRIM2* mutations was negative, indicating this is a rare cause of this phenotype.[226]

Lethal Neonatal Autosomal Recessive Axonal Sensorimotor Polyneuropathy

A single report of an inbred kindred from Northern Mississippi described 13 infants with a severe sensorimotor axonal polyneuropathy causing death before eight months of age.[227] The genetic basis of this disorder is unknown. Prenatal onset was reflected by reduced fetal movement and polyhydramnios. Respiratory failure manifested soon after birth. Weakness was distally predominant, with early facial and bulbar involvement, and contractures were absent. Nerve conduction studies showed a severe sensorimotor axonal neuropathy. Neuropathologic findings included a severe loss of large- and medium-sized myelinated fibers, without features of demyelination. Motor neuron density in the spinal cord was normal.

Early Onset Axonal Neuropathies with CNS Involvement

Giant Axonal Neuropathy

Giant axonal neuropathy (GAN) is due to recessive mutations in the *gigaxonin* (*GAN*) gene,[228] and is characterized by a severe early-onset motor and sensory axonal neuropathy with CNS involvement and characteristic "kinky" hair.[229,230] Onset is usually within the first few years of life.[231,232]

Most children present with delayed development and gait difficulties.[230,232] In some, CNS signs (cerebellar, oculomotor, and upper motor neuron features) are apparent early; in others these features develop later or remain absent during the course of the disease.[232–234] Intellectual disability is common. In some families with confirmed *GAN* mutations kinky hair is absent.[235,236] Associated features include skeletal deformities,[232,234] and less commonly cranial neuropathies[233] and seizures.[237] The clinical course is progressive. Most affected children are wheelchair-dependent by the second decade.[232,236,238]

Nerve conduction studies generally show an axonal sensorimotor neuropathy.[232,234] On MRI periventricular and cerebellar white matter signal abnormalities and/or cerebral, cerebellar, and brainstem atrophy may be present.[232,233] The pathological hallmark of GAN is giant axons (large axonal swellings in thinly myelinated/amyelinated nerve fibers) filled with densely packed structurally normal bundles of neurofilaments (Figure 16.10).[229]

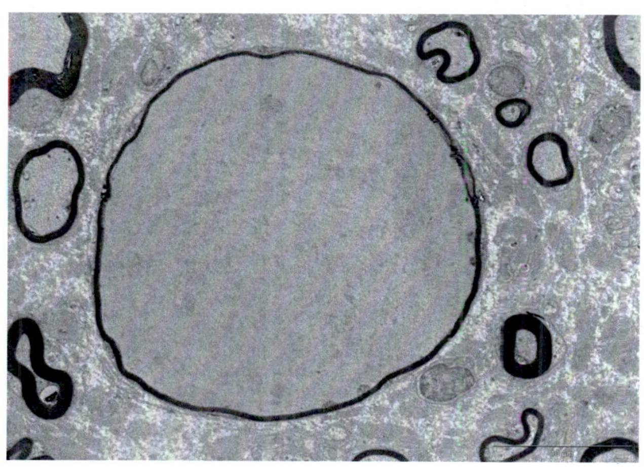

FIGURE 16.10 Sural nerve biopsy from a 6-year-old girl with giant axonal neuropathy. A giant, thinly myelinated and swollen axon is shown. *Reprinted from reference 3 with permission from Wiley.*

Giant axons are not specific for GAN, and are also described in neuropathies due to *SH3TC2*, *NEFL*, and *BAG3* mutations.[113,239,240] *BAG3* mutations are associated with myofibrillar myopathy, but some individuals present with neuropathic features in association with the more typical myopathic manifestations.[240,241] Other pathological features of GAN include variable loss of myelinated nerve fibers, and segmental demyelination and remyelination with onion bulb formations, suggesting a degree of Schwann cell dysfunction.[235]

Gigaxonin is a ubiquitously expressed protein involved in maintenance of microtubule networks. Mutations are thought to cause defective ubiquitin-mediated degradation of microtubule-related proteins, leading to neuronal toxicity and impaired retrograde axonal transport.[242,243]

Hereditary Motor and Sensory Neuropathy with Agenesis of the Corpus Callosum

Hereditary motor and sensory neuropathy with agenesis of the corpus callosum (HMSN/ACC), also known as Andermann syndrome, is an autosomal recessive infantile onset condition characterized by a severe progressive sensorimotor neuropathy, variable degrees of agenesis of the corpus callosum, and cognitive impairment.[244,245] HMSN/ACC is found mainly in the French Canadian population of Quebec.[246,247]

The clinical presentation is fairly homogeneous, with hypotonia, motor delay, and areflexia in the first year of life. Some children walk with support by age two to five years. Others never ambulate. Progressive distal and proximal weakness causes loss of ambulation in later childhood. Cranial nerve involvement may manifest as ptosis, facial weakness, or gaze palsies.[245,248,249] Most affected individuals have intellectual disability. Interestingly, about a third of patients develop an atypical psychosis in late adolescence or early adulthood.[248,250] The average age of death is 33 years.[245]

Median MNCVs are variable (11 to 57 m/s) and CMAPs are low-amplitude. Sensory responses are unobtainable even in the first year of life.[245] On brain imaging complete callosal agenesis is seen in 60% of cases, partial agenesis in 10%, and a normal corpus callosum in 30%.[248] Nerve histopathology shows severe loss of large myelinated fibers, axonal degeneration, and enlarged axons with disproportionately thinned myelin sheaths, suspected to be secondary to previous axoplasmic swelling.[245]

Infantile Neuroaxonal Dystrophy

Classic infantile neuroaxonal dystrophy (INAD) is a severe early-onset neurodegenerative disorder characterized by hypotonia, psychomotor regression, and axonal spheroids on nerve biopsy.[251] Onset is between six months and three years of age, after a period of normal development.[252,253] Affected children rapidly evolve from hypotonia to a spastic tetraparesis, although some children have more prominent features of a neuropathy with amyotrophy and areflexia. Gait ataxia, ophthalmologic abnormalities, and dystonia are frequent. The clinical course is rapidly progressive, with a mean age of death at nine years.[252–254]

Characteristic neurophysiologic findings include a sensorimotor axonal neuropathy, presence of fast rhythms on EEG, and abnormal visual evoked potentials.[251,253] Almost all children have cerebellar atrophy on MRI, along with T2/FLAIR hyperintensities in the cerebellum. Around 50% of children exhibit abnormal iron accumulation in the globus pallidus.[254,255] Prior to identification of the gene, diagnosis was confirmed by the presence of dystrophic axons on electron microscopy of nerves in biopsies of sural nerve, skin, rectum, or conjunctiva.[251,253]

INAD is due to recessive mutations in the *PLA2G6* gene[256,257]; allellic disorders include some cases of neurodegeneration with brain iron accumulation (NBIA2 or atypical neuroaxonal dystrophy) and adult onset parkinsonism-dystonia syndrome.[252,256,258] Recently, a homozygous mutation in the *NALCN* gene was identified in a family with an INAD phenotype with additional facial dysmorphic features.[259]

Congenital Axonal Neuropathy and Encephalopathy

The etiology of these cases is unknown, and they are likely genetically heterogeneous. Affected infants present during early infancy with hypotonia, distal weakness, and areflexia. CNS involvement manifests as seizures, microcephaly, or global developmental delay.[1,260–262] Some subjects have skeletal deformities, dysmorphic features, retinopathy, or cortical blindness. The prognosis is poor, and most children die of respiratory failure. Neurophysiologic and pathologic studies indicate a severe sensorimotor axonal neuropathy. Neuroimaging shows nonspecific white matter abnormalities.

Spinal Muscular Atrophy and Motor Neuronopathies

These conditions may mimic the clinical presentation of infantile neuropathies, and include SMA type 1 and the non 5q infantile SMA variants discussed in Chapter 8. While the primary pathology is motor neuron degeneration, peripheral motor and sensory involvement may also occur,[263–265] and these conditions should form part of the differential diagnosis of infants presenting with congenital hypotonia, areflexia, and contractures[6] (Table 16.2).

Early Onset Axonal Neuropathies Associated with Mitochondrial Disorders

Mitochondrial disorders presenting in infancy may involve the peripheral nervous system, often as part of a multisystem disorder; the neuropathy may be demyelinating or, more commonly, axonal.[6] These disorders include some of the mitochondrial DNA (mtDNA) depletion syndromes, which are characterized by a reduction in cellular mtDNA content, and caused by mutations in nuclear genes responsible for mtDNA duplication or maintenance, e.g. *DGUOK*, *C10orf2*, *TK2*, *POLG1*, and *MPV17* (Tables 16.2 and 16.5). These are discussed in detail in Chapter 19.

Deficiencies of the mitochondrial trifunctional protein (MTP), and less commonly long-chain 3-hydroxyacyl-CoA dehydrogenase (LCHAD, a part of the MTP complex), are also associated with peripheral neuropathy. These conditions usually present in the neonatal or infantile periods with hypoketotic hypoglycemia, but MTP deficiency occasionally presents with hypotonia, weakness, and failure to thrive.[266–268]

Immunologically Mediated Neuropathies

Guillain-Barré syndrome (GBS) and chronic inflammatory demyelinating polyneuropathy (CIDP) may present in infancy, although both are rare.

Guillain-Barré Syndrome

In a cohort of 293 children with GBS, only 1.7% presented at less than 1 year of age.[269] Both neonatal and infantile onset GBS have been described,[270–273] including a case report of GBS affecting both a mother and her newborn child.[274] Generalized hypotonia, limb weakness (often affecting the lower more than the upper limbs), and absent reflexes are frequent findings. Pain, presenting as irritability, is common and may lead to misdiagnosis.[272] Most reported cases have neurophysiological findings consistent with acute inflammatory demyelinating polyneuropathy (AIDP).[271,272,274] Recovery appears good with or without treatment with intravenous immunoglobulin.[271,273,274] For a detailed description of childhood GBS the reader is referred to Chapter 20.

Chronic Inflammatory Demyelinating Polyneuropathy

CIDP (discussed in Chapter 21) may be difficult to distinguish from the inherited demyelinating neuropathies, especially during infancy; its chronic course may mimic inherited neuropathies, and a preceding normal clinical history may be difficult to establish with certainty. Congenital, neonatal, and infantile cases of CIDP have been reported.[270,275–278] Developmental delay, hypotonia, progressive weakness, and areflexia have been described, and some infants have bulbar and respiratory compromise. Nerve biopsy showing inflammatory cell infiltrates assisted in the diagnosis of some cases[276–278]; in others a relapsing course, or subsequent recovery, only later confirmed the diagnosis of CIDP.[278,279]

CONCLUSION

The congenital and infantile onset neuropathies represent a rare but heterogeneous and complex group of disorders. Most have a genetic basis, and although much progress has been made in determining the genetic etiology of these conditions, at least half of children presenting in the first year of life remain without a specific diagnosis. Important clinical, neurophysiologic, and neuropathologic clues to aid the diagnosis and management of the demyelinating and axonal neuropathies in this age group have been presented. While treatment remains largely symptomatic, targeted disease-specific therapies may become available in the future.

ACKNOWLEDGMENTS

The authors wish to thank Professor Catriona McLean (Neuropathology Department, The Alfred Hospital, Melbourne, Australia) for assistance with Figure 16.4, and Professor Jean-Michel Vallat (Service de Neurologie, Centre de Référence Neuropathies périphériques rares, CHU Limoges, France) for assistance with Figures 16.9 and 16.10. EMY is funded by a National Health and Medical Research Council of Australia (NHMRC) Early Career Fellowship. JB is funded by the Association Belge contre les Maladies Neuromusculaires (ABMM) and the European Community's Seventh Framework Programme (FP7/2007-2013) under grant agreement number 2012-305121, 'Integrated European'omics research project for diagnosis and therapy in rare neuromuscular and neurodegenerative diseases (NEUROMICS).

REFERENCES

1. Wilmshurst JM, Pollard JD, Nicholson G, Antony J, Ouvrier R. Peripheral neuropathies of infancy. *Dev Med Child Neurol* 2003;**45**:408–14.
2. Baets J, Deconinck T, De Vriendt E, Zimon M, Yperzeele L, Van Hoorenbeeck K, et al. Genetic spectrum of hereditary neuropathies with onset in the first year of life. *Brain* 2011;**134**:2664–76.
3. Yiu EM, Ryan MM. Demyelinating prenatal and infantile developmental neuropathies. *J Peripher Nerv Syst* 2012;**17**:32–52.
4. Parano E, Uncini A, De Vivo DC, Lovelace RE. Electrophysiologic correlates of peripheral nervous system maturation in infancy and childhood. *J Child Neurol* 1993;**8**:336–8.
5. Miller RG, Kuntz NL. Nerve conduction studies in infants and children. *J Child Neurol* 1986;**1**:19–26.
6. Yiu EM, Ryan MM. Genetic axonal neuropathies and neuronopathies of pre-natal and infantile onset. *J Peripher Nerv Syst* 2012;**17**:285–300.

7. Dyck PJ, Lambert EH. Lower motor and primary sensory neuron diseases with peroneal muscular atrophy. I. Neurologic, genetic, and electrophysiologic findings in hereditary polyneuropathies. *Arch Neurol* 1968;**18**:603–18.
8. Ouvrier RA, McLeod JG, Conchin TE. The hypertrophic forms of hereditary motor and sensory neuropathy. A study of hypertrophic Charcot-Marie-Tooth disease (HMSN type I) and Dejerine-Sottas disease (HMSN type III) in childhood. *Brain* 1987;**110** (Pt 1):121–48.
9. Hayasaka K, Himoro M, Sawaishi Y, Nanao K, Takahashi T, Takada G, et al. De novo mutation of the myelin P0 gene in Dejerine-Sottas disease (hereditary motor and sensory neuropathy type III). *Nat Genet* 1993;**5**:266–8.
10. Roa BB, Dyck PJ, Marks HG, Chance PF, Lupski JR. Dejerine-Sottas syndrome associated with point mutation in the peripheral myelin protein 22 (PMP22) gene. *Nat Genet* 1993;**5**:269–73.
11. Warner LE, Mancias P, Butler IJ, McDonald CM, Keppen L, Koob KG, et al. Mutations in the early growth response 2 (EGR2) gene are associated with hereditary myelinopathies. *Nat Genet* 1998;**18**:382–4.
12. Gabreels-Festen A. Dejerine-Sottas syndrome grown to maturity: overview of genetic and morphological heterogeneity and follow-up of 25 patients. *J Anat* 2002;**200**:341–56.
13. Benstead TJ, Kuntz NL, Miller RG, Daube JR. The electrophysiologic profile of Dejerine-Sottas disease (HMSN III). *Muscle Nerve* 1990;**13**:586–92.
14. Gabreels-Festen A, Gabreels F. Hereditary demyelinating motor and sensory neuropathy. *Brain Pathol* 1993;**3**:135–46.
15. Charnas L, Trapp B, Griffin J. Congenital absence of peripheral myelin: abnormal Schwann cell development causes lethal arthrogryposis multiplex congenita. *Neurology* 1988;**38**:966–74.
16. Phillips JP, Warner LE, Lupski JR, Garg BP. Congenital hypomyelinating neuropathy: two patients with long-term follow-up. *Pediatr Neurol* 1999;**20**:226–32.
17. Schroder JM. Neuropathology of Charcot-Marie-Tooth and related disorders. *Neuromolecular Med* 2006;**8**:23–42.
18. Boylan KB, Ferriero DM, Greco CM, Sheldon RA, Dew M. Congenital hypomyelination neuropathy with arthrogryposis multiplex congenita. *Ann Neurol* 1992;**31**:337–40.
19. Hahn JS, Henry M, Hudgins L, Madan A. Congenital hypomyelination neuropathy in a newborn infant: unusual cause of diaphragmatic and vocal cord paralyses. *Pediatrics* 2001;**108**:E95.
20. McMillan HJ, Santagata S, Shapiro F, Batish SD, Couchon L, Donnelly S, et al. Novel MPZ mutations and congenital hypomyelinating neuropathy. *Neuromuscul Disord* 2010;**20**:725–9.
21. Kasman M, Bernstein L, Schulman S. Chronic polyradiculoneuropathy of infancy. A report of three cases with familial incidence. *Neurology* 1976;**26**:565–73.
22. Smit LS, Roofthooft D, van Ruissen F, Baas F, van Doorn PA. Congenital hypomyelinating neuropathy, a long term follow-up study in an affected family. *Neuromuscul Disord* 2008;**18**:59–62.
23. Warner LE, Hilz MJ, Appel SH, Killian JM, Kolodry EH, Karpati G, et al. Clinical phenotypes of different MPZ (P0) mutations may include Charcot-Marie-Tooth type 1B, Dejerine-Sottas, and congenital hypomyelination. *Neuron* 1996;**17**:451–60.
24. Simonati A, Fabrizi GM, Pasquinelli A, Taioli F, Cavallaro T, Morbin M, et al. Congenital hypomyelination neuropathy with Ser72Leu substitution in PMP22. *Neuromuscul Disord* 1999;**9**:257–61.
25. Boerkoel CF, Takashima H, Bacino CA, Daentl D, Lupski JR. EGR2 mutation R359W causes a spectrum of Dejerine-Sottas neuropathy. *Neurogenetics* 2001;**3**:153–7.
26. Lupski JR, de Oca-Luna RM, Slaugenhaupt S, Pentao L, Guzzetta V, Trask BJ, et al. DNA duplication associated with Charcot-Marie-Tooth disease type 1A. *Cell* 1991;**66**:219–32.
27. Raeymaekers P, Timmerman V, Nelis E, De Jonghe P, Hoogendijk JE, Baas F, et al. Duplication in chromosome 17p11.2 in Charcot-Marie-Tooth neuropathy type 1a (CMT 1a). The HMSN Collaborative Research Group. *Neuromuscul Disord* 1991;**1**:93–7.
28. Chance PF, Alderson MK, Leppig KA, Lensch MW, Matsunami N, Smith B, et al. DNA deletion associated with hereditary neuropathy with liability to pressure palsies. *Cell* 1993;**72**:143–51.
29. Ionasescu VV, Ionasescu R, Searby C, Neahring R. Dejerine-Sottas disease with de novo dominant point mutation of the PMP22 gene. *Neurology* 1995;**45**:1766–7.
30. Fabrizi GM, Simonati A, Taioli F, Cavallaro T, Ferrarini M, Rigatelli F, et al. PMP22 related congenital hypomyelination neuropathy. *J Neurol Neurosurg Psychiatry* 2001;**70**:123–6.
31. Tyson J, Ellis D, Fairbrother U, King RH, Muntoni F, Jacobs J, et al. Hereditary demyelinating neuropathy of infancy. A genetically complex syndrome. *Brain* 1997;**120**(Pt 1):47–63.
32. Marques Jr. W, Thomas PK, Sweeney MG, Carr L, Wood NW. Dejerine-Sottas neuropathy and PMP22 point mutations: a new base pair substitution and a possible "hot spot" on Ser72. *Ann Neurol* 1998;**43**:680–3.
33. Ikegami T, Ikeda H, Aoyama M, Matsuki T, Imota T, Fukuuchi Y, et al. Novel mutations of the peripheral myelin protein 22 gene in two pedigrees with Dejerine-Sottas disease. *Hum Genet* 1998;**102**:294–8.
34. Ionasescu VV, Searby C, Greenberg SA. Dejerine-Sottas disease with sensorineural hearing loss, nystagmus, and peripheral facial nerve weakness: de novo dominant point mutation of the PMP22 gene. *J Med Genet* 1996;**33**:1048–9.
35. Russo M, Laura M, Polke JM, Davis MB, Blake J, Brandner S, et al. Variable phenotypes are associated with PMP22 missense mutations. *Neuromuscul Disord* 2011;**21**:106–14.
36. Jen J, Baloh RH, Ishiyama A, Baloh RW. Dejerine-Sottas syndrome and vestibular loss due to a point mutation in the PMP22 gene. *J Neurol Sci* 2005;**237**:21–4.
37. Marques Jr. W, Neto JM, Barreira AA. Dejerine-Sottas' neuropathy caused by the missense mutation PMP22 Ser72Leu. *Acta Neurol Scand* 2004;**110**:196–9.
38. Parman Y, Plante-Bordeneuve V, Guiochon-Mantel A, Eraksoy M, Said G. Recessive inheritance of a new point mutation of the PMP22 gene in Dejerine-Sottas disease. *Ann Neurol* 1999;**45**:518–22.
39. Abe A, Nakamura K, Kato M, Numakura C, Honma T, Seiwa C, et al. Compound heterozygous PMP22 deletion mutations causing severe Charcot-Marie-Tooth disease type 1. *J Hum Genet* 2010;**55**:771–3.
40. Al-Thihli K, Rudkin T, Carson N, Poulin C, Melancon S, Der Kaloustian VM. Compound heterozygous deletions of PMP22 causing severe Charcot-Marie-Tooth disease of the Dejerine-Sottas disease phenotype. *Am J Med Genet A* 2008;**146A**:2412–16.
41. Sturtz FG, Latour P, Mocquard Y, Cruz S, Fenoll B, LeFur JM, et al. Clinical and electrophysiological phenotype of a homozygously

duplicated Charcot-Marie-Tooth (type 1A) disease. *Eur Neurol* 1997;**38**:26–30.

42. Kaku DA, Parry GJ, Malamut R, Lupski JR, Garcia CA. Nerve conduction studies in Charcot-Marie-Tooth polyneuropathy associated with a segmental duplication of chromosome 17. *Neurology* 1993;**43**:1806–8.
43. LeGuern E, Gouider R, Mabin D, Tardieu S, Birouk N, Parent P, et al. Patients homozygous for the 17p11.2 duplication in Charcot-Marie-Tooth type 1A disease. *Ann Neurol* 1997;**41**:104–8.
44. Snipes GJ, Suter U, Welcher AA, Shooter EM. Characterization of a novel peripheral nervous system myelin protein (PMP-22/SR13). *J Cell Biol* 1992;**117**:225–38.
45. Haney C, Snipes GJ, Shooter EM, Suter U, Garcia C, Griffin JW, et al. Ultrastructural distribution of PMP22 in Charcot-Marie-Tooth disease type 1A. *J Neuropathol Exp Neurol* 1996;**55**:290–9.
46. Colby J, Nicholson R, Dickson KM, Orfali W, Naef R, Suter U, et al. PMP22 carrying the trembler or trembler-J mutation is intracellularly retained in myelinating Schwann cells. *Neurobiol Dis* 2000;**7**:561–73.
47. Khajavi M, Shiga K, Wiszniewski W, He F, Shaw CA, Yan J, et al. Oral curcumin mitigates the clinical and neuropathologic phenotype of the Trembler-J mouse: a potential therapy for inherited neuropathy. *Am J Hum Genet* 2007;**81**:438–53.
48. Shy ME. Peripheral neuropathies caused by mutations in the myelin protein zero. *J Neurol Sci* 2006;**242**:55–66.
49. Shy ME, Jani A, Krajewski K, Grandis M, Lewis RA, Li J, et al. Phenotypic clustering in MPZ mutations. *Brain* 2004;**127**:371–84.
50. Mandich P, Mancardi GL, Varese A, Soriani S, Di Maria E, Bellone E, et al. Congenital hypomyelination due to myelin protein zero Q215X mutation. *Ann Neurol* 1999;**45**:676–8.
51. Szigeti K, Saifi GM, Armstrong D, Belmont JW, Miller G, Lupski JR. Disturbance of muscle fiber differentiation in congenital hypomyelinating neuropathy caused by a novel myelin protein zero mutation. *Ann Neurol* 2003;**54**:398–402.
52. Fabrizi GM, Cavallaro T, Morbin M, Simonati A, Taioli F, Rizzuto N. Novel mutation of the P0 extracellular domain causes a Dejerine-Sottas syndrome. *J Neurol Neurosurg Psychiatry* 1999;**66**:386–9.
53. Bird TD, Kraft GH, Lipe HP, Kenney KL, Sumi SM. Clinical and pathological phenotype of the original family with Charcot-Marie-Tooth type 1B: a 20-year study. *Ann Neurol* 1997;**41**:463–9.
54. Gabreels-Festen AA, Hoogendijk JE, Meijerink PH, Gabreels FJ, Bolhuis PA, van Beersum S, et al. Two divergent types of nerve pathology in patients with different P0 mutations in Charcot-Marie-Tooth disease. *Neurology* 1996;**47**:761–5.
55. Ikegami T, Nicholson G, Ikeda H, Ishida A, Johnston H, Wise G, et al. A novel homozygous mutation of the myelin Po gene producing Dejerine-Sottas disease (hereditary motor and sensory neuropathy type III). *Biochem Biophys Res Commun* 1996;**222**:107–10.
56. Filbin MT, Zhang K, Li W, Gao Y. Characterization of the effect on adhesion of different mutations in myelin P0 protein. *Ann NY Acad Sci* 1999;**883**:160–7.
57. Khajavi M, Inoue K, Wiszniewski W, Ohyama T, Snipes GJ, Lupski JR. Curcumin treatment abrogates endoplasmic reticulum retention and aggregation-induced apoptosis associated with neuropathy-causing myelin protein zero-truncating mutants. *Am J Hum Genet* 2005;**77**:841–50.
58. Grandis M, Vigo T, Passalacqua M, Jain M, Scazzola S, La Padula V, et al. Different cellular and molecular mechanisms for early and late-onset myelin protein zero mutations. *Hum Mol Genet* 2008;**17**:1877–89.
59. Boerkoel CF, Takashima H, Garcia CA, Olney RK, Johnson J, Berry K, et al. Charcot-Marie-Tooth disease and related neuropathies: mutation distribution and genotype-phenotype correlation. *Ann Neurol* 2002;**51**:190–201.
60. Timmerman V, De Jonghe P, Ceuterick C, De Vriendt E, Lofgren A, Nelis E, et al. Novel missense mutation in the early growth response 2 gene associated with Dejerine-Sottas syndrome phenotype. *Neurology* 1999;**52**:1827–32.
61. Vandenberghe N, Upadhyaya M, Gatignol A, Boutrand L, Boucherat M, Chazot G, et al. Frequency of mutations in the early growth response 2 gene associated with peripheral demyelinating neuropathies. *J Med Genet* 2002;**39**:e81.
62. Numakura C, Shirahata E, Yamashita S, Kanai M, Kijima K, Matsuki T, et al. Screening of the early growth response 2 gene in Japanese patients with Charcot-Marie-Tooth disease type 1. *J Neurol Sci* 2003;**210**:61–4.
63. Szigeti K, Wiszniewski W, Saifi GM, Sherman DL, Sule N, Adesina AM, et al. Functional, histopathologic and natural history study of neuropathy associated with EGR2 mutations. *Neurogenetics* 2007;**8**:257–62.
64. Briani C, Taioli F, Lucchetta M, Bombardi R, Fabrizi GM. Adult onset Charcot-Marie-Tooth disease type 1D with an Arg381Cys mutation of EGR2. *Muscle Nerve* 2010;**41**:888–9.
65. Pareyson D, Taroni F, Botti S, Morbin M, Baratta S, Lauria G, et al. Cranial nerve involvement in CMT disease type 1 due to early growth response 2 gene mutation. *Neurology* 2000;**54**:1696–8
66. Nagarajan R, Svaren J, Le N, Araki T, Watson M, Milbrandt J. EGR2 mutations in inherited neuropathies dominant-negatively inhibit myelin gene expression. *Neuron* 2001;**30**:355–68.
67. Musso M, Balestra P, Bellone E, Cassandrini D, Di Maria E, Doria LL, et al. The D355V mutation decreases EGR2 binding to an element within the Cx32 promoter. *Neurobiol Dis* 2001;**8**:700–6.
68. Jang SW, Svaren J. Induction of myelin protein zero by early growth response 2 through upstream and intragenic elements. *J Biol Chem* 2009;**284**:20111–20.
69. Warner LE, Svaren J, Milbrandt J, Lupski JR. Functional consequences of mutations in the early growth response 2 gene (EGR2) correlate with severity of human myelinopathies. *Hum Mol Genet* 1999;**8**:1245–51.
70. Mersiyanova IV, Perepelov AV, Polyakov AV, Sitnikov VF, Dadali EL, Oparin RB, et al. A new variant of Charcot-Marie-Tooth disease type 2 is probably the result of a mutation in the neurofilament-light gene. *Am J Hum Genet* 2000;**67**:37–46.
71. De Jonghe P, Mersivanova I, Nelis E, Del Favero J, Martin JJ, Van Broeckhoven C, et al. Further evidence that neurofilament light chain gene mutations can cause Charcot-Marie-Tooth disease type 2E. *Ann Neurol* 2001;**49**:245–9.
72. Jordanova A, De Jonghe P, Boerkoel CF, Takashima H, De Vriendt E, Ceuterick C, et al. Mutations in the neurofilament light chain gene (NEFL) cause early onset severe Charcot-Marie-Tooth disease. *Brain* 2003;**126**:590–7.
73. Abe A, Numakura C, Saito K, Koide H, Oka N, Honma A, et al. Neurofilament light chain polypeptide gene mutations in Charcot-Marie-Tooth disease: nonsense mutation probably causes a recessive phenotype. *J Hum Genet* 2009;**54**:94–7.

74. Yum SW, Zhang J, Mo K, Li J, Scherer SS. A novel recessive Nefl mutation causes a severe, early-onset axonal neuropathy. *Ann Neurol* 2009;**66**:759–70.
75. Fabrizi GM, Cavallaro T, Angiari C, Cabrini I, Taioli F, Malerba G, et al. Charcot-Marie-Tooth disease type 2E, a disorder of the cytoskeleton. *Brain* 2007;**130**:394–403.
76. Miltenberger-Miltenyi G, Janecke AR, Wanschitz JV, Timmerman V, Windpassinger C, Auer-Grumbach M. Clinical and electrophysiological features in Charcot-Marie-Tooth disease with mutations in the NEFL gene. *Arch Neurol* 2007;**64**:966–70.
77. Lee MK, Cleveland DW. Neuronal intermediate filaments. *Annu Rev Neurosci* 1996;**19**:187–217.
78. Brownlees J, Ackerley S, Grierson AJ, Jacobsen NJ, Shea K, Anderton BH, et al. Charcot-Marie-Tooth disease neurofilament mutations disrupt neurofilament assembly and axonal transport. *Hum Mol Genet* 2002;**11**:2837–44.
79. Sasaki T, Gotow T, Shiozaki M, Sakaue F, Saito T, Julien JP, et al. Aggregate formation and phosphorylation of neurofilament-L Pro22 Charcot-Marie-Tooth disease mutants. *Hum Mol Genet* 2006;**15**:943–52.
80. Fabrizi GM, Cavallaro T, Angiari C, Bertolasi L, Cabrini I, Ferrarini M, et al. Giant axon and neurofilament accumulation in Charcot-Marie-Tooth disease type 2E. *Neurology* 2004;**62**:1429–31.
81. Baxter RV, Ben Othmane K, Rochelle JM, Stajich JE, Hulette C, Dew-Knight S, et al. Ganglioside-induced differentiation-associated protein-1 is mutant in Charcot-Marie-Tooth disease type 4A/8q21. *Nat Genet* 2002;**30**:21–2.
82. Cuesta A, Pedrola L, Sevilla T, Garcia-Planells J, Chumillas MJ, Mayordomo F, et al. The gene encoding ganglioside-induced differentiation-associated protein 1 is mutated in axonal Charcot-Marie-Tooth type 4A disease. *Nat Genet* 2002;**30**:22–5.
83. Senderek J, Bergmann C, Ramaekers VT, Nelis E, Bernert G, Makowski A, et al. Mutations in the ganglioside-induced differentiation-associated protein-1 (GDAP1) gene in intermediate type autosomal recessive Charcot-Marie-Tooth neuropathy. *Brain* 2003;**126**:642–9.
84. Nelis E, Erdem S, Van Den Bergh PY, Belpaire-Dethiou MC, Ceuterick C, Van Gerwen V, et al. Mutations in GDAP1: autosomal recessive CMT with demyelination and axonopathy. *Neurology* 2002;**59**:1865–72.
85. Sevilla T, Cuesta A, Chumillas MJ, Mayordomo F, Pedrola L, Palau F, et al. Clinical, electrophysiological and morphological findings of Charcot-Marie-Tooth neuropathy with vocal cord palsy and mutations in the GDAP1 gene. *Brain* 2003;**126**:2023–33.
86. Sevilla T, Jaijo T, Nauffal D, Collado D, Chumillas MJ, Vilchez JJ, et al. Vocal cord paresis and diaphragmatic dysfunction are severe and frequent symptoms of GDAP1-associated neuropathy. *Brain* 2008;**131**:3051–61.
87. Birouk N, Azzedine H, Dubourg O, Muriel MP, Benomar A, Hamadouche T, et al. Phenotypical features of a Moroccan family with autosomal recessive Charcot-Marie-Tooth disease associated with the S194X mutation in the GDAP1 gene. *Arch Neurol* 2003;**60**:598–604.
88. Boerkoel CF, Takashima H, Nakagawa M, Izumo S, Armstrong D, Butler I, et al. CMT4A: identification of a Hispanic GDAP1 founder mutation. *Ann Neurol* 2003;**53**:400–5.
89. Claramunt R, Pedrola L, Sevilla T, Lopez de Munain A, Berciano J, Cuesta A, et al. Genetics of Charcot-Marie-Tooth disease type 4A: mutations, inheritance, phenotypic variability, and founder effect. *J Med Genet* 2005;**42**:358–65.
90. De Sandre-Giovannoli A, Chaouch M, Boccaccio I, Bernard R, Delague V, Grid D, et al. Phenotypic and genetic exploration of severe demyelinating and secondary axonal neuropathies resulting from GDAP1 nonsense and splicing mutations. *J Med Genet* 2003;**40**:e87.
91. Pedrola L, Espert A, Wu X, Claramunt R, Shy ME, Palau F. GDAP1, the protein causing Charcot-Marie-Tooth disease type 4A, is expressed in neurons and is associated with mitochondria. *Hum Mol Genet* 2005;**14**:1087–94.
92. Niemann A, Ruegg M, La Padula V, Schenone A, Suter U. Ganglioside-induced differentiation associated protein 1 is a regulator of the mitochondrial network: new implications for Charcot-Marie-Tooth disease. *J Cell Biol* 2005;**170**:1067–78.
93. Niemann A, Wagner KM, Ruegg M, Suter U. GDAP1 mutations differ in their effects on mitochondrial dynamics and apoptosis depending on the mode of inheritance. *Neurobiol Dis* 2009;**36**:509–20.
94. Bolino A, Lonie LJ, Zimmer M, Boerkoel CF, Takashima H, Monaco AP, et al. Denaturing high-performance liquid chromatography of the myotubularin-related 2 gene (MTMR2) in unrelated patients with Charcot-Marie-Tooth disease suggests a low frequency of mutation in inherited neuropathy. *Neurogenetics* 2001;**3**:107–9.
95. Quattrone A, Gambardella A, Bono F, Aguglia U, Bolino A, Bruni AC, et al. Autosomal recessive hereditary motor and sensory neuropathy with focally folded myelin sheaths: clinical, electrophysiologic, and genetic aspects of a large family. *Neurology* 1996;**46**:1318–24.
96. Houlden H, King RH, Wood NW, Thomas PK, Reilly MM. Mutations in the 5′ region of the myotubularin-related protein 2 (MTMR2) gene in autosomal recessive hereditary neuropathy with focally folded myelin. *Brain* 2001;**124**:907–15.
97. Ohnishi A, Murai Y, Ikeda M, Fujita T, Furuya H, Kuroiwa Y. Autosomal recessive motor and sensory neuropathy with excessive myelin outfolding. *Muscle Nerve* 1989;**12**:568–75.
98. Salih MA, Maisonobe T, Kabiraj M, al Rayess M, al-Turaiki MH, Akbar M, et al. Autosomal recessive hereditary neuropathy with focally folded myelin sheaths and linked to chromosome 11q23: a distinct and homogeneous entity. *Neuromuscul Disord* 2000;**10**:10–15.
99. Laporte J, Bedez F, Bolino A, Mandel JL. Myotubularins, a large disease-associated family of cooperating catalytically active and inactive phosphoinositides phosphatases. *Hum Mol Genet* 2003;**12** (Spec. No. 2):R285–92.
100. Berger P, Berger I, Schaffitzel C, Tersar K, Volkmer B, Suter U. Multi-level regulation of myotubularin-related protein-2 phosphatase activity by myotubularin-related protein-13/set-binding factor-2. *Hum Mol Genet* 2006;**15**:569–79.
101. Robinson FL, Dixon JE. The phosphoinositide-3-phosphatase MTMR2 associates with MTMR13, a membrane-associated pseudophosphatase also mutated in type 4B Charcot-Marie-Tooth disease. *J Biol Chem* 2005;**280**:31699–707.
102. Bolino A, Bolis A, Previtali SC, Dina G, Bussini S, Dati G, et al. Disruption of Mtmr2 produces CMT4B1-like neuropathy with

myelin outfolding and impaired spermatogenesis. *J Cell Biol* 2004;**167**:711–21.

103. Bolis A, Zordan P, Coviello S, Bolino A. Myotubularin-related (MTMR) phospholipid phosphatase proteins in the peripheral nervous system. *Mol Neurobiol* 2007;**35**:308–16.

104. Senderek J, Bergmann C, Weber S, Ketelsen UP, Schorle H, Rudnik-Schoneborn S, et al. Mutation of the SBF2 gene, encoding a novel member of the myotubularin family, in Charcot-Marie-Tooth neuropathy type 4B2/11p15. *Hum Mol Genet* 2003;**12**:349–56.

105. Azzedine H, Bolino A, Taieb T, Birouk N, Di Duca M, Bouhouche A, et al. Mutations in MTMR13, a new pseudophosphatase homologue of MTMR2 and Sbf1, in two families with an autosomal recessive demyelinating form of Charcot-Marie-Tooth disease associated with early-onset glaucoma. *Am J Hum Genet* 2003;**72**:1141–53.

106. Kiwaki T, Umehara F, Takashima H, Nakagawa M, Kamimura K, Kashio N. Hereditary motor and sensory neuropathy with myelin folding and juvenile onset glaucoma. *Neurology* 2000;**55**:392–7.

107. Hirano R, Takashima H, Umehara F, Arimura H, Michizono K, Okamoto Y, et al. SET binding factor 2 (SBF2) mutation causes CMT4B with juvenile onset glaucoma. *Neurology* 2004;**63**:577–80.

108. Nakhro K, Park JM, Hong YB, Park JH, Nam SH, Yoon BR, et al. SET binding factor 1 (SBF1) mutation causes Charcot-Marie-Tooth disease type 4B3. *Neurology* 2013;**81**:165–73.

109. Kim SA, Vacratsis PO, Firestein R, Cleary ML, Dixon JE. Regulation of myotubularin-related (MTMR)2 phosphatidylinositol phosphatase by MTMR5, a catalytically inactive phosphatase. *Proc Natl Acad Sci USA* 2003;**100**:4492–7.

110. Senderek J, Bergmann C, Stendel C, Kirfel J, Verpoorten N, De Jonghe P, et al. Mutations in a gene encoding a novel SH3/TPR domain protein cause autosomal recessive Charcot-Marie-Tooth type 4C neuropathy. *Am J Hum Genet* 2003;**73**:1106–19.

111. Azzedine H, Bontoux L, LeGuern E. Charcot-Marie-Tooth Neuropathy Type 4C. In: Pagon R, Bird T, Dolan C, Stephens K, editors. *GeneReviews at GeneTests: Medical Genetics Information Resource (database online)*. Seattle: Copyright, University of Washington; 1997–2011. Available at: <http://www.genetests.org> Updated July 6, 2010.

112. Kessali M, Zemmouri R, Guilbot A, Maisonobe T, Brice A, LeGuern E, et al. A clinical, electrophysiologic, neuropathologic, and genetic study of two large Algerian families with an autosomal recessive demyelinating form of Charcot-Marie-Tooth disease. *Neurology* 1997;**48**:867–73.

113. Azzedine H, Ravise N, Verny C, Gabreels-Festen A, Lammens M, Grid D, et al. Spine deformities in Charcot-Marie-Tooth 4C caused by SH3TC2 gene mutations. *Neurology* 2006;**67**:602–6.

114. Colomer J, Gooding R, Angelicheva D, King RH, Guillen-Navarro E, Parman Y, et al. Clinical spectrum of CMT4C disease in patients homozygous for the p.Arg1109X mutation in SH3TC2. *Neuromuscul Disord* 2006;**16**:449–53.

115. Gooding R, Colomer J, King R, Angelicheva D, Marns L, Parman Y, et al. A novel Gypsy founder mutation, p.Arg1109X in the CMT4C gene, causes variable peripheral neuropathy phenotypes. *J Med Genet* 2005;**42**(e69).

116. Gosselin I, Thiffault I, Tetreault M, Chau V, Dicaire MJ, Loisel L, et al. Founder SH3TC2 mutations are responsible for a CMT4C French-Canadians cluster. *Neuromuscul Disord* 2008;**18**:483–92.

117. Houlden H, Laura M, Ginsberg L, Jungbluth H, Robb SA, Blake J, et al. The phenotype of Charcot-Marie-Tooth disease type 4C due to SH3TC2 mutations and possible predisposition to an inflammatory neuropathy. *Neuromuscul Disord* 2009;**19**:264–9.

118. Gabreels-Festen A, van Beersum S, Eshuis L, LeGuern E, Gabreels F, van Engelen B, et al. Study on the gene and phenotypic characterisation of autosomal recessive demyelinating motor and sensory neuropathy (Charcot-Marie-Tooth disease) with a gene locus on chromosome 5q23-q33. *J Neurol Neurosurg Psychiatry* 1999;**66**:569–74.

119. Arnaud E, Zenker J, de Preux Charles AS, Stendel C, Roos A, Medard JJ, et al. SH3TC2/KIAA1985 protein is required for proper myelination and the integrity of the node of Ranvier in the peripheral nervous system. *Proc Natl Acad Sci USA* 2009;**106**:17528–33.

120. Stendel C, Roos A, Kleine H, Arnaud E, Ozcelik M, Sidiropoulos PN, et al. SH3TC2, a protein mutant in Charcot-Marie-Tooth neuropathy, links peripheral nerve myelination to endosomal recycling. *Brain* 2010;**133**:2462–74.

121. Lupo V, Galindo MI, Martinez-Rubio D, Sevilla T, Vilchez JJ, Palau F, et al. Missense mutations in the SH3TC2 protein causing Charcot-Marie-Tooth disease type 4C affect its localization in the plasma membrane and endocytic pathway. *Hum Mol Genet* 2009;**18**:4603–14.

122. Delague V, Bareil C, Tuffery S, Bouvagnet P, Chouery E, Koussa S, et al. Mapping of a new locus for autosomal recessive demyelinating Charcot-Marie-Tooth disease to 19q13.1-13.3 in a large consanguineous Lebanese family: exclusion of MAG as a candidate gene. *Am J Hum Genet* 2000;**67**:236–43.

123. Boerkoel CF, Takashima H, Stankiewicz P, Garcia CA, Leber SM, Rhee-Morris L, et al. Periaxin mutations cause recessive Dejerine-Sottas neuropathy. *Am J Hum Genet* 2001;**68**:325–33.

124. Marchesi C, Milani M, Morbin M, Cesani M, Lauria G, Scaioli V, et al. Four novel cases of periaxin-related neuropathy and review of the literature. *Neurology* 2010;**75**:1830–8.

125. Guilbot A, Williams A, Ravise N, Verny C, Brice A, Sherman DL, et al. A mutation in periaxin is responsible for CMT4F, an autosomal recessive form of Charcot-Marie-Tooth disease. *Hum Mol Genet* 2001;**10**:415–21.

126. Takashima H, Boerkoel CF, De Jonghe P, Ceuterick C, Martin JJ, Voit T, et al. Periaxin mutations cause a broad spectrum of demyelinating neuropathies. *Ann Neurol* 2002;**51**:709–15.

127. Parman Y, Battaloglu E, Baris I, Bilir B, Poyraz M, Bissar-Tadmouri N, et al. Clinicopathological and genetic study of early-onset demyelinating neuropathy. *Brain* 2004;**127**:2540–50.

128. Dytrych L, Sherman DL, Gillespie CS, Brophy PJ. Two PDZ domain proteins encoded by the murine periaxin gene are the result of alternative intron retention and are differentially targeted in Schwann cells. *J Biol Chem* 1998;**273**:5794–800.

129. Scherer SS, Xu YT, Bannerman PG, Sherman DL, Brophy PJ. Periaxin expression in myelinating Schwann cells: modulation by axon-glial interactions and polarized localization during development. *Development* 1995;**121**:4265–73.

130. Court FA, Sherman DL, Pratt T, Garry EM, Ribchester RR, Cottrell DF, et al. Restricted growth of Schwann cells lacking

Cajal bands slows conduction in myelinated nerves. *Nature* 2004;**431**:191–5.
131. Gillespie CS, Sherman DL, Fleetwood-Walker SM, Cottrell DF, Tait S, Garry EM, et al. Peripheral demyelination and neuropathic pain behavior in periaxin-deficient mice. *Neuron* 2000;**26**:523–31.
132. De Sandre-Giovannoli A, Delague V, Hamadouche T, Chaouch M, Krahn M, Boccaccio I, et al. Homozygosity mapping of autosomal recessive demyelinating Charcot-Marie-Tooth neuropathy (CMT4H) to a novel locus on chromosome 12p11.21-q13.11. *J Med Genet* 2005;**42**:260–5.
133. Delague V, Jacquier A, Hamadouche T, Poitelon Y, Baudot C, Boccaccio I, et al. Mutations in FGD4 encoding the Rho GDP/GTP exchange factor FRABIN cause autosomal recessive Charcot-Marie-Tooth type 4H. *Am J Hum Genet* 2007;**81**:1–16.
134. Stendel C, Roos A, Deconinck T, Pereira J, Castagner F, Niemann A, et al. Peripheral nerve demyelination caused by a mutant Rho GTPase guanine nucleotide exchange factor, frabin/FGD4. *Am J Hum Genet* 2007;**81**:158–64.
135. Houlden H, Hammans S, Katifi H, Reilly MM. A novel Frabin (FGD4) nonsense mutation p.R275X associated with phenotypic variability in CMT4H. *Neurology* 2009;**72**:617–20.
136. Fabrizi GM, Taioli F, Cavallaro T, Ferrari S, Bertolasi L, Casarotto M, et al. Further evidence that mutations in FGD4/frabin cause Charcot-Marie-Tooth disease type 4H. *Neurology* 2009;**72**:1160–4.
137. Ikeda W, Nakanishi H, Takekuni K, Itoh S, Takai Y. Identification of splicing variants of Frabin with partly different functions and tissue distribution. *Biochem Biophys Res Commun* 2001;**286**:1066–72.
138. Chow CY, Zhang Y, Dowling JJ, Jin N, Adamska M, Shiga K, et al. Mutation of FIG4 causes neurodegeneration in the pale tremor mouse and patients with CMT4J. *Nature* 2007;**448**:68–72.
139. Nicholson G, Lenk GM, Reddel SW, Grant AE, Towne CF, Ferguson CJ, et al. Distinctive genetic and clinical features of CMT4J: a severe neuropathy caused by mutations in the PI(3,5)P2 phosphatase FIG4. *Brain* 2011;**134**:1959–71.
140. Zhang X, Chow CY, Sahenk Z, Shy ME, Meisler MH, Li J. Mutation of FIG4 causes a rapidly progressive, asymmetric neuronal degeneration. *Brain* 2008;**131**:1990–2001.
141. Vaccari I, Dina G, Tronchere H, Kaufman E, Chicanne G, Cerri F, et al. Genetic interaction between MTMR2 and FIG4 phospholipid phosphatases involved in Charcot-Marie-Tooth neuropathies. *PLOS Genetics* 2011;**7**:e1002319.
142. Bergoffen J, Scherer SS, Wang S, Scott MO, Bone LJ, Paul DL, et al. Connexin mutations in X-linked Charcot-Marie-Tooth disease. *Science* 1993;**262**:2039–42.
143. Yiu EM, Geevasinga N, Nicholson GA, Fagan ER, Ryan MM, Ouvrier RA. A retrospective review of X-linked Charcot-Marie-Tooth disease in childhood. *Neurology* 2011;**76**:461–6.
144. Dubourg O, Tardieu S, Birouk N, Gouider R, Leger JM, Maisonobe T, et al. Clinical, electrophysiological and molecular genetic characteristics of 93 patients with X-linked Charcot-Marie-Tooth disease. *Brain* 2001;**124**:1958–67.
145. Liang GS, de Miguel M, Gomez-Hernandez JM, Glass JD, Scherer SS, Mintz M, et al. Severe neuropathy with leaky connexin32 hemichannels. *Ann Neurol* 2005;**57**:749–54.
146. Nicholson G, Nash J. Intermediate nerve conduction velocities define X-linked Charcot-Marie-Tooth neuropathy families. *Neurology* 1993;**43**:2558–64.
147. Hahn AF, Ainsworth PJ, Bolton CF, Bilbao JM, Vallat JM. Pathological findings in the x-linked form of Charcot-Marie-Tooth disease: a morphometric and ultrastructural analysis. *Acta Neuropathol* 2001;**101**:129–39.
148. Birouk N, LeGuern E, Maisonobe T, Rouger H, Gouider R, Tardieu S, et al. X-linked Charcot-Marie-Tooth disease with connexin 32 mutations: clinical and electrophysiologic study. *Neurology* 1998;**50**:1074–82.
149. Ionasescu VV, Trofatter J, Haines JL, Summers AM, Ionasescu R, Searby C, et al. X-linked recessive Charcot-Marie-Tooth neuropathy: clinical and genetic study. *Muscle Nerve* 1992;**15**:368–73.
150. Cowchock FS, Duckett SW, Streletz LJ, Graziani LJ, Jackson LG. X-linked motor-sensory neuropathy type-II with deafness and mental retardation: a new disorder. *Am J Med Genet* 1985;**20**:307–15.
151. Schiffmann R, van der Knaap MS. Invited article: an MRI-based approach to the diagnosis of white matter disorders. *Neurology* 2009;**72**:750–9.
152. MacFaul R, Cavanagh N, Lake BD, Stephens R, Whitfield AE. Metachromatic leucodystrophy: review of 38 cases. *Arch Dis Child* 1982;**57**:168–75.
153. Cameron CL, Kang PB, Burns TM, Darras BT, Jones Jr. HR. Multifocal slowing of nerve conduction in metachromatic leukodystrophy. *Muscle Nerve* 2004;**29**:531–6.
154. Siddiqi ZA, Sanders DB, Massey JM. Peripheral neuropathy in Krabbe disease: electrodiagnostic findings. *Neurology* 2006;**67**:263–7.
155. Haberlandt E, Scholl-Burgi S, Neuberger J, Felber S, Gotwald T, Sauter R, et al. Peripheral neuropathy as the sole initial finding in three children with infantile metachromatic leukodystrophy. *Eur J Paediatr Neurol* 2009;**13**:257–60.
156. De Silva KL, Pearce J. Neuropathy of metachromatic leucodystrophy. *J Neurol Neurosurg Psychiatry* 1973;**36**:30–3.
157. Korn-Lubetzki I, Dor-Wollman T, Soffer D, Raas-Rothschild A, Hurvitz H, Nevo Y. Early peripheral nervous system manifestations of infantile Krabbe disease. *Pediatr Neurol* 2003;**28**:115–18.
158. Zara F, Biancheri R, Bruno C, Bordo L, Assereto S, Gazzerro E, et al. Deficiency of hyccin, a newly identified membrane protein, causes hypomyelination and congenital cataract. *Nat Genet* 2006;**38**:1111–13.
159. Biancheri R, Zara F, Bruno C, Rossi A, Bordo L, Gazzerro E, et al. Phenotypic characterization of hypomyelination and congenital cataract. *Ann Neurol* 2007;**62**:121–7.
160. Traverso M, Yuregir OO, Mimouni-Bloch A, Rossi A, Aslan H, Gazzerro E, et al. Hypomyelination and congenital cataract: identification of novel mutations in two unrelated families. *Eur J Paediatr Neurol* 2013;**17**:108–11.
161. Ugur SA, Tolun A. A deletion in DRCTNNB1A associated with hypomyelination and juvenile onset cataract. *Eur J Hum Genet* 2008;**16**:261–4.
162. Rossi A, Biancheri R, Zara F, Bruno C, Uziel G, van der Knaap MS, et al. Hypomyelination and congenital cataract: neuroimaging features of a novel inherited white matter disorder. *AJNR Am J Neuroradiol* 2008;**29**:301–5.

163. Gazzerro E, Baldassari S, Giacomini C, Musante V, Fruscione F, La Padula V, et al. Hyccin, the molecule mutated in the leukodystrophy hypomyelination and congenital cataract (HCC), is a neuronal protein. *PLOS ONE* 2012;**7**:e32180.
164. Pingault V, Bondurand N, Kuhlbrodt K, Goerich DE, Prehu MO, Puliti A, et al. SOX10 mutations in patients with Waardenburg-Hirschsprung disease. *Nat Genet* 1998;**18**:171–3.
165. Bondurand N, Dastot-Le Moal F, Stanchina L, Collot N, Baral V, Marlin S, et al. Deletions at the SOX10 gene locus cause Waardenburg syndrome types 2 and 4. *Am J Hum Genet* 2007;**81**:1169–85.
166. Pingault V, Bodereau V, Baral V, Marcos S, Watanabe Y, Chaoui A, et al. Loss-of-function mutations in SOX10 cause Kallmann syndrome with deafness. *Am J Hum Genet* 2013;**92**:707–24.
167. Inoue K, Tanabe Y, Lupski JR. Myelin deficiencies in both the central and the peripheral nervous systems associated with a SOX10 mutation. *Ann Neurol* 1999;**46**:313–18.
168. Inoue K, Khajavi M, Ohyama T, Hirabayashi S, Wilson J, Reggin JD, et al. Molecular mechanism for distinct neurological phenotypes conveyed by allelic truncating mutations. *Nat Genet* 2004;**36**:361–9.
169. Touraine RL, Attie-Bitach T, Manceau E, Korsch E, Sarda P, Pingault V, et al. Neurological phenotype in Waardenburg syndrome type 4 correlates with novel SOX10 truncating mutations and expression in developing brain. *Am J Hum Genet* 2000;**66**:1496–503.
170. Inoue K, Shilo K, Boerkoel CF, Crowe C, Sawady J, Lupski JR, et al. Congenital hypomyelinating neuropathy, central dysmyelination, and Waardenburg-Hirschsprung disease: phenotypes linked by SOX10 mutation. *Ann Neurol* 2002;**52**:836–42.
171. Pingault V, Guiochon-Mantel A, Bondurand N, Faure C, Lacroix C, Lyonnet S, et al. Peripheral neuropathy with hypomyelination, chronic intestinal pseudo-obstruction and deafness: a developmental "neural crest syndrome" related to a SOX10 mutation. *Ann Neurol* 2000;**48**:671–6.
172. Verheij JB, Sival DA, van der Hoeven JH, Vos YJ, Meiners LC, Brouwer OF, et al. Shah-Waardenburg syndrome and PCWH associated with SOX10 mutations: a case report and review of the literature. *Eur J Paediatr Neurol* 2006;**10**:11–17.
173. Kuhlbrodt K, Herbarth B, Sock E, Hermans-Borgmeyer I, Wegner M. Sox10, a novel transcriptional modulator in glial cells. *J Neurosci* 1998;**18**:237–50.
174. Britsch S, Goerich DE, Riethmacher D, Peirano RI, Rossner M, Nave KA, et al. The transcription factor Sox10 is a key regulator of peripheral glial development. *Genes Dev* 2001;**15**:66–78.
175. Nance MA, Berry SA. Cockayne syndrome: review of 140 cases. *Am J Med Genet* 1992;**42**:68–84.
176. Neilan EG. Cockayne syndrome. In: Pagon RA, Bird TC, Dolan CR, Stephens K, editors. *GeneReviews at GeneTests: Medical Genetics Information Resource (database online)*. Seattle: Copyright, University of Washington; 1997–2011. Available at <http://www.genetests.org> Updated Mar 07 2006.
177. Vos A, Gabreels-Festen A, Joosten E, Gabreels F, Renier W, Mullaart R. The neuropathy of Cockayne syndrome. *Acta Neuropathol* 1983;**61**:153–6.
178. Schenone A, Rolando S, Ferrari M, Romagnoli P, Tabaton M, Mancardi GL. Peripheral neuropathy in Cockayne syndrome. *Ital J Neurol Sci* 1986;**7**:447–52.
179. Varon R, Gooding R, Steglich C, Marns L, Tang H, Angelicheva D, et al. Partial deficiency of the C-terminal-domain phosphatase of RNA polymerase II is associated with congenital cataracts facial dysmorphism neuropathy syndrome. *Nat Genet* 2003;**35**:185–9.
180. Mullner-Eidenbock A, Moser E, Klebermass N, Amon M, Walter MC, Lochmuller H, et al. Ocular features of the congenital cataracts facial dysmorphism neuropathy syndrome. *Ophthalmology* 2004;**111**:1415–23.
181. Tournev I, Kalaydjieva L, Youl B, Ishpekova B, Guergueltcheva V, Kamenov O, et al. Congenital cataracts facial dysmorphism neuropathy syndrome, a novel complex genetic disease in Balkan Gypsies: clinical and electrophysiological observations. *Ann Neurol* 1999;**45**:742–50.
182. Kalaydjieva L, Lochmuller H, Tournev I, Baas F, Beres J, Colomer J, et al. 125th ENMC International Workshop: Neuromuscular disorders in the Roma (Gypsy) population, 23–25 April 2004, Naarden, the Netherlands. *Neuromuscul Disord* 2005;**15**:65–71.
183. Tournev I, King RH, Workman J, Nourallah M, Muddle JR, Kalaydjieva L, et al. Peripheral nerve abnormalities in the congenital cataracts facial dysmorphism neuropathy (CCFDN) syndrome. *Acta Neuropathol* 1999;**98**:165–70.
184. Shabo G, Scheffer H, Cruysberg JR, Lammens M, Pasman JW, Spruit M, et al. Congenital cataract facial dysmorphism neuropathy syndrome: a clinically recognizable entity. *Pediatric Neurology* 2005;**33**:277–9.
185. Zuchner S, Mersiyanova IV, Muglia M, Bissar-Tadmouri N, Rochelle J, Dadali EL, et al. Mutations in the mitochondrial GTPase mitofusin 2 cause Charcot-Marie-Tooth neuropathy type 2A. *Nat Genet* 2004;**36**:449–51.
186. Verhoeven K, Claeys KG, Zuchner S, Schroder JM, Weis J, Ceuterick C, et al. MFN2 mutation distribution and genotype/phenotype correlation in Charcot-Marie-Tooth type 2 *Brain* 2006;**129**:2093–102.
187. Chung KW, Kim SB, Park KD, Choi KG, Lee JH, Eun HW, et al. Early onset severe and late-onset mild Charcot-Marie-Tooth disease with mitofusin 2 (MFN2) mutations. *Brain* 2006;**129**:2103–18.
188. Ouvrier RA, McLeod JG, Morgan GJ, Wise GA, Conchin TE. Hereditary motor and sensory neuropathy of neuronal type with onset in early childhood. *J Neurol Sci* 1981;**51**:181–97.
189. Gabreels-Festen AA, Joosten EM, Gabreels FJ, Jennekens FG, Gooskens RH, Stegeman DF. Hereditary motor and sensory neuropathy of neuronal type with onset in early childhood. *Brain* 1991;**114**:1855–70.
190. Ouvrier R, Grew S. Mechanisms of disease and clinical features of mutations of the gene for mitofusin 2: an important cause of hereditary peripheral neuropathy with striking clinical variability in children and adults. *Dev Med Child Neurol* 2010;**52**:328–30.
191. Feely SM, Laura M, Siskind CE, Sottile S Davis M, Gibbons VS, et al. MFN2 mutations cause severe phenotypes in most patients with CMT2A. *Neurology* 2011;**76**:1690–6.
192. Zuchner S, De Jonghe P, Jordanova A, Claeys KG, Guergueltcheva V, Cherninkova S, et al. Axonal neuropathy with optic atrophy is caused by mutations in mitofusin 2. *Ann Neurol* 2006;**59**:276–81.

193. Brockmann K, Dreha-Kulaczewski S, Dechent P, Bonnemann C, Helms G, Kyllerman M, et al. Cerebral involvement in axonal Charcot-Marie-Tooth neuropathy caused by mitofusin2 mutations. *J Neurol* 2008;**255**:1049–58.
194. Vallat JM, Ouvrier RA, Pollard JD, Magdelaine C, Zhu D, Nicholson GA, et al. Histopathological findings in hereditary motor and sensory neuropathy of axonal type with onset in early childhood associated with mitofusin 2 mutations. *J Neuropathol Exp Neurol* 2008;**67**:1097–102.
195. Baloh RH, Schmidt RE, Pestronk A, Milbrandt J. Altered axonal mitochondrial transport in the pathogenesis of Charcot-Marie-Tooth disease from mitofusin 2 mutations. *J Neurosci* 2007;**27**:422–30.
196. Chen H, Chomyn A, Chan DC. Disruption of fusion results in mitochondrial heterogeneity and dysfunction. *J Biol Chem* 2005;**280**:26185–92.
197. de Brito OM, Scorrano L. Mitofusin 2 tethers endoplasmic reticulum to mitochondria. *Nature* 2008;**456**:605–10.
198. Chen H, Vermulst M, Wang YE, Chomyn A, Prolla TA, McCaffery JM, et al. Mitochondrial fusion is required for mtDNA stability in skeletal muscle and tolerance of mtDNA mutations. *Cell* 2010;**141**:280–9.
199. Nicholson GA, Magdelaine C, Zhu D, Grew S, Ryan MM, Sturtz F, et al. Severe early-onset axonal neuropathy with homozygous and compound heterozygous MFN2 mutations. *Neurology* 2008;**70**:1678–81.
200. Lawson VH, Graham BV, Flanigan KM. Clinical and electrophysiologic features of CMT2A with mutations in the mitofusin 2 gene. *Neurology* 2005;**65**:197–204.
201. Auer-Grumbach M, Olschewski A, Papic L, Kremer H, McEntagart ME, Uhrig S, et al. Alterations in the ankyrin domain of TRPV4 cause congenital distal SMA, scapuloperoneal SMA and HMSN2C. *Nat Genet* 2010;**42**:160–4.
202. Deng HX, Klein CJ, Yan J, Shi Y, Wu Y, Fecto F, et al. Scapuloperoneal spinal muscular atrophy and CMT2C are allelic disorders caused by alterations in TRPV4. *Nat Genet* 2010;**42**:165–9.
203. Deng HX, Klein CJ, Yan J, Shi Y, Wu Y, Fecto F, et al. Mutations in TRPV4 cause Charcot-Marie-Tooth disease type 2C. *Nat Genet* 2010;**42**:170–4.
204. Dyck PJ, Litchy WJ, Minnerath S, Bird TD, Chance PF, Schaid DJ, et al. Hereditary motor and sensory neuropathy with diaphragm and vocal cord paresis. *Ann Neurol* 1994;**35**:608–15.
205. DeLong R, Siddique T. A large New England kindred with autosomal dominant neurogenic scapuloperoneal amyotrophy with unique features. *Arch Neurol* 1992;**49**:905–8.
206. Fleury P, Hageman G. A dominantly inherited lower motor neuron disorder presenting at birth with associated arthrogryposis. *J Neurol Neurosurg Psychiatry* 1985;**48**:1037–48.
207. Zimon M, Baets J, Auer-Grumbach M, Berciano J, Garcia A, Lopez-Laso E, et al. Dominant mutations in the cation channel gene transient receptor potential vanilloid 4 cause an unusual spectrum of neuropathies. *Brain* 2010;**133**:1798–809.
208. Rock MJ, Prenen J, Funari VA, Funari TL, Merriman B, Nelson SF, et al. Gain-of-function mutations in TRPV4 cause autosomal dominant brachyolmia. *Nat Genet* 2008;**40**:999–1003.
209. Krakow D, Vriens J, Camacho N, Luong P, Deixler H, Funari TL, et al. Mutations in the gene encoding the calcium-permeable ion channel TRPV4 produce spondylometaphyseal dysplasia, Kozlowski type and metatropic dysplasia. *Am J Hum Genet* 2009;**84**:307–15.
210. Nishimura G, Dai J, Lausch E, Unger S, Megarbane A, Kitoh H, et al. Spondylo-epiphyseal dysplasia, Maroteaux type (pseudo-Morquio syndrome type 2), and parastremmatic dysplasia are caused by TRPV4 mutations. *Am J Med Genet A* 2010;**152A**:1443–9.
211. Chen DH, Sul Y, Weiss M, Hillel A, Lipe H, Wolff J, et al. CMT2C with vocal cord paresis associated with short stature and mutations in the TRPV4 gene. *Neurology* 2010;**75**:1968–75.
212. Mercuri E. Congenital form of spinal muscular atrophy predominantly affecting the lower limbs: a clinical and muscle MRI study. *Neuromuscul Disord* 2004;**14**:125–9.
213. Cho TJ, Matsumoto K, Fano V, Dai J, Kim OH, Chae JH, et al. TRPV4-pathy manifesting both skeletal dysplasia and peripheral neuropathy: A report of three patients. *Am J Med Genet A* 2012;**158A**:795–802.
214. Reddel S, Ouvrier R, Nicholson G, Dierick I, Irobi J, Timmerman V, et al. Autosomal dominant congenital spinal muscular atrophy—A possible developmental deficiency of motor neurones? *Neuromuscul Disord* 2008;**18**:530–5.
215. Frijns CJ, Van Deutekom J, Frants RR, Jennekens FG. Dominant congenital benign spinal muscular atrophy. *Muscle Nerve* 1994;**17**:192–7.
216. Santoro L, Manganelli F, Di Maio L, Barbieri F, Carella M, D'Adamo P, et al. Charcot-Marie-Tooth disease type 2C: a distinct genetic entity. Clinical and molecular characterization of the first European family. *Neuromuscul Disord* 2002;**12**:399–404.
217. McEntagart ME, Reid SL, Irrthum A, Douglas JB, Eyre KE, Donaghy MJ, et al. Confirmation of a hereditary motor and sensory neuropathy IIC locus at chromosome 12q23-q24. *Ann Neurol* 2005;**57**:293–7.
218. Plant TD, Strotmann R. TRPV4: A multifunctional nonselective cation channel with complex regulation. In: Liedtke WB, Heller S, editors. *TRP Ion Channel Function in Sensory Transduction and Cellular Signaling Cascades*. Boca Raton (FL): CRC Press; 2007.
219. Liedtke W. Molecular mechanisms of TRPV4-mediated neural signaling. *Ann NY Acad Sci* 2008;**1144**:42–52.
220. Weedon MN, Hastings R, Caswell R, Xie W, Paszkiewicz K, Antoniadi T, et al. Exome sequencing identifies a DYNC1H1 mutation in a large pedigree with dominant axonal Charcot-Marie-Tooth disease. *Am J Hum Genet* 2011;**89**:308–12.
221. Tsurusaki Y, Saitoh S, Tomizawa K, Sudo A, Asahina N, Shiraishi H, et al. A DYNC1H1 mutation causes a dominant spinal muscular atrophy with lower extremity predominance. *Neurogenetics* 2012;**13**:327–32.
222. Harms MB, Ori-McKenney KM, Scoto M, Tuck EP, Bell S, Ma D, et al. Mutations in the tail domain of DYNC1H1 cause dominant spinal muscular atrophy. *Neurology* 2012;**78**:1714–20.
223. Poirier K, Lebrun N, Broix L, Tian G, Saillour Y, Boscheron C, et al. Mutations in TUBG1, DYNC1H1, KIF5C and KIF2A cause malformations of cortical development and microcephaly. *Nat Genet* 2013;**45**:639–47.
224. Willemsen MH, Vissers LE, Willemsen MA, van Bon BW, Kroes T, de Ligt J, et al. Mutations in DYNC1H1 cause severe intellectual disability with neuronal migration defects. *J Med Genet* 2012;**49**:179–83.

225. Pfister KK, Shah PR, Hummerich H, Russ A, Cotton J, Annuar AA, et al. Genetic analysis of the cytoplasmic dynein subunit families. *PLOS Genet* 2006;**2**:e1.

226. Ylikallio E, Poyhonen R, Zimon M, De Vriendt E, Hilander T, Paetau A, et al. Deficiency of the E3 ubiquitin ligase TRIM2 in early-onset axonal neuropathy. *Hum Mol Genet* 2013;**22**:2975−83.

227. Vedanarayanan VV, Smith S, Subramony SH, Bock GO, Evans OB. Lethal neonatal autosomal recessive axonal sensorimotor polyneuropathy. *Muscle Nerve* 1998;**21**:1473−7.

228. Bomont P, Cavalier L, Blondeau F, Ben Hamida C, Belal S, Tazir M, et al. The gene encoding gigaxonin, a new member of the cytoskeletal BTB/kelch repeat family, is mutated in giant axonal neuropathy. *Nat Genet* 2000;**26**:370−4.

229. Berg BO, Rosenberg SH, Asbury AK. Giant axonal neuropathy. *Pediatrics* 1972;**49**:894−9.

230. Bruno C, Bertini E, Federico A, Tonoli E, Lispi ML, Cassandrini D, et al. Clinical and molecular findings in patients with giant axonal neuropathy (GAN). *Neurology* 2004;**62**:13−16.

231. Kuhlenbaumer G, Young P, Oberwittler C, Hunermund G, Schirmacher A, Domschke K, et al. Giant axonal neuropathy (GAN): case report and two novel mutations in the gigaxonin gene. *Neurology* 2002;**58**:1273−6.

232. Koop O, Schirmacher A, Nelis E, Timmerman V, De Jonghe P, Ringelstein B, et al. Genotype-phenotype analysis in patients with giant axonal neuropathy (GAN). *Neuromuscul Disord* 2007;**17**: 624−30.

233. Demir E, Bomont P, Erdem S, Cavalier L, Demirci M, Kose G, et al. Giant axonal neuropathy: clinical and genetic study in six cases. *J Neurol Neurosurg Psychiatry* 2005;**76**:825−32.

234. Tazir M, Nouioua S, Magy L, Huehne K, Assami S, Urtizberea A, et al. Phenotypic variability in giant axonal neuropathy. *Neuromuscul Disord* 2009;**19**:270−4.

235. Ben Hamida M, Hentati F, Ben Hamida C. Giant axonal neuropathy with inherited multisystem degeneration in a Tunisian kindred. *Neurology* 1990;**40**:245−50.

236. Roth LA, Johnson-Kerner BL, Marra JD, Lamarca NH, Sproule DM. The absence of curly hair is associated with a milder phenotype in Giant Axonal Neuropathy. *Neuromuscul Disord* 2013.

237. Houlden H, Groves M, Miedzybrodzka Z, Roper H, Willis T, Winer J, et al. New mutations, genotype phenotype studies and manifesting carriers in giant axonal neuropathy. *J Neurol Neurosurg Psychiatry* 2007;**78**:1267−70.

238. Bomont P, Ioos C, Yalcinkaya C, Korinthenberg R, Vallat JM, Assami S, et al. Identification of seven novel mutations in the GAN gene. *Hum Mutat* 2003;**21**:446.

239. Fabrizi GM, Cavallaro T, Angiari C, Cabrini I, Taioli F, Malerba G, et al. Charcot-Marie-Tooth disease type 2E, a disorder of the cytoskeleton. *Brain* 2007;**130**:394−403.

240. Jaffer F, Murphy SM, Scoto M, Healy E, Rossor AM, Brandner S, et al. BAG3 mutations: another cause of giant axonal neuropathy. *J Peripher Nerv Syst* 2012;**17**:210−16.

241. Selcen D, Muntoni F, Burton BK, Pegoraro E, Sewry C, Bite AV, et al. Mutation in BAG3 causes severe dominant childhood muscular dystrophy. *Ann Neurol* 2009;**65**:83−9.

242. Allen E, Ding J, Wang W, Pramanik S, Chou J, Yau V. Gigaxonin-controlled degradation of MAP1B light chain is critical to neuronal survival. *Nature* 2005;**438**:224−8.

243. Ding J, Allen E, Wang W, Valle A, Wu C, Nardine T, et al. Gene targeting of GAN in mouse causes a toxic accumulation of microtubule-associated protein 8 and impaired retrograde axonal transport. *Hum Mol Genet* 2006;**15**:1451−63.

244. De Braekeleer M, Dallaire A, Mathieu J. Genetic epidemiology of sensorimotor polyneuropathy with or without agenesis of the corpus callosum in northeastern Quebec. *Human Genetics* 1993;**91**:223−7.

245. Dupre N, Howard HC, Mathieu J, Karpati G, Vanasse M, Bouchard JP, et al. Hereditary motor and sensory neuropathy with agenesis of the corpus callosum. *Ann Neurol* 2003;**54** 9−18.

246. Howard HC, Mount DB, Rochefort D, Byun N, Dupre N, Lu J, et al. The K-Cl cotransporter KCC3 is mutant in a severe peripheral neuropathy associated with agenesis of the corpus callosum. *Nat Genet* 2002;**32**:384−92.

247. Salin-Cantegrel A, Riviere JB, Dupre N, Charron FM, Shekarabi M, Karemera L, et al. Distal truncation of KCC3 in non-French Canadian HMSN/ACC families. *Neurology* 2007;**69**: 1350−5.

248. Mathieu J, Bedard F, Prevost C, Langevin P. [Motor and sensory neuropathies with or without agenesis of the corpus callosum: a radiological study of 64 cases]. *Can J Neurol Sci* 1990;**17**:103−8.

249. Larbrisseau A, Vanasse M, Brochu P, Jasmin G. The Andermann syndrome: agenesis of the corpus callosum associated with mental retardation and progressive sensorimotor neuronopathy. *Can J Neurol Sci* 1984;**11**:257−61.

250. Filteau MJ, Pourcher E, Bouchard RH, Baruch P, Mathieu J, Bedard F, et al. Corpus callosum agenesis and psychosis in Andermann syndrome. *Arch Neurol* 1991;**48**:1275−80.

251. Aicardi J, Castelein P. Infantile neuroaxonal dystrophy. *Brain* 1979;**102**:727−48.

252. Kurian MA, Morgan NV, MacPherson L, Foster K, Peake D, Gupta R, et al. Phenotypic spectrum of neurodegeneration associated with mutations in the PLA2G6 gene (PLAN). *Neurology* 2008;**70**:1623−9.

253. Nardocci N, Zorzi G, Farina L, Binelli S, Scaioli W, Ciano C, et al. Infantile neuroaxonal dystrophy: clinical spectrum and diagnostic criteria. *Neurology* 1999;**52**:1472−8.

254. Gregory A, Westaway SK, Holm IE, Kotzbauer PT, Hogarth P, Sonek S, et al. Neurodegeneration associated with genetic defects in phospholipase A(2). *Neurology* 2008;**71**:1402−9.

255. Farina L, Nardocci N, Bruzzone MG, D'Incerti L, Zorzi G, Verga L, et al. Infantile neuroaxonal dystrophy: neuroradiological studies in 11 patients. *Neuroradiology* 1999;**41**:376−80

256. Morgan NV, Westaway SK, Morton JE, Gregory A, Gissen P, Sonek S, et al. PLA2G6, encoding a phospholipase A2, is mutated in neurodegenerative disorders with high brain iron. *Nat Genet* 2006;**38**:752−4.

257. Khateeb S, Flusser H, Ofir R, Shelef I, Narkis G, Vardi G, et al. PLA2G6 mutation underlies infantile neuroaxonal dystrophy. *Am J Hum Genet* 2006;**79**:942−8.

258. Paisan-Ruiz C, Bhatia KP, Li A, Hernandez D, Davis M, Wood NW, et al. Characterization of PLA2G6 as a locus for dystonia-parkinsonism. *Ann Neurol* 2009;**65**:19−23.

259. Koroglu C, Seven M, Tolun A. Recessive truncating NALCN mutation in infantile neuroaxonal dystrophy with facial dysmorphism. *J Med Genet* 2013;**50**:515−20.

260. Chau V, Clement JF, Robitaille Y, D'Anjou G, Vanasse M. Congenital axonal neuropathy and encephalopathy. *Pediatr Neurol* 2008;**38**:261–6.
261. Guzzetta F, Ferriere G. Congenital neuropathy with prevailing axonal changes. A clinical and histological report. *Acta Neuropathol* 1985;**68**:185–90.
262. Mizuno T, Sasaki M, Komaki H, Sakuma H, Saito Y, Nakagawa E, et al. A case of congenital axonal neuropathy associated with West syndrome. *Brain Dev* 2011;**33**:692–6.
263. Korinthenberg R, Sauer M, Ketelsen UP, Hanemann CO, Stoll G, Graf M, et al. Congenital axonal neuropathy caused by deletions in the spinal muscular atrophy region. *Ann Neurol* 1997;**42**:364–8.
264. Rudnik-Schoneborn S, Sztriha L, Aithala GR, Houge G, Laegreid LM, Seeger J, et al. Extended phenotype of pontocerebellar hypoplasia with infantile spinal muscular atrophy. *Am J Med Genet A* 2003;**117A**:10–17.
265. Wilmshurst JM, Bye A, Rittey C, Adams C, Hahn AF, Ramsay D, et al. Severe infantile axonal neuropathy with respiratory failure. *Muscle Nerve* 2001;**24**:760–8.
266. den Boer ME, Dionisi-Vici C, Chakrapani A, van Thuijl AO, Wanders RJ, Wijburg FA. Mitochondrial trifunctional protein deficiency: a severe fatty acid oxidation disorder with cardiac and neurologic involvement. *J Pediatr* 2003;**142**:684–9.
267. den Boer ME, Wanders RJ, Morris AA, Heymans LIJ, Wijburg HS, Long-chain FA. 3-hydroxyacyl-CoA dehydrogenase deficiency: clinical presentation and follow-up of 50 patients. *Pediatrics* 2002;**109**:99–104.
268. Spiekerkoetter U, Bennett MJ, Ben-Zeev B, Strauss AW, Tein I. Peripheral neuropathy, episodic myoglobinuria, and respiratory failure in deficiency of the mitochondrial trifunctional protein. *Muscle Nerve* 2004;**29**:66–72.
269. Tang J, Dai Y, Li M, Cheng M, Hong S, Jiang L, et al. Guillain-Barré syndrome in Chinese children: a retrospective analysis. *Pediatr Neurol* 2011;**45**:233–7.
270. Jackson AH, Baquis GD, Shah BL. Congenital Guillain-Barré syndrome. *J Child Neurol* 1996;**11**:407–10.
271. al-Qudah AA, Shahar E, Logan WJ, Murphy EG. Neonatal Guillain-Barré syndrome. *Pediatr Neurol* 1988;**4**:255–6.
272. Roodbol J, de Wit MC, Walgaard C, de Hoog M, Catsman-Berrevoets CE, Jacobs BC, et al. Recognizing Guillain-Barré syndrome in preschool children. *Neurology* 2011;**76**:807–10.
273. Vasconcelos A, Abecasis F, Monteiro R, Camilo C, Vieira M, de Carvalho M, et al. A 3-month-old baby with H1N1 and Guillain-Barré syndrome. *BMJ Case Rep* 2012;2012. Available from: http://dx.doi.org/10.1136/bcr.12.2011.5462.
274. Luijckx GJ, Vles J, de Baets M, Buchwald B, Troost J. Guillain-Barré syndrome in mother and newborn child. *Lancet* 1997;**349**:27.
275. Herguner MO, Incecik F, Altunbasak S. Cyclosporin treatment in three children with chronic inflammatory demyelinating neuropathy. *Pediatr Neurol* 2009;**41**:223–5.
276. Majumdar A, Hartley L, Manzur AY, King RH, Orrell RW, Muntoni F. A case of severe congenital chronic inflammatory demyelinating polyneuropathy with complete spontaneous remission. *Neuromuscul Disord* 2004;**14**:818–21.
277. Pearce J, Pitt M, Martinez A. A neonatal diagnosis of congenital chronic inflammatory demyelinating polyneuropathy. *Dev Med Child Neurol* 2005;**47**:489–92.
278. Plante-Bordeneuve V, Parman Y, Guiochon-Mantel A, Alj Y, Deymeer F, Serdaroglu P, et al. The range of chronic demyelinating neuropathy of infancy: a clinico-pathological and genetic study of 15 unrelated cases. *J Neurol* 2001;**248**:795–803.
279. Levy BK, Fenton GA, Loaiza S, Hayat GR. Unexpected recovery in a newborn with severe hypomyelinating neuropathy. *Pediatr Neurol* 1997;**16**:245–8.

Chapter 17

Hereditary Neuropathies in Late Childhood and Adolescence

Kathryn M. Brennan and Michael E. Shy

DEFINITION AND NOMENCLATURE

This chapter will focus on those hereditary neuropathies that can present in late childhood and adolescence. Inherited neuropathies are often referred to collectively as Charcot-Marie-Tooth (CMT) disease, an eponym used in recognition of the three men who initially described the disorder.[1,2] CMT is, however, an umbrella term that encompasses a wide variety of inherited sensory and/or motor neuropathies. These inherited neuropathies also include other closely related neuropathies such as hereditary neuropathy with liability to pressure palsy (HNPP); distal hereditary motor neuropathy (dHMN), also known as distal spinal muscular atrophy (dSMA); hereditary sensory and autonomic neuropathy (HSAN/HSN); and neuropathies associated with inherited metabolic disease (see Table 17.1). HSAN/HSN will be discussed in Chapter 18 and neuropathies associated with inherited metabolic disease will be discussed in Chapter 19.

The first step is to determine whether the patient has a genetic neuropathy. Where there is an affected parent, either an autosomal dominant (AD) or X-linked (if there is no definite male-to-male transmission) inheritance is likely. If there are multiple affected siblings and/or consanguineous parents, then autosomal recessive (AR) inheritance is likely. However, recognizing CMT can be challenging, particularly where there is no family history or if families are small. Factors pointing to an inherited neuropathy in such circumstances include presentations in infancy, slow progression, the presence of foot deformities, and the lack of positive sensory symptoms (dysesthesias, paresthesias) in the presence of clear sensory signs. Many patients may have undergone foot surgery in childhood or report difficulties with sports at school

CMT is the most common inherited neurological condition with a prevalence of around 1:2500.[3] A heterogeneous disorder, CMT is divided into subtypes based on the pattern of inheritance and also by neurophysiological studies. Subtypes include AD demyelinating (CMT 1), AD axonal (CMT 2), AR (CMT 4), and X-linked (CMTX).[4] CMT 1 typically has slow nerve conduction velocities (less than 38 m/s in the upper extremities) and pathological evidence of a hypertrophic demyelinating neuropathy, whereas CMT 2 has relatively normal nerve conduction velocities with evidence of axonal degeneration.[5]

As most forms of CMT have both motor and sensory involvement, CMT 1 and 2 are often classified as hereditary motor and sensory neuropathy (HMSN) type I or II. Neurophysiology studies have also identified another group, CMT Intermediate, with nerve conduction demonstrating intermediate velocities (less than and greater than 38 m/s). Originally severe forms of CMT with uninformative family histories were thought to be AR in inheritance and were classified as HMSN III or termed Dejerine-Sottas syndrome.[6] Advances in molecular technology have led to a slight revision of this classification with most of the patients in the AR HMSN III group demonstrated to have spontaneous mutations in dominantly inherited genes. The term HMSN III is no longer used; however, Dejerine-Sottas syndrome is still used to describe severely affected infants with CMT and typically very slow nerve conduction velocities.

The discovery of causal genes has led to the modification of the classification of CMT further to include the gene. The groups (e.g. CMT1, CMT, CMTX, as outlined in Table 17.1) remain but letters have been added to incorporate the specific gene that causes the disorder. Each type of CMT is now subdivided according to the specific genetic cause of the neuropathy (see Table 17.2). For example, the most common form of CMT1, termed CMT1A, is caused by a duplication of a fragment of chromosome 17 containing the peripheral myelin protein 22-kDa (*PMP22*) gene, while CMT1B is caused by mutations in the myelin protein zero (*MPZ*) gene (see later).

B.T. Darras, H. Royden Jones, Jr., M.M. Ryan & D.C. De Vivo (Eds): Neuromuscular Disorders of Infancy, Childhood and Adolescence, Second edition.
DOI: http://dx.doi.org/10.1016/B978-0-12-417044-5.00017-2

TABLE 17.1 A Basic Classification of CMT and Inherited Neuropathies

CMT Subtype	Inheritance Pattern	Neurophysiology
CMT1	AD	NCV <38 m/s
CMT2 CMTX	AD X-linked	NCV >38 m/s & axonal Slow/intermediate velocities
CMT4 CMT5 CMT6	AR AD AD	Variable NCV >38 m/s & axonal NCV >38 m/s & axonal
CMT Intermediate	AD or AR	NCV < & > 38 m/s
Dejerine-Sottas	AD or AR	NCV <15 m/s & axonal
HSAN/HSN dHMN/dSMA	AD or AR AD or AR	NCV >38 m/s & axonal – sensory only NCV >38 m/s & axonal – motor only

Abbreviations: AD, autosomal dominant; AR, autosomal recessive; HSAN/HSN, hereditary sensory and autonomic neuropathy/hereditary sensory neuropathy; dHMN, distal hereditary motor neuropathy; dSMA, distal spinal muscular atrophy.

Currently, mutations in more than 70 genes have been identified as causes of inherited neuropathies.

EPIDEMIOLOGY

The prevalence of CMT is about 1 in 2500, without ethnic predisposition.[3] The 17p11.2 duplication causing CMT1A accounts for 60% to 70% of CMT1 patients, CMT1X for approximately 20% of CMT1 cases, and CMT1B for less than 10% of CMT1 cases. Only around 25% of patients with CMT2 receive a genetic diagnosis with the majority accounted for by CMT2A.[7,8] The prevalence of HNPP is not well known, but this subtype probably represents around 6% of the total CMT population.[17]

PATHOBIOLOGY

A common feature of most genes mutated in CMT is the role they play in maintaining the structure or function of cellular components of the peripheral nervous system, Schwann cells, and the axons of the peripheral nerve (Figure 17.1).

The first genes identified to cause CMT express proteins that are essential for compact myelin (peripheral myelin protein 22 [*PMP22*] and myelin protein zero [*MPZ*]) and noncompact myelin (gap junction protein beta 1 [*GJB1*]),[9] and their altered expressions cause demyelination or dysmyelination. A novel concept derived from the identification of PMP22 duplication as the basic pathomechanism in CMT 1A is that of gene-protein dosage. It has become clear that correct stoichiometry of *PMP22* is necessary to maintain compact myelin integrity. Too much *PMP22* (i.e. with a duplication) causes CMT1A; too little (as seen with a deletion) causes hereditary neuropathy with liability to pressure palsies.[10] Abnormal expression of *MPZ* also causes demyelination, although in this case it is usually due to point mutations in the *MPZ* gene.

An important biologic feature common to both neurons and Schwann cells is their highly specialized and polarized cellular architecture.[11] Although the polarization of neurons is a well-recognized feature of these cells, with their axons extending more than 1 m in humans, Schwann cells are also very polarized because their membranes have to expand while they concentrically wrap around axons. To overcome the long distances between the cell nucleus and the more distal segments of the membrane, Schwann cells have areas of noncompact myelin rich in gap junctions that provide a radial pathway directly to the layers of myelin sheath. Connexin 32 (Cx32), the protein expressed by the *GJB1* gene, is the main component of gap junctions in the myelin of Schwann cells and this may explain, at least in part, why *GJB1* mutations cause CMT1X.[12] The high polarization of neurons and Schwann cells may also explain why mutations in ubiquitously expressed genes, such as mitofusin 2 (*MFN2*), ganglioside-induced differentiation-associated protein 1 (*GDAP1*), or glycyl-tRNA synthetase (*GARS*), cause preferential dysfunction of the peripheral nervous system. The length-dependent neuropathy commonly found in patients with CMT seems to support this hypothesis that distal peripheral axons are especially susceptible to disruptions in organelle and metabolite axonal transport.

Schwann cells and axons interact at multiple points along the peripheral nerve, including the adaxonal (opposing the axon) membrane, paranodal myelin loops, microvilli, and juxtaparanodal basal lamina. These interactions are mutually beneficial, providing trophic support to the axon and myelinating cues to the Schwann cell. An example of this important interaction is the occurrence of secondary axonal degeneration in all forms of demyelinating CMT. This axonal degeneration is deemed to occur as a consequence of ineffective Schwann cell support to the axon and is actually more directly related to the clinical functional impairment than the demyelination itself.[13]

Several recent studies have demonstrated a susceptibility of Schwann cells to mutations yielding misfolded proteins, as seen in certain *PMP22*[14] and *MPZ*[15,16] point mutations. Misfolded proteins may accumulate in the endoplasmic reticulum (ER) of Schwann cells inducing a transitory unfolded protein response (UPR), a series of cellular events that helps the ER to cope with the increased metabolic demand caused by retention of the misfolded protein. This, in turn, causes downregulation of the myelination program genes and dedifferentiation of Schwann cells, a toxic gain of function that worsens with the

TABLE 17.2 Classification of Charcot-Marie-Tooth Disease Subtypes

Type	Gene/Locus	Specific Phenotype
Autosomal dominant		
CMT1 (AD CMT1) CMT1A	Dup 17p (*PMP22*)	Classic CMT1
CMT1B	*MPZ*	CMT1/DSN/CHN/intermediate/CMT2
CMT1C	*LITAF*	Classic CMT1
CMT1D	*EGR2*	Classic CMT1/DSN/CHN
CMT1E	*PMP22* (point mutation)	Classic CMT1/DSN/CHN
CMT1F	*NEFL*	CMT2 but can have slow MCVs in CMT1 range +/− early onset severe disease Classic CMT1/DSN/CHN
Hereditary neuropathy with liability to pressure palsies (HNPP)		
HNPP	Del 17p (*PMP-22*)	Typical HNPP
	PMP-22 (point mutation)	Typical HNPP
X linked CMT1 (CMT1X)		
CMT1X	*GJB1*	Intermediate +/− patchy MCVs/male MCVs, female MCVs
Autosomal recessive demyelinating (CMT4)		
CMT4A	*GDAP1*	CMT1 or CMT2 usually early onset and severe/vocal cord and diaphragm paralysis described/rare AD CMT2 families described
CMT4B1	*MTMR2*	Severe CMT1/facial/bulbar/focally folded myelin
CMT4B2	*MTMR13*	Severe CMT1/glaucoma/focally folded myelin
CMT4C	*KIAA1985* (*SH3TC2*)	Severe CMT1/scoliosis/cytoplasmic expansions
CMT4D (HMSNL)	*NDRG1*	Severe CMT1/gypsy/deafness/tongue atrophy
CMT4E	*EGR2*	Classic CMT1/DSN/CHN
CMT4F	*PRX*	CMT1/more sensory/focally folded myelin
CMT4H	*FGD4*	CMT1
CMT4J	*FIG4*	CMT1
CCFDN	*CTDP1*	CMT1/gypsy/cataracts/dysmorphic features
HMSN Russe	10q22−q23	CMT1
CMT1	*PMP22* (point mutation)	Classic CMT1/DSN/CHN/HNPP
CMT1	*MPZ*	CMT1/DSN/CHN/intermediate/CMT2
Autosomal dominant CMT2 (AD CMT 2)		
CMT2A	*KIF1Bb*	Classic CMT2
CMT2A	*MFN 2*	CMT2/usually severe/optic atrophy
CMT2B	*RAB7*	CMT2 with predominant sensory involvement and sensory complications
CMT2C	*TRPV4*	CMT2 with vocal cord and respiratory involvement
CMT2D	*GARS*	CMT2 with predominant hand wasting/weakness or dHMN-V
CMT2E	*NEFL*	CMT2 but can have slow MCVs in CMT1 range +/2 early onset severe disease
CMT2F	*HSP27* (*HSPB1*)	Classic CMT2 or dHMN-II
CMT2G	12q12-q13.3	Classic CMT2
CMT2L	*HSP22* (*HSPB8*)	Classic CMT2 or dHMN-II

(*Continued*)

TABLE 17.2 (Continued)

Type	Gene/Locus	Specific Phenotype
CMT2M	*DNM2*	CMT2 with infantile onset
CMTN	*AARS*	CMT2/dHMN
CMTO	*DYNC1H1*	CMT2 with proximal involvement
CMT2	*MPZ*	CMT1/DSN/CHN/intermediate/CMT2
CMT2 (HMSNP)	3q13.1	CMT2 with proximal involvement
Autosomal recessive CMT2 (also called CMT4)		
AR CMT2A	*LMNA*	CMT2 proximal involvement and rapid progression described; also causes muscular dystrophy/cardiomyopathy/lipodystrophy
AR CMT2B	*MED25*	Typical CMT2
AR CMT2	*GDAP1*	CMT1 or CMT2 usually early onset and severe/vocal cord and diaphragm paralysis described/rare AD CMT2 families described
Intermediate CMT (DI-CMT)		
DI-CMTA	10q24.1-25.1	Typical CMT
DI-CMTB	*DNM2*	Typical CMT
DI-CMTC	*YARS*	Typical CMT
CMT RIA	*GDAP1*	Onset in early childhood
CMT RIB	*KARS*	± CNS features
CMT RIC	*PLEKHG5*	Typical CMT
Maternal inheritance		
Mitochondrial	*MT-ATP6*	Typical CMT2 or dHMN
Hereditary neuralgic amyotrophy (HNA)		
HNA	*SEPT9*	Recurrent neuralgic amyotrophy

Abbreviations: AD, autosomal dominant; AR, autosomal recessive; CHN, congenital hypomyelinating neuropathy; CMT, Charcot-Marie-Tooth; DI, dominant intermediate; RI, recessive intermediate; CTDP1, CTD phosphatase subunit 1; Del, deletion; DMN2, dynamin 2; DSN, Dejerine-Sottas neuropathy; Dup, duplication; EGR2, early growth response 2; FGD4, FYVE, RhoGEF and PH domain containing 4; FIG4, FIG 4 homologue; GARS, glycyl-tRNA synthetase; GDAP1, ganglioside induced differentiation associated protein 1; GJB1, gap junction protein beta1; HNPP, hereditary neuropathy with liability to pressure palsies; HSP22, heat shock 22-kDa protein 8; HSP27, heat shock 27-kDa protein 1; KIF1Bb, kinesin family member 1B-b; LITAF, lipopolysaccharide induced tumor necrosis factor; LMNA, lamin A/C; MCV, motor conduction velocity; MFN2, mitofusin 2; MPZ, myelin protein zero; MTMR2, myotubularin related protein 2; MTMR13, myotubularin related protein 13; NDRG1, N-myc downstream regulated gene 1; NEFL, neurofilament, light polypeptide 68 kDa; PMP22, peripheral myelin protein 22; PRX, periaxin; RAB7, member RAS oncogene family; SEPT9, septin 9; SH3TC2, SH3 domain and tetratricopeptide repeats 2; YARS, tyrosyl-tRNA synthetase.
Adapted from Reilly and Shy.[49] From BMJ Publishing Group Ltd, with permission.

demyelination and is potentially amenable to therapeutic intervention.[17,18]

At a pathological level, dysmyelination, demyelination, remyelination, and axonal loss are characteristic features of the various demyelinating forms of CMT1. In Dejerine-Sottas neuropathy, myelin may never have formed normally, which is referred to as dysmyelination. In CMT1, onion bulbs of concentric Schwann cell lamellae are usually present on nerve biopsies, with loss of both small- and large-diameter myelinated fibers and sometimes axons. Focal, sausage-like thickenings of the myelin sheath (tomacula) are characteristic of HNPP but may also be found in other forms of CMT1, particularly CMT1B. In CMT1, as aforementioned, disability typically correlates better with secondary axonal degeneration than with demyelination itself, thereby demonstrating the importance of Schwann cell–axonal interactions in demyelinating disease.

CLINICAL MANIFESTATIONS

Despite phenotypic variability, the typical clinical course of CMT1 and CMT2 patients includes normal development

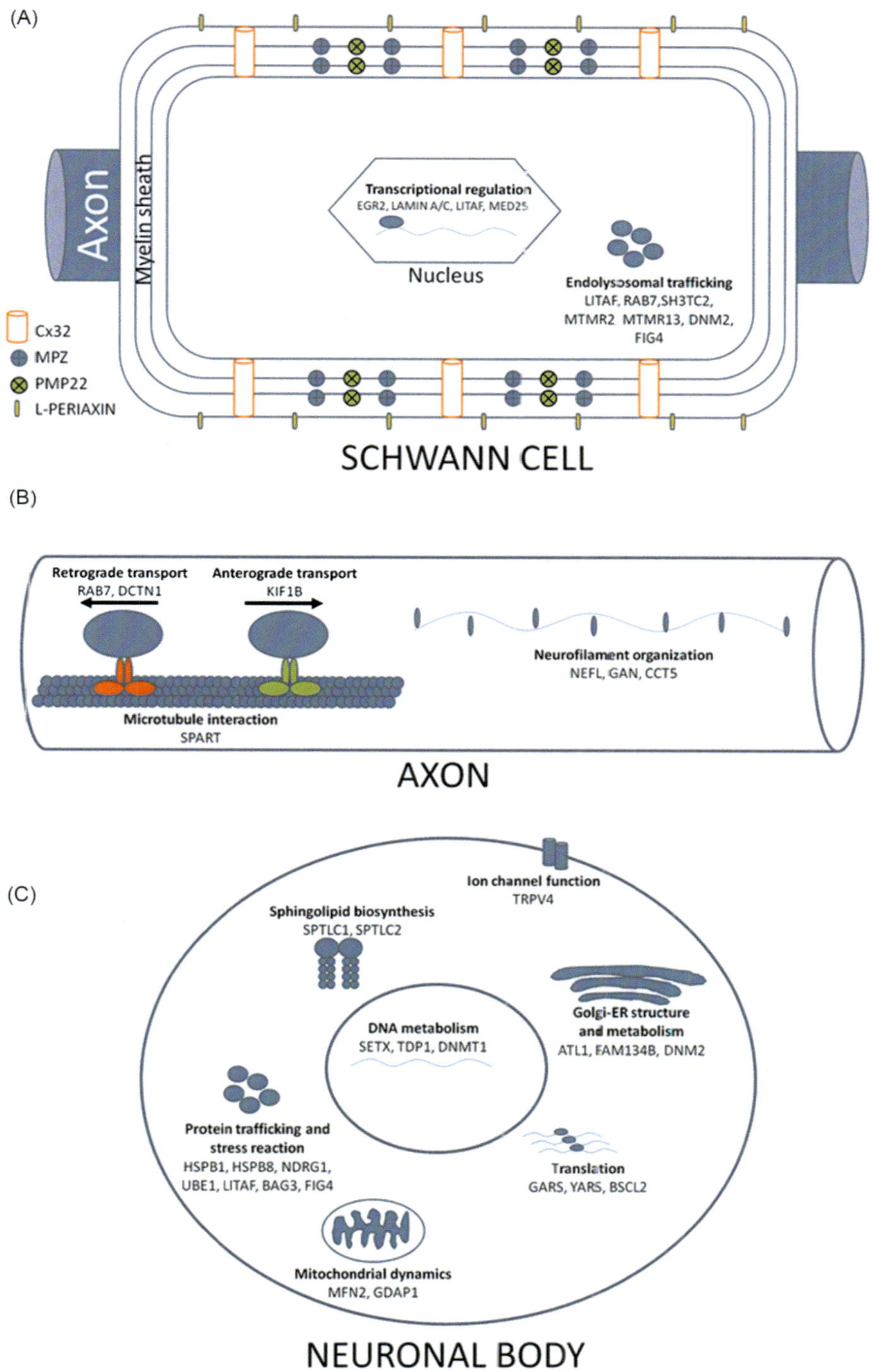

FIGURE 17.1 A neuron: its axon and Schwann cell with the major genes associated with Charcot-Marie-Tooth disease with their respective function and cellular compartment. *From Saporta MA & Shy ME[129] reproduced with permission from Elsevier.*

before weakness and sensory loss appearing gradually within the first two decades of life. This is often referred to as the "classical phenotype." Affected children are often slow runners and have difficulty with activities that require balance (e.g. skating, walking along a log across a stream). Ankle-foot orthotics (AFOs) are frequently required by the third decade. Fine movements of the hands for activities such as turning a key or using buttons and zippers may be impaired, but the hands are rarely as affected as the feet. Most patients remain ambulatory throughout life and have a normal lifespan.

A minority of CMT patients have a more severe phenotype with delayed motor milestones and onset in infancy, termed "Dejerine-Sottas neuropathy." Especially severe cases are classified as congenital hypomyelination if myelination appears to be disrupted during embryologic development. Many patients have *de novo* autosomal dominant disorders, and the term Dejerine-Sottas neuropathy is currently used primarily to denote severe early-onset clinical phenotypes regardless of the inheritance pattern.

Patients with distal hereditary motor neuropathies (dHMN) sometimes have mild sensory abnormalities, and patients with hereditary sensory and autonomic neuropathies (HSAN/HSN) usually have some weakness. Indeed, the same mutation in the same gene can cause both CMT and dHMN within the same family; for example, glycyl-tRNA synthetase or *GARS* mutations can cause both CMT2D and hereditary motor neuropathy type V.[19,20]

NEUROPHYSIOLOGY

Nerve conduction velocity testing will distinguish between demyelinating and axonal neuropathies. In clinical practice, around 60% of CMT patients have demyelinating CMT (CMT1) and around 20% have axonal CMT (CMT2).[7] Neurophysiology is also useful to detect sensory involvement that often is unreported by patients. It can also be helpful with the further classification of axonal inherited neuropathies into mixed motor and sensory axonal neuropathy (CMT2), pure motor axonal neuropathy (dHMN), and pure sensory axonal neuropathy (HSAN/HSN).

Most CMT1 patients, particularly those with CMT1A, have a uniformly slow nerve conduction velocity of about 20 m/s. However, asymmetrical slowing, which is characteristic of HNPP, may be found in patients with missense mutations in *PMP22, MPZ, EGR2*, and *GJB1*.[21–23] With *GJB1* mutations, the asymmetry is particularly common in women with the disorder. It is important to try to recognize these cases because if the disorder is in fact acquired, it raises the possibility of a superimposed inflammatory neuropathy, which can benefit from immunosuppressive therapy.[22–25] Virtually all forms of CMT1 have axonal loss as well as demyelination. In cases where the motor and sensory nerve responses are absent, it is worthwhile to look at proximal nerve conduction (e.g. axillary nerve latency) to fully investigate the possibility of a severe demyelinating rather than an axonal pathology such as is found in CMT2.

GENETIC TESTING AND DIAGNOSTIC STRATEGIES

Molecular testing, performed after the family history, neurologic examination, and neurophysiologic testing, have suggested the probable candidate genes (GeneClinics—available at www.geneclinics.org), is the "gold standard" for the diagnosis of inherited neuropathies. Strategies for focused genetic testing based on inheritance and clinical phenotype have been in place for over 10 years. The choice of genes to be tested should reflect the specific population tested. For example, in Northern European and Northern American populations, most patients have dominant inheritance even if the cases are sporadic. AR CMT is observed in less than 10% of these cases. However, in populations with high rates of consanguinity, AR CMT is likely to be much more common, even reaching levels of 40%.

An algorithm exists based on the North American population that utilizes clinical information, neurophysiological data, and inheritance patterns to guide the clinician on which genes to consider testing.[7] Flow charts incorporating the phenotype, illustrated in Figures 17.2 and 17.3, guide the clinician on appropriate testing strategies. Using such tools, most CMT clinics reach a genetic diagnosis in around two-thirds of their total CMT population.[7,8] Importantly, around 90% of patients found to have a genetic diagnosis are found to have changes in one of only four genes: *PMP22, GJβ1, MPZ*, and *MFN2*. Therefore the flow charts emphasize the need for testing these particular genes first.

Very Slow MNCV (Less Than or Equal To 15 m/s)

Many patients in this group had an onset in infancy with delayed walking. However, most patients with very slow MNCV and who walked by 15 months of age have CMT1A; therefore, in this scenario, testing for *PMP22* duplications is warranted in the child or adolescent. If CMT1A testing is negative in this group, the next most common cause of CMT is CMT1B. If this is also negative, then testing for more rare forms of CMT may be indicated.

Slow MNCV (Greater Than 15 and Less Than or Equal To 35 m/s)

This represents the largest group seen in a typical CMT clinic with most patients giving a history of walking on time with onset of symptoms in the first or second decade. Approximately 89% of this group had CMT1A, so again testing should begin with analysis for *PMP22* duplication. This test should be performed even in the absence of a family history as 10% of cases present with apparently *de novo* mutations.[26] CMT1X is the next most common form of CMT in this group but should only be investigated if there is no male-to-male transmission. CMT1X is much more likely if the patient has intermediately slowed NCV (see the following). CMT1B is a much less likely possibility in this scenario, with testing only warranted if CMT1A or CMT1X is negative or if there is evidence of male-to-male transmission.

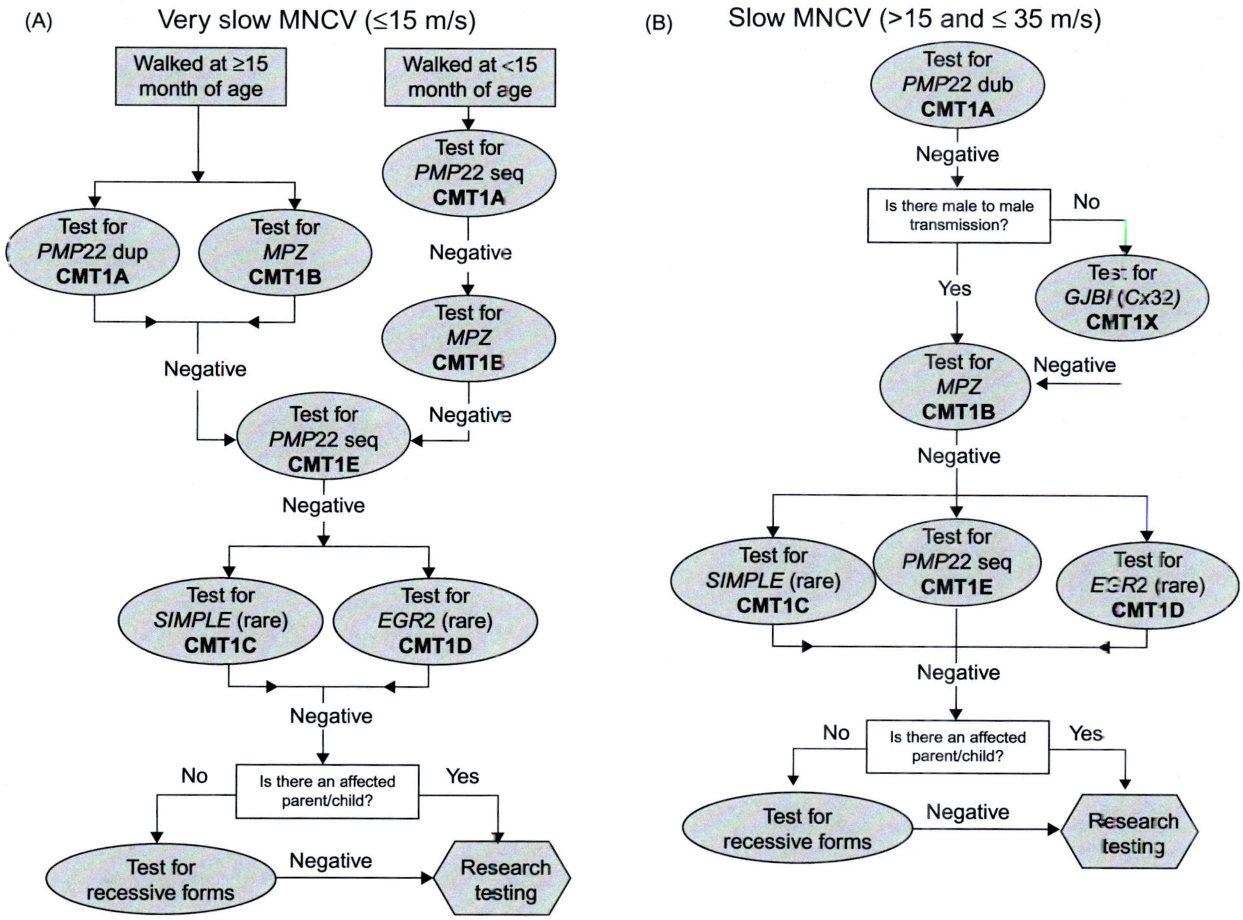

FIGURE 17.2 Algorithm for the genetic diagnosis of patients with Charcot-Marie-Tooth disease and very slow (**A**) or slow (**B**) upper extremity motor nerve conduction velocities. Dup, duplication; Seq, sequencing. *From Saporta et al.*[7] *with permission.*

Intermediate MNCV (Greater Than 35 and Less Than 45 m/s)

Intermediate conductions in a classical CMT clinical picture are typically caused by CMT1X or later onset forms of CMT1B. The first step is to identify an accurate family history. Where there is no male-to-male transmission, test for CMT1X as this represented almost 80% of such cases in one large CMT clinic.[7] If this testing is negative or if there is male-to-male transmission, then CMTIB should be investigated.

Axonal MNCV (Greater Than 45 m/s or Unobtainable CMAP)

With a classical CMT picture with onset by the second decade, the two major genetic considerations are CMT1X (if there is no male-to-male transmission) and CMT1B. The most common cause of axonal CMT, CMT2A, will have presented clinically in infancy or early childhood. If these tests are negative, clinical findings can help guide the clinician to the diagnostic genetic test (e.g. severe involvement of upper extremities should prompt the consideration of CMT2D).

While a detailed review of the pros and cons for genetic testing is beyond the scope of this chapter, we think it is reasonable to provide some information about how to pursue genetic testing.[27] Clearly, not every patient with a genetic neuropathy wants or needs testing to identify the genetic cause of their disease. The ultimate decision rests with the patient and their parents (when the patient is under the age of 18 years). Reasons that patients and their families give for obtaining testing include identifying the inheritance pattern of their CMT, making family planning decisions, and obtaining knowledge about the cause and natural history of their form of CMT. As yet, natural history data is only available for CMT1A and CMT1X, which can provide guidance for prognosis recognizing that there can be phenotypic variability within these subtypes.[28,29] Patients with other forms of CMT frequently choose to undergo genetic testing to contribute to the natural history data collection for other patients with the

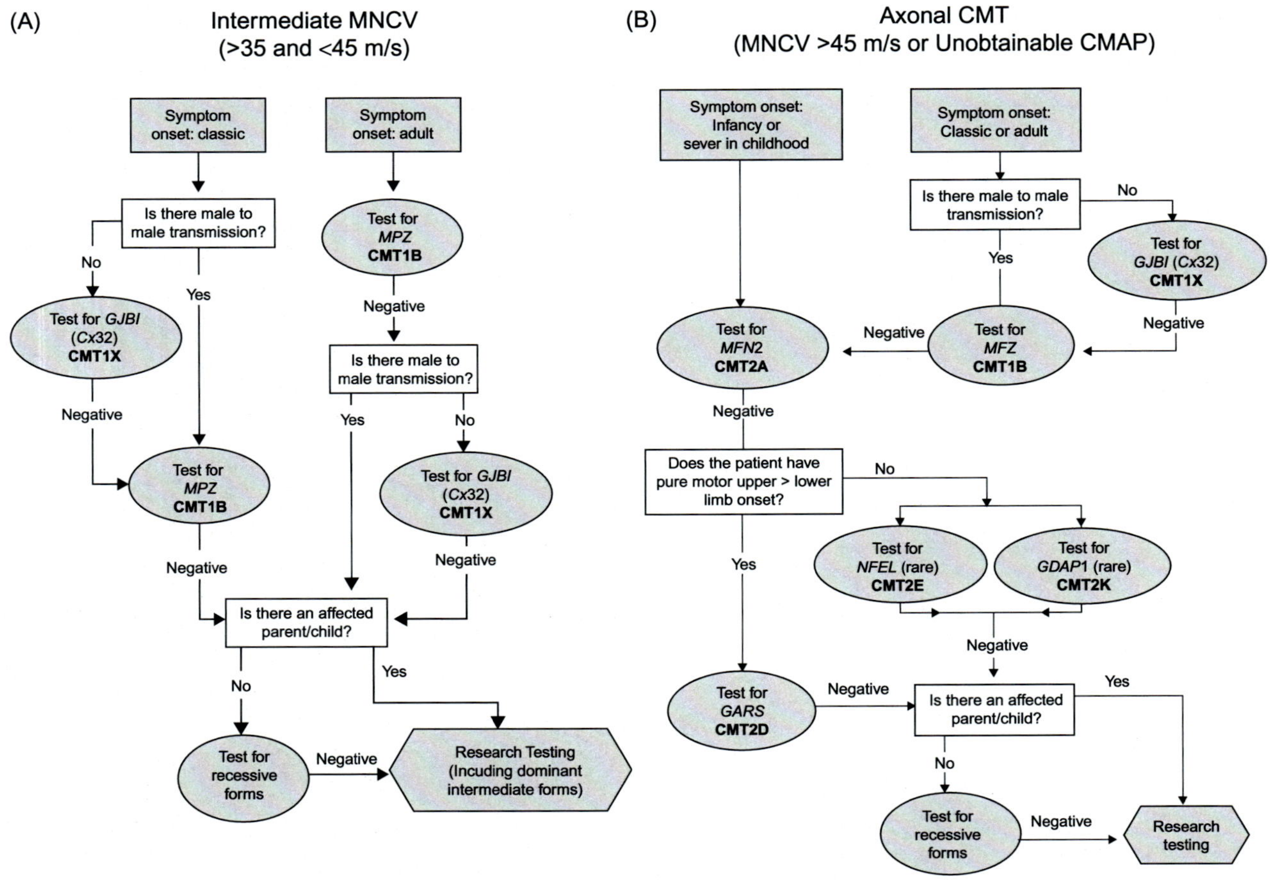

FIGURE 17.3 Algorithm for the genetic diagnosis of patients with Charcot-Marie-Tooth disease and intermediate (**A**) or normal (**B**) upper extremity motor nerve conduction velocities. *From Saporta et al.*[10] *with permission.*

same subtype. Some decline testing for reasons including the high cost of commercial testing and fears of discrimination in school or the workplace or in obtaining health insurance. Because there are currently no medications to reverse any form of CMT, some decide against testing. We maintain that it is always the patient's (or parents'/guardian's) decision whether or not to pursue genetic testing.

Once genetic testing has been done in a patient, other family members usually do not need genetic testing but can be identified by clinical evaluation with neurophysiology. We do not typically test patients for multiple genetic causes of CMT simultaneously, although our group has identified 11 patients out of nearly 1000 patients with multiple genetic causes of CMT.[7] It is our current policy to only consider genetic testing in clinically affected family members if their phenotype is atypical for the type of CMT in the family. In addition, we do not test asymptomatic minors with a family history of CMT, either electrophysiologically or by genetic testing, owing to the chance for increased psychological harm to the child.[30] We do routinely perform limited nerve conduction studies (NCSs), though not needle electromyogram (EMG), on symptomatic children with CMT. Because changes such as slowing are uniform, testing of a single nerve is often adequate to guide genetic testing or determine whether a symptomatic child is affected in a family with CMT.

The landscape of genetic testing is rapidly changing with the introduction of next generation sequencing (NGS) into clinical practice in some areas, which enables sequencing of either the entire genome or exome (only the protein coding sequences) in days.[31] These new testing methods are not without fault, however; for example, the X-chromosome and many GC-rich regions are poorly covered by whole genome sequencing (WGS) and whole exome sequencing (WES). That said, even Sanger sequencing, the so-called gold standard of molecular techniques, can miss small indels (insertions or deletions of DNA) that have been picked up by NGS. Disease specific panels where multiple genes are tested via NGS are a future viable option and are in development. Here, only the coding regions of known CMT genes are tested. However, panels with a high number of genes being tested will require careful interpretation. Many genes such as *periaxin* and *MFN2* have a high number of polymorphisms that are not pathogenic.

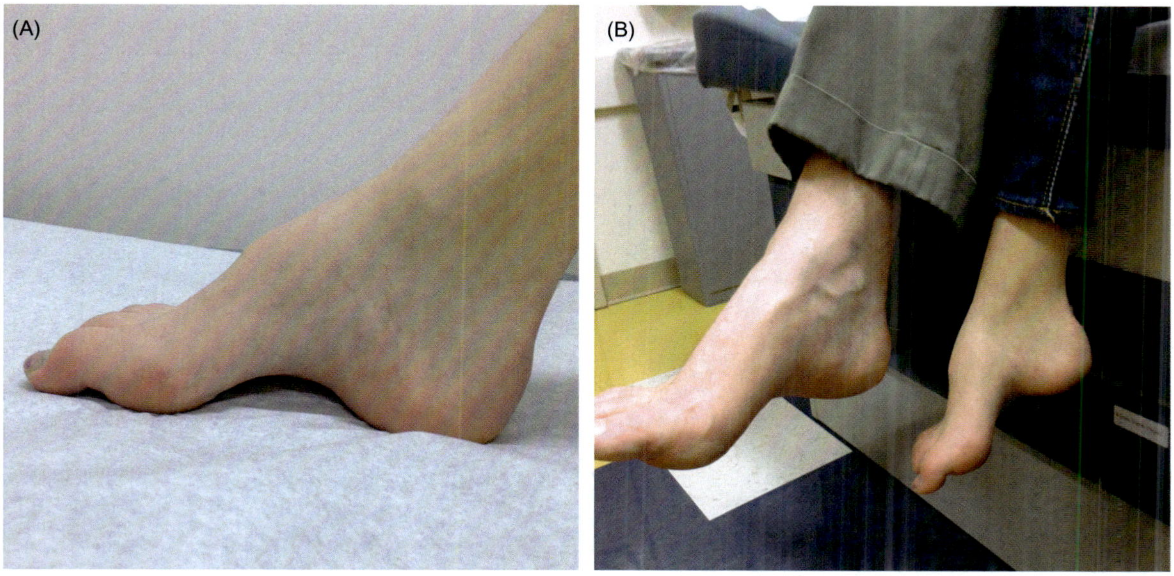

FIGURE 17.4 (**A**) High arches (pes cavus) and hammertoes in a 14-year-old female with CMT1. Patient walked before the age of 15 months. Nerve conduction velocities in the upper limbs ranged from 12.7 to 23.7 m/s and in the lower limbs from 10.3 to 15.4 m/s. Sensory nerve action potentials could not be elicited in the median, ulnar, sural, or superficial peroneal nerves. (**B**) Patient's father also has pes cavus but of milder severity. Multiple members on father's side of the family, including his mother, are affected by CMT with definite male-to-male transmission in the pedigree. Patient in (**A**) is currently being tested for the *PMP22* duplication (CMT1A). If this testing is negative, she will be tested for CMT1B (*MPZ*).

In addition, as more genes are being discovered in the pathogenesis of CMT, this will require modification of existing panels that will likely contribute to cost.

All things considered, it is likely that in the near future, the sequential Sanger sequencing of genes according to phenotypes as described above will no longer be the most cost-effective way of performing genetic testing. Currently, genetic testing is in a transition phase with both traditional Sanger sequencing and NGS technology being used by most dedicated CMT clinics, and whole exome sequencing is not far behind. It is our current practice to perform whole exome sequencing on patients for whom the above algorithm (using traditional techniques such as Sanger sequencing or MLPA for *PMP22* duplication) has failed to identify a causal mutation. A review of these techniques is beyond the scope of this chapter, but a reference is provided.[31] Ensuring that genetic testing is focused and cost effective is important for patients and their families.

SPECIFIC FORMS OF CMT

We will mention the phrase "classical phenotype" throughout the next section. This term is based on the original description by Harding and Thomas.[4,5] Affected patients with the classical phenotype begin walking on time, usually between 12 and 15 months, and develop weakness or sensory loss in the first two decades of life. Impairment slowly increases thereafter. The most common phenotypes seen in this population in particular (i.e. late childhood and adolescence) are CMT1A and CMT1X (especially in males). We will next discuss the various specific forms of CMT with emphasis on those that present in late childhood/adolescence.

CMT1: Autosomal Dominant Demyelinating Neuropathies

CMT1 is the most common subtype of CMT and is caused primarily by mutations in four genes. These genes are essential to formation of the myelin sheath and also Schwann cell function. Two forms in particular, CMT1A and CMT1C, will present with a classical phenotype.

CMT1A

This is the most common form of CMT in most populations, affecting 55% of genetically determined CMT and 36% of all CMT.[7] Accordingly, it is the most common form of CMT1, accounting for 80 to 90% of CMT1 cases, and will be the most commonly observed CMT neuropathy in late childhood and adolescence Patients usually present with the "classical CMT phenotype," which includes lower limb symptoms (difficulty walking/foot deformity) with onset in the first two decades of life accompanied by distal weakness, atrophy, sensory loss, hyporeflexia, and foot deformities such as high arches (pes cavus) and hammertoes (Figure 17.4). Patients with

CASE EXAMPLE 17.1

A 25-year-old man with no family history of neuropathy had been weak since infancy. He was able to stand independently by 3 years of age but was never able to run normally and always had an abnormal gait. He is currently only able to walk if wearing ankle-foot orthotics. He also has pronounced weakness with fine movements of his fingers and is unable to button his clothes, cut his own food or perform activities such as turning a key in his front door. His neurological function has been relatively stable since his teenage years. Nerve conduction studies showed markedly slowed NCVs (<10 m/s) in his upper extremities; NCVs in his legs were unobtainable at routine recording sites. Compound muscle action potentials were significantly reduced in the arms and absent in the legs. Sensory nerve action potentials were absent in the arms and legs. Genetic testing revealed an Arg98Cys mutation in MPZ (myelin protein zero) leading to a diagnosis of severe CMT1B.

Comment

In North America, if one has a genetically diagnosable form of CMT, it is likely that the causal mutation is in 1 of 4 genes (*PMP22, MPZ, GJB1,* or *MFN2*) unless the family history strongly suggests an autosomal recessive inheritance pattern (multiple affected siblings with no parents affected). CMT1A, the most common form of CMT, typically has an NCV of around 20 m/s in the arms and a classical CMT phenotype with normal early milestones and gradual weakness developing in the first two decades of life. Delayed early milestones and NCV <10 m/s are suggestive of an early onset form of CMT1B. *GJB1* mutations causing CMT1X typically have intermediately slowed NCV (35–45 m/s) with an X-linked inheritance. MFN2 mutations cause the most common form of CMT2. Another group of patients with CMT1B often present symptoms in adulthood, with intermediate to normal NCV.

the "classical phenotype" usually will walk on time.[7] Lifespan is not affected. While patients frequently require ankle-foot orthotics, they rarely require wheelchairs for ambulation. Only very rarely will CMT1A patients present with a Dejerine-Sottas phenotype. Conduction velocities from the median and ulnar nerves are below 38 m/s and are typically in the 20 s range. The sensory action potentials are either reduced or absent. Nerve biopsy is not required, but, if performed, reveals demyelination and onion bulb formation. CMT1A is an autosomal dominant condition, so most patients will report a family history, but there is a *de novo* mutation rate of 10%.[32] Therefore, children and adolescents without a family history with ulnar MNCV under 35 m/s should first be screened for CMT1A before proceeding with other genetic testing.[7] CMT1A is caused by a 1.4 Mb duplication on chromosome 17p11.2 in the region that carries the *PMP22* gene.[33] It is the extra copy of *PMP22* that causes the neuropathy.

CMT1B

This the fourth most common type of CMT[7] and clinically is associated with two distinct phenotypes: (1) an early infantile onset severe phenotype with delayed walking and NCV <10 m/s, often referred to as Dejerine-Sottas disease; or (2) a much later, milder phenotype with onset at around age 40 and NCV of around 40 m/s. CMT1B can, however, cause the "classical CMT phenotype," doing so in about 15% of total CMT1B cases.[7] CMT1B is caused by point mutations in the myelin protein zero (*MPZ*) gene on chromosome 1q22-23.[34] *MPZ* is the major component of peripheral nerve myelin, comprising at least 50% of the protein. See Case Example 17.1.

CMT1C

CMT1C is clinically indistinguishable from CMT1A, presenting in the second decade with the "classical phenotype" of distal wasting and weakness, sensory loss, reduced reflexes, and NCVs ranging from 17 to 26 m/s.[35] It is a rare cause of CMT, accounting for <1% of the CMT population. CMT1C is caused by mutations in the lipopolysaccharide-induced tumor necrosis factor (TNF)-α factor (*LITAF*) gene, also known as SIMPLE.[36]

CMT1D

The CMT1D phenotype is severe: patients typically present in infancy and may have hypomyelination (hypotonia, delayed motor milestones, MNCV < 10 m/s).[37] In addition, scoliosis and cranial nerve involvement may be features.[38,39] This is a rare phenotype of CMT and is caused by mutations in *EGR2*, a member of the early growth response family. *EGR2* is a Schwann cell transcription factor that binds DNA along three zinc finger domains and is thought to regulate the expression of myelin genes such as MPZ and MBP.[37] Autosomal recessive inheritance has also been described with *EGR2* mutations (known as CMT4E) causing a very similar clinical phenotype.

CMT1E

CMT1E is caused by point mutations in the *PMP22* gene. Often, but not always, these patients have an earlier onset with a more severe phenotype sometimes associated with deafness, compared to CMT1A patients who simply have a duplication in the *PMP22* gene. Onset within the first 2 years of life with a delay in walking is not uncommon; however, the clinical phenotype is variable and onset may be later.[40] The disease severity depends on the particular PMP22 mutation; some cases can be very mild or even resemble HNPP, particularly if they induce processes such as nonsense-mediated decay, in which truncated proteins are degraded by the cell.

CMT1F

See CMT2E.

HNPP

Point mutations in the *PMP22* gene can also be associated with another hereditary neuropathy called "hereditary neuropathy with liability to pressure palsies," or HNPP. More commonly, however, HNPP is caused by a deletion of the 1.4 Mb stretch of chromosome 17p11.2.[41] It is this same stretch of chromosome that is duplicated in CMT1A. HNPP is the third most common type of CMT and is an AD disorder with patients typically presenting with recurrent neuropathies occurring at sites of entrapments[7,42] with an age of onset similar to that in CMT1A,[43] although one larger study suggests >50% of HNPP has an earlier onset presenting typically with delay in ambulation.[7] The literature suggests a wide range of onset ranging from early childhood right through to adulthood. Sometimes motor NCVs are normal but are slowed around sites of entrapment such as fibular head, carpal tunnel, and medial epicondyle.[44] Electrophysiological clues to the diagnosis include marked slowing of the ulnar nerve around the elbow and the peroneal nerve around the knee when they are combined with prolonged distal motor latencies. Motor nerve conduction velocities can be normal or minimally slow.[45] Clinically, patients present with a history of recurrent focal mononeuropathies that are transient, lasting anything from hours to days or weeks and occasionally longer. Many patients develop a predominantly length-dependent large fiber sensory neuropathy as they age.

CMTX: X-linked CMT

CMTIX is the main X-linked form of CMT, although other X-linked forms exist that are typically components of syndromes involving more than the peripheral nervous system. CMT1X is caused by mutations in the gap junction beta 1 or *GJB1* gene, which encodes the protein connexin32 (Cx32).[12] This is the second most common form of CMT, accounting for around 10 to 15% of CMT.[7] Despite the fact that there are a huge number of mutations described in *GJB1*, the phenotype in males is virtually the same, including those with entire gene deletions, suggesting that mutations lead to a loss of function of the protein.[28] The majority of males will start to have symptoms in childhood, but around 20% will have a later onset.[7]

Males are typically more severely affected and tend to have marked atrophy of the intrinsic hand muscles and all compartments of the calf muscles. Males with CMT1X often display a "split hand syndrome" with abductor pollicis brevis being weaker and more wasted than first dorsal interosseous. Episodes of transient stroke-like symptoms have been rarely reported for males with CMT1X, usually following a stressor. Presumably these events occur because Cx32 is also expressed in oligodendrocytes in the CNS.[46]

Females with *GJB1* will usually have abnormalities on neurological examination or nerve conduction testing.[47] Typically females group into two distinct phenotypes. Two-thirds have mild symptoms of CMT and progress slowly. One-third, however, have a more moderate neurology and clinically resemble affected males with CMTX.[47] It has been assumed that the cause of the difference in phenotype observed in women is due to X-inactivation, but there is no evidence as yet to support this theory.[47,48] Typically males with CMTX have MNCVs between 25 and 45 m/s, whereas women usually have MNCVs greater than 35 m/s.[7] Therefore, the presence of near normal to intermediate MNCVs in females and slow to intermediate MNCVs in males in families with an absence of male-to-male transmission should prompt the clinician to consider CMTX.

CMT2: Autosomal Dominant Axonal Neuropathies

Basic nerve conduction studies suggest the diagnosis of CMT2, with patients demonstrating NCVs of >38 m/s defining this group.[5] It is at this stage that clinical and electrophysiological data enable axonal neuropathies to be classified further into pure or predominant sensory form (HSAN/HSN), the mixed motor and sensory form (CMT2), or the pure motor form (dHMN). CMT2 presents clinically in one of three distinct clinical phenotypes.[49] The first is the "classical CMT phenotype" with lower limb symptoms (difficulty walking/foot deformity) with onset in the first two decades of life, although much later ages of onset can be seen than with CMT1. Profound sensory impairment characterizes the second phenotype, and selective initial involvement of the upper limbs is seen in the third phenotype. It is more challenging to obtain a genetic diagnosis in CMT2 as there are many genes known to cause axonal CMT (see the Mutation Database of Inherited Peripheral Neuropathies [IPNMDB][50]), with each gene accountable for only a very small proportion of cases, with the notable exception of *MFN2*.

As disease-causing CMT genes account for the minority of CMT2 patients (around 35%), the overwhelming majority of CMT2 patients still await an exact genetic diagnosis. Because the number of patients with CMT2 with a genetic diagnosis is currently small, it is difficult to be accurate about exact genotype-phenotype correlations and in particular with regard to age of onset. However, with the advent of Next Generation Sequencing, it is certain that the number of genetic causes of CMT2 will significantly increase over the next few years.

CMT2A

This is the most common type of CMT2, accounting for about 20 to 25% of patients with CMT2.[51] CMT2A often has a severe phenotype with onset in infancy or early childhood, with many patients requiring a wheelchair to ambulate by the age of 20. CMT2A is caused by mutations in the nuclear encoded mitofusin 2 or *MFN2* gene.[52] This GTPase is a component of the outer mitochondrial membrane and is essential for fusion of mitochondria to each other or to the endoplasmic reticulum.[53,54] Often, these patients have unrecordable CMAP amplitudes and NCS even in the upper extremities, leading to the suggestion to test for MFN2 mutations as the initial genetic test in such patients.[7] There are a high number of polymorphisms encountered in MFN2 genetic testing, so care must be taken when ascertaining if mutations are indeed disease-causing. Most disease-causing mutations are in the GTPase domain, coiled-coil domains, or in other evolutionary conserved regions of the protein.[51] Rarely, some patients present later in life (during childhood, adolescence, or even adulthood) with a milder phenotype. Some have additional clinical features such as optic atrophy (known as CMT5)[55] or pyramidal signs (CMT6)[56] as part of their disease.

CMT2B

CMT2B is a rare disorder characterized by mild to moderate sensory loss that typically leads to ulcerations, infections, and sometimes amputations of toes.[57–59] Spontaneous pain is absent. Additionally, patients with CMT2B usually have the typical motor symptoms of CMT. On occasion, the sensory loss can be severe and patients can be clinically indistinguishable from HSAN type 1. Nerve conduction studies usually show intermediately slowed or normal nerve conduction velocities and reduced amplitudes. Absent responses typically follow a length dependent form in both motor and sensory nerves.[57–59] Age of onset is typically from adolescence onwards with foot deformity and sensory loss being the first signs.[57] CMT2B is caused by mutations in the *RAB7* gene, located at chromosome 3q13-q22.59. RAB7 belongs to the Rab family of Ras-related GTPases that are essential for regulation of intracellular membrane vesicle led transport.[60]

CMT2C

CMT2C is characterized by a predominantly motor neuropathy with vocal cord and diaphragm paralysis.[61] Other distinctive clinical features may include bilateral sensorineural hearing loss, raspy voice, and impaired bladder control with or without incontinence.[62,63] It is allelic with spondylometaphyseal dysplasia, metatropic dysplasia, and brachyolmia, and some patients may have features of an overlapping phenotype such as scoliosis, short stature, and/or scapular winging with their neuropathy.[63–65] The age of onset is often in the first decade of life, but adolescent and adult onset forms have been reported with potentially marked intrafamilial variability.[63] CMT2C is caused by mutations in the *TRPV4* gene found at chromosome 12q23-24.[63–66] TRPV4 is part of a large family of cation selective channels involved in the transduction of many sensory stimuli ranging from capsaicin to mechanical force.

CMT2D

CMT2D is a predominantly motor disorder that usually presents in the second decade. The earliest manifestations are often transient cramping or pain in the hands after exposure to the cold.[19] Often, but not always, there is a distinct clinical phenotype with more pronounced weakness and wasting of the distal upper extremities compared to the lower extremities. Patients may have a "split hand" appearance with wasting more obvious in the first dorsal interosseous and thenar eminence muscles compared to the hypothenar eminence muscles.[19,20] This is in contrast to males with CMT1X where abductor pollicis brevis is weaker and more wasted than the first dorsal interosseous. CMT2D is caused by mutations in the glycyl-tRNA synthetase or *GARS* gene.[20] *GARS* is involved in protein synthesis, specifically catalyzing aminoacetylation of glycyl-tRNA, which is required for the insertion of glycine into proteins.[67] GARS is a ubiquitously expressed protein required for translation in all cells. How mutations in *GARS* lead to neuropathy is not fully understood, but the mechanism may involve the mutation leading to a toxic gain of function.[68,69] Sensory findings are usually mild and may even be absent. In these situations, the disorder has been classified as distal hereditary motor neuropathy (dHMN) type V. Since the discovery of *GARS* involvement with CMT2D, mutations have been described in several other aminoacyl synthetase genes, with motor involvement and intermediate slowing of nerve conduction velocities being a predominant feature of many these clinical phenotypes.

CMT2E

CMT2E is caused by mutations in the *NEFL* (neurofilament light) gene.[70] Confusingly for the clinician, NCVs can be in either the axonal or demyelinating range.[71–73] Typically, those patients with slowed conduction present early either in infancy with delayed motor milestones or in late childhood/early adolescence. However, the age of onset ranges from infancy and childhood, typically with severe impairment, to adulthood with mild impairment. Weakness and wasting starts in the legs, with all sensory modalities reported to be affected in the early/childhood/adolescent onset cases.[73] Neurofilaments are important cytoskeletal components of the axon,[74] and therefore patients with NEFL mutations are referred as having an axonal neuropathy regardless of the NCVs. As aforementioned, some patients present with a demyelinating phenotype and they are sometimes referred to as CMT1F.

CMT2F

CMT2F is caused by mutations in the *HSPB1* gene, a member of the distal heat shock protein (HSP)-27 superfamily, and is also known as HSP27.[75] The heat shock proteins are a group of ubiquitously expressed chaperone proteins classified according to their molecular weight. This is a rare cause of CMT. Most patients with mutations in this gene have a distal hereditary motor neuropathy (dHMN) phenotype with nerve conduction velocities in the axonal range.[76] Sensory findings may or may not be present. Typically, this is a distal length dependent neuropathy with slow progression proximally of symptoms.[77] The age of onset is typically late adolescence to early adulthood.[76]

CMT2L

CMT2L is caused by mutations in the *HSPB8* gene, also a member of the HSP superfamily, and is also known as HSP22.[78,79] Typically, this is a distal axonal motor neuropathy known as dHMN type II[80] but sensory involvement can occur, in which case the correct diagnosis is CMT2L.[79] The age of onset is usually from mid-adolescence to the fourth decade,[79] with predominantly distal lower limb weakness and wasting although proximal involvement has been reported in a few cases. Overall, this is also a rare cause of CMT.

CMT2K

CMT2K is caused by mutations in the ganglioside induced differentiation protein 1 or *GDAP1* gene.[81] An autosomal recessive form also exists, CMT4A, which is more common than the autosomal dominant form, CMT2K. While the recessive form, CMT4A, may be quite severe, the dominant form, CMT2K, typically results in a much milder phenotype with onset ranging from adolescence to even adulthood.[82,83] Interestingly, Sivera et al. found that ankle dorsiflexion and plantar flexion were consistently affected to a similar degree in a Spanish cohort of patients.[82] A second recent study by this group found that *GDAP1* mutations were found to account for 14.7% of their cases of CMT2, suggesting that this may be related to the high levels of consanguinity in certain groups in southern Spain. This again emphasizes the importance of considering a patient's ethnic background in guiding the diagnostic work-up.[84]

CMT2M

CMT2M is caused by mutations in *DNM2* and is allelic with centronuclear myopathy as well as the dominant intermediate CMT, CMTDIB. The onset of the neuropathy in reported patients is typically in infancy.[85]

CMT2O

A heterozygous mutation in cytoplasmic dynein heavy chain 1 (*DYNC1H1*) has also been identified in a family with dHMN or CMTO using exome sequencing. The patients in this series presented early in childhood with delayed motor milestones or in the second decade with motor/gait difficulties. Here proximal weakness was prominent. Dynein is a retrograde cytoskeletal motor protein[86] with mutations in the murine model leading to defects in axonal transport.[87] Recently, mutations in the bicaudal D homolog 2 (BICD2) gene have been shown to cause a similar phenotype, demonstrating the importance of retrograde axonal transport in the pathogenesis of motor disorders.[88]

CMT4: Autosomal Recessive Neuropathies

CMT4 includes all forms of autosomal recessive CMT and are typically associated with demyelinating conduction velocities on electrophysiological testing. CMT4 should be considered especially where multiple family members in one generation are affected or if there is a history of consanguinity. Most of these recessive forms of CMT present earlier than late childhood or adolescence and therefore are only briefly mentioned.

CMT4A

As mentioned above, CMT4A is caused by mutations in the *GDAP1* gene.[81] GDAP1 belongs to a subfamily of glutathione-S-transferases and is located in the outer membrane of the mitochondria, where it acts as a fission or fragmentation factor.[89] Affected individuals have a severe early childhood onset with weakness starting in the distal lower extremities and affecting the distal upper extremities within the first decade of life.[90] Progression to wheelchair is not uncommon and there may be additional clinical features such as vocal cord paralysis or hoarseness.[91]

CMT4B1

This neuropathy is characterized by an onset in early childhood with symmetrical weakness in all four limbs, with patients typically becoming wheelchair dependent by adulthood.[92] Additional clinical features of scoliosis and diaphragmatic and facial weakness may be present.[93,94] CMT4B1 is caused by mutations in myotubularin-related protein 2 (*MTMR2*), located at chromosome 11q22.[95]

CMT4B2

CMT4B2 is caused by mutations in *SBF2*, a myotubularin-related protein also known as *MTMR13*, located at chromosome 11p15.[96,97] The age of onset here is later than CMT4B1 but symptoms have begun typically by age 5,[97] with nerve conduction velocities around or below 20 m/s. MTMR2 and MTMR13 interact with each other in regulating phosphorylation of phosphoinositides (PIP), which is presumably why mutations in these two genes cause a common phenotype.

CMT4C

CMT4C is a demyelinating neuropathy with a slightly later onset than other AR CMTs with childhood-adolescent onsets reported.[98,99] Scoliosis is a particularly common feature and may precede other findings. Occasional additional clinical findings include movement disorders (ataxia and tremors),[100] facial and bulbar weakness,[98] sensorineural deafness,[101] and respiratory insufficiency.[99] It is caused by mutations in the SH3 domain and tetratricopeptide repeat domain 2 (*SH3TC2*) gene.[99] It is the most common of the AR CMTs in North American and Northern European populations.[10,11,102]

CMT4F

CMT4F is caused by mutations in the *periaxin* gene.[103,104] Characteristically, patients have demyelinating conductions with a severe early onset neuropathy, and often have predominant sensory findings on examination, including sensory ataxia.[103,105] There is a particularly large number of polymorphisms in periaxin, and care should be taken to ensure that mutations, either homozygous or compound heterozygous, segregate with the neuropathy in a family and that the phenotype is appropriate before concluding that mutations are disease-causing.

CMT4J

CMT4J is caused by mutations in the *FIG4* gene.[106] Again, the onset here is early childhood with a severe motor phenotype, which is sometimes asymmetric.[107] Rapid progression to a wheelchair in later adulthood has been described for patients who were only mildly affected in their first two decades of life, however. Nerve conductions are usually slow in patients with CMT4J, although the subsequent rapid progression appears to result from damage to the axon in this ubiquitously expressed gene. FIG4 functions in the same PIP pathway described for MTMR2/MTMR13.

CMT4: Autosomal Recessive Axonal Neuropathies Also Known as AR-CMT2

A small number of genes have been found to cause axonal forms of CMT4, also known as AR-CMT2, with lamin A/C most commonly presenting in the second decade.

A: Lamin A/C

Most patients with lamin A/C or *LMNA* mutations will present in adolescence with a severe CMT phenotype, including proximal muscle involvement, although some have a milder phenotype. Lamin A/C mutations have been associated with a wide variety of clinical syndromes including partial lipodystrophy, cardiac conduction defects, and Emery-Dreifuss phenotype muscular dystrophy.

Intermediate CMT

In some families, nerve conduction velocities overlap the axonal and demyelinating definitions. This is termed intermediate CMT and can be inherited in a dominant, recessive, or X-linked manner (see Table 17.2). The most common causes of intermediate CMT are, in order of frequency, CMT1X and CMT1B, and both have been discussed previously. The other intermediate CMTs appear to be rare. It is worth mentioning, however, that *DNM2* mutations cause a dominant intermediate form, CMT, known as CMTDIB, that presents in the first and second decade. *DNM2* mutations are also associated with CMT2M.[108] In addition, mutations in the tRNA synthetase genes such as *GARS*, *YARS*, *AARS*, and mutations in the distal heat shock proteins, also frequently present with intermediate nerve conduction velocities.

TABLE 17.3 Classification of the Distal Hereditary Motor Neuropathies

Type	Inheritance	Gene/Locus	Specific Phenotype
HMN I	AD	*HSPB1/HSPB8/GARS/ DYNCIH1*	Juvenile onset dHMN
HMN II	AD	*HSPB1/ HSPB8*	Adult onset typical dHMN/CMT2F or dHMN/CMT2L
HMN III	AR	11q13	Early onset, slowly progressive
HMN IV	AR	11q13	Juvenile onset, diaphragmatic involvement
HMN V	AD	*GARS*	Upper limb onset, slowly progressive/CMT2D
HMN V	AD	*BSCL2*	Upper limb onset, +/2 spasticity lower limbs/Silver syndrome
HMN VI	AR	*IGHMBP2*	Spinal muscle atrophy with respiratory distress (SMARD1), infantile onset respiratory distress
HMN VIIA	AD	2q14	Adult onset, vocal cord paralysis
HMN VIIB	AD	*DCTN1/TRPV4*	Adult onset/vocal cord paralysis/facial weakness
HMN/ALS4	AD	*SETX*	Early onset, pyramidal signs
HMN-J	AR	9p21.1-p12	Juvenile onset, pyramidal features, Jerash origin.
Congenital distal SMA	AD	*TRPV4*	Antenatal onset, arthrogryposis

Abbreviations: AD, autosomal dominant; AR, autosomal recessive; HSPB1, heat-shock protein B1; HSPB8, heat-shock protein B8; GARS, glycyl-tRNA synthetase; BSCL2, Berardinelli-Seip congenital lipodystrophy 2 (seipin); DYNC1H1, cytoplasmic dynein heavy chain 1; IGHMBP2, immunoglobulin mu binding protein 2; DCTN1, dynactin 1; TRPV4, transient receptor vallinoic 4; SETX, sentaxin; SMA, spinal muscular atrophy.
Adapted from Reilly and Shy.[49] From BMJ Publishing Group Ltd, with permission.

dHMNs: Distal Hereditary Motor Neuropathies

The dHMNs are very similar to CMT in that they are inherited length-dependent, slowly progressive neuropathies with onset usually starting in the first two decades.[109] Clinically, however, the dHMNs are defined by being exclusively motor in nature. However, many forms of dHMN have minor sensory abnormalities and there is a degree of overlap between CMT2 and dHMN where the same mutation in the same gene can cause both phenotypes within a family and be defined as either CMT or a dHMN as discussed above. Bulbar involvement in dHMN apart from the recurrent laryngeal nerve is rare. Clinical examination confirms distal weakness and wasting with reduced or absent reflexes. Neurophysiology testing reveals reduced motor amplitude potentials with no sensory abnormalities, and EMG testing may reveal a predominantly distal pattern of denervation. Table 17.3 outlines this group of disorders. For many of the dHMNs, however, the exact genetic mutation remains to be elucidated.

dHMNs: Juvenile Onset

The juvenile onset dHMNs include the dominantly inherited dHMN type I and type V and the recessively inherited HMN type IV and HMN-J.

AD dHMN I

dHMN type I is typically caused by mutations in *HSPB1* and *B8*, with mutations in both genes producing a clinically indistinguishable picture[79] of a length-dependent distal motor neuropathy. *HSPB1* and *B8* mutations are also called CMT2F and CMT2L, respectively, when sensory loss is present. How these ubiquitously expressed chaperone protein mutations lead to neuropathy has not been elucidated as yet. While *GARS* mutations typically produce an upper limb dominant pattern of weakness, some patients clearly present with a lower limb onset occurring in the second decade of life.[110] A heterozygous mutation in the cytoplasmic dynein heavy chain 1 (*DYNC1H1*) has also been identified in a family with dHMN or CMTO using exome sequencing. The patients in this series presented early in childhood with delayed motor milestones or in the second decade with motor/gait difficulties. Proximal weakness was prominent. Dynein and BICD2 mutations are retrograde cytoskeletal motor protein[86,88] with mutations in the murine model leading to defects in axonal transport[87]

AR dHMN IV

The exact mutation here is unknown and this subtype of dHMN has only been reported in one large consanguineous Lebanese family and linked to chromosome 11q13.

The age of onset varied from infancy (with variable presence of diaphragmatic palsy) to early adulthood. Most patients presented with a distal length-dependent pattern of limb weakness.

AD dHMN V

GARS mutations have already been discussed in the section on CMT2D. The dHMN phenotype seen with GARS mutations is clinically indistinguishable from CMT2D with the exception that there is no sensory involvement in dHMN, although overlap can occur.

The Berardinelli-Seip congenital lipodystrophy type 2 (*BSCL2*) gene is so named because it was first identified as a cause of congenital lipodystrophy.[111] However, mutations in this gene also cause a spectrum of clinical neurological phenotypes including upper-limb onset dHMN, Silver syndrome (a combination of spasticity in the lower limbs with distal upper extremity weakness and wasting), and hereditary spastic paraparesis. The largest series reported a variable age of onset from 6 to 66; however, the mean age at onset for males in particular was 19.[112] Patients mostly presented with unilateral wasting of thenar eminence and/or dorsal interosseous of the dominant hand. While reflexes were normal in the upper extremities, they were brisk in the lower extremities in the majority of affected individuals. The *BSCL2* gene encodes seipin, an endoplasmic reticulum based glycoprotein[113] with mutations presumed to lead to a toxic gain of function originating from activation of the unfolded protein response.

AR dHMN-J

Another autosomal recessive form has been described in seven consanguineous families from the Jerash region of Jordan and hence named dHMN-J.[114] The age of onset observed here is between 6 and 10 with weakness and wasting of the lower limbs with pyramidal features such as brisk reflexes. Characteristically, there is a relatively quick progression to involve the upper extremities within 2 years. The exact mutation has not been delineated; however, this form has been linked to chromosome 9p21.1-p12.

DIFFERENTIAL DIAGNOSIS

Inherited neuropathies must be distinguished from acquired neuropathies (see Chapters 20–24). Additionally, genetic disorders of the CNS such as hereditary spastic paraplegia or leukodystrophies (Chapters 9 and 19) may mimic inherited neuropathies by causing length-dependent weakness, sensory loss, and foot deformities such as pes cavus; these patients will frequently have upper motor neuron signs, such as increased reflexes or Babinski's signs, and do not have neurophysiologic evidence of neuropathy.

TREATMENT STRATEGIES

Despite the great improvement in our biologic understanding of inherited neuropathies, derived mostly from developments in molecular biology and transgenic animal models in the last 25 years, there is still no treatment available for any type of CMT. Physical therapy, occupational therapy, and a few orthopedic procedures are still the cornerstone of CMT treatment. A detailed family history and often examination of family members are required for prognosis and genetic counseling.

A dedicated, multidisciplinary rehabilitation team can significantly contribute to the management of patients with CMT and improve functionality and quality of life. Physical therapy strategies to maintain muscle strength and tone, prevent muscle contractures, and improve balance are a common need for most patients with CMT. Orthotics are also an important component of treating these patients, providing support and improving balance for ambulation. Occupational therapy focused on developing tools and strategies to help patients with activities of daily living will benefit patients with CMT, especially those with hand weakness. Tendon lengthening and tendon transfers can benefit a subset of CMT patients with muscle contractures and tendon shortening and patients with significant weakness in functionally relevant muscles, respectively; however, the optimal timing of such procedures is still controversial. Foot surgery is sometimes offered to correct inverted feet, pes cavus, and hammertoes. This surgical intervention may improve walking, alleviate pain over pressure points, and prevent plantar ulcers. However, foot surgery is generally unnecessary and does not improve weakness and sensory loss.

Reducing the expression of *PMP22* in Schwann cells (hence treating the overexpression of *PMP22*) is a biologic strategy being tested to treat CMT1A. High-dose ascorbic acid (vitamin C) was shown to decrease *PMP22* levels and symptoms in mice with CMT1A, so that they were able to stay on a rotating rod longer, cross a beam more rapidly, and grip for longer than untreated mice.[115] Several studies have been performed in humans with CMT1A, testing different doses of vitamin C (1–4 g/d) for up to 2 years. Unfortunately, all studies failed to meet their primary outcome measures and did not show a significant effect on phenotype.[116–118] Progesterone antagonists have also been shown to decrease *PMP22* expression in a rat model of CMT1A, improving the phenotype (specifically, the axonal loss seen during disease progression).[119,120] Unfortunately, onapristone, the compound shown to have therapeutic effects in this study, is toxic to humans.

Efforts to develop bioequivalent compounds with a better safety profile are ongoing.

Recent studies have demonstrated the role of ER accumulation of misfolded proteins and UPR activation in the pathogenesis of several animal models of CMT associated with point mutations in myelin-related genes, including *PMP22*[14] and *MPZ*.[15,16] Furthermore, treatment with an agent that relieves ER stress (curcumin) improved the phenotype of both models.[17,18] Therefore, compounds that either relieve ER stress or reduced UPR activation are promising therapeutic strategies to treat patients with mutations that cause misfolded proteins to accumulate in the ER of Schwann cells.

Treatment strategies for axonal forms of CMT have not been as easily identified as for demyelinating forms. Recently, histone deacetylase-6 inhibitors have been shown to correct axonal transport defects in a mouse model of CMT2F associated with point mutations in the *HSPB1* gene, rescuing the axonal loss and clinical phenotype of these mice.[121] It remains to be shown whether this same strategy could be useful in other forms of axonal CMT, but correcting axonal transport defects may be a common treatment option for most of these CMT types.

Two new technologies recently developed hold enormous potential in the search for compounds to treat CMT: cellular reprogramming and high-throughput drug screening. Cellular reprogramming is a technique that allows the generation of specific cell types (including stem cell-like cells, neurons, and glia) by genetically modifying readily available somatic cells such as fibroblasts or lymphocytes.[122,123] Using this technology, researchers are able to generate unlimited supplies of patient-specific cell lines for use in mechanistic studies and drug development.[124] These patient specific cells lines will be particularly useful when combined with high-throughput screening of drug libraries containing thousands of compounds. In these highly automated platforms, the process of identifying compounds capable of correcting certain disease-related cell phenotypes is streamlined, allowing for a faster target selection of compounds to be tested in phase I animal studies. The use of patient-derived human cells offer the theoretical advantage of a more translational platform, which could facilitate the process of moving from phase I studies to human clinical trials. Whether this is actually true remains to be proven. A recent study using cellular reprogramming successfully generated human neural crest progenitors derived from a patient with HSAN type III.[125] These cells are the precursors of sensory and autonomic neurons, the cell types most affected by this condition. Interestingly, patient-derived neural crest precursors expressed very low levels of normal inhibitor of kappa light polypeptide gene enhancer in B cells, kinase complex-associated protein (*IKBKAP*) transcript, while also displaying marked defects in neuronal differentiation and migration. The investigators were also able to find compounds that rescued this phenotype, at least partially, validating this platform for drug discovery in inherited neuropathies.

Vital to clinical trials in inherited neuropathy is the establishment of solid outcome measures. This can be challenging with neuropathies that progress slowly over many years. To this end, a validated scoring system based on symptoms, signs, and neurophysiological data has been developed for adults with excellent inter- and intra-observer correlations and that can detect changes over 1 year.[29,126] A more sensitive pediatric score exists for those patients between the ages of 3 and 21 that takes into account growth related changes and should prove useful in future clinical trials.[127] Sequential yearly MRI of limb muscles looking at progressive replacement of muscle with fatty tissue may also prove to be a valuable outcome measurement and is currently under investigation in a cohort of CMT1A patients.[128]

SUMMARY

Although CMT is a genetically heterogeneous condition, it is often possible to determine the type of CMT a person has by distinguishing characteristics. The prevalence of the various mutations, inheritance pattern, nerve conductions, and age of onset should be taken into account when deciding what genetic testing should be ordered. New genes causing CMT continue to be found, prevalence continues to be studied, and recommendations for testing will continue to evolve over time. Our increasing understanding of biologic processes involved in CMT has offered new therapeutic targets for drug development and new tools recently developed hold the promise of even faster drug discovery in CMT.

REFERENCES

1. Charcot JM, Marie P. Sur une forme particuliére d atrophie musculaire progressive souvent familiale débutant par les pieds et les jambes et atteignant plus tard les mains. *Rev Méd Paris* 1886;**6**:97–138.
2. Tooth HH. *The Peroneal Type of Progressive Muscular Atrophy*. London: Lewis; 1886.
3. Skre H. Genetic and clinical aspects of Charcot-Marie-Tooth's disease. *Clin Genet* 1974;**6**:98–118.
4. Harding AE, Thomas PK. Genetic aspects of hereditary motor and sensory neuropathy (types I and II). *J Med Genet* 1980 **17**:329–36.
5. Harding AE, Thomas PK. The clinical features of hereditary motor and sensory neuropathy types I and II. *Brain* 1980;**103**:259–80.
6. Dejerine J, Sottas J. Sur la nevrite interstitielle, hypertrophique et progressive de l'enfance. *C R Soc Biol* 1893;**45** 63–96.
7. Saporta AS, Sottile SL, Miller LJ, Feely SM, Siskind CE, Shy ME. Charcot-Marie-Tooth disease subtypes and genetic testing strategies *Ann Neurol* 2011;**69**:22–33.

8. Murphy SM, Laura M, Fawcett K, Pandraud A, Liu YT, Davidson GL, et al. Charcot-Marie-Tooth disease: frequency of genetic subtypes and guidelines for genetic testing. *J Neurol Neurosurg Psychiatry* 2012;**83**:706–10.
9. Trapp B, Pfeiffer S, Anitei A. Cell biology of myelin assembly. In: Lazzarini R, editor. *Myelin Biology and Disorders*. San Diego: Elsevier Academic Press; 2004. p. 29–56.
10. Niemann A, Berger P, Suter U. Pathomechanisms of mutant proteins in Charcot-Marie-Tooth disease. *Neuromolecular Med* 2006;**8**:217–42.
11. Li J. Hypothesis of double polarization. *J Neurol Sci* 2008;**275**:33–6.
12. Bergoffen J, Scherer SS, Wang S, Scott MO, Bone LJ, Paul DL, et al. Connexin mutations in X-linked Charcot-Marie-Tooth disease. *Science* 1993;**262**:2039–42.
13. Krajewski KM, Lewis RA, Fuerst DR, Turansky C, Hinderer SR, Garbern J, et al. Neurological dysfunction and axonal degeneration in Charcot-Marie-Tooth disease type 1A. *Brain* 2000;**123**(Pt 7): 1516–27.
14. Colby J, Nicholson R, Dickson KM, Orfali W, Naef R, Suter U, et al. PMP22 carrying the trembler or trembler-J mutation is intracellularly retained in myelinating Schwann cells. *Neurobiol Dis* 2000;**7**:561–73.
15. Pennuto M, Tinelli E, Malaguti M, Del Carro U, D'Antonio M, Ron D, et al. Ablation of the UPR-mediator CHOP restores motor function and reduces demyelination in Charcot-Marie-Tooth 1B mice. *Neuron* 2008;**57**:393–405.
16. Saporta MA, Shy BR, Patzko A, Bai Y, Pennuto M, Ferri C, et al. MpzR98C arrests Schwann cell development in a mouse model of early-onset Charcot-Marie-Tooth disease type 1B. *Brain* 2012;**135**: 2032–47.
17. Khajavi M, Shiga K, Wiszniewski W, He F, Shaw CA, Yan J, et al. Oral curcumin mitigates the clinical and neuropathologic phenotype of the Trembler-J mouse: a potential therapy for inherited neuropathy. *Am J Hum Genet* 2007;**81**:438–53.
18. Patzko A, Bai Y, Saporta MA, Katona I, Wu X, Vizzuso D, et al. Curcumin derivatives promote Schwann cell differentiation and improve neuropathy in R98C CMT1B mice. *Brain* 2012;**135**:3551–66.
19. Sivakumar K, Kyriakides T, Puls I, Nicholson GA, Funalot B, Antonellis A, et al. Phenotypic spectrum of disorders associated with glycyl-tRNA synthetase mutations. *Brain* 2005;**128**:2304–14.
20. Antonellis A, Ellsworth RE, Sambuughin N, Puls I, Abel A, Lee-Lin SQ, et al. Glycyl tRNA synthetase mutations in Charcot-Marie-Tooth disease type 2D and distal spinal muscular atrophy type V. *Am J Hum Genet* 2003;**72**:1293–9.
21. Houlden H, King R, Blake J, Groves M, Love S, Woodward C, et al. Clinical, pathological and genetic characterization of hereditary sensory and autonomic neuropathy type 1 (HSAN I). *Brain* 2006;**129**:411–25.
22. Cottenie E, Menezes MP, Rossor AM, Morrow JM, Yousry TA, Dick DJ, et al. Rapidly progressive asymmetrical weakness in Charcot-Marie-Tooth disease type 4J resembles chronic inflammatory demyelinating polyneuropathy. *Neuromuscul Disord* 2013;**23**:399–403.
23. Michell AW, Laura M, Blake J, Lunn MP, Cox A, Gibbons VS, et al. GJB1 gene mutations in suspected inflammatory demyelinating neuropathies not responding to treatment. *J Neurol Neurosurg Psychiatry* 2009;**80**:699–700.
24. Houlden H, Laura M, Ginsberg L, Jungbluth H, Robb SA, Blake J, et al. The phenotype of Charcot-Marie-Tooth disease type 4C due to SH3TC2 mutations and possible predisposition to an inflammatory neuropathy. *Neuromuscul Disord* 2009;**19**:264–9.
25. Ginsberg L, Malik O, Kenton AR, Sharp D, Muddle JR, Davis MB, et al. Coexistent hereditary and inflammatory neuropathy. *Brain* 2004;**127**:193–202.
26. Nelis E, Van Broeckhoven C, De Jonghe P, Lofgren A, Vandenberghe A, Latour P, et al. Estimation of the mutation frequencies in Charcot-Marie-Tooth disease type 1 and hereditary neuropathy with liability to pressure palsies: a European collaborative study. *Eur J Hum Genet* 1996;**4**:25–33.
27. Krajewski KM, Shy ME. Genetic testing in neuromuscular disease. *Neurol Clin* 2004;**22**:481–508.
28. Shy ME, Siskind C, Swan ER, Krajewski KM, Doherty T, Fuerst DR, et al. CMT1X phenotypes represent loss of GJB1 gene function. *Neurology* 2007;**68**:849–55.
29. Shy ME, Chen L, Swan ER, Taube R, Krajewski KM, Herrmann D, et al. Neuropathy progression in Charcot-Marie-Tooth disease type 1A. *Neurology* 2008;**70**:378–83.
30. Points to consider: ethical, legal, and psychosocial implications of genetic testing in children and adolescents. American Society of Human Genetics Board of Directors, American College of Medical Genetics Board of Directors. *Am J Hum Genet* 1995;**57**:1233–41.
31. Rossor AM, Polke JM, Houlden H, Reilly MM. Clinical implications of genetic advances in Charcot-Marie-Tooth disease. *Nat Rev Neurol* 2013;**9**:562–71.
32. Blair IP, Nash J, Gordon MJ, Nicholson GA. Prevalence and origin of de novo duplications in Charcot-Marie-Tooth disease type 1A: first report of a de novo duplication with a maternal origin. *Am J Hum Genet* 1996;**58**:472–6.
33. Lupski JR, de Oca-Luna RM, Slaugenhaupt S, Pentao L, Guzzetta V, Trask BJ, et al. DNA duplication associated with Charcot-Marie-Tooth disease type 1A. *Cell* 1991;**66**:219–32.
34. Hayasaka K, Himoro M, Sato W, Takada G, Uyemura K, Shimizu N, et al. Charcot-Marie-Tooth neuropathy type 1B is associated with mutations of the myelin P0 gene. *Nat Genet* 1993;**5**:31–4.
35. Saifi GM, Szigeti K, Wiszniewski W, Shy ME, Krajewski K, Hausmanowa-Petrusewicz I, et al. SIMPLE mutations in Charcot-Marie-Tooth disease and the potential role of its protein product in protein degradation. *Hum Mutat* 2005;**25**:372–83.
36. Street VA, Bennett CL, Goldy JD, Shirk AJ, Kleopa KA, Tempel BL, et al. Mutation of a putative protein degradation gene LITAF/SIMPLE in Charcot-Marie-Tooth disease 1C. *Neurology* 2003;**60**:22–6.
37. Warner LE, Mancias P, Butler IJ, McDonald CM, Keppen L, Koob KG, et al. Mutations in the early growth response 2 (EGR2) gene are associated with hereditary myelinopathies. *Nat Genet* 1998;**18**:382–4.
38. Vandenberghe N, Upadhyaya M, Gatignol A, Boutrand L, Boucherat M, Chazot G, et al. Frequency of mutations in the early growth response 2 gene associated with peripheral demyelinating neuropathies. *J Med Genet* 2002;**39**:e81.
39. Pareyson D, Taroni F, Botti S, Morbin M, Baratta S, Lauria G, et al. Cranial nerve involvement in CMT disease type 1 due to early growth response 2 gene mutation. *Neurology* 2000;**54**:1696–8.
40. Russo M, Laura M, Polke JM, Davis MB, Blake J, Brandner S, et al. Variable phenotypes are associated with PMP22 missense mutations. *Neuromuscul Disord* 2011;**21**:106–14.
41. Chance PF, Alderson MK, Leppig KA, Lensch MW, Matsunami N, Smith B, et al. DNA deletion associated with hereditary neuropathy with liability to pressure palsies. *Cell* 1993;**72**:143–51.

42. Stogbauer F, Young P, Kuhlenbaumer G, De Jonghe P, Timmerman V. Hereditary recurrent focal neuropathies: clinical and molecular features. *Neurology* 2000;**54**:546–51.
43. Kim SM, Chung KW, Choi BO, Yoon ES, Choi JY, Park KD, et al. Hereditary neuropathy with liability to pressure palsies (HNPP) patients of Korean ancestry with chromosome 17p11.2-p12 deletion. *Exp Mol Med* 2004;**36**:28–35.
44. Andersson PB, Yuen E, Parko K, So YT. Electrodiagnostic features of hereditary neuropathy with liability to pressure palsies. *Neurology* 2000;**54**:40–4.
45. Li J, Krajewski K, Shy ME, Lewis RA. Hereditary neuropathy with liability to pressure palsy: the electrophysiology fits the name. *Neurology* 2002;**58**:1769–73.
46. Siskind C, Feely SM, Bernes S, Shy ME, Garbern JY. Persistent CNS dysfunction in a boy with CMT1X. *J Neurol Sci* 2009;**279**:109–13.
47. Siskind CE, Murphy SM, Ovens R, Polke J, Reilly MM, Shy ME. Phenotype expression in women with CMT1X. *J Peripher Nerv Syst* 2011;**16**:102–7.
48. Murphy SM, Ovens R, Polke J, Siskind CE, Laura M, Bull K, et al. X inactivation in females with X-linked Charcot-Marie-Tooth disease. *Neuromuscul Disord* 2012;**22**:617–21.
49. Reilly MM, Shy ME. Diagnosis and new treatments in genetic neuropathies. *J Neurol Neurosurg Psychiatry* 2009;**80**:1304–14.
50. Mutation Database of Inherited Peripheral Neuropathies (IPNMDB). Human Genome Variation Society. <http://www.molgen.ua.ac.be/cmtmutations/> [accessed from 14.06.2014].
51. Feely SM, Laura M, Siskind CE, Sottile S, Davis M, Gibbons VS, et al. MFN2 mutations cause severe phenotypes in most patients with CMT2A. *Neurology* 2011;**76**:1690–6.
52. Zuchner S, Mersiyanova IV, Muglia M, Bissar-Tadmouri N, Rochelle J, Dadali EL, et al. Mutations in the mitochondrial GTPase mitofusin 2 cause Charcot-Marie-Tooth neuropathy type 2A. *Nat Genet* 2004;**36**:449–51.
53. Rojo M, Legros F, Chateau D, Lombes A. Membrane topology and mitochondrial targeting of mitofusins, ubiquitous mammalian homologs of the transmembrane GTPase Fzo. *J Cell Sci* 2002;**115**:1663–74.
54. de Brito OM, Scorrano L. Mitofusin 2 tethers endoplasmic reticulum to mitochondria. *Nature* 2008;**456**:605–10.
55. Zuchner S, De Jonghe P, Jordanova A, Claeys KG, Guergueltcheva V, Cherninkova S, et al. Axonal neuropathy with optic atrophy is caused by mutations in mitofusin 2. *Ann Neurol* 2006;**59**:276–81.
56. Zhu D, Kennerson ML, Walizada G, Zuchner S, Vance JM, Nicholson GA. Charcot-Marie-Tooth with pyramidal signs is genetically heterogeneous: families with and without MFN2 mutations. *Neurology* 2005;**65**:496–7.
57. Auer-Grumbach M, De Jonghe P, Wagner K, Verhoeven K, Hartung HP, Timmerman V. Phenotype-genotype correlations in a CMT2B family with refined 3q13-q22 locus. *Neurology* 2000;**55**:1552–7.
58. De Jonghe P, Timmerman V, FitzPatrick D, Spoelders P, Martin JJ, Van Broeckhoven C. Mutilating neuropathic ulcerations in a chromosome 3q13-q22 linked Charcot-Marie-Tooth disease type 2B family. *J Neurol Neurosurg Psychiatry* 1997;**62**:570–3.
59. Kwon JM, Elliott JL, Yee WC, Ivanovich J, Scavarda NJ, Moolsintong PJ, et al. Assignment of a second Charcot-Marie-Tooth type II locus to chromosome 3q. *Am J Hum Genet* 1995;**57**:853–8.
60. Verhoeven K, De Jonghe P, Coen K, Verpoorten N, Auer-Grumbach M, Kwon JM, et al. Mutations in the small GTP-ase late endosomal protein RAB7 cause Charcot-Marie-Tooth type 2B neuropathy. *Am J Hum Genet* 2003;**72**:722–7.
61. Santoro L, Manganelli F, Di Maio L, Barbieri F, Carella M, D'Adamo P, et al. Charcot-Marie-Tooth disease type 2C: a distinct genetic entity. Clinical and molecular characterization of the first European family. *Neuromuscul Disord* 2002;**12**:399–404.
62. Dyck PJ, Litchy WJ, Minnerath S, Bird TD, Chance PF, Schaid DJ, et al. Hereditary motor and sensory neuropathy with diaphragm and vocal cord paresis. *Ann Neurol* 1994;**35**:608–15.
63. Landoure G, Zdebik AA, Martinez TL, Burnett BG, Stanescu HC, Inada H, et al. Mutations in TRPV4 cause Charcot-Marie-Tooth disease type 2C. *Nat Genet* 2010;**42**:170–4.
64. Chen DH, Sul Y, Weiss M, Hillel A, Lipe H, Wolff J, et al. CMT2C with vocal cord paresis associated with short stature and mutations in the TRPV4 gene. *Neurology* 2010;**75**:1968–75.
65. Klein CJ, Cunningham JM, Atkinson EJ, Schaid DJ, Hebbring SJ, Anderson SA, et al. The gene for HMSN2C maps to 12q23-24: a region of neuromuscular disorders. *Neurology* 2003;**60**:1151–6.
66. Deng HX, Klein CJ, Yan J, Shi Y, Wu Y, Fecto F, et al. Scapuloperoneal spinal muscular atrophy and CMT2C are allelic disorders caused by alterations in TRPV4. *Nat Genet* 2010;**42**:165–9.
67. Ge Q, Trieu EP, Targoff IN. Primary structure and functional expression of human Glycyl-tRNA synthetase, an autoantigen in myositis. *J Biol Chem* 1994;**269**:28790–7.
68. Motley WW, Talbot K, Fischbeck KH. GARS axonopathy: not every neuron's cup of tRNA. *Trends Neurosci* 2010;**33**:59–66.
69. Motley WW, Seburn KL, Nawaz MH, Miers KE, Cheng J, Antonellis A, et al. Charcot-Marie-Tooth-linked mutant GARS is toxic to peripheral neurons independent of wild-type GARS levels. *PLOS Genet* 2011;**7**:e1002399.
70. Mersiyanova IV, Perepelov AV, Polyakov AV, Sitnikov VF, Dadali EL, Oparin RB, et al. A new variant of Charcot-Marie-Tooth disease type 2 is probably the result of a mutation in the neurofilament-light gene. *Am J Hum Genet* 2000;**67**:37–46.
71. Georgiou DM, Zidar J, Korosec M, Middleton LT, Kyriakides T Christodoulou K. A novel NF-L mutation Pro22Ser is associated with CMT2 in a large Slovenian family. *Neurogenetics* 2002;**4**:93–6.
72. Yoshihara T, Yamamoto M, Hattori N, Misu K, Mori K, Koike H, et al. Identification of novel sequence variants in the neurofilament-light gene in a Japanese population: analysis of Charcot-Marie-Tooth disease patients and normal individuals. *J Peripher Nerv Syst* 2002;**7**:221–4.
73. Jordanova A, De Jonghe P, Boerkoel CF, Takashima H, De Vriendt E, Ceuterick C, et al. Mutations in the neurofilament light chain gene (NEFL) cause early onset severe Charcot-Marie-Tooth disease. *Brain* 2003;**126**:590–7.
74. Friede RL, Samorajski T. Axon caliber related to neurofilaments and microtubules in sciatic nerve fibers of rats and mice. *Anat Rec* 1970;**167**:379–87.
75. Evgrafov OV, Mersiyanova I, Irobi J, Van Den Bosch L, Dierick I, Leung CL, et al. Mutant small heat-shock protein 27 causes axonal Charcot-Marie-Tooth disease and distal hereditary motor neuropathy. *Nat Genet* 2004;**36**:602–6.
76. Chung KW, Kim SB, Cho SY, Hwang SJ, Park SW, Kang SH, et al. Distal hereditary motor neuropathy in Korean patients with a small heat shock protein 27 mutation. *Exp Mol Med* 2008;**40**:304–12.

77. Houlden H, Laura M, Wavrant-De Vrieze F, Blake J, Wood N, Reilly MM. Mutations in the HSP27 (HSPB1) gene cause dominant, recessive, and sporadic distal HMN/CMT type 2. *Neurology* 2008;**71**:1660–8.

78. Tang BS, Zhao GH, Luo W, Xia K, Cai F, Pan Q, et al. Small heat-shock protein 22 mutated in autosomal dominant Charcot-Marie-Tooth disease type 2L. *Hum Genet* 2005;**116**:222–4.

79. Tang BS, Luo W, Xia K, Xiao JF, Jiang H, Shen L, et al. A new locus for autosomal dominant Charcot-Marie-Tooth disease type 2 (CMT2L) maps to chromosome 12q24. *Hum Genet* 2004;**114**: 527–33.

80. Irobi J, Van Impe K, Seeman P, Jordanova A, Dierick I, Verpoorten N, et al. Hot-spot residue in small heat-shock protein 22 causes distal motor neuropathy. *Nat Genet* 2004;**36**:597–601.

81. Baxter RV, Ben Othmane K, Rochelle JM, Stajich JE, Hulette C, Dew-Knight S, et al. Ganglioside-induced differentiation-associated protein-1 is mutant in Charcot-Marie-Tooth disease type 4A/8q21. *Nat Genet* 2002;**30**:21–2.

82. Sivera R, Espinos C, Vilchez JJ, Mas F, Martinez-Rubio D, Chumillas MJ, et al. Phenotypical features of the p.R120W mutation in the GDAP1 gene causing autosomal dominant Charcot-Marie-Tooth disease. *J Peripher Nerv Syst* 2010;**15**:334–44.

83. Crimella C, Tonelli A, Airoldi G, Baschirotto C, D'Angelo MG, Bonato S, et al. The GST domain of GDAP1 is a frequent target of mutations in the dominant form of axonal Charcot Marie Tooth type 2K. *J Med Genet* 2010;**47**:712–16.

84. Sivera R, Sevilla T, Vilchez JJ, Martinez-Rubio D, Chumillas MJ, Vazquez JF, et al. Charcot-Marie-Tooth disease: genetic and clinical spectrum in a Spanish clinical series. *Neurology* 2013;**81**:1617–25.

85. Susman RD, Quijano-Roy S, Yang N, Webster R, Clarke NF, Dowling J, et al. Expanding the clinical, pathological and MRI phenotype of DNM2-related centronuclear myopathy. *Neuromuscul Disord* 2010;**20**:229–37.

86. Hook P, Vallee RB. The dynein family at a glance. *J Cell Sci* 2006;**119**:4369–71.

87. Bilsland LG, Sahai E, Kelly G, Golding M, Greensmith L, Schiavo G. Deficits in axonal transport precede ALS symptoms in vivo. *Proc Natl Acad Sci USA* 2010;**107**:20523–8.

88. Oates EC, Rossor AM, Hafezparast M, Gonzalez M, Speziani F, MacArthur DG, et al. Mutations in BICD2 cause dominant congenital spinal muscular atrophy and hereditary spastic paraplegia. *Am J Hum Genet* 2013;**92**:965–73.

89. Niemann A, Ruegg M, La Padula V, Schenone A, Suter U. Ganglioside-induced differentiation associated protein 1 is a regulator of the mitochondrial network: new implications for Charcot-Marie-Tooth disease. *J Cell Biol* 2005;**170**:1067–78.

90. Senderek J, Bergmann C, Ramaekers VT, Nelis E, Bernert G, Makowski A, et al. Mutations in the ganglioside-induced differentiation-associated protein-1 (GDAP1) gene in intermediate type autosomal recessive Charcot-Marie-Tooth neuropathy. *Brain* 2003;**126**:642–9.

91. Sevilla T, Cuesta A, Chumillas MJ, Mayordomo F, Pedrola L, Palau F, et al. Clinical, electrophysiological and morphological findings of Charcot-Marie-Tooth neuropathy with vocal cord palsy and mutations in the GDAP1 gene. *Brain* 2003;**126**:2023–33.

92. Parman Y, Battaloglu E, Baris I, Bilir B, Poyraz M, Bissar-Tadmouri N, et al. Clinicopathological and genetic study of early-onset demyelinating neuropathy. *Brain* 2004;**127**:2540–50.

93. Tyson J, Ellis D, Fairbrother U, King RH, Muntoni F, Jacobs J, et al. Hereditary demyelinating neuropathy of infancy. A genetically complex syndrome. *Brain* 1997;**120**(Pt 1):47–63.

94. Verny C, Ravise N, Leutenegger AL, Pouplard F, Dubourg O, Tardieu S, et al. Coincidence of two genetic forms of Charcot-Marie-Tooth disease in a single family. *Neurology* 2004;**63**:1527–9.

95. Bolino A, Muglia M, Conforti FL, LeGuern E, Salih MA, Georgiou DM, et al. Charcot-Marie-Tooth type 4B is caused by mutations in the gene encoding myotubularin-related protein-2. *Nat Genet* 2000;**25**:17–19.

96. Azzedine H, Bolino A, Taieb T, Birouk N, Di Duca M, Bouhouche A, et al. Mutations in MTMR13, a new pseudophosphatase homologue of MTMR2 and Sbf1, in two families with an autosomal recessive demyelinating form of Charcot-Marie-Tooth disease associated with early-onset glaucoma. *Am J Hum Genet* 2003;**72**:1141–53.

97. Senderek J, Bergmann C, Weber S, Ketelsen UP, Schorle H, Rudnik-Schoneborn S, et al. Mutation of the SBF2 gene, encoding a novel member of the myotubularin family, in Charcot-Marie-Tooth neuropathy type 4B2/11p15. *Hum Mol Genet* 2003;**12**:349–56.

98. Azzedine H, Ravise N, Verny C, Gabreels-Festen A, Lammens M, Grid D, et al. Spine deformities in Charcot-Marie-Tooth 4C caused by SH3TC2 gene mutations. *Neurology* 2006;**67**:602–6.

99. Senderek J, Bergmann C, Stendel C, Kirfel J, Verpoorten N, De Jonghe P, et al. Mutations in a gene encoding a novel SH3/TPR domain protein cause autosomal recessive Charcot-Marie-Tooth type 4C neuropathy. *Am J Hum Genet* 2003;**73**:1106–19.

100. Gooding R, Colomer J, King R, Angelicheva D, Marns L, Parman Y, et al. A novel Gypsy founder mutation, p.Arg1109X in the CMT4C gene, causes variable peripheral neuropathy phenotypes. *J Med Genet* 2005;**42**(e69).

101. Colomer J, Gooding R, Angelicheva D, King RH, Guillen-Navarro E, Parman Y, et al. Clinical spectrum of CMT4C disease in patients homozygous for the p.Arg1109X mutation in SH3TC2. *Neuromuscul Disord* 2006;**16**:449–53.

102. Lassuthova P, Mazanec R, Vondracek P, Siskova D, Haberlova J, Sabova J, et al. High frequency of SH3TC2 mutations in Czech HMSN I patients. *Clin Genet* 2011;**80**:334–45.

103. Boerkoel CF, Takashima H, Stankiewicz P, Garcia CA, Leber SM, Rhee-Morris L, et al. Periaxin mutations cause recessive Dejerine-Sottas neuropathy. *Am J Hum Genet* 2001;**68**:325–33.

104. Guilbot A, Williams A, Ravise N, Verny C, Brice A, Sherman DL, et al. A mutation in periaxin is responsible for CMT4F, an autosomal recessive form of Charcot-Marie-Tooth disease. *Hum Mol Genet* 2001;**10**:415–21.

105. Marchesi C, Milani M, Morbin M, Cesani M, Lauria G, Scaioli V, et al. Four novel cases of periaxin-related neuropathy and review of the literature. *Neurology* 2010;**75**:1830–8.

106. Chow CY, Zhang Y, Dowling JJ, Jin N, Adamska M, Shiga K, et al. Mutation of FIG4 causes neurodegeneration in the pale tremor mouse and patients with CMT4J. *Nature* 2007;**448**:68–72.

107. Zhang X, Chow CY, Sahenk Z, Shy ME, Meisler MH, Li J. Mutation of FIG4 causes a rapidly progressive, asymmetric neuronal degeneration. *Brain* 2008;**131**:1990–2001.

108. Zuchner S, Noureddine M, Kennerson M, Verhoeven K, Claeys K, De Jonghe P, et al. Mutations in the pleckstrin homology domain of dynamin 2 cause dominant intermediate Charcot-Marie-Tooth disease. *Nat Genet* 2005;**37**:289–94.

109. Rossor AM, Kalmar B, Greensmith L, Reilly MM. The distal hereditary motor neuropathies. *J Neurol Neurosurg Psychiatry* 2012;**83**:6–14.

110. Dubourg O, Azzedine H, Yaou RB, Pouget J, Barois A, Meininger V, et al. The G526R glycyl-tRNA synthetase gene mutation in distal hereditary motor neuropathy type V. *Neurology* 2006;**66**:1721–6.

111. Agarwal AK, Simha V, Oral EA, Moran SA, Gorden P, O'Rahilly S, et al. Phenotypic and genetic heterogeneity in congenital generalized lipodystrophy. *J Clin Endocrinol Metab* 2003;**88**:4840–7.

112. Auer-Grumbach M, Schlotter-Weigel B, Lochmuller H, Strobl-Wildemann G, Auer-Grumbach P, Fischer R, et al. Phenotypes of the N88S Berardinelli-Seip congenital lipodystrophy 2 mutation. *Ann Neurol* 2005;**57**:415–24.

113. Ito D, Suzuki N. Seipinopathy: a novel endoplasmic reticulum stress-associated disease. *Brain* 2009;**132**:8–15.

114. Christodoulou K, Zamba E, Tsingis M, Mubaidin A, Horani K, Abu-Sheik S, et al. A novel form of distal hereditary motor neuropathy maps to chromosome 9p21.1-p12. *Ann Neurol* 2000;**48**:877–84.

115. Passage E, Norreel JC, Noack-Fraissignes P, Sanguedolce V, Pizant J, Thirion X, et al. Ascorbic acid treatment corrects the phenotype of a mouse model of Charcot-Marie-Tooth disease. *Nat Med* 2004;**10**:396–401.

116. Verhamme C, de Haan RJ, Vermeulen M, Baas F, de Visser M, van Schaik IN. Oral high dose ascorbic acid treatment for one year did not improve myelination in young CMT1a patients: a proof of principle pilot study. *J Peripher Nerv Syst* 2009;**14**: 152–3.

117. Burns J, Ouvrier RA, Yiu EM, Joseph PD, Kornberg AJ, Fahey MC, et al. Ascorbic acid for Charcot-Marie-Tooth disease type 1A in children: a randomised, double-blind, placebo-controlled, safety and efficacy trial. *Lancet Neurol* 2009;**8**:537–44.

118. Micallef J, Attarian S, Dubourg O, Gonnaud PM, Hogrel JY, Stojkovic T, et al. Effect of ascorbic acid in patients with Charcot-Marie-Tooth disease type 1A: a multicentre, randomised, double-blind, placebo-controlled trial. *Lancet Neurol* 2009;**8**:1103–10.

119. Sereda MW, Meyer zu Horste G, Suter U, Uzma N, Nave KA. Therapeutic administration of progesterone antagonist in a model of Charcot-Marie-Tooth disease (CMT-1A). *Nat Med* 2003;**9**:1533–7.

120. Meyer zu Horste G, Prukop T, Liebetanz D, Mobius W, Nave KA, Sereda MW. Antiprogesterone therapy uncouples axonal loss from demyelination in a transgenic rat model of CMT1A neuropathy. *Ann Neurol* 2007;**61**:61–72.

121. d'Ydewalle C, Krishnan J, Chiheb DM, Van Damme P, Irobi J, Kozikowski AP, et al. HDAC6 inhibitors reverse axonal loss in a mouse model of mutant HSPB1-induced Charcot-Marie-Tooth disease. *Nat Med* 2011;**17**:968–74.

122. Takahashi K, Yamanaka S. Induction of pluripotent stem cells from mouse embryonic and adult fibroblast cultures by defined factors. *Cell* 2006;**126**:663–76.

123. Dimos JT, Rodolfa KT, Niakan KK, Weisenthal LM, Mitsumoto H, Chung W, et al. Induced pluripotent stem cells generated from patients with ALS can be differentiated into motor neurons. *Science* 2008;**321**:1218–21.

124. Saporta MA, Grskovic M, Dimos JT. Induced pluripotent stem cells in the study of neurological diseases. *Stem Cell Res Ther* 2011;**2**:37.

125. Lee G, Papapetrou EP, Kim H, Chambers SM, Tomishima MJ, Fasano CA, et al. Modelling pathogenesis and treatment of familial dysautonomia using patient-specific iPSCs. *Nature* 2009;**461**:402–6.

126. Murphy SM, Herrmann DN, McDermott MP, Scherer SS, Shy ME, Reilly MM, et al. Reliability of the CMT neuropathy score (second version) in Charcot-Marie-Tooth disease. *J Peripher Nerv Syst* 2011;**16**:191–8.

127. Burns J, Ouvrier R, Estilow T, Shy R, Laura M, Pallant JF, et al. Validation of the Charcot-Marie-Tooth disease pediatric scale as an outcome measure of disability. *Ann Neurol* 2012;**71**:642–52.

128. Sinclair CD, Morrow JM, Miranda MA, Davagnanam I, Cowley PC, Mehta H, et al. Skeletal muscle MRI magnetisation transfer ratio reflects clinical severity in peripheral neuropathies. *J Neurol Neurosurg Psychiatry* 2012;**83**:29–32.

129. Saporta MA, Shy ME. Peripheral neuropathies. In: Zigmond MJ, editor. *Neurobiology of brain disorders*. Amsterdam, Netherlands: Elsevier; 2014. p. 167–88.

Chapter 18

Hereditary Sensory and Autonomic Neuropathies

Felicia B. Axelrod and Horacio Kaufmann

INTRODUCTION

The hereditary sensory and autonomic neuropathies (HSANs) are a group of rare disorders caused by different genetic mutations, each affecting specific aspects of development, function, or survival of small myelinated and unmyelinated neurons resulting in variable phenotypic expression. Affected patients have impaired pain and temperature perception of different severity and varying autonomic abnormalities.[1–3] These sensory deficits frequently lead to unnoticed acral ulcerations that can become infected, resulting in osteomyelitis and autoamputations, which are perhaps the most dramatic feature of patients with HSANs.[1–5] The sensory and autonomic neurons have common embryonic origin in the multipotent neural crest cells. Their differentiation and commitment to function in the mature nervous system require exposure to growth factors along the migratory route and then within the target tissue.[6,7] Differentiation will determine the neuron's ability to produce specific neurotransmitters. HSANs causing mutations affect these processes at different steps.

The HSANs have been classified by Dyck and Ohta as types I to V on the basis of age at onset, mode of inheritance, and the predominant clinical features.[8] More recently, the identification of a number of disease-causing mutations in ten HSAN disease-causing genes,[9] a number that will likely continue to increase, has allowed a genotype-based classification that is still evolving (Table 18.1). Many published cases lacked genetic confirmation, and there is still controversy over terminology and classification.[1–3] HSAN-causing mutations affect key proteins involved in the development of nerves such as the nerve growth factor receptor in HSAN 4 or other proteins, such as *IKBKAP* in HSAN 3, whose functions are less well understood. For each HSAN type, penetrance is complete but there can be marked variability in phenotypic expression.

HSAN TYPE I

Molecular Genetics

HSAN I, also known as hereditary sensory radicular neuropathy,[10] is an autosomal dominant disorder that usually presents between the second and fifth decades, but congenital or childhood onset has been reported.[5] Several causative mutations have been identified in five genes: *SPTLC1*, *SPLTC2*, *ATL1*, *RAB7*, and *DNMT1*.[3,5,11,12]

SPTLC1 and *SPTLC2* encode subunits 1 and 2 of serine palmitoyltransferase (SPT),[13] an enzyme that catalyzes the biosynthesis of serine and palmitoyl CoA. Upregulation of SPT is suggested to play a role in apoptosis. Gain of function mutations in the *SPTLC1* and *SPTLC2* genes lead to the formation of atypical deoxysphingoid bases that cannot be converted to complex sphingolipids or degraded, resulting in their intracellular accumulation with pronounced neurotoxic effects on neurite formation and neurofilament structure.[13] The single patient reported with congenital presentation and very severe phenotype had a *de novo* missense mutation in *SPTLC1* (c.992C > T; p.Ser331Phe) that was not present in either parent.[5]

Four missense mutations in *RAB7* have been identified that are associated with the phenotype of profound sensory loss and motor weakness, now termed Charcot-Marie-Tooth type 2B (CMT2B) disease.[12] *RAB7* mutants impair epidermal growth factor (EGF) receptor trafficking.[12] The resulting downregulation of EGF receptor dependent nuclear transcription impedes normal axon outgrowth and peripheral innervation and results in neurodegeneration.

B.T. Darras, H. Royden Jones, Jr., M.M. Ryan & D.C. De Vivo (Eds): Neuromuscular Disorders of Infancy, Childhood and Adolescence, Second edition.
DOI: http://dx.doi.org/10.1016/B978-0-12-417044-5.00018-4

TABLE 18.1 HSAN Pattern of Inheritance and Associated Genes

Type	Other Names	Inheritance	Genes	OMIM
I	Hereditary sensory radicular neuropathy	Autosomal dominant	*SPTLC1, SPLTC2* *ATL1* *RAB7* *DNMT1*	162400 613640 613708 600882 126375
IIA IIB IIC	Congenital insensitivity to pain	Autosomal recessive	*WNK1* *FAM134B* *KIF1A*	201300 613115 614213
III	Familial dysautonomia (FD)/Riley Day	Autosomal recessive	*IKBKAP*	223900
IV	Congenital insensitivity to pain with anhidrosis (CIPA)	Autosomal recessive	*TrK A/NGF R*	256800
V	Congenital insensitivity to pain	Autosomal recessive	*NGF B*	608654

Pathophysiology

Neurophysiology studies show a sensory axonal neuropathy, but in many individuals there is also electrical evidence of demyelination.[13] The disease process affects the axons and cell bodies of primary sensory neurons in the dorsal root ganglia and motor neurons in the anterior horns of the spinal cord. Sural nerve biopsies and postmortem findings showed distal axonal degeneration.[14] Although there is loss of unmyelinated, small myelinated, and large myelinated fibers, the small fibers are affected to a greater degree and there is subsequent dorsal root ganglion cell loss.[14] Dorsal columns in the spinal cord are diminished in size. Dorsal roots are small, but motor roots are normal. Primary sensory neurons show degenerative changes, and residual nodules are prominent, especially in sacral and lumbar sensory ganglia.

Clinical Features

Hick's original clinical description in 1922 was of an autosomal dominant progressive disease in an English family. The features described included perforating ulcers of the feet, shooting pains on the body, and deafness, with clinical signs of dissociated sensory loss and areflexia in the feet.[15] The same family was later reported by Denny-Brown in 1951, when he termed the disorder as hereditary sensory radicular neuropathy.[10] With increased awareness of the disorder and discovery of various mutations, it was appreciated that there can be great variability in both penetrance and age of onset. Patients with *SPTLC1* and *SPTLC2* are phenotypically indistinguishable. The usual sensory findings include prominent loss of pain and temperature sensation that is particularly marked in the lower limbs, causing ulcerations, infected calluses, and bony deformities of the feet. Intermittent lancinating pains occur in some kinships.[1–3] Touch and pressure sensations are lost to a lesser extent. Distal tendon reflexes are lost, and there is moderate weakness. The autonomic disturbances are usually limited to hypohidrosis, which roughly matches the distribution of sensory deficits. Blood pressure control, and sphincter and sexual functions are normal. Intelligence may be mildly impaired. A study by Houlden et al. showed that there is clinical heterogeneity both within and between families, suggesting the influence of other genetic and acquired factors.[11] There is also a report of congenital onset in a French Gypsy patient with a *de novo* missense mutation in *SPTLC1* resulting in an unusually severe phenotype associated with severe growth and mental retardation, hypotonia, gastroesophageal reflux, and vocal cord paralysis.[5] Nerve conduction studies showed absent sensory and motor responses in the upper and lower limbs in this patient.

ATL1

Patients with HSAN type I caused by atlastin-GTPase 1 (*ATL1*) mutations may have additional symptoms that include trophic skin and nail changes. Patients with mutations in DNA (cytosine-5-)-methyltransferase 1 (*DNMT1*), have dementia and hearing loss.[3]

RAB7

Because mutations in *RAB7* also cause distal weakness, the disease is also called CMT2B. Affected patients are indistinguishable from those with HSAN.[5,12] In addition, the sensory loss is more generalized, affecting vibration, touch, and proprioception, as well as pain and temperature.[3] Patients with *RAB7* mutations are also more likely to present in adolescence.[5,12]

HSAN TYPE II

Molecular Genetics

HSAN type II is also termed congenital insensitivity to pain. There are at least three causative gene mutations resulting in subtype designations 2A, 2B, and 2C. Although mutations in different genes are involved, the clinical features of HSAN type II are similar.

HSAN 2A

The affected *WNK1* gene encodes the serine-threonine protein kinase WNK1, an osmotic sensor that regulates sodium, chloride, and potassium homeostasis and can change membrane excitability. It is normally expressed in the central and peripheral nervous systems, with particularly high levels in dorsal root ganglia and sciatic nerves as well as increased expression in sensory axons as compared to motor.[16] All mutations in *WNK1* are presumed to result in loss of function as they cause truncations in the transcript. WNK1 is known to decrease the cell-surface expression of TRpV4, a nonselective cation channel that is involved in thermal and mechanical nociception.[17] It has been suggested that a loss of function mutation could increase expression of this receptor, raising the threshold for sensitivity to pain as well as cold and heat detection.

HSAN 2B

The function of the affected *FAM134B* is poorly understood. It is known to be a component of the Golgi matrix and as such has a role in cellular structure. It has been suggested that the *FAM134B* protein is involved in neuronal survival, particularly nociceptive neurons, and that nonfunctional *FAM134B* protein would result in apoptosis of neuronal cells.[18] As *FAM134B* is predominantly expressed in sensory and autonomic ganglia, it would explain the clinical phenotype seen in HSAN type II patients.[18]

There may be genes other than the *WNK1* or *FAM134B* genes that also cause the HSAN type II phenotype since there are some individuals who have all clinical features of HSAN type II but fail to demonstrate mutations in either gene. Many of these individuals also manifest sensorineural hearing loss.[19]

HSAN 2C

Recently, a third mutation has been found to cause the HSAN2 phenotype. Erlich et al. described three brothers from an inbred Palestinian family with ulcero-mutilating sensory neuropathy and distal sensory loss along with distal motor involvement resulting in spastic paraplegia.[20] Loss of function mutations in the *KIF1A* gene were identified via exome sequencing. *KIF1A* encodes KIF1A, a member of the kinesin superfamily that transports membranous organelles and macromolecules along microtubules.[20] In axons, precursors of synaptic vesicles are transported anterogradely by KIF1A.

Pathophysiology

No cutaneous sensory receptors or nerve fibers are seen.[14,19] Within the sural nerve there is severe depletion of myelinated axons and a lesser loss of nonmyelinated fibers.[19] Neurophysiological evaluation reveals elevated vibratory and thermal thresholds at the hands and feet. Typically, nerve conduction velocities cannot be recorded as there is absence of the sensory nerve action potentials. Motor nerve conduction velocities are at or slightly below the normal limit and compound motor action potentials show slightly reduced amplitudes. EEG and electromyographic (EMG) studies are normal, but rate-dependent changes in brainstem auditory evoked potentials are increased indicating immature pathways.[1,2]

Catecholaminergic sympathetic fibers were demonstrated by aldehyde-induced fluorescence.[19]

Clinical Features

HSAN type II manifests in infancy or early childhood. Affected patients have profound sensory loss and frequent gastrointestinal and respiratory problems.[1–3] The neonatal course is characterized by severe feeding problems and frequent apnea. Gastroesophageal reflux and vomiting occur commonly. Other autonomic disturbances such as erythematous blotching of the skin and episodic hyperhidrosis have been observed, especially in the *FAM134B*-related HSAN (HSAN 2B), but not postural hypotension.[18] Tearing is frequently delayed but is eventually normal.

All sensory modalities are affected, including pain, temperature, and position senses. Unintentional self-injury is common and may begin with eruption of primary teeth. Taste sensation is diminished and lingual fungiform papillae are hypotrophic. Corneal and gag reflexes are diminished. Deep tendon reflexes are decreased. Painless fractures and acral ulcers can develop and repeated injury can lead to Charcot joints (Figure 18.1). Generalized hypotonia can delay attainment of developmental milestones. The severe hypotonia may contribute to scoliosis. Growth is normal. Cognitive function varies greatly. Some patients are mildly retarded with expressive aphasia, whereas the subset with sensorineural hearing loss has normal intelligence.

HSAN TYPE III

Molecular Genetics

HSAN type III was originally described by Conrad Riley and Richard Day and was renamed by Riley himself as

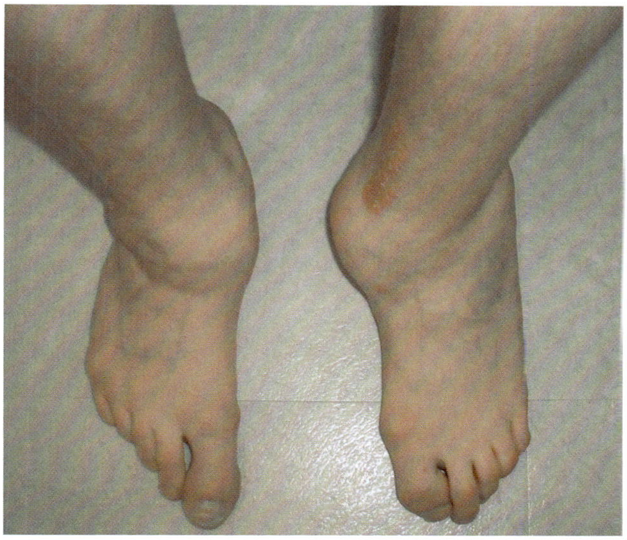

FIGURE 18.1 Orthopedic sequelae in HSAN type II. Repeated multiple fractures of weight bearing joints have resulted in bilateral Charcot joints of ankles.

familial dysautonomia (FD).[21,22] HSAN III is an autosomal recessive disease that appears to almost exclusively affect children with Eastern European Jewish parents.[23] Three causative mutations have been reported in the *IKBKAP* gene, but 99.5% of HSAN III patients are homozygous for a splice-site mutation, a T to C change at base pair 6 of intron 20.[24] The other two mutations, R696P and P914L, are missense mutations and thus far are always paired with the common mutation.[24,25] The carrier frequency of the most common HSAN III mutation among the Ashkenazi Jews of North America has been reported to be between 1 in 27 and 1 in 32.

The missplice mutation, IVS20 + 6T_C, causes a tissue-specific decrease in splicing efficiency with variable skipping of exon 20 in the *IKBKAP* message resulting in a truncated protein product, IKAP or ELP1.[24,26,27] ELP1 is a scaffold protein that is required to assemble the RNA polymerase II elongator complex, a key mediator of transcriptional elongation.[27] Many of its target genes are required for migration and maturation of the nervous system.[27–29] The IVS20 + 6T_C mutation does not cause complete loss of function, but neuronal tissues seem most severely affected as they primarily express mutant mRNA, somatic tissues express roughly equal levels of normal and mutant mRNA, and lymphoblast cell lines primarily express the normal product.[24,26]

Pathophysiology

The mutation responsible for HSAN III results in an abnormal IKAP protein, now known as ELP1. How abnormal ELP1 results in the phenotype is unknown. As ELP1 is part of normal elongator transcription function, most studies have focused on identifying genes that are downregulated in the absence of a functional elongator complex, especially those that might regulate cell motility, development, differentiation, and survival.[27–31] Because elongator may also facilitate acetylation of α-tubulin in the cytoplasm, decreased ELP1 might result in defective cytoskeletal organization.[29]

Sensory Nervous System

Consistent with the theory of incomplete development or early cell death, pathologic studies have shown decreased numbers of unmyelinated and small myelinated neurons in the peripheral nervous system.[1,2,19,23,32] Dorsal root ganglia are small and have markedly reduced numbers of neurons.[33] The sural nerve is hypoplastic with markedly reduced numbers of nonmyelinated and small myelinated fibers, but there is also depletion of large-myelinated axons. Lissauer's tracts and dorsal columns are diminished in size.[32–35] Lingual sensory axons and submucosal neurons are diminished in number. Circumvallate papillae are hypoplastic and bear few taste buds. Fungiform papillae are absent and filiform papillae are rudimentary.[36]

Autonomic Nervous System

Recent work from our laboratory indicates that the main autonomic defect in patients with HSAN III is in the afferent (sensory) neurons that relay incoming information from the arterial baroreceptors.[37] Despite the sympathetic ganglia being one third normal size with about one tenth the normal neuronal population,[38] efferent sympathetic function is intact and can be activated to raise blood pressure dramatically.[37,39] Preganglionic sympathetic neurons in the spinal cord are reduced to about half the normal population.[38] Surviving neurons tend to be slightly larger and have markedly increased tyrosine hydroxylase levels.[40] Despite fewer remaining sympathetic terminals on vessels of skin, and kidney,[41] the plasma concentration of norepinephrine, the sympathetic neurotransmitter, is normal.[39,42–44]

Parasympathetic ganglia are variably affected.[45] The ciliary ganglion is normal or marginally depleted of neurons, but the sphenopalatine ganglion has less than one tenth the normal population.[45]

Patients are hypersensitive to sympathomimetics and parasympathomimetics.[46,47] Small doses of clonidine can cause profound falls in blood pressure. Doses of norepinephrine that would be below threshold in normal subjects cause hypertension. The pupil is supersensitive to constriction by methacholine but not to dilation by epinephrine.[47]

Excretion of breakdown products of epinephrine and norepinephrine is decreased, but those of dopamine are

CASE EXAMPLE 18.1

A 5-year-old girl presented for evaluation with short stature and failure to thrive. Her parents were Ashkenazi Jewish. She was breech presentation and delivered by C-section at 37 weeks. After delivery, she had bilateral pneumothoraces. She was also hypotonic and had absent myotatic reflexes. She failed to breastfeed and had gastroesophageal reflux and recurrent pneumonia. She had nocturnal choking episodes. Poor oral coordination with aspiration was confirmed on swallow study. She was not able to meet caloric or fluid needs and was diagnosed with failure to thrive at age 2. She breath-held and the mother described decerebrate posturing. Developmental milestones were delayed. Words were noted at 18 months. Physical therapy and occupational therapy were added. By age 5, she was verbal and had a history of tongue ulcers and pneumonias. Her parents could not recall her crying with tears. She had just been toilet trained.

On examination, she was fair, very thin, and small for age. Facial features were normal. She made good eye contact. Her tongue appeared smooth. Neurological exam revealed mild hypotonia with absent deep tendon and corneal reflexes. Gait was clumsy. Skin blotching, hyperhidrosis, and hypertension were observed when she was upset and subsided when relaxed. The gene mutation was found on testing.

Comment

As this child's case illustrates, feeding problems that are due to uncoordinated swallow occur early, compromise nutrition, and produce respiratory infections. The neurological examination showed hypotonia, depressed tendon reflexes, and gait ataxia. Erythematous skin blotching and sweating when upset are signs of sympathetic arousal and occur because of catecholamine release. Breath-holding spells show lack of hypoxic ventilatory drive. Although prenatal diagnosis is available, testing is not always done and children with this disorder are still being born.

eliminated in increased amounts.[48,49] During emotional arousal, sympathetic activity is unrestrained and dopamine and norepinephrine rise markedly.[44] Blood pressure falls when upright because of a failure to activate sympathetic neurons and increase norepinephrine and vascular resistance when needed.[39,49,50]

Clinical Features

There is considerable variation in some of the clinical features of HSAN III. Signs of the disorder are present from birth and some neurological features like gait and vision slowly deteriorate with age.[51]

Somatic Sensory and Motor Systems

In HSAN III, the peripheral sensory abnormalities involve primarily the nociceptive functions, pain and temperature perceptions which are transmitted by Aδ (smaller myelinated) and C (unmyelinated) fibers. Thermal discrimination is more profoundly affected than pain and there are varying degrees of insensitivity.[51] Typically, bone and skin pain are diminished but not absent, and sensitivity to visceral pain is present. Patients have decreased pain perception in the distribution of the trigeminal nerve, diminished corneal reflexes, and deficient taste, especially for sweet, which is due to the absence of fungiform papillae on the tip of the tongue.[36] Defects in taste correlate with a decreased number of neurons in the geniculate ganglion.[52]

Tendon reflexes are decreased or absent and the gait is ataxic, at least partly explained by loss of functional muscle spindles and impaired proprioception.[51,53,54]

Autonomic System

There is widespread involvement of afferent autonomic neurons resulting in dysfunction of multiple organ system.

Gastrointestinal

Feeding difficulties manifesting as poor suck, uncoordinated swallowing, aerophagia, and frequent misdirection of food or liquid into the trachea[1,2,23] are frequently the earliest clinical signs in a baby with FD (see Case Example 18.1). Recurrent misdirection, especially of liquids, and frequent gastroesophageal reflux put the patient at risk for aspiration and chronic lung disease. Severe cases are usually managed with a combined fundoplication and gastrostomy. Protracted episodes of nausea and vomiting, termed dysautonomic or hyperadrenergic crisis, are characteristic for patients with HSAN III. Circulating catecholamines, including dopamine, rise during these crises.[39,44] Nausea and retching are likely caused by activation of dopamine receptors in the chemoreceptor trigger zone.[44] Other physical signs associated with a crisis include agitation, tachycardia, hypertension, blotchy erythema, and diaphoresis, all signs of exagerated sympathetic activation. In some patients, these crises are daily morning events whereas in others, they occur only with extreme physical or emotional stress.

Respiratory

Repeated aspiration pneumonias and abnormal ventilatory drive to hypoxia and hypercapnea result in a particular type of chronic lung disease.[55] Additional cases may be further complicated by a component of restrictive lung disease and neuromuscular weakness.

Minute ventilation fails to increase in response to hypoxia (at high altitude, during air travel, or during a pneumonia).[55,56] Severe hypoxemia can cause hypotension, bradyarrhythmia, and syncope. Agitated patients are able to hold their breath without the drive to breathe producing cyanosis, syncope, decerebrate posturing, and seizures. Diving or underwater swimming can be hazardous. Tachypnea does not accompany respiratory infections. Frequently, abnormal spinal curvature further compromises respiratory function decreasing chest compliance, and it is unclear if corrective spine surgery adds to the restrictive component of the lung disease.[56,57]

Cardiovascular

Cardiovascular lability as a result of afferent baroreflex failure is characteristic for this disorder.[37] Patients with FD are unable to modulate sympathetic activity and to release vasopressin by baroreflex-mediated stimuli, and they exhibit marked sympathetic activation during cognitive arousal. Thus, patients can exhibit both profound and rapid postural hypotension without compensatory tachycardia with change of position as well as extreme hypertension and tachycardia when emotionally and physically stressed.[37,49,50] Although postural hypotension can be pronounced, patients appear to tolerate extremely low blood pressures well. Symptoms are exaggerated and syncope is more apt to appear with cutaneous vasodilatation (hot weather), reductions in venous return (straining), or intravascular volume depletion (dehydration or anemia). Hypertension frequently occurs when the individual is supine, is responding to emotional stress or visceral pain, or is experiencing a dysautonomic crisis. Hypertension may contribute to target organ damage, particularly left ventricular hypertrophy and chronic renal failure.[50,58–60]

Blotchy erythema occurring during eating or emotional excitement is further evidence of abnormal vasomotor control. The extremities may intermittently become cold, red, and mottled. Inappropriate temperature response to infection or environmental changes can result in either hypothermia or hyperthermia.[1,2,23]

The Eye

Absence of tearing (alacrima) with emotional crying is a cardinal feature of this disorder. Baseline eye moisture varies among affected individuals and is further compromised by hypoesthesia resulting in a decreased blink frequency with rapid evaporation of baseline moisture. Corneal de-epithelialization, ulceration, and scarring with opacity often occur. The pupil responds appropriately to light and accommodation, suggesting intact parasympathetic pathways. Supersensitivity to infused methacholine may be caused by reduced tears resulting in higher local concentration of pharmacologic agents.[47] Recently FD patients have been shown to have a specific type of optic neuropathy with predominant loss of papillomacular nerve fibers.[61]

Cognition and Development

Developmental milestones are commonly delayed. Intelligence is usually within low normal ranges and learning disabilities are common.[62] There is a general tendency for better verbal than motor performance. Memory is usually intact but executive planning skills are frequently impaired and there is a tendency to concrete thinking with problems in extrapolation. Anxiety and obsessive-compulsive repetitive and oppositional behavior disorders appear to occur more frequently. Puberty is usually delayed in both sexes. Average age at menarche is 15 years (normal 12.8 years). Nevertheless, primary and secondary sexual characteristics do develop, and males and females have demonstrated reproductive capability. Offspring of affected individuals and their nonaffected partners have been healthy obligatory carriers of the mutation.[1,2,23]

Musculoskeletal

Individuals with FD tend to have similarities in facial characteristics that may be related to poor oromotor coordination as these characteristics are not noted at birth but become more distinguishable with age.[63] Small jaws, mild dental crowding, and malocclusions are frequent. In addition, the upper lip tends to be thin and straight especially with smiling. Somatic growth is poor, the average adult height being about 5 feet. Growth hormone levels are usually normal.[23] The high incidence of juvenile scoliosis further compromises eventual height.[56,57] By 10 years of age, 85% of patients have scoliosis (Figure 18.2).

Survival and Prognosis

With supportive measures, about half of the patients now reach adulthood.[64] The major problems in the adult population are progressive optic atrophy, worsening ataxia, and development of renal dysfunction.[59,61] Chronic kidney disease may be a consequence of cardiovascular lability.[58–60] By age 25, nearly one-half of all patients with FD will have reached stage 3 chronic kidney disease.[59] Dialysis and renal transplants may be required.[60]

HSAN TYPE IV

Molecular Genetics

HSAN type IV, or congenital insensitivity to pain with anhidrosis (CIPA), is caused by loss of function mutations

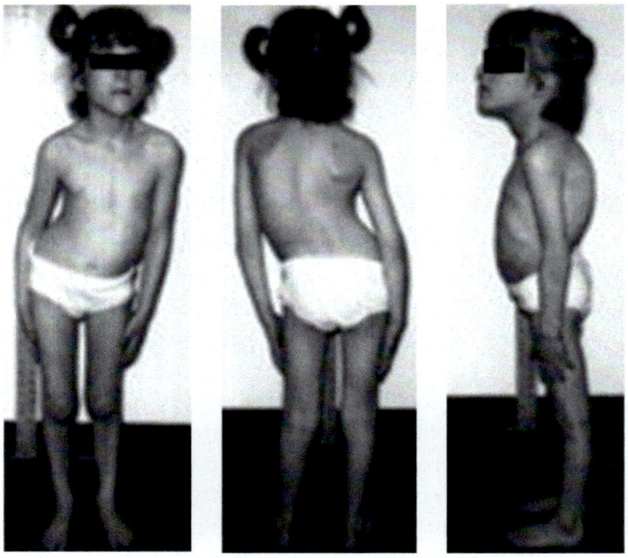

FIGURE 18.2 Severe kyphoscoliosis in HSAN type III. Left thoracic juvenile scoliosis is manifested by age 8 years in this patient.

in the *NTRK1* gene located on chromosome 1 (1q21-q22).[1,2,4,65] The *NTRK1* (also known as *TRKA*) gene encodes TrkA, receptor tyrosine kinase type 1, which is necessary for normal binding of NGF. TrkA is one of three tyrosine kinases involved in neurotrophin signaling. Upon binding its ligand β-NGF, TrkA is autophosphorylated and activates downstream signaling pathways that support neurite outgrowth as well as survival of sympathetic ganglia neurons, nociceptive sensory neurons in dorsal root ganglia, and cholinergic neurons of the basal forebrain.[60] If binding is impeded as a result of the impaired receptor function, then NGF dependent neurons, including nociceptive sensory and sympathetic neurons, fail to develop during embryogenesis. The β-NGF-TrkA signaling pathway is also active in the immune response,[65,66] which may explain the observed slow healing and chronic inflammation manifested by many HSAN type IV patients.

The pattern of inheritance follows an autosomal recessive pattern. There is no particular ethnic distribution for this disorder, but one-half of reported cases have occurred in consanguineous marriages. Since Indo's original report in 1996,[67] over 40 different loss of function mutations have been identified, and it has been suggested that the various mutations may correlate with phenotypic expression.[1,2,4]

Pathophysiology

Patients with HSAN type IV have an even more profound decrease in unmyelinated neurons than HSAN type III patients.[8,19,68] There is a virtual absence of unmyelinated fibers in peripheral nerves, in Lissauer's tract from the spinal cord, and in dorsal root ganglia. The overall size of the sural nerve is markedly reduced. Skin biopsies of HSAN IV patients revealed deficient C and Aδ fibers in the epidermis and absent or hypoplastic dermal sweat glands without innervation.[69,70]

Clinical Features

Clinical features are present from birth with extensive involvement of ectodermal structures including skin, bone, and nervous system.[71,72] Although penetrance is complete, phenotypic expression varies and may be related to the site of the mutation on the NGF receptor.[1,65]

Sensory Nervous System

In HSAN type IV patients, the insensitivity to pain is profound and affects cranial nerves and visceral sensation. Self-mutilation is frequent and can be the result of lip biting, corneal scarring, and autoamputation of digits. Ectodermal tissues skin and bone heal poorly. Large weight-bearing joints are susceptible to repeated trauma resulting in Charcot joints and osteomyelitis.[71,72] Temperature sensation (i.e. perception of hot and cold) is decreased or absent. Deep tendon reflexes are intact, position sense is preserved, and patients are not ataxic. Although hypotonia may be present in the early years, developmental milestones are usually normal to only mildly delayed. In contrast to patients with type III, emotional tearing is normal, but hypoalgesia can result in corneal damage.[73]

Autonomic Nervous System

HSAN type IV is characterized by anhidrosis secondary to impaired sudomotor sympathetic outflow. Anhidrosis involves the trunk and upper extremities in all cases; other areas of the body are variably affected.[65] The anhidrosis causes episodic fevers and extreme hyperpyrexia, which is usually the earliest sign of the disorder (see Case Example 18.2). Anhidrosis also contributes to the thick and calloused appearance of the skin with lichenification of palms, dystrophic nails, and areas of hypotrichosis on the scalp (Figure 18.3). There is no acrocyanosis. Cardiovascular responses are not well studied, but are assumed to be normal as syncope is rare.[1] Gastrointestinal dysmotility is infrequent; vomiting is not a feature. Insensitivity to hypoxia and hypercapnea has not been noted.

Cognition

Some type of intellectual disability is a consistent feature of HSAN type IV patients. Most have severe learning and behavioral problems. Impulsivity, hyperactivity, and

CASE EXAMPLE 18.2

A 13-month-old girl was referred with a history of unexplained high fevers and repeated traumatic injuries that healed poorly. She was the second child of Palestinian parents living in the West Bank, who were not aware of being relations. The father reported having a paternal first cousin with an HSAN. The girl was a full term normal weight infant from an uneventful pregnancy and delivery. Postdelivery, she was noted to be hypotonic and hypothermic (94°F). In addition, she had periods of sleepiness. She was treated for meningitis with 14 days of antibiotics although MRI, spinal tap, CT scans, and blood cultures were all negative. She was discharged by 14 days of life. Within the first month of life, she was hospitalized for hyperpyrexia of unknown origin. When her teeth appeared, she developed tongue ulcers. Her parents needed to cover her hands at night to stop her biting and injuring her hands. They could not remember her ever sweating.

On initial examination, she was noted to be a thin, fair-haired child who was an appropriate size for her age. She was responsive and smiled. She had signs of self-injury. The tongue was misshapen and scarred. Her skin was generally dry, especially the palmar surfaces of the hands. Tears were present but corneal reflexes were absent. She was very hypotonic and had lax joints. She was able to stand with support and had prespeech sounds. Deep tendon reflexes were present. An intradermal histamine test was performed and did not produce an axon flare. Blood pressure determinations in supine position and when held upright were stable.

Comment

This case shows that the earliest sign of HSAN type IV can be unexplained fever due to anhidrosis. Insensitivity to pain was also noted early with tongue biting, ulcers, and other painless injuries. Fractures are common once children begin walking. In contrast to HSAN type III patients, this girl never exhibited gastrointestinal problems or respiratory problems and her tearing was normal. The diagnosis was suspected because another family member was affected, but in many cases the diagnosis should be considered if a child demonstrates diminished response to noxious stimuli and heat intolerance. Diagnosis was confirmed by identification of loss of function mutation in the *NTRK1* gene, suggesting a developmental defect of small fiber neurons.

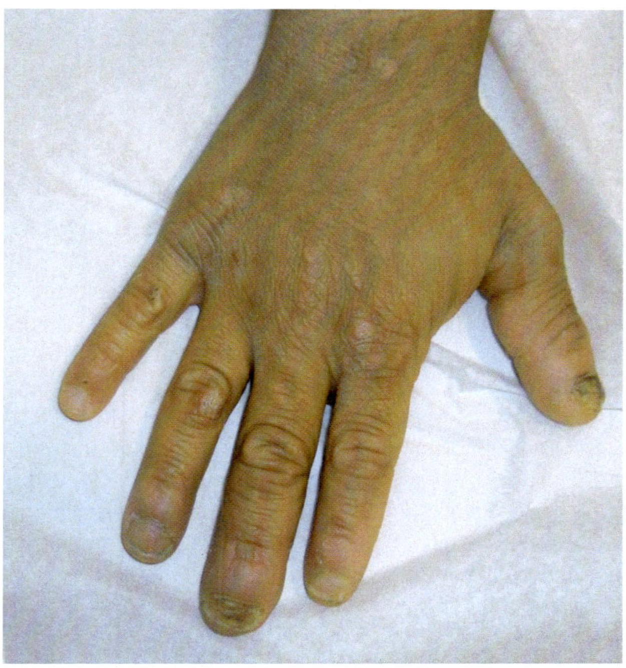

FIGURE 18.3 Hand of a patient with HSAN type IV. The skin is dry and multiple scars are noted due to inadvertent injuries. The tip of the third finger has been autoamputated.

reckless behaviors are common. We have noted a few patients with only borderline reduction in IQ, but most HSAN IV children are unable to stay in mainstream education.

HSAN TYPE V

Molecular Genetics

HSAN type V is an autosomal recessive neuropathy caused by a mutation in the *NGF* gene, which encodes β-NGF.[1,3–5,74] Interestingly, heterozygous carriers can present in adulthood with a mild HSAN phenotype.[75] β-NGF is a neurotrophin protein that aids in regulation of neuronal survival, development, and function. It also mediates inflammatory and immune responses after tissue injury.[76] It remains unclear how the functions of β-NGF are actually altered, but it is assumed that the mutations alter NGF signaling and efficiency, leading to a neuropathy mainly affecting nociception. The mutations can be missense or frameshift, and have a more severe phenotype in general patients with frameshift mutations than in those with the missense mutation.[77]

Pathophysiology

Nerve conduction studies are normal. In the sural nerve, unmyelinated fibers are severely reduced and small diameter myelinated fibers are moderately diminished.[1,8,78,79]

Clinical Features

Clinically, patients with HSAN type V closely resemble those with HSAN type IV. The distinguishing clinical features are degree of sweating and intellect. Patients with

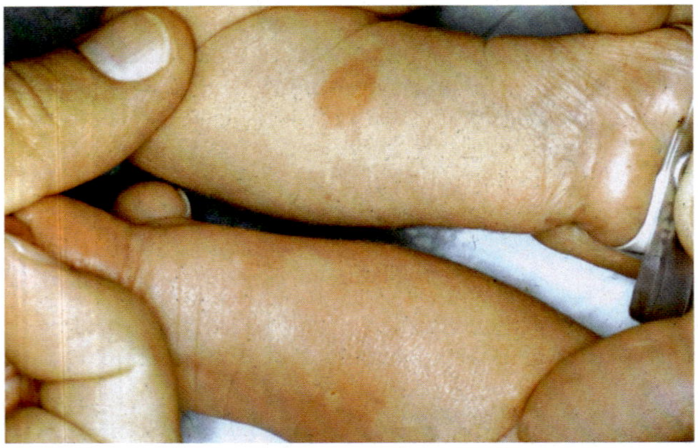

FIGURE 18.4 Histamine test. The top forearm displays the abnormal reaction to intradermal histamine that is characteristic for HSAN. Only a narrow areola surrounds the wheal. Lower forearm displays normal reaction to histamine test with diffuse axon flare around a central wheal.

type V have variable degrees of hypohidrosis rather than anhidrosis, and mental retardation is not characteristic. Onset is usually evident at birth, and the course of the disease does not appear to be progressive. Clinical diagnosis is based on documentation of loss of pain perception and absent corneal reflexes with otherwise normal sensory function, including preserved temperature and touch perception. Axon reflex and tendon reflexes are intact. Early manifestations include acral ulcers, painless fractures, and other trophic injuries. The loss of pain is suspected by lack of discomfort with cold exposure of one extremity, heat pain stimulation, or electrical stimulation with high intensities during nerve conduction studies.[80]

DIAGNOSIS OF AN HSAN

Phenotypic expression of each HSAN varies because each is caused by a different genetic error affecting a specific aspect of small fiber development, function, or survival. Classification of the HSANs is still evolving. Clinical descriptions of apparently new cases of HSANs periodically appear in the literature that still await genotype identification.[1,19,81,82] There are ongoing efforts to correlate genotype with phenotype to provide correct diagnosis and to gain insight into the factors that disrupt normal development of the sensory and autonomic nervous systems.[2–5] With the exception of HSAN type I, which is a dominant trait usually presenting in the second decade, the other HSANs are autosomal recessive and are frequently expressed at birth.

Clinical suspicion of an HSAN is based on recognizing that the patient has early sensory and autonomic dysfunction. The characteristic shared by all HSANs is the failure of intradermal injection of histamine phosphate to elicit a normal axon flare response.[1,2,83] As sensory fibers are diminished, arteriolar dilation cannot be mediated via usual antidromic conduction (Figure 18.4).[1,2]

The characteristic sensory deficits are decreased pain and temperature perception putting the young child at high risk for inadvertent injury. One of the earliest clinical signs of insensitivity to pain are ventral tongue ulcerations or lip chewing that coincides with the eruption of primary teeth. Lack of corneal sensitivity causes decreased blink frequency so that even if the baseline eye moisture is adequate, decreased blink frequency can result in relatively dry eye syndromes and development of keratitis.[61,72] Early clinical signs of autonomic dysfunction can include feeding difficulties with impaired swallowing leading to repeated aspiration pneumonias, labile body temperatures, and erratic blood pressure. Because all affected HSAN individuals fail to produce an axon flare after intradermal histamine, assessment of other clinical signs and symptoms is necessary to distinguish between the various HSAN types (Table 18.2). Some of the HSANs have unique features, such as alacrima in patients with HSAN type III and anhidrosis in patients with HSAN type IV. Because of extreme variability in phenotypic expression, however, clinical criteria are not sufficient and molecular diagnosis is required. For HSAN type III, DNA diagnosis is commercially available; for the other HSANs, molecular confirmation is more difficult and relies either on a research facility or private genetic testing companies.

THERAPEUTIC INTERVENTIONS

Morbidity and mortality related to the various HSANs are substantial, and all have reduced life expectancy.[1,2,58,60] Hopefully, gene identification may eventually yield more specific treatments for the HSANs. Targeted gene-specific agents are currently being investigated for HSAN type III, and clinical trials are under way.[79,80,84,85] At present, most therapeutic interventions for the HSAN disorders are supportive modalities. Treatment of the problems caused by sensory deficits can be utilized by almost all HSAN patients. The autonomic perturbations manifested by each of the HSANs vary, so treatment of autonomic dysfunction is disease specific.

Management of Decreased Sensation

Decreased pain sensation and propensity to self-injury require constant supervision against inadvertent trauma. Swelling and skin discoloration may be the only clues to serious injuries without symptoms. The ankles, knees, and fingers are particularly susceptible.[71,72] Fracture setting casts require vigilance as pressure ulcers may not be recognized. For the child who is biting lips or developing a tongue ulcer, early smoothing of the sharp saw-edged

TABLE 18.2 Affected Neurological Functions in the Five HSANs

Neurologic Function	HSAN Type I (Hereditary Radicular Neuropathy)	HSAN Type II (Congenital Sensory Neuropathy)	HSAN Type III (Familial Dysautonomia)	HSAN Type IV (Congenital Sensory Neuropathy with Anhidrosis)	HSAN Type V (Congenital Sensory Neuropathy without Anhidrosis)
Cranial Nerves					
Corneal (V)	Normal	Diminished	Diminished/Absent	Absent	Normal/Diminished
Gag (IX,X)	Normal	Diminished	Diminished/[closed up solidus] Normal	Normal	Normal
Taste (XII)	Normal	Reduced	Reduced	Normal	Normal
Motor					
Gait	Normal	Normal	Sensory ataxia	Normal	Normal
Sensory					
Pain	Distally absent	Absent	Reduced	Absent	Absent
Temperature	Distally absent	Absent	Reduced	Reduced	Normal
Touch	Reduced/Normal	Reduced	Normal	Reduced	Normal
Position Sense	Reduced	Normal/ Reduced	Normal/Reduced late	Normal	Normal
Vibration	Normal	Normal/ Reduced	Normal/Reduced late	Normal	Normal
Axon flare	Distally absent	Absent	Absent	Absent	Absent
Reflexes					
Tendon	Decreased or Normal	Decreased	Absent	Decreased/[closed up solidus] Normal	Normal
Superficial	Normal	Decreased/ Normal	Decreased/Normal	Decreased/Normal	Normal
Autonomic					
Tears	Normal	Normal	No emotional overflow	Normal	Normal
Sweating	Distally decreased	Increased/ Normal	Increased (with stress)	Decreased	Normal
Cardiovascular function	Not affected	Skin blotching ± acrocyanosis	Postural hypotension, Episodic hypertension, skin blotching± acrocyanosis	Occasional blotching	Not affected
Gastrointestinal symptoms	Normal	Occasional dysphagia	Dysphagia, vomiting, gastroesophageal reflux	Occasional dysphagia	Normal
Cognition					
Cognition	Mild decrease or Normal	Normal, increased OCD, increased anxiety	Normal (about 5% PDD), frequent learning disabilities, increased OCD, increased anxiety	Decreased, severe learning disabilities, frequent hyperactivity	Normal

primary incisors may suffice and help avoid dental extraction.[63]

Anesthetic corneas need to be protected. Corneal complications have been decreasing with regular use of artificial tear solutions. Moisture chamber spectacle attachments on glasses or goggles help maintain eye moisture and protect the eye from wind and foreign bodies. Tarsorrhaphy of the medial or lateral part of the palpebral fissure has been reserved for unresponsive and chronic situations. Cautery of the tear ducts increases baseline moisture and is beneficial in promoting corneal healing. Scleral lenses also have been used to provide protection and help maintain moisture.

Management of Autonomic Dysfunction

Anhidrosis

Hyperpyrexia can be fatal. Overheating should be avoided and early use of antipyretics is advised. Treatment includes the use of cool baths and sponging as well as rubbing of the trunk and extremities. Use of cooling vests have been beneficial in decreasing risk of heat exhaustion and hyperpyrexia.

Abnormal Swallowing

Gastrostomy may be required to ensure adequate caloric and fluid intake and to prevent dehydration or aspirations caused by misdirected swallows. If gastroesophageal reflux is also present, it is recommended that patients have fundoplication with gastrostomy.[23] For salivary incontinence due to poor swallowing of oral secretions, periodic use of glycopyrrolate can be used. Some patients have responded well to injections of Botox into the salivary glands.

Respiratory Dysfunction

For those who have chronic lung disease, appropriate chest therapy should be instituted. Sleep studies are recommended to detect hypoventilation and apneas. If sleep disordered breathing is documented, noninvasive ventilation during sleep is indicated.

Parents of patients with HSAN type III who exhibit insensitivity to hypoxia should be warned about the need for appropriate precautions with underwater swimming, air travel, and visits to high altitudes, which includes having supplementary oxygen available in potential hypoxic environments. Breath-holding episodes may be misinterpreted as a seizure disorder.

Hyperadrenergic Crises

These episodes are unique to HSAN type III patients and respond poorly to standard antiemetics. Traditionally, the episodes are treated with centrally acting medications such as the benzodiazepine diazepam, along with clonidine, an alpha2-agonist that inhibits sympathetic outflow, but their use is limited by the unwanted side effects of sedation and hypotension. Newer agents like pregabalin are only moderately effective.[86] Carbidopa, an inhibitor of DOPA decarboxylase that blocks the synthesis of dopamine outside the brain, was shown to be effective in a recent double-blind crossover study.[44]

Blood Pressure Lability

Afferent baroreflex failure is unique to HSAN type III and manifests as extreme blood pressure lability. Elevation of the head of the bed as well as relaxation techniques are advised for hypertension when supine. Postural hypotension can usually be managed by good hydration, increased salt intake, and physical counter maneuvers to promote venous return. The alpha-adrenergic agonist midodrine produces a short-lived pressor effect and can be taken prior to exercise or prolonged postural stress. When blood pressure is persistently low, hypoxemia, hypovolemia, anemia, or severe infection should be suspected.

Emotional Lability and High Levels of Anxiety

In addition to the labile autonomic nervous system and deficiencies in sensory stimulation, cognition and control of emotions are frequently affected in the HSANs. Intelligence is lower in patients than in their unaffected siblings and learning disabilities are common, so special education is often required. Use of stimulant medications should be avoided in HSAN type III and have not been shown to be useful in controlling hyperactivity in HSAN IV.

REFERENCES

1. Axelrod FB. Autonomic and sensory disorders. In: 5th ed. Rimoin DL, Connoe JM, Pyeritz RE, Korf BR, editors. *Principles and Practice of Medical Genetics*, vol. 3. Philadelphia: Churchill Livingstone; 2007. pp. 2802–16.
2. Axelrod FB. Hereditary sensory and autonomic neuropathies: familial dysautonomia and other HSANs. *Clin Auton Res* 2002;**12** (Suppl. 1):2–14.
3. Rotthier A, Baets J, Timmerman V, Janssens K. Mechanisms of disease in hereditary sensory and autonomic neuropathies. *Nat Rev Neurol* 2012;**8**:73–85.
4. Auer-Grumbach M, Mauko B, Auer-Grumbach P, Pieber. Molecular genetics of hereditary sensory neuropathies. *Neuromolecular Med* 2006;**8**:147–58.
5. Rotthier A, Baets J, De Vriendt E, Jacobs A, Auer-Grumbach M, Levy N, et al. Genes for hereditary sensory and autonomic neuropathies: a genotype-phenotype correlation. *Brain* 2009;**132**:2699–711.
6. Edlund T, Jessell TM. Progression from extrinsic to intrinsic signaling in cell fate specifications: A view from the nervous system. *Cell* 1999;**96**:211–24.
7. Goridis C, Brunet JF. Transcriptional control of neurotransmitter phenotype. *Curr Opin Neurobiol* 1999;**9**:47–53.
8. Dyck PJ. Neuronal atrophy and degeneration predominantly affecting peripheral sensory neurons. In: Dyck PJ, Griffin JVV, Low PA, Poduslo JF, editors. *Peripheral Neuropathy*. 3rd ed. Philadelphia: WB Saunders; 1993. pp. 1065–93.
9. Verhoeven K, Timmerman V, Mauko B, Pieber TR, De Jonghe P, Auer-Grumbach M. Recent advances in hereditary sensory and autonomic neuropathies. *Curr Opin Neurol.* 2006;**19**:474–80.

10. Denny-Brown D. Hereditary sensory radicular neuropathy. *J Neurol Neurosurg Psychiatry* 1951;**14**:237–52.
11. Houlden H, King R, Blake J, Groves M, Love S, Woodward C, et al. Clinical, pathological and genetic characterization of hereditary sensory and autonomic neuropathy type 1 (HSAN I). *Brain* 2006;**129**:411–25.
12. BasuRay S, Mukherjee S, Romero EG, Seaman MN, Wandinger-Ness A. Rab7 mutants associated with Charcot-Marie-Tooth disease cause delayed growth factor receptor transport and altered endosomal and nuclear signaling. *J Biol Chem* 2013;**288**(2):1135–49.
13. Penno A, Reilly MM, Houlden H, Laura M, Rentsch K, Niederkofler V, et al. Hereditary sensory neuropathy type 1 is caused by the accumulation of two neurotoxic sphingolipids. *J Biol Chem* 2010;**285**:11178–87.
14. Dyck PJ. Histologic measurements and fine structures of biopsied sural nerve: Normal, and in peroneal muscular atrophy, hypertrophic neuropathy, and congenital sensory neuropathy. *Mayo Clinic Proc* 1966;**41**:742–74.
15. Hicks EP. Hereditary perforating ulcer of the foot. *Lancet* 1922;**1**:319–21.
16. Shekarabi M, Girard N, Riviere JB, Dion P, Houle M, Toulouse A, et al. Mutations in the nervous system-specific HSN2 exon of WNK1 cause hereditary sensory neuropathy type II. *J Clin Invest* 2008;**118**:2496–505.
17. Fu Y, Suabramanya A, Rozansky D, Cohen DM. WNK kinases influence TRPV4 channel function and localization. *Am J Physiol Renal Physiol* 2006;**290**:F1305–14.
18. Kurth I, Pamminger T, Hennings JC, Soehendra D, Huebner AK, Rotthier A, et al. Mutations in FAM134B, encoding a newly identified Golgi protein, cause severe sensory and autonomic neuropathy. *Nat Genet* 2009;**41**:1179–81.
19. Axelrod FB, Pearson J. Congenital sensory neuropathies. Diagnostic distinction from familial dysautonomia. *Am J Dis Child* 1984;**138**:947–54.
20. Erlich Y, Edvardson S, Hodges E, Zenvirt S, Thekkat P, Shaag A, et al. Exome sequencing and disease-network analysis of a single family implicate a mutation in KIF1A in hereditary spastic paraparesis. *Genome Res* 2011;**21**:658–64.
21. Riley CM, Day RL, Greely D, Langford WS. Central autonomic dysfunction with defective lacrimation. *Pediatrics* 1949;**3**:468–77.
22. Riley CM. Familial dysautonomia. *Adv Pediatr* 1957;**9**:157–90.
23. Axelrod FB. Familial dysautonomia (Invited Review). *Muscle Nerve* 2004;**29**:352–63.
24. Slaugenhaupt SA, Blumenfeld A, Gill SP, Leyne M, Mull J, Cuajungco MP, et al. Tissue-specific expression of a splicing mutation in the IKBKAP gene causes familial dysautonomia. *Am J Hum Genet* 2001;**68**:598–605.
25. Leyne M, Mull J, Gill SP, Cuajungco MP, Oddoux C, Blumenfeld A, et al. Identification of the first non-Jewish mutation in Familial Dysautonomia. *Am J Med Genet* 2003;**118A**:305–8.
26. Cuajungco MP, Leyne M, Mull J, Gill SP, Lu W, Zagzag D, et al. Tissue-specific reduction in splicing efficiency of IKBKAP due to the major mutation associated with familial dysautonomia. *Am J Hum Genet* 2003;**72**:749–58.
27. Close P, Hawkes N, Cornez I, Creppe C, Lambert CA, Rogister B, et al. Transcription impairment and cell migration defects in elongator-depleted cells: implication for familial dysautonomia. *Mol Cell* 2006;**22**:521–31.
28. Lee G, Papapetrou EP, Kim H, Chambers SM, Tomishima MJ, Fasano CA, et al. Modelling pathogenesis and treatment of familial dysautonomia using patient-specific iPSCs. *Nature* 2009;**461**:402–6.
29. Winkler GS, Kristjuhan A, Endjument-Bromage H, Tempst P, Svejstrup JQ. Elongator is a histone H3 and H4 acetyltransferase important for normal histone acetylation levels in vivo. *Proc Natl Acad Sci USA* 2002;**99**:3517–22.
30. Cheishvilli D, Maayan C, Smith Y, Ast G, Razin. IKAP/hELP1 deficiency in the cerebrum of familial dysautonomia patients results in down regulation of genes involved in oligodendrocyte differentiation and in myelination. *Hum Mol Genet* 2007;**16**:2097–104.
31. Creppe C, Malinouskaya L, Volvert ML, Gillard M, Close P, Malaise O, et al. Elongator controls the migration and differentiation of cortical neurons through acetylation of alpha-tubulin. *Cell* 2009;**136**(3):551–64.
32. Pearson J, Axelrod F, Dancis J. Current concepts of dysautonomia: Neuropathological defects. *Ann NY Acad Sci* 1974;**228**:288–300.
33. Pearson J, Pytel B, Grover-Johnson N, Axelrod FB, Dancis J. Quantitative studies of dorsal root ganglia and neuropathologic observations on spinal cords in familial dysautonomia. *J Neurol Sci* 1978;**35**:77–92.
34. Aguayo A, Nair CPV, Bray GM. Peripheral nerve abnormalities in the Riley-Day syndrome. *Arch Neurol* 1971;**24**:106–16.
35. Pearson J, Dancis J, Axelrod F, Grover N. The sural nerve in familial dysautonomia. *J Neuropathol Exp Neurol* 1975;**34**:413–24.
36. Pearson J, Finegold M, Buzilovich G. The tongue and taste in familial dysautonomia. *Pediatrics* 1970;**45**:739–45.
37. Norcliffe-Kaufmann L, Axelrod F, Kaufmann H. Afferent baroreflex failure in familial dysautonomia. *Neurology* 2010;**75**:1904–11.
38. Pearson J, Pytel B. Quantitative studies of sympathetic ganglia and spinal cord intermedio-lateral gray columns in familial dysautonomia. *J Neurol Sci* 1978;**39**:47–59.
39. Norcliffe-Kaufmann LJ, Axelrod FB, Kaufmann H. Cyclic vomiting associated with excessive dopamine in Riley-Day syndrome. *J Clin Gastroenterol* 2012;**47**:136–8.
40. Pearson J, Brandeis L, Goldstein M. Tyrosine hydroxylase immunoreactivity. *Science* 1979;**206**:71–2.
41. Grover-Johnson N, Pearson J. Deficient vascular innervation in familial dysautonomia, an explanation for vasomotor instability. *Neuropathol Appl Neurobiol* 1976;**2**:217–24.
42. Axelrod FB, Goldstein DS, Holmes C, Berlin D, Kopin IJ. Pattern of plasma levels of catecholamines in familial dysautonomia. *Clin Auton Res* 1996;**6**:205–9.
43. Goldstein DS, Holmes C, Axelrod FB. Plasma catechols in familial dysautonomia: a long-term follow-up study. *Neurochem Res* 2008;**33**(9):1889–93.
44. Norcliffe-Kaufmann LJ, Martinez J, Axelrod FB, Kaufmann H. Hyperdominergic crises in familial dysautonomia: A randomized trial of carbidopa. *Neurology* 2013;**80**(17):1611–17.
45. Pearson J, Pytel B. Quantitative studies of ciliary and sphenopalatine ganglia in familial dysautonomia. *J Neurol Sci* 1978;**39**:123–30.
46. Smith AA, Hirsch JI, Dancis J. Responses to infused methacholine in familial dysautonomia. *Pediatrics* 1965;**36**:225–30.

47. Smith A, Dancis J, Breinin G. Ocular responses to autonomic drugs in familial dysautonomia. *Invest Ophthalmol Vis Sci* 1965;**4**:358–61.
48. Goldstein M, Dancis J, Freedman L, Axelrod F. Serum dopamine-beta-hydroxylase in normal subjects and patients with familial dysautonomia. *Nature* 1972;**236**:310–11.
49. Ziegler MG, Lake RC, Kopin IJ. Deficient sympathetic nervous response in familial dysautonomia. *N Engl J Med* 1976;**294**:630–3.
50. Axelrod FB, Glickstein JS, Weider J, Gluck MC, Friedman D. The effects of postural change and exercise on renal haemodynamics in familial dysautonomia. *Clin Auton Res* 1993;**3**:195–200.
51. Tokita N, Sekhar HK, Sachs M, Daly JF. Familial dysautonomia (Riley-Day syndrome). Temporal bone findings and otolaryngological manifestations. *Ann Otol Rhinol Laryngol Suppl* 1978;**87**:1–12.
52. Macefield VG, Norcliffe-Kaufmann L, Axelrod FB, Kaufmann H. Cardiac-locked bursts of muscle sympathetic nerve activity are absent in familial dysautonomia. *J Physiol* 2013;**591**:689.
53. Macefield VG, Norcliffe-Kaufmann LJ, Axelrod FB, Kaufmann H. Relationship between proprioception at the knee joint and gait ataxia in HSAN III. *Mov Disord* 2013;**28**(6):823–7.
54. Edelman NH, Cherniack NS, Lahiri S, Richards E, Fishman AP. The effects of abnormal sympathetic nervous function upon the ventilatory response to hypoxia. *J Clin Invest* 1970;**41**:1153–65.
55. Bernardi L, Hilz M, Stemper B, Passino C, Welsch G, Axelrod FB. Respiratory and cerebrovascular responses to hypoxia and hypercapnia in familial dysautonomia. *Am J Respir Crit Care Med* 2002;**167**:141–9.
56. Rubery PT, Spielman JH, Hester P, Axelrod FB, Burke SW, Levine DB. Scoliosis in familial dysautonomia. *J Bone Joint Surg* 1995;**77**:1362–9.
57. Hayek S, Laplaza J, Axelrod FB, Burke SW. Spinal deformity in familial dysautonomia: Prevalence and results of brace management. *J Bone Joint Surg (Am)* 2000;**82**:1558–62.
58. Norcliffe-Kaufmann L, Axelrod FB, Kaufmann H. Developmental abnormalities, blood pressure variability and renal disease in Riley Day syndrome. *J Hum Hypertens* 2013;**27**:51–6.
59. Elkayam L, Matalon A, Tseng CH, Axelrod FB. The prevalence and severity of renal disease in familial dysautonomia. *Am J Kidney Dis* 2006;**48**:780–6.
60. Rekhtman Y, Bomback AS, Nash MA, Cohen SD, Matalon A, Jan DM, et al. Renal transplantation in familial dysautonomia: report of two cases and review of the literature. *Clin J Am Soc Nephrol* 2010;**5**(9):1676–80.
61. Mendoza-Santiesteban CE, Hedges III TR, Norcliffe-Kaufmann L, Warren F, Reddy S, Axelrod FB, et al. Clinical neuro-ophthalmic findings in familial dysautonomia. *Neuroophthalmology* 2012;**32**:23–6.
62. Welton W, Clayson D, Axelrod FB, Levine DB. Intellectual development and familial dysautonomia. *Pediatrics* 1979;**63**:708–12.
63. Mass E. A review of the oro-dento-facial characteristics of hereditary sensory and autonomic neuropathy type III (familial dysautonomia). *Spec Care Dentist* 2012;**32**(1):15–20.
64. Axelrod FB, Goldberg JD, Ye XY, Maayan C. Survival in familial dysautonomia: Impact of early intervention. *J Pediatr* 2002;**141**:518–23.
65. Indo Y. Genetics of congenital insensitivity to pain with anhidrosis (CIPA) or hereditary sensory and autonomic neuropathy type IV. *Clin Auton Res* 2002;**12**(Suppl.):20–32.
66. Levi-Montalcini R, Skaper SD, Dal TR, Petrelli L, Leon A. Nerve growth factor from neurotrophin to neurokine. *Trends Neurosci* 1996;**19**:514–20.
67. Indo Y, Tsuruta M, Hayashida Y, Karim MA, Ohta K, Kawano T, et al. Mutations in the TRKA/NGF receptor gene in patients with congenital insensitivity to pain with anhidrosis. *Nat Genet* 1996;**13**:485–8.
68. Goebel HH, Veit S, Dyck PJ. Confirmation of virtual unmyelinated fiber absence in hereditary sensory neuropathy type IV. *J Neuropathol Exp Neurol* 1980;**39**:670–5.
69. Nolano M, Crisci C, Santoro L. Absent innervation of skin and sweat glands in congenital insensitivity to pain with anhidrosis. *Clin Neurophysiol* 2000;**111**:1596–601.
70. Langer J, Goebel HH, Veit S. Eccrine sweat glands are not innervated in hereditary sensory neuropathy type IV. An electron microscopic study. *Acta Neuropath* 1981;**54**:199–202.
71. Bar-On E, Weigl D, Parvari R, Katz K, Weitz R, Steinberg T. Congenital insensitivity to pain: Orthopedic manifestations. *J Bone Joint Surg [Br]* 2002;**84-B**:252–7.
72. Feldman DS, Ruchelsman DE, Spencer DB, Straight JJ, Schweitzer ME, Axelrod FB. Peripheral arthropathy in hereditary sensory and autonomic neuropathy types III and IV. *J Ped Ortho* 2009;**29**(1):91–7.
73. Amano S, Fukuoda S, Sushi T, Honda N, Ideta R, Ochiai M, et al. Ocular manifestations of congenital insensitivity to pain with anhidrosis. *Am J Ophthalmol* 2006;**141**:472–7.
74. Einarsdottir E, Carlsson A, Minde J, Toolanen G, Svensson O, Solders G, et al. A mutation in the nerve growth factor beta gene (NGFB) causes loss of pain perception. *Hum Mol Genet* 2004;**13**:799–805.
75. Minde J, Toolanen G, Andersson T, Nennesmo I, Remahl IN, Svensson O, et al. Familial insensitivity to pain (HSAN V) and a mutation in the NGFB gene. A neurophysiological and pathological study. *Muscle Nerve* 2004;**30**(6):752–60.
76. Nicol GD, Vasko MR. Unraveling the story of NGF-mediated sensitization of nociceptive sensory neurons: ON or OFF the Trks? *Mol Interv* 2007;**7**:26–41.
77. Carvalho OP, Thornton GK, Hertecant J, Houlden H, Nicholas AK, Cox JC, et al. A novel NGF mutation clarifies the molecular mechanism and extends the phenotypic spectrum of the HSAN5 neuropathy. *J Med Genet* 2011;**48**:131–5.
78. Low PA, Burke WJ, McLeod JG. Congenital sensory neuropathy with selective loss of small myelinated fibers. *Ann Neurol* 1978;**3**(2):179–82.
79. Dyck PJ, Mellinger JF, Reagan TJ, Horowitz SJ, McDonald JW, Litchy WJ, et al. Not "indifference to pain" but varieties of hereditary sensory and autonomic neuropathy. *Brain* 1983;**106**:373–90.
80. Hilz MJ. Assessment and evaluation of hereditary sensory and autonomic neuropathies with autonomic and neurophysiological examinations. *Clin Auton Res* 2002;**12**(Suppl. 1):33–43.
81. Axelrod FB, Pearson J, Tepperberg J, Ackerman B. Congenital sensory neuropathy with skeletal dysplasia. *J Pediatr* 1983;**102**:727–30.
82. Ørbeck H, Oftedal G. Familial dysautonomia in a non-Jewish child. *Acta Paediatr Scand* 1977;**66**:777–81.
83. Smith AA, Dancis J. Response to intradermal histamine in familial dysautonomia—a diagnostic test. *J Pediatr* 1963;**63**:889–94.
84. Slaugenhaupt SA, Mull J, Leyne M, Cuajungco MP, Gill SP, Hims M, et al. Rescue of a human mRNA splicing defect by the plant cytokinin kinetin. *Hum Mol Genet* 2004;**13**(4):429–36.
85. Axelrod FB, Liebes L, Gold-von Simson G, Mendoza S, Mull J, Leyne M, et al. Kinetin Improves IKBKAP mRNA Splicing in Patients With Familial Dysautonomia. *Pediatr Res* 2011;**70**:480–3.
86. Axelrod FB, Berlin D. Pregabalin: a new approach to treatment of the dysautonomic crisis. *Pediatrics* 2009;**124**:743–6.

Chapter 19

Peripheral Neuropathy in Inherited Metabolic Disease

Marc C. Patterson and Alan K. Percy

INTRODUCTION

Peripheral neuropathy is associated with numerous inherited metabolic diseases, such as Fabry disease (trihexosylceramide lipidosis) where it may be the presenting complaint, or adrenomyeloneuropathy, in which it may play a relatively minor role in contrast to the more prominent and progressive spastic paraparesis. In the mucopolysaccharidoses, neuropathy may represent a complication of the underlying storage disease, such as an entrapment neuropathy. This chapter describes the specific inherited metabolic disorders in which peripheral neuropathies are common. While the biochemical and molecular bases of these disorders are well established, effective treatment is quite elusive without specific therapies directed at the primary metabolic defect. In some cases, pharmacologic interventions may be helpful.

The inherited metabolic diseases in which a peripheral neuropathy is prominent may be separated into several categories (Table 19.1) including storage diseases, disorders of amino acid and fatty acid metabolism, and mitochondrial cytopathies.

LYSOSOMAL STORAGE DISEASES

Sphingolipidoses

Peripheral neuropathy is an integral component of several disorders of sphingolipid metabolism including Krabbe disease, metachromatic leukodystrophy, and Fabry disease. Histologic evidence of lipid storage in peripheral nerve also occurs in rapidly progressive forms of GM1 and GM2 gangliosidoses, infantile Gaucher disease, Niemann-Pick disease, and Farber disease, but in these disorders nerve involvement is not clinically evident, most likely due to rapid central nervous system progression. These conditions will therefore not be discussed further. Involvement of the peripheral and autonomic nervous systems has been recently recognized in chronic GM2 gangliosidosis.[1] This rather uncommon form of the disease typically presents in late childhood or adolescence as ataxia, dystonia, motor neuron disease, and dementia. Nonetheless, peripheral nerve disease may be sufficient to produce clinical symptomatology and objective evidence of diffuse sensorimotor neuropathy with reduced NCV and amplitude (Table 19.2).

Krabbe Disease: Globoid Cell Leukodystrophy (Galactosylceramide Lipidosis)

Clinical Characteristics

Globoid leukodystrophy (GLD), described in 1916 by the Danish physician Knud Krabbe,[2] has four clinical presentations, based upon age of onset.[3–5] The *early infantile* form (3–6 months of age) presents with episodes of rigidity and opisthotonus, often in response to auditory, tactile, or visual stimuli; progressive psychomotor retardation; feeding difficulties; febrile episodes; spastic paraparesis; and optic atrophy. Results from an international registry demonstrated abnormalities in most children with early infantile form; 3-year survival is 14%.[6] The *late infantile* form (6 months to 3 years) presents with ataxia, generalized hypertonicity, psychomotor regression, and visual failure. The combined incidence of these two forms is estimated at 1:100,000 to 1:200,000. The *juvenile* form (3–10 years of age) presents with dystonia, visual loss, and progressive spastic paraparesis, and the *adult* form with progressive spastic paraparesis, dementia, and optic atrophy.[3]

Peripheral neuropathy is a prominent feature in the infantile forms, perhaps even before clinical symptoms, but is mild or absent in the juvenile and adult forms. The peripheral neuropathy is associated with elevated levels of

B.T. Darras, H. Royden Jones, Jr., M.M. Ryan & D.C. De Vivo (Eds): Neuromuscular Disorders of Infancy, Childhood and Adolescence, Second edition.
DOI: http://dx.doi.org/10.1016/B978-0-12-417044-5.00019-6

TABLE 19.1 Peripheral Neuropathy in Inherited Metabolic Diseases

Category	Disease	Stored Material	Biochemical Defect	Molecular Defect	Chromosome
1. Lysosomal Storage Diseases					
Sphingolipidoses	Krabbe disease	Galactosylceramide	Galactosylceramide β-galactosidase	*GLAC*	14q25-31
	Metachromatic leukodystrophy	Sulfatide	Sulfatide sulfatase (arylsulfatase A)	*ARSA*	22q13.31-ter
	Fabry disease	Trihexosylceramide	Trihexosylceramide α-galactosidase	*GLA*	Xq22.1
Mucopolysaccharidoses	Multiple (see Table 19.3)	See Table 19.3	See Table 19.3	See Table 19.3	See Table 19.3
Glycoproteinoses	Fucosidosis	Oligosaccharides	α-fucosidase		1p34.1-36.1
	α and β mannosidosis	Oligosaccharides	α and β mannosidase	FUCS1 *MAN2B1*, (α) *MANBA*, (β) *NAGA*	α-19p13.2-q12 β-4
	Sialidosis type II	Oligosaccharides	Sialidase	*NEU*	6p21.3
	Schindler disease	Oligosaccharides	N-acetylgalactosaminidase	*AGB*	22q13.1-13.2
2. Peroxisomal Diseases	Adrenomyeloneuropathy	Very long chain fatty acids	VLCFA CoA synthetase	*ALDP*	X128
	Refsum disease	Phytanic acid	Phytanic acid oxidase	*PHYH*	10p
	Hyperoxaluria	Calcium oxalate	Alanineglyoxylate aminotransferase	*AGT*	2q36-37
3. Lipids	Cerebrotendinous xanthomatosis	Cholestanol Cholesterol esters	Cholesterol 27-hydroxylase	CTX	2q35
	Tangier disease		Unknown	*ABCA1*	
	Abetalipoproteinemia		Unknown	*APOB, PCSK9, MTP*	
4. Glycogen Storage Disease	Adult polyglucosan body disease	Glycogen	Amylo-1, 4-1, 6 transglucosidase	GBE1	3p12
5. Amino Acids	Tyrosinemia type I	Tyrosine/succinyl-acetone	Fumarylaceto-acetate hydrolase	FAH	15
6. Mitochondrial Diseases	LCHAD	3-Hydroxy dicarboxylic aciduria	Trifunctional protein and long-chain 3-hydroxy-acyl-CoA dehydrogenase	HADHA/HADHB	2p23.3
	Leigh	Lactate/pyruvate	Multiple enzymes		Multiple nucleotide
	NARP	Lactate/pyruvate		ATP synthase subunit 6	8993 (mtDNA)
7. Other	Acute intermittent porphyria	δ-Aminolevulinic acid	Porphobilinogen deaminase	PBGD	11q24.1-q24.2
			δ-aminolevulinic acid dehydratase	*ALAD*	9q34

TABLE 19.2 Peripheral Neuropathic Features of Inherited Metabolic Diseases

Category	Disease	Neuropathic Features
1. Lysosomal Storage Diseases		
Sphingolipidoses	Krabbe disease	Schwann cell inclusions
		Segmental demyelination
	Metachromatic leukodystrophy	Schwann cell sulfatide (metachromatic) accumulation
		Segmental demyelination
	Fabry disease	Lamellar inclusions in perineurial cells
		Axonal loss
Mucopolysaccharidoses	Multiple (see Table 19.3)	Entrapment neuropathy
Glycoproteinoses	Fucosidosis	Entrapment neuropathy
	Mannosidosis	Axonal inclusion
	Sialidosis type II	Schwann cell vacuoles
		Axonal inclusions
	Schindler disease	Axonal spheroids
2. Peroxisomal Diseases	Adrenomyeloneuropathy	Schwann cell inclusions
		Segmental demyelination
	Refsum disease	Schwann cell inclusions
		Onion-bulb formation
		Segmental demyelination
	Hyperoxaluria	Axonal degeneration
		Segmental demyelination
3. Lipids	Cerebrotendinous xanthomatosis	Axonal loss
		Schwann cell vacuoles
	Tangier disease	Axonal loss
		Schwann cell vacuoles
	Abetalipoproteinemia	Axonal degeneration
		Segmental demyelination
4. Glycogen Storage Disease	Adult polyglucosan body disease	Schwann cell inclusions (polyglucosan bodies)
5. Amino Acids	Tyrosinemia type I	Axonal loss
		Segmental demyelination
6. Mitochondrial Diseases	LCHAD	Axonal loss
		Segmental demyelination
	Leigh	Segmental demyelination
	NARP	Segmental demyelination
7. Other	Acute intermittent	Axonal degeneration
	porphyria	Segmental demyelination

cerebrospinal fluid protein (up to 400 mg/dL) and reduced nerve conduction velocities (NCV = 10 m/s or lower). Distal latencies may be mildly prolonged. Compound motor action potential (CMAP) amplitudes may be mildly to moderately diminished.[7] Electromyography (EMG) is normal. An infant with prenatal diagnosis demonstrated slowed NCVs at 7 weeks.[8] Patients with the late-onset forms rarely show clinical evidence of peripheral nerve involvement, and their cerebrospinal fluid protein levels are normal or only mildly increased, but NCVs are decreased in most instances. Both axonal and demyelinating patterns have been reported in late onset Krabbe disease[9]; the latter occurs in about 80% of those who exhibit peripheral neuropathy (about 60% of adults).[10]

Pathology

The main histopathological changes in the central nervous system are extensive demyelination, gliosis, and the presence of characteristic multinucleated cells, referred to as globoid cells. These phagocytic cells can be produced experimentally in rats by injecting galactosylceramide into the spleen or brain white matter.[11] Peripheral nerves demonstrate segmental demyelination and remyelination with marked reduction in the density of myelinated fibers and a relative predominance of small thinly myelinated fibers (Table 19.2). Sural nerve biopsies demonstrate Schwann cell cytoplasmic inclusions similar to those in the brain; however, classic globoid cells are not found.[7,12]

Biochemistry

GLD is due to deficiency of the lysosomal enzyme galactosylceramidase (galactosylceramide-β-galactosidase),[13] which metabolizes galactosylceramide to ceramide and galactose as well as galactosylsphingosine (psychosine) to sphingosine and galactose. Diagnosis is made by measurement of GALC activity in white blood cells, cultured skin fibroblasts, amniocytes, or chorionic villus cells.[13] Psychosine accumulation in patients with GLD is a key factor in the pathogenesis of the disease.[14] Animal model studies suggest an axonopathy related to psychosine.[15]

Molecular Defect

GLD is an autosomal recessive disorder in which the gene *GALC* maps to 14q25-31. One hundred and forty-seven disease-causing mutations are known.[16] The most common, a 502T/del, is present in 40% of GLD patients of European origin. The second most frequent mutation, a C > T change at nucleotide 1538, is found in approximately 10% of patients with the infantile phenotype. Mutation analysis is the most reliable method of carrier identification.

Management

Hematopoietic stem cell transplantation improves survival in babies with Krabbe disease treated within the first few weeks of life,[17] but its benefits may be transient.[18] Hematopoietic stem cell transplantation (HSCT) can be effective in the juvenile form of the disease but does not improve nerve conduction velocities.[19]

Metachromatic Leukodystrophy (Sulfatide Lipidosis)

Clinical Characteristics

Metachromatic leukodystrophy (MLD) presents in late infancy (most commonly), adolescence or adulthood.[20–22] Onset is between 1 and 3 years of age with loss of previously acquired motor skills such as ambulation, followed later by ataxia, dysarthria and dysphagia, and impairment of cognitive function, vision, and hearing. The deficits are progressive, leading to death after several years. Occasionally, MLD presents in infancy as peripheral neuropathy without overt CNS findings[23]; cranial neuropathy (esotropia and facial weakness) has also been reported as a presenting feature, associated with cranial nerve enhancement on MR imaging.[24] Neurologic examination shows evidence of both central nervous and peripheral nerve dysfunction, with hypotonia and diminished muscle stretch reflexes and positive dorsal plantar responses, ataxia, dystonic movements, and nystagmus. The juvenile form is similar to the late infantile form but with slower progression. In the adult form, behavioral and cognitive changes may precede motor changes by years or even decades. The frontal lobe syndrome in adult MLD may include disinhibition, impulsivity, poor judgment, emotional lability, social inappropriateness, and poor attention span.[25]

Peripheral nerve involvement is a prominent feature in all forms of MLD, the combined peripheral and central nervous system involvement being an important diagnostic clue. NCVs are significantly decreased (<10 m/second) and distal latencies may be prolonged. Sural nerve biopsy shows metachromatic staining. In juvenile metachromatic leukodystrophy, diminished NCVs may prompt evaluation for combined central and peripheral nervous system demyelination.[26] A rapidly progressive ascending paralysis has been described in an 18-month-old girl with MLD.[27] Respiratory failure and ptosis responded to intravenous immunoglobulin, but limb weakness did not, suggesting that the child had an immune-mediated neuropathy superimposed on the chronic neuropathy of MLD.

Pathology

The pathology of MLD consists primarily of demyelination and metachromatic granule deposition in the central and peripheral nervous system (Figure 19.1). These changes can also be seen in the kidney, urine (urinary sulfatide excretion is increased), and other tissues. Viewed by electron microscopy the metachromatic granules present a hexagonal honeycomb pattern, and are referred to as tuffstone bodies. Peripheral nerves show segmental demyelination, with metachromatic granules singly or in clusters in Schwann cells (Table 19.2, Figure 19.2), in endoneurial macrophages, and to a lesser extent in the Remak cells associated with unmyelinated nerve fibers.[28]

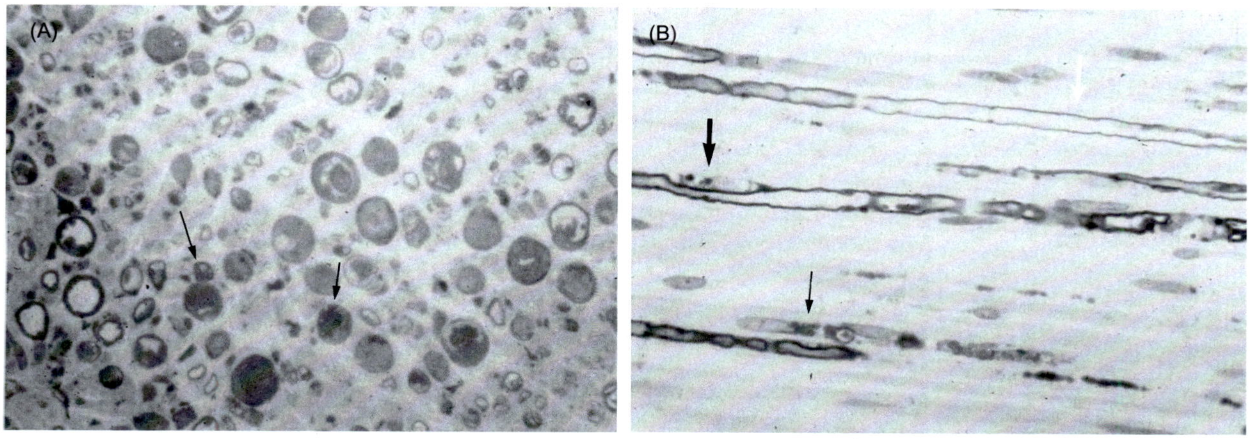

FIGURE 19.1 (A) Metachromatic leukodystrophy: Cross-section of sural nerve demonstrating metachromatic material (arrows) and irregular myelin sheaths. Methylene blue. ×400. (B) Metachromatic leukodystrophy: Longitudinal section of sural nerve demonstrating metachromatic material within Schwann cell cytoplasm (black arrows) and narrowing of myelin sheath (white arrow). Cresyl violet. ×400.

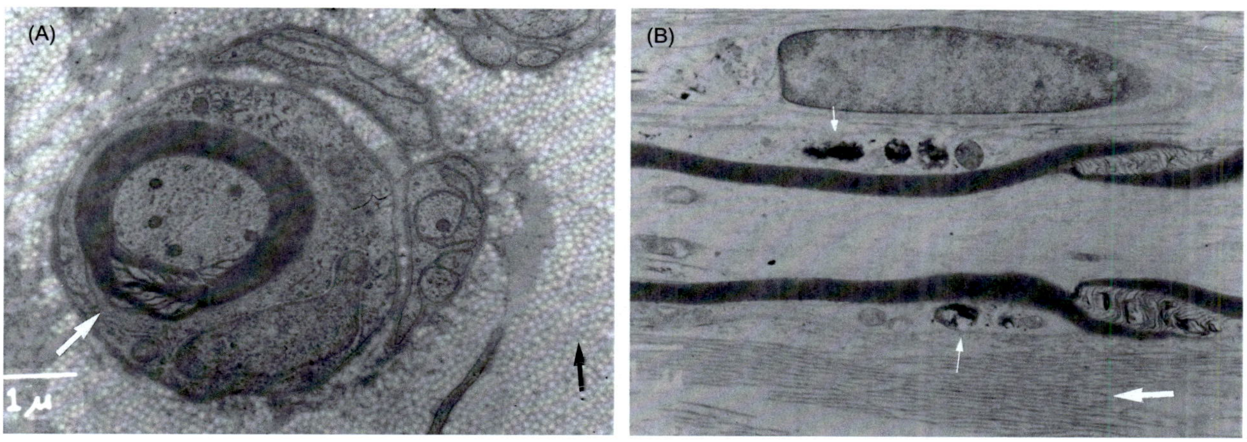

FIGURE 19.2 (A) Metachromatic leukodystrophy: Electron micrograph of sural nerve in cross-section, demonstrating excess interstitial collagen (black arrow) and remyelination (white arrow). (B) Metachromatic leukodystrophy: Electron micrograph of sural nerve in longitudinal section demonstrating granular inclusions in Schwann cell cytoplasm (small white arrows) and excess interstitial collagen (large white arrow). ×12,000.

In addition to the expected metachromatic granules revealed by cresyl violet staining, one study of 40 cases of nerve biopsies in MLD showed accumulation of orthochromatic granules in perivascular macrophages, suggesting storage of multiple species of glycosphingolipids.[29]

Biochemistry

The key abnormality in MLD is the accumulation of sulfatide (galactosylceramide-3-sulfate), the metachromasia due to the sulfate group. Deficiency of the lysosomal enzyme arylsulfatase-A (ASA), which cleaves sulfatide to galactosylceramide and sulfate, is the most common cause of sulfatide accumulation, with an estimated incidence of 1:40,000. Diagnosis of ASA deficiency can be determined in white blood cells, cultured skin fibroblasts, amniocytes, or chorion villus cells. Great caution must be used to distinguish MLD from a condition referred to as pseudoarylsulfatase deficiency (PSD).[30] PSD is a common and entirely benign condition in which ASA activity is reduced, at times to levels nearly as low as in MLD. Persons with PSD do not display any of the clinical manifestations of MLD. Distinction between PSD and MLD is crucial in prenatal studies and in asymptomatic infants or adults. Molecular analysis[30] and measurement of urinary sulfatide levels aid the distinction. Urinary sulfatide levels are increased in MLD but not PSD.[31] Less commonly, sulfatide accumulation is due to multiple sulfatase deficiency, in which at least seven sulfated lipids or mucopolysaccharides accumulate.[32] The third and least common cause is a deficiency of saposin B, one of the sphingolipid activator proteins, which is required for the biological activity of ASA.[33] Mutations in the PSAP gene, which encodes saposin B, are associated with a wide variety of phenotypes.[34]

Molecular Defect

All forms of MLD are transmitted as autosomal recessive traits, the gene, *ASA*, mapping to 22q13, with more than 150 mutations identified.[35] Mutations that abolish function completely cause the severe late infantile phenotype. Missense mutations, in which up to 3% to 5% of normal activity is maintained, tend to be associated with the adult form of the disease. Individuals who are heterozygous for a null mutation and one of the milder missense mutations often have the juvenile phenotype of MLD.[36]

Management

Bone marrow transplantation (BMT) can stabilize or slow the rate of progression of central nervous system dysfunction, and stabilize peripheral nerve involvement in patients with the juvenile, but not the early infantile presentation of MLD.[37] BMT is recommended only for patients with relatively mild nervous system involvement, or for asymptomatic patients identified by testing of at-risk relatives, with care taken to differentiate MLD from PSD. The procedure is not recommended for patients with advanced involvement, as it is associated with a high risk of additional worsening of the disability. Gene therapy is under investigation in cultured cells and in animal models. Enzyme replacement therapy is currently in human trials, after promising results in animal models.[38]

Cholecystectomy is occasionally required in those in whom the accumulated sulfatide has formed gallstones.

Fabry Disease (Trihexosylceramide Lipidosis)

Clinical Characteristics

Trihexosylceramide lipidosis (angiokeratoma corporis diffusum universale, Fabry disease) was described independently in 1893 by Fabry and Anderson.[39] Among the sphingolipid storage diseases, it is the only X-linked disorder. The principal clinical manifestations involve skin, heart, kidney, and peripheral nerve.

The hallmark of Fabry disease is its cutaneous manifestation—*angiokeratoma*—violaceous angiectatic lesions over the genitalia, thighs, buttock, back, and lower abdomen. Similar lesions may be seen in the oral mucosa. Angiokeratomata have also been described in fucosidosis, β-mannosidosis, galactosialidosis, and Schindler disease, and are not specific for Fabry disease. Fabry disease has its onset most commonly in adolescence or early adulthood. Despite significant chronic medical issues, survival is often quite prolonged. Ultimately, renal or cardiac involvement will determine longevity. Cardiac manifestations include left ventricular hypertrophy, valvular dysfunction (especially mitral insufficiency), and conduction abnormalities.[40,41] Death usually occurs in the fourth or fifth decades.

Chronic renal disease poses the major clinical problem. As such, Fabry disease had been regarded as the best candidate among the sphingolipidoses for direct therapeutic intervention. However, improvement of renal function with dialysis or renal transplantation has not reversed other aspects of the disease.[42]

The most common neurological problem is a painful peripheral neuropathy,[43,44] which generally involves the distal extremities (acroparesthesias). Clinical features include both severe crises of lancinating pain and chronic, unremitting pain. The painful crises may be precipitated by exercise, fatigue, stress, or exposure to sunshine or hot weather. Decreased sweating (hypohidrosis) leads to temperature intolerance and untoward responses to sun exposure.

Other common features are corneal and lenticular opacifications and retinal vasculopathies. CNS vaso-occlusive events may be noted, and Fabry disease should be considered in the differential diagnosis of stroke in the young individual.[45,46] Cranial imaging often demonstrates changes consistent with stroke, even the absence of a history of stroke-like events. Occasionally, lymphedema and episodic diarrhea may be significant problems. Both are attributed to lipid storage.

Female carriers (heterozygotes) of trihexosylceramide lipidosis may develop signs similar to those in affected males, although typically much later in adulthood.[47,48] Corneal opacifications are the most common signs, being present in nearly three-fourths of carriers. Other features are less prevalent but equally severe to those seen in men.

Cardiac Variant A variant form of Fabry disease is limited predominantly to cardiac involvement with onset typically after age 40, without other features of Fabry disease except for mild proteinuria. The investigation of unexplained proteinuria in these males led to the recognition of this milder variant.

Pathology

Lipid materials accumulate in vascular smooth muscle and endothelium, kidney, cornea, peripheral nerves, and in neurons of the autonomic nervous system, both central and peripheral. Electron microscopy of peripheral nerves reveals lamellar inclusions in perineurial cells. Similar lamellar inclusions can be demonstrated in vascular smooth muscle. The degree of peripheral nerve involvement can be easily assessed by analysis of biopsy specimens[43] (Table 19.2). A decrease in small peripheral sensory neurons may be noted.[49]

Biochemistry

Deficient α-galactosidase (trihexosylceramide α-galactosidase) activity is the basis for Fabry disease.[50] Trihexosylceramide is an important membrane sphingolipid in kidney and the first intermediate in the degradation of globoside, the prominent sphingolipid of red cell membranes. In Fabry disease, trihexosylceramide accumulates in kidney, liver, and lung from 30 to 300 times normal levels. A lesser degree of digalactosylceramide accumulation is also seen, primarily in the kidney. Diagnosis can be made in leukocytes and cultured skin fibroblasts, and numerous mutations are described.[51–53]

Management

The painful and disabling dysesthesias of hands and feet may be treated with carbamazepine, phenytoin, gabapentin, or pregabalin.[54] The systemic clinical manifestations, including neurological problems, arise from vaso-occlusion due to sphingolipid deposition in vascular endothelial cells, and may be difficult to treat. Renal transplantation may be required for end-stage kidney disease, but does not usually result in overall clinical and biochemical improvement.[42]

Enzyme replacement therapy (ERT) was approved for use in Fabry disease in Europe in 2001 and in the USA in 2003; the response to therapy is variable. ERT appears to be least effective in individuals with advanced disease.[46] ERT reduces neuropathic pain, improves the detection threshold for cold and warm sensation in the hand and foot, and leads to improvement in sweating and heat tolerance, but not does not fully reverse peripheral nerve involvement in Fabry disease.[44] The extent of renal impairment correlates with small fiber response to ERT; i.e. patients with more severe renal impairment have a more severe neuropathy.[55] The deacylated product, trihexosylsphingosine, has been suggested as a useful biomarker following ERT for Fabry disease.[56,57] Newborn screening in Fabry disease suggests a high frequency of under-diagnosis of later onset forms that could be treated effectively if diagnosed.[58–61]

Mucopolysaccharidoses

The mucopolysaccharidoses (Table 19.3) are lysosomal storage diseases due to deficiencies of specific lysosomal enzymes responsible for the degradation of heparan sulfate, dermatan sulfate, or keratan sulfate, singly or in combination. Lack of lysosomal enzymes activity specific to each disorder leads to tissue accumulation and urinary excretion of mucopolysaccharides.[62]

The mucopolysaccharidoses have many common clinical features including hepatosplenomegaly, deposition of mucopolysaccharide in arterial smooth muscle with resultant coronary insufficiency and compromise of the aortic wall, and accumulation in heart valves producing both insufficiency and stenosis.[63] Varying degrees of skeletal involvement (dysostosis multiplex) and joint contractures with limitation of motion occur in all but Morquio disease. Mixed or sensorineural deafness occurs in all except Sanfilippo disease. Corneal clouding occurs to varying degrees in these disorders, but may be absent or seen only by slit lamp in Hunter disease. Pigmentary retinopathy is occasionally seen in Scheie, Sanfilippo, and Hunter diseases.

The most frequent neurologic manifestation is psychomotor retardation. This is present, albeit with varying severity, in Hurler, Hunter, and Sanfilippo diseases, the result, most likely, of mucopolysaccharide and lipid deposition in brain cells. Thickening of the meninges secondary to mucopolysaccharide storage may result in hydrocephalus, particularly in Hurler and Hunter disease. Meningeal thickening in the cervical region may result in myelopathy secondary to spinal cord compression and nerve root compromise.

Peripheral neuropathy may develop from nerve entrapment as the result of mucopolysaccharide accumulation in adjacent connective tissue (Table 19.2). Median nerve compression in the carpal tunnel is the most common neuropathic manifestation. Carpal tunnel compression has been noted prominently in Scheie (MPS I S), Hunter (MPS II), and Maroteaux-Lamy (MPS VI) diseases.[64] Mucopolysaccharidoses are the most common cause of carpal tunnel syndrome in childhood, usually presenting with impaired fine motor function.[65] One study found that 43 of 45 boys with MPS II had carpal tunnel syndrome; it was bilateral in all cases.[66] The profound cognitive impairment in Scheie or Hunter disease may delay recognition of peripheral nerve symptoms until involvement is advanced. Surgical management is generally effective.

The original classification of the mucopolysaccharidoses as distinct disorders—Hurler, Hunter, Sanfilippo, Morquio, Scheie, Maroteaux-Lamy, and β-glucuronidase deficiency—derived from their variable clinical features. Subsequent biochemical and molecular studies indicated that Hurler and Scheie are allelic,[67] caused by different mutations of the same gene. In contrast, Sanfilippo and Morquio diseases have multiple subtypes caused by mutations in different genes. Enzyme replacement therapy delivered by intravenous infusion has been approved in the USA for MPS types I, II, and VI; it ameliorates somatic manifestations, but does not influence nervous system disease.[68] Several lines of evidence support early initiation of ERT in this population to minimize the consequences of irreversible tissue damage.[69] Hematopoietic stem cell transplantation stabilizes cognitive decline and prolongs survival in MPS type I, but not in other MPS disorders; animal studies suggest that substrate reduction therapy and gene therapy[70] are promising alternative strategies for MPS disorders with no, or currently inadequate, treatment.[68,71]

Glycoproteinoses

The glycoproteinoses include fucosidosis, α- and β-mannosidosis, the sialidoses, and aspartylglucosaminuria,[72] and resemble the other lysosomal hydrolase deficiency diseases in many respects (Table 19.4). These

TABLE 19.3 Mucopolysaccharidoses

Disease	Common Name	Relevant Metabolite	Metabolic Defect	Clinical Features
MPS I H	Hurler	Dermatan sulfate Heparan sulfate	α-L-Iduronidase	Severe physical and cognitive difficulties; early death
MPS I S	Scheie	Dermatan sulfate Heparan sulfate	α-L-Iduronidase	Severe physical difficulties; normal mentation and survival
MPS I H/S	Hurler-Scheie	Dermatan sulfate Heparan sulfate	α-L-Iduronidase	Intermediate
MPS II	Hunter	Dermatan sulfate Heparan sulfate	Iduronate-2-sulfatase	Similar to Hurler, somewhat milder
MPS III A	Sanfilippo A	Heparan sulfate	Heparan-N-sulfatase	Severe cognitive difficulties; mild physical changes
MPS III B	Sanfilippo B	Heparan sulfate	N-Acetyl-α-glucosaminidase	Severe cognitive difficulties; mild physical changes
MPS III C	Sanfilippo C	Heparan sulfate	Acetyl-CoA: α-glucosaminide N-acetyltransferase	Severe cognitive difficulties; mild physical changes
MPS III D	Sanfilippo D	Heparan sulfate	N-Acetyl-α-glucosamine-6-sulfatase	Severe cognitive difficulties; mild physical changes
MPS IV A	Morquio A	Keratan sulfate Chondroitin 6-sulfate	N-Acetylgalactosamine-6-sulfatase; galactose-6-sulfate sulfatase	Severe skeletal features; normal mentation
MPS IV B	Morquio B	Keratan sulfate	β-Galactosidase	Milder skeletal features
MPS VI	Maroteaux-Lamy	Dermatan sulfate	Arylsulfatase B	Severe skeletal features; normal mentation
MPS VII	Sly	Dermatan sulfate Heparan sulfate Chondroitin 4-, 6-sulfate	β-Glucuronidase	Severe skeletal features; moderate cognitive impairment

glycoproteinoses are characterized by the inability to degrade the oligosaccharide portion of glycoproteins, resulting in the excretion of large quantities of oligosaccharides in urine. Significant clinical heterogeneity exists within the specific entities, each of which is inherited in an autosomal recessive fashion. Prenatal diagnosis is possible, but specific treatment is lacking.

Fucosidosis, mannosidosis, type II sialidosis, and infantile sialic acid storage disease have many features in common with the mucopolysaccharidoses. As such, entrapment neuropathies may be noted. In addition, inclusions have been noted in peripheral nerves and sympathetic neurons in type II sialidosis (Table 19.2).

Cherry Red Spot Myoclonus Syndrome (Sialidosis I)

Debilitating myoclonus occurs in the second or third decade in cherry red spot myoclonus syndrome and is associated with nonpigmentary macular degeneration secondary to lysosomal neuraminidase deficiency. Sural nerve biopsy may demonstrate segmental demyelination with Schwann cells containing indeterminate abnormal storage material within membrane bound vacuoles.[73]

Schindler Disease

Clinical Characteristics

Peripheral nerve involvement in α-N-acetylgalactosaminidase deficiency (Schindler disease) is much more marked than in the other glycoproteinoses.[74] The hallmark of Schindler disease is a central and peripheral axonopathy.

Schindler disease is a very rare disorder, described in only a small number of individuals. An acute form has its onset from 9 to 15 months as an exaggerated startle response after normal early infancy. Psychomotor deterioration is rapid, leading to marked spasticity, cortical blindness, and myoclonic epilepsy. The EEG shows

TABLE 19.4 Glycoproteinoses

Disease	Relevant Metabolite	Metabolic Defect	Clinical Features
Fucosidosis	Fucose-containing glycoproteins and oligosaccharides	α-fucosidase	Coarse facies Dysostosis multiplex Hearing loss Psychomotor decline
α-Mannosidosis	Mannose-containing glycoproteins and oligosaccharides	α-mannosidase	Coarse facies Dysostosis multiplex Hearing loss Psychomotor decline
β-Mannosidosis	Mannosyl-β-N-acetyl-glucosamine	β-mannosidase	Recurrent infections Psychomotor decline Peripheral neuropathy in delayed-onset forms
Sialidoses			
Type I	Sialic acid-containing oligosaccharides	Sialidase	Progressive myoclonus Visual loss Cherry-red maculae Ataxia
Type II	Sialic-acid containing oligosaccharides	Sialidase	Coarse facies Contractures of major joints Mild dysostosis multiplex Cherry-red maculae Psychomotor decline Peripheral neuropathy
Sialic acid storage diseases	Sialic acid	Sialic acid lysosomal transport	Coarse facies Dysostosis multiplex Growth retardation Psychomotor decline
Galactosialidosis	Sialic-acid containing oligosaccharides	Protective protein/Cathepsin A	Similar to type II sialidosis
Aspartylglucosaminuria	Aspartylglucosamine	Aspartylglucosaminidase	Coarse facies Mild dysostosis multiplex Cognitive impairment
N-acetylgalactos-aminidase deficiency (Schindler disease)	Sialic acid-containing and asialoglycoproteins and oligosaccharides	α-N-acetylgalactosaminidase	Psychomotor decline Myoclonic epilepsy Cortical blindness

background slowing and multifocal spike and spike and wave activity. The VER, ABR, and SSEP are reduced, whereas NCV and ERG are normal.

A chronic form features mild cognitive impairment and slow progression with dilated, tortuous retinal vessels and cutaneous angiokeratoma, most prominent over the torso.[75]

Pathology

An axonal neuropathy is prominent, resembling that of infantile neuroaxonal dystrophy.[76] Axonal spheroids are present in peripheral nerve axons and in axons ranging from cerebral cortex to the myenteric plexus. Electron microscopy revealed membranous cytoplasmic bodies in these tissues. α-N-acetylgalactosaminidase B deficiency, which establishes the diagnosis, may be measured in leucocytes or cultured skin fibroblasts.[76,77] This enzyme is identical to its isomer, α-galactosidase A, the deficient enzyme in Fabry disease.

The only three patients described with Schindler disease and neuroaxonal dystrophy were all related; this observation, and the wide range of phenotypes associated with α-N-acetylgalactosaminidase B deficiency, has led some authors to propose that children with enzyme deficiency and the neurological phenotype also have mutations in the calcium-independent phospholipase A_2 gene, *PLA2G6*, which is the cause of infantile neuroaxonal dystrophy.[78]

PEROXISOMAL STORAGE DISEASES

Peripheral neuropathy has been described mainly in three peroxisomal storage diseases, Refsum disease, adrenomyeloneuropathy, and hyperoxaluria (Table 19.2). Refsum disease is quite responsive to dietary therapy when early diagnosis is established. Hyperoxaluria may be treated with combined liver-renal transplantation.

Phytanic Acid Lipidosis (Refsum Disease; Heredopathia Atactica Polyneuritiformis)

Clinical Characteristics

Refsum disease, or heredopathia atactica polyneuritiformis, was first described by Sigvald Refsum in Norway.[79] The primary clinical features are the triad of retinitis pigmentosa, chronic polyneuropathy, and cerebellar ataxia. Most patients have additional clinical features such as sensorineural hearing loss, anosmia, skin changes, skeletal malformations, and cardiac abnormalities.[80] Clinical manifestations are observed before age 10 years in 43% and before age 20 in 75% of patients. Refsum disease must be distinguished clearly from infantile Refsum disease, which is a disorder of peroxisome biogenesis, presenting in infancy with a wide range of clinical manifestations.[81] Refsum disease is restricted to those individuals with clinical manifestations of heredopathia atactica polyneuritiformis and the biochemical and gene defects described below.

Peripheral neuropathy is a prominent part of the symptomatology, but may be preceded for several years by retinal degeneration. The neuropathy is symmetric and usually affects the distal parts of the lower limbs with muscle atrophy, weakness, and foot drop. The upper extremities are usually also involved. Over the course of the illness, muscle weakness can become widespread and disabling. The trunk musculature is also commonly involved. Sensory disturbances are frequent, usually involving vibration and position sense, but cutaneous hypesthesia of the glove and stocking type may also occur. The peripheral nerves (ulnar, radial, peroneal, and great auricular) may be palpably enlarged and firm. Motor and sensory conduction velocities are markedly reduced. The level of cerebrospinal fluid protein is increased.

Pigmentary retinal degeneration is often the first clinical manifestation of Refsum disease, night blindness often being the only clinical manifestation in children and young adults. Electroretinograms (ERG) demonstrate a reduction or complete absence of rod and cone responses even at that time. Concentric visual field constriction gradually develops, so that only tubular vision remains. It is important to consider Refsum disease in all patients with retinitis pigmentosa. Other ocular abnormalities that may be seen in Refsum disease include cataracts, optic atrophy, and vitreous opacities.

Other clinical manifestations include *hyposmia* or *anosmia*, which is an early clinical sign.[82] *Progressive loss of hearing* is often an early symptom and may eventually become almost complete. *Cerebellar ataxia* was originally considered part of the principal clinical triad of Refsum disease,[79] but reevaluation by Skjeldal et al.[83] concluded that the unsteadiness of gait is attributable mainly to the neuropathy, with clinically detectable signs of cerebellar involvement found in less than a third of individuals. *Cardiac abnormalities* are common, and include cardiac enlargement, heart failure, conduction disturbances, and electrocardiographic changes. Cardiac failure is the most common cause of sudden death in Refsum disease. *Dry skin* and *ichthyosis*, related to phytanic acid accumulation, occur to a variable extent.

Pathology

Pathologically, lipid storage is evident in the leptomeninges, the ependyma, the choroid epithelium, the liver, and the kidneys. The peripheral nerves are enlarged by a hypertrophic neuropathy with onion bulbs and segmental demyelination. Electron microscopy shows the frequent presence of lipid granules in the Schwann cell cytoplasm. The central nervous system may show diffuse myelin destruction around the inferior olivary nuclei and in the brain stem and dentate nucleus.[84] The retina shows an almost complete loss of photoreceptors, thinning of the inner nuclear layer, and reduction in the number of ganglion cells.

Biochemistry

The principal biochemical abnormality is the accumulation of phytanic acid, a branched chain fatty acid, 3,7,11,15-tetramethylhexadecanoic acid, which is toxic to cells and is exclusively of dietary origin.[85–87]

Phytanic acid levels are increased in the plasma.[88] Due to the presence of the α-methyl group, phytanic acid cannot be degraded by the beta-oxidation mechanism used for most fatty acids, and an alpha oxidation reaction that results in the oxidation of phytanic acid to pristanic acid (2,6,10,14-tetramethylpentadecanoic acid) is the first step in phytanic acid degradation.

The enzymatic defect in Refsum disease, phytanoyl-CoA hydroxylase deficiency, has been localized to the peroxisome.[89] Phytanic acid accumulates in several peroxisomal disorders other than Refsum disease, including rhizomelic chondrodysplasia punctata and deficiency of peroxisomal bifunctional enzyme[90] and the disorders of peroxisome biogenesis, infantile Refsum disease, neonatal adrenoleukodystrophy, and Zellweger syndrome.[91] Determination of the ratio of phytanic to pristanic acid and the measurement of other peroxisomal functions aid the distinction between Refsum disease and these other disorders. In Refsum disease, biochemical defects are usually confined to phytanic acid, except for what appears to be an allelic disorder in which pipecolic acid levels are also increased.[92] Plasma phytanic acid levels are elevated, while the levels of pristanic acid, which is below the metabolic block, are low or undetectable.[88]

Molecular Defect

Refsum disease is inherited as an autosomal recessive trait, the gene encoding phytanoyl-CoA hydroxylase, *PHYH*, mapping to 10p.[89,93] A subgroup of affected individuals (about 10% of patients with clinical and biochemical features of Refsum disease), do not have mutations in *PHYH*, but rather in *PEX7*, which encodes the peroxin 7 receptor protein, an important component of the peroxisomal protein import system.[94,95] A disorder with features mimicking Refsum disease, but with normal phytanic and pristanic acid levels, has been reported in a consanguineous Norwegian kindred and linked to chromosome 20.[96]

Management

Dietary therapy is based on the knowledge that phytanic acid, which is derived exclusively from dietary intake, is the toxic agent in this disease.[85] Reduction of dietary phytanic acid intake lowers plasma levels of phytanic acid over some months.[80] This goal can be accomplished by restricting the intake of products from ruminant animals and fish. The contribution of phytanic acid from vegetables is negligible. Brown et al. have measured the phytanic acid content of 151 foods and provide practical guidelines on how to provide a nutritionally adequate and acceptable diet.[97] Dietary therapy improves peripheral nerve function[85] and appears to stabilize central nervous system deficits.[98,99] Early, presymptomatic diagnosis combined with prompt institution of dietary therapy may prevent many disease sequelae. Very high plasma levels of phytanic acid may produce toxic and life-threatening symptoms, including quadriparesis and cardiac dysfunction, necessitating the use of plasmapheresis.[100]

Peripheral Neuropathy in Disorders of Peroxisome Biogenesis

Peripheral neuropathy may also be a presenting symptom in disorders of peroxisome biogenesis (Table 19.2). Baumgartner et al. (case 26 in[101]) described a woman who developed normally until 11 years of age and then developed ataxia and gait difficulties. EMG and nerve conduction studies displayed axonal sensory polyneuropathy. Slight elevation of plasma phytanic acid levels was found and Refsum disease was diagnosed. However, at age 21 years, her condition worsened and she developed dementia and evidence of severe and progressive central nervous system involvement. Subsequent studies revealed a disorder of peroxisome biogenesis.

X-linked Adrenoleukodystrophy

X-linked adrenoleukodystrophy (X-ALD) is a peroxisomal disorder that mainly affects the central nervous system white matter and axons, the adrenal cortex, and the testis,[102] with an incidence of 1:17,000.[103,104]

Phenotypic expression varies widely, ranging from the rapidly progressive childhood cerebral form that often leads to total disability and death by 10 years of age, to the somewhat milder adult form that presents in young adulthood with a paraparesis that progresses over decades and is referred to as adrenomyeloneuropathy (AMN). Some males with the same gene defect have adrenocortical insufficiency (Addison disease) without apparent nervous system involvement. At least 70% of males with X-ALD have adrenocortical insufficiency. Approximately 50% of female heterozygotes develop a milder AMN-like syndrome in middle age or later. This form is often unrecognized.[105]

The principal biochemical abnormality is the accumulation of saturated very long chain fatty acids (VLCFA) in tissues and plasma. Measurement of plasma VLCFA is the most common diagnostic assay.[106] The defective gene maps to Xq28, encoding a peroxisomal membrane protein[107] referred to as ALDP or ABCD 1 More than 690 nonrecurrent mutations have been identified and are accessible at the X-ALD database website (www.x-ald.nl).

The childhood cerebral and the AMN phenotypes co-occur, often within the same family. The rapidly progressive cerebral phenotypes are associated with an inflammatory demyelinating response that may be mediated by cytokines and autoimmune mechanisms.[108,109] AMN, in contrast, presents as a noninflammatory distal axonopathy that involves the spinal cord most severely, with the gracile tracts involved most in the cervical cord, and the corticospinal tracts in the lumbar region.[110] Some degree of peripheral nerve involvement occurs in AMN and in heterozygous women. It is generally not a feature in childhood X-ALD.

Clinical Characteristics of AMN

The typical AMN patient presents with stiffness or clumsiness in the legs in the third or fourth decade, with inexorable progression over the next several decades to severe spastic paraparesis. Voiding abnormalities are seen early in the disease, as are impaired vibration and position sense with abnormal somatosensory and brainstem auditory evoked responses.

Several large studies of peripheral nerve function in AMN have been reported, describing abnormalities more commonly in the legs than in the arms.[111,112] Almost 60% of X-ALD heterozygotes develop peripheral neuropathy.[113] Abnormalities represent either a mixture of axonal loss and multifocal demyelination or primary axonopathy; 26% of individuals partially fulfilled the neurophysiological criteria of primary demyelination. Restuccia et al.[114] demonstrated abnormalities of motor and sensory evoked responses in 12 of 19 heterozygotes. Approximately 30% of AMN patients develop cerebral involvement. When this occurs, the disorder may progress rapidly.

Pathology

The most severe pathological changes in AMN affect the spinal cord, are bilateral, and are usually symmetrical. Ascending tract (gracile and dorsal spinocerebellar) degeneration is worse in the upper cord (further from the cell of origin), and descending tract degeneration (pyramidal) is worse in the lower cord (furthest from the cell of origin). Axonal loss is equal to or greater than myelin loss, and controversy exists about the degree of peripheral nerve involvement as peripheral nerve lesions are variable and mild compared with the myelopathy. Spinal nerve roots and ganglia, sciatic, popliteal, and ulnar nerves have been reported as unremarkable, whereas sural and peroneal nerves reveal loss of large and small diameter myelinated fibers.

Management

It is essential that adrenocortical function be monitored in all male X-ALD patients and treated by replacement therapy in order to prevent death or morbidity from adrenal crisis.[115] Boys and adolescents in the early stages of cerebral involvement can benefit from bone marrow transplantation, which may stabilize cerebral lesion long-term,[116] but which is a high-risk procedure.[117–120] The favorable effect of BMT in these patients appears to be on the brain inflammatory response. It is not known whether BMT has a favorable effect on the noninflammatory myelopathy in AMN. Cyclophosphamide therapy is ineffective.[121] Dietary therapy with a 4:1 mixture of glyceryl trioleate and glyceryl trierucate can normalize the plasma VLCFA levels in patients with X-ALD, but favorable clinical effects have not been demonstrated.[122]

Hyperoxaluria

Hyperoxaluria may be primary, due to an inherited metabolic disease, or secondary to environmental factors including diet, drugs,[123] or coincident renal disease.[124]

Hyperoxaluria resulting from an inherited metabolic defect is rather rare, arising from deficiency of the liver peroxisomal enzyme, alanine-glyoxylate aminotransferase (AGT), or the cytosolic enzyme, D-glycerate dehydrogenase/glyoxylate reductase (DGDH/GR). As such, these disorders are classified as primary hyperoxaluria type 1 (PH1) due to AGT deficiency,[125] type 2 (PH2) due to DGDH/GR deficiency,[126] and type 3 (PH3) associated with defects in the liver-specific mitochondrial enzyme 4-hydroxy-2-oxoglutarate aldolase (HOGA).[127] Only PH1 has systemic features, including peripheral neuropathy, due to oxalate storage in multiple tissues, whereas PH2 is restricted to renal involvement, and PH3 has the mildest phenotype.[128]

PH1 has its onset before age 5 in most instances, presenting with hematuria or renal colic secondary to urolithiasis.[129] Ultimately, oxalate (calcium oxalate) accumulation leads to chronic renal disease. Oxalate accumulation is widespread including bone, eye, blood vessels, brain, and peripheral nerves.

Peripheral nerve involvement includes evidence of both axonal degeneration and segmental demyelination.[130–133] Although painful paresthesias and muscle weakness have been described, clinical manifestations of peripheral neuropathy tend to be minor.

As effective therapy for PH1 is now feasible, appropriate diagnostic methodologies should be employed to distinguish PH1 from PH2 and PH3, as well as from secondary hyperoxaluria.[134] AGT is a liver-specific peroxisomal enzyme. In the past, definitive diagnosis required liver biopsy. The distinction from PH2 was made by measuring DGDH activity in leucocytes.[135] Now, direct molecular analysis of the genes responsible for PH permits rapid, noninvasive diagnosis.[128]

Renal transplantation alone does not prevent ongoing oxalate accumulation, as the deficient enzyme, AGT, is present only in liver peroxisomes. In order to reverse systemic complications, and thus provide effective therapy for PH1, combined kidney-liver transplantation is required.[129,136] Vitamin B6 therapy is effective in lowering oxalate levels in patients with specific mutations causing PH1.[137]

DISEASES OF LIPIDS AND LIPOPROTEINS

The disorders of lipids and lipoproteins associated with peripheral neuropathy in childhood include cholestanol lipidosis (cerebrotendinous xanthomatosis), Tangier disease, and abetalipoproteinemia (Table 19.2). As effective treatment is possible for cholestanol lipidosis and for abetalipoproteinemia, early diagnosis is important.

Cholestanol Lipidosis (Cerebrotendinous Xanthomatosis)

Cholestanol lipidosis (cerebrotendinous xanthomatosis, CTX), a rather rare autosomal recessive disorder, results from defective cholesterol 27-hydroxylase activity.[138]

As a consequence, cholesterol is converted via a side-reaction to cholestanol (dihydrocholesterol). Clinical manifestations begin in late childhood or adolescence and include cataracts, xanthomata over extensor tendons, especially the Achilles tendon, and slowly progressive spasticity, cerebellar ataxia, and dementia. These xanthomata, which contain cholesterol and cholestanol, develop over tendons and within brain white matter, particularly in the cerebellum. Atherosclerotic changes are prominent throughout the cardiovascular system. Psychiatric disturbance (hallucinations, delusions, or other manifestations of schizophrenia) may precede other neurologic features.[139,140] Survival to age 40–60 is usual. Death is usually secondary to CNS sequelae or myocardial infarction. Cranial CT and MRI reveal diffuse low-density lesions of white matter as well as cerebral and cerebellar atrophy.[141]

Although peripheral nerve pathology is common, clinical signs are mild.[142] Nerve conduction velocities are reduced. Sural nerve biopsy demonstrates segmental demyelination and remyelination. In addition to demyelination, peripheral nerve pathology reveals axonal degeneration. An Italian study of 35 CTX patients found that 26 had peripheral neuropathy; there was no correlation between peripheral nerve involvement or other clinical or biochemical data.[143]

Serum cholesterol is normal, but cholestanol levels in serum, CSF, and red cells are increased.[144] Cholesterol 27-hydroxylase activity is deficient in cultured skin fibroblasts,[145] although measurement of bile acid intermediates may be an easier screening assessment.[146] Several mutations have been identified in the gene for CTX.[147]

Management

Treatment consists of dietary chenodeoxycholic acid (CDCA) or cholic acid, to reduce plasma cholestanol levels via feedback inhibition of the rate-limiting reaction of cholesterol synthesis, hydroxymethylglutaryl Co-A (HMG-CoA) reductase.[148] Cholic acid appears to produce fewer side effects and is less expensive. This dietary modification stabilizes or improves neurological involvement, normalizing the EEG and correcting peripheral neuropathic findings (NCVs) as well as other evoked response abnormalities (ABR and VER). The Italian study referenced above confirmed that CDCA treatment improved conduction velocities, regardless of duration of treatment, but that amplitudes of motor and sensory action potentials were unchanged.[143] HMG-CoA reductase inhibitors (e.g. lovastatin) will also reduce cholestanol, but long-term benefit is unclear.[149,150]

Tangier Disease

Tangier disease is an autosomal recessive disorder associated with very deficient plasma high-density lipoprotein (HDL); prominent cholesterol ester storage in liver, spleen, bone marrow lymph nodes, and tonsils; and a significant, predominantly sensory polyneuropathy.[151]

About 100 patients with Tangier disease have been described.[152] Inheritance of the clinical phenotype is autosomal recessive, but the biochemical phenotype is autosomal co-dominant.[152]

A progressive polyneuropathy is usually the presenting sign of Tangier disease in adults; younger children usually present with enlarged orange tonsils. The neuropathy has its onset in childhood or adolescence.[153,154] However, diagnosis is often delayed or confused due to asymmetric involvement or a syringomyelia-like syndrome with a disassociated sensory loss primarily affecting small myelinated or unmyelinated sensory fibers that subserve pain and temperature modalities.[155] A waxing and waning course suggestive of mononeuritis multiplex is also described. Clinical signs of neuropathy are mainly sensory, ranging from pain and paresthesias to diminished pain and temperature sensation or analgesia. Slowing of nerve conduction velocities is not invariable. Sural nerve biopsy typically reveals Schwann cell vacuolation (cholesterol and cholesterol esters) in unmyelinated fibers, and severe axonal loss in advanced disease.[156]

Other clinical features include hepatosplenomegaly, thrombocytopenia, and corneal opacification. The tonsils are typically orange in color, reflecting profound storage of cholesterol esters. Coronary artery disease is a later feature but may be quite significant. Otherwise, survival can be quite long.

Diagnosis is suspected in patients with low or absent plasma HDL, low total plasma cholesterol, 1% to 2% of normal plasma apolipoprotein AI levels,[157] and 5% to 10% of normal plasma apolipoprotein AII levels.[153,158] Tangier disease is caused by mutations in the ATP-Binding Cassette transporter A1 (*ABCA1*) gene, which encodes the membrane transporter ABCA1. This transporter functions in the initial phase of reverse cholesterol transport, whereby free cholesterol effluxing from peripheral cells is transferred to lipid-poor apoA-I.[152] There is no known disease modifying therapy; raising HDL cholesterol levels has not proven clinically effective.

Abetalipoproteinemia (Bassen-Kornzweig Disease)

A family of autosomal recessive disorders, characterized by very low levels of apolipoprotein B (apoB) and LDL cholesterol, have been described in association with mutations in the apoB, proprotein convertase subtilisin/kexin type 9 (*PCSK9*), and *MTP* genes, causing either abetalipoproteinemia or hypobetalipoproteinemia. A combined form of hypolipoproteinemia is caused by mutations in the *ANGPTL3* gene.[159,160]

Abetalipoproteinemia (ABL; Bassen-Kornzweig disease) is the prototype of these diseases. Its incidence at birth is estimated at 1:1,000,000.[161] ABL is a relatively benign condition if it is detected early and treated promptly. Abetaliproteinemia is characterized by absence of plasma apolipoprotein B and low levels of plasma cholesterol, triglycerides, and fat-soluble vitamins, particularly vitamin E, as well as absence of plasma VLDL, LDL, and chylomicrons.[162]

Clinical involvement is apparent from birth due to poor feeding and weight gain and prominent gastrointestinal problems (nausea, vomiting, abdominal distension, and fatty stools) secondary to fat malabsorption.[163] The presence of acanthocytes (spiked red cells), detectable on blood smears from the newborn period, may herald the diagnosis.

During childhood, a sensory neuropathy may develop. This involves all sensory modalities and includes reduced or absent muscle stretch reflexes. In addition, more widespread neurologic signs emerge, which reflect involvement of spinal cord posterior columns and spinocerebellar pathways. A degenerative retinopathy (retinitis pigmentosa) initially causes night blindness, which can progress to complete loss of vision.[164,165]

Motor CVs are largely preserved whereas sensory CVs are quite prolonged.[166] Peripheral nerve histology demonstrates both axonal degeneration and demyelination, although the primary pathology appears to be axonal loss.[167]

Vitamin E deficiency is the likely pathogenetic mechanism in abetalipoproteinemia. Perhaps the most compelling evidence supporting this mechanism is the virtual prevention of neurologic signs by vitamin E (100 mg/kg/d) replacement.[168,169] Vitamin A (20–30 IU/kg/daily or every other day) replacement obviates the retinopathy, but caution must be taken to avoid vitamin A toxicity. Recent data suggests that even early vitamin E supplementation may not prevent retinopathy; guidelines for management have been published.[170] Dietary fat intake should also be reduced, although the abnormalities of serum lipids cannot be completely corrected. Death before age 40 may occur.

Abetaliproteinemia should be distinguished from neuroacanthocytosis (chorea-acanthocytosis; Levine-Critchley syndrome), in which plasma lipoprotein levels are normal despite the presence of acanthocytes.[171–173] The formation of acanthocytes in neuroacanthocytosis likely results from abnormal Lyn activation, which leads to impaired interactions between the red cell membrane and the cytoskeleton.[174] Neurologic involvement occurs during adulthood and consists mainly of basal ganglia dysfunction (chorea, tics, dyskinesia, dystonia, and rigidity). Peripheral nerve involvement, which is uncommon, consists of axonal degeneration with normal nerve conduction velocities; the condition may mimic motor neuron disease.[175] Neuroacanthocytosis is caused by loss of function mutations in the vacuolar protein sorting 13 homolog A (*VPS13A*) gene, which encodes chorein. This protein interacts with beta-adducin and beta-actin, which localize to the major sites of neuropathology in this disorder.[176] No treatment is available.

GLYCOGEN STORAGE DISEASE

Although involvement of peripheral nerve has been described in type II glycogen storage disease (Pompe disease), clinical expression is not noted. However, an adult variant of type IV glycogen storage disease (Anderson disease) does include peripheral neuropathy as a prominent clinical component. While this is principally an adult-onset condition, similar neuropathic changes may be noted in the more typical childhood form of this disease.[177]

Adult Polyglucosan Disease

Adult polyglucosan disease, which is a variant of type IV glycogen storage disease (Anderson disease; brancher deficiency; amylo-1,4-1,6-transglucosidase deficiency) is characterized by slowly progressive involvement of both the CNS and PNS. Onset usually occurs after age 40.

Clinical Characteristics

An international study of 50 patients found that 100% of patients had neurogenic bladder, 90% spastic paraparesis, and 90% axonal neuropathy.[178] The mean ages of onset for these findings were 51, 63, and 70 years respectively. Onset in infancy with delayed walking, however, has been described.[179] Seizures, ataxia, an ALS-like picture, and movement disorder occur much less commonly. Polyglucosan bodies are noted throughout the nervous system. Cranial MRI reveals white matter hyperintensity on T2 and fluid-attenuated inversion recovery sequences predominantly in the periventricular regions, the posterior limb of the internal capsule, the external capsule, and the pyramidal tracts and medial lemnisci of the pons and medulla. All subjects showed atrophy of the medulla and spine.[178] MCVs are reduced. No specific therapy is available.

Only one gene, *GBE1*, has been associated with this disease. The *GBE1* c.1076A > C mutation is frequent in the Ashkenazi Jewish population, with a carrier frequency estimated at 1:34.5[180]

About one-third of patients with decreased debrancher enzyme activity harbor only one *GBE1* mutation. It is likely that such individuals have large gene rearrangements, mutations in noncoding regions, or mutations in so far unrecognized genes.[179]

The more typical form of type IV glycogen storage disease has its onset within the first 3 to 6 months of life as failure to thrive, hepatosplenomegaly, and liver failure with cirrhosis. Psychomotor delay, hypotonia, weakness, and muscle atrophy accompanied by absent or decreased muscle stretch reflexes are prominent.[181] Clinical evidence of peripheral neuropathy is lacking secondary to the relatively rapid course and cognitive delay. A severe cardiomyopathy has been described.[182] Type IV is a multisystem disease with death by age 5, secondary to progressive cirrhosis and portal hypertension. Further, a congenital form has significant involvement of both muscle and Schwann cells.[183]

A more benign form presents at 2 years of age as hepatomegaly and elevated liver enzyme activity. No neurologic abnormalities are noted, and liver disease is not progressive.[184,185]

Pathology

Inclusion (polyglucosan) bodies are found in muscle, heart, liver, and brain in both adult polyglucosan disease and the typical childhood form. Polyglucosan bodies are prominent in subcortical cerebral white matter, as well as in a perivascular distribution in gray matter, cerebellum, spinal cord, and peripheral nerve (Table 19.2).

Biochemistry

Brancher enzyme deficiency, encoded by the *GBE1* gene, located on chromosome 3p12, results in the synthesis of unbranched, elongated glycogen chains resembling the structure of starch or amylopectin. Brancher enzyme deficiency can be detected in leukocytes, erythrocytes, cultured skin fibroblasts, and cultured amniotic cells or chorionic villi.[186]

AMINO ACID DISEASE

Tyrosinemia

Tyrosinemia type I, or hepatorenal tyrosinemia, is a rare, autosomal recessive disease often presenting in early infancy as liver dysfunction or failure, in which case rapid progression to death may occur.[187]

Clinical features include failure to thrive, vomiting, diarrhea, and hepatosplenomegaly. Liver failure is associated with generalized edema, GI bleeding, and ascites whereas liver dysfunction is accompanied by hyperbilirubinemia and hypoproteinemia. Cirrhosis and hepatocellular carcinoma are common, with the latter occurring as early as age 1 year. Renal tubular dysfunction is also noted in association with generalized aminoaciduria, glucosuria, proteinuria, and excessive excretion of tyrosine. Vitamin D-resistant renal rickets is a constant feature. The precise mechanism of hepatorenal toxicity is unknown at present. However, in addition to tyrosine, succinylacetone, a product of tyrosine degradation, is also excreted in urine. Succinylacetone may be the responsible component leading to hepatorenal toxicity. Excess tyrosine and succinylacetone excretion result from deficiency of the enzyme that is responsible for tyrosinemia type I. This enzyme, fumarylacetoacetate hydrolase (FAH), has been mapped to chromosome 15.

The principal neurologic manifestation of tyrosinemia type I is painful peripheral neuropathic crises resembling those of acute intermittent porphyria, as described below[188] (Table 19.2). These crises may occur in early childhood and can be recurrent.[189] As with hepatorenal toxicity, succinylacetone appears to be the offending agent. Succinylacetone inhibits δ-aminolevulinic acid hydratase, leading to accumulation of δ-aminolevulinic acid (ALA). ALA is the likely neurotoxin in acute intermittent porphyria. It has been shown to block peripheral myelin formation in tissue culture.[190] Psychomotor impairments are variable and likely result from acute hepatic encephalopathy.

Management

Treatment consists of dietary restriction of phenylalanine and tyrosine. However, dietary modification alone does not appear to prevent the development of cirrhosis, reduce the risk of hepatocellular carcinoma, or obviate neuropathic crises. Since 1991, nitisinone [NTBC, 2-(2-nitro-4-trifluoromethylbenzoyl)-1,3-cyclohexane-dione)], a potent inhibitor of 4-hydroxyphenylpyruvate dioxygenase (pHPPD), a key enzyme in tyrosine catabolism, has proven very effective in protecting children against liver disease including hepatocellular carcinoma. It also decreases or prevents neuropathic crises and renal tubular dysfunction.[191] Inhibition of tyrosine degradation blocks production of the offending metabolite, succinylacetone. Ninety percent of treated children respond to treatment. Serum α-fetoprotein levels, which are markedly elevated without treatment, are substantially reduced with nitisinone, providing a useful marker of therapeutic efficacy and the presence of hepatocellular cancer. Nitisinone itself may produce corneal lesions. Thus, careful follow-up is required. Liver transplantation may be required in nonresponders or those treated by dietary restriction alone. The management of tyrosinemia type I has recently been reviewed.[192]

BOX 19.1 Diseases of Fatty Acid β-oxidation

- Acyl-CoA Dehydrogenase Deficiency Diseases
  - Very Long-Chain Acyl-CoA Dehydrogenase (VLCAD) Deficiency
  - (Long-Chain Acyl-CoA Dehydrogenase (LCAD) Deficiency)
  - Medium-Chain Acyl-CoA Dehydrogenase (MCAD) Deficiency
  - Short-Chain Acyl-CoA Dehydrogenase (SCAD) Deficiency
  - Electron Transfer Flavoprotein (ETF) Deficiency
  - ETF-Coenzyme Q Oxidoreductase (ETF-QO) Deficiency
- Hydroxyacyl-CoA Dehydrogenase Deficiency Diseases
  - Long-Chain 3-Hydroxyacyl-CoA Dehydrogenase (LCHAD) Deficiency
  - Short-Chain 3-Hydroxyacyl-CoA Dehydrogenase (SCHAD) Deficiency

MITOCHONDRIAL DISEASES

Mitochondrial diseases associated with peripheral neuropathy in childhood include disorders of fatty acid β-oxidation (Box 19.1), the pyruvate dehydrogenase complex, and the electron transport chain. The specific disorders are mitochondrial trifunctional protein and long chain hydroxyacyl-CoA dehydrogenase deficiencies, Leigh disease, and neuropathy (Table 19.2), ataxia, and retinitis pigmentosa (NARP).

Mitochondrial Trifunctional Protein and Long-Chain 3-Hydroxy-Acyl-CoA Dehydrogenase Deficiencies

3-Hydroxy-acyl-CoA dehydrogenase is part of the mitochondrial trifunctional protein (MTP) complex, the third enzyme in very long chain fatty acid β-oxidation.[193] The other components are 2-enoyl-CoA hydratase and 3-oxoacyl-CoA thiolase. MTP and long-chain 3-hydroxyacyl-CoA dehydrogenase (LCHAD) deficiencies occur in early infancy[194,195] and are characterized by lethargy and coma, hypoglycemia, hypoketonemia, and hyperammonemia with acidosis. Hypotonia and hepatomegaly with lipid storage are noted. Pigmentary retinopathy, cardiomyopathy, and progressive sensorimotor neuropathy are typical.[196] Peripheral nerve histology reveals both axonal loss and demyelination with predominant changes in sensory fibers (Table 19.2).

3-Hydroxydicarboxylic aciduria is usually present. Diagnosis can be confirmed by measuring LCHAD in cultured skin fibroblasts. It is possible that some children with LCHAD deficiency actually have MTP deficiency.

TABLE 19.5 Leigh Disease: Metabolic Defects

Enzyme	Mutation Site	Mutation
Pyruvate dehydrogenase complex (PDHC)	Nuclear DNA	Chromosome 7
PDHC E1α subunit	Nuclear DNA	Chromosome X
Succinate dehydrogenase	Nuclear DNA	
NADH: Ubiquinone oxidoreductase (Complex I)	Nuclear DNA	NDUFS8
Cytochrome C Oxidase (Complex IV)	Nuclear DNA	COX – chromosome 9q
Mitochondrial ATP synthase (Complex V)	Mitochondrial DNA	Nucleotide 8993
Transfer RNAs	Mitochondrial DNA	Multiple

Management

Treatment requires frequent feeding, avoidance of fasting, and alteration of fat intake by substitution of MCT for long-chain fatty acids. Docosahexaenoic acid replacement appears to improve the sensorimotor neuropathy.[197] Death in infancy was common before the institution of newborn screening,[195] which has reduced morbidity and mortality from this disorder.[198]

Leigh Disease (Subacute Necrotizing Encephalopathy)

Leigh disease typically has its onset in early infancy or childhood, and rarely in adulthood.[199] It features multisystem neurologic manifestations that may be diagnostically challenging. These may include lethargy or coma, swallowing and feeding difficulty, and hypotonia. Some manifestations are more system-specific, including signs of ataxia and intention tremor, involuntary movements, peripheral neuropathy, external ophthalmoplegia, loss of vision, impaired hearing, vascular-type headaches, and seizures.[200] A retrospective study of 130 patients from eight European centers found that 80% of patients presented by 2 years of age, most with motor (hypotonia, abnormal reflexes, and ataxia), and/or ocular findings.[201]

Enzyme deficiencies (Table 19.5) may involve mutations of nuclear or mitochondrial DNA (mtDNA). Most affected children have involvement of either the pyruvate dehydrogenase complex (PDHC), specifically the E1α subunit,[202,203] or the electron transport complexes,[200,204–207] particularly cytochrome c oxidase (COX).[208–210] As such, the inheritance pattern of Leigh disease is heterogeneous, involving both Mendelian and mitochondrial mechanisms. As the E1α subunit is encoded on the X-chromosome, both X-linked and autosomal recessive Mendelian inheritance occur with Leigh disease. PDHC deficiency is usually evident in early infancy, whereas COX deficiencies are usually associated with later infantile onset.[211] In many instances, a specific enzymatic defect is not identified.

Clinical Characteristics

Leigh disease usually presents during the preschool years. Its variable clinical features may include impaired motor and cognitive development, feeding and swallowing difficulties, vomiting, weakness, external ophthalmoplegia, optic atrophy and visual loss, seizures, ataxia, dystonia, and peripheral neuropathy. Brainstem dysfunction is virtually always present. In infancy, the clinical picture is dominated by lethargy or coma, feeding and swallowing difficulties, and hypotonia. Lactic acidosis may be significant. Death occurs within 6 to 24 months of onset.

When children initially become symptomatic during the preschool period, from 2 to 5 years, growth failure, external ophthalmoplegia, movement disorders (dystonia, choreoathetosis, or myoclonus),[212] ataxia and intention tremor, and vascular-type headaches are more prominent. Lactic acidosis may or may not be seen. Later onset is associated with survival of 5 to 10 years. The dominance of this multisystem picture may obscure the presence of peripheral neuropathic signs including hypotonia, weakness, and reduced or absent muscle stretch reflexes. Nerve conduction velocities are often significantly reduced to the demyelinating range.

Brainstem dysfunction, including cranial nerve dysfunction and respiratory irregularities, is typically present in both age groups. Cardiac instability and hypertension reflect autonomic system involvement.[213] Acute clinical events may be triggered by acute infections or other illness. Recurrent exacerbations may result in a step-wise decline, because recovery from each exacerbation may be less than complete. Cranial MRI is superior to cranial CT in depicting specific abnormalities. Typical findings are found most frequently in the thalamus, basal ganglia and brainstem. Involvement of the putamen is virtually uniform. The ABR is typically abnormal.[214]

Pathology

Brain pathology is similar to that of Wernicke encephalopathy, with the exception that the mammillary bodies are generally spared in Leigh disease. On the other hand, the substantia nigra, which is usually unaffected in Wernicke disease, is frequently affected in Leigh disease. Striking neuronal necrosis is seen in deep gray matter, especially putamen, globus pallidus, the periaqueductal gray, and dentate nuclei of the cerebellum. Necrotizing changes are also located symmetrically throughout the entire brainstem, including the tectum and tegmentum of the midbrain, pons, and rostral medulla. Similar lesions are often present in the hypothalamus, thalamus, and caudate. The most devastated areas are the globus pallidus, red nucleus, substantia nigra, and the dorsal tegmentum of the midbrain and pons. Peripheral nerve pathology, when present, consists of segmental demyelination (Table 19.2).

Biochemistry

Leigh disease is caused by several abnormalities in energy metabolism including PDHC, the electron transport chain,[206] and other enzymes related to oxidative phosphorylation. Lactic acidosis is frequently detectable during acute events, although an elevated lactate may only be found in CSF.[215,216] Cytochrome c oxidase deficiency has been identified in brain, skeletal muscle, kidney, cardiac muscle, liver, and cultured skin fibroblasts

Leigh disease is caused by a mutation in the mitochondrial DNA (mtDNA) gene, encoding ATP synthase subunit 6 (complex V) at nucleotide 8993. The clinical presentation relates to the percentage of mutant DNA in the affected child. Early onset Leigh disease occurs when this is at high levels (>95%).[217] Where the percentage of mutant mtDNA is <75%, the clinical phenotype is that of NARP or neuropathy, ataxia, and retinitis pigmentosa syndrome (see below). Intermediate levels may produce a later onset Leigh disease.[218]

Management

A variety of vitamin and cofactor replacement strategies have been trialed but effective therapy is lacking. Some benefit may be seen from a "mitochondrial cocktail" in individual patients.[219] A small open-label study of EPI-743, a para-benzoquinone targeting repletion of reduced intracellular glutathione, showed promising results in reversing disease progression in children with Leigh disease.[220] mTOR inhibition alleviated markers of mitochondrial disease in a murine model, raising the possibility of using these agents in human disease.[221,222] Supportive care, including home-ventilator programs, may be quite effective.

Neuropathy, Ataxia, and Retinitis Pigmentosa

As described above, a clinical continuum between Leigh disease and neuropathy, ataxia, and retinitis pigmentosa (NARP) is based on the degree of mutant mtDNA at nucleotide 8993. NARP usually appears in adolescence or young adults, but may be seen in early childhood. NARP is characterized by sensory demyelinating neuropathy (Table 19.2), ataxia, and retinitis pigmentosa. Proximal weakness and cognitive decline may also occur. The findings may evolve over many decades.[223] The phenotype is quite variable, reflecting the extent of mutant mtDNA. Ragged red fibers (RRFs) are not seen. Peripheral nerve histology typically reveals segmental demyelination.

Biochemistry

Serum lactate is normal. Most cases of NARP are due to a point mutation at nucleotide 8993 (either T > C or T > G) in the mtDNA gene encoding subunit 6 of ATP synthase (complex V), but other mutations have been reported.[224,225] The degree of clinical involvement reflects variable percentages of mutant mtDNA.[226] When the percentage of mutant mtDNA is <75%, the NARP phenotype is seen.[205,227,228] When mutant mtDNA exceeds 95%, the clinical pattern is that of Leigh disease. Levels of mutant mtDNA between 75% and 95% result in an intermediate clinical picture that may include vascular-type headache, cognitive impairment, and retinitis pigmentosa. Recent studies have found that mutation load alone is not absolutely predictive of outcome, since other variations in the mitochondrial DNA and environmental triggers also play a role in the expression of this phenotype.[229]

OTHER INHERITED METABOLIC DISEASES

Acute Intermittent Porphyria

Acute intermittent porphyria (AIP), unlike the other disorders in this chapter, is inherited in an autosomal dominant fashion.[230] AIP is characterized by intermittent, recurrent episodes of pain that is mainly visceral, but may include peripheral neuropathic signs in the form of pain and subsequent weakness.[231–234] The pain consists mainly of paresthesias, dysesthesias, and hypesthesias. Weakness is typically proximal, and greater in the upper than lower extremities. Occasionally, weakness may involve facial muscles and those involved in breathing, chewing, and swallowing.

Peripheral neuropathic changes in AIP include segmental demyelination and axonal degeneration, but are typically seen only in individuals with recurrent clinical events.[235,236] Bolton (personal communication, 1992) performed nerve conduction studies (NCS) on a child with porphyria and noted a pure axonal motor neuropathy. The results were similar to the NCS and EMG findings reported by Albers et al., who demonstrated a similar neuropathy in eight patients with quadriparesis caused by acute porphyria.[237]

Autonomic neuropathy is common, featuring hypertension[238] and postural hypotension,[239] as well as tachycardia, and gastrointestinal and bladder dysfunction. These typically occur only during exacerbations.

Colicky pain, sometimes occurring as a severe abdominal crisis, is associated with nausea, vomiting, and constipation. Despite this clinical appearance of an acute surgical abdomen, palpation of the belly does not elicit any pain or tenderness. In addition, central nervous system involvement sometimes occurs and is characterized by severe anxiety, insomnia, hallucinations, and aggressive or violent behavior. Seizures are also occasionally witnessed.

AIP is due to deficiency of porphobilinogen (PBG) deaminase (also known as hydroxymethylbilane synthase, HMBS), a key enzyme in porphyrin synthesis. The acute attacks that typify AIP may be precipitated by a number of exogenous factors. These include a number of drugs or medications such as alcohol, sulfonamides, and chloroquine, and various antiepileptic agents such as barbiturates, phenytoin, carbamazepine, valproic acid, and the succinimides.[240] Infection, inadequate nutrition, and sex hormones may also trigger attacks.

AIP has its onset after puberty in most instances, perhaps explaining the important role of sex hormones. Nevertheless, onset in childhood is well documented.[241,242] PBG deaminase deficiency is relatively common in the general population. The prevalence in Sweden, for example, is about 1:10,000. However, 10% or fewer of individuals with PBG deaminase deficiency are symptomatic. This often occurs with exposure to one of the above noted potential precipitating factors. These medications or alcohol either induce δ-aminolevulinic acid (ALA) synthetase or inhibit PBG deaminase. As a result, ALA and PBG excretion in urine increases dramatically during these attacks. Asymptomatic individuals also excrete excess amounts of ALA and PBG.

Diagnosis is established by measuring PBG deaminase activity in erythrocytes,[243] leukocytes,[244] or cultured skin fibroblasts.[245] However, RBC PBG deaminase activity may be normal in 10% to 15% of cases, and definitive diagnosis rests on demonstration of a mutation in the PBGD gene.[230]

Management

Treatment is based largely on nutritional advice and prevention of exposure to the exogenous agents known to provoke acute attacks. The treatment of acute attacks includes high carbohydrate oral intake and, if involvement is severe, parenteral dextrose. Hematin or heme arginate can suppress ALA and PBG excretion; delivery of heme in this fashion is the most effective means of suppressing ALA and porphobilinogen excretion. This treatment is sometimes associated with phlebitis or thrombophlebitis.[230]

REFERENCES

1. Salman MS, Clarke JT, Midroni G, Waxman MB. Peripheral and autonomic nervous system involvement in chronic GM2-gangliosidosis. *J Inherit Metab Dis* 2001;**24**:65–71.
2. Krabbe K. A new infantile form of diffuse brain sclerosis. *Brain* 1916;**39**:74–114.
3. Kolodny E. Globoid leukodystrophies. In: Moser HW, editor. *Neurodystrophies and Neurolipidoses: Handbook of Clinical Neurology*. Amsterdam: Elsevier; 1991. pp. 187–210.
4. Ohno K. Kunihiko Suzuki and sphingolipidoses. *J Biochem* 2011;**150**:597–605.
5. Wenger DA, Escolar ML, Luzi P, Rafi MA. Krabbe disease (globoid cell leukodystrophy). In: Valle D, Beaudet AL, Vogelstein B, Kinzler KW, Antonarakis SE, Ballabio A, et al., editors. *OMMBID: The Online Metabolic and Molecular Bases of Inherited Diseases*. New York: McGraw HIll; 2013 [cited 29 May 2014]; Available from: http://ommbid.mhmedical.com/content.aspx?bookid=474&Sectionid=45374150
6. Duffner PK, Barczykowski A, Jalal K, Yan L, Kay DM, Carter RL. Early infantile Krabbe disease: results of the World-Wide Krabbe Registry. *Pediatr Neurol* 2011;**45**:141–8.
7. Dunn HG, Lake BD, Dolman CL, Wilson J. The neuropathy of Krabbe's infantile cerebral sclerosis (globoid cell leucodystrophy). *Brain* 1969;**92**:329–44.
8. Lieberman JS, Oshtory M, Taylor RG, Dreyfus PM. Perinatal neuropathy as an early manifestation of Krabbe's disease. *Arch Neurol* 1980;**37**:446–7.
9. Malandrini A, D'Eramo C, Palmeri S, Gaudiano C, Gambelli S, Sicurelli F, et al. Peripheral neuropathy in late-onset Krabbe disease: report of three cases. *Neurol Sci* 2013;**34**:79–83.
10. Debs R, Froissart R, Aubourg P, Papeix C, Douillard C, Degos B, et al. Krabbe disease in adults: phenotypic and genotypic update from a series of 11 cases and a review. *J Inherit Metab Dis* 2013;**36**:859–68.
11. Austin JH, Lehfeldt D. Studies in Globoid (Krabbe) Leukodystrophy. 3. Significance of experimentally-produced globoid-like elements in rat white matter and spleen. *J Neuropathol Exp Neurol* 1965;**24**:265–89.
12. Bischoff A, Ulrich J. Peripheral neuropathy in globoid cell leukodystrophy (Krabbe's disease). Ultrastructural and histochemical findings. *Brain* 1969;**92**:861–70.
13. Suzuki K, Suzuki Y. Galactosylceramide lipidosis: globoid-cell leukodystrophy (Krabbe disease). In: Scriver CR, Sly WS, Childs B, Beaudet AL, Valle D, Kinzler KW, et al., editors. *The Metabolic Bases of Inherited Disease*. 8th ed. New York: McGraw Hill; 1989. pp. 1699–720.
14. Suzuki K. Twenty five years of the "psychosine hypothesis": a personal perspective of its history and present status. *Neurochem Res* 1998;**23**:251–9.
15. Castelvetri LC, Givogri MI, Zhu H, Smith B, Lopez-Rosas A, Qiu X, et al. Axonopathy is a compounding factor in the pathogenesis of Krabbe disease. *Acta Neuropathol* 2011;**122**:35–48.
16. Wenger DA, Luzi P, Rafi MA. Krabbe disease: are certain mutations disease-causing only when specific polymorphisms are present or when inherited in trans with specific second mutations? *Mol Genet Metab* 2014;**111**:307–8.
17. Escolar ML, Poe MD, Provenzale JM, Richards KC, Allison J, Wood S, et al. Transplantation of umbilical-cord blood in babies with infantile Krabbe's disease. *N Engl J Med* 2005;**352**:2069–81.
18. Siddiqi ZA, Sanders DB, Massey JM. Peripheral neuropathy in Krabbe disease: effect of hematopoietic stem cell transplantation. *Neurology* 2006;**67**:268–72.
19. Krivit W, Shapiro EG, Peters C, Wagner JE, Cornu G, Kurtzberg J, et al. Hematopoietic stem-cell transplantation in globoid-cell leukodystrophy. *N Engl J Med* 1998;**338**:1119–26.
20. Kolodny E, Fluharty AL. Metachromatic leukodystrophy and multiple sulfatase deficiency: sulfatide lipidosis. In: Scriver CR, Sly WS, Childs B, Beaudet AL, Valle D, Kinzler KW, et al., editors. *The Metabolic and Molecular Bases of Inherited Disease*. 8th ed. New York: McGraw Hill; 1995. pp. 2693–739.
21. Gieselmann V, Krageloh-Mann I. Metachromatic leukodystrophy—an update. *Neuropediatrics* 2010;**41**:1–6.
22. Von Figura K, Gieselmann V, Jaeken J. Metachromatic leukodystrophy. In: Valle D, Beaudet AL, Vogelstein B, Kinzler KW, Antonarakis SE, Ballabio A, et al., editors. *OMMBID: The Online Metabolic and Molecular Bases of Inherited Diseases*. New York: McGraw Hill; 2013 [cited 29 May 2014]; Available from: http://ommbid.mhmedical.com/content.aspx?bookid=474&Sectionid=45374151.
23. Haberlandt E, Scholl-Burgi S, Neuberger J, Felber S, Gotwald T, Sauter R, et al. Peripheral neuropathy as the sole initial finding in three children with infantile metachromatic leukodystrophy. *Eur J Paediatr Neurol* 2009;**13**:257–60.
24. Singh RK, Leshner RT, Kadom N, Vanderver AL. Isolated cranial nerve enhancement in metachromatic leukodystrophy. *Pediatr Neurol* 2009;**40**:380–2.
25. Shapiro EG, Lockman LA, Knopman D, Krivit W. Characteristics of the dementia in late-onset metachromatic leukodystrophy. *Neurology* 1994;**44**:662–5.
26. Lewis RA, Sumner AJ, Shy ME. Electrophysiological features of inherited demyelinating neuropathies: A reappraisal in the era of molecular diagnosis. *Muscle Nerve* 2000;**23**:1472–87.
27. Dubey R, Chakrabarty B, Gulati S, Sharma MC, Deopujari S, Baheti N, et al. Leukodystrophy presenting as acute-onset polyradiculoneuropathy. *Pediatr Neurol* 2014;**50**:616–18.
28. Martin JJ, Ceuterick C, Mercelis R, Joris C. Pathology of peripheral nerves in metachromatic leucodystrophy. A comparative study of ten cases. *J Neurol Sci* 1982;**53**:95–112.
29. Bindu PS, Mahadevan A, Taly AB, Christopher R, Gayathri N, Shankar SK. Peripheral neuropathy in metachromatic leucodystrophy. A study of 40 cases from south India. *J Neurol Neurosurg Psychiatry* 2005;**76**:1698–701.
30. Gieselmann V, Fluharty AL, Tonnesen T, Von Figura K. Mutations in the arylsulfatase A pseudodeficiency allele causing metachromatic leukodystrophy. *Am J Hum Genet* 1991;**49** 407–13.
31. Natowicz MR, Prence EM, Chaturvedi P, Newburg DS. Urine sulfatides and the diagnosis of metachromatic leukodystrophy. *Clin Chem* 1996;**42**:232–8.
32. Recksiek M, Selmer T, Dierks T, Schmidt B, von Figura K. Sulfatases trapping of the sulfated enzyme intermediate by substituting the active site formylglycine. *J Biol Chem* 1998;**273**:6096–103.
33. Schlote W, Harzer K, Christomanou H, Paton BC, Kustermann-Kuhn B, Schmid B, et al. Sphingolipid activator protein 1 deficiency in metachromatic leucodystrophy with normal arylsulphatase A activity. A clinical, morphological, biochemical, and immunological study. *Eur J Pediatr* 1991;**150**:584–91.
34. Siri L, Rossi A, Lanza F, Mazzotti R, Costa A, Stroppiano M, et al. A novel homozygous splicing mutation in PSAP gene causes

metachromatic leukodystrophy in two Moroccan brothers. *Neurogenetics* 2014;**15**:101–6.

35. Luzi P, Rafi MA, Rao HZ, Wenger DA. Sixteen novel mutations in the arylsulfatase A gene causing metachromatic leukodystrophy. *Gene* 2013;**530**:323–8.
36. Gieselmann V, Matzner U, Hess B, Lullmann-Rauch R, Coenen R, Hartmann D, et al. Metachromatic leukodystrophy: molecular genetics and an animal model. *J Inherit Metab Dis* 1998;**21**:564–74.
37. Martin HR, Poe MD, Provenzale JM, Kurtzberg J, Mendizabal A, Escolar ML. Neurodevelopmental outcomes of umbilical cord blood transplantation in metachromatic leukodystrophy. *Biol Blood Marrow Transplant* 2013;**19**:616–24.
38. Patil SA, Maegawa GH. Developing therapeutic approaches for metachromatic leukodystrophy. *Drug Des Dev Ther* 2013;**7**:729–45.
39. Desnick RJ, Ioannou YA, Eng CM. Alpha-glactosidase A deficiency: Fabry disease. In: Valle D, Beaudet AL, Vogelstein B, Kinzler KW, Antonarakis SE, Ballabio A, et al., editors. *OMMBID: The Online Metabolic and Molecular Bases of Inherited Diseases*. New York: McGraw Hill; 2013 [cited 29 May 2014]; Available from: http://ommbid.mhmedical.com/content.aspx?bookid=474&Sectionid=45374153.
40. Roudebush CP, Foerster JM, Bing OH. The abbreviated PR interval of Fabry's disease. *N Engl J Med* 1973;**289**:357–8.
41. Desnick RJ, Blieden LC, Sharp HL, Hofschire PJ, Moller JH. Cardiac valvular anomalies in Fabry disease. Clinical, morphologic, and biochemical studies. *Circulation* 1976;**54**:818–25.
42. Buhler FR, Thiel G, Dubach UC, Enderlin F, Gloor F, Tholen H. Kidney transplantation in Fabry's disease. *Br Med J* 1973;**3**:28–9.
43. Scott LJ, Griffin JW, Luciano C, Barton NW, Banerjee T, Crawford T, et al. Quantitative analysis of epidermal innervation in Fabry disease. *Neurology* 1999;**52**:1249–54.
44. Schiffmann R. Neuropathy and Fabry disease: pathogenesis and enzyme replacement therapy. *Acta Neurol Belg* 2006;**106**:61–5.
45. Maisey DN, Cosh JA. Basilar artery aneurysm and Anderson-Fabry disease. *J Neurol Neurosurg Psychiatry* 1980;**43**:85–7.
46. Bersano A, Lanfranconi S, Valcarenghi C, Bresolin N, Micieli G, Baron P. Neurological features of Fabry disease: clinical, pathophysiological aspects and therapy. *Acta Neurol Scand* 2012;**126**:77–97.
47. Whybra C, Kampmann C, Willers I, Davies J, Winchester B, Kriegsmann J, et al. Anderson-Fabry disease: clinical manifestations of disease in female heterozygotes. *J Inherit Metab Dis* 2001;**24**:715–24.
48. Dobrovolny R, Dvorakova L, Ledvinova J, Magage S, Bultas J, Lubanda JC, et al. Relationship between X-inactivation and clinical involvement in Fabry heterozygotes. Eleven novel mutations in the alpha-galactosidase A gene in the Czech and Slovak population. *J Mol Med (Berl)* 2005;**83**:647–54.
49. Ohnishi A, Dyek P. Loss of small peripheral sensory neurons in Fabry's disease. *Arch Neurol* 1974;**31**:120–7.
50. Brady RO, Gal AE, Bradley RM, Martensson E, Warshaw AL, Laster L. Enzymatic defect in Fabry's disease. Ceramidetrihexosidase deficiency. *N Engl J Med* 1967;**276**:1163–7.
51. Eng CM, Desnick RJ. Molecular basis of Fabry disease: mutations and polymorphisms in the human alpha-galactosidase A gene. *Hum Mutat* 1994;**3**:103–11.
52. Okumiya T, Ishii S, Kase R, Kamei S, Sakuraba H, Suzuki Y. Alpha-galactosidase gene mutations in Fabry disease: heterogeneous expressions of mutant enzyme proteins. *Hum Genet* 1995;**95**:557–61.
53. Davies JP, Eng CM, Hill JA, Malcolm S, MacDermot K, Winchester B, et al. Fabry disease: fourteen alpha-galactosidase A mutations in unrelated families from the United Kingdom and other European countries. *Eur J Hum Genet* 1996;**4**:219–24.
54. Lockman LA, Hunninghake DB, Krivit W, Desnick RJ. Relief of pain of Fabry's disease by diphenylhydantoin. *Neurology* 1973;**23**:871–5.
55. Uceyler N, He L, Schonfeld D, Kahn AK, Reiners K, Hilz MJ, et al. Small fibers in Fabry disease: baseline and follow-up data under enzyme replacement therapy. *J Peripher Nerv Syst* 2011;**16**:304–14.
56. Aerts JM, Groener JE, Kuiper S, Donker-Koopman WE, Strijland A, Ottenhoff R, et al. Elevated globotriaosylsphingosine is a hallmark of Fabry disease. *Proc Natl Acad Sci USA* 2008;**105**:2812–17.
57. Togawa T, Kodama T, Suzuki T, Sugawara K, Tsukimura T, Ohashi T, et al. Plasma globotriaosylsphingosine as a biomarker of Fabry disease. *Mol Genet Metab* 2010;**100**:257–61.
58. Spada M, Pagliardini S, Yasuda M, Tukel T, Thiagarajan G, Sakuraba H, et al. High incidence of later-onset fabry disease revealed by newborn screening. *Am J Hum Genet* 2006;**79**:31–40.
59. Hwu WL, Chien YH, Lee NC, Chiang SC, Dobrovolny R, Huang AC, et al. Newborn screening for Fabry disease in Taiwan reveals a high incidence of the later-onset GLA mutation c.936 + 919G > A (IVS4 + 919G > A). *Hum Mutat* 2009;**30**:1397–405.
60. Chien YH, Lee NC, Chiang SC, Desnick RJ, Hwu WL. Fabry disease: incidence of the common later-onset alpha-galactosidase A IVS4 + 919G > A—mutation in Taiwanese newborns—superiority of DNA-based to enzyme-based newborn screening for common mutations. *Mol Med* 2012;**18**:780–4.
61. Inoue T, Hattori K, Ihara K, Ishii A, Nakamura K, Hirose S. Newborn screening for Fabry disease in Japan: prevalence and genotypes of Fabry disease in a pilot study. *J Hum Genet* 2013;**58**:548–52.
62. Neufeld EF, Muenzer J. The mucopolysaccharidoses. In: Valle D, Beaudet AL, Vogelstein B, Kinzler KW, Antonarakis SE, Ballabio A, et al., editors. *OMMBID: The Online Metabolic and Molecular Bases of Inherited Diseases*. New York: McGraw Hill; 2013 [cited 29 May 2014]; Available from: http://ommbid.mhmedical.com/content.aspx?bookid=474&Sectionid=45374136.
63. Gross DM, Williams JC, Caprioli C, Dominguez B, Howell RR. Echocardiographic abnormalities in the mucopolysaccharide storage diseases. *Am J Cardiol* 1988;**61**:170–6.
64. Pronicka E, Tylki-Szymanska A, Kwast O, Chmielik J, Maciejko D, Cedro A. Carpal tunnel syndrome in children with mucopolysaccharidoses: needs for surgical tendons and median nerve release. *J Ment Defic Res* 1988;**32**(Pt 1):79–82.
65. Van Meir N, De Smet L. Carpal tunnel syndrome in children. *Acta Orthop Belg* 2003;**69**:387–95.
66. Kwon JY, Ko K, Sohn YB, Kim SJ, Park SW, Kim SH, et al. High prevalence of carpal tunnel syndrome in children with mucopolysaccharidosis type II (Hunter syndrome). *Am J Med Genet A* 2011;**155A**:1329–35.
67. Wiesmann U, Neufeld EF. Scheie and Hurler syndromes: apparent identity of the biochemical defect. *Science* 1970;**169**:72–4.
68. Noh H, Lee JI. Current and potential therapeutic strategies for mucopolysaccharidoses. *J Clin Pharm Ther* 2014;**39**:215–24.
69. Muenzer J. Early initiation of enzyme replacement therapy for the mucopolysaccharidoses. *Mol Genet Metab* 2014;**111**:63–72.

70. Ferla R, Claudiani P, Cotugno G, Saccone P, De Leonibus E, Auricchio A. Similar therapeutic efficacy between a single administration of gene therapy and multiple administrations of recombinant enzyme in a mouse model of lysosomal storage disease. *Hum Gene Ther* 2014;**25**:609–18.
71. Anson DS, McIntyre C, Byers S. Therapies for neurological disease in the mucopolysaccharidoses. *Curr Gene Ther* 2011;**11**:132–43.
72. Thomas GH. Disorders of glycoprotein degradation: alpha-mannosidosis, beta-mannosidosis, fucosidosis, and sialidosis. In: Valle D, Beaudet AL, Vogelstein B, Kinzler KW, Antonarakis SE, Ballabio A, et al., editors. *OMMBID: The Online Metabolic and Molecular Bases of Inherited Diseases*. New York: McGraw Hill; 2013 [cited 29 May 2014]; Available from: http://ommbid.mhmedical.com/content.aspx?bookid=474&Sectionid=45374141.
73. Steinman L, Tharp BR, Dorfman LJ, Forno LS, Sogg RL, Kelts KA, et al. Peripheral neuropathy in the cherry-red spot-myoclonus syndrome (sialidosis type I). *Ann Neurol* 1980;**7**:450–6.
74. Desnick RJ, Schindler D. Alpha-N-acetylgalactosaminidase deficiency: Schindler disease. In: Valle D, Beaudet AL, Vogelstein B, Kinzler KW, Antonarakis SE, Ballabio A, et al., editors. *OMMBID: The Online Metabolic and Molecular Bases of Inherited Diseases*. New York: McGraw Hill; 2013 [cited 29 May 2014]; Available from: http://ommbid.mhmedical.com/content.aspx?bookid=474&Sectionid=45374140.
75. Chabas A, Coll MJ, Aparicio M, Rodriguez Diaz E. Mild phenotypic expression of alpha-N-acetylgalactosaminidase deficiency in two adult siblings. *J Inherit Metab Dis* 1994;**17**:724–31.
76. Wolfe DE, Schindler D, Desnick RJ. Neuroaxonal dystrophy in infantile alpha-N-acetylgalactosaminidase deficiency. *J Neurol Sci* 1995;**132**:44–56.
77. Kanzaki T, Wang AM, Desnick RJ. Lysosomal alpha-N-acetylgalactosaminidase deficiency, the enzymatic defect in angiokeratoma corporis diffusum with glycopeptiduria. *J Clin Invest* 1991;**88**:707–11.
78. Westaway SK, Gregory A, Hayflick SJ. Mutations in PLA2G6 and the riddle of Schindler disease. *J Med Genet* 2007;**44**:e64.
79. Refsum S. Heredopathia atactica polyneuritiformis. *Acta Psychiatr Scand* 1946;**38**:1–303.
80. Skjeldal OH. Heredopathia atactica polyneuritiformis, Refsum disease. In: Moser HW, editor. *Handbook of Clinical Neurology*. Amsterdam: Elsevier; 1996. pp. 485–503.
81. Torvik A, Torp S, Kase BF, Ek J, Skjeldal O, Stokke O. Infantile Refsum's disease: a generalized peroxisomal disorder. Case report with postmortem examination. *J Neurol Sci* 1988;**85**:39–53.
82. Gibberd FB, Billimoria JD, Goldman JM, Clemens ME, Evans R, Whitelaw MN, et al. Heredopathia atactica polyneuritiformis: Refsum's disease. *Acta Neurol Scand* 1985;**72**:1–17.
83. Skjeldal OH, Stokke O, Refsum S, Norseth J, Petit H. Clinical and biochemical heterogeneity in conditions with phytanic acid accumulation. *J Neurol Sci* 1987;**77**:87–96.
84. Cammermeyer J. Refsum's disease: neurological aspects. In: Vinken PJ, Bruyn GW, editors. *Handbook of Clinical Neurology*. Amsterdam: North Holland Publishing Company; 1975. pp. 231–57.
85. Steinberg D. Refsum disease. In: Scriver CR, Sly WS, Childs B, Beaudet AL, Valle D, Kinzler KW, et al., editors. *The Metabolic and Molecular Bases of Human Disease*. 8th ed. New York: McGraw Hill; 1995. pp. 2351–69.
86. Powers JM, Kenjarski TP, Moser AB, Moser HW. Cerebellar atrophy in chronic rhizomelic chondrodysplasia punctata: a potential role for phytanic acid and calcium in the death of its Purkinje cells. *Acta Neuropathol* 1999;**98**:129–34.
87. Wanders RA, Jakobs C, Skjeldal OH. Refsum disease. In: Valle D, Beaudet AL, Vogelstein B, Kinzler KW, Antonarakis SE, Ballabio A, et al., editors. *OMMBID: The Online Metabolic and Molecular Bases of Inherited Diseases*. New York: McGraw Hill; 2013 [cited 29 May 2014]; Available from: http://ommbid.mhmedical.com/content.aspx?bookid=474&Sectionid=45374131.
88. Verhoeven NM, Kulik W, van den Heuvel CM, Jakobs C. Pre- and postnatal diagnosis of peroxisomal disorders using stable-isotope dilution gas chromatography–mass spectrometry. *J Inherit Metab Dis* 1995;**18**(Suppl. 1):45–60.
89. Jansen GA, Ofman R, Ferdinandusse S, Ijlst L, Muijsers AO, Skjeldal OH, et al. Refsum disease is caused by mutations in the phytanoyl-CoA hydroxylase gene. *Nat Genet* 1997;**17**:190–3.
90. van Grunsven EG, van Berkel E, Ijlst L, Vreken P, de Klerk JB, Adamski J, et al. Peroxisomal D-hydroxyacyl-CoA dehydrogenase deficiency: resolution of the enzyme defect and its molecular basis in bifunctional protein deficiency. *Proc Natl Acad Sci USA* 1998;**95**:2128–33.
91. Lazarow PB, Moser HW. Disorders of peroxisome biogenesis. In: Scriver CR, Sly WS, Childs B, Beaudet AL, Valle D, Kinzler KW, et al., editors. *The Metabolic and Molecular Bases of Human Disease*. 8th ed. New York: McGraw Hill; 1995. pp. 2128–33.
92. Nadal N, Rolland MO, Tranchant C, Reutenauer L, Gyapay G, Warter JM, et al. Localization of Refsum disease with increased pipecolic acidaemia to chromosome 10p by homozygosity mapping and carrier testing in a single nuclear family. *Hum Mol Genet* 1995;**4**:1963–6.
93. Mihalik SJ, Morrell JC, Kim D, Sacksteder KA, Watkins PA, Gould SJ. Identification of PAHX, a Refsum disease gene. *Nat Genet* 1997;**17**:185–9.
94. van den Brink DM, Brites P, Haasjes J, Wierzbicki AS, Mitchell J, Lambert-Hamill M, et al. Identification of PEX7 as the second gene involved in Refsum disease. *Am J Hum Genet* 2003;**72**:471–7.
95. Jansen GA, Waterham HR, Wanders RJ. Molecular basis of Refsum disease: sequence variations in phytanoyl-CoA hydroxylase (PHYH) and the PTS2 receptor (PEX7). *Hum Mutat* 2004;**23**:209–18.
96. Fiskerstrand T, Knappskog P, Majewski J, Wanders RJ, Boman H, Bindoff LA. A novel Refsum-like disorder that maps to chromosome 20. *Neurology* 2009;**72**:20–7.
97. Brown PJ, Mei G, Gibberd FB, Burston D, Mayne PD, McClinchy JE, et al. Diet and Refsum disease. The determination of phytanic acid and phytol in certain foods and the application of this knowledge to the choice of suitable convenience foods for patients with Refsum's disease. *J Hum Nutr Diet* 1993;**6**:295–305.
98. Refsum S. Heredopathia atactica polyneuritiformis phytanic-acid storage disease, Refsum's disease: a biochemically well-defined disease with a specific dietary treatment. *Arch Neurol* 1981;**38**:605–6.
99. Baldwin EJ, Gibberd FB, Harley C, Sidey MC, Feher MD, Wierzbicki AS. The effectiveness of long-term dietary therapy in the treatment of adult Refsum disease. *J Neurol Neurosurg Psychiatry* 2010;**81**:954–7.

100. Lundberg A, Lilja LG, Lundberg PO, Try K. Heredopathia atactica polyneuritiformis (Refsum's disease). Experiences of dietary treatment and plasmapheresis. *Eur Neurol* 1972;**8**:309–24.
101. Baumgartner MR, Poll-The BT, Verhoeven NM, Jakobs C, Espeel M, Roels F, et al. Clinical approach to inherited peroxisomal disorders: a series of 27 patients. *Ann Neurol* 1998;**44**:720–30.
102. Moser HW. Genotype-phenotype correlations in peroxisomal disorders. *Dev Brain Dysfunct* 1997;**10**:292.
103. Bezman L, Moser AB, Raymond GV, Rinaldo P, Watkins PA, Smith KD, et al. Adrenoleukodystrophy: incidence, new mutation rate, and results of extended family screening. *Ann Neurol* 2001;**49**:512–17.
104. Moser HW, Smith KD, Watkins PA, Powers JM. X-linked adrenoleukodystrophy. In: Valle D, Beaudet AL, Vogelstein B, Kinzler KW, Antonarakis SE, Ballabio A, et al., editors. *OMBID: The Online Metabolic and Molecular Bases of Inherited Diseases*. New York: McGraw Hill; 2013 [cited 29 May 2014]; Available from: http://ommbid.mhmedical.com/content.aspx?bookid=474&Sectionid=45374130.
105. Jangouk P, Zackowski KM, Naidu S, Raymond GV. Adrenoleukodystrophy in female heterozygotes: underrecognized and undertreated. *Mol Genet Metab* 2012;**105**:180–5.
106. Moser AB, Kreiter N, Bezman L, Lu S, Raymond GV, Naidu S, et al. Plasma very long chain fatty acids in 3,000 peroxisome disease patients and 29,000 controls. *Ann Neurol* 1999;**45**:100–10.
107. Mosser J, Douar AM, Sarde CO, Kioschis P, Feil R, Moser H, et al. Putative X-linked adrenoleukodystrophy gene shares unexpected homology with ABC transporters. *Nature* 1993;**361**:726–30.
108. Schluter A, Espinosa L, Fourcade S, Galino J, Lopez E, Ilieva E, et al. Functional genomic analysis unravels a metabolic-inflammatory interplay in adrenoleukodystrophy. *Hum Mol Genet* 2012;**21**:1062–77.
109. Singh I, Pujol A. Pathomechanisms underlying X-adrenoleukodystrophy: a three-hit hypothesis. *Brain Pathol* 2010;**20**:838–44.
110. Powers JM, DeCiero DP, Ito M, Moser AB, Moser HW. Adrenomyeloneuropathy: a neuropathologic review featuring its noninflammatory myelopathy. *J Neuropathol Exp Neurol* 2000;**59**:89–102.
111. Chaudhry V, Moser HW, Cornblath DR. Nerve conduction studies in adrenomyeloneuropathy. *J Neurol Neurosurg Psychiatry* 1996;**61**:181–5.
112. van Geel BM, Koelman JH, Barth PG, Ongerboer de Visser BW. Peripheral nerve abnormalities in adrenomyeloneuropathy: a clinical and electrodiagnostic study. *Neurology* 1996;**46**:112–18.
113. Engelen M, Barbier M, Dijkstra IM, Schur R, de Bie RM, Verhamme C, et al. X-linked adrenoleukodystrophy in women: a cross-sectional cohort study. *Brain* 2014;**137**:693–706.
114. Restuccia D, Di Lazzaro V, Valeriani M, Oliviero A, Le Pera D, Barba C, et al. Neurophysiologic follow-up of long-term dietary treatment in adult-onset adrenoleukodystrophy. *Neurology* 1999;**52**:810–16.
115. Polgreen LE, Chahla S, Miller W, Rothman S, Tolar J, Kivisto T, et al. Early diagnosis of cerebral X-linked adrenoleukodystrophy in boys with Addison's disease improves survival and neurological outcomes. *Eur J Pediatr* 2011;**170**:1049–54.
116. Shapiro E, Krivit W, Lockman L, Jambaque I, Peters C, Cowan M, et al. Long-term effect of bone-marrow transplantation for childhood-onset cerebral X-linked adrenoleukodystrophy. *Lancet* 2000;**356**:713–18.
117. Ferrer I, Aubourg P, Pujol A. General aspects and neuropathology of X-linked adrenoleukodystrophy. *Brain Pathol* 2010;**20**:817–30.
118. Berger J, Pujol A, Aubourg P, Forss-Petter S. Current and future pharmacological treatment strategies in X-linked adrenoleukodystrophy. *Brain Pathol* 2010;**20**:845–56.
119. Cartier N, Aubourg P. Hematopoietic stem cell transplantation and hematopoietic stem cell gene therapy in X-linked adrenoleukodystrophy. *Brain Pathol* 2010;**20**:857–62.
120. Miller WP, Rothman SM, Nascene D, Kivisto T, DeFor TE, Ziegler RS, et al. Outcomes after allogeneic hematopoietic cell transplantation for childhood cerebral adrenoleukodystrophy: the largest single-institution cohort report. *Blood* 2011;**118**:1971–8.
121. Horvath GA, Eichler F, Poskitt K, Stockler-Ipsiroglu S. Failure of repeated cyclophosphamide pulse therapy in childhood cerebral X-linked adrenoleukodystrophy. *Neuropediatrics* 2012;**43**:48–52.
122. van Geel BM, Assies J, Haverkort EB, Koelman JH, Verbeeten Jr. B, Wanders RJ, et al. Progression of abnormalities in adrenomyeloneuropathy and neurologically asymptomatic X-linked adrenoleukodystrophy despite treatment with "Lorenzo's oil". *J Neurol Neurosurg Psychiatry* 1999;**67**:290–9.
123. Kwan TK, Chadban SJ, McKenzie PR, Saunders JR. Acute oxalate nephropathy secondary to orlistat-induced enteric hyperoxaluria. *Nephrology* 2013;**18**:241–2.
124. Danpure CJ. Primary hyperoxaluria. In: Valle D, Beaudet AL, Vogelstein B, Kinzler KW, Antonarakis SE, Ballabio A, et al., editors. *OMMBID: The Online Metabolic and Molecular Bases of Inherited Diseases*. New York: McGraw Hill; 2013 [cited 29 May 2014]; Available from: http://ommbid.mhmedical.com/content.aspx?bookid = 474&Sectionid = 45374132.
125. Danpure CJ. Peroxisomal alanine:glyoxylate aminotransferase and prenatal diagnosis of primary hyperoxaluria type 1. *Lancet* 1986;**2**:1168.
126. Danpure CJ, Jennings PR, Mistry J, Chalmers RA, McKerrell RE, Blakemore WF, et al. Enzymological characterization of a feline analogue of primary hyperoxaluria type 2: a model for the human disease. *J Inherit Metab Dis* 1989;**12**:403–14.
127. Belostotsky R, Seboun E, Idelson GH, Milliner DS, Becker-Cohen R, Rinat C, et al. Mutations in DHDPSL are responsible for primary hyperoxaluria type III. *Am J Hum Genet* 2010;**87**:392–9.
128. Cochat P, Rumsby G. Primary hyperoxaluria. *N Engl J Med* 2013;**369**:649–58.
129. Danpure CJ. Molecular and clinical heterogeneity in primary hyperoxaluria type 1. *Am J Kidney Dis* 1991;**17**:366–9.
130. Hall BM, Walsh JC, Horvath JS, Lytton DG. Peripheral neuropathy complicating primary hyperoxaluria. *J Neurol Sci* 1976;**29**:343–9.
131. Bilbao JM, Berry H, Marotta J, Ross RC. Peripheral neuropathy in oxalosis. A case report with electron microscopic observations. *Can J Neurol Sci* 1976;**3**:63–7.
132. Moorhead PJ, Cooper DJ, Timperley WR. Progressive peripheral neuropathy in patient with primary hyperoxaluria. *Br Med J* 1975;**2**:312–13.
133. Anderson B, Adams QM. Facial-auditory nerve oxalosis. *Am J Med* 1990;**88**:87–8.
134. Danpure CJ, Jennings PR. Further studies on the activity and subcellular distribution of alanine:glyoxylate aminotransferase in the

livers of patients with primary hyperoxaluria type 1. *Clin Sci* 1988;**75**:315–22.

135. Chalmers RA, Tracey BM, Mistry J, Griffiths KD, Green A, Winterborn MH. L-Glyceric aciduria (primary hyperoxaluria type 2) in siblings in two unrelated families. *J Inherit Metab Dis* 1984;**7**(Suppl. 2):133–4.

136. Nair P, Al-Otaibi T, Nampoory N, Al-Qabandi W, Said T, Halim MA, et al. Combined liver and kidney transplantation in primary hyperoxaluria: a report of three cases and review of the literature *Saudi J Kidney Dis Transpl* 2013;**24**:969–75.

137. Hoyer-Kuhn H, Kohbrok S, Volland R, Franklin J, Hero B, Beck BB, et al. Vitamin b6 in primary hyperoxaluria I: first prospective trial after 40 years of practice. *Clin J Am Soc Nephrol* 2014;**9**:468–77.

138. Bjorkhem I, Boberg K, Leitersdorf E. Inborn errors in bile acid biosynthesis and storage of sterols other than cholesterol. In: Valle D, Beaudet AL, Vogelstein B, Kinzler KW, Antonarakis SE, Ballabio A, et al., editors. *OMMBID: The Online Metabolic and Molecular Bases of Inherited Diseases*. New York: McGraw Hill; 2013 [cited May 29 2014]; Available from: http://ommbid.mhmedical.com/content.aspx?bookid=474&Sectionid=45374119.

139. Walterfang M, Bonnot O, Mocellin R, Velakoulis D. The neuropsychiatry of inborn errors of metabolism. *J Inherit Metab Dis* 2013;**36**:687–702.

140. Bonnot O, Klunemann HH, Sedel F, Tordjman S, Cohen D, Walterfang M. Diagnostic and treatment implications of psychosis secondary to treatable metabolic disorders in adults: a systematic review. *Orphanet J Rare Dis* 2014;**9**:65.

141. Berginer VM, Berginer J, Korczyn AD, Tadmor R. Magnetic resonance imaging in cerebrotendinous xanthomatosis: a prospective clinical and neuroradiological study. *J Neurol Sci* 1994;**122**: 102–8.

142. Argov Z, Soffer D, Eisenberg S, Zimmerman Y. Chronic demyelinating peripheral neuropathy in cerebrotendinous xanthomatosis. *Ann Neurol* 1986;**20**:89–91.

143. Ginanneschi F, Mignarri A, Mondelli M, Gallus GN, Del Puppo M, Giorgi S, et al. Polyneuropathy in cerebrotendinous xanthomatosis and response to treatment with chenodeoxycholic acid. *J Neurol* 2013;**260**:268–74.

144. Salen G, Berginer V, Shore V, Horak I, Horak E, Tint GS, et al. Increased concentrations of cholestanol and apolipoprotein B in the cerebrospinal fluid of patients with cerebrotendinous xanthomatosis. Effect of chenodeoxycholic acid. *N Engl J Med* 1987;**316**:1233–8.

145. Setoguchi T, Salen G, Tint GS, Mosbach EH. A biochemical abnormality in cerebrotendinous xanthomatosis. Impairment of bile acid biosynthesis associated with incomplete degradation of the cholesterol side chain. *J Clin Invest* 1974;**53**:1393–401.

146. Koopman BJ, Wolthers BG, van der Molen JC, van der Slik W, Waterreus RJ, van Spreeken A. Cerebrotendinous xanthomatosis: a review of biochemical findings of the patient population in The Netherlands. *J Inherit Metab Dis* 1988;**11**:56–75.

147. Leitersdorf E, Safadi R, Meiner V, Reshef A, Bjorkhem I, Friedlander Y, et al. Cerebrotendinous xanthomatosis in the Israeli Druze: molecular genetics and phenotypic characteristics. *Am J Hum Genet* 1994;**55**:907–15.

148. Berginer VM, Salen G, Shefer S. Long-term treatment of cerebrotendinous xanthomatosis with chenodeoxycholic acid. *N Engl J Med* 1984;**311**:1649–52.

149. Peynet J, Laurent A, De Liege P, Lecoz P, Gambert P, Legrand A, et al. Cerebrotendinous xanthomatosis: treatments with simvastatin, lovastatin, and chenodeoxycholic acid in 3 siblings. *Neurology* 1991;**41**:434–6.

150. Bjorkhem I, Hansson M. Cerebrotendinous xanthomatosis: an inborn error in bile acid synthesis with defined mutations but still a challenge. *Biochem Biophys Res Commun* 2010;**396**:46–9.

151. Assmann G, Von Eckardstein A, Bryan Brewer HH. Familial Analphalipoproteinemia: Tangier disease. In: Valle D, Beaudet AL, Vogelstein B, Kinzler KW, Antonarakis SE, Ballabio A, et al., editors. *OMMBID: The Online Metabolic and Molecular Bases of Inherited Diseases*. New York: McGraw Hill; 2013 [cited 29 May 2014]; Available from: http://ommbid.mhmedical.com/content.aspx?bookid=474&Sectionid=45374118.

152. Puntoni M, Sbrana F, Bigazzi F, Sampietro T. Tangier disease: epidemiology, pathophysiology, and management. *Am J Cardiovasc Drugs* 2012;**12**:303–11.

153. Assmann GA, Von Eckardstein A, Brewer Jr HB. Familial high density lipoprotein deficiency: Tangier disease. In: Scriver CR, Sly WS, Childs B, Beaudet AL, Valle D, Kinzler KW, et al., editors. *The Metabolic and Molecular Bases of Inherited Disease*. 8th ed. New York: McGraw Hill; 2001. pp. 2937–60.

154. Schaefer EJ. Clinical, biochemical, and genetic features in familial disorders of high density lipoprotein deficiency. *Arteriosclerosis* 1984;**4**:303–22.

155. Zyss J, Behin A, Couvert P, Bouhour F, Sassolas A, Kolev I, et al. Clinical and electrophysiological characteristics of neuropathy associated with Tangier disease. *J Neurol* 2012 **259**:1222–6.

156. Pollock M, Nukada H, Frith RW, Simcock JP, Allpress S. Peripheral neuropathy in Tangier disease. *Brain* 1983;**106**(Pt 4):911–28.

157. Schaefer EJ, Blum CB, Levy RI, Jenkins LL, Alaupovic P, Foster DM, et al. Metabolism of high-density lipoprotein apolipoproteins in Tangier disease. *N Engl J Med* 1978;**299**:905–10.

158. Assmann G, Herbert PN, Fredrickson DS, Forte T. Isolation and characterization of an abnormal high density lipoprotein in Tangier Disease. *J Clin Invest* 1977;**60**:242–52.

159. Welty FK. Hypobetalipoproteinemia and abetalipoproteinemia. *Curr Opin Lipidol* 2014;**25**:161–8.

160. Hooper AJ, Burnett JR. Update on primary hypobetalipoproteinemia. *Curr Atheroscler Rep* 2014;**16**:423.

161. Burnett JR, Bell DA, Hooper AJ, Hegele RA. Clinical utility gene card for: Familial Hypobetalipoproteinaemia (APOB). *Eur J Hum Genet* 2012;**20**(8).

162. Kane JP, Havel RJ. Disorders of the biogenesis and secretion of lipoproteins containing the B apolipoproteins. In: Valle D, Beaudet AL, Vogelstein B, Kinzler KW, Antonarakis SE, Ballabio A, et al., editors. *OMMBID: The Online Metabolic and Molecular Bases of Inherited Diseases*. New York: McGraw Hill; 2013 [cited 29 May 2014]; Available from: http://ommbid.mhmedical.com/content.aspx?bookid=474&Sectionid=45374110.

163. Weinstein MA, Pearson KD, Agus SG. Abetalipoproteinemia. *Radiology* 1973;**108**:269–73.

164. Sperling MA, Hengstenberg F, Yunis E, Kenny FM, Drash AL. Abetalipoproteinemia: metabolic, endocrine, and electron-microscopic investigations. *Pediatrics* 1971;**48**:91–102.

165. Cogan DG, Rodrigues M, Chu FC, Schaefer EJ. Ocular abnormalities in abetalipoproteinemia. A clinicopathologic correlation. *Ophthalmology* 1984;**91**:991–8.

166. Brin MF, Pedley TA, Lovelace RE, Emerson RG, Gouras P, MacKay C, et al. Electrophysiologic features of abetalipoproteinemia: functional consequences of vitamin E deficiency. *Neurology* 1986;**36**:669–73.
167. Wichman A, Buchthal F, Pezeshkpour GH, Gregg RE. Peripheral neuropathy in abetalipoproteinemia. *Neurology* 1985;**35**:1279–89.
168. Muller DP, Lloyd JK. Effect of large oral doses of vitamin E on the neurological sequelae of patients with abetalipoproteinemia. *Ann NY Acad Sci* 1982;**393**:133–44.
169. Runge P, Muller DP, McAllister J, Calver D, Lloyd JK, Taylor D. Oral vitamin E supplements can prevent the retinopathy of abetalipoproteinaemia. *Br J Ophthalmol* 1986;**70**:166–73.
170. Lee J, Hegele RA. Abetalipoproteinemia and homozygous hypobetalipoproteinemia: a framework for diagnosis and management. *J Inherit Metab Dis* 2014;**37**:333–9.
171. Aminoff MJ. Acanthocytosis and neurological disease. *Brain* 1972;**95**:749–60.
172. Bird TD, Cederbaum S, Valey RW, Stahl WL. Familial degeneration of the basal ganglia with acanthocytosis: a clinical, neuropathological, and neurochemical study. *Ann Neurol* 1978;**3**:253–8.
173. Gross KB, Skrivanek JA, Carlson KC, Kaufman DM. Familial amyotrophic chorea with acanthocytosis. New clinical and laboratory investigations. *Arch Neurol* 1985;**42**:753–6.
174. De Franceschi L, Bosman GJ, Mohandas N. Abnormal red cell features associated with hereditary neurodegenerative disorders: the neuroacanthocytosis syndromes. *Curr Opin Hematol* 2014;**21**:201–9.
175. Neutel D, Miltenberger-Miltenyi G, Silva I, de Carvalho M. Chorea-acanthocytosis presenting as motor neuron disease. *Muscle Nerve* 2012;**45**:293–5.
176. Shiokawa N, Nakamura M, Sameshima M, Deguchi A, Hayashi T, Sasaki N, et al. Chorein, the protein responsible for chorea-acanthocytosis, interacts with beta-adducin and beta-actin. *Biochem Biophys Res Comm* 2013;**441**:96–101.
177. Chen Y, Kishnani PS, Koeberl D. Glycogen storage diseases. In: Valle D, Beaudet AL, Vogelstein B, Kinzler KW, Antonarakis SE, Ballabio A, et al., editors. *OMMBID: The Online Metabolic and Molecular Bases of Inherited Diseases*. New York: McGraw Hill; 2013 [cited 29 May 2014]; Available from: http://ommbid.mhmedical.com/content.aspx?bookid=474&Sectionid=45374051.
178. Mochel F, Schiffmann R, Steenweg ME, Akman HO, Wallace M, Sedel F, et al. Adult polyglucosan body disease: Natural History and Key Magnetic Resonance Imaging Findings. *Ann Neurol* 2012;**72**:433–41.
179. Bathgate D, Wigley R, Gorman G, Horvath R, Chinnery PF. Childhood presentation of "adult" polyglucosan body disease: normal GBE1 sequence with no glycogen branching enzyme activity. *Ann Neurol* 2013;**73**:317–18.
180. Hussain A, Armistead J, Gushulak L, Kruck C, Pind S, Triggs-Raine B, et al. The adult polyglucosan body disease mutation GBE1 c.1076A > C occurs at high frequency in persons of Ashkenazi Jewish background. *BioChem Biophys Res Commun* 2012;**426**:286–8.
181. McMaster KR, Powers JM, Hennigar Jr. GR, Wohltmann HJ, Farr Jr. GH. Nervous system involvement in type IV glycogenosis. *Arch Pathol Lab Med* 1979;**103**:105–11.
182. Servidei S, Riepe RE, Langston C, Tani LY, Bricker JT, Crisp-Lindgren N, et al. Severe cardiopathy in branching enzyme deficiency. *J Pediatr* 1987;**111**:51–6.
183. Nolte KW, Janecke AR, Vorgerd M, Weis J, Schroder JM. Congenital type IV glycogenosis: the spectrum of pleomorphic polyglucosan bodies in muscle, nerve, and spinal cord with two novel mutations in the GBE1 gene. *Acta Neuropathol* 2008;**116**:491–506.
184. Guerra AS, van Diggelen OP, Carneiro F, Tsou RM, Simoes S, Santos NT. A juvenile variant of glycogenosis IV (Andersen disease). *Eur J Pediatr* 1986;**145**:179–81.
185. Greene HL, Brown BI, McClenathan DT, Agostini Jr. RM, Taylor SR. A new variant of type IV glycogenosis: deficiency of branching enzyme activity without apparent progressive liver disease. *Hepatology* 1988;**8**:302–6.
186. Shin YS, Steiguber H, Klemm P, Endres W, Schwab O, Wolff G. Branching enzyme in erythrocytes. Detection of type IV glycogenosis homozygotes and heterozygotes. *J Inherit Metab Dis* 1988;**11**(Suppl. 2):252–4.
187. Mitchell GA. Hypertyrosinemia. In: Valle D, Beaudet AL, Vogelstein B, Kinzler KW, Antonarakis SE, Ballabio A, et al., editors. *OMMBID: The Online Metabolic and Molecular Bases of Inherited Diseases*. New York: McGraw Hill; 2013 [cited 29 May 2014]; Available from: http://ommbid.mhmedical.com/content.aspx?bookid=474&Sectionid=45374064.
188. Mitchell G, Larochelle J, Lambert M, Michaud J, Grenier A, Ogier H, et al. Neurologic crises in hereditary tyrosinemia. *N Engl J Med* 1990;**322**:432–7.
189. Gibbs TC, Payan J, Brett EM, Lindstedt S, Holme E, Clayton PT. Peripheral neuropathy as the presenting feature of tyrosinaemia type I and effectively treated with an inhibitor of 4-hydroxyphenylpyruvate dioxygenase. *J Neurol Neurosurg Psychiatry* 1993;**56**:1129–32.
190. Felitsyn N, McLeod C, Shroads AL, Stacpoole PW, Notterpek L. The heme precursor delta-aminolevulinate blocks peripheral myelin formation. *J Neurochem* 2008;**106**:2068–79.
191. Holme E, Lindstedt S. Tyrosinaemia type I and NTBC (2-(2-nitro-4-trifluoromethylbenzoyl)-1,3-cyclohexanedione). *J Inherit Metab Dis* 1998;**21**:507–17.
192. de Laet C, Dionisi-Vici C, Leonard JV, McKiernan P, Mitchell G, Monti L, et al. Recommendations for the management of tyrosinaemia type 1. *Orphanet J Rare Dis* 2013;**8**:8.
193. Roe CR, Ding J. Mitochondrial fatty acid oxidation disorders. In: Valle D, Beaudet AL, Vogelstein B, Kinzler KW, Antonarakis SE, Ballabio A, et al., editors. *OMMBID: The Online Metabolic and Molecular Bases of Inherited Diseases*. New York: McGraw Hill; 2013 [cited 29 May 2014]; Available from: http://ommbid.mhmedical.com/content.aspx?bookid=474&Sectionid=45374093.
194. Tyni T, Pihko H. Long-chain 3-hydroxyacyl-CoA dehydrogenase deficiency. *Acta Paediatr* 1999;**88**:237–45.
195. Spiekerkoetter U. Mitochondrial fatty acid oxidation disorders: clinical presentation of long-chain fatty acid oxidation defects before and after newborn screening. *J Inherit Metab Dis* 2010;**33**:527–32.
196. Ibdah JA, Tein I, Dionisi-Vici C, Bennett MJ, IJlst L, Gibson B, et al. Mild trifunctional protein deficiency is associated with progressive neuropathy and myopathy and suggests a novel genotype-phenotype correlation. *J Clin Invest* 1998;**102**:1193–9.
197. Tein I, Vajsar J, MacMillan L, Sherwood WG. Long-chain L-3-hydroxyacyl-coenzyme A dehydrogenase deficiency neuropathy: response to cod liver oil. *Neurology* 1999;**52**:640–3.

198. Sykut-Cegielska J, Gradowska W, Piekutowska-Abramczuk D, Andresen BS, Olsen RK, Oltarzewski M, et al. Urgent metabolic service improves survival in long-chain 3-hydroxyacyl-CoA dehydrogenase (LCHAD) deficiency detected by symptomatic identification and pilot newborn screening. *J Inherit Metab Dis* 2011;**34**:185–95.
199. Robinson BH. Lactic acidemia: disorders of pyruvate carboxylase and pyruvate dehydrogenase. In: Valle D, Beaudet AL, Vogelstein B, Kinzler KW, Antonarakis SE, Ballabio A, et al., editors. *OMMBID: The Online Metabolic and Molecular Bases of Inherited Diseases*. New York: McGraw Hill; 2013 [cited 29 May 2014]; Available from: http://ommbid.mhmedical.com/content.aspx?bookid=474&Sectionid=45374092.
200. Rahman S, Blok RB, Dahl HH, Danks DM, Kirby DM, Chow CW, et al. Leigh syndrome: clinical features and biochemical and DNA abnormalities. *Ann Neurol* 1996;**39**:343–51.
201. Sofou K, De Coo IF, Isohanni P, Ostergaard E, Naess K, De Meirleir L, et al. A multicenter study on Leigh syndrome: disease course and predictors of survival. *Orphanet J Rare Dis* 2014;**9**:52.
202. Kretzschmar HA, DeArmond SJ, Koch TK, Patel MS, Newth CJ, Schmidt KA, et al. Pyruvate dehydrogenase complex deficiency as a cause of subacute necrotizing encephalopathy (Leigh disease). *Pediatrics* 1987;**79**:370–3.
203. Matthews PM, Brown RM, Otero L, Marchington D, Leonard JV, Brown GK. Neurodevelopmental abnormalities and lactic acidosis in a girl with a 20-bp deletion in the X-linked pyruvate dehydrogenase E1 alpha subunit gene. *Neurology* 1993;**43**:2025–30.
204. de Vries DD, van Engelen BG, Gabreels FJ, Ruitenbeek W, van Oost BA. A second missense mutation in the mitochondrial ATPase 6 gene in Leigh's syndrome. *Ann Neurol* 1993;**34**:410–12.
205. Santorelli FM, Shanske S, Macaya A, DeVivo DC, DiMauro S. The mutation at nt 8993 of mitochondrial DNA is a common cause of Leigh's syndrome. *Ann Neurol* 1993;**34**:827–34.
206. Loeffen J, Smeitink J, Triepels R, Smeets R, Schuelke M, Sengers R, et al. The first nuclear-encoded complex I mutation in a patient with Leigh syndrome. *Am J Hum Genet* 1998;**63**:1598–608.
207. Triepels RH, van den Heuvel LP, Loeffen JL, Buskens CA, Smeets RJ, Rubio Gozalbo ME, et al. Leigh syndrome associated with a mutation in the NDUFS7 (PSST) nuclear encoded subunit of complex I. *Ann Neurol* 1999;**45**:787–90.
208. DiMauro S, De Vivo DC. Genetic heterogeneity in Leigh syndrome. *Ann Neurol* 1996;**40**:5–7.
209. Tiranti V, Hoertnagel K, Carrozzo R, Galimberti C, Munaro M, Granatiero M, et al. Mutations of SURF-1 in Leigh disease associated with cytochrome c oxidase deficiency. *Am J Hum Genet* 1998;**63**:1609–21.
210. Zhu Z, Yao J, Johns T, Fu K, De Bie I, Macmillan C, et al. SURF1, encoding a factor involved in the biogenesis of cytochrome c oxidase, is mutated in Leigh syndrome. *Nat Genet* 1998;**20**:337–43.
211. Morin C, Dube J, Robinson BH, Lacroix J, Michaud J, De Braekeleer M, et al. Stroke-like episodes in autosomal recessive cytochrome oxidase deficiency. *Ann Neurol* 1999;**45**:389–92.
212. Macaya A, Munell F, Burke RE, De Vivo DC. Disorders of movement in Leigh syndrome. *Neuropediatrics* 1993;**24**:60–7.
213. Narita T, Yamano T, Ohno M, Takano T, Ito R, Shimada M. Hypertension in Leigh syndrome—a case report. *Neuropediatrics* 1998;**29**:265–7.
214. Taylor MJ, Robinson BH. Evoked potentials in children with oxidative metabolic defects leading to Leigh syndrome. *Pediatr Neurol* 1992;**8**:25–9.
215. Miyabayashi S, Ito T, Narisawa K, Iinuma K, Tada K. Biochemical study in 28 children with lactic acidosis, in relation to Leigh's encephalomyelopathy. *Eur J Pediatr* 1985;**143**:278–83.
216. Stacpoole PW, Bunch ST, Neiberger RE, Perkins LA, Quisling R, Hutson AD, et al. The importance of cerebrospinal fluid lactate in the evaluation of congenital lactic acidosis. *J Pediatr* 1999;**134**:99–102.
217. Tatuch Y, Christodoulou J, Feigenbaum A, Clarke JT, Wherret J, Smith C, et al. Heteroplasmic mtDNA mutation (T—G) at 8993 can cause Leigh disease when the percentage of abnormal mtDNA is high. *Am J Hum Genet* 1992;**50**:852–8.
218. Nagashima T, Mori M, Katayama K, Nunomura M, Nishihara H, Hiraga H, et al. Adult Leigh syndrome with mitochondrial DNA mutation at 8993. *Acta Neuropathol* 1999;**97**:416–22.
219. Mermigkis C, Bouloukaki I, Mastorodemos V, Plaitakis A, Alogdianakis V, Siafakas N, et al. Medical treatment with thiamine, coenzyme Q, vitamins E and C, and carnitine improved obstructive sleep apnea in an adult case of Leigh disease. *Sleep Breath* 2013;**17**:1129–35.
220. Martinelli D, Catteruccia M, Piemonte F, Pastore A, Tozzi G, Dionisi-Vici C, et al. EPI-743 reverses the progression of the pediatric mitochondrial disease—genetically defined Leigh Syndrome. *Mol Genet Metab* 2012;**107**:383–8.
221. Geach T. Neurometabolic disease: Treating mitochondrial diseases with mTOR inhibitors—a potential treatment for Leigh syndrome? *Nat Rev Neurol* 2014;**10**:2.
222. Johnson SC, Yanos ME, Kayser EB, Quintana A, Sangesland M, Castanza A, et al. mTOR inhibition alleviates mitochondrial disease in a mouse model of Leigh syndrome. *Science* 2013;**342**:1524–8.
223. Rawle MJ, Larner AJ. NARP syndrome: A 20-year follow-up. *Case Rep Neurol* 2013;**5**:204–7.
224. Lopez-Gallardo E, Solano A, Herrero-Martin MD, Martinez-Romero I, Castano-Perez MD, Andreu AL, et al. NARP syndrome in a patient harbouring an insertion in the MT-ATP6 gene that results in a truncated protein. *J Med Genet* 2009;**46**:64–7.
225. Duno M, Wibrand F, Baggesen K, Rosenberg T, Kjaer N, Frederiksen AL. A novel mitochondrial mutation m.8989G > C associated with neuropathy, ataxia, retinitis pigmentosa—the NARP syndrome. *Gene* 2013;**515**:372–5.
226. Makela-Bengs P, Suomalainen A, Majander A, Rapola J, Kalimo H, Nuutila A, et al. Correlation between the clinical symptoms and the proportion of mitochondrial DNA carrying the 8993 point mutation in the NARP syndrome. *Pediatr Res* 1995;**37**:634–9.
227. Santorelli FM, Mak SC, Vazquez-Memije ME, Shanske S, Kranz-Eble P, Jain KD, et al. Clinical heterogeneity associated with the mitochondrial DNA T8993C point mutation. *Pediatr Res* 1996;**39**:914–17.
228. Santorelli FM, Tanji K, Shanske S, DiMauro S. Heterogeneous clinical presentation of the mtDNA NARP/T8993G mutation. *Neurology* 1997;**49**:270–3.
229. Kara B, Arikan M, Maras H, Abaci N, Cakiris A, Ustek D. Whole mitochondrial genome analysis of a family with NARP/MILS caused by m.8993T > C mutation in the MT-ATP6 gene. *Mol Genet Metab* 2012;**107**:389–93.

230. Anderson KE, Sassa S, Bishop DF, Desnick RJ. Disorders of heme biosynthesis: X-linked sideroblastic anemia and the porphyrias. In: Valle D, Beaudet AL, Vogelstein B, Kinzler KW, Antonarakis SE, Ballabio A, et al., editors. *OMMBID: The Online Metabolic and Molecular Bases of Inherited Diseases*. New York: 2013.
231. Waldenstrom J. The porphyrias as inborn errors of metabolism. *Am J Med* 1957;**22**:758–73.
232. Goldberg A. Acute intermittent porphyria: a study of 50 cases. *Q J Med* 1959;**28**:183–209.
233. Stein JA, Tschudy DP. Acute intermittent porphyria. A clinical and biochemical study of 46 patients. *Medicine* 1970;**49**:1–16.
234. Pischik E, Kauppinen R. Neurological manifestations of acute intermittent porphyria. *Cell Mol Biol (Noisy-le-grand)* 2009;**55**:72–83.
235. Sorensen AW, With TK. Persistent pareses after porphyric attacks. *S Afr Med J* 1971;101–3.
236. Poser CM, Edwards K. Transient monoparesis in acute intermittent porphyria. *Arch Neurol* 1978;**35**:550.
237. Albers JW, Robertson Jr. WC, Daube JR. Electrodiagnostic findings in acute porphyric neuropathy. *Muscle Nerve* 1978;**1**:292–6.
238. Allen SC, Rees GA. A previous history of acute intermittent porphyria as a complication of obstetric anaesthesia. *Br J Anaesth* 1980;**52**:835–8.
239. Sim M, Hudon R. Acute intermittent porphyria associated with postural hypotension. *Can Med Assoc J* 1979;**121**:845–6.
240. Kappas A, Sassa S, Galbraith RA, Drummond GS, Nordman Y. The porphyrias. In: Scriver CR, Beaudet AL, Sly WS, Valle D, editors. *The Metabolic and Molecular Bases of Human Disease*. 7th ed. New York: McGraw Hill; 1995. pp. 2103–59.
241. Barclay N. Acute intermittent porphyria in childhood. A neglected diagnosis? *Arch Dis Child* 1974;**49**:404–6.
242. Becker DM, Kramer S. The neurological manifestations of porphyria: a review. *Medicine* 1977;**56**:411–23.
243. Mustajoki P, Tenhunen R. Variant of acute intermittent porphyria with normal erythrocyte uroporphyrinogen-I-synthase activity. *Eur J Clin Invest* 1985;**15**:281–4.
244. Sassa S, Zalar GL, Kappas A. Studies in porphyria. VII. Induction of uroporphyrinogen-I synthase and expression of the gene defect of acute intermittent porphyria in mitogen-stimulated human lymphocytes. *J Clin Invest* 1978;**61**:499–508.
245. Sassa S, Solish G, Levere RD, Kappas A. Studies in porphyria. IV. Expression of the gene defect of acute intermittent porphyria in cultured human skin fibroblasts and amniotic cells: prenatal diagnosis of the porphyric trait. *J Exp Med* 1975;**142**: 722–31.

Chapter 20

Acute Polyneuropathies

H. Royden Jones, Jr., Padraic J. Grattan-Smith, and Monique M. Ryan

INTRODUCTION

The acute polyneuropathies are a group of acquired, generally immune-mediated or infectious disorders causing dysfunction and sometimes degeneration of peripheral nerves, spinal sensory and motor nerve roots, and occasionally the cranial nerves. The most common of these conditions are the Guillain-Barré syndrome and its variants. This chapter will also cover the presentation and treatment of tick paralysis.

GUILLAIN-BARRÉ SYNDROME

Guillain-Barré syndrome (GBS) is an acute inflammatory polyneuropathy that is most often characterized by rapidly progressive, essentially symmetric weakness and areflexia in a previously well child.[1,2] Weakness is usually distally predominant, at least at onset, but an initially proximal pattern of weakness is seen in 15–20% of children, and cranial nerve involvement is common in pediatric GBS. Distal paresthesias and neuropathic pain are prominent in many affected children. The neurologic deficit of GBS progresses over a few days to a month. The clinical diagnosis of GBS is confirmed by the findings of cytoalbuminologic dissociation (elevation of the cerebrospinal fluid (CSF) protein without concomitant CSF pleocytosis), and by neurophysiologic abnormalities suggesting an acute (usually demyelinating) neuropathy. These abnormalities are not always present in the early stages of the illness. Diagnostic criteria for GBS have been suggested but do not encompass the full spectrum of this disorder, so that the diagnosis is based on consistent clinical, laboratory, and neurophysiologic findings with exclusion of alternate diagnoses (Box 20.1).[1]

In the post-polio era, GBS is the most common cause of acute flaccid paralysis in childhood, with an annual incidence of 0.34–1.34 cases per 100,000 persons aged 18 years or less. GBS is slightly more common in males. It occurs worldwide, affecting all age groups and ethnicities, with an increase in frequency in each decade of life.[3–5] Rare cases develop *in utero* or in the neonatal period. Variation in the incidence of GBS in different populations, and the occasional reports of familial cases of GBS, may reflect variable genetic susceptibility or geographic exposure to causative pathogens.[6,7]

Clinical Findings

In 50–70% of cases, GBS develops 2–4 weeks after a prodromal gastroenteritic or respiratory illness, such as those caused by Epstein-Barr virus, cytomegalovirus, *Mycoplasma pneumoniae*, hepatitis, varicella, and other herpesviruses.[8–18] *Campylobacter jejuni* is the most common infectious agent associated with axonal forms of GBS[2,16–18] while cytomegalovirus infection causes a form of GBS with prominent sensory symptoms and involvement of the cranial nerves.[10] The incidence of GBS after *C. jejuni* infection, however, is only 0.25–0.65/1000 cases. Occasional cases follow immunization or surgery, or develop in association with other illnesses, but childhood GBS has not been proven to result from vaccination against poliomyelitis, tetanus, or measles.[19–21] Uncommonly, GBS may be the presenting manifestation of primary central nervous system or systemic lymphoma.[22,23]

GBS can occur at any age. Occasional congenital and neonatal cases have been described, with vertical transmission between an affected mother and fetus also having been reported on several occasions.[24–26] As a treatable cause of neonatal hypotonia, GBS should therefore be included in the differential diagnosis of the "floppy infant." Approximately one-third of childhood cases occur before the age of 3 years.[27]

Infants with GBS present with hypotonia and decreased movement, and sometimes with respiratory insufficiency. Irritability may be prominent. Diagnosis of infantile GBS is often delayed.[28]

B.T. Darras, H. Royden Jones, Jr., M.M. Ryan & D.C. De Vivo (Eds): Neuromuscular Disorders of Infancy, Childhood and Adolescence, Second edition.
DOI: http://dx.doi.org/10.1016/B978-0-12-417044-5.00020-2

BOX 20.1 Diagnostic Criteria for Guillain-Barré Syndrome

Development of essentially symmetric limb weakness in arms and legs—may start in legs only
Cranial nerve involvement
Loss or decrease in muscle stretch reflexes within 1 week of onset
Autonomic instability
Pain in neck, back, or extremities—often poorly localized
Progression of these features over several days to 4 weeks
Paresthesias of the hands and feet

Features casting doubt on the diagnosis of GBS:

Persistent asymmetry of weakness
Identifiable sensory level
Prominent and persisting bladder or bowel dysfunction
More than 50 mononuclear cells/mm^3 in CSF
Polymorphonuclear cells in CSF

Laboratory abnormalities that support the diagnosis:

Elevation of CSF protein >45 mg/dL within 3 weeks of onset
Neurophysiologic abnormalities consistent with an acute inflammatory polyneuropathy in at least two limbs:

1. Slowing of motor and sensory nerve conduction (<80% of lower limit of normal values for age)
2. Conduction block or temporal dispersion of compound muscle action potentials (CMAPs)
3. Increased distal latencies
4. Abnormalities of F waves (loss, impersistence, dispersion)
5. Criteria for axonal forms include lack of neurophysiologic evidence of demyelination, with loss or dispersion of CMAP or sensory nerve action potential amplitudes to at least <80% of lower limit of normal values for age

Modified from Asbury and Cornblath.[1]

Older children usually complain of weakness and fatigue, paresthesias and pain, and have difficulty walking, rising from the floor, and climbing stairs.[9,28–33] Weakness is generally symmetrical, starting in the lower extremities and then ascending into the upper extremities over days to weeks, and remaining distally predominant throughout the course of the illness.[9,27,28] In a significant proportion (15–20%) of cases, however, weakness is primarily proximal. Cranial nerve involvement is more common in pediatric than adult GBS. Facial weakness and ophthalmoplegia are seen in up to 45% of affected children. Internal ophthalmoplegia is rare in GBS and should prompt consideration of tick paralysis, diphtheria, and botulism.[9,27,28]

Neck, back, buttock, or leg pain, which is presumed to result from nerve root and peripheral nerve inflammation, is the initial manifestation in as many as 50% of children with GBS.[8,9,29,34–39] This pain is often poorly localized and may cause marked irritability.[40] Recognition of this symptom is especially important in young children who may be unable to describe weakness, and in whom pain and refusal to weight-bear is often initially thought to result from orthopedic or rheumatologic causes.[41] The severity of pain tends to broadly correspond to the degree of weakness.[8]

As many as 50% of children with GBS have ataxia resulting from weakness and sensory loss ("deafferentation") rather than from cerebellar involvement.[9,28,29,42] Rare cases of pediatric GBS present as the Miller Fisher variant of ataxia, ophthalmoplegia, and areflexia without peripheral weakness.[43,44]

Respiratory muscle weakness in GBS is usually slowly progressive and tends to correlate with the degree of limb muscle weakness. Very rare cases of pediatric GBS present with acute-onset respiratory failure, which is life-threatening and may be associated with complete flaccid paralysis.[45,46] Such severe and acute cases should prompt consideration of an underlying genetic neuropathy.[47]

Meningism, headache, drowsiness, and irritability are seen occasionally in children presenting with a "pseudo-encephalopathic" form of GBS that mimics acute meningitis or meningoencephalitis.[9,40] There are several reports of locked-in syndrome in children with severe GBS presenting with quadriparesis, areflexia, respiratory failure, and cranial nerve dysfunction. Neurophysiologic studies show inexcitable peripheral nerves. The prognosis of such cases is poor.[45,48]

By definition GBS has a rapid course, reaching a maximal deficit within 4 weeks, although as many as 80% of children reach their clinical nadir within 2 weeks of onset.[9] At their nadir 60% of children cannot walk, and 24% lose functional movement of their arms.[8]

Physical examination reveals weakness, which ascends from the distal lower extremities over days–weeks, and loss of the muscle stretch reflexes. These changes may be quite patchy and can be asymmetric, especially at onset. Weakness usually begins in the legs and progresses rostrally, but can be proximally predominant at onset. The deep tendon reflexes are usually lost within the first week of the illness, but are occasionally preserved throughout its course.[9,28,29] Rare cases of hyperreflexia in GBS are usually associated with the acute motor axonal variant of this condition.[49]

Sensory loss is rarely prominent in childhood GBS, but can be identified on detailed examination in about 40% of cases.[9] Identification of a sensory level is not compatible with GBS and should prompt immediate consideration of spinal cord pathology.

Involvement of the autonomic nervous system is seen in 25–50% of children with GBS, usually manifesting as blood pressure instability, sinus tachycardia, pupillary abnormalities, and abnormalities of sweating.[8,9,27,43,50] Autonomic instability and urinary incontinence are more common in atypical or focal forms of pediatric GBS.[43] Recent reports of hypertensive encephalopathy and posterior reversible encephalopathy as initial manifestations of GBS highlight the importance of considering this condition as part of the differential diagnosis of any acute dysautonomia.[51,52]

Sphincter dysfunction (urinary retention, urinary and fecal incontinence) is occasionally seen in children with GBS but is far more common in transverse myelitis, tumors, and other spinal cord lesions.[9] Although urinary retention occurs in 10–15% of children early in the course of GBS, persistent bladder or bowel involvement are indications for spinal imaging.[28]

Clinical Variants of Guillain-Barré Syndrome

The various clinical subtypes of GBS are defined by their differential involvement of motor and sensory axons of peripheral nerves (Table 20.1). In most populations, the most common form of GBS is acute inflammatory demyelinating polyneuropathy, which in most series accounts for about 75% of cases, while acute motor axonal neuropathy most often occurs in outbreaks associated with *Campylobacter jejuni* infection.[2,16,18]

Acute Inflammatory Demyelinating Polyneuropathy

Acute inflammatory demyelinating polyneuropathy (AIDP) is the classic form of the Guillain-Barré syndrome and its most common subtype in Western nations.[2,4,27] This acute immune-mediated neuropathy is thought to reflect injury, mediated by activated T cells and antibody responses, to Schwann cell and myelin epitopes in spinal roots and peripheral nerves. The antibody responses are poorly understood, as AIDP is only infrequently linked to specific antibodies against peripheral nerve gangliosides and glycolipids. The pathologic correlates of AIDP include inflammatory infiltrates in affected nerve segments, with concomitant macrophage-mediated vesicular demyelination along the course of sensory and motor axons. The neurophysiologic correlates include patchy, multifocal slowing of nerve conduction, conduction block, and loss of late responses.[27,28,53,54] The clinical deficits resulting from these processes are determined by the severity of conduction block and the degree of concomitant axonal degeneration.

Acute Motor Axonal Neuropathy

The acute motor axonal neuropathy (AMAN) form of GBS was originally described in southern Mexico, and subsequently found to underpin epidemics of acute flaccid paralysis following *Campylobacter jejuni* enteritis in Chinese children. *C. jejuni* expresses lipo-oligosaccharides which resemble the carbohydrate component of gangliosides. AMAN mimics AIDP clinically, but the fulminant motor deficits of AMAN are unassociated with sensory deficits. Neurophysiologic studies in AMAN show normal speed of peripheral nerve conduction, decreased amplitude and dispersion of compound motor action potential (CMAP) amplitudes with normal sensory responses, and electromyographic evidence of denervation.[2,18,19] Pathology studies confirm the discrete involvement of peripheral motor axons, with sparing of dorsal spinal nerve roots and sensory axons.[55–57] Nerve histology shows inflammatory mononuclear infiltrates with macrophage-mediated phagocytosis of motor axons. This inflammation is often associated with development of IgG antibodies against the GM1 and/or

TABLE 20.1 Clinical Variants of Guillain-Barré Syndrome in Childhood

Clinical Syndrome	Relative Frequency	IgG Antiganglioside Antibody Association(s)
Acute inflammatory demyelinating polyneuropathy	Common	GM1 (minority)
Acute motor axonal neuropathy	Common	GM1, GD1a
Acute motor and sensory axonal neuropathy	Uncommon	GM1, GD1a
Miller Fisher syndrome	Uncommon	GQ1b, GT1a
Pharyngeal-cervical-brachial variant	Rare	GT1a, GQ1b, GD1a
Polyneuritis cranialis	Rare	GQ1b, GT1a
Acute (ataxic) sensory neuropathy	Very rare	GQ1b, GT1a
Acute pandysautonomia	Very rare	
Acute ophthalmoparesis	Very rare	GQ1b, GT1a

GD1a gangliosides.[2,58] Although the peripheral nerve axon is the primary immunologic target in this disorder, its clinical course and tempo of recovery are often very similar to AIDP, suggesting that most nerve injury is either due to antibody-mediated paranodal or very distal nerve injury, and hence quickly repaired.[56]

Acute Motor-Sensory Axonal Neuropathy

The acute motor-sensory axonal neuropathy (AMSAN) variant is a rare, severe form of GBS. Like AMAN, the clinical presentation of AMSAN may be indistinguishable from AIDP, but neurophysiologic studies show pure axonal changes (diminished CMAP and sensory nerve action potential amplitudes, with relative preservation of motor and sensory nerve conduction velocities). In severe cases peripheral nerve responses may not be recordable. There is evidence of denervation on EMG.[59–61] The histopathologic features reflect severe degeneration of motor and sensory axons.[62] Differentiation of AMSAN from AIDP is contingent on the neurophysiologic findings, and is important because children with the former generally have a more severe deficit with a slower, often incomplete recovery.

Miller Fisher Syndrome and the Pharyngo-Cervical-Brachial Variant of GBS

Although generally attributed to C. Miller Fisher, this clinical syndrome of ophthalmoparesis, ataxia, and areflexia was initially recognized as a GBS variant by Georges Guillain himself in 1916.[44] Bulbar weakness may also be present, and children initially presenting with these findings often later evolve truncal and limb weakness.[63,64] Neurophysiologic testing may show loss of late responses (F waves and/or H responses), and there is often slowing of peripheral nerve conduction or evidence of an axonal neuropathy.[65] Formes fruste of Miller Fisher syndrome (MFS) include isolated acute ptosis or ophthalmoplegia, with or without ataxia,[66] while Bickerstaff's brainstem encephalitis may present as MFS with additional long-tract signs attributable to central nervous system involvement. Most patients eventually recover completely. MFS is often associated with circulating IgG and IgM antibodies to the GQ1b and/or GT1a gangliosides, glycolipids expressed on the axolemmal membrane of cranial nerves.[2,58] The CSF protein is generally normal in the first week but elevated by the second or third week of the illness.[67]

The pharyngo-cervical-brachial variant of GBS also has a regional predominance, but can be associated with late evolution of more generalized weakness, disseminated changes on nerve conduction studies, and antibodies to GT1a or/or GQ1b.[58] In practice these clinical syndromes overlap quite significantly. All tend to recover quickly.

Polyneuritis Cranialis

This uncommon form of GBS most often presents with bilateral facial neuropathies, with or without dysphagia and dysphonia, but with little (if any) limb weakness, although this may develop later in the disease course.[9,29,65,68–70] The muscle stretch reflexes are diminished or absent. Polyneuritis cranialis is also associated with prominent sensory symptoms and a more severe and prolonged clinical course. Cerebrospinal fluid examination discloses typical cytoalbuminologic dissociation, while nerve conduction studies show a demyelinating neuropathy. Multiple cranial nerve involvement may be associated with serologic evidence of antibody response to gangliosides GT1a and GQ1b. Polyneuritis cranialis is common in GBS after cytomegalovirus infection.[10]

Occasional cases of pediatric GBS are limited to acute pandysautonomia, pure sensory loss, or acute ataxic neuropathy. Diagnosis is often delayed in such cases, in which nerve conduction studies often but not invariably show evidence of a more generalized neuropathy. CSF examination may show cytoalbuminologic dissociation.[51,52,71–74]

Pathophysiology

GBS is likely an autoimmune disorder resulting from T-cell activation and production of antibodies directed at antigenic proteins of the peripheral nerves.[2,13,28] Infectious agents such as Epstein-Barr virus, cytomegalovirus, *Mycoplasma pneumoniae*, and *Campylobacter jejuni*, or stressors such as immunizations, surgery, or parturition may trigger formation of these antibodies.[13,18] Although most such antibodies are directed at myelin proteins, in some cases axonal moieties may be the primary target of immune-mediated nerve injury.[2,9,54] Thus, while GBS has traditionally also been known as acute inflammatory demyelinating polyneuropathy (AIDP), the primary pathologic process of this condition is in some cases axonal injury, as in the GBS variants acute motor axonal neuropathy (AMAN) and acute motor and sensory axonal neuropathy (AMSAN).[1,2,9]

The immunologic mechanisms of nerve injury underlying GBS have been extensively studied but remain incompletely understood. The peripheral nervous system is usually protected from circulating macromolecules by a blood-nerve barrier, maintained at the level of endothelial cell tight junctions within capillary beds. This barrier blocks access of proteins and lymphocytes to the brain and peripheral nervous system, and stops antibody binding to peripheral nerve antigens.

The first step in the process underlying GBS is breakdown of this immunologic tolerance, which may occur after an infection or other immune-activating event as a result of molecular mimicry, superantigen mechanisms, or cytokine activation. The blood-nerve barrier is relatively deficient at the level of the nerve roots and distal motor nerve terminals, rendering these sites particularly vulnerable to immune-mediated peripheral nerve injury.

The next step involves antigen recognition by T cell receptors, and antigen processing through major histocompatibility complexes. Activated T cells gain access to their target autoantigens by crossing the blood-nerve barrier and migrating into the endoneurium via adhesion molecules on endoneurial endothelial cell walls. In concert with macrophages, these activated T cells react against autoantigens on Schwann cell or axonal constituents. Simultaneous activation of complement is triggered by antibody binding to Schwann cells and deposition of activated complement components.[75–78]

In AIDP, these inflammatory infiltrates trigger segmental demyelination of motor and sensory nerve fibers, often in association with secondary axonal degeneration, which is generally first apparent at the level of the spinal roots and distal nerve endings where the blood-nerve barrier is relatively deficient. The target antigens on Schwann cells in AIDP are essentially unknown. Vesicular myelin degeneration in GBS may be discrete or diffuse, and may affect the peripheral nerves at any point from their origin in the spinal cord to the neuromuscular junction. The weakness of GBS results from conduction block and concomitant or primary axonal injury in affected motor nerves. Pain and paresthesiae are the clinical correlates of sensory nerve involvement. Panel A in Figure 20.1 shows a schema of the possible pathogenesis of antibody-mediated Guillian-Barré syndrome.

In AMAN, IgG antibodies and activated complement bind to the motor nerve axolemma at the nodes of Ranvier, with subsequent formation of the membrane attack complex, axonal injury and degeneration, without focal demyelination or lymphocytic infiltrates (Panel B, Figure 20.1).[2,57,75–78] This may impair nerve conduction by blocking sodium axonal channels or by triggering complement-mediated disruption of nodal and paranodal sodium channel clusters. Similar changes are seen in both motor and sensor nerve fibers in AMSAN.

Many investigators have tried to correlate specific antibody-ganglioside responses to symptoms and signs in discrete GBS variants (Table 20.1). Gangliosides (ceramide-sugar complexes containing sialic acid and an oligosaccharide core) are common constituents in and on the surface of neuronal and axolemmal membranes, but they are relatively rare on the surface of Schwann cells and myelin.[79] Not surprisingly, antibody responses against specific gangliosides appear most robust in those GBS variants that particularly target peripheral nerve axons. In most series, the most commonly identified autoantibodies are those against GM1 gangliosides, which are often seen in patients with AMAN and AMSAN.[80–83] AMAN has also been linked with an antibody response to ganglioside GD1a, and less commonly to other ganglioside species.[80,83] The Miller Fisher syndrome is associated with IgG1 antibodies to the ganglioside GQ1b.[18,68,69] Antibody responses are not specific to the various GBS clinical and neurophysiologic sub-types.[84–86]

Molecular mimicry has been postulated to underlie the link between antecedent infections and other stressors, and the development of specific antibody responses in GBS. Specific surface components of infectious agents, such as lipo-oligosaccharide in the outer membrane of *C. jejuni*, resemble specific gangliosides, GM1 and GD1a.[14,16,18] A cross-reacting immune response may underlie some cases of GBS, with regional patterns of neurologic deficit reflecting varying ganglioside expression in different neural tissues.[18,55,69,81–83]

B-cell mediated antibody responses are not the sole basis of GBS. Experimental allergic neuritis (EAN), a murine model of GBS, can be induced by immunization using peripheral myelin, whole constituent myelin proteins, or even protein fragments, or by passive transfer of activated cells sensitized against these proteins. Segmental demyelination in EAN is more severe when it occurs in response to both activated T cells and targeting antibodies, suggesting that both T cell- and B cell-mediated immunity are important in this disease model.[87] Pro-inflammatory cytokine levels are also often diffusely increased in GBS, but their exact role in disease pathogenesis remains unclear.[88,89]

The current evidence favors a pathogenic mechanism for GBS in which antibodies and activated T cells act in concert to coordinate a macrophage-mediated attack on specific components of the peripheral nerve that may include myelin, Schwann cells, motor and/or sensory axons. Nerve injury may be increased or lessened by local and circulating pro- or anti-inflammatory cytokines. Immune targeting in this process can be very focal, as in specific pure sensory, motor, or dysautonomic forms of GBS, or may result in diffuse injury to sensory and motor axons and their accompanying Schwann cells and myelin.

Investigations

There is no single investigation that can confirm or disprove the diagnosis of GBS, particularly early in its course. The diagnosis is contingent on supporting evidence from the clinical examination, lumbar puncture, neurophysiologic, radiologic and/or ancillary investigations, with exclusion of alternative diagnoses by the same means.

Lumbar puncture is generally undertaken in cases of suspected GBS, to rule out infectious conditions and identify the typical findings of GBS. These include elevation of the CSF protein (reflecting nerve root demyelination) without evidence of active infection (lack of CSF

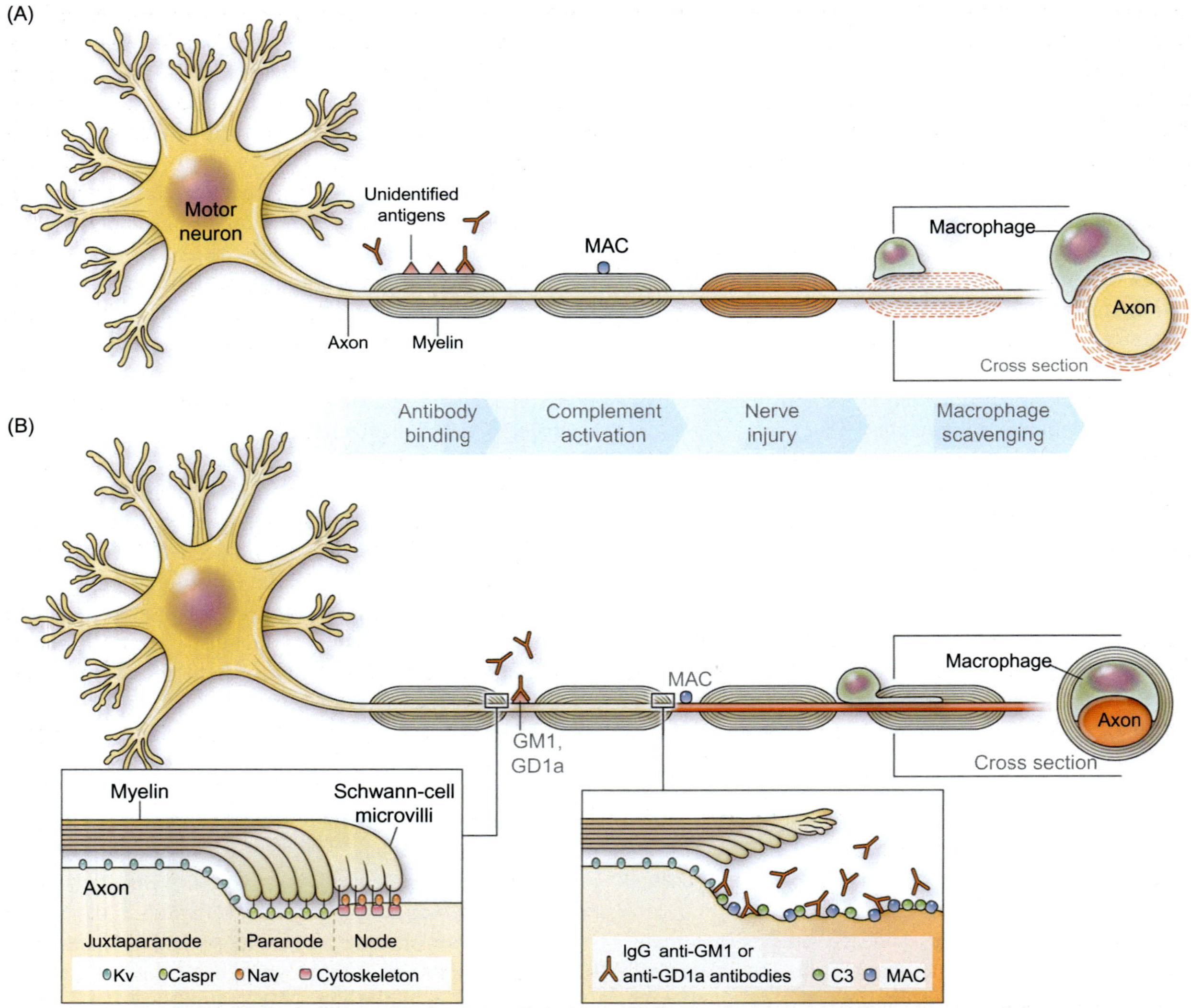

FIGURE 20.1 Possible immunopathogenesis of the Guillain-Barré syndrome. Panel **A** shows the putative immunopathogenesis of acute inflammatory demyelinating polyneuropathy. Autoantibodies may bind to myelin antigens, activating complement and triggering formation of the membrane-attack complex (MAC) on the outer surface of Schwann cells. Vesicular myelin degeneration is followed by removal of myelin debris by macrophages. Panel **B** shows the proposed immunopathogenesis of acute motor axonal neuropathy (AMAN). Myelinated axons are divided into four functional regions: the nodes of Ranvier, paranodes, juxtaparanodes, and internodes. Myelin gangliosides GM1 and GD1a are strongly expressed at the nodes of Ranvier, where the voltage-gated sodium (Nav) channels are localized. Contactin-associated protein (Caspr) and voltage-gated potassium (Kv) channels are sited at the paranodes and juxtaparanodes. In AMAN, IgG anti-GM1 or anti-GD1a autoantibodies bind to the nodal axolemma, leading to MAC formation, complement deposition, loss of Nav clusters and detachment of paranodal myelin. This leads to failure of nerve conduction and axonal degeneration. *From Yuki and Hartung*[2] *with permission.*

pleocytosis). This finding, known as *cytoalbuminologic dissociation*, is not always present in GBS; an increase in the CSF protein may not be seen in the first 48 hours of symptoms, and the CSF protein sometimes remains normal for as long as a week.[1,2,9,27] The CSF protein level generally correlates poorly with disease severity. CSF microscopy in GBS generally reveals fewer than 5 leukocytes/mm^3, but occasional patients have 10–50 mononuclear cells/mm^3.[9,28,33] Significant CSF pleocytosis (>50 leukocytes/mm^3) is very atypical of childhood GBS, and should prompt consideration of other disorders such as poliomyelitis, enterovirus 71, and HIV infection.[27,28,90–92] As CSF pleocytosis is not invariable in myelitis, or even encephalitis, the CSF cell count may not reliably differentiate between GBS and other conditions. The CSF is usually normal in tick paralysis and botulism.

Neurophysiologic studies (nerve conduction studies and electromyography) confirm the clinical diagnosis of GBS and characterize the underlying process as primarily demyelinating or axonal. Peripheral nerve demyelination is reflected in slowing of nerve conduction, which tends to be more profound in younger children.[9,29] Because these changes are patchy in GBS, testing of multiple nerves characteristically reveals variable increases in distal latencies, slowing of nerve conduction, conduction block, and dispersion of motor responses. The late responses (F and H waves), which test both distal and more proximal segments of the peripheral nerves, are often more sensitive in identifying mild slowing of nerve conduction.[9,27] Both severe demyelination and axonal injury cause loss of amplitude of motor and sensory responses. In the pure axonal forms of GBS, conduction velocities are preserved despite loss of amplitude of motor responses. Electromyography reveals a loss of motor unit action potentials (decreased recruitment) early in the illness. Changes of acute denervation (fibrillation potentials and positive sharp waves) are seen only in cases with significant axonal injury (whether primary or secondary), and are not usually apparent until 7–10 days into the disease course, although they may be seen earlier in very small children.[9,27,29]

Electrodiagnostic finding are, to some extent, predictive of outcome in pediatric GBS.[9,27,28,31] In primarily axonal forms of GBS, a severe reduction in CMAP amplitudes generally predicts a prolonged, and potentially incomplete, recovery.[33,56,57] If the underlying process is acute peripheral nerve demyelination, this often primarily affects proximal nerve roots and the terminal segments of motor nerves, and can be accompanied by conduction block, predominantly as the nerve reaches its termination in the intramuscular regions. In some cases this results in unrecordable motor responses, and it may not be possible, on neurophysiologic criteria, to distinguish this situation from extensive axonal degeneration. The prognostic implications of severely reduced CMAP amplitudes must therefore be interpreted with caution. When peripheral responses are unrecordable, a search for more proximal responses (such as from stimulation of the phrenic or axillary nerves) may enable identification of a primarily demyelinating process. Electromyographic evidence of acute denervation is always suggestive of severe axonal injury and a less good prognosis.[9,27,28,30–33]

Performance of a detailed neurophysiologic study enables diagnosis of GBS in as many as 90% of cases during the first week of symptoms.[27] Changes are virtually universal by the second week of illness, by which time a definitive diagnosis can almost always be made. Electrodiagnostic studies are uncomfortable, however, and can be technically challenging in small children; therefore, they should be performed only by individuals with appropriate pediatric expertise.

Neuroimaging has become a valuable adjunct to neurophysiologic studies in suspected GBS. Post-gadolinium enhancement of the peripheral nerve roots and cauda equina may be seen on spinal magnetic resonance imaging in children with GBS.[93,94] Similar findings are often also seen in the cranial nerves of patients with clinical evidence of cranial neuropathy.[95] These findings are seen in as many as 95% of patients with GBS, but they are not specific to this disorder. Similar changes can be seen after lumbar puncture and in chronic genetic and inflammatory neuropathies.[96] Spinal magnetic resonance imaging (MRI) should be undertaken as a matter of urgency in children presenting with severe back pain, a sensory level, or prominent sphincter dysfunction, in order to exclude transverse myelitis or spinal cord compression.

Nerve biopsy is virtually never required for diagnosis of pediatric GBS but, where performed, typically shows mononuclear endoneurial inflammatory infiltrates, variable degrees of Wallerian degeneration, and macrophage-mediated segmental demyelination.[62,84]

Antiganglioside antibodies are identifiable in about 50% of patients with GBS.[58,76,78] There is a degree of ganglioside specificity to GBS subtype, but there is overlap between these subtypes and the ganglioside antibody subtypes identifiable in individual subjects (Table 20.1) Where available, testing for these antibodies can be very useful in confirming the clinical diagnosis, particularly in patients with atypical presentations or formes fruste of GBS.

Differential Diagnosis

The differential diagnosis of childhood GBS includes disorders involving the central nervous system, peripheral nerves, neuromuscular junction, nerve, and muscle (Box 20.2). See Case Example 20.1.

As many as 50% of children with GBS have significant ataxia, which may relate to proximal weakness causing clumsiness and frequent falls, or to sensory ataxia from decreased afferent input.[9,28,31] These findings raise the possibility of acute cerebellar ataxia, which can be distinguished from GBS in that it is not associated with weakness or loss of the muscle stretch reflexes.[42]

Transverse myelitis and acute disseminated encephalomyelitis (ADEM) are syndromes of central nervous system demyelination that share with GBS an autoimmune pathogenesis and tendency to develop after a viral illness or other prodrome. Occasional cases of concurrent ADEM and GBS or GBS and transverse myelitis are reported, in which instance there is a case for combined treatment with corticosteroids and IVIg or plasma exchange.[8,97,98] There can be considerable overlap in the clinical presentation of acute myelopathy and early GBS. Both can be associated with back and/or leg pain, and weakness initially confined to the lower extremities.[99,100] A sensory level may be difficult to identify in the young child, and reflexes may be depressed early in the course of the acute

BOX 20.2 Differential Diagnosis of Childhood Guillain-Barré Syndrome

Spinal cord lesions

Brainstem encephalitis, transverse myelitis, epidural abscess, vascular malformations, cord infarction or compression, fibrocartilaginous embolism, diskitis, trauma, congenital anomalies, tumors

Anterior horn cell disorders

Poliomyelitis, other enteroviral myelitides, West Nile virus, Hopkins' syndrome

Peripheral neuropathies

Toxic: Vincristine, other chemotherapeutics, lead, other heavy metals, organophosphates, glue sniffing, buckthorn wild cherry, fish toxins
Infectious: Lyme disease, HIV, diphtheria
Inborn errors of metabolism: Leigh disease, other mitochondrial neuropathies, Tangier disease, Fabry disease, porphyria, Refsum's disease, tyrosinemia
Critical illness polyneuropathy
Inflammatory: Chronic inflammatory demyelinating polyneuropathy, vasculitic neuropathies
Nutritional: Vitamin B1 deficiency

Disorders of neuromuscular transmission

Myasthenia gravis, botulism, tick paralysis, hypercalcemia, hypermagnesemia, Lambert-Eaton syndrome

Myopathies

Viral myositis, dermatomyositis, critical illness myopathy, periodic paralyses

CASE EXAMPLE 20.1

A previously well 2-year-old girl presented with refusal to walk, irritability, and sleep disturbance. On examination she had neck and back stiffness. She cried constantly and was reluctant to weight-bear. Formal strength testing could not be undertaken, but her deep tendon reflexes were absent. She was admitted to hospital with a putative diagnosis of viral illness. Over the following couple of days she developed obvious limb weakness without respiratory or swallowing difficulties. Magnetic resonance imaging of the brain was normal, while a contrast-enhanced MRI of the spine showed enhancement of the cauda equina and proximal nerve roots. Nerve conduction studies showed patchy slowing of motor and sensory nerve conduction, with conduction block and loss of F-waves.

Comment

Meningism is common in children with Guillain-Barré syndrome, which in young children often presents with pain and irritability. This sort of pain syndrome can mimic synovitis or even meningoencephalitis. A careful neurologic examination is important in all irritable unwell children with poorly differentiated symptoms.

spinal cord syndrome. Sphincter dysfunction is common with spinal cord lesions, but relatively rare (and usually transient) in GBS.[28] Transverse myelitis may be associated with marked elevation of the CSF protein and significant CSF pleocytosis.[99] Neurophysiologic studies are usually normal in transverse myelitis, but there may be loss of late responses (F and H waves) corresponding with the affected spinal level(s).[27]

Poliomyelitis has been largely eradicated in countries with widespread immunization programs, but continues to surface in nonimmunized groups and immunocompromised individuals. Where the attenuated oral vaccine is employed, cases of vaccine-associated paralytic polio can follow mutation of the attenuated live virus to a wild-type phenotype causing paralytic disease. The risk of paralytic poliomyelitis after immunization with the oral live-attenuated Sabin vaccine is 1 in 2.5 million vaccinations.[101] This disorder will hopefully no longer be seen now that most countries have transitioned to immunization with the inactivated Salk vaccine. Other enteroviruses can also cause acute focal, asymmetric limb weakness, usually in association with fever and pain, which can mimic GBS. Probably the most common single species is enterovirus 71, which has been responsible for epidemic outbreaks with a relatively high incidence of neurologic complications among infected children.[90–92] In addition to direct viral invasion of the central nervous system causing meningoencephalitis or poliomyelitis, this strain of enterovirus appears to be highly immunogenic, resulting in postinfectious disorders such as acute cerebellar ataxia and GBS.[90] CSF analysis shows a polymorphonuclear pleocytosis, while nerve conduction studies reflect acute denervation without demyelination. Neuroimaging may show striking preferential involvement of the anterior horn. A definitive diagnosis is achieved by serologic testing.

Other acute neuropathies can be distinguished from GBS based on the clinical setting or associated findings. Children with mitochondrial diseases such as pyruvate dehydrogenase deficiency can present with an acute illness resembling GBS.[102,103] In such cases an inborn error of metabolism may be suggested by pre-existing developmental delay or epilepsy, the presence of systemic lactic acidosis, or a suggestive family history. Nerve conduction studies in such cases may be normal or may show a demyelinating or axonal neuropathy.

Guillain-Barré syndrome can occur in subjects with underlying inherited neuropathies such as Charcot-Marie-Tooth disease. In such cases the diagnosis may be known prior to development of an acute deterioration raising the question of a superimposed inflammatory process, or may be suspected when a child with GBS recovers less quickly and completely than would be expected.[104] An underlying inherited neuropathy should be suspected in very severe cases of GBS,[47] or where multiple family members develop an apparently acquired neuropathy.[105,106]

Neuromuscular transmission disorders are sometimes confused with GBS. In small children infantile botulism should be considered, particularly where there is early constipation, involvement of the bulbar muscles, and internal and/or external ophthalmoplegia. Infant botulism is generally seen in the first 36 weeks of life.[107] The CSF examination is normal in botulism, while the neurophysiologic examination reveals incremental responses on rapid repetitive nerve stimulation, with florid electromyographic changes (abundant fibrillation potentials and low-amplitude polyphasic motor unit action potentials).[108] Myasthenia gravis tends to have a slower time course than GBS, and can generally be distinguished from a neuropathy because of the findings of ptosis, fluctuating ophthalmoparesis, and prominent proximal weakness.[109] Tick paralysis typically affects young children between 2 and 5 years of age, and causes an ascending quadriparesis with internal and external ophthalmoplegia and early bulbar involvement (see below). When appendicular muscles are involved, children are frequently areflexic. On nerve conduction studies, tick paralysis causes loss of amplitude of the motor responses. The CSF is usually normal in this condition.[110]

Acute myopathies are uncommon in childhood. Children with inflammatory myopathies may complain of myalgia and weakness, with muscle tenderness on palpation. The serum creatine kinase is usually markedly elevated in children with significant weakness due to an acute myositis, and mildly elevated in dermatomyositis, which tends to have a more indolent course and is often associated with skin changes. Periodic paralysis may present in childhood with weakness, hypo- or areflexia with normal level of consciousness, and a normal sensory examination. Bouts of paralysis are usually relatively brief and self-limited, resolving over hours, and may be associated with unusually high or low serum potassium values.

Management

On diagnosis, all children with GBS should be admitted to hospital for monitoring. Although weakness and hypotonia may be relatively mild at onset, the potential for sudden, sometimes fatal respiratory or autonomic compromise should always be anticipated in this condition.[27] Children with mild GBS who are able to ambulate unassisted are usually treated expectantly. Those with rapid clinical progression, loss of the ability to walk, or significant bulbar or respiratory compromise should receive specific immunomodulatory treatment, with the aims of amelioration of disease severity and acceleration of recovery. Many pediatric clinicians would have a lower threshold for treating infants with immunosuppressive therapies because of the greater difficulty in assessing clinical progression in very small children.

Vigilant supportive therapy is vital in GBS and includes monitoring for respiratory and autonomic complications of this disorder, in addition to pain management and prevention of complications of immobility (constipation, pressure areas, contractures, and renal calculi). The vital signs and respiratory capacity should be closely monitored. Children should be admitted to a pediatric intensive care unit if they have a rapidly progressive course, inability to flex the neck or arms, flaccid limb weakness, or significant bulbar, respiratory, or autonomic dysfunction.[111] Intubation and mechanical ventilation should be considered when vital capacity falls below 15 mL/kg of body weight, arterial pO_2 falls below 70 mmHg, or there is significant fatigue.[111] As many as 15–20% of children require ventilatory support for acute GBS.[8,9,28,112] The need for mechanical ventilation relates to the severity of the limb and axial weakness, and is also predicted by severe hypotension.[112]

Although there have been no prospective placebo-controlled randomized trials of these therapies in pediatric GBS, a number of prospective and retrospective studies have suggested that treatment with plasma exchange (PE) or intravenous immunoglobulin (IVIg) hastens recovery of independent ambulation in GBS, and shortens requirement for hospital admission.[8,9,113–120] Both therapies appear most effective when given within 2 weeks of onset of symptoms. Neither therapy has been proven to improve long-term outcome in this disorder, although both improve outcome at 12 months after presentation. Both are thought to act by neutralizing antibody-mediated peripheral nerve dysfunction, and are probably most effective when administered within 7 days of symptom onset. Specifically, intravenous human immunoglobulin may act by binding pathogenic antibodies, downregulation of B cell-mediated antibody production and accelerated antibody catabolism, and complement inhibition. Other possible mechanisms of action include blockade of activated T cell receptors, augmentation of suppressor T cell activity, or inhibition of lymphocyte proliferation.[116] Plasmapheresis reduces levels of circulating autoantibodies, and may reduce levels of circulating proinflammatory cytokines or cell adhesion molecules.[117]

Intravenous immunoglobulin is generally preferred for treatment of childhood GBS because of administrative ease and a preferable side-effect profile.[115,119,120] A total dose of 2g/kg IVIg is given over 2 to 5 days and is generally well tolerated at all ages.[8,9,28] Most centers give IVIg over 2 days, although two small studies have suggested a slightly greater risk of relapse with this regimen as opposed to a 5-day course.[119,121] Possible side-effects of immunoglobulin include fever, headache, nausea and vomiting, myalgia, allergic reactions, hypercoagulability, and transmission of bloodborne infections. Comparison of therapeutic regimens has been limited by differences between treatment groups, including disease severity and time to treatment, but a meta-analysis of such trials that have been conducted to date suggests that IVIg is as effective as plasma exchange in accelerating recovery from GBS.[116] A clinical response to IVIg is usually apparent within 3 to 7 days. Intravenous immunoglobulin is the agent of choice in patients with proven antibody-mediated GBS.[2]

Plasma exchange (PE), or plasmapheresis, can be a safe and effective treatment of GBS for children weighing more than 10 kg, and is usually performed 4 to 6 times on an alternate-day schedule, to a total of 250 ml/kg (or triple-volume) exchange. Adverse effects of plasma exchange may include hypotension, infection, and hypercoagulability. A recent meta-analysis of treatment trials with plasmapheresis for GBS suggested that, overall, adults treated for GBS with PE have a slightly higher risk of relapse in the first year after treatment than those who are untreated, but 12 months after presentation are more likely to have recovered fully.[117] PE has been shown in several small studies of pediatric GBS to decrease time to independent walking.[114,122–124] There are few comparative trials of PE and IVIg in childhood GBS. A recent Egyptian study reported that PE might be superior to IVIg in very severe pediatric GBS necessitating ventilatory support, although this benefit was limited to time required for mechanical ventilation, and not to length of intensive care unit stay or short-term neurological outcome.[121]

The effects of treatment with both PE and IVIg are commonly delayed for several days, during which interval clinical progression may occur. In severe cases, or patients whose condition deteriorates a few weeks after treatment ("treatment-related fluctuations"), repeated courses of plasmapheresis or IVIg may be beneficial.[125,126] Sequential or combined administration of these treatments has not been studied in childhood but does not show additional benefit in adults with severe GBS.[116,127] A current focus of research interest is individual variability in immunoglobulin pharmacokinetics, which may result in very variable IgG levels after immunoglobulin infusions and hence variable responsiveness to treatment.[128] An ongoing double-blind randomized trial is examining whether there is a subpopulation of GBS patients with low IgG increments after infusions who would benefit from a second course of therapy.[129]

Alternative techniques of immunoglobulin ultrafiltration, such as CSF filtration and immunoabsorption, have been reported in small adult series but not yet trialed in children with GBS.[130,131]

Corticosteroid therapy was previously a mainstay of treatment for GBS, but has now been shown to be ineffective when used alone or in combination with IVIg. Steroid therapy may in fact slow recovery from GBS, and may be complicated by glucose intolerance and hypertension.[132] Corticosteroids may, however, sometimes have a place in the treatment of severe radicular or neuropathic pain in GBS.[38]

Blood pressure lability and arrhythmias can be life-threatening in very unwell children with GBS.[50–52,112] Hypertension should be treated only when symptomatic or extreme, as there may be marked sensitivity to antihypertensive agents. Pacing devices are rarely required for children with severe bradycardia. Occasionally patients with severe GBS require treatment for gastroparesis, ileus, and urinary retention.

Pain is commonly underrecognized and undertreated in childhood GBS. Nonsteroidal anti-inflammatory medications and agents used to treat neuropathic pain (gabapentin, pregabalin, and amitriptyline) are often ineffective early in the disease course, but should be considered early, with rapid dose increases as tolerated. Corticosteroids may be effective in alleviating pain, although they do not otherwise ameliorate the course of GBS.[38] Opioid analgesics are often required for adequate analgesia.[36] Pain control is a paramount consideration even though narcotic analgesia may contribute to respiratory depression and constipation.

Physiotherapy should be initiated immediately upon control of pain in childhood GBS, and continued throughout the recovery period, to maintain skin integrity and joint range of motion, and prevent contractures. Careful attention should also be paid to nutrition, as the consequences of inadequate caloric intake include accelerated muscle catabolism.

Immobilization hypercalcemia is occasionally seen in children with severe GBS, and when severe may require treatment with calcitonin and bisphosphonates.[133]

Outcome

Children with GBS typically have a shorter clinical course and more complete recovery than adults with this disorder.[8,9,27,28] As many as 60% of children become nonambulant during their illness.[8,111] Fifteen to 20% require ventilatory support, but respiratory failure does not predict a persisting deficit.[8,9,27,28] Most children reach their clinical nadir within 2 weeks, and recovery begins soon thereafter. In most cases there is minimal residual impairment by 1 to 4 months from onset.[8,9,30,111] Ultimately more than 90% of children recover fully, while a small minority have mild weakness (most commonly of the ankle dorsiflexors) but are able to walk unaided.[8,9,43] Children with AIDP generally recover more quickly than children affected by AMAN or AMSAN.[36,43,112]

Electrodiagnostic markers of severe axonal injury do not invariably predict poor outcome in pediatric GBS.[8,9] Similarly, development of antibodies to GM1, GD1b, GD1a, and/or GT1a is not invariably associated with poor prognosis.[78] Outcomes are less good in very young children and children who are very weak at presentation, quadriparetic on day 10, require ventilator support, or have inexcitable motor nerves.[43,133–137]

Mortality in childhood GBS is low—of the order of 1–2%—and generally results from respiratory failure.[28] The importance of zealous respiratory, cardiovascular, and autonomic monitoring in all cases of GBS cannot be overemphasized. Many deaths are due to potentially preventable respiratory complications. Autonomic instability is a predictor of fatal cardiac arrhythmias in GBS.[111] Relapses occur in about 4% of cases of childhood GBS, and are generally responsive to immunomodulatory treatment.[8] Occasionally children go on to have recurrent weakness and are ultimately diagnosed with chronic inflammatory demyelinating polyneuropathy.[8,9]

Future Directions

Various aspects of the natural history and response to therapy in GBS are poorly understood. Individual susceptibility to development of forms of GBS (e.g. to AMAN after *C. jejuni* infection) is likely underpinned by as yet undefined genetic and other host factors, including immune-response polymorphisms. These are an active area of research interest.[138,139]

An international multicenter collaborative study, the IGOS 1000 trial, is currently underway. This study aims to better characterize clinical, neurophysiologic, therapeutic, pharmacokinetic, immunological, and genetic features of children and adults with GBS, with a goal to identify predictors of course, prognosis, and response to therapy.[140] An important aspect of such studies is both better definition of the natural history of this clinically heterogeneous condition and the establishment of outcome measures by which to assess the efficacy of new treatments.[141] Research into pediatric GBS is particularly challenging in that it must address these issues in the context of the maturational and growth-related changes in peripheral nerve function in childhood.

TICK PARALYSIS

Ticks are arachnids and live by sucking blood from mammals, birds, and reptiles. They have been regarded as malevolent creatures since antiquity. Pliny (AD 77) described the tick as "most disgusting...living on blood with its head always fixed and swelling; being one of those animals which has no exit for its food, it bursts with over-repletion and dies from actual nourishment."[142] Only mosquitoes transmit more infectious diseases than ticks. Human infections transmitted by ticks include those caused by protozoa (babesiosis), rickettsiae (Rocky Mountain spotted fever, Queensland tick typhus), spirochetes (Lyme disease), bacteria (ehrlichiosis, tularemia), and viruses (Russian spring-summer encephalitis, Colorado tick fever, Crimean-Congo hemorrhagic fever). Ticks may also provoke hypersensitivity reactions including anaphylaxis.[143]

Ticks of medical importance are divided into two families: Argasidae (soft ticks) and Ixodidae (hard ticks).[144] Argasid ticks have a soft, leathery body and feed for short periods (usually 5 to 25 minutes), while ixodid ticks are characterized by a hard body plate and feed for as long as 11 days.[145]

Tick paralysis was probably first recorded in 1824, by the Australian explorer Hovell. The first medical descriptions of generalized paralysis in humans came from the USA, Canada, and Australia in 1912.[146–148]

Epidemiology

Generalized tick paralysis is uncommon in humans. In an Australian study of acute flaccid paralysis in children under 15 years of age, of 143 cases reported over a 4-year period, only 3 were due to tick paralysis.[149] Apparent clustering of cases in some reports suggests that a geographic variability in toxin potency may result from variable virulence of some toxin strains.[150]

Tick paralysis mainly affects children aged 1 to 5 years. Girls are more commonly affected, perhaps because long hair makes detection of a tick embedded in the scalp more difficult.[110,151–153] Adults may be protected by their larger body mass and because they appreciate the presence of the tick before it becomes fully engorged. In most large series around 80% of cases, and most fatalities, occur in children younger than 16 years of age.[110,151–153]

Pathogenesis

A number of characteristics contribute to the success of ticks as parasites and disease vectors.[154] They have a relatively long life span, living for years. Their salivary glands secrete multiple hemostatic and anti-inflammatory agents, enabling attachment and feeding, and are also the source of the paralysis toxin. Most species of ticks do not paralyze their host; there are 217 species of *Ixodes*, but only 7 cause paralysis. Paralysis usually develops as the tick becomes fully engorged, and the host may die after the tick has fully fed and detached itself.

In North America, tick paralysis in humans is usually caused by *Dermacentor andersoni* or *Dermacentor variabilis*.[151] *Ixodes holocyclus* is responsible for most cases in Australia. It seems to be the most potent of all the world's paralyzing ticks and has produced paralysis in a number of animal species as well as humans.[155] It is the adult female tick that causes paralysis, and these are most abundant in spring and early summer, which is the most common time for tick paralysis. When a tick comes in contact with a child, it must first attach itself, then neutralize the host defenses (inflammatory and cell-mediated immunity, and the local hemostatic systems), and finally prepare itself for a massive increase in body size as blood is sucked from the host. The scalp behind the ear, and other sites on the head and neck, are preferred site for ticks causing paralysis (Figure 20.2). Once the tick is firmly attached, the tick begins feeding, its buccal canal becoming a common duct for intake of host blood and output of tick saliva.[157] Feeding is spasmodic for several days, while the cuticle growth necessary for rapid expansion of body size is completed. The rapid feeding phase then commences. Comparatively huge volumes of blood are taken up over a 12 to 24 hour period, with the body weight of the tick increasing 100 to 200 times. This phase of rapid feeding and intense salivation coincides with the time of peak production of toxin.[157]

There is a delay of several days from the time of attachment of the tick to the onset of symptoms. It seems likely that one component of this delay is the time taken for the tick to secrete significant quantities of toxin. The circulating toxin may also need to undergo a process of binding and internalization before intracellular poisoning occurs.[158] The ongoing clinical deterioration often seen after the removal of *I. holocyclus* may be due to the continued action of internalized toxin.

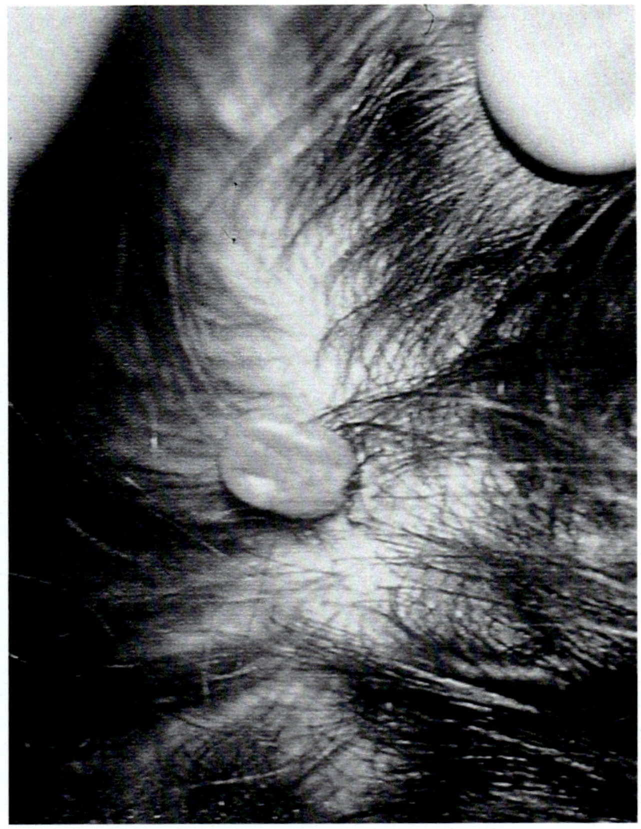

FIGURE 20.2 Tick bite. Engorged female tick (*Dermacentor variabilis*) in the scalp of a 6-year-old girl with ascending paralysis, ataxia, areflexia, bulbar signs, and hypoventilation. *From Felz et al.,[156] with permission.*

Pathophysiology of Tick Paralysis

The pathophysiology of tick paralysis is poorly understood. A number of studies have examined the site of action of the toxin of *Dermacentor andersoni*. There appears to be a failure of liberation or synthesis of acetylcholine (ACh) at peripheral nerve terminals, secondary to toxin action on slower-conducting, small diameter motor and sensory fibers.[159,160] Animal studies of paralysis induced by *I. holocyclus* toxin have shown a marked reduction in muscle compound action potentials (CMAPs) evoked by nerve stimulation. Nerve conduction velocities (NCVs) and nerve action potentials, and responses to fast and slow repetitive nerve stimulation, were normal.[110,161,162] There was normal spontaneous release of ACh at rest, but there was temperature-dependent inhibition of evoked ACh release from nerve terminals after nerve stimulation. There was no change in the postsynaptic sensitivity to ACh, and direct muscle stimulation produced a normal contraction even when the nerves were unresponsive.[156,161–164] These studies suggest that the site of action of the toxin is at or near the neuromuscular junction: either a presynaptic abnormality of the excitation-secretion mechanism, possibly due to reduced calcium availability, or failure of distal fibers to conduct action potentials to their terminals, with the toxin interrupting sodium flux across axonal membranes.[156] Krishnan and colleagues used sophisticated nerve excitability techniques in studying a 45-year-old man with localized paralysis of the brachial plexus, and suggested that the terminal motor neuron was the likely site of the pathology, possibly due to a blockade in nodal sodium channels.[165]

The toxins of *D. variabilis* and *I. holocyclus* may act at different sites or in different ways. For example, the toxin of *I. holocyclus* may become internalized, resulting in a prolonged period of action, whereas that of *D. variabilis* might reversibly attach to sodium receptors on axonal membranes, with rapid loss of action once the tick is removed.[156] Alternatively, these toxins can have multiple sites of action; snake venoms, for example, can cause both presynaptic and postsynaptic neuromuscular blockade.

Clinical Features

The paralysis caused both by North American and Australian ticks is preceded by several days of fatigue, anorexia, and nonspecific head and body pains, followed by unsteadiness of gait. In some cases ataxia initially seems more prominent than weakness, but there is then development of an ascending symmetrical paralysis and areflexia, which can take more than 24 hours to fully develop. There may be bulbar and facial weakness. Occasionally cases with only focal (face or arm weakness) are reported,[166–168] and some children present only with acute cerebellar signs or bulbar weakness.[169,170] There is no sensory loss. Tick paralysis evolves slowly, but it can be deadly.[146,151] If the tick is not removed or respiratory support not provided, affected children may die of respiratory failure.

The paralytic syndromes caused by North American and Australian ticks differ. Internal and external ophthalmoplegia are common with paralysis caused by *I. holocyclus*.[110] Pupillary abnormalities are rare in paralysis caused by North American ticks, although they may cause development of nystagmus.[146,171,172] The most important intercontinental difference, however, is in the rate of recovery after tick removal. Removal of the North American ticks usually produces dramatic changes: "improvement in 1 hour and recovery within a day." Most patients recover completely within a few days of tick removal.[156,162,172] In contrast, with Australian ticks paralysis often progresses for as long as 48 hours after the tick has been removed or has dropped off. Although this deterioration is part of the natural history of *I. holocyclus* paralysis, the presence of other undetected ticks should always be considered. Recovery from *I. holocyclus* may take some weeks.[110] Once the paralysis has resolved, a complete recovery is expected.

Although tick toxin seems to mainly act on peripheral nerves, there is human and animal evidence of autonomic and possibly central nervous system involvement. Autonomic disturbance with heart rate and blood pressure instability mimicking Guillain-Barré syndrome have been reported in a 57-year-old female.[173] Pupillary involvement is seen with Australian tick paralysis.[161] Vomiting is occasionally reported and mental state changes are apparent in some children with tick paralysis.[110] More unusual complications of tick paralysis in humans include myocarditis[174] and myositis.[175,176]

Case Example 20.2 emphasizes the importance of a careful search for a tick in any child with ascending paralysis.

Differential Diagnosis

The main differential diagnosis of tick paralysis is GBS, which can present very similarly. Tick paralysis must always be considered in endemic areas, particularly in the spring and summer months. Careful examination

CASE EXAMPLE 20.2

A previously-well 23-month-old girl presented with a 24-hour history of lethargy, vomiting, and poor fluid intake. She had stopped talking, had become unsteady, and was unable to walk without assistance.

On examination, she was drowsy, irritable, and mildly dehydrated. Her right pupil was dilated and reacted sluggishly to light. There were no other focal neurologic signs and her general examination was normal. The full blood count and biochemical profile were normal but she had a metabolic acidosis with pH 7.3, carbon dioxide 23 mmHg, bicarbonate 11 mmol/L, and base excess of 14. Her creatine kinase was elevated at 452 U/L (range 0–200 U/L). The cerebrospinal fluid, electrocardiograph, cerebral magnetic resonance imaging, and chest X-ray were normal.

She was thought to be suffering from encephalitis. She was rehydrated and given intravenous cefotaxime, but her condition deteriorated overnight. The following day, her pupils were mid-size and nonreactive to light. She had bilateral ptosis and complete paralysis of the ocular movements, with facial weakness and a poor gag reflex. She was unable to lift her limbs against gravity. She was areflexic, with no Babinski sign. Sensory testing was normal. A member of the nursing staff discovered a tick behind her right ear. The tick was carefully removed, using forceps to the mouth parts, and identified as *I. holocyclus*.

Over the following days she received several doses of tick antitoxin but her condition continued to worsen. She required intubation for airway protection. Two days later some pupillary responses were seen. Her spontaneous movement improved, but she remained ventilator-dependent for another week. By day 16 she could stand briefly and feed herself. By day 20, she was walking unaided, and 4 months later she had returned to normal.

Adapted from Grattan-Smith et al.,[110] with permission.

(including combing of the hair) may prevent a large number of unnecessary investigations and incorrect treatment. In Australia, early involvement of the pupils is an important clue that the diagnosis may be tick paralysis rather than GBS. Fixed, dilated pupils associated with ascending paralysis have also been described following snake bite[177] and blue-ringed octopus bite,[178] in which the evolution of signs is much more rapid than in tick paralysis. Infant botulism can produce a similar picture but is almost always seen in the first year of life.[107] Other causes of evolving weakness such as progressive spinal cord compression, transverse myelitis, myasthenia gravis, poliomyelitis, various forms of periodic paralysis, and organophosphate or other poisoning should also be considered in the differential diagnosis.

With rapid intercontinental travel, and the relatively long time between attachment and the development of symptoms, examples of tick paralysis can appear in unexpected locations, as in the example of the 59-year-old Japanese man presenting with paralysis due to *Ixodes holocyclus* after returning home from a visit to Australia.[179]

Clinical Neurophysiology

The neurophysiologic study of children with tick paralysis is technically challenging. The child is usually unwell and very weak, and these studies are often performed in the difficult environment of an intensive care unit, rather than a neurophysiology laboratory. As a result, only a small number of reports have been published, and in most only rudimentary studies have been performed.

There have been seven North American reports of nerve conduction studies in tick paralysis.[156,162,171,172,180–182] Each was limited to a single patient, and in 3 patients only 1 nerve was studied. The tick was identified as *D. variabilis* in 4 cases. A constant finding in these studies was a marked reduction in size of the CMAPs, often with prolongation of the distal latencies and mild slowing of motor nerve conduction. Repeat studies show reversibility of these changes once children have recovered. Sensory responses are normal. Repetitive motor nerve stimulation shows variable effects but is generally normal.[162,172,180–182] In one interesting, somewhat atypical case in which the child had more than 50 ticks on his back, serial nerve conduction studies showed severe changes with persistently low-amplitude CMAPs and changes of denervation on EMG.[181]

The largest Australian study of the neurophysiologic findings in tick paralysis in humans is that of Grattan-Smith and colleagues, who described 6 pediatric cases.[110] Almost all nerves studied showed very low CMAP amplitudes. Motor nerve conduction was normal or showed mild slowing of nerve conduction. Repetitive motor nerve stimulation, at high and low rates, was normal, as were the sensory studies. Attempts to cool the limbs were technically complicated, and no definitive response was seen.

Therapy

The first step in treatment is finding the tick and then removing it. The most common site of attachment is in the scalp behind an ear. Not uncommonly, a swelling is noted but it may not be readily identified as an engorged tick, and there are many reports in the literature in which tick discovery is delayed or even post-mortem.[110,183] There may be more than one tick, and a careful search of body crevices should be made, including the auditory meatus, nasal cleft, vulva, and nose.[146,184]

Many methods of tick removal have been advocated, but traditional methods such as the application of a lighted match, alcohol, or petroleum jelly are not effective.[185] It may be difficult to remove the tick intact and without local trauma. It has been recommended that the tick be grasped close to the skin with curved forceps and removed with steady pressure, but there are concerns that this may lead to rapid dispersal of toxin and exacerbation of weakness. Use of a topical pyrethrin-based insecticidal spray may be a useful initial step.[186]

In North America, removal of the tick is usually followed by rapid recovery and no further treatment is required. After removal of *I. holocyclus*, in a child showing any signs of weakness it should be assumed that the child's condition will worsen and that the child may require respiratory support. The child should be observed in hospital for at least 48 hours. Antitoxin, a hyperimmune serum prepared from dogs, is an effective treatment for animals paralyzed by *I. holocyclus*.[155] In humans it has been used only in severely ill patients, because of the significant risk of acute reactions and serum sickness. The antitoxin has appeared effective in some,[110,153] but not all cases.[184,187,188] Variation in efficacy of the antitoxin could be the result of different toxins secreted by different ticks, but it is also possible that the toxin may have become internalized and inaccessible to the circulating antitoxin.

REFERENCES

1. Asbury AK, Cornblath DR. Assessment of current diagnostic criteria for Guillain-Barré syndrome. *Ann Neurol* 1990;**27**(Suppl.):S21–4.
2. Yuki N, Hartung HP. Guillain-Barré syndrome. *N Engl J Med* 2012;**366**:2294–304.
3. McGrogan A, Madle GC, Seaman HE, de Vries CS. The epidemiology of Guillain-Barré syndrome worldwide. A systematic literature review. *Neuroepidemiology* 2009;**32**:150–63.
4. Hughes RA, Rees JH. Clinical and epidemiologic features of Guillain-Barré syndrome. *J Infect Dis* 1997;**176**(Suppl. 2):S92–8.

5. Sejvar JJ, Baughman AL, Wise M, Morgan OW. Population incidence of Guillain-Barré syndrome: a systematic review and meta-analysis. *Neuroepidemiology* 2011;**36**:123–33.
6. Geleijns K, Brouwer BA, Jacobs BC, Houwing-Duistermaat JJ, van Duijn CM, van Doorn PA. The occurrence of Guillain-Barré syndrome within families. *Neurology* 2004;**63**:1747–50.
7. Rees JH, Vaughan RW, Kondeatis E, Hughes RA. HLA-class II alleles in Guillain-Barré syndrome and Miller Fisher syndrome and their association with preceding *Campylobacter jejuni* infection. *J Neuroimmunol* 1995;**62**:53–7.
8. Korinthenberg R, Schessl J, Kirschner J. Clinical presentation and course of childhood Guillain-Barré syndrome: a prospective multicentre study. *Neuropediatrics* 2007;**38**:10–17.
9. Bradshaw DY, Jones Jr. HR. Guillain-Barré syndrome in children: clinical course, electrodiagnosis, and prognosis. *Muscle Nerve* 1992;**15**:500–6.
10. Visser LH, van der Meche FG, Meulstee J, Rothbarth PP, Jacobs BC, Schmitz PI, et al. Cytomegalovirus infection and Guillain-Barré syndrome: the clinical, electrophysiologic, and prognostic features. Dutch Guillain-Barré Study Group. *Neurology* 1996;**47**:668–73.
11. Boucquey D, Sindic CJ, Lamy M, Delmee M, Tomasi JP, Laterre EC. Clinical and serological studies in a series of 45 patients with Guillain-Barré syndrome. *J Neurol Sci* 1991;**104**:56–63.
12. Grose C, Henle W, Henle G, Feorino PM. Primary Epstein-Barr-virus infections in acute neurologic diseases. *N Engl J Med* 1975;**292**:392–5.
13. Jacobs BC, Rothbarth PH, van der Meche FG, Herbrink P, Schmitz PI, de Klerk MA, et al. The spectrum of antecedent infections in Guillain-Barré syndrome: a case-control study. *Neurology* 1998;**51**:1110–15.
14. Kusunoki S, Chiba A, Hitoshi S, Takizawa H, Kanazawa I. Anti-Gal-C antibody in autoimmune neuropathies subsequent to mycoplasma infection. *Muscle Nerve* 1995;**18**:409–13.
15. Meseguer MA, Aparicio M, Calvo A, de Blas G, Lorenzo G. Mycoplasma pneumoniae antigen detection in Guillain-Barré syndrome. *Eur J Pediatr* 1998;**157**:1034.
16. Rees JH, Soudain SE, Gregson NA, Hughes RA. *Campylobacter jejuni* infection and Guillain-Barré syndrome. *N Engl J Med* 1995;**333**:1374–9.
17. Poropatich KO, Walker CL, Black RE. Quantifying the association between *Campylobacter* infection and Guillain-Barré syndrome: a systematic review. *J Health Popul Nutr* 2010;**28**:545–52.
18. Yuki N. Pathogenesis of Guillain-Barré and Miller Fisher syndromes subsequent to *Campylobacter jejuni* enteritis. *Jpn J Infect Dis* 1999;**52**:99–105.
19. Haber P, Sejvar J, Mikaeloff Y, DeStefano F. Vaccines and Guillain-Barré syndrome. *Drug Saf* 2009;**32**:309–23.
20. Tuttle J, Chen RT, Rantala H, Cherry JD, Rhodes PH, Hadler S. The risk of Guillain-Barré syndrome after tetanus-toxoid-containing vaccines in adults and children in the United States. *Am J Public Health* 1997;**87**:2045–8.
21. da Silveira CM, Salisbury DM, de Quadros CA. Measles vaccination and Guillain-Barré syndrome. *Lancet* 1997;**349**:14–16.
22. Toren A, Mandel M, Shahar E, Rimmoni E, Roizin H, Neuman Y. Primary central nervous system Burkitt's lymphoma presenting as Guillain-Barré syndrome. *Med Pediatr Oncol* 1994;**23**:372–5.
23. Gucuyener K, Keskil S, Baykaner MK, Bilir E, Oguz A, Ceviker N. Co-incidence of Guillain-Barré syndrome and spinal cord compression in non-Hodgkin lymphoma. *Neuropediatrics* 1994;**25**:36–8.
24. Jackson AH, Baquis GD, Shah BL. Congenital Guillain-Barré syndrome. *J Child Neurol* 1996;**11**:407–10.
25. Rolfs A, Bolik A. Guillain-Barré syndrome in pregnancy: reflections on immunopathogenesis. *Acta Neurol Scand* 1994;**89**:400–2.
26. Luijckx GJ, Vles J, de Baets M, Buchwald B, Troost J. Guillain-Barré syndrome in mother and newborn child. *Lancet* 1997;**349**:27.
27. Delanoe C, Sebire G, Landrieu P, Huault G, Metral S. Acute inflammatory demyelinating polyradiculopathy in children: clinical and electrodiagnostic studies. *Ann Neurol* 1998;**44**:350–6.
28. Jones HR. Childhood Guillain-Barré syndrome: clinical presentation, diagnosis, and therapy. *J Child Neurol* 1996;**11**:4–12.
29. Jones Jr. HR. Guillain-Barré syndrome: perspectives with infants and children. *Semin Pediatr Neurol* 2000;**7**:91–102.
30. Sladky JT. Immune neuropathies in childhood. *Baillieres Clin Neurol* 1996;**5**:233–44.
31. Rantala H, Uhari M, Niemela M. Occurrence, clinical manifestations, and prognosis of Guillain-Barré syndrome. *Arch Dis Child* 1991;**66**:706–8, discussion 8–9.
32. Kleyweg RP, van der Meche FG, Loonen MC, de Jonge J, Knip B. The natural history of the Guillain-Barré syndrome in 18 children and 50 adults. *J Neurol Neurosurg Psychiatry* 1989;**52**:853–6.
33. Korinthenberg R, Monting JS. Natural history and treatment effects in Guillain-Barré syndrome: a multicentre study. *Arch Dis Child* 1996;**74**:281–7.
34. Khatri A, Pearlstein L. Pain in Guillain-Barré syndrome. *Neurology* 1997;**49**:1474.
35. Mikati MA, DeLong GR. Childhood Guillain-Barré syndrome masquerading as a protracted pain syndrome. *Arch Neurol* 1985;**42**:839.
36. Moulin DE, Hagen N, Feasby TE, Amireh R, Hahn A. Pain in Guillain-Barré syndrome. *Neurology* 1997;**48**:328–31.
37. Nguyen DK, Agenarioti-Belanger S, Vanasse M. Pain and the Guillain-Barré syndrome in children under 6 years old. *J Pediatr* 1999;**134**:773–6.
38. Ropper AH, Shahani BT. Pain in Guillain-Barré syndrome. *Arch Neurol* 1984;**41**:511–14.
39. Roca B, Mentero A, Simon E. Pain and opioid analgesics in Guillain-Barré syndrome. *Neurology* 1998;**51**:924.
40. Bradshaw DY, Jones HR. Pseudomeningoencephalitic presentation of pediatric Guillain-Barré syndrome. *J Child Neurol* 2001;**16**:505–8.
41. Roodbol J, de Wit MC, Walgaard C, de Hoog M, Catsman-Berrevoets CE, Jacobs BC. Recognizing Guillain-Barré syndrome in preschool children. *Neurology* 2011;**76**:807–10.
42. Gieron-Korthals MA, Westberry KR, Emmanuel PJ. Acute childhood ataxia: 10-year experience. *J Child Neurol* 1994;**9**:381–4.
43. Lin JJ, Hsia SH, Wang HS, Lyu RK, Chou ML, Hung PC, et al Clinical variants of Guillain-Barré syndrome in children. *Pediatr Neurol* 2012;**47**:91–6.
44. Fisher M. An unusual variant of acute idiopathic polyneuritis (syndrome of ophthalmoplegia, ataxia and areflexia). *N Engl J Med* 1956;**255**:57–65.
45. Medici C, Gonzalez G, Cerisola A, Scavone C. Locked-in syndrome in three children with Guillain-Barré syndrome. *Pediatr Neurol* 2011;**45**:125–8.
46. Larsen A, Tobias JD. Landry-Guillain-Barré syndrome presenting with symptoms of upper airway obstruction. *Pediatr Emerg Care* 1994;**10**:347–8.

47. Munch C, Epplen JT, Meins M, Meyer R, Weber JR, Meyer T. Severe Guillain-Barré syndrome associated with chromosome 17p11.2–12 duplication. *Muscle Nerve* 2008;**37**:256–8.
48. Bakshi N, Maselli RA, Gospe Jr. SM, Ellis WG, McDonald C, Mandler RN. Fulminant demyelinating neuropathy mimicking cerebral death. *Muscle Nerve* 1997;**20**:1595–7.
49. Yuki N, Kokubun N, Kuwabara S, Sekiguchi Y, Ito M, Odaka M, et al. Guillain-Barré syndrome associated with normal or exaggerated tendon reflexes. *J Neurol* 2012;**259**:1181–90.
50. Dimario Jr. FJ, Edwards C. Autonomic dysfunction in childhood Guillain-Barré syndrome. *J Child Neurol* 2012;**27**:581–6.
51. Aleyadhy AA, Hassan GM. Hypertensive encephalopathy as the initial manifestation of Guillain-Barré syndrome in a 7-year-old girl. *Neurosciences (Riyadh)* 2013;**18**:163–5.
52. Sutter R, Mengiardi B, Lyrer P, Czaplinski A. Posterior reversible encephalopathy as the initial manifestation of a Guillain-Barré syndrome. *Neuromuscul Disord* 2009;**19**:709–10.
53. Brown WF, Snow R. Patterns and severity of conduction abnormalities in Guillain-Barré syndrome. *J Neurol Neurosurg Psychiatry* 1991;**54**:768–74.
54. Ropper AH, Wijdicks EF, Shahani BT. Electrodiagnostic abnormalities in 113 consecutive patients with Guillain-Barré syndrome. *Arch Neurol* 1990;**47**:881–7.
55. Hafer-Macko C, Hsieh ST, Li CY, Ho TW, Sheikh K, Cornblath DR, et al. Acute motor axonal neuropathy: an antibody-mediated attack on axolemma. *Ann Neurol* 1996;**40**:635–44.
56. Ho TW, Hsieh ST, Nachamkin I, Willison HJ, Sheikh K, Kiehlbauch J, et al. Motor nerve terminal degeneration provides a potential mechanism for rapid recovery in acute motor axonal neuropathy after *Campylobacter* infection. *Neurology* 1997;**48**: 717–24.
57. Lu JL, Sheikh KA, Wu HS, Zhang J, Jiang ZF, Cornblath DR, et al. Physiologic-pathologic correlation in Guillain-Barré syndrome in children. *Neurology* 2000;**54**:33–9.
58. Ho TW, Willison HJ, Nachamkin I, Li CY, Veitch J, Ung H, et al. Anti-GD1a antibody is associated with axonal but not demyelinating forms of Guillain-Barré syndrome. *Ann Neurol* 1999;**45**: 168–73.
59. Ho TW, Li CY, Cornblath DR, Gao CY, Asbury AK, Griffin JW, et al. Patterns of recovery in the Guillain-Barré syndromes. *Neurology* 1997;**48**:695–700.
60. Alam TA, Chaudhry V, Cornblath DR. Electrophysiological studies in the Guillain-Barré syndrome: distinguishing subtypes by published criteria. *Muscle Nerve* 1998;**21**:1275–9.
61. Yuki N, Kuwabara S, Koga M, Hirata K. Acute motor axonal neuropathy and acute motor-sensory axonal neuropathy share a common immunological profile. *J Neurol Sci* 1999;**168**:121–6.
62. Massaro ME, Rodriguez EC, Pociecha J, Arroyo HA, Sacolitti M, Taratuto AL, et al. Nerve biopsy in children with severe Guillain-Barré syndrome and inexcitable motor nerves. *Neurology* 1998;**51**:394–8.
63. Ter Bruggen JP, van der Meche FG, de Jager AE, Polman CH. Ophthalmoplegic and lower cranial nerve variants merge into each other and into classical Guillain-Barré syndrome. *Muscle Nerve* 1998;**21**:239–42.
64. Ropper AH. Miller Fisher syndrome and other acute variants of Guillain-Barré syndrome. *Baillieres Clin Neurol* 1994;**3**:95–106.
65. Wong V. A neurophysiological study in children with Miller Fisher syndrome and Guillain-Barré syndrome. *Brain Dev* 1997;**19**: 197–204.
66. Lee SH, Lim GH, Kim JS, Oh SY, Kim JK, Cha JK, et al. Acute ophthalmoplegia (without ataxia) associated with anti-GQ1b antibody. *Neurology* 2008;**71**:426–9.
67. Nishimoto Y, Odaka M, Hirata K, Yuki N. Usefulness of anti-GQ1b IgG antibody testing in Fisher syndrome compared with cerebrospinal fluid examination. *J Neuroimmunol* 2004;**148**:200–5.
68. Willison HJ, O'Hanlon GM. The immunopathogenesis of Miller Fisher syndrome. *J Neuroimmunol* 1999;**100**:3–12.
69. Chiba A, Kusunoki S, Obata H, Machinami R, Kanazawa I. Serum anti-GQ1b IgG antibody is associated with ophthalmoplegia in Miller Fisher syndrome and Guillain-Barré syndrome: clinical and immunohistochemical studies. *Neurology* 1993;**43**:1911–17.
70. Yuki N, Susuki K, Hirata K. Ataxic Guillain-Barré syndrome with anti-GQ1b antibody: relation to Miller Fisher syndrome. *Neurology* 2000;**54**:1851–3.
71. Ropper AH. The Guillain-Barré syndrome. *N Engl J Med* 1992; **326**:1130–6.
72. Jackman NL, Klig JE. Lower extremity pain in a three year old: manifestations of Guillain-Barré syndrome. *Pediatr Emerg Care* 1998;**14**:272–4.
73. Ito M, Matsuno K, Sakumoto Y, Hirata K, Yuki N. Ataxic Guillain-Barré syndrome and acute sensory ataxic neuropathy form a continuous spectrum. *J Neurol Neurosurg Psychiatry* 2011;**82**:294–9.
74. Yuki N, Ang CW, Koga M, Jacobs BC, van Doorn PA, Hirata K, et al. Clinical features and response to treatment in Guillain-Barré syndrome associated with antibodies to GM1b ganglioside. *Ann Neurol* 2000;**47**:314–21.
75. Hughes RA, Hadden RD, Gregson NA, Smith KJ. Pathogenesis of Guillain-Barré syndrome. *J Neuroimmunol* 1999;**100**:74–97.
76. Willison HJ, Yuki N. Peripheral neuropathies and anti-glycolipid antibodies. *Brain* 2002;**125**:2591–625.
77. Yuki N, Kuwabara S. Axonal Guillain-Barré syndrome: carbohydrate mimicry and pathophysiology. *J Peripher Nerv Syst* 2007; **12**:238–49.
78. Schessl J, Koga M, Funakoshi K, Kirschner J, Muellges W, Weishaupt A, et al. Prospective study on anti-ganglioside antibodies in childhood Guillain-Barré syndrome. *Arch Dis Child* 2007; **92**:48–52.
79. Chiba A, Kusunoki S, Obata H, Machinami R, Kanazawa I. Ganglioside composition of the human cranial nerves, with special reference to pathophysiology of Miller Fisher syndrome. *Brain Res* 1997;**745**:32–6.
80. Sekiguchi Y, Uncini A, Yuki N, Misawa S, Notturno F, Nasu S, et al. Antiganglioside antibodies are associated with axonal Guillain-Barré syndrome: a Japanese-Italian collaborative study. *J Neurol Neurosurg Psychiatry* 2012;**83**:23–8.
81. Oomes PG, Jacobs BC, Hazenberg MP, Banffer JR, van der Meche FG. Anti-GM1 IgG antibodies and *Campylobacter* bacteria in Guillain-Barré syndrome: evidence of molecular mimicry. *Ann Neurol* 1995;**38**:170–5.
82. Rees JH, Gregson NA, Hughes RA. Anti-ganglioside GM1 antibodies in Guillain-Barré syndrome and their relationship to *Campylobacter jejuni* infection. *Ann Neurol* 1995;**38**:809–16.

83. Kuwabara S, Yuki N, Koga M, Hattori T, Matsuura D, Miyake M, et al. IgG anti-GM1 antibody is associated with reversible conduction failure and axonal degeneration in Guillain-Barré syndrome. *Ann Neurol* 1998;**44**:202–8.
84. Asbury AK, Arnason BG, Adams RD. The inflammatory lesion in idiopathic polyneuritis. Its role in pathogenesis. *Med (Baltimore)* 1969;**48**:173–215.
85. Griffin JW, Li CY, Macko C, Ho TW, Hsieh ST, Xue P, et al. Early nodal changes in the acute motor axonal neuropathy pattern of the Guillain-Barré syndrome. *J Neurocytol* 1996;**25**:33–51.
86. Griffin JW, Li CY, Ho TW, Tian M, Gao CY, Xue P, et al. Pathology of the motor-sensory axonal Guillain-Barré syndrome. *Ann Neurol* 1996;**39**:17–28.
87. Hahn AF, Feasby TE, Wilkie L, Lovgren D. Antigalactocerebroside antibody increases demyelination in adoptive transfer experimental allergic neuritis. *Muscle Nerve* 1993;**16**:1174–80.
88. Koski CL. Mechanisms of Schwann cell damage in inflammatory neuropathy. *J Infect Dis* 1997;**176**(Suppl. 2):S169–72.
89. Lisak RP, Skundric D, Bealmear B, Ragheb S. The role of cytokines in Schwann cell damage, protection, and repair. *J Infect Dis* 1997;**176**(Suppl. 2):S173–9.
90. Takimoto S, Waldman EA, Moreira RC, Kok F, Pinheiro Fde P, Saes SG, et al. Enterovirus 71 infection and acute neurological disease among children in Brazil (1988–1990). *Trans R Soc Trop Med Hyg* 1998;**92**:25–8.
91. Alexander Jr. JP, Baden L, Pallansch MA, Anderson LJ. Enterovirus 71 infections and neurologic disease—United States, 1977–1991. *J Infect Dis* 1994;**169**:905–8.
92. McMinn P, Stratov I, Nagarajan L, Davis S. Neurological manifestations of enterovirus 71 infection in children during an outbreak of hand, foot, and mouth disease in Western Australia. *Clin Infect Dis* 2001;**32**:236–42.
93. Mulkey SB, Glasier CM, El-Nabbout B, Walters WD, Ionita C, McCarthy MH, et al. Nerve root enhancement on spinal MRI in pediatric Guillain-Barré syndrome. *Pediatr Neurol* 2010;**43**:263–9.
94. Yikilmaz A, Doganay S, Gumus H, Per H, Kumandas S, Coskun A. Magnetic resonance imaging of childhood Guillain-Barré syndrome. *Childs Nerv Syst* 2010;**26**:1103–8.
95. Zuccoli G, Panigrahy A, Bailey A, Fitz C. Redefining the Guillain-Barré spectrum in children: neuroimaging findings of cranial nerve involvement. *AJNR Am J Neuroradiol* 2011;**32**:639–42.
96. Iwata F, Utsumi Y. MR imaging in Guillain-Barré syndrome. *Pediatr Radiol* 1997;**27**:36–8.
97. Howell KB, Wanigasinghe J, Leventer RJ, Ryan MM. Concomitant transverse myelitis and acute motor axonal neuropathy in an adolescent. *Pediatr Neurol* 2007;**37**:378–81.
98. Mohammed RR, Jan MM. Co-morbid Guillain-Barré syndrome and acute disseminated encephalomyelitis. *Neurosciences (Riyadh)* 2013;**18**:166–8.
99. Knebusch M, Strassburg HM, Reiners K. Acute transverse myelitis in childhood: nine cases and review of the literature. *Dev Med Child Neurol* 1998;**40**:631–9.
100. Yiu EM, Kornberg AJ, Ryan MM, Coleman LT, Mackay MT. Acute transverse myelitis and acute disseminated encephalomyelitis in childhood: spectrum or separate entities? *J Child Neurol* 2009;**24**:287–96.
101. David WS, Doyle JJ. Acute infantile weakness: a case of vaccine-associated poliomyelitis. *Muscle Nerve* 1997;**20**:747–9.
102. Strassburg HM, Koch J, Mayr J, Sperl W, Boltshauser E. Acute flaccid paralysis as initial symptom in 4 patients with novel E1alpha mutations of the pyruvate dehydrogenase complex. *Neuropediatrics* 2006;**37**:137–41.
103. Singhi P, De Meirleir L, Lissens W, Singhi S, Saini AG. Pyruvate dehydrogenase-e1alpha deficiency presenting as recurrent demyelination: an unusual presentation and a novel mutation. *JIMD Rep* 2013;**10**:107–11.
104. Desurkar A, Lin JP, Mills K, Al-Sarraj S, Jan W, Jungbluth H, et al. Charcot-Marie-Tooth (CMT) disease 1A with superimposed inflammatory polyneuropathy in children. *Neuropediatrics* 2009; **40**:85–8.
105. Ginsberg L, Malik O, Kenton AR, Sharp D, Muddle JR, Davis MB, et al. Coexistent hereditary and inflammatory neuropathy. *Brain* 2004;**127**:193–202.
106. Verny C, Prundean A, Nicolas G, Pautot V, Maugin D, Levade T, et al. Refsum's disease may mimic familial Guillain Barré syndrome. *Neuromuscul Disord* 2006;**16**:805–8.
107. Underwood K, Rubin S, Deakers T, Newth C. Infant botulism: a 30-year experience spanning the introduction of botulism immune globulin intravenous in the intensive care unit at Childrens Hospital Los Angeles. *Pediatrics* 2007;**120**:e1380–5.
108. Cornblath DR, Sladky JT, Sumner AJ. Clinical electrophysiology of infantile botulism. *Muscle Nerve* 1983;**6**:448–52.
109. Castro D, Derisavifard S, Anderson M, Greene M, Iannaccone S. Juvenile myasthenia gravis: a twenty-year experience. *J Clin Neuromuscul Dis* 2013;**14**:95–102.
110. Grattan-Smith PJ, Morris JG, Johnston HM, Yiannikas C, Malik R, Russell R, et al. Clinical and neurophysiological features of tick paralysis. *Brain* 1997;**120**(Pt 11):1975–87.
111. Rantala H, Uhari M, Cherry JD, Shields WD. Risk factors of respiratory failure in children with Guillain-Barré syndrome. *Pediatr Neurol* 1995;**13**:289–92.
112. Hu MH, Chen CM, Lin KL, Wang HS, Hsia SH, Chou ML, et al. Risk factors of respiratory failure in children with Guillain-Barré syndrome. *Pediatr Neonatol* 2012;**53**:295–9.
113. Sladky JT. Guillain-Barré syndrome in children. *J Child Neurol* 2004;**19**:191–200.
114. Lamont PJ, Johnston HM, Berdoukas VA. Plasmapheresis in children with Guillain-Barré syndrome. *Neurology* 1991;**41**:1928–31.
115. Vajsar J, Sloane A, Wood E, Murphy EG. Plasmapheresis vs intravenous immunoglobulin treatment in childhood Guillain-Barré syndrome. *Arch Pediatr Adolesc Med* 1994;**148**:1210–12.
116. Hughes RA, Swan AV, van Doorn PA. Intravenous immunoglobulin for Guillain-Barré syndrome. *Cochrane Database Syst Rev* 2012;**7**: CD002063.
117. Raphael JC, Chevret S, Hughes RA, Annane D. Plasma exchange for Guillain-Barré syndrome. *Cochrane Database Syst Rev* 2012;**7**: CD001798.
118. Plasmapheresis and acute Guillain-Barré syndrome. The Guillain-Barré syndrome Study Group. *Neurology* 1985;**35**:1096–104.
119. Korinthenberg R, Schessl J, Kirschner J, Monting JS. Intravenously administered immunoglobulin in the treatment of childhood Guillain-Barré syndrome: a randomized trial. *Pediatrics* 2005;**116**:8–14.

120. Gurses N, Uysal S, Cetinkaya F, Islek I, Kalayci AG. Intravenous immunoglobulin treatment in children with Guillain-Barré syndrome. *Scand J Infect Dis* 1995;**27**:241–3.
121. El-Bayoumi MA, El-Refaey AM, Abdelkader AM, El-Assmy MM, Alwakeel AA, El-Tahan HM. Comparison of intravenous immunoglobulin and plasma exchange in treatment of mechanically ventilated children with Guillain Barré syndrome: a randomized study. *Crit Care* 2011;**15**:R164.
122. Epstein MA, Sladky JT. The role of plasmapheresis in childhood Guillain-Barré syndrome. *Ann Neurol* 1990;**28**:65–9.
123. Khatri BO, Flamini JR, Baruah JK, Dobyns WB, Konkol RJ. Plasmapheresis with acute inflammatory polyneuropathy. *Pediatr Neurol* 1990;**6**:17–19.
124. Jansen PW, Perkin RM, Ashwal S. Guillain-Barré syndrome in childhood: natural course and efficacy of plasmapheresis. *Pediatr Neurol* 1993;**9**:16–20.
125. Farcas P, Avnun L, Frisher S, Herishanu YO, Wirguin I. Efficacy of repeated intravenous immunoglobulin in severe unresponsive Guillain-Barré syndrome. *Lancet* 1997;**350**:1747.
126. van der Meche FG, Schmitz PI. A randomized trial comparing intravenous immune globulin and plasma exchange in Guillain-Barré syndrome. Dutch Guillain-Barré Study Group. *N Engl J Med* 1992;**326**:1123–9.
127. Hughes RA, Wijdicks EF, Barohn R, Benson E, Cornblath DR, Hahn AF, et al. Practice parameter: immunotherapy for Guillain-Barré syndrome: report of the Quality Standards Subcommittee of the American Academy of Neurology. *Neurology* 2003;**61**:736–40.
128. Kuitwaard K, de Gelder J, Tio-Gillen AP, Hop WC, van Gelder T, van Toorenenbergen AW, et al. Pharmacokinetics of intravenous immunoglobulin and outcome in Guillain-Barré syndrome. *Ann Neurol* 2009;**66**:597–603.
129. Walhgaard C, Jacobs BC, Cornblath DR, Steyerberg EW, van Doorn PA, for the Dutch GBS Study group. Second IVIg dose for Guillain-Barré syndrome with poor prognosis, a double-blind randomised controlled trial. 2011 *Meeting of the Peripheral Nerve Society*, Potomac, Maryland [poster].
130. Wollinsky KH, Hulser PJ, Brinkmeier H, Aulkemeyer P, Bossenecker W, Huber-Hartmann KH, et al. CSF filtration is an effective treatment of Guillain-Barré syndrome: a randomized clinical trial. *Neurology* 2001;**57**:774–80.
131. Haupt WF, Rosenow F, van der Ven C, Borberg H, Pawlik G. Sequential treatment of Guillain-Barré syndrome with extracorporeal elimination and intravenous immunoglobulin. *J Neurol Sci* 1996;**137**:145–9.
132. Hughes RA, van Doorn PA. Corticosteroids for Guillain-Barré syndrome. *Cochrane Database Syst Rev* 2012;**8**: CD001446.
133. Go T. Low-dose oral etidronate therapy for immobilization hypercalcaemia associated with Guillain-Barré syndrome. *Acta Paediatr* 2001;**90**:1202–4.
134. Hung PL, Chang WN, Huang LT, Huang SC, Chang YC, Chang CJ, et al. A clinical and electrophysiologic survey of childhood Guillain-Barré syndrome. *Pediatr Neurol* 2004;**30**:86–91.
135. Vajsar J, Fehlings D, Stephens D. Long-term outcome in children with Guillain-Barré syndrome. *J Pediatr* 2003;**142**:305–9.
136. Ammache Z, Afifi AK, Brown CK, Kimura J. Childhood Guillain-Barré syndrome: clinical and electrophysiologic features predictive of outcome. *J Child Neurol* 2001;**16**:477–83.
137. Ortiz-Corredor F, Pena-Preciado M, Diaz-Ruiz J. Motor recovery after Guillain-Barré syndrome in childhood. *Disabil Rehabil* 2007;**29**:883–9.
138. Geleijns K, Roos A, Houwing-Duistermaat JJ, van Rijs W, Tio-Gillen AP, Laman JD, et al. Mannose-binding lectin contributes to the severity of Guillain-Barré syndrome. *J Immunol* 2006;**177**:4211–17.
139. Geleijns K, Emonts M, Laman JD, van Rijs W, van Doorn PA, Hermans PW, et al. Genetic polymorphisms of macrophage-mediators in Guillain-Barré syndrome. *J Neuroimmunol* 2007;**190**:127–30.
140. van Doorn PA. Diagnosis, treatment and prognosis of Guillain-Barré syndrome (GBS). *Presse Med* 2013;**42**:e193–201.
141. van Nes SI, Faber CG, Merkies IS. Outcome measures in immune-mediated neuropathies: the need to standardize their use and to understand the clinimetric essentials. *J Peripher Nerv Syst* 2008;**13**:136–47.
142. Arthur DR. *Ticks and Disease*. Oxford: Pergamon Press; 1962.
143. Banfield JF. Tick bites in man. *Med J Aust* 1966;**2**:600–1.
144. Sonenshine DE. The mouthparts and foregut: capitulum, pharynx and oesophagus. In: Sonenshine DE, Roe RM, editors. *Biology of Ticks*. New York: Oxford University Press; 1991. pp. 119–40.
145. Bowman AS, Dillwith JW, Sauer JR. Tick salivary prostaglandins: Presence, origin and significance. *Parasitol Today* 1996; **12**:388–96.
146. Temple IU. Acute ascending paralysis, or tick paralysis. *Med Sentinel* 1912;**20**:507–14.
147. Todd JL. Tick bite in British Columbia. *Can Med Assoc J* 1912;**2**:1118–19.
148. Cleland JB. Injuries and diseases of man in Australia attributable to animals (excepting insects). *Aust Med Gazette* 1912;**32**:295–9.
149. Morris AM, Elliott EJ, D'Souza RM, Antony J, Kennett M, Longbottom H. Acute flaccid paralysis in Australian children. *J Paediatr Child Health* 2003;**39**:22–6.
150. Lysyk TJ. Tick paralysis caused by *Dermacentor andersoni* (Acari: Ixodidae) is a heritable trait. *J Med Entomol* 2010; **47**:210–14.
151. Schmitt N, Bowmer EJ, Gregson JD. Tick paralysis in British Columbia. *Can Med Assoc J* 1969;**100**:417–21.
152. Dworkin MS, Shoemaker PC, Anderson DE. Tick paralysis: 33 human cases in Washington State, 1946–1996. *Clin Infect Dis* 1999;**29**:1435–9.
153. Pearn J. The clinical features of tick bite. *Med J Aust* 1977; **2**:313–18.
154. Stone BF. Tick paralysis, particularly involving *Ixodes holocyclus* and other *Ixodes* species. In: Harris KF, editor. *Advances in Disease Vector Research*. New York: Springer-Verlag; 1988. pp. 61–85.
155. Malik R, Farrow BR. Tick paralysis in North America and Australia. *Vet Clin North Am Small Anim Pract* 1991;**21**:157–71.
156. Felz MW, Smith CD, Swift TR. A six-year-old girl with tick paralysis. *N Engl J Med* 2000;**342**:90–4.
157. Kemp DH, Stone BF, Binnington KC. Tick attachment and feeding: role of the mouthparts, feeding apparatus, salivary gland secretions and the host response. In: Obenchain FD, Galun R, editors. *The Physiology of Ticks*. Oxford: Pergamon Press; 1982. pp. 119–67.

158. Simpson LL. Botulinum toxin: a deadly poison sheds its negative image. *Ann Intern Med* 1996;**125**:616–17.
159. Murnaghan MF. Nerve fiber conduction block in tick paralysis. *Rev Can Biol* 1961;**20**:19–24.
160. Emmons P, McLennan H. Failure of acetylcholine release in tick paralysis. *Nature* 1959;**183**:474–5.
161. Cooper BJ. *Studies on the Pathogenesis of Tick Paralysis Ph.D. thesis.*. Sydney, Australia: The University of Sydney; 1976.
162. Swift TR, Ignacio OJ. Tick paralysis: electrophysiologic studies. *Neurology* 1975;**25**:1130–3.
163. Murnaghan MF. Neuro-anatomical site in tick paralysis. *Nature* 1958;**181**:131.
164. Rose I. Evidence of a neuromuscular block in tick paralysis. *Nature* 1956;**178**:95–6.
165. Krishnan AV, Lin CS, Reddel SW, McGrath R, Kiernan MC. Conduction block and impaired axonal function in tick paralysis. *Muscle Nerve* 2009;**40**:358–62.
166. Engin A, Elaldi N, Bolayir E, Dokmetas I, Bakir M. Tick paralysis with atypical presentation: isolated, reversible involvement of the upper trunk of brachial plexus. *Emerg Med J* 2006;**23**:e42.
167. Prakairungthong S, Limviriyakul S, Chongvisal S. Temporal bone histopathology case of the month: Acute unilateral facial paralysis in child from intra-aural hard tick. *Otol Neurotol* 2012;**33**:e81–2.
168. Patil MM, Walikar BN, Kalyanshettar SS, Patil SV. Tick induced facial palsy. *Indian Pediatr* 2012;**49**:57–8.
169. Gorman RJ, Snead OC. Tick paralysis in three children. The diversity of neurologic presentations. *Clin Pediatr (Phila)* 1978; **17**:249–51.
170. Daugherty RJ, Posner JC, Henretig FM, McHugh LA, Tan CG. Tick paralysis: atypical presentation, unusual location. *Pediatr Emerg Care* 2005;**21**:677–80.
171. DeBusk FL, O'Connor S. Tick toxicosis. *Pediatrics* 1972;**50**: 328–9.
172. Haller JS, Fabara JA. Tick paralysis. Case report with emphasis on neurological toxicity. *Am J Dis Child* 1972;**124**:915–17.
173. Crawford P, Mitchell D. Tick paralysis as a cause of autonomic dysfunction in a 57-year-old female. *South Med J* 2009;**102**: 190–2.
174. Garrettson LK. Poisoning. In: Pellock JM, Myer EC, editors. *Neurological Emergencies in Infancy and Childhood.* Philadelphia: Harper and Row; 1984. pp. 187.
175. Pearn JH. A case of tick paralysis with myocarditis. *Med J Aust* 1966;**1**:629–30.
176. Boffey GC, Paterson DC. Creatine phosphokinase elevation in a case of tick paralysis. *Can Med Assoc J* 1973;**108**:866–8.
177. Sutherland SK. Genus Oxyuranus Kinghorn, the Taipan and the small-scaled or fierce snake. In: Sutherland SK, editor. *Australian Animal Toxins: the Creatures, Their Toxins and Care of the Poisoned Patient.* Melbourne: Oxford University Press; 1983. pp. 110–27.
178. Sutherland SK. Ticks. In: Sutherland SK, editor. *Australian Animal Toxins: the Creatures, Their Toxins and Care of the Poisoned Patient.* Melbourne: Oxford University Press; 1983. pp. 299–315.
179. Inokuma H, Takahata H, Fournier PE, Brouqui P, Raoult D, Okuda M, et al. Tick paralysis by *Ixodes holocyclus* in a Japanese traveler returning from Australia. *Ann N Y Acad Sci* 2003;**990**:357–8.
180. Donat JR, Donat JF. Tick paralysis with persistent weakness and electromyographic abnormalities. *Arch Neurol* 1981;**38**:59–61.
181. Cherington M, Synder RD. Tick paralysis. Neurophysiologic studies. *N Engl J Med* 1968;**278**:95–7.
182. Morris 3rd HH. Tick paralysis: electrophysiologic measurements. *South Med J* 1977;**70**:121–2.
183. Rose I. A review of tick paralysis. *Can Med Assoc J* 1954;**70**: 175–6.
184. Hamilton DG. Tick paralysis: a dangerous disease in children. *Med J Aust* 1940;**2**:759–65.
185. Needham GR. Evaluation of five popular methods for tick removal. *Pediatrics* 1985;**75**:997–1002.
186. Stone BF. Tick paralysis: a suggestion. *Aust Vet Pract* 1990;**20**: 38–9.
187. Tomkins J. The treatment of tick bite in Australia. *Ann Gen Pract* 1963;**8**:92–5.
188. Cameron RJ, Rowley MP. Tick paralysis requires prolonged observation. *Med J Aust* 1999;**171**:334–5.

Chapter 21

Chronic Inflammatory Demyelinating Polyradiculoneuropathy

Hugh J. McMillan and Peter B. Kang

EPIDEMIOLOGY

Chronic inflammatory demyelinating polyradiculoneuropathy (CIDP) comprises a small but important subset of chronic childhood polyneuropathies. One series identified childhood CIDP in 11/125 (9%) of all children with a chronic polyneuropathy confirmed by sural nerve biopsy.[1] A second series reported CIDP in 35/249 (14%) of all children diagnosed with a sensorimotor polyneuropathy.[2]

CIDP can present from infancy[3] to late adulthood. The mean age of childhood CIDP onset is 8 to 9 years,[4,5] but it can appear at any age. No gender predilection is observed.[4] The overall annual incidence of CIDP (all ages) is estimated to be 0.15 per 100,000 per year.[6] By comparison the annual incidence of acute inflammatory demyelinating polyradiculoneuropathy, or Guillain-Barré syndrome, has been estimated to be 1.3 per 100,000 per year (range 0.4 to 4.0)[7] reflecting a 10-fold higher annual incidence of the acute form of the disease. The overall prevalence of CIDP is estimated at 1–2 per 100,000[6,8] with disease prevalence increasing with advancing age. An Australian study reported the prevalence of childhood CIDP to be 0.5 per 100,000, increasing more than 12-fold to 6.7 per 100,000 among 70- to 79-year-old adults.[6]

PATHOGENESIS

CIDP is an autoimmune disease resulting from a loss of immunologic self-tolerance. Cellular and humoral factors both contribute to autoimmune attack as autoreactive T and B lymphocytes, antibodies, and complement target myelinated peripheral nerves.[9] Molecular mimicry is believed to be the mechanism underlying the autoimmune attack in CIDP.[10,11] Molecular mimicry occurs when an individual is exposed to a new exogenous antigen (i.e. viral or bacterial infection) or when a normal endogenous intracellular protein becomes exposed to the immune system as a result of trauma, cancer, or a degenerative disease. The antigenic determinant known as the epitope is identified as foreign by the host's immune system and targeted for cellular and humoral immune attack. While this process is essential for normal immune functioning, in circumstances where the foreign epitope resembles normal host tissue it can lead to an ongoing autoimmune attack redirected toward the host's own tissue. In CIDP, the autoimmune attack is directed toward peripheral nerves and nerve roots. Only about one-third of children with CIDP can point to an inciting or antecedent event in the months prior to disease onset.[12–14] Even among children with an acute inflammatory neuropathy such as Guillain-Barré syndrome, where less time has passed between potential preceding infection and disease onset, only two-thirds of patients recall prior infectious symptoms.[15] Nevertheless, there are multiple levels of evidence to support an autoimmune basis for CIDP regardless of the initiating event.

Animal models of experimental allergic neuritis show a strong clinical and pathological resemblance to human CIDP.[16] By injecting components of peripheral nerve myelin such as myelin protein zero (MPZ) or peripheral myelin protein 22 (PMP22) into the blood of healthy animals, a syndrome of chronic neuritis can be induced that is clinically and pathologically analogous to human CIDP.[17] Affected animals show chronic inflammatory changes in their endoneurium as well as onion bulb formation similar to that seen in sural nerves of humans with CIDP. Such experiments were important in confirming an autoimmune basis for CIDP, redirecting attention away from a potential infectious cause. Later studies confirmed that passive transfer of serum or purified IgG from human CIDP patients to healthy rats can induce conduction block and demyelination.[18] Antibody-mediated disease offers an

B.T. Darras, H. Royden Jones, Jr., M.M. Ryan & D.C. De Vivo (Eds): Neuromuscular Disorders of Infancy, Childhood and Adolescence, Second edition.
DOI: http://dx.doi.org/10.1016/B978-0-12-417044-5.00021-4

explanation for specific types of inflammatory nerve disease such as Miller Fisher syndrome (MFS). This subtype of Guillain-Barré syndrome is characterized by ataxia, areflexia, and ophthalmoplegia and shows a very strong association with anti-GQ1b antibodies, which opsonize the GQ1b protein that is highly expressed on motor nerves innervating extraocular muscles.[19] Unlike MFS, there has been no convincing antibody link to CIDP. Fewer than 10% of all CIDP patients have anti-ganglioside, anti-sulfatide, or other autoantibodies.[19] Although an antibody-mediated attack does not appear to be the main immune process in patients with CIDP, the link with an autoimmune process is clear. T-lymphocytes remain important in the pathophysiology of CIDP. Sural nerve biopsies of adult CIDP patients show a correlation between the abundance of T lymphocytes in the endoneurium and the severity of demyelination.[20]

Finally, support for autoimmunity in CIDP is supported by response of patients to immunomodulating therapies such as intravenous immunogloblins (IVIg), plasma exchange (PLEX), and corticosteroids, which will be described in greater detail later in this chapter.

CLINICAL FEATURES

Three main areas must be considered by the clinician evaluating a child with possible chronic inflammatory demyelinating polyradiculoneuropathy. The first is to determine if the child exhibits the necessary clinical features to meet diagnostic criteria for CIDP. The second is to confirm the duration of symptoms that are reported. Finally, it is important to ascertain if the child demonstrates any clinical features potentially consistent with one of the many CIDP mimics, raising clinical suspicion of an alternative diagnosis and possibly excluding CIDP.

Children with CIDP demonstrate clinical signs and symptoms consistent with a sensorimotor polyneuropathy. Generalized muscle weakness and reduced or absent deep tendon reflexes are the two mandatory clinical criteria required for diagnosis of CIDP[21] (Box 21.1). Children will often have evidence of both proximal and distal muscle weakness, owing to the patchy immune attack along the entire length of the peripheral nerves. Weakness in CIDP patients is typically symmetrical; significant asymmetry should raise suspicion for alternative diagnoses. Hyporeflexia or areflexia is an early clinical feature of an inflammatory polyneuropathy and reflects involvement of both sensory afferents and—to a lesser degree—motor efferent fibers in this disease.

Gait impairment and/or clumsiness are the most common presenting features of CIDP,[24] particularly in younger children. Parents will often become concerned about their child's increasing clumsiness, frequent falls, and inability to run or get up from the floor without assistance. Our recently published cohort of childhood CIDP noted gait difficulty as an initial symptom in 29/30 (97%) of patients.[4] All 30 patients demonstrated some degree of muscle weakness, but the severity of weakness was typically moderate, with only 23% being nonambulatory. Gait ataxia results from a combination of muscle weakness and sensory loss due to involvement of large myelinated sensory afferents conveying proprioceptive input. Decreased or absent deep tendon reflexes in an ataxic child provide a valuable clinical clue directing attention toward a peripheral nerve disorder, and away from central nervous system disorders such as a posterior fossa tumor or a central demyelinating disease.

Sensory symptoms such as paresthesia and hypesthesia are seen in about 65–90% of children with CIDP,[10,13,24] reflecting involvement of small fiber sensory afferents. Since sensory symptoms are subjective, they can be challenging to evaluate in some young children, and they are not included in the diagnostic criteria. There are also rare reports of childhood CIDP presenting with isolated motor nerve involvement.[25,26] While clinical and electrodiagnostic evidence of sensory nerve involvement reassures clinicians that the lesion localizes best to the peripheral nerve, this evidence is not required for diagnosis of CIDP.

Back pain and limb pain are less common but are reported in <30% of children with CIDP. In adults, CIDP has been reported to cause nerve root inflammation causing nonspecific backache, massive nerve root enlargement has caused radiculopathy-like symptoms, and leg pain with walking has been has been attributed to a spinal cord stenosis-like syndrome related to massive enlargement of the cauda equina.[27] Awareness that these symptoms can be associated with CIDP is important for correct diagnosis, institution of immunomodulatory therapy, and avoidance of unnecessary surgical interventions. Magnetic resonance imaging in CIDP is discussed in detail below.

Cranial nerve involvement, manifesting as facial or oculobulbar weakness, is seen in only 10–15% of children with CIDP.[12,28] Dysautonomia and respiratory failure is exceedingly rare in children and adult CIDP patients.[4,29,30]

Temporal progression of CIDP symptoms provides additional clinical information that can help differentiate this chronic inflammatory polyneuropathy from its acute counterpart, Guillain-Barré syndrome (GBS). Diagnostic criteria stipulate that the progression of muscle weakness (i.e. time from initial symptom onset until time of maximal clinical deficit) in CIDP must be over at least 4 weeks. In situations where patients demonstrate initial rapid symptom onset (i.e. a GBS-like presentation), this must be followed by a relapsing or protracted disease course lasting at least 12 months (Box 21.1). Accurate differentiation of CIDP from GBS is important to ensure correct management with regard to the choice and duration of immunomodulatory therapy.

BOX 21.1 Diagnostic Criteria for Childhood Chronic Inflammatory Demyelinating Polyradiculoneuropathy

Mandatory Clinical Criteria

- Progression of muscle weakness in proximal and distal muscles of upper and lower extremities over at least 4 weeks, or rapid progression (Guillain-Barré-like presentation) followed by relapsing or protracted course (>1 year)
- Areflexia or hyporeflexia

Major Laboratory Features

Electrophysiologic Criteria

- Must demonstrate at least 3 of 4 major criteria in motor nerves ***or*** 2 major and 2 supportive

Major Criteria

- Conduction block or abnormal temporal dispersion in one or more motor nerves at sites not prone to compression:
  - Conduction block[a] = ≥50% drop in negative peak area or peak-to-peak amplitude of proximal compound muscle action potential (CMAP) if duration of negative peak of proximal CMAP is <130% of distal CMAP duration
  - Temporal dispersion[a] = abnormal if duration of negative peak of proximal CMAP is >130% of distal CMAP duration
- Reduction in conduction velocity (CV) in 2 or more nerves: <75% of the mean −2 standard deviations (SD) CV value for age
- Prolonged distal latency (DL) in 2 or more nerves: >130% of mean +2 SD DL value for age
- Absent F waves or prolonged F-wave minimal latency (ML) in 2 or more nerves: >130% of mean +2 SD F-wave ML for age (F-wave should include a minimum of 10 trials)

Supportive Criteria

(When conduction block is absent, the following abnormal electrophysiologic parameters are indicative of nonuniform slowing and thus of an acquired neuropathy):

- Abnormal median sensory nerve action potential (SNAP) while the sural nerve SNAP is normal
- Abnormally low terminal latency index[b]
- Side-to-side comparison of motor CVs showing a difference of >10 m/sec between nerves

Cerebrospinal Fluid (CSF) Criteria

- Protein >35 mg/dL (i.e. >0.35 g/L)
- Cell count <10 cells/mm^3

Nerve Biopsy Features

- Predominant features of demyelination

Exclusion Criteria

- Clinical features or history of a hereditary neuropathy, other disease, or exposure to drugs or toxins known to cause peripheral neuropathy
- Laboratory findings (including nerve biopsy or genetic testing) that show evidence of a cause other than CIDP
- Electrodiagnostic features of abnormal neuromuscular transmission, myopathy, or anterior horn cell disease

Diagnostic Criteria (must meet all exclusion criteria)

Confirmed CIDP

- Mandatory clinical features
- Electrodiagnostic and CSF features

Possible CIDP

- Mandatory clinical features
- One of three laboratory findings

a. Conduction block and temporal dispersion can only be assessed in nerves with distal CMAP amplitude >1 mV.
b. Terminal latency index of motor nerve (TLI) is calculated as follows: the terminal distance (TD) between the stimulation site and recording electrode (in millimeters) divided by the conduction velocity (CV) of the proximal nerve segment (in m/sec) and distal motor latency (in msec). Expressed as a mathematical formula: $TLI = TD / (CV \times DML)$. [Shahani (1979)[22] and Uncini (1991)[23]].
Reference: Nevo and Topaloglu (2002).[21]

Most children with CIDP present with an initial insidious or chronic symptom-onset. We recently reported a large cohort of children with CIDP, in which 60% of patients demonstrated at least an 8 week period between their initial symptoms and the time of maximal clinical deficit.[4] The mean time between symptom onset and maximal clinical deficit was 7.9 months for this group. A smaller group (20%) of patients demonstrated subacute onset of disease, which was defined as between 4 and 8 weeks between initial symptom onset and maximal clinical deficit. Subacute onset of CIDP has been well described in prior studies.[13] Combining the subacute and chronic subgroups, 80% of childhood CIDP patients demonstrate at least a 4 week period between their initial symptoms and the time of maximal clinical deficit.

A third group of children with CIDP (20% patients) initially presented with acute-onset disease, where less than 4 weeks elapsed between initial symptom and maximal clinical deficit.[4] One of these children presented with rapidly progressive weakness over several days, becoming quadriplegic and requiring intubation and mechanical ventilation. The proportion of our acute-onset group was similar to that of a prior group of pediatric and adult CIDP patients, wherein 16% of individuals eventually found to have CIDP presented with an initial acute onset of disease.[13] CIDP patients with acute-onset disease (A-CIDP) are initially difficult to distinguish from patients with demyelinating Guillain-Barré syndrome (acute inflammatory demyelinating polyradiculoneuropathy, AIDP). However, unlike AIDP patients, children who eventually go on to be diagnosed with CIDP must demonstrate persistent and/or relapsing disease with symptoms persisting for longer than 12 months.[4,13]

Adult studies have looked for features distinguishing the 16–20% of acute-onset CIDP patients from the 10% of GBS patients who may show transient weakness or treatment-related fluctuations after treatment with intravenous immunoglobulin (IVIg) or plasma exchange (PLEX).[31] Adults with AIDP show less prominent sensory symptoms and a higher incidence of autonomic dysregulation, facial weakness, preceding infection, and need for mechanical ventilation compared to the acute-onset CIDP group.[29] Another study found that adults with AIDP generally showed a more rapid time to maximal clinical deficit (mean = 8.5 days) compared to the acute-onset CIDP group (mean = 16.5 days), as well as a shorter time to symptom exacerbation and a higher likelihood of needing mechanical ventilation.[30]

Childhood CIDP has a relapsing-remitting or polyphasic course in 60–70% patients, with the remaining 30–40% of children demonstrating a protracted monophasic disease.[4,14,32,33]

The final area of inquiry is to ensure that the child does not demonstrate any clinical features that could be consistent with one of the many CIDP mimics which will be discussed in the next section.

DIFFERENTIAL DIAGNOSIS

There are numerous features on clinical history and/or physical examination that should raise concern for a potential diagnosis other than CIDP. These include: (1) sensory level or unequivocal sphincter disturbance; (2) self-mutilation; (3) marked asymmetry of symptoms; (4) early or predominant oculobulbar symptoms; (5) drug or toxin exposure; (6) central nervous system involvement; and/or (7) early onset and/or family history of polyneuropathy (Table 21.1).

Sensory findings in CIDP are typically length-dependent. A clear sensory level should raise suspicion of a lesion within the spinal cord. Another diagnostic clue may be brisk reflexes or extensor plantar responses, indicating corticospinal tract involvement. Caution is required, however, in the patient who shows an abrupt clinical deterioration (i.e. an acute upon chronic decline) since deep tendon reflexes may be temporarily depressed in a patient with spinal shock. The extensor plantar response should point toward a central or spinal lesion in such cases. Similarly, while children with CIDP may exhibit urinary or fecal incontinence due to an inability to reach the toilet in time, this typically occurs later in the disease course. Incontinence is not an early or predominant feature in CIDP, and should raise concern about a potential spinal cord lesion.

Self-mutilation can be seen with some hereditary sensory neuropathies (e.g. hereditary sensory and autonomic neuropathy (HSAN)). Patients with spinal cord lesions (i.e. syrinx or hydromyelia) may also develop painless burns, ulcerations, lacerations, or even fractures. While patients with CIDP may demonstrate decreased pain sensation, particularly in the feet, this again should not be an isolated or predominant finding.

Marked asymmetry of symptoms should raise suspicion of motor neuron diseases such as Hirayama disease, poliomyelitis, Hopkins syndrome, or other polio-like illnesses, as

TABLE 21.1 Differential Diagnosis of Childhood CIDP and Clinical Mimics of CIDP (Chronic progressive weakness and/or paresthesia)

Disease	Differentiating Clinical Features
Intracranial or spinal cord tumor	Sensory level, sphincter disturbance
Spinal muscular atrophy	Muscle weakness/wasting with intact sensation
Hereditary sensory and autonomic neuropathy	Painless ulceration, laceration, self-mutilation
Vasculitic neuropathy	Asymmetrical, painful
Gasoline, glue inhalation abuse	History of substance abuse
Charcot-Marie-Tooth disease	Family history of polyneuropathy, pes cavus
Myasthenia gravis	Predominant oculobulbar weakness
Peroxisomal disorders (Refsum disease, adrenomyeloneuropathy)	Prominent cerebellar ataxia, retinitis pigmentosa
Leukodystrophies	Central nervous system (CNS) involvement
Krabbe disease	
Metachromatic leukodystrophy	
Mitochondrial disease	CNS and/or other end-organ involvement
Cerebrotendinous xanthomatosis	CNS, tendon xanthomata, cataracts

well as other forms of inflammatory nerve disease including vasculitic neuropathies. CIDP is generally a symmetrical disease of peripheral nerves. While mild side-to-side differences can be seen in clinical and electrodiagnostic abnormalities, marked asymmetry is atypical.

Although cranial nerve involvement is seen in 10–15% of children with CIDP,[12,28] early or predominant oculobulbar symptoms are less typical of this disease. Early ocular weakness (i.e. ophthalmoplegia, ptosis) is common with disorders of neuromuscular transmission such as congenital myasthenic syndrome or juvenile myasthenia gravis, but can also be seen with mitochondrial diseases and congenital myopathies.

Prior or concomitant drug or toxin exposure should raise suspicion for a potential CIDP mimic. Glue or gasoline inhalation can give rise to a sensorimotor polyneuropathy with demyelinating features.[34]

Central nervous system involvement raises suspicion of metabolic conditions such as globoid cell leukodystrophy (Krabbe disease) or metachromatic leukodystrophy, rare diseases inherited in an autosomal recessive manner. Due to their inheritance patterns, there may be no family history of these conditions. Affected children demonstrate progressive irritability as well as cognitive and language delay or regression due to central and peripheral nerve demyelination. Symptoms of neuropathy may predate central nervous system involvement in some patients with leukodystrophies.[35,36] Peroxisomal diseases such as Refsum disease or adrenoleukodystrophy may also be associated with central and peripheral nerve demyelination. Peroxisomal diseases can present from infancy to early adulthood[37] and can show progressive patterns of abnormality on electrodiagnostic testing, that could be mistaken for CIDP.[38] Cockayne syndrome may show a resemblance to CIDP on electrodiagnostic grounds; however, the distinctive clinical phenotype—distinctive facial features (deep set eyes, progerioid appearance), microcephaly, developmental delay and failure to thrive—should alert the clinician to this genetic disorder.[39] Mitochondrial disease can show clinical and electrodiagnostic overlap with CIDP.[40]

Lastly, in very young children, or those with a family history of neuropathy, hereditary neuropathies such as Charcot-Marie-Tooth (CMT) disease should be considered. Both CMT and CIDP cause chronically progressive motor and sensory deficits. Although hereditary neuropathies such as CMT1A typically show uniform conduction slowing,[41] occasional patients with CMT1B[42] or CMTX[43] demonstrate features of nonuniform slowing, including temporal dispersion and conduction block (Table 21.2).

TABLE 21.2 Electrodiagnostic Mimics of CIDP

Disease	Reference
Charcot-Marie-Tooth disease 1 A (*PMP22* duplication)[a]	Kaku et al.[41]
Charcot-Marie-Tooth disease 1B (*MPZ* mutation)[a]	Street et al.[42]
Charcot-Marie-Tooth disease X (*GJB1* mutation)[a]	Ryan and Jones[43]; Tabaraud et al.[44]
Hereditary neuropathy with liability to pressure palsy (*PMP22* deletion or point mutation)[a]	Sellman and Mayer[45]
Peroxisomal disorders (e.g. Refsum, adrenomyeloneuropathy)[a]	Chaudhry et al.[46]
Mitochondrial disease[a]	Rusanen et al.[40]
Solvent inhalation abuse (i.e. gasoline, glue sniffing)[a]	Burns et al.[47]
Medication-induced	
Amiodarone[a]	Fraser et al.[48]
Tacrolimus[a]	Wilson et al.[49]
Suramin	Chaudhry et al.[50]
Cerebrotendinous xanthomatosis	Argov et al.[51]
Cockayne syndrome	Moosa and Dubowitz[52]
Krabbe disease (globoid cell leukodystrophy)	Marks et al.[35]; Tullu et al.[53]
Metachromatic leukodystrophy[a]	Nevo[36]; Cameron et al.[54]
Niemann-Pick disease type B	Wasserstein et al.[55]

[a]*Nonuniform slowing of conduction velocity (i.e. conduction block, temporal dispersion) may be seen.*

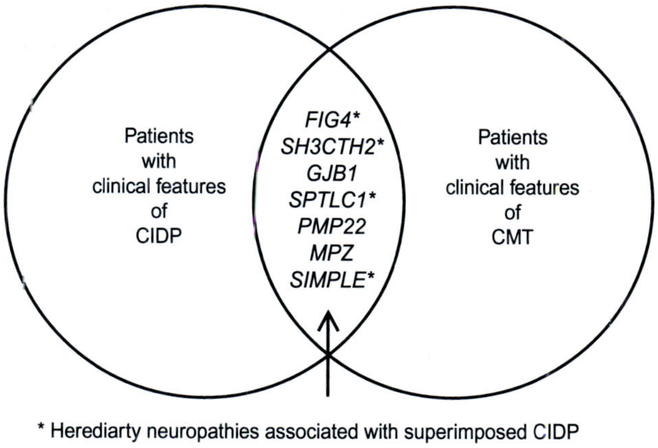

FIGURE 21.1 There can be significant clinical overlap between chronic inflammatory demyelinating polyneuropathy (CIDP) and Charcot-Marie-Tooth (CMT) disease. A small subset of genetic neuropathies has been clearly associated with a predisposition to CIDP (*).

Similar findings are also seen occasionally in children with metachromatic or globoid cell dystrophies.[54] Hereditary neuropathy with liability to pressure palsies (HNPP) is most commonly associated with a 1.5 Mb deletion at chromosome 17p11, a region that contains the *PMP22* gene.[56] The gene deletion results in inadequate peripheral myelin protein expression, rendering peripheral nerves more susceptible to injury after minor trauma or compression.[57] HNPP patients typically present in early adulthood with a painless mononeuropathy usually involving the peroneal or ulnar nerve. Less commonly patients develop a painless brachial plexopathy or a progressive polyneuropathy.[58] Electrodiagnostic testing demonstrates evidence of both generalized and segmental demyelination in all symptomatic and most asymptomatic HNPP patients over 15 years old.[58] Although the clinical phenotype infrequently overlaps CIDP, the electrodiagnostic findings can show similarities between these two disorders. CMTX can also show significant clinical and neurophysiologic overlap with CIDP.[42,43] Reports of CMT and CIDP coexisting in the same patient raise concern that some CIDP patients who demonstrate only a transient response to immunosuppressant therapy could be affected by two overlapping or concomitant diseases (Figure 21.1). One such report is of a 26-year-old man who walked at 26 months of age and had progressive weakness from early childhood, necessitating wheelchair use by age 13 years. At 22 years old he suffered a left femoral fracture. Around that time he developed severe, progressive weakness of his left upper extremity, clinically mimicking an inflammatory neuropathy, which left his hand functionally useless within 2 years. He was found to have compound heterozygous mutations in *FIG4* consistent with the diagnosis of CMT type 2J.[59]

Another report is of a woman with normal early neurodevelopmental milestones, who was diagnosed with kyphoscoliosis at 4 years old and from age 14 developed progressive muscle weakness and sensory loss. She was found to have CMT due to *SH3TC2* gene mutations (CMT type 4C). She was diagnosed with ulcerative colitis at age 24 years old and treated with corticosteroids. Upon withdrawal of corticosteroid therapy she developed marked progressive muscle weakness. She had an acellular CSF with a markedly elevated CSF protein (1.06 g/L, normal <0.45 g/L). Her sural nerve biopsy revealed classic onion bulb formations, with CD4 cell staining showing T-lymphocyte infiltration supporting a superimposed inflammatory disease. She was diagnosed with CIDP superimposed upon CMT. She initially showed a partial response to IVIG which later became ineffective, and failed to respond to corticosteroids and azathioprine.[60] These examples are important for clinicians caring for CIDP patients who demonstrate a transient or inadequate clinical response to multiple immunomodulatory therapies. There are other cases of patients with inherited neuropathies due to *GJB1*,[61] *SPTLC1*[62] and *SIMPLE*[63] gene mutations who were initially diagnosed with an inflammatory polyneuropathy.

Molecular mimicry has been discussed as the most accepted etiology of inflammatory neuropathy. Patients with inherited polyneuropathy are particularly susceptible to this process as their chronic axonal degeneration, with accumulation of myelin and cellular debris, can presumably act as a focus to sensitize the immune system to further autoimmune attack.

LABORATORY STUDIES

An elevated CSF protein (>35 mg/dL or >0.35 g/L) without pleocytosis (<10 leukocytes per mm^3) is known as albuminocytologic dissociation. This finding is observed in 87–100% of childhood CIDP patients (see Case Example 21.1).[4,12,14,28,64] Elevated CSF protein is certainly not specific to CIDP and can be seen in children with congenital hypomyelinating neuropathy,[65] Charcot-Marie-Tooth disease or leukodystrophies such as Krabbe disease.[53,64] Infrequently children with CIDP have a normal CSF protein, as seen in Case Example 21.2. In such circumstances clinicians must look to other tests to ensure that diagnostic criteria are met. A CSF leukocyte count greater than 10 cells/mm^3 is atypical of childhood CIDP, and should prompt consideration of alternative diagnoses such as demyelinating disorders or infectious myelitis (i.e. human immunodeficiency virus or Lyme disease).

CASE EXAMPLE 21.1

A 14-year-old boy presented with a 5-month history of gait difficulty and clumsiness. He described increased difficulty walking in snow and on ice and reported falling more frequently. He remained able to climb stairs and rise from the floor without difficulty. He had bilateral foot paresthesia below the ankles. His upper extremities were unaffected, and he had no symptoms of bowel or bladder dysfunction. His symptoms had improved slightly by the time he was evaluated by a neurologist. On examination his muscle bulk and strength were normal. He had decreased sensation to pin-prick and fine touch below the ankles and decreased vibration to his great toes, and was areflexic. No pes cavus was apparent. Nerve conduction studies identified a moderate sensorimotor polyneuropathy with demyelinating features. Based upon the clinical and electrodiagnostic testing, he was diagnosed with a resolving Guillain-Barré syndrome due to his normal strength. Over the ensuing months, his symptoms remained stable but he did not return to baseline. He was unable to use a skateboard over the summer months. Twelve months after initial symptom onset, he had an upper respiratory tract infection, followed within two weeks by a marked deterioration in strength. He could no longer run or jump. He began falling more frequently, although he remained independent with dressing and walking, could climb stairs, grasp a pen, and write normally. His lower limb paresthesia returned, and he described numbness extending up to both knees. His past medical history was significant only for stable scoliosis, and the family history was noncontributory. His physical examination revealed normal growth parameters and normal cranial nerve examination. Muscle strength testing (MRC scale) revealed mild distal upper limb weakness: abductor pollicus brevis 4+, first dorsal interosseous 4+ as well as lower extremity weakness: tibialis anterior 4+, extensor hallicus longus 4. Proximal muscle strength remained full. The deep tendon reflexes were absent. Sensory testing revealed decreased sensation below the mid-thigh and absent below the ankles, with decreased pin-prick sensation in all fingers. Vibration sensation was absent at the toes, ankles, and knees. Proprioception was absent in the toes. Cold sensation was intact. No dysmetria or tremor was apparent. Repeat nerve conduction studies (12 months after symptom onset) revealed a severe sensorimotor polyneuropathy with demyelinating features. Sensory responses were absent at his bilateral median, ulnar, peroneal, and sural nerves. His right radial sensory nerve amplitude was low at 2.3 mcV (normal >11 mcV). Motor nerve studies revealed prolonged distal latencies, slowed conduction velocity, and absent late responses, although these findings did not meet criteria for conduction block or temporal dispersion. Concentric needle electromyography (EMG) of his left medial gastrocnemius revealed mild (1 +) fibrillation potentials and positive sharp waves as well as an increase in polyphasic and long duration motor units. His left tibialis anterior revealed an increase in polyphasic and long duration units. Spinal fluid analysis revealed a marked elevation of cerebrospinal fluid (CSF) protein 1.33 g/L (normal 0.15 to 0.60 g/L), with normal CSF leukocytes (3 cells/mm^3), erythrocytes (1 cells/mm^3), and glucose (2.7 mmol/L). A spinal MRI with gadolinium 12 months after symptom onset revealed diffuse hypertrophy and enhancement of the nerve roots in the cervical and lumbar spine bilaterally (Figure 21.2). The nerve root thickening was symmetric with no nodular appearance noted. There was enhancement of cranial nerves VII and VIII in the internal auditory canal. Based upon the findings of weakness and areflexia persisting >12 months, the nerve condition studies and CSF results, he was diagnosed with CIDP (Table 21.3). Nerve biopsy was not performed. He was treated with IVIg (1 g/kg) × 2 days followed by maintenance dosing of IVIg 1g/kg once per month. He demonstrated progressive improvement in the first few months of IVIg therapy. After 3 months of IVIg his distal muscle strength returned to normal, although he remained areflexic and reported decreased sensation distally. After 6 months of IVIg, knee and ankle reflexes could be elicited (1+), and sensation was almost normal. The interval between IVIG treatments was gradually increased. After 12 months, therapy was discontinued, subsequent to which he has remained well and has returned to all activities including trampoline, snowboarding, and skating.

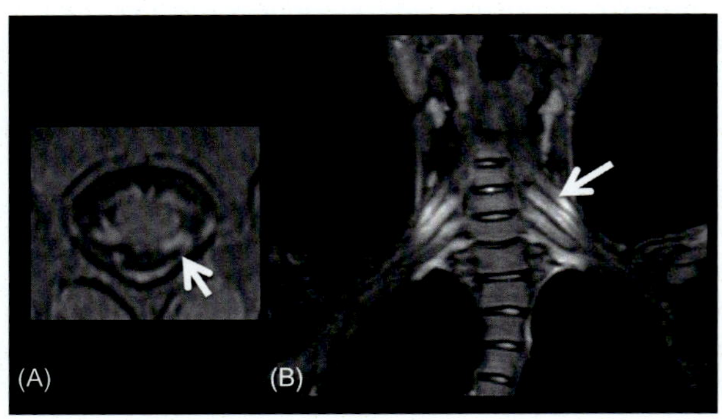

FIGURE 21.2 MR imaging of the spine in a 14-year-old boy (Case Example 21.1) with CIDP, performed 12 months after initial symptom-onset. **(A)** Axial T1-weighted image with gadolinium contrast illustrates nerve root thickening and enhancement. **(B)** Coronal STIR images reveal smooth thickening of nerve roots in the cervical spine. Thickening and enhancement was also noted in cranial nerves VII and VIII and lumbar nerve roots (not shown). Image **(B)** is previously published.[88]

(Continued)

CASE EXAMPLE 21.1 (Continued)

TABLE 21.3 Motor Nerve Conduction Studies of a 14-year-old Boy (Case Example 21.1) with CIDP

Site	NR	Distal Onset Latency (ms)	Amplitude (mV)	Conduction Velocity (m/sec)	F-Response
Left Median Motor (at abductor pollicis brevis; APB)					
Wrist		**7.7**	7.7	**26**	**NR**
Elbow		16.9	4.5		
Right Median Motor (at abductor pollicis brevis; APB)					
Wrist		**8.5**	6.8	**26**	**NR**
Elbow		18.2	5.2		
Left Ulnar Motor (at adductor digiti minimi; ADM)					
Wrist		**5.9**	**2.6**	**30**	**NR**
B. Elbow		12.9	2.4	**28**	
A. Elbow		17.2	2.4		
Right Ulnar Motor (at adductor digiti minimi; ADM)					
Wrist		**6.0**	**4.2**	**25**	**NR**
B. Elbow		14.6	3.1	29	
A. Elbow		18.8	3.2		
Left Peroneal Motor (at extensor digitorum brevis; EDB)					
Ankle	**NR**				
B. Fib					
Poplt					
Right Peroneal Motor (at extensor digitorum brevis; EDB)					
Ankle		**13.9**	**0.2**	**18**	
B. Fib		33.8	0.2	11	
Poplt		39.5	0.2		
Left Tibial Motor (at abductor hallucis; AH)					
Ankle	**NR**				
Knee					
Right Tibial Motor (at abductor hallucis; AH)					
Ankle		**28.8**	**0.1**	**23**	
Knee		47.5	0.1		

Legend: NR = no response.
Abnormal values shown in bold.
Sensory responses were absent at the following sites: bilateral median nerve, bilateral ulnar nerve, right superficial peroneal and bilateral sural nerves.

Biochemical testing does not offer any additional benefit except to rule out other metabolic diseases as described above. Some adults with CIDP have elevated titers of serum antibodies to ganglioside and sulfatides,[66] tubulin,[67] Schwann cells[68] and peripheral myelin protein 22,[69] but these autoantibodies are present only in a subset of CIDP patients. As their clinical significance and predictive value remain unknown, they are not routinely ordered at most centers and are not included in CIDP diagnostic criteria.

CASE EXAMPLE 21.2

A 2.5-year-old previously healthy boy was admitted to hospital with refusal to walk. Two weeks earlier, he had had an upper respiratory tract infection with low grade fever, mild cough, and rhinorrhea. One week prior to hospitalization he began limping and falling, and complained of leg pain. He remained in diapers, with no apparent change in his bowel or bladder function. His past medical history was significant only for an idiopathic left CN IV palsy (prior MRI brain was normal) which was being patched and followed clinically. He was a developmentally normal boy with an expansive vocabulary. On admission he was irritable. His cranial nerve examination was remarkable only for a partial left CN IV palsy. He had symmetric weakness of the proximal and distal arm and leg muscles. He was areflexic. Sensory testing could not be reliably performed due to age and poor cooperation. There was no pes cavus. CSF studies revealed a normal protein and white blood cell count. Nerve conduction studies identified a severe sensorimotor polyneuropathy. Sensory responses could not be elicited from the median, ulnar, superficial peroneal, or sural nerves. Motor responses showed severe prolongation of latencies and decreased conduction velocities (5 m/sec), with low amplitude compound motor action potentials (CMAPs). Given the severe reduction of CMAP amplitudes it was not possible to comment upon the presence or absence of non-uniform slowing (temporal dispersion, conduction block), or on the absence of late responses (F-wave, H-reflex).

The boy was diagnosed with Guillain-Barré syndrome and treated with IVIg (1 g/kg × 2 days) with subsequent clinical improvement. When discharged home a week later, he was able to walk. When examined 6 weeks later, however, he had deteriorated; he was walking only short distances and demonstrated a broad-based gait, proximal and distal muscle weakness, and diffuse areflexia. Further history from his mother revealed that, although he had achieved his initial gross motor milestones appropriately, he had always seemed weaker and tired more quickly than his peers. Given this additional history of a more longstanding problem, the differential diagnosis was changed from GBS to either CIDP or a hereditary polyneuropathy such as Charcot-Marie-Tooth disease. Repeat CSF and neurophysiologic studies were unchanged. Concentric needle EMG revealed active and chronic changes of denervation in tibialis anterior. An MRI scan of the brain and spine showed no evidence of nerve root thickening or enhancement. Given that the electrodiagnostic testing and CSF studies could not confirm the diagnosis of CIDP in this boy, he underwent a sural nerve biopsy with the goal of attaining a more definite diagnosis prior to embarking upon prolonged treatment with immunomodulating drugs. His sural biopsy was consistent with an inflammatory neuropathy (Figure 21.3). The toluidine blue preparation showed a decrease in the number of small myelinated and large myelinated axons.

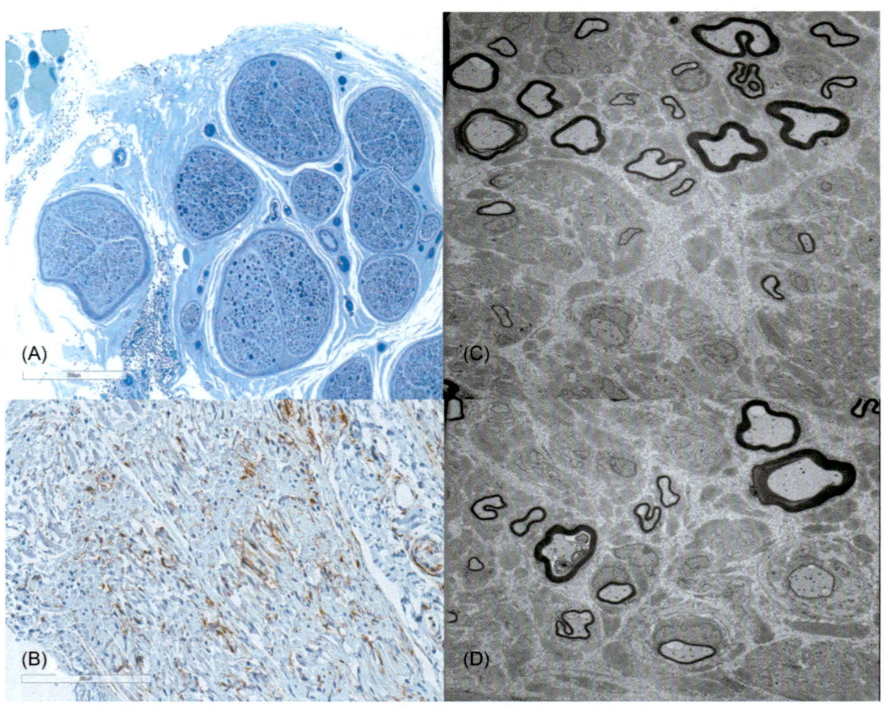

FIGURE 21.3 Sural nerve biopsy (Case Example 21.2) showing features of CIDP. (**A**) Toluidine blue stained semi-thin section: Variable numbers of myelinated fibers within different nerve fascicles. (**B**) Leukocyte common antigen immunohistochemistry showing inflammatory cells in endoneurium and epineurium. (**C**) Ultrastructural views (electron microscopy) showing unequal loss of myelinated fibers; severe drop-out of axons is seen in the lower portion of the figure while axons in the upper portion are relatively preserved but only thinly myelinated. (**D**) Two axons with thinner myelin sheath and one axon (lower right) showing early hypertrophic changes (i.e. onion bulb formation). *Photo credit: Dr. Jean Michaud, Department of Pathology, Children's Hospital of Eastern Ontario.*

(*Continued*)

CASE EXAMPLE 21.2 (Continued)

Immunohistochemistry staining for CD45 (leukocyte common antigen) and CD68 (macrophages, monocytes) revealed mild positivity in the endoneurium, epineurium, and perineurium, with lymphocytes within the endoneurium and around small blood vessels. Electron microscopy revealed variable fascicular loss of myelinated axons. Some axons had thin myelin sheaths indicating remyelination. Early hypertrophic changes were also noted. The presence of inflammatory cells and, more importantly, the patchy nature of the myelin loss supported an acquired inflammatory neuropathy (i.e. CIDP) and not a hereditary neuropathy (i.e. CMT), in which myelin loss would be more uniform and hypertrophic changes more widespread. Based upon the nerve biopsy findings, IVIg therapy was reinstituted. He received initial IVIg 1g/kg × 2 doses followed by monthly IVIg 1 g/kg, and his symptoms completely resolved within 6 months. The boy is now 6 months off IVIg therapy with no recurrence noted.

ELECTROPHYSIOLOGIC TESTING

Nerve conduction studies (NCS) and electromyography (EMG) are important electrophysiological tests to confirm peripheral nerve involvement and rule out the many diseases that can mimic CIDP (Table 21.1). NCS are also an important part of the diagnostic criteria of CIDP.

Clinicians should attempt to study as many peripheral nerves as possible in the upper and lower extremities. Since the electrophysiological findings in CIDP are patchy, it is not unusual for them to vary slightly from limb to limb, and they are often worse in longer nerves (i.e. the lower extremities). Diagnosis can be problematic in cases where no sensory responses can be obtained and compound motor action potential (CMAP) amplitudes are absent or extremely low. In situations such as these, studying motor nerves in the upper limb can be very important to permit complete electrodiagnostic assessment and ascertain whether or not CIDP diagnostic criteria are met.

Peripheral nerve myelination begins before birth and progresses until around 3 to 5 years old, at which age children attain adult values for conduction velocity.[70,71] Normal values for age should be used in young children, to avoid erroneously diagnosing a demyelinating disease.[72]

There are four major electrophysiological criteria for childhood CIDP. To meet NCS criteria for diagnosis, children must demonstrate evidence of at least three of four major criteria affecting motor nerves: (1) nonuniform slowing; (2) abnormal conduction velocity; (3) prolonged distal latency; and/or (4) abnormal late responses. Alternatively children meet diagnostic criteria if they have two of the major criteria as well as two of the following minor criteria: (1) abnormal median nerve sensory nerve action potential (SNAP) amplitude while the sural nerve SNAP remains normal; (2) abnormal minimal latency index; (3) differential conduction velocity in upper and lower limbs (Box 21.1).

Nerve conduction slows in all demyelinating neuropathies. Nonuniform slowing of conduction, indicating segmental demyelination, is one of the hallmarks of acquired, inflammatory neuropathies. In acquired peripheral nerve diseases, including CIDP, the severity of demyelination varies along the length of each axon, and from one axon and another. As a result, the speed of conduction begins to vary between one large myelinated axon and another, giving rise to temporal dispersion, which has a "drawn out appearance" on a CMAP waveform. As demyelination progresses, the normal saltatory conduction eventually fails along some axons (i.e. conduction block) resulting in a drop between distal and proximal CMAP amplitudes. Conduction block of an intact motor or sensory axon prevents the transmission of its impulses and produces motor or sensory deficits.

Nonuniform slowing including conduction block and temporal dispersion (Figure 21.4) is highly suggestive, but not pathognomonic, of acquired polyneuropathies.[73] Although hereditary neuropathies including CMT1A typically show uniform conduction slowing,[41] occasional patients with CMT1B[42] or CMTX[43] demonstrate features of nonuniform slowing, including temporal dispersion and conduction block (Table 21.2). Similar findings are also seen in occasional children with late infantile or early juvenile forms of metachromatic or globoid cell dystrophy.[54] A large series of 127 patients with CMT1 found that only 19 of 360 nerve segments studied (5.3%) showed a reduction of >50% amplitude on proximal versus distal stimulation.[41] A separate series initially reported conduction block in the median motor nerve of 9/37 (24%) of patients with CMT1 (conduction block defined as ≥50% drop in CMAP amplitude without an increase in temporal dispersion >15%)[74]; however, the authors questioned whether supramaximal stimulation was attained (upon reexamination of three of these subjects with higher degrees of stimulation, only one still demonstrated conduction block). Higher degrees of stimulation are often required in inherited demyelinating neuropathies, and this case series illustrates the importance of using supramaximal stimulation. Caution must also be taken at common sites of nerve entrapment or compression, including the common peroneal nerve between the

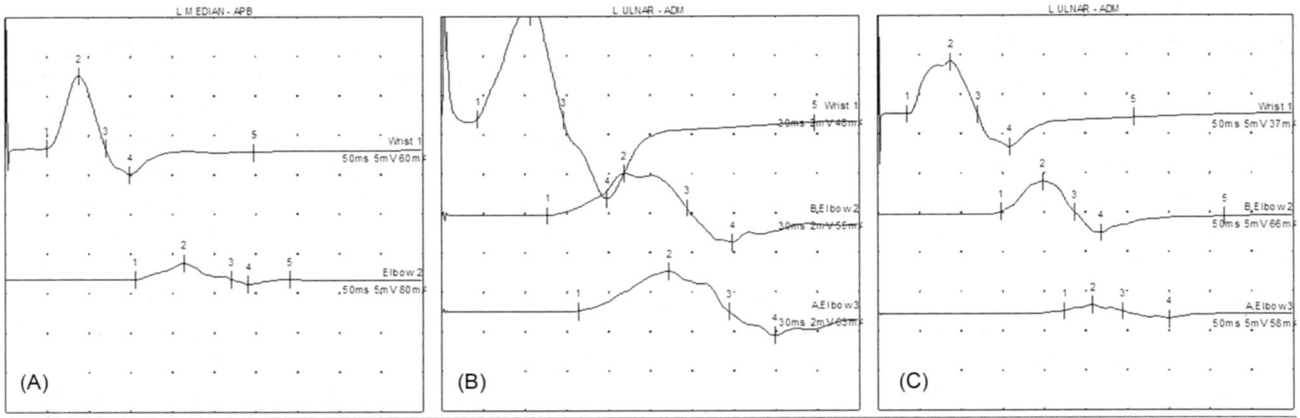

FIGURE 21.4 Nerve conduction studies in children with CIDP showing nonuniform slowing. From left to right: (A) Left median motor nerve (at APB) response in a 16-year-old female with CIDP showing conduction block. Compound motor action potential (CMAP) amplitude at wrist (distal stimulation site) was 9.3 mV compared to the amplitude at the antecubital fossa (proximal stimulation site), which was 2.1 mV (77% amplitude drop between the elbow and the wrist). Conduction velocity 22 m/sec (normal >50 m/sec). (B) Left ulnar nerve motor response (at ADM) in a 6-year-old female with CIDP showing temporal dispersion. Proximal CMAP duration is 166% longer compared to distal CMAP duration at the wrist. Ulnar nerve CMAP amplitudes are: 6.6 mV (at wrist); 2.1 mV (below elbow) and 2.0 mV (above elbow). Although the CMAP amplitude drops by 75%, conduction block cannot be diagnosed when temporal dispersion is also present. Conduction velocity is 23 m/sec (normal >50 m/sec). (C) Left ulnar nerve motor response (at ADM) in same 16-year-old female with CIDP as in (B) showing conduction block across the elbow. Ulnar nerve CMAP amplitudes are: 6.8 mV (at wrist); 3.8 mV (below elbow) and 0.8 mV (above elbow); 79% amplitude drop across the elbow. Conduction block must be interpreted with extreme caution at sites where nerve compression is commonly seen.

ankle and the fibular head, the median nerve between the wrist and elbow, or the ulnar nerve between the wrist and elbow. Hereditary neuropathy with liability to pressure palsies (HNPP) is another inherited neuropathy that commonly shows electrophysiological evidence of nonuniform slowing, although this typically is seen at common nerve compression sites.[45,58] Uncini identified conduction block in 25% of HNPP patients (defining it as a >20% drop in amplitude), and in 6% of patients where conduction block was defined as a >50% drop in amplitude.[75]

Definitions of conduction block have changed with time. The initial American Academy of Neurology (AAN) CIDP research criteria required at least a 20% amplitude drop in negative peak area or peak-to-peak amplitude. A change of less than 15% in duration between proximal and distal site stimulation was required to meet criteria for conduction block.[76,77] Defined pediatric criteria are even more stringent.[21] The electrodiagnostic criteria for conduction block for childhood CIDP demand at least a 50% amplitude drop in negative peak area or the peak-to-peak amplitude of the proximal and distal compound muscle action potentials (CMAPs), if the duration of the proximal CMAP duration is <130% of the distal CMAP duration. Temporal dispersion is noted when the proximal CMAP duration is >130% of the distal CMAP duration. When evaluating for nonuniform slowing, three conditions must be met. First, motor nerves must be evaluated at sites that are not prone to compression, in order to avoid misdiagnosis of patients with HNPP. Second, the distal CMAP amplitude must exceed 1 mV. Third, supramaximal stimulation must be used.

Slowing of nerve conduction is a second electrophysiological criterion of CIDP. Children must demonstrate a reduction in conduction velocity in two or more motor nerves, to <75% of the mean (−2 standard deviations) for their age-matched normal values.[72]

Prolongation of distal latencies is a third criterion for CIDP. The distal latency is the time (in milliseconds) between stimulation of a motor nerve and the arrival of the first sign of the action potential at the muscle innervated by that nerve. Distal latency provides information about signal conduction in the most distal segment of a motor nerve. Since the distal nerve segment is tortuous and involves the neuromuscular junction, conduction velocity is not easily calculated over this segment. Distal latency provides information analogous to conduction velocity. A distal latency is prolonged when it is >130% of the mean normal value (+2 SD) in 2 or more motor nerves.

Late responses, including F-waves, provide additional information fulfilling the fourth criteria of CIDP. F waves assess the most proximal segments of nerves and nerve roots. Absent or prolonged F-wave responses are an important indicator of proximal nerve and/or nerve root pathology, which is frequently present in CIDP. Criteria for abnormal response are met when F-waves are absent, or the minimal F-wave latency is prolonged in two or more nerves by >130% of the mean for age (+2 SD). F-wave assessments should include a minimum of 10 stimulations at each site, with M waves recorded with each stimulation. Care must be taken in patients with low CMAP amplitudes, as F responses may not be present regardless of whether or not nerve roots are involved.

Minor electrodiagnostic criteria which can be used in the electrophysiological diagnosis of CIDP are aimed at evaluating patchy or unequal findings. These include: (1) an abnormal median sensory nerve action potential (SNAP) with a normal sural SNAP; and (2) motor nerve conduction velocities varying >10 m/sec from side to side for a specific nerve. Both indicate an unequal distribution of demyelination that can be seen in CIDP. The last minor criterion is the abnormally low terminal latency index (TLI), which measures how well the most distal segments of peripheral nerves are conducting. TLI is calculated as follows: the terminal distance (TD) or the distance between the site of stimulation and the recording electrode (mm) divided by the conduction velocity (CV) of the more proximal nerve segment (m/sec) times the distal motor latency (msec). Expressed as a mathematical formula, this is $TLI = TD / (CV \times DML)$. TLI is generally very low in patients with demyelinating diseases because of their poor distal conduction.[22,23]

These electrophysiological criteria must be interpreted within the correct clinical context, and additional biochemical or even pathological studies may be required when results are uncertain. It must be emphasized that the diagnosis of CIDP should not be made on electrodiagnostic criteria alone if the clinical presentation does not support the diagnosis. In addition to hereditary polyneuropathies, which can show overlapping features on electrodiagnostic testing, several metabolic diseases may also show similar patterns of abnormality (Table 21.2). Peroxisomal diseases including Refsum disease and adrenomyeloneuropathy can show evidence of multifocal demyelination on NCS, although axonal loss is usually more predominant.[46,78] Milder phenotypes of peroxisomal dysfunction also show multifocal demyelination that can mimic inflammatory polyneuropathies.[38]

NERVE BIOPSY

Nerve biopsy is not required for diagnosis of childhood CIDP. In most cases, a definite diagnosis of childhood CIDP can be made upon clinical, electrodiagnostic, and biochemical grounds (i.e. Case Example 21.1). In our recent cohort of 30 childhood CIDP patients treated at Boston Children's Hospital, only 5/30 (16%) underwent nerve biopsy. However, there are cases where a nerve biopsy can be helpful to confirm or strengthen the diagnosis of CIDP prior to embarking upon lengthy immunomodulation therapy, which can be costly, inconvenient, and potentially put the patient at unnecessary risk. Case Example 21.2 is an excellent example; this patient was found to have severe sensorimotor polyneuropathy with such a degree of CMAP amplitude reduction in all motor nerves that it was not possible to reliably define the underlying pathology as demyelinating or axonal. He had absent sensory responses and a severe reduction of amplitude of motor responses. Moreover, his CSF protein levels were normal and MRI of his spinal nerve roots did not reveal any enhancement or hypertrophy. On this basis it was not possible to differentiate a chronic inflammatory polyneuropathy from a hereditary polyneuropathy and a sural nerve biopsy was performed to differentiate between the two (Figure 21.3).

The sural nerve biopsy of our patient supported the diagnosis of CIDP, since it demonstrated two of the hallmark pathologic features of this disease: (1) patchy, non-uniform or segmental demyelination and; (2) mononuclear cell infiltration of involved peripheral nerves. In contrast, patients with hereditary neuropathies (CMT1A) typically demonstrate more uniform demyelination and more uniform and pronounced hypertrophic changes (i.e. onion bulb formations). Our patient also demonstrated evidence of early and mild hypertrophic changes (i.e. onion bulb formations) as well as evidence of an overall reduction of myelinated fiber density and remyelinating fibers. Some patients may also demonstrate subperineurial edema and axonal degeneration.[79,80]

The most characteristic pathologic finding in CIDP is macrophage-mediated stripping of myelin from axons. Onion bulb formations are composed chiefly of circumferentially directed processes of Schwann cells, proliferating around axons during continued cycles of demyelination and remyelination. Autopsy studies of adult CIDP patients have shown that the inflammatory process is multifocal and predominantly proximal (i.e. spinal nerve roots and proximal nerve trunks), although the lesions extend throughout the peripheral nervous system.[81] Inflammatory infiltrates observed on sural nerve biopsy are predominantly T cells and macrophages.[20,82–84] B-cell lymphocytes are uncommon. Macrophages are more common in the endoneurium, and T cells are predominantly in the epineurium.[20] CD8-positive T cells (cytotoxic-suppressor phenotype) are more common than CD4-positive T cells (helper-inducer phenotype).[20] In CIDP, the inflammatory infiltrates are perivascular and do not invade vessel walls. Nerve biopsies are not invariably abnormal in patients with CIDP, since inflammation and demyelination can be patchy.[79] Similarly, not all pathological abnormalities are demonstrated on all biopsies. Where the nerve biopsy is normal in a patient with suspected CIDP, the clinician should ask him or herself if the biopsy site was appropriate: specifically, if the patient demonstrated clinical or neurophysiologic evidence of abnormality at that location.

Another situation where a clinician should contemplate a nerve biopsy is in a patient with significant asymmetry to his/her symptoms, with or without pain, raising concern regarding the possibility of a vasculitic neuropathy. Symptoms or a preexisting diagnosis of connective tissue disease should also raise suspicion in such patients. Although vasculitic neuropathy is more common in adults, it is well described in adolescent patients with lupus and other connective tissue diseases.[85]

The sural nerve is typically biopsied in cases of suspected CIDP, as it is purely sensory, easily accessed, and usually involved. The sural nerve provides sensory innervation to the lateral aspect of the ankle and foot. Since it does not carry motor fibers, the patient will not experience increased weakness post-biopsy, as would be the case with a mixed nerve. Patients and/or parents should be counseled about the potential for loss of sensation along the lateral aspect of the foot as well as the fact that pain can be a common problem in the days or weeks post-biopsy. Infection and poor incision healing are potential risks. Although the patient will have a scar at the incision site near the ankle, this usually becomes less noticeable with time.

NEUROIMAGING

Magnetic resonance imaging (MRI) can assist in diagnosis of CIDP, by identifying nerve root and/or cauda equina thickening, as well as gadolinium enhancement of these structures. Such MRI findings have been listed as supportive tests in recently revised adult CIDP diagnostic criteria.[86] However, these findings, when present, must be interpreted within clinical context and in conjunction with other diagnostic criteria.

Nerve root thickening and gadolinium enhancement are seen in approximately 40% of children[4] and 60% of adults with CIDP.[87] This means that most children with CIDP do not demonstrate nerve root abnormalities on MR imaging. Thus, the absence of this finding does not offer any reassurance that the child does not have CIDP. When children with CIDP do demonstrate MRI changes, their nerve roots can appear diffusely abnormal, extending as smooth or fusiform thickening to the more distal nerve roots and brachial or lumbar plexus.[88] Nerve root enlargement can be massive in some adult CIDP patients. Lumbosacral nerve root enlargement can mimic lumbar stenosis[89] and in some cases has produced symptoms of spinal cord compression.[27] Cranial nerves can also appear thickened and enhanced[90] and there are reports of adult patients with CIDP undergoing biopsy of a cranial nerve due to concern for a possible acoustic neuroma.[27]

The presence of nerve root thickening is by no means unique to CIDP. Homogeneous nerve root enlargement is also reported in patients with Charcot-Marie-Tooth disease,[91,92] Guillain-Barré syndrome[93,94] and infection causes (HIV, CMV).[95] Nodular nerve root thickening can be seen in neurofibromatosis-related nerve sheath tumors,[96] neoplasias (e.g. neurolymphomatosis)[97] and granulomatous disease.[98] Gadolinium enhancement of nerve roots may be more sensitive at identifying inflammatory polyneuropathies like CIDP and may be more discriminating than nerve hypertrophy in differentiating it from disorders like CMT1. Gadolinium enhancement of the cauda equina was reported during a blind MRI interpretation in 11 of 16 (70%) CIDP patients and in none of 13 controls, which included 5 patients with CMT1.[99] However, nerve root enhancement (without thickening) can also be seen in adult patients when the MRI of the spine with gadolinium is completed after lumbar puncture has already been performed. In patients with symptoms of a postlumbar puncture or postural headache, dilation of the anterior internal vertebral plexus on contrast-enhanced sequences can be mistaken for anterior nerve root thickening.[100]

TREATMENT

Childhood CIDP can be successfully treated with intravenous immunoglobulins (IVIg), oral corticosteroids, or plasmapheresis (PLEX). Since continuous therapy is typically required for months or years, the treating physician must decide whether the potential benefit of the treatment outweighs its likely adverse effects. We advocate treating children with disabling symptoms (i.e. unable to play sports or complete age-appropriate self-care tasks, and/or with difficulty walking), if those symptoms are progressive or prolonged. On rare occasions, children with CIDP may have mild and resolving symptoms at the time of first assessment; these children must be followed closely since many will demonstrate chronic symptoms and/or disease relapse at a later time. Predetermined outcome measures should be monitored closely to assess the response to treatment or to follow untreated patients. Muscle strength and sensory loss may be reasonable clinical end-points in older children or adolescents but can be difficult to objectively measure in young children. The deep tendon reflexes are of questionable value as an end-point since hyporeflexia or areflexia are often an early clinical feature of this disease and one of the later symptoms to recover when disease remission does occur. The modified Rankin disability score has been used by many researchers as a measure of disability associated with CIDP and to monitor response to treatment.[4,28,32,33,64] Other disability scores have been used in other pediatric CIDP studies.[5] The modified Rankin score was initially devised to quantify clinical deficit in patients after stroke.[101] This functional scale ranges from 0 to 6 as follows: 0 = asymptomatic; 1 = mild symptoms that do not interfere with work, school, or extracurricular activity; 2 = slight disability (i.e. child has given up one or more activities) but is able to perform all age-appropriate personal care (i.e. dressing, eating) and complex tasks (i.e. handwriting, age-appropriate food preparation); 3 = moderate symptoms (i.e. child is still able to walk independently (may require cane or walker) but requires assistance for age-appropriate tasks (see above); 4 = moderate-to-severe symptoms (i.e. child is unable to walk (carried by parent and/or wheelchair is required) and unable to perform age-appropriate care); 5 = severe disability (i.e. patient is bed-ridden and requires constant nursing care), may require intubation and mechanical ventilation; 6 = death. Although each functional category is quite broad, the scale nevertheless provides a means of evaluation of disease progression or remission. The chronic inflammatory demyelinating polyneuropathy disease activity status scale has also been proposed as a serial measure of CIDP activity,[102] and may be a better measure of response to therapy.[103]

Repeat NCS can be useful in some children with CIDP. However, we recommend at least several months between repeat neurophysiologic studies unless there is an acute or subacute clinical decline and/or the clinician wishes to have repeat NCS values immediately prior to instituting or changing an immunomodulation therapy. Although clinical recovery can be seen over weeks to months in some patients, it is not unusual for improvements in NCS to take longer. Many patients have resolution or near resolution of clinical symptoms and signs but still show slow conduction velocity and prolonged distal latencies. Thus, treatment decisions should not be based solely on these parameters.[28]

Intravenous Immunoglobulin (IVIg) in Childhood CIDP

Most children with CIDP receive IVIg as first-line therapy at our center. The rationale for its use as initial immunomodulating treatment is based upon its favorable side-effect profile, excellent efficacy, and ease of administration. Approximately 80% of children with CIDP show a good response to IVIg monotherapy. Many can be weaned off this medication after months of regular administration.[4] Like most pediatric centers, we recommend the administration of IVIg 1 g/kg on 2 consecutive days. Treatment should be continued at 4-week intervals with the patient receiving an additional 1 g/kg (single dose) at that time. As the patient's clinical symptoms resolve, we recommend gradually increasing the interval between IVIg doses; such a taper typically takes 6–12 months or even longer. After the dosing interval is increased to every 6 to 8 weeks, we recommend discontinuing IVIg therapy to see if the patient remains in remission. Although there is evidence from randomized trials that IVIg is beneficial in pediatric GBS[104,105] the evidence for its efficacy in pediatric CIDP has been derived from pediatric retrospective cohort studies[4,5,28,32,33,64] and extrapolation from adult CIDP data. Several large randomized control studies of adult CIDP patients have demonstrated the efficacy of IVIg versus placebo[106–111] as well as its efficacy compared to other immunomodulating therapies including plasma exchange (PLEX)[112] and corticosteroids.[113] The mechanism of action of IVIg may include the blocking effects of some autoantibodies,[114] an overall downregulation of T-cell activity, and endogenous antibody and cytokine production causing relative immune quiescence. The main limitations of IVIg use are availability and cost. IVIg is generally very well tolerated in children and adults. Potential side effects can include fever, malaise, hypotension, headache, vomiting, vasomotor or allergic symptoms (i.e. skin flushing, chills, chest tightness, wheezing). These symptoms are generally noted during the IVIg infusion or within the hours following the completion of therapy.[115,116] Aseptic meningitis occurs occasionally.[117,118] Premedication with acetaminophen and diphenhydramine and/or switching to a different IVIg preparation may be useful in reducing some IVIg-related side effects. For patients who have experienced severe post-IVIg headache and/or aseptic meningitis who require repeat dosing, a small dose of oral daily prednisone (×5 days) can help reduce the incidence and severity of these symptoms. IVIg must be used cautiously in patients at risk of volume overload (i.e. patients with cardiac or renal disease) due to its high protein content and tendency to increase intravascular volume. In such situations it may be advisable to follow an administration schedule more typically used in adults, namely a smaller volume of 0.4 g/kg given over 5 consecutive days. Severe anaphylactic reactions may occur, particularly in patients with IgA deficiency if a non-IgA depleted formulation is used. The risk of viral transmission (hepatitis B or C, HIV) is negligible. Thromboembolic events and renal failure are occasionally reported in adult patients treated with IVIg.[119,120]

Corticosteroid Therapy in Childhood CIDP

Corticosteroids, including prednisone or prednisolone, are very effective therapies for childhood CIDP. When the use of first-line corticosteroids was evaluated across many retrospective pediatric CIDP studies it was found overall to be efficacious in 84% of children.[3–5,14,23,28 32,33,121,122] Corticosteroid use was more common in earlier reports, with more recent studies favoring IVIg as a first-line agent due to its more favorable side-effect profile. A randomized control study of adult CIDP patients found oral prednisolone to be equally efficacious as IVIg therapy

6 weeks after therapy initiation.[113] Corticosteroids have also been demonstrated to improve symptoms in adult CIDP patients relative to no treatment.[123] Corticosteroids exert a broad immunosuppressive effect.[124] The advantages of corticosteroid therapy include ease of administration and low cost. The disadvantages relate to the numerous adverse effects frequently associated with its long-term use.[125] These side effects are particularly worrisome for children. Risks can include bone demineralization and ensuing vertebral and long-bone fractures, osteonecrosis of the hip, decreased linear growth and a reduction in final adult height attainment, behavior problems, and weight gain. Other side effects can include hypertension, hyperglycemia, altered immune function, skin striae, and acne, as well as cataracts. There are rare reports of children with CIDP who have demonstrated clinical deterioration associated with high-dose IV methylprednisone.[126]

The route of administration and dosing of corticosteroids have varied from one study to another, but most studies report an initial dose of 1–2 mg/kg/day, tapering over 6–12 months if a positive response to therapy is seen.[28,64,127] Most pediatric data support oral corticosteroid use. Some patients have been treated with an initial intravenous pulse dose of corticosteroids.[4,28] For patients requiring chronic corticosteroid use we recommend that vitamin D and calcium intake be monitored and optimized, growth velocity be monitored, and that bone densitometry studies be completed at least annually to ensure that bone mineralization remains within an acceptable range. Some children with CIDP will demonstrate signs of disease relapse with weaning of corticosteroid therapy, necessitating an increase in the dosing of this medication and/or consideration of other steroid-sparing immunosuppressive agents or use of other immune modulating drugs.

Plasma Exchange in Childhood CIDP

Plasma exchange (PLEX) may have a role for a select subset of children with CIDP. PLEX is used less frequently in children with CIDP compared to IVIg or corticosteroids, primarily because of concerns about technical factors and possible adverse effects, particularly in younger patients who correspondingly have a smaller blood volume. PLEX requires secure large bore venous access, which for many children necessitates central line insertion. Potential adverse effects can include hypotension, electrolyte imbalance, and coagulopathy. Chronic PLEX can be associated with anemia, hypocalcemia, and hypogammaglobulinemia. There are reports of children with CIDP showing a transient response to PLEX[14] while others have reported a sustained benefit comparable to that seen with IVIg.[12] In our experience, older children and adolescents have a higher likelihood of benefiting from and tolerating PLEX since venous access becomes a frequent problem in younger children.[4] There are good data from adult CIDP studies indicating that treatment with PLEX shows similar efficacy to IVIg.[112]

Other Immunosuppressant Medications in Childhood CIDP

Data pertaining to the use of other corticosteroid-sparing agents in childhood CIDP are limited. No prospective studies are available and data are limited to a few reports in retrospective case series. We have successfully treated three patients with azathioprine and one with mycophenolate mofetil. Other studies have reported some success.[4] Overall, azathioprine has been used most widely in childhood CIDP although there are also reports of the occasional child receiving cyclophosphamide or cyclosporine.[5,28,32] More recently, rituximab has been trialed in several cases of refractory CIDP with good effect.[103] In adults, a relatively short prospective study of azathioprine failed to demonstrate efficacy,[128] possibly because the trial was too short.[129] Mycophenolate mofetil has some effect in adult CIDP.[130] Methotrexate,[131] cyclosporine,[132] and interferon beta-1a[133,134] have not proven beneficial in adult CIDP. More novel therapies such as TNF-alpha (etanercept)[135] and anti-CD20 antibodies (rituximab)[32,136] have been reported to offer benefit, but have not yet been used widely, particularly in childhood CIDP.

PROGNOSIS

The long-term outcome of childhood CIDP appears to be quite good in most children, especially with appropriate immunotherapy.

In a recent study at Boston Children's Hospital, long-term data were available for 23/30 children with CIDP followed for a mean of 3.8 years (range 5 months to 13 years). Follow-up modified Rankin scores (MRS) had improved in all but three children. The mean most recent follow-up MRS score in the remaining 20/23 children in the Boston Children's Hospital series was 0.5, improved from a mean initial MRS score of 2.8. Nine children (45%) were off all immunosuppressant medication at the time of their last clinical follow-up. No clinical deficit was seen on clinical examination of 11/20 children. Minimal, primarily distal focal weakness was observed in 9/20 (45%) children. These very mild deficits were insufficient to cause functional disability or restrict school or extracurricular activities (i.e. MRS = 1). Only 1 patient had residual clinical deficits causing restriction of activities (MRS = 2).

Combined long-term outcome in the CIDP patients seen at the Boston Children's Hospital was compared with that in four prior series.[28,32,33,64] The overall mean MR score improved from 3.7 at presentation to 0.7 at last follow-up. Our series, and two others, reported favorable long-term outcome of childhood CIDP,[28,33] whereas two other case series showed mild-to-moderate clinical deficits in some children with CIDP.[32,64] One additional study reported favorable long-term outcome with childhood CIDP although this study used a different functional rating scale.[5] McCombe et al. reported a favorable prognosis in childhood CIDP with 7 of 8 patients free of disability and 1 patient with minor disability at follow-up.[13] Overall, the prognosis of childhood CIDP is generally quite favorable.

REFERENCES

1. Ouvrier RA, McLeod JG. Chronic peripheral neuropathy in childhood: an overview. *Aust Paediatr J* 1988;**24**(Suppl. 1):30–2.
2. Sladky JT. Immune neuropathies in childhood. *Baillieres Clin Neurol* 1996;**5**(1):233–44.
3. Sladky JT, Brown MJ, Berman PH. Chronic inflammatory demyelinating polyneuropathy of infancy: a corticosteroid-responsive disorder. *Ann Neurol* 1986;**20**(1):76–81.
4. McMillan HJ, Kang PB, Jones HR, Darras BT. Childhood chronic inflammatory demyelinating polyradiculoneuropathy: combined analysis of a large cohort and eleven published series. *Neuromuscul Disord* 2013;**23**(2):103–11.
5. Korinthenberg R. Chronic inflammatory demyelinating polyradiculoneuropathy in children and their response to treatment. *Neuropediatrics* 1999;**30**(4):190–6.
6. McLeod JG, Pollard JD, Macaskill P, Mohamed A, Spring P, Khurana V. Prevalence of chronic inflammatory demyelinating polyneuropathy in New South Wales, Australia. *Ann Neurol* 1999;**46**(6):910–13.
7. Hughes RA, Rees JH. Clinical and epidemiologic features of Guillain-Barré syndrome. *J Infect Dis* 1997;**176**(Suppl. 2):S92–8.
8. Lunn MP, Manji H, Choudhary PP, Hughes RA, Thomas PK. Chronic inflammatory demyelinating polyradiculoneuropathy: a prevalence study in south east England. *J Neurol Neurosurg Psychiatry* 1999;**66**(5):677–80.
9. Dalakas MC. Basic aspects of neuroimmunology as they relate to immunotherapeutic targets: present and future prospects. *Ann Neurol* 1995;**37**(Suppl. 1):S2–13.
10. Kieseier BC, Dalakas MC, Hartung HP. Immune mechanisms in chronic inflammatory demyelinating neuropathy. *Neurology* 2002;**59**(12 Suppl. 6):S7–12.
11. Weiss MD, Luciano CA, Semino-Mora C, Dalakas MC, Quarles RH. Molecular mimicry in chronic inflammatory demyelinating polyneuropathy and melanoma. *Neurology* 1998;**51**(6):1738–41.
12. Simmons Z, Wald JJ, Albers JW. Chronic inflammatory demyelinating polyradiculoneuropathy in children: I. Presentation, electrodiagnostic studies, and initial clinical course, with comparison to adults. *Muscle Nerve* 1997;**20**(8):1008–15.
13. McCombe PA, Pollard JD, McLeod JG. Chronic inflammatory demyelinating polyradiculoneuropathy. A clinical and electrophysiological study of 92 cases. *Brain* 1987;**110**:1617–30.
14. Nevo Y, Pestronk A, Kornberg AJ, Connolly AM, Yee WC, Iqbal I, et al. Childhood chronic inflammatory demyelinating neuropathies: clinical course and long-term follow-up. *Neurology* 1996;**47**(1):98–102.
15. Oomes PG, Jacobs BC, Hazenberg MP, Banffer JR, van der Meche FG. Anti-GM1 IgG antibodies and Campylobacter bacteria in Guillain-Barré syndrome: evidence of molecular mimicry. *Ann Neurol* 1995;**38**(2):170–5.
16. Waksman BH, Adams RD. Allergic neuritis: an experimental disease of rabbits induced by the injection of peripheral nervous tissue and adjuvants. *J Exp Med* 1955;**102**(2):213–36.
17. Brostoff SW, Levit S, Powers JM. Induction of experimental allergic neuritis with a peptide from myelin P2 basic protein. *Nature* 1977;**268**(5622):752–3.
18. Yan WX, Taylor J, Andrias-Kauba S, Pollard JD. Passive transfer of demyelination by serum or IgG from chronic inflammatory demyelinating polyneuropathy patients. *Ann Neurol* 2000;**47**(6):765–75.
19. Chiba A, Kusunoki S, Obata H, Machinami R, Kanazawa I. Ganglioside composition of the human cranial nerves, with special reference to pathophysiology of Miller Fisher syndrome. *Brain Res* 1997;**745**(1–2):32–6.
20. Matsumuro K, Izumo S, Umehara F, Osame M. Chronic inflammatory demyelinating polyneuropathy: histological and immunopathological studies on biopsied sural nerves. *J Neurol Sci* 1994;**127**(2):170–8.
21. Nevo Y, Topaloglu H. 88th ENMC international workshop: childhood chronic inflammatory demyelinating polyneuropathy (including revised diagnostic criteria), Naarden, The Netherlands, December 8–10, 2000. *Neuromuscul Disord* 2002;**12**(2):195–200.
22. Shahani BT, Young RR, Potts F, Maccabee P. Terminal latency index (TLI) and late response studies in motor neuron disease (MND), peripheral neuropathies and entrapment syndromes (Abstract). *Acta Neurol Scand* 1979;**73**(Suppl.):118.
23. Uncini A, Parano E, Lange DJ, De Vivo DC, Lovelace RE. Chronic inflammatory demyelinating polyneuropathy in childhood: clinical and electrophysiological features. *Childs Nerv Syst* 1991;**7**(4):191–6.
24. Dyck PJ, Lais AC, Ohta M, Bastron JA, Okazaki H, Groover RV. Chronic inflammatory polyradiculoneuropathy. *Mayo Clin Proc* 1975;**50**(11):621–37.
25. Sinno DD, Darras BT, Yamout BI, Rebeiz JG, Mikati MA. Motor variant of chronic inflammatory demyelinating polyneuropathy in a child. *Pediatr Neurol* 2008;**38**(6):426–9.
26. Sabatelli M, Madia F, Mignogna T, Lippi G, Quaranta L, Tonali P. Pure motor chronic inflammatory demyelinating polyneuropathy. *J Neurol* 2001;**248**(9):772–7.
27. Schady W, Goulding PJ, Lecky BR, King RH, Smith CM. Massive nerve root enlargement in chronic inflammatory demyelinating polyneuropathy. *J Neurol Neurosurg Psychiatry* 1996;**61**(6):636–40.
28. Ryan MM, Grattan-Smith PJ, Procopis PG, Morgan G, Ouvrier RA. Childhood chronic inflammatory demyelinating polyneuropathy: clinical course and long-term outcome. *Neuromuscul Disord* 2000;**10**(6):398–406.

29. Dionne A, Nicolle MW, Hahn AF. Clinical and electrophysiological parameters distinguishing acute-onset chronic inflammatory demyelinating polyneuropathy from acute inflammatory demyelinating polyneuropathy. *Muscle Nerve* 2010;**41**(2):202–7.
30. Ruts L, Drenthen J, Jacobs BC, van Doorn PA. Distinguishing acute-onset CIDP from fluctuating Guillain-Barré syndrome: a prospective study. *Neurology* 2010;**74**(21):1680–6.
31. Kleyweg RP, van der Meche FG. Treatment related fluctuations in Guillain-Barré syndrome after high-dose immunoglobulins or plasma-exchange. *J Neurol Neurosurg Psychiatry* 1991;**54**(11):957–60.
32. Rossignol E, D'Anjou G, Lapointe N, Haddad E, Vanasse M. Evolution and treatment of childhood chronic inflammatory polyneuropathy. *Pediatr Neurol* 2007;**36**(2):88–94.
33. Simmons Z, Wald JJ, Albers JW. Chronic inflammatory demyelinating polyradiculoneuropathy in children: II. Long-term follow-up, with comparison to adults. *Muscle Nerve* 1997;**20**(12):1569–75.
34. Rossignol E, D'Anjou G, Lapointe N, Haddad E, Vanasse M. Toxic polyneuropathy due to glue sniffing. *J Neurol Neurosurg Psychiatry* 1974;**37**(7):848–53.
35. Marks HG, Scavina MT, Kolodny EH, Palmieri M, Childs J. Krabbe's disease presenting as a peripheral neuropathy. *Muscle Nerve* 1997;**20**(8):1024–8.
36. Nevo Y, Pestronk A, Lopate G, Carroll SL. Neuropathy of metachromatic leukodystrophy: improvement with immunomodulation. *Pediatr Neurol* 1996;**15**(3):237–9.
37. Moser HW, Moser AB, Smith KD, Bergin A, Borel J, Shankroff J, et al. Adrenoleukodystrophy: phenotypic variability and implications for therapy. *J Inherit Metab Dis* 1992;**15**(4):645–64.
38. McMillan HJ, Worthylake T, Schwartzentruber J, Gottlieb CC, Lawrence SE, Mackenzie A. Specific combination of compound heterozygous mutations in 17beta-hydroxysteroid dehydrogenase type 4 (HSD17B4) defines a new subtype of D-bifunctional protein deficiency. *Orphanet J Rare Dis* 2012;**7**:90.
39. Ozdirim E, Topcu M, Ozon A, Cila A. Cockayne syndrome: review of 25 cases. *Pediatr Neurol* 1996;**15**(4):312–16.
40. Rusanen H, Majamaa K, Tolonen U, Remes AM, Myllyla R, Hassinen IE. Demyelinating polyneuropathy in a patient with the tRNA(Leu)(UUR) mutation at base pair 3243 of the mitochondrial DNA. *Neurology* 1995;**45**(6):1188–92.
41. Kaku DA, Parry GJ, Malamut R, Lupski JR, Garcia CA. Uniform slowing of conduction velocities in Charcot-Marie-Tooth polyneuropathy type 1. *Neurology* 1993;**43**(12):2664–7.
42. Street VA, Meekins G, Lipe HP, Seltzer WK, Carter GT, Kraft GH, et al. Charcot-Marie-Tooth neuropathy: clinical phenotypes of four novel mutations in the MPZ and Cx 32 genes. *Neuromuscul Disord* 2002;**12**(7–8):643–50.
43. Ryan MM, Jones Jr. HR. CMTX mimicking childhood chronic inflammatory demyelinating neuropathy with tremor. *Muscle Nerve* 2005;**31**(4):528–30.
44. Tabaraud F, Lagrange E, Sindou P, Vandenberghe A, Levy N, Vallat JM. Demyelinating X-linked Charcot-Marie-Tooth disease: unusual electrophysiological findings. *Muscle Nerve* 1999;**22**(10):1442–7.
45. Sellman MS, Mayer RF. Conduction block in hereditary neuropathy with susceptibility to pressure palsies. *Muscle Nerve* 1987;**10**(7):621–5.
46. Chaudhry V, Moser HW, Cornblath DR. Nerve conduction studies in adrenomyeloneuropathy. *J Neurol Neurosurg Psychiatry* 1996;**61**(2):181–5.
47. Burns TM, Shneker BV, Juel VC. Gasoline sniffing multifocal neuropathy. *Pediatr Neurol* 2001;**25**(5):419–21.
48. Fraser AG, McQueen IN, Watt AH, Stephens MR. Peripheral neuropathy during long-term high-dose amiodarone therapy. *J Neurol Neurosurg Psychiatry.* 1985;**48**(6):576–8.
49. Wilson JR, Conwit RA, Eidelman BH, Starzl T, Abu-Elmagd K. Sensorimotor neuropathy resembling CIDP in patients receiving FK506. *Muscle Nerve.* 1994;**17**(5):528–32.
50. Chaudhry V, Eisenberger MA, Sinibaldi VJ, Sheikh K, Griffin JW, Cornblath DR. A prospective study of suramin-induced peripheral neuropathy. *Brain* 1996;**119**(Pt 6):2039–52.
51. Argov Z, Soffer D, Eisenberg S, Zimmerman Y. Chronic demyelinating peripheral neuropathy in cerebrotendinous xanthomatosis. *Ann Neurol* 1986;**20**(1):89–91.
52. Moosa A, Dubowitz V. Peripheral neuropathy in Cockayne syndrome. *Arch Dis Child.* 1970;**45**(243):674–7.
53. Tullu MS, Muranjan MN, Kondurkar PP, Bharucha BA. Krabbe disease—clinical profile. *Indian Pediatr* 2000;**37**(9):939–46.
54. Cameron CL, Kang PB, Burns TM, Darras BT, Jones Jr. HR. Multifocal slowing of nerve conduction in metachromatic leukodystrophy. *Muscle Nerve* 2004;**29**(4):531–6.
55. Wasserstein MP, Aron A, Brodie SE, Simonaro C, Desnick RJ, McGovern MM. Acid sphingomyelinase deficiency: prevalence and characterization of an immediate phenotype of Niemann-Pick disease. *J Pediatr.* 2006;**149**(4):554–9.
56. Chance PF, Alderson MK, Leppig KA, Lensch MW, Matsunami N, Smith B, et al. DNA deletion associated with hereditary neuropathy with liability to pressure palsies. *Cell* 1993;**72**(1):143–51.
57. Shy ME, Garbern JY, Kamholz J. Hereditary motor and sensory neuropathies: a biological perspective. *Lancet Neurol* 2002;**1**(2):110–18.
58. Mouton P, Tardieu S, Gouider R, Birouk N, Maisonobe T, Dubourg O, et al. Spectrum of clinical and electrophysiologic features in HNPP patients with the 17p11.2 deletion. *Neurology* 1999;**52**(7):1440–6.
59. Cottenie E, Menezes MP, Rossor AM, Morrow JM, Yousry TA, Dick DJ, et al. Rapidly progressive asymmetrical weakness in Charcot-Marie-Tooth disease type 4J resembles chronic inflammatory demyelinating polyneuropathy. *Neuromuscul Disord* 2013;**23**(5):399–403.
60. Houlden H, Laura M, Ginsberg L, Jungbluth H, Robb SA, Blake J, et al. The phenotype of Charcot-Marie-Tooth disease type 4C due to SH3TC2 mutations and possible predisposition to an inflammatory neuropathy. *Neuromuscul Disord* 2009;**19**(4):264–9.
61. Michell AW, Laura M, Blake J, Lunn MP, Cox A, Gibbons VS, et al. GJB1 gene mutations in suspected inflammatory demyelinating neuropathies not responding to treatment. *J Neurol Neurosurg Psychiatry* 2009;**80**(6):699–700.
62. Houlden H, King R, Blake J, Groves M, Love S, Woodward C, et al. Clinical, pathological and genetic characterization of hereditary sensory and autonomic neuropathy type 1 (HSAN I). *Brain* 2006;**129**(Pt 2):411–25.
63. Bennett CL, Shirk AJ, Huynh HM, Street VA, Nelis E, Van Maldergem L, et al. SIMPLE mutation in demyelinating

neuropathy and distribution in sciatic nerve. *Ann Neurol* 2004;**55**(5):713–20.

64. Hattori N, Ichimura M, Aoki S, Nagamatsu M, Yasuda T, Kumazawa K, et al. Clinicopathological features of chronic inflammatory demyelinating polyradiculoneuropathy in childhood. *J Neurol Sci* 1998;**154**(1):66–71.

65. McMillan HJ, Santagata S, Shapiro F, Batish SD, Couchon L, Donnelly S, et al. Novel MPZ mutations and congenital hypomyelinating neuropathy. *Neuromuscul Disord* 2010;**20**(11):725–9.

66. Fredman P, Vedeler CA, Nyland H, Aarli JA, Svennerholm L. Antibodies in sera from patients with inflammatory demyelinating polyradiculoneuropathy react with ganglioside LM1 and sulphatide of peripheral nerve myelin. *J Neurol* 1991;**238**(2):75–9.

67. Connolly AM, Pestronk A, Mehta S, Yee WC, Green BJ, Fellin C, et al. Serum IgM monoclonal autoantibody binding to the 301 to 314 amino acid epitope of beta-tubulin: clinical association with slowly progressive demyelinating polyneuropathy. *Neurology* 1997;**48**(1):243–8.

68. van Schaik IN, Vermeulen M, van Doorn PA, Brand A. Anti-GM1 antibodies in patients with chronic inflammatory demyelinating polyneuropathy (CIDP) treated with intravenous immunoglobulin (IVIg). *J Neuroimmunol* 1994;**54**(1–2):109–15.

69. Gabriel CM, Gregson NA, Hughes RA. Anti-PMP22 antibodies in patients with inflammatory neuropathy. *J Neuroimmunol* 2000;**104**(2):139–46.

70. Thomas JE, Lambert EH. Ulnar nerve conduction velocity and H-reflex in infants and children. *J Appl Physiol* 1960;**15**:1–9.

71. Wagner AL, Buchthal F. Motor and sensory conduction in infancy and childhood: reappraisal. *Dev Med Child Neurol* 1972;**14**(2):189–216.

72. Pitt M. Maturational changes vis-a-vis neurophysiology markers and the development of peripheral nerves. In: Holmes GL, Moshe S, Jones HR, editors. *Clinical Neurophysiology of Infancy, Childhood, and Adolescence*. Philadelphia: Elsevier; 2006. pp. 146–67.

73. Lewis RA, Sumner AJ. The electrodiagnostic distinctions between chronic familial and acquired demyelinative neuropathies. *Neurology* 1982;**32**(6):592–6.

74. Hoogendijk JE, de Visser M, Bour LJ, Jennekens FG, Ongerboer BW. Conduction block in hereditary motor and sensory neuropathy type I. *Muscle Nerve* 1992;**15**(4):520–1, author reply 3.

75. Uncini A, Di Guglielmo G, Di Muzio A, Gambi D, Sabatelli M, Mignogna T, et al. Differential electrophysiological features of neuropathies associated with 17p11.2 deletion and duplication. *Muscle Nerve* 1995;**18**(6):628–35.

76. Research criteria for diagnosis of chronic inflammatory demyelinating polyneuropathy (CIDP). Report from an Ad Hoc Subcommittee of the American Academy of Neurology AIDS Task Force. *Neurology* 1991;**41**(5):617–18.

77. Saperstein DS, Katz JS, Amato AA, Barohn RJ. Clinical spectrum of chronic acquired demyelinating polyneuropathies. *Muscle Nerve* 2001;**24**(3):311–24.

78. van Geel BM, Koelman JH, Barth PG, Ongerboer de Visser BW. Peripheral nerve abnormalities in adrenomyeloneuropathy: a clinical and electrodiagnostic study. *Neurology* 1996;**46**(1):112–18.

79. Prineas JW, McLeod JG. Chronic relapsing polyneuritis. *J Neurol Sci* 1976;**27**(4):427–58.

80. Vital A, Vital C, Brechenmacher C, Fontan D, Castaing Y. Chronic inflammatory demyelinating polyneuropathy in childhood: ultrastructural features of peripheral nerve biopsies in four cases. *Eur J Pediatr* 1990;**149**(9):654–8.

81. Dyck PJ, Prineas JW, Pollard JD. Chronic inflammatory demyelinating polyradiculoneuropathy. In: Dyck PJ, Thomas PK, editors. *Peripheral Neuropathy*. 3rd ed. Philadelphia: Saunders; 1993. pp. 1498–517.

82. Pollard JD, McCombe PA, Baverstock J, Gatenby PA, McLeod JG. Class II antigen expression and T lymphocyte subsets in chronic inflammatory demyelinating polyneuropathy. *J Neuroimmunol* 1986;**13**(2):123–34.

83. Krendel DA, Parks HP, Anthony DC, St Clair MB, Graham DG. Sural nerve biopsy in chronic inflammatory demyelinating polyradiculoneuropathy. *Muscle Nerve* 1989;**12**(4):257–64.

84. Cornblath DR, Griffin DE, Welch D, Griffin JW, McArthur JC. Quantitative analysis of endoneurial T-cells in human sural nerve biopsies. *J Neuroimmunol* 1990;**26**(2):113–18.

85. Ryan MM, Tilton A, De Girolami U, Darras BT, Jones Jr. HR. Paediatric mononeuritis multiplex: a report of three cases and review of the literature. *Neuromuscul Disord* 2003;**13**(9):751–6.

86. Van den Bergh PY, Hadden RD, Bouche P, Cornblath DR, Hahn A, Illa I, et al. European Federation of Neurological Societies/Peripheral Nerve Society guideline on management of chronic inflammatory demyelinating polyradiculoneuropathy: report of a joint task force of the European Federation of Neurological Societies and the Peripheral Nerve Society—first revision. *Eur J Neurol* 2010;**17**(3):356–63.

87. Duggins AJ, McLeod JG, Pollard JD, Davies L, Yang F, Thompson EO, et al. Spinal root and plexus hypertrophy in chronic inflammatory demyelinating polyneuropathy. *Brain* 1999;**122**(Pt 7):1383–90.

88. McMillan HJ, Miller E. Cranial nerve hypertrophy in pediatric chronic inflammatory demyelinating polyradiculoneuropathy. *Pediatr Radiol* 2010;**40**(Suppl. 1):S176.

89. Ginsberg L, Platts AD, Thomas PK. Chronic inflammatory demyelinating polyneuropathy mimicking a lumbar spinal stenosis syndrome. *J Neurol Neurosurg Psychiatry* 1995;**59**(2):189–91.

90. Duarte J, Martinez AC, Rodriguez F, Mendoza A, Sempere AP, Claveria LE. Hypertrophy of multiple cranial nerves and spinal roots in chronic inflammatory demyelinating neuropathy. *J Neurol Neurosurg Psychiatry* 1999;**67**(5):685–7.

91. Haubrich C, Krings T, Senderek J, Zuchner S, Schroder JM, Noth J, et al. Hypertrophic nerve roots in a case of Roussy-Levy syndrome. *Neuroradiology* 2002;**44**(11):933–7.

92. Choi SK, Bowers RP, Buckthal PE. MR imaging in hypertrophic neuropathy: a case of hereditary motor and sensory neuropathy, type I (Charcot-Marie-Tooth). *Clin Imaging* 1990;**14**(3):204–7.

93. Gorson KC, Ropper AH, Muriello MA, Blair R. Prospective evaluation of MRI lumbosacral nerve root enhancement in acute Guillain-Barré syndrome. *Neurology* 1996;**47**(3):813–17.

94. Yikilmaz A, Doganay S, Gumus H, Per H, Kumandas S, Coskun A. Magnetic resonance imaging of childhood Guillain-Barré syndrome. *Childs Nerv Syst* 2010;**26**(8):1103–8.

95. Borgstein BJ, Koster PA, Portegies P, Peeters FL. Myelography in patients with acquired immuno deficiency syndrome. Indications and results. *Neuroradiology* 1989;**31**(4):326–30.

96. Lin J, Martel W. Cross-sectional imaging of peripheral nerve sheath tumors: characteristic signs on CT, MR imaging, and sonography. *AJR Am J Roentgenol* 2001;**176**(1):75–82.

97. Grisariu S, Avni B, Batchelor TT, van den Bent MJ, Bokstein F, Schiff D, et al. Neurolymphomatosis: an International Primary CNS Lymphoma Collaborative Group report. *Blood* 2010;**115**(24):5005–11.
98. Smith JK, Matheus MG, Castillo M. Imaging manifestations of neurosarcoidosis. *AJR Am J Roentgenol* 2004;**182**(2):289–95.
99. Midroni G, de Tilly LN, Gray B, Vajsar J. MRI of the cauda equina in CIDP: clinical correlations. *J Neurol Sci* 1999;**170**(1):36–44.
100. Yousry I, Forderreuther S, Moriggl B, Holtmannspotter M, Naidich TP, Straube A, et al. Cervical MR imaging in postural headache: MR signs and pathophysiological implications. *AJNR Am J Neuroradiol* 2001;**22**(7):1239–50.
101. van Swieten JC, Koudstaal PJ, Visser MC, Schouten HJ, van Gijn J. Interobserver agreement for the assessment of handicap in stroke patients. *Stroke* 1988;**19**(5):604–7.
102. Gorson KC, van Schaik IN, Merkies IS, Lewis RA, Barohn RJ, Koski CL, et al. Chronic inflammatory demyelinating polyneuropathy disease activity status: recommendations for clinical research standards and use in clinical practice. *J Peripher Nerv Syst* 2010;**15**(4):326–33.
103. Ware TL, Kornberg AJ, Rodriguez-Casero MV, Ryan MM. Childhood chronic inflammatory demyelinating polyneuropathy: an overview of 10 cases in the modern era. *J Child Neurol* 2014;**29**(1):43–8.
104. Gurses N, Uysal S, Cetinkaya F, Islek I, Kalayci AG. Intravenous immunoglobulin treatment in children with Guillain-Barré syndrome. *Scand J Infect Dis* 1995;**27**(3):241–3.
105. Korinthenberg R, Schessl J, Kirschner J, Monting JS. Intravenously administered immunoglobulin in the treatment of childhood Guillain-Barré syndrome: a randomized trial. *Pediatrics* 2005;**116**(1):8–14.
106. Eftimov F, Winer JB, Vermeulen M, de Haan R, van Schaik IN. Intravenous immunoglobulin for chronic inflammatory demyelinating polyradiculoneuropathy. *Cochrane Database Syst Rev* 2009;(1):CD001797
107. Hahn AF, Bolton CF, Pillay N, Chalk C, Benstead T, Bril V. Plasma-exchange therapy in chronic inflammatory demyelinating polyneuropathy. A double-blind, sham-controlled, cross-over study. *Brain* 1996;**119**(Pt 4):1055–66.
108. Hughes RA, Donofrio P, Bril V, Dalakas MC, Deng C, Hanna K, et al. Intravenous immune globulin (10% caprylate-chromatography purified) for the treatment of chronic inflammatory demyelinating polyradiculoneuropathy (ICE study): a randomised placebo-controlled trial. *Lancet Neurol* 2008;**7**(2):136–44.
109. Thompson N, Choudhary P, Hughes RA, Quinlivan RM. A novel trial design to study the effect of intravenous immunoglobulin in chronic inflammatory demyelinating polyradiculoneuropathy. *J Neurol* 1996;**243**(3):280–5.
110. Mendell JR, Barohn RJ, Freimer ML, Kissel JT, King W, Nagaraja HN. Randomized controlled trial of IVIg in untreated chronic inflammatory demyelinating polyradiculoneuropathy. *Neurology* 2001;**56**(4):445–9.
111. Vermeulen M, van Doorn PA, Brand A, Strengers PF, Jennekens FG, Busch HF. Intravenous immunoglobulin treatment in patients with chronic inflammatory demyelinating polyneuropathy: a double blind, placebo controlled study. *J Neurol Neurosurg Psychiatry* 1993;**56**(1):36–9.
112. Dyck PJ, Litchy WJ, Kratz KM, Suarez GA, Low PA, Pineda AA, et al. A plasma exchange versus immune globulin infusion trial in chronic inflammatory demyelinating polyradiculoneuropathy. *Ann Neurol* 1994;**36**(6):838–45.
113. Hughes R, Bensa S, Willison H, Van den Bergh P, Comi G, Illa I. Randomized controlled trial of intravenous immunoglobulin versus oral prednisolone in chronic inflammatory demyelinating polyradiculoneuropathy. *Ann Neurol* 2001;**50**(2):195–201.
114. Buchwald B, Ahangari R, Weishaupt A, Toyka KV. Intravenous immunoglobulins neutralize blocking antibodies in Guillain-Barré syndrome. *Ann Neurol* 2002;**51**(6):673–80.
115. Thornton CA, Ballow M. Safety of intravenous immunoglobulin. *Arch Neurol* 1993;**50**(2):135–6.
116. Brannagan III TH, Nagle KJ, Lange DJ, Rowland LP. Complications of intravenous immune globulin treatment in neurologic disease. *Neurology* 1996;**47**(3):674–7.
117. Casteels-Van Daele M, Wijndaele L, Brock P, Kruger M, Gillis P. Aseptic meningitis associated with high dose intravenous immunoglobulin therapy. *J Neurol Neurosurg Psychiatry* 1992;**55**(10):980–1.
118. Sekul EA, Cupler EJ, Dalakas MC. Aseptic meningitis associated with high-dose intravenous immunoglobulin therapy: frequency and risk factors. *Ann Intern Med* 1994;**121**(4):259–62.
119. Dalakas MC. High-dose intravenous immunoglobulin and serum viscosity: risk of precipitating thromboembolic events. *Neurology* 1994;**44**(2):223–6.
120. Cayco AV, Perazella MA, Hayslett JP. Renal insufficiency after intravenous immune globulin therapy: a report of two cases and an analysis of the literature. *J Am Soc Nephrol* 1997;**8**(11):1788–94.
121. Rodriguez-Casero MV, Shield LK, Kornberg AJ. Subacute inflammatory demyelinating polyneuropathy in children. *Neurology* 2005;**64**(10):1786–8.
122. Colan RV, Snead III OC, Oh SJ, Benton Jr. JW. Steroid-responsive polyneuropathy with subacute onset in childhood. *J Pediatr* 1980;**97**(3):374–7.
123. Casteels-Van Daele M, Wijndaele L, Brock P, Kruger M, Gillis P, et al. Prednisone improves chronic inflammatory demyelinating polyradiculoneuropathy more than no treatment. *Ann Neurol* 1982;**11**(2):136–41.
124. Fauci AS, Dale DC, Balow JE. Glucocorticosteroid therapy: mechanisms of action and clinical considerations. *Ann Intern Med* 1976;**84**(3):304–15.
125. Bromberg MB, Carter O. Corticosteroid use in the treatment of neuromuscular disorders: empirical and evidence-based data. *Muscle Nerve* 2004;**30**(1):20–37.
126. Rostasy KM, Diepold K, Buckard J, Brockmann K, Wilken B, Hanefeld F. Progressive muscle weakness after high-dose steroids in two children with CIDP. *Pediatr Neurol* 2003;**29**(3):236–8.
127. Vedanarayanan VV, Kandt RS, Lewis Jr. DV, DeLong GR. Chronic inflammatory demyelinating polyradiculoneuropathy of childhood: treatment with high-dose intravenous immunoglobulin. *Neurology* 1991;**41**(6):828–30.
128. Dyck PJ, O'Brien P, Swanson C, Low P, Daube J. Combined azathioprine and prednisone in chronic inflammatory-demyelinating polyneuropathy. *Neurology* 1985;**35**(8):1173–6.

129. De Sousa EA, Brannagan III TH. Diagnosis and treatment of chronic inflammatory demyelinating polyneuropathy. *Curr Treat Options Neurol* 2006;**8**(2):91–103.

130. Gorson KC, Amato AA, Ropper AH. Efficacy of mycophenolate mofetil in patients with chronic immune demyelinating polyneuropathy. *Neurology* 2004;**63**(4):715–17.

131. Randomised controlled trial of methotrexate for chronic inflammatory demyelinating polyradiculoneuropathy (RMC trial): a pilot, multicentre study. *Lancet Neurol* 2009;**8**(2):158–64.

132. Barnett MH, Pollard JD, Davies L, McLeod JG. Cyclosporin A in resistant chronic inflammatory demyelinating polyradiculoneuropathy. *Muscle Nerve* 1998;**21**(4):454–60.

133. Hadden RD, Sharrack B, Bensa S, Soudain SE, Hughes RA. Randomized trial of interferon beta-1a in chronic inflammatory demyelinating polyradiculoneuropathy. *Neurology* 1999;**53**(1):57–61.

134. Hughes RA, Gorson KC, Cros D, Griffin J, Pollard J, Vallat JM, et al. Intramuscular interferon beta-1a in chronic inflammatory demyelinating polyradiculoneuropathy. *Neurology* 2010;**74**(8):651–7.

135. Chin RL, Sherman WH, Sander HW, Hays AP, Latov N. Etanercept (Enbrel) therapy for chronic inflammatory demyelinating polyneuropathy. *J Neurol Sci* 2003;**210**(1–2):19–21.

136. Gorson KC, Natarajan N, Ropper AH, Weinstein R. Rituximab treatment in patients with IVIg-dependent immune polyneuropathy: a prospective pilot trial. *Muscle Nerve* 2007;**35**(1):66–9.

Chapter 22

Neuropathies Secondary to Systemic Disorders

Jo M. Wilmshurst and Robert A. Ouvrier

INTRODUCTION

There is a clear distinction between a peripheral neuropathy that is clinically significant and which is a presenting feature of the underlying condition (e.g. those associated with diabetes mellitus, lymphoma, and uremia) and a peripheral neuropathy that is asymptomatic and detected only by careful neurologic examination and electrophysiologic studies. Peripheral polyneuropathy is frequently asymptomatic in children with systemic disorders. This chapter reviews some of the relatively uncommon but clinically significant polyneuropathies that may be encountered in pediatric patients with various systemic illnesses. A summary of these neuropathies is provided in Table 22.1. Neuropathies seen in response to infections and metabolic and neurodegenerative disorders are also part of systemic processes, but these are described in other chapters.

CRITICAL ILLNESS POLYNEUROPATHY

Critical illness polyneuropathy (CIP) occurs in patients with sepsis, asthma, organ transplantation, and multiple organ failure as part of a systemic inflammatory response syndrome (see Chapter 50). Initially reported in adults with severe systemic illnesses in intensive care units,[1] it is often first recognized when it proves difficult to wean a child from mechanical ventilation.[2] The accompanying signs of generalized weakness, muscle wasting, and depressed or absent reflexes may be difficult to detect or misinterpreted in the intensive care setting. Onset may manifest as early as within the first week of illness, with a range of 4 to 26 days documented.[2,3] The differential diagnosis includes paraparesis or tetraparesis secondary to spinal cord lesions, prolonged neuromuscular blockade, steroid- or relaxant-induced myopathy, acute necrotizing myopathy, hypophosphatemia, toxic and thiamine-deficiency neuropathies, asthma-amyotrophy (Hopkins') syndrome, and Guillain-Barré syndrome (Box 22.1). A similar condition may occasionally follow severe burns.[4] A careful and systematic investigation includes testing of the serum creatine kinase, which is normal in most cases, magnesium and phosphate levels, neurophysiologic studies (nerve conduction studies, electromyography [EMG], and occasionally electroencephalography), and relevant imaging modalities. If diagnostic doubt persists, nerve and/or muscle biopsy sometimes provide diagnostic information. In typical cases of critical illness polyneuropathy, nerve conduction studies and histopathologic observations confirm widespread axonal degeneration with extensive muscle denervation.

CIP and critical illness myopathy are on a continuum of neuromuscular disorders. While far more data are available on affected adult patients, the limited studies in the pediatric age group suggest that the condition causes significant morbidity in critically ill children.[2] Treatment is mainly supportive, but early correction of abnormal osmolar states, glycemic levels, and electrolyte imbalance is considered important, as well as early mobilization.[5,6] Recovery, where reported, is spontaneous and of variable timing.[2] See Case Example 22.1.

RENAL DISORDERS

Chronic Renal Failure

The neuropathy of chronic renal failure is typically an axonal degenerative type, which in most children is asymptomatic.[7,8] Nerve conduction studies in long-term hemodialysis-dependent children generally show normal or only mildly impaired values.[9] The evaluation of the "H" reflex has been reported as a useful indicator of clinically asymptomatic uremic polyneuropathy, identifying

B.T. Darras, H. Royden Jones, Jr., M.M. Ryan & D.C. De Vivo (Eds): Neuromuscular Disorders of Infancy, Childhood and Adolescence, Second edition.
DOI: http://dx.doi.org/10.1016/B978-0-12-417044-5.00022-6

TABLE 22.1 Peripheral Neuropathies Associated with Systemic Diseases

Systemic Disease	Predominant Neuropathy Type[a]	Onset and Course	Pathology
Metabolic Disorders			
Uremia[b]	Sensory (sensorimotor)	Chronic (acute)	AD, SD
Hypoglycemia	Motor	Chronic	AD
Endocrine Disorders			
Hypothyroidism	Sensory	Chronic	AD
Acromegaly	Sensory	Chronic	AD
Diabetes mellitus[b]	Sensory, motor	Chronic	AD, SD
Malignancies and Reticuloses			
Carcinoma	Sensory, sensorimotor	Subacute or chronic	AD (SD in acute relapsing type)
Lymphoma[b]	Sensorimotor	Acute, subacute	AD, SD (acute relapsing type)
Chronic lymphatic leukemia	Sensorimotor	Acute (subacute)	AD
Deficiency States			
Vitamin B12[b]	Sensory	Chronic	AD
Thiamine[b]	Sensorimotor	Chronic	AD
Folic acid	Sensory	Chronic	AD
Vitamin E[b]	Sensory	Chronic	AD
Connective Tissue Disorders			
Rheumatoid arthritis[b]	Sensory, mononeuritis	Chronic, acute	SD, AD
Polyarteritis nodosa[b]	Mononeuritis	Acute, chronic	AD, SD
Systemic lupus erythematosus[b]	Sensory, motor	Chronic, acute	AD, SD
Miscellaneous			
Chronic liver disease[b]	Sensory (sensorimotor)	Chronic	AD, SD
Primary biliary cirrhosis[b]	Sensory	Chronic	AD
Viral hepatitis	Sensory or sensorimotor	Acute	SD
Celiac disease[b]	Sensorimotor	Chronic	AD
Chronic obstructive lung disease	Sensory (sensorimotor)	Chronic	AD
Sarcoidosis[b]	Sensorimotor, motor	Acute or chronic	AD
Amyloidosis[b]	Sensorimotor	Chronic	AD
Paraproteinemias and Dysproteinemias			
Multiple myeloma	Sensorimotor (sensory)	Chronic	AD
Cryoglobulinemia	Sensorimotor	Chronic	AD
Macroglobulinemia	Sensorimotor	Subacute or chronic	AD, SD
Monoclonal Gammopathy			
IgA	Sensorimotor	Chronic	AD
IgG	Sensorimotor	Chronic	AD, SD
IgM	Sensorimotor	Chronic	SD

[a]*Words in parentheses indicate less common types.*
[b]*Reported in childhood.*
AD = axonal degeneration; Ig = immunoglobulin; SD = segmental demyelination.

this complication in almost 60% of one series of children with chronic renal failure.[10]

This neuropathy usually remains stable or improves with treatment; if not, increasing the frequency and duration of dialysis may result in improvement. Rapid improvement usually follows successful renal transplantation, even in patients with long-standing neuropathy.

ENDOCRINE DISORDERS

Diabetes Mellitus

Diabetic neuropathy may manifest as a polyneuropathy, focal neuropathy, or autonomic neuropathy.[11] The latter is the most common and has the most serious side effects, related to unawareness of hypoglycemia and cardiovascular dysfunction. Children with a length-dependent polyneuropathy suffer distal sensory loss (numbness) and paraesthesiae; they may also have pain manifesting as burning, aching, and electric-sharp sensations. Weakness, clumsiness, loss of balance, and falls may be associated with distal anhidrosis, postprandial bloating, constipation, diarrhea, and lack of awareness of hypoglycemia. Glove and stocking sensory loss, loss of deep tendon reflexes, distal weakness, and even foot ulcers may occur.

BOX 22.1 Differential Diagnosis of Critical Illness Polyneuropathy

- Spinal cord lesions
- Prolonged neuromuscular blockade
- Steroid or relaxant-induced myopathy
- Acute necrotizing myopathy
- Hypophosphatemia, hypermagnesemia
- Toxic and thiamine-deficiency neuropathies
- Asthma-amyotrophy (Hopkins') syndrome
- Guillain-Barré syndrome

Between 10% and 68% of children with diabetes suffer from neuropathies.[11,12] In one large cohort of 146 children (under 18 years of age) with diabetes, 27.4% had peripheral neuropathy.[13] Of this affected group, 62.5% had subclinical disease, a minority being clinically affected.[13]

Although patients with diabetic neuropathy may not complain of it, their physical examination may reveal variable degrees of sensory loss. In one study, 30.8% of subjects had numbness and about 7% to 10% showed large myelinated nerve fiber dysfunction, while only 1.4% had impairment in pain or temperature sensation.[13] Nerve conduction studies most often show abnormalities of the peroneal and sural nerves.[13–15]

Large studies have suggested that neuropathy can develop in young children with short illness duration and despite good glycemic control.[16] As such, the concept of the duration of illness as a key disease-inducing factor has not been established; it has been suggested that the nerve damage can occur acutely and rapidly at the onset of the diabetes, subsequent to which the neuropathy develops more slowly or may even plateau.[13] The best way to prevent this complication is still strict blood glucose control.[13]

Since the neuropathy can progress and ultimately affect quality of life, but is often initially subclinical, regular

CASE EXAMPLE 22.1 Critical Illness Polyneuropathy after Septic Shock

An 18-month-old boy with Down syndrome, global developmental delay, epilepsy, and cortical blindness was admitted for septicemia and multiorgan failure. He had previously had recurrent aspiration pneumonia necessitating a fundoplication and gastrostomy, and a urinary tract infection. He received antibiotics (ceftriaxone, vancomycin, and gentamicin) as well as corticosteroids. He required massive inotropic support (adrenaline, dopamine, and dobutamine), platelets, fresh frozen plasma, and colloid infusions to stabilize his condition. Various metabolic and electrolyte problems were corrected. Hypoglycemic episodes developed, requiring glucose infusions.

Two weeks after admission, he developed an adult respiratory distress-like syndrome and was started on prednisone (2 mg/kg/day). He was slowly weaned from the ventilator but continued to have episodes of central apnea and hypercapnea. Alveolar lavage showed persistent *Pseudomonas* colonization. Ceftazidime and gentamicin were administered. Recurrent seizures necessitated treatment with multiple anticonvulsants. He developed generalized weakness with minimal movements limited to his left arm. The muscle stretch reflexes were absent. A cranial computed tomography scan demonstrated generalized cortical atrophy with diffuse white matter changes compatible with a hypoxic insult. An electroencephalogram was nondiagnostic. Cervical spine radiographs were unremarkable. Based on his stormy clinical course, the persistent difficulty weaning him from the ventilator, and his associated apneic spells during sleep, a diagnosis of critical illness polyneuropathy was considered.

Motor nerve conduction studies demonstrated findings consistent with an axonal polyneuropathy with markedly reduced compound muscle action potential amplitudes on all motor studies in the upper limbs, and an absent common peroneal motor response. Needle EMG of the tibialis anterior and biceps brachii muscles demonstrated fibrillations.

Although he had some improvement, he was still bedridden owing to weakness 3 months after the onset of his illness.

Comment

This is a fairly typical course for a child with critical illness polyneuropathy.

nerve conduction studies are recommended in order to monitor for development of this serious disease complication.[17]

Multiple Endocrine Neoplasia Type 2B

Multiple endocrine neoplasia (MEN) type 2B is a rare, dominantly inherited syndrome accounting for 5% of all cases of MEN 2, characterized by medullary thyroid carcinoma, pheochromocytoma, ganglioneuromatosis, and a variety of skeletal and connective tissue abnormalities.[18] MEN 2B cases usually carry either an M918T or A883T mutation of the RET (REarranged during Transfection) oncogene.[18,19] Affected patients have an arresting appearance because of a diffuse, thick, fleshy enlargement of the everted lips, apparent eversion of the eyelids, and a marfanoid habitus.[18] Other oral manifestations include a high arched palate and small, whitish-yellow, firm, protruding neuromatous nodules on the tongue. Fibular atrophy, pes cavus, scoliosis, and proximal or distal muscle atrophy are common.[20]

A polyneuropathy, with mild weakness of ankle dorsiflexion or occasionally of the intrinsic hand muscles, may be the presenting manifestation of this disorder.[21] Nerve conduction studies demonstrate mild abnormalities of motor and sensory nerve conduction. Chronic denervation is apparent on needle EMG.[21,22]

Sural nerve biopsy reveals moderate loss of small- and large-diameter myelinated fibers. At autopsy, unusual plaques of hyperplastic interlacing bands of Schwann cells and myelinated fibers overlie the posterior columns of the spinal cord.[22] Autonomic nerves are also involved and are responsible for the severe constipation or diarrhea frequently seen in these patients.[22,23]

Early diagnosis is extremely important because of the high risk of malignancy. Medullary thyroid carcinoma is almost always present at the time of diagnosis.[18]

VITAMIN DEFICIENCY STATES

Vitamin B1 (Thiamine) Deficiency

Childhood thiamine deficiency, or beriberi neuropathy, is rare in resource-equipped settings except in special circumstances, such as prolonged parenteral nutrition without adequate vitamin supplementation. Such complications, related to inadequate thiamine intake or uptake, are reported in preterm infants, children after gastric surgery, and patients undergoing allogeneic hematopoietic stem cell transplantation.[24–26] Infantile encephalitic beriberi (IEBB) is a rare form of thiamine deficiency reported in "Third-world" countries in the breast-fed infants of thiamine-deficient mothers.[27] IEBB appears suddenly, after an episode of gastroenteritis, with cardiac symptoms, neck stiffness, and acute peripheral neuritis. IEBB can mimic Leigh syndrome, with life-threatening respiratory and central nervous system symptoms. There is a dramatic response to thiamine. Epidemic spastic paraparesis (konzo) is a neurological condition associated with the consumption of cassava roots and minimal protein consumption.[28] The condition is prevalent in Nigeria, Tanzania, Sierra Leone, Mozambique, Central African Republic, and the Democratic Republic of the Congo. It occurs in children aged 4 to 12 years and in young women, causing asymmetrical spastic paraparesis with marked sensory polyneuropathy and ataxia. There is overlap with the features of dry beriberi. There is a strong case that the condition is, at least in part, the result of thiamine deficiency resulting from the overconsumption of cassava roots.

Nerve conduction studies show a mainly sensory neuropathy with a superimposed acute motor axonopathy.[29] Pathologically, peripheral nerves demonstrate the typical changes of large-fiber-predominant axonal degeneration and subperineurial edema.[30] A prompt response to therapy is reported in most cases.[29]

Vitamin B2 (Riboflavin) Deficiency

The Brown-Vialetto-Van Laere syndrome is a rare neurodegenerative disorder that usually causes death from respiratory disease in early childhood.[31] Fazio-Londe syndrome is considered the same disease entity (but without deafness).[31] Affected patients have a rapidly progressive axonal sensorimotor neuropathy manifesting with sensory ataxia, severe weakness of the upper limbs and axial muscles often with distinctly preserved strength in the lower limbs, bulbar palsy, hearing loss, optic atrophy, and respiratory insufficiency.[31] Presentations are protean and include a severe infantile form, a slowly progressive sensorimotor neuropathy with prominent ataxia, resembling Friedreich ataxia and a slowly progressive ponto-bulbar palsy, often evidenced by tongue fasciculations (Brown-Vialetto-van Laere syndrome). Sensorineural deafness and optic atrophy are common associations resulting in a form of the optico-acoustic neuropathy syndrome. Patients with these conditions have been found to carry mutations in the *SLC52A2* and *SLC52A3* genes, which encode two of the three known riboflavin transporter genes (RFVT).[31–33] These mutations cause reduced riboflavin uptake and transporter protein expression.[32] High-dose riboflavin supplementation in affected patients results in significant and sustained clinical and biochemical improvements.[31,32]

Vitamin B6 (Pyridoxine) Deficiency

Vitamin B6 supplementation is recommended for patients treated with isoniazid, to avoid development of a painful sensory neuropathy.[34] However, there are a few reports whereby pyridoxine therapy itself resulted in the development of a dorsal root ganglionopathy in adults but not children.[35]

Vitamin B12 (Cobalamin) Deficiency

Although vitamin B12 deficiency anemia in childhood due to familial malabsorption of cobalamin or inadequate dietary intake is well reported,[36] descriptions of peripheral neuropathy in association with the recognized neurologic complications in children are infrequent.[37–39] Children from resource-poor countries with dietary insufficiency, malabsorption, and malnutrition may have neurological sequelae inclusive of neuropathy.[37,38] The clinical, pathologic, and electrophysiologic features of the neuropathy due to vitamin B12 deficiency are better described in adults.[40]

Vitamin E Deficiency

A progressive neurologic syndrome secondary to deficient vitamin E absorption occurs in children with longstanding obstructive liver disease, chronic intestinal malabsorption, and cystic fibrosis.[41–43] (See Case Example 22.2.) Children suffering from protein-energy malnutrition are also at risk of vitamin E deficiency.[44] Vitamin E deficiency is a major factor in the pathogenesis of the polyneuropathy associated with abetalipoproteinemia.[45] Affected patients may develop nystagmus, gaze paresis, retinitis pigmentosa, an ataxic gait, and areflexia, and have impaired position and vibration awareness.[43,46]

In rare cases, vitamin E malabsorption appears to be an isolated defect, a condition known as ataxia with vitamin E deficiency (AVED) or familial isolated vitamin E deficiency related to mutations in the alpha-tocopherol transfer protein gene (*TTP1*).[46,47] Vitamin E has antioxidant properties and modulates glutamate neurotoxicity. The defect in AVED impairs incorporation of vitamin E into plasma low-density lipoproteins. Subsequently, although vitamin E is absorbed in the intestine, normal recycling of vitamin E, which is dependent on the alpha-tocopherol transfer protein, is deficient. Thus, vitamin E is rapidly eliminated and levels are low.

AVED is characterized by progressive ataxia, dysarthria, weakness, areflexia, and proprioceptive loss with undetectable or very low serum vitamin E levels, in the absence of hypolipidemia or fat malabsorption. Onset of symptoms is usually between 3 and 13 years of age. In contrast to patients with vitamin E deficiency secondary to other gastrointestinal disorders, patients with AVED do not usually have retinopathy, ophthalmoplegia, or abnormal peripheral sensory conduction.[46,48] Extensor plantar responses, tremor, and dystonic movements may be present.[49] Abnormal posterior column function is confirmed by somatosensory evoked potential studies.[50] Later, some children are found to have alterations of sensory conduction.[49]

Nerve conduction studies typically show a sensory neuronopathy with normal motor conduction and absent or markedly reduced sensory nerve action potentials (SNAPs).[51] Sural nerve biopsy from a 22-year-old man with isolated vitamin E deficiency revealed loss of large myelinated fibers and evidence of remyelination.[52] This case study differed from others in that it suggested a demyelinating process rather than axonal disease, as is described in other reports.[49]

Autopsy findings include loss of large myelinated fibers and axonal degeneration in the peripheral nerves, loss of dorsal root ganglion cells, posterior column degeneration, and swollen dystrophic axons (spheroids) and astrocytosis in the gracile and cuneate nuclei, posterior columns, and Clarke's column.[53]

Dietary vitamin E supplementation can lead to clinical and electrophysiological recovery of sensory conduction and evoked potentials.[52]

MALIGNANCIES AND RETICULOSES

The peripheral nervous system is commonly involved in patients with cancer because of direct infiltration, compression, paraneoplastic syndromes, the complications of treatment, and a number of unpredictable complications such as infections and the impact of poor nutrition.[54–56]

CASE EXAMPLE 22.2 Vitamin E Deficiency Due to Chronic Cholestasis

A female child became jaundiced soon after birth and was eventually diagnosed with Alagille's syndrome with chronic cholestasis. At 8 years of age, she was noted to be walking on her toes and a tendon-lengthening procedure was performed.

From the age of 10 years she fell frequently, developed a squint, and had trouble straightening her arms. There was evidence of night blindness.

At the age of 13 years, she had tigroid pigmentation of the peripheral retina. There was limitation of abduction and elevation of both eyes. Her horizontal extraocular movements were jerky. There was slight dysarthria, and mild ataxia on finger-nose-finger and tandem gait testing. Vibration sense and muscle stretch reflexes were absent. The plantar responses were equivocal.

Motor nerve conduction velocities were 52 and 42 m/s in the median and common peroneal nerves, respectively. Sensory nerve conduction velocities were normal in the median, ulnar, and sural nerves, but the amplitudes of the responses were slightly low at 10, 7, and 6 mV, respectively. The serum vitamin E level was 0.5 units (normal range, 12–36 units).

The child was treated with vitamin E (2000 units/day). Twelve months later, there was slight improvement in her ataxia and an increase in the amplitude of the sensory nerve action potentials from the median and ulnar nerves.

Complications related to direct infiltration and the toxic effects of cancer therapies are addressed in Chapters 12 and 23. Most data are from adult studies. When drug toxicity and local neoplastic infiltration are excluded as factors, peripheral neuropathies in the context of childhood malignancies are rare. A review of 96 long-term survivors of childhood malignancy found no clinical or electrophysiologic evidence of neuropathy in the 49 patients who had not received vincristine.[57] In contrast, there was a high rate of areflexia and sensory conduction velocity abnormalities in the 47 children treated with vincristine, although these children were generally not clinically compromised.

Lymphoma

About 5% of adults with lymphoma develop a peripheral neuropathy.[54] This association is extremely rare in childhood. As in carcinomatous neuropathy, two main clinical types are recognized: a sensory neuropathy and a sensorimotor neuropathy.

Sensory Neuropathy

Pure sensory neuropathy is considerably less common in lymphoma than in carcinoma.[57] The clinical features of paresthesia, dysesthesia, pain, and sensory ataxia are similar to those seen in carcinoma.

Sensorimotor Neuropathy

Acute polyneuropathy of the Guillain-Barré type is occasionally found in association with lymphoma, particularly Hodgkin's disease. Various reports have described the challenge of differentiation between acute inflammatory polyradiculoneuropathy (AIDP) and vincristine toxicity in children with acute lymphoblastic leukemia, non-Hodgkin's lymphoma, and Hodgkin's lymphoma[58–60] who develop a predominantly lower limb sensorimotor neuropathy.[58,59] This contrasts with other malignant tumors, which have no association with Guillain-Barré syndrome. The distinction can be difficult. The polyneuropathy probably occurs as a result of the underlying disturbances of immune function in patients with diseases of the lymphoreticular system. Children with AIDP may have absent F waves if studied early in their course, and may respond well to intravenous immunoglobulin.[59] Occasionally children develop more chronic demyelinating or axonal neuropathies in association with lymphoma.[61] Malignant infiltration of nerve roots and peripheral nerves occurs more commonly in lymphoma than in carcinoma.

Paraneoplastic Neuropathies

Paraneoplastic neurological syndromes affect about 1% of patients with cancer. In adults, these syndromes often cause peripheral neuropathies mediated by antibodies to either intracellular (Hu-D and CV2-CRMP5) or cell membrane antigens (voltage-gated calcium channel, LGI and CASPR2 proteins).[55,56,62] The malignancies most commonly associated with paraneoplastic peripheral neuropathy, such as small cell lung cancer, are much more common in adults.[62–64] Children with paraneoplastic neurologic syndromes are more likely to present with limbic encephalitis than neuropathy,[63] but occasional cases are reported.[65,66] A 13-year-old boy with Hodgkin's disease developed an acute polyneuropathy and autoimmune hemolytic anemia.[65] Sural nerve biopsy and post-mortem examination demonstrated axonal degeneration in the nerves and dorsal funiculus, without metastatic involvement or cellular inflammation in the nerves. A 14-year-old boy suffered from a chronic demyelinating neuropathy during the 14-month period before he was diagnosed with Hodgkin's disease.[66] This case raises the issue of whether cases of apparent AIDP and CIDP in association with lymphoreticular malignancies might represent paraneoplastic conditions.[58,66,67]

Graft-versus-Host Disease

Acute immune-mediated neuropathies can arise as a complication of allogeneic hematopoietic stem cell transplantation in both children and adults.[68–72] All variants of GBS are reported in the context of graft-versus-host disease (GVHD), often after preceding infections.[69] Recognition of an acquired immune-mediated process through the usual evaluation (clinical, CSF, and neurophysiologic) is important, as patients usually respond well to parenteral immunoglobulins. Chronic inflammatory neuropathy following bone marrow transplantation complicated by severe graft-versus-host disease has also been reported in a 7-month-old girl.[73]

Although children are at risk of peripheral neuropathy as part of a graft-versus-host response, other causative or contributory factors (such as toxins, infections, and paraneoplastic processes) must also be considered and excluded.[74,75]

Neurofibromatosis

The autosomal dominant condition neurofibromatosis type 1 (NF1) is associated with benign nerve-sheath tumors (Figure 22.1), which undergo malignant change in about 8% to 13% of cases.[76] This malignant transformation is associated pathologically with mutations in the *p53* and *INK4a* genes, and aberrant signaling in the Notch pathway.[56] Malignant peripheral nerve-sheath tumors (MPNST) are invasive sarcomas and carry a poor prognosis.[56,77] A study of 123 Chinese children with NF1 detected MPNST in 3 patients (2.4%).[77] Affected individuals usually present with severe pain and rapid growth of the soft-tissue lesion.[56] Various serum markers have been suggested to aid early detection of risk of malignant change.[76]

Sickle Cell Disease

Sickle cell disease is prevalent throughout Africa. Children with sickle cell disease are predisposed to development of peripheral neuropathy related to lead intoxication.[78] Lead intoxication remains a hazard in many resource-poor countries, and should be considered in children with sickle cell disease presenting with neuropathy. More recent reports describe sensory neuropathy developing in adolescents and adults with sickle cell disease, but this is not generally seen in younger children.[79]

CONNECTIVE TISSUE DISORDERS

Rheumatoid Arthritis

Polyneuropathy and entrapment neuropathies are fairly frequent in adults with rheumatoid arthritis, but uncommon in children with this condition.[80] There are isolated reports of carpal tunnel syndrome in patients with juvenile rheumatoid arthritis.[81,82] In one series of 99 patients with juvenile rheumatoid arthritis, 4 children were found to have slowing of nerve conduction in individual nerves. Clinical symptoms of mononeuropathy were not evident, and conduction in other nerves was normal.[83] Another report described a 14-year-old boy with a 10-year history of juvenile rheumatoid arthritis who developed progressive muscle weakness and hyporeflexia over a 4-month period. Denervation was evident on EMG, but motor nerve conduction studies were normal.[7]

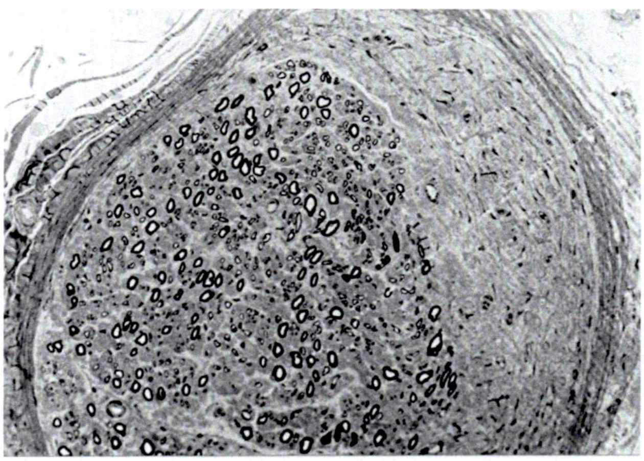

FIGURE 22.1 Transverse section of a nerve fascicle from a patient with neurofibromatosis type 1 (LM, toluidine blue, magnification ×200). The subperineurial space is increased, containing neurofibromatous tissue.

Systemic Lupus Erythematosus

Reports of peripheral neuropathy in children with systemic lupus erythematosus (SLE) are uncommon, but symptomatic or subclinical nerve involvement may be present in as many as 15% of children with this condition.[84–89] Neuropathy—presenting with foot drop, or pain—may be the presenting symptom of this condition,[90] but most children develop neuropathy a few years after the diagnosis of SLE (range 1–5 years).[85] Affected children commonly have neuropsychiatric manifestations, which can complicate diagnosis.[90] Three main clinical presentations are described: mononeuritis multiplex, an acute sensorimotor neuropathy similar to GBS (see Case Example 22.3), and a distal sensory neuropathy. There is some overlap of the latter two types. Nerve conduction studies typically reflect sensorimotor axonal pathology, which may be slowly progressive and fluctuate over time.[91] Sural nerve biopsies demonstrate axonal degeneration of both myelinated and unmyelinated fibers, often but not always with evidence of vasculitis.[92,93] Various mechanisms are proposed for the pathogenesis of the peripheral neuropathy in SLE, including vasculitis of the vasa nervorum, endoneurial immune complex deposition, and antibody-mediated damage to components of neural tissue.[85,93] A role for antiphospholipid antibodies has

CASE EXAMPLE 22.3 Mononeuritis Multiplex in a Teenager with Systemic Lupus Erythematosus

A 16-year-old girl was followed at Boston Children's Hospital with dialysis-dependent proliferative membranous glomerulonephritis secondary to SLE. She had associated hypertension, pericardial effusion, and seizures. She was noncompliant with medication and therefore not a candidate for renal transplantation.

One month before her neuromuscular evaluation, she suddenly developed left hand numbness; shortly thereafter, an acute left foot-drop occurred, followed some weeks later by numbness of her right hand. She became incapacitated when an acute right foot-drop developed the next day. On admission, her neurologic examination demonstrated a left median and ulnar neuropathy as well as bilateral, asymmetrical peroneal neuropathies.

Nerve conduction studies showed an absent left peroneal compound muscle action potential with low-amplitude right peroneal and left median motor responses. Nerve conduction velocities and latencies were normal. Both sural and the left median sensory responses were absent. The left ulnar, radial, and right median motor responses were normal. Needle EMG demonstrated active denervation with a decreased number of motor unit potentials in an asymmetrical distribution appropriate to her clinical examination.

A sural nerve biopsy showed evidence of a vasculitis. Treatment was begun with high-dose corticosteroids.

also been suggested.[84,85,88,94] Wallerian degeneration of nerve fibers results from ischemic infarction due to occlusion of blood vessels by leukocytoclastic vasculitis.[95,96] Treatment may include standard immunomodulation and intravenous immunoglobulin.[97] Most studies recommend regular nerve conduction studies to screen children with SLE for early evidence of subclinical neuropathy.[84,90]

Mononeuritis multiplex in relation to vasculitides is reported in pediatric patients with SLE, Churg-Strauss syndrome, and polyarteritis nodosa.[98–104] Sjögren syndrome is an autoimmune exocrinopathy that primarily affects the salivary and lacrimal glands, and that causes a sensory neuropathy thought to be due to either a vasculitis or ganglionitis of the dorsal root.[105] Most cases are in adults, but occasional pediatric cases are reported.[105]

MISCELLANEOUS SYSTEMIC DISORDERS

Chronic Idiopathic Intestinal Pseudo-Obstruction

Intestinal pseudo-obstruction can present at all ages, including the neonatal period, and is occasionally associated with motor neuropathy. Neurogenic causes include various forms of neuronal dysplasia in the myenteric plexus.[106,107] Other etiologies include Duchenne muscular dystrophy, myotonic dystrophy, mitochondrial cytopathy, myxedema, porphyria, scleroderma, amyloidosis, and Chagas disease.[108,109] The disorder is often familial, referred to as familial visceral neuropathy.[107] Non-intestinal abnormalities, including pupillary abnormalities, may be present.[110–112] An overlap with familial visceral myopathy has been noted.[106,107] Electrophysiological studies usually show demyelinating changes, but axonal pathology is also reported.[107] Inheritance is debated and probably variable.[107]

Another condition combining bowel dysmotility and peripheral neuropathy is Waardenburg-Shah syndrome, which is allelic to Waardenburg syndromes type 2E and type 4C and occurs as a result of recessive mutations in the *SOX10* gene.[113] Patients affected by this neurocristopathy present early in life with Hirschsprung disease, Waardenburg syndrome, central dysmyelination, and a hypomyelinating peripheral neuropathy.[113–119] Sural nerve biopsy from an affected infant showed hypoplasia and severe reduction in the number of axons.[116]

Mitochondrial neurogastrointestinal encephalomyopathy (MNGIE) is a rare multisystemic recessive disorder consisting of progressive external ophthalmoplegia, gastrointestinal dysmotility, cachexia, peripheral neuropathy, and leukoencephalopathy caused by mutations in the *TMMP* gene, encoding for thymidine phosphorylase (see Chapter 41).[120] The condition begins in childhood or adolescence.[121] The peripheral neuropathy is usually demyelinating but occasionally mixed or predominantly axonal.[120] Diagnosis of MNGIE is often delayed.[120]

Celiac disease is a gluten-sensitive enteropathy, which is occasionally associated with an axonal or demyelinating sensory or sensorimotor peripheral neuropathy.[122–124] Recurrent Guillain-Barré syndrome was reported in another case and thought due to a process of molecular mimicry.[125] A 3-year-old child with celiac disease who developed a progressive polyneuropathy unresponsive to a gluten-free diet, vitamin E, and folic acid supplementation underwent a sural nerve biopsy that showed marked loss of myelinated fibers without evidence of regeneration.[126] Distal involvement and features of nerve pathology were consistent with a “dying-back” mechanism underlying this nerve injury. The lack of response to a gluten-free diet suggested that direct toxicity of gliadin was not implicated.[126]

Sarcoidosis

Peripheral neuropathy may be an isolated manifestation of sarcoidosis or part of the more generalized systemic manifestations of the disease.[127] Facial nerve palsy is the most common neurologic manifestation of the condition, especially in affected adults.[127] More generalized neuropathies are typically asymmetric, and may take the form of mononeuritis multiplex or Guillain-Barré syndrome.[127] An unusual feature is sensory loss over large areas of the trunk, which may be associated with dysesthesia or pain.[128]

The combination of cranial neuropathy with generalized chronic neuropathy is highly suggestive of sarcoidosis. The diagnosis is often made clinically when there are associated features of hilar lymphadenopathy (Figure 22.2), uveitis, parotitis, and erythema nodosum.[129] Pediatric cases are uncommon, but children as young as 13 years old with peripheral neuropathy are described, both such patients presenting with a Guillain-Barré syndrome.[129,130]

Nerve conduction studies may be normal or show only mild degrees of slowing of conduction.[128] Sarcoid granulomas may be seen on muscle or sural nerve biopsy.[131] The cerebrospinal fluid may have a slight increase in cells and an elevated total protein.[128]

The natural history of sarcoid neuropathy is uncertain, although the prognosis is better than in central nervous system sarcoidosis. It is difficult to be confident of the value of steroid therapy, although it is usually given.

Amyloidosis

The amyloidoses are a group of diseases characterized by extracellular tissue deposition of fibrils composed of low-molecular-weight subunits of a variety of proteins. The site of deposition depends on the type of amyloidosis, which can be acquired or hereditary, localized or systemic. At least 25 different precursors of amyloid fibrils

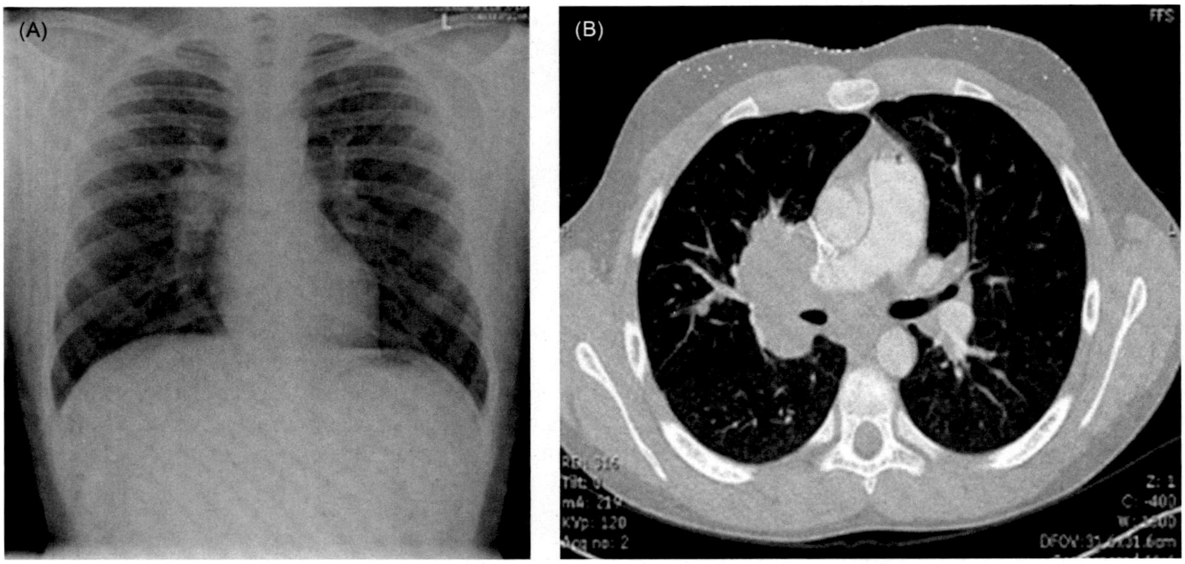

FIGURE 22.2 (**A**) Chest X-ray demonstrating right hilar lymphadenopathy in a 13-year-old boy with sarcoidosis.[129] (**B**) Axial CT image showing mass encasing right main bronchus from a 13-year-old boy with sarcoidosis.[129] *Reprinted with permission from Wurzel et al.*[129]

are recognized.[132] The most common form of amyloidosis in pediatrics is reactive AA amyloidosis due to chronic inflammatory diseases such as familial Mediterranean fever (FMF).[132–134] A 14-year-old girl with familial Mediterranean fever was reported with recurrent facial nerve palsy, but this was considered related to vasculitis rather than amyloid deposition.[135] Peripheral neuropathy is not seen in reactive (AA, secondary) amyloidosis or in most of the inherited amyloidoses characterized by renal, hepatic, or cardiac deposition.[136]

The hereditary forms of amyloidosis with associated neuropathy (previously termed familial amyloidotic polyneuropathy [FAP]), occur in patients with hereditary transthyretin amyloidosis, hereditary apoA-I amyloidosis caused by the glycine 26 arginine (Gly26Arg) mutation and hereditary gelsolin (Gel) amyloidosis.[136] These dominantly inherited disorders are almost exclusively seen in adulthood.[137,138]

Most subjects present with numbness and pain in the extremities. Because of loss of pain and temperature sense, some patients experience painless burns on the hands and feet; they also later complain of weakness. Autonomic dysfunction is frequently present, with patients complaining of symptoms of postural hypotension (lightheadedness, dizziness, and syncope), impotence, and bladder disturbances.[136]

On neurologic examination there is distal wasting and weakness with hyporeflexia or areflexia, but the sensory signs are predominant. There is frequently dissociated sensory loss, with pain and temperature sense being disproportionately affected. Most patients have postural hypotension and impairment of sweating.[136]

On EMG this is primarily a small-fiber sensory neuropathy. On other occasions, absent or reduced sensory nerve action potentials are found, with normal or only mildly impaired motor conduction parameters. Sural nerve biopsy characteristically reveals a severe loss of both small-diameter myelinated fibers and unmyelinated fibers. Amyloid is deposited around capillaries in the endoneurium, epineurium, and perineurium. Cerebrospinal fluid protein levels are frequently elevated, and there is occasionally a pleocytosis. The neuropathy is relentlessly progressive and unaffected by medical treatment.[136]

REFERENCES

1. Zochodne DW, Bolton CF, Wells GA, Gilbert JJ, Hahn AF, Brown JD, et al. Critical illness polyneuropathy. A complication of sepsis and multiple organ failure. *Brain* 1987;**110**(Pt 4):819–41.
2. Williams S, Horrocks IA, Ouvrier RA, Gillis J, Ryan MM. Critical illness polyneuropathy and myopathy in pediatric intensive care: A review. *Pediatr Crit Care Med* 2007;**8**:18–22.
3. Banwell BL, Mildner RJ, Hassall AC, Becker LE, Vajsar J, Shemie SD. Muscle weakness in critically ill children. *Neurology* 2003;**61**: 1779–82.
4. Tamam Y, Tamam C, Tamam B, Ustundag M, Orak M, Tasdemir N. Peripheral neuropathy after burn injury. *Eur Rev Med Pharmacol Sci* 2013;**17**(Suppl. 1):107–11.
5. Gheith O, Al Otaibi T, Halim M, Said T, Nair P, Balaha M, et al. Successful management of critical illness polyneuropathy and myopathy in renal transplant recipients. *Exp Clin Transplant* 2012;**10**:62–6.
6. Hermans G, De Jonghe B, Bruyninckx F, Van den Berghe G. Interventions for preventing critical illness polyneuropathy and critical illness myopathy. *Cochrane Database Syst Rev* 2009;**1**:CD006832.

7. Evans OB. Polyneuropathy in childhood. *Pediatrics* 1979;**64**: 96–105.
8. Elzouki A, Carroll J, Butinar D, Moosa A. Improved neurological outcome in children with chronic renal disease from infancy. *Pediatr Nephrol* 1994;**8**:205–10.
9. de Beaufort CE, Andre JL, Heimans JJ, van der Eerden HA, van Diemen NG, Duc ML, et al. Peripheral nerve function in children with end-stage renal failure. *Pediatr Nephrol* 1989;**3**:175–8.
10. Mendoza-Guevara L, Cervantes A, Aguilar-Kitsu A, Rendon E, Morales A, Rodriguez F, et al. "H" reflex as a measure of subclinical uremic polyneuropathy in children with chronic renal failure. *Adv Perit Dial* 1997;**13**:285–90.
11. Trotta D, Verrotti A, Salladini C, Chiarelli F. Diabetic neuropathy in children and adolescents. *Pediatr Diabetes* 2004;**5**:44–57.
12. Bao XH, Wong V, Wang Q, Low LC. Prevalence of peripheral neuropathy with insulin-dependent diabetes mellitus. *Pediatr Neurol* 1999;**20**:204–9.
13. Hasani N, Khosrawi S, Hashemipour M, Haghighatiyan M, Javdan Z, Taheri MH, et al. Prevalence and related risk-factors of peripheral neuropathy in children with insulin-dependent diabetes mellitus. *J Res Med Sci* 2013;**18**:132–6.
14. Karsidag S, Morali S, Sargin M, Salman S, Karsidag K, Us O. The electrophysiological findings of subclinical neuropathy in patients with recently diagnosed type 1 diabetes mellitus. *Diabetes Res Clin Pract* 2005;**67**:211–19.
15. Meh D, Denislic M. Subclinical neuropathy in type I diabetic children. *Electroencephalogr Clin Neurophysiol* 1998;**109** 274–80.
16. Moser JT, Langdon DR, Finkel RS, Ratcliffe SJ, Foley LR, Andrews-Rearson ML, et al. The evaluation of peripheral neuropathy in youth with type 1 diabetes. *Diabetes Res Clin Pract* 2013; **100**:e3–6.
17. Holiner I, Haslinger V, Lutschg J, Muller G, Barbarini DS, Fussenegger J, et al. Validity of the neurological examination in diagnosing diabetic peripheral neuropathy. *Pediatr Neurol* 2013; **49**:171–7.
18. Moline J, Eng C. Multiple endocrine neoplasia type 2: an overview. *Genet Med* 2011;**13**:755–64.
19. Nakao KT, Usui T, Ikeda M, Mori Y, Yamamoto T, Kawashima ST, et al. Novel tandem germline RET proto-oncogene mutations in a patient with multiple endocrine neoplasia type 2B: report of a case and a literature review of tandem RET mutations with in silico analysis. *Head Neck* 2013;**35**:E363–8.
20. Carney JA, Bianco Jr. AJ, Sizemore GW, Hayles AB. Multiple endocrine neoplasia with skeletal manifestations. *J Bone Joint Surg Am* 1981;**63**:405–10.
21. Ramos-Levi AM, Diaz-Perez A, Sobrido MJ, Pinearo-Hermida S, Blanco-Arias P, Cabezas-Agricola JM, et al. Axonal neuropathy, long limbs and bumpy tongue: think of MEN2B. *Muscle Nerve* 2012;**46**:961–4.
22. Dyck PJ, Carney JA, Sizemore GW, Okazaki H, Brimijoin WS, Lambert EH. Multiple endocrine neoplasia, type 2b: phenotype recognition; neurological features and their pathological basis. *Ann Neurol* 1979;**6**:302–14.
23. Mass E, Lapidot M, Gadoth N. Case report: multiple endocrine neoplasia type 2B misdiagnosed as familial dysautonomia. *Eur J Paediatr Dent* 2005;**6**:48–50.
24. Oguz SS, Ergenekon E, Tumer L, Koc E, Turan O, Onal E, et al. A rare case of severe lactic acidosis in a preterm infant: lack of thiamine during total parenteral nutrition. *J Pediatr Endocrinol Metab* 2011;**24**:843–5.
25. Masumoto K, Esumi G, Teshiba R, Nagata K, Nakatsuji T, Nishimoto Y, et al. Need for thiamine in peripheral parenteral nutrition after abdominal surgery in children. *JPEN J Parenter Enteral Nutr* 2009;**33**:417–22.
26. Han JW, Lim S, Shin HS, Park HJ, Jung WJ, Kwon SY, et al. Two cases of Wernicke's encephalopathy in young age patients receiving allogeneic hematopoietic stem cell transplantation. *Yonsei Med J* 2012;**53**:1049–53.
27. Rao SN, Mani S, Madap K, Kumar MV, Singh L, Chandak GR. High prevalence of infantile encephalitic beriberi with overlapping features of Leigh's disease. *J Trop Pediatr* 2008;**54**:328–32.
28. Adamolekun B. Neurological disorders associated with cassava diet: a review of putative etiological mechanisms. *Metab Brain Dis* 2011;**26**:79–85.
29. Ishibashi S, Yokota T, Shiojiri T, Matunaga T, Tanaka H, Nishina K, et al. Reversible acute axonal polyneuropathy associated with Wernicke-Korsakoff syndrome: impaired physiological nerve conduction due to thiamine deficiency? *J Neurol Neurosurg Psychiatry* 2003;**74**:674–6.
30. Koike H, Iijima M, Sugiura M, Mori K, Hattori N, Ito H, et al. Alcoholic neuropathy is clinicopathologically distinct from thiamine-deficiency neuropathy. *Ann Neurol* 2003;**54**:19–29.
31. Bosch AM, Stroek K, Abeling NG, Waterham HR, Ijlst L, Wanders RJ. The Brown-Vialetto-Van Laere and Fazio Londe syndrome revisited: natural history, genetics, treatment and future perspectives. *Orphanet J Rare Dis* 2012;**7**:83.
32. Johnson JO, Gibbs JR, Megarbane A, Urtizberea JA, Hernandez DG, Foley AR, et al. Exome sequencing reveals riboflavin transporter mutations as a cause of motor neuron disease. *Brain* 2012;**135**:2875–82.
33. Ciccolella M, Corti S, Catteruccia M, Petrini S, Tozzi G, Rizza T, et al. Riboflavin transporter 3 involvement in infantile Brown-Vialetto-Van Laere disease: two novel mutations. *J Med Genet* 2013;**50**:104–7.
34. Snider Jr. DE. Pyridoxine supplementation during isoniazid therapy. *Tubercle* 1980;**61**:191–6.
35. Baxter P. Pyridoxine-dependent and pyridoxine-responsive seizures. *Dev Med Child Neurol* 2001;**43**:416–20.
36. MacLean Jr. WC, Graham GG. Vegetarianism in children. *Am J Dis Child* 1980;**134**:513–19.
37. Conway SP, Gillies DR, Littlewood JM. Vitamin B12 neuropathy in a 6 year old. *Arch Dis Child* 1984;**59**:575–6.
38. van Loon M, Postels DG, Heikens GT, Molyneux E. Severe pernicious anaemia in an 8-year-old African girl. *Ann Trop Paedictr* 2009;**29**:231–4.
39. Grattan-Smith PJ, Wilcken B, Procopis PG, Wise GA. The neurological syndrome of infantile cobalamin deficiency: developmental regression and involuntary movements. *Mov Disord* 1997;**12**:39–46.
40. McCombe PA, McLeod JG. The peripheral neuropathy of vitamin B12 deficiency. *J Neurol Sci* 1984;**66**:117–26.
41. Chakrabarty B, Kabra SK, Gulati S, Toteja GS, Lodha R, Kabra M, et al. Peripheral neuropathy in cystic fibrosis: a prevalence study. *J Cyst Fibros* 2013;**12**:754–60.
42. Ueda N, Suzuki Y, Rino Y, Takahashi T, Imada T, Takanashi Y, et al. Correlation between neurological dysfunction with vitamin E deficiency and gastrectomy. *J Neurol Sci* 2009;**287**:216–20.

43. Muller DP. Vitamin E and neurological function. *Mol Nutr Food Res* 2010;**54**:710–18.
44. Kalra V, Grover JK, Ahuja GK, Rathi S, Gulati S, Kalra N. Vitamin E administration and reversal of neurological deficits in protein-energy malnutrition. *J Trop Pediatr* 2001;**47**:39–45.
45. Hentati F, El-Euch G, Bouhlal Y, Amouri R. Ataxia with vitamin E deficiency and abetalipoproteinemia. *Handb Clin Neurol* 2012; **103**:295–305.
46. Di Donato I, Bianchi S, Federico A. Ataxia with vitamin E deficiency: update of molecular diagnosis. *Neurol Sci* 2010;**31**:511–15.
47. Fusco C, Frattini D, Pisani F, Gellera C, Della Giustina E. Isolated vitamin E deficiency mimicking distal hereditary motor neuropathy in a 13-year-old boy. *J Child Neurol* 2008;**23**:1328–30.
48. Hentati A, Deng HX, Hung WY, Nayer M, Ahmed MS, He X, et al. Human alpha-tocopherol transfer protein: gene structure and mutations in familial vitamin E deficiency. *Ann Neurol* 1996;**39**: 295–300.
49. Jackson CE, Amato AA, Barohn RJ. Isolated vitamin E deficiency. *Muscle Nerve* 1996;**19**:1161–5.
50. Muller DP, Goss-Sampson MA. Neurochemical, neurophysiological, and neuropathological studies in vitamin E deficiency. *Crit Rev Neurobiol* 1990;**5**:239–63.
51. Harding AE, Muller DP, Thomas PK, Willison HJ. Spinocerebellar degeneration secondary to chronic intestinal malabsorption: a vitamin E deficiency syndrome. *Ann Neurol* 1982;**12**:419–24.
52. Puri V, Chaudhry N, Tatke M, Prakash V. Isolated vitamin E deficiency with demyelinating neuropathy. *Muscle Nerve* 2005;**32**: 230–5.
53. Guggenheim MA, Ringel SP, Silverman A, Grabert BE. Progressive neuromuscular disease in children with chronic cholestasis and vitamin E deficiency: diagnosis and treatment with alpha tocopherol. *J Pediatr* 1982;**100**:51–8.
54. Hughes RA, Britton T, Richards M. Effects of lymphoma on the peripheral nervous system. *J R Soc Med* 1994;**87**:526–30.
55. Antoine JC, Camdessanche JP. Treatment options in paraneoplastic disorders of the peripheral nervous system. *Curr Treat Options Neurol* 2013;**15**:210–23.
56. Antoine JC, Camdessanche JP. Peripheral nervous system involvement in patients with cancer. *Lancet Neurol* 2007;**6**:75–86.
57. Lowitzsch K, Gutjahr P, Ottes H. Clinical and neurophysiological findings in 47 long-term survivors of childhood malignancies treated with various doses of vincristine. In: Canal N, Pozza G, editors. *Peripheral Neuropathies*. Amsterdam: North-Holland; 1978. pp. 459–66.
58. Bahl A, Chakrabarty B, Gulati S, Raju KN, Raja A, Bakhshi S. Acute onset flaccid quadriparesis in pediatric non-Hodgkin lymphoma: vincristine induced or Guillain-Barré syndrome? *Pediatr Blood Cancer* 2010;**55**:1234–5.
59. Brigo F, Balter R, Marradi P, Ferlisi M, Zaccaron A, Fiaschi A, et al. Vincristine-related neuropathy versus acute inflammatory demyelinating polyradiculoneuropathy in children with acute lymphoblastic leukemia. *J Child Neurol* 2012;**27**:867–74.
60. Apjok E, Marosi A, Magyarosy E. Guillain-Barré syndrome in patients treated for Hodgkin disease. *Orv Hetil* 2003;**144**:1039–40.
61. McLeod JG, Walsh JC. Peripheral neuropathy associated with lymphomas and other reticuloses. In: Dyck PJ, Thomas PK, Lambert EH, Bunge R, editors. *Peripheral Neuropathy*. 2nd ed. Philadelphia: WB Saunders; 1984. pp. 2192–203.
62. Dalmau J, Graus F, Rosenblum MK, Posner JB. Anti-Hu-associated paraneoplastic encephalomyelitis/sensory neuronopathy. A clinical study of 71 patients. *Medicine (Baltimore)* 1992;**71**: 59–72.
63. Honnorat J, Didelot A, Karantoni E, Ville D, Ducray F, Lambert L, et al. Autoimmune limbic encephalopathy and anti-Hu antibodies in children without cancer. *Neurology* 2013;**80**:2226–32.
64. Briani C, Vitaliani R, Grisold W, Honnorat J, Graus F, Antoine JC, et al. Spectrum of paraneoplastic disease associated with lymphoma. *Neurology* 2011;**76**:705–10.
65. Kurczynski TW, Choudhury AA, Horwitz SJ, Roessmann U, Gross S. Remote effect of malignancy on the nervous system in children. *Dev Med Child Neurol* 1980;**22**:205–22.
66. Rowland LP, Schneck SA. Neuromuscular disorders associated with malignant neoplastic disease. *J Chronic Dis* 1963;**16**:777–95.
67. Vallat JM, De Mascarel HA, Bordessoule D, Jauberteau MO, Tabaraud F, Gelot A, et al. Non-Hodgkin malignant lymphomas and peripheral neuropathies—13 cases. *Brain* 1995;**118**(Pt 5): 1233–45.
68. Bulsara KR, Baron PW, Tuttle-Newhall JE, Clavien PA, Morgenlander J. Guillain-Barré syndrome in organ and bone marrow transplant patients. *Transplantation* 2001;**71**:1169–72.
69. Wen PY, Alyea EP, Simon D, Herbst RS, Soiffer RJ, Antin JH. Guillain-Barré syndrome following allogeneic bone marrow transplantation. *Neurology* 1997;**49**:1711–14.
70. Rabinstein AA, Wijdicks EF. Weaning from the ventilator using BiPAP in myasthenia gravis. *Muscle Nerve* 2003;**27**:252–3.
71. Rodriguez V, Kuehnle I, Heslop HE, Khan S, Krance RA. Guillain-Barré syndrome after allogeneic hematopoietic stem cell transplantation. *Bone Marrow Transplant* 2002;**29**:515–17.
72. Ostronoff F, Perales MA, Stubblefield MD, Hsu KC. Rituximab-responsive Guillain-Barré syndrome following allogeneic hematopoietic SCT. *Bone Marrow Transplant* 2008;**42**:71–2.
73. Adams C, August CS, Maguire H, Sladky JT. Neuromuscular complications of bone marrow transplantation. *Pediatr Neurol* 1995;**12**: 58–61.
74. Lavoratore SR, Navarro OM, Grunebaum E, Ali M, Koo A, Schechter T, et al. Cyclosporine-induced pain syndrome in a child undergoing hematopoietic stem cell transplant. *Ann Pharmacother* 2009;**43**:767–71.
75. Onozawa M, Hashino S, Haseyama Y, Hirayama Y, Iizuka S, Ishida T, et al. Incidence and risk of postherpetic neuralgia after varicella zoster virus infection in hematopoietic cell transplantation recipients: Hokkaido Hematology Study Group. *Biol Blood Marrow Transplant* 2009;**15**:724–9.
76. Park SJ, Sawitzki B, Kluwe L, Mautner VF, Holtkamp N, Kurtz A. Serum biomarkers for neurofibromatosis type 1 and early detection of malignant peripheral nerve-sheath tumors. *BMC Med* 2013;**11**:109.
77. Cheuk DK, Chiang AK, Ha SY, Chan GC. Malignancies in Chinese patients with neurofibromatosis type 1. *Hong Kong Med J* 2013;**19**:42–9.
78. Imbus CE, Warner J, Smith E, Pegelow CH, Allen JP, Powars DR. Peripheral neuropathy in lead-intoxicated sickle cell patients. *Muscle Nerve* 1978;**1**:168–71.
79. Kehinde MO, Temiye EO, Danesi MA. Neurological complications of sickle cell anemia in Nigerian Africans—a case-control study. *J Natl Med Assoc* 2008;**100**:394–9.

80. Kaymak B, Bilginer Y, Ozcakar L, Topaloglu R. Effect of wrist involvement on median nerve electrophysiology in juvenile idiopathic arthritis. *Acta Reumatol Port* 2009;**34**:614–17.
81. Unal O, Ozcakar L, Cetin A, Kaymak B. Severe bilateral carpal tunnel syndrome in juvenile chronic arthritis. *Pediatr Neurol* 2003;**29**:345–8.
82. De Smet L, Wouters C. Severe carpal tunnel syndrome in a patient with juvenile idiopathic arthritis due to proximal migration of hypertrophic lumbrical muscles. *Clin Rheumatol* 2004;**23**:552–4.
83. Puusa A, Lang HA, Makela AL. Nerve conduction velocity in juvenile rheumatoid arthritis. *Acta Neurol Scand* 1986;**73** 145–50.
84. Bhowmik A, Banerjee P. Mononeuritis multiplex complicating systemic lupus erythematosus. *Indian Pediatr* 2012;**49**:581–2.
85. Harel L, Mukamel M, Brik R, Blau H, Straussberg R. Peripheral neuropathy in pediatric systemic lupus erythematosus. *Pediatr Neurol* 2002;**27**:53–6.
86. Loh WF, Hussain IM, Soffiah A, Lim YN. Neurological manifestations of children with systemic lupus erythematosus. *Med J Malaysia* 2000;**55**:459–63.
87. Ryan M. Peripheral neuropathy in pediatric systemic lupus erythematosus. *Pediatr Neurol* 2003;**28**:236, author reply.
88. Steinlin MI, Blaser SI, Gilday DL, Eddy AA, Logan WJ, Laxer RM, et al. Neurologic manifestations of pediatric systemic lupus erythematosus. *Pediatr Neurol* 1995;**13**:191–7.
89. Sibbitt Jr. WL, Brandt JR, Johnson CR, Maldonado ME, Patel SR, Ford CC, et al. The incidence and prevalence of neuropsychiatric syndromes in pediatric onset systemic lupus erythematosus. *J Rheumatol* 2002;**29**:1536–42.
90. Sridhar AV, Gosalakkal J, Pye IF, Houtman P. Peripheral neuropathy—a rare complication of paediatric systemic lupus erythematosus. *Eur J Paediatr Neurol* 2004;**8**:253–6.
91. Omdal R, Loseth S, Torbergsen T, Koldingsnes W, Husby G, Mellgren SI. Peripheral neuropathy in systemic lupus erythematosus—a longitudinal study. *Acta Neurol Scand* 2001;**103**:386–91.
92. Goransson LG, Tjensvoll AB, Herigstad A, Mellgren SI, Omdal R. Small-diameter nerve fiber neuropathy in systemic lupus erythematosus. *Arch Neurol* 2006;**63**:401–4.
93. McCombe PA, McLeod JG, Pollard JD, Guo YP, Ingall TJ. Peripheral sensorimotor and autonomic neuropathy associated with systemic lupus erythematosus. Clinical, pathological and immunological features. *Brain* 1987;**110**(Pt 2):533–49.
94. Seaman DE, Londino Jr. AV, Kwoh CK, Medsger Jr. TA, Manzi S. Antiphospholipid antibodies in pediatric systemic lupus erythematosus. *Pediatrics* 1995;**96**:1040–5.
95. Scheinberg L. Polyneuritis in systemic lupus erythematosus; review of the literature and report of a case. *N Engl J Med* 1956; **255**:416–21.
96. Bódi I, Váradi P, Pokorny G, Engelhardt J, Dibo G, Vecsei L, et al. Polyneuropathy with endoneurial immune complex deposition as the first manifestation of systemic lupus erythematosus. *Acta Neuropathol* 1998;**96**:297–300.
97. Levy Y, Uziel Y, Zandman GG, Amital H, Sherer Y, Langevitz P, et al. Intravenous immunoglobulins in peripheral neuropathy associated with vasculitis. *Ann Rheum Dis* 2003;**62**:1221–3.
98. Guillevin L, Cohen P, Gayraud M, Lhote F, Jarrousse B, Casassus P. Churg-Strauss syndrome. Clinical study and long-term follow-up of 96 patients. *Medicine (Baltimore)* 1999;**78**:26–37.
99. Ford FR. *Diseases of the Nervous System in Infancy, Childhood and Adolescence*. 5th ed. Springfield, IL: C. C. Thomas; 1966.
100. Draaisma JM, Fiselier TJ, Mullaart RA. Mononeuritis multiplex in a child with cutaneous polyarteritis. *Neuropediatrics* 1992;**23**: 28–9.
101. Wang SJ, Yang YH, Lin YT, Tsai MJ, Chiang BL. Childhood Churg-Strauss syndrome: report of a case. *J Microbiol Immunol Infect* 2000;**33**:263–6.
102. Penas PF, Porras JI, Fraga J, Bernis C, Sarria C, Dauden E. Microscopic polyangiitis. A systemic vasculitis with a positive P-ANCA. *Br J Dermatol* 1996;**134**:542–7.
103. Ryan MM, Tilton A, De Girolami U, Darras BT, Jones Jr. HR. Paediatric mononeuritis multiplex: a report of three cases and review of the literature. *Neuromuscul Disord* 2003;**13**:751–6.
104. Andrade R, Moya Machado A, Gomez Conde S, Garcia Espinosa A, Molina Diaz M, Machado Rojas A. [Neuropathies due to vasculitis in infancy]. *Rev Neurol* 2004;**38**:619–24.
105. Kumon K, Satake A, Mizumoto M, Kobayashi I, Ishikawa N. A case of sensory neuropathy associated with childhood Sjögren syndrome. *Eur J Pediatr* 2000;**159**:630–1.
106. Kapur RP. Neuronal dysplasia: a controversial pathological correlate of intestinal pseudo-obstruction. *Am J Med Genet A* 2003; **122A**:287–93.
107. Roper EC, Gibson A, McAlindon ME, Williams LH, Cook JA, Kandler RH, et al. Familial visceral neuropathy: a defined entity? *Am J Med Genet A* 2005;**137A**:249–54.
108. Garcia Aroca J, Sanz N, Alonso JL, de Mingo L, Rollan V. Intestinal pseudo-obstruction secondary to systemic neuropathies and myopathies. *Cir Pediatr* 1994;**7**:115–20.
109. Byrne WJ, Cipel L, Euler AR, Halpin TC, Ament ME. Chronic idiopathic intestinal pseudo-obstruction syndrome in children—clinical characteristics and prognosis. *J Pediatr* 1977;**90**:585–9.
110. Arrindell EL, Trobe JD, Sieving PA, Barnett JL. Pupillary and electroretinographic abnormalities in a family with neuronal intranuclear hyaline inclusion disease. *Arch Ophthalmol* 1991;**109**: 373–8.
111. Barnett JL, McDonnell WM, Appelman HD, Dobbins WO. Familial visceral neuropathy with neuronal intranuclear inclusions: diagnosis by rectal biopsy. *Gastroenterology* 1992;**102**: 684–91.
112. Faber J, Fich A, Steinberg A, Steiner I, Granot E, Alon I, et al. Familial intestinal pseudoobstruction dominated by a progressive neurologic disease at a young age. *Gastroenterology* 1987;**92**: 786–90.
113. Chaoui A, Watanabe Y, Touraine R, Baral V, Goossens M, Pingault V, et al. Identification and functional analysis of SOX10 missense mutations in different subtypes of Waardenburg syndrome. *Hum Mutat* 2011;**32**:1436–49.
114. Inoue K, Shilo K, Boerkoel CF, Crowe C, Sawady J, Lupski JR, et al. Congenital hypomyelinating neuropathy, central dysmyelination, and Waardenburg-Hirschsprung disease: phenotypes linked by SOX10 mutation. *Ann Neurol* 2002;**52**:836–42.
115. Mahmoudi A, Rami M, Khattala K, Elmadi A, Afifi MA, Youssef B. Shah-waardenburg syndrome. *Pan Afr Med J* 2013;**14**:60.
116. Parthey K, Kornhuber M, Kunze C, Wand D, Nolte KW, Nikolin S, et al. SOX10 mutation with peripheral amyelination and developmental disturbance of axons. *Muscle Nerve* 2012;**45**: 284–90.

117. Shimotake T, Tanaka S, Fukui R, Makino S, Maruyama R. Neuroglial disorders of central and peripheral nervous systems in a patient with Hirschsprung's disease carrying allelic SOX10 truncating mutation. *J Pediatr Surg* 2007;**42**:725–31.
118. Unzicker A, Pingault V, Meyer T, Rauthe S, Schutz A, Kunzmann S. A novel SOX10 mutation in a patient with PCWH who developed hypoxic-ischemic encephalopathy after *E. coli* sepsis. *Eur J Pediatr* 2011;**170**:1475–80.
119. Verheij JB, Sival DA, van der Hoeven JH, Vos YJ, Meiners LC, Brouwer OF, et al. Shah-Waardenburg syndrome and PCWH associated with SOX10 mutations: a case report and review of the literature. *Eur J Paediatr Neurol* 2006;**10**:11–17.
120. Garone C, Tadesse S, Hirano M. Clinical and genetic spectrum of mitochondrial neurogastrointestinal encephalomyopathy. *Brain* 2011;**134**:3326–32.
121. Libernini L, Lupis C, Mastrangelo M, Carrozzo R, Santorelli FM, Inghilleri M, et al. Mitochondrial neurogastrointestinal encephalomyopathy: novel pathogenic mutations in thymidine phosphorylase gene in two Italian brothers. *Neuropediatrics* 2012; **43**:201–8.
122. Lionetti E, Francavilla R, Pavone P, Pavone L, Francavilla T, Pulvirenti A, et al. The neurology of coeliac disease in childhood: what is the evidence? A systematic review and meta-analysis. *Dev Med Child Neurol* 2010;**52**:700–7.
123. Cakir D, Tosun A, Polat M, Celebisoy N, Gokben S, Aydogdu S, et al. Subclinical neurological abnormalities in children with celiac disease receiving a gluten-free diet. *J Pediatr Gastroenterol Nutr* 2007;**45**:366–9.
124. Ruggieri M, Incorpora G, Polizzi A, Parano E, Spina M, Pavone P. Low prevalence of neurologic and psychiatric manifestations in children with gluten sensitivity. *J Pediatr* 2008;**152**:244–9.
125. Gupta V, Kohli A. Celiac disease associated with recurrent Guillain Barré syndrome. *Indian Pediatr* 2010;**47**:797–8.
126. Simonati A, Battistella PA, Guariso G, Clementi M, Rizzuto N. Coeliac disease associated with peripheral neuropathy in a child: a case report. *Neuropediatrics* 1998;**29**:155–8.
127. Sharma OP. Neurosarcoidosis: a personal perspective based on the study of 37 patients. *Chest* 1997;**112**:220–8.
128. Matthews WB. Sarcoid neuropathy. In: Dyck PJ, Thomas PK, Lambert EH, Bunge R, editors. *Peripheral Neuropathy*. 2nd ed. Philadelphia: Saunders; 1984. pp. 2018–26.
129. Wurzel DF, Steinfort DP, Massie J, Ryan MM, Irving LB, Ranganathan SC. Paralysis and a perihilar protuberance: an unusual presentation of sarcoidosis in a child. *Pediatr Pulmonol* 2009;**44**:410–14.
130. Le Luyer B, Devaux AM, Dailly R, Ensel P. [Polyradiculoneuritis as a manifestation of childhood sarcoidosis]. *Arch Fr Pediatr* 1983;**40**:175–8.
131. Nemni R, Galassi G, Cohen M, Hays AP, Gould R, Singh N, et al. Symmetric sarcoid polyneuropathy: analysis of a sural nerve biopsy. *Neurology* 1981;**31**:1217–23.
132. Bilginer Y, Akpolat T, Ozen S. Renal amyloidosis in children. *Pediatr Nephrol* 2011;**26**:1215–27.
133. Oner A, Erdogan O, Demircin G, Bulbul M, Memis L. Efficacy of colchicine therapy in amyloid nephropathy of familial Mediterranean fever. *Pediatr Nephrol* 2003;**18**:521–6.
134. Lane T, Loeffler JM, Rowczenio DM, Gilbertson JA, Bybee A, Russell TL, et al. AA amyloidosis complicating the hereditary periodic fever syndromes. *Arthritis Rheum* 2013;**65**:1116–21.
135. Yilmaz U, Gulez N, Cubukcu D, Guzel O, Akinci G, Ozturk A. Recurrent peripheral facial palsy in a child with familial Mediterranean fever. *Pediatr Neurol* 2013;**49**:289–91.
136. Benson MD, Kincaid JC. The molecular biology and clinical features of amyloid neuropathy. *Muscle Nerve* 2007;**36**:411–23.
137. Munar-Ques M, Saraiva MJ, Viader-Farre C, Zabay-Becerril JM, Mulet-Ferrer J. Genetic epidemiology of familial amyloid polyneuropathy in the Balearic Islands (Spain). *Amyloid* 2005;**12**: 54–61.
138. Soares M, Buxbaum J, Sirugo G, Coelho T, Sousa A, Kastner D, et al. Genetic anticipation in Portuguese kindreds with familial amyloidotic polyneuropathy is unlikely to be caused by triplet repeat expansions. *Hum Genet* 1999;**104**:480–5.

Chapter 23

Toxic Neuropathies

David Pleasure

INTRODUCTION

In a sense, all peripheral nervous system (PNS) disorders caused by toxic molecules are toxic neuropathies. Even normal PNS proteins can have deleterious effects if overexpressed. For example, Charcot-Marie-Tooth syndrome type 1A (hereditary motor and sensory neuropathy type 1A) results from excess synthesis by Schwann cells of the myelin protein PMP22, caused by inheritance of a duplicated chromosomal segment within which PMP22 is encoded. Excessive deposition in the PNS of proteins produced by neoplastic cells elsewhere in the body (e.g. immunoglobulin light chains) can also damage the PNS. Neuropathies caused by bacterial proteins, such as diphtheritic neuropathy, as well as cases of Guillain-Barré syndrome associated with immunologic responses to bacterial glycolipids, also fall within this broad definition of a toxic neuropathy. Traditionally, however, the term "toxic neuropathy" is limited to PNS disorders caused by exposure to exogenous poisons or neurotoxic drugs. Furthermore, toxic disorders that primarily injure the perikarya of motor neurons or perturb neurotransmitter release at neuromuscular junctions are not generally classified as toxic neuropathies. To avoid overlap with other chapters, the discussion in this chapter is limited to disorders caused by environmental toxins and drugs that affect primarily Schwann cells, axons, and sensory and autonomic neurons.

GENERAL FEATURES OF TOXIC NEUROPATHIES

Clinical Features

Toxins and drugs with the potential to damage the PNS of infants and children usually have similar but often better-defined deleterious effects in adults. Arsenic neuropathy is a good example. Some neurotoxins also have the potential to damage the PNS *in utero* (e.g. poisoning by organic mercury).[1,2] As in adults, most toxic childhood polyneuropathies present with evidence of length-dependent axonopathy. Typically, these toxic substances affect both small- and large-fiber sensory modalities.

Some toxic compounds, however, specifically affect other types of nerve fibers or nerve cells, providing important exceptions to the general rule that toxins damage primarily sensory nerves. Pyridoxine (vitamin B6) and the platinum-containing anticancer drugs (e.g. cisplatin) are examples of toxins that exclusively damage dorsal root ganglion neurons, leading to large-fiber sensory loss and sensory ataxia.[3–7] In contrast, N-hexane inhalation produces almost entirely motor deficits.[8,9] An acute sensorimotor demyelinating polyneuropathy may be caused by the ingestion of buckthorn "fruit": clinically, this is sometimes hard to distinguish from idiopathic Guillain-Barré syndrome.[10] Children poisoned by thallium initially present with pain and autonomic dysfunction.[11] An encephalopathy associated with autonomic dysfunction occurs in children poisoned by organophosphates.[12] In both thallium and organophosphate poisoning, sensory polyneuropathy usually becomes evident only as these early encephalopathic symptoms abate.

Pathophysiology

An appealing hypothesis is that length-dependent toxic axonopathies are a direct consequence of impaired axoplasmic transport of essential molecules from the neuronal perikaryon distally, or of end organ-derived trophic molecules proximally. Colchicine, paclitaxel (Taxol), and the vinca alkaloids can all induce symmetrical distal polyneuropathy and are known to perturb turnover of microtubules, which are an essential factor in axoplasmic transport. Studies of neuropathies caused by acrylamide, hexane, and the rodenticide Vacor also suggest perturbation of axoplasmic flow.[13–15] It remains unclear,

B.T. Darras, H. Royden Jones, Jr., M.M. Ryan & D.C. De Vivo (Eds): Neuromuscular Disorders of Infancy, Childhood and Adolescence, Second edition.
DOI: http://dx.doi.org/10.1016/B978-0-12-417044-5.00023-8

however, whether the impairment of axoplasmic transport is caused by these various toxic agents or whether this occurs secondary to the axonal degeneration itself.

The similarities in clinical, electrophysiologic, and pathologic features of cisplatin and pyridoxine sensory neuronopathies suggest common underlying pathophysiologic mechanisms. In both instances, the toxin evidently penetrates the blood-nerve barrier to reach dorsal root ganglion neurons; this likely causes the death of large sensory neurons by chemical cross-linking of macromolecules (DNA in the case of cisplatin, probably proteins in the case of pyridoxine).

The neuropathies caused by chloroquine and amiodarone appear to result from the lipophilic nature and resistance to degradation of these molecules. These properties cause the drugs to accumulate within Schwann cells as membrane-bound inclusions. This may be the basis for the Schwann cell metabolic dysfunction, eventually leading to segmental demyelination.[16,17]

Diagnosis

A comprehensive history is the most essential tool for correctly diagnosing children with toxic neuropathies. Which, if any, medications has the child been prescribed or did he or she have access to in the home? Is there a suggestion of inadvertent or voluntary hydrocarbon inhalation? Has an exterminator visited the home recently? Have friends or family members had similar symptoms? Recognition of a drug-induced neuropathy is frequently delayed by the presence of an intercurrent illness. For example, is the abrupt onset of a painful neuropathy in a child receiving treatment for human immunodeficiency virus caused by the virus *per se*, or is it secondary to the purine or pyrimidine analogue with which the child is being treated?

The general physical examination may also yield important clues, such as hair loss in thallium poisoning and transverse white Mee's lines on the nails in either arsenic or thallium poisoning. The neurologic examination establishes the distribution of sensory deficits (e.g. stocking-glove versus asymmetrical) and the modalities affected (large or small sensory, motor, autonomic). Obtaining reliable results from a sensory examination is challenging in children but is of particular importance, given their inability to describe their symptoms precisely. It is often desirable in young children to test perception of cold, rather than pinprick, to detect abnormalities in small-fiber sensory perception. Electrophysiologic testing is helpful in documenting the distribution of involvement, detecting subclinical damage and muscle denervation, and demonstrating whether the disorder affects axons, myelin sheaths, or both through measurements of conduction velocity and action potential amplitude.

DNA studies are useful for diagnosing the genetic neuropathies. Occasionally one may find that both genetic and toxic mechanisms are active in the same patient. Identification of a genetic neuropathy therefore does not exclude the remote possibility of a superimposed toxic or inflammatory process leading to an acute exacerbation of the underlying genetic process. This has been particularly well defined with the combination of vincristine superimposed on children with occult Charcot-Marie-Tooth disease type 1.[18–21] Blood, urine, and hair can be analyzed for heavy metals. Although nerve biopsy often shows axonal balloons in patients with hydrocarbon-induced neuropathy or membrane-bound bodies in endoneurial cells in neuropathies caused by Vacor, amiodarone, or chloroquine, useful information is seldom obtained from nerve biopsy in patients with toxic neuropathies.

Treatment

The primary treatment for toxic neuropathies is withdrawal of the offending toxin. In most instances, this is sufficient to stop the progression of deficits, and recovery follows. However, the "coasting phenomenon," in which the disease worsens for weeks after cessation of exposure, is well recognized, particularly in patients exposed to hydrocarbons, cisplatin, or pyridoxine. Protection of deafferented limbs from trauma, range-of-motion physical therapy, and motor and gait reconditioning are frequently useful, and adequate suppression of pain is also important. Specific therapies are available in a few cases (e.g. pyridoxine supplementation for children at risk of isoniazid neuropathy; chelation of arsenic with penicillamine or BAL; treatment with atropine and pyridine 2-aldoxime during the early stages of organophosphate poisoning).

SELECTED TOXIC NEUROPATHIES

Antineoplastic Agents

Cisplatin and Other Platin Compounds. These drugs are used for the treatment of solid tumors but often cause sensory neuronopathy. Patients present with paresthesias and evidence of impairment of large-fiber sensory modalities, including sensory ataxia. Diminished pain and temperature perception, if present, are generally mild. Distal muscle stretch reflexes are usually depressed, and generalized areflexia may occur. Autonomic disturbances are infrequent, and motor function is spared. Symptoms and signs may progress for months after cessation of therapy. Electrophysiologic studies show reduced or absent sensory nerve action potentials, normal motor potentials, and no evidence of muscle denervation.[4–6]

Vincristine and Other Vinca Alkaloids. These drugs, which are included in several chemotherapy protocols for childhood cancer, interfere with microtubule assembly. Vincristine toxicity is the most common cause of neuropathy in children with malignancies. Many children receiving vinca alkaloids have reduced ankle reflexes and mild diminution in distal extremity sensation. Some may also show evidence of autonomic dysfunction, such as urinary retention or orthostatic hypotension. One clinical review of 96 survivors of vincristine-treated childhood malignancies demonstrated that 47 (49%) had areflexia associated with decreased motor and sensory amplitudes on nerve conduction studies (NCS), consistent with an axonopathy. Needle electromyogram (EMG) may show muscle denervation. Although symptoms and signs of the axonal neuropathy typically dissipate gradually after cessation of exposure, some children develop substantial persistent weakness, sometimes with bilateral facial paresis. This is especially of concern with vincristine toxicity if the child has an unsuspected hereditary neuropathy.[18–21]

Paclitaxel. This drug is used occasionally to treat solid tumors in children. It is frequently associated with burning dysesthesias secondary to a dose-dependent, predominantly sensory, axonal polyneuropathy.[22,23]

Antibacterial, Antiprotozoal, and Antiretroviral Drugs

Chloroquine. This drug is used for malaria prophylaxis and for treatment of connective tissue diseases. Neuropathy may appear months or years after institution of therapy and may be accompanied by proximal weakness. Electrophysiologic testing may show evidence of both segmental demyelination and axonopathy; in those with proximal weakness, there may be evidence of myopathy as well. Membrane-bound inclusions similar to those described in amiodarone toxicity are seen in nerve biopsies of patients with chloroquine neuropathy.[17]

Metronidazole. This antibiotic is used to treat amebiasis and other protozoan diseases, and may also be useful in anaerobic bacterial infections. Children treated with metronidazole may develop a mild, reversible, predominantly sensory axonal polyneuropathy.[24]

Isoniazid. Children treated for tuberculosis with isoniazid occasionally develop a mild distal sensory axonal polyneuropathy. Isoniazid is an inhibitor of pyridoxal phosphokinase, and isoniazid neuropathy can usually be prevented by supplementation with pyridoxine.[25]

Dapsone. This drug is used to treat leprosy and dermatitis herpetiformis. Occasionally, patients develop clinical and electrophysiologic evidence of a primary motor polyneuropathy.[26]

Nitrofurantoin. Polyneuropathy caused by this antibacterial drug can be acute in onset and is most apt to occur in children with impaired renal function. It affects both sensory and motor axons. Distinguishing the toxic effects of nitrofurantoin from those of uremic toxins can be difficult.[27]

Antiretroviral drugs. Chronic administration of a variety of purine and pyrimidine analogues, including dideoxycytidine (ddC), didanosine (ddI), and zidovudine (ZDV), can elicit either a rapid or gradual onset of a painful, length-dependent, small-fiber polyneuropathy, typically without autonomic deficits. Electrophysiology shows diminished sensory action potentials and normal motor action potentials. Symptoms and signs may worsen for a month or more after cessation of therapy.[28,29]

Miscellaneous Medications

Amiodarone. This medication is used for the treatment of arrhythmias. As many as 10% of those who take it for a year or more develop a primary sensory polyneuropathy. Electrophysiologic testing shows evidence of segmental demyelination, but there may also be muscle denervation, indicating an axonal component. Nerve biopsy shows accumulation of intracellular, osmiophilic, membrane-bound inclusions in Schwann cells, perineurium, and endoneurium.[16]

Phenytoin. Children treated chronically with this anticonvulsant occasionally develop reduced muscle stretch reflexes in the legs, and some suggest that patients may rarely develop distal sensory impairment. However, they do not have a clinically relevant polyneuropathy.

Colchicine. This drug is occasionally used in children for the treatment of inflammatory disorders. Prolonged therapy may elicit an axonal sensory polyneuropathy. Patients may also develop proximal weakness and an elevated serum creatine kinase level; in such cases, muscle biopsy shows a vacuolar myopathy.[30]

Thalidomide. Indications are increasing for the use of this drug to treat inflammatory and neoplastic diseases. When children receive thalidomide, it is reasonable to expect that some of them, like many adults, will develop a motor, sensory, and autonomic polyneuropathy, which in some cases may worsen even after discontinuation of treatment.[31,32] At Boston Children's Hospital, preliminary data from the EMG laboratory suggest that this medication may lead to a significant polyneuropathy in at least some treated children after approximately 2 to 3 years of therapy. In adults, serial electrophysiologic monitoring has proved valuable in monitoring therapy to prevent this complication.[31,32]

Pyridoxine (Vitamin B6). Megadoses of pyridoxine (e.g. 25–50 mg/kg per day) are occasionally used in children with pyridoxine-dependent epilepsies, but this is uncommon now that pyridoxal-5-phosphate has largely superseded pyridoxine for these conditions. When given in very large doses, pyridoxine often elicits large-fiber

CASE EXAMPLE 23.1

A 12-year-old boy from a depressed industrial town was referred for evaluation of progressive weakness over a 2-week period. The initial history and general physical examination were unrevealing. Neurologic examination demonstrated areflexia and distal weakness equal to proximal weakness in his legs, with normal sensation and normal strength in the arms. The cranial nerves, cerebellar function, and mentation were normal.

Spinal fluid showed a normal protein level and no cells. Electrophysiologic testing demonstrated a 30% to 40% slowing in nerve conduction velocities in the legs, lesser slowing in the arms, no conduction block, and no evidence of muscle denervation.

A day after admission to the hospital, the young man told his nurse that he and his friends had been sniffing glue from paper bags for several months. His weakness progressed for several weeks while he was on the rehabilitation service of the hospital, and then gradually subsided. His neurologic function was normal when last seen, a year later.

Comment

Children and particularly adolescents are prone to experimentation with a variety of potentially toxic agents. Glue sniffing is one of the best-known sources of an occult PNS toxin, and the astute physician needs to keep this in mind. It is also important that family and friends be questioned individually when attempting to establish the source of the neuropathy. An empathetic listener, as illustrated by the nurse in this instance, can often gain important diagnostic information.

The results of the clinical neurophysiologic studies described in Case Example 23.1 emphasize the need to consider possible toxin exposure in children presenting with a subacute demyelinating polyneuropathy resembling Guillain-Barré syndrome or CIDP. However, most toxic polyneuropathies are associated primarily with axon loss.

sensory defects and sensory ataxia. Lower doses, as little as 200 mg/day in adults, may also cause sensory deficits after prolonged therapy. Electrophysiologic examination shows diminished or absent sensory nerve action potentials. The primary site of pyridoxine toxicity is the dorsal root ganglia. The considerable degree of eventual recovery in most patients after cessation of pyridoxine exposure suggests that this drug, like cisplatin, causes a reversible functional impairment of some sensory neurons, as well as death of others.[3,33]

Dichloroacetate. Dichloroacetate is a potent lactate-lowering agent that inhibits the pyruvate dehydrogenase complex and has been shown to relieve symptoms in mitochondrial myopathy, encephalopathy, lactic acidosis, and stroke-like episodes (MELAS) caused by 3243A > G mutants (also discussed in Chapter 41). A recent large, randomized, double-blind, placebo-controlled cross-over trial of dichloroacetate, however, was terminated because of peripheral nerve toxicity. All subjects treated with dichloroacetate had to be taken off treatment because of development or worsening of a length-dependent axonal neuropathy that was felt to overshadow assessment of any benefit from the medication.[34]

Hexane and Methyl-N-Butyl Ketone. Exposure of children to these hydrocarbons is most commonly the result of voluntary inhalation, so-called glue sniffing (see Case Example 23.1). Glue-sniffing neuropathy is characterized by predominant proximal as well as distal weakness. Distal paresthesias may be the first evidence of toxicity. This adolescent addiction should be considered in the differential diagnosis of any subacute polyneuropathy, especially chronic inflammatory demyelinating polyradiculoneuropathy (CIDP). Weakness may continue to worsen for months following cessation of glue sniffing or other exposure to these solvents, but then gradually abates.[8,9,35–37] Occasionally, glue or gasoline sniffing can give rise to a mononeuritis multiplex-like clinical picture. The N-hexane component of glue is also found in certain commercial settings, particularly the shoe industry, where this toxin is a component of cements, and occupational exposure can lead to development of generalized weakness with areflexia, nausea, anorexia, and weight loss.[35]

In the setting of glue or gasoline sniffing, NCS often show prolonged distal latencies, moderate slowing of motor conduction, conduction block, and absent F-waves.[29,30] Active denervation may be seen on needle EMG, suggesting a concomitant axonal as well as demyelinating process.[8,37]

Sural nerve biopsy in children with glue-sniffing neuropathy classically shows giant axons with axonal masses and concomitant paranodal demyelination (Figure 23.1). These axonal swellings are filled with neurofilaments surrounded by thin myelin, similar to the findings of giant axonal neuropathy.[9,37,38]

Insecticides, Rodenticides, and Heavy Metals

Organophosphates. Insecticides containing organophosphates inhibit acetylcholinesterase activity, producing excessive muscarinic and nicotinic cholinergic activity. Acute exposure usually causes diffuse cholinergic effects, characterized by increased salivation, lacrimation, miosis, sweating, brady- or tachyarrhythmias, pulmonary edema, muscle fasciculations, and muscle twitching. Mental status changes may follow, with delirium and even coma. Neuropathy usually appears later. A distal, symmetrical, predominantly motor polyneuropathy may

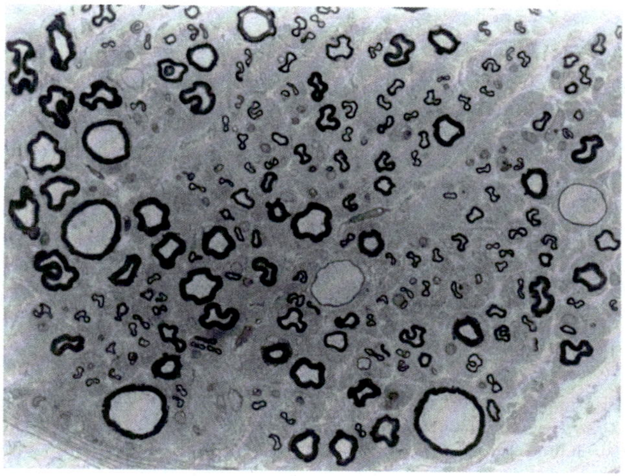

FIGURE 23.1 Glue sniffing neuropathy. Sural nerve biopsy with characteristic peripheral nerve changes; note the marked variation in diameter of myelinated fibers, with disproportionately thin myelin sheaths surrounding some very large axons. On electron microscopy, the large axons were packed with neurofilaments. *Used with permission from Ouvrier et al.*[38]

CASE EXAMPLE 23.2

A 3-year-old boy was brought to the hospital because of abdominal pain and vomiting. The initial workup, including a neurologic examination, was unrevealing, with the exception of a sinus tachycardia. Several days later, after his abdominal symptoms subsided, he was noted to be losing scalp hair.

Repeat neurologic examination showed diminished temperature sensation distally in his legs and general hyporeflexia. He cried vigorously when his feet or hands were stroked.

Urine thallium excretion was markedly elevated. His parents then recalled having obtained a rodenticide powder from a neighbor and using some of it in their bedroom closet. Over the next several months, the child's hair regrew, but his distal sensory deficits did not fully disappear for more than a year.

Comment

Here the clinical finding of depilation provided the important clue to the eventual laboratory diagnosis.

occur subacutely in patients exposed to organophosphate agricultural insecticides.

Treatment with atropine and pyridine 2-aldoxime can be lifesaving during the acute phase of organophosphate poisoning. A typical axonal polyneuropathy typically develops 1 to 3 weeks after the acute cholinergic crisis, and may mimic Guillain-Barré syndrome.[12,39]

Vacor (N-3-pyridylmethyl-N′-p-nitrophenyl urea). This lipophilic toxin is a component of some rodenticide preparations. Inadvertent ingestion causes an autonomic polyneuropathy. This may be preceded by acute diabetic ketoacidosis, as a consequence of the additional toxicity of this molecule to the pancreatic islet cells.[40]

Thallium. This metal is used primarily as a rodenticide or a depilatory agent in women. Acute ingestion of thallium may mimic arsenic poisoning, with prominent autonomic dysfunction including severe gastrointestinal distress, polyneuropathy, and even Mee's lines.[11,41] Children may also present initially with acute encephalopathy. After one adolescent ingested thallium in a suicide attempt, compound muscle action potentials could not be recorded distally, but a proximal median nerve study demonstrated a motor conduction velocity of 20 m/s, suggesting a demyelinating effect.[42] As these features subside, hair loss becomes evident, and a dysesthetic polyneuropathy develops. Electrophysiologic studies demonstrate length-dependent axonopathy, and the diagnosis can be made by measuring thallium content of the urine or blood.[11] See Case Example 23.2.

Arsenic. After acute exposure to this heavy metal, children may develop gastrointestinal symptoms and may become encephalopathic. A dysesthetic axonal polyneuropathy develops days to weeks after these initial symptoms subside. Occasionally, adolescents with arsenic poisoning present with a rapidly progressive motor and sensory polyneuropathy mimicking either Guillain-Barré syndrome or CIDP.[43,44] The diagnosis can be confirmed by elevated levels of arsenic in hair, nails, or urine. Treatment with penicillamine or BAL may be useful in reducing the severity of neuropathy, though this is controversial.

Lead. Lead neuropathy is exceedingly rare in childhood, especially in comparison to pica encephalopathy, which usually presents with encephalopathy if acute and with learning and behavioral abnormalities if chronic.[45,46] On rare occasions, lead poisoning causes focal motor neuropathy in children. Unlike in adults, children's legs may be affected instead of, or in addition to, the extensor muscles of the forearms. Electrophysiologic studies show evidence of demyelination, but an axonopathy may also develop in chronic lead poisoning. Blood studies typically show microcytic anemia, basophilic stippling of red blood cells, and elevated lead levels; hyperuricemia may also be demonstrated.

One well-documented case is significant because levels of lead in both blood and urine were normal.[46] This 3-year-old experienced two episodes of nausea, vomiting, thigh pain, and recurrent subacute motor neuropathy, with a history of 2 years of pica. The child was anemic. Urinary levels of coproporphyrins were positive, and δ-aminolevulinic acid levels were 9 times normal. NCS were not performed, but the EMG demonstrated denervation. Calcium disodium edetate treatment resulted in a 1000-fold increase in lead excretion. Within 6 months, the child had fully recovered. The same authors reported lead-induced neuropathies in 8 other children,[46] each with a different clinical picture. In general, there was an acute motor greater than sensory neuropathy, associated with a history of pica, as well as anemia, usually with basophilic stippling of erythrocytes.

Mercury. Very rarely, inorganic mercury causes a peripheral neuropathy in children.[41,47–49] Children exposed to mercury may demonstrate incoordination, weakness, abdominal pain, lethargy, acrodynia, and anorexia. Electrophysiologic studies may show evidence of axonopathy or sensory neuronopathy. Intra-uterine exposure to mercury may damage both the central and peripheral nervous systems.[1]

Poisonous Plants

Karwinskia humboldtiana **(Buckthorn, Wild Cherry).** Buckthorn is a poisonous "fruit" found on shrubs in northern Mexico, the southwestern United States, Central America, Colombia, and the Caribbean countries. If an unsuspecting child accidentally ingests this toxic fruit, an acute symmetrical motor polyneuropathy mimicking Guillain-Barré syndrome may develop within 1 to 15 days.[10,50] The cerebrospinal fluid protein is usually normal. If the intoxication is severe, a bulbar paralysis and a potentially fatal outcome are possible. Some of the most severely affected children also have diarrhea and vomiting. Motor nerve conduction velocities are decreased to 40% to 60% of normal, and there is a mild loss of amplitude of the compound muscle action potentials.[10,50] Sural nerve biopsy demonstrates segmental demyelination.

Buckthorn toxin targets the metabolic activity of the Schwann cell, with subsequent myelin disruption. This toxin is now more widely recognized as an important consideration when a child native to these geographic areas presents with acute flaccid paralysis. The ease of diagnosis is related to the availability of a blood immunoassay test for *Karwinskia humboldtiana* toxin.

CONCLUSION

The list of environmental agents that can cause neuropathy in children is virtually endless; some of the better recognized peripheral nerve toxins are mentioned in this chapter. Toxic neuropathy must always be considered in the differential diagnosis of subacute or chronic symmetrical distal sensory polyneuropathy. Primarily motor toxic neuropathic syndromes are also encountered. A careful history is often crucial in making the diagnosis.

Avoidance of further exposure is the primary treatment for most toxic neuropathies. Specific therapies are available in a few instances, and symptomatic measures, including pain relief and physical therapy, are often of great value. In future, pharmacogenetic analysis to quantify individual susceptibilities to neurotoxic agents can be expected to play an increasing role in preventing toxic neuropathies.

REFERENCES

1. Takeuchi T, Eto K, Oyanag S, Miyajima H. Ultrastructural changes of human sural nerves in the neuropathy induced by intrauterine methylmercury poisoning (so-called fetal Minamata disease). *Virchows Arch B Cell Pathol* 1978;**27**:137–54.
2. Carman KB, Tutkun E, Yilmaz H, Dilber C, Dalkiran T, Cakir B, et al. Acute mercury poisoning among children in two provinces of Turkey. *Eur J Pediatr* 2013;**172**:821–7.
3. Schaumburg H, Kaplan J, Windebank A, Vick N, Rasmus S, Pleasure D, et al. Sensory neuropathy from pyridoxine abuse. A new megavitamin syndrome. *N Engl J Med* 1983;**309**:445–8.
4. Krarup-Hansen A, Rietz B, Krarup C, Heydorn K, Rorth M, Schmalbruch H. Histology and platinum content of sensory ganglia and sural nerves in patients treated with cisplatin and carboplatin: an autopsy study. *Neuropathol Appl Neurobiol* 1999;**25**:29–40.
5. Bissett D, Kunkeler L, Zwanenburg L, Paul J, Gray C, Swan IR, et al. Long-term sequelae of treatment for testicular germ cell tumours. *Br J Cancer* 1990;**62**:655–9.
6. Spunt SL, Freeman 3rd BB, Billups CA, McPherson V, Khan RB, Pratt CB, et al. Phase I clinical trial of oxaliplatin in children and adolescents with refractory solid tumors. *J Clin Oncol* 2007;**25**:2274–80.
7. Windebank AJ, Grisold W. Chemotherapy-induced neuropathy. *J Peripher Nerv Syst* 2008;**13**:27–46.
8. Towfighi J, Gonatas NK, Pleasure D, Cooper HS, McCree L. Glue sniffer's neuropathy. *Neurology* 1976;**26**:238–43.
9. Smith AG, Albers JW. n-Hexane neuropathy due to rubber cement sniffing. *Muscle Nerve* 1997;**20**:1445–50.
10. Martinez HR, Bermudez MV, Rangel-Guerra RA, de Leon Flores L. Clinical diagnosis in *Karwinskia humboldtiana* polyneuropathy. *J Neurol Sci* 1998;**154**:49–54.
11. Bank WJ, Pleasure DE, Suzuki K, Nigro M, Katz R. Thallium poisoning. *Arch Neurol* 1972;**26**:456–64.
12. Aiuto LA, Pavlakis SG, Boxer RA. Life-threatening organophosphate-induced delayed polyneuropathy in a child after accidental chlorpyrifos ingestion. *J Pediatr* 1993;**122**:658–60.
13. Pleasure DE, Mishler KC, Engel WK. Axonal transport of proteins in experimental neuropathies. *Science* 1969;**166**:524–5.
14. Watson DF, Griffin JW. Vacor neuropathy: ultrastructural and axonal transport studies. *J Neuropathol Exp Neurol* 1987;**46**:96–108.
15. LaPointe NE, Morfini G, Brady ST, Feinstein SC, Wilson L, Jordan MA. Effects of eribulin, vincristine, paclitaxel and ixabepilone on fast axonal transport and kinesin-1 driven microtubule gliding: implications for chemotherapy-induced peripheral neuropathy. *Neurotoxicology* 2013;**37**:231–9.
16. Jacobs JM, Costa-Jussa FR. The pathology of amiodarone neurotoxicity. II. Peripheral neuropathy in man. *Brain* 1985;**108**:753–69.
17. Tegner R, Tome FM, Godeau P, Lhermitte F, Fardeau M. Morphological study of peripheral nerve changes induced by chloroquine treatment. *Acta Neuropathol* 1988;**75**:253–60.

18. Weimer LH, Podwall D. Medication-induced exacerbation of neuropathy in Charcot Marie Tooth disease. *J Neurol Sci* 2006;**242**:47–54.
19. Nakamura T, Hashiguchi A, Suzuki S, Uozumi K, Tokunaga S, Takashima H. Vincristine exacerbates asymptomatic Charcot-Marie-tooth disease with a novel EGR2 mutation. *Neurogenetics* 2012;**13**:77–82.
20. Casey EB, Jellife AM, Le Quesne PM, Millett YL. Vincristine neuropathy. Clinical and electrophysiological observations. *Brain* 1973;**96**:69–86.
21. Hogan-Dann CM, Fellmeth WG, McGuire SA, Kiley VA. Polyneuropathy following vincristine therapy in two patients with Charcot-Marie-Tooth syndrome. *JAMA* 1984;**252**:2862–3.
22. Lipton RB, Apfel SC, Dutcher JP, Rosenberg R, Kaplan J, Berger A, et al. Taxol produces a predominantly sensory neuropathy. *Neurology* 1989;**39**:368–73.
23. Postma TJ, Vermorken JB, Liefting AJ, Pinedo HM, Heimans JJ. Paclitaxel-induced neuropathy. *Ann Oncol* 1995;**6**:489–94.
24. Duffy LF, Daum F, Fisher SE, Selman J, Vishnubhakat SM, Aiges HW. Peripheral neuropathy in Crohn's disease patients treated with metronidazole. *Gastroenterology* 1985;**88**:681–4.
25. Ochoa J. Isoniazid neuropathy in man: quantitative electron microscope study. *Brain* 1970;**93**:831–50.
26. Rhodes LE, Coleman MD, Lewis-Jones MS. Dapsone-induced motor peripheral neuropathy in pemphigus foliaceus. *Clin Exp Dermatol* 1995;**20**:155–6.
27. Yiannikas C, Pollard JD, McLeod JG. Nitrofurantoin neuropathy. *Aust NZ J Med* 1981;**11**:400–5.
28. Pardo CA, McArthur JC, Griffin JW. HIV neuropathy: insights in the pathology of HIV peripheral nerve disease. *J Peripher Nerv Syst* 2001;**6**:21–7.
29. Verma A. Epidemiology and clinical features of HIV-1 associated neuropathies. *J Peripher Nerv Syst* 2001;**6**:8–13.
30. Kuncl RW, Cornblath DR, Avila O, Duncan G. Electrodiagnosis of human colchicine myoneuropathy. *Muscle Nerve* 1989;**12**:360–4.
31. Calabrese L, Fleischer AB. Thalidomide: current and potential clinical applications. *Am J Med* 2000;**108**:487–95.
32. Fleming FJ, Vytopil M, Chaitow J, Jones Jr. HR, Darras BT, Ryan MM. Thalidomide neuropathy in childhood. *Neuromuscul Disord* 2005;**15**:172–6.
33. McLachlan RS, Brown WF. Pyridoxine dependent epilepsy with iatrogenic sensory neuronopathy. *Can J Neurol Sci* 1995;**22**:50–1.
34. Kaufmann P, Engelstad K, Wei Y, Jhung S, Sano MC, Shungu DC, et al. Dichloroacetate causes toxic neuropathy in MELAS: a randomized, controlled clinical trial. *Neurology* 2006;**66**:324–30.
35. Rizzuto N, Terzian H, Galiazzo-Rizzuto S. Toxic polyneuropathies in Italy due to leather cement poisoning in shoe industries. A light- and electron-microscopic study. *J Neurol Sci* 1977;**31**:343–54.
36. Kuwabara S, Nakajima M, Tsuboi Y, Hirayama K. Multifocal conduction block in n-hexane neuropathy. *Muscle Nerve* 1993;**16**: 1416–7.
37. Korobkin R, Asbury AK, Sumner AJ, Nielsen SL. Glue-sniffing neuropathy. *Arch Neurol* 1975;**32**:158–62.
38. Ouvrier RA, McLeod JG, Pollard JD. Toxic neuropathies. In: Ouvrier R, McLeod JG, Pollard JD, editors. *Peripheral Neuropathy in Childhood*. 2nd ed. London: MacKeith Press for the International Child Neurology Association; 1999. pp. 207–8.
39. Jenkins RB, Toole JF. Polyneuropathy following exposure to insecticides. Two cases of polyneuropathy with albuminocytologic dissociation in the spinal fluid following exposure to DDD and aldrin and DDT and endrin. *Arch Intern Med* 1964;**113**:691–5.
40. LeWitt PA. The neurotoxicity of the rat poison Vacor. A clinical study of 12 cases. *N Engl J Med* 1980;**302**:73–7.
41. Windebank AJ, McCall JT, Dyck PJ. Metal neuropathy. In: Dyck PJ, Thomas PK, Lambert EH, Bunge R, editors. *Peripheral Neuropathy*. 3rd ed. Philadelphia: Saunders; 1993. pp. 1561–3.
42. Davis LE, Standefer JC, Kornfeld M, Abercrombie DM, Butler C. Acute thallium poisoning: toxicological and morphological studies of the nervous system. *Ann Neurol* 1981;**10**:38–44.
43. Donofrio PD, Wilbourn AJ, Albers JW, Rogers L, Salanga V, Greenberg HS. Acute arsenic intoxication presenting as Guillain-Barré-like syndrome. *Muscle Nerve* 1987;**10**:114–20.
44. Le Quesne PM, McLeod JG. Peripheral neuropathy following a single exposure to arsenic. Clinical course in four patients with electrophysiological and histological studies. *J Neurol Sci* 1977;**32**:437–51.
45. Feldman RG, Hayes MK, Younes R, Aldrich FD. Lead neuropathy in adults and children. *Arch Neurol* 1977;**34**:481–8.
46. Seto DS, Freeman JM. Lead neuropathy in childhood. *Am J Dis Child* 1964;**107**:337–42.
47. Ross AT. Mercuric polyneuropathy with albumino-cytologic dissociation and eosinophilia. *JAMA* 1964;**188**:830–1.
48. Swaiman KF, Flagler DG. Mercury poisoning with central and peripheral nervous system involvement treated with penicillamine. *Pediatrics* 1971;**48**:639–42.
49. Agocs MM, Etzel RA, Parrish RG, Paschal DC, Campagna PR, Cohen DS, et al. Mercury exposure from interior latex paint. *N Engl J Med* 1990;**323**:1096–101.
50. Calderon-Gonzalez R, Rizzi-Hernandez H. Buckthorn polyneuropathy. *N Engl J Med* 1967;**277**:69–71.

Chapter 24

Hansen's Disease (Leprosy): Leprous Neuropathy

Wilson Marques, Jr.

INTRODUCTION

Leprosy is a chronic infectious disease caused by *Mycobacterium leprae* (*M. leprae*), an acid-fast, Gram-positive, rod-shaped, obligate intracellular parasite which has the unique propriety of systematic invasion and multiplication inside peripheral nerves. Although mostly known due to the resulting skin manifestations, most of the feared consequences of leprosy result from nerve damage. Although known from time immemorial, it is still a poorly understood disease, affecting thousands of people in poor countries and constituting a diagnostic problem in most wealthy countries.

The introduction of multidrug therapy in 1981 dramatically changed leprosy epidemiology around the world. The 2013 Global Leprosy Update[1] reported that the number of new cases detected in 2012 was 232,857, 95% of these originating from only 16 countries, led by India (134,752), Brazil (33,303) and Indonesia (18,994). Grade 2 disability was already present at diagnosis in 14,049 cases, attesting to late diagnosis. The proportion of children among new cases ranged from 0.6% in Argentina to 41.3% in Micronesia, indicating that transmission is still very active in several world regions.

Leprosy affects people of any age, from infancy to elderly, and of both sexes, although in many areas men are mostly affected (in a proportion of 2:1). The main portal of exit of *M. leprae* is the nasal mucosa, which can eliminate up to 1 million viable organisms in a single day. Similarly the main portal of entry seems to be the nasal mucosa, although it seems that skin may also be an alternative, via abrasions and cuts followed by hematological dissemination. *M. leprae* bacilli preferentially invade and multiply inside nonmyelinating Schwann cells in the coldest regions of the body, as the ideal temperature for *M. leprae* is around 30°C.[2] The incubation period varies from 6 months to 20 years. *M. leprae* has a very long generation time (~13 days).[3] Approximately 95% of the population is naturally resistant and does not develop infection after exposure. Among those who are susceptible, those in the "tuberculoid" pole have a strong cell-mediated immunity and delayed hypersensitivity response,[4] effectively controlling the infection and not transmitting the disease. These patients develop localized manifestations, with only one or a few well-demarcated skin and nerve lesions. There is a severe granulomatous response that may completely destroy the nerve and the bacilli. At the other extreme, the "lepromatous" patients have no cellular immune response to *M. leprae*, developing slowly progressive and disseminated skin and nerve injury, and being highly infectious.[4] Most patients [borderline (BB), borderline-tuberculoid (BT), and borderline lepromatous (BL)] have reactions to infection between these extremes. These subjects may develop a more severe disease, especially those in the BT group, in whom the immune response cannot prevent bacterial dissemination, but which is then sufficient to destroy *M. leprae* bacilli and infected tissue.

SKIN MANIFESTATIONS

The typical skin manifestations are hypopigmented macules, papules, or nodules, which sometimes have a reddish or copper-colored appearance. These are most frequently localized in the trunk and abdomen, although the limbs may also be affected. Thermoalgesic and tactile sensation are usually compromised in these areas, which may also show anhidrosis and alopecia. Plantar ulcers and other trophic disturbances usually develop late in the course of the disease in those with significant sensory loss.[5]

B.T. Darras, H. Royden Jones, Jr., M.M. Ryan & D.C. De Vivo (Eds): Neuromuscular Disorders of Infancy, Childhood and Adolescence, Second edition.
DOI: http://dx.doi.org/10.1016/B978-0-12-417044-5.00024-X

LEPROSY NEUROPATHY

Virtually all patients with leprosy have a neuropathy, as Schwann cells are the primary target of the bacilli. As a result, worldwide, leprosy is still one of the most common treatable neuropathies. Although the number of new cases has decreased significantly over the years, treated patients may be left with severe and permanent nerve damage, may develop a progressive neuropathy or even manifest clinically a previously asymptomatic neuropathy, situations that may also result in severe and irreversible nerve destruction. Early detection and early treatment are essential to control leprosy dissemination and avoid permanent nerve injury.[1] The time from the first clinical manifestation to diagnosis may be as long as 1.8 years.[6] This interval may even be longer in countries where leprosy is rarely seen, in pure neural leprosy and in countries where leprosy still has a heavy stigma.

Characteristics of Leprosy Neuropathy

Asymmetric sensory loss is the hallmark of leprosy neuropathy. Nerve hypertrophy is also often present, although not always detected on examination.[5,7] Nerve ultrasonography is effective in detecting nerve hypertrophy and should be considered in patients with idiopathic asymmetric neuropathy.[8]

There are some characteristics which together are extremely suggestive of leprosy neuropathy:[2,5]

1. There are two patterns of nerve involvement in leprosy. *Dermal nerve injury* does not follow a specific nerve territory distribution, and affects thermal, pain, tactile, and autonomic fibers of the skin in a mosaic distribution.[9] *Nerve trunk injury* compromises sensation and muscle strength in the territory of specific nerves. Superposition of both patterns results in a very suggestive clinical presentation.
2. *M. leprae* has a predilection to involve initially the dermal nerves and the unmyelinated and small myelinated nerve fibers of the nerve trunks. Consequently, abnormalities of proprioception, vibration, motor function, and reflexes are usually seen late in the disease course.
3. *M. leprae* has a predilection to invade and proliferate in nerves localized in the coolest areas of the body, such as the elbow, dorsal regions of the forearm, dorsal hand, knees, lateral aspects of the legs, earlobes, and malar region, resulting in a temperature-dependent distribution, in clear contrast to the length-dependent pattern observed in most polyneuropathies. Even in advanced disease, a careful sensory examination will demonstrate that the warmest body parts have intact (or at least partially preserved) sensation. The confluent neuropathy seen in advanced cases may superficially simulate a length-dependent symmetrical polyneuropathy, but a careful examination will not reveal a glove-and-stocking distribution but rather a very asymmetrical pattern when involvement of the coolest and the warmest regions of the lower limbs, upper limbs, trunk, and face are compared.[2]
4. The preference of *M. leprae* to invade and multiply in the coldest and superficial regions probably explains the preferential sites of involvement of the nerve trunks in leprosy, resulting in focal nerve damage. The ulnar nerve is typically involved 10–15 cm above the olecranon and 2–4 cm proximal to the wrist, the median nerve 5–7 cm proximal to the carpal tunnel, the radial nerve 4–6 cm proximal to the elbow, the fibular nerve at the head of the fibula, and the posterior tibial nerve at the ankle and above the gastrocnemius muscle. The facial branches are asymmetrically involved. Affected nerve segments may be enlarged and tender. Neurophysiologic studies usually show focal slowing of nerve conduction at these sites with distal axonal degeneration.
5. Although leprosy neuropathy mainly results in negative sensory and motor manifestations, positive sensory symptoms—such as neuropathic pain—may be debilitating.[7]
6. The final pattern of nerve injury depends on the balance between body temperature and the immunological competence of the host, which is at least partially genetically determined. Individuals who are highly resistant to the bacilli (tuberculoid pole of leprosy) will have a mononeuropathy or a restricted mononeuritis multiplex, with involvement of only a few nerves, which will usually be severely damaged. In contrast, in subjects with no resistance to *M. leprae* (lepromatous and borderline-lepromatous patients), there will be a diffuse and slowly progressive neuropathy with late nerve destruction. In these patients, early diagnosis and treatment is essential to avoid a diffuse and incapacitating neuropathy. Those patients with intermediate resistance to *M. leprae* (borderline leprosy) may present with a less discrete neuropathy, as their immune response to *M. leprae* is not strong enough to prevent dissemination of the bacilli but is sufficient to mount a marked local response resulting in severe damage to the affected nerves.

Intradermal sensory loss in leprosy neuropathy is most often apparent at the elbows, knees, and ears. The ulnar, superficial peroneal, radial, median, and sural are the most frequently involved sensory nerves. Weakness is most severe in the distributions of the ulnar, median, peroneal, and tibial nerves. Subjects in the tuberculoid pole generally present with a mononeuropathy or mononeuritis multiplex, while those in the lepromatous pole have a more diffuse mononeuritis multiplex, with both nerve trunk and dermal involvement.

The absence of protective sensation predisposes affected individuals to soft tissue injuries and other trophic manifestations. As proprioception and motor function are involved late in the disease course, injury to insensate areas may not be recognized as being due to neuropathy, especially in children and those with low educational background. Trophic ulcers are most commonly plantar, but ulcers may also develop over the knees, hands, ears, nose, and eyes. Persisting infection may be complicated by bone resorption.

Some clinical variants of leprosy neuropathy have been described but most are unconfirmed, do not affect children, or are just a variation of the classic phenotype. Silent neuritis is progressive neural damage that occurs in the absence of symptoms.[10]

Lepromatous reactions are episodes of acute change in the immunological host status, usually (but not always) in response to treatment.[7] The episodes may result in acute, severe nerve damage.[2] The reversal reaction (type 1 reaction) occurs mainly in subjects with borderline and tuberculoid leprosy, due to an increase in cell-mediated response. There is effective destruction of the bacilli, but this response may also damage the peripheral nerve. Erythema nodosum leprosum (type 2 reaction) occurs in lepromatous and borderline lepromatous patients, usually in association with death of bacilli. It is a systemic reaction, caused by a vasculitis affecting medium and small arteries and arterioles, which should be promptly treated to avoid permanent damage.

DIAGNOSIS

The gold standard of leprosy diagnosis is identification of *M. leprae* in slit-skin smears or biopsies from skin, nerves, or other affected tissues, but this is not always feasible, particularly in pure neural leprosy.[3] Nerve conduction studies may be extremely helpful in defining the neuropathy characteristics. The most frequent electrophysiological patterns are predominantly sensory or sensory-motor mononeuritis multiplex, frequently accompanied by focal slowing of nerve conduction if the nerves are studied at the segments where they are preferentially invaded. This pattern of neurophysiologic abnormality is extremely rare in childhood. Nerve ultrasonography may also reveal abnormalities in affected regions.[8] Nerve biopsy may be very helpful in patients with pure neural leprosy,[5] even when bacilli are not detected. In lepromatous patients, there is a striking asymmetry of nerve involvement; some fascicles seem to be preserved while others are totally destroyed. An inflammatory reaction is present at the epineurium and perineurium, and

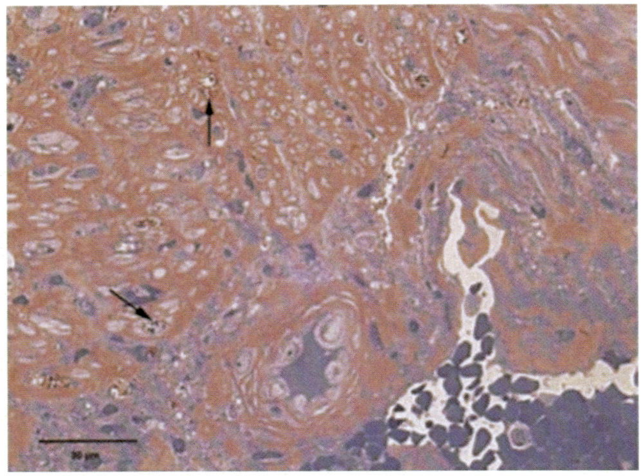

FIGURE 24.1 Endoneurial area of a sural nerve biopsy from a lepromatous leprosy patient. Transverse section of an Epon 812®-embedded semithin (0.5 μm) section showing complete loss of myelin fibers and increase in endoneurial collagen (red stain). Large numbers of *Mycobacterium leprae* bacilli are seen as clusters or "globi" within Schwann cells. Two "globi" are indicated by arrows. Azure stain. Bar: 30 μm. [About azure stain: alcohol-acid resistant stains (Fite-Faraco or Ziehl-Neelsen) are used for identification of bacilli in paraffin-embedded histological sections. In semithin (0.5 to 1.5 μm thick) sections azure stain is used more frequently because the conventional alcohol-acid resistant stains fail to identify bacilli. Bacilli can be identified in semithin sections of more than 1000× magnification on oil embedded slides using optical microscopy. Azure stains the *Mycobacterium leprae* bacillus very well when used to stain semithin sections of peripheral nerve fascicles of lepromatous leprosy patients.]

some vessels show a lymphocytic vasculitis. *M. leprae* bacilli are usually abundant (Figure 24.1). Fibrosis may be a prominent picture. Teased fiber preparations show demyelination and axonal loss. In the tuberculoid pole, there is usually severe nerve destruction with epithelioid-cell granulomas and lymphocytic infiltration. *M leprae* bacilli are rarely seen. Phenolic glycolipid-1 (PGL-1) is a unique glycolipid seen in *M. leprae* which binds to the Schwann cell basement membrane and induces demyelination. Serological tests for anti PGL-1 antibodies are helpful in lepromatous leprosy, but not in paucibacillar patients. In patients where the *M. leprae* bacillus is not identified, diagnosis may be based on the epidemiological, clinical, and electrophysiological characteristics, and confirmed by a response to treatment.

PATHOPHYSIOLOGY OF NERVE DAMAGE

Leprosy neuropathy is a neurodegenerative infectious disease, the primary cause of which is invasion of Schwann cells by *M. leprae.*[11] Schwann cells protect the bacilli against the host immune response, allowing bacterial multiplication which results in neural damage and allows dissemination to other Schwann cells and other tissues.[11]

M. leprae has a specific molecular affinity to Schwann cells via laminin-2 and α-dystroglycan receptors, but preferentially invades unmyelinated Schwann cells, as compact myelin seems to be a physical and physiological barrier to the bacilli.[11] At the early stages of the disease, attachment to myelinating Schwann cells induces a contact-dependent demyelination, which is mediated by ErbB2 receptor tyrosine kinase signaling.[12]

Nerve destruction usually becomes significant only years after infection, and results from complex (not yet fully understood) mechanisms involving not only the infection but also the host immune response and a multifactorial inflammatory reaction.[4,13] Nerve edema caused by the inflammatory response may contribute to nerve damage, especially in entrapment sites. The type of leprosy an infected person develops depends essentially upon his/her cell-mediated immune response.[14] In tuberculoid leprosy, the granulomatous host response seems to destroy the nerve structure,[4] while in lepromatous leprosy macrophages and Schwann cells are heavily loaded by bacilli which slowly multiply inside these cells, protected against the host immune response, which in this situation can be effective only when the loaded cell loses its integrity, releasing the bacteria. This results in a slowly chronic inflammation with intraneural fibrosis and progressive nerve destruction.[4]

TREATMENT

Patients should be treated as early as possible, in order to avoid permanent nerve damage, but this may still occur before, during, and after treatment.[3,13] The World Health Organization recommends 6 months of multidrug therapy (monthly rifampicin, daily dapsone) for paucibacillary (≤5 skin lesions) patients and 12 months of multidrug therapy (monthly rifampicin, daily dapsone and clofazimine) for multibacillary patients (>5 skin lesions). Relapses are very rare. The use of dapsone should be closely monitored, as it can cause an axonal motor neuropathy. Chemoprophylaxis seems to prevent the disease in contacts, but its use has not yet been systematized.[13] Corticosteroids should be used for acute neuritis. The type 1 reaction should be treated with prednisone and type 2 reactions with thalidomide (which may also cause a sensory and painful motor polyneuropathy).[10] Patients with neuropathic pain should be treated symptomatically.

REFERENCES

1. World Health Organization (WHO). Weekly Epidemiological Record No. 35, 365–390, 30 August 2013.
2. Sabin TD, Swift TR, Jacobson RR. Leprosy. In: Dyck PJ, Thomas PK, editors. *Peripheral Neuropathy*. 4th ed. Philadelphia: Elsevier Saunders; 2005. pp. 2081–108.
3. Goulart IMB, Goulart LR. Leprosy: diagnostic and control challenges for a worldwide disease. *Arch Dermatol Res* 2008;**300**:269–90.
4. Scollard DM, Adams LB, Gillis TP, Krahenbuhl JL, Truman RW, Willians DL. The continuing challenges of leprosy. *Clin Microbiol Rev* 2006;**19**(2):338–81.
5. Said G. Neuropathies due to Lyme disease, leprosy and Chagas disease. In: Asbury AK, Thomas PK, editors. *Peripheral Nerve Disorders*. Oxford: Butterworth-Heinemann; 1995. pp. 205–22.
6. Lockwood DN, Reid AJ. The diagnosis of leprosy is delayed in United Kingdom. *QJM* 2001;**94**:207–12.
7. Nascimento OJM. Leprosy neuropathy: clinical presentations. *Arq Neuropsiquiatr* 2013;**71**(9-B):661–6.
8. Elias Jr J, Nogueira-Barbosa MH, Feltrin LT, Furini RB, Foss NT, Marques Jr W, et al. Role of ulnar nerve sonography in leprosy neuropathy with electrophysiologic correlation. *Ultrasound Med* 2009;**28**:1201–9.
9. de Faria CR, Silva IM. Electromyographic diagnosis in leprosy. *Arq Neuropsiquiat* 1990;**48**:403–13.
10. Ooi WW, Srinivasan J. Leprosy and the peripheral nervous system: basic and clinical aspects. *Muscle Nerve* 2004;**30**:393–409.
11. Rambukkana A. Usage of signaling in neurodegeneration and regeneration of peripheral nerves in leprosy. *Prog Neurobiol* 2010;**91**:102–7.
12. Rambukkana A, Zanazzi G, Tapinos N, Salzer JL. Contact-dependent demyelination by *Mycobacteriom leprae*-induced demyelination: a model for early nerve degeneration. *Curr Opin Immunol* 2002;**16**:511–18.
13. Scollard DM. The biology of nerve injury in leprosy. *Lepr Rev* 2008;**79**:242–53.
14. Rodrigues LC, Lockwood DNJ. Leprosy now: epidemiology, progress, challenges, and research gaps. *Lancet Infect Dis* 2011;**11**: 464–70.

Part V

Neuromuscular Junction Disorders

Chapter 25

Acquired Presynaptic Neuromuscular Junction Disorders

Infant Botulism and Lambert-Eaton Myasthenic Syndrome

Ai Sakonju and Thomas O. Crawford

INFANT BOTULISM

Botulism is derived from the Latin *botulus*, for sausage. Van Ermengem in 1897 isolated the spore-forming anaerobe *Bacillus botulinum* from the spleen of a victim and the contaminated ham involved.[1] This toxigenic form of botulism is the best known, and before the advent of modern food processing, was the most common form of botulism in the developed world. There are three other forms of botulism in addition to food-borne toxin ingestion: wound, infant, "adult onset infant botulism," and iatrogenic.

It would have been impossible to recognize infant botulism in an earlier era of high infant mortality, and thus it was not recognized until 1976.[2] This form of human botulism is caused by *Clostridium botulinum* colonization within the flora of the susceptible infant colon and subsequent production of toxin. Infant botulism is now the most common form of human botulism in the developed world.[3]

Clinical Concerns

Typical patients are healthy full-term infants with painless constipation for days to months before onset of weakness. This weakness, poor feeding, a paucity of movement, and seeming lethargy are the most common initial concerns. At presentation, three cardinal features are recognized: symmetrical weakness of bulbar, face, and neck more than appendicular musculature; a nonirritable, awake sensorium; and the absence of fever. Bulbar weakness and poor head support are seen in all patients; ophthalmoparesis and pupillary sluggishness or dilatation are frequent and helpful findings.[4] The face is typically expressionless, drooling is present, and a high-pitched, mewing cry develops that is recognized as characteristic by clinicians in areas of high prevalence. Some infants have a suggestion of fatigability, manifesting as bursts of movement amid a too-immobile restful state. Muscle stretch reflexes are often absent, but their presence does not rule out the diagnosis. Many infants progress rapidly to require assisted ventilation, usually initiated out of concern for a secure airway.

Once infant botulism is recognized and treated, patients with the classic syndrome may need ventilatory assistance or airway support for many months, with a mean of 3 weeks.[5] Patients typically recover in the reverse order of their symptoms' appearance, with limb movements reappearing before the infant has a competent airway. In exceptional cases, relapse may occur well into the course of recovery.[6]

Epidemiology

The median age of affected infants is 10 weeks, with 95% being younger than 6 months old. The vast majority of patients are between 1 week and 11 months old, but rare adult cases of "infant botulism" can arise in the setting of severe gastrointestinal illness.[7–9] There are case reports of infant botulism starting with symptoms at 38 hours and 54 hours of life due to different forms of *Clostridia* such as type F.[10,11]

Although infant botulism is widespread, having been recognized in all regions of the United States[12] and four continents,[3] clusters of increased incidence in California,[13] Utah,[14] and the suburban Philadelphia region[15] are the best documented. This may be related to soil type and alkalinity. In both Pennsylvania and Utah, many patients were known to have been exposed to nearby construction

B.T. Darras, H. Royden Jones, Jr., M.M. Ryan & D.C. De Vivo (Eds): Neuromuscular Disorders of Infancy, Childhood and Adolescence, Second edition.
DOI: http://dx.doi.org/10.1016/B978-0-12-417044-5.00025-1

or agricultural soil conditions, with wind and alkaline soil noted in the Utah group.[14] There is some suggestion of seasonal variation in susceptibility, with the greatest incidence in the summer and fall when there is no snow and conditions favor spore dissemination,[16] but cases do occur year-round in more temperate climates. Whether breast feeding confers a susceptibility risk or protective benefit is controversial. In southeastern Pennsylvania, a case-controlled epidemiologic survey found that breastfeeding[15] was a major factor in relative risk. One report showing cases occurring across all major racial and ethnic groups in California also found that breastfeeding seemed to slow the onset of symptoms, and those who were breastfed had a later age of onset of symptoms (median age of onset 116 days) compared to bottle-fed infants (median age of symptom onset 66 days).[3,17]

The pathogenic *C. botulinum* organisms are ubiquitous. *C. botulinum* is comprised of four phylogenetically distinct species of bacteria that form botulinum neurotoxin, which accounts for the multiple forms of botulinum toxin designated types A through G. Infant botulism is associated almost exclusively with toxin type A or B. The distribution of human disease parallels the regional soil distribution, with type A being common in the Rocky Mountain states and type B infection in the Great Plains states and to the east.[16] Unlike the food-borne toxigenic form of human botulism, epidemic outbreaks of infant botulism are not seen. There are no reported cases of disease by fecal-oral contamination from other affected infants, and the affected infant is typically the only ill member of the family. Taken together, these facts suggest that human exposure to *C. botulinum* spores is a fairly common event, and colonization causing symptomatic weakness reflects features of individual susceptibility more than an effect of random exposure to the organism. Exposure's partial role in incidence is suggested, however, by an increased relative risk following dietary exposure to honey (odds ratio 9.8) or corn syrup (odds ratio 5.2),[18] in which spores may be found. In Europe, as many as 59% of infant botulism cases have been linked to honey with ongoing case reports.[19–21] Nonetheless, before the widely distributed American Academy of Pediatrics cautionary recommendation, exposure to honey explained only 16% to 30% of cases in the USA.[18,22]

There are hints that risk factors change with the age of the infant. In older infants, a history of long-standing infrequent stooling, breastfeeding, and honey or corn syrup exposure is correlated with disease. In infants younger than 2 months of age, only rural residence with presumed aerosolized soil exposure is a risk factor.[16,18] The fact that normal intestinal microflora changes with the introduction of formula or solid foods likely plays a role in the ability of *C. botulinum* to colonize the infant's colon. This is supported by the fact that in a majority of the Pennsylvania cases, the infants had first been introduced to non-breast milk foods within 4 weeks of the onset of disease.[23] Although breastfeeding is a risk factor for infant botulism, it appears to provide some protection against the most serious, acute presentation of sudden infant death.[24]

Given these features of epidemiology, it is hard to conceive how any program of prevention could be successful. Infant botulism is a rare disease, and the evidence suggests that host features are more substantial risk factors than exposure. Although breastfeeding is a known risk factor, the clear benefits of breastfeeding outweigh the tiny risk of contracting infant botulism. The single addressable public health factor has already been dealt with—the recommended exclusion of honey and corn syrup from the diet of infants younger than 1 year. More important is a program of education for primary practitioners who, with enhanced awareness and continued vigilance, are best situated to prevent the complications of infant botulism through early diagnosis.

Pathogenesis

Infant botulism is caused by colonization of the large intestine by toxin-forming *C. botulinum* or related organisms.[25,26] *C. botulinum* is not a normal constituent of gut microflora.[27] Apparently, its absence is due to competition from other microorganisms: adult germ-free mice support large-intestine colonization, but immediately upon exposure to other normal flora, the colonization is cleared.[28] An important exception that may have relevance for human infant botulism is seen during early postnatal development of mice: normal pups of 7 to 13 days can sustain enteric colonization, but older and younger mice cannot.[29] Interestingly, like *C. botulinum*, other *Clostridium* species are excluded from the gut of breastfed infants, but the addition of any other food supplement results in the occasional appearance of colonization.[29] This may explain the increased incidence of breastfeeding in older affected infants[23,27] who, with even a small amount of supplementation from other food sources, may become more susceptible to colonization with *C. botulinum*.[30]

Clostridia are obligatory anaerobic, Gram-positive, spore-forming rods. In general, they grow best in high pH media, although a combination of temperature, osmolarity, redox potential, the presence of food preservatives, and competing microorganisms affects growth in an interrelated fashion.[12] Each strain produces a single toxin type, although the genes for multiple toxins sometimes exist within a strain, and occasionally a single strain produces two types of toxin.[31] The *C. botulinum* species is an aggregate of multiple strains, likely the result of convergent evolutionary branches that have in common the production of botulinum toxins. Since this classification of species was developed, other *Clostridium* species have been found to produce botulinum toxin, thus demonstrating the difficulty of representing the complexity of bacterial phylogeny with a single nomenclature.[32]

Botulinum toxin is the most toxic substance known. The lethal ingested dose is estimated to be less than 1 nanogram/kg body weight.[33] Although abundant toxin is produced in the colon in infant botulism, apparently only a minute amount is absorbed. In the human infant, *C. botulinum* behaves more like the normal flora than an enteric pathogen; it does not invade the colonic mucosa, and there is no direct cytotoxic effect. The colon must therefore be a good barrier to toxin uptake, but more proximal regions of the gastrointestinal tract are likely not barriers to absorption, given the expression of symptoms with only minute toxin ingestion. The fact that many infants with symptomatic botulism have a history of chronic constipation[18] suggests that diminished intestinal motility is a cause as well as a consequence of infant botulism. These infants may be at higher risk for colonization in the first place, they may be more likely to absorb the toxin once it is made by reflux through the length of the colon and the ileocecal junction, or both possibilities may be true. Breastfeeding is associated with increased colonic motility, which may explain the later onset and less severe expression, though not the increased incidence, of infant botulism in breastfed infants.

The various exquisitely tailored botulinum toxins are likely the consequence of complex pathways of convergent evolution. Seven different serotypes have evolved, designated by the letters A through G. The botulinum toxins are similar in action to tetanus toxin, although the targeted cell differs: tetanus toxin affects Renshaw cells of the spinal cord while botulinum toxin alters the secretive exocytosis of motor neurons of the brain stem and spinal cord. Botulinum toxin is synthesized initially as a large 150-kDa protein that is autocleaved into two fragments bonded by disulfide links. The larger 100-kDa heavy-chain fragment of each of the botulinum toxins is responsible for binding to a variety of specific gangliosides that characterize the axon terminus of cholinergic motor neurons.[34] Once internalized by endosomic uptake, the light chain separates and escapes the endosome into the cytoplasm of the presynaptic terminal. Here, a zinc-dependent protease function of the light chain targets one of three proteins essential to the sequence of synaptic vesicle docking and exocytosis. These targeted proteins are enzymatically cleaved at a site distinct for each toxin type. Types A and E toxin cleave a protein known as SNAP 25. Types B, D, F, and G (as well as the short chain of the tetanus toxin) cleave a critical vesicle-associated membrane protein (VAMP, also known as synaptobrevin). Botulinum toxin type C cleaves the syntaxin molecule.[35–40] The consequence of inactivation of any one of these crucial docking proteins is an inability to release quanta of acetylcholine from the presynaptic terminal in response to an action potential.[41] The remarkable potency of botulinum toxin stems from this combination of high-affinity uptake at presynaptic terminals, mediated by the 100-kDa fragment, and highly efficient enzymatic protease activity mediated by the 50-kDa fragment.

Recovery from this intoxication requires the spontaneous degradation of the toxin and the synthesis and transport of new SNAP 25, VAMP, or syntaxin protein from the neuronal perikaryon. Morphologically, recovery is associated with new sprouting from the presynaptic terminal in the neuromuscular junction; whether this is a necessary feature of recovery is not known. While the muscle is denervated, there is spread of the acetylcholine receptor from the site of the original junction, but with recovery of synaptic quantal release, the regular arrangement of the synapse is reestablished. Subtle differences in the frequency of spontaneous quantal release of acetylcholine are apparent in the different forms of botulinum toxin.[42] With type B cleavage of VAMP only, the stimulated release of acetylcholine quanta is inhibited. In contrast, with type A cleavage of SNAP 25, both stimulated and spontaneous quantal release of acetylcholine are inhibited. This difference may be associated with more significant denervation responses in muscle affected by type A toxin, with greater alteration of resting membrane potential and frequency of abnormal spontaneous activity. These differences may be reflected in single-fiber electrophysiologic studies.[42]

Differential Diagnosis and Evaluation

The differential for subacute onset of hypotonia may be very broad, but due to the age at presentation the bias may lean towards infectious etiologies. Clinical suspicion for botulism must remain high in order to proceed toward appropriate evaluations in a timely manner. Actual clinical mimics were described by a review of 681 cases of infant botulism from 1992 to 2005 in which 32 patients (4.7%) met the clinical diagnosis of infant botulism but were not laboratory confirmed and subsequently were given a different diagnosis. Of those without laboratory confirmation, 28% had no other diagnosis and were felt to have clinical botulism. The remainder 23 patients fell into five diagnostic categories including metabolic disorders (25%, including glutaric aciduria type 1, maple syrup urine disease, Leigh's syndrome), miscellaneous (22%, including GBS variants, central demyelinating disease, cerebral infarct), SMA type 1 (16%), and infectious diseases (9%, RSV, enterovirus encephalitis)[43] (Box 25 1).

Evaluations may include cerebrospinal fluid examination, but a spinal tap may be dangerous as there may be impending respiratory compromise. Other overlapping

BOX 25.1 Infantile Botulism Differential Diagnosis

- Inborn errors of metabolism
- Peripheral neuropathies with cranial nerve involvement: Guillain-Barré syndrome variants such as Pharyngeal-cervical-brachial variant, "Miller-Fisher syndrome," or Bickerstaff encephalitis
- Poliomyelitis
- Central demyelination or infarction
- Spinal muscular atrophy type 1
- Infections: viral encephalitides, RSV infection
- Congenital myasthenic syndromes
- Organophosphate poisoning or acquired electrolyte disturbances

Abbreviation: RSV = Respiratory syncytial virus.

syndromes include electrolyte disturbances including hypo- or hypercalcemia, hypernatremia, hypo- or hyperkalemia, hypo- or hypermagnesemia, and hypoglycemia. The prospect of an inborn error of metabolism should be considered in very young infants who become lethargic after the initiation of enteral feedings, as well as in any infant newly challenged by fasting or intercurrent infection. In very young infants in whom the possibility of an inborn error is high, initial serum and urine samples should be saved for later, more definitive analysis, if indicated.

Other rare causes of weakness in previously healthy infants include organophosphate poisoning, which presents with additional signs of cholinergic excess. Poisoning with heavy metals may cause a chronic or acute flaccid state in infants. Now exceptionally rare, infantile poliomyelitis generally accompanies vaccination with live attenuated virus, although transmission from contact with another recently vaccinated infant can be a source of the virus. Vaccine-associated poliomyelitis usually involves back-mutation to a virulent state in the type 3 polio serotype. Alternatively, it can affect some infants with unsuspected immune deficiency in a more chronic fashion. The syndrome of infantile polio is quite different from that of infant botulism however, with recent diarrhea, a low fever on presentation, irritability, and aseptic meningitis evident. The weakness of infantile polio may or may not be asymmetrical during the acute phase. Another possible but unusual diagnosis is Guillain-Barré syndrome (GBS), which is exceptionally rare in infants, (see Chapter 23). GBS should be distinguished by electrophysiologic criteria and elevation of cerebrospinal fluid protein as the disease progresses.

One of the most difficult diagnoses to distinguish from infant botulism is an unrecognized heritable disorder of the neuromuscular junction, known as a congenital myasthenic syndrome. These infants present with increased weakness precipitated by some form of metabolic stress. Like infants with botulism, such infants characteristically have ocular involvement (though rarely pupillary sluggishness) and bulbar weakness greater than limb weakness. The use of edrophonium in the diagnosis can be confusing because infants with botulism may be transiently stronger. With the congenital myasthenic syndromes, the history is often one of a more indolent process, though this can overlap with milder, chronic forms of infant botulism. The two disorders are best distinguished by electrophysiologic studies and recovery of toxin or organism from the stool.

Nearly all infants suspected of having infant botulism must first be considered to have a serious systemic or central nervous system infection; appropriate cultures should be obtained, and prophylactic antibiotics administered. Early laboratory tests also need to include determination of glucose, electrolytes, calcium, magnesium, acid-base status, anion gap, and ammonia levels. Intravenous glucose can be started, but enteral feedings should be held pending results. Plasma and urine specimens for toxicology and metabolic studies must be obtained in case their evaluation is warranted by later developments.

The diagnosis can be made most rapidly by electrophysiologic studies. In the most carefully documented study, an incremental response to rapid repetitive stimulation (20 or 50 Hz), found in 92% of patients, was the most sensitive and specific finding (Figure 25.1). Slower rates of repetitive stimulation produce variable results.[44] Concentric needle electromyography is also sensitive but less specific, with many patients demonstrating short-duration, low-amplitude motor unit potentials and abnormal spontaneous activity.[4,44] Several technical problems can undermine the sensitivity of the conventional rapid repetitive stimulation test for incremental compound muscle action potential (CMAP) amplitude,[45] including inadequate immobilization of the limb, studying a nerve-muscle combination that has a very low CMAP amplitude below a physiologic threshold allowing increment, or studying a muscle that already has a normal CMAP amplitude and cannot increment further. One frequent error is to unwittingly study a nerve-muscle combination that has been transiently incremented by the baby's exercising the muscle during the set-up for rapid repetitive stimulation. Concentric needle electromyographic studies demonstrate frequent brief, small-amplitude, polyphasic, voluntary motor unit potentials reflecting denervation of a portion of the array of muscle fibers innervated by each motor axon. This finding, along with the frequent presence of abnormal spontaneous fibrillation and positive sharp-wave potentials, is also seen in other disorders of infants and is not specific for infant botulism.

Studies of jitter with a stimulated single-fiber electrode may have enhanced sensitivity over conventional rapid repetitive stimulation. This method is technically easy in weak infants and requires a much lower-level stimulus, making it less painful. A positive result is

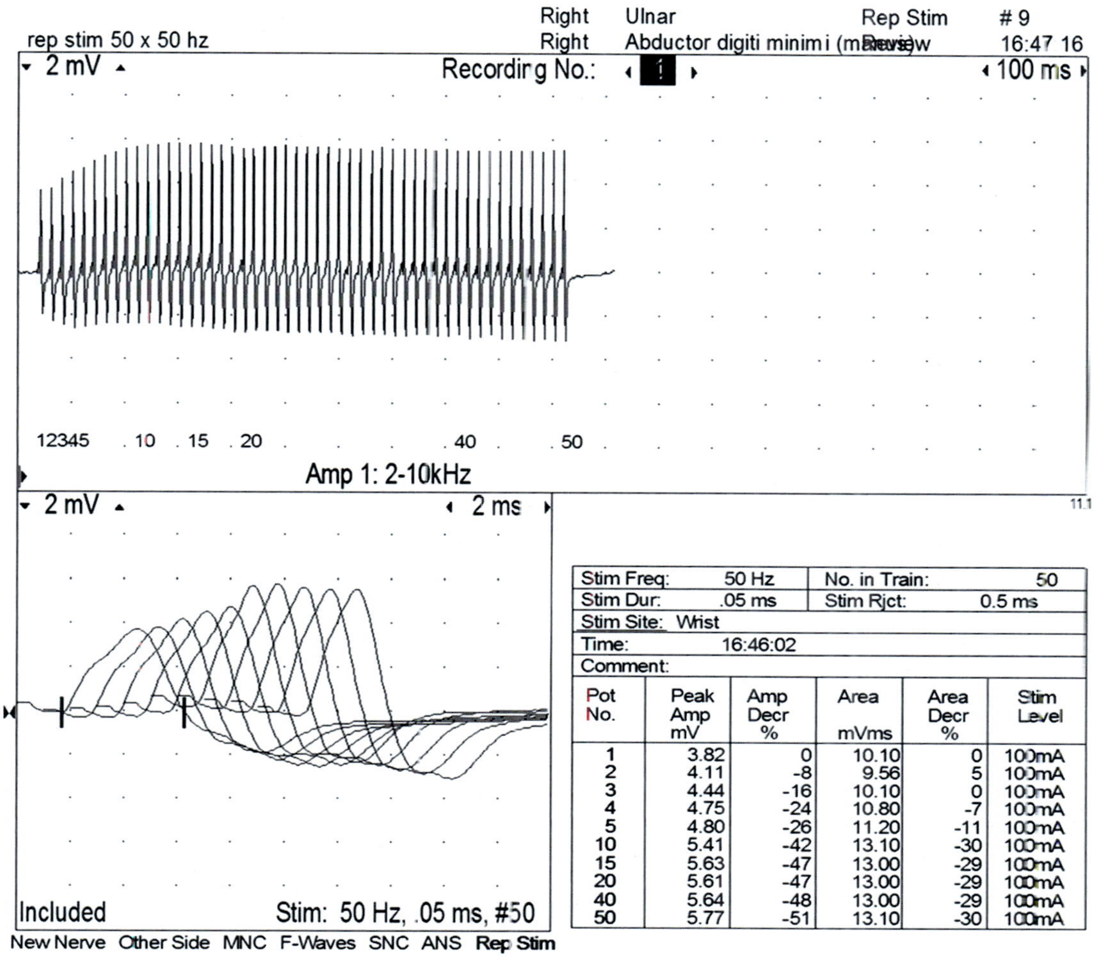

FIGURE 25.1 Tracing derived from first author's patient with confirmed type A infantile botulism. Recording of the right abductor digiti minimi (ADM) muscle with 50 Hz repetitive nerve stimulation shows >50% facilitation of the baseline compound muscle action potential (CMAP).

reported when abnormal jitter between the stimulus artifact and an isolated single muscle fiber potential decreases with an increasing stimulation rate, similar to that expected with Lambert-Eaton myasthenic syndrome. Autoimmune myasthenia and other postsynaptic disorders would be expected to have increasing jitter with an increased stimulation rate.

Most infants have both toxin and organism recoverable from the stool, although local logistical concerns may favor one test over the other. Because affected babies have a reduced stooling frequency, the fluid recovered from a gentle 20-mL saline enema contains enough material to test for the toxin. The blood is usually, but not always, negative for toxin. Stool can remain positive for culture and toxin for a long time, though others report that it is clear within 1 month of diagnosis.[46] Polymerase chain reaction techniques have been developed[47] that may be incrementally more sensitive but are not yet in routine use.

Therapy

Prior to the introduction of human botulism immune globulin (Baby-BIG), approved by the Food and Drug Administration in October 2003, treatment was largely supportive with respiratory and nutritional care. Baby-BIG is comprised of immunoglobulin isolated from donor plasma immunized with pentavalent botulinum toxoid (ABCDE).[48] A 5-year randomized placebo-controlled trial treating with Baby-BIG vs. IVIg showed a significant reduction in morbidity with reduction in length of hospital stay and duration of intubation by a mean of 3 weeks. This study was followed by another 5-year open-label study and a 30-year retrospective review[49] that recapitulated findings. Mean intensive care stay was reduced by 3.2 weeks, mechanical ventilation by 2.6 weeks, and tube feeding by 6.4 weeks. This resulted in significant cost reduction as well. Although the cost of BIG is high (approximately $45,000 in 2005), the overall savings in hospital stay and morbidity were far more significant with an estimated $34.2 million in hospital charges that were avoided.[48] BIG has been demonstrated to be of benefit in infant botulism when given within the first 3 days after admission to the hospital.[50] Equine-derived botulinum antitoxin is not used in infants because of the high incidence of rash and anaphylaxis and the potential for lifelong sensitization to horse serum products. A Cochrane review also supports BIG as the only medical treatment that showed significant effect compared to 3,4-diaminopyridine (3,4 DAP), plasma exchange, equine antitoxin, and guanidine.[51]

Baby BIG is available worldwide by the California Department of Health Services Infant Botulism Prevention Program (telephone, 1-510-231-7600; up to date as of 8/2014). Because of the necessity for quick treatment, BIG can be administered on the strength of a classic history and physical examination (details at the California Department of Health Infant Botulism Prevention and Treatment Program website[52]), with laboratory confirmation of the diagnosis and botulinum toxin type to follow.

Other than administration of BIG, primary management consists of respiratory and nutritional management. At-risk infants should be watched carefully in the intensive care unit and intubated early when signs of airway compromise, bulbar dysfunction, or respiratory insufficiency arise. Infants who require an artificial airway do so for weeks and may acquire a respiratory virus that can prolong their ventilator requirement. Most now agree that despite this long-term requirement, tracheostomy is rarely indicated[5] because normal lung compliance and a weak chest wall permit full ventilation even with a slightly small, uncuffed tube that has some degree of air leak. Under such conditions, few patients develop airway strictures, even with long periods of intubation.[53]

Because infants often recover in the reverse order of the original progression, the clinician may be faced with the odd circumstance of an infant intubated for weakness who has good arm and leg muscle power. Some have advocated the return of head stability as a predictor of adequate airway competence.[54] Others wait for the maximum inspiratory pressure to exceed -25 cm H_2O and the return of protective airway reflexes.[5] In either case, it is just as important to follow the infant's airway after extubation, during recovery from the disease, as it is when the child first presents. Throughout the illness, the capacity for enhanced fatigue of airway and respiratory muscles is a greater concern than is baseline weakness.

Given the severe constipation, concern about enteral feedings led many to give infants parenteral alimentation in the past. Many infants do not tolerate bolus feedings by nasogastric tube, but most can be adequately nourished by continuous nasogastric feedings, beginning slowly, as early as 3 days after presentation.[5] Concern about gastroesophageal reflux with aspiration has prompted some to use the nasojejunal route for tube feeding.[55] Caloric requirements in the absence of physical activity are significantly reduced.

Many infants experience altered autonomic function, including sudden unexplained alterations in heart rate, blood pressure, or skin color. These are usually mild, transient, or both and rarely require specific therapy.[15] Some of these changes are undoubtedly due to supersensitivity of cholinergic autonomic terminals after functional denervation by the toxin. Urinary retention may require intermittent catheterization to prevent bladder infections. Some very young infants develop the syndrome of inappropriate antidiuretic hormone secretion, especially early in the disease, just after being placed on positive-pressure ventilators. This should respond to fluid restriction and usually resolves spontaneously over several days to a week. Left unrecognized or untreated, the low sodium values that result can cause serious, anticonvulsant-refractory seizures.

Although many infants are given antibiotics immediately upon presentation because of the possibility of unrecognized sepsis, further treatment with antibiotics has not been successful and might be harmful, either by perpetuating *C. botulinum* colonization of the gut with the alteration of normal enteric flora, or by increasing the release of botulinum toxin, which is in high concentration within the organism. In particular, treatment with aminoglycoside antibiotics may be contraindicated because of the potential for increased neuromuscular blockade.[56] Giving specific foods such as yogurt in an attempt to alter the normal colonic flora has not been tried systematically, but the results with individual patients have not been striking. Although colonic enemas may theoretically reduce the enteric burden of toxin, there is concern that they may also increase absorption. Because recovery from infant botulism is very good once the diagnosis is made and appropriate supportive care is initiated, there is little enthusiasm for these treatments that may carry some risk. Treatment with acetylcholine esterase blockers produces mixed results.

Although some infants may worsen during convalescence, relapse after full recovery is not seen.[6] Some infants may harbor toxin in the colon well after recovery, suggesting that some degree of circulating humeral antibody aids in absorption of residual toxin taken up systemically.[13,23]

Very importantly, the parents need to be assured of the probable good outcome for their infant, despite the prospect of a prolonged intensive care hospitalization. Parents should be encouraged early on to establish a regular routine that includes time away from the hospital. This will enable them to better tolerate the personal stress associated with a prolonged intensive care stay.

LAMBERT-EATON MYASTHENIC SYNDROME

An acquired immune-mediated presynaptic neuromuscular junction disorder, Lambert-Eaton myasthenic syndrome (LEMS) is typically divided into two forms based on whether there is an underlying neoplasm or not; that is, paraneoplastic and nonparaneoplastic, respectively. In children, the associated neoplasms have included leukemia, neuroblastoma and Wilms' tumor, whereas in adults it is classically associated with small cell lung cancer as first described in 1956.[57] Transient neonatal LEMS has been described as well and may be due to placental crossover of maternal antibodies causing hypotonia at birth lasting only for a few weeks and requiring only supportive care.[58] Adult LEMS occurs in approximately 0.5 per million.[58] Although the incidence is unknown, pediatric LEMS is much rarer than infantile botulism, with less than two dozen LEMS cases reported in the literature to date. The classic syndrome occurs in adults as an immune-mediated disorder associated with P/Q-type voltage-gated calcium channel antibodies found in 85–90% of patients.[59]

Clinical Concerns

LEMS presents as a clinical triad of proximal muscle weakness, hyporeflexia, and autonomic symptoms. This is essentially confirmed in a review of nearly a dozen reported cases of pediatric LEMS: progressive proximal muscle weakness was the initial symptom in the majority, and ptosis in 3 of 12 reported children.[60] Ptosis may be a potentially serious sign since, in this review, the three patients with ptosis were found to have malignancy. Other initial presenting features included ophthalmoparesis, bulbar weakness with dysarthria, dysphagia, fatigable weakness, dysautonomia with constipation, dry mouth, and nonspecific pain including muscle aches.[60] In adults, approximately 62% had extraocular or bulbar symptoms.[61] Autonomic symptoms at onset may be missed in young patients with limited ability to express these types of symptoms.

On examination there may be signs of proximal muscle weakness including waddling gait, facial weakness, and eye movement abnormalities or ptosis. In particular, a characteristic feature is the potentiation of muscle stretch reflexes after maximal voluntary contraction in proximally weak patients with hypoactive or absent reflexes at baseline. One technique is to extend the knee against resistance for about 10 seconds and then test the patellar reflex which should be amplified in the classic patient. There may also be amplification of strength in a patient after brief maximal voluntary contraction that can be misinterpreted as "functional weakness," suggesting the incorrect diagnosis of conversion disorder.

Diagnostic Evaluations

The differential diagnosis for proximal muscle weakness in children is broad. The two main immune mediated diseases that need to be distinguished from LEMS include GBS and myasthenia gravis. In contrast to LEMS, GBS will have associated sensory symptoms and distal involvement. Also, with myasthenia gravis, the typical progression is cranial to caudal, whereas LEMS typically presents in a caudal to cranial fashion with lower extremity weakness initially. Further, acetylcholine receptor antibodies can be obtained to help distinguish autoimmune myasthenia gravis from LEMS.

Even rarer than autoimmune myasthenia gravis are the congenital myasthenic syndromes with presynaptic defects. In general, onset of symptoms spans from infancy to childhood and diagnosis requires specific genetic testing as well as electrophysiologic studies, the results of which may be hard to distinguish from acquired LEMS. These patients may present similarly to myopathy patients with severe hypotonia as their predominant feature, and without sophisticated electrophysiologic and other studies, as described in Chapter 26, they may be misdiagnosed.

LEMS patients can also present similarly to patients with congenital myopathy with normal CK. The combination of normal CK and low CMAPs in a patient with proximal weakness and no significant muscle atrophy should prompt neuromuscular junction and exercise testing, and thus may prevent an unnecessary muscle biopsy.

However, there is no single test alone that confirms the diagnosis and thus LEMS may be under-reported in children. The reliability of elevated voltage-gated calcium channel (VGCC) antibody testing is unknown in pediatric LEMS. In one review, 5 of 11 pediatric patients underwent testing and 4 of them were positive for VGCC antibodies.[62]

Electrodiagnostic testing is usually helpful but the diagnosis must be suspected in order to obtain the best results and increase the sensitivity of the procedure. Early on, testing may be normal in individuals with LEMS, but on repeat nerve conduction testing the baseline CMAPs may become lower with disease progression. Once low CMAPs are obtained, the differential diagnosis generally includes neuropathy and, in the right clinical setting, a neuromuscular junction defect may be suspected; at this point, specialized exercise testing, repetitive nerve conduction studies, and sometimes single fiber EMG must be requested, as they are not always part of standard EMG testing. Single-fiber electromyography can be done to detect a neuromuscular transmission defect that may be technically difficult to detect in very young children. This can be done using stimulated single fiber EMG as described in the botulism section of this chapter.

Most importantly, a characteristic feature associated with LEMS is exercise testing when reduced CMAPs are

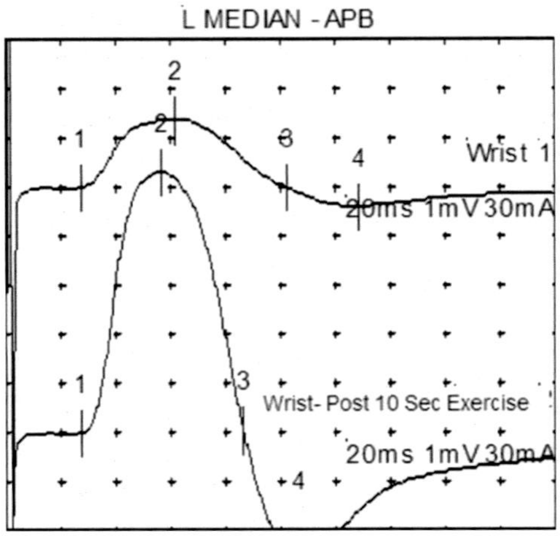

FIGURE 25.2 Left median motor nerve compound muscle action potentials (CMAPs) recorded over the abductor pollicis brevis (APB) muscle. Top trace shows baseline CMAP. Bottom trace shows 276% facilitation of the CMAP 10 seconds after maximal voluntary exercise. *Reprinted with permission from Elsevier: Hajjar et al. (2014).*[60]

obtained at baseline. Supportive of LEMS diagnosis is the finding of post-exercise CMAP amplitude increase (facilitation) of >25% from baseline. A baseline CMAP is obtained and then the patient is asked to provide 10 seconds of maximum voluntary contraction of the respective muscle. In LEMS, a repeat CMAP soon after the exercise should reveal significant facilitation (Figure 25.2). More specific testing is obtained next with repetitive motor nerve stimulation.

Fast and slow repetitive nerve stimulation will both be abnormal in LEMS patients; however, fast rates or high-frequency stimulation at >10 Hz proves to be the diagnostic test of choice. Repetitive stimulation at low rates such as 3 Hz is usually obtained first and may show a >10% decrement similar to myasthenia gravis patients, thereby confirming a neuromuscular transmission defect. However, the important finding as mentioned above is the increment with high-frequency stimulation at 20–50 Hz after 10 to 20 seconds of brief electrical 10 Hz stimulation to electrically exercise the muscle or less painfully in the cooperative patient by exercising the muscle being examined. An increment of >60% is diagnostic of LEMS with 97% sensitivity and 99% specificity, and excludes myasthenia gravis.[63]

Facilitation of at least 25% may be acceptable for a diagnosis of LEMS according to the American Association of Electrodiagnostic Medicine.[64] In 11 of 11 children who underwent low-frequency repetitive nerve stimulation (1–3 Hz), there was a >10% decrement indicating a neuromuscular junction defect.[60] High-frequency (10–50 Hz) repetitive nerve stimulation in 8 of 8 pediatric LEMS patients resulted in an increment of 124% to 500% from baseline with the same number also showing 60% to 1000% facilitation after 10 to 30 seconds maximal voluntary effort.[59] The increment may be similar to the one seen in previously described infantile botulism because both are presynaptic disorders of the neuromuscular junction. The distinction between the two is based primarily on the clinical presentation and the age of patients with these conditions.

Needle EMG reveals small myopathic motor units without fibrillation potentials. This may help distinguish patients with LEMS from a neurogenic disorder such as spinal muscular atrophy. Muscle biopsy has been done in a few patients and is generally nondiagnostic with or without nonspecific findings.

Overall, when the clinical presentation is associated with the presence of anti-VGCC antibodies and post-exercise CMAP amplitude facilitation along with repetitive motor nerve stimulation findings as described above, the diagnosis of LEMS is almost definite.

Pathogenesis

There are several lines of evidence that LEMS is an autoimmune disease mediated by anti-VGCC antibodies. Passive transfer of patients' immunoglobulin induces similar pathophysiology in mice.[64] Patients have improved clinical response to plasma exchange[65] and up to 90% of patients with LEMS have detectable P/Q type anti-VGCC antibodies.[66] Pediatric LEMS was first described in 1974, associated with childhood leukemia, but the first pediatric patients with a positive VGCC antibody titer were initially reported in 2002.[67,68] Muscle biopsies are occasionally performed in LEMS patients, but unless specific neuromuscular junction studies are done on a research basis, no clear pathology is seen.

The putative mechanism for aberrant presynaptic neurotransmission is due to blockage of acetylcholine release into the synapse by interference via voltage-gated calcium channel antibodies, resulting in reduced quantal release of neurotransmitter detected by low-frequency repetitive stimulation, which in turn results in a CMAP decrement. Higher rates of electrical stimulation or maximal voluntary contraction essentially induce release of acetylcholine into the neuromuscular junction and an incremental electrical response. This incremental response after high-frequency stimulation is seen in botulism as well and is a characteristic feature of presynaptic neuromuscular junction disorders.

Therapy

The initial treatment that was used was guanidine,[69] which increases neurotransmitter release at the neuromuscular junction but was retracted due to the risk for bone marrow suppression and renal failure. In the following decade, a similarly acting compound 4-AP (4-aminopyridine) that increases acetylcholine release via voltage gated potassium channels was used to treat patients with LEMS but was limited due to the increased risk for seizures. A related compound that crosses the blood-brain barrier less readily is 3,4 DAP, which is now considered the treatment of choice. Few randomized placebo controlled trials (RCTs) have been done, partly due to the rarity of the disease. A recent Cochrane review described five RCTs, four of which included 3,4 DAP and one, IVIg.[70] Overall, these trials showed that muscle strength was improved after 3,4 DAP administration by testing myometry and other scores of muscle strength and by repeating CMAP testing after medication was administered.[71–75] IVIg dosing 2 g/kg over 2 days in the single RCT also resulted in improved muscle strength testing by myometry.[76] Muscle strength scores and CMAP amplitudes showed improvement in both groups, 3,4 DAP and IVIg, over a period of days or for up to 8 weeks, respectively. Cost-benefit analysis of using IVIg or 3,4 DAP depended on availability, and both may be costly therapies.

Pyridostigmine is also often used in conjunction with 3,4 DAP for symptomatic treatment. No RCTs have been done involving oral immunosuppressive medications such as prednisolone, azathioprine, or cyclosporine, although there are case reports describing good results in pediatric patients with long-term follow-up.[60,77,78]

In those with paraneoplastic syndrome, treatment of the malignancy generally results in modification if not resolution of LEMS symptoms. Therefore, once the diagnosis is confirmed, the next step for the patient is to undergo rigorous and periodic surveillance screening for malignancy for at least 2 years as suggested by the adult literature.[79] However, in children, prognosis may be better than adult onset LEMS; the prevalence of neoplasm is less frequent in childhood LEMS where only 3 of 12 reported patients in the literature were found to have a neoplasm (lymphoproliferative disorders, Wilms' tumor, neuroblastoma), compared to 60% of adult LEMS patients.[60,79,80]

SUMMARY

As LEMS is a treatable disorder with potential for an occult neoplasm and debilitating progression, this disorder should remain a diagnostic consideration in patients with proximal muscle weakness. It remains a challenging disease to diagnose but patients may display significant improvement in quality of life that makes this disorder worth understanding.

REFERENCES

1. Van Ermengem E. Ueber einen neuen anaeroben Bacillus und seine Beziehungen zum Botulismus. *Z Hyg Infekt* 1897;**26**:1–56.
2. Pickett J, Berg B, Chaplin E, Brunstetter-Shafer MA. Syndrome of botulism in infancy: clinical and electrophysiologic study. *N Engl J Med* 1976;**295**:770–2.
3. Arnon SS. Infant botulism: anticipating the second decade. *J Infect Dis* 1986;**154**:201–6.
4. Clay SA, Ramseyer JC, Fishman LS, Sedgwick RP. Acute infantile motor unit disorder. Infantile botulism? *Arch Neurol* 1977;**34**:236–43.
5. Schreiner MS, Field E, Ruddy R. Infant botulism: a review of 12 years' experience at the Children's Hospital of Philadelphia. *Pediatrics* 1991;**87**:159–65.
6. Glauser TA, Maguire HC, Sladky JT. Relapse of infant botulism. *Ann Neurol* 1990;**28**:187–9.
7. Bartlett JC. Infant botulism in adults. *N Engl J Med* 1986;**315**:254–5.
8. Griffin PM, Hatheway CL, Rosenbaum RB, Sokolow R. Endogenous antibody production to botulinum toxin in an adult with intestinal colonization botulism and underlying Crohn's disease. *J Infect Dis* 1997;**175**:633–7.
9. McCroskey LM, Hatheway CL. Laboratory findings in four cases of adult botulism suggest colonization of the intestinal tract. *J Clin Microbiol* 1988;**26**:1052–4.
10. Barash JR, Tang TW, Arnon SS. First case of infant botulism caused by *Clostridium baratii* type F in California. *J Clin Microbiol* 2005;**43**:4280–2.
11. Keet CA, Fox CK, Margeta M, Marco E, Shane AL, Dearmond SJ, et al. Infant botulism, type F, presenting at 54 hours of life. *Pediatr Neurol* 2005;**32**:193–6.
12. Centers for Disease Control and Prevention (CDC). *Botulism in the United States, 1899–1996. Handbook for Epidemiologists, Clinicans and Laboratory Workers.* Atlanta, GA: CDC; 1998.
13. Arnon SS, Midura TF, Clay SA, Wood RM, Chin J. Infant botulism. Epidemiological, clinical, and laboratory aspects. *JAMA* 1977;**237**:1946–51.
14. Thompson JA, Glasgow LA, Warpinski JR, Olson C. Infant botulism: clinical spectrum and epidemiology. *Pediatrics* 1980;**66**:936–42.
15. Long SS. Epidemiologic study of infant botulism in Pennsylvania: report of the Infant Botulism Study group. *Pediatrics* 1985;**75**:928–34.
16. Arnon SS, Damus K, Chin J. Infant botulism: epidemiology and relation to sudden infant death syndrome. *Epidemiol Rev* 1981;**3**:45–66
17. Morris Jr. JG, Snyder JD, Wilson R, Feldman RA. Infant botulism in the United States: an epidemiologic study of cases occurring outside of California. *Am J Public Health* 1983;**73** 1385–8.
18. Spika JS, Shaffer N, Hargrett-Bean N, Collin S, MacDonald KL, Blake PA. Risk factors for infant botulism in the United States. *Am J Dis Child* 1989;**143**:828–32.
19. Aureli P, Franciosa G, Fenicia L. Infant botulism and honey in Europe: a commentary. *Pediatr Infect Dis J* 2002;**21**:866–8.
20. Hoarau G, Pelloux I, Gayot A, Wroblewski I, Popoff MR, Mazuet C, et al. Two cases of type a infant botulism in Grenoble, France: no honey for infants. *Eur J Pediatr* 2012;**171** 589–91.
21. Ramroop S, Williams B, Vora S, Moshal K. Infant botulism and botulism immune globulin in the UK: a case series of four infants. *Arch Dis Child* 2012;**97**:459–60.
22. Arnon SS, Midura TF, Damus K, Thompson B, Wood RM, Chin J. Honey and other environmental risk factors for infant botulism. *J Pediatr* 1979;**94**:331–6.
23. Long SS, Gajewski JL, Brown LW, Gilligan PH. Clinical, laboratory, and environmental features of infant botulism in Southeastern Pennsylvania. *Pediatrics* 1985;**75**:935–41.

24. Arnon SS, Damus K, Thompson B, Midura TF, Chin J. Protective role of human milk against sudden death from infant botulism. *J Pediatr* 1982;**100**:568–73.
25. Mills DC, Arnon SS. The large intestine as the site of *Clostridium botulinum* colonization in human infant botulism. *J Infect Dis* 1987;**156**:997–8.
26. Suen JC, Hatheway CL, Steigerwalt AG, Brenner DJ. Genetic confirmation of identities of neurotoxigenic *Clostridium baratii* and *Clostridium butyricum* implicated as agents of infant botulism. *J Clin Microbiol* 1988;**26**:2191–2.
27. Arnon SS. Infant botulism. In: Borriello SP, editor. Clostridia *in Gastrointestinal Disease*. Boca Raton, FL: CRC Press; 1985. pp. 40–55.
28. Burr DH, Sugiyama H. Susceptibility to enteric botulinum colonization of antibiotic-treated adult mice. *Infect Immun* 1982;**36**: 103–6.
29. George WL, Finegold SM. *Clostridia* in the human gastrointestinal flora. In: Borriello SP, editor. Clostridia *in gastrointestinal disease*. Boca Raton, FL: CRC Press; 1985. pp. 2–37.
30. Stark PL, Lee A. The microbial ecology of the large bowel of breast-fed and formula-fed infants during the first year of life. *J Med Microbiol* 1982;**15**:189–203.
31. Franciosa G, Ferreira JL, Hatheway CL. Detection of type A, B, and E botulism neurotoxin genes in *Clostridium botulinum* and other *Clostridium* species by PCR: evidence of unexpressed type B toxin genes in type A toxigenic organisms. *J Clin Microbiol* 1994;**32**:1911–17.
32. Arnon SS. Botulism as an intestinal toxemia. In: Blaser MJ, Smith PD, Greenberg HB, Guerrant RL, Ravdin JI, editors. *Infections of the Gastrointestinal Tract*. New York: Raven Press; 1995. pp. 257–71.
33. Morton HE. *The toxicity of* Clostridium botulinum *type A toxin for various species of animals, including man*. Philadelphia: The Institute for Cooperative Research, University of Pennsylvania; 1961.
34. Black JD, Dolly JO. Interaction of 125I-labeled botulinum neurotoxins with nerve terminals. I. Ultrastructural autoradiographic localization and quantitation of distinct membrane acceptors for types A and B on motor nerves. *J Cell Biol* 1986;**103**:521–34.
35. Huttner WB. Cell biology. Snappy exocytoxins. *Nature* 1993;**365**: 104–5.
36. Schiavo G, Benfenati F, Poulain B, Rossetto O, de Laureto PP, DasGupta BR, et al. Tetanus and botulinum-B neurotoxins block neurotransmitter release by proteolytic cleavage of synaptobrevin. *Nature* 1992;**359**:832–5.
37. Blasi J, Chapman ER, Link E, Binz T, Yamasaki S, De Camilli P, et al. Botulinum neurotoxin A selectively cleaves the synaptic protein SNAP-25. *Nature* 1993;**365**:160–3.
38. Blasi J, Chapman ER, Yamasaki S, Binz T, Niemann H, Jahn R. Botulinum neurotoxin C1 blocks neurotransmitter release by means of cleaving HPC-1/syntaxin. *EMBO J* 1993;**12**:4821–8.
39. Schiavo G, Rossetto O, Catsicas S, de Laureto PP, DasGupta BR, Benfenati F. Identification of the nerve terminal targets of botulinum neurotoxin serotypes A, D, and E. *J Biol Chem* 1993;**268**: 23784–7.
40. Schiavo G, Shone CC, Rossetto O, Alexander FC, Montecucco C. Botulinum neurotoxin serotype F is a zinc endopeptidase specific for VAMP/synaptobrevin. *J Biol Chem* 1993;**268**:11516–19.
41. Simpson LL. Molecular pharmacology of botulinum toxin and tetanus toxin. *Annu Rev Pharmacol Toxicol* 1986;**26**:427–53.
42. Mandler RN, Maselli RA. Stimulated single-fiber electromyography in wound botulism. *Muscle Nerve* 1996;**19**:1171–3.
43. Francisco AM, Arnon SS. Clinical mimics of infant botulism. *Pediatrics* 2007;**119**:826–8.
44. Cornblath DR, Sladky JT, Sumner AJ. Clinical electrophysiology of infantile botulism. *Muscle Nerve* 1983;**6**:448–52.
45. Graf WD, Hays RM, Astley SJ, Mendelman PM. Electrodiagnosis reliability in the diagnosis of infant botulism. *J Pediatr* 1992;**120**:747–9.
46. Hatheway CL, McCroskey LM. Examination of feces and serum for diagnosis of infant botulism in 336 patients. *J Clin Microbiol* 1987;**25**:2334–8.
47. Szabo EA, Pemberton JM, Gibson AM, Eyles MJ, Desmarchelier PM. Polymerase chain reaction for detection of *Clostridium botulinum* types A, B and E in food, soil and infant faeces. *J Appl Bacteriol* 1994;**76**:539–45.
48. California Department of Health Services. *Summary basis of approval: botulism immune globulin intravenous (human) (BIG-IV)*. California: California Department of Health Services; 2003.
49. Underwood K, Rubin S, Deakers T, Newth C. Infant botulism: a 30-year experience spanning the introduction of botulism immune globulin intravenous in the intensive care unit at Childrens Hospital Los Angeles. *Pediatrics* 2007;**120**:e1380–5.
50. Arnon SS. Infant botulism. In: Feigen RD, Cherry JD, editors. *Textbook of Pediatric Infectious Diseases*. 4th ed. Philadelphia: Saunders; 1998. pp. 1570–7.
51. Chalk C, Benstead TJ, Keezer M. Medical treatment for botulism. *Cochrane Database Syst Rev* 2011;**2**:CD008123.
52. Infant Botulism Prevention and Treatment Program [Internet]. California Department of Health. [cited 3/10/2014]. Available at <http://www.infantbotulism.org/>.
53. Wohl DL, Tucker JA. Infant botulism: considerations for airway management. *Laryngoscope* 1992;**102**:1251–4.
54. L'Hommedieu C, Polin RA. Progression of clinical signs in severe infant botulism. Therapeutic implications. *Clin Pediatr (Phila)* 1981;**20**:90–5.
55. L'Hommedieu C, Stough R, Brown L, Kettrick R, Polin R. Infant botulism in the age of botulism immune globulin. *Neurology* 2005;**64**:2029–32.
56. L'Hommedieu C, Stough R, Brown L, Kettrick R, Polin R. Potentiation of neuromuscular weakness in infant botulism by aminoglycosides. *J Pediatr* 1979;**95**:1065–70.
57. Lambert EH, Eaton LM, Rooke ED. Defect of neuromuscular conduction associated with malignant neoplasms. *Am J Physiol* 1956;**187**:612–13.
58. Reuner U, Kamin G, Ramantani G, Reichmann H, Dinger J. Transient neonatal lambert-eaton syndrome. *J Neurol* 2008;**255**: 1827–8.
59. Titulaer MJ, Lang B, Verschuuren JJ. Lambert-Eaton myasthenic syndrome: From clinical characteristics to therapeutic strategies. *Lancet Neurol* 2011;**10**:1098–107.
60. Hajjar M, Markowitz J, Darras BT, Kissel JT, Srinivasan J, Jones HR. Lambert-Eaton syndrome, an unrecognized treatable pediatric neuromuscular disorder: three patients and literature review. *Pediatr Neurol* 2014;**50**:11–17.
61. O'Neill JH, Murray NM, Newsom-Davis J. The Lambert-Eaton myasthenic syndrome. *A review of 50 cases. Brain* 1988; **111**(Pt 3):577–96.

62. Morgan-Followell B, de Los Reyes E. Child neurology: diagnosis of Lambert-Eaton myasthenic syndrome in children. *Neurology* 2013;**80**:e220–2.
63. Oh SJ, Kurokawa K, Claussen GC, Ryan Jr. HF. Electrophysiological diagnostic criteria of Lambert-Eaton myasthenic syndrome. *Muscle Nerve* 2005;**32**:515–20.
64. AAEM Quality Assurance Committee. American Association of Electrodiagnostic Medicine. Literature review of the usefulness of repetitive nerve stimulation and single fiber EMG in the electrodiagnostic evaluation of patients with suspected myasthenia gravis or Lambert-Eaton myasthenic syndrome. *Muscle Nerve* 2001;**24**:1239–47.
65. Lang B, Newsom-Davis J, Wray D, Vincent A, Murray N. Autoimmune aetiology for myasthenic (Eaton-Lambert) syndrome. *Lancet* 1981;**2**:224–6.
66. Motomura M, Lang B, Johnston I, Palace J, Vincent A, Newsom-Davis J. Incidence of serum anti-P/O-type and anti-N-type calcium channel autoantibodies in the Lambert-Eaton myasthenic syndrome. *J Neurol Sci* 1997;**147**:35–42.
67. Shapira Y, Cividalli G, Szabo G, Rozin R, Russell A. A myasthenic syndrome in childhood leukemia. *Dev Med Child Neurol* 1974;**16**:668–71.
68. Tsao CY, Mendell JR, Friemer ML, Kissel JT. Lambert-Eaton myasthenic syndrome in children. *J Child Neurol* 2002;**17**:74–6.
69. Lambert EH. Defects of neuromuscular transmission in syndromes other than myasthenia gravis. *Ann NY Acad Sci* 1966;**135**:367–84.
70. Keogh M, Sedehizadeh S, Maddison P. Treatment for Lambert-Eaton myasthenic syndrome. *Cochrane Database Syst Rev* 2011 **2**:CD003279.
71. McEvoy KE, Windebank AJ, Daube JR. 3,4-diaminopyridine in Lambert-Eaton myasthenic syndrome. *Ann Neurol* 1988;**24**:122.
72. McEvoy KM, Windebank AJ, Daube JR, Low PA. 3 4-Diaminopyridine in the treatment of Lambert-Eaton myasthenic syndrome. *N Engl J Med* 1989;**321**:1567–71.
73. Oh SJ, Claussen GG, Hatanaka Y, Morgan MB. 3,4-Diaminopyridine is more effective than placebo in a randomized, double-blind, cross-over drug study in LEMS. *Muscle Nerve* 2009;**40**:795–800.
74. Sanders DB, Massey JM, Sanders LL, Edwards LJ. A randomized trial of 3,4-diaminopyridine in Lambert-Eaton myasthenic syndrome. *Neurology* 2000;**54**:603–7.
75. Wirtz PW, Verschuuren JJ, van Dijk JG, de Kam ML, Schoemaker RC, van Hasselt JG, et al. Efficacy of 3,4-diaminopyridine and pyridostigmine in the treatment of Lambert-Eaton myasthenic syndrome: a randomized, double-blind, placebo-controlled, crossover study. *Clin Pharmacol Ther* 2009;**86**:44–8.
76. Bain PG, Motomura M, Newsom-Davis J, Misbah SA, Chapel HM, Lee ML, et al. Effects of intravenous immunoglobulin on muscle weakness and calcium-channel autoantibodies in the Lambert-Eaton myasthenic syndrome. *Neurology* 1996;**47**:678–83.
77. Kostera-Pruszczyk A, Ryniewicz B, Rowinska-Marcinska K, Dutkiewicz M, Kaminska A. Lambert-Eaton myasthenic syndrome in childhood. *Eur J Paediatr Neurol* 2009;**13**:194–6.
78. Portaro S, Parisi D, Polizzi A, Ruggieri M, Andreetta F, Bernasconi P, et al. Long-term follow-up in infantile-onset Lambert-Eaton Myasthenic Syndrome. *J Child Neurol* 2013. [Epub ahead of press].
79. Titulaer MJ, Soffietti R, Dalmau J, Gilhus NE, Giometto B, Graus F, et al. Screening for tumours in paraneoplastic syndromes: report of an EFNS task force. *Eur J Neurol* 2011;**18** 19–e3.
80. Wirtz PW, Smallegange TM, Wintzen AR, Verschuuren JJ. Differences in clinical features between the Lambert-Eaton myasthenic syndrome with and without cancer: an analysis of 227 published cases. *Clin Neurol Neurosurg* 2002;**104**:359–63.

Chapter 26

Congenital Myasthenic Syndromes

Andrew G. Engel

INTRODUCTION

Congenital myasthenic syndromes (CMS) are heterogeneous disorders in which the safety margin of neuromuscular transmission is compromised by one or more specific mechanisms. These syndromes are neither new nor uncommon. They were recognized by Rothbart, who in 1937 described four brothers with myasthenia younger than 2 years of age in one family,[1] and by Sarah Bundey, who in 1972 collected 97 familial cases with onset in early childhood.[2] After the discovery of the autoimmune origins of myasthenia gravis in the 1970s and of the Lambert-Eaton syndrome in the 1980s, it became apparent that myasthenic disorders occurring in a familial or congenital setting must have a different etiology. Since then, clinical, ultrastructural, *in vitro* microelectrode, and molecular genetic studies of CMS patients have revealed a phenotypically and genetically heterogeneous group of disorders. To date, no fewer than 19 CMS disease genes have been identified. Most disease proteins reside in presynaptic, synaptic basal lamina, or postsynaptic endplate (EP) compartments, but recently it has become apparent that proteins distributed in many tissues, such as those providing cytoskeletal support or subserving glycosylation of nascent peptides, are also CMS targets (Figure 26.1).

DIAGNOSIS

Generic Diagnosis

A generic diagnosis of a CMS can be made on clinical grounds from a history of fatigable weakness involving ocular, bulbar, and limb muscles since infancy or early childhood, a history of similarly affected relatives, a decremental electromyographic response, and negative tests for antibodies against the acetylcholine receptor (AChR) and the muscle specific receptor tyrosine kinase (MuSK). In some CMS, however, the onset is delayed, the electromyography (EMG) abnormalities are not present in all muscles or are present only intermittently, and the weakness has a restricted distribution. Box 26.1 lists the differential diagnosis of CMS.

Genetic Diagnosis

A direct approach to diagnosis is Sanger sequencing based on available phenotypic clues (see Box 26.2). However, many CMS patients lack phenotypic clues. In that case mutation analysis can be based on frequencies of heretofore identified mutations in different proteins as shown in Table 26.1. In large kinships, linkage analysis followed by Sanger sequencing of genes at the candidate chromosomal locus can be helpful. The use of microarray chips designed to detect variants in currently identified CMS disease genes will likely increase in the near future. Exome sequencing has now become a powerful tool for obtaining a genetic diagnosis when other approaches fail. However, this approach presently captures close to only ~85% of the entire exome and reads a smaller proportion of the exome at a high coverage. It also misses changes in noncoding DNA and large deletions or duplications, is less efficient at detecting dominant than recessive mutations, and is still very expensive. DNA from both parents, and whenever possible from other affected family members, should be analyzed. The bioinformatics analysis of the enormous amount of generated data remains challenging, and the putative pathogenic mutations must be confirmed by Sanger sequencing. The effects of novel nontruncating mutations can be assessed by available software programs, but the accuracy of these tools is not currently validated. Ideally, pathogenicity of novel variants should be evaluated by expression studies. Finally, pathogenic mutations may reside in novel genes whose function is yet unknown. Sequencing the whole genome is also feasible but the interpretation is even more expensive and complicated than that of exome sequencing.[4]

B.T. Darras, H. Royden Jones, Jr., M.M. Ryan & D.C. De Vivo (Eds): Neuromuscular Disorders of Infancy, Childhood and Adolescence, Second edition.
DOI: http://dx.doi.org/10.1016/B978-0-12-417044-5.00026-3

In some CMS, and especially in novel syndromes, additional studies are required to elucidate the pathogenesis and to recommend appropriate therapy. Such studies include estimation of the number of AChRs per EP; histochemical, immunocytochemical, electron microscopy analysis of EP morphology; and electrophysiology assessment of EP function *in vitro*. The last includes microelectrode recording of EP potentials (EPPs) and currents, and single-channel patch-clamp recordings of currents flowing through single AChR channels.

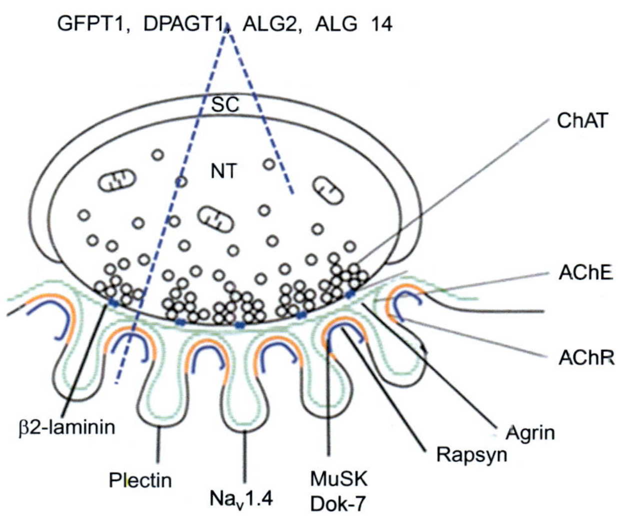

FIGURE 26.1 Schematic diagram of an EP indicating locations of presynaptic, synaptic, and postsynaptic CMS disease proteins. Green line, synaptic basal lamina; red line, AChR on crests of the junctional folds; blue line, MuSK and Dok-7 closely associated with AChR. GFPT1, DPAGT1, ALG2, and ALG14 proteins are expressed at both pre- and postsynaptic sites.

CLASSIFICATION

The CMS can be classified according to inheritance, the clinical phenotype, the disease gene or protein, or the site of the defect (presynaptic, synaptic, or postsynaptic). Classification by inheritance is simple but not sufficiently informative: the slow-channel syndromes are caused by autosomal dominant gain-of-function mutations: all other CMS recognized to date are caused by autosomal recessive loss-of-function mutations. A classification according to the clinical phenotype, which includes the EMG findings, has merits because it can point to some syndromes (see Box 26.2), but it is still unsatisfactory because different presynaptic and postsynaptic syndromes cannot be reliably differentiated by clinical and EMG criteria. Classification by an altered EP-specific protein or gene is unsatisfactory because the disease gene of some CMS remains unidentified, and because different mutations of a given gene can cause different clinical phenotypes. A classification according to the site of the defect distinguishes among presynaptic, synaptic basal lamina, and postsynaptic syndromes and, by necessity, requires morphologic and sometimes *in vitro* electrophysiologic studies, but it is still not entirely satisfactory because disease genes that regulate EP development and maintenance are expressed at multiple anatomic sites.

The classification shown in Table 26.1 is based on the site of the defect and the effect on EP development and maintenance in 351 patients investigated at the Mayo Clinic. It shows that postsynaptic CMS are more frequent than presynaptic or synaptic CMS, and that most postsynaptic CMS are caused by deficiency or kinetic abnormality of the AChR. The distribution of the different types of CMS in any series could be biased by ease of identification of a given type of CMS by clinical criteria. Conversely, patients whose disease closely mimics a type of muscular dystrophy, congenital myopathy, or spinal

BOX 26.1 The Differential Diagnosis of Congenital Myasthenic Syndromes

Neonatal Period, Infancy, Childhood

- Spinal muscular atrophy
- Congenital myopathies (central core disease, nemaline myopathy, myotubular myopathy)
- Congenital muscular dystrophies
- Limb-girdle muscular dystrophy
- Infantile myotonic dystrophy
- Mitochondrial myopathy
- Brain stem anomaly
- Möbius syndrome
- Infantile botulism
- Seropositive and seronegative forms autoimmune myasthenia gravis[a]

Older Patients

- Limb girdle or facioscapulohumeral dystrophy
- Seropositive and seronegative forms of autoimmune myasthenia gravis
- Mitochondrial myopathy
- Chronic fatigue syndrome
- Motor neuron disease
- Peripheral neuropathy[b]
- Syringomyelia[b]
- Radial nerve palsy[b]

a. Not reported in the first year of life.
b. This diagnosis was made in some slow-channel CMS patients.

BOX 26.2 Clinical Clues Pointing to a Specific of Congenital Myasthenic Syndromes

Endplate AChR Deficiency Due to Low-Expressor Mutations in AChR Subunit Gene
- Ptosis, oculoparesis, weakness of limb muscles

Endplate acetylcholinesterase deficiency
- Repetitive CMAPs
- Refractoriness to cholinesterase inhibitors
- Delayed pupillary light reflexes in some cases
- Unresponsive to/worsened by cholinergic agonists; improved by ephedrine or albuterol

Slow-channel congenital myasthenic syndrome
- Repetitive CMAPs
- Selectively severe involvement of cervical and wrist and finger extensor muscles in most cases
- Dominant inheritance in most cases
- Worsened by cholinergic agonists; improved by long-lived AChR channel blockers

Endplate choline acetyltransferase deficiency
- Recurrent apneic episodes, spontaneous or with fever, vomiting, or excitement
- Variable myasthenic symptoms between acute episodes
- Stimulation at 10 Hz for 5 min causes marked decrease of the CMAP amplitude followed by slow recovery over 10 minutes. Rapid recovery in 2 to 3 minutes is not diagnostic
- EMG decrement at 2 Hz can be absent at rest but appears after stimulation at 10 Hz for 5 min, then disappears slowly over 10 minutes. Rapid recovery in 2 to 3 minutes is not specific

Rapsyn deficiency
- Multiple congenital joint contractures in one fourth of patients
- Increased weakness and respiratory insufficiency precipitated by intercurrent infections
- EMG decrement can be mild or absent
- Ophthalmoparesis in ~25%; strabismus relatively common

Dok-7 myasthenia
- Predominantly limb-girdle and axial distribution of weakness; mild facial weakness, and ptosis are common; normal ocular ductions in most patients
- Significant bulbar muscles involvement in some patients
- Worsened by cholinesterase inhibitors; responds to ephedrine or albuterol
- Can present with stridor and vocal cord paralysis in neonates and infants

GFPT1 myasthenia
- Predominantly limb-girdle and axial distribution of weakness.
- CK elevation in about one fourth
- Tubular aggregates in muscle in most patients; autophagic myopathy in some patients
- Improved by cholinergic agonists

DPAGT1 myasthenia
- Limb-girdle or diffuse weakness; ocular ductions spared
- Can be associated with congenital malformations and cognitive deficits
- Abnormal isolelectric focusing pattern of serum transferrin in some patients
- Tubular aggregates in muscle in most patients; autophagic myopathy in some patients
- Improved by cholinesterase agonists

Laminin β2 myasthenia
- Nephrotic syndrome, ocular abnormalities (Pierson syndrome)
- Refractoriness to cholinesterase inhibitors

Plectin deficiency myasthenia
- Epidermolysis bullosa simplex
- Refractoriness to cholinesterase inhibitors

Na-channel myasthenia
- Recurrent apneic episodes
- Suprathreshold EPPs fail to generate muscle fiber action potentials

Centronuclear myopathy-myasthenia
- High proportion of centronucleated muscle fibers as the predominant pathologic alteration

There are no specific phenotypic clues to the diagnosis of the fast-channel CMS and most cases of rapsyn deficiency.

muscular atrophy, may not be referred for further studies and may well remain undiagnosed. Also, identification and awareness of a given clinical phenotype (e.g. the slow-channel syndrome or EP AChE deficiency) promote identification of more patients of the same phenotype. The frequency of given types of CMS may also vary in different geographic areas and in different ethnic groups. In some closed communities the frequency of a given CMS is increased by a founder effect or by consanguineous marriages.

PRESYNAPTIC CONGENITAL MYASTHENIC SYNDROME

Endplate Choline Acetyltransferase Deficiency (Congenital Myasthenic Syndrome with Episodic Apnea)

Choline acetyltransferase (ChAT) catalyzes the resynthesis of ACh by transfer of an acetyl group from acetyl-CoA to choline in cholinergic neurons. Pathogenic mutations, alone

TABLE 26.1 Classification of Congenital Myasthenic Syndrome Based on Index Patients Investigated at the Mayo Clinic between 1988 and 2013

	Index Cases at Mayo Clinic	%
Presynaptic (5.6%)		
• Choline acetyltransferase deficiency	18	5.1
• Paucity of synaptic vesicles and reduced quantal release[a]	1	0.3
• Congenital Lambert-Eaton-like[a]	1	0.3
Synaptic basal lamina (13.3%)		
• Endplate acetylcholinesterase deficiency	45	13
• Laminin β2 deficiency	1	0.3
Postsynaptic (52%)		
• Primary kinetic defect ± AChR deficiency	62	18
• Primary AChR deficiency ± minor kinetic defect	118	33
• Na channel myasthenia	1	0.3
• Plectin deficiency	2	0.6
Defects in endplate development and maintenance (29%)[b]		
• Agrin deficiency	1	0.3
• MuSK deficiency	1	0.3
• Rapsyn deficiency	51	14.5
• Dok-7 myasthenia	35	10
• GFPT1 myasthenia	11	3.1
• DPAGT1 myasthenia	2	0.6
• Myasthenic syndrome with centronuclear myopathy[a]	1	0.3
Total	**351**	**100**

[a]*No gene defect detected.*
[b]*Mutations in* ALG2 *and* ALG14 *were identified at other medical centers.*[3]

or in combination, alter the expression, catalytic efficiency, or structural stability of the enzyme.[5,6] When neuronal impulse flow is increased, as during exercise, the decreased rate of ACh synthesis progressively depletes the ACh content of the synaptic vesicles, and hence decreases the amplitude of the EPP and of the compound muscle action potential (CMAP). A useful clinical test for this disease consists of monitoring the EMG decrement and the CMAP amplitude during 10 Hz simulation over 5 minutes and for 10 to 15 minutes after stimulation. A nonspecific decrease below 50% of the potential from the baseline followed by return to the baseline in 2 to 3 minutes occurs in different CMS. In contrast, in patients with ChAT deficiency the recovery occurs slowly over 5 to 15 minutes. Therefore, the CMAP must be monitored for 10 to 15 minutes after 5 Hz stimulation.

Some patients present with hypotonia, bulbar paralysis, and apnea at birth (Figure 26.2). Others are normal at birth and develop apneic attacks during infancy or childhood precipitated by infection, excitement, or no apparent cause.[6–10] In some children an acute attack is followed by respiratory insufficiency that lasts for weeks.[11] Few patients are apneic, respirator dependent, and paralyzed since birth,[6] and some develop cerebral atrophy after episodes of hypoxemia.[6,10] Others improve with age, but still have variable ptosis, ophthalmoparesis, fatigable weakness, and recurrent cyanotic episodes, and some complain only of mild to moderately severe fatigable weakness. The symptoms are worsened by exposure to cold because this further reduces the catalytic efficiency of the mutant enzyme.[3]

Mutations with severe kinetic effects are located at or near the active-site tunnel or the substrate binding site of the enzyme, or exert their effect allosterically. Some mutations have no kinetic effects and express well but impair the thermal stability of ChAT by introducing a

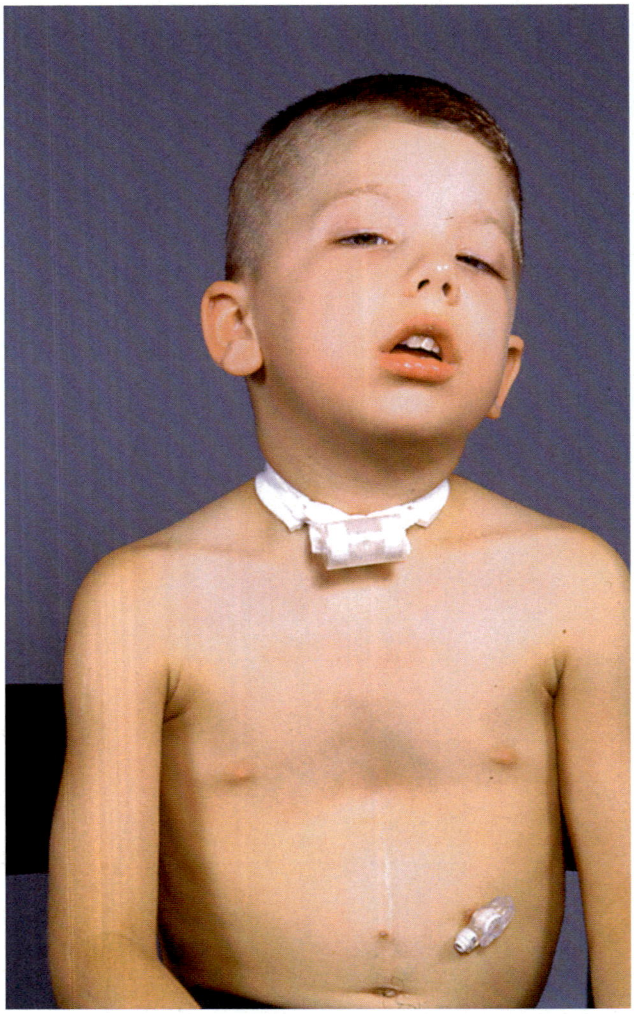

FIGURE 26.2 Patient with congenital myasthenic syndrome caused by a defect in choline acetyltransferase. Note ptosis, esotropia, hyperactive frontalis muscle, facial weakness, tracheostomy, and percutaneous gastrostomy.

proline residue into an α-helix.[6] Genotype–phenotype correlations are hindered in patients with biallelic mutations. Interestingly, however, in three very severely affected patients with lifelong apnea, permanent paralysis, and failure to respond to pyridostigmine, one of the two mutations was near the active-site of ChAT.

Treatment consists of prophylactic therapy with pyridostigmine. The parents should be provided with an inflatable rescue bag and a fitted mask, instructed in the intramuscular injection of neostigmine methylsulfate, and advised to install an apnea monitor in their home.

Paucity of Synaptic Vesicles Associated with Reduced Quantal Release

The clinical features closely mimic moderately severe to severe autoimmune myasthenia gravis, but the onset is at birth or in early infancy and tests for anti-AChR antibodies are negative. A specific diagnosis requires morphologic and electrophysiologic studies of the EP in a muscle biopsy specimen intact from origin to insertion. These studies reveal no EP AChR deficiency. A presynaptic defect is indicated by a severe decrease (to ~20% of normal) in the number of ACh quanta (m) released by nerve impulse. The decrease in m is due to a decrease in the number of readily releasable quanta (n), associated with a comparable decrease (to ~20% of normal) in the numerical density of the synaptic vesicles. The putative defect resides in the synthesis or axonal transport of vesicle precursors from the anterior horn cell to the nerve terminal, or is related to impaired recycling of the synaptic vesicles.[12] Treatment consists of AChE inhibitors.

CMS Resembling the Lambert-Eaton Myasthenic Syndrome

In one young child reported with this syndrome, the CMAP amplitude was abnormally small but facilitated several-fold on tetanic stimulation, and the symptoms were improved by guanidine.[13] The defect likely resides in a subunit of the presynaptic voltage-gated calcium channel or in a component of the synaptic vesicle release complex.

SYNAPTIC BASAL LAMINA ASSOCIATED CMS

EP Acetylcholinesterase Deficiency

The EP species of acetylcholinesterase (AChE) is an asymmetric enzyme composed of catalytic subunits encoded by $ACHE_T$ and a collagenic structural subunit encoded by *COLQ* that anchors the enzyme in the synaptic basal lamina. No spontaneous mutations have been observed in $AChE_T$. The ColQ (collagen Q) protein is composed of three identical strands. The N-terminal residues of each strand bind a catalytic homotetramer. Mutations in this domain prevent ColQ association with the catalytic subunits and abolish enzyme activity. Frameshift or nonsense mutations in the collagenic midsection of ColQ produce a single-stranded insertion incompetent enzyme. Mutations of critical residues in the globular C-terminal region of ColQ prevent ColQ insertion into the synaptic basal lamina or hinder the triple-helical assembly of the ColQ strands (Figure 26.3).[14,15]

Patients typically present in the neonatal period or infancy with apnea and generalized weakness that persist through life (Figure 26.4A). In some patients the ocular ductions are spared and in some the pupillary light reflex is delayed. Milder cases associated with missense mutations in the C-terminal domain of ColQ may present later in childhood.[16] The edrophonium test is negative and pyridostigmine

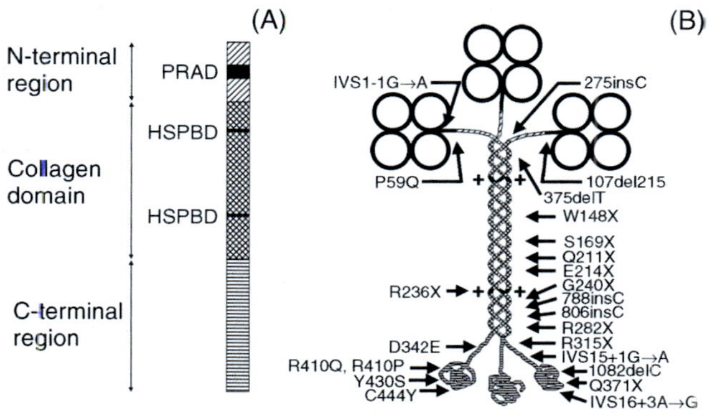

FIGURE 26.3 Schematic diagram showing domains of a ColQ strand, components of the asymmetric species of AChE, and observed mutations. The mutations appear in each ColQ domain.

causes increased secretions but has no effect on the weakness. Absence of AChE from the EP prolongs the lifetime of ACh in the synaptic cleft because each ACh binds multiple AChRs before leaving the synaptic space by diffusion. This prolongs the duration of the miniature EP potential (MEPP) and EPP, and when the EPP outlasts the absolute refractory period of the muscle fiber, it generates a second (or repetitive) muscle fiber action potential, reflected by a repetitive compound muscle action potential (CMAP) that is unaffected by edrophonium (Figure 26.4B). Cytochemical localization of AChE shows that the synaptic space is devoid of reaction product (Figure 26.4C and D).

Neuromuscular transmission is compromised by smallness of the nerve terminals and their encasement by Schwann cells, which reduces the number of releasable

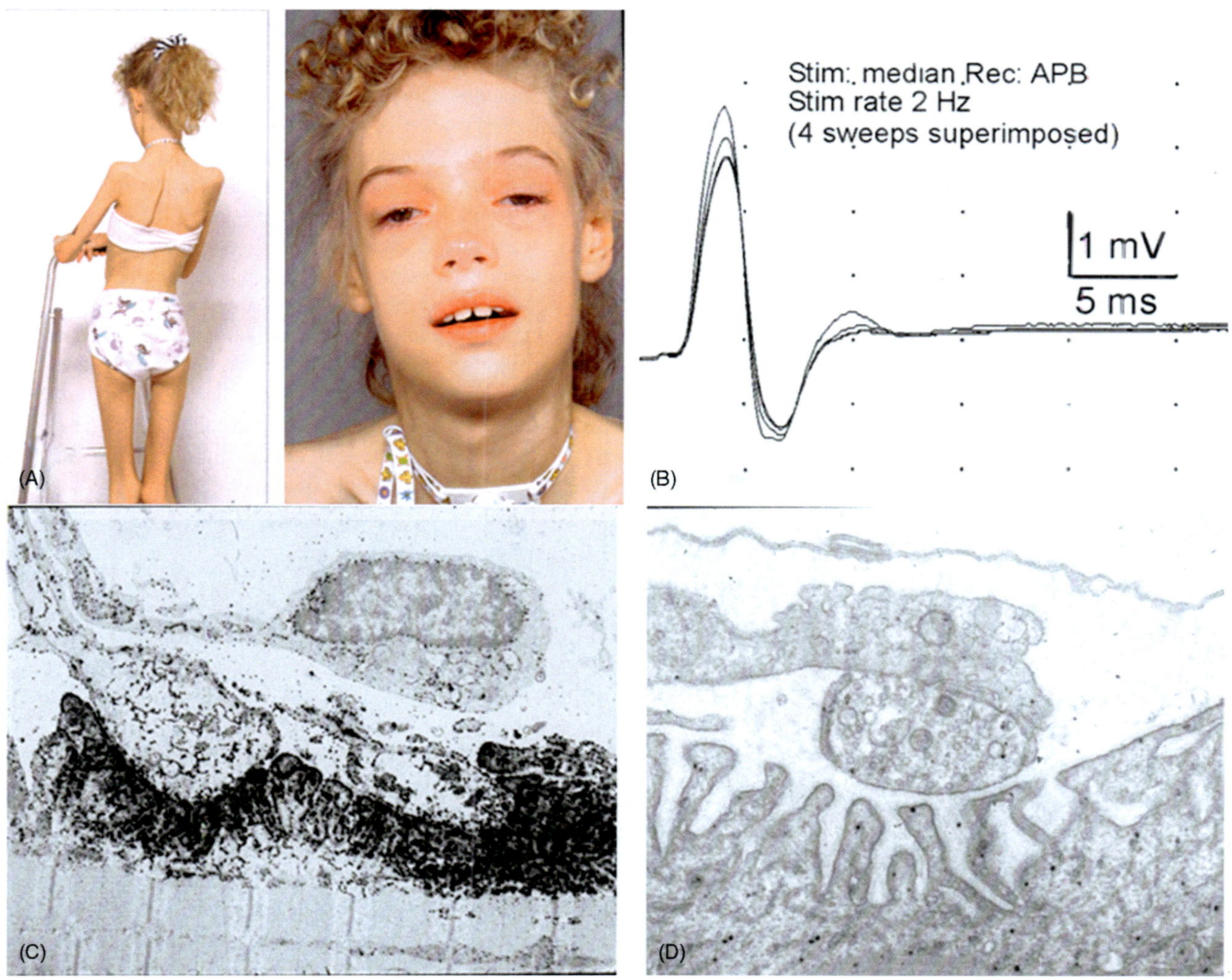

FIGURE 26.4 EP AChE deficiency. (**A**) Eleven-year-old girl with EP AChE deficiency. Note ptosis, hyperactive frontalis muscle, tracheostomy, ophthalmoparesis, scoliosis, and small muscle bulk. The patient can only stand with support. (**B**) Repetitive CMAP. (**C** and **D**) Electron cytochemical localization of AChE at control and patient EPs. At control EP the dark reaction product fills the synaptic space. At patient EP there is no reaction product in the synaptic space.

quanta, by an EP myopathy from cholinergic overactivity, and by desensitization and depolarization block of AChR at physiologic rates of stimulation.[17,18] Smallness of nerve terminals and their encasement by Schwann cells are likely compensatory to protect the EP from overexposure to ACh.

Therapy is still unsatisfactory, but ephedrine[16,19,20] and albuterol[21] were noted to have a gradually developing beneficial effect.

CMS Associated with β2-Laminin Deficiency

β2-Laminin, encoded by *LAMB2*, is a component of the basal lamina of different tissues and is highly expressed in kidney, eye, and the neuromuscular junction. Synaptic β2-laminin governs the appropriate alignment of the axon terminal with the postsynaptic region and, hence, pre- and postsynaptic trophic interactions. Defects in β2-laminin result in Pierson syndrome with renal and ocular malformations. A patient carrying heteroallelic missense and frameshift mutations in *LAMB2* had Pierson syndrome associated with ocular, respiratory, and proximal limb muscle weakness.[22] The renal defect was corrected by renal transplant at age 15 months. *In vitro* microelectrode studies revealed decreased quantal release by nerve impulse and a reduced MEPP amplitude. Electron microscopy showed abnormally small nerve terminals often encased by Schwann cells, accounting for the decreased quantal release. The synaptic space was widened and the junctional folds were simplified, accounting for the decreased MEPP amplitude.

POSTSYNAPTIC CMS

Most postsynaptic CMS stem from molecular defects in the muscle form of the nicotinic AChR. We therefore review the structure and kinetic properties of this species of AChR.

The Muscle Form of the Nicotinic Acetylcholine Receptor

The muscle form of nicotinic AChR is a pentameric transmembrane macromolecule with a subunit composition of $\alpha_2\beta\delta\varepsilon$ at adult EPs, and $\alpha_2\beta\delta\gamma$ at fetal EPs and extrajunctional sites (Figure 26.5). The genes encoding the α, δ, and γ subunits are at different loci on chromosome 2q; the genes encoding the β and ε subunits are at different loci on chromosome 17p. Neuronal AChRs are also pentameric, composed of moieties of two α ($\alpha_2-\alpha_9$) and three β ($\beta_2-\beta_4$) subunits, or exist as α homopentamers. Both muscle and neuronal AChRs are members of a superfamily of ligand-gated ion channels that also includes receptors for 5-hydroxytryptamine, γ-aminobutyric acid, and glycine. At the normal human EP AChR is concentrated on the crests of the junctional folds.

The five subunits of AChR are organized like barrel staves around a central cation-permeable channel (Figure 26.5). Each subunit has an N-terminal extracellular domain that comprises approximately 50% of the primary sequence, four putative transmembrane domains (M1−M4), and a small C-terminal extracellular domain. M2, which lines the ion channel, forms an α-helix interrupted by a short stretch of β-sheet. C-terminal residues of the M1 domain may also contribute to the channel lumen in the resting state. The transmembrane domains are connected by an extracellular M2/M3 linker and by intracellular M1/M2 and M3/M4 linkers. The M3/M4 linker forms a long cytoplasmic loop that likely serves as an attachment site for cytoskeletal elements, contains phosphorylation sites that may be important for desensitization, and harbors residues that keep the gating properties of the ion channel consistent.[23,24] Each AChR has two ACh-binding pockets positioned at the α/ε (or α/γ) and α/δ interfaces. Residues contributing to the binding pocket appear on three peptide loops on the α subunit and four peptide loops of the δ, ε, and γ subunits. Crystallography studies indicate that each ACh-binding pocket is open to the extracellular space.[25]

Two ACh molecules (A) bind to the receptor (R) with association rates k_{+1} and k_{+2} and dissociate from it with rates k_{-2} and k_{-1}. The double-liganded receptor (A2R) opens with rate β, and the open receptor (A2R*) closes with rate α. A channel opening episode, or opening burst, consists of one or more openings interrupted by brief closures due to oscillations of the diliganded receptor between the open and closed state. The above scheme predicts that the mean duration of channel-opening episode bursts is approximately $(1 + \beta/k_{-2})/\alpha$. This value also corresponds to the decay time constant of the miniature EP current (MEPC) evoked by a single quantum of ACh.

Slow-Channel Syndrome

This syndrome arises from dominant gain of function mutations that occur in extracellular or transmembrane domains of each AChR subunit. The mutations enhance the agonist affinity or the gating efficiency of the receptor.[26] Either mechanism prolongs the single-channel currents, EP currents, and EPPs. As in EP AChE deficiency, when the length of the EPP exceeds the absolute refractory period of the muscle fiber, it triggers a repetitive CMAP (Figure 26.6A−C), but unlike in AChE deficiency, the repetitive response is accentuated by edrophonium. Also, at physiologic rates of stimulation, each prolonged EPP arises in the wake of the preceding EPP causing a progressive depolarization block of the postsynaptic membrane. The prolonged EP currents and leakiness of the ion channels even in the closed state result in Ca^{2+} overloading of the postsynaptic region[27] (Figure 26.6 D and E) and an EP myopathy (Figure 26.6F). The safety margin of

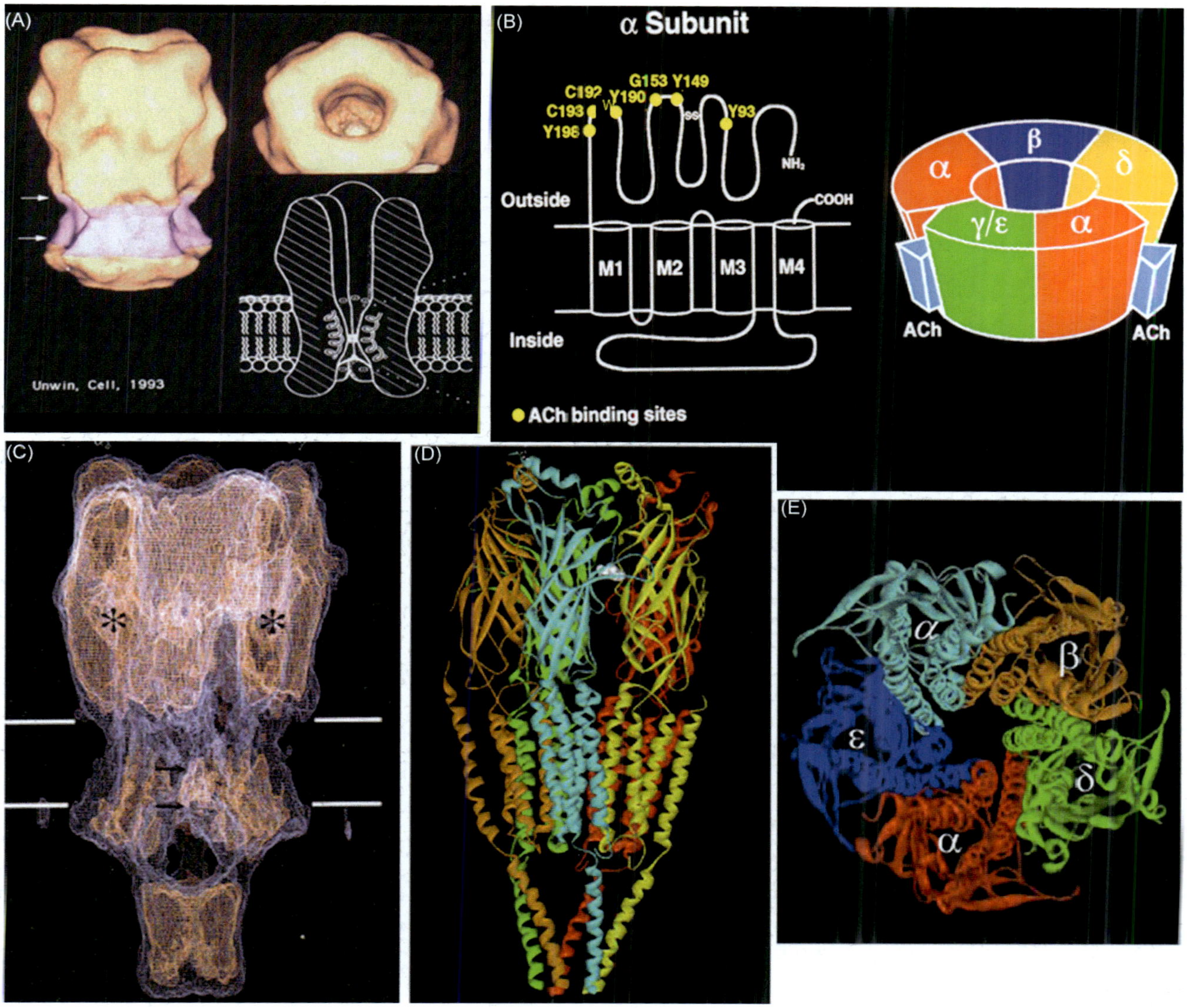

FIGURE 26.5 Schematic diagrams: (**A**) AChR in lipid bilayer showing the central pore of the channel. (**B**) Folding pattern of the α subunit; residues indicated on three peptide loops are implicated in governing affinity for ACh and cross-section of AChR at the level of the ACh binding pocket. Note circular arrangement of AChR subunits and ACh binding pockets appear at the α/ε or α/γ and at the α/δ subunit interfaces. (**C**) Cryo-electron microscopic structure of Torpedo AChR at 7.5 μm resolution. (**D** and **E**) Structure of the Torpedo AChR as determined at 4 Å resolution (PDB2BG9). For each subunit, the extracellular domain contains predominantly β-sheets, the transmembrane domains contain four α helices and the cytoplasmic domain contains an α helix and unresolved structure.

neuromuscular transmission is compromised by a depolarization block, by desensitization of the AChR during physiologic activity, and by the endplate myopathy that causes loss of AChR from the junctional folds and alters the EP geometry.

The syndrome usually presents in the first decade, but severely affected patients present in the neonatal period. There is selectively severe involvement of the cervical, scapular, and dorsal forearm muscles (Figure 26.7). The ocular muscles are usually spared. Some patients have mild asymmetric ptosis. The syndrome is refractory to, or worsened by, cholinergic agonists but is improved by long-lived open-channel blockers of the AChR channel, such as quinidine,[28] quinine, or fluoxetine.[29–31]

Fast-Channel Syndrome

This syndrome is caused by a recessive fast-channel mutation of one allele accompanied by a null or low-expressor mutation of the second allele, or rarely by another fast-channel mutation on the second allele.[26] It is the physiological opposite of the slow-channel syndrome

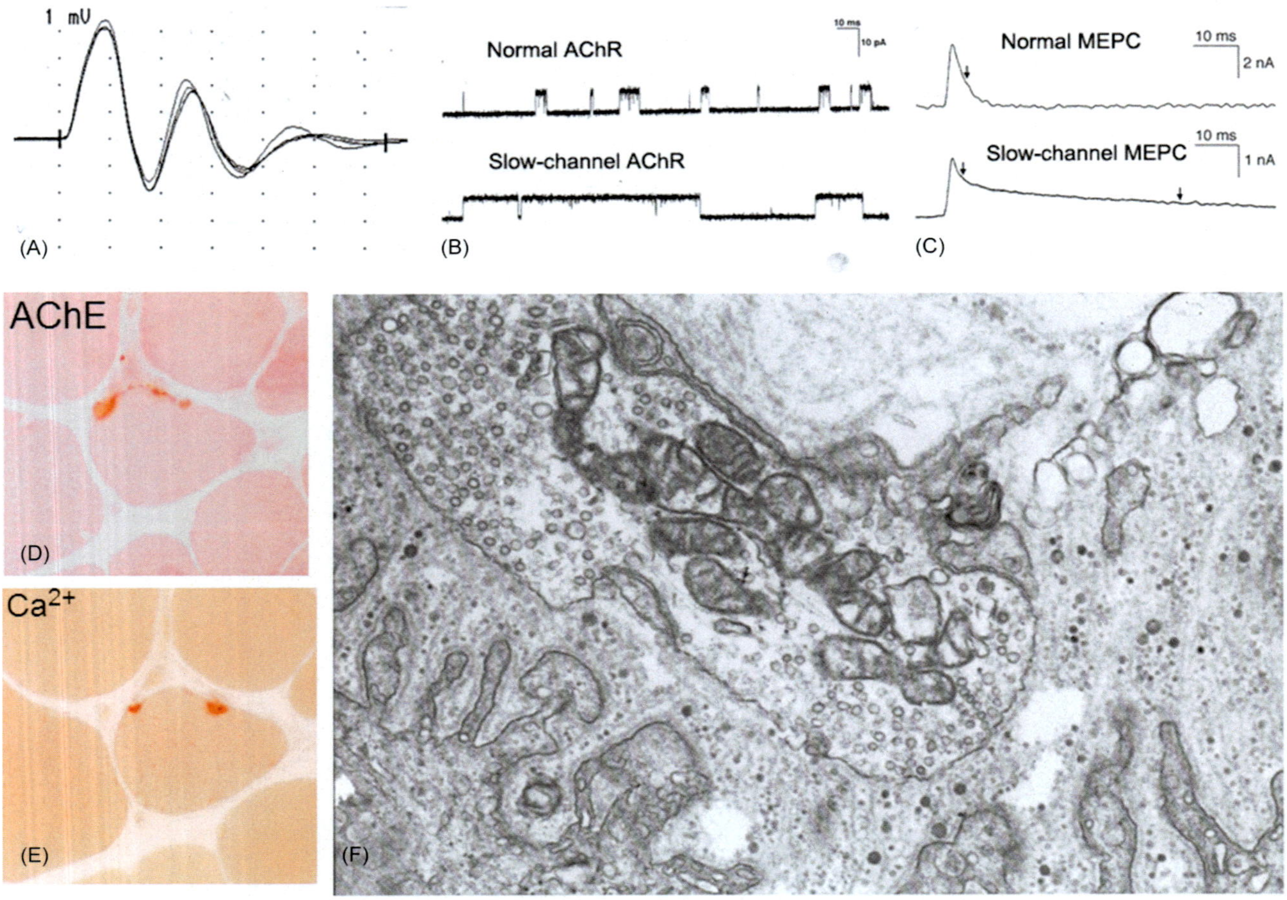

FIGURE 26.6 Slow-channel syndrome. Repetitive compound muscle potentials (**A**) are generated by prolonged AChR channel opening events (**B**) that generate abnormally prolonged endplate currents (**C**). The slow-channel MEPC decays biexponentially owing to presence of both wild-type and mutant channels at the endplate. The prolonged endplate currents as well as leakiness of the receptor in the resting state result in Ca^{2+} overloading of the postsynaptic region (**D** and **E**) and in an EP myopathy (**F**). *(Reproduced from Engel et al.[26])*

in that it decreases affinity for ACh, or reduces gating efficiency, or it can destabilize channel kinetics or act by a combination of these factors.[23,24,32–37] Consequently, the EP currents decay abnormally fast, the channel activation episodes are abnormally brief, and the amplitude of the synaptic currents and potentials is reduced owing to a reduced probability that the AChR channel is opened by physiologic concentrations of ACh (Figure 26.8A and B). The most mutations reside in the extracellular domain of AChR where they hinder agonist binding to gating (Figure 26.8D) or decrease affinity for ACh (Figure 26.8E). Mutations in the C-terminal region of the M3–M4 linker result in mode-switching kinetics causing an excessive number of brief channel openings and a milder phenotype (Figure 26.8F).

There are no clinical or EMG clues pointing to the specific diagnosis of a fast-channel syndrome. Consequently, in vitro microelectrode studies of muscle specimens intact from origin to insertion are required to detect the kinetic abnormality. However, if a mutation is identified in the disease gene, the position of the mutation in the structural model of the receptor may inform on its kinetic consequences.[37]

The clinical consequences are severe when the pathogenic mutation affects an agonist binding site[37] or affects coupling of agonist binding to gating (Figure 26.8D),[33] or is positioned in a transmembrane domain of the receptor (Figure 26.8E),[34] and less severe when it is in the long cytoplasmic loop of the ε subunit where it affects the fidelity of gating[23,34] (Figure 26.8F). Therapy consists of 3,4-diaminopyridine (3,4-DAP), which increases the number of ACh quanta released by nerve impulse, and pyridostigmine, which increases the number of AChRs activated by each quantum.

Primary Acetylcholine Receptor Deficiency

CMS with severe EP AChR deficiency result from different types of homozygous or more frequently heterozygous

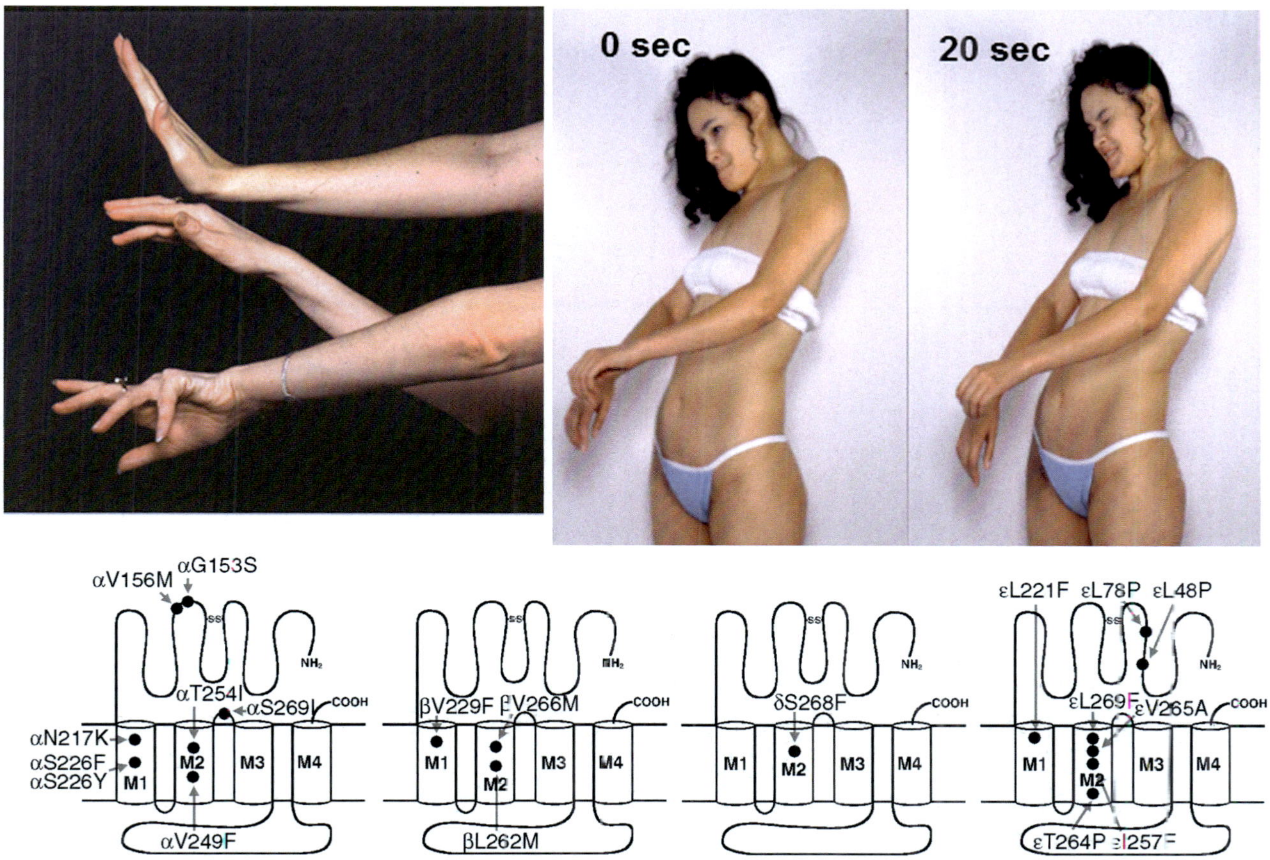

FIGURE 26.7 Slow-channel syndrome patients and mutations. **Left upper panel**: Patient harboring an Leu269Phe mutation in the M2 domain of the AChR ε subunit cannot fully extend her fingers and wrists as shown by examiner with sleeves. **Right upper panels**: patient harboring a Gly153Ser mutation at the binding site of the AChR α subunit is unable to sustain even partial elevation of her right arm for 20 sec. Note atrophic forearm muscles and the lordotic stance that worsens after 20 sec. **Lower panel**: identified slow-channel mutations in different AChR subunits.

recessive mutations in the AChR α, β, δ, and ε subunit genes. The mutations are concentrated in the ε subunit (Figure 26.9A and B). The main reason for this is that expression of the fetal type γ subunit, although at a low level, partially compensates for absence of the ε subunit (Figure 26.9D),[23,38,39] whereas patients harboring null mutations in subunits other than ε might not survive for lack of a substituting subunit (Figure 26.9C). Ptosis and oculoparesis is present in nearly all patients.

Some mutations cause premature termination of the translational chain. These mutations are frameshifting,[39–43] occur at a splice site,[40,42] produce a stop codon directly,[39] or arise from a chromosomal microdeletion.[44] An important mutation in this group is ε1369delG because it results in loss of a C-terminal Cys470 residue crucial to maturation and surface expression of the adult receptor.[45] Thus any mutation that truncates the ε subunit upstream of Cys470 abrogates expression of the ε subunit.

Structural studies reveal multiple small EP regions over an extended span of the muscle fiber and attenuated expression of AChR on the junctional folds. The structural integrity of the junctional folds is maintained but some EP regions are simplified. The MEPP amplitude is reduced whereas quantal release is normal or even higher than normal. The duration of the EP currents is mildly increased owing to expression of the fetal γ-A.

A generic diagnosis of a CMS is readily made from the clinical and EMG findings, but muscle biopsy studies are needed to demonstrate the EP AChR deficiency by cytochemical or electron microscopy visualization of α-bungarotoxin labeled EPs (Figure 26.10), and *in vitro* microelectrode studies are required to prove or disprove an associated kinetic abnormality of AChR.

Therapy consists of AChE inhibitors in combination with 3,4-DAP. However, only one-third of the patients with severe EP AChR deficiency derive significant

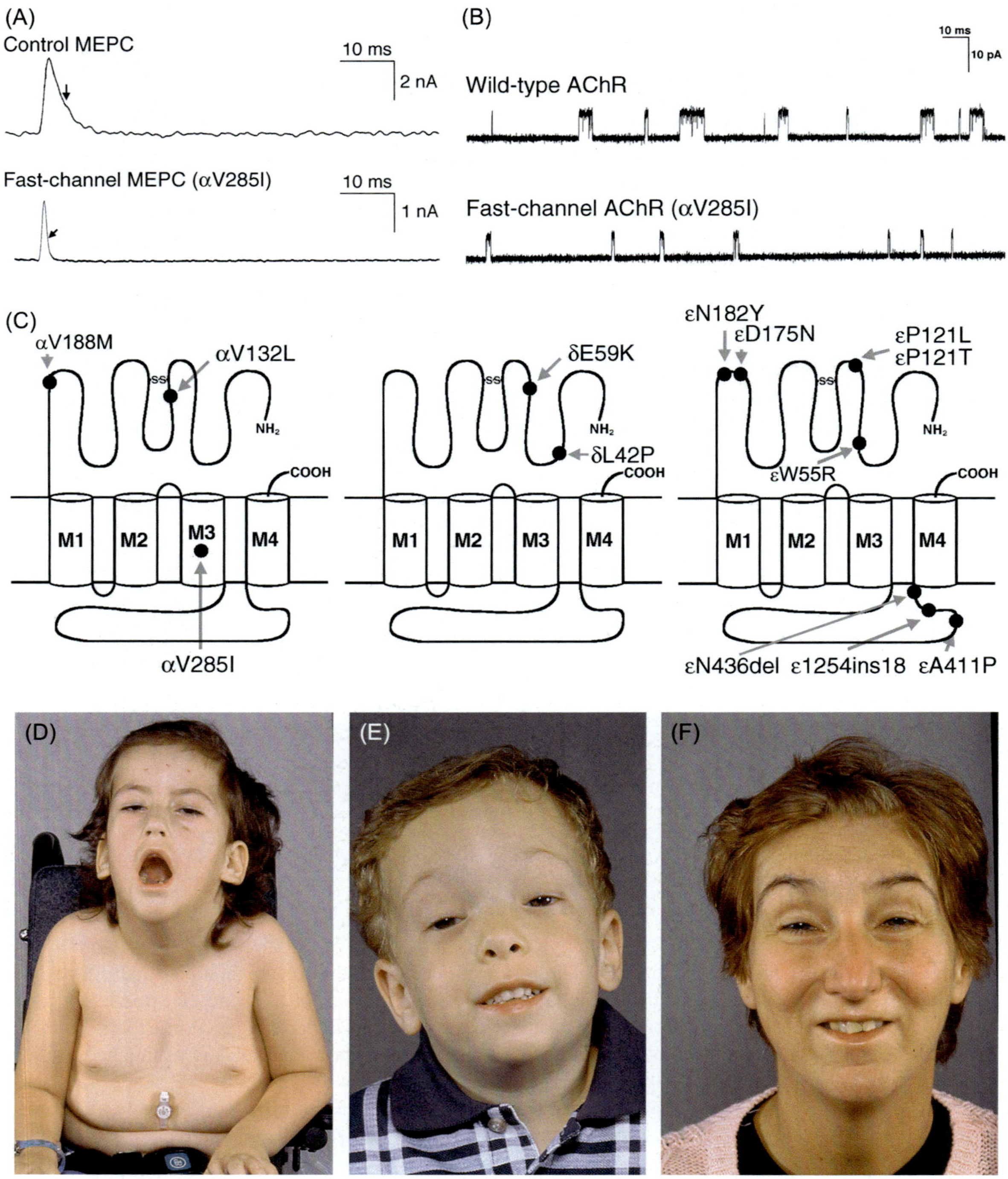

FIGURE 26.8 Fast-channel syndrome. (**A**) MEPCs and (**B**) single channel currents in a control subject and in a fast-channel patient who harbors a Val285Ile mutation in the AChR α subunit. (**C**) Schematic diagram showing positions of identified fast-channel mutations. (**D**) Severely affected patient harbors a Val132Leu mutation in the Cys-loop of the AChR α subunit that hinders coupling of agonist binding to gating. (**E**) A moderately severely affected patient with a Pro121Leu mutation in the AChR ε subunit that reduces affinity for ACh in the open-channel state and thereby the channel opening rate and gating efficiency. (**F**) A mildly affected patients whose phenotype is determined by a 1254ins18 mutation in the long cytoplasmic loop of the ε subunit that causes mode-switching channel kinetics. *(Panels A–D are reproduced from Engel AG,* Congenital Myasthenic Syndromes in Myasthenia Gravis and Myasthenic Syndromes, *2nd ed., edited by Engel AG, Oxford University Press, 2012.)*

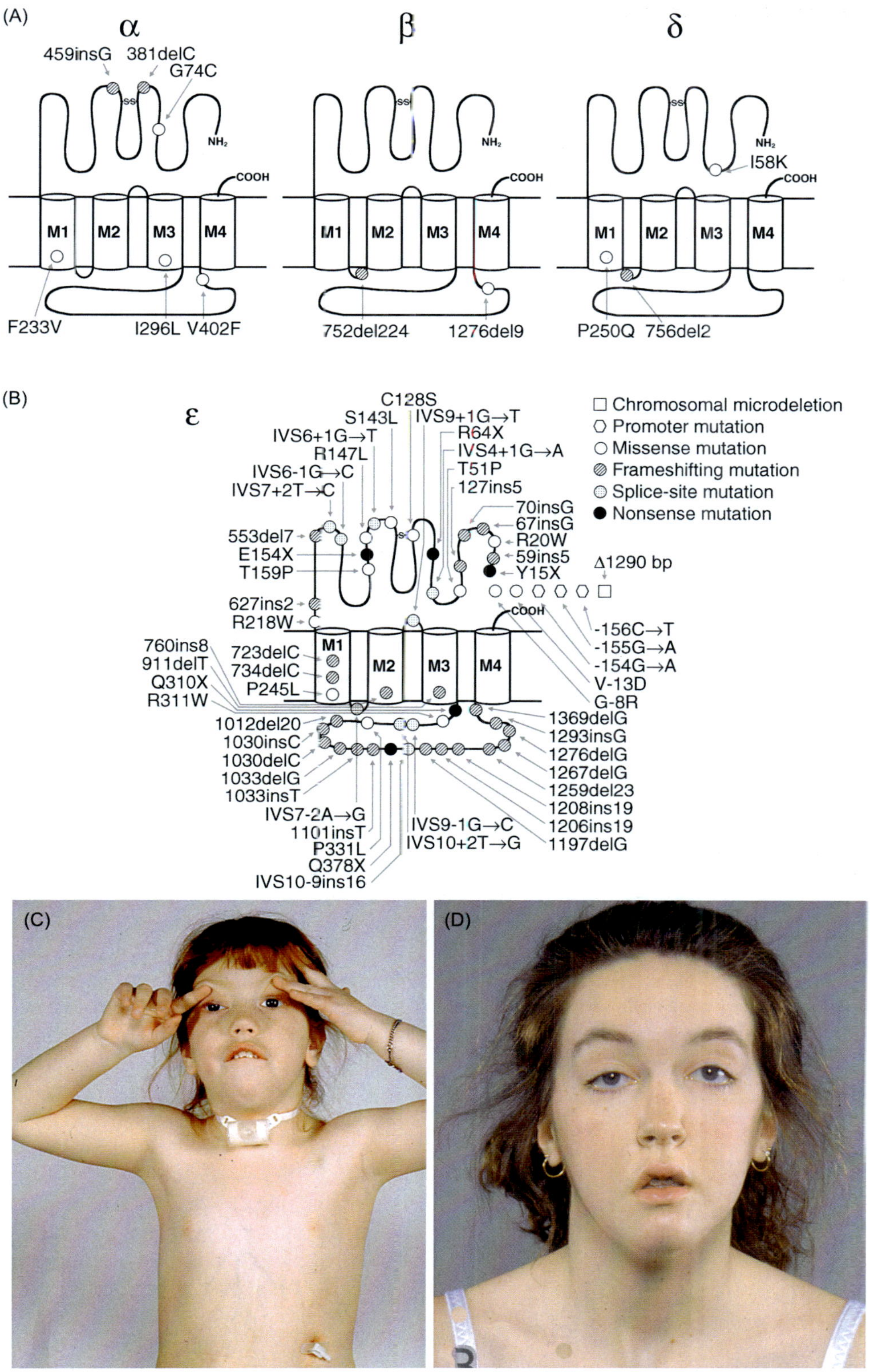

FIGURE 26.9 Low expressor mutations of AChR. (**A** and **B**) Schematic diagram of reported low-expressor and null mutations in the AChR α, β, δ, and ε subunits. Most mutations appear in the ε subunit and especially in the long cytoplasmic loop between M3 and M4. Square indicates a chromosomal microdeletion; hexagons are promoter mutations; open circles are missense mutations; closed circles are nonsense mutations; shaded circles are frame-shifting mutations; dotted circles are splice-site mutations. The most likely consequence of a splice-site mutation is skipping of a flanking exon; therefore the splice-site mutations point to N-terminal codons of the predicted skipped exons. (**C**) A severely affected 4-year-old girl with a heteroallelic low expressor and a null mutation in the AChR β subunit. She lifts her eyelids with her hands. Also note lordotic stance, tracheostomy, and G-tube. At age 6 years she died of acute respiratory failure. (**D**) A less severely affected patient is homozygous ε 553del7 mutation that truncates the extracellular domain of the ε subunit. Note asymmetric ptosis, facial paresis, and mild esotropia. Single channel recordings from this patient's EPs revealed only low-amplitude long-duration openings characteristic of AChR harboring the γ instead of the ε subunit. *(Panels A–D are reproduced from Engel AG,* Congenital Myasthenic Syndromes in Myasthenia Gravis and Myasthenic Syndromes, *2nd ed., edited by Engel AG, Oxford University Press, 2012.)*

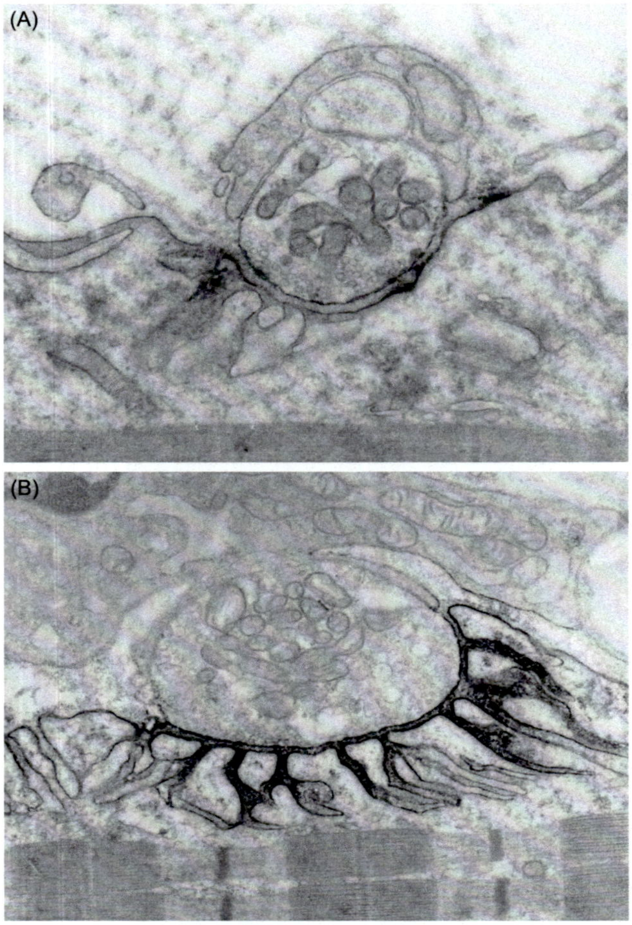

FIGURE 26.10 AChR localization with peroxidase-labeled α-bungarotoxin. (**A**) Markedly attenuated and patchy expression of AChR (dark reaction product) on the postsynaptic membrane of a patient harboring two low-expressor mutations in the AChR ε subunit. (**B**) AChR expression at a control EP. *(Panel A is reproduced from Engel AG,* Congenital Myasthenic Syndromes in Myasthenia Gravis and Myasthenic Syndromes, *2nd ed., edited by Engel AG, Oxford University Press, 2012.)*

benefit from the additional use of 3,4-DAP.45 Recently, albuterol was found to be effective in patients responding poorly to pyridostigmine and 3,4-DAP.[46]

Prenatal CMS Caused by Mutations in AChR Subunits and Other EP Specific Proteins

The first identified prenatal myasthenic syndrome was traced to mutations in the fetal AChR γ subunit. In humans, AChR harboring the fetal γ subunit appears on myotubes around the ninth developmental week and becomes concentrated at nascent nerve-muscle junctions around the sixteenth developmental week. Subsequently, the γ subunit is replaced by the adult ε subunit and is no longer present at fetal EPs after the thirty-first developmental week.[47] Thus pathogenic mutations of the γ subunit result in hypomotility *in utero* mostly during the sixteenth and thirty-first developmental week. The clinical consequences at birth are multiple joint contractures, small muscle bulk, multiple pterygia (webbing of the neck, axilla, elbows, fingers, or popliteal fossa), fixed flexion contractures of the fingers (camptodactyly), rocker-bottom feet with prominent heels, and characteristic facies with mild ptosis and a small mouth with downturned corners. Myasthenic symptoms are absent after birth because by then the normal adult ε subunit is expressed at the EPs.[47,48] More recently, lethal fetal akinesia syndromes arising from biallelic null mutations in the AChR α, β, and δ subunits as well as in rapsyn[49,50] and Dok-7[51] were also identified.

CMS Caused by Plectin Deficiency

Plectin, encoded by *PLEC*, is a highly conserved and ubiquitously expressed intermediate filament-linking protein. By virtue of its tissue and organelle-specific isoforms, it links cytoskeletal filaments to target organelles in different tissues.[52–54] It is concentrated at sites of mechanical stress, such as the postsynaptic membrane covering the junctional folds, the sarcolemma, Z-disks in skeletal muscle, hemidesmosomes in skin, and intercalated disks in cardiac muscle. In skeletal muscle, multiple alternatively spliced transcripts link intermediate filaments to the outer nuclear membrane, the outer mitochondrial membrane, Z disks, and sarcolemmal costameres.[54] Mutations in plectin result in epidermolysis bullosa simplex (EBS), a progressive muscular dystrophy in many patients, and a myasthenic syndrome in some patients.[55–57] In two patients investigated by the author, microelectrode studies of intercostal muscle EPs showed low-amplitude MEPPs. Morphologic studies revealed clustering of nuclei (Figure 26.11A), endomysial fibrosis (Figure 26.11B) and focal calcium accumulation in some fibers, as in Duchenne dystrophy (Figure 26.11C). The synaptic contacts were arrayed over an extended span of the fibers (Figure 26.11D). EM studies showed degenerating muscle fiber organelles, plasma membrane defects accounting for calcium accumulation in the fibers, and extensive degeneration of the junctional folds (Figure 26.11 E and F), all attributable to lack of cytoskeletal support.[55] Interestingly, a recent study identified a homozygous deletion mutation in *PLEC* isoform 1f that only caused limb-girdle muscular dystrophy.[58]

Na-Channel Myasthenia

Only one patient with this syndrome has been identified to date. This was the case of a normokalemic patient who had abrupt attacks of respiratory and bulbar paralysis since birth lasting from 3 to 30 minutes. Intracellular microelectrode studies of neuromuscular transmission revealed that

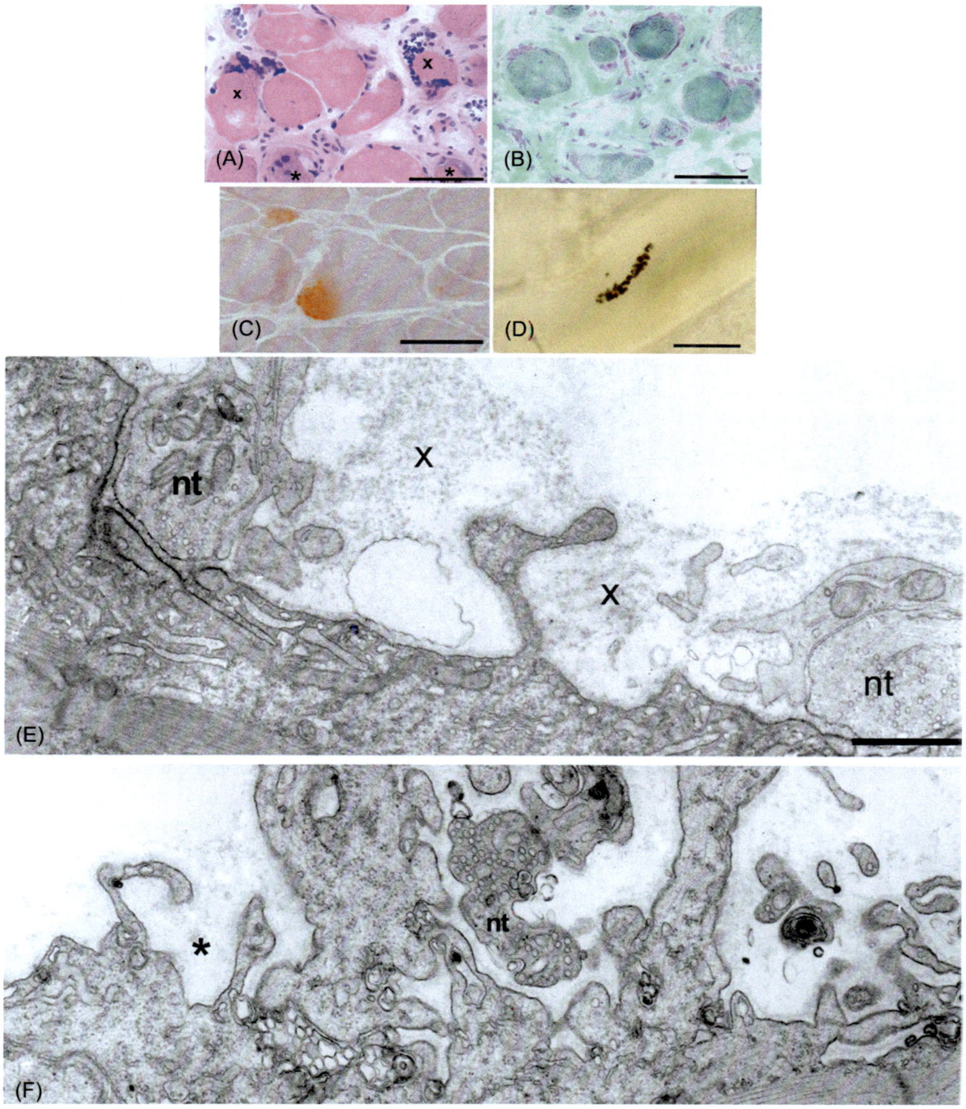

FIGURE 26.11 Plectin deficiency. (**A**) Clusters of dislocated nuclei. (**B**) Endomysial fibrosis. (**C**) Focal calcium overloading of the muscle fibers revealed by the Alizarin-red stain. (**D**) Cholinesterase reaction reveals multiple small synaptic contacts arrayed over an extended length of the muscle fiber. (**E**) The imaged EP regions show partial occupancy of the postsynaptic region by two nerve terminals (nt) and remnants of degenerate folds (x). Dark reaction product on postsynaptic membrane shows AChR localization with peroxidase labeled α-bungarotoxin. (**F**) Small nerve terminal occupies only part of a highly simplified postsynaptic region. Asterisk indicates remnants of basal lamina investing remnants of preexisting folds. *(Reproduced from Selcen et al.*[55]*)*

suprathreshold EPPs failed to generate muscle action potentials pointing to Nav1.4, encoded by *SCN4A*, as the culprit. EP structure and Nav1.4 expression at the EPs were normal, but *SCN4A* harbored two mutations (S246L in the S4/S5 linker in domain I and V1442E in S4/S5 linker in domain IV). Recombinant V1442E-sodium channels expressed in HEK cells showed marked enhancement of fast inactivation close to the resting potential and enhanced use-dependent inactivation on high-frequency stimulation; S246L had only minor kinetic effects and is likely a benign mutation. The safety margin in this congenital myasthenic syndrome is impaired because a large fraction of the postsynaptic Nav1.4 channels are inexcitable in the resting state.[59]

CMS Caused by Defects in Endplate Development or Maintenance

To date, mutations in proteins essential for EP development and maintenance have been detected in MuSK, agrin, Dok-7, rapsyn, and four enzymes subserving protein glycosylation, namely GFPT1, DAPGT1, ALG2, and ALG14. Agrin, secreted into the synaptic space by the nerve terminal, binds to the lipoprotein-related protein LRP4 in the postsynaptic membrane. The agrin-LRP4 complex binds to and activates MuSK. Binding of LRP4 to MuSK enhances MuSK phosphorylation and leads to clustering of LRP4 and MuSK. Activated MuSK in concert with postsynaptic Dok-7 and other postsynaptic proteins acts on rapsyn to concentrate AChR in the postsynaptic membrane, increase synapse specific gene expression by postsynaptic nuclei, and promote postsynaptic differentiation. Clustered LRP4, in turn, promotes differentiation motor axons. The agrin-LRP4-MusK-Dok-7 signaling system is also essential for maintaining the structure of the adult neuromuscular junction (reviewed by Burden et al.).[60]

CMS Caused by Mutations in Rapsyn

Rapsyn, under the influence of agrin, LRP4, MuSK, and Dok-7, concentrates AChR in the postsynaptic membrane and links it to the subsynaptic cytoskeleton through dystroglycan.[61–65] Mutations have now been detected in the entire open reading frame and promoter region of *RAPSN.* Nearly all Indo-Europeans harbor a common N88K mutation in *RAPSN*[66] (Figure 26.12A–C). Rapsyn mutations impede rapsyn self-association, hinder rapsyn binding to AChR, interfere with agrin- and LRP4-mediated clustering of AChR in the postsynaptic membrane, or decrease rapsyn expression.[67,69] There are no consistent genotype-phenotype correlations except for near-Eastern Jewish patients harboring a homozygous E-box mutation ($-38A > G$) have masticatory and facial muscle weakness, ptosis, prognathism, and hypernasal speech. Cervical, axial and limb muscles are usually spared.[70]

In most patient, myasthenic symptoms present at birth or infancy; in a few they present in the second or third decade.[71] Arthrogryposis at birth and other congenital malformations occur in nearly one-third.[67,68,71] Respiratory infections or other febrile illnesses precipitate increased weakness and respiratory crises and can cause an anoxic encephalopathy.[67,71–73] Mutations in the open reading frame of *RAPSN* result in clinical features that resemble those of autoimmune myasthenia except that ophthalmoparesis was reported as uncommon,[71] but 9 out of 39 of our patients had constant or episodic ophthalmoparesis[68] and most had eyelid ptosis. Faciobulbar and cervical muscle weakness is common. The weakness is more severe in proximal than distal weakness. Out-of-proportion weakness of the anterior tibial muscle has been noted in the late-onset cases.[71]

The morphologic features of the EP and the factors that impair the safety margin of neuromuscular transmission are like those in primary AChR deficiency. The synaptic contacts are multiple and dispersed over the fiber surface (Figure 26.12D). The EP AChR deficiency is relatively mild in most patients and the postsynaptic regions are less well differentiated than in patients harboring low-expressor mutations in AChR subunits (Figure 26.12E). In some patients, single-fiber EMG is required to demonstrate a defect in neuromuscular transmission.

Most patients respond well to AChE inhibitors; some derive additional benefit from 3,4-DAP,[72] and some benefit from the added use of ephedrine or albuterol.

CMS Caused by Mutations in Agrin

Two reports describe a CMS caused by defects in agrin. In one report, two siblings with eyelid ptosis but normal ocular ductions had only mild weakness of the facial and hip-girdle muscle.[74] They carry a homozygous missense mutation (G1709R) in the laminin G-like 2 domain required for MuSK activation. AChR and agrin expression at the EP were normal. Structural studies showed misshapen EPs with postsynaptic gutters partially denuded of nerve endings and formation of new EP regions. The postsynaptic regions were preserved. Expression studies revealed no effect on MuSK activation by agrin. Forced expression of a mutant agrin minigene in mouse muscle induced changes similar to those at patient EPs. Thus, the observed mutation perturbs the maintenance of the EP without preventing agrin initiating postsynaptic differentiation. Cholinesterase inhibitors and 3,4-DAP were ineffective.

The second report describes a severe CMS caused by N-terminal nonsense (Q353X) and missense (V1727F) mutations located at the N-terminal mutation in the second laminin G-like domain of agrin.[75] Expression studies revealed that the nonsense mutation abolished agrin expression and the missense mutation markedly reduced AChR clustering in C2 muscle cells. The synaptic contacts were dispersed and fragmented, the postsynaptic regions were simplified, and the nerve terminals were reduced in size. In addition degenerative changes were present in the functional folds space and in the junctional cytoplasm. The index patient failed to respond to pyridostigmine and 3,4-DAP but responded partially to ephedrine.

CMS Caused by Mutations in MuSK

This CMS has been reported three times. In one kinship, two siblings harbored heterozygous frame-shifting (c.220insC) and a missense (V790M) mutation that decreased the expression and stability of MuSK and reduced AChR aggregation. The EPs consisted of multiple small regions linked by nerve sprouts. The EP AChR content was 45% of normal.[76] Overexpression of the missense mutation at mouse EPs reproduced the findings at patient EPs.[77]

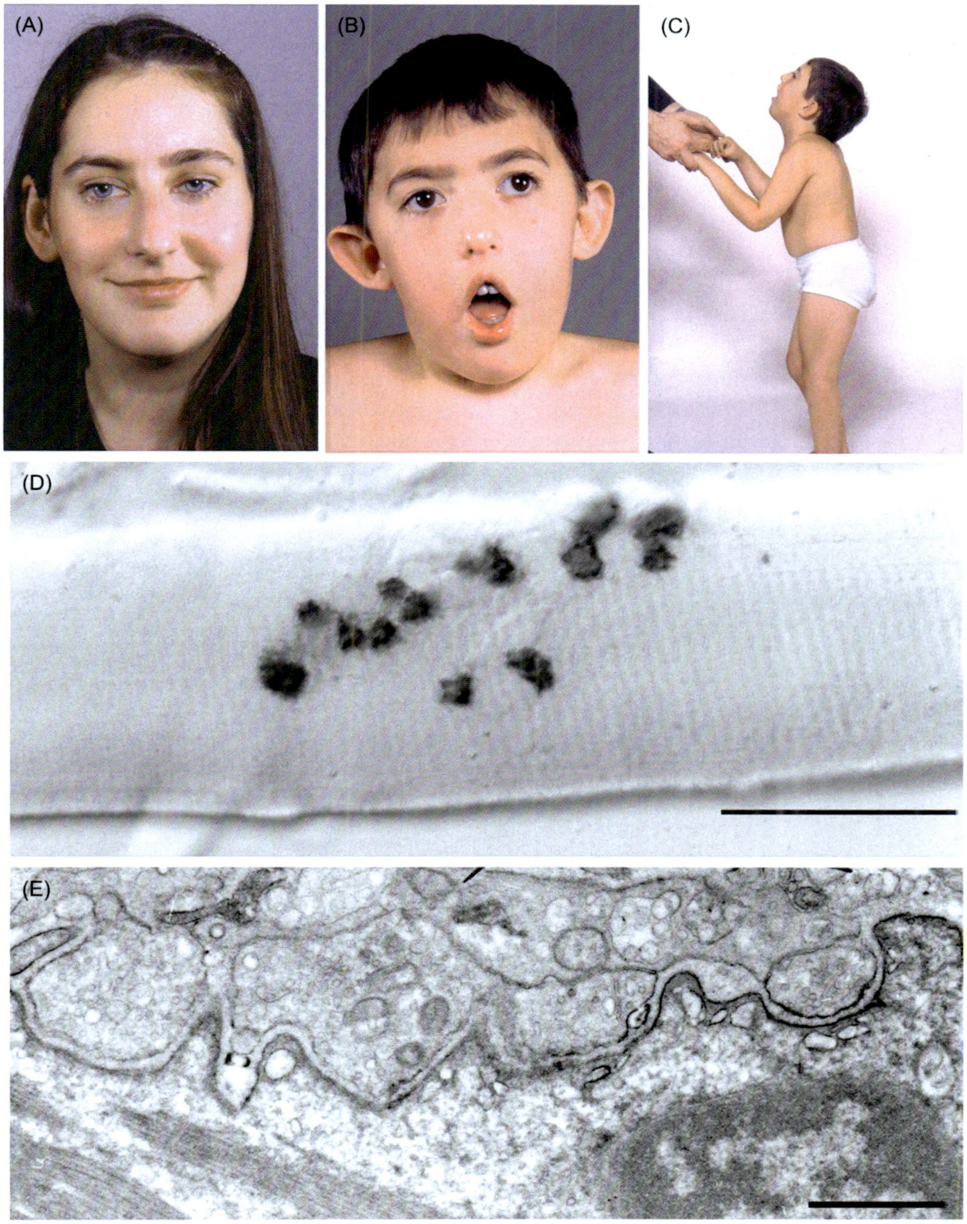

FIGURE 26.12 Rapsyn deficiency. (A) A 27-year-old woman and a 6-year-old boy (B and C) homozygous for the N88K rapsyn mutation. The 27-year-old woman shows only mild eyelid ptosis and mild fatigable weakness on exertion. The 6-year-old boy can only stand with support. Note small stature, short neck, low set ears, and contractures at the elbow and knee. (D) Small cholinesterase reactive EP regions are dispersed over an extended length of the muscle fiber. (E) Multiple small nerve terminals overlie highly simplified postsynaptic regions with no junctional folds. The distribution of AChR on the postsynaptic membrane, visualized with peroxidase-labeled α-bungarotoxin, is patchy. Bar 50 micrometers in (D) and 1 micrometer in (E). *(Panels A and B are reproduced from Ohno et al.[67] and panels D and E from Milone et al.[68])*

A second report describes heterozygous M605I and A727V mutations in a woman with severe myasthenic symptoms since early life that improved after puberty but worsened after menstrual periods. The synaptic response to ACh was reduced to 30% and quantal release to 50% of normal. The synaptic contacts were small and electron microscopy showed simplified postsynaptic regions with sparse secondary synaptic clefts. The patient failed to respond to pyridostigmine, ephedrine, or 3,4-DAP but responded partially to albuterol.[78]

A third kinship harbored a homozygous P344R missense mutation in the cysteine-rich extracellular domain of MuSK.[79] The clinical course was progressive. Low doses of pyridostigmine and 3,4-DAP led to gradual improvement; ephedrine or higher doses of pyridostigmine were not tolerated.

CMS Caused by Mutations in Dok-7

After the discovery that Dok-7 is a muscle-intrinsic activator of MuSK,[64] numerous CMS-related mutations were identified in *DOK7*.[80–84] All had limb-girdle weakness with lesser face, jaw, or neck muscle weakness. The EPs were small and the synaptic response to ACh was half-normal.[85] A subsequent study of 16 patients showed a wider clinical spectrum (Figure 26.13).[82] All patients experienced short-term fatigability on exertion. Neonatal onset was evident in 8, severe bulbar weakness in 5, and significant oculoparesis in 3. The clinical course was mild to severe. Type 1 fiber preponderance, type 2 fiber atrophy, isolated necrotic and regenerating fibers, and pleomorphic oxidative enzyme decreases or target formations were variable associated features. All EPs consisted of one to multiple small synaptic contacts. Impaired maintenance of the EPs was indicated by ongoing destruction and remodeling of the EPs (Figure 26.14). Neuromuscular transmission was compromised by decreased quantal release and reduced response to released ACh. Each patient harbored two heteroallelic mutations in *DOK7*. No consistent phenotype-genotype correlations emerged.[86] Importantly, this CMS is worsened by cholinergic agonists but responds well over a period of time to ephedrine[87] and albuterol.[86]

CMS Caused by Congenital Defects in Glycosylation (CDG)

Glycosylation increases solubility, folding, stability, assembly, and intracellular transport of nascent peptides. O-glycosylation occurs in the Golgi apparatus with addition of sugar residues to hydroxyl groups of serine and threonine; N-glycosylation occurs in the endoplasmic reticulum in sequential reactions that decorate amino group of asparagine with a core glycan composed of 2 glucose, 9 mannose, and 2 N-acetylglucosamine (GlcNAc) residues. To date, four enzymes subserving glycosylation are known to be associated with a CMS: GFPT1 (glutamine fructose-6-phosphate transaminase),[88,89] DPAGT1 (dolichyl-phosphate [UDP-N-acetylglucosamine] N-acetylglucosaminephosphotransferase 1),[90,91]

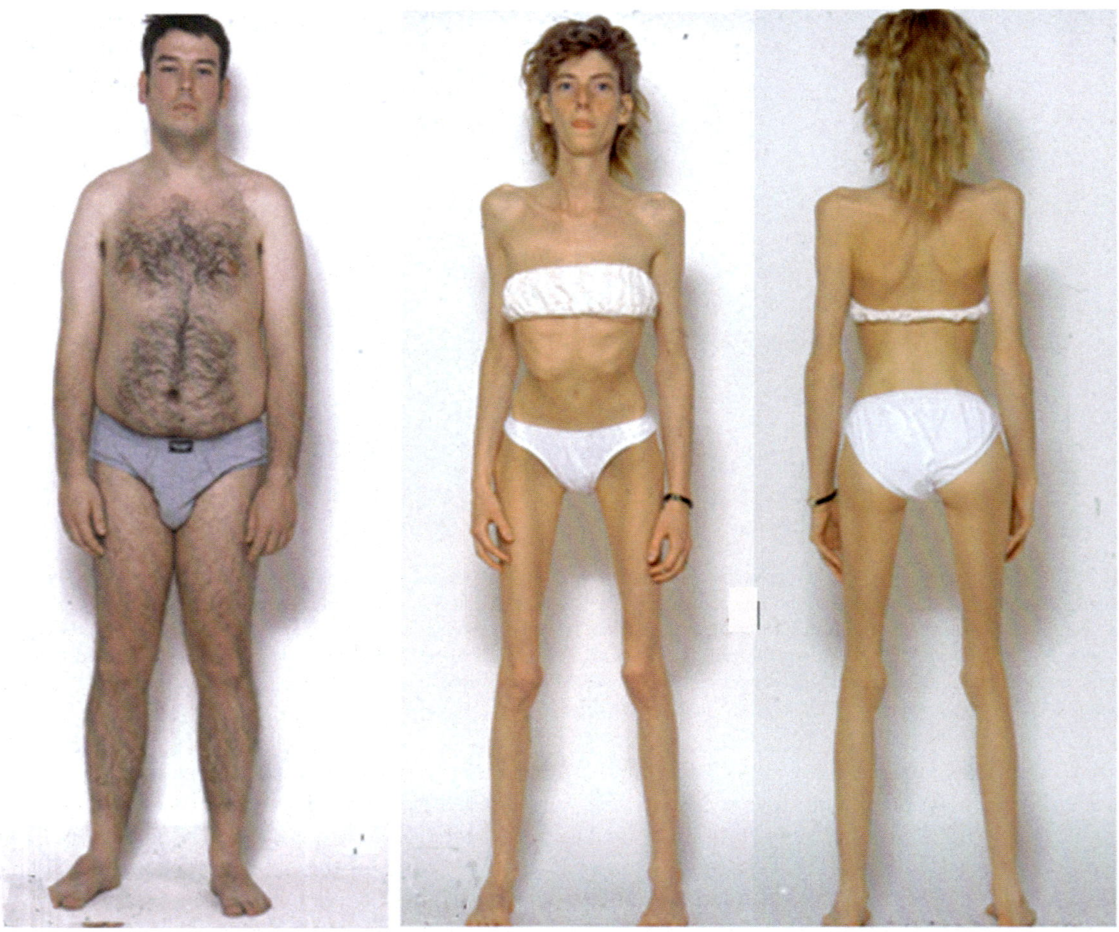

FIGURE 26.13 Phenotypic variability of Dok-7 myasthenia. Patient on left has mild weakness and atrophy of limb girdle muscles, and mild eyelid ptosis. Patient on right has severe diffuse weakness and atrophy of limb and axial muscles. *(Reproduced from Selcen et al.[82])*

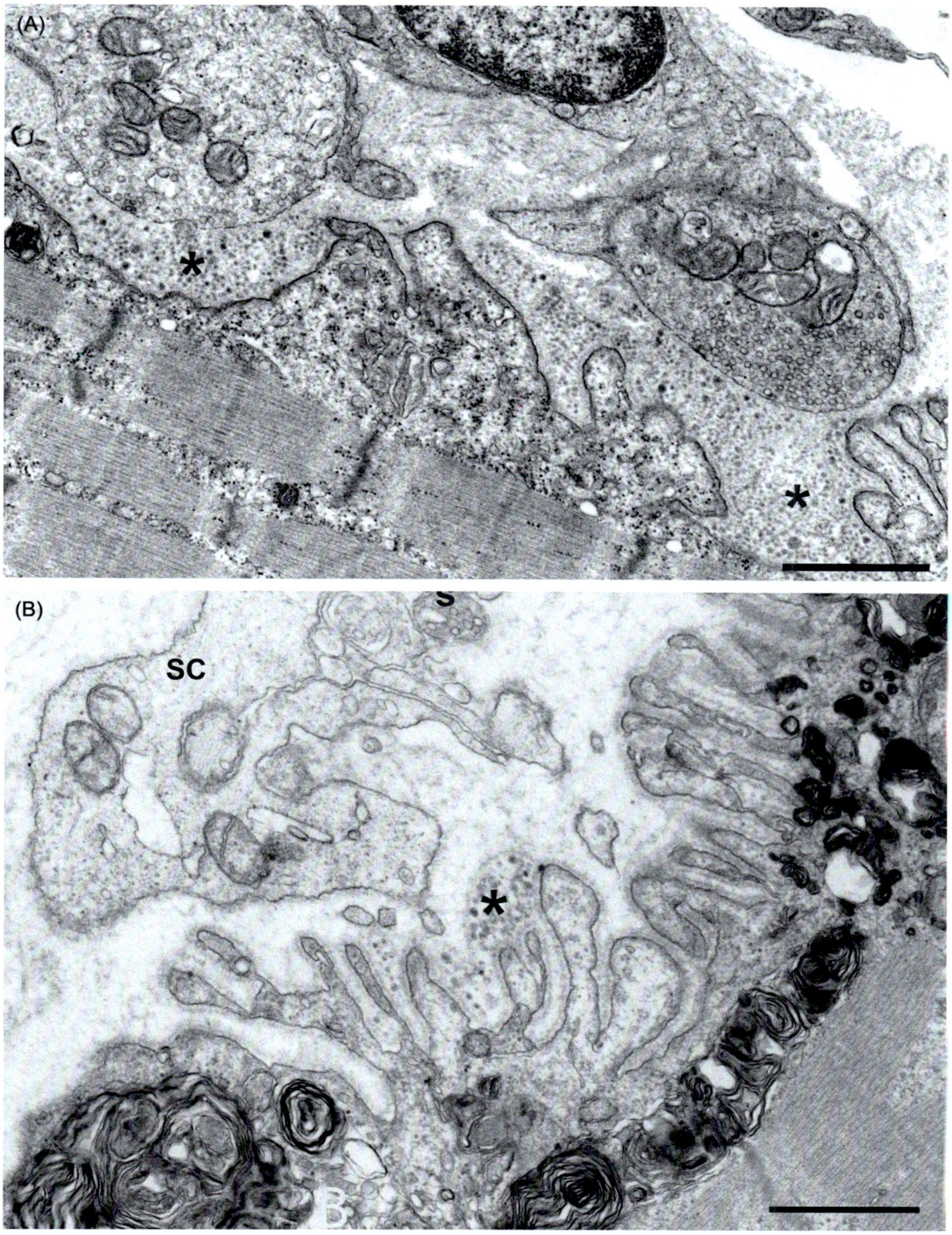

FIGURE 26.14 Dok-7 myasthenia. (**A**) Extensive degeneration of the junctional folds results in a simplified postsynaptic membrane and widening of the synaptic space. This predicts a decreased synaptic response to ACh due to loss of AChR from the destroyed folds, loss of ACh by diffusion from the widened synaptic space, and decreased input resistance of the remaining simple folds. (**B**) A highly abnormal EP region devoid of nerve terminal. Some junctional folds are degenerating (asterisk). The subsynaptic sarcoplasm harbors large myeloid structures. Nerve sprouts (s) surrounded by Schwann cell (SC) appear above the junction. *(Reproduced from Selcen et al.[82])*

ALG2 (alpha-1,3-mannosyltransferase), and ALG14 (UDP-N-acetylglucosaminyltransferase subunit).[3]

CMS Caused by Mutations in GFPT1

GFPT1 controls the flux of glucose into the hexosamine pathway, and thus formation of hexosamine products and availability of precursors for N- and O-linked protein glycosylation. A CMS caused by defects in GFPT1 causing limb-girdle muscle weakness was reported in 2011 in 16 patients.[88] The effects of the different mutations on EP fine structure and the extent to which they alter parameters of neuromuscular transmission were not determined. A subsequent report of 11 patients extended the phenotypic spectrum of the disease and described *in vitro* electrophysiologic and ultrastructural features of the

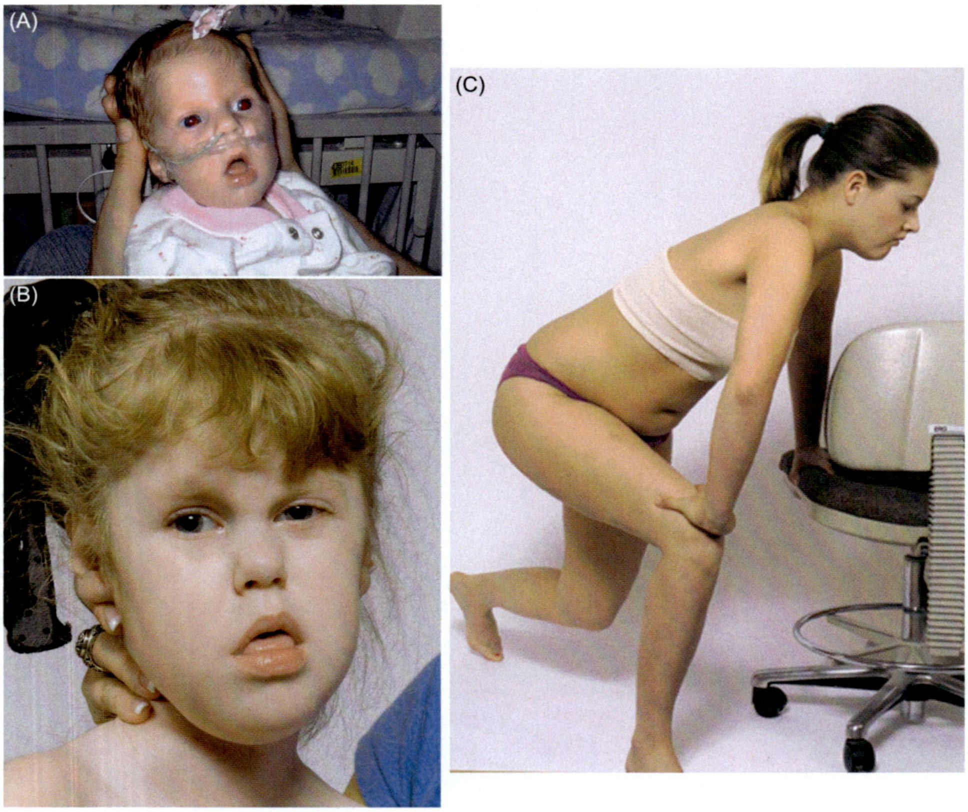

FIGURE 26.15 GFPT1 myasthenia. (**A** and **B**) Patient with deletion of muscle specific exon of GFPT1 during infancy and at age 4 years. Note facial diplegia, open mouth, protruding tongue, lack of head control, asymmetric ptosis, and strabismus. (**C**) Patient harboring a p.Arg545Pro mutation and a c.*22C > A mutation has only mild limb-girdle weakness at age 12 years. *(Reproduced by permission from Selcen et al.[89])*

disease.[89] In 10 of 11 patients the weakness was slowly progressive and involved distal arm and anterior leg as well as limb girdle muscles. One patient with mutations disrupting the muscle-specific exon of GFPT1 never moved *in utero*, was arthrogrypotic at birth, and subsequently developed a cardiomyopathy and has remained bedfast and tube-fed (Figure 26.15).

Histochemical studies in 9 less affected patients revealed tubular aggregates in 6 and rimmed vacuoles in 3. Electron microscopy demonstrated abnormally small EP regions and poorly developed junctional folds (Figure 26.16A and B). Microelectrode studies of intercostal muscle EPs in 3 of 5 patients indicated a reduced synaptic response to ACh.

The patient whose mutations disrupt the muscle-specific exon has a severe myopathy with numerous dilated and degenerating vesicular profiles, autophagic vacuoles, and bizarre apoptotic nuclei[89] (Figure 26.16C and D).

That many EPs in this CMS are underdeveloped likely stems from hypoglycosylation and altered function of EP-specific glycoproteins, such as MuSK, agrin, and dystroglycans. However, it is unclear why neuromuscular transmission and muscle fiber architecture are selectively affected when glycosylated proteins are widely distributed in many tissues and organs.

CMS Caused by Mutations in DPAGT1

DPAGT1 catalyzes the first committed step of N-linked protein glycosylation. A defect in DAPGT1 predicts failure of asparagine glycosylation in multiple proteins distributed throughout the organism, but in the first 5 patients with mutations in *DAPGT1*, only neuromuscular transmission was selectively affected.[90] The adverse effect on neuromuscular transmission was attributed to reduced expression of AChR at the EP, but direct EP studies were not performed

A subsequent study of two siblings (Pts 1 and 3) and of another patient with DPAGT1 deficiency (Pt 2) extended the phenotypic spectrum of the disease.[91] They have moderately severe to severe weakness (Figure 26.17), are cognitively challenged, and Pt 3 also has autistic features. Histochemical studies reveal mild EP AChR deficiency, fiber type disproportion, small tubular aggregates, autophagic vacuoles, and ovoid synaptic contacts. Ultrastructural studies demonstrate few degenerating EPs, small pre- and postsynaptic regions with poorly developed junctional folds, and an autophagic vacuolar myopathy (Figure 26.18A–I). Microelectrode studies demonstrate a ~50% decrease in evoked quantal release, postsynaptic response to ACh, and EP AChR content. Immunoblots of muscle extracts with two different antibodies

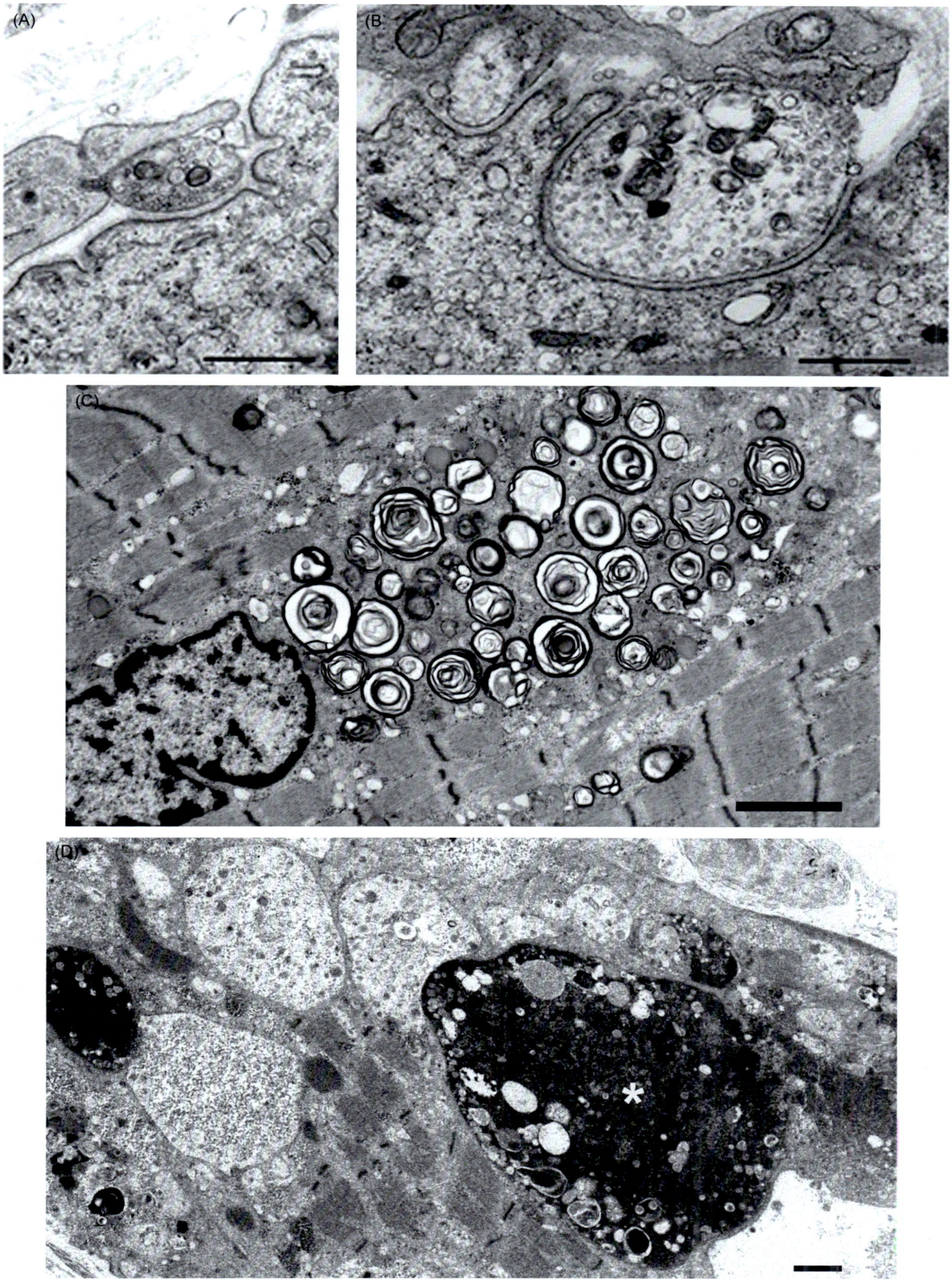

FIGURE 26.16 Ultrastructural alterations in GFPT1 myasthenia. (**A** and **B**) Hypoplastic EPs with small nerve terminals and simplified junctional folds. (**C** and **D**) Degenerating muscle fibers in patient lacking the muscle specific GFPT1 exon. In (**C**), note accumulation of dilated and degenerating vesicular profiles near a nucleus. In (**D**), note multiple autophagic vacuoles containing debris and apoptotic nuclei, one of which is large and bizarre (asterisk). *(Reproduced by permission from Selcen et al.*[89]*)*

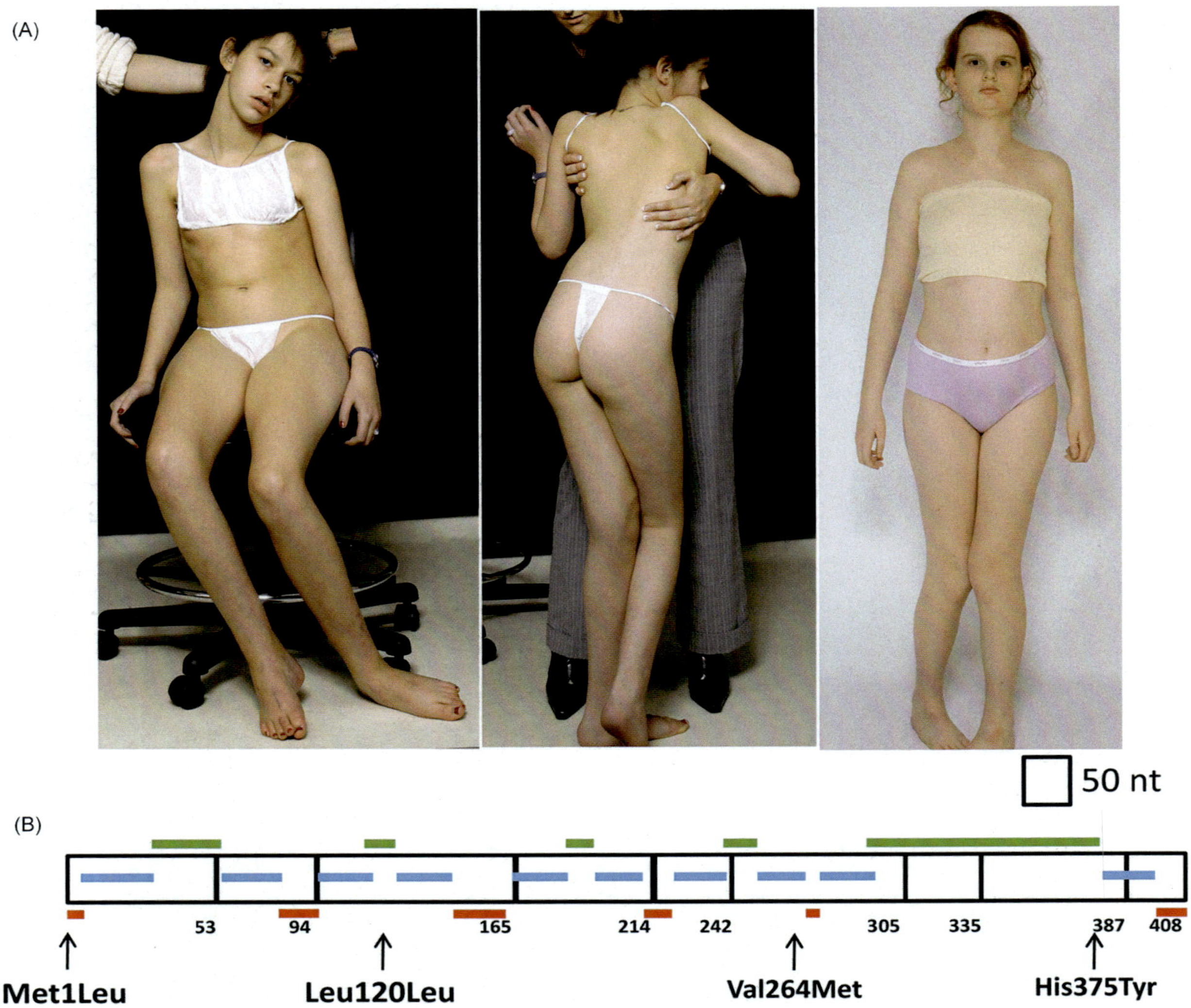

FIGURE 26.17 DPAGT1 myasthenia. (A) Patient 1 (left panels) is unable to sit upright or stand without support. Patient 2 (right panel) shows reduced muscle bulk, lordotic stance, internally rotated thighs causing cubitus valgus and in-toeing stance. (B) Scaled linear model of DPAGT1 with color coded luminal (red), transmembrane (blue), and cytoplasmic (green) sites, and identified mutations.

indicate reduced to absent protein glycosylation. Patient 1 responds poorly to pyridostigmine and 3,4-DAP; Patient 2 responds partially to pyridostigmine and albuterol.

Mutations in ALG2 and ALG14

ALG2 catalyzes the second and third committed steps of N-glycosylation. In one kinship, four affected siblings were homozygous for an insertion/deletion mutation; and another patient was homozygous for a low-expressor V68G mutation. ALG14 forms a multiglycosyltransferase complex with ALG13 and DPAGT1 and catalyzes the first committed step of N-glycosylation. In one family two affected siblings carried heteroallelic P65L and V68G mutations. EP ultrastructure and parameters of neuromuscular transmission were not investigated.[3]

CMS Associated with Centronuclear Myopathy

Centronuclear myopathies are clinically and genetically heterogeneous congenital disorders in which the predominant pathologic alteration is centralization of the muscle fiber nuclei. The implicated disease proteins/genes are myotubularin (*MTM1*), dynamin 2 (*DNM2*), amphiphysin 2 (*BIN1*), and the ryanodine receptor (*RYR1*).[92] Features suggesting a myasthenic disorder, ptosis, ophthalmoparesis, abnormal fatigability, decremental EMG response, or abnormally increased jitter were observed in clinically and genetically different patients (reviewed by Vogt et al.),[49] but the pathogenesis of the putative myasthenic disorder was not elucidated.

We investigated a 39-year-old man with centronuclear myopathy and a myasthenic syndrome. His early motor

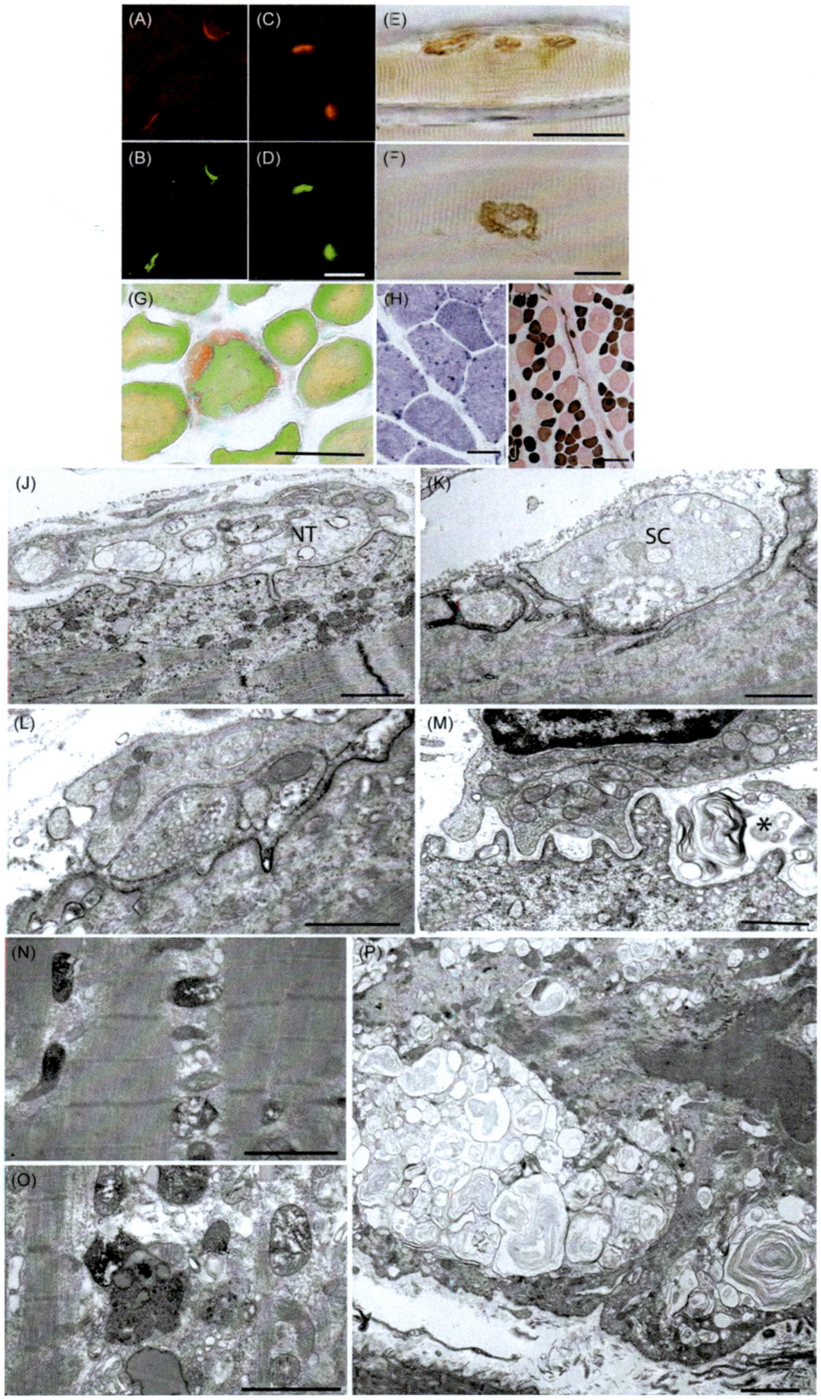

FIGURE 26.18 Structural observations in DPAGT1 myasthenia. (**A–D**) Paired fluorescence localization of AChR (red color) and AChE (green color) at patient (**A** and **C**) and control EPs (**B** and **D**). Note attenuated reactivity for AChR and AChE at patient EPs. (**E** and **F**) Face-on views of AChE reactive EPs. In panel (**E**), single muscle fiber displays three ovoid-shaped EPs. EP in panel (**F**) is composed of multiple small round regions. (**G**) Muscle fiber harboring acid-phosphatase-positive autophagic vacuoles. (**H**) Type 2 muscle fibers display punctate NADH-dehydrogenase-reactive tubular aggregates. (**I**) Fiber-type disproportion (ATPase reaction, pH 4.3). Examination of EP ultrastructure reveals degenerating nerve terminal (NT) in (**J**), degenerating Schwann cell (SC) in (**K**), degenerating junctional folds (asterisk) in (**M**), and highly simplified postsynaptic regions (**J–L**). Localization of AChR with peroxidase-labeled α-bgt reveals patchy and focally attenuated expression of AChR on the postsynaptic membrane (**K, L**). Examination of muscle fiber ultrastructure shows degenerating organelles (**N, O**), loss of myofibrils (**O**), and large autophagic vacuoles and degraded myofibrils (**P**).

BOX 26.3 Pharmacotherapy of the CMS

- ChAT deficiency: Pyridostigmine
- AChE deficiency: Albuterol or ephedrine. **Avoid pyridostigmine.**
- Simple AChR deficiency: Pyridostigmine; 3,4-DAP also helps in 30–50%; addition of albuterol may benefit patients refractory to pyridostigmine and 3,4-DAP.
- Slow-channel CMS: Quinidine, quinine, or fluoxetine (long-lived open channel blockers of AChR channel). **Avoid pyridostigmine.**
- Fast-channel CMS: Pyridostigmine and 3,4-DAP.
- Rapsyn deficiency: Pyridostigmine, 3,4-DAP, albuterol.
- Na-channel myasthenia: Pyridostigmine and acetazolamide.
- Plectin deficiency: Refractory to pyridostigmine; 3,4-DAP may help.
- Dok-7 myasthenia: Albuterol or ephedrine. **Avoid pyridostigmine.**
- Agrin deficiency: No response to pyridostigmine or 3,4DAP in one patient,[74] some response to pyridostigmine in another patient.[75]
- MuSK deficiency: Variable responses to low doses of pyridostigmine or to 3,4-DAP or albuterol.[76,78,79]
- GFPT1, DAPGT1, ALG2, and ALG14 myasthenia: Pyridostigmine, 3,4-DAP.

development was normal but he could never run. He had mild ptosis when tired and moderately severe fatigable limb-girdle weakness since his teens. He had a 19–35% EMG decrement in different muscles and responded partially to pyridostigmine. Serologic tests for AChR and MuSK antibodies were negative. No mutations were detected in *MTM1*, *DNM2*, *BIN1*, and *RYR1*.

Intercostal muscle EP studies demonstrated formation of new EP regions on individual fibers. AChE expression was normal but AChR expression was mildly reduced. Quantitative electron microscopy revealed simplified postsynaptic regions, with normal nerve terminal size and normal synaptic vesicle density. The MEPP amplitude was decreased to 60% of normal. Quantal release by nerve impulse (m) was reduced to 40% of normal due to a decreased number of quanta available for release (n). The probability of quantal release (p) was normal. The safety margin of neuromuscular transmission in this patient is compromised by decreased quantal release by nerve impulse and, to a lesser extent, by simplification of the postsynaptic regions and the mild AChR deficiency.

CURRENTLY AVAILABLE THERAPIES

Agents that benefit one type of CMS can be ineffective or harmful in another type. For example, slow-channel syndromes are worsened whereas fast-channel syndromes are improved by cholinergic agonists. Similarly, patients harboring mutations in DOK7 are worsened by cholinergic agonists but are improved by adrenergic agonists. Therefore it is essential that a molecular diagnosis should inform the choice of therapy. Also, different mutations can have different functional consequences, so patients harboring different mutations in a given gene may respond differently to a given form of therapy. Finally, the cholinergic agonists pyridostigmine and 3,4-DAP exert their effects as soon as the medication is absorbed whereas the beneficial effects of fluoxetine, quinidine, quinine, ephedrine, and albuterol develop more slowly over days, weeks, or months. Box 26.3 summarizes therapy of the currently recognized CMS.

REFERENCES

1. Rothbart HB. Myasthenia gravis. Familial occurrence. *JAMA* 1937; **108**:715–17.
2. Bundey S. A genetic study of infantile and juvenile myasthenia gravis. *J Neurol Neurosurg Psychiatry* 1972;**35**:41–51.
3. Cossins J, Belaya K, Hicks D, Salih MA, Finlayson S, Carboni N, et al. Congenital myasthenic syndromes due to mutations in ALG2 and ALG14. *Brain* 2013;**136**:944–56.
4. Kaiser J. Human genetics. Affordable 'exomes' fill gaps in a catalog of rare diseases. *Science* 2010;**330**:903.
5. Ohno K, Tsujino A, Brengman JM, Harper CM, Bajzer Z, Udd B, et al. Choline acetyltransferase mutations cause myasthenic syndrome associated with episodic apnea in humans. *Proc Natl Acad Sci USA* 2001;**98**:2017–22.
6. Shen XM, Crawford TO, Brengman J, Acsadi G, Iannaconne S, Karaca E, et al. Functional consequences and structural interpretation of mutations of human choline acetyltransferase. *Hum Mutat* 2011;**32**:1259–67.
7. Byring RF, Pihko H, Tsujino A, Shen XM, Gustafsson B, Hackman P, et al. Congenital myasthenic syndrome associated with episodic apnea and sudden infant death. *Neuromuscul Disord* 2002;**12**:548–53.
8. Maselli RA, Chen D, Mo D, Bowe C, Fenton G, Wollmann RL. Choline acetyltransferase mutations in myasthenic syndrome due to deficient acetylcholine resynthesis. *Muscle Nerve* 2003;**27**:180–7.
9. Mallory LA, Shaw JG, Burgess SL, Estrella E, Nurko S, Burpee TM, et al. Congenital myasthenic syndrome with episodic apnea. *Pediatr Neurol* 2009;**41**:42–5.
10. Schara U, Christen HJ, Durmus H, Hietala M, Krabetz K, Rodolico C, et al. Long-term follow-up in patients with congenital myasthenic syndrome due to CHAT mutations. *Eur J Paediatr Neurol* 2010;**14**:326–33.
11. Kraner S, Laufenberg I, Strassburg HM, Sieb JP, Steinlein OK. Congenital myasthenic syndrome with episodic apnea in patients homozygous for a CHAT missense mutation. *Arch Neurol* 2003;**60**: 761–3.
12. Kraner S, Laufenberg I, Strassburg HM, Sieb JP, Steinlein OK. Congenital myasthenic syndrome with paucity of synaptic vesicles and reduced quantal release [abstract]. *Neurology* 1990;**40**:278.

13. Bady B, Chauplannaz G, Carrier H. Congenital Lambert-Eaton myasthenic syndrome. *J Neurol Neurosurg Psychiatry* 1987;**50**:476–8.
14. Ohno K, Brengman J, Tsujino A, Engel AG. Human endplate acetylcholinesterase deficiency caused by mutations in the collagen-like tail subunit (ColQ) of the asymmetric enzyme. *Proc Natl Acad Sci USA* 1998;**95**:9654–9.
15. Kimbell LM, Ohno K, Engel AG, Rotundo RL. C-terminal and heparin-binding domains of collagenic tail subunit are both essential for anchoring acetylcholinesterase at the synapse. *J Biol Chem* 2004;**279**:10997–1005.
16. Bestue-Cardiel M, Saenz de Cabezon-Alvarez A, Capablo-Liesa JL, Lopez-Pison J, Pena-Segura JL, Martin-Martinez J, et al. Congenital endplate acetylcholinesterase deficiency responsive to ephedrine. *Neurology* 2005;**65**:144–6.
17. Engel AG, Lambert EH, Gomez MR. A new myasthenic syndrome with end-plate acetylcholinesterase deficiency, small nerve terminals, and reduced acetylcholine release. *Ann Neurol* 1977;**1**:315–30.
18. Hutchinson DO, Walls TJ, Nakano S, Camp S, Taylor P, Harper CM, et al. Congenital endplate acetylcholinesterase deficiency. *Brain* 1993;**116**(Pt 3):633–53.
19. Brengman JM, Capablo-Liesa JL, Lopez-Pison J, Pena-Segura J, Benaventa-Aguilar I, Sadeh M, et al. Ephedrine treatment of seven patients with congenital endplate acetylcholinesterase deficiency. *Neuromuscul Disord* 2006;**16**:S129.
20. Mihaylova V, Muller JS, Vilchez JJ, Salih MA, Kabiraj MM, D'Amico A, et al. Clinical and molecular genetic findings in COLQ-mutant congenital myasthenic syndromes. *Brain* 2008;**131**:747–59.
21. Liewluck T, Selcen D, Engel AG. Beneficial effects of albuterol in congenital endplate acetylcholinesterase deficiency and Dok-7 myasthenia. *Muscle Nerve* 2011;**44**:789–94.
22. Maselli RA, Ng JJ, Anderson JA, Cagney O, Arredondo J, Williams C, et al. Mutations in LAMB2 causing a severe form of synaptic congenital myasthenic syndrome. *J Med Genet* 2009;**46**:203–8.
23. Milone M, Wang HL, Ohno K, Prince R, Fukudome T, Shen XM, et al. Mode switching kinetics produced by a naturally occurring mutation in the cytoplasmic loop of the human acetylcholine receptor epsilon subunit. *Neuron* 1998;**20**:575–88.
24. Wang HL, Ohno K, Milone M, Brengman JM, Evoli A, Batocchi AP, et al. Fundamental gating mechanism of nicotinic receptor channel revealed by mutation causing a congenital myasthenic syndrome. *J Gen Physiol* 2000;**116**:449–62.
25. Brejc K, van Dijk WJ, Klaassen RV, Schuurmans M, van Der Oost J, Smit AB, et al. Crystal structure of an ACh-binding protein reveals the ligand-binding domain of nicotinic receptors. *Nature* 2001;**411**:269–76.
26. Engel AG, Ohno K, Sine SM. Sleuthing molecular targets for neurological diseases at the neuromuscular junction. *Nat Rev Neurosci* 2003;**4**:339–52.
27. Ohno K, Hutchinson DO, Milone M, Brengman JM, Bouzat C, Sine SM, et al. Congenital myasthenic syndrome caused by prolonged acetylcholine receptor channel openings due to a mutation in the M2 domain of the epsilon subunit. *Proc Natl Acad Sci USA* 1995;**92**:758–62.
28. Harper CM, Engel AG. Quinidine sulfate therapy for the slow-channel congenital myasthenic syndrome. *Ann Neurol* 1998;**43**:480–4.
29. Harper CM, Fukodome T, Engel AG. Treatment of slow-channel congenital myasthenic syndrome with fluoxetine. *Neurology* 2003;**60**:1710–13.
30. Colomer J, Muller JS, Vernet A, Nascimento A, Pons M, Gonzalez V, et al. Long-term improvement of slow-channel congenital myasthenic syndrome with fluoxetine. *Neuromuscul Disord* 2006;**16**:329–33.
31. Chaouch A, Muller JS, Guergueltcheva V, Dusl M, Schara U, Rakocevic-Stojanovic V, et al. A retrospective clinical study of the treatment of slow-channel congenital myasthenic syndrome. *J Neurol* 2012;**259**:474–81.
32. Ohno K, Wang HL, Milone M, Bren N, Brengman JM, Nakano S, et al. Congenital myasthenic syndrome caused by decreased agonist binding affinity due to a mutation in the acetylcholine receptor epsilon subunit. *Neuron* 1996;**17**:157–70.
33. Shen XM, Ohno K, Tsujino A, Brengman JM, Gingold M, Sine SM, et al. Mutation causing severe myasthenia reveals functional asymmetry of AChR signature cystine loops in agonist binding and gating. *J Clin Invest* 2003;**111**:497–505.
34. Wang HL, Milone M, Ohno K, Shen XM, Tsujino A, Batocchi AP, et al. Acetylcholine receptor M3 domain: stereochemical and volume contributions to channel gating. *Nat Neurosci* 1999;**2**:226–33.
35. Sine SM, Shen XM, Wang HL, Ohno K, Lee WY, Tsujino A, et al. Naturally occurring mutations at the acetylcholine receptor binding site independently alter ACh binding and channel gating. *J Gen Physiol* 2002;**120**:483–96.
36. Shen XM, Brengman JM, Sine SM, Engel AG. Myasthenic syndrome AChRalpha C-loop mutant disrupts initiation of channel gating. *J Clin Invest* 2012;**122**:2613–21.
37. Shen XM, Brengman JM, Edvardson S, Sine SM, Engel AG. Highly fatal fast-channel syndrome caused by AChR epsilon subunit mutation at the agonist binding site. *Neurology* 2012;**79**:449–54.
38. Engel AG, Ohno K, Bouzat C, Sine SM, Griggs RC. End-plate acetylcholine receptor deficiency due to nonsense mutations in the epsilon subunit. *Ann Neurol* 1996;**40**:810–17.
39. Ohno K, Quiram PA, Milone M, Wang HL, Harper MC, Pruitt JN, et al. Congenital myasthenic syndromes due to heteroallelic nonsense/missense mutations in the acetylcholine receptor epsilon subunit gene: identification and functional characterization of six new mutations. *Hum Mol Genet* 1997;**6**:753–66.
40. Ohno K, Anlar B, Ozdirim E, Brengman JM, DeBleecker JL, Engel AG. Myasthenic syndromes in Turkish kinships due to mutations in the acetylcholine receptor. *Ann Neurol* 1998;**44**:234–41.
41. Abicht A, Stucka R, Karcagi V, Herczegfalvi A, Horvath R, Mortier W, et al. A common mutation (epsilon1267delG) in congenital myasthenic patients of Gypsy ethnic origin. *Neurology* 1999;**53**:1564–9.
42. Middleton L, Ohno K, Christodoulou K, Brengman J, Milone M, Neocleous V, et al. Chromosome 17p-linked myasthenias stem from defects in the acetylcholine receptor epsilon-subunit gene. *Neurology* 1999;**53**:1076–82.
43. Croxen R, Newland C, Betty M, Vincent A, Newsom-Davis J, Beeson D. Novel functional epsilon-subunit polypeptide generated by a single nucleotide deletion in acetylcholine receptor deficiency congenital myasthenic syndrome. *Ann Neurol* 1999;**46**:639–47.
44. Abicht A, Stucka R, Schmidt C, Briguet A, Hopfner S, Song IH, et al. A newly identified chromosomal microdeletion and an

N-box mutation of the AChR epsilon gene cause a congenital myasthenic syndrome. *Brain* 2002;**125**:1005–13.

45. Ealing J, Webster R, Brownlow S, Abdelgany A, Oosterhuis H, Muntoni F, et al. Mutations in congenital myasthenic syndromes reveal an epsilon subunit C-terminal cysteine, C470, crucial for maturation and surface expression of adult AChR. *Hum Mol Genet* 2002;**11**:3087–96.
46. Sadeh M, Shen XM, Engel AG. Beneficial effect of albuterol in congenital myasthenic syndrome with epsilon-subunit mutations. *Muscle Nerve* 2011;**44**:289–91.
47. Hesselmans LF, Jennekens FG, Van den Oord CJ, Veldman H, Vincent A. Development of innervation of skeletal muscle fibers in man: relation to acetylcholine receptors. *Anat Rec* 1993;**236**: 553–62.
48. Morgan NV, Brueton LA, Cox P, Greally MT, Tolmie J, Pasha S, et al. Mutations in the embryonal subunit of the acetylcholine receptor (CHRNG) cause lethal and Escobar variants of multiple pterygium syndrome. *Am J Hum Genet* 2006;**79**:390–5.
49. Vogt J, Harrison BJ, Spearman H, Cossins J, Vermeer S, ten Cate LN, et al. Mutation analysis of CHRNA1, CHRNB1, CHRND, and RAPSN genes in multiple pterygium syndrome/fetal akinesia patients. *Am J Hum Genet* 2008;**82**:222–7.
50. Michalk A, Stricker S, Becker J, Rupps R, Pantzar T, Miertus J, et al. Acetylcholine receptor pathway mutations explain various fetal akinesia deformation sequence disorders. *Am J Hum Genet* 2008;**82**: 464–76.
51. Vogt J, Morgan NV, Marton T, Maxwell S, Harrison BJ, Beeson D, et al. Germline mutation in DOK7 associated with fetal akinesia deformation sequence. *J Med Genet* 2009;**46**:338–40.
52. Elliott CE, Becker B, Oehler S, Castanon MJ, Hauptmann R, Wiche G. Plectin transcript diversity: identification and tissue distribution of variants with distinct first coding exons and rodless isoforms. *Genomics* 1997;**42**:115–25.
53. Fuchs P, Zorer M, Rezniczek GA, Spazierer D, Oehler S, Castanon MJ, et al. Unusual 5′ transcript complexity of plectin isoforms: novel tissue-specific exons modulate actin binding activity. *Hum Mol Genet* 1999;**8**:2461–72.
54. Konieczny P, Fuchs P, Reipert S, Kunz WS, Zeold A, Fischer I, et al. Myofiber integrity depends on desmin network targeting to Z-disks and costameres via distinct plectin isoforms. *J Cell Biol* 2008;**181**: 667–81.
55. Selcen D, Juel VC, Hobson-Webb LD, Smith EC, Stickler DE, Bite AV, et al. Myasthenic syndrome caused by plectinopathy. *Neurology* 2011;**76**:327–36.
56. Forrest K, Mellerio JE, Robb S, Dopping-Hepenstal PJ, McGrath JA, Liu L, et al. Congenital muscular dystrophy, myasthenic symptoms and epidermolysis bullosa simplex (EBS) associated with mutations in the PLEC1 gene encoding plectin. *Neuromuscul Disord* 2010;**20**:709–11.
57. Maselli RA, Arredondo J, Cagney O, Mozaffar T, Skinner S, Yousif S, et al. Congenital myasthenic syndrome associated with epidermolysis bullosa caused by homozygous mutations in PLEC1 and CHRNE. *Clin Genet* 2011;**80**:444–51.
58. Gundesli H, Talim B, Korkusuz P, Balci-Hayta B, Cirak S, Akarsu NA, et al. Mutation in exon 1f of PLEC, leading to disruption of plectin isoform 1f, causes autosomal-recessive limb-girdle muscular dystrophy. *Am J Hum Genet* 2010;**87**: 834–41.
59. Tsujino A, Maertens C, Ohno K, Shen XM, Fukuda T, Harper CM, et al. Myasthenic syndrome caused by mutation of the SCN4A sodium channel. *Proc Natl Acad Sci USA* 2003;**100**:7377–82.
60. Burden SJ, Yumoto N, Zhang W. The role of MuSK in synapse formation and neuromuscular disease. *Cold Spring Harb Perspect Biol* 2013;**5**:a009167.
61. Froehner SC, Luetje CW, Scotland PB, Patrick J. The postsynaptic 43K protein clusters muscle nicotinic acetylcholine receptors in *Xenopus* oocytes. *Neuron* 1990;**5**:403–10.
62. Cartaud A, Coutant S, Petrucci TC, Cartaud J. Evidence for in situ and in vitro association between beta-dystroglycan and the subsynaptic 43K rapsyn protein. Consequence for acetylcholine receptor clustering at the synapse. *J Biol Chem* 1998;**273**:11321–6.
63. Mittaud P, Camilleri AA, Willmann R, Erb-Vogtli S, Burden SJ, Fuhrer C. A single pulse of agrin triggers a pathway that acts to cluster acetylcholine receptors. *Mol Cell Biol* 2004;**24**:7841–54.
64. Okada K, Inoue A, Okada M, Murata Y, Kakuta S, Jigami T, et al. The muscle protein Dok-7 is essential for neuromuscular synaptogenesis. *Science* 2006;**312**:1802–5.
65. Zhang B, Luo S, Wang Q, Suzuki T, Xiong WC, Mei L. LRP4 serves as a coreceptor of agrin. *Neuron* 2008;**60**:285–97.
66. Muller JS, Mildner G, Muller-Felber W, Schara U, Krampfl K, Petersen B, et al. Rapsyn N88K is a frequent cause of congenital myasthenic syndromes in European patients. *Neurology* 2003;**60**:1805–10.
67. Ohno K, Engel AG, Shen XM, Selcen D, Brengman J, Harper CM, et al. Rapsyn mutations in humans cause endplate acetylcholine-receptor deficiency and myasthenic syndrome. *Am J Hum Genet* 2002;**70**:875–85.
68. Milone M, Shen XM, Selcen D, Ohno K, Brengman J, Iannaccone ST, et al. Myasthenic syndrome due to defects in rapsyn: Clinical and molecular findings in 39 patients. *Neurology* 2009;**73**:228–35.
69. Cossins J, Burke G, Maxwell S, Spearman H, Man S, Kuks J, et al. Diverse molecular mechanisms involved in AChR deficiency due to rapsyn mutations. *Brain* 2006;**129**:2773–83.
70. Ohno K, Sadeh M, Blatt I, Brengman JM, Engel AG. E-box mutations in the RAPSN promoter region in eight cases with congenital myasthenic syndrome. *Hum Mol Genet* 2003;**12**:739–48.
71. Burke G, Cossins J, Maxwell S, Owens G, Vincent A, Robb S, et al. Rapsyn mutations in hereditary myasthenia: distinct early- and late-onset phenotypes. *Neurology* 2003;**61**:826–8.
72. Banwell BL, Ohno K, Sieb JP, Engel AG. Novel truncating RAPSN mutations causing congenital myasthenic syndrome responsive to 3,4-diaminopyridine. *Neuromuscul Disord* 2004;**14**:202–7.
73. Skeie GO, Aurlien H, Muller JS, Lochmuller H, Norgard G, Bindoff LA. Unusual features in a boy with the rapsyn N88K mutation. *Neurology* 2006;**67**:2262–3.
74. Huze C, Bauche S, Richard P, Chevessier F, Goillot E, Gaudon K, et al. Identification of an agrin mutation that causes congenital myasthenia and affects synapse function. *Am J Hum Genet* 2009; **85**:155–67.
75. Maselli RA, Fernandez JM, Arredondo J, Navarro C, Ngo M, Beeson D, et al. LG2 agrin mutation causing severe congenital myasthenic syndrome mimics functional characteristics of non-neural (z-) agrin. *Hum Genet* 2012;**131**:1123–35.
76. Chevessier F, Faraut B, Ravel-Chapuis A, Richard P, Gaudon K, Bauche S, et al. MUSK, a new target for mutations causing congenital myasthenic syndrome. *Hum Mol Genet* 2004;**13**:3229–40.

77. Chevessier F, Girard E, Molgo J, Bartling S, Koenig J, Hantai D, et al. A mouse model for congenital myasthenic syndrome due to MuSK mutations reveals defects in structure and function of neuromuscular junctions. *Hum Mol Genet* 2008;**17**:3577–95.
78. Maselli RA, Arredondo J, Cagney O, Ng JJ, Anderson JA, Williams C, et al. Mutations in MUSK causing congenital myasthenic syndrome impair MuSK-Dok-7 interaction. *Hum Mol Genet* 2010;**19**:2370–9.
79. Mihaylova V, Salih MA, Mukhtar MM, Abuzeid HA, El-Sadig SM, von der Hagen M, et al. Refinement of the clinical phenotype in musk-related congenital myasthenic syndromes. *Neurology* 2009;**73**:1926–8.
80. Beeson D, Higuchi O, Palace J, Cossins J, Spearman H, Maxwell S, et al. Dok-7 mutations underlie a neuromuscular junction synaptopathy. *Science* 2006;**313**:1975–8.
81. Muller JS, Herczegfalvi A, Vilchez JJ, Colomer J, Bachinski LL, Mihaylova V, et al. Phenotypical spectrum of DOK7 mutations in congenital myasthenic syndromes. *Brain* 2007;**130**:1497–506.
82. Selcen D, Milone M, Shen XM, Harper CM, Stans AA, Wieben ED, et al. Dok-7 myasthenia: phenotypic and molecular genetic studies in 16 patients. *Ann Neurol* 2008;**64**:71–87.
83. Anderson JA, Ng JJ, Bowe C, McDonald C, Richman DP, Wollmann RL, et al. Variable phenotypes associated with mutations in DOK7. *Muscle Nerve* 2008;**37**:448–56.
84. Ben Ammar A, Petit F, Alexandri N, Gaudon K, Bauche S, Rouche A, et al. Phenotype genotype analysis in 15 patients presenting a congenital myasthenic syndrome due to mutations in DOK7. *J Neurol* 2010;**257**:754–66.
85. Slater CR, Fawcett PR, Walls TJ, Lyons PR, Bailey SJ, Beeson D, et al. Pre- and post-synaptic abnormalities associated with impaired neuromuscular transmission in a group of patients with 'limb-girdle myasthenia'. *Brain* 2006;**129**:2061–76.
86. Selcen D, Milone M, Shen XM, Brengman JM, Harper CM, Engel AG. Dok-7 myasthenia: clinical spectrum, endplate electrophysiology and morphology, 12 novel DNA rearrangements, and genotype-phenotype relations in a Mayo cohort of 13 patients. *Neurology* 2007;**68**:A106–7.
87. Schara U, Barisic N, Deschauer M, Lindberg C, Straub V, Strigl-Pill N, et al. Ephedrine therapy in eight patients with congenital myasthenic syndrome due to DOK7 mutations. *Neuromuscul Disord* 2009;**19**:828–32.
88. Senderek J, Muller JS, Dusl M, Strom TM, Guergueltcheva V, Diepolder I, et al. Hexosamine biosynthetic pathway mutations cause neuromuscular transmission defect. *Am J Hum Genet* 2011;**88**:162–72.
89. Selcen D, Shen XM, Milone M, Brengman J, Ohno K, Deymeer F, et al. GFPT1-myasthenia: clinical, structural, and electrophysiologic heterogeneity. *Neurology* 2013;**81**:370–8.
90. Belaya K, Finlayson S, Slater CR, Cossins J, Liu WW, Maxwell S, et al. Mutations in DPAGT1 cause a limb-girdle congenital myasthenic syndrome with tubular aggregates. *Am J Hum Genet* 2012;**91**:193–201.
91. Selcen D, Shen XM, Li Y, Wieben E, Engel AG. Congenital myasthenic syndrome, autophagic myopathy, and cognitive dysfunction caused by mutations in DPAGT1. *Ann Neurol* 2012;**72**:S128.
92. Romero NB. Centronuclear myopathies: a widening concept. *Neuromuscul Disord* 2010;**20**:223–8.

Chapter 27

Juvenile and Neonatal Myasthenia Gravis

Peter B. Kang, Wendy K.M. Liew, Maryam Oskoui, and Angela Vincent

BACKGROUND

Juvenile myasthenia gravis is a major category of neuromuscular junction disorders that shares many pathophysiologic and clinical features with adult autoimmune myasthenia gravis, yet displays distinct demographic patterns, clinical features, and therapeutic challenges.[1] These considerations underscore the need to discuss this disease separately from its adult counterpart.

Juvenile myasthenia gravis affects a broad range of ages within the pediatric population and has a worldwide geographic distribution. In North American and European populations, juvenile myasthenia gravis tends to be rare compared to adult myasthenia gravis, having been estimated to comprise 10% to 20% of myasthenia cases overall.[2–4] The incidence of juvenile myasthenia gravis is estimated to be 0.9 to 2.0 new cases per million per year in Canada,[5] and 1.5 new cases per million children per year in the United Kingdom.[6] Although incidence and prevalence studies are rare, the disease appears to be more common in some parts of Asia. One series from China found that 50% of myasthenia gravis patients were less than 15 years old,[7] while another series from Hong Kong found that 39.3% of myasthenia gravis cases were of juvenile onset.[8,9] Gender ratios in juvenile myasthenia gravis are either balanced[10] or show a slight female predominance[11,12] prior to puberty, whereas affected females are much more common than males in adolescence, mimicking the gender ratio in adults.[12–14]

Epidemiological data regarding neonatal myasthenia gravis are hard to come by and mild cases may not be recognized. Overall, it appears to be relatively rare[9,15] and the incidence may be declining as maternal myasthenia gravis is better controlled. One study found a 30% incidence of neonatal myasthenia gravis among infants born to mothers with known cases of myasthenia gravis.[16] Even rarer are the more severe variants: the fetal akinesia sequence and the fetal acetylcholine inactivation syndrome (see Chapter 7).

PATHOPHYSIOLOGY

As in the adult form of the disease, juvenile myasthenia gravis is caused by antibodies attacking the postsynaptic membrane at the neuromuscular junction. Neonatal myasthenia gravis is somewhat different in etiology compared to juvenile myasthenia gravis, as the former is caused by transplacental transmission of maternal antibodies. Presynaptic autoimmune injuries typically cause a slightly different disease, Lambert-Eaton myasthenic syndrome, which has been reported in childhood[17,18] but is exceedingly rare in this population. Genetic mutations that impair the function of the neuromuscular junction lead to congenital myasthenic syndrome (see Chapter 26).

The most common pathogenic antibody binds to the acetylcholine receptor (AChR), which is a key component of the synaptic signaling mechanism. Acetylcholine, the primary neurotransmitter at the neuromuscular junction, is released into the synapse in response to depolarization of the presynaptic motor nerve membrane with its accompanying influx of calcium ions. The acetylcholine diffuses rapidly across the synapse and binds to the acetylcholine receptors, triggering a depolarization of the postsynaptic membrane and a cascade of depolarizations that culminate in contraction of the muscle fiber (Figure 27.1A). All of this occurs in a blink of an eye.

The acetylcholine receptor is a pentameric complex that exists in two forms: $\alpha_2\beta\delta\gamma$ in fetal and denervated muscle, and $\alpha_2\beta\delta\varepsilon$ in adult muscle. Most antibodies bind to the acetylcholine receptor and are detected by a radioimmunoprecipitation assay that is widely available. These antibodies bind mainly to the α and γ subunits, but since the γ subunit is not present in mature muscle, the less common antibodies to ε subunits can be important in some patients. These antibodies activate the complement cascade and can lead to destruction of the postsynaptic membrane and loss of the high density of acetylcholine receptors. The antibodies can also cause internalization of the acetylcholine receptors, without damage, because

B.T. Darras, H. Royden Jones, Jr., M.M. Ryan & D.C. De Vivo (Eds): Neuromuscular Disorders of Infancy, Childhood and Adolescence, Second edition.
DOI: http://dx.doi.org/10.1016/B978-0-12-417044-5.00027-5

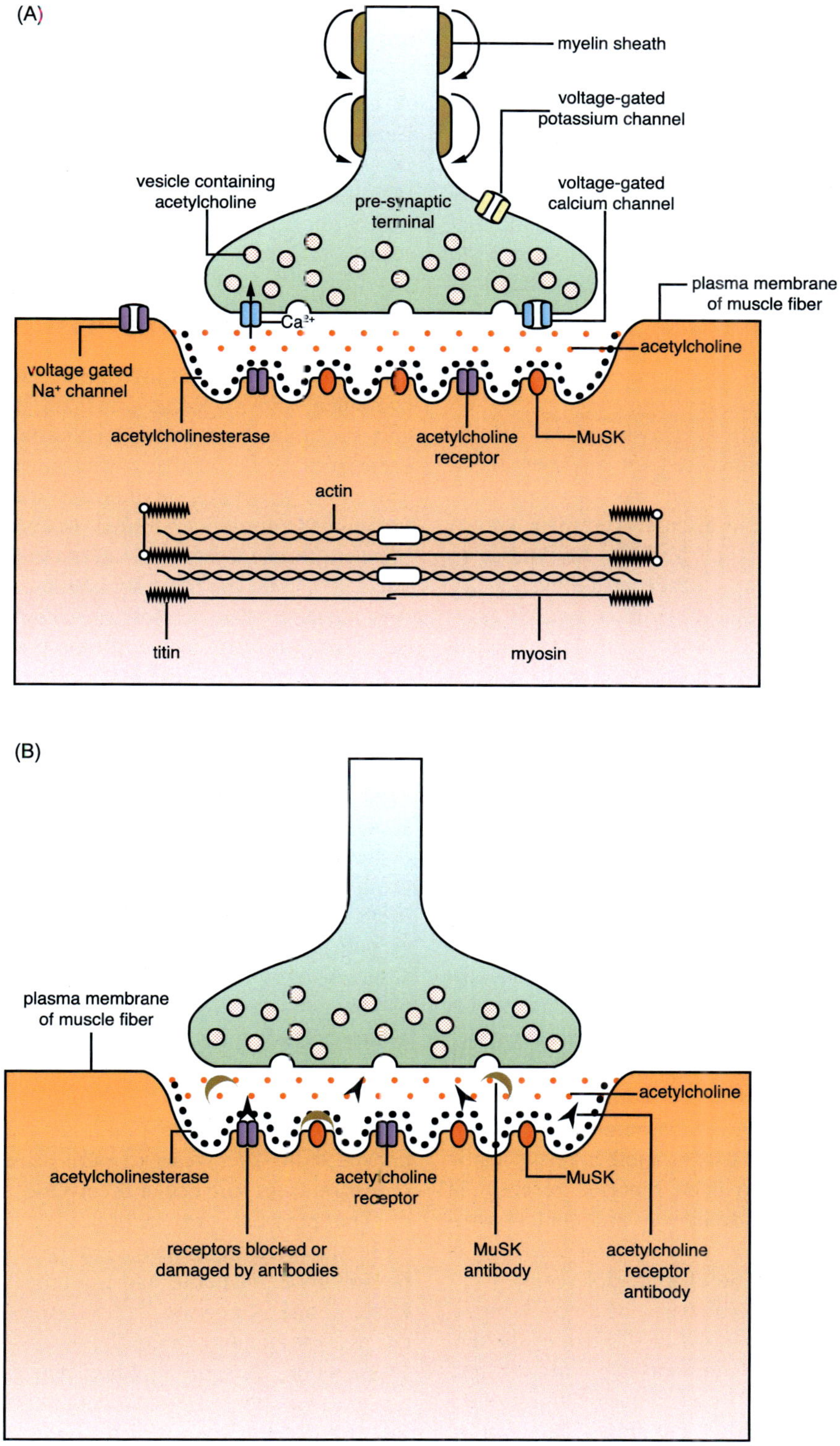

FIGURE 27.1 Schematic diagrams of (**A**), a normal neuromuscular junction illustrating locations of the presynaptic vesicles, acetylcholine within the synapse, and the postsynaptic acetylcholine receptors, as well as (**B**), a neuromuscular junction affected by myasthenia gravis, with antibodies that interfere with the binding and/or structure of the acetylcholine receptors and muscle-specific kinase (MuSK) (note that antibodies to both proteins are not typically present in the same individual, though this phenomenon has been reported on occasion).

being divalent they can cross-link adjacent acetylcholine receptors via binding to the two acetylcholine α subunits; however, this process may not be very important in myasthenia gravis as it is accompanied, in experimental animals, by an increased synthesis of acetylcholine receptors. More rarely, blocking antibodies prevent the normal ion channel function of the acetylcholine receptors. The end result of any of these three mechanisms is impairment of neuromuscular transmission (Figure 27.1B), leading to myasthenia gravis (the original Greek term translates as "muscle weakness severe"). It should be noted that most antibody tests are designed to help diagnosis and not to make accurate and standardized measurements for evaluation of long-term treatment efficacy.

In the 15% of adult myasthenia gravis patients who do not have elevated acetylcholine receptor antibody titers, other antibody-mediated attacks on the neuromuscular junction are likely to cause the disease. The main known target is muscle specific kinase (MuSK). Antibodies to MuSK tend to be elevated in the absence of elevations of acetylcholine receptor antibodies. Myasthenia gravis associated with MuSK antibodies tends to be less common in northern Europe or Canada. MuSK antibody positive juvenile myasthenia gravis is rare, but has been reported on multiple occasions, in children as early as 14 months.[19] Patients with MuSK antibody-associated myasthenia gravis often do not respond to pyridostigmine,[19] so it is important to identify the immunological basis of individual cases whenever possible as a guide to potential treatments.

Titin and ryanodine receptors are intracellular muscle proteins and the antibodies that bind to them serve as biomarkers for the presence of thymoma in adults younger than 60 years. They are often, though not always, accompanied by elevations in acetylcholine receptor antibodies in adults. Children and adolescents with juvenile myasthenia gravis occasionally have isolated elevations of this antibody, but the association with thymoma appears to be reduced or absent in this age group based on a small number of cases.[20] Antibodies to LRP4, another postsynaptic membrane protein, have also been reported recently in myasthenia gravis,[21] but their frequency and clinical associations with the disease have not been characterized in detail to date, especially in children and adolescents.

Thymomas are found in some cases of juvenile myasthenia gravis, just as they are in the adult form of the disease, but are much less common in children and adolescents than in adults. This association with thymoma places myasthenia gravis within the broad category of paraneoplastic diseases. Histologic examinations of thymus tissue obtained from children with juvenile myasthenia gravis who have undergone thymectomy have found a high incidence of lymphoid follicular hyperplasia.[20,22,23]

Some evidence suggests a link between Epstein Barr virus (EBV) and some autoimmune diseases. An early hint that this may be the case for myasthenia gravis was the discovery of the presence of Epstein Barr virus genome in thymus tissue derived from a few affected patients.[24] A more recent study found evidence for active Epstein Barr virus infection in all 17 thymuses obtained from myasthenia gravis patients, compared to none of 6 control thymuses.[25] However, 2 subsequent studies found no evidence for such an association,[26,27] casting doubt on this theory. This proposed association remains controversial.

CLINICAL PRESENTATION

Nearly all patients with juvenile myasthenia gravis demonstrate ocular involvement, principally the levator palpebrae and/or extraocular muscles, leading to ptosis, diplopia, and sometimes frank ophthalmoparesis or ophthalmoplegia.[28] The ocular symptoms are often the first or primary manifestation of the disease at all ages. In rare cases, ocular symptoms are absent at presentation, but in those cases, bulbar or vocal symptoms are typically present, and ocular symptoms may develop later. Dysphagia, dysarthria, nasal speech, extremity weakness, and respiratory distress may occur in conjunction with the ocular symptoms. It is rare for myasthenia gravis to present with extremity fatigue in the absence of cranial nerve-innervated muscle involvement.

The bimodal anatomical distribution of involvement in most cases of myasthenia gravis has yielded its subcategorization into pure ocular and generalized variants. Both the ocular and generalized forms of myasthenia gravis may occur at any age. Ocular myasthenia gravis is more common in children than in adults[7] and tends to occur more often in prepubertal children, whereas generalized myasthenia gravis is more common in postpubertal adolescents.[5,12,29–31] Most patients who develop generalized myasthenia gravis display symptoms and signs of generalization soon after onset; thus, it is usually possible to assign individual patients to one of these anatomical subcategories early in the course of the disease. However, in rare cases, generalization may occur unexpectedly after years of pure ocular symptoms and/or remission from pure ocular involvement.

The physical examination of a child or adolescent with suspected myasthenia gravis is critical. Ptosis may be unilateral, bilateral, and asymmetric (Figure 27.2), or bilateral and symmetric.[15,32] When bilateral and symmetric, the ptosis may be difficult to discern on physical examination, as some children may appear to have mild ptosis at baseline. In such situations, it is helpful to examine any older photos of the child that are available for comparison. When unilateral or asymmetric, the ptosis is usually more obvious. One finding that can help verify the presence of abnormal ptosis is the curtain sign. To elicit the curtain sign, the examiner should lift each eyelid in turn. The affected side(s) will drop when the contralateral eyelid is raised. Ophthalmoparesis may be detected

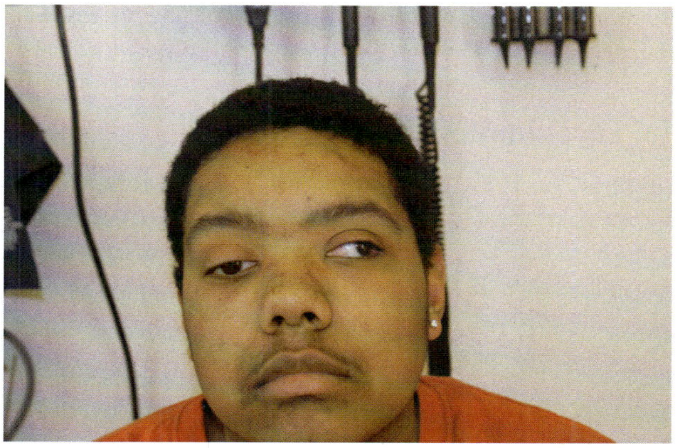

FIGURE 27.2 Asymmetric ptosis and ophthalmoparesis. Note lack of movement of the right eye when patient was asked to look to his left.

by lack of conjugate gaze in one or more directions, and is supported by the patient's report of subjective diplopia in those directions (Figure 27.2). Quantification of misalignments is most accurately performed by a neuro-ophthalmologist. Ptosis and/or ophthalmoparesis may become more obvious in response to a maneuver to fatigue the eyes, such as sustained upgaze for 1 minute. The ice pack test is used by some practitioners in the clinic or at the bedside; clinical signs of myasthenia gravis should improve in response to cold temperatures. Neck flexion strength is especially useful to assess while examining a patient with suspected myasthenia gravis; however, it is important to remember that children under the age of approximately 10 years tend to have apparent neck flexion weakness at baseline. While examining the extremities, some practitioners attempt to elicit fatigability in the deltoid muscle by asking the patient to abduct and adduct the arms repeatedly. When basic pulmonary function testing is available, measurement of vital capacity and/or negative inspiratory force is useful to assess respiratory function. When such equipment is not readily available, a good bedside assessment of vital capacity can be performed by asking the patient to take a deep breath and count numbers out loud as high as possible on one breath. Adolescents and adults should be able to count beyond 30 on one breath.

In all suspected cases of myasthenia gravis, serum thyroid studies are indicated, as symptoms of thyroid disorders may mimic those of myasthenia gravis, and thyroid disorders are sometimes present concurrently with myasthenia gravis.

A differential diagnosis should be considered in the appropriate settings. For pure ocular presentations, alternative diagnoses may include ptosis due to hypoplasia of the levator palpebrae muscles. The Marcus Gunn jaw winking phenomenon also occurs at times in children, and should be suspected when the ptosis occurs during chewing movements. When generalized symptoms are present, the possibility of Lambert-Eaton myasthenic syndrome should be considered, though this diagnosis is exceedingly rare in children compared to adults. Botulism may sometimes be a diagnostic consideration. A typical history for infant botulism includes living in an endemic geographic area (Pennsylvania, Utah, and California have high rates of infant botulism in the United States), a family member who works in the construction industry (spores are often stirred during construction projects), and constipation in addition to weakness. In older children and adolescents, rare cases of botulism may occur primarily as a result of ingesting botulinum toxin in improperly canned foods. Clues on physical examination that may indicate the presence of botulism include internal as well as external ophthalmoplegia (sluggish pupillary reactions in addition to eye movement abnormalities) and flaccid paralysis; tendon reflexes may be preserved. When generalized weakness appears to be chronic, the diagnosis of congenital myasthenic syndrome is a consideration (see Chapter 26), as well as other inherited disorders associated with ptosis including myotonic dystrophy (see Chapter 37) and congenital myopathies (see Chapter 28). Serum creatine kinase levels and electromyography may help sort out this differential diagnosis.

DIAGNOSTIC TESTING

Juvenile myasthenia gravis, like adult myasthenia gravis, is ultimately a clinical diagnosis. A strongly suggestive history and physical examination may be sufficient to make the diagnosis; however, in a number of cases the clinical presentation is not clear-cut, and it is usually helpful to have positive results on one or more standard diagnostic tests. It is important to remember that some of these diagnostic tests do not always have high sensitivities, especially in cases of ocular myasthenia gravis.

Antibody Testing

Antibody testing is usually highly specific for myasthenia gravis. In the appropriate clinical setting, positive antibody titers confirm the diagnosis, and further testing may only be indicated if the clinical presentation is atypical. Antibody testing only involves serum testing and, especially in the case of acetylcholine receptor antibodies, is relatively inexpensive and straightforward.[33] Thus, this test is easy to obtain in children. Some laboratories distinguish between binding, blocking, and modulating antibodies, but the binding antibody titer is usually sufficient and the other two are not universally available.

As in adults, children with generalized juvenile myasthenia gravis are more likely to have elevated acetylcholine receptor antibodies than children with pure ocular juvenile myasthenia gravis.[5] Because the acetylcholine receptor antibody was the first and most common antibody to be associated with myasthenia gravis, patients who have elevated acetylcholine receptor antibody titers have traditionally been termed "seropositive," with the remainder categorized as "seronegative," even if they are found to have elevated titers of other antibodies.[34–37] Seropositive adult patients tend to have more severe clinical courses,[38,39] though the antibody titers do not correlate with clinical status or history of thymectomy.[40] In pediatrics, seronegative myasthenia gravis is more common before puberty.[14,29] In adults, a significant proportion of seronegative myasthenia patients have been shown to have elevated antibodies to muscle-specific kinase (MuSK), ranging from a majority in the first report[41] to a lower proportion in other reports from the Netherlands,[42] Thailand,[43] and Sri Lanka.[44] In the United States, MuSK antibody testing is expensive and elevated MuSK antibody titers are rare in seronegative juvenile myasthenia gravis,[5,45] but can be identified in some generalized cases.[46] A prepubertal child with MuSK antibody positive juvenile myasthenia gravis has been reported to present with ocular symptoms, only developing a brief period of generalized symptoms 2 years after onset.[47] There are rare reports of patients, including an adolescent, found to have elevated titers of both acetylcholine receptor antibodies and MuSK antibodies.[48] It is important to identify the occasional patients with elevations in both types of antibodies, as the acetylcholine receptor antibody elevation is typically low and less relevant than the elevation in the MuSK antibody.

Among the rare instances when striated muscle antibody titers are elevated in juvenile myasthenia gravis, they are sometimes associated with acetylcholine receptor antibody elevations and sometimes are found in isolation. When any antibody titer is borderline or mildly elevated, or if the clinical presentation is discordant with the antibody titers, repeat testing may help clarify the situation, though any intervening immunotherapies may reduce the sensitivity of further rounds of testing. Antibody titers do not generally correlate with the clinical status of affected individuals, though some reports suggest that a correlation may exist for MuSK antibody positive cases.[19,49,50]

The Ice Test

The ice test is a traditional bedside test in which ice is placed on the eyes of a patient with ptosis who is suspected of having myasthenia gravis. A positive result is determined when cooling for approximately 2 minutes results in a 2 mm or greater improvement in the ptosis. The ice test has been determined in numerous adult studies to have a reasonably high sensitivity and specificity, but should not be relied on alone for diagnostic confirmation.[51] The ice test has not been studied extensively in juvenile myasthenia gravis, but its use has been documented in children,[52] and one study that included adults and children over the age of 9 years found that the ice test had reasonably good sensitivity but low specificity.[53]

The Edrophonium Test

Edrophonium is a short-acting acetylcholinesterase inhibitor that is used only for diagnostic purposes; the half-life is so short that this medication has no therapeutic value. Tensilon is an old brand name for edrophonium, and the diagnostic test that makes use of this medication is often still called the Tensilon test, but given the multiple brands currently in use, it would be more accurate to call it the edrophonium test. The test is most useful when there is a clear physical sign such as unilateral ptosis that can be objectively observed for a response to this medication. It is best if this testing is performed in a monitored setting (i.e. with cardiac telemetry) where advanced cardiorespiratory support is available in case of bradycardia or asystole, which in rare instances occurs during this testing, especially when higher doses are used. An intravenous line should be placed, with a 3-way stopcock attached close to the catheter. One port should be attached to a syringe of normal saline flush, and the other to a syringe of edrophonium. At its typical 10 mg/mL concentration, edrophonium should either be drawn up in a 1 mL tuberculin syringe or diluted for use in a larger syringe. There is some conflict in the literature regarding the optimal dose of edrophonium in infants, with one report recommending 0.1 mg only,[54] and another documenting 0.1 mg/kg.[55] A pharmacokinetic study of edrophonium that included infant data suggests that 0.1 mg/kg is needed to counteract neuromuscular blockade. It may be reasonable to start with a dose of 0.1 mg as a test dose, and then administer 0.1 mg/kg (with a maximum of 0.5 mg) once or twice. For children older than a year who are below 35 kg in weight, the test dose should be 0.5 mg, followed by subsequent 1 mg doses several minutes apart to a maximum total dose of 5 mg. For children and adolescents who are above 35 kg, adult dosing ranges may be used, i.e. a test dose of 1 mg and subsequent doses of 1 to 2 mg each administered several minutes apart to a maximum total dose of 10 mg. The physical sign in question (e.g. ptosis, dramatic ophthalmoparesis, nasal speech) should be monitored before and after each dose to determine if there is a transient improvement. The heart rate should be tracked throughout, and dosing paused if there is a substantial dip. If bradycardia ensues, atropine is an effective antidote, and should be administered intravenously immediately. When the parents consent, photos taken before and after the test may be useful to document any possible response (Figure 27.3). There are many instances when the results of this testing are equivocal, so the parents should be cautioned ahead of time that the findings may not yield a clear diagnosis.

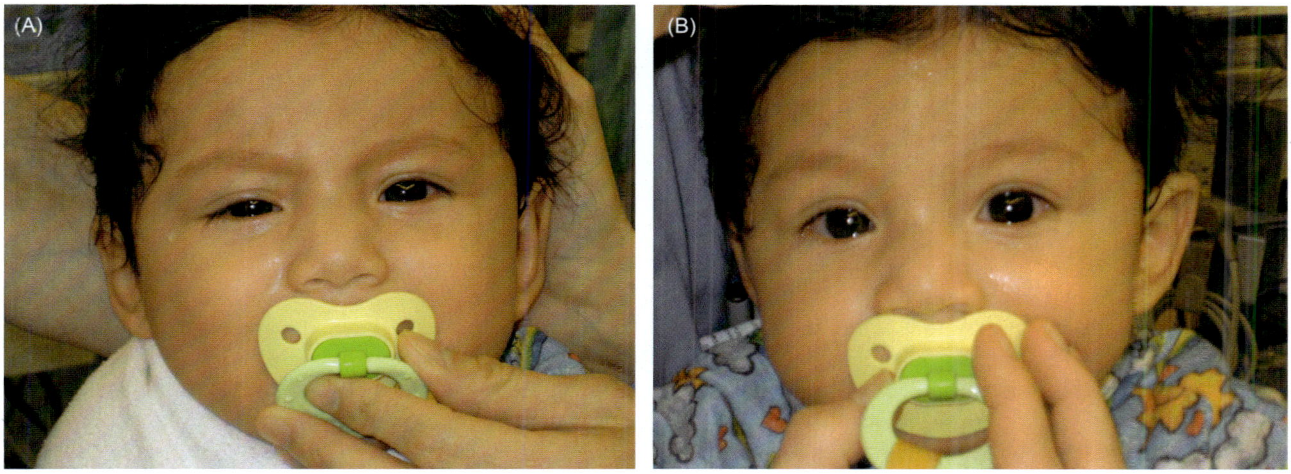

FIGURE 27.3 Photos illustrating a positive edrophonium test in a child presenting with bilateral ptosis. (A) The bilateral ptosis that was present prior to the administration of edrophonium. (B) The rapid but transient resolution of ptosis after the dose of edrophonium was given.

Repetitive Stimulation

Repetitive stimulation testing is an adaptation of a motor nerve conduction study and is a useful diagnostic test, especially when the results of other diagnostic tests are negative, equivocal, or unavailable. As the term implies, the electromyography machine is programmed to deliver a series of electrical stimuli to a motor nerve at fixed intervals, and the compound motor action potential is recorded at a muscle innervated by that nerve. Common sites of stimulation include the ulnar nerve, spinal accessory nerve, and facial nerve, recording at the abductor digiti minimi, trapezius, and nasalis muscles, respectively, though any nerve and corresponding muscle that are amenable to motor nerve conduction studies may potentially be tested. Low frequency repetitive stimulation is administered at a frequency lower than 10 Hz, typically 2 or 3 Hz, whereas high frequency repetitive stimulation is administered at a higher frequency than 10 Hz, typically 20 to 50 Hz. The 10 Hz threshold corresponds to the 100 milliseconds that are needed after each stimulation for calcium ions to be effluxed from the nerve terminal to their original extracellular space, repolarizing the presynaptic membrane.

On low frequency repetitive stimulation, a decrement of 10% or greater in the compound motor action potential amplitude by the fourth stimulus is abnormal, indicating the presence of a defect in neuromuscular junction transmission. It is important to remember that decrements on low frequency stimulation cannot distinguish among different disorders of the neuromuscular junction. Low frequency testing is first obtained at baseline. When testing adults, it is customary to ask the patient to contract the muscle under study for a full minute, then perform further rounds of low frequency repetitive stimulation. In some instances, fatiguing the muscle in this manner brings out or accentuates the decrement. However, a study in adults indicates that post-exercise low frequency repetitive stimulation only produces an incremental increase in the diagnostic yield of this testing overall[56]; this finding, combined with the tenuous nature of children's cooperation with these studies, suggests that post-exercise testing should be considered on a case-by-case basis in children and adolescents.

If a presynaptic disorder such as Lambert Eaton myasthenic syndrome or botulism is suspected, high frequency stimulation testing should be considered, as an increment in the compound motor action potential amplitude during prolonged high frequency stimulation indicates the presence of a presynaptic disorder. However, high frequency repetitive stimulation is painful for the child or adolescent, and needs to be performed with adequate analgesia, especially in younger children. Opioid analgesia is recommended, as milder analgesics such as acetaminophen and ibuprofen are inadequate, and topical lidocaine preparations are similarly ineffective under these circumstances. When adequate analgesia is not available, other clues on routine testing may suggest the presence of a presynaptic disorder. For example, a low baseline compound motor action potential amplitude with normal sensory nerve action potentials may be found in presynaptic disorders, as well as abnormal spontaneous activity and small, myopathic motor units on needle electromyography. This pattern of findings may also be consistent with a myopathic disorder; the results of low frequency repetitive stimulation would help distinguish between the two possibilities, though a low baseline compound motor action potential amplitude may make it difficult to interpret the repetitive stimulation testing in some patients.

As a general rule, repetitive stimulation at any frequency is more technique-dependent than routine motor nerve conduction studies, and artifactual findings are common, especially in some children who have difficulty holding still for this testing. Thus, it is important to attempt to replicate abnormalities in multiple nerves whenever possible before concluding that there is electrophysiologic evidence for a disorder of neuromuscular transmission.

Single Fiber Electromyography

Single fiber electromyography is a sophisticated adaptation of needle electromyography that determines the jitter for a set of individual muscle fibers within a motor unit. The jitter is the variability in the latency of neuromuscular transmission for individual muscle fibers; increased variability, or jitter, may be consistent with a range of neuromuscular disorders, including disorders of neuromuscular junction transmission. Single fiber electromyography is highly sensitive for myasthenia gravis, but like repetitive stimulation is not entirely specific for this disorder; thus, other evidence should be assembled in addition to abnormal jitter to make the diagnosis of myasthenia gravis. For single fiber electromyography to work, the patient needs to be cooperative and to be able to contract individual motor units voluntarily. The needle electrodes used are the same as for routine needle electromyography. Muscles that are commonly tested include the extensor digitorum brevis, frontalis, and orbicularis oculi. Traditional single fiber electromyography is not practical for most children, especially as it cannot be performed under sedation or general anesthesia, though some adolescents may be able to cooperate enough to produce a technically adequate study.

A variant of single fiber electromyography, stimulated single fiber electromyography, is much more useful in children and adolescents than traditional single fiber electromyography.[57] In this technique, a monopolar needle electrode is inserted close to the nerve innervating the muscle under study, and the recording electrode may be a standard concentric bipolar needle inserted into the muscle itself. The monopolar needle acts as a stimulating electrode, and at low stimulus currents, activates individual motor units. Thus, aside from holding still during the study, this technique requires no further cooperation on the part of the patient, so stimulated single fiber electromyography may be performed in an awake or sedated patient.

Regardless of whether traditional single fiber or stimulated single fiber electromyography is employed, the data generated are similar. The motor unit potentials of a pair of individual muscle fibers within a motor unit are identified on the screen, and a set of these potentials (typically 100) are collected in rapid succession. The differences in latencies for the potentials representing the pair of muscle fibers is then averaged, yielding a measure of the variability of these differences, called the jitter. Increased jitter indicates the presence of a neuromuscular disorder, in some cases myasthenia gravis, and in some cases other diseases. Normal values for jitter on stimulated single fiber electromyography have been published.[58]

In selected cases, when all or most of the diagnostic testing noted above is negative or unavailable, a therapeutic trial of pyridostigmine may be considered. However, the use of such a trial for diagnostic purposes may be unreliable, and it is important for expectations to be discussed clearly with the family in advance, and for the family to express an understanding of the side effects. One particular concern revolves around the possibility of a placebo effect. Thus, in assessing the response to such a trial, monitoring of objective physical examination findings should be performed whenever possible.

Imaging

An imaging study of the thorax to assess for thymoma is important to consider in cases of confirmed juvenile myasthenia gravis. Though rare in childhood and adolescence, thymomas have been reported on multiple occasions in this age group. Thorax imaging is critical for any child or adolescent with generalized juvenile myasthenia gravis. For ocular juvenile myasthenia gravis, thorax imaging should also be considered, although the yield is much lower in this situation. When choosing an imaging modality to use, three considerations are important. One is diagnostic yield. Both CT and MRI scans have good resolution and diagnostic yield, whereas chest X-rays are not sensitive enough to rely on alone. A second consideration is the need for sedation. MRIs may require sedation, especially in younger children, adding to its already considerable cost. CT scans typically do not require sedation, though exceptions do arise. A third consideration is radiation exposure. CT scans generate significant radiation exposure, whereas MRI scans do not generate any. The radiation exposure issue becomes especially important when a thymoma is identified and a patient will need serial imaging after a thymectomy to monitor for a recurrence of thymoma.

MEDICAL THERAPIES

General Approach to Treating Juvenile Myasthenia Gravis

Medical therapies for juvenile myasthenia gravis include drugs that increase the availability of the acetylcholine neurotransmitter, as well as immunomodulatory and immunosuppressive drugs.[59] Immunomodulation and immunosuppression are more powerful approaches to medical therapy. However, immunomodulation and immunosuppression are typically more invasive and/or carry greater side effect burdens.[60,61] As in adults, pediatric patients and their families should be counseled regarding medications that may exacerbate myasthenia gravis.[62]

Among the standard medical therapies for juvenile myasthenia gravis, it is reasonable to start with pyridostigmine except in the rare cases where a child presents in myasthenic crisis. But for those children who do not respond adequately to pyridostigmine, a complicated decision awaits the family and neurologist. Plasmapheresis, intravenous immunoglobulins, and corticosteroids are all efficacious in the treatment of myasthenia gravis, and this has been documented in the adult literature.[63–68] Corticosteroids are easy to administer and are low in cost compared to plasmapheresis and intravenous immunoglobulins. However, chronic corticosteroid therapy is significantly more toxic in children and adolescents than in adults, as noted previously. It is thus worth considering plasmapheresis and/or intravenous immunoglobulins before corticosteroids.

Acetylcholinesterase Inhibition

Pyridostigmine is a synthetic small molecule ($C_9H_{13}N_2O_2$) that was patented in 1951, and is often known by the brand name Mestinon. It is an acetylcholinesterase inhibitor, thus reducing the natural enzymatic metabolism of acetylcholine at the neuromuscular junction. The end effect is an increased amount of acetylcholine available to bind to postsynaptic receptors. Its potency is typically mild, especially in comparison with certain forms of immunomodulation. However, most patients tolerate the medication well, and some patients with mild symptoms may be adequately controlled on pyridostigmine monotherapy, so it is usually the first therapy used in juvenile myasthenia gravis. An exception to this approach involves MuSK antibody positive cases, which often do not respond well to this medication.[19] Many patients experience no side effects at all. Some will develop diarrhea and less commonly other cholinergic side effects. Occasionally, somnolence will occur. A typical starting dose for children and adolescents is 0.5 to 1.0 mg/kg/dose, given 3 times a day, with a maximum starting dose of 60 mg/dose. The usual maximum for children and adolescents is 7 mg/kg/day, though some patients may need higher doses, such as confirmed cases of certain subtypes of congenital myasthenic syndrome. The 7 mg/kg/day maximum rarely needs to be exceeded in juvenile myasthenia gravis, and the risk of cholinergic side effects increases as the dose escalates. For adolescents and adults, some recommend a maximum absolute dose of 300 mg/day, though patients do exceed that level in certain cases.

Plasmapheresis

Plasmapheresis (also known as plasma exchange) is a therapeutic procedure in which whole blood is removed from the circulation, proteins such as antibodies and clotting factors are filtered from the plasma, and then the whole filtered blood is returned to the circulation. In most cases, the therapeutic targets are the antibodies rather than the clotting factors. It is effective in children and adolescents with juvenile myasthenia gravis, just as it is in adults with autoimmune myasthenia gravis. Traditionally, plasmapheresis has been used primarily to optimize a patient's clinical status prior to thymectomy[69,70] and also to treat myasthenic crisis. However, it can also be used as maintenance therapy for ambulatory children and adolescents, with one report indicating a more consistent response to this intervention compared to intravenous immunoglobulin.[20] Some practical considerations in the implementation of this therapy are specific to children and adolescents. Chief among them is the issue of venous access. Some adolescents and older children may be able to accommodate the two large intravenous lines needed for peripheral plasmapheresis. However, many cannot, and thus the risks and benefits of having central venous access for this therapy will need to be considered and discussed with the families in those cases. Traditional plasmapheresis catheters and Broviac lines have double lumens and can easily accommodate plasmapheresis. Pheresis catheters are only meant to stay in for a brief period of time, whereas a Broviac may stay in longer. These lines carry a risk of infection, especially in situations where they remain in place after the patient is discharged (more likely with Broviac lines than pheresis catheters) and proper line care does not occur at home. The filtered antibodies are typically regenerated by the immune system in a matter of weeks, so plasmapheresis is either used in acute settings, such as myasthenic crisis, or repeated at regular intervals for those on maintenance therapy. Regimens vary widely depending on the individual patient's response.

Intravenous Immunoglobulins

Intravenous immunoglobulin (IVIg) therapy consists of infusions of immunoglobulins pooled from multiple blood donors. Only standard venous access is required for these infusions, so most pediatric patients, including infants, are able to receive this treatment via peripheral venous access. IVIg has been reported to be effective in juvenile myasthenia gravis.[71] In the past, allergic reactions could be severe and sometimes life-threatening, but the more severe reactions were found to be due to IgA deficiency; thus, many current IVIg formulations are depleted of IgA during processing. The most common side effects include headache, nausea, and vomiting. Some patients develop an aseptic meningitis, in which these common symptoms become more severe than usual, often persisting several days after treatment. As these meningitides resolve spontaneously with supportive treatment, it is usually not necessary to diagnose them with lumbar punctures, especially as the occasional post-lumbar puncture headache would only make matters worse. Rare cases of renal failure have been associated with IVIg therapy, but those have generally occurred in adult patients. The standard initial IVIg treatment regimen consists of 1 gram/kg infused intravenously $\times$ 2 consecutive days. IVIg is eliminated from the circulation over a period of several weeks, with a half-life estimated to be 21 days. Thus, patients with chronic diseases such as myasthenia gravis may need a regular schedule of recurring treatments. Factors that may limit access to IVIg for certain patients include cost and the availability of sufficient numbers of blood donors. Shortages of IVIg have occurred in recent years, and the cost is extremely expensive, depending on the actual dose ordered.

Corticosteroids

Corticosteroids have traditionally played a central role in the treatment of autoimmune disorders in children, including juvenile myasthenia gravis.[11,72] They are inexpensive, convenient to take in either oral or intravenous form, and have a long track record, so the side effect profile is well known. Corticosteroids are known to be effective in myasthenia gravis.[73] The problem with their use revolves around the extensive list of side effects, including several that have significant long-term implications. The long-term toxicities are more consequential for children and adolescents than in adults, so steroids should be used more judiciously in the pediatric population than in adults. For example, the linear growth deceleration that occurs in some patients who are receiving chronic corticosteroid therapy is deeply disturbing to many patients and families, but is only relevant to children and adolescents who have not reached their full adult height. The weight gain and Cushingoid facies that may occur with corticosteroid therapy are troubling to both children and adults, but adolescents in particular who are often already sensitive about their weight are typically very worried about the prospect of embarking on a course of corticosteroids that may last months or even several years. The bone demineralization that occurs in many patients receiving corticosteroids is more consequential in children and adolescents as they are at an age when bone growth is typically increasing. There are numerous other side effects of corticosteroids that are well documented elsewhere, and are as concerning in children and adolescents as in adults.

Steroid-sparing Agents

Azathioprine is a commonly used steroid-sparing immunosuppressive agent for adults with myasthenia gravis.[74] Its mechanism of action is inhibition of purine metabolism, resulting in impairment of DNA and RNA replication. The traditional indication for its use is to facilitate weaning of steroids.[75] It has also been used in children and adolescents with juvenile myasthenia gravis, with good outcomes noted in combination with other therapies.[29,30,76,77] Azathioprine is generally well tolerated, though side effects may include myelosuppression and some liver toxicity. There is a slight risk of malignancy reported, though this seems to be a negligible concern for relatively short courses of therapy.[78] It has become increasingly recognized that certain variants of *TPMT*, the gene encoding thiopurine methyltransferase, can lead to deficient activity of that enzyme, leading to a higher risk of severe myelosuppression in patients with those variants who receive azathioprine. Thus, it is important to test patients, including children and adolescents, for thiopurine methyltransferase activity and/or genetic variants in the *TPMT* gene prior to initiation of azathioprine therapy so that dosing can be calibrated appropriately.[79] Baseline and ongoing monitoring of blood counts and liver function tests should also be routine in patients receiving azathioprine.

Cyclosporine is another steroid-sparing immunosuppressant medication that is a standard option for the treatment of adults with myasthenia gravis,[80–82] though it is not used frequently in some countries such as the United Kingdom. It inhibits T-cell activation by dampening the production and release of interleukin 2. Its track record in juvenile myasthenia gravis is somewhat limited, but it has been used in that setting as well.[29]

Tacrolimus is a macrolide that inhibits T-cell activation by blocking the dephosphorylation of nuclear factor of activated T cells (NF-AT). It is another immunosuppressive medication that has the potential to help wean steroids in the setting of myasthenia gravis. Tacrolimus seems to be beneficial in some cases of adult myasthenia gravis,[83–85] and it is also used for various indications in children, but not much is known about its potential as a therapy for juvenile myasthenia gravis. One recent case report suggests that it may be helpful in some children with ocular juvenile myasthenia gravis.[86]

Mycophenolate mofetil inhibits purine synthesis and is another steroid-sparing immunosuppressant that is used for the treatment of myasthenia gravis. Its history in this setting is brief but somewhat turbulent. Mycophenolate mofetil began to be used for myasthenia gravis based on case reports,[87] small open label studies,[88] and retrospective case series[89] suggesting that it was quite effective in some adults and adolescents. However, when two major randomized, controlled trials were performed, they showed no statistically significant treatment effect, causing confusion among many neurologists and patients.[90,91] It is possible that these trials did not show a clear treatment effect due to the concomitant use of steroids in both arms of the studies. A subsequent retrospective study detected another possible explanation for the discrepancy, finding that the benefits of mycophenolate mofetil may only begin to emerge in a quantifiable manner after 6 months.[92] Mycophenolate mofetil continues to be used in some affected children and adolescents based on retrospective data,[89] but has not been systematically studied for this age group.

SURGICAL THERAPIES

The potential beneficial effect of thymectomy on patients with myasthenia gravis was first described in 1939 by Blalock, who performed this procedure on a 21-year-old woman with myasthenia gravis who had a thymoma.[93] The thymectomy was performed to treat the thymoma, without expectation that the natural history of the myasthenia gravis would be altered. This, in fact, is what happened, and the woman's myasthenia symptoms improved after the procedure. A few years later, Blalock and others performed thymectomies on a series of patients with myasthenia gravis who did not have thymomas, and documented that a number of them experienced improvements in their myasthenia symptoms.

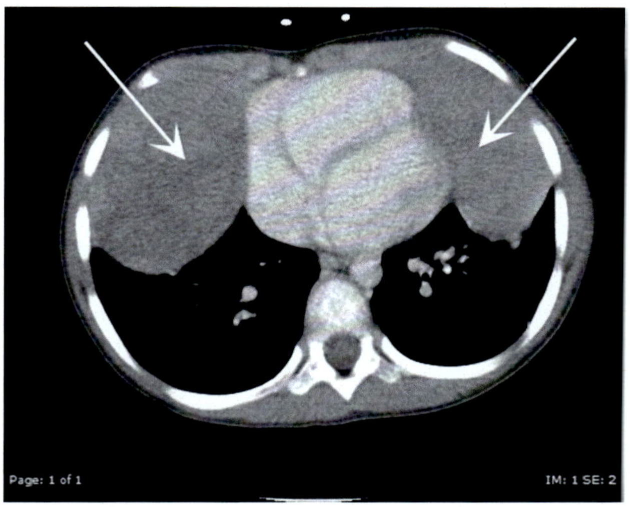

FIGURE 27.4 An axial CT image of the thorax in a 5-year-old boy with generalized juvenile myasthenia gravis, illustrating a thymoma in the shape of a butterfly draping over the heart and compressing the lungs (arrows). The thymoma was subsequently resected.

It is indisputable that the presence of a thymoma in the setting of myasthenia gravis at any age is an absolute indication for thymectomy (Figure 27.4). Thymomas are less common in juvenile myasthenia gravis compared to adult myasthenia gravis,[30,32,76,94,95] but they do occur on occasion.[7,9] Though many investigators and clinicians believe that thymectomy is also helpful for the treatment of myasthenia gravis in the absence of a thymoma, it has been difficult to acquire unequivocal evidence for this. Despite efforts to orchestrate randomized, prospective studies of the efficacy of thymectomy in patients with myasthenia gravis in the absence of thymoma, the currently published literature consists primarily of retrospective series in adults. Using comparison groups or natural history controls, a number of these studies found a higher remission rate in adult patients who received thymectomy versus those who did not, and thus thymectomy is recommended as a therapeutic option in these situations.[96] A randomized prospective trial of thymectomy in adults with myasthenia gravis who have elevated acetylcholine receptor antibody titers is in process, and results are expected to be available in 2015. Given the relative rarity of juvenile myasthenia gravis versus adult myasthenia gravis, it is not feasible to generate large patient cohorts, especially at single sites. However, numerous retrospective pediatric studies indicate that thymectomy is also potentially beneficial in juvenile myasthenia gravis (see Case Example 27.1).[3,12,20,22,23,32,78,95,97–99]

NEONATAL MYASTHENIA GRAVIS AND THE FETAL ACETYLCHOLINE RECEPTOR INACTIVATION SYNDROME

In early development, fetal acetylcholine receptors are widely distributed throughout the myotubular surface.

CASE EXAMPLE 27.1

A previously well 17-year-old adolescent female presented with a 1-month history of intermittent ptosis, diplopia, nasal speech, and fatigue. These symptoms tended to be worse later in the day. She could not recall any antecedent illnesses. Diagnostic evaluation demonstrated markedly elevated acetylcholine receptor binding antibodies and a 25% decrement in compound muscle action potential amplitude on 2 Hz repetitive stimulation of the accessory nerve. An MRI of the thorax demonstrated a normal thymus.

The response to pyridostigmine was initially promising, but within 2 weeks the symptoms recurred. Her peripheral venous access was good, and thus a course of plasmapheresis was started, with excellent response. However, after 4 months, she was still dependent on a frequent schedule of plasmapheresis, and her veins were becoming more difficult to access. She was thus referred for a surgical consultation and underwent thymectomy via a transsternal approach without complications. Thymus pathology demonstrated lymphoid follicular hyperplasia, but no frank thymoma was found. Two months after thymectomy, her clinical status had stabilized to the point where she began lengthening the intervals between plasma exchanges, and within a year, she was tapered off all plasmapheresis. She remained stable at her last follow-up visit 3 years post-thymectomy, remaining on a modest dose of pyridostigmine.

Clustering of these acetylcholine receptors occurs at sites of contact with the motor nerve terminal, where released agrin activates a MuSK molecule to cluster acetylcholine receptors through a cytoplasmic scaffolding protein called rapsyn. Bidirectional signalling between the muscle and nerve modulates the maturation of the postsynaptic membrane.[100] In the few fetal human muscles examined, the fetal acetylcholine receptor is present until 33 weeks gestation, when the γ subunit is replaced by the adult ε subunit.[101] Experimental studies suggest that the fetal acetylcholine receptor is crucial to the orderly innervation pattern of skeletal muscle.[102]

The effect of circulating maternal acetylcholine receptor antibodies on newborns of myasthenic mothers varies from asymptomatic in the majority (85–90%), transient neonatal myasthenia (10–15%), to rare variants of fetal akinesia sequence and the fetal acetylcholine receptor inactivation syndrome.[103] The effect on the fetus has been shown to be correlated with higher titers of fetal acetylcholine receptor antibodies in the mother, with no correlation between disease severity in the mother and the infant.[104] Indeed, asymptomatic mothers with high fetal acetylcholine receptor antibody levels have been reported with affected infants, and the risk to subsequent infants is similar with each pregnancy. Although the majority of newborns of affected mothers will be asymptomatic, about 10% to 15% will have transient symptoms of neonatal myasthenia gravis.[105] Newborns develop symptoms of muscle weakness within a few hours of life, which include in decreasing order of frequency: feeding difficulties, generalized weakness and hypotonia, respiratory difficulty, weak cry, facial paresis, and ptosis. Management is symptomatic, and infants improve as the circulating maternal antibodies are naturally cleared over 2 to 4 weeks (see Case Example 27.2). When the classical form of neonatal myasthenia gravis is managed properly with supportive care, the outcome is universally favorable.

Maternal antibodies to the fetal γ subunit of the acetylcholine receptor can competitively block neurotransmission in the fetus and prevent normal neuromuscular junction maturation causing lack of fetal movements that lead to arthrogryposis multiplex congenita (see Chapter 7). Maternal titers of fetal acetylcholine receptor antibody correlate with clinical severity in the neonate, whereas the clinical status of the mother does not.[104]

The timing and magnitude of such circulating antibodies affecting the fetal subunit of the acetylcholine receptor during gestation will carry different consequences to the fetus. The more severe and less frequent variants of neonatal myasthenia gravis include the fetal akinesia sequence and the fetal acetylcholine inactivation syndrome, which are part of a clinical continuum. Over 40 cases of fetal akinesia sequence have been described in children of myasthenic mothers, most with fixed joint contractures, and with craniofacial anomalies (hypertelorism, micrognathia, low-set ears), limb anomalies, pulmonary hypoplasia (from decreased fetal breathing), short umbilical cord, and polyhydramnios (from decreased fetal swallowing).[106,107] There have been over a dozen recognized cases of the fetal acetylcholine receptor inactivation syndrome, with facial diplegia, highly arched palate, velopharyngeal incompetence, conductive hearing loss, and cryptorchidism.[103,108] These cases have all been reported in infants of myasthenic mothers who have high fetal acetylcholine receptor antibody titers, with a recurrence rate

CASE EXAMPLE 27.2

A term baby boy was transferred to the neonatal intensive care unit at 5 hours of life for respiratory distress. He was born to a $G_3P_1A_2$ mother who had 2 previous spontaneous abortions in the first trimester. His older brother had a history of feeding difficulties at birth, not being able to breastfeed or bottle feed effectively due to poor suck and swallow function. This brother improved spontaneously after a week of gavage feeding. The mother had no other known past medical history, and family history was otherwise negative for neuromuscular disorders.

On arrival in the neonatal intensive care unit, the infant was tachypneic at 60 breaths per minute with an oxygen saturation of 94% on room air. He had facial diplegia, with inability to fully close his eyes or mouth. Rooting and sucking reflexes were present but reduced, as was his Moro response. The gag reflex was diminished. Axial and appendicular hypotonia was present, with intact tendon reflexes. The lower extremities were in a frog leg position, with minimal spontaneous movements. He was unable to breastfeed or bottle feed, and required nasogastric tube feedings. He required BiPAP support for 48 hours, after which his respiratory status remained stable on room air.

Laboratory investigations were normal including a complete blood count, cerebrospinal fluid analysis, and serum electrolytes (including glucose and calcium). A chest X-ray and brain MRI were normal. Intravenous ampicillin and gentamicin were administered for 48 hours until bacterial cultures were confirmed to be negative. On further examination of the mother, it was noted that she had variable ptosis that worsened towards the evening. She was otherwise asymptomatic. Serum acetylcholine receptor antibody titers were elevated in both the baby and the mother (>1000 nm/L in the infant, with a normal range <0.4 nm/L). After a course of treatment with intravenous immunoglobulins and gavage feeds, the infant improved by 5 weeks of age.

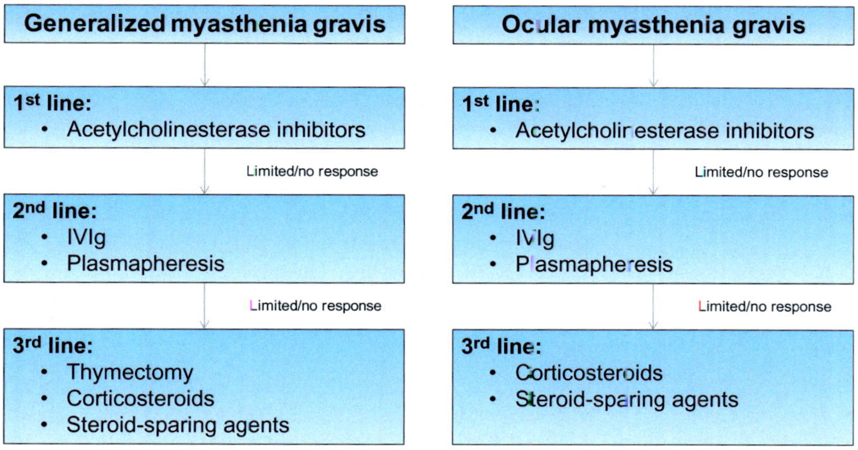

FIGURE 27.5 A treatment algorithm for generalized juvenile myasthenia gravis, illustrating the different tiers of interventions typically used in this disorder. Note that acetylcholinesterase inhibitors are not effective in most cases of MuSK antibody-associated disease, and thus this tier should be bypassed in that subset of patients.

near 100%. Improved outcome in future pregnancies has been reported with immunomodulatory treatment during pregnancy.[103,109]

The persistent weakness of selective muscle groups in these children suggests a differential vulnerability during fetal development. The distribution and density of the fetal acetylcholine receptors may be another factor in the developmental susceptibility of the midline muscles involved in oro-bucco-lingual function.

PROGNOSIS

Overall, complete clinical remission in adult and juvenile myasthenia gravis is difficult to attain. One study that included adults and children reported a 22% remission rate 1 year after onset.[110] However, juvenile myasthenia gravis generally has a better prognosis, with higher reported rates of remission and improvement,[76] especially for children and adolescents with the generalized form who underwent early thymectomy[78] and those with ocular juvenile myasthenia gravis.[9] The pure ocular form of juvenile myasthenia gravis is not life-threatening (assuming that generalization does not occur). There is a variable response to medical treatments, and thymectomy is generally not recommended for ocular juvenile myasthenia gravis. The generalized form of juvenile myasthenia gravis may be quite serious, depending on the degree of weakness and respiratory involvement. Myasthenic crises are emergencies requiring intensive care support, as in adults. Some patients with generalized juvenile myasthenia gravis undergo spontaneous remission, but a number of retrospective studies focusing on pediatric populations indicate that remission is facilitated by thymectomy.

The outcome in the classical form of neonatal myasthenia gravis is excellent, assuming that appropriate respiratory and nutritional support is provided. The more severe and rare variants of neonatal myasthenia gravis, however, are associated with numerous long-term complications.

CONCLUSIONS

Though important studies have been performed on juvenile myasthenia gravis, the quantity and depth of research on this disease is far exceeded by the body of work on adult autoimmune myasthenia gravis, due in part to the relative scarcity of the juvenile form in many populations. The existing literature indicates that there are important differences between juvenile and adult autoimmune myasthenia gravis with respect to demographics, clinical presentations, and therapeutic decision-making. To characterize these issues with greater precision and to optimize the care of children and adolescents with juvenile myasthenia gravis, further research is needed. The establishment of multicenter research consortia and the infrastructure to conduct prospective clinical studies should be considered by major pediatric facilities. Based on the data available, a treatment algorithm may be proposed for generalized juvenile myasthenia gravis that takes into account the distinct considerations needed in children and adolescents (Figure 27.5).

Traditionally, neonatal myasthenia gravis has been regarded as a relatively benign disease with few to no long-term consequences when properly cared for. More severe variants associated with long-term medical consequences have been described, including the fetal acetylcholine receptor inactivation syndrome. Thus, these variants should be considered in the differential diagnosis of children who have unexplained muscle weakness in the setting of distinct clinical features known to be associated with these forms of neonatal myasthenia gravis.

REFERENCES

1. Chiang LM, Darras BT, Kang PB. Juvenile myasthenia gravis. *Muscle Nerve* 2009;**39**:423–31.
2. Osserman KE, Genkins G. Studies in myasthenia gravis: review of a twenty-year experience in over 1200 patients. *Mt Sinai J Med* 1971;**38**:497–537.

3. Szobor A, Mattyus A, Molnar J. Myasthenia gravis in childhood and adolescence. Report on 209 patients and review of the literature. *Acta Paediatr Hung* 1988;**29**:299–312.
4. Phillips II LH, Torner JC, Anderson MS, Cox GM. The epidemiology of myasthenia gravis in central and western Virginia. *Neurology* 1992;**42**:1888–93.
5. VanderPluym J, Vajsar J, Jacob FD, Mah JK, Grenier D, Kolski H. Clinical characteristics of pediatric myasthenia: a surveillance study. *Pediatrics* 2013;**132**:e939–44.
6. Parr JR, Andrew MJ, Finnis M, Beeson D, Vincent A, Jayawant S. How common is childhood myasthenia? The UK incidence and prevalence of autoimmune and congenital myasthenia. *Arch Dis Child* 2014;**99**:539–42.
7. Zhang X, Yang M, Xu J, Zhang M, Lang B, Wang W, et al. Clinical and serological study of myasthenia gravis in HuBei Province, China. *J Neurol Neurosurg Psychiatry* 2007;**78**:386–90.
8. Yu YL, Hawkins BR, Ip MS, Wong V, Woo E. Myasthenia gravis in Hong Kong Chinese. 1. Epidemiology and adult disease. *Acta Neurol Scand* 1992;**86**:113–19.
9. Wong V, Hawkins BR, Yu YL. Myasthenia gravis in Hong Kong Chinese. 2. Paediatric disease. *Acta Neurol Scand* 1992;**86**:68–72.
10. Haliloglu G, Anlar B, Aysun S, Topcu M, Topaloglu H, Turanli G, et al. Gender prevalence in childhood multiple sclerosis and myasthenia gravis. *J Child Neurol* 2002;**17**:390–2.
11. Morita MP, Gabbai AA, Oliveira AS, Penn AS. Myasthenia gravis in children: analysis of 18 patients. *Arq Neuro-psiquiatr* 2001;**59**:681–5.
12. Batocchi AP, Evoli A, Palmisani MT, Lo Monaco M, Bartoccioni M, Tonali P. Early-onset myasthenia gravis: clinical characteristics and response to therapy. *Eur J Pediatr* 1990;**150**:66–8.
13. Andrews PI, Massey JM, Howard Jr. JF, Sanders DB. Race, sex, and puberty influence onset, severity, and outcome in juvenile myasthenia gravis. *Neurology* 1994;**44**:1208–14.
14. Andrews PI, Massey JM, Sanders DB. Acetylcholine receptor antibodies in juvenile myasthenia gravis. *Neurology* 1993;**43**:977–82.
15. Gamio S, Garcia-Erro M, Vaccarezza MM, Minella JA. Myasthenia gravis in childhood. *Binocul Vis Strabismus Q* 2004;**19**:223–31.
16. Djelmis J, Sostarko M, Mayer D, Ivanisevic M. Myasthenia gravis in pregnancy: report on 69 cases. *Eur J Obstet Gynecol Reprod Biol* 2002;**104**:21–5.
17. Morgan-Followell B, de Los Reyes E. Child neurology: diagnosis of Lambert-Eaton myasthenic syndrome in children. *Neurology* 2013;**80**:e220–2.
18. Kostera-Pruszczyk A, Ryniewicz B, Rowinska-Marcinska K, Dutkiewicz M, Kaminska A. Lambert-Eaton myasthenic syndrome in childhood. *Eur J Paediatr Neurol* 2009;**13**:194–6.
19. Yilmaz S, Gokben S, Serdaroglu G, Akcay A. Early-onset acquired myasthenia gravis secondary to anti-muscle-specific kinase autoantibodies. *J Child Neurol* 2014;**29**:108–10.
20. Liew WK, Powell CA, Sloan SR, Shamberger RC, Weldon CB, Darras BT, et al. A comparison of plasmapheresis and IVIG as maintenance therapies for juvenile myasthenia gravis. *JAMA Neurol* 2014;**71**:575–80.
21. Higuchi O, Hamuro J, Motomura M, Yamanashi Y. Autoantibodies to low-density lipoprotein receptor-related protein 4 in myasthenia gravis. *Ann Neurol* 2011;**69**:418–22.
22. Adams C, Theodorescu D, Murphy EG, Shandling B. Thymectomy in juvenile myasthenia gravis. *J Child Neurol* 1990;**5**:215–18.
23. Rodriguez M, Gomez MR, Howard Jr. FM, Taylor WF. Myasthenia gravis in children: long-term follow-up. *Ann Neurol* 1983;**13**:504–10.
24. McGuire LJ, Huang DP, Teoh R, Arnold M, Wong K, Lee JC. Epstein-Barr virus genome in thymoma and thymic lymphoid hyperplasia. *Am J Pathol* 1988;**131**:385–90.
25. Cavalcante P, Serafini B, Rosicarelli B, Maggi L, Barberis M, Antozzi C, et al. Epstein-Barr virus persistence and reactivation in myasthenia gravis thymus. *Ann Neurol* 2010;**67**:726–38.
26. Kakalacheva K, Maurer MA, Tackenberg B, Munz C, Willcox N, Lunemann JD. Intrathymic Epstein-Barr virus infection is not a prominent feature of myasthenia gravis. *Ann Neurol* 2011;**70**:508–14.
27. Meyer M, Hols AK, Liersch B, Leistner R, Gellert K, Schalke B, et al. Lack of evidence for Epstein-Barr virus infection in myasthenia gravis thymus. *Ann Neurol* 2011;**70**:515–18.
28. Jayawant S, Parr J, Vincent A. Autoimmune myasthenia gravis. *Handb Clin Neurol* 2013;**113**:1465–8.
29. Evoli A, Batocchi AP, Bartoccioni E, Lino MM, Minisci C, Tonali P. Juvenile myasthenia gravis with prepubertal onset. *Neuromuscul Disord: NMD* 1998;**8**:561–7.
30. Lindner A, Schalke B, Toyka KV. Outcome in juvenile-onset myasthenia gravis: a retrospective study with long-term follow-up of 79 patients. *J Neurol* 1997;**244**:515–20.
31. Heckmann JM, Hansen P, Van Toorn R, Lubbe E, Janse van Rensburg E, Wilmshurst JM. The characteristics of juvenile myasthenia gravis among South Africans. *S Afr Med J* 2012;**102**:532–6.
32. Millichap JG, Dodge PR. Diagnosis and treatment of myasthenia gravis in infancy, childhood, and adolescence: a study of 51 patients. *Neurology* 1960;**10**:1007–14.
33. Furukawa S, Akazawa S, Furukawa Y, Kamo I, Satoyoshi E, Hayashi K. A practical enzyme immunoassay for anti-acetylcholine receptor antibodies in myasthenia gravis. *J Neuroimmunol* 1984;**6**:397–409.
34. Evoli A, Batocchi AP, Lo Monaco M, Servidei S, Padua L, Majolini L, et al. Clinical heterogeneity of seronegative myasthenia gravis. *Neuromuscul Disord* 1996;**6**:155–61.
35. Verma PK, Oger JJ. Seronegative generalized myasthenia gravis: low frequency of thymic pathology. *Neurology* 1992;**42**:586–9.
36. Yamamoto T, Vincent A, Ciulla TA, Lang B, Johnston I, Newsom-Davis J. Seronegative myasthenia gravis: a plasma factor inhibiting agonist-induced acetylcholine receptor function copurifies with IgM. *Ann Neurol* 1991;**30**:550–7.
37. Vincent A, Li Z, Hart A, Barrett-Jolley R, Yamamoto T, Burges J, et al. Seronegative myasthenia gravis. Evidence for plasma factor (s) interfering with acetylcholine receptor function. *Ann NY Acad Sci* 1993;**681**:529–38.
38. Romi F, Aarli JA, Gilhus NE. Seronegative myasthenia gravis: disease severity and prognosis. *Eur J Neurol* 2005;**12**:413–18.
39. Romi F, Gilhus NE, Aarli JA. Myasthenia gravis: disease severity and prognosis. *Acta Neurol Scand Suppl* 2006;**183**:24–5.
40. Seybold ME, Baergen RN, Nave B, Lindstrom JM. Anti-acetylcholine-receptor antibody concentrations after thymectomy in patients with myasthenia gravis. *Br Med J* 1978;**2**:1051–3.
41. Hoch W, McConville J, Helms S, Newsom-Davis J, Melms A, Vincent A. Auto-antibodies to the receptor tyrosine kinase MuSK in patients with myasthenia gravis without acetylcholine receptor antibodies. *Nat Med* 2001;**7**:365–8.

42. Niks EH, Kuks JB, Verschuuren JJ. Epidemiology of myasthenia gravis with anti-muscle specific kinase antibodies in The Netherlands. *J Neurol Neurosurg Psychiatry* 2007;**78**:417–18.
43. Witoonpanich R, Dejthevaporn C, Pulkes T, Tunlayadechanont S, Boonkongchuen P, Pongpakdee S, et al. Prevalence, clinical features and treatment outcomes of patients with myasthenia gravis positive for antibodies to muscle-specific kinase in Thailand. *J Clin Neurosci* 2013;**20**:707–9.
44. Chang T, Gunaratne P, Gamage R, Riffsy MT, Vincent A. MuSK-antibody-positive myasthenia gravis in a South Asian population *J Neurol Sci* 2009;**284**:33–5.
45. Anlar B, Vincent A. Antibodies against muscle-specific kinase in juvenile myasthenia gravis. *Neuropediatrics* 2003;**34**:110–11.
46. Guptill JT, Sanders DB, Evoli A. Anti-MuSK antibody myasthenia gravis: clinical findings and response to treatment in two large cohorts. *Muscle Nerve* 2011;**44**:36–40.
47. Gungor-Tuncer O, Orhan EK, Yilmaz V, Parman Y, Oflazer F, Saruhan-Direskeneli G, et al. Prepubertal anti-Musk positive myasthenia gravis with long remission. *Neuromuscul Disord* 2014;**24**: 36–9.
48. Kostera-Pruszczyk A, Kwiecinski H. Juvenile seropositive myasthenia gravis with anti-MuSK antibody after thymectomy. *J Neurol* 2009;**256**:1780–1.
49. Ohta K, Shigemoto K, Fujinami A, Maruyama N, Konishi T, Ohta M. Clinical and experimental features of MuSK antibody positive MG in Japan. *Eur J Neurol* 2007;**14**:1029–34.
50. Anlar B, Yilmaz V, Saruhan-Direskeneli G. Long remission in muscle-specific kinase antibody-positive juvenile myasthenia. *Pediatr Neurol* 2009;**40**:455–6.
51. Fakiri MO, Tavy DL, Hama-Amin AD, Wirtz PW. Accuracy of the ice test in the diagnosis of myasthenia gravis in patients with ptosis. *Muscle Nerve* 2013;**48**:902–4.
52. Bodack MI. Ptosis and cranial nerve IV palsy reveal juvenile myasthenia gravis. *Optometry* 2009;**80**:342–9.
53. Mittal MK, Barohn RJ, Pasnoor M, McVey A, Herbelin L, Whittaker T, et al. Ocular myasthenia gravis in an academic neuro-ophthalmology clinic: clinical features and therapeutic response. *J Clin Neuromuscul Dis* 2011;**13**:46–52.
54. Fenichel GM. Clinical syndromes of myasthenia in infancy and childhood. A review. *Arch Neurol* 1978;**35**:97–103.
55. Hays RM, Michaud LJ. Neonatal myasthenia gravis: specific advantages of repetitive stimulation over edrophonium testing. *Pediatr Neurol* 1988;**4**:245–7.
56. Rubin DI, Hentschel K. Is exercise necessary with repetitive nerve stimulation in evaluating patients with suspected myasthenia gravis? *Muscle Nerve* 2007;**35**:103–6.
57. Pitt M. Neurophysiological strategies for the diagnosis of disorders of the neuromuscular junction in children. *Dev Med Child Neurol* 2008;**50**:328–33.
58. Stalberg EV, Sanders DB. Jitter recordings with concentric needle electrodes. *Muscle Nerve* 2009;**40**:331–9.
59. Skeie GO, Apostolski S, Evoli A, Gilhus NE, Illa I, Harms L, et al. Guidelines for treatment of autoimmune neuromuscular transmission disorders. *Eur J Neurol* 2010;**17**:893–902.
60. McMillan HJ, Darras BT, Kang PB. Autoimmune neuromuscular disorders in childhood. *Curr Treatment Options Neurol* 2011;**13**:590–607.
61. Liew WK, Kang PB. Update on juvenile myasthenia gravis. *Curr Opin Pediatr* 2013;**25**:694–700.
62. Adams SL, Mathews J, Grammer LC. Drugs that may exacerbate myasthenia gravis. *Ann Emerg Med* 1984;**13**:532–8.
63. Gajdos P, Chevret S. Treatment of myasthenia gravis acute exacerbations with intravenous immunoglobulin. *Ann N Y Acad Sci* 2008;**1132**:271–5.
64. Gajdos P, Chevret S, Toyka K. Intravenous immunoglobulin for myasthenia gravis. *Cochrane Database of Syst Rev* 2008;**1:** CD002277.
65. Barth D, Nabavi Nouri M, Ng E, Nwe P, Bril V. Comparison of IVIg and PLEX in patients with myasthenia gravis. *Neurology* 2011;**76**:2017–23.
66. Jensen P, Bril V. A comparison of the effectiveness of intravenous immunoglobulin and plasma exchange as preoperative therapy of myasthenia gravis. *J Clin Neuromuscul Dis* 2008;**9**:352–5.
67. Mandawat A, Kaminski HJ, Cutter G, Katirji B, Alshekhlee A. Comparative analysis of therapeutic options used for myasthenia gravis. *Ann Neurol* 2010;**68**:797–805.
68. Zinman L, Ng E, Bril V. IV immunoglobulin in patients with myasthenia gravis: a randomized controlled trial. *Neurology* 2007;**68**:837–41.
69. Nagayasu T, Yamayoshi T, Matsumoto K, Ide N, Hashizume S, Nomura M, et al. Beneficial effects of plasmapheresis before thymectomy on the outcome in myasthenia gravis. *Jpn J Thorac Cardiovasc Surg* 2005;**53**:2–7.
70. Sarkar BK, Sengupta P, Sarkar UN. Surgical outcome in thymic tumors with myasthenia gravis after plasmapheresis – a comparative study. *Interact Cardiovasc Thorac Surg* 2008;**7**:1007–10.
71. Selcen D, Dabrowski ER, Michon AM, Nigro MA. High-dose intravenous immunoglobulin therapy in juvenile myasthenia gravis. *Pediatr Neurol* 2000;**22**:40–3.
72. Snead III OC, Benton JW, Dwyer D, Morley BJ, Kemp GE, Bradley RJ, et al. Juvenile myasthenia gravis. *Neurology* 1980;**30**: 732–9.
73. Schneider-Gold C, Gajdos P, Toyka KV, Hohlfeld RR Corticosteroids for myasthenia gravis. *Cochrane Database Syst Rev* 2005;**2:** CD002828.
74. Sathasivam S. Steroids and immunosuppressant drugs in myasthenia gravis. *Nat Clin Pract Neurol* 2008;**4**:317–27.
75. Palace J, Newsom-Davis J, Lecky BA. Randomized double-blind trial of prednisolone alone or with azathioprine in myasthenia gravis. Myasthenia Gravis Study Group. *Neurology* 1998;**50**: 1778–83.
76. Ashraf VV, Taly AB, Veerendrakumar M, Rao S. Myasthenia gravis in children: a longitudinal study. *Acta Neurol Scand* 2006;**114**:119–23.
77. Raksadawan N, Kankirawatana P, Balankura K, Prateepratana P, Sangruchi T, Atchaneeyasakul LO. Childhood onset myasthenia gravis. *J Med Assoc Thai* 2002;**85**(Suppl. 2):S769–77.
78. Seybold ME. Thymectomy in childhood myasthenia gravis. *Ann NY Acad Sci* 1998;**841**:731–41.
79. Relling MV, Gardner EE, Sandborn WJ, Schmiegelow K, Pui CH, Yee SW, et al. Clinical Pharmacogenetics Implementation Consortium guidelines for thiopurine methyltransferase genotype and thiopurine dosing. *Clin Pharmacol Ther* 2011;**89**:387–91.
80. Bonifati DM, Angelini C. Long-term cyclosporine treatment in a group of severe myasthenia gravis patients. *J Neurol* 1997;**244**:542–7.

81. Lavrnic D, Vujic A, Rakocevic-Stojanovic V, Stevic Z, Basta I, Pavlovic S, et al. Cyclosporine in the treatment of myasthenia gravis. *Acta Neurol Scand* 2005;**111**:247–52.
82. Tindall RS, Phillips JT, Rollins JA, Wells L, Hall K. A clinical therapeutic trial of cyclosporine in myasthenia gravis. *Ann NY Acad Sci* 1993;**681**:539–51.
83. Nagane Y, Utsugisawa K, Obara D, Kondoh R, Terayama Y. Efficacy of low-dose FK506 in the treatment of Myasthenia gravis – a randomized pilot study. *Eur Neurol* 2005;**53**:146–50.
84. Ponseti JM, Gamez J, Azem J, Lopez-Cano M, Vilallonga R, Armengol M. Tacrolimus for myasthenia gravis: a clinical study of 212 patients. *Ann NY Acad Sci* 2008;**1132**:254–63.
85. Yoshikawa H, Kiuchi T, Saida T, Takamori M. Randomised, double-blind, placebo-controlled study of tacrolimus in myasthenia gravis. *J Neurol Neurosurg Psychiatry* 2011;**82**:970–7.
86. Mori T, Mori K, Suzue M, Ito H, Kagami S. Effective treatment of a 13-year-old boy with steroid-dependent ocular myasthenia gravis using tacrolimus. *Brain Dev* 2013;**35**:445–8.
87. Meriggioli MN, Rowin J. Treatment of myasthenia gravis with mycophenolate mofetil: a case report. *Muscle Nerve* 2000;**23**:1287–9.
88. Ciafaloni E, Massey JM, Tucker-Lipscomb B, Sanders DB. Mycophenolate mofetil for myasthenia gravis: an open-label pilot study. *Neurology* 2001;**56**:97–9.
89. Meriggioli MN, Ciafaloni E, Al-Hayk KA, Rowin J, Tucker-Lipscomb B, Massey JM, et al. Mycophenolate mofetil for myasthenia gravis: an analysis of efficacy, safety, and tolerability. *Neurology* 2003;**61**:1438–40.
90. Muscle Study Group. A trial of mycophenolate mofetil with prednisone as initial immunotherapy in myasthenia gravis. *Neurology* 2008;**71**:394–9.
91. Sanders DB, Hart IK, Mantegazza R, Shukla SS, Siddiqi ZA, De Baets MH, et al. An international, phase III, randomized trial of mycophenolate mofetil in myasthenia gravis. *Neurology* 2008;**71**:400–6.
92. Hehir MK, Burns TM, Alpers J, Conaway MR, Sawa M, Sanders DB. Mycophenolate mofetil in AChR-antibody-positive myasthenia gravis: outcomes in 102 patients. *Muscle Nerve* 2010;**41**:593–8.
93. Blalock A, Mason MF, Morgan HJ, Riven SS. Myasthenia gravis and tumors of the thymic region: Report of a case in which the tumor was removed. *Ann Surg* 1939;**110**:544–61.
94. Mullaney P, Vajsar J, Smith R, Buncic JR. The natural history and ophthalmic involvement in childhood myasthenia gravis at the hospital for sick children. *Ophthalmology* 2000;**107**:504–10.
95. Youssef S. Thymectomy for myasthenia gravis in children. *J Pediatr Surg* 1983;**18**:537–41.
96. Gronseth GS, Barohn RJ. Practice parameter: thymectomy for autoimmune myasthenia gravis (an evidence-based review): report of the Quality Standards Subcommittee of the American Academy of Neurology. *Neurology* 2000;**55**:7–15.
97. Tracy MM, McRae W, Millichap JG. Graded response to thymectomy in children with myasthenia gravis. *J Child Neurol* 2009;**24**:454–9.
98. Castro D, Derisavifard S, Anderson M, Greene M, Iannaccone S. Juvenile myasthenia gravis: a twenty-year experience. *J Clin Neuromuscul Dis* 2013;**14**:95–102.
99. Heng HS, Lim M, Absoud M, Austin C, Clarke D, Wraige E, et al. Outcome of children with acetylcholine receptor (AChR) antibody positive juvenile myasthenia gravis following thymectomy. *Neuromuscul Disord* 2014;**24**:25–30.
100. Sanes JR, Lichtman JW. Development of the vertebrate neuromuscular junction. *Annu Rev Neurosci* 1999;**22**:389–442.
101. Missias AC, Chu GC, Klocke BJ, Sanes JR, Merlie JP. Maturation of the acetylcholine receptor in skeletal muscle: regulation of the AChR gamma-to-epsilon switch. *Dev Biol* 1996;**179**:223–38.
102. Koenen M, Peter C, Villarroel A, Witzemann V, Sakmann B. Acetylcholine receptor channel subtype directs the innervation pattern of skeletal muscle. *EMBO Rep* 2005;**6**:570–6.
103. Oskoui M, Jacobson L, Chung WK, Haddad J, Vincent A, Kaufmann P, et al. Fetal acetylcholine receptor inactivation syndrome and maternal myasthenia gravis. *Neurology* 2008;**71**:2010–12.
104. Vernet-der Garabedian B, Lacokova M, Eymard B, Morel E, Faltin M, Zajac J, et al. Association of neonatal myasthenia gravis with antibodies against the fetal acetylcholine receptor. *J Clin Invest* 1994;**94**:555–9.
105. Greer M, Schotland M. Myasthenia gravis in the newborn. *Pediatrics* 1960;**26**:101–8.
106. Riemersma S, Vincent A, Beeson D, Newland C, Hawke S, Vernet-der Garabedian B, et al. Association of arthrogryposis multiplex congenita with maternal antibodies inhibiting fetal acetylcholine receptor function. *J Clin Invest* 1996;**98**:2358–63.
107. Polizzi A, Huson SM, Vincent A. Teratogen update: maternal myasthenia gravis as a cause of congenital arthrogryposis. *Teratology* 2000;**62**:332–41.
108. D'Amico A, Bertini E, Bianco F, Papacci P, Jacobson L, Vincent A, et al. Fetal acetylcholine receptor inactivation syndrome and maternal myasthenia gravis: a case report. *Neuromuscul Disord* 2012;**22**:546–8.
109. Carr SR, Gilchrist JM, Abuelo DN, Clark D. Treatment of antenatal myasthenia gravis. *Obstet Gynecol* 1991;**78**:485–9.
110. Beghi E, Antozzi C, Batocchi AP, Cornelio F, Cosi V, Evoli A, et al. Prognosis of myasthenia gravis: a multicenter follow-up study of 844 patients. *J Neurol Sci* 1991;**106**:213–20.

Part VI

Myopathies

Chapter 28

Congenital and Other Structural Myopathies

James J. Dowling, Kathryn N. North, Hans H. Goebel, and Alan H. Beggs

The congenital myopathies are a heterogeneous group of neuromuscular conditions defined by distinctive morphologic abnormalities in skeletal muscle (Figure 28.1). These disorders typically present in infancy or early childhood with hypotonia and muscle weakness, and are usually nonprogressive, or only slowly progressive with age. There is wide variation in clinical severity within each group and marked clinical overlap with other neuromuscular disorders, including the muscular dystrophies, congenital myasthenic syndromes, metabolic myopathies, and spinal muscular atrophy. Therefore, accurate diagnosis requires the combined evaluation of the clinical data, the electromyogram, and histopathologic findings. The congenital myopathies usually have a genetic basis, although for some morphologic entities only sporadic cases have been reported (see later discussion).

Clinically, the congenital myopathies share a number of common features: generalized weakness, hypotonia and hyporeflexia, poor muscle bulk, and dysmorphic features secondary to the myopathy (e.g. pectus carinatum, scoliosis, foot deformities, a high arched palate, and elongated facies) (Figure 28.2). The pathologic process appears to exclusively affect the striated muscle, and thus congenital myopathies may be distinguished clinically from neuromuscular conditions with multisystem involvement, such as facioscapulohumeral muscular dystrophy, which is associated with hearing defects or retinal vascular pathology, myotonic dystrophy with central nervous system involvement, cataracts and cardiac conduction defects, and a subset of patients with congenital muscular dystrophies (CMD) associated with brain pathology (e.g. Walker Warburg syndrome, muscle-eye-brain disease, Fukuyama CMD). Some of the important clinical features that distinguish congenital myopathies from other similar diagnoses and that delineate specific congenital myopathy subtypes are presented in Table 28.1 (for clinical signs in infancy) and Table 28.2 (for clinical features in older children).

The first report of a congenital myopathy was by Shy and Magee in 1956.[1] They described a hereditary, nonprogressive disorder that subsequently became known as central core disease. Prior to this, the spinal muscular atrophies were the only well-defined neuromuscular disorders of infancy and childhood. Most patients with congenital myopathies were confusingly classified under such descriptive terms as "amyotonia congenita" or "benign congenital hypotonia." In 1969, Dubowitz clarified the classification of these disorders with his delineation of the "new myopathies," later termed congenital myopathies.[2] The pathologic changes associated with congenital myopathies appear to originate within the myofiber itself. The structural diversity of the striated muscle fiber allows for an enormous spectrum of myopathologic features rarely encountered in pathology of any other mammalian cell. Damaging forces from outside the muscle fiber, such as inflammation or denervation, rarely result in the morphologic changes associated with congenital myopathies. Exceptions include target formation in denervation, nemaline bodies secondary to tenotomy, and tubulofilamentous aggregates in inclusion body myositis.

The congenital myopathies are rare disorders, although precise epidemiologic data of frequency and geographic distribution are available in only limited circumstances. A recent comprehensive examination of prevalence was performed using a well-delineated pediatric population in a small but ethnically diverse region of the United States (southeastern Michigan).[3] These data estimated the point prevalence of myopathies to be 1:26,000, with core myopathies the most common histopathologic subtype and *RYR1*-related myopathies the most prevalent genetic subtype. Additional information comes from several case series. In one series of 250 patients with severe neonatal

B.T. Darras, H. Royden Jones, Jr., M.M. Ryan & D.C. De Vivo (Eds): Neuromuscular Disorders of Infancy, Childhood and Adolescence, Second edition.
DOI: http://dx.doi.org/10.1016/B978-0-12-417044-5.00028-7

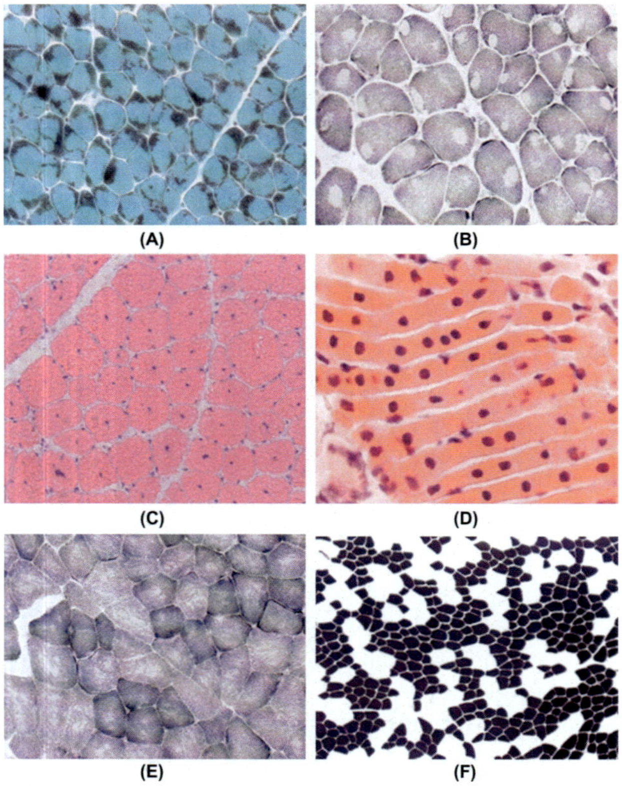

FIGURE 28.1 Pathological features that define the major subtypes of congenital myopathy. **(A)** *Nemaline rods:* Biopsy from patient with nemaline myopathy with a dominant mutation in the *ACTA1* gene, showing clusters of purple staining rods at the periphery of most fibers and some internal within fibers (Gomori trichrome). **(B)** *Central cores:* Biopsy of the quadriceps from a 3-year-old patient with central core disease with a dominant mutation in the ryanodine receptor gene showing mild variation in fiber size (fiber diameter range 15–65 mm), fiber type uniformity and numerous cores of varying size centrally or peripherally (oxidative enzyme stain SDH). **(C)** *Central nuclei:* Quadriceps biopsy from a 28-year-old patient with autosomal dominant centronuclear myopathy due to a *DNM2* mutation. The biopsy demonstrated small type 1 fibers and centrally placed nuclei in the majority of fibers (H&E). **(D)** *Central nuclei (longitudinal section):* Quadriceps biopsy from a case of X-linked myotubular myopathy aged 8 months showing large central nuclei. Note the widely spaced nuclei which affects the number seen in transverse section. Most fibers are less than 10 mm in diameter. **(E)** *Multiminicores:* Areas in both fiber types of varying size and number devoid of oxidative enzyme stain in a quadriceps biopsy from a case of "multiminicore disease" aged 11 years with recessive mutations in the *SEPN1* gene (NADH-TR). **(F)** *Congenital fiber type disproportion with mutation in ACTA1:* The only apparent pathology in this case was the small size of the dark-staining type 1 fibers and type 1 fiber predominance (ATPase preincubated at pH 4.3). Fiber diameter 25–70 mm. *Reprinted with permission from North et al.*[4] *© 2014 Elsevier.*

hypotonia, 34 of the 250 (14%) were diagnosed as having a congenital myopathy.[5] Eight cases (3% of the total) had nemaline myopathy, which is perhaps the most common type presenting in infancy. A more recent series of 66 consecutive patients seen at one center in the United Kingdom identified over half (54%) with core myopathies, 17% with nemaline myopathies, and 13% with centronuclear myopathies, together accounting for the majority of diagnoses.[6]

Over the past three decades, the number of morphologically distinct congenital myopathies has grown rapidly, paralleling the advent and refinement of myopathologic techniques such as electron microscopy, enzyme histochemistry and immunocytochemistry, and more recently, molecular genetic techniques. Recent efforts have focused on defining and classifying the distinct and "real" disease entities. The current clinicopathological classification is based on the specific structural abnormalities seen on light and electron microscopy (Box 28.1), although improved genetic diagnoses are revealing extensive genetic heterogeneity within morphological subtypes, and considerable overlap between subtypes sharing common genetic etiologies.[4] Certain congenital myopathies are well defined clinically, morphologically, and genetically (Table 28.3). These classical congenital myopathies are nemaline myopathy, core myopathies (encompassing central core disease and multiminicore myopathy), the centronuclear myopathies (including X-linked myotubular myopathy), and congenital fiber type disproportion and aggregation myopathies (including the myofibrillar, or desmin-related, myopathies and the autophagic vacuolar myopathies) and will be reviewed in detail (Figure 28.1). There are several more rare forms of congenital myopathy, many of which have been described in only a very small number of patients. Some of these entities are now recognized as variant forms of the more common subtypes (for example, cap myopathy is likely an allelic variant of nemaline myopathy), while some continue to be unique and poorly understood conditions. A list of the rare myopathies not discussed further in this chapter, along with some of their distinguishing features, is presented in Table 28.4.

While the congenital myopathy subtypes share many common clinical features, some specific signs and symptoms may be present to help distinguish them (Tables 28.1 and 28.2). For example, ophthalmoparesis is common in centronuclear myopathies and in recessive RYR1-related myopathies,[7] but far less common in nemaline myopathy.[8] Foot drop may be observed in nemaline myopathy,[9] *DNM2*-related centronuclear myopathy,[10] and *MYH7* related myopathies.[11] Specific disease features are provided in greater detail in the specific subsections following.

Most commonly, congenital myopathies are distinguished by their features on muscle biopsy (Figure 28.1).[4] Routine laboratory studies are typically not helpful, though serum creatine phosphokinase (CPK) levels are useful in that very high levels (>1000) are more typically associated with muscular dystrophies (though there are rare exceptions). Adjunct studies, in addition to clinical features, that can aid in the differential diagnosis include

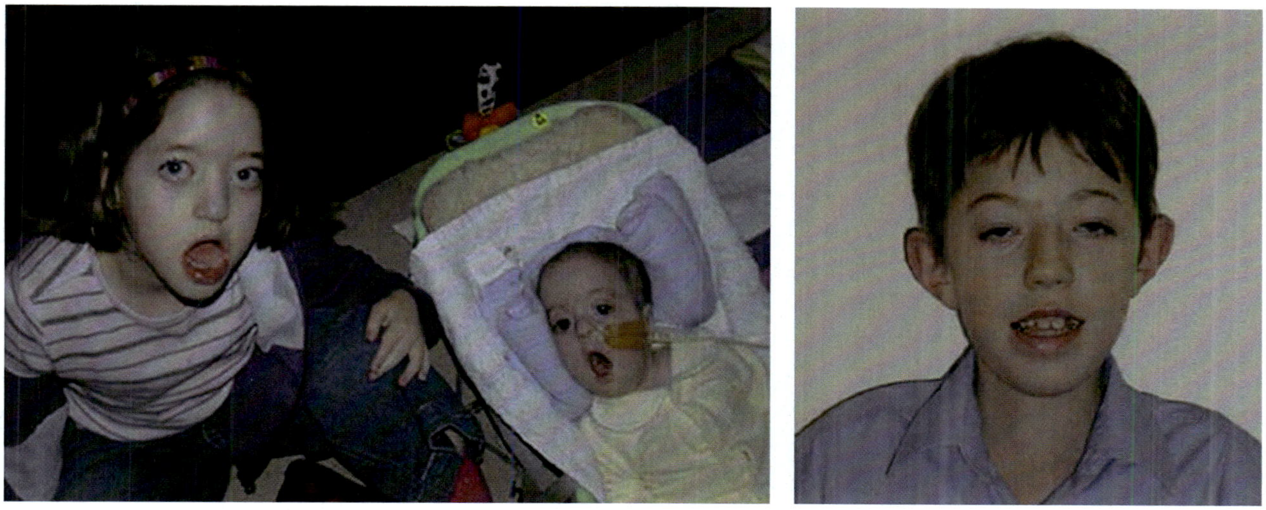

FIGURE 28.2 Facial involvement in congenital myopathies. (A) Pronounced facial weakness, particularly affecting the lower face and mouth resulting in craniofacial dysmorphism ("myopathic facies") in sisters aged 6 years and 3 months with autosomal recessive nemaline myopathy (likely due to nebulin gene (*NEB*) mutation). (B) Ptosis and ophthalmoplegia in a patient with *DNM2*-related centronuclear myopathy at age 9 years. *Reprinted with permission from North et al.[4] © 2014 Elsevier.*

TABLE 28.1 Clinical Clues Particularly Suggestive of Specific Diagnosis in Congenital Myopathies: Newborn and Infant <2 Years

Clinical Features	Congenital Myopathy	Differential Diagnosis
Facial weakness	NM, CNM (*MTM1, RYR1, DNM2*)	DM1, CMS (rapsyn)
Ophthalmoplegia	CNM (*MTM1, RYR1, DNM2*), MmD (*RYR1*)	CMS, mitochondrial
Ptosis	CNM (*MTM1, RYR1, DNM2*), MmD, CCD	CMS, DM1
Facial dysmorphism (long face, dolichocephaly, high arched palate)	NM, CNM (*MTM1*, severe *DNM2*), severe *RYR1*	DM1
Bulbar weakness (sucking/swallowing)	NM, CNM (*MTM1*), severe *RYR1*	CMS, DM1, PWS, SMA
Severe respiratory involvement at birth	NM, CNM (*MTM1*), severe *RYR1*	DM1, SMA 0, CMS, Pompe
Predominant axial hypotonia	*RYR1, SEPN1*	*LMNA*
Severe congenital hypotonia	NM, *MTM1, RYR1*	DM1, PWS, Down Syndrome
Orthopedic deformities	*RYR1*, NM	COL6, CMS
Hip dislocation	*RYR1*	COL6
Fetal akinesia/severe arthrogryposis	NM (*ACTA1, NEB*), severe *RYR1, KLHL40*	CMS, SMA 0, CHS
Club feet	NM, *RYR1*	CMS, DM1, CHS

Abbreviations: NM, nemaline myopathy; DM1, myotonic dystrophy type 1; CMS, congenital myasthenic syndromes; CNM, centronuclear myopathy; mito, mitochondrial myopathy; MmD, multiminicore disease; CCD, central core disease; PWS, Prader-Willi syndrome; SMA, spinal muscular atrophy; Pompe Pompe disease; COL6, Ullrich congenital muscular dystrophy.
Please note that the table is meant to indicate if a clinical finding is a particular clue to one of the congenital myopathies. Specific clinical findings can occur in the other forms of congenital myopathy at lower frequency.
Reprinted with permission from North et al.[4] © 2014 Elsevier.

TABLE 28.2 Clinical Clues Suggestive of Specific Diagnosis in Congenital Myopathies: Older Child

Clinical Features	Congenital Myopathies	Differential Diagnoses
Scoliosis	*SEPN1*, *RYR1*, NM	COL6, *LAMA2*
Rigid spine	*SEPN1*, *RYR1*	
Cardiomyopathy	*TTN*, *MYH7*, rarely *ACTA1*	Pompe disease
Foot drop/pes cavus	NM (*NEB*, *TPM3*, *TPM2*), *DNM2*, *MYH7*	Peripheral neuropathy
Malignant hyperthermia	CCD, MmD and CNM (*RYR1* only)	
Respiratory involvement and axial involvement out of proportion to skeletal muscle weakness	*SEPN1*, NM (*NEB*, *TPM3*, *ACTA1*)	*LMNA*, CMS, Pompe disease

Abbreviations: NM, nemaline myopathy; COL6, collagen VI associated myopathy; CCD, central core disease; CMS, congenital myasthenic syndrome.
Please note that the table is meant to indicate if a clinical finding is a particular clue to one of the congenital myopathies. Each clinical finding may occur in the other forms at lower frequency.
Reprinted with permission from North et al.[4]

BOX 28.1 Classification of the Congenital and Other Structural Myopathies

Structured Types

Abnormalities of Sarcomeres

1. Central core disease
2. Multiminicore disease
3. Myofibrillar lysis myopathy/hyaline body myopathy
4. Actinopathy
5. Cap disease
6. Trilaminar fiber myopathy
7. Broad A-band myopathy

Z-disk Abnormalities

1. Nemaline/rod myopathy
2. Desmin-related/myofibrillar myopathies
   Cytoplasmic body myopathy
   Spheroid body myopathy
   Mallory body-like myopathy
   Sarcoplasmic body myopathy
   Granulofilamentous myopathy

Nuclear Abnormalities

1. Myotubular myopathy
   Centronuclear myopathy
2. Intranuclear rod myopathy

Abnormal Inclusions

1. Fingerprint body myopathy
2. Zebra body myopathy
3. Reducing body myopathy
4. Cylindrical spirals myopathy

Abnormalities of Organelles

1. Sarcotubular myopathy
   Tubular aggregate myopathy
2. Lysosome-related vacuolar myopathies

Miscellaneous

1. Minimal change myopathy
2. Nonspecific congenital myopathies in syndromes

Mixed Myopathies

1. Rod-core myopathy
2. Cytoplasmic body–reducing body myopathy
3. Multiminicore–centronuclear myopathy

Unstructured Types

1. Congenital fiber type disproportion
2. Microfiber myopathy
3. Type 1 fiber predominance
4. Type 1 fiber uniformity
5. Type 1 myofiber hypotrophy
6. Type 2A fiber uniformity and smallness
7. Type 2 muscle fiber hypoplasia

especially muscle MRI (see Figure 28.3 and Table 28.5).[15] Electromyography may be helpful in distinguishing congenital myopathies from congenital myasthenic syndromes but not in delineating myopathy subtypes. In the majority of cases, a genetic diagnosis can be achieved, and ultimately represents the defining characteristic for a given patient. More than 15 genes have now been discovered. It is estimated that approximately 25–50% of cases remain genetically unsolved, though this number is rapidly dwindling with the application of new genomic technologies.

NEMALINE MYOPATHY

Nemaline myopathy is characterized by the presence of electron dense rod-shaped structures in the muscle fibers. The first descriptions of the disorder are generally attributed to Shy and colleagues and Conen and colleagues in

TABLE 28.3 Congenital and Structural Myopathies and Their Identified Gene Loci

Subtype	Gene	OMIM Designtion	Chromosome Location	Inheritance Pattern
Nemaline Myopathy	*ACTA1*	NEM3	1q42.1	AD, AR
	CFL2	NEM7	14q13.1	AR
	KBTBD13	NEM6	15q22.31	AD
	KLHL40	NEM8	3p33.1	AR
	KLHL41	NEM9	2q31.1	AR
	NEB	NEM2	2q23.3	AR
	RYR1		19q13.2	AR
	TNNT1	NEM5	19q13.42	AR
	TPM2	NEM4	9p13.3	AD
	TPM3	NEM1	1q21.3	AD, AR
Cap Disease (NM Variant)	*ACTA1*		1q42.1	AD
	TPM2	CAPM2	9p13.3	AD
	TPM3	CAPM1	1q21.3	AD
Zebra Body Myopathy	*ACTA1*		1q42.1	AD
Hyaline Body Myopathy (Myosin Storage Myopathy)	*MYH7*		14q11.2	AD
Core-Rod Myopathy	*KBTBD13*		15q22.31	AD
	NEB		2q23.3	AR
	RYR1		19q13.2	AD, AR
	TPM2		9p13.3	AD
Central Core Disease	*RYR1*	CCD	19q13.2	AD, AR
Multiminicore Disease	*RYR1*		19q13.2	AR
	SEPN1		1p36.11	AR
	MYH7		14q11.2	AD
Centronuclear Myopathy	*BIN1*	CNM2	2q14.3	AR
	CCDC78	CNM4	16p13.3	AD
	DNM2	CNM1	19p13.2	AD
	MTM1	CNMX	Xq28	XL
	MYF6	CNM3	12q21.31	AD
	RYR1		19q13.2	AR
	TTN		2q31.2	AR
	SPEG	CNM5	2q35	AR
Congenital Fiber Type Disproportion	*ACTA1*		1q42.1	AD
	MYH7		14q11.2	AD
	RYR1		19q13.2	AR
	SEPN1		1p36.11	AR
	TPM2		9p13.3	AD
	TPM3		1q21.3	AD

(Continued)

TABLE 28.3 (Continued)

Subtype	Gene	OMIM Designtion	Chromosome Location	Inheritance Pattern
Autophagic Vacuolar Myopathies	*VMA21*	MEAX	Xq28	XL
	LAMP2	Danon Disease	Xq24	XL
	GAA	Pompe Disease	17q25.3	AR
Myofibrillar Myopathies	*DES*	MFM1	2q35	AD, AR
	CRYAB	MFM2	11q23.1	AD
	MYOT	MFM3	5q31.2	AD
	ZASP	MFM4	10q23.2	AD
	FLNC	MFM5	7q32.1	AD
	BAG3	MFM6	10q26.11	AD

TABLE 28.4 Rare Structural Myopathies

Rare Myopathy	Histopathology	Clinical Features	Genetics
Cylindrical spirals myopathy (Figure 28.20)	Parallel spiraled lamellae as cylindrical bodies	Minimal weakness, myotonia, cramps	Dominant familial cases
Tubular aggregate myopathy (Figure 28.21)	Tubular aggregates in hexagonal patterns	Myalgias, mild proximal weakness, cramps	*STIM1* *ORAI1*
Fingerprint body myopathy	Fingerprint bodies, type I fiber hypotrophy	Hypotonia, kyphosis, delayed ambulation	Familial cases
Hexagonal crystalloid body myopathy	Hexagonally cross-linked tubular arrays, caveolin 3-positive	Post-exercise myalgias	Familial cases

1963.[16,17] However, it is now recognized that a patient with congenital myopathy and rod bodies was described by Dr. Douglas Reye, a pathologist in Sydney, Australia, in 1958.[18,19] The term *nemaline* was first applied by Shy and colleagues[16] because of the threadlike appearance of the rod bodies (the Greek word *nema* means "thread"). Nemaline myopathy is a rare disorder with an estimated incidence of 0.02 per 1000 live births.[20]

Classification

The existing classification of nemaline myopathy is based on age of onset and severity of motor and respiratory involvement with severe congenital, intermediate congenital, typical congenital, childhood-onset, and adult-onset forms, as reviewed by North and associates.[21–23] However, there is marked overlap between each of these groups. With the recent confirmation of both autosomal dominant and recessive forms of nemaline myopathy, and the discovery of at least nine genetic loci for the disorder (see later discussion), further molecular studies may eventually result in a meaningful classification that permits prediction of prognosis or determination of the mode of inheritance in singleton cases.

The *severe neonatal form* presents at birth with severe hypotonia and muscle weakness, little spontaneous movement, difficulties with sucking and swallowing, gastroesophageal reflux, and respiratory insufficiency. Decreased fetal movements and polyhydramnios may complicate the pregnancy, and death *in utero* associated with fetal akinesia has been described.[24] Dilated cardiomyopathy occurs infrequently. Arthrogryposis may be a presenting feature in some individuals, and can be quite severe. Death due to respiratory insufficiency or recurrent pneumonia is common during the first weeks or months of life. However, even patients with severe floppiness and lack of spontaneous respiration

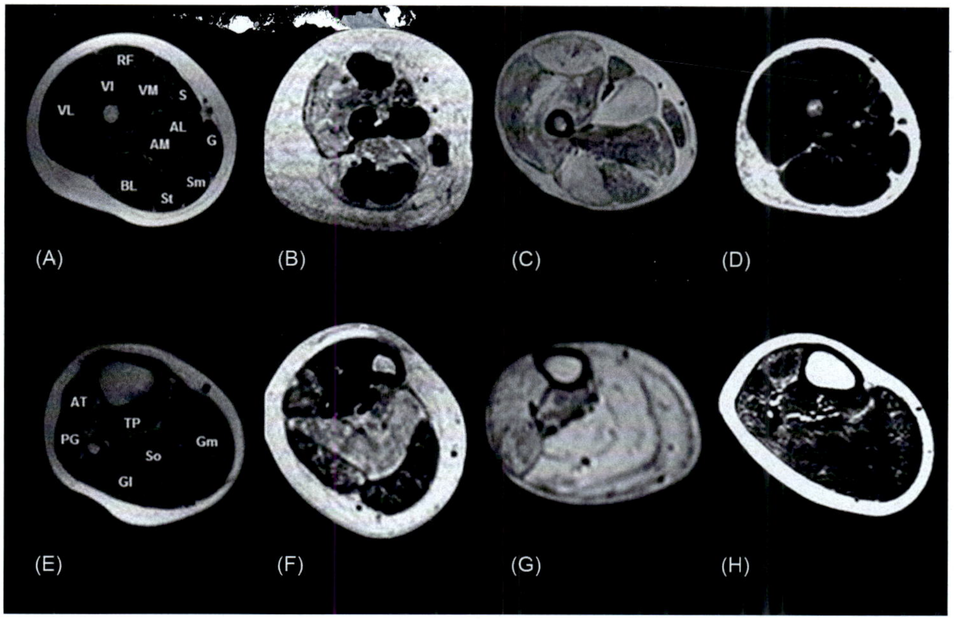

FIGURE 28.3 Muscle magnetic resonance imaging (MRI) in the diagnosis of the congenital myopathies: T1-weighted images, transverse sections, through the thigh (**A–D**) and the lower leg (**E–H**) in a normal control (**A, E**) and patients with *RYR1*-related dominant central core disease (CCD) (**B, F**), *DNM2*-related dominant centronuclear myopathy (CNM) (**C, G**) and *NEB*-related recessive nemaline myopathy (NM) (**D, H**). In *RYR1*-related dominant CCD (**B, F**), there is marked involvement of the thigh with relative sparing of rectus femoris (RF), adductor longus (AL), gracilis (G) and hamstring muscles. In the lower leg, soleus (So), gastrocnemius lateralis (Gl) and peroneal group (PG) are markedly affected whereas tibialis anterior (AT), tibialis posterior (TP), and gastrocnemius medialis (Gm) are relatively spared. In *DNM2*-related dominant centronuclear myopathy (CNM) (**C, G**), the thigh is diffusely affected with prominent involvement of rectus femoris (RF) and adductor longus (AL) compared to other muscle groups. In the lower leg, there is diffuse involvement of all muscle groups with relative sparing of the tibialis posterior (TP). In *NEB*-related recessive nemaline myopathy (NM) (**D, H**), the thigh is spared whereas the lower leg shows early involvement of the tibialis anterior (AT) and, to a lesser extent, soleus (So), in keeping with the clinical finding of early distal involvement in NM. *Composite image from Jungbluth H, et al., 2004[12]; Jungbluth H et al., 2004[13]; and Schessl J et al., 2007.[14] Reprinted with permission from North et al.[1] © 2014 Elsevier.*

at birth have been known to survive, some of them with little residual disability.[25–27]

The *mild congenital* or *classical* form of nemaline myopathy usually presents at birth or during the first year of life with hypotonia, weakness, and feeding difficulties; however, the severity of muscle involvement is often less marked than in the severe neonatal form. Some cases present later with delayed attainment of motor milestones, waddling gait, or speech abnormalities. There is commonly distal involvement in addition to the proximal muscle weakness, and some patients have initially been thought to have peroneal paresis because of their foot drop. The respiratory muscles are usually involved, although hypoventilation may not be clinically obvious; cardiac involvement is rare. The course of the disease is often static or only very slowly progressive and most patients will be able to lead an active life. Others may experience deterioration during the prepubertal period of rapid growth and some will start using a wheelchair at this time. See Case Example 28.1.

The *childhood-onset form* of nemaline myopathy was first described in a large autosomal dominant Australian kindred by Laing and colleagues.[9] Early motor development was normal, with the development of symmetrical weakness of ankle dorsiflexion in the late first or early second decade. Weakness was slowly progressive with eventual involvement of all ankle movements, and more proximal limb musculature. Two older family members were wheelchair-bound by age 40 years. Cardiac muscle was spared.

The *late-onset* or *adult-onset* form of nemaline myopathy is heterogeneous in terms of clinical presentation and disease progression. Patients with a true isolated adult-onset form of nemaline myopathy have no family history and no preceding symptoms, and onset of proximal and distal weakness occurs in the third to sixth decades. Respiratory and cardiac involvement occurs in the minority of cases, but in these patients the disease often follows a clearly progressive course.

Clinical Features

The cardinal features of nemaline myopathy are generalized weakness and hypotonia. Muscle weakness is usually most severe in the face, the neck flexors, and the proximal limb muscles. In some patients there is an additional

TABLE 28.5 Lower Extremity Muscle MRI Findings in Congenital Myopathy Subtypes

Gene	Muscles																			
	AL	AM	BF	GSL	GSM	GMa	GMi	GRA	PER	PT	PTB	RF	SAT	SMM	ST	SOL	TA	VIM	VL	VM
ACTA1	S	A				A		S	A			S	A		A		A			
BIN1				S	A				A		S					S	A			
DNM2	A		A		A		A	S		S			S	A		A		A		
MTM1	S	A						S		A		S								A
MYH7					S				A							A	A			
NEB					A						S					A	A			
RYR1		A				A		S				S	A			A			A	
SEPN1	S			A	A					A		S	A							
TPM2		A	A					S				S	S	A						

Abbreviations: A, affected; S, spared; AL, adductor longus; AM, adductor magnus; BF, biceps femoris; GSL, lateral gastrocnemius; GSM, medial gastrocnemius; GMa, gluteus maximus; GMi, gluteus minimus; GRA, gracilis; PER, peroneal; PT, posterior thigh; PTB, posterior tibialis; RF, rectus femoris; SAT, sartorius; SMM, semimembranosus; ST, semitendinosus; SOL, soleus; TA, tibialis anterior; VIM, vastus intermedius; VL, vastus lateralis; VM, vastus medialis.

CASE EXAMPLE 28.1

MLL is the first child of a nonconsanguineous Caucasian couple with no family history of neuromuscular disease. She presented at birth with generalized weakness, and particularly bulbar weakness resulting in poor suck and swallow. She had characteristic lower facial weakness (Figure 28.4). During infancy, frequent aspiration led to recurrent pneumonias. Because of her bulbar weakness, MLL required gastrostomy tube feeding until the age of 2.5 years. Through intensive therapy, her swallowing improved, eventually allowing for independent feeding. However, she has had persistent and profuse drooling, which has required frequent suctioning throughout her life.

At 6 months of age, MLL began physical, occupational, and speech therapies. She was never able to lift her head while supine and could never crawl. Her first steps were taken at 13.5 months and independent walking was achieved at 15 months. She has subsequently maintained the ability to walk (though with reduced exercise tolerance), but has never been able to run.

At 4 years of age, MLL was fitted with a prosthetic lift to treat her immobile soft palate and hypernasal speech. At 13 years of age, a pharyngeal flap procedure was performed to improve velopharyngeal incompetency. At 13.5 years of age, she developed scoliosis that was treated with bracing. To date, she has not required scoliosis surgery. Also, breathing assistance has never been required. Of note, MLL has advanced intellectual ability confirmed through standardized neuropsychological assessment.

At her most recent examination (age 16 years), she had marked lower facial weakness with preserved eye movements. She was able to walk independently, but used a mobility scooter for long distances. She was able to jump with assistance but not run. She had a positive Gowers' maneuver.

Muscle biopsy at 5 years of age demonstrated nemaline bodies on light and electron microscopy. Sequencing of the nebulin (*NEB)* gene at 14 years of age revealed a novel heterozygous variant (c.3252_3255 + 3delTGACGTA). The variant is predicted to abolish the splice donor site in intron 32. A *NEB* array was also performed and identified a 2.7 kb deletion encompassing exon 77. The approximate genomic locations are 152,469,299 in intron 77 and 152,471,995 in intron 76. Parental studies showed that the two mutations were inherited from different parents, and thus present in *trans*. Together, the clinical findings, muscle biopsy, and genetic testing results are consistent with a diagnosis of nemaline myopathy due to compound heterozygous mutation in *NEB*. As both parents are carriers and the condition is autosomal recessive, the risk of recurrence for future pregnancies is 25%.

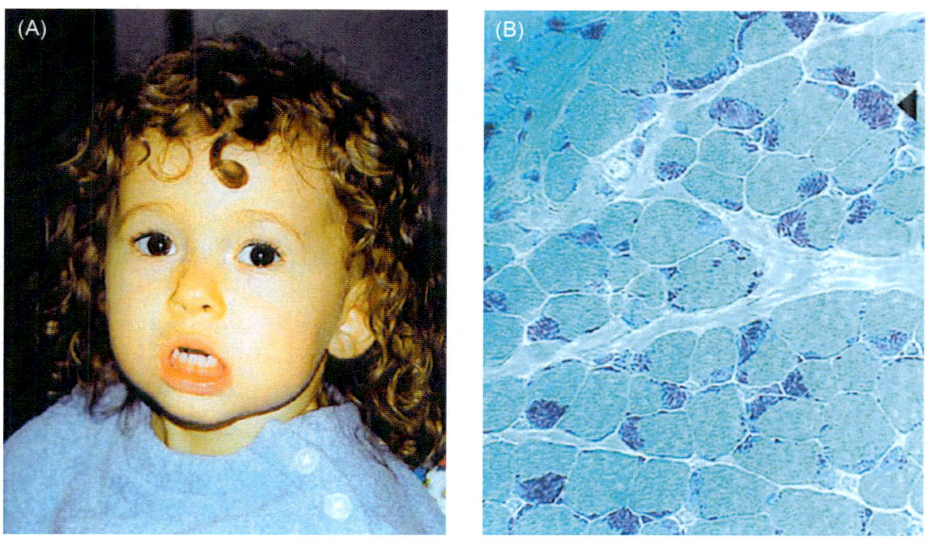

FIGURE 28.4 Case example of nemaline myopathy. (**A**) Photograph of patient MLL at age 3 The classical facial appearance, particularly of lower facial weakness, of NM is depicted. (**B**) Photomicrograph of the muscle biopsy from patient MLL. Abundant nemaline rods (arrowhead) are present, confirming the histopathologic diagnosis of nemaline myopathy. Stain = modified Gomori trichrome.

distal involvement. The extraocular muscles are most often spared. In congenital forms of nemaline myopathy, the face is often elongated and expressionless, the upper lip tented, and the palate high-arched (Figures 28.2A and 28.4). The gait is usually waddling. The build is slender but muscle bulk is not necessarily reduced, especially in young children. The spine is hyperlordotic and sometimes rigid. In some patients, chest deformity is evident at birth. Tendon reflexes are depressed or absent. Gross motor development is slow, whereas fine motor activity is normal. Many patients have hypermobile joints, and joint contractures commonly develop with time. Severe arthrogryposis is not a characteristic feature of the classical or later onset forms of nemaline myopathy, but is more common (and may be quite severe) among patients with the severe neonatal form of the disease.[23,28–32]

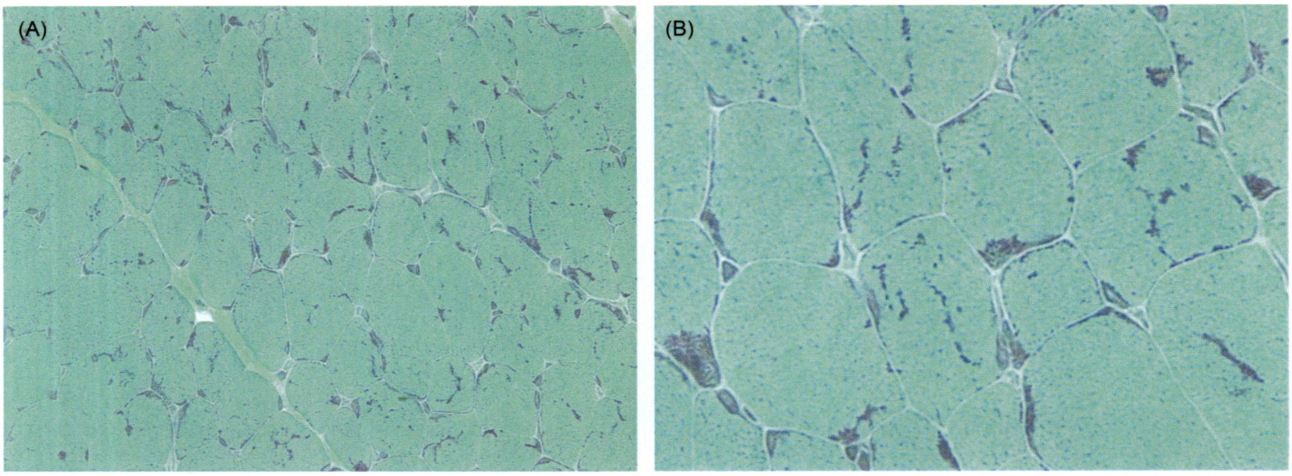

FIGURE 28.5 Nemaline myopathy. Light microscopy (Gomori trichrome stain) at low power (**A**) and high power (**B**). Multiple rods are present in most fibers. Note aggregation of rods under the sarcolemmal membrane.

Respiratory problems are a common feature of congenital nemaline myopathy, not only in the neonatal period but throughout life. The degree of skeletal muscle weakness does not necessarily reflect the degree of respiratory muscle involvement. Even if asymptomatic, most patients will show restricted respiratory capacity on testing. Sleep studies are an important part of the management plan, as patients are at risk of insidious nocturnal hypoxia even in the absence of morning symptoms, and several patients have experienced sudden respiratory failure.[23,27,33,34] Infants with the congenital onset form of nemaline myopathy commonly have feeding difficulties, and older patients may present with isolated swallowing difficulties. Cardiac contractility is usually normal in congenital nemaline myopathy, but cardiac involvement, particularly dilated cardiomyopathy, may rarely occur.[35] The central nervous system is not usually affected in nemaline myopathy and intelligence is usually normal.[27]

Laboratory Investigations

Laboratory examinations are of little help in making the diagnosis of nemaline myopathy. The serum CK concentration is usually normal or only slightly elevated (up to five times higher than normal). EMG may be normal in young patients and mild cases but usually shows polyphasic motor unit potentials with small amplitude, and a full interference pattern during weak effort (early recruitment). In addition to these "myopathic" features, electromyographic signs often interpreted as neurogenic (large motor unit potentials with discrete pattern on full effort, and abnormal jitter) may develop with time, especially in distal muscles. However, most patients will have normal findings on examination of peripheral nerves, including normal conduction velocities.[36] Ultrasonography often shows abnormally high echogenicity in affected muscles, computed tomography (CT) will show low density of muscles with preservation of volume, and magnetic resonance imaging (MRI) will commonly reveal fatty infiltration of the muscle tissue. Muscle MRI (particularly T1 imaging) is increasingly used as an adjunct diagnostic modality, and certain nemaline myopathy subtypes (particularly due to *NEB* mutations) have relatively specific patterns of selective muscle involvement (Figure 28.3 and Table 28.5).[12,37]

Pathologic Features

The pathologic changes of all genetic forms of nemaline myopathy are much the same irrespective of the severity of the clinical manifestations or the age of onset.[38,39] The diagnostic hallmark of these disorders is the presence of distinct rod-like inclusions, so-called nemaline bodies, in the skeletal muscle fibers of affected patients (Figure 28.1A, Figure 28.4, and Figure 28.5). The rods are not visible on hematoxylin and eosin (H&E) staining, but appear as red or purple structures against the blue-green myofibrillar background with the modified Gomori trichrome stain. The distribution of rods within myofibers may be random, but they show a tendency to cluster under the sarcolemma and around nuclei. The proportion of fibers containing rods varies from case to case and from muscle to muscle, and the number of rods in a muscle specimen does not appear to correlate with disease severity.[39] Replacement of muscle fibers by fat and fibrous tissue may be seen in advanced cases, but necrotic and regenerating fibers are uncommon. Inflammation is not a feature. There may be internal nuclei and occasional fiber splitting.

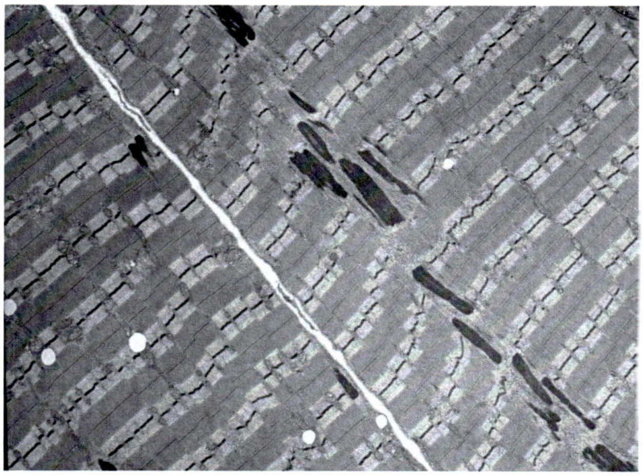

FIGURE 28.6 Nemaline myopathy. Electron microscopy of nemaline bodies. The rods have the same electron density as the Z-lines of adjacent sarcomeres (×28,350).

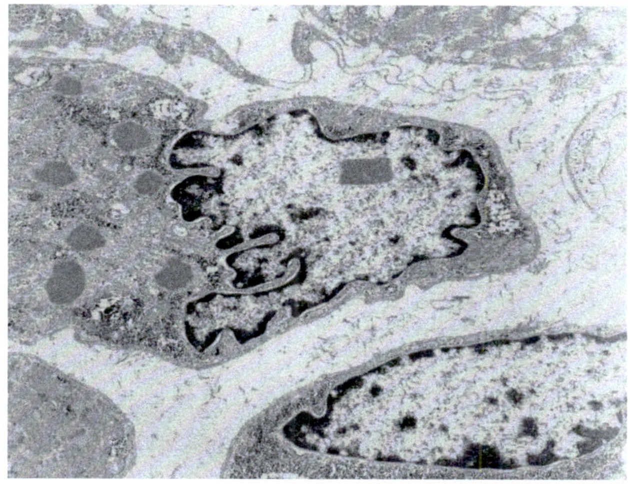

FIGURE 28.7 Nemaline myopathy. Electron micrograph showing intranuclear rod in a single nucleus. Rods are also present in the adjacent muscle fiber, where there is loss of the regular sarcomeric structure (×36,000).

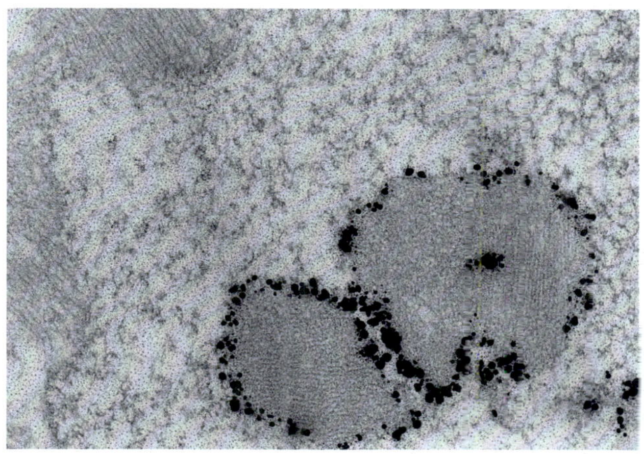

FIGURE 28.8 Nemaline myopathy. Intranuclear rods are marked by a criss-cross pattern and are immunoelectron microscopically labeled with an antibody against α-actinin enhanced by additional silver impregnation (×67,500).

The rods are considered to be derived from the lateral expansion of the Z-line, based on their structural continuity with Z-lines and electron density and criss-cross pattern on electron microscopy (EM), and positive staining with antibodies to α-actinin-2 and α-actinin-3 which are major components of the Z-line in skeletal muscle (Figure 28.6).[40,41] Focal disruption of the myofibrillar pattern is often seen on EM, and there may be accumulations of thin filaments in areas devoid of sarcomeric structure.[42] Predominance of type 1 fibers is a common feature of nemaline myopathy, and some patients show exclusively type 1 fibers or poor fiber type differentiation. In many biopsies, the mean diameter of the type 1 fibers is smaller than that of the type 2 fibers. Fiber type 1 predominance and atrophy, i.e. fiber type disproportion, tend to become more prominent with age, and there is often a complete deficiency of type 2B (fast glycolytic) fibers.[43,44] Rods are usually present in both type 1 and 2 fibers, though are restricted to type 1 fibers in patients with *TPM3* mutations. Biopsies from patients with severe forms of the disease and extensive myofibrillar disruption often have numerous small rods, and sometimes these rods are visible only on electron microscopy.[32,39]

Whereas nemaline bodies typically occur in the sarcoplasm of the muscle fiber, occasional instances of intranuclear nemaline bodies (Figures 28.7 and 28.8) have been observed in muscle biopsies from patients with severe neonatal myopathy, often caused by *ACTA1* mutations[29,45] and in some adult-onset cases (so-called SLONM or sporadic late-onset nemaline myopathy), presumably reflecting an autoimmune condition with a progressive course.[46,47] Intranuclear rods are generally considered to be associated with more severe muscle involvement and a worse prognosis. Rarely, a combination of rods and cores (described below) are noted, and are described as core-rod myopathy. Genes associated with this finding include *ACTA1*, *TPM2*, *NEB*, *RYR1*, and *KBTBD13*.

Two rare variants of nemaline myopathy are cap myopathy and zebra body myopathy. Cap myopathy is defined by well-delineated areas on muscle biopsy (Figure 28.9), which are eosinophilic on hematoxylin and eosin stain (giving the appearance of caps on the myofiber).[48] Cap myopathy has been observed most commonly in patients with *TPM2* and *TPM3* mutations[49–53] and

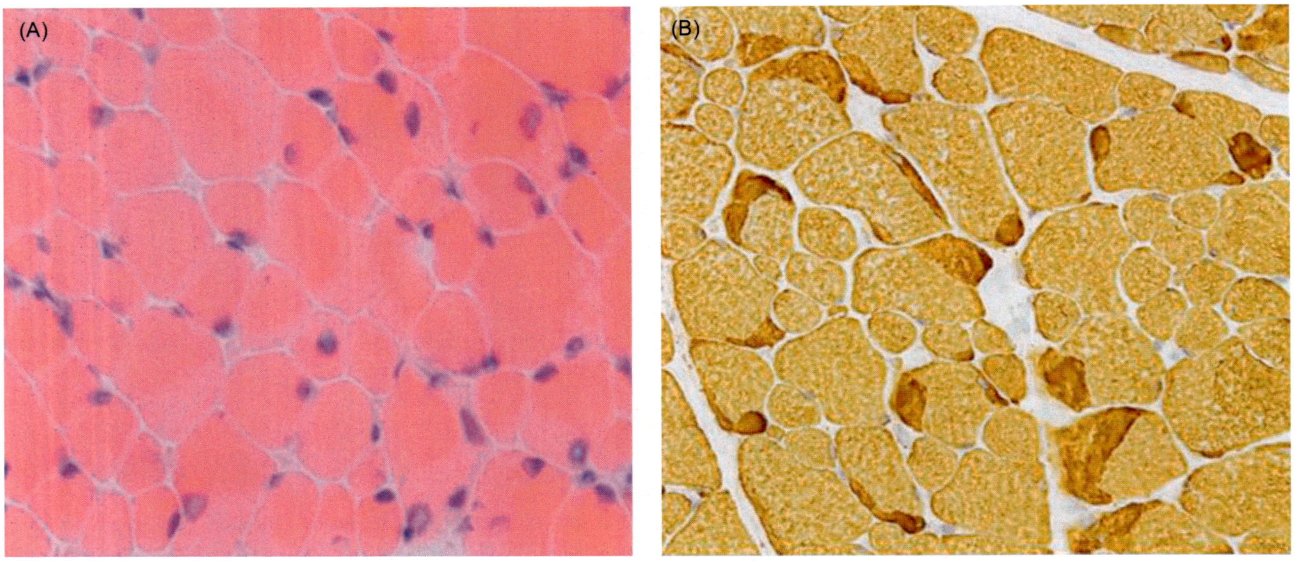

FIGURE 28.9 Cap myopathy. (**A**) Hematoxylin and eosin stain of a muscle biopsy with cap myopathy. Note the pale and granular areas that correspond to the caps. (**B**) Immunostaining with anti-myosin antibodies highlights the caps.

reported once in a patient with an *ACTA1* mutation.[54] Zebra bodies are extremely rarely encountered, and have been observed in association with nemaline bodies, suggesting that they may represent a variant form of the condition.[55,56] The identification of *ACTA1* mutations in some individuals with zebra body myopathy strengthens the hypothesis that zebra body myopathy is likely a variant of NM.[57]

Involvement of cardiac muscle is unusual in nemaline myopathy, although occasional cases have been reported of dilated cardiomyopathy, most often in the setting of *ACTA1* mutations.[58–64] There is a single report of cardiac dysfunction with *TPM2* mutation as well. However, in most autopsy studies, nemaline bodies are not present in heart muscle, supporting the assumption that most individuals with NM are not at risk of cardiac disease. Atypical cases have been documented with rods present in both the diaphragm and cardiac muscle, but these are exceptional.[65,66]

Differential Diagnosis

Nemaline bodies may occur as a secondary phenomenon in a number of conditions, including tenotomized rat muscle,[67] mitochondrial myopathy,[68,69] and human immunodeficiency virus (HIV) infection.[70] Combinations of histologic abnormalities characteristic for different congenital myopathies, particularly cores, have been encountered within families, sometimes even in the same patient.[71,72] It is not justified, therefore, to make the diagnosis of hereditary nemaline myopathy solely on the basis of pathology in the absence of a typical clinical picture.

Genetics

Nemaline myopathy is genetically heterogeneous. Mutations in nine different genes have been identified in a subset of nemaline myopathy patients: α-tropomyosin slow (*TPM3*),[73] nebulin (*NEB*),[74] skeletal muscle α-actin (*ACTA1*),[75] β-tropomyosin (*TPM2*),[76] *KBTBD13* (exclusively in core-rod myopathy, discussed later),[77] *CFL2*,[78–80] *KLHL40*,[32] *KLHL41*,[81] and troponin T (*TNNT1*)[80,83] (see Table 28.3). Five of these genes encode protein components of the thin filaments of muscle fibers, while the others likely participate as regulators of the degradation/turnover of the thin filament apparatus.[84] Because of this, nemaline myopathy is broadly considered a disorder of the thin filament. There are additional cases of nemaline myopathy that do not link to any of the identified loci, suggesting further genetic heterogeneity. The mode of inheritance of nemaline myopathy is variable, with autosomal recessive and autosomal dominant families, as well as sporadic cases. All mutations identified to date in the *NEB* gene are inherited in an autosomal recessive fashion.[74,85] Both autosomal dominant and autosomal recessive families have been identified with mutations in *TPM3*[50,73] and *ACTA1*.[75,86] Many of the mutations identified in the skeletal muscle α-actin gene represent new dominant mutations appearing as sporadic cases. The proportion of new mutations and the incidence of germline mosaicism in patients with nemaline myopathy are unknown.

As yet, no definitive clinical or pathologic markers for the various genetic forms of nemaline myopathy have been identified, although a clue to the presence of *TPM3* mutations may be the selective involvement of type 1 fibers in these patients.[39,87] Congenital and childhood-onset forms of nemaline myopathy have been described in

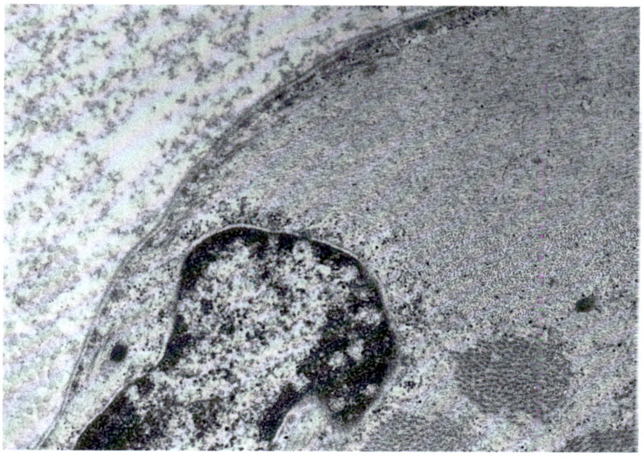

FIGURE 28.10 Actinopathy. A large perinuclear subsarcolemmal aggregate of thin, actin filaments. These aggregates may occur in association with typical nemaline myopathy (×36,400).

association with mutations in the *TPM3*[50] and *ACTA1*[86] genes, while only severe cases have been uncovered with mutations in *KLHL40*.[32] Patients with mutations in the *NEB* gene were initially thought to largely fall into the typical congenital category, but recent studies have identified cases with a variety of clinical presentations including the neonatal, severe form of the disease.[85,88,89] Detailed pathologic studies may provide morphologic clues to guide mutation analysis. For example, α-tropomyosinSLOW (*TPM3)* is expressed only in type 1 (slow) fibers, and fiber atrophy and nemaline bodies in patients with the *TPM3* mutation occur preferentially in type 1 fibers.[22,39,87,90,91] Similarly, abnormal accumulation of actin filaments,[92] and abnormal distribution of actin staining[93] have been observed in patients with *ACTA1* mutations whose clinical presentations can range from the most severe, with neonatal lethal presentations associated with fetal akinesia, to mild congenital forms compatible with full lifespans and reproductive fitness (Figure 28.10).[86,94] The number of patients that have been studied in detail is still too small to conclude whether these findings are specific. Of note, *ACTA1* mutations are most commonly encountered in dominantly inherited disease[95] and *NEB* mutations are the most common recessive cause.[85] Of the remaining genes, *KLHL40*, *TPM3*, and *TPM2* are the next most frequently encountered,[81,96] while *CFL2, TNNT1*, and *KLHL41* mutations appear extremely rare.[78–83] *KLHL40* mutations are particularly common in severe autosomal recessive NM. The del7lys mutation in *TPM2* is a highly recurrent dominant mutation described in more than 50 individuals.[97,98]

CORE MYOPATHIES

Core myopathies are defined by the presence, in myofibers, of regions devoid of oxidative staining, which may appear as large longitudinally oriented central cores in central core disease (CCD) (Figure 28.1, B and E), or multiple smaller "minicores" in multiminicore myopathy.[99] Cores seen in association with other characteristic findings, such as nemaline rods, can lead to hybrid diagnoses, often termed core-rod myopathies, but genetic analysis may allow delineation based on the underlying etiology; for example, there are nemaline myopathies with cores associated with mutations of *ACTA1*,[95] *NEB*,[100] *TPM2*, or *KBTBD13*[77] and CCD or minicores with nemaline bodies caused by mutations of the skeletal muscle ryanodine receptor, *RYR1*.[101–104] The "pure core myopathies," CCD and multiminicore myopathy, are most often associated with mutations in *RYR1* (see Case Example 28.2), though it is not the only cause. *SEPN1* mutations are the next most frequently encountered, and are associated with minicore myopathy without ophthalmoplegia. This subsection will discuss both types of core myopathy and associated conditions.

CCD was the first congenital myopathy defined on the basis of specific morphologic changes in skeletal muscle. As noted earlier, the disorder was first described by Shy and Magee in 1956,[1] and the term "central core disease" coined by Greenfield, Cornman, and Shy in 1958.[105] The characteristic ultrastructural abnormality of CCD is the presence, in the majority of muscle fibers, of centrally placed cores. A core is a circumscribed lesion of disorganized myofibrils with absent mitochondria and absent staining for oxidative enzymes, phosphorylase, and glycogen. The important and potentially fatal association of CCD with malignant hyperthermia was first noted by Denborough and colleagues in 1973.[106] CCD is a rare condition, but its true incidence is not known.

Clinical Features

The original case report of Shy and Magee[1] exemplifies the typical clinical manifestations of CCD.[107] The disorder usually presents in infancy with weakness and hypotonia. Motor milestones are delayed; many patients do not walk until 3 to 4 years of age. Muscle weakness is symmetrical and mild; active movement of all muscle groups against gravity and resistance is usually present, and most patients achieve ambulation. Weakness is more severe in the lower limbs and preferentially affects proximal musculature. Usually patients are only mildly disabled, and weakness is nonprogressive or only slowly progressive. However, a subset of individuals may present with severe neonatal weakness, arthrogryposis, and respiratory failure. These individuals usually do not walk independently, and often continue to require respiratory support.[107,108]

Patients with CCD often have poor muscle bulk, but muscle atrophy is not conspicuous. Fasciculations are

CASE EXAMPLE 28.2

KL is the first child of a nonconsanguineous Caucasian couple with no family history of neuromuscular disease. During the pregnancy, decreased fetal movements were noted. Fetal hydrops was identified on ultrasound but later resolved. KL was born at 38 weeks by Caesarean section for a breech position. Birth weight was 3.055 kilograms. At birth he had mild respiratory distress, hypotonia, severe weakness, and multiple joint contractures (bilateral clubfeet and bilateral knee and wrist contractures), as well as bilateral hip dislocations.

Soon after early infancy, he developed kyphoscoliosis, which first required bracing and eventually (by age 5) surgical correction. He never achieved the ability to sit, crawl, or walk. He did not, however, have significant eye muscle or bulbar weakness. In addition, he could feed adequately despite having significantly reduced body mass.

At 9 years of age, KL had mild lower facial weakness, proximal greater than distal weakness, decreased muscle bulk, hypotonia, hip and knee contractures, and absent deep tendon reflexes. He was nonambulatory and had severe restrictive lung disease, but did not require continuous respiratory support.

KL had a comprehensive diagnostic workup in the first years of life. His karyotype was normal. He had two muscle biopsies. The first, done in infancy, showed only atrophy. The second biopsy, done after 1 year of age, showed central cores (Figure 28.11A and B), increased adipose tissue, atrophy, and endomysial fibrosis. Based on the second biopsy, sequencing of the entire coding sequence of the ryanodine (receptor gene *RYR1*) was performed. A novel variant was identified, c.14759C>T (p.Thr4920Ile), in a conserved amino acid located in close proximity to several well-established central core disease-associated mutations. Together, the clinical findings and diagnostic testing results are consistent with a diagnosis of central core disease due to *RYR1* mutation. Parental *RYR1* testing was not performed. However, this variant is likely *de novo*, with a probable risk of recurrence of <1%.

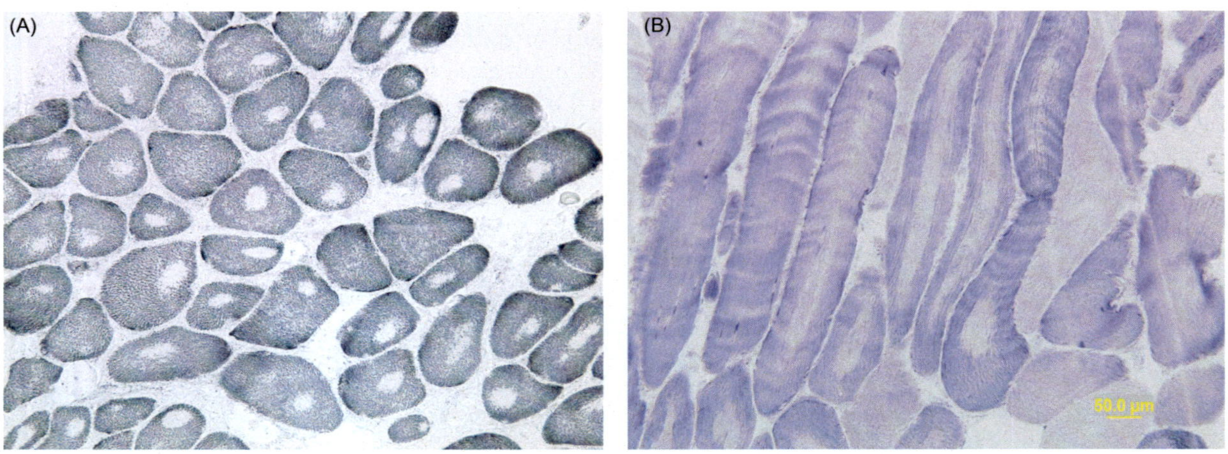

FIGURE 28.11 Central core disease (Case Example 28.2). Transverse (**A**) and longitudinal (**B**) sections stained for NADH show absence of enzyme activity in single large central cores.

absent, and deep tendon reflexes are reduced or absent. Mild weakness of facial and neck muscles may occur, but the extraocular muscles are usually spared. Muscle cramps were a predominant feature in one family, and many patients complain of muscle pain or stiffness on moderate exertion.[109,110] Musculoskeletal deformities such as kyphoscoliosis, congenital dislocation of the hip, pes cavus, pes planus, and thoracic deformities occur frequently.[1,108] The severity of the deformity often does not correlate with the degree of muscle weakness; in some individuals, skeletal deformities are the sole manifestation of disease. Significant respiratory insufficiency is unusual, cardiac abnormalities occur extremely rarely, and intellectual performance is not impaired.[110]

Although the majority of patients reported display the "typical" clinical phenotype, central cores may be found in individuals who are asymptomatic, in patients with mild weakness or skeletal deformity, and in patients with a raised CK as their only manifestation of the disorder.[110,111] As mentioned above, a small subset of patients with CCD can have more severe muscle weakness and motor impairment.[110,112,113] Amburgey et al. describe two cases in a larger series that typify this severe presentation.[114] Both children had weakness, hypotonia, and arthrogryposis noted at birth. Neither achieved independent ambulation, and both had respiratory difficulties and severe musculoskeletal complications (including severe scoliosis requiring corrective surgery in the first decade of life). Muscle biopsy in both cases demonstrated typical central cores. Mutation analysis revealed *de novo* heterozygous sequence changes in the CCD hotspot area of *RYR1*.

Central Core Disease and Malignant Hyperthermia

CCD is associated with an increased risk of malignant hyperthermia (MH).[106,110,115] MH is an autosomal dominant disorder characterized by an increase in skeletal muscle metabolism in response to certain inhalational anesthetics (particularly halothane) and depolarizing muscle relaxants (particularly succinylcholine).[116,117] It can also be seen in response to extreme heat or exercise in susceptible individuals. The association between CCD and MH is complex. Most patients with MH have histologically normal muscle, and less than 30% of patients with CCD have MH susceptibilty.[118] Both CCD and MH are most commonly caused by mutations in the *RYR1* gene. As such, MH is considered a disorder of calcium metabolism; this is further confirmed by the fact that the other known genetic cause of MH is mutations in the L-type calcium channel.[119]

In anesthesia-related MH, the triggering agent increases the sarcoplasmic calcium concentration, resulting in an increased production of heat. The major symptom is generalized hyperthermia, which may be fatal if untreated. Despite general recognition among anesthesiologists, MH is still considered to be one of the most frequent causes of death during anesthesia. At present, susceptibility to MH can be diagnosed only by means of the pharmacologic *in vitro* contracture test and/or by demonstration of one of the proven MH mutations in *RYR1*.[120,121] Triggering anesthetic agents should be avoided in susceptible individuals. As the first exposure to the trigger substances elicits an event in only 50% of MH-susceptible patients, a previous history of tolerance of halothane or succinylcholine does not ensure that these anesthetic agents can be used safely in the future.

Two unique MH-related syndromes are well described and worth noting. The first is King-Denborough syndrome, which is characterized by the triad of myopathy (including mild proximal weakness and a distinctive gait), dysmorphic features (including hypertelorism and plagiocephaly), and susceptibility to MH.[106] Muscle biopsy may be normal or may demonstrate nonspecific myopathic changes. All genetically solved cases to date are due to *RYR1* mutations.[122] However, some individuals with the King-Denborough phenotype are *RYR1* mutation negative, indicating genetic variability. Also, the possibility that in some instances King-Denborough syndrome is due to multiple simultaneous gene mutations (i.e. double trouble) has not been excluded. The other syndrome is Native American myopathy (NAM), which has to date only been identified in Lumbee Native Americans in North Carolina (USA).[123] Like King-Denborough syndrome, this condition is characterized by mild facial dysmorphism, skeletal abnormalities, and mild extremity weakness. All individuals with NAM identified to date harbor recessive mutations in *STAC3*.[124] STAC3 is a protein associated with the triad that serves as a regulator of excitation-contraction coupling.

Laboratory Investigations

In patients with CCD the serum CK is normal or slightly raised, and the EMG is usually normal or myopathic with short-duration, small-amplitude polyphasic motor unit potentials. The average number of fibers innervated by a single anterior horn cell is increased, resulting in abnormal single-fiber EMG studies.[125] Muscle ultrasound may also be used to demonstrate increased echogenicity associated with the primary myopathic process.[113] Muscle MRI is useful for diagnostic assessment, as individuals with CCD due to *RYR1* mutations typically have selective sparing of rectus femoris (Figure 28.3 and Table 28.5).[15,126]

Pathologic Features

The diagnostic feature of muscle from patients with CCD is a single, well-circumscribed circular region in the center of most type 1 fibers in the biopsy (see Figures 28.1B, 28.11A and B). These "central cores" are devoid of mitochondria and sarcoplasmic reticulum, demonstrate altered T-tubule systems, and are deficient in oxidative enzymes and phosphorylase activity. In longitudinal sections, the cores extend the entire length of the muscle fiber. The cores are best seen on sections stained for oxidative enzyme activity (succinate dehydrogenase [SDH], cytochrome-c-oxidase [COX], or nicotinamide adenine dinucleotide [NADH] dehydrogenase reacted sections). The noncore areas of the fiber stain normally, and the absence or marked reduction of oxidative enzyme activity in the cores demarcates them sharply from the rest of the muscle fiber. The cores may be overlooked on routine sections with H&E staining (Figure 28.12).

Although the presence of cores is the dominant pathologic feature in the biopsy, there is often associated variability in fiber size and an increase in the number of internal nuclei. Dystrophic features such as necrosis, regenerating fibers, and endomysial fibrosis are not typically seen. In the majority of cases there is type 1 fiber predominance and, sometimes, type 1 fiber uniformity, even without cores, and the cores selectively involve type 1 fibers. Immunocytochemical studies have demonstrated overexpression of actin, and the presence of actin-binding proteins α-actinin and gelsolin. The desmin network is

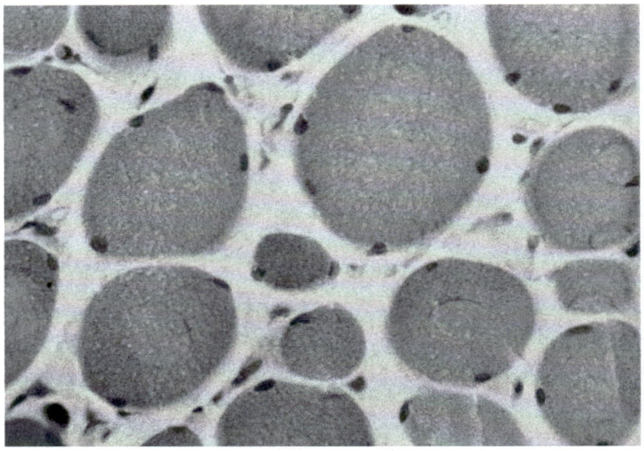

FIGURE 28.12 Central core disease. The cores are not apparent on hematoxylin and eosin staining.

markedly modified in CCD, with excessive staining at the edge of the cores or abnormal accumulation of desmin within the cores.[127] Interesting coexistence of cores and rods (so-called core-rod myopathy) has been described in a number of families.[22,128] Mutations in the *RYR1* gene have been identified in two large kindreds with core-rod myopathy.[101,102]

On EM examination, the cores contain densely packed and disorganized myofibrils. Central cores have been divided into two types, depending on whether the myofibrillar organization is maintained. In *structured cores,* the basic sarcomeric architecture is maintained, although the sarcomeres are out of register with adjacent normal fibrils, as well as each other, and sarcomere length is reduced. The myofibrils are more densely packed and the Z-lines may be thickened and wavy. In *unstructured cores*, the sarcomeric structure and normal banding pattern are lost, with large areas of Z-line streaming. Structured and unstructured cores may coexist within the same biopsy specimen. As predicted from histochemical staining, the cores are devoid of mitochondria and there is decreased glycogen. Sarcoplasmic reticulum and T-tubules are visible within cores, but their anatomy is distorted.[129]

Genetics

Both CCD and MH are known to be genetically heterogeneous, though the vast majority of CCD patients (>90%) have mutations in *RYR1*.[107] Most commonly these are dominant or *de novo* dominant mutations, with *de novo* mutations occurring more commonly in the severe cases. Occasionally, recessive mutations in *RYR1* will yield CCD pathology. Mutations throughout the *RYR1* gene have been uncovered in patients with CCD, though there may be a clustering of mutations in the C terminal transmembrane domain (i.e. hotspot 3).[130] Of note is the fact that some missense *RYR1* mutations are associated either with MH alone or else with frank myopathy/CCD. For example, an identical mutation in *RYR1* (Arg163Cys) resulted in CCD and MH in one family and MH alone in an unrelated pedigree.[131] As yet there is no clear explanation for the phenotypic heterogeneity within families affected by CCD or MH.

Rare patients with CCD and clinical myopathy have been observed with mutations in β-myosin heavy chain gene (*MYH7*),[132] and central cores have additionally been found in patients with *MYH7* mutations and a purely cardiomyopathic presentation.[133] Distal weakness at onset is often a clinical clue to *MYH7* mutations (particularly a "dropped" big toe as seen in Laing distal myopathy due to *MYH7* mutations), though patients may also manifest an axial myopathy; muscle pathology tends to more commonly show minicores or atypical cores in these individuals.[11] There are cases of CCD with no *RYR1* or *MYH7* mutation, but the additional candidate genes have not yet been identified.

RYR1-*related Myopathies*

CCD is only one of the histopathologic phenotypes within the broader grouping of myopathies due to *RYR1* mutation (or *RYR1*-related myopathies). Overall *RYR1*-related myopathies appear to be one of the commonest genetic types of congenital myopathy.[3] While dominant/*de novo* *RYR1* mutations are nearly always associated with CCD, with a few cases reported also of core-rod myopathy, patients with recessive mutations can have a broad range of biopsy findings. These include multiminicore myopathy,[134,135] centronuclear myopathy,[136] and congenital fiber type disproportion,[137,138] all of which are discussed in greater detail in subsequent subsections of this chapter. There has even been an association with dystrophic biopsy changes and *RYR1* mutations.[139] Such patients have a very severe clinical picture. Of note, a little more than half of recessive cases have cores on biopsy.[114]

Patients with recessive *RYR1* mutations typically have a more severe presentation than those with dominant mutations. In addition to extremity weakness, ophthalmoplegia is commonly encountered, and is a distinguishing feature compared to CCD and to minicore myopathy caused by *SEPN1* mutations.[140,141] Recessive mutations are found throughout *RYR1*, without obvious clustering mutations.[107,114] Patients may have two missense mutations or one missense mutation and one hypomorphic allele (frameshift, splice site mutation, etc.). The presence of a hypomorphic allele is correlated with a more severe clinical presentation.[114] Interestingly, missense mutations associated with CCD, when present in the heterozygous state, are often found along with a second mutation in recessive *RYR1* myopathies. This creates very complex genetic pedigrees, where one, or two or more mutations segregate through the family.[107] The complexity is furthered by the variable expressivity seen with the condition,[142] and by the rare instances where mutations are associated with reduced penetrance.

The association between dominant/*de novo* mutations in *RYR1* and MH, as already detailed, is well described. The relationship between patients with recessive *RYR1* mutations and MH is less certain. MH reactions have been documented in a minority of cases of recessive disease. MH precautions are, however, recommended in all cases of *RYR1*-related myopathies unless proven otherwise.

Several atypical clinical presentations have been described in the setting of *RYR1* mutations. These include nonanesthesia MH-like reactions, exertional myalgias, and exercise-related rhabdomyolysis.[143] Other presentations include rigid spine myopathy, late onset axial myopathy,[144,145] periodic paralysis,[134] and ophthalmoplegia with facial weakness (either in the presence or, less commonly, the absence of extremity weakness).[146]

MULTIMINICORE DISEASE

Ten years after the description of CCD, Engel and colleagues reported a patient with multiple small cores within muscle fibers, later designated as multicores.[147] Subsequently, many different terms have been used to describe the same ultrastructural appearance, including "minicores," "focal loss of cross-striation," "target-like lesions," and "miniature cores."[99,148] The term multiminicore disease (MmD) is now preferred.[149]

Clinical Features

MmD due to *RYR1* mutations is characterized by extremity weakness, often with prominent axial involvement, plus ophthalmoplegia.[141] This section will focus on minicore myopathy not caused by *RYR1* mutations, which is usually not associated with ophthalmoplegia and is most commonly associated with mutations of *SEPN1*.[150] It usually has its onset with weakness in infancy or early childhood, although occasional cases with juvenile and adult onset are described.[151–154] Facial and bulbar weakness may be present though are not prominent; limb weakness is usually proximal and may be associated with contractures or the development of scoliosis and respiratory compromise. Axial weakness is often the dominant clinical feature, and can include early head drop and severe, progressive scoliosis. Patients typically have a thin body habitus with muscle atrophy/hypotrophy. One variant of presentation is rigid spine muscular dystrophy, and biopsies in this setting may have dystrophic features with or without minicores.[155]

The clinical course is usually static or only slowly progressive during early childhood, but may be associated with decline in adolescence or young adulthood.[150] This is particularly the case for spine curvature and for respiratory function. Some individuals with minicore myopathy progress to respirator-dependent ventilatory failure while still ambulant. More commonly, patients have nocturnal hypoventilation and require respiratory support only at night. Ptosis and ophthalmoplegia occur rarely. Cardiac defects and cardiomyopathy are exceptional features.[156,157] A reaction to anesthesia has been reported in a single case,[158] but MH susceptibility is not a common feature of minicore not due to *RYR1* mutation. The EMG is nonspecifically myopathic, and the serum CK is normal or only mildly elevated. MRI of the muscle can be quite helpful, particularly for distinguishing *SEPN1* mutations from other causes of rigid spine myopathy or axial myopathy.[141] Specifically, there can be selective and prominent involvement of the sartorius muscle on thigh T1 imaging.

Pathologic Features

The pathologic hallmark of MmD is the presence of multiple foci of sarcomeric disruption including smearing or streaming of the Z-disks, concomitant absence of mitochondria, and the presence of abnormal components of the sarcotubular system, such as misaligned triads (Figure 28.1E, Figure 28.13A and B). Their size and, particularly, their number can best be demonstrated in semithin plastic sections. Enzyme histochemical preparations, including NADH, cytochrome-C oxidase (COX), and succinate dehydrogenase (SDH), may show multiple defects with ill-defined borders, the latter finding quite unlike the sharp demarcation of central or eccentric cores seen in CCD. The multiminicores are present in both type 1 and type 2 muscle fibers and contain desmin, α-actinin, and αB-crystallin, as well as heat-shock protein, dystrophin, and filamin C.[127,159,160] There is frequently type 1 fiber predominance and hypotrophy.

Genetics

Non-RYR1 MmD is most commonly caused by recessive mutations in the selenoprotein-N (*SEPN1*) gene.[161] *SEPN1* mutations are also associated with rigid spine muscular dystrophy, with or without cores,[155] desmin-related myopathy with Mallory body-like inclusions,[162] and congenital fiber type disproportion.[163] There is additional genetic variability among cases of MmD. *MYH7* mutations have been described in patients with minicore myopathy,[164] as have mutations in the giant sarcomeric protein titin (TTN, see below). In both instances, cardiomyopathy may be a complicating feature of the skeletal myopathy.

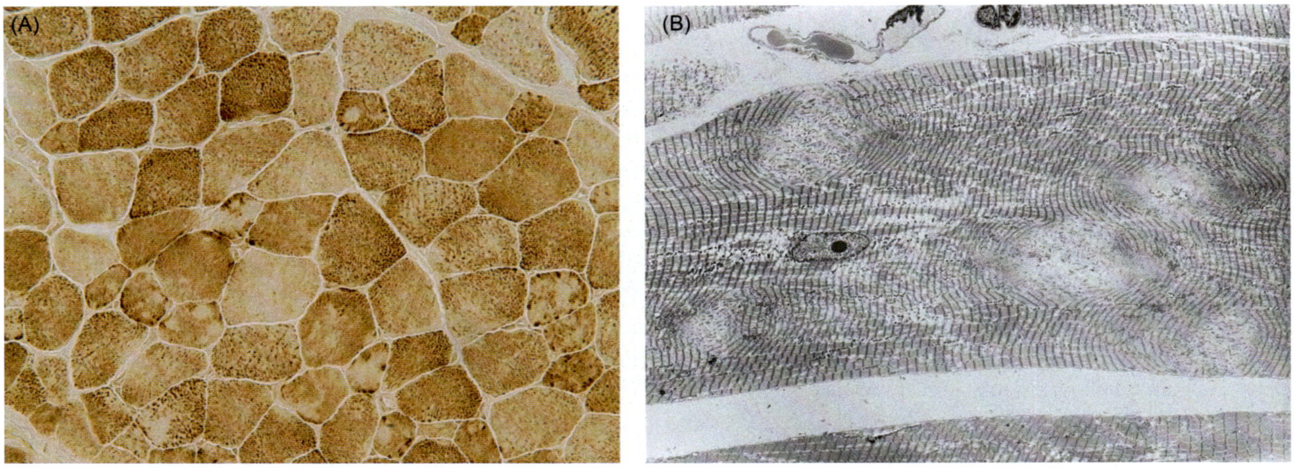

FIGURE 28.13 Multi-minicore myopathy. (**A**) Transverse section stained with NADH demonstrates foci of decreased oxidative enzyme activity throughout the muscle fiber. (**B**) Low-power electron microscopic view of longitudinal muscle sections demonstrates focal loss of cross striations, corresponding to areas of decreased mitochondrial enzyme activity.

Multiminicores are not confined to minicore myopathy. Multiminicores can occur as a nonspecific feature in other neuromuscular conditions, such as type III glycogenosis,[165] short-chain acyl-CoA dehydrogenase deficiency,[166] a complex cerebroretinomuscular syndrome,[167] and Marfan syndrome.[159] Sometimes precise classification of a patient's neuromuscular condition may be difficult when core lesions are neither of the central nor multiminicore type, but are intermediate in appearance or represent a combination of all forms. In CCD, multiminicores may occur as an associated feature. However, central cores tend to run along the entire long axis of the muscle fiber, distinguishing them from abundant fusing multiminicores.

Internal nuclei may appear within muscle fibers in MmD and may indicate another nosologic variant combining central nuclei and multiminicores.[71,159,168] Such patients and their muscle specimens are categorized either as having centronuclear myopathy with multiminicores, or multiminicore disease with central nuclei. These features have been described in patients with *RYR1* mutations and in patients with recessive titin (*TTN*) mutations.[136,169] Titinopathies are described in further detail in a separate subsection.

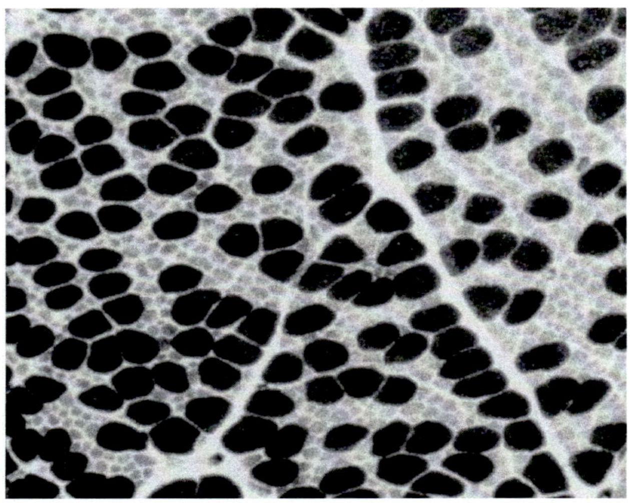

FIGURE 28.14 Congenital fiber type disproportion. Myosin ATPase stain at pH 9.4 demonstrating small type 1 fibers (pale) compared to type 2 fibers (dark stain).

CONGENITAL FIBER TYPE DISPROPORTION

Congenital fiber type disproportion (CFTD) refers to a particular histologic pattern with multiple etiologies.[137] The characteristic finding on light microscopy is the discrepancy in size between type 1 muscle fibers (slow twitch), which are significantly smaller than type 2 fibers (fast twitch) (Figure 28.1F and Figure 28.14).[170] The term was first used by Brooke, who in 1973 described a series of 14 children with this finding on muscle biopsy.[171] Most of the patients were hypotonic from birth and had diffuse weakness that improved with age. The majority had multiple joint contractures, short stature, a long thin face, a high-arched palate, and a tented upper lip. Many had relatives that were similarly affected. From these findings it was suggested that CFTD was a marker for a discrete disease entity, and other clinicians have supported this conclusion.[172,173] Over the last two decades it has become clear that the pathologic picture of CFTD can be associated with diverse clinical presentations and occurs as a secondary phenomenon in a variety of skeletal and neuromuscular disorders including congenital myotonic dystrophy, spondyloepimetaphyseal dysplasia, spinal muscular atrophy, Krabbe leukodystrophy, adrenoleukodystrophy, and nemaline myopathy.[174] The largest well-characterized clinical series to date was reported by Iannaccone and associates,[174] who analyzed 37 patients with the pathologic features of CFTD. In 15 (41%) patients, they were able to diagnose an alternative primary neuromuscular disorder. In six patients CFTD appeared to develop as a result of perinatal asphyxiae. In the remaining 16 patients (43%) an alternative neuromuscular condition could not be diagnosed. These cases were categorized into four groups: those with clinical features reminiscent of a congenital myopathy (3 patients), those with clinical and histologic signs of congenital muscular dystrophy (4 patients), and those with significant developmental delay suggestive of primary central nervous system disease (4 patients). In the remaining fourth group (5 patients), the hypotonia and mild motor delay slowly resolved with age, and it was postulated that the CFTD reflected a delay in muscle maturation.

The consensus has been that once the known etiologies of CFTD have been excluded, there remains a subset of patients in which CFTD is the primary pathologic condition. Recent discoveries of the genetic basis for many of these cases has confirmed the supposition that CFTD is a useful diagnostic classification, although in each instance, the responsible genes are also implicated in other well-defined myopathies, illustrating that fiber type disproportion is a relatively nonspecific response to certain insults to skeletal muscle. The most common cause of CFTD is mutations of the *TPM3* gene encoding slow alpha-tropomyosin.[87,91,175] Given that *TPM3* is a type 1 fiber-specific protein, it is satisfying to know that its mutation preferentially affects those fibers, and in fact, the more rare patients with *TPM3* mutation and nemaline myopathy also have selective involvement of their type 1 fibers.[39,87] Mutations of *RYR1* have also been identified as a relatively common cause of CFTD, although the mechanism leading to fiber type disproportion is unknown.[138] Rare mutations of *ACTA1*[176] and *SEPN1*[163] have also been identified in patients with CFTD. Most recently, mutations of the *LMNA* gene, which otherwise causes several forms of muscular dystrophy and Charcot-Marie-Tooth disease, have been identified in several Japanese patients with CFTD,[177] raising the point that such patients may represent a subset of individuals with CFTD at risk for developing cardiomyopathy.

CENTRONUCLEAR MYOPATHIES

The centronuclear myopathies (CNMs) encompass a group of conditions characterized by a congenital myopathy in association with significantly increased frequency of central nuclei in skeletal muscle myofibers (Figure 28.1C and D), in the absence of other diagnostic abnormalities such as nemaline bodies, cores, or other such structural defects.[178–181] Subtypes include X-linked, autosomal recessive, and autosomal dominant forms, with clinical severity ranging from neonatal lethality to adult onset slowly progressive disease. An emerging unifying feature for several of these types is defective excitation-contraction coupling, secondary to abnormalities in structure or function of the terminal cisternae of the sarcoplasmic reticulum or proteins localized at the triad.[182]

X-linked myotubular myopathy (XLMTM) is defined pathologically by the presence of centrally placed nuclei in the majority of extrafusal muscle fibers (see Case Example 28.3). These fibers are often abnormally small in diameter. The disorder was originally called *myotubular myopathy* because the muscle fibers with centrally placed nuclei resemble the myotubes present during normal fetal muscle development. Although the terms *myotubular* and *centronuclear* may be used interchangeably, some authors suggest that the term *myotubular myopathy* should be restricted to the X-linked form of the disorder (see later discussion).[5,178]

Centronuclear myopathy was first described in 1966 by Spiro and colleagues[183]; these first cases were characterized clinically by early childhood-onset weakness affecting the extraocular, facial, neck, and limb muscles, with a slowly progressive course. Since the original description, there have been many patients described with similar pathologic features but with marked clinical heterogeneity, with some patients presenting later in life. Those that survive into late childhood typically have muscle biopsies with larger, more mature-appearing myofibers that no longer resemble myotubes, but retain the characteristic central nucleation. Consequently, these cases came to be called *centronuclear myopathy*. Due to the wide variability in the age of onset and clinical severity, the degree of extraocular muscle involvement, and mode of inheritance, the disorder is now commonly divided into three groups based on clinical and genetic features[180]: the severe X-linked recessive form with prenatal or neonatal onset, the autosomal recessive form with onset in infancy or childhood, and the autosomal dominant form, which is typically mild with late onset but can present with a severe phenotype related to *de novo* mutations in early childhood. All forms of myotubular (centronuclear) myopathy are rare, though as a group it is probably as common as other subtypes of congenital myopathy.

Clinical Features

The X-linked recessive form, XLMTM, is the best defined, clinically and genetically (Figure 28.15A).[178,184] Onset is commonly *in utero*, and the pregnancy is

CASE EXAMPLE 28.3

Mr. and Mrs. F. are unrelated individuals of Italian/Indian descent. They are both healthy with normal strength and muscle bulk and no evidence of myotonia. Their first male infant was delivered postterm after a normal pregnancy. The delivery was difficult with a breech presentation, fetal distress, and meconium-stained amniotic fluid. The infant was hypotonic at delivery and severely asphyxiated, with Apgar scores of 1 and 1 at 1 and 5 minutes, respectively. He was ventilated from birth, but his course was complicated by bilateral pneumothoraces, Gram-negative sepsis, and ongoing severe hypoxia. Thin ribs were noted on chest X-ray. He died on day 5.

At postmortem examination he was described as being in a "squatting position" with generalized muscle wasting (proximal and distal), and a bell-shaped chest. He had long fingers and toes and normal genitalia. The brain and spinal cord were normal. Postmortem histologic examination of skeletal muscle showed variation in fiber size with many small fibers with central nuclei and perinuclear clearing. Cardiac muscle was normal.

Subsequent to this Mrs. F. had a needle biopsy of her left gastrocnemius, which showed mild myopathic features with a small number of abnormally large fibers, some of which contained numerous internal nuclei. Her EMG was normal. A provisional diagnosis for the infant of type 1 fiber hypotrophy with central nuclei was made, and Mrs. F. was thought to have a less severe form. At that time the possibility of X-linked recessive and autosomal dominant inheritance were discussed with Mr. and Mrs. F., but they did not attend formal genetic counseling or further neurologic follow-up. Mrs. F. has three brothers, all of whom are well. There is no other significant family history.

In 1988, Mr. and Mrs. F. had a normal female infant. JF, a male infant, was born in 1992. The pregnancy was normal to 37 weeks, when decreased fetal movements were noted. He was delivered by Cesarean section and at delivery was noted to be hypotonic with poor respiratory effort. His Apgar scores were 3 and 6 at 1 and 5 minutes, respectively, and his birth weight and head circumference were between the 50th and 90th percentiles. He required intubation and was ventilated for 5 hours prior to extubation into head-box oxygen.

On initial examination, he was hypotonic with a "squatting posture" (flexed at the hips and knees with ankle dorsiflexion) and had contractures of the knee joints bilaterally. He had little spontaneous movement, his respirations were shallow, and he had poor muscle bulk. He cried with painful stimuli but had little facial expression, and his cry was weak. Reflexes were absent. He had a high-arched palate, true arachnodactyly of the fingers and toes, and bilaterally undescended testes. The rest of his examination was unremarkable. At 24 hours of age, JF had episodes of bradycardia with decreased oxygen saturation and required reintubation. On chest X-ray, thin ribs and right upper lobe collapse were present. Head ultrasound was normal.

Open muscle biopsy from the right deltoid and right quadriceps was performed under local anesthetic on day 15. Histopathologic examination of the specimens showed increased fiber size variability, and many of the fibers contained central nuclei (Figure 28.15B), some of which were surrounded by a clear zone in the ATPase stains. On electron microscopy there was central accumulation of mitochondria and glycogen with occasional redundant folds of basal lamina due to atrophy. These changes were consistent with myotubular myopathy.

JF remained hypotonic, but his level of activity and respiratory effort increased, and he was extubated on day 23 of life. By day 34 he was able to take all feedings orally, although he still required suction to cope with pharyngeal secretions. He was discharged from hospital on day 36. At 4 months of age, JF remained severely hypotonic with little active movement. He required suction to cope with his pharyngeal secretions. He was feeding well but failing to thrive (weight at 4 months was 3.305 kg). He died of respiratory insufficiency at 8 months of age.

Once the myotubularin gene was identified, DNA from JF and his mother was sent for mutation analysis and a mutation has been identified which could be used for prenatal diagnosis and carrier detection in the female sibling. The abnormalities on muscle biopsy in Mrs. F. were thought to represent carrier status for X-linked myotubular myopathy. There was a 50% carrier risk for her female children and a 50% risk of being affected for each male infant. Mr. and Mrs. F. elected not to have additional children.

complicated by polyhydramnios and reduced fetal movements. There may be a history of miscarriages and neonatal deaths of male infants in the maternal line. Affected males usually present at birth with severe floppiness and weakness, facial diplegia, and difficulties with respiration and feeding. Additional features include thin ribs, contractures of the hips and knees, puffy eyelids, ophthalmoplegia, and cryptorchidism. Cardiac muscle is spared and, in the absence of significant hypoxia, cognitive development is normal.

The long-term prognosis of XLMTM, as reported in the literature, is extremely poor, with death occurring during infancy or the first year of life from respiratory failure in 25–50% of cases.[185,186] Many affected males lack spontaneous antigravity movement and fail to achieve independent respiration. As a result, the management of affected patients is sometimes conservative, with withdrawal of ventilatory support once the diagnosis is made.

However, cases have been reported of children who survived for months or years.[187–189] Furthermore, one

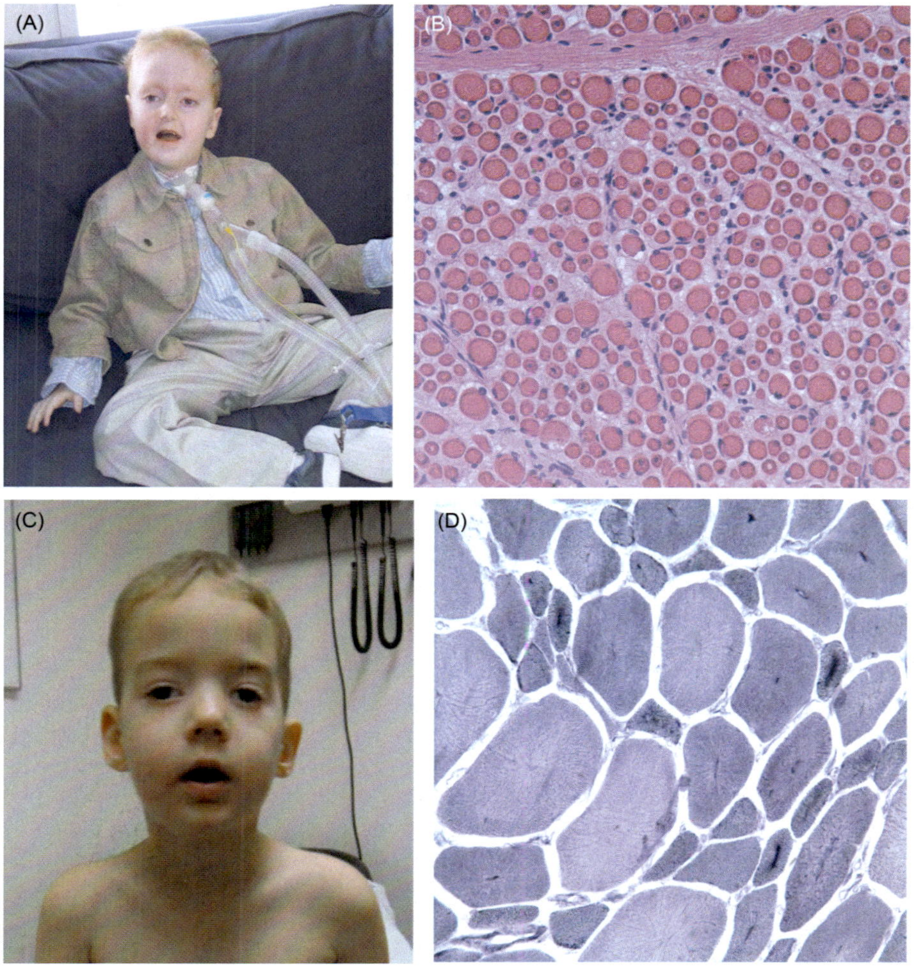

FIGURE 28.15 Clinical and pathological aspects of CNM. (A) 4-year-old boy with XLMTM illustrating generalized weakness and myopathic facies. (B) H&E stained muscle from a newborn with XLMTM. Note extremely small size of myofibers relative to size of central nuclei, characteristic of myotubes. (C) 4-year-old boy with autosomal dominant CNM due to a *DNM2* mutation. (D) NADH stained muscle from a young adult with *DNM2* mutation. Note characteristic radial arrays in some myofibers, sometimes referred to as "spokes on a wheel."

study of 55 genetically defined patients[185] suggests that prognosis of XLMTM may not be as poor as previously reported: 64% of patients survived to at least 1 year of age, and approximately three-fourths of these patients were alive at the time of reporting (age range up to 27 years); 80% of living patients were completely or partially ventilator-dependent, and the muscle weakness appeared nonprogressive. Seven patients were classified as "mild" with only slightly delayed milestones and no need for ventilatory support beyond the newborn period. Herman and colleagues[185] did not identify any prognostic indicators, although their raw data suggest that some missense mutations may be associated with the milder phenotype. Of interest is the finding that several patients had additional medical problems, including pyloric stenosis (4 patients), spherocytosis (2), gallstones (4), nephrocalcinosis (2), vitamin K-responsive bleeding diathesis (2), and abnormal liver function tests (6). Three patients had fatal abdominal hemorrhages secondary to hepatic peliosis; 40% had height greater than the 90th percentile with rapid linear growth, advanced bone age, and premature adrenarche. All these features may be part of the XLMTM phenotype, and suggest that the function of the myotubularin protein, which is defective in these cases, is not restricted to muscle cells, but may also be important in the signaling pathways controlling growth and differentiation in the liver, red blood cells, and adrenal cortex (see later discussion). This study highlights the need for careful genetic and prognostic counseling for patients with XLMTM and their families.

The autosomal recessive type of CNM is relatively rare (as compared to myotubular myopathy), and usually presents in infancy or early childhood with respiratory distress, hypotonia, a weak cry, and difficulty sucking. Ophthalmoplegia, ptosis, and facial diplegia are common. The clinical course is characterized by delay in motor milestones, slowly progressing weakness, and development of scoliosis; by adolescence many patients are confined to a wheelchair.[190,191] Not infrequently, individuals with recessive CNM will present with clinical pictures nearly indistinguishable from the X-linked form making genetic diagnosis essential, especially in the case of male births. Patients with autosomal recessive CNM usually have a typical course that includes moderate to severe weakness and hypotonia. However, both genetic and clinical heterogeneity among this subgroup makes generalizations difficult; for example, patients with *BIN1* mutation may have particularly severe and progressive courses,[192] those with *TTN* mutations may be at risk of developing cardiomyopathy, while recessive *RYR1* mutations may place the patient at particular risk for MH.

The autosomal dominant and *de novo* dominant forms have a much broader range of presentation.[193] Originally described primarily as presenting in adolescence and adulthood with slowly progressive weakness, there are now abundant reports of presentations in the first years of life (at least for those with *de novo DNM2* mutations).[10] Weakness is typically in a limb-girdle pattern, though there can be distal predominance with patients with *DNM2* mutations.[194] Facial and ocular muscles are usually affected, and involvement often parallels age of onset, with early onset cases associated with ptosis and complete ophthalmoplegia (Figures 28.2B and 28.15C). Most patients are ventilator independent, though there can be a phase of progressive respiratory decline that may require noninvasive ventilation but rarely necessitates invasive support. Scoliosis is common, and can be progressive in the teen years.[195,196]

Laboratory Investigations

Ancillary investigations often are not useful in making the diagnosis, making genetic testing a critical component of the clinical workup.[181] The CK is normal or only mildly elevated, the EMG usually shows a nonspecific myopathic pattern, and nerve conduction velocities are normal. Muscle MRI patterns are less well defined for XLMTM and for autosomal recessive CNM, but may reveal a specific pattern of involvement in the dominant cases due to *DNM2* mutation.[14,197] Specifically, there is prominent involvement of the soleus and the tibialis anterior in the lower leg.

Pathologic Features

The cardinal pathologic feature of centronuclear myopathy is central nuclei in >25% of the muscle fibers, without obvious evidence of dystrophy.[178] There can be a great deal of variability, both in the number of fibers with central nuclei as well as the number of central nuclei within a single fiber. There can also be a range of other changes, particularly in the cases that are not due to *MTM1* or *DNM2* mutation. For XLMTM, the characteristic histologic feature is the predominance of type 1 fibers with centrally placed nuclei or a central clear zone corresponding to space between adjacent nuclei just outside their plane of section (Figures 28.1D and 28.15B).[198] Most fibers are significantly smaller and rounder than those of age-matched control subjects, and resemble fetal myotubes. In longitudinal sections, long chains of nuclei are seen in the center of the fibers. Unlike other congenital myopathies, particularly CCD, the pathologic diagnosis can be made on H&E staining of formalin-fixed, paraffin-embedded tissue. There is usually an aggregation of mitochondria in the center of the muscle fiber, associated with dense oxidative enzyme staining and a lack of staining with myosin ATPase (Figure 28.16). The remainder of the muscle fiber is structurally normal, although there may be a mild increase in perimysial fibrous connective tissue. There is usually a marked type 1 fiber predominance and fiber type disproportion with atrophy or hypotrophy of type 1 fibers. The muscle changes are widespread and are found in all muscles, including the diaphragm and intercostal muscles. Obligate carriers of the XLMTM are usually asymptomatic, but approximately 50% may have their carrier status verified by muscle biopsy. Genetic testing is most informative.[181,190,191]

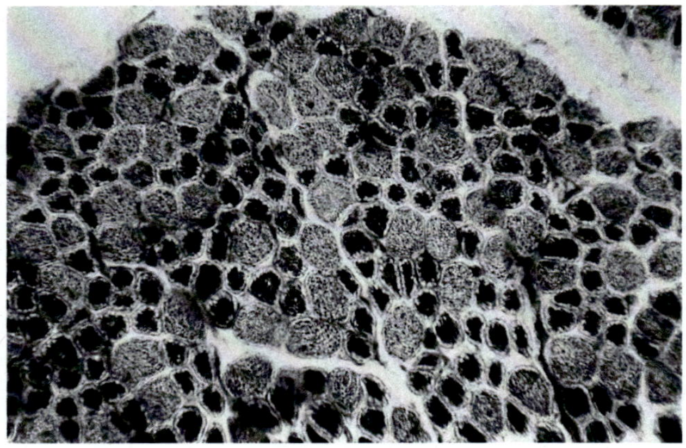

FIGURE 28.16 Muscle biopsy from a male infant with X-linked myotubular myopathy stained with nicotinamide adenine dinucleotide (NADH). Note characteristic central aggregation of staining with NADH in myotube-like fibers.

In surviving or clinically less affected XLMTM patients, peculiar muscle fibers, so-called necklace fibers, have been identified with a ring of increased histochemical oxidative enzyme activity beneath, but separated from the sarcolemma,[199,200] but they have also been found associated with *DNM2* mutations.[201] Another peculiar feature are "radiant strands" in oxidative enzyme preparations seen in *DNM2* CNM (Figure 28.15D), but less numerous in *BIN1* CNM.[202] In *TTN*-related CNM, the internal nuclei are often accompanied by disorganized myofibrillar material that may be detected either by immunostaining for actin/desmin or else by electron microscopy.[169] Interestingly, in *RYR1*-related CNM, cores are rarely if ever observed.[136]

Differential Diagnosis

Because of its histologic similarities with XLMTM, congenital myotonic dystrophy must be excluded by other than histologic means (DNA analysis, EMG in the mother).[4] In myotonic dystrophy, the involved fibers within the same fascicle appear arrested at all stages of maturation, whereas in XLMTM all the fibers are morphologically relatively uniform, except for the few fibers that have a mature appearance.

In some instances it may be difficult to distinguish between XLMTM and the autosomal forms of centronuclear myopathy on the basis of the clinical picture and muscle biopsy. In fact, a number of cases have been diagnosed initially as XLMTM based on congenital weakness and neonatal muscle biopsy findings, only to be reclassified as autosomal centronuclear myopathy based on clinical course, evolution of the biopsy and, most significantly, by negative gene testing for *MTM1*.[203] In other words, in the absence of gene mutation data and an extensive family history, the differentiation between XLMTM and autosomal centronuclear myopathy can be difficult.

Genetics

To date, mutations in five genes have been definitively associated with centronuclear myopathy (Table 28.3). The most common and extensively studied are mutations of the gene for myotubularin (*MTM1*), which cause the X-linked form of the disease.[204] Over 470 mutations have been identified in *MTM1*, with most presumably causing loss of expression.[205] There are a few recurring missense mutations, notably p.R69C, which can cause either severe/classic disease or more mild symptomatology.[189] Mutations in dynamin-2 (*DNM2*) are the second most common cause of CNM, and the only recognized dominant cause of disease.[10,206] Mutations in the middle domain of *DNM2* are typically associated with a milder presentation, while mutations in the PH domain are more likely to cause severe disease.

For recessive disease, mutations have been identified in *BIN1*,[207] *RYR1*,[136] and *TTN*,[169] with *RYR1* likely the most common cause within this group. *RYR1*-related CNM is often quite severe, though it is associated with significant variability. Multiple individuals with a phenotype resembling the X-linked form have been found to harbor recessive *RYR1* mutations. *BIN1* mutations cause a fairly classic early onset presentation (as mentioned above), while recessive *TTN* mutations presenting as CNM tend to be associated with milder disease and may also lead to cardiomyopathy.[169] There are clearly additional genetic loci that cause CNM, as highlighted by a recent study using whole exome sequencing on a cohort of 29 individuals, in which only a fraction of individuals had a clear documented mutation in the known CNM genes.[169] Of note, sequence variants in *MTMR14*, a gene similar in function to *MTM1*, have been reported in two individuals with CNM.[208] It is unclear if these variants are the true cause of disease, or instead are acting as modifiers of a second site gene mutation. Further identification of *MTMR14* cases will be necessary to understand the association between this gene and CNM.

At present, the specific reason why mutations in the known genes associated with CNM cause the unique pathologic feature of the disease (i.e. central nuclei) is unclear. However, a primary mechanism of pathology that underlies weakness in patients with mutations in many of these genes appears to be defective excitation-contraction coupling. The *MTM1*,[209–211] *DNM2*,[212,213] and *BIN1*[214] gene products all localize to the triad where they play important roles in establishing proper membrane curvature and structures, while the ryanodine receptor itself is a critical component of the calcium channels necessary for excitation-contraction coupling.[136]

TITINOPATHIES

Titin (*TTN*) is "giant" sarcomeric protein that is one of the largest genes and proteins in the human genome. Heterozygous/dominant mutations were first identified in a rare adult myopathy called tibial muscular dystrophy.[215–217] Historically, interrogation of the *TTN* gene for mutations in childhood myopathies has been limited by its large size and the difficulties sequencing the gene. However, massively parallel sequencing approaches have led to the ability to comprehensively evaluate *TTN* in a range of childhood myopathies.[218] The result has been identification of recessive *TTN* mutations in an increasing number of clinical contexts.

As mentioned above, *TTN* has been identified as a relative common cause of centronuclear myopathy [169] Patients with CNM and *TTN* mutations appear to have a relatively nondescript clinical presentation (diffuse weakness without eye muscle involvement), and have a mutation pattern of two truncating alleles. The pathology is often reported as centronuclear myopathy, but myofibers in many biopsies may contain multiple internal nuclei and some dystrophic features, allowing differentiation from the other forms of CNM characterized by a single central nucleus. Ferreiro and colleagues have described four families with recessive *TTN* mutations and core myopathy plus cardiac disease.[219] The clinical phenotypes in these patients range from arthrogryposis, without progression, to an Emery-Dreifuss-like presentation. Muscle biopsies show minicores and central nuclei. The *TTN* mutations in each patient from this cohort are a combination of missense and truncating. Udd and colleagues described a cohort of individuals with childhood-onset disease and generalized weakness with a more dystrophic biopsy pattern.[220] The biopsies also included rimmed vacuoles in most instances. The mutation pattern was similar to the Ferreiro study, with compound heterozygous mutations featuring one missense and one truncating/loss of expression mutation.

In summary, recessive mutations in *TTN* are associated with a range of clinical pictures and biopsy findings. Identifying *TTN* mutations is important because, unlike most congenital myopathies, there is a significant risk of cardiomyopathy. A challenge with *TTN* is the fact that the gene is highly polymorphic and determining sequencing variant pathogenicity can be very challenging. One feature that unifies the existing studies is that in the firmly documented cases there is supporting evidence for a change of *TTN* expression. This has come in the form of RNA analysis (showing splicing alterations by certain mutations) or protein analysis (Western blot or immunostaining showing reduction in titin expression).

MYOFIBRILLAR MYOPATHIES

The myofibrillar myopathies are a clinically and genetically heterogeneous group of disorders defined by the presence of focal myofibrillar disarray and desmin-positive myofibrillar aggregates on muscle biopsy.[221] The definition of myofibrillar myopathies has evolved through the application of immunohistochemistry to muscle biopsy specimens, and the demonstration of accumulated desmin, the muscle fiber–specific intermediate filament, within muscle fibers. Nosologically, the desmin-related myopathies can be divided into three groups, based on morphologic and clinical criteria: the inclusion body type, with sarcoplasmic, cytoplasmic, or spheroid bodies; the granulofilamentous type; and a third group with Mallory body-like inclusions.[222] Because desmin is not the only abnormal material that accumulates, and because the common denominator is myofibrillar degradation, the noncommittal term of *myofibrillar myopathy* has been proposed as a substitute for desmin storage myopathy and desmin-related myopathy.[223,224]

The myofibrillar myopathies are genetically heterogeneous, with identified disease-causing mutations in *ZASP*, *BAG3*, αB crystallin, myotilin, filamin C and in desmin itself (see later). In addition, there are likely additional gene mutations that cause MFM, as the known genetic loci account for approximately half of all cases. This group of disorders is rare, although the true incidence is not known. Sporadic and familial patients have been recorded worldwide.

Clinical Features

Myofibrillar myopathies are primarily a disorder of adulthood, but their age of onset ranges from early childhood to middle age, and does include a severe, progressive childhood presentation. Weakness is most commonly distal, or may be more diffuse, and bulbar or respiratory involvement occurs in some families. Cardiac symptoms may dominate the clinical picture, particularly hypertrophic cardiomyopathy or atrioventricular conduction block.[225] Lens opacities and neuropathic features have also been reported in some families.

Subgroups of patients can be identified on the basis of clinical and pathologic findings. Many patients and families with the inclusion bodies or granulofilamentous inclusions demonstrate late onset of symptoms and preferential involvement of distal muscles,[226] although early-onset weakness has been described.[227–229] Cardiomyopathy is a frequent finding in the granulofilamentous type, and may be severe, necessitating heart transplantation in adults or even children.[227] Mallory body-like inclusions tend to be associated with a more severe clinical phenotype with onset of muscle weakness in childhood, scoliosis, respiratory muscle involvement, and autosomal recessive inheritance.[222,230]

Laboratory Investigations

The serum CK is either normal or only mildly elevated. In children, EMG may demonstrate mild, nonspecific "myopathic" features. Electrocardiography and echocardiography are essential to identify associated cardiomyopathy and cardiac conduction defects.

Pathologic Features

The common pathologic feature in this group of myopathies is the activation of a degradative process that primarily affects the myofibrils. The lytic lesions become depleted of actin, α-actinin, and occasionally of titin, nebulin, and myosin. Compacted and degraded myofibrillar components accumulate in hyaline spheroidal structures. Desmin accumulates in some, but not all, abnormal fiber regions (Figure 28.17), together with many other proteins, such as lamin B, αB crystallin, gelsolin, ubiquitin, α1-antichymotrypsin, dystrophin, large deposits of congophilic amyloid material, neural cell adhesion molecule (NCAM, a marker of regeneration),[231–233] and vimentin.[223,234,235]

Similar protein aggregates were found in autosomal-recessive desminopathy[236] and αB crystallinopathy.[237,238]

The inclusion bodies, which characterize the desmin-related or myofibrillar myopathies, are termed cytoplasmic bodies, sarcoplasmic bodies, or spheroid bodies (Figure 28.18A). They consist of intermediate filaments and electron-dense granular material. These two components are often present in a separate fashion, granular material forming a core surrounded by a halo of filaments similar to intraneuronal Lewy bodies in Parkinson disease

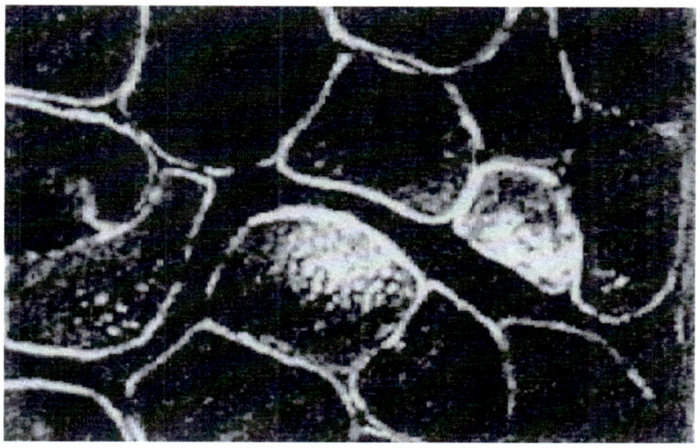

FIGURE 28.17 Desmin-related/myofibrillar myopathy. Immunocytochemical staining with antibody to desmin demonstrating accumulation of desmin within individual fibers under light microscopy.

and other neurodegenerative diseases of the brain, or Rosenthal fibers, a feature of abnormal astrocytes. In granulofilamentous myopathy, the granular and filamentous components appear more admixed and scattered across the myofiber, and they also amass beneath the sarcolemma (Figure 28.18B). At the light microscopic level, the granulofilamentous material appears as patches or plaques differing from the often well-delineated inclusion bodies. Some patients have demonstrated both inclusion bodies and granulofilamentous material on biopsy.[239]

More refined ultrastructural studies[240] disclosed that granulofilamentous material predominated in desminopathy and αB crystallinopathy, but was less obvious in other MFM. Filamentous bundles were frequent in ZASPopathy, tubulofilamentous aggregates in myotilinopathy and ZASPopathy, sarcoplasmic and intranuclear rods in ZASPopathy, myotilinopathy, plectinopathy[241] and filaminopathy-rod.[242] These EM findings may facilitate molecular analysis.

The pathogenetic mechanisms underlying the myofibrillar myopathies are not yet known. However, it is likely that the mutant proteins, for example desmin and αB crystallin, are biophysically different from the wild type proteins and act as a "seed" for other proteins to aggregate. Thus, the desmin-related myopathies appear to be "protein surplus myopathies,"[243] in contrast to the concept in many muscular dystrophies in which disease-specific proteins are absent or reduced, as with dystrophin and sarcoglycans.

Genetics

Desminopathies

Mutations in the desmin gene (true desminopathy) have been identified in a subgroup of patients with myofibrillar myopathy.[244] Missense mutations affecting one allele, and with autosomal dominant inheritance, are associated with late-onset disease. Onset is typically in the 20s. Weakness is often distally predominant. Cardiomyopathy can be a prominent feature, and may be present in childhood and be the dominating feature of pediatric onset disease. Recessively inherited mutations affecting both alleles tend to be associated with earlier and more severe myopathy.[225,245,246]

BAG3opathies

Mutations in *BAG3* account for the majority of pediatric onset cases of myofibrillar myopathy. It was first described in three patients with childhood onset and progressive weakness.[247] All had cardiomyopathy and developed respiratory failure, and two also showed evidence of rigid spine. *De novo* heterozygous mutations of *BAG3* were found in all three. Since this initial description, several more patients have been identified with *BAG3* mutation and a similar progressive myopathy.[248,249] Serum CK levels are typically elevated (3–15 ×), and neurophysiologic studies can show evidence of a peripheral neuropathy. Interestingly, *BAG3* mutations have now been found as a cause of giant axonal neuropathy.[250] Of note, while mutations are typically *de novo*, parental mosaicism has been uncovered in at least one case.

AlphaB Crystallinopathies

AlphaB crystallinopathies typically present in late adulthood and are due to heterozygous mutations in the *CRYAB* gene.[251] However, a severe infantile presentation has been described. In particular, recessive *CRYAB* mutation was uncovered in Canadian aboriginal peoples with fatal infantile muscular dystrophy.[237] This condition involves progressive limb and axial weakness, peculiar muscle stiffness, respiratory failure and, usually, death in the first year of life. A similarly severe myopathy has been described in one additional non-Canadian patient.[238]

HYALINE BODY MYOPATHY

Hyaline body myopathy (HBM) (also called myosin storage myopathy) was originally reported as "familial myopathy with probable lysis of myofibrils in type-I fibers"[252] in two young siblings who were floppy from birth with delayed motor development. These original patients are now in their mid-30s and are wheelchair-bound and prone to develop pulmonary infections (DeMyer, Indiana University Medical Center: personal communication, 1997). The disorder was renamed when the characteristic morphologic features in the muscle fiber became known as hyaline bodies.[45,253,254] The condition is rare, with fewer than 20 patients reported.

Clinical Features

HBM does not have any specific clinical features apart from muscle weakness, which may be proximal or generalized. Onset may be in infancy or childhood,[252] although adult-onset cases have been described,[254] presenting as scapuloperoneal syndrome. Cardiac involvement has been reported in at least one case.[255] CK may be normal or only slightly elevated, and the EMG may show myopathic features.

Pathologic Features

The characteristic structural lesions, the hyaline bodies (Figure 28.19A), are large areas devoid of sarcomeres, or areas of "myofibrillar lysis."[252] Actually, the areas are not real "inclusion bodies" but rather plaques or patches of extrasarcomeric finely granular material largely devoid of organelles (Figure 28.19B), which amasses in the periphery of the muscle fibers. The myofibrillar ATPases are active within these lesions, so that fibers are positive on ATPase reaction after acid preincubation.[253,256] Immunocytochemically, hyaline bodies contain myosin, especially of the slow type, but apparently no other proteins, with an increase in desmin staining at the periphery.[253] Based on this observation, this myopathy is also referred to as myosin storage myopathy. They are negative for several antibodies such as α-actinin, spectrin, vimentin, dystrophin, and ubiquitin.[254] Ultrastructurally, the hyaline bodies consist of a finely granular material. The granularity is less electron-dense than that in granulofilamentous material, and granules are smaller than glycogen granules. Hyaline bodies are not membrane bound, and are sharply demarcated from adjacent sarcomeres.

Hyaline bodies occur in 10–30% of muscle fibers.[253,256] They occur predominantly in type 1 fibers, and there may be associated fiber type disproportion, as well as type 1 fiber predominance.[252–254,256]

Genetics

To date, all genetically solved cases of HBM are due to heterozygous missense mutations in the slow skeletal myosin (*MYH7*) gene.[257] Both *de novo* and autosomal dominantly inherited cases with *MYH7* mutations have been reported,[258–260] and there is a single case of a homozygous *MYH7* mutation associated with both HBM and cardiomyopathy.[255] HBM thus adds to the growing group of disorders associated with *MYH7* mutations. This group additionally includes Laing distal myopathy and core myopathy, as well as dilated cardiomyopathy.[261]

AUTOPHAGIC VACUOLAR MYOPATHIES

Autophagic vacuolar myopathies are a heterogeneous group of conditions united by the observation of accumulated lysosomes on muscle biopsy.[262] In normal muscle, lysosomes are scantly visualized. In autophagic vacuolar myopathies, there is aggregation of basophilic material on H and E staining, which corresponds to pathologically appearing lysosomes. The prototypical vacuolar myopathy is Pompe disease, or acid maltase deficiency. In Pompe disease, the muscle biopsy contains vacuoles with amorphous material that, based on PAS staining, correspond to accumulated glycogen. The clinical features and treatment of Pompe disease are discussed elsewhere in this book. The other autophagic vacuolar myopathies fall broadly into two categories based on the characterization of the vacuoles. One has autophagic

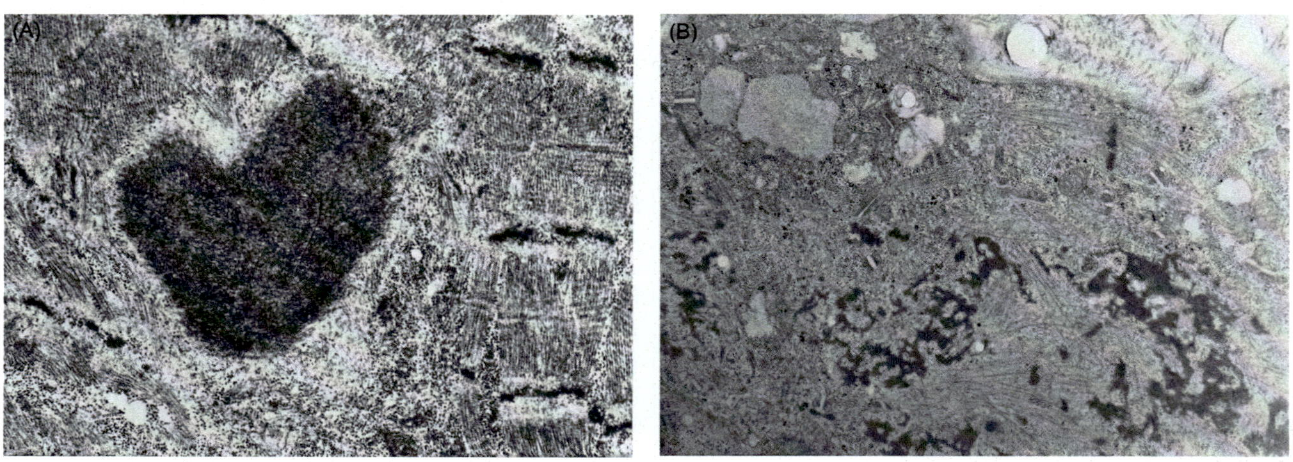

FIGURE 28.18 Desmin-related/myofibrillar myopathy. Ultrastructure of desmin aggregation within muscle fibers: (**A**) A cytoplasmic body consists of an electron-dense core and is surrounded by loosely packed filaments (×35,000). (**B**) Granulofilamentous material accumulated in the subsarcolemmal region of the muscle fiber (×21,375).

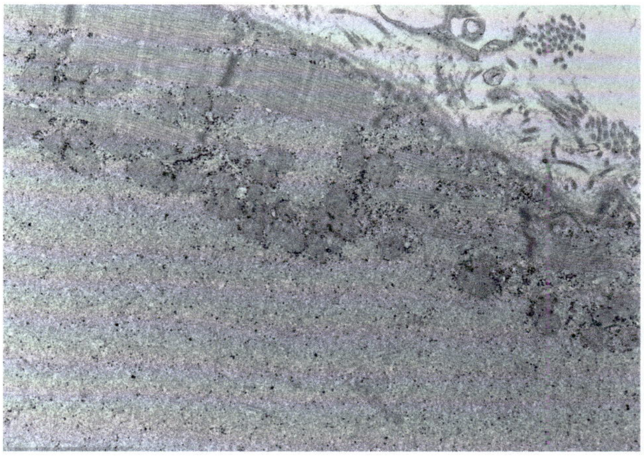

FIGURE 28.19 Hyaline body myopathy. (**A**) Sharply limited pale areas within muscle fibers represent "hyaline" bodies (modified Gomori trichrome stain). (**B**) A large, sharply demarcated area devoid of sarcomeres and organelles is filled with finely granular material, hyaline body (×25,000).

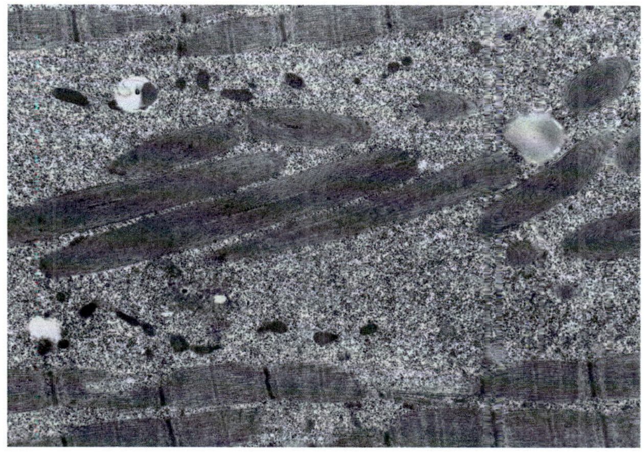

FIGURE 28.20 Cylindrical spirals myopathy. Electron microscopy demonstrating a group of cylindrical spirals cut transversely, obliquely, and longitudinally (×15,390).

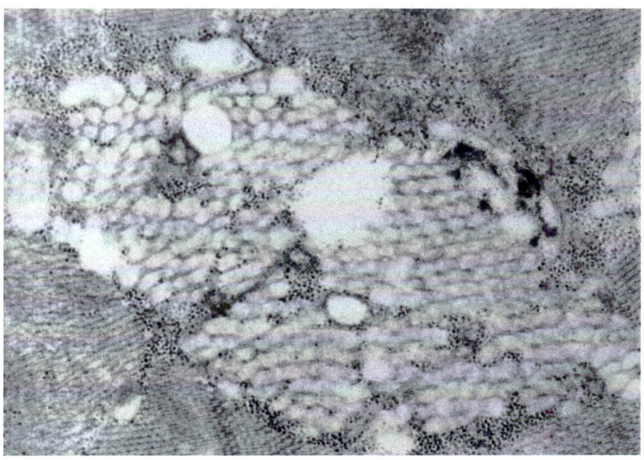

FIGURE 28.21 Myopathy with tubular aggregates. The aggregate of tubules is focally distributed in the individual fibers and lies adjacent to normal myofibrils.

vacuoles with sarcomeric features (AVSF), while the other has rimmed vacuoles. Myopathies with rimmed vacuoles, which include distal myopathy with rimmed vacuoles (DMRV) or Nonaka myopathy and hereditary inclusion body myopathy,[263,264] are not typically encountered in the pediatric setting. Myopathies with AVSFs, conversely, can present with symptoms in childhood, and include Danon disease and X-linked vacuolar myopathy with excessive autophagy (XMEA).

Danon disease is a vacuolar myopathy that presents with the triad of cardiomyopathy, muscle weakness, and mild mental retardation.[265] Diagnosis is usually made in the context of severe cardiomyopathy, which dominates the clinical picture. Diagnosis is established by cardiac or muscle biopsy, where features of a vacuolar myopathy are observed, and immunostaining for LAMP2 is absent. Danon disease is an X-linked disorder caused by mutations in *LAMP2*. The disease primarily affects boys, though women also manifest symptoms, particularly cardiomyopathy. The skeletal myopathy is mild, and usually does not impair ambulation. Weakness is shoulder girdle and neck predominant. The serum CK is elevated, often >1000. Death occurs from cardiac failure, with the mean age of death in males of 19 years, and in females, 40.

XMEA is most typically a myopathy of adulthood.[259] Biopsies from XMEA patients contain vacuoles with sarcomeric features. This X-linked disorder is caused by mutations in *VMA21*. VMA21 is a chaperone that stabilizes the ATPase complex of the lysosome. There are also reports of cases with similar biopsy findings presenting in infancy. One is called infantile autophagic vacuolar myopathy,[266] which has been described in two cases and is associated with severe congenital weakness and early death. Another is X-linked congenital autophagic vacuolar myopathy.[259] The relationship between these two conditions and XMEA is uncertain, but it is possible one or both represent severe presentations of VMA21-associated vacuolar myopathy.

TREATMENT OF CONGENITAL MYOPATHIES

There are currently no curative treatments for patients with congenital myopathy. Nevertheless, a multidisciplinary approach to the management of the individual patient will greatly improve quality of life and can influence survival.[267] The therapeutic approach to this group of patients should be considered in the following four areas:

1. Prevention: genetic counseling and prenatal diagnosis.
2. Monitoring: developing a plan for the prospective detection of complications.

3. Risk management: special care during surgical procedures, anesthesia, and childbirth.
4. Symptomatic therapy and rehabilitation: physical and occupational therapy, orthopedic intervention, respiratory care, and management of feeding difficulties.

Prevention: Genetic Counseling and Prenatal Diagnosis

Genetic counseling and prediction of genetic risk for future offspring for families with congenital myopathies can be challenging, even in those disorders in which disease genes have been identified and molecular genetic testing is available. For example, nemaline myopathy is the best characterized of the myopathies at the molecular level, but is genetically heterogeneous. Five disease genes have been identified, but mutations have been identified in only a minority of cases, and it is likely that a number of other disease genes can cause the same clinical and pathologic phenotype. In addition, for those genes that have been identified, inheritance may be autosomal dominant, sporadic (new dominant), or autosomal recessive; and an unaffected parent may have more than one affected offspring as a result of mosaicism for a disease-causing mutation in their germ cells (ova or sperm; so-called *germline mosaicism*). Therefore, definitive genetic counseling and prenatal diagnosis can be offered only when a disease-causing mutation has been identified and prenatal diagnosis is performed on tissue from the unborn fetus (by chorionic villus biopsy or amniocentesis). In some cases, linkage analysis may also be possible for prenatal diagnosis, but with the recent development of next generation sequencing-based whole genome and whole exome sequencing, as well as targeted next-generation panels covering dozens or even hundreds of genes, rapid and specific tests for the genetic basis for many congenital myopathies are now becoming available, and will likely revolutionize the diagnosis and detection of these disorders.

When only one person in a family is affected by a specific congenital myopathy, determining the mode of inheritance can be a problem. Clinical evaluation of both parents is necessary to rule out minor muscle weakness. Muscle biopsy of parents may be useful, but interpretation of abnormal findings can be difficult. For example, if a clinically healthy parent has abnormalities on muscle biopsy, it may not be possible to distinguish whether they are a manifesting heterozygote for a recessive gene, or whether the disorder is dominant and they have no clinical manifestations (incomplete penetrance). If one parent shows overt disease clinically and typical histologic abnormalities on muscle biopsy, but the other parent is healthy and shows normal findings on muscle biopsy, the likely mode of inheritance is autosomal dominant. If both parents are clinically healthy and show no abnormality on muscle biopsy, dominant transmission from one of the parents is unlikely, leaving the possibility of a new dominant mutation in the child (with minimal risk to future offspring), germline mosaicism in one of the parents (with low but significant risk to future offspring), or recessive inheritance with a 25% recurrence risk. As the molecular genetics of these disorders are clarified, some of these diagnostic issues may be resolved.

Monitoring

The congenital myopathies are clinically heterogeneous in terms of their age of onset, rate of progression, and the severity of respiratory, cardiac, bulbar, and skeletal muscle involvement. The mainstay of therapy for patients with congenital myopathy is *early detection* of disease manifestations and complications, and early referral to a *multidisciplinary team* of health care professionals for ongoing assessment and intervention. A coordinating physician who then makes the appropriate referrals must assess the clinical status and therapeutic needs of each patient at regular intervals. The majority of congenital myopathy patients will require *at least* ongoing input from a physiotherapist and occupational therapist, plus possibly a speech pathologist, regular monitoring of pulmonary function, a baseline sleep study, and screening for scoliosis.

Risk Management

Surgical Procedures and Anesthetic Risks

CCD is associated with an increased risk of MH. However, because the diagnosis is unknown in many patients undergoing muscle biopsy, MH precautions should be taken prior to a definitive diagnosis. MH has not been clearly associated with other congenital myopathies, although one report describes three children with nemaline myopathy in whom the heart rate decreased during induction of anesthesia for cardiac surgery, and body temperature increased during or after surgery.[268]

In general, patients with congenital myopathy tolerate surgical procedures and general anesthetics well. However, respiratory complications may occur. In our series of 128 patients with nemaline myopathy,[23] five patients had unexpected postoperative respiratory failure that necessitated prolonged ventilatory support in three patients and resulted in the death of another. Preoperative assessment of respiratory status and performance of serial pulmonary function tests may be useful to guide timing of surgical intervention. In addition, prolonged immobility may exacerbate or worsen muscle weakness, so patients should be mobilized as soon as possible after a surgical procedure.

Obstetric Complications

In general, adult females with congenital myopathies tolerate childbirth well, but there is a high incidence of obstetric complications associated with the delivery of a child with congenital myopathies presenting in the newborn period. Problems include polyhydramnios, decreased fetal movements, and delivery complicated by an abnormal presentation, fetal distress, failure to progress, or prematurity.

Symptomatic Therapy and Rehabilitation

The main factors influencing prognosis in patients with congenital myopathy are respiratory capacity and the development of scoliosis.[33,34] Thus, the monitoring of respiratory function and the spine are essential elements in the ongoing care of these patients.

Respiratory Care

Respiratory problems are a common feature of congenital myopathies and are the primary cause of death, not only in the neonatal period but also throughout life. Respiratory failure can occur at any age, and the degree of skeletal muscle weakness does not necessarily reflect the degree of respiratory muscle involvement. Respiratory compromise occurs secondary to involvement of the intercostal muscles and diaphragm in combination with weakening of the abdominal muscles or progressive scoliosis. Bulbar involvement increases the risk of aspiration, and poor nutritional status may increase susceptibility to respiratory infection.

Most patients will show restriction of their respiratory capacity on testing, even if they are symptom-free. In addition, patients with congenital myopathy run a great risk of insidious nocturnal hypoxia. Symptoms include sleep disturbance, nightmares, morning headache, daytime tiredness, and weight loss. However, nocturnal hypoxia may occur even in the absence of morning symptoms, and several patients have experienced sudden respiratory failure.[33,34,187]

All patients with congenital myopathy should have at least baseline evaluation of their respiratory status, with the frequency of re-evaluation depending on the patient's clinical status. Investigations include lung function testing (vital capacity, FEV_1, and maximal inspiratory and expiratory pressures), oximetry, and evaluation of CO_2 during waking and sleep and an assessment of bulbar function. Postural drainage, regular chest physiotherapy, and a manually assisted cough may assist respiratory toilet in patients with bulbar weakness, reduced vital capacity, and recurrent aspiration. Respiratory infections should be treated early and aggressively, including intravenous antibiotics. The patient and the family should be provided with information and education concerning the possibility of respiratory insufficiency and the options for home mechanical ventilation. In addition, they should meet with a physician and other patients who have experience with home mechanical ventilation so that informed choices can be made, if and when ventilation is necessary.[269]

Feeding Difficulties

Feeding intolerance is common in patients with congenital myopathy presenting in the newborn period. In addition, bulbar dysfunction can result in chewing and swallowing difficulties, recurrent aspiration, gastroesophageal reflux, and poor nutritional status at any age. Feeding problems in the newborn period often necessitate commencement of gavage feeds. Approximately half of these infants will improve and will come to tolerate bottle feedings and soft solids.[23] However, insertion of a gastrostomy tube, possibly with fundoplication, should be considered if problems persist after the first few months of life. Those children with relatively mild neonatal symptoms tend to improve with increasing age, ultimately tolerating a full oral diet with few complications.

Joint Contractures and Scoliosis

The physiotherapist and orthopedic surgeon play an essential role in the ongoing care of children with congenital myopathy. A regular program of muscle stretches and exercise will help to prevent or minimize joint contractures and promote cardiorespiratory health. Ideally, the majority of such therapies should take place at home and become integrated into the child's day-to-day activities. Orthotics, splinting, and serial plaster casting may be necessary, particularly in the treatment of tight Achilles tendons. Surgical release is indicated for contractures that do not respond to aggressive physiotherapy—particularly if such surgery will result in improved function, or will facilitate activities of daily living.

All patients with congenital myopathy should be monitored for the development of scoliosis, kyphosis, or kyphoscoliosis. Once a curve is detected, the child should be referred to an orthopedic surgeon and have serial spinal X-rays. Progressive spinal deformity can lead to pain and difficulty in standing, walking, sitting, or balancing. There may be deterioration in motor function and independence. More importantly, respiratory function may be further compromised with the development of restrictive lung disease. Treatment options include bracing and spinal fusion (arthrodesis). Surgery is indicated if the curve is progressing, pulmonary function is impaired, and spinal fusion is unlikely to impair motor function.

Exercise

The ultimate cause of physical impairment in neuromuscular disorders is loss of normally functioning muscle fibers. Submaximal strengthening exercises with frequent rest periods are recommended for patients with neuromuscular disorders, although exercising to the point of exhaustion should be avoided (reviewed in Kilmer and McDonald[270]). Regular, enjoyable, low-impact activities, such as cycling or swimming, are ideal, with supervision to monitor for excessive fatigue or increased weakness. Although no direct medical benefits have been scientifically demonstrated in children with muscle disorders, regular exercise promotes cardiovascular well-being and general fitness. Any exercise program should also include regular stretches to help maintain joint range of motion. Participation in regular group activities can be beneficial in terms of improved socialization, self-esteem, and independence, limiting isolation and loneliness.[271]

NOVEL THERAPEUTIC STRATEGIES

While no therapies have yet been clinically proven to ameliorate symptoms in congenital myopathies, several treatments have been tried in a small series of patients. The medication with perhaps the most exposure is the acetylcholinesterase inhibitor pyridostigmine. Mainly due to the clinical overlap between congenital myasthenic syndrome and centronuclear myopathy, this medication has been tried in subgroups of CNM patients.[272] Positive benefit has been reported in individual cases of *DNM2*-related CNM, *MTM1*-related myotubular myopathy, and in CNM without genetic diagnosis. The most dramatic report is of an individual with myotubular myopathy who regained ambulation after starting pyridostigmine. The potential efficacy of drugs that target the NMJ in CNM is supported by preclinical studies in the zebrafish and mouse, which have demonstrated neuromuscular junction abnormalities in both *DNM2* and *MTM1* models, and improvement in motor function with pyridostigmine exposure.[267,273]

Another therapy that has shown benefit in small case series is L-tyrosine for nemaline myopathy. Ryan, North and colleagues reported experience with five patients who received L-tyrosine and had clinical improvements in drooling (with significant reduction) and in fatigue (with increased energy levels).[274] This medication is also supported by preclinical data, in which tyrosine has been shown to improve motor function in a mouse model of nemaline myopathy due to *ACTA1* mutations.[275]

A few other medications have been reported in single or small case series. There is one case of dantrolene improving motor function in an adult patient with CCD due to heterozygous *RYR1* mutation.[276] A single case with improvement in muscle strength and respiratory function after treatment with albuterol and exercise was recorded for a 9 year old with central core disease and heterozygous *RYR1* mutation. However, this individual had progression of joint contractures during the treatment period. Finally, 13 patients with core myopathy were treated for 6 months with salbutamol by Mercuri and colleagues[277]; 10/13 completed the study, and all had significant increases in strength, respiratory measurements, and functional ability. The overall conclusion regarding off-label treatment with medications such as salbutamol, pyridostigmine, and L-tyrosine is that the case studies are promising but that randomized, double-blind, placebo controlled trials are necessary to truly establish their efficacy and before they can be recommended as standard therapies.

DRUG DISCOVERY

A recent explosion in model system development for congenital myopathies[278–280] has greatly aided the development of exciting potential therapies for these currently untreatable conditions. Perhaps the most exciting preclinical development has been the observation that gene or protein replacement therapies for myotubular myopathy are able to ameliorate the disease process.[281,282] This has been shown in the mouse model for protein replacement therapy[282] and in both the mouse and canine models for gene therapy.[281] Gene editing strategies[283] are being considered for several congenital myopathies, and offer a novel treatment strategy for dominant diseases such as *DNM2*-related CNM and *RYR1*-related central core disease.[284] Antioxidant therapy using the potent antioxidant N-acetylcysteine has shown great promise in preclinical models and in patient cells for *SEPN1* and *RYR1* mutations.[285] Rigorous testing of these potential therapies in a clinical trial setting is planned.

REFERENCES

1. Magee KR, Shy GM. A new congenital non-progressive myopathy. *Brain* 1956;**79**(4):610–21.
2. Dubowitz V. The "new" myopathies. *Neuropediatrics* 1969;**1**: 137–48.
3. Amburgey K, McNamara N, Bennett LR, McCormick ME, Acsadi G, Dowling JJ. Prevalence of congenital myopathies in a representative pediatric United States population. *Ann Neurol* 2011;**70**(4):662–5.
4. North KN, Wang CH, Clarke N, Jungbluth H, Vainzof M, Dowling JJ, et al. Approach to the diagnosis of congenital myopathies. *Neuromuscul Disord: NMD* 2014;**24**(2):97–116.
5. Fardeau M, Tome F. Congenital myopathies. In: Engel AG, Franzini-Armstrong C, editors. *Myology*. 2nd ed. New York: McGraw-Hill; 1994. pp. 1487–533.

6. Maggi L, Scoto M, Cirak S, Robb SA, Klein A, Lillis S, et al Congenital myopathies—clinical features and frequency of individual subtypes diagnosed over a 5-year period in the United Kingdom. *Neuromuscul Disord: NMD* 2013;**23**(3):195–205.
7. Bevilacqua JA, Monnier N, Bitoun M, Eymard B, Ferreiro A, Monges S, et al. Recessive RYR1 mutations cause unusual congenital myopathy with prominent nuclear internalization and large areas of myofibrillar disorganization. *Neuropathol Appl Neurobiol* 2011;**37**(3):271–84.
8. Wallgren-Pettersson C. Nemaline and myotubular myopathies. *Semin Pediatr Neurol* 2002;**9**(2):132–44.
9. Laing NG, Majda BT, Akkari PA, Layton MG, Mulley JC, Phillips H, et al. Assignment of a gene (NEMI) for autosomal dominant nemaline myopathy to chromosome I. *Am J Hum Genet* 1992;**50**(3):576–83.
10. Bohm J, Biancalana V, Dechene ET, Bitoun M, Pierson CR, Schaefer E, et al. Mutation spectrum in the large GTPase dynamin 2, and genotype-phenotype correlation in autosomal dominant centronuclear myopathy. *Hum Mutat* 2012;**33**(6):949–59.
11. Lamont PJ, Wallefeld W, Hilton-Jones D, Udd B, Argov Z, Barboi AC, et al. Novel mutations widen the phenotypic spectrum of slow skeletal/beta-cardiac myosin (MYH7) distal myopathy. *Hum Mutat* 2014;**35**(7):868–79.
12. Jungbluth H, Sewry CA, Counsell S, Allsop J, Chattopadhyay A, Mercuri E, et al. Magnetic resonance imaging of muscle in nemaline myopathy. *Neuromuscul Disord: NMD* 2004;**14**(12):779–84.
13. Jungbluth H, et al. Magnetic resonance imaging of muscle in congenital myopathies associated with RYR1 mutations. *Neuromuscul Disord* 2004;**14**:785–90.
14. Schessl J, Medne L, Hu Y, Zou Y, Brown MJ, Huse JT, et al. MRI in DNM2-related centronuclear myopathy: evidence for highly selective muscle involvement. *Neuromuscul Disord: NMD* 2007;**17**(1):28–32.
15. Quijano-Roy S, Carlier RY, Fischer D. Muscle imaging in congenital myopathies. *Semin Pediatr Neurol* 2011;**18**(4):221–9.
16. Shy GM, Engel WK, Somers JE, Wanko T. Nemaline myopathy. A new congenital myopathy. *Brain* 1963;**86**:793–810.
17. Conen PE, Murphy EG, Donohue WL. Light and electron microscopic studies of "Myogranules" in a child with hypotonia and muscle weakness. *Can Med Assoc J* 1963;**89**:983–6.
18. Goebel HH, Lenard HG. Congenital myopathies. *Clin Neurol Neurosurg* 1992;**18**:331–67.
19. Schnell C, Kan A, North KN. "An artefact gone awry": identification of the first case of nemaline myopathy by Dr R.D.K. Reye. *Neuromuscul Disord: NMD* 2000;**10**(4–5):307–12.
20. Wallgren-Petterson C. *Congenital Nemaline Myopathy: A Longitudinal Study*. Helsinki: University of Helsinki; 1990.
21. North KN, Laing NG, Wallgren-Pettersson C. Nemaline myopathy: current concepts. The ENMC International Consortium and Nemaline Myopathy. *J Med Genet* 1997;**34**(9):705–13.
22. Wallgren-Pettersson C, Beggs AH, Laing NG. 51st ENMC International Workshop: Nemaline Myopathy. 13–15 June 1997, Naarden, The Netherlands. *Neuromuscul Disord: NMD* 1998;**8**(1):53–6.
23. Ryan MM, Schnell C, Strickland CD, Shield LK, Morgan G, Iannaccone ST, et al. Nemaline myopathy: a clinical study of 143 cases. *Ann Neurol* 2001;**50**(3):312–20.
24. Lammens M, Moerman P, Fryns JP, Lemmens F, van de Kamp GM, Goemans N, et al. Fetal akinesia sequence caused by nemaline myopathy. *Neuropediatrics* 1997;**28**(2):116–19.
25. Banwell BL, Singh NC, Ramsay DA. Prolonged survival in neonatal nemaline rod myopathy. *Pediatr Neurol* 1994;**10**(4):335–7.
26. Roig M, Hernandez MA, Salcedo S. Survival from symptomatic nemaline myopathy in the newborn period. *Pediatr Neurosci* 1987;**13**(2):95–7.
27. Wallgren-Pettersson C. Congenital nemaline myopathy. A clinical follow-up of twelve patients. *J Neurol Sci* 1989;**89**(1):1–14.
28. Bucher HU, Boltshauser E, Briner J. Neonatal nemaline myopathy presenting with multiple joint contractures. *Eur J Pediatr* 1985;**144**(3):288–90.
29. Rifai Z, Kazee AM, Kamp C, Griggs RC. Intranuclear rods in severe congenital nemaline myopathy. *Neurology* 1993;**43**(11):2372–7.
30. Schmalbruch H, Kamieniecka Z, Arroe M. Early fatal nemaline myopathy: case report and review. *Dev Med Child Neurol* 1987;**29**(6):800–4.
31. Wallgren-Pettersson C, Laing N. 40th ENMC Sponsored International Workshop: Nemaline Myopathy. 2–4 February 1996, Naarden, The Netherlands. *Neuromuscul Disord: NMD* 1996;**6**(5):389–91.
32. Ravenscroft G, Miyatake S, Lehtokari VL, Todd EJ, Vornanen P, Yau KS, et al. Mutations in KLHL40 are a frequent cause of severe autosomal-recessive nemaline myopathy. *Am J Hum Genet* 2013;**93**(1):6–18.
33. Howard RS, Wiles CM, Hirsch NP, Spencer GT. Respiratory involvement in primary muscle disorders: assessment and management. *Q J Med* 1993;**86**(3):175–89.
34. Sasaki M, Yoneyama H, Nonaka I. Respiratory muscle involvement in nemaline myopathy. *Pediatr Neurol* 1990;**6**(6):425–7.
35. Airenne AL, Wallgren-Pettersson C. Normal cardiac contractility in patients with congenital nemaline myopathy. *Neuropediatrics* 1988;**19**(3):115–17.
36. Wallgren-Pettersson C, Sainio K, Salmi T. Electromyography in congenital nemaline myopathy. *Muscle Nerve* 1989;**12**(7):587–93.
37. Jarraya M, Quijano-Roy S, Monnier N, Behin A, Avila-Smirnov D, Romero NB, et al. Whole-Body muscle MRI in a series of patients with congenital myopathy related to TPM2 gene mutations. *Neuromuscul Disord: NMD* 2012;**22**(Suppl 2):S137–47.
38. Shimomura C, Nonaka I. Nemaline myopathy: comparative muscle histochemistry in the severe neonatal, moderate congenital, and adult-onset forms. *Pediatr Neurol* 1989;**5**(1):25–31.
39. Ryan MM, Ilkovski B, Strickland CD, Schnell C, Sanoudou D, Midgett C, et al. Clinical course correlates poorly with muscle pathology in nemaline myopathy. *Neurology* 2003;**60**(4):665–73.
40. Jockusch BM, Veldman H, Griffiths GW, van Oost BA, Jennekens FG. Immunofluorescence microscopy of a myopathy. alpha-Actinin is a major constituent of nemaline rods. *Exp Cell Res* 1980;**127**(2):409–20.
41. Wallgren-Pettersson C, Jasani B, Newman GR, Morris GE, Jones S, Singhrao S, et al. Alpha-actinin in nemaline bodies in congenital nemaline myopathy: immunological confirmation by light and electron microscopy. *Neuromuscul Disord: NMD* 1995;**5**(2):93–104.
42. Goebel HH, Piirsoo A, Warlo I, Schofer O, Kehr S, Gaude M. Infantile intranuclear rod myopathy. *J Child Neurol* 1997;**12**(1):22–30.
43. Miike T, Ohtani Y, Tamari H, Ishitsu T, Une Y. Muscle fiber type transformation in nemaline myopathy and congenital fiber type disproportion. *Brain Dev* 1986;**8**(5):526–32.

44. Volpe P, Damiani E, Margreth A, Pellegrini G, Scarlato G. Fast to slow change of myosin in nemaline myopathy: electrophoretic and immunologic evidence. *Neurology* 1982;**32**(1):37–41.
45. Barohn RJ, Jackson CE, Kagan-Hallet KS. Neonatal nemaline myopathy with abundant intranuclear rods. *Neuromuscul Disord: NMD* 1994;**4**(5–6):513–20.
46. Paulus W, Peiffer J, Becker I, Roggendorf W, Schumm F. Adult-onset rod disease with abundant intranuclear rods. *J Neurol* 1988;**235**(6):343–7.
47. Keller CE, Hays AP, Rowland LP, Moghadaszadeh B, Beggs AH, Bhagat G. Adult-onset nemaline myopathy and monoclonal gammopathy. *Arch Neurol* 2006;**63**(1):132–4.
48. Fidzianska A, Badurska B, Ryniewicz B, Dembek I. "Cap disease": new congenital myopathy. *Neurology* 1981;**31**(9):1113–20.
49. Schreckenbach T, Schroder JM, Voit T, Abicht A, Neuen-Jacob E, Roos A, et al. Novel TPM3 mutation in a family with cap myopathy and review of the literature. *Neuromuscul Disord: NMD* 2014;**24**(2):117–24.
50. Marttila M, Lehtokari VL, Marston S, Nyman TA, Barnerias C, Beggs AH, et al. Mutation update and genotype-phenotype correlations of novel and previously described mutations in TPM2 and TPM3 causing congenital myopathies. *Hum Mutat* 2014;**35**(7):779–90.
51. De Paula AM, Franques J, Fernandez C, Monnier N, Lunardi J, Pellissier JF, et al. A TPM3 mutation causing cap myopathy. *Neuromuscul Disord: NMD* 2009;**19**(10):685–8.
52. Clarke NF, Domazetovska A, Waddell L, Kornberg A, McLean C, North KN. Cap disease due to mutation of the beta-tropomyosin gene (TPM2). *Neuromuscul Disord: NMD* 2009;**19**(5):348–51.
53. Lehtokari VL, Ceuterick-de Groote C, de Jonghe P, Marttila M, Laing NG, Pelin K, et al. Cap disease caused by heterozygous deletion of the beta-tropomyosin gene TPM2. *Neuromuscul Disord: NMD* 2007;**17**(6):433–42.
54. Hung RM, Yoon G, Hawkins CE, Halliday W, Biggar D, Vajsar J. Cap myopathy caused by a mutation of the skeletal alpha-actin gene ACTA1. *Neuromuscul Disord: NMD* 2010;**20**(4):238–40.
55. Reyes MG, Goldbarg H, Fresco K, Bouffard A. Zebra body myopathy: a second case of ultrastructurally distinct congenital myopathy. *J Child Neurol* 1987;**2**(4):307–10.
56. Lake BD, Wilson J. Zebra body myopathy. Clinical, histochemical and ultrastructural studies. *J Neurol Sci* 1975;**24**(4):437–46.
57. Sewry CA, Holton J, Dick DJ, Jacques TS, Muntoni F, Hanna M. Zebra body myopathy resolved. *Neuromuscul Disord: NMD* 2009;**19**(8–9):637–8.
58. D'Amico A, Graziano C, Pacileo G, Petrini S, Nowak KJ, Boldrini R, et al. Fatal hypertrophic cardiomyopathy and nemaline myopathy associated with ACTA1 K336E mutation. *Neuromuscul Disord: NMD* 2006;**16**(9–10):548–52.
59. Nakajima M, Shima Y, Kumasaka S, Kuwabara K, Migita M, Fukunaga Y. An infant with congenital nemaline myopathy and hypertrophic cardiomyopathy. *J Nippon Med Sch* 2008;**75**(6):350–3.
60. Kim SY, Park YE, Kim HS, Lee CH, Yang DH, Kim DS. Nemaline myopathy and non-fatal hypertrophic cardiomyopathy caused by a novel ACTA1 E239K mutation. *J Neurol Sci* 2011;**307**(1–2):171–3.
61. Nagata R, Kamimura D, Suzuki Y, Saito T, Toyama H, Dejima T, et al. A case of nemaline myopathy with associated dilated cardiomyopathy and respiratory failure. *Int Heart J* 2011;**52**(6):401–5.
62. Mir A, Lemler M, Ramaciotti C, Blalock S, Ikemba C. Hypertrophic cardiomyopathy in a neonate associated with nemaline myopathy. *Congenit Heart Dis* 2012;**7**(4):E37–41.
63. Taglia A, D'Ambrosio P, Palladino A, Politano L. On a case of respiratory failure due to diaphragmatic paralysis and dilated cardiomyopathy in a patient with nemaline myopathy. *Acta Myol* 2012;**31**(3):201–3.
64. Gatayama R, Ueno K, Nakamura H, Yanagi S, Ueda H, Yamagishi H, et al. Nemaline myopathy with dilated cardiomyopathy in childhood. *Pediatrics* 2013;**131**(6):e1986–90.
65. Bergmann M, Kamarampaka M, Kuchelmeister K, Klein H, Koch H. Nemaline myopathy: two autopsy reports. *Childs Nerv Syst* 1995;**11**(10):610–15.
66. Ishibashi-Ueda H, Imakita M, Yutani C, Takahashi S, Yazawa K, Kamiya T, et al. Congenital nemaline myopathy with dilated cardiomyopathy: an autopsy study. *Hum Pathol* 1990;**21**(1):77–82.
67. Karpati G, Carpenter S, Eisen AA. Experimental core-like lesions and nemaline rods. A correlative morphological and physiological study. *Arch Neurol* 1972;**27**(3):237–51.
68. Fukunaga H, Osame M, Igata A. A case of nemaline myopathy with ophthalmoplegia and mitochondrial abnormalities. *J Neurol Sci* 1980;**46**(2):169–77.
69. Kornfeld M. Mixed nemaline-mitochondrial "myopathy". *Acta Neuropathol* 1980;**51**(3):185–9.
70. Simpson DM, Bender AN. Human immunodeficiency virus-associated myopathy: analysis of 11 patients. *Ann Neurol* 1988;**24**(1):79–84.
71. Fitzsimons RB, McLeod JG. Myopathy with pathological features of both centronuclear myopathy and multicore disease. *J Neurol Sci* 1982;**57**(2–3):395–405.
72. Seitz RJ, Toyka KV, Wechsler W. Adult-onset mixed myopathy with nemaline rods, minicores, and central cores: a muscle disorder mimicking polymyositis. *J Neurol* 1984;**231**(3):103–8.
73. Laing NG, Wilton SD, Akkari PA, Dorosz S, Boundy K, Kneebone C, et al. A mutation in the alpha tropomyosin gene TPM3 associated with autosomal dominant nemaline myopathy. *Nat Genet* 1995;**9**(1):75–9.
74. Pelin K, Hilpela P, Donner K, Sewry C, Akkari PA, Wilton SD, et al. Mutations in the nebulin gene associated with autosomal recessive nemaline myopathy. *Proc Natl Acad Sci USA* 1999;**96**(5):2305–10.
75. Nowak KJ, Wattanasirichaigoon D, Goebel HH, Wilce M, Pelin K, Donner K, et al. Mutations in the skeletal muscle alpha-actin gene in patients with actin myopathy and nemaline myopathy. *Nat Genet* 1999;**23**(2):208–12.
76. Donner K, Ollikainen M, Ridanpaa M, Christen HJ, Goebel HH, de Visser M, et al. Mutations in the beta-tropomyosin (TPM2) gene—a rare cause of nemaline myopathy. *Neuromuscul Disord: NMD* 2002;**12**(2):151–8.
77. Sambuughin N, Yau KS, Olive M, Duff RM, Bayarsaikhan M, Lu S, et al. Dominant mutations in KBTBD13, a member of the BTB/Kelch family, cause nemaline myopathy with cores. *Am J Hum Genet* 2010;**87**(6):842–7.
78. Agrawal PB, Greenleaf RS, Tomczak KK, Lehtokari VL, Wallgren-Pettersson C, Wallefeld W, et al. Nemaline myopathy with minicores caused by mutation of the CFL2 gene encoding the skeletal muscle actin-binding protein, cofilin-2. *Am J Hum Genet* 2007;**80**(1):162–7.

79. Ockeloen CW, Gilhuis HJ, Pfundt R, Kamsteeg EJ, Agrawal PB, Beggs AH, et al. Congenital myopathy caused by a novel missense mutation in the CFL2 gene. *Neuromuscul Disord: NMD* 2012;**22**(7):632–9.
80. Ong RW, Alsaman A, Selcen D, Arabshahi A, Yau KS, Ravenscroft G, et al. Novel cofilin-2 (CFL2) four base pair deletion causing nemaline myopathy. *J Neurol Neurosurg Psychiatry* 2014;**85**(9):1058–60.
81. Gupta VA, Ravenscroft G, Shaheen R, Todd EJ, Swanson LC, Shiina M, et al. Identification of KLHL41 mutations implicates BTB-Kelch-mediated ubiquitination as an alternate pathway to myofibrillar disruption in nemaline myopathy. *Am J Hum Genet* 2013;**93**(6):1108–17.
82. Johnston JJ, Kelley RI, Crawford TO, Morton DH, Agarwala R, Koch T, et al. A novel nemaline myopathy in the Amish caused by a mutation in troponin T1. *Am J Hum Genet* 2000;**67**(4):814–21.
83. van der Pol WL, Leijenaar JF, Spliet WG, Lavrijsen SW, Jansen NJ, Braun KP, et al. Nemaline myopathy caused by TNNT1 mutations in a Dutch pedigree. *Mol Genet Genomic Med* 2014;**2**(2):134–7.
84. Gupta VA, Beggs AH. Kelch proteins: emerging roles in skeletal muscle development and diseases. *Skelet Muscle* 2014;**4**:11.
85. Lehtokari VL, Pelin K, Sandbacka M, Ranta S, Donner K, Muntoni F, et al. Identification of 45 novel mutations in the nebulin gene associated with autosomal recessive nemaline myopathy. *Hum Mutat* 2006;**27**(9):946–56.
86. Agrawal PB, Strickland CD, Midgett C, Morales A, Newburger DE, Poulos MA, et al. Heterogeneity of nemaline myopathy cases with skeletal muscle alpha-actin gene mutations. *Ann Neurol* 2004;**56**(1):86–96.
87. Lawlor MW, Dechene ET, Roumm E, Geggel AS, Moghadaszadeh B, Beggs AH. Mutations of tropomyosin 3 (TPM3) are common and associated with type 1 myofiber hypotrophy in congenital fiber type disproportion. *Hum Mutat* 2010;**31**(2):176–83.
88. Lehtokari VL, Greenleaf RS, DeChene ET, Kellinsalmi M, Pelin K, Laing NG, et al. The exon 55 deletion in the nebulin gene—one single founder mutation with world-wide occurrence. *Neuromuscul Disord: NMD* 2009;**19**(3):179–81.
89. Lawlor MW, Ottenheijm CA, Lehtokari VL, Cho K, Pelin K, Wallgren-Pettersson C, et al. Novel mutations in NEB cause abnormal nebulin expression and markedly impaired muscle force generation in severe nemaline myopathy. *Skelet Muscle* 2011;**1**(1):23.
90. Tan P, Briner J, Boltshauser E, Davis MR, Wilton SD, North K, et al. Homozygosity for a nonsense mutation in the alpha-tropomyosin slow gene TPM3 in a patient with severe infantile nemaline myopathy. *Neuromuscul Disord: NMD* 1999;**9**(8):573–9.
91. Clarke NF, Kolski H, Dye DE, Lim E, Smith RL, Patel R, et al. Mutations in TPM3 are a common cause of congenital fiber type disproportion. *Ann Neurol* 2008;**63**(3):329–37.
92. Goebel HH, Anderson JR, Hubner C, Oexle K, Warlo I. Congenital myopathy with excess of thin myofilaments. *Neuromuscul Disord: NMD* 1997;**7**(3):160–8.
93. Ilkovski B, Cooper ST, Nowak K, Ryan MM, Yang N, Schnell C, et al. Nemaline myopathy caused by mutations in the muscle alpha-skeletal-actin gene. *Am J Hum Genet* 2001;**68**(6):1333–43.
94. Stenzel W, Prokop S, Kress W, Huppmann S, Loui A, Sarioglu NM, et al. Fetal akinesia caused by a novel actin filament aggregate myopathy skeletal muscle actin gene (ACTA1) mutation. *Neuromuscul Disord: NMD* 2010;**20**(8):531–3.
95. Laing NG, Dye DE, Wallgren-Pettersson C, Richard G, Monnier N, Lillis S, et al. Mutations and polymorphisms of the skeletal muscle alpha-actin gene (ACTA1). *Hum Mutat* 2009;**30**(9):1267–77.
96. Citirak G, Witting N, Duno M, Werlauff U, Petri H, Vissing J. Frequency and phenotype of patients carrying TPM2 and TPM3 gene mutations in a cohort of 94 patients with congenital myopathy. *Neuromuscul Disord NMD* 2014;**24**(4):325–30.
97. Davidson AE, Siddiqui FM, Lopez MA, Lunt P, Carlson HA, Moore BE, et al. Novel deletion of lysine 7 expands the clinical, histopathological and genetic spectrum of TPM2-related myopathies. *Brain* 2013;**136**(Pt 2):508–21.
98. Mokbel N, Ilkovski B, Kreissl M, Memo M, Jeffries CM, Marttila M, et al. K7del is a common TPM2 gene mutation associated with nemaline myopathy and raised myofibre calcium sensitivity. *Brain* 2013;**136**(Pt 2):494–507.
99. Jungbluth H, Sewry CA, Muntoni F. Core myopathies. *Semin Pediatr Neurol* 2011;**18**(4):239–49.
100. Romero NB, Lehtokari VL, Quijano-Roy S, Monnier N, Claeys KG, Carlier RY, et al. Core-rod myopathy caused by mutations in the nebulin gene. *Neurology* 2009;**73**(14):1159–61.
101. Monnier N, Romero NB, Lerale J, Nivoche Y, Qi D, MacLennan DH, et al. An autosomal dominant congenital myopathy with cores and rods is associated with a neomutation in the RYR1 gene encoding the skeletal muscle ryanodine receptor. *Hum Mol Genet* 2000;**9**(18):2599–608.
102. Scacheri PC, Hoffman EP, Fratkin JD, Semino-Mora C, Senchak A, Davis MR, et al. A novel ryanodine receptor gene mutation causing both cores and rods in congenital myopathy. *Neurology* 2000;**55**(11):1689–96.
103. von der Hagen M, Kress W, Hahn G, Brocke KS, Mitzscherling P, Huebner A, et al. Novel RYR1 missense mutation causes core rod myopathy. *Eur J Neurol* 2008;**15**(4):e31–2.
104. Hernandez-Lain A, Husson I, Monnier N, Farnoux C, Brochier G, Lacene E, et al. De novo RYR1 heterozygous mutation (I4898T) causing lethal core-rod myopathy in twins. *Eur J Med Genet* 2011;**54**(1):29–33.
105. Greenfield JG, Cornman T, Shy GM. The prognostic value of the muscle biopsy in the floppy infant. *Brain* 1958;**81**(4):461–84.
106. Denborough MA, Dennett X, Anderson RM. Central-core disease and malignant hyperpyrexia. *Br Med J* 1973;**1**(5848):272–3.
107. Klein A, Lillis S, Munteanu I, Scoto M, Zhou H, Quinlivan R, et al. Clinical and genetic findings in a large cohort of patients with ryanodine receptor 1 gene-associated myopathies. *Hum Mutat* 2012;**33**(6):981–8.
108. Jungbluth H. Central core disease. *Orphanet J Rare Dis* 2007;**2**:25.
109. Bethlem J, van Gool J, Hulsmann WC, Meijer AE. Familial non-progressive myopathy with muscle cramps after exercise. A new disease associated with cores in the muscle fibres. *Brain* 1966;**89**(3):569–88.
110. Shuaib A, Paasuke RT, Brownell KW. Central core disease. Clinical features in 13 patients. *Medicine (Baltimore)* 1987;**66**(5):389–96.

111. Byrne E, Blumbergs PC, Hallpike JF. Central core disease. Study of a family with five affected generations. *J Neurol Sci* 1982;**53**(1):77–83.
112. Lynch PJ, Tong J, Lehane M, Mallet A, Giblin L, Heffron JJ, et al. A mutation in the transmembrane/luminal domain of the ryanodine receptor is associated with abnormal Ca^{2+} release channel function and severe central core disease. *Proc Natl Acad Sci USA* 1999;**96**(7):4164–9.
113. Manzur AY, Sewry CA, Ziprin J, Dubowitz V, Muntoni F. A severe clinical and pathological variant of central core disease with possible autosomal recessive inheritance. *Neuromuscul Disord: NMD* 1998;**8**(7):467–73.
114. Amburgey K, Bailey A, Hwang JH, Tarnopolsky MA, Bonnemann CG, Medne L, et al. Genotype-phenotype correlations in recessive RYR1-related myopathies. *Orphanet J Rare Dis* 2013;**8**:117.
115. Wedel DJ. Malignant hyperthermia and neuromuscular disease. *Neuromuscul Disord: NMD* 1992;**2**(3):157–64.
116. Rosenberg H, Davis M, James D, Pollock N, Stowell K. Malignant hyperthermia. *Orphanet J Rare Dis* 2007;**2**:21.
117. Malicdan MCV, Nishino I. Central Core Disease. In: Pagon RA, Adam MP, Ardinger HH, Bird TD, Dolan CR, Fong CT, et al., editors. *GeneReviews(R)*. Seattle (WA)1993.
118. Krivosic-Horber R, Krivosic I. Central core disease associated with malignant hyperthermia sensitivity. *Presse Med* 1989;**18**(16):828–31.
119. Monnier N, Procaccio V, Stieglitz P, Lunardi J. Malignant-hyperthermia susceptibility is associated with a mutation of the alpha 1-subunit of the human dihydropyridine-sensitive L-type voltage-dependent calcium-channel receptor in skeletal muscle. *Am J Hum Genet* 1997;**60**(6):1316–25.
120. Group EMH. A protocol for the invesitgation of malignant hyperpyrexia (MH) susceptibility. *Br J Anaesth* 1984;**56**:1267–9.
121. Robinson R, Carpenter D, Shaw MA, Halsall J, Hopkins P. Mutations in RYR1 in malignant hyperthermia and central core disease. *Hum Mutat* 2006;**27**(10):977–89.
122. Dowling JJ, Lillis S, Amburgey K, Zhou H, Al-Sarraj S, Buk SJ, et al. King-Denborough syndrome with and without mutations in the skeletal muscle ryanodine receptor (RYR1) gene. *Neuromuscul Disord: NMD* 2011;**21**(6):420–7.
123. Stamm DS, Aylsworth AS, Stajich JM, Kahler SG, Thorne LB, Speer MC, et al. Native American myopathy: congenital myopathy with cleft palate, skeletal anomalies, and susceptibility to malignant hyperthermia. *Am J Med Genet A* 2008;**146A**(14):1832–41.
124. Horstick EJ, Linsley JW, Dowling JJ, Hauser MA, McDonald KK, Ashley-Koch A, et al. Stac3 is a component of the excitation-contraction coupling machinery and mutated in Native American myopathy. *Nat Commun* 2013;**4**:1952.
125. Cruz Martinez A, Ferrer MT, Lopez-Terradas JM, Pascual-Castroviejo I, Mingo P. Single fibre electromyography in central core disease. *J Neurol Neurosurg Psychiatry* 1979;**42**(7):662–7.
126. Klein A, Jungbluth H, Clement E, Lillis S, Abbs S, Munot P, et al. Muscle magnetic resonance imaging in congenital myopathies due to ryanodine receptor type 1 gene mutations. *Arch Neurol* 2011;**68**(9):1171–9.
127. De Bleecker JL, Ertl BB, Engel AG. Patterns of abnormal protein expression in target formations and unstructured cores. *Neuromuscul Disord: NMD* 1996;**6**(5):339–49.
128. Thomas C. Nemaline rod and central core disease: a coexisting Z-band myopathy. *Muscle Nerve* 1997;**20**(7):893–6.
129. Hayashi K, Miller RG, Brownell AK. Central core disease: ultrastructure of the sarcoplasmic reticulum and T-tubules. *Muscle Nerve* 1989;**12**(2):95–102.
130. Monnier N, Romero NB, Lerale J, Landrieu P, Nivoche Y, Fardeau M, et al. Familial and sporadic forms of central core disease are associated with mutations in the C-terminal domain of the skeletal muscle ryanodine receptor. *Hum Mol Genet* 2001;**10**(22):2581–92.
131. Quane KA, Healy JM, Keating KE, Manning BM, Couch FJ, Palmucci LM, et al. Mutations in the ryanodine receptor gene in central core disease and malignant hyperthermia. *Nat Genet* 1993;**5**(1):51–5.
132. Clarke NF, Amburgey K, Teener J, Camelo-Piragua S, Kesari A, Punetha J, et al. A novel mutation expands the genetic and clinical spectrum of MYH7-related myopathies. *Neuromuscul Disord: NMD* 2013;**23**(5):432–6.
133. Fananapazir L, Dalakas MC, Cyran F, Cohn G, Epstein ND. Missense mutations in the beta-myosin heavy-chain gene cause central core disease in hypertrophic cardiomyopathy. *Proc Natl Acad Sci USA* 1993;**90**(9):3993–7.
134. Zhou H, Lillis S, Loy RE, Ghassemi F, Rose MR, Norwood F, et al. Multi-minicore disease and atypical periodic paralysis associated with novel mutations in the skeletal muscle ryanodine receptor (RYR1) gene. *Neuromuscul Disord: NMD* 2010;**20**(3):166–73.
135. Duarte ST, Oliveira J, Santos R, Pereira P, Barroso C, Conceicao I, et al. Dominant and recessive RYR1 mutations in adults with core lesions and mild muscle symptoms. *Muscle Nerve* 2011;**44**(1):102–8.
136. Wilmshurst JM, Lillis S, Zhou H, Pillay K, Henderson H, Kress W, et al. RYR1 mutations are a common cause of congenital myopathies with central nuclei. *Ann Neurol* 2010;**68**(5):717–26.
137. DeChene ET, Kang PB, Beggs AH. Congenital fiber-type disproportion. In: Pagon RA, Adam MP, Bird TD, Dolan CR, Fong CT, Smith RJH, et al., editors. *GeneReviews(R)*. Seattle (WA)1993.
138. Clarke NF, Waddell LB, Cooper ST, Perry M, Smith RL, Kornberg AJ, et al. Recessive mutations in RYR1 are a common cause of congenital fiber type disproportion. *Hum Mutat* 2010;**31**(7):E1544–50.
139. Bharucha-Goebel DX, Santi M, Medne L, Zukosky K, Dastgir J, Shieh PB, et al. Severe congenital RYR1-associated myopathy: the expanding clinicopathologic and genetic spectrum. *Neurology* 2013;**80**(17):1584–9.
140. Jungbluth H, Zhou H, Hartley L, Halliger-Keller B, Messina S, Longman C, et al. Minicore myopathy with ophthalmoplegia caused by mutations in the ryanodine receptor type 1 gene. *Neurology* 2005;**65**(12):1930–5.
141. Jungbluth H. Multi-minicore Disease. *Orphanet J Rare Dis* 2007;**2**:31.
142. Attali R, Aharoni S, Treves S, Rokach O, Becker Cohen M, Fellig Y, et al. Variable myopathic presentation in a single family with novel skeletal RYR1 mutation. *PLOS ONE* 2013;**8**(7):e69296.

143. Dlamini N, Voermans NC, Lillis S, Stewart K, Kamsteeg EJ, Drost G, et al. Mutations in RYR1 are a common cause of exertional myalgia and rhabdomyolysis. *Neuromuscul Disord: NMD* 2013;**23**(7):540–8.

144. Jungbluth H, Lillis S, Zhou H, Abbs S, Sewry C, Swash M, et al Late-onset axial myopathy with cores due to a novel heterozygous dominant mutation in the skeletal muscle ryanodine receptor (RYR1) gene. *Neuromuscul Disord: NMD* 2009;**19**(5):344–7.

145. Loseth S, Voermans NC, Torbergsen T, Lillis S, Jonsrud C, Lindal S, et al. A novel late-onset axial myopathy associated with mutations in the skeletal muscle ryanodine receptor (RYR1) gene. *J Neurol* 2013;**260**(6):1504–10.

146. Shaaban S, Ramos-Platt L, Gilles FH, Chan WM, Andrews C, De Girolami U, et al. RYR1 mutations as a cause of ophthalmoplegia, facial weakness, and malignant hyperthermia. *JAMA Ophthalmol* 2013;**131**(12):1532–40.

147. Engel AG, Gomez MR, Groover RV. Multicore disease. A recently recognized congenital myopathy associated with multifocal degeneration of muscle fibers. *Mayo Clin Proc* 1971;**46**(10):666–81.

148. Beggs AH, Agrawal PB. Multiminicore disease. In: Pagon RA, Adam MP, Ardinger HH, Bird TD, Dolan CR, Fong CT, et al., editors. *GeneReviews(R)*. Seattle (WA)1993.

149. Ferreiro A, Fardeau M. 80th ENMC International Workshop on Multi-Minicore Disease: 1st International MmD Workshop. 12–13th May, 2000, Soestduinen, The Netherlands. *Neuromuscul Disord: NMD* 2002;**12**(1):60–8.

150. Scoto M, Cirak S, Mein R, Feng L, Manzur AY, Robb S, et al. SEPN1-related myopathies: clinical course in a large cohort of patients. *Neurology* 2011;**76**(24):2073–8.

151. Bonnette H, Roelofs R, Olson WH. Multicore disease: report of a case with onset in middle age. *Neurology* 1974;**24**(11):1039–44.

152. Zeman AZ, Dick DJ, Anderson JR, Watkin SW, Smith IE, Shneerson JM. Multicore myopathy presenting in adulthood with respiratory failure. *Muscle Nerve* 1997;**20**(3):367–9.

153. Ferreiro A, Estournet B, Chateau D, Romero NB, Laroche C, Odent S, et al. Multi-minicore disease—searching for boundaries: phenotype analysis of 38 cases. *Ann Neurol* 2000;**48**(5):745–57.

154. Jungbluth H, Sewry C, Brown SC, Manzur AY, Mercuri E, Bushby K, et al. Minicore myopathy in children: a clinical and histopathological study of 19 cases. *Neuromuscul Disord: NMD* 2000;**10**(4–5):264–73.

155. Moghadaszadeh B, Petit N, Jaillard C, Brockington M, Quijano Roy S, Merlini L, et al. Mutations in SEPN1 cause congenital muscular dystrophy with spinal rigidity and restrictive respiratory syndrome. *Nat Genet* 2001;**29**(1):17–18.

156. Davis DG, Nelson KR, Markesbery WR. Congenital myopathy and cardiomyopathy with identical ultrastructural changes. *Arch Neurol* 1990;**47**(10):1141–4.

157. Shuaib A, Martin JM, Mitchell LB, Brownell AK. Multicore myopathy: not always a benign entity. *Can J Neurol Sci* 1988;**15**(1):10–14.

158. Koch BM, Bertorini TE, Eng GD, Boehm R. Severe multicore disease associated with reaction to anesthesia. *Arch Neurol* 1985;**42**(12):1204–6.

159. Pages M, Echenne B, Pages AM, Dimeglio A, Sires A. Multicore disease and Marfan's syndrome: a case report. *Eur Neurol* 1985;**24**(3):170–5.

160. Bonnemann CG, Thompson TG, van der Ven PF, Goebel HH, Warlo I, Vollmers B, et al. Filamin C accumulation is a strong but nonspecific immunohistochemical marker of core formation in muscle. *J Neurol Sci* 2003;**206**(1):71–8.

161. Ferreiro A, Quijano-Roy S, Pichereau C, Moghadaszadeh B, Goemans N, Bonnemann C, et al. Mutations of the selenoprotein N gene, which is implicated in rigid spine muscular dystrophy, cause the classical phenotype of multiminicore disease: reassessing the nosology of early-onset myopathies. *Am J Hum Genet* 2002;**71**(4):739–49.

162. Ferreiro A, Ceuterick-de Groote C, Marks JJ, Goemans N, Schreiber G, Hanefeld F, et al. Desmin-related myopathy with Mallory body-like inclusions is caused by mutations of the selenoprotein N gene. *Ann Neurol* 2004;**55**(5):676–86.

163. Clarke NF, Kidson W, Quijano-Roy S, Estournet B, Ferreiro A, Guicheney P, et al. SEPN1: associated with congenital fiber-type disproportion and insulin resistance. *Ann Neurol* 2006;**59**(3):546–52.

164. Cullup T, Lamont PJ, Cirak S, Damian MS, Wallefeld W, Gooding R, et al. Mutations in MYH7 cause Multi-minicore Disease (MmD) with variable cardiac involvement. *Neuromuscul Disord: NMD* 2012;**22**(12):1096–104.

165. Pellissier JF, de Barsy T, Faugere MC, Rebuffel P. Type III glycogenosis with multicore structures. *Muscle Nerve* 1979;**2**(2):124–32.

166. Tein I, Haslam RH, Rhead WJ, Bennett MJ, Becker LE, Vockley J. Short-chain acyl-CoA dehydrogenase deficiency: a cause of ophthalmoplegia and multicore myopathy. *Neurology* 1999;**52**(2):366–72.

167. Avoni P, Monari L, Carelli V, Carcangiu R, Barboni P, Donati C, et al. Congenital encephalomyopathy with epilepsy, chorioretinitis, basal ganglia involvement, and muscle minicores. *Ann Neurol* 2000;**47**(3):395–9.

168. Goebel HH, Meinck HM, Reinecke M, Schimrigk K, Mielke U. Centronuclear myopathy with special consideration of the adult form. *Eur Neurol* 1984;**23**(6):425–34.

169. Ceyhan-Birsoy O, Agrawal PB, Hidalgo C, Schmitz-Abe K, DeChene ET, Swanson LC, et al. Recessive truncating titin gene, TTN, mutations presenting as centronuclear myopathy. *Neurology* 2013;**81**(14):1205–14.

170. Brooke MH, Engel WK. The histographic analysis of human muscle biopsies with regard to fiber types. 4. Children's biopsies. *Neurology* 1969;**19**(6):591–605.

171. Brooke MH. A neuromuscular disease characterized by fiber type disproportion. In: Kakulas BA, editor. *Clinical Studies in Myology*. Excerpta Medica; 1973. pp. 147–59

172. Cavanagh NP, Lake BD, McMeniman P. Congenital fibre type disproportion myopathy. A histological diagnosis with an uncertain clinical outlook. *Arch Dis Child* 1979;**54**(10):735–43.

173. Kinoshita M, Satoyoshi E, Kumagai M. Familial type I fiber atrophy. *J Neurol Sci* 1975;**25**(1):11–17.

174. Iannaccone ST, Bove KE, Vogler CA, Buchino JJ. Type 1 fiber size disproportion: morphometric data from 37 children with myopathic, neuropathic, or idiopathic hypotonia. *Pediatr Pathol* 1987;**7**(4):395–419.

175. Munot P, Lashley D, Jungbluth H, Feng L, Pitt M, Robb SA, et al Congenital fibre type disproportion associated with mutations in the tropomyosin 3 (TPM3) gene mimicking congenital myasthenia. *Neuromuscul Disord* 2010;**20**(12):796–800.

176. Laing NG, Clarke NF, Dye DE, Liyanage K, Walker KR, Kobayashi Y, et al. Actin mutations are one cause of congenital fibre type disproportion. *Ann Neurol* 2004;**56**(5):689–94.
177. Kajino S, Ishihara K, Goto K, Ishigaki K, Noguchi S, Nonaka I, et al. Congenital fiber type disproportion myopathy caused by LMNA mutations. *J Neurol Sci* 2014;**340**(1–2):94–8.
178. Pierson CR, Tomczak K, Agrawal P, Moghadaszadeh B, Beggs AH. X-linked myotubular and centronuclear myopathies. *J Neuropathol Exp Neurol* 2005;**64**(7):555–64.
179. Jungbluth H, Wallgren-Pettersson C, Laporte J. Centronuclear (myotubular) myopathy. *Orphanet J Rare Dis* 2008;**3**:26.
180. Romero NB, Bitoun M. Centronuclear myopathies. *Semin Pediatr Neurol* 2011;**18**(4):250–6.
181. Biancalana V, Beggs AH, Das S, Jungbluth H, Kress W, Nishino I, et al. Clinical utility gene card for: Centronuclear and myotubular myopathies. *Eur J Hum Genet* 2012;**20**:10.
182. Cowling BS, Toussaint A, Muller J, Laporte J. Defective membrane remodeling in neuromuscular diseases: insights from animal models. *PLoS Genet* 2012;**8**(4):e1002595.
183. Spiro AJ, Shy GM, Gonatas NK. Myotubular myopathy. Persistence of fetal muscle in an adolescent boy. *Arch Neurol* 1966;**14**(1):1–14.
184. Das S, Dowling J, Pierson CR. X-Linked centronuclear myopathy. In: Pagon RA, Adam MP, Bird TD, Dolan CR, Fong CT, Smith RJH, et al., editors. *GeneReviews(R)*. Seattle (WA)1993.
185. Herman GE, Finegold M, Zhao W, de Gouyon B, Metzenberg A. Medical complications in long-term survivors with X-linked myotubular myopathy. *J Pediatr* 1999;**134**(2):206–14.
186. McEntagart M, Parsons G, Buj-Bello A, Biancalana V, Fenton I, Little M, et al. Genotype-phenotype correlations in X-linked myotubular myopathy. *Neuromuscul Disord: NMD* 2002;**12**(10):939–46.
187. Heckmatt JZ, Sewry CA, Hodes D, Dubowitz V. Congenital centronuclear (myotubular) myopathy. A clinical, pathological and genetic study in eight children. *Brain* 1985;**108**(Pt 4):941–64.
188. Barth PG, Dubowitz V. X-linked myotubular myopathy—a long-term follow-up study. *Eur J Paediatr Neurol* 1998;**2**(1):49–56.
189. Pierson CR, Agrawal PB, Blasko J, Beggs AH. Myofiber size correlates with MTM1 mutation type and outcome in X-linked myotubular myopathy. *Neuromuscul Disord: NMD* 2007;**17**(7):562–8.
190. Wallgren-Petterson C, Clarke A. Myotubular myopathy. In: Emery AEH, editor. *Neuromuscular Disorders: Clinical and Molecular Genetics*. Chichester: Wiley; 1998. pp. 263–76.
191. Wallgren-Pettersson C, Thomas NS. Report on the 20th ENMC sponsored international workshop: myotubular/centronuclear myopathy. *Neuromuscul Disord: NMD* 1994;**4**(1):71–4.
192. Mejaddam AY, Nennesmo I, Sejersen T. Severe phenotype of a patient with autosomal recessive centronuclear myopathy due to a BIN1 mutation. *Acta Myol* 2009;**28**(3):91–3.
193. Jeannet PY, Bassez G, Eymard B, Laforet P, Urtizberea JA, Rouche A, et al. Clinical and histologic findings in autosomal centronuclear myopathy. *Neurology* 2004;**62**(9):1484–90.
194. Fischer D, Herasse M, Bitoun M, Barragan-Campos HM, Chiras J, Laforet P, et al. Characterization of the muscle involvement in dynamin 2-related centronuclear myopathy. *Brain* 2006;**129**(Pt 6):1463–9.
195. McLeod JG, Baker Wde C, Lethlean AK, Shorey CD. Centronuclear myopathy with autosomal dominant inheritance. *J Neurol Sci* 1972;**15**(4):375–87.
196. Wallgren-Pettersson C, Clarke A, Samson F, Fardeau M, Dubowitz V, Moser H, et al. The myotubular myopathies: differential diagnosis of the X linked recessive, autosomal dominant, and autosomal recessive forms and present state of DNA studies. *J Med Genet* 1995;**32**(9):673–9.
197. Catteruccia M, Fattori F, Codemo V, Ruggiero L, Maggi L, Tasca G, et al. Centronuclear myopathy related to dynamin 2 mutations: clinical, morphological, muscle imaging and genetic features of an Italian cohort. *Neuromuscul Disord: NMD* 2013;**23**(3):229–38.
198. Shichiji M, Biancalana V, Fardeau M, Hogrel JY, Osawa M, Laporte J, et al. Extensive morphological and immunohistochemical characterization in myotubular myopathy. *Brain Behav* 2013;**3**(4):476–86.
199. Bevilacqua JA, Bitoun M, Biancalana V, Oldfors A, Stoltenburg G, Claeys KG, et al. "Necklace" fibers, a new histological marker of late-onset MTM1-related centronuclear myopathy. *Acta Neuropathol* 2009;**117**(3):283–91.
200. Gurgel-Giannetti J, Zanoteli E, de Castro Concentino EL, Abath Neto O, Pesquero JB, Reed UC, et al. Necklace fibers as histopathological marker in a patient with severe form of X-linked myotubular myopathy. *Neuromuscul Disord: NMD* 2012;**22**(6):541–5.
201. Liewluck T, Lovell TL, Bite AV, Engel AG. Sporadic centronuclear myopathy with muscle pseudohypertrophy, neutropenia, and necklace fibers due to a DNM2 mutation. *Neuromuscul Disord: NMD* 2010;**20**(12):801–4.
202. Hanisch F, Muller T, Dietz A, Bitoun M, Kress W, Weis J, et al. Phenotype variability and histopathological findings in centronuclear myopathy due to DNM2 mutations. *J Neurol* 2011;**258**(6):1085–90.
203. Wockel L, Ketelsen UP, Stotter M, Laule S, Meyermann R, Bornemann A. Abundant minute myotubes in a patient who later developed centronuclear myopathy. *Acta Neuropathol* 1998;**95**(5):547–51.
204. Laporte J, Hu LJ, Kretz C, Mandel JL, Kioschis P, Coy JF, et al. A gene mutated in X-linked myotubular myopathy defines a new putative tyrosine phosphatase family conserved in yeast. *Nat Genet* 1996;**13**(2):175–82.
205. Oliveira J, Oliveira ME, Kress W, Taipa R, Pires MM, Hilbert P, et al. Expanding the MTM1 mutational spectrum: novel variants including the first multi-exonic duplication and development of a locus-specific database. *Eur J Hum Genet: EJHG* 2013;**21**(5):540–9.
206. Bitoun M, Maugenre S, Jeannet PY, Lacene E, Ferrer X, Laforet P, et al. Mutations in dynamin 2 cause dominant centronuclear myopathy. *Nat Genet* 2005;**37**(11):1207–9.
207. Nicot AS, Toussaint A, Tosch V, Kretz C, Wallgren-Pettersson C, Iwarsson E, et al. Mutations in amphiphysin 2 (BIN1) disrupt interaction with dynamin 2 and cause autosomal recessive centronuclear myopathy. *Nat Genet* 2007;**39**(9):1134–9.
208. Tosch V, Rohde HM, Tronchere H, Zanoteli E, Monroy N, Kretz C, et al. A novel PtdIns3P and PtdIns(3,5)P2 phosphatase with an inactivating variant in centronuclear myopathy. *Hum Mol Genet* 2006;**15**(21):3098–106.

209. Al-Qusairi L, Weiss N, Toussaint A, Berbey C, Messaddeq N, Kretz C, et al. T-tubule disorganization and defective excitation-contraction coupling in muscle fibers lacking myotubularin lipid phosphatase. *Proc Natl Acad Sci USA* 2009;**106**(44):18763–8.
210. Dowling JJ, Vreede AP, Low SE, Gibbs EM, Kuwada JY, Bonnemann CG, et al. Loss of myotubularin function results in T-tubule disorganization in zebrafish and human myotubular myopathy. *PLoS Genet* 2009;**5**(2):e1000372.
211. Al-Qusairi L, Laporte J. T-tubule biogenesis and triad formation in skeletal muscle and implication in human diseases. *Skelet Muscle* 2011;**1**(1):26.
212. Cowling BS, Toussaint A, Amoasii L, Koebel P, Ferry A, Davignon L, et al. Increased expression of wild-type or a centronuclear myopathy mutant of dynamin 2 in skeletal muscle of adult mice leads to structural defects and muscle weakness. *Am J Pathol* 2011;**178**(5):2224–35.
213. Gibbs EM, Davidson AE, Telfer WR, Feldman EL, Dowling JJ. The myopathy-causing mutation DNM2-S619L leads to defective tubulation in vitro and in developing zebrafish. *Dis Model Mech* 2014;**7**(1):157–61.
214. Fugier C, Klein AF, Hammer C, Vassilopoulos S, Ivarsson Y, Toussaint A, et al. Misregulated alternative splicing of BIN1 is associated with T tubule alterations and muscle weakness in myotonic dystrophy. *Nat Med* 2011;**17**(6):720–5.
215. Hackman P, Vihola A, Haravuori H, Marchand S, Sarparanta J, De Seze J, et al. Tibial muscular dystrophy is a titinopathy caused by mutations in TTN, the gene encoding the giant skeletal-muscle protein titin. *Am J Hum Genet* 2002;**71**(3):492–500.
216. Udd B, Partanen J, Halonen P, Falck B, Hakamies L, Heikkila H, et al. Tibial muscular dystrophy. Late adult-onset distal myopathy in 66 Finnish patients. *Arch Neurol* 1993;**50**(6):604–8.
217. Hackman P, Marchand S, Sarparanta J, Vihola A, Penisson-Besnier I, Eymard B, et al. Truncating mutations in C-terminal titin may cause more severe tibial muscular dystrophy (TMD). *Neuromuscul Disord : NMD* 2008;**18**(12):922–8.
218. Dowling JJ. Titin and centronuclear myopathy: The tip of the iceberg for TTN-ic mutations? *Neurology* 2013;**81**(14):1189–90.
219. Chauveau C, Bonnemann CG, Julien C, Kho AL, Marks H, Talim B, et al. Recessive TTN truncating mutations define novel forms of core myopathy with heart disease. *Hum Mol Genet* 2014;**23**(4):980–91.
220. Evila A, Vihola A, Sarparanta J, Raheem O, Palmio J, Sandell S, et al. Atypical phenotypes in titinopathies explained by second titin mutations. *Ann Neurol* 2014;**75**(2):230–40.
221. Selcen D, Engel AG. Myofibrillar myopathy. In: Pagon RA, Adam MP, Bird TD, Dolan CR, Fong CT, Smith RJH, et al., editors. *GeneReviews(R)*. Seattle (WA), 1993.
222. Fidzianska A, Goebel HH, Osborn M, Lenard HG, Osse G, Langenbeck U. Mallory body-like inclusions in a hereditary congenital neuromuscular disease. *Muscle Nerve* 1983;**6**(3):195–200.
223. De Bleecker JL, Engel AG, Ertl BB. Myofibrillar myopathy with abnormal foci of desmin positivity. II. Immunocytochemical analysis reveals accumulation of multiple other proteins. *J Neuropathol Exp Neurol* 1996;**55**(5):563–77.
224. Nakano S, Engel AG, Waclawik AJ, Emslie-Smith AM, Busis NA. Myofibrillar myopathy with abnormal foci of desmin positivity. I. Light and electron microscopy analysis of 10 cases. *J Neuropathol Exp Neurol* 1996;**55**(5):549–62.
225. Dalakas MC, Park KY, Semino-Mora C, Lee HS, Sivakumar K, Goldfarb LG. Desmin myopathy, a skeletal myopathy with cardiomyopathy caused by mutations in the desmin gene. *N Engl J Med* 2000;**342**(11):770–80.
226. Goebel HH, Muller J, Gillen HW, Merritt AD. Autosomal dominant "spheroid body myopathy." *Muscle Nerve* 1978;**1**(1):14–26.
227. Prelle A, Moggio M, Comi GP, Gallanti A, Checcarelli N, Bresolin N, et al. Congenital myopathy associated with abnormal accumulation of desmin and dystrophin. *Neuromuscul Disord: NMD* 1992;**2**(3):169–75.
228. Sabatelli M, Bertini E, Ricci E, Salviati G, Magi S, Papacci M, et al. Peripheral neuropathy with giant axons and cardiomyopathy associated with desmin type intermediate filaments in skeletal muscle. *J Neurol Sci* 1992;**109**(1):1–10.
229. Vajsar J, Becker LE, Freedom RM, Murphy EG. Familial desminopathy: myopathy with accumulation of desmin-type intermediate filaments. *J Neurol Neurosurg Psychiatry* 1993;**56**(6):644–8.
230. Fidzianska A, Ryniewicz B, Barcikowska M, Goebel HH. A new familial congenital myopathy in children with desmin and dystrophin reacting plaques. *J Neurol Sci* 1995;**131**(1):88–95.
231. Osborn M, Goebel HH. The cytoplasmic bodies in a congenital myopathy can be stained with antibodies to desmin, the muscle-specific intermediate filament protein. *Acta Neuropathol* 1983;**62**(1–2):149–52.
232. Rappaport L, Contard F, Samuel JL, Delcayre C, Marotte F, Tome F, et al. Storage of phosphorylated desmin in a familial myopathy. *FEBS Lett* 1988;**231**(2):421–5.
233. Caron A, Chapon F, Berthelin C, Viader F, Lechevalier B. Inclusions in familial cytoplasmic body myopathy are stained by anti-dystrophin antibodies. *Neuromuscul Disord: NMD* 1993;**3**(5–6):541–6.
234. Helliwell TR, Green AR, Green A, Edwards RH. Hereditary distal myopathy with granulo-filamentous cytoplasmic inclusions containing desmin, dystrophin and vimentin. *J Neurol Sci* 1994;**124**(2):174–87.
235. Goebel HH, Voit T, Warlo I, Jacobs K, Johannsen U, Muller CR. Immunohistologic and electron microscopic abnormalities of desmin and dystrophin in familial cardiomyopathy and myopathy. *Rev Neurol* 1994;**150**(6–7):452–9.
236. Pinol-Ripoll G, Shatunov A, Cabello A, Larrode P, de la Puerta I, Pelegrin J, et al. Severe infantile-onset cardiomyopathy associated with a homozygous deletion in desmin. *Neuromuscul Disord: NMD* 2009;**19**(6):418–22.
237. Del Bigio MR, Chudley AE, Sarnat HB, Campbell C, Goobie S, Chodirker BN, et al. Infantile muscular dystrophy in Canadian aboriginals is an alphaB-crystallinopathy. *Ann Neurol* 2011;**69**(5):866–71.
238. Forrest KM, Al-Sarraj S, Sewry C, Buk S, Tan SV, Pitt M, et al. Infantile onset myofibrillar myopathy due to recessive CRYAB mutations. *Neuromuscul Disord: NMD* 2011;**21**(1):37–40.
239. Cameron CH, Mirakhur M, Allen IV. Desmin myopathy with cardiomyopathy. *Acta Neuropathol* 1995;**89**(6):560–6.
240. Olive M, Odgerel Z, Martinez A, Poza JJ, Bragado FG, Zabalza RJ, et al. Clinical and myopathological evaluation of early- and late-onset subtypes of myofibrillar myopathy. *Neuromuscul Disord: NMD* 2011;**21**(8):533–42.

241. Winter L, Wiche G. The many faces of plectin and plectinopathies: pathology and mechanisms. *Acta Neuropathol* 2013;**125**(1):77–93.
242. Furst DO, Goldfarb LG, Kley RA, Vorgerd M, Olive M, van der Ven PF. Filamin C-related myopathies: pathology and mechanisms. *Acta Neuropathol* 2013;**125**(1):33–46.
243. Goebel HH, Borchert A. Protein surplus myopathies and other rare congenital myopathies. *Semin Pediatr Neurol* 2002;**9**(2):160–70.
244. Goldfarb LG, Park KY, Cervenakova L, Gorokhova S, Lee HS, Vasconcelos O, et al. Missense mutations in desmin associated with familial cardiac and skeletal myopathy. *Nat Genet* 1998;**19**(4):402–3.
245. Munoz-Marmol AM, Strasser G, Isamat M, Coulombe PA, Yang Y, Roca X, et al. A dysfunctional desmin mutation in a patient with severe generalized myopathy. *Proc Natl Acad Sci USA* 1998;**95**(19):11312–17.
246. Sjoberg G, Saavedra-Matiz CA, Rosen DR, Wijsman EM, Borg K, Horowitz SH, et al. A missense mutation in the desmin rod domain is associated with autosomal dominant distal myopathy, and exerts a dominant negative effect on filament formation. *Hum Mol Genet* 1999;**8**(12):2191–8.
247. Selcen D, Muntoni F, Burton BK, Pegoraro E, Sewry C, Bite AV, et al. Mutation in BAG3 causes severe dominant childhood muscular dystrophy. *Ann Neurol* 2009;**65**(1):83–9.
248. Odgerel Z, Sarkozy A, Lee HS, McKenna C, Rankin J, Straub V, et al. Inheritance patterns and phenotypic features of myofibrillar myopathy associated with a BAG3 mutation. *Neuromuscul Disord: NMD* 2010;**20**(7):438–42.
249. Lee HC, Cherk SW, Chan SK, Wong S, Tong TW, Ho WS, et al. BAG3-related myofibrillar myopathy in a Chinese family. *Clin Genet* 2012;**81**(4):394–8.
250. Jaffer F, Murphy SM, Scoto M, Healy E, Rossor AM, Brandner S, et al. BAG3 mutations: another cause of giant axonal neuropathy. *J Peripher Nerv Syst* 2012;**17**(2):210–16.
251. Vicart P, Caron A, Guicheney P, Li Z, Prevost MC, Faure A, et al. A missense mutation in the alphaB-crystallin chaperone gene causes a desmin-related myopathy. *Nat Genet* 1998;**20**(1):92–5.
252. Cancilla PA, Kalyanaraman K, Verity MA, Munsat T, Pearson CM. Familial myopathy with probable lysis of myofibrils in type I fibers. *Neurology* 1971;**21**(6):579–85.
253. Ceuterick C, Martin JJ, Martens C. Hyaline bodies in skeletal muscle of a patient with a mild chronic nonprogressive congenital myopathy. *Clin Neuropathol* 1993;**12**(2):79–83.
254. Masuzugawa S, Kuzuhara S, Narita Y, Naito Y, Taniguchi A, Ibi T. Autosomal dominant hyaline body myopathy presenting as scapuloperoneal syndrome: clinical features and muscle pathology. *Neurology* 1997;**48**(1):253–7.
255. Tajsharghi H, Oldfors A, Swash M. Myosin storage myopathy with cardiomyopathy. *Neuromuscul Disord: NMD* 2007;**17**(9–10):725.
256. Barohn RJ, Brumback RA, Mendell JR. Hyaline body myopathy. *Neuromuscul Disord: NMD* 1994;**4**(3):257–62.
257. Bohlega S, Abu-Amero SN, Wakil SM, Carroll P, Al-Amr R, Lach B, et al. Mutation of the slow myosin heavy chain rod domain underlies hyaline body myopathy. *Neurology* 2004;**62**(9):1518–21.
258. Tajsharghi H, Thornell LE, Lindberg C, Lindvall B, Henriksson KG, Oldfors A. Myosin storage myopathy associated with a heterozygous missense mutation in MYH7. *Ann Neurol* 2003;**54**(4):494–500.
259. Laing NG, Ceuterick-de Groote C, Dye DE, Liyanage K, Duff RM, Dubois B, et al. Myosin storage myopathy: slow skeletal myosin (MYH7) mutation in two isolated cases. *Neurology* 2005;**64**(3):527–9.
260. Shingde MV, Spring PJ, Maxwell A, Wills EJ, Harper CG, Dye DE, et al. Myosin storage (hyaline body) myopathy: a case report. *Neuromuscul Disord: NMD* 2006;**16**(12):882–6.
261. Tajsharghi H, Oldfors A. Myosinopathies: pathology and mechanisms. *Acta Neuropathol* 2013;**125**(1):3–18.
262. Nishino I. Autophagic vacuolar myopathy. *Semin Pediatr Neurol* 2006;**13**(2):90–5.
263. Malicdan MC, Noguchi S, Nishino I. Recent advances in distal myopathy with rimmed vacuoles (DMRV) or hIBM: treatment perspectives. *Curr Opin Neurol* 2008;**21**(5):596–600.
264. Park YE, Kim HS, Choi ES, Shin JH, Kim SY, Son EH, et al. Limb-girdle phenotype is frequent in patients with myopathy associated with GNE mutations. *J Neurol Sci* 2012;**321**(1–2):77–81.
265. Yang Z, Vatta M. Danon disease as a cause of autophagic vacuolar myopathy. *Congenit Heart Dis* 2007;**2**(6):404–9.
266. Yamamoto A, Morisawa Y, Verloes A, Murakami N, Hirano M, Nonaka I, et al. Infantile autophagic vacuolar myopathy is distinct from Danon disease. *Neurology* 2001;**57**(5):903–5.
267. Wang CH, Dowling JJ, North K, Schroth MK, Sejersen T, Shapiro F, et al. Consensus statement on standard of care for congenital myopathies. *J Child Neurol* 2012;**27**(3):363–82.
268. Asai T, Fujise K, Uchida M. Anaesthesia for cardiac surgery in children with nemaline myopathy. *Anaesthesia* 1992;**47**(5):405–8.
269. Rutgers M, Lucassen H, Kesteren RV, Leger P. Respiratory insufficiency and ventilatory support. 39th ENMC International Workshop, Naarden, The Netherlands, 26–28 January 1996. European Consortium on Chronic Respiratory Insufficiency. *Neuromuscul Disord: NMD* 1996;**6**(6):431–5.
270. Kilmer DD, McDonald CM. Childhood progressive neuromuscular disease. In: Goldberg B, editor. *Sports and Exercise in Children with Chronic Health Conditions*. Champaign, IL: Human Kinetics; 1995. pp. 109–21.
271. Goebel HH, Anderson JR. Structural congenital myopathies (excluding nemaline myopathy, myotubular myopathy and desminopathies): 56th European Neuromuscular Centre (ENMC) sponsored International Workshop. December 12–14, 1997, Naarden, The Netherlands. *Neuromuscul Disord: NMD* 1999;**9**(1):50–7.
272. Robb SA, Sewry CA, Dowling JJ, Feng L, Cullup T, Lillis S, et al. Impaired neuromuscular transmission and response to acetylcholinesterase inhibitors in centronuclear myopathies. *Neuromuscul Disord: NMD* 2011;**21**(6):379–86.
273. Gibbs EM, Clarke NF, Rose K, Oates EC, Webster R, Feldman EL, et al. Neuromuscular junction abnormalities in DNM2-related centronuclear myopathy. *J Mol Med* 2013;**91**(6):727–37.
274. Ryan MM, Sy C, Rudge S, Ellaway C, Ketteridge D, Roddick LG, et al. Dietary L-tyrosine supplementation in nemaline myopathy. *J Child Neurol* 2008;**23**(6):609–13.

275. Nguyen MA, Joya JE, Kee AJ, Domazetovska A, Yang N, Hook JW, et al. Hypertrophy and dietary tyrosine ameliorate the phenotypes of a mouse model of severe nemaline myopathy. *Brain* 2011;**134**(Pt 12):3516–29.
276. Michalek-Sauberer A, Gilly H. Prophylactic use of dantrolene in a patient with central core disease. *Anesth Analg* 1998;**86**(4):915–16.
277. Messina S, Hartley L, Main M, Kinali M, Jungbluth H, Muntoni F, Mercuri E. Pilot trial of salbutamol in central core and multi-minicore diseases. *Neuropediatrics* 2004;**35**(5):262–6.
278. Telfer WR, Nelson DD, Waugh T, Brooks SV, Dowling JJ. Neb: a zebrafish model of nemaline myopathy due to nebulin mutation *Dis Model Mech* 2012;**5**(3):389–96.
279. Ottenheijm CA, Buck D, de Winter JM, Ferrara C, Piroddi N, Tesi C, et al. Deleting exon 55 from the nebulin gene induces severe muscle weakness in a mouse model for nemaline myopathy. *Brain* 2013;**136**(Pt 6):1718–31.
280. Smith LL, Gupta VA, Beggs AH. Bridging integrator 1 (Bin1) deficiency in zebrafish results in centronuclear myopathy. *Hum Mol Genet* 2014;**23**(13):3566–78.
281. Childers MK, Joubert R, Poulard K, Moal C, Grange RW, Doering JA, et al. Gene therapy prolongs survival and restores function in murine and canine models of myotubular myopathy. *Sci Transl Med* 2014;**6**(220): 220ra10.
282. Lawlor MW, Armstrong D, Viola MG, Widrick JJ, Meng H, Grange RW, et al. Enzyme replacement therapy rescues weakness and improves muscle pathology in mice with X-linked myotubular myopathy. *Hum Mol Genet* 2013;**22**(8):1525–38.
283. Cathomen T, Ehl S. Translating the genomic revolution—targeted genome editing in primates. *N Engl J Med* 2014;**370**(24): 2342–5.
284. Jungbluth H, Wallgren-Pettersson C, Laporte JF. 198th ENMC International Workshop: 7th Workshop on Centronuclear (Myotubular) myopathies, 31st May–2nd June 2013, Naarden, The Netherlands. *Neuromuscul Disord: NMD* 2013;**23**(12):1033–43.
285. Jungbluth H, Muntoni F, Ferreiro A. 150th ENMC International Workshop: Core Myopathies, 9–11th March 2007, Naarden, The Netherlands. *Neuromuscul Disord: NMD* 2008;**18**(12): 989–96.

Chapter 29

Congenital Muscular Dystrophies

Eugenio Mercuri and Francesco Muntoni

INTRODUCTION

The identification of several new genes responsible for different forms of congenital muscular dystrophies (CMD) has dramatically expanded our knowledge of the spectrum of these conditions and their pathogenesis, allowing a better characterization of the individual forms and of the specific pathomechanisms underlying them.

Definition of CMD

Classically, the term *congenital muscular dystrophy* includes a group of genetically, clinically, and biochemically distinct entities sharing clinical and pathological features such as early presentation of weakness and hypotonia and dystrophic features on muscle biopsy.[1,2] The term *congenital* implies origin of the pathological process during fetal life, which distinguishes these variants from later onset, postnatal conditions such as Duchenne muscular dystrophy (DMD) and limb girdle muscular dystrophies (LGMDs). It is nevertheless appreciated that the onset of the *clinical* symptoms might be delayed as not all affected children have frank symptoms at birth, but they will start developing weakness, contractures, and delayed motor milestones in the first year of life.

This "expanded" definition can be confusing, however, when assessing children in whom allelic mutations result in a broad spectrum of clinical syndromes, ranging from severe CMDs with clear prenatal onset to mild LGMD variants with onset in childhood, adolescence, or adulthood. This is particularly confusing when the severity of their conditions overlaps between CMD and LGMD and in the presence of central nervous system involvement, when delayed motor milestones could be attributed to muscle weakness, or global developmental delay, or a combination of the two. This is best exemplified by mutations in genes responsible for secondary dystroglycanopathy, such as *FKRP, FKTN, POMT1, POMT2*, or *ISPD*. Many authors have therefore followed a pragmatic approach that includes in the CMD definition children with onset of weakness prenatally or before ambulation is acquired, while LGMD, in contrast, refers to all other cases with later onset weakness, with overt clinical signs after ambulation has been acquired at an appropriate age.[3]

The definition of CMD should also take into account that some disorders classically labeled as CMD do not always have the typical dystrophic pattern but only milder myopathic changes, at least in the muscle available for the routine diagnostic procedures. Indeed, in some conditions such as rigid spine syndrome or rigid spine muscular dystrophy 1 (RSMD1), a phase of "myopathic" changes can precede dystrophic abnormalities, or some muscles (such as the paraspinal muscles) may have more severe involvement than others.

A comprehensive definition of CMD could therefore be that of a heterogeneous group of genetically, clinically, and biochemically distinct entities sharing clinical and pathological features, with onset of clinical signs at birth or typically in the first 12 months. In some variants such as Ullrich CMD and SEPN1-related myopathies, the onset can be delayed even beyond the end of the first year. Muscle biopsy reveals a dystrophic or myopathic pattern without other diagnostic features suggestive of a structural congenital myopathy or a metabolic disorder.

Classification

The first attempt at a rational classification of CMD was proposed in 1994 by the International Consortium on CMD, which recommended recognizing forms of CMD with structural brain changes from a classical or "pure" CMD without structural brain abnormalities.[1] The original cornerstone in the CMD classification came after the identification of the deficiency of the extracellular matrix protein merosin (more specifically the laminin α2 chain of the merosin trimer) in a subset of CMD cases,[4,5]

B.T. Darras, H. Royden Jones, Jr., M.M. Ryan & D.C. De Vivo (Eds): Neuromuscular Disorders of Infancy, Childhood and Adolescence, Second edition.
DOI: http://dx.doi.org/10.1016/B978-0-12-417044-5.00029-9

followed shortly thereafter by the isolation of the responsible gene (*LAMA2*, encoding for the laminin α2 chain of the merosin trimer) on chromosome 6.[5] This allowed the subdivision of the "pure" form into two groups, merosin-deficient and merosin-positive CMD, according to the presence or deficiency of merosin. The term *merosin-positive* CMD then became commonly used to describe variants other than merosin-deficient CMD; however, this term is no longer used as the majority of patients in this category now have well described conditions, with specific patterns of abnormal protein expression and/or with specific primary genetic defects. The most common subtypes in this category are the Ullrich phenotype with distal laxity, RSMD1, and LMNA-CMD, well-recognized variants in which the genetic basis has been identified.

Another major breakthrough came in the last decade following the discovery that various forms of CMD are characterized by a profound depletion of the glycosylation of α-dystroglycan. The underlying genetic defects in these disorders, which are collectively described as *dystroglycanopathies*, are mutations in known or putative glycosyltransferase enzymes, which are involved in the process of constructing an unusual glycostructure on α-dystroglycan.[6]

In the last few years, CMDs have been classified according to combined clinical, genetic and pathological approaches[3]:

1. Forms of CMD due to mutations in genes encoding for structural proteins of the basal lamina or extracellular matrix or receptors for extracellular matrix proteins. This category includes mutations in the collagen 6 genes, laminin α2 (merosin), in integrin α7 and the more recent variant due to integrin α9 gene deficiency. Mutations in *DAG1* (dystroglycan 1 gene) have very recently also been described and clearly belong to this subgroup.[7]
2. Forms secondary to genes encoding for putative or confirmed glycosyltransferases that affect the glycosylation of α-dystroglycan. These include Fukuyama CMD, muscle-eye-brain (MEB) disease, Walker-Warburg syndrome (WWS), MDC1C and MDC1D, and other phenotypes associated with mutations in one of the 15 known genes.
3. Defects of nuclear envelope proteins (LMNA and nesprin).
4. Defects of proteins with thus far unknown function localized in the endoplasmic reticulum, which include the form of CMD with rigid spine syndrome secondary to mutations in the *SEPN1* gene.
5. CMD with mitochondrial structural abnormalities (CMDmt).

Table 29.1 provides a list of the known genes responsible for CMD and of the phenotypes associated with mutations in each of them. The locations of a number of proteins described in this chapter are shown in Figure 29.1. In this chapter, we will specifically concentrate on new findings related to diagnosis, clinical management, and differential diagnosis of the most common forms of CMD.

MEROSIN-DEFICIENT CMD

In a recent UK survey, merosin-deficient CMD (MDC1A) accounted for 22% of all the molecularly defined CMD subtypes, with a lower frequency than that reported in the early studies reporting this.[8] Similar figures have been reported in the USA[9] and in a recent Italian population study (Mercuri, personal communication). Patients with absent merosin invariably present at birth or in the first few weeks of life with hypotonia, weakness, and, in some cases, contractures. The maximal motor ability achieved is sitting or, in rare instances, walking with support. Weakness generally affects upper limbs more severely than lower limbs. There are often contractures such as flexion deformity at the hips, knees, elbows, and ankles, followed by rigidity and scoliosis of the spine. Another clinical sign is the progressive limitation of eye movements and more specifically of the upward gaze,[10] a feature typically noted in the second decade of life.

Respiratory function is invariably reduced in the first decade and night time hypoventilation is frequently observed in early childhood.[11] Feeding problems are also frequent.[12]

Diagnosis is generally performed after the onset of clinical signs, as the muscle biopsy shows a classical dystrophic picture and immunofluorescence techniques can readily demonstrate the reduction or absence of laminin α2 chain. In most cases, the protein is totally absent or only present in traces. A severe inflammatory infiltrate can sometimes be seen, and this explains why in the past some of these children received a diagnosis of infantile inflammatory myopathy.

Since a secondary, partial merosin deficiency can also be found in other forms of CMD, such as dystroglycanopathies, the diagnosis of merosin-deficient CMD should be genetically confirmed by studying the laminin α2 chain (*LAMA2*) gene, mapped to chromosome 6q22-23,[13] especially in cases in which there is residual protein expression.

In patients with mutations in the *LAMA2* gene and partial merosin deficiency, the severity of the phenotype is often milder compared to patients with absent merosin. Patients with partial primary merosin deficiency usually have normal motor milestones in the first years of life, and the onset of symptoms may be delayed and occur in the second decade[13]; these patients are also much less likely to require enteral feeding and ventilatory support.[11]

TABLE 29.1 Classification of Congenital Muscular Dystrophies: Clinical, Biochemical, and Genetic Characteristics

Biochemical Defect	Locus	Gene	Disease Phenotype(s)
Extracellular Matrix Proteins	6q22-23	*LAMA2*	Primary merosin deficiency (MDC1A)
	21q22.3 2q37	*COL6A1* *COL6A2* *COL6A3*	Ullrich CMD
External Sarcolemmal Proteins	12q13	*ITGA7*	Integrin α7-related CMD
	3p23-21	*ITGA9*	Integrin α9-related CMD
Dystroglycan and Glycosyltransferase Enzymes	9q34.1	*POMT1*	WWS, MEB, CMD with cerebellar involvement, CMD with mental retardation and microcephaly
	1q32-34	*POMGnT1*	WWS, MEB, CMD with cerebellar involvement
	14q24.3	*POMT2*	WWS, MEB, CMD with cerebellar involvement, CMD with mental retardation and microcephaly
	19q13.3	*FKRP*	WWS, MEB, CMD with cerebellar involvement, CMD with mental retardation and microcephaly, CMD with no mental retardation and normal brain MRI
	9q31	*FCMD*	Fukuyama CMD
	22q12.3-13.1	*LARGE*	WWS, MEB, white matter changes
	1q12-q21	*DPM2/DPM3*	CMD with mental retardation and severe epilepsy
	7p21.2	*ISPD*	WWS, LGMD
	3p22.1	*GTDC2*	WWS
	11q13.2	*B3GALNT2*	WWS, MEB
	3p21.23	*GMPPB*	CMD with mental retardation and severe epilepsy, LGMD
	3p21	*DAG1*	Primary dystroglycanopathy, LGMD with early onset and mental retardation, normal brain MRI
	8p11.21	*SGK196*	MEB
	1q42	–	MDC1B
Endoplasmic Reticulum Protein	1p35-36	*SEPN1*	CMD with spinal rigidity (RSMD1)
Nuclear Envelope Proteins	6q25	*SYNE1 (nesprin 1)*	CMD with adducted thumbs
	1q21.2	*LMNA*	Congenital laminopathy
Sarcolemmal and Mitochondrial Membrane Protein	22q13	*CHKB*	Mitochondrial CMD (CMDmt)

The classification is based on combined clinical, genetic, and pathological data.
Abbreviations: CMD, congenital muscular dystrophy; CMDmt, CMD with mitochondrial structural abnormalities; LGMD, limb girdle muscular dystrophy; MEB, muscle eye brain; MRI, magnetic resonance imaging; RSMD1, rigid spine muscular dystrophy 1; WWS, Walker-Warburg syndrome; MDC, muscular dystrophy congenital.

Patients with both complete or partial absence of merosin have diffuse white matter changes on brain magnetic resonance imaging (MRI) affecting both hemispheres (Figure 29.2A), with sparing of the internal capsule, corpus callosum, basal ganglia, thalami, and cerebellum.[14,15] These changes are best visualized on MRI after 6 months.[16] The changes on brain MRI are thought to be the result of an increase in water content rather than being a sign of demyelination and are generally not associated with clinical signs of CNS involvement. While until recently these changes were thought to be specific of MDC1A, the same pattern has also been found in patients with dystroglycanopathies who also often have secondary partial merosin deficiency on muscle biopsy. In these

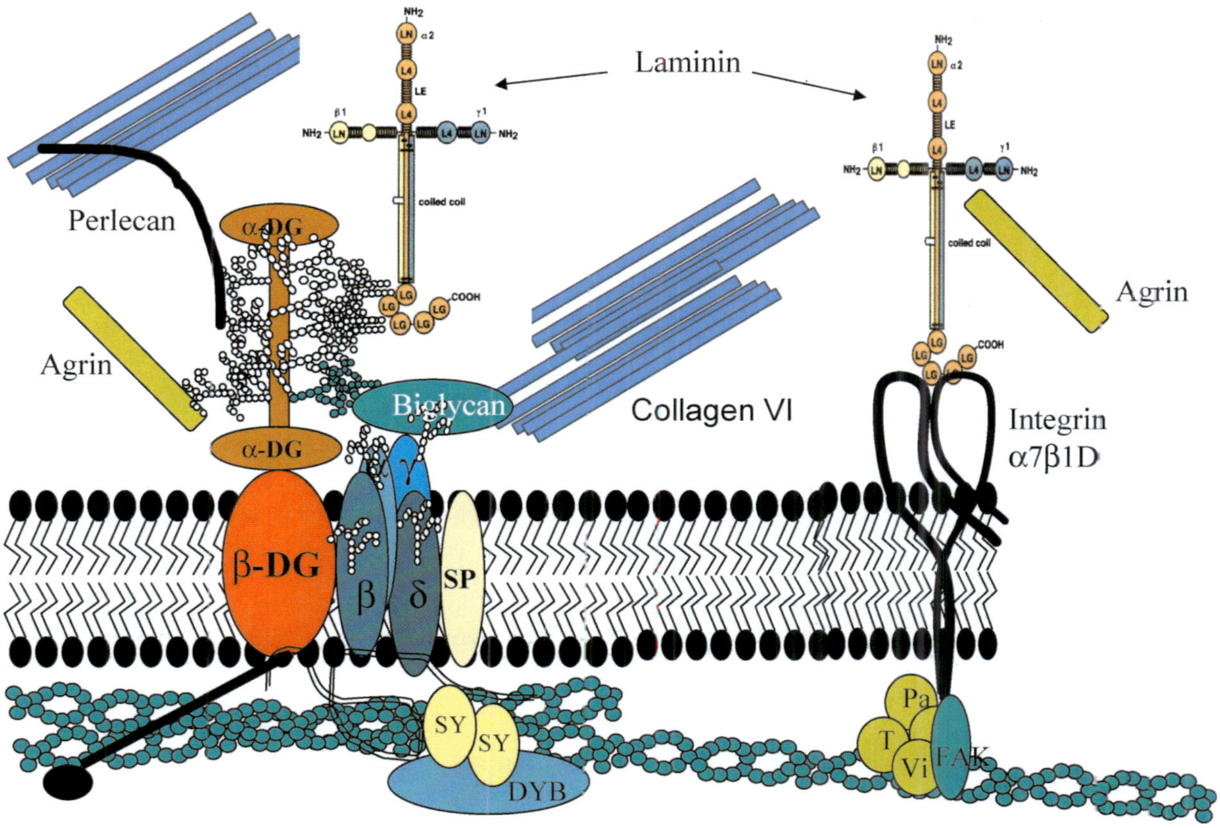

FIGURE 29.1 The dystrophin-associated protein complex. This schematic shows the location of key proteins involved in the extracellular matrix-dystroglycan-dystrophin axis; a number of these proteins are described in this chapter, in particular α-dystroglycan (α-DG), the laminin α2 chain of the merosin trimer, the integrin complex, and collagen VI. *Adapted with permission from Muntoni F and Voit T, 2004.*[2]

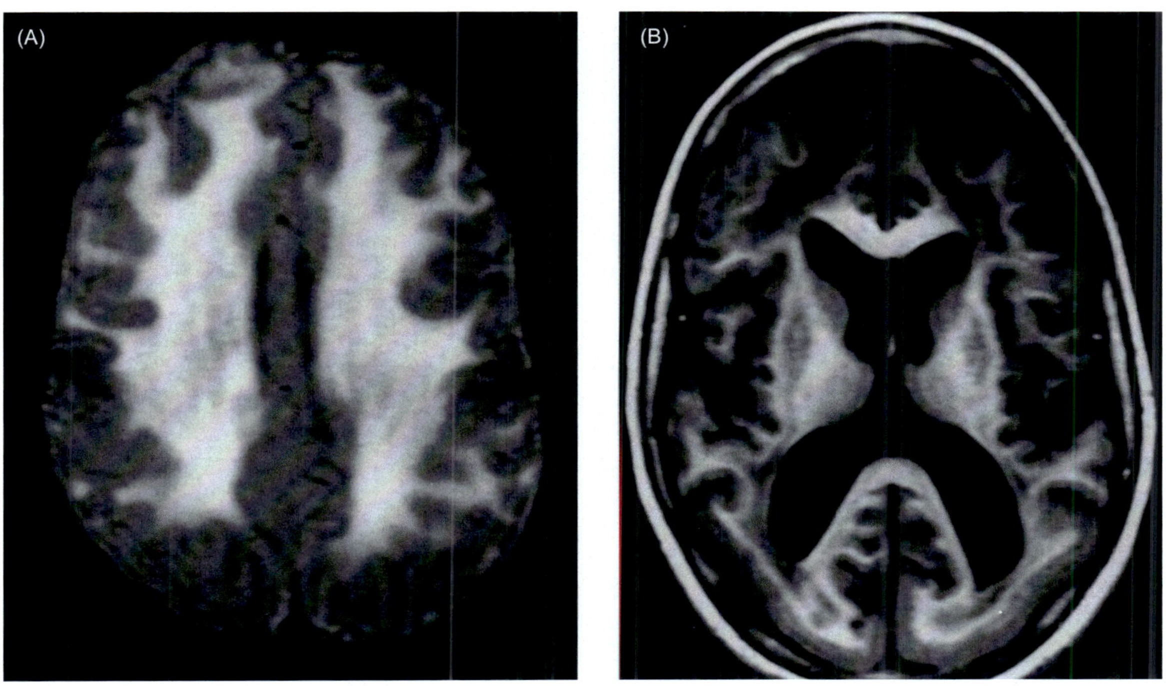

FIGURE 29.2 Merosin-deficient CMD. (**A**) Transverse brain MRI section showing diffuse white matter hyperintensities involving both hemispheres (T2-weighted image). (**B**) Transverse brain MRI section showing bilateral occipital cortical dysplasia and also diffuse white matter hyperintensities in both hemispheres (T1-weighted image). MRI, magnetic resonance imaging. *Adapted with permission from Mercuri et al., 1999.*[14]

cases, genetic confirmation is needed for differential diagnosis. While the white matter abnormalities are essentially invariably found, we recently came across an individual with mild LGMD due to *LAMA2* mutations in whom the brain MRI did not show the typical changes.

Cognitive and visual function are generally normal with the exception of the small percentage (~<10%) of cases with associated structural brain changes affecting the occipital regions (Figure 29.2B).[14,17] The molecular explanation for this rare but consistent complication in MDC1A is not known.

DYSTROGLYCANOPATHIES

The term *dystroglycanopathies* includes a genetically heterogeneous group of muscle disorders, with hypoglycosylated α-dystroglycan (ADG) on muscle biopsy, frequently associated with concomitant central nervous system pathology and more rarely with eye pathology. Mutations in dystroglycan itself have been very recently reported but are rare, hence almost invariably patients in this group of conditions will have mutations in genes involved in the glycosylation of ADG rather than in dystroglycan (*DAG1*) itself. A more appropriate terminology is therefore of secondary dystroglycanopathies for all conditions in which *DAG1* is genetically intact.

So far, defects in a total of 15 putative and demonstrated glycosyltransferases or accessory proteins of glycosyltransferases have been shown to cause a dystroglycanopathy phenotype, with several new genes reported in the last two years[18–20] (Table 29.1). The main pathway related to defects in glycosylation of ADG relates to O-mannosylation; however, recently a few cases with mutations in genes involved also in the N-glycosylation pathways have been reported also.[19,21]

Patients with primary or secondary dystroglycanopathies have common features, such as predominant involvement of the upper limbs and markedly elevated CK. The spectrum of clinical severity ranges from forms with severe structural muscle, eye, and brain involvement (WWS), muscle-eye-brain disease (MEB), and Fukuyama congenital muscular dystrophy (FCMD), to mild cases with late onset and no brain or eye involvement. The classification of these forms has changed over the years in order to take into account not only the complexity of genetic defects but also the heterogeneity of the phenotypes. While the early studies following the identification of the first genes involved mainly focused on distinct known phenotypes, namely MEB, WWS, and FCMD, or on their absence in patients with normal brain MRI scans, it soon became obvious that the spectrum of CNS involvement is much more complex. Recent brain MRI studies have highlighted that a number of patients have posterior fossa lesions, such as cerebellar dysplasia or hypoplasia, occurring in isolation or variably associated with brainstem and pons abnormalities, suggesting that infratentorial structures have a particular susceptibility to the underlying disease process in the majority of the dystroglycanopathies with brain involvement. We have proposed a simplified clinical classification system comprising seven broad phenotypic categories (five for the CMD forms),[22] with patients subdivided according to the presence/absence and degree of structural and functional brain involvement, and motor functional abilities (Tables 29.2 and 29.3).

Recent studies on large cohorts have demonstrated that, although the involvement of each of these genes was originally reported in patients with a distinct phenotype, mutations in individual genes have subsequently been associated with different phenotypes and, conversely, the same phenotype has been reported in association with many, if not all, of the known genes.[22,23] This suggests that the identity of the defective gene cannot be predicted from the clinical phenotype. The genetic heterogeneity of these forms has been further suggested by the evidence that not all dystroglycanopathy patients have mutations in the known genes, thus suggesting that novel genes are also involved in the pathogenesis of this group of disorders.[22,23] While precise figures on the percentage of patients in whom a final molecular diagnosis cannot be reached are not available due to the very recent identification of novel genes, we estimate it to be approximately 20% to 30% of cases now.

New entries in the Online Mendelian Inheritance in Man (OMIM) database have recently been created, subdividing the group of muscular dystrophies with deficit of dystroglycan or muscular dystrophy-dystroglycanopathy (MDDG) into three broad phenotypic groups denoted type A, B, and C, each including some of the categories proposed by Godfrey et al. (Table 29.3).[22]

Type A includes the phenotypes with brain and eye abnormalities and the most severe phenotypes: WWS, WWS-like, MEB, and FCMD-like. The brain abnormalities include cortical malformations such as lissencephaly type 2 or cobblestone lissencephaly, pachygyria, polymicrogyria, cerebellar hypoplasia (vermis/hemispheres), cerebellar cysts or other signs of cerebellar dysplasia, white matter hyperintensities on MRI, flat pons and brain stem, ventricular dilatation, absent corpus callosum and/or cerebellar vermis, and others (Figure 29.3A, B). Ocular involvement can be variable, including malformation of the anterior chamber and persistent primary vitreous in WWS, micophthalmia, buphthalmos, progressive myopia, retinal degeneration and cataracts in WWS or MEB, or simple myopia without structural changes in FCMD. The prognosis in type A phenotypes ranges from death in early infancy to survival beyond the age of 3 years (Table 29.2).

TABLE 29.2 Dystroglycanopathy Core Phenotypic Categories for Congenital Muscular Dystrophies

Phenotypic Classifications	Description
Onset	
Further subdivisions based on degree of structural or functional brain involvement	
Congenital onset weakness	
WWS/WWS-like	Onset is prenatal or at birth. Patients have severe structural brain abnormalities including complete agyria or severe lissencephaly with only rudimentary cortical folding, marked hydrocephalus, severe cerebellar involvement, and complete or partial absence of the corpus callosum. Eye abnormalities including congenital cataracts, microphthalmia, and buphthalmos are common. Motor development is typically severely impaired with most patients attaining head control only.
MEB/FCMD-like	These categories are merged due to the overlapping phenotypic features. Included in this group are patients with CMD and structural brain defects less severe than those seen with WWS. MRI findings include pachygyria with preferential frontoparietal involvement, polymicrogyria, cerebellar hypoplasia, cerebellar dysplasia, and frequent flattening of the pons and brainstem. Eye abnormalities are often present in this group and include congenital glaucoma, progressive myopia, retinal atrophy and juvenile cataracts. Individuals may rarely acquire the ability to walk, although this is delayed. Patients typically speak only a few meaningful words.
CMD CRB (CMD with cerebellar involvement)	CMD with mental retardation and MRI evidence of cerebellar involvement as the only structural brain abnormality. Cerebellar abnormalities may include cysts, hypoplasia, or dysplasia. Cerebellar cysts in the absence of supratentorial white matter or cortical involvement have been encountered most commonly in patients with defects in *FKRP* or *POMGNT1*.
CMD MR (CMD with mental retardation)	CMD with mental retardation and a structurally normal brain. This category includes patients with isolated microcephaly or minor white matter changes or entirely normal brain on MRI.
CMD no MR (CMD with no mental retardation)	Patients within this group have no evidence of abnormal cognitive development. This category includes patients who have not had neuroimaging performed.

Abbreviations: WWS, Walker-Warburg syndrome; MEB, muscle-eye-brain; CMD, congenital muscular dystrophy; FCMD, Fukuyama congenital muscular dystrophy; CRB, cerebellar; MR, mental retardation.
Adapted with permission from Godfrey, 2011.[6]

Type B includes cases of CMD with posterior fossa/cerebellar (CRB) abnormalities or normal MRI (CMD CRB, CMD MR, and CMD no MR), hence mixing together patients with and without MR (mental retardation), which has some shortcomings from a clinical perspective and prognostic counseling (Table 29.3).

Type C includes the limb-girdle muscular dystrophy (LGMD) forms that are not discussed in this chapter (see Chapter 41).

Each of these clinical categories is further subdivided according to the patient's gene defect.

The broad spectrum of phenotypes secondary to mutations in a single gene can well be exemplified by the FKTN, FKRP, and ISPD genes: mutations in each one of these genes can give rise to severe Walker-Warburg or MEB syndromes, to intermediate phenotypes such as those with cerebellar cysts or normal brain MRI, or to the mild forms of limb-girdle muscular dystrophies (LGMD2I in the case of *FKRP*). The large series of patients reported,[22,23] however, have highlighted that even if the identity of the defective gene cannot be always predicted from the clinical phenotype, there are some hints that make the primary involvement of one specific gene more probable.

1. WWS has been described in association with mutations in all the known genes, including the most recent and still relatively uncommon genes.
2. MEB is more often found with *POMGnT1, POMT1* and *POMT2, ISPD*, and *FKRP*.
3. Prominent cerebellar cysts are very common with *POMGnT1, ISPD*, and *FKRP* mutations (when the brain is structurally affected); they are also seen in *GMPPB* and *B3GALNT2* but are rarely seen in *POMT1* and *POMT2*.[24]
4. The phenotype with mental retardation, either with normal brain MRI or with cerebellar involvement and microcephaly, are more often associated with *POMT1* and *POMT2* mutations, but are also associated with *ISPD*.
5. The phenotype with normal brain MRI and no mental retardation is more often associated with *FKRP*.
6. Cataracts in ambulant patients are more commonly associated with *GMPPB*.

TABLE 29.3 Phenotypes Associated with Mutations in Causative Dystroglycanopathy Genes

Dystroglycanopathy core phenotypic categories						
Congenital muscular dystrophy (CMD)						
		WWS/ WWS-like	MEB/ FCMD-like	CMD CRB	CMD MR	CMD no MR
OMIM phenotypic classifications assigned in 2010: Muscular dystrophy-dystroglycanopathy (MDDG)						
Causative genes	TYPE	A- Congenital with brain and eye abnormalities		B- Congenital with / without mental retardation		
POMT1	-1	—	—	—	—	—
POMT2	-2	—	—	—	—	
POMGNT1	-3	—	—	—		
FKTN	-4	—	—			—
FKRP	-5	—	—	—	— (MDC1C)	—
LARGE	-6	—	—		— (MDC1D)	
DPM2				—		
DPM3	CDG-1o					
		WWS/ WWS-like	MEB/ FCMD-like	CMD CRB	CMD MR	CMD no MR

CMD, congenital muscular dystrophy; MEB, muscle eye brain; WWS, Walker-Warburg syndrome; FCMD, Fukuyama congenital muscular dystrophy; CRB, cerebellar; MR, mental retardation; CDG-1o, congenital disorder of glycosylation type 1o.
Adapted with permission from Godfrey et al.[6]

CMD WITH COLLAGEN VI DEFICIENCY (ULLRICH DISEASE)

This CMD form, also known as Ullrich disease, and its milder allelic form, Bethlem myopathy, are due to deficiency of collagen type VI,[25] an extracellular matrix component (Figure 29.1). This form appears to be the most common form of CMD[8,26] and is characterized by an unusual combination of marked distal laxity, contractures of the proximal joints, skin changes, and normal intelligence.

In Ullrich CMD, serum CK is often normal or only mildly elevated.[27] Muscle biopsy can show a variable pathology ranging from myopathic changes (always in the presence of neonatal myosin positive fibers) to clearly dystrophic patterns, with abundant adipose tissue. A marked deficiency or absence of collagen VI is a clear marker of

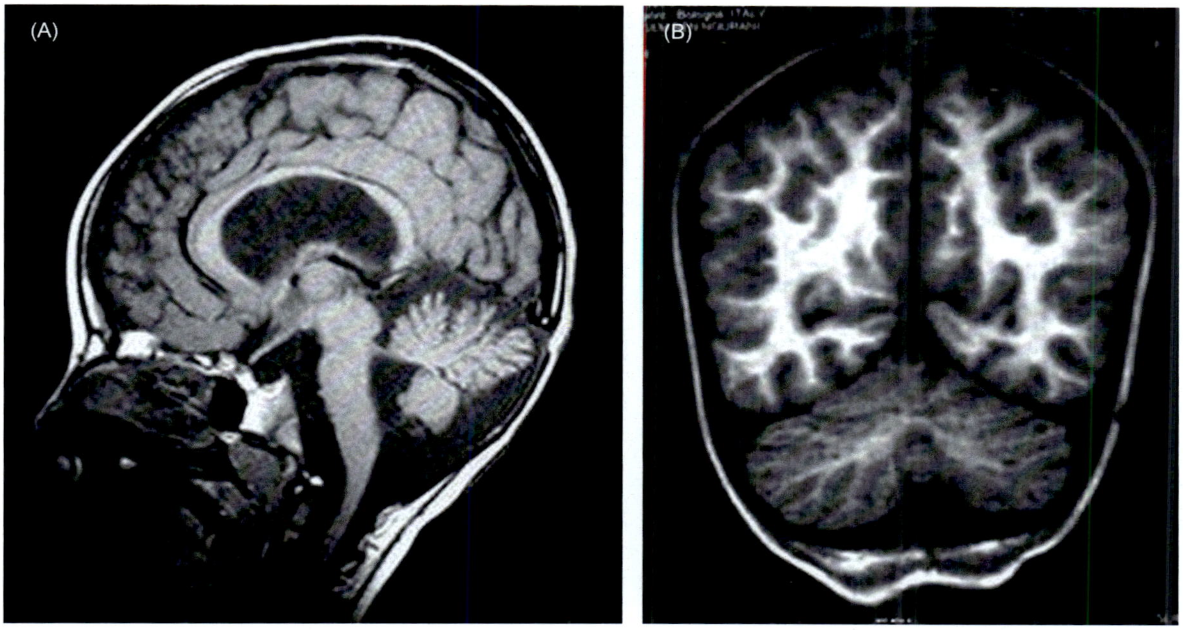

FIGURE 29.3 (A) Enlarged cisterna magna and mild cerebellar vermis hypoplasia in a 16-month-old child with *POMT2* mutations (sagittal T1-weighted brain MR image). (B) Cerebellar cysts in a 7-year-old with *POMT2* mutations (coronal T1-weighted MR image). MR, magnetic resonance. *Adapted with permission from Messina et al., 2008.*[24]

Ullrich CMD, but in some patients the deficiency is more subtle and affects the basal lamina expression of collagen VI in muscle, not its interstitial deposition. In these cases, careful double labeling with another marker of basal lamina such as perlecan or collagen V is usually diagnostic[23]; additional confirmatory data on abnormal expression in fibroblasts cultures can also be performed, and quantitative techniques using flow cytometry to assess objectively collagen VI expression have recently been described.[29]

The diagnosis needs to be confirmed by genetic analysis. Collagen VI (Figure 29.1) is composed of three α chains encoded by three genes: *COL6A1* and *COL6A2* on chromosome 21q22 or *COL6A3* on chromosome 2q37.

Following the identification of mutations in collagen VI, there was a straightforward distinction between Ullrich CMD, initially thought to be due to recessive mutations only, and its allelic disorder, Bethlem myopathy, related to dominant mutations. In the last few years, it has become obvious not only that both disorders can be due to both recessive and dominant mutations but also that the phenotypes of the two forms are not always easily distinguishable,[30–35] as there are reports of patients with Bethlem myopathy having early onset and a more severe phenotype of that typically reported in Bethlem myopathy.[27,36]

Patients at the severe end of the clinical spectrum, conforming to the original description of Ullrich congenital muscular dystrophy, show proximal contractures associated with distal laxity (Figure 29.4), rigidity of the spine, and early respiratory failure. Contractures and hypotonia, often associated with torticollis, hip dysplasia, and congenital kyphosis, can be present at birth, but delayed milestones can also be the presenting sign, and the mean age at onset of symptoms has been reported to be around 12 months.[26,37,38] The severe muscle weakness often prevents ambulation, and although a proportion of these patients can acquire the ability to walk independently, this is eventually lost. A longitudinal observation in Ullrich CMD showed that approximately 70% of patients became constant wheelchair users at a mean age

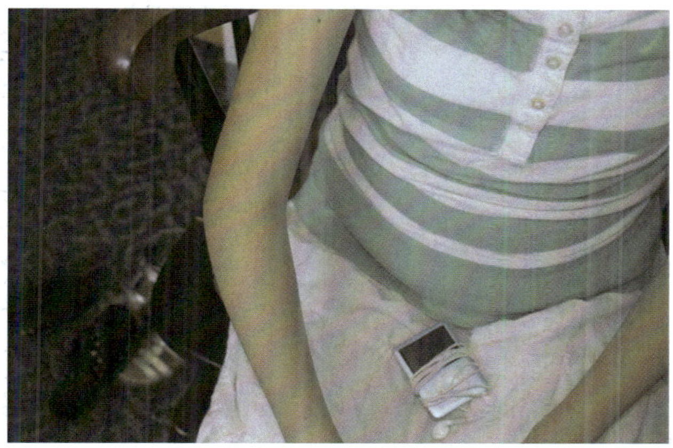

FIGURE 29.4 CMD with collagen VI deficiency (Ullrich disease). Note proximal contracture at the elbow. This 12-year-old female patient had significant laxity of hands and feet. *Courtesy of Basil T. Darras, MD, Boston Children's Hospital.*

of 11.1 years (SD 4.8 years), with another 20% only being able to ambulate indoors with assistance.[37]

On examination, there is a significant laxity of hands (Figure 29.5) and feet and often feet deformities with protrusion of the calcanei (Figure 29.6A). The distal laxity is typically associated with proximal contractures (Figure 29.4), which become more marked with time. Other clinical signs are the tendency to develop follicular hyperkeratosis (keratosis pilaris) (Figure 29.6B), hypertrophic scars, and keloids (Figure 29.6C).[39] All patients develop rigidity of the spine, often associated with scoliosis, irrespective of whether they have achieved independent ambulation. By the end of the first decade, there is often failure to thrive and respiratory compromise. In the severe Ullrich CMD, there is a cumulative decline in forced vital capacity of 2.6% per year ($p < 0.0001$).

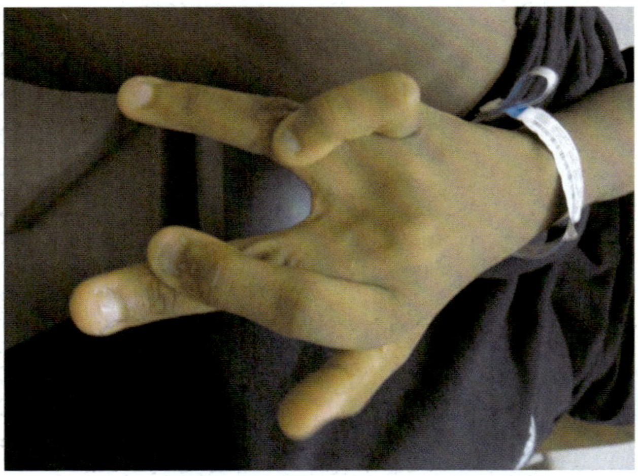

FIGURE 29.5 Laxity of hands/fingers in a child with Ullrich congenital muscular dystrophy. *Courtesy of Basil T. Darras, MD, Boston Children's Hospital.*

At the other end of the spectrum, patients with Bethlem myopathy have a more stable clinical course with only mild progression of weakness, a typical pattern of contractures, and little or no respiratory involvement.

In between these two phenotypes, however, there are a number of patients in whom walking with/without assistance is achieved and who have an intermediate phenotype, as they have a milder motor phenotype than the severe Ullrich CMD (of intermediate severity between Ullrich CMD and Bethlem myopathy) but demonstrate a continuous decline in pulmonary function with a cumulative decline in forced vital capacity of 2.3% per year.[27]

None of the patients with declining pulmonary function attained the ability to hop or run; these patients were therefore categorized as intermediate collagen VI-related myopathy to reflect the more guarded outcome and need for more stringent surveillance when compared to Bethlem myopathy.

SEPN1-RELATED CMD

Mutations in the *selenoprotein N1* (*SEPN1*) gene located on chromosome 1p36-13[40] have been reported in patients with minimal dystrophic changes, and labeled as one of the CMD forms (referred to as rigid spine syndrome type 1 [RSMD1]).[41] The clinical phenotype is characterized by weakness and rigidity of the spine with early respiratory insufficiency, a slow course of motor impairment, and

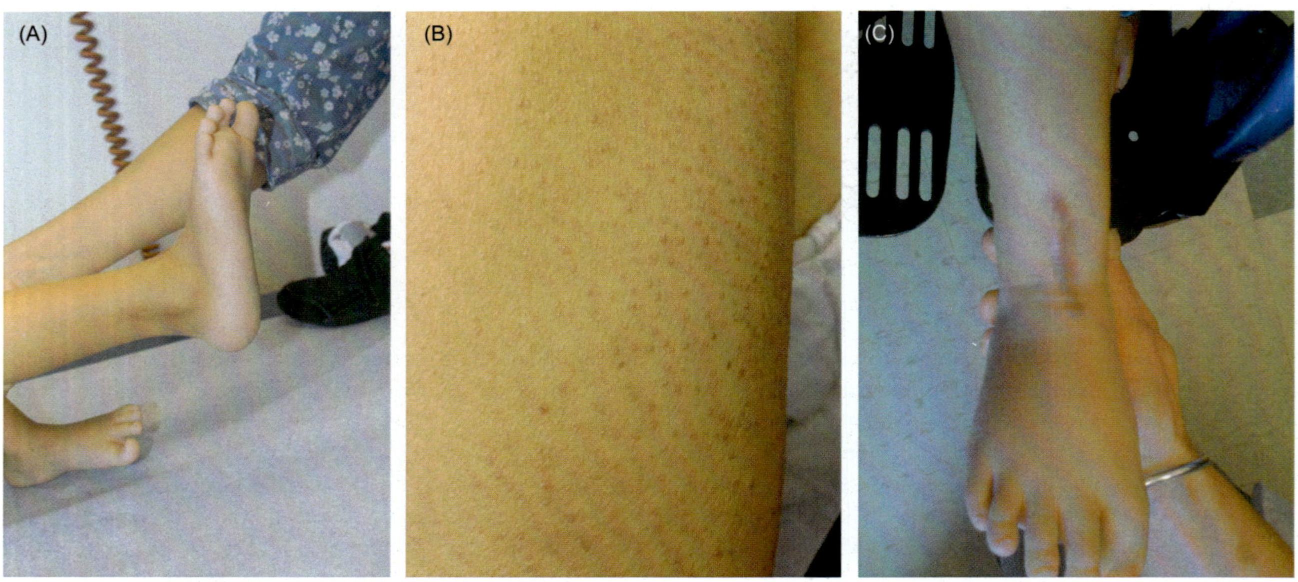

FIGURE 29.6 Ullrich CMD. Note the prominence of the calcaneus (**A**), keratosis pilaris (**B**), and a hypertrophic scar (keloid) (**C**) in this child with a homozygous *COL6A1* gene recessive mutation. *Courtesy of Basil T. Darras, MD, Boston Children's Hospital.*

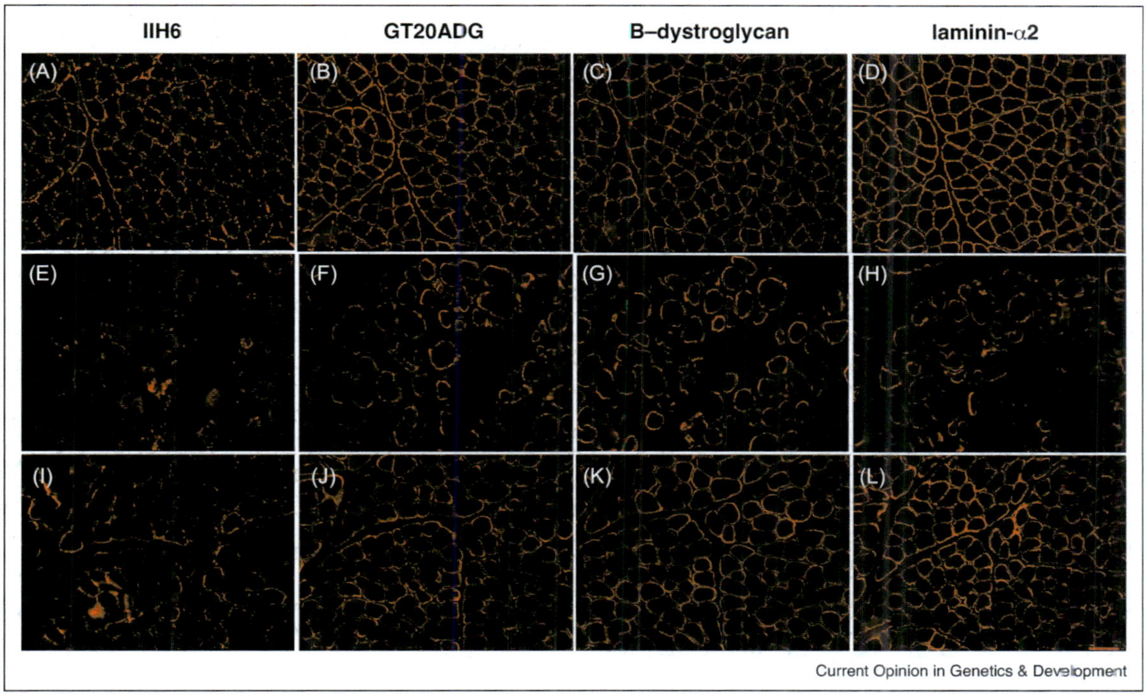

FIGURE 29.7 Skeletal muscle in dystroglycanopathies. The pathological feature common to all these disorders is the reduced glycosylation (hypoglycosylation) of α-dystroglycan shown as reduction of immunoreactivity to IIH6 (an antibody specific for α-dystroglycan glycan moieties) (**A**, **E**, **I**) in transverse sections of skeletal muscle from two patients with mutations in FKRP (**E**–**H**) and LARGE (**I**–**L**) genes. Control normal muscle stains well with all antibodies (**A**–**D**). Note the retention of immunoreactivity to GT20ADG (an antibody that recognizes the core α-dystroglycan peptide) (**B**, **F**, **J**) and β-dystroglycan (**C**, **G**, **K**). Laminin-α2 labeling shows secondary reduction of immunoreactivity related to the reduced capacity of hypoglycosylated α-dystroglycan to bind its extracellular matrix ligands (**D**, **H**, **L**). FKRP, Fukutin-related protein. *Adapted with permission from Godfrey et al., 2011.*[6]

limited evidence of muscle damage, as indicated by serum CK levels.[42] This phenotype has subsequently also been found in association with multiminicores, congenital fiber-type disproportion, and myofibrillar-like myopathy with Mallory body-like inclusions.[40,43–45]

Mutations in the *SEPN1* gene[40] have been reported in association with all these individual pathological findings with a significant clinical overlap and no obvious correlation between genotype and clinical course or pathological findings.[43,46,47] In most patients with mutations in *SEPN1*, the onset of clinical signs occurs in the first year with delayed milestones even though in a number of cases the early signs are often underestimated and patients are seen by a neurologist after the age of 1 year. Joint contractures are frequent but often are not the presenting sign. With few exceptions, ambulation is achieved and maintained over the years. Scoliosis and respiratory insufficiency are almost invariable findings in the second decade, with the possibility of early signs from the first decade.

LAMIN A/C-RELATED CMD

Following the identification of mutations in the lamin A/C gene (*LMNA*) in patients with the dominant form of Emery-Dreifuss muscular dystrophy,[48] it has become obvious that mutations in this gene could also be found in children with early onset muscular dystrophy with severe axial weakness and wasting of cervicoaxial muscles, "dropped head" syndrome phenotype, often associated with feeding and respiratory involvement.[49–51]

DIFFERENTIAL DIAGNOSIS

In most cases, a combination of clinical, pathological, and MRI features can help to select the most appropriate genetic analysis. Severe presentation at birth and very high CK are generally found in patients with merosin-deficient CMD or with dystroglycanopathies.

Merosin-deficient CMD and dystroglycanopathies also share a more marked involvement of upper limbs and relative hypertrophy of lower limbs. Muscle biopsy will confirm the dystrophic features and will help to differentiate the forms with primary merosin deficiency from those with more marked reduction of α-dystroglycan (Figure 29.7). Brain MRI often provides additional information and will facilitate differential diagnosis among the individual forms (see previous paragraph).

Mild to moderately elevated CK levels are often found in *LMNA*-related CMD. In contrast, normal or mildly elevated CK levels are generally found in Ullrich or *SEPN1* myopathies or in integrin α7, and are generally associated with children having delayed milestones and only occasionally very early severe onset.

THERAPEUTIC APPROACHES

A better understanding of the mechanisms underlying the individual CMD forms has allowed, for the first time, the development of a number of possible therapeutic approaches for specific forms of CMD that currently are mostly still undergoing preclinical testing.

1. In MDC1A, several studies using mouse models are being used to assess different therapeutic approaches such as the use of mini-agrin, a minigene derived from agrin, a protein with binding activity that may help to restore the connection between α-DG and the basement membrane lost as a consequence of laminin α2 deficiency.[52] Other ongoing pharmacological approaches have focused on inhibitors of apoptosis, such as omigapil or doxycycline, which have been studied in the MDC1A mouse model,[53,54] or on using drugs such as losartan that, by decreasing the activation of transforming growth factor (TGF)-β, a promoter of fibrosis,[55] may result in reduction of fibrosis.
2. In dystroglycanopathies, it has been demonstrated that in myodystrophy (*Large^{myd}*) mice, *LARGE* gene transfer using adenovirus can induce glycosylation of hypoglycosylated α-DG and improve muscle pathology.[56] Several groups, including ours, are involved in high throughput analysis of compounds potentially capable of hyperglycosylating α-dystroglycan.[57] However, recent data from our laboratory indicates that the timing and extent of α-dystroglycan hyperglycosylation induced by *LARGE* overexpression needs careful further evaluation as it could potentially also be detrimental if not carefully delivered.[58]
3. In Ullrich CMD, work in the mouse model of collagen VI deficiency has suggested that drugs involved in inhibiting the inappropriate opening of the permeability transition pores, such as cyclosporine A, may reverse the mitochondrial damage and decrease apoptosis.[59,60] Other antiapoptotic agents, such as the cyclophilin D inhibitor D-MeAla3-EtVal4-cyclosporine (Debio 025) have also been investigated.[60] More recently, defective autophagy was demonstrated in the animal model of collagen VI deficiency,[61,62] with evidence that the reactivation of the autophagic process can improve muscle pathology in the affected mice.
4. In SEPN1, recent studies have identified the role in the physiological redox-related calcium homeostasis and cell protection against oxidative stress,[63] suggesting a potential role of antioxidants such as N-acetylcysteine (NAC) in experimental therapeutic approaches of SEPN1-related myopathies.[64]

ACKNOWLEDGMENTS

The authors wish to thank Dr Basil T Darras, Boston Children's Hospital, for helpful comments and for providing clinical photographs. Francesco Muntoni acknowledges the contribution of the Muscular Dystrophy Campaign and the National Specialist Commissioning Team for the work on congenital muscular dystrophies.

REFERENCES

1. Dubowitz V. 22nd ENMC sponsored workshop on congenital muscular dystrophy held in Baarn, The Netherlands, 14–16 May 1993. *Neuromuscul Disord* 1994;**4**:75–81.
2. Muntoni F, Voit T. The congenital muscular dystrophies in 2004: a century of exciting progress. *Neuromuscul Disord* 2004;**14**: 635–49.
3. Mercuri E, Muntoni F. The ever-expanding spectrum of congenital muscular dystrophies. *Ann Neurol* 2012;**72**:9–17.
4. Tome FM, Evangelista T, Leclerc A, Sunada Y, Manole E, Estournet B, et al. Congenital muscular dystrophy with merosin deficiency. *C R Acad Sci III* 1994;**317**:351–7.
5. Helbling-Leclerc A, Zhang X, Topaloglu H, Cruaud C, Tesson F, Weissenbach J, et al. Mutations in the laminin alpha 2-chain gene (LAMA2) cause merosin-deficient congenital muscular dystrophy. *Nat Genet* 1995;**11**:216–18.
6. Godfrey C, Foley AR, Clement E, Muntoni F. Dystroglycanopathies: coming into focus. *Curr Opin Genet Dev* 2011;**21**:278–285.
7. Hara Y, Balci-Hayta B, Yoshida-Moriguchi T, Kanagawa M, de Bernabe DBV, Gundesli H, et al. A dystroglycan mutation associated with limb-girdle muscular dystrophy. *N Engl J Med* 2011;**364**: 939–46.
8. Clement EM, Feng L, Mein R, Sewry CA, Robb SA, Manzur AY, et al. Relative frequency of congenital muscular dystrophy subtypes: analysis of the UK diagnostic service 2001–2008. *Neuromuscul Disord* 2012;**22**:522–7.
9. Bonnemann CG. Congenital muscular dystrophy. In: Squire LR, editor. *Encyclopedia of Neuroscience*. Oxford: Academic Press; 2009. pp. 67–74.
10. Philpot J, Muntoni F. Limitation of eye movement in merosin-deficient congenital muscular dystrophy. *Lancet* 1999;**353**:297–8.
11. Geranmayeh F, Clement E, Feng LH, Sewry C, Pagan J, Mein R, et al. Genotype-phenotype correlation in a large population of muscular dystrophy patients with LAMA2 mutations. *Neuromuscul Disord* 2010;**20**:241–50.
12. Philpot J, Bagnall A, King C, Dubowitz V, Muntoni F. Feeding problems in merosin deficient congenital muscular dystrophy. *Arch Dis Child* 1999;**80**:542–7.
13. Hillaire D, Leclerc A, Faure S, Topaloglu H, Chiannilkulchai N, Guicheney P, et al. Localization of merosin-negative congenital muscular dystrophy to chromosome 6q2 by homozygosity mapping. *Hum Mol Genet* 1994;**3**:1657–61.
14. Mercuri E, Gruter-Andrew J, Philpot J, Sewry C, Counsell S, Henderson S, et al. Cognitive abilities in children with congenital

muscular dystrophy: correlation with brain MRI and merosin status. *Neuromuscul Disord* 1999;**9**:383–7.
15. Philpot J, Cowan F, Pennock J, Sewry C, Dubowitz V, Bydder G, et al. Merosin-deficient congenital muscular dystrophy: the spectrum of brain involvement on magnetic resonance imaging. *Neuromuscul Disord* 1999;**9**:81–5.
16. Mercuri E, Rutherford M, De Vile C, Counsell S, Sewry C, Brown S, et al. Early white matter changes on brain magnetic resonance imaging in a newborn affected by merosin-deficient congenital muscular dystrophy. *Neuromuscul Disord* 2001;**11**:297–9.
17. Messina S, Bruno C, Moroni I, Pegoraro E, D'Amico A, Biancheri R, et al. Congenital muscular dystrophies with cognitive impairment. A population study. *Neurology* 2011;**75**:898–903.
18. Willer T, Lee H, Lommel M, Yoshida-Moriguchi T, de Bernabe DB, Venzke D, et al. ISPD loss-of-function mutations disrupt dystroglycan O-mannosylation and cause Walker-Warburg syndrome. *Nat Genet* 2012;**44**:575–80.
19. Carss KJ, Stevens E, Foley AR, Cirak S, Riemersma M, Torelli S, et al. Mutations in GDP-mannose pyrophosphorylase B cause congenital and limb-girdle muscular dystrophies associated with hypoglycosylation of alpha-dystroglycan. *Am J Hum Genet* 2013;**93**:29–41.
20. Cirak S, Foley AR, Herrmann R, Willer T, Yau S, Stevens E, et al. ISPD gene mutations are a common cause of congenital and limb-girdle muscular dystrophies. *Brain* 2013;**136**:269–81.
21. Barone R, Aiello C, Race V, Morava E, Foulquier F, Riemersma M, et al. DPM2-CDG: a muscular dystrophy-dystroglycanopathy syndrome with severe epilepsy. *Ann Neurol* 2012;**72**:550–8.
22. Godfrey C, Clement E, Mein R, Brockington M, Smith J, Talim B, et al. Refining genotype phenotype correlations in muscular dystrophies with defective glycosylation of dystroglycan. *Brain* 2007;**130**:2725–35.
23 Mercuri E, Messina S, Bruno C, Mora M, Pegoraro E, Comi GP, et al. Congenital muscular dystrophies with defective glycosylation of dystroglycan: a population study. *Neurology* 2009;**72**: 1802–9.
24. Messina S, Mora M, Pegoraro E, Pini A, Mongini T, D'Amico A, et al. POMT1 and POMT mutations in CMD patients: a multicentric Italian study. *Neuromusc Disord* 2008;**18**:565–71.
25. Camacho-Vanegas O, Bertini E, Zhang RZ, Petrini S, Minosse C, Sabatelli P, et al. Ullrich scleroatonic muscular dystrophy is caused by recessive mutations in collagen type VI. *Proc Natl Acad Sci USA* 2001;**98**:7516–21.
26. Mercuri E, Yuva Y, Brown SC, Brockington M, Kinali M, Jungbluth H, et al. Collagen VI involvement in Ullrich syndrome: a clinical, genetic, and immunohistochemical study. *Neurology* 2002;**58**:1354–9.
27. Foley AR, Quijano-Roy S, Collins J, Straub V, McCallum M, Deconinck N, et al. Natural history of pulmonary function in collagen VI-related myopathies. *Brain* 2013;**136**:3625–33.
28. Hicks D, Lampe AK, Barresi R, Charlton R, Fiorillo C, Bonnemann CG, et al. A refined diagnostic algorithm for Bethlem myopathy. *Neurology* 2008;**70**:1192–9.
29. Kim J, Jimenez-Mallebrera C, Foley AR, Fernandez-Fuente M, Brown SC, Torelli S, et al. Flow cytometry analysis: a quantitative method for collagen VI deficiency screening. *Neuromuscul Disord* 2012;**22**(2):138–48.
30. Baker NL, Morgelin M, Peat R, Goemans N, North KN, Bateman JF, et al. Dominant collagen VI mutations are a common cause of Ullrich congenital muscular dystrophy. *Hum Mol Genet* 2005;**14**:279–93.
31. Brinas L, Richard P, Quijano-Roy S, Gartioux C, Ledeuil C, Lacene E, et al. Early onset collagen VI myopathies: Genetic and clinical correlations. *Ann Neurol* 2010;**68**:511–20.
32. Foley AR, Hu Y, Zou Y, Columbus A, Shoffner J, Dunn DM, et al. Autosomal recessive inheritance of classic Bethlem myopathy. *Neuromuscul Disord* 2009;**19**:813–17.
33. Giusti B, Lucarini L, Pietroni V, Lucioli S, Bandinelli B, Sabatelli P, et al. Dominant and recessive COL6A1 mutations in Ullrich scleroatonic muscular dystrophy. *Ann Neurol* 2005;**58**:400–10.
34. Lampe AK, Bushby KM. Collagen VI related muscle disorders. *J Med Genet* 2005;**42**:673–85.
35. Gualandi F, Urciuolo A, Martoni E, Sabatelli P, Squarzoni S, Bovolenta M, et al. Autosomal recessive Bethlem myopathy. *Neurology* 2009;**73**:1883–91.
36. Pepe G, Lucarini L, Zhang RZ, Pan TC, Giusti B, Quijano-Roy S, et al. COL6A1 genomic deletions in Bethlem myopathy and Ullrich muscular dystrophy. *Ann Neurol* 2006;**59**:190–5.
37. Nadeau A, Kinali M, Main M, Jimenez-Mallebrera C, Aloysius A, Clement E, et al. Natural history of Ullrich congenital muscular dystrophy. *Neurology* 2009;**73**:25–31.
38. Demir E, Sabatelli P, Allamand V, Ferreiro A, Moghadaszadeh B, Makrelouf M, et al. Mutations in COL6A3 cause severe and mild phenotypes of Ullrich congenital muscular dystrophy. *Am J Hum Genet* 2002;**70**:1446–58.
39. Nadeau A, Muntoni F. Skin changes in Ullrich congenital muscular dystrophy. *Neuromuscul Disord* 2008;**18**:982.
40. Moghadaszadeh B, Petit N, Jaillard C, Brockington M, Roy SQ, Merlini L, et al. Mutations in SEPN1 cause congenital muscular dystrophy with spinal rigidity and restrictive respiratory syndrome. *Nat Genet* 2001;**29**:17–18.
41. Moghadaszadeh B, Topaloglu H, Merlini L, Muntoni F, Estournet B, Sewry C, et al. Genetic heterogeneity of congenital muscular dystrophy with rigid spine syndrome. *Neuromuscul Disord* 1999;**9**:376–82.
42. Flanigan KM, Kerr L, Bromberg MB, Leonard C, Tsuruda J, Zhang P, et al. Congenital muscular dystrophy with rigid spine syndrome: a clinical, pathological, radiological, and genetic study. *Ann Neurol* 2000;**47**:152–61.
43. Ferreiro A, Quijano-Roy S, Pichereau C, Moghadaszadeh B, Goemans N, Bonnemann C, et al. Mutations of the selenoprotein N gene, which is implicated in rigid spine muscular dystrophy, cause the classical phenotype of multiminicore disease: reassessing the nosology of early-onset myopathies. *Am J Hum Genet* 2002;**71**:739–49.
44. Ferreiro A, Ceuterick-de Groote C, Marks JJ, Goemans N, Schreiber G, Hanefeld F, et al. Desmin-related myopathy with Mallory body-like inclusions is caused by mutations of the selenoprotein N gene. *Ann Neurol* 2004;**55**:676–86.
45. Clarke NF, Kidson W, Quijano-Roy S, Estournet B, Ferreiro A, Guicheney P, et al. SEPN1: associated with congenital fiber-type disproportion and insulin resistance. *Ann Neurol* 2006;**59**:546–52.
46. Schara U, Kress W, Bonnemann CG, Breitbach-Faller N, Korenke CG, Schreiber G, et al. The phenotype and long-term follow-up in 11 patients with juvenile selenoprotein N1-related myopathy. *Eur J Paediatr Neurol* 2008;**12**:224–30.

47. Mercuri E, Talim B, Moghadaszadeh B, Petit N, Brockington M, Counsell S, et al. Clinical and imaging findings in six cases of congenital muscular dystrophy with rigid spine syndrome linked to chromosome 1p (RSMD1). *Neuromuscul Disord* 2002;**12**:631–8.
48. Bonne G, Mercuri E, Muchir A, Urtizberea A, Becane HM, Recan D, et al. Clinical and molecular genetic spectrum of autosomal dominant Emery-Dreifuss muscular dystrophy due to mutations of the lamin A/C gene. *Ann Neurol* 2000;**48**:170–80.
49. Mercuri E, Manzur AY, Jungbluth H, Bonne G, Muchir A, Sewry C, et al. Early and severe presentation of autosomal dominant Emery-Dreifuss muscular dystrophy (EMD2). *Neurology* 2000;**54**:1704–5.
50. Mercuri E, Poppe M, Quinlivan R, Messina S, Kinali M, Demay L, et al. Extreme variability of phenotype in patients with an identical missense mutation in the lamin A/C gene: from congenital onset with severe phenotype to milder classic Emery-Dreifuss variant. *Arch Neurol* 2004;**61**:690–4.
51. Quijano-Roy S, Mbieleu B, Bonnemann CG, Jeannet PY, Colomer J, Clarke NF, et al. De novo LMNA mutations cause a new form of congenital muscular dystrophy. *Ann Neurol* 2008;**64**:177–86.
52. Moll J, Barzaghi P, Lin S, Bezakova G, Lochmuller H, Engvall E, et al. An agrin minigene rescues dystrophic symptoms in a mouse model for congenital muscular dystrophy. *Nature* 2001;**413**:302–7.
53. Erb M, Meinen S, Barzaghi P, Sumanovski LT, Courdier-Fruh I, Ruegg MA, et al. Omigapil ameliorates the pathology of muscle dystrophy caused by laminin-alpha2 deficiency. *J Pharmacol Exp Ther* 2009;**331**:787–95.
54. Girgenrath M, Beermann ML, Vishnudas VK, Homma S, Miller JB. Pathology is alleviated by doxycycline in a laminin-alpha2-null model of congenital muscular dystrophy. *Ann Neurol* 2009;**65**: 47–56.
55. Meinen S, Shuo L, Ruegg MA. Treatment approaches in laminin-α2-deficient congenital muscular dystrophy (MDC1A). *Neuromuscul Disord* 2009;**19**:543.
56. Barresi R, Michele DE, Kanagawa M, Harper HA, Dovico SA, Satz JS, et al. LARGE can functionally bypass alpha-dystroglycan glycosylation defects in distinct congenital muscular dystrophies. *Nat Med* 2004;**10**:696–703.
57. Brockington M, Torelli S, Sharp PS, Liu K, Cirak S, Brown SC, et al. Transgenic overexpression of LARGE induces alpha-dystroglycan hyperglycosylation in skeletal and cardiac muscle. *PLOS ONE* 2010;**5**:e14434.
58. Whitmore C, Fernandez-Fuente M, Booler H, Parr C, Kavishwar M, Ashraf A, et al. The transgenic expression of LARGE exacerbates the muscle phenotype of dystroglycanopathy mice. *Hum Mol Genet* 2013;**23**(7):1842–55.
59. Irwin WA, Bergamin N, Sabatelli P, Reggiani C, Megighian A, Merlini L, et al. Mitochondrial dysfunction and apoptosis in myopathic mice with collagen VI deficiency. *Nat Genet* 2003;**35**:367–71.
60. Tiepolo T, Angelin A, Palma E, Sabatelli P, Merlini L, Nicolosi L, et al. The cyclophilin inhibitor Debio 025 normalizes mitochondrial function, muscle apoptosis and ultrastructural defects in Col6a1-/- myopathic mice. *Br J Pharmacol* 2009;**157**:1045–52.
61. Grumati P, Coletto L, Sabatelli P, Cescon M, Angelin A, Bertaggia E, et al. Autophagy is defective in collagen VI muscular dystrophies, and its reactivation rescues myofiber degeneration. *Nat Med* 2010;**16**:1313–20.
62. Grumati P, Coletto L, Sandri M, Bonaldo P. Autophagy induction rescues muscular dystrophy. *Autophagy* 2011;**7**:426–8.
63. Arbogast S, Beuvin M, Fraysse B, Zhou H, Muntoni F, Ferreiro A. Oxidative stress in SEPN1-related myopathy: from pathophysiology to treatment. *Ann Neurol* 2009;**65**:677–86.
64. Arbogast S, Ferreiro A. Selenoproteins and protection against oxidative stress: selenoprotein N as a novel player at the crossroads of redox signaling and calcium homeostasis. *Antioxid Redox Signal* 2010;**12**:893–904.

Chapter 30

Dystrophinopathies

Basil T. Darras, Caroline C. Menache-Starobinski, Veronica Hinton, and Louis M. Kunkel

INTRODUCTION

Dystrophinopathies are X-linked recessive diseases related to dystrophin deficiency; the major ones are Duchenne muscular dystrophy (DMD), the most common form of muscular dystrophy, and Becker muscular dystrophy (BMD). BMD has a presentation similar to that of DMD but a relatively milder clinical course. In addition to DMD and BMD, there is an intermediate group of muscular dystrophy patients with either mild DMD or severe BMD, who are known as "Outliers."

Other dystrophinopathies, occurring at a lower incidence, include:

- DMD-associated dilated cardiomyopathy
- muscle cramps with myoglobinuria
- asymptomatic elevation of muscle enzymes
- manifesting DMD/BMD carrier females.

Historical Background

Duchenne's description of DMD was published in the 1861 edition of his book on the electrical stimulation of muscle, but John Little and Edward Meryon had first reported this entity 8–9 years earlier. In fact, in 1852, Meryon[1] described very accurately the clinical course of the disease in four brothers and also the pathological findings on microscopic examination of muscle. In 1853, Little described two brothers with "abnormal increase of bulk of muscle combined with contraction and adipose degeneration," who were unable to walk after the age of 11 years.[2]

In 1861, Duchenne described the disease as "hypertrophic paraplegia of infancy due to a cerebral cause"[3] but, by 1868, he recognized that it was muscular in origin and was the first one to propose criteria for diagnosis.[4] He also used a small harpoon to perform muscle biopsies and described the replacement of muscle by fat and connective tissue.

The occurrence of a more benign form of X-linked muscular dystrophy was first proposed in 1955 by Becker and Kiener[5] and Walton[6]; the disorder was named Becker muscular dystrophy after several families were reported with similar features. It was not clear at that time whether DMD and BMD were genetically distinct disorders, or allelic forms of the same genetic entity.

In the early 1980s, advances in cytogenetics and molecular genetics led to mapping of the gene responsible for DMD in band 21 of the short arm of the X chromosome (Xp21). First, a number of female patients with muscular dystrophy were reported who, on cytogenetic analysis, had X;autosome translocations (e.g., X;11, X;21); the translocation breakpoint in the X chromosome was consistently at Xp21.[7,8] Second, using restriction fragment length polymorphisms (RFLPs) and linkage analysis, cloned DNA segments were linked to the DMD locus on the short arm of the X chromosome, supporting the Xp localization of the *DMD* gene.[9,10] Third, in 1985, Francke et al. at Yale University detected a microscopically observable deletion at Xp21 on the short arm of the X chromosome in a male patient ("BB"); this patient suffered from Duchenne muscular dystrophy, chronic granulomatous disease, retinitis pigmentosa, and the McLeod syndrome.[11] Subsequently, in 1985, by using a competitive hybridization strategy, Kunkel et al. at Boston Children's Hospital isolated specific DNA fragments absent from the genome of patient "BB"[12]; molecular analysis of a large number of DNAs from DMD patients showed that one of these fragments/clones was deleted in 6.5% of DMD patients.[13] A 200 kilobase (kb) chromosomal walk was initiated from this clone which, in 1986, led to the isolation of an exon of the *DMD* gene and partial cloning of a muscle-specific transcript.[14] Using a different strategy, Worton's group at the University of Toronto isolated a breakpoint junction clone from the DNA of a female patient with an X;21 translocation[15]; with starting point this junction clone, in 1987 a 180 kb walk led to the identification of a second exon of

B.T. Darras, H. Royden Jones, Jr., M.M. Ryan & D.C. De Vivo (Eds): Neuromuscular Disorders of Infancy, Childhood and Adolescence, Second edition.
DOI: http://dx.doi.org/10.1016/B978-0-12-417044-5.00030-5

the *DMD* gene and more of the cDNA of the 14 kb transcript.[16] In 1987–88, the complete cDNA sequence of the *DMD* gene was cloned and sequenced by Kunkel's group and the protein product was named dystrophin.[17–19] Dystrophin was localized in the sarcolemma of muscle fibers and was shown to be almost completely absent in DMD and decreased in BMD.[20–22] These major scientific advances have made diagnosis and genetic counseling far more accurate and have allowed reliable prenatal testing; they have also paved the way for the discovery of effective therapies.

ETIOLOGY AND PATHOGENESIS

All dystrophinopathies are allelic conditions, resulting from different mutations of the *DMD* gene.

DMD Gene

DMD and titin are the largest genes yet identified in humans. The *DMD* gene, with a 14 kb transcript and 79 exons, spans approximately 2.2 megabases on the short arm of the X chromosome (Xp21.3-p21.2), or about 1% of the entire X chromosome.[14,19,23] The introns of the gene are large, with some near the 5′ third of the gene and between exons 44 and 45 being nearly 200 kb. The genomic (exonic and intronic) sequence of the *DMD* gene can be obtained from the National Center for Biotechnology Information (NCBI) / Genbank web site pages and also from the Leiden Muscular Dystrophy web pages. The *DMD* gene product dystrophin is recognized on Western blots of human skeletal muscle proteins using antidystrophin antibodies.[24] With the use of immunocytochemistry, dystrophin has been localized to the cytoplasmic side of the plasma membrane of muscle fibers. Dystrophin has four functional domains: the amino-terminal or actin-binding domain, the rod domain, the cysteine-rich domain and the carboxy-terminal domain.

- The amino-terminal domain (exons 1–8) binds to actin through three high-affinity actin-binding sites.
- The rod domain (exons 9–63) is large, comprised of 24 homologous "spectrin-like" repeats forming an α-helical structure. The repeats are interrupted by two nonhelical regions known as "hinges," which are thought to confer flexibility on the rod domain during muscle contraction.
- The cysteine-rich domain (exons 64–69) is a region of the *DMD* gene near the carboxy-terminus that seems to be the region of dystroglycan binding.
- The carboxy-terminal domain (exons 70–79) binds with a protein complex that links the cytoskeleton with membrane proteins that in turn bind with proteins in the extracellular matrix.

Therefore, dystrophin is part of a large, tightly associated glycoprotein complex containing other proteins, the so-called dystrophin associated proteins (DAP) (Figure 30.1).

The *DMD* gene can generate several cell type-specific isoforms of dystrophin of different molecular weight, each driven by a distinct promoter.[25–31] These promoters drive the transcription of the dystrophin gene from their own first exon and give rise to dystrophin isoforms of various molecular weights (Table 30.1) (Figure 30.2A, B).[32] The brain, muscle, and Purkinje cell dystrophins are predicted to have a molecular weight of 427 kilodaltons (kDa), whereas the glial/general[33] and Schwann cell[34] dystrophins are encoded by carboxy-terminal transcripts and have predicted molecular weights of 71 and 116 kDa; they are designated as B/Dp427, M/Dp427, P/Dp427, G/Dp71, and S/Dp116 dystrophins, respectively. The dystrophins of retina (R/Dp260)[35] and brain/kidney (B-K/Dp140)[31] have been described also (Figure 30.2B).

Other dystrophin variants (i.e. alternatively spliced transcripts) result from alternative splicing of exons 71, 74, 78, and 79 and are expressed in muscle, brain and heart tissue; alternative splicing occurs in regions involved in binding of dystrophin to the DAP complex and, therefore, may regulate their association in various tissues.[27,36] Muscle dystrophin is expressed in skeletal (all muscle fiber types), smooth, and cardiac muscles, and also in the outer plexiform layer of the retina. The expression of this and probably other isoforms is developmentally regulated; dystrophin is first detected in human fetal muscle at 9 weeks of gestation and its expression is increased as myoblasts differentiate into multinucleated myotubes. The highest expression of brain dystrophin is in the neocortex and hippocampus; Purkinje cell dystrophin comprises most or nearly all of cerebellar dystrophin.[37–39] Dp71 is expressed in glial cells, viscera, and mature cardiac and fetal skeletal muscle but not in mature skeletal muscle.[29,33,40–42] While muscle dystrophin is detected on the plasma membrane of skeletal muscle fibers, it appears to be particularly abundant in neuromuscular and myotendinous junctions.[21,43,44]

The different isoforms must be interacting with different proteins in the various tissues in which they are expressed. In DMD, the most important of these isoforms are the three full-length, large molecular weight (427 kDa) dystrophins, present in skeletal and smooth muscle (muscle type, Dp427), in cerebral cortex and hippocampus (brain type, Dp427), and in Purkinje cells (Purkinje type, Dp427).[45]

The Dystrophin Associated Protein Complex

As discussed above, the dystrophin-associated proteins together with dystrophin form a complex known as the dystrophin associated protein complex (DAP), which,

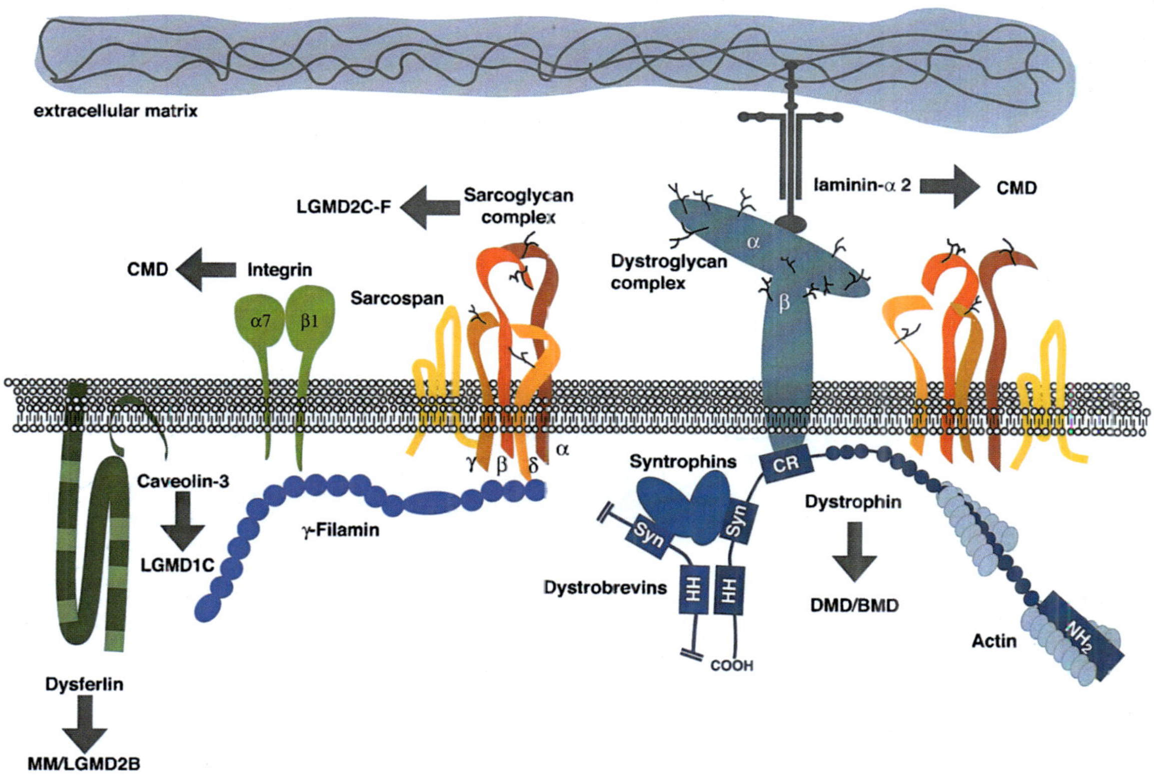

FIGURE 30.1 The dystrophin associated protein complex and other proteins known to be involved in muscular dystrophy. Arrows indicate the protein components mutated in various muscular dystrophies. Alterations in the dystrophin gene cause Duchenne muscular dystrophy (DMD) and Becker muscular dystrophy (BMD). Mutations in the sarcoglycan proteins, caveolin-3 and dysferlin, lead to forms of limb girdle muscular dystrophy (LGMD). Laminin α2-chain is mutated in a subtype of congenital muscular dystrophy (CMD) without structural brain anomalies, as is α7-integrin. Abbreviations: COOH, carboxy-terminal domain; CR, cysteine-rich domain; HH, two helices of the coiled-coil domain; MM, Miyoshi myopathy; NH_2, amino-terminal domain; Syn, syntrophin-binding domain. *Courtesy of K. O'Brien, C. Bönnemann, and L. Kunkel, Boston Children's Hospital, Boston, MA, USA.*

TABLE 30.1 Major Dystrophin Isoforms

Name	Location of Promoter	Tissue Expression
Brain, B/Dp427[a]	− 90 kb Exon 1[b]	Cerebral cortex
		Hippocampal cortex, amygdala
Muscle, M/Dp427	− 140 bases Exon 1[b]	Skeletal, smooth muscle
		Cardiac muscle, retina
Purkinje cell, P/Dp427	Intron 1[c]	Purkinje cells
Retinal, R/Dp260	Intron 29	Retina
Brain-Kidney, B-K/Dp140	Intron 44	Brain/kidney
Schwann cell, S/Dp116	Intron 55	Schwann cells
Glial/general, G/Dp71	Intron 62	Glial cells, viscera, mature cardiac muscle
		Fetal but not mature skeletal muscle

[a]*Numbers in this column denote the molecular weight in kilodaltons of the respective dystrophin isoforms.*
[b]*Exon 1 of muscle dystrophin (Dp427M in Figure 30.2A).*
[c]*Intron between the muscle promoter-exon1 (Dp427M) and the second exon (common to all three full-length dystrophins).*

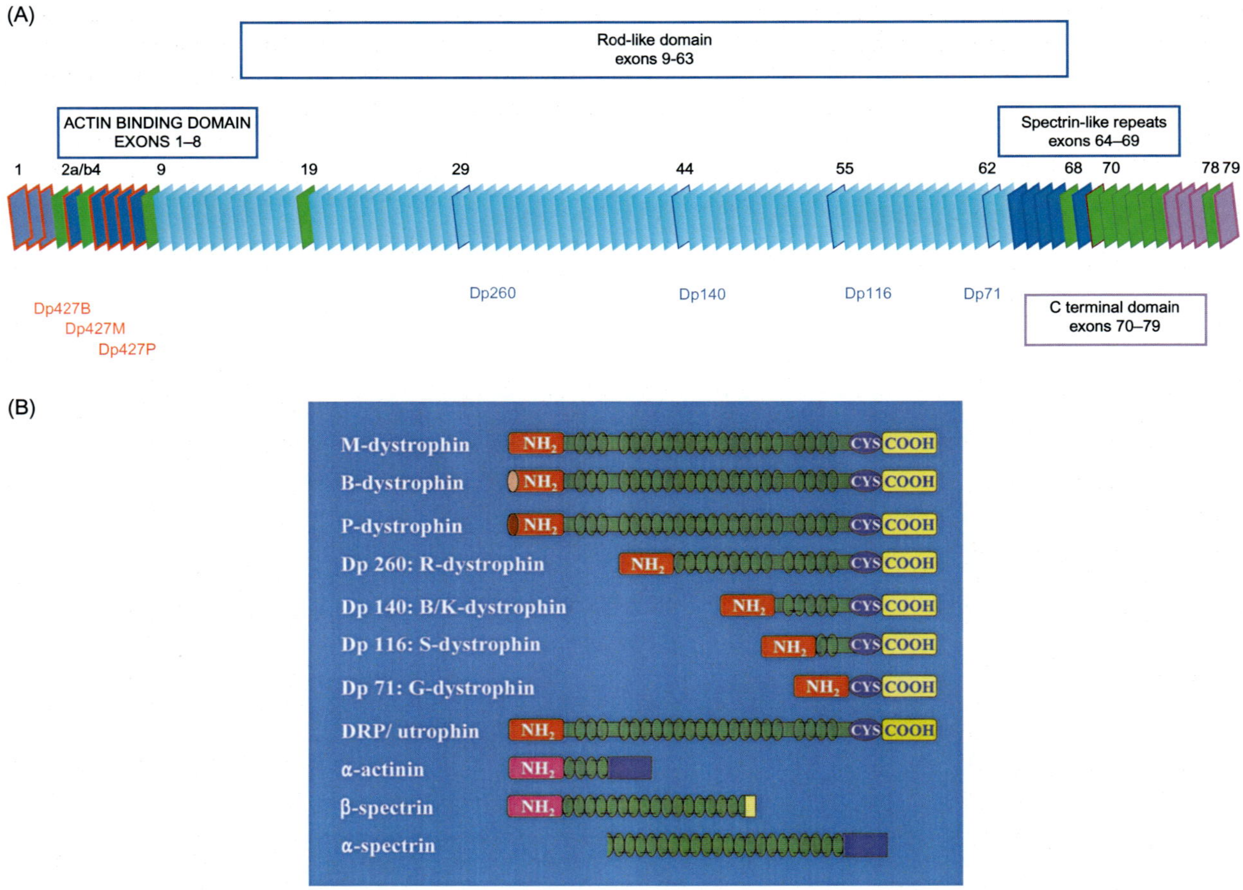

FIGURE 30.2 (**A**) Dystrophin gene structure and promoters. The schematic diagram shows 7 distinct promoters, which drive the transcription of the dystrophin gene from their own first exon and give rise to dystrophin isoforms of various molecular weights. It also shows a schematic map of 79 exons of the dystrophin gene, protein domains (boxes), plus the three 5′ promoter-specific first exons in red (Dp427B-M-D). Lines in blue indicate the 3′ promoters and the first exons of isoforms: Dp260 (retina), Dp140 (brain/kidney), Dp116 (Schwann cell), Dp71 (general). Exons in green are alternatively spliced or skipped. (**B**) Schematic diagram of three full-length dystrophins and smaller isoforms. The actin-binding domain (red/pink) is shared among a number of the various dystrophins, utrophin, α-actinin, and β-spectrin. The internal rod-domain (green) consists of homologous repeats of ~109 amino acids. The carboxy-terminus (yellow) is homologous only to utrophin (gold). Abbreviations: M, muscle; B, brain; P, Purkinje cell; R, retina; B/K, brain/kidney; S, Schwann cell; G, general or glial; DRP, dystrophin related protein; NH_2, amino-terminus; COOH, carboxy-terminus; CYS, cysteine-rich domain. *(A) is from Ferlini et al.*[23] *Used with permission.*

acting as a bridge, connects the intracellular cytoskeletal actin to the basal lamina through the extracellular matrix (Figure 30.1).[23] The DAP complex is believed to stabilize the sarcolemmal membrane during contraction and relaxation.[46,47] The extracellular matrix is a network situated between the sarcolemma and the extracellular basal lamina and is composed mainly of laminin, collagen, fibronectin, and proteoglycans. The DAP complex will be discussed herein briefly and in detail in Chapter 34 on Limb-girdle Muscular Dystrophies.

The DAP complex is composed of three groups of proteins:[47,48]

1. Dystroglycan group
   α-dystroglycan
   β-dystroglycan.
2. Sarcoglycan group
   α-sarcoglycan (or adhalin)
   β-sarcoglycan
   γ-sarcoglycan
   δ-sarcoglycan
   sarcospan.
3. Syntrophin/dystrobrevin group
   α-syntrophin
   β1-syntrophin
   β2-syntrophin
   α-dystrobrevin
   β-dystrobrevin.

The dystroglycan subcomplex functions as a receptor of certain extracellular matrix proteins. Indeed, α-dystroglycan is an extracellular protein that binds to β-dystroglycan, which is located in the sarcolemma, and to laminin-α2 in the extracellular matrix. β-dystroglycan is a transmembranous protein that binds intracellularly to dystrophin (at the cysteine-rich domain and the first half of the carboxy-terminal domain).

The "dystrophin-axis" (actin-dystrophin-dystroglycan-laminin) serves a very important function in creating a link between the sarcolemma and the extracellular basal lamina, and thus forms the structural basis for mechanical stabilization of the sarcolemma during contraction and relaxation of muscle fibers.[45]

Laminins are a group of proteins situated in the basal lamina and extending into the extracellular matrix. They are each a heterotrimer molecule composed of 1 heavy chain (α2) and 2 light chains (β1 or β2 and γ1).[49] Several α, β and γ chains have been identified, and their various combinations result in different types of laminin. Merosin is a term that encompasses all laminins that contain an α2 heavy chain. Laminin α2-chain mutations have been described in a subset of patients with congenital muscular dystrophy (discussed in Chapter 29). In the sarcolemma, several proteins have been considered as receptors for laminins, including dystroglycans (described above), integrins (α7-β1 integrin), and lectins.[45]

The sarcoglycan group is comprised of a number of sarcolemmal transmembrane proteins (α-, β-, γ-, δ-sarcoglycans and sarcospan) and is described as the side arm of the dystrophin axis because of its lateral linkage with the dystrophin-dystroglycan complex.[50] However, the exact details of the dystrophin-sarcoglycan complex association have not yet been elucidated.[45] Deficiency of the various sarcoglycan complex proteins is responsible for several limb-girdle muscular dystrophies, known as sarcoglycanopathies (discussed in detail in Chapter 34). A muscle-specific form of filamin, termed filamin 2 (FLN2, FLNC or γ) was identified as a γ- and δ-sarcoglycan interacting protein.[51] FLN2 is located in two intracellular pools, with about 97% of the cellular content contained in the cytoplasm and approximately 3% localized at the sarcolemmal membrane. Interestingly, in LGMD and DMD patients the membrane content of FLN2 increases to about 30%, which suggests that FLN2 binds to membrane-bound proteins other than sarcoglycans, possibly β1-integrin; the increase of FLN2 at the membrane in patients with LGMD and DMD and its diminished translocation intracellularly to bind F-actin also imply an essential role in maintaining membrane integrity.[51]

The second group of dystrophin-associated proteins contains the intracellular syntrophins and the dystrobrevins. The syntrophin subcomplex, made of 3 intracytoplasmic proteins (α-, β1-, and β2-syntrophin), also binds to the carboxy-terminus of dystrophin. In addition, the syntrophins have been shown to bind neuronal nitric oxide synthase (nNOS).[52] Dystrobrevin is another intracytoplasmic protein complex, comprised of α- and β-dystrobrevin, that binds to syntrophins and dystrophin. The dystrobrevins, in addition to binding to dystrophin, are highly homologous to the dystrophin carboxy-terminal domain. The quantity of dystrobrevins is reduced in both DMD and patients with sarcoglycanopathies, probably on a secondary basis.[45]

DMD Gene Mutations

Of the DMD/BMD mutations identified thus far, most are deletions, detected in approximately 50–65% of males with DMD and 65–70% of males with BMD.[53–56] Partial gene duplications have also been reported in a small percentage of patients with DMD and BMD (about 5–10%).[57–59] Deletions or duplications involve 1 or more exons of the *DMD* gene. In both DMD and BMD, partial deletions and duplications cluster in two recombination hot spots, one proximal at the 5′ end of the gene in the region of exons 3–7 (nearly 30%) and one more distal, comprising exons 44–53 (nearly 70%), with additional deletions throughout the rest of the gene.[23] The hot spots seem to occur in the regions of the gene where the introns are long; for example, as discussed earlier, the intron 44 between exons 44 and 45 is extremely large; the genomic breakpoints of the 5′ hotspot lie within introns 2 and 7. In the remaining 25–35% of DMD patients without detectable deletions or duplications, and in about 20–30% of males with BMD, the molecular lesions represent single base changes, small deletions or insertions, or splicing errors.[59–63]

Nonsense mutations occur more commonly in DMD than in BMD, in the range of 20–25% of DMD cases compared with fewer than 5% of BMD cases.[58,59] In both DMD and BMD, a substantial portion of sequence changes are splice site mutations and small insertion/deletion mutations (in/del). In neither DMD nor BMD are missense mutations a common cause. Atypical mutations (deep intronic, those rarely occurring in the 5′ and 3′ untranslated regions) account for less than 1% of all *DMD* gene mutations.[23]

Published studies have failed to reveal any apparent correlation between the size of dystrophin gene deletions and the severity and progression of the DMD/BMD phenotype.[64] The molecular basis of Duchenne versus Becker muscular dystrophy seems to be related to the disruption or preservation of the amino acid translational reading frame by the deletion mutations (Figure 30.3). In other words, during the synthesis of mature mRNA, the joining ends of the exons (after splicing of introns) have to be in phase in order to maintain the correct translational open reading frame. A deletion that juxtaposes exons that shift the translational reading frame (out-of-frame deletion) usually results in unstable mRNA and eventually in the

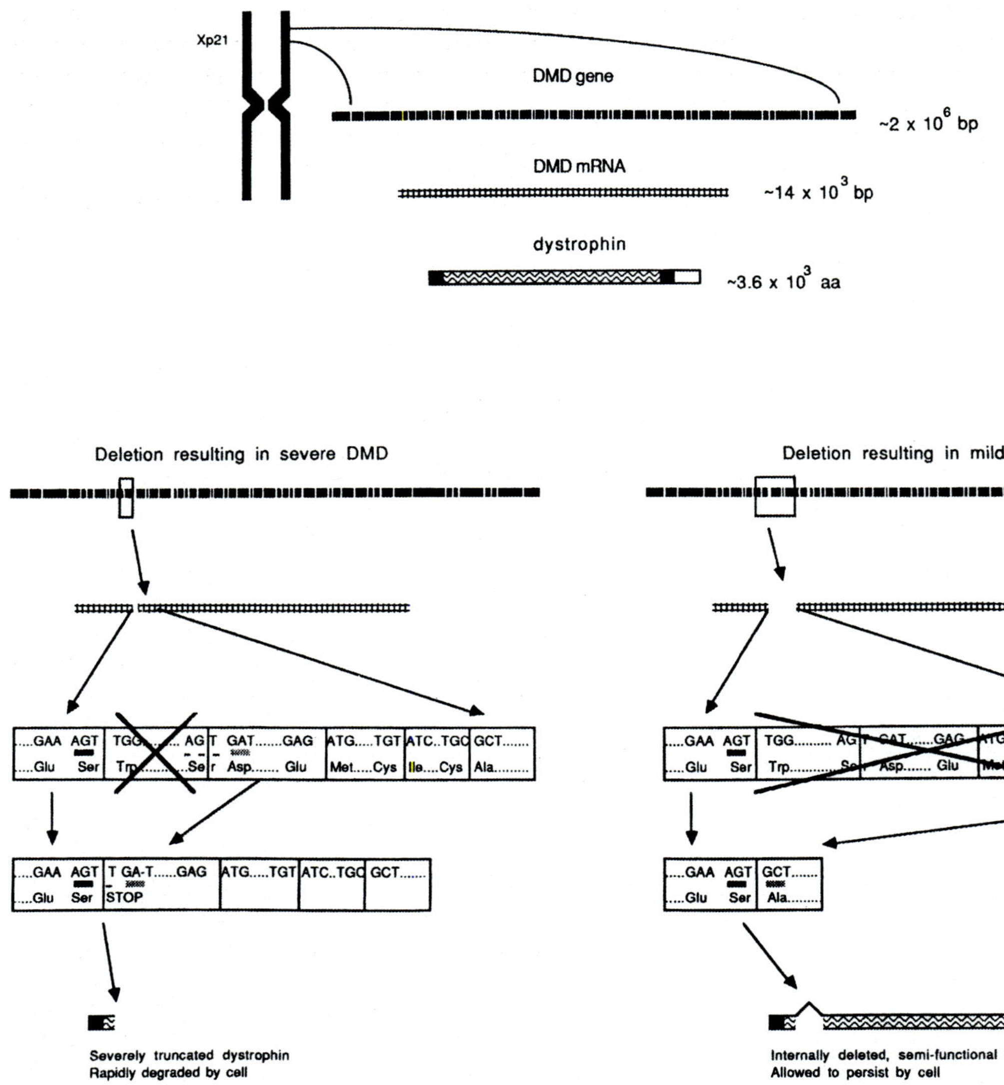

FIGURE 30.3 Molecular basis for Duchenne muscular dystrophy (DMD) versus Becker muscular dystrophy (BMD) phenotypes. The top of the figure indicates the Xp21 map position of DMD/BMD locus and also approximates the size of the *DMD* gene, its transcription product (mRNA), and dystrophin protein. The bottom of the figure shows how a small deletion can lead to a severe DMD phenotype, in contrast to a large one, which yields a milder BMD phenotype. On the left, deletion of single exon (crossed-out box) disrupts the translational reading frame because exons flanking the deletion are out of phase (out-of-frame deletion). Because mRNA can be translated only in triplet codons, T-GAT... will be converted to TGA-T..., which creates a nonsense mutation (TGA codon is translation stop), leading to severely truncated dystrophin protein. On the right, a relatively larger deletion contains integral number of triplet codons and therefore preserves the reading frame (in-frame deletion). The resulting protein is internally deleted and probably semifunctional, accounting for the milder BMD phenotype. *From Hoffman and Kunkel LM.*[65] *Reprinted by permission of Cell Press; © Cell Press.*

production of a severely truncated dystrophin molecule, which is rapidly degraded in the cell and, thus, leads to a more severe DMD phenotype.[66] This phenomenon is known as nonsense mediated decay and aims to eliminate the pathological allele to avoid dominant negative effects.[67] Other mutations disrupting the reading frame include stop (nonsense) mutations, some splicing mutations, and duplications. A deletion which juxtaposes exons that preserve the translational reading frame (in-frame deletion) will result in an internally deleted but semifunctional dystrophin protein (with intact amino- and carboxy-termini), which can persist at some quantity in the cell and thus result in a milder BMD phenotype. Other "in-frame" mutations include duplications, some splicing mutations and most non-truncating single base changes. However, exceptions to this "reading frame hypothesis" occur in approximately 8% of cases[53]; they were initially reported by Malhotra et al.[68] and subsequently were noted by others as well.[69–73]

When in-frame deletions in BMD patients are compared, as a general rule some regions of the dystrophin protein seem more critical than others; deletions in the regions coding for the amino-terminal or carboxy-terminal domains of dystrophin result in more severe phenotypes than mutations affecting the rod domain[69,74]; nonetheless, exceptions to this rule do occur also[73] and are discussed below. Exceptions most frequently occur with deletions of exons 3 to 7 or exon 45; the exceptions are also detailed below. Although these mutations are out-of-frame they can result in BMD, DMD, or an intermediate phenotype.[72,75] Exon-skipping events or activation of new cryptic translational start sites may create situations in which apparently out-of-frame deletions behave as in-frame deletions or vice versa.[76] Alternatively, posttranscriptional mechanisms (e.g. ribosomal frame shifting or reinitiation) may help restore the mRNA reading frame in BMD patients with out-of-frame 3 to 7 exon deletions.[72,77] In addition, variations in the severity of the phenotype among patients with similar mutations (e.g. deletion of exon 45) have been reported, supporting the contribution of other factors in the phenotypic expression of these mutations.[69,78] Furthermore, very large deletions or deletions in protein-binding domains (e.g. cysteine-rich domain), even if in-frame, can result in a severe phenotype.[70,71]

Exceptions to the Reading-Frame Rule: DMD Patients with In-Frame Deletions

Large in-frame deletions in the 5′ region that extend into the mid rod domain (e.g. exons 3–31, 3–25, 4–41, or 4–18) result in severe phenotype.[79] In contrast, large deletions in the rod domain that do not involve the 5′ putative actin binding site of dystrophin result in a mild BMD phenotype.[80–83] However, small in-frame deletions, such as the deletion of exons 3–13, which disrupt the 5′ actin binding domain, are associated with a DMD phenotype.[84] These exceptions underscore the functional relevance of the 5′ actin binding domain.

Exceptions to the Reading-Frame Rule: BMD Patients with Out-of-Frame Deletions

Frameshift deletions or duplications involving the 5′ end of the gene (exons 3–7, 5–7, 3–6) or further downstream (exons 51, 49–50, 47–52, 44, or 45) have been described in BMD patients. The carboxy-terminus is not disrupted in these patients and some dystrophin is produced because of exon skipping that occurs via alternative splicing.[85] The exon skipping events probably create larger mRNA deletions which presumably are in frame and hence functional.[86,87] In BMD patients with out-of-frame deletions involving exons 3–7, it has been hypothesized that an alternative translation initiation site within exon 8 becomes activated.[88]

Exon skipping methodologies using antisense oligonucleotides are currently being investigated as potential therapies in patients with DMD.

Point Mutations

Certain missense or nonsense mutations may occur within exonic splicing enhancers (ESEs) and induce exon skipping with variable results on the phenotype.[89] For example, a nonsense mutation within exon 27 disrupts an ESE sequence and induces partial skipping of the exon which results in a BMD phenotype.[90]

Nonsense mutations within DMD muscle isoform exon 1 have been associated with an extremely mild BMD phenotype with loss of ambulation in the seventh decade. It seems that internal initiation of translation at two AUG codons within exon 6 ameliorates the phenotype.[91]

Dystrophin

Dystrophin is a rod-shaped molecule of 427 kDa and can easily be detected on Western blots (immunoblots) of 100 μg of total muscle protein derived from a small portion of a muscle biopsy, by utilizing antidystrophin antibodies to the amino-terminus, rod domain, and carboxy-terminus. The quantity and quality of dystrophin can be evaluated either visually or by using densitometry. If the 427 kDa dystrophin protein is normal in size and amount, the diagnosis of DMD or BMD can be excluded. More than 99% of DMD patients display complete or almost complete absence of dystrophin in skeletal muscle biopsy specimens.[24] Most patients with BMD (about 85%) have dystrophin of abnormal molecular weight, either smaller (80%) or larger (5%), in gene deletion or gene duplication cases, respectively; the dystrophin amount is reduced to normal. About 15% of the BMD patients, however, have normal-sized protein of reduced quantity. The Western blot test is very specific, because patients with neuromuscular diseases other than DMD/BMD have normal dystrophin. Thus, dystrophin immunoblotting, in conjunction with muscle biopsy immunostaining and other clinical and laboratory parameters, can be used to help predict the severity of the evolving muscular dystrophy phenotype (Table 30.2).[92]

Pathogenesis

The functional features of dystrophin are not fully understood[93] and probably differ in the various isoforms present

TABLE 30.2 Genetic, Clinical, and Pathological Features of the Dystrophinopathies

Type	Gene Location	Protein	Inheritance	Clinical Features	Pathology
Duchenne	Xp21	Dystrophin	XR	Onset: 2–5 yr Pseudohypertrophy Diminished I.Q. Cardiac involvement Rapid decline Wheelchair confinement: 11–13 yr or earlier Death: 15–30 yr	Severe dystrophic changes Complete/almost total absence of dystrophin by immunohistochemistry[a] Dystrophin 0–5% of normal quantity[b] by Western blot[a]
Intermediate "Outliers"	Xp21	Dystrophin	XR	Intermediate severity Wheelchair confinement: 13–16 yr	Dystrophin 5–20% of normal quantity by Western blot of muscle protein
Becker	Xp21	Dystrophin	XR	Onset: 5–20 yr or later More benign course Wheelchair confinement: after 16 yr	Less marked changes Normal appearing or reduced intensity ± patchy dystrophin staining by immunohistochemistry Normal or abnormal molecular weight[c] dystrophin, quantity >20% by Western blot

[a]*Uses monoclonal antibodies to the carboxy-terminus, amino-terminus, and mid-rod domain (6–10 antibody) of dystrophin.*
[b]*The quantity of dystrophin is expressed as a percentage of control values (standardized versus myosin post transfer with Coommasie stain).*
[c]*Normal molecular weight is 427 kDa.*
Abbreviations: XR, X-linked recessive; I.Q., intelligence quotient.

in different tissues.[45] In skeletal muscle the primary role of dystrophin is probably mechanical stabilization of the plasma membrane.[94] Dystrophin-deficient muscle fibers of the *mdx* mouse exhibit an increased susceptibility to contraction-induced sarcolemmal rupture.[94] Dystrophin may allow the membrane to fold during contraction-relaxation of muscle fibers[95]; it may also function as a stabilizer of membrane-spanning molecules such as sarcoglycans. Therefore, the clinical manifestations of dystrophin deficiency may be related to the loss of stability of sarcoglycans or of other DAP complex glycoprotein functions.[45] Furthermore, because the sarcolemmal expression of the DAP complex proteins depends on the presence of normal dystrophin, there is almost always significant secondary reduction in the amount of DAP glycoproteins[96] in the muscle tissue of patients with DMD.

Histopathologic findings and animal experiments support the notion that a decreased amount of dystrophin results in a mechanically weakened plasma membrane,[97,98] which is prone to focal tears during contractile activity.[94] This allows then a massive influx of extracellular calcium, which activates proteolytic enzymes like calpains and thus leads gradually over time to necrosis of muscle fibers.[99,100] This theory could explain why dystrophinopathies are progressive diseases; it could also serve as an explanation for the fact that small-caliber fibers (i.e. fibers of extraocular muscles) are relatively spared, as the relative mechanical stress per unit surface-membrane area in small-caliber muscle fibers is less than that exerted in larger fibers.[101] The report by Kimura et al.[102] seems to illustrate this theory. They describe a patient having both DMD and X-linked recessive myotubular myopathy (MTM1) whose muscle biopsy demonstrated changes typical of MTM1 only, and lacked the typical myonecrosis of DMD. They hypothesize that not only the immaturity and small size of many fibers, but also the inactivity of muscle fibers due to MTM1 protected these fibers from necrosis.

In *mdx* mice, despite the increased susceptibility to contraction-induced membrane damage, the severity of muscular dystrophy is quite mild. Several explanations have been postulated.

1. Utrophin expression in muscle, which can compensate physiologically for the absence of dystrophin in *mdx* mice.[103]

2. Sialic acid N-glycosylneuraminic acid (Neu5Gc) expression in murine but not in human skeletal and cardiac muscle. It is incorporated in muscle glycoproteins and glycolipids and presumably ameliorates disease expression in *mdx* mice.[104]
3. Progressive loss of muscle stem cell function due to shorter telomeres in humans compared to murine muscle stem cells. This may lead to inability to repair the muscle injury induced by dystrophin deficiency in humans.[105]

Novel pathogenetic mechanisms have been described such as impaired signaling, intermittent ischemia of skeletal and cardiac muscle due to vascular smooth muscle dysfunction, and defective membrane repair and resealing of damaged muscle fibers.[106] In particular, filamin proteins (FLN1 and FLN2) have not only been found to be essential in membrane mechanoprotection during force application but also seem to be involved in signal transduction associated with cell adhesion and differentiation and, perhaps, membrane shape change. Thus, the interaction between FLN2 and the sarcoglycans implies a dynamic role of the DAP complex in muscle cell function (i.e. signal transduction) in addition to its originally believed structural role in the maintenance of a link between actin, the sarcolemma, and the extracellular matrix.[51]

Sarcolemmal nNOS catalyzes the production of nitric oxide and it is secured to the sarcolemma by dystrophin, as part of the DAP complex.[107] Nitric oxide mediates vasodilation and thus increases blood flow into muscles, which prevents early muscle fatigue with exercise.[108,109] Dystrophin deficiency results in loss of muscle nNOS[107] and hence in exercise-induced muscle fatigue.[110]

Utrophin

Utrophin or dystrophin related protein (DRP), a protein that resembles dystrophin in its amino acid sequence, structural organization, and size (395 kDa) (Figure 30.2B), is the product of a dystrophin homologous gene mapped to chromosome 6q24. It is present in the cellular membranes of most well-differentiated tissues, intramuscular nerves and blood vessels, Schwann cells, and uterine smooth muscle fibers, and also in myoblasts and myotubes. In skeletal muscle fibers, its expression is confined to the neuromuscular and myotendinous junctions, whereas dystrophin is present throughout the plasma membrane. At the neuromuscular junction, utrophin is detected near the acetylcholine receptor on the crests of the junctional folds, whereas dystrophin is concentrated in the troughs of the junctional folds.[111,112] During fetal development, utrophin is also expressed in the extrajunctional sarcolemma but at birth it is replaced by dystrophin in this location.[113,114] As is dystrophin, utrophin is also associated with a glycoprotein complex, similar to the DAP one.[115] Experimental data in the dystrophin deficient *mdx* mouse suggest that upregulation of utrophin may be able to compensate for dystrophin deficiency.[116] However, it appears that utrophin expression is already upregulated in humans with DMD, resulting in its presence in the extrajunctional sarcolemma.[115,117,118] Despite that, modest upregulation of utrophin expression may modify substantially the natural course of DMD. In humans, it would be reasonable to speculate that the spontaneous utrophin upregulation that occurs in DMD is a mitigating mechanism and that in its absence, the rate of progression of the disease would be even faster. This is supported by a report of a young patient with DMD, who had a very severe clinical course, and whose muscle biopsy revealed complete absence of utrophin in addition to absent dystrophin.[119] Also, dystrophin-utrophin double knockout (dko) mice exhibit multiple systemic degenerative changes in addition to premature muscle degeneration.[120] Strategies designed to identify small molecules that could activate the utrophin promoter in humans and thus upregulation of utrophin expression in dystophic muscle, may lead to the discovery of therapies for DMD.[121] Compounds such as biglycan and SMTG1100 upregulate utrophin and reduce the dystrophic pathology and/or symptoms in the *mdx* mouse.[122,123]

Germline Mosaicism

Dystrophinopathies are familial diseases; however, sporadic cases do occur in one-third of cases, resulting from a spontaneous *de novo* mutation in a single germ cell of the mother or grandparent. It can also occur in the proband's dystrophin gene early in postzygotic embryo development; in that case the mutation will be present in some, but not all, cells of the body, leading to "somatic mosaicism." If the mother does not have a detectable mutation in DNA derived from her white blood cells but has more than one affected offspring, that may be related to the phenomenon of "germline mosaicism" (i.e. the mutation is present in the germ cells only and is not detectable in a blood sample)[124,125]; the latter accounts for 15 to 20% of new DMD mutations.[126,127] This phenomenon is probably due to a mutation at a mitotic stage of germ cell (egg) lineage development, which results in a subpopulation of mutated eggs; the germ cells derived from the normal progenitor cell(s) will not carry the mutation. The earlier the mutation event occurs in the stages of germ cell lineage development, the higher will be the percentage of mutated eggs. If "germline mosaicism" occurs in a female, her ovaries will contain a mosaic of normal and mutated eggs and thus she will be at substantial risk for having more than one affected male offspring. If the germline mutation occurs in a male, his female offspring are at significant risk for becoming carriers.[125] As a matter of fact, because of the possibility of germline mosaicism, if the mother of an affected male is found not to be a carrier by genetic analysis of her lymphocyte DNA, her risk of having another affected son could be 7–10%[128]; depending on the degree of mosaicism, in individual cases, this risk could be much higher.

Dystrophinopathies in Females

Almost all patients with dystrophinopathies are male, but rarely females may manifest mild or severe symptoms. During early embryonic development, random inactivation of one of the two X chromosomes occurs (lyonization), leaving active 50% of the maternally derived chromosomes and 50% of the paternally derived ones.[129] Thus, if the maternal X chromosome carries a mutated dystrophin gene, 50% of the remaining nuclei have an active normal dystrophin gene and thus will produce enough dystrophin to prevent muscle necrosis. In the case of nonrandom X chromosome inactivation, less than 50% of the nuclei may have the normal dystrophin gene in an active state, thus resulting in dystrophin deficiency and clinical manifestations in females.[130] Skewed X chromosome inactivation or uneven lyonization has been described in one of monozygotic twins.[130] Twinning could predispose to nonrandom lyonization by reducing the size of the muscle cell population at the time of X chromosome inactivation (i.e. the fewer the cells when lyonization occurs, the greater the chance that the normal X chromosome will be inactivated); alternatively, splitting of the embryonic mass after lyonization could, by chance, result in two embryos with significantly different proportions of active paternal and maternal X chromosomes.[130] Symptomatic females with X-autosomal translocations with breakpoints at Xp21.3-21.2 within the dystrophin gene, resulting in preferential inactivation of the normal chromosome, have also been described.[7,8] In Turner syndrome (karyotype XO) or Turner mosaic (XO/XX or XO/XX/XXX) syndrome, if the X chromosome carries a mutated dystrophin gene, the female child will be clinically affected also. Rarely, uniparental disomy for the entire X chromosome[131] or partial uniparental disomy for the Xp21 region could result in a fully symptomatic female with a DMD or BMD phenotype.

INCIDENCE AND PREVALENCE

DMD is the most common form of muscular dystrophy, with an incidence of about 1 in 5000 live male births.[132] The mutation rate is about one-third of the incidence of the disease in males, or approximately 1 in 10,000.[133–135] The reported incidence of BMD has varied from about 1 in 18,000[136,137] to 1 in 31,000 male births.[138] The birth incidence of DMD in northern England has been reported as 1 in 5618 live male births, and that of BMD as 1 in 18,450 live male births.[136] In Nova Scotia, the overall annual incidence of DMD was 1 in 4500 males born between 1968 and 2008[139]; in Wales, a newborn screening program revealed an incidence of 1 in 5,867 live males births.[132] The prevalence of DMD/BMD among males (5 to 24 years old) in four states (Arizona, Colorado, Iowa, and western New York) in the United States ranged from 1.3 to 1.8 per 10,000.[140]

CLINICAL PRESENTATION

Great heterogeneity in the clinical features and course of the various dystrophinopathies has been observed, creating a spectrum ranging from very mild to very severe presentations. The severe end of the spectrum includes Duchenne/Becker muscular dystrophy and the "Outliers" or intermediate phenotype when skeletal muscle is primarily affected (Table 30.2), and DMD-associated dilated cardiomyopathy (DCM) when the heart is the organ primarily affected.[141–144] The phenotypes of muscle cramps with myoglobinuria and asymptomatic increase in serum concentration of muscle enzymes represent the mild end of the dystrophinopathy spectrum in males. DMD/BMD carrier females can be totally asymptomatic or can manifest mild to severe symptoms.

Table 30.2 summarizes the main genetic, clinical, and pathological features of the dystrophinopathies; the main distinction between DMD and BMD is made by the age of wheelchair dependency, which is less than 13 years in DMD and beyond 16 years in BMD. Patients who become wheelchair-dependent between 13 years and 16 years are classified as "Outliers" or as exhibiting an intermediate phenotype. The mild end of the BMD spectrum has been extended by several investigators to include patients with an increase in serum concentration of creatine kinase (CK) and abnormal dystrophin on muscle biopsy, but with "subclinical" skeletal muscle involvement.[145,146] The distinction between BMD and DMD-associated DCM is hard to make in these atypical patients when they develop severe cardiomyopathy.[143]

Duchenne Muscular Dystrophy

Clinical Features

In children with DMD, although there is histologic and laboratory evidence (elevated serum CK) of myopathy at birth, clinical manifestations are usually absent during infancy. Short stature, however, may be seen even in presymptomatic children with DMD; at birth, height and weight are normal, but during early childhood, growth curves fall off the normal percentiles for age.[147] During the first two years of life, some affected boys exhibit mildly delayed gross motor development presenting primarily as delayed sitting, standing, and independent walking. The mean age of walking is approximately 18 months (range 12–24 months). Parents of children with DMD usually identify the following first symptoms: gross-motor delay (42%), delay in walking (20%), gait difficulties, including persistent toe-walking and flat feet (30%), learning problems (5%), and speech delay (3%).[148] Many parents also notice frequent falls and calf enlargement.[149] Weakness is usually noted between 2 and 3 years of age; in some cases, however, signs of weakness may be delayed and become apparent after the age of 3 years. Weakness selectively affects proximal limb muscles before distal and the lower extremities before the upper ones, and presents usually as difficulty with running, jumping, standing up from a squatting position, and climbing stairs.[150] The mean age at diagnosis in males with negative family history for DMD is approximately 4 years 10 months, with a range of 16 months to 8 years.[151,152] Between 3 and 6 years of age, a broad-based waddling gait, exaggerated lumbar lordosis, and calf enlargement are usually observed. Early on, patients may complain of leg pains. In arising from a supine position on the floor, affected boys turn their face to the floor, spread their legs and use their hands to climb up their thighs to an upright position (Gowers' sign) (Figure 30.4A,B), thus compensating for weak pelvic girdle muscles. Neck flexor weakness often goes unnoticed but it does occur at all stages of the disease; when lying on his back, even at an early stage, a child with DMD is unable to lift his head against gravity. This sign distinguishes boys with DMD from patients with milder phenotypes, such as "Outliers" and BMD, who, at least early on, exhibit preservation of neck flexor strength.[153]

Physical examination shows enlargement of the calf muscles and, in some instances, of the quadriceps, gluteal, deltoid, infraspinatus, and rarely masseter muscles. The calf muscles feel firm and rubbery to palpation. The enlargement, early in the course of the disease, is the result of true hypertrophy, however later on is related to replacement of the muscle fibers by fat and connective tissue (pseudohypertrophy) (Figure 30.5A,B, Figure 30.6).[153] With the exception of the sternocleidomastoids, cranial-nerve innervated muscles and also the levator ani and the external sphincter muscles remain essentially unaffected.[154,155] The external ocular muscles are always spared; the tongue, however, often enlarges. The ankle plantar flexors (gastrocnemius and soleus) and invertors (tibialis posterior) remain very strong until late in the course of the disease while the tibialis anterior and peroneal muscles become weak gradually, resulting in heel cord contractures and toe walking; eventually, this

FIGURE 30.4 Gowers' sign in a boy with Duchenne muscular dystrophy. This series of drawings, from left to right, illustrates the various components of a classic Gowers' maneuver. In the upper panel (A), the patient uses hand support on the floor, initially bilateral then unilateral, and hand support on the thighs, either unilateral or bilateral, in order to attain the standing position. The lower panel (B) shows the "butt first" maneuver, which evolves into a classic "one-handed" or "two-handed" Gowers' sign. *Drawings courtesy of Mrs. Julianna Wittrédi, Geneva, Switzerland.*

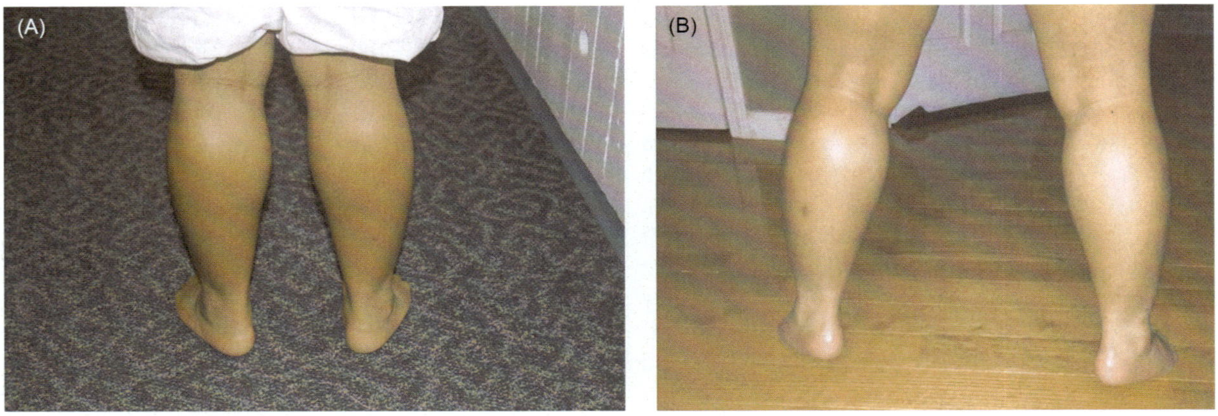

FIGURE 30.5 A boy with Duchenne muscular dystrophy, at the ages of 8 years (**A**) and 11.5 years (**B**). Note enlargement of gastrocnemii muscles bilaterally, known as "pseudohypertrophy." Also note the progression in foot position from plantigrade (**A**) to mild equinovarus (**B**).

imbalance in muscle strength will lead to bilateral equinovarous foot deformities (Figure 30.5B). Neck flexors are weaker than extensors, biceps and triceps are weaker than deltoid, wrist extensors are more involved than flexors, and the quadriceps more than hamstrings.[156]

Between 3 and 6 years of age there may be some evidence of transient improvement, known as the "honeymoon phase" of muscular dystrophy; this is related to the fact that the normal motor developmental progression of the child outpaces disease evolution, but this phase is followed gradually by relentless deterioration. Between age 6 and 11 years, muscle strength decreases almost linearly but the ability to climb stairs, stand from a supine position, climb stairs with rails, go to all fours from supine, and walk a short distance (750 cm) declines rather rapidly in the mentioned order.[157] Loss of function, therefore, does not correlate well with loss of strength of selected muscle groups.[157] When evaluated using the Vignos scale (1, normal function; 9, nonambulatory), most DMD patients remain in functional grade 2 for a long time and then lose function very fast, over a period of 2 or 3 years.[157,158] The speed at which function is lost, however, may vary substantially among children with DMD. When patients are evaluated with the 6-minute walk test (6MWT), children with DMD younger than 7 years of age may show stability or increase in distance walked (6MWD) over about one year despite progressive muscular impairment.[159] Baseline 6MWD and age ($\geq$7 years) are strong predictors of $\geq$10% worsening in 6MWD and decline in ambulation in general. In a recent study, baseline 6MWD of $<$350 meters was predictive of greater functional decline, with loss of ambulation seen only in those patients with a baseline 6MWD $<$325 meters.[159] Patients $\geq$7 years of age usually decline, but some may remain stable or even show improvement in their 6MWD over 48 weeks (Figure 30.7). As the disease progresses, deep tendon reflexes diminish or cannot be elicited; before the age of 10 years, the triceps, biceps, and knee reflexes become hard to elicit in approximately 50% of patients. The brachioradialis stretch reflex remains active longer, with the ankle reflex being elicited in one-third of patients even during the last phase of the disease.[160]

The development of joint contractures is another almost constant clinical feature of DMD. Between age 6 and 10 years, the majority of DMD patients (70%) develop contractures of the heel cords, iliotibial bands, and hip joints, causing toe walking and limitation of hip flexion.[158] By age 8 years, knee, wrist extensor and elbow contractures appear, and correlate with decreasing ambulation.[153] Shoulder contractures occur only late in the disease course. In most patients, the weakness leads to dependency on long leg braces for ambulation at around age 10 years[153] and wheelchair confinement by approximately the age of 11–13 years (Figure 30.6).[161] Weakness of muscles in the upper extremities increases gradually, and wheelchair-dependent patients can perform only minor movements involving their hand and forearm muscles. Muscles of all extremities and trunk become atrophic once ambulation is lost. Wheelchair-bound children tend then to develop scoliosis due to weakness of paraspinal muscles, and gradually their pulmonary function deteriorates related to the weakening of respiratory muscles and probably the increasing scoliosis. The progressive weakness of respiratory muscles is assessed by pulmonary function testing, which shows gradual decline in vital and total lung capacities beginning at the age of 8 or 9 years.[162,163] With advancing disease, carbon dioxide retention and anoxemia may occur even without infection; nighttime carbon dioxide retention usually presents as morning headaches upon awakening.

The majority of patients die during the period of late teens to late twenties, more frequently from respiratory

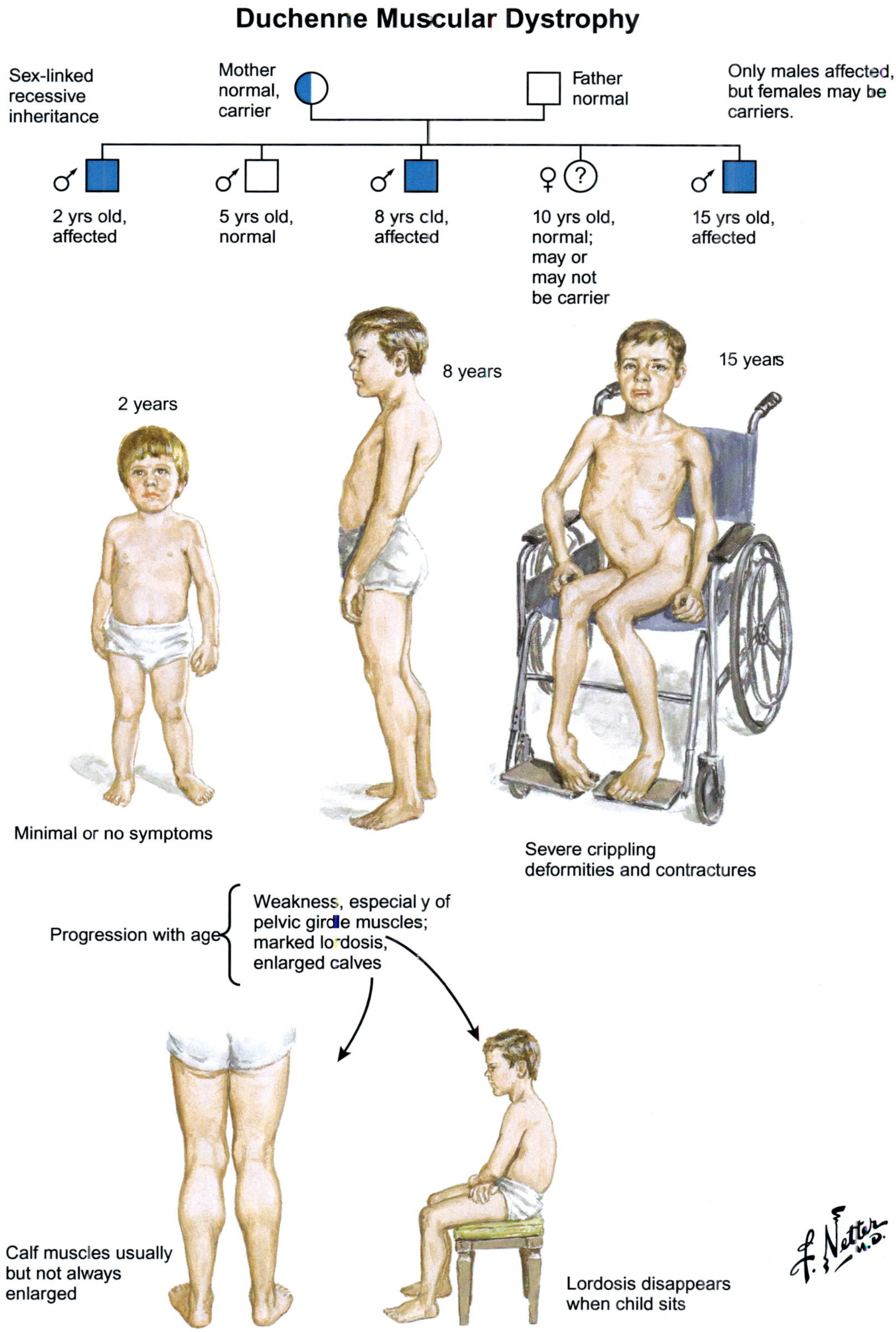

FIGURE 30.6 Symptomatic progression of Duchenne muscular dystrophy. *From Jones HR, Burns TM, Aminoff MJ, Pomeroy SL. Nervous System: Part II, Spinal Cord and Peripheral Motor and Sensory Systems. Vol 7 in Netter FH. The Netter Collection of Medical Illustrations, 2nd Ed. Elsevier: 2013, p 265, plate 12-9. All rights reserved.*

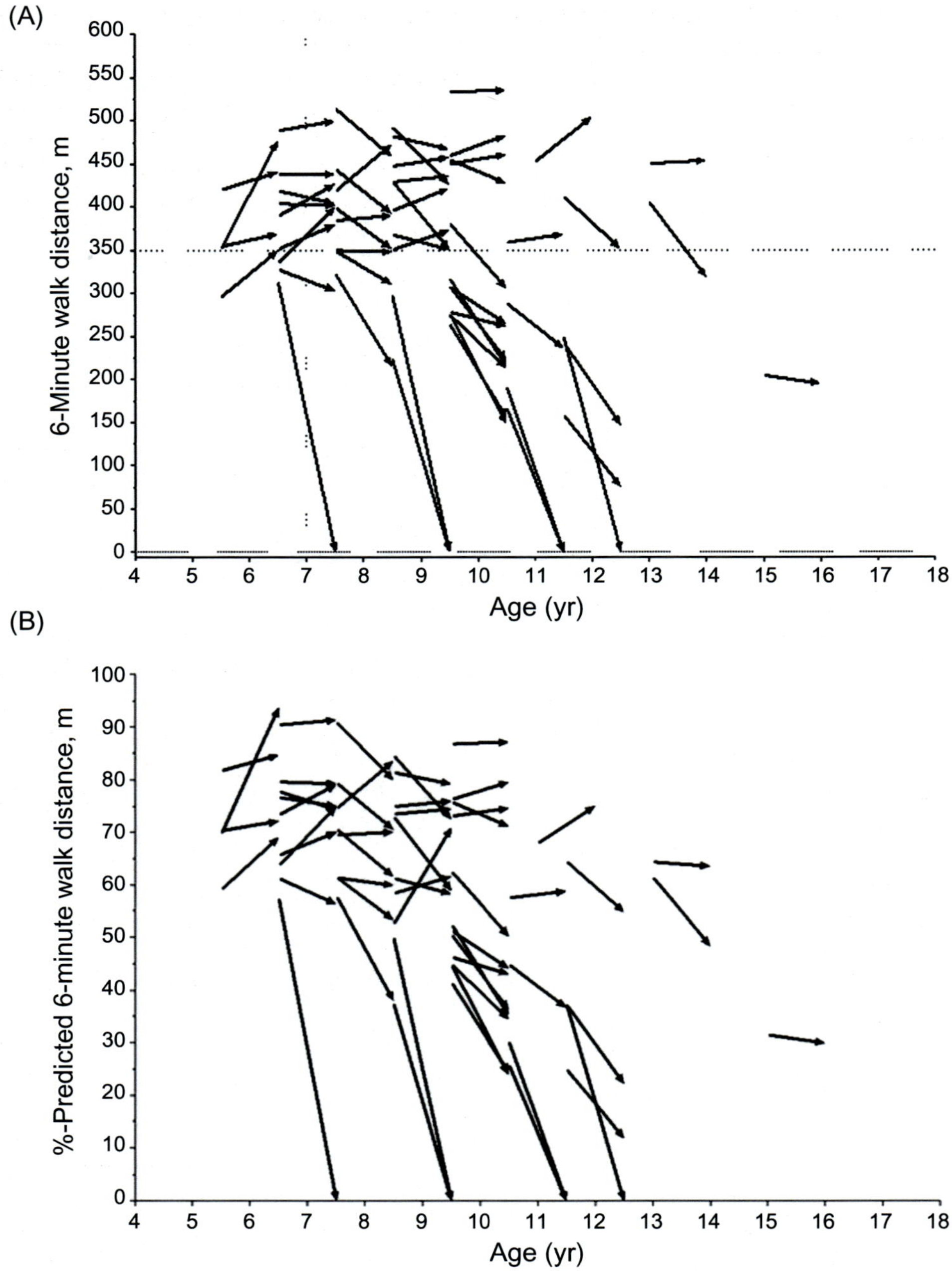

FIGURE 30.7 (**A**) shows change in 6-minute walk distance (6MWD) over 48 weeks in a cohort of 57 placebo-treated DMD patients. (**B**) shows change in percent predicted 6MWD value calculated using age and height. Note the tendency of the 6MWD values to decline in most patients ≥7 years of age and to improve or be stable in patients younger than 7 years of age. *Reproduced with permission from McDonald et al.*[159] *Courtesy of Allen Reha, MS, PTC Therapeutics, South Plainfield, NJ, USA.*

failure with or without an associated pulmonary infection, but also from cardiac failure secondary to progressive cardiomyopathy.[164] However, survival into the thirties has been observed in patients receiving ventilatory support.[164] In some cases, the immediate cause of death is not apparent. A recent study of patients with molecularly confirmed diagnoses has determined a median survival of 24.0 years, with ventilated patients reaching a median survival of 27.0 years; for unventilated patients, the median survival was 19.0 years.[165] In another study, the overall age for cardiac deaths was 19.6 years (range 13.4–27.5); for respiratory patients without ventilatory support, the age at death was 17.7 years (range 11.6–27.5) but increased to 27.9 years (range 23–38.6) in patients who received mechanical ventilation.[164] In a cohort of patients having both spinal surgery and nocturnal ventilation, the median survival was 30 years.[166]

Other Organ Involvement

Heart, brain, and smooth muscle are also involved in the disease process, most likely due to the lack of dystrophin expression in these organs.

Heart

As a result of evolving cardiomyopathy, 90% of patients with DMD exhibit abnormalities in their electrocardiogram (ECG)[167,168] (increased R/S amplitude ratio in lead V_1, deep Q waves in left [V_5, V_6] precordial leads, increased QT dispersion). Intra-atrial conduction disturbances, sinus tachycardia, or other sinus arrhythmias are more frequent than atrioventricular conduction defects and infranodal/ventricular abnormalities. The incidence of cardiomyopathy increases gradually in teenage years, with about one-third of patients being affected by age 14 years, one-half by age 18 years, and all patients after age 18 years.[169] In a series of 78 steroid-naive DMD patients less than 6 years of age, ECG abnormalities were identified in 78% but only 1 echocardiogram was abnormal.[170]

Echocardiography shows diminished contractility of the posterobasal ventricular wall and adjacent left ventricular myocardium. Subclinical or clinical cardiac involvement occurs in about 90% of patients with DMD/BMD, but it is the cause of death in only 20% of individuals with DMD.[171] The majority of patients with DMD remain free of cardiovascular symptomatology until late in the disease course, probably due to their inability to exercise, which may mask the cardiac symptoms.[126] In the late stages of the disease, congestive heart failure and arrhythmias may develop, especially during intercurrent infections; in very rare cases, congestive heart failure dominates the picture and can be the immediate cause of death without marked compromise of respiratory function.[153] Pericardial effusion and cardiac tamponade as well as myocardial inflammation precipitating heart failure have been described in patients with DMD.[172,173]

Brain

There is wide variation in intellectual performance across individuals diagnosed with dystrophinopathy. As a group, IQ scores are shifted down from the normative population about one standard deviation, yet the frequency distribution does not differ significantly, such that most affected boys score within the average range and about a third score in the intellectually delayed range.[174]

Recent work has demonstrated that lower scores on cognitive tests are evident early in life. Among infants with dystrophinopathy, developmental quotients on standardized tests are lower than expected, with performance on cognitive, language, and motor tasks shifted down about one standard deviation.[175] Further, parents of affected children report a greater number of delayed language and motor developmental milestones than in unaffected siblings.[176] Preschoolers with dystrophinopathy struggle with language delays and generalized cognitive and adaptive skills deficits.[177,178] As such, the cognitive aspect of the phenotype occurs prior to obvious physical limitations of the disease.

Most research has examined cognitive skills among school-aged children with dystrophinopathy. Physical disability, overall level of intellectual function, environmental background and age variables may all influence results. Nonetheless, some findings appear to be consistent across studies. Most studies have found that individuals with dystrophinopathy have compromised verbal and reading skills and limited immediate verbal memory.[179–188] There are also data indicating poor executive functions, attention, and visuospatial memory,[189–191] although findings in these areas are less consistent across studies. What is very clear is that there is a cognitive phenotype associated with dystrophinopathy that impacts general academic performance and overall quality of life. The extent of the cognitive impairment is variable, yet all affected individuals are at an increased risk for cognitive difficulties.

There is also a behavioral phenotype associated with the disorder. Children with dystrophinopathy are at increased risk for poor social skills, as well as limited attention and difficulties with mental flexibility and executive control.[191–194] Some of this may be related to the language difficulties many boys have. These characteristics in exaggerated form underlie the diagnosis of autistic spectrum disorder, and among boys with dystrophinopathy the incidence of autistic spectrum disorders is higher than in the general population, with rates estimated to range from 3 to 18 percent.[193,195,196] Additionally, as boys age and experience the physical and emotional challenges associated with the disorder, the risk for reactive depressive and anxiety behaviors increases.[193,197] Children in supportive environments where their illness is discussed openly and their concerns are addressed may develop without experiencing symptoms of depression or anxiety.[198–200]

The cognitive phenotype appears to be related to mutation position, such that deletions in the second half of the gene are associated with more severe intellectual impairment.[201,202] There is a lack of consensus on whether mutation position is associated with specific cognitive profiles, although one report suggests that it is.[203]

In the multidisciplinary care of all affected children, recommendations include a neuropsychological evaluation at diagnosis and/or prior to the start of school, as well as working with the child's school to implement an individualized education program to ensure that the child receives any necessary educational support.[204] Further, individual neuropsychological assessment should be provided at any time when the child may be experiencing educational or psychosocial difficulties. Identification of the problems can determine the appropriate interventions needed to ensure an optimal quality of life for each affected individual.

A neuropathologic abnormality underlying the cognitive deficit in DMD has not been found.[153] Autopsy studies of the CNS in patients with DMD have not revealed any consistent and/or specific pathological changes.[205,206] Wingeier et al.[186] evaluated 25 boys with genetically confirmed DMD and found no clear relationship between mutation and specific aspects of cognitive function, or overall performance on standard measures of cognitive abilities. However, the investigators did note that a lack of dystrophin brain distal isoform Dp140 was associated with greater impairments overall; this observation confirms findings of a previous study suggesting that dystrophin deletions involving Dp140 are associated with intellectual impairment.[202] In males with mutations affecting Dp140, mild intellectual disability is significantly more common; also, most males with mutations involving isoform Dp71 are cognitively disabled.[207,208]

Dp71 and Dp140 are shorter isoforms of dystrophin and are highly expressed in fetal brain with gradual increase from the embryonic stage to adult. Dp71 is abundant in the hippocampus and some layers of the cerebral cortex and is sublocalized in synaptic membranes, microsomes, synaptic vesicles, and mitochondria. The location of the mutation seems to correlate with full-scale IQ (FSIQ) values; in a study of 62 boys with DMD, mutations affecting the Dp140 isoform 5′ UTR correlated less with FSIQ than did mutations affecting the Dp140 promoter or coding regions.[208] The risk of cognitive deficit increases with cumulative loss of isoforms expressed in the central nervous system.[208] Despite all this data, the precise role of brain dystrophins in the pathogenesis of cognitive impairment remains uncertain.

Smooth Muscle

Degeneration of gastrointestinal tract smooth muscle in DMD patients can lead to important and even life-threatening complications, such as intestinal hypomotility, also known as intestinal pseudo-obstruction, and acute gastroparesis; it may present with acute gastric dilatation, vomiting, and abdominal pain and distension.[153,209,210] Delayed gastric emptying has been documented in 8 of 11 DMD patients.[209] Degeneration of the outer, longitudinal, smooth-muscle wall of the stomach has been documented pathologically; this is probably related to dystrophin deficiency in smooth muscle.

Skeleton

It has been demonstrated that osteoporosis in boys with DMD begins to develop early, while they are still ambulating[211]; it is more severe in the lower extremities, and may lead to frequent fractures that, sometimes in older children, expedite the loss of ambulation. Unfortunately, fractures involving the arms and legs are frequent and occur in the setting of falls[212]; in a series of 378 DMD patients (ages 1 to 25 years), 79 (21%) had experienced fractures.[213]

Progressive scoliosis develops in most nonambulant patients with DMD because of paraspinal muscle weakness.[214] Whether the severity of scoliosis and its correction by spinal fusion determine the rate of respiratory function decline has been debated in the literature.[215,216]

Adverse Anesthetic Reactions

When halothane and succinylcholine are used for general anesthesia, children with DMD may present with malignant hyperthermia-like episodes, acute hyperkalemia in the absence of hyperthermia, profound increase of serum CK over the baseline levels, and also with signs of acute cardiac failure related to their chronic cardiomyopathy.[217–220] When compared to classic malignant hyperthermia episodes, hyperthermia and rigidity are less common in DMD but myoglobinuria is more frequent[221] (see also Chapter 45 on malignant hyperthermia). It is possible that the increased amount of calcium in the muscle fibers of DMD patients makes them more susceptible to changes in ion fluxes across the plasma membrane triggered by the anesthetic agents. The risk of adverse reactions can be reduced substantially by not using inhalation anesthetics and by inducing muscle relaxation with nondepolarizing agents instead of succinylcholine.[126]

Becker Muscular Dystrophy

In Becker dystrophy the age of onset of symptoms is usually later, between the ages of 5 and 20 years, sometimes even in the third or fourth decade or later. The mean age of onset is approximately 12 years.[222–225] Patients with clinical symptoms before the age of 8 years are sometimes difficult to distinguish clinically from DMD patients.[226] By the age of 10 years, almost 50% of the patients are symptomatic and by the age of 20 years almost 90% have clinical signs of the disease.[226] The degree of clinical involvement is milder but the pattern of muscle wasting is similar to the one noted in DMD. Pelvic girdle and thigh muscles are involved first and calf muscle (pseudo)hypertrophy occurs early in most but not all patients.[223,227,228] Tibialis anterior and peroneal muscle groups are less affected. Shoulder girdle weakness develops later after the onset of proximal lower extremity weakness. In a series of

67 patients, lower extremity weakness was noted first at the mean age of 11 years, and upper extremity weakness at the mean age of 31 years.[225] Calf muscles, forearm and hand intrinsics remain strong until the late stage of the disease. Calf pain upon or following exercise was reported as a presenting symptom in 28% and as a complaint at some point during the course of the disease in 91% of all patients. Only 2 of the 67 patients presented with myoglobinuria. Thus, muscle pain is very common in patients with BMD but myoglobinuria occurs infrequently.[229,230] As discussed earlier, in BMD, as in the intermediate type of dystrophinopathy ("Outliers"), there is relative preservation of neck flexor muscle strength; neck flexors become weak later on in the course of the disease. In patients with BMD, facial muscles are usually preserved. Contractures are less likely to develop, and scoliosis is also less common, but it may develop after the patient with BMD is confined to a wheelchair.

Cardiac disease and cognitive impairment are not as common or as severe as in the Duchenne variety, and gastrointestinal symptoms are essentially absent. Rarely, however, cardiac symptoms can be severe in patients with only mild skeletal muscle weakness[92,143,231–233] or can precede muscle weakness by several years.[234,235] BMD patients with deletions involving the amino-terminal domain are more likely to experience early-onset cardiomyopathy (mid-20s), whereas those with deletions of the rod domain and hinge 3 experience a later onset (mid-40s).[236] Although most patients with BMD have IQs within the normal range, as a group mean IQ is shifted down slightly from that for the general population. Patients with BMD typically remain ambulatory beyond the age of 16 years and into adult life; the mean age of BMD patients when they lose ambulation is in the fourth decade.[222,224–226,237] They usually survive beyond the age of 30 years,[161] with death from respiratory insufficiency or cardiomyopathy/cor pulmonale usually occurring between 30 and 60 years.[45] Cardiac involvement is the cause of death in 50% of patients with BMD.[171] Mean age at death is in the mid-40s.[151,225] When compared to unaffected brothers, fertility is about 70% in the average BMD patient but below 10% in severely affected individuals.[222,225]

The mild end of the spectrum includes men with onset of symptoms after the age of 40 years who remain ambulatory even into their 60s.[238–240] Mildly affected patients with abnormal *DMD* gene molecular genetic testing and/or dystrophin immunostaining of muscle biopsies have been classified as having (1) "benign" skeletal myopathy with subclinical findings consisting of muscle weakness in the pelvic and/or shoulder girdle with or without associated cardiomyopathy, or (2) BMD with "subclinical" skeletal muscle involvement but with elevated serum CK concentration, calf enlargement, muscle aches and cramps, and myoglobinuria upon exertion.[145,146]

"Outliers" or Intermediate Phenotype

The intermediate group of patients with mild DMD or severe BMD (called "Outliers") presents with symptoms intermediate between DMD and BMD. They usually remain ambulatory beyond the age of 13 years but become wheelchair dependent before the age of 16 years. This subgroup of patients was identified during the natural history of DMD study.[241] They can be differentiated from DMD early in life by the preservation of their ability to flex their neck against gravity or against resistance. DMD patients lack this ability even before the age of 5 years, essentially throughout their entire life.[153] Cardiomyopathy and cognitive impairment are also part of the presentation of "Outliers."[153]

DMD-associated Dilated Cardiomyopathy

In 1987 a five-generation, 63-member family was described who had dilated cardiomyopathy, but no evidence of skeletal myopathy.[242] Towbin et al.[243] showed linkage of DMD-associated dilated cardiomyopathy (DCM) to the dystrophin locus at Xp21 in this family and one other. Subsequent reports defined DMD-associated DCM as a progressive and fatal type of heart disease that presents in the second or third decade of life with congestive heart failure in patients who have little or no skeletal muscle weakness.[141,243–245] Males present in their teens and twenties, the disease course is rapidly progressive, and associated ventricular arrhythmias are common. Female carriers develop mild cardiomyopathy in the fourth or fifth decade and exhibit slow progression. The only abnormal biochemical finding is an elevation in serum CK. It has been demonstrated that in patients with the most severe cardiac phenotype, the cardiac muscle is usually unable to produce dystrophin, while skeletal muscle is unaffected.[246,247]

DMD-associated DCM may be the presenting and main feature in individuals with BMD who have little or no clinical skeletal muscle weakness. Some classify such patients as "subclinical" or "benign" BMD,[146] whereas others may classify such patients as having DMD-associated DCM with increased serum CK activity.[144] In one study of 28 patients with "subclinical" and "benign" BMD ranging in age from 6 to 48 years, 19 (68%) had myocardial involvement, although only two were symptomatic.[146] In another study of 21 patients ranging in age from 3 to 63 years (mean age 40 years), 33% had cardiac failure despite relatively mild skeletal muscle findings.[248]

Different mutations in the dystrophin gene cause selective absence of dystrophin in heart muscle.[126,144] In several patients with dystrophin-deficient DMD-associated DCM, cardiomyopathy is caused by mutations in the *DMD* gene that affect the muscle promoter (P_M)

and the first muscle-specific exon (M1) via deletion of P_M and M1 or altered mRNA splicing, resulting in no dystrophin transcripts (M isoform) being produced in cardiac muscle[141,249,250]; however, two alternative promoters that are normally only active in the brain (P_B) and Purkinje cells (P_P) are active in the skeletal muscle, resulting in sufficient dystrophin expression to avoid skeletal muscle symptoms.[251] When cardiac dystrophin was tested in two muscle promoter/exon 1-deletion cases, its level was undetectable, whereas skeletal muscle dystrophin levels were normal or reduced to as low as 30%.[142,247,252,253] The exact mechanism for this difference in transcriptional control between cardiac and skeletal muscle remains unclear. A review of all the reported families with DMD-associated DCM points out that mutations involving the 5′ end of the gene result in more severe cardiomyopathy than mutations situated in the spectrin-like region (centered around exons 48–49).[246] With mutations involving the 5′ end of the dystrophin gene, the exclusive cardiac involvement may also be related to a difference in RNA splicing regulation between heart and skeletal muscle. Also, DMD-associated DCM may also be caused by alteration of epitopes in a region of the protein of particular functional importance to cardiac muscle[254] or possibly by mutations in hypothetical cardiac-specific exons. Further, cardiac conduction abnormalities in patients with dystrophinopathies may be related to reduced expression of cardiac sodium channel NA(v)1.5 due to dystrophin deficiency.[255]

The report of a mutation resulting either in late onset cardiomyopathy, or in a Becker phenotype in the same family, raises the possibility of modification of the phenotypic expression of a specific mutation by epigenetic factors.[256]

Muscle Cramps with Myoglobinuria

Exercise-induced myalgias and cramps without skeletal muscle weakness, starting in childhood, have been demonstrated to be caused by mutations in the dystrophin gene of affected members in reported families.[257–259] In one of the families,[257] nine affected males had high resting serum CK levels, and hypertrophic calves but no clinical evidence of skeletal muscle weakness; all experienced muscle cramps triggered by normal activity, and some reported exercise-induced myalgia and cramps. DNA testing showed a deletion in the first third of the dystrophin gene involving at least exons 10–22. This deletion would be expected to remove the first few repeat units in the central rod-like domain of the protein. Western blot analysis revealed dystrophin of reduced molecular weight but no reduction in the amount of the protein present. One of the authors (BTD) has routinely tested patients presenting with muscle cramps, pains and/or myoglobinuria for dystrophin deficiency (via deletion analysis and in selected cases via muscle biopsy Western blot assay) and has identified only two patients in almost 24 years of neuromuscular practice. One of these two patients, however, had mild to moderate fixed proximal weakness in conjunction with exercise-induced myalgias, cramps, and microscopic myoglobinuria, and carried the diagnosis of BMD. Another family with cardiomyopathy and muscle cramps without clinical weakness proved to have a mutation in the dystrophin gene of the affected members.[258] Although the duration of the follow-up of some of these patients has not been long enough to rule out late-onset skeletal muscular weakness, these cases suggest that dystrophinopathies can sometimes present with signs and symptoms of a metabolic myopathy without permanent or fixed muscle weakness. Myoglobinuria without muscle weakness has also been linked to dystrophin gene mutations in several patients and has been associated with cardiomyopathy and exertional cramps[258] and/or exercise-induced myalgias.[229,257] As in the previously mentioned cases, a longer follow-up period would be necessary to definitely rule out late-onset skeletal muscle weakness in some of these patients[229,257]; it is certain, however, that dystrophin deficiency induced myalgias may persist indefinitely without progression to weakness.

Two molecular regions on the dystrophin gene have been linked to the phenotype of cramps and myalgias: exons 10–44[69,257] and exons 45–52,[229,258] both in the rod-like domain.[259] Some of the patients in these studies have also had associated fixed weakness.[69]

Asymptomatic Elevation of Muscle Enzymes

The incidental finding of an asymptomatic 11-year-old child with elevated CK and a deletion in the region of exons 33–45, together with the essential absence of mutations in the same region in symptomatic patients, has raised the possibility of mutations in the rod domain of the dystrophin gene causing a very mild or even normal phenotype with asymptomatic elevation of muscle enzymes.[69] Also, the "reading frame" theory predicts more BMD than DMD cases assuming the deletion mutations are evenly distributed throughout the gene; nevertheless, the incidence of DMD in the population is much higher than that of BMD. This suggests that many "in-frame" deletions in the rod domain of the dystrophin gene may not be detected because the individuals bearing them are either normal or only minimally symptomatic[69,260,261]; therefore, most probably they will not be seen in neuromuscular clinics or elsewhere, resulting in an underestimation of the prevalence of asymptomatic cases due to an ascertainment bias. It is also possible that some "in-frame" deletions are not detected because they result in non-DMD/non-BMD clinical features.

As discussed before, large "in-frame" deletions of the rod domain may result in severe presentation.[79] A nonsense mutation, truncating the carboxy-terminal domain of dystrophin, has also resulted in an asymptomatic dystrophinopathy in an 8-year-old boy; mRNA analysis failed to implicate an alternative splicing mechanism.[262]

Manifesting DMD/BMD Carrier Females

Carriers are usually free of symptoms but may have mildly increased serum CK and usually mild calf hypertrophy. Two studies have shown that 2.5–7.8% of DMD and BMD carriers exhibit various clinical abnormalities.[263,264] It should be noted, however, that one study[263] was done before the discovery of the dystrophin gene, and in the other[264] the ascertainment of carrier status was based on pedigree and CK analysis. More recent studies cite an incidence of clinical symptoms between 8% and 22% of female carriers of DMD mutations.[265] Manifesting carriers (MCs) can present with mild but on occasion with severe myopathy of the limb-girdle type or even DMD/BMD.[135,266] Despite the recruitment of dystrophin-positive satellite cells, severely affected females do not improve over time. If anything, their weakness increases with time due to depletion of satellite cells and thus exhaustion of the regenerative capacity of the muscle similar to that seen in DMD patients. Cardiac involvement is usually subclinical; a cross-sectional study of a Dutch cohort demonstrated that 8% of definite *DMD* gene mutation carriers had DCM, versus none in definite BMD carriers. Only 38% of the studied population of definite carriers aged 18 to 60 years had a completely normal heart investigation. The remaining had subclinical ECG or echocardiographic abnormalities (Table 30.3).[267,268] Some carriers, however, can manifest severe cardiac symptoms.[269] Longitudinal follow-up of the same Dutch cohort of 99 DMD/BMD carriers over a period of 7 to 10.6 years identified 11 cases of DCM (10 DMD, 1 BMD); 1 patient died of heart failure at age 57 years.[270] In a different group of 23 proven carriers aged 6.2 to 15.9 years, none had cardiac abnormalities.[271]

TABLE 30.3 Symptoms and Signs and Cardiac Abnormalities in DMD and BMD Carriers

Clinical Features	DMD Carriers	BMD Carriers
Myalgia/cramps	5%	5%
Muscle weakness[a]	19%	14%
No symptoms/signs	76%	81%
Left ventricular dilation[b]	19%	16%
Dilated cardiomyopathy[b]	8%	0

[a]*Mild to moderate weakness.*
[b]*These patients had no cardiac symptoms or clinical signs. Cardiac status was assessed by electrocardiography and echocardiography.*
Data derived from Hoogerwaard et al.[267]

Criteria supporting the diagnosis of MCs include a positive family history for DMD or BMD in male patients; heterozygosity for a dystrophin gene mutation; cytogenetic abnormalities (X:autosome translocations with breakpoints at Xp21 being the most common); carrier status proven via linkage analysis; persistently elevated serum CK levels, usually above 1000 IU/L, and myopathic findings in their muscle biopsies in the setting of a positive family history and/or a positive DNA test for a mutation. Thus, a positive family history for DMD or BMD is not always required for the diagnosis of MCs ("isolated" MCs). In a large study of girls and women with neuromuscular complaints, about 10% of the isolated cases of hyperCKemic myopathy were found by dystrophin analysis to have a dystrophinopathy as the cause of their disease.[272] Their muscle biopsies showed a mosaic of dystrophin-positive and dystrophin-negative fibers (Figure 30.8E, F). Females with a disease-causing *DMD* gene mutation may be symptomatic also if they have Turner syndrome (i.e. complete or partial absence of an X chromosome), X chromosome uniparental disomy, male pseudohermaphroditism due to a mutation in the androgen receptor gene,[274] or nonrandom X chromosome inactivation. The inactivation status of a female's X chromosomes can be measured in leukocytes or muscle tissue; however, some studies have shown no consistent correlation between X chromosome inactivation (XCI) ratios in leukocytes and serum CK activity, severity of clinical manifestations, or the proportion of dystrophin-negative fibers detected in muscle biopsies.[129,275] In another study of 7 proven MCs, the XCI pattern was skewed toward nonrandom (greater than 80:20) in the 4 out of 7 with deletions or duplications but was random in all 3 with nonsense mutations.[276] In contrast, Pegoraro et al. showed that more than 90% of heterozygous females with skewed XCI (defined as equal or greater than 75:25) in leukocytes developed mild, moderate, or severe symptoms.[277] The occurrence of both highly skewed[278] and completely random XCI patterns in MCs suggests lack of clear correlation possibly related to: (i) different patterns of XCI in lymphocyte DNA and muscle tissue in certain patients, or (ii) difference in methylation status between the DMD locus and the androgen receptor gene locus, which is usually tested in methylation assays.[276]

Monozygotic female twins with *DMD* gene mutations typically show discordance in manifesting the disease due to skewed X chromosome inactivation in the

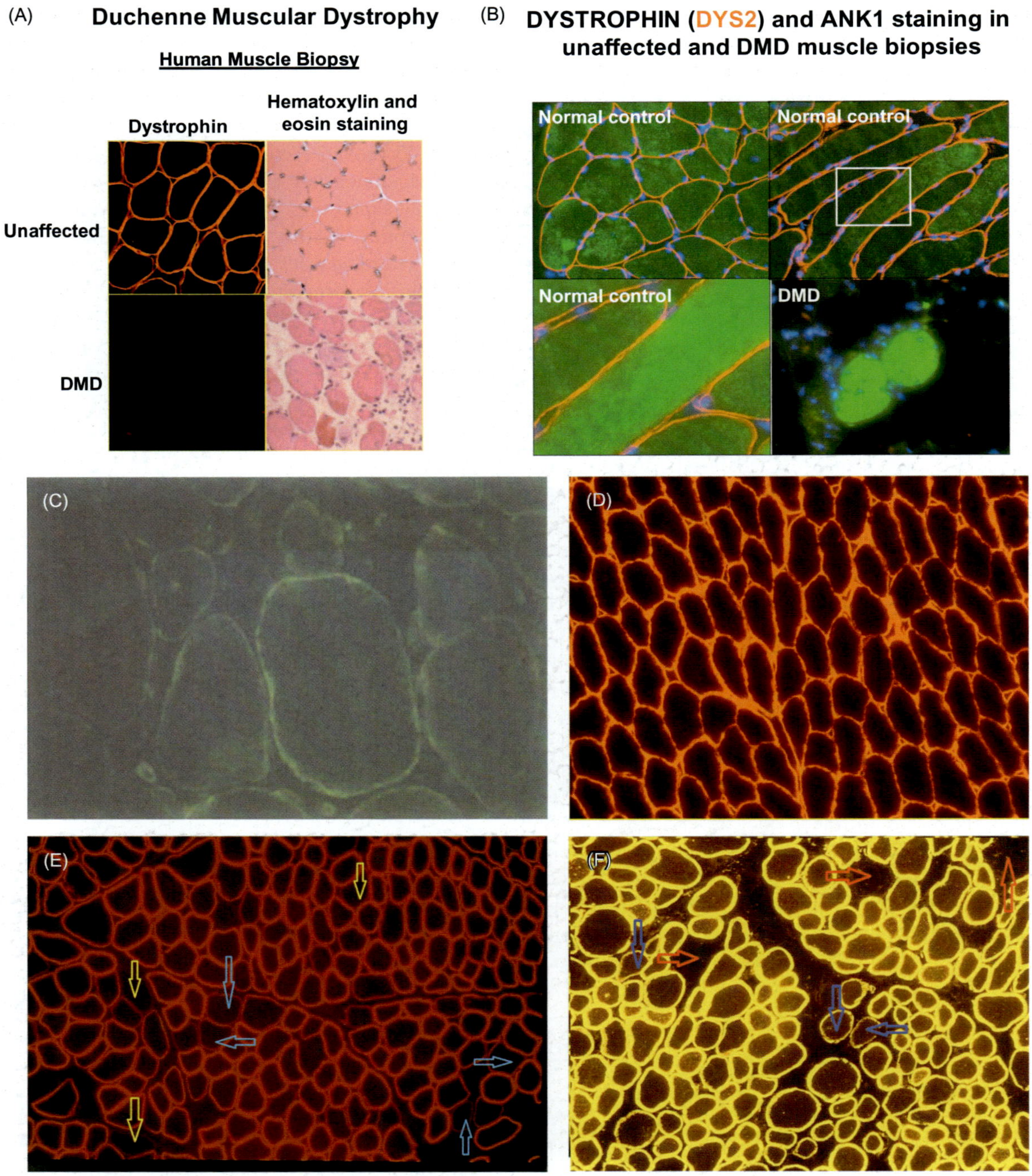

FIGURE 30.8 (**A, B**) Immunostaining of frozen sections of human skeletal muscle biopsies from unaffected controls and Duchenne muscular dystrophy (DMD) patients with antidystrophin antibodies using indirect immunofluorescence. **A** also displays hematoxylin and eosin-stained sections corresponding to unaffected and DMD biopsies, respectively. There is complete absence of immunofluorescence at the sarcolemma in a muscle section from a DMD patient, compared with the homogeneous staining of the plasma membrane in normal, unaffected muscle (**A, B**). In **B**, muscle biopsies have been immunostained with a combination of antidystrophin (DYS2) and ankyrin 1 (ANK1) antibodies; the latter stain the sarcomere green. (**C**) Immunofluorescence of frozen section of muscle tissue from a Becker muscular dystrophy (BMD) patient immunostained with antidystrophin antibodies. Note the discontinuous and weak labeling in all muscle fibers in BMD. (**D**) Immunostaining of muscle biopsy from normal female control. (**E, F**) Immunostaining of muscle biopsies from symptomatic females with antidystrophin antibodies. Note the mosaic pattern with complete loss of immunofluorescence in certain fibers (yellow, red arrows) and partial loss in others (blue arrows). *A and B courtesy of Matthew Alexander, PhD, Division of Genetics, Boston Children's Hospital. A has been reprinted from* Molecular Genetics and Metabolism, *Vol 74, O'Brien KF and Kunkel LM, Dystrophin and muscular dystrophy: past, present, and future, pages 75–88, Copyright © 2001, with permission from Elsevier. C has been reproduced from Muntoni et al.,*[273] *with permission from BMJ Publishing Group, Ltd. D, E, and F courtesy of Dr. Hart Lidov, Department of Pathology, Boston Children's Hospital.*

manifesting twin[279,280] (discussed earlier in the subsection "Dystrophinopathies in Females"). Mutations in both DMD alleles in MCs have also been described.[276,281]

DIFFERENTIAL DIAGNOSIS

Limb-girdle Muscular Dystrophies

In childhood, the clinical presentation of certain limb-girdle muscular dystrophies (LGMDs) resembles DMD or BMD except for the absence of cognitive deficits, the less frequent occurrence of cardiomyopathy, and the less marked calf hypertrophy. LGMDs comprise a wide clinical spectrum associated with elevated levels of serum CK and thus, at times, are difficult to distinguish from a sporadic case of DMD or BMD. LGMDs are transmitted in an autosomal recessive or dominant fashion; thus, if there is a clear evidence for autosomal transmission, a dystrophinopathy can easily be excluded. If not, negative genetic testing for a *DMD* gene mutation will exclude a dystrophinopathy and raise the possibility of an LGMD. LGMD2I is the type of autosomal recessive LGMD that closely resembles DMD/BMD because of the frequent respiratory and cardiac involvement; it is caused by mutation in *FKRP*, the gene encoding fukutin-related protein.[282] Details on the diagnostic approach to LGMDs can be found in Chapter 2, "Genetics of Neuromuscular Disorders," and also in Chapter 34 on LGMDs.

If symptoms of limb-girdle dystrophy are noted in a girl, one should always consider the rare possibility of a manifesting DMD or BMD carrier.

Acid Maltase Deficiency, Spinal Muscular Atrophy

Acid maltase deficiency (AMD) and spinal muscular atrophy (SMA) may both present in childhood, before the age of 5 years, with proximal muscle weakness and sometimes calf muscle hypertrophy; however, the serum level of CK (usually less than 1000 IU/L in SMA, an increase in AMD smaller than the increase in DMD), the muscle biopsy findings (i.e. glycogen storage in AMD), and electromyography will easily permit the distinction from DMD.[126] SMA is caused by mutations in the *SMN* gene (chromosomal locus 5q11-q13); molecular genetic testing is easily available to confirm the diagnosis of SMA.

Emery-Dreifuss Muscular Dystrophy

Emery-Dreifuss muscular dystrophy is another X-linked or autosomal dominant or recessive muscular dystrophy[283,284] that can at times be confused with BMD at the onset of the symptoms. However, the lack of pseudohypertrophy, the mild CK elevation, the selective involvement of biceps and triceps muscles, and the early occurrence of elbow and heel contractures will allow one to differentiate them clinically; genetic testing for Emery-Dreifuss muscular dystrophy is available (emerin, lamin A/C, *FHL1* and other gene mutation analysis) to confirm the diagnosis.

Congenital Muscular Dystrophies

As mentioned previously, in very rare instances dystrophinopathies can present similarly to congenital muscular dystrophies[285,286] with neonatal weakness, hypotonia, and subsequent developmental delay. The differential diagnosis can be made clinically by the early onset, the modestly increased CK level (albeit maybe markedly elevated in congenital muscular dystrophies also), and the frequent facial involvement and contractures seen in congenital muscular dystrophies. Genetic testing and muscle biopsy immunostaining may also be helpful in differentiating the two entities.

Mitochondrial DNA Depletion Syndrome

Mitochondrial DNA (mtDNA) depletion syndrome (MDS) is an autosomal recessive mitochondrial cytopathy presenting with a wide spectrum of clinical manifestations, such as weakness, hypotonia, developmental delay, and/or seizures, and categorized as myopathic, encephalomyopathic, hepatocerebral, or multisystemic forms.[287] Patients with MDS may have involvement of single or multiple organs, including the central and/or peripheral nervous system, muscle, liver, and kidney. Vu et al.[288] described five patients with MDS; age at onset ranged from 3 months to 5 years, and two patients had laboratory and clinical features suggestive of dystrophinopathy. CK levels in these two patients were high (4229/6000 IU/L) and muscle biopsies in all five showed abundant ragged-red fibers. Thus, mtDNA depletion should be considered in children with weakness and high CK values, particularly if associated with multiorgan involvement (e.g. liver, brain, peripheral nerves).

Dermatomyositis

Dermatomyositis before the appearance of the skin rash can be mistaken for a dystrophinopathy, but the acute or subacute onset of symptoms and the muscle MRI and/or biopsy findings usually permit easy distinction from DMD or BMD.

Elevation of Liver Enzymes

The liver enzymes alanine aminotransferase (ALT) and aspartate aminotransferase (AST) are also present in muscle and thus are elevated in DMD and BMD.[289] Therefore, an unexplained persistent elevation of liver enzymes should always alert the clinician to the possibility of an underlying muscular dystrophy; this is true even in females, who could prove to be symptomatic carriers, asymptomatic carriers of DMD/BMD, rarely symptomatic patients with DMD/BMD (e.g. if they have Turner syndrome), or patients with other muscular dystrophies.[290] In these instances, measurement of serum CK, which is a more specific muscle enzyme, should always be made to prove the myogenic origin of the elevated liver enzymes. See Case Example 30.1.

Dilated Cardiomyopathy

Finally, a dystrophinopathy should be considered in all cases of dilated cardiomyopathy (DCM). DCM can be sporadic or familial. In a large series of patients, one-third of cases had sporadic DCM and two-thirds had familial DCM. Familial DCM can include patients with autosomal dominant (56%) or autosomal recessive (16%) inheritance, X-linked recessive due to *DMD* gene mutation (10%), autosomal dominant with subclinical muscle disease (7.7%), DCM with conduction defects (2.6%), and unclassified (7.7%).[291,292]

Other X-linked infantile cardiomyopathies caused by mutations in the *G4.5* gene (chromosomal locus Xq28) include Barth syndrome, X-linked endocardial fibroelastosis, left ventricular noncompaction (LVNC), and severe X-linked infantile cardiomyopathy.[293] The *G4.5* gene encodes a novel protein family called "tafazzins."[294]

DIAGNOSTIC TESTING

Serum Muscle Enzymes

Before the age of 5 years, the serum creatine kinase (CK) levels are usually 20–200 times the upper limit of normal in BMD, and 50–200 times in DMD.[153,241] Thus, CK values in the order of 10,000–50,000 IU/L are not unusual in DMD/BMD. In a child with DMD, during the first 3 years of life, the serum CK concentration is always more than 10 times the upper limit of normal (Table 30.4); if it is less than that, the diagnosis should be

CASE EXAMPLE 30.1 Duchenne Muscular Dystrophy

A 2-year-old boy was referred to our hospital for evaluation of hepatomegaly and elevated liver function tests. He was well until 1 month before admission, when he was hospitalized elsewhere for gastroenteritis and dehydration. At that time his liver was felt to be enlarged and liver function tests were obtained. The AST was noted to peak at over 1000 U and an ALT was in the 500s, although the bilirubin, alkaline phosphatase, and hepatitis screens were normal. The patient improved clinically but was referred to our hospital for further evaluation of the enlarged liver and abnormal laboratory values.

Initial examination was notable for a temperature of 38.3°C, but otherwise normal vital signs. The liver was down 3 cm, with a 10-cm span, and a spleen tip was not palpable. Height, weight, and head circumference were appropriate for age, and the rest of the examination was initially felt to be unremarkable. Liver function tests were repeated and revealed an AST of 123 U, an ALT of 253 U, and LDH of 353 U, with normal bilirubin and alkaline phosphatase clotting studies.

Question and Comment

What is the diagnosis and what further laboratory tests might be of help?

The differential diagnosis at this point was that of hepatomegaly in an otherwise previously healthy male. While infectious etiologies need to be considered, particularly viral etiologies, one also needs to evaluate the possibility of storage diseases, infiltrative processes, or toxins (although the history was noncontributory for this possibility). An ultrasound was performed and showed a borderline enlarged liver and normal spleen.

Just when it appeared that the child might remain a diagnostic dilemma, a CPK returned at 5880 U, and it became clear that muscle rather than liver was making the significant contribution to the enzyme elevations. In fact, on repeat careful neurologic examination, the calves appeared slightly larger and firmer than one might expect. Moreover, strength was slightly diminished in the deltoids and iliopsoas muscles, although the patient did not have a classical Gowers' sign, or have a gait disturbance. A subsequent muscle biopsy, however, confirmed the diagnosis of Duchenne muscular dystrophy. Western blot analysis of muscle protein showed almost complete absence of dystrophin (less than 5% of normal). Also, dystrophin was not detectable by immunohistochemistry in muscle biopsy tissue sections. Dystrophin gene deletion analysis by multiplex PCR showed an out-of-frame deletion involving exons 41, 42, and 43.

We mention this case because usually an elevated AST with ALT point to liver pathology, although, as demonstrated here, a muscle disorder can also present this way. It is probable that without his viral gastroenteritis prompting initial studies, the diagnosis of muscular dystrophy (especially with a negative family history and normal developmental milestones to date) would have gone undetected for at least several more months. Of note, a check of mother's CPK was normal, although the normal finding is often seen in female carriers when they reach adulthood. In fact, analysis of gene dosage indicated that she carried the same deletion in the dystrophin gene known to occur in her son; therefore, she is carrier.

TABLE 30.4 Serum Creatine Phosphokinase (CK) Concentration in the Dystrophinopathies

	Percent of Patients	Serum CK Level
Males		
DMD	100%	>10 × ULN
BMD	100%	>5 × ULN
DMD-associated DCM	Most patients[a]	Increased
Female Carriers		
DMD	~50%	2–10 × ULN
BMD	~30%	2–10 × ULN

[a]*Normal CK levels have been reported in DMD-associated DCM.*
Abbreviations: CK, creatine phosphokinase; ULN, upper limit of normal; DMD, Duchenne muscular dystrophy; BMD, Becker muscular dystrophy; DCM, dilated cardiomyopathy.
Modified from Darras et al.[295]

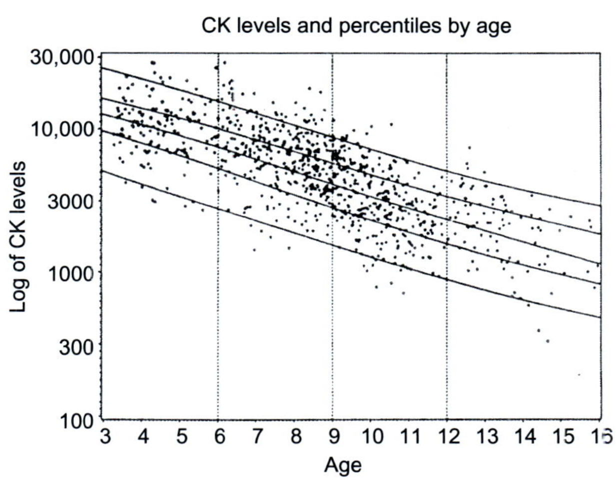

FIGURE 30.9 In Duchenne muscular dystrophy, the concentration of creatine kinase (CK) declines gradually with advancing age, at a rate of about 20% per year. *From Brooke et al.[241] Reprinted by permission of John Wiley & Sons, Inc.*

questioned. CK levels in DMD are elevated even during the first year of life when the child is totally asymptomatic, and they peak at approximately the age of 2–3 years.[296] The concentration of CK, however, tends to decline gradually with advancing age, at a rate of about 20% per year[241] (Figure 30.9); the decline is related to the progressive elimination of dystrophic muscle fibers, which are the source of the elevated serum CK levels.[24,297] Although on average the values are lower than in DMD, serum CK concentration can be markedly increased in Becker dystrophy and, therefore, cannot be used as a way of differentiating between the two types of dystrophy (Table 30.4). At the age of 20 years, in a series of 52 BMD patients, the mean serum CK concentration was increased 35-fold[225]; the highest elevation, however, is observed in the first 10 years of life and, as in DMD, it declines with advancing age and disease progression.[222,224,225] In patients with mild BMD, or BMD patients with mild skeletal muscle involvement but early cardiomyopathy,[258,298] and also in the ones with muscle pains, cramps upon exertion but minimal baseline weakness,[258] serum CK concentration is increased from 3 to 20 times the upper limit of normal.

Blood aldolase, a tetrameric enzyme, is similarly increased in both BMD and DMD. It is a glycolytic enzyme found in almost all tissues, but predominantly in skeletal muscle, liver, brain, and erythrocytes; aldolase levels, therefore, are not as specific or as sensitive for the diagnosis of muscle disease as are CK levels. Hemolysis of a blood specimen is the most common cause of spurious aldolase elevation in patients with normal serum CK and no liver disease. Serum aminotransferases, AST and ALT, are also present in muscle and, therefore, they can also be elevated in DMD and BMD. The mechanism of their increased levels is not completely understood, but is thought to be related to leakage through defective muscle membranes.[290] Their values tend to follow a course similar to that of CK levels, with a decrease in the late stages of DMD/BMD; although AST and ALT are somewhat less reliably linked to disease stage than the serum CK level,[299–301] they tend to be highest in ambulant boys with DMD.[289] Serum AST activity is the least organ specific, due to its presence in liver, kidney, pancreas, lung and erythrocytes, in addition to heart and skeletal muscle ALT is more liver specific due to its increased concentration in liver compared to muscle and other organs. Other enzymes increased in DMD, BMD, and other dystrophinopathies are pyruvate kinase, glucose phosphate isomerase, enolase, carbonic anhydrase III, and lactate dehydrogenase (LDH). LDH, another tetrameric enzyme, is the only one of this last group of enzymes that can be assayed easily by clinical laboratories, but it lacks specificity and thus utility in the diagnosis of DMD/BMD; it is found in almost every tissue in the body and as a result is increased in a variety of human diseases. CK is present only in skeletal muscle, heart and brain, but not in other organs; thus, it is the most useful and cost-effective enzyme assayed in clinical neuromuscular practice.[290]

CK levels are usually increased in DMD-associated DCM, but normal levels have been reported.[291] In DMD and BMD female carriers, serum CK concentration is increased 2–10 times the upper limit of normal in approximately 50% and 30% of carriers respectively (Table 30.4).[267,302] In a study of 129 DMD/BMD carriers, the mean serum CK level was 306 IU/L (range: 48–1860 IU/L; upper limit of normal: 193 IU/L). The difference in mean serum CK concentration between asymptomatic and symptomatic carriers was not statistically significant.[135,267,296,302] Thus, a normal CK level does not exclude DMD or BMD carrier status in a female. In fact, serum CK levels vary widely in DMD/BMD carriers, with the mean serum CK concentration significantly increased in carriers younger than 20 years compared with the concentration in those older than 20 years, in whom CK levels tend to stabilize at lower abnormal values or normalize,[275] hence the clinical practice of checking CK levels earlier in life (preferably during the first few years) in prospective female carriers.

CK is a dimer molecule with three distinct isoenzyme forms—known as MM, MB, and BB—which can be identified electrophoretically. It consists of muscle-specific (M) and brain-specific (B) monomers.[303] Skeletal muscle has the highest concentration of CK of any tissue in the human body. Normal mature skeletal muscle contains more than 99% MM with small amounts of MB isoenzyme.[304,305] The BB isoenzyme is abundant in brain tissue. Cardiac tissue has the highest concentration of CK-MB isoenzyme (approximately 20% of cardiac CK). In DMD and other dystrophinopathies, the concentration of the MB fraction increases, probably due to the increased number of regenerating muscle fibers in these conditions[306–309] which revert to the fetal state of increased MB isoenzyme production.[310] When the serum CK values are extremely high, as occurs early in the course of DMD and BMD, the serum concentration of CK-MB isoenzyme is so much increased that it could not originate exclusively from the heart; nonetheless, at least a fraction of serum CK-MB in patients with DMD/BMD probably does emanate from the heart.

Other Serum and Urine Abnormalities

Myoglobinemia is frequent in patients with DMD and BMD,[311–313] with the highest serum myoglobin concentration detected in patients with the highest CK concentration and the highest level of physical activity. In ambulatory patients, wide fluctuations are noted, which correlate well with the amount of physical activity. As DMD patients gradually lose the ability to ambulate, serum myoglobin levels decline. In normal individuals, mild exercise does not change significantly serum CK or myoglobin levels. After prolonged and intensive exercise, levels of serum myoglobin peak within 2 hours and CK levels after 10 to 18 hours; myoglobin concentration returns toward normal when serum CK reaches maximal values.[313] In DMD patients, after mild physical exertion, serum CK levels increase significantly within 4 hours (as much as four-fold) following an earlier and proportionate increase in serum myoglobin concentration.[313] This suggests that in patients with DMD, even mild exercise can induce muscle fiber membrane damage. Nevertheless, in the DMD group, barring anesthesia-induced rhabdomyolysis, macroscopic myoglobinuria is highly unusual, but it does occur infrequently in BMD patients.[229,230]

Electromyography

Electromyography (EMG) is used very rarely in the diagnosis of dystrophinopathies. EMG is useful for distinguishing a myopathic process from a neurogenic disorder, but the findings in cases of dystrophinopathies are nonspecific and occur in all myogenic disorders. These findings include short-duration, low-amplitude, polyphasic, rapidly recruited motor unit potentials. As the disease progresses, the interference pattern becomes incomplete because of reduced recruitment, and eventually the muscle becomes electrically silent.

Muscle Biopsy

Histopathological Findings

By light microscopy, the muscle biopsy demonstrates degeneration, regeneration, isolated "opaque" hypertrophic fibers, and significant replacement of muscle by fat and connective tissue (Figure 30.10A, C, D). The degenerating necrotic fibers are recognized on trichrome staining by their green-blue color (instead of the normal deep blue staining) and their glassy or homogeneous cytoplasm (Figure 30.10C). Histochemical reactions for oxidative enzymes and/or myofibrillar ATPase are either reduced or lost. Most necrotic fibers are invaded by mononuclear cells, a few have no invading cells within their cytoplasm, and one-fifth are essentially replaced by invading mononuclear cells.[314] Increased variability in muscle fiber size is also observed, with larger than normal fibers in younger patients followed, with progression of the disease, by the appearance of numerous muscle fibers having a size smaller than normal.[315] Regenerating fibers are clusters of small fibers with basophilic cytoplasm and large vesicular nuclei with prominent nucleoli. Regenerating fibers express the fetal isoform of myosin heavy chain[316] and utrophin throughout the sarcolemma.[111,317] By myofibrillar ATPase staining, they are type 2C fibers, thus the increased proportion of type 2C fibers in DMD.[318,319]

"Opaque" fibers are also known as hypercontracted, hyaline, or large dark fibers (Figure 30.10D). Their origin is unclear but they could be produced at the time of the muscle biopsy and be an artifact of tissue removal,[320] perhaps enhanced by the fragility of the plasma membrane; thus, their significance in DMD has been disputed. Experimentally, hypercontraction develops when defects

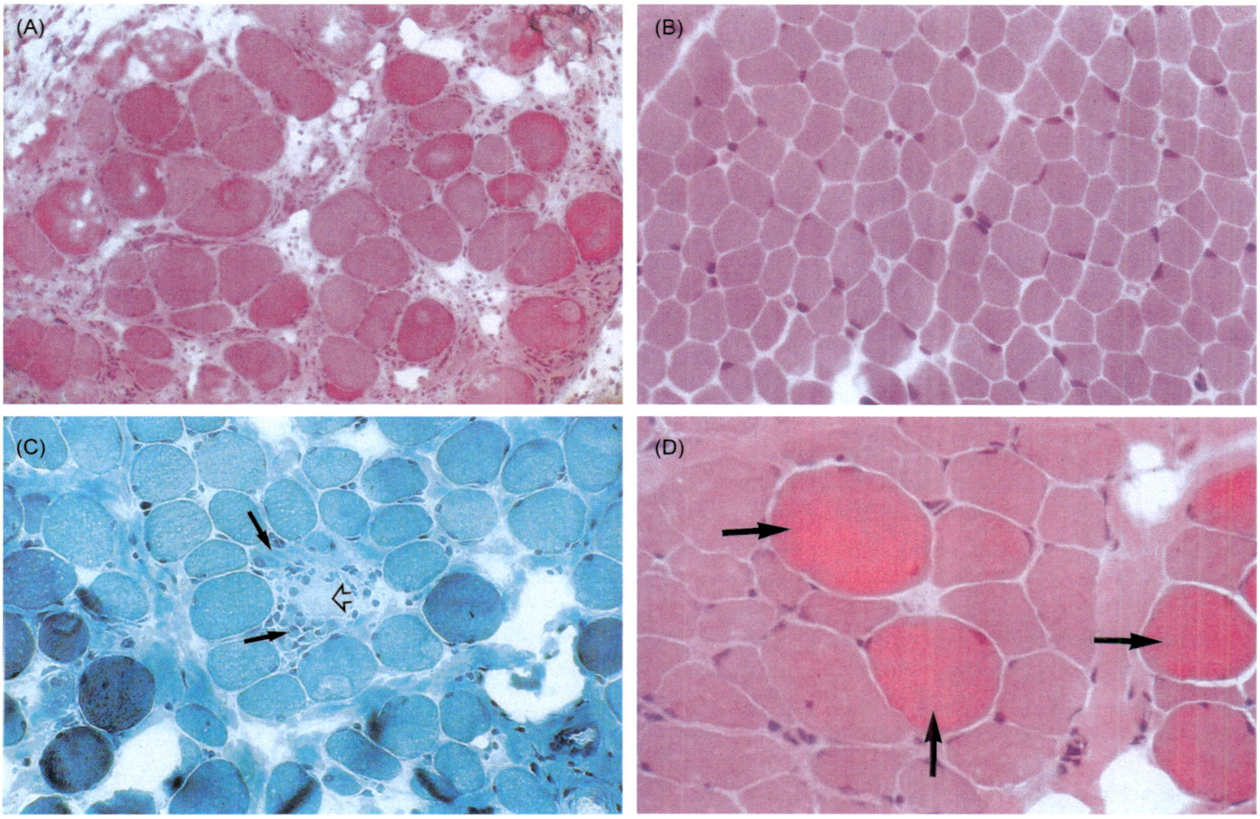

FIGURE 30.10 Muscle biopsy of Duchenne muscular dystrophy (**A**, **C**, **D**) and normal control (**B**). By light microscopy, the biopsy demonstrates marked variability in fiber size, degenerating and regenerating fibers, "opaque" hypertrophic fibers, and replacement of muscle by fat and connective tissue (most findings shown in **A** and **C**). Note the lack of these findings in the normal muscle biopsy (**B**). **C** shows show small regenerating muscle fibers (arrows) and a necrotic fiber invaded by mononuclear cells (open arrow). In **D**, arrows indicate the dark-staining "opaque" or hypercontracted large fibers. (**A**, **B**, **D**, hematoxylin-eosin; **C**: modified trichrome.)

in plasma membrane allow the influx of calcium-rich extracellular fluid. Most hypercontracted fibers in DMD are overloaded with calcium[321] and show diminished amounts of desmin, α-actinin,[320] and other intracellular proteins, supporting the hypothesis of calcium-activated protease induced segmental muscle fiber damage. In longitudinal and serial transverse sections of muscle tissue from patients with DMD, some hypercontracted fibers exhibit a necrotic region in between the hypercontracted segments.[322] The hypercontracted nonnecrotic segments stain dark pink with eosin-hematoxylin and dark blue with trichrome stains; they are overreactive for oxidative enzyme (NADH) and glycogen (PAS) stains. Opaque fibers are not specific for DMD, as they are also seen in muscle biopsies from patients with BMD, congenital muscular dystrophy, inflammatory myopathies, and denervation atrophy.[321,323] Central nuclei are seen in 2–4% of the fibers in DMD and 7–25% of the fibers in BMD.[319] Rimmed vacuoles have been identified in BMD muscle; they seem to represent autophagic vacuoles and probably reflect chronicity of the disease.[324]

Inflammatory cell infiltrates are seen in the perimysium, endomysium, and perivascular spaces and consist mostly of cytotoxic T cells and macrophages.[325] Another striking characteristic is the development of endomysial and perimysial fibrosis with disease progression. In the neonatal period the connective tissue distribution is normal but, progressively, fibrosis and fatty tissue accumulation replace the degenerating muscle fibers; in the late stage of the disease only a few muscle fibers remain, separated by large zones of connective tissue and fat. Type III collagen is the main component of the fibrous connective tissue.[326] The differences in the microscopic appearance of muscle in DMD versus BMD correlate well with the severity of the disease, with fewer necrotic, hypercontracted, and regenerating fibers seen in milder BMD phenotypes. Variation in fiber size, mild endomysial fibrosis, and central nuclei in only a small percentage of the muscle fibers may be the only findings in mild cases of BMD.

In DMD, electron microscopy shows plasma membrane defects in nonnecrotic muscle fibers with an intact basal lamina over the defects.[327,328] The membrane defects are followed later by morphologic evidence of segmental fiber necrosis with hypercontracted myofibrils, scattered degenerating mitochondria, tubular profiles, and eventually rarefaction of the myofibrils. Although interesting, at this time electron microscopy of muscle biopsies is not performed routinely by clinical laboratories. It has been superseded by more useful and specific tests such as immunostaining and Western blot analysis.

Dystrophin Immunostaining

In muscle biopsies derived from DMD patients, there is markedly reduced or no detectable staining of the sarcolemma in almost all muscle fibers using antidystrophin antibodies (Table 30.2) (Figure 30.8A, B). The amount of detectable dystrophin immunoreactivity depends on the amount of dystrophin present in the sarcolemma but also on the type and titer of the antidystrophin antibodies utilized in the immunocytochemical analysis. Currently, the commercially available antibodies recognize epitopes in the rod-domain, and the amino-terminal and carboxy-terminal regions of the dystrophin protein.[329,330] In about 80% of DMD patients, dystrophin expression is detected in less than 1% of fibers, which are called "revertant" fibers. It is believed that the presence of dystrophin in these fibers is due to a somatic mutation in a second region of the gene that suppresses the original "out-of-frame" mutation and thus restores the reading frame[331,332]; the exact timing of this somatic reversion/suppression is not known. In approximately 60% of DMD patients, immunostaining of muscle cryosections with either the rod-domain or the amino-terminal antibodies shows faint sarcolemmal immunoreactivity not seen with carboxy-terminal antibodies.[329,330,333] A monoclonal antibody against the carboxy-terminus detected dystrophin in all 30 BMD and 24 control skeletal muscles but not in 21 DMD specimens.[85] Therefore, the carboxy-terminal antibodies provide one of the best reagents for differentiating between DMD and BMD muscle biopsies; nonetheless, see also the discussion below about the need to use at least two different antibodies. Exon-specific monoclonal antibodies are better suited for the characterization of altered dystrophins in DMD/BMD; however, they are not used routinely in clinical practice.[334]

In BMD patients, partial staining of the sarcolemma is observed (Table 30.2) (Figure 30.8C); in a few BMD cases, however, the pattern of dystrophin immunostaining is indistinguishable from that seen in normal biopsies.[225,273,330,333,335,336] When the immunostaining is reduced, the reduction may be seen uniformly in all fibers in a cryosection or may vary from fiber to fiber or be patchy within single fibers. When the epitopes recognized by a specific antibody are deleted due to a corresponding deletion mutation, membrane immunostaining is not detected, thus the need to immunostain muscle biopsies with at least two different antibodies.[82,85,337,338] In patients with other neuromuscular diseases and normal females, there is homogeneous staining of the plasma membrane (Figure 30.8D).

As discussed above, this test appears useful in identifying sporadic cases of symptomatic or asymptomatic female DMD carriers with high serum CK levels in families without a male proband or in families with no detectable mutation.[339] Thus, symptomatic and asymptomatic DMD carriers with elevated CK values may exhibit a characteristic mosaic pattern of dystrophin immunostaining (Figure 30.8E, F).[339–345] This mosaic pattern is more common in symptomatic DMD carriers[269,345] but is less commonly seen in asymptomatic DMD carriers.[346,347] Attenuated staining of a few muscle fibers has been reported also in some BMD carriers.[343] A negative (normal) result nonetheless does not exclude carrier status, due to the possibility of selective loss of dystrophin-negative fibers, with advancing age.

Western Blot Analysis

Immunoblot or Western blot analysis can be used for dystrophin quantitation and also for assessing the size (molecular weight) of the residual dystrophin protein. Theoretically, epitope-specific antibodies could also detect the regions of the protein molecule that are lost because of deletion mutations in the *DMD* gene. As with muscle biopsy immunostaining, multiple antibodies directed against different epitopes of dystrophin (e.g. amino-terminus, rod-domain, carboxy-terminus) need to be used to avoid erroneous results. Also, because the analysis is quantitative, a reference or internal standard is required to indicate the amount of protein present in the muscle specimen being tested; degradation of the specimen, related to poor handling, will lead to reduction in the amount of all proteins including the internal standard. An internal standard used in most laboratories is myosin; the intensity of the myosin signal can be used to estimate the amount of protein loaded in each lane and thus normalize the visual or densitometric estimate of dystrophin on the Western blot.[24,70,92,329]

In DMD, there is significant reduction in the amount of dystrophin, or absence of dystrophin (0–5% of normal control values), when carboxy-terminal antibodies are used for its detection on a Western blot (Figure 30.11).[92] This occurs because most DMD mutations (out-of-frame deletions or nonsense mutations) create stop codons that cause premature termination of translation and thus result in severely truncated dystrophin molecules. When rod-domain or amino-terminal antibodies are used for the Western blot analysis, amounts as high as 25% of normal

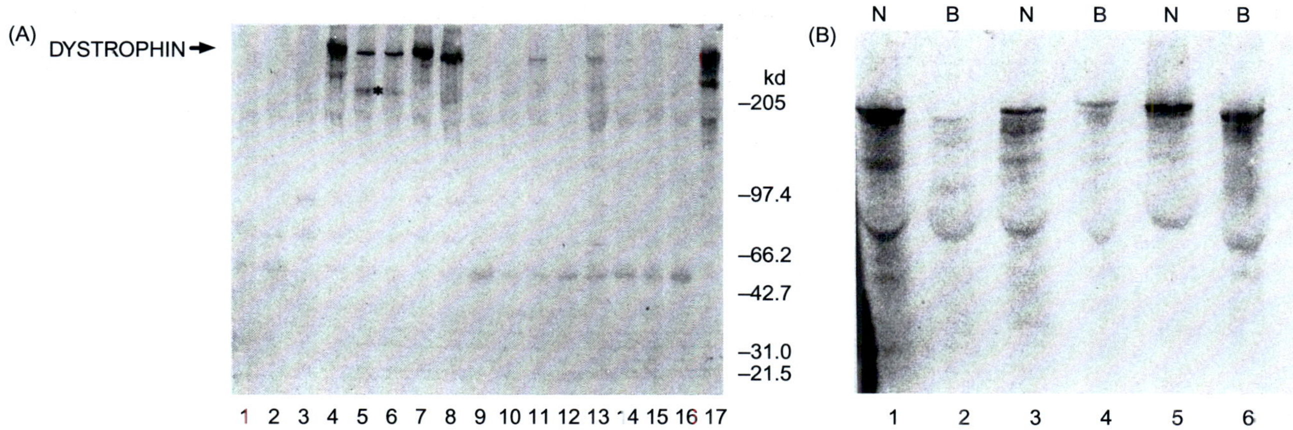

FIGURE 30.11 Dystrophin assay by Western blot. In **A**, complete or almost complete absence of dystrophin is displayed in lanes 1, 2, 3, 9, 10, 12, 13, 14, 15, and 16, corresponding to muscle biopsy specimens from patients with Duchenne muscular dystrophy (DMD). Lanes 4, 5, 6, and 7 in **A**, and 2, 4, and 6 in **B** display the amount (reduced in **A**) and size (altered, smaller or larger, in **B**) of dystrophin in patients with Becker muscular dystrophy (BMD). Lane 11 corresponds to an intermediate type of dystrophy (mild DMD or severe BMD). Lane 17 in **A** and lanes 1, 3, and 5 in **B** show dystrophin of both normal size and quantity detected in muscle specimens from normal individuals. *From Hoffman et al.*[24] *Reprinted with permission of the New England Journal of Medicine.*

may be detected in about 50% of DMD patients; the dystrophin molecule, however, is smaller than normal.[329,335,336,348] In a study of 100 patients with dystrophinopathy, all patients missing single exon 44 or 45 had some detectable dystrophin[348] with rod-domain and amino-terminal antibodies.

In BMD, Western blot analysis with either carboxy-terminal, or rod-domain, or amino-terminal antibodies shows reduced or normal quantity (20%–100%) of dystrophin of altered size or reduced quantity (20%–50% of normal) of normal protein size: decreased molecular weight (with deletion mutations) in about 80% of the cases, increased molecular weight (with duplications) in about 5%, and normal size but reduced amount in approximately 15% of BMD patients (Figure 30.11).[65,225,329,336,348–350] The size of the dystrophin protein in BMD cases is inversely proportional to the size of in-frame deletion mutations (i.e. smaller protein in larger deletions)[225]; in exceptional cases, nonetheless, where very large deletions remove the epitopes recognized by a specific antibody, dystrophin may be undetectable by a Western blot assay. This may create diagnostic confusion, especially if the patient's phenotype is very mild.[82] In patients with a deletion starting in exon 45 and extending towards exons 47, 48, 49, or 54, an additional, smaller than normal "unique" 210-to-230-kDa degradation fragment may be detected with amino-terminal and rod-domain but not with carboxy-terminal antibodies[92,225]; this fragment probably results from proteolytic cleavage of dystrophin at the deletion junctions.[350]

In a study of 97 patients with possible BMD, Hoffman et al.[92] detected a correlation between amount of dystrophin and severity of the phenotype: patients with a 3–10% of normal quantity, regardless of size, conformed to a severe Becker or intermediate phenotype and those with 20% or above had mild or moderate BMD. Most patients with abnormal sized dystrophin but none of those with normal sized molecule had dystrophin quantities exceeding 40% of normal control samples. Therefore quantities of normal sized dystrophin greater than 40% may be enough to prevent clinical manifestations, a hypothesis substantiated in mice when dystrophin is transgenically expressed at different levels.[351] In another study, dystrophin levels as low as 30% of normal were found sufficient to prevent muscular dystrophy in humans.[247] Various quantitative studies have shown overlap in the relative amount of dystrophin in patients with severe (DMD), intermediate, or milder (BMD) phenotypes, thus suggesting that differential diagnosis of disease severity based entirely on dystrophin quantitation may be inaccurate.[329] Other studies have shown also a loose correlation between the amount of dystrophin detected on a Western blot and the age at loss of ambulation.[77,352] The carboxy-terminal antibody differentiates very well DMD from BMD; however, it tends to estimate poorly the amount of residual dystrophin.[329]

In most clinical laboratories, quantitation of muscle dystrophin via Western blot analysis is performed using antibodies against the carboxy-terminus and the mid-rod domain (6-10 antibody); presumably, it is more than 99% accurate. Nonetheless, less than 1% of patients with BMD lack detectable muscle dystrophin due to the specificity of the antibodies being used. The reference ranges shown in Table 30.2 and Table 30.5 are the ones currently

TABLE 30.5 Clinical Correlation of Western Blot Analysis Results

Clinical Phenotype	Dystrophin Protein Size	Dystrophin Quantity (% of Normal)[a]
Duchenne	Normal or abnormal size	0–5%
Intermediate or severe Becker	Normal or abnormal size	5–20%
Mild/Moderate Becker	Normal size	20–50%
	Abnormal size	20–100%

[a]The reference ranges shown in this table are those currently used by clinical laboratories and reflect approximate and reconciled data from the literature.

TABLE 30.6 Summary of Molecular Genetic Testing Used in Duchenne Muscular Dystrophy and Becker Muscular Dystrophy

Gene Symbol	Test Method	Mutations Detected	Mutation Detection Rate by Test Method[a]			
			DMD		BMD	
			Affected Males	Heterozygous Females[g]	Affected Males	Heterozygous Females[g]
DMD	Deletion/ duplication analysis[b]	Deletion of one or more exons or the whole gene	~50–65%	~50–65%	~65–70%	Unknown
		Duplication of one or more exons	~5–10%	~5–10%	~5–10%	Unknown
	Sequence analysis/ mutation scanning[c]	Point mutations (sequence variants)[d]	~20–35 %[e]	~25–35%[f]	~20–30%[e]	Unknown

[a]The ability of the test method used to detect a mutation that is present in the indicated gene. The mutation detection rates reflect approximate values.
[b]Testing that identifies deletions/duplications not readily detectable by sequence analysis of genomic DNA; a variety of methods including quantitative PCR, long-range PCR, multiplex ligation-dependent probe amplification (MLPA), or targeted array GH (gene/segment-specific) may be used. A full array GH analysis that detects deletions/duplications across the genome may also include this gene/segment.
[c]Sequence analysis and mutation scanning of the entire gene can have similar detection frequencies; however, detection rates for mutation scanning may vary considerably between laboratories based on specific protocol used.
[d]Examples of mutations detected by sequence analysis may include small intragenic deletions/insertions and missense, nonsense, and splice site mutations.
[e]Lack of amplification by PCR prior to sequence analysis can suggest a putative deletion of one or more exons or the entire X-linked gene in a male; confirmation may require additional testing by deletion/duplication analysis.
[f]Sequence analysis of genomic DNA cannot detect deletion of one or more exons or the entire X-linked gene in a heterozygous female.
[g]The mutation spectrum for females heterozygous for a DMD mutation resulting in BMD has not been well documented, but is presumed to be similar to the spectrum for males with BMD. Similarly, there are fewer data about females heterozygous for a DMD mutation resulting in DMD, but it is also presumed to be similar to that seen in males with DMD.
Modified from Darras et al.[295]

used by clinical laboratories and reflect approximate and reconciled data from the literature cited above. In one of the author's (BTD) clinical experience, these reference ranges correlate quite well with the clinical findings most of the time but not always; in some cases, the Western blot results do not match the patient's phenotype. Given the reported exceptions and the overlap between the phenotypes, test results should be interpreted with caution and always in the context of, and correlated with, the patient's clinical presentation. Quantitation of dystrophin in female carriers is not useful in clinical practice because of the wide range of values and significant overlap with normal values.

Molecular Genetic Testing

DMD is the only gene in which mutations are known to cause DMD, BMD, and DMD-associated DCM. The following data regarding frequency of *DMD* mutations refer to individuals with DMD or BMD. Data are insufficient to estimate the percentage of individuals with DMD-associated DCM who have detectable *DMD* mutations. Three broad types of mutations are identified: deletions, duplications, and point mutations (sequence variants).

Deletions of one or more exons account for approximately 60–70% of mutations in individuals with DMD or BMD (Table 30.6).[54–56,59]

Duplications, which may lead to in-frame or out-of-frame transcripts, account for the disease-causing mutations in approximately 5%−10% of males with DMD and BMD.[57−59] Duplications may be slightly more common in BMD; rates not higher than 10% in BMD have been found in most studies, including those using newer techniques such as multiplex ligation probe amplification (MLPA),[59] but one smaller series found duplications in 14/75 (19%) of males with BMD.[353] Among individuals whose mutations were not identified by deletion or point mutation screens, duplications were detected in 87%.[57]

Point mutations include small deletions or insertions, single-base changes, and splicing mutations. Approximately 25%−35% of mutations in patients with DMD and about 20%−30% in BMD are accounted for by point mutations.[59−63,354] Nonsense mutations occur more commonly in DMD, in the range of 20−25% of cases as compared to less than 5% of cases in BMD.[58,59] In both DMD and BMD, splice site mutations and small insertion/deletion mutations (in/del) are a substantial proportion of sequence changes. Missense mutations are not a common cause of either DMD or BMD.

The rate of individuals with a dystrophinopathy for whom no molecular diagnosis can be established appears slightly higher in BMD compared to DMD. However, this finding is likely secondary to the inclusion of phenotypes similar to the dystrophinopathies but not caused by mutations in *DMD*.[353]

Clinical Testing

Deletions or duplications of *DMD* in probands and carrier females are detected by two types of analysis: multiplex ligation probe amplification (MLPA),[355] and array genomic hybridization (aGH), also called chromosome microarray (CMA).[356−358] The latter technology offers an advantage over MLPA because it can detect intronic rearrangements. Up to 7% of persons with a dystrophinopathy do not have coding region mutations[55]; the fraction of these that have intronic rearrangements of clinical significance is unknown. aGH or CMA does not detect intronic point mutations. MLPA and aGH/CMA have largely supplanted methods used in the past to detect deletions or duplications, including multiplex PCR,[359] Southern blotting,[360] and FISH (with probes covering *DMD* exons 3−6, 8, 12, 13, 17, 19, 32−34, 43−48, 50, 51, and 60).

Point mutations (i.e. small deletions or insertions, single-base changes, and splicing mutations) are detected by sequence analysis or mutation scanning. Sequencing of the entire *DMD* gene can be performed by traditional PCR and Sanger sequencing, or by more automated methods such as universal long PCR combined with massive pyrosequencing.[361] Sequencing of the entire gene is often necessary to detect rare or private mutations.

Single-base changes can be detected by mutation scanning methods such as denaturing high performance liquid chromatography (dHPLC) or newer methods such as single-condition amplification internal primer sequencing (SCAIP)[58,61] and high-resolution melting curve analysis.[362] These methods provide such detection at a reasonable cost compared with the previously high costs of sequencing. However, both the time required for direct sequencing and its costs have decreased significantly due to advances in technology, and therefore only a few clinical laboratories still perform mutation scanning for individuals in whom a deletion or duplication has not been identified.

Testing Strategies in Specific Populations

Because no other phenotypes are known to be associated with mutations in *DMD*, the following strategies will help to confirm a diagnosis in the indicated populations. Although prenatal and preimplantation genetic diagnoses for at-risk pregnancies are possible, they require prior identification of the disease-causing mutation in the family through one of the following means.

In males with an elevated serum CK concentration whose clinical findings suggest a dystrophinopathy, first perform *DMD* molecular genetic testing using deletion/duplication analysis. If a mutation is not identified, perform sequence analysis/mutation scanning. If a disease-causing *DMD* mutation is identified, the diagnosis of a dystrophinopathy is established, but the distinction between DMD and BMD can remain difficult in some cases. For example, deletion of exons 3−7 has been found in DMD and BMD patients, and is the most extensively investigated deletion associated with both phenotypes.[363] As discussed in the section on *DMD* gene mutations, the reading frame rule can help address the issue of DMD versus BMD without the need for a muscle biopsy. The "rule" states that the more severe DMD phenotype correlates with mutations that alter the reading frame (out-of-frame) while the milder BMD phenotype correlates with mutations that do not alter the reading frame (in-frame deletions/duplications).[66] This determination can distinguish between the DMD and BMD phenotypes with 91% to 92% accuracy in young children who are simplex cases (i.e. a single occurrence in a family).[363] It remains very important to correlate clinical features with molecular test results because exceptions to the reading frame rule have been documented to occur at a rate below 10%,[363] although other studies suggest that the rate of exception may be higher with a BMD phenotype for both deletions and duplications.[59,353] If no disease-causing *DMD* mutation is identified via the above methods in individuals with suspected DMD or BMD, skeletal muscle biopsy is warranted for Western blot and immunohistochemistry studies of dystrophin. However, the clinician may choose to order molecular genetic testing for LGMDs before performing a muscle biopsy[282] because of the phenotypic resemblance between DMD/BMD and certain limb-girdle muscular dystrophies (LGMDs) such as LGMD2I.

A different testing strategy is used to establish the diagnosis in a boy who has DMD and one or more other X-linked disorders. Other X-linked disorders in these contiguous gene deletions include retinitis pigmentosa, chronic granulomatous disease, and McLeod red cell phenotype[11] or glycerol kinase deficiency and adrenal hypoplasia.[364] The diagnosis of a contiguous gene deletion can be confirmed by CMA studies to look for deletions involving Xp21.2, or by FISH analysis with probes (covering *GK* and *NR0B1* in addition to exons in *DMD*) to look for rearrangements involving Xp21.2.

The testing strategy to establish the diagnosis in a girl with classic DMD includes identification of the genetic mechanism underlying this rare occurrence. These genetic mechanisms (and the tests to identify them) include: deletion involving Xp21.2 (CMA); X chromosome rearrangement involving Xp21.2 or complete absence of an X chromosome, also known as Turner syndrome (cytogenetic studies); uniparental disomy (UPD) of the X chromosome (UPD studies); compound heterozygosity for two *DMD* mutations[276] (deletion/duplication analysis and/or sequence analysis); and nonrandom X chromosome inactivation (XCI). As discussed, XCI pattern-phenotype correlation has limitations.

Several strategies can be used when performing carrier testing for at-risk female relatives. Carriers are heterozygotes for this X-linked disorder and may develop clinical findings related to the disorder. When the proband's *DMD* mutation is known to be a deletion or duplication, any of the methods for deletion/duplication analysis that can detect the proband's mutation can be used in the at-risk female; when the proband is known to have a point mutation, sequence analysis of the gene region in which the mutation has occurred can be used in the at-risk female. If an affected male is not available for testing, the strategy consists of molecular genetic testing of the at-risk female first by deletion/duplication analysis, and then, if no mutation is identified, by sequence analysis.

The strategy in at-risk females when the proband's *DMD* mutation is not known and *DMD* mutation is not identified in a carrier female is to offer linkage analysis. Performing a muscle biopsy for dystrophin immunostaining is another option, particularly if linkage analysis is not available or the family is not informative. Testing by linkage analysis is not available to families in which there is a single affected male, but to families who have more than one affected male with an unequivocal diagnosis of DMD/BMD/DMD-associated DCM. Linkage studies are based on accurate clinical diagnosis of DMD/BMD/DMD-associated DCM in the affected family members and accurate understanding of the genetic relationships in the family. Family members must be available and willing to be tested in order for linkage analysis to be successful. The markers used for linkage in DMD/BMD/DMD-associated DCM are highly informative and lay both within and flanking the *DMD* locus; therefore, they can be used in most families with more than one case of DMD/BMD/DMD-associated DCM.[365] However, it should be noted that the large size of *DMD* leads to an appreciable risk of recombination. The gene itself may span a genetic distance of 12 centimorgans[366]; thus, multiple recombination events among different members of a family may complicate the interpretation of a linkage study.

Finally, the following is a testing strategy in families with a dystrophinopathy with only one affected member, in order to identify the origin of a *de novo* mutation for genetic counseling purposes. The proband's mutation can often be identified by performing deletion/duplication analysis or sequencing in conjunction with linkage analysis. The haplotype associated with the mutated *DMD* allele in the affected individual can be tracked back through the mother and, if necessary, through maternal grandparents to identify the individual in whom the mutation originated.[367]

Genetic Counseling

The dystrophinopathies are inherited in an X-linked manner. The father of an affected male will not have the disease and will not be a carrier of a *DMD* mutation. A woman who has an affected son and one other affected relative in the maternal line is an obligate heterozygote (i.e. a carrier of the *DMD* mutation). A woman who has more than one affected son and no other family history of a dystrophinopathy has either a germline mutation or germline mosaicism (see previous subsection "Germline Mosaicism").

When the proband is the only affected family member, one of several genetic mechanisms is responsible.[368] First, a *de novo DMD* mutation can result when the mutation occurred in the egg at the time of the proband's conception; in this instance, the mutation is present in every cell of the proband's body, and the proband's mother does not have a *DMD* mutation. Second, a *de novo DMD* mutation can occur after conception; in this instance, the mutation is present in some but not all cells of the proband's body (somatic mosaicism) and the mother does not have a *DMD* mutation. Third, the proband's mother can have a *de novo DMD* mutation via one of four mechanisms: (1) the mutation occurred in the egg or sperm at the time of her conception (germline mutation present in every cell of her body and detectable in DNA extracted from leukocytes); (2) the mutation is present in some but not all cells of her body (somatic mosaicism detectable or not detectable in DNA extracted from leukocytes); (3) the mutation is present in her egg cells only (germline mosaicism not detectable in DNA extracted from a blood sample, empiric risk 15–20%, with each offspring having an increased risk of inheriting the *DMD* mutation[127,369]); or (4) the proband's mother has inherited a *DMD* mutation from her mother who is a carrier, from her mother or her father who has somatic mosaicism, or from her mother or her father who has germline mosaicism. Molecular genetic testing combined with linkage analysis can often determine the point of origin of a *de novo* mutation, which will help determine which branches of the family are at risk for dystrophinopathies.

The risk to sibs of a proband depends on the carrier status of the mother. If the mother of the proband has the family-specific *DMD* mutation, the chance of transmitting it in each pregnancy is 50%. Males who inherit the mutation will be affected; females who inherit the mutation will be carriers. If the mother does not have the family-specific *DMD* mutation detectable in her DNA, it is possible that the proband has a *de novo* mutation. However, because the incidence of germline mosaicism in mothers is 15%–20%, the sibs of a proband are at increased risk of inheriting the family-specific *DMD* mutation. If the mother has concomitant somatic and germline mosaicism, the risk to sibs of inheriting the family-specific *DMD* mutation may be higher than if the mother has germline mosaicism only.[127]

Males with DMD usually die before reproductive age or are too debilitated to reproduce. If they were to reproduce, all male offspring would be unaffected and all female offspring would be carriers. Male probands with BMD and DMD-associated DCM will produce daughters who are carriers; none of the sons will inherit their father's *DMD* mutation.

The proband's maternal grandmother, maternal aunts, and their offspring may be at risk of being carriers or being affected, depending on their gender, family relationship, and the carrier status of the proband's mother. Carrier testing is clinically available for at-risk females. Females who are identified as carriers of a *DMD* mutation need to be informed of their risk for DMD-associated DCM, as well as the recommended surveillance. BMD and DMD-associated DCM are sometimes observed in the same family.[256] Thus, the entire spectrum of possible muscle disease should be considered when obtaining a family history and providing genetic counseling.

Optimally, before pregnancy families have determined genetic risk, clarified carrier status, and investigated the availability of prenatal testing. Young adults who are affected, who are carriers, or who are at risk of being carriers will benefit from counseling regarding reproductive options and the potential risks to offspring.

Prenatal Testing

Prenatal testing is possible for pregnancies of women who are carriers, if the *DMD* mutation has been identified in a family member or if linkage has been established. The process first involves determination of fetal sex in one of two ways: at approximately 10 to 12 weeks' gestation by karyotype or specialized studies to identify the sex chromosomes from cells obtained by chorionic villus sampling (CVS), or at approximately 15–18 weeks' gestation by amniocentesis. When the karyotype is 46, XY, DNA is then extracted from fetal cells and analyzed for the known disease-causing mutation, or using the linkage previously established. When the causative *DMD* mutation is not known or genetic linkage has not been established, *in utero* fetal muscle biopsy has been used in the prenatal diagnosis of DMD,[370] but it is not a routine procedure. Preimplantation genetic diagnosis may be available for families in which the disease-causing mutation has been identified.

The history of molecular diagnostic testing in DMD and the impact of new techniques such as CMA analysis and noninvasive prenatal diagnosis methods are reviewed in detail by Raymond et al.[371] Banking of the DNA of an affected individual may contribute to future advances in testing methodology and our understanding of genes, mutations, and diseases, particularly when the individual's disease-causing mutation has not been identified.

TREATMENT AND MANAGEMENT

The clinical management and treatment of dystrophinopathies, including molecular therapies, are discussed in Chapters 31 and 51.

REFERENCES

1. Meryon E. On granular and fatty degeneration of the voluntary muscles. *Med Chir Trans* 1852;**35**:73.
2. Little W. *On the nature and treatment of the deformities of the human frame: being a course of lectures delivered at the Royal Orthopaedic Hospital in 1843: With numerous notes and additions, to the present time*. London: Longman, Brown, Green and Longmans; 1853. pp. 14–16.
3. Duchenne de Boulogne GBA. *De l'electrisation Localisee et de Son Application a la Pathologie et a la Therapeutique*. 2nd ed. Paris: Bailiere & Fils; 1861.
4. Duchenne de Boulogne GBA. Recherches sur la paralysie musculaire pseudo-hypertrophique, ou paralysie myo-sclerosique. Reprint In: Wilkins RH, Brody IA, editors. *Neurological Classics*. New York: Johnson; 1973. pp. 59–67.
5. Becker PE, Kiener F. Eine neue X-chromosomale muskeldystrophie. *Arch Psychiatr Z Neurol* 1955;**193**:427.
6. Walton JN. On the inheritance of muscular dystrophy. *Ann Hum Genet* 1955;**20**:1.

7. Greenstein RM, Reardon MP, Chan TS, Middleton AB, Mulivor RA, Greene AE, et al. An (X;11) translocation in a girl with Duchenne muscular dystrophy. Repository identification No. GM1695. *Cytogenet Cell Genet* 1980;**27**(4):268.
8. Boyd Y, Buckle V, Holt S, Munro E, Hunter D, Craig I. Muscular dystrophy in girls with X; autosome translocations. *J Med Genet* 1986;**23**(6):484–90.
9. Murray JM, Davies KE, Harper PS, Meredith L, Mueller CR, Williamson R. Linkage relationship of a cloned DNA sequence on the short arm of the X chromosome to Duchenne muscular dystrophy. *Nature* 1982;**300**(5887):69–71.
10. Davies KE, Pearson PL, Harper PS, Murray JM, O'Brien T, Sarfarazi M, et al. Linkage analysis of two cloned DNA sequences flanking the Duchenne muscular dystrophy locus on the short arm of the human X chromosome. *Nucleic Acids Res* 1983;**11**(8):2303–12.
11. Francke U, Ochs HD, de Martinville B, Giacalone J, Lindgren V, Disteche C, et al. Minor Xp21 chromosome deletion in a male associated with expression of Duchenne muscular dystrophy, chronic granulomatous disease, retinitis pigmentosa, and McLeod syndrome. *Am J Hum Genet* 1985;**37**(2):250–67.
12. Kunkel LM, Monaco AP, Middlesworth W, Ochs HD, Latt SA. Specific cloning of DNA fragments absent from the DNA of a male patient with an X chromosome deletion. *Proc Natl Acad Sci USA* 1985;**82**(14):4778–82.
13. Kunkel LM. Analysis of deletions in DNA from patients with Becker and Duchenne muscular dystrophy. *Nature* 1986;73–7.
14. Monaco AP, Neve RL, Colleti-Feener C, Bertelson CJ, Kurnit DM, Kunkel LM. Isolation of candidate cDNAs for portions of the Duchenne muscular dystrophy gene. *Nature* 1986;**323**:646–50.
15. Ray PN, Belfall B, Duff C, Logan C, Kean V, Thompson MW, et al. Cloning of the breakpoint of an X;21 translocation associated with Duchenne muscular dystrophy. *Nature* 1985;**318**(6047):672–5.
16. Burghes AH, Logan C, Hu X, Belfall B, Worton RG, Ray PH. A cDNA clone from the Duchenne/Becker muscular dystrophy gene. *Nature* 1987;**328**:434–7.
17. Hoffman EP, Brown Jr. RH, Kunkel LM. Dystrophin: the protein product of the Duchenne muscular dystrophy locus. *Cell* 1987;**51**(6):919–28.
18. Koenig M, Hoffman EP, Bertelson CJ, Monaco AP, Feener C, Kunkel LM. Complete cloning of the Duchenne muscular dystrophy (DMD) cDNA and preliminary genomic organization of the DMD gene in normal and affected individuals. *Cell* 1987;**50**:509–17.
19. Koenig M, Monaco AP, Kunkel LM. The complete sequence of dystrophin predicts a rod-shaped cytoskeletal protein. *Cell* 1988;**53**(2):219–26.
20. Arahata K, Ishiura S, Ishiguro T, Tsukahara T, Suhara Y, Eguchi C, et al. Immunostaining of skeletal and cardiac muscle surface membrane with antibody against Duchenne muscular dystrophy peptide. *Nature* 1988;**333**(6176):861–3.
21. Bonilla E, Samitt CE, Miranda AF, Hays AP, Salviati G, DiMauro S, et al. Duchenne muscular dystrophy: deficiency of dystrophin at the muscle cell surface. *Cell* 1988;**54**(4):447–52.
22. Zubrzycka-Gaarn EE, Bulman DE, Karpati G, Burghes AH, Belfall B, Klamut HJ, et al. The Duchenne muscular dystrophy gene product is localized in sarcolemma of human skeletal muscle. *Nature* 1988;**333**:466–9.
23. Ferlini A, Neri M, Gualandi F. The medical genetics of dystrophinopathies: molecular genetic diagnosis and its impact on clinical practice. *Neuromuscul Disord: NMD* 2013;**23**(1):4–14.
24. Hoffman EP, Fischbeck KH, Brown RH, Johnson M, Medori R, Loike JD, et al. Characterization of dystrophin in muscle-biopsy specimens from patients with Duchenne's or Becker's muscular dystrophy. *N Engl J Med* 1988;**318**(21):1363–8.
25. Nudel U, Robzyk K, Yaffe D. Expression of the putative Duchenne muscular dystrophy gene in differentiated myogenic cell cultures and in the brain. *Nature* 1988;**331**(6157):635–8.
26. Nudel U, Zuk D, Einat P, Zeelon E, Levy Z, Neuman S, et al. Duchenne muscular dystrophy gene product is not identical in muscle and brain. *Nature* 1989;**337**(6202):76–8.
27. Feener CA, Koenig M, Kunkel LM. Alternative splicing of human dystrophin mRNA generates isoforms at the carboxy terminus. *Nature* 1989;**338**(6215):509–11.
28. Boyce FM, Beggs AH, Feener C, Kunkel LM. Dystrophin is transcribed in brain from a distant upstream promoter. *Proc Natl Acad Sci USA* 1991;**88**(4):1276–80.
29. Lederfein D, Levy Z, Augier N, Mornet D, Morris G, Fuchs O, et al. A 71-kilodalton protein is a major product of the Duchenne muscular dystrophy gene in brain and other nonmuscle tissues. *Proc Natl Acad Sci USA* 1992;**89**(12):5346–50.
30. Ahn AH, Kunkel LM. The structural and functional diversity of dystrophin. *Nature Genet* 1993;**3**:283–9.
31. Lidov HG, Selig S, Kunkel LM. Dp140: a novel 140 kDa CNS transcript from the dystrophin locus. *Hum Mol Genet* 1995;**4**(3):329–35.
32. Makover A, Zuk D, Breakstone J, Yaffe D, Nudel U. Brain-type and muscle-type promoters of the dystrophin gene differ greatly in structure. *Neuromuscul Disord* 1991;**1**(1):39–45.
33. Rapaport D, Fuchs O, Nudel U, Yaffe D. Expression of the Duchenne muscular dystrophy gene products in embryonic stem cells and their differentiated derivatives. *J Biol Chem* 1992;**267**(30):21289–92.
34. Byers TJ, Lidov HG, Kunkel LM. An alternative dystrophin transcript specific to peripheral nerve. *Nat Genet* 1993;**4**(1):77–81.
35. D'Souza VN, Nguyen TM, Morris GE, Karges W, Pillers DA, Ray PN. A novel dystrophin isoform is required for normal retinal electrophysiology. *Hum Mol Genet* 1995;**4**(5):837–42.
36. Bies RD, Phelps SF, Cortez MD, Roberts R, Caskey CT, Chamberlain JS. Human and murine dystrophin mRNA transcripts are differentially expressed during skeletal muscle, heart, and brain development. *Nucleic Acids Res* 1992;**20**(7):1725–31.
37. Lidov HG, Byers TJ, Watkins SC, Kunkel LM. Localization of dystrophin to postsynaptic regions of central nervous system cortical neurons. *Nature* 1990;**348**(6303):725–8.
38. Gorecki DC, Monaco AP, Derry JM, Walker AP, Barnard EA, Barnard PJ. Expression of four alternative dystrophin transcripts in brain regions regulated by different promoters. *Hum Mol Genet* 1992;**1**(7):505–10.
39. Lidov HG, Byers TJ, Kunkel LM. The distribution of dystrophin in the murine central nervous system: an immunocytochemical study. *Neuroscience* 1993;**54**(1):167–87.
40. Bar S, Barnea E, Levy Z, Neuman S, Yaffe D, Nudel U. A novel product of the Duchenne muscular dystrophy gene which greatly differs from the known isoforms in its structure and tissue distribution. *Biochem J* 1990;**272**(2):557–60.

41. Rapaport D, Lederfein D, den Dunnen JT, Grootscholten PM, Van Ommen GJ, Fuchs O, et al. Characterization and cell type distribution of a novel, major transcript of the Duchenne muscular dystrophy gene. *Differentiation* 1992;**49**(3):187–93.
42. Hugnot JP, Gilgenkrantz H, Vincent N, Chafey P, Morris GE, Monaco AP, et al. Distal transcript of the dystrophin gene initiated from an alternative first exon and encoding a 75-kDa protein widely distributed in nonmuscle tissues. *Proc Natl Acad Sci USA* 1992;**89**(16):7506–10.
43. Byers TJ, Kunkel LM, Watkins SC. The subcellular distribution of dystrophin in mouse skeletal, cardiac, and smooth muscle. *J Cell Biol* 1991;**115**(2):411–21.
44. Zubrzycka-Gaarn EE, Hutter OF, Karpati G, Klamut HJ, Bulman DE, Hodges RS, et al. Dystrophin is tightly associated with the sarcolemma of mammalian skeletal muscle fibers. *Exp Cell Res* 1991;**192**(1):278–88.
45. Schapira AHV, Griggs RC, editors. *Muscle Diseases*. Boston: Butterworth Heinemann; 1999.
46. Campbell KP, Kahl SD. Association of dystrophin and an integral membrane glycoprotein. *Nature* 1989;**338**(6212):259–62.
47. Ozawa E, Yoshida M, Hagiwara Y, Suzuki A, Mizuno Y, Noguchi S. Dystrophin-associated proteins in muscular dystrophy. *Hum Mol Genet* 1990;**4**:11711–16.
48. Ervasti JM, Campbell KP. Membrane organization of the dystrophin-glycoprotein complex. *Cell* 1991;**66**:1121–31.
49. Wewer UM, Durkin ME, Zhang X, Laursen H, Nielsen NH, Towfighi J, et al. Laminin beta 2 chain and adhalin deficiency in the skeletal muscle of Walker-Warburg syndrome (cerebro-ocular dysplasia-muscular dystrophy). *Neurology* 1995;**45**:2099–101.
50. Suzuki A, Yoshida M, Yamamoto H, Ozawa E. Glycoprotein-binding site of dystrophin is confined to the cysteine-rich domain and the first half of the carboxy-terminal domain. *FEBS Lett* 1992;**308**(2):154–60.
51. Thompson TG, Chan YM, Hack AA, Brosius M, Rajala M, Lidov HG, et al. Filamin 2 (FLN2): A muscle-specific sarcoglycan interacting protein. *J Cell Biol* 2000;**148**(1):115–26.
52. Hillier BJ, Christopherson KS, Prehoda KE, Bredt DS, Lim WA. Unexpected modes of PDZ domain scaffolding revealed by structure of nNOS-syntrophin complex. *Science* 1999;**284**(5415):812–15.
53. Koenig M, Beggs AH, Moyer M, Scherpf S, Heindrich K, Bettecken T, et al. The molecular basis for Duchenne versus Becker muscular dystrophy: correlation of severity with type of deletion. *Am J Hum Genet* 1989;**45**(4):498–506.
54. Yan J, Feng J, Buzin CH, Scaringe W, Liu Q, Mendell JR, et al. Three-tiered noninvasive diagnosis in 96% of patients with Duchenne muscular dystrophy (DMD). *Hum Mutat* 2004;**23**(2):203–4.
55. Dent KM, Dunn DM, von Niederhausern AC, Aoyagi AT, Kerr L, Bromberg MB, et al. Improved molecular diagnosis of dystrophinopathies in an unselected clinical cohort. *Am J Med Genet A* 2005;**134**(3):295–8.
56. Prior TW, Bridgeman SJ. Experience and strategy for the molecular testing of Duchenne muscular dystrophy. *J Mol Diagn* 2005;**7**(3):317–26.
57. White SJ, Aartsma-Rus A, Flanigan KM, Weiss RB, Kneppers AL, Lalic T, et al. Duplications in the DMD gene. *Hum Mutat* 2006;**27**(9):938–45.
58. Flanigan KM, Dunn DM, von Niederhausern A, Soltanzadeh P, Gappmaier E, Howard MT, et al. Mutational spectrum of DMD mutations in dystrophinopathy patients: application of modern diagnostic techniques to a large cohort. *Hum Mutat* 2009;**30**(12):1657–66.
59. Takeshima Y, Yagi M, Okizuka Y, Awano H, Zhang Z, Yamauchi Y, et al. Mutation spectrum of the dystrophin gene in 442 Duchenne/Becker muscular dystrophy cases from one Japanese referral center. *J Hum Genet* 2010;**55**(6):379–88.
60. Dolinsky LC, de Moura-Neto RS, Falcao-Conceicao DN. DGGE analysis as a tool to identify point mutations, de novo mutations and carriers of the dystrophin gene. *Neuromuscul Disord* 2002;**12**(9):845–8.
61. Flanigan KM, von Niederhausern A, Dunn DM, Alder J, Mendell JR, Weiss RB. Rapid direct sequence analysis of the dystrophin gene. *Am J Hum Genet* 2003;**72**(4):931–9.
62. Hofstra RM, Mulder IM, Vossen R, de Koning-Gans PA, Kraak M, Ginjaar IB, et al. DGGE-based whole-gene mutation scanning of the dystrophin gene in Duchenne and Becker muscular dystrophy patients. *Hum Mutat* 2004;**23**(1):57–66.
63. Bennett RR, den Dunnen J, O'Brien KF, Darras BT, Kunkel LM. Detection of mutations in the dystrophin gene via automated DHPLC screening and direct sequencing. *BMC Genet* 2001;**2**(1):17.
64. Darras BT, Blattner P, Harper JF, Spiro AJ, Alter S, Francke U. Intragenic deletions in 21 Duchenne muscular dystrophy (DMD)/Becker muscular dystrophy (BMD) families studied with the dystrophin cDNA: location of breakpoints on HindIII and BglII exon-containing fragment maps, meiotic and mitotic origin of the mutations. *Am J Hum Genet* 1988;**43**(5):620–9.
65. Hoffman EP, Kunkel LM. Dystrophin abnormalities in Duchenne/Becker muscular dystrophy. *Neuron* 1989;**2**(1):1019–29.
66. Monaco AP, Bertelson CJ, Liechti-Gallati S, Moser H, Kunkel LM. An explanation for the phenotypic differences between patients bearing partial deletions of the DMD locus. *Genomics* 1988;**2**(1):90–5.
67. Muhlemann O, Eberle AB, Stalder L, Zamudio Orozco R. Recognition and elimination of nonsense mRNA. *Biochim Biophys Acta* 2008;**1779**(9):538–49.
68. Malhotra SB, Hart KA, Klamut HJ, Thomas NS, Bodrug SE, Burghes AH, et al. Frame-shift deletions in patients with Duchenne and Becker muscular dystrophy. *Science* 1988;**242**(4879):755–9.
69. Beggs AH, Hoffman EP, Snyder JR, Arahata K, Specht L, Shapiro F, et al. Exploring the molecular basis for variability among patients with Becker muscular dystrophy: dystrophin gene and protein studies. *Am J Hum Genet* 1991;**49**(1):54–67.
70. Hoffman EP, Garcia CA, Chamberlain JS, Angelini C, Lupski JR, Fenwick R. Is the carboxyl-terminus of dystrophin required for membrane association? A novel, severe case of Duchenne muscular dystrophy. *Ann Neurol* 1991;**30**(4):605–10.
71. Bies RD, Caskey CT, Fenwick R. An intact cysteine-rich domain is required for dystrophin function. *J Clin Invest* 1992;**90**(2):666–72.
72. Gangopadhyay SB, Sherratt TG, Heckmatt JZ, Dubowitz V, Miller G, Shokeir M, et al. Dystrophin in frameshift deletion patients with Becker muscular dystrophy. *Am J Hum Genet* 1992;**51**(3):562–70.
73. Helliwell TR, Ellis JM, Mountford RC, Appleton RE, Morris GE. A truncated dystrophin lacking the C-terminal domains is localized at the muscle membrane. *Am J Hum Genet* 1992;**50**(3):508–14.
74. Baumbach LL, Chamberlain JS, Ward PA, Farwell NJ, Caskey CT. Molecular and clinical correlations of deletions leading to Duchenne and Becker muscular dystrophies. *Neurology* 1989;**39**(4):465–74.

75. Winnard AV, Klein CJ, Coovert DD, Prior T, Papp A, Snyder P, et al. Characterization of translational frame exception patients in Duchenne/Becker muscular dystrophy. *Hum Mol Genet* 1993;**2**(6):737–44.
76. Chelly J, Gilgenkrantz H, Lambert M, Hamard G, Chafey P, Recan D, et al. Effect of dystrophin gene deletions on mRNA levels and processing in Duchenne and Becker muscular dystrophies. *Cell* 1990;**63**(6):1239–48.
77. Bushby KM. Genetic and clinical correlations of Xp21 muscular dystrophy. *J Inherit Metab Dis* 1992;**15**(4):551–64.
78. Prior TW, Bartolo C, Papp AC, Snyder PJ, Sedra MS, Burghes AH, et al. Dystrophin expression in a Duchenne muscular dystrophy patient with a frame shift deletion. *Neurology* 1997;**48**(2):486–8.
79. Nevo Y, Muntoni F, Sewry C, Legum C, Kutai M, Harel S, et al. Large in-frame deletions of the rod-shaped domain of the dystrophin gene resulting in severe phenotype. *Isr Med Assoc J* 2003;**5**(2):94–7.
80. Love DR, Flint TJ, Marsden RF, Bloomfield JF, Daniels RJ, Forrest SM, et al. Characterization of deletions in the dystrophin gene giving mild phenotypes. *Am J Med Genet* 1990;**37**(1):136–42.
81. Bushby KM, Gardner-Medwin D, Nicholson LV, Johnson MA, Haggerty ID, Cleghorn NJ, et al. The clinical, genetic and dystrophin characteristics of Becker muscular dystrophy. II. Correlation of phenotype with genetic and protein abnormalities. *J Neurol* 1993;**240**(2):105–12.
82. England SB, Nicholson LV, Johnson MA, Forrest SM, Love DR, Zubrzycka-Gaarn EE, et al. Very mild muscular dystrophy associated with the deletion of 46% of dystrophin. *Nature* 1990;**343**(6254):180–2.
83. Love DR, Flint TJ, Genet SA, Middleton-Price HR, Davies KE. Becker muscular dystrophy patient with a large intragenic dystrophin deletion: implications for functional minigenes and gene therapy. *J Med Genet* 1991;**28**(12):860–4.
84. Muntoni F, Gobbi P, Sewry C, Sherratt T, Taylor J, Sandhu SK, et al. Deletions in the 5′ region of dystrophin and resulting phenotypes. *J Med Genet* 1994;**31**(11):843–7.
85. Arahata K, Beggs AH, Honda H, Ito S, Ishiura S, Tsukahara T, et al. Preservation of the C-terminus of dystrophin molecule in the skeletal muscle from Becker muscular dystrophy. *J Neurol Sci* 1991;**101**(2):148–56.
86. Nicholson LV, Bushby KM, Johnson MA, den Dunnen JT, Ginjaar IB, van Ommen GJ. Predicted and observed sizes of dystrophin in some patients with gene deletions that disrupt the open reading frame. *J Med Genet* 1992;**29**(12):892–6.
87. Patria SY, Alimsardjono H, Nishio H, Takeshima Y, Nakamura H, Matsuo M. A case of Becker muscular dystrophy resulting from the skipping of four contiguous exons (71–74) of the dystrophin gene during mRNA maturation. *Proc Assoc Am Physicians* 1996;**108**(4):308–14.
88. Winnard AV, Mendell JR, Prior TW, Florence J, Burghes AH. Frameshift deletions of exons 3–7 and revertant fibers in Duchenne muscular dystrophy: mechanisms of dystrophin production. *Am J Hum Genet* 1995;**56**(1):158–66.
89. Ginjaar IB, Kneppers AL, v d Meulen JD, Anderson LV, Bremmer-Bout M, van Deutekom JC, et al. Dystrophin nonsense mutation induces different levels of exon 29 skipping and leads to variable phenotypes within one BMD family. *Eur J Hum Genet: EJHG* 2000;**8**(10):793–6.
90. Shiga N, Takeshima Y, Sakamoto H, Inoue K, Yokota Y, Yokoyama M, et al. Disruption of the splicing enhancer sequence within exon 27 of the dystrophin gene by a nonsense mutation induces partial skipping of the exon and is responsible for Becker muscular dystrophy. *J Clin Invest* 1997;**100**(9):2204–10.
91. Gurvich OL, Maiti B, Weiss RB, Aggarwal G, Howard MT, Flanigan KM. DMD exon 1 truncating point mutations: amelioration of phenotype by alternative translation initiation in exon 6. *Hum Mutat* 2009;**30**(4):633–40.
92. Hoffman EP, Kunkel LM, Angelini C, Clarke A, Johnson M, Harris JB. Improved diagnosis of Becker muscular dystrophy by dystrophin testing. *Neurology* 1989;**39**(8):1011–17.
93. Karpati G. Recent developments in the biology of dystrophin and related molecules. *Curr Opin Neurol Neurosurg* 1992;**5**(5):615–21.
94. Petrof BJ, Shrager JB, Stedman HH, Kelly AM, Sweeney HL. Dystrophin protects the sarcolemma from stresses developed during muscle contraction. *Proc Natl Acad Sci USA* 1993;**90**(8):3710–14.
95. Hutter OF. The membrane hypothesis of Duchenne muscular dystrophy: quest for functional evidence. *J Inherit Metab Dis* 1992;**15**(4):565–77.
96. Matsumura K, Tome FM, Ionasescu V, Ervasti JM, Anderson RD, Romero NB, et al. Deficiency of dystrophin-associated proteins in Duchenne muscular dystrophy patients lacking COOH-terminal domains of dystrophin. *J Clin Invest* 1993;**92**(2):866–71.
97. Menke A, Jockusch H. Decreased osmotic stability of dystrophin-less muscle cells from the mdx mouse. *Nature* 1991;**349**(6304):69–71.
98. Matsumura K, Campbell KP. Dystrophin-glycoprotein complex: its role in the molecular pathogenesis of muscular dystrophies. *Muscle Nerve* 1994;**17**(1):2–15.
99. Carpenter S, Karpati G. Segmental necrosis and its demarcation in experimental micropuncture injury of skeletal muscle fibers. *J Neuropathol Exp Neurol* 1989;**48**(2):154–70.
100. Spencer MJ, Croall DE, Tidball JG. Calpains are activated in necrotic fibers from mdx dystrophic mice. *J Biol Chem* 1995;**270**(18):10909–14.
101. Karpati G, Carpenter S. Small-caliber skeletal muscle fibers do not suffer deleterious consequences of dystrophic gene expression. *Am J Med Genet* 1986;**25**(4):653–8.
102. Kimura S, Sugino S, Ohtani Y, Matsukura M, Nishino I, Ikezawa M, et al. Muscle fiber immaturity and inactivity reduce myonecrosis in Duchenne muscular dystrophy. *Ann Neurol* 1998;**44**(6):967–71.
103. Grady RM, Teng H, Nichol MC, Cunningham JC, Wilkinson RS, Sanes JR. Skeletal and cardiac myopathies in mice lacking utrophin and dystrophin: a model for Duchenne muscular dystrophy. *Cell* 1997;**90**(4):729–38.
104. Chandrasekharan K, Yoon JH, Xu Y, deVries S, Camboni M, Janssen PM, et al. A human-specific deletion in mouse Cmah increases disease severity in the mdx model of Duchenne muscular dystrophy. *Sci Transl Med* 2010;**2**(42):42ra54.
105. Sacco A, Mourkioti F, Tran R, Choi J, Llewellyn M, Kraft P, et al. Short telomeres and stem cell exhaustion model Duchenne muscular dystrophy in mdx/mTR mice. *Cell* 2010;**143**(7):1059–71.
106. Cohn RD, Campbell KP. Molecular basis of muscular dystrophies. *Muscle Nerve* 2000;**23**(10):1456–71.

107. Lai Y, Thomas GD, Yue Y, Yang HT, Li D, Long C, et al. Dystrophins carrying spectrin-like repeats 16 and 17 anchor nNOS to the sarcolemma and enhance exercise performance in a mouse model of muscular dystrophy. *J Clin Invest* 2009;**119**(3):624–35.
108. Percival JM, Anderson KN, Gregorevic P, Chamberlain JS, Froehner SC. Functional deficits in nNOSmu-deficient skeletal muscle: myopathy in nNOS knockout mice. *PLOS ONE* 2008;**3**(10):e3387.
109. Kobayashi YM, Rader EP, Crawford RW, Iyengar NK, Thedens DR, Faulkner JA, et al. Sarcolemma-localized nNOS is required to maintain activity after mild exercise. *Nature* 2008;**456**(7221):511–15.
110. Heydemann A, McNally E. NO more muscle fatigue. *J Clin Invest* 2009;**119**(3):448–50.
111. Khurana TS, Watkins SC, Chafey P, Chelly J, Tome FM, Fardeau M, et al. Immunolocalization and developmental expression of dystrophin related protein in skeletal muscle. *Neuromuscul Disord* 1991;**1**(3):185–94.
112. Ohlendieck K, Ervasti JM, Matsumura K, Kahl SD, Leveille CJ, Campbell KP. Dystrophin-related protein is localized to neuromuscular junctions of adult skeletal muscle. *Neuron* 1991;**7**(3):499–508.
113. Tinsley JM, Blake DJ, Zuellig RA, Davies KE. Increasing complexity of the dystrophin-associated protein complex. *Proc Natl Acad Sci USA* 1994;**91**(18):8307–13.
114. Blake DJ, Tinsley JM, Davies KE. Utrophin: a structural and functional comparison to dystrophin. *Brain Pathol* 1996;**6**(1):37–47.
115. Matsumura K, Ervasti JM, Ohlendieck K, Kahl SD, Campbell KP. Association of dystrophin-related protein with dystrophin-associated proteins in mdx mouse muscle. *Nature* 1992;**360**(6404):588–91.
116. Tinsley JM, Potter AC, Phelps SR, Fisher R, Trickett JI, Davies KE. Amelioration of the dystrophic phenotype of mdx mice using a truncated utrophin transgene. *Nature* 1996;**384**(6607):349–53.
117. Love DR, Byth BC, Tinsley JM, Blake DJ, Davies KE. Dystrophin and dystrophin-related proteins: a review of protein and RNA studies. *Neuromuscul Disord* 1993;**3**(1):5–21.
118. Sewry CA, Sansome A, Matsumura K, Campbell KP, Dubowitz V. Deficiency of the 50 kDa dystrophin-associated glycoprotein and abnormal expression of utrophin in two south Asian cousins with variable expression of severe childhood autosomal recessive muscular dystrophy. *Neuromuscul Disord* 1994;**4**(2):121–9.
119. Chevron MP, Echenne B, Demaille J. Absence of dystrophin and utrophin in a boy with severe muscular dystrophy [letter]. *N Engl J Med* 1994;**331**(17):1162–3.
120. Isaac C, Wright A, Usas A, Li H, Tang Y, Mu X, et al. Dystrophin and utrophin "double knockout" dystrophic mice exhibit a spectrum of degenerative musculoskeletal abnormalities. *J Orthop Res* 2013;**31**(3):343–9.
121. Moorwood C, Khurana TS. Duchenne muscular dystrophy drug discovery—the application of utrophin promoter activation screening. *Expert Opin Drug Discov* 2013;**8**(5):569–81.
122. Amenta AR, Yilmaz A, Bogdanovich S, McKechnie BA, Abedi M, Khurana TS, et al. Biglycan recruits utrophin to the sarcolemma and counters dystrophic pathology in mdx mice. *Proc Natl Acad Sci USA* 2011;**108**(2):762–7.
123. Tinsley JM, Fairclough RJ, Storer R, Wilkes FJ, Potter AC, Squire SE, et al. Daily treatment with SMTC1100, a novel small molecule utrophin upregulator, dramatically reduces the dystrophic symptoms in the mdx mouse. *PLOS ONE* 2011;**6**(5):e19189.
124. Bakker E, Van Broeckhoven C, Bonten EJ, van de Vooren MJ, Veenema H, Van Hul W, et al. Germline mosaicism and Duchenne muscular dystrophy mutations. *Nature* 1987;**329**(6139):554–6.
125. Darras BT, Francke U. A partial deletion of the muscular dystrophy gene transmitted twice by an unaffected male. *Nature* 1987;**329**(6139):556–8.
126. Engel AG, Franzini-Armstrong C, editors. *Myology*. 2nd ed. New York: McGraw-Hill, Inc; 1994.
127. van Essen AJ, Mulder IM, van der Vlies P, van der Hout AH, Buys CH, Hofstra RM, et al. Detection of point mutation in dystrophin gene reveals somatic and germline mosaicism in the mother of a patient with Duchenne muscular dystrophy. *Am J Med Genet A* 2003;**118**(3):296–8.
128. Van Essen AJ, Abbs S, Bauget M, et al. Parental origin and germline mosaicism of deletions and duplications of the dystrophin gene: a European study. *Hum Genet* 1992;**8**:249–57.
129. Matthews PM, Benjamin D, Van Bakel I, Squier MV, Nicholson LV, Sewry C, et al. Muscle X-inactivation patterns and dystrophin expression in Duchenne muscular dystrophy carriers. *Neuromuscul Disord* 1995;**5**(3):209–20.
130. Pena SD, Karpati G, Carpenter S, Fraser FC. The clinical consequences of X-chromosome inactivation: Duchenne muscular dystrophy in one of monozygotic twins. *J Neurol Sci* 1987;**79**(3):337–44.
131. Quan F, Janas J, Toth-Fejel S, Johnson DB, Wolford JK, Popovich BW. Uniparental disomy of the entire X chromosome in a female with Duchenne muscular dystrophy. *Am J Hum Genet* 1997;**60**(1):160–5.
132. Moat SJ, Bradley DM, Salmon R, Clarke A, Hartley L. Newborn bloodspot screening for Duchenne muscular dystrophy: 21 years experience in Wales (UK). *Eur J Hum Genet: EJHG* 2013;**21**(10):1049–53.
133. Gardner-Medwin D. Mutation rate in Duchenne type of muscular dystrophy. *J Med Genet* 1970;**7**(4):334–7.
134. Emery AE. Duchenne muscular dystrophy. Genetic aspects, carrier detection and antenatal diagnosis. *Br Med Bull* 1980;**36**(2):117–22.
135. Moser H. Duchenne muscular dystrophy: pathogenetic aspects and genetic prevention. *Hum Genet* 1984;**66**(1):17–40.
136. Bushby KM, Thambyayah M, Gardner-Medwin D. Prevalence and incidence of Becker muscular dystrophy. *Lancet* 1991;**337**(8748):1022–4.
137. Mostacciuolo ML, Lombardi A, Cambissa V, Danieli GA, Angelini C. Population data on benign and severe forms of X-linked muscular dystrophy. *Hum Genet* 1987;**75**(3):217–20.
138. Nakagawa M, Nakahara K, Yoshidome H, Suehara M, Higuchi I, Fujiyama J, et al. Epidemiology of progressive muscular dystrophy in Okinawa, Japan. Classification with molecular biological techniques. *Neuroepidemiology* 1991;**10**(4):185–91.
139. Dooley J, Gordon KE, Dodds L, MacSween J. Duchenne muscular dystrophy: a 30-year population-based incidence study. *Clin Pediatr (Phila)* 2010;**49**(2):177–9.
140. Prevalence of Duchenne/Becker muscular dystrophy among males aged 5–24 years—four states, 2007. *MMWR* Morbidity and mortality weekly report 2009;**58**(40):1119–22.

141. Muntoni F, Cau M, Ganau A, Congiu R, Arvedi G, Mateddu A, et al. Brief report: deletion of the dystrophin muscle-promoter region associated with X-linked dilated cardiomyopathy. *N Engl J Med* 1993;**329**(13):921–5.
142. Beggs AH. Dystrophinopathy, the expanding phenotype. Dystrophin abnormalities in X-linked dilated cardiomyopathy [editorial]. *Circulation* 1997;**95**(10):2344–7.
143. Cox GF, Kunkel LM. Dystrophies and heart disease. *Curr Opin Cardiol* 1997;**12**(3):329–43.
144. Towbin JA. The role of cytoskeletal proteins in cardiomyopathies. *Curr Opin Cell Biol* 1998;**10**(1):131–9.
145. Angelini C, Fanin M, Pegoraro E, Freda MP, Cadaldini M, Martinello F. Clinical-molecular correlation in 104 mild X-linked muscular dystrophy patients: characterization of sub-clinical phenotypes. *Neuromuscul Disord* 1994;**4**(4):349–58.
146. Melacini P, Fanin M, Danieli GA, Villanova C, Martinello F, Miorin M, et al. Myocardial involvement is very frequent among patients affected with subclinical Becker's muscular dystrophy. *Circulation* 1996;**94**(12):3168–75.
147. Eiholzer U, Boltshauser E, Frey D, Molinari L, Zachmann M. Short stature: a common feature in Duchenne muscular dystrophy. *Eur J Pediatr* 1988;**147**(6):602–5.
148. Marshall PD, Galasko CS. No improvement in delay in diagnosis of Duchenne muscular dystrophy [letter]. *Lancet* 1995;**345**(8949):590–1.
149. Firth M, Gardner-Medwin D, Hosking G, Wilkinson E. Interviews with parents of boys suffering from Duchenne muscular dystrophy. *Dev Med Child Neurol* 1983;**25**(4):466–71.
150. Li QX, Yang H, Zhang N, Xiao B, Bi FF, Li J. Clinical and pathological features of 50 children with Duchenne's muscular dystrophy. *Zhongguo Dang Dai Er Ke Za Zhi* 2012;**14**(10):746–50.
151. Bushby KM. The limb-girdle muscular dystrophies-multiple genes, multiple mechanisms. *Hum Mol Genet* 1999;**8**(10):1875–82.
152. Zalaudek I, Bonelli RM, Koltringer P, Reisecker F, Wagner K. Early diagnosis in Duchenne muscular dystrophy [letter]. *Lancet* 1999;**353**(9168):1975.
153. Griggs RC, Mendell JR, Miller RG. *Evaluation and Treatment of Myopathies*. Philadelphia: F.A. Davis Company; 1995.
154. Ohtake T, Takahashi R, Nagashima T, Hirose K, Tanabe H. Sparing of the perineal muscles in muscular dystrophies [letter]. *J Neurol* 1989;**236**(4):252.
155. Kaminski HJ, al-Hakim M, Leigh RJ, Katirji MB, Ruff RL. Extraocular muscles are spared in advanced Duchenne dystrophy. *Ann Neurol* 1992;**32**(4):586–8.
156. Gardner-Medwin D. Clinical features and classification of the muscular dystrophies. *Br Med Bull* 1980;**36**(2):109–15.
157. Allsop KG, Ziter FA. Loss of strength and functional decline in Duchenne's dystrophy. *Arch Neurol* 1981;**38**(7):406–11.
158. Brooke MH, Griggs RC, Mendell JR, Fenichel GM, Shumate JB. The natural history of Duchenne muscular dystrophy: a caveat for therapeutic trials. *Trans Am Neurol Assoc* 1981;**106**:195–9.
159. McDonald CM, Henricson EK, Abresch RT, Florence JM, Eagle M, Gappmaier E, et al. The 6-minute walk test and other endpoints in Duchenne muscular dystrophy: longitudinal natural history observations over 48 weeks from a multicenter study. *Muscle Nerve* 2013;**48**(3):343–56.
160. Perlstein MA. Deep-tendon reflexes in pseudohypertrophic muscular dystrophy: rate and order of loss. *JAMA* 1965;**193**:540.
161. Dubowitz V. *Muscle Biopsy: A Practical Approach*. London: Bailliere Tindall; 1985. pp. 289–339.
162. Inkley SR, Oldenburg FC, Vignos Jr. PJ. Pulmonary function in Duchenne muscular dystrophy related to stage of disease. *Am J Med* 1974;**56**(3):297–306.
163. Khirani S, Ramirez A, Aubertin G, Boule M, Chemouny C, Forin V, et al. Respiratory muscle decline in Duchenne muscular dystrophy. *Pediatr Pulmonol* 2013. Available from: http://dx.doi.org/10.1002/ppul.22847.
164. Passamano L, Taglia A, Palladino A, Viggiano E, D'Ambrosio P, Scutifero M, et al. Improvement of survival in Duchenne Muscular Dystrophy: retrospective analysis of 835 patients. *Acta Myol* 2012;**31**(2):121–5.
165. Rall S, Grimm T. Survival in Duchenne muscular dystrophy. *Acta Myol* 2012;**31**(2):117–20.
166. Eagle M, Bourke J, Bullock R, Gibson M, Mehta J, Giddings D, et al. Managing Duchenne muscular dystrophy—the additive effect of spinal surgery and home nocturnal ventilation in improving survival. *Neuromuscul Disord: NMD* 2007;**17**(6):470–5.
167. Farah MG, Evans EB, Vignos Jr. PJ. Echocardiographic evaluation of left ventricular function in Duchenne's muscular dystrophy. *Am J Med* 1980;**69**(2):248–54.
168. Finsterer J, Stollberger C. The heart in human dystrophinopathies. *Cardiology* 2003;**99**(1):1–19.
169. Nigro G, Comi LI, Politano L, Bain RJ. The incidence and evolution of cardiomyopathy in Duchenne muscular dystrophy. *Int J Cardiol* 1990;**26**(3):271–7.
170. James J, Kinnett K, Wang Y, Ittenbach RF, Benson DW, Cripe L. Electrocardiographic abnormalities in very young Duchenne muscular dystrophy patients precede the onset of cardiac dysfunction. *Neuromuscul Disord: NMD* 2011;**21**(7):462–7.
171. Hermans MC, Pinto YM, Merkies IS, de Die-Smulders CE, Crijns HJ, Faber CG. Hereditary muscular dystrophies and the heart. *Neuromuscul Disord: NMD* 2010;**20**(8):479–92.
172. Lin JJ, Hwang MS, Hsia SH, Chung HT, Chang YS, Lin KL. Pericardial effusion with cardiac tamponade as a cardiac manifestation of Duchenne muscular dystrophy. *Muscle Nerve* 2009;**40**(3):476–80.
173. Mavrogeni S, Papavasiliou A, Spargias K, Constandoulakis P, Papadopoulos G, Karanasios E, et al. Myocardial inflammation in Duchenne Muscular Dystrophy as a precipitating factor for heart failure: a prospective study. *BMC Neurol* 2010;**10**:33.
174. Cotton S, Voudouris NJ, Greenwood KM. Intelligence and Duchenne muscular dystrophy: full-scale, verbal, and performance intelligence quotients. *Dev Med Child Neurol* 2001;**43**(7):497–501.
175. Pane M, Scalise R, Berardinelli A, D'Angelo G, Ricotti V, Alfieri P, et al. Early neurodevelopmental assessment in Duchenne muscular dystrophy. *Neuromuscul Disord: NMD* 2013;**23**(6):451–5.
176. Cyrulnik SE, Fee RJ, De Vivo DC, Goldstein E, Hinton VJ. Delayed developmental language milestones in children with Duchenne's muscular dystrophy. *J Pediatr* 2007;**150**(5):474–8.
177. Smith RA, Sibert JR, Harper PS. Early development of boys with Duchenne muscular dystrophy. *Dev Med Child Neurol* 1990;**32**(6):519–27.
178. Cyrulnik SE, Fee RJ, Batchelder A, Kiefel J, Goldstein E, Hinton VJ. Cognitive and adaptive deficits in young children with Duchenne muscular dystrophy (DMD). *J Int Neuropsychol Soc* 2008;**14**(5):853–61.

179. Billard C, Gillet P, Barthez M, Hommet C, Bertrand P. Reading ability and processing in Duchenne muscular dystrophy and spinal muscular atrophy. *Dev Med Child Neurol* 1998;**40**(1):12–20.

180. Whelan TB. Neuropsychological performance of children with Duchenne muscular dystrophy and spinal muscle atrophy. *Dev Med Child Neurol* 1987;**29**(2):212–20.

181. Hinton VJ, De Vivo DC, Nereo NE, Goldstein E, Stern Y. Poor verbal working memory across intellectual level in boys with Duchenne dystrophy. *Neurology* 2000;**54**(11):2127–32.

182. Hinton VJ, De Vivo DC, Nereo NE, Goldstein E, Stern Y. Selective deficits in verbal working memory associated with a known genetic etiology: the neuropsychological profile of Duchenne muscular dystrophy. *J Int Neuropsychol Soc* 2001;**7**(1):45–54.

183. Hinton VJ, De Vivo DC, Fee R, Goldstein E, Stern Y. Investigation of poor academic achievement in children with Duchenne muscular dystrophy. *Learn Disabil Res Pract* 2004;**19**(3):146–54.

184. Hinton VJ, Fee RJ, Goldstein EM, De Vivo DC. Verbal and memory skills in males with Duchenne muscular dystrophy. *Dev Med Child Neurol* 2007;**49**(2):123–8.

185. Hendriksen JG, Vles JS. Are males with Duchenne muscular dystrophy at risk for reading disabilities? *Pediatr Neurol* 2006;**34**(4):296–300.

186. Wingeier K, Giger E, Strozzi S, Kreis R, Joncourt F, Conrad B, et al. Neuropsychological impairments and the impact of dystrophin mutations on general cognitive functioning of patients with Duchenne muscular dystrophy. *J Clin Neurosci* 2011;**18**(1):90–5.

187. Dorman C, Hurley AD, D'Avignon J. Language and learning disorders of older boys with Duchenne muscular dystrophy. *Dev Med Child Neurol* 1988;**30**(3):316–27.

188. Lorusso ML, Civati F, Molteni M, Turconi AC, Bresolin N, D'Angelo MG. Specific profiles of neurocognitive and reading functions in a sample of 42 Italian boys with Duchenne Muscular Dystrophy. *Child Neuropsychol* 2013;**19**(4):350–69.

189. de Moura MC, Do Valle LE, Resende MB, Pinto KO. Visuospatial attention disturbance in Duchenne muscular dystrophy. *Dev Med Child Neurol* 2010;**52**(2):e10–15.

190. Mento G, Tarantino V, Bisiacchi PS. The neuropsychological profile of infantile Duchenne muscular dystrophy. *Clin Neuropsychol* 2011;**25**(8):1359–77.

191. Donders J, Taneja C. Neurobehavioral characteristics of children with Duchenne muscular dystrophy. *Child Neuropsychol* 2009;**15**(3):295–304.

192. Hendriksen JG, Vles JS. Neuropsychiatric disorders in males with Duchenne muscular dystrophy: frequency rate of attention-deficit hyperactivity disorder (ADHD), autism spectrum disorder, and obsessive-compulsive disorder. *J Child Neurol* 2008;**23**(5):477–81.

193. Darke J, Bushby K, Le Couteur A, McConachie H. Survey of behaviour problems in children with neuromuscular diseases. *Eur J Paediatr Neurol* 2006;**10**(3):129–34.

194. Hinton VJ, Nereo NE, Fee RJ, Cyrulnik SE. Social behavior problems in boys with Duchenne muscular dystrophy. *J Dev Behav Pediatr* 2006;**27**(6):470–6.

195. Wu JY, Kuban KC, Allred E, Shapiro F, Darras BT. Association of Duchenne muscular dystrophy with autism spectrum disorder. *J Child Neurol* 2005;**20**(10):790–5.

196. Hinton VJ, Cyrulnik SE, Fee RJ, Batchelder A, Kiefel JM, Goldstein EM, et al. Association of autistic spectrum disorders with dystrophinopathies. *Pediatr Neurol* 2009;**41**(5):339–46.

197. Fitzpatrick C, Barry C, Garvey C. Psychiatric disorder among boys with Duchenne muscular dystrophy. *Dev Med Child Neurol* 1986;**28**(5):589–95.

198. Fee RJ, Hinton VJ. Resilience in children diagnosed with a chronic neuromuscular disorder. *J Dev Behav Pediatr* 2011;**32**(9):644–50.

199. Elsenbruch S, Schmid J, Lutz S, Geers B, Schara U. Self-reported quality of life and depressive symptoms in children, adolescents, and adults with Duchenne muscular dystrophy: a cross-sectional survey study. *Neuropediatrics* 2013;**44**(5):257–64.

200. Rahbek J, Werge B, Madsen A, Marquardt J, Steffensen BF, Jeppesen J. Adult life with Duchenne muscular dystrophy: observations among an emerging and unforeseen patient population. *Pediatr Rehabil* 2005;**8**(1):17–28.

201. Moizard MP, Billard C, Toutain A, Berret F, Marmin N, Moraine C. Are Dp71 and Dp140 brain dystrophin isoforms related to cognitive impairment in Duchenne muscular dystrophy? *Am J Med Genet* 1998;**80**(1):32–41.

202. Felisari G, Martinelli Boneschi F, Bardoni A, Sironi M, Comi GP, Robotti M, et al. Loss of Dp140 dystrophin isoform and intellectual impairment in Duchenne dystrophy. *Neurology* 2000;**55**(4):559–64.

203. D'Angelo MG, Lorusso ML, Civati F, Comi GP, Magri F, Del Bo R, et al. Neurocognitive profiles in Duchenne muscular dystrophy and gene mutation site. *Pediatric neurology* 2011;**45**(5):292–9.

204. Bushby K, Finkel R, Birnkrant DJ, Case LE, Clemens PR, Cripe L, et al. Diagnosis and management of Duchenne muscular dystrophy, part 1: diagnosis, and pharmacological and psychosocial management. *Lancet Neurol* 2010;**9**(1):77–93.

205. Rosman NP, Kakulas BA. Mental deficiency associated with muscular dystrophy. A neuropathological study. *Brain* 1966;**89**(4):769–88.

206. Dubowitz V, Crome L. The central nervous system in Duchenne muscular dystrophy. *Brain* 1969;**92**(4):805–8.

207. Daoud F, Angeard N, Demerre B, Martie I, Benyaou R, Leturcq F, et al. Analysis of Dp71 contribution in the severity of mental retardation through comparison of Duchenne and Becker patients differing by mutation consequences on Dp71 expression. *Hum Mol Genet* 2009;**18**(20):3779–94.

208. Taylor PJ, Betts GA, Maroulis S, Gilissen C, Pedersen RL, Mowat DR, et al. Dystrophin gene mutation location and the risk of cognitive impairment in Duchenne muscular dystrophy. *PLOS ONE* 2010;**5**(1):e8803.

209. Barohn RJ, Levine EJ, Olson JO, Mendell JR. Gastric hypomotility in Duchenne's muscular dystrophy. *N Engl J Med* 1988;**319**(1):151–8.

210. Chung BC, Park HJ, Yoon SB, Lee HW, Kim KW, Lee SI, et al. Acute gastroparesis in Duchenne's muscular dystrophy. *Yonsei Med J* 1998;**39**(2):175–9.

211. Larson CM, Henderson RC. Bone mineral density and fractures in boys with Duchenne muscular dystrophy. *J Pediatr Orthop* 2000;**20**(1):71–4.

212. Parker AE, Robb SA, Chambers J, Davidson AC, Evans K, O'Dowd J, et al. Analysis of an adult Duchenne muscular dystrophy population. *QJM* 2005;**98**(10):729–36.

213. McDonald DG, Kinali M, Gallagher AC, Mercuri E, Muntoni F, Roper H, et al. Fracture prevalence in Duchenne muscular dystrophy. *Dev Med Child Neurol* 2002;**44**(10):695–8.
214. Hsu JD, Quinlivan R. Scoliosis in Duchenne muscular dystrophy (DMD). *Neuromuscul Disord: NMD* 2013;**23**(8):611–17.
215. Velasco MV, Colin AA, Zurakowski D, Darras BT, Shapiro F. Posterior spinal fusion for scoliosis in Duchenne muscular dystrophy diminishes the rate of respiratory decline. *Spine* 2007;**32**(4):459–65.
216. Alexander WM, Smith M, Freeman BJ, Sutherland LM, Kennedy JD, Cundy PJ. The effect of posterior spinal fusion on respiratory function in Duchenne muscular dystrophy. *Eur Spine J* 2013;**22**(2):411–16.
217. Karpati G, Watters GV. Adverse anaesthetic reactions in Duchenne dystrophy. *Excerpta Medica ICS* 1980;**527**:206.
218. Rosenberg H, Heiman-Patterson T. Duchenne's muscular dystrophy and malignant hyperthermia: another warning [letter]. *Anesthesiology* 1983;**59**(4):362.
219. Sethna NF, Rockoff MA, Worthen HM, Rosnow JM. Anesthesia-related complications in children with Duchenne muscular dystrophy. *Anesthesiology* 1988;**68**(3):462–5.
220. Nathan A, Ganesh A, Godinez RI, Nicolson SC, Greeley WJ. Hyperkalemic cardiac arrest after cardiopulmonary bypass in a child with unsuspected Duchenne muscular dystrophy. *Anesth Analg* 2005;**100**(3):672–4.
221. Willner J, Nakagawa M, Wood D. Drug-induced fiber necrosis in Duchenne dystrophy. *Ital J Neurol Sci* 1984;**5**(Suppl. 3):117.
222. Emery AE, Skinner R. Clinical studies in benign (Becker type) X-linked muscular dystrophy. *Clin Genet* 1976;**10**(4):189–201.
223. Ringel SP, Carroll JE, Schold SC. The spectrum of mild X-linked recessive muscular dystrophy. *Arch Neurol* 1977;**34**(7):408–16.
224. Bradley WG, Jones MZ, Mussini JM, Fawcett PR. Becker-type muscular dystrophy. *Muscle Nerve* 1978;**1**(2):111–32.
225. Bushby KM, Gardner-Medwin D. The clinical, genetic and dystrophin characteristics of Becker muscular dystrophy. I. Natural history. *J Neurol* 1993;**240**(2):98–104.
226. Specht L, Shapiro F. Boys with Duchenne and Becker muscular dystrophy are clinically indistinguishable at presentation [abstr]. *Ann Neurol* 1990;**28**:443.
227. Emery AEH. Muscular dystrophy, benign X-linked (Becker). In: Vinken PJ, Bruyn GW, editors. *Handbook of Clinical Neurology*. Amsterdam: North Holland; 1982. pp. 86–8.
228. de Visser M, Bakker E, Defesche JC, Bolhuis PA, van Ommen GJ. An unusual variant of Becker muscular dystrophy. *Ann Neurol* 1990;**27**(5):578–81.
229. Doriguzzi C, Palmucci L, Mongini T, Chiado-Piat L, Restagno G, Ferrone M. Exercise intolerance and recurrent myoglobinuria as the only expression of Xp21 Becker type muscular dystrophy. *J Neurol* 1993;**240**(5):269–71.
230. Samaha FJ, Quinlan JG. Myalgia and cramps: dystrophinopathy with wide-ranging laboratory findings. *J Child Neurol* 1996;**11**(1):21–4.
231. Yazawa M, Ikeda S, Owa M, Haruta S, Yanagisawa N, Tanaka E, et al. A family of Becker's progressive muscular dystrophy with severe cardiomyopathy. *Eur Neurol* 1987;**27**(1):13–19.
232. Casazza F, Brambilla G, Salvato A, Morandi L, Gronda E, Bonacina E. Cardiac transplantation in Becker muscular dystrophy. *J Neurol* 1988;**235**(8):496–8.
233. Donofrio PD, Challa VR, Hackshaw BT, Mills SA, Cordell AR. Cardiac transplantation in a patient with muscular dystrophy and cardiomyopathy. *Arch Neurol* 1989;**46**(6):705–7.
234. Finsterer J, Bittner RE, Grimm M. Cardiac involvement in Becker's muscular dystrophy, necessitating heart transplantation, 6 years before apparent skeletal muscle involvement. *Neuromuscul Disord* 1999;**9**(8):598–600.
235. Doo KH, Ryu HW, Kim SS, Lim BC, Hwang H, Kim KJ, et al. A case of Becker muscular dystrophy with early manifestation of cardiomyopathy. *Korean J Pediatr* 2012;**55**(9):350–3.
236. Kaspar RW, Allen HD, Ray WC, Alvarez CE, Kissel JT, Pestronk A, et al. Analysis of dystrophin deletion mutations predicts age of cardiomyopathy onset in Becker muscular dystrophy. *Circ Cardiovasc Genet* 2009;**2**(6):544–51.
237. Norman AM, Thomas NS, Kingston HM, Harper PS. Becker muscular dystrophy: correlation of deletion type with clinical severity. *J Med Genet* 1990;**27**(4):236–9.
238. Heald A, Anderson LV, Bushby KM, Shaw PJ. Becker muscular dystrophy with onset after 60 years. *Neurology* 1994;**44**(12):2388–90.
239. Yazaki M, Yoshida K, Nakamura A, Koyama J, Nanba T, Ohori N, et al. Clinical characteristics of aged Becker muscular dystrophy patients with onset after 30 years. *Eur Neurol* 1999;**42**(3):145–9.
240. Bosone I, Bortolotto S, Mongini T, Doriguzzi C, Chiado-Piat L, Ugo I, et al. Late onset and very mild course of Xp21 Becker type muscular dystrophy. *Clin Neuropathol* 2001;**20**(5):196–9.
241. Brooke MH, Fenichel GM, Griggs RC, Mendell JR, Moxley R, Miller JP, et al. Clinical investigation in Duchenne dystrophy: 2. Determination of the "power" of therapeutic trials based on the natural history. *Muscle Nerve* 1983;**6**(2):91–103.
242. Berko BA, Swift M. X-linked dilated cardiomyopathy. *N Engl J Med* 1987;**316**(19):1186–91.
243. Towbin JA, Hejtmancik JF, Brink P, Gelb B, Zhu XM, Chamberlain JS, et al. X-linked dilated cardiomyopathy. Molecular genetic evidence of linkage to the Duchenne muscular dystrophy (dystrophin) gene at the Xp21 locus. *Circulation* 1993;**87**(6):1854–65.
244. Palmucci L, Doriguzzi C, Mongini T, Chiado-Piat L, Restagno G, Carbonara A, et al. Dilating cardiomyopathy as the expression of Xp21 Becker type muscular dystrophy. *J Neurol Sci* 1992;**111**(2):218–21.
245. Yoshida K, Ikeda S, Nakamura A, Kagoshima M, Takeda S, Shoji S, et al. Molecular analysis of the Duchenne muscular dystrophy gene in patients with Becker muscular dystrophy presenting with dilated cardiomyopathy. *Muscle Nerve* 1993;**16**(11):1161–6.
246. Ferlini A, Sewry C, Melis MA, Mateddu A, Muntoni F. X-linked dilated cardiomyopathy and the dystrophin gene. *Neuromuscul Disord* 1999;**9**(5):339–46.
247. Neri M, Torelli S, Brown S, Ugo I, Sabatelli P, Merlini L, et al. Dystrophin levels as low as 30% are sufficient to avoid muscular dystrophy in the human. *Neuromuscul Disord: NMD* 2007;**17**(11–12):913–18.
248. Saito M, Kawai H, Akaike M, Adachi K, Nishida Y, Saito S. Cardiac dysfunction with Becker muscular dystrophy. *Am Heart J* 1996;**132**(3):642–7.
249. Milasin J, Muntoni F, Severini GM, Bartoloni L, Vatta M, Krajinovic M, et al. A point mutation in the 5′ splice site of the

dystrophin gene first intron responsible for X-linked dilated cardiomyoathy. *Hum Mol Genet* 1996;**5**:73–9.

250. Yoshida K, Nakamura A, Yazaki M, Ikeda S, Takeda S. Insertional mutation by transposable element, L1, in the DMD gene results in X-linked dilated cardiomyopathy. *Hum Mol Genet* 1998;**7**(7):1129–32.

251. Neri M, Valli E, Alfano G, Bovolenta M, Spitali P, Rapezzi C, et al. The absence of dystrophin brain isoform expression in healthy human heart ventricles explains the pathogenesis of 5′. X-linked dilated cardiomyopathy. *BMC Med Genet* 2012;**13**:20.

252. Muntoni F, Melis MA, Ganau A, Dubowitz V. Transcription of the dystrophin gene in normal tissues and in skeletal muscle of a family witih X-linked dilated cardiomyopathy. *Am J Hum Genet* 1995;**56**:151–7.

253. Muntoni F, Wilson L, Marrosu G, Marrosu MG, Cianchetti C, Mestroni L, et al. A mutation in the dystrophin gene selectively affecting dystrophin expression in the heart. *J Clin Invest* 1995;**96**:693–9.

254. Ortiz-Lopez R, Li H, Su J, Goytia V, Towbin JA. Evidence for a dystrophin missense mutation as a cause of X-linked dilated cardiomyopathy. *Circulation* 1997;**95**(10):2434–40.

255. Gavillet B, Rougier JS, Domenighetti AA, Behar R, Boixel C, Ruchat P, et al. Cardiac sodium channel Nav1.5 is regulated by a multiprotein complex composed of syntrophins and dystrophin. *Circ Res* 2006;**99**(4):407–14.

256. Palmucci L, Mongini T, Chiado-Piat L, Doriguzzi C, Fubini A. Dystrophinopathy expressing as either cardiomyopathy or Becker dystrophy in the same family. *Neurology* 2000;**54**(2):529–30.

257. Gospe Jr. SM, Lazaro RP, Lava NS, Grootscholten PM, Scott MO, Fischbeck KH. Familial X-linked myalgia and cramps: a nonprogressive myopathy associated with a deletion in the dystrophin gene. *Neurology* 1989;**39**(10):1277–80.

258. Gold R, Kress W, Meurers B, Meng G, Reichmann H, Muller CR. Brief communication: Becker muscular dystrophy: detection of unusual disease courses by combined approach to dystrophin analysis. *Muscle Nerve* 1992;**15**:214–18.

259. Figarella-Branger D, Baeta Machado AM, Putzu GA, Malzac P, Voelckel MA, Pellissier JF. Exertional rhabdomyolysis and exercise intolerance revealing dystrophinopathies. *Acta Neuropathol* 1997;**94**(1):48–53.

260. Melis MA, Cau M, Muntoni F, Mateddu A, Galanello R, Boccone L, et al. Elevation of serum creatine kinase as the only manifestation of an intragenic deletion of the dystrophin gene in three unrelated families. *Eur J Paediatr Neurol* 1998;**2**(5):255–61.

261. Saengpattrachai M, Ray PN, Hawkins CE, Berzen A, Banwell BL. Grandpa and I have dystrophinopathy?: approach to asymptomatic hyperCKemia. *Pediatr Neurol* 2006;**35**(2):145–9.

262. Suminaga R, Takeshima Y, Wada H, Yagi M, Matsuo M. C-terminal truncated dystrophin identified in skeletal muscle of an asymptomatic boy with a novel nonsense mutation of the dystrophin gene. *Pediatr Res* 2004;**56**(5):739–43.

263. Moser H, Emery AE. The manifesting carrier in Duchenne muscular dystrophy. *Clin Genet* 1974;**5**(4):271–84.

264. Norman A, Harper P. A survey of manifesting carriers of Duchenne and Becker muscular dystrophy in Wales. *Clin Genet* 1989;**36**(1):31–7.

265. Juan-Mateu J, Rodriguez MJ, Nascimento A, Jimenez-Mallebrera C, Gonzalez-Quereda L, Rivas E, et al. Prognostic value of X-chromosome inactivation in symptomatic female carriers of dystrophinopathy. *Orphanet J Rare Dis* 2012;**7**:82.

266. Olson BJ, Fenichel GM. Progressive muscle disease in a young woman with family history of Duchenne's muscular dystrophy. *Arch Neurol* 1982;**39**(6):378–80.

267. Hoogerwaard EM, Bakker E, Ippel PF, Oosterwijk JC, Majoor-Krakauer DF, Leschot NJ, et al. Signs and symptoms of Duchenne muscular dystrophy and Becker muscular dystrophy among carriers in The Netherlands: a cohort study. *Lancet* 1999;**353**(9170):2116–19.

268. Hoogerwaard EM, van der Wouw PA, Wilde AA, Bakker E, Ippel PF, Oosterwijk JC, et al. Cardiac involvement in carriers of Duchenne and Becker muscular dystrophy. *Neuromuscul Disord* 1999;**9**(5):347–51.

269. Kamakura K, Kawai M, Arahata K, Koizumi H, Watanabe K, Sugita H. A manifesting carrier of Duchenne muscular dystrophy with severe myocardial symptoms. *J Neurol* 1990;**237**(8):483–5.

270. Schade van Westrum SM, Hoogerwaard EM, Dekker L, Standaar TS, Bakker E, Ippel PF, et al. Cardiac abnormalities in a follow-up study on carriers of Duchenne and Becker muscular dystrophy. *Neurology* 2011;**77**(1):62–6.

271. Nolan MA, Jones OD, Pedersen RL, Johnston HM. Cardiac assessment in childhood carriers of Duchenne and Becker muscular dystrophies. *Neuromuscul Disord* 2003;**13**(2):129–32.

272. Hoffman EP, Arahata K, Minetti C, Bonilla E, Rowland LP. Dystrophinopathy in isolated cases of myopathy in females. *Neurology* 1992;**42**(5):967–75.

273. Muntoni F, Mateddu A, Cianchetti C, Marrosu MG, Clerk A, Cau M, et al. Dystrophin analysis using a panel of anti-dystrophin antibodies in Duchenne and Becker muscular dystrophy. *J Neurol Neurosurg Psychiatry* 1993;**56**(1):26–31.

274. Katayama Y, Tran VK, Hoan NT, Zhang Z, Goji K, Yagi M, et al. Co-occurrence of mutations in both dystrophin- and androgen-receptor genes is a novel cause of female Duchenne muscular dystrophy. *Hum Genet* 2006;**119**(5):516–19.

275. Sumita DR, Vainzof M, Campiotto S, Cerqueira AM, Canovas M, Otto PA, et al. Absence of correlation between skewed X inactivation in blood and serum creatine-kinase levels in Duchenne/Becker female carriers. *Am J Med Genet* 1998;**80**(4):356–61.

276. Soltanzadeh P, Friez MJ, Dunn D, von Niederhausern A, Gurvich OL, Swoboda KJ, et al. Clinical and genetic characterization of manifesting carriers of DMD mutations. *Neuromuscul Disord NMD* 2010;**20**(8):499–504.

277. Pegoraro E, Schimke RN, Arahata K, Hayashi Y, Stern H, Marks H, et al. Detection of new paternal dystrophin gene mutations in isolated cases of dystrophinopathy in females. *Am J Hum Genet* 1994;**54**(6):989–1003.

278. Viggiano E, Picillo E, Cirillo A, Politano L. Comparison of X-chromosome inactivation in Duchenne muscle/myocardium-manifesting carriers, non-manifesting carriers and related daughters. *Clin Genet* 2013;**84**(3):265–70.

279. Lupski JR, Garcia CA, Zoghbi HY, Hoffman EP, Fenwick RG. Discordance of muscular dystrophy in monozygotic female twins: evidence supporting asymmetric splitting of the inner cell mass in a manifesting carrier of Duchenne dystrophy. *Am J Med Genet* 1991;**40**(3):354–64.

280. Zneimer SM, Schneider NR, Richards CS. In situ hybridization shows direct evidence of skewed X inactivation in one of

monozygotic twin females manifesting Duchenne muscular dystrophy. *Am J Med Genet* 1993;**45**(5):601–5.

281. Fujii K, Minami N, Hayashi Y, Nishino I, Nonaka I, Tanabe Y, et al. Homozygous female Becker muscular dystrophy. *Am J Med Genet A* 2009;**149A**(5):1052–5.
282. Schwartz M, Hertz JM, Sveen ML, Vissing J. LGMD2I presenting with a characteristic Duchenne or Becker muscular dystrophy phenotype. *Neurology* 2005;**64**(9):1635–7.
283. Zacharias AS, Wagener ME, Warren ST, Hopkins LC. Emery-Dreifuss muscular dystrophy. *Semin Neurol* 1999;**19**(1):67–79.
284. Raffaele Di Barletta M, Ricci E, Galluzzi G, Tonali P, Mora M, Morandi L, et al. Different mutations in the LMNA gene cause autosomal dominant and autosomal recessive Emery-Dreifuss muscular dystrophy. *Am J Hum Genet* 2000;**66**(4):1407–12.
285. Arikawa E, Ishihara T, Nonaka I, Sugita H, Arahata K. Immunocytochemical analysis of dystrophin in congenital muscular dystrophy. *J Neurol Sci* 1991;**105**(1):79–87.
286. Prelle A, Medori R, Moggio M, Chan HW, Gallanti A, Scarlato G, et al. Dystrophin deficiency in a case of congenital myopathy. *J Neurol* 1992;**239**(2):76–8.
287. Chanprasert S, Wang J, Weng SW, Enns GM, Boue DR, Wong BL, et al. Molecular and clinical characterization of the myopathic form of mitochondrial DNA depletion syndrome caused by mutations in the thymidine kinase (TK2) gene. *Mol Genet Metab* 2013;**110**(1–2):153–61.
288. Vu TH, Sciacco M, Tanji K, Nichter C, Bonilla E, Chatkupt S, et al. Clinical manifestations of mitochondrial DNA depletion. *Neurology* 1998;**50**(6):1783–90.
289. McMillan HJ, Gregas M, Darras BT, Kang PB. Serum transaminase levels in boys with Duchenne and Becker muscular dystrophy. *Pediatrics* 2011;**127**(1):e132–6.
290. Morse RP, Rosman NP. Diagnosis of occult muscular dystrophy: importance of the "chance" finding of elevated serum aminotransferase activities. *J Pediatr* 1993;**122**(2):254–6.
291. Mestroni L, Rocco C, Gregori D, Sinagra G, Di Lenarda A, Miocic S, et al. Familial dilated cardiomyopathy: evidence for genetic and phenotypic heterogeneity. Heart Muscle Disease Study Group. *J Am Coll Cardiol* 1999;**34**(1):181–90.
292. Watkins H, Ashrafian H, Redwood C. Inherited cardiomyopathies. *N Engl J Med* 2011;**364**(17):1643–56.
293. D'Adamo P, Fassone L, Gedeon A, Janssen EA, Bione S, Bolhuis PA, et al. The X-linked gene G4.5 is responsible for different infantile dilated cardiomyopathies. *Am J Hum Genet* 1997;**61**(4):862–7.
294. Yen TY, Hwu WL, Chien YH, Wu MH, Lin MT, Tsao LY, et al. Acute metabolic decompensation and sudden death in Barth syndrome: report of a family and a literature review. *Eur J Pediatr* 2008;**167**(8):941–4.
295. Darras BT, Miller DT, Urion DK. Dystrophinopathies. 2000 Sep 5 [Updated 2011 Nov 23]. In: Pagon RA, Adam MP, Bird TD, et al., editors. *GeneReviews™ [Internet]*. Seattle (WA): University of Washington, Seattle; 1993–2013. Available from: <http://www.ncbi.nlm.nih.gov/books/NBK1119/>.
296. Rosalki SB. Serum enzymes in disease of skeletal muscle. *Clin Lab Med* 1989;**9**(4):767–81.
297. Zatz M, Rapaport D, Vainzof M, Passos-Bueno MR, Bortolini ER, Pavanello Rde C, et al. Serum creatine-kinase (CK) and pyruvate-kinase (PK) activities in Duchenne (DMD) as compared with Becker (BMD) muscular dystrophy. *J Neurol Sci* 1991;**102**(2):190–6.
298. Kuhn E, Fiehn W, Schroder JM, Assmus H, Wagner A. Early myocardial disease and cramping myalgia in Becker-type muscular dystrophy: a kindred. *Neurology* 1979;**29**(8):1144–9.
299. Pearson CM. Serum enzymes in muscular dystrophy and certain other muscular and neuromuscular diseases. I. Serum glutamic oxalacetic transaminase. *N Engl J Med* 1957;**256**:1069–75.
300. Fowler WM, Pearson CM. Diagnostic and prognostic significance of serum enzymes. I. Muscular dystrophy. *Arch Phys Med* 1964;**45**:117–24.
301. Munsat TL, Baloh R, Pearson CM, Fowler Jr. W. Serum enzyme alterations in neuromuscular disorders. *JAMA* 1973;**226**(13):1536–43.
302. Griggs RC, Mendell JR, Brooke MH, Fenichel GM, Miller JP, Province M, et al. Clinical investigation in Duchenne dystrophy: V. Use of creatine kinase and pyruvate kinase in carrier detection. *Muscle Nerve* 1985;**8**(1):60–7.
303. Dawson DM, Fine IH. Creatin kinase in human tissues. *Arch Neurol* 1967;**16**(2):175–80.
304. Goto I, Nagamine M, Katsuki S. Creatine phosphokinase isozymes in muscles. Human fetus and patients. *Arch Neurol* 1969;**20**(4):422–9.
305. Somer H, Dubowitz V, Donner M. Creatine kinase isoenzymes in neuromuscular diseases. *J Neurol Sci* 1976;**29**(2–4):129–36.
306. Somer H, Donner M, Murros J, Konttinen A. A serum isozyme study in muscular dystrophy. Particular reference to creatine kinase, aspartate aminotransferase, and lactic acid dehydrogenase isozymes. *Arch Neurol* 1973;**29**(5):343–5.
307. Goedde HW, Christ I, Benkmann HG, Beckmann R, Lang H. Creatine kinase isoenzyme patterns in Duchenne muscular dystrophy. *Clin Genet* 1978;**14**(5):257–60.
308. Takahashi K, Shutta K, Matsuo B, Takai T, Takao H, Imura H. Serum creatine kinase isoenzymes in Duchenne muscular dystrophy. *Clin Chim Acta* 1977;**75**(3):435–42.
309. Tzvetanova E. Serum creatine kinase isoenzymes in progressive muscular dystrophy. *Enzyme* 1978;**23**(4):238–45.
310. Wolf PL. Abnormalities in serum enzymes in skeletal muscle diseases. *Am J Clin Pathol* 1991;**95**(3):293–6.
311. Ando T, Shimizu T, Kato T, Ohsawa M, Fukuyama Y. Myoglobinemia in children with progressive muscular dystrophy. *Clin Chim Acta* 1978;**85**(1):17–22.
312. Kagen LJ, Moussavi S, Miller SL, Tsairis P. Serum myoglobin in muscular dystrophy. *Muscle Nerve* 1980;**3**(3):221–6.
313. Florence JM, Fox PT, Planer GJ, Brooke MH. Activity, creatine kinase, and myoglobin in Duchenne muscular dystrophy: a clue to etiology? *Neurology* 1985;**35**(5):758–61.
314. Engel AG, Arahata K. Monoclonal antibody analysis of mononuclear cells in myopathies. II: Phenotypes of autoinvasive cells in polymyositis and inclusion body myositis. *Ann Neurol* 1984;**16**(2):209–15.
315. Bell CD, Conen PE. Change in fiber size in Duchenne muscular dystrophy. *Neurology* 1967;**17**(9):902–13.
316. Schiaffino S, Gorza L, Dones I, Cornelio F, Sartore S. Fetal myosin immunoreactivity in human dystrophic muscle. *Muscle Nerve* 1986;**9**(1):51–8.
317. Karpati G, Carpenter S, Morris GE, Davies KE, Guerin C, Holland P. Localization and quantitation of the chromosome

6-encoded dystrophin-related protein in normal and pathological human muscle. *J Neuropathol Exp Neurol* 1993;**52**(2):119–28.

318. Watkins SC, Cullen MJ. Histochemical fibre typing and ultrastructure of the small fibres in Duchenne muscular dystrophy. *Neuropathol Appl Neurobiol* 1985;**11**(6):447–60.
319. Kaido M, Arahata K, Hoffman EP, Nonaka I, Sugita H. Muscle histology in Becker muscular dystrophy. *Muscle Nerve* 1991;**14**(11):1067–73.
320. Uchino M, Araki S, Miike T. Correlative study of the incidence of opaque, necrotic and regenerative fibers in Duchenne dystrophy. *Acta Neuropathol* 1988;**75**(3):308–12.
321. Bodensteiner JB, Engel AG. Intracellular calcium accumulation in Duchenne dystrophy and other myopathies: a study of 567,000 muscle fibers in 114 biopsies. *Neurology* 1978;**28**(5):439–46.
322. Nonaka I, Sugita H. Muscle pathology of Duchenne dystrophy, with particular reference to "opaque fibers". *Adv Neurol Sci* 1980;**24**:718.
323. Cornelio F, Dones I. Muscle fiber degeneration and necrosis in muscular dystrophy and other muscle diseases: cytochemical and immunocytochemical data. *Ann Neurol* 1984;**16**(6):694–701.
324. Momma K, Noguchi S, Malicdan MC, Hayashi YK, Minami N, Kamakura K, et al. Rimmed vacuoles in Becker muscular dystrophy have similar features with inclusion myopathies. *PLOS ONE* 2012;**7**(12):e52002.
325. Arahata K, Engel AG. Monoclonal antibody analysis of mononuclear cells in myopathies. I: Quantitation of subsets according to diagnosis and sites of accumulation and demonstration and counts of muscle fibers invaded by T cells. *Ann Neurol* 1984;**16**(2):193–208.
326. Stephens HR, Duance VC, Dunn MJ, Bailey AJ, Dubowitz V. Collagen types in neuromuscular diseases. *J Neurol Sci* 1982;**53**(1):45–62.
327. Mokri B, Engel AG. Duchenne dystrophy: electron microscopic findings pointing to a basic or early abnormality in the plasma membrane of the muscle fiber. *Neurology* 1975;**25**(12):1111–20
328. Carpenter S, Karpati G. Duchenne muscular dystrophy: plasma membrane loss initiates muscle cell necrosis unless it is repaired. *Brain* 1979;**102**(1):147–61.
329. Bulman DE, Murphy EG, Zubrzycka-Gaarn EE, Worton RG, Ray PN. Differentiation of Duchenne and Becker muscular dystrophy phenotypes with amino- and carboxy-terminal antisera specific for dystrophin. *Am J Hum Genet* 1991;**48**(2):295–304.
330. Voit T, Stuettgen P, Cremer M, Goebel HH. Dystrophin as a diagnostic marker in Duchenne and Becker muscular dystrophy. Correlation of immunofluorescence and Western blot. *Neuropediatrics* 1991;**22**(3):152–62.
331. Winnard AV, Jia-Hsu Y, Gibbs RA, Mendell JR, Burghes AH. Identification of a 2 base pair nonsense mutation causing a cryptic splice site in a DMD patient. *Hum Mol Genet* 1992;**1**(8):645–6.
332. Klein CJ, Coovert DD, Bulman DE, Ray PN, Mendell JR, Burghes AH. Somatic reversion/suppression in Duchenne muscular dystrophy (DMD): evidence supporting a frame-restoring mechanism in rare dystrophin-positive fibers. *Am J Hum Genet* 1992;**50**(5):950–9.
333. Nicholson LV, Johnson MA, Gardner-Medwin D, Bhattacharya S, Harris JB. Heterogeneity of dystrophin expression in patients with Duchenne and Becker muscular dystrophy. *Acta Neuropathol* 1990;**80**(3):239–50.
334. Nguyen TM, Morris GE. Use of epitope libraries to identify exon-specific monoclonal antibodies for characterization of altered dystrophins in muscular dystrophy. *Am J Hum Genet* 1993;**52**(6):1057–66.
335. Yamamoto H, Brengman JM, McCormick DJ, Nishino H, Engel AG. Dystrophin deficiency examined by combined PCR, immunoblot and immunostain analysis [abstr]. *Neurology* 1993;**43**(Suppl.):A293–4.
336. Nicholson LV, Johnson MA, Bushby KM, Gardner-Medwin D, Curtis A, Ginjaar IB, et al. Integrated study of 100 patients with Xp21 linked muscular dystrophy using clinical, genetic, immunochemical, and histopathological data. Part 1. Trends across the clinical groups. *J Med Genet* 1993;**30**(9):728–36.
337. Ikeya K, Saito K, Hayashi K, Tanaka H, Hagiwara Y, Yoshida M, et al. Molecular genetic and immunological analysis of dystrophin of a young patient with X-linked muscular dystrophy. *Am J Med Genet* 1992;**43**(3):580–7.
338. Gold R, Kress W, Reichmann H, Muller CR. The use of monoclonal antibodies in diagnostic tests for Becker and Duchenne muscular dystrophy. *J Neurol* 1993;**240**(1):21–4.
339. Hoogerwaard EM, Ginjaar IB, Bakker E, de Visser M. Dystrophin analysis in carriers of Duchenne and Becker muscular dystrophy. *Neurology* 2005;**65**(12):1984–6.
340. Bonilla E, Schmidt B, Samitt CE, Miranda AF, Hays AP, de Oliveira AB, et al. Normal and dystrophin-deficient muscle fibers in carriers of the gene for Duchenne muscular dystrophy. *Am J Pathol* 1988;**133**(3):440–5.
341. Arahata K, Ishihara T, Kamakura K, Tsukahara T, Ishiura S, Baba C, et al. Mosaic expression of dystrophin in symptomatic carriers of Duchenne's muscular dystrophy. *N Engl J Med* 1989;**320**(3):138–42.
342. Morandi L, Mora M, Gussoni E, Tedeschi S, Cornelio F. Dystrophin analysis in Duchenne and Becker muscular dystrophy carriers: correlation with intracellular calcium and albumin. *Ann Neurol* 1990;**28**(5):674–9.
343. Vainzof M, Pavanello RC, Pavanello I, Tsanaclis AM, Levy JA, Passos-Bueno MR, et al. Dystrophin immunofluorescence pattern in manifesting and asymptomatic carriers of Duchenne's and Becker muscular dystrophies of different ages. *Neuromuscul Disord* 1991;**1**(3):177–83.
344. Muntoni F, Mateddu A, Marrosu MG, Cau M, Congiu R, Melis MA et al. Variable dystrophin expression in different muscles of a Duchenne muscular dystrophy carrier. *Clin Genet* 1992;**42**(1):35–8.
345. Mercier S, Toutain A, Toussaint A, Raynaud M, de Barace C, Marcorelles P, et al. Genetic and clinical specificity of 26 symptomatic carriers for dystrophinopathies at pediatric age. *Eur J Hum Genet: EJHG* 2013;**21**(8):855–63.
346. Clerk A, Rodillo E, Heckmatt JZ, Dubowitz V, Strong PN, Sewry CA. Characterisation of dystrophin in carriers of Duchenne muscular dystrophy. *J Neurol Sci* 1991;**102**(2):197–205.
347. Bernier FP, Greenberg CR, Halliday WC, Wrogemann K. Single-blind study of dystrophin staining in carriers of Duchenne muscular dystrophy. *Can J Neurol Sci* 1993;**20**(1):44–7.
348. Nicholson LV, Johnson MA, Bushby KM, Gardner-Medwin D, Curtis A, Ginjaar IB, et al. Integrated study of 100 patients with Xp21 linked muscular dystrophy using clinical, genetic, immunochemical, and histopathological data. Part 2. Correlations within individual patients. *J Med Genet* 1993;**30**(9):737–44

349. Arikawa E, Hoffman EP, Kaido M, Nonaka I, Sugita H, Arahata K. The frequency of patients with dystrophin abnormalities in a limb-girdle patient population. *Neurology* 1991;**41**(9):1491–6.
350. Beggs AH, Hoffman EP, Kunkel LM. Additional dystrophin fragment in Becker muscular dystrophy may result from proteolytic cleavage at deletion junctions. *Am J Med Genet* 1992;**44**(3):378–81.
351. Cox GA, Cole NM, Matsumura K, Phelps SF, Hauschka SD, Campbell KP, et al. Overexpression of dystrophin in transgenic mdx mice eliminates dystrophic symptoms without toxicity. *Nature* 1993;**364**:673–5.
352. Nicholson LV, Johnson MA, Bushby KM, Gardner-Medwin D, Curtis A, Ginjaar IB, et al. Integrated study of 100 patients with Xp21 linked muscular dystrophy using clinical, genetic, immunochemical, and histopathological data. Part 3. Differential diagnosis and prognosis. *J Med Genet* 1993;**30**(9):745–51.
353. Kesari A, Pirra LN, Bremadesam L, McIntyre O, Gordon E, Dubrovsky AL, et al. Integrated DNA, cDNA, and protein studies in Becker muscular dystrophy show high exception to the reading frame rule. *Hum Mutat* 2008;**29**(5):728–37.
354. Mendell JR, Buzin CH, Feng J, Yan J, Serrano C, Sangani DS, et al. Diagnosis of Duchenne dystrophy by enhanced detection of small mutations. *Neurology* 2001;**57**(4):645–50.
355. Gatta V, Scarciolla O, Gaspari AR, Palka C, De Angelis MV, Di Muzio A, et al. Identification of deletions and duplications of the DMD gene in affected males and carrier females by multiple ligation probe amplification (MLPA). *Hum Genet* 2005;**117**(1):92–8.
356. Bovolenta M, Neri M, Fini S, Fabris M, Trabanelli C, Venturoli A, et al. A novel custom high density-comparative genomic hybridization array detects common rearrangements as well as deep intronic mutations in dystrophinopathies. *BMC Genomics* 2008;**9**:572.
357. Hegde MR, Chin EL, Mulle JG, Okou DT, Warren ST, Zwick ME. Microarray-based mutation detection in the dystrophin gene. *Hum Mutat* 2008;**29**(9):1091–9.
358. del Gaudio D, Yang Y, Boggs BA, Schmitt ES, Lee JA, Sahoo T, et al. Molecular diagnosis of Duchenne/Becker muscular dystrophy: enhanced detection of dystrophin gene rearrangements by oligonucleotide array-comparative genomic hybridization. *Hum Mutat* 2008;**29**(9):1100–7.
359. Multicenter Study Group. Diagnosis of Duchenne and Becker muscular dystrophies by polymerase chain reaction. A multicenter study. *JAMA* 1992;**267**(19):2609–15.
360. Darras BT, Koenig M, Kunkel LM, Francke U. Direct method for prenatal diagnosis and carrier detection in Duchenne/Becker muscular dystrophy using the entire dystrophin cDNA. *Am J Med Genet* 1988;**29**(3):713–26.
361. Bonnal RJ, Severgnini M, Castaldi A, Bordoni R, Iacono M, Trimarco A, et al. Reliable resequencing of the human dystrophin locus by universal long polymerase chain reaction and massive pyrosequencing. *Anal Biochem* 2010;**406**(2):176–84.
362. Almomani R, van der Stoep N, Bakker E, den Dunnen JT, Breuning MH, Ginjaar IB. Rapid and cost effective detection of small mutations in the DMD gene by high resolution melting curve analysis. *Neuromuscul Disord: NMD* 2009;**19**(6):383–90.
363. Aartsma-Rus A, Van Deutekom JC, Fokkema IF, Van Ommen GJ, Den Dunnen JT. Entries in the Leiden Duchenne muscular dystrophy mutation database: an overview of mutation types and paradoxical cases that confirm the reading-frame rule. *Muscle Nerve* 2006;**34**(2):135–44.
364. Darras BT, Francke U. Myopathy in complex glycerol kinase deficiency patients is due to 3′ deletions of the dystrophin gene. *Am J Hum Genet* 1988;**43**(2):126–30.
365. Kim UK, Chae JJ, Lee SH, Lee CC, Namkoong Y. Molecular diagnosis of Duchenne/Becker muscular dystrophy by polymerase chain reaction and microsatellite analysis. *Mol Cells* 2002;**13**(3):385–8.
366. Abbs S, Roberts RG, Mathew CG, Bentley DR, Bobrow M. Accurate assessment of intragenic recombination frequency within the Duchenne muscular dystrophy gene. *Genomics* 1990;**7**(4):602–6.
367. Ferreiro V, Szijan I, Giliberto F. Detection of germline mosaicism in two Duchenne muscular dystrophy families using polymorphic dinucleotide (CA)n repeat loci within the dystrophin gene. *Mol Diagn* 2004;**8**(2):115–21.
368. van Essen AJ, Kneppers AL, van der Hout AH, Scheffer H, Ginjaar IB, ten Kate LP, et al. The clinical and molecular genetic approach to Duchenne and Becker muscular dystrophy: an updated protocol. *J Med Genet* 1997;**34**(10):805–12.
369. van Essen AJ, Busch HF, te Meerman GJ, ten Kate LP. Birth and population prevalence of Duchenne muscular dystrophy in The Netherlands. *Hum Genet* 1992;**88**(3):258–66.
370. Ladwig D, Mowat D, Tobias V, Taylor PJ, Buckley MF, McNally G, et al. In utero fetal muscle biopsy in the diagnosis of Duchenne muscular dystrophy. *Aust NZ J Obstet Gynaecol* 2002;**42**(1):79–82.
371. Raymond FL, Whittaker J, Jenkins L, Lench N, Chitty LS. Molecular prenatal diagnosis: the impact of modern technologies. *Prenat Diagn* 2010;**30**(7):674–81.

Chapter 31

Clinical Management of Dystrophinopathies: A Systematic Approach

Michela Guglieri and Katharine Bushby

INTRODUCTION

This chapter reviews the current management recommendations for the skeletal muscle predominant disorders caused by mutations in the dystrophin gene: Duchenne muscular dystrophy (DMD), Becker muscular dystrophy (BMD), and manifesting carriers of dystrophinopathy. Duchenne muscular dystrophy is the most common form of dystrophinopathy, affecting around 1 in 5000 male births.[1] The incidence of Becker muscular dystrophy is around 1 in 14,000.[2,3] Figures for manifesting carrier status are variable depending on the definition of this condition, which can be highly heterogeneous, ranging from minor symptoms to full-blown disease as seen in males. However, around 10% of female carriers of a dystrophin mutation have some kind of symptoms at some stage in their lives. Rarely, male and female patients have an isolated cardiomyopathy without any skeletal muscle manifestations at all. The management of these cases depends on the nature of their cardiac disease, and they are not discussed in detail here.

Diagnosis is dealt with in detail in Chapter 30. The age of diagnosis of DMD remains a crucial factor to improve patient management and ensure appropriate genetic counselling.[4,5] Only by recognizing the primary manifestations and promptly identifying possible complications of this disorder we can ensure optimal treatment.[6,7] Although still controversial, recent data suggest that early initiation of steroid treatment could be associated with better long-term outcomes.[8] Moreover, it is intuitive to think that any interventions that potentially modify the natural history of the condition may be of greater benefit in younger children with less damaged muscles. Early diagnosis will therefore potentially allow early access to investigational drugs.

While newborn screening has been suggested to reduce the diagnostic delay in DMD, this is not widely available and raises ethical issues due to the current lack of a curative treatment.[9–11] Earlier consideration of DMD in the differential diagnosis of boys with motor or speech developmental delay may expedite diagnosis and improve management and quality of life of these patients.

DUCHENNE MUSCULAR DYSTROPHY

Duchenne muscular dystrophy remains a devastating diagnosis with inexorable and relatively predictable progression and development of complications. Internationally generated standardized recommendations for the diagnosis and management of DMD were published in 2010[12,13] and are also available in a simplified version for families in many languages.[14] As outlined in the 2010 publications,[12,13] there are many areas where patient care can be dramatically improved. The standards of care recognize that DMD is a multisystemic and progressive condition needing a multidisciplinary approach, timed and adapted to disease stage-specific needs. The natural course of DMD can be subdivided into five major stages: presymptomatic, early ambulatory, late ambulatory, early and late nonambulatory phases. Although the age of onset and duration of each stage can vary from patient to patient and may be affected by pharmacological and nonpharmacological interventions, each boy with DMD passes through each of these stages sequentially. A stage-specific approach therefore addresses specific needs and areas of interventions better than one based on age alone.

B.T. Darras, H. Royden Jones, Jr., M.M. Ryan & D.C. De Vivo (Eds): Neuromuscular Disorders of Infancy, Childhood and Adolescence, Second edition.
DOI: http://dx.doi.org/10.1016/B978-0-12-417044-5.00031-7

A comprehensive approach to management must be multidisciplinary and should be coordinated with the family by a neuromuscular specialist with a genetic, pediatric, or neurological background. Although a multidisciplinary approach is required throughout the lifespan, depending on the stage of the condition, the relative involvement of genetic counselors, physiotherapist, cardiologist, respiratory clinician, orthopedic surgeon, endocrinologist, dietician, gastroenterologist, psychologist, and speech and language therapist will vary (see Table 31.1). Management should be anticipatory, preventing complications wherever possible. Access to advice on social care issues is also imperative at all stages. Ideally, social and medical issues should be addressed in a holistic manner within a recognized center of excellence. The prolonged survival of DMD patients has highlighted the need to adapt a proactive approach to transition services and the engagement of appropriate adult-based specialists to assume the patient ongoing care into adulthood.

Although there are no specific treatments currently available for DMD, the application of the care recommendations to include corticosteroids, respiratory, cardiac, orthopedic, and rehabilitative interventions has contributed to a "new natural history" altering the characteristic progression of the condition.[15–19] A recent comprehensive review of the burden of illness in DMD across the USA, the UK, Italy, and Germany showed that, while annual direct health costs are approximately ten times the national average for a comparable age group, these are just a small proportion of the overall burden of the disease.[20] Although the generation of an international publication on standards of care has significantly improved the diagnosis and management of DMD worldwide, variable and inconsistent health care is still reported in different countries and even among different specialized centers within the same country[20–22]

Assessments

Patients should be seen by the neuromuscular specialist at least once every six months. The aim of these assessments is to recognize manifestations of the condition, prevent and promptly address disease complications, inform decisions about therapeutic and physiotherapy interventions, and monitor response to treatments. More frequent evaluations might be required in transitional phases, such as initiation of steroid treatment, in the late ambulatory and early non-ambulatory phases. A three- and six-month follow-up visit should be offered after initiation of corticosteroid treatment to closely monitor individual response and side effects.

Assessments should include medical and physiotherapy examinations. The involvement of a psychologist, a social care advisor, and an occupational therapist at each assessment ensures early interventions for specific issues raised by the patient and family (Table 31.2).

Questions about family history should be included in the patient's first evaluation and at each follow-up, to identify subjects at risk of developing the condition or of being carriers and to ensure appropriate counseling and cardiac surveillance for female carriers.

Vital signs and physical examination are of particular importance for subjects on corticosteroids to monitor side effects.

Physiotherapy assessments should monitor muscle strength and function, posture, gait, and range of joint motion in a standardized manner to allow comparison over time. Scales and assessments commonly used in DMD are listed in Table 31.3. Essential requirements for any evaluation scale are feasibility, safety, reliability, reproducibility, and sensitivity to change. These assessments should be used to guide steroid dosing, physiotherapy interventions, and needs for aids and appliances. In a clinical setting, important factors are also ease of use, familiarity of clinical staff with the tests, restrictions related to clinic setting and timing, and the ways in which the results of the testing will influence patient care.

For clinical trials, a wider range of measures may be mandated, depending on the study design, to demonstrate reliability, validity, and responsiveness to interventions. Although the aim of these assessments is different in the two situations, some consistency across the two settings is desirable in order to ensure clinical relevance of the measures used in trials.

Manual muscle testing (Medical Research Council [MRC] scale) is used clinically to assess muscle strength in the limbs, monitor disease progression, and evaluate response to treatment. Advantages of the MRC scale include its ease of use, as it does not require any equipment and can easily be performed at any clinic. However, the assessment is highly evaluator-sensitive and should be performed only by clinicians with experience in neuromuscular diseases to ensure reliability.[23] Quantitative muscle testing (QMT) has been shown to be a valid and reliable outcome measure in DMD trials,[24,25] but quantitative measurement systems are expensive and available only in a limited number of centers.

The North Star Ambulatory Assessment (NSAA) is a widely used 17-item rating scale specifically developed by the North Star Clinical Network for Paediatric Neuromuscular Disease for children with DMD.[26,27] The scale assesses 17 motor skills (Table 31.2) considered both clinically meaningful and sensitive to change,[28] with scores summed to generate a total ambulatory function score (higher scores indicate better motor function).[29] The assessment can be completed in approximately 10 minutes and requires minimal equipment, making it both feasible and practical.[30] As long as adequate training is provided, the scale is reliable and reproducible.[31]

TABLE 31.1 Care Recommendations Based on Disease Stages

	Stage 1: Pre-symptomatic	Stage 2: Early Ambulatory	Stage 3: Late Ambulatory	Stage 4: Early Nonambulatory	Stage 5: Late Nonambulatory
Diagnostics	High suspicion in boys with motor and/or speech developmental delay (CK testing); Diagnostic testing (genetic testing); Genetic counseling		–	–	–
Neuromuscular Specialist assessment	Involve the multidisciplinary team Ensure immunizations schedule	Evaluation of disease status (muscle strength and function, cognitive and psychological issues) Consider initiation of corticosteroid treatment; Evaluate treatment response and monitor side effects	Monitoring of disease progression, response to treatment, development of complications Management of steroid-related side effects		
Treatment consideration (corticosteroids)	Start discussion about effects of corticosteroids	Initiation of treatment recommended during the plateau phase (age 2–5)	Dose and regime to be adjusted balancing growth and side effects	Continuation of treatment to preserve upper limb function, reduce progression of scoliosis, and delay decline in respiratory and cardiac function Considering dose reduction (0.5 mg/kg); Initiation of corticosteroids in this phase is dubious	Carefully evaluate benefits and side effect profile
Rehabilitation management	Encourage adequate exercise and activity	Continue adequate exercise and activity; Preventive interventions to maintain muscle and joint movements (orthoses); Encourage daily stretching exercises	Previous recommendation to be continued; Provision of adequate wheelchair to allow maximal independency and social participation; Establish adaptation at school and at home	Continue previous recommendations; Ensure correct posture in sitting to prevent development of spinal deformities; Provide adequate aids and devices to guarantee independency in daily activities	
Orthopedic Management	Orthopedic interventions rarely needed		Consideration of surgical correction of contractures (Achilles tendons) if necessary	Monitoring spinal deformities	Monitor spinal deformities and consider surgical interventions (spinal fusion) if necessary; Consider surgical correction of joint contractures (e.g., foot, hip) for wheelchair positioning and pain control

(*Continued*)

TABLE 31.1 (Continued)

	Stage 1: Pre-symptomatic	Stage 2: Early Ambulatory	Stage 3: Late Ambulatory	Stage 4: Early Nonambulatory	Stage 5: Late Nonambulatory
Cardiac management	ECG and echocardiogram at diagnosis to exclude dilated cardiomyopathy	Cardiac assessment every 2 years until age of 10, annually thereafter	Continue cardiac assessments to monitor development of early signs of cardiac dysfunction	Increased risk of cardiomyopathy with age Consider treatment when $FS < 28\%$ and/or $EF < 55\%$	Severe cardiomyopathy often asymptomatic; Establish standard heart failure interventions with deterioration of heart function
Respiratory Management	Normal respiratory function	Monitor respiratory function to establish compliance with the techniques; Ensure immunization schedule with 23-polivalent pneumococcal vaccine and influenza vaccines	Low risk of respiratory problems; Monitor progress; Ensure annual influenza vaccine	Increased risk of respiratory impairment; Monitor cough peak and early signs of nocturnal hypoventilation	Increased risk of respiratory complications; Establish respiratory physiotherapy and consider cough assist devices; Promptly trigger respiratory investigations and interventions (e.g. antibiotics)
Gastrointestinal management	Monitor for normal growth for age	Monitor weight; Consider gastro protection with corticosteroid treatment		Monitor weight; Nutritional assessment for over/underweight	Monitor for possible dysphagia and aspiration pneumonia
Psychological management	Family support; Early evaluation and interventions for speech developmental delay, learning difficulties, and behavioral problems	Assess educational needs and establish interventions at school; Monitoring and management of behavioral issues Support coping strategies		Promote independence and social development; Monitor coping strategies; Consider interventions (psychological support)	Promote independence and social participation; Transition planning to adult service; Consider pharmacological treatment if necessary

TABLE 31.2 Assessments in DMD: Standard Recommendations

Assessments	Frequency (Minimum)	Aims
Neuromuscular specialist	Every 6 months	• Evaluation of disease status and progression (muscle strength, function, and range of movements) • Anticipatory planning of future developments and prevention of complications (e.g. scoliosis) • Ensuring immunizations schedule (e.g. varicella zoster; pneumococcal vaccine; flu jab) • Planning treatment interventions (e.g. steroid therapy), efficacy, and side effect management • Evaluation of psychological issues, daily activity, and functioning; support in place
Physiotherapist	Every 4 months	• Assessment of disease progression and complications (e.g. joint contractures; spinal deformities) • Trigger interventions for management of complications (e.g. orthoses; referral to orthopedic surgeon) • Advice about stretching exercises (to be done locally 4–6 times per week by parents or local physiotherapist at home or at school) and monitoring of progress
Cardiac specialist	At diagnosis, every 2 years until age of 10, annually thereafter	• ECG and Echocardiogram to evaluate dyskinesia, LV dysfunction and dilated cardiomyopathy • Prompt initiation of treatment (ACE inhibitors +/– beta blockers) with early signs of cardiac dysfunction
Pulmonary evaluation	Every 6 months	• Evaluation of respiratory function (FVC; cough peak flow) to allow timely prevention and management of complications • Ensuring immunization schedule (pneumococcal vaccine; flu jab) • Trigger further respiratory investigations (overnight pulse oximetry; hemogas analysis) • Trigger respiratory interventions (respiratory physiotherapy; cough assist machine; NIV; tracheostomy)
Gastrointestinal assessment	Every 6 months	• Growth and weight monitoring • Nutritional assessment for over/underweight • Assessment of dysphagia
Speech and language therapist	As indicated	• Evaluation of speech developmental delay and establishment of prompt interventions • Assessment of dysphagia
Rehabilitation and orthoses	As indicated	• Evaluation of orthoses (AFOs; KAFOs; standing frame) • Evaluation of wheelchairs to ensure independency and correct posture
Orthopedic surgeon	As indicated	• Evaluation of surgical options for joint contractures (e.g. Achilles tendons; hips) • Monitoring for scoliosis and interventions (e.g. spinal fusion)
Psychologist	As indicated	• Provision of family support • Early evaluation and timely interventions for speech development; learning and behavioral issues • Evaluation of coping strategies; promoting independency and social development
Occupational therapist	As indicated	• Evaluation of upper limb assistive devices and orthoses; housing adaptations

Timed function tests (i.e. time to climb stairs, run 10 meters, rise from the floor) are also clinically relevant endpoints that have been shown to be sensitive, reliable, and strong predictors of decline in ambulation in DMD.[28,32,33] The 10 m run and rise from floor are included in the NSAA.

NSAA and timed function testing have shown good correlation with clinically meaningful outcomes for ambulatory DMD patients, and can therefore be used as predictive factors of motor decline. A total NSAA score higher than 18 infers a lower risk of losing ambulation within the following 2 years. Similar correlations have been found for times to rise from the floor of less than 7 seconds, and run 10m in less than 10 seconds.[34] Inability to rise from the floor predicts loss of ambulation within 12 months in 40% of cases.[33]

Other scales, including the Motor Function Measure,[35] Hammersmith Motor Ability Scale,[36] Vignos[37] and Brooke scales[38] have been largely replaced by the NSAA or are not in current use in clinical practice due to their lower sensitivity and length of the assessment.

Based on the assumption that gait abnormalities are a major disease manifestation with great importance to patients and families, McDonald et al.[39] developed a

TABLE 31.3 Physiotherapy Scales and Assessments

	Scale	Advantages	Disadvantages
Muscle strength testing	Manual muscle testing	Ambulatory and nonambulatory stages; easy to perform in clinic	Evaluator sensitive; value of the testing uncertain in late nonambulatory patients
	Quantitative myometry	Reliable	Requirement of specific equipment; length of the assessment
Muscle function testing	North Star Ambulatory Assessment (NSAA)	Reliable and reproducible if trained personnel; clinically meaningful; minimal equipment required	Limited to ambulatory patients
	Motor Function Measure (MFM)	It includes function related to activities of daily living	Poor sensibility; not specific for DMD
	Brooke upper extremity scale	Easy to perform	Poor sensitivity, especially for late ambulatory stage
	Egen Klassification (EK) functional assessment	It can be performed in nonambulatory patients	Poor reliability and sensitivity
	Hammersmith motor scale	It can be performed in ambulatory and nonambulatory patients	Poor sensibility
	Performance of the Upper Limb (PUL)	It can be performed across stages (from ambulatory to late nonambulatory phases)	Not fully validated yet
Timed functional testing	6 minute walking test (6MWT)	Integrated measure of ambulatory function; reliable and reproducible if trained personnel	Difficult to perform in a clinical setting due to space and equipment requirements; limited to ambulatory patients
	Time to rise from the floor	Clinically meaningful; easy to perform in clinic	Limited to ambulatory patients
	Time to run 10 meters	Clinically meaningful; easy to perform in clinic	Limited to ambulatory patients
	Time to climb 4 stairs	Clinically meaningful; easy to perform in clinic	Requirement of standardized stairs; limited to ambulatory patients
	Time to descend 4 stairs	Clinically meaningful; easy to perform in clinic	Requirement of standardized stairs; limited to ambulatory patients
Range of motion	Goniometry	Minimal equipment required	Poor sensitivity

modified 6-minute Walk Test (6MWT) for DMD boys. This integrated measure of ambulatory function in DMD is influenced by lower limb strength, joint contractures, fatigability, and cardiorespiratory dysfunction.[33,40,41] In recent years, the 6MWT has been used as primary outcome measure in several clinical trials involving ambulatory DMD patients. Natural history and clinical trial data suggest a good correlation of the 6-minute walk distance (6MWD) with age and disease progression as measured by patient-reported outcomes and timed functional tests.[33,40] A threshold 6MWD below 50–55% predicted for age, height, and weight, or approximately 325 meters, is predictive of more rapid decline in motor function with a 40% risk of loss of ambulation over the subsequent year. The 6MWT requires a 25-m length, quiet straight corridor, which limits its applicability in routine clinical practice.

All of these assessments are focused on lower limb muscle strength and are therefore applicable only during the ambulatory stages of the disease. Moreover, the complexity of some of these scales requires patient compliance, limiting their application in very young children. The importance of outcome measures able to reflect functional abilities across different stages of DMD has been recently emphasized by advocacy groups and regulatory agencies.[42–44] With this aim, a number of upper limb functional tests have been developed to evaluate muscle weakness, manual abilities, and dexterity in DMD, and to explore the impact of upper limb involvement on daily activities such as transfers, feeding, or washing.[43,44] These scales are still under development but are likely to provide a useful addition to the currently available assessments by allowing follow-up across the disease spectrum.

Meanwhile there is some experience with other measures in the nonambulant population, such as the (mainly patient-reported) Egen Klassifikation (EK) scale.

International collaborations have been developed over the last few years to find a consensus on outcome measures and assessment in DMD, with an aim to collect consistent natural history data, monitor disease progression, and assess the impact of therapies. Moreover, standardized and validated assessment of muscle strength and motor function are essential for assessment of efficacy of new investigational drugs. Until an agreement on standardized assessment is reached, training and experience of the assessor and consistency in the tests performed within an individual clinic are essential to allow monitoring of disease progression.[12]

Pharmacological Interventions

Corticosteroids are the only pharmacological treatment to have been proven effective in DMD. Corticosteroids improve muscle strength and function in DMD their prescriptions in routine clinical practice has dramatically changed the natural history of the disease.

Several additional pharmacological approaches have been developed over the last 10 years to modify the underlying pathology of DMD; none has yet been proven effective, but several remain under investigation in clinical trials.

Corticosteroids

Although the precise mechanism of action of corticosteroids in DMD remains unknown, it is now well recognized that corticosteroids slow the natural decline in muscle strength and function in DMD.[45] Several studies have shown that corticosteroids improve muscle strength and reduce the progression of weakness, preserving important functions such as independent walking and ability to stand from supine, rise from a chair, and climb stairs.[46,47] In the longer term, corticosteroids also improve upper limb function by prolonging the ability to raise hands to the mouth and respiratory function, maintaining respiratory muscle strength and possibly reducing ventilatory dependency in later stages.[48,49]

Corticosteroids also slow the development of scoliosis, leading to a significant reduction in the number of patients needing to undergo spinal surgery.[50] It remains unclear whether corticosteroids truly prevent scoliosis or simply delay its onset by prolonging ambulation and maintaining axial and trunk muscle strength,[51] but increasing experience in the adult age group suggests that the protection against the development of scoliosis is long-lasting.

Recent studies support previous evidence from cohort observations that corticosteroids impact cardiac function, reducing the development of cardiomyopathy and slowing progression of left ventricular dysfunction, further reducing mortality.[52–54] The cardioprotective mechanism of corticosteroids remains unclear. Although this may relate to reduction of inflammation and fibrosis in the heart, such changes are not seen in steroid-treated *mdx* mice.[55,56] The improvement in cardiac function following steroid therapy might simply represent a secondary effect from improved skeletal muscle and respiratory function and reduced risk of scoliosis.[53]

Although corticosteroids have been prescribed in DMD for over 20 years, there is still clear equipoise about their use in routine clinical practice, due mainly to their significant side effects. Moreover, it is anecdotally known that some DMD boys do not respond to corticosteroids, although how often patients are truly nonresponders is unknown. Treatment with corticosteroids should be routinely offered unless there are substantial contraindications such as major pre-existing risk factors for steroid side effects or poor family compliance.[12] As risk of developing side effects and effectiveness seem to be independent responses, discontinuation of steroid therapy is recommended if there is no improvement or stabilization of muscle strength and function within 6 to 12 months of initiation of treatment. No predictive factors for steroid nonresponsiveness have yet been identified, but several groups are currently investigating biomarkers and genetic modifiers that might also be useful in predicting response to steroids and minimizing side effects.

Initiation of Corticosteroid Therapy

Exactly when corticosteroids should be initiated in DMD is still debated.[57] Current guidelines recommend commencing treatment during the plateau phase of the child's motor development, when there is no longer ongoing acquisition of new motor skills, but before the decline and development of muscle weakness. In the early stages of DMD, a "honeymoon phase" is often described, during which children continue to acquire motor skills due to growth and improved motor coordination. Recent systematic studies in young children show that this is a misnomer, because even at a young age there is an increasing disparity in motor performance between patients with DMD and their peers, even though they are gaining skills.[58] The plateau phase can be identified by assessment and parental report when children are no longer gaining skills but not yet deteriorating. This typically occurs around the age of 4, and rarely continues beyond age 5–6. This should not be confused with the plateau in motor performance that can be seen later with the use of steroids.

Concerns about the long-term side effects of corticosteroids have, to date, limited their prescriptions while new skills are still being achieved, and certainly in children younger than 2; equally, starting treatment during the decline phase is not optimal as this is probably less effective.[12] Recent data suggest that boys beginning corticosteroids before the age of 5 have a trend to a slower functional decline compared to boys starting treatment later.[59] Some studies also suggest that earlier treatment—even before the age of 4—might better preserve strength in the long term, possibly prolonging ambulation.[8] Concerns regarding the increased risk of long-term side effects have so far prevented large-scale early prescription of corticosteroids, but further study would be valuable to better define long-term outcomes in larger cohorts of patients.

Current recommendations are for steroid initiation between 3 and 5 years of age, but their applicability is limited by the fact that in most countries, the average age of diagnosis is about 4 years.[4,5]

Steroid Formulations, Doses, and Regimes

A number of different steroid dosage regimes are in use for DMD worldwide. The main formulations used are prednisolone/prednisone and deflazacort. Typical starting doses are 0.75mg/kg/day for prednisolone/prednisone and 0.9mg/kg/day for deflazacort.[60] The use of higher daily doses (1.5 mg/kg/day) is not associated with greater benefits,[61] while doses of 0.3 mg/kg/day are less effective. Daily, alternate day, 10 days on/off, and high dose weekend regimes all have their advocates,[21,46,62–64] and there are distinct regional differences in how steroids are prescribed.

Prednisolone/prednisone represents the first choice in several countries due to low cost and availability in tablet and liquid formulations, which might be preferable in younger children. Deflazacort is more expensive and is not currently approved for use in DMD in all countries (e.g. USA), though its use is relatively widespread.

In ambulant boys, the dose is usually increased according to growth, taking into account changes in strength and development of side effects, which should be regularly monitored (Figure 31.1). Doses are usually increased according to weight when there is a decline in motor performance, and if side effects are acceptable to the child and family.[12] The aim is to reach the minimum effective dose; therefore, doses can be maintained at the initial levels where strength remains stable.

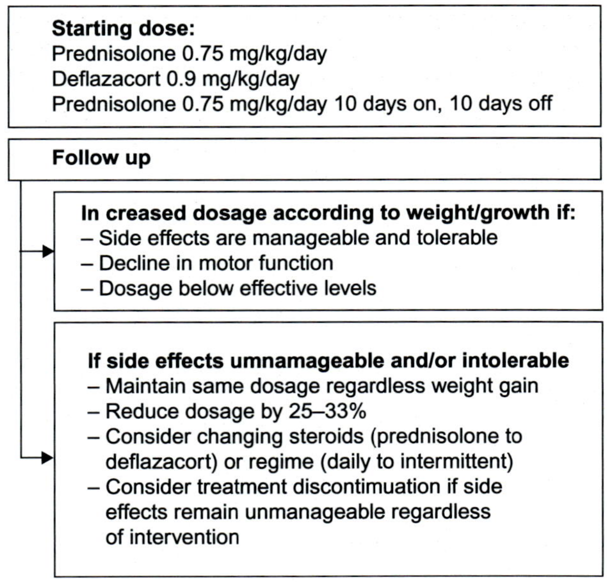

FIGURE 31.1 Corticosteroids: dosage management.

Regimes—Equipoise Since the initial publications on the use of corticosteroids in DMD in 1974, concerns about the side effects of daily regimens have led to the development of alternative regimens trying either to give a lower dosage of corticosteroids or to allow "steroid holidays." This has led to lack of consistency of dosage and regimes in different countries and centers.[21,62] Few large randomized studies on corticosteroids in DMD have been published so far; therefore, comparison between different regimes or different corticosteroids is difficult.[44] Prednisolone and deflazacort have shown similar effect on muscle strength as well as respiratory and cardiac function. Compared to prednisolone, deflazacort seems to be associated with less weight gain but more frequent development of cataracts.[65,66]

Daily administration seems to be more effective than alternate day treatment.[67] An intermittent regime of 10 days on followed by 10 days off steroids seems less effective in preserving muscle strength and function but may cause fewer side effects (weight gain, behavioral problems, hypertension, and vertebral fractures) compared to daily treatment.[59] A weekend high-dose regime was suggested to minimize side effects. A double-blind clinical trial comparing high-dose weekend prednisone (10 mg/kg/week) with standard daily regime (0.75 mg/kg/day) over 12 months showed that high-dose weekend prednisone was well tolerated, equally effective in preserving muscle function, and possibly had less negative impact on growth and weight gain[68,69]; however, the long-term outcomes of these regimens remain unclear.

Unfortunately, today, the choice of glucocorticoid and regime often depends on availability, cost, and personal experience of the clinician prescribing the treatment, rather than clinically based evidence. The lack of a standard regimen may result in patients being treated suboptimally or exposed to higher risk of side effects, and may confound efficacy evaluation in clinical trials of new therapeutic approaches. An international clinical trial funded by the National Institutes of Health (FOR-DMD: Finding the Optimum Regimen of Corticosteroids for DMD) is currently comparing different steroid regimes in a young (4–7 years) steroid-naïve DMD population with respect to efficacy, treatment satisfaction, side effects, tolerability, and quality of life (ClinicalTrials.gov #NCT01603407). The study is designed to generate the data that will allow standardization of treatment of DMD, ensuring harmonization of care, improving management, and preparing a population for current and future studies with new investigational drugs.[70]

Nonambulant Patients

The use of corticosteroids in nonambulatory patients is controversial. Further research is needed to investigate the long-term effects of late corticosteroid treatment, especially in view of the increasing adult population with DMD. In nonambulatory patients, the potential benefits of ongoing steroid therapy (preservation of respiratory and cardiac function and prevention of scoliosis) must be carefully balanced against the risk of side effects, in particular in fractures and obesity, which can negatively impact cardiorespiratory and upper limb function.

A recent study showed that older boys continuing corticosteroid treatment following loss of independent ambulation are more likely to preserve muscle function in the lower and upper limbs compared to those who discontinued treatment.[71] Maintenance of the ability to self-feed improves independence and health-related quality of life. The probable cardioprotective effects of corticosteroids are also important at this stage.[52,54]

Together, these data support continuation of treatment after the loss of ambulation in patients who have used corticosteroids while ambulatory. No controlled or long-term data are available on the effect of commencing treatment in patients who are already nonambulatory. Stabilization of respiratory function in early nonambulatory patients has been suggested in small and short-term studies; presence of cardiomyopathy does not represent a contraindication to corticosteroid therapy, but the impact of treatment in patients with severe left ventricular dysfunction is unclear. Currently, the prescription of corticosteroids in steroid-naïve nonambulatory cases is based on the experience and judgment of the clinician.

Administration

Corticosteroids are usually administered in the morning, with breakfast, to mimic the natural diurnal variation in endogenous steroid production. Although evening administration is sometimes tried in children who develop behavioral issues (emotional lability, hyperactivity, and temper) in the hours after steroid administration, evening administration may cause insomnia. Corticosteroids should never be taken on an empty stomach due to the risk of gastrointestinal complications. Sudden discontinuation of corticosteroids must be avoided as it can cause acute adrenal failure. Although the risk seems to be related to duration and cumulative dose of steroids,[72] all children who have received systemic glucocorticoid therapy at a dose ≥0.3mg/kg daily for more than three weeks should be considered at risk of adrenal insufficiency. A slow, tapered reduction must therefore be planned if discontinuation is required.

Side Effects

Corticosteroids have a well-documented range of side effects, not all of which have been reported in the DMD population. The most common adverse events reported in DMD are weight gain, height restriction, behavioral problems, Cushingoid change in facial appearance, loss of bone mass, and delayed puberty. Hypertension, acne, cataracts, and diabetes have been reported but do not appear to be so frequent in this population.

Side effects remain the main cause of steroid dosage modification and discontinuation in DMD.[45] Weight gain and behavioral problems are the most common causes for treatment discontinuation. Other side effects of corticosteroids, such as Cushingoid facies, hirsutism, delayed puberty, and gastrointestinal problems, do not usually lead to discontinuation of treatment.[45] Measures to prevent steroid-related side effects should be put in place before commencing treatment and reinforced at each clinic visit. Dietary and behavioral advice are of particular importance in this patient population, given the inherent risk of developing these complications as part of DMD regardless of concomitant treatment with steroids (see Table 31.4).

There are few absolute indications to discontinue steroids due to side effects. The decision to stop steroid therapy must be undertaken with full discussion of the family's and child's perceptions of the tolerability of the side effects, and the pros and cons of dosage modification Dosage can be maintained (i.e. not increased for weight gain) when a side effect is present but manageable. If dose reduction is recommended, this should be slow and progressive. Deflazacort is an alternative to prednisolone in patients at risk of obesity or who develop weight gain in spite of careful dietary control. Changing from a daily to an intermittent regime may be considered when a side effect persists despite dose reduction, but the risk of adrenal failure should be considered. Recommended interventions for the most common steroid side effects are summarized in Table 31.4.

Recent research has focused on development of new steroid molecules with the same efficacy but fewer side effects than those currently available.[73,74] Until new and better corticosteroids are available, proactive management of steroid therapy should be provided by specialized neuromuscular clinicians with appropriate expertise to balance adverse events and response to treatment.

TABLE 31.4 Management of Steroid Related Side Effects

Side Effect	Recommended Monitoring	Prophylaxis and Management	Dose Change Recommendations and Other Therapy
Weight gain	Monitor weight at each clinic visit	Reinforce dietary advice pre-steroids and at each visit; Refer to a dietician for at risk patients and in case of weight gain; Consider deflazacort as first treatment option in at risk children	Consider maintaining same dosage regardless growth or reduce dosage if annual BMI increases regardless dietary advice; Consider changing to deflazacort
Cushingoid features	Physical examination at each visit	Dose or regime changes might be considered if unacceptable for the child/family	Consider dosage changes rarely required and to be considered only if distressful for the child/family
Behavioral problems	Evaluate history of behavioral problems prior to starting treatment Forewarn parents	Provide behavioral management advice pre-steroids and at each visit; Consider referral for behavioral health or a child psychologist	Consider changing steroid administration to later in the day Consider maintaining same dosage; Regardless of growth or reduce dosage if behavioral problems remain disruptive for the child/family
Growth retardation	Monitor height at each clinic visit	Refer for endocrine consultation if growth plateau	Consider dosage changes or treatment discontinuation if distressful for the child/family
Delayed puberty	Monitor Tanner stage as indicated per patient's age	Refer to an Endocrinologist for possible treatment	Consider testosterone treatment
Bone demineralization and increased fracture risk	Evaluate fracture history; arrange annual DEXA scan; advise parents about calcium and vitamin D intake; maintain high suspicion for paucisymptomatic vertebral fractures	Treat vitamin D deficiency/insufficiency; Refer to an Endocrinologist for possible treatment	Consider bisphosphonate in case of vertebral fractures
GERD and peptic ulcer disease	Enquire about symptoms; advise parents to promptly report symptoms; perform an abdominal examination at each clinic visit	Avoid NSAIDs; Seek gastroenterologist opinion if symptoms persist regardless of treatment	Prescribe ranitidine or Proton-pump inhibitors if symptomatic; Dosage modification or treatment discontinuation might be required in case of confirmed peptic ulcer disease
Immune/adrenal suppression	Avoid live vaccines; advise parents of risk of serious infection Advise parents to inform all medical personnel that the child is on steroids; ensure that steroids are not stopped abruptly for more than 24 hours	Provide parents with an "I am on steroids" card; Obtain varicella zoster immunization before starting steroids	Give intravenous corticosteroids if nothing by mouth; Consider intravenous stress-dose steroid coverage for surgery or major illness
Hirsutism	Forewarn parents	–	Dose or regime changes are usually not necessary
Cataracts	Annual eye check	Refer for ophthalmologist consultation if evidence of cataract; Surgical correction might be required if symptomatic	Dose or regime changes are usually not necessary
Acne, tinea, warts	Forewarn parents	More frequent in teenager	Treat as indicated; Dose or regime changes are usually not necessary
Hypertension	BP monitoring at each clinic visit	Provide dietary advice (reduction of salt intake, weight control) if BP > 90%	Consider treatment (ACE inhibitors and/or beta blockers) if high BP persist > 99th centile; Dose or regime changes are usually not necessary
Glucose intolerance	Urine dip stick for glucose at each clinic visit	Provide dietary advice pre-steroids and at each visit; Check for fasting or postprandial glucose levels if glycosuria is confirmed; Seek endocrine consultation	Consider dosage changes or treatment discontinuation if diagnosis of diabetes is confirmed

Weight Gain Excessive weight gain was apparent in some patients with DMD even before corticosteroids were regularly prescribed for this condition. Overweight and obese states are frequently induced by corticosteroids but conversely are exacerbated by reduced mobility, which steroid use can counteract.[46] Excess weight gain impacts mobility, especially activities requiring antigravity movement, such as getting up from the floor or climbing stairs. Obesity also negatively impacts respiratory and cardiac function. Sleep apnea can represent a severe complication of excess weight, especially in the early nonambulatory phases, and even in the absence of compromised respiratory function.

Dietary advice, including the likelihood of increased appetite with steroid use with an emphasis on the reduced calorie intake required for people with reduced mobility, should be provided to all families before commencing corticosteroids, and reinforced at each clinical appointment.[75] Referral to a dietician might be indicated at the outset for some families, but should be considered for any child showing excessive weight gain. Reduction in steroid doses should be considered if excessive weight gain continues regardless of nutritional interventions.[76]

Medications, such as metformin, have been suggested in cases of severe obesity and associated glucose intolerance. Although short-term weight loss or slower weight gained has been reported,[77] the data currently available are insufficient to justify routine use of metformin in DMD.

Behavioral Problems Children with DMD may have behavioral problems with or without steroid therapy, which can cause or exacerbate these issues. Externalizing behavior (inattentive, impulsive, and overactive; defiant, oppositional, and rule-breaking) and internalizing behavior (passive/withdrawn and anxious/depressive) are the most common mental health problems in children.[77] Physical restriction, medical issues, dependency and social needs, and the life-limiting nature of the disease further contribute to the emotional impact of DMD.[78] Boys with DMD also show a higher prevalence of cognitive and learning problems and neuropsychiatric disorders, which can cause further distress to patients and their families.

Behavioral problems are a common but poorly characterized adverse effect of corticosteroids, which can limit their prescription in at-risk cases or cause treatment discontinuation. The most commonly reported steroid-related behavior issues are temper tantrums, emotional lability, hyperactivity and aggressiveness. Very rarely, boys develop psychosis, which is an absolute indication for tapering and potentially stopping steroid treatment. Behavioral side effects are possibly more common in boys on daily rather than intermittent regimes,[59] but there are limited data on the comparative impact on behavior of different steroid formulations and regimes.

Guidelines on basic behavioral support should be provided to all subjects prior to starting corticosteroids, and reinforced at each follow-up assessment. Evening administration might help in controlling behavioral problems at school but significant behavioral changes or alterations that affect daily life or distress the patient or his family, might require dosage modification.

Growth Adult height is influenced by a combination of endocrine, genetic, and nutritional factors, all of which can be affected in DMD.[76] Short stature is reported in boys with DMD even without corticosteroids, though it is further aggravated by steroid therapy.[79] Short stature may be associated with better motor outcomes in DMD,[80] but height restriction can have a significant psychological impact on boys. Growth should be measured at each clinic appointment. Standing height should be measured in ambulatory patients and indirect measures of height (e.g. arm span or ulnar length) made in nonambulatory patients.

Daily corticosteroid regimes seem to be associated with more severe growth restriction than are intermittent regimes,[59] but data on the variable impact of different steroids and regimes are limited. Treatment with growth hormone may slightly increase growth velocity in boys with glucocorticoid-induced growth failure, but its use in boys with DMD has been limited because of its potential side effects (intracranial hypertension, insulin resistance, and the possible negative impact of excessive growth on muscle strength and scoliosis).[81–83] Prospective studies are required to determine risk and benefits before its applicability in clinical practice.

The impact of alternative approaches (e.g. IGF-1) on muscle strength and growth are under investigation[82] (ClinicalTrials.gov #NCT01207908). Further information is needed about the impact on growth of treatments currently prescribed for other indications, such as testosterone (see section on delayed puberty).

Bone Health The increased fracture risk in DMD results from a combination of factors including long-term exposure to corticosteroids, immobility, and changes in body fat composition.[84] Corticosteroids have a negative impact on bone, arresting bone growth and inducing osteoporosis through demineralization of trabecular bone.[85] Reduced bone age and steroid-induced delayed puberty further impact bone density in DMD boys. As corticosteroids affect trabecular bone, treated boys have an increased risk of vertebral fractures but do not show a higher incidence of long-bone fractures compared to steroid naïve boys.[86–88] Reduced muscle strength and physical activity may also contribute to bone pathology and lower bone density in this population, even without corticosteroid therapy.[89] Thus, preservation of muscle function in treated patients might at least partially attenuate the negative effect of a corticosteroid on bone.[85]

Vertebral fractures in DMD are often asymptomatic, possibly because of corticosteroid-induced analgesia, and hence they often remain undiagnosed. In the long term, these fractures can cause chronic pain and may impact upon muscle function and ambulation.

There is no consensus regarding screening and treatment of bone health in DMD, although several workshops have called for standardized recommendations for prophylaxis and treatment of osteoporosis and fractures in DMD.[90,91] Due to the frequency of short stature in DMD, height-adjusted bone mineral density (BMD) values should be used to evaluate bone density and fracture risk in DMD boys instead of age-adjusted values.[89,92] Adequate dietary calcium intake is important and supplements should be considered in deficiency cases. Vitamin D levels should be maintained within normal ranges and deficiencies corrected with oral supplementation[89,93] according to country-specific pediatric recommendations.

Treatment with bisphosphonates is recommended in patients with symptomatic vertebral fractures to improve spinal bone density and control pain. There is currently no consensus on the use of bisphosphonates for prophylaxis of vertebral fracture in DMD boys on long-term steroid treatment.[94–98] Further longitudinal data are needed to clarify the long-term benefits and side effects of bisphosphonates in DMD. The effect of different corticosteroid regimes on bone health and fracture risk also needs further investigations.[59]

Gastrointestinal (GI) Problems GI side effects of corticosteroids are not particularly common in DMD, but peptic ulcer should be considered in the differential diagnosis of a boy with DMD on corticosteroids who develops acute abdominal pain. Abdominal examination should be part of the routine assessment, and advice on avoidance of nonsteroidal anti-inflammatory drugs (NSAIDs) should be provided to boys on ongoing steroid therapy. Symptoms or signs of gastritis or gastroesophageal reflux (GERD) should prompt initiation of treatment with proton-pump inhibitors and/or H2-antagonists.

Delayed Puberty Long-term corticosteroid therapy can result in absent, delayed, or arrested puberty through the inhibition of the hypothalamic-pituitary axis and induction of testosterone deficiency.[75] Delayed puberty has a negative impact on bone health, growth, and emotional maturity, and can cause psychological distress.

Replacement therapy with testosterone can be considered by the age of 14, or when raised as an issue by the family, and it is usually well tolerated. Testosterone induces virilization, growth, and gains in bone density, but its effect in DMD can be attenuated due to the underlying muscle condition and concomitant treatment. Inducing puberty also increases libido, and there may be challenges for patients and families in addressing issues around sexuality. Effects and expectations should therefore be carefully discussed with the patient by an endocrinologist with experience with DMD.[76]

Immune System Live vaccines should not be administered to subjects on long-term treatment with corticosteroids; therefore, immunization status should be evaluated according to local clinical practice before commencing treatment. This also applies to flu prophylaxis, for which an inactivated vaccine should be used. Complications related to increased susceptibility to infection are infrequent in patients with DMD on long-term steroids; however, vigilance should be maintained.

Emergencies Management of patients on long-term treatment with corticosteroids in case of emergencies is an area that needs to be carefully addressed. According to local practice, patients should be provided with an information card stating that they are on corticosteroids to alert medical professionals who might be involved in the patient's care in case of acute illness, fracture, serious infection, and need for surgery and general anesthesia. Due to the risk of stress-induced adrenal insufficiency, administration of extra doses of corticosteroids should be considered. The avoidance of missed doses also needs to be stressed, as well as options for parenteral treatment in the case of prolonged vomiting. In the meantime, parents need to be reminded at all clinic visits that there is no room for complacency in patients who have been taking steroids long term and that the fact of long-term steroid treatment should be brought to the attention of any medical professional coming into contact with the child.

Other Pharmaceutical Interventions

Numerous supplements (coenzyme Q10, carnitine, glutamine, arginine), anti-inflammatory and antioxidants (Protandim, green tea extract, fish oil, vitamin E) have been studied in DMD over the last few years. Few have been tested in randomized double-blind studies, and limited information is available regarding their efficacy and side-effect profile. In the absence of supportive data, these medications are not routinely prescribed.

Cardiac Management

Although the natural history of cardiac disease in dystrophinopathies is poorly characterized, all patients with dystrophinopathies are at risk of developing cardiac dysfunction.

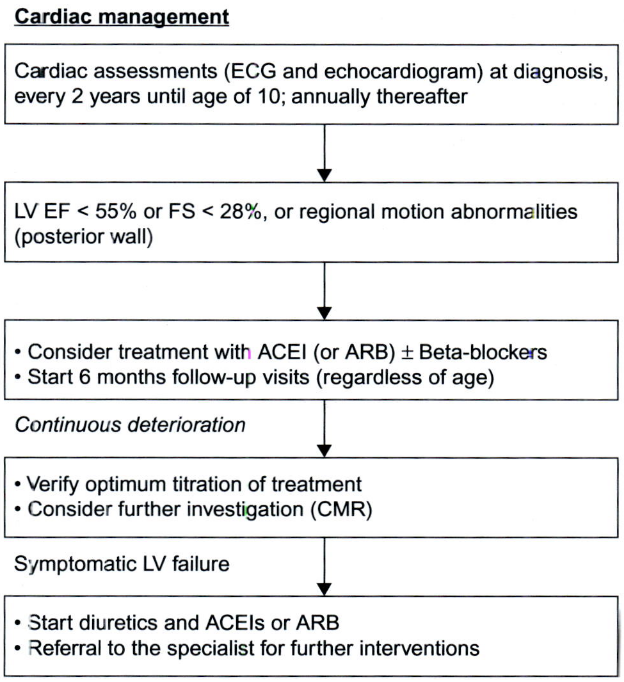

FIGURE 31.2 Cardiac management.

Virtually all boys with DMD have evidence of cardiac impairment by the age of 18. Cardiomyopathy in DMD is due to progressive replacement of myocardial fibers by connective tissue, which seems to first and primarily involve the posterobasal and lateral left ventricular regions.[99,100] This ultimately results in dilated cardiomyopathy with left ventricular impairment and/or cardiac arrhythmia.[101,102]

Cardiac symptoms are often unrecognized until very late stages in this population, due to the diminished muscle strength and reduced physical activity. Moreover, DMD patients often develop cardiac and respiratory dysfunction in parallel; therefore, non-specific symptoms such as dyspnea, orthopnea, difficulties with secretions, and poor sleep may reflect a primarily cardiac or pulmonary problem.[103] Following implementation of respiratory care and introduction of ventilatory support, dilated cardiomyopathy has been the major cause of death in DMD, significantly shortening life expectancy.[15,104,105]

Although cardiomyopathy is progressive in DMD and the incidence of LV dysfunction increases with age, the evolution of cardiomyopathy can be independent of the severity of skeletal and respiratory muscle involvement.[105] Early-onset cardiomyopathy occurs in about 25% of DMD boys.[106] Cardiac assessment (echocardiogram and ECG) should therefore be initiated at diagnosis or at latest by age 6, with follow-up assessments every two years until the age of 10 and yearly thereafter. Closer follow-up is recommended following detection of cardiomyopathy (Figure 31.2). Further studies are required to better understand Duchenne cardiomyopathy, and to optimize treatment and prophylaxis.

Abnormal conduction patterns on ECG, including deep Q waves in leads I, aVL, V5, and V6, are commonly observed in young patients with DMD and are of unknown clinical significance.[107] Sinus tachycardia is also often observed in young patients; however, attention should be paid to new onset of sinus tachycardia, which could reflect early systolic dysfunction.

Echocardiograms can show epicardial thinning, areas of dyskinesis and akinesis, and reduced left ventricular function measures (ejection fraction [EF] and fractional shortening [FS]).[100] With progression of the cardiac involvement, patients can show loss of circadian rhythm, resting tachycardia, and more serious arrhythmias (atrial fibrillation, ventricular tachycardia, and ventricular fibrillation).[108]

Cardiovascular magnetic resonance imaging (CMRI) is emerging as a new tool to identify early abnormalities of ventricular size, function, and tissue characteristics in Duchenne cardiomyopathy.[109] Gadolinium CMRI studies have identified myocardial fibrosis in patients with normal echocardiogram parameters and EF.[110] Cost and poor tolerability of the CMRI studies limit their routine application in DMD. Further studies are required to understand the clinical significance of these early abnormalities and long-term outcome of early interventions based on these findings.

Several studies have shown that ACE inhibitors (ACEIs) slow deterioration of cardiomyopathy in DMD, supporting their early prescription to prevent cardiac dysfunction.[18,111] Although type 1 angiotensin II receptor blockers (ARBs) have not been largely used in DMD, a recent double-blind study showed similar efficacy of lisinopril and losartan in treatment of DMD-associated dilated cardiomyopathy over a year of treatment,[112] suggesting that ARBs represent a valid alternative in case of poor tolerability of ACEIs.[103]

The combination of beta blockers and ACEIs seems to lead to initial improvement of LV function, while ACEIs alone causes only a stabilization of the fractional shortening. Although there are no evidence-based guidelines about the optimal time to start treatment in DMD, data from other diseases support initiation of treatment when there is any early evidence of cardiac involvement, and certainly when the EF drops below 55% and FS is less than 28%.[113,114] A double-blind placebo controlled study is currently investigating the impact of combined prophylactic treatment with ACEIs and beta blockers in DMD (controlled-trials.com #ISRCTN50395346).[18]

Heart failure in DMD patients should be treated as per published guidelines.[12,17,115–118] The benefit of implantable defibrillators and/or biventricular pacing in DMD has not been established, and clinical experience with these devices in DMD is very limited. Successful implantation of left ventricular assistance devices in two young patients, one male and one female, with DMD with end-stage dilated cardiomyopathy was recently reported,[119] but use of these sorts of interventions in DMD should be considered only by cardiologists with extensive experience in the field.

Cardiac transplantation is an established intervention in patients with Becker muscular dystrophy in whom medical treatment has failed,[120] but its place in DMD is not clear.

Despite the clear guidance provided in existing care guidelines, recent cross-sectional analyses of multi-institutional, international DMD clinical data[21,121] have reported that cardiac diagnostic tests are still underutilized, suggesting that DMD cardiomyopathy is frequently undertreated. In view of the impact of cardiac dysfunction on survival, and the availability of effective treatment, early and close cardiac monitoring is an important medical obligation.

Respiratory Management

Patients with DMD are at risk of respiratory complications due to progressive loss of respiratory muscle strength. The progressive weakness of respiratory muscles exposes patients to recurrent chest infections, sleep-disordered breathing, and ultimately respiratory failure. Pulmonary function in DMD follows the progression of the condition, with an initial improvement phase associated with age-related growth (which is enhanced by steroid treatment), followed by a plateau and finally a decline phase, which usually develops in the nonambulatory stages.[122,123] During the ambulatory stages, respiratory complications are infrequent. With the deterioration of respiratory muscle function, the ability to clear the airway from secretions becomes critical. Respiratory muscle weakness and impaired cough increase the risk of chest infections, potentially leading to episodes of atelectasis and respiratory insufficiency.[124]

Regular screening of respiratory function should be initiated from the early stages of the disease to establish a baseline and allow familiarity with the equipment; more intensive pulmonary care becomes essential after the loss of independent ambulation, when patients develop progressive respiratory muscle weakness with ineffective cough and hypoventilation.

Prior to the introduction of ventilatory support, pulmonary complications and acute respiratory failure were the cause of death in 90% of patients with DMD.[123,125] Improved respiratory care and initiation of ventilatory assistance when necessary, significantly improve survival in DMD, potentially doubling life expectancy.[16,126–128] A proactive approach by an experienced team to allow early identification of symptoms and signs of respiratory dysfunction should be an integral part of the care for DMD (Figure 31.3). Although guidelines for respiratory management in DMD are well established,[13,129,130] there are worrying studies showing that respiratory management, timing, and frequency of assessments and interventions vary greatly in different centers and countries.[21,131]

Lung function tests should be performed at least annually during the ambulatory phase and twice a year in nonambulatory patients or when signs of respiratory involvement are detected (forced vital capacity $<80\%$). Measurements should include forced vital capacity (FVC) in sitting and supine, and peak expiratory flow during cough (PCF).[13] Maximal inspiratory and expiratory pressure (MIP and MEP) are more accurate measures of cough strength in DMD patients and may help identify early signs of compromised respiratory function,[132,133] but FVC is well understood and correlates with progression to respiratory failure.[134]

Prophylactic measures include immunization with 23-valent pneumococcal polysaccharide vaccination for patients older than 2 years, and annual immunization with trivalent inactivated influenza vaccine, especially in nonambulatory patients and those with compromised respiratory function.[13] Formulations using inactive virus should be used in patients on corticosteroid treatment due to their potential immunosuppression.

Acute chest infections should be promptly treated with antibiotics. Manual and mechanically assisted cough should be considered in presence of abundant respiratory secretions and in any patients with compromised PCF. Regular lung volume recruitment (LVR) maneuvers (breath-stacking) slow pulmonary function decline in DMD patients[135] and should be offered as part of respiratory physiotherapy to patients with reduced PCF.

Mechanical insufflation-exsufflation (MI-E) using a mechanical cough assistance device (Cough Assist) is well tolerated and induces improvements in PCF, CO_2 clearance, and respiratory muscle performance.[136–138] Cough assist devices should therefore be considered in patients with compromised respiratory function, and their regular use even outside acute episodes of infection should be considered as part of respiratory prophylaxis.

Hypercapnic respiratory insufficiency is an inevitable complication in the disease progression of patients with DMD, and usually evolves from sleeping disorders without hypercapnia to sleep hypercapnia to diurnal hypercapnia.[139]

Respiratory insufficiency often develops insidiously in DMD and symptoms can remain mild and unspecific, especially in patients with reduced mobility.[139] Particular attention should be paid to early symptoms, including increased frequency of infections, morning headaches, decreased appetite, weight loss, sleep disturbances, reduced energy levels and daytime sleepiness. Oxygen therapy should be used with caution in patients with neuromuscular diseases

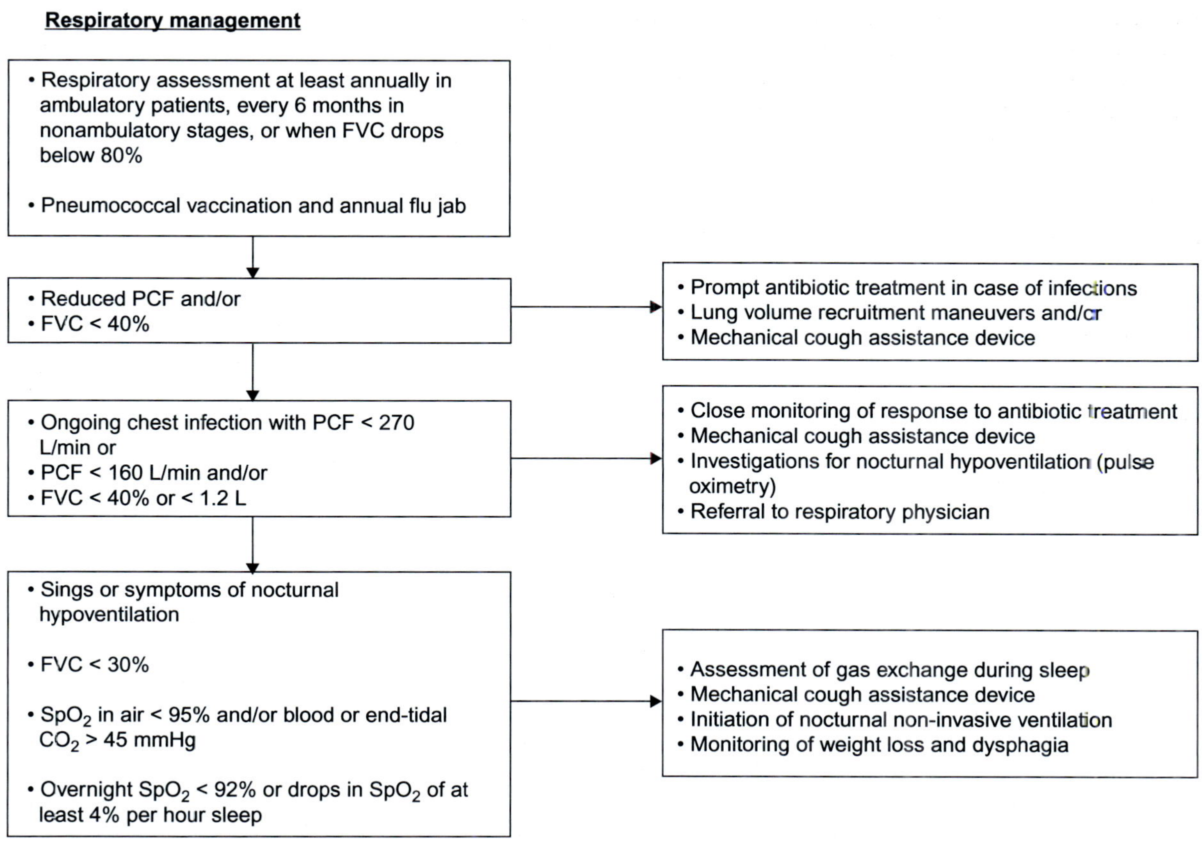

FIGURE 31.3 Respiratory management.

at risk of respiratory compromise, as it might impair central respiratory drive and exacerbate hypercapnia.[140–142]

Although the absolute values for intervention are not universally agreed upon, an FVC < 1 liter or <40% of the predicted value per patient's age and height is predictive of nocturnal hypercapnia, and should therefore prompt overnight pulse oximetry or sleep studies to exclude episodes of nocturnal desaturation and hypercapnia.[134,139] If not already involved, at this stage a physician and therapist with skill in the initiation and management of noninvasive ventilation and associated interfaces should be brought into the care team of the patient. Interventions to correct overnight hypercapnia delay the development of daytime hypercapnia, improving quality of life and survival.[143] There is evidence showing that after development of daytime hypercapnia, mean survival is less than a year if ventilatory support is not offered.[144] Noninvasive intermittent positive pressure ventilation (NIV) initiated at night represents the first choice in case of chronic respiratory insufficiency and has been shown to be effective. Once support of coughing and/or breathing is instituted, respiratory assessments are recommended every 3 to 6 months.

NIV is usually well tolerated and remains the first choice to assist ventilation when support is required overnight only. With further progression of respiratory muscle weakness, additional interventions may be indicated. Prolonged NIV is often associated with inefficient cough, due to the involvement of expiratory muscles, and therefore manual and instrumental cough assistance is often indispensable. In some cases, masks can become uncomfortable, causing sores, while large leaks may lead to conjunctivitis. Daytime ventilatory support via a mouthpiece complicates efforts to eat.[145] Progression to a tracheostomy is advocated in many countries where it is reported to be very well tolerated and associated with an excellent quality of life, but in other countries it is rarely or never used.[146–149] Disadvantages of tracheostomy include frequent respiratory infections, increased bronchial secretions and tracheal lesions,[150,151] stenosis and rare occurrence of fistulas. Correct timing of tracheostomy placement is crucial to avoid complications during the surgical procedure.

Both permanent NIV and tracheostomy increase patient dependence, possibly necessitating 24-hour care. The final choice between invasive versus noninvasive ventilatory support should be made by the patient, ideally electively, but in practice is often dependent on local expertise and treatment bias.

Physiotherapy Input and Interventions

The muscle weakness in DMD leads to progressive loss of strength and function. Physical therapy therefore remains an essential part of the patient management.

Reduced range of movements and joint contractures in DMD result not only from limited mobility with inability to actively move a joint and static positioning, but also from primary fibrotic changes in muscles and muscle imbalance around the joints.[152–155] Physical therapy should be in place throughout all stages of DMD and should be tailored to individual needs, stages of the disease, response to therapy, and tolerance. Maintenance of good range of joint mobility will prolong ambulation; in later stages, it will allow functional positioning, avoid development of fixed joint deformities, prevent pain, and preserve skin integrity.[156–158]

Effective stretching is achieved with multiple interventions including active, active-assisted and passive stretching, orthoses and, where indicated, standing devices.[159–161] A proactive and preventative approach is essential, as joint contractures are difficult to treat when already present. Physiotherapy assessments should be provided every 3 to 4 months to ensure early identification of issues and establish prompt interventions. Active and/or passive stretching should be part of the patient's daily routine and should be performed ideally daily and at least 4 to 6 times per week at home or at school. Stretching exercises should focus on the ankles, knees, and hips in the early stages of the disease, and should be directed to the upper extremities during the nonambulatory phases. Ankle-foot orthoses (AFOs) help to prevent and control progressive equinus contractures.[159,162,163] They should be custom-made for optimum alignment and comfort. Ambulatory patients should wear AFOs at night, from an early age, to improve compliance and prevent development of fixed contractures. Serial casting might be required to reach correct joint positioning before prescribing AFOs.[164] AFOs should be worn at daytime by full-time wheelchair users, but their daytime use is not recommended in ambulatory boys, who usually do not have the strength to walk safely in them.

Dynamic stretching devices (contracture control devices, CCDs) can be prescribed in cases of severe Achilles tendon contractures and/ or poor tolerability to AFOs. CCDs provide effective stretching,[165] but comparative data are required to demonstrate their relative benefit (as compared to standard AFOs) to justify their higher cost.

Knee-ankle foot orthoses (KAFOs) are used in some centers to allow standing and limited ambulation in the late ambulatory and early nonambulatory stages. These require careful custom-molding and patient compliance, but when tolerated, they prolong useful standing and ambulation, and are effective in delaying or preventing the onset of scoliosis and other secondary complications of prolonged wheelchair use. Standing devices can be useful in early nonambulatory patients to manage contractures and delay the development of scoliosis.[166]

The indication, timing and outcomes of surgical correction of joint contractures are still debated. Prophylactic surgical interventions to release Achilles and hamstring tendons have been suggested, but there is little data suggesting long-term benefit. In the ambulatory phase, surgery can be suggested when walking is affected by joint contractures with otherwise preserved limb muscle strength. If incorrectly timed, however, these interventions can lead to postsurgery immobilization and consequent earlier loss of ambulation. In the late nonambulatory patients, joint pain is a major indication for surgical correction of contractures and bone deformities.[167]

A wheelchair should be offered when reduced mobility affects the patient's independence and participation in social life. Young children can be provided with a lightweight manual wheelchair for long-distance mobility due to their reduced endurance, and an electric wheelchair should be considered as soon as functional ambulation declines, to ensure independence and quality of life. Correct posture is essential to avoid the development of contractures and bone deformities, including scoliosis. Custom seating and provision of an appropriate headrest, solid seat and back, lateral trunk support, and power tilt and recline should be included as mandatory adaptations for full-time wheelchair users. As muscle weakness progress over time and patients often lose the ability to walk prepuberty, wheelchairs should be regularly reviewed and adapted to patient growth, mobility, and concomitant complications.

Few orthoses are currently available to assist upper limb function and maintain independence with the progression of muscle involvement.[168] Hand splints can be prescribed in late nonambulatory stages to maintain functional hand and finger range of movements. Powered mobile arm supports seem to increase independence, improve confidence and improve ability to participate in social life,[169] but experience with these devices is very limited.

The effect of exercise in neuromuscular diseases remains controversial, and no guidelines on optimum exercises and physical activity for patients with DMD are currently available. Most data are based on studies in animal models that show contraction-induced muscle damage, suggesting a negative impact of such exercise in patients with muscular dystrophy. However, recent studies have shown a possible effect of exercise on utrophin expression, potentially impacting the pathophysiology of dystrophinopathies.[170] Randomized studies in patients are limited due to difficulties in controlling and quantifying the exercise and ensuring standardized compliance. Several factors, including joint contractures and respiratory and cardiac function, interfere with the ability to perform exercises, making standardized protocols difficult to design. Whole-body vibration training and cycling training have recently been suggested for DMD. Small studies conducted with both approaches have shown that training was well tolerated, but their long-term impact on muscle function requires further research.[171,172]

Given the lack of evidence-based guidelines, current recommendations suggest regular physiotherapy and gentle, submaximal exercise such as swimming, to avoid disused muscle atrophy and secondary complications of inactivity, including joint contractures and weight gain. Resistance exercise is not recommended, and strenuous activity such as trampolining should be avoided due to the risk of myoglobinuria, a complication that seems to be more common in DMD since the widespread use of steroids.[173]

Spinal Management

According to natural history data, DMD boys who are not treated with corticosteroids have a 75–90% chance of developing progressive scoliosis.[174] The risk of scoliosis increases after loss of independent ambulation, but some patients develop spinal curves even when still walking, especially in the presence of asymmetric contractures of the Achilles tendons.[175] Corticosteroids reduce the risk or delay the development of spinal deformities[44]; however, they increase frequency of fractures, and more evidence is required to better understand the long-term effects of vertebral fractures on scoliosis.[176]

Spinal deformities can cause pain and impair respiratory and cardiac function, upper limb movement, and seated posture. The spine should be regularly assessed in patients with DMD. Prophylactic interventions include monitoring of any spinal curve, adequate seating and spinal support in the wheelchair to ensure spinal and pelvic symmetry. In ambulatory patients, monitoring is usually provided by clinical observation; however, back pain, especially following a fall or in patients on corticosteroid therapy, should warrant radiography to exclude vertebral fractures.

A spinal assessment, including anteroposterior and lateral full-spine radiograph, should be arranged in the early nonambulatory phase and repeated on detection of new spinal curves on examination.

A spinal surgeon with experience in neuromuscular diseases should be involved in the management of DMD patients from their early nonambulatory phase, to establish adequate follow up and timed interventions. Spinal braces are often poorly tolerated and cannot prevent the development or the progression of scoliosis in DMD[176,177]; however, they may be considered for patients who cannot undergo spinal surgery and are indicated in some patients to maintain comfort and posture.[13] Spinal fusion should be considered in nonambulatory patients with severe or progressive spinal deformity, to control pain and preserve respiratory and cardiac function. The optimal timing of surgery is still debated. Correction of spinal deformities should be performed before onset of major cardiac and respiratory dysfunction, and early surgery appears to result in better long-term outcomes.[178–181] A multidisciplinary team should be involved in the planning of spinal surgery to ensure a comprehensive preoperative evaluation including consideration of anesthetic management, cardiopulmonary care, changes in steroid therapy, and options for postoperative rehabilitation.[182,183]

Fracture Management

Long bone fractures are common in DMD due to the reduced mobility, concomitant osteopenia and tendency to falls.[87] Conservative versus surgical treatment of long bone fractures should be evaluated case by case.[184] In ambulatory patients, early mobilization and rehabilitation are essential to ensure that independent ambulation is not lost. Internal fixation should therefore be considered if surgical intervention can facilitate prompt rehabilitation, but the type of fracture and anesthetic risk usually dictate the more suitable approach. In all cases, physiotherapy input should be provided promptly after the fracture to maintain general mobility.[13]

Nutritional, Swallowing, and Gastrointestinal Management

Patients with DMD are at risk of both undernutrition and being overweight/obese.[185,186] This risk differs from patient to patient and at different ages and stages of the condition. Obesity is often induced by corticosteroid treatment and exacerbated by reduced mobility. Undernutrition is more common in the later stages of the condition and often has multifactorial causes, including physical dependency, poor appetite due to respiratory and cardiac impairment, fatigable chewing, impaired swallowing (Figure 31.4), and psychological issues. Similarly, undernutrition causes fatigability, affects mobility, and predisposes to complications such as infections, and respiratory and cardiac dysfunction.[187]

Weight gain or loss should be monitored in patients at all stages of the disease. Unexplained weight loss or gain should warrant referral to a dietician and nutritionist. The diet should be evaluated to ensure correct intake of calories, proteins, fluids, calcium, and vitamins. Multivitamins or other supplements and high calorie drinks are not part of the routine care but can be prescribed if indicated.

With progression of the condition, patients can develop chewing difficulties and dysphagia primarily due to compromised oral phase, with insufficient or effortful chewing. In later stages, pharyngeal weakness can lead to incomplete clearance, leaving pharyngeal residue and increasing risk of aspiration.[188,189] The onset of swallowing difficulties can be gradual, and they often remain under-recognized by

Alerting symptoms
- Recurrent cough when eating
- Episodes of choking
- Tiredness in chewing
- Increased time to complete a meal
- Loss of weight
- Recurrent or severe, treatment-resistant chest infections

Interventions

Step 1

- Consider possible causes (respiratory or cardiac dysfunction, swallowing problems)
- Monitor weight
- Refer to a dietician
- Prescribe calorie supplements

Step 2

- Refer to a speech and language therapist
- Evaluate PEG insertion

FIGURE 31.4 Swallowing problems.

patients.[190] Dysphagia can lead to aspiration pneumonia, dehydration, and emaciation.[188,191,192] Specific questions about nutrition and swallowing should therefore be raised at each clinical evaluation of nonambulatory patients.[193] Referral to a speech and language therapist and a videofluoroscopic study of swallowing need to be considered in any patient presenting with weight loss, recurrent cough when eating, prolonged time to complete a meal, episodes of choking, or treatment-resistant chest infections.

Gastrostomy feeds should be offered when weight loss continues despite compensatory strategies such as high calorie oral supplementation with thickened textures, after aspiration pneumonia, and in patients with progressive deterioration of respiratory function. Gastrostomy feeds are preferred to nasogastric tubes for long-term enteral feeding, are usually well tolerated, and can be used easily at home.[194–196] Nausea and vomiting are the most frequent side effects of gastric tube placement in this patient population, often necessitating adaptation of feed volumes and timing.[197] The risks and benefits of parenteral feeding should be discussed carefully with the patients and the family, and gastrostomy placement needs to be timed carefully to balance the patient's needs and anesthetic risk. Placement of a gastric tube during acute episodes and emergency situations should be avoided.[198]

Surgical and Emergency Care Considerations

Due to its multisystem involvement, several factors need to be taken into consideration approaching a DMD patient in an emergency room. The advances in the standards of care, prolonged life expectancy and improved quality of life should be taken into account and a proactive approach should always be considered. The underlying diagnosis and concomitant medications, including corticosteroid treatment, should be evaluated in addition to cardiac, respiratory, and nutritional status.

Long-term corticosteroid use can mask infections and reduce response to stress. Steroid-induced gastric ulceration can have high mortality risk if not promptly suspected.

Respiratory, cardiac, and nutritional issues can rapidly progress in compromised patients and can both hide and aggravate the underlying cause of the acute primary medical event. The use of oxygen without respiratory support should be limited due to the risk of hypercapnia. The respiratory team should be involved promptly for all patients who are already using ventilation.

Undernutrition can further affect the response to stress and infections; alternative feeding might therefore need to be considered during acute medical events, especially in patients with previous history of compromised swallowing and inadequate caloric intake.

Anesthetic issues need to be considered at all times if surgery or sedation is needed: surgical procedures in this population require careful planning and a multispecialty approach.[199] Whenever possible, surgery should be planned in a full-service hospital involving a team with experience with neuromuscular patients. Anesthetic risks in DMD patients include potentially fatal reactions to anesthesia, hypoventilation, atelectasis, heart failure and cardiac dysrhythmias, and difficulty weaning from mechanical ventilation. Close evaluation of cardiac and respiratory status and concomitant medication should be part of the pre-operative assessment and intraoperative management, taking into consideration the impact of hypotension or major fluid shifts.[200] Anesthetics should be carefully chosen; exposure to inhalational anesthetic agents (e.g. halothane and isoflurane) and depolarizing muscle relaxants such as suxamethonium chloride can cause rhabdomyolysis, malignant hyperthermia-like reactions, and other life-threatening reactions.[201–203] Cases of fat embolism syndrome (FES) following low-energy traumas have been reported in patients with DMD.[204,205] Whether corticosteroids increase the risk of FES remains unclear, although a possible role of adrenal insufficiency has been hypothesized.[204] Patients should be reminded to use a seatbelt at all times while in a wheelchair, and about precautions during transfers to limit risk of falls and fractures. Although FES remains a clinical diagnosis of exclusion as investigations, including chest CT scans, are often negative, early suspicion should be kept in any patient with unexpected acute neurological or respiratory symptoms following a fracture or minor injuries to improve the outcomes. Treatment is primarily supportive, with maintenance of oxygenation and ventilation.[206,207]

Cognitive and Speech Management

Nonprogressive cognitive impairment is common in boys and young men with DMD, while global developmental delay or speech disorders can be presenting features of this condition.[5,208–210] There is also a well-documented association between DMD and social interaction deficits, including attention-deficiency hyperactivity, obsessive-compulsive disorders, and autistic spectrum disorders.[211]

Cognitive impairment in DMD is often unrecognized, leading to inadequate management and poor or delayed interventions.[58] Because it is not progressive, early treatment may significantly improve neurocognitive development. An assessment of educational needs should be offered to all affected children, to promptly identify learning difficulties and areas of need.

Psychological and Social Issues

Psychosocial dysfunction in DMD derives from primary biological deficits in specific behavioral and cognitive skills including learning difficulties, speech and language delay, short-term memory deficiency, and autistic spectrum disorders, but also as a consequence of physical disability leading to social isolation, social withdrawal, and reduced access to social activities. For some families, the stress caused by psychosocial problems might exceed the stress associated with the physical disability.[212]

Of crucial importance from the earliest stage following diagnosis is provision for families of a point of contact, an individual acting as the care coordinator to meet the family's information needs, facilitate communication with clinicians, coordinate appointments, and proactively identify and intervene on social problems. Psychological and social support should be considered for families at the time around diagnosis and should be part of the medical care of all patients with DMD throughout their life, also considering psychological needs of the parents and siblings.[12]

Needs and interventions change with the disease progression and therefore should be customized to the age of the subject, the stage of the condition, and individual requirements. Crucial times include the time of diagnosis, starting school, and times of functional transition (e.g. loss of independent ambulation, need for ventilator support). The transition from childhood to adulthood represents a critical stage for these patients, and additional support might be required around this time.

Increasing awareness and knowledge of DMD at school and educating peers about the condition are essential steps to improve quality of life of these children. When available, social activities such as participation in adapted sports groups should be encouraged. A social service professional can help access financial and other social supports, while occupational therapists can identify adaptations and equipment needed to facilitate family care, maintain independence, reduce discomfort, and improve quality of life.

Pharmacological interventions should be considered in patients with moderate or severe psychiatric symptoms. Their prescription should follow standard recommendations and guidelines, taking into account disease-specific considerations, in particular cardiac and respiratory dysfunction, concomitant medications, weight gain secondary to steroid treatment, and behavioral problems.

Adults with DMD

Although no curative treatment is yet available, prolonged survival in DMD has led to a new adult population with DMD with specific multidisciplinary needs.[99] However, DMD is still often considered a pediatric condition, and the absence of guidelines for management of adults with DMD is becoming a major issue in several countries.[20,213,214] Pediatric care is usually family-oriented, and the transition to adult services can be disorientating, demotivating, and potentially distressing, especially as this transition often happens at a time of medical deterioration. There is a concern that poor transition services may result in lower quality of life, worse prognosis, and higher mortality.[215]

Adolescence is often a personally challenging process; this becomes even more complex for patients with a chronic disease such as DMD.[213] Increased physical dependency, separation from parental care, social isolation, sexuality, continuous loss of expectation, end of life care, and the prospect of premature death are problematic issues for this patient population. Transition should aim to encourage independence and self-confidence through social and physical support; ensure participation in social and community activities to avoid isolation and promote satisfaction; guarantee psychological support for acceptance and coping; provide medical and psychosocial information to allow discussion and informed decision about care planning and end of life. Ensuring the patient's involvement in the management of their care, their relationships with clinicians and other professionals, and in decisions about their health and treatment is important but may be difficult, as adults with DMD often remain in the care of their parents due to their physical disability and dependency.

The life-limiting nature of DMD requires discussions about end-of-life interventions and treatment with the patient, the family, and the clinicians involved in the care. In the later stages of the condition, a palliative care team should be involved in the design of a care plan, in order to clarify with the patient and family their treatment goals and expectations. A management plan should be in place in case of critical events when difficult medical decisions might need to be made rapidly, in order to reduce distress and ensure treatment and interventions as desired by the patient.[99]

OTHER DYSTROPHINOPATHIES: BECKER MUSCULAR DYSTROPHY AND MANIFESTING CARRIERS OF DMD/BMD

There are no consensus guidelines for management of patients with the milder dystrophinopathies (Becker muscular dystrophy [BMD] and manifesting carriers of DMD/BMD). Little has been published so far on the natural history of these conditions and their response to treatment. The current recommendations are based on the assumption that these conditions are milder phenocopies of DMD, such that recommendations for assessment and care are extrapolated from those for DMD. However, it remains unclear whether these patients require more specific management considerations; this has led to inconsistencies in the care provided to these patients.

The diagnosis of BMD may be made at any age, though the mean age of diagnosis is around 11 years. Patients with BMD may present in early childhood with developmental delay; in such cases, distinction from DMD can be initially challenging. The finding of an in-frame deletion or other nonframeshifting mutation of the dystrophin gene is supportive of a diagnosis of BMD; however, in these cases, it may be useful to supplement the genetic diagnosis with dystrophin estimation on muscle biopsy. Early presentation may not necessarily predict a poor prognosis, but may reflect the concomitant learning difficulties often seen in BMD. Another frequent age at presentation in BMD is in the early teenage years when muscle aches and pains are common, particularly on exercise, and associated in some patients with myoglobinuria. Other patients are not diagnosed until adult life. Muscle weakness is progressive but often only slowly so, and the mean age that patients with BMD need to permanently use a wheelchair is around 40 years. Life expectancy is typically not reduced.

The diagnosis of manifesting carriers of DMD/BMD may be made at any age. For every 100 boys with DMD or BMD, there are probably fewer than 5 affected girls. Familial cases are easily suspected and diagnosed through CK and genetic testing, while the diagnosis of new cases is often reached only after muscle biopsy; dystrophin gene screening by MLPA should be considered in all girls with developmental delay and a high CK. In patients presenting later, the diagnosis may be suggested by motor problems, cardiomyopathy, learning problems, or a combination of all three.

A behavioral/learning difficulty phenotype similar to that seen in DMD can be a major problem for patients with BMD and manifesting carriers. This may even be the presenting problem, and this is likely to be increasingly recognized with the more widespread use of CGH array in place of cytogenetic analysis for patients with learning difficulties.[216]

Clinical assessments are usually recommended annually in BMD patients and DMD/BMD manifesting carriers, although 6-month follow-ups might be required in more severely affected patients. Physiotherapy assessments should be part of the routine clinical evaluation of these patients, to allow early recognition of muscle symptoms and avoid delay in the treatment of muscle complications such as joint contractures.

The use of corticosteroids in BMD and manifesting DMD/BMD carriers is controversial. Observational studies and single case reports suggest possible benefits in muscle strength and function in BMD,[217–219] while indications for steroid therapy in manifesting carriers of DMD/BMD are unclear. Benefits versus side effects need to be balanced, taking into consideration the patient's age, severity of muscle involvement and associated complications such as cardiac impairment, and concomitant medical conditions (e.g. hypertension, diabetes). The risk of complications in adults on long-term treatment with corticosteroids should be carefully pondered.

Cardiac management and interventions remain very similar to those for DMD. Both BMD patients and manifesting carriers are at risk of cardiomyopathy at all stages of their condition, regardless of the severity of muscle involvement. In these patients, dilated cardiomyopathy might represent the presenting symptom of the condition and in some cases may remain the only clinical manifestation of the dystrophinopathy. Cardiac follow-up is therefore mandatory for all patients. BMD patients are usually monitored once every 1 to 2 years, while manifesting carriers are offered cardiac reviews at diagnosis and every 3 to 5 years thereafter. Whether more frequent assessments would lead to earlier identification of cardiac impairment and be associated with a better outcome needs to be clarified. Treatment should be commenced promptly following detection of cardiac impairment or deterioration of left ventricular function on echocardiogram, with the same guidelines as for DMD-associated cardiomyopathy. Heart transplant is a realistic and successful option for these patients, especially when severe cardiac dysfunction is the prominent clinical feature, with relatively preserved skeletal muscle function.

Respiratory and spine assessment should be part of the clinical evaluation. Respiratory dysfunction and spinal deformities are less common and tend to correlate with the severity of muscle weakness. Management and interventions do not differ from those for DMD. Like patients with DMD, patients with BMD and DMD/BMD manifesting carriers have an increased anesthetic risk that may be further aggravated by concomitant cardiac or respiratory complications.[203]

CONCLUSIONS

The "incurable" nature of the dystrophinopathies has for many years led to a certain nihilism in the approach to their management, but in recent times improved clinical care has markedly increased life expectancy and quality of life in patients affected by these conditions. This has concentrated efforts on a more comprehensive and proactive approach to care, as a baseline for the novel therapies for these conditions that are currently approaching or in clinical trials. It is to be hoped that the next challenge in management of these conditions will be the systematic integration of new disease-modifying therapies into the management paradigm, with even better clinical outcomes.

REFERENCES

1. Mendell JR, Shilling C, Leslie ND, Flanigan KM, al-Dahhak R, Gastier-Foster J, et al. Evidence-based path to newborn screening for Duchenne muscular dystrophy. *Ann Neurol* 2012;**71**:304–13.
2. Norwood FL, Harling C, Chinnery PF, Eagle M, Bushby K, Straub V. Prevalence of genetic muscle disease in Northern England: in-depth analysis of a muscle clinic population. *Brain* 2009;**132**: 3175–86.
3. Emery AE. Population frequencies of inherited neuromuscular diseases—a world survey. *Neuromuscul Disord* 1991;**1**:19–29.
4. Van Ruiten HJA, Straub V, Bushby K, Guglieri M. Improving recognition of Duchenne muscular dystrophy—a retrospective case note review. *Arch Dis Child* 2014. [submitted].
5. Ciafaloni E, Fox DJ, Pandya S, Westfield CP, Puzhankara S, Romitti PA, et al. Delayed diagnosis in Duchenne muscular dystrophy: data from the Muscular Dystrophy Surveillance, Tracking, and Research Network (MD STARnet). *J Pediatr* 2009;**155**:380–5.
6. Parsons EP, Clarke AJ, Bradley DM. Developmental progress in Duchenne muscular dystrophy: lessons for earlier detection. *Eur J Paediatr Neurol* 2004;**8**:145–53.
7. Cyrus A, Street N, Quary S, Kable J, Kenneson A, Fernhoff P. Clinic-based infant screening for Duchenne muscular dystrophy a feasibility study. *PLoS Curr* 2012. e4f99c5654147a.
8. Merlini L, Gennari M, Malaspina E, Cecconi I, Armaroli A, Gnudi S, et al. Early corticosteroid treatment in 4 Duchenne muscular dystrophy patients: 14-year follow-up. *Muscle Nerve* 2012;**45**:796–802.
9. Moat SJ, Rees D, King L, Ifederu A, Harvey K, Hall K, et al. Newborn blood spot screening for sickle cell disease by using tandem mass spectrometry: implementation of a protocol to identify only the disease states of sickle cell disease. *Clin Chem* 2014;**60**:373–80.
10. Ellis JA, Vroom E, Muntoni F. 195th ENMC international workshop: newborn screening for duchenne muscular dystrophy 14–16th December, 2012, Naarden, The Netherlands. *Neuromuscul Disord* 2013;**23**:682–9.
11. Parsons EP, Clarke AJ, Hood K, Lycett E, Bradley DM. Newborn screening for Duchenne muscular dystrophy: a psychosocial study. *Arch Dis Child Fetal Neonatal Ed* 2002;**86**:F91–5.
12. Bushby K, Finkel R, Birnkrant DJ, Case LE, Clemens PR, Cripe L, et al. Diagnosis and management of Duchenne muscular dystrophy, part 1: diagnosis, and pharmacological and psychosocial management. *Lancet Neurol* 2010;**9**:77–93.
13. Bushby K, Finkel R, Birnkrant DJ, Case LE, Clemens PR, Cripe L, et al. Diagnosis and management of Duchenne muscular dystrophy, part 2: implementation of multidisciplinary care. *Lancet Neurol* 2010;**9**:177–89.
14. Treat-NMD Neuromuscular Network. Available from: <http://www.treat-nmd.eu/care/dmd/family-guide/translations/> [accessed 24.06.14].
15. Goemans N, Buyse G. Current treatment and management of dystrophinopathies. *Curr Treat Options Neurol* 2014;**16**:287.
16. Eagle M, Baudouin SV, Chandler C, Giddings DR, Bullock R, Bushby K. Survival in Duchenne muscular dystrophy: improvements in life expectancy since 1967 and the impact of home nocturnal ventilation. *Neuromuscul Disord* 2002;**12**:926–9.
17. Bushby K, Muntoni F, Bourke JP. 107th ENMC international workshop: the management of cardiac involvement in muscular dystrophy and myotonic dystrophy. 7th–9th June 2002, Naarden, the Netherlands. *Neuromuscul Disord* 2003;**13**:166–72.
18. Duboc D, Meune C, Lerebours G, Devaux JY, Vaksmann G, Becane HM. Effect of perindopril on the onset and progression of left ventricular dysfunction in Duchenne muscular dystrophy. *J Am Coll Cardiol* 2005;**45**:855–7.
19. Bushby K, Connor E. Clinical outcome measures for trials in Duchenne muscular dystrophy: report from International Working Group meetings. *Clin Investig (Lond)* 2011;**1**:1217–35.
20. Landfeldt E, Lindgren P, Bell CF, Schmitt C, Guglieri M, Straub V, et al. The burden of Duchenne muscular dystrophy: an international, cross-sectional study. *Neurology* 2014;**83**:529–36.
21. Landfeldt E, Lindgren P, Bell CF, Schmitt C, Guglieri M, Straub V, et al. Mapping real-world medical management against international care recommendations in 770 patients with Duchenne muscular dystrophy: an international, cross-sectional study. 2014 [submitted].
22. Care-NMD International. Available from: <http://www.treat-nmd.eu/care/dmd/family-guide/translations/> [Accessed 24.06.14].
23. Florence JM, Pandya S, King WM, Robison JD, Baty J, Miller JP, et al. Intrarater reliability of manual muscle test (Medical Research Council scale) grades in Duchenne's muscular dystrophy. *Phys Ther* 1992;**72**:115–22, discussion 22–6.
24. Mayhew JE, Florence JM, Mayhew TP, Henricson EK, Leshner RT, McCarter RJ, et al. Reliable surrogate outcome measures in multicenter clinical trials of Duchenne muscular dystrophy. *Muscle Nerve* 2007;**35**:36–42.
25. Lerario A, Bonfiglio S, Sormani M, Tettamanti A, Marktel S, Napolitano S, et al. Quantitative muscle strength assessment in Duchenne muscular dystrophy: longitudinal study and correlation with functional measures. *BMC Neurol* 2012;**12**:91.
26. Scott E, Eagle M, Main M, Sheehan J. The north star ambulatory assessment [poster]. *Dev Med Child Neurol* 2006;**48**:27.
27. Scott E, Eagle M, Mayhew A, Freeman J, Main M, Sheehan J, et al. Development of a functional assessment scale for ambulatory boys with Duchenne muscular dystrophy. *Physiother Res Int* 2012;**17**:101–9.
28. Mayhew AG, Cano SJ, Scott E, Eagle M, Bushby K, Manzur A, et al. Detecting meaningful change using the North Star

Ambulatory Assessment in Duchenne muscular dystrophy. *Dev Med Child Neurol* 2013;**55**:1046–52.

29. Mayhew A, Cano S, Scott E, Eagle M, Bushby K, Muntoni F. Moving towards meaningful measurement: Rasch analysis of the North Star Ambulatory Assessment in Duchenne muscular dystrophy. *Dev Med Child Neurol* 2011;**53**:535–42.
30. Mazzone ES, Messina S, Vasco G, Main M, Eagle M, D'Amico A, et al. Reliability of the North Star Ambulatory Assessment in a multicentric setting. *Neuromuscul Disord* 2009;**19**:458–61.
31. Mazzone E, Martinelli D, Berardinelli A, Messina S, D'Amico A, Vasco G, et al. North Star Ambulatory Assessment, 6-minute walk test and timed items in ambulant boys with Duchenne muscular dystrophy. *Neuromuscul Disord* 2010;**20**:712–16.
32. Florence JM, Pandya S, King WM, Robison JD, Signore LC, Wentzell M, et al. Clinical trials in Duchenne dystrophy. Standardization and reliability of evaluation procedures. *Phys Ther* 1984;**64**:41–5.
33. McDonald CM, Henricson EK, Abresch RT, Florence J, Eagle M, Gappmaier E, et al. The 6-minute walk test and other clinical endpoints in Duchenne muscular dystrophy: reliability, concurrent validity, and minimal clinically important differences from a multicenter study. *Muscle Nerve* 2013;**48**:357–68.
34. Berard C, Payan C, Hodgkinson I, Fermanian J. A motor function measure for neuromuscular diseases. Construction and validation study. *Neuromuscul Disord* 2005;**15**:463–70.
35. Mazzone ES, Pane M, Sormani MP, Scalise R, Berardinelli A, Messina S, et al. 24 month longitudinal data in ambulant boys with Duchenne muscular dystrophy. *PLOS ONE* 2013;**8**:e52512.
36. Scott OM, Hyde SA, Goddard C, Dubowitz V. Quantitation of muscle function in children: a prospective study in Duchenne muscular dystrophy. *Muscle Nerve* 1982;**5**:291–301.
37. Vignos Jr. PJ, Spencer Jr. GE, Archibald KC. Management of progressive muscular dystrophy in childhood. *JAMA* 1963;**184**: 89–96.
38. Brooke MH, Griggs RC, Mendell JR, Fenichel GM, Shumate JB, Pellegrino RJ. Clinical trial in Duchenne dystrophy. I. The design of the protocol. *Muscle Nerve* 1981;**4**:186–97.
39. McDonald CM, Henricson EK, Han JJ, Abresch RT, Nicorici A, Elfring GL, et al. The 6-minute walk test as a new outcome measure in Duchenne muscular dystrophy. *Muscle Nerve* 2010;**41**: 500–10.
40. Henricson E, Abresch R, Han JJ, Nicorici A, Goude Keller E, de Bie E, et al. The 6-minute walk test and person-reported outcomes in boys with Duchenne muscular dystrophy and typically developing controls: longitudinal comparisons and clinically-meaningful changes over one year. *PLoS Curr* 2013 July 8;5.
41. Goemans N, Klingels K, van den Hauwe M, Boons S, Verstraete L, Peeters C, et al. Six-minute walk test: reference values and prediction equation in healthy boys aged 5 to 12 years. *PLOS ONE* 2013;**8**:e84120.
42. Pane M, Mazzone ES, Fanelli L, De Sanctis R, Bianco F, Sivo S, et al. Reliability of the Performance of Upper Limb assessment in Duchenne muscular dystrophy. *Neuromuscul Disord* 2014;**24**: 201–6.
43. Mayhew A, Mazzone ES, Eagle M, Duong T, Ash M, Decostre V, et al. Development of the Performance of the Upper Limb module for Duchenne muscular dystrophy. *Dev Med Child Neurol* 2013; **55**:1038–45.
44. Mazzone ES, Vasco G, Palermo C, Bianco F, Galluccio C, Ricotti V, et al. A critical review of functional assessment tools for upper limbs in Duchenne muscular dystrophy. *Dev Med Child Neurol* 2012;**54**:879–85.
45. Moxley 3rd RT, Pandya S, Ciafaloni E, Fox DJ, Campbell K. Change in natural history of Duchenne muscular dystrophy with long-term corticosteroid treatment: implications for management. *J Child Neurol* 2010;**25**:1116–29.
46. Manzur AY, Kuntzer T, Pike M, Swan A. Glucocorticoid corticosteroids for Duchenne muscular dystrophy. *Cochrane Database Syst Rev* 2008;**1**:CD003725.
47. Henricson EK, Abresch RT, Cnaan A, Hu F, Duong T, Arrieta A, et al. The cooperative international neuromuscular research group Duchenne natural history study: glucocorticoid treatment preserves clinically meaningful functional milestones and reduces rate of disease progression as measured by manual muscle testing and other commonly used clinical trial outcome measures. *Muscle Nerve* 2013;**48**:55–67.
48. Daftary AS, Crisanti M, Kalra M, Wong B, Amin R. Effect of long-term steroids on cough efficiency and respiratory muscle strength in patients with Duchenne muscular dystrophy. *Pediatrics* 2007;**119**:e320–4.
49. Bach JR, Martinez D, Saulat B. Duchenne muscular dystrophy: the effect of glucocorticoids on ventilator use and ambulation. *Am J Phys Med Rehabil* 2010;**89**:620–4.
50. Lebel DE, Corston JA, McAdam LC, Biggar WD, Alman BA. Glucocorticoid treatment for the prevention of scoliosis in children with Duchenne muscular dystrophy: long-term follow-up. *J Bone Joint Surg Am* 2013;**95**:1057–61.
51. King WM, Ruttencutter R, Nagaraja HN, Matkovic V, Landoll J, Hoyle C, et al. Orthopedic outcomes of long-term daily corticosteroid treatment in Duchenne muscular dystrophy. *Neurology* 2007;**68**:1607–13.
52. Markham LW, Kinnett K, Wong BL, Benson DW, Cripe LH. Corticosteroid treatment retards development of ventricular dysfunction in Duchenne muscular dystrophy. *Neuromuscul Disord* 2008;**18**:365–70.
53. Barber BJ, Andrews JG, Lu Z, West NA, Meaney FJ, Price ET, et al. Oral corticosteroids and onset of cardiomyopathy in Duchenne muscular dystrophy. *J Pediatr* 2013;**163**:1080–4.e1.
54. Schram G, Fournier A, Leduc H, Dahdah N, Therien J, Vanasse M, et al. All-cause mortality and cardiovascular outcomes with prophylactic steroid therapy in Duchenne muscular dystrophy. *J Am Coll Cardiol* 2013;**61**:948–54.
55. Janssen PM, Murray JD, Schill KE, Rastogi N, Schultz EJ, Tran T, et al. Prednisolone attenuates improvement of cardiac and skeletal contractile function and histopathology by lisinopril and spironolactone in the mdx mouse model of Duchenne muscular dystrophy. *PLOS ONE* 2014;**9**:e88360.
56. Bauer R, Straub V, Blain A, Bushby K, MacGowan GA. Contrasting effects of steroids and angiotensin-converting-enzyme inhibitors in a mouse model of dystrophin-deficient cardiomyopathy. *Eur J Heart Fail* 2009;**11**:463–71.
57. McDonald CM, Han JJ, Mah JK, Carter GT. Corticosteroids and Duchenne muscular dystrophy: does earlier treatment really matter? *Muscle Nerve* 2012;**45**:777–9.
58. Connolly AM, Florence JM, Cradock MM, Malkus EC, Schierbecker JR, Siener CA, et al. Motor and cognitive

assessment of infants and young boys with Duchenne Muscular Dystrophy: results from the Muscular Dystrophy Association DMD Clinical Research Network. *Neuromuscul Disord* 2013;**23**: 529–39.

59. Ricotti V, Ridout DA, Scott E, Quinlivan R, Robb SA, Manzur AY, et al. Long-term benefits and adverse effects of intermittent versus daily glucocorticoids in boys with Duchenne muscular dystrophy. *J Neurol Neurosurg Psychiatry* 2013;**84**:698–705.
60. Griggs RC, Moxley 3rd RT, Mendell JR, Fenichel GM, Brooke MH, Pestronk A, et al. Prednisone in Duchenne dystrophy. A randomized controlled trial defining the time course and dose response. Clinical Investigation of Duchenne Dystrophy Group. *Arch Neurol* 1991;**48**:383–8.
61. Mendell JR, Moxley RT, Griggs RC, Brooke MH, Fenichel GM, Miller JP, et al. Randomized, double-blind six-month trial of prednisone in Duchenne's muscular dystrophy. *N Engl J Med* 1989; **320**:1592–7.
62. Griggs RC, Herr BE, Reha A, Elfring G, Atkinson L, Cwik V, et al. Corticosteroids in Duchenne muscular dystrophy: major variations in practice. *Muscle Nerve* 2013;**48**:27–31.
63. Bushby K, Muntoni F, Urtizberea A, Hughes R, Griggs R. Report on the 124th ENMC International Workshop. Treatment of Duchenne muscular dystrophy; defining the gold standards of management in the use of corticosteroids. 2–4 April 2004, Naarden, The Netherlands. *Neuromuscul Disord* 2004;**14**:526–34.
64. Moxley 3rd RT, Ashwal S, Pandya S, Connolly A, Florence J, Mathews K, et al. Practice parameter: corticosteroid treatment of Duchenne dystrophy: report of the Quality Standards Subcommittee of the American Academy of Neurology and the Practice Committee of the Child Neurology Society. *Neurology* 2005;**64**:13–20.
65. Biggar WD, Gingras M, Fehlings DL, Harris VA, Steele CA. Deflazacort treatment of Duchenne muscular dystrophy. *J Pediatr* 2001;**138**:45–50.
66. Reitter B. Deflazacort vs. prednisone in Duchenne muscular dystrophy: trends of an ongoing study. *Brain Dev* 1995; **17**(Suppl.):39–43.
67. Fenichel GM, Mendell JR, Moxley 3rd RT, Griggs RC, Brooke MH, Miller JP, et al. A comparison of daily and alternate-day prednisone therapy in the treatment of Duchenne muscular dystrophy. *Arch Neurol* 1991;**48**:575–9.
68. Escolar DM, Hache LP, Clemens PR, Cnaan A, McDonald CM, Viswanathan V, et al. Randomized, blinded trial of weekend vs daily prednisone in Duchenne muscular dystrophy. *Neurology* 2011;**77**:444–52.
69. Moxley 3rd RT, Pandya S. Weekend high-dosage prednisone: a new option for treatment of Duchenne muscular dystrophy. *Neurology* 2011;**77**:416–17.
70. Bushby K, Griggs R. 145th ENMC International Workshop: planning for an International Trial of Steroid Dosage Regimes in DMD (FOR DMD), 22–24th October 2006, Naarden, The Netherlands. *Neuromuscul Disord* 2007;**17**:423–8.
71. McDonald CM, Henricson EK, Abresch RT, Han JJ, Escolar DM, Florence JM, et al. The cooperative international neuromuscular research group Duchenne natural history study—a longitudinal investigation in the era of glucocorticoid therapy: design of protocol and the methods used. *Muscle Nerve* 2013;**48**:32–54.
72. Sacre K, Dehoux M, Chauveheid MP, Chauchard M, Lidove O, Roussel R, et al. Pituitary-adrenal function after prolonged glucocorticoid therapy for systemic inflammatory disorders: an observational study. *J Clin Endocrinol Metab* 2013;**98**: 3199–205.
73. Reeves EK, Hoffman EP, Nagaraju K, Damsker JM, McCall JM. VBP15: preclinical characterization of a novel anti-inflammatory delta 9,11 steroid. *Bioorg Med Chem* 2013;**21**:2241–9.
74. Heier CR, Damsker JM, Yu Q, Dillingham BC, Huynh T, Van der Meulen JH, et al. VBP15, a novel anti-inflammatory and membrane-stabilizer, improves muscular dystrophy without side effects. *EMBO Mol Med* 2013;**5**:1569–85.
75. Bianchi ML, Biggar D, Bushby K, Rogol AD, Rutter MM, Tseng B. Endocrine aspects of Duchenne muscular dystrophy. *Neuromuscul Disord* 2011;**21**:298–303.
76. Leung DG, Germain-Lee EL, Denger BE, Wagner KR. Report on the Second Endocrine Aspects of Duchenne Muscular Dystrophy Conference December 1–2, 2010, Baltimore, Maryland, USA. *Neuromuscul Disord* 2011;**21**:594–601.
77. Rutter MM, Wong BL, Rose SR, et al. Metformin improves weight gain and body mass index in Duchenne muscular dystrophy [poster]. In: 14th International Congress of the World Muscle Society: Geneva, Switzerland; 2009.
78. Hendriksen JG, Poysky JT, Schrans DG, Schouten EG, Aldenkamp AP, Vles JS. Psychosocial adjustment in males with Duchenne muscular dystrophy: psychometric properties and clinical utility of a parent-report questionnaire. *J Pediatr Psychol* 2009;**34**:69–78.
79. Nagel BH, Mortier W, Elmlinger M, Wollmann HA, Schmitt K, Ranke MB. Short stature in Duchenne muscular dystrophy: a study of 34 patients. *Acta Paediatr* 1999;**88**:62–5.
80. Bodor M, McDonald CM. Why short stature is beneficial in Duchenne muscular dystrophy. *Muscle Nerve* 2013;**48**:336–42.
81. Rutter MM, Collins J, Rose SR, Woo JG, Sucharew H, Sawnani H, et al. Growth hormone treatment in boys with Duchenne muscular dystrophy and glucocorticoid-induced growth failure. *Neuromuscul Disord* 2012;**22**:1046–56.
82. Cittadini A, Ines Comi L, Longobardi S, Rocco Petretta V, Casaburi C, Passamano L, et al. A preliminary randomized study of growth hormone administration in Becker and Duchenne muscular dystrophies. *Eur Heart J* 2003;**24**:664–72.
83. Frank GR, Smith RE. Effective growth hormone therapy in a growth hormone deficient patient with Duchenne muscular dystropy without evidence of acceleration of the dystrophic process. *J Pediatr Endocrinol Metab* 2001;**14**:211–14.
84. Joyce NC, Hache LP, Clemens PR. Bone health and associated metabolic complications in neuromuscular diseases. *Phys Med Rehabil Clin N Am* 2012;**23**:773–99.
85. Manolagas SC, Weinstein RS. New developments in the pathogenesis and treatment of steroid-induced osteoporosis. *J Bone Miner Res* 1999;**14**:1061–6.
86. Bianchi ML, Mazzanti A, Galbiati E, Saraifoger S, Dubini A, Cornelio F, et al. Bone mineral density and bone metabolism in Duchenne muscular dystrophy. *Osteoporos Int* 2003;**14**:761–7.
87. McDonald DG, Kinali M, Gallagher AC, Mercuri E, Muntoni F, Roper H, et al. Fracture prevalence in Duchenne muscular dystrophy. *Dev Med Child Neurol* 2002;**44**:695–8.
88. Talim B, Malaguti C, Gnudi S, Politano L, Merlini L. Vertebral compression in Duchenne muscular dystrophy following deflazacort. *Neuromuscul Disord* 2002;**12**:294–5.

89. Mayo AL, Craven BC, McAdam LC, Biggar WD. Bone health in boys with Duchenne Muscular Dystrophy on long-term daily deflazacort therapy. *Neuromuscul Disord* 2012;**22**:1040–5.
90. Quinlivan R, Roper H, Davie M, Shaw NJ, McDonagh J, Bushby K. Report of a Muscular Dystrophy Campaign funded workshop. Birmingham, UK, January 16th 2004. Osteoporosis in Duchenne muscular dystrophy; its prevalence, treatment and prevention. *Neuromuscul Disord* 2005;**15**:72–9.
91. Biggar WD, Bachrach LK, Henderson RC, Kalkwarf H, Plotkin H, Wong BL. Bone health in Duchenne muscular dystrophy: a workshop report from the meeting in Cincinnati, Ohio, July 8, 2004. *Neuromuscul Disord* 2005;**15**:80–5.
92. Zemel BS, Kalkwarf HJ, Gilsanz V, Lappe JM, Oberfield S, Shepherd JA, et al. Revised reference curves for bone mineral content and areal bone mineral density according to age and sex for black and non-black children: results of the bone mineral density in childhood study. *J Clin Endocrinol Metab* 2011;**96**: 3160–9.
93. Bianchi ML, Morandi L, Andreucci E, Vai S, Frasunkiewicz J, Cottafava R. Low bone density and bone metabolism alterations in Duchenne muscular dystrophy: response to calcium and vitamin D treatment. *Osteoporos Int* 2011;**22**:529–39.
94. Quinlivan R, Shaw N, Bushby K. 170th ENMC International Workshop: bone protection for corticosteroid treated Duchenne muscular dystrophy. 27–29 November 2009, Naarden, The Netherlands. *Neuromuscul Disord* 2010;**20**:761–9.
95. Sbrocchi AM, Rauch F, Jacob P, McCormick A, McMillan HJ, Matzinger MA, et al. The use of intravenous bisphosphonate therapy to treat vertebral fractures due to osteoporosis among boys with Duchenne muscular dystrophy. *Osteoporos Int* 2012;**23**: 2703–11.
96. Houston C, Mathews K, Shibli-Rahhal A. Bone density and alendronate effects in Duchenne muscular dystrophy patients. *Muscle Nerve* 2014;**49**:506–11.
97. Hawker GA, Ridout R, Harris VA, Chase CC, Fielding LJ, Biggar WD. Alendronate in the treatment of low bone mass in steroid-treated boys with Duchennes muscular dystrophy. *Arch Phys Med Rehabil* 2005;**86**:284–8.
98. Palomo AE, Ballester HMJ, Marquez de La Plata MA, Medina Cano E, Carmona VRM. Alendronate treatment of osteoporosis secondary to Duchenne muscular dystrophy. *An Pediatr (Barc)* 2011;**74**:122–5.
99. Parker AE, Robb SA, Chambers J, Davidson AC, Evans K, O'Dowd J, et al. Analysis of an adult Duchenne muscular dystrophy population. *QJM* 2005;**98**:729–36.
100. Kaspar RW, Allen HD, Montanaro F. Current understanding and management of dilated cardiomyopathy in Duchenne and Becker muscular dystrophy. *J Am Acad Nurse Pract* 2009;**21**:241–9.
101. de Kermadec JM, Becane HM, Chenard A, Tertrain F, Weiss Y. Prevalence of left ventricular systolic dysfunction in Duchenne muscular dystrophy: an echocardiographic study. *Am Heart J* 1994;**127**:618–23.
102. Chenard AA, Becane HM, Tertrain F, de Kermadec JM, Weiss YA. Ventricular arrhythmia in Duchenne muscular dystrophy: prevalence, significance and prognosis. *Neuromuscul Disord* 1993;**3**:201–6.
103. Romfh A, McNally EM. Cardiac assessment in Duchenne and Becker muscular dystrophies. *Curr Heart Fail Rep* 2010;**7**:212–18.
104. Spurney CF. Cardiomyopathy of Duchenne muscular dystrophy: current understanding and future directions. *Muscle Nerve* 2011;**44**:8–19.
105. Kieny P, Chollet S, Delalande P, Le Fort M, Magot A, Pereon Y, et al. Evolution of life expectancy of patients with Duchenne muscular dystrophy at AFM Yolaine de Kepper centre between 1981 and 2011. *Ann Phys Rehabil Med* 2013;**56**:443–54.
106. Nigro G, Comi LI, Politano L, Bain RJ. The incidence and evolution of cardiomyopathy in Duchenne muscular dystrophy. *Int J Cardiol* 1990;**26**:271–7.
107. Finsterer J, Stollberger C. The heart in human dystrophinopathies. *Cardiology* 2003;**99**:1–19.
108. Corrado G, Lissoni A, Beretta S, Terenghi L, Tadeo G, Foglia-Manzillo G, et al. Prognostic value of electrocardiograms, ventricular late potentials, ventricular arrhythmias, and left ventricular systolic dysfunction in patients with Duchenne muscular dystrophy. *Am J Cardiol* 2002;**89**:838–41.
109. Puchalski MD, Williams RV, Askovich B, Sower CT, Hor KH, Su JT, et al. Late gadolinium enhancement: precursor to cardiomyopathy in Duchenne muscular dystrophy? *Int J Cardiovasc Imaging* 2009;**25**:57–63.
110. Silva MC, Meira ZM, Giannetti JG, da Silva MM, Campos AF, Barbosa MM, et al. Myocardial delayed enhancement by magnetic resonance imaging in patients with muscular dystrophy. *J Am Coll Cardiol* 2007;**49**:1874–9.
111. Ramaciotti C, Heistein LC, Coursey M, Lemler MS, Eapen RS, Iannaccone ST, et al. Left ventricular function and response to enalapril in patients with Duchenne muscular dystrophy during the second decade of life. *Am J Cardiol* 2006;**98**:825–7.
112. Allen HD, Flanigan KM, Thrush PT, Dvorchik I, Yin H, Canter C, et al. A randomized, double-blind trial of lisinopril and losartan for the treatment of cardiomyopathy in Duchenne muscular dystrophy. *PLoS Curr* 2013;5.
113. Viollet L, Thrush PT, Flanigan KM, Mendell JR, Allen HD. Effects of angiotensin-converting enzyme inhibitors and/or beta blockers on the cardiomyopathy in Duchenne muscular dystrophy. *Am J Cardiol* 2012;**110**:98–102.
114. Jefferies JL, Eidem BW, Belmont JW, Craigen WJ, Ware SM, Fernbach SD, et al. Genetic predictors and remodeling of dilated cardiomyopathy in muscular dystrophy. *Circulation* 2005;**112**:2799–804.
115. Hunt SA, Abraham WT, Chin MH, Feldman AM, Francis GS, Ganiats TG, et al. ACC/AHA 2005 Guideline Update for the Diagnosis and Management of Chronic Heart Failure in the Adult: a report of the American College of Cardiology/American Heart Association Task Force on Practice Guidelines (Writing Committee to Update the 2001 Guidelines for the Evaluation and Management of Heart Failure): developed in collaboration with the American College of Chest Physicians and the International Society for Heart and Lung Transplantation: endorsed by the Heart Rhythm Society. *Circulation* 2005;**112**:e154–235.
116. Saito T, Matsumura T, Miyai I, Nozaki S, Shinno S. Carvedilol effectiveness for left ventricular-insufficient patients with Duchenne muscular dystrophy. *Rinsho Shinkeigaku* 2001;**41**:691–4.
117. Ishikawa Y, Bach JR, Minami R. Cardioprotection for Duchenne's muscular dystrophy. *Am Heart J* 1999;**137**:895–902.
118. Shaddy RE, Tani LY, Gidding SS, Pahl E, Orsmond GS, Gilbert EM, et al. Beta-blocker treatment of dilated cardiomyopathy with

congestive heart failure in children: a multi-institutional experience. *J Heart Lung Transplant* 1999;**18**:269–74.

119. Ryan TD, Jefferies JL, Sawnani H, Wong BL, Gardner A, Del Corral M, et al. Implantation of the HeartMate II and HeartWare left ventricular assist devices in patients with Duchenne muscular dystrophy: lessons learned from the first applications. *ASAIO J* 2014;**60**:246–8.
120. Wu RS, Gupta S, Brown RN, Yancy CW, Wald JW, Kaiser P, et al. Clinical outcomes after cardiac transplantation in muscular dystrophy patients. *J Heart Lung Transplant* 2010;**29**:432–8.
121. Spurney C, Shimizu R, Hache LP, Kolski H, Gordish-Dressman H, Clemens PR. CINRG Duchenne Natural History Study demonstrates insufficient diagnosis and treatment of cardiomyopathy in Duchenne muscular dystrophy. *Muscle Nerve* 2014;**50**:250–6.
122. Gayraud J, Ramonatxo M, Rivier F, Humberclaude V, Petrof B, Matecki S. Ventilatory parameters and maximal respiratory pressure changes with age in Duchenne muscular dystrophy patients. *Pediatr Pulmonol* 2010;**45**:552–9.
123. Inkley SR, Oldenburg FC, Vignos Jr. PJ. Pulmonary function in Duchenne muscular dystrophy related to stage of disease. *Am J Med* 1974;**56**:297–306.
124. Dohna-Schwake C, Ragette R, Teschler H, Voit T, Mellies U. Predictors of severe chest infections in pediatric neuromuscular disorders. *Neuromuscul Disord* 2006;**16**:325–8.
125. Vignos Jr. PJ. Respiratory function and pulmonary infection in Duchenne muscular dystrophy. *Isr J Med Sci* 1977;**13**:207–14.
126. Passamano L, Taglia A, Palladino A, Viggiano E, D'Ambrosio P, Scutifero M, et al. Improvement of survival in Duchenne Muscular Dystrophy: retrospective analysis of 835 patients. *Acta Myol* 2012;**31**:121–5.
127. Gomez-Merino E, Bach JR. Duchenne muscular dystrophy: prolongation of life by noninvasive ventilation and mechanically assisted coughing. *Am J Phys Med Rehabil* 2002;**81**:411–15.
128. Mellies U, Ragette R, Dohna Schwake C, Boehm H, Voit T, Teschler H. Long-term noninvasive ventilation in children and adolescents with neuromuscular disorders. *Eur Respir J* 2003;**22**:631–6.
129. Finder JD, Birnkrant D, Carl J, Farber HJ, Gozal D, Iannaccone ST, et al. Respiratory care of the patient with Duchenne muscular dystrophy: ATS consensus statement. *Am J Respir Crit Care Med* 2004;**170**:456–65.
130. Birnkrant DJ, Bushby KM, Amin RS, Bach JR, Benditt JO, Eagle M, et al. The respiratory management of patients with Duchenne muscular dystrophy: a DMD Care Considerations working group specialty article. *Pediatr Pulmonol* 2010;**45**:739–48.
131. Katz SL, McKim D, Hoey L, Barrowman N, Kherani T, Kovesi T, et al. Respiratory management strategies for Duchenne muscular dystrophy: practice variation amongst Canadian subspecialists. *Pediatr Pulmonol* 2013;**48**:59–66.
132. Szeinberg A, Tabachnik E, Rashed N, McLaughlin FJ, England S, Bryan CA, et al. Cough capacity in patients with muscular dystrophy. *Chest* 1988;**94**:1232–5.
133. Kang SW, Kang YS, Sohn HS, Park JH, Moon JH. Respiratory muscle strength and cough capacity in patients with Duchenne muscular dystrophy. *Yonsei Med J* 2006;**47**:184–90.
134. Phillips MF, Quinlivan RC, Edwards RH, Calverley PM. Changes in spirometry over time as a prognostic marker in patients with Duchenne muscular dystrophy. *Am J Respir Crit Care Med* 2001;**164**:2191–4.
135. McKim DA, Katz SL, Barrowman N, Ni A, LeBlanc C. Lung volume recruitment slows pulmonary function decline in Duchenne muscular dystrophy. *Arch Phys Med Rehabil* 2012;**93**:1117–22.
136. Bach JR. Mechanical insufflation-exsufflation. Comparison of peak expiratory flows with manually assisted and unassisted coughing techniques. *Chest* 1993;**104**:1553–62.
137. Chatwin M, Ross E, Hart N, Nickol AH, Polkey MI, Simonds AK. Cough augmentation with mechanical insufflation/exsufflation in patients with neuromuscular weakness. *Eur Respir J* 2003;**21**:502–8.
138. Fauroux B, Guillemot N, Aubertin G, Nathan N, Labit A, Clement A, et al. Physiologic benefits of mechanical insufflation-exsufflation in children with neuromuscular diseases. *Chest* 2008;**133**:161–8.
139. Toussaint M, Steens M, Soudon P. Lung function accurately predicts hypercapnia in patients with Duchenne muscular dystrophy. *Chest* 2007;**131**:368–75.
140. Smith PE, Edwards RH, Calverley PM. Oxygen treatment of sleep hypoxaemia in Duchenne muscular dystrophy. *Thorax* 1989;**44**:997–1001.
141. Niranjan V, Bach JR. Noninvasive management of pediatric neuromuscular ventilatory failure. *Crit Care Med* 1998;**26**:2061–5.
142. Piastra M, Antonelli M, Caresta E, Chiaretti A, Polidori G, Conti G. Noninvasive ventilation in childhood acute neuromuscular respiratory failure: a pilot study. *Respiration* 2006;**73**:791–8
143. Simonds AK, Muntoni F, Heather S, Fielding S. Impact of nasal ventilation on survival in hypercapnic Duchenne muscular dystrophy. *Thorax* 1998;**53**:949–52.
144. Ward S, Chatwin M, Heather S, Simonds AK. Randomised controlled trial of non-invasive ventilation (NIV) for nocturnal hypoventilation in neuromuscular and chest wall disease patients with daytime normocapnia. *Thorax* 2005;**60**:1019–24.
145. Toussaint M, Steens M, Wasteels G, Soudon P. Diurnal ventilation via mouthpiece: survival in end-stage Duchenne patients. *Eur Respir J* 2006;**28**:549–55.
146. Bach JR, Martinez D. Duchenne muscular dystrophy: continuous noninvasive ventilatory support prolongs survival. *Respir Care* 2011;**56**:744–50.
147. Rideau Y, Politano L. Research against incurability. Treatment of lethal neuromuscular diseases focused on Duchenne Muscular Dystrophy. *Acta Myol* 2004;**23**:163–78.
148. Soudon P, Steens M, Toussaint M. A comparison of invasive versus noninvasive full-time mechanical ventilation in Duchenne muscular dystrophy. *Chron Respir Dis* 2008;**5**:87–93.
149. McKim DA, Griller N, LeBlanc C, Woolnough A, King J. Twenty-four hour noninvasive ventilation in Duchenne muscular dystrophy: a safe alternative to tracheostomy. *Can Respir J* 2013;**20**:e5–9.
150. Toussaint M, Chatwin M, Soudon P. Mechanical ventilation in Duchenne patients with chronic respiratory insufficiency: clinical implications of 20 years published experience. *Chron Respir Dis* 2007;**4**:167–77.
151. Saito T, Sawabata N, Matsumura T, Nozaki S, Fujimura H, Shinno S. Tracheo-arterial fistula in tracheostomy patients with Duchenne muscular dystrophy. *Brain Dev* 2006;**28**:223–7.
152. Brooke MH, Fenichel GM, Griggs RC, Mendell JR, Moxley R, Florence J, et al. Duchenne muscular dystrophy: patterns of

clinical progression and effects of supportive therapy. *Neurology* 1989;**39**:475–81.
153. Sutherland DH, Olshen R, Cooper L, Wyatt M, Leach J, Mubarak S, et al. The pathomechanics of gait in Duchenne muscular dystrophy. *Dev Med Child Neurol* 1981;**23**:3–22.
154. Johnson ER, Fowler Jr. WM, Lieberman JS. Contractures in neuromuscular disease. *Arch Phys Med Rehabil* 1992;**73**:807–10.
155. Hsu JD, Furumasu J. Gait and posture changes in the Duchenne muscular dystrophy child. *Clin Orthop Relat Res* 1993;122–5.
156. McDonald CM, Abresch RT, Carter GT, Fowler Jr. WM, Johnson ER, Kilmer DD, et al. Profiles of neuromuscular diseases. Duchenne muscular dystrophy. *Am J Phys Med Rehabil* 1995;**74**: S70–92.
157. Fowler Jr. WM. Rehabilitation management of muscular dystrophy and related disorders: II. Comprehensive care. *Arch Phys Med Rehabil* 1982;**63**:322–8.
158. Vignos Jr. PJ. Physical models of rehabilitation in neuromuscular disease. *Muscle Nerve* 1983;**6**:323–38.
159. Hyde SA, FlLytrup I, Glent S, Kroksmark AK, Salling B, Steffensen BF, et al. A randomized comparative study of two methods for controlling Tendo Achilles contracture in Duchenne muscular dystrophy. *Neuromuscul Disord* 2000;**10**:257–63.
160. Scott OM, Hyde SA, Goddard C, Dubowitz V. Prevention of deformity in Duchenne muscular dystrophy. A prospective study of passive stretching and splintage. *Physiotherapy* 1981;**67**:177–80.
161. McDonald CM. Limb contractures in progressive neuromuscular disease and the role of stretching, orthotics, and surgery. *Phys Med Rehabil Clin N Am* 1998;**9**:187–211.
162. Siegel IM. Plastic-molded knee-ankle-foot orthoses in the treatment of Duchenne muscular dystrophy. *Arch Phys Med Rehabil* 1975;**56**:322.
163. Bakker JP, De Groot IJ, De Jong BA, Van Tol-De Jager MA, Lankhorst GJ. Prescription pattern for orthoses in The Netherlands: use and experience in the ambulatory phase of Duchenne muscular dystrophy. *Disabil Rehabil* 1997;**19**:318–25.
164. Glanzman AM, Flickinger JM, Dholakia KH, Bonnemann CG, Finkel RS. Serial casting for the management of ankle contracture in Duchenne muscular dystrophy. *Pediatr Phys Ther* 2011;**23**: 275–9.
165. Farmer SE, Woollam PJ, Patrick JH, Roberts AP, Bromwich W. Dynamic orthoses in the management of joint contracture. *J Bone Joint Surg Br* 2005;**87**:291–5.
166. Miller G, Dunn N. An outline of the management and prognosis of Duchenne muscular dystrophy in Western Australia. *Aust Paediatr J* 1982;**18**:277–82.
167. Scher DM, Mubarak SJ. Surgical prevention of foot deformity in patients with Duchenne muscular dystrophy. *J Pediatr Orthop* 2002;**22**:384–91.
168. Dunning AG, Herder JL. A review of assistive devices for arm balancing. *IEEE Int Conf Rehabil Robot* 2013;**2013**:6650485.
169. Kumar A, Phillips MF. Use of powered mobile arm supports by people with neuromuscular conditions. *J Rehabil Res Dev* 2013;**50**:61–70.
170. Gordon BS, Lowe DA, Kostek MC. Exercise increases utrophin protein expression in the mdx mouse model of duchenne muscular dystrophy. *Muscle Nerve* 2014 Jun;**49**(6):915–18.
171. Vry J, Schubert IJ, Semler O, Haug V, Schonau E, Kirschner J. Whole-body vibration training in children with Duchenne muscular dystrophy and spinal muscular atrophy. *Eur J Paediatr Neurol* 2014;**18**:140–9.
172. Jansen M, van Alfen N, Geurts AC, de Groot IJ. Assisted bicycle training delays functional deterioration in boys with Duchenne muscular dystrophy: the randomized controlled trial "no use is disuse". *Neurorehabil Neural Repair* 2013;**27**:816–27.
173. Garrood P, Eagle M, Jardine PE, Bushby K, Straub V. Myoglobinuria in boys with Duchenne muscular dystrophy on corticosteroid therapy. *Neuromuscul Disord* 2008;**18**:71–3.
174. Kinali M, Messina S, Mercuri E, Lehovsky J, Edge G, Manzur AY, et al. Management of scoliosis in Duchenne muscular dystrophy: a large 10-year retrospective study. *Dev Med Child Neurol* 2006;**48**:513–18.
175. Cambridge W, Drennan JC. Scoliosis associated with Duchenne muscular dystrophy. *J Pediatr Orthop* 1987;**7**:436–40.
176. Harvey A, Baker L, Williams K. Non-surgical prevention and management of scoliosis for children with Duchenne muscular dystrophy: what is the evidence? *J Paediatr Child Health* 2013. Available from: http://dx.doi.org/doi:10.1111/jpc.12177.
177. Heller KD, Forst R, Forst J, Hengstler K. Scoliosis in Duchenne muscular dystrophy: aspects of orthotic treatment. *Prosthet Orthot Int* 1997;**21**:202–9.
178. Suk KS, Lee BH, Lee HM, Moon SH, Choi YC, Shin DE, et al. Functional outcomes in Duchenne muscular dystrophy scoliosis: comparison of the differences between surgical and nonsurgical treatment. *J Bone Joint Surg Am* 2014;**96**:409–15.
179. Cervellati S, Bettini N, Moscato M, Gusella A, Dema E, Maresi R. Surgical treatment of spinal deformities in Duchenne muscular dystrophy: a long term follow-up study. *Eur Spine J* 2004;**13**: 441–8.
180. Gayet LE, Duport G, Pries P. Flexible and semi-early vertebral instrumentation in surgical treatment of Duchenne muscular dystrophy scoliosis. *Eur J Orthop Surg Traumatol* 1999;**9**:223–31.
181. Arun R, Srinivas S, Mehdian SM. Scoliosis in Duchenne's muscular dystrophy: a changing trend in surgical management: a historical surgical outcome study comparing sublaminar, hybrid and pedicle screw instrumentation systems. *Eur Spine J* 2010;**19**:376–83.
182. Mullender M, Blom N, De Kleuver M, Fock J, Hitters W, Horemans A, et al. A Dutch guideline for the treatment of scoliosis in neuromuscular disorders. *Scoliosis* 2008;**3**:14.
183. Hsu JD, Quinlivan R. Scoliosis in Duchenne muscular dystrophy (DMD). *Neuromuscul Disord* 2013;**23**:611–17.
184. Hsu JD, Garcia-Ariz M. Fracture of the femur in the Duchenne muscular dystrophy patient. *J Pediatr Orthop* 1981;**1**:203–7.
185. Pane M, Vasta I, Messina S, Sorleti D, Aloysius A, Sciarra F, et al. Feeding problems and weight gain in Duchenne muscular dystrophy. *Eur J Paediatr Neurol* 2006;**10**:231–6.
186. Willig TN, Carlier L, Legrand M, Riviere H, Navarro J. Nutritional assessment in Duchenne muscular dystrophy. *Dev Med Child Neurol* 1993;**35**:1074–82.
187. Shimizu-Fujiwara M, Komaki H, Nakagawa E, Mori-Yoshimura M, Oya Y, Fujisaki T, et al. Decreased resting energy expenditure in patients with Duchenne muscular dystrophy. *Brain Dev* 2012;**34**:206–12.
188. Aloysius A, Born P, Kinali M, Davis T, Pane M, Mercuri E. Swallowing difficulties in Duchenne muscular dystrophy: indications for feeding assessment and outcome of videofluoroscopic swallow studies. *Eur J Paediatr Neurol* 2008;**12**:239–45.

189. van den Engel-Hoek L, Erasmus CE, Hendriks JC, Geurts AC, Klein WM, Pillen S, et al. Oral muscles are progressively affected in Duchenne muscular dystrophy: implications for dysphagia treatment. *J Neurol* 2013;**260**:1295–303.
190. Shinonaga C, Fukuda M, Suzuki Y, Higaki T, Ishida Y, Ishii E, et al. Evaluation of swallowing function in Duchenne muscular dystrophy. *Dev Med Child Neurol* 2008;**50**:478–80.
191. Hanayama K, Liu M, Higuchi Y, Fujiwara T, Tsuji T, Hase K, et al. Dysphagia in patients with Duchenne muscular dystrophy evaluated with a questionnaire and videofluorography. *Disabil Rehabil* 2008;**30**:517–22.
192. Nozaki S, Umaki Y, Sugishita S, Tatara K, Adachi K, Shinno S. Videofluorographic assessment of swallowing function in patients with Duchenne muscular dystrophy. *Rinsho Shinkeigaku* 2007;**47**:407–12.
193. Archer SK, Garrod R, Hart N, Miller S. Dysphagia in Duchenne muscular dystrophy assessed by validated questionnaire. *Int J Lang Commun Disord* 2013;**48**:240–6.
194. Grant JP. Percutaneous endoscopic gastrostomy. Initial placement by single endoscopic technique and long-term follow-up. *Ann Surg* 1993;**217**:168–74.
195. Gauderer MW. Percutaneous endoscopic gastrostomy: a 10-year experience with 220 children. *J Pediatr Surg* 1991;**26**:288–92, discussion 92–4.
196. Martigne L, Seguy D, Pellegrini N, Orlikowski D, Cuisset JM, Carpentier A, et al. Efficacy and tolerance of gastrostomy feeding in Duchenne muscular dystrophy. *Clin Nutr* 2010;**29**:60–4.
197. Mizuno T, Komaki H, Sasaki M, Takanoha S, Kuroda K, Kon K, et al. Efficacy and tolerance of gastrostomy feeding in Japanese muscular dystrophy patients. *Brain Dev* 2012;**34**:756–62.
198. Zickler RW, Barbagiovanni JT, Swan KG. A simplified open gastrostomy under local anesthesia. *Am Surg* 2001;**67**:806–8.
199. Cripe LH, Tobias JD. Cardiac considerations in the operative management of the patient with Duchenne or Becker muscular dystrophy. *Paediatr Anaesth* 2013;**23**:777–84.
200. Muenster T, Mueller C, Forst J, Huber H, Schmitt HJ. Anaesthetic management in patients with Duchenne muscular dystrophy undergoing orthopaedic surgery: a review of 232 cases. *Eur J Anaesthesiol* 2012;**29**:489–94.
201. Hayes J, Veyckemans F, Bissonnette B. Duchenne muscular dystrophy: an old anesthesia problem revisited. *Paediatr Anaesth* 2008;**18**:100–6.
202. Yemen TA, McClain C. Muscular dystrophy, anesthesia and the safety of inhalational agents revisited; again. *Paediatr Anaesth* 2006;**16**:105–8.
203. Segura LG, Lorenz JD, Weingarten TN, Scavonetto F, Bojanic K, Selcen D, et al. Anesthesia and Duchenne or Becker muscular dystrophy: review of 117 anesthetic exposures. *Paediatr Anaesth* 2013;**23**:855–64.
204. McAdam LC, Rastogi A, Macleod K, Douglas Biggar W. Fat Embolism Syndrome following minor trauma in Duchenne muscular dystrophy. *Neuromuscul Disord* 2012;**22**:1035–9.
205. Medeiros MO, Behrend C, King W, Sanders J, Kissel J, Ciafaloni E. Fat embolism syndrome in patients with Duchenne muscular dystrophy. *Neurology* 2013;**80**:1350–2.
206. Pender ES, Pollack Jr. CV, Evans OB. Fat embolism syndrome in a child with muscular dystrophy. *J Emerg Med* 1992;**10**:705–11.
207. Shaikh N. Emergency management of fat embolism syndrome. *J Emerg Trauma Shock* 2009;**2**:29–33.
208. Cotton S, Voudouris NJ, Greenwood KM. Intelligence and Duchenne muscular dystrophy: full-scale, verbal, and performance intelligence quotients. *Dev Med Child Neurol* 2001;**43**:497–501.
209. Hinton VJ, Fee RJ, Goldstein EM, De Vivo DC. Verbal and memory skills in males with Duchenne muscular dystrophy. *Dev Med Child Neurol* 2007;**49**:123–8.
210. D'Angelo MG, Lorusso ML, Civati F, Comi GP, Magri F, Del Bo R, et al. Neurocognitive profiles in Duchenne muscular dystrophy and gene mutation site. *Pediatr Neurol* 2011;**45**:292–9.
211. Hendriksen JG, Vles JS. Neuropsychiatric disorders in males with Duchenne muscular dystrophy: frequency rate of attention-deficit hyperactivity disorder (ADHD), autism spectrum disorder, and obsessive-compulsive disorder. *J Child Neurol* 2008;**23**:477–81.
212. Nereo NE, Fee RJ, Hinton VJ. Parental stress in mothers of boys with Duchenne muscular dystrophy. *J Pediatr Psychol* 2003;**28**:473–84.
213. Schrans DG, Abbott D, Peay HL, Pangalila RF, Vroom E, Goemans N, et al. Transition in Duchenne muscular dystrophy: An expert meeting report and description of transition needs in an emergent patient population: (Parent Project Muscular Dystrophy Transition Expert Meeting 17–18 June 2011, Amsterdam, The Netherlands). *Neuromuscul Disord* 2013;**23**:283–6.
214. Abbott D, Carpenter J, Bushby K. Transition to adulthood for young men with Duchenne muscular dystrophy: research from the UK. *Neuromuscul Disord* 2012;**22**:445–6.
215. Wedgwood N, Llewellyn G, Honey A, Schneider J. The transition of adolescents with chronic health conditions from paediatric to adult services. Australian Research Alliance for Children and Youth (ARACY); 2008 [updated 2008 February 20; cited May 08 2014]; Available from: <http://www.academia.edu/719601/The_Transition_of_Adolescents_with_Chronic_Health_Conditions_from_Paediatric_to_Adult_Services>.
216. del Gaudio D, Yang Y, Boggs BA, Schmitt ES, Lee JA, Sahoo T, et al. Molecular diagnosis of Duchenne/Becker muscular dystrophy: enhanced detection of dystrophin gene rearrangements by oligonucleotide array–comparative genomic hybridization. *Hum Mutat* 2008;**29**:1100–7.
217. Johnsen SD. Prednisone therapy in Becker's muscular dystrophy. *J Child Neurol* 2001;**16**:870–1.
218. Higuchi I, Nakamura K, Nakagawa M, Nakamura N, Usuki F, Inose M, et al. Steroid-responsive myalgia in a patient with Becker muscular dystrophy. *J Neurol Sci* 1993;**115**:219–22.
219. Backman E, Henriksson KG. Low-dose prednisolone treatment in Duchenne and Becker muscular dystrophy. *Neuromuscul Disord* 1995;**5**:233–41.

Chapter 32

Facioscapulohumeral Dystrophy

Wendy K.M. Liew, Silvère M. van der Maarel, and Rabi Tawil

INTRODUCTION

Facioscapulohumeral muscular dystrophy (FSHD) is the third most common muscular dystrophy after dystrophinopathies and myotonic muscular dystrophy. It affects 1 in 15,000 to 20,000 individuals with an estimated prevalence of 4 to 7 per 100,000.[1] FSHD was first described by Landouzy and Dejerine in 1884.[2] A study in 1952 looking at six generations of a single family from Utah further described the clinical manifestations of FSHD.[3]

FSHD is a slowly progressive form of muscular dystrophy with an autosomal dominant inheritance in most cases that is caused by a contraction of the macrosatellite repeat array D4Z4 in the subtelomere of chromosome 4q (FSHD1).[4,5] In approximately 5% of patients with FSHD (FSHD2), there is no contracted D4Z4 repeat array in chromosome 4, but these patients most often have a mutation in *SMCHD1*.[6–9] Although these two forms of FSHD, FSHD1 and FSHD2, are genetically distinct, they are identical in their clinical presentation and disease progression.[7] The age of onset is variable with a 95% penetrance by the age of 20.[10]

Diagnostic criteria were first developed by the Facioscapulohumeral Consortium at the International Conference on the Cause and Treatment of FSHD in 1997 (Box 32.1). Following that, similar diagnostic criteria were formulated by a European Neuromuscular Centre workshop.[11] These clinical diagnostic criteria for FSHD are now superseded by molecular genetic testing since diagnostic DNA testing for the D4Z4 contraction on chromosome 4q is readily available.[12]

CLINICAL FEATURES

Patients with FSHD classically present with asymmetric weakness in the face, shoulders, limb girdle, and abdomen. There is a selective sparing of the masticatory, lingual, and extraocular muscles in the facial weakness. Progression of disease is slow and there is a variable spectrum of disease severity. Some individuals remain asymptomatic or minimally symptomatic throughout their lives, with 20% of patients becoming wheelchair dependent and significantly disabled. There are reports that suggest that several factors may affect the prognosis and clinical severity of FSHD. Women appear to be less affected compared to men,[1,11,13] although this was not noted in patients with FSHD2.[7] Severely affected cases also seem to originate more often through new mutations or more commonly transmitted through maternal than through paternal lines.[13] Some reports suggest that there may also be an element of clinical anticipation in FSHD, which predicts increasing severity with each subsequent generation;[14–16] however, this was not confirmed by subsequent studies.

The initial manifestation in the majority of patients is asymptomatic facial weakness affecting the orbicularis oculi and orbicularis oris. Later on, difficulty in eye closure (Figure 32.1A) and the inability to smile, whistle, drink from a straw, or blow up balloons is noted. Patients also develop pouting of the lips (bouche de tapir), a relatively expressionless face, and the characteristic dimpling lateral to the angles of the mouth, which become deeper when the patient smiles or tries to bare the teeth (Figure 32.1B). The onset is usually insidious, with a slow progression of muscle weakness involving the scapular fixators (latissimus dorsi, lower trapezius, rhomboids, and serratus anterior), humeral, truncal, and lower extremities.

Weakness of scapular fixators together with the atrophic sternocostal head of the pectoralis major muscles results in an upward slope of the axillary folds. The deltoid muscles are relatively spared early on in the disease; even as the disease progresses, they are less affected compared to the other shoulder girdle muscles. This gives the "humeral" appearance seen in FSHD. Patients typically report difficulty in raising their arms. When arm

B.T. Darras, H. Royden Jones, Jr., M.M. Ryan & D.C. De Vivo (Eds): Neuromuscular Disorders of Infancy, Childhood and Adolescence, Second edition.
DOI: http://dx.doi.org/10.1016/B978-0-12-417044-5.00032-9

BOX 32.1 Diagnostic Criteria for Facioscapulohumeral Dystrophy

Inclusion

Weakness of face or scapular stabilizers (in familial cases, facial weakness is present in >90% of affected individuals)
Scapular stabilizer weakness greater than hip-girdle weakness (applicable in mild to moderate cases)
Autosomal dominant inheritance in familial cases

Exclusion

Extraocular or pharyngeal muscle weakness
Prominent and diffuse elbow contractures
Cardiomyopathy
Distal symmetrical sensory loss
Dermatomyositic rash or signs of an alternative diagnosis
Electromyographic evidence of myotonia or neurogenic potentials

Supportive Features

Asymmetry of muscle weakness
Descending sequence of involvement
Early, often partial, abdominal muscle weakness (positive Beevor's sign)
Sparing of deltoid muscles
Typical shoulder profile: straight clavicles, forward sloping of shoulders
Relative sparing of neck flexors
Selective weakness of wrist extensors in distal upper extremities
Sparing of calf muscles
High-frequency hearing loss
Retinal vasculopathy

Source: Facioscapulohumeral Consortium at the International Conference on the Cause and Treatment of Facioscapulohumeral Dystrophy, Boston, 1997.

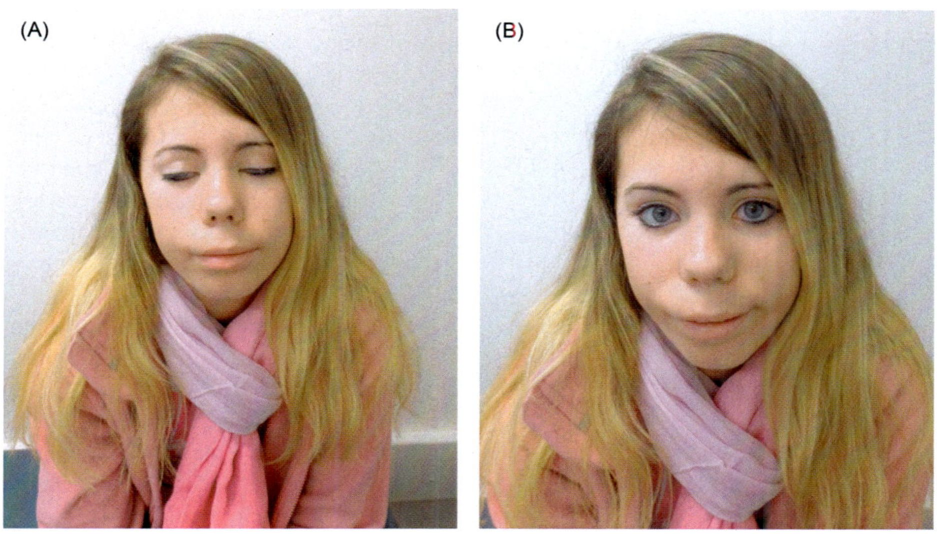

FIGURE 32.1 A patient with FSHD. (**A**) illustrates the inability to bury the eyelashes upon eye closure and (**B**) demonstrates the overt facial weakness, horizontal smile, and the dimpling on either side of the face upon attempting to smile.

abduction is attempted, the scapulae ride upwards and over the trapezius, resulting in a hypertrophic appearance (Figure 32.2). Scapular winging is a common feature. In the early stages, scapular winging can be made more apparent with the abduction and forward movement of the arms (Figure 32.3). Patients can be asked to push against the wall, which activates the serratus anterior and brings out the scapular winging. While the supraspinatus and infraspinatus muscles are still preserved early on in the disease process, patients have little difficulty abducting their arms up to approximately 25 degrees, as this movement utilizes the supraspinatus muscle. Mechanically fixing the scapula in place allows the spared deltoid muscles to complete the remaining abduction movement to the horizontal position. Abduction of the arm beyond that requires rotation and fixation of the scapula to the chest wall.

Truncal and abdominal wall weakness results in the lumbar lordosis and protuberant abdomen commonly seen in patients with FSHD. The upper abdominal wall muscles, which are usually stronger than those of the lower abdominal wall, cause an upward movement of the umbilicus during neck flexion in a patient lying in a supine position. This is called the Beevor's sign and is fairly specific for

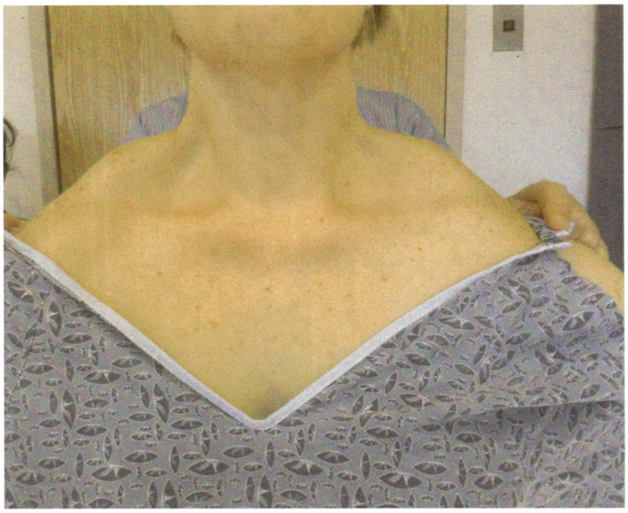

FIGURE 32.2 This figure illustrates the hypertrophic appearance of the trapezius, which becomes apparent when shoulder abduction is attempted.

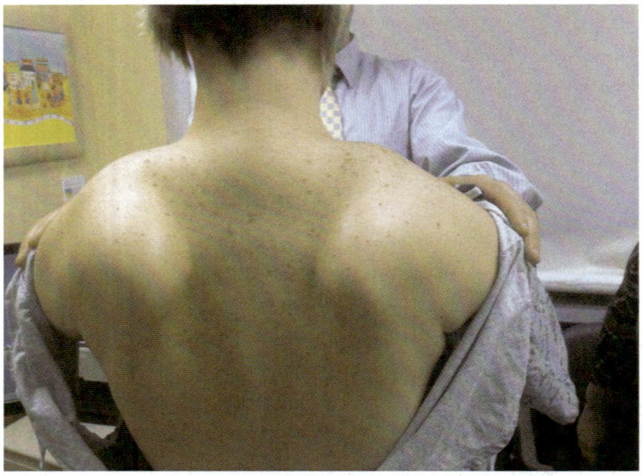

FIGURE 32.3 Scapular winging is seen in this patient with FSHD, with the scapulae rising upwards and laterally, and jutting out posteriorly of the inferomedial corners.

FSHD.[17] In a prospective case-control study, the Beevor's sign was found in 90% of FSHD patients.[17]

The peroneal muscles are often affected in FSHD, resulting in foot drop. Mild tendon contractures and pseudohypertrophy of the muscles, particularly the extensor digitorum brevis, may be present. In contrast, severe contractures are usually not found. In advanced stages of the disease, the hip girdle muscles may also be affected, in some cases more so than the shoulder girdle muscles. For patients presenting only in this late stage of the disease, it may be difficult to make a clinical distinction between FSHD and limb girdle muscular dystrophy.[18]

EXTRAMUSCULAR MANIFESTATIONS

Involvement of other systems is reported in FSHD. These include hearing loss, retinal disease, cardiac arrhythmia, cognitive impairment, and seizures. Sensorineural hearing loss is described in both infantile and typical FSHD patients. Screening audiometry is abnormal in up to 75% of affected individuals,[19,20] and initially presents as high-frequency hearing loss with progression to involvement of all frequencies. In some patients, FSHD-associated hearing loss can remain stable over long periods of time.[21] Symptomatic hearing loss, however, is seen almost exclusively in patients with large D4Z4 deletions.

Minimal retinal vascular abnormalities are seen in 50% to 75% of patients.[22] Capillary telangiectasia, microaneurysms, capillary closure, and protein exudation have been described. In rare cases, retinal detachment and visual loss (Coat's syndrome) can occur.[23] In a study performed by Fitzsimons et al., 56 out of 75 patients with a molecular diagnosis of FSHD had evidence of retinal capillary abnormalities as demonstrated by fluorescein angiography.[22] In some of these patients, the retinal vascular disease seems to have developed early, even before any overt muscle weakness was present. Photocoagulation of capillary telangiectasia, if done early, may be effective in preventing progression to Coats' syndrome. There does not appear to be any clear association between the extent of the retinal vasculopathy and the severity of the myopathy.[22] As in hearing loss, retinal vasculopathy is largely asymptomatic except in patients with very large D4Z4 deletions.

Minimally symptomatic or asymptomatic cardiac arrhythmias, specifically atrial arrhythmias, are reported in approximately 5% of patients.[24–28] Generally, symptomatic respiratory compromise in FSHD is rare, with evidence of restrictive lung disease on pulmonary function testing in about 10% of patients.[29] In a study of FSHD patients in the Dutch population, respiratory insufficiency requiring ventilatory support was seen in 1% of their patients.[30] These patients were typically wheelchair dependent and had severe scoliosis and lumbar lordosis. Cognitive impairment and seizures are rarely reported in early onset FSHD.[31,32]

INFANTILE FSHD

The infantile form of FSHD was first described by Brooke in 1977.[33] This is a more severe and rapidly progressive form of FSHD seen in approximately 4% of the total FSHD population.[31,33] Infantile FSHD is defined as the presence of signs or symptoms of facial weakness before 5 years of age, or the development of signs or symptoms of shoulder girdle weakness before 10 years of age.[31] Symptomatic sensorineural hearing loss and retinal vasculopathy is almost exclusively associated with infantile FSHD or patients with large D4Z4 deletions. In addition, there seems to be a correlation between clinical severity and the very short D4Z4 repeat array in chromosome 4q35.[34] Children with a very small number of chromosome 4q35 repeats also tend to have seizures, cognitive disabilities, and severe sensorineural hearing loss.[35,36] Similarly to adult patients with FSHD, cardiac and respiratory complications are rare.

Inheritance is often sporadic in nature. Although a positive family history is rare, infantile FSHD patients with a *de novo* mutation frequently have an unaffected carrier parent with somatic mosaicism of the mutation. Klinge et al. have reported a cohort of 7 patients who satisfy the clinical criteria of infantile FSHD.[33] Age of onset was between infancy and 4.5 years of age. Although gross motor milestones were appropriate in this cohort, the authors reported early development of facial weakness, with rapid progression to involve the shoulder and hip girdle muscles. There is usually severe hyperlordosis of the lumbar spine (Figure 32.4), exaggerated forward pelvic tilt, and hyperextension of the knees on walking. Most patients lose ambulation and become wheelchair dependent before their teens. While there is significant heterogeneity in the FSHD phenotype, there is less variability in the infantile form. However, there have been rare case reports describing children with mild limb weakness who remained ambulant despite severe facial weakness.[37,38]

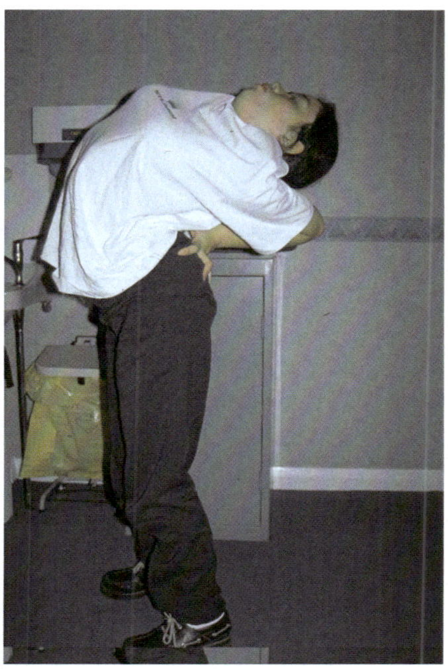

FIGURE 32.4 Severe hyperlordosis seen in infantile FSHD. *Reprinted from Klinge et al., 2006[33] with permission from Elsevier.*

In infantile FSHD, disease progression may not follow the classical adult FSHD phenotype of gradual descending asymmetric muscle weakness. In contrast, weakness in the pelvic girdle has been reported to precede shoulder girdle weakness, with both symptoms occurring within 1.5 to 2 years. Development of muscle weakness also appeared to be symmetrical.

FSHD VARIANTS

Linkage to the FSHD 4q35 locus has been established in patients presenting with phenotypes other than the classic constellation of signs typical of FSHD (including facial weakness and asymmetric involvement of the shoulder, hip girdle, and distal muscles). No clear associations have been made between these atypical phenotypes and the size of the deletion on chromosome 4q35.[39]

SCAPULOHUMERAL DYSTROPHY PHENOTYPE WITH FACIAL SPARING

In a retrospective study by Felice et al. conducted over a 7-year period, 15% of genetically confirmed FSHD patients were found to have the absence of facial weakness. These patients otherwise had the characteristic features of FSHD, such as limb girdle weakness and scapular winging.[40,41] Other phenotypes that have been reported included asymmetric, localized limb weakness of the shoulder, thigh, calf, or foot.[39,42–44]

DIFFERENTIAL DIAGNOSIS

The clinical distinction between FSHD and several neuromuscular conditions can sometimes be difficult to make as the clinical distribution of muscle weakness can also be seen in scapuloperoneal syndromes, limb girdle muscular dystrophy, polymyositis,[45] mitochondrial myopathies,[46] congenital myopathies,[47] inclusion body myositis, and autosomal recessive inclusion body myopathy, as well as juvenile and adult forms of acid maltase deficiency. Definitive diagnosis in these instances will require further investigations such as muscle biopsy or molecular testing.

LABORATORY TESTS

Serum creatine kinase (CK) may be elevated up to five times the upper limit of normal in symptomatic individuals. In asymptomatic patients, serum CK is usually normal. The electromyogram typically demonstrates a generalized myopathic picture with polyphasic motor units of short duration and low amplitude. Occasionally, fibrillations and positive sharp waves may also be seen. Motor and sensory nerve conduction studies are usually normal. Muscle biopsy usually shows nonspecific myopathic changes such as variability in fiber size. Inflammatory cellular infiltrates can be seen in up to 40% of FSHD patients and may be extensive in some. Definitive diagnosis is made with molecular genetic testing.

MANAGEMENT

To date, there are no effective treatments or interventions known to slow the progression of FSHD or reverse muscle weakness in patients. Management in this group of patients is mainly supportive. The mainstay of management involves physical therapy, pain control, and orthopedic interventions, as well as surveillance and interventions targeting FSHD-specific complications. While mildly symptomatic patients may require infrequent follow-up and minimal interventions, patients with infantile FSHD will often benefit from close monitoring in a multidisciplinary care clinic, with input from various specialties.

Physical Therapy and Rehabilitation

Patients will benefit from a rehabilitation consult addressing their functional limitations, as well as an assessment of their gait, posture, and need for assistive devices or orthoses. An individualized exercise plan including stretching, aerobics, and resistive training can then be recommended based on each patient's needs. Current evidence has shown that regular aerobic exercise for 30 minutes at least three times per week can improve cardiovascular fitness and strength in FSHD patients.[48–50] In patients who are unable to participate in aerobic exercises, moderate resistance training can be used instead.

Orthopedic/Surgical Interventions

One common and major functional limitation faced by patients with FSHD is the inability to elevate the arm to or above shoulder level, which often impacts activities of daily living such as combing one's hair or reaching for high objects. Scapulodesis, the fixation of scapula to the posterior part of the thorax, has been shown to improve upper arm mobility and range of motion.[51–54] This intervention can also improve the appearance of the shoulder in patients with significant scapular winging. Patients should have reasonable upper arm strength at baseline before this procedure is considered. Potential surgical complications include breakage of the hardware used to fix the scapula and, rarely, brachial plexus injuries.[55] These complications may inadvertently result in pain and a loss of function.

Assistive devices such as ankle-foot orthoses may be helpful in patients with foot drop.

Pain

Pain and fatigue is a common complaint in FSHD patients.[56–58] The etiologies appear to be multifactorial in origin and should be addressed appropriately. A combination of approaches using physical therapy, analgesic medications, and energy conservation strategies can be used in the management of pain and fatigue.[50,59]

Respiratory Function

Although significant respiratory insufficiency is rare and occurs in less than 1% of FSHD patients, all patients with moderate to severe FSHD should undergo routine screening for symptoms of nocturnal hypoventilation.[60] Recommendations made by an expert panel include yearly forced vital capacity (FVC) for all patients who are wheelchair dependent, have pelvic girdle weakness and superimposed pulmonary disease, and have significant spinal or chest wall deformities. Surveillance with a supine and sitting FVC is also recommended for any patient prior to any surgical intervention that requires general anesthesia or conscious sedation.

Retinal Vascular Disease and Hearing Loss

All patients should be seen by an ophthalmologist at the time of diagnosis to assess for the presence of retinal vascular disease. In the absence of retinal disease, no further ophthalmological follow-up is warranted unless the patient develops visual symptoms. In children with infantile onset FSHD, annual follow-up is still recommended until the child is able to verbalize and report visual symptoms. More frequent ophthalmological surveillance has been suggested for patients with D4Z4 repeat array fragments of <15 kb in size.[61]

Sensorineural hearing loss has been frequently associated with infantile FSHD patients, and if left undetected, may result in delay of speech and language development; therefore, preschool children should be routinely screened with hearing tests.[21] For children of school-going age, an audiogram is not warranted if they have had normal hearing screens in school and if there are no ongoing concerns regarding their speech and language development. Asymptomatic FSHD adults do not need routine hearing assessments.[60]

Surveillance during Pregnancy

There is limited data regarding pregnancy outcomes in patients with FSHD. In general, patients have good pregnancy outcomes, although there have been a few reports demonstrating an increased incidence of operative and preterm deliveries.[62,63] There may also be persistent worsening of motor function in approximately 25% of pregnant women. Based on this evidence, the following recommendations have been made: (1) pregnant women with FSHD should plan their delivery in a center that can provide intensive perinatal care, and (2) evaluation by a pulmonologist may be warranted, especially in those who have impaired baseline respiratory function.

PROGNOSIS

FSHD is typically a slowly progressive neuromuscular disorder, often resulting in a gradual decline of muscle strength and motor function. Although there may be some stagnation of clinical progression in some patients, there is usually no reversibility of the muscle weakness. In a natural history study of FSHD patients, patients experience a 1% to 4% per year decline in composite strength scores, with either no change or a slow progression in disease using upper and lower extremity functional severity scales.[64,65] One study showed that 10% to 20% of patients require a wheelchair at some stage.[66] Clinical severity appears to correlate with smaller number of D4Z4 repeats[15,67] and with an early onset and longer disease duration.[64] (See Case Example 32.1.) A retrospective analysis of 313 symptomatic patients with FSHD confirmed these findings.[68] The authors found that patients with small D4Z4 repeat arrays had the greatest risk for early wheelchair use in their first three decades, compared to a gradual age-dependent increase in risk observed in the other patients.[68] The prevalence and risk of other functional impairments such as difficulties in word pronunciation, using arms for activities of daily living, being on disability, or having modifications at work, appear to be higher in patients with a larger contraction, although this was not statistically significant. Another study looking at the quality of life in FSHD using a health questionnaire found correlations between the clinical severity of disease and its impact on quality of life.[69] Lifespan in FSHD patients is generally unaffected.

Over the last decade, there have been increasing advances in the understanding of the pathophysiology of FSHD. There is hope that these advances may lead to the identification of possible therapeutic targets in the coming years. In addition, efforts are being made in the establishment of patient registries, together with validated biomarkers of disease severity and progression and clinical outcome measures, to facilitate patient enrollment into future clinical trials as well as to reliably measure therapeutic responses.[70]

GENETICS

There are two forms of FSHD that can be recognized by DNA analysis, which can be performed on peripheral blood mononuclear cells (Figure 32.5). The common form of FSHD1 is linked to a genetic lesion in the tip of the long arm of chromosome 4.[71,72] This 4q subtelomere contains a polymorphic repeat array, D4Z4, which consists of 11 to 100 repeat units 3.3 kb in size.[4,73] Patients with FSHD1 have an array of 1 to 10 D4Z4 units, which is considered diagnostic for the disease.[4,5] At least one D4Z4 unit seems necessary for the development of FSHD since monosomy of 4q subtelomere is not associated with FSHD.[74] There is a rough and inverse relationship between the onset and clinical severity, and the size of the pathogenic array. Individuals with 1 to 3 D4Z4 units are typically at the severe end of the disease spectrum and are usually sporadic cases. In families typically carrying arrays of 4 to 10 units, the clinical variation can range from asymptomatic to severely affected and wheelchair dependent.[4,14,15,75] Only about half of the contractions of the D4Z4 repeat array to a size of 1 to 10 units are pathogenic as the array needs to reside on an FSHD-permissive genetic background of chromosome 4 that contains a polymorphic polyadenylation signal for *DUX4*, a gene embedded in each D4Z4 unit and encoding a germline transcription factor.[76–79]

There is a copy of the D4Z4 repeat array in the subtelomere of chromosome 10q but, with few exceptions, contractions of this array does not cause FSHD because of the absence of a *DUX4* polyadenylation signal on chromosome 10q.[79–81] In somatic cells, such as skeletal muscle, the D4Z4 repeat array is epigenetically repressed, leading to the silencing of the *DUX4* gene. Upon repeat

CASE EXAMPLE 32.1

Brouwer et al. described a 17-year-old boy with early-onset FSHD.[32] He reportedly had always slept with his eyes open. At age 3 years, he became hyperactive, was clumsy, and had speech delay. At age 7 years, facial and shoulder girdle weakness, scapular winging, increased lumbar lordosis, and waddling gait were noted.

Serum creatine kinase (CK) was mildly raised at 683 U/L (normal, <100 U/L). Electromyogram was normal. Muscle biopsy demonstrated myopathic features. Audiometry demonstrated right-sided hearing loss of 60 dB at 4000 Hz.

His condition progressed, and at age 12 years he had difficulty standing up from a chair. He also developed bilateral foot drop.

His parents and elder sister had no symptoms or signs of FSHD. DNA analysis with probe p13E-11 showed a small EcoRI fragment of 13.5 kb. This small fragment was not found in other family members.

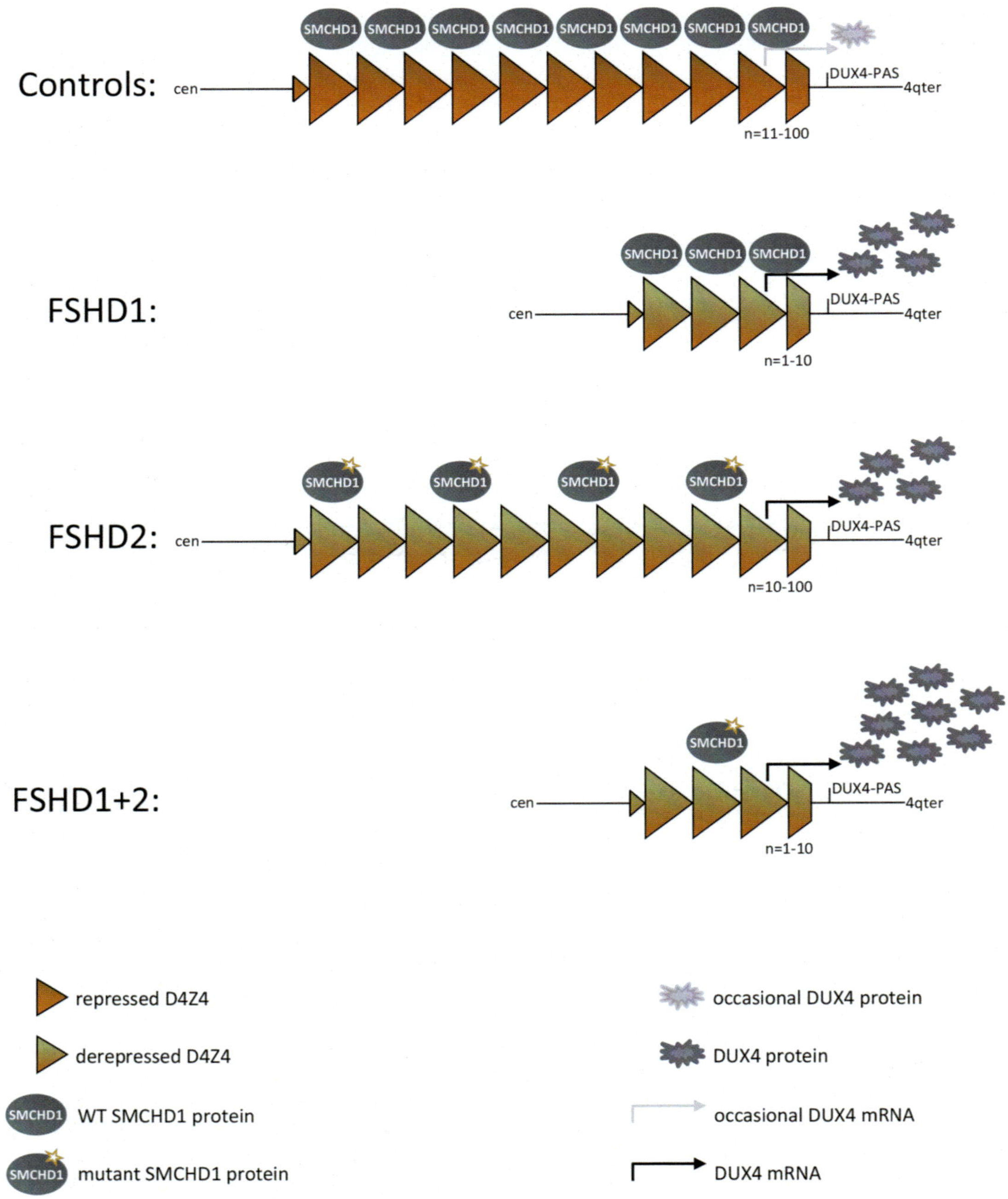

FIGURE 32.5 FSHD is characterized by a greater likelihood of expression of the germline transcription factor DUX4 in skeletal muscle. Normally, the D4Z4 repeat array contains 11–100 units of D4Z4, each encoding a copy of the *DUX4* gene. These long D4Z4 arrays have a repressive chromatin structure and only on FSHD-permissive chromosomes that contain a *DUX4* polyadenylation signal (DUX4-PAS), sporadically, traces of DUX4 expression can be observed. In the common FSHD1 form, a contraction of the D4Z4 repeat array to a size of 1–10 units results in a less repressive chromatin structure and a greater likelihood of DUX4 expression in skeletal muscle. SMCHD1 is a chromatin modifier involved in the establishment and maintenance of a repressed D4Z4 chromatin structure in somatic cells. Mutations in *SMCHD1* in FSHD2 also lead to a less repressive D4Z4 chromatin structure and an increase in DUX4 expression. In some families, FSHD1 and FSHD2 can cosegregate and the combination of both genetic defects leads to greater levels of DUX4 expression and a more severe phenotype.

contraction, this epigenetic silencing is partly released, causing the derepression of the *DUX4* gene that, in combination with the presence of a *DUX4* polyadenylation signal, results in a greater likelihood of transcription of stable DUX4 mRNA and the molecular hallmark of the disease of sporadic myonuclei expressing abundant amounts of DUX4 protein.[6,79,82]

The second form of FSHD that can be genetically confirmed is FSHD2. Patients with FSHD2 are clinically indistinguishable from those with FSHD1 but do not have a chromosome 4 with a contracted D4Z4 repeat array.[7–9] However, while the epigenetic changes of the D4Z4 repeat array in FSHD1 patients are confined to the contracted array, similar epigenetic changes in FSHD2 patients are much more widespread and can be observed on all four normal-sized arrays on chromosomes 4 and 10. The partial release of epigenetic repression of D4Z4 and derepression of *DUX4* in a majority of cases of FSHD2 is caused by heterozygous mutations in the structural maintenance of chromosomes flexible hinge domain containing 1 (*SMCHD1*) gene on chromosome 18.[83] Thus, FSHD2 is caused by the digenic inheritance of a chromosome 4 containing a *DUX4* polyadenylation signal and a mutation in *SMCHD1*. SMCHD1 is a chromatin modifier that directly binds to the D4Z4 repeat to maintain a repressed chromatin state in somatic cells, and its reduced activity in FSHD2 patients leads to the derepression of *DUX4* transcription and, similarly to FSHD1 patients, the molecular hallmark of this disease with sporadic myonuclei expressing DUX4 protein.[6,83] SMCHD1 is not only causal to FSHD2 but is also a modifier of disease severity in FSHD1 families in whom a combination of an upper FSHD1-sized D4Z4 repeat array of 9 to 10 units and a mutation in *SMCHD1* causes a more severe phenotype and enhanced levels of DUX4.[84]

DUX4 is a germline transcription factor with a very restricted expression pattern. It is expressed in the germline and early stem cells but gets silenced in somatic cells. Its ectopic expression in FSHD muscle leads to the activation of germline and early stem cell programs, the activation of specific classes of retrotransposons, and suppression of the innate immune system.[6,85,86]

Although over recent years several lines of support have been presented for this uniform molecular mechanism of D4Z4 derepression and ectopic DUX4 expression that unifies FSHD1 and FSHD2, the epigenetic changes of the 4q subtelomere and the consequences thereof are complex and not fully understood. It cannot be excluded that, in addition to *DUX4*, the regulation of other genes located in the vicinity of the D4Z4 repeat, such as *FRG1*, *FRG2*, and *FAT1*, are also contributing to the pathology. Moreover, not all FSHD2 patients can be explained by mutations in *SMCHD1*, suggesting that other genes are involved.[87–89]

REFERENCES

1. Padberg GW, Frants RR, Brouwer OF, Wijmenga C, Bakker E, Sandkuijl LA. Facioscapulohumeral muscular dystrophy in the Dutch population. *Muscle Nerve Suppl* 1995;**2**:S81–4.
2. Landouzy L, Dejerine J. De la myopathie atrophique progressive (myopathie héréditaire, débutant dans l'enfance par la face, sans altération du systéme nerveux). *C R Acad Sci Gen (Paris)* 1884;**98**:53–5.
3. Tyler FH, Stephens FE. Studies in disorders of muscle. II. Clinical manifestations and inheritance of facioscapulohumeral dystrophy in a large family. *Ann Intern Med* 1950;**32**:640–60.
4. van Deutekom JC, Wijmenga C, van Tienhoven EA, Gruter AM, Hewitt JE, Padberg GW, et al. FSHD associated DNA rearrangements are due to deletions of integral copies of a 3.2 kb tandemly repeated unit. *Hum Mol Genet* 1993;**2**:2037–42.
5. Lunt PW. 44th ENMC International Workshop: Facioscapulohumeral Muscular Dystrophy: Molecular Studies. 19–21 July 1996, Naarden, The Netherlands. *Neuromuscul Disord* 1998;**8**:126–30.
6. Snider L, Geng LN, Lemmers RJ, Kyba M, Ware CB, Nelson AM, et al. Facioscapulohumeral dystrophy: incomplete suppression of a retrotransposed gene. *PLoS Genet* 2010;**6**:e1001181.
7. de Greef JC, Lemmers RJ, Camano P, Day JW, Sacconi S, Dunand M, et al. Clinical features of facioscapulohumeral muscular dystrophy 2. *Neurology* 2010;**75**:1548–54.
8. de Greef JC, Lemmers RJ, van Engelen BG, Sacconi S, Venance SL, Frants RR, et al. Common epigenetic changes of D4Z4 in contraction-dependent and contraction-independent FSHD. *Hum Mutat* 2009;**30**:1449–59.
9. van Overveld PG, Lemmers RJ, Sandkuijl LA, Enthoven L, Winokur ST, Bakels F, et al. Hypomethylation of D4Z4 in 4q-linked and non-4q-linked facioscapulohumeral muscular dystrophy. *Nat Genet* 2003;**35**:315–17.
10. Lunt PW, Compston DA, Harper PS. Estimation of age dependent penetrance in facioscapulohumeral muscular dystrophy by minimising ascertainment bias. *J Med Genet* 1989;**26**:755–60.
11. Padberg GW, Lunt PW, Koch M, Fardeau M. Diagnostic criteria for facioscapulohumeral muscular dystrophy. *Neuromuscul Disord* 1991;**1**:231–4.
12. Lemmers RJ, O'Shea S, Padberg GW, Lunt PW, van der Maarel SM. Best practice guidelines on genetic diagnostics of Facioscapulohumeral muscular dystrophy: Workshop 9th June 2010, LUMC, Leiden, The Netherlands. *Neuromuscul Disord* 2012;**22**:463–70.
13. Zatz M, Marie SK, Cerqueira A, Vainzof M, Pavanello RC, Passos-Bueno MR. The facioscapulohumeral muscular dystrophy (FSHD1) gene affects males more severely and more frequently than females. *Am J Med Genet* 1998;**77**:155–61.
14. Tawil R, Forrester J, Griggs RC, Mendell J, Kissel J, McDermott M, et al. Evidence for anticipation and association of deletion size with severity in facioscapulohumeral muscular dystrophy. The FSH-DY Group. *Ann Neurol* 1996;**39**:744–8.
15. Lunt PW, Jardine PE, Koch MC, Maynard J, Osborn M, Williams M, et al. Correlation between fragment size at D4F104S1 and age at onset or at wheelchair use, with a possible generational effect, accounts for much phenotypic variation in 4q35-facioscapulohumeral muscular dystrophy (FSHD). *Hum Mol Genet* 1995;**4**:951–8.
16. Zatz M, Marie SK, Passos-Bueno MR, Vainzof M, Campiotto S, Cerqueira A, et al. High proportion of new mutations and possible

anticipation in Brazilian facioscapulohumeral muscular dystrophy families. *Am J Hum Genet* 1995;**56**:99–105.
17. Awerbuch GI, Nigro MA, Wishnow R. Beevor's sign and facioscapulohumeral dystrophy. *Arch Neurol* 1990;**47**:1208–9.
18. Tawil R, Van Der Maarel SM. Facioscapulohumeral muscular dystrophy. *Muscle Nerve* 2006;**34**:1–15.
19. Brouwer OF, Padberg GW, Ruys CJ, Brand R, de Laat JA, Grote JJ. Hearing loss in facioscapulohumeral muscular dystrophy. *Neurology* 1991;**41**:1878–81.
20. Padberg GW, Brouwer OF, de Keizer RJ, Dijkman G, Wijmenga C, Grote JJ, et al. On the significance of retinal vascular disease and hearing loss in facioscapulohumeral muscular dystrophy. *Muscle Nerve Suppl* 1995;**2**:S73–80.
21. Darras BT, Tawil R. Predicting hearing loss in facioscapulohumeral muscular dystrophy. *Neurology* 2013;**81**:1370–1.
22. Fitzsimons RB, Gurwin EB, Bird AC. Retinal vascular abnormalities in facioscapulohumeral muscular dystrophy. A general association with genetic and therapeutic implications. *Brain* 1987;**110**(Pt 3):631–48.
23. Desai UR, Sabates FN. Long-term follow-up of facioscapulohumeral muscular dystrophy and Coats' disease. *Am J Ophthalmol* 1990;**110**:568–9.
24. Stevenson WG, Perloff JK, Weiss JN, Anderson TL. Facioscapulohumeral muscular dystrophy: evidence for selective, genetic electrophysiologic cardiac involvement. *J Am Coll Cardiol* 1990;**15**:292–9.
25. Marca GD, Frusciante R, Scatena M, Dittoni S, Testani E, Vollono C, et al. Heart rate variability in facioscapulohumeral muscular dystrophy. *Funct Neurol* 2010;**25**:211–16.
26. Trevisan CP, Pastorello E, Armani M, Angelini C, Nante G, Tomelleri G, et al. Facioscapulohumeral muscular dystrophy and occurrence of heart arrhythmia. *Eur Neurol* 2006;**56**:1–5.
27. Topaloglu H, Celiker A, Ozme S, Gogus S, Gucuyener K, Erdinc OO, et al. Facioscapulohumeral syndrome with cardiomyopathy. *Postgrad Med J* 1990;**66**:1088–9.
28. Laforet P, de Toma C, Eymard B, Becane HM, Jeanpierre M, Fardeau M, et al. Cardiac involvement in genetically confirmed facioscapulohumeral muscular dystrophy. *Neurology* 1998;**51**:1454–6.
29. Scully MA, Eichinger KJ, Donlin-Smith CM, Tawil R, Statland JM. Restrictive lung involvement in facioscapulohumeral muscular dystrophy. *Muscle Nerve Epub* 2014. Available from: http://dx.doi.org/10.1002/mus.24218.
30. Wohlgemuth M, van der Kooi EL, van Kesteren RG, van der Maarel SM, Padberg GW. Ventilatory support in facioscapulohumeral muscular dystrophy. *Neurology* 2004;**63**:176–8.
31. Brouwer OF, Padberg GW, Wijmenga C, Frants RR. Facioscapulohumeral muscular dystrophy in early childhood. *Arch Neurol* 1994;**51**:387–94.
32. Brouwer OF, Padberg GW, Bakker E, Wijmenga C, Frants RR. Early onset facioscapulohumeral muscular dystrophy. *Muscle Nerve Suppl* 1995;**2**:S67–72.
33. Klinge L, Eagle M, Haggerty ID, Roberts CE, Straub V, Bushby KM. Severe phenotype in infantile facioscapulohumeral muscular dystrophy. *Neuromuscul Disord* 2006;**16**:553–8.
34. Chen TH, Lai YH, Lee PL, Hsu JH, Goto K, Hayashi YK, et al. Infantile facioscapulohumeral muscular dystrophy revisited: Expansion of clinical phenotypes in patients with a very short EcoRI fragment. *Neuromuscul Disord* 2013;**23**:298–305.
35. Bindoff LA, Mjellem N, Sommerfelt K, Krossnes BK, Roberts F, Krohn J, et al. Severe fascioscapulohumeral muscular dystrophy presenting with Coats' disease and mental retardation. *Neuromuscul Disord* 2006;**16**:559–63.
36. Funakoshi M, Goto K, Arahata K. Epilepsy and mental retardation in a subset of early onset 4q35-facioscapulohumeral muscular dystrophy. *Neurology* 1998;**50**:1791–4.
37. Okinaga A, Matsuoka T, Umeda J, Yanagihara I, Inui K, Nagai T, et al. Early-onset facioscapulohumeral muscular dystrophy: two case reports. *Brain Dev* 1997;**19**:563–7.
38. Felice KJ, Jones JM, Conway SR. Facioscapulohumeral dystrophy presenting as infantile facial diplegia and late-onset limb-girdle myopathy in members of the same family. *Muscle Nerve* 2005;**32**:368–72.
39. Krasnianski M, Eger K, Neudecker S, Jakubiczka S, Zierz S. Atypical phenotypes in patients with facioscapulohumeral muscular dystrophy 4q35 deletion. *Arch Neurol* 2003;**60**:1421–5.
40. Felice KJ, North WA, Moore SA, Mathews KD. FSH dystrophy 4q35 deletion in patients presenting with facial-sparing scapular myopathy. *Neurology* 2000;**54**:1927–31.
41. Jardine PE, Upadhyaya M, Maynard J, Harper P, Lunt PW. A scapular onset muscular dystrophy without facial involvement: possible allelism with facioscapulohumeral muscular dystrophy. *Neuromuscul Disord* 1994;**4**:477–82.
42. van der Kooi AJ, Visser MC, Rosenberg N, van den Berg-Vos R, Wokke JH, Bakker E, et al. Extension of the clinical range of facioscapulohumeral dystrophy: report of six cases. *J Neurol Neurosurg Psychiatry* 2000;**69**:114–16.
43. Felice KJ, Moore SA. Unusual clinical presentations in patients harboring the facioscapulohumeral dystrophy 4q35 deletion. *Muscle Nerve* 2001;**24**:352–6.
44. Hassan A, Jones Jr. LK, Milone M, Kumar N. Focal and other unusual presentations of facioscapulohumeral muscular dystrophy. *Muscle Nerve* 2012;**46**:421–5.
45. Bates D, Stevens JC, Hudgson P. "Polymyositis" with involvement of facial and distal musculature. One form of the fascioscapulohumeral syndrome? *J Neurol Sci* 1973;**19**:105–8.
46. Slipetz DM, Aprille JR, Goodyer PR, Rozen R. Deficiency of complex III of the mitochondrial respiratory chain in a patient with facioscapulohumeral disease. *Am J Hum Genet* 1991;**48**:502–10.
47. Serratrice G, Pellissier JF, Bes A, Arne-Bes MC. [A case of centronuclear myopathy in adults with facio-scapulo-peroneal topography]. *Rev Neurol (Paris)* 1987;**143**:693–7.
48. Milner-Brown HS, Miller RG. Muscle strengthening through high-resistance weight training in patients with neuromuscular disorders. *Arch Phys Med Rehabil* 1988;**69**:14–19.
49. van der Kooi EL, Vogels OJ, van Asseldonk RJ, Lindeman E, Hendriks JC, Wohlgemuth M, et al. Strength training and albuterol in facioscapulohumeral muscular dystrophy. *Neurology* 2004;**63**:702–8.
50. Olsen DB, Orngreen MC, Vissing J. Aerobic training improves exercise performance in facioscapulohumeral muscular dystrophy. *Neurology* 2005;**64**:1064–6.
51. Mummery CJ, Copeland SA, Rose MR. Scapular fixation in muscular dystrophy. *Cochrane Database Syst Rev* 2003;**3**:CD003278.
52. Rhee YG, Ha JH. Long-term results of scapulothoracic arthrodesis of facioscapulohumeral muscular dystrophy. *J Shoulder Elbow Surg* 2006;**15**:445–50.

53. Giannini S, Faldini C, Pagkrati S, Grandi G, Digennaro V, Luciani D, et al. Fixation of winged scapula in facioscapulohumeral muscular dystrophy. *Clin Med Res* 2007;**5**:155–62.
54. Diab M, Darras BT, Shapiro F. Scapulothoracic fusion for facioscapulohumeral muscular dystrophy. *J Bone Joint Surg Am* 2005;**87**:2267–75.
55. Wolfe GI, Young PK, Nations SP, Burkhead WZ, McVey AL, Barohn RJ. Brachial plexopathy following thoracoscapular fusion in facioscapulohumeral muscular dystrophy. *Neurology* 2005;**64**:572–3.
56. Jensen MP, Hoffman AJ, Stoelb BL, Abresch RT, Carter GT, McDonald CM. Chronic pain in persons with myotonic dystrophy and facioscapulohumeral dystrophy. *Arch Phys Med Rehabil* 2008;**89**:320–8.
57. Bushby KM, Pollitt C, Johnson MA, Rogers MT, Chinnery PF. Muscle pain as a prominent feature of facioscapulohumeral muscular dystrophy (FSHD): four illustrative case reports. *Neuromuscul Disord* 1998;**8**:574–9.
58. Miro J, Gertz KJ, Carter GT, Jensen MP. Pain location and intensity impacts function in persons with myotonic type 1 and facioscapulohumeral dystrophy with chronic pain. *Muscle Nerve* 2014;**49**:900–5.
59. Smith AE, McMullen K, Jensen MP, Carter GT, Molton IR. Symptom burden in persons with myotonic and facioscapulohumeral muscular dystrophy. *Am J Phys Med Rehabil* 2014;**93**:387–95.
60. Tawil R, van der Maarel S, Padberg GW, van Engelen BG. 171st ENMC international workshop: Standards of care and management of facioscapulohumeral muscular dystrophy. *Neuromuscul Disord* 2010;**20**:471–5.
61. Statland JM, Sacconi S, Farmakidis C, Donlin-Smith CM, Chung M, Tawil R. Coats syndrome in facioscapulohumeral dystrophy type 1: frequency and D4Z4 contraction size. *Neurology* 2013;**80**:1247–50.
62. Ciafaloni E, Pressman EK, Loi AM, Smirnow AM, Guntrum DJ, Dilek N, et al. Pregnancy and birth outcomes in women with facioscapulohumeral muscular dystrophy. *Neurology* 2006;**67**:1887–9.
63. Rudnik-Schoneborn S, Glauner B, Rohrig D, Zerres K. Obstetric aspects in women with facioscapulohumeral muscular dystrophy, limb-girdle muscular dystrophy, and congenital myopathies. *Arch Neurol* 1997;**54**:888–94.
64. A prospective, quantitative study of the natural history of facioscapulohumeral muscular dystrophy (FSHD): implications for therapeutic trials. The FSH-DY Group. *Neurology* 1997;48:38–46.
65. Stubgen JP, Stipp A. Facioscapulohumeral muscular dystrophy: a prospective study of weakness and functional impairment. *J Neurol* 2010;**257**:1457–64.
66. Lunt PW, Jardine PE, Koch M, Maynard J, Osborn M, Williams M, et al. Phenotypic-genotypic correlation will assist genetic counseling in 4q35-facioscapulohumeral muscular dystrophy. *Muscle Nerve Suppl* 1995;**2**:S103–9.
67. Ricci E, Galluzzi G, Deidda G, Cacurri S, Colantoni L, Merico B, et al. Progress in the molecular diagnosis of facioscapulohumeral muscular dystrophy and correlation between the number of KpnI repeats at the 4q35 locus and clinical phenotype. *Ann Neurol* 1999;**45**:751–7.
68. Statland JM, Tawil R. Risk of functional impairment in Facioscapulohumeral muscular dystrophy. *Muscle Nerve* 2014;**49**:520–7.
69. Padua L, Aprile I, Frusciante R, Iannaccone E, Rossi M, Renna R, et al. Quality of life and pain in patients with facioscapulohumeral muscular dystrophy. *Muscle Nerve* 2009;**40**:200–5.
70. Tawil R, Shaw DW, van der Maarel SM, Tapscott SJ. Clinical trial preparedness in facioscapulohumeral dystrophy: outcome measures and patient access: 8–9 April 2013, Leiden, The Netherlands. *Neuromuscul Disord* 2014;**24**:79–85.
71. Wijmenga C, Brouwer OF, Padberg GW, Frants RR. Transmission of de-novo mutation associated with facioscapulohumeral muscular dystrophy. *Lancet* 1992;**340**:985–6.
72. Wijmenga C, Hewitt JE, Sandkuijl LA, Clark LN, Wright TJ, Dauwerse HG, et al. Chromosome 4q DNA rearrangements associated with facioscapulohumeral muscular dystrophy. *Nat Genet* 1992;**2**:26–30.
73. Hewitt JE, Lyle R, Clark LN, Valleley EM, Wright TJ, Wijmenga C, et al. Analysis of the tandem repeat locus D4Z4 associated with facioscapulohumeral muscular dystrophy. *Hum Mol Genet* 1994;**3**:1287–95.
74. Tupler R, Berardinelli A, Barbierato L, Frants R, Hewitt JE, Lanzi G, et al. Monosomy of distal 4q does not cause facioscapulohumeral muscular dystrophy. *J Med Genet* 1996;**33**:366–70.
75. Scionti I, Greco F, Ricci G, Govi M, Arashiro P, Vercelli L, et al. Large-scale population analysis challenges the current criteria for the molecular diagnosis of fascioscapulohumeral muscular dystrophy. *Am J Hum Genet* 2012;**90**:628–35.
76. Snider L, Asawachaicharn A, Tyler AE, Geng LN, Petek LM, Maves L, et al. RNA transcripts, miRNA-sized fragments and proteins produced from D4Z4 units: new candidates for the pathophysiology of facioscapulohumeral dystrophy. *Hum Mol Genet* 2009;**18**:2414–30.
77. Leidenroth A, Hewitt JE. A family history of DUX4: phylogenetic analysis of DUXA, B, C and Duxbl reveals the ancestral DUX gene. *BMC Evol Biol* 2010;**10**:364.
78. Dixit M, Ansseau E, Tassin A, Winokur S, Shi R, Qian H, et al. DUX4, a candidate gene of facioscapulohumeral muscular dystrophy, encodes a transcriptional activator of PITX1. *Proc Natl Acad Sci USA* 2007;**104**:18157–62.
79. Lemmers RJ, van der Vliet PJ, Klooster R, Sacconi S, Camano P, Dauwerse JG, et al. A unifying genetic model for facioscapulohumeral muscular dystrophy. *Science* 2010;**329**:1650–3.
80. Bakker E, Wijmenga C, Vossen RH, Padberg GW, Hewitt J, van der Wielen M, et al. The FSHD-linked locus D4F104S1 (p13E-11) on 4q35 has a homologue on 10qter. *Muscle Nerve Suppl* 1995;**2**:S39–44.
81. Deidda G, Cacurri S, Grisanti P, Vigneti E, Piazzo N, Felicetti L. Physical mapping evidence for a duplicated region on chromosome 10qter showing high homology with the facioscapulohumeral muscular dystrophy locus on chromosome 4qter. *Eur J Hum Genet* 1995;**3**:155–67.
82. Jones TI, Chen JC, Rahimov F, Homma S, Arashiro P, Beermann ML, et al. Facioscapulohumeral muscular dystrophy family studies of DUX4 expression: evidence for disease modifiers and a quantitative model of pathogenesis. *Hum Mol Genet* 2012;**21**:4419–30.
83. Lemmers RJ, Tawil R, Petek LM, Balog J, Block GJ, Santen GW, et al. Digenic inheritance of an SMCHD1 mutation and an FSHD-permissive D4Z4 allele causes facioscapulohumeral muscular dystrophy type 2. *Nat Genet* 2012;**44**:1370–4.

84. Sacconi S, Lemmers RJ, Balog J, van der Vliet PJ, Lahaut P, van Nieuwenhuizen MP, et al. The FSHD2 gene SMCHD1 is a modifier of disease severity in families affected by FSHD1. *Am J Hum Genet* 2013;**93**:744–51.

85. Geng LN, Yao Z, Snider L, Fong AP, Cech JN, Young JM, et al. DUX4 activates germline genes, retroelements, and immune mediators: implications for facioscapulohumeral dystrophy. *Dev Cell* 2012;**22**:38–51.

86. Young JM, Whiddon JL, Yao Z, Kasinathan B, Snider L, Geng LN, et al. DUX4 binding to retroelements creates promoters that are active in FSHD muscle and testis. *PLoS Genet* 2013;**9**: e1003947.

87. Gabellini D, Green MR, Tupler R. Inappropriate gene activation in FSHD: a repressor complex binds a chromosomal repeat deleted in dystrophic muscle. *Cell* 2002;**110**:339–48.

88. Caruso N, Herberth B, Bartoli M, Puppo F, Dumonceaux J, Zimmermann A, et al. Deregulation of the protocadherin gene FAT1 alters muscle shapes: implications for the pathogenesis of facioscapulohumeral dystrophy. *PLoS Genet* 2013;**9**:e1003550.

89. Rijkers T, Deidda G, van Koningsbruggen S, van Geel M, Lemmers RJ, van Deutekom JC, et al. FRG2, an FSHD candidate gene, is transcriptionally upregulated in differentiating primary myoblast cultures of FSHD patients. *J Med Genet* 2004;**41**: 826–36.

Chapter 33

Distal Myopathies

Bjarne Udd

EARLY ONSET DISTAL MYOPATHIES

Laing Distal Myopathy (LDM, MPD1)

Clinical Phenotype

Symptoms of ankle stiffness are usually observed in infancy before the child starts to walk. However, ankle dorsiflexion weakness may be very mild and remain unnoticed until later. Weakness is marked in ankle dorsiflexor, toe extensor, and neck flexor muscles. Later in the disease course finger extensors and proximal limb muscles are also affected, but most patients remain ambulatory throughout their lifetime. Disease severity ranges from the "hanging big toe" sign in the 30s to the more generalized weakness in childhood with scoliosis and ankle contractures. Cardiomyopathy may occur in a minority of patients and some cardiac monitoring is advised. LDM is an autosomal dominant disorder, but many patients appear in clinic as sporadic cases. Since the first report in 1995,[1] many families have been identified in different populations.[2–6]

Serum creatine kinase (CK) level is normal or mildly elevated (up to three-fold). Electromyography (EMG) shows short, brief myopathic potentials, and muscle biopsy reveals moderate myopathy with fiber type disproportion. Vacuoles are rare but cores and core-like changes are frequent. Muscle MRI is very informative and shows fatty degeneration consistently in anterior compartment muscles of the lower legs. Later in the disease course, medial gastrocnemius and proximal thigh muscles may be involved.

The exact prevalence is not known; however, in some well-studied smaller populations such as Western Australia, Finland, Norway, and Valencia, three to four different mutations in each have been identified, which suggests that the overall prevalence may be 1–5 per 1 million.

Molecular Genetics

The disease was first linked to chromosome 14q11[1] in an Australian family and later the causative mutations in *MYH7* were identified.[7] The gene encodes slow beta myosin heavy chain, which is the main myosin isoform in type 1 slow skeletal muscle fibers and in cardiac muscle fibers. Myosin molecules are hexameres and consist of two large heavy chain and four small light chain subunits. To date some 30 different MYH7 mutations have been identified. Many of them are recurring hotspot mutations and many are *de novo* mutations in the sporadic patients. The vast majority of mutations are located in the tail region of the MYH7 molecule between amino acids 1500 and 1855. The mutations are predicted to break the regularity of the heptad coiled-coil structure of the rod domain. Mutations in other parts of the protein may cause cardiomyopathy or hyaline body myopathy.[7]

Diagnostics

The key for diagnosis is the clinical and muscle imaging phenotype, although the disease is still confused with Charcot-Marie-Tooth. Lack of high amplitude motor unit potentials on EMG in the affected muscles, despite the chronic condition, should indicate a myopathic process. Final diagnosis requires molecular genetic verification of mutated MYH7.

Therapy

Ankle orthotic devices are useful in severe foot drop situations. Tendon release in the ankles and scoliosis surgery may be needed. For the disabled child, multidisciplinary rehabilitation measures should be provided.

B.T. Darras, H. Royden Jones, Jr., M.M. Ryan & D.C. De Vivo (Eds): Neuromuscular Disorders of Infancy, Childhood and Adolescence, Second edition.
DOI: http://dx.doi.org/10.1016/B978-0-12-417044-5.00033-0

Distal Nebulin Myopathy

Clinical Phenotype

Patients with a similar early-onset ankle dorsiflexion weakness but without a dominant family history should also be considered with respect to nebulin mutations. Anterior compartment muscles of the lower legs and, later, extensors of the fingers and hands are predominantly affected. Neck flexor and mild facial weakness may occur. The progression is very mild and adult patients do not have major disability. There are no respiratory difficulties, no dysphagia, and no cardiomyopathy with the disease.[8]

Muscle MRI shows selective fatty degeneration in the anterior compartment muscles of the lower legs, first in the tibialis anterior. EMG is myopathic and CK is normal or mildly elevated. Muscle biopsy in the affected muscles may be confusing with scattered and grouped atrophic fibers, easily interpreted as neurogenic pathology. The biopsy findings in proximal muscles are very mild with nonspecific increase of fiber size variation. Electron microscopy may or may not show a few small dense rods associated with Z-disks.

Molecular Genetics

Mutations in the nebulin gene are known to cause autosomal recessive congenital nemaline myopathy.[9] The distal phenotype without nemaline rod pathology on light microscopy is explained by the fact that nemaline myopathy patients have more disrupting mutations, and the distal phenotype without nemaline rods is caused by missense mutations on both chromosomes.[8] Sequencing of the large nebulin gene has not been available for diagnostic purposes, but the introduction of next generation sequencing capacity for diagnostic use will probably increase the number of patients identified with this disease.

Distal Myopathy Caused by Mutations in KLHL9

Clinical Phenotype

So far, only one dominant German family has been described. Symptoms started between ages 8 and 16 years with ankle dorsiflexion weakness, and patients developed ankle contractures.[10] In the disease course, finger weakness and intrinsic hand muscles were affected. Distal sensory loss was also reported by clinical examination, and progression was slow.

Serum CK levels were elevated (200–1400 U/L). Nerve conduction studies were usually normal, although a few increased distal motor latencies were recorded and sural sensory action potential was absent in one. EMG findings were also confusing with fibrillations and neurogenic motor unit potentials in distal muscles. MRI of muscles showed degenerative atrophic changes of the tibialis anterior, gastrocnemius, and soleus muscles with relative preservation of the tibialis posterior and peroneus longus muscles. In the thigh, the semimembranosus and biceps femoris muscles as well as the vastus intermedius muscle were more severely affected. The lateral and medial vastus muscles, gracilis, and sartorius were preserved. Gastrocnemius muscle biopsy showed fiber size variation, internal nuclei, and increased connective tissue. Biopsy of the sural nerve did not show axon loss or demyelination.

Molecular Genetics

A missense mutation p.L95F in the *KLHL9* gene encoding a kelch-like homolog 9 protein segregated with the disease phenotype.[10] KLHL9 is involved in a ubiquitin ligase complex with Cul3, and the mutation is thought to impair crucial interactions.

Distal Caveolinopathy

Dominant mutations in caveolin-3 cause different phenotypes: LGMD1C, hyperCKemia, rippling muscle disease, and a novel form of distal myopathy in a sporadic 12-year-old patient with hyperCKemia, and atrophy with moderate weakness of small muscles in hands and feet.[11] EMG was myopathic in hand muscles, and muscle biopsy showed mild myopathic changes. Caveolin-3 and dysferlin were reduced on immunohistochemistry with normal labeling of dystrophin, merosin, α-sarcoglycan, and β-dystroglycan. Caveolin-3 was absent on blots, whereas dysferlin was normal. A previously reported mutation, *R27Q*, was found in the patient.

JUVENILE ONSET DISTAL MYOPATHIES

Miyoshi Distal Myopathy

Clinical Phenotype

Symptoms of calf muscle weakness usually begin between 15 and 25 years of age, which means that a minority of patients will be referred to a pediatric neurologist. Exercise myalgia and swelling in the calves with high CK levels may occur many years earlier. At onset of weakness, calf muscles become atrophic and ankle reflexes are lost.[12] The anterior compartment muscles are usually spared at early stages and become involved later. Calf atrophy is symmetric although asymmetry has been reported, as well as early ankle dorsiflexion weakness.[13]

Proximal muscles in legs and arms get weak 5–10 years later, and the two phenotypes in dysferlinopathy, Miyoshi myopathy, and LGMD2B cannot be distinguished at the later stage.

Miyoshi myopathy shows very high CK levels, in the range of 20 to 150 times normal levels. EMG shows low myopathic motor unit potentials and early recruitment. Muscle biopsy shows fiber necrosis, and the severely affected gastrocnemius muscle may show "end-stage" changes with widespread fibrosis, fatty replacement, and loss of most muscle tissue. Inflammatory cell infiltrates are common.

Molecular Genetics

Recessive mutations in dysferlin are the cause of Miyoshi myopathy.[14,15] Diagnosis can be established by loss of protein on immunohistochemistry and Western blotting of muscle biopsy, subsequently finalized by molecular genetic mutation detection. Dysferlin normally localizes to the sarcolemma and is absent in patients with Miyoshi myopathy. Immunohistochemistry is not always straightforward and some cytoplasmic labeling may occur, whereas Western blotting is more reliable.

Dysferlin interacts with caveolin-3 but not with the dystrophin associated protein complex.[16] The protein contains C2 domains which bind calcium and trigger signal transduction involved in membrane trafficking events. Changes in the sarcolemma have been reported in muscle and a central role for dysferlin in membrane repair mechanisms has been reported.[17] The natural mouse model for dysferlin deficiency (SJL-Dysf) has spontaneous mutations in the fourth C2 domain.[18] This and other dysferlinopathy models develop muscular dystrophy and are useful in the development of therapeutic rescue strategies.

Distal Anoctaminopathy

Clinical Phenotype

Patients with a Miyoshi-like phenotype without dysferlinopathy may have the recently identified distal anoctaminopathy disease.[19,20] Common early symptoms are painful and sometimes large calf muscles in the late teens combined with high CK values, 40–100 times the upper normal limit. HyperCKemia as an incidental finding may occur even years before, which means that very high CK in an asymptomatic individual after age 10 could be caused by anoctaminopathy. After the early phase of myalgia, the calf muscles will be atrophic and frequently asymmetric. EMG is myopathic and muscle biopsy shows scattered necrotic fibers without generalized increase of MHC class1 expression.[19,20]

Molecular Genetics

Affected patients have recessive mutations in the *ANO-5* gene coding for a transmembrane anoctamin 5 protein with the structure of a chloride channel. The more common phenotype of anoctaminopathy is later onset proximal muscle weakness, LGMD2L, without clear phenotype-genotype correlations. Females tend to have a milder phenotype compared to males with similar mutations.[19,20] To date no clear understanding of the subcellular localization of the protein or protein changes in the patients has been established. A defect in membrane repair similar to dysferlinopathy has been proposed but not confirmed.[21]

REFERENCES

1. Laing NG, Laing BA, Meredith C, Wilton SD, Robbins P, Honeyman K, et al. Autosomal dominant distal myopathy: linkage to chromosome 14. *Am J Hum Genet* 1995;**56**:422–7.
2. Voit T, Kutz P, Leube B, Neuen-Jacob E, Schröder JM, Cavallotti D, et al. Autosomal dominant distal myopathy: further evidence of a chromosome 14 locus. *Neuromuscul Disord* 2001;**11**:11–19.
3. Zimprich F, Djamshidian A, Hainfellner JA, Budka H, Zeitlhofer J. An autosomal dominant early adult-onset distal muscular dystrophy. *Muscle Nerve* 2000;**23**:1876–9.
4. Lamont PJ, Udd B, Mastaglia FL, de Visser M Hedera P, Voit T, et al. Laing early onset distal myopathy: slow myosin defect with variable abnormalities on muscle biopsy. *J Neurol Neurosurg Psychiatry* 2006;**77**:208–15.
5. Muelas N, Hackman P, Luque H, Garcés-Sánchez M, Azorín I, Suominen T, et al. *MYH7* gene tail mutation causing myopathic profiles beyond Laing distal myopathy. *Neurology* 2010;**75**:732–41.
6. Dubourg O, Maisonobe T, Behin A, Suominen T, Raheem O, Penttilä S, et al. A novel MYH7 mutation occurring independently in French and Norwegian Laing distal myopathy families and de novo in one Finnish patient. *J Neurol* 2011;**258**:1157–63.
7. Meredith C, Herrmann R, Parry C, Liyanage K, Dye DE, Durling HJ, et al. Mutations in the slow skeletal muscle fiber myosin heavy chain gene (MYH7) cause Laing early-onset distal myopathy (MPD1). *Am J Hum Genet.* 2004;**75**:703–8.
8. Wallgren-Pettersson C, Lehtokari V-L, Kalimo H, Paetau A, Nuutinen E, Hackman P, et al. Distal myopathy caused by homozygous missense mutations in the nebulin gene. *Brain* 2007;**130**:1465–76.
9. Wallgren-Pettersson C, Pelin K, Nowak KJ, Muntoni F, Romero NB, Goebel HH, et al. Genotype-phenotype correlations in nemaline myopathy caused by mutations in the genes for nebulin and skeletal muscle alpha-actin. *Neuromuscul Disord* 2004;**14**:461–70.
10. Cirak S, von Deimling F, Sachdev S, Errington WJ, Herrmann R, Bönnemann C, et al. Kelch-like homologue 9 mutation is associated with an early onset autosomal dominant distal myopathy. *Brain* 2010;**133**:2123–35.
11. Tateyama M, Aoki M, Nishino I, Hayashi YK, Sekiguchi S, Shiga Y, et al. Mutation in the caveolin-3 gene causes a peculiar form of distal myopathy. *Neurology* 2002;**58**:323–5.

12. Miyoshi K, Kawai H, Iwasa M, Kusaka K, Nishino H. Autosomal recessive distal muscular dystrophy as a new type of progressive muscular dystrophy: Seventeen cases in eight families, including an autopsied case. *Brain* 1986;**109**:31–54.
13. Illa I, Serrano-Munuera C, Gallardo E, Lasa A, Rojas-García R, Palmer J, et al. Distal anterior compartment myopathy: a dysferlin mutation causing a new muscular dystrophy phenotype. *Ann Neurol* 2001;**49**:130–4.
14. Liu J, Aoki M, Illa I, Wu C, Fardeau M, Angelini C, et al. Dysferlin, a novel skeletal muscle gene, is mutated in Miyoshi myopathy and limb girdle muscular dystrophy. *Nat Genet* 1998;**20**:31–6.
15. Anderson LV, Davison K, Moss JA, Young C, Cullen MJ, Walsh J, et al. Dysferlin is a plasma membrane protein and is expressed early in human development. *Hum Mol Genet* 1999;**8**:855–61.
16. Matsuda C, Hayashi YK, Ogawa M, Aoki M, Murayama K, Nishino I, et al. The sarcolemmal proteins dysferlin and caveolin-3 interact in skeletal muscle. *Hum Mol Genet* 2001;**10**:1761–6.
17. Bansal D, Campbell KP. Dysferlin and the plasma membrane repair in muscular dystrophy. *Trends Cell Biol* 2004;**14**:206–13.
18. Bittner RE, Anderson LV, Burkhardt E, Bashir R, Vafiadaki E, Ivanova S, et al. Dysferlin deletion in SJL mice (SJL-Dysf) defines a natural model for limb girdle muscular dystrophy 2B. *Nat Genet* 1999;**23**:141–2.
19. Bolduc V, Marlow G, Boycott KM, Saleki K, Inoue H, Kroon J, et al. Recessive mutations in the putative calcium-activated chloride channel Anoctamin 5 cause proximal LGMD2L and distal MMD3 muscular dystrophies. *Am J Hum Genet* 2010;**86**: 213–21.
20. Penttila S, Palmio J, Suominen T, Raheem O, Evilä A, Muelas Gomez N, et al. Eight new mutations and the expanding phenotype variability in muscular dystrophy caused by ANO5. *Neurology* 2012;**78**:897–903.
21. Jaiswal JK, Marlow G, Summerill G, Mahjneh I, Mueller S, Hill M, et al. Patients with a non-dysferlin Miyoshi myopathy have a novel membrane repair defect. *Traffic* 2007;**8**:77–88.

Chapter 34

Limb-girdle Muscular Dystrophies

Payam Mohassel and Carsten G. Bönnemann

INTRODUCTION

Historical Background and Definition

The limb-girdle muscular dystrophies (LGMDs) are a diverse and heterogeneous group of disorders within the broader field of genetic muscle disease. The term LGMD was first formally introduced in the 1950s to include the category of a larger group of patients with muscle weakness that could not be recognized under the major muscular dystrophy groups identified at the time, such as X-linked Duchenne muscular dystrophy (DMD) or facioscapulohumeral dystrophy (FSHD). In a classic paper, Walton and Nattrass delineated the clinical phenotype: childhood onset of proximal weakness, slow progression, and autosomal recessive inheritance in the majority of cases.[1]

Subsequently, a number of diverse conditions including acquired inflammatory, metabolic, mitochondrial, congenital/structural, toxic, paraneoplastic, and endocrine myopathies, and even neurogenic diseases such as spinal muscular atrophy (SMA) were found to present with a similar clinical phenotype, complicating and challenging the clinical usefulness of LGMDs as an entity. Some authors use the term *limb-girdle syndrome* to describe the clinical picture prior to elucidation of a pathophysiological or genetic cause.[2,3]

LGMD is defined as a muscular dystrophy presenting with predominantly proximal weakness, sparing facial, extraocular, and distal extremity muscles (at least early in the course of the disease).[4] Based on this definition, the muscle biopsy is of great importance for inclusion into this group and also to evaluate for other causes of the limb-girdle syndrome. The muscle biopsy typically shows dystrophic features, including degeneration and regeneration, increased internalized nuclei, fiber size variability, increased endomysial fibrosis, and fatty replacement. However, just mild, nonspecific myopathic changes may also be seen (in milder cases, earlier in the disease or in less affected muscles) and still be consistent with the diagnosis of LGMD.[5]

Exact epidemiologic estimates for LGMDs are difficult to ascertain and vary depending on different populations, some owing to founder effects (e.g. the Amish, Libyan Jews, a Reunion Island community, among others).[6–9] Using strict diagnostic criteria, some estimate an overall prevalence of 6 to 8 per million for autosomal recessive and sporadic cases.[10] A more recent study in a population in Northern England estimated the overall prevalence of LGMD at 2.27 per 100,000.[11] Autosomal recessive LGMDs are more common than their autosomal dominant counterparts and will be discussed first in the chapter.

Genetic Nomenclature

The nomenclature first proposed by the European Neuromuscular Center (ENMC) workshop on LGMD[4] has simply designated all autosomal dominant LGMD as LGMD 1A, 1B, 1C, and so on and autosomal recessive LGMD as LGMD 2A, 2B, 2C, and so on. The letters indicate separate LGMDs by the order of the gene product identified or linked to a specific locus. The current most common LGMDs are summarized in Table 34.1.

As discussed in this chapter, a purely genetic nomenclature is an oversimplification and may have outlived its usefulness. Several of the genes and proteins implicated in LGMDs can result in very divergent phenotypes, e.g. congenital muscular dystrophy, Emery-Dreifuss muscular dystrophy (EDMD), cardiomyopathy, neuropathy, or even lipodystrophy for *LMNA*. In addition, other muscular dystrophies may also present with an LGMD-like phenotype. As a result, a nomenclature based on proteins or cellular physiology informed by the clinical phenotype may be more useful.

Nonetheless, we will still be mentioning both nomenclatures, as the primary goal of this chapter is to characterize the most common limb-girdle muscular

B.T. Darras, H. Royden Jones, Jr., M.M. Ryan & D.C. De Vivo (Eds): Neuromuscular Disorders of Infancy, Childhood and Adolescence, Second edition.
DOI: http://dx.doi.org/10.1016/B978-0-12-417044-5.00034-2

TABLE 34.1 Classification of Limb-girdle Muscular Dystrophies

LGMD Genetic Classification and Alternative Name	Locus	Gene Symbol	Protein Product
Autosomal Dominant			
LGMD 1A (Myotilinopathy)	5q31.2	*TTID, MYOT*	Myotilin
LGMD 1B (Laminopathy)	1q22	*LMNA*	lamin A/C
LGMD 1C (Caveolinopathy)	3p25.3	*CAV3*	Caveolin-3
LGMD 1D (Desminopathy)	6q23	*DES*	Desmin
LGMD 1E	7q36.3	*DNAJB6*	DNAJB6
Autosomal Recessive			
LGMD 2A (Calpainopathy)	15q15.1	*CAPN3*	Calpain-3
LGMD 2B (Dysferlinopathy)	2p13.2	*DYSF*	Dysferlin
LGMD 2C (γ-Sarcoglycanopathy)	13q12	*SCCG*	γ-Sarcoglycan
LGMD 2D (α-Sarcoglycanopathy)	17q21	*SGCA*	α-Sarcoglycan
LGMD 2E (β-Sarcoglycanopathy)	4q12	*SGCB*	β-Sarcoglycan
LGMD 2F (δ-Sarcoglycanopathy)	5q33-34	*SGCD*	δ-Sarcoglycan
LGMD 2G (Telethoninopathy)	17q12	*TCAP*	Telethonin
LGMD 2H	9q33.1	*TRIM32*	TRIM32
LGMD 2I	19q13.3	*FKRP*	FKRP
LGMD 2J (titinopathy)	2q31.2	*TTN*	Titin
LGMD 2K	9q34.13	*POMT1*	POMT1
LGMD 2L (Anoctaminopathy)	11p14.3	*ANO5*	Anoctamin-5
LGMD 2M	9q31	*Fukutin*	Fukutin
LGMD 2N	14q24	*POMT2*	POMT2
LGMD 2O	1p32	*POMGnT1*	POMGnT1
LGMD 2P	3p21	*POMGnT1*	DAG1
LGMD 2T	3p21.31	*GMPPB*	GMPPB

Abbreviations: LGMD, Limb-girdle muscular dystrophy; TRIM32, tripartite-motif containing 32; FKRP, Fukutin related protein; POMT1, protein-O-mannosyltransferase 1; POMT2, protein-O-mannosyltransferase 2; POMGnT1, protein O-linked mannose N-acetylglucosaminyltransferase 1; DAG1, dystroglycan; GMPPB, GDP-Mannose pyrophosphorylase B.

dystrophies in a way that is useful for clinical practice and provide a rational approach to the workup of patients with LGMD.

AUTOSOMAL RECESSIVE LIMB-GIRDLE MUSCULAR DYSTROPHIES

Classic autosomal recessive LGMDs include the sarcoglycanopathies (LGMD 2C–F), calpainopathy (LGMD 2A), and dysferlinopathy (LGMD 2B). This order reflects their average age of onset from younger to older. Two additional more common autosomal recessive LGMDs are attributed to mutations in the α-dystroglycanopathy gene fukutin-related protein (*FKRP*) (LGMD 2I), with a broad range of age of onset, and anoctamin-5 (LGMD 2L), a primarily adult-onset disease that resembles dysferlinopathy. There are a number of additional recessive LGMD forms, stemming from mutations of other α-dystroglycanopathy genes as well as others that we will touch upon briefly. In this section, we will discuss the clinical features and pathophysiologic mechanisms of major autosomal recessive LGMDs. Understanding the role of protein products of these genes and their functional associations in muscle has provided a window to study pathophysiologic mechanisms of LGMDs with the goal of identifying therapeutic targets.

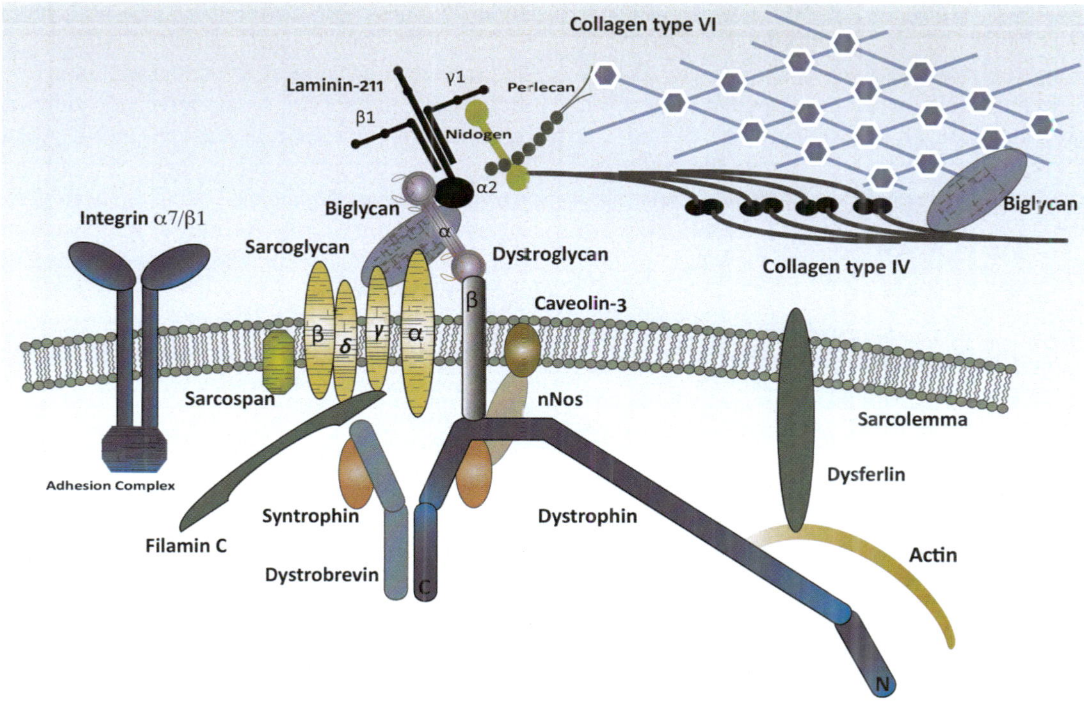

FIGURE 34.1 Schematic depiction of the complex of dystrophin-associated proteins and their association with the extracellular matrix.

Sarcoglycanopathies (LGMD 2C–F)

Sarcoglycanopathies were first described as autosomal recessive disorders resembling Duchenne muscular dystrophy (DMD).[12–14] In addition to the autosomal recessive pattern of inheritance, the immunohistochemical presence of dystrophin on muscle biopsy differentiated sarcoglycanopathies from DMD.[15] While initially referred to as SCARMD (for **s**evere **c**hildhood **a**utosomal **r**ecessive **m**uscular **d**ystrophy), many patients with milder presentations were subsequently identified, and the term SCARMD is no longer used widely.

The proportion of patients with sarcoglycanopathies among patients with muscular dystrophy depends on age, clinical severity, and the population studied; on average, they are found in approximately 20% to 25% of all patients with muscular dystrophy.[16,17] However, sarcoglycanopathies comprise roughly 50% to 60% of the more severe LGMDs as opposed to 10% to 20% of the milder forms.[18–20] Thus, a patient with an early onset, severe muscular dystrophy has a higher likelihood of having a sarcoglycanopathy compared to a patient with a juvenile onset, milder phenotype.

Pathophysiology

Dystrophin-associated proteins

Dystrophin, a large molecule located under the sarcolemma of myocytes, is mutated in DMD and BMD as described in Chapter 30. Its N-terminus and rod domain interact with the cytoskeleton while its C-terminus interacts with several intracellular and transmembrane proteins, known as the dystrophin-associated proteins (DAPs)[21–23] (Figure 34.1).

Intracellular DAPs The intracellular DAPs include the dystrobrevins and the syntrophins. Dystrobrevins share a similar structure to the C-terminal domain of dystrophin, with α-helical coiled-coils[24] mediating their association.[21] In addition, dystrobrevins interact with the sarcoglycan complex and neuronal nitric oxide synthase (nNOS).[25] Both dystrophins and dystrobrevins bind syntrophins. To date, mutations in dystrobrevins or syntrophins have not been associated with muscle disease in humans. However, α-dystrobrevin deficient mice show evidence of a muscular dystrophy.[26,27]

Transmembrane DAPs: Dystroglycan complex The dystroglycan complex includes two proteins: β-dystroglycan, a transmembrane protein, binds the C-terminal region of dystrophin intracellularly and interacts with α-dystroglycan extracellularly.[28] α-Dystroglycan, a heavily glycosylated protein, connects the dystrophin-β-dystroglycan axis of proteins to the G domain laminin α2, the heavy chain of laminin 211 (merosin).[29] Laminin α2 (*LAMA2*) mutations are one of the most common causes of congenital muscular dystrophy (see Chapter 29). In essence, the dystroglycan complex structurally links the intracellular dystrophin-cytoskeleton complex to the extracellular matrix.[30–32] Mutations in dystroglycan itself are extremely rare, as they are usually incompatible with development. However, abnormal glycosylation of α-dystroglycan caused by mutations in a series of proven and putative glycosyltransferases, as well as cooperating components, can cause congenital muscular dystrophy or autosomal recessive LGMD, specifically LGMD2I[33] (see the following).

Transmembrane DAPs: Sarcoglycan complex The sarcoglycan complex in skeletal muscle includes four proteins: α-, β- , γ-, and δ-sarcoglycan.[34] α-Sarcoglycan (50kDa) and γ-sarcoglycan (35kDa) are encoded on chromosome 17q21[35,36] and 13q12,[37] respectively, and are almost exclusively expressed in skeletal muscle, though γ- and δ-sarcoglycan are expressed in smooth muscle as well. β-Sarcoglycan (43kDa) and δ-sarcoglycan (35kDa) are encoded on chromosome 4q12[38,39] and 5q33,[40] respectively, and have more widespread expression.

Two additional widely expressed sarcoglycan-like molecules, ε- and ζ-sarcoglycan, have also been identified.[41,42] While no human diseases are linked to ζ-sarcoglycan mutations, some patients with an autosomal dominant myoclonus-dystonia syndrome carry heterozygous ε-sarcoglycan mutations.[43–45]

The exact nature of the association of the sarcoglycan complex with dystrophin is not known; however, associations with dystroglycans, synaptobrevin, dystrobrevin, and nNOS have been shown.[25,34,46] γ-Filamin/filamin C, an actin binding protein, is an intracellular molecule that is known to interact with the sarcoglycan complex, an association that theoretically provides an additional point of interaction with the cytoskeleton.[47] Biglycan provides an extracellular link between the sarcoglycan complex and dystroglycan.[48] Recently, muscle specific aquaporin-4, a water channel, was shown to interact with the sarcoglycan complex via α1-syntrophin.[49,50] Sarcospan, which is severely reduced in sarcoglycanopathies, is another protein that is intimately associated with the sarcoglycan complex.[51,52] Sarcospan mutations are not linked to muscle disease, however, and its function remains elusive. These complex associations likely stabilize the DAPs and provide a structural and functional link to intracellular signals through different pathways, but the exact relevance of these interactions to muscle disease is yet to be identified.

Genetics and Mutations

Mutations in all four major sarcoglycan genes cause four genetically separate forms of autosomal recessive LGMD with almost indistinguishable clinical phenotypes.[35,37–40] Although all four disorders appear to have a worldwide distribution, there are certain regional differences owing primarily to founder mutations (e.g. a high proportion of γ-sarcoglycanopathy in North Africa and the Mediterranean because of a founder mutation).[13,53] Of the sarcoglycan mutations, α-sarcoglycanopathies are likely the most common type in Europe, North America, and Brazil, accounting for more than 50% of genetically proven sarcoglycanopathies.[17,18,54,55] These are followed in varying proportion by β- and γ-sarcoglycan mutations, with δ-sarcoglycan as the least commonly mutated sarcoglycan in most series.

In α- and β-sarcoglycanopathy, in particular, there are high proportions of missense mutations, the majority of which are located in the respective extracellular domains. Approximately one-third of the α-sarcoglycan mutations map to the cadherin-like domains, suggesting a role for calcium-mediated protein interactions in the pathogenesis process.[56] Among them, the single mutation *Arg77Cys* is the most frequent.[57] Many other mutations are also identified, however, and genotype/phenotype correlations are difficult to evaluate because mutations often occur in compound heterozygous states. Even with homozygous mutations, there can be considerable phenotypic variability. Truncating or nonsense mutations likely give rise to a more severe phenotype, whereas missense mutations are much more variable.[55,58]

Four β-sarcoglycan missense and nonsense mutations are known to occur. In the Indiana Amish, there is a founder mutation, *Thr151Arg*, associated with a variable phenotype.[39] There is a relative clustering of mutations in the immediately extracellular domain, encoded on exon 3 of the gene.[59,60] Similar to α-sarcoglycan, nonsense or truncating mutations lead to a more severe phenotype, but a higher proportion of missense mutations in β-sarcoglycan cause a phenotype at least as severe, particularly when the mutations are predicted to disrupt the secondary structure in the domain immediately following the transmembrane stretch.[61] This may reflect the central position of β-sarcoglycan within the complex.[62]

The most common mutation in γ-sarcoglycan is the frame-shifting deletion of a single thymidine, *del521T*, which originated in North Africa as a founder mutation and is readily found in Mediterranean countries or migrants from that area.[63] As a result, γ-sarcoglycanopathy is the most common form of sarcoglycanopathy in these countries. Phenotypes may vary significantly even within the same families. Another founder mutation, *Cys283Tyr*, commonly found in the Romany (Gypsy) population in Europe, results in a consistently severe phenotype despite being a missense mutation.[64–68]

Mutations in δ-sarcoglycan are the least common overall but are found commonly in Brazil, where the disorder was originally described.[40] A single nucleotide deletion, nonsense mutations, and missense mutations have all been reported with rather severe phenotypes.[69–71] Dilated cardiomyopathy has been seen in some families with heterozygous mutations in the δ-sarcoglycan gene, with or without skeletal muscle weakness. It has been hypothesized that these mutations may act in a dominant-negative way on the sarcoglycan complex as a whole, interfering with its function in cardiomyocytes.[72]

Most milder sarcoglycanopathies are related to α-sarcoglycan mutations, including almost asymptomatic patients with high CK levels and lordosis only.[73,74] β- and γ-sarcoglycan mutations have a higher proportion of severe early childhood cases though significant intrafamilial variability can be seen.[75] δ-Sarcoglycan mutations are generally more severe.

Clinical Features

Clinical features of the four genetically distinct sarcoglycanopathies overlap significantly, with little difference among the different types. They predominantly affect young children with a median age of onset around 6 to 8 years with a broad range from even younger onset to adult onset disease or almost asymptomatic mutation carriers.[75] Early motor milestones are usually normal, but there may be mild delay or toe walking in some children. Some children are slower than their peers in physical activities. (See Case Example 34.1.)

Time course and distribution of motor symptoms

The first symptoms generally relate to pelvic muscle weakness, evidenced by a waddling gait, which limits activities such as running, getting up from the floor, or climbing stairs. Primary toe walking may be present in some children. Muscle cramps, pain, and exercise intolerance with or without myoglobinuria can also occur, the so-called "pseudometabolic presentation."[58] The distribution of weakness is reminiscent of dystrophinopathies: the glutei and adductors are more involved than the quadriceps, sartorius, and gracilis. However, unlike dystrophinopathies, anterior and posterior compartments of the thigh may be equally affected. Shoulder girdle weakness follows. The deltoid, infraspinatus, and biceps muscles are involved early in the disease. Scapular weakness tends to be more pronounced compared to dystrophinopathies (Figure 34.2). Facial and extraocular muscles are spared. Late in the disease, distal muscles may be involved, starting in the anterior tibial compartment. Very mild facial involvement with a "transverse smile" may also be seen. In later stages—and similar to DMD—neck flexor weakness may occur. The weakness is quite rapidly progressive, with loss of strength towards the end of the first decade similar to DMD in the early onset cases, although there is more clinical variability compared to DMD. Loss of independent ambulation may occur around 12 to 16 years of age, although there again is considerable variability. CK levels are elevated 10- to 100-fold early in the course of the disease, but tend to decrease as weakness progresses.[58,60,77]

Cardiac features

Cardiac involvement clearly occurs in sarcoglycanopathies, though symptomatic dilated cardiomyopathy is only seen in a minority of patients.[78,79] In subjects with cardiomyopathy, however, cardiac failure or sudden cardiac

CASE EXAMPLE 34.1 Sarcoglycanopathy

Two siblings were evaluated for weakness and elevated CK. In the older sister, now 9.5 years old, an elevated CK was incidentally noted during blood work for gastroenteritis at the age of 7. She had a normal developmental history; she walked at age 12 months and was able to run. There was some reported fatigability but no other symptoms of muscle disease. Her CK levels were persistently elevated. On examination, she had normal eye movements and facial strength, with a minimal lumbar lordosis and mild winging of the scapulae. No contractures were present. Formal strength testing identified very mild weakness of neck flexion and the proximal upper and lower extremities (Medical Research Council [MRC] grade 4 to 4 + /5), and quadriceps greater than hamstring weakness. She arose from the floor slowly, but without a Gowers' sign.

Her brother, now 7 years of age, had delayed early motor milestones, sitting without support at 12 months, and walking independently at 22 months of age. His CK was also persistently elevated. He has always had difficulty in physically keeping up with his peers. He had to roll onto one side to arise from the floor. There was no reported muscle pain, cramping, or discoloration of the urine. The family history was unremarkable; there was no parental consanguinity.

His physical examination showed lumbar lordosis and mild scapular winging, without any contractures. Cranial nerve function is also normal. There was marked neck flexor weakness and proximal upper extremity weakness (MRC grade 4/5, triceps more affected than biceps). There was no facial weakness. In the lower extremities, the proximal muscles were somewhat weaker, in the 4- to 4/5 range, with hip abduction greater than adduction weakness and knee flexion greater than extension. He arose from the floor by rolling to his side and using a modified Gowers' maneuver. He ran slowly.

Muscle biopsy in the younger brother showed dystrophic muscle with normal dystrophin immunoreactivity and complete absence of immunoreactivity for all four sarcoglycan proteins. Genetic testing showed two missense compound heterozygous mutations in the β-sarcoglycan gene (Figure 34.2).

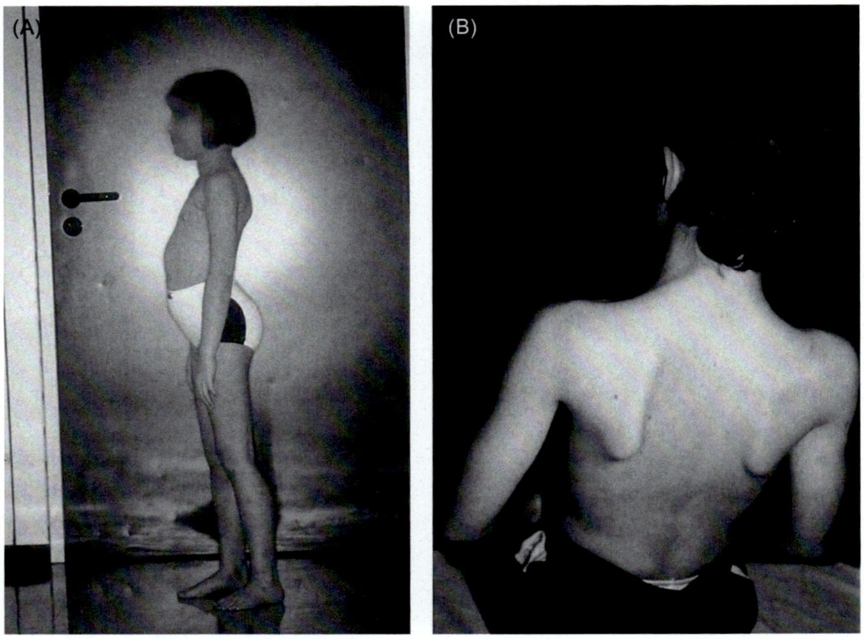

FIGURE 34.2 Two patients with γ-sarcoglycanopathy (LGMD 2C). (**A**) In the 9-year-old girl, note pronounced lordosis and calf hypertrophy. (**B**) In the 11-year-old boy, note prominent scapular winging and hypertrophy of the triceps.

death may occur and cardiac transplantation may become necessary. Dilated cardiomyopathy may be more common in γ- and δ-sarcoglycanopathies but can be seen in all sarcoglycanopathies.[80] Subclinical cardiac involvement is more frequently seen on electrocardiography and echocardiography.[80,81]

The pathogenesis of cardiomyopathy in sarcoglycanopathies is complex. Animal studies suggest that, in addition to abnormalities of the cardiomyocytes themselves, it may be partially related to smooth muscle dysfunction due to the disruption of the sarcoglycan complex in coronary arteries.[82,83] This finding may have implications for therapy. For example, calcium channel blockers favorably influence development of cardiomyopathy in δ-sarcoglycan deficient mice.[84] On the other hand, some δ-sarcoglycan mutations can cause a dilated cardiomyopathy without skeletal muscle disease. In these patients, no evidence of coronary pathology is found.[72] Carriers of δ-sarcoglycan mutations may develop late onset cardiac disease and should be monitored carefully, similarly to carriers of dystrophin mutations.[85]

Pulmonary features

Symptomatic respiratory failure is not an early feature of sarcoglycanopathies. However, similarly to DMD, some patients with severe early onset muscle weakness develop severe restrictive lung disease once nonambulant. Mild to moderate restrictive lung disease, on the other hand, affects the majority of patients.[80] Patients with α- or γ-sarcoglycanopathy may have more severe respiratory disease, suggesting a specific role for the different subtypes of sarcoglycans in different skeletal muscles.

Contractures and other signs and symptoms

Other signs in some LGMDs include calf hypertrophy and macroglossia. Achilles tendon shortening and lumbar lordosis may occur early in the course, so toe walking can be an early manifestation of the condition. Later, as the disease progresses, more contractures involving the hip flexors, lateral tractus, and knee flexors may develop. Progressive scoliosis may worsen respiratory compromise later in the disease. Unlike DMD, sarcoglycanopathies do not cause intellectual impairment.[58]

Diagnosis

Sarcoglycanopathies can be suspected based on clinical grounds, in particular in a young patient with muscular dystrophy of Duchenne-like severity. The probability of an autosomal recessive LGMD in boys with a Duchenne-like phenotype is 6% to 8%.[86,87] However, this probability obviously increases if the patient is a girl, or a boy with negative genetic studies of the dystrophin gene.

Definitive histological diagnosis requires examination of the muscle biopsy specimen, including immunohistochemical stains with antibodies against sarcoglycans and dystrophin. Histology usually shows marked degeneration and regeneration and severely dystrophic muscle. Dystrophin immunoreactivity is expected to be normal, though it can be reduced similarly to BMD or female

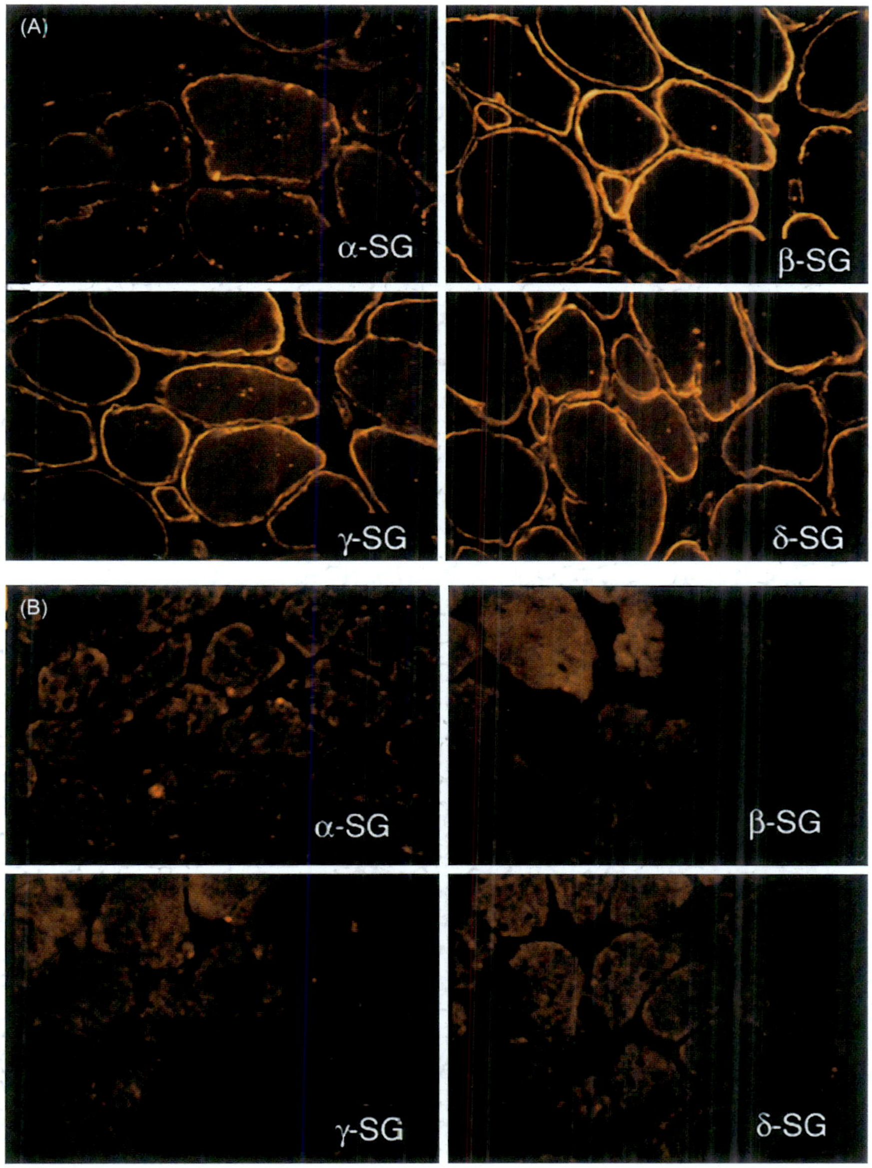

FIGURE 34.3 Sarcoglycan immunohistochemistry. (**A**) α-Sarcoglycanopathy; note prominent reduction of α-sarcoglycan with preservation of the other sarcoglycans. (**B**) β-Sarcoglycanopathy; note complete loss of membrane location for all four sarcoglycans.

carriers of dystrophin mutation.[88,89] Western blot analysis of muscle usually shows dystrophin with a normal molecular weight and quantities within 10% of normal.[90]

The pattern of *in situ* sarcoglycan immunoreactivity in sarcoglycanopathies is rather complex. The entire sarcoglycan complex tends to be affected as a unit, but in some cases the pattern of immunoreactivity of some components may appear more normal. For example, in α- and γ-sarcoglycanopathy, immunoreactivity of mutated proteins is reduced while all others may be preserved or only mildly affected.[91] In contrast, in β- and δ-sarcoglycanopathies, the entire complex is severely reduced or completely absent.[69,90,92,93] Thus, these patterns can be diagnostically helpful (Figure 34.3).

Residual expression of α-sarcoglycan in some cases of missense mutations may correlate with milder clinical phenotypes, but this is not always true.[58,74,94] For example, even complete absence of the protein has been reported with mild phenotypes in γ-sarcoglycanopathy.[63]

Nevertheless, immunohistochemical stains can guide molecular genetic testing for the absent proteins in question in a stepwise manner.

Treatment

No specific therapies are available for sarcoglycanopathies, but unlike dystrophinopathies, sarcoglycanopathies are more amenable to gene therapy owing to the fact that cDNA for sarcoglycans are small enough to be suitable for package in viral vectors and delivery.[95–99] Feasibility of local delivery of adeno-associated virus containing the α-sarcoglycan gene, with long-lasting expression of the wild-type gene without adverse effects, has been shown.[100] Vascular delivery of the vector with systemic delivery remains an area of active investigation. Similarly, a phase I trial of adeno-associated virus γ-sarcoglycan gene therapy was recently conducted with promising proof of concept, though its widespread clinical use awaits larger studies and longer follow-up.[101]

Anecdotal reports advocate the use of steroids in sarcoglycanopathies similar to that in DMD.[102] However, no controlled studies have examined these effects systematically to date. Nifedipine was used in an animal model of δ-sarcoglycan deficient mice, and showed a positive effect on the course of cardiomyopathy attributed to coronary smooth muscle pathology, but the use of such agents in humans is not yet reported.[84] Principles of respiratory and orthopedic management are similar to those for DMD (see Chapters 31, 44, 52, and 53).

LGMD 2I (Fukutin-related Protein Mutation)

Fukutin-related protein (*FKRP*) belongs to the growing list of genes underlying the group of α-dystroglycanopathies, characterized by abnormal O-mannosyl glycosylation of α-dystroglycan, which more typically are associated with forms of congenital muscular dystrophy (CMD, discussed in Chapter 29). Mutations in *FKRP* cause an extremely wide spectrum of phenotypes ranging from severe Walker-Warburg syndrome, transitional phenotypes of CMD with variable central nervous system involvement or normal brain imaging, to less severe LGMDs resembling the dystrophinopathies, which are referred to as LGMD 2I.[103]

Pathophysiology and Genetics

This form, designated as LGMD 2I, was first identified and mapped in Tunisia to chromosome 19q13.3.[104] FKRP was later identified based on its homology to fukutin, the protein deficient in Fukuyama congenital muscular dystrophy.[33] It is thought to be involved in glycosylation of α-dystroglycan, in particular of its LARGE glycolepitope, which is variably reduced in muscle by immunohistochemistry and Western blot in patients with LGMD 2I.[33] These findings are also replicated in the mouse model of this disease.[105,106]

The mutations identified so far include missense as well as null mutations. Most mutations are contained in exon 4; as a result, mutation analysis is not complicated.[33,104,107–109] A missense mutation, *Leu276Ile*, is seen in almost all LGMD patients and results in a relatively milder phenotype in the homozygous state, compared to the compound heterozygous state with other, more severe mutations.[33,110]

Clinical Features

The LGMD phenotype of FKRP mutations can present as a DMD phenocopy, with early loss of ambulation (before 10 years of age in some cases).[111] The pattern of muscle weakness is very similar to DMD, showing a pronounced predilection for axial muscles, neck flexors, and proximal limb muscles with a prominent lordosis. In some patients, shoulder girdle muscles may be weaker than those of the pelvic girdle, with atrophy of the deltoid and pectoralis muscles. There can be tongue, calf, and brachioradialis hypertrophy. The hamstrings are generally more involved than the quadriceps. Mild facial weakness can also be seen. Scapular winging is uncommon.[112–115] There is significant variability; patients homozygous for *Leu276Ile* may have a milder phenotype more similar to BMD, and there are adult-onset forms of the disorder. CK levels are high, 10 to 30 times the upper limit of normal.

Clinically significant dilated cardiomyopathy can occur even before loss of ambulation, while limb girdle weakness may still be relatively mild, and in occasional patients can be rapidly progressive. Respiratory failure necessitating nocturnal ventilatory support, on the other hand, typically manifests after loss of ambulation and should be evaluated and treated appropriately.[114,116–118]

Diagnosis

LGMD 2I may be suspected on clinical grounds, based on features such as the peculiar hypertrophy of brachioradialis and tongue in some patients. Muscle biopsy immunohistochemical analysis usually shows prominently reduced α-dystroglycans and a mild reduction of laminin α2.[103] While these are not specific, they can be helpful in prompting genetic analysis of FKRP gene in the right clinical setting.

Treatment

No specific treatments are available at this time. Similar to other muscular dystrophies causing cardiomyopathy and respiratory failure, the recognition, evaluation, and treatment of dilated cardiomyopathy and potential need for nocturnal ventilatory support is of utmost importance.

Other α-Dystroglycanopathies

Numerous other gene mutations may result in hypoglycosylation of α-dystroglycan with an LGMD phenotype. These include protein-O-mannosyl transferase 1 and 2 (*POMT1* and *POMT2*), causing LGMD 2K and LGMD 2N, respectively; protein-O-mannose 1,2-N-acetylglucosaminyltransferase 1 *(POMGnT1)*, causing LGMD 2O; *fukutin*, causing LGMD 2M; and *DAG1*, causing LGMD 2P. These conditions are also phenotypically heterogeneous.

Fukutin mutations, typically associated with a CMD, have also been reported to cause a mild LGMD without intellectual disability.[112,119] A mutation in *POMGnT1* has also been reported to cause a mild LGMD phenotype with onset at 12 years of age and normal intellect.[120] In contrast, LGMD phenotypes associated with *POMT1* and *POMT2* mutations are almost always accompanied by intellectual disability or central nervous system dysfunction.[112,121–124] In these patients, muscle weakness remains mild and most are ambulatory into adulthood. Primary mutations in *LARGE* have not been associated with a LGMD phenotype, but a patient with LGMD, intellectual disability, and a mutation in *DAG1* was described.[125] *DAG1* encodes dystroglycan, which interacts with and is glycosylated by *LARGE.* Similar to the other secondary dystroglycanopathies discussed above, this mutation leads to aberrant posttranslational glycosylation, underscoring the importance of this process in normal muscle membrane function.

Mutations in the isoprenoid synthase domain containing *(ISPD)* gene and GDP-mannose pyrophosphorylase B (*GMPPB)*, both of which are involved in glycosylation of α-dystroglycan, have recently been identified in patients with a LGMD phenotype. *ISPD* mutations, previously shown to cause WWS and CMD phenotypes, can cause a mild LGMD pattern of weakness. In addition to the expected cardiac and pulmonary involvement similar to other dystroglycanopathies, these patients may also have infratentorial CNS involvement with cerebellar atrophy and/or oculomotor apraxia.[126] *GMPPB* mutations also cause a wide range of abnormalities including LGMD. Patients may also have epilepsy, cataracts, and developmental abnormalities of the posterior fossa structures.[127]

Calpainopathy (LGMD 2A)

Background and Epidemiology

Calpainopathy, or LGMD 2A, one of the most common LGMDs, was the first autosomal recessive LGMD localized by linkage analysis. The first clinical description of this juvenile onset muscular dystrophy can be attributed to Wilhelm Erb in 1880s.[128] Genetic linkage analysis in families of the Reunion Island off the coast of Africa identified the locus on chromosome 15q15.[129] LGMD 2A is likely the most common juvenile onset form of LGMD and may account for 40% to 50% of all cases of LGMD.[130]

Pathophysiology, Genetics, and Mutations

CAPN3, the gene mutated in LGMD 2A,[131] encodes calpain-3, a calcium-activated protease that localizes to the cytoplasm and nuclei of the cells. It has a nuclear localization signal where it is thought to help in processing of transcription factors relevant to muscle.[132,133] It has been thought to stabilize NF-kB, which activates anti-apoptotic genes, and inhibit its degradation. Presence of apoptotic cell death in muscle cells of patients with calpainopathy further supports this mechanism.[134,135] In addition, calpain-3 interacts with titin, a large sarcomeric protein.[136,137] Additional studies propose a role for calpain-3 in membrane repair via its interactions with the dysferlin complex.[138,139]

A large number of mutations in calpain-3 have been identified. These include (more commonly) missense mutations and (less commonly) large deletions and truncations.[130] In general, truncating mutations tend to be associated with more severe phenotypes, though there are no strong genotype/phenotype correlations as a whole.[7,140] There are no specific hotspots for mutations in the gene.[141]

Clinical Features

Calpainopathy has a characteristic clinical phenotype, although atypical presentations do occur. The age of onset is slightly later than sarcoglycanopathies, between 8 and 15 years of age (range 2 to 40 years). Early milestones are generally normal, but some children are reported to be weaker than peers in physical activities. Primary toe walking and delayed walking are also reported in some individuals. Exercise-induced myalgias or muscle stiffness may predate muscle weakness.[7,142–147] See Case Example 34.2.

Time course and distribution of motor symptoms

Similarly to other LGMDs, the first symptoms generally relate to pelvic girdle muscle weakness limiting activities such as climbing stairs, running quickly, or getting up from the floor. Later, shoulder girdle and arm weakness become apparent. Minimal facial weakness is reported very late in the disease. The facial, extraocular, and pharyngeal muscles are otherwise spared.[7,145,148–150]

The pattern of weakness in the lower extremities is characteristic and involves the gluteus maximus and hip adductor muscles. The hip flexors and hamstrings are involved to a lesser degree (hamstrings are weaker

CASE EXAMPLE 34.2

A 14-year-old boy presented with proximal weakness. He was born after an uncomplicated pregnancy and delivery with normal early motor development. Problems were first noted at 12 years of age with difficulty in skiing, jumping, and climbing stairs. His mother also had noted increased tendency to toe-walk and decreased muscle mass in the legs and around the shoulders. Over the next 1.5 years, he developed increasing difficulty getting up from the floor and lifting heavy objects, without muscle cramping or myalgias. The family history was unremarkable and there was no parental consanguinity.

On examination, his cranial nerves were normal without facial weakness or external ophthalmoplegia. There was pronounced atrophy of the proximal upper and lower extremities, with scapular winging. On formal strength testing, there was weakness of the periscapular muscles and proximal upper extremity muscles (biceps greater than triceps) with normal distal strength. In the lower extremities, the hip flexors and glutei were MRC grade 2 to 3/5; the quadriceps and hip abductors were clearly stronger than hamstrings and hip adductors. There was a mild lumbar lordosis and prominent Achilles contractures.

The CK was elevated, at 2584 U/L. A muscle biopsy showed dystrophic muscle with normal immunoreactivity for dystrophin, sarcoglycans, and dysferlin. The Western blot analysis showed a complete absence of calpain-3. Genetic analysis showed two compound heterozygous missense mutations in the *CAPN3* gene.

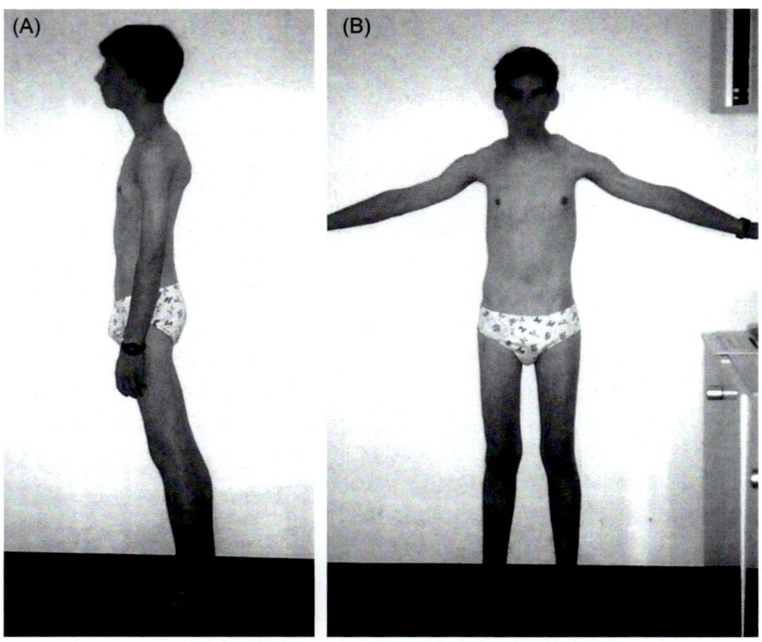

FIGURE 34.4 Patient with calpainopathy (LGMD 2A), a 14-year-old boy. (**A**) Note atrophic muscles and prominent contractures of the Achilles tendons. There is no calf hypertrophy. (**B**) Note prominent high-riding scapulae.

compared to the quadriceps), and the hip abductors generally have relatively better strength. Distal anterior leg muscles are affected later in the course of the disease. Upper extremity weakness is characterized by early and striking weakness of the shoulder muscles including the latissimus dorsi, rhomboids, serratus anterior, and pectoralis major muscles, resulting in scapular winging and high-riding scapulae. To a lesser degree, deltoid, biceps, and brachioradialis muscles weaken but triceps is comparatively spared. There may be significant asymmetry early on, particularly in the upper extremities. In contrast to DMD and the sarcoglycanopathies, neck flexors are preserved. The abdominal muscles show notable laxity. Another characteristic feature is that calpainopathy usually results in an *atrophic* muscular dystrophy; unlike DMD and sarcoglycanopathies, calf hypertrophy is seen only exceptionally and transiently, and most patients have an asthenic habitus (Figure 34.4).[151]

Variations occur: some patients develop early Achilles and elbow contractures, similar to EDMD; others may have a rapidly progressive course reminiscent of DMD and indistinguishable from it on clinical grounds.

CK concentrations in the serum are significantly elevated early in the disease course (5–20 times the upper limit of normal), decreasing as it progresses, but on average tend to be lower than those seen in DMD and the sarcoglycanopathies.

The course of the disease is progressive with interindividual variability.[146] Loss of independent ambulation typically occurs around 14 years of age or later (range 5−39 years).[7]

Cardiac features

No primary cardiac involvement has been attributed to *CAPN3* mutations, and the protein is normally not expressed in cardiac myocytes. However, reports of right bundle branch block in two brothers with *CAPN3* mutations and a 23-year-old with ventricular failure and *CAPN3* mutations are noted in the literature.[152,153]

Pulmonary features

Respiratory failure is a rare event, only occurring in severe and very advanced disease. Mild restrictive lung disease evidenced by reduced forced vital capacity is quite prevalent and usually asymptomatic.[148]

Contractures and other signs and symptoms

Contractures appear in the Achilles tendons first, and may progressively involve the elbows and the spine. In some patients, contractures may be severe earlier in the course, mimicking Emery-Dreifuss muscular dystrophy. Early lordosis may be seen.[154,155] Scapular winging and high riding scapulae may be asymmetric, which may raise FSHD as a differential diagnostic consideration.

Diagnosis

As mentioned above, the clinical features may be strongly suggestive of calpainopathy, but further studies are needed for confirmation.[156,157] Muscle imaging may reveal the characteristic selective involvement of the posterior compartment of the thigh and the hip adductors,[154,158] but their reliability in differentiation of the LGMDs is not entirely clear.[159]

Muscle biopsy may show evidence of degeneration, regeneration, and necrosis. In addition to dystrophic features, type 1 fibers may appear lobulated on the oxidative stains such as NADH. Also, eosinophilic myositis has now been reported in several cases of *CAPN3* mutations.[136,160,161]

Direct immunohistochemistry for calpain-3 with the current antibodies is an unreliable test, but Western blot analysis can be informative. In 80% of patients with known calpain-3 mutations, calpain-3 level is either reduced or absent in muscle biopsy immunoblots. Normal studies in the other 20% of cases suggest that loss-of-function mutations may affect calpain-3 function without affecting protein quantities.[162,163] Functional biochemical assays for calpain-3 can be used to supplement immunoblotting in detection of loss-of-function mutations.[157,164,165] Of note, calpain-3 may be secondarily reduced in cases of primary dysferlin mutation.[166,167]

Direct genetic testing may help in diagnosis of LGMD 2A cases where there is clinical suspicion and normal biochemical and histochemical tests, or when muscle biopsy is not available.[152]

Treatment

No specific therapies are currently available for calpainopathy. Findings of eosinophilic inflammation in the muscle in some LGMD2A patients has raised interest in use of immunosuppressive therapies, but no studies have addressed the utility of this approach.[168] In mouse models of calpainopathy, adenovirus mediated gene transfer has been successful, suggesting a potential approach for future clinical trials.[169]

Dysferlinopathy (LGMD 2B)

The protein affected in LGMD type 2B is dysferlin, which is encoded by *DYSF*, a gene on chromosome 2p.[170−172] *DYSF* mutations are associated with a number of distinct clinical phenotypes: a typical LGMD, a distal muscular dystrophy known as Miyoshi myopathy,[173] and a third phenotype with early anterior tibial compartment weakness.[174,175] Even in the limb-girdle phenotype, the phenotypic spectrum is broad.[176−180] Most patients first develop weakness in late adolescence and early adulthood. However, some patients develop a congenital myopathy[181] and some remain only mildly affected well into adulthood.[182]

Pathophysiology

Dysferlin, a 230-kDa transmembrane protein, is a plasma membrane-based protein. Its intracellular domains are considerably larger than the small extracellular domain and contain six C2 domains.[183] Dysferlin is expressed in skeletal muscle early in limb development (5−6 weeks of gestation).[184] It interacts with caveolin-3,[185,186] and its C2 domain may interact with calcium and phospholipids,[187] suggesting a role in membrane repair.[188−190] Unusually high CK levels in patients with dysferlinopathy further support this postulate.

Inflammation may also play a role in pathogenesis of dysferlinopathy. Prominent inflammation on muscle biopsy of patients with dysferlinopathy supports this notion. Additional studies have shown aberrant recruitment of inflammatory cells and abnormal monocyte adhesion in cellular and animal models of dysferlinopathy.[191,192]

Recent studies show that dysferlin expression is enriched in the T-tubule membranes, and may play a role in mediating calcium homeostasis and maintenance of normal T-tubule structure and function, especially after muscle injury.[193] Furthermore, in a mouse model of dysferlinopathy, extracellular calcium depletion or blockade of L-type calcium channels with diltiazem reduces muscle injury, necrosis, and inflammation, suggesting a possible therapeutic role for calcium channel blockade in dysferlinopathy.[194]

Further studies on the intersection of the divergent roles of dysferlin in myofiber physiology may shed light on the pathogenesis of the disease and provide avenues for future therapeutic interventions.

Genetics and Mutations

Different mutations including missense and splicing mutations, deletions, and truncations have been identified in dysferlinopathies.[195–199] Interestingly, all three phenotypes—the limb-girdle phenotype, the Miyoshi myopathy, and the anterior tibial presentation—have been reported with the same mutations, some within the same family, suggesting a genetic or epigenetic factor resulting in the different clinical phenotypes.[175,200,201]

Clinical Features

Dysferlinopathy generally presents later than other recessive LGMDs, regardless of its phenotype. Weakness typically starts around 18 to 20 years of age, but congenital forms and mild adult onset disease have also been reported.[144,179,180,202] Early motor milestones are usually normal and some patients have been competitive athletes. Weakness is generally slowly progressive, and ambulation is usually preserved until late in the disease, typically into the fourth decade.[177]

Time course and distribution of motor symptoms

In the limb-girdle form of dysferlinopathy (LGMD 2B), weakness is initially of pelvifemoral distribution, particularly affects the quadriceps, and later involves the arms. However, on targeted examination, mild gastrocnemius and soleus muscle wasting and weakness are present from early in the disease course, causing early inability to walk on toes even in the LGMD phenotype[203] (Figure 34.5A). This unique feature, which can be demonstrated on muscle imaging studies, can be diagnostically helpful[204,205] (Figure 34.5B). Arm weakness is usually mild at first, involving the distal biceps. The deltoid and periscapular muscles are usually spared. As a result, in contrast to sarcoglycanopathies and calpainopathy, scapular winging is not a usual feature. As in other LGMDs, the facial and extraocular muscles are preserved. In the Miyoshi phenotype, gastrocnemius and soleus muscles are affected first, manifesting as the typical inability to walk on the toes.[173] In the upper extremities, the muscles of the forearms are quite weak, yet the intrinsic muscles of the hand are spared. The proximal muscles are affected as the disease progresses. Other phenotypes (i.e. those affecting the anterior leg compartment) are also described.[206] Variable proximal, distal, or concurrent proximal and distal onset disease can be seen within a single family.

CK levels tend to be very high during the early and active phase of the disease (20–150 times the upper limit of normal); in the presymptomatic patients, it may only be mildly elevated.

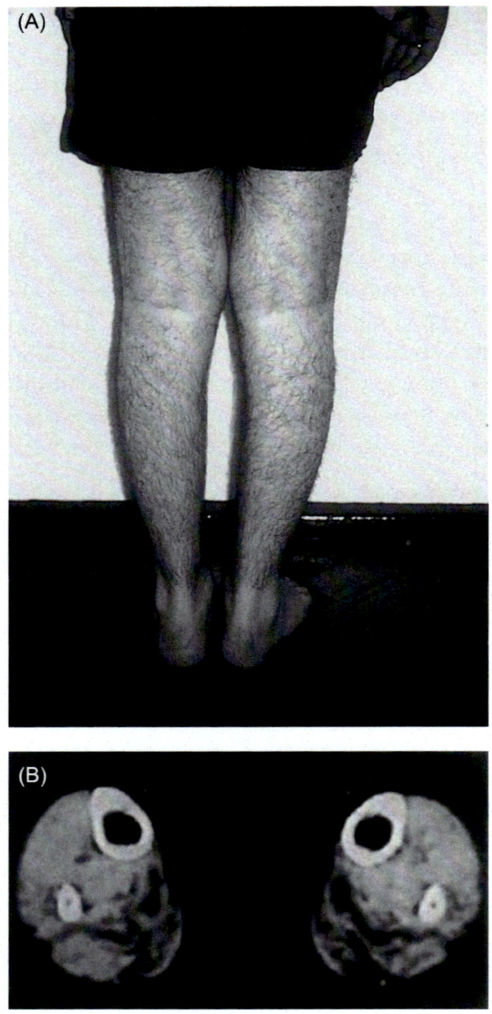

FIGURE 34.5 (**A**) Patient with LGMD 2B; note atrophic calf muscles. (**B**) Computed tomographic scan through the calf in a patient with LGMD 2B; note obvious involvement of the gastrocnemius muscle with extensive hypodense areas. *(**A**) Courtesy of K.M.D. Bushby, Newcastle upon Tyne; (**B**) Courtesy of I. Mahjneh, Oulu, Finland.*

Cardiac and pulmonary features

Dysferlinopathy does not generally cause significant cardiac or pulmonary disease, and in their absence life expectancy is not reduced.

Contractures and other signs and symptoms

Contractures are not a usual feature of dysferlinopathy, but are occasionally reported in patients with early anterior compartment weakness.[207]

Diagnosis

The characteristic clinical presentation of dysferlinopathy includes the following features: age of onset around 18 to 20 years of age, highly elevated CK levels, proximal lower extremity weakness in conjunction with inability to walk on toes, relative sparing of the periscapular muscles, and early gastrocnemius muscle involvement on muscle imaging. However, there can be significant phenotypic variation, even among family members.

Muscle biopsy usually shows dystrophic features, but inflammatory changes may be prominent and can delay accurate diagnosis by suggesting myositis.[208–210] Amyloid positive deposits in intramuscular vessels on Congo red stain have also been reported though are not specific.[211,212] Immunostains usually show absent dysferlin.[179] There may be secondary reduction of and/or staining abnormalities in dysferlin in other dystrophic conditions such as calpainopathy (with normal appearing dysferlin Western blot).[213] Similarly, calpain levels may be secondarily reduced or absent in dysferlinopathy.[166,167] Western blot analysis is a more specific diagnostic test[214] that can also be attempted on blood-derived macrophages.[215,216] Dystrophin and DAPs are usually normal.

Mutation analysis is the most specific test but can be labor intensive with conventional methods given the large size of the gene (55 exons).[217] There are no clear mutation hot spots.

Treatment

No specific treatments are available as yet. Judicious use of orthotics can provide help in maintenance of ambulation in many patients, especially those with distally predominant weakness.

Anoctaminopathy (LGMD 2L)

Recessive mutations in *ANO5* on chromosome 11, the gene encoding anoctamin-5, have recently been linked with an adult-onset limb-girdle muscular dystrophy. Some subjects have quadriceps atrophy and others a distal myopathy clinically very similar to dysferlinopathy.[218–222] The median age of onset is in the fourth decade.

Pathophysiology, Genetics, and Mutations

Anoctamin-5 is a 107-kDa protein that was first identified as a causative mutation of gnathodiaphyseal dysplasia before being implicated in muscular dystrophy.[223,224] Although highly expressed in skeletal muscle, cardiac myocytes, and bone, its function is still poorly understood. Anoctamin-5 shares significant homology to a family of calcium-activated chloride channels, but its potential role in membrane integrity and function is debated.[221,224]

The most common mutation is a truncating mutation in exon 5, but compound heterozygous missense mutations have also been reported. There is relative phenotypic homogeneity in patients with homozygous truncating mutations.[219,221,222,225]

Clinical Features

The median age of onset is in the adult years, usually in the fourth decade. Some patients are completely normal in their youth, able to run marathons and participate in sports. The first noticeable symptoms are usually difficulty walking long distances, climbing stairs, and, in some patients, difficulty toe-walking. As mentioned previously, some patients develop a distal onset of the disease similar to Miyoshi myopathy, but as the disease progresses proximal muscles also become involved.

Weakness is noted in the hip muscles as well as knee flexors, knee extensors, and plantar flexors in the lower extremities. In the upper extremities (which are usually considerably less affected) there is preferential weakness of biceps, brachioradialis, and triceps muscles. There is quadriceps atrophy, especially involving the vastus medialis muscle. Other atrophied muscles include medial gastrocnemius, hamstrings, biceps, and brachioradialis. Significant asymmetry in the degree of involvement may be seen. CK levels are moderately elevated (median range of 4000 to 5000 U/L). The disease progresses very slowly and most patients remain ambulatory well into adulthood, in particular in women, who may remain relatively asymptomatic.[220]

Muscle MRI may show asymmetric fatty replacement of semitendinosus, semimembranosus, and adductor muscles with preservation of gracilis and sartorius. The lateral lower leg compartment is generally preserved. Changes in the quadriceps muscle are noted later in the disease[226,227] (Figure 34.6).

Muscle biopsy usually shows nonspecific dystrophic changes with normal immunoreactivity for other proteins such as dysferlin and calpain-3. Some have reported amyloid deposits in intramuscular vessels, similar to some cases of dysferlinopathy.[228,229]

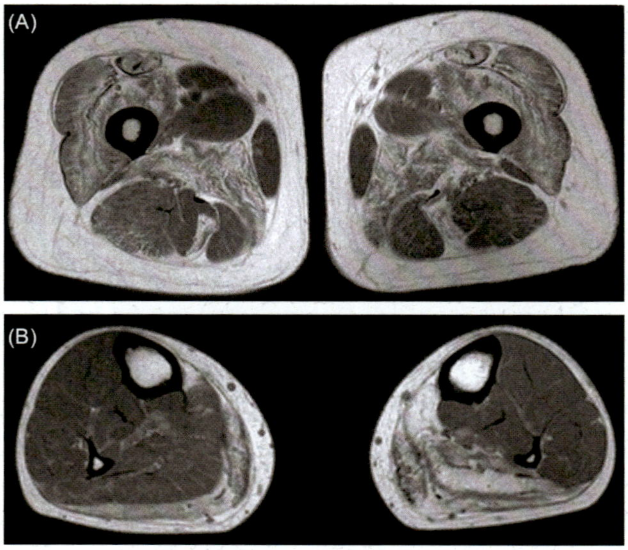

FIGURE 34.6 Muscle magnetic resonance images of a patient with LGMD 2L/anoctaminopathy. (A) Involvement of the quadriceps and hamstring muscles. (B) Asymmetric involvement of the medial gastrocnemius muscle in the same patient.

Cardiac and pulmonary features

There are mixed reports of mild cardiac disease in patients with *ANO5* mutations. While some cohorts report no cardiac disease,[220,230,231] others report mild arrhythmias, such as premature ventricular contractions, or a subclinical dilated cardiomyopathy.[225,232] Clinically significant pulmonary disease is not reported.

Other Rare Autosomal Recessive Limb-girdle Muscular Dystrophies

Telethoninopathy (LGMD 2G)

Pathophysiology and genetics

Telethonin, the mutated protein in LGMD 2G, is a sarcomeric protein mapping to chromosome 17q11.[233–235] Telethonin is a small protein found in cardiac and skeletal muscle, localizing to the Z-disc.[236] It binds to the N-terminal of titin and effectively increases its mechanical resistance.[237] It is phosphorylated by the serine kinase domain of titin, and as a result is also referred to as T-cap.[238] Some mutations identified are in the C-terminus, the site of titin phosphorylation, highlighting the important functional role of telethonin-titin interactions in skeletal muscle.[235]

Clinical features

Based on the studies of a few families of Brazilian origin and rare single cases, most patients experience symptoms at a mean age of 12 to 13 years.[233,235,239] Similarly to other LGMDs, patients show difficulty with walking, running, and climbing stairs. Uniquely, the majority of the initially reported patients developed prominent distal compartment lower extremity weakness and foot drop. Loss of ambulation occurs in the third or fourth decade. Most patients with this LGMD have no cardiac disease. Pronounced calf hypertrophy and highly elevated CKs of 10- to 30-fold are common. Significant intrafamilial variability is observed.

Diagnosis

The clinical phenotype is heterogeneous. The early anterior tibial compartment involvement, with foot drop, can be a guide to diagnosis. Muscle biopsy may show rimmed vacuoles, a nonspecific feature in distal myopathies.[240] Immunohistochemical staining, and Western blot analysis, may show reduced telethonin immunoreactivity.[235,241] However, mutation analysis of this relatively small gene is the easiest way to secure a diagnosis.

Treatment

No specific treatment is available at this time. Like other small genes, telethonin may be more amenable to gene therapy with cDNA delivery with virus vectors.

LGMD 2H

LGMD 2H was noted with high prevalence in the Hutterites of Canada. Genetic linkage studies in this population located the gene to chromosome 9q33.[242] Later, mutations in the *TRIM32* gene, coding for a tripartite motif-containing protein predicted to be involved in protein degradation by ubiquitinating proteins for proteosomal degradation, was identified.[243,244] Mutations in the same gene may result in a sarcotubular myopathy.[245–247] Based on mouse studies, TRIM32 may also play an active role in skeletal muscle regeneration, maintenance of satellite cells, and myoblast differentiation.[248,249]

Affected subjects show significant clinical variability but generally develop proximal muscle weakness and wasting in the lower extremities. Later, trapezius and deltoid weakness, as well as some distal weakness in the anterior tibial group and brachioradialis, may be seen. Mild facial weakness is another feature of this disease. The age of onset of symptoms is variable, ranging from 8 to 27 years. The initial symptoms include exercise-induced muscle weakness, fatigue, and myalgia. Serum CKs are only slightly raised. Compared to the other autosomal recessive LGMDs, most patients have a slower rate of progression. Muscle biopsy usually shows dystrophic muscle without any distinctive features. Electrocardiography may show evidence of mild cardiac disease but no significant pulmonary disease is reported.[250,251]

Titinopathy (LGMD 2J)

Titin is a sarcomeric protein encoded by *TTN* gene on chromosome 2q31.2. *TTN* mutations were first identified in Finnish families with an autosomal dominant distal myopathy known as tibial muscular dystrophy.[252] In some of these families, another muscle disease phenotype with a limb-girdle pattern of weakness emerged and was linked to the same gene (LGMD 2J).[253,254]

Several other phenotypes, including dilated cardiomyopathy[255] and adult-onset hereditary myopathy with early respiratory failure, have since been linked to autosomal dominant *TTN* mutations.[256,257] Recently, compound heterozygous mutations in *TTN* were found in multiple patients with centronuclear congenital myopathy.[258] However, the limb-girdle autosomal recessive phenotype is quite rare and has only been described in the Finnish families with high prevalence of *TTN* mutations.

AUTOSOMAL DOMINANT LIMB-GIRDLE MUSCULAR DYSTROPHIES

Autosomal dominant LGMDs are generally less common, constituting only about 10% of LGMDs. They occur later, and in comparison to autosomal recessive LGMDs generally have a milder course. Commonly, a positive family history is evident; however, *de novo* mutations and germline mosaicism account for some patients without an obvious family history.[10] Unlike most autosomal recessive LGMDs, which constitute the most frequent phenotype of the disease, autosomal dominant LGMDs can be thought of as the less frequent variants of other genetic muscle diseases.

In this section, we will divide autosomal dominant LGMDs into two groups: one without cardiac involvement (LGMD 1A, 1C, and 1E) and another with cardiac involvement (LGMD 1B). Of note, LGMD 1D was originally thought to be an autosomal dominant LGMD with cardiac involvement that genetically linked to chromosome 6q22. Further studies since then have shown the mutation to be on chromosome 2, involving desmin as part of a myofibrillar myopathy.[259]

Autosomal Dominant LGMDs without Cardiac Involvement

Myotilinopathy (LGMD 1A)

This was the first autosomal dominant girdle muscular dystrophy to be linked to a specific genetic locus. It was first described in a family with late-onset LGMD.[260] Later on, the gene was identified to encode myotilin (myofibrillar protein with titin-like Ig domains), a sarcomeric protein.[261] Mutation of a sarcomeric protein was later seen in another LGMD (LGMD 2G as described above) and provides a novel pathophysiologic concept in muscular dystrophy. Myotilin mutations have also been implicated in myofibrillar myopathy 3, a distal, adult-onset, autosomal dominant myopathy, which in fact is the more common clinical presentation of myotilin mutations.[262]

Pathophysiology and mutations

Myotilin, a 57-kDa protein, is mainly found in skeletal and cardiac muscle. It contains two immunoglobulin-like domains in its C-terminus. Its N-terminus contains no known functional domains. It localizes to the Z-disc and sarcolemma and forms functional homodimers.[263] Myotilin directly binds F-actin, α-actinin (an actin cross-linking protein), and filamins. It also prevents actin disassembly induced by latrunculin.[264] Based on these interactions, myotilin may provide a functional and structural link between the cytoskeleton and the sarcolemma via filamins, and thus in particular to the sarcoglycan complex.[265]

Two missense mutations in the N-terminus, *Ser55Phe* and *Thr57Ile*, have been identified in the majority of the cases of LGMD 1A.[261,262,266] The functional significance of these mutations in myotilin protein-protein interactions and Z-disc structure is a matter of active investigation but does not seem to disrupt its binding to α-actinin or filamin.[261]

Clinical features

Onset is usually in adulthood, with a median onset age in the late 20s. The disease is usually slowly progressive and involves proximal leg and arm weakness As the disease progresses, weakness can involve distal muscles, but it almost never leads to loss of ambulation. Apparent anticipation of the weakness may be seen.[267] Most patients have a particular nasal speech quality and dysarthria.[266,268] Achilles tendon contractures may be seen.[269] CK values may be raised 1.6- to 10-fold the upper limit of normal. Cardiac and pulmonary involvement is not a typical feature of this disease.

Diagnosis

The peculiar clinical feature of nasal speech quality in the setting of an autosomal dominant LGMD with mildly raised CK values suggests this diagnosis on clinical grounds. Muscle biopsy shows dystrophic features and in some cases autophagic vacuoles; Z-line streaming and disorganization as well as rod-shaped accumulations similar to nemaline myopathies have been reported on electron microscopy.[261] Immunohistochemistry and Western blot analysis usually show normal levels of myotilin.

Ultimately, mutation analysis would be necessary to arrive at the specific diagnosis.

Treatment

Again, no specific therapies are available at this time.

Caveolinopathy (LGMD 1C)

Pathophysiology and genetics

Mutations in caveolin-3, the muscle specific form of caveolin, result in LGMD 1C as well as other phenotypes.[270,271] Caveolae, localized to the cytoplasmic surface, contribute to sarcolemmal integrity, regulate ion channels, and play a role in signal transduction, among other roles.[272] Caveolins are membrane-based proteins that span the membrane and include large intracellular domains. They form homodimers, which subsequently oligomerize to form membrane-based specialized lipid rafts.[273] These lipid rafts bring receptors and their signal transduction apparatus in close proximity, facilitating signal transduction.

In the muscle, caveolin-3 is expressed throughout the T-tubule system.[274] It also interacts with dystrophin-sarcoglycan complex of proteins by binding the intracellular portion of β-dystroglycan.[275,276]

Most mutations resulting in the limb-girdle phenotype are missense or microdeletions, and act in a dominant negative fashion.[277,278] This is supported by the significant reduction of caveolin-3 immunoreactivity in muscle biopsy specimens of heterozygous patients.[270,275] *De novo* mutations without a positive family history have also been described.[279]

Clinical features

Caveolin-3 mutations generally result in four different muscle disease phenotypes: LGMD, rippling muscle disease, asymptomatic CK elevation, and a distal myopathy. There is significant overlap of the symptoms in different patients. There is also significant intrafamilial variability without a typical clinical phenotype. Some patients develop weakness in early childhood while others remain asymptomatic. Myalgias and cramping with exertion have been reported.[275] Some patients have calf hypertrophy.[280] While other caveolin mutations have been implicated in cardiac conduction problems such as long QT syndrome,[281] cardiac involvement is not a typical feature of LGMD 1C. Pulmonary disease is also uncommon.

Diagnosis

As described above, there is significant phenotypic variability of caveolin-3 mutations. However, in the case of an apparently autosomal dominant LGMD, muscle biopsy and immunohistochemical analysis usually show significantly reduced caveolin-3 immunoreactivity due to the dominant negative effects. In some patients with caveolin-3 mutations, dysferlin may be mislocalized with apparently reduced immunoreactivity on muscle biopsy, but normal Western blot levels.[185] Ultimately, genetic confirmation is necessary in suggestive cases. *De novo* mutations necessitate consideration of caveolin-3 as a diagnosis even in the absence of clear family history.

Treatment

No specific treatments are known, but supportive management can be helpful to patients.

LGMD 1E

This particular muscular dystrophy has been genetically linked to chromosome 7q in several families.[282] Most patients develop weakness in the third and fourth decade, more pronounced in the lower extremities but also involving the proximal shoulder girdle muscles. CK levels are only mildly elevated (1.3- to 3-fold) in most cases but can be elevated up to 10-fold. Mild dysphagia is also noted in some patients.[283] Muscle biopsy usually shows dystrophic muscle. Electron microscopy may show Z-line streaming, disorganization, and autophagic vacuoles.[284]

Whole exome sequencing has recently identified three different mutations on chromosome 7q36.6 in some patients with LGMD 1E.[283,285] The mechanism of pathogenesis of these mutations in muscle is unknown.

Autosomal Dominant LGMDs with Cardiac Involvement

Autosomal dominant LGMDs with cardiac involvement have a close relationship to autosomal dominant EDMD, which is discussed elsewhere in this book (Chapter 35). The LGMD phenotype is discussed in following sections. For a more detailed discussion of the pathophysiology and genetics of lamin A/C mutations, the reader is referred to the chapter dedicated to EDMD.

Laminopathy (LGMD 1B)

This LGMD is allelic with autosomal dominant EDMD. It was originally described in Dutch families with an LGMD profile with prominent cardiac involvement.[286] Subsequently, linkage to chromosome 1q was established, a region where AD-EDMD had its locus as well.[287] The nuclear intermediate filament lamin A/C was ultimately found to be the causative mutation in both disorders.[288,289]

Pathophysiology and genetics

Lamin A/C is an intermediate filament, lining the inner nuclear membrane, part of the nuclear envelope.[290] The transcript from the gene is spliced differentially to give rise to the two different forms, lamin A and lamin C. These molecules bind to other inner nuclear membrane proteins such as emerin, which is mutated in X-linked EDMD, but their exact function is not fully known. Anchoring of chromatin to the nuclear membrane has been suggested.[291,292] Other possible roles include functions in DNA replication, nuclear pore formation, nuclear material transport, and influencing gene regulation by changing nuclear infrastructure.[293,294]

Mutations described in lamin A/C are mostly missense mutations but include some single codon deletions, nonsense mutations, and splice mutations.[295] A substantial portion of patients have *de novo* mutations, as many as 76% in one series.[296]

Several different phenotypes have been linked to lamin A/C mutations: autosomal dominant EDMD,[297] LGMD 1B, a form of congenital muscular dystrophy, dilated cardiomyopathy,[298] lipodystrophies,[299] peripheral neuropathy,[300] and premature aging.[301,302] It is possible that different mutations on different domains of the protein may result in different clinical phenotypes but this remains to be shown unequivocally.[303]

Clinical features

Whereas the EDMD and CMD forms have onset in earlier childhood, the LGMD presentation of laminopathy usually occurs in the late teens and early adulthood with a relatively wide range extending all the way to the fourth decade.[286] Most patients develop weakness in the pelvic girdle muscles, with difficulty getting up from the floor and with subsequent development of arm weakness. Retrospectively, some patients report a history of waddling gait, lordosis, and difficulties with running, starting in childhood. Mild facial weakness and mild anterior tibial compartment weakness may be seen in some patients.[295] Progression is rather slow but appears to be fully penetrant by age 45. Biceps and Achilles tendon contractures may develop, though they are not typically as severe as in EDMD.[286] CK values are usually mildly elevated, 1.5- to 3-fold.

Cardiac features

Cardiac problems may first manifest in the third and fourth decade with first-degree atrioventricular (AV) block that progresses to complete AV block. Similarly to the EDMD and CMD presentations of *LMNA* mutations, sudden cardiac death is a prominent risk and may be its first and only manifestation.[304] Dilated cardiomyopathy only rarely occurs.[305–307] The risk of developing cardiac conductive problems increases with increasing age and as a result careful cardiac studies including echocardiography, as well as 24-hour electrocardiography for evaluation of subclinical cardiac features, are important.

Diagnosis

The diagnosis is based on suggestive clinical features, the cardiac manifestations, the presence of contractures, and autosomal dominant hereditary pattern in the family. Muscle biopsy evaluation is not particularly helpful as lamin A/C immunoreactivity appears normal even in the presence of mutations; however, it is helpful to rule out other causes of LGMD. In the absence of family history, diagnosis purely based on clinical features is virtually impossible. Thus, direct molecular genetic analysis should be initiated when LGMD 1B is suspected.

Treatment

No specific treatments are available. However, careful comprehensive cardiac evaluation and prospective monitoring is important to evaluate for the need for intracardiac defibrillator placement to prevent fatal cardiac arrhythmias.[308]

OTHER MUSCULAR DYSTROPHIES THAT MAY PRESENT AS LGMD

As described in the preceding text, the limb-girdle muscular dystrophies have a broad phenotypic spectrum with significant overlap. It is not surprising that other major muscular dystrophies may present with a similar phenotype and should be considered in the differential diagnosis. As discussed in the introduction section, the *limb-girdle syndrome* has a number of causes. Thus, any patient with weakness in a limb-girdle distribution should be evaluated for acquired inflammatory myopathies, metabolic myopathies such as acid maltase deficiency, mitochondrial disease, spinal muscular atrophy, endocrine myopathies, and congenital myopathies. Once these are evaluated, the possibility of other genetic myopathies should be considered. This section will focus on other genetic myopathies and dystrophies that have a different typical presentation but in some cases may manifest with an LGMD-like pattern.

Autosomal Recessive Conditions Presenting as LGMD

Partial Laminin $\alpha 2$ Deficiency

This condition is allelic to primary laminin $\alpha 2$ deficiency, the form of congenital muscular dystrophy related to severe deficiency of laminin α2, the heavy chain of the heterotrimer laminin211 (merosin). Primary mutations in the *LAMA2* gene on chromosome 6q are responsible. In case of complete deficiency, the clinical phenotype is that of a severe congenital muscular dystrophy (discussed in Chapter 29) and independent walking is almost never achieved. MRI of the brain may show supratentorial white matter T2 hyperintensities, and some children may develop seizures.[309]

The clinical presentation, however, may be that of a milder congenital muscular dystrophy or may resemble LGMD with proximal weakness only, in which case walking may be achieved. Joint contractures can be prominent and resemble EDMD. In these cases, only a partial deficiency of laminin α2 on muscle biopsy specimen is seen, most clearly with an antibody against the 300 kDa fragment.[310] Western blot analysis is a more sensitive measure and helpful in such cases. Patients with partial laminin α2 deficiency almost always show subtle alterations of white matter signal on T2-weighted MRIs.[311] Patients with the LGMD phenotype usually have a missense mutation or an in-frame deletion in compound heterozygosity with a null mutation.[312,313]

X-linked Recessive Conditions Presenting as LGMD

Some patients with emerin mutations, the protein mutated in classical X-linked EDMD, may present with an LGMD-like presentation without cardiac involvement.[314] These patients may develop proximal weakness in early childhood without significant contractures at the outset. Immunohistochemistry is helpful as most patients with emerin mutations show abnormal emerin immunoreactivity in muscle, lymphocytes, buccal mucosa cells, or even skin fibroblasts.

Autosomal Dominant Conditions that May Present as LGMD

Facioscapulohumeral Dystrophy

Facioscapulohumeral dystrophy (FSHD) is probably the single most important entity to consider in patients with an LGMD-like presentation, especially those with an autosomal dominant pattern of inheritance. A significant portion of patients evaluated for autosomal dominant LGMD are found to have FSHD.[315,316] Some patients with FSHD have a predominantly limb-girdle pattern of weakness with minimal or subtle facial involvement that is easily overlooked.[317] Therefore, a high index of clinical suspicion is prudent.

Myotonic Dystrophy Types 1 and 2

Myotonic dystrophy (DM) is one of the most common muscular dystrophies. Patients with myotonic dystrophy, especially DM type 2, may present with a limb-girdle pattern of muscle weakness. Affected family members may have subclinical disease. Careful clinical examination of the patient and their family members as well as EMG will usually be revealing. Molecular genetic testing should be pursued in the appropriate clinical setting.[318]

Collagen VI Related Dystrophies

Collagen VI related disorders constitute a spectrum of severities from the (milder) Bethlem myopathy to the (more severe) Ullrich congenital muscular dystrophy. Bethlem myopathy is the most important to consider in the differential diagnosis of LGMD. Bethlem myopathy is an autosomal dominant (and rarely also recessive) disorder caused by mutations in one of three genes coding for the alpha chains of collagen type VI, *COL6A1* and *A2* on chromosome 21q22.3[319] or *COL6A3* on chromosome 2q37.[320] Missense mutations and in-frame exon skipping mutations are the most common causes of the disease, likely due to a dominant negative effect on formation of heterotrimers. Recessive and *de novo* dominant mutations in collagen VI underlie the more severe phenotype Ullrich congenital muscular dystrophy.[321]

Ullrich CMD is characterized by severe weakness with joint hyperlaxity and contractures at birth.[322] More severely affected patients may not achieve the ability to walk due to both severe weakness and contractures, but most achieve that ability, albeit with some delay.[323] Weakness is only gradually progressive and worsening contractures likely cause more severe loss of function over time. Respiratory insufficiency, which may begin before loss of independent ambulation, also worsens with aging.[322] As a result, treatment with noninvasive bilevel positive airway pressure ventilation may be required. There are also patients presenting with phenotypes of intermediate severity between Bethlem and Ullrich, with independent ambulation into late teenage years and young adulthood.[324] Typically patients of this severity are not able to jump or run, while the typical coexistence between distal joint hyperlaxity and more proximal contractures is evident, with prominent elbow and shoulder contractures in the upper extremity. Patients in this intermediate phenotype also experience progressive respiratory insufficiency, often requiring nighttime noninvasive ventilator support while remaining ambulant.[325]

Diagnosis is suspected based on congenital weakness and concomitant severe contractures and joint laxity. Skin findings with prominent keratosis pilaris of the extensor surfaces of the arms and legs and abnormal keloid formation may aid in diagnosis. Muscle imaging by MRI or

CASE EXAMPLE 34.3 Bethlem Myopathy

A mother presented with her 3-year-old daughter for evaluation of an apparent familial muscle disorder. The 33-year-old mother was born with torticollis, requiring surgical intervention at 6 months of age. She had always had difficulty keeping up at school sports and could never run very fast. At about age 10, she could no longer participate in sports and began developing contractures of the Achilles tendons, elbows, and finger flexors. Weakness has been more pronounced but slowly progressive over the years.

On examination, there were no cranial nerve abnormalities. The neck flexors, proximal muscles, biceps, and triceps were MRC grade 4/5, the distal upper extremity muscles were at 5-/5, and the proximal lower extremity muscles were at 4- to 4/5. Quadriceps, hamstrings, and foot eversion and dorsiflexion were rated at 4+/5. Mild contractures of the neck flexors and pronounced contractures of the elbows and finger flexors were noted. The fingers could not be straightened with the wrists dorsiflexed. There were prominent bilateral plantar flexor contractures. Hypertrophic scars were also noted. The skin was otherwise normal. The serum CK level, echocardiography, and electrocardiography were normal.

The daughter was born at full term after a normal pregnancy, and was found at birth to have mild hip dysplasia and congenital torticollis. She walked at 17 months of age. At age 3 years, she had difficulty getting up from the floor and climbing stairs, ran slowly, and had a waddling gait.

On examination, she had a mild torticollis, with proximal greater than distal muscular hypotonia and weakness. Muscle bulk was reduced and there was mild scapular winging. The distal joints were markedly hyperextensible, but she had Achilles tendon and elbow contractures. There was a lumbar lordosis without scoliosis. She walked with a positive Trendelenburg sign and arose from the floor using a Gowers' maneuver. She was unable to run.

Further family history revealed that the maternal uncle and grandfather were also affected by weakness and contractures to a variable extent.

Muscle biopsy in the daughter revealed nonspecific myopathic changes and occasional degenerating and regenerating fibers. Collagen VI immunohistochemistry was normal. Muscle biopsy in the mother showed nonspecific myopathic changes with slight reduction in collagen type VI immunoreactivity.

ultrasound in the collagen VI related dystrophies may show a characteristic pattern of muscle degeneration beginning adjacent to the fascia, resulting in the typical rimming of muscles and the "central cloud" in the rectus femoris along its central fascia.[326–328] Molecular testing is similar for all collagen VI related phenotypes, in that dominant as well as recessive mechanisms in all three genes have to be considered (see the following).

Most patients with Bethlem myopathy develop symptoms in childhood; some may have congenital weakness/hypotonia, torticollis, or foot deformities. (See Case Example 34.3.) However, age of onset is usually variable and may be in adulthood (Figure 34.7). Weakness is usually more predominant in the proximal muscles, although distal weakness may also be present. Early on, there may be no contractures; on the contrary, hyperlaxity of joints may be present. Contractures usually develop at the end of the first decade, initially involving the deep finger flexors and proximal joints. When these are absent or minor, the phenotype can resemble LGMD (Figure 34.8). As the contractures worsen over time, patients may lose their ability to ambulate. Approximately half of the patients are dependent on assisted mobility by age 60.[329,330] There may be prominent keloid formation as a sign of skin involvement.[331] While cardiac disease is not a typical feature of this disease, respiratory compromise does occur in particular in more severe cases, and needs to be monitored closely.[325]

Diagnosis is usually secured by identifying mutations in one of the three genes encoding one of the alpha chains of collagen type VI. Although mutations in the Bethlem group are more commonly dominant, recessive mutations have also been described. Immunohistochemistry on the muscle biopsy specimen is difficult in the Bethlem group, as there may only be a subtle deficiency of collagen VI in the basement membrane, detection of which requires double labeling techniques. Fibroblast cultures from skin biopsies can be used to evaluate collagen VI production and for identifying mutations on RNA.[332,333] Sanger sequencing, as well as next-generation sequencing, is available to screen for mutations in the collagen type VI genes and arrive at a specific diagnosis in the appropriate clinical setting.

SUMMARY AND APPROACH TO PATIENTS

As discussed in the preceding text, there is significant genetic and phenotypic variability of the different LGMDs. As a result, and similarly to other genetic diseases, a careful family history and even examination of family members who may or may not be aware of their own symptoms can play a big role in narrowing down the differential diagnosis. For example, in the case of autosomal dominant LGMDs, a careful review of the family history for cases of sudden cardiac death or other cardiac problems becomes crucial and may have significant implications for the care of the patient.

In addition, certain diagnostically helpful clinical features need to be evaluated fully.

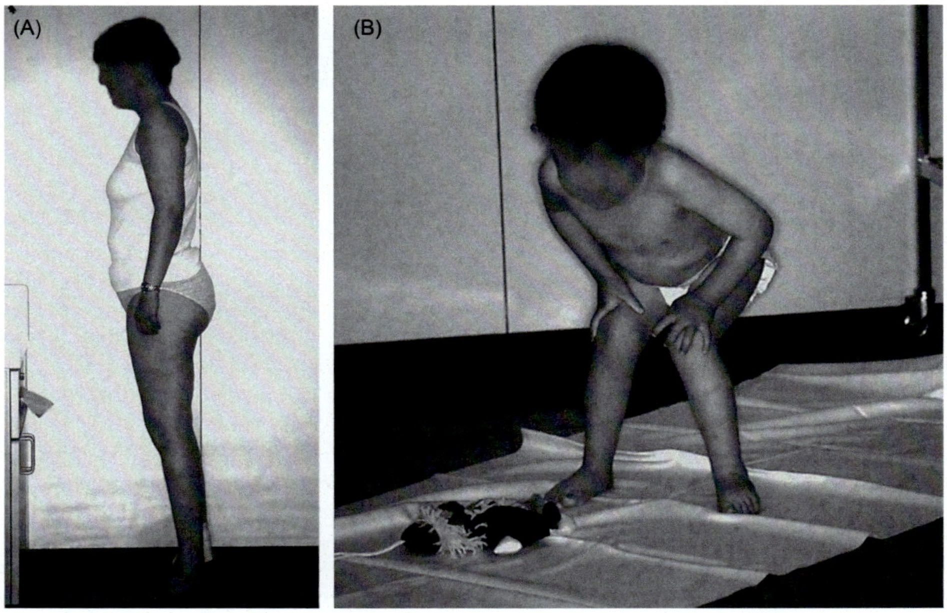

FIGURE 34.7 Bethlem myopathy. (A) 38-year-old mother and (B) her 2.5-year-old daughter. In the mother there are contractures of the neck extensors, pectoralis muscles, elbows, finger flexors, and Achilles tendons (note bandage from recent tendon release surgery). In the daughter there are no contractures except for torticollis; she has diffuse but more proximally pronounced weakness (she is using a Gowers' maneuver to get up from the floor).

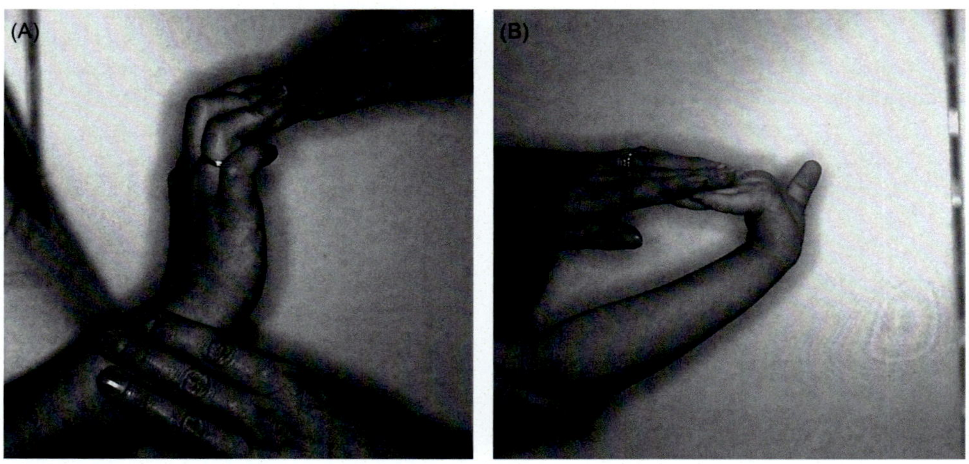

FIGURE 34.8 Bethlem myopathy. (A) Typical deep finger-flexor contractures in an adult with Bethlem myopathy. (B) Marked distal laxity in a young child with Bethlem myopathy.

1. Age of onset and speed of progression may provide an important clue in cases of suspected LGMD. Earlier onset and more severe course with elevated CK are suggestive of mutations in the sarcoglycan complex or FKRP. In cases with later childhood onset symptoms, calpainopathy becomes more likely. If symptoms first appear in adolescence or early adulthood, dysferlinopathy should be suspected, especially if CK levels are extremely elevated. Mild adult-onset symptoms with significantly elevated CK may suggest anoctaminopathy, or myotilinopathy if the CK is close to normal. However, as discussed, there is significant phenotypic variability of presentation of each of these disorders, and onset of the disease should only be used as a general guideline.
2. The distribution of muscle weakness, presence or absence of contractures, or muscle groups that are spared can be important clues. Obviously, these

TABLE 34.2 Characteristic Clinical and Biopsy Findings in Autosomal Dominant Limb-girdle Muscular Dystrophies

	Helpful Clinical Features	Immunohistochemistry and Western Blot
LGMD 1A (Myotilinopathy)	– Late onset and milder disease – Nasal speech quality – Early contractures of the Achilles tendons	Usually normal
LGMD 1B (Laminopathy)	– Contractures mimic EDMD but can be minimal – Conductive cardiac disease and AV block – Early childhood onset possible – *De novo* mutations common	Usually normal
LGMD 1C (Caveolinopathy)	– Highly variable – May present with isolated elevated CK, with no/minimal weakness	Majority of cases show reduced or absent protein

Abbreviations: LGMD, Limb-girdle muscular dystrophy; EDMD, Emery-Dreifuss muscular dystrophy; AV, atrioventricular.

TABLE 34.3 Characteristic Clinical and Biopsy Findings in Autosomal Recessive Limb-girdle Muscular Dystrophies

	Helpful Clinical Features	Immunohistochemistry and Western Blot
LGMD 2A (Calpainopathy)	– Juvenile onset (~10 yrs) – Early periscapular weakness – Muscle wasting and atrophy – Hamstrings weakness – Sparing of the hip abductors	– IHC is not available but WB shows reduction in most cases – Secondary reduction in dysferlin – Eosinophilic myositis is reported
LGMD 2B (Dysferlinopathy)	– Late adolescent onset (~18 yrs) – Can present with proximal and/or distal weakness – Early gastrocnemius involvement, esp. on MRI – Periscapular muscles spared – CK levels extremely high	– Possible by IHC but WB is more specific – Secondary reduction in calpain – Intramuscular vessels may show amyloid deposits
LGMD 2C–F (Sarcoglycanopathies)	– Early childhood onset (6–8 yrs) – Similar pattern to DMD/BMD except early scapular weakness – Calf hypertrophy very common – CK levels are very high – Cardiomyopathy in a subset	– IHC and WB possible – Pattern of absence of all sarcoglycan proteins may be helpful
LGMD 2G (Telethoninopathy)	– Variable clinical phenotype – Early anterior tibial weakness and foot drop	– IHC and WB possible
LGMD 2H	– Variable clinical phenotype – LE > UE weakness – Trapezius and deltoid affected	– Not possible
LGMD 2I	– Typically mild but early childhood onset is seen – Weakness in the neck flexors and axial muscles – UE > LE weakness – Brachioradialis, tongue, and calf hypertrophy – Cardiopulmonary disease possible	– IHC may show reduced α-dystroglycan and laminin α2 but not specific
LGMD 2L (Anoctaminopathy)	– Adult onset – Knee flexor and extensor weakness – Medial gastrocnemius, hamstring, and quadriceps atrophy, esp. vastus medialis	– Not possible – Intramuscular vessels may show amyloid deposits

Abbreviations: LGMD, Limb-girdle muscular dystrophy; WB, Western blot; IHC, Immunohistochemistry; DMD, Duchenne muscular dystrophy; BMD, Becker muscular dystrophy; UE, upper extremity; LE, lower extremity.

features are more informative earlier in the disease, before diffuse weakness occurs, and may not be feasible in very young children. Ancillary radiologic imaging such as MRI can be helpful in defining the distribution of myopathy. For a summary of the specific clinical features in the commonly encountered LGMDs, please refer to Table 34.2 and Table 34.3.

3. Presence or absence of cardiac features can be an important clue, especially in autosomal dominant LGMDs, and should be evaluated exhaustively due to its important clinical repercussions.

Muscle Biopsy and Protein Studies

Muscle biopsy can be an important step in evaluation of patients with the limb girdle syndrome. As discussed previously, it is usually a critical part of the diagnostic evaluation for the mimickers of a genetic LGMD. Therefore, prior to categorization of a patient in the LGMD group, a muscle biopsy is usually obtained.

In LGMD patients, muscle biopsy with routine histochemical stains usually shows dystrophic features with fiber size variability, increased endomysial fibrosis, fatty replacement, and myofiber degeneration and regeneration. Intermyofibrillary architecture can be disrupted, showing moth-eaten, lobulated, or whorled fibers. For example, a predominance of lobulated fibers of the type I histochemical type is seen in calpainopathy. Special attention should be paid to evaluate for the presence of ragged red fibers, glycogen accumulation on the periodic acid-Schiff stain, or findings suggestive of an acquired inflammatory myopathy. The presence of rimmed vacuoles may suggest myotilinopathy or telethoninopathy, but prominent vacuoles or inclusion bodies may suggest an alternative diagnosis to LGMD.

Immunohistochemistry in addition to Western blot analysis can be helpful in many cases. While assessment against many antibodies can be done, these methods are at best semiquantitative and should be interpreted with caution. Major patterns of immunohistochemical reduction of expression in different LGMDs are summarized in Tables 34.2 and 34.3.

Molecular and Genetic Testing

Ultimately, it is preferable that all diagnoses of genetic LGMD be confirmed by molecular genetics when available. The availability of whole exome sequencing, and simultaneous sequencing of multiple genes at relatively low cost using next-generation technology, has recently provided an alternative to traditional diagnostic methods. However, some mutations with large deletions or duplications are not identified using this approach. In addition, variations of unknown significance complicate the interpretation of the results in the absence of functional data. As a result, these approaches should be used as a complement to the clinical features, and molecular and clinical features carefully correlated. It is always important to adequately consider acquired muscle disease, as its treatment and genetic implications are so different.

REFERENCES

1. Walton JN, Nattrass FJ. On the classification, natural history and treatment of the myopathies. *Brain* 1954;**77**:169–231.
2. Shields R. Limb girdle syndromes. In: Engel A, Franzini-Armstrong C, editors. *Myology*. New York: McGraw-Hill; 1994. pp. 1258–74.
3. Walton J, Gardner-Medwin D. The muscular dystrophies. In: Walton J, editor. *Disorders of Voluntary Muscle*. Edinburgh: Churchill Livingstone; 1991. pp. 519–68.
4. Bushby KM, Beckmann JS. The 105th ENMC sponsored workshop: pathogenesis in the non-sarcoglycan limb-girdle muscular dystrophies, Naarden, April 12–14, 2002. *Neuromuscul Disord* 2003;**13**:80–90.
5. Dubowitz V. The muscular distrophies and allied disorders II: Limb-girdle muscular dystrophies. In: Dubowitz V, Sewry C, Oldfors A, Lane R, editors. *Muscle Biopsy: A Practical Approach*. London: Saunders Elsevier; 2013. pp. 276–301.
6. Yates JR, Emery AE. A population study of adult onset limb-girdle muscular dystrophy. *J Med Genet* 1985;**22**:250–7.
7. Fardeau M, Hillaire D, Mignard C, Feingold N, Feingold J, Mignard D, et al. Juvenile limb-girdle muscular dystrophy. Clinical, histopathological and genetic data from a small community living in the Reunion Island. *Brain* 1996;**119**(Pt 1):295–308.
8. Argov Z, Sadeh M, Mazor K, Soffer D, Kahana E, Eisenberg I, et al. Muscular dystrophy due to dysferlin deficiency in Libyan Jews. Clinical and genetic features. *Brain* 2000;**123**(Pt 6):1229–37.
9. Jackson CE, Strehler DA. Limb-girdle muscular dystrophy: clinical manifestations and detection of preclinical disease. *Pediatrics* 1968;**41**:495–502.
10. van der Kooi AJ, Barth PG, Busch HF, de Haan R, Ginjaar HB, van Essen AJ, et al. The clinical spectrum of limb girdle muscular dystrophy. A survey in The Netherlands. *Brain* 1996;**119** (Pt 5):1471–80.
11. Norwood FL, Harling C, Chinnery PF, Eagle M, Bushby K, Straub V. Prevalence of genetic muscle disease in Northern England: in-depth analysis of a muscle clinic population. *Brain* 2009;**132**:3175–86.
12. Matsumura K, Tomé FM, Collin H, Azibi K, Chaouch M, Kaplan JC, et al. Deficiency of the 50K dystrophin-associated glycoprotein in severe childhood autosomal recessive muscular dystrophy. *Nature* 1992;**359**:320–2.
13. Azibi K, Bachner L, Beckmann JS, Matsumura K, Hamouda E, Chaouch M, et al. Severe childhood autosomal recessive muscular dystrophy with the deficiency of the 50 kDa dystrophin-associated glycoprotein maps to chromosome 13q12. *Hum Mol Genet* 1993;**2**:1423–8.
14. Fardeau M, Matsumura K, Tomé FM, Collin H, Leturcq F, Kaplan JC, et al. Deficiency of the 50 kDa dystrophin associated glycoprotein (adhalin) in severe autosomal recessive muscular dystrophies

in children native from European countries. *C R Acad Sci III* 1993;**316**:799–804.

15. Matsumura K, Campbell KP. Deficiency of dystrophin-associated proteins: a common mechanism leading to muscle cell necrosis in severe childhood muscular dystrophies. *Neuromuscul Disord* 1993;**3**:109–18.
16. Ginjaar HB, van der Kooi AJ, Ceelie H, Kneppers AL, van Meegen M, Barth PG, et al. Sarcoglycanopathies in Dutch patients with autosomal recessive limb girdle muscular dystrophy. *J Neurol* 2000;**247**:524–9.
17. Vainzof M, Passos-Bueno MR, Pavanello RC, Marie SK, Oliveira AS, Zatz M. Sarcoglycanopathies are responsible for 68% of severe autosomal recessive limb-girdle muscular dystrophy in the Brazilian population. *J Neurol Sci* 1999;**164**:44–9.
18. Fanin M, Duggan DJ, Mostacciuolo ML, Martinello F, Freda MF, Sorarù G, et al. Genetic epidemiology of muscular dystrophies resulting from sarcoglycan gene mutations. *J Med Genet* 1997;**34**:973–7.
19. Dinçer P, Leturcq F, Richard I, Piccolo F, Yalnizoglu D, de Toma C, et al. A biochemical, genetic, and clinical survey of autosomal recessive limb girdle muscular dystrophies in Turkey. *Ann Neurol* 1997;**42**:222–9.
20. Duggan DJ, Gorospe JR, Fanin M, Hoffman EP, Angelini C. Mutations in the sarcoglycan genes in patients with myopathy. *N Engl J Med* 1997;**336**:618–24.
21. Waite A, Tinsley CL, Locke M, Blake DJ. The neurobiology of the dystrophin-associated glycoprotein complex. *Ann Med* 2009;**41**:344–59.
22. Tinsley JM, Blake DJ, Pearce M, Knight AE, Kendrick-Jones J, Davies KE. Dystrophin and related proteins. *Curr Opin Genet Dev* 1993;**3**:484–90.
23. Ozawa E, Yoshida M, Suzuki A, Mizuno Y, Hagiwara Y, Noguchi S. Dystrophin-associated proteins in muscular dystrophy. *Hum Mol Genet* 1995;(4 Spec No):1711–16.
24. Sadoulet-Puccio HM, Rajala M, Kunkel LM. Dystrobrevin and dystrophin: an interaction through coiled-coil motifs. *Proc Natl Acad Sci USA* 1997;**94**:12413–18.
25. Sandonà D, Betto R. Sarcoglycanopathies: molecular pathogenesis and therapeutic prospects. *Expert Rev Mol Med* 2009;**11**:e28.
26. Grady RM, Grange RW, Lau KS, Maimone MM, Nichol MC, Stull JT, et al. Role for alpha-dystrobrevin in the pathogenesis of dystrophin-dependent muscular dystrophies. *Nat Cell Biol* 1999;**1**:215–20.
27. Grady RM, Akaaboune M, Cohen AL, Maimone MM, Lichtman JW, Sanes JR. Tyrosine-phosphorylated and nonphosphorylated isoforms of alpha-dystrobrevin: roles in skeletal muscle and its neuromuscular and myotendinous junctions. *J Cell Biol* 2003;**160**:741–52.
28. Jung D, Yang B, Meyer J, Chamberlain JS, Campbell KP. Identification and characterization of the dystrophin anchoring site on beta-dystroglycan. *J Biol Chem* 1995;**270**:27305–10.
29. Ervasti JM, Campbell KP. A role for the dystrophin-glycoprotein complex as a transmembrane linker between laminin and actin. *J Cell Biol* 1993;**122**:809–23.
30. Henry MD, Campbell KP. Dystroglycan: an extracellular matrix receptor linked to the cytoskeleton. *Curr Opin Cell Biol* 1996;**8**: 625–31.
31. Henry MD, Campbell KP. Dystroglycan inside and out. *Curr Opin Cell Biol* 1999;**11**:602–7.
32. Moore CJ, Winder SJ. The inside and out of dystroglycan post-translational modification. *Neuromuscul Disord* 2012;**22**:959–65.
33. Brockington M, Yuva Y, Prandini P, Brown SC, Torelli S, Benson MA, et al. Mutations in the fukutin-related protein gene (FKRP) identify limb girdle muscular dystrophy 2I as a milder allelic variant of congenital muscular dystrophy MDC1C. *Hum Mol Genet* 2001;**10**:2851–9.
34. Ozawa E, Mizuno Y, Hagiwara Y, Sasaoka T, Yoshida M. Molecular and cell biology of the sarcoglycan complex. *Muscle Nerve* 2005;**32**:563–76.
35. Roberds SL, Leturcq F, Allamand V, Piccolo F, Jeanpierre M, Anderson RD, et al. Missense mutations in the adhalin gene linked to autosomal recessive muscular dystrophy. *Cell* 1994;**78**:625–33.
36. McNally EM, Yoshida M, Mizuno Y, Ozawa E, Kunkel LM. Human adhalin is alternatively spliced and the gene is located on chromosome 17q21. *Proc Natl Acad Sci USA* 1994;**91**:9690–4.
37. Noguchi S, McNally EM, Ben Othmane K, Hagiwara Y, Mizuno Y, Yoshida M, et al. Mutations in the dystrophin-associated protein gamma-sarcoglycan in chromosome 13 muscular dystrophy. *Science* 1995;**270**:819–22.
38. Bönnemann CG, Modi R, Noguchi S, Mizuno Y, Yoshida M, Gussoni E, et al. Beta-sarcoglycan (A3b) mutations cause autosomal recessive muscular dystrophy with loss of the sarcoglycan complex. *Nat Genet* 1995;**11**:266–73.
39. Lim LE, Duclos F, Broux O, Bourg N, Sunada Y, Allamand V, et al. Beta-sarcoglycan: characterization and role in limb-girdle muscular dystrophy linked to 4q12. *Nat Genet* 1995;**11**:257–65.
40. Nigro V, de Sá Moreira E, Piluso G, Vainzof M, Belsito A, Politano L, et al. Autosomal recessive limb-girdle muscular dystrophy, LGMD2F, is caused by a mutation in the delta-sarcoglycan gene. *Nat Genet* 1996;**14**:195–8.
41. McNally EM, Ly CT, Kunkel LM. Human epsilon-sarcoglycan is highly related to alpha-sarcoglycan (adhalin), the limb girdle muscular dystrophy 2D gene. *FEBS Lett* 1998;**422**:27–32.
42. Wheeler MT, Zarnegar S, McNally EM. Zeta-sarcoglycan, a novel component of the sarcoglycan complex, is reduced in muscular dystrophy. *Hum Mol Genet* 2002;**11**:2147–54.
43. Valente EM, Misbahuddin A, Brancati F, Placzek MR, Garavaglia B, Salvi S, et al. Analysis of the epsilon-sarcoglycan gene in familial and sporadic myoclonus-dystonia: evidence for genetic heterogeneity. *Mov Disord* 2003;**18**:1047–51.
44. Valente EM, Edwards MJ, Mir P, DiGiorgio A, Salvi S, Davis M, et al. The epsilon-sarcoglycan gene in myoclonic syndromes. *Neurology* 2005;**64**:737–9.
45. Szubiga M, Rudzińska M, Bik-Multanowski M, Pietrzyk JJ, Szczudlik A. A novel conserved mutation in SGCE gene in 3 unrelated patients with classical phenotype myoclonus-dystonia syndrome. *Neurol Res* 2013;**35**:659–62.
46. Miller G, Wang EL, Nassar KL, Peter AK, Crosbie RH. Structural and functional analysis of the sarcoglycan-sarcospan subcomplex. *Exp Cell Res* 2007;**313**:639–51.
47. Thompson TG, Chan YM, Hack AA, Brosius M, Rajala M, Lidov HG, et al. Filamin 2 (FLN2): A muscle-specific sarcoglycan interacting protein. *J Cell Biol* 2000;**148**:115–26.
48. Rafii MS, Hagiwara H, Mercado ML, Seo NS, Xu T, Dugan T, et al. Biglycan binds to alpha- and gamma-sarcoglycan and regulates their expression during development. *J Cell Physiol* 2006;**209**: 439–47.

49. Assereto S, Mastrototaro M, Stringara S, Gazzerro E, Broda P, Nicchia GP, et al. Aquaporin-4 expression is severely reduced in human sarcoglycanopathies and dysferlinopathies. *Cell Cycle* 2008;**7**:2199–207.
50. Crosbie RH, Dovico SA, Flanagan JD, Chamberlain JS, Ownby CL, Campbell KP. Characterization of aquaporin-4 in muscle and muscular dystrophy. *FASEB J* 2002;**16**:943–9.
51. Crosbie RH, Heighway J, Venzke DP, Lee JC, Campbell KP. Sarcospan, the 25-kDa transmembrane component of the dystrophin-glycoprotein complex. *J Biol Chem* 1997;**272**:31221–4.
52. Crosbie RH, Lebakken CS, Holt KH, Venzke DP, Straub V, Lee JC, et al. Membrane targeting and stabilization of sarcospan is mediated by the sarcoglycan subcomplex. *J Cell Biol* 1999;**145**:153–65.
53. Ben Hamida M, Ben Hamida C, Zouari M, Belal S, Hentati F. Limb-girdle muscular dystrophy 2C: clinical aspects. *Neuromuscul Disord* 1996;**6**:493–4.
54. Passos-Bueno MR, Vainzof M, Moreira ES, Zatz M. Seven autosomal recessive limb-girdle muscular dystrophies in the Brazilian population: from LGMD2A to LGMD2G. *Am J Med Genet* 1999;**82**:392–8.
55. Trabelsi M, Kavian N, Daoud F, Commere V, Deburgrave N, Beugnet C, et al. Revised spectrum of mutations in sarcoglycanopathies. *Eur J Hum Genet* 2008;**16**:793–803.
56. Dickens NJ, Beatson S, Ponting CP. Cadherin-like domains in alpha-dystroglycan, alpha/epsilon-sarcoglycan and yeast and bacterial proteins. *Curr Biol* 2002;**12**:R197–9.
57. Carrié A, Piccolo F, Leturcq F, de Toma C, Azibi K, Beldjord C, et al. Mutational diversity and hot spots in the alpha-sarcoglycan gene in autosomal recessive muscular dystrophy (LGMD2D). *J Med Genet* 1997;**34**:470–5.
58. Eymard B, Romero NB, Leturcq F, Piccolo F, Carrié A, Jeanpierre M, et al. Primary adhalinopathy (alpha-sarcoglycanopathy): clinical, pathologic, and genetic correlation in 20 patients with autosomal recessive muscular dystrophy. *Neurology* 1997;**48**:1227–34.
59. Bönnemann CG, Wong J, Ben Hamida C, Hamida MB, Hentati F, Kunkel LM. LGMD 2E in Tunisia is caused by a homozygous missense mutation in beta-sarcoglycan exon 3. *Neuromuscul Disord* 1998;**8**:193–7.
60. Bönnemann CG, Passos-Bueno MR, McNally EM, Vainzof M, de Sá Moreira E, Marie SK, et al. Genomic screening for beta-sarcoglycan gene mutations: missense mutations may cause severe limb-girdle muscular dystrophy type 2E (LGMD 2E). *Hum Mol Genet* 1996;**5**:1953–61.
61. Balci B, Wilichowski E, Haliloğlu G, Talim B, Aurino S, Kremer E, et al. Beta-sarcoglycan gene mutations in Turkey. *Acta Myol* 2004;**23**:154–8.
62. Hack AA, Lam MY, Cordier L, Shoturma DI, Ly CT, Hadhazy MA, et al. Differential requirement for individual sarcoglycans and dystrophin in the assembly and function of the dystrophin-glycoprotein complex. *J Cell Sci* 2000;**113**(Pt 14):2535–44.
63. McNally EM, Passos-Bueno MR, Bönnemann CG, Vainzof M, de Sá Moreira E, Lidov HG, et al. Mild and severe muscular dystrophy caused by a single gamma-sarcoglycan mutation. *Am J Hum Genet* 1996;**59**:1040–7.
64. Georgieva B, Todorova A, Tournev I, Mitev V, Kremensky I. C283Y gamma-sarcoglycan gene mutation in the Bulgarian Roma (Gypsy) population: prevalence study and carrier screening in a high-risk community. *Clin Genet* 2004;**66**:467–72.
65. Lasa A, Piccolo F, de Diego C, Jeanpierre M, Colomer J, Rodríguez MJ, et al. Severe limb girdle muscular dystrophy in Spanish gypsies: further evidence for a founder mutation in the gamma-sarcoglycan gene. *Eur J Hum Genet* 1998;**6**:396–9.
66. Merlini L, Kaplan JC, Navarro C, Barois A, Bonneau D, Brasa J, et al. Homogeneous phenotype of the gypsy limb-girdle MD with the gamma-sarcoglycan C283Y mutation. *Neurology* 2000;**54**:1075–9.
67. Piccolo F, Jeanpierre M, Leturcq F, Dodé C, Azibi K, Toutain A, et al. A founder mutation in the gamma-sarcoglycan gene of gypsies possibly predating their migration out of India. *Hum Mol Genet* 1996;**5**:2019–22.
68. Todorova A, Ashikov A, Beltcheva O, Tournev I, Kremensky I. C283Y mutation and other C-terminal nucleotide changes in the gamma-sarcoglycan gene in the Bulgarian Gypsy population. *Hum Mutat* 1999;**14**:40–4.
69. Dinçer P, Bönnemann CG, Erdir Aker O, Akçoren Z, Nigro V, Kunkel LM, et al. A homozygous nonsense mutation in delta-sarcoglycan exon 3 in a case of LGMD2F. *Neuromuscul Disord* 2000;**10**:247–50.
70. Duggan DJ, Manchester D, Stears KP, Mathews DJ, Hart C, Hoffman EP. Mutations in the delta-sarcoglycan gene are a rare cause of autosomal recessive limb-girdle muscular dystrophy (LGMD2). *Neurogenetics* 1997;**1**:49–58.
71. Moreira ES, Vainzof M, Marie SK, Nigro V, Zatz M, Passos-Bueno MR. A first missense mutation in the delta sarcoglycan gene associated with a severe phenotype and frequency of limb-girdle muscular dystrophy type 2F (LGMD2F) in Brazilian sarcoglycanopathies. *J Med Genet* 1998;**35**:951–3.
72. Tsubata S, Bowles KR, Vatta M, Zintz C, Titus J, Muhonen L, et al. Mutations in the human delta-sarcoglycan gene in familial and sporadic dilated cardiomyopathy. *J Clin Invest* 2000;**106**:655–62.
73. Angelini C, Fanin M, Menegazzo E, Freda MP, Duggan DJ, Hoffman EP. Homozygous alpha-sarcoglycan mutation in two siblings: one asymptomatic and one steroid-responsive mild limb-girdle muscular dystrophy patient. *Muscle Nerve* 1998;**21**:769–75.
74. Angelini C, Fanin M, Freda MP, Duggan DJ, Siciliano G, Hoffman EP. The clinical spectrum of sarcoglycanopathies. *Neurology* 1999;**52**:176–9.
75. van der Kooi AJ, de Visser M, van Meegen M, Ginjaar HB, van Essen AJ, Jennekens FG, et al. A novel gamma-sarcoglycan mutation causing childhood onset, slowly progressive limb girdle muscular dystrophy. *Neuromuscul Disord* 1998;**8**:305–8.
76. Beckmann JS, Brown RH, Muntoni F, Urtizberea A, Bonnemann C, Bushby KM. 66th/67th ENMC sponsored international workshop: The limb-girdle muscular dystrophies, 26–28 March 1999, Naarden, The Netherlands. *Neuromuscul Disord* 1999;**9**:436–45.
77. Norwood F, de Visser M, Eymard B, Lochmüller H, Bushby K, Force EGT. EFNS guideline on diagnosis and management of limb girdle muscular dystrophies. *Eur J Neurol* 2007;**14**:1305–12.
78. Fanin M, Melacini P, Boito C, Pegoraro E, Angelini C. LGMD2E patients risk developing dilated cardiomyopathy. *Neuromuscul Disord* 2003;**13**:303–9.
79. Melacini P, Fanin M, Duggan DJ, Freda MP, Berardinelli A, Danieli GA, et al. Heart involvement in muscular dystrophies due to sarcoglycan gene mutations. *Muscle Nerve* 1999;**22**:473–9.
80. Politano L, Nigro V, Passamano L, Petretta V, Comi LI, Papparella S, et al. Evaluation of cardiac and respiratory involvement in sarcoglycanopathies. *Neuromuscul Disord* 2001;**11**:178–85.

81. Townsend D, Yasuda S, McNally E, Metzger JM. Distinct pathophysiological mechanisms of cardiomyopathy in hearts lacking dystrophin or the sarcoglycan complex. *FASEB J* 2011;**25**:3106–14.
82. Coral-Vazquez R, Cohn RD, Moore SA, Hill JA, Weiss RM, et al. Disruption of the sarcoglycan-sarcospan complex in vascular smooth muscle: a novel mechanism for cardiomyopathy and muscular dystrophy. *Cell* 1999;**98**:465–74.
83. Wheeler MT, Allikian MJ, Heydemann A, Hadhazy M, Zarnegar S, McNally EM. Smooth muscle cell-extrinsic vascular spasm arises from cardiomyocyte degeneration in sarcoglycan-deficient cardiomyopathy. *J Clin Invest* 2004;**113**:668–75.
84. Cohn RD, Durbeej M, Moore SA, Coral-Vazquez R, Prouty S, Campbell KP. Prevention of cardiomyopathy in mouse models lacking the smooth muscle sarcoglycan-sarcospan complex. *J Clin Invest* 2001;**107**:R1–7.
85. Ordoñez-Razo RM, Garrido-Garduño MH, Pérez-Martínez RA, Ruiz VM, Herrera-Tepatlán E, Rodríguez-Cruz M, et al. A delta-sarcoglycan gene polymorphism as a risk factor for hypertrophic cardiomyopathy. *Genet Test Mol Biomarkers* 2012;**16**:855–8.
86. Zatz M, Passos-Bueno MR, Rapaport D. Estimate of the proportion of Duchenne muscular dystrophy with autosomal recessive inheritance. *Am J Med Genet* 1989;**32**:407–10.
87. Stec I, Kress W, Meng G, Müller B, Müller CR, Grimm T. Estimate of severe autosomal recessive limb-girdle muscular dystrophy (LGMD2C, LGMD2D) among sporadic muscular dystrophy males: a study of 415 familes. *J Med Genet* 1995;**32**:930–3.
88. Vainzof M, Moreira ES, Ferraz G, Passos-Bueno MR, Marie SK, Zatz M. Further evidence for the organisation of the four sarcoglycans proteins within the dystrophin-glycoprotein complex. *Eur J Hum Genet* 1999;**7**:251–4.
89. Draviam R, Billington L, Senchak A, Hoffman EP, Watkins SC. Confocal analysis of the dystrophin protein complex in muscular dystrophy. *Muscle Nerve* 2001;**24**:262–72.
90. Vainzof M, Passos-Bueno MR, Canovas M, Moreira ES, Pavanello RC, Marie SK, et al. The sarcoglycan complex in the six autosomal recessive limb-girdle muscular dystrophies. *Hum Mol Genet* 1996;**5**:1963–9.
91. Higuchi I, Kawai H, Umaki Y, Kawajiri M, Adachi K, Fukunaga H, et al. Different manners of sarcoglycan expression in genetically proven alpha-sarcoglycan deficiency and gamma-sarcoglycan deficiency. *Acta Neuropathol* 1998;**96**:202–6.
92. Draviam RA, Shand SH, Watkins SC. The beta-delta-core of sarcoglycan is essential for deposition at the plasma membrane. *Muscle Nerve* 2006;**34**:691–701.
93. Nigro V, Piluso G, Belsito A, Politano L, Puca AA, Papparella S, et al. Identification of a novel sarcoglycan gene at 5q33 encoding a sarcolemmal 35 kDa glycoprotein. *Hum Mol Genet* 1996;**5**:1179–86.
94. Klinge L, Dekomien G, Aboumousa A, Charlton R, Epplen JT, Barresi R, et al. Sarcoglycanopathies: can muscle immunoanalysis predict the genotype? *Neuromuscul Disord* 2008;**18**:934–41.
95. Duclos F, Straub V, Moore SA, Venzke DP, Hrstka RF, Crosbie RH, et al. Progressive muscular dystrophy in alpha-sarcoglycan-deficient mice. *J Cell Biol* 1998;**142**:1461–71.
96. Greelish JP, Su LT, Lankford EB, Burkman JM, Chen H, Konig SK, et al. Stable restoration of the sarcoglycan complex in dystrophic muscle perfused with histamine and a recombinant adeno-associated viral vector. *Nat Med* 1999;**5**:439–43.
97. Li J, Dressman D, Tsao YP, Sakamoto A, Hoffman EP, Xiao X. rAAV vector-mediated sarcogylcan gene transfer in a hamster model for limb girdle muscular dystrophy. *Gene Ther* 1999;**6**:74–82.
98. Cordier L, Hack AA, Scott MO, Barton-Davis ER, Gao G, Wilson JM, et al. Rescue of skeletal muscles of gamma-sarcoglycan-deficient mice with adeno-associated virus-mediated gene transfer. *Mol Ther* 2000;**1**:119–29.
99. Holt KH, Lim LE, Straub V, Venzke DP, Duclos F, Anderson RD, et al. Functional rescue of the sarcoglycan complex in the BIO 14.6 hamster using delta-sarcoglycan gene transfer. *Mol Cell* 1998;**1**:841–8.
100. Mendell JR, Rodino-Klapac LR, Rosales XQ, Coley BD, Galloway G, Lewis S, et al. Sustained alpha-sarcoglycan gene expression after gene transfer in limb-girdle muscular dystrophy, type 2D. *Ann Neurol* 2010;**68**:629–38.
101. Herson S, Hentati F, Rigolet A, Behin A, Romero NB, Leturcq F, et al. A phase I trial of adeno-associated virus serotype 1-γ-sarcoglycan gene therapy for limb girdle muscular dystrophy type 2C. *Brain* 2012;**135**:483–92.
102. Connolly AM, Pestronk A, Mehta S, Al-Lozi M. Primary alpha-sarcoglycan deficiency responsive to immunosuppression over three years. *Muscle Nerve* 1998;**21**:1549–53.
103. Bushby K, Anderson LV, Pollitt C, Naom I, Muntoni F, Bindoff L. Abnormal merosin in adults. A new form of late onset muscular dystrophy not linked to chromosome 6q2. *Brain* 1998;**121** (Pt 4):581–8.
104. Driss A, Amouri R, Ben Hamida C, Souilem S, Gouider-Khouja N, Ben Hamida M, et al. A new locus for autosomal recessive limb-girdle muscular dystrophy in a large consanguineous Tunisian family maps to chromosome 19q13.3. *Neuromuscul Disord* 2000;**10**:240–6.
105. Holzfeind PJ, Grewal PK, Reitsamer HA, Kechvar J, Lassmann H, Hoeger H, et al. Skeletal, cardiac and tongue muscle pathology, defective retinal transmission, and neuronal migration defects in the Large(myd) mouse defines a natural model for glycosylation-deficient muscle – eye – brain disorders. *Hum Mol Genet* 2002;**11**:2673–87.
106. Grewal PK, Holzfeind PJ, Bittner RE, Hewitt JE. Mutant glycosyltransferase and altered glycosylation of alpha-dystroglycan in the myodystrophy mouse. *Nat Genet* 2001;**28**:151–4.
107. Driss A, Noguchi S, Amouri R, Kefi M, Sasaki T, Sugie K, et al. Fukutin-related protein gene mutated in the original kindred limb-girdle MD 2I. *Neurology* 2003;**60**:1341–4.
108. de Paula F, Vieira N, Starling A, Yamamoto LU, Lima B, de Cássia Pavanello R, et al. Asymptomatic carriers for homozygous novel mutations in the FKRP gene: the other end of the spectrum. *Eur J Hum Genet* 2003;**11**:923–30.
109. Harel T, Goldberg Y, Shalev SA, Chervinski I, Ofir R, Birk OS. Limb-girdle muscular dystrophy 2I: phenotypic variability within a large consanguineous Bedouin family associated with a novel FKRP mutation. *Eur J Hum Genet* 2004;**12**:38–43.
110. Sveen ML, Schwartz M, Vissing J. High prevalence and phenotype-genotype correlations of limb girdle muscular dystrophy type 2I in Denmark. *Ann Neurol* 2006;**59**:808–15.
111. Poppe M, Cree L, Bourke J, Eagle M, Anderson LV, Birchall D, et al. The phenotype of limb-girdle muscular dystrophy type 2I. *Neurology* 2003;**60**:1246–51.

112. Godfrey C, Clement E, Mein R, Brockington M, Smith J, Talim B, et al. Refining genotype phenotype correlations in muscular dystrophies with defective glycosylation of dystroglycan. *Brain* 2007;**130**:2725–35.
113. Mercuri E, Brockington M, Straub V, Quijano-Roy S, Yuva Y, Herrmann R, et al. Phenotypic spectrum associated with mutations in the fukutin-related protein gene. *Ann Neurol* 2003;**53**:537–42.
114. Bourteel H, Vermersch P, Cuisset JM, Maurage CA, Laforet P, Richard P, et al. Clinical and mutational spectrum of limb-girdle muscular dystrophy type 2I in 11 French patients. *J Neurol Neurosurg Psychiatry* 2009;**80**:1405–8.
115. Boito CA, Melacini P, Vianello A, Prandini P, Gavassini BF, Bagattin A, et al. Clinical and molecular characterization of patients with limb-girdle muscular dystrophy type 2I. *Arch Neurol* 2005;**62**:1894–9.
116. Margeta M, Connolly AM, Winder TL, Pestronk A, Moore SA. Cardiac pathology exceeds skeletal muscle pathology in two cases of limb-girdle muscular dystrophy type 2I. *Muscle Nerve* 2009;**40**:883–9.
117. Kefi M, Amouri R, Chabrak S, Mechmeche R, Hentati F. Variable cardiac involvement in Tunisian siblings harboring FKRP gene mutations. *Neuropediatrics* 2008;**39**:113–15.
118. Poppe M, Bourke J, Eagle M, Frosk P, Wrogemann K, Greenberg C, et al. Cardiac and respiratory failure in limb-girdle muscular dystrophy 2I. *Ann Neurol* 2004;**56**:738–41.
119. Godfrey C, Escolar D, Brockington M, Clement EM, Mein R, Jimenez-Mallebrera C, et al. Fukutin gene mutations in steroid-responsive limb girdle muscular dystrophy. *Ann Neurol* 2006;**60**:603–10.
120. Clement EM, Godfrey C, Tan J, Brockington M, Torelli S, Feng L, et al. Mild POMGnT1 mutations underlie a novel limb-girdle muscular dystrophy variant. *Arch Neurol* 2008;**65**:137–41.
121. Saredi S, Gibertini S, Ardissone A, Fusco I, Zanotti S, Blasevich F, et al. A fourth case of POMT2-related limb girdle muscle dystrophy with mild reduction of α-dystroglycan glycosylation. *Eur J Paediatr Neurol* 2014;**18**:404–8.
122. Biancheri R, Falace A, Tessa A, Pedemonte M, Scapolan S, Cassandrini D, et al. POMT2 gene mutation in limb-girdle muscular dystrophy with inflammatory changes. *Biochem Biophys Res Commun* 2007;**363**:1033–7.
123. Bello L, Melacini P, Pezzani R, D'Amico A, Piva L, Leonardi E, et al. Cardiomyopathy in patients with POMT1-related congenital and limb-girdle muscular dystrophy. *Eur J Hum Genet* 2012;**20**:1234–9.
124. Balci B, Uyanik G, Dincer P, Gross C, Willer T, Talim B, et al. An autosomal recessive limb girdle muscular dystrophy (LGMD2) with mild mental retardation is allelic to Walker-Warburg syndrome (WWS) caused by a mutation in the POMT1 gene. *Neuromuscul Disord* 2005;**15**:271–5.
125. Hara Y, Balci-Hayta B, Yoshida-Moriguchi T, Kanagawa M, Beltrán-Valero de Bernabé D, Gündeşli H, et al. A dystroglycan mutation associated with limb-girdle muscular dystrophy. *N Engl J Med* 2011;**364**:939–46.
126. Cirak S, Foley AR, Herrmann R, Willer T, Yau S, Stevens E, et al. ISPD gene mutations are a common cause of congenital and limb-girdle muscular dystrophies. *Brain* 2013;**136**:269–81.
127. Carss KJ, Stevens E, Foley AR, Cirak S, Riemersma M, Torelli S, et al. Mutations in GDP-mannose pyrophosphorylase B cause congenital and limb-girdle muscular dystrophies associated with hypoglycosylation of α-dystroglycan. *Am J Hum Genet* 2013;**93**:29–41.
128. Erb W. Über die "juvenile Form" der progressiven Muskelatrophie und ihre Beziehung zur sogenannten Pseudohypertrohpie der Muskeln. *Dusch Archiv Klin Med* 1884;467–519.
129. Beckmann JS, Richard I, Hillaire D, Broux O, Antignac C, Bois E, et al. A gene for limb-girdle muscular dystrophy maps to chromosome 15 by linkage. *C R Acad Sci III* 1991;**312**:141–8.
130. Richard I, Roudaut C, Saenz A, Pogue R, Grimbergen JE, Anderson LV, et al. Calpainopathy—a survey of mutations and polymorphisms. *Am J Hum Genet* 1999;**64**:1524–40.
131. Richard I, Broux O, Allamand V, Fougerousse F, Chiannilkulchai N, Bourg N, et al. Mutations in the proteolytic enzyme calpain 3 cause limb-girdle muscular dystrophy type 2 A. *Cell* 1995;**81**:27–40.
132. Baghdiguian S, Martin M, Richard I, Pons F, Astier C, Bourg N, et al. Calpain 3 deficiency is associated with myonuclear apoptosis and profound perturbation of the IkappaB alpha/NF-kappaB pathway in limb-girdle muscular dystrophy type 2A. *Nat Med* 1999;**5**:503–11.
133. Garnham CP, Hanna RA, Chou JS, Low KE, Gourlay K, Campbell RL, et al. Limb-girdle muscular dystrophy type 2A can result from accelerated autoproteolytic inactivation of calpain 3. *Biochemistry* 2009;**48**:3457–67.
134. Baghdiguian S, Richard I, Martin M, Coopman P, Beckmann JS, Mangeat P, et al. Pathophysiology of limb girdle muscular dystrophy type 2A: hypothesis and new insights into the IkappaBalpha/NF-kappaB survival pathway in skeletal muscle. *J Mol Med (Berl)* 2001;**79**:254–61.
135. Hauerslev S, Sveen ML, Duno M, Angelini C, Vissing J, Krag TO. Calpain 3 is important for muscle regeneration: evidence from patients with limb girdle muscular dystrophies. *BMC Musculoskelet Disord* 2012;**13**:43.
136. Gallardo E, Saenz A, Illa I. Limb-girdle muscular dystrophy 2A. *Handb Clin Neurol* 2011;**101**:97–110.
137. Sorimachi H, Ono Y, Suzuki K. Skeletal muscle-specific calpain, p94, and connectin/titin: their physiological functions and relationship to limb-girdle muscular dystrophy type 2A. *Adv Exp Med Biol* 2000;**481**:383–95, discussion 95–7.
138. Huang Y, de Morree A, van Remoortere A, Bushby K, Frants RR, den Dunnen JT, et al. Calpain 3 is a modulator of the dysferlin protein complex in skeletal muscle. *Hum Mol Genet* 2008;**17**:1855–66.
139. Richard I, Brenguier L, Dinçer P, Roudaut C, Bady B, Burgunder JM, et al. Multiple independent molecular etiology for limb-girdle muscular dystrophy type 2A patients from various geographical origins. *Am J Hum Genet* 1997;**60**:1128–38.
140. Saenz A, Leturcq F, Cobo AM, Poza JJ, Ferrer X, Otaegui D, et al. LGMD2A: genotype-phenotype correlations based on a large mutational survey on the calpain 3 gene. *Brain* 2005;**128**:732–42.
141. Pratt VM, Jackson CE, Wallace DC, Gurley DS, Feit A, Feldman GL. DNA studies of limb-girdle muscular dystrophy type 2A in the Amish exclude a modifying mitochondrial gene and show no evidence for a modifying nuclear gene. *Am J Hum Genet* 1997;**61**:231–3.
142. Kawai H, Akaike M, Kunishige M, Inui T, Adachi K, Kimura C, et al. Clinical, pathological, and genetic features of limb-girdle muscular dystrophy type 2A with new calpain 3 gene mutations in

seven patients from three Japanese families. *Muscle Nerve* 1998;**21**:1493–501.

143. Passos-Bueno MR, Moreira ES, Marie SK, Bashir R, Vasquez L, Love DR, et al. Main clinical features of the three mapped autosomal recessive limb-girdle muscular dystrophies and estimated proportion of each form in 13 Brazilian families. *J Med Genet* 1996;**33**:97–102.

144. Pathak P, Sharma MC, Sarkar C, Jha P, Suri V, Mohd H, et al Limb girdle muscular dystrophy type 2A in India: a study based on semi-quantitative protein analysis, with clinical and histopathological correlation. *Neurol India* 2010;**58**:549–54.

145. Pollitt C, Anderson LV, Pogue R, Davison K, Pyle A, Bushby KM. The phenotype of calpainopathy: diagnosis based on a multidisciplinary approach. *Neuromuscul Disord* 2001;**11**:287–96.

146. Urtasun M, Saenz A, Roudaut C, Poza JJ, Urtizberea JA, Cobo AM, et al. Limb-girdle muscular dystrophy in Guipuzcoa (Basque Country, Spain). *Brain* 1998;**121**(Pt 9):1735–47.

147. Pénisson-Besnier I, Richard I, Dubas F, Beckmann JS, Fardeau M. Pseudometabolic expression and phenotypic variability of calpain deficiency in two siblings. *Muscle Nerve* 1998;**21**:1078–80.

148. Fardeau M, Eymard B, Mignard C, Tomé FM, Richard I, Beckmann JS. Chromosome 15-linked limb-girdle muscular dystrophy: clinical phenotypes in Reunion Island and French metropolitan communities. *Neuromuscul Disord* 1996;**6**:447–53.

149. Bushby KM. Making sense of the limb-girdle muscular dystrophies. *Brain* 1999;**122**(Pt 8):1403–20.

150. Laval SH, Bushby KM. Limb-girdle muscular dystrophies—from genetics to molecular pathology. *Neuropathol Appl Neurobiol* 2004;**30**:91–105.

151. Bushby K. Diagnosis and management of the limb girdle muscular dystrophies. *Pract Neurol* 2009;**9**:314–23.

152. Chou FL, Angelini C, Daentl D, Garcia C, Greco C, Hausmanowa-Petrusewicz I, et al. Calpain III mutation analysis of a heterogeneous limb-girdle muscular dystrophy population. *Neurology* 1999;**52**:1015–20.

153. Okere A, Reddy SS, Gupta S, Shinnar M. A cardiomyopathy in a patient with limb girdle muscular dystrophy type 2A. *Circ Heart Fail* 2013;**6**:e12–13.

154. Mercuri E, Bushby K, Ricci E, Birchall D, Pane M, Kinali M, et al. Muscle MRI findings in patients with limb girdle muscular dystrophy with calpain 3 deficiency (LGMD2A) and early contractures. *Neuromuscul Disord* 2005;**15**:164–71.

155. Häffner K, Speer A, Hübner C, Voit T, Oexle K. A small inframe deletion within the protease domain of muscle-specific calpain, p94 causes early-onset limb-girdle muscular dystrophy 2A. *Hum Mutat* 1998;(Suppl. 1):S298–300.

156. Groen EJ, Charlton R, Barresi R, Anderson LV, Eagle M, Hudson J, et al. Analysis of the UK diagnostic strategy for limb girdle muscular dystrophy 2A. *Brain* 2007;**130**:3237–49.

157. Fanin M, Nascimbeni AC, Tasca E, Angelini C. How to tackle the diagnosis of limb-girdle muscular dystrophy 2A. *Eur J Hum Genet* 2009;**17**:598–603.

158. Angelini C, Nardetto L, Borsato C, Padoan R, Fanin M, Nascimbeni AC, et al. The clinical course of calpainopathy (LGMD2A) and dysferlinopathy (LGMD2B). *Neurol Res* 2010;**32**:41–6.

159. ten Dam L, van der Kooi AJ, van Wattingen M, de Haan RJ, de Visser M. Reliability and accuracy of skeletal muscle imaging in limb-girdle muscular dystrophies. *Neurology* 2012;**79**:1716–23.

160. Amato AA. Adults with eosinophilic myositis and calpain-3 mutations. *Neurology* 2008;**70**:730–1.

161. Krahn M, Lopez de Munain A, Streichenberger N, Bernard R, Pecheux C, Testard H, et al. CAPN3 mutations in patients with idiopathic eosinophilic myositis. *Ann Neurol* 2006;**59**:905–11.

162. Talim B, Ognibene A, Mattioli E, Richard I, Anderson LV, Merlini L. Normal calpain expression in genetically confirmed limb-girdle muscular dystrophy type 2A. *Neurology* 2001;**56**:692–3.

163. Fanin M, Nascimbeni AC, Fulizio L, Trevisan CP, Meznaric-Petrusa M, Angelini C. Loss of calpain-3 autocatalytic activity in LGMD2A patients with normal protein expression. *Am J Pathol* 2003;**163**:1929–36.

164. Fanin M, Fulizio L, Nascimbeni AC, Spinazzi M, Piluso G, Ventriglia VM, et al. Molecular diagnosis in LGMD2A: mutation analysis or protein testing? *Hum Mutat* 2004;**24**:52–62.

165. Fanin M, Nascimbeni AC, Angelini C. Screening of calpain-3 autolytic activity in LGMD muscle: a functional map of CAPN3 gene mutations. *J Med Genet* 2007;**44**:38–43.

166. Anderson LV, Harrison RM, Pogue R, Vafiadaki E, Pollitt C, Davison K, et al. Secondary reduction in calpain 3 expression in patients with limb girdle muscular dystrophy type 2B and Miyoshi myopathy (primary dysferlinopathies). *Neuromuscul Disord* 2000;**10**:553–9.

167. Hermanova M, Zapletalova E, Sedlackova J, Chrobakova T, Letocha O, Kroupova I, et al. Analysis of histopathologic and molecular pathologic findings in Czech LGMD2A patients. *Muscle Nerve* 2006;**33**:424–32.

168. Oflazer PS, Gundesli H, Zorludemir S, Sabuncu T, Dincer P. Eosinophilic myositis in calpainopathy: could immunosuppression of the eosinophilic myositis alter the early natural course of the dystrophic disease? *Neuromuscul Disord* 2009;**19**:261–3.

169. Danièle N, Richard I, Bartoli M. Ins and outs of therapy in limb girdle muscular dystrophies. *Int J Biochem Cell Biol* 2007;**39**:1608–24.

170. Bashir R, Strachan T, Keers S, Stephenson A, Mahjneh I, Marconi G, et al. A gene for autosomal recessive limb-girdle muscular dystrophy maps to chromosome 2p. *Hum Mol Genet* 1994;**3**:455–7.

171. Bueno MR, Moreira ES, Vainzof M, Chamberlain J, Marie SK, Pereira L, et al. A common missense mutation in the adhalin gene in three unrelated Brazilian families with a relatively mild form of autosomal recessive limb-girdle muscular dystrophy. *Hum Mol Genet* 1995;**4**:1163–7.

172. Passos-Bueno MR, Bashir R, Moreira ES, Vainzof M, Marie SK, Vasquez L, et al. Confirmation of the 2p locus for the mild autosomal recessive limb-girdle muscular dystrophy gene (LGMD2B) in three families allows refinement of the candidate region. *Genomics* 1995;**27**:192–5.

173. Miyoshi K, Kawai H, Iwasa M, Kusaka K, Nishino H. Autosomal recessive distal muscular dystrophy as a new type of progressive muscular dystrophy. Seventeen cases in eight families including an autopsied case. *Brain* 1986;**109**(Pt 1):31–54.

174. Illarioshkin SN, Ivanova-Smolenskaya IA, Tanaka H, Vereshchagin NV, Markova ED, Poleshchuk VV, et al. Clinical and molecular analysis of a large family with three distinct phenotypes of progressive muscular dystrophy. *Brain* 1996;**119** (Pt 6):1895–909.

175. Vilchez JJ, Gallano P, Gallardo E, Lasa A, Rojas-Garcia R, Freixas A, et al. Identification of a novel founder mutation in the DYSF gene causing clinical variability in the Spanish population. *Arch Neurol* 2005;**62**:1256–9.
176. Bushby KM. Dysferlin and muscular dystrophy. *Acta Neurol Belg* 2000;**100**:142–5.
177. Mahjneh I, Marconi G, Bushby K, Anderson LV, Tolvanen-Mahjneh H, Somer H. Dysferlinopathy (LGMD2B): a 23-year follow-up study of 10 patients homozygous for the same frameshifting dysferlin mutations. *Neuromuscul Disord* 2001;**11**:20–6.
178. Nakagawa M, Matsuzaki T, Suehara M, Kanzato N, Takashima H, Higuchi I, et al. Phenotypic variation in a large Japanese family with Miyoshi myopathy with nonsense mutation in exon 19 of dysferlin gene. *J Neurol Sci* 2001;**184**:15–19.
179. Nalini A, Gayathri N. Dysferlinopathy: a clinical and histopathological study of 28 patients from India. *Neurol India* 2008;**56**:379–85 discussion, 86–7.
180. Nguyen K, Bassez G, Krahn M, Bernard R, Laforet P, Labelle V, et al. Phenotypic study in 40 patients with dysferlin gene mutations: high frequency of atypical phenotypes. *Arch Neurol* 2007;**64**:1176–82.
181. Paradas C, Gonzalez-Quereda L, De Luna N, Gallardo E, Garcia-Consuegra I, Gomez H, et al. A new phenotype of dysferlinopathy with congenital onset. *Neuromuscul Disord* 2009;**19**:21–5.
182. Paradas C, Llauger J, Diaz-Manera J, Rojas-Garcia R, De Luna N, Iturriaga C, et al. Redefining dysferlinopathy phenotypes based on clinical findings and muscle imaging studies. *Neurology* 2010;**75**:316–23.
183. Bashir R, Britton S, Strachan T, Keers S, Vafiadaki E, Lako M, et al. A gene related to *Caenorhabditis elegans* spermatogenesis factor fer-1 is mutated in limb-girdle muscular dystrophy type 2B. *Nat Genet* 1998;**20**:37–42.
184. Anderson LV, Davison K, Moss JA, Young C, Cullen MJ, Walsh J, et al. Dysferlin is a plasma membrane protein and is expressed early in human development. *Hum Mol Genet* 1999;**8**:855–61.
185. Hernandez-Deviez DJ, Martin S, Laval SH, Lo HP, Cooper ST, North KN, et al. Aberrant dysferlin trafficking in cells lacking caveolin or expressing dystrophy mutants of caveolin-3. *Hum Mol Genet* 2006;**15**:129–42.
186. Hernandez-Deviez DJ, Howes MT, Laval SH, Bushby K, Hancock JF, Parton RG. Caveolin regulates endocytosis of the muscle repair protein, dysferlin. *J Biol Chem* 2008; **283**:6476–88.
187. Rizo J, Südhof TC. C2-domains, structure and function of a universal Ca2+-binding domain. *J Biol Chem* 1998;**273**:15879–82.
188. Bansal D, Miyake K, Vogel SS, Groh S, Chen CC, Williamson R, et al. Defective membrane repair in dysferlin-deficient muscular dystrophy. *Nature* 2003;**423**:168–72.
189. Bansal D, Campbell KP. Dysferlin and the plasma membrane repair in muscular dystrophy. *Trends Cell Biol* 2004;**14**:206–13.
190. Cai C, Weisleder N, Ko JK, Komazaki S, Sunada Y, Nishi M, et al. Membrane repair defects in muscular dystrophy are linked to altered interaction between MG53, caveolin-3, and dysferlin. *J Biol Chem* 2009;**284**:15894–902.
191. Chiu YH, Hornsey MA, Klinge L, Jorgensen LH, Laval SH, Charlton R, et al. Attenuated muscle regeneration is a key factor in dysferlin-deficient muscular dystrophy. *Hum Mol Genet* 2009;**18**:1976–89.
192. de Morree A, Flix B, Bagaric I, Wang J, van den Boogaard M, Grand Moursel L, et al. Dysferlin regulates cell adhesion in human monocytes. *J Biol Chem* 2013;**288**:14147–57.
193. Klinge L, Harris J, Sewry C, Charlton R, Anderson L, Laval S, et al. Dysferlin associates with the developing T-tubule system in rodent and human skeletal muscle. *Muscle Nerve* 2010;**41**:166–73.
194. Kerr JP, Ziman AP, Mueller AL, Muriel JM, Kleinhans-Welte E, Gumerson JD, et al. Dysferlin stabilizes stress-induced Ca2+ signaling in the transverse tubule membrane. *Proc Natl Acad Sci USA* 2013;**110**:20831–6.
195. Cagliani R, Fortunato F, Giorda R, Rodolico C, Bonaglia MC, Sironi M, et al. Molecular analysis of LGMD-2B and MM patients: identification of novel DYSF mutations and possible founder effect in the Italian population. *Neuromuscul Disord* 2003;**13**:788–95.
196. Kawabe K, Goto K, Nishino I, Angelini C, Hayashi YK. Dysferlin mutation analysis in a group of Italian patients with limb-girdle muscular dystrophy and Miyoshi myopathy. *Eur J Neurol* 2004;**11**:657–61.
197. Krahn M, Beroud C, Labelle V, Nguyen K, Bernard R, Bassez G, et al. Analysis of the DYSF mutational spectrum in a large cohort of patients. *Hum Mutat* 2009;**30**:E345–75.
198. Illarioshkin SN, Ivanova-Smolenskaya IA, Greenberg CR, Nylen E, Sukhorukov VS, Poleshchuk VV, et al. Identical dysferlin mutation in limb-girdle muscular dystrophy type 2B and distal myopathy. *Neurology* 2000;**55**:1931–3.
199. McNally EM, Ly CT, Rosenmann H, Mitrani Rosenbaum S, Jiang W, Anderson LV, et al. Splicing mutation in dysferlin produces limb-girdle muscular dystrophy with inflammation. *Am J Med Genet* 2000;**91**:305–12.
200. Weiler T, Greenberg CR, Nylen E, Halliday W, Morgan K, Eggertson D, et al. Limb-girdle muscular dystrophy and Miyoshi myopathy in an aboriginal Canadian kindred map to LGMD2B and segregate with the same haplotype. *Am J Hum Genet* 1996;**59**:872–8.
201. Weiler T, Bashir R, Anderson LV, Davison K, Moss JA, Britton S, et al. Identical mutation in patients with limb girdle muscular dystrophy type 2B or Miyoshi myopathy suggests a role for modifier gene(s). *Hum Mol Genet* 1999;**8**:871–7.
202. Linssen WH, de Voogt WG, Krahn M, Bernard R, Levy N, Wokke JH, et al. Long-term follow-up study on patients with Miyoshi phenotype of distal muscular dystrophy. *Eur J Neurol* 2013;**20**:968–74.
203. Mahjneh I, Passos-Bueno MR, Zatz M, Vainzof M, Marconi G, Nashef L, et al. The phenotype of chromosome 2p-linked limb-girdle muscular dystrophy. *Neuromuscul Disord* 1996;**6**:483–90.
204. Kesper K, Kornblum C, Reimann J, Lutterbey G, Schröder R, Wattjes MP. Pattern of skeletal muscle involvement in primary dysferlinopathies: a whole-body 3.0-T magnetic resonance imaging study. *Acta Neurol Scand* 2009;**120**:111–18.
205. Brummer D, Walter MC, Palmbach M, Knirsch U, Karitzky J, Tomczak R, et al. Long-term MRI and clinical follow-up of symptomatic and presymptomatic carriers of dysferlin gene mutations. *Acta Myol* 2005;**24**:6–16.
206. Illa I, Serrano-Munuera C, Gallardo E, Lasa A, Rojas-Garcia R, Palmer J, et al. Distal anterior compartment myopathy: a dysferlin mutation causing a new muscular dystrophy phenotype. *Ann Neurol* 2001;**49**:130–4.

207. Saito H, Suzuki N, Ishiguro H, Hirota K, Itoyama Y, Takahashi T, et al. Distal anterior compartment myopathy with early ankle contractures. *Muscle Nerve* 2007;**36**:525–7.
208. Gayathri N, Alefia R, Nalini A, Yasha TC, Anita M, Santosh V, et al. Dysferlinopathy: spectrum of pathological changes in skeletal muscle tissue. *Indian J Pathol Microbiol* 2011;**54**:350–4.
209. Confalonieri P, Oliva L, Andreetta F, Lorenzoni R, Dassi P, Mariani E, et al. Muscle inflammation and MHC class I upregulation in muscular dystrophy with lack of dysferlin: an immunopathological study. *J Neuroimmunol* 2003;**142**:130–6.
210. Gallardo E, Rojas-Garcia R, de Luna N, Pou A, Brown Jr. RH, Illa I. Inflammation in dysferlin myopathy: immunohistochemical characterization of 13 patients. *Neurology* 2001;**57**:2136–8.
211. Rosales XQ, Gastier-Foster JM, Lewis S, Vinod M, Thrush DL, Astbury C, et al. Novel diagnostic features of dysferlinopathies. *Muscle Nerve* 2010;**42**:14–21.
212. Spuler S, Carl M, Zabojszcza J, Straub V, Bushby K, Moore SA, et al. Dysferlin-deficient muscular dystrophy features amyloidosis. *Ann Neurol* 2008;**63**:323–8.
213. Vainzof M, Anderson LV, McNally EM, Davis DB, Faulkner G, Valle G, et al. Dysferlin protein analysis in limb-girdle muscular dystrophies. *J Mol Neurosci* 2001;**17**:71–80.
214. Piccolo F, Moore SA, Ford GC, Campbell KP. Intracellular accumulation and reduced sarcolemmal expression of dysferlin in limb-girdle muscular dystrophies. *Ann Neurol* 2000;**48**:902–12.
215. De Luna N, Freixas A, Gallano P, Caselles L, Rojas-Garcia R, Paradas C, et al. Dysferlin expression in monocytes: a source of mRNA for mutation analysis. *Neuromuscul Disord* 2007;**17**:69–76.
216. Gallardo E, de Luna N, Diaz-Manera J, Rojas-Garcia R, Gonzalez-Quereda L, Flix B, et al. Comparison of dysferlin expression in human skeletal muscle with that in monocytes for the diagnosis of dysferlin myopathy. *PLOS ONE* 2011;**6**:e29061
217. Aoki M, Liu J, Richard I, Bashir R, Britton S, Keers SM, et al. Genomic organization of the dysferlin gene and novel mutations in Miyoshi myopathy. *Neurology* 2001;**57**:271–8.
218. Jarry J, Rioux MF, Bolduc V, Robitaille Y, Khoury V, Thiffault I, et al. A novel autosomal recessive limb-girdle muscular dystrophy with quadriceps atrophy maps to 11p13-p12. *Brain* 2007;**130**:368–80.
219. Penttilä S, Palmio J, Suominen T, Raheem O, Evilä A, Muelas Gomez N, et al. Eight new mutations and the expanding phenotype variability in muscular dystrophy caused by ANO5. *Neurology* 2012;**78**:897–903.
220. Hicks D, Sarkozy A, Muelas N, Koehler K, Huebner A, Hudson G, et al. A founder mutation in Anoctamin 5 is a major cause of limb-girdle muscular dystrophy. *Brain* 2011;**134**:171–82.
221. Bolduc V, Marlow G, Boycott KM, Saleki K, Inoue H, Kroon J, et al. Recessive mutations in the putative calcium-activated chloride channel Anoctamin 5 cause proximal LGMD2 L and distal MMD3 muscular dystrophies. *Am J Hum Genet* 2010;**86**:213–21.
222. van der Kooi AJ, Ten Dam L, Frankhuizen WS, Straathof CS, van Doorn PA, de Visser M, et al. ANO5 mutations in the Dutch limb girdle muscular dystrophy population. *Neuromuscul Disord* 2013;**23**:456–60.
223. Mizuta K, Tsutsumi S, Inoue H, Sakamoto Y, Miyatake K, Miyawaki K, et al. Molecular characterization of GDD1/TMEM16E, the gene product responsible for autosomal dominant gnathodiaphyseal dysplasia. *Biochem Biophys Res Commun* 2007;**357**:126–32.
224. Marconi C, Brunamonti Binello P, Badiali G, Caci E, Cusano R, Garibaldi J, et al. A novel missense mutation in ANO5/TMEM16E is causative for gnathodiaphyseal dyplasia in a large Italian pedigree. *Eur J Hum Genet* 2013;**21**:613–19.
225. Witting N, Duno M, Petri H, Krag T, Bundgaard H, Kober L, et al. Anoctamin 5 muscular dystrophy in Denmark: prevalence, genotypes, phenotypes, cardiac findings, and muscle protein expression. *J Neurol* 2013;**260**:2084–93.
226. Mahjneh I, Bashir R, Kiuru-Enari S, Linssen W, Lamminen A, Visser M. Selective pattern of muscle involvement seen in distal muscular dystrophy associated with anoctamin 5 mutations: a follow-up muscle MRI study. *Neuromuscul Disord* 2012;**22** (Suppl. 2):S130–6.
227. Sarkozy A, Deschauer M, Carlier RY, Schrank B, Seeger J, Walter MC, et al. Muscle MRI findings in limb girdle muscular dystrophy type 2L. *Neuromuscul Disord* 2012;**22**(Suppl. 2): S122–9.
228. Liewluck T, Winder TL, Dimberg EL, Crum BA, Heppelmann CJ, Wang Y, et al. ANO5-muscular dystrophy: clinical, pathological and molecular findings. *Eur J Neurol* 2013;**20**:1383–9.
229. Milone M, Liewluck T, Winder TL, Pianosi PT. Amyloidosis and exercise intolerance in ANO5 muscular dystrophy. *Neuromuscul Disord* 2012;**22**:13–15.
230. Sarkozy A, Hicks D, Hudson J, Laval SH, Barresi R, Hilton-Jones D, et al. ANO5 gene analysis in a large cohort of patients with anoctaminopathy: confirmation of male prevalence and high occurrence of the common exon 5 gene mutation. *Hum Mutat* 2013;**34**: 1111–18.
231. Magri F, Del Bo R, D'Angelo MG, Sciacco M, Gandossini S, Govoni A, et al. Frequency and characterisation of anoctamin 5 mutations in a cohort of Italian limb-girdle muscular dystrophy patients. *Neuromuscul Disord* 2012;**22**:934–43.
232. Wahbi K, Béhin A, Bécane HM, Leturcq F, Cossée M, Laforêt P et al. Dilated cardiomyopathy in patients with mutations in anoctamin 5. *Int J Cardiol* 2013;**168**:76–9.
233. Moreira ES, Vainzof M, Marie SK, Sertié AL, Zatz M, Passos-Bueno MR. The seventh form of autosomal recessive limb-girdle muscular dystrophy is mapped to 17q11-12. *Am J Hum Genet* 1997;**61**:151–9.
234. Valle G, Faulkner G, De Antoni A, Pacchioni B, Pallavicini A, Pandolfo D, et al. Telethonin, a novel sarcomeric protein of heart and skeletal muscle. *FEBS Lett* 1997;**415**:163–8.
235. Moreira ES, Wiltshire TJ, Faulkner G, Nilforoushan A, Vainzof M, Suzuki OT, et al. Limb-girdle muscular dystrophy type 2G is caused by mutations in the gene encoding the sarcomeric protein telethonin. *Nat Genet* 2000;**24**:163–6.
236. Mues A, van der Ven PF, Young P, Fürst DO, Gautel M. Two immunoglobulin-like domains of the Z-disc portion of titin interact in a conformation-dependent way with telethonin. *FEBS Lett* 1998;**428**:111–14.
237. Bertz M, Wilmanns M, Rief M. The titin-telethonin complex is a directed, superstable molecular bond in the muscle Z-disk. *Proc Natl Acad Sci USA* 2009;**106**:13307–10.
238. Mayans O, van der Ven PF, Wilm M, Mues A, Young P, Fürst DO, et al. Structural basis for activation of the titin kinase domain during myofibrillogenesis. *Nature* 1998;**395**:863–9.

239. Zatz M, de Paula F, Starling A, Vainzof M. The 10 autosomal recessive limb-girdle muscular dystrophies. *Neuromuscul Disord* 2003;**13**:532–44.

240. Nonaka I. Distal myopathies. *Curr Opin Neurol* 1999;**12**:493–9.

241. Vainzof M, Moreira ES, Suzuki OT, Faulkner G, Valle G, Beggs AH, et al. Telethonin protein expression in neuromuscular disorders. *Biochim Biophys Acta* 2002;**1588**:33–40.

242. Weiler T, Greenberg CR, Zelinski T, Nylen E, Coghlan G, Crumley MJ, et al. A gene for autosomal recessive limb-girdle muscular dystrophy in Manitoba Hutterites maps to chromosome region 9q31-q33: evidence for another limb-girdle muscular dystrophy locus. *Am J Hum Genet* 1998;**63**:140–7.

243. Frosk P, Weiler T, Nylen E, Sudha T, Greenberg CR, Morgan K, et al. Limb-girdle muscular dystrophy type 2H associated with mutation in TRIM32, a putative E3-ubiquitin-ligase gene. *Am J Hum Genet* 2002;**70**:663–72.

244. Kudryashova E, Kudryashov D, Kramerova I, Spencer MJ. Trim32 is a ubiquitin ligase mutated in limb girdle muscular dystrophy type 2H that binds to skeletal muscle myosin and ubiquitinates actin. *J Mol Biol* 2005;**354**:413–24.

245. Liewluck T, Tracy JA, Sorenson EJ, Engel AG. Scapuloperoneal muscular dystrophy phenotype due to TRIM32-sarcotubular myopathy in South Dakota Hutterite. *Neuromuscul Disord* 2013;**23**:133–8.

246. Borg K, Stucka R, Locke M, Melin E, Ahlberg G, Klutzny U, et al. Intragenic deletion of TRIM32 in compound heterozygotes with sarcotubular myopathy/LGMD2H. *Hum Mutat* 2009;**30**:E831–44.

247. Schoser BG, Frosk P, Engel AG, Klutzny U, Lochmüller H, Wrogemann K. Commonality of TRIM32 mutation in causing sarcotubular myopathy and LGMD2H. *Ann Neurol* 2005;**57**:591–5.

248. Kudryashova E, Kramerova I, Spencer MJ. Satellite cell senescence underlies myopathy in a mouse model of limb-girdle muscular dystrophy 2H. *J Clin Invest* 2012;**122**:1764–76.

249. Nicklas S, Otto A, Wu X, Miller P, Stelzer S, Wen Y, et al. TRIM32 regulates skeletal muscle stem cell differentiation and is necessary for normal adult muscle regeneration. *PLOS ONE* 2012;**7**:e30445.

250. Saccone V, Palmieri M, Passamano L, Piluso G, Meroni G, Politano L, et al. Mutations that impair interaction properties of TRIM32 associated with limb-girdle muscular dystrophy 2H. *Hum Mutat* 2008;**29**:240–7.

251. Shieh PB, Kudryashova E, Spencer MJ. Limb-girdle muscular dystrophy 2H and the role of TRIM32. *Handb Clin Neurol* 2011;**101**:125–33.

252. Udd B, Partanen J, Halonen P, Falck B, Hakamies L, Heikkilä H, et al. Tibial muscular dystrophy. Late adult-onset distal myopathy in 66 Finnish patients. *Arch Neurol* 1993;**50**:604–8.

253. Udd B. Limb-girdle type muscular dystrophy in a large family with distal myopathy: homozygous manifestation of a dominant gene? *J Med Genet* 1992;**29**:383–9.

254. Udd B, Kääriänen H, Somer H. Muscular dystrophy with separate clinical phenotypes in a large family. *Muscle Nerve* 1991;**14**:1050–8.

255. Itoh-Satoh M, Hayashi T, Nishi H, Koga Y, Arimura T, Koyanagi T, et al. Titin mutations as the molecular basis for dilated cardiomyopathy. *Biochem Biophys Res Commun* 2002;**291**:385–93.

256. Hedberg C, Melberg A, Dahlbom K, Oldfors A. Hereditary myopathy with early respiratory failure is caused by mutations in the titin FN3 119 domain. *Brain* 2014;**137**:e271.

257. Pfeffer G, Barresi R, Wilson IJ, Hardy SA, Griffin H, Hudson J, et al. Titin founder mutation is a common cause of myofibrillar myopathy with early respiratory failure. *J Neurol Neurosurg Psychiatry* 2014;**85**:331–8.

258. Ceyhan-Birsoy O, Agrawal PB, Hidalgo C, Schmitz-Abe K, DeChene ET, Swanson LC, et al. Recessive truncating titin gene, TTN, mutations presenting as centronuclear myopathy. *Neurology* 2013;**81**:1205–14.

259. Greenberg SA, Salajegheh M, Judge DP, Feldman MW, Kuncl RW, Waldon Z, et al. Etiology of limb girdle muscular dystrophy 1D/1E determined by laser capture microdissection proteomics. *Ann Neurol* 2012;**71**:141–5.

260. Speer MC, Yamaoka LH, Gilchrist JH, Gaskell CP, Stajich JM, Vance JM, et al. Confirmation of genetic heterogeneity in limb-girdle muscular dystrophy: linkage of an autosomal dominant form to chromosome 5q. *Am J Hum Genet* 1992;**50**:1211–17.

261. Hauser MA, Horrigan SK, Salmikangas P, Torian UM, Viles KD, Dancel R, et al. Myotilin is mutated in limb girdle muscular dystrophy 1A. *Hum Mol Genet* 2000;**9**:2141–7.

262. Selcen D, Engel AG. Mutations in myotilin cause myofibrillar myopathy. *Neurology* 2004;**62**:1363–71.

263. Salmikangas P, Mykkänen OM, Grönholm M, Heiska L, Kere J, Carpén O. Myotilin, a novel sarcomeric protein with two Ig-like domains, is encoded by a candidate gene for limb-girdle muscular dystrophy. *Hum Mol Genet* 1999;**8**:1329–36.

264. Salmikangas P, van der Ven PF, Lalowski M, Taivainen A, Zhao F, Suila H, et al. Myotilin, the limb-girdle muscular dystrophy 1A (LGMD1A) protein, cross-links actin filaments and controls sarcomere assembly. *Hum Mol Genet* 2003;**12**:189–203.

265. van der Ven PF, Wiesner S, Salmikangas P, Auerbach D, Himmel M, Kempa S, et al. Indications for a novel muscular dystrophy pathway. gamma-filamin, the muscle-specific filamin isoform, interacts with myotilin. *J Cell Biol* 2000;**151**:235–48.

266. Hauser MA, Conde CB, Kowaljow V, Zeppa G, Taratuto AL, Torian UM, et al. Myotilin mutation found in second pedigree with LGMD1A. *Am J Hum Genet* 2002;**71**:1428–32.

267. Gamez J, Navarro C, Andreu AL, Fernandez JM, Palenzuela L, Tejeira S, et al. Autosomal dominant limb-girdle muscular dystrophy: a large kindred with evidence for anticipation. *Neurology* 2001;**56**:450–4.

268. Feit H, Silbergleit A, Schneider LB, Gutierrez JA, Fitoussi RP, Réyès C, et al. Vocal cord and pharyngeal weakness with autosomal dominant distal myopathy: clinical description and gene localization to 5q31. *Am J Hum Genet* 1998;**63**:1732–42.

269. Gilchrist JM, Pericak-Vance M, Silverman L, Roses AD. Clinical and genetic investigation in autosomal dominant limb-girdle muscular dystrophy. *Neurology* 1988;**38**:5–9.

270. Minetti C, Sotgia F, Bruno C, Scartezzini P, Broda P, Bado M, et al. Mutations in the caveolin-3 gene cause autosomal dominant limb-girdle muscular dystrophy. *Nat Genet* 1998;**18**:365–8.

271. McNally EM, de Sá Moreira E, Duggan DJ, Bönnemann CG, Lisanti MP, Lidov HG, et al. Caveolin-3 in muscular dystrophy. *Hum Mol Genet* 1998;**7**:871–7.

272. Parton RG, del Pozo MA. Caveolae as plasma membrane sensors, protectors and organizers. *Nat Rev Mol Cell Biol* 2013;**14**:98–112.

273. Okamoto T, Schlegel A, Scherer PE, Lisanti MP. Caveolins, a family of scaffolding proteins for organizing "preassembled signaling complexes" at the plasma membrane. *J Biol Chem* 1998;**273**:5419–22.
274. Minetti C, Bado M, Broda P, Sotgia F, Bruno C, Galbiati F, et al. Impairment of caveolae formation and T-system disorganization in human muscular dystrophy with caveolin-3 deficiency. *Am J Pathol* 2002;**160**:265–70.
275. Herrmann R, Straub V, Blank M, Kutzick C, Franke N, Jacob EN, et al. Dissociation of the dystroglycan complex in caveolin-3-deficient limb girdle muscular dystrophy. *Hum Mol Genet* 2000;**9**:2335–40.
276. Sotgia F, Lee JK, Das K, Bedford M, Petrucci TC, Macioce P, et al. Caveolin-3 directly interacts with the C-terminal tail of beta -dystroglycan. Identification of a central WW-like domain within caveolin family members. *J Biol Chem* 2000;**275**:38048–58.
277. Galbiati F, Volonte D, Minetti C, Chu JB, Lisanti MP. Phenotypic behavior of caveolin-3 mutations that cause autosomal dominant limb girdle muscular dystrophy (LGMD-1C). Retention of LGMD-1C caveolin-3 mutants within the Golgi complex. *J Biol Chem* 1999;**274**:25632–41.
278. Cagliani R, Bresolin N, Prelle A, Gallanti A, Fortunato F, Sironi M, et al. A CAV3 microdeletion differentially affects skeletal muscle and myocardium. *Neurology* 2003;**61**:1513–19.
279. Carbone I, Bruno C, Sotgia F, Bado M, Broda P, Masetti E, et al. Mutation in the CAV3 gene causes partial caveolin-3 deficiency and hyperCKemia. *Neurology* 2000;**54**:1373–6.
280. González-Pérez P, Gallano P, González-Quereda L, Rivas-Infante E, Teijeira S, Navarro C, et al. Phenotypic variability in a Spanish family with a Caveolin-3 mutation. *J Neurol Sci* 2009;**276**:95–8
281. Hayashi T, Arimura T, Ueda K, Shibata H, Hohda S, Takahashi M, et al. Identification and functional analysis of a caveolin-3 mutation associated with familial hypertrophic cardiomyopathy. *Biochem Biophys Res Commun* 2004;**313**:178–84.
282. Speer MC, Vance JM, Grubber JM, Lennon Graham F, Stajich JM, Viles KD, et al. Identification of a new autosomal dominant limb-girdle muscular dystrophy locus on chromosome 7. *Am J Hum Genet* 1999;**64**:556–62.
283. Harms MB, Sommerville RB, Allred P, Bell S, Ma D, Cooper P, et al. Exome sequencing reveals DNAJB6 mutations in dominantly-inherited myopathy. *Ann Neurol* 2012;**71**:407–16.
284. Hackman P, Sandell S, Sarparanta J, Luque H, Huovinen S, Palmio J, et al. Four new Finnish families with LGMD1D; refinement of the clinical phenotype and the linked 7q36 locus. *Neuromuscul Disord* 2011;**21**:338–44.
285. Sarparanta J, Jonson PH, Golzio C, Sandell S, Luque H, Screen M, et al. Mutations affecting the cytoplasmic functions of the co-chaperone DNAJB6 cause limb-girdle muscular dystrophy. *Nat Genet* 2012;**44**(450–5):S1–2.
286. van der Kooi AJ, Ledderhof TM, de Voogt WG, Res CJ, Bouwsma G, Troost D, et al. A newly recognized autosomal dominant limb girdle muscular dystrophy with cardiac involvement. *Ann Neurol* 1996;**39**:636–42.
287. van der Kooi AJ, van Meegen M, Ledderhof TM, McNally EM, de Visser M, Bolhuis PA. Genetic localization of a newly recognized autosomal dominant limb-girdle muscular dystrophy with cardiac involvement (LGMD1B) to chromosome 1q11-21. *Am J Hum Genet* 1997;**60**:891–5.
288. Mercuri E, Poppe M, Quinlivan R, Messina S, Kinali M, Demay L, et al. Extreme variability of phenotype in patients with an identical missense mutation in the lamin A/C gene: from congenital onset with severe phenotype to milder classic Emery-Dreifuss variant. *Arch Neurol* 2004;**61**:690–4.
289. Bonne G, Di Barletta MR, Varnous S, Bécane HM, Hammouda EH, Merlini L, et al. Mutations in the gene encoding lamin A/C cause autosomal dominant Emery-Dreifuss muscular dystrophy. *Nat Genet* 1999;**21**:285–8.
290. Worman HJ, Foisner R. The nuclear envelope from basic biology to therapy. *Biochem Soc Trans* 2010;**38**:253–6.
291. Andrés V, González JM. Role of A-type lamins in signaling, transcription, and chromatin organization. *J Cell Biol* 2009;**187**:945–57.
292. Maraldi NM, Lattanzi G, Capanni C, Columbaro M, Merlini L, Mattioli E, et al. Nuclear envelope proteins and chromatin arrangement: a pathogenic mechanism for laminopathies. *Eur J Histochem* 2006;**50**:1–8.
293. Maraldi NM, Capanni C, Cenni V, Fini M, Lattanzi G. Laminopathies and lamin-associated signaling pathways. *J Cell Biochem* 2011;**112**:979–92.
294. Cacciottolo M, Numitone G, Aurino S, Caserta IR, Fanin M, Politano L, et al. Muscular dystrophy with marked dysferlin deficiency is consistently caused by primary dysferlin gene mutations. *Eur J Hum Genet* 2011;**19**:974–80.
295. Benedetti S, Menditto I, Degano M, Rodolico C, Merlini L, D'Amico A, et al. Phenotypic clustering of lamin A/C mutations in neuromuscular patients. *Neurology* 2007;**69**:1285–92.
296. Bonne G, Mercuri E, Muchir A, Urtizberea A, Bécane HM, Recan D, et al. Clinical and molecular genetic spectrum of autosomal dominant Emery-Dreifuss muscular dystrophy due to mutations of the lamin A/C gene. *Ann Neurol* 2000 **48**:170–80.
297. Raffaele Di Barletta M, Ricci E, Galluzzi G, Tonali P, Mora M, Morandi L, et al. Different mutations in the LMNA gene cause autosomal dominant and autosomal recessive Emery-Dreifuss muscular dystrophy. *Am J Hum Genet* 2000;**66**:1407–12.
298. Brodsky GL, Muntoni F, Miocic S, Sinagra G, Sewry C, Mestroni L. Lamin A/C gene mutation associated with dilated cardiomyopathy with variable skeletal muscle involvement. *Circulation* 2000;**101**:473–6.
299. Shackleton S, Lloyd DJ, Jackson SN, Evans R, Niermeijer MF, Singh BM, et al. LMNA, encoding lamin A/C, is mutated in partial lipodystrophy. *Nat Genet* 2000;**24**:153–6.
300. De Sandre-Giovannoli A, Chaouch M, Kozlov S, Vallat JM, Tazir M, Kassouri N, et al. Homozygous defects in LMNA, encoding lamin A/C nuclear-envelope proteins, cause autosomal recessive axonal neuropathy in human (Charcot-Marie-Tooth disorder type 2) and mouse. *Am J Hum Genet* 2002;**70**:726–36.
301. De Sandre-Giovannoli A, Bernard R, Cau P, Navarro C, Amiel J, Boccaccio I, et al. Lamin a truncation in Hutchinson-Gilford progeria. *Science* 2003;**300**:2055.
302. De Sandre-Giovannoli A, Lévy N. Altered splicing in prelamin A-associated premature aging phenotypes. *Prog Mol Subcell Biol* 2006;**44**:199–232.
303. Scharner J, Lu HC, Fraternali F, Ellis JA, Zammit PS. Mapping disease-related missense mutations in the immunoglobulin-like fold domain of lamin A/C reveals novel genotype-phenotype associations for laminopathies. *Proteins* 2014;**82**:904–15.

304. Bécane HM, Bonne G, Varnous S, Muchir A, Ortega V, Hammouda EH, et al. High incidence of sudden death with conduction system and myocardial disease due to lamins A and C gene mutation. *Pacing Clin Electrophysiol* 2000;**23**:1661–6.

305. Al-Saaidi R, Rasmussen TB, Palmfeldt J, Nissen PH, Beqqali A, Hansen J, et al. The LMNA mutation p.Arg321Ter associated with dilated cardiomyopathy leads to reduced expression and a skewed ratio of lamin A and lamin C proteins. *Exp Cell Res* 2013;**319**:3010–19.

306. Chen W, Huo J, Ma A, Bai L, Liu P. A novel mutation of the LMNA gene in a family with dilated cardiomyopathy, conduction system disease, and sudden cardiac death of young females. *Mol Cell Biochem* 2013;**382**:307–11.

307. Sims-Williams HP, Nye HJ, Walker PR. Dilated cardiomyopathy and skeletal myopathy: presenting features of a laminopathy. *BMJ Case Rep* 2013. Jan 17;2013.

308. Meune C, Van Berlo JH, Anselme F, Bonne G, Pinto YM, Duboc D. Primary prevention of sudden death in patients with lamin A/C gene mutations. *N Engl J Med* 2006;**354**:209–10.

309. Vainzof M, Marie SK, Reed UC, Schwartzman JS, Pavanello RC, Passos-Bueno MR, et al. Deficiency of merosin (laminin M or alpha 2) in congenital muscular dystrophy associated with cerebral white matter alterations. *Neuropediatrics* 1995; **26**:293–7.

310. Herrmann R, Straub V, Meyer K, Kahn T, Wagner M, Voit T. Congenital muscular dystrophy with laminin alpha 2 chain deficiency: identification of a new intermediate phenotype and correlation of clinical findings to muscle immunohistochemistry. *Eur J Pediatr* 1996;**155**:968–76.

311. Rajakulendran S, Parton M, Holton JL, Hanna MG. Clinical and pathological heterogeneity in late-onset partial merosin deficiency. *Muscle Nerve* 2011;**44**:590–3.

312. Tezak Z, Prandini P, Boscaro M, Marin A, Devaney J, Marino M, et al. Clinical and molecular study in congenital muscular dystrophy with partial laminin alpha 2 (LAMA2) deficiency. *Hum Mutat* 2003;**21**:103–11.

313. Naom I, D'Alessandro M, Sewry CA, Philpot J, Manzur AY, Dubowitz V, et al. Laminin alpha 2-chain gene mutations in two siblings presenting with limb-girdle muscular dystrophy. *Neuromuscul Disord* 1998;**8**:495–501.

314. Muntoni F, Lichtarowicz-Krynska EJ, Sewry CA, Manilal S, Recan D, Llense S, et al. Early presentation of X-linked Emery-Dreifuss muscular dystrophy resembling limb-girdle muscular dystrophy. *Neuromuscul Disord* 1998;**8**:72–6.

315. Felice KJ, Moore SA. Unusual clinical presentations in patients harboring the facioscapulohumeral dystrophy 4q35 deletion. *Muscle Nerve* 2001;**24**:352–6.

316. Krasnianski M, Eger K, Neudecker S, Jakubiczka S, Zierz S. Atypical phenotypes in patients with facioscapulohumeral muscular dystrophy 4q35 deletion. *Arch Neurol* 2003;**60**:1421–5.

317. Tawil R. Facioscapulohumeral muscular dystrophy. *Neurotherapeutics* 2008;**5**:601–6.

318. Udd B, Krahe R. The myotonic dystrophies: molecular, clinical, and therapeutic challenges. *Lancet Neurol* 2012;**11**:891–905.

319. Jöbsis GJ, Keizers H, Vreijling JP, de Visser M, Speer MC, Wolterman RA, et al. Type VI collagen mutations in Bethlem myopathy, an autosomal dominant myopathy with contractures. *Nat Genet* 1996;**14**:113–15.

320. Pan TC, Zhang RZ, Pericak-Vance MA, Tandan R, Fries T, Stajich JM, et al. Missense mutation in a von Willebrand factor type A domain of the alpha 3(VI) collagen gene (COL6A3) in a family with Bethlem myopathy. *Hum Mol Genet* 1998;**7**:807–12.

321. Camacho Vanegas O, Bertini E, Zhang RZ, Petrini S, Minosse C, Sabatelli P, et al. Ullrich scleroatonic muscular dystrophy is caused by recessive mutations in collagen type VI. *Proc Natl Acad Sci USA* 2001;**98**:7516–21.

322. Nadeau A, Kinali M, Main M, Jimenez-Mallebrera C, Aloysius A, Clement E, et al. Natural history of Ullrich congenital muscular dystrophy. *Neurology* 2009;**73**:25–31.

323. Briñas L, Richard P, Quijano-Roy S, Gartioux C, Ledeuil C, Lacène E, et al. Early onset collagen VI myopathies: Genetic and clinical correlations. *Ann Neurol* 2010;**68**:511–20.

324. Bonnemann CG. The collagen VI-related myopathies: muscle meets its matrix. *Nat Rev Neurol* 2011;**7**:379–90.

325. Foley AR, Quijano-Roy S, Collins J, Straub V, McCallum M, Deconinck N, et al. Natural history of pulmonary function in collagen VI-related myopathies. *Brain* 2013;**136**:3625–33.

326. Bönnemann CG, Brockmann K, Hanefeld F. Muscle ultrasound in Bethlem myopathy. *Neuropediatrics* 2003;**34**:335–6.

327. Mercuri E, Pichiecchio A, Allsop J, Messina S, Pane M, Muntoni F. Muscle MRI in inherited neuromuscular disorders: past, present, and future. *J Magn Reson Imaging* 2007;**25**:433–40.

328. Deconinck N, Dion E, Ben Yaou R, Ferreiro A, Eymard B, Briñas L, et al. Differentiating Emery-Dreifuss muscular dystrophy and collagen VI-related myopathies using a specific CT scanner pattern. *Neuromuscul Disord* 2010;**20**:517–23.

329. Jöbsis GJ, Boers JM, Barth PG, de Visser M. Bethlem myopathy: a slowly progressive congenital muscular dystrophy with contractures. *Brain* 1999;**122**(Pt 4):649–55.

330. Merlini L, Morandi L, Granata C, Ballestrazzi A. Bethlem myopathy: early-onset benign autosomal dominant myopathy with contractures. Description of two new families. *Neuromuscul Disord* 1994;**4**:503–11.

331. Collins J, Foley AR, Straub V, Bönnemann CG. Spontaneous keloid formation in patients with Bethlem myopathy. *Neurology* 2012;**79**:2158.

332. Hicks D, Lampe AK, Barresi R, Charlton R, Fiorillo C, Bonnemann CG, et al. A refined diagnostic algorithm for Bethlem myopathy. *Neurology* 2008;**70**:1192–9.

333. Pepe G, Bertini E, Giusti B, Brunelli T, Comeglio P, Saitta B, et al. A novel de novo mutation in the triple helix of the COL6A3 gene in a two-generation Italian family affected by Bethlem myopathy. A diagnostic approach in the mutations' screening of type VI collagen. *Neuromuscul Disord* 1999;**9**:264–71.

Chapter 35

Emery-Dreifuss Muscular Dystrophy: Nuclear Envelopathies

Hacer Durmus, Piraye Serdaroglu-Oflazer, and Feza Deymeer

INTRODUCTION

Emery-Dreifuss muscular dystrophy (EDMD) is a rare inherited disorder presenting in childhood or adolescence with relatively benign neuromuscular features and potentially fatal cardiac involvement. It is characterized by the triad of (1) early contractures of the Achilles tendons, elbows, and posterior cervical muscles; (2) slowly progressive muscle weakness and wasting, with humeroperoneal distribution in the early stages; and (3) cardiomyopathy with conduction defects.[1,2] The disease is transmitted usually as an X-linked recessive trait,[3] rarely as an autosomal dominant (AD) trait,[4,5] and exceptionally as an autosomal recessive (AR)[6] or X-linked dominant trait.[7] First clinically defined almost half a century ago,[1] the progress made in the last 20 years, with several genes discovered to be associated with the EDMD phenotype, has provided tremendous insight into its pathogenesis. Strikingly, all these genes encode nuclear envelope proteins, so that it may be appropriate to classify EDMD under the rubric "nuclear envelopathies."[8]

HISTORICAL BACKGROUND

In the early 1900s, Cestan and LeJonne[9] and later Schenk and Mathias[10] reported patients with muscular dystrophy and early contractures. In 1961, Dreifuss and Hogan[11] described a large family from Virginia with "an unusual type of benign X-linked muscular dystrophy." Contractures and cardiac abnormalities were not noted at the time, and the authors proposed that the slowly and rapidly progressive muscular dystrophies be grouped under an all-embracing term such as X-chromosomal or Duchenne-type muscular dystrophy. A reassessment of the same family a few years later by Emery and Dreifuss[1,12] made it clear that they were facing a unique nosological entity. In particular, they drew attention to the presence of contractures and the absence of pseudohypertrophy as being different from Becker muscular dystrophy. Cardiac involvement was considered an essential part of the clinical picture. Thus, the symptoms of EDMD had been delineated. The report of Emery and Dreifuss was followed by a period of confusion, when cases with similar phenotypes were published under different designations.[13–15] Rowland and coworkers[16] suggested the use of the eponym Emery-Dreifuss muscular dystrophy for all these cases, based on the remarkably constant features of the clinical phenotype.

Several linkage studies pointed the localization of the gene to the long arm of the X chromosome.[17–19] The gene responsible for X chromosome-linked EDMD (X-EDMD), *STA* or *EMD*, and the gene product, emerin, were identified in 1994.[3] While the X-linked form was clinically well established in the literature, several reports of families showing male-to-male inheritance made it clear that an autosomal dominant form (AD-EDMD) with essentially the same clinical features also existed.[20–22] The gene for the AD and AR forms, *LMNA* on 1q21, encoding lamin A/C, was identified in 1999.[4]

About 50% of EDMD patients do not carry mutations in *EMD* or *LMNA*, suggesting the existence of other causative genes. Recently, mutations in *SYNE1* and *SYNE2*, encoding nesprin 1 and nesprin 2 proteins, were also found to be associated with the AD form of EDMD.[5] The latest additions to the genes causing EDMD were *FHL1* encoding four-and-a-half-LIM protein 1 (FHL1)[7] and *TMEM43* encoding LUMA.[23]

B.T. Darras, H. Royden Jones, Jr., M.M. Ryan & D.C. De Vivo (Eds): Neuromuscular Disorders of Infancy, Childhood and Adolescence, Second edition.
DOI: http://dx.doi.org/10.1016/B978-0-12-417044-5.00035-4

MOLECULAR BIOLOGY

X-linked EDMD

Mutations in *EMD* result in X-linked recessive EDMD (EDMD 1).[3] The gene is located on the long arm of the X chromosome at Xq28 and contains 6 exons. The product of *EMD*, emerin, is a 254-amino acid conserved nuclear envelope (NE) protein.

Recently, mutations in *FHL1* encoding FHL1 protein were found to be associated with X-linked dominant EDMD phenotype myopathy with hypertrophic cardiomyopathy (EDMD 6).[7] FHL1 is highly expressed in skeletal and cardiac muscles. Mutations in *FHL1* also cause X-linked dominant scapuloperoneal myopathy,[24] X-linked myopathy with postural muscle atrophy,[25] and reducing body myopathy.[26]

See Case Example 35.1 for an example of X-linked recessive EDMD.

AD and AR EDMD

The gene for AD-EDMD (EDMD 2) and AR-EDMD (AR-EDMD, EDMD 3) is *LMNA* on chromosome 1q11−q23.[4] *LMNA* is composed of 12 exons encoding lamin A and lamin C, two A-type lamins produced as a result of alternative RNA splicing.

LMNA mutations were found to be associated with a great variety of conditions that affect striated muscle, adipose tissue, peripheral nerves, or multiple systems with signs of accelerated aging.[30] They are collectively called "laminopathies" and include limb girdle muscular dystrophy 1B,[31] congenital muscular dystrophy,[32] Charcot-Marie-Tooth neuropathy 2,[33] dilated cardiomyopathy 1 A,[34] Dunnigan type partial lypodystrophy,[35] mandibuloacral dysplasia,[36] Hutchinson-Gilford progeria syndrome,[37] atypical Werner syndrome,[38] lipoatrophy with diabetes, hepatic steatosis, hypertrophic cardiomyopathy, leukomelanodermic papules syndrome,[39] and tight skin contracture syndrome.[40]

Additional genes encoding NE proteins such as *SYNE1* (EDMD 4) encoding nesprin 1 and *SYNE2* (EDMD 5) encoding nesprin 2 have also been found to cause the AD form of EDMD.[5] More recently, another NE protein, LUMA, encoded by *TMEM43*, was implicated in an AD-EDMD like phenotype (EDMD 7).[23]

Case Example 35.2 gives an example of autosomal EDMD.

PATHOGENESIS

The proteins that play a key role in the pathogenesis of EDMD are located in the nucleus, in contrast to the proteins involved in most other muscular dystrophies that are associated with the sarcolemma.

A definition of the NE and a brief summary of the localization and interrelationships of the nuclear proteins involved in the pathogenesis of EDMD are necessary to have some idea of the complex.[30,42−45] The NE acts as a barrier separating the nucleus from the cytoplasm and is essential for the maintenance of the nuclear shape and its integrity. It consists of two membranes (inner and outer) and a dense network of nuclear intermediate filaments

CASE EXAMPLE 35.1 X-linked Recessive Emery-Dreifuss Muscular Dystrophy

The propositus, a 54-year-old man, was one of a pair of monozygotic twins.[27] He was admitted to the hospital because of chest pain and was found to have atrial arrest, for which a pacemaker was inserted. Elbow contractures had been noted before age 3 years, and he had been toe-walking since early childhood. Mild weakness, which showed very little progression over the years, had been noticed around 10 years of age. He had had a cerebrovascular accident with left hemiparesis at age 31, at which time a slow heart rate (40 beats/minute) was noted. He had been on anticoagulation since then.

Neurologic examination showed a left hemiparesis with mild left facial weakness. The right side was taken into consideration in reporting the muscle weakness. Moderate weakness of scapulohumeroperoneal muscles and mild weakness of the proximal lower extremity muscles were noted. Contractures were present at the elbows, wrists, knees, and ankles. The neck could not be fully flexed. Deep tendon reflexes were absent. A left Babinski sign was present.

Serum creatine kinase (CK) was mildly elevated. EMG was predominantly myopathic, combined with neurogenic features. Muscle biopsy showed minimal changes, consisting of fiber size variation and increased perimysial connective tissue. Cardiac evaluation revealed atrial paralysis.

His twin brother had a similar neuromuscular picture. He had had a cerebrovascular accident with aphasia and right hemiparesis at age 41. Cardiac evaluation revealed atrial paralysis, for which a pacemaker was inserted. Of the two affected maternal nephews, the elder, aged 32, had neck, elbow, wrist, and ankle contractures and mild humeral weakness. He also had atrial paralysis. The younger nephew, aged 13 years, had a similar examination, except that he lacked neck contractures and cardiac involvement.

In this family with X-linked recessive inheritance, limitation of elbow extension was noted in very early childhood, followed by toe-walking in all the patients. Of note was the fact that the youngest patient had not yet developed the full phenotype at age 13 years; he had no neck contractures and no cardiac involvement. The mutation in this family was a splice donor mutation 421 G>A in intron 2 of *EMD*.[28,29]

CASE EXAMPLE 35.2 Autosomal Emery-Dreifuss Muscular Dystrophy

The proband is an 11-year-old girl. She walked at 1 year of age and developed symptoms of muscle weakness and an awkward gait at 3 years. By age 6 years, she had bilateral elbow contractures (10 degrees), a stiff gait, a positive Gowers' sign, and mild right knee and ankle contractures. Lower extremity involvement was asymmetrical. At age 8 years, right heel cord lengthening and right hamstring lengthening at the knee were performed for the 20-degree equinus and flexion contractures, respectively. These procedures were subsequently done to relieve left-sided contractures. At age 11 years, elbow contractures had increased to 30 degrees. Her neck still had a full range of forward flexion, but early thoracolumbar paraspinal muscle tightness developed, limiting forward spinal flexion. Cardiac evaluations, which included physical examination, electrocardiogram (ECG), and two-dimensional echocardiogram, were normal at 9 and 11 years of age. Head and spinal cord magnetic resonance imaging was normal. Serial serum CK levels were consistently elevated (280 IU/L; normal, 4–150 IU/L). Three quadriceps muscle biopsies done at 5, 8, and 11 years of age showed progressive worsening of myocyte degeneration with fibrofatty infiltrates.

The proband's parents and an older female sibling were evaluated by physical examination, 12-lead ECG, and echocardiogram. In each, these evaluations were normal. A serum CK level in her father was normal (153 IU/L; normal, 41–500 IU/L).

Based on the proband's clinical diagnosis—probable autosomal EDMD—direct sequencing of *LMNA* was undertaken. DNA samples from the proband revealed a G > A transition at nucleotide 1072 in exon 6, which replaces the normal glutamic acid residue at position 358 (GAG) with lysine (AAG) (designated Glu358Lys). No sample derived from other family members or 200 normal control samples contained this G > A transition.

A second sequence variant was identified in exon 11 of the proband's DNA. A C > A transition was present at nucleotide 1871 that replaces arginine 624 (CGC) with histidine (CAC) (designated Arg624His). Analyses of DNA from the proband's father also demonstrated the adenosine nucleotide at position 1871. DNA from the proband's sister, mother, and 200 normal control samples did not exhibit this transition.

It seems that the compound heterozygous mutations account for the particularly severe phenotype in the proband, and it has been hypothesized that the Arg624His amino acid change may exert a synergistic or modifier effect on disease expression. The absence of pathology in the father suggests that Arg624His functions either as a dominant mutation with incomplete penetrance or as a recessive allele.[41]

formed by A-type and B-type lamins. Lamins A and C are the major isoforms of A-type lamins.

The NE contains over 100 different proteins, the functions of most of which still remain unexplored. Three groups of NE proteins relevant to the pathogenesis of EDMD are LEM-domain proteins (emerin and others), SUN-domain proteins (SUN1 and SUN2), and KASH-domain proteins (nesprins 1, 2, 3, and others). Emerin is an integral membrane protein localizing to the inner membrane. It binds to lamins (most importantly to the A-type lamins) and to BAF (barrier-to-autointegration factor), which is a highly conserved protein essential for chromosome segregation, cell cycle progression, and post-mitotic nuclear assembly.[46] This trio (emerin-lamins-BAF) forms a major component of the NE-associated nucleoskeletal structure known as "nuclear lamina."[43] In addition to their role in maintaining the mechanical stability of the NE throughout the phases of the cell cycle, emerin and lamins interact with many proteins that regulate DNA synthesis, chromatin organization, gene transcription, and cell differentiation.[43]

Emerin also binds directly to SUN-domain and KASH-domain proteins. SUN-domain proteins, which span the inner nuclear membrane, bridge to KASH-domain proteins, which span the outer nuclear membrane. SUN- and KASH-domain proteins form the LINC-complex (linker of nucleoskeleton and cytoskeleton), which is proposed to form a mechanical link between the nucleoskeleton and cytoskeleton.

As can be seen, the proteins causing EDMD, emerin, lamin A/C, nesprin 1, and nesprin 2, interact with each other.[43] LUMA is also a binding partner of emerin. The interactions of FHL-1 with NE proteins are not yet known. FHL1 is localized to the sarcomere, the sarcolemma, and the nucleus, and has important roles in maintaining the stability of the sarcolemma, myofibrillar assembly, and transcriptional regulation. Absence or reduced level of FHL1 protein may cause delayed myotube formation.

Disrupted function of these proteins is the key pathophysiological mechanism in EDMD. Accumulating data suggest that EDMD might be caused by the uncoupling of the nucleoskeleton and cytoskeleton because of damaged emerin-nesprin-lamin interactions.[5] However, how the mutations cause disease is still far from being elucidated. It is particularly puzzling that these ubiquitous proteins lead to tissue-specific disorders.[47] Proposed hypotheses on pathogenesis include NE defects affecting nuclear stiffness and increased susceptibility to mechanical stress, NE defects as determinants of altered nucleocytoplasmic interplay, altered cell cycle control, alterations of the nuclear morphology affecting chromatin rearrangements, disruption of the process of skeletal muscle regeneration, altered DNA repair due to oxidative stress, and increased susceptibility to apoptosis.[48,49]

CLINICAL FEATURES

Neuromuscular Features

The distinguishing feature of the neuromuscular picture is the early development of flexion contractures of the arm and leg muscles and extension contractures of the cervical muscles. The arms are held in a semiflexed position and cannot be extended (see Figure 35.1). The feet are in an equinus position. The neck cannot be flexed onto the sternum, although extension is possible (see Figure 35.2). There may also be contractures at the knees, wrists (see Figure 35.3), and fingers. In contrast to other neuromuscular diseases in which contractures develop when weakness is advanced, contractures in EDMD develop before any significant weakness occurs, and their severity remains out of proportion to the degree of weakness throughout the course of the disease. Contractures are "in the direction of the weakest joint-moving muscle."[50] Muscle weakness is mild and slowly progressive. Muscle wasting is most pronounced in the biceps brachii and posterior leg muscles (see Figure 35.4). Patients rarely lose ambulation. All forms of EDMD are clinically similar but not identical.[51] The severity of the symptoms can show striking variability even within the same family.[51,52]

Onset is usually in childhood, when the child is noted to have difficulty with ambulation, usually because of toe-walking due to flexion contractures of the leg

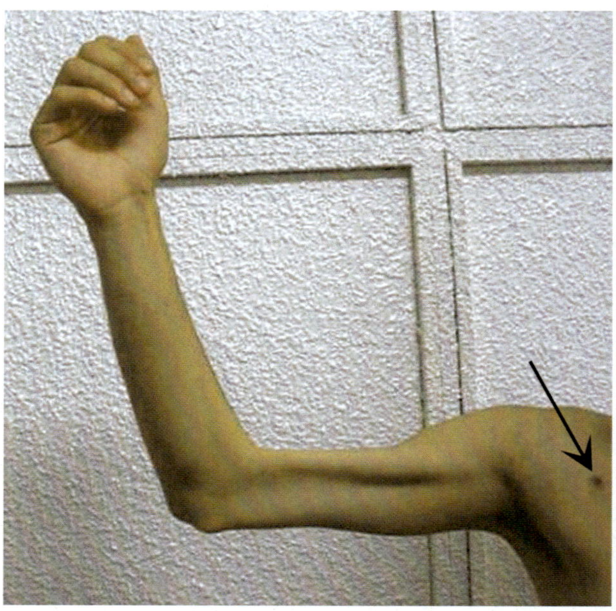

FIGURE 35.1 Flexion contracture of arm muscles at the elbow and marked humeral muscle wasting with relative sparing of shoulder muscles in Emery-Dreifuss muscular dystrophy. The arm cannot be fully extended at the elbow and is maintained in a semiflexed position. The pacemaker insertion site is evident (arrow).

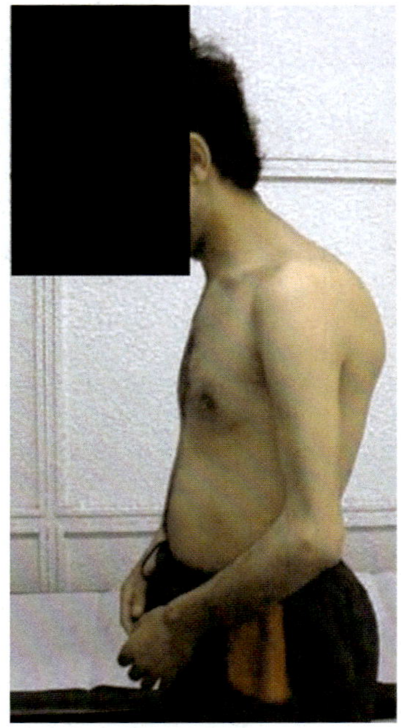

FIGURE 35.2 Extension contracture of posterior cervical muscles in X-linked Emery-Dreifuss muscular dystrophy. Note the significant limitation of neck flexion (the patient was asked to flex his neck fully onto the sternum).

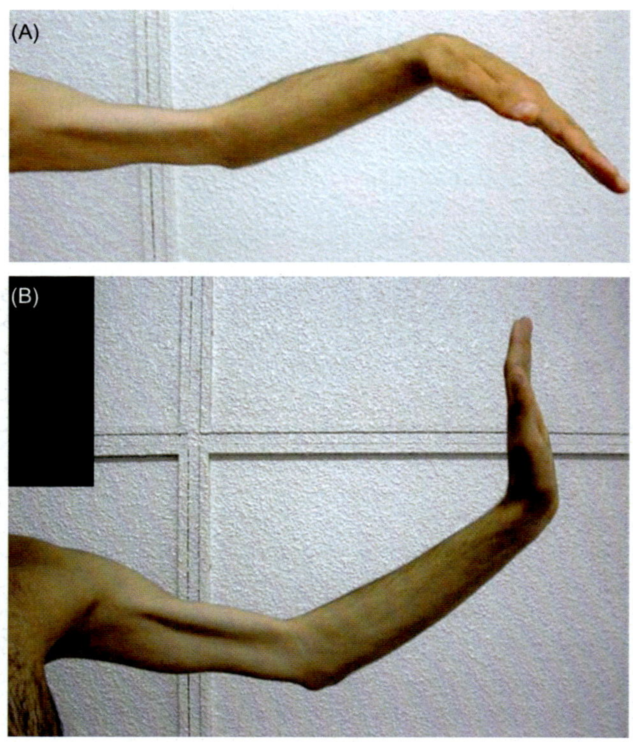

FIGURE 35.3 Extension contracture of forearm muscles at the wrist. The patient (the same one shown in Figure 35.2) is unable to flex his wrist beyond a few degrees (**A**), although he can fully extend it (**B**).

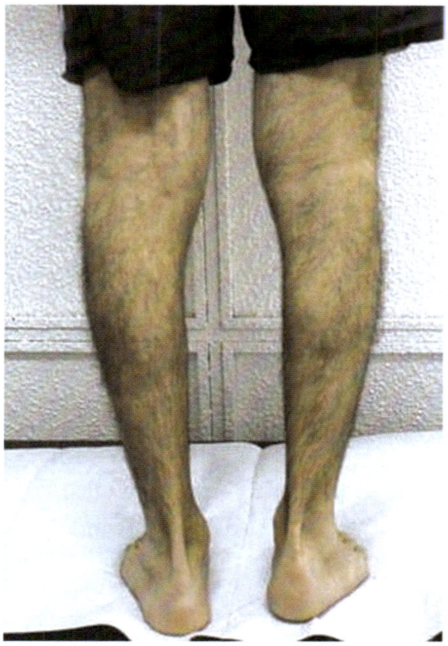

FIGURE 35.4 Atrophy of posterior calf muscles, particularly of the medial gastrocnemius muscles.

muscles. The problem is sometimes noticed as soon as the child starts to walk. Limitation of elbow extension is usually noticed later in the first decade; however, this is the first symptom in some patients.[13,15,27] Limitation of neck flexion is usually noted in the second and rarely in the first decade.[14] Muscle weakness starts in the first or the second decade. Mental retardation or even mild cognitive delay is not a feature of the disease.[1]

Examination reveals flexion contractures at the elbows (see Figure 35.1), shortening of the Achilles tendons, and extension contracture of the cervical muscles (see Figure 35.2). Mild weakness and atrophy of biceps brachii, triceps (see Figure 35.1), and posterior calf muscles (see Figure 35.4) are noted, although more widespread involvement may be seen. Upper extremity distal muscles can sometimes be weak,[27] and distal contractures may occur (see Figure 35.3). Mild facial weakness may be present. Deep tendon reflexes are reduced or absent. Pseudohypertrophy is usually not seen. Scoliosis is rarely observed.

Cardiac Features

Atrial conduction defects are the most common cardiac abnormalities in EDMD 1. First-degree atrioventricular block, sinus bradycardia, and supraventricular tachycardias may be early signs of cardiac involvement.[53–55] They may progress to atrial flutter, atrial fibrillation (see Figure 35.5), and atrial standstill. The onset of cardiac abnormalities usually occurs in the third decade in EDMD. Earlier onset of severe involvement is rare.[53,55,56] A recent study reported serious ventricular arrhythmia in a young patient with EDMD 1.[57] Cardiac involvement is more aggressive in EDMD 2;[58] ventricular arrythmias occur earlier and lead to dilated cardiomyopathy.[59,60] Sudden death can occur, more frequently in EDMD 2, but also in EDMD 1. There is no correlation between the severity of the neuromuscular involvement and the degree of cardiac abnormality.[53,54,61]

DIFFERENTIAL DIAGNOSIS

EDMD first needs to be differentiated from other neuromuscular diseases presenting with contractures in the extremities and rigidity at cervical and dorsolumbar spine.[62] They include congenital muscular dystrophies (CMD), congenital myopathies (such as central core and centronuclear congenital myopathies), and rigid spine syndrome (RSS).[62] None of these disorders shows cardiac conduction defects or blocks.

One might be led to underestimate this differentiation, thinking that EDMD has prominent cardiac conduction defects in contrast to all the other conditions. This is not at all helpful when one encounters a child with a rigid spine because the cardiac defects in EDMD become apparent at a much later age. Thus, differentiation is done on other grounds in early childhood.[63]

There are several forms of CMD, and they present within the first year of life with severe hypotonia accompanied by possible contractures, very high CK, and dystrophic muscle biopsy findings. Symptoms related to the central nervous system might be a part of the clinical picture. Most of them are inherited in an autosomal recessive pattern. Congenital myopathies have distinctive muscle biopsy findings.

The most difficult differential diagnosis is that from RSS, as evidenced by cases previously diagnosed as RSS that have turned out to be EDMD after the advent of molecular biology.[53,64,65] Early and severe scoliosis in the first decade is in favor of RSS and should prompt search for selenoprotein 2 gene mutation. Severe respiratory involvement is usually present in RSS but is exceptional in EDMD.[62]

GENOTYPE-PHENOTYPE CORRELATION

At present, mutations in six genes (*EMD*, *LMNA*, *SYNE1*, *SYNE2*, *FHL1*, and *TMEM43*) are associated with the EDMD phenotype. They cause very similar phenotypes, difficult to distinguish from each other.[66]

The majority of EMD mutations are null mutations with no emerin expression. Missense mutations with reduced expression of emerin are very rare.[67] There is remarkable intra- and interfamilial variation regarding age

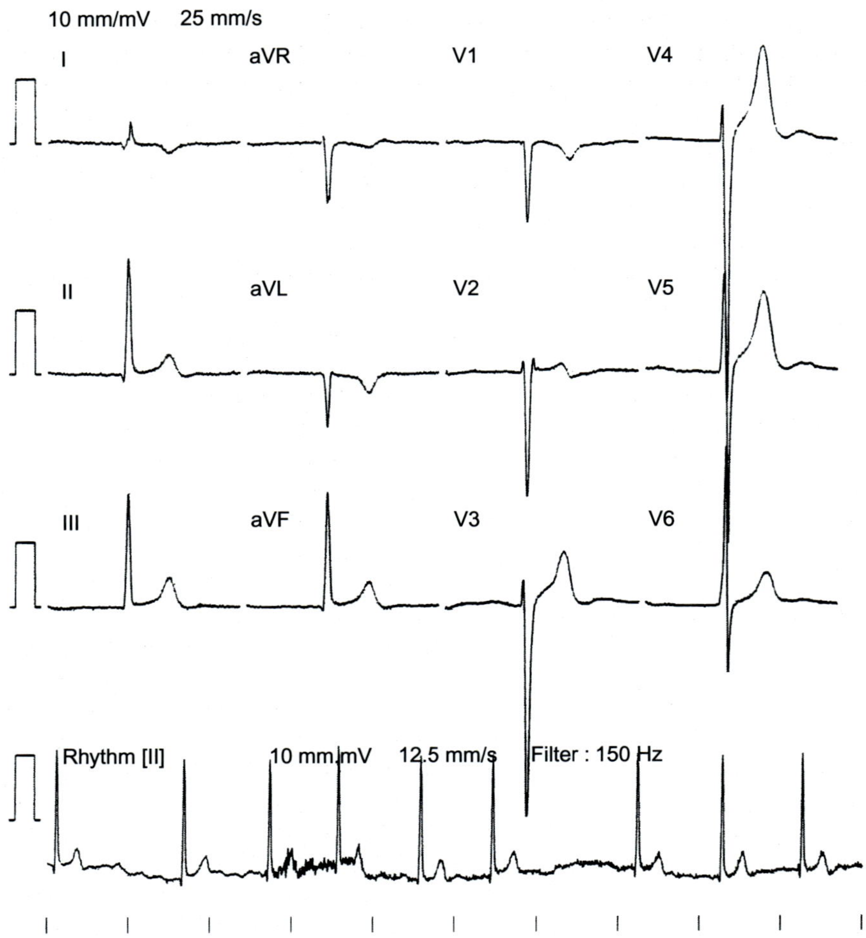

FIGURE 35.5 Electrocardiogram of a 25-year-old patient with X-linked Emery-Dreifuss muscular dystrophy. The main features are a heart rate of 52 beats/minute, irregular rhythm, atrial fibrillation, and incomplete right bundle branch block.

of onset, severity, and course[68] of neuromuscular and cardiac manifestations, independent of each other. This variation is more pronounced than in other dystrophies. It could be speculated that some modifier gene yet unexplored might have an important role in the modulation of disease severity in nuclear envelopathies.[69]

LMNA mutations are mostly missense mutations.[41,51] No correlation has been found between the location of the mutation and the clinical phenotype, including severity.[51,70] In addition to the typical EDMD phenotype, severe phenotypes with very early onset (second year of life) weakness have been reported.[51,70]

EDMD 2 is phenotypically very similar to EDMD 1, particularly with respect to its neuromuscular features.[51,71] Onset of the weakness might be earlier and more severe in EDMD 2 and might precede contractures, which are usually the presenting symptoms in EDMD 1. Loss of ambulation might be seen in EDMD 2, which is very unlikely in EDMD 1.[72] Furthermore, patients with EDMD 2 have more severe and progressive weakness of the biceps brachii compared with EDMD 1.[51] Calf hypertrophy might be more common in EDMD 2.[73]

The major distinguishing phenotypical difference of EDMD 2 from EDMD 1 concerns the nature of the cardiac involvement, whereas conduction disturbances, especially atrial arrhythmias, are frequent and the major cardiac problem in EDMD 1, dilation of the left ventricle, cardiac failure, and ventricular tachyarrhythmias appear to be more common in EDMD 2.[58,71] Cardiac involvement in EDMD 2 is more severe compared to EDMD 1 patients.[71] Isolated cardiac involvement is more frequently found in EDMD 2.

LMNA mutations rarely cause AR phenotypes. In a family with homozygous mutations in the two alleles in the proband and heterozygous mutations in the carriers, the patient had severe neuromuscular symptoms with confinement to a wheelchair, but she had no cardiac involvement even at age 40.[6]

A more severe phenotype of EDMD with unusual disease progression rate is reported in individuals who are carrying mutations in *EMD* and the desmin gene at the same time.[74]

FHL1 mutations cause a similar phenotype to EDMD 1. *FHL1* mutations should be considered in patients with X-linked inheritance pattern, especially if hypertrophic cardiomyopathy is present.[26]

SYNE1, *SYNE2*, and *TMEM43* mutations are very rare and it is too early to try to make phenotype-genotype correlations in patients with these mutations.

LABORATORY INVESTIGATIONS

Serum Creatine Kinase

The CK level can be normal or mildly to moderately elevated.[53] A level as high as 10 times the upper limit of normal can occasionally be seen,[53] but it does not usually get as high as that seen in Duchenne or Becker muscular dystrophy. Data from several families suggest that CK is normal at birth, rises after age 1 year, peaks in childhood and adolescence, and declines thereafter toward normal values.[15,75]

CK is not useful in identifying carriers. Obligate carriers are likely to have normal CK levels[15,75,76]; the level is high in only a small percentage.[1,17]

Electromyography

The EMG is myopathic in most patients. In some studies, both myopathic and neurogenic features were present,[13,16,76] but other patients from the same families showed clear-cut myopathic features.[13,76] Hopkins and coworkers[76] pointed out that classic findings of myopathy are not always present in chronic dystrophies, and that confusing EMG findings reflect the mildness and long duration of the myopathic process in EDMD. A quantitative EMG study concluded that the findings were compatible with myopathy and attributed the presence of long duration motor unit potentials to increased variability of muscle fiber size and reorganization of the motor unit in a slowly progressive process.[77] Fortunately, with the advent of immunohistochemistry and DNA testing, EMG has become less important as a diagnostic procedure; it may support the myopathic nature of the process, but EMG is not required if the immunohistochemical or genetic tests are diagnostic of EDMD.[78]

Histopathology and Immunohistochemistry

X-linked and autosomal forms have similar histological changes. Quadriceps biopsies showed mild myopathic changes including fiber size variation and increased internal nuclei. Necrotic and regenerating fibers and increased connective tissue were rarely found.[79] Severe fibrosis was observed in biceps muscle biopsies of X-linked recessive EDMD patients.[53,80] In an autopsy study of a 45-year-old man with X-linked recessive EDMD,[13] macroscopic evaluation disclosed triceps, deeper portions of biceps brachii, and calf muscles to be white and fibrotic. Microscopically, gastrocnemius muscle was almost totally replaced by, and biceps brachii had a substantial increase in, fibrofatty connective tissue. Generalized or focal fibrosis was also present in cardiac muscle biopsies.[13,53,76]

Electron microscopy of muscle showed structural abnormalities in myonuclei both in AD-EDMD[79,81–83] and X-linked recessive EDMD.[82] Furthermore, chromatin changes were observed in the satellite cells in EDMD 2[81] implying the presence of defective regeneration.[84]

Unexpected pathological findings were also reported. In one study, inflammatory changes were found in the muscle biopsies of some children (≤2 years of age) with *LMNA* mutations.[85] A few patients with EDMD 1 had rimmed vacuoles and electron microscopic tubulofilamentous structures reminiscent of inclusion body myositis.[86]

Absence of emerin in X-linked EDMD can be demonstrated by immunohistochemistry in muscle, skin, buccal cells, and lymphocytes[87–89] (see Figure 35.6). No mutations were found in *EMD* in patients who were emerin-positive, confirming its diagnostic value.[71] Emerin determination[89] can also be useful in identifying female carriers; mosaic expression of emerin is observed in carriers of null mutations.[90,91] Immunohistochemistry is not useful in the diagnosis of EDMD 2.

Immunohistochemistry and immunoblotting for FHL1 showed a strong reduction in FHL1 protein expression in the muscles of patients with EDMD, suggesting that these might be helpful before genetic testing.[7]

Radiological Studies of the Extremities

MR and CT studies showed a selective involvement of muscles in EDMD with particular involvement of posterior calf muscles. Of these muscles, the medial head of the gastrocnemius was more severely affected in EDMD 2 while soleus was the more severely affected muscle in EDMD 1.[92] At the thigh level, patients with EDMD 1 had mild involvement while vastus lateralis and intermedius were moderately to severely involved in patients with EDMD 2.[92] The sparing of rectus femoris was emphasized.[93] Of note was the presence of similar radiological involvement of posterior leg muscles in different phenotypes with *LMNA* mutations such as LGMD1B and isolated dilated cardiomyopathy[94] and asymptomatic carriers of *EMD* mutations.[57] MR of the extremities was also

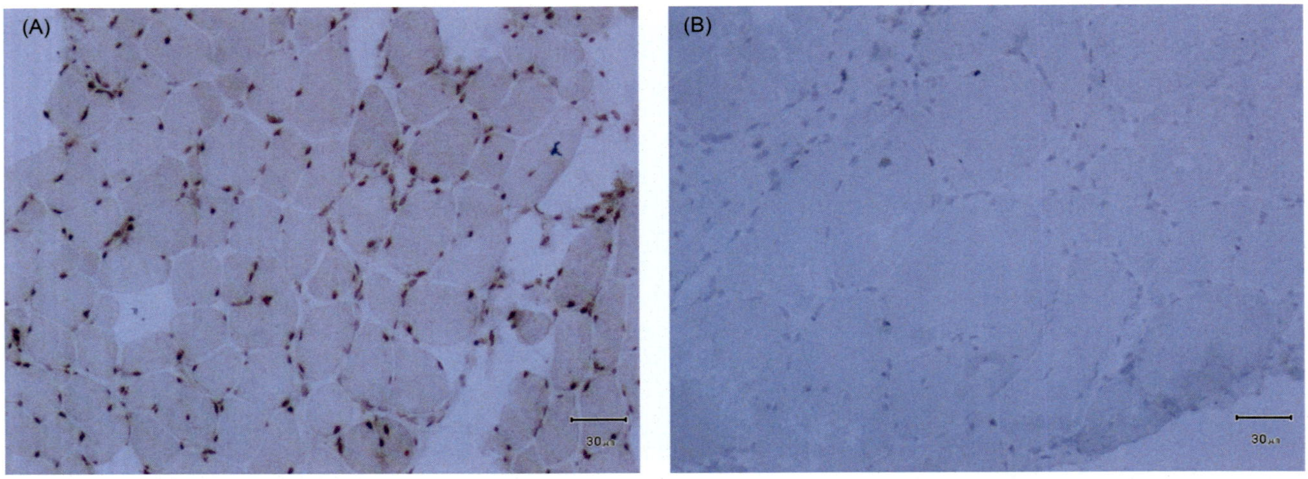

FIGURE 35.6 Muscle immunohistochemistry. Emerin staining in normal muscle (**A**), no emerin staining in nuclei in EDMD muscle (**B**).

found to be helpful in differentiating EDMD from other disorders with rigidity and contractures.[95,96]

Genetic Testing

The choice of the gene to be screened for mutations depends on the sex of the child and the family history. In boys with a phenotype suggestive of EDMD, a family history showing X-linked inheritance, or both, *EMD* should be screened first for mutations. If the test is negative, *FHL1* should be screened.

Patients with AD inheritance pattern should be screened first for mutations in *LMNA*. Mutations in *SYNE1* and *SYNE2* are also known to cause AD-EDMD-like phenotypes, but it should be kept in mind that these are very rare. Almost half of the causative mutations of AD inherited EDMD are still unknown. If the initial screening of *LMNA* is negative, whole exome sequencing might be helpful and even more cost effective for identifying the causative mutation.[97]

In sporadic cases, a negative search for *EMD* mutations should be followed by a search for *LMNA* mutations in boys. Since the occurrence of symptomatic X-linked EDMD is unlikely in girls, one should go directly to a search for *LMNA* mutations.

In centers where DNA testing is not available, immunohistochemistry performed on a muscle biopsy specimen can establish the diagnosis of X-linked recessive EDMD if it shows the absence of emerin. If the diagnosis is made by immunohistochemistry, it is not necessary to proceed to DNA analysis unless prenatal diagnosis is a consideration. It should be noted, however, that genetic testing is less invasive than muscle biopsy, probably less expensive, and far more reliable. Nevertheless, a muscle biopsy may have to be performed before any genetic testing if the cause of the child's muscle weakness is not clear and the family history is negative for EDMD.

Cardiac Investigations

ECG is the most important test in detecting arrhythmias in EDMD[54] and should be performed periodically because cardiac involvement may not be present at an early age. If there is any abnormality on the routine ECG, yearly 24-hour ECG monitoring is mandatory.[54] Echocardiography is necessary to evaluate the extent of cardiomyopathy.

Female carriers are at a substantial risk of having cardiac conduction defects. Sudden death, stroke, and heart attack at an early age have been reported.[56,98] Female carriers should thus have regular cardiac evaluations, especially as they get older.

THERAPY

The most important aspect of management is the timely detection of cardiac conduction defects and the implantation of a pacemaker,[99] or preferably implantable cardioconverter-defibrillator (ICD),[100] which can be lifesaving. However, the appearance of dilated cardiomyopathy and ventricular arrhythmia despite adequate pacemaker function[54] emphasizes the seriousness of the cardiac involvement and may necessitate the performance of cardiac transplantation.[101]

Physiotherapy may be helpful. Surgical release of heel and elbow contractures may improve mobility,[61] and heel cord lengthening in particular may be warranted, because the long-term prognosis for ambulation is good.[102]

The small size of the gene and the fact that it need only be delivered to the heart are reasons to be optimistic about gene therapy in the future.[66,98,103] Increased activity of extracellular signal-regulated kinase (ERK) 1/2 was demonstrated in the heart of an autosomal EDMD mouse model; blocking its action improved cardiac function. Thus, selumetinib, which blocks mitogen-activated protein kinase/extracellular signal-regulated kinase kinase 1/2 that activates ERK 1/2, appears to be a promising drug.[104,105]

PROGNOSIS

Progression of neuromuscular symptoms is slow, and the symptoms seem to stabilize at about age 20 years.[61] Patients usually remain ambulatory throughout their lives

In a review of the literature, more than half of the reported patients died suddenly, most of them before age 50.[72] The outlook improves with therapeutic measures taken against the conduction defects and the dilated cardiomyopathy.[50]

ACKNOWLEDGMENTS

We thank Dr. Basil T. Darras for contributing Case Example 35.2.

REFERENCES

1. Emery AE, Dreifuss FE. Unusual type of benign X-linked muscular dystrophy. *J Neurol Neurosurg Psychiatry* 1966;**29**:338–42.
2. Maraldi NM, Merlini L. Emery-Dreifuss muscular dystrophy. In: Engel AG, Franzini-Armstrong C, editors. *Myology*. New York: McGraw-Hill; 2004, pp. 1027–37.
3. Bione S, Maestrini E, Rivella S, Mancini M, Regis S, Romeo G, et al. Identification of a novel X-linked gene responsible for Emery-Dreifuss muscular dystrophy. *Nat Genet* 1994;**8**:323–7.
4. Bonne G, Di Barletta MR, Varnous S, Becane HM, Hammouda EH, Merlini L, et al. Mutations in the gene encoding lamin A/C cause autosomal dominant Emery-Dreifuss muscular dystrophy. *Nat Genet* 1999;**21**:285–8.
5. Zhang Q, Bethmann C, Worth NF, Davies JD, Wasner C, Feuer A, et al. Nesprin-1 and -2 are involved in the pathogenesis of Emery Dreifuss muscular dystrophy and are critical for nuclear envelope integrity. *Hum Mol Genet* 2007;**16**:2816–33.
6. Di Barletta RM, Ricci E, Galluzzi G, Tonali P, Mora M, Morandi L, et al. Different mutations in the LMNA gene cause autosomal dominant and autosomal recessive Emery-Dreifuss muscular dystrophy. *Am J Hum Genet* 2000;**66**:1407–12.
7. Gueneau L, Bertrand AT, Jais JP, Salih MA, Stojkovic T, Wehnert M, et al. Mutations of the FHL1 gene cause Emery-Dreifuss muscular dystrophy. *Am J Hum Genet* 2009;**85**:338–53.
8. Nagano A, Arahata K. Nuclear envelope proteins and associated diseases. *Curr Opin Neurol* 2000;**13**:533–9.
9. Cestan R, LeJonne NJ. Une myopathie avec retractions familiales. *Nouvelle Iconographie Salpetriere* 1902;**15**:38–52.
10. Schenk P, Mathias E. Zur Kasuistik der Dystrophia musculorum progressiva retrahens. *Klin Wochenschr* 1920;**57**:557–8.
11. Dreifuss FE, Hogan GR. Survival in x-chromosomal muscular dystrophy. *Neurology* 1961;**11**:734–7.
12. Emery AE. Impact commentaries. Unusual type of benign X linked muscular dystrophy. *J Neurol Neurosurg Psychiatry* 2013;**84**:4–5.
13. Thomas PK, Calne DB, Elliott CF. X-linked scapuloperoneal syndrome. *J Neurol Neurosurg Psychiatry* 1972;**35**:208–15.
14. Mawatari S, Katayama K. Scapuloperoneal muscular atrophy with cardiopathy. An X-linked recessive trait. *Arch Neurol* 1973;**28**:55–9.
15. Johnston AW, McKay E. X linked muscular dystrophy with contractures. *J Med Genet* 1986;**23**:591–5.
16. Rowland LP, Fetell M, Olarte M, Hays A, Singh N, Wanat FE. Emery-Dreifuss muscular dystrophy. *Ann Neurol* 1979;**5**:111–17.
17. Hodgson S, Boswinkel E, Cole C, Walker A, Dubowitz V, Granata C, et al. A linkage study of Emery-Dreifuss muscular dystrophy. *Hum Genet* 1986;**74**:409–16.
18. Yates JR, Affara NA, Jamieson DM, Ferguson-Smith MA, Hausmanowa-Petrusewicz I, Zaremba J, et al. Emery-Dreifuss muscular dystrophy: localisation to Xq27.3----qter confirmed by linkage to the factor VIII gene. *J Med Genet* 1986;**23**:587–90.
19. Thomas NS, Williams H, Elsas LJ, Hopkins LC, Sarfarazi M, Harper PS. Localisation of the gene for Emery-Dreifuss muscular dystrophy to the distal long arm of the X chromosome. *J Med Genet* 1986;**23**:596–8.
20. Fenichel GM, Sul YC, Kilroy AW, Blouin R. An autosomal-dominant dystrophy with humeropelvic distribution and cardiomyopathy. *Neurology* 1982;**32**:1399–401.
21. Serratrice G, Pouget J. Emery-Dreifuss disease or syndrome of amyotrophy with early contractures and secondary disorders of cardiac conduction with variable heredity. *Rev Neurol (Paris)* 1986;**142**:766–70.
22. Yates JR. 43rd ENMC International Workshop on Emery-Dreifuss Muscular Dystrophy, 22 June 1996, Naarden, The Netherlands. *Neuromuscul Disord* 1997;**7**:67–9.
23. Liang WC, Mitsuhashi H, Keduka E, Nonaka I, Noguchi S, Nishino I, et al. TMEM43 mutations in Emery-Dreifuss muscular dystrophy-related myopathy. *Ann Neurol* 2011;**69**:1005–13.
24. Quinzii CM, Vu TH, Min KC, Tanji K, Barral S, Grewal RP, et al. X-linked dominant scapuloperoneal myopathy is due to a mutation in the gene encoding four-and-a-half-LIM protein 1. *Am J Hum Genet* 2008;**82**:208–13.
25. Windpassinger C, Schoser B, Straub V, Hochmeister S, Noor A, Lohberger B, et al. An X-linked myopathy with postural muscle atrophy and generalized hypertrophy, termed XMPMA, is caused by mutations in FHL1. *Am J Hum Genet* 2008;**82**:88–99.
26. Schessl J, Feldkirchner S, Kubny C, Schoser B. Reducing body myopathy and other FHL1-related muscular disorders. *Semin Pediatr Neurol* 2011;**18**:257–63.
27. Deymeer F, Oge AE, Bayindir C, Kaymaz C, Nisanci Y, Adalet K, et al. Emery-Dreifuss muscular dystrophy with unusual features. *Muscle Nerve* 1993;**16**:1359–65.
28. Yates JR, Warner JP, Smith JA, Deymeer F, Azulay JP, Hausmanowa-Petrusewicz I, et al. Emery-Dreifuss muscular dystrophy: linkage to markers in distal Xq28. *J Med Genet* 1993;**30**:108–11.
29. Yates JR, Bagshaw J, Aksmanovic VM, Coomber E, McMahon R, Whittaker JL, et al. Genotype-phenotype analysis in X-linked Emery-Dreifuss muscular dystrophy and identification of a missense mutation associated with a milder phenotype. *Neuromuscul Disord* 1999;**9**:159–65.
30. Worman HJ. Nuclear lamins and laminopathies. *J Pathol* 2012;**226**:316–25.

31. Muchir A, Bonne G, van der Kooi AJ, van Meegen M, Baas F, Bolhuis PA, et al. Identification of mutations in the gene encoding lamins A/C in autosomal dominant limb girdle muscular dystrophy with atrioventricular conduction disturbances (LGMD1B). *Hum Mol Genet* 2000;**9**:1453–9.
32. Mercuri E, Poppe M, Quinlivan R, Messina S, Kinali M, Demay L, et al. Extreme variability of phenotype in patients with an identical missense mutation in the lamin A/C gene: from congenital onset with severe phenotype to milder classic Emery-Dreifuss variant. *Arch Neurol* 2004;**61**:690–4.
33. De Sandre-Giovannoli A, Chaouch M, Kozlov S, Vallat JM, Tazir M, Kassouri N, et al. Homozygous defects in LMNA, encoding lamin A/C nuclear-envelope proteins, cause autosomal recessive axonal neuropathy in human (Charcot-Marie-Tooth disorder type 2) and mouse. *Am J Hum Genet* 2002;**70**:726–36.
34. Fatkin D, MacRae C, Sasaki T, Wolff MR, Porcu M, Frenneaux M, et al. Missense mutations in the rod domain of the lamin A/C gene as causes of dilated cardiomyopathy and conduction-system disease. *N Engl J Med* 1999;**341**:1715–24.
35. Shackleton S, Lloyd DJ, Jackson SN, Evans R, Niermeijer MF, Singh BM, et al. LMNA, encoding lamin A/C, is mutated in partial lipodystrophy. *Nat Genet* 2000;**24**:153–6.
36. Novelli G, Muchir A, Sangiuolo F, Helbling-Leclerc A, D'Apice MR, Massart C, et al. Mandibuloacral dysplasia is caused by a mutation in LMNA-encoding lamin A/C. *Am J Hum Genet* 2002;**71**:426–31.
37. De Sandre-Giovannoli A, Bernard R, Cau P, Navarro C, Amiel J, Boccaccio I, et al. Lamin a truncation in Hutchinson-Gilford progeria. *Science* 2003;**300**:2055.
38. Chen L, Lee L, Kudlow BA, Dos Santos HG, Sletvold O, Shafeghati Y, et al. LMNA mutations in atypical Werner's syndrome. *Lancet* 2003;**362**:440–5.
39. Caux F, Dubosclard E, Lascols O, Buendia B, Chazouilleres O, Cohen A, et al. A new clinical condition linked to a novel mutation in lamins A and C with generalized lipoatrophy, insulin-resistant diabetes, disseminated leukomelanodermic papules, liver steatosis, and cardiomyopathy. *J Clin Endocrinol Metab* 2003;**88**:1006–13.
40. Navarro CL, De Sandre-Giovannoli A, Bernard R, Boccaccio I, Boyer A, Genevieve D, et al. Lamin A and ZMPSTE24 (FACE-1) defects cause nuclear disorganization and identify restrictive dermopathy as a lethal neonatal laminopathy. *Hum Mol Genet* 2004;**13**:2493–503.
41. Brown CA, Lanning RW, McKinney KQ, Salvino AR, Cherniske E, Crowe CA, et al. Novel and recurrent mutations in lamin A/C in patients with Emery-Dreifuss muscular dystrophy. *Am J Med Genet* 2001;**102**:359–67.
42. Margalit A, Brachner A, Gotzmann J, Foisner R, Gruenbaum Y. Barrier-to-autointegration factor—a BAFfling little protein. *Trends Cell Biol* 2007;**17**:202–8.
43. Berk JM, Tifft KE, Wilson KL. The nuclear envelope LEM-domain protein emerin. *Nucleus* 2013;**4**:298–314.
44. Dubinska-Magiera M, Zaremba-Czogalla M, Rzepecki R. Muscle development, regeneration and laminopathies: how lamins or lamina-associated proteins can contribute to muscle development, regeneration and disease. *Cell Mol Life Sci* 2013;**70**:2713–41.
45. Butin-Israeli V, Adam SA, Goldman AE, Goldman RD. Nuclear lamin functions and disease. *Trends Genet* 2012;**28**:464–71.
46. Haraguchi T, Koujin T, Osakada H, Kojidani T, Mori C, Masuda H, et al. Nuclear localization of barrier-to-autointegration factor is correlated with progression of S phase in human cells. *J Cell Sci* 2007;**120**:1967–77.
47. Gomez-Cavazos JS, Hetzer MW. Outfits for different occasions: tissue-specific roles of Nuclear Envelope proteins. *Curr Opin Cell Biol* 2012;**24**:775–83.
48. Politano L, Carboni N, Madej-Pilarczyk A, Marchel M, Nigro G, Fidziaoska A, et al. Advances in basic and clinical research in laminopathies. *Acta Myol* 2013;**32**:18–22.
49. Lattanzi G, Marmiroli S, Facchini A, Maraldi NM. Nuclear damages and oxidative stress: new perspectives for laminopathies. *Eur J Histochem* 2012;**56**:e45.
50. Merlini L. Emery-Dreifuss muscular dystrophy. In: Katirji B, Kaminski HJ, Preston DC, Ruff RL, Shapiro BE, editors. *Neuromuscular Disorders in Clinical Practice*. Boston: Butterworth-Heinemann; 2002, pp. 1106–13.
51. Bonne G, Mercuri E, Muchir A, Urtizberea A, Becane HM, Recan D, et al. Clinical and molecular genetic spectrum of autosomal dominant Emery-Dreifuss muscular dystrophy due to mutations of the lamin A/C gene. *Ann Neurol* 2000;**48**:170–80.
52. Goldblatt J, Schram LJ, Wallis G, Oswald A, Beighton P. Emery-Dreifuss syndrome and X-linked muscular dystrophy with contractures: evidence for homogeneity. *Clin Genet* 1989;**35**:1–4.
53. Voit T, Krogmann O, Lenard HG, Neuen-Jacob E, Wechsler W, Goebel HH, et al. Emery-Dreifuss muscular dystrophy: disease spectrum and differential diagnosis. *Neuropediatrics* 1988;**19**:62–71.
54. Bialer MG, McDaniel NL, Kelly TE. Progression of cardiac disease in Emery-Dreifuss muscular dystrophy. *Clin Cardiol* 1991;**14**:411–16.
55. Buckley AE, Dean J, Mahy IR. Cardiac involvement in Emery Dreifuss muscular dystrophy: a case series. *Heart* 1999;**82**:105–8.
56. Dickey RP, Ziter FA, Smith RA. Emery-Dreifuss muscular dystrophy. *J Pediatr* 1984;**104**:555–9.
57. Carboni N, Mura M, Mercuri E, Marrosu G, Manzi RC, Cocco E, et al. Cardiac and muscle imaging findings in a family with X-linked Emery-Dreifuss muscular dystrophy. *Neuromuscul Disord* 2012;**22**:152–8.
58. Taylor MR, Fain PR, Sinagra G, Robinson ML, Robertson AD, Carniel E, et al. Natural history of dilated cardiomyopathy due to lamin A/C gene mutations. *J Am Coll Cardiol* 2003;**41**:771–80.
59. Becane HM, Bonne G, Varnous S, Muchir A, Ortega V, Hammouda EH, et al. High incidence of sudden death with conduction system and myocardial disease due to lamins A and C gene mutation. *Pacing Clin Electrophysiol* 2000;**23**:1661–6.
60. Cattin ME, Muchir A, Bonne G. 'State-of-the-heart' of cardiac laminopathies. *Curr Opin Cardiol* 2013;**28**:297–304.
61. Waters DD, Nutter DO, Hopkins LC, Dorney ER. Cardiac features of an unusual X-linked humeroperoneal neuromuscular disease. *N Engl J Med* 1975;**293**:1017–22.
62. Merlini L, Granata C, Ballestrazzi A, Marini ML. Rigid spine syndrome and rigid spine sign in myopathies. *J Child Neurol* 1989;**4**:274–82.
63. Bertini E, Marini R, Sabetta G, Palmieri GP, Spagnoli LG, Vaccario ML, et al. The spectrum of the so-called rigid spine syndrome: nosological considerations and report of three female cases. *J Neurol* 1986;**233**:248–53.

64. Goebel HH, Lenard HG, Gorke W, Kunze K. Fibre type disproportion in the rigid spine syndrome. *Neuropadiatrie* 1977;**8**:467–77.
65. Kubo S, Tsukahara T, Takemitsu M, Yoon KB, Utsumi H, Nonaka I, et al. Presence of emerinopathy in cases of rigid spine syndrome. *Neuromuscul Disord* 1998;**8**:502–7.
66. Koch AJ, Holaska JM. Emerin in Health and Disease. *Semin Cell Dev Biol* 2014;**29**:95–106.
67. Tsuchiya Y, Arahata K. Emery-Dreifuss syndrome. *Curr Opin Neurol* 1997;**10**:421–5.
68. Hoeltzenbein M, Karow T, Zeller JA, Warzok R, Wulff K, Zschiesche M, et al. Severe clinical expression in X-linked Emery-Dreifuss muscular dystrophy. *Neuromuscul Disord* 1999;**9**:166–70.
69. Granger B, Gueneau L, Drouin-Garraud V, Pedergnana V, Gagnon F, Ben Yaou R, et al. Modifier locus of the skeletal muscle involvement in Emery-Dreifuss muscular dystrophy. *Hum Genet* 2011;**129**:149–59.
70. Bonne G, Capeau J, De Visser M, Duboc D, Merlini L, Morris GE, et al. 82nd ENMC international workshop, 5th international Emery-Dreifuss muscular dystrophy (EDMD) workshop, 1st Workshop of the MYO-CLUSTER project EUROMEN (European muscle envelope nucleopathies), 15–16 September 2000, Naarden, The Netherlands. *Neuromuscul Disord* 2002;**12**:187–94.
71. Vytopil M, Benedetti S, Ricci E, Galluzzi G, Dello Russo A, Merlini L, et al. Mutation analysis of the lamin A/C gene (LMNA) among patients with different cardiomuscular phenotypes. *J Med Genet* 2003;**40**:e132.
72. Merlini L, Granata C, Dominici P, Bonfiglioli S. Emery-Dreifuss muscular dystrophy: report of five cases in a family and review of the literature. *Muscle Nerve* 1986;**9**:481–5.
73. Astejada MN, Goto K, Nagano A, Ura S, Noguchi S, Nonaka I, et al. Emerinopathy and laminopathy clinical, pathological and molecular features of muscular dystrophy with nuclear envelopathy in Japan. *Acta Myol* 2007;**26**:159–64.
74. Muntoni F, Bonne G, Goldfarb LG, Mercuri E, Piercy RJ, Burke M, et al. Disease severity in dominant Emery Dreifuss is increased by mutations in both emerin and desmin proteins. *Brain* 2006;**129**:1260–8.
75. Bialer MG, Bruns DE, Kelly TE. Muscle enzymes and isoenzymes in Emery-Dreifuss muscular dystrophy. *Clin Chem* 1990;**36**:427–30.
76. Hopkins LC, Jackson JA, Elsas LJ. Emery-Dreifuss humeroperoneal muscular dystrophy: an X-linked myopathy with unusual contractures and bradycardia. *Ann Neurol* 1981;**10**:230–7.
77. Rowinska-Marcinska K, Szmidt-Salkowska E, Fidzianska A, Zalewska E, Dorobek M, Karwanska A, et al. Atypical motor unit potentials in Emery-Dreifuss muscular dystrophy (EDMD). *Clin Neurophysiol* 2005;**116**:2520–7.
78. Darras BT, Jones HR. Diagnosis of pediatric neuromuscular disorders in the era of DNA analysis. *Pediatr Neurol* 2000;**23**:289–300.
79. Sewry CA, Brown SC, Mercuri E, Bonne G, Feng L, Camici G, et al. Skeletal muscle pathology in autosomal dominant Emery-Dreifuss muscular dystrophy with lamin A/C mutations. *Neuropathol Appl Neurobiol* 2001;**27**:281–90.
80. Serratrice G, Guastalla B, Pellissier JF. Emery-Dreifuss disease. A discussion of 2 new cases. *Rev Med Interne* 1988;**9**:365–8.
81. Park YE, Hayashi YK, Goto K, Komaki H, Hayashi Y, Inuzuka T, et al. Nuclear changes in skeletal muscle extend to satellite cells in autosomal dominant Emery-Dreifuss muscular dystrophy/limb-girdle muscular dystrophy 1B. *Neuromuscul Disord* 2009;**19**:29–36.
82. Fidzianska A, Hausmanowa-Petrusewicz I. Architectural abnormalities in muscle nuclei. Ultrastructural differences between X-linked and autosomal dominant forms of EDMD. *J Neurol Sci* 2003;**210**:47–51.
83. Sabatelli P, Lattanzi G, Ognibene A, Columbaro M, Capanni C, Merlini L, et al. Nuclear alterations in autosomal-dominant Emery-Dreifuss muscular dystrophy. *Muscle Nerve* 2001;**24**:826–9.
84. Gnocchi VF, Ellis JA, Zammit PS. Does satellite cell dysfunction contribute to disease progression in Emery-Dreifuss muscular dystrophy? *Biochem Soc Trans* 2008;**36**:1344–9.
85. Komaki H, Hayashi YK, Tsuburaya R, Sugie K, Kato M, Nagai T, et al. Inflammatory changes in infantile-onset LMNA-associated myopathy. *Neuromuscul Disord* 2011;**21**:563–8.
86. Fidzianska A, Rowinska-Marcinska K, Hausmanowa-Petrusewicz I. Coexistence of X-linked recessive Emery-Dreifuss muscular dystrophy with inclusion body myositis-like morphology. *Acta Neuropathol* 2004;**107**:197–203.
87. Nagano A, Koga R, Ogawa M, Kurano Y, Kawada J, Okada R, et al. Emerin deficiency at the nuclear membrane in patients with Emery-Dreifuss muscular dystrophy. *Nat Genet* 1996;**12**:254–9.
88. Manilal S, Nguyen TM, Sewry CA, Morris GE. The Emery-Dreifuss muscular dystrophy protein, emerin, is a nuclear membrane protein. *Hum Mol Genet* 1996;**5**:801–8.
89. Sabatelli P, Squarzoni S, Petrini S, Capanni C, Ognibene A, Cartegni L, et al. Oral exfoliative cytology for the non-invasive diagnosis in X-linked Emery-Dreifuss muscular dystrophy patients and carriers. *Neuromuscul Disord* 1998;**8**:67–71.
90. Manilal S, Sewry CA, Man N, Muntoni F, Morris GE. Diagnosis of X-linked Emery-Dreifuss muscular dystrophy by protein analysis of leucocytes and skin with monoclonal antibodies. *Neuromuscul Disord* 1997;**7**:63–6.
91. Mora M, Cartegni L, Di Blasi C, Barresi R, Bione S, Raffaele di Barletta M, et al. X-linked Emery-Dreifuss muscular dystrophy can be diagnosed from skin biopsy or blood sample. *Ann Neurol* 1997;**42**:249–53.
92. Mercuri E, Counsell S, Allsop J, Jungbluth H, Kinali M, Bonne G, et al. Selective muscle involvement on magnetic resonance imaging in autosomal dominant Emery-Dreifuss muscular dystrophy. *Neuropediatrics* 2002;**33**:10–14.
93. Merlini L. Selectivity of muscle sparing in Emery-Dreifuss muscular dystrophy. *Neuromuscul Disord* 2009;**19**:500–1.
94. Carboni N, Mura M, Marrosu G, Cocco E, Marini S, Solla E, et al. Muscle imaging analogies in a cohort of patients with different clinical phenotypes caused by LMNA gene mutations. *Muscle Nerve* 2010;**41**:458–63.
95. Deconinck N, Dion E, Ben Yaou R, Ferreiro A, Eymard B, Brinas L, et al. Differentiating Emery-Dreifuss muscular dystrophy and collagen VI-related myopathies using a specific CT scanner pattern. *Neuromuscul Disord* 2010;**20**:517–23.
96. Mercuri E, Clements E, Offiah A, Pichiecchio A, Vasco G, Bianco F, et al. Muscle magnetic resonance imaging involvement in muscular dystrophies with rigidity of the spine. *Ann Neurol* 2010;**67**:201–8.
97. Jimenez-Escrig A, Gobernado I, Garcia-Villanueva M, Sanchez-Herranz A. Autosomal recessive Emery-Dreifuss muscular dystrophy caused by a novel mutation (R225Q) in the lamin A/C gene identified by exome sequencing. *Muscle Nerve* 2012;**45**:605–10.

98. Fishbein MC, Siegel RJ, Thompson CE, Hopkins LC. Sudden death of a carrier of X-linked Emery-Dreifuss muscular dystrophy. *Ann Intern Med* 1993;**119**:900–5.
99. Emery AE. Emery-Dreifuss muscular dystrophy—a 40 year retrospective. *Neuromuscul Disord* 2000;**10**:228–32.
100. Golzio PG, Chiribiri A, Gaita F. 'Unexpected' sudden death avoided by implantable cardioverter defibrillator in Emery Dreifuss patient. *Europace* 2007;**9**:1158–60.
101. Dell'Amore A, Botta L, Martin Suarez S, Lo Forte A, Mikus E, Camurri N, et al. Heart transplantation in patients with Emery-Dreifuss muscular dystrophy: case reports. *Transplant Proc* 2007;**39**:3538–40.
102. Shapiro F, Specht L. Orthopedic deformities in Emery-Dreifuss muscular dystrophy. *J Pediatr Orthop* 1991;**11**:336–40.
103. Bione S, Small K, Aksmanovic VM, D'Urso M, Ciccodicola A, Merlini L, et al. Identification of new mutations in the Emery-Dreifuss muscular dystrophy gene and evidence for genetic heterogeneity of the disease. *Hum Mol Genet* 1995;**4**:1859–63.
104. Worman HJ, Fong LG, Muchir A, Young SG. Laminopathies and the long strange trip from basic cell biology to therapy. *J Clin Invest* 2009;**119**:1825–36.
105. Muchir A, Kim YJ, Reilly SA, Wu W, Choi JC, Worman HJ. Inhibition of extracellular signal-regulated kinase 1/2 signaling has beneficial effects on skeletal muscle in a mouse model of Emery-Dreifuss muscular dystrophy caused by lamin A/C gene mutation. *Skelet Muscle* 2013;**3**:17.

Chapter 36

Myopathies of Systemic Disease

Thornton B.A. Mason and Darryl C. De Vivo

INTRODUCTION

When evaluating a child with myopathy, the clinician must not only determine the pattern and extent of muscle weakness but also the cause. In addition to the many primary disorders of muscle, there are many systemic diseases that produce myopathies. These generalized disorders include endocrine abnormalities, renal disorders, infections, medication-related effects, connective tissue diseases, and eosinophilic syndromes. In this chapter, we detail the disorders that may have a significant impact on skeletal muscle function in childhood.

HYPERTHYROIDISM

Thyrotoxic myopathy symptoms are primarily those associated with a chronic proximal myopathy, and patients may complain especially of mild, symmetrical hip flexion weakness. Bulbar or respiratory involvement is rare. Pectoral muscles are more severely affected than the pelvic girdle, and muscle wasting may be noted on examination. Occasionally, distal muscles may also be affected. Despite muscle weakness and wasting in thyrotoxicosis, serum creatine kinase (CK) concentrations are normal or decreased, probably because of increased clearance. Serum myoglobin concentrations are also normal. Findings on muscle biopsy are nonspecific, both by routine histopathology and ultrastructural evaluation. Histochemical staining may demonstrate atrophy of both type I and type II fibers; in some cases, more prominent atrophy of type II fibers is found. Additionally, some studies describe scattered fiber necrosis and inflammatory infiltrates. Patients with thyrotoxic myopathy may respond satisfactorily to antithyroid medications alone and not require additional therapy, such as steroids.[1,2]

Thyrotoxic periodic paralysis (TPP) is a common cause of sudden-onset muscle weakness in young Asian and Latin American adults that rarely occurs in Caucasians. In North America, TPP occurs in approximately 0.1% to 0.2% of the hyperthyroid population, but is roughly 10-fold more common in Asian populations. Males are more commonly affected than females. Almost all cases occur sporadically, without a history of familial periodic paralysis.[3] Recently, several mutations in *KCNJ18*, the gene coding for Kir2.6 (an inwardly rectifying potassium channel), have been shown to be associated with susceptibility to TPP[4] in individuals from the USA, Brazil, and France. Because these mutations were not common in Hong Kong Chinese populations, further studies in Chinese populations demonstrated a new TTP susceptibility locus at 17q24.3,[5] near *KCNJ2* (also known as *KIR2.1*). Importantly, Kir2.6 forms functional heterotetramers with Kir2.1.[6]

TPP begins typically in the early morning, often following a day of strenuous exercise; the limbs are primarily involved, and speech is usually spared. TPP clinically resembles idiopathic familial periodic paralysis, except for a greater tendency toward dysrhythmias, cardiac arrest, and respiratory failure. Management includes admission for acute paralysis, cardiac monitoring, correction of hypokalemia, and initiation of antithyroid medication. Episodes of weakness always cease when a euthyroid state is achieved. Short-term use of β-blockers may be indicated. Strenuous exercise and high-carbohydrate diets should be avoided.[3]

A series of four patients with hyperthyroidism due to Graves' disease described abnormal increases in serum CK concentration that occurred during antithyroid medication treatment. Three of the four patients developed myalgia and muscle cramps. While the cause is not established, a rapid reduction of thyroid hormones in skeletal muscle may produce the muscle symptoms and increased CK level. If myalgias, muscle cramps, or increased CK concentration occurs a few months after the initiation of medical treatment of Graves' disease, levothyroxine might be added to the regimen rather than discontinuing the antithyroid agents.[7]

B.T. Darras, H. Royden Jones, Jr., M.M. Ryan & D.C. De Vivo (Eds): Neuromuscular Disorders of Infancy, Childhood and Adolescence, Second edition.
DOI: http://dx.doi.org/10.1016/B978-0-12-417044-5.00036-6

HYPOTHYROIDISM

Myopathy is a significant complication of hypothyroidism, with an estimated incidence of 30% to 80% among patients with myxedema. In general, the severity of the myopathy parallels the duration and degree of hypothyroidism.[8] Findings in hypothyroid myopathy include muscle pain and cramps, proximal weakness, and slow myotatic reflexes.[9] Serum CK concentration is characteristically increased, and in some cases may be an order of magnitude greater than control levels (i.e. comparable to levels usually associated with muscular dystrophy). With medical treatment for hypothyroidism, CK levels always return to normal. Electromyography (EMG) studies commonly demonstrate myopathic patterns. Magnetic resonance spectroscopy of hypothyroid muscle at rest shows low intracellular pH and delayed glycogen breakdown during exercise. Muscle biopsy specimens from patients with hypothyroid myopathy reveal changes including type I fiber predominance, selective atrophy of type II fibers, and increased numbers of internal nuclei. A fraction of type I fibers may contain core-like structures with reduced nicotinamide adenine dinucleotide reductase activity.[10,11] Ultrastructural changes include myofilament loss, accumulations of glycogen and lipid, increased vacuoles, abnormal mitochondria, Z-line streaming, dilated sarcoplasmic reticulum, and increased lipofuscin pigment and lysosomes.[8,12,13]

Hypothyroidism affects muscle by impacting glycogenolysis (with reduced acid maltase activity) and mitochondrial oxidation.[10] In patients with subclinical hypothyroidism (elevated circulating thyroid-stimulating hormone levels in the face of normal free thyroid hormone levels), incremental, submaximal exercise results in venous lactate values that are significantly higher than in controls; blood pyruvate concentrations are not elevated. These results suggest that there is *in vivo* functional impairment of mitochondrial function in patients with subclinical hypothyroidism.[12] Rarely, hypothyroid myopathy may be associated with exercise-induced myalgias and recurrent rhabdomyolysis (the latter being supported by increased serum CK level and muscle fiber necrosis on biopsy). Exercise-induced myalgia may be an early symptom of hypothyroid myopathy.[8]

Moderate to severe hypothyroidism during infancy can produce myopathy in association with multiple other features, resulting in the Kocher-Debré-Sémélaigne syndrome. Examination findings may include coarse facial features; protruding tongue; hoarse cry; persistence of an open fontanelle; abdominal distention; umbilical hernia; and cool, mottled, or dry skin (see Case Example 36.1). Marked hypertrophy of calf, thigh, deltoid, and forearm muscles gives a "herculean" appearance; involved muscles have a firm consistency and are prominently weak. Myotatic reflexes may display delayed relaxation.[14,15] Immediate treatment of hypothyroidism in infancy is critical to limit cognitive and developmental deficits.

The term Hoffman's syndrome refers to the rare association of hypothyroidism with increased muscle volume/hypertrophy, weakness, and slowness of movement in adolescents and adults.[1] Muscle atrophy is much more common in older patients with hypothyroidism than with hypertrophy. While hypertrophy in Hoffman's syndrome resembles the pseudohypertrophy of muscular dystrophy, it occurs in the setting of normal serum aspartate aminotransferase levels and responds briskly to thyroid therapy.[16]

PARATHYROID DISEASE

Hypoparathyroidism can be caused by altered parathyroid gland function, surgical resection of parathyroid glands, infiltrative diseases, and irradiation. The clinical features of hypoparathyroidism are linked to hypocalcemia. At the neuromuscular junction, hypocalcemia is associated with hyperexcitability and may result in laryngospasm, tetany, muscle cramps, and carpopedal spasm. Myopathy with increased serum CK has been reported in patients with hypoparathyroidism, but is not common; there appears to be an inverse relationship between serum calcium and serum CK. The etiology of CK elevation in hypoparathyroidism with myopathy is unclear, but could be related to hypocalcemia-induced changes in the sarcolemma, or less likely tetani.[17] Muscle biopsy findings in such cases have been unremarkable or nonspecific; type II fiber atrophy, focal myofibrillar degeneration, and perinuclear accumulation of mitochondria have been reported.[18,19] Some patients with myopathy have been reported to have a concurrent rash (face, legs).[17,18]

The pathogenesis of primary hyperparathyroid myopathies is thought to be a combination of altered energy metabolism/utilization and parathyroid-induced muscle breakdown. Phosphorus and calcium levels do not correlate with the severity of neuromuscular symptoms in hyperparathyroidism, CK levels are only slightly elevated or normal, and muscle biopsies typically show only nonspecific findings.[20] Ischemic calcific myopathy has been reported as a complication of azotemic hyperparathyroidism, but is rare.[21] If surgical management is not pursued, cinacalcet therapy can normalize serum calcium levels and promote resolution of muscle weakness over weeks to months.[20]

CUSHING'S SYNDROME

The diagnosis of Cushing's myopathy should be considered in patients with proximal muscle weakness, myopathic EMG, and normal serum CK and aspartate aminotransferase levels.[22] EMG findings in myopathy of

CASE EXAMPLE 36.1

S.R. is a 2.5-year-old girl who was brought to the clinic with a history of motor and language delay. She was born at 42 weeks' gestation with Apgar scores of 9 and 9. Her mother reported the child had mild feeding difficulties as a newborn, and the baby had an especially hoarse cry. More recently, the girl has appeared apathetic, clumsy, and slow. On examination, the patient had coarse facial features; a large, protruding tongue; and prominent abdominal distention (with an umbilical hernia; Figure 36.1A and B). Her anterior fontanelle was open and readily palpable. Her skin was very dry and cool (Figure 36.1C). Multiple muscles appeared hypertrophied and were firm on palpation; these included the deltoid, thigh, and calf muscles. She had difficulty rising from a seated position, and had an unsteady, waddling gait. Myotatic reflexes were abnormal with delayed relaxation. Thyroid function studies revealed that total T_4, T_3, and free thyroxine were all depressed, while thyroid-stimulating hormone level was markedly elevated. These laboratory findings confirmed the suspicion of primary hypothyroidism. Her muscle weakness improved remarkably with thyroxine replacement, and her pseudoathletic appearance gradually softened (Figure 36.1D and E). She was enrolled in an early intervention program.

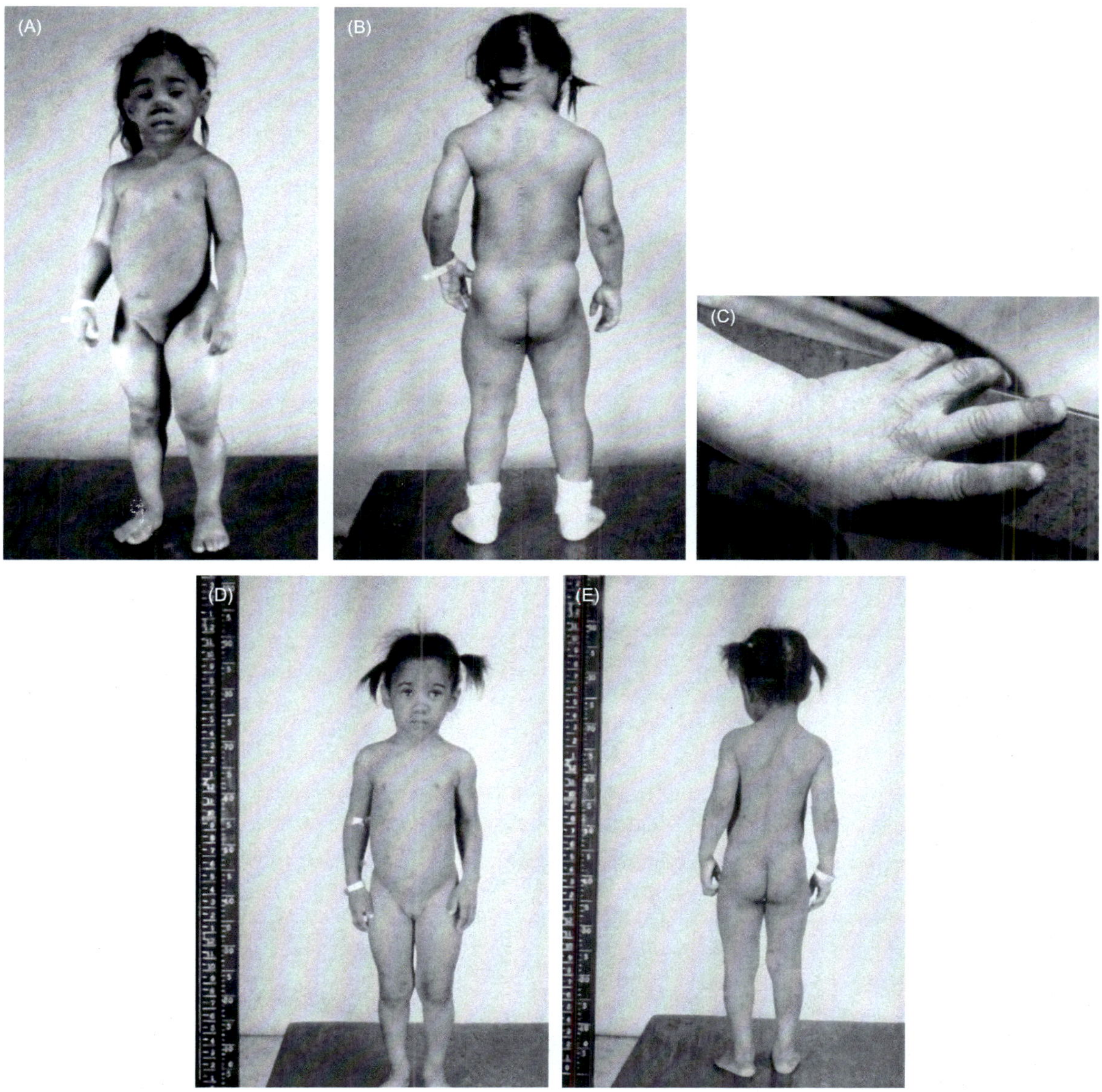

FIGURE 36.1 Hypothyroid myopathy (Case Example 36.1). (**A** and **B**) Note the coarse facial features, diminished facial expression, and protruding abdomen and buttocks. (**C**) The skin is xerotic and relatively pale. The fine wrinkling imparts a parchment-like quality to the skin. (**D** and **E**) Following treatment, there is increased alertness and facial animation. The abdomen, buttocks, and limb musculature appear normal, and the skin is less pale and sallow.

Cushing's syndrome share features seen in inflammatory myopathies and include low-amplitude, short-duration motor unit potentials that are recruited in increased numbers with minimal effort. Some patients may also have abnormal insertional activity, with brief runs of positive waves or fibrillation potentials.[23] Atrophy of type II muscle fibers on biopsy analysis may offer a clue to the diagnosis, although such a finding is nonspecific. Type II fiber atrophy is also found in exogenous steroid myopathy, as well as the endocrine myopathies of thyrotoxicosis, myxedema, and osteomalacia. Nonspecific ultrastructural findings include excessive glycogen accumulation, myofilament loss, large mitochondria, and excess lipofuscin.[24]

Once hypercortisolism is suspected, the diagnosis is further supported by increased morning serum cortisol levels, increased 24-hour urinary free cortisol levels, and the failure of suppression of endogenous cortisol with dexamethasone. Several months after definitive treatment (e.g. excision of an adrenal adenoma), muscle strength increases, blood pressure normalizes, and excess weight decreases.[22]

UREMIA

Muscle weakness in uremia is prominent in the lower extremities, especially the proximal muscles. Biopsy studies show moderate muscle fiber atrophy, especially type II fibers. Ultrastructurally, severe yet nonspecific degenerative changes are found. Multiple factors have been implicated in the myopathy of chronic renal failure: (1) abnormalities in vitamin D metabolism; (2) various uremic toxins; (3) excess parathyroid hormone; (4) malnutrition and reduced energy uptake; (5) carnitine depletion and impaired energy uptake; (6) abnormalities in cellular energetics; (7) impaired protein synthesis and amino acid metabolism; (8) hypophosphatemia and diminished intracellular phosphorus stores; and (9) impaired biochemical integrity of the sarcolemma.[25]

OSTEOMALACIA

Osteomalacia is known to be associated with symmetrical weakness and wasting of the proximal limb muscles, especially the legs. Myotatic reflexes and sensation are intact. The serum alkaline phosphatase level is always elevated, but CK level is often normal; muscle biopsy may prove unremarkable. With appropriate treatment, such as the administration of vitamin D, the weakness is reversible. Because neuromuscular abnormalities may precede osteoarticular findings in some infants with vitamin D-deficiency rickets, osteomalacia should be considered in young children with muscle weakness.[26]

HYPOKALEMIA

Paralytic myopathy is a recognized feature of primary aldosteronism, with a high incidence reported among Asian populations. Patients with myopathy have low serum potassium levels, and the greater the degree of hypokalemia, the more severe the muscle weakness. Limb muscles are most commonly involved, but facial and neck muscles may also be affected. An associated feature may be hypertension, related to the marked sodium retention associated with the disorder. Mechanistically, hypokalemia results in transient membrane depolarization and consequently renders muscle fibers electrically unexcitable. In addition, hypokalemia can cause a cellular energy crisis because potassium is required to maintain intracellular muscle energy stores through phosphorylation of creatine and adenosine diphosphate.

The degree of hypokalemia found in patients with primary aldosteronism is in part related to sodium intake. The restriction of sodium leads to potassium retention and improves hypokalemic manifestations, while sodium excess promotes renal potassium wastage and can produce profound hypokalemia. In Taiwan, average daily salt intake is greater than 14 grams. One may therefore propose that the higher incidences of thyrotoxic hypokalemic periodic paralysis and aldosterone hypokalemic paralysis reported among the Asian population are dietary rather than ethnic or genetically induced.[27]

Although hypokalemic periodic paralysis and hypokalemic myopathy may both present with paralysis, they are distinct diseases. Hypokalemic periodic paralysis is caused by mutations in the dihydropyridine receptor.[28] In hypokalemic periodic paralysis, skeletal muscle is overloaded with glycogen and rhabdomyolysis is absent. On the other hand, in hypokalemic myopathy, muscle glycogen is initially absent, necrosis is prominent, muscle-derived enzyme levels are elevated, and recovery requires a prolonged course of potassium supplementation. Hypokalemic myopathy may occur as a result of injudicious use of diuretics, laxatives, treatment with mineralocorticoids, amphotericin B, and lithium, and also in compulsive licorice eaters. Laxative and diuretic abuse can be prominent in patients with anorexia nervosa.[29,30] The periodic paralyses are described in Chapter 38.

Licorice

Licorice contains glycyrrhizin as a major ingredient. It has been used as a Chinese medication and as a sweetener in other medications, candies, foods, and drinks in the United States and Europe. Hypokalemia, metabolic alkalosis, and hypertension due to ingestion of licorice in large amounts has been termed licorice-induced pseudoaldosteronism; low levels of plasma renin activity and aldosterone, altered renal function, cardiac rhythm disturbances, and myopathy accompany these findings.[31,32] In the 1950s, licorice was included as a sweetener in preparations of p-aminosalicylic acid, which was used in the treatment of pulmonary tuberculosis in children. Consequently, children and young adults suffered from glycyrrhizin-induced hypokalemic myopathy (GIHM). Factors that aggravate the risk of developing GIMH are those that may worsen hypokalemia: diuretics, diarrhea, and alcoholism. Potassium levels are likely to be less than 2.0 mEq/L, and CK level is elevated often over 10-fold; blood aldosterone level is normal or decreased. Muscle biopsy findings in GIMH include scattered necrotic fibers, patterns of phagocytosis and regeneration, and cellular infiltration around necrotic fibers and vessels without angiitis.[29]

Intoxication with glycyrrhizin may give rise to rhabdomyolysis. This generally begins as a rapidly progressive muscle deficit, usually with proximal weakness of the limbs, myalgias, and cramps. More severe cases may result in acute renal failure. Rhabdomyolysis associated with glycyrrhizin ingestion correlates with potassium levels around 1.5 mEq/L.[32]

The active form of glycyrrhizin, glycyrrhetinic acid, has a mineralocorticoid-like effect and resembles desoxycorticosterone structurally. Glycyrrhetinic acid either exerts a mineralocorticoid effect directly on the kidney or increases the effect of aldosterone indirectly. Consequently, K^+ excretion is elevated following enhanced sodium resorption in the distal tubules, leading to hypokalemia. Patients are generally completely cured with careful K^+ repletion; cardiac monitoring is indicated because of risks for cardiac arrhythmias associated with severe hypokalemia.[29] Hypokalemia-related findings on electrocardiography include prominent U waves and prolonged QT intervals. Chronic hypokalemia from licorice ingestion can cause renal tubule damage, resulting in hypokalemia that persists following courses of oral/intravenous repletion.[30]

Renal Tubular Acidosis

Renal tubular acidosis can produce hypokalemia, which in turn may lead to recurrent hypokalemic paralysis. The presence of failure to thrive and episodic hypokalemia since early childhood suggest distal renal tubular acidosis. To help establish the diagnosis, urinary pH should be measured in the early morning and following oral ammonium chloride administration. Inappropriately alkaline urine (pH $>$ 6) in a patient with systemic acidosis is consistent with a diagnosis of renal tubular acidosis.[33]

Reports include a child with repeated episodes of hypokalemic muscle weakness and an associated incomplete proximal renal tubular acidosis; this scenario was determined to be the initial presentation of Wilson's disease. Management included potassium supplementation, D-penicillamine chelation, and pyridoxine. Because both renal and neurologic aspects of Wilson disease can be effectively treated, it is vitally important that the condition is recognized. Wilson disease should be suspected in patients with recurrent hypokalemic paralysis associated with renal abnormalities of unknown origin.[34]

Gitelman syndrome is a disorder characterized by renal hypocalcemia and hypomagnesemia in addition to features found in Bartter syndrome. Bartter syndrome includes metabolic alkalosis, hypokalemia, hyperreninemia, hyperaldosteronism, and hyperplasia of the juxtaglomerular apparatus. These two syndromes are distinguishable on the basis of urinary calcium levels: hypocalciuria in Gitelman syndrome and normocalciuria or hypercalciuria in Bartter syndrome.[35] As compared with Bartter syndrome, Gitelman syndrome generally has milder clinical symptoms, and is often diagnosed in late childhood or adulthood. Patients with Gitelman syndrome are often asymptomatic, and growth failure is rare. Studies support a defect in Gitelman syndrome that affects the thiazide-sensitive Na-Cl transporter in the distal tubule. Hypokalemic rhabdomyolysis has been reported in Bartter and Gitelman syndromes.[36–38] A 12-year-old boy with Gitelman syndrome has been reported with a myopathy; findings included weakness in the legs greater than the arms, serum potassium 1.6 mmol/L, CK 9700 U/L, and low-amplitude polyphasic motor unit potentials. Muscle biopsy revealed focal myocyte necrosis and central vacuolar degeneration of myocytes. Treatment options for Gitelman syndrome myopathy include magnesium supplementation alone, as well as indomethacin, potassium-sparing diuretics, and sodium/potassium supplements.[39]

Hypokalemic myopathy in childhood and young adulthood may also be caused by an autosomal recessive 17α-hydroxylase/17,20-lyase deficiency. With these enzymes absent, 17α-hydroxyprednisolone and 17α-hydroxyprogesterone, as well as C_{12} steroid precursors of androgens and estrogens, fail to form. Because of an impaired production of cortisol, adrenocorticotropic hormone levels are increased; in turn, mineralocorticoids are overproduced. Patients then go on to develop hypertension, hypokalemia, and suppression of the renin-angiotensin system. Moreover, depletion of the sex hormones causes sexual infantilism and primary amenorrhea in females and failure of virilization in males.[40]

HYPERNATREMIA

Marked hypernatremia may be seen in the setting of hypothalamic lesions, which produce a defective thirst drive and impaired antidiuretic hormone levels. In addition to altered states of consciousness, such as lethargy and delirium, hypernatremia may also be associated with tenderness and weakness of proximal muscles. EMG results are consistent with a myopathic pattern. After correction of serum sodium, muscle tenderness and weakness disappear, CK values are restored to a normal range, and EMG findings become unremarkable. The mechanism of muscle weakness in hypernatremia has not been established, but may be caused by a depletion of intracellular energy stores (e.g. reduced adenosine triphosphate because of overactive Na-K pump consumption).[41]

PRIMARY HYPERPARATHYROIDISM

This disorder may present as a clinical syndrome of easy fatigability and muscle weakness, especially involving the proximal leg muscles. Other common features include glucose intolerance and insulin resistance. Myotatic reflexes are preserved or hyperactive. Patients may also exhibit tremors of the tongue, minor sensory defects, and abnormalities of phonation, gait, station, and mentation. EMG findings are mixed, with some patients showing action potentials of abnormally short duration and low amplitude whereas others show high-amplitude, long-duration polyphasic potentials. Atrophy of both type I and type II fibers is seen on muscle biopsy. As a rule, serum CK and transaminase levels are normal. In addition to the deleterious effects of the associated hypercalcemia, parathyroid hormone itself in abnormally high concentrations may be toxic to muscle fibers.[42]

HUMAN IMMUNODEFICIENCY VIRUS

Human immunodeficiency virus (HIV) myopathy may be present early in HIV infection, but more commonly complicates full-blown acquired immunodeficiency syndrome. HIV myopathy occurs almost exclusively in adults. HIV myopathy begins with the subacute onset of proximal, symmetrical muscle weakness that may be associated with wasting of the legs more often than the arms. Myalgias can be seen, and serum CK level is increased as much as 10 to 15 times normal.[43] EMG findings include spontaneous activity with fibrillations and positive sharp waves, and brief, low-amplitude, polyphasic units. Nerve conduction studies are either normal or may suggest a mild, axonal sensory neuropathy. The muscle biopsy shows inflammatory cells, chiefly lymphocytes and macrophages, in a perivascular, endomysial, or perimysial distribution, surrounding healthy muscle fibers; degenerating and necrotic muscle fibers may also be present. There is an increase in connective tissue in more severe cases. Rarely, scattered cytoplasmic bodies and nemaline (rod) bodies can be seen. If some angulated fibers are also present, there may be a coexisting axonal neuropathy. Ultrastructurally, there is disorganization of the myofibrillar structure and macrophages, lymphocytes, and plasma cells.[44]

Other factors may contribute to muscle disease in HIV infection. Although rare, infection of skeletal muscle with opportunistic organisms can occur in the course of disseminated infection in older patients. Such organisms include *Cryptococcus neoformans*, *Toxoplasma gondii*, *Mycobacterium avium-intracellulare*, and *Mycobacterium tuberculosis*. Further myotoxic factors include vitamin deficiencies; induction of type II fiber atrophy due to disuse, prolonged immobilization, or muscle wasting related to bacterial toxins and systemic sepsis; and myotoxicity from the assortment of drugs used to treat underlying illnesses. Increased levels of cachexin and tumor necrosis factor are found in patients with acquired immunodeficiency syndrome and may contribute to muscle wasting syndrome by interfering with aspects of lipid metabolism. HIV wasting syndrome, then, in most cases is not a true myopathy, but probably is due to metabolic or nutritional factors, or both.[44,45]

Pyomyositis is the most common myopathy in children who are HIV positive. Findings suggestive of pyomyositis include low-grade fever (even without leukocytosis), localized pain and swelling in a large muscle group, and elevated creatine kinase level.[46] An enhancing lesion, often with a fluid density, may be demonstrated by ultrasound, magnetic resonance imaging, or computed tomography scan with contrast. *Staphylococcus aureus* is implicated, or more rarely gram-negative organisms; a single case of an 11-year-old boy with perinatally acquired HIV and pyomyositis due to *Streptococcus agalactiae* has been reported.[47] Risk factors for pyomyositis include exercise-induced trauma, underlying muscle abnormalities, or hematogenous spread of a bacterial infection; HIV-associated defects of neutrophil function may contribute to the disease.[44]

Zidovudine (AZT) is capable of inducing a myopathy with depletion of mitochondrial DNA, secondary to its mitochondrial toxicity. AZT inhibits γ-DNA polymerase, which is found solely in the mitochondrial matrix, and produces termination of DNA strands. AZT myopathy may occur in as many as 30% of adults treated with long-term (>1 year), high-dose (>1200 mg daily) AZT. The clinical features of AZT myopathy include proximal muscle weakness; myalgias, especially in the calves and thighs, that are often exacerbated with exercise; and elevated serum CK levels (that may also worsen with exercise). EMG studies demonstrate myopathic units like those observed in HIV myopathy. Because AZT myopathy is clinically similar to HIV myopathy, the only clinical method of differentiating between them is to stop AZT and monitor for clinical improvement. Such improvement includes decreased muscle pain, increased strength, and reduced serum CK levels. On biopsy, the most prominent characteristic of AZT myopathy is the presence of "ragged-red" fibers on trichrome-stained preparations, indicating abnormalities in muscle mitochondria. Electron microscopy studies have confirmed that abnormal mitochondria are present and show further that they are ubiquitous, even in areas appearing normal by light microscopy. Following the discontinuation of AZT, these muscle changes are partially or fully reversible.[44,48]

LYME DISEASE

Lyme disease, caused by *Borrelia burgdorferi*, is the most common vector-borne disease in the United States. The clinical features of Lyme disease in children are similar to those in adults; however, the long-term complications of the illness in the pediatric population appear to be more rare.[49] Lyme disease is commonly associated with muscle stiffness and nonspecific myalgia early in its course; less often, myositis is noted, which is characteristically localized and may occur concomitant with or up to a year after erythema migrans lesions. Myositis is demonstrated by EMG or biopsy typically near an involved joint or localized neuropathy. Although myopathic patterns by electromyography can be found, they are not mandatory for the diagnosis; neuropathic patterns are secondary to *Borrelia*-associated neuropathy. Histopathologic examination reveals features of interstitial myositis or focal nodular myositis, with a cellular infiltrate that is predominantly T cells (CD4+ > CD8+). B cells are relatively rarely noted.[50] Muscle fiber degeneration is minimal. Although not successfully cultured from muscle, *Borrelia* spirochetes are detected in biopsy specimens by silver or immunogold-silver staining. Positive PCR analysis of fluid from involved muscle tissue at biopsy can provide additional diagnostic support.[51] Nuclear medicine imaging (gallium 67 uptake) may be useful in documenting potential areas of localized myositis in Lyme disease. With appropriate antibiotic therapy, a good clinical response is expected.[52]

SEPSIS/INTENSIVE CARE UNIT MYOPATHIES

Thick Filament Myopathy

Thick filament myopathy is a specific syndrome that occurs in children or adults in the setting of status asthmaticus. Patients receive obligate endotracheal intubation and are placed on ventilatory support. Often for days or longer, patients are given high-dose corticosteroids to control the asthma and neuromuscular blocking agents to facilitate mechanical ventilation. On subsequent attempts to wean the ventilatory support, affected patients are found to have severe neuromuscular ventilatory insufficiency and decreased limb strength. Ophthalmoplegia may also be present, and CK levels are often significantly elevated.[53] Nerve conduction studies are normal, and EMG findings support a myopathy. On muscle biopsy, there is a loss of structure in the central region of muscle fibers. This loss of structure has been shown by electron microscopy to result from the destruction of thick myosin filaments.[54] Recovery usually occurs rapidly.[55]

Cachectic Myopathy

Cachectic myopathy, catabolic myopathy, and disuse atrophy are cited as complications of critical illness, such as sepsis or severe trauma.[56] Although these disorders cause muscle weakness and wasting, they are poorly defined clinically. EMG/nerve conduction studies and plasma CK values are normal. Muscle biopsies are fully normal or show nonspecific type II fiber atrophy.[55]

Acute Necrotizing Myopathy of Intensive Care

Acute necrotizing myopathy of intensive care may result from a wide variety of chemical exposures, infections, and other insults. As a result, the incidence of cases from this category of myopathy would be expected to be increased in critical care units with a high frequency of infection, trauma, and use of various medications. Cases described include severe weakness, highly elevated CK values, EMG studies consistent with severe weakness, and muscle biopsies showing widespread muscle fiber necrosis.[55,57]

Hopkins' Syndrome

Hopkins' syndrome is a poliomyelitis-type acute paralytic illness in children that occurs approximately 4 to 7 days after an acute asthma episode.[58] The syndrome does not appear to be related to the severity of the preceding asthma attack.[59] Clinical features include rapid-onset muscle weakness leading to flaccid limb paralysis, with absent myotatic reflexes in the involved regions. When patients suffer weakness in more than one extremity, the weakness is asymmetrical.[60] Sensation is preserved, and indeed, myalgia in the paralyzed limb has been frequently reported.[61] These clinical features, along with electrophysiologic investigations (absent motor action potentials, normal sensory nerve conduction studies, denervation changes on electromyography) support an anterior horn cell disease. Moreover, magnetic resonance imaging signal changes in the anterior horn of the ipsilateral cervical spinal cord have been reported in a boy with flaccid left arm paralysis.[62] There may be a mild cerebrospinal fluid pleocytosis (usually less than 50 cells/mL) and mildly elevated cerebrospinal fluid protein level; initially normal cerebrospinal fluid may later manifest these changes. Surprisingly, a causal (putative viral) organism has not been identified. Affected children have been previously immunized against polio, and no single asthma medication appears to be implicated. The prognosis is uniformly poor, inasmuch as affected patients later develop marked limb amyotrophy and remain selectively handicapped.[60] Given the rarity of the syndrome, no single treatment has emerged as clearly effective, but pulse therapy with methylprednisolone, plasmapheresis, and intravenous immunoglobulin therapy have all been tried.[59]

Neutral Lipid Storage Disease

Neutral lipid storage disease results from mutations in *CGI-58* (the gene coding for alpha/beta-hydrolase domain containing protein, ABHD5) or *PNPLA2* (encoding adipose triglyceride lipase, ATGL).[63] These disorders are marked by triglyceride-containing droplets in the cytoplasm of neutrophils (Jordans' bodies), as well as other tissues such as skin, muscle, and bone marrow. Chanarin-Dorfman syndrome is a condition characterized by lipid storage and ichthyosis, often with hepatomegaly, mild myopathy, and sometimes hearing loss, ophthalmologic symptoms, short stature, mild intellectual disability, microcephaly, and intestinal involvement. The manifestations of patients with *PNPLA2* mutations include myopathy (often with proximal weakness predominating in the upper and lower extremities); additional symptoms can include cardiomyopathy, diabetes, and hepatomegaly, while skin is unaffected. Rhabdomyolysis is rare, and disease progression is slow.[64,65] It appears that hyperCKemia is the early sign of neutral lipid storage disease in childhood and precedes clinical overt myopathy.[63]

MEDICATION-RELATED MYOPATHIES

Corticosteroids

Corticosteroid myopathy is likely the most common drug-induced myopathy. There are two distinct types of corticosteroid-induced myopathies: acute and chronic. Acute corticosteroid myopathy occurs less often than the chronic variety. The acute variety is usually encountered in the setting of high-dose systemic corticosteroid treatment (e.g. intravenous methylprednisolone or dexamethasone).[66] Approximately a week after high doses of corticosteroids have been started, patients often present with generalized proximal and distal weakness, including the respiratory musculature. There is support for a direct correlation between the occurrence of an acute corticosteroid myopathy and the type or total dose of corticosteroids administered. In addition, combining corticosteroids and an aminosteroid-based nondepolarizing neuromuscular blocking agent appears to increase the risk of developing an acute myopathy (see earlier discussion of thick filament myopathy).[67,68] Examples of these blockers include pancuronium, vecuronium, pipecuronium, and atracurium. In most cases of acute corticosteroid myopathy, serum CK levels are increased and may be accompanied by myoglobinuria. Muscle biopsies show atrophy of all fiber types as well as focal and diffuse necrosis. After corticosteroid therapy is discontinued, recovery is slow, extending over 6 months or beyond.

Chronic corticosteroid myopathy is the more frequently encountered type. Fluorinated corticosteroids (e.g. triamcinolone, betamethasone) appear to be more often implicated in myopathy than nonfluorinated corticosteroids (hydroxycortisone, prednisone). Patients develop progressive proximal weakness with inability to rise from low chairs, climb stairs, and brush the hair. Wasting of the thighs and proximal arm muscles may accompany the weakness. Face, hand, and sphincter muscles are relatively spared. Myalgia is mild if present at all. Serum CK levels in chronic corticosteroid myopathy are generally normal; if elevated, another type of myopathy should be considered. Urinary excretion of creatine and 3-methylhistidine is increased. EMG studies are often normal, but decreased compound muscle action potentials may be found. On muscle biopsy, there is selective atrophy of type IIb, fast-twitch glycolytic muscle fibers. The mechanism of corticosteroid myopathy may be related to the inhibition of mRNA synthesis for muscle proteins, as well as an increase in the rate of protein breakdown (myofibrillar proteins and muscle extracellular matrix proteins). Activation of two proteolytic systems (ubiquitin proteasome and lysosomal systems) is thought to be mediated through upregulation of several genes involved in atrophy: *FOXO*, *Atrogin-1*, and *MuRF-1*.[69] Chronic corticosteroid myopathy is usually reversible over months if the drug can be withdrawn or the dose decreased. In some patients, a confusing scenario can arise when the medication is tapered. Approximately a week after the dosage is lowered, the patient may complain of diffuse aching and stiffness in the muscle. These symptoms are due to corticosteroid withdrawal and will improve over the next several weeks, only to recur when the dose is again lowered. The risk and extent of corticosteroid myopathy may be lessened by a regular exercise program.[70] See also Case Example 36.2.

Amphotericin B

Amphotericin B use is associated with a myopathy, likely because of hypokalemia induced by the effect of medication on renal function. The incidence of hypokalemia in the setting of amphotericin B therapy is estimated at 75% to 90%; since the distal tubulopathy caused by amphotericin

CASE EXAMPLE 36.2

G.B. is a 12-year-old male who was evaluated in consultation, at the request of the pediatric pulmonary service, for severe muscle weakness and inability to walk. He had a history of severe asthma, with four previous hospital admissions. Three weeks previously, he was taken to the hospital's emergency department in respiratory distress, with marked intercostal retractions and prominent accessory respiratory muscle use. Despite multiple bronchodilator therapies, he had worsening gas exchange and was admitted to the intensive care unit in status asthmaticus. There he was intubated, and begun on high-dose intravenous methylprednisolone and a theophylline drip. Because of agitation, he was given benzodiazepines and several doses of vecuronium. His intensive care unit course was further complicated by a concurrent pneumonia. Because he was difficult to wean from the ventilator, he received a total of 15 days of ventilatory support and continued on steroids before extubation. Following transfer to the floor, his asthma was well controlled, but he was found to be unusually weak. On examination, he had diffuse muscle weakness on confrontational testing (3–4/5); he was unable to rise from a seated position and could take only a few steps independently. Serum CK level was mildly elevated. Prior to this hospitalization, he had normal strength. A diagnosis of corticosteroid myopathy was made based on history and examination. His steroids were tapered; physical therapy was recommended and initiated in the hospital. The patient showed gradual but steady improvement of strength over the next 3 months, as documented by his outpatient therapist and by follow-up neurologic evaluation.

B can persist for up to 6 months following its withdrawal,[71] the risk for potential myopathy exceeds the duration of therapy. Clinical aspects include severe muscle pain, weakness, swelling, and areflexia. Cranial muscles are spared in this myopathy. Histology of muscle biopsies in amphotericin B myopathy shows degeneration, segmental necrosis, and vacuoles. After adequate potassium replacement, recovery is typically complete; however, symptoms from rhabdomyolysis may take longer to resolve.[70]

Didanosine

Didanosine is a purine nucleoside analogue that is active against HIV-1 and HIV-2 infections. During phase I clinical trials, myopathy or arthralgia was reported in up to 4% of adult patients. In pediatric patients, reports included muscle atrophy (8%), myalgia (9%), and decreased strength (6%). Serum CK levels were elevated in most cases. With discontinuation of didanosine, these symptoms were reversible.[70]

Chloroquine and Hydroxychloroquine

Both chloroquine and hydroxychloroquine are used for treatment of malaria and rheumatologic disease. These drugs are known to cause myopathy or peripheral neuropathy, depending on the dosage regimen and duration of use. Hydroxychloroquine appears to have a lower toxicity than chloroquine, perhaps because of lower tissue drug levels and faster metabolism. Typically, patients develop proximal muscle weakness and a moderate increase in CK level. Although the precise mechanism of myotoxicity is unknown, lysosomal damage with inhibition of lysosomal enzyme activity likely plays a role; consequently, a large number of vacuoles accumulate filled with lipids or other cytoplasmic degradation products. Electromyography demonstrates fibrillations and myopathic motor unit potentials. On muscle biopsy, there is a vacuolar degeneration particularly affecting type I muscle fibers; the curvilinear bodies seen with electron microscopy are specific for hydroxychloroquine-associated myopathy.[72] Discontinuation of the drug will reverse the myopathy, but improvement will require many months.[70]

Clofibrate

Clofibrate is a branched-chain fatty acid ester that is employed as a hypolipidemic agent. The myopathy that is associated with clofibrate use can develop soon after starting the medication or following months of drug therapy; a dosage increase may precipitate the onset. Clofibrate myopathy is manifested clinically by muscle pain and proximal weakness. Serum CK levels are usually significantly elevated in patients with myopathic symptoms, but can also be elevated in asymptomatic patients receiving the medication. EMG may demonstrate normal compound muscle action potentials in mildly symptomatic patients or show frank myopathic changes when obvious weakness is present. Muscle biopsies may show scattered muscle fiber necrosis. Following withdrawal of clofibrate, there is gradual recovery over days to months; the CK levels may decrease prior to strength improving.[70,73]

Epsilon-aminocaproic Acid

Epsilon-aminocaproic acid is an antifibrinolytic agent used to treat patients with hereditary angioneurotic edema and subarachnoid hemorrhage. Patients treated for a period of at least 4 weeks may develop myalgia in the shoulder and hip regions that is accompanied by proximal muscle weakness. Serum CK level is usually markedly elevated. EMG demonstrates a reduction in the amplitude and duration of the compound muscle action potentials; profuse fibrillation potentials and positive sharp waves are seen in clinically affected muscles. On biopsy, there is fiber necrosis with selective type I fiber involvement. Following discontinuation of epsilon-aminocaproic acid, symptoms resolve over weeks to months.[70,74]

Labetalol

Labetalol, a competitive antagonist of alpha$_1$- and beta-adrenergic receptors, is used to treat hypertension and pheochromocytoma. Two children placed on labetalol for the management of renal disease and hypertension developed a severe generalized myopathy. These patients had proximal muscle weakness and markedly increased CK levels; muscle biopsy was consistent with a necrotic myopathy. In the months following the discontinuation of labetalol, both patients regained normal muscle strength, although one patient had persistent CK level elevation that returned to normal over the next year.[75]

Lovastatin

Lovastatin reduces the blood concentration of low-density lipoproteins through its potent inhibition of 3-hydroxy-3-methylglutaryl coenzyme A reductase. Lovastatin myopathy includes features such as muscle pain, cramps, proximal muscle weakness, and myotonia. Approximately 0.2% of patients treated with lovastatin alone for 48 months develop such symptoms. Serum CK levels exceed 2000 U/L. EMG studies demonstrate normal compound muscle action potentials in mildly affected patients or definite reductions in compound muscle action potential duration and amplitude when significant weakness is present. On muscle biopsy, there is often fiber necrosis, mononuclear cell infiltration with phagocytosis, and regeneration. Although infrequent, this necrotizing myopathy associated with lovastatin is important to recognize so that the medication can be promptly withdrawn, because life-threatening hemolysis can occur.[74] Within 30 days after discontinuing the drug, improvement generally occurs. Concurrent use of erythromycin, cyclosporine, nicotinic acid, or gemfibrozil with lovastatin may increase the toxicity of lovastatin. Patients with renal insufficiency and severe hepatobiliary dysfunction are at increased risk for developing lovastatin myopathy.[70] The authors have one patient with lovastatin-associated myopathy who harbored a mitochondrial DNA point mutation (A3243G) and recovered on high-dose coenzyme Q_{10}. Other statins are also associated with a range of muscle pain and injury that vary in frequency and severity. *SLCO1B1* affects the transport of statins into hepatocytes, and the rs4149056 variant of this gene is significantly linked to simvastatin-associated myopathy.[76]

Other Medications

Niacin can cause a myopathy, but only in patients taking more than 1 gram per day. These patients develop myalgias, proximal muscle weakness, and elevated CK levels; myotactic reflexes are preserved. After discontinuing niacin, the symptoms resolve rapidly, with a complete recovery in a month.[77] Quinacrine, used in the treatment of giardiasis, may cause a myopathy that includes proximal muscle pain and weakness and a cardiomyopathy. The CK level is often mildly elevated. Findings on muscle biopsy may include small groups of necrotic fibers with accompanying fibroblast proliferation and limited lymphocyte infiltration. As a rule, gradual recovery occurs over a period of weeks or months after withdrawal of quinacrine.[70]

INFLAMMATORY MYOPATHIES ASSOCIATED WITH SYSTEMIC DISORDERS

Myositis is a general term used to indicate inflammation of muscle. The cause may be known (bacterial, viral, or parasitic) or unknown (idiopathic). Idiopathic inflammatory myopathies include polymyositis, dermatomyositis (see Case Example 36.3), and inclusion body myositis. Together, the inflammatory myopathies form a heterogeneous group of subacute to chronic muscle diseases that share a common and distinguishing feature: muscle degeneration mediated through inflammatory processes.[78]

Inflammatory Myopathies and Malignancies

Inflammatory myopathies and malignancies have an established association, but the incidence of malignancy varies according to the type of myopathy and the patient age group. Although dermatomyositis is more common in children than adults (see Figure 36.5), polymyositis is extremely rare in childhood. In adults with dermatomyositis, the reported frequency of malignancy is 6% to 45%; in adults with polymyositis, the association with malignancy is less prominent, with a reported frequency of 0% to 28%.[79] In contrast, childhood dermatomyositis and polymyositis are generally not associated with malignancy, and the few cases that are reported in the literature have unusual clinical courses.[80] Muscle weakness usually improves significantly following removal of the tumor or chemotherapy, whereas tumor regrowth is often marked by the reappearance of muscle weakness or increased serum muscle enzyme levels. A large variety of tumors are associated with inflammatory myopathies; the most common are stomach and ovarian cancers. In Asian adults, nasopharyngeal carcinoma is frequently implicated.[78,81] Inflammatory myopathies may occur as a result of autoantibodies to tumor antigens that cross-react with muscle

CASE EXAMPLE 36.3

S.G. is a 13-year-old male who presented with a 3-month history of difficulty climbing stairs and proximal muscle pain. His father reported that he frequently complained of feeling tired and was not his usual self (more irritable and socially isolated). On examination, the patient appeared miserable, with facial flushing and a light blue-purple hue around his eyes. An erythematous rash was also present on his neck, anterior chest, and on his elbows and dorsa of the hands bilaterally (Figure 36.2A, B, and D). Pinpoint telangiectasis of the nail beds could be appreciated with the aid of a lens. Proximal muscle strength was 4/5 at the shoulders and hips bilaterally, with an element of give-way weakness and restricted movement secondary to pain (Figure 36.2C). Serum studies revealed an increased erythrocyte sedimentation rate of 100, and an elevated CK level of 450 IU. Muscle biopsy from the right deltoid demonstrated atrophy of both type I and type II fibers in the peripheral regions of many fascicles. The muscle nuclei in multiple fibers, especially in the perifascicular zones, appeared large, internally located, and hyperchromatic. In the septa separating muscle fibers, foci of lymphatic cell accumulation were seen (Figure 36.3A and B). Because his physical findings and evaluation were consistent with childhood dermatomyositis, a course of daily then every-other-day prednisone was initiated; after 6 months, and following the normalization of the sedimentation rate and CK values, his steroid therapy was slowly tapered. The course was highly effective in improving his strength and demeanor (Figure 36.4).

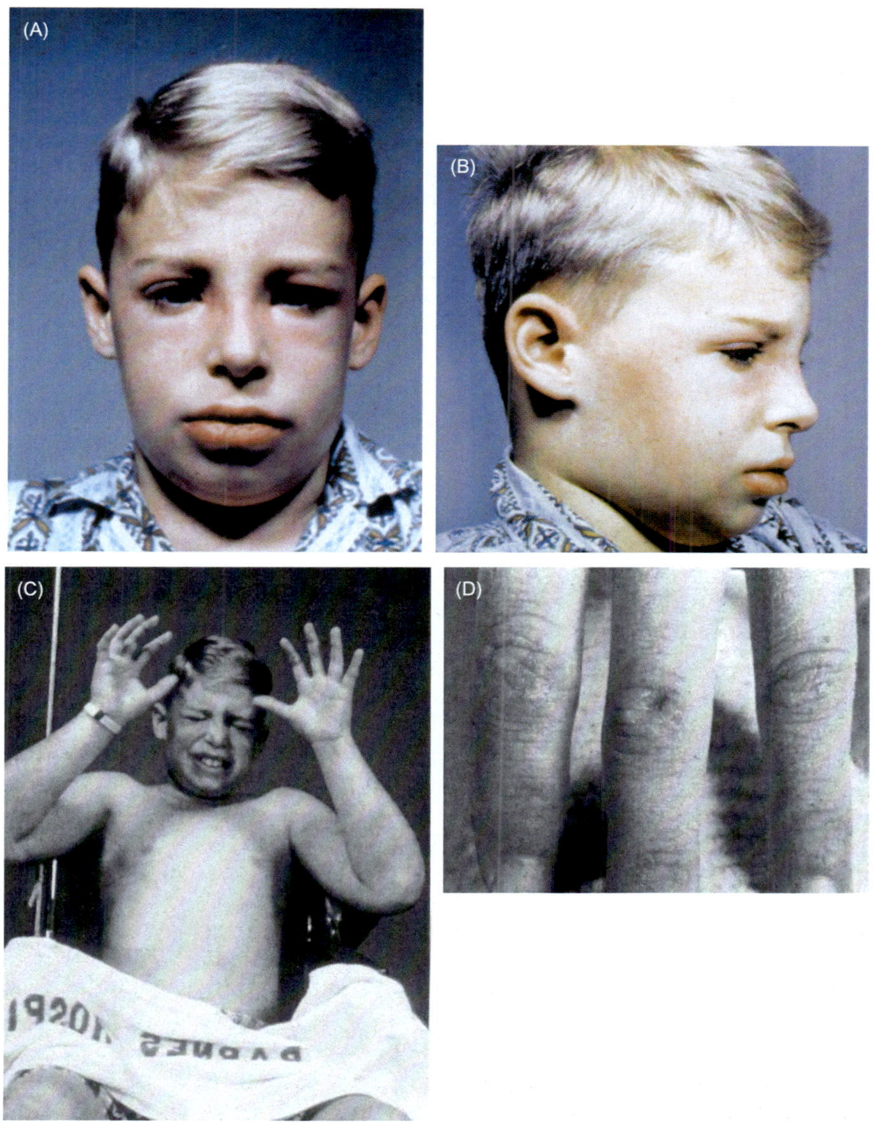

FIGURE 36.2 Dermatomyositis (Case Example 36.3). **(A and B)** Note the erythematous rash in the periorbital regions bilaterally and the facial edema. **(C)** Limited abduction is demonstrated at the shoulders secondary to muscle pain. **(D)** Violaceous flat-topped papules (Gottron's papules) are present on the dorsal interphalangeal joints.

(Continued)

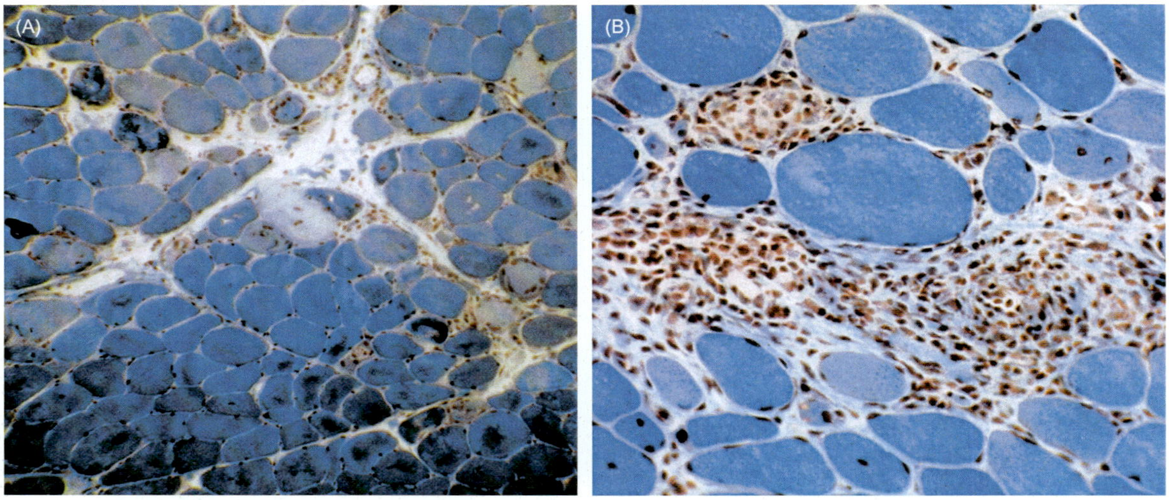

FIGURE 36.3 Muscle biopsy in dermatomyositis. (**A**) Low-power view of muscle biopsy specimen showing pale, necrotic muscle fibers, as well as angulated muscle fibers associated with concurrent inflammatory nerve injury. A prominent inflammatory cell infiltrate (red) is apparent (modified trichrome). (**B**) Higher power view of muscle biopsy specimen demonstrating the perivascular/connective tissue location of the inflammation. *(Modified trichrome).*

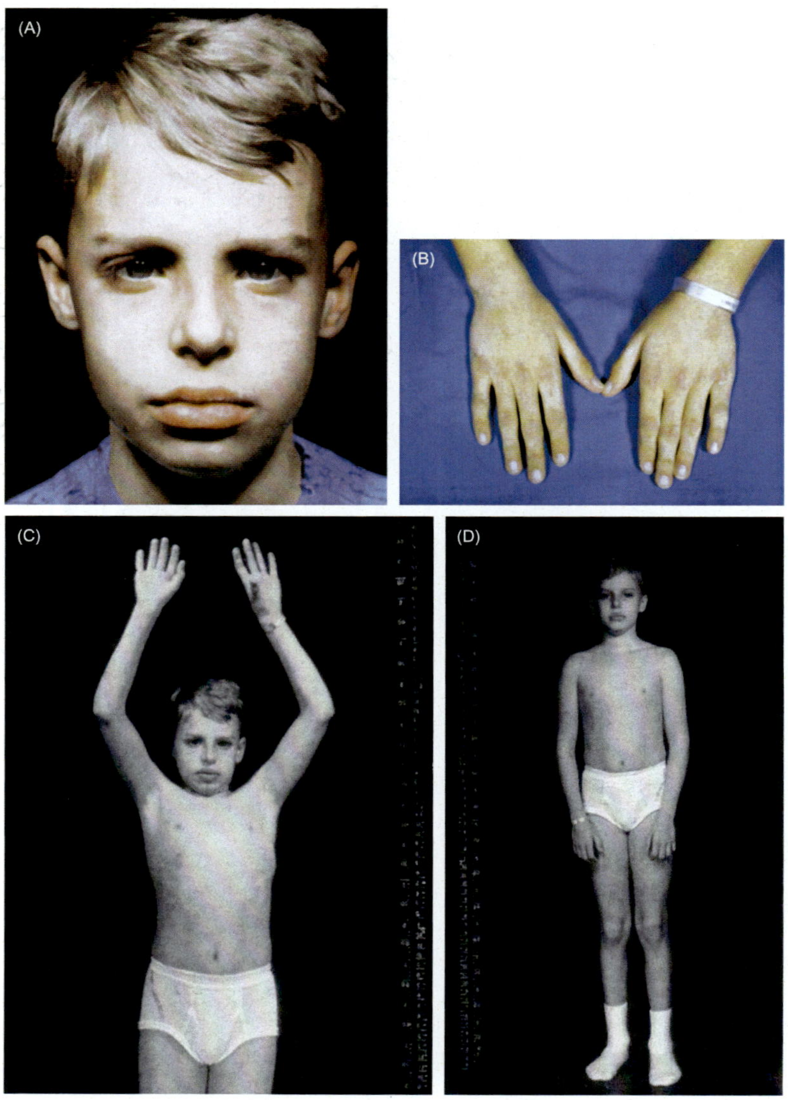

FIGURE 36.4 Improvement of dermatomyositis following steroid therapy (Case Example 36.3). (**A**) There is resolution of the periorbital edema following the prednisone treatment course. (**B**) Some erythema persists in the dorsa of the hands bilaterally, especially over the metacarpal-phalangeal joints. (**C**) Improved strength and loss of muscle pain are noted following treatment. (**D**) Ambulation and stance are normal following treatment.

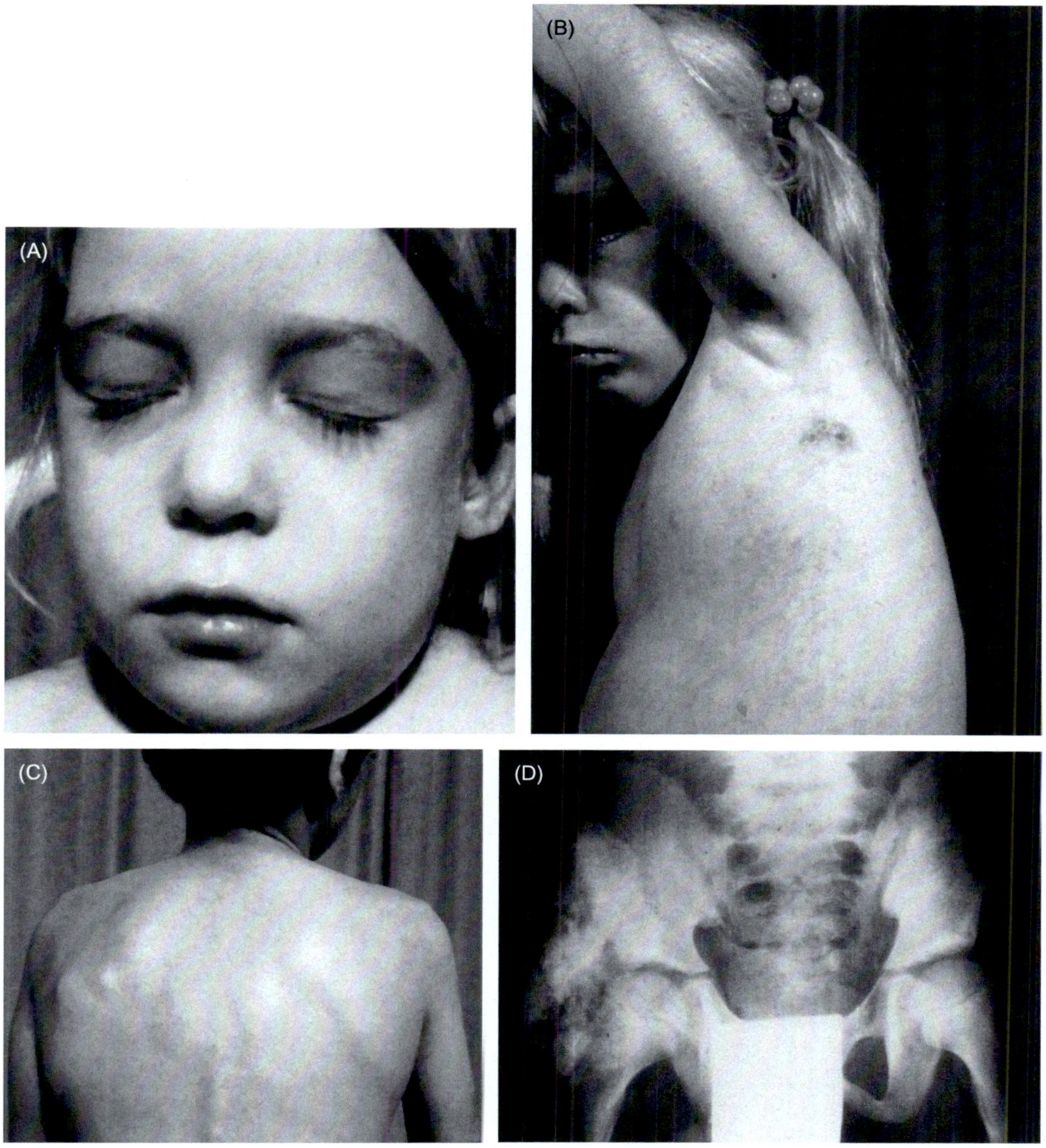

FIGURE 36.5 Further clinical manifestations of dermatomyositis. (**A**) A characteristic purplish-red heliotrope erythema of the eyelids and cheeks is seen in a girl with dermatomyositis. (**B**) Cutaneous calcinosis of the left lateral chest, with extrusion and ulceration of the skin (**C**) An example of severe calcinosis in childhood dermatomyositis, involving a large suprascapular region. (**D**) Radiographic evidence of calcinosis of the right hip in chronic dermatomyositis, seen here as patchy radiodense deposits.

antigens in a fraction or of toxic factors elaborated by tumors that produce muscle fiber inflammation and necrosis. The effectiveness of steroid therapy is unclear, inasmuch as responses appear to vary greatly between patients.[78,82]

Rheumatoid Disease and Myopathy

Skeletal muscle from patients with rheumatoid arthritis shows at least three patterns of infiltration by inflammatory cells: polymyositis, focal nodular myositis, and arteritis at the light microscopy and ultrastructural levels.[83] It should be noted that none of these patterns is specific for rheumatoid disease, and the relationship between these patterns is unclear.[84] Typically in arterial inflammation, there is a predominance of mononuclear elements. The smaller arterial vessels tend to be most severely affected; there is frequent luminal narrowing, and fibrinoid necrosis may be present.[85] Polymyositis is known to occur in the setting of rheumatoid disease, and in many cases is morphologically indistinguishable from primary polymyositis; changes include the infiltration by mononuclear inflammatory cells, widespread fiber necrosis and regeneration, and later endomysial connective tissue proliferation. Focal nodular myositis may occur in a number of disorders, but is especially prominent in patients with active, chronic rheumatoid disease. The characteristic lesion in focal nodular myositis is a small, nodular 1- to 2-mm accumulation of plasma cells and lymphocytes adjacent to an arteriole, with or without coincident damage to regional muscle fibers.[84] It should be emphasized that impressive muscle weakness in patients with rheumatoid arthritis may be secondary to arthritis, pain, medications (e.g. steroids), inactivity, or peripheral neuropathy; a recent quantitative controlled EMG study found an increased prevalence of neurogenic but not myogenic changes in patients with rheumatoid arthritis compared with controls.[86]

Scleroderma–Myositis

Between 5% and 7% of patients with scleroderma also have myositis. Usually these patients have mild weakness, with normal or slightly increased serum CK level. Muscle biopsy features include perimysial inflammation and fibrosis, atrophy of type II fibers, and vasculopathy typical of scleroderma. Patients with scleroderma-myositis have one of the two forms of scleroderma: CREST syndrome (calcinosis, Raynaud's phenomenon, esophageal disease, sclerodactyly, and cutaneous telangiectasis) or progressive systemic sclerosis. Children with scleroderma-myositis show less visceral involvement (in areas such as esophagus, lung, and heart) than do adult patients.[87] Immunosuppressive treatment is usually effective in these conditions. Scleroderma-myositis may be associated with anti-PM-scleroderma (PM-1 or PM-Scl) antibodies, but these antibodies are also found in idiopathic myositis and rheumatoid arthritis.[78]

Systemic Lupus Erythematosus–Myositis

Either dermatomyositis or polymyositis can occur in the course of systemic lupus erythematosus (SLE), but the incidence is low. Generally, patients with SLE and myositis have moderate to severe, proximal, symmetrical muscle weakness associated with myalgia; lupus myositis is often as severe as primary myositis.[88] Raynaud's phenomenon and dysphagia may be present. SLE can be associated with a number of autoimmune disorders, including idiopathic thrombocytopenia, thrombotic thrombocytopenic purpura, hemolytic anemia, vasculitis, progressive systemic sclerosis, autoimmune neutropenia, myasthenia gravis, sarcoidosis, and glomerulonephritis. Antibodies to native DNA are highly specific in SLE; antibodies to denatured DNA are also found but are less specific. Up to 50% of SLE patients have anti-RNP antibodies, including anti-Sm antibodies (specific for SLE) and anti-U1 snRNP antibodies (also present in mixed connective tissue disease [MCTD]).[78] The most common clinical course for patients with lupus myositis is a relapsing and remitting one. Corticosteroids are used as a first-line therapeutic agent, often with an additional immunosuppressive agent such as azothiaprine.[88]

Sjögren's Syndrome–Myositis

The major manifestations of Sjögren's syndrome (SS) are dryness of the mouth, mucous membranes, and eyes, as well as chronic arthritis. In severe forms, lack of secretion may involve the entire gastrointestinal tract, respiratory tract, vagina, and skin. Myopathy occurs very frequently and is usually severe, affecting proximal muscles symmetrically. Nasal speech and dysphagia may be present. Muscle biopsies in SS–myositis show inflammation of muscle and sometimes an associated vasculitis. Up to 40% of patients with SS develop tubulointerstitial nephritis leading to distal renal tubular acidosis; proximal muscle weakness may complicate osteomalacia due to distal renal tubular acidosis in SS and has been reported to mimic polymyositis.[89] SS may overlap with SLE, polyarteritis, primary biliary cirrhosis, chronic active hepatitis, and lymphoproliferative hyperplasia. Quantitative immunoglobulin levels may demonstrate increased IgG and IgM; antinuclear antibodies, specific for ribonucleoprotein antigens SS-A (Ro) and SS-B (La), and anti-γ-globulin are present in up to 90% of patients.[78]

Mixed Connective Tissue Disease–Myositis

MCTD is associated with titers of anti-U_1 snRNP antibodies. The clinical features of MCTD include those of SLE, progressive systemic sclerosis, and dermatomyositis that usually appear following one another rather than simultaneously. Approximately 35% to 72% of patients with MCTD also have myositis. Muscle symptoms include mild weakness and myalgia. CK levels are usually in the normal range. The pathologic changes seen on muscle biopsy resemble those found in dermatomyositis.[78]

Churg-Strauss Syndrome

Churg-Strauss syndrome is characterized by the triad of eosinophilia, asthma, and granulomatous vasculitis. Frequent but minor manifestations include arthralgia and arthritis. Typical regions of clinical involvement include the upper and lower respiratory tracts, peripheral nerves, skin, and heart.[90] Rarely, patients with Churg-Strauss syndrome may manifest features of polymyositis, with muscle biopsy showing focal interstitial infiltration by eosinophils and necrotizing vasculitis. A rapid response to prednisone has been described, with resolution of myalgia and weakness complaints within a week.[91]

EOSINOPHILIC MYOPATHIC SYNDROMES

Idiopathic Eosinophilic Myopathy

Idiopathic eosinophilic myopathy may be subdivided into three types: focal eosinophilic myositis, eosinophilic polymyositis, and eosinophilic perimyositis. Deep eosinophilic infiltration of muscle with invasion of muscle fibers is seen in focal eosinophilic myositis.[92,93] The overlying skin and fascia are normal. In focal eosinophilic myositis, there is typically evidence of peripheral blood eosinophilia as well as features of myositis, such as abnormal electromyography and elevated CK levels. The legs are most often affected; however, a pseudotumor of the sternocleidomastoid from focal eosinophilic myositis has been described in a 14-year-old boy.[94] The cause of focal eosinophilic myositis is unknown; some cases appear following trauma or excessive alcohol consumption. Many patients improve spontaneously in a few weeks following presentation; treatment options include corticosteroid therapy or surgical excision of localized soft tissue masses. Relapses commonly occur, and may be noted several years after the initial presentation.[93]

In eosinophilic polymyositis, eosinophils deeply and diffusely infiltrate muscle, and marked perivenular eosinophilic cuffing is seen. As compared with focal eosinophilic myositis, muscle fiber invasion is not commonly noted, but necrosis may occur in regions of dense infiltration.[95] The clinical presentation of eosinophilic polymyositis is distinct from focal eosinophilic myositis. Muscle swelling, myalgia, and proximal weakness occur in the setting of a systemic illness with pericarditis, myocarditis, and heart failure that may be complicated by conduction block or atrial fibrillation.[93,96] Often, patients have a history of Raynaud's phenomenon and weight loss; skin involvement includes erythema, subcutaneous induration, or angioedema. Because muscle damage is widespread, CK levels are generally very high. Corticosteroid therapy is effective, and may need to be continued on a prolonged course.[93]

In eosinophilic perimyositis, the eosinophilic infiltrate is detected primarily in the fascia and superficial perimysium. Muscle fibers usually appear normal, and fiber necrosis is rare. Typically, muscle tenderness mainly affects the legs with localized induration of the calves or thighs. In contrast to findings in focal eosinophilic myositis and eosinophilic polymyositis, peripheral blood eosinophilia is not usually seen, and CK levels are generally normal.[97] Spontaneous recovery is the rule.[93]

Eosinophilic Myopathy and Parasitic Infection

Parasitic infections with *Taenia solium* and *Trichinella spiralis* commonly involve focal eosinophilic infiltration of muscle. When muscle biopsy demonstrates the causative organism, the diagnosis is readily apparent. Treatment is aimed at eliminating the parasitic infection.

Toxic Oil Syndrome

In 1981, an epidemic of what became known as toxic oil syndrome (TOS) occurred in Spain. Investigations determined that TOS was caused by ingestion of a mixture of oils including denatured rapeseed oil, contaminated with aniline, that was originally intended only for industrial use. The oil was refined and sold in bulk as olive oil by itinerant salesmen. TOS affected more than 20,000 people and resulted ultimately in the deaths of more than 300. Once the oil was withdrawn, the incidence of TOS sharply dropped.[98] In the early phase of TOS, patients suffered fever, pulmonary edema, severe myalgia, eosinophilia, skin rash, malaise, and generalized lymphadenopathy. Intermediate phase symptoms included hepatic disease, subcutaneous edema, arthritis, and diarrhea. Although the majority recovered, approximately one fourth of patients developed a later phase of severe myalgia, muscle atrophy, polyneuropathy, joint contractures, and weight loss. Chronic findings included scleroderma-like skin thickening, muscle cramps, and Raynaud's phenomenon.[93,98] Muscle biopsies revealed evidence of an inflammatory myopathy, with infiltrates of eosinophils, lymphocytes, neutrophils, and histiocytes; these infiltrates primarily involved the fascia, epimysium, and perimysium. Muscle necrosis was uncommon. More chronic muscle findings included denervation atrophy with endomysial fibrosis. Thus, TOS shares histologic features with idiopathic eosinophilic perimyositis and clinical features with eosinophilia-myalgia syndrome, as described in the next section.[93,99]

Eosinophilia-Myalgia Syndrome

In 1989, eosinophilia-myalgia syndrome (EMS) was reported as an outbreak primarily in the United States but also in Canada, the United Kingdom, and Germany.[99,100] Through epidemiologic studies, EMS was found to be strongly associated with ingestion of contaminated L-tryptophan from a single bulk manufacturer.[101–103] L-tryptophan is an essential amino acid that has been sold as a dietary supplement; because it is metabolized to serotonin, it has been used to treat depression, insomnia, and other disorders.[103] The specific components in the tryptophan preparations that caused EMS have yet to be characterized.[99] With the recall of over-the-counter preparations of L-tryptophan in November 1989, the incidence of this potentially fatal disorder fell dramatically.[102] In total, more than 1500 cases of EMS have occurred, with at least 37 deaths associated with this syndrome.[104]

The primary features of EMS are acute, severe myalgia and peripheral eosinophilia. The most frequent multisystemic features include arthritis/arthralgia, erythematous macular rashes, dyspnea or cough, peripheral edema, increased serum aldolase levels, and elevated liver enzyme concentrations.[100,102] During the 2 to 3 months following onset of EMS, scleroderma-like cutaneous induration and peripheral neuromuscular disease occur; features may include myopathy, sensory or sensorimotor neuropathy, mononeuropathy multiplex, and a Guillain-Barré-like syndrome. Other potentially life-threatening features include myocarditis and pulmonary hypertension. Muscle biopsy specimens show perimysial and epimysial inflammatory infiltrates composed of histiocytes, lymphocytes, and occasional eosinophils and plasma cells.[105,106] When inflammation has been found to affect myofibers, there is an associated inflammation of the endomysium at the periphery of muscle fascicles. Muscle fiber necrosis is apparently uncommon. Additional inflammation involves muscle spindles, the walls of some small blood vessels, the perineurium of small nerve twigs, and fibrous septae of subcutaneous adipose tissue.[105] Muscle spindle inflammation may be an important contributing factor in chronic myalgia and muscle weakness.[93] No specific treatment has been used, but some patients reported improvement with oral corticosteroid therapy. Within 2 years after onset of EMS, the majority of patients noted improvement or resolution of most symptoms. Long-term functional disability may be associated with chronic neuropathy, myopathy, cutaneous induration, or neurocognitive dysfunction.[100,107]

REFERENCES

1. Kendall-Taylor P, Turnbull DM. Endocrine myopathies. *Br Med J (Clin Res Ed)* 1983;**287**:705–8.
2. Hardiman O, Molloy F, Brett F, Farrell M. Inflammatory myopathy in thyrotoxicosis. *Neurology* 1997;**48**:339–41.
3. Kodali VR, Jeffcote B, Clague RB. Thyrotoxic periodic paralysis: a case report and review of the literature. *J Emerg Med* 1999;**17**:43–5.
4. Ryan DP, da Silva MR, Soong TW, Fontaine B, Donaldson MR, Kung AW, et al. Mutations in potassium channel Kir2.6 cause susceptibility to thyrotoxic hypokalemic periodic paralysis. *Cell* 2010;**140**:88–98.
5. Cheung CL, Lau KS, Ho AY, Lee KK, Tiu SC, Lau EY, et al. Genome-wide association study identifies a susceptibility locus for thyrotoxic periodic paralysis at 17q24.3. *Nat Genet* 2012;**44**:1026–9.
6. Lin SH, Huang CL. Mechanism of thyrotoxic periodic paralysis. *J Am Soc Nephrol* 2012;**23**:985–8.
7. Suzuki S, Ichikawa K, Nagai M, Mikoshiba M, Mori J, Kaneko A, et al. Elevation of serum creatine kinase during treatment with antithyroid drugs in patients with hyperthyroidism due to Graves disease. A novel side effect of antithyroid drugs. *Arch Intern Med* 1997;**157**:693–6.
8. Lochmuller H, Reimers CD, Fischer P, Heuss D, Muller-Hocker J, Pongratz DE. Exercise-induced myalgia in hypothyroidism. *Clin Investig* 1993;**71**:999–1001.
9. Duyff RF, Van den Bosch J, Laman DM, van Loon BJ, Linssen WH. Neuromuscular findings in thyroid dysfunction: a prospective clinical and electrodiagnostic study. *J Neurol Neurosurg Psychiatry* 2000;**68**:750–5.
10. Rodolico C, Toscano A, Benvenga S, Mazzeo A, Bartolone S, Bartolone L, et al. Myopathy as the persistently isolated symptomatology of primary autoimmune hypothyroidism. *Thyroid* 1998; **8**:1033–8.
11. Modi G. Cores in hypothyroid myopathy: a clinical, histological and immunofluorescence study. *J Neurol Sci* 2000;**175**:28–32.
12. Monzani F, Caraccio N, Siciliano G, Manca L, Murri L, Ferrannini E. Clinical and biochemical features of muscle dysfunction in subclinical hypothyroidism. *J Clin Endocrinol Metab* 1997;**82**:3315–18.
13. Khaleeli AA, Gohil K, McPhail G, Round JM, Edwards RH. Muscle morphology and metabolism in hypothyroid myopathy: effects of treatment. *J Clin Pathol* 1983;**36**:519–26.
14. Virmani A, Gambhir A, Iyer PU. Kocher Debre Semelaigne syndrome mimicking primary muscle disease. *Indian Pediatr* 1990;**27**:88–9.
15. Virtanen M. Manifestations of congenital hypothyroidism during the 1st week of life. *Eur J Pediatr* 1988;**147**:270–4.
16. Chopra BK, Singh G, Puri A. Hoffmann's syndrome (hypothyroid myopathy). *J Assoc Physicians India* 1972;**20**:387–90.
17. Syriou V, Kolitsa A, Pantazi L, Pikazis D. Hypoparathyroidism in a patient presenting with severe myopathy and skin rash. Case report and review of the literature. *Hormones (Athens)* 2005;**4**:161–4.
18. Barber J, Butler RC, Davie MW, Sewry CA. Hypoparathyroidism presenting as myopathy with raised creatine kinase. *Rheumatology (Oxford)* 2001;**40**:1417–18.
19. Policepatil SM, Caplan RH, Dolan M. Hypocalcemic myopathy secondary to hypoparathyroidism. *WMJ* 2012;**111**:173–5.
20. Rymanowski JV, Twydell PT. Treatable dropped head syndrome in hyperparathyroidism. *Muscle Nerve* 2009;**39**:409–10.
21. De Luca GC, Eggers SD. A rare complication of azotemic hyperparathyroidism: ischemic calcific myopathy. *Neurology* 2010;**75**:1942.
22. Lacomis D, Chad DA, Aronin N, Smith TW. The myopathy of Cushing's syndrome. *Muscle Nerve* 1993;**16**:880–1.
23. Olafsson E, Jones Jr. HR, Guay AT, Thomas CB. Myopathy of endogenous Cushing's syndrome: a review of the clinical and electromyographic features in 8 patients. *Muscle Nerve* 1994;**17**:692–3.
24. Khaleeli AA, Edwards RH, Gohil K, McPhail G, Rennie MJ, Round J, et al. Corticosteroid myopathy: a clinical and pathological study. *Clin Endocrinol (Oxf)* 1983;**18**:155–66.
25. Brautbar N. Skeletal myopathy in uremia: abnormal energy metabolism. *Kidney Int Suppl* 1983;**16**:S81–6.
26. Torres CF, Forbes GB. Muscle weakness in infants with vitamin D deficiency rickets [abstract]. *Ann Neurol* 1984;**16**:406.
27. Huang YY, Hsu BR, Tsai JS. Paralytic myopathy—a leading clinical presentation for primary aldosteronism in Taiwan. *J Clin Endocrinol Metab* 1996;**81**:4038–41.

28. Ptacek LJ, Tawil R, Griggs RC, Engel AG, Layzer RB, Kwiecinski H, et al. Dihydropyridine receptor mutations cause hypokalemic periodic paralysis. *Cell* 1994;**77**:863–8.
29. Shintani S, Murase H, Tsukagoshi H, Shiigai T. Glycyrrhizin (licorice)-induced hypokalemic myopathy. Report of 2 cases and review of the literature. *Eur Neurol* 1992;**32**:44–51.
30. Ishikawa S, Kato M, Tokuda T, Momoi H, Sekijima Y, Higuchi M, et al. Licorice-induced hypokalemic myopathy and hypokalemic renal tubular damage in anorexia nervosa. *Int J Eat Disord* 1999;**26**:111–14.
31. Conn JW, Rovner DR, Cohen EL. Licorice-induced pseudoaldosteronism. Hypertension, hypokalemia, aldosteronopenia, and suppressed plasma renin activity. *JAMA* 1968;**205**:492–6.
32. Barrella M, Lauria G, Quatrale R, Paolino E. Hypokalemic rhabdomyolysis associated with liquorice ingestion: report of an atypical case. *Ital J Neurol Sci* 1997;**18**:217–20.
33. Bagga A, Dutta S. Hypokalemic muscle paralysis. *Indian Pediatr* 1994;**31**:1297–8.
34. Chu CC, Huang CC, Chu NS. Recurrent hypokalemic muscle weakness as an initial manifestation of Wilson's disease. *Nephron* 1996;**73**:477–9.
35. Bettinelli A, Bianchetti MG, Girardin E, Caringella A, Cecconi M, Appiani AC, et al. Use of calcium excretion values to distinguish two forms of primary renal tubular hypokalemic alkalosis: Bartter and Gitelman syndromes. *J Pediatr* 1992;**120**:38–43.
36. Bierbach H, Bohl J, Goldner HJ, Majdandzic J. Hypokalemic rhabdomyolysis associated with Bartter's syndrome. *Klin Wochenschr* 1983;**61**:183–6.
37. van der Loos TL. Profound hypokalemia with rhabdomyolysis in Bartter's syndrome. *JAMA* 1986;**256**:2518.
38. Nishihara G, Higashi H, Matsuo S, Yasunaga C, Sakemi T, Nakamoto M. Acute renal failure due to hypokalemic rhabdomyolysis in Gitelman's syndrome. *Clin Nephrol* 1998;**50**:330–2.
39. Tang NL, Hui J, To KF, Ng HK, Hjelm NM, Fok TF. Severe hypokalemic myopathy in Gitelman's syndrome. *Muscle Nerve* 1999;**22**:545–7.
40. Satoh J, Kuroda Y, Nawata H, Yanase T. Molecular basis of hypokalemic myopathy caused by 17alpha-hydroxylase/17,20-lyase deficiency. *Neurology* 1998;**51**:1748–51.
41. Hiromatsu K, Kobayashi T, Fujii N, Itoyama Y, Goto I, Murakami J. Hypernatremic myopathy. *J Neurol Sci* 1994;**122**:144–7.
42. Knochel JP. Neuromuscular manifestations of electrolyte disorders. *Am J Med* 1982;**72**:521–35.
43. Engel AG, Emslie-Smith AM. Inflammatory myopathies. *Curr Opin Neurol Neurosurg* 1989;**2**:695–700.
44. Dalakas MC. Retroviruses and inflammatory myopathies in humans and primates. *Baillieres Clin Neurol* 1993;**2**:659–91.
45. Miro O, Pedrol E, Cebrian M, Masanes F, Casademont J, Mallolas J, et al. Skeletal muscle studies in patients with HIV-related wasting syndrome. *J Neurol Sci* 1997;**150**:153–9.
46. Schwartzman WA, Lambertus MW, Kennedy CA, Goetz MB. Staphylococcal pyomyositis in patients infected by the human immunodeficiency virus. *Am J Med* 1991;**90**:595–600.
47. Hull R, Gay H, Giles H, Nowicki M. *Streptococcus agalactiae* myositis in a child with perinatally acquired human immunodeficiency virus. *South Med J* 2008;**101**:317–19.
48. Chalmers AC, Greco CM, Miller RG. Prognosis in AZT myopathy. *Neurology* 1991;**41**:1181–4.
49. Shapiro ED. Lyme disease in children. *Am J Med* 1995;**98**:69S–73S.
50. Reimers CD, de Koning J, Neubert U, Preac-Mursic V, Koster JG, Muller-Felber W, et al. *Borrelia burgdorferi* myositis: report of eight patients. *J Neurol* 1993;**240**:278–83.
51. Holmgren AR, Matteson EL. Lyme myositis. *Arthritis Rheum* 2006;**54**:2697–700.
52. Ilowite NT. Muscle, reticuloendothelial, and late skin manifestations of Lyme disease. *Am J Med* 1995;**98**:63S–68S.
53. Sitwell LD, Weinshenker BG, Monpetit V, Reid D. Complete ophthalmoplegia as a complication of acute corticosteroid- and pancuronium-associated myopathy. *Neurology* 1991;**41**:921–2.
54. Danon MJ, Carpenter S. Myopathy with thick filament (myosin) loss following prolonged paralysis with vecuronium during steroid treatment. *Muscle Nerve* 1991;**14**:1131–9.
55. Bolton CF. Sepsis and the systemic inflammatory response syndrome: neuromuscular manifestations. *Crit Care Med* 1996;**24**:1408–16.
56. Clowes Jr. GH, George BC, Villee Jr. CA, Saravis CA. Muscle proteolysis induced by a circulating peptide in patients with sepsis or trauma. *N Engl J Med* 1983;**308**:545–52.
57. Zochodne DW, Ramsay DA, Saly V, Shelley S, Moffatt S. Acute necrotizing myopathy of intensive care: electrophysiological studies. *Muscle Nerve* 1994;**17**:285–92.
58. Hopkins IJ. A new syndrome: poliomyelitis-like illness associated with acute asthma in childhood. *Aust Paediatr J* 1974;**10**:273–6.
59. Yeung SC, Antonio G, Ip KS. Flaccid paralysis of the limbs after an asthmatic attack. *Pediatr Neurol* 2010;**42**:133–6.
60. Shahar EM, Hwang PA, Niesen CE, Murphy EG. Poliomyelitis-like paralysis during recovery from acute bronchial asthma: possible etiology and risk factors. *Pediatrics* 1991;**88**:276–9.
61. Nihei K, Naitoh H, Ikeda K. Poliomyelitis-like syndrome following asthmatic attack (Hopkins syndrome). *Pediatr Neurol* 1987;**3**:166–8.
62. Arita J, Nakae Y, Matsushima H, Maekawa K. Hopkins syndrome: T2-weighted high intensity of anterior horn on spinal MR imaging. *Pediatr Neurol* 1995;**13**:263–5.
63. Fiorillo C, Brisca G, Cassandrini D, Scapolan S, Astrea G, Valle M, et al. Subclinical myopathy in a child with neutral lipid storage disease and mutations in the PNPLA2 gene. *Biochem Biophys Res Commun* 2013;**430**:241–4.
64. Fischer J, Lefevre C, Morava E, Mussini JM, Laforet P, Negre-Salvayre A, et al. The gene encoding adipose triglyceride lipase (PNPLA2) is mutated in neutral lipid storage disease with myopathy. *Nat Genet* 2007;**39**:28–30.
65. Reilich P, Horvath R, Krause S, Schramm N, Turnbull DM, Trenell M, et al. The phenotypic spectrum of neutral lipid storage myopathy due to mutations in the PNPLA2 gene. *J Neurol* 2011;**258**:1987–97.
66. Williams TJ, O'Hehir RE, Czarny D, Horne M, Bowes G. Acute myopathy in severe acute asthma treated with intravenously administered corticosteroids. *Am Rev Respir Dis* 1988;**137**:460–3.
67. Op de Coul AA, Lambregts PC, Koeman J, van Puyenbroek MJ, Ter Laak HJ, Gabreels-Festen AA. Neuromuscular complications in patients given Pavulon (pancuronium bromide) during artificial ventilation. *Clin Neurol Neurosurg* 1985;**87**:17–22.

68. Hirano M, Ott BR, Raps EC, Minetti C, Lennihan L, Libbey NP, et al. Acute quadriplegic myopathy: a complication of treatment with steroids, nondepolarizing blocking agents, or both. *Neurology* 1992;**42**:2082–7.
69. Schakman O, Kalista S, Barbe C, Loumaye A, Thissen JP. Glucocorticoid-induced skeletal muscle atrophy. *Int J Biochem Cell Biol* 2013;**45**:2163–72.
70. Shultz CE, Kincaid JC. Drug induced myopathies. In: Biller J, editor. *Iatrogenic Neurology*. Boston: Butterworth-Heineman; 1998. pp. 305–17.
71. Lucas da Silva PS, Iglesias SB, Waisberg J. Hypokalemic rhabdomyolysis in a child due to amphotericin B therapy. *Eur J Pediatr* 2007;**166**:169–71.
72. Kwon JB, Kleiner A, Ishida K, Godown J, Ciafaloni E, Looney Jr. RJ. Hydroxychloroquine-induced myopathy. *J Clin Rheumatol* 2010;**16**:28–31.
73. Hodel C. Myopathy and rhabdomyolysis with lipid-lowering drugs. *Toxicol Lett* 2002;**128**:159–68.
74. Victor M, Sieb JP. Myopathies due to drugs, toxins, and nutritional deficiency. In: Engel AG, Franzini-Armstrong C, editors. *Myology*. 2nd ed. New York: McGraw-Hill; 1994. pp. 1697–725.
75. Willis JK, Tilton AH, Harkin JC, Boineau FG. Reversible myopathy due to labetalol. *Pediatr Neurol* 1990;**6**:275–6.
76. Brunham LR, Lansberg PJ, Zhang L, Miao F, Carter C, Hovingh GK, et al. Differential effect of the rs4149056 variant in SLCO1B1 on myopathy associated with simvastatin and atorvastatin. *Pharmacogenomics J* 2012;**12**:233–7.
77. Gharavi AG, Diamond JA, Smith DA, Phillips RA. Niacin-induced myopathy. *Am J Cardiol* 1994;**74**:841–2.
78. Mantegazza R, Bernasconi P, Confalonieri P, Cornelio F. Inflammatory myopathies and systemic disorders: a review of immunopathogenetic mechanisms and clinical features. *J Neurol* 1997;**244**:277–87.
79. Callen JP. Myositis and malignancy. *Curr Opin Rheumatol* 1994;**6**:590–4.
80. Sherry DD, Haas JE, Milstein JM. Childhood polymyositis as a paraneoplastic phenomenon. *Pediatr Neurol* 1993;**9**:155–6.
81. Koh ET, Seow A, Ong B, Ratnagopal P, Tjia H, Chng HH. Adult onset polymyositis/dermatomyositis: clinical and laboratory features and treatment response in 75 patients. *Ann Rheum Dis* 1993;**52**:857–61.
82. Engel AG, Hohlfeld R, Banker BQ. Inflammatory myopathies, the polymyositis and dermatomyositis syndromes. In: Engel AG, Franzini-Armstrong C, editors. *Myology*. New York: McGraw-Hill; 1994. pp. 1335–83.
83. de Palma L, Chillemi C, Albanelli S, Rapali S, Bertoni-Freddari C. Muscle involvement in rheumatoid arthritis: an ultrastructural study. *Ultrastruct Pathol* 2000;**24**:151–6.
84. Kim RC, Collins GH. The neuropathology of rheumatoid disease. *Hum Pathol* 1981;**12**:5–15.
85. Magyar E, Talerman A, Mohacsy J, Wouters HW, de Bruijn WC. Muscle changes in rheumatoid arthritis. A review of the literature with a study of 100 cases. *Virchows Arch A Pathol Anat Histol* 1977;**373**:267–78.
86. Bekkelund SI, Torbergsen T, Husby G, Mellgren SI. Myopathy and neuropathy in rheumatoid arthritis. A quantitative controlled electromyographic study. *J Rheumatol* 1999;**26**:2348–51.
87. Jablonska S, Blaszczyk M. Scleromyositis: a scleroderma/polymyositis overlap syndrome. *Clin Rheumatol* 1998;**17**:465–7.
88. Garton MJ, Isenberg DA. Clinical features of lupus myositis versus idiopathic myositis: a review of 30 cases. *Br J Rheumatol* 1997;**36**:1067–74.
89. Zimhony O, Sthoeger Z, Ben David D, Bar Khayim Y, Geltner D. Sjögren's syndrome presenting as hypokalemic paralysis due to distal renal tubular acidosis. *J Rheumatol* 1995;**22**:2366–8.
90. Lanham JG, Elkon KB, Pusey CD, Hughes GR. Systemic vasculitis with asthma and eosinophilia: a clinical approach to the Churg-Strauss syndrome. *Medicine (Baltimore)* 1984;**63**:65–81.
91. De Vlam K, De Keyser F, Goemaere S, Praet M, Veys EM. Churg-Strauss syndrome presenting as polymyositis. *Clin Exp Rheumatol* 1995;**13**:505–7.
92. Hall FC, Krausz T, Walport MJ. Idiopathic eosinophilic myositis. *QJM* 1995;**88**:581–6.
93. Pickering MC, Walport MJ. Eosinophilic myopathic syndromes. *Curr Opin Rheumatol* 1998;**10**:504–10.
94. Agrawal BL, Giesen PC. Eosinophilic myositis. An unusual cause of pseudotumor and eosinophilia. *JAMA* 1981;**246**:70–1.
95. Strosberg JM, Lava NS. Polymyositis with eosinophilia and mononeuropathy multiplex. *J Rheumatol* 1981;**8**:523–6.
96. Stark RJ. Eosinophilic polymyositis. *Arch Neurol* 1979;**36**:721–2.
97. Trueb RM, Becker-Wegerich P, Hafner J, Wuthrich B, Burg G. Relapsing eosinophilic perimyositis. *Br J Dermatol* 1995;**133**:109–14.
98. Alonso-Ruiz A, Calabozo M, Perez-Ruiz F, Mancebo L. Toxic oil syndrome. A long-term follow-up of a cohort of 332 patients. *Medicine (Baltimore)* 1993;**72**:285–95.
99. Williamson BL, Tomlinson AJ, Hurth KM, Posada de la Paz M, Gleich GJ, Naylor S. Rapid HPLC screening method for contaminants found in implicated L-tryptophan associated with eosinophilia myalgia syndrome and adulterated rapeseed oil associated with toxic oil syndrome. *Biomed Chromatogr* 1998;**12**:255–61.
100. Kaufman LD. The eosinophilia-myalgia syndrome: current concepts and future directions. *Clin Exp Rheumatol* 1992;**10**:87–91.
101. Hertzman PA, Blevins WL, Mayer J, Greenfield B, Ting M, Gleich GJ. Association of the eosinophilia-myalgia syndrome with the ingestion of tryptophan. *N Engl J Med* 1990;**322**:869–73.
102. Swygert LA, Maes EF, Sewell LE, Miller L, Falk H, Kilbourne EM. Eosinophilia-myalgia syndrome. Results of national surveillance. *JAMA* 1990;**264**:1698–703.
103. Belongia EA, Hedberg CW, Gleich GJ, White KE, Mayeno AN, Loegering DA, et al. An investigation of the cause of the eosinophilia-myalgia syndrome associated with tryptophan use. *N Engl J Med* 1990;**323**:357–65.
104. Mayeno AN, Gleich GJ. Eosinophilia-myalgia syndrome and tryptophan production: a cautionary tale. *Trends Biotechnol* 1994;**12**:346–52.
105. Herrick MK, Chang Y, Horoupian DS, Lombard CM, Adornato BT. L-tryptophan and the eosinophilia-myalgia syndrome: pathologic findings in eight patients. *Hum Pathol* 1991;**22**:12–21.
106. Verity MA, Bulpitt KJ, Paulus HE. Neuromuscular manifestations of L-tryptophan-associated eosinophilia-myalgia syndrome: a histomorphologic analysis of 14 patients. *Hum Pathol* 1991;**22**:3–11.
107. Hertzman PA, Clauw DJ, Kaufman LD, Varga J, Silver RM, Thacker HL, et al. The eosinophilia-myalgia syndrome: status of 205 patients and results of treatment 2 years after onset. *Ann Intern Med* 1995;**122**:851–5.

Chapter 37

Myotonic Dystrophy

Richard T. Moxley, Emma Ciafaloni, and Debra Guntrum

CLINICAL PRESENTATION AND PHENOTYPES

Brief Overview of History and Epidemiology

In 1909, Hans Steinert (Leipzig, Germany) described the typical pattern of muscle weakness and myotonia in adult-onset myotonic dystrophy type 1 (DM1). Over subsequent years, clinical studies demonstrated dominant inheritance and a wide spectrum of variable multisystem manifestations (skeletal and smooth muscle, heart, brain, eyes, respiratory, gastrointestinal, immune, and endocrine systems).[1–4] In 1992, researchers discovered that DM1 results from an unstable trinucleotide repeat expansion (CTG) in the dystrophia myotonica-protein kinase (*DMPK*) gene located on chromosome 19q13.3.[5–7] This discovery stimulated widespread DNA testing of clinically diagnosed patients felt to have Steinert's disease. Interestingly, in 1994, clinical researchers in the United States and Germany identified kindreds of patients with dominantly inherited myotonia, cataracts, and proximal weakness without an abnormal expansion of the CTG repeat in the *DMPK* gene and without linkage to hereditary myotonic disorders caused by mutations of the skeletal muscle chloride or sodium channels.[8,9] They proposed two names for this newly identified disorder: myotonic dystrophy type 2 (DM2)[8] and proximal myotonic myopathy (PROMM).[9] In 1998, genetic mapping linked both disorders to chromosome 3q21,[10] and in 2001, researchers discovered that DM2/PROMM resulted from an unstable CCTG repeat expansion in intron 1 of the *CNBP/ZNF9* gene at locus 3q21.3.[11]

The relatively recent discoveries of the assays to identify the mutations responsible for DM1 and DM2 create a challenge to determining the exact prevalence of DM1 and DM2. DNA confirmation of DM1 has only become widely available since the mid 1990s,[5–7] and for DM2 since 2003.[10–12] Prior estimates of the prevalence of myotonic dystrophy relied solely upon clinical diagnosis. Study populations primarily involved symptomatic adults and probably included a mixture of patients with DM1 and DM2. Patients with minimal or no symptoms are unlikely to have participated. Within these limitations, prior studies report a range of prevalence for myotonic dystrophy of 5 to 20 per 100,000.[1] There are clusters of higher prevalence in the Basque region of Spain, northern Sweden,[13] Istria region of Croatia,[14] and the Saguenay-Lac St. Jean region of Quebec, Canada (158–189 per 100,000).[15–17] Few studies of the prevalence of congenital myotonic dystrophy type 1 (CDM) and childhood DM1 are available. However, a recent nationwide study of CDM in Canada calculated the prevalence to be approximately 2.1/100,000.[18] This estimate is lower than estimates in previous non-population-based studies in Sweden, Spain, and Britain, and probably reflects a stricter definition of CDM. The definition used for CDM in the Canadian study was: symptoms of DM1 < 30 days after birth (hypotonia, feeding and/or respiratory difficulty, required hospitalization > 72 hours) and positive DNA testing.[18] Other reports classifying children with positive DNA testing for DM1 have defined CDM as an infant < 1 year of age with symptoms consistent with DM1; childhood DM1 (ChDM) as a child between 1–10 years of age with symptoms; and older teens and adults as the adult group.[2] More epidemiological studies are necessary in DM1 to establish the rate of appearance and progression of disease manifestations, especially in children.

Until recently no estimates of prevalence for DM2 were available. In 2011, clinical researchers in Finland published findings indicating a higher prevalence of DM2 in Europeans than most clinicians suspected. In a large study of over 5000 anonymous blood donors, they observed a higher prevalence for DM2 than DM1.[19] The higher prevalence of DM2 identified in this Finnish population emphasizes the wide spectrum of phenotypes and points out the diagnostic challenge that exists due to the

B.T. Darras, H. Royden Jones, Jr., M.M. Ryan & D.C. De Vivo (Eds): Neuromuscular Disorders of Infancy, Childhood and Adolescence, Second edition.
DOI: http://dx.doi.org/10.1016/B978-0-12-417044-5.00037-8

TABLE 37.1 Summary of Genetics and Core Clinical Features of Myotonic Dystrophy (DM1 and DM2) in Mild to Moderately Affected Adult Patients (mother of congenital DM1 patient or late teenage onset DM1 patient as well as the father or mother of childhood onset DM1 patient may have these features)

	Adult Onset Myotonic Dystrophy Type 1 (DM1)	Adult Onset Myotonic Dystrophy Type 2 (DM2)
Genetics		
Inheritance	Autosomal dominant	Autosomal dominant
Anticipation	Pronounced	Exceptionally rare
Congenital form	Yes	No
Chromosome	19q13.3	3q21.3
Locus	*DMPK*	*CNBP*
Expansion mutation	$(CTG)_n$	$(CCTG)_n$
Location of the expansion	3′ untranslated region	Intron 1
Core features: ***Mild to moderately affected patients***		
Clinical myotonia	• Usually occurs in the distal limb (grip and intrinsic hand muscles); • Can manifest in the tongue, jaw, eye, and smooth muscles; • Percussion myotonia in thenar and forearm muscles often present; • Myotonia may be absent in some patients.	• Not as prominent as in DM1; • Often first appears intermittently with hand grip; • May be limited mainly to percussion of wrist extensor muscles; • May be absent in some patients.
Muscle weakness	• Forearm finger flexors; foot dorsi- and plantar flexors; and facial muscles; • Ptosis; • Laryngeal (dysarthria); • Occasional extraocular weakness; • Weakness of shoulder girdle and proximal upper limb, neck (flexors > extensors), and knee flexors and extensors more prominent in later stages as disease progresses.	• Involvement of distal and facial muscles is usually absent; • Initial weakness is in proximal hip girdle and neck (flexors > extensors) muscles; • Occasional mild ptosis less common than in DM1; • Calf muscle hypertrophy is common.
Muscle wasting	• Significant atrophy is typically apparent in patients affected moderately; • Anterior neck; • Masseter and temporalis; • Distal forearm; • Intrinsic hand; • Distal lower extremity.	• Significant muscle atrophy is typically mild and involves anterior neck and hip and shoulder girdle.
Cataracts	• Iridescent, posterior capsular opacities, typically before age 50 years.	• Similar in appearance to DM1 but less commonly seen on initial exam.

variation in different symptoms that occur in DM2. For example, clinical findings such as myotonia on physical examination or even on electromyographic testing may be absent, limiting the ability of myotonia screening alone as a means to identify individuals with DM2.[20,21] Whole blood DNA testing in individuals at known risk for DM1 or DM2 or in individuals presenting with one of the common manifestations (Tables 37.1 and 37.2) is needed to determine more exactly the actual prevalence of individuals with abnormal repeat expansions of DMPK and *CNBP/ZNF9* genes.

Mode of Inheritance and Prevalence

DM1 and DM2 are autosomal dominant. Anticipation (earlier onset of more severe symptoms in successive generations) occurs primarily in DM1 (see Figure 37.1).[1,2,22,23] Congenital myotonic dystrophy

TABLE 37.2 Summary of Multisystem Manifestations of Myotonic Dystrophy (DM1 and DM2) in Mild to Moderately Affected Adult Patients

	Adult Onset Myotonic Dystrophy Type 1 (DM1)	Adult Onset Myotonic Dystrophy Type 2 (DM2)
Multisystem manifestations: ***Mild to moderately affected patients***		
Brain	• Visual spatial deficits; • Reduced executive function (trouble organizing & staying on task, reduced goal directed action); • Apathy.	• Similar visual-spatial and executive function deficits to those present in DM1.
Heart	• Conduction defects, particularly heart block and arrhythmias.	• Significant disturbances in conduction much less common than in DM1.
Respiratory	• Reduced forced vital capacity; • Obstructive and central sleep apnea; • Weak cough; • Microatelectasis; • Frequent respiratory infections.	• Obstructive sleep apnea.
Anesthesia	• Increased frequency of respiratory insufficiency/ failure following general anesthesia; • Paradoxical reaction to depolarizing muscle relaxant medications; • Hypersensitivity to opioids and sedative medications; • Neuroleptic malignant syndrome; • Failed respiratory wean; • Arrhythmias.	• Limited information is available to determine if there is a significant and increased risk of general anesthesia; • Recommend careful monitoring in postoperative period until more information is published.
Hypersomnia and fatigue	• Excessive daytime sleepiness; • Obstructive and central sleep apnea; • CNS and muscle-related fatigue.	• Excessive daytime sleepiness is not as prominent as in DM1; • Obstructive sleep apnea; • CNS and muscle related fatigue.
Endocrine	• Gonadal insufficiency; • Low testosterone; • Erectile dysfunction; • Insulin resistance; • Hyperlipidemia; • Hypothyroidism; • Decreased growth hormone release.	• Gonadal insufficiency; • Low testosterone; • Erectile dysfunction; • Insulin resistance; • Hyperlipidemia; • Hypothyroidism.
Gastrointestinal	• Dysphagia; • Gastroesophageal reflux; • Gastroparesis; • Small intestinal dysmotility (diarrhea, abdominal cramps, bloating, pain); • Colonic dysmotility (pain, "spastic colon," constipation); • Pseudo-obstruction; • Anorectal dysfunction (anal incontinence, constipation, fecal impaction).	• Limited information is available; • Manifestations often less severe than DM1 and can include: • Dysphagia; • Gastroesophageal reflux; • Abdominal pain; • Constipation.
Muscle pain	• Posterior neck, trapezius, anterior lateral chest wall, and upper and lower back; • Can worsen with exercise and cold temperature.	• Often a major symptom, especially the arms and upper and lower back; • Fluctuates in duration, location, and intensity; • Can worsen with exercise and cold temperature; • Aches and stiffness; • Pain on muscle palpitation.
Pregnancy	• Vital capacity declines and shortness of breath occurs often in the 3rd trimester; • Frequent miscarriage; • Failed maintenance of pregnancy; • Polyhydramnios; • Prolonged 1st stage of labor; • Placenta previa; • Postpartum hemorrhage;	• Limited information is available to determine if there is significant risk of complications during pregnancy and delivery; • Weakness and stiffness may worsen during pregnancy and improve following delivery.

(*Continued*)

TABLE 37.2 (Continued)

	Adult Onset Myotonic Dystrophy Type 1 (DM1)	Adult Onset Myotonic Dystrophy Type 2 (DM2)
Liver	• Elevated liver enzymes, especially gamma glutamyl transferase; • Decreased to low normal albumin; • Gallstones and diffuse fatty liver.	• Similar to DM1.
Immune System	• Decreased immunoglobulins, primarily IgG and IgM.	• Similar to DM1.
Eye	• Cataract; • Ptosis; • Exposure keratitis; • Retinal pigmentary abnormalities; • Decreased intraocular pressure; • Abnormal dark adaptation; • Epiretinal membrane; • Oculomotor abnormalities.	• Cataract.
Pilomatricoma	• Asymptomatic, slow growing, firm, superficial mass (0.15–4.0 cm) usually in the head/neck region.	• Not reported.

(CDM) occurs almost exclusively through maternal transmission (see Figures 37.2 and 37.3). In contrast, both mothers and fathers with DM1 can have children with childhood onset of DM1.[1,2] However, only 5–10% of DM1 patients have onset before their mid-teens.[1,2] DM2 occurs neither in infancy nor childhood, although teenage onset of myotonic symptoms can occur in some DM2 patients carrying concomitant mutations in the gene for the skeletal muscle chloride channel.[24]

Clinical Presentation and Phenotypes

Tables 37.1 and 37.2 summarize the genetics, core clinical features, and multisystem manifestations typically observed in the initial evaluation of adults with either mild or moderate signs and symptoms of DM1 and DM2. Figure 37.4 outlines the workup and Table 37.3 presents general guidelines for patient care. Childhood and infant onset symptoms occur only in DM1. Figures 37.5 and 37.6 outline the evaluation and workup of DM1 patients who present in infancy and childhood. Table 37.4 summarizes the clinical problems they often have as well as their management.

Myotonic Dystrophy Type 1 (Steinert's Disease): Age of Onset and Clinical Presentation

DM1 can present from birth to late adult life. Teenage or later onset is typical for 90% of individuals with DM1, and in this phenotype one of the core features (Table 37.1), such as myotonia, is often the initial sign or symptom.[1,3,4] However, one of the multisystem manifestations (Table 37.2) can be the first complaint. In the absence of an established family history of DM1, care providers must have a high degree of suspicion in order to avoid overlooking the diagnosis of DM1. Because of the variable nature of these multisystem features, such as cataracts,[1,3] gastrointestinal complaints,[25–30] and sleep disturbances,[31–37] care providers often do not consider DM1 as the cause and a delay in diagnosis occurs. Early onset in infants or children often shortens the time for diagnostic testing for DM1. Symptoms of CDM appear in the newborn period, can be life threatening, and manifestations include generalized weakness and hypotonia, respiratory failure/insufficiency, feeding difficulty, clubfoot deformity, increased risk of germinal matrix/intracerebral hemorrhage, and eventration of the diaphragm[1–4,18,38–40] (Table 37.4, Figure 37.5). Childhood myotonic dystrophy (ChDM) may be more difficult to diagnose if there is no diagnosis of an affected parent or no positive family history of DM1[1–4] (Table 37.4, Figure 37.6). ChDM is often defined as having onset of DM symptoms <10 years, an uneventful prenatal and neonatal history, normal development within the first year of life, and increasing manifestations in the toddler years.[41] ChDM typically causes intellectual deficiency, difficulty with speech or hearing, delayed bowel-bladder training, clumsiness, or rarely cardiac dysrhythmia or postoperative apnea.[2,39,42–49]

The challenge in reaching a rapid diagnosis of DM1 often comes in thinking of this disease. DM1, especially in adults and mild cases in childhood, has a wide

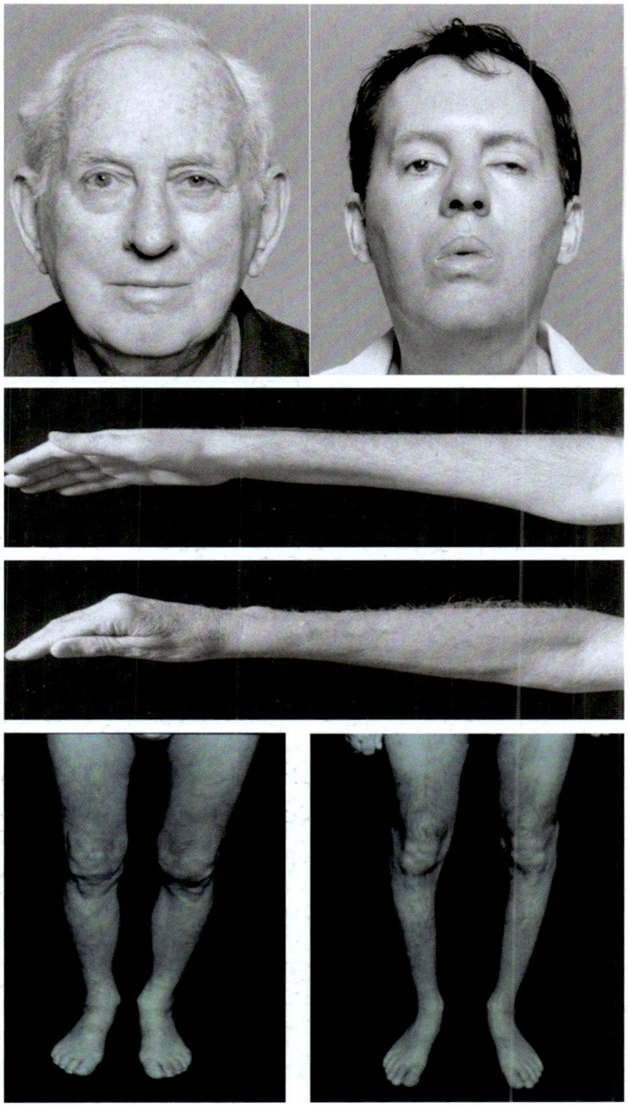

FIGURE 37.1 The uppermost panel of this three panel figure shows photographs of a father and son both afflicted with myotonic dystrophy type 1 (DM1). The left photo is of the father (age 77 years) and the right is of the son (age 41 years). Both patients have had cataract surgery, right eye lens implant for the father and left eye for the son. Ptosis, bifacial weakness, wasting of the sternocleidomastoid, temporalis, and masseter muscles as well as frontal balding is apparent in the son and difficult to detect in the father. The middle panel shows the forearms of both father and son. There is greater severity of forearm muscle wasting (forearm flexor compartment being more wasted than forearm extensors) in the son (lower forearm photo) compared to the father (upper forearm photo). The bottom panel shows the lower extremities of the father (left) and son (right). The son shows more wasting of the leg musculature. The shaved areas on the anterior lateral regions of both thighs in the son represent the sites for needle muscle biopsies that were obtained in association with a research study in which the patient participated. *Reprinted from Moxley (1992)*[3] *with permission from Elsevier.*

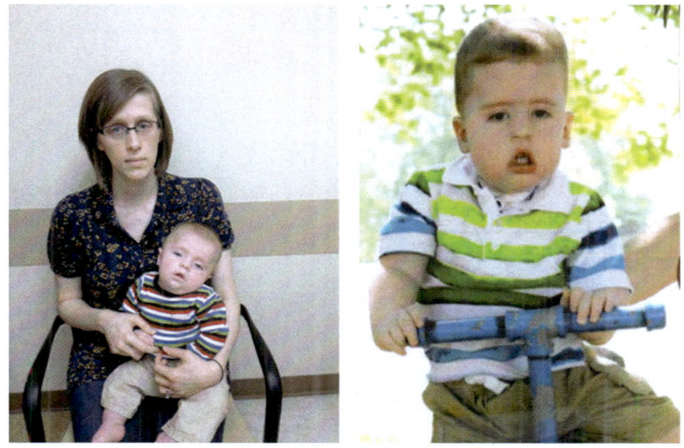

FIGURE 37.2 Mother and her son are the same individuals described in Case Example 37.1 in the text (left panel). At the time of this photograph, she was 27 years old and he was 8 months of age. The information in Case Example 37.1 also includes discussions of video clips of myotonia in the mother and floppiness in the infant. The right panel shows a 4 year old with congenital DM1 riding his tricycle This child is described in Case Example 37.2 in the text.

FIGURE 37.3 Mother, who has mild to moderate myotonic dystrophy type 1, with her two daughters, both of whom have congenital DM1. At the time of the photograph, the mother was 34 years old and the daughters were 9 and 5 years old. The mother is now 49 years old, performs all activities of daily living well, cares for the daughters, has mild facial and distal limb weakness, and has left bundle branch block on her electrocardiogram. The older daughter is 24 years old, ambulatory, goes horseback riding each week, has no difficulty with feeding, lives with her parents, performs activities of daily living, and participates in housekeeping activities. Her electrocardiogram is normal. The younger, more severely affected daughter with severe CDM is 20 years old, still has a tracheostomy, and receives G-tube feeding and night-time ventilation. She is ambulatory, using ankle-foot orthoses during the day, and is off ventilator support. She climbs stairs and rides a bike. She attends a day program and receives weekly physical therapy, occupational therapy, and speech therapy. The younger daughter has regular follow-up care with pulmonary, cardiology, and otolaryngology consultants. Both daughters along with their mother have regular neuromuscular physician evaluations. *Reproduced from Moxley and Meola (2000),*[4] *with permission from S. Karger AG.*

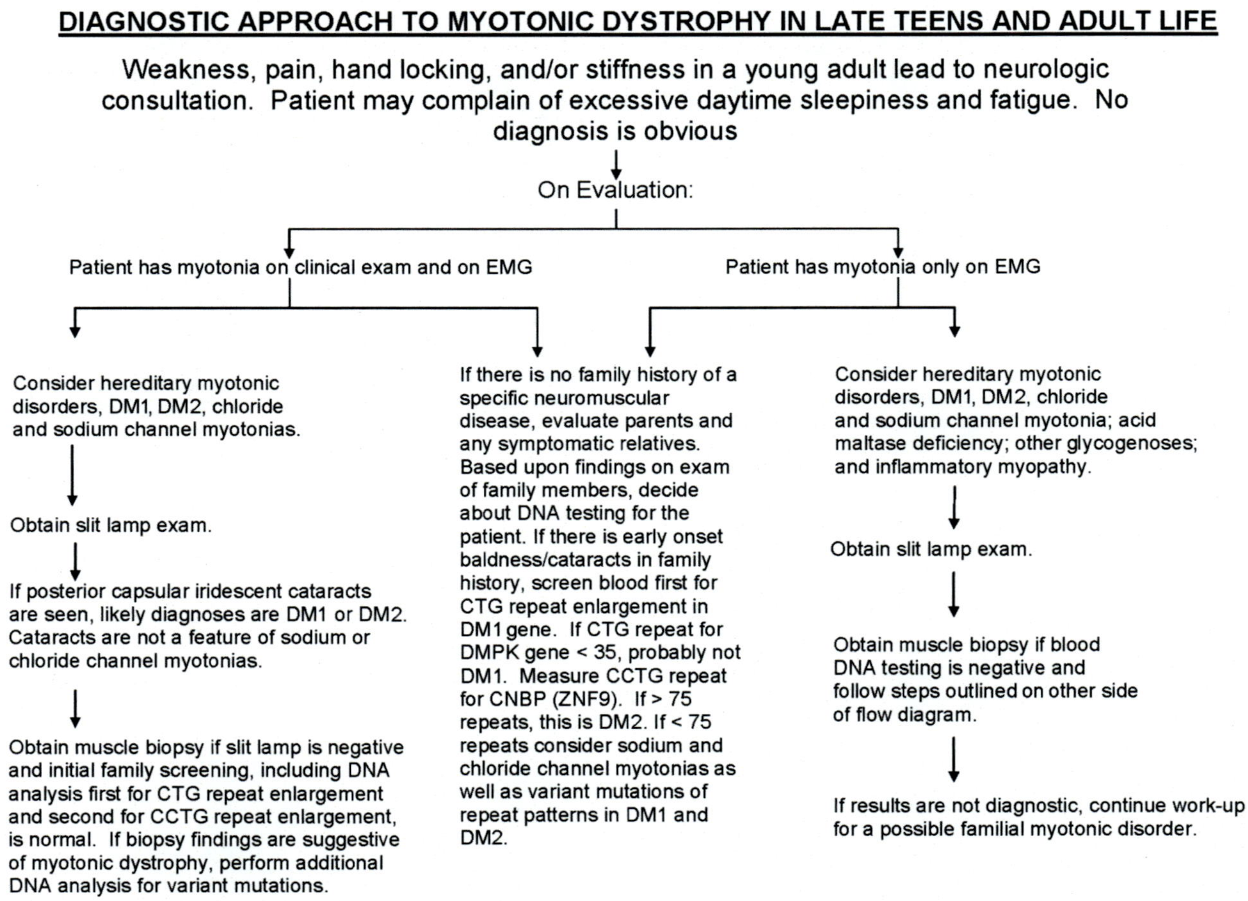

FIGURE 37.4 Diagnostic approach to myotonic dystrophy in late teens and adult life.

spectrum of clinical manifestations (gastrointestinal, brain, heart, eyes), many not clearly related to skeletal muscle. They can present in isolation, staggered in time of onset and severity, and separate from the subsequent onset of skeletal muscle signs and symptoms. DM1 can present in late life with only cataracts, baldness, or occasionally with heart block. Weakness and myotonia may be absent or very mild. In the second and third decades, DM1 can present with variable grip myotonia, trouble holding items, ankle weakness, or intermittent slurring of speech. Mild weakness of the face, long finger flexors, intrinsic hand, and foot dorsiflexor muscles is usually present, as is myotonia. The myotonia typically occurs following a strong grip or after percussion of the thenar and forearm wrist extensor muscles. An affected parent of an infant or child with DM1 may have only mild disease manifestations. These clinical findings may be subtle, for example in an undiagnosed pregnant mother with DM1; they may only be detected on a focused clinical evaluation to detect myotonia after she gives birth to a baby with severe CDM. Not uncommonly the mother has a history of decreased fetal movements and polyhydramnios followed by failed labor and an urgent cesarean section. Placenta previa and a history of miscarriage are common.[1–4] Because of the marked variation in severity and spectrum of manifestations of DM1, birth of an infant with CDM often comes with a sudden impact. Abruptly, three generations of a family become aware that they are affected by a dominantly inherited, progressive disease that at present has no curative treatment. Family counseling and emotional support to the mother and father as well as emergent medical treatment for the infant become essential responsibilities for care. Grip and percussion myotonia are not present in infancy and early childhood in DM1, and are sometimes difficult to detect on clinical examination in adults with marked wasting of forearm and intrinsic hand muscles.[1–4] Dysarthria, difficulty swallowing, gastrointestinal dysmotility (frequent bowel movements, intermittent constipation), hypersomnia, cognitive-behavioral problems, sleep apnea, decreased vital capacity, and cardiac conduction disturbances develop over subsequent years.[1–4,50] Further longitudinal

TABLE 37.3 General Guidelines for Care in Adults with Myotonic Dystrophy (DM1 and DM2)

System	Evaluation	Approaches to Care
Skeletal muscle	• Serial monitoring strength, timed function testing, and grip myotonia; • Evaluate balance/risk of falls and refer to PT/OT as necessary; • Evaluate for assistive devices, scooter, wheelchair as required.	• Antimyotonia treatment as needed; • Ankle-foot orthoses, cane, walker, other assistive devices.
Brain	• Psychological, educational, and counseling evaluations as needed; • Structural imaging as required; • Routinely assess for sleep disturbances and respiratory insufficiency.	• Psychological, educational, and other counseling treatment and services; • CNS medications (e.g. stimulants) as necessary under close supervision of care providers.
Heart	• Yearly electrocardiograms; • Cardiology consultation for symptomatic patients and long-term follow-up care.	• Prompt treatment of heart block or atrial tachyarrhythmias with implantable device for ablation therapy; • Medications as appropriate for left ventricular dysfunction (ACE inhibitors/angiotensin receptor blockers and beta blockers).
Respiratory	• Serial monitoring of sitting and supine respiratory function, including forced vital capacity; • Polysomnography and pulmonary medicine consultation as required.	• Yearly immunizations; • Noninvasive or invasive ventilation as required; • Serial evaluation by pulmonary medicine and sleep consultation as required.
Anesthesia	• Before elective surgery, have anesthesia consultation and pulmonary medicine evaluation; • ECG reviewed by cardiology consult; • Discuss known risks and any previous anesthesia-related problems.	• Use of regional anesthesia over general when appropriate; • Use of nondepolarizing muscle relaxants; • Reduce use of opioids; • In general anesthesia, protection of the airway and minimizing aspiration, careful cardiac monitoring, and extensive postoperative monitoring (at least 24 hours).
Hypersomnia and Fatigue	• Polysomnograms; metabolic-endocrine screens; • Psychological, educational, and sleep consultant evaluations.	• If obstructive sleep apnea or respiratory insufficiency found, use of continuous positive airway pressure (CPAP) or bilevel positive airway pressure (BiPAP); • Use of CNS stimulants (modafinil, methylphenidate, caffeine).
Endocrine	• Symptomatic assessment of testosterone deficiency; • Yearly lipid profile, thyroid screening, diabetes screening; • Monitor sleep disturbances.	• Hormone replacement as required; • Dietary intervention; • Medications for lipid and glucose control as needed; • Treatment for sleep disturbances as required.
Gastrointestinal	• Swallowing function evaluation; • Metabolic and endocrine screens; • Dietician, gastrointestinal consultations; • Careful assessment of bloating and signs of pseudo-obstruction.	• Gastroesophageal reflux may be treated with avoidance of late night meals, elevation of the head of the bed, and medications (proton pump inhibitors); • Constipation, diarrhea, abdominal pain, and bloating may be treated with modifying the diet to small, low fat meals; • Surgery as appropriate for gallbladder disease; • Use of cholestyramine may help alleviate diarrhea; • Small intestine and colon dysmotility may respond to antimyotonia medications.
Muscle pain	• Psychological, educational, and pain counseling; • Pain treatment consultants as required.	• Antimyotonia agents; NSAIDs; • Avoidance of opioids; • Aquatic therapy; • Other pain medication as coordinated between pain treatment consult and primary care provider.

(Continued)

TABLE 37.3 (Continued)

System	Evaluation	Approaches to Care
Pregnancy	• Obtain obstetrician and genetic consultation prior to pregnancy as appropriate; • Discuss possible complications; • Coordinate monitoring of pregnancy with other care providers, including a neonatal pediatric/ neonatology specialist; • Closely monitor respiratory function during the 3rd trimester.	• During delivery, monitor mother's ECG; • Use regional anesthesia; • Notify consultants of mother's status and request urgent evaluations as necessary.
Liver	• Monitor elevated liver enzymes serially; • Monitor for signs of gallbladder inflammation.	• Surgery to remove gallbladder as required.
Immune system	• Low IgG level noted.	• No specific treatment required.
Eye	• Cataract monitored serially; • Ptosis monitored and effect on vision assessed; • Exposure keratitis requires serial monitoring; • Retinal pigmentary changes monitored by ophthalmology; • Epiretinal membrane and oculomotor abnormalities monitored serially by ophthalmology consultant.	• Cataract removal as necessary; • Treat conservatively but consider surgery if vision impaired; • Sleep-related eye care, eye drops, salve, patches, ophthalmology consultations as needed; • Treatment directed by ophthalmology consultant; • Treatment directed by ophthalmology consultant.
Pilomatricoma	• Monitor size of mass and search for other subcutaneous masses, refer for dermatological and surgical consultations.	• Care directed by dermatologic and surgical consultants.

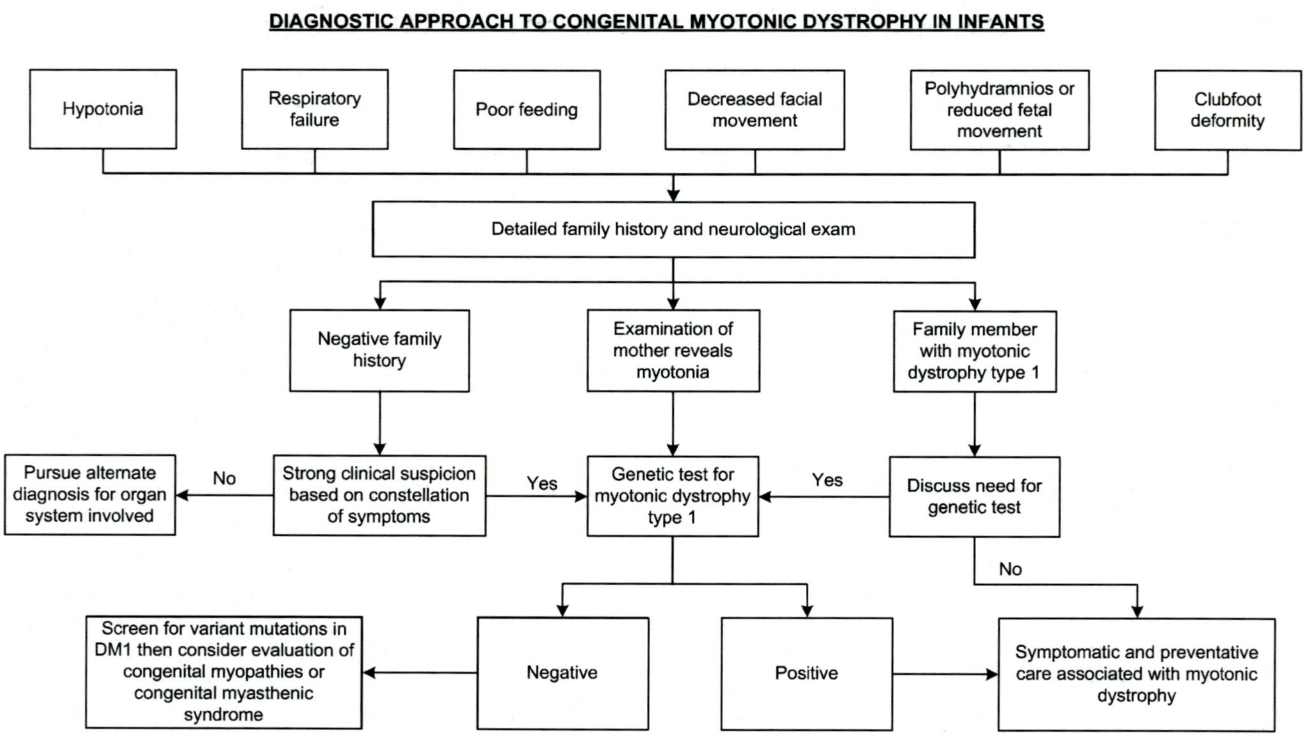

FIGURE 37.5 Evaluation of infant with decreased facial movement and one or more of these other findings: suspect congenital myotonic dystrophy.

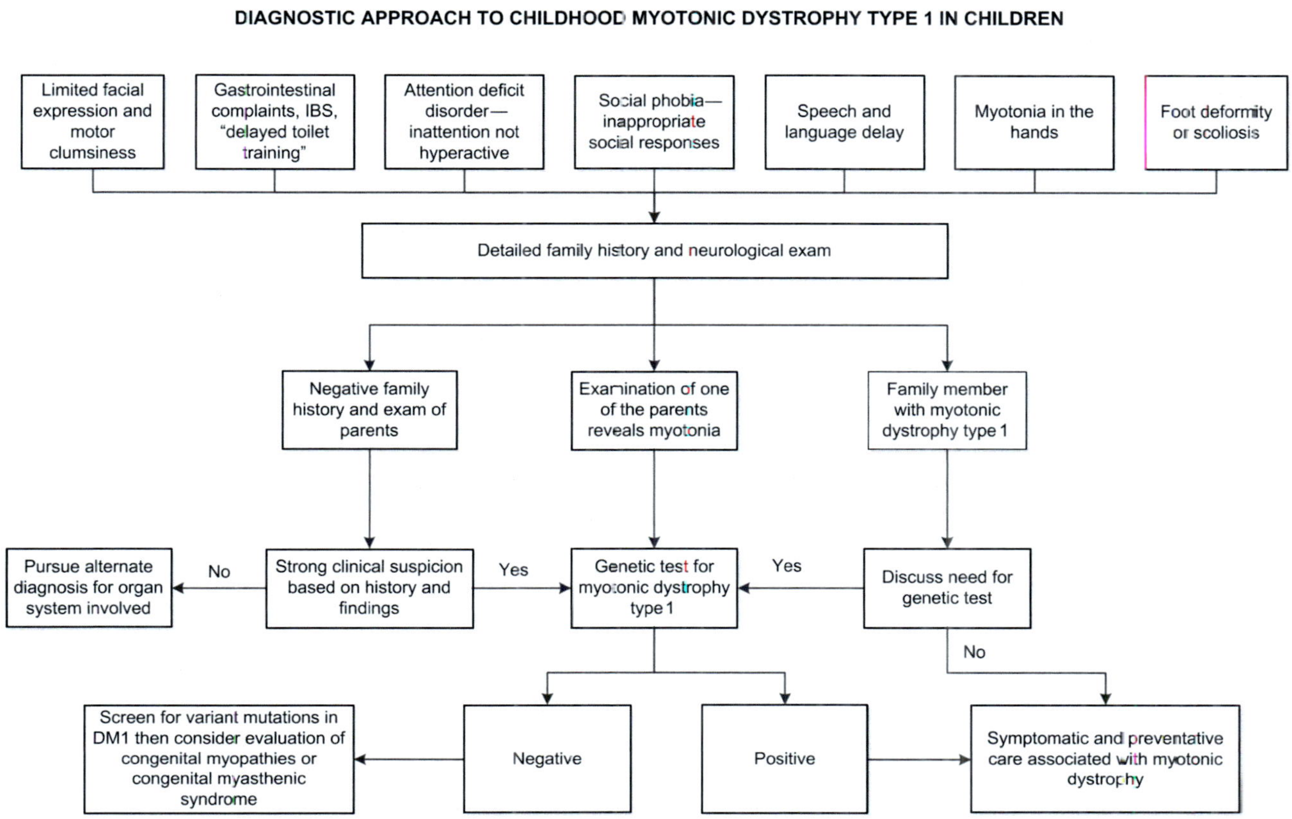

FIGURE 37.6 Evaluation of child with developmental and/or cognitive delay with one or more of these findings: suspect childhood myotonic dystrophy type 1.

studies of the rate of progression of the wide spectrum of the clinical manifestations of the different phenotypes in DM1, especially in children, are needed to supplement the limited number of prior studies.[51–56] Such studies will help extend and enhance current standards of care for DM1.[57–59] Our present estimates of mortality and end of life events are also limited.

In CDM, estimated mortality ranges between 16% and 41% in the neonatal period.[18] Contributors to reduced survival include respiratory failure and elective withdrawal of care. In a study of 838 adults with DM1 in Quebec from 1985 to 2010, there were 321 deaths with a median age of death of 56 years.[16] This study showed a significant difference between median age of death between childhood onset (median = 48.5 years), adult onset (median = 54 years), and mild phenotypes (median = 73 years). Another study of 406 DM1 patients in the United States reported 118 deaths with a median age of 54 years,[60] the majority caused by respiratory and cardiac complications, similar to previous reports.[15,61,62] Going forward, it will be important to document how mortality rates for DM1 change over the next decade as care providers implement improvements in the treatment of respiratory and cardiac manifestations.

Myotonic Dystrophy Type 2 (Proximal Myotonic Myopathy): Age of Onset and Clinical Presentation

DM2 presents in adult life, usually with myotonic stiffness or weakness (proximal legs or long finger flexors).[8,9,12,63,64] Tables 37.1 and 37.2 summarize the core features and multisystem manifestations of DM2, while Table 37.3 provides guidelines for care. Frequently, the myotonia fluctuates, being prominent only on "bad days."[12,63–65] Muscle pain similarly varies in severity over days to weeks, the cause for which is unknown.[12,65–67] It typically appears to be independent of exercise and unrelated to the severity of myotonia on clinical examination. In the initial stages of DM2, there is often mild weakness of long finger flexors, thigh flexors, neck flexors, and hip extensors along with grip myotonia.[12,65] The myotonia is most apparent following percussion of forearm extensor and thenar muscles. The myotonia that occurs after a powerful grip often has a jerky, tremulous quality that differs from the grip myotonia observed in DM1. Patients with this jerky myotonia frequently have brisk tendon reflexes and often have mild sustention tremor.

TABLE 37.4 Summary of Clinical Problems of Congenital Myotonic Dystrophy in Infants and Childhood Onset (1–10 Years Old) Myotonic Dystrophy Type 1 and Their Management

Clinical Problem	Congenital Myotonic Dystrophy (CDM)	Childhood Onset Myotonic Dystrophy Type 1
Respiratory insufficiency or failure	Neonatal intensive care, ventilator support, chest radiograph to check on elevation of diaphragm; may require long-term ventilator support via tracheostomy or noninvasive nasal ventilation; long-term coordination of care with pulmonary medicine is essential; standard care as necessary for pulmonary immaturity; need to monitor throughout infancy and childhood for aspiration pneumonitis and obstructive sleep apnea.	Occasionally respiratory insufficiency is a problem and usually related to obstructive sleep apnea. Otolaryngology consultation and evaluation for tonsil removal as well as sleep study are needed along with pulmonary medicine consultation. Older children may develop hypersomnia and/or other sleep disturbances. May require polysomnography and stimulant medication.
Feeding problems and aspiration	Frequent monitoring of respiratory status and serial checks of chest radiography; daily assessment of swallowing and monitoring for recurrent aspiration; care in neonatal intensive care unit or highly skilled nursing environment is needed; evaluate esophageal function and cough if problems persist; may require feeding gastrostomy on a long-term basis. Aggressive persistent care of atelectatic localized collapse is necessary to re-expand those areas of lung.	Children with significant facial weakness with or without high arched palate often have trouble chewing, swallowing, and speaking. Speech therapy, swallowing function evaluation, and monitoring for signs of aspiration are needed. Consider pulmonary medicine consultation.
Risk of anoxia and cerebral hemorrhage	Neonatal intensive care; serial cerebral ultrasound and/or brain MRI evaluation.	Cerebral hemorrhage is not a feature of childhood onset DM1.
General delay in development and cognitive function	Accurate evaluation of contributory factors, especially muscular and cerebral components. Long-term global cognitive deficit is common.	Attention deficit, inappropriate social behavior, autistic features, as well as learning disability can occur in isolation or combinations; requires individualized care program.
Speech problems	Anticipate early; distinguish palatal weakness, hearing difficulty, and other causes from global cognitive deficit; may need speech therapy and training in sign language, often accompanied by marked facial weakness and dysphagia. Long-term treatment typically needed and coordinated with school program over time.	May occur in children with marked facial weakness and/or high arched palate. Speech therapy is helpful.
Recurrent otitis and hearing difficulty	Active management of otitis; awareness of later onset neural hearing loss; requires continued monitoring throughout infancy and childhood.	Similar approach as with CDM; poor facial expression–mobility often delays detection of hearing loss.
Talipes-foot problems	Early diagnosis and active correction; splinting; surgery, if indicated.	See Table 37.3. Adult care.
Abdominal pain and bowel problems	Delayed bowel and bladder training is typical throughout early and mid-childhood; usually requires a specific management program coordinated with kindergarten and school programs.	Older children and teenagers often develop "irritable bowel syndrome"; role of smooth muscle and myotonia is poorly understood; response to medications is variable and not established; may require pediatric gastroenterology evaluation; trial of antimyotonia therapy deserves consideration.
School decisions	Not an issue during infancy and very early childhood. Becomes important thereafter. May need special classroom setting and vocational education in later school years; requires full assessment of functional capabilities, awareness of facial immobility and hearing and speech defects that exaggerate cognitive limitations; physical disability over time is typically not severe and function usually improves throughout childhood and early teens.	Decisions about special classroom requirements, counseling, serial neuropsychological evaluations, behavioral management, and medications for attention deficits depend upon careful evaluation of the spectrum of problems—their severity in each individual patient. Sleep disturbance and fatigue are often coexisting complaints that require evaluation and treatment.
Restriction of physical activity	Activity should not be restricted based solely upon the diagnosis of DM1. Individualized exercise programs are often necessary. Yearly ECG is useful to identify signs of cardiac conduction disturbances. Arrhythmias are relatively infrequent in infants and children.	As with CDM. Tailor physical activity to the physical limitations in each patient and other non-DM1 related medical problems. Monitor ECG yearly and obtain cardiology evaluation if potential cardiovascular symptoms occur.

The prominent facial weakness and the wasting of facial, forearm, and distal leg muscles that are typically seen in DM1 are not observed in DM2. Even in the late stages, typical DM2 patients have relatively mild muscle wasting with the occasional exception of the gluteal and proximal thigh muscles. Occasionally, patients with DM2 have calf muscle hypertrophy.[8,9,12,63] Early onset of cataracts that require surgery is common. Later in life, patients may present with complaints of difficulty climbing stairs or arising from a squat, often attributed to "arthritis" or "old age." Myotonia and muscle pain are less prominent when onset is late. In contrast to DM1, respiratory failure and cardiac conduction disturbances are less common in DM2.[1,63,65] Difficulty swallowing, gastrointestinal dysfunction, and cognitive disturbances do occur in DM2, but their frequency seems less than in DM1.[63–65,68] Additional investigation of the natural history of DM2 is necessary to establish the actual frequency of these complaints and their relationship, if any, to the severity of muscle weakness and/or myotonia. There are no large-scale studies evaluating the disease manifestations that contribute to mortality in DM2. Future research is necessary to determine if the causes of death in DM2 differ significantly from those of the general population.

MOLECULAR GENETICS

DM1: Normal Gene Product

Table 37.1 summarizes information on the dystrophia myotonica protein kinase *(DMPK)* gene and the associated mutation that leads to DM1. The protein encoded by this gene is a serine-threonine kinase closely related to other kinases that interact with members of the Rho family of small GTPases. Substrates for this enzyme include myogenin, the beta-subunit of the L-type calcium channels, and phospholemman. The serine-threonine kinase appears to function normally in patients with DM1. The 3′ untranslated region of the *DMPK* gene in normal leukocyte DNA contains 5 to 37 copies of a CTG trinucleotide. This CTG repeat is unstable in DM1 and may range in length from 50 to 4000 repeats in the circulating blood of different patients. In asymptomatic or "premutation carriers" of the unstable CTG repeat in the *DMPK* gene, the expansion size is 30 to 50 repeats.[23,69] The repeat size becomes unstable once it exceeds 50 CTG repeats.[23]

DM2: Normal Gene Product

Table 37.1 summarizes information on the *CNBP* gene (often termed the zinc finger 9 protein gene, *ZNF9* gene) that is responsible for DM2. *CNBP* encodes cellular nucleic acid-binding protein. Its role in humans is not clearly understood, but it influences embryonic development. It appears to function normally in DM2, in which there is an unstable expansion of the CCTG repeat in intron 1 of the *CNBP* gene. The normal CCTG repeat structure is approximately 10 to 20 repeats. In carriers of the premutation, there are 22 to 33 uninterrupted CCTG repeats. These premutation carriers are asymptomatic and are unlikely ever to show symptoms. Symptomatic DM2 patients have CCTG repeat lengths ranging from 75 to 11,000 repeats (mean 5000 CCTG repeats)[23,69]

DM1 and DM2: Genotype-Phenotype Correlations

In DM1, there is a rough correlation between the length of the abnormal CTG expansion in circulating leukocyte DNA and the severity of disease, with exceptions occurring frequently in clinical practice.[1,23,53–56,59] Age of onset of DM1 shows a correlation. Individuals having 50 to 100 CTG repeats in leukocyte DNA typically develop mild symptoms later in life.[23,53,55,56,69] Patients with 150 to 1000 CTG repeats often have manifestations typical of adult onset DM1 as summarized in Tables 37.1 and 37.2. Individuals with over 1000 repeats are likely to have congenital or childhood forms of DM1, although a number of these patients have CTG expansions that range between 150 and 1000 repeats.[38,43–45]

Published reports of genotype-phenotype correlations have made the assumption that most DM1 patients have continuous uninterrupted abnormally expanded tracts of CTG repeats in the 3′ untranslated region of the *DMPK* gene; however, recent reports point out the limitations in this assumption.[23,70–72] The most recent report estimates that approximately 4.8% of DM1 patients have CCG interruptions within the 3′-end of the CTG motif at the *DMPK* gene locus.[70] This study used bidirectional triplet primed PCR (TP-PCR) and sequencing to identify the variant expansions.[23,70] Of the 5 patients identified, 3 tested negative for an abnormal CTG expansion using a standard DNA analysis of the *DMPK* gene by routine long-range PCR.[70] More investigations of variant mutations affecting the CTG repeat motif of the *DMPK* gene locus are necessary before making firm conclusions about their contribution to phenotypic variation in DM1. However, these findings support the use of bidirectional TP-PCR in the detection of "variant" DM1 CTG repeat expansions and its inclusion in the routine diagnostic DNA testing to screen for DM1.[23]

In DM2 the abnormal length of the CCTG repeat in circulating blood spans a wider range of size than in DM1 and does not correlate with the clinical severity.[23,69]

Both DM2 and DM1 show somatic instability of the mutant alleles. The size of the repeat expansion increases over time, across generations, and shows somatic mosaicism, even in post-mitotic tissue.[69] For example, CTG repeat lengths in skeletal muscle, heart, and brain of DM1 patients often are 5- to 10-fold longer than in their circulating leukocyte DNA.[69,73–76] Intergenerational repeat variations and somatic mosaicism result from contractions and expansions due to defects in DNA replication, DNA repair, and recombination.[77–81] Studies from population-based mathematical modeling of DM1 mutant alleles from leukocyte DNA demonstrate a bias toward expansion. This bias appears to be due to expansion and contractions that are coupled to DNA repair or transcription rather than dependent on DNA replication.[81]

Disease Mechanisms for DM1 and DM2

DM1 and DM2 are RNA-mediated Diseases

While the fundamental mechanisms that account for the abnormal expansion of the CTG repeats in the *DMPK* gene and CCTG repeats in the *CNBP* gene that cause DM1 and DM2 remain elusive, strong evidence supports the hypothesis that both disorders are RNA-mediated diseases with many similarities in their molecular pathomechanisms. The comments that follow come largely from two selected reviews that describe the disease mechanisms of DM in more detail.[69,82]

DM1 results from the abnormally expanded CTG repeat in exon 15. Expression of the mutated allele produces mRNA containing expanded CUG repeats. The mutant RNA containing the expanded CUG repeats forms a double-stranded hairpin loop that is highly stable and retained in the nucleus.

An expansion of the CCTG repeat in intron 1 of the *CNBP* gene causes DM2. As occurs with the normal-sized CCUG repeat, editing of mutant pre-mRNA removes the expanded CCUG repeat from the mutant *CNBP* pre-mRNA allele. Expanded CCUG repeats are absent from the *CNBP* mRNA. However, the expanded CCUG remnant of intron 1 is highly resistant to degradation due to formation of a double-stranded hairpin structure similar to that adopted by the expanded CUG repeat in mutant DM1 mRNA.

In both DM1 and DM2, the abnormally expanded pathogenic RNA remains in the nucleus, where it binds nuclear regulatory proteins, especially proteins in the muscleblind-like (MBNL) family such as MBNL1, and forms ribonuclear inclusions. Sequestration of MBNL1 by this expanded mutant RNA occurs in both DM1 and DM2 skeletal muscle and other target tissues, hampering the normal function of MBNL1 and leading to misregulation of alternative splicing of a large number of pre-mRNAs.[69,82] A number of studies in patients with DM1 and DM2, as well as studies in animal models, support the toxic RNA gain-of-function model summarized above.[69,82] However, several other factors contribute to complex pathomechanisms that likely participate in one or both forms of myotonic dystrophy. These factors include repeat-associated non-ATG translation (RAN),[83–85] bidirectional transcription,[71,80,85–87] aberrant DNA methylation,[88,89] microRNA (miRNA) dysregulation,[90–93] abnormal regulation of protein translation,[94] and other alterations.[69,95] Figure 37.7 provides a schematic of the different currently hypothesized pathomechanisms.[69]

Human Studies Supporting RNA-mediated Pathomechanism for DM1 and DM2

Skeletal muscle, heart, and brain from patients with DM1 and DM2 have abnormal ribonuclear inclusions containing mutant mRNA with abnormally expanded CUG repeats that sequester nuclear regulatory proteins, specifically MBNL1, MBNL2, and CELF (CUG BP).[69,82] Studies in skeletal muscle of DM1 patients demonstrate toxic effects of the mutant mRNA on the regulation of alternative splicing of pre-mRNAs of the insulin receptor[96] and skeletal chloride channel.[97–100] The defect in splicing leads to an imbalance in the synthesis of the correct isoforms of the insulin receptor and chloride channel with an excess of fetal and a deficiency of adult isoforms.[69,82] This imbalance results primarily from sequestration of MBNL1 causing a preponderance of the less responsive neonatal isoform of the insulin receptor, leading to insulin resistance and in the overproduction of the neonatal isoform of the skeletal muscle chloride channel causing a marked reduction in chloride conductance and myotonia.[97–100]

Animal Models

Researchers have developed a number of cell and animal models for DM1 and DM2 since the discoveries of the *DMPK* and *ZNF9* expansion mutations.[69,101–104] While there are no animal models that recapitulate the full spectrum of disease manifestations of DM1 or DM2, the available model systems play a major role in our understanding of these disorders.[69,101–104] Cell models demonstrate that CTG and CCTG repeats accumulate in intranuclear RNA foci together with an assortment of cellular factors including MBNL1, MBNL2, hnRNP H, and Sp1. Mouse models include transgenic mice expressing either ubiquitously expressed or tissue-specific constitutive, as well as inducible, CTG and CCTG repeats.[69,101–104] Knockout models of Mbnl1 and Mbnl2 recapitulate many features of the DM-associated spliceopathy and other characteristic disease manifestations including myotonia, subcapsular ocular cataracts, and cardiac arrhythmias observed in patients. Another key pathological feature in patients with DM1, severe muscle

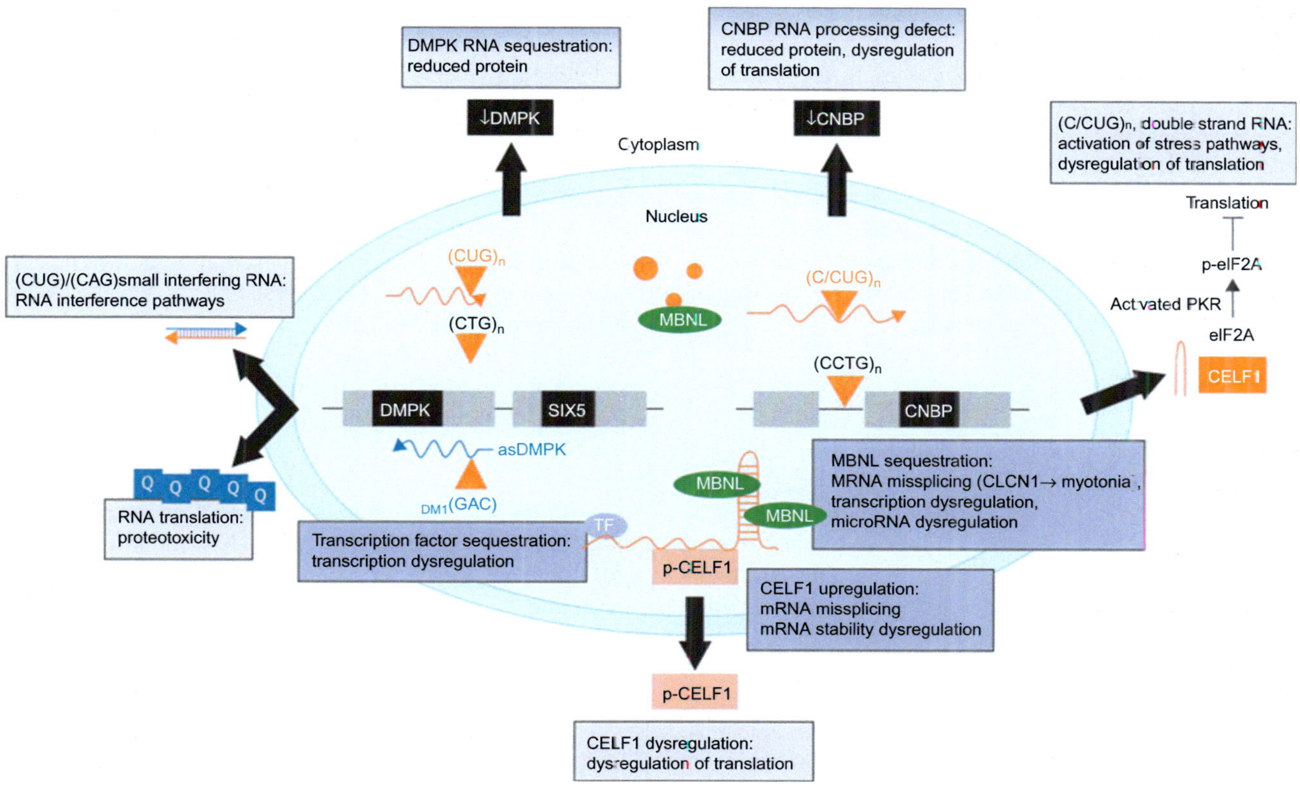

FIGURE 37.7 Postulated pathological mechanisms underlying myotonic dystrophy types 1 and 2. Most mechanisms are consistent with the prevailing toxic RNA gain-of-function model and affect multiple cellular aspects in the nucleus (blue boxes), the cytoplasm (grey boxes), or both (blue and grey boxes), including reduced protein function of the mutated genes, missplicing, dysregulation of transcription, protein translation and turnover, and activation of cellular stress pathways. Both mutant $(CTG)_n$ and $(CCTG)_n$ expansions in noncoding regions of *DMPK* and *CNBP* give rise to C/CUG-containing transcripts (red) that form stable secondary structures detectable as RNA foci (red circles) in the nucleus and the cytoplasm, and result in reduced amounts of DMPK and CNBP protein because of DMPK mRNA sequestration or a CNBP pre-mRNA processing defect. Members of the muscleblind-like (MBNL) (green) protein family, such as MBNL1, are sequestered in ribonuclear foci leading to loss of function and dysregulation of MBNL splice and transcription targets and microRNA metabolism. Hyperphosphorylation of nuclear and cytoplasmic CELF1 (p-CELF1; salmon) by several protein kinases, including RNA-dependent protein kinase (PKR), leads to stabilization of different isoforms, which in turn affects alternative splicing, translation, and protein turnover. Inactivation of the translation initiation factor eIF2A leads to general attenuation of translation. Sequestration of transcription factors and other nuclear factors also contributes to dysregulation of gene expression. Generation of small interfering RNA from sense (s) and antisense (as) DMPK transcripts might activate RNA interference pathways that lead to wider dysregulation of mRNA and protein amounts. sDMPK and asDMPK transcripts are subject to repeat-associated non-ATG translation in the polyglutamine (poly$[Q]_n$) reading frame and possibly others, giving rise to toxic homopolymeric polypeptides that accumulate in the cytoplasm. Although some mechanisms have only been reported in myotonic dystrophy type 1, theoretically they could also operate in type 2 disease. TF = transcription factor. *Reproduced from Udd and Krahe (2012),[69] with permission from Elsevier.*

wasting, has been observed in inducible CTG960 and CELF1 overexpression mice.[103] See recent reviews for more information describing animal and cell model systems as well as future treatments.[69,101–105]

Current Human Research

Much of the current research involving patients with DM1 and DM2 focuses upon the identification of reliable, feasible measures of disease progression that are meaningful to patients and are appropriate for use in future therapeutic trials (see ClinicalTrials.gov online database). These measures include assessments of: (i) strength using manual muscle testing and quantitative myometry; (ii) muscle mass (whole body and regional); (iii) myotonia (quantitative isometric grip relaxation, voluntary hand opening time, needle electromyography); (iv) timed functional tests (6-minute walk distance, time to travel 30 feet); (v) forced vital capacity; (vi) whole body dual energy X-ray absorptiometry DEXA; (vii) patient reported quality of life measures; and (viii) tissue biomarkers. One very recent study of patients with DM1 and DM2 included a cohort of 50 DM1 patients in whom investigators found 42 defects in alternative splicing of specific pre-mRNAs in small needle biopsy specimens of their tibialis anterior muscles. Twenty of these splicing defects showed graded changes that correlated with the degree of weakness of foot dorsiflexion.[106] Monitoring of defects in alternative splicing offers promise as a surrogate measure to assess early therapeutic response in future treatment trials in DM1. The promising results of treatment with antisense oligonucleotides in the mouse model of DM1[105,107] have accelerated plans to pursue their use in patients, and the methodology of small needle biopsy analysis of alterations in splicing, especially alterations in the skeletal muscle chloride channel along with assessments of myotonia, may prove very useful in detecting an early beneficial therapeutic response.

DIAGNOSTIC APPROACH

The major problem in making the diagnosis of DM1 and DM2 is the tendency of clinicians to ignore these entities in their practice. DM1 can present as a floppy infant or as a child with delayed motor and cognitive development (see Figures 37.5 and 37.6). There are long lists that focus on more common causes for these presentations than DM1. If there is a low diagnostic suspicion, the clinician may not probe the family history sufficiently and may not examine the parents of the patient using a focused approach (see Figure 37.4) to reveal characteristic features of DM1 (Tables 37.1 and 37.2). The failure to consider DM1 is the principal contributor to the diagnostic odyssey and delays in DNA testing, family counseling, and appropriate management and treatment.[23,50,57–69,108]

DM1 and DM2 present primarily in adult life. The core clinical features in Table 37.1 and the multisystem manifestations in Table 37.2 can occur in isolation or combination and usually develop gradually.[1–4,50–59] Muscle stiffness, muscle pain, intermittent trouble with clear speech and swallowing, dropping items, gait instability, and trouble seeing are frequent initial symptoms.[1–4] Early onset cataracts (typically, iridescent posterior capsular cataracts) and cardiac arrhythmias are examples of multisystem manifestations that can occur early in DM1 and DM2. In contrast, patients with other hereditary myotonic disorders, such as chloride or sodium channel myotonias, do not have cardiac arrhythmias or posterior capsular cataracts. DM1 patients occasionally present with prolonged apnea or delayed-onset apnea following general anesthesia. These complications can occur in individuals with only mild clinical signs of DM1. The risk of complications after general anesthesia in DM2 patients appears to be low, but until more long-term data are available, it is prudent to monitor both DM2 and DM1 patients closely during and 24 hours following general anesthesia.[4,12,65,109–111]

Figure 37.4 shows a diagnostic flow chart for identifying DM1 and DM2 in an adult with muscle weakness, stiffness, or pain but without clear diagnostic findings on clinical evaluation. The current gold standard to establish the diagnosis is DNA testing to identify the abnormally expanded CTG repeat in the *DMPK* gene for DM1 and the abnormally expanded CCTG repeat in the *CNBP* gene for DM2[23,69] (Table 37.1; see previous section, "Molecular Genetics"). In a small percentage of DM1 patients evaluated with current standard DNA testing, affected individuals have a false negative DNA test due to an interruption in the tract of CTG repeats by other nucleotide repeats (see previous subsection "DM1 and DM2: Genotype-Phenotype Correlations"). These DM1 patients have an abnormally expanded total CTG repeat length that requires the use of bidirectional TP-PCR to detect the "variant" interruptions of the CTG repeat expansion.[23] Figure 37.4 points out that if a patient suspected of having DM1 or DM2 has negative DNA testing, performance of a muscle biopsy may help establish the diagnosis of another myotonic disorder. However, if the muscle biopsy findings are typical for DM1 (increased central nuclei, atrophy of type-I fibers, ringed fibers, subsarcolemmal masses) or DM2 (very small fibers, type-II nuclear clumps, type-II fiber atrophy), further genetic analysis, including bidirectional TP-PCR, may provide evidence of a mutation variant.[70]

As indicated previously, the diagnosis of CDM or ChDM (Figures 37.5 and 37.6) often involves the discovery that three or more generations of the family have a progressive muscle disease for which there is no curative therapy. However, DM1 has a number of important supportive and anticipatory treatments, and the next portion of this chapter outlines many of these approaches (Tables 37.3 and 37.4). Case Examples 37.1 and 37.2 highlight the clinical features and management related to treatment of severe and very severe CDM, the least frequent form of DM1.

MANAGEMENT

Table 37.2 summarizes the multisystem manifestations. Table 37.3 outlines the general guidelines for management of patients with DM1 and DM2. Table 37.4 summarizes management for infants and children with DM1. Detailed discussion of each disease manifestation is beyond the scope of this review. More specific information about guidelines for care is available in textbooks focused primarily on myotonic dystrophy[1,2,50] and recent evidence-based guidelines.[57,59]

Standards of Care

Symptomatic treatment remains the mainstay of therapy for DM1 and DM2. Tables 37.3 and 37.4 outline the general approach to evaluation and approaches to care of several multisystem manifestations. The references cited below are to accompany the information in these

CASE EXAMPLE 37.1 A Case of Congenital Myotonic Dystrophy (CDM) with Transient Respiratory Failure

Maternal Background

The mother's diagnosis of myotonic dystrophy type 1 (DM1) occurred at six weeks' gestation. Care providers suspected DM1 based upon positive family history and her clinical manifestations. Her brother had been diagnosed with DM1 one year previously. She recalled that at 16 years of age, she had developed "locking up" of her hands and jaw muscles intermittently. These manifestations continued. Her DNA testing was positive for the *DMPK* gene mutation. She had 270 CTG repeats in her mutated *DMPK* allele (normal 3−35 CTG repeats). Following the diagnosis of DM1, she and her husband received genetic counseling. They decided not to have prenatal testing. Neuromuscular consultation first occurred at 32 weeks' gestation. On examination at that time she had mild bilateral ptosis, thin face, weakness of long finger flexors, grip and percussion myotonia. (See video clips that demonstrate her percussion and grip myotonia as well as her facial appearance. The infant is also shown.) She complained of muscle stiffness and emphasized having a history of intermittent chronic gastrointestinal problems, diarrhea alternating with constipation. Her electrocardiogram showed incomplete right bundle branch block, asymptomatic bradycardia, and a PR interval duration of 174 msec (normal <210 msec).

Pregnancy

Ultrasound of the mother revealed polyhydramnios, and she reported decreased fetal movements. Care providers suspected that the baby had CDM. As a result, the mother received gestational care from a high-risk obstetrics practice. Care providers established plans to deliver the baby at a tertiary medical center with a neonatal intensive care unit (NICU).

Delivery and Immediate Care

The baby presented breech, and he was delivered by Caesarean section at 32.4 weeks' gestation. The baby boy had APGAR scores of 4 and 8. Care providers transferred him urgently to the NICU, where he spent the next 6 weeks. An initial trial of noninvasive ventilation proved ineffective. Acute respiratory failure developed. The infant required intubation and positive pressure ventilation. Eventually he was weaned from the ventilator, transitioned for a period of time to continuous positive airway pressure support, subsequently placed on high flow nasal cannula respiratory care, and then discontinued from respiratory support. DNA testing of the baby was positive for DM1 with the mutated allele having 1020 CTG repeats.

Initial Home Care

The baby received total parenteral nutrition during his NICU stay and initially at home. He also underwent apnea/bradycardia monitoring at home. After several weeks of care at home, the baby was transitioned to nasal gastric feeds and eventually to feeding by mouth. He underwent a video swallow evaluation at 7 months of age and had moderate to severe oropharyngeal dysphagia with aspiration. He was placed on a thickened liquid diet. On examination he had an elongated face, tented upper lip, decreased truncal and appendicular tone, head lag, and frog leg posture when supine (see video of floppy infant). He had reduced tendon reflexes and a decreased range of motion of the hips, knees, and ankles. He underwent treatment with serial casting for contractures of the ankles and over several weeks, treatment was transitioned to ankle foot orthoses.

Follow-up Neuromuscular Evaluation at 8 Months of Age

The baby continued to have decreased truncal and appendicular tone, prominent head lag, frog leg posture when supine, elongated face, tented upper lip, and weak cry. He was unable to lift his head when prone and was unable to roll or sit independently. He continued to have decreased tendon reflexes and decreased range of motion at the hips, knees, and ankles. Treatment during home care had included physical therapy twice weekly, occupational therapy once a week, and speech therapy once a week.

Treatment Recommendations Following 8-Month Examination

Recommendations included the continuation of the current schedule of physical therapy, occupational therapy, and speech therapy; continuation of close monitoring of swallowing function; and continuation of frequent follow-up assessments of respiratory function by Pediatric Pulmonary Group and serial evaluations by Developmental Pediatrics and Neuromuscular consultants as well as regular visits to his attending pediatrician.

tables and provide more details related to guidelines for care involving skeletal muscle complaints,[51,112,113] musculoskeletal,[114,115] myotonia,[116−120] heart,[121−127] eye other than cataracts,[128−133] brain,[44,58,134,135] respiratory,[32,40,136] sleep disturbances and fatigue,[33,136−138] endocrine dysfunction,[3,4,139,140] gastrointestinal,[25−29,141] muscle pain,[66,67,142] prenatal and preimplantation genetic diagnosis,[143−145] and pregnancy-delivery.[146−148] There are also recent publications reporting typical abnormalities in routine blood tests in DM1[149] and DM2.[150]

Therapies under Investigation

As of January 2013, the US National Institutes of Health (NIH) database ClinicalTrials.gov listed three treatment trials (NCT01225614, NCT01530841, and NCT01406873) actively recruiting DM1 patients. One is a multicenter study of DM1 patients with respiratory insufficiency randomized to receive early initiation of nighttime noninvasive bilevel pressure ventilation compared to usual annual clinical monitoring. The primary outcome is the number of pulmonary complications

CASE EXAMPLE 37.2 A Case of Congenital Myotonic Dystrophy with Persisting Respiratory Failure

Maternal Background

The mother of the patient had DNA-confirmed diagnosis of DM1 with 431 CTG repeats at the *DMPK* gene locus in the mutated allele. Her father died at 45 years of age from cardiac complications of DM1 and her paternal aunt had symptoms of DM1. Initial symptoms in the mother of the patient included hand stiffness, grip myotonia, drooping of her eyelids, and excessive sleeping.

Pregnancy, Delivery, and Immediate Care

Ultrasound demonstrated polyhydramnios during pregnancy. The mother did not detect fetal movements. She had premature rupture of the membranes at 33 weeks' gestation. The baby boy presented breech and delivery was by emergency Caesarean section. He had serial APGAR scores of 2, 3, and 5. He was blue, floppy, and without spontaneous respiratory effort. He was intubated, placed on positive pressure ventilator support, and transferred to the NICU, where he remained for 4 months.

DNA analysis of the baby's blood showed 1683 CTG repeats in his mutated *DMPK* gene. He had poor suck and required gastrostomy feeding. Numerous attempts to wean him from ventilator support failed, and he received a tracheostomy with continuous positive pressure respiratory support. During his 4-month hospitalization on the NICU, he had generalized hypotonia, bilateral talipes foot deformities requiring bracing therapy, weak grasp, frog leg appearance, markedly decreased voluntary movement of his legs, and moderate power in his upper extremity movements. Hearing testing also revealed significant hearing loss.

Follow-up Neuromuscular Evaluation at Age 4 Years

The patient received multidisciplinary follow-up care during late infancy and early childhood. At 4 years of age, he remained fully dependent upon gastrostomy feeding. He was able to breathe on room air around his tracheostomy during the day and received continuous positive pressure ventilation at night. On examination, he had a long face, droopy lower lip, head lag, decreased muscle tone, and absent deep tendon reflexes. He could sit independently for brief periods of time and was able to stand with support, and he used a standing frame. He could hold crayons with a fist grip. He spoke in one syllable words and communicated using an iPad. He was not toilet trained. He received physical therapy, occupational therapy, speech therapy, and special education services at home. He also received home nursing to assist with his care.

Treatment Recommendations for the CDM 4-Year-Old after Follow-up Evaluation

The treatment plan included continuation of physical therapy, occupational therapy, and speech therapy along with special education services. Continued close monitoring of pulmonary and cardiac function was also necessary.

Treatment Recommendations and Current Evaluation of the Mother

Primary complaints of the mother include excessive daytime sleepiness, sleep apnea, recurrent gastroesophageal reflux, and ptosis. On examination, the mother had prominent facial weakness with weak closure of her eyes and lips, ptosis, high-arched palate, and moderate dysarthria. Strength testing revealed grade 4 out of 5 weakness of proximal arm and leg muscles and 3–4 out of 5 weakness distal muscles of the forearms, legs, and hands. She had prominent grip myotonia and percussion myotonia of thenar and forearm wrist extensor muscles.

Treatment recommendations for the mother included yearly monitoring of cardiac status with electrocardiogram and other testing requested by cardiology consultants; treatment of her sleep apnea coordinated through her internist and pulmonary consultant; treatment of her hypersomnia with stimulant medication monitored by her internist and pulmonary consultant; yearly ophthalmology evaluation to monitor cataract development and other potential eye problems related to weakness of eye closure; and yearly routine gynecologic care for health screening and cancer screening.

---

observed over 5 years. The second is a single center randomized trial of DM1 patients with respiratory insufficiency comparing nocturnal home ventilation with bilevel pressure support with average volume assured pressure support (AVAPS) to nocturnal home ventilation without AVAPS. The primary outcome is to compare differences at day 7 on arterial pCO_2 under ventilation. The third is a single center double-blind randomized 6-month trial of mexiletine 150 mg 3 times daily compared to placebo. The primary outcome is the change in distance walked in 6 minutes and the degree of grip myotonia comparing baseline to 6 months of treatment.

Preclinical experimental therapies for myotonic dystrophy involve animal models and have focused on the use of antisense oligonucleotides (ASOs), small interfering RNAs, small molecules, ribozymes, and engineered small nuclear RNAs.[69,100–105,151–153]

Especially promising are ASO therapies (Figure 37.8). ASOs are short synthetic analogues of DNA or RNA that bind to target RNA and act by blocking RNA function or by cleaving RNA, for example through the enzyme RNase H.[153] In 2009, two studies demonstrated the efficacy of ASO treatments in disrupting mutant RNA and reversing disease manifestations in a mouse model of DM1.[154,155] Newer approaches using ASO therapies focus on degrading the mutant RNA. Two studies have shown promising results to target and cleave toxic RNA via the enzyme RNase H in mouse models of

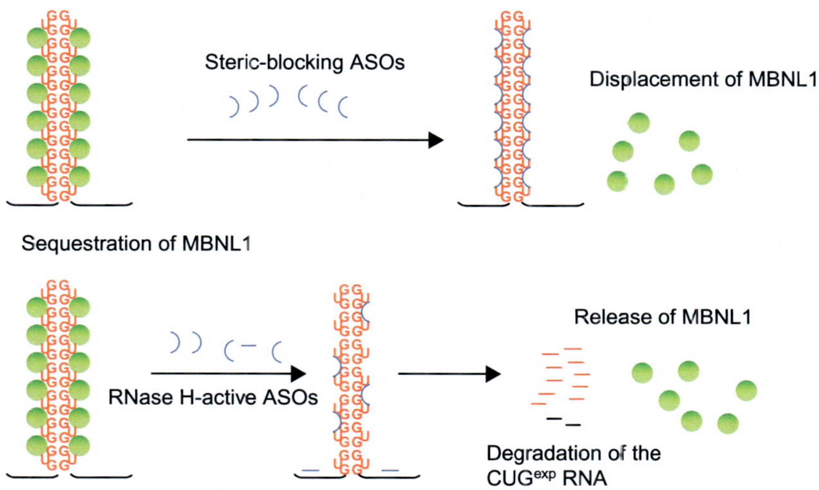

FIGURE 37.8 Application of antisense oligonucleotides (ASOs) to eliminate RNA toxicity in myotonic dystrophy type 1 (DM1). Steric blocking ASOs bind to CUG^{exp} RNA with high affinity and disrupt the RNA–MBNL1 (muscleblind-like 1) interactions, thereby displacing MBNL1 from the toxic RNA and reversing some features of DM1. In contrast, RNase H-active ASOs bind to the CUG^{exp}RNA and directly induce the degradation of the toxic RNA, resulting in release of MBNL1, restoration of its normal function, and reversal of disease features. *Reprinted with permission from Gao and Cooper (2013).*[105] *The publisher for this copyrighted material is Mary Ann Liebert, Inc., publishers.*

DM1.[107,151] One used a combination ASO therapy, one to block and one to degrade (through RNase H) toxic RNA.[151] The ASOs were delivered by intramuscular injection, had high efficacy and potency, but resulted in local muscle damage.[151] The other study used subcutaneous ASO injections (twice weekly for 4 weeks) and showed very encouraging results.[107] After a total of 8 injections, DM1 mice showed release of MBNL proteins from nuclear aggregates, correction of measured RNA splicing defects, disappearance of myotonia, and reversal of all these alterations even 1 year after cessation of treatment.[107] These results offer considerable promise that more effective treatment for DM1 is close at hand, but challenges remain. Researchers need to assure that ASO therapies go to the tissues primarily affected in DM1 (skeletal muscle, brain, and heart), and that there are no serious off-target effects. There is also an urgent need for clinical researchers to develop and validate the most appropriate measures to use in treatment trials to assess the progression of each of the different disease manifestations, particularly those manifestations that are most important to patients. Partnership between care providers and patients will become even more important going forward.

Several questions will need to be asked by patients and care providers with the development of promising new treatments. How do we manage expectations about the distribution of these promising treatments to care providers and their patients? Which patients are "first in line" to participate in early phase clinical trials? Will treatments be safe and available for children? Do we need to have further discussions of genetic testing of at-risk family members and newborn screening? The future for treatment of DM1 and DM2 requires shared responsibilities and holds great promise for success.

REFERENCES

1. Harper PS. *Myotonic Dystrophy*. 3rd ed. London: W. B. Saunders Company; 2001.
2. Echenne B, Bassez G. Congenital and infantile myotonic dystrophy. *Handb Clin Neurol* 2013;**113**:1387–93.
3. Moxley RT. Myotonic muscular dystrophy. In: Rowland LP DiMauro S, editors. *Handbook of Clinical Neurology*. New York Elsevier Science Publishers; 1992. pp. 209–59.
4. Moxley III RT, Meola G. Myotonic dystrophy. In: Deymeer F, editor. *Neuromuscular Diseases: From Basic Mechanisms to Clinical Management*. Basel: S. Karger AG; 2000. pp. 61–78.
5. Brook JD, McCurrach ME, Harley HG, Buckler AJ, Church D, Aburatani H, et al. Molecular basis of myotonic dystrophy: expansion of a trinucleotide (CTG) repeat at the 3′ end of a transcript encoding a protein kinase family member. *Cell* 1992;**68**:799–808.
6. Fu YH, Pizzuti A, Fenwick Jr. RG, King J, Rajnarayan S, Dunne PW, et al. An unstable triplet repeat in a gene related to myotonic muscular dystrophy. *Science* 1992;**255**:1256–8.
7. Mahadevan M, Tsilfidis C, Sabourin L, Shutler G, Amemiya C, Jansen G, et al. Myotonic dystrophy mutation: an unstable CTG repeat in the 3′ untranslated region of the gene. *Science* 1992;**255**:1253–5.
8. Thornton CA, Griggs RC, Moxley III RT. Myotonic dystrophy with no trinucleotide repeat expansion. *Ann Neurol* 1994;**35**:269–72.
9. Ricker K, Koch MC, Lehmann-Horn F, Pongratz D, Otto M, Heine R, et al. Proximal myotonic myopathy: a new dominant disorder with myotonia, muscle weakness, and cataracts. *Neurology* 1994;**44**:1448–52.
10. Ranum LP, Rasmussen PF, Benzow KA, Koob MD, Day JW. Genetic mapping of a second myotonic dystrophy locus. *Nat Genet* 1998;**19**:196–8.
11. Liquori CL, Ricker K, Moseley ML, Jacobsen JF, Kress W, Naylor SL, et al. Myotonic dystrophy type 2 caused by a CCTG expansion in intron 1 of ZNF9. *Science* 2001;**293**:864–7.
12. Day JW, Ricker K, Jacobsen JF, Rasmussen LJ, Dick KA, Kress W, et al. Myotonic dystrophy type 2: molecular, diagnostic and clinical spectrum. *Neurology* 2003;**60**:657–64.

13. Rolander A, Floderus S. Dystrophia myotonica in the Norbotten district. *Sven Lakartidn* 1961;**58**:648–52.
14. Medica I, Logar N, Mileta DL, Peterlin B. Genealogical study of myotonic dystrophy in Istria (Croatia). *Ann Genet* 2004;**47**:139–46.
15. Mathieu J, Allard P, Potvin L, Prevost C, Begin P, et al. A 10-year study of mortality in a cohort of patients with myotonic dystrophy. *Neurology* 1999;**52**:1658–62.
16. Mathieu J, Prevost C. Epidemiological surveillance of myotonic dystrophy type 1: a 25-year population-based study. *Neuromuscul Disord* 2012;**22**:974–9.
17. Mathieu J, De Braekeleer M, Prevost C. Genealogical reconstruction of myotonic dystrophy in the Saguenay-Lac-Saint-Jean area (Quebec, Canada). *Neurology* 1990;**40**:839–42.
18. Campbell C, Levin S, Siu VM, Venance S, Jacob P. Congenital myotonic dystrophy: Canadian population-based surveillance study. *J Pediatr* 2013;**163**(120–5):e1–3.
19. Suominen T, Bachinski LL, Auvinen S, Hackman P, Baggerly KA, Angelini C, et al. Population frequency of myotonic dystrophy: higher than expected frequency of myotonic dystrophy type 2 (DM2) mutation in Finland. *Eur J Hum Genet* 2011;**19**:776–82.
20. Milone M, Batish SD, Daube JR. Myotonic dystrophy type 2 with focal asymmetric muscle weakness and no electrical myotonia. *Muscle Nerve* 2009;**39**:383–5.
21. Young NP, Daube JR, Sorenson EJ, Milone M. Absent, unrecognized, and minimal myotonic discharges in myotonic dystrophy type 2. *Muscle Nerve* 2010;**41**:758–62.
22. Harper PS, Harley HG, Reardon W, Shaw DJ. Anticipation in myotonic dystrophy: new light on an old problem. *Am J Hum Genet* 1992;**51**:10–16.
23. Kamsteeg EJ, Kress W, Catalli C, Hertz JM, Witsch-Baumgartner M, Buckley MF, et al. Best practice guidelines and recommendations on the molecular diagnosis of myotonic dystrophy types 1 and 2. *Eur J Hum Genet* 2012;**20**:1203–8.
24. Cardani R, Giagnacovo M, Botta A, Rinaldi F, Morgante A, Udd B, et al. Co-segregation of DM2 with a recessive CLCN1 mutation in juvenile onset of myotonic dystrophy type 2. *J Neurol* 2012;**259**:2090–9.
25. Bellini M, Biagi S, Stasi C, Costa F, Mumolo MG, Ricchiuti A, et al. Gastrointestinal manifestations in myotonic muscular dystrophy. *World J Gastroenterol* 2006;**12**:1821–8.
26. Garrett JM, DuBose Jr. TD, Jackson JE, Norman JR. Esophageal and pulmonary disturbances in myotonia dystrophica. *Arch Intern Med* 1969;**123**:26–32.
27. Ronnblom A, Forsberg H, Danielsson A. Gastrointestinal symptoms in myotonic dystrophy. *Scand J Gastroenterol* 1996;**31**:654–7.
28. Degraeuwe J, Van Laecke E, De Muynck M, Van Biervliet S, Vande Velde S, Van Winckel M. Faecal incontinence due to atrophy of the anal sphincter in myotonic dystrophy: a case report. *Acta Gastroenterol Belg* 2011;**74**:88–90.
29. Sjogreen L, Tulinius M, Kiliaridis S, Lohmander A. The effect of lip strengthening exercises in children and adolescents with myotonic dystrophy type 1. *Int J Pediatr Otorhinolaryngol* 2010;**74**:1126–34.
30. Prendergast P, Magalhaes S, Campbell C. Congenital myotonic dystrophy in a national registry. *Paediatr Child Health* 2010;**15**:514–18.
31. Bianchi ML, Losurdo A, Di Blasi C, Santoro M, Masciullo M, Conte G, et al. Prevalence and clinical correlates of sleep disordered breathing in myotonic dystrophy types 1 and 2. *Sleep Breath* 2013. [Epub ahead of print].
32. Monteiro R, Bento J, Goncalves MR, Pinto T, Winck JC. Genetics correlates with lung function and nocturnal ventilation in myotonic dystrophy. *Sleep Breath* 2013;**17**:1087–92.
33. Laberge L, Gagnon C, Dauvilliers Y. Daytime sleepiness and myotonic dystrophy. *Curr Neurol Neurosci Rep* 2013;**13**:340.
34. Quera Salva MA, Blumen M, Jacquette A, Durand MC, Andre S, De Villiers M, et al. Sleep disorders in childhood-onset myotonic dystrophy type 1. *Neuromuscul Disord* 2006;**16**:564–70.
35. Laberge L, Begin P, Dauvilliers Y, Beaudry M, Laforte M, Jean S, et al. A polysomnographic study of daytime sleepiness in myotonic dystrophy type 1. *J Neurol Neurosurg Psychiatry* 2009;**80**:642–6.
36. Kalkman JS, Schillings ML, van der Werf SP, Padberg GW, Zwarts MJ, van Engelen BG, et al. Experienced fatigue in facioscapulohumeral dystrophy, myotonic dystrophy, and HMSN-I. *J Neurol Neurosurg Psychiatry* 2005;**76**:1406–9.
37. Ciafaloni E, Mignot E, Sansone V, Hilbert JE, Lin L, Lin X, et al. The hypocretin neurotransmission system in myotonic dystrophy type 1. *Neurology* 2008;**70**:226–30.
38. Echenne B, Rideau A, Roubertie A, Sebire G, Rivier F, Lemieux B. Myotonic dystrophy type I in childhood: Long-term evolution in patients surviving the neonatal period. *Eur J Paediatr Neurol* 2008;**12**:210–23.
39. Harper PS. Congenital myotonic dystrophy in Britain. I. Clinical aspects. *Arch Dis Child* 1975;**50**:505–13.
40. Campbell C, Sherlock R, Jacob P, Blayney M. Congenital myotonic dystrophy: assisted ventilation duration and outcome. *Pediatrics* 2004;**113**:811–16.
41. Koch MC, Grimm T, Harley HG, Harper PS. Genetic risks for children of women with myotonic dystrophy. *Am J Hum Genet* 1991;**48**:1084–91.
42. Douniol M, Jacquette A, Cohen D, Bodeau N, Rachidi L, Angeard N, et al. Psychiatric and cognitive phenotype of childhood myotonic dystrophy type 1. *Dev Med Child Neurol* 2012;**54**:905–11.
43. Angeard N, Jacquette A, Gargiulo M, Radvanyi H, Moutier S, Eymard B, et al. A new window on neurocognitive dysfunction in the childhood form of myotonic dystrophy type 1 (DM1). *Neuromuscul Disord* 2011;**21**:468–76.
44. Ekstrom AB, Hakenas-Plate L, Tulinius M, Wentz E. Cognition and adaptive skills in myotonic dystrophy type 1: a study of 55 individuals with congenital and childhood forms. *Dev Med Child Neurol* 2009;**51**:982–90.
45. Ekstrom AB, Hakenas-Plate L, Samuelsson L, Tulinius M, Wentz E. Autism spectrum conditions in myotonic dystrophy type 1: a study on 57 individuals with congenital and childhood forms. *Am J Med Genet B Neuropsychiatr Genet* 2008;**147B**:918–26.
46. Johnson NE, Luebbe E, Eastwood E, Chin N, Moxley III RT, Heatwole CR. The impact of congenital and childhood myotonic dystrophy on quality of life: A qualitative study of associated symptoms. *J Child Neurol* 2013;**29**:983–6.
47. Sjogreen L, Engvall M, Ekstrom AB, Lohmander A, Kiliaridis S, Tulinius M. Orofacial dysfunction in children and adolescents with myotonic dystrophy. *Dev Med Child Neurol* 2007;**49**:18–22.
48. Engvall M, Sjogreen L, Kjellberg H, Robertson A, Sundell S, Kiliaridis S. Oral health in children and adolescents with myotonic dystrophy. *Eur J Oral Sci* 2007;**115**:192–7.

49. Sinclair JL, Reed PW. Risk factors for perioperative adverse events in children with myotonic dystrophy. *Paediatr Anaesth* 2009;**19**:740–7.
50. Harper PS, Engelen BG, Eymard B, Wilcox DE. *Myotonic Dystrophy: Present Management, Future Therapy*. New York: Oxford University Press; 2004.
51. Johnson ER, Abresch RT, Carter GT, Kilmer DD, Fowler Jr. WM, Sigford BJ, et al. Profiles of neuromuscular diseases. Myotonic dystrophy. *Am J Phys Med Rehabil* 1995;**74**:S104–16.
52. Dahlbom K, Ahlstrom G, Barany M, Kihlgren A, Gunnarsson LG. Muscular dystrophy in adults: a five-year follow-up. *Scand J Rehabil Med* 1999;**31**:178–84.
53. Marchini C, Lonigro R, Verriello L, Pellizzari L, Bergonzi P, Damante G. Correlations between individual clinical manifestations and CTG repeat amplification in myotonic dystrophy. *Clin Genet* 2000;**57**:74–82.
54. Mathieu J, Boivin H, Richards CL. Quantitative motor assessment in myotonic dystrophy. *Can J Neurol Sci* 2003;**30**:129–36.
55. Kroksmark AK, Ekstrom AB, Bjorck E, Tulinius M. Myotonic dystrophy: muscle involvement in relation to disease type and size of expanded CTG-repeat sequence. *Dev Med Child Neurol* 2005;**47**:478–85.
56. Arsenault ME, Prevost C, Lescault A, Laberge C, Puymirat J, Mathieu J. Clinical characteristics of myotonic dystrophy type 1 patients with small CTG expansions. *Neurology* 2006;**66**:1248–50.
57. Turner C, Hilton-Jones D, Lochmuller H, Hanna MG. MRC Centre for Neuromuscular Diseases, 1st (1st December 2010) and 2nd (2nd May 2012) Myotonic Dystrophy Workshops, London, UK and the Myotonic Dystrophy Standards of Care and National Registry Meeting, Newcastle, UK July 2011. *Neuromuscul Disord* 2013;**23**:1069–80.
58. Gagnon C, Chouinard MC, Laberge L, Veillette S, Begin P, Breton R, et al. Health supervision and anticipatory guidance in adult myotonic dystrophy type 1. *Neuromuscul Disord* 2010;**20**:847–51.
59. Ashizawa T, Moxley R, Day J, Groh W, Puymirat J, Cwik V, et al. Evidence-based guideline: evaluation and management of myotonic dystrophy. *Neurology* 2014. [in press].
60. Groh WJ, Groh MR, Shen C, Monckton DG, Bodkin CL, Pascuzzi RM. Survival and CTG repeat expansion in adults with myotonic dystrophy type 1. *Muscle Nerve* 2011;**43**:648–51.
61. Mladenovic J, Pekmezovic T, Todorovic S, Rakocevic-Stojanovic V, Savic D, Romac S, et al. Survival and mortality of myotonic dystrophy type 1 (Steinert's disease) in the population of Belgrade. *Eur J Neurol* 2006;**13**:451–4.
62. de Die-Smulders CE, Howeler CJ, Thijs C, Mirandolle JF, Anten HB, Smeets HJ, et al. Age and causes of death in adult-onset myotonic dystrophy. *Brain* 1998;**121**(Pt 8):1557–63.
63. Meola G, Moxley III RT. Myotonic dystrophy type 2 and related myotonic disorders. *J Neurol* 2004;**251**:1173–82.
64. Meola G, Sansone V, Marinou K, Cotelli M, Moxley III RT, Thornton CA, et al. Proximal myotonic myopathy: a syndrome with a favourable prognosis? *J Neurol Sci* 2002;**193**:89–96.
65. Udd B, Meola G, Krahe R, Wansink DG, Bassez G, Kress W, et al. Myotonic dystrophy type 2 (DM2) and related disorders: Report of the 180th ENMC workshop including guidelines on diagnostics and management 3–5 December 2010, Naarden, The Netherlands. *Neuromuscul Disord* 2011;**21**:443–50.
66. Suokas KI, Haanpaa M, Kautiainen H, Udd B, Hietaharju AJ. Pain in patients with myotonic dystrophy type 2: a postal survey in Finland. *Muscle Nerve* 2012;**45**:70–4.
67. George A, Schneider-Gold C, Zier S, Reiners K, Sommer C. Musculoskeletal pain in patients with myotonic dystrophy type 2. *Arch Neurol* 2004;**61**:1938–42.
68. Moxley RT, Hilbert JE, Meola G. The myotonic dystrophies. In: Rosenberg RN, Pascual JM, editors. *Rosenberg's Molecular and Genetic Basis of Neurological and Psychiatric Disease*. 5th ed. San Diego: Academic Press; 2014.
69. Udd B, Krahe R. The myotonic dystrophies: molecular, clinical, and therapeutic challenges. *Lancet Neurol* 2012;**11**:891–905.
70. Santoro M, Masciullo M, Pietrobono R, Conte G, Modoni A, Bianchi ML, et al. Molecular, clinical, and muscle studies in myotonic dystrophy type 1 (DM1) associated with novel variant CCG expansions. *J Neurol* 2013;**260**:1245–57.
71. Braida C, Stefanatos RK, Adam B, Mahajan N, Smeets HJ, Niel F, et al. Variant CCG and GGC repeats within the CTG expansion dramatically modify mutational dynamics and likely contribute toward unusual symptoms in some myotonic dystrophy type 1 patients. *Hum Mol Genet* 2010;**19**:1399–412.
72. Musova Z, Mazanec R, Krepelova A, Ehler E, Vales J, Jaklova R, et al. Highly unstable sequence interruptions of the CTG repeat in the myotonic dystrophy gene. *Am J Med Genet A* 2009;**149A**:1365–74.
73. Ashizawa T, Dubel JR, Harati Y. Somatic instability of CTG repeat in myotonic dystrophy. *Neurology* 1993;**43**:2674–8.
74. Martorell L, Monckton DG, Gamez J, Johnson KJ, Gich I, Lopez de Munain A, et al. Progression of somatic CTG repeat length heterogeneity in the blood cells of myotonic dystrophy patients. *Hum Mol Genet* 1998;**7**:307–12.
75. Anvret M, Ahlberg G, Grandell U, Hedberg B, Johnson K, Edstrom L. Larger expansions of the CTG repeat in muscle compared to lymphocytes from patients with myotonic dystrophy. *Hum Mol Genet* 1993;**2**:1397–400.
76. Thornton CA, Johnson K, Moxley III RT. Myotonic dystrophy patients have larger CTG expansions in skeletal muscle than in leukocytes. *Ann Neurol* 1994;**35**:104–7.
77. Goula AV, Stys A, Chan JP, Trottier Y, Festenstein R, Merienne K. Transcription elongation and tissue-specific somatic CAG instability. *PLoS Genet* 2012;**8**:e1003051.
78. Liu G, Chen X, Leffak M. Oligodeoxynucleotide binding to (CTG)*(CAG) microsatellite repeats inhibits replication fork stalling, hairpin formation, and genome instability. *Mol Cell Biol* 2013;**33**:571–81.
79. Kurosaki T, Ueda S, Ishida T, Abe K, Ohno K, Matsuura T. The unstable CCTG repeat responsible for myotonic dystrophy type 2 originates from an AluSx element insertion into an early primate genome. *PLoS One* 2012;**7**:e38379.
80. Nakamori M, Pearson CE, Thornton CA. Bidirectional transcription stimulates expansion and contraction of expanded (CTG)*(CAG) repeats. *Hum Mol Genet* 2011;**20**:580–8.
81. Higham CF, Morales F, Cobbold CA, Haydon DT, Monckton DG. High levels of somatic DNA diversity at the myotonic dystrophy type 1 locus are driven by ultra-frequent expansion and contraction mutations. *Hum Mol Genet* 2012;**21**:2450–63.
82. Wheeler TM, Thornton CA. Myotonic dystrophy: RNA-mediated muscle disease. *Curr Opin Neurol* 2007;**20**:572–6.

83. Cleary JD, Ranum LP. Repeat-associated non-ATG (RAN) translation in neurological disease. *Hum Mol Genet* 2013;**22**: R45–51.
84. Pearson CE. Repeat associated non-ATG translation initiation: one DNA, two transcripts, seven reading frames, potentially nine toxic entities! *PLoS Genet* 2011;**7**:e1002018.
85. Zu T, Gibbens B, Doty NS, Gomes-Pereira M, Huguet A, Stone MD, et al. Non-ATG-initiated translation directed by microsatellite expansions. *Proc Natl Acad Sci USA* 2011;**108**:260–5.
86. Moseley ML, Zu T, Ikeda Y, Gao W, Mosemiller AK, Daughters RS, et al. Bidirectional expression of CUG and CAG expansion transcripts and intranuclear polyglutamine inclusions in spinocerebellar ataxia type 8. *Nat Genet* 2006;**38**:758–69.
87. Huguet A, Medja F, Nicole A, Vignaud A, Guiraud-Dogan C, Ferry A, et al. Molecular, physiological, and motor performance defects in DMSXL mice carrying >1,000 CTG repeats from the human DM1 locus. *PLoS Genet* 2012;**8**:e1003043.
88. Lopez Castel A, Nakamori M, Tome S, Chitayat D, Gourdon G, Thornton CA, et al. Expanded CTG repeat demarcates a boundary for abnormal CpG methylation in myotonic dystrophy patient tissues. *Hum Mol Genet* 2011;**20**:1–15.
89. Lopez Castel A, Nakamori M, Thornton CA, Pearson CE. Identification of restriction endonucleases sensitive to 5-cytosine methylation at non-CpG sites, including expanded (CAG)n/(CTG) n repeats. *Epigenetics* 2011;**6**:416–20.
90. Furling D. Misregulation of alternative splicing and microRNA processing in DM1 pathogenesis. *Rinsho Shinkeigaku* 2012;**52**:1018–22.
91. Greco S, Perfetti A, Fasanaro P, Cardani R, Capogrossi MC, Meola G, et al. Deregulated microRNAs in myotonic dystrophy type 2. *PLOS ONE* 2012;**7**:e39732.
92. Rau F, Freyermuth F, Fugier C, Villemin JP, Fischer MC, Jost B, et al. Misregulation of miR-1 processing is associated with heart defects in myotonic dystrophy. *Nat Struct Mol Biol* 2011;**18**:840–5.
93. Perbellini R, Greco S, Sarra-Ferraris G, Cardani R, Capogrossi MC, Meola G, et al. Dysregulation and cellular mislocalization of specific miRNAs in myotonic dystrophy type 1. *Neuromuscul Disord* 2011;**21**:81–8.
94. Meola G, Jones K, Wei C, Timchenko LT. Dysfunction of protein homeostasis in myotonic dystrophies. *Histol Histopathol* 2013;**28**:1089–98.
95. Ravel-Chapuis A, Belanger G, Yadava RS, Mahadevan MS, DesGroseillers L, Cote J, et al. The RNA-binding protein Staufen1 is increased in DM1 skeletal muscle and promotes alternative pre-mRNA splicing. *J Cell Biol* 2012;**196**:699–712.
96. Echeverria GV, Cooper TA. Muscleblind-like 1 activates insulin receptor exon 11 inclusion by enhancing U2AF65 binding and splicing of the upstream intron. *Nucleic Acids Res* 2014;**42**:1893–903.
97. Mankodi A, Takahashi MP, Jiang H, Beck CL, Bowers WJ, Moxley RT, et al. Expanded CUG repeats trigger aberrant splicing of ClC-1 chloride channel pre-mRNA and hyperexcitability of skeletal muscle in myotonic dystrophy. *Mol Cell* 2002;**10**:35–44.
98. Charlet BN, Savkur RS, Singh G, Philips AV, Grice EA, Cooper TA. Loss of the muscle-specific chloride channel in type 1 myotonic dystrophy due to misregulated alternative splicing. *Mol Cell* 2002;**10**:45–53.
99. Lueck JD, Mankodi A, Swanson MS, Thornton CA, Dirksen RT. Muscle chloride channel dysfunction in two mouse models of myotonic dystrophy. *J Gen Physiol* 2007;**129**:79–94.
100. Wheeler TM, Lueck JD, Swanson MS, Dirksen RT, Thornton CA. Correction of ClC-1 splicing eliminates chloride channelopathy and myotonia in mouse models of myotonic dystrophy. *J Clin Invest* 2007;**117**:3952–7.
101. Sicot G, Gomes-Pereira M. RNA toxicity in human disease and animal models: from the uncovering of a new mechanism to the development of promising therapies. *Biochim Biophys Acta* 2013;**1832**:1390–409.
102. Gomes-Pereira M, Cooper TA, Gourdon G. Myotonic dystrophy mouse models: towards rational therapy development. *Trends Mol Med* 2011;**17**:506–17.
103. Ward AJ, Rimer M, Killian JM, Dowling JJ, Cooper TA. CUGBP1 overexpression in mouse skeletal muscle reproduces features of myotonic dystrophy type 1. *Hum Mol Genet* 2010;**19**:3614–22.
104. Sicot G, Gourdon G, Gomes-Pereira M. Myotonic dystrophy, when simple repeats reveal complex pathogenic entities: new findings and future challenges. *Hum Mol Genet* 2011;**20**: R116–23.
105. Gao Z, Cooper TA. Antisense oligonucleotides: rising stars in eliminating RNA toxicity in myotonic dystrophy. *Hum Gene Ther* 2013;**24**:499–507.
106. Nakamori M, Sobczak K, Puwanant A, Welle S, Eichinger K, Pandya S, et al. Splicing biomarkers of disease severity in myotonic dystrophy. *Ann Neurol* 2013;**74**:862–72.
107. Wheeler TM, Leger AJ, Pandey SK, MacLeod AR, Nakamori M, Cheng SH, et al. Targeting nuclear RNA for in vivo correction of myotonic dystrophy. *Nature* 2012;**488**:111–15.
108. Hilbert JE, Ashizawa T, Day JW, Luebbe EA, Martens WB, McDermott MP, et al. Diagnostic odyssey of patients with myotonic dystrophy. *J Neurol* 2013;**260**:2497–504.
109. Veyckemans F, Scholtes JL. Myotonic dystrophies type 1 and 2: anesthetic care. *Paediatr Anaesth* 2013;**23**:794–803.
110. Campbell N, Brandom B, Day JW, Moxley III RT. *Practical suggestions for the anesthetic management of a myotonic dystrophy patient. Myotonic Dystrophy Foundation Toolkit*. Menlo Park, CA: Myotonic Dystrophy Foundation; 2013. pp. 73–80.
111. Kirzinger L, Schmidt A, Kornblum C, Schneider-Gold C, Kress W, Schoser B. Side effects of anesthesia in DM2 as compared to DM1: a comparative retrospective study. *Eur J Neurol* 2010;**17**: 842–5.
112. Wiles CM, Busse ME, Sampson CM, Rogers MT, Fenton-May J, van Deursen R. Falls and stumbles in myotonic dystrophy. *J Neurol Neurosurg Psychiatry* 2006;**77**:393–6.
113. Voet NB, van der Kooi EL, Riphagen II, Lindeman E, van Engelen BG, Geurts ACh. Strength training and aerobic exercise training for muscle disease. *Cochrane Database Syst Rev* 2013;**7**: CD003907.
114. Canavese F, Sussman MD. Orthopaedic manifestations of congenital myotonic dystrophy during childhood and adolescence. *J Pediatr Orthop* 2009;**29**:208–13.
115. Schilling L, Forst R, Forst J, Fujak A. Orthopaedic disorders in myotonic dystrophy type 1: descriptive clinical study of 21 patients. *BMC Musculoskelet Disord* 2013;**14**:338.
116. Logigian EL, Martens WB, Moxley RT 4th, McDermott MP, Dilek N, Wiegner AW, et al. Mexiletine is an effective

antimyotonia treatment in myotonic dystrophy type 1. *Neurology* 2010;**74**:1441–8.

117. Logigian EL, Blood CL, Dilek N, Martens WB, Moxley RT 4th, Wiegner AW, et al. Quantitative analysis of the "warm-up" phenomenon in myotonic dystrophy type 1. *Muscle Nerve* 2005;**32**:35–42.

118. Trip J, Drost G, van Engelen BG, Faber CG. Drug treatment for myotonia. *Cochrane Database Syst Rev* 2006;**1**:CD004762.

119. de Swart BJ, van Engelen BG, Maassen BA. Warming up improves speech production in patients with adult onset myotonic dystrophy. *J Commun Disord* 2007;**40**:185–95.

120. Heatwole CR, Statland JM, Logigian EL. The diagnosis and treatment of myotonic disorders. *Muscle Nerve* 2013;**47**:632–48.

121. Groh WJ, Bhakta D. Arrhythmia management in myotonic dystrophy type 1. *JAMA* 2012;**308**:337–8.

122. Wahbi K, Meune C, Porcher R, Becane HM, Lazarus A, Laforet P, et al. Electrophysiological study with prophylactic pacing and survival in adults with myotonic dystrophy and conduction system disease. *JAMA* 2012;**307**:1292–301.

123. Bhakta D, Shen C, Kron J, Epstein AE, Pascuzzi RM, Groh WJ. Pacemaker and implantable cardioverter-defibrillator use in a US myotonic dystrophy type 1 population. *J Cardiovasc Electrophysiol* 2011;**22**:1369–75.

124. Ha AH, Tarnopolsky MA, Bergstra TG, Nair GM, Al-Qubbany A, Healey JS. Predictors of atrio-ventricular conduction disease, long-term outcomes in patients with myotonic dystrophy types I and II. *Pacing Clin Electrophysiol* 2012;**35**:1262–9.

125. Facenda-Lorenzo M, Hernandez-Afonso J, Rodriguez-Esteban M, de Leon-Hernandez JC, Grillo-Perez JJ. Cardiac manifestations in myotonic dystrophy type 1 patients followed using a standard protocol in a specialized unit. *Rev Esp Cardiol* 2013;**66**:193–7.

126. Brembilla-Perrot B, Schwartz J, Huttin O, Frikha Z, Sellal JM, Sadoul N, et al. Atrial flutter or fibrillation is the most frequent and life-threatening arrhythmia in myotonic dystrophy. *Pacing Clin Electrophysiol* 2014;**37**:329–35.

127. Turillazzi E, Neri M, Riezzo I, Di Paolo M, Evangelisti L, Fineschi V. Cardiac fibrosis, arrhythmia and sudden death in myotonic dystrophy type 1: could TGF-ss1 improve the predictive accuracy of patients at risk, opening new therapeutic challenges? *Int J Cardiol* 2013;**168**:4976–8.

128. Kersten HM, Roxburgh RH, Child N, Polkinghorne PJ, Frampton C, Danesh-Meyer HV. Epiretinal membrane: a treatable cause of visual disability in myotonic dystrophy type 1. *J Neurol* 2014;**261**:37–44.

129. Aring E, Ekstrom AB, Tulinius M, Sjostrom A. Ocular motor function in relation to gross motor function in congenital and childhood myotonic dystrophy type 1. *Acta Ophthalmol* 2012;**90**:369–74.

130. Rosa N, Lanza M, Borrelli M, De Bernardo M, Palladino A, Di Gregorio MG, et al. Low intraocular pressure resulting from ciliary body detachment in patients with myotonic dystrophy. *Ophthalmology* 2011;**118**:260–4.

131. Ekstrom AB, Tulinius M, Sjostrom A, Aring E. Visual function in congenital and childhood myotonic dystrophy type 1. *Ophthalmology* 2010;**117**:976–82.

132. Sarks J, Penfold P, Liu H, Sarks S, Killingsworth M, Horowitz G. Retinal changes in myotonic dystrophy: a clinicomorphological study. *Aust NZ J Ophthalmol* 1985;**13**:19–36.

133. Hayasaka S, Kiyosawa M, Katsumata S, Honda M, Takase S, Mizuno K. Ciliary and retinal changes in myotonic dystrophy. *Arch Ophthalmol* 1984;**102**:88–93.

134. Meola G, Sansone V. Cerebral involvement in myotonic dystrophies. *Muscle Nerve* 2007;**36**:294–306.

135. Cup EH, Kinebanian A, Satink T, Pieterse AJ, Hendricks HT, Oostendorp RA, et al. Living with myotonic dystrophy; what can be learned from couples? A qualitative study. *BMC Neurol* 2011;**11**:86.

136. Bhat S, Sander HW, Grewal RP, Chokroverty S. Sleep disordered breathing and other sleep dysfunction in myotonic dystrophy type 2. *Sleep Med* 2012;**13**:1207–8.

137. Annane D, Moore DH, Barnes PR, Miller RG. Psychostimulants for hypersomnia (excessive daytime sleepiness) in myotonic dystrophy. *Cochrane Database Syst Rev* 2006;**3**:CD003218.

138. Shepard P, Lam EM, St Louis EK, Dominik J. Sleep disturbances in myotonic dystrophy type 2. *Eur Neurol* 2012;**68**:377–80.

139. Cruz Guzman Odel R, Chavez Garcia AL, Rodriguez-Cruz M. Muscular dystrophies at different ages: metabolic and endocrine alterations. *Int J Endocrinol* 2012. 2012:485376.

140. Orngreen MC, Arlien-Soborg P, Duno M, Hertz JM, Vissing J. Endocrine function in 97 patients with myotonic dystrophy type 1. *J Neurol* 2012;**259**:912–20.

141. Tieleman AA, van Vliet J, Jansen JB, van der Kooi AJ, Borm GF van Engelen BG. Gastrointestinal involvement is frequent in Myotonic Dystrophy type 2. *Neuromuscul Disord* 2008;**18**:646–9.

142. Martorell L, Cobo AM, Baiget M, Naudo M, Poza JJ, Parra J. Prenatal diagnosis in myotonic dystrophy type 1. Thirteen years of experience: implications for reproductive counselling in DM1 families. *Prenat Diagn* 2007;**27**:68–72.

143. Verpoest W, De Rademaeker M, Sermon K, De Rycke M, Seneca S, Papanikolaou E, et al. Real and expected delivery rates of patients with myotonic dystrophy undergoing intracytoplasmic sperm injection and preimplantation genetic diagnosis. *Hum Reprod* 2008;**23**:1654–60.

144. De Rademaeker M, Verpoest W, De Rycke M, Seneca S, Sermon K, Desmyttere S, et al. Preimplantation genetic diagnosis for myotonic dystrophy type 1: upon request to child. *Eur J Hum Genet* 2009;**17**:1403–10.

145. Dechanet C, Castelli C, Reyftmann L, Coubes C, Hamamah S, Hedon B, et al. Myotonic dystrophy type 1 and PGD: ovarian stimulation response and correlation analysis between ovarian reserve and genotype. *Reprod Biomed Online* 2010;**20**:610–18.

146. Awater C, Zerres K, Rudnik-Schoneborn S. Pregnancy course and outcome in women with hereditary neuromuscular disorders: comparison of obstetric risks in 178 patients. *Eur J Obstet Gynecol Reprod Biol* 2012;**162**:153–9.

147. Rudnik-Schoneborn S, Schneider-Gold C, Raabe U, Kress W, Zerres K, Schoser BG. Outcome and effect of pregnancy in myotonic dystrophy type 2. *Neurology* 2006;**66**:579–80.

148. Rudnik-Schoneborn S, Zerres K. Outcome in pregnancies complicated by myotonic dystrophy: a study of 31 patients and review of the literature. *Eur J Obstet Gynecol Reprod Biol* 2004;**114**:44–53.

149. Heatwole CR, Miller J, Martens B, Moxley III RT. Laboratory abnormalities in ambulatory patients with myotonic dystrophy type 1. *Arch Neurol* 2006;**63**:1149–53.

150. Heatwole C, Johnson N, Goldberg B, Martens W, Moxley III R. Laboratory abnormalities in patients with myotonic dystrophy type 2. *Arch Neurol* 2011;**68**:1180–4.
151. Lee JE, Bennett CF, Cooper TA. RNase H-mediated degradation of toxic RNA in myotonic dystrophy type 1. *Proc Natl Acad Sci USA* 2012;**109**:4221–6.
152. Gonzalez-Barriga A, Mulders SA, van de Giessen J, Hooijer JD, Bijl S, van Kessel ID, et al. Design and analysis of effects of triplet repeat oligonucleotides in cell models for myotonic dystrophy. *Mol Ther Nucleic Acids* 2013;**2**:e81.
153. Bennett CF, Swayze EE. RNA targeting therapeutics: molecular mechanisms of antisense oligonucleotides as a therapeutic platform. *Annu Rev Pharmacol Toxicol* 2010;**50**:259–93.
154. Mulders SA, van den Broek WJ, Wheeler TM, Croes HJ, van Kuik-Romeijn P, de Kimpe SJ, et al. Triplet-repeat oligonucleotide-mediated reversal of RNA toxicity in myotonic dystrophy. *Proc Natl Acad Sci USA* 2009;**106**:13915–20.
155. Wheeler TM, Sobczak K, Lueck JD, Osborne RJ, Lin X, Dirksen RT, et al. Reversal of RNA dominance by displacement of protein sequestered on triplet repeat RNA. *Science* 2009;**325**:336–9.

Chapter 38

Muscle Channelopathies: Myotonias and Periodic Paralyses

Karin Jurkat-Rott, Reinhardt Rüdel, and Frank Lehmann-Horn

INTRODUCTION

Defective ion channels may play a causal role in disease pathogenesis. This implication was first concluded from the observation of an abnormal ion conductance in muscle fibers biopsied from myotonic goats.[1] In humans, a similar conclusion was reached for patients with paramyotonia congenita (PC) and with periodic paralyses.[2,3] The term "ion channelopathies" was then coined in the 1990s,[4] and defined for disorders that are caused by malfunction or altered regulation of ion channel proteins. Channelopathies, therefore, can be either hereditary or acquired (the latter usually caused by auto-antibodies).

Hereditary channelopathies can be categorized in those affecting the motor endplate (congenital myasthenic syndromes), the sarcolemma (myotonias and periodic paralyses), and excitation-contraction coupling (malignant hyperthermia, central core myopathy, and multiminicore myopathy). The channel defects result in changes of excitability that one would expect to be constantly present. Fortunately, this is not the case. Clinical symptoms mainly appear episodically, provoked by an out-of-the-normal situation, a so-called trigger.[5] Compensatory mechanisms such as the normalization of the serum potassium level often allow an episode to cross over to complete remission. In addition to the episodes, progressive manifestations with muscular degeneration are present in a substantial percentage of patients.

As many disease mechanisms have been elucidated by functional characterization on the molecular level, the channelopathies are regarded as model disorders. Molecular genetics identify the underlying mutations and thus often enable one to predict the course of the diseases and to assign an individual therapy for the patient. They also allow one to calculate the patient's risk of having affected offspring.

ELECTROPHYSIOLOGY OF CLINICAL WEAKNESS AND PARALYSIS

The excitability of a muscle fiber, and thus its ability to generate force and contraction, depends on a *high* resting potential across its sarcolemma (−80 mV inside with respect to outside). Such fibers are assuming the P1 state in the nomenclature of Jurkat-Rott et al.,[6] and they make up the majority of the fibers in a healthy muscle. Owing to the complicated voltage dependence of the membrane conductance on the membrane potential, muscle fibers can also assume a *low* stable resting potential of about −60 mV. Such "depolarized" fibers are said to be in the P2 state. Their major functional difference is that they cannot be excited and, therefore, are "paralyzed." In a healthy muscle, fibers in the P2 state make up only a small minority. Mutations in the genes encoding the various ion channels responsible for the membrane potential may influence the electrical stability so that fibers can shift from the P1 state to the P2 state, for short time intervals but also for very long durations. The number of excitable fibers is then very much reduced.

MYOTONIAS

Clinical Presentation of Cl^- Channel Myotonias Becker and Thomsen

The major symptom of the myotonias is a generalized muscle stiffness that appears particularly following strong and/or sudden muscle contractions (e.g. when a patient gets scared). The stiffness is caused by a series of involuntary action potentials that can occur in practically all skeletal muscle fibers. The muscles are often hypertrophic, in contrast to what is seen in most other muscle diseases. The involuntary action potentials are caused by one (or

B.T. Darras, H. Royden Jones, Jr., M.M. Ryan & D.C. De Vivo (Eds): Neuromuscular Disorders of Infancy, Childhood and Adolescence, Second edition.
DOI: http://dx.doi.org/10.1016/B978-0-12-417044-5.00038-X

two in the case of recessive Becker myotonia) of a great number of mutations in genes encoding muscular ion channels, which all result in increased excitability of the sarcolemma. Depending on the type of channels affected, chloride and sodium channel myotonias are distinguished.[7]

There are two forms of *chloride channel myotonia*, the rare dominant *Thomsen myotonia* and the more common recessive *Becker myotonia*. Both forms are characterized by generalized myotonia and the so-called warm-up phenomenon, a reduction of myotonia in the course of repeated contractions. A typical feature of myotonia would be the inability of let-go after a hearty handshake (Figure 38.1). Such myotonia during closure of the fist is most pronounced when the muscles have remained relaxed for at least 10 minutes. Another typical sign is the "lid-lag" phenomenon, where after a brief look upward, upon a brisk look downward the eyelids do not immediately follow the eyeballs due to muscle stiffness. Gentle tapping on a relaxed muscle may elicit a "myotonic reaction," a local muscle contraction called "percussion myotonia." A distinct myotonia may be followed by a "transient weakness" of the same muscle. This weakness hampers the patient more than the stiffness.[8,9] Hypertrophy is more pronounced with the Becker form than with the Thomsen, whereby the glutei, thighs, and calves are most affected (Figure 38.2). Paradoxically, the power produced by these muscles is less than expected. In severe cases, the hyperactivity will result in shortened muscles leading to *pes equinus* with consequent lordosis and in reduced extension of the elbow and wrist joints. Males are usually more affected than females. Females

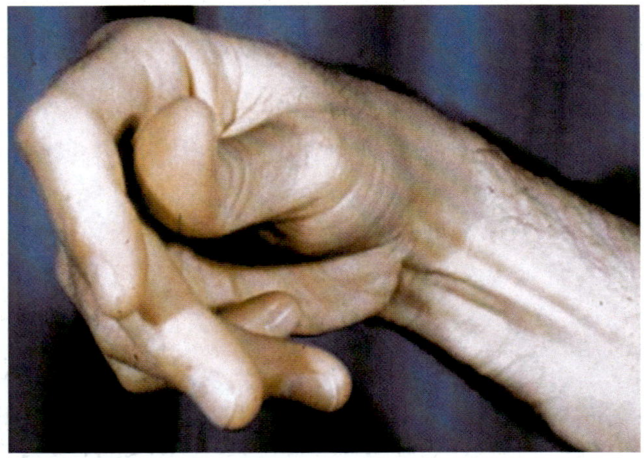

FIGURE 38.1 Myotonic stiffness of the hand of a patient with recessive myotonia (Becker). The patient was asked to rest his fingers for 5 minutes, and then to close the fist with full strength for 3 seconds, and then to stretch the fingers. Complete opening of the hand lasted more than 10 seconds. *Reproduced from Lehmann-Horne et al.*[13] *with permission.*

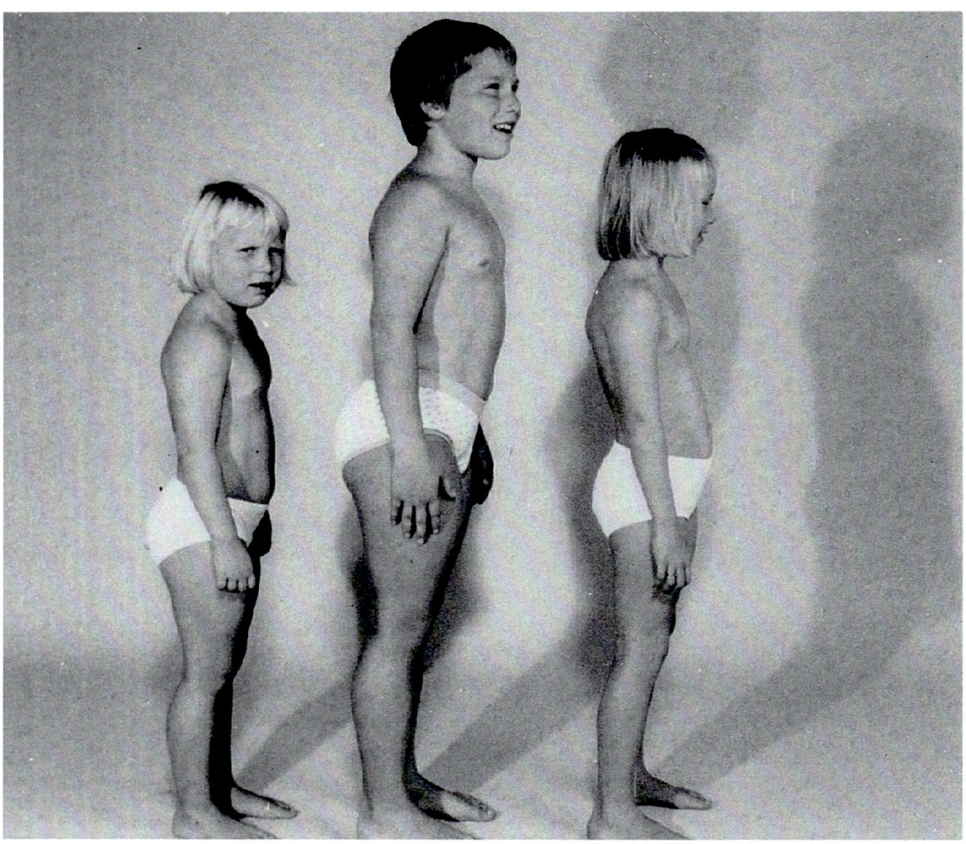

FIGURE 38.2 Three siblings from a family with generalized myotonia (Becker). The girl on the left and the boy are affected, while the girl on the right is not. Note the different posture in the patients, and particularly the hypertrophy of the thighs and upper arms. *Reproduced from Lehmann-Horne et al.*[13] *with permission.*

often develop myotonic symptoms only when hypothyroid or pregnant. In Becker myotonia, myotonic stiffness is usually first detected within the rather long time span between first school enrollment and the third decade of life. During adulthood, the severity of myotonia remains constant; life expectancy is normal. Becker patients have reduced manual skills, which should be given consideration when occupational choices are made.

Clinical Presentation of Na^+ Channel Myotonias Fluctuans and Permanens

Sodium channel myotonia is also called potassium-sensitive myotonia because the myotonic stiffness is increased after oral intake of potassium (e.g. 1 h after 1–2 potassium tablets). Depending on the mutation, the symptoms may occur in a wide range of severity. The most benign form is called *myotonia fluctuans*,[10] and the most severe form *myotonia permanens*.[11,12] With myotonia fluctuans, work-induced stiffness does occur, but usually with a delay of about an hour. It then lasts for about 1 to 2 hours. The typical warm-up phenomenon is often masked by this delay but may be observed with repeated contractions and relaxations of the extremities. Patients with myotonia permanens may be rendered more or less immobile by the continual activity in many of their muscles. Diagnostic tests administering oral potassium should not be performed in these patients because of the danger that stiff respiratory muscles may lead to respiratory insufficiency.[13] Patients with sodium channel myotonia tend to suffer from cramping when their muscles are being stressed.[14]

For a clinical discrimination between chloride and sodium channel myotonia, one may use the fact that chloride channel myotonia shows warm-up of the eyelid muscles, whereas sodium channel mutations cause paradoxical eyelid myotonia.[13,15,16] Further specification is possible by investigating the reactions to cold and potassium. For instance, after cooling of an eye with an ice bag for 10 minutes, a forceful closure of the eye may lead to the inability to open for many seconds. Creatine kinase (CK) values of five times normal may be observed in sodium channel myotonia, whereas in chloride channel myotonia they are at the most a little increased. Some patients, in particular those with sodium-channel myotonia, suffer from muscle pain that increases with the amount of muscle work. Myotonia patients deserve special attention before and during narcosis. In particular, if the myotonia is not readily recognizable as in hyperkalemic periodic paralysis and myotonia fluctuans, the use of potassium and other depolarizing agents such as suxamethonium and cholinesterase inhibitors can provoke the myotonic reaction and thus compromise intubation and respiration. This can induce life-threatening incidents.[17]

Myotonias: Diagnostic Approach

Typical signs of all types of myotonia are bursts of discharges in the EMG that are elicited by the movement of the needle or by tapping on the muscle (Figure 38.3). In addition, sodium channel myotonias show long-lasting series of fibrillation potentials. As with myasthenia gravis at 3 Hz, repetitive nerve stimulation at 10 Hz yields a decrement in chloride channel myotonia. In contrast to myasthenia, however, the decrement begins later, is more pronounced, is not antagonized by Tensilon, and runs parallel with the described transient weakness, i.e. the amplitude recovers with continuing stimulation.[18] No decrement is found in sodium channel myotonia.

The most severe type of sodium channel myotonia is characterized by persistent and severe myotonia, and is therefore called myotonia permanens. Molecular biology has revealed that this condition is caused by specific mutations (Figure 38.4) in the *SCN4A* gene product, $Na_v1.4$.[11,12,19,20] The continuous electrical myotonia leads to a generalized muscle hypertrophy that also involves muscles of the face, neck, and shoulders. If the myotonia is aggravated by intake of potassium-rich food or by exercise, ventilation can be impaired by laryngospasm and stiffness of the thoracic muscles. Children are particularly at risk of suffering acute hypoventilation leading to cyanosis and unconsciousness (Figure 38.5). This has led to confusion with epileptic seizures and resulted in treatment with anticonvulsants. As most anticonvulsants block sodium channels, such therapy is effective although the diagnosis is wrong. Severely affected patients would probably not survive without continuous treatment. One patient was misdiagnosed as having the "myogenic type" of Schwartz-Jampel syndrome.[21] Electrophysiological studies revealed impaired sodium channel inactivation.[22] Finally, molecular genetics showed an *SCN4A* mutation.[13] A further indication of the severity of myotonia permanens is that all patients except one have been sporadic. Due to the severity of the disease, ingestion of potassium or exposure to cold may cause further worsening and should be avoided.

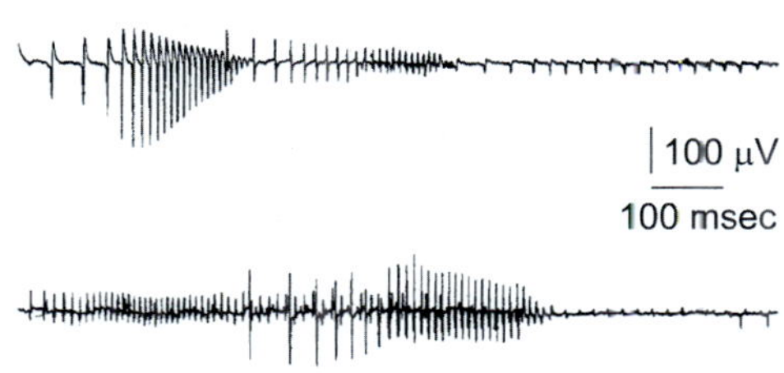

FIGURE 38.3 Myotonic discharges in a patient with chloride channel myotonia that were elicited by the movement of the EMG needle. The potentials show the typical modulation in frequency and amplitude.

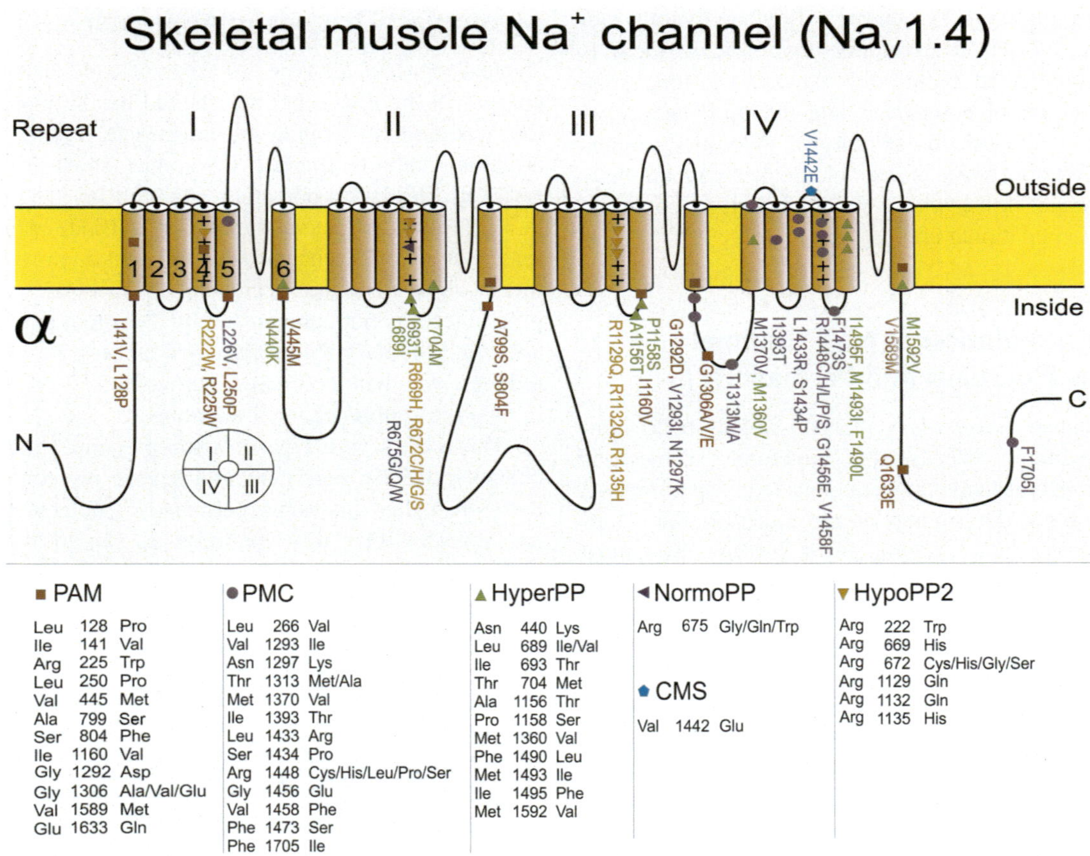

FIGURE 38.4 Schematic of the alpha subunit of the voltage-gated sodium channel $Na_V1.4$ indicating all known mutations responsible for potassium-sensitive myotonia (PAM), paramyotonia (PMC) and hyperkalemic (HyperPP), normokalemic (NormoPP), and hypokalemic periodic paralysis type 2 (HypoPP2). The subunit consists of 4 almost homolog domains (repeats I–IV), formed by 6 transmembraneous alpha-helical segments (1–6). Within the membrane, the 4 domains are so arranged that they form a central pore with segments 5 and 6 lining its wall. The segments 4 contain several positively charged amino acids (arginine and lysine), which are responsible for the voltage sensitivity of the channel. The intracellular loop between domains II and IV represents the channel's inactivation gate. Also, domain IV is important for the inactivation process. All the HypoPP mutations replace the arginines within the voltage sensors by neutral amino acids (which are given in the 3-letter code, e.g. Arg for arginine).

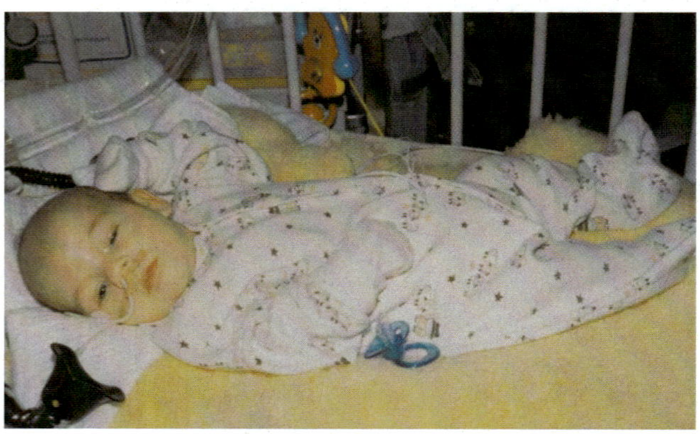

FIGURE 38.5 Male infant with myotonia permanens with a *de novo* $Na_V1.4$-G1306E sodium channel mutation. Many times, ventilation was impaired by laryngospasm and stiffness of the thoracic muscles. The remittent hypoventilation led to cyanosis and unconsciousness and demanded frequent oxygen supply.

Myotonias: Differential Diagnoses

In childhood, recessive or sporadic myotonia may resemble Schwartz-Jampel syndrome (SJS), a rare autosomal recessive disorder characterized by permanent myotonia, myopia, and osteochondrodysplasia including reduced stature, kyphoscoliosis, bending of the diaphyses, and irregular epiphyses.[23] SJS has also been referred to as chondrodystrophic myotonia due to the presence of myotonia, particularly in the face and thighs. Facial characteristics include blepharospasm, myokymic twitching at the chin, and puckered lips (Figure 38.6). Warm-up, the typical phenomenon of chloride channel myotonia, is minimal or lacking. The thigh muscles are often hypertrophic, but the shoulder girdle muscles may be atrophic. Joint contractures, thoracic deformities, and other skeletal abnormalities such as platyspondylosis can be present. Most patients are small in stature due to slow postnatal growth.

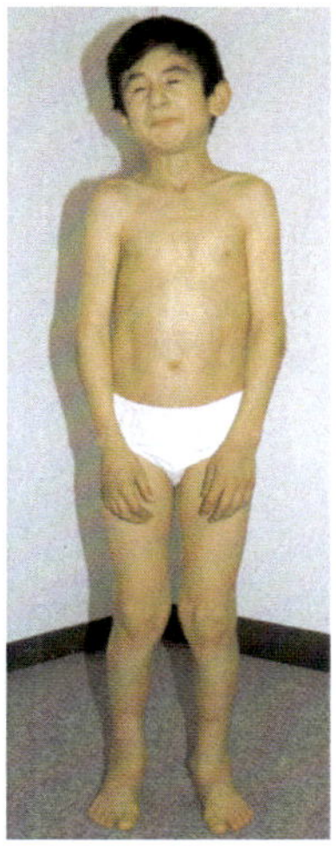

FIGURE 38.6 Schwartz-Jampel syndrome in an 11-year-old boy with marked myopia. Note the blepharospasm, low-set ears, puckered lips, and myokymia visible on the chin, and bending of the diaphyses.

Frequent additional symptoms are myopia, cataract, strabismus, and nystagmus. Other findings can consist of low-set ears, receding chin, laryngeal stridor, high-pitched and forced voice, and high-arched palate. Most reports describe patients under 10 years of age. Little information is available about the prognosis, but few reports of older children suggest a nonprogressive or stationary course. The EMG shows persistent spontaneous activity, particularly in the face and thigh muscles, described as pseudomyotonic or as complex repetitive discharges. This activity is diminished by curare.

SJS is caused by missense and splice-site mutations in *HSPG2* at 1p34–p36, the gene encoding perlecan.[24] Perlecan co-assembles with dystroglycan, and this complex serves as a cell surface receptor for acetylcholine esterase at the neuromuscular junction. The structural importance of perlecan for cartilage and bone development can be attributed in part to the binding partners of perlecan: (1) structural proteins (i.e. collagen type IV, laminins, etc.) protecting the cartilage extracellular matrix, and (2) growth factors such as FGF-7 or FGF-BP influencing chondrocyte activity.[25] EMG on a genetic mouse model confirmed the peripheral nerve as the origin of the spontaneous activity.[26] Perlecan deficiency can be confirmed in skin and muscle samples immunocytochemically.

In adults, when diagnosing myotonia patients one should always think of the myotonic dystrophies (DM), in particular if they have cataracts as in DM1 or atypical breast pain as in DM2.

Myotonias: Therapeutic Interventions

Systematic studies for the therapy of the myotonias do not exist and, therefore, recommendations are not supported by evidence-based data.[27–29] In cases of benign myotonia no medication is required because patients learn to handle their disorder. Patients with sodium channel myotonia should avoid strenuous ventures such as hiking in the mountains. In cases of severe myotonia where medication seems appropriate, as in myotonia permanens or in cases of Becker myotonia, antiarrhythmic drugs have the best effect. By virtue of their "use-dependent" action, they block in essence the pathologic after-activity without affecting the excitability *per se*. Their therapeutic window is narrow, however. They do not affect the chloride channels but do improve chloride channel myotonia, although not as effectively as with sodium channel myotonia. For a long time the Ib-antiarrhythmic drug mexiletine was the drug of choice,[30,31] but recently it has been taken off the market (with the exception of Japan and the United States) because of low cost-effectiveness. As alternatives, the Ic-antiarrhythmic drugs flecainide and propafenone are available.[32] Electro- and echocardiogram should be recorded before administration to avoid a heart bundle block. During use of one of the above antiarrhythmics, patients with heart insufficiency or arrhythmias should have no prolongation of the QRS-complex by more than 20% and the QT time should be no longer than 500 ms. The absolute QTc time should remain stable. Regular cardiologic controls are recommended with all patients. The control of serum potassium values and the monitoring of potential cardiac or central nervous side effects, as well as the avoidance of dehydration, will permit a lifelong treatment, even with beginning in childhood. In the case of cardiac disturbances, one may sidestep to the well-compatible antiepileptic drug lamotrigine. Its most frequent adverse effects are of neurosensory type (diplopia, dizziness) and sometimes exacerbation of epilepsy attacks. Severe cutaneous symptoms were exceptionally observed. Carbamazepine and phenytoin should not be used because of their low antimyotonic action.

Myotonias: Pathophysiology

The mutations responsible for the myotonias are located in the genes *CLCN1* and *SCN4A*, encoding the muscular chloride and sodium channels, respectively. In both forms of chloride channel myotonia, Becker and Thomsen, the mutations lead to decreased activity of the chloride channels in the sarcolemma, which reduces the chloride conductance of the latter. Physiologically the chloride conductance of a muscle fiber comes up to 80% of the total membrane conductance at rest. This resting conductance is passive ("ohmic"), and its rather high value is very important for a stabilization of the muscle fiber's resting potential of the P1 state close to −80 mV. When the chloride conductance is lower than 30% of the total membrane conductance, the muscle fibers still stay

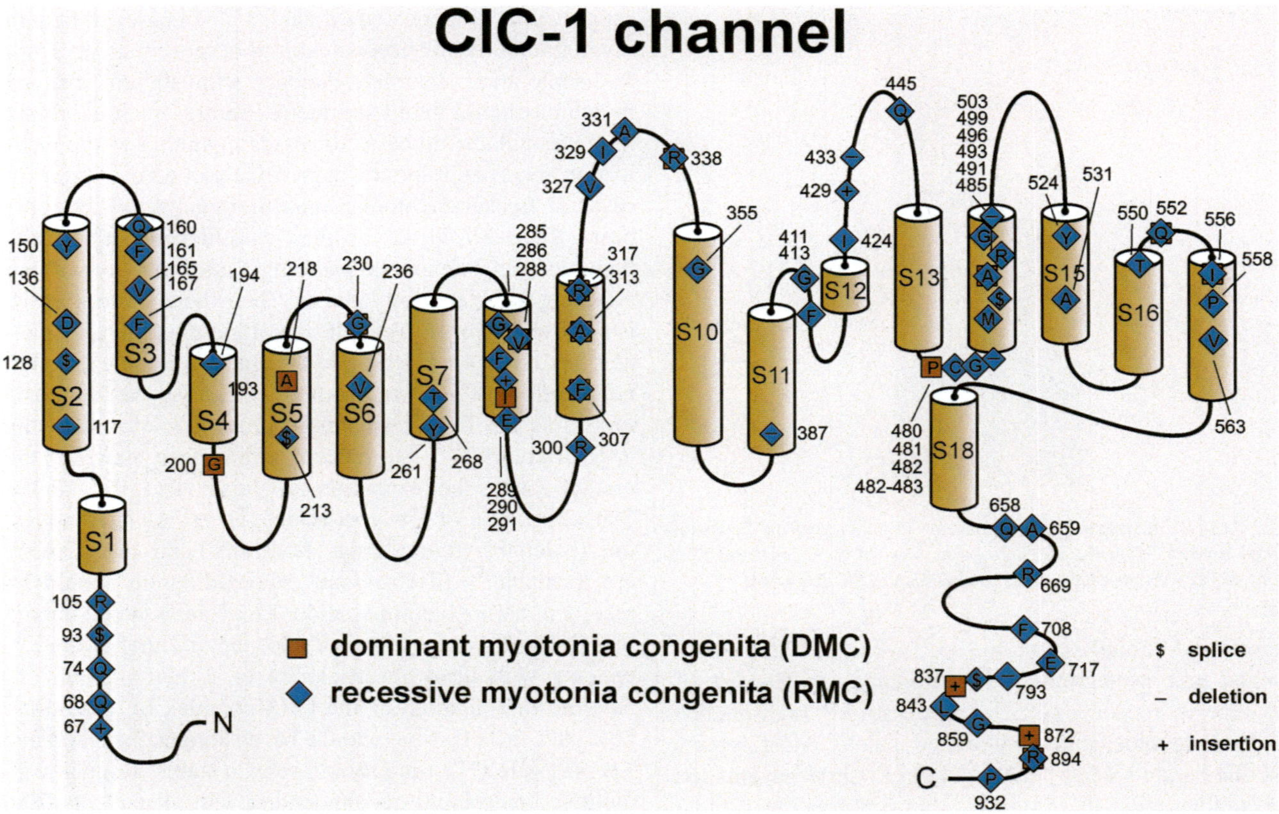

FIGURE 38.7 Schematic of the voltage-gated ClC-1 monomer. The functional channel is an antiparallel assembled homodimer. Symbols are used for the mutations leading to either dominant or recessive myotonia congenita. The amino acids at which substitutions occur are indicated by 1-letter abbreviations and numbered according to the protein sequence. Dominant mutations regard areas involved in the assembly of the proteins.

favorably in the P1 state at rest, but they become hyperexcitable. Clinically, this results in myotonia.[33] During the state of transient weakness, an increased number of muscle fibers are believed to stay in the P2 state at the end of a myotonic series of action potentials.

Mutations of the muscle chloride channel gene with dominant and recessive mode of inheritance are responsible for Becker and Thomsen type myotonia, respectively (Figure 38.7). Functionally, the approximately 10 dominant mutations exert a dominant-negative effect on the homodimeric channel complex as shown by co-expression studies, meaning that mutant/mutant and mutant/wild-type complexes are malfunctional. The most common feature of the resulting chloride currents is a shift of the activation threshold towards less negative membrane potentials almost out of the physiological range.[34,35] As a consequence of this, the chloride conductance is drastically reduced in the vicinity of the resting membrane potential (Figure 38.8). Interestingly, both testosterone and progesterone rapidly and reversibly exert a similar effect on chloride conductance.[36] The approximately 100 recessive mutations do not functionally hinder the associated subunit. This explains why two mutant alleles are required to

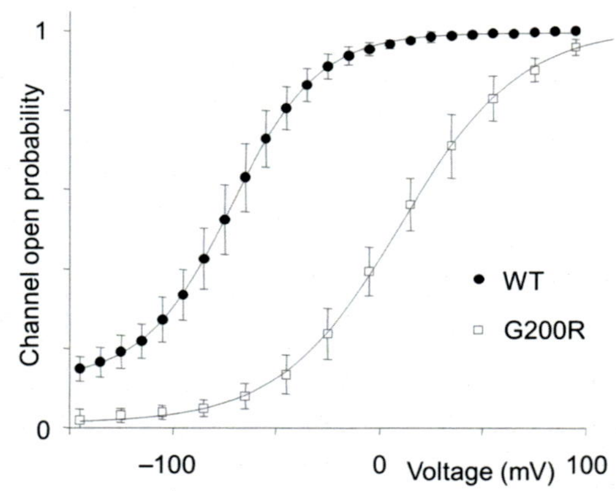

FIGURE 38.8 Voltage dependence of relative open probability of wild-type and G200R channels. The mutant curve is shifted to the right, indicating a reduced opening probability over a physiological range of potentials. Mutations causing such voltage shifts have dominant effects. *Modified after Wagner et al. (1998).*[35]

reduce chloride conductance sufficiently for myotonia to develop clinically in Becker myotonia. Heterozygous carriers of a recessive mutation are healthy but may exhibit some myotonic runs in the EMG.

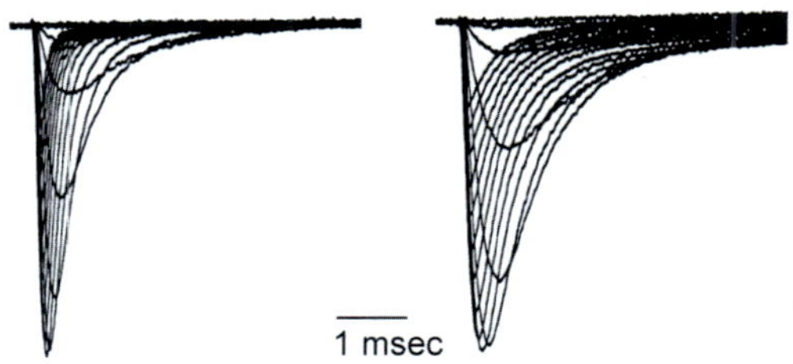

FIGURE 38.9 Example of faulty inactivation of mutant $Na_v1.4$ channels expressed in HEK293 cells. Sodium currents from normal (left) and mutant channels (right) were measured by use of the patch-clamp technique and normalized to equal size for better comparison of the time course. Note the slower inactivation and the persistent current of the mutant channels.

Dominant mutations in the gene encoding the muscular voltage-dependent sodium channel $Na_v1.4$ are responsible for sodium-channel myotonia (Figure 38.4).[10,12,37] The electrophysiologic findings consist of a small persistent current and a mildly slowed current decay (Figure 38.9). The fast inactivation is always accompanied by accelerated recovery from the inactivated state, which reduces the refractory time of the channels and contributes to the generation of repetitive action potentials. This pattern may be explained by a destabilization of the inactivated channel state. That is, the channel does not "like" to enter this state and tries to "escape" (e.g. due to a reduced binding affinity between the "latch bar and the catch"). In agreement with the phenotypic severity of three mutations at the same site, the electrophysiological alterations are minor for G1306A (myotonia fluctuans), moderate for G1306V (intermediate myotonia), and most pronounced for G1306E (myotonia permanens).[12,19,38] As the persistent sodium current and the associated membrane depolarization are small, the main clinical feature is myotonia while weakness does not occur. In addition to dominant missense mutations, aberrant splicing has been reported to cause sodium-channel myotonia.[39]

Myotonias: Molecular Genetics

The prevalence of Thomsen myotonia is now estimated at approximately 1:400,000,[13] i.e. much lower than the 1:23,000 of the premolecular era.[40] This is owing to the fact that about 20% of families with "Thomsen" myotonia are now identified as having sodium-channel mutations that result in a different disease with very similar symptomatology. Other "Thomsen" families were found to have Becker myotonia with pseudodominant inheritance. In contrast to the decreased prevalence of Thomsen myotonia, the prevalence of Becker myotonia is now thought to be 1:24,000,[13,41] much higher than Becker's original estimate of 1:50,000,[40] suggesting a prevalence of 1:78 for the occurrence of a single recessive *CLCN1* mutation. This high prevalence explains the high probability of pseudodominant chloride channel myotonia. Another surprise was the observation that most myotonia permanens patients harbor a *de novo* $Na_v1.4$ mutation. The fact that the parents are not affected can lead to the clinical diagnosis of Becker myotonia.

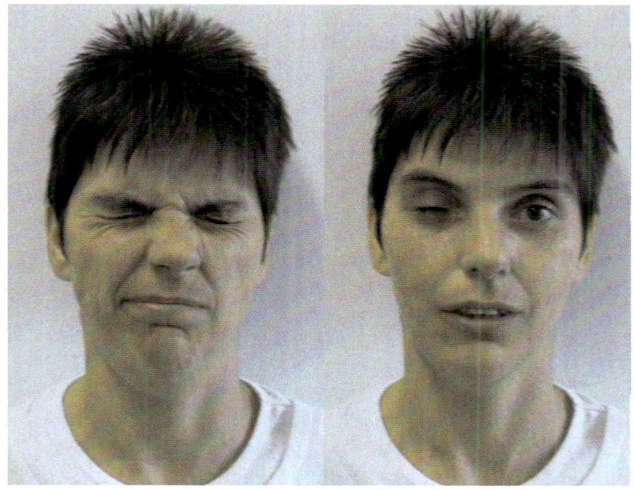

FIGURE 38.10 Local cooling elicits myotonia in a patient with paramyotonia congenita. After the patient's right eye was cooled for 10 min, she was asked to close her eyes forcefully (left) and then open them fast (right). The cooled lid remained closed for some time, initially due to myotonia, later due to flaccid weakness (ptosis).

PARAMYOTONIA

Clinical Presentation

Paramyotonia is a sodium channel myotonia that, depending on the specific mutation, may or may not also show characteristics of hyperkalemic periodic paralysis. Typically, the myotonic stiffness is negligible in warm muscles, but may assume very cumbersome degrees during exposure of the muscles to cold. A slight cold draft may suffice to "freeze" the musculature (Figure 38.10). The warm-up phenomenon might be present in some muscles; however, repeated contractions usually rather increase the muscle stiffness ("paradoxical myotonia"). With extended duration of cooling, the stiffness tends to give way to weakness, which can then last for hours, even after warming the muscles. The mentioned periods of generalized paralysis experienced by patients with certain mutations may occur at rest and in a warm environment.

Paramyotonia: Diagnostic Approach

In patients without myotonia in a warm environment, electrical discharges in the EMG may be absent at normal or increased temperature, but occasional myotonic runs do occur. In patients with myotonia in a warm environment, there are myotonic bursts and myotonic after-activity reveals a warm-up phenomenon. Upon cooling, a fibrillation-like spontaneous EMG activity develops consistently that is maximal at a muscle temperature of about 29°C. This spontaneous activity decreases with a further drop in temperature, and voluntary activity almost disappears at 24°C when the muscle becomes paralyzed. Consequently, CMAP amplitudes in paramyotonia patients decrease upon cooling corresponding to the developing weakness.

Paramyotonia: Differential Diagnoses

Brody disease is a rare inherited myopathy due to reduced SERCA1 Ca-ATPase activity caused by recessive *ATP2A1* mutations. The reduced calcium reuptake into the sarcoplasmic reticulum causes delayed muscle relaxation and silent contractures. The disease has mostly been diagnosed by measurement of SERCA1 activity. Since mutation analysis became available in 1996,[42] it has appeared that not all patients with reduced SERCA1 activity indeed have *ATP2A1* mutations, and a distinction between Brody disease (with *ATP2A1* mutations) and Brody syndrome (without *ATP2A1* mutations) was proposed in 2012.[43]

Paramyotonia: Therapeutic Interventions

With paramyotonia patients, a short-term therapy (e.g. for 2 to 3 days before exposure to cold) is preferable to a long-term treatment. The same is true for sodium channel myotonia and extreme bodily stress. Patients with hyperkalemic periodic paralysis should avoid strong bodily stress and should have frequent and smaller meals. They should also avoid potassium-rich food (dried fruit, bananas, apricots, cauliflower, leguminous plants, etc.). For prophylaxis, medication with thiazide diuretics or carbonic anhydrase inhibitors is recommended.[44] While both types of drugs stimulate excretion of potassium, the latter do so only to a small extent and can therefore be used for the hypokalemic form as well (see below). Beta-sympathomimetic drugs like salbutamol, when taken at the onset of an episode, may ease the symptom rather quickly. Also, the simultaneous application of glucose and insulin may lower serum potassium rather quickly.

Paramyotonia: Pathophysiology

Most paramyotonia mutations are situated in protein parts relevant for channel inactivation, in the inactivation gate itself (i.e. the intracellular loop connecting domains III and IV like T1313M), in the outermost arginine of the voltage sensor in domain IV (R1448H/C/S/P),[45,46] in intracellular S4−S5 loops of domain III or IV (e.g. F1473S),[47] or in the C-terminus (F1705I) (Figure 38.4).[38] During cooling to 27°C *in vitro*, paramyotonia muscle fibers slowly depolarize from −85 mV to about −40 mV whereas normal muscle fibers depolarize by not more than 5 mV. The depolarization is associated with a long-lasting burst of action potentials, which ends when the membrane potential approximates values of −55 mV.[2,46] At this voltage, most genetically normal sodium channels are inactivated and therefore the muscle fibers become inexcitable and paralyzed. Functional expression of the mutant channels revealed slowed fast inactivation and accelerated recovery from the inactivated state and an uncoupling of *fast* inactivation from activation (Figure 38.9).[5,45] As *slow* sodium channel inactivation should also be incomplete to maintain depolarization-induced paralysis,[48] several groups examined the effects of temperature on slow inactivation of the mutant channels.[49−51] The results were not uniform and were difficult to interpret since entry into slow inactivation was already changed by the strikingly slowed fast inactivation. Bouts of long-lasting weakness in paramyotonia can be explained with an ongoing membrane depolarization in the majority of muscle fibers (i.e. a preference of fibers to stay in the P2 state). Further sustained depolarization of paramyotonic fibers to potentials positive to −40 mV induces a calcium release in the absence of an action potential, leading to an electrically silent muscle contraction, a contracture. Highly specific for PC is ^{23}Na MRI, which can visualize the myoplasmic sodium accumulation that is related to the membrane depolarization and correlates well with the grade of paresis (Figure 38.11).

Paramyotonia: Molecular Genetics

Paramyotonia is considered an extremely rare disorder, though little epidemiological work has been done. Prevalence is generally higher in Europe-derived populations and lower among Asians. Epidemiological estimates provided for the German population suggest a prevalence of between 1:350,000 and 1:180,000.[53,54] It should be noted, however, that paramyotonia patients are not uniformly distributed across Germany. Most affected families with the $Na_v1.4$-R1448H mutation stem from the Bielefeld-Ravensberg area in Northwest Germany, where a founder effect seems to be responsible for most cases.[53,54] The prevalence in that area is estimated at 1:6,000.

PERIODIC PARALYSES

Clinical Presentation of Hyperkalemic, Normokalemic and Hypokalemic Periodic Paralyses

Although genes that encode channels conducting ions other than potassium are mutated in all the periodic

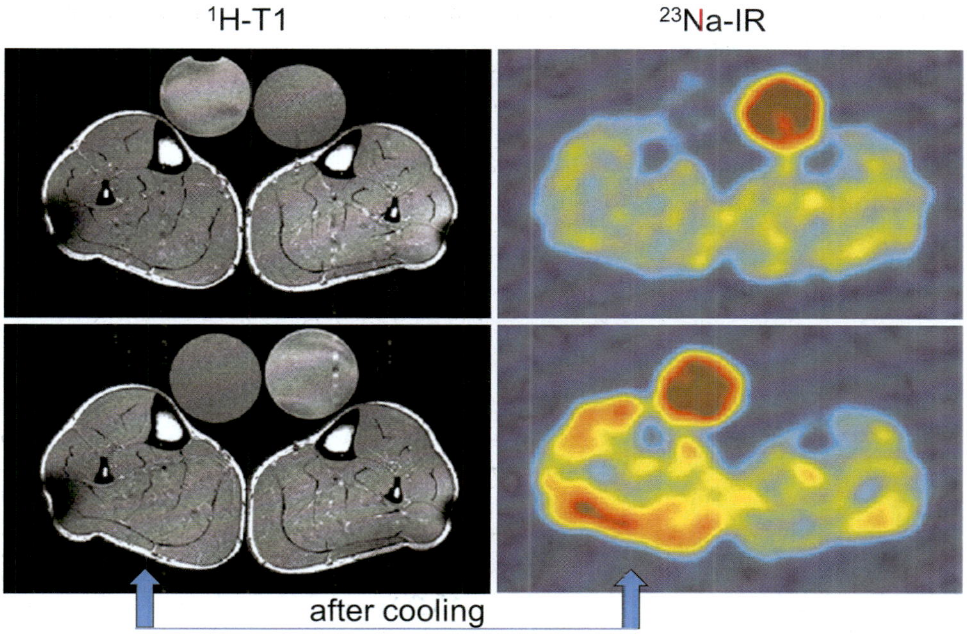

FIGURE 38.11 ¹H-MR and ²³Na-MR images of a PC patient before and after cooling of the lower leg (arrow). While the T1-weighted ¹H-MR signal is unaltered, the ²³Na inversion recovery (Na-IR) signal is markedly increased in the muscles of the cooled leg (blue: low intensity, red: high intensity). *Modified after Nagel et al.*[52]

paralyses, with the exception of the Andersen-Tawil syndrome (see below), the level of serum potassium plays a decisive role in the symptoms, namely by triggering attacks in the hyperkalemic and normokalemic forms and relieving paralysis in the hypokalemic form. The situation with glucose is similar because, in combination with insulin, it shifts potassium from the extracellular to the intracellular space and thus may likewise have triggering or therapeutic effects. The same is true for cortisone and drugs that modify the serum potassium level. In contrast, bodily and psychic stress, cold, infections, vaccinations, operations, and drugs always act adversely in all forms of channelopathies. Muscle biopsy is indicated only if molecular genetics do not provide the diagnosis or when there is reasonable suspicion of additional neurogenic or inflammatory processes. Because of possible dangerous situations of hypoglycemia or hypokalemia, diagnostic tests using glucose/insulin loading should be performed only in exceptional cases. Stand-by anesthesia is required.

The typical symptom of the periodic paralyses is the experience of episodes of localized or generalized flaccid paralysis that may last for a few minutes to several days. Respiratory and vocal muscles may be affected. Muscle reflexes might be reduced during an episode. At increased age of patients, a permanent weakness may additionally develop and result in muscular dystrophy. *Hyperkalemic, normokalemic* and *hypokalemic periodic paralyses* are distinguished depending on the serum potassium during a full-fledged episode. The diagnosis between these variants is not always easy, as the result of a single measurement of serum potassium may be misleading. Additional information received from the determination of triggers and the duration of an episode as well as from the reaction to the administration of potassium, sodium, or glucose may be helpful. Attacks of hyperkalemic periodic paralysis typically begin in childhood and are usually brief, being of shorter duration (minutes to hours) than attacks of hypokalemic periodic paralysis. Episodes of hyper- and normokalemic, but not of hypokalemic, periodic paralyses may be accompanied by myotonia. Between episodes, myotonia might be detectable only in the EMG, but may be present at the beginning of an episode.[55]

Symptoms set in during childhood or youth and only rarely during early adulthood. Frequency and severity of the paralytic episodes decrease from 50 years of age onward, but about 50% of cases then suffer from a chronic progressive muscular weakness with increased CK values, myopathic EMG changes, and histological alterations similar to a vacuolar myopathy or a limb-girdle dystrophy. Cases with progressive atrophies have also been described.[56] Females with incomplete penetrance for the paralytic attacks may well present with such a myopathy at a later age. In addition to the prevailing medical problems, there are psychosocial and judicial questions that arise during the course of these diseases (occupational suitability, military service, degree of disability, possibility of therapy). Therefore, advocacy groups, social services, psychotherapists, and paramedical services may be of essential help. Moreover, stress caused by an operation may lead to hypokalemia and may trigger a paralytic attack in patients with hypokalemic periodic paralysis. Development of hypokalemia during general anesthesia may evoke local or generalized muscle weakness in all forms of periodic paralysis. Therefore, measures to retain body temperature should always obviate hypothermia of the patient during the operation.

Periodic Paralyses: Andersen-Tawil Syndrome

This is a very rare disease, characterized not only by episodes of paralysis but also by life-threatening ventricular arrhythmias. During paralysis, the serum potassium level may be increased, normal, or reduced.[57] Accordingly, the response to oral potassium is unpredictable. Andersen-Tawil syndrome is characterized by a clinical triad that includes periodic paralysis, long QT interval with ventricular arrhythmias, and dysmorphic physical findings. The ECG should be monitored at rest as well as in long-term observations. Some of these patients show skeletal abnormalities of the face, hands, and/or feet (Figure 38.12). The dysmorphic features include low-set ears, hypertelorism, broad nasal bridge, retro- or micrognathia, short stature, and clinodactyly. Diagnosis is difficult in some families because the triad of findings is sometimes incomplete in different family members. Some patients have only long QT syndrome while others may not have episodes of paralysis. The dysmorphic features may be accompanied by a mild fixed muscle weakness that resembles a congenital myopathy, but this myopathy is usually nonprogressive. Mutations of the Kir2.1 potassium channel, an inward rectifier expressed in skeletal and cardiac muscle, are causative of the disorder.[58] Kir2.1 channels are essential for maintaining the highly negative resting membrane potential of muscle fibers and accelerating the repolarization phase of the cardiac action potential. The mutations mediate loss of channel function by haplo-insufficiency or by dominant-negative effects on the wild-type allele and may lead to long-lasting depolarization and membrane inexcitability. The prevalence is estimated at <1:100,000.

Thyrotoxic Hypokalemic Periodic Paralysis

Thyrotoxic hypokalemic periodic paralysis resembles familial hypokalemic periodic paralysis with respect to changes in serum and urinary electrolytes during attacks and in its response to glucose, insulin, and rest after exertion. It differs from familial hypokalemic periodic paralysis as it occurs only in the course of hyperthyroidism, even though the latter does not need to be clinically obvious. While Graves' disease (hyperthyroidism only) shows a 5:1 female to male predominance with a prevalence of 2% in the general population, this ratio is about 1:6 in thyrotoxic hypokalemic periodic paralysis, at least in Asian patients. Statistically, the incidence of periodic paralysis in Asian men with Graves' disease has been estimated at between 13% and 24%.[13] More than 75% of the cases occur in Asians, suggesting a predisposing racial factor. The attacks occur much more frequently in summer than in winter. A geographical component is not likely, because Chinese and Japanese immigrants in North or South America have the same disease frequency as in their country of origin. Reports of cases in Caucasians and Blacks indicate that the disease rarely occurs in non-Asians.[59]

The onset of disease is usually after the age of 20 years. Forty-five percent of patients develop the syndrome in the third decade of life, another 35% in the fourth, and the rest in the fifth. The thyrotoxicosis precedes or appears simultaneously with the periodic paralysis in more than 80% of patients,[60] but the thyrotoxic signs are relatively mild at the time of the initial attack (no palpitations, goiter, or exophthalmos). Sudden paralytic attacks of proximal limb muscles are typical after strenuous exercise or at rest following high-carbohydrate meals in the evening or during the night. Hypokalemia is also typical during the attacks. The serum potassium falls to levels below 3.5 mM in 80% of the patients. In some patients, it may be as low as 1.2 mM and cause life-threatening arrhythmias or sinoatrial block. As the hypokalemia is the result of an insulin-induced shift of potassium from the extracellular to the intracellular space of the muscles, potassium is released

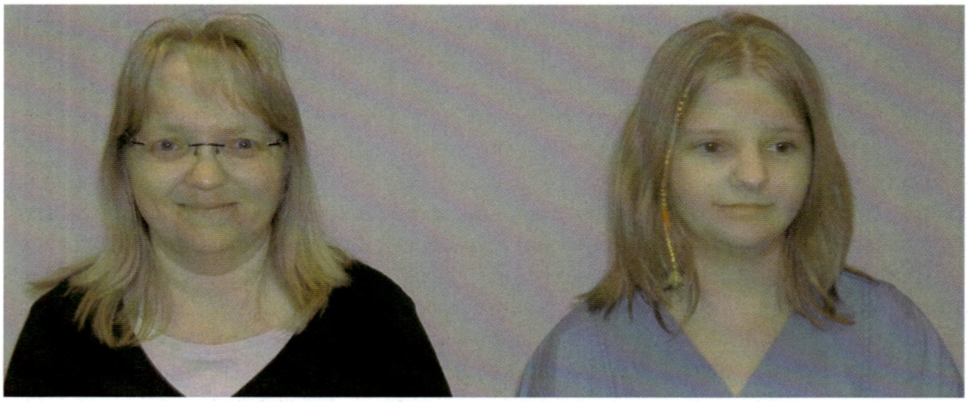

FIGURE 38.12 Mother and daughter with Andersen's syndrome. Note the hypertelorism and the broad bridge of the nose.

from muscle at the end of an attack, which causes rebound hyperkalemia. During an attack, both the arrhythmia and the acute paralytic attack are relieved by administration of potassium. The paralytic attacks cease when the euthyroid state is restored, but recurrence of the hyperthyroid state causes recurrence of the paralytic attacks.

KCNJ18, a gene encoding the inwardly rectifying potassium channel Kir2.6, was mutated in 0% to 33% of patients of various ethnic origins.[61] Kir2.6 is transcriptionally regulated by thyroid hormone.

Periodic Paralyses: Diagnostic Approach

Dyskalemia during a weakness episode, frequently preceded by a known trigger and return of strength coincident with normalization of serum potassium or positive genetic testing, are unequivocal criteria. Auxiliary findings include prevention of episodes with either increasing or decreasing serum potassium or prophylactic medication; ictally reduced stretch reflexes, voluntary EMG activity, or cMAP amplitude; and electrically silent muscle. Interictally, a late decrement in the long exercise test supports the diagnosis of periodic paralyses. Electrical myotonia is evidence for the diagnosis of hyperkalemic periodic paralysis or paramyotonia. To decide on the appropriate therapy, it does not suffice to clarify the form of dyskalemia; one should also search for an ictal or interictal edema and for dystrophy using MRI (Figure 38.13).[6,62] The ischiocrural musculature is particularly likely to be affected by fatty degeneration.[63]

Periodic Paralyses: Differential Diagnoses

The most important differential diagnosis is episodic ataxia. Episodic ataxias (EAs) are characterized by episodic spells of cerebellar ataxia that can be triggered by stress, startle, or heavy exertion such as exercise. Symptoms can first appear in infancy. There is a phenotypic overlap with migraine, spinocerebellar ataxia, and epilepsy. EA1 is associated with myokymia (neuromyotonia)—i.e. continuous muscle movement—and usually presents with paroxysmal truncal and limb ataxia and dysarthria lasting seconds to minutes. Nystagmus is absent. Typically, episodes are triggered by strong emotion or exercise and last seconds to minutes. The syndrome usually presents in childhood or adolescence and often improves spontaneously in the third decade. About 10% of patients also have epilepsy. Inheritance is autosomal dominant. Approximately 20 mutations have been described, almost all of which are missense mutations of the *KCNA1* gene that encodes the voltage-gated potassium channel $K_v1.1$. Most involve highly conserved amino acids such as those in the transmembrane segments. If the functional changes mainly show a slowing of the time course of activation, the phenotype may be primarily neuromyotonia without ataxia; if the threshold of activation is shifted or the current reduced, the ataxia is more prominent. Reduced penetrance can occur.

EA2 is caused by mutations of *CACNA1A*, the gene encoding the neuronal voltage-gated P/Q-type calcium channel $\alpha1$ subunit, $Ca_v2.1$. The ataxia lasts longer and mild interictal nystagmus and ataxia are present. Vertigo, nausea, and vomiting precede the episodes in over half of the patients. Over 50% have migraine as well. For diagnosis, interictal gaze-evoked nystagmus with features typical of rebound nystagmus may be elicited. Spontaneous vertical nystagmus, particularly downbeat nystagmus, is seen in ~30% of cases. Penetrance is 80 to 90%. Acetazolamide and 4-aminopyridine are effective in controlling or reducing the frequency and severity of attacks. More than 50 $Ca_v2.1$ EA2 mutations have been described of which the majority represent nonsense mutations leading to premature truncations of the protein with loss of function. The prevalence has been estimated at lower than 1:100,000 population.

Patients having events of paralysis may suffer from dyskalemia (e.g. accompanying Conn's, Addison's, or Bartter's syndrome). In contrast to genuine periodic paralysis, these conditions are accompanied by long-term pathologic alterations of muscle force and serum potassium. Of course, a so-far compensated state can be decompensated by an acute additional change in serum potassium, and thus lead to an episode of paralysis.

Periodic Paralyses: Therapeutic Interventions

Therapeutic studies of the periodic paralyses have so far only been performed with dichlorphenamide,[64] a carbonic anhydrase inhibitor no longer available. With normokalemic periodic paralysis patients, avoidance of dyskalemia and medication with carbonic anhydrase inhibitors is a suitable prophylaxis against paralytic episodes. Patients with hypokalemic periodic paralysis should avoid food rich in carbohydrates and strong bodily work. They should eat food rich in potassium (see above) and low in sodium. To avoid paralytic episodes, medication with carbonic anhydrase inhibitors like acetazolamide and/or the aldosterone antagonist eplerenone is recommended (Figure 38.13). Potassium saving diuretics like triamterene or amiloride would also be suitable but in certain countries they are only available in combination with potassium-excreting diuretics; this, of course, counteracts their effect. The dose of the drugs should be slowly increased (e.g. in adults, eplerenone should start with 25 mg/d and be increased to 50 mg/d in the second week; the dose will need adjustment in children based on weight). Safety has been established in children younger than 18 years of age.[65] A combination of carbonic

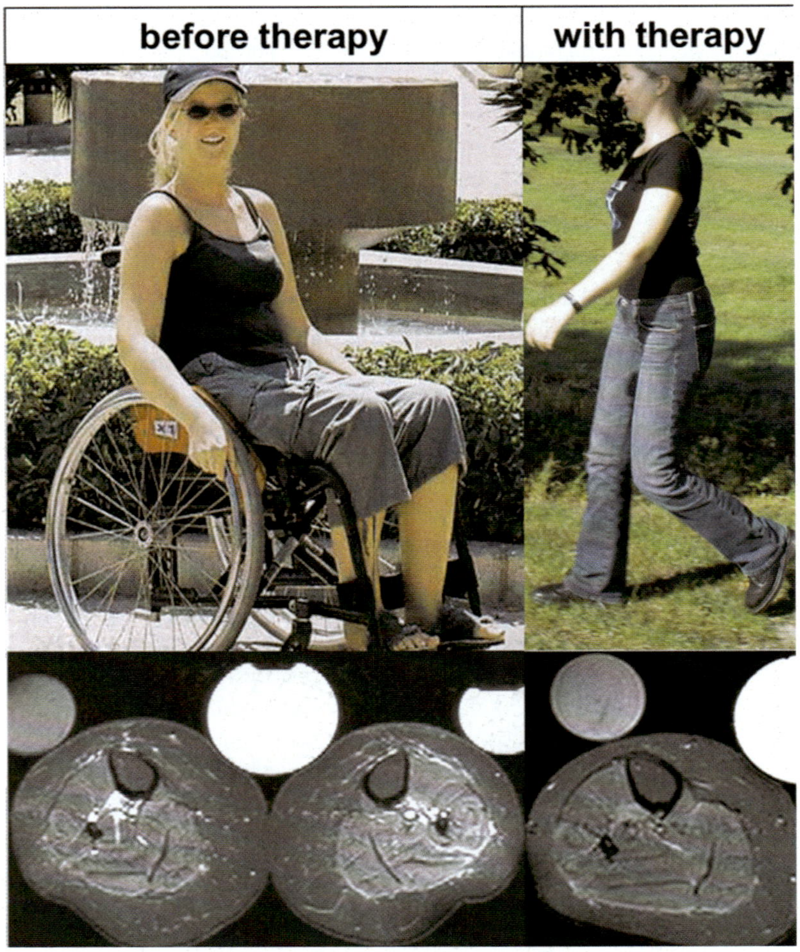

FIGURE 38.13 Patient with wheelchair-boundness since age of 14 years (left upper panel). At age of 24, the diagnosis of hypokalemic periodic paralysis was made genetically. After administration of an aldosterone antagonist and physical therapy, the patient was able to walk and to jog (right upper panel). Moreover, the hypotrophic and edematous muscles (left lower panel—both lower legs) became normotrophic and the edema was washed out by the medication (right lower panel—a lower leg). The edema causes the bright signal in the fat-suppressed ^{1}H MRI. This example shows that the weakness is reversible as long as the muscles have not undergone fatty degeneration.

anhydrase inhibitors, potassium-sparing diuretics, and potassium, each at low dosage, may reduce the side effects of the drugs. With this combination, the possibility of hyperkalemia with muscular weakness and cardiac arrhythmias exists, which persists until the hyperkalemia is reversed. The danger of this possibility happening is, however, much lower than feared by many physicians, because most of the patients present even in the interval with serum potassium values at the lower norm, and the potential of eplerenone to cause hyperkalemia is extremely low in contrast to the formerly used spironolactone. In fact, it turns out to be rather difficult to raise their serum potassium level to a persisting 4.8 to 5.2 mM, values that would suffice to let them experience less muscle weakness, less muscle pain, and a reduced number of paralytic episodes.

In some patients, carbonic anhydrase inhibitors have an opposite effect: for reasons unknown they lead to an increase in the number and severity of paralytic attacks.[66] Aldosterone antagonists and potassium-sparing diuretics are helpful in these cases.[6] In an acute episode, at least, potassium should be administered. The aforementioned drugs are also effective in antagonizing the lasting muscle weakness, provided muscle tissue is still preserved.[6] Since it has been shown that potassium may repolarize the depolarized muscle fibers of patients with hypokalemic periodic paralysis, potassium should not only be given during a paralytic episode, but also as a prophylactic measure. This may sound paradoxical, but often it works! Best suited are slow-retard potassium tablets.[6] A thyrotoxic hypokalemic periodic paralysis is best cured by treating the hyperthyroidism. In an acute case, potassium and carbonic anhydrase inhibitors may be used.[13] Some experts also advocate beta-blockers.

In Andersen-Tawil syndrome, carbonic anhydrase inhibitors can be used as prophylactic means against paralytic attacks. Extrasystoles should not be treated at any cost; often the arrhythmias disappear with tachycardia, as with moderate bodily stress or fever. A treatment with blockers of the sodium or calcium channels or with beta-blockers is common but often not effective. Earlier investigators reported imipramine as useful. Amiodarone should be given only in severe cases because of its substantial side effects. For the long-QT syndrome, this drug is even contraindicated.[67] In cases of arrhythmogenic syncope, a pacemaker or a combined pacemaker/defibrillator may be indicated. All drugs that prolong the QT time are strongly contraindicated.

Periodic Paralyses: Pathophysiology

Dominant mutations in the gene encoding the muscular voltage-dependent sodium channel $Na_V1.4$ are responsible for *hyperkalemic periodic paralysis* (Figure 38.4).[68,69] The mutant channels persistently conduct sodium currents that generate repetitive action potentials. As in paramyotonia, the progressive membrane depolarization inactivates the wild-type sodium channels and turns the hyperexcitability into hypoexcitability, thereby causing flaccid weakness (Figure 38.9).

Hypokalemic periodic paralysis is caused by mutations in the protein segments containing the voltage sensors in either of two very homolog muscular ion channels: the sodium channel $Na_V1.4$ and the L-type calcium channel $Ca_V1.1$ (encoded by *CACNA1S*).[70–72] In either channel, the mutations generate additional pores in the voltage sensor segments.[73] These pores conduct aberrant sodium currents that induce the depolarized P2 state in the affected fibers (Figure 38.14).[6] Determination of the specific mutant allows one to predict the minima of serum potassium levels adopted by a particular patient. These minima determine the severity of future paralytic attacks and of secondary cardiac disturbances. They are also responsible for the course of the progressive myopathy. If the muscle fibers stay in the P2 state for a very long time, sodium ions will accumulate in the intracellular space, and by virtue of osmotic forces, this will be followed by accumulation of intracellular water.[6] This disequilibrium plus the electrical inexcitability leads to a progressive myopathy that may develop vacuoles or may histologically look like a dystrophy.

Periodic Paralyses: Molecular Genetics

The periodic paralyses are rare diseases with dominant inheritance. The prevalence of all forms taken together is about 1:50,000. The *hyperkalemic form* is transmitted with full penetrance, a male-to-female ratio of 1:1, and a prevalence of 1:200,000.[13] Patients without interictal myotonia are much more prone to develop progressive myopathy and permanent weakness than those with myotonia. This is obvious in individuals with the most common T704M mutation: about half of them show myotonia in the EMG, and these are the individuals who do not develop a permanent myopathy. The second most frequent mutation, M1592V, is always associated with myotonia in the EMG, and patients have never been reported to suffer from permanent myopathy.

The *hypokalemic form* is the most frequent, with a prevalence of 1:100,000. Hypokalemia causes a sustained membrane depolarization that reduces membrane excitability.[75] The disease is transmitted with reduced penetrance in women (the male to female ratio is 3 or 4 to 1). The most common mutations for hypokalemic periodic paralysis type 1 (HypoPP-1) are the $Ca_V1.1$ substitutions R528H and R1239H, both of which are situated in voltage sensors. Because of the large omega current, both lead to progressive myopathy.[6] Also, the rare R528G and R1239G mutations are severe mutations that can cause respiratory

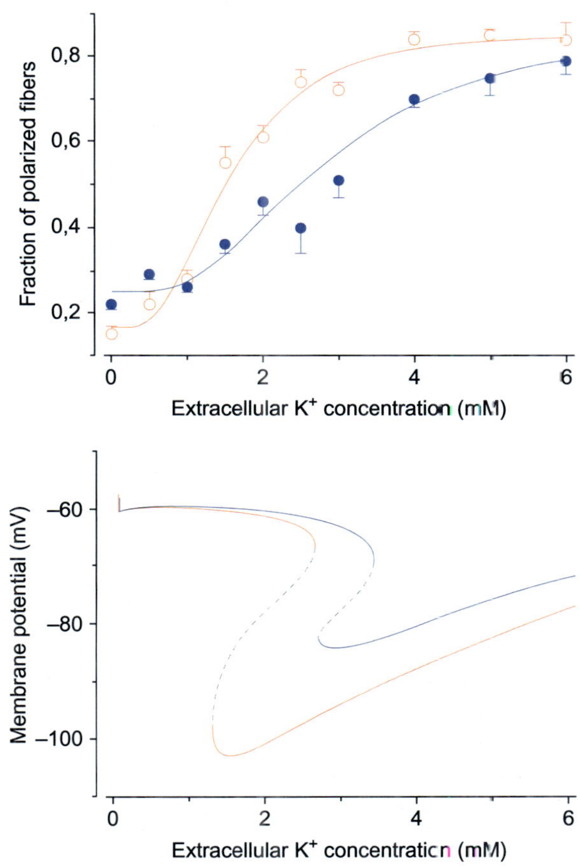

FIGURE 38.14 Bistable membrane potentials and pathogenesis of hypokalemic periodic paralysis. *Upper panel*: Fraction of rat muscle fibers with normal membrane potentials (P1) after reduction of extracellular K^+ to various values, without (red) and with (blue) a Na^+ ionophore mimicking the omega pore. Curves represent fits of a log-normal cumulative distribution function. *Lower panel*: Bifurcation diagram of a mathematical model of skeletal muscle with the extracellular K^+ concentration as control parameter. The *red* curve shows the normal muscle behavior, i.e. a reduction of the extracellular K^+ concentration causes instability of the resting potential at a pathologically low K^+ concentration of about 1 mM from which a sudden transition to P2 of −58 mV occurs (paradoxical depolarization). In the presence of an omega pore (*blue*), the membrane instability already occurs at a mild K^+ reduction. *Modified after Jurkat-Rott et al.*[74]

insufficiency and probably progressive myopathy.[76,77] Other $Ca_v1.1$ mutations also affect voltage sensors, such as R897S and R900G,[78,79] while two other $Ca_v1.1$ mutations (V876E, H916Q) reported to cause HypoPP are not located in voltage sensors and must have a different pathogenesis.[80,81] The $Na_v1.4$ mutations that cause HypoPP-2 are rare. All affect voltage sensors (Figure 38.4).

The name *normokalemic periodic paralysis* was originally given to a variant described in the 1960s. The disorder resembled the hyperkalemic form in many aspects; the only real differences were the lack of increase in the concentration of serum potassium even during serious attacks, and the lack of a beneficial effect of glucose administration. The existence as a nosologic entity was questioned because of the potassium sensitivity of the patients and the identification of the most frequent hyperkalemic periodic paralysis mutations T704M or M1592V in such families, including the original family.

In 2004, Vicart and colleagues reported a potassium-sensitive type of periodic paralysis with normokalemia and episodes of weakness reminiscent of those in both hyperkalemic (initiation of an attack by potassium) and hypokalemic forms (duration of attacks).[82] This phenotype, also named normokalemic periodic paralysis, is caused by *SCN4A* mutations at deeper locations of the voltage sensor of domain II at codon 675 (Figure 38.4). Functionally, R675 mutations generate a leak current with a reversed voltage dependence compared to mutations causing HypoPP-2 since this site is exposed to the extracellular space at stronger depolarizations. Future studies will show whether normokalemic periodic paralysis is a separate clinical entity. The diagnostics are as described for the two more common forms of the disease.

ACKNOWLEDGMENTS

We thank the Else-Kröner-Fresenius-Stiftung (2010-A27), the BMBF (IonoNeurOnet), and the nonprofit Hertie Foundation for support afforded to Karin Jurkat-Rott and Frank Lehmann-Horn. Frank Lehmann-Horn is endowed Senior Research Professor for Neurosciences of the Hertie Foundation.

REFERENCES

1. Bryant SH. Cable properties of external intercostal muscle fibres from myotonic and nonmyotonic goats. *J Physiol* 1969;**204**:539–50.
2. Lehmann-Horn F, Rudel R, Dengler R, Lorkovic H, Haass A, Ricker K. Membrane defects in paramyotonia congenita with and without myotonia in a warm environment. *Muscle Nerve* 1981;**4**:396–406.
3. Lehmann-Horn F, Rudel R, Ricker K, Lorkovic H, Dengler R, Hopf HC. Two cases of adynamia episodica hereditaria: in vitro investigation of muscle cell membrane and contraction parameters. *Muscle Nerve* 1983;**6**:113–21.
4. Hoffman EP, Lehmann-Horn F, Rudel R. Overexcited or inactive: ion channels in muscle disease. *Cell* 1995;**80**:681–6.
5. Lehmann-Horn F, Jurkat-Rott K. Voltage-gated ion channels and hereditary disease. *Physiol Rev* 1999;**79**:1317–72.
6. Jurkat-Rott K, Weber MA, Fauler M, Guo XH, Holzherr BD, Paczulla A, et al. K+-dependent paradoxical membrane depolarization and Na+ overload, major and reversible contributors to weakness by ion channel leaks. *Proc Natl Acad Sci USA* 2009;**106**:4036–41.
7. Jurkat-Rott K, Lehmann-Horn F. Muscle channelopathies and critical points in functional and genetic studies. *J Clin Invest* 2005;**115**:2000–9.
8. Deymeer F, Cakirkaya S, Serdaroglu P, Schleithoff L, Lehmann-Horn F, Rudel R, et al. Transient weakness and compound muscle action potential decrement in myotonia congenita. *Muscle Nerve* 1998;**21**:1334–7.
9. Rudel R, Ricker K, Lehmann-Horn F. Transient weakness and altered membrane characteristic in recessive generalized myotonia (Becker). *Muscle Nerve* 1988;**11**:202–11.
10. Ricker K, Moxley III RT, Heine R, Lehmann-Horn F. Myotonia fluctuans. A third type of muscle sodium channel disease. *Arch Neurol* 1994;**51**:1095–102.
11. Kubota T, Kinoshita M, Sasaki R, Aoike F, Takahashi MP, Sakoda S, et al. New mutation of the Na channel in the severe form of potassium-aggravated myotonia. *Muscle Nerve* 2009;**39**:666–73.
12. Lerche H, Heine R, Pika U, George Jr. AL, Mitrovic N, Browatzki M, et al. Human sodium channel myotonia: slowed channel inactivation due to substitutions for a glycine within the III-IV linker. *J Physiol* 1993;**470**:13–22.
13. Lehmann-Horn F, Rüdel R, Jurkat-Rott K. Nondystrophic myotonias and periodic paralyses. In: Engel AG, Franzini-Armstrong C, editors. *Myology*. 3rd ed. New York: McGraw-Hill; 2004. pp. 1257–300.
14. Orrell RW, Jurkat-Rott K, Lehmann-Horn F, Lane RJ. Familial cramp due to potassium-aggravated myotonia. *J Neurol Neurosurg Psychiatry* 1998;**65**:569–72.
15. Stunnenberg BC, Ginjaar HB, Trip J, Faber CG, van Engelen BG, Drost G. Isolated eyelid closure myotonia in two families with sodium channel myotonia. *Neurogenetics* 2010;**11**:257–60.
16. Trip J, Drost G, Ginjaar HB, Nieman FH, van der Kooi AJ, de Visser M, et al. Redefining the clinical phenotypes of non-dystrophic myotonic syndromes. *J Neurol Neurosurg Psychiatry* 2009;**80**:647–52.
17. Klingler W, Lehmann-Horn F, Jurkat-Rott K. Complications of anaesthesia in neuromuscular disorders. *Neuromuscul Disord* 2005;**15**:195–206.
18. Dupre N, Chrestian N, Bouchard JP, Rossignol E, Brunet D, Sternberg D, et al. Clinical, electrophysiologic, and genetic study of non-dystrophic myotonia in French-Canadians. *Neuromuscul Disord* 2009;**19**:330–4.
19. Mitrovic N, George Jr. AL, Heine R, Wagner S, Pika U, Hartlaub U, et al. K(+)-aggravated myotonia: destabilization of the inactivated state of the human muscle Na+ channel by the V1589M mutation. *J Physiol* 1994;**478**(Pt 3):395–402.
20. Gay S, Dupuis D, Faivre L, Masurel-Paulet A, Labenne M, Colombani M, et al. Severe neonatal non-dystrophic myotonia

secondary to a novel mutation of the voltage-gated sodium channel (SCN4A) gene. *Am J Med Genet A* 2008;**146**:380–3.

21. Spaans F, Theunissen P, Reekers AD, Smit L, Veldman H. Schwartz-Jampel syndrome: I. Clinical, electromyographic, and histologic studies. *Muscle Nerve* 1990;**13**:516–27.
22. Lehmann-Horn F, Iaizzo PA, Franke C, Hatt H, Spaans F. Schwartz-Jampel syndrome: II. Na+ channel defect causes myotonia. *Muscle Nerve* 1990;**13**:528–35.
23. Schwartz O, Jampel RS. Congenital blepharophimosis associated with a unique generalized myopathy. *Arch Ophthalmol* 1962;**68**:52–7.
24. Nicole S, Davoine CS, Topaloglu H, Cattolico L, Barral D, Beighton P, et al. Perlecan, the major proteoglycan of basement membranes, is altered in patients with Schwartz-Jampel syndrome (chondrodystrophic myotonia). *Nat Genet* 2000;**26**:480–3.
25. Forsberg E, Kjellen L. Heparan sulfate: lessons from knockout mice. *J Clin Invest* 2001;**108**:175–80.
26. Echaniz-Laguna A, Rene F, Marcel C, Bangratz M, Fontaine B, Loeffler JP, et al. Electrophysiological studies in a mouse model of Schwartz-Jampel syndrome demonstrate muscle fiber hyperactivity of peripheral nerve origin. *Muscle Nerve* 2009;**40**:55–61.
27. Lehmann-Horn F, Jurkat-Rott K, Rudel R. Diagnostics and therapy of muscle channelopathies—Guidelines of the Ulm Muscle Centre. *Acta Myol* 2008;**27**:98–113.
28. Sansone V, Meola G, Links TP, Panzeri M, Rose MR. Treatment for periodic paralysis. *Cochrane Database Syst Rev* 2008;**1**:CD005045.
29. Trip J, Drost G, van Engelen BG, Faber CG. Drug treatment for myotonia. *Cochrane Database Syst Rev* 2006;**1**:CD004762.
30. Heatwole CR, Moxley III RT. The nondystrophic myotonias. *Neurotherapeutics* 2007;**4**:238–51.
31. Mohammadi B, Jurkat-Rott K, Alekov A, Dengler R, Bufler J, Lehmann-Horn F. Preferred mexiletine block of human sodium channels with IVS4 mutations and its pH-dependence. *Pharmacogenet Genomics* 2005;**15**:235–44.
32. Alfonsi E, Merlo IM, Tonini M, Ravaglia S, Brugnoni R, Gozzini A, et al. Efficacy of propafenone in paramyotonia congenita. *Neurology* 2007;**68**:1080–1.
33. Palade PT, Barchi RL. On the inhibition of muscle membrane chloride conductance by aromatic carboxylic acids. *J Gen Physiol* 1977;**69**:879–96.
34. Pusch M, Steinmeyer K, Koch MC, Jentsch TJ. Mutations in dominant human myotonia congenita drastically alter the voltage dependence of the CIC-1 chloride channel. *Neuron* 1995;**15**:1455–63.
35. Wagner S, Deymeer F, Kurz LL, Benz S, Schleithoff L, Lehmann-Horn F, et al. The dominant chloride channel mutant G200R causing fluctuating myotonia: clinical findings, electrophysiology, and channel pathology. *Muscle Nerve* 1998;**21**:1122–8.
36. Fialho D, Kullmann DM, Hanna MG, Schorge S. Non-genomic effects of sex hormones on CLC-1 may contribute to gender differences in myotonia congenita. *Neuromuscul Disord* 2008;**18**:869–72.
37. Heine R, Pika U, Lehmann-Horn F. A novel SCN4A mutation causing myotonia aggravated by cold and potassium. *Hum Mol Genet* 1993;**2**:1349–53.
38. Wu FF, Gordon E, Hoffman EP, Cannon SC. A C-terminal skeletal muscle sodium channel mutation associated with myotonia disrupts fast inactivation. *J Physiol* 2005;**565**:371–80.
39. Kubota T, Roca X, Kimura T, Kokunai Y, Nishino I, Sakoda S, et al. A mutation in a rare type of intron in a sodium-channel gene results in aberrant splicing and causes myotonia. *Hum Mutat* 2011;**32**:773–82.
40. Becker PE. *Myotonia Congenita and Syndromes Associated with Myotonia*. Stuttgart: Thieme Press, Inc.; 1977.
41. Baumann P, Myllyla VV, Leisti J. Myotonia congenita in northern Finland: an epidemiological and genetic study. *J Med Genet* 1998; **35**:293–6.
42. Odermatt A, Taschner PE, Khanna VK, Busch HF, Karpati G, Jablecki CK, et al. Mutations in the gene-encoding SERCA1, the fast-twitch skeletal muscle sarcoplasmic reticulum Ca2+ ATPase, are associated with Brody disease. *Nat Genet* 1996;**14**: 191–4.
43. Voermans NC, Laan AE, Oosterhof A, van Kuppevelt TH, Drost G, Lammens M, et al. Brody syndrome: a clinically heterogeneous entity distinct from Brody disease: a review of literature and a cross-sectional clinical study in 17 patients. *Neuromuscul Disord* 2012;**22**:944–54.
44. Jurkat-Rott K, Lehmann-Horn F. Genotype-phenotype correlation and therapeutic rationale in hyperkalemic periodic paralysis. *Neurotherapeutics* 2007;**4**:216–24.
45. Chahine M, George Jr. AL, Zhou M, Ji S, Sun W, Barchi RL, et al. Sodium channel mutations in paramyotonia congenita uncouple inactivation from activation. *Neuron* 1994;**12**:281–94
46. Lerche H, Mitrovic N, Dubowitz V, Lehmann-Horn F. Paramyotonia congenita: the R1448P Na+ channel mutation in adult human skeletal muscle. *Ann Neurol* 1996;**39**:599–608.
47. Fleischhauer R, Mitrovic N, Deymeer F, Lehmann-Horn F, Lerche H. Effects of temperature and mexiletine on the F1473S Na+ channel mutation causing paramyotonia congenita. *Pflugers Arch* 1998;**436**:757–65.
48. Ruff RL. Slow Na+ channel inactivation must be disrupted to evoke prolonged depolarization-induced paralysis. *Biophys J* 1994; **66**:542.
49. Carle T, Fournier E, Sternberg D, Fontaine B, Tabti N. Cold-induced disruption of Na+ channel slow inactivation underlies paralysis in highly thermosensitive paramyotonia. *J Physiol* 2009;**587**:1705–14.
50. Ruff RL. Effects of temperature on slow and fast inactivation of rat skeletal muscle Na(+) channels. *Am J Physiol* 1999;**277**:C937–47.
51. Webb J, Cannon SC. Cold-induced defects of sodium channel gating in atypical periodic paralysis plus myotonia. *Neurology* 2008; **70**:755–61.
52. Nagel AM, Amarteifio E, Lehmann-Horn F, Jurkat-Rott K, Semmler W, Schad L, et al. 3 Tesla sodium inversion recovery magnetic resonance imaging allows for improved visualization of intracellular sodium content changes in muscular channelopathies. *Invest Radiol.* 2011;**46**(12):759–66.
53. Becker PE. *Paramyotonia congenita (Eulenburg). Fortschritte Der Allgemeinen Und Klinischen Humangenetik (Advances in Human Genetics)*. Stuttgart: Thieme Press, Inc.; 1970.
54. Meyer-Kleine C, Otto M, Zoll B, Koch MC. Molecular and genetic characterisation of German families with paramyotonia congenita and demonstration of founder effect in the Ravensberg families. *Hum Genet* 1994;**93**:707–10.
55. Jurkat-Rott K, Lehmann-Horn F. *Hyperkalemic periodic paralysis type 1*. Seattle: University of Washington; 2003 [updated 2011 May 31; cited 2013 December 1]; Available from: <http://www.genetests.org>.

56. Meyer T, Jurkat-Rott K, Huebner A, Lehmann-Horn F, Linke P, Van Landeghem F, et al. Progressive muscle atrophy with hypokalemic periodic paralysis and calcium channel mutation. *Muscle Nerve* 2008;**37**:120–4.
57. Chun TU, Epstein MR, Dick II M, Andelfinger G, Ballester L, Vanoye CG, et al. Polymorphic ventricular tachycardia and KCNJ2 mutations. *Heart Rhythm* 2004;**1**:235–41.
58. Plaster NM, Tawil R, Tristani-Firouzi M, Canun S, Bendahhou S, Tsunoda A, et al. Mutations in Kir2.1 cause the developmental and episodic electrical phenotypes of Andersen's syndrome. *Cell* 2001; **105**:511–19.
59. Reisin RC, Martinez O, Moran M, Rovira M, Roccatagliata G, Pardal A, et al. Thyrotoxic periodic paralysis in Caucasians. Report of 8 cases. *Neurologia* 2000;**15**:222–5.
60. Engel AG. Thyroid function and periodic paralysis. *Am J Med* 1961;**30**:327–33.
61. Ryan DP, da Silva MR, Soong TW, Fontaine B, Donaldson MR, Kung AW, et al. Mutations in potassium channel Kir2.6 cause susceptibility to thyrotoxic hypokalemic periodic paralysis. *Cell* 2010;**140**:88–98.
62. Weber MA, Nielles-Vallespin S, Essig M, Jurkat-Rott K, Kauczor HU, Lehmann-Horn F. Muscle Na + channelopathies: MRI detects intracellular 23Na accumulation during episodic weakness. *Neurology* 2006;**67**:1151–8.
63. Links TP, Smit AJ, Molenaar WM, Zwarts MJ, Oosterhuis HJ. Familial hypokalemic periodic paralysis. Clinical, diagnostic and therapeutic aspects. *J Neurol Sci* 1994;**122**:33–43.
64. Tawil R, McDermott MP, Brown Jr. R, Shapiro BC, Ptacek LJ, McManis PG, et al. Randomized trials of dichlorphenamide in the periodic paralyses. Working Group on Periodic Paralysis. *Ann Neurol* 2000;**47**:46–53.
65. Li JS, Flynn JT, Portman R, Davis I, Ogawa M, Shi H, et al. The efficacy and safety of the novel aldosterone antagonist eplerenone in children with hypertension: a randomized, double-blind, dose-response study. *J Pediatr* 2010;**157**:282–7.
66. Ikeda K, Iwasaki Y, Kinoshita M, Yabuki D, Igarashi O, Ichikawa Y, et al. Acetazolamide-induced muscle weakness in hypokalemic periodic paralysis. *Intern Med* 2002;**41**:743–5.
67. Junker J, Haverkamp W, Schulze-Bahr E, Eckardt L, Paulus W, Kiefer R. Amiodarone and acetazolamide for the treatment of genetically confirmed severe Andersen syndrome. *Neurology* 2002; **59**:466.
68. Fontaine B, Khurana TS, Hoffman EP, Bruns GA, Haines JL, Trofatter JA, et al. Hyperkalemic periodic paralysis and the adult muscle sodium channel alpha-subunit gene. *Science* 1990;**250**: 1000–2.
69. Rojas CV, Wang JZ, Schwartz LS, Hoffman EP, Powell BR, Brown Jr RH, et al. A Met-to-Val mutation in the skeletal muscle Na+ channel alpha-subunit in hyperkalaemic periodic paralysis. *Nature* 1991;**354**:387–9.
70. Fontaine B, Vale-Santos J, Jurkat-Rott K, Reboul J, Plassart E, Rime CS, et al. Mapping of the hypokalaemic periodic paralysis (HypoPP) locus to chromosome 1q31-32 in three European families. *Nat Genet* 1994;**6**:267–72.
71. Jurkat-Rott K, Lehmann-Horn F, Elbaz A, Heine R, Gregg RG, Hogan K, et al. A calcium channel mutation causing hypokalemic periodic paralysis. *Hum Mol Genet* 1994;**3**:1415–19.
72. Jurkat-Rott K, Mitrovic N, Hang C, Kouzmekine A, Iaizzo P, Herzog J, et al. Voltage-sensor sodium channel mutations cause hypokalemic periodic paralysis type 2 by enhanced inactivation and reduced current. *Proc Natl Acad Sci USA* 2000;**97**:9549–54.
73. Jurkat-Rott K, Holzherr B, Fauler M, Lehmann-Horn F. Sodium channelopathies of skeletal muscle result from gain or loss of function. *Pflugers Arch* 2010;**460**:239–48.
74. Jurkat-Rott K, Groome J, Lehmann-Horn F. Pathophysiological role of omega pore current in channelopathies. *Front Pharmacol* 2012;**3**:112.
75. Rudel R, Lehmann-Horn F, Ricker K, Kuther G. Hypokalemic periodic paralysis: in vitro investigation of muscle fiber membrane parameters. *Muscle Nerve* 1984;**7**:110–20.
76. Ptacek LJ, Tawil R, Griggs RC, Engel AG, Layzer RB, Kwiecinski H, et al. Dihydropyridine receptor mutations cause hypokalemic periodic paralysis. *Cell* 1994;**77**:863–8.
77. Kil TH, Kim JB. Severe respiratory phenotype caused by a de novo Arg528Gly mutation in the CACNA1S gene in a patient with hypokalemic periodic paralysis. *Eur J Paediatr Neurol* 2010;**14**: 278–81.
78. Chabrier S, Monnier N, Lunardi J. Early onset of hypokalaemic periodic paralysis caused by a novel mutation of the CACNA1S gene. *J Med Genet* 2008;**45**:686–8.
79. Hirano M, Kokunai Y, Nagai A, Nakamura Y, Saigoh K, Kusunoki S, et al. A novel mutation in the calcium channel gene in a family with hypokalemic periodic paralysis. *J Neurol Sci* 2011;**309**:9–11.
80. Ke T, Gomez CR, Mateus HE, Castano JA, Wang QK. Novel CACNA1S mutation causes autosomal dominant hypokalemic periodic paralysis in a South American family. *J Hum Genet* 2009; **54**:660–4.
81. Li FF, Li QQ, Tan ZX, Zhang SY, Liu J, Zhao EY, et al. A novel mutation in CACNA1S gene associated with hypokalemic periodic paralysis which has a gender difference in the penetrance. *J Mol Neurosci* 2012;**46**:378–83.
82. Vicart S, Sternberg D, Fournier E, Ochsner F, Laforet P, Kuntzer T, et al. New mutations of SCN4A cause a potassium-sensitive normokalemic periodic paralysis. *Neurology* 2004;**63**:2120–7.

Chapter 39

Glycogen Storage Diseases of Muscle

Hasan O. Akman, Anders Oldfors, and Salvatore DiMauro

INTRODUCTION AND BACKGROUND

Metabolic myopathies are clinical disorders in which defects of adenosine triphosphate (ATP) production cause muscle dysfunction. A review of muscle metabolism is beyond the scope of this chapter and readers are referred to other reviews for further details.[1–3] Nevertheless, a brief description of the biochemical pathways supporting the energy requirement of skeletal muscle is necessary to understand the symptoms of patients with metabolic myopathies, including disorders of glycogen metabolism.

Skeletal muscle is highly energy dependent and therefore vulnerable to disorders of energy metabolism. The hydrolysis of ATP to adenosine diphosphate (ADP) and inorganic phosphate (Pi) is the immediate and essential source of energy for contraction and relaxation. In order to regenerate ATP, skeletal muscle utilizes different substrates (high-energy phosphate compounds, glucose and glycogen, and free fatty acids [FFAs]), and different metabolic pathways (the creatine kinase (CK) reaction; anaerobic glycolysis; the β-oxidation spiral; oxidative phosphorylation).

Resting muscle utilizes predominantly FFAs[1] (see Chapter 40). In working muscle, the bioenergetic pathway is dictated by the type, intensity, and duration of exercise.[2]

Anaerobic glycolysis and the creatine kinase reaction, the major anaerobic sources of ATP, are important fuel sources in ischemic or isometric exercise, when muscle blood flow and oxygen delivery to exercising muscles are interrupted or reduced, and during extremely intense physical activity, when the energy demand is above the anaerobic threshold. During isotonic exercise, such as walking or running, ATP is regenerated by high-energy phosphate compounds during the first several minutes, after which carbohydrate oxidation increases and glycogen becomes the most important fuel. With prolonged exercise, glucose and glycogen stores are depleted, and there is a metabolic switch to fatty acid metabolism. After a longer period of exercise, FFAs are the predominant sources of ATP. Except for anaerobic glycolysis, the final common pathway for ATP production is the mitochondrial respiratory chain (oxidative phosphorylation; see Chapter 41) (Figure 39.1).

Defects in ATP production cause clinical disorders characterized by muscle symptoms, which will depend on the specific metabolic pathway that is impaired. For example, defects of glycolysis will be evident after short intervals of moderate to intense exercise, whereas fatty acid metabolism disorders will cause symptoms after prolonged exercise.

This chapter reviews those disorders of glycogen metabolism (glycogenoses) that affect skeletal muscle alone, or in combination with other tissues (Figure 39.2). These disorders include three enzyme defects involving nonlysosomal glycogenolysis and one affecting lysosomal glycogenolysis, seven defects of glycolysis, and three of glycogen synthesis.

Glycogenoses cause two main clinical syndromes[3,5,6]:

1. Exercise intolerance, with cramps, and myoglobinuria: these are generally associated with defects of glycogenolysis (phosphorylase kinase and myophosphorylase deficiencies) and glycolysis (phosphoglucomutase, phosphofructokinase, aldolase, beta-enolase, phosphoglycerate kinase, phosphoglycerate mutase, and lactate dehydrogenase deficiencies).
2. Progressive weakness involving limb and trunk muscles: this is characteristic of defects in the glycogenosynthetic pathway (glycogenin and brancher enzyme deficiency) and in the lysosomal glycogenolytic system (acid maltase deficiency).

However, there are exceptions: (1) Defects of debrancher, a glycogenolytic enzyme, usually cause weakness rather than cramps and myoglobinuria; (2) patients with myophosphorylase or phosphofructokinase deficiency often develop fixed weakness later in life; (3) some patients with deficiencies of myophosphorylase, phosphorylase kinase, or phosphofructokinase present with

B.T. Darras, H. Royden Jones, Jr., M.M. Ryan & D.C. De Vivo (Eds): Neuromuscular Disorders of Infancy, Childhood and Adolescence, Second edition.
DOI: http://dx.doi.org/10.1016/B978-0-12-417044-5.00039-1

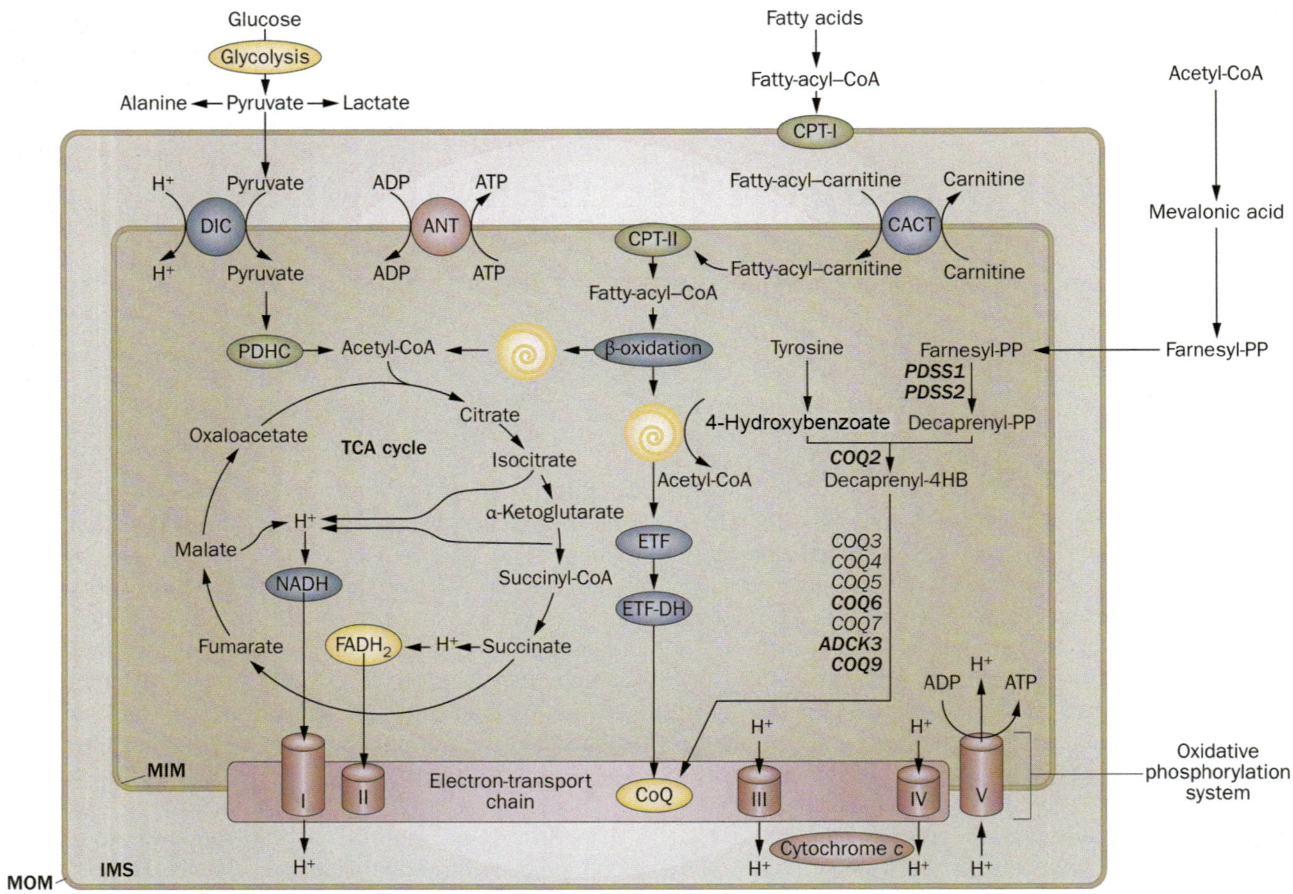

FIGURE 39.1 Schematic view of cellular metabolism showing the metabolic "fuels" ultimately utilized for the production of ATP in the mitochondrial respiratory chain. Glucose, derived from glycogen breakdown, is converted to pyruvate through the glycolytic pathway, which is located in the cytoplasm. In conditions of anaerobiosis, pyruvate is converted to lactate and alanine. In aerobic conditions, pyruvate is transported into mitochondria through the dicarboxylate carrier (DIC) and oxidized by the pyruvate dehydrogenase complex (PDHC) to acetyl-CoA. Long-chain fatty acids are carried into the mitochondrial through a complex transport mechanism that includes two carnitine palmitoyltransferases (CPT-I and CPT-II), a transport molecule (L-carnitine), and a translocase (carnitine-acylcarnitine translocase, CACT), resulting in the formation of acetyl-CoA, which is the main anaplerotic substrate of the tricarboxylic acid cycle (TCA, Krebs cycle). Electrons derived from dehydrogenases in the Krebs cycle and in the beta-oxidation spirals flow down the electron carrier chain, composed of four complexes (I–IV) and two small carriers (coenzyme Q10 [CoQ10] and cytochrome c). The electron carrier chain—together with ATP synthase (complex V)—comprises the respiratory chain. *(Reproduced with permission from DiMauro S and Schon EA[4]).*

weakness, without cramps and myoglobinuria; and (4) conversely, exercise intolerance characterizes a defect of glycogen synthesis, glycogen synthetase deficiency.

All glycogenoses are autosomal recessive, except phosphoglycerate kinase, which is X-linked, and phosphorylase kinase, which may be either. All of the genes encoding these enzymes have been cloned, their chromosomes have been localized, and many molecular genetic defects have been identified.

FOREARM ISCHEMIC EXERCISE TEST

The forearm ischemic exercise test (FIET) is a useful tool to corroborate the diagnosis of defects in the glycogenolytic or glycolytic pathways. The test measures basal and post-exercise venous lactate and ammonia levels. In normal adults, lactate increases 3- to 5-fold over baseline with a peak at 1 to 2 minutes post-exercise, whereas ammonia increases 5- to 10-fold. Patients with myophosphorylase, phosphofructokinase, or distal glycolytic enzyme deficiencies generally show absent or inadequate (less than 1.5-fold) increases of lactate and exaggerated rises in ammonia. Conversely, a normal lactate response with impaired ammonia production is characteristic of myoadenylate deaminase deficiency.[7]

However, the test has several limitations. The rise of venous lactate depends on the subject's ability and willingness to exercise vigorously. The simultaneous measurement of ammonia levels is therefore important in order to confirm the adequate effort of the patient.

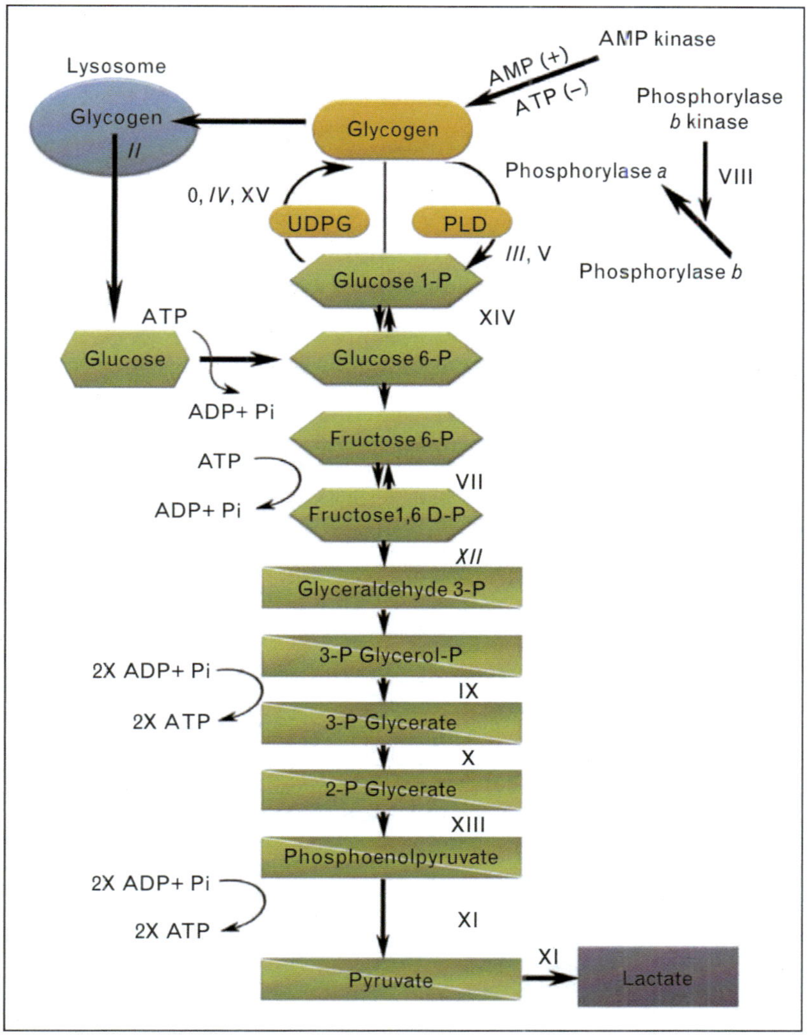

FIGURE 39.2 Scheme of glycogen metabolism and glycolysis. Roman numerals indicate enzymes whose deficiencies are associated with muscle glycogenoses: II, acid maltase (GAA, Pompe disease); III, debrancher (GDE, Cori-Forbes disease); IV, brancher (GBE, Andersen disease); V, myophosphorylase (PYGM, McArdle disease); VII, phosphofructokinase (PFK, Tarui disease); VIII, phosphorylase kinase (PHK); IX, phosphoglycerate kinase (PGK); X, phosphoglycerate mutase (PGAM); XI, lactate dehydrogenase (LDH); XII, aldolase A; XIII, beta-enolase (ENO3); XIV, phosphoglucomutase (PGM); XV, glycogenin (GYG1); 0, glycogen synthase (GYS1). UDPG, uridine diphosphate glucose; PLD, phosphorylase-limit dextrin; AMP, adenosine monophosphate; ATP, adenosine triphosphate; P, inorganic phosphate. *(Reproduced with permission from Oldfors and DiMauro.[5])*

Furthermore, affected individuals can develop muscle cramping and prolonged contractures,[8] and should stop exercising at the first appearance of cramping to reduce the risk of muscle necrosis. Because of these reasons, the test is rarely performed in children.

To avoid pain and contractures, the test has been performed in 30 children with suspected muscle glycogenoses under semi-ischemic conditions, allowing systolic blood flow.[9] Using this modification, all subjects were able to exercise for the whole minute without problems, and two patients with a defect in the glycolytic pathway failed to adequately increase venous lactate but did not experience contractures during exercise.

Furthermore, it has been shown that maximal intermittent handgrip exercise for one minute without ischemia provides a stimulus strong enough to elicit maximal glycolytic flux.[10] Ischemic and non-ischemic tests were performed in eight patients with McArdle disease and in eight matched healthy control subjects. The non-ischemic test had the same diagnostic power for McArdle disease as the ischemic test but, unlike the ischemic test, did not cause muscle cramps or pain to the patients. Therefore, a non-ischemic forearm exercise test is an easy, safe, and well-tolerated tool in the diagnostic investigation of disorders of muscle carbohydrate metabolism.

DEFECTS OF GLYCOGENOLYSIS

Phosphorylase Kinase Deficiency

Phosphorylase b kinase (PhK) is a regulatory enzyme in the activation cascade of glycogenolysis. By phosphorylating and thus activating glycogen phosphorylase, PhK stimulates glycogen degradation in response to various neural and hormonal signals.

The enzyme is a decahexameric protein composed of four subunits (α,β,γ,δ). Subunits α and β are regulatory, γ is catalytic, and δ is identical to calmodulin.[11,12] Two isoforms are known for subunit α (muscle and liver, α_M and α_L)[13] and for subunit γ (muscle and testis, γ_M and γ_T).[14] The genes encoding α_M (*PHKA1*) and α_L (*PHKA2*) are both on the X chromosome, whereas the β (*PHKB*), γ_M (*PHKG1*), and γ_T (*PHKG2*) genes are autosomal.[15]

PhK deficiency is associated with four different phenotypes, distinguished by tissue involvement (liver, muscle, liver and muscle, heart), and mode of inheritance (X-linked or autosomal).[11]

1. Liver disease, the most frequent type, is usually a benign condition of infancy or childhood, characterized by hepatomegaly, growth retardation, delayed motor development, and fasting hypoglycemia. Inheritance is X-linked recessive, due to mutations in *PHKA2*. The enzyme defect is also expressed in erythrocytes and leukocytes, but not in muscle.
2. Liver and muscle disease is characterized by hepatomegaly, usually resolving with age, and nonprogressive myopathy. Transmission appears to be autosomal recessive in most cases and due to mutations in *PHKB*.[16] The enzyme defect is expressed in muscle and liver.
3. Myopathy, inherited as an autosomal recessive or as an X-linked recessive trait, has been reported in about 20 cases.[17] The clinical picture consists of exercise intolerance, with cramps, myalgia, and weakness of exercising muscle. Myoglobinuria after strenuous exercise has been described in less than 50% of the cases. Onset is usually in childhood or adolescence. Although fixed weakness was only present in a few cases, one of the patients reported by Clemens and associates,[18] a 58-year-old man, had progressive weakness starting at age 46 and affecting distal more than proximal muscles. In most patients, serum CK level is variably increased.

   The FIET is often normal or only mildly altered, and electromyography (EMG) can be normal or show nonspecific myopathic changes. Muscle biopsies may be normal or show subsarcolemmal accumulation of glycogen. Although PhK activity in muscle homogenate is absent or markedly decreased, the generally mild presentation of this myopathy has raised the question of whether this is indeed a disease or merely a metabolic variant.[19]
4. Cardiomyopathy has been described in a few infants and is well defined in terms of clinical pathology and biochemistry. The clinical phenotype is uniformly severe and survival has not exceeded months. The isolated involvement of the heart was a riddle because there is no cardiac isoform of PhK. Now we know that the PhK deficiency in these cases is secondary to *de novo* mutations in the gene (*PRKAG2*) encoding the γ2 subunit of the AMP-activated protein kinase (AMPK).[11,20]

Myophosphorylase Deficiency

Muscle phosphorylase (PYGM) initiates glycogen breakdown by removing 1,4-glucosyl residues phosphorolytically from the outer branches of the glycogen molecule, with liberation of glucose-1-phosphate: this goes on until the peripheral chains are shortened to about four glucosyl units. The degraded polysaccharide is known as phosphorylase-limit dextrin (PLD), and its short peripheral chains are removed by the debranching enzyme. In muscle, phosphorylase exists in two forms: a more active, phosphorylated *a* form, and a less active, dephosphorylated *b* form. The phosphorylation of myophosphorylase is catalyzed by PhK.

Human phosphorylase has three isozymes, muscle (PYGM), liver (PYGL), and brain/heart (PYGB), encoded by different genes that have been mapped on different chromosomes.[21–23] Normal adult human muscle expresses only the muscle isozyme, whereas heart and brain express, in different amounts, both the brain and the muscle isoforms.[24] In particular, heart contains the muscle isozyme (13%), the brain isozyme (58%), and a hybrid of the two isozymes (29%). Brain contains the muscle isozyme (8%), the brain isozyme (64%), and a hybrid of the two (28%), whereas liver contains exclusively the liver isozyme. Therefore, patients with the muscle isozyme deficiency must also have partial defects in heart and brain; however, symptoms of heart or brain involvement have not been reported.

A defect of the muscle isoform underlies glycogenosis type V (McArdle disease), whereas phosphorylase deficiency in liver is referred to as glycogenosis type VI (Hers' disease).

McArdle disease is characterized by exercise intolerance with premature fatigue, myalgia, and cramps (see Case Example 39.1). Symptoms usually occur during brief and intense exercise (pushing or lifting heavy objects) or during sustained, but less intense exercise (climbing stairs).[3,25,26] Most patients experience the so-called second-wind phenomenon: if they rest briefly at the onset of symptoms, they can resume exercising without discomfort. Myoglobinuria occurs in about 50% of patients and half of these develop renal failure. Although exercise intolerance is present from childhood, cramps and myoglobinuria develop later in life, and the diagnosis is usually established in young adult life.

Although the clinical phenotype is rather uniform, a few variants have been reported, including a mild form with excessive tiredness and poor stamina,[3] a late-onset form with fixed weakness in the fifth to sixth decade without cramps or myoglobinuria,[27] and a rare fatal infantile form characterized by weakness, severe respiratory insufficiency, and early death.[28,29] Congenital myopathy and mental retardation were reported in a 4-year-old boy.[30]

CASE EXAMPLE 39.1

A 16-year-old Caucasian boy presented with exercise intolerance since early childhood. Any attempt to perform vigorous exercise caused painful shortening of the muscles that were engaged in the activity (e.g. arms and shoulder muscles after shoveling snow), which resolved with rest. He also noticed that, if he paused at the first appearance of muscle stiffness, he was able to resume activity with more ease. He never noted pigmenturia.

Physical and neurologic examinations were unremarkable. Resting serum CK level was slightly elevated (468 U/L, normal <150). A FIET caused no increase of venous lactate above baseline, but normal rise in ammonia levels. Electrophysiologic studies were noncontributory. A muscle biopsy specimen (vastus lateralis) showed excessive variation of fiber size, subsarcolemmal deposits of glycogen by periodic acid–Schiff (PAS) staining, and negative histochemical reaction for phosphorylase. Biochemical analysis showed complete absence of phosphorylase activity whereas activities for phosphofructokinase and other glycolytic enzymes were normal. Molecular studies of genomic DNA extracted from blood showed that the patient was homozygous and his parents heterozygous for the common R50X nonsense mutation.

Comment

The symptoms in this young man (cramps induced by vigorous exercise) are typical—but certainly not specific—of McArdle disease. The description of a second wind phenomenon bolstered the diagnosis of McArdle[25] and suggested looking for the common PYGM mutation in blood cells, which confirmed the diagnosis already established by muscle histochemistry and biochemistry.

Laboratory tests show increased resting serum CK, flat venous lactate response to ischemic exercise, and myopathic EMG. The diagnosis is confirmed by the histochemical or biochemical documentation of phosphorylase deficiency in muscle and by molecular analysis of DNA. The presence of subsarcolemmal or intermyofibrillar deposits of glycogen gives muscle a vacuolar appearance (Figure 39.3A,B).

In patients with myophosphorylase deficiency, phosphorylase activity is normal in erythrocytes, leukocytes, platelets, and fibroblasts. Therefore, the diagnosis relies on measurement of phosphorylase activity in muscle or on genetic testing.

The histochemical reaction for phosphorylase is a valuable diagnostic tool because in most patients muscle fibers have no enzyme activity (Figure 39.3D,E). However, the phosphorylase reaction can be positive in rare cases with some residual enzyme activity and in biopsy specimens with abundant regenerating fibers. This false-positive reaction, which is due to the expression of PYGMB, may be misleading in muscle biopsy specimens taken too soon after an episode of myoglobinuria, when regeneration is at its most active.

The myophosphorylase gene has been cloned, assigned to chromosome 11q1,[22] and its genomic structure has been described.[31] The disease is transmitted as an autosomal recessive trait, but pseudo-dominant transmission has been documented in a few families. The molecular heterogeneity of myophosphorylase deficiency is striking: in the 20 years from the description of the first molecular defects,[32,33] more than 100 different mutations have been reported.[26] The most common mutation in North America and Northern Europe is the R50X. This has allowed the use of molecular genetic analysis in leukocytes for diagnostic purposes, thus circumventing the need for a muscle biopsy.[34] However, as leukocytes are increasingly used for diagnosis, it becomes important to keep in mind the relative frequency of distinct mutations in different ethnic groups. For example, the R50X mutation has never been observed in Japan, where the most common mutation appears to be a 3-bp deletion, TTC at codon 708/709.[35] The genotype/phenotype relationship in McArdle disease remains unclear. For example, the common R49X mutation was also found in an infant with the fatal variant,[32] in a child who died of sudden infant death syndrome,[36] and in two virtually asymptomatic children with persistent and unusually high levels of serum CK.[37,38]

The exercise intolerance in McArdle disease is due to two main mechanisms. First, the block of anaerobic glycolysis deprives muscle of the energy required for isometric exercise. Second, the block of aerobic glycogen utilization and the consequent shortage of pyruvate and acetyl-CoA impairs dynamic exercise above a certain intensity (about 50% VO_{2max}). The impairment of oxidative phosphorylation is evidenced by the decrease in oxygen extraction and maximum oxygen uptake documented in patients with myophosphorylase deficiency, which can at least partially be restored by intravenous glucose infusion.[39]

There is no specific therapy for McArdle disease, but regular moderate aerobic training is effective because it favors alternate fuel delivery and utilization.[40] In cases with residual phosphorylase activity, vitamin B6 should be tried because the overall body stores of pyridoxal phosphate (PLP) are depleted in McArdle disease due to the frequent lack of enzyme protein (to which PLP is bound[41–43]). Patients should be warned about the risks of strenuous exercise and advised to seek medical attention at the first appearance of pigmenturia, especially if accompanied by oliguria. Sucrose ingestion before exercise is beneficial but may lead to weight gain.[44]

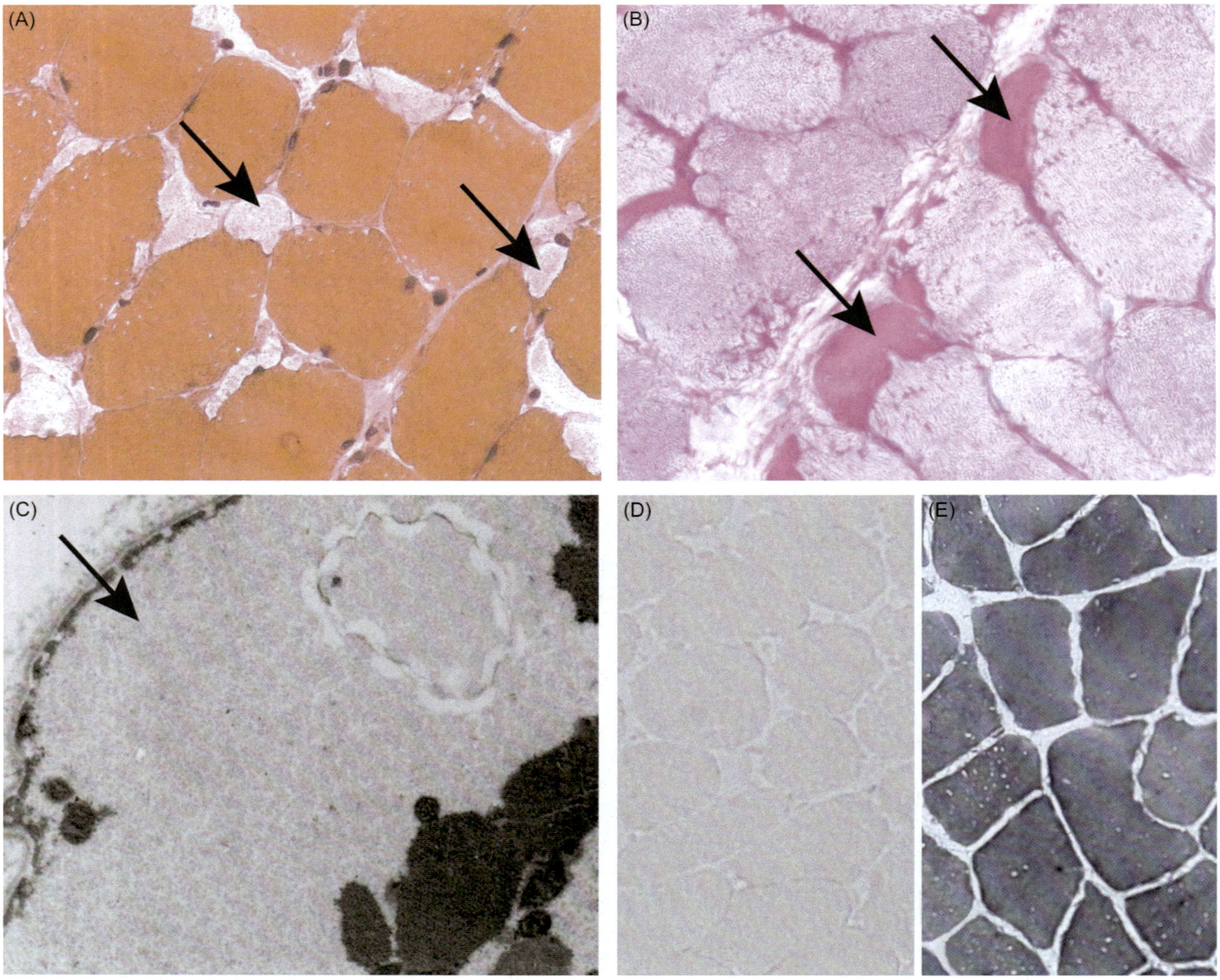

FIGURE 39.3 Muscle biopsy from a patient with McArdle disease (GSD V). (**A**) Staining with hematoxylin-eosin (H&E) reveals subsarcolemmal vacuoles (arrows). (**B**) The vacuoles contain glycogen, as shown by the periodic acid-Schiff (PAS) stain. (**C**) Electron microscopy (EM) documents the presence of glycogen beta particles in the vacuoles. (**D**) Negative histochemical reaction for phosphorylase in the muscle of the patient. (**E**) Positive histochemical reaction in a control muscle.

The two spontaneous animal models of McArdle disease—Charolais cattle[45] and Merino sheep[46]—are not practical for experimentation. However, a knock-in mouse model of the R50X mutation recapitulates faithfully the human disease[47] and will provide valuable information on pathophysiology and experimental therapy.

Debranching Deficiency

Debranching enzyme is a monomeric protein encoded by a single gene and expressed ubiquitously in all tissues. It has two independent catalytic activities: transferase and α-glucosidase. After phosphorylase has shortened the peripheral chains of glycogen to about four glucosyl units, these residual stubs are removed by the debranching enzyme in two steps, catalyzed by an oligo-1,4-1,4-glucantransferase and by an amylo-1,6-glucosidase. The transferase and glucosidase activities can be assessed separately or together by using appropriate substrates.

Debrancher deficiency (glycogenosis type III) typically presents as a childhood-onset liver disease with hepatomegaly, growth failure, fasting hypoglycemia, and, less frequently, hypoglycemic seizures.[48] Liver symptoms improve with age and often resolve completely around puberty.

Although clinical cardiopathy is infrequent, cardiac involvement is demonstrable by laboratory tests in most patients with myopathy[48–50] and a child was reported to have died of cardiac failure at 4 years of age.[51]

Myopathy is not always manifest in infants or children, who may show hypotonia, delayed motor milestones, or mild weakness.[48] Myopathy often appears in adult life, long after liver symptoms have remitted or in patients without any history of hepatopathy.[52–56] However, in 7 of 22 patients reviewed by Cornelio and colleagues[55] onset was in childhood and the diagnosis was facilitated by the coexistence of liver disease. Two of these children had exercise intolerance, cramps, and premature fatigue, and five had diffuse weakness and wasting, growth retardation, and delayed motor development.

Adult-onset myopathies can be predominantly distal or generalized.[52,54] Patients with distal myopathy develop wasting of distal leg and intrinsic hand muscles, often leading to the diagnosis of motor neuron disease or peripheral neuropathy: the "mixed" myopathic and neurogenic EMG pattern and the often slowed nerve conduction velocity suggests that weakness in these patients may have a neurogenic component.[54,57] Patients with generalized myopathy tend to have more severe weakness, often affecting respiratory muscles.[58]

It may seem paradoxical that the typical presentation of debrancher deficiency is so different from that of McArdle disease, although the two defective enzymes act sequentially in the first step of glycogenolysis. However, cycle ergometry with or without glucose infusion in six young patients with debrancher deficiency and no or only mild proximal weakness has documented that peak oxygen consumption was below normal and glucose improved work capacity by lowering heart rate and increasing peak work rate.[59] Thus, even before developing overt weakness, patients with debrancher deficiency have exercise intolerance, not unlike patients with McArdle disease, albeit to a much lower extent.

Muscle biopsy specimens typically show large PAS-positive vacuoles containing diastase-digestible glycogen under the sarcolemma and between myofibrils (Figure 39.4A,B). Ultrastructurally, the vacuoles are collections of free and apparently normal glycogen β-particles (Figure 39.4C).

The gene encoding human debrancher (*AGL*) has been cloned, sequenced,[60,61] and assigned to chromosome 1p2,[62] and several different mutations have been identified.[48]

Although there is no specific therapy for debrancher deficiency, various interventions, including dietary regimens, liver transplantation, physical therapy, and precautions with anesthesia and pregnancy, are discussed in a review article.[48]

A Pediatrician's Perspective

The disorders of glycogenolysis described earlier, and especially those affecting muscle, tend to affect adults more than children. However, the following exceptions are worthy of note.

The "liver and muscle" variant of PhK deficiency is typically seen in childhood. An interesting phenomenon is illustrated by a large family that we reported in 1985.[63] Seven male children had an X-linked disorder characterized by hepatomegaly, hypotonia, and weakness in childhood (but resolving with age), delay in growth and sexual maturation, and gouty arthritis. Glycogen storage was severe in liver but also present in muscle, both histochemically and biochemically. We considered PhK deficiency a likely diagnosis, but found normal enzyme activities in both liver and muscle. In 1997, molecular analysis in the propositus by Drs. Jan Hendrickx and Patrick Willems (University of Groningen, the Netherlands) revealed a pathogenic mutation in the gene encoding the liver α-subunit of PhK (*PHKA2*). As these authors have

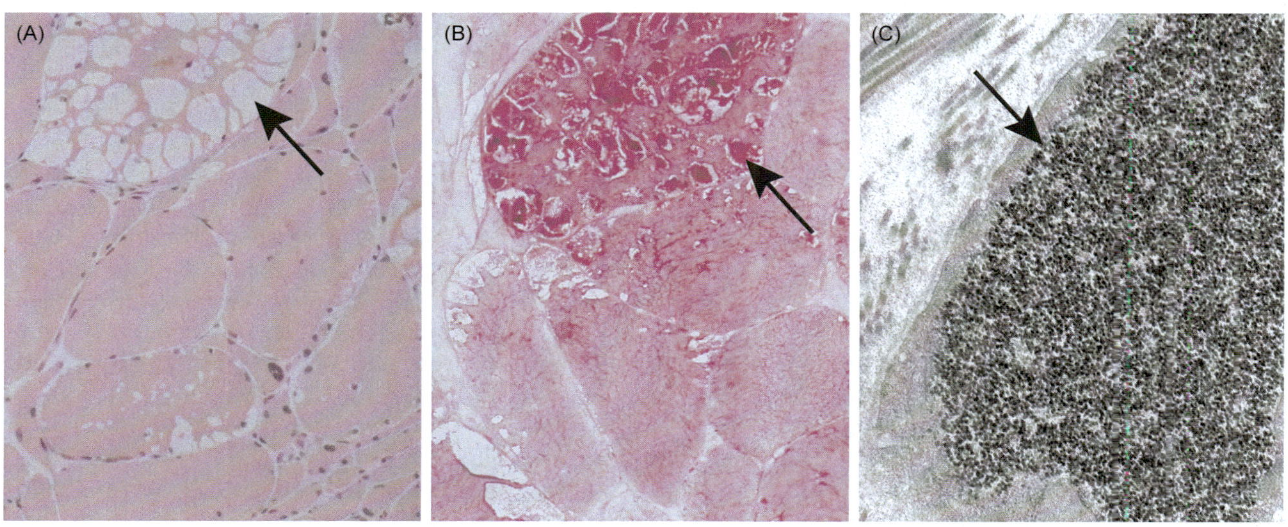

FIGURE 39.4 Muscle biopsy from a patient with debrancher enzyme (DBE) deficiency (GSD III). (**A**) H&E reveals large vacuoles in muscle fibers (arrow), and the PAS stain (**B**) shows that they are filled with glycogen with apparently normal ultrastructural beta-particle structure (**C**).

documented,[64] there are two subgroups of X-linked glycogenosis, XLGI and XLGII, distinguished by the fact that PhK deficiency is demonstrable in liver from patients with XLGI but not in liver from patients with XLGII. Mutational analysis has shown that molecular defects in XLGI affect the stability of the protein whereas those in XLGII affect the regulation of enzyme activity, which probably explains the normal PhK activity that we had observed in two tissues. Although hypotonia in these children can be attributed to liver dysfunction, weakness and glycogen storage in muscle are more difficult to explain. Irrespective of the exact pathogenesis, mutations in *PHKA2* should be considered in the differential diagnosis of children with liver and muscle glycogenosis, even in the absence of enzyme deficiency.

Myophosphorylase deficiency (McArdle disease) is typically a disease of young adults and is usually diagnosed in children only when older siblings are affected.[65,66] In addition to the rare fatal infantile variant, pediatricians should also remember that early onset of symptoms, including myoglobinuria, can be due to genetic "double-trouble"—that is, the coexistence of homozygous mutations in the myophosphorylase gene and in the gene encoding myoadenylate deaminase.[67]

DEFECTS OF GLYCOLYSIS

Phosphoglucomutase Deficiency

Phosphoglucomutase (PGM) deficiency has been described thus far in a single patient, a 34-year-old man who had two episodes of exercise-induced myoglobinuria.[68] FIET caused only moderate increase in venous lactate and the muscle biopsy showed mild increase of glycogen. In keeping with the site of the block, cycle ergometry showed that glucose infusion decreased the abnormal heart rate and increased peak work rate, just as in McArdle disease.[69] Interestingly, PGM deficiency has also been associated with a congenital disorder of glycosylation (CDG) in a patient with exercise intolerance, increased serum CK, growth hormone deficiency, and first branchial arch syndrome.[70]

Phosphofructokinase Deficiency

Phosphofructokinase (PFK) is a tetrameric enzyme composed of three distinct subunits, muscle (M), liver (L), and platelet (P), which are variably expressed in different tissues.[6] Mature human muscle contains exclusively the M isoform, whereas erythrocytes express both M and L subunits.

In the typical form of PFK deficiency (glycogenosis type VII, Tarui disease), there is a total lack of activity in muscle and partial enzyme deficiency in erythrocytes. The clinical presentation is indistinguishable from myophosphorylase deficiency and is characterized by intolerance to vigorous exercise, with cramps and myoglobinuria.[6,71] However, PFK-M deficiency may also cause compensated hemolytic anemia, jaundice, and gouty arthritis due to "myogenic hyperuricemia."[72] In contrast to McArdle disease, patients with PFK deficiency do not experience a second wind[73] and their exercise intolerance appears to worsen, rather than improve, with high carbohydrate intake.[74]

Atypical clinical presentations include hemolytic anemia without myopathy and late-onset fixed weakness.[75–79] A strikingly different clinical presentation includes myopathy in infancy or early childhood, with respiratory failure and death before 2 years of age. This variant has been reported in several children.[6] Although they all had severe myopathy with PFK deficiency, no molecular defect was documented and the heterogeneity of clinical and biochemical data suggests that we may be dealing with different and possibly diverse molecular etiologies.

In patients with typical muscle disease, laboratory investigations reveal elevated resting levels of serum CK, bilirubin, and uric acid. Reticulocytosis is common, but anemia is not. The FIET is positive but nonspecific. EMG may be normal or show myopathic pattern with irritative features.

Muscle biopsy shows accumulation of subsarcolemmal and intermyofibrillar glycogen. A specific stain for phosphofructokinase allows histochemical diagnosis,[80] which can be confirmed biochemically in muscle homogenate. In most cases, careful observation of the muscle biopsy specimen reveals the presence of an abnormal polysaccharide, which stains intensely with PAS, but is not digested by diastase (Figure 39.5A).[76,81] Ultrastructurally, this abnormal glycogen consists of finely granular and filamentous material, similar to the amylopectin-like polysaccharide that accumulates in branching enzyme deficiency, that is, polyglucosan (Figure 39.5B). The deposit of polyglucosan can be explained by the accumulation in muscle of glucose-6-phosphate (G6P), a metabolite upstream of the glycolytic block and an activator of glycogen synthase (GS). This skews the delicate ratio between GS and the glycogen branching enzyme (GBE) in favor of linear synthesis, thus leading to polyglucosan formation.[76,81,82]

The gene encoding PFK-M has been assigned to chromosome 1[83] and the full-length cDNA and the structure of genomic DNA have been reported.[84] Several pathogenic mutations have been identified in Ashkenazi Jewish individuals, who account for most of the patients seen in the United States,[85,86] and about 20 mutations have been identified worldwide.[87]

Pathogenic mutations in *PFKM* cause virtual absence of enzyme activity in skeletal muscle and partial PFK deficiency (about 50%) in erythrocytes. Tissues such as liver and platelets express predominantly or exclusively

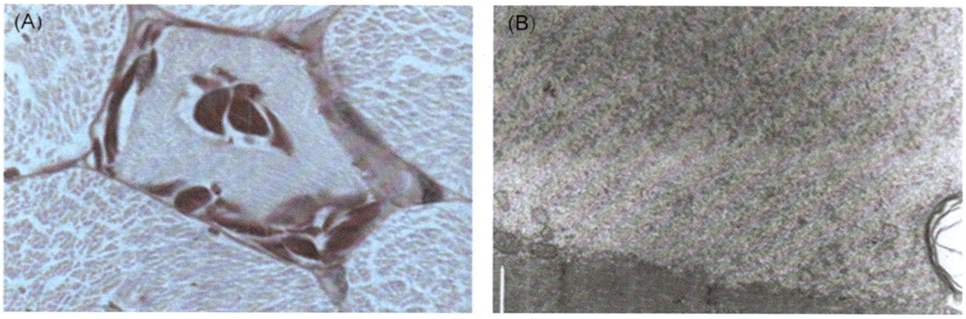

FIGURE 39.5 Muscle biopsy from a patient with phosphofructokinase (PFK) deficiency (GSD VII). (A) PAS stain shows PAS-positive and diastase-resistant material (polyglucosan) under the sarcolemma of one fiber. (B) The material has the finely granulo-fibrillar structure of polyglucosan.

the non-M PFK subunits and are not affected. The lack of clinical cardiomyopathy or encephalopathy is more difficult to explain because PFK-M accounts for over 90% of heart PFK and 50% of brain PFK.[88] Maybe, we should pay closer attention to the heart function of patients with PFK deficiency, as at least one patient had, besides the typical myopathic symptoms, "idiopathic" mild hypertrophic cardiomyopathy and paroxysmal atrial fibrillation.[87]

There is no specific therapy. Glucose administration is not only useless because glucose enters the glycolytic pathway upstream of the metabolic block, but is in fact detrimental because it lowers the blood concentration of alternative fuels, such as fatty acids and ketone bodies.[74]

A 2-year-old boy with the infantile variant of PFK deficiency benefited from a ketogenic diet:[89] there was clear improvement in strength, EMG features, and electroencephalogram (EEG) pattern. Unfortunately, the child worsened suddenly at 35 months of age and died of pneumonia. Still, a ketogenic diet ought to be considered in children with the more severe infantile form of PFK deficiency.

Aldolase Deficiency

Aldolase (ALD) deficiency has been reported in two children with transfusion-requiring nonspherocytic hemolytic anemia, muscle weakness, exercise-induced myalgia, and elevated serum CK.[90,91] Muscle symptoms became more evident and CK levels were especially high during febrile illnesses. One patient was alive at 4½ years[90] and the other died at 4 years during an episode of myoglobinuria and hyperkalemia.[91] Both children were compound heterozygous for mutations in *ALDOA*, the gene that encodes aldolase A, the only isozyme present in skeletal muscle and erythrocytes. The vulnerability of patients to fever may be explained by the abnormal thermolability of the mutant enzymes.

Phosphoglycerate Kinase Deficiency

In the second stage of the glycolytic pathway, phosphoglycerate kinase (PGK) catalyzes the transfer of the acylphosphate group of 1,3-diphosphoglycerate to ADP with formation of 3-phosphoglycerate and ATP.

Primary isolated myopathy is not a common manifestation of PGK deficiency, an X-linked recessive disorder most commonly (9 of 33 patients) presenting as nonspherocytic hemolytic anemia and central nervous system disorder (seizures, mental retardation).[92] However, a purely myopathic presentation is equivalent (9 of 33 patients), while isolated blood dyscrasia was reported in 6 patients, and the association of myopathy and CNS dysfunction in 4 patients. However, we may have underestimated this latter group: after describing one patient with myopathy and juvenile Parkinsonism,[93] we found that a second unrelated patient with apparently isolated myopathy[93] but harboring the same mutation in *PGK1* also developed Parkinsonism a year later.[94] The lesson here is that we should follow patients with PGK deficiency for a few years before concluding that they have only an isolated myopathy.

Nonetheless, isolated myopathy has been reported in a few patients.[95–98] These patients complained of exercise intolerance, with cramps and myoglobinuria. Resting serum CK level was inconsistently increased. FIET caused no or inadequate venous lactate increase. EMG was normal. Muscle biopsy showed mild diffuse increase of PAS-positive material.

Human PGK (PGK1) is a monomeric enzyme encoded by a single gene located on chromosome Xq13 and expressed ubiquitously.[99] A testicular isozyme (PGK2) is encoded by a gene on chromosome 19.[100]

The monomeric nature of PGK1 makes it difficult to explain the wide spectrum of clinical phenotypes in PGK deficiency. Different amounts of residual activity in different tissues do not fully explain the clinical heterogeneity. Lack of myoglobinuria in patients with severe hemolytic anemia and encephalopathy may be due to their inability to exercise strenuously, but it is more difficult to explain why patients with pure myopathy do not manifest blood dyscrasia or CNS involvement.

Phosphoglycerate Mutase Deficiency

Phosphoglycerate mutase (PGAM) catalyzes one of the terminal steps of the glycolytic pathway, the interconversion of 2-phosphoglycerate and 3-phosphoglycerate.

Human PGAM requires 2,3-biphosphoglycerate as a cofactor and is a dimeric enzyme containing, in different tissues, different proportions of a slow-migrating muscle isozyme (MM), a fast migrating brain isozyme (BB), and an intermediate hybrid form (MB). Normal mature muscle contains predominantly the MM homodimer.[101] Liver, kidney, and brain contain mainly the BB form, and heart contains all three dimer types, MM, BB, and MB. The only other tissue containing substantial amounts of the M subunit is sperm, but there is no evidence of cardiomyopathy or infertility in patients with PGAM deficiency.

The full-length cDNA encoding the M-subunit has been cloned,[102] and the entire gene encoding PGAM-M has been isolated, sequenced,[103] and mapped to chromosome 7p12–7p13.[104,105]

About 15 patients have been reported, most of them African Americans.[106,107] All African American patients harbor one common nonsense mutation (W78X), suggesting a founder effect.[107] Different mutations have been found in Italian families,[108,109] in a Japanese family,[110] and in a Pakistani patient[107,111] Despite the abundance of PGAM in muscle, we and others have observed exercise intolerance in heterozygous carriers.[107,110,112]

All patients had intolerance to intense exercise, with myalgia and cramps, and about half of them had recurrent myoglobinuria. Serum CK level was increased between attacks. FIET caused decreased but not absent venous lactate response. EMG and nerve conduction were normal. Muscle biopsy was normal in most patients, but diffuse or patchy glycogen accumulation was seen in a few cases (Figure 39.6), and about one-third of patients had tubular aggregates, a finding not shared by any other glycolytic disorder and whose pathogenesis remains obscure.

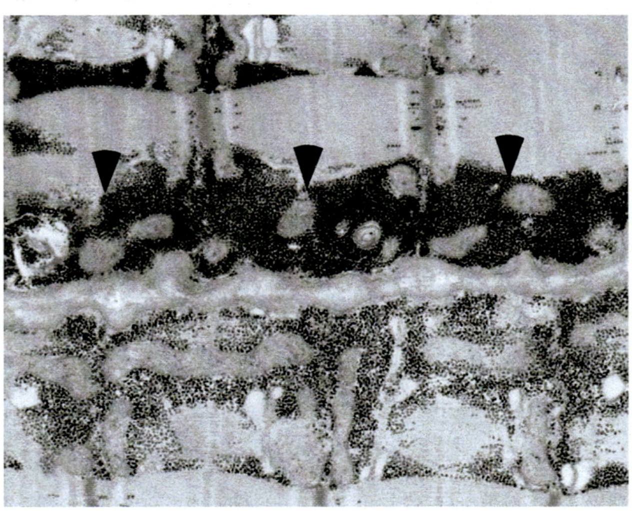

FIGURE 39.6 Phosphoglycerate mutase deficiency (glycogenosis type X). An electron micrograph shows focal excess of electron-dense (dark) glycogen particles in a subsarcolemmal zone (arrowheads) and within the cytoplasm. A similar ultrastructural pattern of glycogen excess is found in all disorders of glycogenolysis and glycolysis.

Cycle exercise responses in two patients were markedly different from those seen in McArdle disease: the PGAM patients had virtually normal cycle exercise and oxidative capacity, no second wind, and no improvement of their exercise capacity with lipid or lactate supplementation.[113]

Beta-enolase Deficiency

Like PGM deficiency, beta-enolase deficiency has thus far been reported in a single patient.[114] This 47-year-old man had adult-onset exercise intolerance and generalized weakness but no myoglobinuria. There was no rise of venous lactate with FIET, muscle ultrastructure showed subsarcolemmal deposits of glycogen, and muscle biochemistry revealed an isolated defect of enolase activity, which was 5% of normal. This is because over 90% of muscle beta-enolase is encoded by *ENO3*, and the patient harbored two missense mutations in this gene.

Lactate Dehydrogenase Deficiency

Lactate dehydrogenase (LDH) is a tetrameric enzyme composed of two subunits, M (or A), and H (or B), resulting in five isozymes, the two homotetramers M4 and H4, and three hybrid forms. LDH-M predominates in muscle, while LDH-H predominates in heart and in other tissues.

Muscle LDH deficiency (glycogenosis type XI) has been documented in six Japanese families[115–119] and in two Caucasian patients.[120] The clinical features of this disorder are similar to those of other glycogenoses caused by partial or complete blocks of glycolysis. Some peculiar laboratory and clinical features, however, can offer useful diagnostic clues. During episodes of myoglobinuria, the predictably sky-high levels of serum CK contrast with low levels of serum LDH. FIET causes poor rise of lactate but excessive rise of pyruvate. A characteristic seasonally recurrent erythematosquamous skin rash has been observed in some patients.[121–123]

Muscle biopsy shows nonspecific myopathic changes,[120] and biochemical study of muscle homogenate reveals a deficit of LDH activity. The sequence of the LDH-M cDNA[124] and the organization of the chromosomal gene[125] have been reported, and the gene has been assigned to chromosome 11p15.4.[126] At least seven different mutations have been reported in patients with LDH-M deficiency.[127]

A Pediatrician's Perspective

The infantile presentation of PFK deficiency is perplexing because myopathy, often manifesting as arthrogryposis multiplex congenita, is associated with signs of multisystem involvement, including seizures, cortical blindness, corneal opacifications, and cardiopathy.[77,128–131] Curiously, none of the infantile cases had evidence of hemolytic anemia, which is difficult to explain. Although the involvement of heart and brain could be attributed to the high proportion of the M-subunit in these tissues, this does not explain why typical adult patients do not have either cardiopathy or encephalopathy.

A more likely possibility is that infantile PFK deficiency may not be due to a molecular defect of *PFKM*, which has never been documented, but to the defect of another gene, possibly encoding an activator of PFK. Irrespective of the precise cause, ketogenic diet may be useful in these children, based on the experience with one patient.[89]

The defects of terminal glycolysis (PGK, PGAM, and LDH deficiencies) tend to affect adults more than children. However, aldolase A deficiency did affect children and crises were precipitated by febrile illnesses.

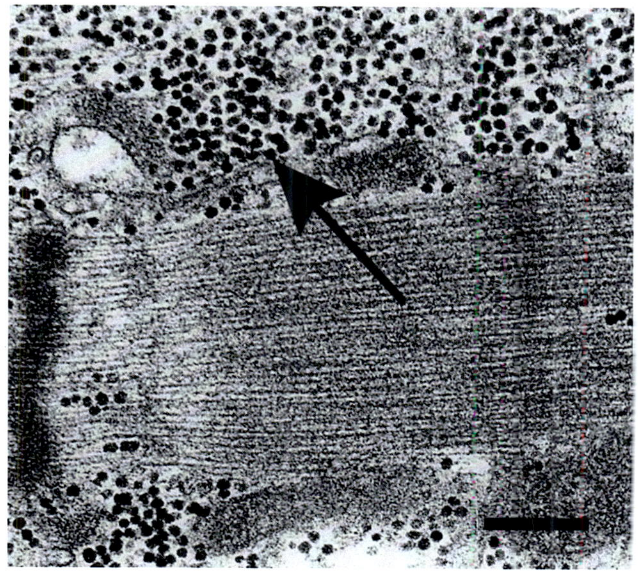

FIGURE 39.7 Electron micrograph demonstrating the normal spherical appearance of glycogen beta-particles (arrow) in skeletal muscle (bar + 0.2 μm).

DEFECTS OF GLYCOGEN SYNTHESIS

Until recently, our knowledge of disorders due to defects of glycogen synthesis was limited to glycogen branching enzyme (GBE) deficiency. However, two novel defects of glycogen synthesis in muscle have been discovered, affecting glycogen synthase (GYS1) and glycogenin (GYG1). Because these new glycogenoses result in the *absence* rather than the *storage* of glycogen, they have been called "glycogenoses type zero (GSD 0)."[132] Before describing them, a brief description of glycogen synthesis is in order. It has been known since the 1970s that the richly branched structure of glycogen, which is crucial to its spherical shape (evident in ultrastructural images of glycogen beta particles (Figure 39.7) and to its solubility in the cytoplasm, depends on the synergistic and delicately balanced action of two enzymes. Glycogen synthase (GYS) strands together glucosyl units one to the other in a linear fashion through α-1,4-glycosidic links. When a branch is about 12 glucosyl units long, the branching enzyme (GBE) has to intervene, in two steps. First, it detaches a short linear "twig" of about 4 glucosyl units from a linear chain. Second, it reattaches it in an α-1,6-glycosidic link to the side of a long linear chain, thus starting a new branch that GYS will elongate, and so on. However, one basic question remained unanswered: how does a molecule of glycogen start? What is the "condensation point" that puts together the first glucose molecules? In 1984, William Whelan identified a self-glycosylating dimeric enzyme, glycogenin (GYG), that fits the bill.[133] While defects of the liver isozyme of GYS have been known for many years and attributed to mutations in *GYS2*, defects in the muscle/heart isozyme of GYS (gene *GYS1*) and in the muscle isozyme of glycogenin (gene *GYG1*) are relatively recent acquisitions.[134,135]

Muscle Glycogenin Deficiency

The only reported patient with muscle glycogenin (GYG1) deficiency was diagnosed at age 27 in dramatic circumstances: after exercising, he experienced dizziness and palpitations, went into ventricular fibrillation, and was saved by emergency cardiac defibrillation.[135] In retrospect, he had been unable to keep up with his peers during childhood because of shortness of breath and mild weakness. At age 19, he was dismissed from military duty because of muscle atrophy and EKG abnormalities (an echocardiogram was normal). Family history revealed Charcot Marie Tooth (CMT) type 2 on the paternal side. Neurological examination showed diffuse mild weakness sparing hip and thigh muscles. Serum CK was normal and needle EMG showed slight myopathic changes. The patient was equipped with a cardioverter-defibrillator.

Muscle biopsy showed extreme pallor with the periodic acid-Schiff (PAS) stain, indicating profound glycogen depletion (Figure 39.8A). There was marked type I fiber predominance and evidence of mitochondrial proliferation. The endomyocardial biopsy showed large centrally located vacuoles containing PAS-positive material that was digested by diastase (Figure 39.8B). Despite the histochemical appearance, ultrastructural examination showed no intravacuolar glycogen but unstructured material together with small vesicles, tubular structures, mitochondria, and lipid droplets (Figure 39.8C). Molecular

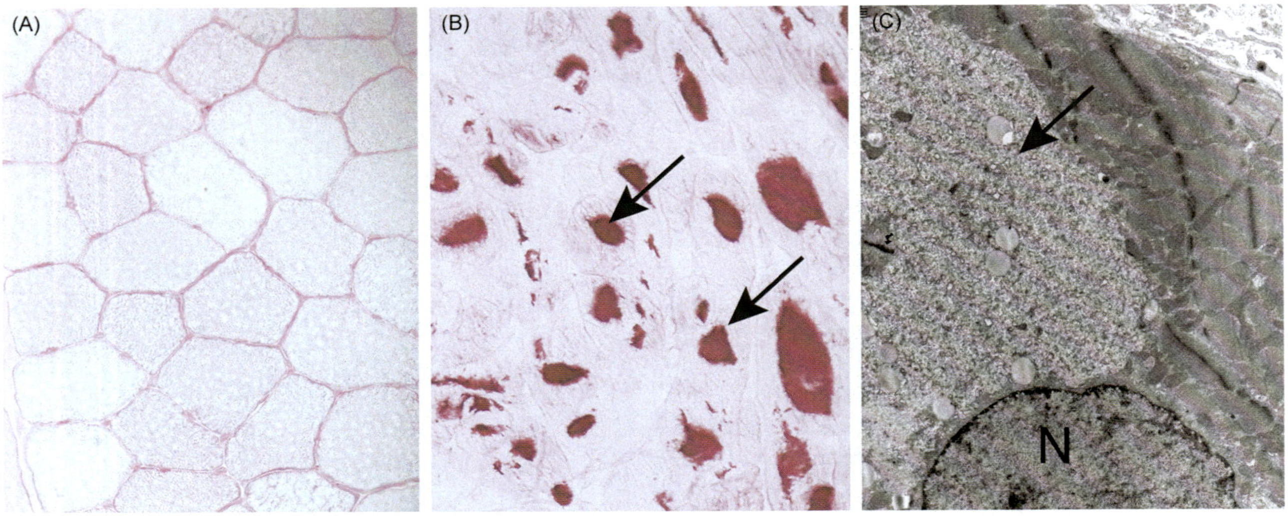

FIGURE 39.8 Muscle biopsy from a patient with glycogenin (GYG1) deficiency (GSD XV). (**A**) The PAS stain shows glycogen depletion in muscle fibers. (**B**) An endomyocardial biopsy from the same patient demonstrates PAS-positive inclusions in cardiomyocytes (see (B)) (arrows). (**C**) By EM, these inclusions (see (B)) (arrow) have polymorphic features, including fat droplets and small vesicular and filamentous structures, but virtually no normal glycogen.

genetic analysis revealed one nonsense and one missense mutation in *GYG1*: the missense mutation impaired the autoglucosylation of GYG1, which is essential for the priming of glycogen synthesis.[135]

Muscle Glycogen Synthase Deficiency

Three siblings and two unrelated patients have been described with muscle glycogen synthase (GYS1) deficiency, which involves both skeletal and cardiac muscle.[134,136,137] In two of the three siblings, who were born of consanguineous parents, the myopathy was characterized by exercise intolerance noted in childhood, with trouble keeping up with their peers and premature fatigue.[134] The cardiomyopathy was more severe and resulted in a fatal cardiac arrest during exercise in the oldest sibling at 10 years of age. Cardiac involvement was also documented in the other two children, who needed cardioprotective medication. At 4 years of age, the oldest sibling had also had tonic-clonic seizures and received an EEG diagnosis of partial complex epilepsy: however, brain MRI and MRS were normal and the EEG normalized by 6 years of age. In the two younger children, skeletal muscle biopsies stained with PAS showed virtual absence of glycogen in all fibers (Figure 39.9A), which was confirmed by electron microscopy (EM) (Figure 39.9B,C). There was also a marked predominance of type I muscle fibers and mitochondrial proliferation. Similarly, examination of postmortem cardiac muscle from the oldest patient with hypertrophic cardiomyopathy showed myocyte hypertrophy and lack of glycogen, whereas the content of glycogen in her liver appeared normal. In all three patients, molecular analysis documented a homozygous nonsense mutation in the *GYS1* gene.

One other patient, an 8-year-old boy born to consanguineous parents, had a clinical history that was tragically similar to that of the oldest sibling described above: while horsing around at school, he suffered a cardiac arrest and died. There was a history of exercise intolerance. Postmortem examination of skeletal muscle was initially suggestive of a mitochondrial myopathy, with subsarcolemmal accumulation of mitochondria and predominance of type I fibers. However, there was also glycogen depletion and molecular genetic analysis identified a homozygous 2 bp deletion in exon 2 of *GYS1*.[138]

The fourth patient was an 11-year-old Japanese girl who, aside from two tonic-clonic seizures (one at age 2 and one at 4), started having repeated episodes of syncope following exercise. Typically, exercise caused myalgia and weakness, making her squat down. This was followed by loss of consciousness for a few hours.[137] Interestingly, FIET failed to raise venous lactate. Surprisingly, however, except for "a mild ischemic finding on exercise electrocardiography," extensive cardiologic evaluation yielded normal results. Neurological examination showed proximal limb weakness and muscle MRI showed fatty degeneration in glutei and thigh muscles. Muscle biopsy showed glycogen depletion in muscle fibers but not in the endomysium by PAS stain, and lack of histochemical phosphorylase reaction (which depends on the presence of endogenous glycogen). Also, this patient had mitochondrial proliferation. Biochemical analysis of muscle documented the paucity of glycogen (3% of the normal mean) and of GYS activity (2% of normal).

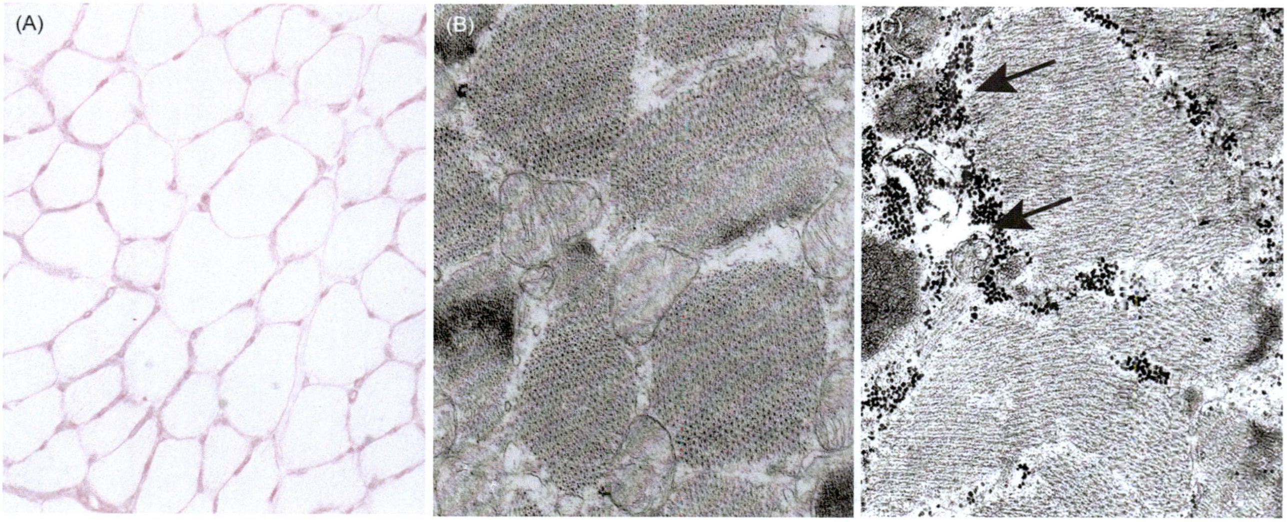

FIGURE 39.9 Muscle biopsy from a patient with glycogen synthase (GYS1) deficiency (GSD 0). (**A**) PAS stain shows obvious glycogen depletion in fibers, which is confirmed by EM (**B**) in contrast to the ultrastructure of a control muscle (**C**).

Glycogen Branching Enzyme Deficiency

The glycogen branching enzyme (GBE) catalyzes the last step in glycogen biosynthesis by attaching a short glucosyl chain (about 7 glucosyl units) in an α-1,6-glucosidic link to a naked peripheral chain of nascent glycogen. The newly added twigs are then elongated by glycogen synthase. Hence, GBE deficiency results in the accumulation of an amylopectin-like polysaccharide with fewer branches and longer outer chains than normal glycogen, which is known as polyglucosan (PG). PG is best recognized in tissues with the PAS stain because it is strongly PAS-positive but largely resistant to diastase digestion (Figure 39.10A,B). Ultrastructurally, the abnormal polysaccharide consists of filamentous and finely granular material, often associated with normal-looking glycogen particles (Figure 39.10C).

GBE deficiency (glycogenosis type IV) is traditionally described as a liver disease with progressive liver cirrhosis leading to hepatic failure and death in early childhood.[139] In fact, this glycogenosis has a very wide spectrum of clinical phenotypes that can affect predominantly the liver, the heart, skeletal muscle, or the brain.[6] Neuromuscular presentations range from congenital or infantile disorders to adult polyglucosan body disease (APBD). See Case Example 39.2.

There are two main infantile presentations. The first is the fetal akinesia deformation sequence (FADS), characterized by intrauterine growth restriction, arthrogryposis multiplex congenita, craniofacial anomalies, pulmonary hypoplasia, and perinatal death.[140] Although the second variant was labeled "congenital,"[140] it should be called "fatal infantile" because it presents at or soon after birth with hypotonia, muscle wasting, neuronal involvement, inconsistent cardiomyopathy, and early death.[140–142] Interestingly, in a series of eight patients with the congenital or infantile neuromuscular form, there was a good correlation between the severity of the phenotype and that of the molecular defect.[140] Detailed neuropathology in a few infants showed PG inclusions in neurons of the basal ganglia and thalamus, and of the pontine and medullar nuclei, and in periaqueductal neurons.[142,143,144] Because the motor neurons of the spinal cord can also be affected, it is not surprising that some of these infants were initially diagnosed as spinal muscular atrophy type 1 (SMA 1) until tissue histochemistry showed PG deposits and molecular diagnosis ruled out mutations in the *SMN1* gene.[141]

In patients with myopathy, serum CK level is variably increased. EMG shows myopathic features, and FIET (when feasible) causes normal rise of venous lactate.

Because PG is ubiquitous, histochemical analysis of fetal tissues can provide prenatal diagnosis: placental biopsies at 25 and 35 weeks of gestation showed unequivocally PG deposits in the extravillous trophoblast in two cases with genetically documented GBE deficiency.[148]

Brain involvement dominates the clinical presentation in adult- or late-onset forms of GBE deficiency, known collectively as "adult polyglucosan body disease" (APBD). Typical APBD affects predominantly individuals of Ashkenazi Jewish descent. After the first case was reported in 1971,[149] Robitaille et al. defined the clinical and neuropathological features of APBD in four patients[150]: they pointed out the "relatively stereotyped chronic neurological disease" characterized by progressive lower and upper motor neuron involvement, marked sensory loss, predominantly in the legs, neurogenic bladder, and, in two of their patients, dementia. They

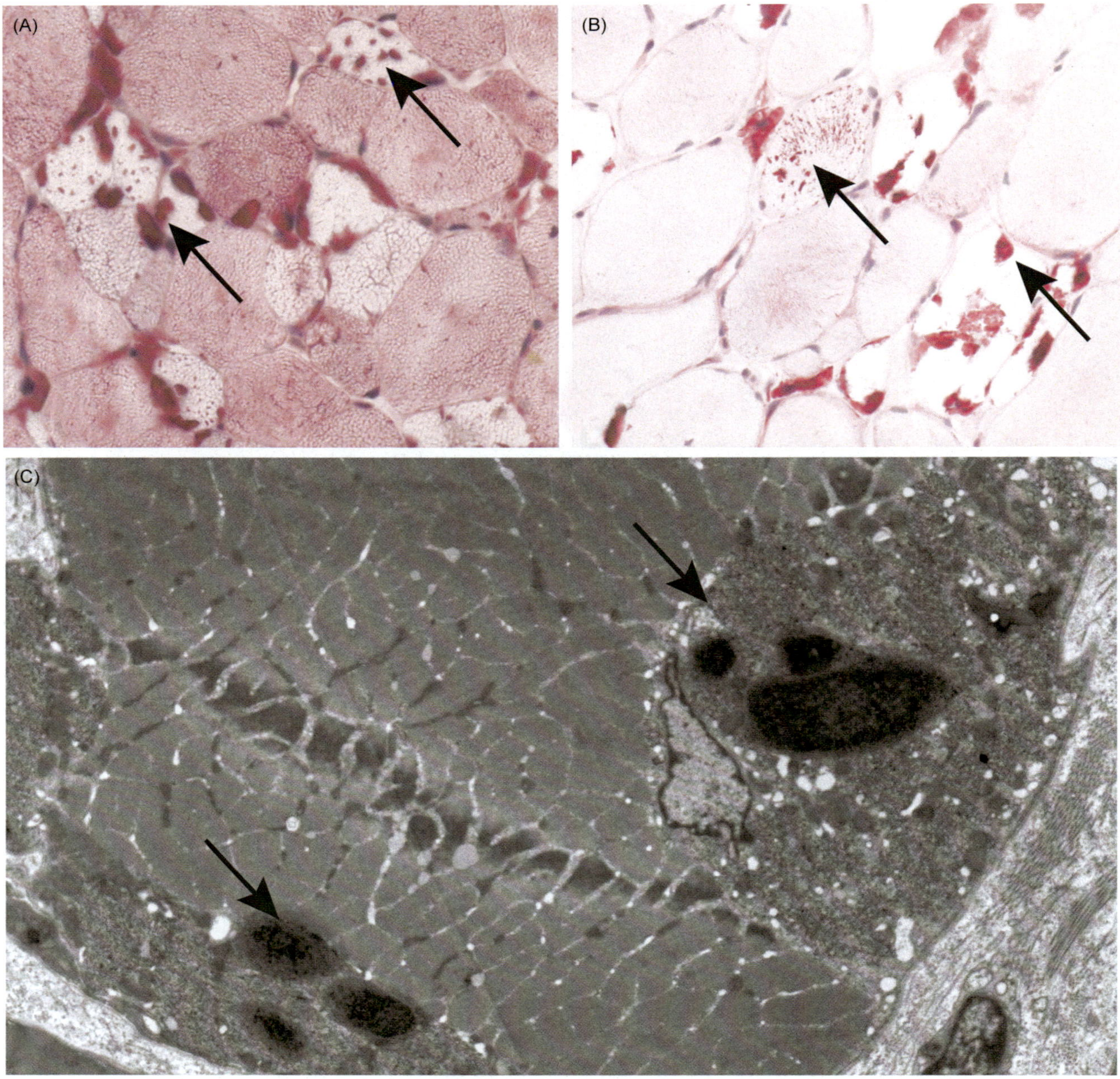

FIGURE 39.10 Muscle biopsy from a patient with branching enzyme (GBE) deficiency (GSD IV). The PAS stain (**A**) shows PAS-positive material that is partly resistant to amylase treatment (**B**). EM documents that the inclusions are compatible with polyglucosan (arrows).

identified the signature neuropathological lesions as structures resembling Lafora bodies, which were abundantly present in both central and peripheral neural processes and in astrocytes but not in perikarya. Thirty-two years later, a multinational study of 50 APBD patients has largely confirmed those early data: the most common clinical finding was neurogenic bladder (thus justifying the dictum that APBD patients see the urologist before seeing the neurologist), followed by spastic paraplegia with vibration loss, and axonal neuropathy.[151] This study confirmed the original observation that dementia is seen in about 50% of patients. The median age at onset was 51 years, the median age of wheelchair dependency was 63 years, and that of death was 70 years.[151] Importantly, this study also highlighted the characteristic MRI abnormalities, with atrophy of the medulla and the spine, and hyperintense white matter abnormalities in T2 sequences affecting mostly the periventricular regions, the posterior limb of the internal capsule, and the pyramidal tracts.[151]

CASE EXAMPLE 39.2

This infant girl was the first child of nonconsanguineous parents. She was born at 33 weeks of gestation after a pregnancy complicated by polyhydramnios. At birth, there was no spontaneous respiratory effort and she was cyanotic and bradycardic. Apgar scores were 1 at 1 minute and 2 at 5 minutes. The baby was intubated and mechanically ventilated. She was markedly hypotonic with absent Moro's and grasp reflexes. There were no tongue fasciculations. There was no arthrogryposis but all tendon reflexes were absent. Muscle tone was diminished and she made minimal spontaneous movements. There was no organomegaly. She was intubated for the entire duration of her life and her hospital course was marked by intermittent episodes of apnea and bradycardia. Electrocardiogram (EKG) and 2D echocardiogram showed no evidence of cardiac dysfunction and liver function studies were unremarkable. The electroencephalogram (EEG) was minimally abnormal, lacking a normal transition to sustained wakefulness, despite stimulation. Electromyogram (EMG) showed positive sharp waves and fibrillations, suggestive of a neurogenic process. The cerebrospinal (CSF) cell count, cultures, protein, glucose, amino acids, lactate and pyruvate levels were all normal. Karyotype. fluorescent *in situ* hybridization and methylation studies for Prader-Willi syndrome, genetic testing for myotonic dystrophy, quantitative serum amino acids, and urinary organic acids were normal.

A quadriceps muscle biopsy showed fibers containing abnormal pale storage material that was strongly PAS-positive and diastase-resistant. There was also increased variation of fiber sizes due to atrophy of some fibers. On day 41 of life, the parents elected to withdraw support and the baby expired immediately after extubation. Autopsy revealed PAS-positive, diastase-resistant globules in liver, heart, skeletal muscle, and neurons of the brain and spinal cord. The polyglucosan was more abundant in motor neurons both of the brainstem and of the spinal cord.

Although typical APBD does not belong, strictly speaking, in a pediatric textbook, it is an important presentation of GBE deficiency and there are reports of APBD patients complaining of exercise intolerance in childhood.[152] More importantly, young adult onset variants of the disorder are being recognized, often in non-Ashkenazi individuals.[153,154] These patients present a diagnostic challenge because they have relapsing-remitting neurological signs and brain MRI lesions simulating multiple sclerosis (MS). Interestingly for pediatricians, liver involvement in childhood either in the patients themselves or in their siblings may give a clue to their diagnosis.[154]

Human GBE is a monomeric enzyme encoded by a single gene, which has been cloned, sequenced, and assigned to chromosome 3p12.[155] Different mutations in the *GBE1* gene have been identified in patients with various clinical presentations.[140] The Y329S mutation is common among Ashkenazi Jewish patients with APBD, undoubtedly the result of a founder effect.

There is no specific therapy for GSD IV but the development of transgenic murine models of both severe infantile and adult forms will facilitate the development of therapeutic modalities.[156]

Other Polyglucosan Diseases

Lafora Progressive Myoclonus Epilepsy

Although Lafora disease (LD) is not traditionally included among the glycogenoses, its pathologic hallmark, the bodies described by Gonzalo Lafora in 1911,[157] are, in fact, polyglucosan (PG) bodies indistinguishable from those that characterize GBE deficiency. This autosomal recessive progressive myoclonus epilepsy (PME) starts in adolescence with increasingly intractable myoclonic, visual, or convulsive seizures and insidious cognitive decline, leading to disinhibited dementia and death (usually in status epilepticus) before 25 years of age.[158]

Besides the brain, PG bodies in LD are found in skeletal and cardiac muscle (Figure 39.11), the liver, and apocrine gland cells of the skin.[160] In the brain, PG bodies are present in all regions and are localized in perikarya (cell bodies) and dendrites, largely sparing neuronal axons[158,160,161]; this is in contrast to the predominant localization of PG bodies in axonal processes and peripheral nerves in APBD (see above).

Also contrary to APBD, the accumulation of PG in LD is not due to a problem with glycogen synthesis. Although the pathogenesis of PG deposition is not completely clear, two mutant genes are responsible for LD, *EPM2A* or *EPM2*, encoding the interacting proteins laforin and malin. Laforin is a phosphatase that eliminates the small amount of phosphate that is normally bound to glycogen.[162] Thus, excessive phosphorylation of glycogen appears to be an important pathogenic step towards PG deposition.[163] The role of malin, a ubiquitin ligase, seems to be that of removing laforin once this has completed its phosphatase function. In the absence of malin, both laforin and its binding to glycogen increase, leading to PG formation.[164]

Although no specific therapy is presently available for LD, there are, as in the case of APBD, valuable transgenic "Lafora mice" that contribute to our understanding of pathogenesis and to the development of therapeutic strategies. One especially interesting and successful approach has been to abrogate glycogen synthesis by removing its main activator, PTG (protein targeting to glycogen),

FIGURE 39.11 Muscle biopsy from a patient with Lafora disease. By EM, several non-membrane-bound vacuoles contain a granulo-fibrillar polysaccharide (denoted as LP for "Lafora polysaccharide") that we now know to be polyglucosan. *(Reproduced with permission from Coleman et al.[159])*

which has resulted in the removal of PG bodies and the resolution of neurodegeneration and myoclonic epilepsy in Lafora mice.[161,165]

Myopathy and Cardiomyopathy Due to Defective Ubiquitin Ligase RBCK1

This new glycogenosis has been described in ten patients from eight families of different ethnic backgrounds.[166] While all patients had childhood or juvenile-onset of proximal limb weakness, eight of them also had rapidly progressive cardiomyopathy, which required cardiac transplantation in four. There was extensive accumulation of PG in muscle (Figure 39.12A,B) and in both skeletal and cardiac muscle in the patients with cardiomyopathy. The stored material had the typical appearance of PG (Figure 39.12C,D).

All patients were either homozygous or compound heterozygous for missense or truncating mutations in *RBCK1*, which encodes a ubiquitin ligase.[166] It should be noted that some patients with mutations in the same gene have a different clinical phenotype, characterized by failure to thrive, chronic autoinflammation, and recurrent episodes of sepsis. These patients also accumulate PG in muscle, heart, and liver.[167] Careful genotype:phenotype correlation analysis will be required to explain the different tissue involvement in these allelic disorders.

Other Polyglucosan Disorders

There are several reports of patients with PG storage disease, either limited to muscle[168,169] or involving muscle and heart,[170] in whom the molecular defect remains to be identified.

A Pediatrician's Perspective

The recent extension of glycogenoses type 0 from liver to muscle and heart offers the neuromuscular pediatrician new diagnostic considerations in children with exercise intolerance and cardiac problems (defects of glycogenin, GYG) or with more severe intermittent exercise-related myalgia associated with syncopal episodes (glycogen synthase, GYS, deficiency). Lack of muscle glycogen, the "high-octane fuel," easily explains several features of GYS deficiency, including the premature fatigue during vigorous activity, the lack of venous lactate increase after ischemic forearm exercise (no glycogen, no lactate), and the "false positive" histochemistry for myophosphorylase (endogenous glycogen is the crucial substrate for this reaction[171]). Two morphological caveats are in order: (i) while muscle pathologists are not likely to miss excessive glycogen in a biopsy, they may overlook lack of glycogen; and (ii) the compensatory mitochondrial proliferation that accompanies both GYG and GYS deficiencies may be severe enough to distract attention from glycogen metabolism and suggest a mitochondrial myopathy instead.

Aside from APBD, most other presentations of brancher deficiency affect infants or children and are surprisingly heterogeneous. Thus, the astute pediatrician should include brancher deficiency in the differential diagnosis of hepatopathy, cardiomyopathy, and apparently isolated congenital or early-onset myopathy. The frequent involvement of the anterior horn cells of the spinal cord may help explain how some infants with congenital or childhood neuromuscular disease are initially diagnosed as having spinal muscular atrophy. After excluding mutations in *SMN1*, muscle histochemistry can easily orient the pediatrician towards GBE deficiency by showing PAS-positive, diastase-resistant deposits of PG. In fact, GBE deficiency is one of three metabolic myopathies that can closely mimic SMA: the other two are mitochondrial diseases, SCO2 deficiency (a form of cytochrome *c* oxidase [COX] deficiency) and TK2 deficiency, an mtDNA depletion syndrome (see Chapter 41).

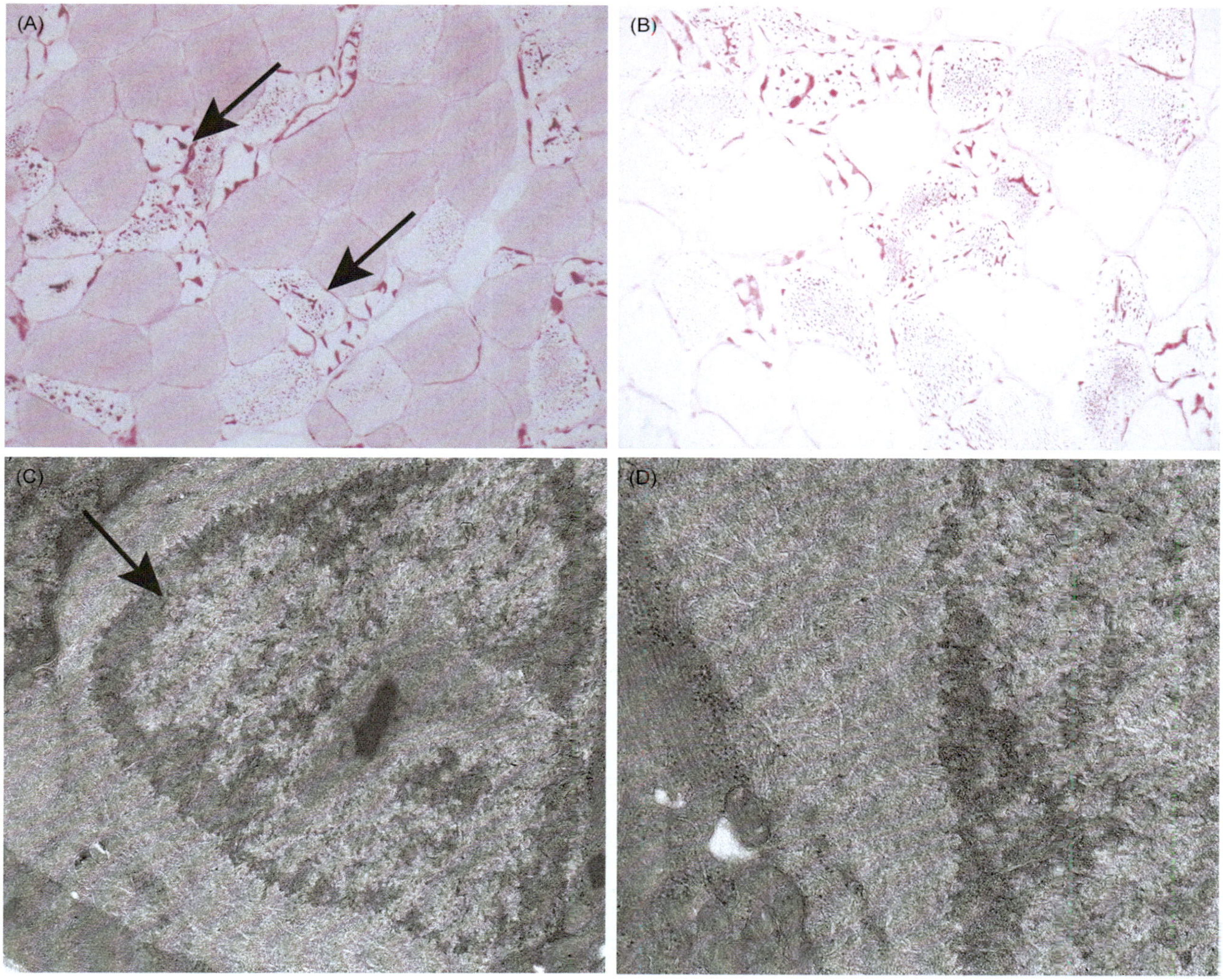

FIGURE 39.12 Muscle biopsy from a patient with RBCK1 deficiency. PAS stain shows PAS-positive inclusins (**A**, arrows) that are only partly digested by amylase treatment (**B**). EM demonstrates that the inclusions are compatible with polyglucosan (**C**, arrow; **D**).

Lafora disease is not primarily a neuromuscular disorder but it is a generalized glycogen storage disease that affects adolescents. Its similarity to APBD despite the different pathogenesis and the existence in both conditions of genetically engineered mouse models that recapitulate clinical features and pathology bode well for the development of novel therapeutic approaches. For example, abolishing glycogen before it becomes the offending polyglucosan, as has been done in Lafora mice,[161,165] appears less extreme an intervention because we know that patients with glycogenoses type 0 tolerate their condition reasonably well.

Finally, it is important to screen for mutations in *RBCK1* in children with polyglucosan myopathy with or without cardiomyopathy because the patients described by Nilsson et al.[166] were diagnostic puzzles for several years and cardiac transplantation is a therapeutic option.

DEFECTS OF LYSOSOMAL GLYCOLYSIS

Acid α-Glucosidase Deficiency

The fatal infantile form of acid maltase (acid α-glucosidase, GAA) deficiency was first described in 1932 by Pompe and Putschar, who called attention to the glycogen storage in the heart.[172,173] In 1963, Hers documented the defect of GAA in liver, heart, and skeletal muscle of infants with "cardiomegalic glycogenosis"[174] and, together with Lejune and Thines-Sempoux, showed that GAA was a lysosomal enzyme.[175] Thus, GAA deficiency became the prototype of inborn lysosomal diseases.[176] GAA catalyzes the acid hydrolysis of α-1,4 and α-1,6 linkages, breaking down glycogen to glucose: the enzyme deficiency results in intralysosomal accumulation of normal glycogen in some or all tissues.

GAA deficiency (glycogenosis type II, Pompe disease) results in two major clinical syndromes, based on age at onset and severity: (i) a generalized and rapidly fatal disease of early infancy,[177] and (ii) a milder neuromuscular disorder beginning in childhood or adulthood.[178,179]

Infantile GAA deficiency (see Case Example 39.3) presents at or soon after birth and is characterized by massive deposition of glycogen in all tissues, especially heart, muscle, and liver. Severe hypotonia and weakness are associated with massive cardiomegaly, hepatomegaly, and, inconsistently, macroglossia. Electrocardiography shows short P-R interval, giant QRS complexes and signs of left ventricular or biventricular hypertrophy. Chest X-ray shows massive cardiomegaly. Cardiac or respiratory failure causes death usually before the first and invariably before the second year of age. In the CNS, anterior horn cells of the spinal cord and neurons of brainstem nuclei are more severely affected than neurons of the cerebral cortex,[180–182] which probably contributes to the flaccid quadriplegia of these infants.

In the childhood form, walking may be delayed and there is slow and progressive weakness of limb-girdle and trunk muscles. Calf hypertrophy may suggest the diagnosis of Duchenne muscular dystrophy in boys. Respiratory muscles are affected early and respiratory failure is the usual cause of death in the second or third decade. There is no cardiomegaly and hepatomegaly is much less common than in infantile GAA deficiency.[183] The adult phenotype begins in the third or fourth decade and is characterized by proximal and truncal muscle weakness. Respiratory muscles are selectively affected and respiratory insufficiency may be the presenting complaint, with morning headache or exertional dyspnea.[184,185] The initial diagnosis in most cases is limb-girdle muscular dystrophy or polymyositis. The common notion that late-onset GAA deficiency is a purely muscular disorder has been challenged by reports of basilary artery and aortic aneurysms,[186,187] urinary and fecal incontinence,[188] and dysphagia. Accordingly, a detailed postmortem study of a late-onset patient documented ultrastructural alterations in smooth muscle of blood vessels, the gastrointestinal tract, and the bladder.[189]

Acid α-glucosidase is virtually absent or markedly reduced in all tissues, not only in the infantile variant but also in the late-onset form.[189,190] In all forms of GAA deficiency, serum CK levels are elevated and EMG shows a myopathic pattern often associated with fibrillation potentials, positive waves, complex repetitive discharges, and myotonic discharges, especially in paraspinal muscles. Whole-body muscle MRI can provide an accurate picture of the muscles involved.[191]

Muscle biopsy shows vacuolar myopathy. In the infantile form, all muscles and all fibers contain many, often confluent vacuoles, resulting in a "lacework" appearance. The vacuoles are lysosomes filled with PAS-positive diastase-digestible material and stain intensely for acid phosphatase (Figure 39.13A,B). Glycogen content is increased up to 10-fold and electron microscopy shows that the polysaccharide is present both within lysosomes and free in the cytoplasm (Figure 39.14). In the adult variant, however, vacuoles are less abundant and biopsies from clinically unaffected muscles may appear normal, despite the marked decrease of GAA activity.

The *GAA* gene has been mapped to chromosome 17q21-q23, and a variety of molecular defects have been described in both infants and adults.[192] In general, there is good correlation between the severity of the mutation and the severity of the clinical phenotype. Thus, deletions and nonsense mutations are usually associated with infantile GAA deficiency, whereas "leaky" mutations, such as the common IVS1(-13T > G) splice site mutation, are associated with the adult-onset variant of the disease.[193,194] Knockout "Pompe mice" are available and have proven very useful in understanding both the pathogenesis of the disease and the mechanism of action of enzyme replacement therapy (ERT).[195]

In a lucid review, where old findings are revisited in the light of present knowledge, Raben et al.[196] point out that Pompe disease is not a purely lysosomal disease because the problem lies upstream of the lysosomes,

CASE EXAMPLE 39.3

A 3-month-old girl was referred because of poor motor development and cardiomegaly, which were first noted during a prior hospital admission at 40 days of age. Born after a normal pregnancy and delivery, she appeared normal during her neonatal period. However, she had poor weight gain and the parents noted head lag and dyspnea. On admission, she had perioral cyanosis, macroglossia, and pseudohypertrophy of her musculature, contrasting with generalized weakness. There was cardiomegaly and hepatomegaly. She had labored breath and decreased spontaneous movements, lying in a "frog position." There was diffuse hypotonia and weakness. She was alert and had a social smile. Electrocardiography showed giant QRS complexes and a short P-R interval, and EMG showed myotonic bursts. Muscle biopsy showed a lacework appearance and the vacuoles were strongly PAS-positive and reacted intensely with the acid phosphatase stain. Biochemical analysis showed a 10-fold increase in glycogen concentration and undetectable acid maltase activity. In the following months, she had repeated respiratory infections and died of cardiorespiratory failure at 6 months of age.

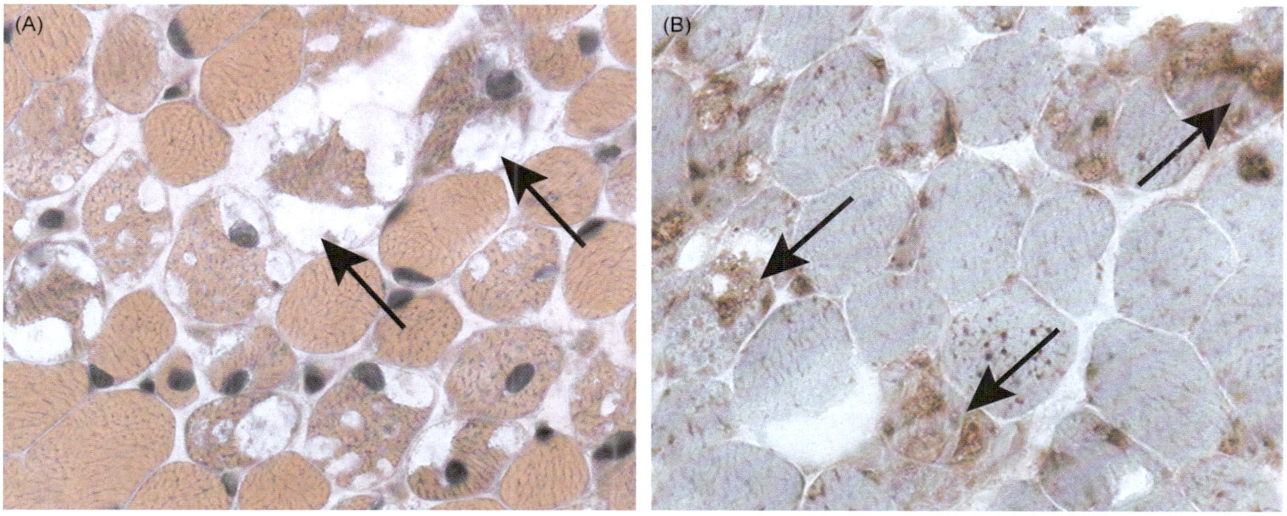

FIGURE 39.13 Muscle biopsy from an infant with acid maltase (GAA) deficiency (Pompe disease, GSD II). H&E shows large vacuoles in all fibers (**A**, arrows) that stain intensely with the acid phosphatase reaction (**B**, arrows).

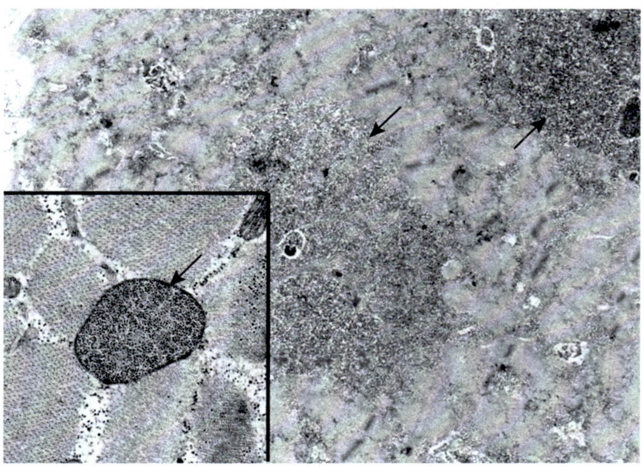

FIGURE 39.14 EM of muscle in infantile Pompe disease, showing large accumulation of free cytoplasmic glycogen (arrows) together with intralysosomal glycogen (inset).

affecting the process of autophagy. In fact, while infantile form muscle biopsies show glycogen-laden lysosomes ("glycogenosomes"), muscle biopsies from late-onset patients show more commonly double-membrane autophagosomes containing not just glycogen but also protein aggregates and worn-out organelles. A compelling scenario for the pathogenesis of Pompe disease, as well as other lysosomal disorders,[197] involves a profound dysfunction of the autophagic process, with accumulation of autophagosomes resulting from defective autophagosome-lysosome fusion.[197,198] In fact, Nishino et al. went as far as stating that "Pompe disease can no longer be viewed simply as a glycogen storage disease," but rather as a problem in handling excessive numbers of autophagosomes.[199]

A central riddle regards the pathogenesis of weakness: is it simply due to the mechanical disarray of the contractile system caused by the presence of glycogenosomes and by cell necrosis due to their rupture, or is it due to an energy defect? The former mechanism may well be true in infantile Pompe disease,[200,201] but does not apply to the late-onset variant. On the other hand, standardized cycle exercise in six patients with late-onset Pompe disease (not on enzyme replacement therapy at the time of the study) showed normal respiratory exchange ratios (RERs), normal fat and carbohydrate oxidation, and no improvement in exercise tolerance upon glucose infusion,[202] thus confirming the conventional wisdom that GAA deficiency does not impair energy metabolism in muscle.

A generally effective enzyme replacement therapy (ERT) with recombinant human GAA (rhGAA) has been widely utilized in the past 6 years[203] and a vast literature has accumulated on the subject.[204] A multicentric, randomized, placebo-controlled trial with alglucosidase alfa, a human rhGAA, was conducted in 90 patients (8 years of age or older) with the late-onset form of Pompe disease.[205] The two primary end points were motor (6-min walk test) and respiratory (percentage of predicted forced vital capacity). Secondary end points were quantitative muscle testing and maximum inspiratory and expiratory pressures. Treatment with rhGAA for 78 weeks was associated with improved walking distance and stabilization of pulmonary function over a period of 18 months. The results were modest but important considering the slow but inexorable progress of the disease.[206]

Because of the different tissue involvement and the presumably different pathogenic mechanism (see above), the response of patients with infantile Pompe disease has been more variable, as illustrated by the first four children treated with rhGAA.[207,208] All four children survived until 4 years or longer, thus meeting the provisional end point of the trial. Three of them were still alive more than 11 years later, but the two who had started therapy late (7 and 8 months of age) were paralytic and ventilator-dependent. The child who had started therapy at 3 months was still able to walk at 11 years of age and went to school, although he had some weakness.[204] The UK experience with 20 infants treated between 2000 and 2009 was equally tattered: overall ventilator-free survival was 35%; 35% died at a median age of 10 months; 30% were alive but ventilator-dependent. The authors wisely concluded that ERT for infants with Pompe disease has to be pondered carefully.[209] In fact, a special issue of the *American Journal of Medical Genetics* dedicated to "The New Era of Pompe Disease" documents how the improved outcome resulting from ERT has created a new and still emerging phenotype for both infantile and late-onset Pompe disease.[210] For example, the prolonged survival of patients with the infantile form and the fact that rhGAA does not cross the blood-brain barrier raise the question of the cognitive outcome and adaptive function of these children.[211,212]

Patients with GAA deficiency may need intermittent or permanent mechanically assisted ventilation.[184] A regimen combining high-protein diet and aerobic exercise has improved strength and respiratory function in some patients[213] but needs to be validated in a large cohort.

A Pediatrician's Perspective

Pompe disease is a well-known devastating disorder of infancy mainly characterized by the association of severe hypotonia and massive cardiomyopathy. It is one important consideration in the differential diagnosis of the "floppy baby syndrome."

While spinal muscular atrophy type 1 (SMA 1) is easily distinguished by the lack of cardiomegaly, a number of metabolic myopathies combine profound weakness with cardiopathy and should be carefully considered. These include one glycogenosis described above, AMPK deficiency (with secondary deficiency of PhK) due to mutations in the *PRKAG* gene.

Primary carnitine deficiency may simulate Pompe disease, although the cardiopathy usually overshadows the myopathy. Muscle biopsy shows lipid rather than glycogen storage. The importance of considering carnitine deficiency is that life-long L-carnitine administration is an effective therapy.

Mitochondrial encephalomyopathies sometimes are accompanied by cardiomyopathy, which can be the dominant feature, as in histiocytoid cardiomyopathy of infancy (see Chapter 41). Lactic acidosis is usually a distinguishing laboratory feature. Myopathy and cardiopathy are also clinical hallmarks of Barth syndrome and Sengers syndrome, both of which, however, are uncommon in infancy.

A major differential consideration is LAMP-2 deficiency (Danon disease), which in fact was initially described as "Lysosomal glycogen storage disease with normal acid maltase."[214] This is an X-linked dominant disorder of autophagy due to mutations in *LAMP-2*,[215] which encodes a structural protein of the lysosomal membrane (lysosome-associated membrane protein 2) and results in myopathy, cardiopathy, and inconsistent mental retardation. Muscle biopsy is of little help because it shows accumulation of vacuoles containing protein aggregates, membranes, and glycogen, similar to those that are seen in late-onset Pompe disease. The presence of cardiopathy in mothers of affected children offers a diagnostic clue. Cardiac transplantation has been helpful in a few patients with LAMP-2 deficiency.

Enzyme replacement therapy (ERT) with recombinant human enzyme (rhGAA) has become the standard of care for patients with both infantile and late-onset Pompe disease, but it is expensive and only partially effective.

Although glycogenoses have been among the first metabolic myopathies to be described, new entities are still being discovered, pathogenic mechanisms are incompletely understood, and effective therapies are still lacking.

REFERENCES

1. DiMauro S, Haller RG. Metabolic myopathies: substrate use defects. In: Schapira AHV, Griggs RC, editors. *Muscle Diseases*. Boston: Butterworth-Heinemann; 1999. pp. 225–49.
2. Haller RG, Vissing J. Functional evaluation of metabolic myopathies. In: Engel AG, Franzini-Armstrong C, editors. *Myology*. New York: McGraw-Hill; 2004. pp. 665–679.
3. DiMauro S, Akman HO, Paradas C. Metabolic myopathies. In: Katirji B, Kamisnski HJ, Ruff RL, editors. *Neuromuscular Disorders in Clinical Practice*. New York: Springer; 2014. pp. 1311–34.
4. DiMauro S, Schon EA. Mitochondrial disorders. In: Waxman SG, editor. *Molecular Neurology*. 1st ed. Burlington MA: Elsevier Academic Press; 2007. pp. 535–51.
5. Oldfors A, DiMauro S. New insights in the field of muscle glycogenoses. *Curr Opin Neurol* 2013;**26**:544–53.
6. DiMauro S, Hays AP, Tsujino S. Nonlysosomal glycogenoses. In: Engel AG, Franzini-Armstrong C, editors. *Myology*. New York: McGraw-Hill; 2004. pp. 1535–58.
7. Fishbein WN, Armbrustmacher VW, Griffin JL. Myoadenylate deaminase deficiency: A new disease of muscle. *Science* 1978;**200**: 545–8.
8. Meinck HM, Goebel HH, Rumpf KW, Kaiser H, Neumann P. The forearm ischaemic work test—hazardous to McArdle patients? *J Neurol Neurosurg Psychiatry* 1982;**45**:1144–6.

9. Bruno C, Bado M, Minetti C, Cordone G. Forearm semi-ischemic exercise test in pediatric patients. *J Child Neurol* 1998;**13**:288–90.
10. Kazemi-Esfarjani P, Skomorowska E, Dysgaard Jensen T, Haller RG, Vissing J. Nonischemic forearm exercise test for McArdle disease. *Ann Neurol* 2002;**52**:153–9.
11. Akman HO, Sampayo JN, Ross FA, Scott JW, Wilson G, Benson L. Fatal infantile cardiac glycogenosis with phosphorylase kinase deficiency and a mutation in the gamma2-subunit of AMP-activated protein kinase. *Pediatr Res* 2007;**62**:499–504.
12. Burwinkel B, Hu B, Schroers A, Clemens PR, Moses SW, Shin YS, et al. Muscle glycogenosis with low phosphorylase kinase activity: mutations in *PHKA1*, *PHKG1* or six other candidate genes explain only a minority of cases. *Eur J Hum Genet* 2003;**11**:516–26.
13. Davidson JJ, Ozcelik T, Hamacher C, Willems PJ, Francke U, Kilimann MW. cDNA cloning of a liver isoform of the phosphorylase kinase alpha subunit and mapping of the gene to Xp22.2-p22.1, the region of human X-linked liver glycogenosis. *Proc Natl Acad Sci USA* 1992;**89**:2096–100.
14. Calalb MB, Fox DT, Hanks SK. Molecular cloning and enzymatic analysis of the rat homolog of PhK-gamma T, an isoform of phosphorylase kinase catalytic subunit. *J Biol Chem* 1992;**267**: 1455–63.
15. Hendrickx J, Lee P, Keating J, Carton D, Sardharwalla IB, Tuchman M, et al. Complete genomic structure and mutational spectrum of PHKA2 in patients with X-linked liver glycogenosis type I and II. *Am J Hum Genet* 1999;**64**:1541–9.
16. Burwinkel B, Maichele AJ, Aagenaes O, Bakker HD, Lerner A, Shin YS, et al. Autosomal glycogenosis of liver and muscle due to phosphorylase kinase deficiency is caused by mutations in the phosphorylase kinase beta subunit (PHKB). *Hum Mol Genet* 1997; **6**:1109–15.
17. Wilkinson DA, Tonin P, Shanske S, Lombes A, Carlson GM, DiMauro S. Clinical and biochemical features of 10 adult patients with muscle phosphorylase kinase deficiency. *Neurology* 1994; **44**:461–6.
18. Clemens PR, Yamamoto M, Engel AG. Adult phosphorylase b kinase deficiency. *Ann Neurol* 1990;**28**:529–38.
19. Preisler N, Orngreen MC, Echaniz-Laguna A, Laforet P, Lonsdorfer-Wolf E, Doutreleau S, et al. Muscle phosphorylase kinase deficiency. A neutral metabolic variant or a disease? *Neurology* 2012;**78**:265–8.
20. Burwinkel B, Scott JW, Buhrer C, van Landeghem FK, Cox GF, Wilson CJ, et al. Fatal congenital heart glycogenosis caused by a recurrent activating R531Q mutation in the gamma2-subunit of AMP-activated protein kinase (*PRKAG2*), not by phosphorylase kinase deficiency. *Am J Hum Genet* 2005;**76**:1034–49.
21. Newgard CB, Nakano K, Hwang PK, Fletterick RJ. Sequence analysis of the cDNA encoding human liver glycogen phosphorylase reveals tissue-specific codon usage. *Proc Natl Acad Sci USA* 1986;**83**:8132–6.
22. Lebo RV, Gorin F, Fletterick RJ, Kao FT, Cheung MC, Bruce BD, et al. High-resolution chromosome sorting and DNA spot-blot analysis assign McArdle's syndrome to chromosome 11. *Science* 1984; **225**:57–9.
23. Newgard CB, Littman DR, van Genderen C, Fletterick RJ. Human brain glycogen phosphorylase: cloning, sequence analysis, chromosomal mapping, tissue expression, and comparison with the human liver and muscle isozymes. *J Biol Chem* 1988;**263**:3850–7.
24. Bresolin N, Miranda AF, Jacobson MP, Lee JH, Capilupi T, DiMauro S. Phosphorylase isoenzymes of human brain. *Neurochem Pathol* 1983;**1**:171–8.
25. Vissing J, Haller RG. A diagnostic cycle test for McArdle's disease. *Ann Neurol* 2004;**54**:539–42.
26. Lucia A, Nogales-Gadea G, Perez M, Martin MA, Andreu AL, Arenas J. McArdle disease: what do neurologists need to know? *Nature Clin Pract Neurol* 2008;**4**:568–77.
27. Quinlivan R, Buckley J, James M, Twist A, Ball S, Duno M, et al. McArdle disease: a clinical review. *J Neurol Neurosurg Psychiatry* 2010;**81**:1182–8.
28. DiMauro S, Hartlage P. Fatal infantile form of muscle phosphorylase deficiency. *Neurology* 1978;**28**:1124–9.
29. Milstein JM, Herron TM, Haas JE. Fatal infantile muscle phosphorylase deficiency. *J Child Neurol* 1989;**4**:186–8.
30. Cornelio F, Bresolin N, DiMauro S, Mora M, Balestrini MR. Congenital myopathy due to phosphorylase deficiency. *Neurology* 1983;**33**:1383–5.
31. Kubisch C, Wicklein EM, Jentsch TJ. Molecular diagnosis of McArdle disease: Revised genomic structure of the myophosphorylase gene and identification of a novel mutation. *Hum Mut* 1998;**12**:27–32.
32. Tsujino S, Shanske S, DiMauro S. Molecular genetic heterogeneity of myophosphorylase deficiency (McArdle's disease). *N Engl J Med* 1993;**329**:241–5.
33. Bartram C, Edwards R, Clague J, Beynon RJ. McArdle's disease: a nonsense mutation in exon 1 of the muscle glycogen phosphorylase gene explains some but not all cases. *Hum Mol Genet* 1993;**2**:1291–3.
34. El-Schahawi M, Tsujino S, Shanske S, DiMauro S. Diagnosis of McArdle's disease by molecular genetic analysis of blood. *Neurology* 1996;**47**:579–80.
35. Sugie H, Sugie Y, Ito M, Fukuda T, Nonaka I, Igarashi Y. Genetic analysis of Japanese patients with myophosphorylase deficiency (McArdle's disease): single-codon deletion in exon 17 is the predominant mutation. *Clin Chim Acta* 1995;**236**:81–6.
36. El-Schahawi M, Bruno C, Tsujino S, Sarrazin AM, Shanske S, LeRoux MG, et al. Sudden infant death syndrome (SIDS) in a family with myophosphorylase deficiency. *Neuromuscul Disord* 1997;**7**:81–3.
37. Bruno C, Bertini E, Santorelli FM, DiMauro S. HyperCKemia as the only sign of McArdle's disease in a child. *J Child Neurol* 2000;**15**:137–8.
38. Gospe SM, El-Schahawi M, Shanske S, Bruno C, DiMauro S, Hoye E, et al. Asymptomatic McArdle's disease associated with hyper-creatine kinase-emia and absence of myophosphorylase. *Neurology* 1998;**51**:1228–9.
39. Haller RG, Lewis SF, Cook JD, Blomqvist CG. Myophosphorylase deficiency impairs muscle oxidative metabolism. *Ann Neurol* 1985;**17**:196–9.
40. Haller RG, Wyrick P, Taivassalo T, Vissing J. Aerobic conditioning: An effective therapy in McArdle's disease. *Ann Neurol* 2006; **59**:922–8.
41. Izumi R, Suzuki N, Kato K, Warita H, Tateyama M, Nakashima I, et al. A case of McArdle disease: efficacy of vitamin B6 on fatigability and impaired glycogenolysis. *Intern Med* 2010;**49**:1623–5.
42. Phoenix J, Hopkins P, Bartram C, Beynon RJ, Quinlivan RCM, Edwards RHT. Effect of vitamin B6 supplementation in McArdle's disease: a strategic case study. *Neuromuscul Disord* 1998;**8**:210–12.

43. Sato S, Ohi T, Nishino I, Sugie H. Confirmation of the efficacy of vitamin B6 supplementation for McArdle disease by follow-up muscle biopsy. *Muscle Nerve* 2012;**45**:436–40.
44. Andersen ST, Haller RG, Vissing J. Effect of oral sucrose shortly before exercise on work capacity in McArdle disease. *Arch Neurol* 2008;**65**:786–9.
45. Tsujino S, Shanske S, Valberg SJ, Cardinet GH, Smith BS, DiMauro S. Cloning of bovine muscle glycogen phosphorylase cDNA and identification of a mutation in cattle with myophosphorylase deficiency, an animal model for McArdle's disease. *Neuromuscul Disord* 1995;**6**:19–26.
46. Tan P, Allen JG, Wilton SD, Akkari PA, Huxtable CR, Laing NG. A splice-site mutation causing ovine McArdle's disease. *Neuromuscul Disord* 1997;**7**:336–42.
47. Nogales-Gadea G, Pinos T, Lucia A, Arenas J, Camara Y, Brull A, et al. Knock-in mice for the R50X mutation in the *PYGM* gene present with McArdle disease. *Brain* 2012;**135**(Pt 7):2048–57.
48. Kishnani P, Austin SL, Arn P, Bali DS, Boney A, Case LE, ACMG, et al. Glycogen storage disease type III diagnosis and management guidelines. *Genet Med* 2010;**12**:446–63.
49. Moses SW, Wanderman KL, Myroz A, Frydman M. Cardiac involvement in glycogen storage disease type III. *Eur J Pediatr* 1989;**148**:764–6.
50. Coleman RA, Winter HS, Wolf B, Gilchrist JM, Chen YT. Glycogen storage disease type III (glycogen debranching enzyme deficiency): correlation of biochemical defects with myopathy and cardiomyopathy. *Ann Intern Med* 1992;**116**:896–900.
51. Miller CG, Alleyne GA, Brooks S. Gross cardiac involvement in glycogen storage disease type III. *Br Heart J* 1972;**34**:862–4.
52. Kiechl S, Kohlendorfer U, Thaler C, Skladal D, Jaksch M, Obermaier-Kusser B, et al. Different clinical aspects of debrancher deficiency myopathy. *J Neurol Neurosurg Psychiatry* 1999;**67**: 364–8.
53. Brunberg JA, McCormick WF, Schochet SS. Type III glycogenosis: An adult with diffuse weakness and muscle wasting. *Arch Neurol* 1971;**25**:171–8.
54. DiMauro S, Hartwig GB, Hays AP, Eastwood AB, Franco R, Olarte M, et al. Debrancher deficiency: neuromuscular disorder in five adults. *Ann Neurol* 1979;**5**:422–36.
55. Cornelio F, Bresolin N, Singer PA, DiMauro S, Rowland LP. Clinical varieties of neuromuscular disease in debrancher deficiency. *Arch Neurol* 1984;**41**:1027–32.
56. Smit G, Fernandes J, Leonard JV, Matthews EE, Moses SW, Odievre M, et al. The long-term outcome of patients with glycogen storage diseases. *J Inherit Metab Dis* 1990;**13**:411–18.
57. Wary C, Nadaj-Pakleza A, Laforet P, Claeys KG, Carlier R, Monnet A, et al. Investigating glycogenosis type III with multiparametric functional NMR imaging and spectroscopy. *Neuromuscul Disord* 2010;**20**:548–58.
58. Kiechl S, Willeit J, Vogel W, Kohlendorfer U, Poewe W. Reversible severe myopathy of respiratory muscles due to adult-onset type III glycogenosis. *Neuromuscul Disord* 1999;**9**:408–10.
59. Preisler N, Pradel A, Husu E, Madsen KL, Becquemin MH, Mollet A, et al. Exercise intolerance in glycogen storage disease type III: weakness or energy deficiency? *Mol Genet Metab* 2013;**109**:14–20.
60. Bao Y, Dawson TL, Chen Y-T. Human glycogen debranching enzyme gene (AGL): Complete structural organization and characterization of the 5′ flanking region. *Genomics* 1996;**38**:155–65.
61. Yang BZ, Ding JH, Enghild JJ, Bao Y, Chen YT. Molecular cloning and nucleotide sequence of cDNA encoding human muscle glycogen debranching enzyme. *J Biol Chem* 1992;**267**:9294–9.
62. Yang-Feng TL, Zheng K, Yu J, Yang BZ, Chen YT, Jao FT. Assignment of the human glycogen debrancher gene to chromosome 1p21. *Genomics* 1992;**13**:931–4.
63. Keating JP, Brown BI, White NH, DiMauro S. X-linked glycogen storage disease. A cause of hypotonia, hyperuricemia, and growth retardation. *Am J Dis Child* 1985;**139**:609–13.
64. Hendickx J, Willems PJ. Genetic deficiencies of the glycogen phosphorylase system. *Hum Genet* 1996;**97**:551–6.
65. Bruno C, Iester A, Bado M, et al. Muscle phosphorylase deficiency in childhood. A case report. *Minerva Pediatr* 1994;**46**:459–62.
66. Kristjansson K, Tsujino S, DiMauro S. Myophosphorylase deficiency. An unusually severe form with myoglobinuria. *J Pediatr* 1994;**125**:409–10.
67. Tsujino S, Shanske S, Carroll JE, et al. Double trouble: Combined myophosphorylase and AMP deaminase deficiency in a child homozygous for nonsense mutations at both loci. *Neuromuscul Disord* 1995;**5**:263–6.
68. Stojkovic T, Vissing J, Petit F, et al. Muscle glycogenosis due to phosphoglucomutase 1 deficiency. *N Engl J Med* 2009;**361**:425–7.
69. Preisler N, Laforet P, Echaniz-Laguna A, et al. Fat and carbohydrate metabolism during exercise in phosphoglucomutase type 1 deficiency. *J Clin Endocr Metab* 2013;**98**:E1235–40.
70. Perez B, Medrano C, Ecay MJ, et al. A novel congenital disorder of glycosylation type without central nervous system involvement caused by mutations in the phosphoglucomutase 1 gene. *J Inherit Metab Dis* 2013;**36**:535–42.
71. Tarui S, Okuno G, Ikua Y, et al. Phosphofructokinase deficiency in skeletal muscle. A new type of glycogenosis. *Biochem Biophys Res Comm* 1965;**19**:517–23.
72. Mineo I, Kono N, Hara N, et al. Myogenic hyperuricemia. A common pathophysiologic feature of glycogenosis types III, V, and VII. *N Engl J Med* 1987;**317**:75–80.
73. Haller RG, Vissing J. No spontaneous second wind in muscle phosphofructokinase deficiency. *Neurology* 2004;**62**:82–6.
74. Haller RG, Lewis SF. Glucose-induced exertional fatigue in muscle phosphofructokinase deficiency. *N Engl J Med* 1991;**324**:364–9.
75. Serratrice G, Monges A, Roux H, et al. Forme myopathique du deficit en phosphofructokinase. *Rev Neurol* 1969;**34**:271–7.
76. Hays AP, Hallett M, Delfs J, et al. Muscle phosphofructokinase deficiency: abnormal polysaccharide in a case of late-onset myopathy. *Neurology* 1981;**31**:1077–86.
77. Danon MJ, Servidei S, DiMauro S, Vora S. Late-onset muscle phosphofructokinase deficiency. *Neurology* 1988;**38**:955–60.
78. Argov Z, Barash V, Soffer D, et al. Late-onset muscular weakness in phosphofructokinase deficiency due to exon5/intron5 junction point mutation: A unique disorder or the natural course of this glycolytic disorder? *Neurology* 1994;**44**:1097–100.
79. Malfatti E, Birouk N, Romero NB, et al. Juvenile-onset permanent weakness in muscle phosphofructokinase deficiency. *J Neurol Sci* 2012;**316**:173–7.
80. Bonilla E, Schotland DL. Histochemical diagnosis of muscle phosphofructokinase deficiency. *Arch Neurol* 1970;**22**:8–12.
81. Agamanolis DP, Askari AD, DiMauro S, et al. Muscle phosphofructokinase deficiency: Two cases with unusual polysaccharide accumulation and immunologically active enzyme protein. *Muscle Nerve* 1980;**3**:456–67.

82. Raben N, Danon MJ, Lu N, et al. Surprises of genetic engineering: a possible model of polyglucosan body disease. *Neurology* 2001;**56**:1739–45.
83. Vora S, Durham S, deMartinville B, et al. Assignment of the human gene for muscle-type phosphofructokinase (PFKM) to chromosome 1 (region cen-q32) using somatic cell hybrids and monoclonal anti-M antibody. *Somatic Cell Genet* 1982;**8**:95–104.
84. Yamasaki T, Nakajima H, Kono N, et al. Structure of the entire human muscle phosphofructokinase-encoding gene: a two-promoter system. *Gene* 1991;**104**:277–82.
85. Raben N, Sherman J, Miller F, et al. A 5′ splice junction mutation leading to exon deletion in an Ashkenazi Jewish family with phosphofructokinase deficiency (Tarui disease). *J Biol Chem* 1993;**268**:4963–7.
86. Sherman JB, Raben N, Nicastri C, et al. Common mutations in the phosphofructokinase-M gene in Ashkenazi Jewish patients with glycogenosis VII—and their population frequency. *Am J Hum Genet* 1994;**55**:305–13.
87. Musumeci O, Bruno C, Mongini T, et al. Clinical features and new molecular findings in muscle phosphofructokinase deficiency (GSD VII). *Neuromuscul Dis* 2012;**22**:325–30.
88. Dunaway GA, Kasten TP, Sebo T, Trapp R. Analysis of the phosphofructokinase subunits and isoenzymes in human tissues. *Biochem J* 1988;**251**:677–83.
89. Swoboda KJ, Specht L, Jones HR, et al. Infantile phosphofructokinase deficiency with arthrogryposis: Clinical benefit of a ketogenic diet. *J Pediatr* 1997;**131**:932–4.
90. Kreuder J, Borkhardt A, Repp R, et al. Brief report: Inherited metabolic myopathy and hemolysis due to a mutation in aldolase A. *N Engl J Med* 1996;**334**:1100–4.
91. Yao DC, Tolan DR, Murray MF, et al. Hemolytic anemia and severe rhabdomyolysis caused by compound heterozygous mutations of the gene for erythrocyte/muscle isozyme of aldolase. *Blood* 2004;**103**:2401–3.
92. Spiegel R, Area Gomez E, Akman HO, et al. Myopathic form of phosphoglycerate kinase (PGK) deficiency: A new case and pathogenic considerations. *Neuromuscul Disord* 2009;**19**:207–11.
93. Sotiriou E, Greene P, Krishna S, et al. Myopathy and Parkinsonism in phosphoglycerate kinase deficiency. *Muscle Nerve* 2010;**41**:707–10.
94. DiMauro S, Spiegel R. Progress and problems in muscle glycogenoses. *Acta Myologica* 2011;**30**:96–102.
95. Rosa R, George C, Fardeau M, et al. A new case of phosphoglycerate kinase deficiency: PGK Creteil associated with rhabdomyolysis and lacking hemolytic anemia. *Blood* 1982;**60**:84–91.
96. DiMauro S, Dalakas M, Miranda AF. Phosphoglycerate kinase deficiency: another cause of recurrent myoglobinuria. *Ann Neurol* 1983;**13**:11–19.
97. Tonin P, Shanske S, Miranda AF, et al. Phosphoglycerate kinase deficiency: Biochemical and molecular genetic studies in a new myopathic variant (PGK Alberta). *Neurology* 1993;**43**:387–91.
98. Hamano T, Mutoh T, Sugie H, et al. Phosphoglycerate kinase deficiency: An adult myopathic form with a novel mutation. *Neurology* 2000;**54**:1188–90.
99. Chen SH, Malcolm LA, Yoshida A, Giblett E. Phosphoglycerate kinase. An X-linked polymorphism in man. *Am J Hum Genet* 1971;**23**:87–91.
100. Willard HF, Gross SJ, Holmes MT, Monroe DL. Regional localization of the phosphoglycerate kinase gene and pseudogene on the human X chromosome and assignment of a related DNA sequence to chromosome 19. *Hum Genet* 1985;**71**:138–43.
101. Omenn GS, Cheung CV. Phosphoglycerate mutase: Isozyme marker for tissue differentiation in man. *Am J Hum Genet* 1974; **26**:393–9.
102. Shanske S, Sakoda S, Hermodson MA, et al. Isolation of a cDNA encoding the muscle-specific subunit of human phosphoglycerate mutase. *J Biol Chem* 1987;**262**:14612–17.
103. Tsujino S, Sakoda S, Mizuno R, et al. Structure of the gene encoding the muscle-specific subunit of human phosphoglycerate mutase. *J Biol Chem* 1989;**264**:15334–7.
104. Edwards YH, Sakoda S, Schon EA, Povey S. The gene for human muscle-specific phosphoglycerate mutase, PGAMM, mapped to chromosome 7 by polymerase chain reaction. *Genomics* 1989; **5**:948–51.
105. Castella-Escola J, Mattei MG, Ojcius DM, et al. In situ mapping of the muscle-specific form of phosphoglycerate mutase gene to human chromosome 7p12–7p13. *Hum Genet* 1990;**84**:210–12.
106. DiMauro S, Miranda AF, Khan S, et al. Human muscle phosphoglycerate mutase deficiency: newly discovered metabolic myopathy. *Science* 1981;**212**:1277–9.
107. Naini A, Toscano A, Musumeci O, et al. Muscle phosphoglycerate mutase deficiency revisited. *Arch Neurol* 2009;**66**:394–8.
108. Toscano A, Tsujino S, Vita G, et al. Molecular basis of muscle phosphoglycerate mutase (PGAM-M) deficiency in the Italian kindred. *Muscle Nerve* 1996;**19**:1134–7.
109. Tonin P, Bruno C, Cassandrini D, et al. Unusual presentation of phosphoglycerate mutase deficiency due to two different mutations in *PGAM-M* gene. *Neuromuscul Disord* 2009;**19**:776–8.
110. Hadjigeorgiou GM, Kawashima N, Bruno C, et al. Manifesting heterozygotes in a Japanese family with a novel mutation in the muscle-specific phosphoglycerate mutase (PGAM-M) gene. *Neuromuscul Disord* 1999;**9**:399–402.
111. Vissing J, Schmalbruch H, Haller RG, Clausen T. Muscle phosphoglycerate mutase deficiency with tubular aggregates: effect of dantrolene. *Ann Neurol* 1999;**46**:274–7.
112. Salameh J, Goyal N, Choudry R, et al. Phosphoglycerate mutase deficiency with tubular aggregates in a patient from Panama. *Muscle Nerve* 2013;**47**:138–40.
113. Vissing J, Quistorff B, Haller RG. Effects of fuels on exercise capacity in muscle phosphoglycerate mutase deficiency. *Arch Neurol* 2005;**62**:1440–3.
114. Comi GP, Fortunato F, Lucchiari S, et al. B-enolase deficiency, a new metabolic myopathy of distal glycolysis. *Ann Neurol* 2001; **50**:202–7.
115. Kanno T, Sudo K, Takeuchi I, et al. Hereditary deficiency of lactate dehydrogenase M-subunit. *Clin Chim Acta* 1980;**108**:267–76.
116. Kanno T, Sudo K, Maekawa M, et al. Lactate dehydrogenase M-subunit deficiency: a new type of hereditary exertional myopathy. *Clin Chim Acta* 1988;**173**:89–98.
117. Maekawa M, Sudo K, Kanno T, Li S. Molecular characterization of genetic mutation in human lactate dehydrogenase-A (M) deficiency. *Biochem Biophys Res Comm* 1990;**168**:677–82.
118. Maekawa M, Sudo K, Li S, Kanno T. Analysis of genetic mutation in human lactate dehydrogenase-A (M) deficiency using DNA conformation polymorphism in combination with polyacrilamide gradient gel and silver staining. *Biochem Biophys Res Comm* 1991;**180**:1083–90.

119. Maekawa M, Sudo K, Kanno T, et al. A novel deletion mutation of lactate dehydrogenase A (M) gene in fifth family with the enzyme deficiency. *Hum Mol Genet* 1994;**3**:825–6.
120. Tsujino S, Shanske S, Brownell A, et al. Molecular genetic studies of muscle lactate dehydrogenase deficiency in white patients. *Ann Neurol* 1994;**36**:661–5.
121. Nazzari G, Crovato F. Annually recurring acroerythema and hereditary lactate dehydrogenase M-subunit deficiency. *J Am Acad Dermatol* 1992;**27**:262–3.
122. Yoshikuni K, Tagami H, Yamada M, et al. Erythematosquamous skin lesions in hereditary lactate dehydrogenase M-subunit deficiency. *Arch Dermatol* 1986;**122**:1420–4.
123. Takayasu S, Fujivara S, Waki T. Hereditary lactate dehydrogenase M-subunit deficiency. Lactate dehydrogenase activity in skin lesions and in hair follicles. *J Am Acad Dermatol* 1991;**24**:339–42.
124. Tsujibo H, Tiano HF, Li SL. Nucleotide sequences of the cDNA and an intronless pseudogene for human lactate dehydrogenase-A isozyme. *Eur J Biochem* 1985;**147**:9–15.
125. Chung FZ, Tsujibo H, Bhattacharya U, et al. Genomic organization of human lactate dehydrogenase-A gene. *Biochemistry* 1985; **231**:537–41.
126. Scrable HJ, Johnson DK, Rinchik EM, Cavenee WK. Rhabdomyosarcoma-associated locus and MYOD1 are synthenic but separate loci on the short arm of human chromosome 11. *Proc Nat Acad Sci USA* 1990;**87**:2182–6.
127. Kanno T, Maekawa M. Lactate dehydrogenase M-subunit deficiency: Clinical features, metabolic background, and genetic heterogeneities. *Muscle Nerve* 1995;(Suppl. 3):S54–60.
128. Guibaud P, Carrier H, Mathieu M, et al. Familial congenital muscular dystrophy caused by phosphofructokinase deficiency. *Arch Fr Pediatr* 1978;**35**:1105–15.
129. Servidei S, Bonilla E, Diedrich RG, et al. Fatal infantile form of muscle phosphofructokinase deficiency. *Neurology* 1986;**36**: 1465–70.
130. Amit R, Bashan N, Abarbanel JM, et al. Fatal familial infantile glycogen storage disease: Multisystem phosphofructokinase deficiency. *Muscle Nerve* 1992;**15**:455–8.
131. Spriggs EL, Marles SL, Lacson A, et al. Long-term survival and normal cognitive development in infantile phosphofructokinase-1 deficiency. *Clin Genet* 1999;**56**:235–7.
132. Orho M, Bosshard NU, Buist NRM, et al. Mutations in the liver glycogen synthase gene in children with hypoglycemia due to glycogen storage disease type 0. *J Clin Invest* 1998;**102**:507–15.
133. Lomako J, Lomako WM, Whelan WJ. Glycogenin: the primer for mammalian and yeast glycogen synthesis. *Biochim Biophys Acta* 2004;**1673**:45–55.
134. Kollberg G, Tulinius M, Gilljam T, et al. Cardiomyopathy and exercise intolerance in muscle glycogen storage disease 0. *N Engl J Med* 2007;**357**:1507–14.
135. Moslemi A-R, Lindberg C, Nilsson J, et al. Glycogenin-1 deficiency and inactivated priming of glycogen synthesis. *N Engl J Med* 2010;**362**:1203–10.
136. Cameron JM, Levandovskiy V, MacKay N, et al. Complex V TMEM70 deficiency results in mitochondrial nucleoid disorganization. *Mitochondrion* 2011;**11**:191–9.
137. Sukigara S, Liang W-C, Komaki H, et al. Muscle glycogen storage disease 0 presenting recurrent syncope with weakness and myalgia. *Neuromuscul Disord* 2012;**22**:162–5.
138. Cameron JM, Levandovskiy V, MacKay N, et al. Identification of a novel mutation in GSY1 (muscle-specific glycogen synthase) resulting in sudden cardiac death, that is diagnosable from skin fibroblasts. *Mol Genet Metab* 2009;**98**:378–82.
139. Chen YT, Burchell A. Glycogen storage disease. In: Scriver CR, Beaudet AL, Sly WS, Valle D, editors. *The Metabolic and Molecular Bases of Inherited Disease*. New York: McGraw-Hill; 1995. pp. 935–65.
140. Bruno C, van Diggelen OP, Cassandrini D, et al. Clinical and genetic heterogeneity of branching enzyme deficiency (glycogenosis type IV). *Neurology* 2004;**63**:1053–8.
141. Tay SKH, Akman HO, Chung WK, et al. Fatal infantile neuromuscular presentation of glycogen storage disease type IV. *Neuromuscul Disord* 2004;**14**:253–60.
142. Janecke AR, Dertinger S, Ketelsen UP, et al. Neonatal type IV glycogen storage disease associated with "null" mutations in glycogen branching enzyme 1. *J Pediatr* 2004;**145**:705–9.
143. Konstantinidou AE, Anninos H, Gyftodimou Y, et al. Neonatal neuromuscular variant of glycogen storage disease type IV: histopathological findings leading to the diagnosis. *Histopathology* 2006;**48**:878–80.
144. Taratuto AL, Akman HO, Saccoliti M, et al. Branching enzyme deficiency/glycogenosis storage disease type IV presenting as a severe congenital hypotonia: Muscle biopsy and autopsy findings, biochemical and molecular studies. *Neuromuscul Disord* 2010;**20**: 783–90.
145. Alegria A, Martins E, Dias M, et al. Glycogen storage disease type IV presenting as hydrops fetalis. *J Inher Metab Dis* 1999; **22**:330–2.
146. Akman HO, Karadimas CL, Gyftodimou Y, et al. Prenatal diagnosis of glycogen storage disease type IV. *Prenat Diagn* 2006;**26**:951–5.
147. Assereto S, van Diggelen OP, Diogo L, et al. Null mutations and lethal congenital form of glycogen storage disease type IV. *Biochem Biophys Res Comm* 2007;**361**:445–50.
148. Konstantinidou A, Anninos H, Dertinger S, et al. Placental involvement in glycogen storage disease type IV. *Placenta* 2008; **29**:378–81.
149. Suzuki K, David E, Kutschman B. Presenile dementia with "Lafora-like" intraneuronal inclusions. *Arch Neurol* 1971;**25**:69–80.
150. Robitaille Y, Carpenter S, Karpati G, DiMauro S. A distinct form of adult polyglucosan body diseases with massive involvement of central and peripheral neuronal processes and astrocytes. *Brain* 1980;**103**:315–36.
151. Mochel F, Schiffmann R, Steenweg M, et al. Adult polyglucosan body disease: Natural history and key MRI findings. *Ann Neurol* 2012;**72**:433–41.
152. Lossos A, Klein CJ, McEvoy KM, Keegan BM. A 63-year-old woman with urinary incontinence and progressive gait disorder. *Neurology* 2009;**72**:1607–13.
153. Billot S, Herve D, Akman HO, et al. Acute but transient neurological deterioration revealing adult polyglucosan body disease. *J Neurol Sci* 2013;**324**:179–82.
154. Paradas C, Akman HO, Ionete C, et al. Branching enzyme deficiency: expanding the clinical spectrum. *JAMA Neurol* 2014;**71**(1): 41–7.
155. Thon VJ, Khalil M, Cannon JF. Isolation of human glycogen branching enzyme cDNAs by screening complementation in yeast. *J Biol Chem* 1993;**268**:7509–13.

156. Akman HO, Sheiko T, Tay SKH, et al. Generation of a novel mouse model that recapitulates early and adult onset glycogenosis type IV. *Hum Mol Genet* 2011;**20**:4430–9.
157. Lafora GR. Uber das Vorkommen amyloider Korperchen in Innern der Ganglienzellen. *Virchows Arch Pathol Anat* 1911;**205**: 295–303.
158. Ramachandran N, Girard J-M, Turnbull J, Minassian B. The autosomal recessively inherited progressive myoclonus epilepsies and their genes. *Epilepsia* 2009;**50**(Suppl. 5):29–36.
159. Coleman DL, Gambetti PL, DiMauro S, Blume RE. Muscle in Lafora disease. *Arch Neurol* 1974;**31**:396–406.
160. Striano P, Ackerley C, Cervasio M, et al. 22-year-old girl with status epilepticus and progressive neurological symptoms. *Brain Pathol* 2009;**19**:727–30.
161. Turnbull J, DePaoli-Roach AA, Zhao X, et al. PTG depletion removes Lafora bodies and rescues the fatal epilepsy of Lafora disease. *PLoS Genet* 2011;**7**(4):e1002037.
162. Tagliabracci VS, Heiss C, Karthik C, et al. Phosphate incorporation during glycogen synthesis and Lafora disease. *Cell Metab* 2011;**13**:274–82.
163. Nitschke F, Wang P, Schmieder P, et al. Hyperphosphorylation of glucose C6 carbons and altered structure of glycogen in the neurodegenerative epilepsy Lafora disease. *Cell Metab* 2013;**17**: 756–67.
164. Tiberia E, Turnbull J, Wang TJ, et al. Increased laforin and laforin binding to glycogen underlie Lafora body formation in malin-deficient Lafora disease. *J Biol Chem* 2012;**287**:25650–9.
165. Pederson BA, Turnbull J, Epp JR, et al. Inhibiting glycogen synthesis prevents Lafora disease in a mouse model. *Ann Neurol* 2013;**74**:297–300.
166. Nilsson J, Schoser BGH, Laforet P, et al. Polyglucosan body myopathy caused by defective ubiquitin ligase RBCK1. *Ann Neurol* 2013;**74**(6):914–19.
167. Boisson B, Laplantine E, Prando C, et al. Immunodeficiency, autoinflammation and amylopectinosis in humans with inherited HOIL-1 and LUBAC deficiency. *Nature Immunol* 2012;**13**: 1178–86.
168. Bornemann A, Besser R, Shin YS, Goebel HH. A mild adult myopathic variant of type IV glycogenosis. *Neuromuscul Disord* 1996;**6**:95–9.
169. Goebel HH, Shin YS, Gullotta F, et al. Adult polyglucosan body myopathy. *J Neuropathol Exp Neurol* 1992;**51**:24–35.
170. Schroder JM, May R, Shin YS, et al. Juvenile hereditary polyglucosan body disease with complete branching enzyme deficiency (type IV glycogenosis). *Acta Neuropathol (Berl)* 1993;**85**:419–30.
171. Takeuchi T, Kuriaki H. Histochemical detection of phosphorylase in animal tissues. *J Histochem Cytochem* 1955;**3**:153–60.
172. Pompe JC. Over Idiopatische Hypertrophie van het Hart. *Ned T Geneesk* 1932;**76**:304–11.
173. Putschar W. Uber angeborene Glykogenspeicherkrankenheit des Herzens: "Thesaurismosis glykogenica". *Beitr Pathol Anat* 1932; **90**:222–3.
174. Hers HG. Alpha-glucosidase deficiency in generalized glycogen storage disease (Pompe's disease). *Biochem J* 1963;**86**:11–16.
175. Lejeune N, Thines-Sempoux D, Hers HG. Tissue fractionation studies: 16. Intracellular distribution and properties of alpha-glucosidases in rat liver. *Biochem J* 1963;**86**:16–21.
176. Hers HG. Inborn lysosomal diseases. *Gastroenterology* 1965;**48**: 625–33.
177. Kishnani PS, Hwu WL, Mandel H, et al. A retrospective, multinational, multicenter study on the natural history of infantile-onset Pompe disease. *J Pediatr* 2006;**148**:671–6.
178. Engel AG, Gomez MR, Seybold ME, Lambert EH. The spectrum and diagnosis of acid maltase deficiency. *Neurology* 1973;**23**:95–105.
179. Karpati G, Carpenter S, Eisen A, et al. The adult form of acid maltase (alpha-1,4-glucosidase) deficiency. *Ann Neurol* 1977;**1**: 276–80.
180. Gambetti PL, DiMauro S, Baker L. Nervous system in Pompe's disease. Ultrastructure and biochemistry. *J Neuropathol Exp Neurol* 1971;**30**:412–30.
181. Martin JJ, DeBarsy T, Van Hoof F. Pompe's disease: an inborn lysosomal disorder with storage of glycogen: a study of brain and striated muscle. *Acta Neuropathol* 1973;**23**:229–44.
182. Engel AG, Hirschhorn R, Huie MI. Acid maltase deficiency. In: Engel AG, Franzini-Armstrong C, editors. *Myology*. New York, NY: McGraw-Hill; 2004. pp. 1559–86.
183. Matsuishi T, Yoshino M, Terasawa K, Nonaka I. Childhood acid maltase deficiency. A clinical, biochemical, and morphologic study of three patients. *Arch Neurol* 1984;**41**:47–52.
184. Mellies U, Lofaso F. Pompe disease: A neuromuscular disease with respiratory muscle involvement. *Respir Med* 2009;**103**:477–84.
185. Amato AA, Leep Hunderfund AN, Selcen D, Keegan BM. A 49-year-old woman with progressive shortness of breath. *Neurology* 2011;**76**:830–6.
186. Kretzchmer HA, Wagner H, Hubner G, et al. Aneurysms and vacuolar degeneration of cerebral arteries in late-onset acid maltase deficiency. *J Neurol Sci* 1990;**98**:169–83.
187. Laforet P, Petiot P, Nicolino M, et al. Dilative arteriopathy and basilar artery dolichoectasia complicating late-onset Pompe disease. *Neurology* 2008;**70**:2063–6.
188. Remiche G, Herbaut A-G, Ronchi D, et al. Incontinence in late-onset Pompe disease: An underdiagnosed treatable condition. *Eur Neurol* 2012;**68**:75–8.
189. Hobson-Webb LD, Proia AD, Thurberg BL, et al. Autopsy findings in late-onset Pompe disease: A case report and systematic review of the literature. *Mol Genet Metab* 2012;**106**:462–9.
190. DiMauro S, Stern LZ, Mehler M, et al. Adult-onset acid maltase deficiency: A postmortem study. *Muscle Nerve* 1978;**1**:27–36.
191. Carlier R-Y, Laforet P, Wary C, et al. Whole-body muscle MRI in 20 patients suffering from late onset Pompe disease: Involvement patterns. *Neuromuscul Disord* 2011;**21**:791–9.
192. Nascimbeni AC, Fanin M, Tasca E, Angelini C. Molecular pathology and enzyme processing in various phenotypes of acid maltase deficiency. *Neurology* 2008;**70**:617–26.
193. Hermans MM, van Leenen D, Kroos M, et al. Twenty-two novel mutations in the lysosomal alpha-glucosidase gene (GAA) underscore the genotype-phenotype correlation in glycogen storage disease type II. *Hum Mutat* 2004;**23**:47–56.
194. Kroos M, Hoogeveen-Westerveld M, Van der Ploeg A, Reuser AJJ. The genotype-phenotype correlation in Pompe disease. *Am J Med Genet* 2012;**160C**:59–68.
195. Raben N, Lu N, Nagaraju K, et al. Conditional tissue-specific expression of the acid alpha-glucosidase (GAA) gene in the GAA knockout mice: implications for therapy. *Hum Mol Genet* 2001;**10**:2039–47.

196. Raben N, Wong A, Ralston E, Myerowitz R. Autophagy and mitochondria in Pompe disease: Nothing is so new as what has long been forgotten. *Am J Med Genet* 2012;**160C**:13–21.

197. Settembre C, Fraldi A, Jahreiss L, et al. A block of autophagy in lysosomal storage disorders. *Hum Mol Genet* 2008;**17**:119–29.

198. Fukuda T, Ewan L, Bauer M, et al. Dysfunction of endocytic and autophagic pathways in a lysosomal storage disease. *Ann Neurol* 2006;**59**:700–8.

199. Malicdan MC, Noguchi S, Nonaka I, et al. Lysosomal myopathies: An excessive build-up in autophagosomes is too much to handle. *Neuromusc Disord* 2008;**18**:521–9.

200. Thurberg BL, Maloney CL, Vaccaro C, et al. Characterization of pre- and post-treatment pathology after enzyme replacement therapy for Pompe disease. *Lab Invest* 2006;**86**:1208–20.

201. Griffin JL. Infantile acid maltase deficiency. I. Muscle fiber destruction after lysosomal rupture. *Virchows Arch B Cell Pathol* 1984;**45**:23–36.

202. Preisler N, Laforet P, Madsen KL, et al. Fat and carbohydrate metabolism during exercise in late-onset Pompe disease. *Mol Genet Metab* 2012;**107**:462–8.

203. Kishnani PS, Corzo D, Nicolino M, et al. Recombinant human acid [alpha]-glucosidase: major clinical benefits in infantile-onset Pompe disease. *Neurology* 2007;**68**:99–109.

204. van der Ploeg AT. Where do we stand in enzyme replacement therapy in Pompe's disease? *Neuromusc Dis* 2010;**20**:733–4.

205. van der Ploeg AT, Clemens PR, Corzo D, et al. A randomized study of alglucosidase alfa in late-onset Pompe's disease. *N Engl J Med* 2010;**362**:1396–406.

206. Hagemans MLC, Janssens ACJW, Winkel LPF, et al. Late-onset Pompe disease primarily affects quality of life in physical health domains. *Neurology* 2004;**63**:1688–92.

207. Van den Hout JMP, Reuser AJJ, de Klerk JBC, et al. Enzyme therapy for Pompe disease with recombinant human alpha-glucoside from rabbit milk. *J Inher Metab Dis* 2001;**24**:266–74.

208. Van den Hout J, Kamphoven JHJ, Winkel LPF, et al. Long-term intravenous treatment of Pompe disease with recombinant human alpha-glucosidase from milk. *Pediatrics* 2004;**113**:e448–57.

209. Chakrapani A, Vellodi A, Robinson P, et al. Treatment of infantile Pompe disease with aglucosidase alpha: the UK experience. *J Inher Metab Dis* 2010;**33**:747–50.

210. Kishnani P, Beckemeyer AA, Mendelsohn NJ. The new era of Pompe disease: advances in the detection, understanding of the phenotypic spectrum, patophysiology, and management. *Am J Med Genet* 2012;**160C**:1–7.

211. Ebbink BJ, Aarsen FK, van Gelder CM, et al. Cognitive outcome of patients with classic infantile Pompe disease receiving enzyme therapy. *Neurology* 2012;**78**:1512–18.

212. van Gelder CM, van Capelle CI, Ebbink RJ, et al. Facial-muscle weakness, speech disorders and dysphagia are common in patients with classic infantile Pompe disease treated with enzyme therapy. *J Inher Metab Dis* 2012;**35**:505–11.

213. Slonim AE, Bulone L, Minikes J, et al. Benign course of glycogen storage disease type IIb in two brothers: nature or nurture? *Muscle Nerve* 2006;**33**:571–4.

214. Danon MJ, Oh SJ, DiMauro S, et al. Lysosomal glycogen storage disease with normal acid maltase. *Neurology* 1981;**31**:51–7.

215. Nishino I, Fu J, Tanji K, et al. Primary LAMP-2 deficiency causes X-linked vacuolar cardiomyopathy and myopathy (Danon disease). *Nature* 2000;**406**:906–10.

Chapter 40

Lipid Storage Myopathies Due to Fatty Acid Oxidation Defects

Ingrid Tein

HISTORICAL BACKGROUND

Defects in fatty acid oxidation (FAO) are an important group of disorders because they are potentially rapidly fatal and a source of major morbidity. They encompass a spectrum of clinical disorders including progressive lipid storage myopathy, recurrent myoglobinuria, neuropathy, progressive cardiomyopathy, recurrent hypoglycemic hypoketotic encephalopathy or Reye-like syndrome, seizures, and mental retardation.[1–15] All of the known conditions are inherited as autosomal recessive diseases, and there often is a family history of sudden unexpected death or SIDS (sudden infant death syndrome) in siblings. Early recognition and prompt institution of therapy and preventative measures, and in certain cases specific therapy, may be life-saving, significantly decreasing long-term morbidity, particularly with respect to central nervous system sequelae.

Engel identified the first genetic defect in FAO in 1970 when he described a skeletal muscle disorder associated with intermittent symptoms and a possible defect in lipid metabolism.[16] DiMauro and DiMauro reported the first enzyme defect in 1973; namely, carnitine palmitoyltransferase (CPT) deficiency, which presented with recurrent myoglobinuria.[1] There are now at least 16 recognized enzyme or transporter defects in FAO, most of which have been diagnosed in the last 25 years. With the significant advances in biomedical technology, there has been a rapid increase in the number of subsequently diagnosed cases, for example of MCAD deficiency.[17,18] The incidence for MCAD deficiency was found to be 1 in 8930 live births in a survey from the Pennsylvania newborn screening program using tandem mass spectrometry, followed by confirmation through molecular analysis for several common mutations.[19] Similarly the incidence of MCADD detected by recent newborn screening using MS/MS and confirmed by mutation analysis was 1/8954 in Denmark.[20] The spectrum of FAOD differs widely among ethnic groups. Incidence calculations from a total of 5,256,999 screened newborns from the USA, Australia, and Germany give a combined incidence of all FAOD of approximately 1:9300.[21]

The delay in recognition of these disorders can be attributed to three major factors. First, FAO does not play a major role in energy production until relatively late in fasting. Thus, affected individuals may remain clinically silent until they are exposed to periods of fasting beyond 12 hours or to prolonged exercise. Acute decompensations may be precipitated by an intercurrent infection with fasting, vomiting, and shivering thermogenesis. Secondly, routine laboratory tests, such as urinary ketones, may not demonstrate the defect in FAO, unless the blood and urine samples are obtained at the time of the acute episode. If they are obtained after the child has received intravenous glucose therapy or has recovered from the acute illness, the defect may be missed. Thirdly, there has been a rapid expansion in the methodologies to identify defects in FAO such as the introduction of acylcarnitine analysis by tandem mass spectrometry in newborn screening; nevertheless there remains a need to unify confirmatory diagnostic procedures such as specific enzyme assays.[22] Mutation analysis may confirm diagnosis and may also help to explain differences in phenotypic expression in patients with the same enzyme defect. Different mutations may have different effects on protein function, as certain missense mutations may result in misfolded but stable proteins that may affect the integrity and function of the FAO enzyme versus pure null mutations that lead to absence of the protein.[22] Gene-gene and environmental factors may also influence phenotypic expression of a given genotype.

An overview will be given of the FAO pathway, its enzymes, and the role of carnitine, as well as a discussion

B.T. Darras, H. Royden Jones, Jr., M.M. Ryan & D.C. De Vivo (Eds): Neuromuscular Disorders of Infancy, Childhood and Adolescence, Second edition.
DOI: http://dx.doi.org/10.1016/B978-0-12-417044-5.00040-8

of normal fasting adaptation in human physiology to lay the groundwork for understanding the pathophysiology of these diseases. This will be followed by the clinical and laboratory features of specific FAO defects, an approach to diagnosis, and current treatments.

PATHOPHYSIOLOGY

Normal Pathway of Fatty Acid Oxidation

The pathway of mitochondrial FAO is outlined in Figure 40.1.[23] During fasting, free fatty acids are liberated from adipocytes and are transported to other tissues either as triglyceride-rich lipoproteins or bound to serum albumin.[24] Triacylglycerols are hydrolyzed outside the cells by lipoprotein lipase to yield free fatty acids (FFAs). The mechanism of FFA entry into cells is not well understood; however, there is kinetic evidence to suggest both a saturable and nonsaturable uptake of FFAs. Once across the plasma membrane, short- (e.g. 4-carbon) and medium-chain (e.g. 8-carbon) fatty acids (less than 10 carbon atoms) are able to cross the outer and inner mitochondrial membranes as free acids to enter the mitochondrial matrix. Here they are then activated by their respective short- and medium-chain acyl-CoA synthetases to their CoA thioesters for ensuant intramitochondrial β-oxidation.

The mitochondrial membrane is impermeable to long-chain fatty acids (e.g. 16-carbon). Therefore, long-chain fatty acids diffuse or are transported to the outer mitochondrial membrane and to the endoplasmic reticulum. At both locations they are activated by conversion to their CoA thioesters. Long-chain acyl-CoA synthetase is a membrane-bound enzyme located in the endoplasmic reticulum and in the outer mitochondrial membrane. This enzyme acts on

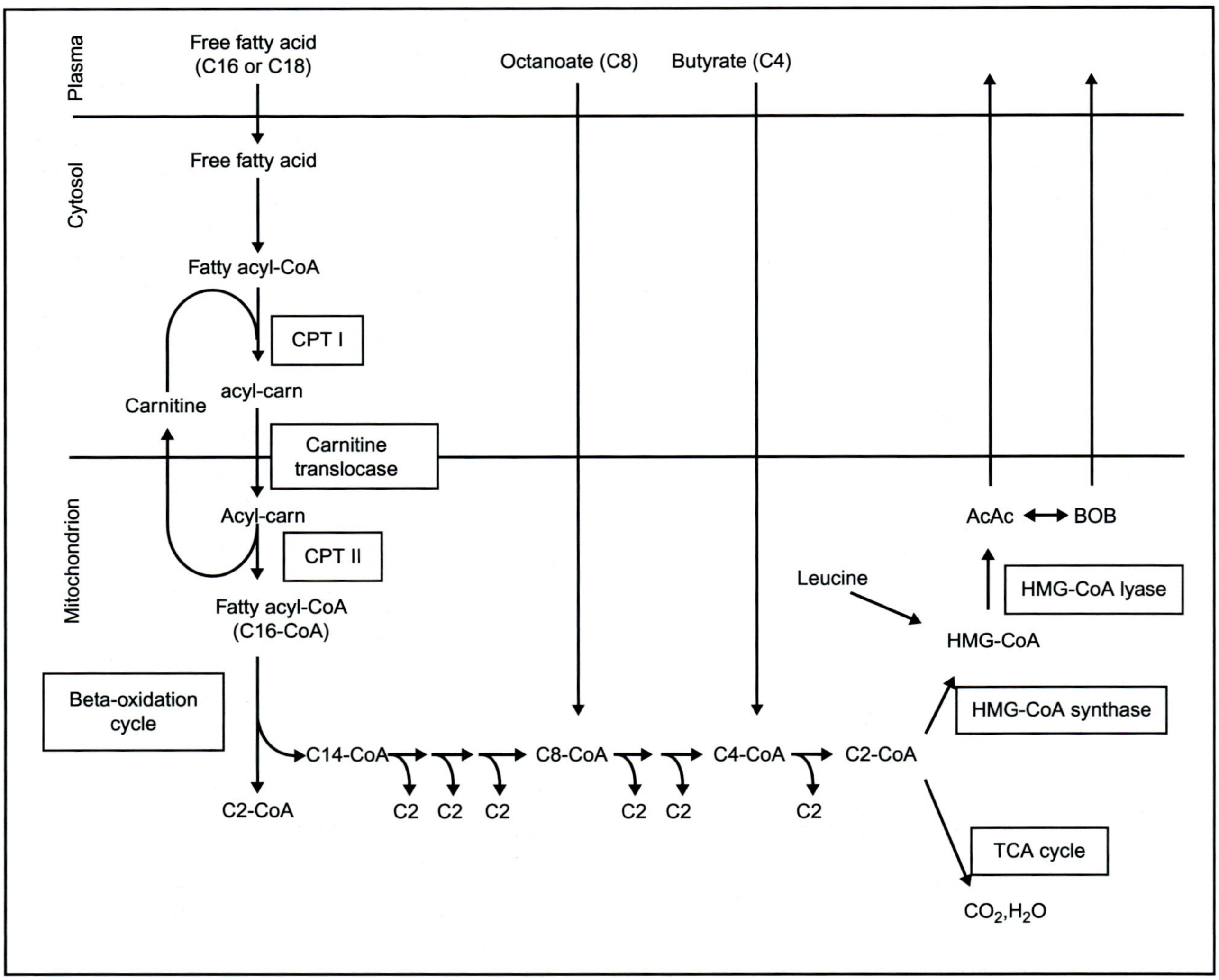

FIGURE 40.1 Pathway of mitochondrial FAO. CoA, coenzyme A; CPT, carnitine palmitoyltransferase; HMG, 3-hydroxy-3-methyl-glutaryl; TCA, tricarboxylic acid. *From Stanley.*[23]

saturated fatty acids containing 10 to 18 carbon atoms and on unsaturated fatty acids containing 16 to 20 carbon atoms.[25] Most of the activated fatty acids are directed toward mitochondrial β-oxidation.[26] The inner mitochondrial membrane is impermeable to CoA and its derivatives, namely fatty acyl-CoA thioesters, which have been formed at the outer mitochondrial membrane. Thus these thioesters cannot directly enter the mitochondrial matrix. Therefore, the long-chain acyl-CoA must first be converted into its acylcarnitine form, e.g. palmitoylcarnitine, with release of free CoA. This is accomplished by the reversible enzyme carnitine palmitoyltransferase I (CPT I), which uses carnitine as a cofactor and is located on the inner side of the outer mitochondrial membrane.[27] The palmitoylcarnitine is then translocated across the inner mitochondrial membrane by carnitine:acylcarnitine translocase[28] which catalyzes a slow unidirectional diffusion of carnitine both in and out of the mitochondrial matrix in addition to a much faster mole-to-mole exchange of acylcarnitine for carnitine, carnitine for carnitine, and acylcarnitine for acylcarnitine. In the mitochondrial matrix, carnitine palmitoyltransferase II (CPT II), which is situated on the inner side of the inner mitochondrial membrane, converts palmitoylcarnitine, in the presence of free CoA, back to palmitoyl-CoA and carnitine.

There are four sequential steps in mitochondrial β-oxidation, which are composed of chain-length specific enzymes (Figure 40.2).[23] It was originally believed that the complete intramitochondrial oxidation of long-chain fatty acids required a minimum of nine enzymes including the three genetically distinct acyl-CoA dehyrogenases (short-, medium-, long-chain)[29] and at least two enoyl-CoA hydratases (crotonase and long-chain enoyl-CoA hydratase),[30] two L-3-hydroxyacyl-CoA dehydrogenases (short- and long-chain)[31] and two 3-ketoacyl-CoA thiolases (acetoacetyl-CoA thiolase and generalized 3-ketoacyl-CoA thiolase).[32] Further enzymatic and protein characterization has revealed the presence of an additional very-long-chain acyl-CoA dehydrogenase enzyme,[33,34] of an acyl-CoA dehydrogenase 9 (ACAD9) that demonstrates maximal activity with unsaturated long-chain acyl-CoAs,[35] and of a trifunctional protein which combines the activities of the long-chain enoyl-CoA hydratase, long-chain-L-3-hydroxyacyl-CoA dehydrogenase, and the long-chain thiolase enzymes.[36–38] These enzymes are chain-length specific.[39] For example, the long-chain acyl-CoA dehydrogenase has specificity for 12–18 carbon fatty acids and the medium-chain acyl-CoA dehydrogenase has specificity for 4–12 carbon fatty acids.[39] Despite a significant overlap of substrate specificity, it appears that ACAD9 and VLCAD are unable to compensate for each other in patients with either deficiency.[35] Studies of the tissue distribution and gene regulation of ACAD9 and VLCAD identify the presence of two independently regulated functional pathways for long-chain fat metabolism, indicating that these two enzymes are likely to be involved in different physiological functions.[35] ACAD9 is required for the assembly of mitochondrial complex I.[40] ACAD9 mutations result in complex I deficiency without disturbing long-chain fatty acid oxidation.[41,42] This strongly contrasts with its evolutionary ancestor VLCAD, which is not required for complex I assembly and clearly plays a role in fatty acid oxidation.[40] The trifunctional protein (TFP) is a heterocomplex of four alpha and four beta

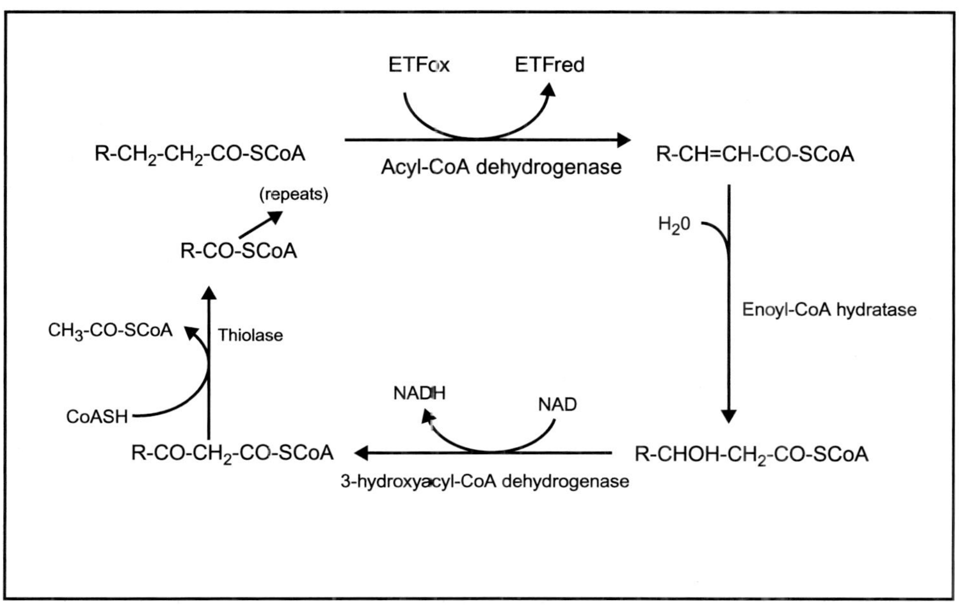

FIGURE 40.2 The four steps of the β-oxidation cycle. CoA, coenzyme A; NAD, nicotinamide-adenine dinucleotide; NADH, reduced form of nicotinamide-adenine dinucleotide. *From Stanley.*[23]

subunits, which are encoded by two nuclear genes. The alpha subunit contains the long-chain enoyl-CoA hydratase and LCHAD activities, and the beta subunit contains the long-chain 3-ketoacyl-CoA thiolase activity.[43] There are two different biochemical phenotypes of TFP deficiency.[44,45] The first includes deficiency of all three enzyme activities of the TFP, with loss of both the alpha and beta subunits by immunoblotting studies. The second has isolated LCHAD deficiency with preservation of the long-chain enoyl-CoA hydratase and 3-ketoacyl-CoA thiolase activities and normal amounts of the alpha and beta subunits.

With each complete cycle, a 2-carbon fragment is cleaved and an acetyl-CoA moiety is released. In most tissues, such as muscle and heart, the acetyl-CoA is oxidized for energy production via the tricarboxylic acid cycle. In liver, about 90% of the hepatic acetyl-CoA is converted into ketones via the coordinated action of acetyl-CoA acetyltransferase, β-hydroxy β-methylglutaryl-CoA synthase, and β-hydroxy β-methylglutaryl-CoA lyase.[46,47] These ketones are then exported for final oxidation by other tissues, such as brain. Ketolysis in extrahepatic mitochondria is mediated by two reversible reactions catalyzed by succinyl-CoA:3-ketoacid-CoA transferase (SCOT) and mitochondrial acetoacetyl-CoA thiolase.[47]

The Regulation of Fatty Acid Oxidation

There are several levels of fatty acid oxidation regulation.[25] The rate of FAO is determined by the availability of fatty acids and by the rate of utilization of β-oxidation products. The concentration of nonesterified fatty acids in plasma is regulated by the hormones glucagon, which stimulates, and insulin, which inhibits the breakdown of triacylglycerols in adipose tissue. Glucagon activates adenylate cyclase leading to an increase in the concentration of cellular cyclic AMP, which in turn activates protein kinase. One of the substrates of protein kinase in adipose tissue is the hormone-sensitive lipase. This lipase is activated by phosphorylation and inactivated by dephosphorylation. Therefore, when the concentration of glucose is low, as in fasting, there is a high glucagon-to-insulin ratio, which results in an increase in plasma nonesterified FFA. These fatty acids will enter cells, where they can be either degraded to acetyl-CoA or incorporated into other lipids. The utilization of fatty acids for either oxidation or lipid synthesis depends both on the nutritional state and on the availability of carbohydrates.

Because of its role in ketogenesis, the regulation of FAO in the liver differs and is more complex than the regulation in heart and skeletal muscle. In the fed state, the liver breaks down carbohydrates to synthesize fatty acids. In contrast, in the fasted animal, FAO, ketogenesis, and gluconeogenesis are more active. McGarry and Foster[48] suggested that the concentration of malonyl-CoA, which is the specific reversible inhibitor of CPT I, determines the rate of FAO. In the fed state, where glucose is converted to fatty acids, the concentration of malonyl-CoA is elevated, thereby leading to an inhibition of CPT I activity. This inhibits the transfer of long-chain fatty acyl residues from CoA to carnitine so that long-chain acylcarnitines cannot be translocated into the mitochondria. Consequently, β-oxidation is depressed. When there is a change from the fed to the fasted state, hepatic metabolism shifts from glucose breakdown to gluconeogenesis, leading to a decrease in fatty acid synthesis. The concentration of malonyl-CoA decreases and the inhibition of CPT I is relieved whereby acylcarnitines are then formed and translocated into mitochondria for β-oxidation and ketogenesis. The cellular concentration of malonyl-CoA is directly related to the activity of acetyl-CoA carboxylase, which is hormonally regulated. In fasting, there is an increase in the glucagon:insulin ratio which causes an increase in cellular cAMP which, in turn, is responsible for the phosphorylation and inactivation of acetyl-CoA carboxylase. As a consequence, the concentration of malonyl-CoA and the rate of fatty acid synthesis decrease, while the rate of β-oxidation increases. A decrease in the glucagon:insulin ratio reverses these effects. Both fatty acid synthesis and FAO are regulated by the glucagon:insulin ratio. A third area that may play a key role in regulating FAO occurs in the mitochondria, where end-product inhibition of proximal β-oxidation by more distal acyl-CoA intermediates prevents excessive accumulation of acyl-CoAs even under conditions of very rapid FAO.[23]

Alternative Pathways for Fatty Acid Metabolism

Under normal circumstances, mitochondrial β-oxidation accounts for the majority of fatty acid metabolism; however, there are other available mechanisms for the oxidation or disposal of fatty acid intermediates. Peroxisomes contain an oxidation pathway which is composed of enzymes that are genetically distinct from the mitochondrial enzymes.[49] The peroxisome may contribute up to 20% of total cellular FAO under conditions of prolonged fasting.[50] When there is an excessive accumulation of acyl-CoAs, they are shunted to microsomes where they undergo omega oxidation. This places a carboxyl group on the methyl-terminal end of the fatty acid and results in the formation of a dicarboxylic acid.[51] Dicarboxylic acids (DCAs) are found in most conditions where the capacity for β-oxidation is exceeded. These include normal fasting, diabetic ketoacidosis, feeding with medium-chain triglycerides, and genetic defects in intramitochondrial β-oxidation. The specific pattern of DCAs found in serum or urine may be diagnostically useful in the identification of specific inborn errors of FAO.

Three additional mechanisms may be important when there is an impairment in mitochondrial β-oxidation. These include the conjugation of acyl groups to glycine and to carnitine and the deacylation of CoA by thioesterases.[52,53] These glycine and carnitine conjugates are able to cross cellular membranes more readily than the respective CoA ester. This prevents excessive accumulation of intracellular acyl-CoAs[54] and is important because acyl-CoA esters inhibit specific enzymes and transporters. Removal of intracellular acyl groups also preserves free CoA for other enzymes which require CoA as a cofactor. Measurement of acylcarnitines and acylglycines is diagnostically useful in the identification of genetic defects in FAO.[55,56]

Central Role of Carnitine Metabolism

Carnitine (beta-hydroxy-gamma-trimethylaminobutyric acid), a water-soluble quartenary amine, has several important intracellular functions[57,58]: (1) it modulates the intramitochondrial acyl-CoA/CoA sulfhydryl ratio in mammalian cells; (2) it serves as an essential cofactor for mitochondrial FAO by transferring long-chain fatty acids as acylcarnitine esters across the inner mitochondrial membrane; (3) it facilitates branched-chain alpha-keto acid oxidation; (4) it shuttles acyl moieties that have been chain-shortened by beta-oxidation out of peroxisomes in the liver; (5) it traps potentially toxic acyl-CoA metabolites that may increase in excess during acute metabolic crises through esterification to carnitine. These metabolites may secondarily impair the citric acid cycle, gluconeogenesis, the urea cycle, and fatty acid oxidation.

In omnivores, approximately 75% of carnitine sources comes from the diet with 25% from endogenous biosynthesis.[59] The principal dietary sources include meat, poultry, fish and dairy products.[60] Approximately 70–80% of dietary carnitine is absorbed in omnivores,[61] whereas in strict vegetarians, endogenous carnitine synthesis provides more than 90% of total available carnitine.[59] There are adequate carnitine concentrations in human milk and most supplemented milk-based formulas to sustain early growth and development. The plasma carnitine concentrations were found to be markedly reduced in term infants fed unsupplemented soy protein-based formulas compared to those in supplemented infants.[62,63] Skeletal muscle is the major tissue reservoir of carnitine, containing over 90% of total body carnitine stores.[59] Under normal conditions, the carnitine concentration in tissues, other than brain, is 20- to 50-fold higher than in plasma and parallels the capacity of the tissue to metabolize fatty acids; human tissue concentrations (nmol/g) are heart (3500–6000) > muscle (2000–4600) > liver (1000–1900) > brain (200–500).[23] Thus, plasmalemmal carnitine uptake occurs across a large concentration gradient which is maintained by a transport system that is generally held to be sodium-gradient and energy-dependent.[57,58,64] The normal serum carnitine concentration is tightly maintained by the renal threshold, which is 40 μmol/L. The kidney is capable of adjusting to wide variations in dietary carnitine as carnitine is not significantly degraded in the body.[65] More than 90% of filtered carnitine is reabsorbed by the kidney at normal physiological plasma carnitine concentrations.[64] Human skeletal muscle, heart, liver, kidney and brain are capable of the biosynthesis of carnitine from methionine and lysine to its immediate precursor gamma-butyrobetaine.[66] However, the final conversion of gamma-butyrobetaine to L-carnitine by gamma-butyrobetaine hydroxylase can only be done in liver, kidney, and brain.[66,67] Gamma-butyrobetaine is thus exported to these tissues for final conversion to L-carnitine. Hepatic gamma-butyrobetaine hydroxylase is developmentally regulated, being approximately 25% of adult activity at birth.[68]

Since 1973, many patients with carnitine deficiency have been described. Originally they were divided into two groups: those with a systemic form characterized by recurrent coma with low carnitine concentrations in serum, liver and muscle versus a muscular form characterized by a progressive lipid storage myopathy in which the serum concentrations were normal and the carnitine deficiency was confined to skeletal muscle. However, with recent advances in biomedical technology, many of these cases were found to be due to intramitochondrial β-oxidation defects with secondary carnitine deficiency.[69] For example, many cases of the systemic form were found to be due to medium-chain acyl-CoA dehydrogenase (MCAD) deficiency. Similarly, certain cases of the myopathic form were attributed to short-chain acyl-CoA dehydrogenase (SCAD) deficiency.[70] The secondary carnitine deficiency disorders can be divided into (1) genetic, (2) acquired, and (3) iatrogenic forms (Box 40.1).[69]

Fasting Adaptation

Fats comprise the most important and efficient fuel for oxidative metabolism. The largest reserve of fuel in the body is comprised of fatty acids stored as adipose tissue triglyceride. As liver glycogen stores are depleted within a few hours of a meal and since there are no reserve stores of protein in the body, fatty acids become the predominant substrate for oxidation quite early in fasting.[23] In adults, approximately 80% of caloric requirements after a 24-hour period of fasting is supplied by fatty acids, which increases to 94% during more prolonged fasting. Fatty acids serve three major functions. First, the partial oxidation of fatty acids by the liver produces ketones (acetoacetate, β-hydroxybutyrate) which are an important auxiliary fuel for almost all tissues and particularly brain, as the blood-brain barrier prevents the direct use of long-chain fatty acids by the brain. This, therefore, provides an important mechanism to spare glucose oxidation

BOX 40.1 Secondary Carnitine Deficiency States and Possible Mechanisms

Genetically Determined Metabolic Errors

- Increased Esterification due to acyl-CoA Accumulation
  - Carnitine acyl-carnitine translocase deficiency
  - Carnitine palmitoyltransferase II deficiency
  - Very long-chain acyl-CoA dehydrogenase deficiency
  - Long-chain acyl-CoA dehydrogenase deficiency
  - Long-chain L-3-hydroxyacyl-CoA dehydrogenase deficiency
  - Trifunctional protein deficiency
  - Medium-chain acyl-CoA dehydrogenase deficiency
  - Short-chain acyl-CoA dehydrogenase deficiency
  - Short-chain L-3-hydroxyacyl-CoA dehydrogenase deficiency
  - Multiple acyl-CoA dehydrogenase (ETF and ETF/Qo) deficiency
  - beta-Hydroxy-beta-methylglutaryl-CoA lyase deficiency
  - Isovaleric acidemia
  - Propionic acidemia
  - Methylmalonic aciduria
  - Glutaryl-CoA dehydrogenase deficiency (glutaric aciduria type I)
  - β-ketothiolase deficiency
- Decreased Biosynthesis
  - Homocystinuria
  - 5-methylene tetrahydrofolate reductase deficiency
  - Adenosine deaminase deficiency
  - Ornithine transcarbamylase heterozygote state
- Increased Urinary Loss
  - Cystinosis
  - Cytochrome oxidase deficiency

Acquired Medical Conditions

- Decreased Biosynthesis[a]
  - Cirrhosis
  - Chronic renal disease
  - Extreme prematurity
- Dietary Deficiency[a]
  - Chronic TPN without carnitine supplementation
  - Malabsorption (cystic fibrosis, short-gut syndrome)
  - Unsupplemented Soybean protein-derived infant formula
- Decreased Body Stores
  - Extreme prematurity
  - Intrauterine growth retardation
  - Infant of carnitine-deficient mother
- Increased Urinary Loss
  - Fanconi syndrome
  - Renal tubular acidosis

Iatrogenic Factors: Drug Therapy

- Increased Esterification and competitive inhibition of carnitine uptake by valproylcarnitine
  - Chronic valproic acid administration
- Impaired Hepatic Biosynthesis
  - Chronic valproic acid administration
- Increased Loss
  - Chronic hemodialysis

[a] often combined factors.
Modified from De Vivo and Tein.[69]

and proteolysis during prolonged fasting.[71] Second, fatty acids serve as a major fuel for cardiac and skeletal muscle. Resting muscle depends mostly on FAO.[72] Depending upon the type, intensity, and duration of exercise, energy in working muscle is derived from either the combination of triglyceride and stored glycogen or the combination of glucose and free fatty acids.[73,74] After 90 minutes, the major fuels are glucose and free fatty acids. During 1–4 hours of mild to moderate prolonged aerobic exercise, free fatty acid uptake by muscle increases by 70% and after 4 hours, free fatty acids are used twice as much as carbohydrate sources. Third, the high rates of hepatic gluconeogenesis and ureagenesis needed for maintaining fasting homeostasis are sustained by the production of ATP, reducing equivalents (nicotinamide adenine dinucleotide, reduced, + H +), and metabolic intermediates (acetyl-coenzyme A) derived from FAO.[39]

Increased Susceptibility of the Child

Infants and young children have an increased risk of problems with fasting adaptation for several reasons.[75–77] First, infants have a larger brain compared with body size which is highly dependent upon glucose and has a high rate of metabolism.[39] Thus, infants and children show an even earlier activation of FAO with hyperketonemia within 12 to 24 hours of fasting. Second, basal energy needs in the infant are high in order to maintain body temperature, given their large ratio of surface area to mass. Their body temperature is maintained by shivering thermogenesis, which is highly dependent upon efficient FAO. Third, there is a lower activity of several key enzymes involved in energy production in the infant compared to the older child and adult, which leads to further impairment of the infant's ability to maintain glucose homeostasis.[78]

CLINICAL AND BIOCHEMICAL FEATURES OF IDENTIFIED DEFECTS

There are now at least 16 identified defects in fatty acid oxidation. They share a number of clinical features which suggest an underlying defect in fatty acid oxidation.

BOX 40.2 Common Features of Fatty Acid Oxidation Disorders

1. Metabolic decompensation during fasting, infection, prolonged exercise, cold exposure, and stress
   a. Decreased oral intake, increased energy expenditure
   b. Progressive obtundation leading to coma
   c. Reye-like syndrome
   d. Sudden infant death syndrome or "near-miss" SIDS
   e. Recurrence, familial occurrence
2. Involvement of fatty acid-dependent tissues
   a. Myalgia, exercise intolerance, myoglobinuria, hypotonia, weakness
   b. Cardiac hypertrophy/dilatation, endocardial fibroelastosis, arrhythmias
   c. Hepatomegaly and hepatic dysfunction
   d. Fatty infiltration of tissues on biopsy or autopsy
3. Hypoketotic hypoglycemia
   a. Elevated serum free fatty acids with serum free fatty acid to ketone body ratio >2:1
   b. Hyperinsulinism and hypopituitarism ruled out (mandatory)
4. Alterations in plasma or tissue carnitine concentration
   a. Decreased total carnitine concentration (10–50% of normal) with increased esterified fraction in intramitochondrial fatty acid oxidation defects
   b. Marked decrease in total carnitine concentration (<5% of normal) with normal esterified fraction in plasmalemmal carnitine transporter defect
   c. CPT I deficiency is exception with high to normal carnitine concentration and low esterified fraction
5. Clinical laboratory abnormal findings
   a. Dicarboxylic aciduria (intramitochondrial FAO defects)
   b. Hyperammonemia
   c. Acidosis
   d. During myoglobinuria, increased serum creatine kinase, hyperkalemia, hyperphosphatemia, hypocalcemia, hyperuricemia and possible increased creatinine (with renal failure)
   e. Increased alanine and aspartate aminotransferase

Modified from Hale and Bennett[39] with permission from Elsevier.

These individual defects can be differentiated on the basis of distinct clinical and biochemical features. In the following sections, the common features will be presented first, followed by a discussion of the differentiating clinical and laboratory features.

Common Features of Fatty Acid Oxidation Disorders

It has been suggested that there are at least four clinical and laboratory features that should lead the clinician to suspect a genetic defect in fatty acid metabolism (Box 40.2).[39] These common features include: (1) acute metabolic decompensation in association with fasting; (2) chronic involvement of tissues highly dependent upon efficient fatty acid oxidation (e.g. muscle, heart, liver); (3) recurrent episodes of hypoketotic hypoglycemia; (4) alterations in the quantity of total carnitine or in the percentage of esterified carnitine in plasma and tissue.

Metabolic Decompensation in Association with Fasting

Episodes of metabolic decompensation occur in affected individuals during conditions that place stress on the FAO pathway for fuel generation, in the context of depleted glycogen and glucose reserves. These stressors include fasting, prolonged exercise (>1 hour of mild to moderate aerobic exercise), infection with vomiting, and cold-induced shivering thermogenesis. Shivering is an involuntary form of muscle activity which depends on long-chain fatty acid oxidation.[79] Thus, ketogenesis is stimulated in normal individuals during cold exposure.[80] Children are most likely to be found comatose in the early-morning hours after an overnight fast. Infections such as a viral illness are frequent precipitants due to the synergistic combination of vomiting and decreased oral intake with shivering thermogenesis. If not recognized and therefore untreated, children may progress to a Reye-like syndrome as seen in medium-chain acyl-CoA dehydrogenase (MCAD) deficiency.[7] Infants and younger children are at greater risk during fasting than older children, because of their limited fasting adaptation capabilities. Thus, prolonged fasting for an infant of less than 1 year of age would be 6 to 8 hours, versus 12 hours for a child between 1 and 4 years of age.[39]

Hypoketotic Hypoglycemia

The accelerated rate of glucose utilization that occurs when fatty acids cannot be used as fuels and ketone bodies are not generated to spare glucose or glycogen stores results in the pattern of hypoketotic hypoglycemia. The increase in the ratio of serum free fatty acids to ketones is another clue to a block in β-oxidation. The normal ratio should be 1:1. If this ratio exceeds 2:1, this would suggest a block in β-oxidation. Problems in fat mobilization, such as hyperinsulinism, are excluded by the increase in serum free fatty acids.

Involvement of Fatty-acid Oxidation Dependent Tissues

Tissues with high energy demands such as skeletal muscle, heart, and liver are dependent upon efficient FAO. With deficient hepatic ketogenesis, glucose becomes the only available fuel and becomes rate-limiting under conditions of FAO stress when glycogen and glucose stores have been depleted. As a result, free fatty acids, which are liberated during fasting and which cannot be metabolized due to the block, may be stored in the cytosol as triglycerides. This may lead to a progressive lipid storage myopathy with weakness as well as a hypertrophic and/or dilatative cardiomyopathy and a fatty liver, which are pathological hallmarks of genetic FAO defects. Increased short- or medium-chain fatty acids and in particular, their dicarboyxlic metabolites from compensatory omega-oxidation, may cause secondary abnormalities including an impairment of gluconeogenesis, beta-oxidation, and the citric acid cycle.[54,81,82] This would further decrease cellular ATP production.

In addition, in the long-chain fatty acid oxidation disorders which may have recurrent episodes of acute muscle breakdown or myoglobinuria (e.g. CPT II, VLCAD, trifunctional enzyme deficiencies), and the accumulation of long-chain fatty acids (LCFAs) and long-chain acylcarnitines may have detergent-like actions on muscle membranes. Excessive amounts of palmitoyl-CoA and palmitoylcarnitine have been shown to have detergent properties on isolated canine myocytic sarcolemmal membranes and to potentiate free radical induced lipid membrane peroxidative injury in ischemia.[83] Long-chain acylcarnitines have also been shown to activate Ca^{2+} channels in cardiac[84,85] and in smooth muscle myocytes[85,86] and thus may potentiate the increase in cytosolic Ca^{2+} associated with arrhythmogenesis as seen in ischemic myocardium.[87]

Alterations in Plasma and Tissue Concentrations of Carnitine

The total carnitine concentration is decreased (<50 % of normal) and the acylcarnitine fraction is increased (>50% esterified; normal = 10–25% in fed state and 30–50% in fasted state) in most cases of intramitochondrial β-oxidation defects.[39] Excessive acyl-CoAs which accumulate proximal to the block, in intramitochondrial β-oxidation defects, may be converted into acylcarnitines by chain-length specific carnitine acyltransferases.[57] When filtered through the kidneys, these acylcarnitines will compete with free carnitine at the renal tubular reabsorptive site. As demonstrated by Stanley et al.[88] in a cultured skin fibroblast model, the longer chain-length acylcarnitines have an increasingly higher affinity for the carnitine transporter than free carnitine, and thus free carnitine will be excreted. This will with time deplete total body and muscle stores of free carnitine. In the plasmalemmal carnitine transporter defect, total carnitine is markedly reduced (e.g. <5% of normal) and the esterified fraction is normal[13] as the transport defect in kidney leads to a decreased renal threshold for carnitine reabsorption.

Additional Laboratory Findings

A number of other biochemical abnormalities may be noted during acute catabolic crises. During the Reye-like syndrome presentation, a modest hyperammonemia (100–200 μmol/L) may be documented accompanied by 3- to 5-fold elevations of liver transaminases.[39] With acute myoglobinuria, there are marked increases in sarcoplasmic enzymes including serum CK which may rise to >100,000 U/L (normal < 250 U/L).[89] There may also be increased serum amino acids (especially taurine), creatinine, potassium, phosphate, and urate.[90] These changes may have deleterious effects on the kidneys and heart, increasing the damage.[91,92] The hyperphosphatemia may lead to secondary hypocalcemia which may be later followed by hypercalcemia because of increased secretion of parathormone and vitamin D, with deposition of calcium in injured muscle followed by release.[92] It is urgent to prevent renal failure. Lactic acidosis may also be noted during the acute catabolic presentations which may reflect poor perfusion or inhibition of critical enzymes such as pyruvate carboxylase by accumulated metabolites.[54] Urine organic acid screening may demonstrate unusual compounds or excessive organic acids which may be diagnostic of a specific block in β-oxidation.

Specific Features of Individual Genetic Defects

With the exception of MCAD, the carnitine (OCTN2) transporter, classic late-onset CPT II, VLCAD, LCHAD, alpha ETF and ETF-CoQ, and HMG-CoA lyase deficiencies, there are fewer than 25 published cases for most of the FAO defects. The clinical and laboratory characteristics that are presented here are based on a summary of these cases; the phenotypes may be further broadened as additional cases are identified. Key clinical features are summarized in Table 40.1.[39]

There are several differentiating features. In the defects involving the transport of fatty acids into the mitochondria (plasmalemmal long-chain fatty acid uptake defect, plasmalemmal carnitine transporter defect OCTN2, CPT I, carnitine-acylcarnitine translocase, CPT II), there is generally not an associated abnormal dicarboxylicaciduria. However, this has been reported in one case of severe infantile CPT II deficiency[94] and on one occasion during an acute metabolic decompensation in a case of infantile CPT I deficiency presenting with recurrent Reye-like syndrome.[95] The carnitine transporter defect, carnitine acylcarnitine translocase deficiency and severe CPT II deficiency present in infancy or early childhood and involve both cardiac and skeletal muscle and recurrent episodes of hypoketotic hypoglycemic encephalopathy. The carnitine transporter OCTN2 defect is

TABLE 40.1 Clinical Features Associated with Specific Genetic Defects of Fatty Acid Oxidation

Deficiency	Fasting Disorder	Tissue Involved	Hypoketotic hypoglycemia	Altered Carnitine	Dicarboxylic acids	Reye-like syndrome	SIDS
LCFAUD	+	L	+	+	NR	NR	NR
OCTN2	+	H, M, L	+	+	NR	+	+
CPT I Liver	+	L, K	+	+	NR[a]	+	NR
TRANS	+	H, L, M, (Mg)	+	+	NR	+	+
CPT II (mild)	±	M, Mg, P	NR	+	NR	NR	NR
CPT II (severe)	+	H, K, M, Mg, L,B	+	+	NR[b]	+	+
VLCAD/LCAD	+	H, M, Mg, L	+	+	+	+	+
Trifunctional/ LCHAD	+	H, M, Mg, L, N, P, R	+	+	+	+	+
Dienoyl CoA reductase	NR	M,D,B, H	NR	+	NR	NR	NR
MCAD	+	(Mg)	+	+	+	+	+
SCAD	+	M, B, D, H	±	+	+	NR	+
SCHAD	+	H, M, Mg, L	+	+	+	NR	+
ETF and ETF/Qo	+	M, H, K, B, D	+	+	+	NR	+
HMG-CoA lyase	+	B, P	+	+	+	+	+ ?

[a]*Reported in one case of infantile CPT 1 deficiency during Reye-like syndrome.*
[b]*Reported in one case of severe infantile CPT II deficiency.*
LCFAUD, long-chain fatty acid uptake defect; OCTN2, plasmalemmal high-affinity carnitine transporter; CPT, carnitine palmitoyltransferase; TRANS, carnitine acylcarnitine translocase; LCAD, long-chain acyl-CoA dehydrogenase; VLCAD, very-long-chain acyl-CoA dehydrogenase; trifunctional, long-chain enoyl-CoA hydratase + long-chain L-3-hydroxyacyl-CoA dehydrogenase + long-chain 3-ketoacyl-CoA thiolase; MCAD, medium-chain acyl-CoA dehydrogenase; SCAD, short-chain acyl-CoA dehydrogenase; SCHAD, short-chain L-3-hydroxyacyl-CoA dehydrogenase; ETF, electron transfer flavoprotein; Qo, coenzyme Q oxidoreductase; HMG, beta-hydroxy-beta-methylglutaryl; B, brain; D, dysmorphic features; H, heart; K, kidney; L, liver; M, muscle; Mg, myoglobinuria; N, neuropathy; P, pancreatitis; R, retinopathy; NR, no case yet reported.
Modified and updated from Hale and Bennett[39] with permission from Elsevier.

characterized by a progressive hypertrophic and/or dilatative cardiomyopathy which is pathologically similar to endocardial fibroelastosis,[96] lipid storage limb girdle myopathy, and recurrent hypoketotic hypoglycemic encephalopathy which is exquisitely responsive to high dose oral L-carnitine supplementation, which reverses the cardiomyopathy and myopathy.[13,88] Renal tubular acidosis has been documented in one case of CPT I deficiency.[97] At this time, the only human defect that has been identified for CPT I deficiency is the hepatic and renal CPT 1A[98] which presents with fasting-induced hepatic failure (Reye-like syndrome) with hypoketotic hypoglycemia. Notably, the P479L variant which was initially described in an Inuit individual[99] has been identified in a large number of Alaskan Inuit and Canadian First Nations individuals.[100] These groups have higher than normal rates of infant deaths and a significant number of newborns identified in the Alaskan newborn screening program have this DNA variant.[101] It is uncertain whether it is a pathological mutation or an adaptive polymorphism, as the amino acid substitution removes any inhibition of CPT 1A by malonyl-CoA[99] theoretically allowing an individual to slowly and continuously generate energy via ketone body production from fatty acids even when not fasted, which may be an adaptive advantage in the Inuit during long winter months when food is scarce and provided there are ample fat stores generated during the summer season.[100] Fasting in the absence of adequate fat stores as in the infant may lead to the high infant mortality rate. One individual with atypical symptoms and homozygosity for P479L had ~20% normal L-CPT1 activity and presented with adult onset recurrent muscle cramps and vomiting with elevated serum CK (2733 U/L).[99] Classic adult CPT II deficiency presents with adolescent onset recurrent episodes of acute myoglobinuria precipitated by prolonged exercise or fasting, in which power between episodes is normal and in which lipid accumulation in muscle is noted only under conditions of fasting and prolonged exercise.[102] Fasting ketogenesis is generally normal in this condition, though it may be delayed. There is no fixed cardiomyopathy, though arrhythmias may occur secondary to hyperkalemia and hypocalcemia during acute myoglobinuric crises. Statistically speaking, classic CPT II deficiency is the most common cause for recurrent myoglobinuria in children and adults.[89,103] However, a number of cases of severe infantile CPT II deficiency have been described which have included recurrent Reye-like syndrome, hepatomegaly with hypoketotic hypoglycemia and elevated liver aminotransferase, and cardiomegaly with cardiac arrhythmias and elevated serum creatine kinase, as well as evidence of lipid storage in heart, skeletal muscle, liver, and kidney.[94,104,105] In these cases the activity of CPT II in cultured skin fibroblasts was <10% of controls in contrast to the 25% residual activity documented in classic CPT II deficiency.[104] A third phenotype is the lethal neonatal form which may have multiple congenital anomalies including CNS malformations that include polymicrogyria, agenesis of the corpus callosum, and ventriculomegaly, seizures, cardiomegaly with hypertrophy, hepatomegaly and enlarged kidneys with renal cysts and neonatal death in which there are null mutations with <5% residual CPT II activity.[106] More than 64 mutations have been described in the literature.[106] Thus the clinical presentation of CPT II deficiency may depend, in part, on the degree of residual CPT II activity.

There is another plasma membrane fatty acid oxidation disorder which involves a defect in the active transport of long-chain free fatty acids across the plasma membrane. This has been described in two young boys who presented with acute liver failure with hyperammonemia, hyperbilirubinemia, coagulopathy and mild encephalopathy with no evidence of cardiac or skeletal myopathy, and who required liver transplantation.[107] One boy had documented hypoketotic hypoglycemia. The uptake of oleic acid (C18:1) in skin fibroblasts from the patient was lower than in controls. The features which were atypical for an FAO defect included the absence of fatty liver with low concentrations of long-chain free fatty acids and elevated carnitine concentrations, due to defective transport of long chain free fatty acids.

There are many common features shared by the intramitochondrial β-oxidation defects (SCAD, MCAD, LCAD, VLCAD, ETF, ETF-CoQ, SCHAD, M/SCHAD, trifunctional protein/LCHAD, MKAT, 2,4,-dienoyl-CoA reductase, HMG-CoA synthase, and HMG-CoA lyase deficiencies), though there are also differentiating characteristics. It should be noted that many cases previously attributed to LCHAD deficiency are now known to be related to a defect in the trifunctional protein (TFP) which combines the activities of the long-chain enoyl-CoA hydratase, long-chain L-3-hydroxyacyl-CoA dehydrogenase, and long-chain 3-ketoacyl-CoA thiolase enzymes[36] although isolated LCHAD deficiency appears to be more common than TFP deficiency.[108] A number of previously diagnosed cases of LCAD deficiency are now believed to be attributable instead to VLCAD deficiency.[109–111] Cardiac involvement (cardiomegaly, cardiomyopathy) is more commonly found in the long-chain defects (e.g. LCAD, VLCAD, trifunctional protein),[3,112,113] but has also been documented in one case of SCHAD deficiency[15] and one case of SCAD deficiency.[114] Cardiomyopathy has been documented in 17 of 24 reported cases of TFP deficiency with presumed or proven marked LCHAD deficiency.[113,115–123]

Chronic skeletal muscle weakness is seen in all of the intramitochondrial β-oxidation defects, but less commonly in MCAD[124] and HMG-CoA lyase deficiency.[125] A unique case of SCAD deficiency has been reported in a 13-year-old girl with progressive external ophthalmoplegia, ptosis, cardiomyopathy, and multicore myopathy.[114] Additional cases of SCAD deficiency with multicore myopathy have now been identified.[126] Recurrent myoglobinuria has been reported with LCAD,[112] VLCAD,[110,127,128] trifunctional protein/LCHAD,[115,129] SCHAD,[15] MCKAT,[130] and MCAD[131] deficiencies. In a long-term follow-up study of 13 patients with VLCAD deficiency, all patients exhibited exercise intolerance and recurrent myoglobinuria triggered by strenuous exercise, fasting, cold, or fever with a mean age at onset of 10 years.[132] Muscle biopsies showed mild lipid accumulation in four patients. Perinatal onset of VLCAD deficiency has been reported in a male neonate with fetal distress, neonatal asphyxia, and transient hypercreatine kinasemia (8400 IU/L) followed by repeated episodes of myoglobinuria one to two times/year during infancy and early childhood.[133] Ohashi et al.[134] showed that immuno-histochemistry is an effective and useful diagnostic method for the detection of VLCAD deficiency, which facilitated identification of 13 new Japanese patients with VLCAD deficiency. Recurrent myoglobinuria has been reported in TFP deficiency beginning at 7 months of age.[135] An episode of myoglobinuria has also been described in an adult with MCAD deficiency.[131]

All of the intramitochondrial enzyme defects demonstrate hepatic dysfunction at the time of acute catabolic decompensations; however, persistent dysfunction has only been documented in trifunctional protein/LCHAD[136] and LCAD/VLCAD[3] deficiencies. Of note, there is increasing evidence that a significant subgroup of women with acute fatty liver of pregnancy (AFLP) or hemolysis, elevated liver enzymes, and low platelets (HELLP) syndrome are heterozygous for LCHAD deficiency and carry affected LCHAD deficient fetuses with a common mutation (G1528C, E474Q) in one or both alleles.[137–140] In AFLP syndrome, the mother may suffer fulminant liver failure and death. In both syndromes there is microvesicular fatty infiltration of the maternal liver. In addition to the recurrent episodes of myoglobinuria, TFP/LCHAD deficiency may present with progressive pigmentary retinopathy[141,142] and motor-sensory neuropathy.[143] Thus, individuals with the myoneuropathic variant of TFP/LCHAD deficiency may mimic the spinal muscular atrophy phenotype. There appear to be three major phenotypes of TFP/LCHAD deficiency with variable overlap; a hepatic form with Reye-like syndrome which is associated with AFLP, a neonatal cardiomyopathic form, and a milder myoneuropathic form. These phenotypes may correlate with specific genotypes and may be a reflection of the severity of the mutation and the presence of additional environmental and genetic factors.[115–123]

Maternal complications in pregnancy are not limited to carriers of LCHAD mutations. Patients with deficiencies of CPT I, MCAD, and SCAD were also born to mothers who developed liver disease during their pregnancies.[144–148] Therefore all babies born following pregnancies that are complicated by AFLP or HELLP syndrome should be considered at risk for having a defect in β-oxidation. A further complication of pregnancy includes placental floor infarction documented in the pregnancy of a fetus with LCHAD deficiency.[145]

SCAD deficiency is a clinically heterogeneous disorder in which the clinical phenotype varies from fatal metabolic decompensations including intermittent metabolic acidosis and neonatal hyperammonemia to infantile onset lipid storage myopathy with failure to thrive, developmental delay, hypotonia, seizures, and hyperreflexia[149] to more subtle later onset progressive myopathy including multicore myopathy[114,126] and to entirely asymptomatic individuals.[150] The characteristic metabolites of ethylmalonic and methylsuccinic acids in SCAD deficiency may also be due to the presence of one of two relatively common variants of SCAD (625G > A and 511C > T) which predispose to excessive ethylmalonic acid production and are also present in normal populations. These common variants may represent disease susceptibility variations when occurring in combination with other genetic and/or environmental factors, especially fever.[151,152] They encode proteins with nearly normal enzymatic activity at physiological conditions *in vitro*; however, SCAD enzyme function is impaired at increased temperature and the tendency to misfold increases under conditions of cellular stress.[153] We have further demonstrated that vulnerability to oxidative stress may contribute to the pathophysiology of SCAD deficiency in patients homozygous for the common variant 625G > A, 625A/319T compound heterozygous patients, and patients homozygous for the pathogenic 319C > T and 1138C > T mutations.[153,154] The c.319C > T *ACADS* gene mutation presents with clinical heterogeneity and appears to be a candidate founder mutation in the Ashkenazim.[126] The challenge with SCAD deficiency is to elucidate whether *ACADS* gene variations are disease-associated, especially when combined with other genetic/cellular/environmental factors such as fever, which may act synergistically.[155]

Multiple acyl-CoA dehydrogenase deficiency or glutaric acidemia type II is due to defects of the electron transfer flavoprotein (ETF) or ETF-ubiquinone oxidoreductase (ETF-QO) which are nuclear encoded proteins through which electrons from flavoprotein acyl-CoA dehydrogenases, dimethylglycine dehydrogenase, and sarcosine dehydrogenase enter the respiratory chain.[156] These patients can be divided into three groups, each consistent within a family and have been designated as (1) neonatal onset with congenital anomalies, (2) neonatal onset without anomalies, and (3) mild or later-onset. The first two groups are sometimes said to have severe multiple acyl-CoA dehydrogenase deficiency (MADD) and the third group to have mild MADD. Neonatal onset patients with congenital anomalies are often premature, presenting within the first 24 to 48 hours of life with severe hypoglycemia, hypotonia, hepatomegaly, and metabolic acidosis and an odor, similar to sweaty feet. The infants may have enlarged kidneys and facial dysmorphism (high forehead, low-set ears, hypertelorism, hypoplastic mid face, etc.), rocker-bottom feet, muscular defects of the anterior abdominal wall, and anomalies of the external genitalia (hypospadias, chordee). Most of these patients die within the first week of life. In others, congenital anomalies are not noted, but renal cysts are found at necropsy. The infants without congenital anomalies usually develop hypotonia, tachypnea, hepatomegaly, hypoglycemia, metabolic acidosis, and a “sweaty foot” odor within the first few days of life, many in the first 24 hours. The few infants with this form who have survived beyond the first week of life due to prompt diagnosis and treatment, have died within a few months, usually with severe cardiomyopathy. A few other infants have been hypoglycemic in the newborn period and later developed typical Reye-like episodes and have survived longer.

Routine laboratory evaluation shows severe metabolic acidosis, mild to moderate hyperammonemia, and severe hypoglycemia without serum or urinary ketones. Serum transaminases may be elevated and prothrombin and partial thromoplastin times may be prolonged. Plasma lactate is usually elevated. There may be hypertrophic cardiomyopathy and renal cysts. Urinary organic acids show various combinations of short-chain volatile acids (e.g. isovaleric, isobutyric, 2-methylbutyric); glutaric, ethylmalonic, 3-hydroxyisovaleric, 2-hydroxyisoglutaric, 5-hydroxyhexaenoic, adipic, suberic, sebacic, and dodecanedioic acids; and isovalerylglycine, isobutyrylglycine, and 2-methylbutyrylglycine. Patients with neonatal onset may have generalized aminoacidemia and aminoaciduria. Plasma carnitine concentrations may be low or normal but acylcarnitine esters are increased in the urine.

Late onset glutaric aciduria type II (GA II) is extremely variable in age at presentation and course of the disease. One infant presented at 7 weeks of age with intermittent episodes of vomiting, hypoglycemia and acidosis, whereas another individual was asymptomatic during childhood and presented in adult life with episodic vomiting, hypoglycemia, hepatomegaly, and proximal myopathy. Some have had progressive lipid storage myopathy with carnitine deficiency. A 12-year-old girl presented with nausea, vomiting, and depression followed by progressive muscle weakness, elevated CK, and lipid storage myopathy which responded well to riboflavin, L-carnitine, and a high calorie, fat-reduced diet.[157] For the riboflavin responsive-MAD deficiency, the challenge is to find the connection between variations in the ETFDH gene and the observed deficiency of a number of different mitochondrial dehydrogenases as well as deficiency of FAD and coenzyme Q10.[155] Isolated myopathic coenzyme Q10 deficiency is caused by mutations in the ETF dehydrogenase gene and can be treated with CoQ10 and riboflavin supplementation.[158] Patients present with exercise intolerance, fatigue, proximal myopathy, and high CK with lipid storage myopathy, subtle signs of mitochondrial myopathy, and severely decreased activities of respiratory chain complexes I and II + III, moderately decreased complex IV, and significantly decreased CoQ10 in muscle. Late-onset GA II and myopathic CoQ10 deficiency may be allelic diseases.

SCHAD deficiency has only been rarely reported. It is also now known as medium- and short-chain L-3-hydroxyacyl-CoA dehydrogenase (M/SCHAD) deficiency, given its broader chain length specificity.[159] Furthermore, it has been argued that it should be called HAD deficiency to distinguish it from the SCHAD enzyme, also known as 17beta-hydroxysteroid dehydrogenase type 10, that is not a member of the HAD family but instead belongs to the short chain dehydrogenase/reductase superfamily that acts on a wide spectrum of substrates, including steroids, cholic acids, and fatty acids, with a preference for short chain methyl-branched acyl-CoAs.[160] Three distinct clinical and biochemical phenotypes have been described. One girl presented with recurrent myoglobinuria, myopathy, cardiomyopathy, and hypoketotic hypoglycemic encephalopathy and died at 16 years of age.[15] SCHAD activity was deficient in muscle but not in cultured skin fibroblasts, though these fibroblasts demonstrated marked microvesicular steatosis when cultured in a glucose-free medium to simulate fasting stress.[161] Other tissues were not examined enzymatically. A second phenotype presented with ketotic hypoglycemia and SCHAD deficiency was described in isolated mitochondria from cultured skin fibroblasts but no other tissues were studied.[162] A third group of patients presented with fatal infantile hepatic involvement and steatosis and SCHAD activity was deficient in liver but normal in muscle and fibroblasts.[163] However, Western blot analysis of SCHAD revealed an identical truncated protein in both liver and muscle from one patient, suggesting that SCHAD is similar in liver and muscle and that the normal activity in muscle may be due to other enzymes with C4 activity. Autopsy findings revealed marked steatosis and a muscle pattern consistent with spinal muscular atrophy in one patient. Mutation analysis of the SCHAD gene in patients with SCHAD deficiency has been inconclusive, suggesting that other proteins in addition to M/SCHAD may be necessary for enzymatic activity and that these proteins may have a tissue-specific distribution. Apparent disease-causing mutations have been described in a patient with fulminant hepatic failure that required transplantation.[164] Hyperinsulinism with hypoglycemia has also been reported in SCHAD deficiency.[165,166] The pathophysiology of hyperinsulinism in M/SCHAD deficiency is not yet fully understood; however, it is becoming increasingly evident that it results from an interaction of the SCHAD protein with glutamate dehydrogenase.[167]

Only one patient with 2,4-dienoyl-CoA reductase deficiency in muscle and liver has been reported, who presented with neonatal hypotonia, microcephaly, a small VSD, and dysmorphism, and who died at 4 months of age of respiratory acidosis with biventricular hypertrophy at autopsy.[168] The clinical significance of the biochemical abnormalities therefore cannot be confirmed until additional patients have been identified. In addition, one patient with medium-chain 3-ketoacyl-coenzyme A thiolase deficiency has been described in a Japanese male neonate who died at 13 days of age after presenting at 2 days of age with vomiting, dehydration, metabolic acidosis, liver dysfunction, and myoglobinuria.[169]

There are several disorders of ketone body production. Deficiency of 3-hydroxy 3-methylglutaryl-CoA lyase (HMG-CoA lyase), which is also active in the metabolism of leucine, presents with hypoketotic hypoglycemia with hyperammonemia and acidosis.[170] Succinyl-coenzyme A:3-ketoacid coenzyme A transferase (SCOT) functions in conjunction with mitochondrial acetoacetyl-coenzyme A thiolase for ketone body utilization (ketolysis) in extra-hepatic tissues.[47] SCOT deficiency presents as persistent ketonuria in the first 1–2 years of life, while acetoacetyl-coenzyme A thiolase deficiency presents with variable clinical symptoms and exaggerated ketoacidosis in response to minor physiological stresses.[171–174] HMG-CoA synthetase deficiency has been reported in a child who presented with fasting hypoketotic hypoglycemia.[175] Remaining challenges in mitochondrial FAO such as MCAD, riboflavin responsive MAD and SCAD deficiencies are summarized well by Gregersen et al.[155]

For illustrative examples, see Case Examples 40.1–40.5.

CASE EXAMPLE 40.1 Carnitine Uptake Defect

A 9-year-old girl was originally born to unrelated asymptomatic parents of East Indian and Caucasian heritage.[13] Her sister presented with congenital hypotonia and muscle atrophy, and at 1 year of age was found to have progressive cardiomyopathy with severe mitral regurgitation. This sibling also had failure to thrive and progressive motor deterioration and died at 3.5 years of age, at which time fatty infiltration of the liver was noted at necropsy.

The index patient was born a footling breech. At 1 month of age, a cardiac murmur was noted. By 14 months, her gross motor milestones plateaued, and over the next 2 years, she developed slowly progressive weakness and recurrent respiratory infections. At 3 years of age, she suffered from congestive heart failure with moderate-to-severe mitral regurgitation. Marked left ventricular dilatation was present on echocardiography. The ECG demonstrated peaked T waves in the anterolateral precordial leads. She was started on digoxin and diuretics. At 4 years 3 months of age, she was found to have very low carnitine concentrations in serum and in muscle. The total muscle carnitine was less than 5% of control values. A muscle biopsy showed lipid storage, but β-oxidation enzymes were normal. Serum creatine kinase was mildly elevated. Serum glucose was normal, as were urinary organic acids. There was no abnormal dicarboxylic aciduria. Oral carnitine was instituted at 1 g three times daily. Within 1 week, appetite, affect, and cardiac function had dramatically improved. By 4 weeks, she had gained weight and had markedly increased exercise tolerance and the ECG showed inversion of the formerly peaked T waves. The predicted left ventricular end systolic and end diastolic dimensions markedly improved from 240% and 205% of normal, respectively, to 175% and 160% of normal. By 4 months, she had good exercise tolerance, and by 7 months of therapy, the predicted left ventricular end systolic and end diastolic dimensions had decreased to 150% and 130% of normal, respectively. At 5 years of age, she had regained gross motor milestones. Her digoxin and diuretics were discontinued by 6.5 years. At 9 years of age, she had an IQ of 140 and was a competitive swimmer at school. On examination, she had a mildly myopathic facies, mild hypotonia, and persistent, although improving, fine motor delay. Her electromyogram was normal. Motor nerve conduction studies revealed borderline abnormalities. Her serum carnitine concentrations were low and she continued to have decreased renal reabsorption of carnitine. Her father's serum carnitine concentrations were normal and her mother's were low normal.

Carnitine uptake studies in the cultured skin fibroblasts of the patient showed minimal carnitine uptake. At a carnitine concentration of 5 μmol/L (K_m value), the mean rate of uptake was 2% of control values. Carnitine uptake studies in the cultured skin fibroblasts of both parents revealed normal K_m values of 5.0 μmol/L (controls = 5.5 ± 0.58) but reduced maximal rates of carnitine uptake with a V_{max} of 1.33 pmol/min per mg protein (controls = 3.4 ± 0.36), which was 40% of control values. The normal K_m values for carnitine binding affinity and reduced V_{max} values for carnitine uptake in the heterozygote parents suggested the presence of a reduced number of normally functioning carnitine transporters. Family history of a previously affected sibling, as well as demonstrated heterozygosity in both parents on carnitine uptake studies, supported the autosomal recessive pattern of inheritance.

CASE EXAMPLE 40.2 Carnitine Palmitoyltransferase II Deficiency—Classic Adult Form

This boy presented at 12 years of age with recurrent episodes of muscle stiffness and pain lasting a few hours to 24 hours at a time following periods of prolonged exercise lasting greater than one hour at a time. On entry into competitive sports at age 15 years, following 2 to 3 hours of uninterrupted exercise, he began experiencing severe episodes of muscle stiffness and pain in his legs and shoulder girdle with subsequent weakness and coca-colored urine (myoglobinuria), which would last from 24 to 48 hours at a time. His serum creatine kinase could be elevated to >20,000 IU/L (normal < 250) and would return to normal in a few days to a few weeks in time, as would his motor power following a period of rest and carbohydrate ingestion. These episodes could also be exacerbated by fasting, cold exposure and infection. Carbohydrate loading prior to exertion would reduce the episodes of pigmenturia. During university exam times with fasting and stress, he also noted that he felt jittery and weak, particularly in the morning prior to breakfast and was immediately improved following carbohydrate ingestion. He had full recovery of power between episodes. He never complained of true muscle cramps and had no second wind phenomenon. With participation in university level basketball and volleyball tournaments, he suffered multiple episodes of pigmenturia in one year. He had no documented episodes of acute renal failure. His brother had an identical clinical presentation with the exception of a probable episode of myoglobinuria at 2 years of age in association with fever and infection. His sister was asymptomatic. Parents were asymptomatic and nonconsanguineous. On neurological examination, the proband was an athletic, muscular young man with full power and no evidence of neuropathy or pigmentary retinopathy. On investigation several months after an episode of myoglobinuria, he had a normal forearm ischemic lactate test, excluding a defect in glycolysis or glycogenolysis. Muscle biopsy of the biceps was essentially normal. There was no evidence of lipid or glycogen accumulation, mitochondrial abnormalities, or necrosis or regeneration. His EKG and echocardiogram were normal. Fibroblast acylcarnitine profile, following incubation with palmitate, showed a slight

(*Continued*)

CASE EXAMPLE 40.2 (Continued)

accumulation of long-chain acylcarnitines. Enzymatic assay in cultured skin fibroblasts revealed normal CPT I activity but a 20% residual activity for CPT II, consistent with mild CPT II deficiency. Following institution of fatty acid oxidation treatment strategies and preventative measures, he had complete cessation of his episodes of myoglobinuria.

Statistically speaking, the most common cause for recurrent myoglobinuria in children and in adults is CPT II deficiency followed by phosphorylase deficiency (type V glycogenosis or McArdle disease).[92] Exercise intolerance may begin in childhood for both conditions. However, overt episodes of myoglobinuria are usually seen in adolescence. These two disorders can be distinguished by clinical history and biochemical testing.[92] In phosphorylase deficiency, there is a history of early onset muscle cramps during the initial stages of high-intensity isometric exercise (e.g. first 10 minutes) or with less intense but sustained dynamic exercise because carbohydrate is then the major source of energy. If the individual continues to exercise at a milder, low-intensity pace, they will then experience a "second wind" phenomena which will allow them to continue to exercise because of hepatic glycogenolysis providing the muscle with glucose as well as a progressive switch-over in muscle metabolism from a primary dependence on carbohydrate to fatty acid utilization for ATP generation. In classic adult-onset CPT II deficiency, individuals will be able to do short-burst, high-intensity exercise for short periods of time. In contrast, they will most frequently complain of muscle stiffness and myalgia following mild to moderate prolonged exercise (e.g. >1 hour) which may occur several hours after the exercise and which may last for several hours or days if an episode of myoglobinuria is precipitated. In CPT II deficiency, episodes may also be precipitated by fasting, infection, cold exposure, and stress. There is no second wind phenomenon. In phosphorylase deficiency there is often a persistent mild elevation in serum creatine kinase between episodes of myoglobinuria and there may be mild fixed proximal greater than distal weakness in approximately one-third of typical cases more commonly seen in older patients. In classic CPT II deficiency, the CK and power typically return to normal between catabolic crises. The "forearm ischemic lactate test" shows a normal three- to five-fold increase in serum lactate in the first 3 minutes in CPT II deficiency and in normal controls. This is paralleled by a similar increase in ammonia. In individuals with a defect in glycolysis/glycogenolysis, there is an insufficient rise in lactate (less than two-fold) and there will be a compensatory increase in ammonia, which also serves to indicate sufficient effort on the part of the individual. On muscle histochemistry with periodic acid-Schiff staining, there may or may not be focal accumulations of glycogen in the subsarcolemmal regions and between myofibrils in phosphorylase deficiency. The phosphorylase stain will show no staining in muscle fibers, though smooth muscle in the walls of intramuscular vessels will stain normally. In classic CPT II deficiency, the biopsy is usually normal following full recovery from episodes of myoglobinuria. However, mild microvesicular steatosis may be seen on Oil-red-O staining, if the individual is stressed with prolonged exercise or fasting prior to the biopsy.

CASE EXAMPLE 40.3 Carnitine Palmitoyltransferase II Deficiency—Infantile Form

This affected boy was born at 3.4 kg after a full-term uncomplicated pregnancy to healthy parents who had two other living children and a past history of a stillborn child.[104] The parents were first cousins. The infant was well until 3 months of age, at which time, following a 3-day history of fever, he was found to be lethargic with respiratory arrest. He suffered partial and generalized seizures which were attributed to severe hypoglycemia (0.2 mmol/L). He had mild hepatomegaly and cardiomegaly (cardiothoracic index 0.57). He suffered cardiac arrhythmias (PVCs, atrioventricular block, and ventricular tachycardia) a few hours following admission to an intensive care unit. Biochemical analysis showed a plasma glucose of 0.7 mmol/L (3.5–6 mmol/L); no plasma or urinary ketones; arterial pH of 7.42; blood lactate of 2.5 mmol/L (0.6–1.9 mmol/L); serum potassium of 6.5 mmol/L (3.5–5.5 mmol/L); aspartate aminotransferase of 167 IU/L (5–25 IU/L); alanine aminotransferase of 91 IU/L (5–25 IU/L); blood ammonia of 75 μmol/L (20–40 μmol/L); fibrinogen of 1 g/L (2–4 g/L); creatine kinase of 1140 IU/L (15–100 IU/L) with MM isozyme of 1100 IU/L (15–100 IU/L) and MB isozyme of 34 IU/L (0–11 IU/L); lactate dehydrogenase of 870 IU/L (138–276 IU/L) with a normal profile of isozymes; and total plasma carnitine of 14 μmol/L (28–71 μmol/L) with esterified/total carnitine ratio of 80% (<30%). The infant recovered after 6 days of symptomatic therapy. L-carnitine administration restored the plasma carnitine to normal, total carnitine of 45 μmol/L and free of 31.5 μmol/L. Withdrawal of carnitine resulted in a fall in plasma carnitine concentration to 16 μmol/L in 1 week. Urinary organic acids, after cessation of L-carnitine, were normal. The infant subsequently had normal growth and development with mild persistent hepatomegaly. At 17 months of age, the child died suddenly after an overnight fast. The oxidation of palmitate was decreased by ~90% in this child's cultured lymphoblasts and skin fibroblasts. CPT I activity in cultured skin fibroblasts was normal in contrast to the CPT II activity, which was reduced by 90% of that of controls.

CASE EXAMPLE 40.4 Medium-chain Acyl-CoA Dehydrogenase (MCAD) Deficiency

A 30-month old girl presented in coma following 24 hours of fever, persistent vomiting, and progressive lethargy.[7] She had normal vital signs but was unresponsive to deep pain and had mild hepatomegaly. Pertinent laboratory values on arrival included a blood glucose of 4 mg/dL, ammonia of 190 μg/dL, SGOT of 34 IU/dL, and SGPT of 21 IU/dL. A small to moderate amount of ketones was found by urine dipstick and a right middle lobe pneumonia was seen on chest X-ray. Within minutes of administering 20 mL of 50% dextrose, she began to respond to commands. After 12 hours of intravenous dextrose infusion and ampicillin, she had recovered completely. Ten months earlier, this child had been treated for a similar but milder episode of vomiting and lethargy without hypoglycemia or hyperammonemia. Family history was notable for two siblings who had died from acute illnesses that had been diagnosed as Reye syndrome. The first, a boy, had presented at 15 months of age with a seizure following 18 hours of vomiting and increasing lethargy. On arrival in the emergency room, he suffered a cardiac arrest and could not be resuscitated. No blood was drawn; however, a postmortem CSF glucose was 2 mg/dL in the absence of pleocytosis or bacteria. Reye syndrome was diagnosed on the basis of fatty degeneration of the liver at necropsy. The second sibling, a 15-month-old girl, suffered vomiting, hypoglycemia, hypotension, and coma 24 hours after uneventful cardiac surgery for repair of atrial and ventricular septal defects and she died within hours of her deterioration. Autopsy revealed marked fatty infiltration of the liver and a diagnosis of Reye syndrome was made. Another 4-year-old brother was healthy. There was no other family history of similar deaths, hypoglycemia, metabolic disease or consanguinity.

Organic acid analysis in the initial urine specimen of the index case revealed large amounts of saturated and unsaturated dicarboxylic acids, primarily adipic, suberic, and sebacic acids, but comparatively low levels of β-hydroxybutyrate and acetoacetate. Total serum carnitine concentration was decreased at 17 μmol/L (normal 40–60 μmol/L). Medium-chain acyl-CoA dehydrogenase deficiency was confirmed by enzyme assay of the patient's mononuclear cells. Following diagnosis, the patient had two additional episodes of lethargy and vomiting with viral syndromes, each of which was treated with and responded quickly to intravenous 10% dextrose. Growth and development remained normal.

CASE EXAMPLE 40.5 Trifunctional Protein Deficiency—The Myoneuropathic Variant

This 16-year-old-boy was born to asymptomatic first-cousin parents following a normal pregnancy and delivery.[129,176] Gross motor milestones were delayed. At age 20 months, he suffered the first of many episodes of profound weakness precipitated by fever, vomiting, or dehydration. His subsequent clinical course was characterized by a slowly progressive limb-girdle myopathy with mild facial weakness and a symmetrical peripheral sensorimotor axonopathy with secondary demyelination. In addition, he suffered recurrent episodes of myoglobinuria with elevations in serum creatine kinase between 20,000 and 50,000 IU/L (up to 5 times per year). These episodes were precipitated by prolonged exertion, infection, cold exposure, fasting, and/or stress. They were intermittently accompanied by somnolence which responded promptly to intravenous glucose administration. Serum total carnitine concentration was 39 μmol/L (35–83 μmol/L) with a free carnitine of 19 μmol/L (24–63 μmol/L) and increased esterified fraction of 50%. Muscle biopsy revealed a mild lipid-accumulative myopathy with intramitochondrial lipid on electron microscopy. Sural nerve biopsy revealed axonal degeneration. During an acute catabolic episode, serum acylcarnitine assay revealed an elevation in long-chain acylcarnitines (C14–18) with 3-hydroxy species. Urinary organic acids revealed a moderate dicarboxylic aciduria with significant 3-hydroxydicarboxylic aciduria. The L-3-hydroxyacyl-CoA dehydrogenase activity measured with the long-chain substrate (C16) in cultured skin fibroblasts and biopsied muscle was significantly reduced. With the introduction of a frequent feeding, high-carbohydrate, low-fat diet and preventive fatty acid oxidation measures at age 7.5 yr, there was a marked reduction in the frequency of myoglobinuric episodes. However, no improvement in power or endurance occurred, and these continued to deteriorate. A trial of prednisone therapy resulted in a significant improvement in his limb-girdle myopathy, which has been sustained over 12 years, as well as a transient improvement in his peripheral neuropathy. The myoglobinuric episodes were markedly reduced.

DIFFERENTIATING LABORATORY FEATURES

The site of biochemical defect may be suggested by the chain-length specificity and the species type of the fatty acid intermediates detected in the serum or urine of affected children. These intermediates require specialized biomedical technology which is generally not available in standard hospital clinical chemistry laboratories and may therefore require referral to specialized metabolic laboratories. Particularly useful are the assessment of plasma total and free carnitine, serum acylcarnitines, urine acylcarnitines, and urine acylglycines and organic acids. It should be emphasized that these intermediates may be absent at times when the child is metabolically stable and

TABLE 40.2 Laboratory Features Associated with Genetic Defects of Fatty Acid Oxidation

Deficiency	Carnitine		Unique or Specific Metabolites		
	Total	Free	Acyl-Carnitine	Acyl-Glycine	Organic Acids
LCFAUD	Low	Low	NR	NR	NR
OCTN2	Very low	Low	NR	NR	NR
CPT I	Normal/ high	High	NR	NR	NR(+)
TRANS	Low	Very low	Long-chain	NR	NR
CPT II (mild)	Low	Very low	Long-chain	NR	NR
CPT II (severe)	Low	Very low	Long-chain	NR	+
VLCAD/LCAD	Low	Low	Long-chain	NR	A, Su, Se
Trifunctional	Low	Low	Long-chain	NR	A, Su, Se; 3-hydroxy intermediates
Dienoyl-CoA reductase	Low	Low	Decadienoyl	NR	NR
MCAD	Low	Low	Octanoyl	Suberyl, hexanoyl, phenprop	A, Su, Se
SCAD	Low	Low	Butyryl	Butyryl	A, Su, Se; ethylmalonic; methylsuccinate
SCHAD	Low	Low	NR	NR	A, Su, Se; 3-hydroxy intermediates
ETF and ETF/Qo	Low	Low	Octanoyl, glutaryl, butyryl, isovaleryl	Suberyl, hexanoyl, butyryl, isovaleryl	A, Su, Se; glutaric; ethylmalonic
HMG-CoA lyase	Low	Low	3-Methylglutaryl	?	3-Hydroxy-3-methyl-glutaric; 3-methyl-glutaconic; 3-methyl-glutaric, A; 3-hydroxy-isovaleric

LCFAUD, long-chain fatty acid uptake defect; OCTN2, plasmalemmal high-affinity carnitine transporter; CPT, carnitine palmitoyltransferase; TRANS, carnitine acylcarnitine translocase; LCAD, long-chain acyl-CoA dehydrogenase; VLCAD, very-long-chain acyl-CoA dehydrogenase; trifunctional, long-chain enoyl-CoA hydratase + long-chain L-3-hydroxyacyl-CoA dehydrogenase + long-chain 3-ketoacyl-CoA thiolase; MCAD, medium-chain acyl-CoA dehydrogenase; SCAD, short-chain acyl-CoA dehydrogenase; SCHAD, short-chain L-3-hydroxyacyl-CoA dehydrogenase; ETF, electron transfer flavoprotein; Qo, coenzyme Q oxidoreductase; HMG, beta-hydroxy-beta-methylglutaryl; NR, No case yet reported; A, Su, Se, adipic, suberic, sebacic acids; phenprop, phenylpropionyl.
Modified from Hale and Bennett[39] with permission from Elsevier.

receiving adequate supplies of glucose, thereby eliminating the stress on the FAO pathway. They are therefore best detected in the serum and urine of children during acute catabolic crises or during times of fasting. The specific intermediates that are characteristic for the individual known defects are presented in Table 40.2.[39]

Carnitine

Fatty acid oxidation disorders are generally associated with a decrease in plasma total carnitine concentration ($<30\,\mu$mol/L, normal 40 to 60 μmol/L). The lowest concentrations are found in the carnitine transporter defect where carnitine values are usually $<5\%$ of control.[88] In the intramitochondrial β-oxidation defects, the plasma total carnitine concentrations vary between 10% and 50% of normal controls. In contrast, the total plasma carnitine concentration may be normal or increased in a child with CPT I deficiency in which the esterification of palmitate to carnitine is defective.[177]

In most FAO defects, with the exception of the OCTN2 and CPT I defects, there is an increase in the ratio of esterified carnitine to total carnitine. This reflects the excessive acyl-CoAs, accumulating proximal to the block in β-oxidation, which become esterified to carnitine. Estimates of the amount of acylcarnitines are based on the difference between the free and total carnitine measurements. Under normal conditions, the esterified carnitine fraction is 10% to 25% of total in the fed state and 30% to 50% of total in the fasted state.[39] When filtered through the kidneys, the increased serum acylcarnitines will compete with free carnitine at the renal tubular reabsorption site. Because the increasingly longer chain-length acylcarnitines have an increasingly higher affinity for the OCTN2 carnitine transporter than free carnitine,[88] free carnitine will be excreted leading to a decrease both in serum free and total carnitine concentrations and an increase in the serum esterified carnitine fraction.

The carnitine esters can be further separated on the basis of the acid insolubility of long-chain acylcarnitine esters.[39] Defects involving long-chain fatty acid oxidation have increased amounts of long-chain acylcarnitines.[173] Further separation and identification of the individual acylcarnitine esters has been facilitated by advances in biomedical technology including the application of fast atom bombardment-tandem mass spectrometry and isotopic exchange high-performance liquid chromatography.[178,179] The identification of specific carnitine esters has been useful in the diagnosis of certain defects (e.g. octanoyl-carnitine in MCAD deficiency).[56] The analysis of serum acylcarnitines may therefore be helpful in screening patients for FAO defects. They are particularly helpful in the diagnosis of long-chain fatty acid oxidation disorders, because they overcome the problem of the renal threshold effect whereby long-chain acylcarnitines are selectively reabsorbed at the renal carnitine transporter site leading to a loss of free carnitine in the urine. They also overcome the problem of the poor solubility of long-chain fatty acids in urine. A novel approach that has been successfully applied to skin fibroblasts for the diagnosis of FAO defects involves probing the β-oxidation pathway by incubating the cells in the presence of a stable isotopically labeled long-chain fatty acid and then analyzing the produced acylcarnitine intermediates by tandem mass spectrometry.[180,181] This method has also been used for successful prenatal diagnosis of MCAD and VLCAD deficiencies in cultured amniocytes.[182]

Electrospray ionization-tandem mass spectrometry (ESI-MS/MS) is increasingly used in newborn screening programs from dried blood spots which can be used to detect FAO disorders, carnitine cycle disorders, and organic acidurias. Stored dried blood spots are also a valuable source for postmortem investigations to elucidate the cause of SIDS. In a systematic study of the stability of acylcarnitines and free carnitine during storage, it was found that at −18°C, acylcarnitines are stable for at least 330 days.[183] If stored for prolonged periods at room temperature (>14 days), acylcarnitines are hydrolyzed to free carnitine and the corresponding fatty acids. The velocity of decay is logarithmic and depends on the acylcarnitine chain length. Short-chain acylcarnitines hydrolyze faster than long-chain acylcarnitines. It was concluded that stored filter cards should only be used for retrospective quantitation of acylcarnitines if appropriate correction for sample decay during storage is applied. Free carnitine was found to increase on storage but could be reliably quantitated under standardized derivatization conditions. OCTN2 deficiency could be reliably diagnosed by examining acylcarnitine profiles.[183]

A novel tandem mass spectrometry method for rapid confirmation of medium- and very-long-chain acyl-CoA dehydrogenase deficiency in newborns has been developed using liquid chromatography tandem mass spectrometry (LC-MS/MS) to measure MCAD- or VLCAD-catalyzed production of enoyl-CoA and hydroxyacyl-CoA in lymphocytes.[184]

Dicarboxylic Acids

Dicarboxylic acids (DCAs) (e.g. adipic, suberic, sebacic acids) are found in all identified intramitochondrial β-oxidation defects.[185] Hale and Bennett[39] point out several limitations to the value of these compounds in the recognition of FAO defects. (i) These DCAs may be seen in children receiving certain formulas containing medium-chain triglycerides or in children who are seriously ill (e.g. with diabetic ketoacidosis)[186] or who are receiving certain medications which interfere with FAO such as valproic acid.[187] It should, however, be noted that in each of the above conditions, the amount of ketone exceeds the amount of DCA whereas in the intramitochondrial FAO defects, the amount of DCA equals or exceeds the amount of ketones when the children are fasting. Furthermore, the pattern of the DCAs is usually easily distinguished, in experienced metabolic laboratories, from the pattern of the pathological DCAs that is associated with FAO defects. (ii) These DCAs are not present when children are not in a catabolic state and are well and eating regularly or are receiving intravenous glucose at rates in excess of normal hepatic glucose production rates, thereby decreasing the dependence on FAO and the production of fatty acid metabolites. (iii) Increased concentrations of DCA in the urine are generally not seen in the disorders involving the transport of fats into the mitochondria. Therefore, an FAO defect can be suspected in the presence of an excess of DCA relative to ketones. However, the absence of DCAs does not necessarily exclude a defect.

The site of defect may be suggested by the organic acid pattern. For example, children with LCAD/LVCAD deficiency excrete primarily medium- and long-chain saturated DCAs in contrast to children with trifunctional protein deficiency, who excrete almost equimolar amounts of the saturated and the 3-hydroxydicarboxylic acids.[188] However, it should be noted that the presence of 3-hydroxy compounds is not specific for trifunctional protein deficiency as they may also be seen in toxic reactions with acetaminophen and with intrinsic liver disease.[189]

Advances in stable-isotope dilution mass spectrometry have improved the ability to detect metabolites in very small quantities in plasma or urine.[190] Acylglycines that are consistently excreted in small quantities in the urine do not appear to have the same limitations as DCAs. Useful glycine metabolites have been identified for several defects including MCAD, SCAD, ETF, and ETF-coenzyme Q oxidoreductase deficiencies. There are increased concentrations of hexanoyl-, suberyl-, and 3-phenyl propionyl glycine in the urine in MCAD deficiency.[191]

DIAGNOSTIC APPROACHES AND SCREENING METHODS

History and Physical Examination

A careful history and clinical examination remain the key to investigation of these patients. The presentation may be either acute and recurrent or more chronic and slowly progressive.[39] The more typical presentation is the acute one in which the child has a history of decreased oral intake during the preceding 24 to 36 hours followed by increasing lethargy and progressive obtundation or coma. A positive family history for SIDS, "near miss" SIDS or Reye syndrome should increase the level of suspicion. Physical examination may reveal myopathy, hypotonia, hepatomegaly, or cardiomyopathy. The initial key investigations in a comatose child should include serum glucose and urine ketone measurements.[39] Determination of urine ketones may be complicated because ill children are often dehydrated and will therefore have concentrated urine. If the blood glucose is above 3.3 mmol/L (60 mg/dL) and if it is accompanied by large amounts of urinary ketones, this tends to rule out an FAO disorder. However, if the blood glucose is <3.3 mmol/L and the urine ketones are trace or small in amount, this would suggest the possibility of an FAO disorder and warrants further investigation. Most importantly and particularly prior to intravenous glucose therapy, samples from the acute presentation should be saved for the determination of serum carnitine total and free, serum acylcarnitines, serum free fatty acids and ketones and urine organic acids, acylglycines and acylcarnitines. Ordinarily, the serum free fatty acid to ketone ratio is 1:1. In the event of a block in FAO, this ratio increases to >2:1 and is therefore a useful initial screen. The serum and urine specimens during the acute episode can also be used to assess the integrity and hormonal regulation of the biochemical pathways involved in glucose homeostasis.

Total Carnitine Measurement

A total plasma carnitine concentration of <30 μmol/L suggests an FAO disorder, but is not diagnostic as there are a number of other genetic and acquired disorders with a secondary carnitine deficiency as previously discussed (Table 40.1).[69] A serum carnitine of 10–50% of normal combined with an increase in the esterified fraction of carnitine (e.g. normal 10–25% fed state; 30–50% fasted state), would suggest an intramitochondrial β-oxidation defect with secondary carnitine deficiency. A total serum carnitine concentration of <5% of normal with a normal esterified fraction, as well as evidence of a progressive cardiomyopathy and myopathy, would suggest a carnitine uptake OCTN2 defect.[13,88] The exception to the rule is CPT I deficiency which presents with recurrent episodes of hypoketotic hypoglycemic encephalopathy or Reye-like syndrome and an increased serum carnitine with a decreased esterified fraction due to the block in palmitate esterification to carnitine.

Urinary Organic Acids

It is critical to obtain urine specimens during the acute catabolic episode prior to intravenous glucose therapy, as the production of fatty acid metabolites diminishes rapidly once normoglycemia is restored. The chain length and species-type of the organic acids may help to identify the specific site of FAO block.

Fasting Studies

A fasting study may be considered for the purposes of distinguishing an FAO defect from other causes of hypoglycemia, if the important diagnostic samples have not been taken during an acute catabolic event. However, it must be strongly emphasized that if a fasting study is to be undertaken, it must be done under very carefully controlled hospital conditions with continuous monitoring and by physicians who are knowledgeable with respect to hypoglycemia, hypopituitarism, hyperinsulinism, and FAO disorders. It is felt by some authorities that fasting studies should *not* be performed in children with FAO disorders, because diagnostic fasting may precipitate an acute metabolic crisis, leading to further morbidity or death. They suggest, instead, that loading tests with carnitine or phenlypropionate can be used to aid in diagnosis. As suggested by Hale and Bennett,[39] there are pros and cons with each method. The advantages of a fasting study are: (1) the duration of fasting tolerance can be determined under carefully controlled conditions which may provide useful information regarding the long-term management of the affected patient and guidelines for prevention of acute catabolic episodes; (2) the full spectrum of abnormal fasting adaptation can be studied through assessment of a number of laboratory parameters including hormonal measurements; (3) the time to precipitation of acute clinical decompensation can be documented. As previously emphasized the cardinal risk is the precipitation of an acute catabolic crisis leading to morbidity and death. It would instead be preferable to collect as much information as possible through the collection of the appropriate samples, during an acute catabolic event that the child has suffered.

The primary advantage of loading tests and the measurement of specific metabolites is their safety. The disadvantages are: (1) they are useful in only certain FAO defects, e.g. MCAD deficiency, and therefore a negative test does not exclude all FAO defects; (2) they do not evaluate the spectrum of fasting adaptation in the individual.

The purpose of the fasting study is to identify the defective metabolic pathway through an analysis of temporal changes in substrates (glucose, free fatty acids, lactate, ketones), metabolites (carnitine, dicarboxylic acids), and relevant hormones (growth hormone, cortisol, insulin). The fast should be terminated when the children demonstrate their first symptoms or have a blood glucose of <3.3 mmol/L (60 mg/dL) or demonstrate a moderate ketonuria. If at this point there is a deficient ketogenic response in the face of a significant dicarboxylic aciduria and a serum free fatty acid:ketone ratio of >2:1, there is strong presumptive evidence for a defect in FAO.

Other Studies

Once there is presumptive evidence for an FAO defect, the clinical picture in combination with serum acylcarnitines, and urinary organic acids and acylglycines may suggest a specific site of defect as well as the chain-length specificity of the defect (e.g. short-, medium-, or long-chain). Further investigations to identify the site of defect are discussed in the following section.

Fatty Acid Oxidation Studies

Provided the defect is expressed in cultured skin fibroblasts, a further useful screening tool would be the measurement of the oxidation rates of [1-^{14}C]-labelled palmitate (C16), octanoate (C8), and butyrate (C4) and deuterium-labeled oleic acid in the fibroblasts to establish the chain-length specificity of the defect.[35,192] Another useful screening test to suggest an underlying defect in FAO is to culture skin fibroblasts in a glucose-free medium, which will simulate fasting and exacerbate microvesicular steatosis in the fibroblasts expressing a genetic defect in FAO.[161] Another method for quantitative acylcarnitine profiling by electrospray ionization-tandem mass spectrometry (ESI-MS-MS) in human skin fibroblasts using unlabelled palmitic acid as substrate has been developed, which has revealed pathognomonic acylcarnitine profiles in a variety of short-, medium-, and long-chain FAO defects.[193] In addition, this method delineates different variants of MCAD deficiency, e.g. mild and classical. A rapid method for acylcarnitine profiling using stable isotopically labeled palmitate and intact mitochondria prepared from fresh biopsied muscle homogenates, which reduces the delay in diagnosis related to tissue culture, has also been described.[194]

Enzymatic Assays

Depending upon the suspected site of defect, a direct enzymatic assay may then be performed for the specific enzyme (Table 40.3).[17,18,33,34,37,38,43–45,109,195–233]

These assays can be performed in cultured skin fibroblasts or in biopsied muscle specimens for CPT I, CACT, CPT II, SCAD, VLCAD, SCHAD, and LCHAD deficiencies. However, it must be emphasized that certain of these defects may be tissue specific. For example, in muscle SCHAD deficiency the defect is expressed in biopsied muscle but not in skin fibroblasts and therefore may be missed unless the affected tissue is examined.[15] The carnitine acylcarnitine translocase[234] and MCAD deficiencies can be measured in cultured skin fibroblasts. Evidence for a defect in ETF or ETF-CoQ relies on a demonstration of a combined deficiency in SCAD, MCAD, LCAD and VLCAD activities. CPT II deficiency has been rapidly diagnosed prenatally by enzyme assay in 10 mg of chorionic villus sampling at the 11th week of gestation in combination with haplotyping using polymorphic markers linked to the CPT II gene.[235]

Uptake Studies

For the carnitine transporter defect, diagnosis is confirmed by *in vitro* studies of carnitine uptake in cultured skin fibroblasts which demonstrate negligible uptake of carnitine in the affected patients, throughout the physiological range (0.1 to 50 μmol/L) of carnitine substrate concentrations (e.g. 2% of uptake at 5 μmol/L of carnitine), thereby precluding the calculation of K_m and V_{max} values.[13,88] This supports the concept that primary carnitine deficiency is due to a defect in the plasmalemmal high-affinity carnitine transporter OCNT2.[236] In one study of heterozygotes, carnitine uptake studies demonstrate normal K_m but reduced V_{max} values of 13% to 44% of control[13] suggesting a decreased number of normally functioning transporters. This is the most sensitive study for detection of the carrier state as serum carnitine in heterozygotes may be normal. Negligible carnitine uptake has also been demonstrated in cultured lymphoblasts[237] and myoblasts[238] of affected children.

TABLE 40.3 Specific Enzymatic and Molecular Characterization of Fatty Acid Oxidation Disorders

Deficiency	Assay	Gene/Chromosome Location	Protein and Molecular Characterization
CUD (OCTN2 transporter)	Carnitine uptake in cultured skin fibroblasts, lymphoblasts, myoblasts	SLC22A5 5q31.1	Wu et al.[195]; Tamai et al.[196]; Lamhonwah and Tein[197]
CPT I, liver	F, M, L	CPT1A 11q13	Esser et al.[198]; Britton et al.[199]
TRANS (CACT)	F	SLC25A20 3p21.31	Indiveri et al.[200,201]; Viggiano et al.[202]
CPT II	F, M, L	CPT2 1p32.3	Woeltje et al.[203]; Finocchiaro et al.[204]; Taroni et al.[205]; Gellera et al.[206]; Verderio et al.[207]
VLCAD	F, M	ACADVL 17p13.1-11.2-	Kelley[33]; Izai et al.[34]; Strauss et al.[208]; Aoyama et al.[109]; Aoyama et al.[209]; Andresen et al.[210]; Andresen et al.[211]
LCAD	F, M, L	ACADL 2q34-q35	Indo et al.[212] Matsubarn et al.[232]
Trifunctional / LCHAD	F, M, L	HADHB/ HADHA 2p23.3	Carpenter et al.[37]; Luo et al.[38]; Uchida et al.[213]; Jackson et al.[44]; Kamijo et al.[43,45]; Sims et al.[214]; Strauss[215]
Dienoyl CoA reductase	M, L	DECR1 8q21.3	Helander et al.[216]
MCAD	F, L, W	ACADM 1p31.1	Kelly et al.[217]; Strauss et al.[218]; Gregersen et al.[18]; Blakemore et al.[17]
SCAD	F, M, L	ACADS 12q24.31	Finocchiaro et al.[219]; Naito et al.[220]; Corydon et al.[221]; Gregersen et al.[222]
SCHAD	F, M, (L)	HADHSC or HADH 4q25	Vredendaal et al.[223]
ETF	F	ETFA 15q24.2-24.3 ETFB 19q13.41	alpha-subunit: Finocchiaro et al.[224]; Indo et al.[225] beta-subunit: Finocchiaro et al.[226,233]
ETF/Qo	F	ETFDH 4q32.1	Goodman et al.[227]
HMG-CoA lyase	F, L, W	HMGCL 1p36.11	Mitchell et al.[228]; Roberts et al.[229]
HMG-CoA synthase	L	HMGCS2 1p13-p12	Bouchard et al.[230]; Mascaro et al. 1995[231]

LCFAUD, long-chain fatty acid uptake defect; CUD, Carnitine uptake defect; CPT, carnitine palmitoyltransferase; TRANS, carnitine acylcarnitine translocase; LCAD, long-chain acyl-CoA dehydrogenase; VLCAD, very-long-chain acyl-CoA dehydrogenase; trifunctional, long-chain enoyl-CoA hydratase + long-chain L-3-hydroxyacyl-CoA dehydrogenase + long-chain 3-ketoacyl-CoA thiolase; MCAD, medium-chain acyl-CoA dehydrogenase; SCAD, short-chain acyl-CoA dehydrogenase; SCHAD, short-chain L-3-hydroxyacyl-CoA dehydrogenase; ETF, electron transfer flavoprotein; Qo, coenzyme Q oxidoreductase; HMG, beta-hydroxy-beta-methylglutaryl; F, fibroblasts, L, liver, M, muscle, W, leukocytes; ND, not done.
Modified and updated from Tein I.[93]

Molecular Studies

Further molecular characterization of the specific defects include Western blotting to determine whether the defects are cross-reacting material positive suggesting a kinetic deficiency or whether they are cross-reacting material negative suggesting decreased production of the affected enzyme. Western blotting has also been used in the determination of the amounts of the alpha and beta subunits of ETF.[239] A number of the genes for the enzymes (e.g. CPT I, CPT II, SCAD, MCAD, VLCAD, SCHAD, TFP) and transporters (OCTN2) have now been cloned (Table 40.3).[17,18,33,34,37,38,43–45,109,200–233] This has led to the discovery of specific mutations resulting in defective enzyme activity, which has given rise to the development of specific molecular probes which can now be used for the precise and rapid detection of certain specific defects in the population. An example would be the gene mutation found in more than 70% of children with MCAD deficiency (A985G point mutation) in the UK population[17] which has facilitated determination of the relative frequency of this mutation in this population. Specific mutations have been documented for CPT II,[205,207] VLCAD,[208,210,211,240] MCAD,[17,18] SCAD,[221,222,241] trifunctional protein,[214,215] and OCTN2 transporter defects,[197,242] among others. The types of mutations may include missense mutations, small amino acid deletions and insertions, truncating mutations, nonsense mutations, out of frame deletions and insertions, and splice site mutations.[243] In SCAD deficiency, ethylmalonic aciduria (EMA) is a biochemical marker though it may be seen in asymptomatic individuals. Two common gene variations 625G > A (G185S) and 5llC > T (R147W) have been identified in the SCAD gene from patients with verified SCAD deficiency but were also found in 14% of the general population, either in homozygous or double heterozygous form.[151] Of the patients with EMA aciduria, 60% were homozygous for 625A, 1% were homozygous for 511T and 8% were double heterozygous with 625A on one allele and 511T on the other. This significant overrepresentation of the variations in SCAD deficient patients versus controls suggests an association and a causal relationship between EMA and the SCAD gene variations. Studies of these variant proteins have indicated that they contribute toward misfolding of the SCAD protein, particularly in the presence of elevated temperature and though not disease-causing themselves, they are susceptibility variations, which together with other genetic, cellular, biochemical or environmental factors, such as fever and starvation, may result in functional SCAD deficiency.[244] The prevalent mutation in CPT II deficiency is a missense mutation S113L, due to a 439C > T substitution[205] which appears to be mild and manifests with adult onset myoglobinuria. Its prevalence in diagnosed cases ranges from 50% to 80% in the USA and Europe.[205,245,246] The most straightforward application of mutation analysis is to confirm diagnosis in suspected patients, particularly in family studies and for prenatal/ preimplantation analysis.

TREATMENT

Due to the small numbers of patients in any given institution, it is difficult to systematically evaluate in a prospective, randomized, double-blind control study any single treatment regimen or to compare different regimens. Recently, treatment recommendations in long-chain FAO defects were made by a consensus panel of 21 FAO investigators based on observations in 75 patients with long-chain FAO defects from 18 metabolic centers in Central Europe.[247] Overall, the mainstay of therapy in FAO disorders is the avoidance of precipitating factors, particularly prolonged fasting. General treatment strategies are discussed in the next sections.

The Avoidance of Precipitating Factors

Precipating factors, such as prolonged fasting, prolonged aerobic exercise >30 minutes, and cold exposure leading to shivering thermogenesis should be avoided. The avoidance of prolonged fasting is key. Suggested maximal fasting periods (during stable metabolic conditions) from a European consensus panel are as follows: neonates→3 hours, also at night; <6 months→4 hours, also at night; >6 months→4 hours at daytime, 6 (−8) hours at night in well-nourished infants; >12 months to 3 years→4 hours at daytime, 10 (−12) hours at night; 4 to 7 years→4 hours during day, 10 (−12) hours at night.[247] These fasting periods apply under healthy, steady-state conditions, when under a normocaloric diet glycogen stores can be built up during the day and used for glucose production during longer fasting intervals at night. In the event of progressive lethargy or obtundation or an inability to take oral feedings due to vomiting, the child should be taken immediately to the emergency room for intravenous glucose therapy. Intravenous glucose should be provided at rates sufficient to prevent fatty acid mobilization (8 to 10 mg/kg/min glucose which, in a young child ≤15 kg, is usually 10% dextrose at one and one-half times maintenance rates, and needs to be adjusted according to weight in older individuals to achieve this same rate of glucose infusion).[39] This should be continued until the child is again able to take oral feedings and the catabolic cascade has been reversed. It is prudent to institute intravenous glucose therapy as soon as there is a concern regarding the status of the child and not to wait until the occurrence of severe symptoms or a drop in blood glucose <3.3 mmol/L, as a delay may lead to serious morbidity or death. Similarly it is wise to avoid prolonged exercise >30 minutes at any given time, as this would be the time period in which there would be increased fat mobilization and increasing dependence upon FAO for ATP production. It is advisable to recommend a high carbohydrate load prior to exercise and a rest period with a repeat carbohydrate load at 15 minutes. The entire period of exercise should not exceed >30 minutes at one given time. It is also important to avoid cold exposure that may lead to shivering thermogenesis, which would again stimulate fatty acid mobilization.

High-carbohydrate, Low-fat Diet

In general, it is advisable to institute a high-carbohydrate, low-fat diet with frequent feedings throughout the day, which would be commensurate with the nutritional needs of the child given his or her age. This goal is best achieved with the aid of a dietitian, aiming toward approximately 60% of calories from carbohydrate sources, 15% from protein, and approximately 25% to 30% from fat. Long-chain fat intake in healthy infants is 40–45% of total energy and in children of school age 30–35%, according to German/Austrian/Swiss Nutrition Societies 2000.[247] Monitoring of essential fatty acid (EFA) levels is important to ensure that the child is receiving adequate EFA, which may require supplementation. In long-chain FAO defects, dietary long-chain fat should be restricted. Augmentation with EFA (at 1–2% of total energy intake) is used to reduce the risk of EFA deficiency.[247–250] Flaxseed, canola, walnut, soy, wheatgerm, or safflower oils can be used. An older child should have three regular meals per day with three equidistantly placed intermeal snacks, including a bedtime snack. In younger children, oral or nasogastric tube administration of an appropriate formula is indicated. In HMG-CoA lyase deficiency a high-carbohydrate, low-fat, low-protein diet with leucine restriction should be implemented.[8]

In symptomatic VLCAD deficiency, long-chain fat content of the diet is recommended to be 25–30% of total energy.[247] Since the predominant clinical symptoms in VLCAD deficient patients are episodic muscle pain and myoglobinuria suggesting muscular energy deficiency, it is recommended that the diet be enriched in medium-chain triglycerides (MCT) to provide 20% of total energy from MCT. Increased MCT (or carbohydrate) prior to more extensive exercise is recommended (e.g. 0.25–0.5 g MCT/kg).

For patients with trifunctional protein deficiency, it is recommended that long-chain fat intake should be as low as possible in both asymptomatic and symptomatic patients.[247] In newborns, a special infant formula low in LCT and high in MCT is important with supplementation of EFA. In view of the high mortality rate in TFP deficiency, dietary treatment in patients detected by newborn screening should be initiated immediately after screening results are available, even before diagnostic confirmation. With institution of solid food, it is suggested that fat content be 25–30% of total energy, with 20–25% as MCT and 5–10% as LCT.

Uncooked Corn Starch

Routine enrichment of the diet with glucose polymers or cornstarch is not recommended, but is part of oral prophylactic and therapeutic emergency management.[247] To delay the onset of fasting overnight, the nightly institution of uncooked corn starch, in doses similar to those used in the treatment of glycogen storage disease (1 to 2 g/kg body weight/day as a single nighttime dose), will prolong the postabsorptive state and delay fasting.[115] Cornstarch provides a sustained release source of glucose, thereby preventing hypoglycemia and lipolysis.[251] Cornstarch is usually initiated at 8 months of age when pancreatic enzymes are first able to function at full capacity for appropriate absorption.[252] The initial recommended doses are generally 1.0 gm/kg/day which can be gradually increased to 1.5 to 2.0 gm/kg/day by age 2 years as needed.[251] However, this may lead to undesirable weight gain.

Specific Measures for Individual Fatty Acid Oxidation Disorders

Carnitine

There is one clear indication for carnitine therapy and this is the carnitine transporter OCTN2 defect resulting from mutations in the *SLC22A5* gene. The carnitine uptake defect is characterized by carnitine-responsive cardiomyopathy with or without weakness, hypoglycemic hypoketotic encephalopathy, and failure to thrive, in which there are very low plasma and tissue concentrations of carnitine (generally <5% of normal), lipid storage in muscle, heart, and liver, and severe renal leak of carnitine with absence of an abnormal dicarboxylicaciduria.[13,88] In this condition there was a dramatic improvement in the cardiomyopathy and myopathy of all 22 patients treated with high dose oral carnitine supplementation within a few weeks and a reduction of heart size toward normal within a few months of therapy. There was reversal of the limb girdle lipid storage myopathy in one patient who underwent muscle biopsy pre- and post-therapy, despite only partial restoration of muscle carnitine concentrations.[13] In addition, three children, with significant failure to thrive prior to therapy, demonstrated a marked improvement in growth after therapy.[13] The impairment in fasting ketogenesis was corrected with carnitine therapy in one patient tested.[236] Eighteen of 19 patients treated with carnitine therapy for periods of 1 to 10 years continue to be healthy.[88] One of these 19 patients was moderately impaired with weakness and cardiomyopathy at age 20, which was thought to be due to poor compliance with the carnitine therapy. Thus, in the carnitine uptake defect, high dose oral L-carnitine supplementation in a dose of 100 mg/kg/day in four divided daily doses is critical and life-saving, significantly reversing the pathology in this otherwise progressive and lethal disease. Some sources recommend up to 300 mg/kg/day of oral carnitine for chronic treatment and intravenous therapy of 100–400 mg/kg/day of carnitine during life-threatening events.[253] Furthermore, if a child is prospectively diagnosed from birth, early carnitine therapy from birth has been shown to prevent development of the clinical phenotype.[242]

In the intramitochondrial β-oxidation defects with secondary carnitine deficiency, the results of carnitine therapy have been highly variable and insufficiently evaluated. The rationale for carnitine supplementation has been to limit the intracellular concentrations of potentially toxic acyl-CoA intermediates within the cell through transesterification and to thereby liberate CoA which is a critical intracellular cofactor.[50] However, there has been no objective prospective study to prove that carnitine administration has a beneficial effect. Stanley et al.[14] studied one patient with MCAD deficiency and demonstrated that the associated carnitine deficiency was not the cause of the defect in FAO, as shown by the lack of effect of carnitine replacement on fasting ketogenesis. After 3 months of oral carnitine therapy, this patient had no increase in plasma ketones and still became ill and hypoglycemic after 14 hours of fasting. Moreover, there is increasing evidence to suggest that in long-chain FAO disorders, carnitine administration may have deleterious effects. In these disorders, there is an accumulation of long-chain acyl-CoAs proximal to the metabolic block, which upon esterification become long-chain acylcarnitines. The long-chain acylcarnitines may have detergent-like actions on muscle membranes. Excessive palmitoyl-CoA and palmitoylcarnitine were shown to have detergent properties on isolated canine myocytic sarcolemmal membranes and to potentiate free radical induced lipid membrane peroxidative injury in ischemia.[83] Long-chain acylcarnitines have also been shown to activate Ca^{2+} channels in cardiac[84,85] and in smooth muscle myocytes[85,86] and thus may potentiate the increase in cytosolic Ca^{2+} associated with arrhythmogenesis as seen in ischemic myocardium.[87]

Medium-chain Triglyceride Oil

MCT oil as a nutritional source would be useful in long-chain FAO disorders as medium-chain fatty acids would circumvent the block in long-chain FAO and facilitate ATP production from the remainder of the patent FAO pathway. MCT oil could be started at a dose of 0.5 gm/kg/day divided in 3 daily doses and could be increased up to 1 or 1.5 gm/kg/day as tolerated. The major side effect is diarrhea. The usefulness of this approach may be limited as excess MCT would ultimately be stored as long-chain fats in adipocytes, which could not be metabolized and would increase the long-chain fat load. The efficacy of this approach, as documented in the literature, has been variable in TFP/LCHAD deficiency ranging from highly beneficial to no significant response and may depend upon the individual case.[129] There have been several reports of successful treatment of cardiomyopathy in VLCAD deficiency.[254,255] When a high percentage of energy from fat is provided by MCT oil, patients are at risk for EFA deficiency and they should be supplemented with EFA (1–2% total energy intake).[250]

Riboflavin

Certain cases of the multiple acyl-CoA dehydrogenase deficiencies (e.g. ETF or ETF-CoQ-linked deficiencies) are responsive to riboflavin supplementation.[256] This may be due to a stabilizing pharmacologic chaperone effect of the riboflavin on a mutant flavin-dependent enzyme, improving ETF conformation and yielding a more stable and active enzyme.[257] The dosages in infants and young children would be approximately 50 mg three times daily and in older children 100 mg three times daily. For the riboflavin responsive-MAD deficiency, the challenge is to find the connection between variations in the ETFDH gene and the observed deficiency of a number of different mitochondrial dehydrogenases as well as deficiency of FAD and coenzyme Q10 (CoQ10).[155] The isolated myopathic form of CoQ10 deficiency is caused by mutations in the ETF dehydrogenase gene and can be treated with CoQ10 and riboflavin.[158]

Specific Therapies for LCHAD/TFP Deficiency

Oral prednisone has been shown to lead to a dramatic reversal of the limb-girdle myopathy and marked reduction in the episodic myoglobinuria in one boy with the myoneuropathic form of LCHAD deficiency.[129] Several children with LCHAD deficiency who had asssociated pigmentary retinopathy were shown to have a deficiency of the (n-3)-polyunsaturated fatty acid docosahexaenoic acid (DHA) and subsequent supplementation with DHA led to some improvement in visual function.[258] Furthermore, the daily oral administration of a cod liver oil extract containing high amounts of DHA led to a marked clinical and electrophysiological recovery of the progressive peripheral sensorimotor axonopathy in one boy with the myoneuropathic form of LCHAD deficiency.[259,260] The recommended dose of DHA supplementation in TFP-complex including LCHAD deficiency by a European consensus panel is 60 mg/day in children weighing <20 kg and 120 mg per day in children >20 kg weight.[247] Higher amounts of DHA were used to reverse the neuropathy.[260]

Triheptanoin

Use of the anaplerotic odd-chain triglyceride, triheptanoic acid, has been reported to be of value in the therapy of long-chain FAO defects through its production of acetyl-CoA for ketone body production and propionyl-CoA as an efficient anaplerotic substrate for the citric acid cycle.[261] In three patients with VLCAD deficiency fed controlled diets in which the fat component was switched from medium-even-chain triglycerides to triheptanoin, this treatment led rapidly to clinical improvement including the resolution of chronic cardiomyopathy, muscle weakness, and myoglobinuria for more than 2 years in one child and of myoglobinuria and weakness in the others. Similarly, in 7 patients with CPT II deficiency who were studied from 7 to 61 months on triheptanoin, there was an elimination of episodic myoglobinuria and a return to normal physical activities including strenuous sports.[262]

Peroxisome Proliferators Activated Receptor (PPAR) Agonists

The PPARs have been shown to upregulate mitochondrial FAO.[263,264] Three different isoforms of these nuclear receptors have been described in mammalian cells, namely PPARalpha, PPARdelta, and PPARgamma, which exhibit distinct tissue distributions, ligand specificities, and physiological functions.[263] Bezafibrate, which is a hypolipidemic agent, is the only fibrate capable of activating the PPARdelta isoform together with PPARalpha.[265,266] It was found to increase palmitate oxidation in fibroblasts from patients with a typical mild myopathic phenotype of CPT II or VLCAD deficiency, but had no effect in cells from severely affected patients in whom the mutations were severe and likely involved residues that were essential for the architecture of the active site.[267] One of the reported adverse effects that should be monitored is possible drug-induced increase in serum CK; however, reports of bezafibrate-induced myoglobinuria are rare and have only been reported in patients with renal insufficiency who tend to accumulate the drug.[268] Djouadi and Bastin[267] recommend that FAO tests in patient fibroblasts may offer a reliable way to assess responsiveness to the drug and with the combination of clinical, biochemical, and molecular analysis, should provide a rational framework for patient stratification in any future clinical bezafibrate trials in CPT II or VLCAD-deficient patients. Bonnefont et al.[269,270] evaluated the efficacy of a 6-month pilot trial of bezafibrate therapy in six adults with mild CPT II deficiency and found a significant increase in muscle palmitoyl L-carnitine oxidation rates and CPT II mRNA expression and a reduction in episodes of myoglobinuria, muscle pain, and physical limitation. Bezafibrate incubated with cultured skin fibroblasts from various FAO disorders including glutaric aciduria type II (GA II), VLCAD, MCAD, CPT II, CACT, and TFP deficiencies, followed by an *in vitro* probe acylcarnitine assay, demonstrated a normalization of the excessive elevation of long- or medium-chain acylcarnitines characteristic for each disorder.[271] Daily bezafibrate in a 2-year-old boy with GA II led to a marked improvement in motor and cognitive skills, with a sustained reduction of C4, C8, C10, and C12 acylcarnitines in blood and normalization of the urine organic acid profile with no adverse events.[271] The practical applications of bezafibrate require future clinical trials. Activation of PPARalpha by clofibrate has been shown to upregulate OCTN2, leading to an increase in intracellular carnitine in animal models.[272,273]

Gene Therapy Approach for VLCAD Deficiency

A gene therapy approach was used by Merritt et al.,[274] who developed a construct containing human VLCAD cDNA under the control of a CMV promoter and demonstrated expression of hVLCAD *in vitro* in transfected human VLCAD deficient fibroblasts which was accompanied by restoration of enzyme activity and correction of the metabolic block *in vitro*.

On a chronic basis, clinical monitoring of individuals with FAO disorders is very important with evaluation of growth, development, neurological exam, and diet and to reinforce the prevention of risk factors and management of acute crises.[275] Yearly ophthalmologic exams are important for long-chain FAO disorders and GA type II. Also for long-chain disorders and OCTN2 defects, EKG and echocardiography are important to monitor cardiac function. Liver ultrasound may be important in LCHAD/TFP deficiency. Biochemical follow-up should include serum carnitine total and free and serum acylcarnitines to monitor metabolic control, and CK.[275] EFA and DHA status should be monitored in long-chain defects. Regular follow-up and education of the child and family is important throughout life to prevent morbidity and mortality.

GENETICS AND PRESYMPTOMATIC RECOGNITION

All known FAO disorders are inherited as autosomal recessive conditions. Thus, both parents of children born with these disorders are obligate carriers of the mutant gene and have a 25% chance of bearing a normal child, 50% chance of bearing a carrier, and a 25% chance of bearing an affected child who manifests the disease. Once an individual has been identified in a given family, this will have significant implications for the other siblings and their management. Screening of affected siblings is important as treatment with preventative measures is relatively simple and without treatment there is significant morbidity and mortality. Although screening of family members, following identification of an affected individual, is of benefit to the family, this approach fails to identify individuals who do not belong to this high-risk group. One approach to screening would be to evaluate populations that are considered to be at risk for an FAO defect such as SIDS, "near miss" SIDS, and Reye syndrome. In a study of more than 100 confirmed cases of MCAD deficiency, almost 20% were originally misdiagnosed as SIDS or "near-miss" SIDS.[39]

Screening approaches that have been considered to date have included analysis of urine or neonatal filter blood-spot cards for identification of specific fatty acid metabolites. Bennett et al.[276] demonstrated that the incidence of dicarboxylic aciduria in an acutely ill pediatric population was about 1 in 12,000 and in most cases the pattern was consistent with a defect in FAO. In another study of 7058 dried postmortem filter-paper blood spot specimens obtained at autopsy from infants with unexplained causes of death using electrospray tandem MS/MS analysis of acylcarnitine profiles, there were 23 cases of MCADD, 9 of VLCADD, 8 of GA II, 6 of CPT II/translocase deficiencies, 4 of severe carnitine deficiency and 4 of LCHAD/TFP deficiency.[277]

Specific molecular probes have been utilized to screen for one of the more common defects, MCAD deficiency. In one study using the specific molecular probe for the gene mutation found in a majority of children with MCAD deficiency (A985G mutation) in the UK, Blakemore et al.[17] examined 410 consecutive neonatal blood-spot cards in a northern English population, and found 6 individuals who were heterozygous and no children who were homozygous for this mutation. This suggested a carrier frequency for this mutation of 1 in 68. This would correspond to a frequency of approximately 1 in 18,500 births for the homozygous MCAD deficient state for this particular mutation. For VLCAD deficiency, the most common long-chain FAO disorder, expanded newborn screening has identified a disease prevalence of up to 1:31,500.[278]

Tandem mass spectrometry (MS/MS) has made screening possible for most FAO defects based upon the profiling of acylcarnitine in blood spots.[279] The inclusion of FAO disorders in newborn screening programs is highly desirable to prevent morbidity and mortality, to implement preventative measures and treatment strategies, and to contain the cost of care of affected patients. This is now run in several countries, allowing the detection of asymptomatic newborns long before the onset of clinical manifestations.[280–282] For example, the incidence of MCAD deficiency was 1 in 8930 live births in a survey from the Pennsylvania newborn screening program using tandem mass spectrometry followed by confirmation through molecular analysis for several common mutations.[19] Implementation of newborn screening has significantly reduced morbidity and mortality of MCAD deficiency and long-chain FAO disorders and has increased disease prevalence figures.[283,284] It has also identified a great number of mildly affected patients who may never develop clinical symptoms throughout life, particularly with the early institution of preventative measures.[285] In MCAD deficiency, the risk of death in the first 72 hours in an Australian series has been estimated to be around 4% with a further approximately 5–7% fatality rate in the first 6 years, but very low subsequent risk in previously undiagnosed patients.[286] The risk of death after diagnosis is very low at any age, with good management. However, neonatal-onset TFP deficiency is still associated with high neonatal lethality.[287] Isolated LCHAD deficiency is treatable, but most patients suffer episodic hypoketotic hypoglycemia and myoglobinuria.[286] VLCAD deficiency poses a risk of death in early infancy, but is generally treatable with good outcome after diagnosis.[286] It has been estimated that about 10–20% of patients diagnosed by newborn screening and treated nevertheless suffer episodic myoglobinuria, though some never become symptomatic.[286] OCTN2 deficiency, formerly lethal, is highly treatable with lifelong high-dose oral carnitine supplementation.[13,88] Carnitine acylcarnitine translocase and neonatal CPT II deficiency have an extremely high neonatal mortality rate.[286]

CONCLUSION

In conclusion, over the past 25 years there has been an exponential expansion in the number of newly identified defects as well as the number of affected children with FAO disorders. This has been due in large part to an increased index of clinical suspicion as well as to the development and application of new biomedical technologies, which have facilitated the diagnosis of these defects. These conditions would, therefore, appear to be far more common than previously recognized. The earlier the identification and institution of specific therapies and appropriate preventative measures, the more significant the reduction in the otherwise significant morbidity and mortality of these diseases and the incidence of acute catabolic decompensations. The opportune collection of samples at the time of an acute catabolic episode is critical to diagnosis and may eliminate the need for more extensive and potentially high-risk investigations at a later date.

ACKNOWLEDGMENTS

This work was supported in part by an Operating Grant from the Heart and Stroke Foundation of Ontario (T-3746) and from the physicians of Ontario through the Physicians' Services Incorporated Foundation.

REFERENCES

1. DiMauro S, DiMauro PM. Muscle carnitine palmitoyltransferase deficiency and myoglobinuria. *Science* 1973;**182**:929–31.
2. Stanley CA, Hale DE, Coates PM, Hall CL, Corkey BE, Yang W, et al. Medium-chain acyl-CoA dehydrogenase deficiency in children with non-ketotic hypoglycemia and low carnitine levels. *Pediatr Res* 1983;**17**:877–84.
3. Hale DE, Batshaw ML, Coates PM, Frerman FE, Goodman SI, Singh I, et al. Long-chain acyl-coenzyme A dehydrogenase deficiency: an inherited cause of non-ketotic hypoglycemia. *Pediatr Res* 1985;**19**:666–71.
4. Duran M, Hofkamp M, Rhead WJ, Saudubray JM, Wadman SK. Sudden child death and "healthy" affected members with medium-chain acyl-coenzyme A dehydrogenase deficiency. *Pediatrcs* 1986; **78**:1052–7.
5. Howat AJ, Bennett MJ, Variend S, Shaw L, Engel PC. Defects of metabolism of fatty acids in the sudden infant death syndrome. *Br Med J* 1985;**290**:1771–3.
6. Amendt BA, Greene C, Sweetman L, Cloherty J, Shih V, Moon A, et al. Short-chain acyl-coenzyme A dehydrogenase deficiency. Clinical and biochemical studies in two patients. *J Clin Invest* 1987;**79**:1303–9.
7. Taubman B, Hale DE, Kelley RI. Familial Reye-like syndrome: a presentation of medium-chain acyl-CoA dehydrogenase deficiency. *Pediatrics* 1987;**79**:382–5.
8. Gibson KM, Breuer J, Nyhan WL. 3-Hydroxy-3-methylglutaryl-coenzyme A lyase deficiency: review of 18 reported patients. *Eur J Pediatr* 1988;**148**:180–6.
9. Demaugre F, Bonnefont JP, Mitchell G, Nguyen-Hoang N, Pelet A, Rimoldi M, et al. Hepatic and muscular presentations of carnitine palmitoyl transferase deficiency: two distinct entities. *Pediatr Res* 1988;**24**:308–11.
10. Goodman SI, Frerman FE. Glutaric acidaemia type II (multiple acyl-CoA dehydrogenation deficiency). *J Inherit Metab Dis* 1984;**7** (Suppl. 1):33–7.
11. Wanders RJ, Duran M, Ijlst L, de Jager JP, van Gennip AH, Jakobs C, et al. Sudden infant death and long-chain 3-hydroxyacyl-CoA dehydrogenase. *Lancet* 1989;**2**:52–3.
12. Hale DE, Thorpe C. Short-chain 3-OH acyl-CoA dehydrogenase deficiency. *Pediatr Res* 1989;**25**:199A.
13. Tein I, De Vivo DC, Bierman F, Pulver P, De Meirleir LJ, Cvitanovic-Sojat L, et al. Impaired skin fibroblast carnitine uptake in primary systemic carnitine deficiency manifested by childhood carnitine-responsive cardiomyopathy. *Pediatr Res* 1990;**28**:247–55.
14. Stanley CA, Hale DE, Coates PM. Medium-chain acyl-CoA dehydrogenase deficiency. In: Tanaka K, Coates PM, editors. *Fatty Acid Oxidation: Clinical, Biochemical and Molecular Aspects*, Vol. 321. New York: Alan R. Liss; 1990. pp. 291–302.
15. Tein I, De Vivo DC, Hale DE, Clarke JTR, Zinman H, Laxer R, et al. Short-chain L-3-hydroxyacyl-CoA dehydrogenase deficiency in muscle: a new cause for recurrent myoglobinuria and encephalopathy. *Ann Neurol* 1991;**30**:415–19.
16. Engel WK, Vick NA, Glueck J, Levy RI. A skeletal muscle disorder associated with intermittent symptoms and a possible defect in lipid metabolism. *N Engl J Med* 1970;**282**:697–704.
17. Blakemore AI, Singleton H, Pollitt RJ, Engel PC, Kolvraa S, Gregersen N, et al. Frequency of the G985 MCAD mutation in the general population. *Lancet* 1991;**337**:298–9.
18. Gregersen N, Andresen BS, Bross P, Winter V, Rudiger N, Engst S, et al. Characterization of a disease-causing Lys329 to Glu mutation in 16 patients with medium-chain acyl-CoA dehydrogenase deficiency. *J Inherited Metab Dis* 1991;**14**:314–16.
19. Ziadeh R, Hoffman EP, Finegold DN, Hoop RC, Brackett JC, Strauss AW, et al. Medium chain acyl-CoA dehydrogenase deficiency in Pennsylvania: neonatal screening shows high incidence and unexpected mutation frequencies. *Pediatr Res* 1995;**37**:675–8.
20. Andresen BS, Lund AM, Hougaard DM, Christensen E, Gahrn B, Christensen M, et al. MCAD deficiency in Denmark. *Mol Genet Metab* 2012;**106**:175–88.
21. Lindner M, Hoffmann GF, Matern D. Newborn screening for disorders of fatty-acid oxidation: experience and recommendations from an expert meeting. *J Inherit Metab Dis* 2010;**33**:521–6.
22. Spiekerkoetter U, Mayatepek E. Update on mitochondrial fatty acid oxidation disorders. *J Inherit Metab Dis* 2010;**33**:467–8.
23. Stanley CA. New genetic defects in mitochondrial fatty acid oxidation and carnitine deficiency. *Adv Pediatr* 1987;**34**:59–88.
24. Stremmel W, Strohmeyer G, Berk PD. Hepatocellular uptake of oleate is energy dependent, sodium linked, and inhibited by an antibody to a hepatocyte plasma membrane fatty acid binding protein. *Proc Natl Acad Sci USA* 1986;**83**:3584–8.
25. Schulz H. Oxidation of fatty acids. In: Vance DE, Vance JE, editors. *Biochemistry of Lipids and Membranes*. Menlo Park, California: Cummings Publishing Co; 1985. pp. 116–42.
26. Mannaerts GP, Debeer LJ, Thomas J, De Schepper PJ. Mitochondrial and peroxisomal fatty acid oxidation in liver homogenates and isolated hepatocytes from control and clofibrate-treated rats. *J Biol Chem* 1979;**254**:4585–95.
27. Murthy MS, Pande SV. Malonyl-CoA binding site and the overt carnitine palmitoyltransferase activity reside on opposite sides of the outer mitochondrial membrane. *Proc Natl Acad Sci USA* 1987;**84**:378–82.
28. Pande SV. A mitochondrial carnitine acylcarnitine translocase system. *Proc Natl Acad Sci USA* 1975;**72**:883–7.
29. Ikeda Y, Dabrowski C, Tanaka K. Separation and properties of five distinct acyl-CoA dehydrogenases from rat liver mitochondria. Identification of a new 2-methyl branched chain acyl-CoA dehydrogenase. *J Biol Chem* 1983;**258**:1066–76.
30. Fong JC, Schulz H. Short-chain and long-chain enoyl-CoA hydratases from pig heart muscle. *Methods Enzymol* 1981;**71**:390–8.
31. El-Fakhri M, Middleton B. The existence of an inner-membrane-bound, long acyl-chain-specific 3-hydroxyacyl-CoA dehydrogenase in mammalian mitochondria. *Biochim Biophys Acta* 1982;**713**:270–9.
32. Middleton B. 3-Ketoacyl-CoA thiolases of mammalian tissues. *Methods Enzymol* 1975;**35**:128–36.
33. Kelley RI. Beta-oxidation of long-chain fatty acids by human fibroblasts: evidence for a novel long-chain acyl-coenzyme A dehydrogenase. *Biochem Biophys Res Commun* 1992;**182**:1002–7.

34. Izai K, Uchida Y, Orii T, Yamamoto S, Hashimoto T. Novel fatty acid beta-oxidation enzymes in rat liver mitochondria. I. Purification and properties of very-long-chain acyl-coenzyme A dehydrogenase. *J Biol Chem* 1992;**267**:1027–33.
35. He M, Rutledge SL, Kelly DR, Palmer CA, Murdoch G, Majumder N, et al. A new genetic disorder in mitochondrial fatty acid beta-oxidation: ACAD9 deficiency. *Am J Hum Genet* 2007;**81**:87–103.
36. Hashimoto T. Peroxisomal and mitochondrial enzymes. In: Coates PM, Tanaka K, editors. *Progress in Clinical and Biological Research. New Developments in Fatty Acid Oxidation*, Vol. 375. New York: Wiley-Liss, Inc; 1992. pp. 19–32.
37. Carpenter K, Pollitt RJ, Middleton B. Human liver long-chain 3-hydroxyacyl-coenzyme A dehydrogenase is a multifunctional membrane-bound beta-oxidation enzyme of mitochondria. *Biochem Biophys Res Commun* 1992;**183**:443–8.
38. Luo MJ, He X-Y, Sprecher H, Schulz H. Purification and characterization of the trifunctional beta-oxidation complex from pig heart mitochondria. *Arch Biochem Biophys* 1993;**304**:266–71.
39. Hale DE, Bennett MJ. Fatty acid oxidation disorders: a new class of metabolic diseases. *J Pediatr* 1992;**121**:1–11.
40. Nouws J, Nijtmans L, Houten SM, van den Brand M, Huynen M, Venselaar H, et al. Acyl-CoA dehydrogenase 9 is required for the biogenesis of oxidative phosphorylation complex I. *Cell Metab* 2010;**12**:283–94.
41. Haack TB, Danhauser K, Haberberger B, Hoser J, Strecker V, Boehm D, et al. Exome sequencing identifies ACAD9 mutations as a cause of complex I deficiency. *Nat Genet* 2010;**42**:1131–4.
42. Nouws J., Wibrand F., van den Brand M., Venselaar H., Duno M., Lund A.M., et al. A patient with complex I deficiency caused by a novel ACAD9 mutation not responding to riboflavin treatment. *JIMD Rep* 2014;**12**:37–45.
43. Kamijo T, Aoyama T, Miyazaki J, Hashimoto T. Molecular cloning of the cDNAs for the subunits of rat mitochondrial fatty acid beta-oxidation multienzyme complex. *J Biol Chem* 1993;**268**: 26452–60.
44. Jackson S, Kler RS, Bartlett K, Briggs H, Bindoff LA, Pourfarzam M, et al. Combined enzyme defect of mitochondrial fatty acid oxidation. *J Clin Invest* 1992;**90**:1219–25.
45. Kamijo T, Wanders RJ, Saudubray JM, Aoyama T, Komiyama A, Hashimoto T. Mitochondrial trifunctional protein deficiency. Catalytic heterogeneity of the mutant enzyme in two patients. *J Clin Invest* 1994;**93**:1740–7.
46. Reed DW, Lane MD. Mitochondrial 3-hydroxy-3-methylglutaryl-CoA synthase from chicken liver. *Methods Enzymol* 1975;**35**:155–60.
47. Kayser MA. Disorders of ketone production and utilization. *Mol Genet Metab* 2006;**87**:281–3.
48. McGarry JD, Foster DW. Regulation of hepatic fatty acid oxidation and ketone body production. *Annu Rev Biochem* 1980;**49**:395–420.
49. Lazarow PB, de Duve C. A fatty acyl-CoA oxidizing system in rat liver peroxisomes; enhancement by clofibrate, a hypolipidemic drug. *Proc Natl Acad Sci USA* 1976;**73**:2043–6.
50. Krahling JB, Gee R, Murphy PA, Kirk JR, Tolbert NE. Comparison of fatty acid oxidation in mitochondria and peroxisomes from rat liver. *Biochem Biophys Res Commun* 1978;**82**:136–41.
51. Preiss B, Bloch K. Omega oxidation of long-chain fatty acids in rat liver. *J Biol Chem* 1964;**239**:85–8.
52. Kølvraa S, Gregersen N. Acyl-CoA: glycine N-acyltransferase: organelle localization and affinity towards straight- and branched-chained acyl-CoA esters in rat liver. *Biochem Med Metab Biol* 1986;**36**:98–105.
53. Roe CR, Millington DS, Maltby DA, Bohan TP, Kahler SG, Chalmers RA. Diagnostic and therapeutic implications of medium-chain acylcarnitines in the medium-chain acyl-CoA dehydrogenase deficiency. *Pediatr Res* 1985;**19**:459–66.
54. Corkey BE, Hale DE, Glennon MC, Kelley RI, Coates PM, Kilpatrick L, et al. Relationship between unusual hepatic acyl coenzyme A profiles and the pathogenesis of Reye syndrome. *J Clin Invest* 1988;**82**:782–8.
55. Baretz BH, Ramsdell HS, Tanaka K. Identification of n-hexanoylglycine in urines from two patients with Jamaican vomiting sickness. *Clin Chim Acta* 1976;**73**:199–202
56. Millington DS. New methods for the analysis of acylcarnitines and acyl-CoA compounds. In: Gaskell SJ, editor. *Mass Spectrometry in Biomedical Research*. New York: Wiley and Sons; 1986. pp. 97–114.
57. Bremer J. Carnitine: metabolism and functions. *Physiol Rev* 1983;**63**:1420–80.
58. Bieber LL. Carnitine. *Annu Rev Biochem* 1988;**57**:261–83.
59. Rebouche CJ. Carnitine function and requirements during the life cycle. *FASEB J* 1992;**6**:3379–86.
60. Rebouche CJ, Engel AG. Kinetic compartmental analysis of carnitine metabolism in the human carnitine deficiency syndromes. Evidence for alterations in tissue carnitine transport. *J Clin Invest* 1984;**73**:857–67.
61. Rebouche CJ, Chenard CA. Metabolic fate of dietary carnitine in human adults: identification and quantification of urinary and fecal metabolites. *J Nutr* 1991;**121**:539–46.
62. Novak M, Monkus EF, Buch M, Lesmes H, Silverio J. The effect of a L-carnitine supplemented soybean formula on the plasma lipids of infants. *Acta Chir Scand* 1983;**517**(Suppl.):149–55.
63. Olson AL, Nelson SE, Rebouche CJ. Low carnitine intake and altered lipid metabolism in infants. *Am J Clin Nutr* 1989;**49**:624–8.
64. Rebouche CJ, Mack DL. Sodium gradient-stimulated transport of L-carnitine into renal brush border membrane vesicles: kinetics, specificity, and regulation by dietary carnitine. *Arch Biochem Biophys* 1984;**235**:393–402.
65. Engel AG, Rebouche CJ, Wilson DM, Glasgow AM, Romshe CA, Cruse RP. Primary systemic carnitine deficiency. II. Renal handling of carnitine. *Neurology* 1981;**31**:819–25.
66. Rebouche CJ, Engel AG. Tissue distribution of carnitine biosynthetic enzymes in man. *Biochim Biophys Acta* 1980;**630**:22–9.
67. Englard S. Hydroxylation of gamma-butyrobetaine to carnitine in human and monkey tissues. *FEBS Lett* 1979;**102**:297–300.
68. Olson AL, Rebouche CJ. Gamma-butyrobetaine hydroxylase activity is not rate limiting for carnitine biosynthesis in the human infant. *J Nutr* 1987;**117**:1024–31.
69. De Vivo DC, Tein I. Primary and secondary disorders of carnitine metabolism. *Int Pediatr* 1990;**5**:134–41.
70. Turnbull DM, Bartlett K, Stevens DL, Alberti KG, Gibson GJ, Johnson MA, et al. Short-chain acyl-CoA dehydrogenase deficiency associated with a lipid-storage myopathy and secondary carnitine deficiency. *N Engl J Med* 1984;**311**:1232–6.
71. Cahill Jr. GF. Starvation in man. *N Engl J Med* 1970;**282**:668–75.

72. Felig P, Wahren J. Fuel homeostasis in exercise. *N Engl J Med* 1975;**293**:1078–84.
73. Gollnick PD, Piehl K, Saltin B. Selective glycogen depletion pattern in human muscle fibres after exercise of varying intensity and at varying pedaling rates. *J Physiol* 1974;**241**:45–57.
74. Essén B. Glycogen depletion of different fibre types in human skeletal muscle during intermittent and continuous exercise. *Acta Physiol Scand* 1978;**103**:446–55.
75. Pildes RS, Patel DA, Nitzan M. Glucose disappearance rate in symptomatic neonatal hypoglycemia. *Pediatr* 1973;**52**:75–82.
76. Bier DM, Leake RD, Haymond MW, Arnold KJ, Gruenke LD, Sperling MA, et al. Measurement of "true" glucose production rates in infancy and childhood with 6,6-dideuteroglucose. *Diabetes* 1977;**26**:1016–23.
77. Aynsley-Green A. Hypoglycaemia in infants and children. *Clin Endocrinol Metab* 1982;**11**:159–94.
78. Pagliara AS, Karl IE, Haymond M, Kipnis DM. Hypoglycemia in infancy and childhood. *J Pediatr* 1973;**82**(Pt 1):365–79.
79. Bell AW, Thompson GE. Free fatty acid oxidation in bovine muscle in vivo: effects of cold exposure and feeding. *Am J Physiol* 1979;**237**:E309–15.
80. Johnson RE, Possmore R, Sargent II F. Multiple factors in experimental human ketosis. *Arch Intern Med* 1961;**107**:43–50.
81. Tonsgard JH, Getz GS. Effect of Reye's syndrome serum on isolated chinchilla liver mitochondria. *J Clin Invest* 1985;**76**:816–25.
82. Tonsgard JH. Serum dicarboxylic acids in patients with Reye syndrome. *J Pediatr* 1986;**109**:440–5.
83. Mak IT, Kramer JH, Weglicki WB. Potentiation of free radical-induced lipid peroxidative injury to sarcolemmal membranes by lipid amphiphiles. *J Biol Chem* 1986;**26**:1153–7.
84. Inoue D, Pappano AJ. L-palmitoylcarnitine and calcium ions act similarly on excitatory ionic currents in avian ventricular muscle. *Circ Res* 1983;**52**:625–34.
85. Spedding M. Activators and inactivators of Ca^{2+} channels: new perspectives. *J Pharmacol* 1985;**16**:319–43.
86. Spedding M, Mir AK. Direct activation of Ca^{2+} channels by palmitoyl carnitine, a putative endogenous ligand. *Br J Pharamacol* 1987;**92**:457–68.
87. Lee HC, Smith N, Mohabir R, Clusin WT. Cytosolic calcium transients from the beating mammalian heart. *Proc Natl Acad Sci USA* 1987;**84**:7793–7.
88. Stanley CA, DeLeeuw S, Coates PM, Vianey-Liaud C, Divry P, Bonnefont JP, et al. Chronic cardiomyopathy and weakness or acute coma in children with a defect in carnitine uptake. *Ann Neurol* 1991;**30**:709–16.
89. Tein I, DiMauro S, De Vivo DC. Recurrent childhood myoglobinuria. *Adv Pediatr* 1990;**37**:77–117.
90. Warren DJ, Leitch AC, Leggett RJ. Hyperuriceamic acute renal failure after epileptic seizures. *Lancet* 1975;**2**:385–7.
91. Knochel JP. Renal injury in muscle disease. In: Suki WN, Eknoyan G, editors. *The Kidney in Systemic Disease*. New York: John Wiley and Sons Inc; 1976. pp. 129–40.
92. Tein I, DiMauro S, Rowland LP. Myoglobinuria. In: Vinken PJ, Bruyn GW, Klawans HL, editors. *Handbook of Clinical Neurology*, vol. 62. Amsterdam: Elsevier Science Publishers B.V.; 1992. pp. 553–93 (18)
93. Tein I. Fatty Acid Oxidation and Associated Defects. *Proceedings of the 47th Annual Meeting of the American Academy of Neurology*, vol. 269. Seattle, WA: Omnipress; 1995. pp. 9–38.
94. Elpeleg ON, Joseph A, Branski A, Christensen E, Holme E, Demaugre F, et al. Recurrent metabolic decompensation in profound carnitine palmitoyltransferase II deficiency. *J Pediatr* 1993;**122**:917–19.
95. Poll-The BT, Duran M, Mousson B, Vianey-Liaud C, Smeitink JAM, Dorland L, et al. Carnitine palmitoyl transferase I deficiency: Is there a diagnostic dicarboxylic aciduria? Published in: *Proceedings of the 30th Annual Symposium of the Society for the Study of Inborn Errors of Metabolism*. Leuven; 1992 Sept. 8–11, p. 128. [Abstract]
96. Tripp ME, Katcher ML, Peters HA, Gilbert EF, Arya S, Hodach RJ, et al. Systemic carnitine deficiency presenting as familial endocardial fibroelastosis: a treatable cardiomyopathy. *N Engl J Med* 1981;**305**:385–90.
97. Falik-Borenstein ZC, Jordan SC, Saudubray JM, Brivet M, Demaugre F, Edmond J, et al. Brief Report: renal tubular acidosis in carnitine palmitoyltransferase type I deficiency. *N Engl J Med* 1992;**327**:24–7.
98. Bennett MJ. Pathophysiology of fatty acid oxidation disorders. *J Inherit Metab Dis* 2010;**33**:533–7.
99. Brown NF, Mullur RS, Subramanian I, Esser V, Bennett MJ, Saudubray JM, et al. Molecular characterization of L-CPT I deficiency in six patients: insights into function of the native enzyme. *J Lipid Res* 2001;**42**:1134–42.
100. Greenberg CR, Dilling LA, Thompson GR, Seargeant LE, Haworth JC, Phillips S, et al. The paradox of the carnitine palmitoyltransferase type Ia P479L variant in Canadian Aboriginal populations. *Mol Genet Metab* 2009;**96**:201–7.
101. Gillingham MB, Banta-Wright SA, Hermerath CA, Skeels MR, Bennett MJ, Narayan SB, et al. CPT1A P479L variant identified in Alaska natives by expanded newborn screening. *J Inherit Metab Dis* 2006;**29**(Suppl. 11). [Abstract 85].
102. DiMauro S, Papadimitriou A. Carnitine palmitoyltransferase deficiency. In: Engel AG, Banker BQ, editors. *Myology*. New York: McGraw-Hill Book Co; 1986. pp. 1697–708.
103. Tonin P, Lewis P, Servidei S, DiMauro S. Metabolic causes of myoglobinuria. *Ann Neurol* 1990;**27**:181–5.
104. Demaugre F, Bonnefont JP, Colonna M, Cepanec C, Leroux JP, Saudubray JM. Infantile form of carnitine palmitoyltransferase II deficiency with hepatomuscular symptoms and sudden death: Physiopathological approach to carnitine palmitoyltransferase II deficiencies. *J Clin Invest* 1991;**87**:859–64.
105. Hug G, Bove KE, Soukup S. Lethal neonatal multiorgan deficiency of carnitine palmitoyltransferase II. *N Engl J Med* 1991;**325**:1862–4.
106. Isackson PJ, Bennett MJ, Lichter-Konecki U, Willis M, Nyhan WL, Sutton VR, et al. CPT2 gene mutations resulting in lethal neonatal or severe infantile carnitine palmitoyltransferase II deficiency. *Mol Genet Metab* 2008;**94**:422–7.
107. Odaib AA, Shneider BL, Bennett MJ, Pober BR, Reyes-Mugica M, Friedman AL, et al. A defect in the transport of long-chain fatty acids associated with acute liver failure. *N Engl J Med* 1998;**339**:1752–7.
108. Rinaldo P, Matern D, Bennett MJ. Fatty acid oxidation disorders. *Annu Rev Physiol* 2002;**64**:477–502.
109. Aoyama T, Souri M, Ushikubo S, Kamijo T, Yamaguchi S, Kelley RI, et al. Purification of human very-long-chain acyl-coenzyme A dehydrogenase and characterization of its deficiency in seven patients. *J Clin Invest* 1995;**95**:2465–73.

110. Ogilvie I, Pourfarzam M, Jackson S, Stockdale C, Bartlett K, Turnbull DM. Very long-chain acyl coenzyme A dehydrogenase deficiency presenting with exercise-induced myoglobinuria. *Neurology* 1994;**44**:467–73.
111. Yamaguchi S, Indo Y, Coates PM, Hashimoto T, Tanaka T. Identification of very-long-chain acyl-CoA dehydrogenase deficiency in three patients previously diagnosed with long-chain acyl-CoA dehydrogenase deficiency. *Pediatr Res* 1993;**34**:111–13.
112. Hale DE, Stanley CA, Coates PM. The long-chain acyl-CoA dehydrogenase deficiency. In: Tanaka K, Coates PM, editors. *Fatty Acid Oxidation: Clinical, Biochemical and Molecular Aspects*. New York: Alan R. Liss; 1990. pp. 303–11.
113. Glasgow AM, Engel AG, Bier DM, Perry LW, Dickie M, Todaro J, et al. Hypoglycemia, hepatic dysfunction, muscle weakness, cardiomyopathy, free carnitine deficiency and long-chain acylcarnitine excess responsive to medium-chain triglyceride diet. *Pediatr Res* 1983;**17**:319–26.
114. Tein I, Haslam RH, Rhead WJ, Bennett MJ, Becker LE, Vockley J. Short-chain acyl-CoA dehydrogenase deficiency: a cause of ophthalmoplegia and multicore myopathy. *Neurology* 1999;**52**:366–72.
115. Dionisi-Vici CD, Burlina AB, Bertini E, Bachmann C, Mazziotta MR, Zacchello F, et al. Progressive neuropathy and recurrent myoglobinuria in a child with long-chain 3-hydroxyacyl-coenzyme A dehydrogenase deficiency. *J Pediatr* 1991;**118**:744–6.
116. Kelley RI, Morton DH. 3-Hydroxyoctanoic aciduria: identification of a new organic acid in the urine of a patient with non-ketotic hypoglycemia. *Clin Chim Acta* 1988;**175**:19–26.
117. Hagengeldt L, von Döbeln U, Holme E, Alm J, Brandberg G, Enocksson E, et al. 3-Hydroxydicarboxylic aciduria—a fatty acid oxidation defect with severe prognosis. *J Pediatr* 1990;**116**:387–92.
118. Rocchiccioli F, Wanders RJ, Aubourg P, Vianey-Liaud C, Ijlst L, Fabre M, et al. Deficiency of long-chain 3-hydroxyacyl-CoA dehydrogenase: a cause of lethal myopathy and cardiomyopathy in early childhood. *Pediatr Res* 1990;**28**:657–62.
119. Duran M, Wanders RJ, deJager JP, Dorland L, Bruinvis L, Ketting D, et al. 3-Hydroxydicarboxylic aciduria due to long-chain 3-hydroxyacyl-coenzyme A dehydrogenase deficiency associated with sudden neonatal death: protective effect of medium-chain triglyceride treatment. *Eur J Pediatr* 1991;**150**:190–5.
120. Jackson S, Bartlett K, Land J, Moxon ER, Pollitt RJ, Leonard JV, et al. Long-chain 3-hydroxyacyl-CoA dehydrogenase deficiency. *Pediatr Res* 1991;**29**:406–11.
121. Wanders RJ, Ijlst L, Duran M, Jakobs C, de Klerk JB, Przyrembel H, et al. Long-chain 3-hydroxyacyl-CoA dehydrogenase deficiency: different clinical expression in three unrelated patients. *J Inherit Metab Dis* 1991;**14**:325–8.
122. Bertini E, Dionisi-Vici C, Garavaglia B, Burlina AB, Sabatelli M, Rimoldi M, et al. Peripheral sensory-motor polyneuropathy, pigmentary retinopathy, and fatal cardiomyopathy in long-chain 3-hydroxy-acyl-CoA dehydrogenase deficiency. *Eur J Pediatr* 1992;**151**:121–6.
123. Ribes A, Riudor E, Navarro C, Boronat M, Marti M, Hale DE. Fatal outcome in a patient with long-chain 3-hydroxyacyl-CoA dehydrogenase deficiency. *J Inherited Metab Dis* 1992;**15**:278–9.
124. Stanley CA, Hale DE, Coates PM. Medium-chain acyl-CoA dehydrogenase deficiency. In: Tanaka K, Coates PM, editors. *Progress in Clinical and Biological Research, vol. 321. Fatty Acid Oxidation. Clinical, Biochemical, and Molecular Aspects*. New York: Alan R. Liss; 1990. pp. 291–302.
125. Gibson EM, Breuer J, Nyhan WL. 3-Hydroxy-3-methylglutaryl-coenzyme A lyase deficiency: review of 18 reported patients. *Eur J Pediatr* 1988;**148**:180–6.
126. Tein I, Elpeleg O, Ben-Zeev B, Korman SH, Lossos A, Lev D, et al. Short-chain acyl-CoA dehydrogenase gene mutation (c.319C > T) presents with clinical heterogeneity and is candidate founder mutation in individuals of Ashkenazi Jewish origin. *Mol Genet Metab* 2008;**93**:179–89.
127. Voermans NC, van Engelen BG, Kluijtmans LA, Stikkelbroeck NM, Hermus AR. Rhabdomyolysis caused by an inherited metabolic disease: very long-chain acyl-CoA dehydrogenase deficiency. *Am J Med* 2006;**119**:176–9.
128. Straussberg R, Harel L, Varsano I, Elpeleg ON, Shamir R, Amir J. Recurrent myoglobinuria as a presenting manifestation of very long chain acyl coenzyme A dehydrogenase deficiency. *Pediatrics* 1997;**99**:894–6.
129. Tein I, Donner EJ, Hale DE, Murphy EG. Clinical and neurophysiologic response of myopathy and neuropathy in long-chain L-3-hydroxyacyl-CoA dehydrogenase deficiency to oral prednisone. *Pediatr Neurology* 1995;**12**:68–76.
130. Kamijo T, Indo Y, Souri M, Aoyama T, Hara T, Yamamoto S, et al. Medium-chain 3-ketoacyl-coenzyme A thiolase deficiency: a new disorder of mitochondrial fatty acid β-oxidation. *Pediatr Res* 1997;**42**:569–76.
131. Ruitenbeek W, Poels PJ, Turnbull DM, Garavaglia B, Chalmers RA, Taylor RW, et al. Rhabdomyolysis and acute encephalopathy in late onset medium chain acyl-CoA dehydrogenase deficiency. *J Neurol Neurosurg Psychiatry* 1995;**58**:209–14.
132. Laforêt P, Acquaviva-Bourdain C, Rigal O, Brivet M, Penisson-Besnier I, Chabrol B, et al. Diagnostic assessment and long-term follow-up of 13 patients with Very Long-Chain Acyl-Coenzyme A dehydrogenase (VLCAD) deficiency. *Neuromuscul Disord* 2009;**19**:324–9.
133. Korematsu S, Kosugi Y, Kumamoto T, Yamaguchi S, Izumi T. Novel mutation of early, perinatal-onset, myopathic-type very-long-chain acyl-CoA dehydrogenase deficiency. *Pediatr Neurol* 2009;**41**:151–3.
134. Ohashi Y, Hasegawa Y, Murayama K, Ogawa M, Hasegawa T, Kawai M, et al. A new diagnostic test for VLCAD deficiency using immunohistochemistry. *Neurology* 2004;**62**:2209–13.
135. Scheuerman O, Wanders RJ, Waterham HR, Dubnov-Raz G, Garty BZ. Mitochondrial trifunctional protein deficiency with recurrent rhabdomyolysis. *Pediatr Neurol* 2009;**40**:465–7.
136. Hagenfeldt L, von Döbeln U, Holme E, Alm J, Brandberg G, Enocksson E, et al. 3-Hydroxydicarboxylic aciduria—a fatty acid oxidation defect with severe prognosis. *J Pediatr* 1990;**116**:387–92.
137. Wilcken B, Leung KC, Hammond J, Kamath R, Leonard JV. Pregnancy and fetal long-chain 3-hydroxyacyl coenzyme A dehydrogenase deficiency. *Lancet* 1993;**341**:407–8.
138. Treem WR, Shoup ME, Hale DE, Bennett MJ, Rinaldo P, Millington DS, et al. Acute fatty liver of pregnancy, hemolysis, elevated liver enzymes, and low platelets syndrome and long chain 3-hydroxyacyl-coenzyme A dehydrogenase deficiency. *Am J Gastroenterol* 1996;**91**:2293–300.

139. Tyni T, Palotie A, Viinikka L, Valannel L, Salo MK, von Dobeln U, et al. Long-chain 3-hydroxyacyl-coenzyme A dehydrogenase deficiency with the G1528C mutation: clinical presentation of thirteen patients. *J Pediatr* 1997;**130**:67–76.
140. Ibdah JA, Zhao M, Hill I, Gibson B, Strauss A. The association between maternal liver disease and the common mutation in mitochondrial trifunctional protein is significant. *Hepatology* 1998;28 [Abstract 316].
141. Tyni T, Paetau A, Strauss AW, Middleton B, Kivelä T. Mitochondrial fatty acid beta-oxidation in the human eye and brain: Implications for retinopathy of long-chain-3-hydroxyacyl-CoA dehydrogenase deficiency. *Pediatr Res* 2004;**56**:744–50.
142. Tyni T, Pihko H, Kivelä T. Ophthalmic pathology in long-chain 3-hydroxyacyl-CoA dehydrogenase deficiency caused by the G1528C mutation. *Curr Eye Res* 1998;**17**:551–9.
143. Spiekerkoetter U, Bennett MJ, Ben-Zeev B, Strauss AW, Tein I. Peripheral neuropathy, episodic myoglobinuria, and respiratory failure in deficiency of the mitochondrial trifunctional protein. *Muscle Nerve* 2004;**29**:66–72.
144. Matern D, Hart P, Murtha AP, Vockley J, Gregersen N, Millington DS, et al. Acute fatty liver of pregnancy associated with short-chain acyl-coenzyme A dehydrogenase deficiency. *J Pediatr* 2001;**138**:585–8.
145. Matern D, Schehata BM, Shekhawa P, Strauss AW, Bennett MJ, Rinaldo P. Placental floor infarction complicating the pregnancy of a fetus with long-chain 3-hydroxyacyl-CoA dehydrogenase (LCHAD) deficiency. *Mol Genet Metab* 2001;**72**:265–8.
146. Innes AM, Seargeant LE, Balachandra K, Roe CR, Wanders RJ, Ruiter JP, et al. Hepatic carnitine palmitoyltransferase I deficiency presenting as maternal illness in pregnancy. *Pediatr Res* 2000;**47**:43–5.
147. Nelson J, Lewis B, Walters B. The HELLP syndrome associated with fetal medium-chain acyl-CoA dehydrogenase deficiency. *J Inherit Metab Dis* 2000;**23**:518–19.
148. Walter JH. Inborn errors of metabolism and pregnancy. *J Inherit Metab Dis* 2000;**23**:229–36.
149. Bhala A, Willi SM, Rinaldo P, Bennett MJ, Schmidt-Sommerfeld E, Hale DE. Clinical and biochemical characterization of short-chain acyl-coenzyme A dehydrogenase deficiency. *J Pediatr* 1995;**126**:910–15.
150. Ribes A, Riudor E, Garavaglia B, Martinez G, Arranz A, Invernizzi F, et al. Mild or absent clinical signs in twin sisters with short-chain acyl-CoA dehydrogenase deficiency. *Eur J Pediatr* 1998;**157**:317–20.
151. Gregersen N, Winter VS, Corydon MJ, Corydon TJ, Rinaldo P, Ribes A, et al. Identification of four new mutations in the short-chain acyl-CoA dehydrogenase (SCAD) gene in two patients: one of the variant alleles, 511C→T, is present at an unexpectedly high frequency in the general population, as was the case for 625G→A, together conferring susceptibility to ethylmalonic aciduria. *Hum Mol Genet* 1998;**7**:619–27.
152. Corydon MJ, Vockley J, Rinaldo P, Rhead WJ, Kjeldsen M, Winter V, et al. Role of common gene variations in the molecular pathogenesis of short-chain acyl-CoA dehydrogenase deficiency. *Pediatr Res* 2001;**49**:18–23.
153. Pedersen CB, Zolkipli Z, Vang S, Palmfeldt J, Kjeldsen M, Stenbroen V, et al. Antioxidant dysfunction: potential risk for neurotoxicity in ethylmalonic aciduria. *J Inherit Metab Dis* 2010;**33**:211–22.
154. Zolkipli Z, Pedersen CB, Lamhonwah AM, Gregersen N, Tein I. Vulnerability to oxidative stress in vitro in pathophysiology of mitochondrial short-chain acyl-CoA dehydrogenase deficiency: response to antioxidants. *PLOS ONE* 2011;**6**:e17534.
155. Gregersen N, Andresen BS, Pedersen CB, Olsen RK, Corydon TJ, Bross P. Mitochondrial fatty acid oxidation defects—remaining challenges. *J Inherit Metab Dis* 2008;**31**:643–57.
156. Frerman FE, Goodman SI. Glutaric aciduria type II and defects of the mitochondrial respiratory chain. In: Scriver CR, Beaudet AL, Sly WS, Valle D, editors. *The Metabolic Basis of Inherited Disease*. New York: McGraw-Hill; 1989. pp. 915–31.
157. Ishii K, Komaki H, Ohkuma A, Nishino I, Nonaka I, Sasaki M. Central nervous system and muscle involvement in an adolescent patient with riboflavin-responsive multiple acyl-CoA dehydrogenase deficiency. *Brain Dev* 2010;**32**:669–72.
158. Gempel K, Topaloglu H, Talim B, Schneiderat P, Schoser BG, Hans VH, et al. The myopathic form of coenzyme Q10 deficiency is caused by mutations in the electron-transferring-flavoprotein dehydrogenase (ETFDH) gene. *Brain* 2007;**130**:2037–44.
159. Kobayashi A, Jiang LL, Hashimoto T. Two mitochondrial 3-hydroxyacyl-CoA dehydrogenases in bovine liver. *J Biochem* 1996;**119**:775–82.
160. Yang SY, He XY, Schulz H. 3-Hydroxyacyl-CoA dehydrogenase and short chain 3-hydroxyacyl-CoA dehydrogenase in human health and disease. *FEBS J* 2005;**272**:4874–83.
161. Renaud DL, Edwards V, Wilson GJ, Tein I. Glucose-free medium exacerbates microvesicular steatosis in cultured skin fibroblasts of genetic defects in fatty acid oxidation. A novel screening test. *J Inherit Metab Dis* 2002;**25**:547–55.
162. Bennett MJ, Weinberger MJ, Kobori JA, Rinaldo P, Burlina AB. Mitochondrial short-chain L-3-hydroxyacyl-coenzyme A dehydrogenase deficiency: a new defect of fatty acid oxidation. *Pediatr Res* 1996;**39**:185–8.
163. Bennett MJ, Spotswood SD, Ross KF, Comfort S, Koonce R, Boriack RL, et al. Fatal hepatic short-chain L-3-hydroxyacyl-coenzyme A dehydrogenase deficiency: clinical, biochemical, and pathological studies on three subjects with this recently identified disorder of mitochondrial beta-oxidation. *Pediatr Dev Pathol* 1999;**2**:337–45.
164. O'Brien LK, Rinaldo P, Sims HF, Alonso EM, Charrow J, Jones PM, et al. Fulminant hepatic failure associated with mutations in the medium and short chain L-3-hydroxyacyl-CoA dehydrogenase gene. *J Inher Metab Dis* 2000;(Suppl.):23 [Abstract 127].
165. Eaton S, Chatziandreou I, Krywawych S, Pen S, Clayton PT, Hussain K. Short-chain 3-hydroxyacyl-CoA dehydrogenase deficiency associated with hyperinsulinism: a novel glucose-fatty acid cycle? *Biochem Soc Trans* 2003;**31**:1137–9.
166. Molven A, Matre GE, Duran M, Wanders RJ, Rishaug U, Njølstad PR, et al. Familial hyperinsulinemic hypoglycemia caused by a defect in the SCHAD enzyme of mitochondrial fatty acid oxidation. *Diabetes* 2004;**53**:221–7.
167. Filling C, Keller B, Hirschberg D, Marschall HU, Jörnvall H, Bennett MJ, et al. Role of short-chain hydroxyacyl CoA dehydrogenases in SCHAD deficiency. *Biochem Biophys Res Commun* 2008;**368**:6–11.

168. Roe CR, Millington DS, Norwood DL, Kodo N, Sprecher H, Mohammed BS, et al. 2,4-Dienoyl-coenzyme A reductase deficiency: a possible new disorder of fatty acid oxidation. *J Clin Invest* 1990;**85**:1703–7.
169. Kamijo T, Indo Y, Souri M, Aoyama T, Hara T, Yamamoto S, et al. Medium-chain 3-ketoacyl-coenzyme A thiolase deficiency: a new disorder of mitochondrial fatty acid beta-oxidation. *Pediatr Res* 1997;**42**:569–76.
170. Mitchell GA, Fukao T. Inborn errors of ketone body metabolism. In: Scriver C, Beaudet AL, Sly W, Valle D, editors. *The Metabolic and Molecular Basis of Inherited Disease*. 8th ed. New York: McGraw-Hill; 2001. pp. 2327–56.
171. Fukao T, Song XQ, Yamaguchi S, Kondo N, Orii T, Matthieu JM, et al. Identification of three novel frameshift mutations (83delAT), 754insCT, and 435 + 1G to A) of mitochondrial acetoacetyl-coenzyme A thiolase gene in two Swiss patients with CRM-negative beta-ketothiolase deficiency. *Hum Mutat* 1997;**9**:277–9.
172. Fukao T, Yamaguchi S, Orii T, Hashimoto T. Molecular basis of beta-ketothiolase deficiency: mutations and polymorphisms in the human mitochondrial acetoacetyl-coenzyme A thiolase gene. *Hum Mutat* 1995;**5**:113–20.
173. Kassovska-Bratinova S, Fukao T, Song XQ, Duncan AM, Chen HS, Robert MF, et al. Succinyl CoA-3-oxoacid CoA transferase (SCOT): human cDNA cloning, human chromosomal mapping to 5p13, and mutation detection in a SCOT-deficient patient. *Am J Hum Genet* 1996;**59**:519–28.
174. Pretorious CJ, Loy Son GG, Bonnici F, Harley EH. Two siblings with episodic ketoacidosis and decreased activity of succinyl-CoA-3-ketoacid CoA-transferase in cultured fibroblasts. *J Inherit Metab Dis* 1996;**19**:296–300.
175. Thompson GN, Hsu BY, Pitt JJ, Treacy E, Stanley CA. Fasting hypoketotic coma in a child with deficiency of mitochondrial 3-hydroxy-3-methylglutaryl-CoA synthase. *N Engl J Med* 1997;**337**:1203–7.
176. Ibdah JA, Tein I, Dionisi-Vici C, Bennett MJ, Ijlst L, Gibson B, et al. Mild trifunctional protein deficiency is associated with progressive neuropathy and myopathy and suggests a novel genotype-phenotype correlation. *J Clin Invest* 1998;**102**:1193–9.
177. Stanley CA, Sunaryo F, Hale DE, Bonnefont JP, Demaugre F, Saudubray JM. Elevated plasma carnitine in the hepatic form of carnitine palmitoyltransferase-1 deficiency. *J Inherited Metab Dis* 1992;**15**:785–9.
178. Millington DS, Terada N, Chace DH, Chen YT, Ding JH, Kodo N, et al. The role of tandem mass spectrometry in the diagnosis of fatty acid oxidation disorders. In: Coates PM, Tanaka K, editors. *Progress in Clinical and Biological Research. New Developments in Fatty Acid Oxidation*, vol. 375. New York: Wiley-Liss, Inc; 1992. pp. 339–54.
179. Kerner J, Beiber LL. A radioisotopic-exchange method for quantitation of short-chain (acid soluble) acylcarnitines. *Anal Biochem* 1983;**134**:459–66.
180. Nada MA, Chace DH, Sprecher H, Roe CR. Investigation of beta-oxidation intermediates in normal and MCAD-deficient human fibroblasts using tandem mass spectrometry. *Biochem Mol Med* 1995;**54**:59–66.
181. Nada MA, Rhead JW, Sprecher H, Schulz H, Roe CR. Evidence for intermediate channelling in mitochondrial beta-oxidation. *J Biol Chem* 1995;**270**:530–5.
182. Nada MA, Vianey-Saban C, Roe CR, Ding JH, Mathieu M, Wappner RS, et al. Prenatal diagnosis of mitochondrial fatty acid oxidation defects. *Prenatal Diagn* 1996;**16**:117–24.
183. Fingerhut R, Ensenauer R, Röschinger W, Arnecke R, Olgemöller B, Roscher AA. Stability of acylcarnitines and free carnitine in dried blood samples: implications for retrospective diagnosis of inborn errors of metabolism and neonatal screening for carnitine transporter deficiency. *Anal Chem* 2009;**81**:3571–5.
184. ter Veld F, Mueller M, Kramer S, Haussmann U, Herebian D, Mayatepek E, et al. A novel tandem mass spectrometry method for rapid confirmation of medium- and very long-chain acyl-CoA dehydrogenase deficiency in newborns. *PLOS ONE* 2009;**4**:e6449.
185. Gregersen N, Kolvraa S, Mortensen PB, Rasmussen K. C6-C10-Dicarboxylic aciduria: biochemical considerations in relation to diagnosis of beta-oxidation defects. *Scand J Clin Lab Invest* 1982;**42**(Suppl. 161):15–27.
186. Mortensen PB, Gregersen N. The biological origin of ketotic dicarboxylic aciduria. II. In vivo and in vitro investigations of the beta-oxidation of C8-C16-dicarboxylic acids in unstarved, starved and diabetic rats. *Biochim Biophys Acta* 1982;**710**:477–84.
187. Mortensen PB. Inhibition of fatty acid oxidation by valproate. *Lancet* 1981;**2**:856–7.
188. Hale DE, Thorpe C, Braat K, Wright JH, Roe CR, Coates PM, et al. The L-3-hydroxyacyl-CoA dehydrogenase deficiency. In: Tanaka K, Coates PM, editors. *Progress in Clinical and Biological Research. Fatty Acid Oxidation: Clinical, Biochemical and Molecular Aspects*, vol. 321. New York: Alan R. Liss; 1990. pp. 503–10.
189. Pollitt RJ. Clinical and biochemical presentations in 20 cases of hydroxydicarboxylic aciduria. In: Tanaka K, Coates PM, editors. *Fatty Acid Oxidation: Clinical, Biochemical and Molecular Aspects*. New York: Alan R. Liss; 1990. pp. 495–502.
190. Gregersen N, Kövlraa S, Mortensen PB. Acyl-CoA: glycine N-acyltransferase: in vitro studies on the glycine conjugation of straight- and branched-chained acyl-CoA esters in human liver. *Biochem Med Metab Biol* 1986;**35**:210–18.
191. Rinaldo P, O'Shea JJ, Coates PM, Hale DE, Stanley CA, Tanaka K. Medium-chain acyl-CoA dehydrogenase deficiency. Diagnosis by stable-isotope dilution measurement of urinary n-hexanoylglycine and 3-phenylpropionylglycine. *N Engl J Med* 1988;**319**:1308–13.
192. Rhead WJ. Screening for inborn errors of fatty acid oxidation in cultured fibroblasts: an overview. In: Tanaka K, Coates PM, editors. *Fatty Acid Oxidation: Clinical, Biochemical and Molecular Aspects*. New York: Alan R. Liss; 1990. pp. 365–82.
193. Okun JG, Kölker S, Schulze A, Kohlmüller D, Olgemöller K, Lindner M, et al. A method for quantitative acylcarnitine profiling in human skin fibroblasts using unlabelled palmitic acid: diagnosis of fatty acid oxidation disorders and differentiation between biochemical phenotypes of MCAD deficiency. *Biochim Biophys Acta* 2002;**1584**:91–8.
194. Tyni T, Pourfarzam M, Turnbull DM. Analysis of mitochondrial fatty acid oxdation intermediates by tandem mass spectrometry from intact mitochondria prepared from homogenates of cultured fibroblasts, skeletal muscle cells, and fresh muscle. *Pediatr Res* 2002;**52**:64–70.
195. Wu X, Prasad PD, Leibach FH, Ganapathy V. cDNA sequence, transport function, and genomic organization of human OCTN2, a

new member of the organic cation transporter family. *Biochem Biophys Res Commun* 1998;**246**:589–95.

196. Tamai I, Ohashi R, Nezu J, Yabuuchi H, Oku A, Shimane M, et al. Molecular and functional identification of sodium ion-dependent, high affinity human carnitine transporter OCTN2. *J Biol Chem* 1998;**273**:20378–82.
197. Lamhonwah AM, Tein I. Carnitine uptake defect: frameshift mutations in the human plasmalemmal carnitine transporter gene. *Biochem Biophys Res Commun* 1998;**252**:396–401.
198. Esser V, Britton CH, Weis BC, Foster DW, McGarry JD. Cloning, sequencing and expression of a cDNA encoding rat liver carnitine palmitoyltransferase I. Direct evidence that a single polypeptide is involved in inhibitor interaction and catalytic function. *J Biol Chem* 1993;**268**:5817–22.
199. Britton CH, Schultz RA, Zhang B, Esser V, Foster DW, McGarry JD. Human liver mitochondrial carnitine palmitoyltransferase I: Characterization of its cDNA and chromosomal localization and partial analysis of the gene. *Proc Natl Acad Sci USA* 1995;**92**(6):1984–8.
200. Indiveri C, Tonazzi A, Palmieri F. Identification and purification of the carnitine carrier from rat liver mitochondria. *Biochim Biophys Acta* 1990;**1020**:81–6.
201. Indiveri C, Tonazzi A, Prezioso G, Palmieri F. Kinetic characterization of the reconstituted carnitine carrier from rat liver mitochondria. *Biochim Biophys Acta* 1991;**1065**:231–8.
202. Viggiano L, Iacobazzi V, Marzella R, Cassano C, Rocchi M, Palmieri F. Assignment of the carnitine/acylcarnitine translocase gene (CACT) to human chromosome band 3p21.31 by in situ hybridization. *Cytogenet Cell Genet* 1997;**79**(1–2):62–3.
203. Woeltje KF, Esser V, Weis BC, Sen A, Cox WF, McPhaul MJ, et al. Cloning, sequencing and expression of a cDNA encoding rat liver mitochondrial carnitine palmitoyltransferase II. *J Biol Chem* 1990;**265**:10720–5.
204. Finocchiaro G, Taroni F, Rocchi M, Martin AL, Colombo I, Tarelli GT, et al. cDNA cloning, sequence analysis, and chromosomal localization of the gene for human carnitine palmitoyltransferase. *Proc Natl Acad Sci USA* 1991;**88**:661–5.
205. Taroni F, Verderio E, Dworzak F, Willems PJ, Cavadini P, DiDonato S. Identification of a common mutation in the carnitine palmitoyltransferase II gene in familial recurrent myoglobinuria patients. *Nat Genet* 1993;**4**:314–20.
206. Gellera C, Verderio E, Floridia G, Finocchiaro G, Montermini L, Cavadini P, et al. Assignment of the human carnitine palmitoyltransferase II gene (CPT1) to chromosome 1p32. *Genomics* 1994;**24**(1):195–7.
207. Verderio E, Cavadini P, Montermini L, Wang H, Lamantea E, Finocchiaro G, et al. Carnitine palmitoyltransferase II deficiency: structure of the gene and characterization of two novel disease-causing mutations. *Hum Mol Genet* 1995;**4**:19–29.
208. Strauss AW, Powell CK, Hale DE, Anderson MM, Ahuja A, Brackett JC, et al. Molecular basis of human mitochondrial very-long-chain acyl-CoA dehydrogenase deficiency causing cardiomyopathy and sudden death in childhood. *Proc Natl Acad Sci USA* 1995;**92**:10496–500.
209. Aoyama T, Souri M, Ueno I, Kamijo T, Yamaguchi S, Rhead WJ, et al. Cloning of human very-long-chain acyl-coenzyme A dehydrogenase and molecular characterization of its deficiency in two patients. *Am J Hum Genet* 1995;**57**:273–83.
210. Andresen BS, Bross P, Vianey-Saban C, Divry P, Zabot MT, Roe CR, et al. Cloning and characterization of human very-long-chain acyl-CoA dehydrogenase cDNA, chromosomal assignment of the gene and identification in four patients of nine different mutations within the VLCAD gene. *Hum Mol Genet* 1996;**5**:461–72.
211. Andresen BS, Vianey-Saban C, Bross P, Divry P, Roe CR, Nada MA, et al. The mutational spectrum in very long-chain acyl-CoA dehydrogenase deficiency. *J Inherit Metab Dis* 1996;**19**:169–72.
212. Indo Y, Yang-Feng T, Glassberg R, Tanaka K. Molecular cloning and nucleotide sequence of cDNAs encoding human long chain acyl-CoA dehydrogenase and assignment of the location of its gene (ACADL) to chromosome 2. *Genomics* 1991;**11**:609–20.
213. Uchida Y, Izai K, Orii T, Hashimoto T. Novel fatty acid beta-oxidation enzymes in rat liver mitochondria. II. Purification and properties of enoyl-coenzyme A (CoA) hydratase/3-hydroxyacyl-CoA dehydrogenase/3-ketoacyl-CoA thiolase trifunctional protein. *J Biol Chem* 1992;**267**(2):1034–41.
214. Sims HF, Brackett JC, Powell CK, Treem WR, Hale DE, Bennett MJ, et al. The molecular basis of pediatric long chain 3-hydroxyacyl-CoA dehydrogenase deficiency associated with maternal acute fatty liver of pregnancy. *Proc Natl Acad Sci USA* 1995;**92**:841–5.
215. Strauss AW. Genotype-phenotype correlations in mitochondrial trifunctional protein deficiency. *Proceedings of the 7th International Congress of Inborn Errors of Metabolism*, 1997, May 21–25. Vienna, Austria [Abstract 16].
216. Helander HM, Koivuranta KT, Horelli-Kuitunen N, Palvimo JJ, Palotie A, Hiltunen JK. Molecular cloning and characterization of the human mitochondrial 2,4-dienoyl-CoA reductase gene (DECR). *Genomics* 1997;**46**(1):112–19.
217. Kelly DP, Kim JJ, Billadello JJ, Hainline BE, Chu TW, Strauss AW. Nucleotide sequence of medium-chain acyl-CoA dehydrogenase mRNA and its expression in enzyme-deficient human tissue. *Proc Natl Acad Sci USA* 1987;**84**:4068–72.
218. Strauss AW, Duran M, Zhang Z, Alpers R, Kelly DP. Molecular analysis of medium chain acyl-CoA dehydrogenase deficiency. In: Tanaka K, Coates PM, editors. *Progress in Clinical and Biological Research. Fatty Acid Oxidation: Clinical, Biochemical and Molecular Aspects*, vol. 321. New York: Alan R. Liss, Inc; 1990. pp. 609–23.
219. Finocchiaro G, Ito M, Tanaka K. Purification and properties of short-chain acyl-CoA, medium-chain acyl-CoA and isovaleryl-CoA dehydrogenases from human liver. *J Biol Chem* 1987;**262**: 7982–9.
220. Naito E, Ozasa H, Ikeda Y, Tanaka K. Molecular cloning and nucleotide sequence of complementary DNAs encoding human short chain acyl-coenzyme A dehydrogenase and the study of the molecular basis of human short chain acyl-Coenzyme A dehydrogenase deficiency. *J Clin Invest* 1989;**83**:1605–13.
221. Corydon MJ, Gregersen N, Lehnert W, Ribes A, Rinaldo P, Kmoch S, et al. Ethylmalonic aciduria is associated with an amino acid variant of short chain acyl-coenzyme A dehydrogenase. *Pediatr Res* 1996;**39**:1059–66.
222. Gregersen N, Winter VS, Corydon MJ, Kristensen J, Ribes A, Martinez G, et al. Characterization of four new mutations in the short-chain acyl-CoA dehydrogenase (SCAD) gene in two patients with SCAD deficiency. *J Inherit Metab Dis* 1996;(Suppl. 1) [Abstract 19].

223. Vredendaal PJC, van den Berg IE, Malingré HE, Stroobants AK, Olde Weghuis DE, Berger R. Human short-chain L-3-hydroxyacyl-CoA dehydrogenase: cloning and characterization of the coding sequence. *Biochem Biophys Res Commun* 1996;**223**: 718–23.
224. Finocchiaro G, Ito M, Ikeda Y, Tanaka K. Molecular cloning and nucleotide sequence of cDNAs encoding the alpha-subunit of human electron transfer flavoprotein. *J Biol Chem* 1988;**263**: 15773–80.
225. Indo Y, Glassberg R, Yokota I, Tanaka K. Molecular characterization of variant alpha-subunit of electron transfer flavoprotein in three patients with glutaric acidemia type II and identification of glycine substitution for valine-157 in the sequence of the precursor, producing an unstable mature protein in a patient. *Am J Hum Genet* 1991;**49**:575–80.
226. Finocchiaro G, Colombo I, Garavaglia B, Gellera C, Valdameri G, Garbuglio N, et al. cDNA cloning and mitochondrial import of the beta-subunit of the human electron-transfer flavoprotein. *Eur J Biochem* 1993;**213**(3):1003–8.
227. Goodman SI, Bemelem KF, Frerman FE. Human cDNA encoding ETF dehydrogenase (ETF:ubiquinone oxidoreductase), and mutations in glutaric acidemia type II. In: Coates PM, Tanaka K, editors. *Progress in Clinical and Biological Research. New Developments in Fatty Acid Oxidation*, vol. 375. New York: Wiley-Liss, Inc; 1992. pp. 567–72.
228. Mitchell GA, Robert MF, Hruz PW, Wang S, Fontaine G, Behnke CE, et al. 3-Hydroxy-3-methylglutaryl coenzyme A lyase (HL). Cloning of the human and chicken liver HL cDNAs and characterization of a mutation causing human HL deficiency. *J Biol Chem* 1993;**25**:4376–81.
229. Roberts JR, Narasimhan C, Hruz PW, Mitchell GA, Miziorko HM. 3-Hydroxy-3-methylglutaryl-CoA lyase: expression and isolation of the recombinant human enzyme and investigation of a mechanism for regulation of enzyme activity. *J Biol Chem* 1994;**269**(27):17841–6.
230. Bouchard L, Robert MF, Vinarov D, Stanley CA, Thompson GN, Morris A, et al. Mitochondrial 3-hydroxy-3-methylglutaryl-CoA synthase deficiency: clinical course and description of causal mutations in two patients. *Pediatr Res* 2001;**49**(3):326–31.
231. Mascaro C, Buesa C, Ortiz JA, Haro D, Hegardt FG. Molecular cloning and tissue expression of human mitochondrial 3-hydroxy-3-methylglutaryl-CoA synthase. *Arch Biochem Biophys* 1995;**317**:385–90.
232. Matsubara Y, Indo Y, Naito E, Ozasa H, Glassberg R, Vockley J, et al. Molecular cloning and nucleotide sequence of cDNAs encoding the precursors of rat long chain acyl-coenzyme A, short chain acyl-coenzyme A, and isovaleryl-coenzyme A dehydrogenases. Sequence homology of four enzymes of the Acyl-CoA dehydrogenase family. *J Biol Chem* 1989;**264**:16321–31.
233. Finocchiaro G, Archidiacono N, Gellera C, Bloisi W, Colombo I. Molecular cloning and chromosomal localization of the beta-subunit of human electron transfer flavoprotein. *Am J Hum Genet* 1989;45:A185.
234. Pande SV, Brivet M, Slama A, Demaugre F, Aufrant C, Saudubray JM. Carnitine-acylcarnitine translocase deficiency with severe hypoglycemia and auriculo ventricular block. Translocase assay in permeabilized fibroblasts. *J Clin Invest* 1993;**91**:1247–52.
235. Vekemans BC, Bonnefont JP, Aupetit J, Royer G, Droin V, Attié-Bitach T. Prenatal diagnosis of carnitine palmitoyltransferase 2 deficiency in chorionic villi: a novel approach. *Prenatal Diag* 2003;**23**:884–7.
236. Treem WR, Stanley CA, Finegold DN, Hale DE, Coates PM. Primary carnitine deficiency due to a failure of carnitine transport in kidney, muscle and fibroblasts. *N Engl J Med* 1988;**319**: 1331–6.
237. Tein I, Xie ZW. The human plasmalemmal carnitine transporter is expressed in cultured lymphoblasts: a new, rapid non-invasive diagnostic method for diagnosis. *Clin Chim Acta* 1996;**252**: 201–4.
238. Pons R, Carrozzo R, Tein I, Walker WF, Addonizio LJ, Rhead W, et al. Deficient muscle carnitine transport in primary carnitine deficiency. *Pediatr Res* 1997;**42**:583–7.
239. Finocchiaro G, Ikeda Y, Ito M, Tanaka K. Biosynthesis, molecular cloning and sequencing of electron transfer flavoprotein. In: Tanaka K, Coates PM, editors. *Progress in Clinical and Biological Research. Fatty Acid Oxidation: Clinical, Biochemical and Molecular Aspects*, vol. 321. New York: Alan R. Liss; 1990. pp. 637–52.
240. Souri M, Aoyama T, Orii K, Yamaguchi S, Hashimoto T. Mutation analysis of very-long-chain acyl-coenzyme A dehydrogenase (VLCAD) deficiency: identification and characterization of mutant VLCAD cDNAs from four patients. *Am J Hum Genet* 1996;**58**:97–106.
241. Naito E, Indo Y, Tanaka K. Identification of two variant short chain acyl-coenzyme A dehydrogenase alleles, each containing a different point mutation in a patient with short chain acyl-coenzyme A dehydrogenase deficiency. *J Clin Invest* 1990;**85**: 1575–82.
242. Lamhonwah AM, Olpin SE, Pollitt RJ, Vianey-Saban C, Divry P, Guffon N, et al. Novel OCTN2 mutations: no genotype-phenotype correlations: early carnitine therapy prevents cardiomyopathy. *Am J Med Genet* 2002;**111**:271–84.
243. Gregersen N, Andresen BS, Corydon MJ, Corydon TJ, Olsen RK, Bolund L, et al. Mutation analysis in mitochondrial fatty acid oxidation defects: Exemplified by acyl-CoA dehydrogenase deficiencies, with special focus on genotype-phenotype relationship. *Hum Mutat* 2001;**18**:169–89.
244. Gregersen N, Andresen BS, Bross P. Prevalent mutations in fatty acid oxidation disorders: diagnostic considerations. *Eur J Pediatr* 2000;**159**(Suppl. 3):S213–18.
245. Kaufmann P, el-Schahawi M, DiMauro S. Carnitine palmitoyl-transferase II deficiency: diagnosis by molecular analysis of blood. *Mol Cell Biochem* 1997;**174**:237–9.
246. Taggart RT, Smail D, Apolito C, Vladutiu GD. Novel mutations associated with carnitine palmitoyltransferase II deficiency. *Hum Mutat* 1999;**13**:210–20.
247. Spiekerkoetter U, Lindner M, Santer R, Grotzke M, Baumgartner MR, Boehles H, et al. Treatment recommendations in long-chain fatty acid oxidation defects: consensus from a workshop. *J Inherit Metab Dis* 2009;**32**:498–505.
248. Gillingham M, Van Calcar S, Ney D, Wolff J, Harding C. Dietary management of long-chain 3-hydroxyacyl-CoA dehydrogenase deficiency (LCHADD). A case report and survey. *J Inherit Metab Dis* 1999;**22**:123–31.

249. Uauy R, Treen M, Hoffman DR. Essential fatty acid metabolism and requirements during development. *Semin Perinatol* 1989;**13**:118–30.

250. Solis JO, Singh RH. Management of fatty acid oxdation disorders: a survey of current treatment strategies. *J Am Diet Assoc* 2002;**102**:1800–3.

251. Fernandes J, Smit G. The glycogen storage diseases. In: Fernandes J, Saudubray J, Van den Berghe C, editors. *Inborn Errors of Metabolic Diseases, Diagnosis and Treatment*. 3rd ed. New York: Springer; 2000. pp. 87–101.

252. Hayde M, Widhalm K. Effects of cornstarch treatment in very young children with type I glycogen storage diseases. *Eur J Pediatr* 1990;**149**:630–3.

253. Laforêt P, Vianey-Saban C. Disorders of muscle lipid metabolism: diagnostic and therapeutic challenges. *Neuromuscul Disord* 2010;**20**:693–700.

254. Brown-Harrison MC, Nada MA, Sprecher H, Vianey-Saban C, Farquhar Jr J, Gilladoga AC, et al. Very long chain acyl-CoA dehydrogenase deficiency: successful treatment of acute cardiomyopathy. *Biochem Mol Med* 1996;**58**:59–65.

255. Cox GF, Souri M, Aoyama T, Rochenmacher S, Varvogli L, Rohr F, et al. Reversal of severe hypertrophic cardiomyopathy and excellent neuropsychologic outcome in very-long-chain acyl-coenzyme A dehydrogenase deficiency. *J Pediatr* 1998;**133**:247–53.

256. Gregersen N, Wintzensen H, Christensen SK, Christensen MF, Brandt NJ, Rasmussen K. C6-C10-dicarboxylic aciduria: investigations of a patient with riboflavin responsive multiple acyl-CoA dehydrogenation defects. *Pediatr Res* 1982;**16**:861–8.

257. Henriques BJ, Rodrigues JV, Olsen RK, Bross P, Gomes CM. Role of flavinylation in a mild variant of multiple acyl-CoA dehydrogenation deficiency: a molecular rationale for the effects of riboflavin supplementation. *J Biol Chem* 2009;**284**:4222–9.

258. Gillingham MB, Mills MD, Vancalcar SC, Verhoeve, JN, Wolff JA, Harding CO. DHA supplementation in children with long-chain 3-hydroxyacyl-CoA dehydrogenase deficiency. *7th International Congress of Inborn Errors of Metabolism*; 1997 May 21–25; Vienna, Austria [Abstract].

259. Tein I, Ibdah JA, Vajsar J, Sherwood WG, MacMillan L, Strauss AW. Molecular basis of LCHAD deficiency causing myopathy, recurrent myoglobinuria and peripheral neuropathy, and marked response of neuropathy to cod liver oil extract. *Proceedings of the 7th International Congress of Inborn Errors of Metabolism*; 1997 May 21–25; Vienna, Austria [Abstract].

260. Tein I, Vajsar J, MacMillan L, Sherwood WG. Long-chain L-3-hydroxyacyl-coenzyme dehydrogenase deficiency neuropathy: response to cod liver oil. *Neurology* 1999;**52**:640–3.

261. Roe CR, Sweetman L, Roe DS, David F, Brunengraber H. Treatment of cardiomyopathy and rhabdomyolysis in long-chain fat oxidation disorders using an anaplerotic odd-chain triglyceride. *J Clin Invest* 2002;**110**:259–69.

262. Roe CR, Yang BZ, Brunengraber H, Roe DS, Wallace M, Garritson BK. Carnitine palmitoyltransferase II deficiency: successful anaplerotic diet therapy. *Neurology* 2008;**71**:260–4.

263. Lee CH, Olson P, Evans RM. Minireview: lipid metabolism, metabolic diseases, and peroxisome proliferator-activated receptors. *Endocrinology* 2003;**144**:2201–7.

264. Feige JN, Gelman L, Michalik L, Desvergne B, Wahli W. From molecular action to physiological outputs: peroxisome proliferator-activated receptors are nuclear receptors at the crossroads of key cellular functions. *Prog Lipid Res* 2006;**45**:120–59.

265. Peters JM, Aoyama T, Burns AM, Gonzalez FJ. Bezafibrate is a dual ligand for PPARalpha and PPARbeta: studies using null mice. *Biochim Biophys Acta* 2003;**1632**:80–9.

266. Tenenbaum A, Motro M, Fisman EZ. Dual and pan-peroxisome proliferator-activated receptors (PPAR) co-agonism: the bezafibrate lessons. *Cardiovasc Diabetol* 2005;**4**:14.

267. Djouadi F, Bastin J. PPARs as therapeutic targets for correction of inborn mitochondrial fatty acid oxidation disorders. *J Inherit Metab Dis* 2008;**31**:217–25.

268. Monk JP, Todd PA. Bezafibrate. A review of its pharmacodynamic and pharmacokinetic properties and therapeutic use in hyperlipidaemia. *Drugs* 1987;**33**:539–59.

269. Bonnefont JP, Bastin J, Behin A, Djouadi F. Bezafibrate for an inborn mitochondrial beta-oxidation defect. *N Engl J Med* 2009;**360**:838–40.

270. Bonnefont JP, Bastin J, Laforêt P, Aubey F, Mogenet A, Romano S, et al. Long-term follow-up of bezafibrate treatment in patients with the myopathic form of carnitine palmitoyltransferase 2 deficiency. *Clin Pharmacol Ther* 2010;**88**:101–8.

271. Yamaguchi S, Li H, Purevsuren J, Yamada K, Furui M, Takahashi T, et al. Bezafibrate can be a new treatment option for mitochondrial fatty acid oxidation disorders: evaluation by in vitro probe acylcarnitine assay. *Mol Genet Metab* 2012;**107**:87–91.

272. Ringseis R, Lüdi S, Hirche F, Eder K. Treatment with pharmacological peroxisome proliferator-activated receptor alpha agonist clofibrate increases intestinal carnitine absorption in rats. *Pharmacol Res* 2008;**58**:58–64.

273. Wen G, Ringseis R, Eder K. Mouse OCTN2 is directly regulated by peroxisome proliferator-activated receptor alpha (PPARalpha) via a PPRE located in the first intron. *Biochem Pharmacol* 2010;**79**:768–76.

274. Merritt II JL, Matern D, Vockley J, Daniels J, Nguyen TV, Schowalter DB. In vitro characterization and in vivo expression of human very-long chain acyl-CoA dehydrogenase. *Mol Genet Metab* 2006;**88**:351–8.

275. Lund AM, Skovby F, Vestergaard H, Christensen M, Christensen E. Clinical and biochemical monitoring of patients with fatty acid oxidation disorders. *J Inherit Metab Dis* 2010;**33**:495–500.

276. Bennett MJ, Worthy E, Pollitt RJ. The incidence and presentation of dicarboxylic aciduria. *J Inherit Metab Dis* 1987;**10**:241–2.

277. Chace DH, DiPerna JC, Mitchell BL, Sgroi B, Hofman LF, Naylor EW. Electrospray tandem mass spectrometry for analysis of acylcarnitines in dried postmortem blood specimens collected at autopsy from infants with unexplained cause of death. *Clin Chem* 2001;**47**:1166–82 [Erratum in: Clin Chem 2001;47:1748].

278. Arnold GL, Van Hove J, Freedenberg D, Strauss A, Longo N, Burton B, et al. A Delphi clinical practice protocol for the management of very long chain acyl-CoA dehydrogenase deficiency. *Mol Genet Metab* 2009;**96**:85–90.

279. Rinaldo P, Matern D. Disorders of fatty acid transport and mitochondrial oxidation: challenges and dilemmas of metabolic evaluation. *Genet Med* 2000;**2**:338–44.

280. Shekhawat PS, Matern D, Strauss AW. Fetal fatty acid oxidation disorders, their effect on maternal health and neonatal outcome: impact of expanded newborn screening on their diagnosis and management. *Pediatr Res* 2005;**57**:78R–86R.

281. Wilcken B. Recent advances in newborn screening. *J Inherit Metab Dis* 2007;**30**:129–33.
282. Wilcken B, Wiley V, Hammond J, Carpenter K. Screening newborns for inborn errors of metabolism by tandem mass spectrometry. *N Engl J Med* 2003;**348**:2304–12.
283. Wilcken B, Haas M, Joy P, Wiley V, Chaplin M, Black C, et al. Outcome of neonatal screening for medium-chain acyl-CoA dehydrogenase deficiency in Australia: a cohort study. *Lancet* 2007; **369**:37–42.
284. Spiekerkoetter U. Mitochondrial fatty acid oxidation disorders: clinical presentation of long-chain fatty acid oxidation defects before and after newborn screening. *J Inherit Metab Dis* 2010;**33**:527–32.
285. Spiekerkoetter U, Lindner M, Santer R, Grotzke M, Baumgartner MR, Boehles H, et al. Management and outcome in 75 individuals with long-chain fatty acid oxidation defects: results from a workshop. *J Inherit Metab Dis* 2009;**32**: 488–97.
286. Wilcken B. Fatty acid oxidation disorders: outcome and long-term prognosis. *J Inherit Metab Dis* 2010;**33**:501–6.
287. Spiekerkoetter U, Mueller M, Cloppenburg E, Motz R, Mayatepek E, Bueltmann B, et al. Intrauterine cardiomyopathy and cardiac mitochondrial proliferation in mitochondrial trifunctional protein (TFP) deficiency. *Mol Genet Metab* 2008;**94**: 428–30.

Chapter 41

Mitochondrial Encephalomyopathies

Darryl C. De Vivo, Carmen Paradas, and Salvatore DiMauro

INTRODUCTION

Historical Background

The history of mitochondria can be traced back at least to 1850 when Kolliker described a regular array of subcellular particles in muscle sarcomeres.[1] Fifty-two years were to pass before Luft and colleagues implicated mitochondria as the cause of a human disease.[2] A 39-year-old woman had florid symptoms of hypermetabolism thought initially to be due to thyroid disease. However, antithyroid medication and partial thyroidectomy were ineffective. Studies of biopsied muscle would prove to be more informative. Isolated mitochondria were loosely coupled *in vitro* by polarographic measurements, and the number of organelles was vastly increased by morphological criteria. Many mitochondria were abnormally large with tightly packed cristae (Figure 41.1). Later, Shy and colleagues would use the terms pleoconial and megaconial when describing similar mitochondrial abnormalities in the biopsied skeletal muscle specimens from some children with limb weakness.[3,4]

The term mitochondrion was coined by Benda in 1898.[5] The specialized function and the unique structure of mitochondria emerged in the 20th century. Keilin described the cytochrome system and Warburg found evidence of a terminal oxidase system.[6,7] He also observed that nicotine adenine dinucleotide behaved as a hydrogen acceptor in oxidation-reduction reactions.[8] Flavoproteins were found to act as intermediary carriers of electrons between the reduced form of nicotine adenine dinucleotide and cytochrome systems. Krebs and Kornberg discovered that phosphorylation in aerobic systems depended on the consumption of oxygen, hence the concept of "oxidative phosphorylation."[9]

Krebs and Kornberg continued their seminal observations in the 1930s, showing that citrate was formed from pyruvate and oxaloacetate.[9] These observations led to the discovery of the tricarboxylic acid cycle (the TCA cycle, now better known as Krebs cycle), which converted pyruvate to carbon dioxide and water. These investigators and others showed that oxidation of intermediates of the tricarboxylic acid cycle was coupled to the generation of adenosine triphosphate.

The importance of the mitochondria as the source of cellular energy was being established by these seminal observations. Additional experiments explained the relationships between fuel combustion to carbon dioxide and water, the generation of reducing equivalents in the Krebs cycle, and the re-oxidation of these equivalents, NADH and $FADH_2$, in the respiratory chain coupled with the synthesis of ATP. The concept of oxidation-phosphorylation was established and the chemiosmotic hypothesis of Mitchell in 1961 proposed the proton-motive force as the primary source of energy for ATP synthesis in complex V.[10] The rudimentary structure of the respiratory chain now was complete with the five enzyme complexes. More critical observations would follow over the next 50 years, characterizing in greater detail the subunit structure of the respiratory chain complexes, the critical shuttling roles of coenzyme Q10 and cytochrome c, and the dual roles of nuclear DNA and mitochondrial DNA in the synthesis of the many subunits that composed the five enzyme complexes.[11]

In parallel with these brilliant biochemical observations, a series of studies was being published describing the ultrastructural characteristics of the mitochondrion.[12–14] Four components of the organelle were characterized: the outer and inner membranes, the intermembranous space, and the inner mitochondrial compartment matrix. A series of studies subsequently demonstrated that the inner mitochondrial membrane was largely impermeable to molecules of all sizes, and special adaptations were necessary for the translocation of metabolites and proteins from the intermembranous space to the mitochondrial matrix. These observations set the stage for an understanding of the complex importation processes that are necessary both to translocate solutes across the inner mitochondrial membrane[15] and to move proteins from the cytoplasmic environment to the intramitochondrial environment.[16,17] Protein translocation is elaborate

B.T. Darras, H. Royden Jones, Jr., M.M. Ryan & D.C. De Vivo (Eds): Neuromuscular Disorders of Infancy, Childhood and Adolescence, Second edition.
DOI: http://dx.doi.org/10.1016/B978-0-12-417044-5.00041-X

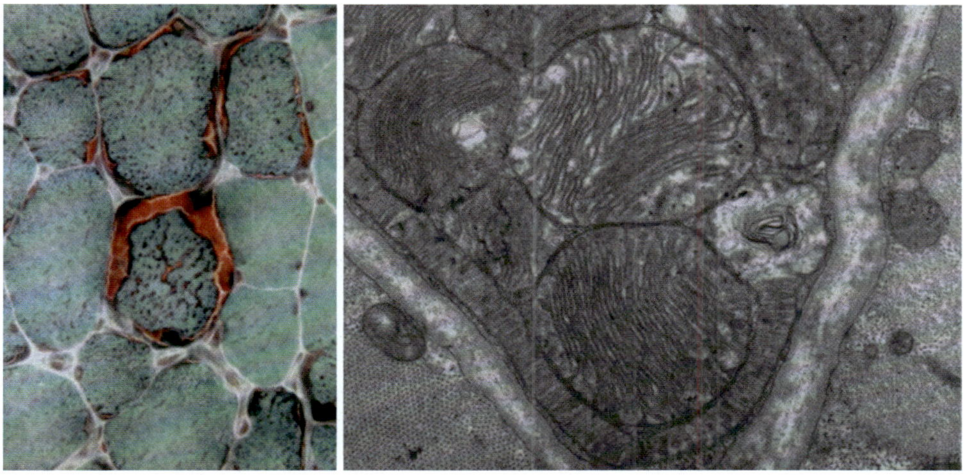

FIGURE 41.1 Muscle biopsy from the second patient with Luft syndrome. In the left panel, histochemistry with the modified Gomori trichrome stain shows two ragged-red fibers. In the right panel, electron microscopic examination shows greatly enlarged mitochondria with tightly packed cristae in one fiber compared to normal-size mitochondria in the adjacent myofiber.

and energy-dependent, requiring the proteins to be unfolded before traversing the mitochondrial outer and inner membranes and then refolded after entering the mitochondrial matrix. The leader sequence that guides the protein through the mitochondrial membrane to the matrix is cleaved by a mitochondrial-specific protease. Only 13 proteins are synthesized within the mitochondrial matrix by protein-coding genes located in the mitochondrial genome.[18]

In 1963, two important observations were made simultaneously and independently. The first observation described the presence of intramitochondrial fibers with DNA characteristics.[19] Twenty-five years would elapse before this observation would take on clinical significance. The second observation was the description of abnormal mitochondria under the light microscope: Engel and Cunningham exploited a modification of the Gomori trichrome stain and found conglomerations of abnormal organelles in the cross section of muscle fibers that they called "ragged-red fibers" (RRF) (Figure 41.1).[20] The RRF almost invariably contained structurally abnormal mitochondria as seen under the electron microscope. These morphological criteria later were displaced by biochemical and genetic criteria for the classification of mitochondrial diseases.

Early morphological observations gradually gave rise to another fundamental biological principle related to mitochondria. Mitochondrial ribosomes resemble bacterial ribosomes under the electron microscope. This observation led Ephrussi and colleagues in 1949 to suggest that mitochondrial morphogenesis was controlled by extra-chromosomal or cytoplasmic genetic factors.[21] The observation of a yeast mutation causing a deficiency of respiratory function and inherited in a non-Mendelian fashion was central to the development of the unique theory put forward by Ephrussi and colleagues. Schatz et al. in 1964 found DNA in yeast mitochondria and this observation led him to propose that the yeast organelle might be self-replicating and relatively independent of the nuclear genome.[22] A series of observations followed and culminated in the report by Fine in 1978 that all mitochondrial DNA was derived from the unfertilized ovum.[23] This hypothesis was validated two years later by Giles and colleagues when they were able to trace the inheritance of a mitochondrial DNA polymorphism in a 33-member three-generation family.[24] The typical mitochondrial DNA cleavage pattern and the atypical mitochondrial DNA cleavage pattern could be traced from one generation to the next, clearly documenting the non-Mendelian pattern of maternal inheritance. Later, it would be shown that clinical conditions related to mitochondrial DNA (mtDNA) mutations were transmitted from the mother to all of her male and female progeny, but only the daughters would be able to pass the condition to succeeding generations. This genetic profile was similar to Mendelian inheritance including autosomal dominant and X-linked patterns, but differed from Mendelian inheritance because both genders were equally affected and there was no evidence of father-to-child transmission. Wallace and colleagues later would add two other important concepts to explain the variable expression of maternally inherited genetic defects: namely, replicative segregation and threshold effect.[25]

The historical stage now was set for the two dramatic publications that appeared in 1988 and that introduced the medical community to the presence of mitochondrial DNA mutations as a new cause of human diseases.[26,27] Holt et al. reported large-scale deletions of mitochondrial DNA in patients with a mitochondrial myopathy[26] and Wallace et al. described a point mutation in the mitochondrial protein-coding gene for subunit 4 of complex I (*ND4*) in a family with Leber's hereditary optic neuropathy (LHON).[27] More than 200 additional mitochondrial DNA point mutations would be described over the next 25 years in addition to numerous mitochondrial DNA large-scale rearrangements.[28] The number of mitochondrial DNA mutations continues to increase, and now exceeds 250 at the time of this chapter revision.

These mutations affect all aspects of the mitochondrial genome, which had been fully characterized and sequenced in 1981 by Anderson and colleagues.[18] The double-stranded molecule contains 16,569 base pairs. Thirty-seven genes are represented including two ribosomal RNAs, 22 transfer RNAs, and 13 protein-coding genes. This molecular machinery allows for the intramitochondrial synthesis of 13 subunits of the respiratory chain. Seven of these subunits are located in complex I, one in complex III, three in complex IV, and two in complex V (Figure 41.2). The simplicity of the mitochondrial genome is shown in Figure 41.2 and is characterized by the circularity, the absence of introns, and the presence of one asymmetric replication origin.

Epidemiology

Epidemiological studies have confirmed the assumption of clinicians that this group of diseases is underestimated. In fact, epidemiological studies limited to mtDNA-related disorders in adults have shown a minimum prevalence of approximately 1 in 5000, making them among the most common genetic disorders and a major burden for society.[29] More interesting to pediatricians, screening of umbilical cord blood for the 10 most common pathogenic mtDNA point mutations revealed that around 1 in 200 infants harbors a mutation.[30] As the prevalence of

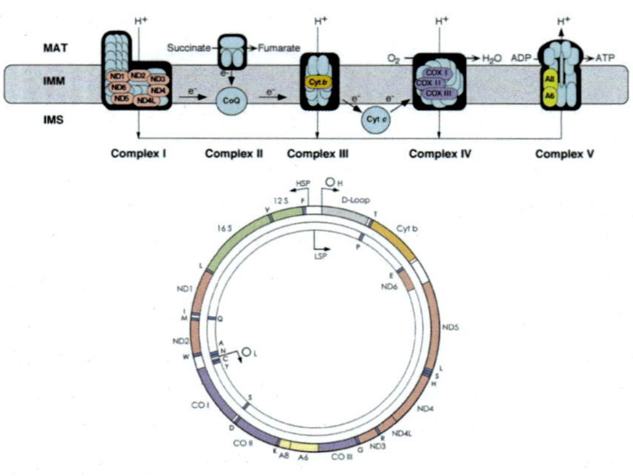

FIGURE 41.2 The mitochondrial respiratory chain (top) and the mitochondrial DNA (bottom). Genes and corresponding gene products are similarly color-coded. ND denotes the subunits of NADH-CoQ oxidoreductase (complex I); Cyt *b*, cytochrome *b*; subunits of cytochrome *c* oxidase (complex IV) are labeled CO in the mitochondrial DNA (mtDNA) scheme and COX in the respiratory chain rendition; A6 and A8 denote subunits 6 and 8 of ATP synthase (complex V). The 22 tRNA genes are identified by one-letter amino acid nomenclature; 12S and 16S denote ribosomal RNAs (rRNAs). O_H and O_L are the origin of heavy- and light-strand replication; HSP and LSP are the promoters of heavy- and light-strand transcription. ADP, adenosine diphosphate; ATP, adenosine triphosphate. *Reproduced with permission from DiMauro S.*[28]

mtDNA-related disorders is far less than 1 in 200, it must be concluded that the mutation load in most carriers remains below the pathogenic threshold (see below). This concept is also in keeping with the results of a supersensitive next-generation sequencing study showing that low-level changes (0.2–2.0% heteroplasmy [the coexistence of mutated and wild-type mtDNAs within a cell—see the following]) are present in blood and skeletal muscle of clinically unaffected individuals, a phenomenon dubbed "universal heteroplasmy."[31] The possibility exists, therefore, that clinically silent mtDNA mutations could expand over time and manifest as *de novo* mitochondrial diseases.

Beginning in 1995, mutations in nuclear genes affecting the respiratory chain[32] were identified at an escalating pace, largely thanks to linkage analysis, homozygosity mapping, monochromosomal transfer techniques, gene complementation studies, and, more recently, whole exome or mito-exome sequencing.[33–35] As Mendelian mitochondrial disorders are more common in infants and children than in adults, as exemplified by Leigh syndrome (LS), next-generation sequencing is particularly important for the molecular diagnosis of pediatric disorders.[36]

Classification

Our earlier classification of mitochondrial diseases was based on the biochemical site of dysfunction,[37] for example substrate transport, substrate utilization, Krebs cycle, respiratory chain, and protein importation. However, this scheme proved inadequate and clumsy when assessing patients with various clinical manifestations. The discovery in 1988 of disease-causing mtDNA mutations and, starting in 1995, the description of myriad diverse Mendelian mitochondrial disorders, forced a revision of the classification, which is today rationally and comprehensively based on molecular criteria.[28]

Keeping in mind that a generally accepted definition of mitochondrial diseases restricts these diseases to defects of the respiratory chain and oxidative phosphorylation (OXPHOS), they fall broadly into two categories: those due to mutations in mtDNA, which cause maternally inherited or sporadic disorders; and those due to mutations in nDNA, which show Mendelian inheritance.

A more practical way of classifying these disorders is to divide them into three groups: mtDNA defects; defects of nDNA that affect the respiratory chain directly or indirectly; and defects of mtDNA maintenance (also known as defects of intergenomic communication) (Table 41.1). We separate out the defects of mtDNA maintenance for two reasons: (i) they involve the mitochondrial genome directly, causing multiple mtDNA deletions or mtDNA depletion; and (ii) although unequivocally Mendelian with regard to inheritance, these disorders share much of the clinical

TABLE 41.1 Defects of the Respiratory Chain

Mitochondrial DNA (mtDNA)-Related Diseases (Maternal or Sporadic)	Nuclear DNA (nDNA)-Related Diseases (Mendelian)
1. Single deletions (sporadic)	1. "direct hits"
2. Point mutations (maternally inherited)	2. "indirect hits"
	3. Defects of mtDNA maintenance (mtDNA depletion and multiple deletions)
	4. Defects of mtDNA translation
	5. Defects of the inner mitochondrial membrane lipid milieu
	6. Defects of mitochondrial dynamics

heterogeneity of primary mtDNA-related diseases because the polyploid mtDNA is involved in both conditions.

Pathogenesis and Genetics

There are two main disturbances of cell function associated with mitochondrial diseases: failure to generate sufficient ATP to carry out cellular work; and failure to neutralize reactive oxygen species (ROS) with resulting damage to structural elements of the cell. All categories of mitochondrial diseases, as shown in Table 41.1, impair ATP production. Clinically, these patients exhibit classical symptoms of energy failure such as weakness, hypotonia, reduced physical stamina, muscle atrophy, heart failure, and limitation of eye movements with ptosis. Developmental processes also may suffer as shown by failure to thrive with short stature, reduced muscle bulk, and deceleration of head growth.

Accumulation of ROS damages tissue membranes, promotes genetic mutations, and contributes to cell death. Symptoms associated with this insidious tissue damage are more difficult to appreciate in a correlative sense but reflect the overall degenerative process. The failing heart, muscular atrophy, and basal ganglia damage likely reflect this microchemical attack on tissue structures, suggesting that the pathogenesis of human mitochondrial diseases directly affecting oxidation-phosphorylation mechanisms is intensified by the combination of energy failure and the accumulation of ROS.[38,39]

Before proceeding with the description of mitochondrial diseases, it may be useful to remind the readers of the *bona fide* mitochondrial diseases that do not belong to this category because they are not due to defects of the respiratory chain and OXPHOS. They belong in three major categories: defects of fatty acid oxidation (FAO); defects of pyruvate metabolism; and defects of the Krebs cycle (except for succinate dehydrogenase [SDH, complex II], which belongs both to the Krebs cycle and to the respiratory chain) (Figure 41.3). FAO defects are discussed in Chapter 40.

The molecular genetics of mitochondrial diseases is complex because it is determined by two genomes, nDNA and mtDNA.[28] The respiratory chain is composed of five complexes containing about 85 subunits, all of which are encoded by nDNA except 13 that are encoded by mtDNA (Figure 41.2). As a result, a defect of the respiratory chain may be genetically determined as an autosomal trait or as a maternal trait. The autosomal conditions may be dominant, recessive, or X-linked traits. Maternally transmitted traits are non-Mendelian and obey the rules of mitochondrial genetics.[41] There are three biological rules that explain mitochondrial genetics. The first is *maternal inheritance*. Sperm mtDNA is actively degraded after fertilization; as a result, all mtDNA is maternally inherited. Mitochondrial DNA mutations, therefore, are transmitted from mother to her male and female progeny, but only her daughters will transmit the mutation to their children.

The second rule is *heteroplasmy*, with the related concept of *threshold effect*. Each mitochondrion contains two to ten mitochondrial DNA copies, and each cell contains hundreds or thousands of mitochondria. Therefore, any given cell may contain as many as several thousand mitochondrial DNA copies. If an mtDNA mutation exists, some of the molecules will harbor the mutation whereas other molecules will be normal. This mixture of mutated and wild-type mtDNAs is referred to as heteroplasmy. Despite recent evidence of "universal heteroplasmy" (see above), for all practical purposes all mtDNAs in normal cells are wild-type, a condition called homoplasmy. An increasing percentage of mutated mtDNAs threatens the biological integrity of the cell until the mutation load reaches a theoretical threshold and the cellular phenotype is transformed from normal to diseased. Pathogenic mtDNA mutations usually are heteroplasmic whereas neutral polymorphisms (nondeleterious mtDNA mutations) are homoplasmic, but there are many exceptions to this

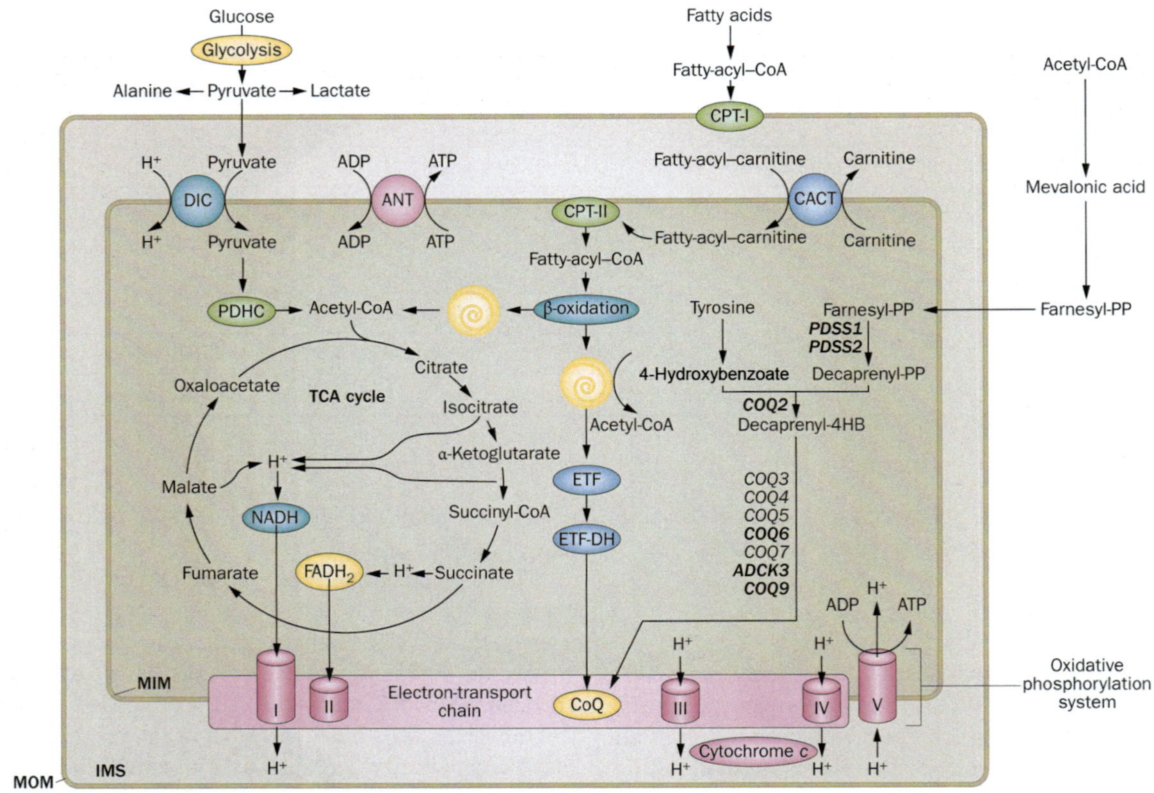

FIGURE 41.3 Schematic view of mitochondrial metabolism. The respiratory chain is highlighted in red. Electrons from dehydrogenases in the Krebs cycle and in beta-oxidation spirals are passed "horizontally" along four protein complexes and two small carriers (the electron-transport chain) that are embedded in the mitochondrial inner membrane (MIM). The electrons travel from complex I (an NADH dehydrogenase) and complex II (a succinate dehydrogenase) to CoQ10 (a small mobile electron carrier also known as ubiquinone), then to complex III (ubiquinone oxidoreductase), cytochrome *c* (another small mobile electron carrier), and complex IV (cytochrome *c* oxidase), ultimately producing water. Concomitant with this horizontal flow of electrons, there is a "vertical" vectorial transport of dehydrogenase-derived protons from the matrix across the MIM into the intermembrane space. This process creates an electrochemical proton gradient across the MIM that is used to drive complex V (F_0F_1 ATPsynthase), a tiny "turbine" that converts ADP to ATP. Conventionally, the five complexes comprise the oxidative phosphorylation system. The biosynthetic pathway of CoQ10, beginning with acetyl-CoA, is shown. ANT, adenine nucleotide translocator; CACT, carnitine-acylcarnitine translocator; CoA, coenzyme A; CoQ10, coenzyme Q10; CPT, carnitine palmitoyltransferase; DIC, dicarboxylate carrier; ETF, electron-transfer flavoprotein; ETF-DH, ETF dehydrogenase; FAD, flavin adenine dinucleotide; IMS, intermembrane space; MIM, mitochondrial inner membrane; MOM, mitochondrial outer membrane; PDHC, pyruvate dehydrogenase complex; TCA, tricarboxylic acid. *Reproduced with permission from DiMauro and Schon.*[40]

rule,[41] notably the three mutations commonly associated with LHON.[42] As the threshold varies inversely with the oxidative demands of the tissue, oxidative-rich tissues are more vulnerable to the presence of pathogenic mtDNA mutations.

The third rule is *mitotic segregation*, which stipulates that the proportion of mutant mtDNAs can shift in daughter cells at division. This stochastic shift in mutated mtDNAs during mitosis can result in segregation of the mutation and is invoked as an explanation for the phenotypic heterogeneity among patients with the same mtDNA point mutation, including siblings. For example, the m.3243A > G mutation, commonly referred to as the MELAS mutation, may produce several different phenotypes, one reflecting the classical MELAS presentation (mitochondrial encephalomyopathy, lactic acidosis, and stroke-like episodes), others reflecting different clinical pictures such as progressive external ophthalmoplegia (PEO), myopathy, cardiomyopathy, and diabetes mellitus and deafness (see Case Example 41.1). Presumably, these phenotypic differences are determined by the segregation of the A3243G point mutation during mitosis, although the influence of modifying mtDNA or nDNA genes cannot be excluded. Other examples of phenotypic heterogeneity and genotypic homogeneity are shown in the morbidity map of the human mitochondrial genome (Figure 41.4). This figure also demonstrates examples of phenotypic homogeneity and genotypic heterogeneity, including the diverse point mutations resulting in Leigh syndrome (LS) or in MELAS.

CASE EXAMPLE 41.1

A 5-year-old girl lost consciousness for 30 minutes after strenuous exercise. She also had multiple episodes of dimmed vision lasting a few minutes. Pregnancy and delivery had been normal and she walked at 1 year. Weakness of the legs had become apparent at 2 years when she had difficulty climbing stairs. At 6 years, episodes of right-sided paresthesias after exertion lasted 30–60 minutes and were followed by blindness. One episode was followed by a generalized tonic-clonic seizure and phenobarbital therapy was started. Postictally, there was hemiparesis that improved after 3 days. A second right focal seizure was followed by a hemiparesis for 5 days. A third generalized seizure was accompanied by blindness, which was thought to be cortical in origin because pupillary reactions to light and fundoscopic examination were normal. Audiogram was normal at age 5 years but showed neurosensory hearing loss at age 9 years. The disorder progressed and at 9 years she was severely demented and deaf, speaking only in short phrases. At 10 years, she died in congestive heart failure. Family history showed that a 17-year-old brother had short stature, mental retardation, and hearing loss. Her older sister was normal and so were her mother and maternal grandmother. On repeated occasions, venous blood obtained without a tourniquet was at least 4 times higher than normal. Brain CT scan at age 7 years showed mild cerebellar atrophy. A muscle biopsy at age 7 years showed numerous COX-positive RRF. Biochemical studies in muscle homogenate showed markedly decreased activities of all respiratory chain complexes except complex II (succinate dehydrogenase), which was actually higher than normal. Citrate synthase activity was increased two-fold. Molecular genetic analysis showed the m.3243A > G mutation in the Leu(UUR) of mtDNA. Mutation loads in blood ranged from 55% in the proposita to 14% in the brother, 59% in the mother, and 15% in the sister. The mutation load in the patient's muscle was 83%. At autopsy, there was mild atrophy of the cerebral hemispheres with countless small cortical infarcts, especially in the occipital lobes and in the posterior portions of the temporal lobes. The cortical lesions were often laminar and involved the three deepest layers.

Comment

This patient was the first reported case of MELAS harboring the "common" MELAS-3243 mutation in mtDNA, and it was revisited in 1992 by Hirano et al.[43] The presentation was typical, with normal initial development followed by recurrent and partially reversible stroke-like episodes, often triggered by seizures. Involvement of the occipital lobe with cortical blindness is common. Maternal inheritance was not obvious phenotypically in this family, but it became evident at the molecular level. The mutational load in blood did not accurately reflect clinical severity, as is often the case. Among easily available tissues, urinary sediment is a better molecular marker of severity. The fact that RRF were COX-positive is a peculiar feature of MELAS among mtDNA-related disorders. The pattern of muscle biochemistry is typical of a mtDNA mutation affecting mitochondrial protein synthesis: increased activities of SDH and citrate synthase (both encoded by nDNA) and variously decreased activities of complexes containing mtDNA-encoded subunits.

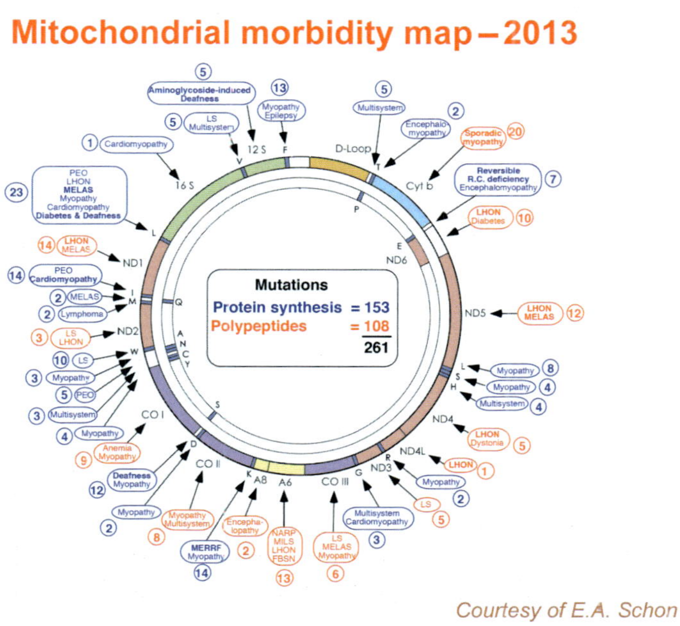

FIGURE 41.4 Morbidity map of the human mtDNA. Disorders caused by mutations in protein-coding genes are shown in red. Disorders caused by mutations in gene controlling protein synthesis are shown in blue. FBSN, familial bilateral striatal necrosis; LHON, Leber hereditary optic neuropathy; LS, Leigh syndrome; MELAS, mitochondrial encephalomyopathy lactic acidosis and stroke-like episodes; MERRF, myoclonus epilepsy and ragged-red fibers; MILS, maternally inherited Leigh syndrome; NARP, neuropathy, ataxia, retinitis pigmentosa; PEO, progressive external ophthalmoplegia. *Reproduced with permission from DiMauro S.*[28]

Clinical Features

There are over 250 pathogenic mtDNA point mutations, most of which affect mitochondrial protein synthesis while a smaller number affect protein-coding genes. Pure myopathies and neuropathies are unusual, but neuromuscular symptoms are common as part of multisystemic phenotypes. Multisystemic disorders are illustrated by Kearns-Sayre syndrome (KSS), Pearson syndrome, MELAS, myoclonus epilepsy and RRF (MERRF), and neuropathy, ataxia, and retinitis pigmentosa/maternally inherited LS (NARP/MILS) (Table 41.2). Two of these diseases (KSS and Pearson syndrome) are due to large-scale deletions; two (MELAS and MERRF) are caused by tRNA mutations that impair mitochondrial protein synthesis globally; and the remaining two diseases (NARP and MILS) are due to mutations in a single protein-coding gene.

Mitochondrial genetics provides some explanation for the clinical heterogeneity of these diseases. For example, the degree of heteroplasmy of the same mutation in the gene encoding ATPase 6 (m.8993T > G) is clearly related to the severity and age at onset of NARP, a disorder of young adults with ~70% mutation load, and MILS, a severe neurodegenerative disease of infancy or young children with mutation loads >90%.[44]

A conundrum that lies at the heart of mitochondrial genetics regards the pathogenesis of typical heteroplasmic

TABLE 41.2 Clinical Features of Some MtDNA-related Diseases

Tissue	Symptom/Sign	Δ-mtDNA		tRNA		ATPase6	
		KSS	Pearson	MERRF	MELAS	NARP	MILS
Central nervous system	Seizures	−	−	+	+	−	+
	Ataxia	+	−	+	+	+	+/−
	Myoclonus	−	−	+	+/−	−	−
	Psychomotor retardation	−	−	−	−	−	+
	Psychomotor regression	+	−	+/−	+	−	−
	Hemiparesis/hemianopia	−	−	−	+	−	−
	Cortical blindness	−	−	−	+	−	−
	Migraine-like headaches	−	−	−	+	−	−
	Dystonia	−	−	−	+	−	+
Peripheral nervous system	Peripheral neuropathy	+/−	−	+/−	+/−	+	−
Muscle	Weakness	+	−	+	+	+	+
	Ophthalmoplegia	+	+/−	−	−	−	−
	Ptosis	+	−	−	−	−	−
Eye	Pigmentory retinopathy	+	−	−	−	+	+/−
	Optic atrophy	−	−	−	−	+/−	+/−
	Cataracts	−	−	−	−	−	−
Blood	Sideroblastic anemia	+/−	+	−	−	−	−
Endocrine	Diabetes mellitus	+/−	−	−	+/−	−	−
	Short stature	+	−	+	+	−	−
	Hypoparathyroidism	+/−	−	−	−	−	−
Heart	Conduction block	+/−	−	−	+/−	−	−
	Cardiomyopathy	+/−	−	−	+/−	−	+/−
Gastrointestinal	Exocrine pancreas dysfunction	+/−	+	−	−	−	−
	Intestinal pseudo-obstruction	−	−	−	−	−	−

KSS = Kearns-Sayre syndrome; MELAS = mitochondrial encephalomyopathy, lactic acidosis, and stroke-like episodes; MERRF = myoclonus epilepsy and ragged-red fibers; MILS = maternally inherited Leigh syndrome; NARP = neuropathy, ataxia, retinitis pigmentosa.

diseases. Simply put, why is KSS different from MELAS and MELAS different from MERRF, when all three conditions result from impaired mitochondrial protein synthesis and ATP production? It would be logical to expect phenotypic homogeneity rather than syndromic individuality and yet—despite occasional overlapping presentations—this is not the case. There is immunohistochemical evidence that the different mutations that characterize the three syndromes have distinct spatial distribution in the brain: the single deletions of KSS are more abundant in the choroid plexus; the most common m.8344A > G MERRF mutation predominates in dentate and olivary nuclei; and the typical m.3443A > G mutation abounds in subpial arterioles.[11] Although these preferential brain localizations do correlate with typical symptomatology (e.g. MELAS is ultimately a mitochondrial angiopathy), the question then becomes what directs the different mutant mtDNAs to selective regions of the brain.

DISEASES DUE TO mtDNA MUTATIONS

Large-Scale Rearrangements of mtDNA

These genetic abnormalities include single large-scale deletions and duplications of mtDNA, i.e. each patient harbors a single species of partially deleted mtDNA. The rearrangements generally are several thousand base pairs in length (Figure 41.5).

Major Deletions and Duplications

Large-scale partial deletions and duplications of mtDNA generally occur spontaneously. More than 200 species of rearranged mtDNAs have been described since the original report by Holt et al. in 1988.[26] Shortly thereafter, investigators established the association of single major deletions with KSS and sporadic PEO.[45–47] The most frequently encountered deletion, termed the "common deletion," spans 4977 base pairs.[48] The common deletion, like most other major deletions, removes both structural and synthetic genes, thus impairing mitochondrial protein synthesis. The following disorders are associated with single large-scale rearrangements of mtDNA.

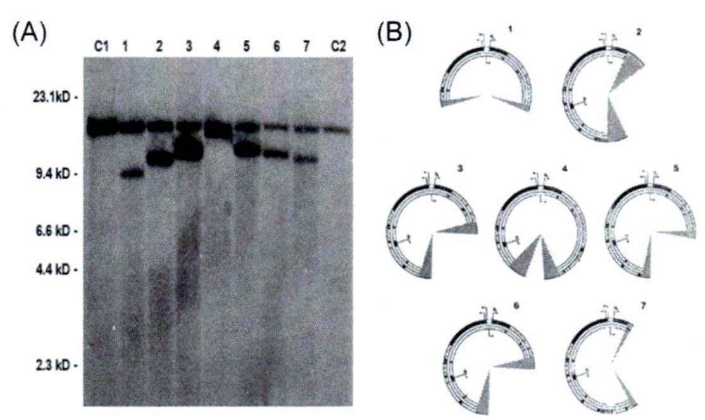

FIGURE 41.5 Southern blot (**A**) of mtDNA in 7 patients with Kearns-Sayre syndrome (KSS) and schematic representation of the single large-scale deletions identified in each of them (**B**). For the Southern blot, total mtDNA (5 μg) was digested with Pvu*II*, electrophoresed through a 0.8% agarose gel, transferred to nitrocellulose, and probed with radiolabeled mtDNA. C1, C2 = control lanes; 1–7 = patient lanes. In **B**, mtDNA maps show the sizes and locations of the deletions. Patients 3 and 6 harbor the same size (4.5 kb) deletion. *Modified with permission from Zeviani et al.*[45]

Kearns-Sayre Syndrome

KSS is defined by onset before age 20 years, progressive ophthalmoplegia, and pigmentary retinopathy. In addition to this triad of symptoms, at least one additional symptom must be present: cardiac conduction block, cerebellar syndrome, or cerebrospinal fluid (CSF) protein concentration >100 mg/dL.[49,50] Patients with KSS exhibit many of the classical signs and symptoms of mitochondrial DNA-related disorders such as short stature, dementia, sensorineural hearing loss, isolated endocrinopathies including diabetes mellitus, hypothyroidism, hypoparathyroidism, and growth hormone deficiency. Surprisingly, seizures are very infrequent and commonly are associated with hypoparathyroidism. Brain MRI shows diffuse white matter abnormalities (Figure 41.6D, KSS). Muscle histology is abnormal, with ragged-red fibers and a mosaic pattern of cytochrome *c* oxidase (COX)-negative fibers. Mitochondrial DNA deletions are easily demonstrated in muscle, and are less likely to be found in blood DNA extracts.

KSS almost always is sporadic and a multicenter retrospective study of 226 families has shown that the risk of a carrier woman to have affected children is very small but finite (1 in 24 births) while disproving the idea that the risk increases with maternal age.[51] There are rare examples of women with the symptoms of KSS who have had children with Pearson's marrow/pancreas syndrome and the same mitochondrial DNA deletion as the mother.[52,53]

Pearson's Syndrome

Pearson's syndrome is a medical disorder affecting infants and causing a severe sideroblastic anemia, pancytopenia and exocrine pancreas dysfunction.[54,55] The hematological disorder often is life-threatening, but those patients who survive later develop the symptoms of KSS.[56,57]

Sporadic PEO with RRF

These patients manifest PEO, ptosis, and pharyngeal and proximal limb weakness, usually beginning in adolescence or young adulthood.[47]

Duplications of mtDNA have been reported in patients with KSS.[58] Maternally inherited duplications have been associated with a syndrome including cerebellar ataxia, proximal renal tubular nephrosis, and diabetes mellitus.[59]

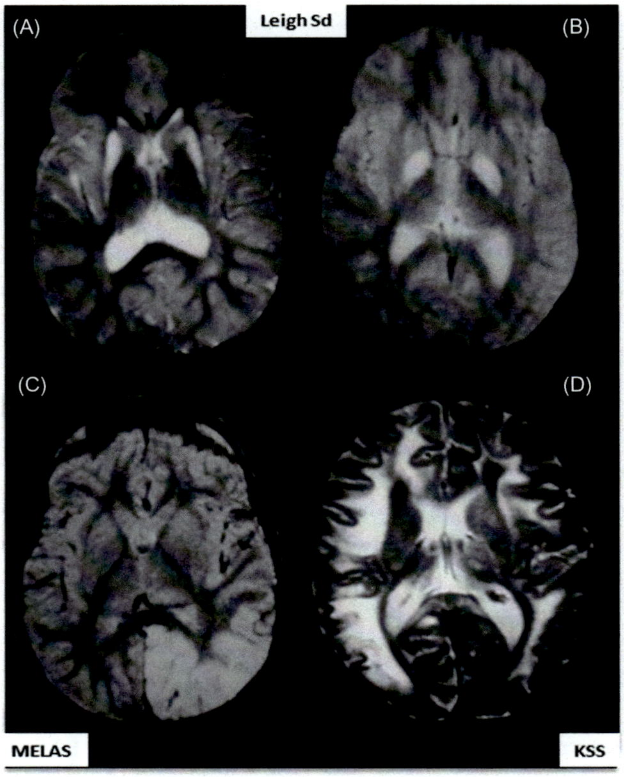

FIGURE 41.6 Characteristic brain MRI abnormalities in mitochondrial encephalomyopathies. These T2-weighted images show bilateral putaminal hyperintensities in a child with cytochrome c oxidase (COX)-associated Leigh syndrome (**A**); bilateral globus pallidus hyperintensities in a child with a familial Leigh syndrome phenotype (**B**); a left posterior cerebral hyperintensity in a child with MELAS (**C**); and bilateral white matter hyperintensities in a child with Kearns-Sayre syndrome (KSS) (**D**).

Brockington et al. have suggested that duplications may be the underlying cause of all maternally transmitted deletions of mitochondrial DNA including the approximately 10 kilobase deletion observed in a family with maternally transmitted diabetes mellitus and deafness.[60]

In summary, sporadic single mtDNA deletions and/or duplications may be associated with the classical phenotype of Kearns-Sayre syndrome and with other phenotypes including progressive external ophthalmoplegia, Pearson's syndrome, and renal tubular nephrosis with diabetes mellitus and ataxia. Limb weakness is evident but may be a less conspicuous clinical finding in these phenotypes.

Point Mutations in mtDNA

Over 250 pathogenic point mutations pepper the small mtDNA molecule (Figure 41.4). We will consider separately those that impair mitochondrial protein synthesis globally (mutations in rRNA and in tRNA genes) and those that affect protein-coding genes.

Point Mutations in rRNA and tRNA Genes

Deleterious mutations in rRNA genes are extremely rare, accounting for only 2% of all mtDNA pathogenic mutations (the coding capacity of the two rRNA genes is 15%). The phenotypes include aminoglycoside-induced deafness and dilated cardiomyopathy.[33,61,62]

The phenotypes associated with point mutations in tRNA genes commonly are multisystemic and include classic clinical syndromes, such as MELAS, MERRF, NARP/MILS (Table 41.2), LHON, and several other disorders.[33,63] Cardiomyopathy and myopathy frequently keep company with encephalopathy in these multisystemic phenotypes. Less frequently, myopathy may dominate the clinical picture. These "pure myopathies" can be explained by at least two mechanisms: (i) somatic mutations that occur *de novo* in the mitochondrial genome of the oocyte or the embryo but affect only myoblasts after germ-layer differentiation, thus sparing other tissues; or (ii) skewed heteroplasmy, a situation in which the tRNA mutation is ubiquitous but the mutation load surpasses the pathogenic threshold only in muscle.[64,65]

MELAS was first described in 1984 as a clinical syndrome manifested by mitochondrial encephalomyopathy with lactic acidosis and stroke-like episodes[66] and probably it is the second most common maternally inherited mitochondrial disease after LHON. Clinically, it is defined by the stroke-like episodes that cause focal brain injury. Seizures are present in nearly 100% of patients and a subacute dementia develops rapidly after the initial clinical presentation. The blood and CSF lactic acid concentrations are elevated and RRF are present in the biopsied muscle tissue. Patients frequently have migraine-like headaches, recurrent vomiting, exercise intolerance, limb weakness and short stature. The stroke-like episodes commonly cause hemiparesis, hemianopia, or cortical blindness, and focal seizures often herald the onset of the stroke-like episodes. The etiology of the stroke-like episodes is uncertain but the prevailing evidence suggests that these episodes represent focal brain injury rather than vaso-occlusive disease. Microscopic studies have documented a mitochondrial angiopathy with increased numbers of enlarged mitochondria that are strongly positive for succinate dehydrogenase in the endothelial cells of both muscle[67] and intracerebral[68–70] small vessels. Diffusion weighted MRI studies have shown evidence for vasogenic rather than cytotoxic edema, consistent with a focal breakdown of the blood-brain barrier (Figure 41.6C, MELAS).[71] This observation suggests a therapeutic role for corticosteroids in the treatment of the acute stroke-like episode, expecting that the corticosteroids would mitigate the vasogenic edema in the same way that corticosteroids benefit the edema associated with brain tumors. There have been no clinical trials to date to examine this speculation.

Phenotypic heterogeneity has been well documented in patients harboring the A3243G point mutation in the $tRNA^{Leu(UUR)}$ gene.[72] These clinical variations may include gastrointestinal features with malabsorption or dysmotility, cardiomyopathy, extrapyramidal syndromes, cluster headaches, pancreatitis, LS, a relatively pure myopathy with PEO,[73] and a number of overlap syndromes. The cardiomyopathy may progress from a hypertrophic to a dilated type, and accessory conduction pathway syndromes may be associated, such as the Wolff-Parkinson-White syndrome.[74]

Dermatological evaluation also has revealed an increased incidence of vitiligo, seborrheic eczema and atopy, and fragility of blood vessels with resulting localized bleeding and ecchymoses. Psychiatric symptomatology is prominent and underreported in patients with the MELAS mutation.[75]

Eighty percent of patients fulfilling the clinical criteria for MELAS harbor the A3243G point mutation. About a dozen other genotypes are associated with the MELAS phenotype (see Table 41.2). Ragged-red fibers are present in the biopsied muscle tissue regardless of whether the mutation involves a synthetic gene or a protein-coding gene. Most of the RRF are COX-positive. Another unusual histological feature is the presence of strongly SDH-reactive blood vessels, the so-called SSRBV.[67,76]

Maternally Inherited PEO

In this genetically determined condition, PEO is the predominant clinical feature.[73] Other less consistent signs include hearing loss, endocrinopathy, heart block, cerebellar ataxia, and pigmentary retinopathy. Lactic acidosis is present but not as frequent or as striking as in patients with MELAS. Muscle biopsy shows RRF with more COX-negative fibers than in typical MELAS.[73] Individual muscle fibers from patients with PEO have shown a higher abundance of mutated mtDNA than the COX-positive RRF in patients with MELAS.

The maternal inheritance pattern distinguishes this PEO phenotype from sporadic PEO associated with single large-scale mtDNA deletions (see above). The most common cause of maternally inherited PEO is the A3243G point mutation that is typically associated with the MELAS phenotype.[47,73] Other mtDNA point mutations associated with this phenotype involve $tRNA^{ILe}$, $tRNA^{Asn}$, $tRNA^{Tyr}$ (see Table 41.2).

Isolated Myopathy without PEO

Although infrequent, pure myopathies have been associated with tRNA mutations: they often affect preferentially respiratory muscles and some have been described in children.[77–80] They can be inherited maternally or be sporadic traits due to *de novo* mutations and the isolated involvement of skeletal muscle is probably due to extremely skewed heteroplasmic distribution of the pathogenic mutation (see above).

An unusual presentation is a severe but spontaneously reversible infantile mitochondrial myopathy characterized biochemically and histochemically by profound deficiency of COX in muscle.[81] A homoplasmic mutation (m.14674T > C in $tRNA^{Glu}$), initially considered a neutral polymorphism, was found in 17 patients from 12 families of various ethnic origins,[82] and the same mutation (or a T > G mutation at the same site) was confirmed in eight Japanese patients.[83] Although the pathogenicity of the mutation was established through high-resolution northern blot analyses of muscle biopsies during the symptomatic phase and through biochemical studies of cybrid cell lines,[82] the mutation *per se* did not explain the tissue specificity or the reversibility of the clinical and biochemical phenotype. It took 4 more years to discover that $tRNA^{Glu}$ has to be 2-thiouridylated by TRMU (5-methylaminomethyl-2-thiouridylate methyltransferase) and that TRMU is downregulated during the symptomatic phase of the disease, thus exacerbating the effects of the $tRNA^{Glu}$ mutation.[84] This is in keeping with previous observations that reversible forms of infantile myopathy and hepatopathy were associated with mutations in TRMU[85,86] and changes the classification of the reversible COX-deficiency myopathy from a primary mtDNA defect to a defect of mtDNA translation[84] (see below).

Myoclonus Epilepsy with Ragged-red Fibers

Myoclonus epilepsy with ragged-red fibers (MERFF) is manifested by myoclonus, ataxia and seizures.[87,88] but it may have protean manifestations.[89] Myoclonus often is the presenting symptom and may be precipitated by action, noise, or photic stimulation. Muscle weakness and exercise intolerance are common.[89] RRF are seen in muscle biopsy and the affected fibers are COX negative.

Additional clinical features have been described in patients and intrafamilial phenotypic variability is recognized.[90] Added clinical features include hearing loss, cognitive impairment, multiple lipomas, neuropathy, and ptosis/ophthalmoplegia.[89] A subtle, slowly progressive dementia may occur in patients, often after several decades, but cognition is relatively spared in contrast to the MELAS phenotype. Phenotypic variability within single families has been described with clinical features ranging from the classic MERRF phenotype to spinocerebellar degeneration, typical Charcot-Marie-Tooth disease, and LS.[91,92]

MERRF was the first well-defined human disease in which maternal inheritance was clearly demonstrated, thus suggesting a defect in mtDNA[93] and the first disorder in which a molecular disorder was associated with epilepsy.[94] Approximately 80% of patients with MERRF have the m.8344A > G point mutation involving the $tRNA^{Lys}$ gene. Other mutations in this gene have also been associated with the MERRF phenotype, including the T8356C and the G8363C point mutations.[95–98] Peculiar presentations of the m.8344A > G mutation include a sporadic case of histiocytoid cardiomyopathy with death at 11 months[99] and a case with myopathy and multiple lipomas but also strokelike episodes and severe gastrointestinal dysfunction, more akin to MELAS than to MERRF.[100] An overlap syndrome of MERRF and KSS was due to a mutation in the $tRNA^{Leu(UUR)}$ gene.[101]

Point Mutations in Protein-Coding Genes

NARP/MILS

These syndromes result from mutations at the same nucleotide (m.8993T > G or, less frequently, m.8993T > C) in the gene encoding the ATPase 6 subunit of complex V.[102] Patients with NARP typically are young adults with the acronymic features whereas maternally related children are likely to have the characteristic symptoms and signs of LS, with developmental delay or arrest, respiratory abnormalities, frequent vomiting, nystagmus, pyramidal signs, bilateral lesions in the basal ganglia and the brainstem, and lactic acidosis (Figure 41.6A,B, Leigh synd). Retinitis pigmentosa distinguishes MILS from other forms of LS and is a useful diagnostic clue.[103] We now know that the different age at onset and severity of NARP and MILS depend on the mutation load: patients with NARP harbor ~70% mutant mtDNAs whereas patients with MILS harbor ~90%.[28,44]

Leber's Hereditary Optic Neuropathy

Leber's hereditary optic neuropathy (LHON) is the most common of the mtDNA-based diseases and the first disorder attributed to a mtDNA point mutation.[27,33] It causes subacute loss of central vision in young adults, predominantly men. LHON is usually due to homoplasmic mutations in one of three genes encoding complex I subunits (m.11778G > A in ND4; m.3460G > A in ND1, or m. 14484T > C in ND6). Despite the homoplasmic nature of these mutations, the retinal ganglion cells are selectively affected and myopathy is not a feature. The predominant involvement of men has been attributed to a protective effect of estrogens in women, an epigenetic effect on the expression of a mtDNA-based disease.[104]

Isolated Myopathies

As of 2000, there were 14 cases of isolated myopathy, with exercise intolerance, weakness, and—often—recurrent episodes of rhabdomyolysis and myoglobinuria associated with mutations in subunits of complex I, III, or IV (Table 3 in[105]). Most of the mutations were in the gene (*MTCYB*) encoding cytochrome *b*[105] (see Case Example 41.2) and all of them were *de novo* somatic changes in sporadic patients. Most patients were adult, although in some of them exercise intolerance had started in childhood. Severe exercise intolerance with or without myoglobinuria and without

CASE EXAMPLE 41.2

A 52-year-old woman had had exercise intolerance since childhood. She remembered rushing to a friend's house at the age of 5 years and arriving there "exhausted and nauseous." These symptoms progressed until she was unable to walk more than one block without experiencing extreme fatigue. She also noted premature fatigue and aching of masticatory muscles when chewing tough meat. Examination at age 39 revealed weakness against resistance in proximal limb muscle. Resting venous lactate concentration was 5 mg/deciliter (normal <2.2). She never noted pigmenturia. Standardized bicycle ergometry with ^{31}P-nuclear magnetic resonance spectroscopy showed a low ratio of phosphocreatine to inorganic phosphate (PCr/Pi) at rest. The ratio decreased further with exercise and, after cessation of exercise, returned to baseline more slowly than normal. Electromyography of limb muscle showed short-duration motor units and polyphasic potentials, suggesting mild myopathy. There was no family history of neuromuscular diseases: the patient's mother had died of cancer at age 69 years and her brother and two children were in good health. Although she reported shortness of breath with exercise, her cardiac function was normal. Serum creatine kinase (CK) concentration was normal, but resting serum lactate concentrations were elevated. A muscle biopsy at age 39 revealed numerous COX-positive RRF and biochemical analysis showed marked isolated complex III deficiency. Sequencing of the gene (*MTCYB*) encoding cytochrome *b* identified a heteroplasmic m14846G > A mutation (p.G34S). Single fiber analysis showed that the mutation was much more abundant in RRF than in normal fibers.

Comment

This patient, as well as four others reported by Andreu et al.,[106] was not considered to have a mtDNA mutation because she had a tissue-specific (as opposed to a multisystemic) disorder, no evidence of maternal inheritance, and an isolated biochemical defect (complex III deficiency). In fact, she is typical of mutations in protein-coding genes of mtDNA and the diagnosis—suggested by extreme exercise intolerance and resting hyperlactacidemia—was confirmed by the muscle biopsy showing COX-positive RRF. The sporadic nature of these myopathies is due to *de novo* somatic mutation affecting exclusively myogenic stem cells after germ-layer differentiation.

evidence of maternal inheritance justifies a muscle biopsy, especially if blood lactate is elevated.

Other Presentations

The most common clinical presentations associated with mtDNA mutations in protein-coding genes have been MELAS and LS. Both conditions have been attributed to the m.13513G > A mutation in *MTND5*[107] while LS has also been associated with diverse mutations in *MTND3*.[108–110] We noted that patients with LS and the common m.10191T > C mutation in *MTND3* tend to have unusually long survival.[111]

Diseases Due to Nuclear DNA Mutations

In the course of evolution, mitochondria have lost much of their bacterial independence and now rely on the nucleus for most of their functions. Therefore, Mendelian mitochondrial disorders are numerous and varied. We can discern at least six disease categories because pathogenic mutations can affect: (1) genes encoding subunits of the respiratory chain (direct hits); (2) genes encoding assembly proteins of respiratory chain complexes (indirect hits); (3) genes needed for mtDNA translation; (4) genes controlling the phospholipid composition of the mitochondrial inner membrane (MIM); (5) genes involved in mitochondrial dynamics. And, as discussed previously, we will consider separately the sixth category, i.e. defects in genes needed for mtDNA integrity or replication ("mtDNA maintenance").

The advent of whole exome sequencing and of the more narrowly targeted mito-exome sequencing[34] has allowed us to sequence in one sweep the approximately 1700 nuclear genes encoding mitochondrial proteins. Mitoexome sequencing has already enriched all six disease categories with new variants,[112–115] and it is to be expected that it may reveal novel disease categories.

Direct Hits

These disorders result from nDNA mutations affecting each of the five respiratory chain complexes. We will review them following the conventional sequential order of the complexes in the respiratory chain but it should be kept in mind that defects in one complex may well have a "ripple effect" because of the supercomplex organization of the respiratory chain.[116]

Complex I

Complex I (NADH-ubiquinone oxidoreductase) is the Goliath of the respiratory chain, containing 45 subunits, of which 38 are encoded by nDNA. Given its size, it is not too surprising that isolated complex I deficiency is the most frequently encountered respiratory chain defect.[117]

Pathogenic mutations have been identified in all of the 14 evolutionarily conserved subunits that comprise the catalytic core. The resulting clinical pictures are remarkably homogeneous and largely correspond to LS. LS is defined neuropathologically (or neuroradiologically) by bilateral symmetrical lesions all along the nervous system, but especially in the basal ganglia, thalamus, brainstem, and cerebellar roof nuclei. Microscopically, there is neuronal loss, proportionate loss of myelin, reactive astrocytosis, and proliferation of cerebral microvessels.

Clinically, these children have psychomotor retardation or regression, respiratory abnormalities, hypotonia, failure to thrive, seizures, dystonia, nystagmus, and optic atrophy. These are the symptoms most commonly encountered by Distelmaier et al. in a study of 15 children with complex I deficiency due to mutations in nDNA-encoded genes,[118] and their findings were reinforced by a review of 130 patients reported in the literature.[119]

Besides LS, four distinct presentations had been attributed to Mendelian complex I deficiency: fatal infantile lactic acidosis (FILA), neonatal cardiomyopathy with lactic acidosis, leukodystrophy with macrocephaly, and hepatopathy with renal tubulopathy.[120,121] However, no specific genotype/phenotype emerged for the different variants and it is more likely that differences in residual activity may account for the apparent differential tissue involvement.

One important clinical feature common to these disorders is that onset is invariably in infancy or childhood but almost never in fetal life.[122] Usually, these children are fine at birth and early in life, although onset of symptoms is followed by a rapid and relentless progression leading to early death. Lactic acid is consistently and markedly elevated both in blood and in the CSF. From a genetic point of view, it is noteworthy that mutations in *NDUFA1* cause X-linked LS,[123] a most unusual hereditary pattern for respiratory chain defects.

Complex II

Complex II (succinate-ubiquinone oxidoreductase) is the smallest multimeric component of the respiratory chain, comprising only 4 subunits. Complex II deficiency is also probably the least common respiratory chain defect.[124] Although only a handful of patients have been described, clinical presentations varied and muscle weakness was usually a consequence of central nervous system involvement.

Four patients had typical LS.[32,125,126] Of these, two children had psychomotor regression and spastic quadriparesis with diffuse leukodystrophy by MRI, and a third child presented in the first year of life with stunted growth and lactic acidosis without neurological symptoms. Interestingly, all three children stabilized or improved after riboflavin administration.[126]

Complex III

Two distinct direct hits for complex III (ubiquinol-cytochrome *c* reductase) have been described, one in a single patient and the other in a large consanguineous family. A deletion in the *UQCRB* gene encoding the ubiquinone-binding protein (QP subunit or subunit VII) was identified in an 8-month-old girl with moderate hepatomegaly, hypoglycemia, metabolic acidosis, and lactic acidosis.[127] At the age of 4 years, the child was normal and the hepatomegaly had regressed. Complex III activity was decreased in liver, lymphocytes, and fibroblasts.

Twenty members of an Israeli-Bedouin family presented with a uniform encephalomyopathy that was severe but compatible with long survival. These patients were normal at birth and for the first few months, then showed psychomotor retardation and developed extrapyramidal signs, including dystonia, athetoid movements, ataxia, and dementia.[128] There was a homozygous mutation in *UQCRQ*.[128]

Coenzyme Q10 Deficiencies

This relatively new group of disorders due to deficiency of Coenzyme Q10 (CoQ10) (ubiquinone) can arguably be included in the "direct hits" because mutations in a cascade of biosynthetic enzymes results in the deficiency of one relatively simple component of the respiratory chain (Figure 41.3). CoQ10 shuttles electrons from complex I and complex II to complex III, acts as an antioxidant, and regulates apoptosis.[129] CoQ10 deficiency was first described in 1989 in two sisters with the clinical triad of mitochondrial and lipid storage myopathy, recurrent muscle breakdown with myoglobinuria, and encephalopathy with seizures, ataxia, and mental retardation.[130] CoQ10 deficiency was also noted in children with spinocerebellar atrophy-like presentations,[131,132] but molecular defects in biosynthetic genes (*PDSS2* and *COQ2*) were not reported until 2006 in two infants with severe infantile mitochondrial encephalomyopathies and glomerular renal involvement.[133,134] Mutations in the same and other biosynthetic genes (*PDSS1*, *COQ6*, *ADCK3*, and *COQ9*) followed[135–139] and five clinical syndromes are now attributed to CoQ10 deficiency: (i) encephalomyopathy; (ii) cerebellar ataxia; (iii) infantile multisystem disorder; (iv) isolated nephropathy; (v) isolated myopathy.[140] It is important to consider primary CoQ10 deficiency in the differential diagnosis of juvenile ataxias and infantile encephalomyopathies because most patients with these disorders benefit from CoQ10 supplementation. (See Case Example 41.3.)

Complex IV

Probably most "direct hits" in nuclear encoded subunits of complex IV (COX) are incompatible with life because years of research have revealed mutations in only one of the 10 nDNA-encoded subunits, COX6B1.[141] Although a sister of the patients had died immediately after birth and

CASE EXAMPLE 41.3

A 33-month-old boy, the son of first cousins, had hypotonia, mild psychomotor delay, and optic atrophy. At 12 months, proteinuria led to a renal biopsy, which showed focal and segmental glomerulosclerosis. His younger sister also had proteinuria. At 18 months, he developed frequent vomiting requiring peritoneal dialysis. He then developed psychomotor regression (loss of the ability to walk or stand unassisted), tremor, and new-onset status epilepticus with focal EEG abnormalities predominant in the left occipital region. Brain MRI showed mild cerebellar atrophy and stroke-like lesions in the left cingulate cortex and subcortical area. Blood and CSF lactate levels were normal. At 22 months, he developed right hemiplegia, myoclonus, and swallowing difficulties. A muscle biopsy showed that many fibers showed increased SDH staining, although they were not ragged-red or COX-deficient. Biochemical analysis of muscle homogenate revealed decreased activities of complex I + III, whereas other complexes were normal. The concentration of CoQ10 was markedly decreased in skeletal muscle (12 μg/g fresh tissue; normal, 32.1 ± 6.7) and in fibroblasts (19 ng/mg protein; normal, 105 ± 14). With oral CoQ10 supplementation (30 mg/kg/day), all neurological manifestations improved dramatically whereas his renal function did not improve and at age 3 years he needed a renal transplantation. He is now 7 years old and his kidney-allograft is functioning normally, but he has neurological sequelae, including cognitive impairment, seizures, and hemiplegia. His sister, who was diagnosed with CoQ10 deficiency and nephrosis at 1 year, before any neurological symptoms developed, was treated with oral CoQ10 and her nephrotic syndrome slowly resolved. Homozygosity mapping followed by candidate gene sequencing revealed a homozygous mutation in *COQ2*, the gene encoding *para*-hydroxybenzoate-polyprenyl transferase, the CoQ10 biosynthetic enzyme that catalyzes the conjugation of the benzoquinone ring with the decaprenyl side chain.

Comment

This child had a severe infantile mitochondrial encephalomyopathy and nephrosis due to primary CoQ10 deficiency and was the first patient in whom a molecular defect of CoQ10 biosynthesis was documented by Quinzii et al.[133] This case is a reminder that the association of infantile encephalomyopathy and nephrosis should raise the suspicion of primary CoQ10 deficiency, which can be documented in skeletal muscle or in cultured skin fibroblasts. These two siblings also remind us that most forms of primary CoQ10 deficiency respond to oral CoQ10 supplementation, albeit not always completely.

the mother had had a second trimester miscarriage, of the two affected brothers one had died at 10 years of age and the other was 8 years old at the time of publication. Both children had had failure to thrive but normal psychomotor development until late childhood, when they developed muscle weakness, cognitive deterioration, visual problems, and lactic acidosis. MRI showed severe cavitating leukodystrophy in both.[141]

Complex V

"Direct hits" of complex V also appear to be very rare, having been reported only in one patient, who harbored a homozygous mutation in the *ATP5E* gene.[142] This girl (patient 3 in[142]), was small at birth and had poor suck, respiratory distress, lactic acidosis, and 3-methylglutaconic aciduria. Although she had recurrent metabolic crises, her clinical course stabilized at 5–6 years and she completed school and was gainfully employed. At the age of 17 years, she had mild ataxia, horizontal nystagmus, exercise intolerance, mixed axonal and demyelinating polyneuropathy, and mild left ventricular hypertrophy. Brain MRI had shown hyperintense lesions in the caudate and lentiform nuclei at age 14, but had normalized by age 17.

Indirect Hits

Even correctly encoded subunits of the various complexes have to be translated, imported into mitochondria, and directed to the MIM. Here, they must assemble with their mtDNA- and nDNA-encoded counterparts and acquire prosthetic groups; some of them multimerize; and most of them must further assemble into supercomplexes. Mutations in genes controlling these pathways are called "indirect hits" because they affect the respiratory chain indirectly.

Complex I

Pathogenic mutations have been reported in 6 assembly subunits[143–148] and in 5 chaperone factors.[149–155] It is safe to predict that many more mutated genes in both groups will be identified through the increased use of whole exome sequencing.[36,156] (See Case Example 41.4.)

The clinical manifestations tend to be more heterogeneous than those associated with "direct hits" in complex I, although there is a good deal of overlap. All the patients had encephalopathy clinically resembling LS but often with leukodystrophy rather than gray matter involvement.[144,145,147] Cardiomyopathy was more common[143,148,149,151,154] and sometimes the dominant feature.[146] Although onset was in infancy or early childhood and early death was common, the course was often prolonged and in some cases fluctuating.[147] Death occurred between 5 and 13 years of age in three patients[145,149,155] and a few patients were alive at ages 4, 5, and 20 years at the times of publication.[147,152,154]

Three siblings in a doubly consanguineous Dutch family and two unrelated patients, who harbored mutations in the *ACAD9* gene, had isolated myopathy with exercise intolerance and lactic acidosis responsive to riboflavin.[153,157]

CASE EXAMPLE 41.4

A 13-year-old boy came to medical attention because of exercise intolerance, weakness, and muscle wasting. He was born after normal pregnancy and delivery but showed psychomotor delay since the first year of life. Neurological examination showed proximal muscle weakness, generalized hypotonia, ataxic gait, bradykinesia and bradylalia, scoliosis, and truncal obesity. A cognitive assessment revealed mild deficit (his IQ was 81). Cardiac function was normal as were both brain MRI and MRS. Laboratory tests showed increased resting plasma lactate and alanine and increased urinary lactate. He also had hypothyroidism. Electromyography was compatible with a myopathic process and a muscle biopsy showed excessive subsarcolemmal staining with the Gomori trichrome stain and with the COX reaction, indicating mitochondrial proliferation. Biochemical analysis of a muscle homogenate showed severe complex I deficiency and moderate complex III deficiency, whereas complex II and citrate synthase values were above normal. Next-generation exome sequencing using a mitochondrial gene library (mito-exome sequencing) revealed a homozygous mutation (c.1240C > T) in the *ACAD* gene, predicted to be deleterious. The patient was treated with riboflavin. He is now 19 years old and has nonprogressive exercise intolerance and weakness.

Comment

This patient, recently reported by Garone et al.,[157] had a mild and rather nonspecific mitochondrial encephalomyopathy with complex I (and less severe complex III) deficiency. Because the trait appeared to be autosomal recessive, we resorted to mito-exome sequencing, which revealed a new mutation in a complex I assembly gene. Mutations in *ACAD9* had been associated with severe and rapidly fatal hypertrophic cardiomyopathy of infancy but also with milder childhood-onset myalgia and exercise intolerance. The beneficial effect of riboflavin, a catalytic cofactor of the ACAD proteins, was documented in cultured skin fibroblasts from the patients with cardiopathy and clinically in the patients with exercise intolerance. This case exemplifies an "indirect hit" to complex I and illustrates the power of whole or mitochondrial exome sequencing in the identification of clinically undefined mitochondrial disorders.

Complex II

Even the small complex II requires assembly factors; and mutations have been described in one of them (*SDHAF1*), which contains a LYR motif and is presumably involved in Fe-S metabolism.[158] After developing normally for the first 6 to 11 months, five children from two consanguineous families developed acute psychomotor regression, impaired growth, spastic quadriplegia, and moderate cognitive decline. One child died at 18 months but the others were alive at ages 6 to 12 years at the time of publication. MRI and positron MRS of the brain showed leukoencephalopathy, sparing U fibers and increased lactate and succinate peaks.[158]

In 1991 and 1993, Haller and coworkers described a young Swedish man with lifelong exercise intolerance, dyspnea, cardiac palpitations, and episodes of myoglobinuria.[159,160] The syndrome was called "mitochondrial myopathy with succinate dehydrogenase and aconitase deficiency" and attributed to altered metabolism of iron-sulfur (Fe-S) cluster proteins, which are prosthetic groups present in complexes I, II, III, and in the Krebs cycle enzyme aconitase. Accordingly, histochemical analysis of muscle showed SDH deficiency, and biochemical analysis showed deficiencies of complex II, complex III, and aconitase. In 2008, homozygosity mapping revealed a single pathogenic mutation in the *ISCU* gene in three Swedish families,[161] and two years later, Kollberg et al.[162] described SDH deficiency and accumulation of iron in muscle as the morphological hallmarks of the disease.

A combined defect of complex II and III in muscle was caused by a homozygous mutation in the *IBA57* gene, which encodes yet another assembly factor for mitochondrial [4Fe-4S] cluster proteins. Clinically, this condition in two siblings was characterized by hypotonia, respiratory insufficiency, arthrogryposis, microcephaly, congenital brain malformations, hyperglycinemia, and perinatal death.[163]

Complex III

The first assembly defect in complex III was identified in 2002 in Finnish infants with an extremely severe syndrome named GRACILE, which summarizes the main symptoms and signs: growth retardation, aminoaciduria, cholestasis, iron overload, and early death (before 5 months of age). The mutated protein, BCS1L, is an ATPase needed for insertion of the Rieske FeS subunit into the complex.[164] Interestingly, these children did not have neurological impairment nor overt complex III deficiency.[165] In contrast, British patients with GRACILE syndrome had both neurological involvement (hypotonia, seizures) and decreased enzyme activity. In even sharper contrast, five Turkish children from four unrelated but consanguineous families developed severe hepatopathy (with complex III deficiency in liver) soon after birth, proximal renal tubulopathy, and encephalopathy compatible with LS in two cases.[166]

Two unrelated patients with mutations in *BC1IL* suffered from severe encephalopathy without visceral involvement: both showed respiratory distress at birth and developed psychomotor delay or regression, spastic quadriparesis, and seizures in one case.[164] Both patients had cerebral atrophy by MRI and, curiously, both had brittle hair. One patient with a novel homozygous *BC1IL* mutation had a mild presentation with infantile onset of hypotonia, psychomotor retardation, coarse facial features, and hirsutism.[167]

The most benign expression of mutations in *BCS1L* is Björnstad syndrome, characterized by congenital sensorineural hearing loss and pili torti, resulting in brittle hair, usually recognized in childhood.[168]

A second assembly factor for complex III is tetratricopeptide repeat 19 (TTC19), a protein of the MIM, where it interacts with complex III.[169] Four Italian patients with homozygous mutations in *TTC19* had severe encephalopathy. Two siblings presented in childhood with neurological symptoms that progressed for years and included mental retardation, ataxia, nystagmus, dysphagia, dysphonia, tremor, dystonia, and hearing loss. The sister was alive but confined to a wheelchair at age 30 and her younger brother was totally incapacitated at age 20. A third patient also had childhood-onset mental retardation, dystonia, and ataxia. At age 19, she was bedridden, on a ventilator, and in a fluctuating comatose state. The fourth patient did not seek medical attention until 42 years of age, when he developed limb weakness and fasciculations, apraxia of gait and hand movements, dysarthria, dystonia, and paraparesis. He died at age 45.

Complex IV

It had long been known that COX (complex IV) deficiency is a common cause of LS,[170–172] but years went by before a mutated gene was associated with LS: the gene (*SURF1*) is essential for correct COX assembly and this was the first example of an "indirect hit" to affect the respiratory chain.[173,174] By 2001, 30 different mutations had been identified.[175]

Mutations in the COX assembly gene *SCO2*, which encodes a metallochaperone involved in mitochondrial copper delivery, cause a much more severe clinical phenotype that combines neonatal hypertrophic cardiomyopathy with encephalopathy,[176] is fatal in the first weeks or months of life, and may be associated with early fetal lethality.[177] The fatal infantile presentation is typical of compound heterozygous patients, who always harbor the "common" E140K mutation.[178] Homozygosity for the same mutation causes delayed onset and longer survival.[179,180]

Importantly, *SCO2* mutations (both in compound heterozygosity and in homozygosity) can simulate spinal muscular atrophy (SMA) and the neurogenic, often SMA-like, histological pattern of the muscle biopsy further complicates the differential diagnosis.[181–183] One postmortem study of the spinal cord showed severe neuronal loss and astrocytosis in the anterior horns.[182] Two distinctive features orient towards the correct diagnosis: clinically, the severe cardiomyopathy; and, morphologically, the lack of COX stain in the muscle biopsy. However, in patients with the clinical picture of SMA but without mutations in the *SMN1* gene, it is important to consider *SCO2* mutations.

Mutations in *SCO1*, a second metallochaperone needed for copper insertion into the COX holocomplex, were first published on in 2000 and reported to affect multiple members of a large family.[184] The child described in that paper was hypotonic and lethargic at birth, had severe metabolic acidosis, and needed respiratory assistance. Liver function tests were severely altered at day 4, followed by hepatomegaly. He developed axial hypotonia, episodes of apnea and bradycardia, and died at 2 months. A liver biopsy obtained immediately postmortem showed swollen hepatocytes with microvesicular steatosis, and a muscle biopsy showed lipid storage. A sibling presented with metabolic acidosis and "severe neurological distress" and died at 5 days.

The report of a single family suggests that *SCO1* mutations are either incompatible with life or very rare, a conclusion supported by several genetic screenings of patients with COX deficiency.[185,186]

Mutations in *COX10*, which encodes a factor involved in the first step of the mitochondrial heme-A biosynthesis, were described in five patients from three families and with different clinical presentations.[187,188] Three siblings from a consanguineous family died at 2, 3, and 5 years of age: the child described in more detail was normal until 18 months, when he developed ataxia, weakness, ptosis, pyramidal signs, and status epilepticus. He had lactic acidosis and increased urinary amino acids, suggesting proximal renal tubulopathy. His younger sister was reported to have similar "progressive neurological deterioration."[187] One of two unrelated children had lactic acidosis, hypotonia, growth retardation, sensorineural hearing loss, transfusion-dependent macrocytic anemia, and biventricular hypertrophic cardiomyopathy, and died at 5 months.[188] The other child had a sudden arrest of her psychomotor development at one and a half months, and developed transfusion-dependent anemia, mild splenomegaly, and hypotonia. Brain MRI showed symmetrical lesions in the basal ganglia typical of LS. She died of central respiratory failure at 4 months.[188]

COX15, like COX10, is involved in heme-A biosynthesis and mutations in *COX15* caused hypotonia, seizures, and lactic acidosis in an infant girl, who had a normal echocardiography at 6 days of age but developed acute biventricular hypertrophy by 22 days and died 2 days later. Postmortem examination revealed massive accumulation of morphologically abnormal mitochondria in the heart.[189,190]

The heart was not affected in a second patient, who presented with typical LS: he was normal at birth and for the first seven months, then developed hypotonia, nystagmus, leg spasticity, progressive microcephaly, and retinopathy. He died of pneumonia a year later. Lactic acid was increased both in blood and in the CSF and brain MRI showed multiple symmetrical lesions in the basal ganglia, dorsal midbrain, cerebral peduncles, and periaqueductal gray matter.[191] A third patient also had a diagnosis of LS, but her course was unusually protracted and, at the time of publication, she was still alive at age 16.[192]

Considering that at least 30 chaperones are needed for the assembly of complex IV in yeast, it is not surprising that the list of indirect hits responsible for COX deficiency in humans keeps getting longer. On the other hand, it is difficult to make general statements on clinical phenotypes because mutations in some genes have been reported in single or very few patients, often siblings from consanguineous marriages. This is the case for a homozygous mutation in *COA5*, which encodes a factor involved in the early steps of COX assembly.[193] Two siblings had onset of cardiomyopathy *in utero* with fetal distress and isolated biventricular hypertrophic cardiomyopathy causing death 8 and 10 days after birth. Postmortem examination of the heart revealed mitochondrial proliferation and lipid storage.

Two siblings from another consanguineous family had a slower but devastating disorder dominated by encephalopathy, with severely delayed psychomotor development, optic atrophy, intractable seizures leading to status epilepticus in the affected boy, and left hemiplegia in the affected girl. At 14 years of age, the girl was unable to stand or walk and had a 20-word vocabulary. The boy was lost to follow-up at the age of 4 years, when he was bedridden and incapable of communicating or moving purposefully.[194] The mutant gene (*KIAA0971*), identified through homozygosity mapping, encodes a protein known as fas-activated serine-threonine kinase domain 2 (FASTKD2) that plays a role in mitochondrial apoptosis rather than in COX assembly.[194] Although COX activity was markedly decreased in muscle, COX and SDH histochemical stains were normal.

Through integrated genomics, based on bioinformatics-generated intersection of DNA, mRNA, and protein data sets, Mootha et al.[195] identified the gene (*LRPPRC*) responsible for the French Canadian type of LS (LSFC) with COX deficiency.[196] Because the LRPPRC (leucine-rich pentatricopeptide repeat cassette) protein has a role in the translation or stability of the mRNA of mtDNA-encoded COX subunits,[197] LSFC might be more appropriately classified among the defects of mtDNA translation (see the following).

Another defect of mitochondrial translation affecting COX, and specifically the COXI subunit, is caused by mutations in the gene *TACO1* (translational activator of COXI).[198] Five children of a consanguineous Turkish family were affected with slowly progressive LS, more severe in girls than in boys.[199] Onset was in childhood or adolescence and progression was slow: all patients were alive at the time of publication and three were in their 20s. The clinical picture was dominated by small stature, mental retardation, dystonia, dysarthria, spasticity with pyramidal signs, and optic atrophy. Lactic acid was inconsistently increased, and MRI of the brain showed bilateral basal ganglia lesions in all patients.

Integrative genomics also facilitated identification of the *ETHE1* gene in children with ethylmalonic encephalomyopathy (EE), an early-onset disorder with microangiopathy, chronic diarrhea, and greatly increased levels of ethylmalonic acid and short-chain acylcarnitines in body fluids.[200] Based on their studies of patients and *Ethe1*-null mice, Valeria Tiranti and Massimo Zeviani introduced a new paradigm of "indirect hit," which can be called "toxic indirect hit." They documented that ETHE1 is a matrix thioesterase and its dysfunction leads to excessive accumulation of sulfide, a powerful COX inhibitor.[201] The clinical presentation has been confirmed and the number of mutations extended in a study of 14 patients with EE.[202]

Complex V

Mutations in two ancillary proteins cause severe infantile disorders. A homozygous missense mutation in *ATP12* (now known as *ATPAF2*) was first reported in 2004 in an infant girl who died at 14 months.[203] She had dysmorphic features, arthrogryposis, hepatomegaly, hypoplastic kidneys, fluctuating lactic acidosis, and—terminally—seizures. Brain MRI showed cortical-subcortical atrophy, dysgenesis of the corpus callosum, and hypoplastic white matter.

A homozygous mutation (c.317-2A > G) in the assembly gene *TMEM70* was identified in 25 infants, of whom 24 were of Roma ethnic origin, with severe multisystem symptoms, lactic acidosis, and 3-methylglutaconic aciduria.[204] The clinical picture was described in detail in a subsequent paper[205] and was characterized by neonatal hypotonia, apneic spells, hypertrophic cardiomyopathy, severe lactic acidosis, and hyperammonemia. Ten patients died within 2 months and the survivors showed hypotonia and psychomotor delay. The cardiomyopathy was nonprogressive and sometimes reversible. The authors suggest that mutations in *TMEM70* should be suspected in critically ill neonates with hypotonia, hypertrophic cardiomyopathy, and the laboratory triad of lactic acidosis, hyperammonemia, and 3-methylglutaric aciduria.[205] The report of more patients has widened the clinical spectrum, including congenital hypertonia with multiple contractures, early-onset cataracts and gastrointestinal problems, and Reye syndrome-like episodes.[206]

Defects of Mitochondrial Protein Importation

A prerequisite for the assembly of any respiratory chain complex is the import of nDNA-encoded subunits from the cytoplasm into mitochondria. Surprisingly, considering the complexity of the mitochondrial importation machinery,[16] only a few diseases belong to this category.

One is the Mohr-Tranebjaerg deafness-dystonia syndrome (DDS), due to mutations in the MIM protein DDP1 (deafness dystonia protein 1) encoded by the gene *DDP/Tim8*.[207] Besides sensorineural deafness and dystonia, patients suffer from spasticity, dysphagia, cognitive deterioration, and cortical blindness.

Mutations in the chaperonin Hsp60 cause an autosomal dominant form of hereditary spastic paraplegia, HSP13,[208] whereas mutations in another chaperonin, DNAJC19, cause early-onset dilated noncompaction cardiomyopathy, anemia, ataxia, testicular dysgenesis, and 3-methylglutaconic aciduria.[209,210]

The *GFER* gene encodes a sulfydryl oxidase (DRS) of the mitochondrial intermembrane space, which is involved in mitochondrial protein import. A homozygous mutation of *GFER* caused congenital cataract, progressive muscle hypotonia, sensorineural hearing loss, and developmental delay in three siblings of a consanguineous family. The pathogenic relationship between this gene and the multiple mtDNA deletions observed in muscle biopsies is not clear.[211]

Defects of mtDNA Translation

Clinically, these disorders came to the attention of pediatricians because they cause severe infantile syndromes characterized by neurological involvement (LS, leukodystrophy, recessive ataxia), hepatocerebral syndrome, or cardiomyopathy, similar to those associated with mtDNA depletion even though mtDNA is normal both qualitatively and quantitatively. Although these disorders are usually characterized by multiple respiratory chain enzyme deficiencies, mutations in some translation factors, such as LRPPRC and TACO1, impair specifically COX activity, as described previously.[195,198]

Translation of mitochondrial mRNAs into the 13 mtDNA-encoded respiratory chain subunits comprises four phases, each requiring multiple ancillary factors, including initiation factors (IF2 and IF3), elongation factors (EF-Tu, EF-Ts, and EF-G1), release factors (eRF1 and ICT1), translational activators (TACO1 and LRP130), and specific tRNA base modifiers (TRMU and PUS1).[212,213]

Examples of specific disorders due to defective mtDNA translation are given in recent reviews.[212,213] We have already described the reversible infantile COX-deficient myopathy due to the combination of a mtDNA $tRNA^{Glu}$ homoplasmic mutation and downregulation of the $tRNA^{Glu}$ base modifier TRMU.[84] Another defect of mtDNA translation with prominent neuromuscular involvement is MLASA (myopathy, lactic acidosis, sideroblastic anemia). This autosomal recessive disorder causes weakness and anemia in childhood and may lead to dependency on blood transfusion.[214,215] Although the selective vulnerability of skeletal muscle and bone marrow remains difficult to explain, one cause of the disease is deficient pseudouridylation of mitochondrial tRNAs due to mutations in the gene (*PUS1*) encoding pseudouridylate synthase 1. A second cause of MLASA is a defect of tyrosyl-tRNA-synthetase due to mutations in the gene *YARS2*.[216,217]

Yet another disease of muscle and the bone marrow that combines mitochondrial myopathy with megaloblastic anemia was recently described in two children and attributed to mutations in the gene *SFAX4*, which encodes sideroflexin 4, a MIM transmembrane protein, probably an iron transporter.[115] As the biochemical defect in these cases involved complex I, this disorder may more appropriately belong to the "indirect hits" (see above).

Defects of the Mitochondrial Inner Membrane Lipid Milieu

The phospholipid component of the mitochondrial inner membrane (MIM) is not just a convenient scaffold for the respiratory chain but it is also an integral functional partner. Increasingly, alterations of the lipid milieu are being associated with mitochondrial encephalomyopathies and—importantly—many of them are neuromuscular disorders.

The first example of these disorders is Barth syndrome, an X-linked myopathy and cardiopathy, with neutropenia and impaired growth.[218] Barth syndrome is due to mutations in the tafazzin (*TAZ*) gene,[219] resulting in deficiency or abnormal composition of cardiolipin, the major phospholipid component of the MIM.[220,221]

In the past few years, four new entities have been described, affecting different phospholipids. Sengers syndrome is similar to Barth syndrome in that: (i) both conditions were described by Dutch pediatricians; (ii) they are both characterized by mitochondrial myopathy and cardiopathy, although in Sengers syndrome congenital cataracts are a distinctive additional sign; and (iii) cardiolipin alterations are involved in both conditions, although in Sengers syndrome they are secondary to mutations in the acylglycerol kinase (*AGK*) gene.[113] The resulting decrease of phosphatidic acid and lysophosphatidic acid in the MIM presumably explains the lack of the adenine nucleotide translocator 1 (ANT1), which was initially considered causative of the disease. The typical clinical triad of myopathy, cardiopathy, and congenital cataracts was further complicated by vascular strokes in one of two families with molecularly documented Sengers syndrome.[222] Furthermore, these patients had extremely high citrate synthase (CS) activity in muscle and CS crystals were identified in muscle mitochondria.

A human congenital muscular dystrophy with giant mitochondria (a megaconial myopathy, according to Shy and Gonatas[4]) displaced to the periphery of the muscle fibers was attributed to mutations in the gene (*CHKB*) encoding choline kinase beta when it was realized that clinical picture and muscle morphology were similar to those of a spontaneously mutant mouse strain with rostrocaudal muscular dystrophy.[223] (See Case Example 41.5.) The human congenital myopathy is compatible with prolonged survival, but all patients have, besides delayed motor development, severe intellectual disability with autistic features and may develop dilated cardiomyopathy.[225]

In 2006, Wortmann and coworkers described four children with the association of 3-**me**thyl**g**lutaconic aciduria, sensorineural **d**eafness, **e**ncephalopathy, and **L**eigh-like features (MEGDEL syndrome). These children had neonatal hypotonia, episodic lactic acidosis and hypoglycemia, recurrent infections, failure to thrive, hearing loss, and neuroradiological evidence of LS. There was also biochemical evidence of respiratory chain dysfunction and persistent 3-methylglutaconic aciduria.[226] Whole exome sequencing in 15 children with MEGDEL syndrome identified mutations in the gene *SERAC1*, which encodes a phospholipid remodeling protein.[227] Alterations in phosphatidylglycerol remodeling lead to altered cardiolipin subspecies and excessive free cholesterol. The discovery of the molecular basis of MEGDEL called attention to the mitochondria-associated endoplasmic reticulum membrane (MAM), a specialized subcellular structure where the endoplasmic reticulum and mitochondria come into close physical and functional association. MAMs are involved in multiple functions, including lipid transport, cholesterol metabolism, calcium signaling, energy metabolism, apoptosis, and mitochondrial dynamics.[228]

Besides the MEGDEL syndrome, other diseases due to MAM dysfunction include the megaconial myopathy due to mutations in *CHKB* mentioned above, and some forms of primary lateral sclerosis due to mutations in *ERLIN2* and of juvenile amyotrophic lateral sclerosis associated with mutations in *SIGMAR1*.[228]

Finally, recurrent rhabdomyolysis with myoglobinuria in children, a syndrome often occurring during febrile illnesses and rarely associated with known inborn errors of metabolism, has been attributed to mutations in *LPIN1*, the gene that encodes the muscle-specific isoform of phosphatidic acid phosphatase.[229]

CASE EXAMPLE 41.5

A 2-year-old boy, the first-born child of a nonconsanguineous African-American couple, was born after a normal pregnancy and delivery. Apgar scores were 8 at 1 minute and 9 at 5 minutes. Soon after he was discharged home, the mother noticed episodes of rhythmic jerking of arms with stiff legs, occurring 3 to 4 times per night, that were labeled "muscle spasms" by the general practitioner. When first seen at age 11 months, the child could roll over both ways, sit without support for 1 minute (if placed), stand for a few seconds holding onto a chair, and speak a few words. At 22 months, he was still not able to sit up and did not utter any intelligible words. At 25.5 months, physical examination showed an irritable and uncooperative child, who fell asleep intermittently. He had a broad flat face, large philtrum, wide nasal bridge, and low-set ears. Neurological examination showed normal cranial nerves but bilateral facial weakness. He was markedly hypotonic and had limb weakness, more pronounced in the legs. Sensory examination appeared normal and there was no ataxia, tremor, or nystagmus. He had episodes of unresponsiveness, with staring, drooling, and slow turning of the head and eyes to one side. Although these appeared to be seizures, the EEG was normal. Normal laboratory examination included serum lactate and pyruvate levels, cardiac evaluation, nerve conduction velocities, and brain auditory evoked responses. Serum creatine kinase (CK) was consistently increased, ranging from 318 to 522 U/L (normal, <176 U/L). Electromyography (EMG) showed polyphasic units with low amplitudes, consistent with myopathy. Brain MRI at age 1 year showed prominent CSF spaces over the frontal convexities and enlargement of the lateral ventricles. Muscle biopsy showed dystrophic features (extreme variation in fiber size, interstitial fibrosis, and necrotic "hyaline" fibers). However, histochemical stains with the modified Gomori trichrome, SDH, and COX revealed greatly enlarged mitochondria, mostly displaced at the periphery of the fibers (Figure 41.7). Electron microscopy confirmed the presence of giant mitochondria with densely packed and whorled cristae.

Comment

This child with congenital myopathy and psychomotor retardation had a muscle biopsy that combined dystrophic features with the presence of giant and displaced mitochondria, what Milton Shy would have called "megaconial myopathy."[4] The peculiar muscle morphology prompted us to sequence the *CHKB* gene encoding choline kinase beta, and we documented a homozygous mutation.[224] Congenital megaconial myopathy is one of the defects of the MIM lipid milieu and—as this case illustrates—it is really an encephalomyopathy. This case also underlies the diagnostic importance of the muscle biopsy, which can direct molecular analysis to the correct gene before more sophisticated (but still expensive) next-generation sequencing techniques are employed.

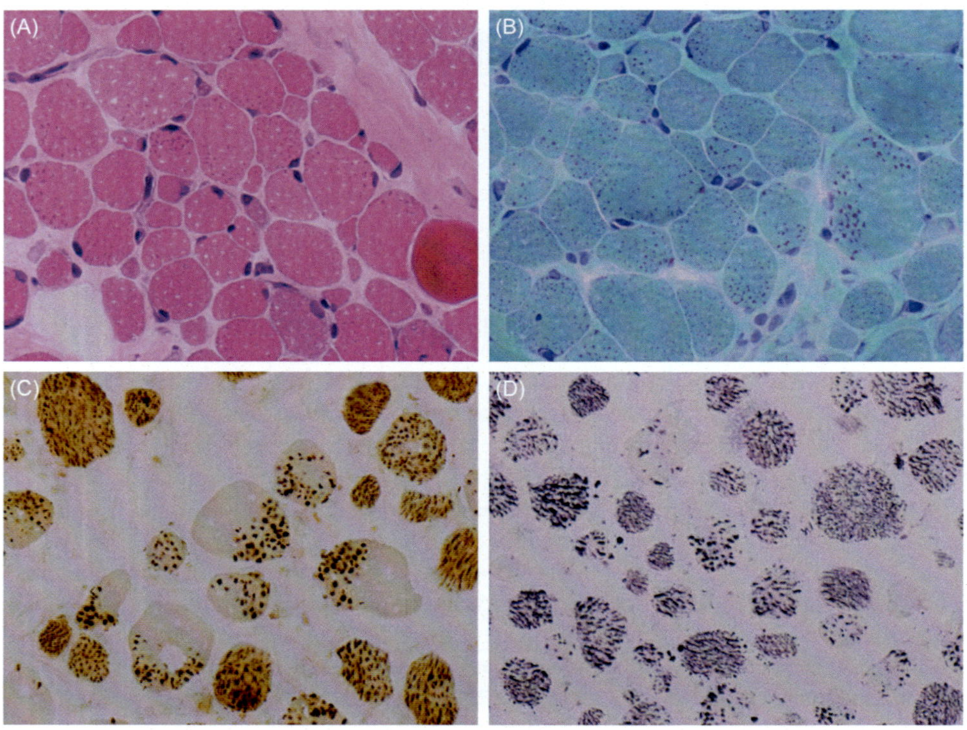

FIGURE 41.7 Muscle biopsy from a child with choline kinase beta (CHKB) deficiency and congenital megaconial myopathy (*Case Example 41.5*). Hematoxylin and eosin (H&E) stain (**A**) shows variation in fiber size, occasional basophilic regenerating fibers, and irregularly distributed areas of basophilic material. The coarse granules were highlighted by the modified Gomori trichrome stain (**B**), the cytochrome *c* oxidase (COX) stain (**C**), and the succinate dehydrogenase (SDH) stain (**D**). Original magnification for all four panels ×200. *Reproduced with permission from Gutierrez-Rios et al.*[224]

Defects of Mitochondrial Dynamics

Faithful to their bacterial origin, mitochondria incessantly move, fuse, and divide, more often than not forming extensive tubular networks that favor a uniform distribution of energy throughout the cell and make possible mitochondrial quality control through mitophagy.[230] Any defect of mitochondrial motility, fusion, or fission results in diseases that tend to affect selectively or predominantly the nervous system.

Mutations in fusion proteins, such as OPA1, MFN2, and GDAP1, result in dominant optic atrophy (DOA) or Charcot-Marie-Tooth (CMT) neuropathies types 2A and 4A.[231–233] However, mutations in *MFN2* and *OPA1* can also cause multisystem disorders with mtDNA depletion and multiple deletions, i.e. a defect of mtDNA maintenance (see the following).

Defects of mitochondrial fission are less common than defects of mitochondrial fusion, but mutations in the gene (*DRP1*) encoding dynamin-related protein 1 have been reported in an infant with mitochondrial and peroxisomal abnormalities, lactic acidosis, and a rapidly fatal encephalopathy with microcephaly, abnormal brain development, and optic atrophy.[234] Homozygosity mapping and mito-exome sequencing allowed the identification of the mutant gene in a child with a clinical presentation also characterized by developmental delay, microcephaly, and optic atrophy.[235] This infant harbored a homozygous mutation in *MFF*, which encodes the mitochondrial fission factor protein. Not surprisingly, mitochondria and peroxisomes in cultured skin fibroblasts from both infants showed excessive tubular pattern.

Defects of mtDNA Maintenance

Nuclear genes are in charge of mtDNA maintenance—that is, the intergenomic dialogue that guarantees mtDNA replication and integrity. When communication between the two genomes goes awry, the diseases that result are characterized either by quantitative mtDNA alterations (mtDNA depletion syndromes) or by qualitative mtDNA alterations (multiple deletions and site-specific point mutations). Actually, we should not say "either or" because we have come to realize that quantitative and qualitative mtDNA alterations often coexist and that mutations in the same gene can cause predominantly mtDNA depletion or predominantly mtDNA multiple deletions (Table 41.3).

TABLE 41.3 Defects of MtDNA Maintenance

Mutated Gene	mtDNA Depletion	Multiple mtDNA Deletions
TK2	Infantile or adult myopathy. SMA phenocopy	Adult AR-PEO
DGUOK	Infantile hepato-cerebral syndrome	Adult myopathy ± PEO
PEO1	Hepato-cerebral syndrome; infantile-onset spinocerebellar ataxia (IOSCA)	Adult AD-PEO-plus
SUCLA2	Infantile encephalomyopathy	
SUCLG1	Infantile encephalomyopathy; methylmalonic aciduria	
RRM2B	Infantile encephalomyopathy	Adult AD/AR-PEO-plus
MPV17	Infantile hepato-cerebral syndrome; Navajo neurohepatopathy (NNH)	Adult AR-PEO-plus
TYMP	Mitochondrial neurogastrointestinal encephalomyopathy (MNGIE)	MNGIE
POLG	Hepato-cerebral syndrome (Alpers)	Adult AD/AR-PEO-plus SANDO; MIRAS
POLG2		Adult AD-PEO
ANT1		Adult AD-PEO-plus
OPA1		Dominant optic atrophy (DOA); PEO-plus
MFN2		DOA-plus
GFER		Congenital cataract, encephalomyopathy

AD-PEO = adult-onset autosomal dominant progressive external ophthalmoplegia; AR-PEO = autosomal recessive progressive external ophthalmoplegia; DOA = dominant optic atrophy; MNGIE = mitochondrial neurogastrointestinal encephalomyopathy; MIRAS = mitochondrial recessive ataxia syndrome; NNH = Navajo neurohepatopathy; PEO = progressive external ophthalmoplegia; SANDO = sensory ataxic neuropathy, dysarthria and ophthalmoparesis.

Defects of mtDNA maintenance can be due to problems with the replication machinery or with the intramitochondrial pool of deoxynucleoside triphosphates (dNTPs), the DNA building blocks (Figure 41.8).

Defects of the Replicating Machinery

The replicative machinery consists of the catalytic subunit of polymerase gamma (gene *POLG*), the accessory subunit POLG2 (gene *POLG2*), and the replicative helicases Twinkle (gene *PEO1*) and DNA2 (gene *DNA2*).

Mutations in *POLG* illustrate how dysfunction of a single gene can cause either mtDNA depletion or mtDNA multiple deletions and how it can give rise to diverse syndromes. Largely based on which of the three *POLG* domains (polymerase, exonuclease, or linker region) harbors the mutation (or mutations), the clinical presentation varies from the severe hepatocerebral disorder known as Alpers-Huttenlocher syndrome[236] to adult-onset autosomal dominant or autosomal recessive PEO (AD-PEO or AR-PEO)[237] to parkinsonism[238] and to other clinical phenotypes, including sensory ataxic neuropathy, dysarthria and ophthalmoparesis (SANDO),[239] or mitochondrial recessive ataxia syndrome (MIRAS).[240] The extreme variety of clinical presentations associated with *POLG* mutations makes this gene an important target to consider by pediatricians dealing with patients presenting complex clinical pictures, often involving seizures or liver dysfunction. Recent reviews on the clinical spectrum of POLG mutations are available[241,242] as is a Human DNA Polymerase Gamma Database.[243] The use of valproate should be avoided in all patients harboring (or suspected to harbor) mutation in *POLG*, as they are prone to develop acute liver toxicity.[244]

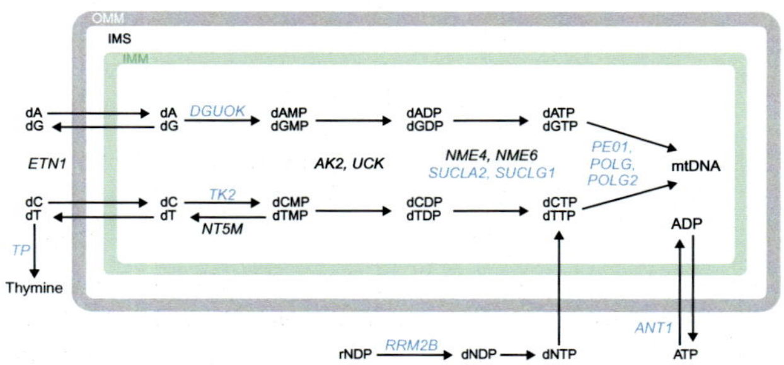

FIGURE 41.8 Schematic representation of the mitochondrial nucleotide pool, showing the genes whose mutations have been associated with multiple mtDNA deletions, those associated with mtDNA depletion, and those associated with both mtDNA multiple deletions and mtDNA depletion. dA, deoxyadenosine; dC, deoxycytosine; dT, deoxythymidine; dG, deoxyguanosine; dAMP, deoxyadenosime monophosphate; dCMP, deoxycytosine monophosphate; dGMP, deoxyguanosine monophosphate; dTMP, deoxythymidine monophosphate; dADP, deoxyadenosine diphosphate; dCDP, deoxycytosine diphosphate; dGDP, deoxyguanosine diphosphate; dTDP, deoxythymidine diphosphate; dATP, deoxyguanosine triphosphate; dCTP, deoxycytosine triphosphate; dGTP, deoxyguanosine triphosphate; dTTP, deoxythymidine triphosphate; *AK2*, adenylate kinase 2; *ANT1*, adenine translocator 1; *dGK* (*DGUOK*) deoxyguanosine kinase; *ENT1*, equilibrative nucleoside transporter 1; *NME4*, nucleoside diphosphate kinase D; *NME6*, nucleoside diphosphate kinase 6; *NT5M*, 5′ nucleotidase, mitochondrial; *PEO1*, mtDNA helicase; *POLG*, polymerase gamma, catalytic form; *POLG2*, polymerase gamma, accessory form; *RRM2B*, ribonucleoside diphosphate reductase subunit M2 B; *TK2*, thymidine kinase 2; *TYMP, (TP)* thymidine phosphorylase; *SUCLA2*, succinate CoA ligase (ADP forming) beta subunit; *SUCLG1*, succinate CoA ligase (GDP forming) alpha subunit; *UCK*, uridine-cytidine kinase. *Reprinted with permission from DiMauro and Schon.*[11]

Mutations in *PEO1* usually cause adult-onset AD-PEO with multiple mtDNA deletions, but can also cause Alpers-Huttenlocher-like autosomal recessive hepatocerebral syndrome with mtDNA liver depletion[245,246] or infantile-onset spinocerebellar ataxia (IOSCA), a disease that is prevalent in the Finnish population and is accompanied by mtDNA depletion in the brain.[247]

The two reported siblings with mutations in the DNA2 helicase suffered from weakness, wasting, exercise intolerance, cramps, and—in one sibling—ptosis and ophthalmoparesis. Symptoms probably started in childhood but brought the patients to medical attention in adult years.[248] Muscle biopsy showed scattered COX-negative RRF and a defect of ATP synthesis.

Defects of the Intramitochondrial Deoxynucleoside Triphosphate (dNTP) Pool

Defects in six enzymes that control the balance of the dNTP pool have been associated primarily with mtDNA depletion[249] and result in four major syndromes (Figure 41.8).

The most important presentation for pediatricians interested in neuromuscular disorders is a myopathic syndrome caused by mutations in thymidine kinase 2 (gene *TK2*).[250] In a recent review of 30 patients, profound infantile and rapidly progressive weakness was the dominant feature, leading to respiratory failure and death in infancy or childhood in 26 patients.[251] Notably, most patients have markedly elevated levels of serum CK, a distinctive unusual lab abnormality in other mitochondrial diseases and a useful clue to the correct diagnosis. Also notably, some patients with *TK2* mutations mimic closely the clinical (and histological) features of SMA.[252,253] In fact, mutations in *TK2*—in addition to mutations in *SCO2* (see above)—ought to be investigated in all patients with the clinical presentation of SMA but without mutations in the *SMN1* gene. (See Case Example 41.6.) However, TK2 mutations can also cause milder myopathy with adult presentation[254] or with early presentation but adult diagnosis[255] or even AR-PEO with multiple mtDNA deletions.[256]

Probably the most common clinical phenotype associated with mtDNA depletion is hepatoencephalopathy,

CASE EXAMPLE 41.6

This boy was born to second-cousin parents after normal pregnancy and delivery. Two older brothers were normal. Early growth and development were normal; he sat unsupported at age 6 months and walked unassisted at 11 months. His cognitive development was normal. At age 2 years, the parents noted difficulty running, tendency to stumble, and decreased physical stamina. At age 2.5 years, his venous lactate was 3.7 mM (normal 0.5-2.2) and CK was 577 IU/L (normal <225). EMG suggested denervation and a first muscle biopsy revealed atrophic and hypertrophic fibers distributed as single fibers or in groups. Histochemical stains for SDH and NADH-dehydrogenase were increased. These findings were considered consistent with SMA. The patient's condition continued to deteriorate involving trunk and limb muscles. At age 3 years, he was unable to walk independently. Neurological examination at age 4 years showed normal cognition, diffuse hypotonia and weakness, inability to stand without assistance, diffusely decreased muscle bulk, and absent or trace tendon reflexes. The tongue was normal without fasciculations. Repeat EMG showed reduced number of fragmented motor units in the vastus medialis and scattered fragmented units in the tibialis anterior and medial gastrocnemius. There was no spontaneous activity and nerve conduction studies were normal. A second muscle biopsy showed RRF and large groups of atrophic fibers, suggesting a mitochondrial myopathy and a superimposed lower motor neuron disorder. At 9 years of age, he needed assisted ventilation. At 11 years, he had severe diffuse wasting, but ocular motility, cognition, and cardiac function were intact. At age 12, he developed bilateral ptosis and he died at age 16 of respiratory failure.

Comment

This child had early-onset and relentlessly progressive weakness and wasting with intact cognition. The initial diagnosis of SMA was supported by the EMG and by the findings of group atrophy in the first muscle biopsy, although the increased histochemical stains for oxidative enzymes were puzzling as was the increased blood level of lactate. At that time sequencing of the *SMN1* gene was not available. However, the second biopsy, showing frank RRF, strongly suggested a primary mitochondrial myopathy, and the diagnosis of SMA was later excluded by genetic analysis. In retrospect, a mitochondrial myopathy with elevated serum CK simulating SMA prompted us to sequence the *TK2* gene, and we documented a homozygous pathogenic mutation.[253] Together with mutations in the COX-assembly gene *SCO2*, mutations in *TK2* often cause clinical phenocopies of SMA. Although the presence of cardiomyopathy is a distinguishing feature of *SCO2* mutations, both genes should be sequenced in patients with features of SMA but without mutations in the *SMN1* gene.

which has been attributed to mutations in the gene (*DGUOK*) encoding deoxyguanosine kinase[257,258] and in that (*MPV17*) encoding a MIM protein of unknown function.[259] However, again, there are exceptional presentations that are of interest to myologists. Thus, *DGUOK* mutations caused a childhood-onset myopathy with mild muscle mtDNA depletion in one patient[260] and a more severe adult-onset mitochondrial myopathy variably associated with PEO, rhabdomyolysis, lower motor neuron syndrome, and mild cognitive impairment in five patients with multiple mtDNA deletions in muscle.[12] Similarly, mutations in *MPV17* have been associated with sensorimotor axonal neuropathy, ptosis, ophthalmoparesis, exercise intolerance, and mitochondrial myopathy with multiple mtDNA deletions (no mtDNA depletion) in one adult man[261] and with neuropathy and leukoencephalopathy in another man with multiple mtDNA deletion in muscle.[262] A severe pediatric disorder prevalent among Navajo children in the southwestern United States and known as Navajo neurohepatopathy (NNH) combines hepatopathy, peripheral neuropathy, corneal anesthesia and scarring, acral mutilation, cerebral leukoencephalopathy, failure to thrive, and recurrent metabolic acidosis.[263] Although long suspected to be an environmental toxic condition, NNH is—in fact—an autosomal recessive disease due to a homozygous mutation in *MPV17*.[264]

The third syndrome associated with mtDNA depletion is a less well defined mitochondrial encephalomyopathy that has been linked to mutations in the enzymes succinyl-CoA synthase (gene *SUCLA2*) and succinate-CoA ligase (gene *SUCLG1*), and in the p53-inducible ribonucleotide reductase (gene *RRM2B*). Hypotonia is often present in patients with *SUCLA2* or *SUCLG1* mutations, but myopathy is overshadowed by brain involvement. On the contrary, mitochondrial myopathy is common in patients with *RRM2B* mutations.[265] In addition, two unrelated families with *RRM2B* mutations had AD-PEO and multiple mtDNA deletions rather than mtDNA depletion.[266] The patients in the first family had only PEO with or without ptosis manifesting in adulthood, whereas the propositus from the second family, besides PEO, also had hearing loss, gait ataxia, depression, and moderate cognitive impairment. Mutations in *RRM2B* are not a rare cause of PEO because screening this gene in 75 patients with molecularly unexplained PEO, COX-negative fibers, and multiple mtDNA deletions, showed that 12 of them (16%) were either compound heterozygous (2 patients with AR-PEO) or heterozygous (10 patients with AD-PEO) for pathogenic mutations.[267]

Mitochondrial neurogastrointestinal encephalomyopathy (MNGIE) is a rare autosomal recessive disease caused by mutations in the *TYMP* gene encoding thymidine phosphorylase (TPase). This peculiar syndrome is characterized by PEO, peripheral neuropathy, gastrointestinal dysmotility leading to severe cachexia, leukoencephalopathy, and mitochondrial myopathy with COX-negative RRF and a combination of multiple mtDNA, mtDNA depletion, and site-specific somatic mtDNA point mutations.[268] Onset occurs in adolescents or young adults and the profound emaciation often leads to the initial diagnosis of anorexia nervosa. Loss of TP activity causes toxic accumulation of the nucleosides thymidine and deoxyuridine that are incorporated by the mitochondrial pyrimidine salvage pathway and leads to dNTP imbalance and mtDNA instability. This otherwise rapidly fatal disorder has been successfully treated in several cases by allogeneic hematopoietic stem cell transplantation (AHSCT).[268]

Other Disorders of mtDNA Maintenance

Although—as described previously—OPA1 is involved in mitochondrial fusion and the signature disease due to *OPA1* mutations is autosomal dominant, optic atrophy (DOA), a syndromic disorder, often dubbed DOA-plus, has emerged, which is characterized more or less sequentially[269] by optic atrophy with visual failure, sensorineural deafness, ataxia, myopathy, axonal sensory-motor polyneuropathy, and PEO.[270–273] Muscle biopsy in these patients shows scattered ragged-blue, COX-negative fibers and multiple mtDNA deletions are demonstrable by long-range PCR. Interestingly, the proportion of COX-negative fibers is much higher in patients with DOA-plus than in those with nonsyndromic DOA. The relationship between this defect of mitochondrial dynamics and altered mitochondrial maintenance is intriguing.

Mutations in another dynamin-like GTPase protein involved in mitochondrial fusion, MFN2, cause—besides CMT type 2A—a syndrome very similar to DOA-plus, with early-onset optic atrophy developing into axonal neuropathy and mitochondrial myopathy. Muscle biopsies show COX-negative RRF and multiple mtDNA deletions.[274]

Finally, and interestingly for myologists, a form of centronuclear myopathy (CNM) has been associated with mutations in the gene encoding dynamin 2 (*DNM2*) and shows—besides the typical features of CNM—mitochondrial changes with COX-negative RRF.[275] Dynamin 2 is a GTPase involved in plasma membrane endocytosis, intracellular membrane trafficking, actin assembly, and centrosome cohesion.

DIAGNOSTIC APPROACH

In a recent review of the mitochondrial diseases,[228] the physician facing the differential diagnosis of this genetically heterogeneous and clinically protean group of disorders was considered a modern-day Theseus, who—after slaying the Minotaur in Crete's labyrinth—found his way out only thanks to Ariadne's thread. If not foolproof Ariadne's threads, there are rules-of-thumb that can help in the differential diagnosis.

Family History

If maternal inheritance is evident, this is a robust clue that we must be dealing with an mtDNA-related disease. However, because of heteroplasmy, maternal inheritance is not always obvious and family history has to be taken meticulously, trying not to ignore soft signs in the maternal lineage, such as short stature, migraines, hearing loss, diabetes, exercise intolerance, and psychiatric disorders (especially depression). On the other hand, we should not forget that disorders due to single mtDNA deletions (KSS, Pearson syndrome, chronic PEO) are almost always sporadic[51] as are diseases due to *de novo* mutations, such as most changes affecting *MTCYB*.[276] Conversely, parental consanguinity suggests autosomal recessive inheritance and homozygosity mapping has helped identify several mutated nuclear genes.

Multisystem Disorders

This feature is common to both mtDNA- and nDNA-related diseases. However, some syndromes are so typically associated with mtDNA mutations (KSS, Pearson syndrome, MELAS, MERRF, NARP, MILS, LHON) or with mutations in specific nuclear genes (mitochondrial neurogastrointestinal encephalomyopathy [MNGIE], SANDO, Alpers-Huttenlocher, DOA) that they provide useful shortcuts to the molecular diagnosis. On the other hand, a diagnosis of LS tells us that we are dealing with a mitochondrial disease, but gives us little information about which genome is involved or to which subgroup of mitochondrial disorders the child belongs.

Three organs are often involved, usually together with the brain and muscle but occasionally in the absence of encephalopathy. Hypertrophic cardiomyopathy can be the presenting or predominant problem. It is often rapidly fatal but it can also be chronic, as in Barth syndrome. Liver involvement is typically associated with mtDNA depletion or with defects of mtDNA translation ("hepatocerebral syndromes") but it can also be seen as part of direct or indirect hits to respiratory chain complexes. Nephropathy typically presents as proximal tubulopathy with Fanconi syndrome, but when it presents as nephrosis with albuminuria, it should raise the diagnostic suspicion of CoQ10 deficiency.

Progressive External Ophthalmoplegia

After excluding myasthenia gravis and oculopharyngeal muscular dystrophy, PEO is a strong clue to mitochondrial dysfunction. Specifically, PEO is commonly associated with primary mtDNA large-scale rearrangements (KSS, chronic PEO) and with nDNA-related defects of mtDNA maintenance resulting in mtDNA deletions or depletion. Sporadic cases suggest single mtDNA deletions whereas autosomal dominant or recessive inheritance suggests disorders of mtDNA maintenance with multiple mtDNA deletions. Maternal inheritance is rare but there are reports of PEO due to mutations in mtDNA tRNA genes, including the common MELAS mutation, m.3243A > G.[73]

The extreme vulnerability of extraocular muscles to mitochondrial dysfunction correlates with their dependence on oxidative metabolism and this, in turn, correlates with their unique function, as both saccadic and smooth pursuit eye movements deal with the same mechanical load throughout life but occur uninterruptedly through the day.[277] In agreement with their sustained and energetically demanding contraction pattern, extraocular muscles are resistant to fatigue,[277] rich in mitochondria, and second only to the heart in vascularization.[278]

A molecular basis for the preferential involvement of extraocular muscles has been offered by histochemical and single fiber studies from 13 patients with PEO, mostly due to single or multiple mtDNA deletions.[279] There were more COX-negative fibers and the mutation threshold for COX deficiency in single fibers was lower in extraocular than in limb muscles, with the exception of one patient with the m.3243A > G MELAS mutation.[279]

Onset

Onset varies widely for mtDNA-related diseases, even within members of the same family, most likely due to varying degrees of heteroplasmy.[44] In most Mendelian disorders except for the defects of mtDNA maintenance, onset is in infancy or early childhood, often following several months of normal development. Prenatal manifestations are uncommon, as shown by the rare occurrence of arthrogryposis and dysmorphisms. In a few instances, fetal cardiopathy is revealed by fetal bradychardia or by echocardiography. Adult onset is the rule in disorders of mtDNA maintenance, most notably multiple mtDNA deletions with PEO, probably because of the overlap of mitochondrial and Mendelian genetics.

Laboratory Exams

Lactic acidosis is a hallmark of all mitochondrial diseases. Although extremely common, it is neither invariably present nor necessarily severe. Increased lactate in the CSF is frequent, and detectable in the brain by ^{1}H-MRS.

Serum CK is only mildly elevated in most mitochondrial disorders except in the mtDNA depletion myopathy due to TK2 deficiency, where increased serum CK serves as a diagnostic red flag.[251]

In 2011, fibroblast growth factor 21 (FGF-21) was introduced as a valuable new serum or plasma biomarker for the detection of mitochondrial disorders, especially those involving skeletal muscle.[280] Although the diagnostic accuracy of FGF-21 was better than that of all conventional biomarkers, including lactate, the test has not yet been widely used, possibly because it requires enzyme-linked immunosorbent assay (ELISA) or simply because of intellectual inertia on the part of the clinical community.

Muscle Biopsy

Although muscle biopsy is an invasive procedure, it remains a gold standard for the diagnosis of mitochondrial diseases, especially those due to primary mtDNA mutations.[281] Systematic analysis of frozen, cross-sectioned muscle with the histochemical reactions for COX, SDH, or both stains superimposed, provides valuable diagnostic clues (Figure 41.9).

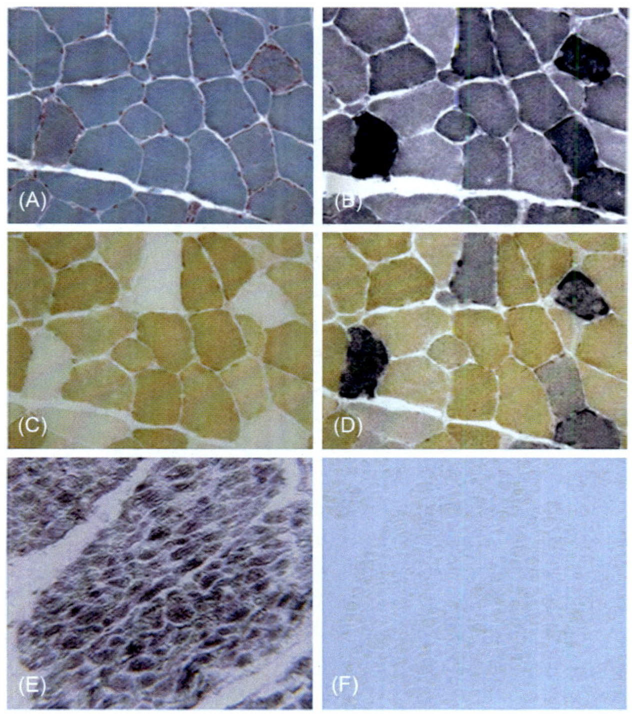

FIGURE 41.9 Serial sections from the muscle biopsy of a patient with MERRF syndrome. With the modified Gomori trichrome stain (**A**), two ragged-red fibers are evident. The same fibers appear *"ragged-blue"* with the succinate dehydrogenase (SDH) stain (**B**); are pale with the cytochrome *c* oxidase (COX) stain (**C**); and appear more or less intensely blue with the combined COX/SDH stain (**D**). In contrast to the checkerboard appearance of the muscle biopsy in an mtDNA-related disease, the biopsy from a patient with mutations in *SCO2*, an nDNA assembly gene of COX, shows normal fiber type-related variable intensity with the SDH stain (**E**) but uniformly lacking reaction with the COX stain (**F**).

It is important to remind the reader that the SDH stain reflects the activity of complex II of the respiratory chain, the only complex that is entirely encoded by nuclear DNA. This means that the SDH stain is unruffled by deleterious mutations of mtDNA while at the same time it is an excellent marker of mitochondrial abundance. On the other hand, the three catalytic subunits of COX are encoded by mtDNA and COX stain is a very good index of mtDNA function. Not only mutations in COX I, COX II, or COX III, but any mtDNA mutation impairing mitochondrial protein synthesis (i.e. large-scale deletions or point mutations in tRNA or rRNA genes) will result in a decrease or lack of COX stain.

Another important reminder is that heteroplasmic mtDNA mutations are not uniformly distributed along syncytial muscle fibers, such that adjacent segmental sections may have widely different amounts of mutant mtDNAs. It follows that cross sections of the muscle biopsy from patients harboring pathogenic mtDNA mutations will show a mosaic pattern of COX-negative and COX-positive fibers.

As mitochondria proliferate in response to energy failure (although the *primum movens* of this phenomenon remains obscure), COX deficiency is often accompanied by excess mitochondria, showing as hyperintense SDH stain (these fibers are dubbed "ragged-blue" because of their analogy with the more famous "ragged-red" fibers revealed with the modified Gomori trichrome stain) (Figure 41.7). Thus, finding a mosaic pattern of normally stained fibers admixed with "ragged-blue" COX-negative fibers in a muscle biopsy is a robust clue to the diagnosis of an mtDNA-related disease affecting mitochondrial protein synthesis (more rarely, affecting one of the three mtDNA-encoded COX subunits). A mosaic pattern of ragged-blue but COX-positive fibers suggests a mutation in a mtDNA gene that encodes a respiratory chain subunit other than COX I, II, or III (for example a complex I subunit or cytochrome *b*).[106]

Heteroplasmy explains not only the mosaic distribution of COX-negative fibers but also the different degree of "COX-negativity," as some fibers have a deficiency rather than a lack of stain. To help reveal the COX-deficient fibers, the combined COX/SDH stain is extremely effective: when the COX and SDH stain are superimposed in normal fibers, the brown COX stain prevails and overshadows the blue SDH stain. However, even a small decrease in COX activity allows the SDH stain to shine through, making the COX-deficient fibers appear blue (Figure 41.9). Nor is this technique limited to skeletal muscle, as it has been successfully applied to reveal COX-deficient motor neurons in patients with ALS,[282] MELAS,[69] and other mtDNA-related diseases.[283]

As expected, in Mendelian disorders, for example COX-deficient LS, muscle histochemistry shows more or less severe but *diffuse* COX deficiency,[284] which is an important diagnostic clue (Figure 41.9). In fact, while routine muscle histochemistry is usually noninformative in LS, COX histochemistry is a useful way to rule in or out one important cause of the disease. There is one important exception to this rule: due to the overlapping features of Mendelian and mitochondrial genetics, muscle biopsies from patients with defects of mtDNA maintenance, and especially from cases of PEO with multiple mtDNA deletions, show the same mosaic pattern of COX-negative ragged-blue fibers as do primary mtDNA mutations. In fact, this morphological pattern, together with a family history suggestive of Mendelian inheritance, is an incentive to look for mutations in mtDNA maintenance genes even in clinically atypical cases.[112,114]

Ultrastructural studies of muscle are usually only confirmatory in that they show mitochondrial proliferation. However, the presence of paracrystalline inclusions, consisting of crystals of CK, reinforces the diagnosis of mitochondrial diseases (although it is still unclear why CK precipitates in pathological conditions); the presence of giant displaced mitochondria suggests problems with mitochondrial dynamics[224]; and abnormal cristae are typical of Barth syndrome.[285]

Biochemistry

Biochemical analysis is usually conducted in frozen muscle tissue or in cultured fibroblasts. A detailed, step-by-step procedure for the measurement of the activities of complexes I–IV has been published[286] and should facilitate comparison of data among laboratories, thus far a thorny issue.

Individual respiratory complex deficiencies are expected to occur in patients with mutations in mtDNA protein-coding genes and with mutations in genes encoding single respiratory chain subunits (direct hits) or assembly genes for specific complexes (indirect hits). While this is largely the case, because the respiratory chain is organized in supercomplexes, a severe defect in one complex is likely to be associated with less severe defects in one or more additional complexes. Thus, for example, mutations in the *MTCYB* gene cause defects of complex III but also of complex I.[276]

Impaired biochemical activities of all complexes containing subunits encoded by mtDNA suggests either mutations in mtDNA genes controlling protein synthesis globally (tRNA or rRNA genes, single deletions) or mutations in nDNA genes controlling mtDNA maintenance (multiple deletions, depletion, defects of mtDNA translation, defects of the MIM lipid milieu).

Some special patterns may be suggestive of specific problems. Combined defects of complex II + III are often associated with primary CoQ10 deficiency whereas combined defects of complexes I, II, and III may be due to defective assembly of the FeS cluster proteins, especially when they are accompanied by aconitase deficiency.[159]

Molecular Studies

Starting in 1995,[32] mutations in nuclear genes affecting the mitochondrial respiratory chain were identified at an escalating pace. These advances were made possible by rapidly evolving technologies, from linkage analysis to either mitoexome sequencing or whole exome sequencing.[34] The advent of whole exome sequencing may affect neurogenetics in the same way that the advent of MRI affected clinical neurology, that is to say, simplifying and—in a sense—banalizing the diagnostic challenge. However, at least in this exciting initial stage, it is hard to overlook the advantages, which are not only practical (diagnosing puzzling patients and providing genetic counseling) but also heuristic (discovering novel mutant genes and novel disease mechanisms).[287] However, it is still possible, more economic, and intellectually satisfactory to target the molecular studies to one, or a few, candidate genes, as indicated in Figure 41.10. As already mentioned, typical mtDNA-related syndromes are usually easily defined at the molecular level through southern blot or Sanger sequencing of specific mtDNA genes. For nDNA-related diseases, the combination of clinical picture, family history, MRI images of the brain, and muscle morphology may orient towards a specific gene defect (e.g. the combination of congenital myopathy, mental retardation, and giant peripherally displaced mitochondria in the muscle biopsy demands sequencing of the *CHKB* gene—see Case Example 41.5). An important, albeit only auxiliary, role of molecular genetics is to reveal the signature changes of mtDNA maintenance defects, mtDNA depletion, multiple mtDNA deletions, or both together. However, if none of the above applies and yet a mitochondrial disease is still strongly suspected, the application of mito-exome sequencing or whole-exome sequencing becomes necessary.

THERAPEUTIC APPROACHES

Although we have defined mitochondrial diseases as defects of the respiratory chain, there is no panacea for respiratory chain dysfunction and no universal therapy for mitochondrial disorders. First, it is important to state the obvious, that incurable diseases are not untreatable diseases and both pharmacological (e.g. antiepileptic drugs) and surgical (e.g. blepharoplasty) remedies are useful in prolonging and improving the life of mitochondrially affected patients.

However, we will briefly review here a few strategies aimed at curing (or even preventing) mitochondrial diseases, some of which have shown promise either *in vitro* or in animals while others have shown encouraging preliminary results in humans.

The most obvious therapeutic approach is to enhance respiratory chain function, thus mitigating both energy crisis (ATP deficit) and oxidative stress (toxic buildup of ROS). The central compound used almost universally in mitochondrial patients is CoQ10, which makes sense considering its pivotal role in electron transport, its antioxidant properties, and its safety even at high doses. Leaving aside some—but not all—patients with primary CoQ10 deficiency who respond dramatically to CoQ10 supplementation,[267] the effect of CoQ10 has been modest and subjective.[288] Two synthetic derivatives of CoQ10 appear more promising. Idebenone, a short-chain benzoquinone, has shown positive results in two studies of LHON patients, a retrospective analysis[289] and a randomized placebo-controlled trial.[290] The second compound, a para-benzoquinone labeled EPI-743, has been tested thus far only in open studies: it reversed vision loss in 4 of 5 LHON patients,[291] showed clinical improvement in 12 children with a variety of mitochondrial disorders,[292] and arrested or reversed disease progression in 13 children with various genetically proven LS.[293]

The second logical therapeutic approach is to eliminate accumulating noxious compounds. To decrease brain lactate in patients with MELAS, we used dichloroacetate (DCA), which keeps pyruvate dehydrogenase (PDH) in the active form and favors lactate oxidation; unfortunately, DCA had unacceptable neurotoxicity and had to be discontinued.[294] A more promising "detoxifying" therapy—despite its inherent risks—appears to be AHSCT aimed at restoring sufficient TPase activity in patients with MNGIE to normalize the circulating toxic levels of thymidine and deoxyuridine.[268] As of 2010, 5 of 11 patients with MNGIE who had undergone AHSCT were alive, and had normal blood TPase activity and virtually undetectable levels of thymidine and deoxyuridine: an international controlled trial is underway.

For mtDNA-related disorders, an obvious—but by no means simple to implement—goal is to shift heteroplasmy, i.e. to lower the mutation load to subthreshold levels. Several strategies have been tried, often with good results in cybrid cell lines. When deprived of glucose and exposed to ketogenic media, cybrids harboring single mtDNA deletions shifted their heteroplasmy level and recovered mitochondrial function, probably through selective mitochondrial autophagy (mitophagy).[295] A genetic approach to heteroplasmic shifting involves using restriction endonucleases as silver bullets to get rid of specific pathogenic mutations.[296] As the dormant stem cells in skeletal muscle (satellite cell) have fewer mtDNA mutations than adult muscle fibers, muscle regeneration after induced myonecrosis has been used to shift heteroplasmy,

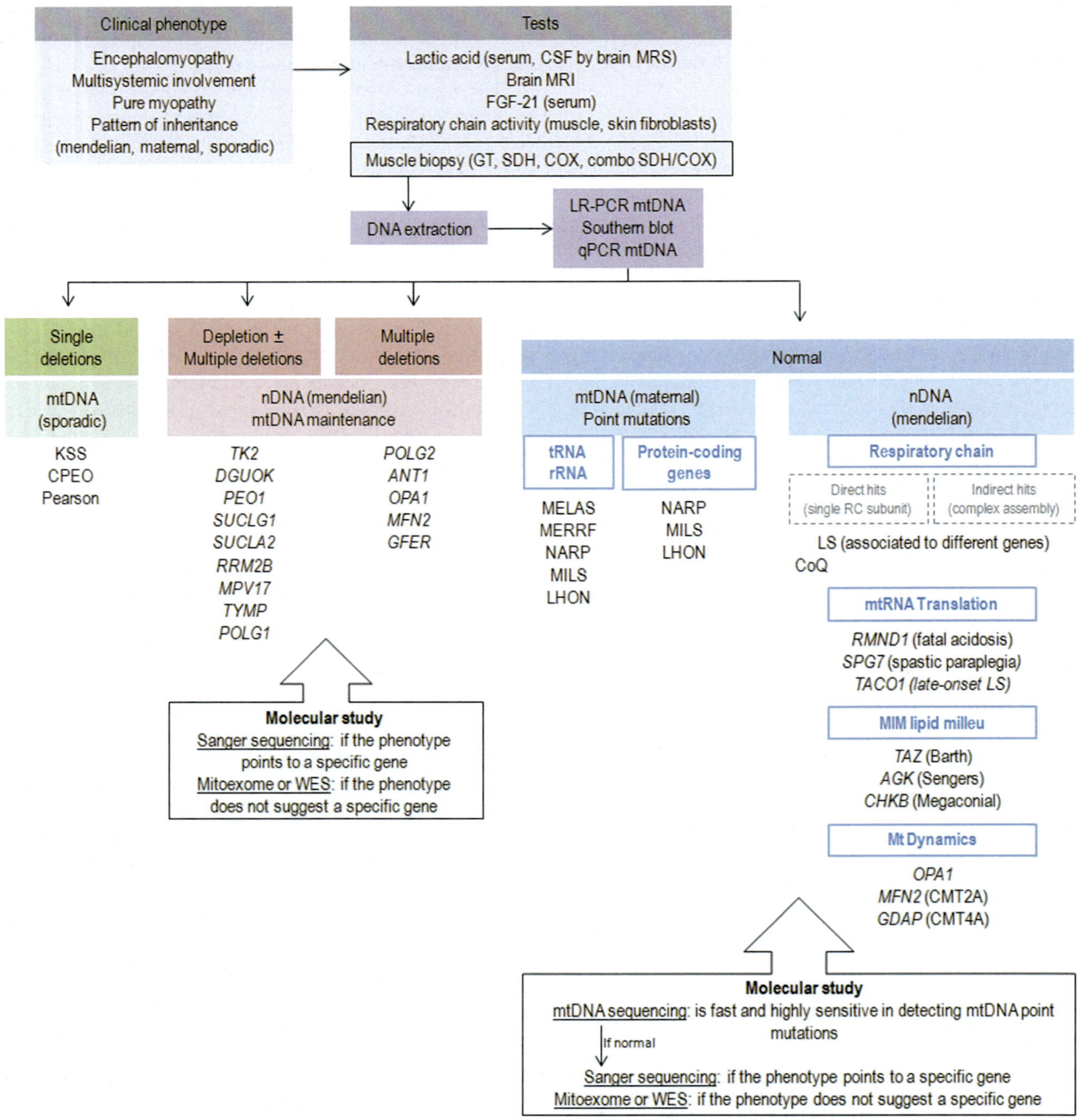

FIGURE 41.10 Algorithm summarizing the main features of mitochondrial diseases and a sequence of laboratory tests and initial molecular studies to sort out mtDNA-related vs. nDNA-related disorders. KSS, Kearns-Sayre syndrome; LHON, Leber's hereditary optic neuropathy; LS, Leigh syndrome; MELAS, mitochondrial encephalomyopathy, lactic acidosis, and stroke-like episodes; MERRF, myoclonus epilepsy and ragged-red fibers; MILS, maternally inherited Leigh syndrome; NARP, neuropathy, ataxia, retinitis pigmentosa; PEO, progressive external ophthalmoplegia.

and with good results, at least in one patient with strabismus.[297]

Mitochondrial dynamics could be exploited therapeutically in two opposite ways. On the one hand, we could enhance mitochondrial fission and favor mitophagy, a natural "quality control" function that sequesters and eliminates dysfunctional mitochondria, perhaps sensing their abnormally low membrane potential.[295,298] On the other hand, enhancing mitochondrial fusion and "networking" would allow complementation of "bad" and "good" mitochondria and normalization of overall mitochondrial function, as shown by the use of mdivi-1 (mitochondrial division inhibitor 1[299]) in human neuronal cells harboring mutations in *PINK1*.[300]

If it is true that mitochondrial proliferation (e.g. the RRF in skeletal muscle) is a futile compensatory attempt, we could try to improve on nature's strategy by enhancing mitochondrial biogenesis through the activation of the transcriptional coactivator PGC-1α (peroxisome proliferator-activated receptor gamma PPARγ coactivator 1α). The advantage over disease-induced mitochondrial proliferation, which favors mutated mtDNAs, is that pharmacological upregulation of mitochondrial biosynthesis increases the number of *all* mtDNAs, thus allowing wild-type genomes to compensate for mutated ones. Encouraging results have been obtained in four mouse models of COX deficiency: one COX10-deficient animal benefited from treatment with bezafibrate, a PPAR panagonist,[301] as did three transgenic mice (with mutations in *Surf1*, *Sco2*, and *Cox15*) treated with the AMP-dependent kinase (AMPK) agonist AICAR, which phosphorylates PGC-1alpha.[302]

For nDNA-related mitochondrial disorders, "classical" gene therapy is—of course—an option, although it carries the well-known "baggage" of this approach, including the choice of appropriate viral or nonviral vectors, delivery to the affected tissues, and potential immunological reactions. Nonetheless, adeno-associated virus-mediated gene transfer has proven successful in two animal models, the *Ant1* mutant mouse that we mentioned above,[303] and in a mouse model of ethylmalonic encephalomyopathy.[304]

For mtDNA-related diseases, many of which are devastating and undiagnosable prenatally (at least as far as their clinical expression will be), the Holy Grail would be to prevent their occurrence altogether via "cytoplasmic transfer." We have long known that this was technically feasible, but recent advances raise the hope that this preventive therapy will soon be available to women carrying mtDNA mutations. The nucleus of an *in vitro*-fertilized oocyte from a carrier is transferred to an enucleated oocyte from a normal donor; the embryo will have the nDNA of the biological parents but the mtDNA of a normal mitochondrial donor. Experiments in primates using spindle-chromosomal complex transfer (ST) have documented that infants were healthy and devoid of maternal mtDNA.[305] The same technique has been applied to fertilized[306] and unfertilized but parthenogenically activated[307] human oocytes and has shown that cells develop into normal blastocysts and contain exclusively donor mtDNA. Even more excitingly, only donor mtDNA was detected after stem cell lines from blastocysts were differentiated into neurons, cardiomyocytes, and beta-cells.[307] Similar results were obtained in the UK after pronuclear transfer in abnormally fertilized human oocytes developed to the blastocyst stage.[308] Despite some minor concerns,[309] the stage seems set for approval of this technique for therapeutic application both in the UK and in the USA.

For now and in the real world, most clinicians prescribe patients with mitochondrial diseases "cocktails" of vitamins and cofactors, whose popularity is based on two factors: (i) their safety makes the doctor feel "Hippocratically correct" (*primum non nocere*) even when their efficacy is doubtful; and (ii) their use is predicated by the two widely accepted pathogenic mechanisms mentioned above, energy shortage and oxidative stress. In the hope of facilitating ATP production by a sluggish respiratory chain, we try to boost electron flux by addition of electron acceptors (CoQ10, vitamin C, vitamin K, succinate). Alternatively, we try to boost the synthesis of a naturally occurring high-energy phosphate compound, phosphocreatine (PC) by administering creatine. We prescribe L-carnitine mostly because plasma free carnitine tends to be lower and esterified carnitine higher than normal, probably reflecting a partial impairment of beta-oxidation. A recent Cochrane review concluded that there is no evidence supporting the use of any such interventions in mitochondrial diseases. The authors advocate the identification of "novel agents to be tried in homogeneous study populations with clinically relevant primary endpoints."[310]

REFERENCES

1. Kolliker A. Einige Bemerkungen uber Endigungen der Hautnerven und den Bau der Muskeln. *Z Wiss Zool* 1856;**8**:311–25.
2. Luft R, Ikkos D, Palmieri G, Ernster L, Afzelius B. A case of severe hypermetabolism of nonthyroid origin with a defect in the maintenance of mitochondrial respiratory control: A correlated clinical, biochemical, and morphological study. *J Clin Invest* 1962;**41**:1776–804.
3. Shy GM, Gonatas NK. Human myopathy with giant abnormal mitochondria. *Science* 1964;**145**:493–6.
4. Shy GM, Gonatas NK. Two childhood myopathies with abnormal mitochondria: I. Megaconial myopathy; II. Pleoconial myopathy. *Brain* 1966;**89**:133–58.
5. Benda C. Weitere Mitteilungen uber die Mitochondria. *Verh Physiolog Ges Berlin* 1898;376–83.
6. Keilin D. Cytochrome, a respiratory pigment common to animals, yeast and higher plants. *Proc R Soc London* 1925;**98**:312.
7. Warburg O, Negelein E. Uber des Absorptions Spectrum des Atmungsferent. *Biochem Z* 1929;**214**:64.
8. Warburg O. *Heavy Metal Prosthetic Groups and Enzyme Action*. London: Oxford University Press; 1949.
9. Krebs HA, Kornberg HL. Energy transformation in living matter. *Ergeb Physiol* 1957;**49**:212.
10. Mitchell P. Coupling of phosphorylation to electron and hydrogen transfer by a chemiosmotic type of mechanism. *Nature* 1961;**191**: 144–8.
11. DiMauro S, Schon EA. Mitochondrial disorders in the nervous system. *Annu Rev Neurosci* 2008;**31**:91–123.

12. Claude A. Fractionation of mammalian liver cells by differential centrifugation. II. Experimental procedure and results. *J Exp Med* 1946;**84**:61–89.
13. Palade GE. An electron microscope study of the mitochondrial structure. *J Histochem Cytochem* 1953;**1**:188–203.
14. Palade GE. Electron microscopy of cytoplasmic structures. In: Gaebler OH, editor. *Enzymes: Units of Biological Structure and Function*. New York: Academic Press; 1956.
15. Palmieri F. Mitochondrial metabolite carrier protein family. *Encyclop Biol Chem* 2013;**3**:149–58.
16. Bolender N, Sickmann A, Wagner R, Meisinger C, Pfanner N. Multiple pathways for sorting mitochondrial precursor proteins. *EMBO J* 2008;**9**:42–9.
17. Schatz G, Dobberstein B. Common principles of protein translocation across membranes. *Science* 1996;**271**:1519–26.
18. Anderson S, Bankier AT, Barrel BG, DeBruijin M, Coulson AR, Drouin J, et al. Sequence and organization of the human mitochondrial genome. *Nature* 1981;**290**:457–65.
19. Nass S, Nass M. Intramitochondrial fibers with DNA characteristics. II. Enzymatic and other hydrolytic treatments. *J Cell Biol* 1963;**19**:613–29.
20. Engel WK, Cunningham CG. Rapid examination of muscle tissue: An improved trichrome stain method for fresh-frozen biopsy sections. *Neurology* 1963;**13**:919–23.
21. Ephrussi B, Hottinguer H, Chimenes AM. Action de l'acriflavine sur les levures. I. La mutation "petite" colonie. *Ann Inst Pasteur* 1949;**76**:351–62.
22. Schatz G, Haslbrunner E, Tuppy H. Deoxyribonucleic acid associated with yeast mitochondria. *Biochem Biophys Res Comm* 1964; **15**:127–32.
23. Fine PE. Mitochondrial inheritance and disease. *Lancet* 1978; **2**:659–62.
24. Giles RE, Blanc H, Cann RM, Wallace DC. Maternal inheritance of human mitochondrial DNA. *Proc Natl Acad Sci USA* 1980; **77**:6715–19.
25. Wallace DC, Zheng XX, Lott MT, Shoffner JM, Hodge JA, Kelley RI, et al. Familial mitochondrial encephalomyopathy (MERRF): genetic, pathophysiological, and biochemical characterization of a mitochondrial DNA disease. *Cell* 1988;**55**:601–10.
26. Holt IJ, Harding AE, Morgan Hughes JA. Deletions of muscle mitochondrial DNA in patients with mitochondrial myopathies. *Nature* 1988;**331**:717–19.
27. Wallace DC, Singh G, Lott MT, Hodge JA, Schurr TG, Lezza A, et al. Mitochondrial DNA mutation associated with Leber's hereditary optic neuropathy. *Science* 1988;**242**:1427–30.
28. DiMauro S. Mitochondrial encephalomyopathies—Fifty years on. *Neurology* 2013;**81**:281–91.
29. Schaefer AM, McFarland R, Blakely E, Langping H, Whittaker RG, Taylor RW, et al. Prevalence of mitochondrial DNA disease in adults. *Ann Neurol* 2008;**63**:35–9.
30. Elliott HR, Samuels DC, Eden JA, Relton CL, Chinnery PF. Pathogenic mitochondrial DNA mutations are common in the general population. *Am J Hum Genet* 2008;**83**:254–60.
31. Payne BAJ, Wilson IJ, Yu-Wai-Man P, Coxhead J, Deehan D, Horvath R, et al. Universal heteroplasmy of human mitochondrial DNA. *Hum Mol Genet* 2013;**22**:384–90.
32. Bourgeron T, Rustin P, Chretien D, Birch-Machin M, Bourgeois M, Viegas-Pequignot E, et al. Mutation of a nuclear succinate dehydrogenase gene results in mitochondrial respiratory chain deficiency. *Nat Genet* 1995;**11**:144–9.
33. Schon EA, DiMauro S, Hirano K-I. Human mitochondrial DNA: roles of inherited and somatic mutations. *Nat Rev Genet* 2012;**13**: 878–90.
34. Calvo S, Compton A, Hershman SG, Lim SC, Lieber DS, Tucker EJ, et al. Molecular diagnosis of infantile mitochondrial disease with targeted next-generation sequencing. *Sci Transl Med* 2012;**4** (118):118ra10.
35. Lieber DS, Calvo SE, Shanahan K, Slate NG, Liu S, Hershman SG, et al. Targeted exome sequencing of suspected mitochondrial disorders. *Neurology* 2013;**80**:1762–79.
36. Calvo SE, Tucker EJ, Compton AG, Kirby DM, Crawford G, Burtt NP, et al. High-throughput, pooled sequencing identifies mutations in NUBPL and FOXRED1 in human complex I deficiency. *Nat Genet* 2010;**42**:851–8.
37. DiMauro S, Bonilla E, Zeviani M, Nakagawa M, De Vivo DC. Mitochondrial myopathies. *Ann Neurol* 1985;**17**:521–38.
38. Quinzii C, Lopez de Munain A, Von-Moltke J, Naini AB, Krishna S, Schuelke M, et al. Respiratory chain dysfunction and oxidative stress correlate with severity of primary CoQ10 deficiency. *FASEB J* 2008;**22**:1874–85.
39. Wallace DC. Pathophysiology of mitochondrial disease as illuminated byanimal models. In: Schapira AHV, DiMauro S, editors. *Mitochondrial Disorders in Neurology,* 2nd Ed. Boston: Butterworth-Heinemann; 2002. pp. 175–212.
40. DiMauro S, Schon EA. Mitochondrial disorders. In: Waxman SG, editor. *Molecular Neurology*. Burlington MA: Elsevier Academic Press; 2007. pp. 535–51.
41. DiMauro S, Davidzon G, Hirano M. A polymorphic polymerase. *Brain* 2006;**126**:1637–9.
42. Carelli V, Barboni P, Sadun AA. Mitochondrial ophthalmology. In: DiMauro S, Hirano M, Schon EA, editors. *Mitochondrial Medicine*. London: Informa Healthcare; 2006. pp. 105–42.
43. Hirano M, Ricci E, Koenigsberger MR, Defendini R, Pavlakis SG, De Vivo DC, et al. MELAS: an original case and clinical criteria for diagnosis. *Neuromusc Disord* 1992;**2**:125–35.
44. Tatuch Y, Christodoulou J, Feigenbaum A, Clarke J, Wherret J, Smith C, et al. Heteroplasmic mtDNA mutation (T > G) at 8993 can cause Leigh disease when the percentage of abnormal mtDNA is high. *Am J Hum Genet* 1992;**50**:852–8.
45. Zeviani M, Moraes CT, DiMauro S. Deletions of mitochondrial DNA in Kearns-Sayre syndrome. *Neurology* 1988; **38**:1339–46.
46. Lestienne P, Ponsot G. Kearns-Sayre syndrome with muscle mitochondrial DNA deletion. *Lancet* 1988;**1**:885.
47. Moraes CT, DiMauro S, Zeviani M, Lombes A, Shanske S, Miranda AF, et al. Mitochondrial DNA deletions in progressive external ophthalmoplegia and Kearns-Sayre syndrome 'see comments. *N Engl J Med* 1989;**320**:1293–9.
48. Schon EA, Rizzuto R, Moraes CT, Nakase H, Zeviani M, DiMauro S. A direct repeat is a hotspot for large-scale deletions of human mitochondrial DNA. *Science* 1989;**244**:346–9.
49. Rowland LP, Hays AP, DiMauro S, De Vivo DC, Behrens M. Diverse clinical disorders associated with morphological abnormalities of mitochondria. In: Scaerlato G, Cerri C, editors. *Mitochondrial Pathology in Muscle Diseases*. Padova, Italy: Piccin; 1983. pp. 141–58.

50. Berenberg RA, Pellock JM, DiMauro S, Rowland LP. Lumping or splitting? "Ophthalmoplegia plus" or Kearns-Sayre syndrome? *Ann Neurol* 1977;**1**:37–43.
51. Chinnery PF, DiMauro S, Shanske S, Schon EA, Zeviani M, Mariotti C, et al. Risk of developing a mitochondrial DNA deletion disorder. *Lancet* 2004;**364**:592–5.
52. Bernes SM, Bacino C, Prezant TR, Pearson MA, Wood TS, Fournier P, et al. Identical mitochondrial DNA deletion in mother with progressive external ophthalmoplegia and son with Pearson marrow-pancreas syndrome. *J Pediatr* 1993;**123**:598–602.
53. Shanske S, Tang Y, Hirano M, Nishigaki Y, Tanji K, Bonilla E, et al. Identical mitochondrial DNA deletion in a woman with ocular myopathy and in her son with Pearson syndrome. *Am J Hum Genet* 2002;**71**:679–83.
54. Pearson HA, Lobel JS, Kocoshis SA, Naiman JL, Windmiller J, Lammi AT, et al. A new syndrome of refractory sideroblastic anemia with vacuolization of marrow precursors and exocrine pancreatic dysfunction. *J Pediatr* 1979;**95**:976–84.
55. Rotig A, Bourgeron T, Chretien D, Rustin P, Munnich A. Spectrum of mitochondrial DNA rearrangements in the Pearson marrow-pancreas syndrome. *Hum Mol Genet* 1995;**4**:1327–30.
56. Larsson NG, Holme B, Kristiansson B. Progressive increase of the mutated mitochondrial DNA fraction in Kearns-Sayre syndrome. *Pediatr Res* 1990;**28**:131–6.
57. McShane MA, Hammans SR, Sweeney M, Holt IJ, Beattie TJ, Brett EM, et al. Pearson syndrome and mitochondrial encephalopathy in a patient with a deletion of mtDNA. *Am J Hum Genet* 1991; **48**:39–42.
58. Poulton J, Deadman ME, Gardiner RM. Duplications of mitochondrial DNA in mitochondrial myopathy. *Lancet* 1989;**1**:236–40.
59. Rotig A, Bessis JL, Romero N, Cormier V, Saudubray JM, Narcy P, et al. Maternally inherited duplication of the mitochondrial genome in a syndrome of proximal tubulopathy, diabetes mellitus, and cerebellar ataxia. *Am J Hum Genet* 1992;**50**:364–70.
60. Brockington M, Sweeney MG, Hammans SR, Morgan-Hughes JA, Harding AE. A tandem duplication in the D-loop of human mitochondrial DNA is associated with deletions in mitochondrial myopathies. *Nat Genet* 1993;**4**:67–71.
61. Prezant TR, Agapian JV, Bohlman MC, Bu X, Oztas S, Qiu W-Q, et al. Mitochondrial ribosomal RNA mutation associated with both antibiotic-induced and non-syndromic deafness. *Nat Genet* 1993; **4**:289–93.
62. Santorelli FM, Tanji K, Manta P, Casali C, Krishna S, Hays AP, et al. Maternally inherited cardiomyopathy: An atypical presentation of the 12S rRNA A1555G mutation. *Am J Hum Genet* 1999;**64**: 295–300.
63. Scaglia F, Wong L-JC. Human mitochondrial transfer tRNAs: role of pathogenic mutation in disease. *Muscle Nerve* 2008;**37**: 150–71.
64. DiMauro S, Bonilla E, Mancuso M, Filosto M, Sacconi S, Salviati L, et al. Mitochondrial myopathies. *Basic Appl Myol* 2003;**13**: 145–55.
65. DiMauro S, Davidzon G. Mitochondrial DNA and disease. *Ann Med* 2005;**37**:222–32.
66. Pavlakis SG, Phillips PC, DiMauro S, De Vivo DC, Rowland LP. Mitochondrial myopathy, encephalopathy, lactic acidosis, and strokelike episodes: a distinctive clinical syndrome. *Ann Neurol* 1984;**16**:481–8.
67. Hasegawa H, Matsuoka T, Goto I, Nonaka I. Strongly succinate dehydrogenase-reactive blood vessels in muscles from patients with mitochondrial myopathy, encephalopathy, lactic acidosis, and stroke-like episodes. *Ann Neurol* 1991;**29**:601–5.
68. Ohama E, Ohara S, Ikuta F, Tanaka K, Nishizawa M, Miyatake T. Mitochondrial angiopathy in cerebral blood vessels of mitochondrial encephalomyopathy. *Acta Neuropath* 1987;**74**:226–33.
69. Betts J, Jaros E, Perry RH, Schaefer AM, Taylor RW, Abdel-All Z, et al. Molecular neuropathology of MELAS: level of heteroplasmy in individual meurones and evidence of extensive vascular involvement. *Neuropath Appl Neurobiol* 2006;**32**:359–73.
70. Tanji K, Kunimatsu T, Vu TH, Bonilla E. Neuropathological features of mitochondrial disorders. *Cell Develop Biol* 2001;**12**: 429–39.
71. Yoneda M, Maeda M, Kimura H, Fujii A, Katayama K, Kuriyama M. Vasogenic edema on MELAS: A serial study with diffusion-weighted MR imaging. *Neurology* 1999;**53**:2182–4.
72. Kaufmann P, Engelstad K, Wei Y, Kulikova R, Oskoui M, Battista V, et al. Protean phenotypic features of the A3243G mitochondrial DNA mutation. *Arch Neurol* 2009;**66**:85–91.
73. Moraes CT, Ciacci F, Silvestri G, Shanske S, Sciacco M, Hirano M, et al. Atypical clinical presentations associated with the MELAS mutation at position 3243 of human mitochondrial DNA. *Neuromusc Disord* 1993;**3**:43–50.
74. Sproule DM, Kaufmann P, Engelstad K, Starc TJ, Hordof AJ, De Vivo DC. Wolff-Parkinson-White syndrome in patients with MELAS. *Arch Neurol* 2007;**64**:1625–7.
75. Kaufmann P, Sano MC, Jhung S, Engelstadt K, De Vivo DC. Psychiatric symptoms are common features of clinical syndromes associated with mitochondrial DNA point mutations. *Neurology* 2002;**58**:A315 [Abstract].
76. Tanji K, Bonilla E. Light microscopic method to visualize mitochondria on tissue sections. *Methods* 2008;**46**:274–80.
77. Goto Y, Tojo M, Tohyama J, Horai S, Nonaka I. A novel point mutation in the mitochondrial tRNALeu(UUR) gene in a family with mitochondrial myopathy. *Ann Neurol* 1992;**31**:672–5.
78. Ogle RF, Christodoulou J, Fagan E, Blok RB, Kirby DM, Seller KL, et al. Mitochondrial myopathy with tRNALeu(UUR) mutation and complex I deficiency responsive to riboflavin. *J Pediatr* 1997; **130**:138–45.
79. Hadjigeorgiou GM, Kim SH, Fischbeck KH, Andreu AL, Berry GT, Bingham P, et al. A new mitochondrial DNA mutation (A3288G) in the TRNALeu(UUR) gene associated with familial myopathy. *J Neurol Sci* 1999;**164**:153–7.
80. Moraes CT, Ciacci F, Bonilla E, Ionasescu V, Schon EA, DiMauro S. A mitochondrial tRNA anticodon swap associated with a muscle disease. *Nat Genet* 1993;**4**:284–8.
81. DiMauro S, Nicholson JF, Hays AP, Eastwood AB, Papadimitriou A, Koenigsberger R, et al. Benign infantile mitochondrial myopathy due to reversible cytochrome c oxidase deficiency. *Ann Neurol* 1983;**14**:226–34.
82. Horvath R, Kemp JP, Tuppen HAL, Hudson G, Oldfors A, Marie SKN, et al. Molecular basis of infantile reversible cytochrome c oxidase deficiency. *Brain* 2009;**132**:3165–74.
83. Mimaki M, Hatakayama H, Komaki H, Yokoyama M, Arai H, Kirino Y, et al. Reversible infantile respiratory chain deficiency: A clinical and molecular study. *Ann Neurol* 2010;**68**: 845–54.

84. Boczonadi V, Smith PC, Pyle A, Gomez-Duran A, Schara U, Tulinius M, et al. Altered 2-thiouridylation impairs mitochondrial translation in reversible infantile respiratory chain deficiency. *Hum Mol Genet* 2013;**22**:4602–15.
85. Uusimaa J, Jungbluth H, Fratter C, Crisponi G, Feng L, Zeviani M, et al. Reversible infantile respiratory chain deficiency is a unique, genetically heterogeneous mitochondrial disease. *J Med Genet* 2011;**48**:660–8.
86. Zeharia A, Shaag A, Pappo O, Mager-Heckel A-M, Saada A, Beinat M, et al. Acute liver failure due to mutations in the TRMU gene. *Am J Hum Genet* 2009;**85**:401–7.
87. Fukuhara N. MERRF: A clinicopathological study. Relationship between myoclonus epilepsies and mitochondrial myopathies. *Rev Neurol* 1991;**147**:476–9.
88. DiMauro S, Hirano M, Kaufmann P, Tanji K, Sano M, Shungu DC, et al. Clinical features and genetics of myoclonic epilepsy with ragged red fibers. In: Fahn S, Frucht SJ, editors. *Myoclonus and Paroxysmal Dyskinesia*. Philadelphia: Lippincott Williams & Wilkins; 2002. pp. 217–29.
89. Mancuso M, Orsucci D, Angelini C, Bertini E, Carelli V, Comi GP, et al. Phenotypic heterogeneity of the 8344A > G mtDNA "MERRF" mutation. *Neurology* 2013;**80**:2049–54.
90. Lombes A, Mendell JR, Nakase H, Barohn RJ, Bonilla E, Zeviani M, et al. Myoclonic epilepsy and ragged-red fibers with cytochrome oxidase deficiency: neuropathology, biochemistry, and molecular genetics. *Ann Neurol* 1989;**26**:20–33.
91. Howell N, Kubacka I, Smith R, Frerman F, Parks JK, Parker WD. Association of the mitochondrial 8344 MERRF mutation with maternally inherited spinocerebellar degeneration and Leigh disease. *Neurology* 1996;**46**:219–22.
92. Santorelli FM, Tanji K, Shanske S, Krishna S, Schmidt RE, Greenwood RS, et al. The mitochondrial DNA A8344G mutation in Leigh syndrome revealed by analysis in paraffin-embedded sections: Revisiting the past. *Ann Neurol* 1998;**44**:962–4.
93. Rosing HS, Hopkins LC, Wallace DC, Epstein CM, Weidenheim K. Maternally inherited mitochondrial myopathy and myoclonic epilepsy. *Ann Neurol* 1985;**17**:228–37.
94. Shoffner JM, Lott MT, Lezza A, Seibel P, Ballinger SW, Wallace DC. Myoclonic epilepsy and ragged-red fiber disease (MERRF) is associated with a mitochondrial DNA tRNALys mutation. *Cell* 1990;**61**:931–7.
95. Silvestri G, Moraes CT, Shanske S, Oh SJ, DiMauro S. A new mtDNA mutation in the tRNALys gene associated with myoclonic epilepsy and ragged-red fibers (MERRF). *Am J Hum Genet* 1992;**51**:1213–17.
96. Shtilbahns A, Shanske S, Goodman S, Sue CM, Bruno C, Johnson TL, et al. G8363A mutation in the mitochondrial DNA transfer ribonucleic acid Lys gene: another cause of Leigh syndrome. *J Child Neurol* 2000;**15**:759–61.
97. Santorelli FM, Mak S-C, El-Schahawi M, Casali C, Shanske S, Baram TZ, et al. Maternally inherited cardiomyopathy and hearing loss associated with a novel mutation in the mitochondrial DNA tRNALys gene (G8363A). *Am J Hum Genet* 1996;**58**: 933–9.
98. Ozawa M, Nishino I, Horai S, Nonaka I, Goto Y-I. Myoclonus epilepsy associated with ragged-red fibers: A G-to-A mutation at nucleotide pair 8363 in mitochondrial tRNALys in two families. *Muscle Nerve* 1997;**20**:271–8.
99. Vallance HD, Jeven G, Wallace DC, Brown MD. A case of sporadic infantile histiocytoid cardiomyopathy caused by the A8344G (MERRF) mitochondrial DNA mutation. *Pediatr Cardiol* 2004;**25**:538–40.
100. Tanji K, Gamez J, Cervera C, Mearin F, Ortega A, de la Torre J, et al. The A8344G mutation in mitochondrial DNA associated with stroke-like episodes and gastrointestinal dysfunction. *Acta Neuropathol* 2003;**105**:69–75.
101. Emmanuele V, Silvers DS, Sotiriou E, Tanji K, DiMauro S, Hirano K-I. MERRF and Kearns-Sayre overlap syndrome due to the mitochondrial DNA m.3291T > C mutation. *Muscle Nerve* 2011;**44**:448–51.
102. Holt IJ, Harding AE, Petty RK, Morgan Hughes JA. A new mitochondrial disease associated with mitochondrial DNA heteroplasmy. *Am J Hum Genet* 1990;**46**:428–33.
103. Santorelli FM, Shanske S, Macaya A, De Vivo DC, DiMauro S. The mutation at nt 8993 of mitochondrial DNA is a common cause of Leigh syndrome. *Ann Neurol* 1993;**34**:827–34.
104. Giordano C, Montopoli M, Perli E, Orlandi M, Fantin M, Ross-Cisneros FN, et al. Oestrogens ameliorate mitochondrial dysfunction in Leber's hereditary optic neuropathy. *Brain* 2011;**134**:220–34.
105. DiMauro S, Andreu AL. Mutations in mtDNA: Are we scraping the bottom of the barrel? *Brain Pathology* 2000;**10**:431–41.
106. Andreu AL, Hanna MG, Reichmann H, Bruno C, Penn AS, Tanji K, et al. Exercise intolerance due to mutations in the cytochrome b gene of mitochondrial DNA. *N Engl J Med* 1999;**341**:1037–44.
107. Shanske S, Coku J, Lu J, Ganesh J, Krishna S, Tanji K, et al. The G13513A mutation in the ND5 gene of mitochondrial DNA as a common cause of MELAS or Leigh syndrome. *Arch Neurol* 2008;**65**:368–72.
108. Sarzi E, Brown MD, Lebon S, Chretien D, Munnich A, Rotig A, et al. A novel mitochondrial DNA mutation in ND3 gene is associated with isolated complex I deficiency causing Leigh syndrome and dystonia. *Am J Med Genet* 2007;**143A**:33–41.
109. McFarland R, Kirby DM, Fowler KJ, Ohtake A, Ryan MT, Amor DJ, et al. De novo mutations in the mitochondrial ND3 gene as a cause of infantile mitochondrial encephalopathy and complex I deficiency. *Ann Neurol* 2004;**55**:58–64.
110. Shanske S, Naini A, Chmait RH, Akman HO, Manshukhani M, Lu J, et al. Mutation in an mtDNA protein-coding gene: prenatal diagnosis aided by fetal muscle biopsy. *J Child Neurol* 2012; **28**:264–8.
111. Levy R, Gutierrez Rios P, Akman HO, De Vivo DC, Sciacco M, DiMauro S. Long survival in patients with Leigh syndrome and the m.10191T > C mutation in MTND3: A case report and review of the literature. *J Child Neurol* 2013. Epub Nov 27. PMID: 24284231.
112. Ronchi D, Garone C, Bordoni A, Gutierrez Rios P, Calvo S, Ripolone M, et al. Next-generation sequencing discloses DGUOK mutations in adult patients with mtDNA multiple deletions. *Brain* 2012;**135**:3404–15.
113. Mayr JA, Haack TB, Graf E, Zimmermann FA, Wieland T, Haberberger B, et al. Lack of mitochondrial protein acylglycerol kinase causes Sengers syndrome. *Am J Hum Genet* 2012;**90**: 314–20.
114. Kornblum C, Nicholls T, Haack TB, Scholer S, Peeva V, Danhauser K, et al. Loss-of-function mutations in MGME1 impair mtDNA replication and cause multisystemic mitochondrial disease. *Nat Genet* 2013;**45**:214–19.

115. Hildick-Smith GJ, Cooney JD, Garone C, Kremer LS, Haack TB, Thon JN, et al. Macrocytic anemia and mitochondriopathy resulting from a defect in sideroflexin 4. *Am J Hum Genet* 2013;**93**: 906–14.
116. Schon EA, Dencher NA. Heavy breathing: energy conversion by mitochondrial respiratory supercomplexes. *Cell Metab* 2009;**9**: 1–3.
117. Smeitink JAM, Sengers R, Trijbels FJM, van den Heuvel L. Human NADH:ubiquinone oxidoreductase. *J Bioenerg Biomembr* 2001;**33**:259–66.
118. Distelmaier F, Koopman WJH, van den Heuvel LP, Rodenburg R, Mayatepek E, Willems PH, et al. Mitochondrial complex I deficiency: from organelle dysfunction to clinical disease. *Brain* 2009;**132**:833–42.
119. Koene S, Rodenburg R, van der Knaap M, Willemsen MAAP, Sperl W, Laugel V, et al. Natural disease course and genotype-phenotype correlations in Complex I deficiency caused by nuclear gene defects: what we learned from 130 cases. *J Inher Metab Dis* 2012;**35**:737–47.
120. Bugiani M, Invernizzi F, Alberio S, Briem E, Lamantea E, Carrara F, et al. Clinical and molecular findings in children with complex I deficiency. *Biochim Biophys Acta* 2004;**1659**:136–47.
121. Janssen RJ, Nijtmans LGJ, Heuvel LP, Smeitink JAM. Mitochondrial complex I: Structure, function and pathology. *J Inher Metab Dis* 2006;**29**:499–515.
122. DiMauro S, Garone C. Metabolic disorders of fetal life: Glycogenoses and mitochondrial defects of the mitochondrial respiratory chain. *Sem Fet Neonat Med* 2011;**16**:181–9.
123. Fernandez-Moreira D, Ugalde C, Smeets R, Rodenburg RJT, Lopez-Laso E, Ruiz-Falco ML, et al. X-linked NDUFA1 gene mutations associated with mitochondrial encephalomyopathy. *Ann Neurol* 2007;**61**:73–83.
124. Rustin P, Rotig A. Inborn errors of complex II—Unusual human mitochondrial diseases. *Biochim Biophys Acta* 2002;**1553**: 117–22.
125. Parfait B, Chretien D, Rotig A, Marsac C, Munnich A, Rustin P. Compound heterozygous mutation in the flavoprotein gene of the respiratory chain complex II in a patient with Leigh syndrome. *Hum Genet* 2000;**106**:236–43.
126. Horvath R, Abicht A, Holinski-Feder E, Gempel K, Prokisch H, Klopstock T, et al. Leigh syndrome caused by mutations in the flavoprotein (Fp) subunit of succinate dehydrogenase (SDHA). *J Neurol Neurosurg Psychiatr* 2006;**77**:74–6.
127. Haut S, Brivet M, Touati G, Rustin P, Lebon S, Garcia-Cazorla A, et al. A deletion in the human QP-C gene causes a complex III deficiency resulting in hypoglycemia and lactic acidosis. *Hum Genet* 2003;**113**:118–22.
128. Barel O, Shorer Z, Flusser H, Ofir R, Narkis G, Finer G, et al. Mitochondrial complex III deficiency associated with a homozygous mutation in UQCRQ. *Am J Hum Genet* 2008;**82**:1211–16.
129. Trevisson E, DiMauro S, Navas P, Salviati L. Coenzyme Q deficiency in muscle. *Curr Opin Neurol* 2011;**24**:440–56.
130. Ogasahara S, Engel AG, Frens D, Mack D. Muscle coenzyme Q deficiency in familial mitochondrial encephalomyopathy. *Proc Nat Acad Sci USA* 1989;**86**:2379–82.
131. Musumeci O, Naini A, Slonim AE, Scavin N, Hadjigeorgiou GL, Krawiecki N, et al. Familial cerebellar ataxia with muscle coenzyme Q10 deficiency. *Neurology* 2001;**56**:849–55.
132. Lamperti C, Naini A, Hirano M, De Vivo DC, Bertini E, Servidei S, et al. Cerebellar ataxia and coenzyme Q10 deficiency. *Neurology* 2003;**60**:1206–8.
133. Quinzii C, Naini A, Salviati L, Trevisson E, Navas P, DiMauro S, et al. A mutation in para-Hydoxybenzoate-polyprenyl transferase (COQ2) causes primary coenzyme Q10 deficiency. *Am J Hum Genet* 2006;**78**:345–9.
134. Lopez LC, Quinzii C, Schuelke M, Kanki T, Naini A, DiMauro S, et al. Leigh syndrome with nephropathy and CoQ10 deficiency due to decaproneyl diphosphate synthase subunit 2 (PDSS2) mutations. *Am J Hum Genet* 2006;**79**:1125–9.
135. Mollet J, Giurgea I, Schlemmer D, Dallner G, Chretien D, Delahodde A, et al. Prenyldiphosphate synthase (PDSS1) and OH-benzoate prenyltransferase (COQ2) mutations in ubiquinone deficiency and oxidative phosphorylation disorders. *J Clin Invest* 2007;**117**:765–72.
136. Heeringa SF, Chernin G, Chaki M, Zhou W, Sloan AJ, Ji Z, et al. COQ6 mutations in human patients produce nephrotic syndrome with sensorineural deafness. *J Clin Invest* 2011;**121**: 2013–24.
137. Mollet J, Delahodde A, Serre V, Chretien D, Schlemmer D, Lombes A, et al. CABC1 gene mutations cause ubiquinone deficiency with cerebellar ataxia and seizures. *Am J Hum Genet* 2008;**82**:623–30.
138. Lagier-Tourenne C, Tazir M, Lopez LC, Quinzii C, Assoum M, Drouot N, et al. ADSK3, an ancestral kinase, is mutated in a form of recessive ataxia associated with coenzyme Q10 deficiency. *Am J Hum Genet* 2008;**82**:661–72.
139. Duncan AJ, Bitner-Glindziez M, Meunier B, Costello H, Hargreaves IP, Lopez LC, et al. A nonsense mutation in COQ9 causes autosomal recessive neonatal-onset primary coenzyme Q10 deficiency: a potentially treatable form of mitochondrial disease. *Am J Hum Genet* 2009;**84**:558–66.
140. Emmanuele V, Lopez LC, Berardo A, Naini A, Tadesse S, Wen B, et al. Heterogeneity of coenzyme Q10 deficiency: Patient study and literature review. *Arch Neurol* 2012;**69**:978–83.
141. Massa V, Fernandez-Vizarra E, Alshahwan S, Bakhsh E, Goffrini P, Ferrero I, et al. Severe infantile encephalomyopathy caused by a mutation in COX6B1, a nucleus-encoded subunit of cytochrome c oxidase. *Am J Hum Genet* 2008;**82**:1281–9.
142. Mayr JA, Havlickova V, Zimmermann F, Magler I, Kaplanova V, Jesina P, et al. Mitochondrial ATP synthase deficiency due to a mutation in the ATP5E gene for the F1 epsilon subunit. *Hum Mol Genet* 2010;**19**:3430–9.
143. Hoefs SJ, Dieteren CEJ, Distelmaier F, RJRJ Janssen, Epplen A, Swarts HGP, et al. NDUFA2 complex I mutation leads to Leigh disease. *Am J Hum Genet* 2008;**82**:1306–15.
144. Saada A, Vogel RO, Hoefs SJ, van den Brand M, Wessels HJ, Willems PH, et al. Mutations in NDUFAF3 (C3ORF60), encoding an NDUFAF4 (C6ORF66)-interacting complex I assembly protein, cause fatal neonatal mitochondrial disease. *Am J Hum Genet* 2009;**84**:718–27.
145. Hoefs SJG, Skjeldal OH, Rodenburg RJ, Nedregaard B, van Kaauwen EPM, Spiekerkotter U, et al. Novel mutations in the NDUFS1 gene cause low residual activities in human complex I deficiencies. *Mol Gen Metab* 2010;**100**:251–6.
146. Fassone E, Taanman J-W, Hargreaves IP, Sebire NJ, Cleary MA, Burch M, et al. Mutations in the mitochondrial complex I

assembly factor NDUFAF1 cause fatal infantilehypertrophic cardiomyopathy. *J Med Genet* 2011;**48**:691–7.

147. Ferreira M, Turraco A, Rizza T, Fattori F, Meschini MC, Castana C, et al. Progressive cavitating leukoencephalopathy associated with respiratory chain complex I deficiency and a novel mutation in NDUFS1. *Neurogenetics* 2011;**12**:9–17.

148. Berger I, Hershkovitz E, Shaag A, Edvardson S, Saada A, Elpeleg O, et al. Mitochondrial complex I deficiency caused by a deleterious NDUFA11 mutation. *Ann Neurol* 2008;**63**:405–8.

149. Ogilvie I, Kennaway NG, Shoubridge EA. A molecular chaperone for mitochondrial complex I assembly is mutated in a progressive encephalopathy. *J Clin Invest* 2005;**115**:2784–92.

150. Sugiana C, Pagliarini DJ, McKenzie M, Kirby DM, Salemi R, Abu-Amero KK, et al. Mutation of C20orf7 distupts complex I assembly and causes lethal neonatal mitochondrial disease. *Am J Hum Genet* 2008;**83**:468–78.

151. Saada A, Edvardson S, Rapoport M, Shaag A, Amry K, Miller C, et al. C6ORF66 is an assembly factor of mitochondrial complex I. *Am J Hum Genet* 2008;**82**:32–8.

152. Dunning CJ, McKenzie M, Sugiana C, Lazarou M, Silke J, Connelly J, et al. Human CIA30 is involved in the early assembly of mitochondrial complex I and mutations in its gene cause disease. *EMBO J* 2007;**26**:3227–37.

153. Gerards M, van den Bosch BJC, Danhauser K, Serre V, van Weeghel M, Wanders RJA, et al. Riboflavin-responsive oxidative phosphorylation complex I deficiency caused by defective ACAD9: new function for an old gene. *Brain* 2011;**134**:210–19.

154. Haack TB, Danhauser K, Haberberger B, Hoser J, Strecker V, Boehm D, et al. Exome sequencing identifies ACAD9 mutations as a cause of complex I deficiency. *Nat Genet* 2010;**42**:1131–4.

155. Saada A, Edvardson S, Shaag A, Chung WK, Segel R, Miller C, et al. Combined OXPHOS complex I and IV defect, due to mutated complex I assembly factor C20ORF7. *J Inher Metab Dis* 2012;**35**:125–31.

156. Pagliarini DJ, Calvo S, Chang B, Sheth SA, Vafai SB, Ong S-E, et al. A mitochondrial protein compendium elucidates complex I disease biology. *Cell* 2008;**134**:112–23.

157. Garone C, Donati AM, Sacchini M, Garcia-Diaz B, Bruno C, Calvo S, et al. Infantile mitochondrial encephalomyopathy due to a novel mutation in ACAD9. *JAMA Neurol* 2013;**70**:1177–9.

158. Ghezzi D, Goffrini P, Uziel G, Horvath R, Klopstock T, Lochmuller H, et al. SDHAF1, encoding a LYR complex II specific assembly factor, is mutated in SDH-defective infantile leukoencephalopathy. *Nat Genet* 2009;**41**:654–6.

159. Haller RG, Henriksson KG, Jorfeldt L, Hultman E, Wibom R, Sahlin K, et al. Deficiency of skeletal muscle succinate dehydrogenase and aconitase. *J Clin Invest* 1991;**88**:1197–206.

160. Hall RE, Henriksson KG, Lewis SF, Haller RG, Kennaway NG. Mitochondrial myopathy with succinate dehydrogenase and aconitase deficiency. Abnormalities of several iron-sulfur proteins. *J Clin Invest* 1993;**92**:2660–6.

161. Mochel F, Knight MA, Tong W-H, Hernandez D, Ayyad K, Taivassalo T, et al. Splice mutation in the iron-sulfur cluster scaffold protein ISCU causes myopathy with exercise intolerance. *Am J Hum Genet* 2008;**82**:652–60.

162. Kollberg G, Melberg A, Holme E, Oldfors A. Transient restoration of succinate dehydrogenase activity after rhabdomyolysis in iron-sulfur cluster deficiency myopathy. *Neuromusc Dis* 2010;**21**:115–20.

163. Bolar NA, Vanlander AV, Wilbrecht C, Van der Aa N, Smet J, De Paepe B, et al. Mutationon of the iron-sulfur cluster assembly gene IBA57 causes severe myopathy and encephalopathy. *Hum Mol Genet* 2013;**22**:2590–602.

164. Fernandez-Vizarra E, Bugiani M, Goffrini P, Carrara F, Farina L, Procopio E, Donati A, et al. Impaired complex III assembly associated with BCS1L gene mutations in isolated mitochondrial encephalomyopathy. *Hum Mol Genet* 2007;**16**:1241–52.

165. Visapaa I, Fellman V, Vesa J, Dasvarma A, Hutton JL, Kumar V, et al. GRACILE syndrome, a lethal metabolic disorder with iron overload, is caused by a point mutation in BCS1L. *Am J Hum Genet* 2002;**71**:863–76.

166. de Lonlay P, Valnot I, Barrientos A, Gorbatyuk M, Tzagoloff A, Taanman J-W, et al. A mutant mitochondrial respiratory chain assembly protein causes complex III deficiency in patients with tubulopathy, encephalopathy and liver failure. *Nat Genet* 2001;**29**:57–60.

167. Blazquez A, Gil-Borlado MC, Moran M, Verdu A, Cazorla-Calleja MR, Martin MA, et al. Infantile mitochondrial encephalomyopathy with unusual phenotype caused by a novel BCS1L mutation in an isolated complex III-deficient patient. *Neuromusc Dis* 2009;**19**:143–6.

168. Hinson JT, Fantin VR, Schonberger J, Breivik N, Siem G, McDonough B, et al. Missense mutations in the BCS1L gene as a cause of the Bjornstad syndrome. *N Engl J Med* 2007;**356**:809–19.

169. Ghezzi D, Arzuffi P, Zordan M, Da Re C, Lamperti C, Benna C, et al. Mutations in TTC19 cause mitochondrial complex III deficiency and neurological impairment in human and flies. *Nat Genet* 2011;**43**:259–63.

170. Willems JL, Monnens L, Trijbels J, Veerkamp JH, Meyer A, vanDam K, et al. Leigh's encephalomyelopathy in a patient with cytochrome c oxidase deficiency in muscle tissue. *Pediatrics* 1977;**60**:850–7.

171. DiMauro S, Servidei S, Zeviani M, DiRocco M, De Vivo DC, DiDonato S, et al. Cytochrome c oxidase deficiency in Leigh syndrome. *Ann Neurol* 1987;**22**:498–506.

172. Van Coster R, Lombes A, De Vivo DC, Chi TL, Dodson WE, Rothman S, et al. Cytochrome c oxidase-associated Leigh syndrome: phenotypic features and pathogenetic speculations. *J Neurol Sci* 1991;**104**:97–111.

173. Zhu Z, Yao J, Johns T, Fu K, De Bie I, Macmillan C, et al. SURF1, encoding a factor involved in the biogenesis of cytochrome c oxidase, is mutated in Leigh syndrome. *Nat Genet* 1998;**20**:337–43.

174. Tiranti V, Hoertnagel K, Carrozzo R, Galimberti C, Munaro M, Granatiero M, et al. Mutations of SURF-1 in Leigh disease associated with cytochrome c oxidase deficiency. *Am J Hum Genet* 1998;**63**:1609–21.

175. Pequignot MO, Dey R, Zeviani M, Tiranti V, Godinot C, Poyau A, et al. Mutations in the SURF1 gene associated with Leigh syndrome and cytochrome c oxidase deficiency. *Hum Mut* 2001;**17**:374–81.

176. Papadopoulou LC, Sue CM, Davidson MM, Tanji K, Nishino I, Sadlock JE, et al. Fatal infantile cardioencephalomyopathy with

COX deficiency and mutations in SCO2, a COX assembly gene. *Nat Genet* 1999;**23**:333–7.

177. Tay SKH, Shanske S, Kaplan P, DiMauro S. Association of mutations in SCO2, a cytochrome c oxidase assembly gene, with early fetal lethality. *Arch Neurol* 2004;**61**:950–2.

178. Leary SC, Mattman A, Wai T, Koehn DC, Clarke LA, Chan S, et al. A hemizygous SCO2 mutation in an early onset rapidly progressive, fatal cardiomyopathy. *Mol Genet Metab* 2006;**89**: 129–33.

179. Jaksch M, Ogilvie I, Yao J, Kortenhaus G, Bresser H-G, Gerbitz K-D, et al. Mutations in SCO2 are associated with a distinct form of hypertrophic cardiomyopathy and cytochrome c oxidase deficiency. *Hum Mol Genet* 2000;**9**:795–801.

180. DiMauro S, Tanji K, Schon EA. The many faces of cytochrome C oxidase deficiency. *Adv Exp Med Biol* 2012;**748**:341–57.

181. Tarnopolsky MA, Bourgeois JM, Fu M-H, Kataeva G, Shah J, Simon DK, et al. Novel SCO2 mutation (G1521A) presenting as a spinal muscular atrophy type I phenotype. *Am J Med Genet* 2004;**125A**:310–14.

182. Salviati L, Sacconi S, Raslan MM, Parton LA, Kronn DF, Braun A, et al. Cytochrome c oxidase deficiency due to a novel SCO2 mutation mimics Werdnig-Hoffman disease. *Arch Neurol* 2002; **59**:862–5.

183. Pronicki M, Kowalski P, Piekutowska-Abramczuk D, Taybert J, Karkucinska-Wieckowska A, Szymanska-Debinska T, et al. A homozygous mutation in the SCO2 gene causes a spinal muscular atrophy like presentation with stridor and respiratory insufficiency. *Eur J Paediatr Neurol* 2010;**14**:253–60.

184. Valnot I, Osmond S, Gigarel N, Mehaye B, Amiel J, Cormier-Daire V, et al. Mutations of the SCO1 gene in mitochondrial cytochrome c oxidase deficiency with neonatal-onset hepatic failure and encephalopathy. *Am J Hum Genet* 2000;**67**:1104–9.

185. Horvath R, Lochmuller H, Stucka R, Yao J, Shoubridge EA, Kim S-H, et al. Characterization of human SCO1 and COX17 genes in mitochondrial cytochrome-c-oxidase deficiency. *Biochem Biophys Res Comm* 2000;**276**:530–2.

186. Sacconi S, Salviati L, Sue CM, Shanske S, Davidson MM, Bonilla E, et al. Mutation screening in patients with isolated cytochrome c oxidase deficiency. *Pediatr Res* 2003;**53**: 224–30.

187. Valnot I, von Kleist-Retzow J-C, Barrientos A, Gorbatyuk M, Taanman J-W, Mehaye B, et al. A mutation in the human heme-A:farnesyltransferase gene (COX 10) causes cytochrome c oxidase deficiency. *Hum Mol Genet* 2000;**9**:1245–9.

188. Antonicka H, Leary SC, Guercin G-H, Agar JN, Horvath R, Kennaway NG, et al. Mutations in COX10 result in a defect in mitochondrial heme A biosynthesis and account for multiple, early-onset clinical phenotypes associated with isolated COX deficiency. *Hum Mol Genet* 2003;**12**:2693–702.

189. Kennaway NG, Carrero-Valenzuela RD, Ewart G, Balan VK, Lightowlers RN, Zhang YZ, et al. Isoforms of mammalian cytochrome c oxidase: correlation with human cytochrome c oxidase deficiency. *Pediatr Res* 1990;**28**:529–35.

190. Antonicka H, Mattman A, Carlson CG, Glerum DM, Hoffbuhr KC, Leary SC, et al. Mutations in COX15 produce a defect in the mitochondrial heme biosynthetic pathway, causing early-onset fatal hypertrophic cardiomyopathy. *Am J Hum Genet* 2003;**72**: 101–14.

191. Oquendo CE, Antonicka H, Shoubridge EA, Reardon W, Brown GK. Functional and genetic studies demonstrate that mutation in the COX15 gene can cause Leigh syndrome. *J Med Genet* 2004;**41**:540–4.

192. Bugiani M, Tiranti V, Farina L, Uziel G, Zeviani M. Novel mutations in COX15 in a long surviving Leigh syndrome patient with cytochrome c oxidase deficiency. *J Med Genet* 2005;**42**(5):e28.

193. Huigsloot M, Nijtmans LGJ, Szklarczyk R, Baars MJH, van den Brand MAM, HendriksFranssen MGM, et al. A mutation in C2orf64 causes impaired cytochrome c oxidase assembly and mitochondrial cardiomyopathy. *Am J Hum Genet* 2011;**88**: 488–93.

194. Ghezzi D, Saada A, D'Adamo P, Fernandez-Vizarra E, Gasparini P, Tiranti V, et al. FASTK2 nonsense mutation in an infantile mitochondrial encephalomyopathy associated with cytochrome c oxidase deficiency. *Am J Hum Genet* 2008;**83**:415–23.

195. Mootha VK, Lepage P, Miller K, Bunkenborg J, Reich M Hjerrild M, et al. Identification of a gene causing human cytochrome c oxidase deficiency by integrative genomics. *Proc Nat Acad Sci USA* 2003;**100**:605–10.

196. Morin C, Mitchell G, Larochelle J, Lambert M, Ogier H, Robinson BH, et al. Clinical, metabolic, and genetic aspects of cytochrome c oxidase deficiency in Saguenay-Lac-Saint-Jean. *Am J Hum Genet* 1993;**53**:488–96.

197. Xu F, Morin C, Mitchell G, Ackerley C, Robinson BH. The role of LRPPRC (leucine-rich pentatricopeptide repeat cassette) gene in cytochrome oxidase assembly: mutation causes lowered levels of COX (cytochrome c oxidase) I and COX III mRNA. *Biochem J* 2004;**382**:331–6.

198. Weraarpachai W, Antonicka H, Sasarman F, Seeger J, Schrank B, Kolesar JE, et al. Mutation in TACO1, encoding a translational activator of COX I, results in cytochrome c oxidase deficiency and late-onset Leigh syndrome. *Nat Genet* 2009;**41**:833–7.

199. Seeger J, Schrank B, Pyle A, Stucka R, Lorcher U, Muller-Ziermann S, et al. Clinical and neuropathological findings in patients with TACO1 mutations. *Neuromuscul Disord* 2010;**20**:720–4.

200. Tiranti V, D'Adamo P, Briem E, Ferrari G, Mineri R, Lamantea E, et al. Ethylmalonic encephalopathy is caused by mutations in ETHE1, a gene encoding a mitochondrial matrix protein. *Am J Hum Genet* 2004;**74**:239–52.

201. Tiranti V, Visconti C, Hildebrandt T, Di Meo I, Mineri R, Tiveron C, et al. Loss of ETHE1, a mitochondrial dioxygenase, causes fatal sulfide toxicity in ethylmalonic encephalopathy. *Nat Med* 2009;**15**:200–5.

202. Mineri R, Rimoldi M, Burlina AB, Koskull S, Perletti C, Heese B, et al. Identification ofnew mutations in the ETHE1 gene in a cohort of 14 patients presenting with ethylmalonic encephalopathy. *J Med Genet* 2008;**45**:473–8.

203. De Meirleir L, Seneca S, Lissens W, De Clercq I, Eyskens F, Gerlo E, et al. Respiratory chain complex V deficiency due to a mutation in the assembly gene ATP12. *J Med Genet* 2004;**41**: 120–4.

204. Cizkova A, Stranecky V, Mayr JA, Tesarova M, Havlikova V, Paul J, et al. TMEM70 mutations cause isolated ATP synthase deficiency and eonatal mitochondrial encephalomyopathy. *Nat Genet* 2009;**40**:1288–90.

205. Honzik T, Tesarova M, Mayr JA, Hansikova H, Jesina P, Bodamer O, et al. Mitochondrial encephalocardiomyopathy with

early neonatal onset due to TMEM70 mutation. *Arch Dis Child* 2010;**95**:296–301.
206. Spiegel R, Khayat M, Shalev SA, Horovitz Y, Mandel H, Hershkovitz E, et al. TMEM70 mutations are a common cause of nuclear encoded ATP synthase assembly defect: further delineation of a new syndrome. *J Med Genet* 2010;**48**:177–82.
207. Roesch K, Curran SP, Tranebjaerg L, Koehler CM. Human deafness dystonia syndrome is caused by a defect in assembly of the DDP1/TIMM8a-TIMM13 complex. *Hum Mol Genet* 2002;**11**:477–86.
208. Hansen JJ, Durr A, Cournu-Rebeix I, Georgopoulos C, Ang D, Nyholm Nielsen M, et al. Hereditary spastic paraplegia SPG13 is associated with a mutation in the gene encoding the mitochondrial chaperonin Hsp60. *Am J Hum Genet* 2002;**70**:1328–32.
209. Davey KM, Parboosingh JS, McLeaod DR, Chan A, Casey R, Ferreira P, et al. Mutation of DNAJC19, a human homologue of yeast inner mitochondrial membrane co-chaperones, causes DCMA syndrome, a novel autosomal recessive Barth syndrome-like condition. *J Med Genet* 2006;**43**:385–93.
210. Ojala T, Polinati P, Manninen T, Hippala A, Rajantie J, Karikoski R, et al. New mutation of mitochondrial DNAJC19 causing dilated and noncompaction cardiomyopathy, anemia, ataxia, and male genital anomalies. *Pediatr Res* 2012;**72**:432–7.
211. Di Fonzo A, Ronchi D, Lodi T, Fassone E, Tigano M, Lamperti C, et al. The mitochondrial disulfide relay system protein GFERis mutated in autosomal-recessive myopathy with cataract and combined respiratory-chain deficiency. *Am J Hum Genet* 2009;**84**: 594–604.
212. Christian BE, Spremulli LL. Mechanism of protein biosynthesis in mammalian mitochondria. *Biochim Biophys Acta* 2012;**1819**: 1035–54.
213. Chrzanowska-Lightowlers ZMA, Horvath R, Lightowlers RN. 175th ENMC international workshop: Mitochondrial protein synthesis in health and disease, 25–27th June, 2010, Naarden, The Netherlands. *Neuromuscul Disord* 2011;**21**:142–7.
214. Bykhovskaya Y, Casas KA, Mengesha E, Inbal A, Fischel-Ghodsian N. Missense mutation in pseudouridine synthase 1 (PUS1) causes mitochondrial myopathy and sideroblastic anemia (MLASA). *Am J Hum Genet* 2004;**74**:1303–8.
215. Fernandez-Vizarra E, Berardinelli A, Valente L, Tiranti V, Zeviani M. Nonsense mutation in pseudouridylate synthase 1 (PUS1) in two brothers affected by myopathy, lactic acidosis and sideroblastic anemia (MLASA). *J Med Genet* 2007;**44**:173–80.
216. Riley LG, Cooper S, Hickey P, Rudinger-Thirion J, McKenzie M, Compton A, et al. Mutation of the mitochondrial tyrosyl-tRNA synthetase gene, YARS, causes myopathy, lactic acidosis, and sideroblastic anemia, MLASA syndrome. *Am J Hum Genet* 2010;**87**:52–9.
217. Shahni R, Wedatilake Y, Cleary MA, Lindley KJ, Sibson KR, Rahman S. A distinct mitochondrial myopathy, lactic acidosis and sideroblastic anemia (MLASA) phenotype associated with YARS2 mutations. *Am J Med Genet* 2013;**161A**:2334–8.
218. Barth PG, Scholte HR, Berden JA, Moorsel V, Luyt-Houwen I, Van T, Veer-Korthof ET, et al. An X-linked mitochondrial disease affecting cardiac muscle, skeletal muscle and neutrophil leucocytes. *J Neurol Sci* 1983;**62**:327–55.
219. Bione S, D'Adamo P, Maestrini E, Gedeon AK, Bolhuis PA, Toniolo D. A novel X-linked gene, G4.5, is responsible for Barth syndrome. *Nat Genet* 1996;**12**:385–9.
220. Schlame M, Ren M. Barth syndrome, a human disorder of cardiolipin metabolism. *FEBS Lett* 2006;**580**:5450–5.
221. Claypool SM, Koehler CM. The complexity of cardiolipin in health and disease. *Trends Biochem Sci* 2012;**37**:32–41.
222. Sriwardena K, MacKey N, Levandovskiy V, Blaser S, Raiman J, Kantor PF, et al. Mitochondrial citrate synthase crystals: Novel finding in Sengers syndrome caused by acylglycerol kinase (ACK) mutations. *Mol Genet Metab* 2013;**108**:40–50.
223. Mitsuhashi S, Ohkuma A, Talim B, Karahashi M, Koumura T, Aoyama C, et al. A congenital muscular dystrophy with mitochondrial structural abnormalities caused by defective de novo phosphatidyl choline biosynthesis. *Am J Hum Genet* 2011;**88**: 845–51.
224. Gutierrez-Rios P, Kalra AA, Wilson JD, Tanji K, Akman HO, Area Gomez E, et al. Congenital megaconial myopathy due to a novel defect in the choline kinase beta (CHKB) gene. *Arch Neurol* 2012;**69**:657–61.
225. Mitsuhashi S, Nishino I. Megaconial congenital muscular dystrophy due to loss-of-function mutations in choline kinase beta. *Curr Opin Neurol* 2013;**26**:536–43.
226. Wortmann S, Rodenburg RJT, Huizing M, Loupatty FJ, de Koning T, Kluijtmans LAJ, et al. Association of 3-methylglutaconic aciduria with sensori-neural deafness, encephalopathy, and Leigh-like syndrome (MEGDEL association) in four patients with a disorder of the oxidative phosphorylation. *Mol Genet Metab* 2006;**88**:47–52.
227. Wortmann RL, Vaz FM, Gardeitchik T, Vissers LELM, Renkema GH, Schuurs-Hoeijmakers JHM, et al. Mutation in the phospholipid remodeling gene SERAC1 impair mitochondrial function and intracellular cholesterol trafficking and cause dystonia and deafness. *Am J Hum Genet* 2012;**44**:797–802.
228. DiMauro S, Schon EA, Carelli V, Hirano M. The clinical maze of mitochondrial neurology. *Nat Rev Neurol* 2013;**9**:429–44.
229. Zeharia A, Shaag A, Houtkooper RH, Hindi T, de Lonlay P, Erez G, et al. Mutations in LPIN cause recurrent childhood myoglobinuria in childhood. *Am J Hum Genet* 2008;**83**: 489–94.
230. Westermann B. Mitochondrial fusion and fission in cell life and death. *Nat Rev Mol Cell Biol* 2010;**11**:872–84.
231. Zuchner S, Mersiyanova IV, Muglia M, Bissar-Tadmouri N, Rochelle J, Dadali EL, et al. Mutations in the mitochondrial GTPase mitofusin 2 cause Charcot-Marie-Tooth neuropathy type 2A. *Nat Genet* 2004;**36**:449–51.
232. Baxter RV, Othmane KB, Rochelle JM, Stajich JE, Hulette C, Dew-Knight S, et al. Ganglioside-induced differentiation-associated protein 1 is mutant in Charcot-Marie-Tooth disease type 4A/8q21. *Nat Genet* 2002;**30**:21–2.
233. Cuesta A, Pedrola L, Sevilla T, Garcia-Planells J, Chumillas MJ, Mayordomo F, et al. The gene encoding ganglioside-induced differentiation-associated protein 1 is mutated in axonal Charcot-Marie-Tooth type 4A disease. *Nat Genet* 2002;**30**:22–5.
234. Waterham HR, Koster J, van Roermund CWT, Mooyer PAW, Wanders RJA, Leonard JV. A lethal defect of mitochondrial and peroxisomal fission. *N Engl J Med* 2007;**356**:1736–41.
235. Shemseldin HE, Alshammari M, Al-Sheddi T, Salih MA, Alkhalidi H, Kentab A, et al. Genomic analysis of mitochondrial diseases in a consanguineous population reveals novel candidate disease genes. *J Med Genet* 2012;**49**:234–41.

236. Naviaux RK, Nguyen KV. POLG mutations associated with Alpers' syndrome and mitochondrial DNA depletion. *Ann Neurol* 2004;**55**:706–12.
237. Van Goethem G, Dermaut B, Lofgren A, Martin J-J, Van Broeckhoven C. Mutation of POLG is associated with progressive external ophthalmoplegia characterized by mtDNA deletions. *Nat Genet* 2001;**28**:211–12.
238. Luoma P, Melberg A, Rinne JO, Kaukonen JA, Nupponen NN, Chalmers RM, et al. Parkinsonism, premature menopause, and mitochondrial DNA polymerase mutations: clinical and molecular genetic study. *Lancet* 2004;**364**:875–82.
239. Milone M, Brunetti-Pierri N, Tang L-Y, Kumar N, Mezei MM, Josephs KA, et al. Sensory ataxic neuropathy with ophthalmoparesis caused by POLG mutations. *Neuromuscul Disord* 2008;**18**: 626–32.
240. Hakonen AH, Davidzon G, Salemi R, Bindoff LA, Van Goethem G, DiMauro S, et al. Abundance of POLG disease mutations in Europe, Australia, New Zealand, and the United States explained by single ancient European founders. *Eur J Hum Genet* 2007; **15**:779–83.
241. Chan SSL, Copeland WC. DNA polymerase gamma and mitochondrial disease: Understanding the consequence of POLG mutations. *Biochim Biophys Acta* 2009;**1787**:312–19.
242. Wong L-JC, Naviaux RK, Brunetti-Pierri N, Zhang Q, Schmitt ES, Truong C, et al. Molecular and clinical genetics of mitochondrial diseases due to POLG mutations. *Hum Mut* 2008;**29**: E150–72.
243. National Institute of Environmental Health Sciences database. National Institutes of Health (NIH). Accessed 2014 from <http://tools.niehs.nih.gov/polg/>.
244. Stewart JD, Horvath R, Baruffini E, Ferrero I, Bulst S, Watkins PB, et al. Polymerase γ gene POLG determines the risk of sodium valproate-induced liver toxicity. *Hepatology* 2010;**52**:1791–6.
245. Hakonen AH, Isohanni P, Paetau A, Herva R, Suomalainen A, Lonqvist T. Recessive twinkle mutations in early onset encephalopathy with mtDNA depletion. *Brain* 2007;**130**:3032–40.
246. Sarzi E, Goffart S, Serre V, Chretien D, Slama A, Munnich A, et al. Twinkle helicase (PEO1) gene mutation causes mitochondrial DNA depletion. *Ann Neurol* 2007;**62**:579–87.
247. Hakonen AH, Goffart S, Marjavaara S, Paetau A, Cooper H, Mattila K, et al. Infantile-onset spinocerebellar recessive ataxia and mitochondrial recessive ataxia syndrome are associated with neuronal complex I defect and mtDNA depletion. *Hum Mol Genet* 2008;**17**:3822–35.
248. Ronchi D, Di Fonzo A, Lin W, Bordoni A, Liu C, Fassone E, et al. Mutations in DNA2 link progressive myopathy to mitochondrial DNA instability. *Am J Hum Genet* 2013;**92**:293–300.
249. Suomalainen A, Isohanni P. Mitochondrial DNA depletion syndromes—Many genes, common mechanisms. *Neuromuscul Disord* 2010;**20**:429–37.
250. Saada A, Shaag A, Mandel H, Nevo Y, Eriksson S, Elpeleg O. Mutant mitochondrial thymidine kinase in mitochondrial DNA depletion myopathy. *Nat Genet* 2001;**29**:342–4.
251. Chanprasert S, Wang J, Weng SW, Enns GM, Boue DR, Wong B, et al. Molecular and clinical characterization of the myopathic form of mitochondrial DNA depletion syndrome caused by mutations in the thyamine kinase (TK2) gene. *Mol Genet Metab* 2013;**110**:153–61.
252. Mancuso M, Salviati L, Sacconi S, Otaegui D, Camano P, Marina A, et al. Mitochondrial DNA depletion. Mutations in thymidine kinase gene with myopathy and SMA. *Neurology* 2002;**59**: 1197–202.
253. Oskoui M, Davidzon G, Pascual J, Erazo R, Gurgel-Giannetti J, Krishna S, et al. Clinical spectrum of mitochondrial DNA depletion due to mutations in the thymidine kinase 2 gene. *Arch Neurol* 2006;**63**:1122–6.
254. Behin A, Jardel C, Claeys KG, Fagart J, Louha M, Romero NB, et al. Adult cases of mitochondrial DNA depletion due to TK2 defect. *Neurology* 2012;**78**:644–8.
255. Paradas C, Gutierrez Rios P, Rivas E, Carbonell P, Hirano M, DiMauro S. TK2 mutation presenting as indolent myopathy. *Neurology* 2013;**80**:504–6.
256. Tyynismaa H, Sun R, Ahola-Erkkila S, Almusa H, Poyhonen R, Korpela M, et al. Thymidine kinase 2 mutations in autosomal recessive progressive external ophthalmoplegia with multiple mitochondrial DNA deletions. *Hum Mol Genet* 2012; **21**:66–75.
257. Mandel H, Szargel R, Labay V, Elpeleg O, Saada A, Shalata A, et al. The deoxyguanosine kinase gene is mutated in individuals with depleted hepatocerebral mitochondrial DNA. *Nat Genet* 2001;**29**:337–41.
258. Dimmock DP, Zhang Q, Dionisi-Vici C, Carrozzo R, Shieh J Tang LY, et al. Clinical and molecular features of mitochondrial DNA depletion due to mutations in deoxyguanosine kinase. *Hum Mut* 2007;**29**:330–45.
259. Spinazzola A, Viscomi C, Fernandez-Vizarra E, Carrara F, D'Adamo P, Calvo S, et al. MPV17 encodes an inner mitochondrial membrane protein and is mutated in infantile hepatic mitochondrial DNA depletion. *Nat Genet* 2006;**38**:570–5.
260. Buchaklian AH, Helbling D, Ware SM, Dimmock DP. Recessive deoxyguanosine kinase deficiency causes juvenile onset mitochondrial myopathy. *Mol Genet Metab* 2012;**107**:92–4.
261. Garone C, Rubio JC, Calvo S, Naini A, Tanji K, DiMauro S, et al. MPV17 mutations causing adult-onset multisystemic disorder with multiple mitochondrial DNA deletions. *Arch Neurol* 2012;**69**:1648–51.
262. Blakely E, Butterworth A, Hadden RDM, Bodi I, He L, McFarland R, et al. MPV17 mutation causes neuropathy and leukoencephalopathy with multiple mtDNA deletions in muscle. *Neuromuscul Disord* 2012;**22**:587–91.
263. Holve S, Hu D, Tyson RW, Sokol RJ. Liver disease in Navajo neuropathy. *J Pediatr* 1999;**135**:482–93.
264. Karadimas CL, Vu TH, Holve SA, Quinzii C, Tanji K, Bonilla E, et al. Navajo neurohepatopathy is caused by a mutation in the MPV17 gene. *Am J Hum Genet* 2006;**79**:544–8.
265. Bornstein B, Area E, Flanigan KM, Ganesh J, Jayakar P, Swoboda KJ, et al. Mitochondrial DNA depletion syndrome due to mutations in the RRM2B gene. *Neuromuscul Disord* 2008; **18**:453–9.
266. Tyynismaa H, Ylikallio E, Patel J, Molnar MJ, Haller RG, Suomalainen A. A heterozygous truncating mutation in RRM2B causes autosomal-dominant progressive external ophthalmoplegia with multiple mtDNA deletions. *Am J Hum Genet* 2009;**85**:290–5.
267. Fratter C, Raman P, Alston CL, Blakeley EL, Craig K, Smith C, et al. RRM2B mutations are frequent in familial PEO with multiple mtDNA deletions. *Neurology* 2011;**76**:2032–4.

268. Hirano M, Garone C, Quinzii C. CoQ10 deficiency and MNGIE: Two treatable mitochondrial disorders. *Biochim Biophys Acta* 2012;**1820**:625–31.
269. Yu-Wai-Man P, Griffiths PG, Gorman GS, Lourenco CM, Wright AF, Auer-Grumbach M, et al. Multi-system neurological disease is common in patients with OPA1 mutations. *Brain* 2010;**133**:771–86.
270. Ranieri M, Del Bo R, Bordoni A, Ronchi D, Colombo I, Riboldi G, et al. Optic atrophy plus phenotype due to mutations in the OPA1 gene: Two more Italian families. *J Neurol Sci* 2012; **315**:146–9.
271. Amati-Bonneau P, Valentino ML, Reynier P, Gallardo ME, Bornstein B, Boissiere A, et al. OPA1 mutations induce mitochondrial DNA instability and optic atrophy "plus" phenotypes. *Brain* 2008;**131**:338–51.
272. Hudson G, Amati-Bonneau P, Blakely E, Stewart JD, He L, Schaefer AM, et al. Mutation of OPA1 causes dominant optic atrophy with external ophthalmoplegia, ataxia, deafness and multiple mitochondrial DNA deletions: a novel disorder of mtDNA maintenance. *Brain* 2008;**131**:329–37.
273. Ferraris S, Clark S, Garelli E, Davidzon G, Moore SA, Kardon RH, et al. Progressive external ophthalmoplegia and vision and hearing loss in a patient with mutations in POLG2 and OPA1. *Arch Neurol* 2008;**65**:125–31.
274. Rouzier C, Bannwarth S, Chaussenot A, Chevrollier A, Verschueren A, Bonello-Palot N, et al. The MFN2 gene is responsible for mitochondrial DNA instability and optic atrophy "plus" phenotype. *Brain* 2012;**136**:23–34.
275. Pirra L, Dubrovsky A, Bitoun M, Guicheney P, Romero NB, Saccolitti M, et al. Ragged red fibres finding in muscle biopsy of dynamin 2-related centronuclear myopathy. *Neuromusc Dis* 2007; **17**:881–2 [Abstract].
276. Emmanuele V, Sotiriou E, Gutierrez Rios P, Ganesh J, Ichord RN, Foley AR, et al. A novel mutation in the mitochondrial DNA cytochrome b gene (MTCYB) in a patient with MELAS syndrome. *J Child Neurol* 2013;**28**:236–42.
277. Fuchs AF, Binder MD. Fatigue resistance of human extraocular muscles. *J Neuropath* 1983;**49**:28–34.
278. Wooten GF, Reis DJ. Blood flow in extraocular muscle of cat. *Arch Neurol* 1978;**26**:350–2.
279. Greaves LC, Yu-Wai-Man P, Blakeley EL, Krishnan KJ, Beadle NE, Kerin J, et al. Mitochondrial DNA defects and selective extraocular muscle involvement in CPEO. *Invest Ophthalmol Vis Sci* 2010;**51**:3340–6.
280. Suomalainen A, Elo JM, Pietilainen KH, Hakonen AH, Sevastianova K, Korpela M, et al. FGF-21 as a biomarker for muscle-manifesting mitochondrial respiratory chain deficiencies: a diagnostic study. *Lancet Neurol* 2011;**10**:806–18.
281. Greaves LC, Reeve AK, Taylor RW, Turnbull D. Mitochondrial DNA and disease. *J Pathol* 2012;**226**:274–86.
282. Borthwick GM, Johnson MA, Ince PG, Shaw PJ, Turnbull DM. Mitochondrial enzyme activity in amyotrophic lateral sclerosis: implications for the role of mitochondria in neuronal cell death. *Ann Neurol* 1999;**46**:787–90.
283. Borthwick GM, Taylor RW, Walls TJ, Tonska K, Taylor GA, Shaw PJ, et al. Motor neuron disease in a patient with a mitochondrial tRNAIle mutation. *Ann Neurol* 2006;**59**:570–4.
284. Sue CM, Karadimas C, Checcarelli N, Tanji K, Papadopoulou LC, Pallotti F, et al. Differential features of patients with mutations in two COX assembly genes, SURF-1 and SCO2. *Ann Neurol* 2000;**47**:589–95.
285. Claypool SM, Boontheung P, McCafferty JM, Loo JA, Koehler CM. The cardiolipin transacylase, tafazzin, associates with two distinct respiratory components providing insight into Barth syndrome. *Mol Biol Cell* 2008;**19**:5143–55.
286. Spinazzi M, Casarin A, Pertegato V, Salviati L, Angelini C. Assessment of mitochondrial respiratory chain enzymatic activities on tissues and cultured cells. *Nat Protoc* 2012;**7**:1235–46.
287. Coppola G, Geschwind DH. Genomic medicine enters the neurology clinic. *Neurology* 2012;**79**:112–14.
288. Glover EI, Martin JJ, Maher A, Thornhill RE, Moran GR, Tarnopolsky MA. A randomized trial of coenzyme Q10 in mitochondrial disorders. *Muscle Nerve* 2010;**42**:739–48.
289. Carelli V, La Morgia C, Valentino ML, Rizzo C, Carbonelli M, De Negri AM, et al. Idebenone treatment in Leber's hereditary optic neuropathy. *Brain* 2011;**134**(Pt 9):e188.
290. Klopstock T, Yu-Wai-Man P, Dimitriadis K, Rouleau J, Heck S, Bailie M, et al. A randomized placebo-controlled trial of idebenone in Leber's hereditary optic neuropathy. *Brain* 2011;**134**: 2677–86.
291. Sadun AA, Chicani CF, Ross-Cisneros FN, Barboni P, Thoolen M, Shrader WD, et al. Effect of EPI-743 on the clinical course of the mitochondrial disease Leber hereditary optic neuropathy. *Arch Neurol* 2012;**69**:331–8.
292. Enns GM, Kinsman SL, Perlman SL, Spicer KM, Abdenur JE, Cohen BH, et al. Initial experience in the treatment of inherited mitochondrial diseases with EPI-743. *Mol Genet Metab* 2012;**105**: 91–102.
293. Martinelli D, Catteruccia M, Piemonte F, Pastore A, Tozzi G, Dionisi-Vici C, et al. EPI-743 reverses the progression of the pediatric mitochondrial disease—genetically defined Leigh symdrome. *Mol Genet Metab* 2012;**107**:383–8.
294. Kaufmann P, Engelstad K, Wei Y-H, Jhung S, Sano MC, Shungu D, Millar WS, et al. Dichloroacetate causes toxic neuropathy in MELAS: A randomized, controlled clinical trial. *Neurology* 2006;**66**:324–30.
295. Gilkerson RW, De Vries RLA, Lebot P, Wikstrom JD, Torgyekes E, Shirhai OS, et al. Mitochondrial autophagy in cells with mtDNA mutations results from the synergistic loss of transmembrane potential and mRORC1 inhibition. *Hum Mol Genet* 2012;**21**:978–90.
296. Tanaka M, Borgeld HJ, Zhang J, Muramatsu S, Gong JS, Yoneda M, et al. Gene therapy for mitochondrial disease by delivering restriction endonuclease SmaI into mitochondria. *J Biomed Sci* 2002;**9**:534–41.
297. Scott AB, Alexander DE, Miller JM. Bupivacaine injection of eye muscles to treat strabismus. *Br J Ophthalmol* 2007;**91**:146–8.
298. Twig G, Hyde B, Shirihai OS. Mitochondrial fusion, fission and autophagy as a quality control axis: The bioenergetic view. *Biochim Biophys Acta* 2008;**1777**:1092–7.
299. Cassidy-Stone A, Chipuk JE, Ingerman E, Song C, Yoo C, Kuwana T, et al. Chemical inhibition of the mitochondrial division dynamin reveals its role in Bax/Bak-dependent mitochondrial outer membrane permeabilization. *Dev Cell* 2008;**14**:193–204.
300. Cui M, Tang X, Christian WV, Yoon Y, Tieu K. Perturbations in mitochondrial dynamics induced by human mutant PINK1 can be rescued by the mitochondrial division inhibitor mdivi-1. *J Biol Chem* 2010;**285**:11740–52.

301. Wenz T, Diaz F, Spiegelman BM, Moraes CT. Activation of the PPAR/PGC-1alpha pathway prevents a bioenergetic deficit and effectively improves a mitochondrial myopathy phenotype. *Cell Metab* 2008;**8**:249–55.
302. Viscomi C, Bottani E, Civiletto G, Cerutti R, Moggio M, Fagiolari G, et al. In vivo correction of COX deficiency by activation of the AMPK/PGC-1alpha axis. *Cell Metab* 2011;**14**:80–90.
303. Flierl A, Chen YT, Coskun PE, Samulski RJ, Wallace DC. Adeno-associated virus-mediated gene transfer of the heart/muscle adenine nucleotide translocator (ANT) in mouse. *Gene Ther* 2005;**12**:570–8.
304. Di Meo I, Auricchio A, Lamperti C, Burlina A, Viscomi C, Zeviani M. Effective AAV-mediated gene therapy in a mouse model of ethylmalonic encephalomyopathy. *EMBO Mol Med* 2012;**4**:1008–14.
305. Tachibana M, Sparman M, Sritanaudomchai H, Ma H, Clepper L, Woodward J, et al. Mitochondrial gene replacement in primate offspring and embryonic stem cells. *Nature* 2009;**461**:367–72.
306. Tachibana M, Amato P, Sparman M, Woodward J, Melguizo Sanchis D, Ma H, et al. Towards germline gene therapy of inherited mitochondrial diseases. *Nature* 2012;**493**:627–31.
307. Paull D, Emmanuele V, Weiss KA, Treff N, Stewart L, Hua H, et al. Nuclear genome transfer in human oocytes eliminates mitochondrial DNA variants. *Nature* 2013;**493**:632–7.
308. Craven L, Tuppen HA, Greggains GD, Harbottle SJ, Murphy JL, Cree LM, et al. Pronuclear transfer in human embryos to prevent transmission of mitochondrial DNA disease. *Nature* 2010;**465**:82–5.
309. Craven L, Elson JL, Irving L, Tuppen HA, Lister LM, Greggains GD, et al. Mitochondrial DNA disease: new options for prevention. *Hum Mol Genet* 2011;**20**:R168–74.
310. Treatment for mitochondrial disorders. Cochrane Database of Systematic Reviews, Issue 4. 2012 [cited 4 July 2011].

Chapter 42

Juvenile Dermatomyositis and Other Inflammatory Myopathies in Children

Lauren M. Pachman

INTRODUCTION

Patients with idiopathic inflammatory myopathies (IIM) manifest a diverse group of syndromes that have in common the loss of muscle function as a disease symptom. The symptoms of inflammatory muscle disease in children can be characterized as either acute (reviewed in Chapter 36 [Case Example 36.3] and Chapter 1) or chronic in nature. Throughout the world, acute muscle complaints are usually associated with bacterial or parasitic infectious agents. In contrast, in North America, acute inflammatory myopathies are more often of viral etiology. With respect to chronic myositis in children, there is considerable variation in disease onset, presentation, and response to therapy. This chapter reviews the spectrum of disease manifestations of the chronic inflammatory myopathies of childhood and their association with specific autoantibodies and immunopathology, as well as our current approaches to medical therapy for the major subgroups of these disorders.

The most common pediatric presentation of the juvenile IIM (JIIM) is juvenile dermatomyositis (JDM).[1] The primary clinical feature of both juvenile dermatomyositis and juvenile polymyositis (JPM) is chronic and progressive weakness of proximal muscles. In JDM, the distinctive skin manifestations (Figure 42.1) and vasculopathy (Figure 42.2) are commonly associated with muscle involvement. In children with JPM, there is no evidence of skin involvement. The child with definite JDM, in addition to the typical rash, has three of the four criteria: elevated serum levels of muscle-derived enzymes, electromyographic (EMG) evidence of inflammatory myopathy, positive muscle biopsy, and proximal muscle weakness (Table 42.1).[2,3] These criteria for diagnosis rely on the selection of the involved inflammatory site for two (EMG, muscle biopsy) of the four criteria. Because the inflammation is quite focal, it is essential to use magnetic resonance imaging (MRI), or other imaging to guide the biopsy site selection.[4] A compatible MRI image is considered by some to be a critical diagnostic criteria.[5] The use of ultrasound for biopsy site selection may also be fruitful.[6] Recent information about serologic characteristics of patients with inflammatory myopathy has led to the recognition of a growing panel of myositis-specific antibodies (MSAs) associated with characteristic patterns of disease onset and course.[7]

The most common of the inflammatory myopathies in children are JDM, JPM, and overlap connective tissue disease syndromes. Children with JDM comprise 85% of this inflammatory myopathy group, compared with about 37% in adults; JPM occurs in 4–8% or less of the total chronic inflammatory myopathies in children, compared with 27% in adults. The third major group of children, 6–12%, fall into the category of connective tissue diseases[8] or overlap syndromes; these children often meet criteria for another autoimmune disease, most commonly systemic sclerosis, with antibody to PM/Scl or Scl-70, or Sjögren syndrome with antibody to SSA/Ro, or SSB/La antigen. There are far fewer cases in which there is highly localized inflammation in muscle (focal or nodular myositis, or orbital or ocular myositis). Finally, there is limited evidence of significant variations in the inflammatory response that can occur in the muscle of the child (proliferative myositis, eosinophilic myositis, granulomatous myositis, cancer-associated myositis), or a necrotizing inflammatory response in response to statins reported in adults. In addition, children may have "dermatomyositis *sine* myositis" where there is no apparent muscle involvement at all.

B.T. Darras, H. Royden Jones, Jr., M.M. Ryan & D.C. De Vivo (Eds): Neuromuscular Disorders of Infancy, Childhood and Adolescence, Second edition.
DOI: http://dx.doi.org/10.1016/B978-0-12-417044-5.00042-1

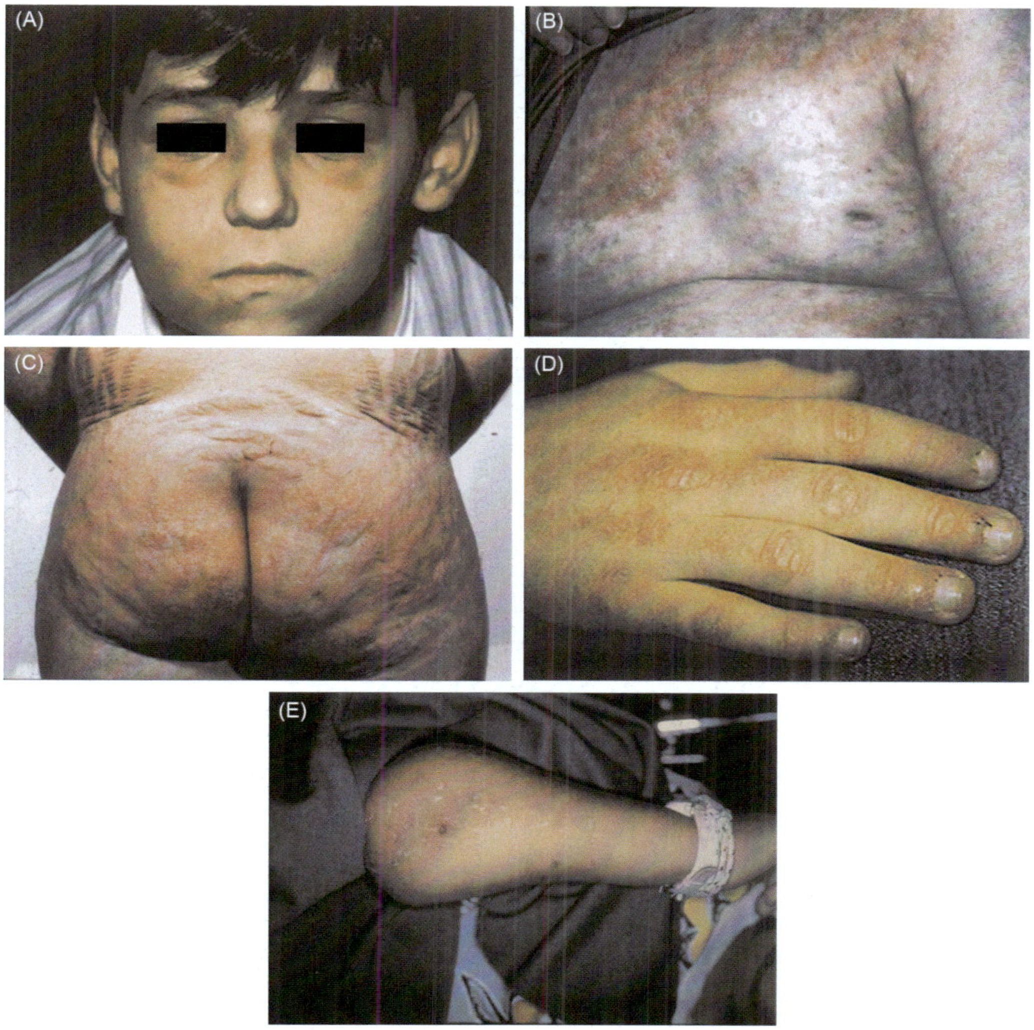

FIGURE 42.1 Cutaneous features of juvenile dermatomyositis. (**A**) The rash of JDM showing the periorbital edema, erythema crossing the bridge of the nose, and swelling of the right cheek. (**B**) Severe diffuse involvement with poikilodermatous changes on the chest. (**C**) Rash over the buttocks can be painful, making sitting uncomfortable. (**D**) The hands of an adolescent with JDM showing severe Gottron's papules, periungual thickening, and dilated nailfold capillary loops. (**E**) Inflammation and secondary infection of an area of dystrophic calcification in a newly diagnosed child with JDM.

THE EPIDEMIOLOGY OF THE MAJOR PEDIATRIC INFLAMMATORY MYOPATHIES

There is a bimodal age distribution for JPM and JDM (at 5 to 9 years of age, with an estimated incidence of 3.7 cases per million per year and at 10 to 14 years of age, with 4.3 pediatric cases per million per year). There is a separate adult peak at 45 to 64 years of age, with 10 cases per million per year) first reported in 1970.[9] In the United States, a National Institutes of Health-sponsored registry of new-onset JDM documented an incidence of JDM of 3.2 cases per million children per year, as determined by the capture and recapture method.[10] Most of the 325 cases occurred in white children (71%), compared with the Hispanic (12%) and African American (9%) populations.[10] In this population, the female-to-male ratio was 2:1, and the mean age of the child at diagnosis was 6.7 for girls, but 7.3 years for boys; 25% of the children were ≤4 years old at disease onset (Figure 42.3). There is controversy concerning the impact of a very young age at disease onset on JDM disease course and severity, as well as tendency to develop calcification.[10–12] Table 42.2 lists some of the major symptoms observed at the diagnosis of JDM in white children and minorities. As seen in Box 42.1, delay in diagnosis and institution of therapy is associated with increased frequency of soft tissue calcification.[11] Table 42.3 displays the first definite symptom of JDM: rash or weakness. Table 42.4 presents some of the systemic manifestations commonly seen in children with

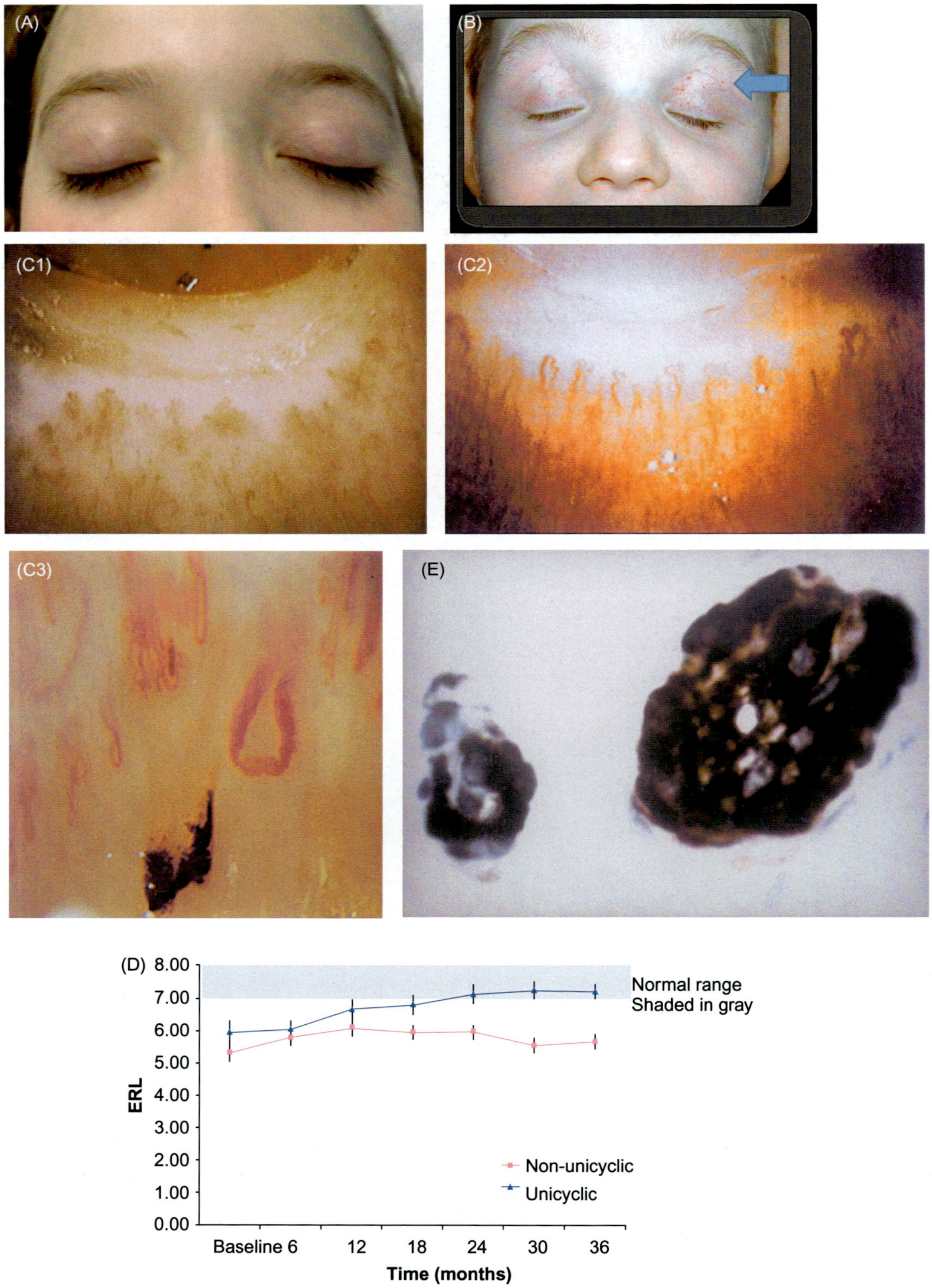

FIGURE 42.2 Vascular features of juvenile dermatomyositis. (**A**) Characteristic capillary dilation on margin of the eyelid of a child with active JDM. Capillary thrombosis is seen, reflecting active intravascular coagulation. (**B**) Telangiectasia (arrow) on the eyelid of a child with chronic severe JDM. (**C1–C3**) The spectrum of changes in the nailfold capillaries in children with untreated JDM. Dropout of nailfold capillary loops are seen with increased periungual avascularity. The dilated capillaries often assume bizarre bush forms. (**D**) Improvement in the number of nailfold capillary end row loops (ERL) over time in a subset of JDM. (**E**) Muscle biopsy of a child with longstanding JDM showing vascular occlusion by smooth muscle proliferation. The lumen of the capillary is further compromised by adhesion of mononuclear cells to the high endothelial lining.

TABLE 42.1 The Modified Diagnostic Criteria for Inflammatory Myopathies

1	Symmetric, often progressive, proximal muscle weakness	
2	Characteristic electromyographic (EMG) triad seen in myositis	Short duration, small, low amplitude potentials
		Fibrillation potentials, seen even at rest
		Bizarre high-frequency repetitive discharges
3	Elevations of serum levels of muscle-associated enzymes	Creatine kinase (CK)
		Aldolase
		Lactate dehydrogenase (LHD)
		Transaminases (ALT/SGPT and AST/SGOT)
4	Evidence of chronic inflammation in muscle biopsy	Necrosis of type I and type II muscle fibers
		Degeneration and regeneration of myofibers with variation in myofiber size
		Interstitial or perivascular mononuclear cells
5	Characteristic rashes of dermatomyositis	Scaly erythematous eruptions over the metacarpal phalangeal or interphalangeal joints (Gottron's papules)
		Periorbital purplish discoloration (heliotrope rash)
		Erythematous scaly rashes over the face, neck (V sign), upper back and arms (shawl sign), and extensor tendons (linear extensor erythema) and other extensor surfaces
6	Other potential criteria	MRI compatible with diagnosis
		Upregulation of class I antigen expression in muscle
		Microvascular damage

Definite Disease: All criteria 1–4, except for dermatomyositis, in which any three of the first four criteria plus the rash are adequate.
Probable Disease: Any 3 of the first 4 criteria, except for dermatomyositis, in which any 2 of the first 4 criteria plus the rash are adequate.
Possible Disease: Any 2 of the first 4 criteria, except for dermatomyositis, in which any one of the first 4 criteria plus the rash is adequate.
Modified with permission from Bohan and Peter (1975).[2,3] (#6 includes other potential criteria currently under discussion.)

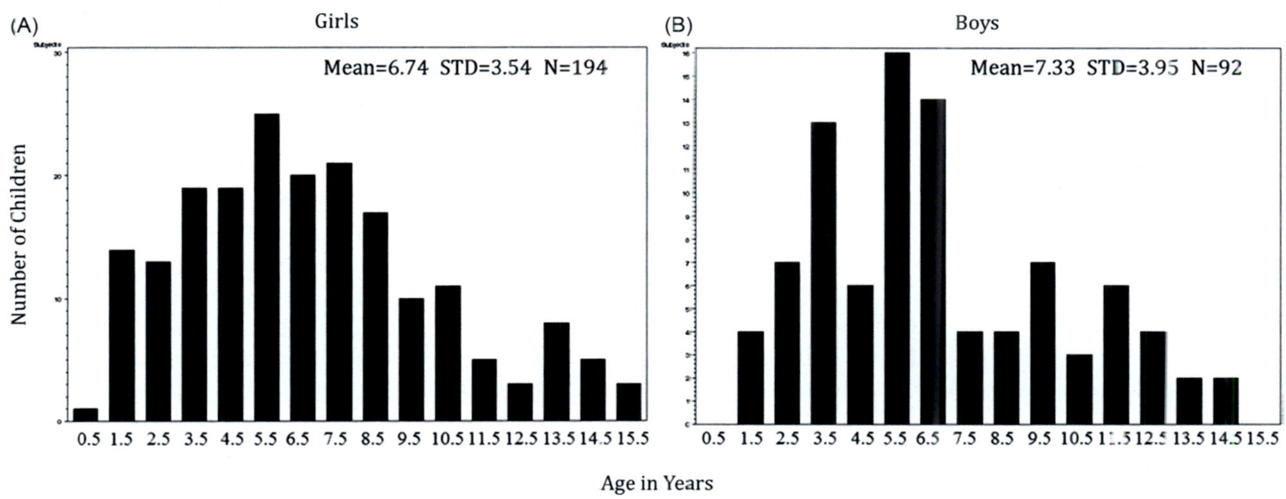

FIGURE 42.3 The age at JDM diagnosis in children with JDM, in panel A for girls and in panel B for boys. NIAMS National Registry (N = 286). *Reproduced with permission from Pachman et al. (2005).*[29]

definite JDM. The mortality rate from JDM has decreased in the past 70 years. Before corticosteroids, there was a 33% mortality rate and a 33% morbidity rate,[12] but both of those outcomes have decreased,[13] with earlier diagnosis and more aggressive therapy (see section on therapy). Factors associated with increased mortality include the

TABLE 42.2 Symptoms at the Diagnosis of JDM

Rash	79 (100)[a]
Weakness	79 (100)
Muscle pain	58 (73)
Fever	51 (65)
Dysphagia	35 (44)
Hoarseness	34 (43)
Abdominal pain	29 (37)
Arthritis	28 (35)
Calcifications	18 (23)
Melena	10 (13)

[a]*Numbers in parentheses indicate percentage of untreated children.*
From Pachman et al. (1998).[11]

BOX 42.1 Duration of Untreated Disease: Association with Dystrophic Calcification

- Calcinosis at diagnosis
  - Total population: 18/79 (23%)
  - Caucasian: 13/59 (22%)
  - Non-Caucasian: 5/20 (25%)
- Median interval between first symptom and diagnosis
  - Without calcinosis: 3.0 months
  - With calcinosis: 4.5 months (p = .04)

Data from Pachman et al. (1998).[11] Used with permission.

TABLE 42.3 The First Symptom (Rash and/or Weakness) of JDM and the Interval of Time (Median in Months) before the Second Symptom

	N	Rash, 1st	Rash-Weakness Interval Median (range)*	Weakness, 1st	Rash-Weakness Interval Median (range)	Rash-Weakness Occurred at Same Time
White	59	30	2.0 (0.2–12.9)	15	2.0 (0.7–9.6)	14
Minority	20	12	3.5 (1.0–9.4)	4	6.0 (4.6–20.0)	4
All	79	42	2.0 (0.2–12.9)	19	3.2 (0.7–20.0)	18

**Median in months.*
From Pachman et al. (1998).[11] Used with permission.

diagnosis of overlap syndrome > JPM > JDM, the presence of aminoacyl-tRNA synthetase autoantibody (particularly anti-alanyl-tRNA synthetase autoantibodies) and accompanying interstitial lung disease (ILD) as well as other autoantibodies: anti-Ku, anti-La, and anti-Sm. Other elements contributing to premature death are older age at illness onset and at diagnosis, dysphagia, and abdominal perforation as the first signs of illness.[13] Cardiac disease has also been identified in the pediatric patients with IIM, exhibiting features similar to those described in studies of adult myositis, encompassing rhythm disturbances, conduction abnormalities, pericardial effusion, left ventricular dysfunction, myocarditis, cardiomyopathy and congestive failure,[14–16] but they do not appear to be associated with early death.[17]

There is controversy about the presence of a seasonal pattern in disease onset of JDM and adults with polymyositis/dermatomyositis (PM/DM), but the data appear to vary by geographic region and by year, defying reproducibility. In the north central region of the United States, children with definite JDM (diagnosed within 4 months of onset) were more likely to have their first symptoms between January and June than at other times in each of 7 years (1974–1980).[18] An investigation of 79 newly diagnosed cases of JDM from all regions of the United States (1989–1992) identified an increased frequency of disease onset in the spring and summer of some years,[11] but a survey of US JDM did not confirm that finding.[10] In the United Kingdom and Ireland, several clusters of children with new symptoms of JDM were identified, the largest of which was in April and May 1992, but the timing of other clusters varied from year to year.[19] When season of birth was investigated, some subgroups of patients with IIM had seasonal birth distributions.[20] Hispanic patients with juvenile-onset IIM had a seasonal birth pattern (mean birth date October 1) significantly different from that of Hispanic controls ($p = 0.002$), who had a uniform birth distribution, and from that of non-Hispanic patients with juvenile-onset IIM ($p < 0.001$), who had a mean birth date of May 2. Juvenile dermatomyositis patients with p155 autoantibody had a birth distribution that differed significantly from that of p155 antibody-negative

TABLE 42.4 Systemic Manifestations of Untreated Children with JDM

1.	General symptoms	Fatigue
		Malaise
		Weight loss
		Raynaud's syndrome
2.	Musculoskeletal features	Muscle weakness, pain, and tenderness
		Arthralgia, usually symmetric and involving the hand joints
		Non-erosive polyarthritis, usually symmetric, mild and responsive to corticosteroids
		Deforming arthropathy of hands is rare, but is associated with subluxation of the thumb interphalangeal joints (floppy thumb sign)
		Carpal tunnel syndrome
3.	Cutaneous findings	Dematomyositis-associated rashes
		Gottron's papules
		Heliotrope rash
		V-sign and shawl sign rashes
		Linear extensor erythema
		Other rashes
		Roughening, scaling, and erythematous fissuring of the palmar and lateral aspects of the fingers (mechanic's hands)
		Photosensitivity
		Periungual abnormalities including telangiectasias and cuticular overgrowth
		Irregular indurated plaques over the fingers with mucin accumulation in the dermis
		Subcutaneous and intradermal calcification, which may ulcerate with secondary infection
		Vasculitis with infarcts and digital ulcers
4.	Pulmonary manifestations	Ventilatory insufficiency caused by respiratory muscle weakness
		Atelectasis
		Aspiration pneumonia in patients with dysphagia
		Interstitial lung disease
		Drug-related pneumonitis
		Opportunistic infections in immunocompromised patients
5.	Cardiac involvement	Myocarditis with arrhythmias and congestive failure
		Cor pulmonale
		Rare pericarditis
6.	Gastrointestinal manifestations	Abnormal pharyngeal and cricopharyngeal function
		Esophageal dysphagia with occasional nasal regurgitation
		Delayed gastric emptying and reflux
		Dysmotility of the small and large intestines
		Vasculitis with infarcts and necrosis of the bowel

Adapted from Pachman and Miller (2001).[361]

juvenile dermatomyositis patients ($p = 0.003$). Juvenile IIM patients with the HLA risk factor allele DRB1*0301 had a birth distribution significantly different from those without the allele ($p = 0.021$). Similar results were observed for juvenile and adult IIM patients with the linked allele DQA1*0501, versus juvenile and adult IIM patients without DQA1*0501, respectively.[20] These data are of importance for they suggest that pre- and perinatal influences may play a role in disease susceptibility.

Other prenatal factors include studies of maternal chimerism[21] in muscle tissue from males with JDM which may contribute to the development of the inflammatory response.[22] Of note, monozygotic twins are equally chimeric because of the shared placenta,[23] but it is still evident that they can be either concordant or discordant for development of IIM.[24]

PATHOGENESIS/GENETIC DATA

Overview

Intensive study of children with newly diagnosed JDM suggests that the symptoms of the disease might follow exposure to infectious agent(s) in genetically susceptible individuals.[25] Disease expression and chronicity appear to be facilitated by genetic risk factors located on chromosome 6, including the tumor necrosis factor alpha (TNFα)-308A allele,[26] which may be exacerbated by exposure to ultraviolet B (UVB) light.[27] Antecedent infection also appears to play a role in initiation of JDM symptoms.[28] A case control study of over 320 US children documented a significant increase in antecedent symptoms in JDM children in the 3 months before diagnosis; the symptoms are primarily respiratory, often with concurrent gastrointestinal symptoms[29]; these observations were supported by those of Canadian investigators.[30]

Infectious Agents and JDM

In addition to temporal and seasonal differences, there may be regional differences in infectious agents identified in association with the first symptoms of JDM. Agents associated with the onset (and on occasion, flare) of JDM include group A beta hemolytic streptococci[31] and the RNA picornaviruses.[18,32,33] Other infections that have been implicated in the pathogenesis of JDM include parvovirus, influenza, parainfluenza, and *Borellia*, as well as parasites, such as *Toxoplasmosis gondii*, but no evidence of related DNA or RNA has been identified in affected JDM muscle.[34] Enteroviral agents are the most tantalizing, when antibody studies are considered. Newly diagnosed children (1974–1980) from the Chicago area were tested and displayed an increased frequency of antibody—both neutralizing (N) and complement fixing (CF)—to the coxsackievirus B antigen.[32] There may be wide variation in different years, because these data were not reproduced by the same laboratory in a study of 20 children with dermatomyositis from the same region with onset of disease in the years 1987 to 1992.[35] In studies searching for genetic evidence of infection, enteroviral RNA was identified in the muscle of UK patients with PM/DM,[33,36] which was not confirmed in a similar patient population living in Sweden.[37] Echovirus has been linked to the development of inflammatory myopathy in children with hypogammaglobulinemia and growth hormone deficiency,[38] which was exacerbated by administration of growth hormone.[39] In contrast, evidence of viral RNA has not been identified in the muscle biopsies from patients with myositis living in the United States, either in Jo-1 positive adults[34] or in newly diagnosed, untreated children with active JDM, who had the specific area of the inflamed muscle identified for biopsy by MRI scan.[35,40] There is also a possible role for molecular mimicry. Evidence supporting this hypothesis is based on a report of a protein present in skeletal myosin (which has homology with the streptococcal type 5M protein), which stimulated peripheral blood mononuclear cells *in vitro*. Blood from children who had "flares" of their dermatomyositis symptoms following streptococcal infections show a higher proliferative response than blood from children without a known antigen exposure.[41] In parallel studies, an epitope specific for rheumatic fever did not stimulate the peripheral blood mononuclear cells from children with JDM.[41]

Early reports of JDM suggested that exposure to *T. gondii* or hepatitis B might initiate the inflammatory myopathy. However, antibody titers to these agents were not increased either in a case control regional study, which tested sera from children with JDM in the mid-1980s[32] or in a national case control study of new onset JDM.[42] Noninfectious agents and exposures currently posited as causative agents in JDM include vaccines (hepatitis B, measles-mumps-rubella vaccine, typhoid, cholera)[43–45] which a recent thoughtful appraisal could not refute.[46] These data contributed to the current ban on the administration of live viral vaccines to children with JDM, irrespective of the level of their disease activity. In addition, drugs such as D-penicillamine,[47] growth hormone,[48] and bone marrow transplants in which graft versus host myositis occurs as a component of immune activation have been invoked as contributing factors to the display of IIM symptoms.[49] Of note, influenza B, which typically causes a short episode of pain in the calf muscles of young children[50] has not been associated with the development of juvenile chronic myositis. Taken together, the above data suggest that the etiology of JDM appears to be multifactorial, and may encompass a specific epitope(s) and environmental conditions in individuals who have a genetic predisposition to develop IIM.

Genetic Data

Several lines of evidence indicate that there is a strong genetic component that contributes to JDM display of symptoms. Although JDM may occur sporadically in more than one family member,[24] the disease has been reported in monozygotic twins in whom muscle-related abnormalities developed 2 weeks after an upper respiratory infection.[51] Proteomic studies of monozygotic twins with autoimmune disease, including JDM, suggests that autoimmune diseases share common pathways.[52] Specific genetic associations have focused on the region controlling autoimmunity on chromosome 6 and previous data have identified DR3*0301 and TNFα in linkage disequilibrium with DQA1*0501 and DQA*0301 in US white patients. The strong association of JDM/DM and JPM/PM with this specific region on chromosome 6 was confirmed by a Genome Wide Association Study (GWAS) of a large population of Western European adult and pediatric patients with these diseases[53] (Figure 42.4). However, this association appears to be modified by race. When 151 white American adults with idiopathic inflammatory myopathy (IIM) were compared with 50 Korean adults with IIM, the linked allele DQA1*0501 and DRB1 alleles sharing the first hypervariable region motif 9EYSTS13 were major genetic risk factors for the development of myositis in Americans. Although both the white and Korean patients had similar distributions of clinical characteristics, autoantibody profiles, and clinical groups, none of the HLA-DRB1 or DQA1 alleles or motifs was a risk factor for IIM in the Korean patients.[54] These data support the currently held assumption that the genetic risk factors for the inflammatory myopathies are multifactorial and appear to be associated with the patient's ethnicity.[55]

When muscle from untreated children with definite JDM was examined, Wt-1 (Wilms Tumor 1) was found to be massively hypomethylated, in contrast to the age-matched healthy controls. Of note, this hypomethylation was also true for children with JIIM who had partial therapy or those with overlap or JPM, suggesting that less muscle wasting may occur in children because they still have the capacity to repair cell damage—controlled by Wt-1.[56] Studies of microRNA (mi-RNA) arrays in untreated JDM muscle disclosed that the mi-12 was downregulated and controlled the inflammatory cytokine cascade via NFkB.[57] In addition, downregulation of mi-12 was associated with increased von Willebrand Factor-antigen (vWF:Ag) in JDM sera, suggesting that this might be one of the inflammatory pathways that contribute to the widespread microvascular vascular damage seen in JDM.[57]

Maternal factors also might be associated with susceptibility to JDM. Boys with JDM appear to be chimeric because 73% of the boys had peripheral blood mononuclear cells (PBMCs) that contained female (XX) cells. These cells were positive for HLA-DQA1*0501 and were activated by the host male cells,[58] which suggests that this antigen (or related MHC antigen) may be an important indicator of predisposing factor(s) for JDM (see the following).

Clinical Features at Diagnosis

The most common physical findings of JDM children at diagnosis are presented in Table 42.2. Less commonly, hepatosplenomegaly and generalized lymphadenopathy are present at diagnosis in about 5% of children, indicating reticuloendothelial system activation. Box 42.1 shows

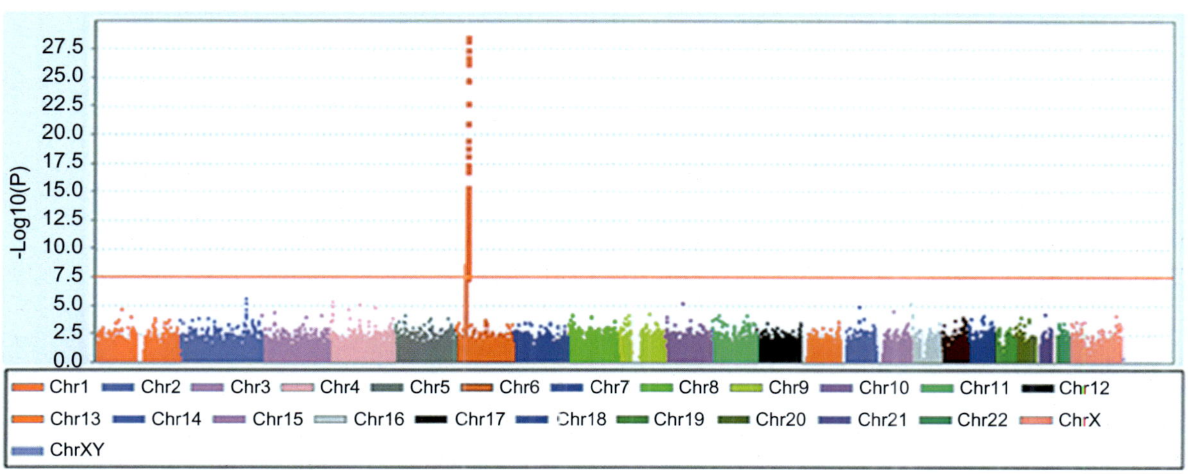

FIGURE 42.4 Genome Wide Association Study (GWAS). Manhattan plot of 1178 patients (European) with dermatomyositis (matched controls = 4724) showing p values for 242,876 SNPs successfully genotyped. The orange line = 7.3 ($p = 5 \times 10^{-8}$) and the blue line = 4 ($p = 1 \times 10^{-4}$). SNP, single nucleotide polymorphism (2). *Reproduced with permission from Miller et al. (2013).*[53]

that delay in diagnosis and therapy is associated with the development of dystrophic calcifications. Table 42.3 presents the first definite symptom of JDM: weakness or rash. The systemic symptoms that may be identified in the various idiopathic inflammatory myopathies are listed in Table 42.4.

Cutaneous Findings

One of the early symptoms of JDM is an intermittent periorbital edema, which may be the only presenting sign.[59] Periorbital erythema and edema (Figure 42.1A), and eyelid capillary vessel dilation may persist long after other signs and symptoms of disease activity have resolved; these are present in the majority of affected children. We have observed a flare in several children indicated only by erythema of the pinna of the ears, heralding the development of full-blown involvement of muscle and skin. As in children with systemic lupus erythematosus (SLE), the erythematous rash may cross the bridge of the nose and is malar in distribution. Sun exposure can precipitate exacerbation of symptoms (the children may sunburn easily) either initially, or once the disease has been diagnosed and treated. When rash is the first definite symptom of JDM, it is first observed in the summer months in 41% of newly diagnosed cases of JDM. Other areas of capillary involvement can be seen on inspection of the roof of the mouth, which shows dilated erythematous capillaries at the junction of the hard and soft palate, as well as swollen and often bleeding erythematous gums.

Proceeding distally, other cutaneous features include the neck, which may have a rash extending to the upper torso (the shawl sign) (Figure 42.1B). The examiner may find involvement of the extensor surfaces of the arms and legs, medial malleoli of the ankles, as well as the buttocks (Figure 42.1C), all of which may occur in the absence of elevated serum concentrations of muscle-derived enzymes. Skin over the knuckles (metacarpophalangeal joints and proximal interphalangeal joints) is often either hypertrophic or pale red (Gottron's sign) and evolves into colorless bands of atrophic skin, which, during active disease, may have a papular, "alligator skin" type of appearance (Gottron's papules) (Figure 42.1D). Pathologic calcifications occur at pressure points, which, if they break through the skin, offer a portal for infection (see Figure 42.1E and C). Partial baldness may result as a consequence of chronic scalp inflammation[60] (Figure 42.5B). The cuticle becomes dry and children frequently try to bite away the thickened skin at the base of the nailbeds. Other cutaneous signs found in JDM include paronychiae, often located on the lateral aspects of the nail of the great toe, which improve with symptomatic

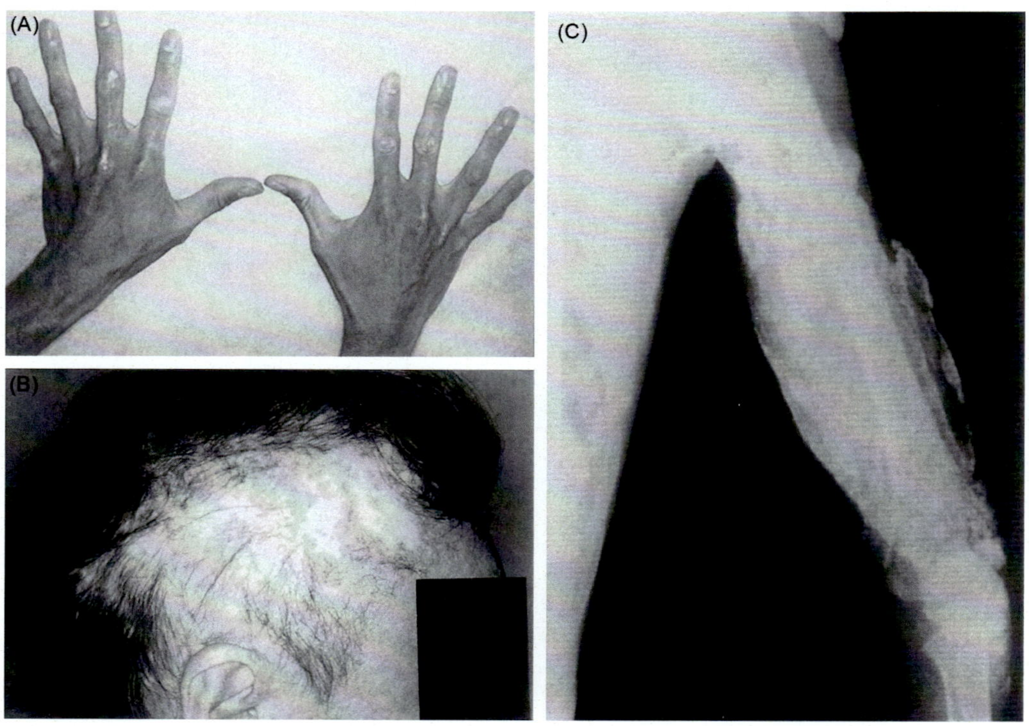

FIGURE 42.5 Features of chronic long-standing disease in JDM. (A) Lipoatrophy showing loss of fat over hand/fingers. (B) Scarring of the scalp following chronic inflammation with extensive hair loss. (C) Extension of pathological calcifications as seen on radiograph of the arm of a child with long-standing active disease.

care and return of sufficient peripheral vascular blood flow. Livedo reticularis (lacy patterning on the skin) is frequent in children with Raynaud's phenomenon and may indicate the presence of a hypercoagulable state as well as the presence of an antiphospholipid antibody.[61] An excellent catalog of cutaneous manifestations in JDM is readily available.[62] Newly diagnosed JDM parents, who were interviewed within 6 months of diagnosis, identified the rash of JDM before proximal muscle weakness in more than 50% of the children. The parents recognized the rash at the same time as the weakness in another 23% of children, and, in the remainder of the JDM children (27%), weakness was the first symptom reported by parents (Table 42.4). Many children present with the amyopathic form of JDM (rash only) much later in their disease course.[63] It is now recognized that anti-MDA5 antibody may be associated with malignancy in adults with amyopathic disease.[64]

The vascular features are one of the hallmarks of JDM. The upper eyelid margins frequently have a fringe of dilated capillaries (Figure 42.2A), some of which become thrombosed in active disease. The rash has a violaceous or heliotrope hue and is often most prominent on the eyelid where small infarctions can occur, as well as in the area of the medial canthi near the eyes in children who have a prominent component of necrotizing vasculopathy. Rarely, children with a chronic disease course can present with eyelid telangiectasia (Figure 42.2B). The spectrum of capillary changes in the nailfold is shown in Figure 42.2C1, 2, and 3. The association with the number of nailfold capillary end row loops and the severity of the rash but not the muscle involvement[65] is graphed in Figure 42.2D and the occlusion of capillaries and arterioles is demonstrated in the muscle biopsy (Figure 42.2E). If the active symptoms of JDM have been present for more than 1 year before treatment, there is greater likelihood that the lipodystrophy will be easy to detect[66] (Figure 42.5A) (see information on chronic course later in this section).

Musculoskeletal Symptoms

Weakness is the first symptom in about 25% of cases (Table 42.3) and is routinely evaluated using the Childhood Myositis Assessment Scale (CMAS).[67] The evaluation of muscle strength is challenging in children, because it requires voluntary cooperation (see later section "Evaluation Tools for Disease Activity and Chronicity"). Proximal muscle weakness may be detected when the child has difficulty with climbing stairs, getting up from a chair, combing hair, or when using their hands to push off from the body in an attempt to stand up from the floor (Gowers' sign). Weakness of the neck flexors, a particularly sensitive indicator of muscular impairment, is often the first area of impairment, and is frequently the last sign of weakness to normalize. Truncal weakness is common and can be demonstrated by asking the child to perform a sit-up (without a counterbalance to the feet) with the arms held in three positions: extended, crossed on the chest, and behind the head. However, even a healthy child younger than 5 years is often not able to clear the scapula; the endurance test of holding first the right arm and then the right leg off the examining table was not achievable by a good proportion of healthy children.[68] More than 60% of children with JDM complain of pain on proximal, but not distal, compression of muscles, which is seen in PM and overlap syndromes, but it is less severe than the muscle pain in bacterial myopathies. Children with spinal muscular atrophy do not usually have muscle pain on compression. Fatigue is a common symptom of many dystrophies (especially Duchenne's and Becker's) and is not useful in differentiating children with JDM. Usually the child with an inflammatory myopathy is more comfortable when the limbs are held in the flexed position, promoting the development of flexion contractures, which may be difficult to reverse.

Cardiorespiratory Abnormalities

In more than half the children with definite JDM the electrocardiogram is abnormal at diagnosis. Asymptomatic conduction abnormalities predominate, with an occasional complete block of the right bundle branch,[69] which usually resolves with control of the inflammatory process. Dilated cardiomyopathy can be a rare presenting complaint[70] and usually resolves when all the indicators of immunologic activation have normalized. Tachycardia is common at diagnosis and when it continues during the disease course it reflects cardiac compromise, which can be detected by an increase in cardiac-specific indicators, such as troponin levels. Because the respiratory musculature in the trunk is almost always involved, most cases of untreated JDM have a restrictive ventilatory defect, accompanied by a decrease in ventilatory capacity.[69] The decrease in ventilatory capacity can also be associated with diminished speech volume. A husky voice or vocal cord nodules, presumably as a result of the stress of trying to be heard, are present in approximately 3% of the over 600 children with JDM followed at our center. A study of long-term follow-up identified restrictive pulmonary impairment, associated with long-term JDM damage in 8% of patients, indicating a need for repetitive pulmonary follow-up in JDM patients.[71]

Pulmonary Fibrosis

Pulmonary fibrosis is more commonly found in individuals who carry an antibody to the t-RNA synthetases, of which anti-Jo-1 is the most common. Children with JDM who have an antibody to Ro, or patients with an overlap syndrome with antibody to U1 RNP,[72] or those with antibody to signal recognition particle (SRP)[73] or anti-alanyl-tRNA synthetase (Pl-12),[74] have a higher frequency of pulmonary fibrosis. They may develop a honeycomb appearance at the base of the lung, identified by high resolution computed tomography scan of the chest, and decreased diffusion of CO_2 and FEV1 on pulmonary function testing. It is critical to identify ILD early in the disease course, for not only is ILD a cause of death in older patients with JDM,[13] but if recognized and treated promptly with compliance in childhood, the fibrosis can be reversed.[74]

Gastrointestinal Involvement

The most severe prognostic indicator is impairment of the flow of secretions associated with decreased esophageal motility. This can be demonstrated by a rehabilitation cookie swallow with radiographic contrast, which shows retention of barium in a widened atonic, pyriform sinus,[75] which can be associated with a "stacked coin" appearance. This anatomic change could allow swallowed liquids to penetrate the adjacent unprotected airway.[76] Esophageal reflux may result in aspiration pneumonia, and appropriate precautions should be taken (e.g. using thickened foods, raising the head of the bed, and giving attention to clearing the respiratory tree). Involvement of the masseter muscles (Figure 42.1A) may be associated with difficulty in chewing as well as with swelling of the cheek(s) and the gradual development (over weeks/months) of buccal fat atrophy characteristic of partial lipodystrophy.

Boys and girls with untreated JDM were shorter and lighter than national norms ($p>0.0005$ for both), perhaps reflecting decreased GI absorption secondary to inflammation, and nonwhite children were weaker than white children ($p>0.0005$). Older children had more dysphagia ($p = 0.017$) and arthritis ($p>0.001$).[77] Vasculopathy can affect any part of the gastrointestinal tract. In severe disease, in addition to weight loss, mucosal ulceration with melena can occur with the possibility of life-threatening perforation. Pneumatosis intestinalis can occur as well, requiring bowel rest and aggressive immunosuppression.[78] In the young child, development of normal speech patterns can be disturbed; soft palate involvement is often revealed by nasal, high-pitched speech (e.g. identified by having the child say the letter "E") and usually resolves with a decrease in the inflammatory component of the myositis. A dilated atonic esophagus can be associated not only with delayed gastric emptying but also with infectious or neoplastic diseases.[76] Smooth muscle dysfunction can also result in decreased lower gastrointestinal motility, making constipation an early and easily ascertainable symptom that responds to a stool modifier, such as Marilax, although chronic intermittent diarrhea has also been reported by the children or their parents as well. If there is a family history of celiac disease or the physician has an increased index of suspicion,[79] then a celiac panel should be obtained, which if positive, should be confirmed by gastrointestinal biopsy, before the children embark on a gluten-free diet because they are thought to have celiac disease.[79]

Genitourinary Function

Severe inflammation can result in massive breakdown of muscle elements as well as primary compromise of the renal parenchyma itself, which may culminate in acute renal failure. This requires prompt hydration with monitoring of renal function. Necrosis of the ureter with calcifications has been reported, involving the middle (iliac) segment, because of the relatively sparse blood supply to this region compared with the upper (lumbar) or lower (pelvic) segments.[80]

Reproductive function may be altered in acute or chronic disease: menses may cease during severe disease, only to resume after the active inflammation is controlled. One study identified delayed menarche with normal cycles and low follicular reserve in JDM patients.[81] The girls with JDM had decreased progesterone levels suggesting an underlying subclinical corpus luteum dysfunction in this disease.[81] Lipodystrophy can result in impairment of fertility in female patients, especially if severe disease onset is before puberty. In addition, JDM girls may have abnormal levels of LH and FSH.[81]

With respect to pregnancy, if the JDM symptoms in the mother are relatively quiescent, then both mother and infant will do well, but a JDM disease flare during pregnancy can be problematic.[82,83] Active JDM symptoms during pregnancy—muscle weakness, rash flare—can usually be managed with the use of modest doses of corticosteroids (both IV and PO), given with periodic IVIg without harm to the fetus. However, pregnancy in a mother with JDM may be associated with increased fetal wastage secondary to antibodies, such as anticardiolipin antibodies, which alter coagulation.[84] Anticardiolipin antibodies have been identified in children with JDM[85] (often with a negative lupus anticoagulant) and can be associated with fetal death, as well as vascular thrombosis.[86] Birth control measures, such as oral contraceptives, should take into account the possibility of anticardiolipin antibodies. The type of birth control used should be discussed with the patient and her OB/GYN physician to diminish the possible risk of increased intravascular thrombosis.

Ophthalmologic Findings

Children with a more severe vasculopathy may display ulceration and infarction of tissue, often in the region of the medial canthus of the eye, which heal and leave a depressed scar. Within the eye, in active inflammation of JDM, transient retinal exudates and "cotton wool" spots may occur after the occlusion of small vessels, leading to intraretinal edema with injury to retinal nerve fibers, optic atrophy, and sustained visual loss. Neovascularization of the retina with spontaneous regression has also been reported.[87] Central retinal vessel occlusion has been associated with anticardiolipin antibody.[88] Disease of conjunctivae vessels can lead to an avascular zone with a potential for infarction. If there is a family history of red-green color blindness, the use of hydroxychloroquine should be avoided. Orbital myositis can be documented by MRI, either targeting a single fast-moving muscle, such as the superior oblique or as part of the spectrum of inflammatory myopathy which responds to corticosteroid administration or other immunosuppressant and may be part of the IgG4 syndrome[89] (see later text). Children treated with steroids should be monitored for glaucoma (especially if there is a positive family history!) and for the development of sublenticular cataracts, which often resolve as the child improves.

Other Disease Manifestations

Vasculopathy involving the central nervous system may be associated with depression or wide mood swings, which may be exacerbated by steroid therapy (see "Psychological Aspects of Juvenile Myositis" below). Very few children with JDM present with Raynaud's phenomenon; these symptoms are more frequently found as isolated phenomena in children with overlap syndromes or scleroderma.

Juvenile Dermatomyositis with a Chronic Course

Lipodystrophy

Three stages of lipodystrophy are now recognized: (1) localized lipodystrophy, occasionally following panniculitis that may be painful; (2) partial lipodystrophy targeting the areas over intense proximal muscle inflammation; and (3) generalized lipodystrophy, in which fat is lost over a larger part of the body, muscle atrophy is prominent, and joint contractures may be present. Several clinical features can provide clues to the duration of disease, for lipodystrophy occurs later in the disease course, reported 4.6 years after first symptom.[66] Loss of subcutaneous fat over the masseter area and over the arms and legs, proximal < distal, gives a ropey appearance to the musculature, with clear definition of muscle groups that are not hypertrophied (Figure 42.5A). In contrast, abdominal muscle tone is diminished, so that the child or young adult has difficulty with sit-ups, even when muscle strength is normal elsewhere. This loss of abdominal tone is accompanied by increased abdominal fat, with the appearance of a "pot belly." Another prominent feature of partial and general lipodystrophy is acanthosis nigricans, or speckled areas of increased pigment, which may be found in the skin folds of the neck, the axilla, behind the knee, or between the buttocks and appears like "dirty skin." An additional phenotypic feature of this illness is coarsening of hair texture, or, if there is chronic inflammation of the scalp, loss of hair in the affected area (Figure 42.5B). If the JDM is still active before puberty, then there may be loss of fertility, with masculinization, reflected by increased facial hair and change in body habitus. The clinical features of lipodystrophy are not rare[90] and occur in about 20% of our JDM population: more than 30% had some evidence of fat atrophy, 22% had increased insulin levels, elevated levels of triglycerides were present in 10%, and three tenths of those tested for glucose tolerance had evidence of insulin resistance.

Dystrophic Calcifications

Calcification in soft tissue is a consequence of the duration and severity of the active symptoms of JDM. Although the number of children with JDM with calcium deposits in the soft tissues has decreased when this disease is diagnosed more promptly and treated aggressively to 14%,[91] the national average is still about 30%[1] and, in South Africa, 70% of children with JDM develop calcinosis.[92] The pathologic calcifications often follow delayed diagnosis or insufficient therapy.[11] In children with the JDM rash only who were not given initial aggressive therapy, myositis often develops at a later date and may be complicated by pathologic calcifications.[93] The calcifications may occur at pressure points, often the elbows or buttocks, but also can be found in fascia surrounding muscle (Figure 42.1E and C), as well as in palmar surface of the digits in children with overlap syndromes, such as Pm/Scl. The calcifications may change in appearance from massive accumulations to become thinner, more pancake-like collections, which then may resolve spontaneously, draining as a white cheesy or serosanguineous exudate containing macrophages[94] and leaving dry, pitted scars (Figure 42.1E). In persistent, active myositis, the calcifications may progress to become a sheath, impairing flexion and function, breaking the barrier of the skin to form a portal for infection. Sepsis is not uncommon, frequently from staphylococci, and is a major contributor to the morbidity and mortality related to this disease. Children with JDM who have calcium deposits have three times higher urinary excretion of γ-carboxyglutamic acid (GLA), a component of the vitamin K-dependent coagulation pathway.[95] An intriguing finding is that children with JDM without pathologic calcifications have GLA excretion that is twice normal, suggesting a prominent role of this protein in the pathophysiology of the disease.[95] Further characterization of the calcifications show that they are not bone-like when studied with Fourier transform infrared spectroscopy,[96] although they do have selective deposition of bone matrix proteins.[97]

Calcification—Differential Diagnosis

It is important to differentiate the etiology of dystrophic calcinosis in JIIM from that in other syndromes such as in heterotopic calcinosis or that following trauma. Heterotopic calcinosis occurs in children with progressive osseous heteroplasia.[98] In these individuals, there may be evidence of both intramembranous and endochondral ossification. In the cases of progressive osseous heterotopia described in the literature, the children presented at birth or within the first 6 months of age with only the cutaneous calcifications.[99] This entity can be differentiated from fibrodysplasia ossificans progressiva, in which congenital malformations of the big toes are present at birth.[100]

Disease Chronicity

Flexion contractures, hyperlipidemia, and cardiovascular disease are a few of the additional consequences of chronic inflammation. The position of comfort in the active phase of acute myositis is the flexed position which, if maintained for too long a period, becomes fixed with loss of both range of motion and of function; calcifications in the flexor tendons compound the problem. ILD is a leading cause of mortality in children with IIM.[13] Short stature, present at diagnosis,[77] is maintained for both boys and girls with JDM once they reach adulthood,[101] when they are at increased risk for atherosclerotic heart disease and the metabolic syndrome,[101] which may share features of the accelerated cardiovascular disease seen in adults with SLE[102] as well as children.[103]

Other Autoimmune Diseases

Many conditions must be considered in the differential diagnosis of inflammatory myopathy in children (Table 42.5). Because myositis is often a component of other autoimmune rheumatic diseases, it is essential to exclude such conditions as SLE, mixed connective tissue disease,[104] chronic arthritis in children (especially systemic onset juvenile rheumatoid arthritis [JRA][105]), the spondyloarthropathies, Sjögren's syndrome, and Kawasaki syndrome. They also may have a myositis specific antibody (MSA) that is often associated with a nonremitting disease course, such as Jo-1, the most common of the antibodies to the t-RNA synthetases[7,106] (Table 42.6). Furthermore, the inflammation can be concentrated in one specific muscle area (see "Other Myopathies" following), as in cases of orbital myositis[107]; it can be a nodular or proliferative process, or it can involve an eosinophilic infiltration of the fascia.[108] Children with muscular dystrophy or a dysferlin deficiency associated with inflammation[109] may present with muscle weakness (most often distal) as well as elevated levels of serum derived. Other conditions must also be excluded before the diagnosis of JDM is established. The endocrinopathies must also be included in the differential diagnosis. We have observed proximal and distal muscle weakness and elevated muscle enzymes in children with hypothyroidism, hyperthyroidism,[110] and hyperparathyroidism.[111]

Laboratory Indicators of Juvenile Dermatomyositis

Magnetic Resonance Imaging

Musculoskeletal inflammation at diagnosis and flare can be identified on magnetic resonance imaging (MRI), which guides the muscle biopsy, EMG, and other serologic assessments. The use of MRI to identify the specific area of often patchy inflammation in the muscle can direct the surgeon to the appropriate site for a muscle biopsy (Figure 42.6). This approach minimizes error in sampling uninvolved areas in this focal disease[112] and can serve, under very special circumstances, as an excellent monitor of a child's response to therapy.[113] The MRI findings may normalize several months later than the muscle enzymes.[114] P31 magnetic spectroscopy has been used for the past few years and gives additional information about the child's muscle strength and performance and has confirmed evidence of defective oxidative phosphorylation in diseased muscle.[115]

Muscle Enzymes

The standard muscle enzymes obtained for diagnosis are creatine kinase (CK), the transaminases aspartate aminotransaminase (AST) and alanine aminotransaminase (ALT), lactate dehydrogenase (LDH), and aldolase.[2] The highest concentration of CK is present in excitable tissues (muscle and nervous system) and occurs in three isoforms, MM, MB, and BB. In skeletal muscle, the MM form predominates (90% to 95%), whereas immature muscle may have increased concentrations of the MB form (20% to 30% in children younger than 1 year of age).[116] AST is found primarily in the heart, liver, skeletal muscle, and kidney, and ALT is also found in skeletal muscle but in much lower concentrations.[117] LDH catalyzes the intraconversion of lactate and pyruvate and is found in most of the tissues of the body.[117] Aldolase is also widely distributed in the tissues of the body, and increased circulating levels of aldolase may be seen not only in myopathies but also in disorders of the liver, the hematologic tissue, and other diseases.[118] The child who has moderately active JDM on muscle biopsy can have an elevated level of aldolase or any one of the other

TABLE 42.5 Differential Diagnosis of JDM

1.	Drug- and toxin-induced myopathies	Corticosteroids
		Ethanol
		Lipid-lowering drugs
		D-Penicillamine
		Colchicine
		Ipecac
		Chloroquine
		Zidovudine (AZT)
		L-Tryptophan (eosinophilia myalgia syndrome)
2.	Endocrine diseases	Hypothyroidism
		Hyperthyroidism
		Acromegaly
		Diabetes mellitus
3.	Neurologic disorders	Amyotrophic lateral sclerosis
		Myasthenia gravis
		Multiple sclerosis
		Guillain-Barré syndrome
		Motor neuron disease
4.	Other connective tissue diseases	Polymyalgia rheumatica
		Rheumatoid arthritis
		Systemic sclerosis
		Tendinitis and overuse syndromes
5.	Metabolic abnormalities	Hypokalemia
		Hypercalcemia
6.	Inherited metabolic defects	Acid maltase deficiency
		Phosphorylase deficiency (McArdle's disease)
		Lipid metabolic defects (carnitine or carnitine palmitoyltransferase deficiency)
		Phosphofructokinase deficiency
7.	Muscular dystrophies	Duchenne's/Becker's
		Facioscapulohumeral
		Limb-girdle
		Myotonic
		Dysferlin deficiency
		Merosin deficiency
8.	Infectious myopathies	Bacterial (pyomyositis)
		Parasitic
		Viral, including human immunodeficiency virus and HTLF-I and -II
9.	Mitochondrial myopathies	

TABLE 42.6 Myositis Specific and Associated Antibodies in JIIM

Serologic Group	Frequency (%)	Clinical Subgroup Association in JIIM	Comments	References
MSA				
Anti-p155/140 (TIF-1-γ)	23–30	JDM and JDM with overlap myositis	Associated with extensive photosensitive skin rashes, including the characteristic rashes of JDM, malar rash, V-sign and shawl-sign rashes, and linear extensor erythema, as well as periungual capillary changes, skin ulceration, and generalized lipodystrophy; patients frequently have a chronic illness course; not associated with cancer-associated myositis, which differs from patients with adult IIM with this autoantibody	7,66,305–308
Anti-MJ (NXP-2)	20–25	Primarily JDM	Frequent muscle cramps, dysphonia, joint contractures, and a monocyclic illness course; this autoantibody group seems to have more severe illness, with some reports of increased calcinosis, frequent muscle atrophy, more frequent hospitalizations and gastrointestinal ulceration	7,307–309
Anti-aminoacyl-tRNA synthetases (Jo-1 & non-Jo-1 synthetases)	2–4	JDM, JPM, and overlap myositis	More common in JPM (9%) and patients with juvenile overlap myositis (8–13%); frequent ILD, arthritis, Raynaud phenomenon, fevers, and mechanic's hands, similar to adults with these autoantibodies; among the MSA phenotypes, this group has the highest mortality caused by ILD	7,141,307,310
Anti-SRP	1	JPM	More common in JPM (18%); seen primarily in African American girls with JPM who have severe to profound proximal and distal muscle weakness, frequent falling episodes, Raynaud phenomenon, very high CK levels, wheelchair use, and a chronic illness course; cardiac disease is also likely associated, and these patients are refractory to several therapies, similarly to patients with adult IIM with anti-SRP autoantibodies	7,73,307
Anti-Mi-2 (NuRD)	2–13	JDM and overlap myositis	JDM with classic cutaneous findings of Gottron's papules, heliotrope rash, and malar rash; predominantly seen in Hispanic patients; mild disease, but children have higher CK levels than adults with this autoantibody	7,307,308,311
Anti-CADM-140 (MDA-5)	Unknown	JDM	Associated with DM and clinically amyopathic DM in adults who frequently have rapidly progressive ILD, cutaneous ulceration, and palmar papules, with a high fatality rate; 7 patients with JDM or amyopathic JDM and ILD have been reported with this autoantibody; to date, most reports are from Japan, and mortality has been high	312–314
Anti-SAE	0.2	JDM	Associated with DM cutaneous manifestations, which predate development of muscle symptoms, an absence of constitutional manifestations, and low frequency of ILD	307,315–317

(Continued)

TABLE 42.6 (Continued)

Serologic Group	Frequency (%)	Clinical Subgroup Association in JIIM	Comments	References
MAA				
Anti-U1-RNP	5–10	JDM, JPM, and overlap myositis	More common in JPM (12%) and overlap myositis (20–27%); associated with arthritis, Raynaud phenomenon, and sclerodactyly	141,307,308,318,319
Anti-Ro	2–6	JDM, JPM, and overlap myositis	More common in overlap myositis (8–15%); little known about the associated clinical features; associated with impaired lung function in 1 report; in adults, patients respond well to prednisone alone; may be seen in association with anti-Jo-1 autoantibodies	141,307,319,320
Anti-PM-Scl	1–4	JDM, JPM, and overlap myositis	More common in JPM (6%) and overlap myositis (10–25%); based on adult cohorts, this autoantibody is associated with Raynaud phenomenon, arthritis, ILD, and esophageal dysmotility; most frequent in Caucasian patients with an HLA DRB1*0301 immunogenetic association	141,307,320,321
Anti-Ku	0.2	JDM, JPM, and overlap myositis	More common in overlap myositis (2%); based on adult cohorts, frequently associated manifestations include arthralgia, Raynaud phenomenon, myalgias, dysphagia, and less commonly, ILD; patients respond well to corticosteroids alone, except those with lung disease	311,322
Other MAAs: anti-U2-, U3-, or U5-RNP; anti-La; anti-Sm; anti-Th	<1	JDM, JPM, and overlap myositis	Each autoantibody is present more frequently in overlap myositis (1–10%); little known, primarily associated with overlap myositis	141,307
Myositis autoantibody negative	28–57	JDM, JPM, and overlap myositis	Relatively mild disease in this subgroup; this group is likely heterogeneous, with several different unrecognized autoantibody groups contained within	141,307

Abbreviations: ILD, interstitial lung disease; JDM, juvenile dermatomyositis; JPM, juvenile polymyositis; MAA, myositis-associated autoantibodies; MDA-5, melanoma differentiation-associated gene 5; MSA, myositis-specific autoantibodies; NuRD, nucleosome remodeling deacetylase complex; SAE, small ubiquitin-like modifier activating enzyme; SRP, signal recognition particle; TIF-1-γ, transcriptional intermediary factor 1γ.
Reproduced with permission from Rider et al. (2013)[1] with minor modifications.

commonly tested muscle derived enzymes, while the others are in the normal range. Furthermore, these muscle enzymes may normalize by 4.5 months after the first JDM symptom, despite continuing disease activity, losing their value as a guide to therapy (Table 42.7).[77]

Electromyogram

Evidence of inflammatory myopathy on EMG is not specific to JDM. Inasmuch as 20% of EMGs not guided by MRI are "negative" or normal,[11] it is prudent to use MRI or ultrasound to identify the specific area of inflammation in the muscle. Once the location of the electrodes has been chosen (not the site of a future biopsy), insertional irritability followed by spontaneous electrical activity at rest is often observed. This pattern can also be seen in the muscular dystrophies and in early acute myositis. Abnormal, early, full recruitment of muscle fibers with moderate effort occurs in about 45% of patients with JDM, and bizarre, high-frequency discharges occur in

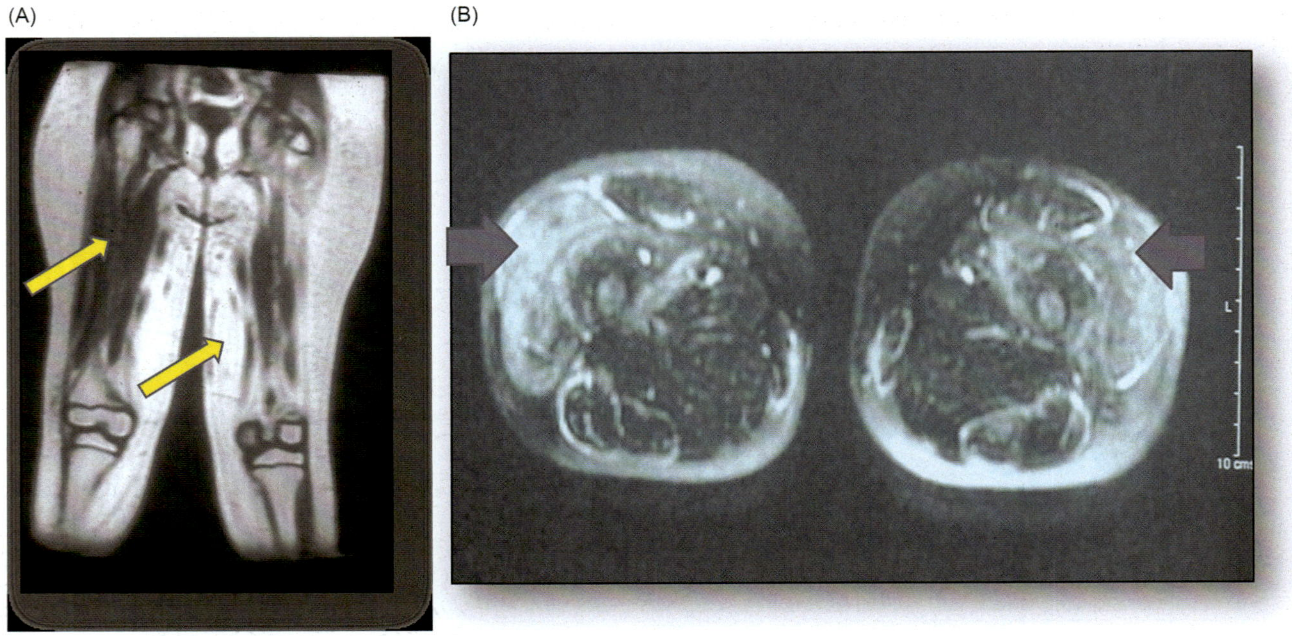

FIGURE 42.6 (**A**) Sagittal view of the MRI of a child positive for U1-RNP. The regions of muscle inflammation (white areas) are asymmetrical; the undirected biopsy of an area subsequently found not to be involved was read as "normal." The MRI directed biopsy (white area) showed characteristic inflammation and muscle fiber damage. (**B**) Magnetic resonance image of a child with early symptoms of JDM and normal muscle enzymes. This T2 weighted image displays the patchy nature of the muscular involvement: the white areas (arrows) show the symmetrical edema of active inflammation surrounded by thickened fascia.

TABLE 42.7 Time-related Normalization of Muscle Derived Enzymes at Diagnosis

Enzyme	Normal (%)	Non-normal	Cutpoint of Untreated JDM in months	p
CK	33 (26%)	93	4.65	< 0.01
LDH	16 (13%)	110	2.53	< 0.01
Aldolase	22 (17%)	104	4.65	< 0.01
SGPT	21 (17%)	105	3.68	< 0.01

From Pachman et al. (2006).[77]

15% to 20% of patients tested. Reduced motor unit activity is seen in Duchenne's muscular dystrophy (DMD) as well as in JIIM. A comprehensive comparison of DMD and JPM showed that the JPM patients more frequently had complex repetitive discharges on electromyography and a complete response to treatment with prednisone or other immunosuppressive agents than dystrophy patients (44% versus 0%).[119] Myasthenia gravis can coexist with an inflammatory myopathy, resulting in a greater degree of instability of motor-unit potential than is found in the uncomplicated inflammatory myopathies.[120] Given the child's variable acceptance of this procedure, we often reserve the EMG testing for the definition of cases that are not straightforward. Study of diagnostic testing of 384 children who were entered into the Children's Arthritis and Rheumatology Research Alliance (CARRA) registry indicated that EMG results were most likely to show non-diagnostic results with 50% of the children requiring an additional diagnostic study to confirm JDM.[121]

Pathology

Muscle Biopsy

The vasculopathy of JDM may occur in the absence of a prominent inflammatory component.[122,123] Capillaries and arteriole damage with loss of the capillary network results in structural change in the nailfold capillary bed as well as in muscle, with a subsequent decrease in the capillary-to-fiber ratio. Vascular compromise and capillary dropout, with perifasicular atrophy of both type I and type II fibers, are seen in the muscle pathology in JDM (Figure 42.7). Multiple satellite cells are frequently found adjacent to atrophic fibers; focal repair occurs concomitantly with fiber atrophy.[124] Low-grade ischemia may also be related to the increased expression of class I and class II major histocompatibility complex gene products.[125] Greater than 50% expression of class I by muscle fibers is now considered as one diagnostic criterion for inflammatory myopathy.[126,127] In contrast, in PM, there appears to be much less primary involvement of vessels; that is, they may be normal in number and structure. Some children with PM have the same histologic features as children with the rash of JDM (perifasicular atrophy, capillary occlusion), which may extend the spectrum of disease pathology for the diagnosis of JDM. Other children with PM and anti-synthetase antibodies, such as SRP, may have areas of necrosis, in comparison with the documentation of apoptotic cell death in JDM muscle, which increases proportionally to the time that the inflammation is untreated.[128]

In JDM, even when the serum levels of muscle-derived enzymes are normal, the MRI-directed muscle biopsy can reveal active foci of inflammation. Children with active inflammation and normal muscle enzymes may also have elevated levels of vWF:AG and/or neopterin, which can be used to guide therapy (see later discussion).

Skin Involvement

Study of the skin in JDM can yield valuable information about disease pathophysiology. Adults with DM and cutaneous ulceration may develop malignancy, which is not true for children with JDM.[129,130] DM patients with "mechanics hands" are at higher risk for ILD. Inflamed DM skin contains an infiltrate composed of C4 and CD8 T cells, plasmacytoid dendritic cells, neutrophils along with basal keratinocyte vacuolar alteration with apoptotic keratinocytes, and mucin deposition.[129] Type 1 interferons are produced by plasmacytoid dendritic cells and genome-wide expression data documents a robust upregulation of IFN-inducible genes in DM skin, as well as evidence of other inflammation pathways, complement activation, and epidermal activation and differentiation.[131] In JDM, mast cells were increased in the skin of the child, compared to their number in the inflamed muscle from the same child obtained by MRI directed biopsy.[130] Unlike in SLE, children with JDM do not display immunoflorescent band-like immunoglobulin deposits at the derma-epidermal junction, which is an important point in differentiating the two similar clinical presentations.

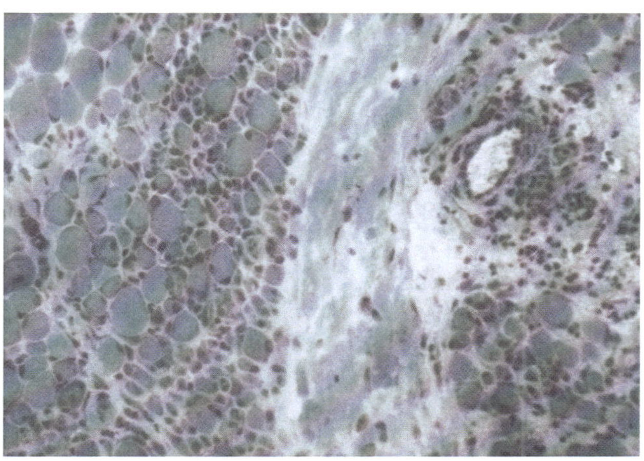

FIGURE 42.7 Muscle histopathology in JDM with 8 months of untreated disease: Hematoxylin and eosin stain shows extensive connective tissue proliferation and fibrosis while muscle fiber size variation is evident. The hallmarks of the disease, perifasicular atrophy and occlusion of capillaries and arterioles, are seen. The active inflammatory infiltrate is composed of primarily mononuclear cells.

Immunologic Data

Humoral Immunity

The complex immune system, consisting of innate and adaptive components, is essential to the pathophysiology of JDM tissue damage, but the complete sequence from immune activation to target damage has not yet been deciphered. Humoral immunity *per se* may be abnormal in a minority of patients with IIM: children with hypogammaglobulinemia may develop inflammatory myopathy associated with echovirus infection,[38] and IgA deficiency has been identified in a few children with undefined inflammatory myopathy.[132] In immunocompetent children, early in the disease course, the IgM and or/or IgG may be elevated.[15]

Myositis Specific Antibodies

As presented in Table 42.6 the evolving serological classification of JIIM has provided insight into some of the factors contributing to clinical myositis disease heterogeneity. Several methods are used to detect myositis-specific antibodies (MSA). For example, ELIZA determinations of MSA may be confounded by false positives[133] as opposed to immunodiffusion, radiolabelling, and immunoprecipitation methods, which yield protein bands of different migratory characteristics.[134] The most common MSA antibody in children with inflammatory myositis is directed against a doublet protein, p155/140.[7] Specific skin sensitivity and patterns of erythema are associated with p155/140 and the children frequently follow a chronic course. This antibody is not associated with malignancy in DM/JDM and differs from the p155/140 antibody directed against TIF-1-γ (antitranscription intermediary factor-1-γ/α (TIF-1-γ/α)) found in adults with DM/PM who harbor a malignancy.[135] Anti-MJ occurs in 20–25% of patients, is directed against nuclear matrix protein 2 (NXP-2),[136] and is associated with a monocyclic course, more severe disease manifestations, more frequent hospitalizations, and calcinosis.[7] Only 2–4% of pediatric IIM patients have antibody to the aminoacyl-tRNA synthetases (Jo-1 and non-Jo-1 synthetases) present both in JPM (9%) and in overlap myositis (13%). These patients have frequent arthritis, and should be rigorously monitored for ILD, which results in the highest mortality of all the MSAs. Another marker of myositis refractory to treatment is anti-SPR (signal recognition protein), primarily found in JPM girls (18%) who are African-American.[73] Anti-M2 (NuRD) is directed against the 218/240 kDa helicase family proteins[137] and is associated with the hallmark cutaneous findings in JDM patients who are predominantly Hispanic and have a milder disease course.[1] Anti-CADM-140 (MDA-5) is common in amyopathic DM patients (i.e. skin findings but no evidence of muscle involvement) who frequently have rapidly progressive ILD, cutaneous ulcerations, and palmar pustules, and who, in Japan, have a high mortality rate.[138] Anti-SAE (small ubiquitin-like modifier activating enzyme) has been identified in children with JDM and adults with skin symptoms preceding muscle involvement and a low frequency of ILD.[139]

Myositis Associated Antibodies

Myositis associated antibodies (MAAs) are primarily found in adults and in children with overlap myositis as well as in JPM and JDM and include the anti-U1-RNP, anti-Ro, anti-PM-Scl, anti-Ku specific antibodies as well as other, less frequently identified MAAs (Table 42.6). Of note, newer, more sensitive methods of antibody detection have indicated that sera containing minute amounts of antibodies to Ro, La, RNP, and Sm can functionally activate the IFN-α pathway in children with JDM,[140] suggesting that these antibodies may actively participate in perpetuating disease pathology. Finally, a large group of children are negative for known MSA or MAA antibodies, and may be either ANA positive or ANA negative. In addition, myositis antibody negative sera may contain precipitin bands against as-yet-unidentified proteins.[141]

Type 1 Interferons

Current concepts in this fast-moving field concerning the driving force of the pediatric inflammatory myopathies center on the massively upregulated type1 interferon response, first identified in muscle of untreated children with active symptoms of JDM of all disease durations as similar to a massive antiviral response[25]; for example MXA (myxovirus A) was upregulated 96-fold (Figure 42.8A). Further study showed the disease duration did impact on many other aspects of gene expression, including vascular remodeling (Figure 42.8B). This type 1 driven interferon response was confirmed by others, both in children and in muscle from adults with DM and PM, but not in adults with IBM.[142,143] There are at least three major types of interferons: type 1 (interferon-α and interferon-β), type 2 (IFN-γ), and type 3, including interferon-λ—which may antagonize IFN-α activity.[144] Recent evidence in adults with DM has identified IFN-β as a strong component of their inflammatory response.[145] Peripheral blood from children and adults with DM was tested which displayed specific upregulation of type 1 interferon induced genes and their proteins.[143,146] However, not all the upregulated genes in the untreated *muscle* were expressed in mRNA from the same child with JDM's *peripheral blood mononuclear cells* (PBMCs) obtained at diagnosis.[147] MxA mRNA expression in JDM PBMC obtained at the initial clinic visit was elevated compared to controls and was positively correlated with Disease Activity Score (DAS) for muscle, but not with DAS for skin, suggesting that damage to skin and muscle in JDM may each have a discrete pathophysiology. Decrease in muscle symptoms was associated with a decrease in PBMC MxA mRNA expression.[147] As type 1 interferons induce many genes including *MXA*, a phase I trial of a monoclonal antibody to type 1 interferon-α was executed, which documented both improved patient symptoms and diminished expression of type 1 interferon genes.[148] The role of TNFα is presented in Figure 42.9. A proposed scheme that integrates some of the environmental and genetic factors in the development of JDM is presented in Figure 42.10.

T Cell Immunity

Children with active untreated JDM may be lymphopenic,[10] and the duration of untreated disease may influence the lymphocyte phenotype. Studies of the cellular composition in the muscle of children with JDM identified CD4, CD8 positive cells and macrophages as well as mast cells—which were more prominent in matched skin samples.[130] Perifasicular atrophy, one of the hallmarks of juvenile dermatomyositis, is characterized by infiltration of CD4 + T cells (dendritic cells) as well as the destruction of micro capillaries which contribute to muscle cell death. The native CD4 + T cells as well as plasmacytoid dendritic cells, which are also CD4 positive, are active producers of IFN-α (see the previous discussion). A possible local source of IFN-β may be immature muscle cells, which appear to be the target of activated DCs expressing IL-1 and IL-23, implicated in T cell polarization. In turn, the local production of IFN-β after TLR-3 activation in the presence of the Th1 cytokine IFN-γ may explain HLA class I overexpression by muscle fibers themselves—a hallmark of inflammatory myopathy.[149]

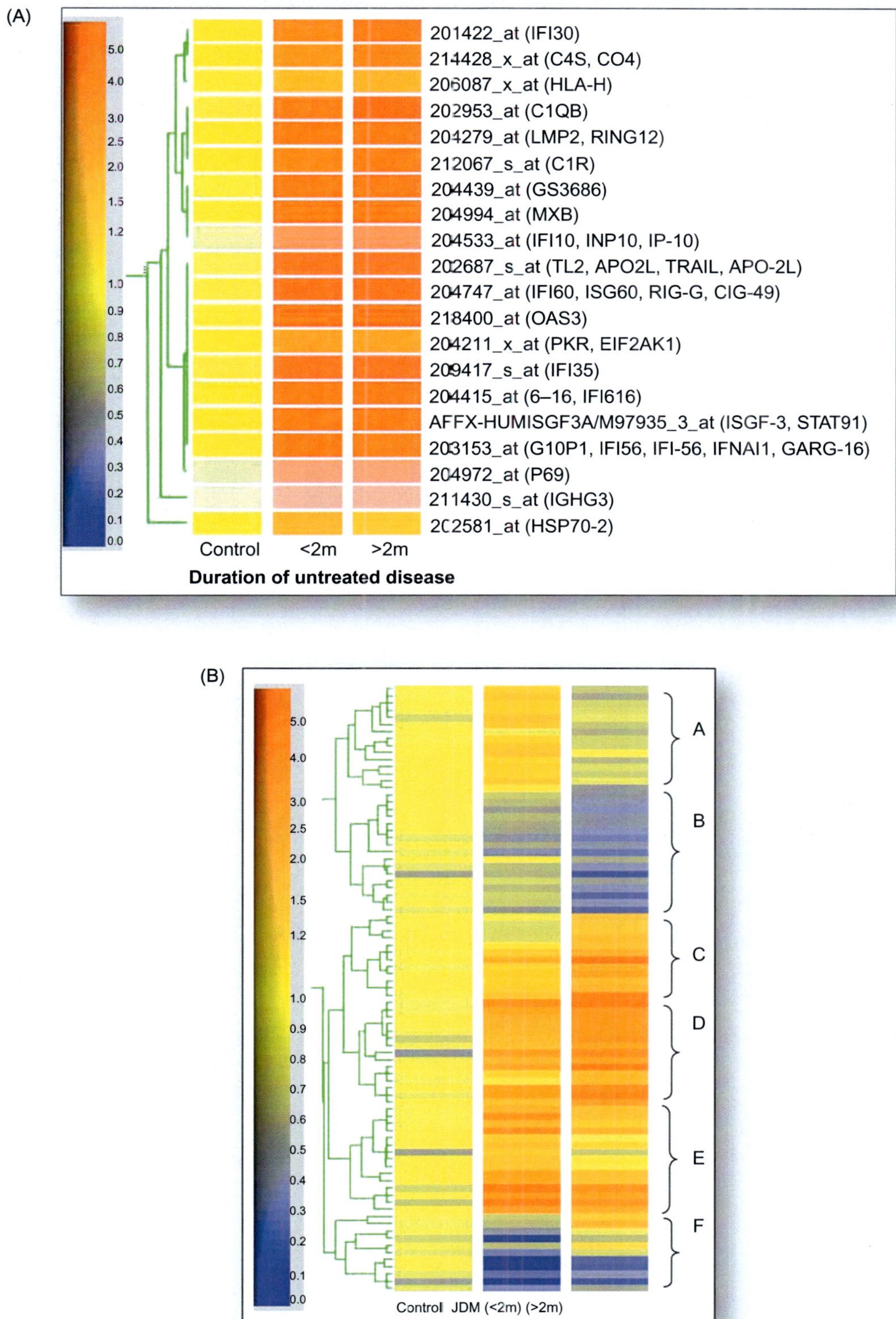

FIGURE 42.8 Gene expression (GE) profile heat maps for type 1 interferon induced genes, comparing data from MRI directed muscle biopsies from (A) untreated children with definite symptoms JDM of 2 months or less duration, with (B) GE profiles from muscle biopsies of untreated children *(Continued)*

with JDM symptoms > 2 months' duration. **B.** Gene tree created by hierarchical clustering based on gene expression profiles. Upregulated genes were clustered into clusters C, D, and F, while downregulated genes were clustered into clusters A, B, and E, in girls with long (≥2 mo) vs. short (<2 mo) duration of active disease. The color codes represent the ratio between each of the JDM groups compared with the age- and sex-matched control samples. The complete list of genes in clusters A–F can be found in supplemental table 1 of Chen et al.[359] Genes involved in vasculature and extracellular matrix remodeling, which is part of angiogenesis, were differentially expressed between patients with long and short duration of active disease. Interestingly, most of these genes cluster in cluster F, in which genes were downregulated more in patients with short duration of the disease (<2 mo). This finding suggests preferential misregulation of the expression of these specific genes in the early phase of the inflammatory disease process. *Reproduced with permission from Chen et al. (2008).*[359]

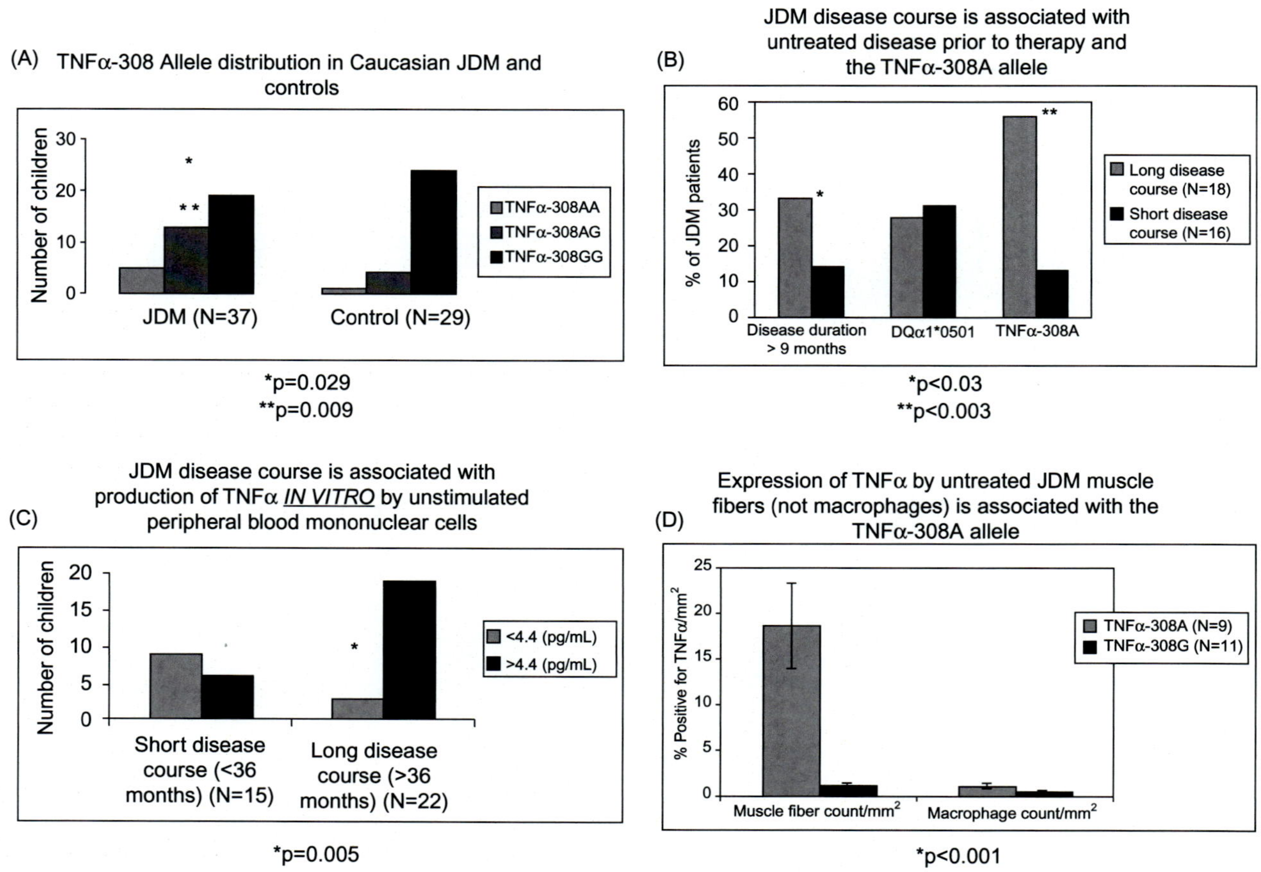

FIGURE 42.9 The association of a long duration of untreated disease and the TNFα-308A allele with a chronic disease course in white children with JDM. **(A)** Increased frequency of the TNFα-308A allele in white children with JDM compared with controls. **(B)** Both a long time period of untreated disease (> 9 months before therapy given) and the TNFα-308A allele contribute to a subsequent chronic disease course in JDM. The DQA1*0501 allele, while increased in children with JDM, appears to be associated with disease susceptibility, not disease chronicity. **(C)** A chronic disease course is associated with increased production of TNFα by peripheral blood mononuclear cells. **(D)** The muscle fiber expression of TNFα on histochemistry in the diagnostic muscle biopsies from untreated JDM is increased in children with the TNFα-308 A allele. Surprisingly, more muscle fibers than macrophages are positive for TNFα, suggesting that this may be an important mechanism of localization of tissue damage. *Reproduced with permission from Pachman et al. (2001).*[360]

Complement

A few studies that included children demonstrated complement activation: the membrane attack complex was localized to the intramuscular microvasculature in 10 of 12 patients[150] and was correlated with the duration of the clinical disease.[151] Complement components and immune complexes appear to participate in the pathophysiology of JDM with evidence of complement activation (C3d) accompanied by increased levels of fibrinopeptide A and vWF:Ag,[152] which are synthesized by and released from damaged endothelial cells[153] (see the following). In children with JDM, decreased levels of C4 protein are directly associated with fewer than normal *4A gene copy*

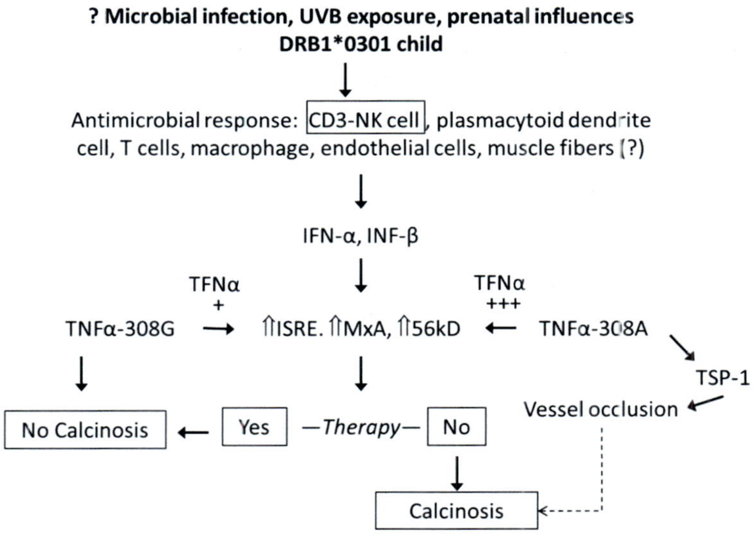

FIGURE 42.10 A proposed scheme for the pathophysiology of JDM. The Major Histocompatibility Locus, DR3 *0301 may serve as a disease susceptibility locus for microbial or other antigenic interactions, while the TNFα-308A allele is associated with increased production of TSP-1 which may play a role in vessel occlusion, as well as a chronic disease course which includes increased pathological calcifications.

number,[154] apparently in linkage disequilibrium with the "autoimmune locus" identified by the GWAS studies.[53]

Cytokines and Chemokines

Investigation of this fast-changing field documents the complex interrelationships of the cytokine network.[155] Many of the type 1 interferon inducible cytokine genes proceed through the NFkB pathway and are pro-inflammatory in nature. A longitudinal investigation of mRNA from adults with DM for analysis of the cytokine-induced gene expression for the signaling pathways of type 1 interferon, TNFα, IL-1β, granulocyte-monocyte colony-stimulating factor, IL-10, and IL-13 demonstrated only a type 1 interferon signature score in 21/24 patients.[146] Similarly, the analysis of children with JDM (MSA specificity not given), showed upregulation of type 1 interferon induced cytokines[156]; analysis of the cytokine pattern was presented as a potential biomarker for disease activity.[157] There is an increased frequency of the A allele in the TNFα-308 promoter region compared with age-matched controls (Figure 42.9A), which is associated with increased production of TNFα protein by peripheral blood mononuclear cells[26] as well as the muscle fibers themselves in untreated children with JDM.[158]

Neopterin

Neopterin is an indicator of macrophage activation and is increased in the serum of untreated children with JDM.[159] Neopterin, a member of the pteridine family, is derived from guanosine triphosphate via guanosine triphosphate cyclohydrolase and is released from macrophages as a consequence of T-cell-dependent interactions[160] involving interferon-γ (IFN-γ). Neopterin levels correlated with the clinical disease activity score in more than 65% of cases.[161] The neopterin levels were not associated with peripheral blood mononuclear cell production of IFN-γ *in vitro* but with the number of macrophages in the muscle biopsy of the child with JDM; they both decrease in response to appropriate therapy. Measurement of urinary neopterin (first voided specimen) showed a good correlation with disease activity.[162]

Acute Phase Reactants

In uncomplicated JM, the usual indicators of an acute-phase reaction, ESR and C-reactive protein, are often within normal limits, even in those children with ILD. However, ESR and C-reactive protein may be elevated in pediatric patients with severe active disease or infected sites of calcinosis, as well as in children with overlap syndromes, where these acute phase reactants are associated arthritis. Of note, if the child is given intravenous IgG therapy, an elevated ESR may be observed.[163] The lymphopenia is not commonly accompanied by an abnormal platelet count, although mild microcytic anemia may be present.

Von Willebrand Factor Antigen

A sensitive indicator of vasculopathy, an elevated von Willebrand factor antigen (vWF:Ag)[164,165] may precede a disease flare (when muscle enzyme data are normal), or it may remain elevated after the enzymes have normalized.[152] Of note, the normal range for this antigen is blood group dependent: people with blood group "O" have lower normal ranges (36–157%) than those associated with other blood group antigens, such as blood group B, for which the normal range is 57% to 241%. In about 65% of children with JDM, levels of vWF:Ag are highly associated with a clinical score of disease activity (DAS).

These clinical indicators of disease activity—the MRI findings, neopterin, and vWF:Ag levels—are easily available, but are imperfect guides to therapy.[166] It is anticipated that accurate measurements of pro-inflammatory immune markers will help characterize the severity of the immunologically mediated inflammatory process, and become more accessible in the future.

Bone Health

Vitamin D Levels

The few studies reporting data from children with JDM at diagnosis comment that the levels of 25-hydroxy vitamin D are often below normal ranges.[167,168] These low values are obtained *before* the children are strongly urged to limit sun exposure, for UVB will exacerbate not only the JDM rash on the exposed skin, but may trigger a full-blown flare in the inflammatory myopathy. It is important to obtain a vitamin D blood level at diagnosis so that appropriate supplementation with 2- or 3-hydroxy-vitamin D can begin concomitantly with the start of corticosteroids, along with supplemental doses of calcium.[169]

Bone Mineral Density

Decreased bone density, frequent in untreated JDM, is measured by DXA, using a "Z score" normalized for age-matched healthy children, and is associated with a longer duration of untreated disease,[170] depressed serum osteocalcin, and 25-hydroxy vitamin D insufficiency.[167,171] This constellation of events places the child at risk for fracture prior to the initiation of glucocorticoid therapy, and this is escalated by the use of corticosteroids. Of note, untreated children with JDM have an elevated RANKL:OPG ratio at the time of diagnosis, resulting in expansion of the number of osteoclasts and activation of the bone resorptive function.[170] This may lead to suppression of normal bone mineral accretion and a subsequent reduction in the lumbar spine bone mineral density (BMD) Z score. Therefore, children with a longer duration of untreated JDM may have lower than normal lumbar spine BMD Z scores. This risk is further augmented by steroid treatment, which impairs calcium absorption, and the child's relative decrease in mobility because of weakness, which promotes calcium mobilization from bone.

COURSE AND THERAPY

Disease Course

The outcome of JDM has greatly improved since the 1960s when one third of the children died, one third were crippled, and one third recovered.[12] Several types of disease courses have been described—monocyclic, recurrent, and continuous[172]—but it may be that the categorization is, in fact, more dichotomous: a single episode of symptoms vs. disease chronicity. The frequency of calcinosis has decreased from <60% of cases[173] to about 20–30% at diagnosis (Box 42.1). Estimates of the average disease duration of JDM vary from 1.5 years to persistent disease over the child's lifespan, and this remains a critical therapeutic problem, when the child appears to have recovered from the initial acute onset of the disease. Follow-up studies of children with myositis cannot yet reflect our current aggressive therapy at diagnosis. For example, one study (mean follow-up, 7.2 years) of 65 children documented calcinosis in 34% (unrelated to gender, therapy, or disease course); 35% were still taking medication 3 or more years after diagnosis; 23% had weakness; 40% still had rash.[174] The concept that a persistent rash predisposes the child to pathologic calcifications was confirmed by other investigators.[96] Late disease recurrence after years of apparent remission has been reported,[175] suggesting the need for periodic monitoring with more sensitive indicators of disease activity. It is difficult to predict outcome at the onset of illness, although the magnitude of the initial serum levels of CK appeared to be a direct correlate of disease severity at the time of observation,[176] and the MSA characteristics may provide some indication of course.[7] Several investigators noted that prognosis is directly related to the degree of vascular involvement.[173,177] Comparison of Korean adults with DM and children with JDM showed that the children had more involvement of neck musculature, had calcifications, and had a better outcome. [178] It is not uncommon for children with JDM to develop another autoimmune disease, such as psoriasis.[179]

Evaluation Tools for Disease Activity and Chronicity

Until recently, there has been little discussion of the development of methods that could distinguish the features of an active inflammatory process from its residue. Although it is clinically held that the active inflammatory process completely resolves in about 33% of the children with JPM and JDM, there are few data, either clinical or histologic, to support this conjecture. One of the operational problems is the lack of standardized tools to assess the child with myopathy before and after therapy, but a funded study is underway to develop an international consensus among experts across the disciplines to evaluate, establish, and validate the criteria for improvement or regression of disease. Furthermore, the impact of the age of the child at disease onset on both the pathophysiology of the disease as well as the degree of potential reversibility has not been defined. In order to address these methodological gaps, several groups have developed criteria for evaluation of the efficacy of therapeutic intervention. A measure of well-being and function, the Child's Health Assessment Questionnaire (CHAQ), initially developed for children with JRA, was tested in children with JDM and identified the extent of disease-associated disability.[180] The CHAQ was found to be reliable in identifying moderate to severe disability in children with JDM.[181] A second larger investigation (115 patients) confirmed these findings and documented the "floor effect" of the CHAQ, that is, that the CHAQ does not accurately determine JDM disease activity (skin involvement, vasculitis) in children who had relatively good muscle function.[174] Most children with inflammatory myopathy have an initial gratifying response to corticosteroids, but these children may continue to have symptoms not related to muscle strength, such as persistence of rash and evidence of vasculopathy. In another study, global assessment, comparing a Likert scale and a visual analogue scale, was used by physician, parent, and patient to determine disease activity and damage indices in 117 children with inflammatory myopathies.[182] These measures appear to be more sensitive to change earlier in the child's illness because there is more rapid change in physical findings at that time, but they are less useful later in the disease course.[182] An evaluation of muscle function, the CMAS, was recently developed and also includes tests of muscle strength as well as tests of muscle endurance for children 4 years of age and older.[181] The total score from this test was positively associated also with assessment of the impact of the disease, the manual muscle test, and the serum CK level.[181] Of note, the CMAS has evolved with minor modifications,[67] and is a useful and fairly reproducible tool. Inasmuch as 25% of children with JDM are 4 years of age or younger at diagnosis,[29] it was important to establish that even healthy children age 4 cannot perform three of the effort dependent maneuvers[68] which are needed to document that the younger child with JDM has fully recovered. A validated method for the rapid assessment of disease activity (the disease activity score, DAS) was developed combining the physician evaluation of the extent and severity of involvement of skin, muscle strength, and function with an estimate of small vessel inflammation as reflected in the palate and nailfold capillaries, and can be used in a child with JDM of any age.[183] This global DAS has been adopted as one of the methods on an international basis to assess disease activity in a child with JDM before and after therapy in addition to the CMAS and manual muscle test.[184] A battery of validated tools to evaluate disease chronicity has been developed and is now commonly employed.[185] There is agreement that chronic damage occurs in some children who have been diagnosed with JDM, for example, persistent painless joint contractures associated with fibrosis; the spectrum of partial lipoatrophy, lipodystrophy, and eventual insulin-resistant diabetes; hypertension; and cardiac disease. Less clear is the status of pathologic calcifications, which can resolve on a rare occasion without therapy or after suppression of chronic inflammation. It is anticipated that with the use of targeted assessment tools, more specific information will be obtained to evaluate the efficacy of not only our current therapies but also new interventions that will be more specifically directed at the pathophysiology of the disease.

Therapy

Development of Consensus Driven Guidelines for Classification and Therapy

Specific investigation of the efficacy of the type, duration, and route of medication administered to children with JDM and the other less common inflammatory myopathies is a continuous process. Several consensus meetings developed guidelines for treatment: one for moderately severe JDM,[186] another for moderate JDM.[187] A concerted assessment of the use of corticosteroids in IIM, as there is in asthma, has not yet been organized.[188] Several factors may confound the existing data. Active vasculopathy, which is often not evaluated, may inhibit oral absorption of the drug compared with the intravenous route.[189] The age of the child may alter drug metabolism, once absorption has occurred, so that younger children may require a higher dose per square meter than their adult counterparts—which has been proven for methotrexate in children,[190] and may apply to other forms of medication. Specific recommendations for therapy have to be considered in this context but are limited because of the lack of uniform assessment of age-dependent patient response during therapy. In addition, although there are some clues as to immunological specificity, these indicators of immunologic activity have not been widely tested to confirm that they have normalized at some point after therapy has been initiated.

Oral Corticosteroids

In general, the treatment of mild uncomplicated JDM consisted of oral prednisone (1 to 2 mg/kg) in conjunction with gastric protection (Table 42.8).[1] The child's linear growth is arrested temporarily if the oral dosage of steroids exceeds 4 mg/m^2. Stunted growth combined with increased weight gain and IIM-associated decreased mobility proves a challenge for the body-image conscious child. For the adolescent, the increased weight gain and the development of steroid-induced acne when taking oral prednisone frequently pose problems with compliance, which can be circumvented to some extent by the intravenous administration of methylprednisolone. In children with microvascular damage which also occurs in the

TABLE 42.8 Medications Used to Treat JDM

Medication	Recommended Regimen	Support & Comments	References
Primary therapy[a]			
Prednisone	Dosage starting at 2 mg/kg/d po; maximum 60 mg/d gradually tapered by 2–4-wk intervals, with a minimum recommended duration of 10–12 mo	Daily prednisone considered the mainstay of treatment of JDM and other JIIM, throughout the illness course, and has led to dramatic improvements in mortality (now less than 2% for JDM) and functional disability	186,187,323
IVMP	30 mg/kg/d IV; maximum 1 g, 3 d in a row at diagnosis, then 2–3 times weekly, guided by laboratory data	Used as an adjunct therapy with daily corticosteroids during the initial treatment of moderate to severe JDM; provides better absorption of corticosteroid, aiding in rapid improvement of disease activity; also valuable for acute or life-threatening manifestations, such as severe dysphagia, interstitial lung disease, myocarditis; may be used later in the illness course for flares or persistent activity	186,189,323,324
Methotrexate	1 mg/kg or 15 mg/m^2 once weekly SQ; max 40 mg weekly	Most commonly used adjunct therapy with steroids; used as first-line therapy for moderate to severe disease; efficacy in combination with steroids is superior to prednisone alone, including a shorter time to inactive disease; methotrexate used in conjunction with steroids from time of diagnosis results in fewer cataracts, less weight gain, and less reduction in growth velocity, and a lower cumulative dose of prednisone	186,325,326
IVIg	2 g/kg IV; max 70 g, every 2 wks × 3, then monthly	Adjunct therapy for moderate to severe disease; usually combined with steroids and methotrexate; used both as an agent for initial treatment of moderately severe disease and for refractory cases; rapid onset of action and lack of immunosuppression are potential benefits; may be helpful in patients with acute, potentially life-threatening manifestations (dysphagia, marked weakness, ulceration), and for treatment of skin disease and calcinosis	186,203,323
Cyclosporine	2.5–7.5 mg/kg/d po divided bid (max 150 mg bid) to maintain trough blood level ≤ 300 ng/mL	Preferred agent in European centers compared with other continents; used for initial treatment but more so for refractory patients; efficacy in combination with steroids is superior to prednisone alone, including a shorter time to inactive disease; however, side effects are greater than with methotrexate	323,325
Therapies for refractory disease[b]			
Mycophenolate mofetil	20–40 mg/kg/d po divided bid or 800–1350 mg/m^2/d	Used to treat severe or refractory disease, with improvements in muscle strength and skin activity and reduction in corticosteroid dose	189,206

(Continued)

TABLE 42.8 (Continued)

Medication	Recommended Regimen	Support & Comments	References
Tacrolimus	Initial dosage of 0.04–0.08 mg/kg/d po divided bid (max 2–3 mg bid), with gradual increase to 0.1–0.25 mg/kg/d divided bid (max 3–6 mg bid) to maintain trough level of 5–10 ng/mL	Used for severe or treatment-refractory disease, particularly with severe weakness, interstitial lung disease, or with anti-synthetase or anti-SRP autoantibodies	209,327
Cyclophosphamide	500–1000 mg/m^2/mo IV	Used for life-threatening disease activity, severe dysphagia, interstitial lung disease, gastrointestinal or cutaneous ulcerations or other forms of vasculitis, or in patients with severe disease activity that is treatment refractory; improvements in muscle strength and function, skin, extramuscular activity, and calcinosis reported	328,329
Rituximab	575 mg/m^2 or 750 mg/m^2 (max 1 g) per dose for 2 weekly infusions	Used for refractory patients who failed to improve on other DMARDs; approximately 80% of treatment-refractory patients have improved in a randomized controlled trial and in open-label studies; onset of improvement lags for about 3 mo after infusion; efficacy more toward muscle rather than skin disease activity; helpful in patients with myositis autoantibodies	217,330–333
Anti-TNFα (infliximab, etanercept)	Infliximab: 3–6 mg/kg IV at weeks 0, 2, 6 and then every 4 wks	Infliximab has best success for JDM in open-label study and small case series of treatment-refractory patients; reported improvement in muscle strength and physical function, muscle enzyme levels, skin rashes, and global activity in >70% of treatment-refractory patients; calcinosis improved in 46%; mixed experience in adult DM/PM, and some reports of disease worsening with anti-TNFα therapy, which may be related to induction of the interferon signature	211,334

[a]*Per the Childhood Arthritis and Rheumatology Research Alliance treatment consensus for moderately severe disease, dietary modifications and dietary consultation if possible (to minimize prednisone weight gain), photoprotective measures, and folic acid are adjunctive primary therapies. Calcium and vitamin D supplementation and referral to physical therapy and/or occupational therapy are also recommended.*[186]
[b]*Combinations of several primary therapies and therapies for refractory disease may also be used.*
Abbreviations: DMARD, disease-modifying anti-rheumatic drug; IV, intravenous; IVIg, intravenous immunoglobulin; IVMP, intravenous methylprednisolone; max, maximum; SQ, subcutaneous; TNFα, tumor necrosis factor α.
Modified from Rider et al. (2013), reproduced with permission.[1]

gastrointestinal tract, absorption of oral prednisone is impaired, providing the rationale for the use of high-dose intermittent methylprednisolone, as shown in Figure 42.11, in which the absorption of 50 mg of prednisolone IV vs. PO was compared in children with JDM and decreased nailfold capillary end row loops.[189]

Additional consequences of high dose oral corticosteroids include gastric distress, ulcers, hypertension, hirsutism, and emotional lability. To decrease these effects, the initial treatment of JDM usually involves a steroid sparing drug—methotrexate. To decrease the impact of corticosteroids, which impair calcium absorption, close attention should be paid to the levels of vitamin D and calcium consumed (see "Bone Health").

Methotrexate

The use of methotrexate early in the disease course at doses of 15 mg/m^2/week has reduced the morbidity of the disease and permitted lower daily steroid dosages with much less weight gain, fewer bone fractures, and improved patient compliance. In active disease, intravenous or subcutaneous administration of methotrexate ensures drug absorption, as opposed to the oral route, and

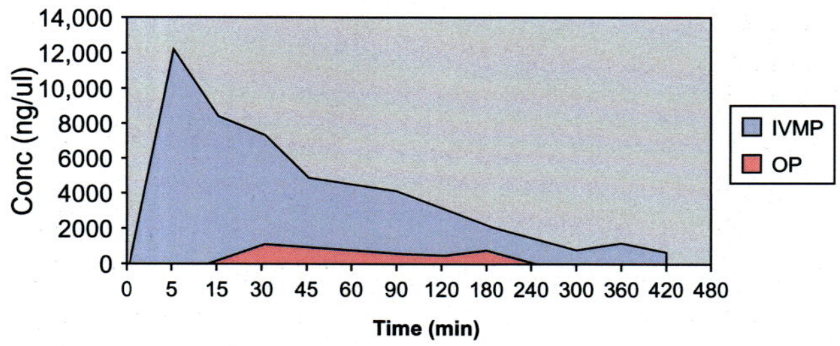

FIGURE 42.11 Prednisolone bioavailability in JDM. IVMP, intravenous methylprednisolone; OP, oral prednisolone. *Modified with permission from Rouster-Stevens et al. (2008).*[189]

the children have less nausea. Complaints of nausea can be circumvented by dividing the dose but giving the full amount in a 24-hour period. With methotrexate, the function of the liver and bone marrow must be monitored at least every 3 months to detect potential toxicity. It is particularly important to inform the adolescent with JDM who is treated with methotrexate to avoid sun exposure (increased sunburn, vitiligo) and ingestion of alcohol (liver toxicity). Loss of hair and oral ulcers, which occur in methotrexate-sensitive patients, can be reduced by daily folic acid, usually at the level of 1 mg/day but increased to 2 mg/day if needed.

High-Dose Intravenous Intermittent (Pulse) Methylprednisolone

The treatment of JDM with intravenous intermittent (pulse) methylprednisolone (IVMP) is now widely accepted,[191–194] once it was recognized that the IVMP administration had to be supplemented with lower daily doses of oral prednisone (as low as 0.5 g/kilo) to achieve disease control.[195] Children with JDM treated with an intravenous course (30 mg/kg, 1 g maximal dose) plus low-dose daily prednisone (0.5 mg/kg/day) versus a standard treatment with oral prednisolone alone had a shorter disease course with respect to persistence of rash (1.5 years compared with 3.9 years) and weakness (1.5 years versus 2.7 years).[196] Furthermore, over the 10 years of follow-up visits, they did not develop either calcinosis or spine compression fractures, although the frequency of cataracts was the same in both groups. Validation of the efficacy of the prompt use of IVMP to prevent calcifications in children with JDM has been published by several groups,[91] including our own.[197] Massive muscle damage with high levels of creatine phosphokinase (CPK) is associated with renal damage, which can be diminished by the prompt use of IV hydration, along with daily IVMP and other immunosuppression, which should continue as long as the CPK is elevated. Most protocols now also involve the use of weekly intravenous methotrexate at a starting dose of 15 mg/m^2, administered in conjunction with high-dose IVMP, especially if there is evidence of dysphagia or difficulty in handling secretions.[186] The frequency of IVMP steroid administration is determined by several factors—the level of the muscle enzymes, the vWF:Ag, the extent of disease activity in the muscle biopsy, and slope of the child's response, using the parameters that are useful for the particular patient. Low-dose oral steroids (0.5 mg/kg/day) are given in the morning on the days that IVMP is not infused. We individualize the protocol for each child and use the immunologic indicators as a guide to therapy. An intensive course of IVMP is employed until there is normalization of the child's specific laboratory data, with gradual reduction in therapy. The use of high-dose IVMP is not without adverse reactions. A prospective log kept over a 5-year period of 213 children with various types of serious rheumatic disease, given over 2622 doses, found that 22% (N = 46) had an adverse reaction, of whom 21 children (47%) had behavioral changes ranging from euphoria to emotional lability.[198] There was one case of anaphylaxis, and adverse reactions to IVMP were closely associated with a record of previous cutaneous allergic eruptions to antibiotics.[198]

Hydroxychloroquine

In children with moderate to severe skin involvement, hydroxychloroquine (HCQ), 5 mg/kg/day),[199] is given, but only if the family history of red-green color blindness is negative. The risk of retinal toxicity increases sharply toward 1% after 5 to 7 years of use, or a cumulative dose of 1000 g, of HCQ; the risk increases further with continued use of the drug.[200] Children with apparent cutaneous involvement alone (amyopathic) should be evaluated for the presence of ILD.[201]

Intravenous Immunoglobulin G (IVIg)

IgG has proven to be a very useful addition to therapy of myositis.[202] In corticosteroid resistant JDM, the use of intravenous immunoglobulin G (IVIg) at 1–2 grams/kilo (70 grams max) as noted in Table 42.8 is usually accompanied by clinical improvement.[203] In children with reactions to IVIg, the dose can be divided over consecutive days. If the child remains severely ill (sometimes despite normalization of muscle enzymes), a thorough evaluation of the immune system may help to guide therapy, for IgG may have been critically depressed by immunotherapy, resulting in frequent infections, accompanied by withholding all immunosuppression except the oral prednisone, and prolonging inflammatory disease activity. If the level of IgG is depressed below the lower limit of normal for age (or 400 mg/dL), then ***replacement doses*** of IVIg should be considered. High-dose intravenous gamma globulin (IVIg) often diminishes the intensity and the distribution of the cutaneous erythema, and improves muscle performance, especially if given early in the disease course.[204]

Treatment of Severe or Resistant IIM

This group is composed of children who present with high levels of muscle enzymes, or, as commonly occurs, their muscle symptoms have responded to the initial corticosteroid regimen, but the rash is still present.

Mycophenolate

The combination of mycophenolate (MMF) in conjunction with other modalities has been successfully used in the treatment of the refractory rashes of JDM, allowing reduction of total corticosteroid dosage.[205,206] If the adolescent girl has passed her menarche, then her urine must be tested at each visit to ensure that she is not pregnant.[1]

Calcinurin inhibitors

These drugs are very effective inhibitors of T cell function, but concomitant renal damage and hypertension are a concern.[207] Tacrolimus has been reported as effective in the treatment of refractory JDM[208] or antisynthetase related ILD,[209] while cyclosporine (CSA) has been used with success in the treatment of myositis in children and some physicians have encouraged the use of calcinurin inhibitors as first-line medications.[210] For both, blood levels are available which helps to monitor compliance/absorption of the drugs. The concurrent use of MMF as a CSA-sparing drug has been proposed in the treatment of liver transplantation to protect renal function, but has not been studied in children with myositis, where renal dysfunction appears to be less of a problem, perhaps related to the lower dosages used (trough CSA blood levels from 80 to 110).

Biologic Agents

Anti-TNFα Agents (Etanercept; Infliximab) Concerning the use of these agents to treat refractory myositis, there are champions [211] as well as naysayers.[212] With the recognition of the strong association of high levels of TNFα with disease activity,[213] in a pilot study the TNFα receptor antagonist etanercept was given to eight children with JDM who had been nonresponsive to IVMP and methotrexate therapy for 36 months or more.[214] In this trial of patients with refractory JDM, use of etanercept did not result in appreciable improvement and some children had worsening of disease. The variance in efficacy may depend on the state of the disease activity in the children; in active JDM, levels of type 1 interferon are high.[215] TNFα may act to suppress IFN-α activity, so that when type 1 interferons are elevated in active JDM, alternatives to the use of TNF inhibitors should be considered.[216]

Rituximab This monoclonal antibody against CD19 depletes B cells, and has been used in a randomized controlled trial. The results indicated that 83% of adult and juvenile myositis patients with refractory disease met the definition of improvement. The role of B cell-depleting therapies in myositis warrants further study, with consideration for a different trial design.[217] Of note, the presence of an antisynthetase and anti-Mi-2 autoantibodies, JDM subset, and lower disease damage strongly predicted clinical improvement in refractory myositis patients.[218]

Other Biologic Agents Abatacept, which targets CTLA-4, has been used successfully to treat arthritis, and was associated with fewer adverse reactions than adalumimab.[219] There are few reports of the use of abatacept to treat children with myositis.[220,221]

Cyclophosphamide

Intravenous cyclophosphamide therapy, starting at 500 mg/m^2 every 3 weeks (after adequate hydration), with Mesna for bladder protection,[222] can be considered Serum levels of IgG must be checked periodically to ensure that they are adequate; if not, replacement therapy (0.4 g/kg every 3 to 4 weeks) is needed to lessen the possibility of recurrent infections. Inasmuch as *Pneumocystis* has been reported in immunosuppressed myositis patients,[223] and is increasingly recognized as a potentially fatal complication of immunosuppression,[224] the concurrent use of trimethoprim sulfate (Bactrim) should be considered when cyclophosphamide is given.

Calcinosis

Currently, there are no proven successful therapies for long-standing calcinosis in children with inactive disease. An MRI study may reveal inflammation associated with calcinosis, permitting more aggressive medical therapy, which may result in regression (and occasionally radiologic resolution) of the calcinosis. Once the myositis is controlled, some of the calcific sites can be carefully appraised. If the active infection is suppressed, the calcinosis can be excised while the child is covered by therapeutic levels of cyclosporin A (trough level 80–110) without recurrence in the incision site, which is a common occurrence in children not taking adequate levels of T cell suppressing agent(s). Other biologic agents include abatacept and rituximab. There is one report of the concurrent use of abatacept and sodium thiosulfate for the treatment of generalized calcifications in a child with chronic JDM and calcifications,[220] and scattered case reports of the successful use of abatacept in adults.[221]

Vitamin D and Bone Health

Previous studies have documented decreased bone density and osteocalcin (an indicator of osteoblastic activity) in untreated active JDM.[171] Protection of bone health can be achieved by starting 2/3-25-OH vitamin D at the same time as the corticosteroids in all children with IIM, along with age-appropriate calcium supplementation and a calcium sufficient diet. In adolescents, who often decrease intake of calcium-containing foods, administration of vitamin D and increasing the calcium intake may aid in calcium absorption from the gastrointestinal tract during steroid therapy (which inhibits calcium absorption). This approach decreases the occurrence of one of the most serious consequences of steroid therapy, osteopenia, which can progress to spinal cord compression or other fractures. The goal is a blood level of 30–40 ng/ml.

Physical Therapy

Exercise Improves JDM Muscle Function

Combined medical treatment to suppress inflammation and physiotherapy (gentle, passive stretching) are required in the early phase of the disease, and more intensive, graded physiotherapy is effective at later stages in the disease, once the inflammation has abated. Muscle metabolism is impaired by the inflammation in JDM,[225] and the children display early fatigue.[226] New data shows that aerobic training is useful to regain muscle function in JDM patients[183] who had impairment of muscle performance,[184] which increases with the duration of chronic symptoms.[185] Nevertheless, return to high-impact sports activities or use of rollercoasters, trampolines, etc. should only be considered once the disease activity is controlled and there is evidence of good/improvement in bone density as well as full recovery of neck flexor stability.

Physical Therapy for Flexion Contractures

An experienced pediatric physical therapist is optimal for children with JDM, and flexion contractures often have decreased bone density and are easily fractured. Serial casting has been successful, similarly to children with cerebral palsy in children with flexion contractures with or without calcinosis.[227] Tendon lengthening and excision of the binding calcified tissue under the cover of documented therapeutic levels of CSA (80–110 trough level) may succeed in restoring range of motion to a compromised child or adolescent.

Psychological Aspects of Juvenile Myositis

Although there is scarce documentation[228,229] of the psychological factors that dominate the child's response to chronic illness, recognition of their importance is critical in achieving a good outcome.[230] The mood swings that occur both as a result of the level of maturation and the administration of corticosteroids can be devastating if not recognized and appropriately treated. It is particularly important to allow the adolescent to participate in decisions affecting therapy and to encourage age-appropriate independence to establish an identity and to master the tools needed for financial independence. For the adolescent who may want to smoke cigarettes, further impairment of lung function is a very serious issue and merits straightforward discussion, as does other high-risk behavior. It is important to treat the adolescent as a separate and competent person: we send the parents out of the room over the age of 12, but invite them back for a summary of medical findings at the end of the visit. We test our female JDM for anticardiolipin antibodies that may guide the selection of appropriate birth control measures, for pregnancy in the face of active disease has a more guarded outcome.[83] The risks of alcohol and other recreational drugs is part of our briefing of both the child with myositis and their family. Quality-of-life studies of children with JDM have shown, not surprisingly, that there is a strong impact of physical symptoms on the patient's perceptions both as a child[231] and as an adult who has had a history of JDM.[232]

Overall, the outcome of JDM has markedly improved, with a reduction in morbidity and mortality. Evidence has been presented of cardiovascular compromise[101,233] but few data are available concerning long-term follow-up. The limited available evidence suggests that children do well in adulthood on socioeconomic and functional assessment.[11]

OTHER LESS COMMON FORMS OF INFLAMMATORY MYOPATHY

Myositis Associated with Connective Tissue Disease

Inflammation in the muscle with localized tenderness and weakness can be a component of most of the connective tissue diseases that have been identified in children. Only 6–12% of children with chronic inflammatory myositis meet the diagnostic criteria for a connective tissue disease other than JDM/PM (Table 42.9) compared with almost 20% of adults. The most frequently observed form is mixed connective tissue disease, in which the myositis can be the presenting complaint with either the concurrent or later development of a high-titer speckled pattern on ANA testing (titer greater than 1:1280) with specificity for U1-RNP, U2-RNP, or U5-RNP.[234] Primary Raynaud's phenomenon is often a presenting complaint accompanied by nailfold capillary abnormalities.[235] Myositis develops in 20% to 90% of children with this antibody,[236] and proximal muscle weakness with a DM-like rash can be identified in about 15% of cases. Cardiac dysfunction has

TABLE 42.9 Spectrum of Juvenile Idiopathic Inflammatory Myopathies

Clinicopathologic Phenotype	Frequency in JIIM (%)	Associations and Comments	References
JDM	85	Most common form of JIIM in children with median age at onset of 7.5 y and female predominance; characterized by Gottron's papules or heliotrope rash, and CD41 T cells and dendritic cells in perivascular distribution in muscle with a type I interferon signature; approximately 50–60% of patients have a chronic course, and 24–40% have a monocyclic course of illness; up to 25% have skin or gastrointestinal ulcerations; calcinosis seen in 25–40% and lipodystrophy in 10%, with a spectrum of other associated manifestations.	307,335–338
Overlap myositis	6–12	Patients meet criteria for another autoimmune disease, most frequently systemic sclerosis, systemic lupus erythematosus, juvenile idiopathic arthritis, localized scleroderma, Sjögren's syndrome, or type I diabetes; Raynaud phenomenon, interstitial lung disease, arthritis, sclerodactyly, and calcinosis are more frequent in this subgroup; mortality is high, often related to lung disease.	141,307,310,335
JPM	4–8	Defined by the absence of the characteristic DM rashes; pathogenesis involves CD81 endomysial infiltration in muscle; patients tend to be older (preteen or teenage), have higher CK levels, and more severe illness onset, with frequent falling episodes as a sign of distal weakness, myalgias, more frequent cardiac involvement, and intermediate mortality.	307,335,339
Amyopathic or hypomyopathic DM	1	DM skin rashes, either without muscle weakness or with subclinical muscle weakness detected only by additional testing (e.g. elevated serum muscle enzymes or abnormal EMG, muscle biopsy, or MRI); approximately 25% of patients develop JDM within 3 y; calcinosis has been reported in only 4%, and cutaneous ulcerations and interstitial lung disease are very rare; none of the children has malignancy, in contrast to adults with clinically amyopathic DM whereby these associated features are common.	340–342
IBM	?	Characterized by slowly progressive proximal and distal weakness, low serum CK level, and rimmed vacuoles on trichrome stain of muscle biopsy, with poor responses to immunosuppressive therapy in people >age 45. IBM is a major clinical subgroup in adults, suggesting a degenerative pathogenesis and/or environmental triggers that are present only in older adults or that have long latency.	256,323,343
Cancer-associated myositis	<1	Malignancy in association with JIIM is so uncommon that routine screening for cancer is generally not warranted, in contrast to adults with DM and PM whereby a malignancy screen is routinely performed because of a paraneoplastic association within 2 y of myositis onset; atypical illness features, such as prominent adenopathy, hepatosplenomegaly, palpable masses, and atypical rashes, have been proposed as part of a profile of indicators that may indicate an underlying associated malignancy that has been observed in most of the 17 reported cases of cancer associated with JIIM; most common cancers in children include lymphoma, leukemia, and solid organ tumors.	298,301,335,344,345
Focal myositis	1–2	Characterized by a subacute, often painful intramuscular swelling or mass in a single muscle group, sometimes preceded by trauma, with histologic features of myositis; most common sites of involvement are the thigh muscles, gastrocnemius, brachialis, and sternocleidomastoid; lesions often spontaneously resolve, but immunosuppressive therapy or surgical resection may be beneficial in refractory cases.	253,254,336

(Continued)

TABLE 42.9 (Continued)

Clinicopathologic Phenotype	Frequency in JIIM (%)	Associations and Comments	References
Orbital myositis	1−2	A focal myositis of the rectus muscles, which presents with ocular pain, headache, diplopia, proptosis, ocular injection, limited ocular mobility, and periorbital edema; children may have atypical presentations, including emesis, anorexia, lethargy, abdominal pain, and weight loss; inflammation of an orbital muscle must be distinguished clinically from thyroid-associated orbitopathy, but the differential diagnosis also includes rhabdomyosarcoma, lymphoproliferative disease, IgG4-related disease, or histiocytosis X; pediatric cases also differ from their adult counterparts in that they are more likely to be bilateral on presentation or associated with uveitis, disc edema, or eosinophilia; orbital myositis may be seen following an infection, such as group A streptococcus, or in the setting of other autoimmune conditions, including inflammatory bowel disease, linear scleroderma, or Kawasaki disease.	263,264,336,346,347
Immune-mediated necrotizing myopathy	1	PM or DM in which the muscle biopsy demonstrates prominent muscle necrosis and little inflammation; thought to be mediated by autoantibodies, including anti-signal recognition particle or anti-3-hydroxy-3-methylglutaryl-coenzyme A reductase.	73,278
Macrophagic myofasciitis	<1	An inflammatory myopathy thought to be related to aluminum-adjuvant vaccines; presents with prominent macrophagic infiltrates and aluminum detectable in the muscle; symptoms include myalgias, arthralgias, muscle weakness, chronic fatigue, and fever; in children, additional features include hypotonia, motor delay with an inability to stand or walk, and failure to thrive; typically, vaccination predates the diagnosis by 2−12 mo, and childhood-onset cases are generally in the first 5 y.	348−351
Eosinophilic myositis	<1	Prominent eosinophilic infiltrates on muscle biopsy, often accompanied by peripheral eosinophilia; may be idiopathic and true polymyositis or focal in nature; may also be drug induced, related to a parasitic infection, or to an underlying hypereosinophilic syndrome; in children, it is particularly important to exclude muscular dystrophies, including calpainopathy, Becker's dystrophy, and gamma sarcoglycanopathy (LGMD2 C), which have been reported to present with eosinophilic myositis.	352−355
Granulomatous myositis	<1	Granulomas prominent in the muscle, which may be idiopathic or seen in association with Wegener vasculitis, sarcoidosis, or tuberculosis.	356,357
Proliferative myositis	Rare	A very rare benign pseudosarcomatous lesion of skeletal muscle; may be diagnosed with fine-needle aspiration cytology, which shows hypercellular and polymorphic lesional cells and distinctive ganglion cell-like cells; most lesions resolve spontaneously within 4 mo; surgical resection is used for the remaining lesions.	279,358
Graft-versus-host myositis	Rare	Polymyositis with weakness, myalgias, and elevated CK, which rarely occurs as part of graft-versus-host disease following stem cell and bone marrow transplantation; responds to prednisone and cyclosporine.	302

Abbreviations: CK, creatine kinase; DM, dermatomyositis; EMG, electromyogram; Ig, immunoglobulin; MRI, magnetic resonance imaging; PM, polymyositis.
Reproduced with permission from Rider et al. (2013)[1] with minor modifications.

been well characterized.[237] Like adults, children may have myositis when they are given the primary diagnosis of SLE, systemic sclerosis, JRA, or Sjögren's syndrome.[234] Alternatively, myositis may occur in the active phase of other connective tissue diseases in the *absence of antibody to U1-RNP*. For example, children with SLE may have an elevated ALT, AST, or LDH on the basis of liver involvement (lupoid hepatitis), but they may also demonstrate myalgia and weakness independent of liver damage. Patients with antisynthetase antibodies, most frequently anti-Jo-1, were found to have a better-than-expected response to infusions of rituximab.[218]

Inflamed muscles may occur in infants and small children with systemic onset JRA, a negative ANA, and no demonstrable antibody to U1-RNP, with or without obvious joint involvement. For example, the child may appear to be inconsolable when held, because pressure on the inflamed muscles can be painful; the myositis associated with JRA can be confirmed by MRI.[238,239] Rituximab is an additional potent therapeutic treatment option for MCTD patients with autoimmune thrombocytopenia refractory to conventional immunosuppressive treatment; the best response may be expected in patients with high titers of anti-platelet antibodies at baseline.[240] In adults with MCTD, myocarditis and coronary artery disease as well as involvement of the small vessels of the myocardium are frequently reported; unfavorable outcomes are associated with pericarditis and pulmonary hypertension.[234]

Myositis Associated with Other Diseases

Kawasaki disease occurs more frequently in young boys, with a mean age of onset of 1.5 years in Japanese children and a slightly older age range in the USA This systemic vasculitis is most symptomatic in the first 2 weeks of illness and the children can complain of both proximal and distal muscle pain as well as dysphagia and dysphonia. They may have a painful arthritis affecting both small and large joints, which has been reported to respond variably to TNFα inhibitors.[241] Myositis has been reported in Behçet's syndrome in both adults[242] and children,[243] and in children with polyarteritis nodosa[244] or Wegener's granulomatosis.[245] Other autoimmune diseases may display a component of myositis, including celiac disease[246] and idiopathic thrombocytopenic purpura.[247] With respect to psoriasis, the symptoms may either precede or follow by several years the symptoms of classical JDM.[179] Myositis is a predominant and early symptom in children with another type of antibody, Pm/Scl or myositis scleroderma overlap, sometimes called scleromyositis.[248] Children with this antibody often present with a severe proximal myositis and severe neck flexor weakness. They may have prominent livedo reticularis, vascular instability, and progressive Raynaud's syndrome accompanied by increasing capillary dropout documented by nailfold capillaroscopy.[249] Children with myositis and Pm/Scl often have good prognosis with a prompt response to corticosteroids, although calcinosis and cuticular overgrowth is a component of this variant. However, we have observed that small joint involvement, with loss of both metacarpal and phalangeal flexion, is common and persistent, accompanied by decreased peripheral subcutaneous tissue. Distal esophageal dysmotility may be manifest by "heartburn" or inability to swallow large pieces of dry or solid food. These children may have cardiac conduction defects with debilitating dysfunction, suggesting that this antibody may be associated with an unremitting course in some. In some children, lung involvement with pulmonary fibrosis is documented by pulmonary function with decreased diffusion of CO_2. Children with Pm/Scl may develop full blown symptoms of scleroderma, and should be monitored for ILD.[250]

Laboratory Testing

In our Myositis Center, laboratory testing at the diagnosis of an inflammatory myopathy includes an ANA and screening for other connective tissue diseases. The ANA is often positive in a titer $\leq$ 1:320 in a speckled pattern (the most frequent pattern identified in the sera from children with myositis), but a proportion of ANAs are negative or homogeneous. Each child is tested for the presence of myositis-specific antibodies (MSA).[134] Serum levels of muscle-derived enzymes are mildly elevated (800 to k0 CPK) in about half of the children with mixed connective tissue disease and can be markedly increased in the presence of diseases that have systemic vasculitis as a major component, such as Wegener's and polyarteritis nodosa.[245] The clinical diagnosis of Wegener's is supplemented by the identification of an antineutrophil cytoplasmic (c-ANCA) or perinuclear (p-ANCA) antibody. Muscle biopsies in children with these systemic conditions show the expected pattern of leukoclastic vasculitis, whereas in Kawasaki disease, the specimens display extensive muscle fiber necrosis and endomysial and perivascular mononuclear infiltrates.

Therapy

There are a range of treatment options for children who have myositis as part of another illness. In diseases in which there is no known specific autoantibody, such as Kawasaki syndrome, the inflammation often diminishes as the primary pathology is controlled, usually by administration of IVIg.[251] In systemic onset JRA, disease "flares" are more likely to occur, which may be accompanied by recurrent myositis. In contrast, the myositis-associated autoantibody such as U1-RNP or Pm/Scl is often an expression of the severity of the underlying disease and frequently responds to concurrent therapy, which may include methotrexate. Induction therapy with cyclophosphamide or Rituxan may be given to children with severe Wegener's disorder, while methotrexate may suffice for more mild disease onset; the myopathy often improves before the renal lesions respond to therapy.[252] In mixed connective tissue disease, arthritis may be a predominant feature, once the initial inflammation is brought under control with immunosuppressive agents, and appears to respond to TNFα receptor antagonists given in doses similar to those dispensed to children with JRA.[8]

Focal Myositis

Focal myositis (FM) presents with localized swelling and pain in a single muscle group, occasionally induced by trauma. Typical muscles involved include those located in the thigh, gastrocnemius, brachialis, and sternocleidomastoid (Table 42.9). Focal myositis described in patients under age 21 was not associated with any other recognized autoimmune disease, and had a distinct pathophysiology in which there was variation in the skeletal muscle myofiber size associated with many angular and rounded atrophic myofibers mixed with myofibers of normal and enlarged diameter.[253] Other characteristics included very dense fibrosis that encompassed both the endomysial and perimysial compartments.[253] Pathologic changes, such as large clusters of nesting muscle fibers surrounded by thick fibrosis, are more characteristic of FM than polymyositis[254] and pyknotic nuclear clumps were suggestive of denervation. Other myopathic changes included basophilic regenerating fibers with large vesicular nuclei and prominent nucleoli, which were often accompanied by degenerating fibers and myophagocytosis.[253] It is also recognized that localized and isolated myositis can be associated with other connective tissue diseases such as localized scleroderma, but the pathologic features differ.[255]

Inclusion Body Myositis

This entity has been reported in children, but as the field of research has evolved these reports may not comply with our current understanding of inclusion body myositis (IBM). New classification criteria for IBM have been developed which include the specification that the patient must be ≥ 45 years of age to be considered for this diagnosis, which automatically excludes children.[256] Further criteria for a diagnosis of clinic-pathologically defined IBM include: duration of symptoms for >12 months; knee extension weakness $\geq$ hip flexion weakness and/or finger flexion weakness $>$ shoulder abduction weakness and a serum CK level no greater than $15\times$ upper limit of normal. On review of the muscle biopsy, all or some of the following may be observed: endomysial inflammatory infiltrate, rimmed vacuoles, protein accumulation, and 15–18 nm filaments.[256] The reported pediatric cases do not appear to meet these current specific criteria.[257–260] A new antibody test for cytosolic 5′-nucleotidase A1 has been identified as highly specific for patients with sporadic IBM, and should help resolve the question of the occurrence of IBM in children.[261]

Orbital Myositis

Pathophysiology and Genetics

Isolated inflammation of the muscles of the eye in children has been primarily documented and reported as individual case reports.[262] This entity, which is relatively uncommon in the population as a whole, and is best documented by MRI, is associated with sarcoidosis, Crohn's disease and ANCA positive vasculitis[263] as well as IgG4 related disease,[264] but can also occur in isolated form. A report of a family with four members with this orbital inflammation suggests that there may be a potential genetic predisposition to this illness, which is usually considered to be sporadic in incidence.[265] Treatment of isolated orbital myositis includes immunosuppression with a range of agents, but the most rapid response is achieved with intravenous high dose methylprednisolone.[266]

Clinical and Laboratory Findings

In orbital myositis, painful diplopia develops as a consequence of enlargement secondary to inflammation of the extraocular muscles. The child initially has severe pain on upward gaze and develops loss of mobility in the field of action of the involved extraocular rectus muscle. The individual rectus muscles most commonly affected are: the lateral, superior, and the medial rectus. Symptoms common to all forms include conjunctival injection and chemosis with local tenderness in the quadrant overlying the involved rectus muscle.[267] Orbital myositis has been reported in patients with both discoid and systemic lupus[268,269] and in Lyme disease.[270] The onset is usually acute and must be differentiated from infectious orbital cellulitis. The other major differential consideration is thyroid eye disease: lid lag, soft tissue involvement, and proptosis are the most common manifestations, whereas restricted eye muscle motility, severe strabismus, and optic neuropathy are quite rare.[271] Unlike orbital myositis, the response to systemic corticosteroids in ocular thyroid disease may not be either prompt or complete, but the results with newer biologic agents appear promising.[272]

Laboratory Testing

The muscle enzymes are often mildly elevated. In the untreated child, there may be a mild peripheral lymphopenia with derangement of lymphocyte subsets.

Therapy

As in JDM, for children with a painful eye, systemic corticosteroids provide prompt relief of the acute symptoms. To achieve this welcome relief, we administer methylprednisolone at 30 mg/kg (1 g maximum dose) intravenously, followed by a lower daily dose of oral prednisolone at 1 mg/kg with gastric protection (PPI). The same precautions to protect bone density (calcium sufficient diet and oral vitamin D) are followed as in JDM. The corticosteroid therapy is slowly decreased once the muscle enzymes and peripheral blood lymphocyte phenotypes have normalized. We have observed no recurrences in children with orbital myositis but the number of patients is too small to generalize. There is a report of response to adalimumab in an adult with Crohn's who had recurrent orbital involvement.[273] A review of adults with recurrences of orbital myositis found that male gender, lack of proptosis, eyelid retraction, horizontal extraocular muscle involvement, multiple or bilateral extraocular muscle involvement, muscle tendon sparing, and lack of response to systemic corticosteroids or nonsteroidal anti-inflammatory agents were associated with recurrent orbital myositis.[274]

Rare Types of Localized Myositis in Children

Immune-Mediated Necrotizing Myopathy

Hyperlipidemia has been increasingly characterized in children with juvenile myositis,[275] prompting the use of statins for control; short-term data for these children appears to show good tolerance.[276] However, in adults, a necrotizing myopathy has been described following statin treatment[277] which led to the identification of autoantibody against the 3-hydroxy-3-methylglutaryl-coenzyme A reductase (HMGCR), the pharmacological target of statins.[278]

Proliferative Myositis: Pathophysiology

These lesions, which are primarily composed of proliferating spindle-shaped cells and uni-nucleated or multinucleated giant ganglion-like cells, can be identified by fine-needle aspiration cytology, and include nodular fasciitis (NF), proliferative fasciitis (PF), and proliferative myositis.[279] Electron microscopy studies suggest that these cells are myofibroblastic in nature and may be derived from pericytes.[280] These lesions usually are identified as a painless mass in the muscle of the extremities, although one case involved the mylohyoid muscle in a 9-year-old child who presented with a neck mass.[281] They appear to resolve within 4 months, or are surgically removed.

Nodular Myositis: Pathophysiology

This rare condition is thought to be a heterogeneous condition in which vascular inflammation is a prominent feature.[282] Several of the reported cases in adults evolve into generalized polymyositis[283] or may be found in graft versus host disease in post-bone marrow transplant patients.[284] Unlike proliferative myositis, these lesions are *painful masses* that occur primarily in the thigh and neck in children. The erythrocyte sedimentation rate and muscle enzyme levels may be elevated. The differential diagnosis of these lesions includes primary tumor of muscle, local muscular abscess or soft tissue infection, osteomyelitis, thrombophlebitis, and diabetic muscular infarction. The course ranges from spontaneous resolution to treatment with immunosuppression, followed by surgical excision for refractory cases.

Eosinophilic Myositis/Fasciitis

Pathophysiology

Eosinophilic fasciitis (EF) is a rare scleroderma-like condition that was first described by Shulman in 1974, and characterized by eosinophilic inflammatory fascial infiltrate which accompanies parasitic infection, such as sarcocystis or other inflammatory disorders.[285] Rarely, it can occur as an isolated observation. In an adult with eosinophilic infiltration of skin and muscle, elevated serum levels of eosinophilic cationic protein and interleukin-5 paralleling disease activity were measured.[286] This entity should be distinguished from the eosinophilic myalgia syndrome and environmental exposure to agents such as rapeseed oil or a contaminant of L-tryptophan that have provoked two epidemics of eosinophilic myositis.[287]

Clinical Findings and Therapy

Painful swelling, usually of the proximal extremity, is seen, although familial painless swelling (angioedema) of the face, hands, and feet has been described.[288] On radiographic imaging, a single muscle may be diffusely enlarged and edematous, with prominent edema in the adjacent fascia.[289] A deep muscle biopsy that includes fascia is mandatory for the diagnosis. [290] The characteristic finding is a striking peripheral eosinophilia, accompanied by eosinophilic infiltrates in the muscle with marked surrounding edema that also involves the fascia (Figure 42.12). This condition is characterized by tightness of the skin, peripheral eosinophilia, and an inflammatory infiltrate of the deep fascia that is composed both of eosinophils and lymphocytes. Children with this appearance should be observed for the development of linear scleroderma.[291] Patients with eosinophilic fasciitis almost always require systemic corticosteroids or other to achieve a good response, but children with EF who have elevated TGF-β may respond to low-dose corticosteroids. [292]

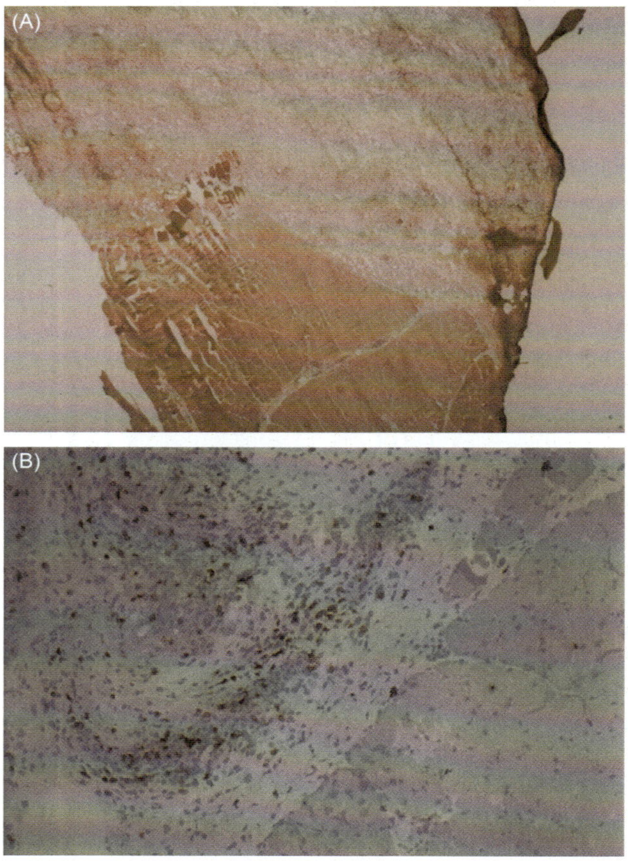

FIGURE 42.12 The muscle biopsy specimen from a child with eosinophilic fasciitis, showing extensive infiltration by eosinophils and lymphocytes of the fascia (**A**) and adjacent muscle (**B**).

Cancer-Associated Myositis

In adults, cancer may be preceded by symptoms of an inflammatory myopathy[293] and is one of the major causes of death.[294,295] Most of the myositis patients with cancer develop an antibody to nuclear matrix protein (NXP2)/TIF-1-λ.[296] Typical cancer-associated myositis (CAM), defined as malignancy occurring within 2 years of diagnosis of myositis, is more common in older age and in men who had dysphagia and lower serum creatine kinase levels.[297] The myopathic process has atypical features, but responds to corticosteroids. In contrast, in children, there are few reports of paraneoplastic myositis.[298] The most frequently encountered malignancies are leukemia and lymphoma, [299] but there are reports of the following as well: hepatocarcinoma, neuroblastoma, small bowel lymphosarcoma, immunoblastic lymph node sarcoma, and ovarian dysgerminoma.[298] Treatment is directed at the primary malignant process as well as the myositis.

Unlike the reports of the high association of malignancy in adults with both DM and PM, there are few case reports of JDM associated with a neoplastic syndrome, as opposed to other pediatric rheumatic disease.[298] This is of importance, for the most common MSA associated with adult DM and PM who develop malignancy is directed against a p155 antigen, TIF-1-γ,[135] while the most common MSA in children with JDM (33%) is to p155/140.[300] However, it is increasingly clear that the various laboratories worldwide have not yet used the same methods (immunodiffusion, immunoprecipitation, ELISA) to identify and characterize the p155/140 protein targets found in sera from both adults and children. Of note, our group reported a 9-year-old girl with symptoms of JDM that resolved with the removal of her ovarian teratoma; she was negative for antibody to p155/140 by both immunoprecipitation[301] and ELISA.

Graft-vs.-Host Myositis

We have reports of polymyositis following either bone marrow transplantation (13 cases of PM/1859 BMT)[302] or administration of allogeneic white cells.[303] Infiltrating CD4 + T cells were identified in the muscle of a patient who developed DM following bone marrow transplantation.[304] These observations provide evidence that activated lymphocytes are a central component of the inflammatory process that results in myositis.

CONCLUSION

Inflammatory myopathy in children is multifactorial. Case Examples 42.1–42.3 illustrate some of the spectrum of the range of inflammatory conditions involving muscle

CASE EXAMPLE 42.1 Juvenile Dermatomyositis

A 4-year-old Caucasian girl had a "sunburn" that persisted on her face despite use of sunscreens. As she became more irritable over the next two months, the parents noted that she began to have trouble getting in and out of bed and asked to be carried on long walks, which she had not done previously. The first physician examined the child and the parents were told that she had eczema, and that the regressive behavior might be associated with the recent birth of a sibling. Three months later, she became constipated and four months after the initial sun exposure, "red dots" appeared over the knuckles on the back of her hand (cf. Figure 42.1D) and she was no longer able to dress herself. The second physician prescribed topical corticosteroids and obtained some laboratory tests. The CPK was 180 (normal range for laboratory), aldolase was 11.2 (normal range <7.8), the AST was 55 (normal range <35) and the ALT was 70 (normal range <45). On the basis of this information, the child was referred to a pediatric rheumatologist, who found an anxious 4-year-old, in the 5th percentile for weight and the 10th percentile for height, blood pressure 90/60, pulse 112. She could not climb up to the examining table.

On our initial examination, the positive findings were: Skin: dilated eyelid capillaries and eyelid erythema (cf. Figure 42.2A), Gottron's papules on hands, with periungual avascularity (cf. Figure 42.2C1). She had a positive shawl sign, and erythema over the elbows, and knees with Gottron's papules there as well. There was a marked head lag, with weak neck flexors, swallowing appeared normal, but her speech had a nasal quality, pronounced on saying the letter "E." Her gums were swollen and bleeding, and the vasculature on the hard palate was erythematous and dilated. She had tenderness on compression of proximal muscles, which were 2+ on muscle testing. She could hold her leg elevated for 15 seconds, and her arm above her head for 20/x seconds (CMAS). She had decreased MCP and PIP flexion of the fingers of both hands, and her ankles were tight (zero degrees dorsiflexion, 20 degrees plantar flexion), as were her hamstrings.

Diagnostic laboratory testing (2 weeks after first blood test): The ANA was positive at 1:160 in a speckled pattern, the myositis specific antibody (MSA) was positive for Mi-2 and p155/140 antigens. Muscle enzymes: CPK = 202; aldolase = 12.0, while the AST and ALT were essentially unchanged and LDH was elevated at 460. The von Willebrand factor antigen, reflecting endothelial cell damage, was elevated for blood group A (normal range = 48–234%) at 380. Her neopterin was elevated at 12.4 (10 is upper limit of normal). An MRI was obtained (cf. Figure 42.5A), depicting patchy symmetrical proximal muscle edema (white area) and bright fascia. The MRI guided muscle biopsy (cf. Figure 42.7) displayed perifasicular atrophy, a mononuclear infiltrate with evidence of apoptosis, but not necrosis, and an unexpected increase in mature and immature collagen, suggesting that the disease was of longer duration than reported, associated with occlusion of capillaries, and capillary drop out. Bone mineral density showed a −1.8 Z score for the lumbar vertebrae. Nailfold capillary studies showed decreased numbers of end row capillary loops (4/7), with moderate dilation and bush formation.

Therapy

After 24 hours, to allow the incision for the biopsy to heal, she was started on 10 mg of omeprazole (Prilosec) and given intravenous methylprednisolone at 30 mg/kg (three days in a row followed by three times/week), until laboratory data normalized and oral prednisolone 0.5 mg/kg on non-IV days. To protect and increase her bone matrix, she was given 400 IU of vitamin D and 500 mg of calcium to take daily, in conjunction with a minimum of three servings of calcium-containing foods daily. IV methotrexate was given at 15 mg/m^2 weekly with 1 g/day folic acid. She was seen every 2 weeks initially, then every 4–6 weeks, and the medication tapered once the von Willebrand factor had normalized a month after the muscle enzymes returned to baseline. After she had a normal eye exam, hydroxychloroquine (5 mg/kg) was given, but the child developed diarrhea on this medication, so it was discontinued. The mother was asked to use sunscreen containing titanium and/or zinc oxide to protect against UVB whenever going outside. Passive stretching was begun with gradual resumption of normal ranges and resistive exercises once the muscle tenderness had abated. She regained normal strength after 3 months of therapy, after which the IVMP was changed to once a week for 2 months, once every other week for 3 doses, and once every 3 weeks for 2 doses and then discontinued when her testing remained normal. The oral prednisolone was maintained on the non-IVMP days. Twelve months later she was only taking weekly subcutaneous methotrexate and folic acid. The methotrexate was tapered over the next 4 months, and the child remained asymptomatic without further medication, but was instructed to continue lifelong use of sunscreen, with avoidance of excessive sun exposure.

---

and fascia that can occur in childhood inflammatory myopathies.

Genetic factors play a critical role, in disease susceptibility and disease chronicity in the children with JDM, as well as in the less common primary inflammations of muscle. We postulate that it is the immune response to (an) as-yet-unknown antigen(s) that initiates the inflammatory response in which the muscle fibers themselves may both potentiate and serve as a target for damage. In summary, the etiology and pathogenesis of JDM, an immunologically mediated vasculopathy, is under intensive investigation. This is an exciting time, because knowledge of the specific factors that play a role in both disease susceptibility and disease severity will lead to more effective and targeted therapies. This information is of vital importance both to children with inflammatory myopathy and to those who provide them with care.

CASE EXAMPLE 42.2 Jo-1 Myositis

At age 10 years this African-American girl had a chronic cough, wheezing, and dyspnea on exertion. She was treated for asthma until the age of 12, when it was realized that she was having trouble climbing stairs, even when her "asthma" was under control with oral prednisone. She developed arthralgia, malar rash, and frontal alopecia. She was referred to a pediatric rheumatologist, who found a mildly dyspneic child (respiration 22), BP 120/78 RAS, P 85. She was obese (80th percentile for weight, 20th percentile for height). Examination of the musculoskeletal system disclosed: sausage fingers with marked decrease in range of motion, increased pad to palm distance on digit flexion and loss of range of motion at the hips, knees, and ankles. Compression of *both* proximal and distal muscles was painful and she had loss of muscle strength on manual muscle testing to a grade 3 − . She could not clear her scapula on performing a sit-up and the Gowers' sign was positive. Cutaneous findings were absent except for the frontal alopecia, malar erythema, and somewhat thickened skin over her small joints. On testing, her CPK was greater than 30,000, the ANA was positive at 1:2560 in a speckled pattern, and her neopterin, which was initially 11, rose to 35 nm/L as her respiratory distress progressed (normal <10). She was negative for antibody to RNP, but was positive for antibody to Jo-1 (t-RNA synthetase) and to Ro.[72] On pulmonary testing, her DLCO% was 40%, and the FEV% was 35%. The initial chest radiograph (Figure 42.13A) progressed over the years until her death at age 23 (Figure 42.13B) due to respiratory failure, secondary to interstitial lung disease, despite aggressive therapy with IVMP, monthly cyclophosphamide and oral corticosteroids.

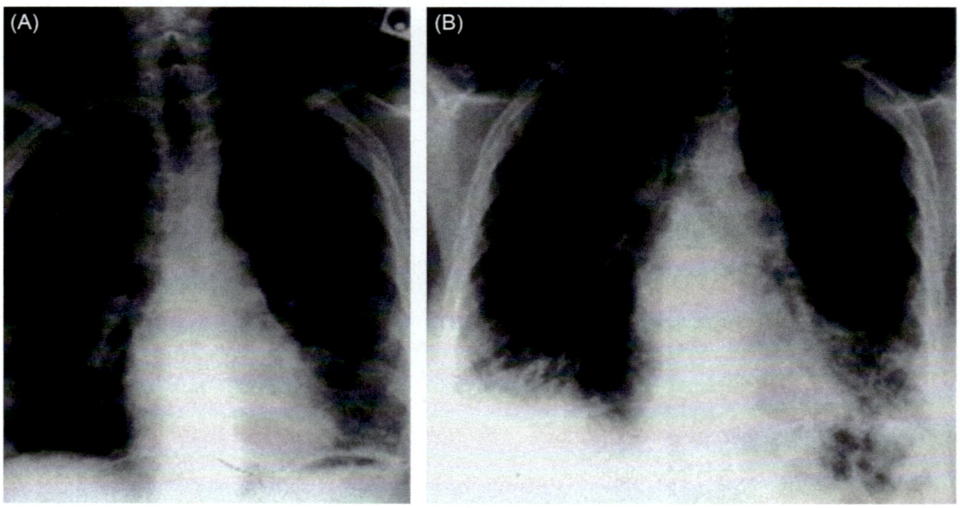

FIGURE 42.13 (**A**) A chest radiograph of the child discussed in Case Example 42.2 at the age of 11 years with muscle weakness, sausage-shaped fingers with painful joints, and a CPK greater than 30,000. (**B**) A chest radiograph of the same patient at the age of 23 years, just before her death, showing extensive pulmonary fibrosis, despite control of the inflammatory myopathy.

CASE EXAMPLE 42.3 Eosinophilic Fasciitis

A 16-year-old Caucasian girl had a 4-month history of tight skin, progressive joint contractures, and eosinophilia. These complaints followed by 5 weeks an illness consisting of tonsillitis and fatigue and serological evidence of elevated titers to Epstein-Barr virus (EVB) despite a negative mono-spot test. She complained of a low-grade fever and swelling of the dorsum of the hands, knees, and ankles. On physical examination, she had a temperature of 101°F and her height and weight were in the 50th percentile and stable. Blood pressure and pulse were within the normal range. She had mild erythema over the cheeks and nasal bridge, and her skin was tight over the hands, elbows, shoulder, and proximal thighs. Laboratory evaluation included leukocyte count 12,000/μL with 38% eosinophils, 2% monocytes, 13% lymphocytes, 42% neutrophils, 5% bands, hemoglobin 12.1 g/dL, platelet count 635,000/μL, sedimentation rate 10 mm/hr. Screening for HLA B27, rheumatoid factor, and antibodies to Scl-70, SS/A, and SS/B were negative. The muscle enzymes were within the normal range. MRI showed increased T2 signal in myofascial distribution, with no increase in signal in the muscle. A biopsy of the involved fascia showed eosinophilic fasciitis (cf. Figure 42.12A and B), which responded to an initial three-day course of IVMP (30 mg/m^2) followed by oral corticosteroids 1 mg/kg (in addition to vitamin D, oral calcium, and at least three servings of calcium-containing food in the diet). She had return of normal ranges within 3 months and the eosinophilia normalized shortly thereafter.

REFERENCES

1. Rider LG, Katz JD, Jones OY. Developments in the classification and treatment of the juvenile idiopathic inflammatory myopathies. *Rheum Dis Clin North Am* 2013;**39**(4):877–904.
2. Bohan A, Peter JB. Polymyositis and dermatomyositis (second of two parts). *N Engl J Med* 1975;**292**(8):403–7.
3. Bohan A, Peter JB. Polymyositis and dermatomyositis (first of two parts). *N Engl J Med* 1975;**292**(7):344–7.
4. Li C, Lu Y, Liu M, Fritz J. Magnetic resonance imaging-guided biopsy of musculoskeletal lesions using open low-field systems. *Top Magn Reson Imaging* 2011;**22**(4):135–41.
5. Brown VE, Pilkington CA, Feldman BM, Davidson JE. An international consensus survey of the diagnostic criteria for juvenile dermatomyositis (JDM). *Rheumatology (Oxford)* 2006;**45**(8): 990–3.
6. Weber MA. Ultrasound in the inflammatory myopathies. *Ann NY Acad Sci* 2009;**1154**:159–70.
7. Rider LG, Shah M, Mamyrova G, Huber AM, Rice MM, Targoff IN, et al. The myositis autoantibody phenotypes of the juvenile idiopathic inflammatory myopathies. *Medicine (Baltimore)* 2013;**92**(4): 223–43.
8. Tsai YY, Yang YH, Yu HH, Wang LC, Lee JH, Chiang BL. Fifteen-year experience of pediatric-onset mixed connective tissue disease. *Clin Rheumatol* 2010;**29**(1):53–8.
9. Medsger Jr. TA, Dawson WN, Masi AT. The epidemiology of polymyositis. *Am J Med* 1970;**48**:715–23.
10. Mendez EP, Lipton RB, Dyer A, Ramsey-Goldman R, Roftcher P, Bowyer S, et al. Incidence of Juvenile Dermatomyositis (JDM) 1995–98: results from the NIAMS Registry. *Arthritis Care Res* 2003;**49**:300–5.
11. Pachman LM, Hayford JR, Chung A, Daugherty CA, Pallansch MA, Fink CW, et al. Juvenile dermatomyositis at diagnosis: clinical characteristics of 79 children. *J Rheumatol* 1998;**25**(6): 1198–204.
12. Bitnum S, Daeschner C, Travis L, Dodge WF, Hopps H. Dermatomyositis. *J Pediatr* 1964;**64**:101–31.
13. Shah M, Mamyrova G, Targoff IN, Huber AM, Malley JD, Rice MM, et al. The clinical phenotypes of the juvenile idiopathic inflammatory myopathies. *Medicine (Baltimore)* 2013;**92**(1): 25–41.
14. Lundberg IE. The heart in dermatomyositis and polymyositis. *Rheumatology (Oxford)* 2006;**45**(Suppl. 4):iv18–21.
15. Pachman LM, Cooke N. Juvenile dermatomyositis: a clinical and immunologic study. *J Pediatr* 1980;**96**(2):226–34.
16. Archard LC, Richardson PJ, Olsen EG, Dubowitz V, Sewry C, Bowles NE. The role of Coxsackie B viruses in the pathogenesis of myocarditis, dilated cardiomyopathy and inflammatory muscle disease. *Biochem Soc Symp* 1987;**53**:51–62.
17. Constantin T, Ponyi A, Orban I, Molnar K, Derfalvi B, Dicso F, et al. National registry of patients with juvenile idiopathic inflammatory myopathies in Hungary—clinical characteristics and disease course of 44 patients with juvenile dermatomyositis. *Autoimmunity* 2006;**39**(3):223–32.
18. Christensen ML, Pachman LM, Maryjowski ML. Antibody to coxsackie-B virus: Increased incidence in sera from children with recently diagnosed juvenile dermatomyositis. *Arthritis Rheum* 1983;**26**:S24.
19. Symmons DPM, Sills JA, Davis SM. The incidence of juvenile dermatomyositis: results from a nation-wide study. *Br J Rheumatol* 1995;**43**:732–6.
20. Vegosen LJ, Weinberg CR, O'Hanlon TP, Targoff IN, Miller FW, Rider LG. Seasonal birth patterns in myositis subgroups suggest an etiologic role of early environmental exposures. *Arthritis Rheum* 2007;**56**(8):2719–28.
21. Artlett CM, Ramos R, Jiminez SA, Patterson K, Miller FW, Rider LG. Chimeric cells of maternal origin in juvenile idiopathic inflammatory myopathies. *Lancet* 2000;**356**(9248):2155–6.
22. Nelson JL. The otherness of self: microchimerism in health and disease. *Trends Immunol* 2012;**33**(8):421–7.
23. Erlich Y. Blood ties: chimerism can mask twin discordance in high-throughput sequencing. *Twin Res Hum Genet* 2011;**14**(2): 137–43.
24. Plamondon S, Dent PB, Reed AM. Familial dermatomyositis. *J Rheumatol* 2000;**27**(12):2031–4.
25. Tezak Z, Hoffman EP, Lutz JL, Fedczyna TO, Stephan DA, Krasnoselska I, et al. Expression profiling in DQA1*0501 children with juvenile dermatomyositis: a novel model of pathogenesis. *J Immunol* 2002;**168**:4154–63.
26. Pachman LM, Liotta-Davis MR, Hong DK, Kinsella TR, Mendez EP, Kinder JM, et al. TNFalpha-308A allele in juvenile dermatomyositis: association with increased production of tumor necrosis factor alpha, disease duration, and pathologic calcifications. *Arthritis Rheum* 2000;**43**(10):2368–77.
27. Shah M, Targoff IN, Rice MM, Miller FW, Rider LG, Childhood Myositis Heterogeneity Collaborative Study Group. Brief report: ultraviolet radiation exposure is associated with clinical and autoantibody phenotypes in juvenile myositis. *Arthritis Rheum* 2013;**65** (7):1934–41.
28. Rider LG, Wu L, Mamyrova G, Targoff IN, Miller FW, Childhood Myositis Heterogeneity Collaborative Study Group. Environmental factors preceding illness onset differ in phenotypes of the juvenile idiopathic inflammatory myopathies. *Rheumatology (Oxford)* 2010;**49**(12):2381–90.
29. Pachman LM, Lipton R, Ramsey-Goldman R, Shamiyeh E, Abbott K, Mendez EP, et al. History of infection before the onset of juvenile dermatomyositis: results from the National Institute of Arthritis and Musculoskeletal and Skin Diseases Research Registry. *Arthritis Rheum* 2005;**53**(2):166–72.
30. Manlhiot C, Liang L, Tran D, Bitnun A, Tyrrell PN, Feldman BM. Assessment of an infectious disease history preceding juvenile dermatomyositis symptom onset. *Rheumatology (Oxford)* 2008;**47**(4): 526–9.
31. Martini A, Ravelli A, Albani S, Viola S, Scotta MS, Magrini U, et al. Recurrent juvenile dermatomyositis and cutaneous necrotizing arteritis with molecular mimicry between streptococcal type 5 M protein and human skeletal myosin. *J Pediatr* 1992;**121**(5 Pt 1): 739–42.
32. Christensen ML, Pachman LM, Schneiderman R, Patel DC, Friedman JM. Prevalence of coxsackie B virus antibodies in patients with juvenile dermatomyositis. *Arthritis Rheum* 1986; **29**:1365.
33. Bowles NE, Olsen EGJ, Richardson PJ, Archard LC. Detection of coxsackie-B-virus-specific RNA sequences in myocardial biopsy samples from patients with myocarditis and dilated cardiomyopathy. *Lancet* 1986;1120–1.

34. Leff RL, Love LA, Miller FW, Greenberg SJ, Klein EA, Dalakas MC, et al. Viruses in idiopathic inflammatory myopathies: absence of candidate viral genomes in muscle. *Lancet* 1992;**339**(8803): 1192–5.
35. Pachman LM, Litt DL, Rowley AH, Hayford JR, Caliendo J, Heller S, et al. Lack of detection of enteroviral RNA or bacterial DNA in magnetic resonance imaging-directed muscle biopsies from twenty children with active untreated juvenile dermatomyositis. *Arthritis Rheum* 1995;**38**(10):1513–18.
36. Yousef GE, Isenberg DA, Mowbray JF. Detection of enterovirus specific RNA sequences in muscle biopsy specimens from patients with adult onset myositis. *Ann Rheum Dis* 1990;**49**(5):310–15.
37. Liljeqvist JA, Bergstrom T, Holmstrom S, Samuelson A, Yousef GE, Waagstein F, et al. Failure to demonstrate enterovirus aetiology in Swedish patients with dilated cardiomyopathy. *J Med Virol* 1993;**39**(1):6–10.
38. Rudge P, Webster AD, Revesz T, Warner T, Espanol T, Cunningham-Rundles C, et al. Encephalomyelitis in primary hypogammaglobulinaemia. *Brain* 1996;**119**(Pt 1):1–15.
39. Wagner DK, Marti GE, Jaffe ES, Straus SE, Nelson DL, Fleisher TA. Lymphocyte analysis in a patient with X-linked agammaglobulinemia and isolated growth hormone deficiency after development of echovirus dermatomyositis and meningoencephalitis. *Int Arch Allergy Appl Immunol* 1989;**89**(2–3):143–8.
40. Rosenberg NL, Rotbart HA, Abzug MJ, Ringel SP, Levin MJ. Evidence for a novel picornavirus in human dermatomyositis. *Ann Neurol* 1989;**26**(2):204–9.
41. Massa M, Costouros N, Mazzoli F, De Benedetti F, La Cava A, Le T, et al. Self epitopes shared between human skeletal myosin and *Streptococcus* pyogenes M5 protein are targets of immune responses in active juvenile dermatomyositis. *Arthritis Rheum* 2002;**46**(11):3015–25.
42. Pachman LM, Hayford JR, Hochberg MC, Pallansch MA, Chung A, Daugherty CD, et al. New-onset juvenile dermatomyositis: comparisons with a healthy cohort and children with juvenile rheumatoid arthritis. *Arthritis Rheum* 1997;**40**(8):1526–33.
43. Cotterill JA, Shapiro H. Dermatomyositis after immunisation. *Lancet* 1978;**2**(8100):1158–9.
44. Gan L, Miller FW. State of the art: what we know about infectious agents and myositis. *Curr Opin Rheumatol* 2011;**23**(6):585–94.
45. Altman A, Szyper-Kravitz M, Shoenfeld Y. HBV vaccine and dermatomyositis: is there an association? *Rheumatol Int* 2008;**28**(6): 609–12.
46. Stubgen JP. A review on the association between inflammatory myopathies and vaccination. *Autoimmun Rev* 2014;**13**(1): 31–9.
47. Dourmishev LA, Stomonjakova SR, Dourmishev AL. D-penicillamine induced polymyositis and morphea in a woman with Hashimoto thyroiditis. *J Eur Acad Dermatol Venereol* 2002;**16**(5): 538–9.
48. Yordam N, Kandemir N, Topaloglu H, Gogus S, Kucukali T. Myositis associated with growth hormone therapy. *J Pediatr* 1994;**125**(4):671.
49. Adams C, August CS, Maguire H, Sladky JT. Neuromuscular complications of bone marrow transplantation. *Pediatr Neurol* 1995;**12**(1): 58–61.
50. Ferrarini A, Lava SA, Simonetti GD, Ramelli GP, Bianchetti MG. Swiss Italian Society of Pediatrics. Influenzavirus B-associated acute benign myalgia cruris: an outbreak report and review of the literature. *Neuromuscul Disord* 2014;**24**(4):342–6.
51. Harati Y, Niakan E, Bergman EW. Childhood dermatomyositis in monozygotic twins. *Neurology* 1986;**36**(5):721–3.
52. O'Hanlon TP, Li Z, Gan L, Gourley MF, Rider LG, Miller FW. Plasma proteomic profiles from disease-discordant monozygotic twins suggest that molecular pathways are shared in multiple systemic autoimmune diseases. *Arthritis Res Ther* 2011;**13**(6):R181.
53. Miller FW, Cooper RG, Vencovsky J, Rider LG, Danko K, Wedderburn LR, et al. Genome-wide association study of dermatomyositis reveals genetic overlap with other autoimmune disorders. *Arthritis Rheum* 2013;**65**(12):3239–47.
54. Rider LG, Shamim E, Okada S, Pandey JP, Targoff IN, O'Hanlon TP, et al. Genetic risk and protective factors for idiopathic inflammatory myopathy in Koreans and American whites: a tale of two loci. *Arthritis Rheum* 1999;**42**(6):1285–90.
55. Mamyrova G, O'Hanlon TP, Monroe JB, Carrick DM, Malley JD, Adams S, et al. Immunogenetic risk and protective factors for juvenile dermatomyositis in Caucasians. *Arthritis Rheum* 2006;**54** (12):3979–87.
56. Wang M, Xie H, Shrestha S, Sredni S, Morgan GA, Pachman LM. Methylation alterations of WT1 and homeobox genes in inflamed muscle biopsy samples from patients with untreated juvenile dermatomyositis suggest self-renewal capacity. *Arthritis Rheum* 2012;**64**(10):3478–85.
57. Xu D, Kachaochana A, Morgan GA, Huang C, Pachman LM. Downregulation of Mir-10a is associated with increased Von Willebrand factor antigen, disease activity score and earlier diagnosis in children with untreated juvenile dermatomyositis. *Arthritis Rheum* 2013;**66**(10 Suppl.):890.
58. Reed AM, McNallan K, Wettstein P, Vehe R, Ober C. Does HLA-dependent chimerism underlie the pathogenesis of juvenile dermatomyositis? *J Immunol* 2004;**172**(8):5041–6.
59. Taban M, Perry JD. Juvenile dermatomyositis presenting with periorbital edema. *Ophthal Plast Reconstr Surg* 2006;**22**(5):393–5.
60. Kasteler JS, Callen JP. Scalp involvement in dermatomyositis. Often overlooked or misdiagnosed. *JAMA* 1994;**272**(24):1939–41.
61. Sherer Y, Livneh A, Levy Y, Shoenfeld Y, Langevitz P. Dermatomyositis and polymyositis associated with the antiphospholipid syndrome—a novel overlap syndrome. *Lupus* 2000;**9**(1): 42–6.
62. Dugan EM, Huber AM, Miller FW, Rider LG. Photoessay of the cutaneous manifestations of the idiopathic inflammatory myopathies. *Dermatol Online J* 2009;**15**(2):1.
63. Tsuchiya H, Tsuno H, Inoue M, Takahashi Y, Yamashita H, Kaneko H, et al. Mycophenolate mofetil therapy for rapidly progressive interstitial lung disease in a patient with clinically amyopathic dermatomyositis. *Mod Rheumatol* 2014;**24**(4): 694–6.
64. Gono T, Sato S, Kawaguchi Y, Kuwana M, Hanaoka M, Katsumata Y, et al. Anti-MDA5 antibody, ferritin and IL-18 are useful for the evaluation of response to treatment in interstitial lung disease with anti-MDA5 antibody-positive dermatomyositis. *Rheumatology* 2012;**51**(9):1563–70.
65. Smith RL, Sundberg J, Shamiyah E, Dyer A, Pachman LM. Skin involvement in juvenile dermatomyositis is associated with loss of end row nailfold capillary loops. *J Rheumatol* 2004;**31**(8): 1644–9.

66. Bingham A, Mamyrova G, Rother KI, Oral E, Cochran E, Premkumar A, et al. Predictors of acquired lipodystrophy in juvenile-onset dermatomyositis and a gradient of severity. *Medicine (Baltimore)* 2008;**87**(2):70–86.
67. Huber AM, Lovell DJ, Pilkington CA, Rennebohm RM, Rider LG. Confusion concerning multiple versions of the Childhood Myositis Assessment Scale. *Arthritis Care Res (Hoboken)* 2014;**66**(4):648.
68. Quinones R, Morgan GA, Amoruso M, Field R, Huang CC, Pachman LM. Lack of achievement of a full score on the childhood myositis assessment scale by healthy four-year-olds and those recovering from juvenile dermatomyositis. *Arthritis Care Res (Hoboken)* 2013;**65**(10):1697–701.
69. Pachman LM, Cooke N. Juvenile dermatomyositis: a clinical and immmunologic study. *J Pediatr* 1980;**96**:226.
70. Cuny C, Eicher JC, Collet E, Chatard C, Chauffert B, Lorcerie B, et al. [Dilated cardiomyopathy disclosing dermatopolymyositis. Management]. *Ann Cardiol Angeiol (Paris)* 1993;**42**(3):155–8.
71. Mathiesen PR, Buchvald F, Nielsen KG, Herlin T, Friis T, Nielsen S. Pulmonary function and autoantibodies in a long-term follow-up of juvenile dermatomyositis patients. *Rheumatology (Oxford)* 2014;**53**(4):644–9.
72. Prestridge A, Morgan G, Ferguson L, Huang CC, Pachman LM. Pulmonary function tests in idiopathic inflammatory myopathy: association with clinical parameters in children. *Arthritis Care Res (Hoboken)* 2013;**65**(9):1424–31.
73. Rouster-Stevens KA, Pachman LM. Autoantibody to signal recognition particle in African American girls with juvenile polymyositis. *J Rheumatol* 2008;**35**(5):927–9.
74. Vega P, Ibarra M, Prestridge A, Pachman LM. Autoantibody to PL-12 (anti-alanyl-tRNA synthetase) in an African American girl with juvenile dermatomyositis and resolution of interstitial lung disease. *J Rheumatol* 2011;**38**(2):394–5.
75. Pachman LM. Juvenile dermatomyositis. Pathophysiology and disease expression. *Pediatr Clin North Am* 1995;**42**(5):1071–98.
76. Laskin BL, Choyke P, Keenan GF, Miller FW, Rider LG. Novel gastrointestinal tract manifestations in juvenile dermatomyositis. *J Pediatr* 1999;**135**(3):371–4.
77. Pachman LM, Abbott K, Sinacore JM, Amoruso L, Dyer A, Lipton R, et al. Duration of illness is an important variable for untreated children with juvenile dermatomyositis. *J Pediatr* 2006;**148**(2):247–53.
78. Berard R, Chedeville G, Saint-Martin C, Scuccimarri R. Benign pneumatosis intestinalis in a patient with juvenile dermatomyositis. *J Rheumatol* 2010;**37**(11):2442–4.
79. Falcini F, Porfirio B, Lionetti P. Juvenile dermatomyositis and celiac disease. *J Rheumatol* 1999;**26**(6):1419–20.
80. Borrelli M, Prado MJ, Cordeiro P, Wroclawski ER, Monteiro Junior J, Kiss MH, et al. Ureteral necrosis in dermatomyositis. *J Urol* 1988;**139**(6):1275–7.
81. Aikawa NE, Sallum AM, Leal MM, Bonfa E, Pereira RM, Silva CA. Menstrual and hormonal alterations in juvenile dermatomyositis. *Clin Exp Rheumatol* 2010;**28**(4):571–5.
82. Kaddour N, Snoussi M, Marzouk S, Frikha F, Jallouli M, Frigui M, et al. [Pregnancy in dermatomyositis and polymyositis]. *Tunisie Medicale* 2012;**90**(8–9):606–12.
83. Madu AE, Omih E, Baguley E, Lindow SW. Juvenile dermatomyositis in pregnancy. *Case Rep Obstet Gynecol* 2013;**2013**:890107.
84. Mintz G. Dermatomyositis. *Rheum Dis Clin North Am* 1989;**15**(2):375–82.
85. Montecucco C, Ravelli A, Caporali R, Viola S, De Gennaro F, Albani S, et al. Autoantibodies in juvenile dermatomyositis. *Clin Exp Rheumatol* 1990;**8**(2):193–6.
86. Willis R, Lakos G, Harris EN. Standardization of antiphospholipid antibody testing—historical perspectives and ongoing initiatives. *Semin Thromb Hemost* 2014;**40**(2):161–2.
87. Fong LP, Yeung J. Spontaneous regression of retinal neovascularisation in juvenile dermatomyositis. *Aust NZ J Ophthalmol* 1990;**18**(1):107–8.
88. Coroi M, Bontas E, Defranceschi M, Bartos D, Dorobantu M. Ocular manifestations of antiphospholipid (Hughes') syndrome—minor features? *Oftalmologia* 2007;**51**(3):16–22.
89. Chen TS, Figueira E, Lau OC, McKelvie PA, Smee RI, Dawes LC, et al. Successful "medical" orbital decompression with adjunctive rituximab for severe visual loss in IgG4-related orbital inflammatory disease with orbital myositis. *Ophthal Plast Reconstr Surg* 2014. PMID: 24481505.
90. Pope E, Janson A, Khambalia A, Feldman B. Childhood acquired lipodystrophy: a retrospective study. *J Am Acad Dermatol* 2006;**55**(6):947–50.
91. Fisler RE, Liang MG, Fuhlbrigge RC, Yalcindag A, Sundel RP. Aggressive management of juvenile dermatomyositis results in improved outcome and decreased incidence of calcinosis. *J Am Acad Dermatol* 2002;**47**(4):505–11.
92. Faller G, Mistry BJ, Tikly M. Juvenile dermatomyositis in South African children is characterised by frequent dystropic calcification: a cross sectional study. *Pediatr Rheumatol Online J* 2014;**12**(1):2.
93. Eisenstein DM, Paller AS, Pachman LM. Juvenile dermatomyositis presenting with rash alone. *Pediatrics* 1997;**100**(3 Pt 1):391–2.
94. Mukamel M, Horev G, Mimouni M. New insight into calcinosis of juvenile dermatomyositis: a study of composition and treatment. *J Pediatr* 2001;**138**(5):763–6.
95. Lian JB, Pachman LM, Gundberg CM, Patridge RE, Maryjowski MC. Gamma-carboxyglutamate excretion and calcinosis in juvenile dermatomyositis. *Arthritis Rheum* 1982;**25**:1094.
96. Pachman LM, Veis A, Stock S, Abbott K, Vicari F, Patel P, et al. Composition of calcifications in children with juvenile dermatomyositis: association with chronic cutaneous inflammation. *Arthritis Rheum* 2006;**54**(10):3345–50.
97. Urganus AL, Zhao YD, Pachman LM. Juvenile dermatomyositis calcifications selectively displayed markers of bone formation. *Arthritis Rheum* 2009;**61**(4):501–8.
98. Shore EM, Kaplan FS. Inherited human diseases of heterotopic bone formation. *Nat Rev Rheumatol* 2010;**6**(9):518–27.
99. Schmidt AH, Vincent KA, Aiona MD. Hemimelic progressive osseous heteroplasia. A case report. *J Bone Joint Surg Am* 1994;**76**(6):907–12.
100. Pignolo RJ, Shore EM, Kaplan FS. Fibrodysplasia ossificans progressiva: clinical and genetic aspects. *Orphanet J Rare Dis* 2011;**6**:80.
101. Eimer MJ, Brickman WJ, Seshadri R, Ramsey-Goldman R, McPherson DD, Smulevitz B, et al. Clinical status and cardiovascular risk profile of adults with a history of juvenile dermatomyositis. *J Pediatr* 2011;**159**(5):795–801.

102. Roman MJ, Shanker BA, Davis A, Lockshin MD, Sammaritano L, Simantov R, et al. Prevalence and correlates of accelerated atherosclerosis in systemic lupus erythematosus. *N Engl J Med* 2003;**349**(25):2399–406.
103. von Scheven E, Bakkaloglu A. What's new in paediatric SLE? *Best Pract Res Clin Rheumatol* 2009;**23**(5):699–708.
104. Ortega-Hernandez OD, Shoenfeld Y. Mixed connective tissue disease: an overview of clinical manifestations, diagnosis and treatment. *Best Pract Res Clin Rheumatol* 2012;**26**(1):61–72.
105. Braidy JF, Poulson JM. Diaphragmatic weakness and myositis associated with systemic juvenile rheumatoid arthritis. *Can Med Assoc J* 1984;**130**(1):47–9.
106. Love LA, Leff RL, Fraser DD, Targoff IN, Dalakas M, Plotz PH, et al. A new approach to the classification of idiopathic inflammatory myopathy: Myositis-specific autoantibodies define useful homogeneous patient groups. *Medicine (Baltimore)* 1991;**70**:360.
107. Scott IU, Siatkowski RM. Idiopathic orbital myositis. *Curr Opin Rheumatol* 1997;**9**(6):504–12.
108. Pinal-Fernandez I, Selva-O'Callaghan A, Grau JM. Diagnosis and classification of eosinophilic fasciitis. *Autoimmun Rev* 2014; **13**(4–5):379–82.
109. Takahashi T, Aoki M, Suzuki N, Tateyama M, Yaginuma C, Sato H, et al. Clinical features and a mutation with late onset of limb girdle muscular dystrophy 2B. *J Neurol Neurosurg Psychiatry* 2013;**84**(4):433–40.
110. Lima JG, Nobrega LH, Nobrega ML, Santos Junior AC, Fernandes Fda C, Mesquita DJ, et al. [Influence of thyroid function in CPK serum levels]. *Arq Bras Endocrinol Metabol* 2012; **56**(3):190–4.
111. Borgia AR, Cavallasca JA, Costa CA, Musuruana JL. Hyperparathyroidism, a forgotten cause of musculoskeletal pain. *Reumatol Clin* 2012;**8**(5):299–301.
112. Pitt AM, Fleckenstein JL, Greenlee Jr. RG, Burns DK, Bryan WW, Haller R. MRI-guided biopsy in inflammatory myopathy: initial results. *Magn Reson Imaging* 1993;**11**(8):1093–9.
113. Malattia C, Damasio MB, Madeo A, Pistorio A, Providenti A, Pederzoli S, et al. Whole-body MRI in the assessment of disease activity in juvenile dermatomyositis. *Ann Rheum Dis* 2014;**73**(6): 1083–90.
114. Huppertz HI, Kaiser WA. Serial magnetic resonance imaging in juvenile dermatomyositis—delayed normalization. *Rheumatol Int* 1994;**4**:127–9.
115. Park JH, Neirmann KJ, Ryder NM, Nelson AE, Das A, Lawton AR, et al. Muscle abnormalities in juvenile dermatomyositis patients: P-31 magnetic resonance spectroscopy studies. *Arthritis Rheum* 2001;**43**(10):2359–67.
116. Larca LJ, Coppola JT, Honig S. Creatine kinase MB isoenzyme in dermatomyositis: a noncardiac source. *Ann Intern Med* 1981; **94**(3):341–3.
117. Wu AH, Perryman MB. Clinical applications of muscle enzymes and proteins. *Curr Opin Rheumatol* 1992;**4**(6):815–20.
118. Rutter WJ, Rajkumar T, Penhoet E, Kochman M, Valentine R. Aldolase variants: structure and physiological significance. *Ann NY Acad Sci* 1968;**151**(1):102–17.
119. Mamyrova G, Katz JD, Jones RV, Targoff IN, Lachenbruch PA, Jones OY, et al. Clinical and laboratory features distinguishing juvenile polymyositis and muscular dystrophy. *Arthritis Care Res (Hoboken)* 2013;**65**(12):1969–75.
120. Paik JJ, Corse AM, Mammen AL. The co-existence of myasthenia gravis in patients with myositis: a case series. *Semin Arthritis Rheum* 2014;**43**(6):792–6.
121. Robinson AB, Hoeltzel MF, Wahezi DM, Becker ML, Kessler E, Schmeling H, et al. Clinical characteristics of children with juvenile dermatomyositis: the Childhood Arthritis and Rheumatology Research Alliance Registry. *Arthritis Care Res (Hoboken)* 2014; **66**(3):404–10.
122. Banker BQ, Victor M. Dermatomyositis (systemic angiopathy) of childhood. *Medicine (Baltimore)* 1966;**45**:261–89.
123. Emslie-Smith AM, Engel AG. Microvascular changes in early and advanced dermatomyositis: a quantitative study. *Ann Neurol* 1990;**27**:343.
124. Woo TR, Rasmussen J, Callen JP. Recurrent photosensitive dermatitis preceding juvenile dermatomyositis. *Pediatr Dermatol* 1985;**2**:207–12.
125. Karpati G, Pouliot Y, Carpenter S. Expression of immunoreactive major histocompatibility complex products in human skeletal muscles. *Ann Neurol* 1988;**23**(1):64–72.
126. van der Pas J, Hengstman GJ, ter Laak HJ, Borm GF, van Engelen BG. Diagnostic value of MHC class I staining in idiopathic inflammatory myopathies. *J Neurol Neurosurg Psychiatry* 2004;**75**(1):136–9.
127. Salaroli R, Baldin E, Papa V, Rinaldi R, Tarantino L, De Giorgi LB, et al. Validity of internal expression of the major histocompatibility complex class I in the diagnosis of inflammatory myopathies. *J Clin Pathol* 2012;**65**(1):14–19.
128. Zhao Y, Fedczyna TO, McVicker V, Caliendo J, Li H, Pachman LM. Apoptosis in the skeletal muscle of untreated children with juvenile dermatomyositis: impact of duration of untreated disease. *Clin Immunol* 2007;**125**(2):165–72.
129. Zaba LC, Fiorentino DF. Skin disease in dermatomyositis. *Curr Opin Rheumatol* 2012;**24**(6):597–601.
130. Shrestha S, Wershil B, Sarwark JF, Niewold TB, Philipp T, Pachman LM. Lesional and nonlesional skin from patients with untreated juvenile dermatomyositis displays increased numbers of mast cells and mature plasmacytoid dendritic cells. *Arthritis Rheum* 2010;**62**(9):2813–22.
131. Wong D, Kea B, Pesich R, Higgs BW, Zhu W, Brown P, et al. Interferon and biologic signatures in dermatomyositis skin: specificity and heterogeneity across diseases. *PLOS ONE* 2012;**7**(1):e29161.
132. Carroll JE, Silverman A, Isobe Y, Brown WR, Kelts KA, Brooke MH. Inflammatory myopathy, IgA deficiency, and intestinal malabsorption. *J Pediatr* 1976;**89**(2):216–19.
133. Tansley SL, Betteridge ZE, McHugh NJ. The diagnostic utility of autoantibodies in adult and juvenile myositis. *Curr Opin Rheumatol* 2013;**25**(6):772–7.
134. Targoff IN. Autoantibodies and their significance in myositis. *Curr Rheumatol Rep* 2008;**10**(4):333–40.
135. Labrador-Horrillo M, Martinez MA, Selva-O'Callaghan A, Trallero-Araguas E, Balada E, Vilardell-Tarres M, et al. Anti-TIF1gamma antibodies (anti-p155) in adult patients with dermatomyositis: comparison of different diagnostic assays. *Ann Rheum Dis* 2012;**71**(6):993–6.
136. Espada G, Maldonado Cocco JA, Fertig N, Oddis CV. Clinical and serologic characterization of an Argentine pediatric myositis cohort: identification of a novel autoantibody (anti-MJ) to a 142-kDa protein. *J Rheumatol* 2009;**36**(11):2547–51.

137. Mimori T, Imura Y, Nakashima R, Yoshifuji H. Autoantibodies in idiopathic inflammatory myopathy: an update on clinical and pathophysiological significance. *Curr Opin Rheumatol* 2007; **19**(6):523–9.

138. Yamasaki Y, Yamada H, Ohkubo M, Yamasaki M, Azuma K, Ogawa H, et al. Longterm survival and associated risk factors in patients with adult-onset idiopathic inflammatory myopathies and amyopathic dermatomyositis: experience in a single institute in Japan. *J Rheumatol* 2011;**38**(8):1636–43.

139. Betteridge ZE, Gunawardena H, Chinoy H, North J, Ollier WE, Cooper RG, et al. Clinical and human leucocyte antigen class II haplotype associations of autoantibodies to small ubiquitin-like modifier enzyme, a dermatomyositis-specific autoantigen target, in UK Caucasian adult-onset myositis. *Ann Rheum Dis* 2009; **68**(10):1621–5.

140. Balboni I, Niewold TB, Morgan G, Limb C, Eloranta ML, Ronnblom L, et al. Interferon-alpha induction and detection of anti-ro, anti-la, anti-sm, and anti-rnp autoantibodies by autoantigen microarray analysis in juvenile dermatomyositis. *Arthritis Rheum* 2013;**65**(9):2424–9.

141. Wedderburn LR, McHugh NJ, Chinoy H, Cooper RG, Salway F, Ollier WE, et al. HLA class II haplotype and autoantibody associations in children with juvenile dermatomyositis and juvenile dermatomyositis-scleroderma overlap. *Rheumatology (Oxford)* 2007;**46**(12):1786–91.

142. Greenberg SA, Sanoudou D, Haslett JN, Kohane IS, Kunkel LM, Beggs AH, et al. Molecular profiles of inflammatory myopathies. *Neurology* 2002;**59**(8):1170–82.

143. Baechler EC, Bilgic H, Reed AM. Type I interferon pathway in adult and juvenile dermatomyositis. *Arthritis Res Ther* 2011; **13**(6):249.

144. Bordi L, Lalle E, Lapa D, Caglioti C, Quartu S, Capobianchi MR, et al. Type III interferon (IFN-lambda) antagonizes the antiviral activity of interferon-alpha in vitro. *J Biol Regul Homeost Agents* 2013;**27**(4):1001–9.

145. Liao AP, Salajegheh M, Nazareno R, Kagan JC, Jubin RG, Greenberg SA. Interferon beta is associated with type 1 interferon-inducible gene expression in dermatomyositis. *Ann Rheum Dis* 2011;**70**(5):831–6.

146. Greenberg SA, Higgs BW, Morehouse C, Walsh RJ, Kong SW, Brohawn P, et al. Relationship between disease activity and type 1 interferon- and other cytokine-inducible gene expression in blood in dermatomyositis and polymyositis. *Genes Immun* 2012; **13**(3):207–13.

147. O'Connor KA, Abbott KA, Sabin B, Kuroda M, Pachman LM. MxA gene expression in juvenile dermatomyositis peripheral blood mononuclear cells: association with muscle involvement. *Clin Immunol* 2006;**120**(3):319–25.

148. Higgs BW, Zhu W, Morehouse C, White WI, Brohawn P, Guo X, et al. A phase 1b clinical trial evaluating sifalimumab, an anti-IFN-alpha monoclonal antibody, shows target neutralisation of a type I IFN signature in blood of dermatomyositis and polymyositis patients. *Ann Rheum Dis* 2014;**73**(1): 256–62.

149. Tournadre A, Lenief V, Eljaafari A, Miossec P. Immature muscle precursors are a source of interferon-beta in myositis: role of Toll-like receptor 3 activation and contribution to HLA class I up-regulation. *Arthritis Rheum* 2012;**64**(2):533–41.

150. Kissel JT, Mendell JR, Rammohan KW. Microvascular deposition of complement membrane attack complex in dermatomyositis. *N Engl J Med* 1986;**314**:329–34.

151. Kissel JT, Halterman RK, Rammohan KW, Mendell JR. The relationship of complement-mediated microvasculopathy to the histologic features and clincial duration of disease in dermatomyositis. *Arch Neurol* 1991;**481991**:26–30.

152. Scott JP, Arroyave C. Activation of complement and coagulation in juvenile dermatomyositis. *Arthritis Rheum* 1987;**30**:572.

153. Yee A, Kretz CA. Von Willebrand factor: form for function. *Semin Thromb Hemost* 2014;**40**(1):17–27.

154. Pachman LM, Lintner KE, Wu YL, Ferguson LJ, Morgan GA, Huang CC, et al. Decreased C4A gene copy numbers in children with juvenile dermatomyositis: association with decreased C4 protein and lower absolute number of CD3 negative CD16/56 + natural killer cells. *Arthritis Rheum* 2012;**64**(Suppl. 10):1948.

155. Tournadre A, Miossec P. Immunoregulation in idiopathic inflammatory myopathies: from dendritic cells to immature regenerating muscle cells. *J Rheumatol* 2013;**40**(2):110–12.

156. Khanna S, Reed AM. Immunopathogenesis of juvenile dermatomyositis. *Muscle Nerve* 2010;**41**(5):581–92.

157. Reed AM, Peterson E, Bilgic H, Ytterberg SR, Amin S, Hein MS, et al. Changes in novel biomarkers of disease activity in juvenile and adult dermatomyositis are sensitive biomarkers of disease course. *Arthritis Rheum* 2012;**64**(12):4078–86.

158. Fedczyna TO, Lutz J, Pachman LM. Expression of TNFalpha by muscle fibers in biopsies from children with untreated juvenile dermatomyositis: association with the TNFalpha-308A allele. *Clin Immunol* 2001;**100**(2):236–9.

159. Ibarra MF, Klein-Gitelman M, Morgan E, Proytcheva M, Sullivan C, Morgan G, et al. Serum neopterin levels as a diagnostic marker of hemophagocytic lymphohistiocytosis syndrome. *Clin Vaccine Immunol* 2011;**18**(4):609–14.

160. Barak M, Merzbach D, Gruener N. Neopterin measured in serum and tissue culture supernates by a competitive enzyme-linked immunosorbant assay. *Clin Chem* 1989;**35**(7):1467–71.

161. De Benedetti F, De Amici M, Aramini L, Ruperto N, Martini A. Correlation of serum neopterin concentrations with disease activity in juvenile dermatomyositis. *Arch Dis Child* 1993;**69**(2): 232–5.

162. Rider LG, Schiffenbauer AS, Zito M, Lim KL, Ahmed A, Zemel LS, et al. Neopterin and quinolinic acid are surrogate measures of disease activity in the juvenile idiopathic inflammatory myopathies. *Clin Chem* 2002;**48**(10):1681–8.

163. Stiehm ER. Adverse effects of human immunoglobulin therapy. *Transfus Med Rev* 2013;**27**(3):171–8.

164. Bloom BJ, Tucker LB, Miller LC, Schaller JG. von Willebrand factor in juvenile dermatomyositis. *J Rheumatol* 1995;**22**(2): 320–5.

165. Guzman J, Petty RE, Malleson PN. Monitoring disease activity in juvenile dermatomyositis: the role of von Willebrand factor and muscle enzymes. *J Rheumatol* 1994;**21**(4):739–43.

166. Pachman LM. Imperfect indications of disease activity in juvenile dermatomyositis. *J Rheumatol* 1995;**22**(2):193–7.

167. Robinson AB, Thierry-Palmer M, Gibson KL, Rabinovich CE. Disease activity, proteinuria, and vitamin D status in children with systemic lupus erythematosus and juvenile dermatomyositis. *J Pediatr* 2012;**160**(2):297–302.

168. Azali P, Barbasso Helmers S, Kockum I, Olsson T, Alfredsson L, Charles PJ, et al. Low serum levels of vitamin D in idiopathic inflammatory myopathies. *Ann Rheum Dis* 2013;**72**(4):512–16.
169. Cohran VC, Griffiths M, Heubi JE. Bone mineral density in children exposed to chronic glucocorticoid therapy. *Clin Pediatr (Phila)* 2008;**47**(5):469–75.
170. Rouster-Stevens KA, Langman CB, Price HE, Seshadri R, Shore RM, Abbott K, et al. RANKL:osteoprotegerin ratio and bone mineral density in children with untreated juvenile dermatomyositis. *Arthritis Rheum* 2007;**56**(3):977–83.
171. Reed A, Haugen M, Pachman LM, Langman CB. Abnormalities in serum osteocalcin values in children with chronic rheumatic diseases. *J Pediatr* 1990;**116**(4):574–80.
172. Spencer CH, Hanson V, Singsen BH, Bernstein BH, Kornreich HK, King KK. Course of treated juvenile dermatomyositis. *J Pediatr* 1984;**105**(3):399–408.
173. Bowyer SL, Blane CE, Sullivan DB. Childhood dermatomyositis: factors predicting functional outcome and development of dystrophic calcification. *J Pediatrics* 1983;**103**:882.
174. Huber AM, Lang B, LeBlanc CM, Birdi N, Bolaria RK, Malleson P, et al. Medium- and long-term functional outcomes in a multicenter cohort of children with juvenile dermatomyositis. *Arthritis Rheum* 2000;**43**(3):541–9.
175. Lovell HB, Lindsley CB. Late recurrence of childhood dermatomyositis. *J Rheumatol* 1986;**13**(4):821–2.
176. Van Rossum MA, Hiemstra I, Prieur AM, Rijkers GT, Kuis W. Juvenile dermato/polymyositis: a retrospective analysis of 33 cases with special focus on initial CPK levels. *Clin Exp Rheumatol* 1994;**12**(3):339–42.
177. Crowe WE, Love KE, Levinson JE, Hilton PK. Clinical and pathogenetic implications of histopathology in childhood polydermatomyositis. *Arthritis Rheum* 1982;**25**:126.
178. Na SJ, Kim SM, Sunwoo IN, Choi YC. Clinical characteristics and outcomes of juvenile and adult dermatomyositis. *J Korean Med Sci* 2009;**24**(4):715–21.
179. Kim NN, Lio PA, Morgan GA, Jarvis JN, Pachman LM. Double trouble: therapeutic challenges in patients with both juvenile dermatomyositis and psoriasis. *Arch Dermatol* 2011;**147**(7):831–5.
180. Feldman BM, Ayling-Campos A, Luy L, Stevens D, Silverman ED, Laxer RM. Measuring disability in juvenile dermatomyositis: validity of the childhood health assessment questionnaire. *J Rheumatol* 1995;**22**(2):326–31.
181. Lovell DJ, Lindsley CB, Rennebohm RM, Ballinger SH, Bowyer SL, Giannini EH, et al. Development of validated disease activity and damage indices for the juvenile idiopathic inflammatory myopathies. II. The Childhood Myositis Assessment Scale (CMAS): a quantitative tool for the evaluation of muscle function. The Juvenile Dermatomyositis Disease Activity Collaborative Study Group. *Arthritis Rheum* 1999;**42**(10):2213–19.
182. Rider LG, Feldman BM, Perez MD, Rennebohm RM, Lindsley CB, Zemel LS, et al. Development of validated disease activity and damage indices for the juvenile idiopathic inflammatory myopathies: I. Physician, parent, and patient global assessments. Juvenile Dermatomyositis Disease Activity Collaborative Study Group. *Arthritis Rheum* 1997;**40**(11):1976–83.
183. Bode RK, Klein-Gitelman MS, Miller ML, Lechman TS, Pachman LM. Disease activity score for children with juvenile dermatomyositis: reliability and validity evidence. *Arthritis Rheum* 2003;**49**(1):7–15.
184. Ruperto N, Pistorio A, Ravelli A, Rider LG, Pilkington C, Oliveira S, et al. The Paediatric Rheumatology International Trials Organisation provisional criteria for the evaluation of response to therapy in juvenile dermatomyositis. *Arthritis Care Res (Hoboken)* 2010;**62**(11):1533–41.
185. Rider LG, Werth VP, Huber AM, Alexanderson H, Rao AP, Ruperto N, et al. Measures of adult and juvenile dermatomyositis, polymyositis, and inclusion body myositis. *Arthritis Care Res (Hoboken)* 2011;**63**(Suppl. 11):S118–57.
186. Huber AM, Giannini EH, Bowyer SL, Kim S, Lang B, Lindsley CB, et al. Protocols for the initial treatment of moderately severe juvenile dermatomyositis: results of a Children's Arthritis and Rheumatology Research Alliance Consensus Conference. *Arthritis Care Res (Hoboken)* 2010;**62**(2):219–25.
187. Huber AM, Robinson AB, Reed AM, Abramson L, Bout-Tabaku S, Carrasco R, et al. Consensus treatments for moderate juvenile dermatomyositis: beyond the first two months. Results of the second Childhood Arthritis and Rheumatology Research Alliance consensus conference. *Arthritis Care Res (Hoboken)* 2012;**64**(4):546–53.
188. Sutherland ER, Busse WW. Designing clinical trials to address the needs of childhood and adult asthma: the National Heart, Lung, and Blood Institute's AsthmaNet. *J Allergy Clin Immunol* 2014;**133**(1):34–8.
189. Rouster-Stevens KA, Gursahaney A, Ngai KL, Daru JA, Pachman LM. Pharmacokinetic study of oral prednisolone compared with intravenous methylprednisolone in patients with juvenile dermatomyositis. *Arthritis Rheum* 2008;**59**(2):222–6.
190. Wallace CA, Sherry DD. Preliminary report of higher dose methotrexate treatment in juvenile rheumatoid arthritis. *J Rheumatol* 1992;**19**(10):1604–7.
191. Laxer RM, Stein LD, Petty RE. Intravenous pulse methylprednisolone treatment of juvenile dermatomyositis. *Arthritis Rheum* 1987;**30**(3):328–34.
192. Miller 3rd JJ. Prolonged use of large intravenous steroid pulses in the rheumatic diseases of children. *Pediatrics* 1980;**65**(5):989–94.
193. Ytterberg SR. Treatment of refractory polymyositis and dermatomyositis. *Curr Rheumatol Rep* 2006;**8**(3):167–73.
194. Ravelli A, Lattanzi B, Consolaro A, Martini A. Glucocorticoids in paediatric rheumatology. *Clin Exp Rheumatol* 2011;**29**(5 Suppl. 68):S148–52.
195. Lang B, Dooley J. Failure of pulse intravenous methylprednisolone treatment in juvenile dermatomyositis. *J Pediatr* 1996;**128**(3):429–32.
196. Pachman LM, Morgan GA, Curran ML, Ferguson LJ, Huang CC. Decreased frequency of dystrophic calcifications in children with juvenile dermatomyositis: a 10 year study. *Arthritis Rheum* 2012;**64**(Suppl. 10):301.
197. Callen AM PL, Hayford J, Chung A, Ramsey-Goldman R. Intermittent high-dose intravenous methylprednisolone (IV pulse) therapy prevents calcinosis and shortens disease course in juvenile dermatomyositis (JDMS). *Arthitis Rheum* 1994;**37**(10):R10.

198. Klein-Gitelman MS, Pachman LM. Intravenous corticosteroids: adverse reactions are more variable than expected in children. *J Rheumatol* 1998;**25**(10):1995–2002.

199. Prasad S, Misra R, Agarwal V, Lawrence A, Aggarwal A. Juvenile dermatomyositis at a tertiary care hospital: is there any change in the last decade? *Int J Rheum Dis* 2013;**16**(5):556–60.

200. Marmor MF, Kellner U, Lai TY, Lyons JS, Mieler WF, American Academy of Ophthalmology. Revised recommendations on screening for chloroquine and hydroxychloroquine retinopathy. *Ophthalmology* 2011;**118**(2):415–22.

201. Femia AN, Vleugels RA, Callen JP. Cutaneous dermatomyositis: an updated review of treatment options and internal associations. *Am J Clin Dermatol* 2013;**14**(4):291–313.

202. Bounfour T, Bouaziz JD, Bezier M, Cordoliani F, Saussine A, Petit A, et al. Clinical efficacy of intravenous immunoglobulins for the treatment of dermatomyositis skin lesions without muscle disease. *J Eur Acad Dermatol Venereol* 2013. PMID: 23906502.

203. Lam CG, Manlhiot C, Pullenayegum EM, Feldman BM. Efficacy of intravenous Ig therapy in juvenile dermatomyositis. *Ann Rheum Dis* 2011;**70**(12):2089–94.

204. Kampylafka EI, Kosmidis ML, Panagiotakos DB, Dalakas M, Moutsopoulos HM, Tzioufas AG. The effect of intravenous immunoglobulin (IVIG) treatment on patients with dermatomyositis: a 4-year follow-up study. *Clin Exp Rheumatol* 2012;**30**(3): 397–401.

205. Rouster-Stevens KA, Morgan GA, Wang D, Pachman LM. Mycophenolate mofetil, a possible therapeutic agent for children with juvenile dermatomyositis. *Arthritis Care Res (Hoboken)* 2010;**62**(10):1446–51.

206. Dagher R, Desjonqueres M, Duquesne A, Quartier P, Bader-Meunier B, Fischbach M, et al. Mycophenolate mofetil in juvenile dermatomyositis: a case series. *Rheumatol Int* 2012; **32**(3):711–16.

207. Azzi JR, Sayegh MH, Mallat SG. Calcineurin inhibitors: 40 years later, can't live without. *J Immunol* 2013;**191**(12):5785–91.

208. Hassan AB, Nikitina-Zake L, Sanjeevi CB, Lundberg IE, Padyukov L. Association of the proinflammatory haplotype (MICA5.1/TNF2/TNFa2/DRB1*03) with polymyositis and dermatomyositis. *Arthritis Rheum* 2004;**50**(3):1013–15.

209. Wilkes MR, Sereika SM, Fertig N, Lucas MR, Oddis CV. Treatment of antisynthetase-associated interstitial lung disease with tacrolimus. *Arthritis Rheum* 2005;**52**(8):2439–46.

210. Labirua-Iturburu A, Selva-O'Callaghan A, Martinez-Gomez X, Trallero-Araguas E, Labrador-Horrillo M, Vilardell-Tarres M. Calcineurin inhibitors in a cohort of patients with antisynthetase-associated interstitial lung disease. *Clin Exp Rheumatol* 2013; **31**(3):436–9.

211. Riley P, McCann LJ, Maillard SM, Woo P, Murray KJ, Pilkington CA. Effectiveness of infliximab in the treatment of refractory juvenile dermatomyositis with calcinosis. *Rheumatology* 2008; **47**(6):877–80.

212. Dastmalchi M, Grundtman C, Alexanderson H, Mavragani CP, Einarsdottir H, Helmers SB, et al. A high incidence of disease flares in an open pilot study of infliximab in patients with refractory inflammatory myopathies. *Ann Rheum Dis* 2008; **67**(12):1670–7.

213. Pachman LM, Liotta-Davis MR, Hong DK, Kinsella TR, Mendez EP, Kinder JM, et al. TNFalpha-308A allele in juvenile dermatomyositis: association with increased production of tumor necrosis factor alpha, disease duration, and pathologic calcifications. *Arthritis Rheum* 2000;**43**(10):2368–77.

214. Rouster-Stevens KA, Ferguson L, Morgan G, Huang CC, Pachman LM. Pilot study of etanercept in patients with refractory juvenile dermatomyositis. *Arthritis Care Res (Hoboken)* 2014; **66**(5):783–7.

215. Niewold TB, Kariuki SN, Morgan GA, Shrestha S, Pachman LM. Elevated serum interferon-alpha activity in juvenile dermatomyositis: associations with disease activity at diagnosis and after thirty-six months of therapy. *Arthritis Rheum* 2009;**60**(6): 1815–24.

216. Palucka AK, Blanck JP, Bennett L, Pascual V, Banchereau J. Cross-regulation of TNF and IFN-alpha in autoimmune diseases. *Proc Natl Acad Sci USA* 2005;**102**(9):3372–7.

217. Oddis CV, Reed AM, Aggarwal R, Rider LG, Ascherman DP, Levesque MC, et al. Rituximab in the treatment of refractory adult and juvenile dermatomyositis and adult polymyositis: a randomized, placebo-phase trial. *Arthritis Rheum* 2013;**65**(2): 314–24.

218. Aggarwal R, Bandos A, Reed AM, Ascherman DP, Barohn RJ, Feldman BM, et al. Predictors of clinical improvement in rituximab-treated refractory adult and juvenile dermatomyositis and adult polymyositis. *Arthritis Rheumatol* 2014;**66**(3):740–9.

219. Schiff M, Weinblatt ME, Valente R, van der Heijde D, Citera G, Elegbe A, et al. Head-to-head comparison of subcutaneous abatacept versus adalimumab for rheumatoid arthritis: two-year efficacy and safety findings from AMPLE trial. *Ann Rheum Dis* 2014;**73**(1):86–94.

220. Arabshahi B, Silverman RA, Jones OY, Rider LG. Abatacept and sodium thiosulfate for treatment of recalcitrant juvenile dermatomyositis complicated by ulceration and calcinosis. *J Pediatr* 2012;**160**(3):520–2.

221. Kerola AM, Kauppi MJ. Abatacept as a successful therapy for myositis-a case-based review. *Clin Rheumatol* 2014. PMID 24493331.

222. Haselberger MB, Schwinghammer TL. Efficacy of mesna for prevention of hemorrhagic cystitis after high-dose cyclophosphamide therapy. *Ann Pharmacother* 1995;**29**(9):918–21.

223. Bachelez H, Schremmer B, Cadranel J, Mouly F, Sarfati C, Agbalika F, et al. Fulminant *Pneumocystis carinii* pneumonia in 4 patients with dermatomyositis. *Arch Intern Med* 1997;**157**(13): 1501–3.

224. Coyle PV, McCaughey C, Nager A, McKenna J, O'Neill H, Feeney SA, et al. Rising incidence of *Pneumocystis jirovecii* pneumonia suggests iatrogenic exposure of immune-compromised patients may be becoming a significant problem. *J Med Microbiol* 2012;**61**(Pt 7):1009–15.

225. Niermann KJ, Olsen NJ, Park JH. Magnesium abnormalities of skeletal muscle in dermatomyositis and juvenile dermatomyositis. *Arthritis Rheum* 2002;**46**(2):475–88.

226. Ratel S, Duche P, Williams CA. Muscle fatigue during high-intensity exercise in children. *Sports Med* 2006;**36**(12):1031–65.

227. Westberry DE, Davids JR, Jacobs JM, Pugh LI, Tanner SL. Effectiveness of serial stretch casting for resistant or recurrent knee flexion contractures following hamstring lengthening in children with cerebral palsy. *J Pediatr Orthop* 2006;**26**(1): 109–14.

228. Patzleiner E, Kirchler E, Mantovan F. [The changed lives of children and adolescents with juvenile dermatomyositis—3 case studies]. *Kinderkrankenschwester* 2011;**30**(10):414–19.
229. Uziel Y, Nemet D, Maimon E, Shalev J, Harris S, Kutz I, et al. Juvenile dermatomyositis: medical and psychosocial team approach. *Clin Exp Rheumatol* 2002;**20**(2):256.
230. Hartley B, Campion-Fuller C. Juvenile dermatomyositis: a Roy nursing perspective. *J Pediatr Nurs* 1994;**9**(3):175–82.
231. Apaz MT, Saad-Magalhaes C, Pistorio A, Ravelli A, de Oliveira SJ, Marcantoni MB, et al. Health-related quality of life of patients with juvenile dermatomyositis: results from the Pediatric Rheumatology International Trials Organisation multinational quality of life cohort study. *Arthritis Rheum* 2009;**61**(4):509–17.
232. Tollisen A, Sanner H, Flato B, Wahl AK. Quality of life in adults with juvenile-onset dermatomyositis: a case-control study. *Arthritis Care Res (Hoboken)* 1964;**64**(7):1020–7.
233. Schwartz T, Sanner H, Husebye T, Flato B, Sjaastad I. Cardiac dysfunction in juvenile dermatomyositis: a case-control study. *Ann Rheum Dis* 2011;**70**(5):766–71.
234. Lundberg IE. The prognosis of mixed connective tissue disease. *Rheum Dis Clin North Am* 2005;**31**(3):535–47, vii–viii
235. Piotto DG, Len CA, Hilario MO, Terreri MT. Nailfold capillaroscopy in children and adolescents with rheumatic diseases. *Rev Bras Reumatol* 2012;**52**(5):722–32.
236. Kotajima L, Aotsuka S, Sumiya M, Yokohari R, Tojo T, Kasukawa R, et al. Clinical features of patients with juvenile onset mixed connective tissue disease; analysis of data collected in a nationwide study in Japan. *J Rheumatol* 1996;**23**:1088–94.
237. Ungprasert P, Wannarong T, Panichsillapakit T, Cheungpasitporn W, Thongprayoon C, Ahmed S, et al. Cardiac involvement in mixed connective tissue disease: a systematic review. *Int J Cardiol* 2014;**171**(3):326–30.
238. Miller ML, Levinson L, Pachman LM, Poznanski A. Abnormal muscle MRI in a patient with systemic juvenile arthritis. *Pediatr Radiol* 1995;**25**(Suppl. 1):S107–8.
239. Gardner-Medwin JM, Irwin G, Johnson K. MRI in juvenile idiopathic arthritis and juvenile dermatomyositis. *Ann NY Acad Sci* 2009;**1154**:52–83.
240. Jovancevic B, Lindholm C, Pullerits R. Anti B-cell therapy against refractory thrombocytopenia in SLE and MCTD patients: long-term follow-up and review of the literature. *Lupus* 2013; **22**(7):664–74.
241. Levin M, Burgner D. Treatment of Kawasaki disease with anti-TNF antibodies. *Lancet* 2014;**383**(9930):1700–3.
242. Jo SE, Kim YJ, Lee KH, Cho SG, Lim MJ, Kwon SR, et al. Muscular involvement of Behcet's disease: ultrasonography, computed tomography, and magnetic resonance imaging findings. *Clin Imaging* 2012;**36**(5):643–6.
243. Lang BA, Laxer RM, Thorner P, Greenberg M, Silverman ED. Pediatric onset of Behcet's syndrome with myositis: case report and literature review illustrating unusual features. *Arthritis Rheum* 1990;**33**:418–25.
244. Bansal NK, Houghton KM. Cutaneous polyarteritis nodosa in childhood: a case report and review of the literature. *Arthritis* 2010;**2010**:687547.
245. Shuhart DT, Torretti DJ, Maksimak JF, George S. Acute myositis as an unusual presentation of Wegener's granulomatosis. *Arch Pediatr Adolesc Med* 1994;**148**:418–25.
246. Selva-O'Callaghan A, Casellas F, de Torres I, Palou E, Grau-Junyent JM, Vilardell-Tarres M. Celiac disease and antibodies associated with celiac disease in patients with inflammatory myopathy. *Muscle Nerve* 2007;**35**(1):49–54.
247. Cooper C, Fairris G, Cotton DW, Steart P, Barth JH. Dermatomyositis associated with idiopathic thrombocytopenia. *Dermatologica* 1986;**172**(3):173–6.
248. Puri V, Agarwal S, Garg T, Chandar R. Importance of cutaneous findings in childhood scleromyositis in Indian scenario. *Indian J Dermatol* 2013;**58**(2):162.
249. Marcus M, Ilyas M, Tolaymat A. Childhood scleromyositis with a negative PM/Scl antibody. *Joint Bone Spine* 2010;**77**(1):73–5.
250. Scalapino K, Arkachaisri T, Lucas M, Fertig N, Helfrich DJ, Londino Jr. AV, et al. Childhood onset systemic sclerosis: classification, clinical and serologic features, and survival in comparison with adult onset disease. *J Rheumatol* 2006;**33**(5):1004–13.
251. Sanchez-Manubens J, Bou R, Anton J. Diagnosis and classification of Kawasaki disease. *J Autoimmun* 2014;**48–49**:113–17.
252. Lutalo PM, D'Cruz DP. Diagnosis and classification of granulomatosis with polyangiitis (aka Wegener's granulomatosis). *J Autoimmun* 2014;**48–49**:94–8.
253. Auerbach A, Fanburg-Smith JC, Wang G, Rushing EJ. Focal myositis: a clinicopathologic study of 115 cases of an intramuscular mass-like reactive process. *Am J Surg Pathol* 2009;**33**(7): 1016–24.
254. Smith AG, Urbanits S, Blaivas M, Grisold W, Russell JW. Clinical and pathologic features of focal myositis. *Muscle Nerve* 2000;**23**(10):1569–75.
255. Zivkovic SA, Freiberg W, Lacomis D, Domsic RT, Medsger TA. Localized scleroderma and regional inflammatory myopathy. *Neuromuscul Disord* 2014;**24**(5):425–30.
256. Rose MR. 188th ENMC International Workshop: Inclusion Body Myositis, 2–4 December 2011, Naarden, The Netherlands. *Neuromuscul Disord* 2013;**23**(12):1044–55.
257. Yunis EJ, Samaha FJ. Inclusion body myositis. *Lab Invest* 1971;**25**(3):240–8.
258. Fidzianska A, Glinka Z. Rimmed vacuoles with beta-amyloid and tau protein deposits in the muscle of children with hereditary myopathy. *Acta Neuropathol* 2006;**112**(2):185–93.
259. Fidzianska A, Kaminska A, Ryniewicz B. Congenital myopathy with tubular aggregates and tubulofilamentous IBM-type inclusions. *Neuropediatrics* 2005;**36**(1):35–9.
260. Fidzianska A, Kaminska A. Congenital myopathy with abundant ring fibres, rimmed vacuoles and inclusion body myositis-type inclusions. *Neuropediatrics* 2003;**34**(1):40–4.
261. Larman HB, Salajegheh M, Nazareno R, Lam T, Sauld J, Steen H, et al. Cytosolic 5′-nucleotidase 1A autoimmunity in sporadic inclusion body myositis. *Ann Neurol* 2013;**73**(3):408–18.
262. Spindle J, Narang S, Purewal B, Shinder R. Atypical idiopathic orbital inflammation in a young girl. *Ophthal Plast Reconstr Surg* 2013;**29**(3):e86–8.
263. Fraser CL, Skalicky SE, Gurbaxani A, McCluskey P. Ocular myositis. *Curr Allergy Asthma Rep* 2013;**13**(3):315–21.
264. Wallace ZS, Deshpande V, Stone JH. Ophthalmic manifestations of IgG4-related disease: single-center experience and literature review. *Semin Arthritis Rheum* 2014;**43**(6):806–17.
265. Maurer I, Zierz S. Recurrent orbital myositis: report of a familial incidence. *Arch Neurol* 1999;**56**(11):1407–9.

266. Kau HC, Kao SC, Peng CH, Hsu WM, Tsai CC. Methylprednisolone pulse therapy in patient with isolated superior oblique myositis. *Eye (Lond)* 2006;**20**(9):1106–9.
267. Pollard ZF. Acute rectus muscle palsy in children as a result of orbital myositis. *J Pediatr* 1996;**128**:230–3.
268. Kono S, Takashima H, Suzuki D, Terada T, Konishi T, Miyajima H. Orbital myositis associated with discoid lupus erythematosus. *Lupus* 2014;**23**(2):220–2.
269. Santosa A, Vasoo S. Orbital myositis as manifestation of systemic lupus erythematosus—a case report. *Postgraduate Med J* 2013; **89**(1047):59.
270. Sauer A, Speeg-Schatz C, Hansmann Y. Two cases of orbital myositis as a rare feature of lyme borreliosis. *Case Rep Infect Dis* 2011;**2011**:372470.
271. Gogakos AI, Boboridis K, Krassas GE. Pediatric aspects in Graves' orbitopathy. *Pediatr Endocrinol Rev* 2010;**7**(Suppl. 2): 234–44.
272. Perez-Moreiras JV, Alvarez-Lopez A, Gomez EC. Treatment of active corticosteroid-resistant Graves' orbitopathy. *Ophthal Plast Reconstr Surg* 2014;**30**(2):162–7.
273. Verma S, Kroeker KI, Fedorak RN. Adalimumab for orbital myositis in a patient with Crohn's disease who discontinued infliximab: a case report and review of the literature. *BMC Gastroenterology* 2013;**13**:59.
274. Mannor GE, Rose GE, Moseley IF, Wright JE. Outcome of orbital myositis. Clinical features associated with recurrence. *Ophthalmology* 1997;**104**(3):409–13.
275. Bingham A, Mamyrova G, Rother KI, Oral E, Cochran E, Premkumar A, et al. Predictors of acquired lipodystrophy in juvenile-onset dermatomyositis and a gradient of severity. *Medicine (Baltimore)* 2008;**87**(2):70–86.
276. Lamaida N, Capuano E, Pinto L, Capuano E, Capuano R, Capuano V. The safety of statins in children. *Acta Paediatrica* 2013;**102**(9):857–62.
277. Hamann PD, Cooper RG, McHugh NJ, Chinoy H. Statin-induced necrotizing myositis—a discrete autoimmune entity within the "statin-induced myopathy spectrum". *Autoimmun Rev* 2013;**12**(12): 1177–81.
278. Mohassel P, Mammen AL. Statin-associated autoimmune myopathy and anti-HMGCR autoantibodies. *Muscle Nerve* 2013;**48**(4): 477–83.
279. Wong NL, Di F. Pseudosarcomatous fasciitis and myositis: diagnosis by fine-needle aspiration cytology. *Am J Clin Pathol* 2009;**132**(6):857–65.
280. el-Jabbour JN, Bennett MH, Burke MM, Lessells A, O'Halloran A. Proliferative myositis. An immunohistochemical and ultrastructural study. *Am J Surg Pathol* 1991;**15**(7):654–9.
281. Mulier S, Stas M, Delabie J, Lateur L, Gysen M, Dal Cin P, et al. Proliferative myositis in a child. *Skeletal Radiol* 1999;**1999**(12): 703–9.
282. Smith CA, Pinals RS. Localized nodular myositis. *J Rheumatol* 1981;**8**(5):815–19.
283. Allen I, Mullally B, Mawhinney H, Sawhney B, McKee P. The nodular form of polymyositis—a possible manifestation of vasculitis. *J Pathol* 1980;**131**(2):183–91.
284. Layfield LJ, Crim J, Gupta D. Fine-needle aspiration findings in nodular myositis: a case report. *Diagn Cytopathol* 2000;**23**(5): 343–7.
285. Arness MK, Brown JD, Dubey JP, Neafie RC, Granstrom DE. An outbreak of acute eosinophilic myositis attributed to human *Sarcocystis* parasitism. *Am J Trop Med Hyg* 1999;**61**(4):548–53.
286. Trueb RM, Lubbe J, Torricelli R, Panizzon RG, Wuthrich B, Burg G. Eosinophilic myositis with eosinophilic cellulitislike skin lesions. Association with increased serum levels of eosinophil cationic protein and interleukin-5. *Arch Dermatol* 1997;**133**(2): 203–6.
287. Hess EV. Are there environmental forms of systemic autoimmune diseases? *Environ Health Perspect* 1999;**107**(Suppl. 5):709–11.
288. Poliak N, Orange JS, Pawel BR, Weiss PF. Eosinophilic fasciitis mimicking angioedema and treatment response to infliximab in a pediatric patient. *Ann Allergy Asthma Immunol* 2011;**106**(5):444–5.
289. Moulton SJ, Kransdorf MJ, Ginsburg WW, Abril A, Persellin S. Eosinophilic fasciitis: spectrum of MRI findings. *Am J Roentgenol* 2005;**184**(3):975–8.
290. Selva-O'Callaghan A, Trallero-Araguas E, Grau JM. Eosinophilic myositis: an updated review. *Autoimmun Rev* 2014;**13**(4–5): 375–8.
291. Balat A, Akinci A, Turgut M, Mizrak B, Aydin A. Eosinophilic fasciitis—progression to linear scleroderma: a case report. *Turkish J Pediatr* 1999;**41**(3):381–5.
292. Dziadzio L, Kelly EA, Panzer SE, Jarjour N, Huttenlocher A. Cytokine abnormalities in a patient with eosinophilic fasciitis. *Ann Allergy Asthma Immunol* 2003;**90**(4):452–5.
293. Aggarwal R, Oddis CV. Paraneoplastic myalgias and myositis. *Rheum DisClin North Am* 2011;**37**(4):607–21.
294. Santo AH, Souza JM, Pinheiro CE, Souza DC, Sato EI. Trends in dermatomyositis- and polymyositis-related mortality in the state of Sao Paulo, Brazil, 1985–2007: multiple cause-of-death analysis. *BMC Public Health* 2010;**10**:597.
295. Ortega C, Limaye V, Chur-Hansen A. Patient perceptions of and experiences with inflammatory myositis. *J Clin Rheumatol* 2010;**16**(7):341–2.
296. Fiorentino DF, Chung LS, Christopher-Stine L, Zaba L, Li S, Mammen AL, et al. Most patients with cancer-associated dermatomyositis have antibodies to nuclear matrix protein NXP-2 or transcription intermediary factor 1gamma. *Arthritis Rheum* 2013;**65**(11):2954–62.
297. Neri R, Simone B, Iacopetti V, Iacopetti G, Pepe P, d'Ascanio A, et al. Cancer-associated myositis: a 35-year retrospective study of a monocentric cohort. *Rheumatol Int* 2013;**34**(4):565–9.
298. Morris P, Dare J. Juvenile dermatomyositis as a paraneoplastic phenomenon: an update. *J Pediatr Hematol Oncol* 2010;**32**(3): 189–91.
299. Millot F, Mechinaud F, Harousseau JL, Mussini JM, Stalder JF. Acute lymphoblastic leukemia associated with dermatomyositis in a child [letter]. *Pediatr Dermatol* 1993;**10**:199–200.
300. Rider LG, Shah M, Mamyrova G, Huber AM, Rice MM, Targoff IN, et al. The myositis autoantibody phenotypes of the juvenile idiopathic inflammatory myopathies. *Medicine (Baltimore)* 2013; **92**(4):223–43.
301. Ibarra M, Chou P, Pachman LM. Ovarian teratoma mimicking features of juvenile dermatomyositis in a child. *Pediatrics* 2011;**128**(5):e1293–6.
302. Stevens AM, Sullivan KM, Nelson JL. Polymyositis as a manifestation of chronic graft-versus-host disease. *Rheumatology (Oxford)* 2003;**42**(1):34–9.

303. Montoro J, Hernandez-Boluda JC, Arbona C, Solano C. Polymyositis after donor lymphocyte infusion. *Int J Hematol* 2012;**96**(3):386–9.
304. Sakamoto K, Imamura T, Niwa F, Komori S, Ishihara Y, Shiga K, et al. Dermatomyositis developed in a recipient of allogeneic BMT; the differentiation of chronic GVHD and autoimmune disease. *Bone Marrow Transplant* 2012;**47**(4):603–4.
305. Trallero-Araguas E, Rodrigo-Pendas JA, Selva-O'Callaghan A, Martinez-Gomez X, Bosch X, Labrador-Horrillo M, et al. Usefulness of anti-p155 autoantibody for diagnosing cancer-associated dermatomyositis: a systematic review and meta-analysis. *Arthritis Rheum* 2012;**64**(2):523–32.
306. Gunawardena H, Wedderburn LR, North J, Betteridge Z, Dunphy J, Chinoy H, et al. Clinical associations of autoantibodies to a p155/140 kDa doublet protein in juvenile dermatomyositis. *Rheumatology (Oxford)* 2008;**47**(3):324–8.
307. Shah M, Mamyrova G, Targoff IN, Huber AM, Malley JD, Rice MM, et al. The clinical phenotypes of the juvenile idiopathic inflammatory myopathies. *Medicine (Baltimore)* 2013;**92**(1):25–41.
308. Espada G, Maldonado Cocco JA, Fertig N, Oddis CV. Clinical and serologic characterization of an Argentine pediatric myositis cohort: identification of a novel autoantibody (anti-MJ) to a 142-kDa protein. *J Rheumatol* 2009;**36**(11):2547–51.
309. Gunawardena H, Wedderburn LR, Chinoy H, Betteridge ZE, North J, Ollier WE, et al. Autoantibodies to a 140-kd protein in juvenile dermatomyositis are associated with calcinosis. *Arthritis Rheum* 2009;**60**(6):1807–14.
310. Huber AM, Mamyrova G, Lachenbruch PA, Lee JA, Katz JD, Targoff IN, et al. Early illness features associated with mortality in the juvenile idiopathic inflammatory myopathies. *Arthritis Care Res (Hoboken)* 2013;**66**(5):732–40.
311. Aikawa NE, Jesus AA, Liphaus BL, Silva CA, Carneiro-Sampaio M, Viana VS, et al. Organ-specific autoantibodies and autoimmune diseases in juvenile systemic lupus erythematosus and juvenile dermatomyositis patients. *Clin Exp Rheumatol* 2012;**30**(1): 126–31.
312. Kobayashi I, Okura Y, Yamada M, Kawamura N, Kuwana M, Ariga T, et al. 5 Antibody is a diagnostic and predictive marker for interstitial lung diseases associated with juvenile dermatomyositis. *J Pediatr* 2011;**158**(4):675–7.
313. Sakurai N, Nagai K, Tsutsumi H, Ichimiya S. Anti-CADM-140 antibody-positive juvenile dermatomyositis with rapidly progressive interstitial lung disease and cardiac involvement. *J Rheumatol* 2011;**38**(5):963–4.
314. Cao H, Pan M, Kang Y, Xia Q, Li X, Zhao X, et al. Clinical manifestations of dermatomyositis and clinically amyopathic dermatomyositis patients with positive expression of anti-melanoma differentiation-associated gene 5 antibody. *Arthritis Care Res (Hoboken)* 2012;**64**(10):1602–10.
315. Fujimoto M, Hamaguchi Y, Kaji K, Matsushita T, Ichimura Y, Kodera M, et al. Myositis-specific anti-155/140 autoantibodies target transcription intermediary factor 1 family proteins. *Arthritis Rheum* 2012;**64**(2):513–22.
316. Fujimoto M, Matsushita T, Hamaguchi Y, Kaji K, Asano Y, Ogawa F, et al. Autoantibodies to small ubiquitin-like modifier activating enzymes in Japanese patients with dermatomyositis: comparison with a UK Caucasian cohort. *Ann Rheum Dis* 2013; **72**(1):151–3.
317. Tarricone E, Ghirardello A, Rampudda M, Bassi N, Punzi L, Doria A. Anti-SAE antibodies in autoimmune myositis: identification by unlabelled protein immunoprecipitation in an Italian patient cohort. *J Immunol Methods* 2012;**384**(1–2):128–34.
318. Vancsa A, Gergely L, Ponyi A, Lakos G, Nemeth J, Szodoray P, et al. Myositis-specific and myositis-associated antibodies in overlap myositis in comparison to primary dermatopolymyositis: relevance for clinical classification: retrospective study of 169 patients. *Joint Bone Spine* 2010;**77**(2):125–30.
319. Koenig M, Fritzler MJ, Targoff IN, Troyanov Y, Senecal JL. Heterogeneity of autoantibodies in 100 patients with autoimmune myositis: insights into clinical features and outcomes. *Arthritis Res Ther* 2007;**9**(4):R78.
320. Marie I, Lahaxe L, Benveniste O, Delavigne K, Adoue D, Mouthon L, et al. Long-term outcome of patients with polymyositis/dermatomyositis and anti-PM-Scl antibody. *Br J Dermatol* 2010;**162**(2):337–44.
321. Mahler M, Raijmakers R. Novel aspects of autoantibodies to the PM/Scl complex: clinical, genetic and diagnostic insights. *Autoimmun Rev* 2007;**6**(7):432–7.
322. Rigolet A, Musset L, Dubourg O, Maisonobe T, Grenier P, Charuel JL, et al. Inflammatory myopathies with anti-Ku antibodies: a prognosis dependent on associated lung disease. *Medicine (Baltimore)* 2012;**91**(2):95–102.
323. Rider LG, Miller FW. Classification and treatment of the juvenile idiopathic inflammatory myopathies. *Rheum Dis Clin North Am* 1997;**23**(3):619–55.
324. Seshadri R, Feldman BM, Ilowite N, Cawkwell G, Pachman LM. The role of aggressive corticosteroid therapy in patients with juvenile dermatomyositis: a propensity score analysis. *Arthritis Rheum* 2008;**59**(7):989–95.
325. Ruperto N, Pistorio A, Oliveira S. A randomized trial in new onset juvenile dermatomyositis: prednisone versus prednisone plus cyclosporine versus prednisone plus methotrexate. *Arthritis Rheum* 2012;**64**(Suppl. 10):S1042–3.
326. Ramanan AV, Campbell-Webster N, Ota S, Parker S, Tran D, Tyrrell PN, et al. The effectiveness of treating juvenile dermatomyositis with methotrexate and aggressively tapered corticosteroids. *Arthritis Rheum* 2005;**52**(11):3570–8.
327. Hassan J, van der Net JJ, van Royen-Kerkhof A. Treatment of refractory juvenile dermatomyositis with tacrolimus. *Clin Rheumatol* 2008;**27**(11):1469–71.
328. Riley P, Maillard SM, Wedderburn LR, Woo P, Murray KJ, Pilkington CA. Intravenous cyclophosphamide pulse therapy in juvenile dermatomyositis. A review of efficacy and safety. *Rheumatology* 2004;**43**(4):491–6.
329. Moraitis E, Arnold K, Pilkington C. Effectiveness of intravenous cyclophosphamide in severe or refractory juvenile dermatomyositis—a national cohort study UK and Ireland [abstract]. *Arthritis Rheum* 2012;**64**(Suppl. 10):S130–1.
330. Aggarwal R, Oddis CV, Bandos A. Effect of B cell depletion therapy with rituximab on myositis associated autoantibody levels in idiopathic inflammatory myopathy. *Arthritis Rheum* 2012;**64**(Suppl. 10):S325.
331. Rios Fernandez R, Callejas Rubio JL, Sanchez Cano D, Saez Moreno JA, Ortego Centeno N. Rituximab in the treatment of dermatomyositis and other inflammatory myopathies. A report of 4 cases and review of the literature. *Clin Exp Rheumatol* 2009; **27**(6):1009–16.

332. Cooper MA, Willingham DL, Brown DE, French AR, Shih FF, White AJ. Rituximab for the treatment of juvenile dermatomyositis: a report of four pediatric patients. *Arthritis Rheum* 2007;**56**(9):3107–11.
333. Chung L, Genovese MC, Fiorentino DF. A pilot trial of rituximab in the treatment of patients with dermatomyositis. *Arch Dermatol* 2007;**143**(6):763–7.
334. Boulter EL, Beard L, Ryder C, Juvenile Dermatomyositis Research Group. Effectiveness of anti-tumor necrosis factor agents in the treatment of refractory juvenile dermatomyositis. *Arthritis Rheum* 2008;**2011**(63):Suppl. 10.
335. Sato JO, Sallum AM, Ferriani VP, Marini R, Sacchetti SB, Okuda EM, et al. A Brazilian registry of juvenile dermatomyositis: onset features and classification of 189 cases. *Clin Exp Rheumatol* 2009;**27**(6):1031–8.
336. McCann LJ, Juggins AD, Maillard SM, Wedderburn LR, Davidson JE, Murray KJ, et al. The Juvenile Dermatomyositis National Registry and Repository (UK and Ireland)—clinical characteristics of children recruited within the first 5 yr. *Rheumatology (Oxford)* 2006;**45**(10):1255–60.
337. Guseinova D, Consolaro A, Trail L, Ferrari C, Pistorio A, Ruperto N, et al. Comparison of clinical features and drug therapies among European and Latin American patients with juvenile dermatomyositis. *Clin Exp Rheumatol* 2011;**29**(1):117–24.
338. Sallum AM, Pivato FC, Doria-Filho U, Aikawa NE, Liphaus BL, Marie SK, et al. Risk factors associated with calcinosis of juvenile dermatomyositis. *J Pediatr (Rio J)* 2008;**84**(1):68–74.
339. Lorenzoni PJ, Scola RH, Kay CS, Prevedello PG, Espindola G, Werneck LC. Idiopathic inflammatory myopathies in childhood: a brief review of 27 cases. *Pediatr Neurol* 2011;**45**(1):17–22.
340. Gerami P, Walling HW, Lewis J, Doughty L, Sontheimer RD. A systematic review of juvenile-onset clinically amyopathic dermatomyositis. *Br J Dermatol* 2007;**157**(4):637–44.
341. Plamondon S, Dent PB. Juvenile amyopathic dermatomyositis: results of a case finding descriptive survey. *J Rheumatol* 2000;**27**(8):2031–4.
342. Mukamel M, Brik R. Amyopathic dermatomyositis in children: a diagnostic and therapeutic dilemma. *J Clin Rheumatol* 2001;**7**(3):191–3.
343. Dimachkie MM, Barohn RJ. Inclusion body myositis. *Curr Neurol Neurosci Rep* 2013;**13**(1):321.
344. Limaye V, Luke C, Tucker G, Hill C, Lester S, Blumbergs P, et al. The incidence and associations of malignancy in a large cohort of patients with biopsy-determined idiopathic inflammatory myositis. *Rheumatol Int* 2013;**33**(4):965–71.
345. Mannion ML, Beukelman T. What is the background incidence of malignancy in children with rheumatic disease? *Curr Rheumatol Rep* 2013;**15**(3):310.
346. Costa RM, Dumitrascu OM, Gordon LK. Orbital myositis: diagnosis and management. *Curr Allergy Asthma Rep* 2009;**9**(4):316–23.
347. Kondolot M, Unal E, Poyrazoglu G, Kara A, Yikilmaz A, Kumandas S, et al. Orbital myositis associated with focal active colitis in a teenage girl. *Childs Nerv Syst* 2012;**28**(4):641–3.
348. Gruis KL, Teener JW, Blaivas M. Pediatric macrophagic myofasciitis associated with motor delay. *Clin Neuropathol* 2006;**25**(4):172–9.
349. Rivas E, Gomez-Arnaiz M, Ricoy JR, Mateos F, Simon R, Garcia-Penas JJ, et al. Macrophagic myofasciitis in childhood: a controversial entity. *Pediatr Neurol* 2005;**33**(5):350–6.
350. Israeli E, Agmom-Levin N, Blank M, Shoenfeld Y. Macrophagic myofasciitis: a vaccine (alum) autoimmune-related disease. *Clin Rev Allergy Immunol* 2011;**41**(2):163–8.
351. Lach B, Cupler EJ. Macrophagic myofasciitis in children is a localized reaction to vaccination. *J Child Neurol* 2008;**23**(6):614–19.
352. Pickering MC, Walport MJ. Eosinophilic myopathic syndromes *Curr Opin Rheumatol* 1998;**10**(6):504–10.
353. Dourmishev LA, Dourmishev AL. Activity of certain drugs in inducing of inflammatory myopathies with cutaneous manifestations. *Expert Opin Drug Saf* 2008;**7**(4):421–33
354. Brown Jr. RH, Amato A. Calpainopathy and eosinophilic myositis. *Ann Neurol* 2006;**59**(6):875–7.
355. Baumeister A, Arber S, Caroni P. Accumulation of muscle ankyrin repeat protein transcript reveals local activation of primary myotube endcompartments during muscle morphogenesis. *J Cell Biol* 1997;**139**(5):1231–42.
356. Hewer E, Goebel HH. Myopathology of non-infectious inflammatory myopathies—the current status. *Pathol Res Pract* 2008;**204**(9):609–23.
357. Le RK, Streichenberger N, Vial C, Petiot P, Feasson L, Bouhour F, et al. Granulomatous myositis: a clinical study of thirteen cases. *Muscle Nerve* 2007;**35**(2):171–7.
358. Talbert RJ, Laor T, Yin H. Proliferative myositis: expanding the differential diagnosis of a soft tissue mass in infancy. *Skeletal Radiol* 2011;**40**(12):1623–7.
359. Chen YW, Shi R, Geraci N, Shrestha S, Gordish-Dressman H, Pachman LM. Duration of chronic inflammation alters gene expression in muscle from untreated girls with juvenile dermatomyositis. *BMC Immunol* 2008;**9**:43.
360. Pachman LM, Fedczyna TO, Lechman TS, Lutz J. Juvenile dermatomyositis: the association of the TNF alpha-308A allele and disease chronicity. *Curr Rheumatol Rep* 2001;**3**(5):379–86.
361. Pachman LM, Miller FW. Idiopathic inflammatory myopathies. In: Austen KF, Frank MM, Atkinson JP, Cantor HI, editors. *Samter's Immunologic Diseases*. 6th ed. Philadelphia: Lippincott Wilkins & Williams; 2001. pp. 94–105.

Part VII

Special Clinical Problems

Chapter 43

Neuromuscular Problems of the Critically Ill Neonate and Child

Basil T. Darras and H. Royden Jones, Jr.

INTRODUCTION

Acute neuromuscular disorders occasionally confront the pediatric intensivist and/or the pediatric neurologist. Many of these are primary disorders of the motor unit that present acutely and may require urgent admission to a neurologic critical care unit. We have chosen to discuss these various processes apropos to two respective age groups, infants and children, because some illnesses are unique to infants and toddlers and because similar disorders can have differing presentations in these two age groups.[1,2] This approach creates somewhat artificial categories, but we feel it provides a practical format for evaluating acute weakness in a child. A third category of acute neuromuscular disorders to be discussed relates to those processes that develop in a child with an acute systemic illness requiring intensive care management and, therefore, concomitantly predisposing the child to another set of acute motor unit disorders.

The differential diagnosis for disorders that present with lesions at each level of the peripheral motor unit is specific to newborns and toddlers. Typical examples seen at or shortly after birth include Werdnig-Hoffman disease (spinal muscular atrophy [SMA], type I), neonatal congenital polyneuropathies, an intrauterine onset of Guillain-Barré syndrome (GBS), and congenital myotonic dystrophy or glycogen/lipid storage myopathies. In contrast, previously healthy infants may develop infantile botulism as well as other unique, genetically determined congenital neuromuscular transmission disorders, and very rarely postvaccine poliomyelitis. Most of these disorders typically present during the first 3–6 months of life, although some may not occur until the infant is almost one year old.

In comparison, the older child and adolescent are predisposed to another set of critical care peripheral motor unit disorders that are usually quite similar to those in the adult patient. These include GBS and myasthenia gravis (MG), which have an increased incidence among older children. However, when GBS and MG affect children, both may vary significantly from their traditional adult presentation. The pseudoencephalopathic form of GBS is a good example. The neuromuscular complications of extended intubation and sepsis affecting either the peripheral nerve, or muscle cell, also occur in children. Careful review of clinical records, by someone as junior as an astute intern, or the nondiagnostic results of an imaging study, often lead clinicians to pursue further diagnostic testing. The differential diagnosis of these various motor unit disorders is also significantly aided by utilizing clinical neurophysiology techniques, particularly nerve conduction studies (NCSs) and electromyography (EMG) or electroencephalography (EEG), as well as cerebrospinal fluid (CSF) analysis. These studies provide the clinician with an objective means to assign a specific pathoanatomic or neurophysiologic site for the child's illness.

Lastly, there is an interesting but relatively uncommon set of neuromuscular disorders that develop in a child predominantly in association with the following three conditions: (1) treatment with a nondepolarizing neuromuscular blocking agent, (2) high-dose systemic glucocorticoid treatment, or (3) sepsis or other severe illness. These processes may independently affect each primary portion of the peripheral motor unit from the anterior horn cell to the peripheral nerve, neuromuscular junction, or muscle fiber. Thus, in comparison to the other disorders discussed in this chapter wherein a primary individual motor unit disorder presents in an acute fashion requiring intensive care management, this set of illnesses is usually serious but self-limited, and somehow secondary to a toxic process related to the underlying illness.[3]

B.T. Darras, H. Royden Jones, Jr., M.M. Ryan & D.C. De Vivo (Eds): Neuromuscular Disorders of Infancy, Childhood and Adolescence, Second edition.
DOI: http://dx.doi.org/10.1016/B978-0-12-417044-5.00043-3

The acute care motor unit disorders discussed in this chapter illustrate a broad anatomic spectrum from the anterior horn cell to the muscle cell. This review is based on our 35-year experience (1979 to 2014) evaluating children, ages newborn through 18 years, at the Boston Children's Hospital (BCH) Neuromuscular Program, intensive care units, and EMG laboratory.

ACUTE NEUROMUSCULAR CRISES IN THE INFANT

Babies may present during the neonatal period with floppy infant syndrome, or acutely with a flaccid paralysis after being healthy during their first few months of life.[4] Some newborns, presenting primarily with an acute respiratory compromise during the immediate neonatal period, are concomitantly observed to be hypotonic, sometimes to the degree of being identified as a floppy baby. A variety of congenital or developmental lesions at any level of the peripheral motor unit may lead to this picture (Box 43.1).

Another subset during the first year of life includes those previously healthy infants who unexpectedly develop a precipitous neuromuscular crisis. Specific clinical mechanisms that require consideration in this instance include a previously unrecognized SMA, Guillain-Barré syndrome, infantile botulism, congenital myasthenic syndromes, or very rarely postvaccine poliomyelitis.

Anterior Horn Cell

The Werdnig-Hoffman form of SMA or SMA type I is one of the most common causes of floppy infant syndrome as well as the most common anterior horn cell disorder of infancy.[1,2] Usually, these infants either present as a floppy baby, or with failure to develop or loss of early motor milestones between 1–3 months of age. Rarely, SMA (classified as type 0 or type IA) may present at birth with profound hypotonia, severe weakness and respiratory failure.[5,6] Eventually some SMA type I infants will require admission to a neonatal intensive care unit (NICU). Their diagnosis is usually made by a combination of EMG and genetic DNA analysis or just DNA analysis.[1] Occasionally, their SMA is not initially recognized and it is not until they are unable to cope with a respiratory illness that they present to the ICU requiring pulmonary support or cannot be weaned from the ventilator. See Case Example 43.1.

Another form of acute anterior horn cell disease, vaccine-associated poliomyelitis (see Case Example 43.2), deserves mention in the differential diagnosis of any infant or child with an acute weakness.[8]

Peripheral Nerve

Some congenital hypomyelinating or axonal polyneuropathies present in the newborn period with a clinical phenotypic appearance similar to that of SMA I, Werdnig-Hoffman disease. These infants may be so ill that they require intensivist support.[1] See Case Example 43.3.

The Guillain-Barré syndrome (GBS) rarely affects newborn infants. It needs consideration in the differential diagnosis of neonatal flaccid paralysis.[13–16] In one instance GBS onset was soon after birth.[13] The baby's mother had an autoimmune disease, chronic ulcerative colitis. During the 30th week of her pregnancy she had observed significantly decreased fetal movements.[13] Her baby was quadriparetic when delivered at 37 weeks with an Apgar score of 6 at 10 minutes. At age 3 days neurologic exam demonstrated severe leg and moderate proximal arm weakness, with generalized absence of muscle stretch reflexes. The CSF protein was normal, 38 mg/dL. In contrast motor NCS demonstrated profound motor NCV slowing between 3 and 15 m/sec, with conduction block and temporal dispersion in many nerves. Needle EMG revealed active denervation in many muscles. He had gradual and complete recovery by age 1 year without any specific treatment.[13]

Another pregnant mother, aged 33, developed severe GBS during the 29th week of her pregnancy.[14] She became tetraplegic and was still respirator dependent when her baby was born at the 38th week. He developed hypotonia, marked respiratory distress, and feeding

BOX 43.1 Conditions Leading to Acute Neuromuscular Crises in the Infant

Anterior Horn Cell
- Spinal muscular atrophy, type 0, type I
- Vaccine-associated poliomyelitis

Peripheral Nerve
- Congenital hypomyelinating or axonal polyneuropathies
- Guillain-Barré syndrome*

Neuromuscular Junction
- Transient neonatal myasthenia gravis
- Congenital myasthenic syndromes
- Infantile botulism
- Magnesium treatment in the mother

Muscle
- Congenital myotonic dystrophy
- Congenital myopathies
- Metabolic/mitochondrial myopathies

* Guillain-Barré syndrome rarely occurs during infancy.

CASE EXAMPLE 43.1 Werdnig-Hoffman Disease Presenting with Respiratory Distress

This previously "healthy" 4-month-old first child was admitted with an acute bronchiolitis and required intubation. Subsequently, when his pulmonary status had significantly improved, his pediatricians had unexpected difficulty weaning him from the respirator. Neurologic consultation demonstrated generalized hypotonia and absent muscle stretch reflexes.

An EMG was performed because of its ability to provide a rapid means to localize the anatomic site for the baby's motor unit abnormality. This demonstrated low amplitude compound muscle action potentials (CMAPs), normal motor conduction velocities (MCVs) and distal latencies (DLs), as well as sensory nerve action potentials (SNAPs). Needle EMG demonstrated a marked decrease in the number of motor unit potentials (MUPs) with widespread active denervation characterized by many fibrillation potentials and positive waves. Subsequent to the EMG, DNA analysis was positive (homozygous deletion of exon 7 of the survival of motor neuron 1 [*SMN1*] gene, which is the most common mutation in SMA patients).

Comment

Infants with SMA are usually easily recognized because of their typical clinical presentation as a floppy baby with an alert facies, absent muscle stretch reflexes, and fasciculations of their tongue. In this instance, DNA analysis of the *SMN1* gene is the appropriate means to make the diagnosis today. Occasionally, however, parents and pediatricians have not recognized a relatively subtle clinical presentation until an acute illness leads to hospitalization. In the intensive care setting, as above, one may need to make a more rapid diagnosis, utilizing EMG, to provide appropriate clinical management as well as counseling of the parents. Rarely, infants with non-5q SMAs may have similar clinical and EMG findings.[7]

CASE EXAMPLE 43.2 Acute Post-immunization Poliomyelitis

A previously healthy 11-week-old boy presented with a 2-day history of fever, irritability, lethargy, and head lag. CT examination of the head and brain was normal. CSF analysis demonstrated 580 white blood cells (WBC), 41% polymorphonuclear leukocytes, 59% lymphocytes; glucose 59 mg/dL and protein 143 mg/dL. Although intravenous antibiotics were initiated, the baby soon developed apneic spells requiring intubation. This infant soon developed a weak cry and asymmetric skeletal muscle weakness particularly sparing left plantar flexor and extraocular muscles.

On the seventh day of his illness an EMG was performed demonstrating an anterior horn cell disease pattern. Subsequently repetitive motor nerve stimulation (RMNS) failed to identify a neuromuscular transmission defect (NMTD).

Review of this baby's immunization records documented that he had received type 3 Sabin live poliomyelitis vaccine 3 weeks prior to onset of his illness. His recovery was poor; he was ventilator dependent more than 2 years later.

Comment

Three cases of acute flaccid paralysis occurring in infants, secondary to immunization-related poliomyelitis, became known to one of the authors (HRJ) within just a few years during the mid-1990s.[8,9] (Goldstein J, personal communication, 1995). All three babies were 3–4 months old, each having received type 3 live polio immunization less than one month earlier. An acute febrile illness preceded a progressive asymmetric extremity weakness, head lag, irritability, and lethargy. A CSF pleocytosis with 100–580 WBC was associated with protein values between 82–143 mg/dL in all and glucose of 49 mg/dL in one.[8] EMG demonstrated classic electrophysiologic evidence of an acute asymmetric anterior horn cell disorder.

Most instances of paralytic poliomyelitis from the attenuated live Sabin oral vaccine occurred among infants less than age 1 year usually subsequent to their first polio vaccine, and within 2 weeks of immunization.[10] Revised CDC guidelines require that the initial immunization protocol utilize the Salk killed vaccine as the primary immunization method for all children.[11] This revised infantile immunization protocol has made postvaccination related poliomyelitis of historical interest in North America. However, the live vaccine is still being utilized in certain countries.

problems 12 days postpartum although he was well at birth. His CSF protein was 243 mg/dL. An EMG was typical for GBS. Both the mother and child had IgG antibodies positive for recent CMV infection. Intravenous immunoglobulin therapy was associated with a complete symptomatic resolution in 2 weeks.[14] One presumes his GBS developed *in utero*.[14]

GBS affected another pregnant woman. Her baby had poor spontaneous ventilation at delivery requiring brief intubation. During the next five days of observation, there were no other signs of GBS.[15] Despite such, both this infant and his mother had high antibody titers to human peripheral nerve myelin glycolipids.[15]

Even without any known intrauterine events, the Guillain-Barré syndrome may occur shortly after birth, as early as age 3 weeks. These previously healthy babies present with an acute, rapidly progressive and often severe hypotonia, with possible respiratory distress, and

CASE EXAMPLE 43.3 Severe Newborn Infantile Demyelinating Polyneuropathy

A 2-week-old old infant was delivered after what the mother perceived as a normal pregnancy with normal fetal movements *in utero*. Since birth, he had a "paucity of spontaneous movements," and a weak cry. His mom stated that he was still able to breastfeed although he did not hold the pacifier tightly. His gag was weak. Fasciculations of the tongue were present. This alert baby was markedly hypotonic with frog leg posture, no spontaneous movements of the legs, wrist drops, weak palmar grasp, and arthrogryposis multiplex. He assumed the inverted U posture when held prone. He was areflexic and the plantar responses were mute. He responded to noxious stimuli applied to all extremities.

The cerebrospinal fluid (CSF) protein was elevated at 190 mg/dL. Sensory NCS failed to demonstrate any median SNAP. Median and peroneal CMAPs were widely dispersed, of very low amplitude (0.05 and 0.02 mV), and had prolonged latencies (3.6 and 8.2 msec), and there were very slow MCVs of 6 and 8 m/sec. EMG demonstrated mild diminution in the number of MUPs which were of slightly longer duration, polyphasic, and normal amplitude and firing at increased frequency. The only areas with occasional fibrillations and positive waves were present in the most distal muscles. These findings were compatible with an acquired, diffuse, predominantly demyelinating, peripheral neuropathy.

Sural nerve biopsy demonstrated hypomyelination with abortive onion bulb formation. These findings did not provide a base to distinguish between an inborn error of myelination or an unusual acquired autoimmune demyelinating peripheral neuropath, such as intrauterine GBS. A trial of prednisone was unsuccessful. Subsequently, progressive respiratory failure ensued. The baby died at age 2 months; an autopsy was declined.

Comment

Any floppy infant presenting with a frog leg posture and tongue fasciculations most likely has SMA type 0 or type I. In contradistinction, this baby's EMG findings of very slow MCVs, with dispersed low amplitude CMAPs, and absent SNAPs, were diagnostic of an acquired demyelinating polyneuropathy and not an abnormality primarily at the motor neuron level. The incidence of peripheral neuropathies was 6 percent in our BCH review of over 100 floppy babies seen in our EMG lab.[1,2] These were more common among the infants who also had arthrogryposis.[2,12] The absence of sensory responses is the most important pediatric EMG clue to the appropriate diagnostic conclusion that the baby has one of these rare neonatal neuropathies. The nerve pathologies included primary axonal, demyelinating, and widespread neuronal degeneration.

feeding difficulties.[17,18] Chronic inflammatory demyelinating polyneuropathy (CIDP) subsequently developed in one 7-week-old infant with acute GBS.[19] An EMG is particularly helpful for the differentiation of GBS from other acute floppy infant syndromes.[17] Absent SNAPs, profound motor conduction slowing, and dispersed CMAPs are the quintessential EMG clues for the diagnosis of neonatal GBS. In our BCH EMG review of more than 100 floppy babies, there was a 6 percent incidence of peripheral neuropathies.[2] Only one of these six infants had EMG findings commensurate with GBS.[2]

Although tick paralysis mimics acute GBS in children, we are unaware of an infantile case. Nonetheless, any baby with an acute onset of generalized muscle weakness needs to be searched for a tick, particularly in the scalp.[20]

Neuromuscular Junction (NMJ)

Various NMJ disorders are occasionally encountered during infancy. Autoimmune maternal associated myasthenia gravis is the most easily recognized and probably most common NMJ pathologic entity occurring in this age group. This relatively short-lived disorder typically occurs at birth in 15–20% of infants whose mothers have myasthenia gravis.[21–23] Rarely, infants born to mothers who had eclampsia and received magnesium treatment also have the potential to develop an acute infantile presynaptic NMJ disorder.[24,25] These two NMTDs only occasionally come to the attention of the pediatric neurologist as pediatricians are well versed in the management of these infants.

There are also a number of relatively rare congenital myasthenic syndromes (CMSs) with a primary pathophysiologic site of difficulty varying between the presynaptic NMJ, the synapse, or the postsynaptic NMJ. An endplate deficiency of choline acetyltransferase is a CMS that predisposes the baby to an acute onset NMJ disorder.[26] Typically, these babies have recurrent episodic apnea. In contrast, it is unusual for other CMS forms to have such an acute presentation.

Infantile botulism is the one other presynaptic NMTD that will also acutely affect previously healthy babies and toddlers (Case Example 43.4). Clinical suspicion, with subsequent EMG, is crucial to making this diagnosis. Because of its somewhat varied presentation we are concerned that possibly infantile botulism is more common than recognized.

Muscle

Congenital myopathies typically present as a floppy infant syndrome but most babies are not ill enough to require NICU monitoring. In addition, only a small subset of myopathies cause acute infantile flaccid paralysis and/or

CASE EXAMPLE 43.4 Infantile Botulism

A previously healthy, 4-month-old baby was admitted to another hospital with a 1-week history of constipation, and subsequently poor feeding, lethargy, a weak cry and dehydration. She had lost her ability to roll over, lift her head, or sit up. A full septic evaluation was unremarkable including CSF examination. After she developed cyanosis with an acute apneic spell, and subsequent increasing lethargy, she was transported to BCH.

Attempt to wean her from the ventilator was unsuccessful. Neurologic examination here demonstrated no extraocular muscle movements, absent corneal responses, and no gag reflex. At this time she still had movement in all extremities, and intact muscle stretch reflexes (MSRs). The initial clinical impression was an "indeterminate encephalopathy." This conclusion was based on nonspecific EEG findings in the right occipital lobe, as her brain MRI, CSF, bacterial and viral cultures, and metabolic evaluation were all normal. During this evaluation, she became increasingly hypotonic and areflexic. Seven days after onset of symptoms and subsequent to further clinical review by a diligent intern, at change of service, an EMG was requested for possible infantile botulism.

The EMG evaluation, in the MICU, demonstrated low amplitude median and peroneal CMAPs (0.56 mV and 0.2 mV, respectively; normal >3.5 mV and 1.6 mV) with normal motor NCV, and distal latencies. Repetitive motor nerve stimulation (RMNS) of the median nerve at 2 Hz demonstrated an 11% decrement, and with 20 Hz stimulation a 160% facilitation. Motor unit potentials were very "myopathic." These EMG findings were consistent with a diagnosis of infantile botulism. Subsequently, a stool sample, as well as an assay of a honey specimen from home, were both positive for type A botulinum toxin. She gradually improved, with supportive treatment, during the next few months.

Comment

Infantile botulism usually has a stereotyped clinical presentation. Typically, a previously healthy infant, between ages 10 days and 12 months, has the acute onset of hypotonia, generalized descending weakness, poor feeding, oculomotor and pupillary abnormalities, dysphagia, and constipation.[27–32] However, one also needs to consider this diagnosis in the setting of unexplained respiratory distress in any baby under 12 months of age. Infantile botulism has been linked to honey ingestion. In its most severe state the presentation may be acute and require ventilatory support. Treatment with human botulinum antitoxin can expedite the infant's recovery.[33,34]

Bacteriologic confirmation usually requires a few weeks, but a mouse bioassay and real-time PCR assays can provide a rapid confirmation of the diagnosis.[35] EMG is the most useful early diagnostic tool. Most babies with infantile botulism have clear-cut evidence of a NMTD. Low rate RMNS, at 2–5 Hz, often but not always demonstrates a decremental response.[36] More rapid RMNS rates of 20–50 Hz provide the primary electrophysiologic technique for diagnosing infantile botulism.[36] A post-tetanic facilitation (PTF) varying between 23% and 313%, with a mean of 73%, was documented in 23 of 25 babies.[36] A longer period of post-tetanic facilitation is also observed. One may not always be able to confirm a diagnosis of infantile botulism with EMG as not all cases have a documented facilitation with RMNS.[36,37] However, in most instances of infantile botulism, babies will have a typical EMG diagnostic triad.[38] This includes: (1) low amplitude CMAPs (beware normal maturation standards for age), (2) post-tetanic facilitation, and (3) absence of post-tetanic exhaustion.[37]

respiratory distress. However, occasionally we are asked to see a NICU baby who is found to have a primary myopathy. See Case Examples 43.5 and 43.6.

ACUTE NEUROMUSCULAR CRISES IN THE TODDLER AND THE OLDER CHILD

A variety of conditions can lead to acute neuromuscular crises in toddlers and older children (see Box 43.2).

Anterior Horn Cell

We are pleased to comment that since the inception of our EMG lab in 1979 at BCH we are not aware of the occurrence of any acute poliomyelitis, beyond infancy. However, one needs to keep this possibility always in mind, as there are some parents who reportedly still decline to have their children appropriately immunized. In addition, with the increased use of air travel and children being adopted, immigrating, or visiting from all parts of the world, one always needs to be alert to the possibility of poliomyelitis in countries with less than ideal immunization standards. However, we do see a few instances of subacute polio in children coming from those nations where standard immunization is not routinely available.[44] The pediatric neurologist must keep poliomyelitis within the differential diagnosis of any child with an acute, particularly asymmetric, flaccid paralysis.

Peripheral Nerve

Today, the Guillain-Barré syndrome (GBS) is the most common cause for an acute generalized paralysis in the older child. A few rare other primary pediatric polyneuropathies occasionally develop. In addition, some of these clinical settings are caused by some unusual neuromuscular transmission defects as noted later in this chapter. In the ICU setting, an EMG is particularly important for

CASE EXAMPLE 43.5 Neonatal Myotonic Dystrophy

Respiratory distress developed soon after birth in this baby girl, with clubfeet. She required intubation for her first month of life. An EMG was requested for evaluation of her generalized hypotonia when she was 2 months old. Motor NCS demonstrated low amplitude CMAPs but other parameters were normal including SNAPs. Profound and prolonged decrescendo myotonic discharges occurred during her EMG. These were associated with rare fibrillation potentials. Her MUPs were normal. The mother had no myotonia detectable by physical examination or on EMG but she recalled that her father had difficulty releasing his grip when driving a car.

Comment

Myotonic dystrophy type 1 (DM1) is one of the few dystrophies that presents as a severely floppy infant. Babies with congenital DM1 may have severe generalized hypotonia requiring intubation at birth for their associated respiratory compromise. Sometimes, the diagnosis is suspected clinically by their classic facies with "tenting" of the mouth that appears like an upside down letter V. Most congenital DM1 infants have a family history of myotonic dystrophy, but it is often unrecognized in the parent. Typically, the mother has a milder form requiring premature cataract removal or has evidence of mild subclinical myotonia and/or distal weakness on clinical exam.[39]

Today when confronted with the typical clinical phenotype, DNA testing for the expansion of the CTG repeat in the *DMPK* gene is very accurate in infants with DM1 as well as their symptomatic or asymptomatic mothers. Myotonia detectable by EMG may be present in the hypotonic newborn infant, as early as age 5 days, although it sometimes is not always evident this early on.[40,41] Occasionally, one finds that a maternal EMG will detect previously unsuspected myotonia.[1] Nowadays, a DNA test will quickly establish the diagnosis.

CASE EXAMPLE 43.6 Metabolic Myopathy: Phosphofructokinase (PFK) Deficiency

A 39-weeks' gestation newborn infant had a history of diminished fetal movement and developed immediate respiratory distress at birth. He had severe generalized and bifacial weakness, a high arched palate, absent muscle stretch reflexes, and marked distal arthrogryposis. At 2 days of age, NCS demonstrated low amplitude to absent CMAPs with otherwise normal motor and sensory conduction parameters. EMG demonstrated very low amplitude, short duration MUPs primarily in proximal muscles typical of a myopathy. Muscle biopsy demonstrated an absence of the glycolytic enzyme PFK. Swoboda[42] suggested that a ketogenic diet might be beneficial. This commenced at age 4 months. He gained some clinical motor improvement through age 2 years; however, he died of other complications later that year.

Comment

Metabolic myopathies rarely present with severe hypotonia and respiratory weakness in the newborn period. A fatal case of phosphorylase deficiency presented at 4 weeks of age with feeding difficulties, tiring easily, and severe respiratory distress. This infant had a rapidly progressive course, becoming very hypotonic with generalized weakness.[43] Muscle biopsy shows an excess of glycogen accumulation in infants with PFK deficiency; the diagnosis is confirmed with enzyme assay and/or genetic analysis.

BOX 43.2 Conditions Leading to Acute Neuromuscular Crises in the Toddler and Older Child

Anterior Horn Cell
- Spinal muscular atrophies (5q and non-5q)
- Poliomyelitis in the nonimmunized child

Peripheral Nerve
- Guillain-Barré syndrome
  - acute demyelinating
  - axonal forms
- Glue sniffing
- Buckthorn wild cherry ingestion
- Insecticides
- Heavy metal poisoning
- Vincristine treatment in patient with CMT1
- Diphtheria
- Acute mononeuritis multiplex

Neuromuscular Junction
- Autoimmune myasthenia gravis
- Congenital myasthenic syndromes
- Tick paralysis

Muscle
- Channelopathies / periodic paralyses
- End-stage muscle disease
- Inflammatory myopathies
- Metabolic myopathies

Other
- Transverse myelitis
- Spinal cord injury

rapid localization of the acute motor unit lesion. It may also prove helpful for making prognostic judgments. Case Examples 43.7 and 43.8 illustrate some of the interesting GBS variants.

An acute vincristine exacerbation of an underlying polyneuropathy may occur if this chemotherapeutic agent is prescribed to children with a mild, subtle, or even unrecognized Charcot-Marie-Tooth type 1 (CMT1) hereditary neuropathy. This toxic neuropathy may be relatively severe when superimposed on this underlying genetic neuropathic process. As sometimes CMT1 is unsuspected early in life, it is important to query the parents before any vincristine therapy is initiated, as to the presence of any personal or family history of CMT1 by asking about

CASE EXAMPLE 43.7 Guillain-Barré Syndrome: Pseudo-encephalopathy, Painful Weakness, and Meningismus

Two weeks prior to evaluation at BCH this previously healthy 5-year-old boy participated in an Audubon field trip to collect insects along the Rhode Island coast. He soon developed a fine papular rash on his face and neck that was followed 24 hours later by headache, nausea, and vomiting during a family sailing cruise. Bilateral pain developed in his knees, posterior thighs, and popliteal fossa 4 days prior to our evaluation. A viral syndrome was diagnosed and the pediatrician treated him symptomatically. Subsequently he lay in his bunk complaining of headaches and occasional vomiting but no fever. He was brought to BCH one week later.

Here he was extremely uncomfortable, groaning and writhing. His conjunctivae were injected, and his neck stiff. Muscle strength was "grossly" normal, and capable of supporting his weight to stand. Sensation seemed intact; however, his muscle stretch reflexes were not elicitable. Brain CT was normal. Subsequent cerebrospinal fluid (CSF) exam demonstrated a protein of 145 mg/dL, glucose 65 mg/dL, and 1 WBC per dL. On neurologic examination the next AM he was listless with bilateral Babinski signs. Muscle strength was still difficult to assess.

Motor NCS were of variably low amplitude with dispersed CMAPs, and multifocal asymmetric prolongation of DLs. Motor NCV varied from 22 m/sec for median to 38 m/sec for ulnar and 50 m/sec for peroneal fibers. Peroneal F waves were absent. No median or ulnar SNAPs were elicited. Needle EMG was normal. This EMG was compatible with an acute demyelinating polyneuropathy typical for GBS.

Comment

With the history of headache and rash, in an agitated child with Babinski signs, an encephalopathic process was the initial diagnostic consideration. After the brain CT was normal, a spinal fluid examination was performed. The high CSF protein led to the consideration of an atypical form of GBS. Other neurologic considerations included a transverse myelitis, an entity that also occasionally mimics GBS; however, there was no evidence of a cord level in this instance. Spinal MRI was therefore not performed. Some individuals with GBS appear to occasionally have an associated, but indeterminate, central nervous system process such as locked-in syndrome, indicated by the presence of acute tetraplegia, areflexia, and cranial nerve involvement despite otherwise typical clinical and EMG findings for GBS. These patients, however, are fully conscious.[45]

Severe pain was the primary presenting symptom of GBS in 21% (5 of 24) children seen at BCH.[46] A significant degree of pain was present in 67% of our GBS cases.[45] At times, as in this instance, this discomfort was so bothersome and led to such irritability, that a number of these children were initially thought to have a meningoencephalopathy.[47]

CASE EXAMPLE 43.8 Axonal Guillain-Barré Syndrome with "Brain Death" Presentation

A 6-year-old boy presented with a 2-week history of headaches and 2 days with difficulty walking. He had had surgery and chemotherapy for a medulloblastoma 8 months previously, and subsequently was feeling fine. His last vincristine, cisplatin, and cyclophosphamide were completed 2 months earlier. Suddenly, he developed acute gasping and shortly thereafter had a respiratory arrest. On admission to the hospital he had no spontaneous movements and he was unresponsive to all noxious stimuli. He had total external ophthalmoplegia and his pupils were fixed and dilated. No muscle stretch reflexes or corneal, oculocephalic, and gag responses were elicitable. Plantar stimulation was mute. Initially he was thought to be "brain dead." His CSF had a protein 167 mg/dL with just 6 WBC. When an EEG was unexpectedly normal, a toxic screen and testing for heavy metals and porphyrins were also unremarkable.

An EMG was first performed on the 5th day of his illness. No motor or sensory responses, or MUPs, were identified. However, fibrillations and positive waves were present in all muscles. Subsequent trial of IVIg therapy did not result in any improvement for a few weeks. Eventually his cranial nerve function slowly recovered followed by respiratory and limb movements. One year later he was able to ambulate with a walker. His arm strength was then normal.

Comment

It is very rare to see fixed dilated pupils with GBS as occurred here. Both diphtheria and botulism are diagnostic considerations with this clinical set particularly when there is predominant bulbar involvement.[48] Myasthenia gravis or poliomyelitis also enters into this differential diagnosis but in these instances the pupillary responses are preserved.

the presence of high arches, weak feet, etc. Testing for the specific genetic defect (a 17p11.2-12 duplication of the PMP22 gene) needs consideration in any child where there is any clinical suspicion of CMT1 prior to treatment with vincristine. One clinical example of such a severe neuropathy occurred in an immuno-compromised child with an unrecognized family history of CMT1.[49]

Prognosis in these pediatric GBS variants is determined by a combination of clinical and/or EMG signs of very significant axonal damage. Often children with GBS have a relatively benign prognosis.[46] In some clinical instances, however, a more prolonged recovery, or even rarely a fatality,[50] may occur. Examples include the child who requires a respirator, those in whom one cannot evoke a CMAP with NCS during an EMG, and those children who have very widespread active denervation on needle EMG. This last group may often require more than a year for reinnervation and subsequent recovery.[48]

The prognosis for recovery in children with GBS is far better than for adults; however, there is no universal agreement on this opinion.[51] Approximately 15 percent of hospitalized children with GBS require mechanical ventilatory support during the course of their admission.[52] It is important to emphasize the need for the intensive-care setting, where it is possible to monitor cardiovascular and autonomic function.[53] In one report, of nine patients who died of Guillain-Barré syndrome, two were children, one of whom died of cardiac arrest; the other had periods of bradycardia and hypotension and became neurologically unresponsive.[50] Fluctuating blood pressure and autonomic instability have been recognized as predictors of a potentially fatal cardiac arrhythmia.[50] Careful monitoring of respiratory function, good management of fluid balance, and avoidance of bed sores and pressure neuropathies are other important areas for care of the child with Guillain-Barré syndrome. A significant cause of death in adults is pulmonary embolism; however, the literature is not clear as to the value of subcutaneous heparin for the prophylaxis of venous thrombosis in a child.[17]

Guidelines as to when to transfer a child with GBS to the ICU and when to proceed with intubation have not been developed. Respiratory failure should be expected in any GBS patient with progressive appendicular or bulbar weakness. In children with GBS cared for on a regular pediatric ward, pulmonary insufficiency may develop insidiously and remain clinically silent without obvious signs of respiratory failure. Hypoxemia may not be detected until the vital capacity has fallen significantly, and hypercarbia is a late finding; therefore, blood gases are not useful in monitoring the diaphragmatic weakness in GBS patients. In older children who can cooperate for pulmonary function testing, bedside measurements of vital capacity (VC) and maximal inspiratory (PI_{max}) and expiratory (PE_{max}) pressures should be carried out every 6 to 8 hours. In the younger uncooperative child, oxygen saturation, especially in sleep, should be monitored by pulse oximetry. It should be noted, however, that a fall in oxygen saturation is a late finding in children receiving oxygen.

Again, practice parameters allowing the prediction of need for mechanical ventilation have not been established in children. However, even in the adult literature the respective guidelines are not unanimous. Box 43.3 describes the indications for a transfer of a child with GBS to the ICU and includes our criteria for elective intubation. Important clinical indicators of impending respiratory failure included bulbar dysfunction and autonomic instability, bilateral facial palsy, and rapid disease progression. The presence of these clinical parameters was associated with increased likelihood of mechanical ventilation. In our experience, however, bilateral facial palsy in children has not been a strong predictor of impending respiratory failure. Analysis of serial respiratory function measurements showed that a 30% reduction in VC, PI_{max} or PE_{max} were associated with subsequent progression to respiratory failure and intubation. Also, the following "threshold" respiratory values—vital capacity less than 20 mL/kg, maximal inspiratory respiratory pressure less than 30 cm of H_2O, and maximal expiratory respiratory pressure less than 40 cm of H_2O—were shown to be highly associated with subsequent progression to respiratory failure. In adults, the use of the so-called 20/30/40 rule allows patients-at-risk to be identified and to be preemptively admitted to the ICU for close monitoring and elective intubation under optimal conditions. In the past, clinical signs of respiratory fatigue and severe bulbar

BOX 43.3 Indications for Transfer of Child with GBS to the Intensive Care Unit at Boston Children's Hospital

1. Rapid disease progression
2. Autonomic instability
3. Bulbar dysfunction
4. Aspiration episode
5. More than 30% reduction in VC from baseline
6. VC less than 20 mL/kg
7. NIF less than 30 cm H_2O
8. Recurrent oxygen desaturation events with values below 90%*

* In a young child not receiving oxygen; however, a fall in oxygen saturation may be a late sign in patients with GBS. In older cooperative children, serial pulmonary function studies are far more reliable predictors of impending respiratory failure than pulse oximetry.
Abbreviations: VC = vital capacity; NIF = negative inspiratory force.
Nos. 5–8 should be considered indications for elective intubation.

weakness with aspiration, or fall of the VC to less than 15 mL/kg or arterial PO_2 values to less than 70 mmHg, had been defined as criteria for intubation.[54] Although these criteria might be accurate in determining the need for intubation, they cannot be considered as practice guidelines for preemptive respiratory management of children with GBS. When serial bedside respiratory function testing shows constant decline of VC and NIF from baseline, the patient should be transferred to the ICU and elective intubation should be considered (Box 43.3). It is also prudent to have young children with oxygen desaturation events with values falling below 90 percent and/or signs of dysautonomia, bulbar dysfunction, or aspiration transferred to the ICU for monitoring and possible intubation. One also needs to be aware of the fact that dysautonomia in GBS patients may also be triggered or accentuated by tracheal manipulation (i.e. intubation, suctioning). Because of the propensity of GBS patients for upper airway collapse, noninvasive methods of respiratory support such as continuous or bilevel positive airway pressure methods have probably a limited role in the respiratory management of severe GBS patients with borderline respiratory values. Nocturnal decompensation is also a common occurrence in patients with severe GBS due to a combination of reduction in vital capacity related to the supine position and also impairment of central respiratory drive; thus the need for close respiratory monitoring during sleep. Immune therapies, plasma exchange and high-dose intravenous immunoglobulin have been shown to be of equal efficacy and are known to expedite recovery from GBS.[55]

Another major immediate diagnostic consideration in the differential diagnosis of GBS relates to excluding an acute spinal cord tumor or transverse myelitis (TM). Children with either lesion may also be admitted to the medical intensive care unit.[56–59] Both of these disease entities may acutely produce a rapidly progressive paralysis, hyporeflexia, and back pain. More commonly spinal cord lesions, and rarely GBS, are associated with early sphincter dysfunction. However, in contrast to spinal lesions, these symptoms are usually transient with GBS.[46] Four children had severe pain, asymmetric lower extremity weakness, and a clear-cut sensory level on careful neurologic examination.[56] MRI was abnormal in 4 of the 8 children evaluated.[56] Each child had a malignant spinal cord tumor.[56] Even imaging studies may cause confusion. One 6-year-old boy developed progressive weakness and areflexia, and his MRI showed transient spinal cord swelling.[57] Both the clinical and EMG findings were commensurate with a GBS diagnosis. If there is enlargement of the spinal cord, swelling with increased signal intensity of the cord, and no contrast enhancement, a diagnosis of GBS is not excluded because, rarely, TM and GBS may occur concurrently.[57,58] TM is the primary spinal cord lesion that may occasionally produce significant confusion in the differential diagnosis of GBS.[59]

It is important to emphasize that although the absence of F-waves is one of the most common early signs of GBS, similar findings are also found with transverse myelitis if the segmental lesion affects the precise spinal level for the legs or arms.[46] EMG is generally normal in this setting; however, if segmental anterior-horn cells are affected, both the CMAPs and F-waves may be affected. This leads to neurophysiologic mimicking of the typical early GBS findings. MR imaging usually confirms the diagnosis.

Mononeuritis Multiplex (MNM)

During our 35-year experience at BCH, we have evaluated just one child with an acute fulminating mononeuritis multiplex affecting a very ill teenager in renal failure (Case Example 43.9). We have seen only two instances of MNM in 35 years at BCH. This limited experience has been reported; two of the three cases were related to systemic vasculitis and one possibly represented a case of nonsystemic vasculitis.[61–63]

Neuromuscular Junction Disorders in the Older Child

The differential diagnosis of acute childhood bulbar and generalized weakness requires consideration of a neuromuscular transmission defect (NMTD). Myasthenia gravis, the Miller-Fisher variant of GBS, and tick paralysis are the primary considerations. When myasthenia gravis occurs in children, it is typically quite similar to its presentation in adults. However, occasionally one may see some clinical variants that lead to an intensive care admission before the diagnosis of myasthenia gravis is even entertained. See Case Example 43.10.

Muscle

Channelopathies

Some of the various channelopathies, particularly hyperkalemic and hypokalemic periodic paralysis, always warrant consideration in the differential diagnosis of acute GBS. As these are autosomal dominant sodium or calcium channelopathies, there is usually a well-documented family history. Clinically the episodes of paralysis are relatively short-lived on most occasions. Routine NCS findings are normal; however, prolonged periods of exercise with EMG monitoring can evoke a diminution in CMAP amplitudes.[67] Needle EMG often demonstrates

CASE EXAMPLE 43.9 Acute Mononeuritis Multiplex

This 16-year-old girl with systemic lupus erythematosus had dialysis dependent, membranous, proliferative glomerulonephritis associated with hypertension, pericardial effusion, and seizures. She was not a renal transplant candidate because of medication noncompliance. A sudden onset of left-hand numbness occurred just one month before we initially evaluated her. She developed an acute left foot drop shortly thereafter followed by an acute numbness in her right hand and an acute right foot drop the very next day. She became incapacitated. Neurologic examination demonstrated a left median and ulnar and bilateral asymmetric peroneal neuropathies.

EMG demonstrated a mononeuritis multiplex. Sural nerve biopsy confirmed the clinical diagnosis of a vasculitis. Treatment with IV Solu-Medrol 30 mg/kg (maximum 1 g) per day for five days followed by PO prednisone 80 mg/d was initiated. Within just a few days, the girl developed an acute gastrointestinal (GI) distress with upper GI bleeding. Gastroscopy demonstrated an active bleeding site. A gastric mucosa biopsy also demonstrated a vasculitis.

Comment

Mononeuritis multiplex (MNM) is very uncommon in children and adolescents. Interestingly, although MNM is commonly seen in diabetics, we are not aware of such complication in the pediatric age group. However, this single case emphasizes that if a child has an underlying connective tissue disorder and presents with an acute mononeuropathy, especially with no evidence of a mechanical cause, the possibility of a systemic vasculitis needs to be considered early on in the differential diagnosis. Today children with renal compromise much less commonly develop a polyneuropathy.[60]

CASE EXAMPLE 43.10 Myasthenia Gravis with Acute Dysphagia, Drooling, and Ptosis

This infant girl was very healthy through age 15 months when she developed drooling, irritability, and inability to finish her bottles. Her pediatrician, noting a red throat, performed a throat swab that proved to be rapid streptococcal negative. During this procedure, she developed acute respiratory distress. After emergent admission to a local hospital she had an otolaryngology evaluation during which she was intubated. She was diagnosed with middle lobe pneumonia and treated with ceftriaxone. Her fever resolved but she had continued difficulty swallowing liquids. While taking oral medication she suddenly became limp and lethargic, sustaining a respiratory arrest. She responded to positive pressure ventilation and corticosteroids. An emergent transfer was made to BCH. Although successfully extubated, she had continued difficulties swallowing. When this child was extubated 4 days later, she was noted to have bilateral ptosis, bifacial weakness, and frequent cough after eating, and was unable to walk.

Tensilon (edrophonium chloride) testing with a dose of 1.6 mgm produced a dramatic improvement in her ptosis and facial appearance. A subsequent EMG, with 3 Hz repetitive motor nerve stimulation, demonstrated 38–45% decrements in the compound muscle action potentials. There was no repetitive component to her CMAPs. Her initial, commercially performed, acetylcholine receptor antibody (AChRAb) was negative; however, it was then definitely positive at the Mayo laboratories. She was treated with Mestinon (pyridostigmine bromide), in gradually increased dosages to 10-mg tid, with gradual but variable improvement of her neuromuscular findings. Interestingly, the ptosis and bifacial weakness were the last to resolve. A thymectomy was considered but with her excellent course it was decided to await a recurrence of her symptoms. The medication was withdrawn during the second six months of her illness. She has been asymptomatic now for a number of years.

Comment

Autoimmune myasthenia gravis (AMG) is a rare but important consideration in the differential diagnosis of either acute respiratory distress or failure to wean from a respirator. This applies to children of any age, even to toddlers, as noted in this instance. The youngest child with non-neonatal AMG that we are aware of was an eleven-month-old.[64] Although most AMG children present with ptosis and/or diplopia,[65] we occasionally see other clinical settings, some of which require urgent intensive care management.[66] This is particularly so when there is any hint of airway compromise. For instance we evaluated one teenager being followed by an otolaryngologist for progressive but variable speech difficulties when suddenly she developed severe dysphagia and increasing respiratory distress.

significant myotonic-like discharges with the hyperkalemic variant.

We do call attention to a case of an 18-month-old, with a previously unrecognized sodium channel myotonia, who presented with respiratory stridor and increasing difficulty breathing, secondary to an acute bronchiolitis. She had trouble opening her eyes that was worsened with crying, as was her stridor. This child had somewhat elevated serum potassium presumably secondary to her increased respiratory effort. These findings were all compatible with a form of paramyotonia. This clinical combination exacerbated her underlying sodium channelopathy with vocal cord myotonia and led to ever-increasing respiratory stridor.

Dermatomyositis

Very occasionally one may witness the acute onset of dermatomyositis in the pediatric age group. More commonly this illness has a slowly ingravescent temporal profile and does not require evaluation in the acute care setting.

CRITICAL ILLNESS NEUROMUSCULAR DISORDERS

The acute weakness or paralysis which may occasionally occur in children with critical illness is sometimes related to conditions involving the peripheral nerve, endplate, or muscle. Children with overwhelming sepsis or status asthmaticus may develop a critical illness neuromuscular syndrome that may mimic a hospital-acquired GBS presenting with a failure to wean from the respirator.[68–70] Some of these syndromes had initially been felt to be secondary to a defect in neuromuscular transmission. Bolton was one of the first to emphasize the need for these various syndromes to be considered by medics when called to the ICU.[69] Subsequent to his initial report, defining the role of the peripheral nerve in some of these critical care failures to wean patients, further careful studies have led to the demonstration that most of these children may have a primary myopathy.[71] These various critical illness neuromuscular disorders may affect each level of the motor unit (Box 43.4). The prevalence, natural history, and clinical significance of these conditions in children remain unknown and/or poorly characterized.[72]

Post-asthmatic Amyotrophy (Hopkins Syndrome)

Hopkins syndrome or post-asthmatic amyotrophy is a poliomyelitis-like illness that occurs following status asthmaticus and primarily in childhood, but it may also occur after puberty.[73] It is characterized by rapidly progressive and permanent monoplegia or diplegia (Table 43.1).[73–81] The etiology of the syndrome remains unknown and, despite its resemblance to poliomyelitis, all 27 reported patients were appropriately immunized against polio (Table 43.2). Other viruses considered as possible etiologies include enteroviruses and HSV type 1 virus. CSF examination in one 7-year-old patient with Hopkins syndrome showed primary herpes simplex virus type 1 infection, supporting the hypothesis of a viral mechanism.[34,85] An outbreak of poliomyelitis-like paralysis indistinguishable from Hopkins syndrome has been described in an epidemic of enterovirus 71[84]; these patients, however, did not have acute asthma. Electrophysiological findings suggest segmental anterior horn cell disease. Magnetic resonance imaging of the cervical spine in Hopkins syndrome has revealed swelling and edema in the region of the clinically involved segments.[78,85] Surgical exploration of the brachial plexus did not show any evidence of pathology.[77] Possible response to IVIg has been reported.[86] Since all patients who have developed Hopkins syndrome received steroids, the possibility of an immunologically mediated mechanism, perhaps an underlying immunodeficiency and/or steroid-mediated susceptibility of patients to invasion of the anterior horn cells by a virus, has also been postulated.[80,81,86]

BOX 43.4 Critical Illness Neuromuscular Disorders

1. Post-asthmatic Amyotrophy (Hopkins syndrome)
2. Critical Illness Polyneuropathy
3. Persistent Neuromuscular Blockade
4. Acute Quadriplegic Myopathy
5. Acute Necrotizing Myopathy of Intensive Care

Critical Illness Polyneuropathy

Initially described by Bolton et al. in 1984,[69] critical illness polyneuropathy (CIP) is a sensorimotor axonal polyneuropathy that develops in association with sepsis and multi-organ failure. Although signs of CIP can usually be seen early in the ICU course,[87] the onset of the neuropathy is difficult to recognize because of the frequently associated encephalopathy, the severity of the underlying illness, and the use of neuromuscular blocking agents (NMBAs) in conjunction with ventilatory support (Table 43.1). It is usually noted when the patient cannot be weaned from the mechanical ventilator. The main clinical findings include distally predominant extremity weakness and loss of muscle stretch reflexes. Sensory loss is hard to demonstrate but is sometimes detectable in patients who can cooperate for a neurologic examination. Cranial nerves are involved only rarely, and therefore, predominant cranial nerve involvement suggests another condition. If the child exhibits facial grimacing in response to rigorous painful stimulation of the distal extremities, implying intact sensory pathways, a myopathic mechanism for the quadriparesis needs to be suspected; nonetheless, the differentiation cannot usually be made clinically.

Critical illness polyneuropathy is a primarily distal axonopathy and its pathogenesis remains uncertain. It is possible that humoral and cellular processes result in disturbed or lack of vascular autoregulation with increased microvascular permeability, endoneural edema, and capillary occlusion.[69,88] Serum CK levels are usually normal or mildly elevated. Electrophysiologic studies show evidence of an axonal neuropathy.[88,89] Nerve conduction studies show reduced amplitude of motor and sensory responses. Significant slowing of nerve conduction or

TABLE 43.1 Differentiating Features of Critical Illness Neuromuscular Disorders

Features	Critical Illness Polyneuropathy	Acute Quadriplegic/ Necrotizing Myopathy	Hopkins Syndrome
Age group	Elderly (reports in children)	Children, adolescents	Young children
Risk factors	Multiorgan failure/sepsis	NMBAs and/or steroids	Status asthmaticus
Timing	>2 weeks	Variable	<2 weeks
Clinical findings	Sensory and motor	Purely motor	Purely motor
Limbs affected	Quadriparesis	Quadriparesis	Monoparesis
Pain	Unusual	Muscular tenderness	Mild
Reflexes	Absent or ↓	↓	Normal in unaffected limbs
Recovery	Good (slow, complete); may be incomplete	Good (slow, complete); may be rapid	Poor
Creatine kinase	< twice upper normal value	Normal to ↑↑↑ *	Normal
Nerve conduction studies	↓ SNAP and CMAP amplitudes	Normal SNAPs; ↓↓ CMAPs	Normal SNAPs
Electromyography	Neuropathic	Myopathic	Localized neuropathic
Muscle excitability	Normal	Absent or ↓↓	Normal
Spinal fluid	Normal	Not reported	Pleocytosis
Muscle biopsy	Grouped atrophy; axonal loss	Type II atrophy; myosin loss	Not reported

**Normal to mildly increased in acute quadriplegic myopathy; significantly increased in acute necrotizing myopathy.*
NMBAs = neuromuscular blocking agents; SNAPs = sensory nerve action potentials; CMAPs = compound muscle action potentials.

TABLE 43.2 Features of 27 Children with Post-asthmatic Amyotrophy*

Age of onset	13 mo – 11 yr
Sex, M/F	19/8
Immunization (polio)	27/27
Days to pareses after asthma	1–11
CSF protein (mild elevation)	6/13 reported
CSF pleocytosis	13/13 reported
Denervation on electromyography	17/17 reported
Slow conduction velocities	4/15
Virus isolation	
Enteroviruses	5/22
HSV type 1 (CSF)	1/22
Permanent paralysis	27/27
Monoplegia	21
Diplegia	4
Hemiplegia	2

**Data derived from cases reported in the literature.*[74–78,80,82,83]
M = Male; F = Female; HSV = herpes simplex virus; CSF = cerebrospinal fluid.

nerve conduction blocks are not expected findings and, if present, suggest other diagnostic possibilities, particularly the Guillain-Barré syndrome. In critical illness polyneuropathy, needle EMG often demonstrates neurogenic motor unit potentials as well as fibrillation potentials and positive sharp waves, in both the extremity musculature and the diaphragm.[88] Phrenic nerve conduction studies show either absent or reduced responses.[90] Nerve biopsy as well as autopsy studies are consistent with axonal degeneration of both motor and sensory fibers without evidence of significant inflammation or demyelination.[88] The differential diagnosis here includes acute Guillain-Barré syndrome, porphyria, botulism, myasthenic crisis, prolonged neuromuscular blockade, and critical illness myopathy.[89] Recovery may be slow and incomplete.

Persistent Neuromuscular Blockade

Neuromuscular blockade following the discontinuation of neuromuscular blocking agents (NMBAs) may persist, resulting in prolonged weakness and inability to wean from the ventilator. This may occur more frequently in children with impaired renal or hepatic function and has been reported with all NMBAs, including pancuronium, vancuronium, and atracurium.[91–93] Serum creatine kinase

(CK) levels are usually normal. Repetitive motor nerve stimulation (RMNS) demonstrates a decremental response of the CMAP, identifying the neuromuscular junction as the anatomic site of this physiologic abnormality. Rarely, however, no motor response is obtained in the child with severe prolonged blockade or in those with acute quadriplegic myopathy (AQM). In children with persistent neuromuscular blockade, the weakness typically lasts for only a few days and should not last for more than a week after the discontinuation of the NMBAs. If the weakness persists beyond seven days, other conditions such as AQM need consideration. Thus the EMG studies must be repeated if the weakness persists, particularly in instances where no motor responses are present on initial evaluation. Normal motor conduction studies with a concomitant decrement on RMNS supports the diagnosis of prolonged blockade. In contrast, the presence of persistently reduced or absent motor responses is compatible with the possible diagnosis of AQM.[93,94]

Acute Quadriplegic Myopathy

Acute quadriplegic myopathy (AQM) was initially reported in 1977 in a woman with status asthmaticus treated with a combination of corticosteroids and NMBAs.[95] A number of other names have been used in the literature to describe this syndrome, such as acute myopathy, critical illness myopathy, critical care myopathy, and acute myopathy with selective loss of myosin filaments.[89] The most common predisposing condition is asthma in conjunction with the use of intravenous corticosteroids, nondepolarizing NMBAs, and aminoglycosides.[94] However, AQM has also been described in patients with sepsis who were not treated with NMBAs or corticosteroids.[96–98] Other predisposing conditions for AQM include muscle unloading, pneumonia, organ transplantation (e.g. heart, liver, lung),[99] hepatic failure, and acidosis.[89] In a study from the Hospital for Sick Children, Toronto, Canada, 8 of the 14 critically ill children with muscle weakness (57%) were solid organ or bone marrow transplant recipients.[100] This study aimed to establish the incidence of muscle weakness in children admitted to pediatric ICU for >24 hours. Fourteen of 830 (1.7%) children had generalized muscle weakness and most had received NMBAs, corticosteroids, or aminoglycoside antibiotics. In 4 of the 5 children who had an EMG performed, the study showed myopathic findings, and in the fifth it was normal. Seven patients underwent NCS. Overall, electrodiagnostic studies were myopathic in 4 children and normal in 2, and showed compressive neuropathies in 1 child, and a mild demyelinative polyneuropathy in 1 patient.[100]

Most patients with AQM are severely affected, having significant weakness or even paralysis, as well as inability to be weaned from the ventilator. In contrast, a few patients may have only mild weakness. Extraocular movements are usually preserved, sensation is intact, and muscle stretch reflexes are decreased proportionate to the decrease in strength. Although rapid recovery is sometimes observed, the usual course of recovery is slow (weeks or months) and may be complete.[71,101] In a series of 8 adults with AQM associated with biopsy-proven loss of thick myofilaments, 6 patients died, 1 patient recovered fully in 1 month, and 1 patient remained quadriparetic at 1 year, probably due to concomitant CNS disease.[102] In another study of 5 children with critical illness neuromuscular disease, both children with critical illness myopathy recovered completely within 3 weeks and 6 months.[103]

In AQM, CK can be normal but is moderately elevated in about 50% of patients.[96,104,105] Nevertheless, in a small number of patients CK levels may be significantly elevated. Muscle biopsy shows loss of myosin ATPase staining which can be localized or be maximal in the center (core-like) of the myofibers; electron microscopy shows loss of myosin-thick filaments.[100,102,104–106] There is relative sparing of actin filaments and Z disks (Figure 43.1). Light microscopy commonly shows small angulated atrophic predominantly type 2 fibers, with basophilic staining of the cytoplasm with hematoxylin and eosin staining. The majority of core-like lesions are seen in type I fibers.[102] Necrosis of myofibers is usually mild but in some cases can be quite severe.

Nerve conduction studies in AQM are usually normal except for diminished CMAP amplitudes.[94,96] Needle EMG often records early recruited low amplitude, short duration myopathic motor units as well as fibrillation potentials.[107] However, many patients may be unable to activate their muscles due to an associated encephalopathy or inability to generate any voluntary activity. The resulting lack of clear EMG findings, combined with the fact that either CIP or AQM may occur in septic patients who have not received corticosteroids or NMBAs, combined with the coexistence of these two syndromes in certain patients, sometimes makes the differentiation of these two syndromes extremely difficult. The presence of normal SNAPs and small CMAPs suggests the presence of AQM, but small or absent SNAPs indicative of CIP do not necessarily exclude AQM because of the possibility of an unrelated neuropathy at baseline (albeit small in children).[108] Also, the finding of low-amplitude CMAPs is not always diagnostic of a polyneuropathy and may indicate AQM.[102] Significant conduction slowing or conduction blocks are not consistent with either CIP or AQM. Myopathic motor unit potentials with an early recruitment pattern are indicative of a myopathic process. However, myopathic potentials may also occur with prolonged neuromuscular blockade from the use of nondepolarizing NMBAs, and thus routine 3 Hz RMNS is required as part of the electrophysiological evaluation of these patients.[89] Direct muscle stimulation may allow the diagnosis of AQM in

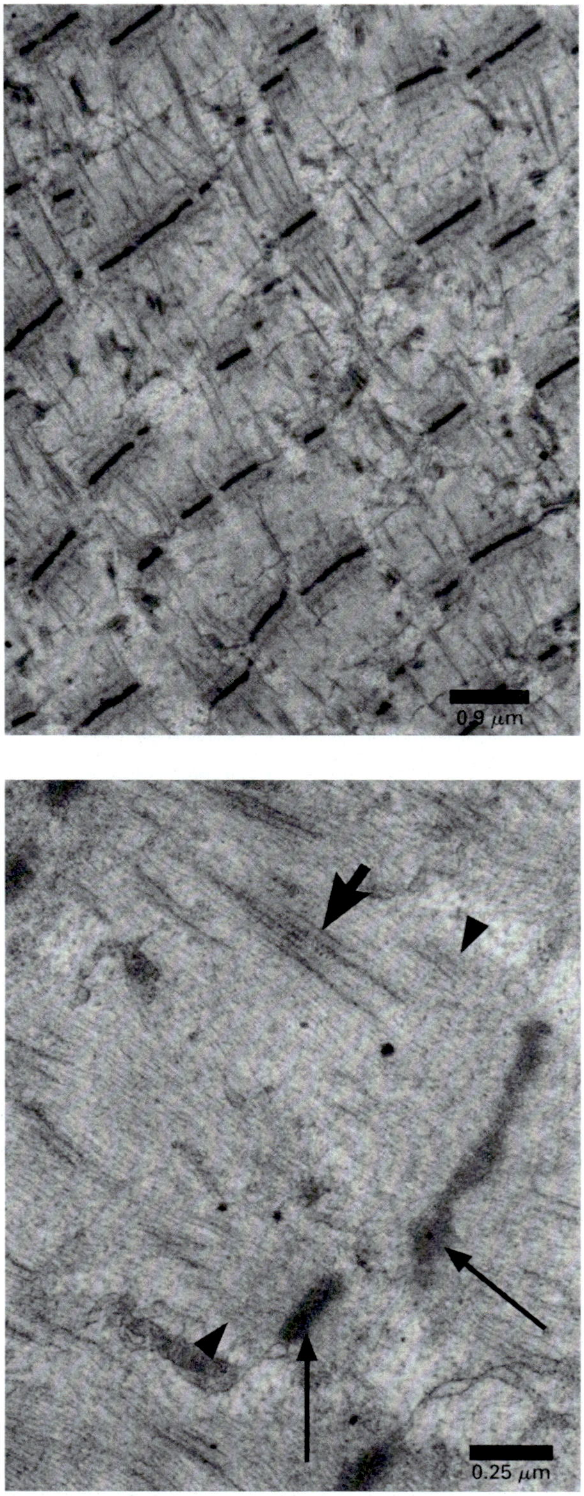

FIGURE 43.1 Electron micrographs show loss of thick filaments and myofibrillar disorganization (top panel, magnification × 11,000). Bottom panel (magnification × 40,500) shows remaining thick (myosin) filaments (short arrow), thin (actin) filaments (arrowheads) and Z bands (long arrows). Note the relative sparing of Z discs and actin filaments. *Reprinted with permission from Sandrock et al. (1997).*[117]

patients with confusing electrophysiological findings, but the test itself can be technically challenging. In very weak patients or patients whose EMG findings are hard to interpret, direct muscle stimulation has been employed to study muscle membrane excitability; the latter is normal in patients with CIP and reduced or absent in patients with AQM. Muscle is reported to be electrically inexcitable in AQM.[96,109] Therefore in patients with AQM, direct muscle stimulation does not lead to action potentials. The membrane excitability does return with clinical recovery of patients. It should be noted that neither author of this chapter has personal experience in the performance of direct muscle stimulation and thus cannot comment on the utility of this test in a clinical setting.

Although muscle biopsy showing the classic pathological findings described above is diagnostic of AQM, this procedure does not always provide a definitive differentiation between AQM and CIP. This is because the characteristic loss of myosin is seen in only a fraction of patients with AQM.

Increased activation of specific proteases such as calpain, a calcium-activated protease, has been found in muscle of patients with AQM. This may be crucial in the pathogenesis of this disorder (Figure 43.2).[97] The expression of ubiquitin and other proteases, as well as DNA fragmentation, was studied in muscle biopsies from three patients with AQM. All of these patients demonstrated significant overexpression of caspases, calpain, cathepsin B, and ubiquitin, and the presence of numerous apoptotic nuclei in over 70 percent of the muscle fibers.[110] This study provides further evidence that proteolytic proteases are important in the pathogenesis of AQM; proteases seem to stimulate apoptosis, but why the thick filaments are preferentially involved remains to be elucidated.[110] Published data generated by the same group provide additional evidence suggesting that stimulation of proaptotic pathways may underlie, at least in part, the pathogenesis of AQM.[111] A more recent study of myofibrillar protein and gene expression supports the important role of transcriptional regulation of myofibrillar protein synthesis in the loss and resynthesis of contractile proteins, myosin in particular, in acute quadriplegic myopathy.[112]

AQM is probably a disorder of protein turnover in muscle. Corticosteroids and other conditions such as sepsis stimulate muscle catabolism, which is further amplified by muscle inactivity secondary initially to neuromuscular blockade and later on to lack of membrane excitability.[113] Conditions such as sepsis and renal or respiratory failure, often associated with significant acidosis, may also result in high levels of cytokines that have a catabolic effect on muscle. The catabolic state usually persists despite treatment of sepsis and/or withdrawal of therapy with corticosteroids and NMBAs; it is possible that the paralysis that

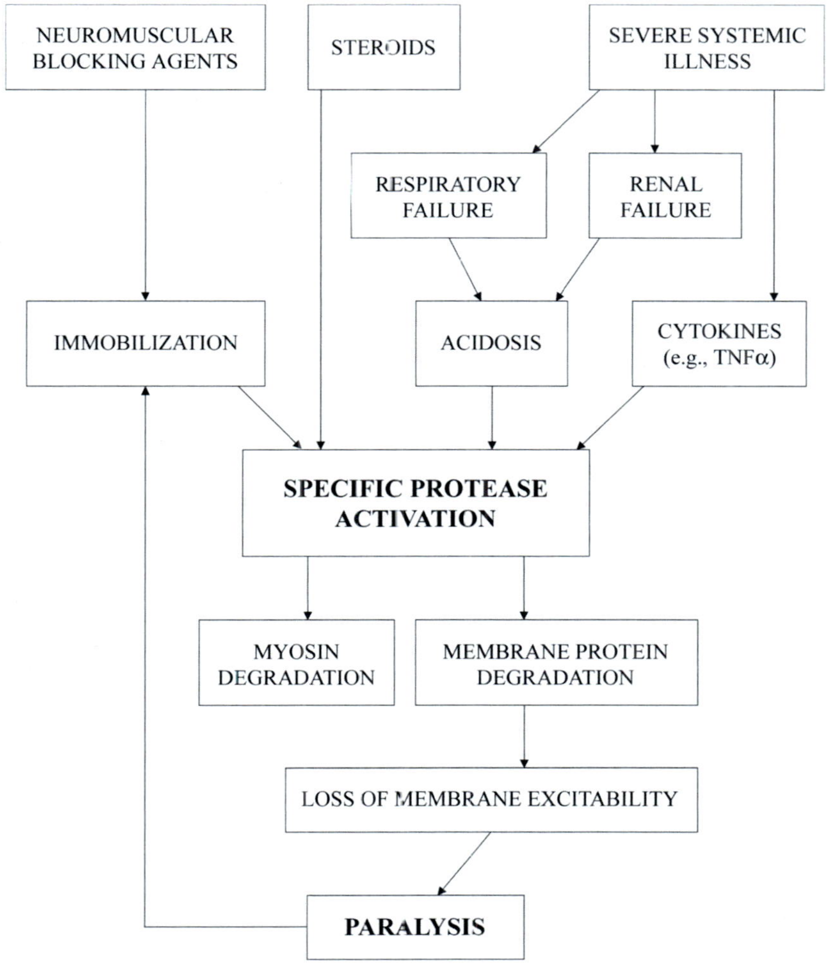

FIGURE 43.2 Risk factors for the development of acute quadriplegic or necrotizing myopathy. *Modified from Salviati L et al.*[101] *Reprinted with permission from: J Child Neurol. 2000;15:63–66. Abbreviation: TNF = tumor necrosis factor.*

results from the lack of membrane excitability prevents muscle from switching to an anabolic state, and thus the recovery from AQM is usually a slow process.[114]

Acute Necrotizing Myopathy of Intensive Care

Critical care myopathies most typically are associated with severe pulmonary disorders. Other clinical settings to consider in this scenario include patients with severe sepsis or those who have had an organ transplant. Most of these patients have received high-dose intravenous corticosteroids in conjunction with low to moderate doses of NMBAs. See Case Example 43.11.

CONCLUSIONS

When infants and children develop an acute flaccid paralysis, there is a clinical differential diagnosis somewhat dissimilar to that considered in the older child. In newborns and during the first 6 months of life, certain unusual syndromes require careful consideration.[116] Infantile botulism is a good example. We question whether this illness may be more common than currently recognized. This is particularly relevant among infants with acute indeterminate respiratory distress. GBS is another example of a possibly underrecognized or not considered illness in the evaluation of the floppy infant syndrome, particularly so with those infants who may have had an intrauterine onset. At times the clinical presentation of some acute pediatric motor unit disorders, such as GBS, mimics central nervous system lesions including various encephalopathies, even brain death, or acute myelopathies such as transverse myelitis or spinal cord tumors. MRI provides an important diagnostic tool for making this differentiation. EMG is also very helpful for defining the presence of a number of primary motor unit lesions. Pediatric neurologic and intensive care colleagues need to

CASE EXAMPLE 43.11 Severe Generalized Weakness after Intubation for Status Asthmaticus

This 12-year-old girl had an acute exacerbation of her asthma and required intubation. She required high-dose corticosteroid therapy and NMBAs. One of her arterial lines became infected, and she developed sepsis. When she was extubated she had severe generalized weakness. Her serum creatine kinase (CK) was markedly elevated with associated myoglobinemia and myoglobinuria. Motor NCVs were essentially normal initially. The median CMAP was mildly diminished in amplitude at 3 mV; however, the peroneal CMAP was normal. Motor unit potentials (MUPs) on needle EMG were myopathic; there were increased percentages of low amplitude, short duration, and at times polyphasic MUPs. Scattered fibrillation potentials were present on needle insertion. A week after the EMG, muscle biopsy demonstrated thick filament fiber loss typical of critical care acute quadriplegic myopathy with extensive areas of panfascicular fiber necrosis. Her recovery was extremely slow.

Comment

These critical care myopathies most typically are associated with severe pulmonary disorders. Other clinical settings to consider in this scenario include patients with severe sepsis or those who have had an organ transplant. Most of these patients have received high-dose intravenous corticosteroids in conjunction with low to moderate doses of NMBAs. As noted in this case, in patients with acute necrotizing myopathy (ANM) of intensive care, muscle biopsy shows panfascicular muscle fiber necrosis, suggesting that ANM may be a severe form of AQM. This pathology contrasts with the multiple pyogenic abscesses typical for pyomyositis, a condition also seen in septic patients but more endemic to the tropics. This is usually associated with significant CK elevation, myoglobinuria, and myopathic EMG findings.[115] Serum CK levels are sometimes, but not uniformly, markedly elevated with AQM. Rarely this laboratory finding is associated with myoglobinemia that may be severe enough to result in acute renal failure.[107,115] The CK level *per se* is not a very helpful tool for making the differentiation between a myopathic or neuropathic lesion. In a study of 14 patients with critical illness myopathy of the AQM variety, only 3 of 14 patients had an elevated CK.[71] With these three instances there was only a mild change, with a maximal value reported to be 331 IU/lt. In another study of 8 adult patients with biopsy-proven AQM, only 4 patients had elevated serum CK levels (417–4770 U/L; normal, up to 295 U/L), but serial CK determinations were not performed in all patients; thus, transient hyperCKemia may not have been detected.[102] The degree of enzyme elevation and disease severity did not correlate. The EMG may exclude a polyneuropathy and, as with this case, often demonstrates typical myopathic findings; as discussed previously, however, in certain cases the EMG findings are hard to interpret. The muscle biopsy defined the typical abnormalities of a primary loss of thick skeletal muscle filaments.

be encouraged to increase their utilization of EMG in some of these critical care, clinical settings. This is especially true when the initial clinical suspicions are not supported by the various investigations.

REFERENCES

1. Jones Jr. HR. Evaluation of the floppy infant. In: Jones Jr. HR, Bolton CF, Harper CM, editors. *Pediatric Clinical Electromyography*. Philadelphia: Lippincott-Raven; 1996. pp. 37–104.
2. David WS, Jones Jr HR. Electromyography and biopsy correlation with suggested protocol for evaluation of the floppy infant. *Muscle Nerve* 1994;**17**:424–30.
3. Darras BT, Jones Jr. HR. Neuromuscular problems of the critically ill neonate and child. *Semin Pediatr Neurol* 2004;**11**:147–68.
4. Jones HRJ, Harmon RL, Bolton CF, Harper CM. An approach to pediatric electromyography. In: Jones HRJ, Bolton CF, Harper CM, editors. *Pediatric Clinical Electromyography. Philadelphia-New York*. Lippincott-Raven; 1996. pp. 1–36.
5. MacLeod MJ, Taylor JE, Lunt PW, Mathew CG, Robb SA. Prenatal onset spinal muscular atrophy. *Eur J Paediatr Neurol* 1999;**3**:65–72.
6. Dubowitz V. Very severe spinal muscular atrophy (SMA type 0): an expanding clinical phenotype. *Eur J Paediatr Neurol* 1999;**3**:49–51.
7. Darras BT. Non-5q spinal muscular atrophies: The alphanumeric soup thickens. *Neurology* 2011;**77**:312–14.
8. David WS, Doyle JJ. Acute infantile weakness: a case of vaccine-associated poliomyelitis. *Muscle Nerve* 1997;**20**:747–9.
9. Beausoleil JL, Nordgren RE, Modlin JF. Vaccine-associated paralytic poliomyelitis. *J Child Neurol* 1994;**9**:334–5.
10. Robbins FC. Eradication of polio in the Americas. *JAMA* 1993;**270**:1857–9.
11. Recommendations of the Advisory Committee on Immunization Practices: revised recommendations for routine poliomyelitis vaccination. *MMWR Morb Mortal Wkly Rep* 1999;48:590.
12. Kang PB, Lidov HG, David WS, Torres A, Anthony DC, Jones HR, et al. Diagnostic value of electromyography and muscle biopsy in arthrogryposis multiplex congenita. *Ann Neurol* 2003;**54**:790–5.
13. Jackson AH, Baquis GD, Shah BL. Congenital Guillain-Barré syndrome. *J Child Neurol* 1996;**11**:407–10.
14. Luijckx GJ, Vles J, de Baets M, Buchwald B, Troost J. Guillain-Barré syndrome in mother and newborn child. *Lancet* 1997; **349**:27.
15. Rolfs A, Bolik A. Guillain-Barré syndrome in pregnancy: reflections on immunopathogenesis. *Acta Neurol Scand* 1994;**89**:400–2.
16. al-Qudah AA, Shahar E, Logan WJ, Murphy EG. Neonatal Guillain-Barré syndrome. *Pediatr Neurol* 1988;**4**:255–6.
17. Jones HR. Childhood Guillain-Barré syndrome: clinical presentation, diagnosis, and therapy. *J Child Neurol* 1996;**11**:4–12.
18. Gilmartin RC, Ch'ien LT. Guillain-Barré syndrome with hydrocephalus in early infancy. *Arch Neurol* 1977;**34**:567–9.

19. Pasternak JF, Fulling K, Nelson J, Prensky AL. An infant with chronic, relapsing polyneuropathy responsive to steroids. *Dev Med Child Neurol* 1982;**24**:504–24.
20. Grattan-Smith PJ, Morris JG, Johnston HM, Yiannikas C, Malik R, Russell R, et al. Clinical and neurophysiological features of tick paralysis. *Brain* 1997;**120**(Pt 11):1975–87.
21. Papazian O. Transient neonatal myasthenia gravis. *J Child Neurol* 1992;**7**:135–41.
22. Lefvert AK, Osterman PO. Newborn infants to myasthenic mothers. A clinical study and an investigation of acetylcholine receptor antibodies in 17 children. *Neurology* 1983;**33**:133–8.
23. O'Carroll P, Bertorini TE, Jacob G, Mitchell CW, Graff J. Transient neonatal myasthenia gravis in a baby born to a mother with new-onset anti-MuSK-mediated myasthenia gravis. *J Clin Neuromuscul Dis* 2009;**11**:69–71.
24. Lipsitz PJ. The clinical and biochemical effects of excess magnesium in the newborn. *Pediatrics* 1971;**47**:501–9.
25. Sokal MM, Koenigsberger MR, Rose JS, Berdon WE, Santulli TV. Neonatal hypermagnesemia and the meconium-plug syndrome. *N Engl J Med* 1972;**286**:823–5.
26. Mallory LA, Shaw JG, Burgess SL, Estrella E, Nurko S, Burpee TM, et al. Congenital myasthenic syndrome with episodic apnea. *Pediatr Neurol* 2009;**41**:42–5.
27. Pickett J, Berg B, Chaplin E, Brunstetter-Shafer MA. Syndrome of botulism in infancy: clinical and electrophysiologic study. *N Engl J Med* 1976;**295**:770–2.
28. Clay SA, Ramseyer JC, Fishman LS, Sedgwick RP. Acute infantile motor unit disorder. Infantile botulism? *Arch Neurol* 1977;**34**:236–43.
29. Hoffman RE, Pincomb BJ, Skeels MR. Type F infant botulism. *Am J Dis Child* 1982;**136**:270–1.
30. Schwartz RH, Eng G. Infant botulism: exacerbation by aminoglycosides. *Am J Dis Child* 1982;**136**:952.
31. Thompson JA, Glasgow LA, Warpinski JR, Olson C. Infant botulism: clinical spectrum and epidemiology. *Pediatrics* 1980;**66**:936–42.
32. King LA, Popoff MR, Mazuet C, Espie E, Vaillant V, de Valk H. Infant botulism in France, 1991–2009. *Arch Pediatr* 2010;**17**:1288–92.
33. May ML, Corkeron MA, Stretton M. Infant botulism in Australia: availability of human botulinum antitoxin for treatment. *Med J Aust* 2010;**193**:614–15.
34. Chalk C, Benstead TJ, Keezer M. Medical treatment for botulism. *Cochrane Database Syst Rev* 2011;**3**:CD008123.
35. Grant KA, Nwarfor I, Mpamugo O, Mithani V, Lister P, Dixon G, et al. Report of two unlinked cases of infant botulism in the UK in October 2007. *J Med Microbiol* 2009;**58**:1601–6.
36. Cornblath DR, Sladky JT, Sumner AJ. Clinical electrophysiology of infantile botulism. *Muscle Nerve* 1983;**6**:448–52.
37. Sheth RD, Lotz BP, Hecox KE, Waclawik AJ. Infantile botulism: pitfalls in electrodiagnosis. *J Child Neurol* 1999;**14**:156–8.
38. Gutierrez AR, Bodensteiner J, Gutmann L. Electrodiagnosis of infantile botulism. *J Child Neurol* 1994;**9**:362–5.
39. Darras BT, Chad DA. Myotonic dystrophy: etiology, clinical features, and diagnosis. In: Shefner JM, Nordli Jr. DR, Dashe JF, editors. Waltham, MA: UpToDate; 2012UpToDate. Available at <http://www.uptodate.com>.
40. Swift TR, Ignacio OJ, Dyken PR. Neonatal dystrophia myotonica. Electrophysiologic studies. *Am J Dis Child* 1975;**129**:734–7.
41. Kuntz NL, Daube JR. *Electrophysiology of Congenital Myotonic Dystrophy*. Rochester, MN: AAEE Course E, American Association of Electromyography and Electrodiagnosis; 1984. 23 pp.
42. Swoboda KJ, Specht L, Jones HR, Shapiro F, DiMauro S, Korson M. Infantile phosphofructokinase deficiency with arthrogryposis: Clinical benefit of a ketogenic diet. *J Pediatr* 1997;**131**:932–4.
43. DiMauro S, Hartlage P. Fatal infantile form of muscle phosphorylase deficiency. *Neurology* 1978;**28**:1124–9.
44. McMillan HJ, Darras BT, Kang PB, Saleh F, Jones HR. Pediatric monomelic amyotrophy: evidence for poliomyelitis in vulnerable populations. *Muscle Nerve* 2009;**40**:860–3.
45. Medici C, Gonzalez G, Cerisola A, Scavone C. Locked-in syndrome in three children with Guillain-Barré syndrome. *Pediatr Neurol* 2011;**45**:125–8.
46. Bradshaw DY, Jones Jr. HR. Guillain-Barré syndrome in children: clinical course, electrodiagnosis, and prognosis. *Muscle Nerve* 1992;**15**:500–6.
47. Bradshaw DY, Jones Jr. HR. Pseudomeningoencephalitic presentation of pediatric Guillain-Barré syndrome. *J Child Neurol* 2001;**16**:505–8.
48. Bakshi N, Maselli RA, Gospe Jr. SM, Ellis WG, McDonald C, Mandler RN. Fulminant demyelinating neuropathy mimicking cerebral death. *Muscle Nerve* 1997;**20**:1595–7.
49. Graf WD, Chance PF, Lensch MW, Eng LJ, Lipe HP, Bird TD. Severe vincristine neuropathy in Charcot-Marie-Tooth disease type 1A. *Cancer* 1996;**77**:1356–62.
50. Honavar M, Tharakan JK, Hughes RA, Leibowitz S, Winer JB. A clinicopathological study of the Guillain-Barré syndrome. Nine cases and literature review. *Brain* 1991;**114**(Pt 3):1245–69.
51. Kleyweg RP, van der Meche FG, Loonen MC, de Jonge J, Knip B. The natural history of the Guillain-Barré syndrome in 18 children and 50 adults. *J Neurol Neurosurg Psychiatry* 1989;**52**:853–6.
52. Sladky JT. Guillain-Barré syndrome. In: Jones HJ, De Vivo DC, Darras BT, editors. *Neuromuscular Disorders of Infancy, Childhood, and Adolescence: A Clinician's Approach*. Philadelphia: Butterworth-Heinemann; 2003. pp. 407–24.
53. Ropper AH. The Guillain-Barré syndrome. *N Engl J Med* 1992;**326**:1130–6.
54. Ropper AH, Kehne SM. Guillain-Barré syndrome: management of respiratory failure. *Neurology* 1985;**35**:1662–5.
55. Lawn ND, Fletcher DD, Henderson RD, Wolter TD, Wijdicks EF. Anticipating mechanical ventilation in Guillain-Barré syndrome. *Arch Neurol* 2001;**58**:893–8.
56. Hesketh E, Eden OB, Gattamaneni HR, Campbell RH, Jenney ME, Lashford L. Spinal cord compression—do we miss it? *Acta Paediatr* 1998;**87**:452–4.
57. Delhaas T, Kamphuis DJ, Witkamp TD. Transitory spinal cord swelling in a 6-year-old boy with Guillain-Barré syndrome. *Pediatr Radiol* 1998;**28**:544–6.
58. Bajaj NP, Rose P, Clifford-Jones R, Hughes PJ. Acute transverse myelitis and Guillain-Barré overlap syndrome with serological evidence for mumps viraemia. *Acta Neurol Scand* 2001;**104**: 239–42.
59. Knebusch M, Strassburg HM, Reiners K. Acute transverse myelitis in childhood: nine cases and review of the literature. *Dev Med Child Neurol* 1998;**40**:631–9.
60. Bolton CF, Young GB. *The Neurologic Complications of Renal Disease*. Boston: Butterworths; 1990.

61. Ryan MM, Tilton A, De Girolami U, Darras BT, Jones Jr. HR. Paediatric mononeuritis multiplex: a report of three cases and review of the literature. *Neuromuscul Disord* 2003;**13**:751–6.
62. Srinivasan J, Ryan MM, Escolar DM, Darras B, Jones HR. Pediatric sciatic neuropathies: a 30-year prospective study. *Neurology* 2011;**76**:976–80.
63. Srinivasan J, Escolar D, Ryan M, Darras B, Jones HR. Pediatric sciatic neuropathies due to unusual vascular causes. *J Child Neurol* 2008;**23**:738–41.
64. Geh GH, Bradbury JA. Ocular myasthenia presenting in an 11-month-old boy. *Eye* 1996;**12**(Pt 2):319–20.
65. Mullaney P, Vajsar J, Smith R, Buncic JR. The natural history and ophthalmic involvement in childhood myasthenia gravis at the Hospital for Sick Children. *Ophthalmology* 2000;**107**:504–10.
66. Chiang LM, Darras BT, Kang PB. Juvenile myasthenia gravis. *Muscle Nerve* 2009;**39**:423–31.
67. Bolton CF. Spinal muscular atrophies of childhood and adolescence. In: Jones HRJ, Bolton CF, Harper CM, editors. *Pediatric Clinical Electromyography*. Philadelphia-New York: Lippincott-Raven; 1996. pp. 105–22.
68. Sheth RD, Pryse-Phillips WE, Riggs JE, Bodensteiner JB. Critical illness neuromuscular disease in children manifested as ventilatory dependence. *J Pediatr* 1995;**126**:259–61.
69. Bolton CF, Gilbert JJ, Hahn AF, Sibbald WJ. Polyneuropathy in critically ill patients. *J Neurol Neurosurg Psychiatry* 1984;**47**:1223–31.
70. Goulden KJ, Dooley JM, Peters S, Ronen GM. Critical illness polyneuropathy: A reversible cause of paralysis in asthmatic children [Abstract]. *Ann Neurol* 1989;**26**:451.
71. Lacomis D, Zochodne DW, Bird SJ. Critical illness myopathy. *Muscle Nerve* 2000;**23**:1785–8.
72. Mohamed A, Ryan MM. Neuromuscular complications of intensive care. *Handb Clin Neurol* 2013;**113**:1481–3.
73. Hopkins IJ, Shield LK. Letter: Poliomyelitis-like illness associated with asthma in childhood. *Lancet* 1974;**1**:760.
74. Shapiro GG, Chapman JT, Pierson WE, Bierman CW. Poliomyelitis-like illness after acute asthma. *J Pediatr* 1979;**94**:767–8.
75. Blomquist HK, Bjorksten B. Poliomyelitis-like illness associated with asthma. *Arch Dis Child* 1980;**55**:61–3.
76. Manson JI, Thong YH. Immunological abnormalities in the syndrome of poliomyelitis-like illness associated with acute bronchial asthma (Hopkins' syndrome). *Arch Dis Child* 1980;**55**:26–32.
77. Nihei K, Naitoh H, Ikeda K. Poliomyelitis-like syndrome following asthmatic attack (Hopkins syndrome). *Pediatr Neurol* 1987;**3**:166–8.
78. Wakamoto H, Morimoto T, Nagao H, Matsuda H. MRI in poliomyelitis-like syndrome. *Pediatr Radiol* 1992;**22**:533–4.
79. Yeung SC, Antonio G, Ip KS. Flaccid paralysis of the limbs after an asthmatic attack. *Pediatr Neurol* 2010;**42**:133–6.
80. Sheth RD, Bolton CF. Neuromuscular complications of sepsis in children. *J Child Neurol* 1995;**10**:346–52.
81. Pavone P, Longo MR, Scalia F, Polosa R, Kira J, Falsaperla R. Recurrent Hopkins' syndrome: a case report and review of the literature. *J Neurol Sci* 2010;**297**:89–91.
82. Shahar EM, Hwang PA, Niesen CE, Murphy EG. Poliomyelitis-like paralysis during recovery from acute bronchial asthma: possible etiology and risk factors. *Pediatrics* 1991;**88**:276–9.
83. Batley R, Johnson EW. Asthmatic amyotrophy. Three cases. *Am J Phys Med Rehabil* 1991;**70**:332–4.
84. Hayward JC, Gillespie SM, Kaplan KM, Packer R, Pallansch M, Plotkin S, et al. Outbreak of poliomyelitis-like paralysis associated with enterovirus 71. *Pediatr Infect Dis J* 1989;**8**: 611–16.
85. Kyllerman MG, Herner S, Bergstrom TB, Ekholm SE. PCR diagnosis of primary herpesvirus type I in poliomyelitis-like paralysis and respiratory tract disease. *Pediatr Neurol* 1993;**9**: 227–9.
86. Cantarín-Extremera V, González-Gutiérrez-Solana L, Ramírez-Orellana M, López-Marín L, Duat-Rodríguez A, Ruíz-Falcó-Rojas ML. Immune-mediated mechanisms in the pathogenesis of Hopkins syndrome. *Pediatr Neurol* 2012;**47**:373–4.
87. Ahlbeck K, Fredriksson K, Rooyackers O, Maback G, Remahl S, Ansved T, et al. Signs of critical illness polyneuropathy and myopathy can be seen early in the ICU course. *Acta Anaesthesiol Scand* 2009;**53**:717–23.
88. Bolton CF, Laverty DA, Brown JD, Witt NJ, Hahn AF, Sibbald WJ. Critically ill polyneuropathy: electrophysiological studies and differentiation from Guillain-Barré syndrome. *J Neurol Neurosurg Psychiatry* 1986;**49**:563–73.
89. Gutmann L. Critical illness neuropathy and myopathy. *Arch Neurol* 1999;**56**:527–8.
90. Hart DE, Rojas LA, Rosario JA, Recalde H, Roman GC. Childhood Guillain-Barré syndrome in Paraguay, 1990 to 1991. *Ann Neurol* 1994;**36**:859–63.
91. O'Connor M, Russell WJ. Muscle strength following anaesthesia with atracurium and pancuronium. *Anaesth Intensive Care* 1988;**16**:255–9.
92. Partridge BL, Abrams JH, Bazemore C, Rubin R. Prolonged neuromuscular blockade after long-term infusion of vecuronium bromide in the intensive care unit. *Crit Care Med* 1990;**18**: 1177–9.
93. Segredo V, Caldwell JE, Matthay MA, Sharma ML, Gruenke LD, Miller RD. Persistent paralysis in critically ill patients after long-term administration of vecuronium. *N Engl J Med* 1992;**327**: 524–8.
94. Bird SJ, Rich MM. Neuromuscular complications of critical illness. *The Neurologist* 2000;**6**:2–11.
95. MacFarlane IA, Rosenthal FD. Severe myopathy after status asthmaticus. *Lancet* 1977;**2**:615.
96. Segredo V, Caldwell JE, Matthay MA, Sharma ML, Gruenke LD, Miller RD. Direct muscle stimulation in acute quadriplegic myopathy. *Muscle Nerve* 1997;**20**:665–73.
97. Showalter CJ, Engel AG. Acute quadriplegic myopathy: analysis of myosin isoforms and evidence for calpain-mediated proteolysis. *Muscle Nerve* 1997;**20**:316–22.
98. Latronico N, Fenzi F, Recupero D, Guarneri B, Tomelleri G, Tonin P, et al. Critical illness myopathy and neuropathy. *Lancet* 1996;**347**:1579–82.
99. Perea M, Picon M, Miro O, Orus J, Roig E, Grau JM. Acute quadriplegic myopathy with loss of thick (myosin) filaments following heart transplantation. *J Heart Lung Transplant* 2001;**20**: 1136–41.
100. Banwell BL, Mildner RJ, Hassall AC, Becker LE, Vajsar J, Shemie SD. Muscle weakness in critically ill children. *Neurology* 2003;**61**:1779–82.
101. Salviati L, Laverda AM, Zancan L, Fanin M, Angelini C, Meznaric-Petrusa M. Acute quadriplegic myopathy in a 17-month-old boy. *J Child Neurol* 2000;**15**:63–6.

102. Sander HW, Golden M, Danon MJ. Quadriplegic areflexic ICU illness: selective thick filament loss and normal nerve histology. *Muscle Nerve* 2002;**26**:499–505.
103. Tabarki B, Coffinieres A, Van Den Bergh P, Huault G, Landrieu P, Sebire G. Critical illness neuromuscular disease: clinical, electrophysiological, and prognostic aspects. *Arch Dis Child* 2002;**86**:103–7.
104. Lacomis D, Petrella JT, Giuliani MJ. Causes of neuromuscular weakness in the intensive care unit: a study of ninety-two patients. *Muscle Nerve* 1998;**21**:610–17.
105. Lacomis D, Giuliani MJ, Van Cott A, Kramer DJ. Acute myopathy of intensive care: clinical, electromyographic, and pathological aspects. *Ann Neurol* 1996;**40**:645–54.
106. Larsson L. Acute quadriplegic myopathy: an acquired "myosinopathy". *Adv Exp Med Biol* 2008;**642**:92–8.
107. Zochodne DW, Ramsay DA, Saly V, Shelley S, Moffatt S. Acute necrotizing myopathy of intensive care: electrophysiological studies. *Muscle Nerve* 1994;**17**:285–92.
108. Primavera A, Abbruzzese M. Case 11-1997: critical-illness myopathy. *N Engl J Med* 1997;**337**:862–3.
109. Rich MM, Teener JW, Raps EC, Schotland DL, Bird SJ. Muscle is electrically inexcitable in acute quadriplegic myopathy. *Neurology* 1996;**46**:731–6.
110. Di Giovanni S, Mirabella M, D'Amico A, Tonali P, Servidei S. Apoptotic features accompany acute quadriplegic myopathy. *Neurology* 2000;**55**:854–8.
111. Di Giovanni S, Molon A, Broccolini A, Melcon G, Mirabella M, Hoffman EP, et al. Constitutive activation of MAPK cascade in acute quadriplegic myopathy. *Ann Neurol* 2004;**55**:195–206.
112. Norman H, Zackrisson H, Hedstrom Y, Andersson P, Nordquist J, Eriksson LI, et al. Myofibrillar protein and gene expression in acute quadriplegic myopathy. *J Neurol Sci* 2009;**285**:28–38.
113. Ochala J, Gustafson AM, Diez ML, Renaud G, Li M, Aare S, et al. Preferential skeletal muscle myosin loss in response to mechanical silencing in a novel rat intensive care unit model: underlying mechanisms. *J Physiol* 2011;**589**:2007–26.
114. Ruff RL. Why do ICU patients become paralyzed? *Ann Neurol* 1998;**43**:154–5.
115. Lannigan R, Austin TW, Vestrup J. Myositis and rhabdomyolysis due to *Staphylococcus aureus* septicemia. *J Infect Dis* 1984;**150**:784.
116. Jones HR, Darras BT. Acute care pediatric electromyography. *Muscle Nerve* 2000;**23**(Suppl.):S53–62.
117. Sandrock AW, Cros DP, Louis DN. Case 11-1997, Case Records of the Massachusetts General Hospital. *N Eng J Med* 1997;**336** (15):1079–88.

Chapter 44

Intensive Care Management, Including Cardiorespiratory Care

William A. Scott and Susan T. Iannaccone

HISTORICAL BACKGROUND

Patients with neuromuscular diseases may present with cardiac dysfunction, arrhythmia, or respiratory failure.[1,2] Information in the literature regarding intensive care for these patients continues to evolve and single institution results have been published.[3,4] Standards of care have been established for neuromuscular diseases.[5–7] Close interaction among the specialties of cardiology, critical care, neurology, and pulmonary medicine are necessary for optimal outcomes. This chapter will provide a brief overview of the management of life-threatening cardiopulmonary complications of pediatric neuromuscular disease.

INTENSIVE CARE

In general, intensive care is not precisely defined. It is often characterized by the location of the patient and the services provided. Guidelines for pediatric transport and intensive care serve as a basis for the specialized care required by patients with neuromuscular disease.[8,9] ICU care includes more frequent and detailed cardiorespiratory assessment, establishment of central venous and arterial access, assisted ventilation, administration of vasoactive medications, and acute therapy for arrhythmia, all in an environment with continuous monitoring and low nurse-to-patient ratios. Specific indications for critical care (Box 44.1) include respiratory failure, heart failure, life threatening arrhythmia, and resuscitated sudden death. The ICU may also be used for recovery following sedation or anesthesia.[10] For the neuromuscular patient, expectant admission is appropriate in the presence of rapidly progressive weakness, markedly reduced vital capacity, bulbar palsy, and autonomic instability.

The care of the critically ill patient begins prior to admission to the ICU and continues after discharge to the inpatient unit. Individuals skilled in pediatric intensive care should transport the critically ill child to the ICU. Ideally, the patient should be stabilized before transport. During appropriate transport, the supportive measures routinely available in the ICU should be available, including oxygen, assisted ventilation, and antiarrhythmic and cardiovascular support medications. The key to successful transport is excellent communication between the referring and accepting facilities as well as the transport team. A pediatric intensivist, pulmonologist, and cardiologist, all of whom are familiar with the complications of neuromuscular disease, should be involved early, before transfer as well as during the hospitalization. Comprehensive discharge planning is necessary to successfully transition from the ICU to an inpatient unit and subsequently home.

ETHICS

The indication for pursuing intensive care support for any patient is a balance between the perceived risks and benefits. In the child with progressive disease, it is often difficult to be certain when aggressive interventions become futile. The patient, depending upon developmental age, and the family must be involved in all decisions regarding extraordinary support measures or withdrawal thereof. Ethics consultation is often appropriate.[11] In a disease known to have progressive deterioration and limited therapies to impact the natural history, the physician should encourage frank discussions about end of life decisions before an acute event occurs.

B.T. Darras, H. Royden Jones, Jr., M.M. Ryan & D.C. De Vivo (Eds): Neuromuscular Disorders of Infancy, Childhood and Adolescence, Second edition.
DOI: http://dx.doi.org/10.1016/B978-0-12-417044-5.00044-5

BOX 44.1 Indications for Critical Care

Cardiac
- Unstable congestive heart failure
- Life-threatening arrhythmias
- Bradycardia requiring cardiac pacing

Respiratory
- Perioperative recovery in presence of chronic respiratory insufficiency
- FiO_2 requirement greater than 50%
- Rising pCO_2 (>70 or with symptoms)
- Markedly reduced vital capacity
- Compromised airway
- Endotracheal intubation

Neurologic
- Rapidly progressive weakness
- Bulbar palsy
- Severe autonomic instability

NORMAL PHYSIOLOGY

Cardiac

A complex control mechanism modulates the pumping action of the heart. Interdependent determinants of systolic function are: preload, the degree to which the ventricle is filled in diastole; afterload, the resistance to forward blood flow; heart rate and contractility, the latter of which describes the ability of the cardiac fibers to shorten. Active relaxation and the resting stiffness of the heart modulate diastolic function.

Normal conduction proceeds from the sinus node to the atria, through the atrioventricular node to the bundle of His, which subsequently divides into the major bundle branches. The electrical impulse utilizes these bundle branches for rapid distribution to the working myocardium of the ventricles. Short- and long-term variation is modulated by the autonomic nervous system and by the local tissue environment including pH, oxygen level, potassium, and calcium.

Respiratory

Ventilation is regulated by the autonomic nervous system, with limited voluntary override. Inspiration is achieved with the contraction of the diaphragm and intercostal muscles, and is dependent upon chest wall compliance and lung volumes. Optimum gas exchange occurs when all alveoli have adequate opening and circulation (ventilation perfusion matching).

Equally important is effective clearance of secretions through an adequate cough, which requires good inspiration, glottic closure, and sufficient contraction of the expiratory muscles to generate high pressure.

PATHOPHYSIOLOGY

Heart Failure

Heart failure results when the pump function of the heart is inadequate to meet the metabolic needs of the body. The clinical signs and symptoms of heart failure are age dependent. In the infant, feeding intolerance, irritability, tachypnea, and tachycardia are most prevalent. The older child and adolescent are more likely to complain of activity intolerance or fatigue. In the patient with skeletal muscle weakness, activity related symptoms may be masked.

Once a specific neuromuscular diagnosis has been established, screening is directed to the associated type of cardiomyopathy. Conversely, if a patient has an unknown neuromuscular disorder, then the presence and the characteristics of a cardiomyopathy may aid in obtaining the correct diagnosis. The two cardiomyopathies associated with neuromuscular diseases that ultimately lead to heart failure are dilated and hypertrophic.

Cardiomyopathy

Dilated cardiomyopathy is the result of decreased contractility and loss of myocardium. Initially dilation is a compensatory mechanism that augments systolic function by increasing the preload. However, left ventricular dilation without increasing wall thickness increases wall stress, and oxygen supply and demand become mismatched. Mechanical stress may accelerate this process. It has been postulated that skeletal muscle weakness may be associated with slower progression of cardiac disease because the heart is never required to respond to exertion or physical stress.[12,13] For example, motor dysfunction progresses more slowly in Becker muscular dystrophy (BMD) than in Duchenne muscular dystrophy (DMD), but symptoms of cardiomyopathy are common in carriers of the DMD gene. Thus, asymptomatic carriers of neuromuscular disease who have little or no limb weakness may present with cardiomyopathy or conduction disturbances.

In DMD, clinically apparent cardiomyopathy is first evident after 10 years of age and increases in incidence with age; it is present in all patients over 18 years of age.[14–16] Cardiac dysfunction typically manifests as left ventricular wall motion abnormalities and dilated cardiomyopathy.[17] Histologic studies indicate that heart muscle from DMD and BMD patients is deficient in dystrophin affecting membrane integrity, calcium movement, and stretch receptor activation. Additional factors contributing to cardiomyopathy may include an abnormal coronary reserve, such that patients are unable to augment myocardial oxygen at times of increased demand.

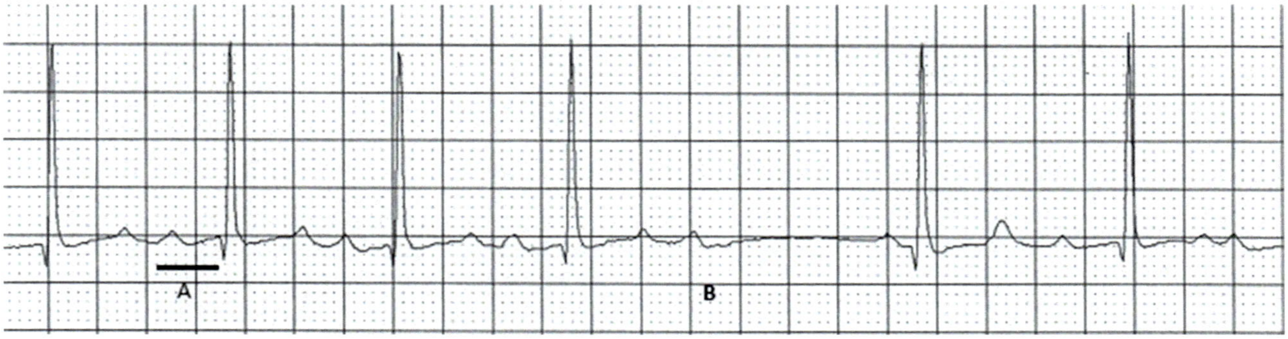

FIGURE 44.1 Cardiac rhythm tracing from a 17-year-old male with myotonic dystrophy that shows two types of heart block. **A.** First degree atrioventricular (AV) block with prolonged PR interval of 250 milliseconds (ms) indicated by solid line (normal < 200 ms). **B.** Second degree AV block indicated by a nonconducted p wave. Both prolonged PR interval and second degree AV block are associated with an increased risk of sudden death in patients with myotonic dystrophy.[20] A pacemaker was implanted and the patient remains asymptomatic from a heart rhythm perspective.

Barth's syndrome presents in infancy and is an example of rapidly fatal dilated cardiomyopathy with skeletal muscle weakness, neutropenia, 3-methylglutaconic aciduria, and abnormal mitochondria.[18]

In *hypertrophic cardiomyopathy*, there is abnormal wall thickness that commonly impairs relaxation and ventricular filling; less commonly there is sufficient hypertrophy to obstruct ventricular emptying. Emery-Dreifuss muscular dystrophy (EDMD) and limb girdle muscular dystrophies, myofibrillar and metabolic myopathies, and Friedreich's ataxia (FA) have been associated with hypertrophic cardiomyopathy. The hypertrophy of the left ventricle in FA appears to correlate with a large number of trinucleotide repeats.[19]

Arrhythmia

The two broad mechanisms of arrhythmia are abnormal impulse formation (automaticity), and abnormal impulse conduction (reentry or block). All clinical arrhythmias involve one or both of these mechanisms. Arrhythmia classification is further subdivided into tachycardia and bradycardia as well as the involved chambers of the heart.

Examples of reentrant tachycardia that are seen in neuromuscular disease include atrioventricular nodal reentrant tachycardia, and atrial flutter. Atrial fibrillation may be due to abnormal automaticity or reentry.

Ventricular tachycardia arises below the level of the AV node and is characterized by a QRS complex that is wider than normal. This also may be due to automatic or reentrant mechanisms.

Bradycardia is due to sinus node dysfunction (abnormal automaticity) or failure of the electrical impulse to conduct to the ventricle (block). Sinus bradycardia has been reported in muscular dystrophy and mitochondrial disease.

Heart block may occur anywhere within the cardiac specialized conductions system. Block can be intermittent, known as second degree (Figure 44.1), or permanent, known as complete or third degree atrioventricular block.

Arrhythmias associated with neuromuscular disease are extensive.[21] Significant arrhythmia frequently coexists with asymptomatic left ventricular dysfunction and wall motion abnormalities. In EDMD, cardiac abnormalities become apparent in the teenage years and are characterized by cardiac conduction defects and infiltration of the myocardium by fibrous and adipose tissue. Usually atrial paralysis is the first manifestation. Subsequent atrioventricular nodal dysfunction may result in progressive heart block and the potential for sudden death. Treatment with ventricular pacing is usually needed and is instituted with the onset of conduction system impairment. Female carriers of EDMD are also at risk of sudden death. Relatives of affected patients should be offered screening with electrocardiography (ECG) and echocardiography.[22–24] Patients with myotonic dystrophy often develop progressive ECG changes without obvious clinical symptoms until the development of complete heart block when, again, there is the risk of sudden death.[25] The severity of the rhythm disturbance seems to correlate with the expansion of unstable repeats. Some clinicians recommend prophylactic placement of a pacemaker in the adult myotonic dystrophy patient.

Sudden death may be arrhythmic, circulatory, or noncardiac. Noncardiac causes are numerous, but include neurologic events, respiratory failure, and trauma. Arrhythmia and congestive heart failure account for the vast majority of sudden deaths in neuromuscular patients. Tachyarrhythmia (ventricular tachycardia or fibrillation) or bradyarrhythmia (atrioventricular block or asystole) may cause sudden death. Complex ventricular ectopy, left ventricular dysfunction, and dilated cardiomyopathy are risk factors for sudden death.[26] Specific risk factors for

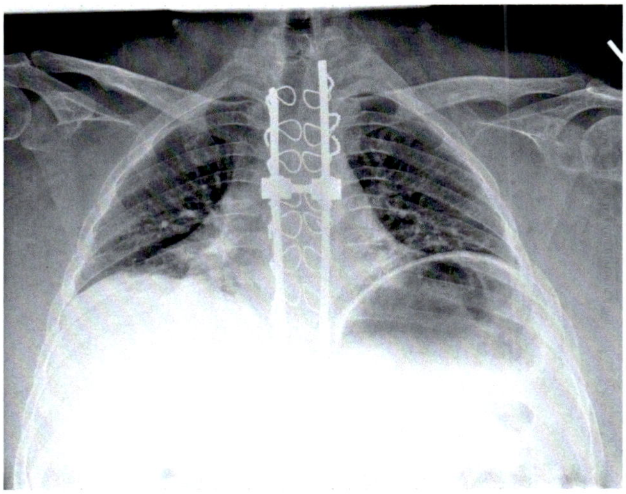

FIGURE 44.2 Chest X-ray from an 18-year-old male with Duchenne muscular dystrophy that demonstrates conditions placing him at increased risk of cardiopulmonary deterioration and the need for intensive cardiopulmonary support. The diaphragms are elevated with loss of air space. There is atelectasis in the right medial lung base as well as gastric distension. Orthopedic hardware is in place for spine stabilization. Cardiac size is difficult to assess with decreased lung volumes, but dysfunction was present on echocardiographic imaging. Cardiorespiratory support included night time noninvasive ventilation and medical management of heart failure.

sudden death in myotonic dystrophy include increased PR interval or QRS duration, abnormal QT intervals, atrial arrhythmia, and a positive family history.[20]

Respiratory Failure

Acute respiratory illness leading to respiratory compromise was the leading cause of unplanned admission to the ICU in children with neuromuscular disease (Figure 44.2).[4] The number of patients placed on home mechanical ventilation has increased dramatically, as has the need for ICU care. For the majority, initiation of mechanical ventilation was unplanned.[27]

There are three primary causes responsible for respiratory failure in neuromuscular diseases.[28,29] Respiratory muscle weakness (pump failure) impairs ventilation.[30] Inefficacy of cough is due to expiratory muscle weakness, upper airway (glottic) muscle weakness, and inspiratory muscle weakness. Glottic muscle weakness may lead to aspiration and pneumonia. Abnormal chest wall and lung compliance contribute to respiratory insufficiency.

Cardiopulmonary Interactions/Pulmonary Hypertension

Pulmonary hypertension may occur as a result of two independent processes. Chronic hypoventilation secondary to weakness in the muscles that control respiration leads to chronic CO_2 retention and elevation of pulmonary arterial pressure. Secondly, cardiac insufficiency leads to an increase in filling pressure in the left ventricle as a compensatory mechanism to augment contractility. This results in an increase in pulmonary venous and, ultimately, pulmonary arterial pressure. These two processes may be additive in their effects on pulmonary pressure.

EVALUATION

Patient History

Historical information is vital to appropriate management. A history of palpitations or syncope is important and mandates further evaluation. Sleep disordered breathing or abnormal pulmonary function testing are also important to note. Prior and current therapy for cardiac and respiratory function should be reviewed in detail.

Physical Examination

Patients with congestive heart failure are typically tachypneic and tachycardic. Blood pressure is usually maintained until heart failure becomes severe or terminal. The precordial activity is usually diminished, but the point of maximal precordial impulse may be displaced secondary to cardiac enlargement. This finding may be obscured by chest wall abnormalities such as occur with scoliosis. There may be a murmur from mitral regurgitation and a gallop is frequently present. Hepatomegaly results from impairment of venous return. Capillary refill time is prolonged. Edema is a very late finding. It should be noted that children who are nonambulatory often develop dependent edema from sitting in a wheelchair for several hours at a time. The diagnosis of dependent edema is confirmed when it is relieved by elevating the child's legs.

Diagnostic Testing

Patients admitted to an ICU routinely have comprehensive laboratory evaluation. Among these, hemoglobin to assess capacity for oxygen delivery, arterial blood gas primarily to evaluate ventilation, and lactate as a measure of overall perfusion are all important first assessments. Brain natriuretic peptide is often evaluated in the presence of congestive heart failure and pulmonary hypertension.

Noninvasive studies on admission include chest X-ray, electrocardiogram, and echocardiogram.

Monitoring

Critically ill patients must be monitored continuously with at least two simultaneous ECG leads. Staff should be skilled in the recognition of rhythm abnormalities as well as the recognition of artifact from patient movement and equipment. All of these data should be archived in a system allowing for simultaneous review for a minimum of 72 hours.

The respiratory rate and the work of breathing supply important information regarding the patient's overall status. Tachypnea is commonly seen in association with respiratory distress. It is also a very early sign of congestive heart failure. Patients with diaphragmatic or chest wall muscle weakness may not manifest the usual signs of respiratory distress and are at high risk for respiratory failure, defined as apnea, or CO_2 retention with intact respiration. Hypoventilation develops insidiously in pediatric neuromuscular patients and can result in "sudden" death because CO_2 retention may cause cardiac arrhythmia. Oxygen supplementation in neuromuscular patients must be followed by frequent monitoring of blood gases. If there is retention of CO_2, the patient requires ventilation rather than O_2 supplementation. Without adequate ventilation, O_2 supplementation depresses inspiration, increases the pCO_2, and worsens respiratory failure.

Noninvasive measurement of blood pressure is adequate for the patient whose condition is stable as it correlates well with simultaneous radial arterial measurements. Complications of noninvasive blood pressure measurement include nerve palsies, particularly from inappropriately placed cuffs that are frequently inflated.

The unstable patient should have arterial pressure continuously monitored using an indwelling catheter. Complications include infection, vascular compromise, and embolization.

Significant deviations from normothermia place additional stress on critically ill patients, particularly neonates and small infants. Additionally, central fever, particularly in conjunction with cool extremities, may be a marker of a low cardiac output state or a significant infection.

Pulse oximetry is a valuable noninvasive tool to monitor the patient's oxygenation. Normal values are 95% to 100%. If any doubts exist regarding accuracy, a blood gas should be obtained.

Blood gas measurements provide pO_2 or dissolved oxygen, pCO_2, and pH as a determination of acid-base status. Venous blood gases are not useful in children with neuromuscular disease because an accurate pCO_2 measure is crucial to their management.

Supplemental Monitoring

Large catheters placed in the central venous circulation allow monitoring of the filling pressure of the right heart and can provide access for frequent laboratory analysis. Moreover, potent vasoactive infusions are more safely administered in the central circulation. A catheter with more than one lumen facilitates simultaneous sampling, pressure monitoring, and medication administration. Complications include infection, bleeding, arterial or nerve injury, thromboembolism, and air embolism.

Pulmonary arterial catheterization provides precise measures of cardiac output and left heart filling pressures. Potential pulmonary complications are significant, which limits its use.

Input and Output

Judicious fluid management is imperative to properly care for the patient with impaired cardiac function. Sufficient circulatory volume must be present to maintain cardiac output, but the primary problem in congestive heart failure is excess fluid retention. Intake and output should be totaled every shift and a running total of the net fluid balance should be maintained. Inputs include all enteral and parenteral fluids including blood products and medication infusions. Outputs include urine, stool, phlebotomy, and those from drainage tubes. Insensible losses include evaporative fluid loss, which are often elevated in patients with fever or receiving assisted ventilation.

Laboratory

The ECG is a standardized recording of the electrical activation and recovery of the heart. The ECG provides an assessment of overall rhythm, cardiac chamber enlargement, and the electrical recovery or repolarization process. Abnormalities associated with cardiomyopathies include left and right ventricular hypertrophy, left and right atrial enlargement, ST segment changes, and arrhythmia. Abnormal Q waves suggest myocardial injury, a very common finding in DMD. Hypertrophy is typical with FA.

Attempts have been made to prognosticate from the ECG. An index to correlate the severity of cardiomyopathy has been proposed.[31] ECG findings correlated with increased risk of death resulting from cardiac failure include decreased R wave amplitude in precordial leads, conduction abnormalities, ventricular ectopy, and sinus tachycardia.[32] Abnormal repolarization is also associated with increased risk of ventricular arrhythmia and sudden death.[33]

Imaging

The chest X-ray often shows cardiomegaly in the presence of cardiomyopathy. The cardiothoracic ratio is the maximum dimension of the heart divided by the thoracic cavity width at the diaphragm on a standard posteroanterior chest film. A cardiothoracic ratio of 0.6 is normal in newborns, whereas 0.45 is normal in older children and adolescents. This measurement may be confounded by kyphoscoliosis as well as lung pathology, including atelectasis (particularly left lower lobe), infiltrate, effusion, or eventration of the diaphragm. There may be passive congestion of the lungs secondary to heart failure. Furthermore, for the intensive care patient, the X-ray can confirm the location of endotracheal tubes, catheters within the circulatory system, and enteral feeding and drainage tubes.

Echocardiography is a standard means to assess cardiac structure and function (Figure 44.3). With chest wall abnormalities, imaging may not be adequate. Better image resolution can be obtained with transesophageal imaging, but requires very heavy sedation or general anesthesia. The echocardiogram provides comprehensive information about the patient's overall anatomy and function and is vital to the assessment of the critically ill patient. It is not routinely used as a tool for daily examination.

Nutrition

Appropriate nutritional support will assist in recovery, prevent infection and decubiti formation, and maintain normal mental status. Whenever possible, the gastrointestinal tract should be used, but neuromuscular patients frequently have bulbar weakness, putting them at risk for aspiration. If the patient cannot be fed orally, then nasogastric, orogastric, or gastrostomy tube feedings are indicated. Enteral feedings are preferred. Parenteral nutrition is reserved for the patient who is intolerant of enteral feeding or when there is evidence of gut hypoperfusion. Parenteral nutrition may be given through a peripheral intravenous line, but the high glucose concentrations (>12.5%) necessary usually require infusion into a central vessel. Lipid emulsions are given to provide fat and fortify calories. The complications of parenteral nutrition include vascular injury and infection. Excessive glucose increases the respiratory quotient (CO_2 production versus

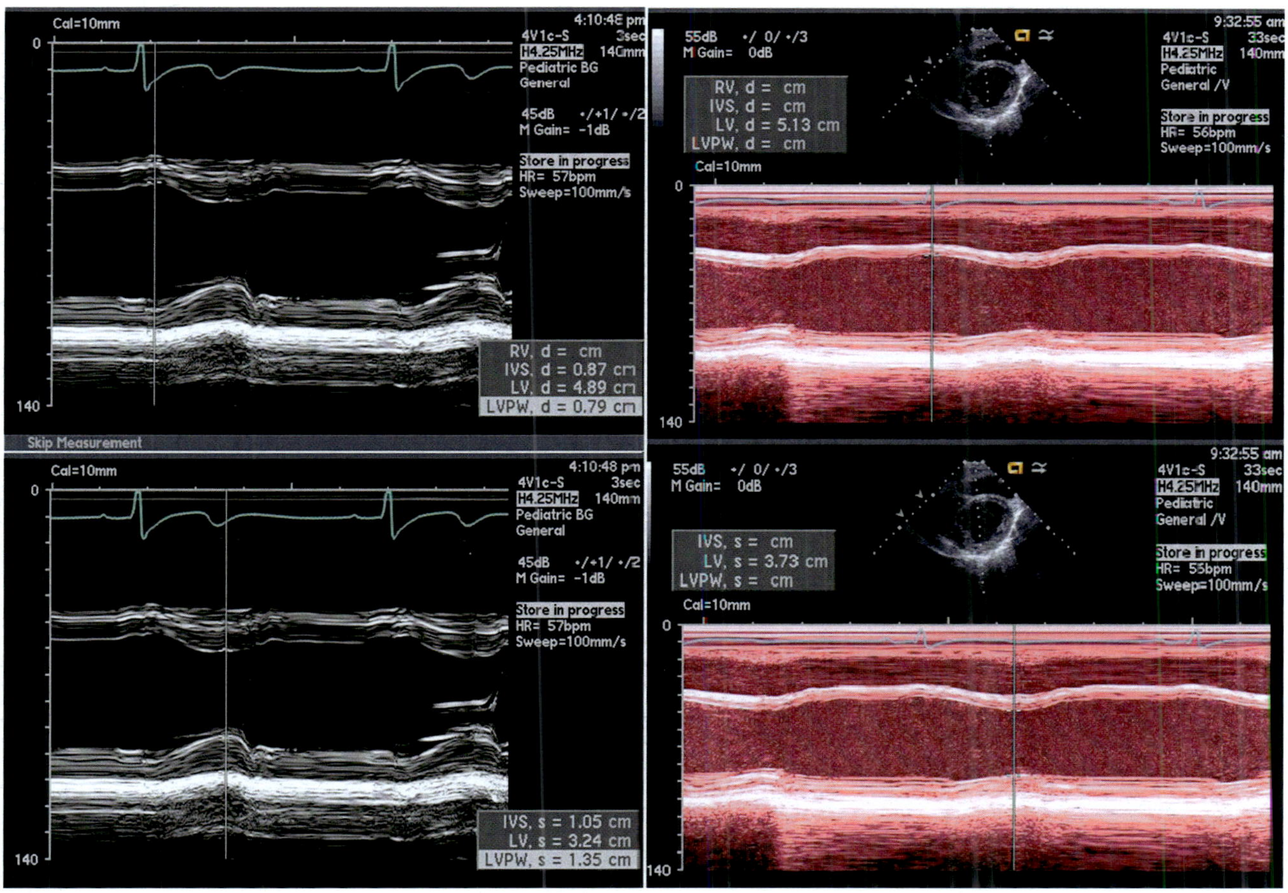

FIGURE 44.3 M-Mode echocardiogram demonstrating left ventricular wall motion. Fractional shortening is the change in cavity dimension during contraction divided by the maximum dimension. **Left:** upper and lower images from a healthy patient illustrate measurements in systole (4.89 cm) and diastole (3.24 cm) that yield a shortening fraction of 33.7% (normal > 29%). **Right:** upper and lower panels show these measurements for a patient with Duchenne muscular dystrophy with dimensions of 5.13 cm in diastole and 3.73 in systole yielding a shortening fraction of 27.3%. The patient was treated with lisinopril with subsequent improvement *(IVS = interventricular septum; LV = left ventricle; LVPW = left ventricular posterior wall; RV = right ventricle).*

oxygen consumption), thereby increasing ventilatory requirements and possibly precipitating or worsening preexisting respiratory failure.

TREATMENT

Pharmacologic

Pharmacologic agents may be used to support circulatory function, to suppress arrhythmia, and to alleviate pain and anxiety.

The circulation can be supported acutely by a number of vasoactive agents (Box 44.2). These agents directly augment cardiac output or blood pressure, or reduce the afterload or work of the heart. Most increase the risk of arrhythmia. They are powerful with very short half-lives and are given as continuous infusions. When high doses or multiple agents are used, an arterial line should be in place. Intravenous diuretics are used acutely to reduce pulmonary and systemic venous congestion.

As the patient stabilizes, a transition to longer acting oral agents is recommended. These include digoxin, oral afterload reduction, blockade of the renin-angiotensin system, β-blockade, and diuretics.[34]

Antiarrhythmic medications (Box 44.3) are primarily used to suppress tachyarrhythmia. Depending on the severity of the rhythm disturbance, many of these agents may be given intravenously. Lidocaine is used as an initial treatment for ventricular arrhythmia. In the past, intravenous amiodarone was reserved for refractory cases, but currently it may be used as a second line drug.

Antiarrhythmic therapy may worsen preexisting conduction abnormalities. Most anti-arrhythmic medications have the potential to worsen conduction, and should be used with care in patients with, or at risk for, heart block. In the event these medications must be used, cardiac pacing must be available.

Occasionally, drugs are required to increase heart rate. In bradycardia, isoproterenol may be used to increase the heart rate. Such treatment may be beneficial to treat atrioventricular block as a temporary measure until pacing can be instituted.

BOX 44.2 Intravenous Drugs to Augment Cardiac Output or Blood Pressure

Increase Contractility
- Dopamine
- Dobutamine
- Epinephrine
- Isoproterenol

Increase Blood Pressure
- Dopamine
- Epinephrine
- Vasopressin
- Phenylephrine

Decrease Afterload
- Milrinone
- Nitroprusside
- Nitroglycerin

Pacemaker

Cardiac pacing is the most reliable therapy for clinically significant bradycardia or atrioventricular block. Temporary pacing can be achieved using a transcutaneous, transesophageal (in neonates and infants), or transvenous electrode system. While providing rate support, loss of coordination for atrioventricular and intraventricular activation can compromise cardiac performance, despite the higher rate. Patients who require prolonged rate support are candidates for permanent pacemaker implantation.

Cardioversion

Arrhythmias that are sufficiently rapid to be life-threatening require immediate intervention. Direct current cardioversion and defibrillation use a large energy electric shock to reset the heart electrically. The usual energy for cardioversion is 0.5 J/kg and the starting energy for defibrillation is 1 J/kg, although initially higher energy may be more effective overall. The patient who is repeatedly at risk for life-threatening arrhythmia whose condition cannot be controlled with pharmacologic agents is a candidate for an implanted cardioverter/defibrillator.

Cardiopulmonary Resuscitation

Cardiopulmonary resuscitation may be complicated in the neuromuscular patient with skeletal abnormalities. Deformity of the chest may preclude effective cardiac compression. Individuals with significant weakness may become ventilator dependent. The use of neuromuscular blocking agents to facilitate intubation may result in hyperkalemia, rhabdomyolysis, and cardiac arrest even in the absence of known cardiac disease.

Mechanical Support

Mechanical support of the circulation (extracorporeal membranous oxygenator or left ventricular assist device) is indicated if there is reversible cardiac or pulmonary impairment despite conventional support with several inotropic agents. In general, these aggressive measures are not used for end-stage cardiomyopathy, except possibly

BOX 44.3 Drugs for Specific Arrhythmias

Supraventricular	Ventricular
Acute (intravenous)	*Acute (intravenous)*
Adenosine	Lidocaine
Diltiazem	Esmolol
Esmolol	Amiodarone
Amiodarone	
	Chronic (oral)
Chronic (oral)	β-Blockers
Digoxin	Dofetilide
β-Blockers	Sotalol
Verapamil	Amiodarone
Flecainide	
Sotalol	
Amiodarone	

as a bridge to cardiac transplantation. Some individuals have proposed the use of ventricular assist devices as "destination therapy" without plans for subsequent transplantation. Complications include bleeding, including intracranial hemorrhage, usually resulting from anticoagulation, renal impairment, infection, and mechanical failure of the circuit.

Cardiac Transplantation

Cardiac transplantation is a viable treatment for severe cardiomyopathy with end stage heart failure. It is therapy of last resort when all other interventions are inadequate. A condition that is likely to be terminal within a few years, other than the heart disease, is a contraindication. Nonetheless, cardiac transplantations have been reported in BMD, Barth's syndrome, Kearns-Sayre syndrome, limb girdle muscular dystrophy, DMD, DMD carriers, and EDMD. These cases are mainly isolated reports. No significant or meaningful follow-up is available.

Respiratory

Respiratory failure frequently requires invasive positive pressure ventilation. Respiratory treatment to mobilize secretions should be optimized and is facilitated with intubation. For patients not on positive pressure ventilation, supplemental O_2 should only be administered when the pCO_2 is known. Continuous positive airway pressure (CPAP) is to be avoided.

Patients who do require intubation must meet appropriate criteria for extubation.[35] Secretions should be minimal and oxygen saturation should be greater than 94% for 12 hours without supplemental oxygen. Continuous noninvasive ventilation is not appropriate immediately post extubation. Those who cannot be extubated successfully are candidates for tracheostomy.

Other Considerations

Patients chronically receiving steroids prior to ICU admission often require supplemental doses to compensate for the stress of their illness. The use of neuromuscular blocking agents to facilitate intubation may result in hyperkalemia, rhabdomyolysis, and cardiac arrest[36] even in the absence of known cardiac disease.[37]

ICU Weakness

Additional mention is warranted regarding patients who acquire weakness in the ICU environment. Almost 2% of children admitted to critical care units develop muscle weakness.[38] Frequently these patients are organ transplant recipients treated with neuromuscular blocking agents, corticosteroids, and aminoglycoside antibiotics. The usual clinical feature is failure to wean from a respirator. Lab evaluations show increased blood concentrations of creatine kinase and histological evidence of myofiber necrosis and depletion of myosin from muscle. Muscle biopsy distinguishes the several types of prolonged weakness that may occur with severe illness requiring intensive care. The condition remains poorly understood.[39]

REFERENCES

1. Melacini P, Fanin M, Danieli GA, Villanova C, Martinello F, Miorin M, et al. Myocardial involvement is very frequent among patients affected with subclinical Becker's muscular dystrophy. *Circulation* 1996;**94**:3168–75.
2. Messina DN, Speer MC, Pericak-Vance MA, McNally EM. Linkage of familial dilated cardiomyopathy with conduction defect and muscular dystrophy to chromosome 6q23. *Am J Hum Genet* 1997;**61**:909–17.
3. Flandreau G, Bourdin G, Leray V, Bayle F, Wallet F, Delannoy B, et al. Management and long-term outcome of patients with chronic neuromuscular disease admitted to the intensive care unit for acute respiratory failure: a single-center retrospective study. *Respir Care* 2011;**56**:953–60.
4. Yates K, Festa M, Gillis J, Waters K, North K. Outcome of children with neuromuscular disease admitted to paediatric intensive care. *Arch Dis Child* 2004;**89**:170–5.
5. Wang CH, Bonnemann CG, Rutkowski A, Sejersen T, Bellini J, Battista V, et al. Consensus statement on standard of care for congenital muscular dystrophies. *J Child Neurol* 2010;**25** 1559–81.
6. Wang CH, Dowling JJ, North K, Schroth MK, Sejersen T, Shapiro F, et al. Consensus statement on standard of care for congenital myopathies. *J Child Neurol* 2012;**27**:363–82.
7. Wang CH, Finkel RS, Bertini ES, Schroth M, Simonds A, Wong B, et al. Consensus statement for standard of care in spinal muscular atrophy. *J Child Neurol* 2007;**22**:1027–49.
8. Woodward GA, Insoft RM, Pearson-Shaver AL, Jaimovich D, Orr RA, Chambliss R, et al. The state of pediatric interfacility transport: consensus of the second National Pediatric and Neonatal Interfacility Transport Medicine Leadership Conference. *Pediatr Emerg Care* 2002;**18**:38–43.

9. Rosenberg DI, Moss MM. American College of Critical Care Medicine of the Society of Critical Care Medicine. Guidelines and levels of care for pediatric intensive care units. *Crit Care Med* 2004;**32**:2117–27.
10. Birnkrant DJ, Panitch HB, Benditt JO, Boitano LJ, Carter ER, Cwik VA, et al. American College of Chest Physicians consensus statement on the respiratory and related management of patients with Duchenne muscular dystrophy undergoing anesthesia or sedation. *Chest* 2007;**132**:1977–86.
11. Geller G, Harrison KL, Rushton CH. Ethical challenges in the care of children and families affected by life-limiting neuromuscular diseases. *J Dev Behav Pediatr* 2012;**33**:548–61.
12. Gnecchi-Ruscone T, Taylor J, Mercuri E, Paternostro G, Pogue R, Bushby K, et al. Cardiomyopathy in Duchenne, Becker, and sarcoglycanopathies: a role for coronary dysfunction? *Muscle Nerve* 1999;**22**:1549–56.
13. Saito M, Kawai H, Akaike M, Adachi K, Nishida Y, Saito S. Cardiac dysfunction with Becker muscular dystrophy. *Am Heart J* 1996;**132**:642–7.
14. Nigro G, Comi LI, Politano L, Bain RJ. The incidence and evolution of cardiomyopathy in Duchenne muscular dystrophy. *Int J Cardiol* 1990;**26**:271–7.
15. Sasaki K, Sakata K, Kachi E, Hirata S, Ishihara T, Ishikawa K. Sequential changes in cardiac structure and function in patients with Duchenne type muscular dystrophy: a two-dimensional echocardiographic study. *Am Heart J* 1998;**135**:937–44.
16. Ishikawa K. Cardiac involvement in progressive muscular dystrophy of the Duchenne type. *Jpn Heart J* 1997;**38**:163–80.
17. Melacini P, Vianello A, Villanova C, Fanin M, Miorin M, Angelini C, et al. Cardiac and respiratory involvement in advanced stage Duchenne muscular dystrophy. *Neuromuscul Disord* 1996;**6**:367–76.
18. Kelley RI, Cheatham JP, Clark BJ, Nigro MA, Powell BR, Sherwood GW, et al. X-linked dilated cardiomyopathy with neutropenia, growth retardation, and 3-methylglutaconic aciduria. *J Pediatr* 1991;**119**:738–47.
19. Isnard R, Kalotka H, Durr A, Cossee M, Schmitt M, Pousset F, et al. Correlation between left ventricular hypertrophy and GAA trinucleotide repeat length in Friedreich's ataxia. *Circulation* 1997;**95**:2247–9.
20. Groh WJ, Groh MR, Saha C, Kincaid JC, Simmons Z, Ciafaloni E, et al. Electrocardiographic abnormalities and sudden death in myotonic dystrophy type 1. *N Engl J Med* 2008;**358**:2688–97.
21. Finsterer J, Stollberger C, Keller H. Arrhythmia-related workup in hereditary myopathies. *J Electrocardiol* 2012;**45**:376–84.
22. Buckley AE, Dean J, Mahy IR. Cardiac involvement in Emery Dreifuss muscular dystrophy: a case series. *Heart* 1999;**82**:105–8.
23. Tsuchiya Y, Arahata K. Emery-Dreifuss syndrome. *Curr Opin Neurol* 1997;**10**:421–5.
24. Rakovec P, Zidar J, Sinkovec M, Zupan I, Brecelj A. Cardiac involvement in Emery-Dreifuss muscular dystrophy: role of a diagnostic pacemaker. *Pacing Clin Electrophysiol* 1995;**18**:1721–4.
25. Dahlbom K, Ahlstrom G, Barany M, Kihlgren A, Gunnarsson LG. Muscular dystrophy in adults: a five-year follow-up. *Scand J Rehabil Med* 1999;**31**:178–84.
26. Chenard AA, Becane HM, Tertrain F, de Kermadec JM, Weiss YA. Ventricular arrhythmia in Duchenne muscular dystrophy: prevalence, significance and prognosis. *Neuromuscul Disord* 1993;**3**:201–6.
27. Paulides FM, Plotz FB, Verweij-van den Oudenrijn LP, van Gestel JP, Kampelmacher MJ. Thirty years of home mechanical ventilation in children: escalating need for pediatric intensive care beds. *Intensive Care Med* 2012;**38**:847–52.
28. Benditt JO. The neuromuscular respiratory system: physiology, pathophysiology, and a respiratory care approach to patients. *Respir Care* 2006;**51**:829–37, discussion 37–9.
29. Benditt JO, Boitano LJ. Pulmonary issues in patients with chronic neuromuscular disease. *Am J Respir Crit Care Med* 2013;**187**:1046–55.
30. Panitch HB. The pathophysiology of respiratory impairment in pediatric neuromuscular diseases. *Pediatrics* 2009;**123**(Suppl. 4): S215–18.
31. Finsterer J, Stollberger C, Keller H, Slany J, Mamoli B. Cardiac involvement in patients with myotonic dystrophy, Becker's muscular dystrophy and mitochondrial myopathy. *Herz* 1997;**22**:96–103.
32. Akita H, Matsuoka S, Kuroda Y. Predictive electrocardiographic score for evaluating prognosis in patients with Duchenne's muscular dystrophy. *Tokushima J Exp Med* 1993;**40**:55–60.
33. Yotsukura M, Yamamoto A, Kajiwara T, Nishimura T, Sakata K, Ishihara T, et al. QT dispersion in patients with Duchenne-type progressive muscular dystrophy. *Am Heart J* 1999;**137**:672–7.
34. Ishikawa Y, Bach JR, Minami R. Cardioprotection for Duchenne's muscular dystrophy. *Am Heart J* 1999;**137**:895–902.
35. Hull J, Aniapravan R, Chan E, Chatwin M, Forton J, Gallagher J, et al. British Thoracic Society guideline for respiratory management of children with neuromuscular weakness. *Thorax* 2012;**67** (Suppl. 1):i1–40.
36. Breucking E, Reimnitz P, Schara U, Mortier W. Anesthetic complications. The incidence of severe anesthetic complications in patients and families with progressive muscular dystrophy of the Duchenne and Becker types. *Anaesthesist* 2000;**49**:187–95.
37. Wu CC, Tseng CS, Shen CH, Yang TC, Chi KP, Ho WM. Succinylcholine-induced cardiac arrest in unsuspected Becker muscular dystrophy—a case report. *Acta Anaesthesiol Sin* 1998;**36**:165–8.
38. Banwell BL, Mildner RJ, Hassall AC, Becker LE, Vajsar J, Shemie SD. Muscle weakness in critically ill children. *Neurology* 2003;**61**:1779–82.
39. Mohamed A, Ryan MM. Neuromuscular complications of intensive care. *Handb Clin Neurol* 2013;**113**:1481–3.

Chapter 45

Malignant Hyperthermia: An Inherited Disorder of Muscle Calcium Metabolism

Werner Klingler and Frank Lehmann-Horn

MALIGNANT HYPERTHERMIA—DEATHS IN GENERAL ANESTHESIA

The name malignant hyperthermia (MH) is derived from the clinical observations of Denborough and Lovell, who analyzed the unexplained deaths of previously asymptomatic patients undergoing general anesthesia in 1960.[1] These patients exhibited symptoms of muscle rigidity and rapidly rising, irrepressible core body temperature; hence the term MH. Within 2 years, Denborough and colleagues realized that young age and blood relationship is a risk factor for the development of the syndrome.[2] An autosomal dominant mode of inheritance was later identified in 1990.[3,4]

In due course, more sophisticated anesthesiological equipment allowed the analysis of signs and symptoms of MH crises in more detail (Table 45.1), the main features of which are rigidity of skeletal muscle, severe acidosis, and inadequate increase in body temperature (Figure 45.1). The majority of reported MH crises indicate that the phenomenon is pharmacologically provoked; crises are triggered by the volatile anesthetic halothane and the depolarizing muscle-relaxant succinylcholine. This predominance is most likely due to the fact that these drugs were common for many years—and still are in economically poorer countries—and that they were not only used frequently, but also regularly in combination. With the exception of xenon and nitrous oxide, MH crises have been observed with all volatile anesthetics (i.e. halothane, sevoflurane, enflurane, isoflurane, and desflurane).[7,8] The preservative 4-chloro-m-cresol (4-CmC) was found to trigger MH and was consequently removed from succinylcholine preparations and other muscle relaxants in the 1980s. 4-CmC is still used as preservative (e.g. for insulin formulations); however, the enclosed dosages are two orders of magnitude below clinical relevance in terms of MH-triggering effects. Hence, modern 4-CmC containing pharmacological formulations can be used safely in MH-disposed individuals.[9]

MUSCULAR HYPERMETABOLISM BY RAMPANT CA^{2+}

In predisposed individuals, application of MH triggers leads to excessive release of Ca^{2+} in skeletal muscle cells. Here, Ca^{2+} is accumulated in intracellular Ca^{2+} stores within the sarcoplasmic reticulum (SR). Ca^{2+} release is regulated by a complex of four dihydropyridine-receptors (DHPR) and the homotetrameric ryanodine receptor type 1 (RyR1), arranged in clusters along the triadic junction between the transverse tubular system and the SR. The voltage sensitive DHPR activates RyR1 predominantly by direct (mechanical) protein-protein interaction and secondarily by promotion of Ca^{2+}-induced Ca^{2+} release (CICR). Volatile anesthetics are direct and potent agonists on RyR1, which is the largest known ion channel (>2MDa, roughly 5000 amino acids) allowing high ion flux and increase in cytoplasmic Ca^{2+} concentration by several orders of magnitude within milliseconds.[10]

Ca^{2+} unmasks actin and leads to force generation by myosin-ATPase. Rampant Ca^{2+} release in MH muscle leads to generalized muscle rigidity, consumption of energy carriers, and stimulation of glycolysis and respiratory enzymes. Accumulation of acidic metabolites, ATP exhaustion, and Ca^{2+} overload finally result in loss of cellular integrity and rhabdomyolysis (Figure 45.1). Clinically, a severe hypermetabolism can be observed including tachypnea, tachycardia, hypoxemia, and rapid increase in end-tidal CO_2 concentrations. Anesthesiologists refer to hot soda lime, which is the CO_2-absorbing component in the breathing circuit of the anesthesia machine. Due to the rigidity of respiratory muscles, blood gas analysis yields a combined

B.T. Darras, H. Royden Jones, Jr., M.M. Ryan & D.C. De Vivo (Eds): Neuromuscular Disorders of Infancy, Childhood and Adolescence, Second edition.
DOI: http://dx.doi.org/10.1016/B978-0-12-417044-5.00045-7

TABLE 45.1 Symptoms and Treatment of a Clinical MH Episode

Malignant Hyperthermia	
Early signs: • increasing $ETCO_2$ • tachycardia • tachypnea • mixed acidosis • masseter spasm/trismus	**Signs that may appear later:** • hyperthermia • muscle rigidity • myoglobinuria • cardiac arrest

Call for Help – Start preparing Dantrolene

Differential diagnoses:

- inadequate anesthesia or analgesia
- insufflation of CO_2
- hypoventilation
- tourniquet ischemia
- used soda lime
- infection, sepsis, external overheating
- hypoxemia
- pheochromocytoma
- thyroid storm

Treatment:

- stop MH triggers
- increase fresh gas flow to 10 L/min (no change of system)
- convert to total intravenous anesthesia
- hyperventilation FiO_2 100%
- preparation of dantrolene 2.5 mg/kg IV bolus (help necessary!)
- rapid administration of dantrolene up to 10 mg/kg until patient stable
- buffer metabolic acidosis / hyperkalemia
- physical cooling
- treat cardiac arrhythmia according to advanced life support guidelines
- intensive care unit, aim at urine output of 2 mL/kg/h,
- blood gas analysis, blood count, liver enzymes, creatine kinase, myoglobin, lactic acid, coagulation
- continue dantrolene 1 mg/kg every 4–6 hours until patient stable for a maximum of 24–36 hours

There is no specific sign for MH. The time course varies significantly. Administration of calcium antagonists can potentiate hyperkalemia. Modified after Harrison et al. (2013).[5]

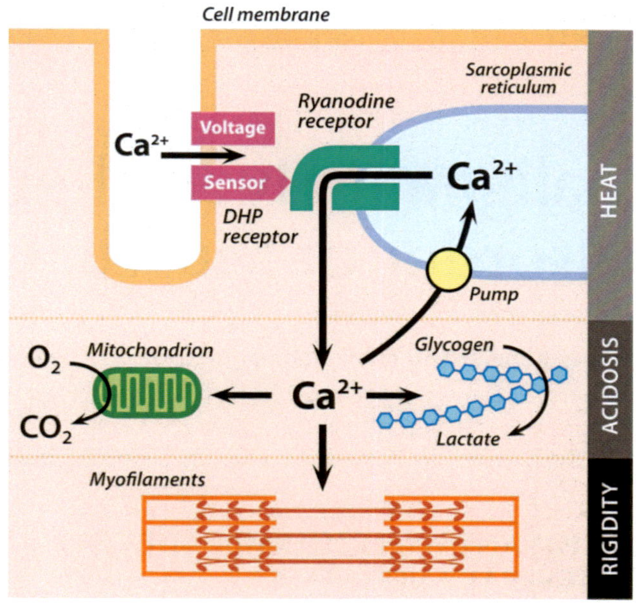

FIGURE 45.1 Pathophysiology of malignant hyperthermia (MH). Cartoon of the key structures of excitation-contraction coupling in the transverse tubule (T-tubule) of skeletal muscle. The dihydropyridine receptor (DHPR) is linked to the ryanodine receptor (RyR1), which is the calcium release channel situated in the membrane of the sarcoplasmic reticulum (SR). The cardinal symptoms of MH are explained by excessive calcium release, and stimulated glycolysis and respiratory chain. *Reprinted with permission from Lehmann-Horn et al. (2011).*[6]

metabolic and respiratory acidosis (i.e. mixed acidosis). Laboratory investigations further show increasing K^+ and creatine phosphokinase levels originating from rhabdomyolysis. The course of an MH crisis may be complicated by renal failure and myoglobinuria, which can appear as "cola-colored" urine, and/or disseminated intravascular coagulation. Cardiac arrhythmia is a phenomenon in MH mainly due to an increase in serum K^+.[11–13]

Membrane depolarization by succinylcholine can boost and rarely induce an MH reaction by stimulation of DHPR and nonspecific Ca^{2+} influx. Nondepolarizing muscle relaxants can be used safely in MH-predisposed individuals. Moreover, precurarization (i.e. pretreatment with nondepolarizing muscle relaxants) may attenuate or even prevent clinical MH reactions.[14,15]

There is no specific symptom pathognomonic for MH because clinical features and time course of MH crises vary markedly.[13,14] In 1994, Larach and colleagues developed a mathematical scoring model to estimate the likelihood of a clinical MH event. Five main processes are assessed: muscle rigidity, muscle breakdown (i.e. rise of the muscle enzyme creatine phosphokinase above 10,000 units/L), acidosis, inadequate increase in body temperature, and the beneficial effect of the specific antidote dantrolene, which also contains mannitol to prevent renal failure.[11]

Treatment of an MH crisis is based on interruption of the positive feedback loop of self-sustaining myoplasmic hypermetabolism. Dantrolene significantly reduces myoplasmic Ca^{2+} overload by various mechanisms, and should be administered immediately after stopping the trigger anesthetics. A clinical MH event can be suspected upon observation of beneficial effects of dantrolene (i.e. reversal of hypermetabolism).

Regarding hypermetabolism and hyperthermia, most interestingly, RyR1 has been identified in B-lymphocytes, where it plays a role in the regulation of specific immune response. In B cells RyR1-liberated Ca^{2+} promotes the release of interleukin 1β (IL-1β), which is an endogenous pyrogen. B-lymphocytes from MH patients show an elevated cellular metabolism and increased IL-1β release after pharmacological challenge with MH-triggering drugs. Action of endogenous pyrogens may explain recrudescence of MH hypermetabolism and even delayed onset, which has been described several hours after anesthesia.[16–18]

ANESTHESIA-RELATED MUSCLE SPASMS

Localized muscle stiffness can present as jaw stiffness (trismus), which may be a threatening condition during induction of general anesthesia because ventilation and endotracheal intubation can be hindered. Anesthesia-related muscle spasm is a nonspecific sign for neuromuscular dysregulation and occurs predominantly in pediatric patients. Independent of underlying disease, masseter spasm is observed in roughly 0.3% of children after administration of the depolarizing muscle relaxant succinylcholine.[19] Compared to other muscles, the masseter has an atypical composition that may explain, in part, the increased sensitivity to succinylcholine. Specific features of the masseter muscles include a highly variable and heterogeneous distribution of myosin isoenzymes, the expression of neonatal and alpha-cardiac myosin heavy chain, and the existence of small motor units.[20] In other words, jaw muscles contain a comparatively high number of motor endplates, where succinylcholine is an agonist on nicotinic acetylcholine receptors (nAChR). nAChR are nonspecific cation channels that depolarize the muscle membrane by influx of Na^+ and, notably, Ca^{2+}.[14] Membrane depolarization initiates excitation-contraction coupling (i.e. Ca^{2+}-release from SR). There are hints that in patients with neuromuscular disorders, most notably MH, the incidence of clinically relevant succinylcholine-induced masseter spasm may be increased.[21,22]

Anesthesia-related localized and generalized muscle spasm is a phenomenon also seen in other conditions, and may complicate artificial ventilation. In disorders of neuromuscular transmission (e.g. denervation, immobilized patients, paralyzed patients, and myasthenia gravis), reference is made to the pathological upregulation of nAChR and expression of abundant extrajunctional nAChR. Enhanced excitability of muscle membranes is also found in inherited channelopathies, such as chloride channel myotonia (myotonia congenita, Thomsen and Becker types) and sodium channel myotonia (paramyotonia). Clinically, the symptoms can mimic an MH event.[23]

AWAKE EPISODES

Outside general anesthesia, individuals carrying the trait for MH are usually asymptomatic. Some patients complain about exercise-dependent myalgias, predominantly in gross muscle such as the thigh. Other observations include heat intolerance and inadequate temperature increase after physical exercise. Case reports describe clinical crises after heavy physical exercise and in febrile conditions.[24,25] Even if there are not enough data for statistical evaluation of incidence, children are obviously more prone to awake MH episodes. Case reports often refer to children, adolescents, or young adults. Crises have been described in hot surroundings, such as a car interior, or during infections with fever.[26–28]

Awake MH episodes are also found in animal models for MH. Most extensively studied is porcine MH, where stress-induced rhabdomyolysis reduces meat quality. The syndrome was named "porcine stress syndrome" and a homozygote mutation in RyR1 was identified. The mode of inheritance is recessive in porcine stress syndrome and dominant in the horse and mouse MH model. In all animal models, nonanesthetic MH crises occur frequently and can be triggered reproducibly by heat or physical exertion. In affected quarter horses, rhabdomyolysis can occur without evident hyperthermia.[29–31]

MH AND CHILDHOOD

The incidence of clinical MH crises as well as awake episodes is more frequent in children (Figure 45.2). Roughly half of the crises occur in children younger than 12.[8] Several reasons may account for this observation. Children have fewer developed compensatory mechanisms for

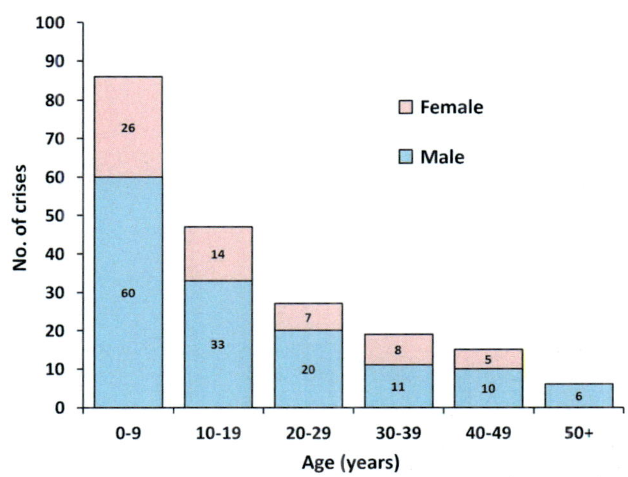

FIGURE 45.2 Age dependence and male predominance. The graph shows age and gender distribution of 200 clinical MH crises evaluated in a European multicenter study. All events were later diagnosed to be proven malignant hyperthermia events by *in vitro* contracture testing on surgically dissected muscle biopsies. Patients were 70% male; 50% of crises occurred in patients younger than 12 years old. 1% of the crises were triggered by succinylcholine, 18% by volatile anesthetics, and 81% by a combination of both. *Figure modified and adapted from: Klingler et al. (2014).*[8]

increased body temperature. Heat tolerance and dissipation is reduced. Generally, children have a higher metabolic rate in skeletal muscle than adults; CPK levels show an age-dependent decline. Fiber type composition is different in children and adults with a shift to fiber type II with increasing age. In other words, children exhibit a relatively high portion of oxidative type I fibers (slow twitching). In the *in vitro* contracture test (IVCT), type I fibers are more sensitive to pharmacological triggers.[32,33] Indeed, other factors of development such as influence of growth hormones and temperature regulation in the central nervous system may contribute to the enhanced vulnerability of skeletal muscle in childhood.

Epidemiological observations in children show that MH-predisposition is associated with anomalies of the musculoskeletal system and connective tissue. Specific diagnoses overrepresented in the MH-predisposed population are various forms of myopathies, strabismus, and scoliosis.[34]

FUNCTIONAL DIAGNOSTICS—*IN VITRO* CONTRACTURE TEST

The *in vitro* contracture test (IVCT) is a robust method to diagnose predisposition to MH. The classification yields MH susceptibility (MHS) or an MH negative (MHN) result. IVCT requires a surgical muscle biopsy at a specialized MH unit. A minimum of four muscle bundles of roughly 20 mm length and 300 mg weight are dissected from human vastus lateralis muscle. To avoid pharmacological interferences, regional or spinal anesthesia is preferred over local anesthesia. The IVCT is a functional test on fresh muscle tissue, so muscle bundles need to be investigated freshly within a maximum of 5 hours; dispatch of specimens to a distant laboratory is impossible. Muscle bundles are suspended and pre-stretched in an organ bath containing physiological Krebs-Ringer solution, pH and temperature are controlled, and the muscle tissue is continuously bubbled with 95% O_2 and 5% CO_2. Force is measured with a highly sensitive force transducer.[14,23] The generated contracture force is used as an indirect marker for myoplasmic Ca^{2+} concentration. Halothane and caffeine, the pharmacological trigger substances, are added to the organ bath in respectively cumulative bolus doses (Figure 45.3). Protocol variants differ depending on the recommendations of the respective North American and European Malignant Hyperthermia Groups. Nonetheless, there is a high concordance between the protocol variants. In Japan, skinned muscle fibers are investigated. In MHS muscle, pharmacologically induced CICR is significantly increased. Although caffeine in the millimolar range is a potent *in vitro* trigger for MH, this is not the case *in vivo* because concentrations differ greatly.[14,35]

At the earliest, IVCT is recommended 3 months after a clinical MH episode. To obtain valid results, recovery of muscle tissue from (partial) rhabdomyolysis and full elimination of pharmaceuticals, most notably dantrolene, is indispensable. Many MH test centers only accept children older than roughly 8 to 10 years because of the required size of the muscle biopsy.[14]

GENETIC DIAGNOSTICS

Estimations for the incidence of clinical MH events range from 1:30,000 in children up to 1:300,000 in adults. Genetic analyses show an autosomal dominant mode of inheritance with an unknown frequency of asymptomatic

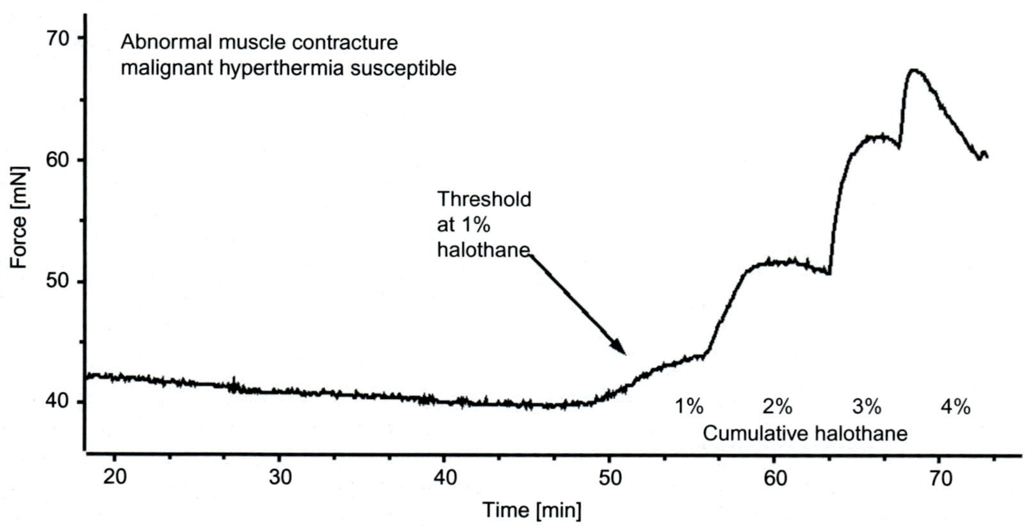

FIGURE 45.3 *In vitro* contracture test (IVCT), original trace. The diagram shows a time versus force diagram of a dissected muscle bundle in an organ bath. The initial pre-stretch is necessary for alignment of contractile proteins. Halothane is added in cumulative doses. The bundle shows a baseline force increase of more than 2 mN at 2% halothane. This response is a pathologic contracture and indicates MH-predisposition.

carriers. Since the discovery of the link between MH susceptibility and the gene encoding RyR1, research efforts have identified more than 300 RyR1 mutations as of 2014. An association with MH is suggested for the majority of these mutations,[3,14,36] many of which are clustered in three hotspot regions of the protein (Figure 45.4). However, causality has only been proven for roughly 10% of these mutations. Certain diagnosis is possible by genetic means according to the guidelines for genetic detection of susceptibility to MH agreed upon by the European Malignant Hyperthermia Group (EMHG).[37,38] The RyR1 gene contains 106 exons; therefore, the search for genetic alterations is laborious and it is difficult to differentiate between polymorphism and functionally relevant mutation. Mutations in DHPR have been identified in singular families.[36,39]

RyR1 is predominantly expressed in skeletal muscle; the level of expression is independent of MH disposition. Isoforms are found in cardiac muscle (RyR2) as well as in the central nervous system (RyR3). Mutations in RyR2 have been linked to ventricular arrhythmia, but general anesthesia including volatile anesthetics and succinylcholine has been described as uneventful.[40]

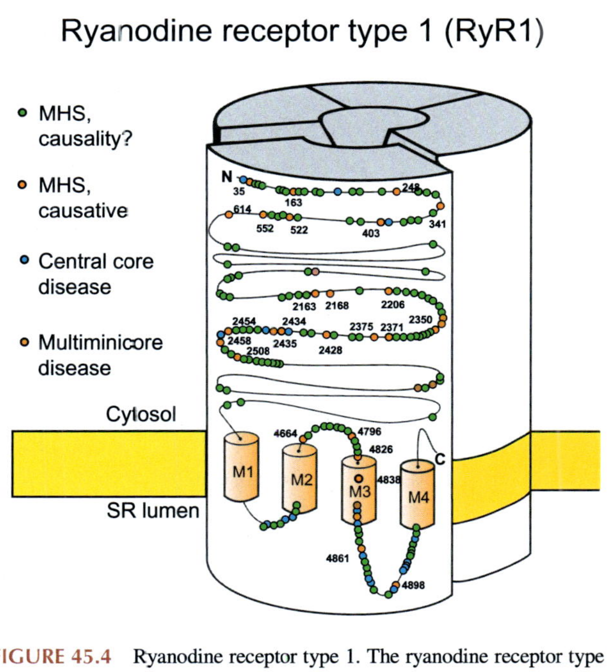

FIGURE 45.4 Ryanodine receptor type 1. The ryanodine receptor type 1 (RyR1), shown here, is a homotetramer composed of more than 5000 amino acids. RyR1 is the largest known human ion channel and is even visible by optical microscopy. RyR1 has four transmembrane domains and serves as the Ca^{2+} release channel in the sarcoplasmic reticulum (SR). The sketch shows a selection of mutations associated to malignant hyperthermia. Mutations in RyR1 are clustered in three hotspot regions. Region 1 is found between amino acids 35 and 614, region 2 between 2163 and 2458, and region 3 between 4664 and 5020. Positions of proven causative mutations are quoted.

Genetic knowledge of MH is still incomplete and does not explain some of the specifics seen in MH, such as age dependence and male preponderance (Figure 45.2). Thus, epigenetic factors seem to contribute to the syndrome and complicate the genetic diagnosis.[31,41,42]

CREATINE PHOSPHOKINASE

Creatine phosphokinase (CPK) is an important enzyme of skeletal muscle energy metabolism, which is usually present only in small concentrations in blood. Recurrent elevated CPK levels in asymptomatic patients can hint toward a subclinical myopathy. In MH-susceptible individuals, CPK is often slightly elevated or in the upper normal range. However, CPK is vulnerable to a number of conditions such as hypo- and hyperthyroidism, soft tissue trauma, medical injections, or strenuous muscle exercise. Therefore, it is advisable to determine CPK several times in intervals of roughly 3 months, in several siblings of the family under suspicion of susceptibility, after sports abstention of roughly 2 weeks, and after exclusion of potential interferences. For evaluation, divergent laboratory methods (meanwhile mostly temperature corrected) and especially the specific age-dependent reference values in children need to be considered. In isolated familial chronic CPK-elevation, an MH diagnostic muscle biopsy is indicated.[14]

HISTOLOGICAL FINDINGS AND CORE FORMATION

Histological examinations in muscle tissue from MH patients show normal findings or mild myopathic changes; therefore, diagnosis of MH is not possible by histological methods. Most commonly, MH muscle exhibits atypical, hypertrophic fibers and type I fiber predominance. The changes are usually disseminated and sometimes accompanied by internal nuclei and myofibrillar necrosis.[43] In some cases, histological examinations may find cores, which are amorphous regions that lack mitochondria and oxidative enzyme activity. There is a strong link or even overlap between MH and core diseases, most notably central core disease (CCD), which presents with muscle weakness of variable degree. Common features are floppy infant syndrome, skeletal anomalies, delayed motor milestones, and generalized muscle hypotonia during adolescence. In adulthood, CCD usually is nonprogressive and functional improvement can occur. Roughly 40% of affected patients are considered asymptomatic in adulthood. Histologically, CCD is characterized by central cores in the muscle fibers (Figure 45.5). A potential explanation of core formation is initial mitochondrial damage due to Ca^{2+} overload and swelling. Neighboring organelles become involved and

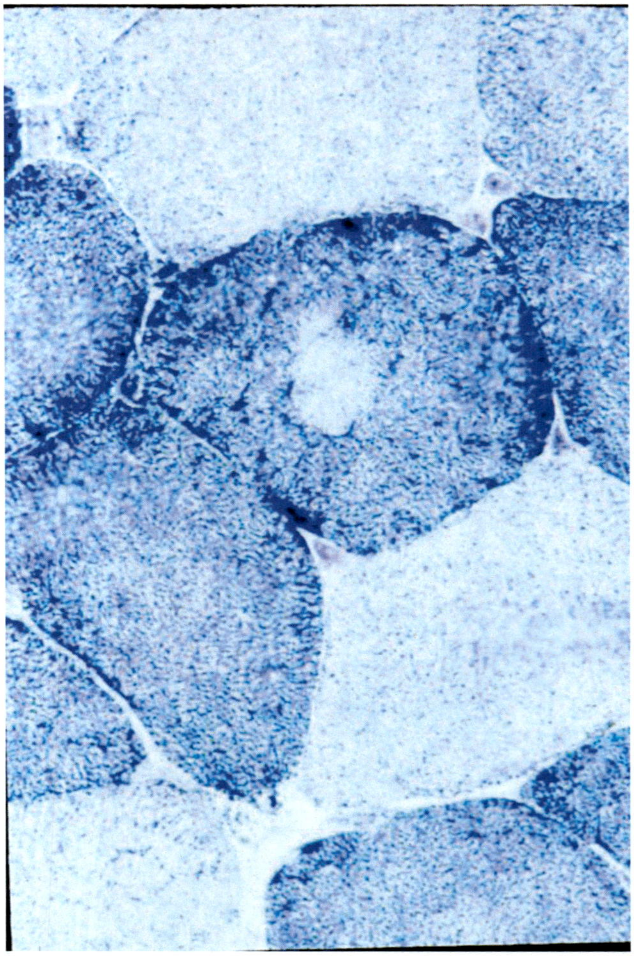

FIGURE 45.5 Central core disease histology. The figure shows a typical histological finding in central core disease (NADH reductase staining, magnification factor 400). The muscle fibers contain unstructured central cores and lack of oxidative enzyme activity indicating loss of mitochondria. [Some of our group published a similar version of this figure in Klingler et al. (2009).[44]]

unstructured areas might result from further progressive damage. Core diseases (multiminicore disease, CCD, nemaline myopathy) show genetic diversity. Presumably, the risk for MH is high when mutations are located in the gene encoding RyR1 (Table 45.2). However, mutations in the C-terminal region of RYR1 lead to "loss of function" of the channel and, in contrast to all other mutations, to reduced Ca^{2+} release. Whether these patients are at risk for clinical MH episodes is unclear.[44]

ASSOCIATED SYNDROMES AND MH-LIKE EPISODES

The specific feature of MH is excessive Ca^{2+} release from the SR leading to pathologic hypermetabolism exhibiting a number of clinical signs and symptoms (Table 45.1). However, hypermetabolic crises are observed in other neuromuscular disorders, which may resemble an MH event despite completely different pathomechanisms. Indeed, in the premolecular era, clinical association with MH was postulated for several neuromuscular diseases and later discarded. For example, an association with MH has been excluded for myotonia congenita (Thomsen and Becker types). In myotonic crises, which may mimic MH-like muscle rigidity, causative therapy is based on inhibition of Na^+ conductance in order to reduce electrical membrane excitability. Furthermore, K^+ in the upper normal range seems to have antimyotonic effects.[23] Volatile anesthetics have not only been used safely in patients suffering from myotonia congenita, but also in mitochondrial myopathy. Mitochondrial disorders have an estimated incidence of 1 in 4000, show a maternal inheritance, and present with various degrees of muscle weakness, cardiomyopathy, and predisposition to epileptic seizures. The most common maternally inherited disorder is the mutation of the mitochondrial tRNA at nucleotide position 3243, which leads to MELAS syndrome (mitochondrial encephalomyopathy, lactic acidosis, and stroke-like episodes). A further mitochondrial disorder depends on nucleotide exchanges within the mitochondrial tRNA and leads to myoclonic epilepsy with ragged red fibers (MERRF) in the muscle biopsy. Other mitochondrial disorders include Leber hereditary optic neuropathy (LHON) and Kearns-Sayre syndrome, for which MH-like events have been described. In contrast to the situation in most other neuromuscular diseases, the use of propofol, most notably for long-term sedation, may be hazardous in patients suffering from mitochondrial myopathies due to the inhibition of oxidative phosphorylation, β-oxidation, and the enzyme carnitine palmitoyltransferase. The pathomechanism also explains why these patients are prone to the development of a propofol infusion syndrome. Skeletal muscle and cerebral and cardiac function are particularly vulnerable to mitochondrial defects. Tight metabolic monitoring (i.e. electrolytes, glucose, lactate, pH) is essential in the perioperative period and metabolism should be supported by the administration of substrates such as glucose or amino acids.[14,45,46]

It is certain that there is an increased susceptibility to MH in patients with core diseases and nemaline myopathy, and it is likely that this is also the case in King-Denborough syndrome.[44] In other myopathies, such as the nondystrophic myotonias, the periodic paralyses, myotonic dystrophy, and the dystrophin-deficient muscular dystrophies, hypermetabolic "MH-like" events have been described (Table 45.2), but there is no consistent triggering agent and it seems likely that the molecular mechanisms responsible differ from those of MH susceptibility.[13,14,47] Several authors also recommend the clarification of MH predisposition in individuals

TABLE 45.2 Syndromes Associated with an Increased Risk of Rhabdomyolysis

Syndrome	RyR1 Mutations
Malignant hyperthermia	>70%
Central core disease	>90%
Nemaline rod myopathy	some cases
Multiminicore disease	some cases
King-Denborough syndrome	some cases
Atypical core diseases	some cases
Hypokalemic periodic paralysis	some cases
Congenital neuromuscular disease with uniform type I fibers	some cases
Samaritan congenital myopathy	some cases
Malignant neuroleptic syndrome	rare cases
Recurrent heat stroke	rare cases
Statin-induced rhabdomyolysis	rare cases
Exercise-induced rhabdomyolysis	rare cases
Amphetamine-induced rhabdomyolysis	not investigated
Mitochondrial myopathies (MELAS, MERF, LHON, Kearns-Sayre syndrome)	no
Myotonia congenita (Becker and Thomson types)	no
Myasthenia gravis, Lambert-Eaton syndrome	no
Myotonic dystrophy	no
Proximal myotonic myopathy	no
Hyperkalemic periodic paralysis	no
Paramyotonia	no
Metabolic myopathies (glycogenoses, lipid storage disorders)	no
Myoadenylate deaminase deficiency	no
Carnitine palmitoyltransferase II deficiency	no
Noonan syndrome	no
Brody myopathy	no

Risk for the development of a true MH crisis depends on the involvement of Ca^{2+} metabolism. This is the case for the majority of patients suffering from CCD, multiminicore disease, or nemaline rod myopathy, where mutations in the RyR1 gene have been documented. The table also shows other syndromes that may present with MH-like hypermetabolism. Potential associations with RyR1 mutations are listed. Interestingly, some cases of sudden infant death have been associated with sequence variants in the cardiac isoform of the RyR (RyR2) gene.

suffering from recurrent heat stroke or exercise-induced rhabdomyolysis, since mutations in RyR1 have been found in a significant portion of these patients.[48] There is weak evidence that MH-like events can occur in enzymopathies/genetic syndromes.[49] Supposedly, multiple factors are involved to create MH-like reactions.

CLINICAL CONSIDERATIONS

Fortunately, perioperative hyperthermia is not MH in most cases. Possible causes for elevated body temperature, especially in children, are warm environment, insolation, heat stroke, hypovolemia, bacterial or viral infection, allergy, hyperthyroidism, side effect of drugs, pheochromocytoma, and other endocrinopathies. Furthermore, preoperative fasting leads to metabolic acidosis in children, which may be mistaken as an MH-associated symptom.[50]

Patient safety is the most prominent concern when dealing with patients with rare diseases. The clinical signs and the emergency management of an MH crisis are summarized in Table 45.1. There is good evidence that checklists improve handling of MH crises independent of professional experience and specialization.[51] Since MH is a potentially life-threatening condition, individuals and siblings at risk should be referred to a rare disease center, equipped with medical emergency bracelets, and advised to inform medical professionals about the condition, most notably in the case of surgical procedures. There are now a multitude of anesthetic drugs at hand that have been proven safe in persons susceptible to MH, allowing the application of regional and general anesthesias in MHS individuals. It is important to note that a person who has passed several general anesthesias without problems may still have MH disposition because an MH crisis is not necessarily triggered in every anesthesia. MH is usually asymptomatic, but it should be mentioned that MHS persons should refrain from extreme physical stress (e.g. long-distance running, triathlon) or exposure to extremely high environmental temperature. The risk for rhabdomyolysis may be increased with drugs altering the sarcolemmal lipid bilayer (e.g. ecstasy, amphetamines, statins, steroids). Further information can be found on the websites of the Malignant Hyperthermia Association of the United States[52] and the EMHG.[53]

In neuromuscular disease, the risk for development of an MH event depends on the involvement of Ca^{2+} conducting proteins. Hundreds of genetic variations have been identified in the gene encoding the sarcoplasmic Ca^{2+} release channel RyR1, most with unclear significance concerning channel function. Presumably, "gain-of-function" mutations are associated with an increased risk for MH. From a clinical point of view and for safety reasons, patients suffering from diseases linked to mutations in RyR1 and/or DHPR should be handled as MHS, even if causality is unknown. Increasing knowledge on genetic and pathophysiologic backgrounds helps to distinguish between classical MH and MH-like events. This discrimination is necessary for optimum management of the patient and, if applicable, of the siblings.

Besides the risk of development of MH or MH-like hypermetabolism, anesthetic risk is mainly determined by respiratory and cardiac complications in patients suffering from myopathies. Cardiac involvement is very frequent in myopathies, although only 10% of patients show clinical signs. It is noteworthy that the severity of cardiac dysfunction does not correlate with the symptoms of the skeletal muscles and that even patients whose cardiac status was fully compensated, suddenly and unexpectedly died. Smooth muscle dysfunction is supposed to cause the increased blood loss in scoliosis surgery, which for example can be twice as high in Duchenne patients as compared to patients with idiopathic scoliosis.

REFERENCES

1. Denborough MA, Lovell RRH. Anaesthetic deaths in a family. *Lancet* 1960;**ii**:45.
2. Denborough MA, Forster JF, Lovell RR, Maplestone PA, Villiers JD. Anaesthetic deaths in a family. *Br J Anaesth* 1962;**34**:395–6.
3. McCarthy TV, Healy JM, Heffron JJ, Lehane M, Deufel T, Lehmann-Horn F, et al. Localization of the malignant hyperthermia susceptibility locus to human chromosome 19q12-13.2. *Nature* 1990;**343**:562–4.
4. MacLennan DH, Duff C, Zorzato F, Fujii J, Phillips M, Korneluk RG, et al. Ryanodine receptor gene is a candidate for predisposition to malignant hyperthermia. *Nature* 1990;**343**:559–61.
5. Harrison TK, Chu LF, Rosenberg H, Goldhaber-Fiebert S. Malignant hyperthermia, in Stanford Anesthesia Cognitive Aid Group: Emergency Manual. Cognitive aids for perioperative critical events. Creative Commons BY-NC-ND 2013. pp. 15–16.
6. Lehmann-Horn F, Klingler W, Jurkat-Rott K. Nonanesthetic malignant hyperthermia. *Anesthesiology* 2011;**115**:915–17.
7. Baur CP, Klingler W, Jurkat-Rott K, Froeba G, Schoch E, Marx T, et al. Xenon does not induce contracture in human malignant hyperthermia muscle. *Br J Anaesth* 2000;**85**:712–16.
8. Klingler W, Heiderich S, Girard T, Gravino E, Heffron JJ, Johannsen S, et al. Functional and genetic characterization of clinical malignant hyperthermia crises: a multi-centre study. *Orphanet J Rare Dis* 2014;**9**:8.
9. Iaizzo PA, Johnson BA, Nagao K, Gallagher WJ. 4-chloro-m-cresol triggers malignant hyperthermia in susceptible swine at doses greatly exceeding those found in drug preparations. *Anesthesiology* 1999;**90**:1723–32.
10. Paolini C, Fessenden JD, Pessah IN, Franzini-Armstrong C. Evidence for conformational coupling between two calcium channels. *Proc Natl Acad Sci USA* 2004;**101**:12748–52.
11. Larach MG, Localio AR, Allen GC, Denborough MA, Ellis FR, Gronert GA, et al. A clinical grading scale to predict malignant hyperthermia susceptibility. *Anesthesiology* 1994;**80**:771–9.
12. Jurkat-Rott K, McCarthy T, Lehmann-Horn F. Genetics and pathogenesis of malignant hyperthermia. *Muscle Nerve* 2000;**23**:4–17.
13. Rosenberg H, Davis M, James D, Pollock N, Stowell K. Malignant hyperthermia. *Orphanet J Rare Dis* 2007;**2**:21.
14. Klingler W, Lehmann-Horn F, Jurkat-Rott K. Complications of anaesthesia in neuromuscular disorders. *Neuromuscul Disord* 2005;**15**:195–206.
15. Hopkins PM. Malignant hyperthermia: pharmacology of triggering. *Br J Anaesth* 2011;**107**:48–56.
16. Morrison AG, Serpell MG. Malignant hyperthermia during prolonged surgery for tumour resection. *Eur J Anaesthesiol* 1998;**15**:114–17.
17. Girard T, Cavagna D, Padovan E, Spagnoli G, Urwyler A, Zorzato F, et al. B-lymphocytes from malignant hyperthermia-susceptible patients have an increased sensitivity to skeletal muscle ryanodine receptor activators. *J Biol Chem* 2001;**276**:48077–82.
18. Zullo A, Klingler W, De Sarno C, Ferrara M, Fortunato G, Perrotta G, et al. Functional characterization of ryanodine receptor (RYR1) sequence variants using a metabolic assay in immortalized B-lymphocytes. *Hum Mutat* 2009;**30**:E575–90.
19. Littleford JA, Patel LR, Bose D, Cameron CB, McKillop C. Masseter muscle spasm in children: implications of continuing the triggering anesthetic. *Anesth Analg* 1991;**72**:151–60.
20. Sciote JJ, Rowlerson AM, Hopper C, Hunt NP. Fibre type classification and myosin isoforms in the human masseter muscle. *J Neurol Sci* 1994;**126**:15–24.
21. Rosenberg H, Fletcher JE. Masseter muscle rigidity and malignant hyperthermia susceptibility. *Anesth Analg* 1986;**65**:161–4.
22. Christian AS, Ellis FR, Halsall PJ. Is there a relationship between masseteric muscle spasm and malignant hyperpyrexia? *Br J Anaesth* 1989;**62**:540–4.
23. Hoppe K, Lehmann-Horn F, Chaiklieng S, Jurkat-Rott K, Adolph O, Klingler W. In vitro muscle contracture investigations on the malignant hyperthermia like episodes in myotonia congenita. *Acta Anaesthesiol Scand* 2013;**57**:1017–23.
24. Wappler F, Fiege M, Steinfath M, Agarwal K, Scholz J, Singh S, et al. Evidence for susceptibility to malignant hyperthermia in patients with exercise-induced rhabdomyolysis. *Anesthesiology* 2001;**94**:95–100.
25. Thomas J, Crowhurst T. Exertional heat stroke, rhabdomyolysis and susceptibility to malignant hyperthermia. *Intern Med J* 2013;**43**:1035–8.
26. Nishio H, Sato T, Fukunishi S, Tamura A, Iwata M, Tsuboi K, et al. Identification of malignant hyperthermia-susceptible ryanodine receptor type 1 gene (RYR1) mutations in a child who died in a car after exposure to a high environmental temperature. *Leg Med (Tokyo)* 2009;**11**:142–3.
27. Groom L, Muldoon SM, Tang ZZ, Brandom BW, Bayarsaikhan M, Bina S, et al. Identical de novo mutation in the type 1 ryanodine

receptor gene associated with fatal, stress-induced malignant hyperthermia in two unrelated families. *Anesthesiology* 2011;**115**:938–45.

28. Lavezzi WA, Capacchione JF, Muldoon SM, Sambuughin N, Bina S, Steele D, et al. Case report: Death in the emergency department: an unrecognized awake malignant hyperthermia-like reaction in a six-year-old. *Anesth Analg* 2013;**116**:420–3.
29. Harrison GG. Pale, soft exudative pork, porcine stress syndrome and malignant hyperpyrexia—an identity? *J S Afr Vet Assoc* 1972;**43**:57–63.
30. Aleman M, Nieto JE, Magdesian KG. Malignant hyperthermia associated with ryanodine receptor 1 (C7360G) mutation in Quarter Horses. *J Vet Intern Med* 2009;**23**:329–34.
31. Yuen B, Boncompagni S, Feng W, Yang T, Lopez JR, Matthaei KI, et al. Mice expressing T4826I-RYR1 are viable but exhibit sex- and genotype-dependent susceptibility to malignant hyperthermia and muscle damage. *FASEB J* 2012;**26**:1311–22.
32. Adnet PJ, Bromberg NL, Haudecoeur G, Krivosic I, Adamantidis MM, Reyford H, et al. Fiber-type caffeine sensitivities in skinned muscle fibers from humans susceptible to malignant hyperthermia. *Anesthesiology* 1993;**78**:168–77.
33. Zukowski F, De Craemer D, Van den Branden C, De Cauwer H, Heytens L, Martin JJ. An image analysis study of vastus lateralis muscle fibers in malignant hyperthermia susceptible patients. *Clin Neuropathol* 1998;**17**:6–11.
34. Li G, Brady JE, Rosenberg H, Sun LS. Excess comorbidities associated with malignant hyperthermia diagnosis in pediatric hospital discharge records. *Paediatr Anaesth* 2011;**21**:958–63.
35. Islander G, Twetman ER. Comparison between the European and North American protocols for diagnosis of malignant hyperthermia susceptibility in humans. *Anesth Analg* 1999;**88**:1155–60.
36. Brandt A, Schleithoff L, Jurkat-Rott K, Klingler W, Baur C, Lehmann-Horn F. Screening of the ryanodine receptor gene in 105 malignant hyperthermia families: novel mutations and concordance with the in vitro contracture test. *Hum Mol Genet* 1999;**8**:2055–62.
37. Urwyler A, Deufel T, McCarthy T, West S. Guidelines for molecular genetic detection of susceptibility to malignant hyperthermia. *Br J Anaesth* 2001;**86**:283–7.
38. Rueffert H, Olthoff D, Deutrich C, Meinecke CD, Froster UG. Mutation screening in the ryanodine receptor 1 gene (RYR1) in patients susceptible to malignant hyperthermia who show definite IVCT results: identification of three novel mutations. *Acta Anaesthesiol Scand* 2002;**46**:692–8.
39. Brandom BW, Bina S, Wong CA, Wallace T, Visoiu M, Isackson PJ, et al. Ryanodine receptor type 1 gene variants in the malignant hyperthermia-susceptible population of the United States. *Anesth Analg* 2013;**116**:1078–86.
40. Swan H, Laitinen PJ, Toivonen L. Volatile anesthetics and succinylcholine in cardiac ryanodine receptor defects. *Anesth Analg* 2004;**99**:435–7.
41. Islander G, Rydenfelt K, Ranklev E, Bodelsson M. Male preponderance of patients testing positive for malignant hyperthermia susceptibility. *Acta Anaesthesiol Scand* 2007;**51**:614–20.
42. Ginz HF, Ruffert H, Levano S, Li Wan PoA, Benthien J, Urwyler A, et al. A mathematical model to improve on phenotyping for molecular genetic research in malignant hyperthermia. *Pharmacogenet Genomics* 2009;**19**:972–8.
43. von Breunig F, Wappler F, Hagel C, von Richthofen V, Fiege M, Weisshorn R, et al. Histomorphologic examination of skeletal muscle preparations does not differentiate between malignant hyperthermia-susceptible and -normal patients. *Anesthesiology* 2004;**100**:789–94.
44. Klingler W, Rueffert H, Lehmann-Horn F, Girard T, Hopkins PM. Core myopathies and risk of malignant hyperthermia. *Anesth Analg* 2009;**109**:1167–73.
45. Finsterer J, Segall L. Drugs interfering with mitochondrial disorders. *Drug Chem Toxicol* 2010;**33**:138–51.
46. Gurrieri C, Kivela JE, Bojanic K, Gavrilova RH, Flick RP, Sprung J, et al. Anesthetic considerations in mitochondrial encephalomyopathy, lactic acidosis, and stroke-like episodes syndrome: a case series. *Can J Anaesth* 2011;**58**:751–63.
47. Sato T, Nishio H, Iwata M, Kentotsuboi, Tamura A, Miyazaki T, et al. Postmortem molecular screening for mutations in ryanodine receptor type 1 (RYR1) gene in psychiatric patients suspected of having died of neuroleptic malignant syndrome. *Forensic Sci Int* 2010;**194**:77–9.
48. Dlamini N, Voermans NC, Lillis S, Stewart K, Kamsteeg EJ, Drost G, et al. Mutations in RYR1 are a common cause of exertional myalgia and rhabdomyolysis. *Neuromuscul Disord* 2013;**23**:540–8.
49. Benca J, Hogan K. Malignant hyperthermia, coexisting disorders, and enzymopathies: risks and management options. *Anesth Analg* 2009;**109**:1049–53.
50. Herlich A. Perioperative temperature elevation: not all hyperthermia is malignant hyperthermia. *Paediatr Anaesth* 2013;**23**:842–50.
51. Arriaga AF, Bader AM, Wong JM, Lipsitz SR, Berry WR, Ziewacz JE, et al. Simulation-based trial of surgical-crisis checklists. *N Engl J Med* 2013;**368**:246–53.
52. Malignant Hyperthermia Association of the United States website. <www.mhaus.org> [Accessed 3.4.14].
53. European Malignant Hyperthermia Group website. <www.emhg.org>. Universität Basel [Accessed 3.4.14].

Chapter 46

Disorders of the Ocular Motor Cranial Nerves and Extraocular Muscles

Monique M. Ryan and Elizabeth C. Engle

GENERAL PRINCIPLES

The extraocular muscles (EOMs) have many anatomic, physiologic, and molecular characteristics distinct from those of other striated muscles. These unique characteristics, which likely developed in response to the specialized demands placed on EOMs, including tonic position, maintaining contractures, conjugate smooth pursuit and saccades, and dysconjugate vergence movements, may account for the often predictable involvement or sparing of extraocular muscle in specific pediatric neuromuscular disorders. This selective involvement of the EOM can assist the clinician in formulating differential diagnoses for neuromuscular conditions.

Extraocular muscle dysfunction in neuromuscular disease typically presents with ptosis, ophthalmoplegia, or incomitant strabismus, or a combination of these symptoms. Ptosis, or blepharoptosis, is drooping of the upper eyelid as a result of dysfunction of the levator palpebrae superioris muscle. Ophthalmoplegia is the inability to move the globe into one or more fields of gaze, and may or may not be accompanied by strabismus. Strabismus is pathologic misalignment of the eyes, resulting in loss of binocular vision. With paralytic (incomitant) strabismus, the angle of deviation of the eyes varies with the direction of gaze. This is the form of strabismus typically associated with neuromuscular disease.

In this chapter, we review isolated ocular motor neuropathies, congenital ptosis, congenital cranial dysinnervation syndromes, and pediatric neuromuscular disorders associated with abnormalities of eye movement. Strabismus syndromes associated with pediatric neuromuscular conditions are summarized in Table 46.1.

THE EXTRAOCULAR LOWER MOTOR UNIT

Four recti and two oblique muscles move the globe, and the levator palpebrae superioris raises the eyelid. These muscles are innervated by the oculomotor (CN III), trochlear (CN IV), and abducens (CN VI) cranial nerves. Axons of the paired oculomotor nuclei (nIII), each composed of contiguous midbrain subnuclei, form two branches. The inferior branch projects uncrossed axons to the medial rectus, inferior rectus, inferior oblique, and pupillary constrictor muscles. The superior branch projects crossed axons to the superior rectus and both crossed and uncrossed axons to the levator palpebrae superioris (LPS) muscle. Notably, the axons innervating the LPS arise from a single midline subnucleus located at the caudal aspect of the oculomotor complex. The paired trochlear nuclei (nIV) in the caudal midbrain send axons across the tectum to exit the brainstem and innervate the contralateral superior oblique muscle. The axons of the paired abducens nuclei (nVI) in the pons innervate the ipsilateral lateral rectus. The anatomy of the ocular cranial nerves and nuclei is depicted schematically in Figures 46.1–46.3 and described in greater detail in the following discussion.

The globe is suspended in the bony orbit by the EOMs, connective tissue fascia, and fat. Because the two orbits point outward at approximately 25 degrees in the anterior-posterior plane, the vertical recti and oblique muscles are not aligned with the primary visual axis. The action of each muscle therefore depends somewhat on the position of the globe at the time of the action. The lateral and medial recti are antagonists in the horizontal plane, abducting and adducting the globe, respectively. The superior and inferior recti are partial antagonists in the vertical plane. The superior rectus primarily elevates and secondarily intorts and adducts the globe, and the inferior rectus primarily depresses and secondarily extorts and adducts the globe. The two obliques are partial antagonists and have greatest effect when the globe is adducted. The superior oblique intorts, depresses, and abducts the

B.T. Darras, H. Royden Jones, Jr., M.M. Ryan & D.C. De Vivo (Eds): Neuromuscular Disorders of Infancy, Childhood and Adolescence, Second edition.
DOI: http://dx.doi.org/10.1016/B978-0-12-417044-5.00046-9

TABLE 46.1 Strabismus in Pediatric Neuromuscular Disorders

Disorder	Ocular Misalignment	Other Visual Abnormalities
Ataxia-telangiectasia	Erratic vertical EOMs	Ocular motor apraxia
		Nystagmus
Cockayne syndrome	Esotropia, exotropia	Nystagmus, pigmented retinal dystrophy, enophthalmos, cataracts, corneal opacities
Mitochondrial Disorders		
MELAS	External ophthalmoplegia	Ptosis
Leigh disease	Horizontal gaze palsy, tonic downgaze deviation, external ophthalmoplegia	Internuclear ophthalmoplegia, dorsal midbrain (Parinaud's) syndrome, ptosis, optic atrophy
Kearns-Sayre syndrome	External ophthalmoplegia	Pigmented retinopathy, ptosis
CPEO	External ophthalmoplegia	Ptosis
Nutritional/Metabolic		
Abetalipoproteinemia with vitamin E deficiency	Convergence insufficiency, upgaze limitation	Internuclear ophthalmoplegia, pigmented retinopathy
Other Disorders		
CIDP	CN III palsy	
Joubert syndrome	Congenital fibrosis syndromes, skew deviation, horizontal tonic gaze deviation, supranuclear EOM deficits	Torsional/pendular nystagmus, ocular motor apraxia, retinal dystrophy, ptosis, colobomata

Abbreviations: CIDP, chronic inflammatory demyelinating polyneuropathy; CN, cranial nerve; CPEO, chronic progressive external ophthalmoplegia; EOMs, extraocular muscles; MELAS, syndrome of mitochondrial encephalopathy, lactic acidosis, and stroke-like episodes.
Source (where not otherwise noted): Brodsky et al. (1996).[1]

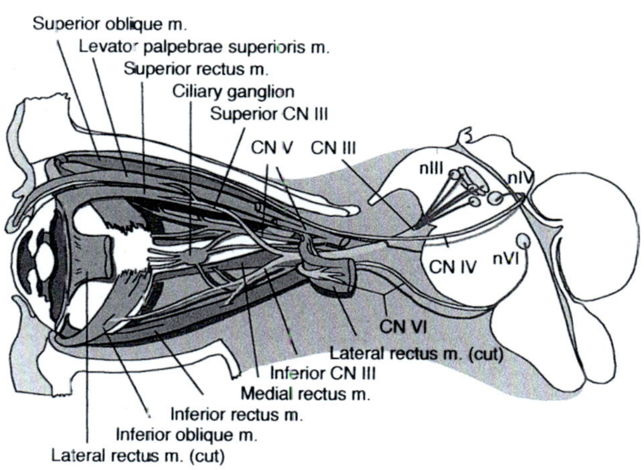

FIGURE 46.1 Schematic lateral view of the brainstem and left orbit in a normal individual. The lateral rectus is cut to expose the contents of the orbit. The cranial nerve nuclei, cranial nerves, and extraocular muscles are labeled. (CN, cranial nerve; n, nucleus.) *From Engle E, 1998,*[2] *with permission from Oxford University Press.*

globe, and the inferior oblique extorts, elevates, and abducts the globe. The extraocular muscles are reciprocally innervated such that when an agonist muscle contracts, its antagonist relaxes, and they are yoked in pairs (i.e. the right lateral rectus and left medial rectus) so that the eyes move together.

The extraocular lower motor unit has a number of unique features allowing a range of pathologic responses that differs from those of other skeletal muscles (reviewed by Porter and Baker, 1996).[3] The histology of EOM is significantly different from that of the skeletal striated muscles, and it appears to be more resistant to pathologic stressors.[4] The architecture of EOM differs from that of limb muscle in at least three ways. EOMs have longer fibrous tendinous insertions than limb muscle. Biopsy specimens may inadvertently be taken from these elongated tendons, resulting in a misdiagnosis of pathologic fibrosis.[5,6] Anatomic and neuroimaging studies have revealed the existence of connective tissue pulleys in the orbit that serve as functional mechanical origins of the four recti muscles. Anterior to these pulleys, the paths of the recti shift with gaze to follow the scleral insertions, whereas posterior to the pulleys, these paths are stable in the orbit.[7,8] The recti and oblique muscles are divided into global and orbital layers (Figure 46.2). The global layer, adjacent to the optic nerve and globe, extends over the entire length of each EOM, while the orbital layer, adjacent to the orbital

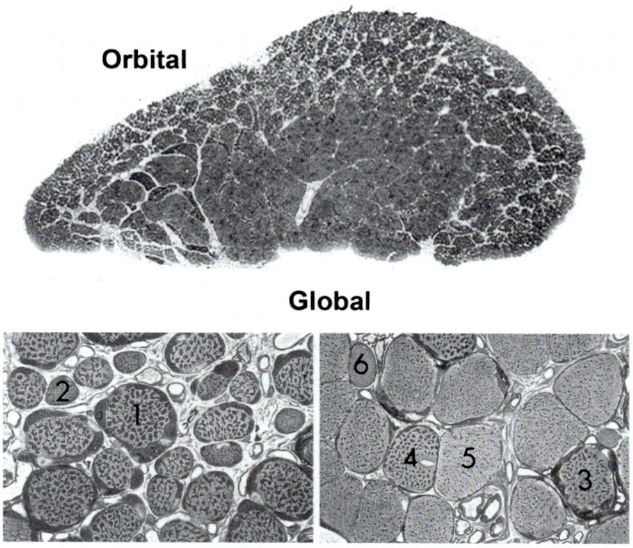

FIGURE 46.2 Light photomicrographs of monkey lateral rectus muscle. **Top:** Low-magnification cross-section of the muscle indicating orbital and global muscle layers. The orbital layer extends around much of the perimeter of the muscle, whereas larger fibers in the global layer fill the central portion of the muscle. **Bottom: (Left)** High-magnification micrograph of the orbital layer illustrating (1) orbital singly innervated and (2) orbital multiply innervated fiber types. **(Right)** High-magnification micrograph of the global layer illustrating the (3) global red singly innervated, (4) global intermediate singly innervated, (5) global white singly innervated, and (6) global multiply innervated fiber types. Original magnification ×21 (Top), ×410 (L and R). *Courtesy of Dr. John Porter, Cleveland.*

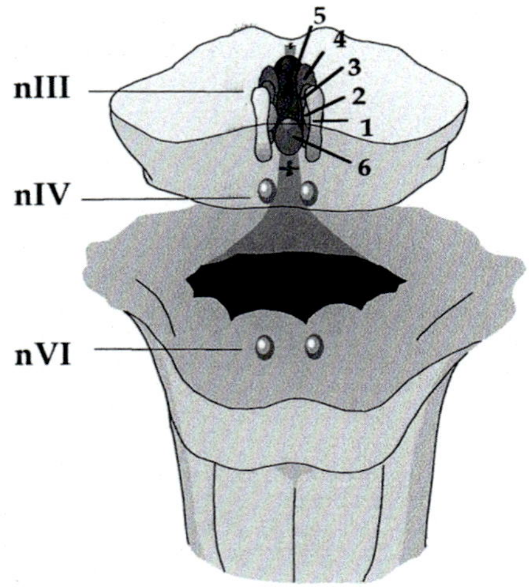

FIGURE 46.3 Schematic anterior view of the human brainstem showing details of the cranial nerve nuclei that innervate the extraocular muscles. (nIII = oculomotor nucleus; nIV = trochlear nucleus; nVI = abducens nucleus.) The nIII subnuclei are shown as follows, with the muscle they innervate in brackets: 1 = ventral lateral [medial rectus]; 2 = medial [superior rectus]; 3 = intermediate lateral [inferior oblique]; 4 = dorsal lateral [inferior rectus]; 5 = central caudal [levator palpebrae superioris]; 6 = Edinger-Westphal [visceral motor]. *From Engle E, 1998,[2] with permission from Oxford University Press.*

walls, is absent in the most anterior aspect of each EOM. The two layers differ from one another in their distribution of fiber types and the richness of their vascular supply, and may have variable insertion points upon the EOM.[8–10]

The morphology of EOM fibers differs from those of limb skeletal muscle in at least two ways. Normal EOM myofibers are rounder and smaller, with greater variability in size, and are surrounded by a greater amount of perimysial and endomysial connective tissue than is limb skeletal muscle.[3,4] EOM fiber typing is different from that of limb skeletal muscle; the traditionally recognized fiber classification schemes cannot be applied to EOM.[9,11] Typing of EOM fibers is based on (1) distribution into global and orbital layers; (2) single versus multiple nerve contacts per fiber; and (3) mitochondrial/oxidative enzyme content. This fiber type scheme identifies six fiber types: orbital singly, orbital multiply, global red singly, global intermediate singly, global pale singly, and global multiply innervated fibers (see Figure 46.2).[9,11] The levator palpebrae superioris differs from other EOMs. This muscle appears to use a ligament as a fulcrum to translate its anterior-posterior line of force into upward movement of the eyelid, has no differentiation of orbital and global layers, and lacks multiply innervated fiber types.[12]

The gene and protein expression profiles of EOM myofibers differ from skeletal limb muscle both during development and in maturity.[13–18] While EOMs contain the normal components of the dystrophin-glycoprotein complex, including dystrophin, dystroglycans, sarcoglycans, syntrophins, dystrobrevin, and merosin, they also retain embryonic myosin,[19,20] fetal acetylcholine receptors,[21–23] and sarcolemmal-wide expression of polysialated neural cell adhesion molecules.[19] These isoforms are more typically associated with developing or regenerating muscle fibers but may be used by EOM as an adaptation for these muscles' normal functional demands.[19]

The molecular and cellular biology of the oculomotor motor neuron is distinct from that of spinal motor neurons. For example, the homeobox gene *PHOX2A*, which is mutated in congenital fibrosis of the extraocular muscles (CFEOM2) (see the following), is essential for development of oculomotor and trochlear motor neurons, but not spinal motor neurons.[24] The alpha motor neurons of nIII, nIV, and nVI form very small motor units, often innervating only three to ten muscle fibers. In addition, a single EOM fiber can be innervated by axons from more than one motor unit (multiply innervated fibers). These motor neurons have an impressively high firing rate, often an order of magnitude higher than that of spinal motor neurons.

EXAMINATION APPROACH TO EXTRAOCULAR MOVEMENT DISORDERS

When interpreting abnormal eye movements or ptosis, or both, several basic approaches may be helpful. Priorities in examination are to:

1. Determine if the abnormality is attributable to a single cranial nerve or to multiple nerves.
2. Determine if the abnormality exists in isolation (e.g. congenital strabismus) or occurs in association with dysfunction of other skeletal muscle groups, syndromic malformations, or features of systemic disease.
3. Determine if the disorder is congenital or acquired. This distinction cannot be made simply by asking whether eye movements appeared normal at birth. The normal development of volitional eye movements such as fixation, visual following, and binocular alignment in the first few months of life may delay recognition of a congenital eye movement disorder. Conversely, acquired deficits may develop insidiously and thus be confused with congenital deficits. Serial examinations over time (including observation of old photographs) and attention to compensatory features (e.g. a large fusion angle suggesting congenital strabismus) may help.
4. Determine whether the dysfunction is static, progressive, recurrent, or resolving.

Examination

Lid Function

Ptosis can be quantified using the following methods:

1. Measure the distance between the upper lid margin and the midcorneal reflex when the globe is in normal primary position. Ptosis is present when this distance is <2 mm or varies by more than 2 mm between the eyes.
2. Measure the amount of the superior portion of the cornea covered by the upper lid when the globe is in primary position; normal is approximately 2 mm. With ptosis, more than 4 mm of cornea is covered.
3. Measure the vertical width of the palpebral fissure; it is normally approximately 9–15 mm.

Pupil

Assess pupil size and reactivity. If anisocoria is present, it should be observed under varied illumination; in Horner syndrome, the anisocoria is more apparent in darkness and a lag in pupillary dilation may be observed, whereas with pupillary constrictor paresis, the difference is more evident under bright light. A relative afferent pupillary defect, indicating optic nerve dysfunction, is detected by swinging a bright light source from one pupil to the other, watching for dilation in the affected eye. Slit lamp examination may identify sectoral irregularities in the iris, an indication of reinnervation.

Ocular Alignment

Assess ocular alignment by the alternate cover test or Maddox rod testing. A patient with a phoria prefers to fixate with one eye and will shift eye position only when the preferred eye is covered; if a tropia is present, fixation will shift when either eye is occluded. The degree of horizontal or vertical misalignment can be quantified using prisms. The Bielschowsky three-step test is valuable for identification of trochlear nerve palsies (see "Trochlear Palsy (CN IV)" section). Information as to whether ocular misalignment is long-standing may be obtained via stereoacuity tests.

Ductions

Assess ductions by having the patient follow a hand-held target to all cardinal positions of gaze; observe the range, speed, and smoothness of these movements in each eye, as well as whether the two eyes move conjugately. The examiner can best observe both eyes simultaneously by fixating on the patient's nose. When there is limitation of movement to a given position of gaze, it may be difficult to distinguish between weakness of an EOM (e.g. lateral rectus) and restriction or overaction (or both) of its antagonist (medial rectus). The speed and smoothness of movement may serve as a clue; sudden slowing toward the end of an excursion suggests restriction of the antagonist. The definitive tests, however, are those of forced duction and active force generation. Forced duction testing is performed by grasping the anesthetized extraocular muscle, tendon, or globe itself, and pulling it through its range of motion; the examiner directly feels any mechanical restriction. In young children, this procedure often requires general anesthesia. Active force generation requires the alert patient's cooperation to attempt eye movements while the examiner grasps muscle, tendon, or globe and senses muscle force.

Horizontal and Vertical Saccades

Assess horizontal and vertical saccades by asking the patient to rapidly switch fixation between the examiner's nose and targets held in each extreme of the visual field. Slow saccades suggest muscle weakness or, in the case of the medially directed saccades, an internuclear ophthalmoplegia; saccades that begin rapidly but slow toward the end of their excursion suggest restriction of the opposing EOM. Repeated saccades may reveal the fatigability associated with myasthenic syndromes. Hyper- or hypometric saccades suggest cerebellar disease. Occasionally, disturbances in smooth pursuit suggest a particular pattern of EOM involvement—for example, the upshoot seen during medial pursuit in Duane's syndrome (see below). Both horizontal and vertical head thrusts take advantage of vestibular input to indicate the true range of motion of the eyes where voluntary maneuvers fail to do so; in addition, they provide helpful information about central control of eye movements. Identification of nystagmus indicates central dysfunction but may also provide an indirect indication of limitation, such as with the abducting nystagmus seen when medial deviation of the opposite eye is limited.

Finally, additional general ophthalmologic testing assists in identifying visual pathway dysfunction that may affect ocular motility. This testing may include assessment of visual acuity at far and at near, color perception, the visual fields, and the funduscopic appearance of the optic nerves and the central and peripheral retina. The latter also may be particularly helpful to identify complex neurologic syndromes of which abnormal EOMs are a single component.

CRANIAL NERVE PALSIES

Oculomotor Palsy (CN III)

Anatomy

The oculomotor nucleus consists of a series of closely associated subnuclei arranged along the dorsal-ventral and rostral-caudal dimensions of the tectum of the midbrain (see Figure 46.3). A midline dorsal and caudal subnucleus provides bilateral innervation to the levator palpebri. Together with the contralateral axons from the superior rectus subnucleus, these axons form the superior division of the nerve. All EOMs other than the superior recti are innervated ipsilaterally via the inferior division of the nerve. Pupillary fibers emerge from the single midline Edinger-Westphal nucleus in the complex, and run superficially within the nerve fascicle to the ipsilateral ciliary muscle and iris sphincter.[1]

Localizing Syndromes

The anatomy of the oculomotor nuclei and nerve produces clinical features helpful in the localization of oculomotor palsies.[1] Third nerve palsies can be localized to the nucleus if they include contralateral involvement of the superior rectus and bilateral involvement of the levator palpebri (and probably also of the medial recti). They may localize to the nerve if ptosis and ophthalmoplegia are ipsilateral, and to only the superior division of the nerve if the ophthalmoplegia is limited to vertical limitations as seen in CFEOM (see the following). Given the close proximity of the oculomotor nucleus to other brainstem structures, acquired nuclear lesions are often associated with other neurologic signs such as somnolence and hemiplegia. Lesions affecting the oculomotor nerve fascicle cause ipsilateral signs because the axons serving the superior recti have already decussated within the nuclear complex.[25] Because the oculomotor nerve fibers pass through the reticular formation, red nucleus, and substantia nigra, a lesion along their course may lead to ipsilateral ataxia (dentatorubrothalamic tract) or contralateral hemitremor, hemichorea, or hemiballismus (red nucleus and substantia nigra). In the interpeduncular cistern, CN III may be subject to compression from aneurysms, although this is very rare in childhood. Early involvement of the pupil in oculomotor palsies points to compressive lesions, as the dorsomedial placement of the pupillary fibers in the nerve fascicle renders them susceptible to compression.[1]

Just before entering the cavernous sinus, the third nerve is susceptible to compression at the free edge of the tentorium and the clivus. Uncal herniation generally causes an ipsilateral, and rarely a contralateral, oculomotor palsy in which the pupil is involved early. Ipsilateral involvement of CN III, CN IV, and CN VI localizes to the cavernous sinus or orbital apex. Involvement of the trigeminal nerve is often more extensive with cavernous sinus lesions, whereas the presence of proptosis or visual loss from optic neuropathy, or both, defines the orbital apex syndrome. In the cavernous sinus, the nerve lies dorsally and deep in the lateral wall, superior to the trochlear nerve; as it enters the superior orbital fissure, it divides into superior and inferior divisions. The superior division innervates the superior rectus and levator palpebrae superioris, whereas the inferior division innervates the inferior and medial recti, inferior oblique, and ciliary ganglion. Thus, an isolated superior division palsy causes globe depression and ptosis, whereas an isolated inferior division palsy results in abduction, elevation, and mydriasis.[26] Lesions at the apex or within the superior orbital fissure typically produce divisional palsies, but because the fibers are segregated within the nerve along much of the fascicular course, more proximal lesions can also produce these findings.

Fundamental Signs and Symptoms

Oculomotor palsies classically present with ptosis, mydriasis, and external ophthalmoplegia. With a complete third nerve palsy, the pupil is large and does not constrict to light or vergence efforts, and near accommodation is impaired. The affected eye is depressed and abducted, and limited in adduction and elevation. Ptosis is severe. Correspondingly, patients complain of ptosis, blurred vision especially at near (due to inability to accommodate sufficiently), and diplopia.

Although the major clinical features are the same in congenital and acquired forms of oculomotor palsy, subtle differences in presentation (e.g. presence of amblyopia, synkinesis, pupillary involvement, or fluctuating symptomatology) may help distinguish the two forms.

Etiology

Third nerve palsies are rare, and less common than fourth or sixth nerve palsies in childhood.[27,28] About ⅔ are partial, and ⅓ complete.[27] As many as 47% of all pediatric oculomotor palsies are congenital.[29] Acquired causes include trauma (12–37%), infection and other inflammation (6–21%), tumor (3–17%), and migraine (3–9%). Aneurysms and other vascular events account for 3% to 11%. The cause remains undetermined less often in children (2–14%) than in adults (25–32%).[27,28,30–33]

Congenital CN III Palsies

Congenital oculomotor palsies in otherwise normal children are thought to relate to perinatal trauma, possibly due to molding of the skull during labor or with forceps use.[31,34] Under these circumstances, the nerve may be compressed against the tentorium by displacement of the temporal lobe or by a diffuse increase in intracranial pressure.[35] In other cases, the nucleus or nerve, or both, fails to develop (see later discussion of CFEOM). Congenital oculomotor palsies may be associated with contralateral hemiplegia,[34,36] or developmental anomalies of the midbrain or cerebellum.[37–39] A frequent clue to the congenital nature of an oculomotor palsy is the presence of aberrant regeneration with synkinesis, which presumably occurs in response to aplasia of the nucleus or disruption of the nerve.[40] Such regeneration may lead to miosis, rather than dilation, of the affected pupil.[30,34] Another common finding is amblyopia, usually in the paretic eye but occasionally in the nonparetic eye.[27,31]

Partial congenital oculomotor palsy may manifest as a divisional palsy or as isolated dysfunction of CN III-innervated muscles, in particular the levator, inferior rectus, inferior oblique, or pupillary constrictor. These disorders likely represent a spectrum of developmental anomalies of the subnuclei and muscles.[6,41–47]

Acquired CN III Palsies

Acquired oculomotor palsies are more often partial than complete, and may be isolated or associated with more generalized neuromuscular processes (Table 46.2). The most common causes of oculomotor palsies are trauma (up to 72%),[29,31,33] neoplasm (5–31%), inflammation or infection (9–25%), migraine (7–19%), and other vascular pathologies (3–13%). As many as 20% of cases are cryptogenic.[28,48,49] Painful ophthalmoplegia often involves more than one cranial nerve, and suggests acquired conditions such as infection, tumor, orbital pseudotumor, hemorrhage, thrombosis, aneurysm, or Tolosa-Hunt syndrome. In contrast, synkinesis or sparing of the pupil suggests a more benign cause. Diplopia in acquired oculomotor palsy is usually oblique in primary position, but varies with the relative degree of weakness in each muscle, and may be obscured if ptosis is severe enough to occlude the paretic eye. Although intorsion due to inferior oblique involvement is not often detected clinically, the patient may perceive a tilted sense of vertical (not always in the paretic eye or in the expected direction).[50]

Acquired partial oculomotor palsies suggest pathology at extremes of the oculomotor nerve: either discrete lesions in the midbrain, involving an oculomotor subnucleus, or focal intraorbital lesions. Nuclear causes include focal metastasis, ischemia, or demyelination, while orbital causes include myasthenia gravis, trauma, tumor, or inflammation.[1,43,51,52]

Recurrent transient oculomotor palsies in childhood are most commonly due to ophthalmoplegic migraine. This uncommon inflammatory cranial neuropathy is much more frequent in children than adults. Onset is generally before age 10. Each episode typically begins with a severe ipsilateral hemicranial headache, with ophthalmoplegia developing within hours to days (rarely up to 14 days), and generally resolving after 3 to 4 days. Deficits rarely persist as long as a month, and synkinesis may develop.[1,53–55] Although a complete oculomotor palsy is most frequent, isolated pupillary involvement can occur, and involvement of the fourth and sixth cranial nerves, oculomotor divisions, and levator palpebrae are also reported in ophthalmoplegic migraine.[55] Rarely, recurrent ophthalmoplegia may occur in the absence of headache, in what may represent a migraine equivalent.[56,57] Magnetic resonance imaging (MRI) studies commonly show enlargement and contrast enhancement of the interpeduncular portion of the oculomotor nerve during, and occasionally following, ophthalmoplegic migraines (Figure 46.4),[53,55,59] while the CSF examination is normal.[55]

Tolosa-Hunt syndrome is an acute syndrome of orbital pain and ophthalmoplegia caused by granulomatous inflammation within the cavernous sinus or orbit, which responds rapidly and dramatically to oral steroids but may recur. Myasthenia gravis should be suspected whenever ptosis or diplopia fluctuate in severity.[52]

Isolated Pupillary Dysfunction

Pediatric CN III palsies occasionally present with isolated pupillary dilatation. This must be distinguished from Adie's tonic pupil and from a contralateral Horner syndrome. Adie's pupil may develop cryptogenically or as a parainfectious, posttraumatic, or migrainous phenomenon, and is often associated with depressed deep tendon

TABLE 46.2 Etiology of Acquired Ocular Motor Palsies in Childhood

Etiology	III Cranial Nerve	IV Cranial Nerve	VI Cranial Nerve
TRAUMA	**25–72%**	**8–50%**	**20–42%**
Diffuse axonal injury	+		
Ischemic	+		
Shearing forces	+	+	+
Subdural hematoma	+		+
Cavernous sinus thrombosis	+	+	+
Postoperative	+	+	+
NEOPLASIA	**5–31%**	**Rare**	**17–39%**
Cavernous sinus			
Meningioma	+	+	+
Schwannoma	+	+	+
Pituitary fossa			
Pituitary adenoma	+	+	+
Craniopharyngioma	+	+	+
Dermoid/epidermoid	+	+	+
Teratoma	+	+	+
Sellar germ cell tumor	+	+	+
Cerebellar			
Astrocytoma		+	+
Medulloblastoma		+	+
Brainstem glioma	+	+	++
Other neoplasms			
Pinealoma		+	+
Meningioma		+	+
Schwannoma		+	+
Acoustic neuroma		+	+
Ependymoma		+	+
Nasopharyngeal carcinoma		+	+
Clivus chordoma			
Infiltrating neoplasms			
Leptomeningeal sarcoma	+	+	+
Lymphoma	+	+	+
Carcinomatous meningitis	+	+	+
Mesencephalic cyst	+		
VASCULAR	**3–13%**		**1–2%**
Subdural hemorrhage	+	+	
Aneurysm	+	+	+
Cavernous hemangioma	+	+	+

(Continued)

TABLE 46.2 (Continued)

Etiology	III Cranial Nerve	IV Cranial Nerve	VI Cranial Nerve
Arteriovenous malformation	+	+	
Carotid-cavernous fistula			+
INFECTION	9–25%		
Meningitis	+	+	+
Encephalitis	+	+	+
Mastoiditis, Gradenigo's syndrome			+
INFLAMMATORY			
Sarcoidosis	+	+	+
Tolosa-Hunt syndrome	+	+	+
Orbital pseudotumor	+	+	+
RHEUMATOLOGIC/AUTO-IMMUNE			
Polyarteritis nodosa	+		
Multiple sclerosis, ADEM	+	+	+
Guillain-Barré syndrome	+	+	+
Sjögren's syndrome		+	
OTHER			
Migraine	7–19%		+
Toxins	+	+	
Increased intracranial pressure		+	+
Thyroid eye disease	+	+	+
Iatrogenic	+	+	
Cryptogenic	<20%	0–21%	9–36%

ADEM = acute disseminated encephalomyelitis.

reflexes.[1,60] Adie's tonic pupil is differentiated from an oculomotor palsy by the presence of light-near dissociation (strong constriction when viewing near targets, with poor reaction to light or accommodation). Slit lamp examination frequently reveals asymmetrical segmental constriction of pupillary fibers, often with vermiform iris movements, once reinnervation has had time to occur. Although there is supersensitivity of the pupil to 0.1% pilocarpine in lesions more than 2 weeks old, this may also be found in CN III palsies.[61]

In its complete form, Horner syndrome consists of ipsilateral ptosis of both upper and lower lids, miosis, and decreased sudomotor function. In congenital Horner syndrome, ocular hypotony may occur and iris heterochromia is common. Lesions along the sympathetic pathway may be secondary to birth or surgical trauma, perinatal infection, the Arnold-Chiari malformation, neuroblastoma, and other tumors.[62,63] Localization may be refined by identification of concurrent involvement of the fourth or sixth nerves (see the following) and by the use of 1% hydroxyamphetamine or 4% cocaine drops to distinguish between involvement of first, second, and third order neurons.

Trochlear Palsy (CN IV)

Trochlear palsy is the most common congenital ocular motor palsy, the most common cause of acquired vertical diplopia, and, overall, the most common isolated cranial nerve palsy.[28,64]

Anatomy

The trochlear is the only cranial nerve to emerge from the dorsal aspect of the brainstem, the only one to cross

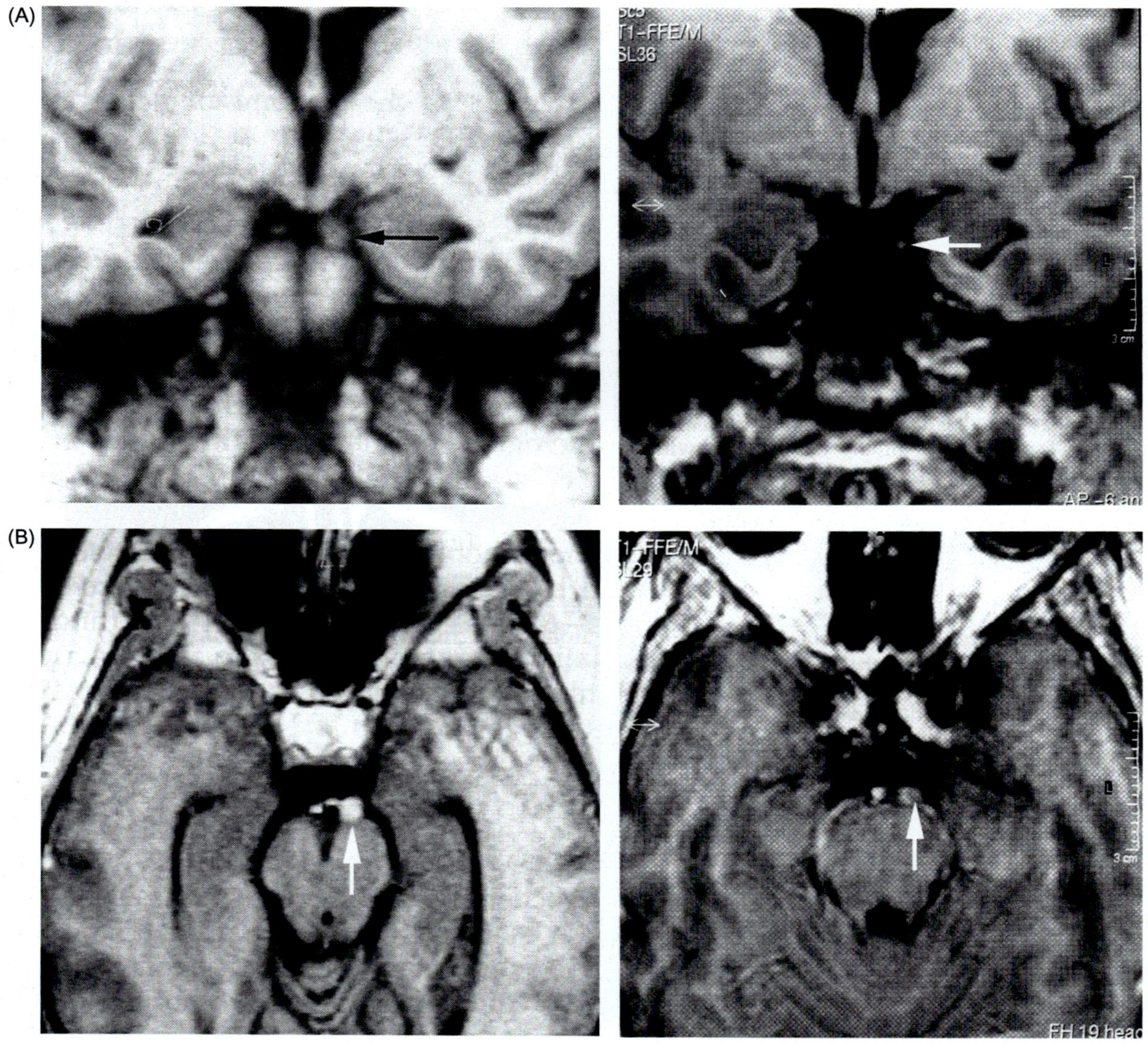

FIGURE 46.4 Ophthalmoplegic migraine. Coronal (**A**) and gadolinium-enhanced axial (**B**) T1-weighted images during (left) and following (right) an episode of ophthalmoplegic migraine in a 13-year-old girl, showing reversible swelling and contrast enhancement of the cisternal portion of the left CN III (arrow). *From Prats et al., 1999, © SAGE Publications.*[58] *Reprinted by permission of SAGE Publications.*

completely, and the longest and thinnest of the ocular motor nerves. The trochlear nucleus is located ventral and lateral to the Sylvian aqueduct at the level of the inferior colliculus, dorsal to the medial longitudinal fasciculus and caudal to the oculomotor nuclear complex. The nerve fascicle leaves the nucleus, courses along the aqueduct, crossing in its roof inferior to the inferior colliculus, and exits the dorsal midbrain contralateral to the nucleus. The cisternal portion travels laterally around the midbrain to its ventral aspect, then passes along the free edge of the tentorium and along the lateral aspect of the clivus to enter the cavernous sinus. Here, it lies in the lateral wall just inferior to the oculomotor nerve, entering the orbit through the superior orbital fissure and running medially across the superior rectus muscle to the superior oblique muscle.[65]

Localizing Syndromes

Associated neurologic signs aid localization of fourth nerve lesions. The dorsal midbrain syndrome, or Parinaud syndrome, may include trochlear palsy with upgaze weakness, downgaze limitation, convergence spasm, convergence-retraction nystagmus, lid retraction, light-near dissociation, or a combination of these symptoms.[1,65] A contralateral Horner syndrome or intranuclear ophthalmoplegia indicates involvement of the fourth nerve within the midbrain.[66] Ipsilateral limb ataxia or contralateral sensory loss may reflect involvement of the cerebellar peduncle or sensory lemniscus, respectively.[67] Concurrent involvement of the third and fourth nerves may localize to the midbrain, the cavernous sinus or orbital apex, where the trigeminal nerve may also be affected.[65]

TABLE 46.3 The Bielschowsky Three-Step Test

Step 1	Step 2		Step 3	
Primary	R gaze	L gaze	R tilt	L tilt
R hypertropia				
RSO		RSO	RSO	
RIR	RIR			RIR
LIO	LIO		LIO	
LSR		LSR		LSR
L hypertropia				
LSO	LSO			LSO
LIR		LIR	LIR	
RIO		RIO		RIO
RSR	RSR		RSR	

Step 1: Determine which eye is higher in primary gaze.
Steps 2 & 3: Determine in which position the deviation is greater. Chart indicates which EOMs may be weak, to yield the observed deviation.
Note: The Three-Step Test can yield false localizing results when more than a single EOM palsy is present, and under a few other unusual circumstances.[a–f]
Abbreviations: EOMs, extraocular muscles; IO, inferior oblique; IR, inferior rectus; L, left; R, right; SO, superior oblique; SR, superior rectus.
[a] *Donahue SP et al., 1999.*[69]
[b] *Ohtsuki H et al., 2000.*[70]
[c] *Kushner BJ, 1989.*[71]
[d] *Kushner BJ, 1981.*[72]
[e] *Getman I and Goldstein JH, 1983.*[73]
[f] *Cartwright MJ and Wyatt DB, 1989.*[74]

Fundamental Signs and Symptoms

In children, fourth nerve palsy often presents with a head tilt away from the affected eye. This helps neutralize the vertical diplopia and rotation of the image from the paretic eye that the patient will perceive with the head erect and gaze in primary position. In addition to the direction of head tilt, the patient's perception in primary position, as well as version testing, will localize the paretic eye. Viewing a horizontal edge, the patient may perceive two images of the edge tilted with respect to one another, and intersecting as if forming an arrow that points to the paretic side. This is less common in children than adults and is not found in congenital trochlear palsies.[50,68]

The Bielschowsky Three-Step Test helps to differentiate between trochlear palsy and its mimickers, such as dissociated vertical deviation, skew deviation, and double elevator palsy (Table 46.3).

Etiology

Most trochlear palsies are congenital (refer to Congenital Cranial Dysinnervation Disorders section below) or arise as a result of trauma. Differentiation between congenital and acquired trochlear palsy may be difficult because the congenital form is often well compensated and thus asymptomatic prior to adolescence or adulthood, when it may decompensate and be mistaken for an acute palsy. Review of photographs from infancy may reveal a long-standing head tilt, suggesting a congenital lesion.[1] Facial asymmetry, with retrusion and upslanting of the mouth on the side of the head tilt, is also indicative of chronicity.[76] Children with congenital trochlear palsy do not generally perceive a tilted image, presumably due to a combination of the head tilt and more complex central physiologic mechanisms.[68] Similarly, amblyopia is rare because the compensatory head tilt enables fusion of the images from both eyes.[1] MRI findings can also be helpful in distinguishing congenital and acquired trochlear palsies[77]. Finally, forced duction testing reveals decreased resistance of the superior oblique in congenital palsies, and normal resistance in acquired forms.

Up to 50% of acquired pediatric trochlear nerve palsies are due to trauma or neurosurgery (see Table 46.2).[28,30,33,78] Orbital fractures may damage the nerve, tendon, trochlea, or the muscle itself.[79,80] Trauma may also cause decompensation of a congenital palsy.[81] Acquired bilateral trochlear palsy is most often due to trauma, and is thought to result from injury at the fascicular decussation. Such cases usually present with a chin-down head position, rather than a head tilt, and with a right hypertropia in left gaze and a left hypertropia in right gaze (alternating adducting hypertropia). A V-pattern esotropia is generally present, and is often large.[1,82]

Inflammatory and infectious causes of trochlear palsies include the Tolosa-Hunt and Guillain-Barré syndromes, *Herpes zoster* ophthalmicus, and, less commonly, sarcoidosis and even tetanus.[83–86] Recurrent CN IV weakness may be seen in the Tolosa-Hunt syndrome, sarcoidosis, ophthalmoplegic migraine (see earlier discussion), and a familial syndrome involving multiple recurrent cranial nerve palsies.

Differential Diagnosis

Ocular motility disturbances, which may be confused with trochlear nerve palsy, include double elevator palsy, dissociated vertical deviation, the ocular tilt reaction, and inferior oblique overaction.[87]

Skew deviation is vertical misalignment due to disruption of supranuclear pathways in the midbrain tegmentum, dorsolateral medulla, cerebellum, or vestibular system.[1] Although most commonly due to trauma, it is also occasionally seen with the Chiari malformation (type II), myelomeningocele, hydrocephalus, or even pseudotumor cerebri.[88] Skew deviation may or may not be comitant, and may alternate periodically over time or with lateral gaze position.[1]

Double elevator palsy consists of an inability to elevate an eye in any horizontal position of gaze, and may be caused by inferior rectus restriction, superior rectus paresis, or supranuclear lesions.[89] The greater elevation of the contralateral eye may be confused with a superior oblique palsy of that eye. In contrast with fourth nerve palsies, this hypertropia is comitant and may be associated with ptosis or pseudoptosis.

In dissociated vertical deviation, each eye deviates in the same vertical direction when it is covered. A head tilt is often present.[1] In true fourth nerve palsy, however, when the affected eye is covered it deviates upward, whereas when the unaffected eye is covered it deviates downward. Other helpful clues to the presence of dissociated vertical deviation include the presence of latent nystagmus or exodeviation, or both.

The ocular tilt reaction consists of a skew deviation, head tilt, and ocular torsion. It may be seen with lesions of the ipsilateral interstitial nucleus of Cajal or a contralateral Wallenberg syndrome.[90]

Weakness of the superior oblique may also be found in myasthenia gravis or thyroid eye disease.[91,92] Finally, torticollis causes a head tilt that may raise suspicion of a superior oblique palsy, but is characterized by resistance to passive tilt of the head in the opposite direction and associated with palpable sternocleidomastoid muscle thickening.[1]

Abducens Palsy (CN VI)

Anatomy

The abducens nuclei lie close to the midline in the caudal paramedian pontine tegmentum, lateral to the medial longitudinal fasciculus and medial to the vestibular nucleus. Each includes two subpopulations of neurons: motor neurons of the abducens nerve and interneurons of the contralateral medial longitudinal fasciculus. The fascicle of the ipsilateral facial nerve wraps around the abducens nucleus. The fascicular portion of CN VI runs ventrally through the paramedian pontine reticular formation, near the trigeminal nerve and superior olivary nucleus, then lateral to the corticospinal tract before exiting from the caudal pons about 1 cm from the midline, turning at a right angle to head rostrally. The nerve crosses the clivus and runs along the basilar artery, penetrating the dura through or above the inferior petrosal sinus. It then turns right, enters the cavernous sinus and then the superior orbital fissure, innervating the lateral rectus muscle as it passes along its medial side. Some sympathetic fibers destined for the ophthalmic branch of the trigeminal nerve join the abducens nerve briefly along its intracavernous portion, accounting for a concurrent Horner syndrome in some cavernous sinus lesions.[93,94]

Localizing Syndromes

Abducens palsies lend themselves well to precise localization, because lesions at various sites result in distinct constellations of symptoms. Nuclear lesions cause an isolated defect in abduction if only the nVI motor neurons are involved. Alternatively, an ipsilateral horizontal gaze palsy will result if the lesion includes the juxtaposed medial longitudinal fasciculus interneurons crossing to the contralateral oculomotor nucleus. A lesion in the dorsolateral pons results in Foville's syndrome—horizontal gaze palsy, ipsilateral facial palsy and analgesia, deafness, and loss of taste in the anterior two thirds of the tongue. A lesion in the ventral pons causes Millard-Gubler syndrome (ipsilateral facial paralysis and contralateral hemiplegia). Pontine lesions may also cause an ipsilateral internuclear ophthalmoplegia or Horner syndrome. Accompanying trigeminal nerve dysfunction, CN VII and CN VIII palsies, nystagmus, and cerebellar signs indicate a lesion at the cerebellopontine angle. Isolated abducens palsies are seen with compression at the clivus, while lesions in the middle fossa may cause facial pain, hypesthesia, or weakness. Those in the cavernous sinus or superior orbital fissure are generally indistinguishable on the basis of clinical signs; in either location, the abducens palsy is frequently accompanied by palsies of CN III, CN IV, and the first branch of the trigeminal nerve, or by an ipsilateral Horner syndrome.[1,64,94]

Fundamental Signs and Symptoms

Because the abducens nerve innervates a single muscle with a single action, the fundamental sign of abducens palsy is straightforward: weakness in abduction of the affected eye. In complete palsies, this causes inability to abduct the eye past the midline. The esotropia is of greater magnitude in the direction of gaze of the affected eye and with distant viewing, and increases when the patient fixates with the affected eye. In mild palsies, the deficit in range may be subtle, and may be identified only by observation of slowed saccades in the direction of the palsy or by formal measurements of ocular alignment using prisms.[1] Children will often present with a head turn toward the affected eye.[95] Another helpful sign in identifying a subtle palsy is medial-beating nystagmus of the contralateral eye on attempted ipsilateral lateral gaze.

Long-standing palsies may result in development of a secondary medial rectus contracture causing a restrictive noncomitant esotropia. In the latter case, ipsilateral saccades will begin with normal rapid speed, then suddenly slow in midcourse, as if "hitting a wall."[1]

Etiology

In contrast to CN III and CN IV palsies, congenital abducens palsies are rare (8–13%) in comparison with acquired forms. Of the acquired forms, the most common causes are trauma (20–42%), neoplasm (18–39%), infection (6–17%), raised intracranial pressure (3–14%), benign/viral (3–18%), and vascular (1–3%). Cryptogenic palsies account for 9% to 36% of cases.[28,33,96,97]

Congenital CN VI Palsies

Transient neonatal abducens palsies probably result from perinatal cranial trauma,[1] and generally resolve within 5 weeks.[98,99] In some infants, a presumed abducens palsy manifests simply as neonatal esotropia.[100] Congenital forms persisting beyond the neonatal period include two forms of congenital cranial dysinnervation disorders (CCDDs), horizontal gaze palsy, and Duane's retraction syndrome. Abducens palsies are also seen in 90% of children with Moebius syndrome (see following).

Acquired CN VI Palsies

Acquired abducens palsies in childhood are commonly due to neoplasms, in which case they arise as a result of tumor infiltration or mass effect, secondary obstructive hydrocephalus, or as a sequela of surgery.[97,101] Tumors frequently associated with sixth nerve palsy are listed in Table 46.2; pontine glioma predominates.

The sixth cranial nerve may be injured by trauma[97] where it crosses the clivus, by direct damage to the petrous bone in basilar skull fractures, or by entrapment in medial orbital fractures.[80,97] The abducens nerve is also vulnerable to traumatic injury during resection of posterior fossa tumors.[1] Trauma may also cause increased intracranial pressure or, rarely, caroticocavernous fistulae. Relatively mild trauma apparently resulting in a sixth nerve palsy should raise suspicion of an occult tumor.[102] Unilateral or bilateral traction injury may result from hydrocephalus (or uncommonly from lumbar puncture).[103,104]

Inflammatory abducens neuropathies (17% of total CN VI palsies) are usually partial.[97] CN VI is affected in as many as 16% of cases of bacterial meningitis.[105] Mastoiditis may directly cause abducens palsies (Gradenigo's syndrome, often also accompanied by intense temporal-parietal pain and ipsilateral facial palsy) or, more commonly, form a nidus for venous thrombosis and subsequent increased intracranial pressure. Otitis media or sinusitis may also be complicated by inflammation of the petrous bone and sixth nerve weakness.[106]

As many as 36% of acquired abducens palsies are idiopathic.[97] Benign isolated complete CN VI palsies may develop acutely, often preceded by a febrile, presumably viral illness.[107] See Case Example 46.1.

Benign recurrent abducens palsy is acute in onset, complete, isolated from other neurologic signs or those of increased intracranial pressure, resolves within 6 to 8 weeks, and usually recurs in the same eye.[108–111] Recurrent abducens palsies have been associated with Epstein-Barr and varicella infections, and with immunizations.[112,113] Less commonly, they are seen with migraine, in association with recurrent CN III weakness, or with vascular anomalies. Recurrent CN VI palsy is more frequent in females and on the left side.[1,107,111]

Increases or decreases in intracranial pressure may result in unilateral or bilateral abducens palsy, presumably caused by nerve compression. Such palsies are usually partial.[1] Aneurysms (usually in the cavernous sinus) rarely cause abducens palsies in childhood.[114]

Differential Diagnosis

Abducens palsy must be distinguished from congenital esotropia, which also presents with esodeviation but is comitant and usually not seen before 6 to 8 weeks of age.[99,100] A horizontal gaze palsy (caused by an ipsilateral lesion to the abducens nucleus or paramedian pontine reticular formation) requires the patient to turn his or her head to see in the affected direction of gaze. This may mimic a unilateral abducens palsy, in which the patient turns his or her head in order to avoid development of diplopia. In abducens palsies, however, diplopia and head turning are abolished by patching of the paretic eye.[95]

Spasm of the near reflex (convergence spasm) may limit abduction and hence mimic bilateral abducens palsies. Frequently caused by head trauma and less commonly seen with Arnold-Chiari malformations or other disorders, it is accentuated with near viewing, and may be identified on the basis of the pupillary constriction that is part of the near reflex.[115]

Lateral rectus weakness is occasionally seen with pediatric myasthenia gravis and thyroid eye disease.[52,116] Orbital pseudotumor and orbital myositis may cause restriction of lateral rectus contraction.[1,117] Restriction or contracture of the medial rectus may mimic or result from CN VI palsies (see earlier discussion).

Unlike cyclic oculomotor palsy, cyclic esotropia, which may follow traumatic abducens palsy, is not due to a true nerve paresis.[118] Affected children alternate between periods of 12, 24, 36, or 48 hours of normal alignment, and periods of similar length during which they manifest esotropia. Despite the intermittent strabismus, a full range of eye movements is preserved.[1,88]

Investigations and Imaging of Cranial Nerve Palsies

Evaluation of ocular motor palsies should include cranial imaging, ideally by magnetic resonance imaging with detailed views of the cavernous sinus, clivus, cranial nerves, and orbit. In children in whom no cause is identified for a persistent acquired ocular motor palsy,

CASE EXAMPLE 46.1

An 8-month-old boy had a 2-week history of intermittent eye crossing. His mother observed that, at times, his eyes appeared normally aligned, but these periods were becoming briefer and less frequent. She also commented that he was increasingly turning his head to the left. Two weeks before the strabismus was noted, he had had an upper respiratory infection with otitis media. Examination revealed a large esotropia in left gaze. Abduction of the left eye was limited to the midline, and fixation alternated between the two eyes. The neurologic examination was otherwise normal. A cranial MRI scan revealed normal ventricles and parenchyma, with enlargement of the extra-axial cerebrospinal fluid spaces. Alternate-day patching was instituted. His incomitant esotropia remained stable over the next 6 months.

Comment

The isolated deficit in abduction and noncomitant esotropia were indicative of a left sixth nerve palsy, which was felt to be benign given the otherwise normal examination and imaging findings. Presentation beyond the first 3 to 4 months of life makes an acquired palsy more likely than a congenital one. Alternating fixation gives the false impression that first one eye, then the other, is turning inward. Weakness appears intermittent at first because the patient is able to overcome the deficit with effort for brief periods. Later the patient minimizes the misalignment by head turning. Patching of the eyes prevents the development of amblyopia. Surgical correction may be considered once the deviation is stable for 6 months.

reimaging after intervals of 1 to 2 years may be indicated in order to exclude an occult mass lesion.[1,49] Magnetic resonance angiography may also be considered for acquired CN III palsy, although intracranial aneurysms and other vascular anomalies are rare in childhood.[48] Magnetic resonance venography may be indicated for CN VI palsy with evidence of increased intracranial pressure.

Lumbar puncture is frequently indicated for both exclusion of central nervous system infection and measurement of intracranial pressure. Other investigations to be considered are the Tensilon test and assessment of forced ductions. Targeted genetic testing should be obtained where appropriate.

Treatment and Prognosis of Cranial Nerve Palsies

General Principles

Three types of disability deserve attention in the management of ocular motor palsies: amblyopia, ocular misalignment, and ptosis. Amblyopia typically results from the child's suppression of the image from one eye in order to avoid diplopia. Occlusion of one eye by ptosis or inaccurate focusing due to failure of accommodation may exacerbate amblyopia and result in strabismus.[119,120] Children are most at risk of developing amblyopia between 6 weeks and 4 years of age. Intermittent patching may prevent both amblyopia and acquired medial rectus contracture. Patching should be continued until the palsy resolves or surgical correction is attempted. Botulinum toxin has been used to weaken selective muscles as an alternative to surgery, but its efficacy is debated.[1,121]

A number of methods for surgical correction of persistent misalignment have been reported. Their success depends both on the degree of residual paresis and on the complex coordination between the various EOMs required to achieve normal conjugate horizontal and vertical movements.[122] In general, the primary goal of strabismus surgery is to allow single binocular vision in primary gaze. The secondary goal is to extend such vision to reading, then to extend it to as wide an angle as possible in any direction from primary gaze, and finally, to improve cosmetic appearance.[1]

Oculomotor Palsy

Surgical procedures to correct adduction deficits entail recession and resection of the horizontal rectus muscles, or division and transfer of the superior oblique tendon.[1,123] To correct vertical misalignment, either of the horizontal recti may be transposed vertically, or recession-resection procedures may be performed on the vertical recti.[124] Surgical correction of severe residual ptosis may best be deferred until ocular alignment has been optimized, because occlusion of the involved eye may be serving a useful function in preventing diplopia. When there is minimal residual levator function, a frontalis suspension may be performed. The long-term prognosis for childhood oculomotor palsy varies with the cause and with the corrective procedure employed. Overall, visual acuity is reduced long term in 50% to 60% of patients, but virtually all of those with congenital oculomotor palsy recover normal acuity. Only 15% recover full motility, and a similar proportion recover stereopsis.[125,126]

Trochlear Palsy

In most cases of nontraumatic acquired trochlear nerve palsy, muscle strength, ocular misalignment, and head tilt gradually improve spontaneously, generally within 3 to 6 months.[1] During this period there is no need for patch occlusion because the patient uses a head tilt to fuse images. Thus, management consists of expectant observation.

Amblyopia is rare; when it does develop, it suggests that the head tilt cannot sufficiently compensate for diplopia, there is another concurrent ocular motility deficit, or that central fusion mechanisms have been disturbed (generally in association with trauma).[127] In some cases, superior oblique myokymia (see separate discussion) may develop.[128] For congenital cases, and acquired cases that do not resolve spontaneously, surgical treatment is usually required.

In contrast to oculomotor palsy, an additional goal is attainment of normal head position, optimal cosmesis, and fusion within a reasonable functional range of eye movements. The choice of procedure depends partly on compensatory responses of other EOMs.[1] Overall, the success rate for restoration of a normal head posture is 75%; for vertical deviation of less than 3 diopters, the success rate is 60%.[129]

Abducens Palsy

Most acquired abducens palsies resolve spontaneously within 3 to 4 months. Prognosis varies with cause and is best in idiopathic cases.[130–132] Younger children have higher rates of permanent strabismus (66%) and amblyopia (20%).[132] Recurrent palsies usually involve the same eye, and resolve within 8 to 12 weeks.[111] Treatment is that of the underlying condition.

Esotropia may persist owing to spread of comitance, incomplete recovery, or acquired contracture of the medial rectus muscle.[1] If residual strabismus is stable for at least 6 months, surgical correction by transposition of the vertical rectus muscles to the lateral rectus or recession of the contralateral medial rectus may be performed.[133]

CONGENITAL PTOSIS SYNDROMES

Blepharoptosis

Eyelid position and movement result from the balance between the opening forces generated by the tonically active levator palpebrae superioris and Muller's muscles, and closing forces generated actively by the (normally quiescent) orbicularis oculi muscle and passively by stretching of the eyelid's ligaments and tendons.[134] Deficiency of levator tonus results in blepharoptosis (ptosis). Ptosis can be congenital or acquired, and can result from upper motor neuron, lower motor neuron, nerve, neuromuscular junction, or primary muscle dysfunction. Congenital ptosis can occur in isolation or in association with other ocular, neuromuscular, or systemic findings. When it occurs with oculomotor nerve palsy, CFEOM, or the Marcus Gunn phenomenon, it is likely neurogenic in origin. When seen with congenital myasthenia or a congenital myopathy, it localizes to the neuromuscular junction or muscle, respectively.

FIGURE 46.5 A mother and two daughters, all affected by isolated congenital ptosis, and an unaffected son. The autosomal dominant trait in this family maps to the chromosome 1 *PTOS1* locus. The mother has moderate right- and mild left-sided ptosis, and both daughters have moderate to severe bilateral ptosis with a compensatory backward tilt to the head. *From Engle et al., 1997,[135] with permission from the University of Chicago Press.*

Isolated Congenital Ptosis

Isolated congenital ptosis (congenital myopathic ptosis or developmental ptosis) is characterized by a deficiency in levator excursion with an elevated or absent lid crease (Figure 46.5). This is unilateral in ~75% of cases. Surgical correction is typically performed between 6 months and 5 years of age, particularly if there is a risk of secondary amblyopia. Levator muscle biopsies at the time of corrective surgery typically reveal a reduction or absence of myofibers and the presence of connective tissue,[136] and some investigators have noted an inverse correlation between the degree of ptosis and the number of residual striated muscle fibers.[137] These observations led to the hypothesis that isolated congenital ptosis results from a myogenic defect. Such biopsy findings, however, could also arise secondary to reduced muscle innervation from maldevelopment of oculomotor caudal central subnucleus motor neurons or their axons. Determining the molecular basis of congenital ptosis should help resolve this issue. Toward this goal, an autosomal dominant *PTOS1* locus (OMIM #178300) on chromosome 1p34.1-p32 (Figure 46.5),[135] an X-linked *PTOS2* locus

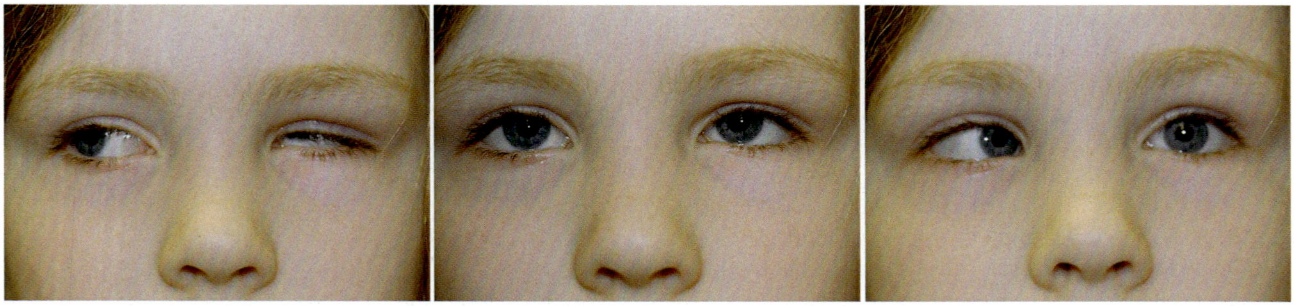

FIGURE 46.6 Patient with Duane's retraction syndrome attempting to look to her right (left photo), straight ahead (middle photo), and to her left (right photo). Her left eye has restricted lateral movement on attempted abduction, and upshoot with narrowing of the palpebral fissure secondary to co-contraction on attempted adduction.

(OMIM #300245) on chromosome Xq24-q27.1,[139] and a translocation disrupting ZFH-4 on chromosome 8[140] have been mapped, but the causative gene mutations not yet reported.

Some individuals with congenital ptosis also have limitation of upgaze. This combination of findings can be neurogenic or myogenic, due either to an error in determining the identity or course of the superior division of the oculomotor nerve that carries axons from nIII to these two muscles, or abnormal development of the levator and superior rectus muscles. These two muscles share a common epimysium and only begin to separate from one another at 7 weeks' gestational age.[141–143]

Blepharophimosis-Ptosis-Epicanthus Inversus Syndrome

Blepharophimosis-ptosis-epicanthus inversus syndrome (BPES, OMIM #110100) is a developmental malformation of the lid and surrounding tissues. Individuals with BPES have ptosis with poor levator function and absent lid crease, palpebral fissures narrowed both horizontally and vertically (blepharophimosis), and a skin fold running inward and upward from the lower lid (epicanthus inversus). BPES can be simplex or can occur as an autosomal dominant trait, in which case the eyelid abnormalities can occur in isolation (BPES type II) or in association with ovarian failure (BPES type I). BPES types I and II map to chromosome 3q23[144–146] and result from mutations in the Forkhead transcription factor, *FOXL2*.[147] *FOXL2* is expressed predominantly in the developing eyelid and perioptic mesenchyme, and in the adult ovary.

CONGENITAL CRANIAL DYSINNERVATION DISORDERS

In 1950, H.W. Brown categorized five types of incomitant strabismus (Duane syndrome, strabismus fixus, vertical retraction syndrome, Brown syndrome, and congenital fibrosis syndrome) under the umbrella term "congenital fibrosis syndromes," based on the observation that they all presented as congenital, nonprogressive restrictive ophthalmoplegia with active limitation and passive restriction of globe movement.[148] The restrictive nature of the ophthalmoplegia with positive forced duction testing and a "tight" feel to the EOMs at surgery, and the finding of connective tissue on surgical biopsies of the EOMs, led Brown and others to propose that these disorders resulted from primary EOM fibrosis. As described in the following discussion, more recent neuropathologic and genetic studies of Duane syndrome and CFEOM have established that a primary defect in motor neuron development accounts for at least a subset of these disorders. Thus, these syndromes have been recategorized as congenital cranial dysinnervation disorders (CCDDs), an umbrella term encompassing maldevelopment of the ocular as well as other cranial nerves.[149–152]

Duane Syndrome

Duane syndrome is the most common of the CCDDs, accounting for 1% to 5% of strabismus cases.[153,154] The affected eye or eyes in individuals with Duane syndrome, also called Duane retraction syndrome (DRS), have limited horizontal gaze and retraction of the globe into the orbit with narrowing of the palpebral fissure on attempted adduction (Figure 46.6). The syndrome was named for Alexander Duane, who published a 1905 paper collating 54 cases.[155] In his series, abduction was virtually absent in 75%, adduction was abnormal in 96%, oblique movements (upshoot, downshoot) occurred on attempted adduction in 57%, and retraction was present in 95% of cases. Many patients had strabismus and abnormal head position in primary gaze. None had accommodation or pupil abnormalities, while associated malformations and family history were not addressed. The preponderance of affected females, unilateral cases, and left eye affection noted by Duane has been borne out by many subsequent studies. Despite the congenital, nonprogressive nature of the disorder, cases are diagnosed equally in infancy, childhood, and adulthood.[156,157]

DRS has been categorized clinically into three types, with types I and III more common than type II.[158] *Type I* is defined as poor abduction with normal or slight limitation of adduction, *type II* as poor adduction with normal or slight limitation of abduction, and *type III* as a combination of poor abduction and adduction. Globe retraction and palpebral fissure narrowing with adduction occurs in all three categories. Many individuals with DRS have strabismus and may maintain a compensatory head turn; esotropia is more common in type I and exotropia in type II. Most patients have good visual acuity; only 10% develop amblyopia. Up to 50% of patients with DRS have additional congenital anomalies, particularly of the skeleton, ear, eye, and kidney.[2,156,159]

Although historically believed to result from a primary myogenic process,[155,160,161] DRS is now accepted as neurogenic in etiology. Postmortem examinations have revealed absence of the abducens nucleus and nerve on the affected side(s), and partial innervation of the lateral rectus muscle(s) by branches from the oculomotor nerve(s).[162,163] Many subsequent magnetic resonance imaging studies have verified this pathology. Electromyographic studies have demonstrated that the globe retraction results from the simultaneous co-contraction of the medial and lateral recti, consistent with the aberrant and paradoxical innervation of these muscles.[164,165]

Multiple genetic causes of DRS have been reported over the last decade, but even in combination these account for a small percent of DRS cases. Among these, *CHN1* mutations result in isolated DRS while *SALL4, HOXA1*, and several chromosomal disorders result in syndromic DRS. In addition, syndromic DRS can result from teratogens, particularly following *in utero* thalidomide exposure between 21 and 26 days' human gestation.[166,167]

Autosomal dominant isolated DRS can result from heterozygous missense mutations in *CHN1*, which encodes alpha2-chimerin, a Rac guanosine triphosphatase-activating protein (RacGAP)[168] Affected individuals have a higher incidence of bilateral DRS and additional vertical eye movement abnormalities compared to individuals with non-*CHN1* DRS.[169–172] Magnetic resonance imaging reveals the anticipated abducens nerve hypoplasia and aberrant lateral rectus innervation, while some affected individuals also have hypoplasia of the oculomotor nerve and oculomotor- and trochlear-innervated muscles.[169,170,172] *CHN1* mutations hyperactivate alpha2-chimarin and lower RacGTP levels. Modeling of these mutations in developing chick and zebrafish oculomotor axons reveal errors in axon growth and guidance.[168,173] *CHN1* mutations are not a common cause of simplex DRS.[174]

Dominant mutations in the transcription factor *SALL4* cause Duane-radial ray syndrome (DRRS), also referred to as Okihiro syndrome, acro-renal-ocular syndrome, or IVIC syndrome.[175–179] *SALL4* mutations cause incompletely penetrant unilateral or bilateral DRS accompanied by radial dysplasia ranging from hypoplasia of the thenar eminence to absent forearm. Deafness, renal anomalies, and imperforate anus can also be co-inherited, while multigene deletions encompassing *SALL4* can result in even more extensive phenotypes.[180] Magnetic resonance imaging of individuals with DRS harboring *SALL4* mutations reveals marked abducens hypoplasia with probable innervation or co-innervation of the lateral rectus muscle by the oculomotor nerve.[181] While several *Sall4* mutant mouse lines have been reported, these have not provided insight into the etiology of DRS in DRRS; *Sall4*$^{-/-}$ embryos die at ~E6.5, and in the one line examined in detail, *Sall4*$^{-/+}$ mice had normal-appearing ocular cranial nuclei and extraocular muscles.[182]

The *HOXA1*-related syndromes result from recessive mutations in the transcription factor *HOXA1*, and include the overlapping Bosley-Salih-Alorainy syndrome (BSAS, OMIM, #601536)[183] and Athabaskan brainstem dysgenesis syndrome (ABDRS).[184] Affected individuals have Duane syndrome type 3 or horizontal gaze palsy with absent abducens nerves, and most have bilateral sensorineural hearing loss caused by an absent cochlea and rudimentary inner ear development. Individuals may also have intellectual disability, autism, moderate-to-severe central hypoventilation, facial weakness, swallowing difficulties, vocal cord paresis, conotruncal heart defects, and/or skull and craniofacial abnormalities.[183–186] *HOXA1* mutations are not, however, a common cause of isolated DRS.[187]

Among a variety of cytogenetic anomalies reported in patients with simplex and syndromic DRS, those in the chromosome 8q12–8q13 region (the DURS1 locus) are most commonly reported. The region has yet to be untangled, but current data suggest that the DURS1 locus could result in DRS by dosage effect in the region of 8q1, through deletion on 8q13 and/or a duplication of 8q12.[188]

Congenital Horizontal Gaze Palsy

Congenital horizontal gaze palsy can occur in isolation but is more frequently found in association with progressive scoliosis or Moebius syndrome, or co-segregating in families with dominant forms of Duane syndrome. Horizontal gaze palsy with progressive scoliosis (HGPPS) is a recessive disorder defined by almost complete limitation of horizontal eye movements with intact vertical gaze, and scoliosis that begins in the first decade of life and is often severe and debilitating.[189] HGPPS results from mutations in the axon guidance receptor *ROBO3*.[190,191] There is failure of axons and cell bodies to

CASE EXAMPLE 46.2

An 11-month-old boy was referred to the neuromuscular clinic for evaluation of ptosis and difficulty moving his eyes. He was born at full term by Caesarean section for cephalopelvic disproportion, following an uneventful pregnancy. He was noted to have bilateral ptosis and mild axial hypotonia at birth. When examined at 11 months of age, he held his head retroflexed and had mild tenting of the upper lip. He had marked bilateral ptosis. In primary position, the eyes were infraducted 30 degrees and esotropic. He had no upgaze, minimal horizontal movement of either eye, and had to fix and follow using his entire head. On attempted upgaze, bilateral convergence occurred. Pupils and fundoscopic examination were normal. He had mild axial hypotonia and an otherwise normal examination. The family history was remarkable for nonprogressive congenital bilateral infraducted ophthalmoplegia and ptosis in his father, grandmother, paternal uncle, and 20 additional members of his extended family. Forced duction testing was positive for marked restriction of globe movement in all fields of gaze. Quadriceps muscle biopsy showed mild fiber type disproportion. Genetic linkage analysis revealed that the family's phenotype mapped to the CFEOM1 locus on chromosome 12 and segregated the most common *KIF21A* mutation, resulting in the R954W amino acid substitution.

Comment

This child demonstrates the classic features of CFEOM1. Over time, his oculomotility disorder did not progress and his mild hypotonia resolved. His visual acuity was monitored closely. He underwent ptosis and strabismus surgeries at several years of age with moderate improvement in primary gaze and head position.

cross the midline of the hindbrain and spinal cord during development, resulting in uncrossed descending corticospinal tracts and ascending sensory tracts.[192] This lack of crossing fibers results in the classic HGPPS midline cleft running rostral-caudal in the hindbrain, which can be seen on magnetic resonance imaging.[189,190,193]

Congenital Fibrosis of the Extraocular Muscles

The diagnosis of congenital fibrosis of the extraocular muscles (CFEOM) refers to forms of CCDDs in which there are limited vertical eye movements with positive forced ductions (vertical concomitant strabismus); most individuals with CFEOM also have ptosis, and many have restricted horizontal movements and strabismus as well. While CFEOM subtypes were initially grouped by clinical presentation, CFEOM phenotypes can overlap and thus genetic classification can be more informative. CFEOM phenotypes resulting from mutations in *KIF21A, PHOX2A, TUBB3*, and *TUBB2B* are now recognized.[24,41,194,195]

Dominant missense mutations in *KIF21A* result primarily in the CFEOM type 1 phenotype (CFEOM1) (see Case Example 46.2); congenital bilateral ptosis, inability to elevate either eye above midline, and typically restricted horizontal gaze; and aberrant residual eye movements (Figure 46.7).[41,196] Human autopsy and MRI data revealed hypoplasia of the superior division of the oculomotor nerve, with absence of the corresponding oculomotor neurons and hypoplasia of the superior rectus and levator palpebrae superioris muscles (Figures 46.8 and 46.9).[6,197] *KIF21A* encodes an anterograde kinesin protein

FIGURE 46.7 Three siblings from a CFEOM1 pedigree who harbor a dominant mutation in *KIF21A*. The two affected children in the foreground have ptosis and infraducted ophthalmoplegia with a compensatory backward head tilt. They have marked restriction of globe movement and positive forced duction testing. The sister in the background is unaffected. *From Engle, et al., 1995,[138] with permission from the University of Chicago Press.*

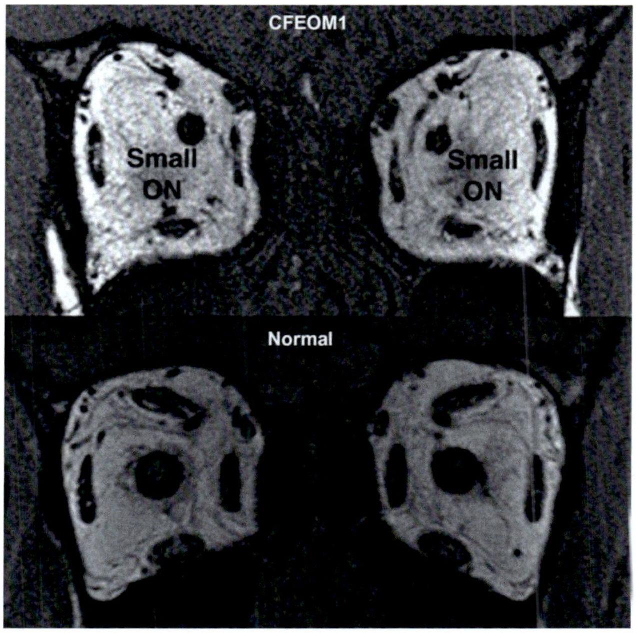

FIGURE 46.8 Quasicoronal MRI of orbits just posterior to the globe–optic nerve junction in a patient with KIF21A-CFEOM1 illustrating a hypoplastic optic nerve and rectus extraocular muscles, in comparison with a normal control subject. *From Demer JL et al., 2005,[75] with permission from the copyright holder, the Association for Research in Vision and Ophthalmology.*

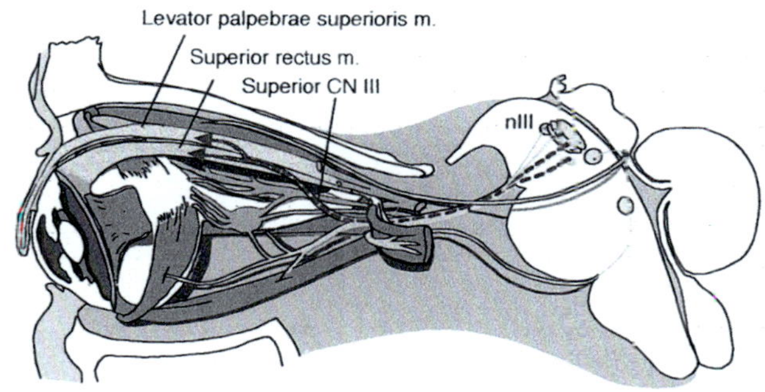

FIGURE 46.9 Schematic lateral view of the brain stem and left orbit demonstrating the neuropathology of CFEOM1. Compared with the normal orbital anatomy depicted in Figure 46.1, the absent medial and central caudal oculomotor (nIII) subnuclei and the corresponding superior division of CN III are depicted as dotted subnuclei and dashed nerves, respectively. The muscles innervated by the superior division of CN III (superior rectus and levator palpebrae superioris) are secondarily abnormal and depicted in light gray.

whose motor domain interacts with and "walks" down microtubules. Wild type Kif21a has been implicated in axonal transport of several cargos and in the inhibition of microtubule growth at the cell cortex *in vitro.*[198–201] It is expressed in most developing and mature neurons in the nervous system, and its expression does not appear to be changed in CFEOM1.[202] *CFEOM1* mutations specifically alter amino acid residues in the motor and third coiled-coil stalk domain of the KIF21A protein.[41,203] These domains interact with one another to autoinhibit Kif21a, and *CFEOM1* mutations attenuate Kif21a autoinhibition and enhance its interaction with microtubules.[42,201] Kif21a knock-in mice harboring the most common *KIF21A* mutation recapitulate the human CFEOM phenotype, and have aberrant stalling and branching of axons within the oculomotor nerve.[42]

A series of recurrent dominant missense mutations in *TUBB3* result in variable phenotypes that correlate with the specific mutation, and include CFEOM1 and CFEOM without bilateral ptosis and/or with some ability to elevate one or both eyes above midline, referred to as CFEOM3 (Figure 46.10). In addition to CFEOM, patients with specific *TUBB3* mutations may have intellectual or social disabilities, facial weakness, vocal cord paralysis, finger contractures, progressive sensorimotor polyneuropathy, Kallmann syndrome, and/or cyclic vomiting.[194,205] In particular, the TUBB3 E410K syndrome may be misdiagnosed as atypical Moebius syndrome.[205] Magnetic resonance imaging reveals oculomotor nerve hypoplasia that can be accompanied by dysgenesis of the corpus callosum, anterior commissure, corticospinal tracts, and/or basal ganglia.[194,205] A knock-in disease mouse model revealed axon guidance defects of both cranial nerves and central axon tracts.[194] *TUBB3* encodes the neuron-specific beta-tubulin isotype III, which is one of multiple beta-tubulin isotypes that heterodimerize with alpha-tubulin isotypes to form microtubules. The disease-associated missense mutations have been shown to impair tubulin heterodimer formation to varying degrees, although folded mutant heterodimers can still polymerize into microtubules. Modeling each CFEOM-causing mutation in yeast tubulin demonstrated that all alter dynamic instability, whereas a subset disrupts the interaction of microtubules with kinesin motors.[194] Similarly, a heterozygous missense mutation in a second beta-tubulin isotype, *TUBB2B*, has been reported to segregate with CFEOM and polymicrogyria.[195] This mutation also incorporates into the microtubule network, where it alters microtubule dynamics and can reduce kinesin localization.[195] The specific molecular relationship between KIF21A-CFEOM1, TUBB3-CFEOM3, and TUBB2B-CFEOM3, and the reason that the developing oculomotor nerve is particularly vulnerable to these disease mutations, remain to be elucidated.

The genetic cause of one recessive form of CFEOM, CFEOM2, has been reported. Patients with CFEOM2 have congenital nonprogressive bilateral ophthalmoplegia and ptosis with the eyes primarily exotropic and vertically midline, and restriction of horizontal and vertical eye movements; they also can have small minimally reactive

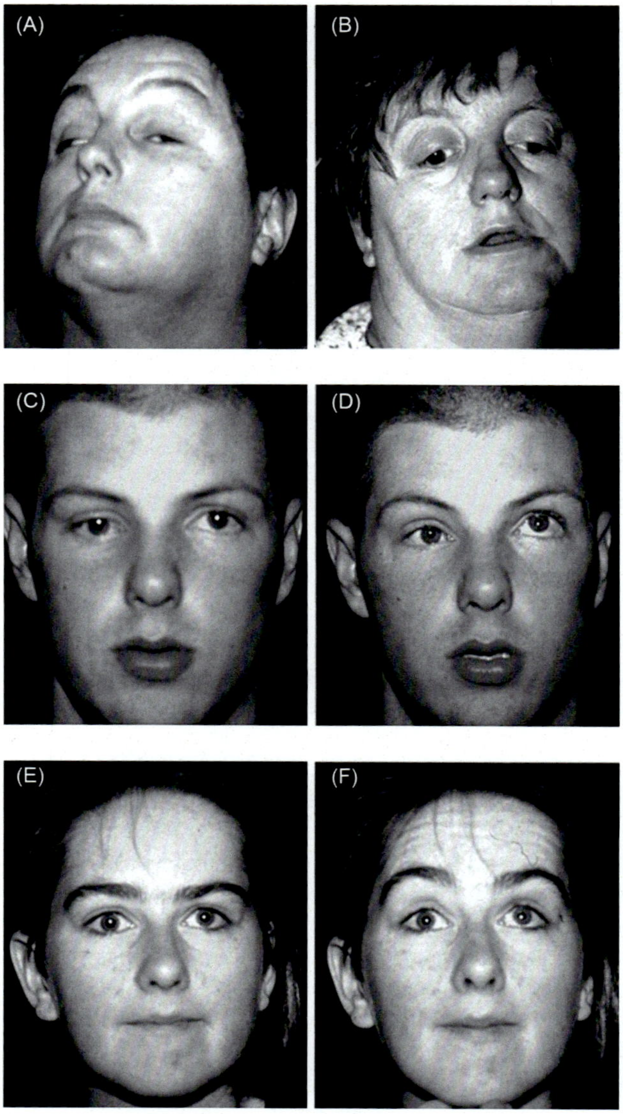

FIGURE 46.10 Members of a family with autosomal dominant CFEOM3 who harbor a dominant mutation in *TUBB3*. The photographs illustrate the spectrum of the CFEOM3 phenotype, which is much broader than that found in CFEOM1 or CFEOM2. Individuals **A** and **B** are severely affected, with bilateral ptosis and a restrictive ophthalmoplegia with the eyes fixed in an infraducted and exotropic position. The individual shown in **C** and **D** is moderately affected with hypotropia of the right eye in primary gaze (**C**) and absent elevation of the right eye on attempted upward gaze (**D**). The individual shown in **E** and **F** is mildly affected, as demonstrated with virtually normal primary gaze (**E**) but bilaterally restricted upgaze (**F**). *From Doherty et al., 1999,*[204] *with permission from the copyright holder, the Association for Research in Vision and Ophthalmology.*

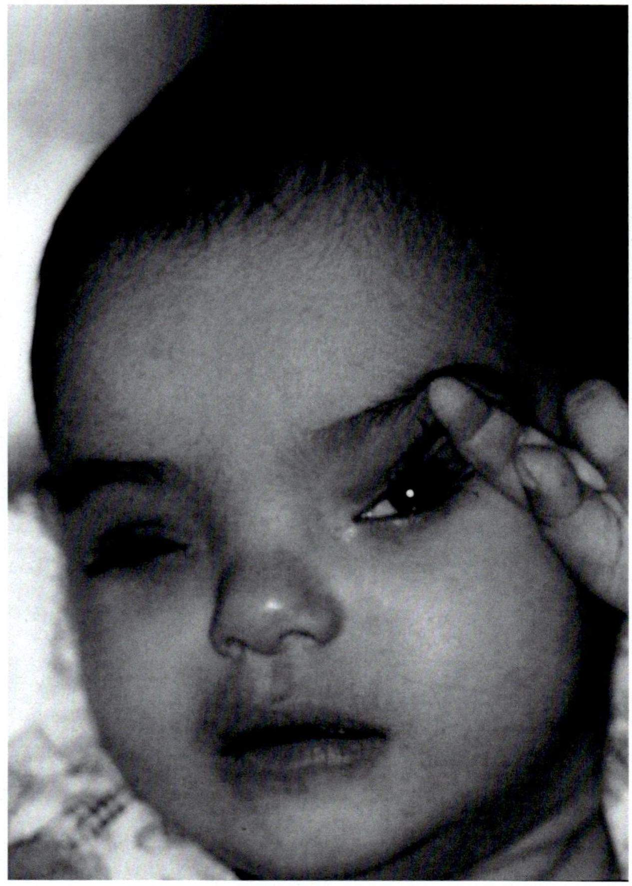

FIGURE 46.11 An affected child from an autosomal recessive CFEOM2 pedigree with a homozygous mutation in *PHOX2A*. The toddler has bilateral ptosis and 60–80 prism-diopter exotropia, right hypertropia, and marked limitation of adduction, elevation, and depression of the globes. He fixes with his exotropic left eye by deviating his head to the right and elevating his left lid with his left index finger. *Reprinted from the American Journal of Human Genetics v. 63, Wang et al., 1998,*[206] *with permission from Elsevier.*

pupils (Figure 46.11).[24,207] CFEOM2 results from recessive loss-of-function mutations in the homeodomain transcription factor *PHOX2A*, which is necessary for the development of oculomotor and trochlear motor neurons in both mouse and zebrafish.[208,209]

In addition to these identified CFEOM disease genes, a recessive locus for CFEOM and postaxial oligodactyly or oligosyndactyly has been mapped to 21qter,[210] and a chromosome 2 to chromosome balanced/unbalanced translocation was reported in a second family.[211] Neither of these genes has been reported.

Congenital Trochlear Palsy and Brown Syndrome

Congenital trochlear palsy (also referred to as congenital fourth nerve palsy or superior oblique palsy) is characterized by hypertropia of the affected eye which increases in adduction and with ipsilateral head tilt. As many as 80% of all congenital ocular motor palsies affect CN IV.[127]

Even so, the incidence of congenital fourth nerve palsy is thought to be underestimated, because it is often asymptomatic or confused with congenital torticollis. Trochlear nerve dysfunction may be isolated or can be associated with oculomotor dysgenesis in CFEOM2,[24] or ocular, oculomotor, auricular, central nervous system, and vertebral abnormalities in the Goldenhar syndrome.[212] Occasional familial cases are described.[213–215]

MR imaging of congenital trochlear palsy reveals hypoplasia of the superior oblique muscle and, in many cases, there is inability to image the trochlear nerve,[216,217] leading to its recent classification as a CCDD. Congenital fourth nerve palsy is rarely inherited, and although *PHOX2A* polymorphisms have been reported in this disorder, these are unlikely to be causative.[218,219]

Congenital Brown syndrome is characterized by absent elevation of the eye in adduction, improved elevation in primary position, and normal or virtually normal elevation in abduction. Forced duction testing reveals a mechanical restriction on attempts to elevate the adducted globe. Affected individuals may also have primary hypotropia, divergence on upgaze, downshoot or widening of the palpebral fissure in adduction, or anomalous head posture.[220,221] Most cases are sporadic. The syndrome can also be acquired, and has been reported in association with trauma and inflammatory disorders. Both congenital and acquired cases can resolve spontaneously or occur intermittently.

Although Brown originally proposed that the syndrome resulted from a congenitally short superior oblique tendon sheath, resulting in the name "superior oblique tendon sheath syndrome," the etiology of Brown syndrome is debated. Proposed etiologies include anomalies of the tendon or trochlea, of the superior oblique muscle, or of the inferior oblique and adjacent structures, and paradoxical innervation of the superior oblique muscle.[222] Interestingly, congenital Brown syndrome can occur with other aberrant innervation syndromes such as Duane syndrome, Marcus Gunn phenomenon, and crocodile tears,[221] and thus it remains possible that at least some cases are also CCDDs.[223,224]

Moebius Syndrome

Moebius syndrome is named for P. J. Möbius, whose 1892 paper categorized the congenital and early childhood cranial nerve palsies into six groups,[225] one of which consisted of six sporadic cases of congenital abducens and facial nerve paralysis. The facial and abduction weakness of Moebius syndrome is congenital and nonprogressive, and usually clinically distinguishable from progressive neuromuscular disorders. Patients with facioscapulohumeral muscular dystrophy,[226] myotonic dystrophy,[227] and unknown progressive neuromuscular disorders,[228] however, are occasionally initially misdiagnosed with Moebius syndrome.

Although the combination of congenital palsies of abducens and facial nerves has been the historical definition of Moebius syndrome, the diagnosis has been applied in a heterogeneous fashion over the years, including in individuals with only isolated facial weakness. Thus, a group of clinicians and researchers met at the 2007 biannual Moebius Syndrome Foundation research meeting and defined the minimum diagnostic criteria for classical Moebius syndrome as "congenital, uni- or bilateral, non-progressive facial weakness and limited abduction of the eye(s)."[229] Applying these minimum diagnostic criteria, Moebius syndrome remains heterogeneous in clinical presentation, and likely in etiology. It frequently occurs in association with additional cranial nerve deficits, orofacial malformations, limb defects, and musculoskeletal, behavioral, and/or cognitive abnormalities.[229–239] Moreover, neuropathological[240–245] and radiological[246–248] reports support multiple disease entities.[249] To improve diagnostic assessment of Moebius syndrome and to optimize genetic analysis, MacKinnon and colleagues have proposed that "full vertical motility" be added to minimum diagnostic criteria for Moebius syndrome.[250]

Classic Moebius syndrome is typically sporadic. Chromosomal anomalies in several sporadic patients defined the *MBS1* locus on 13q12.2-q13.[251,252] In addition, however, patients diagnosed with Moebius syndrome have been found to harbor mutations in *HOXA1* (see previous discussion), *HOXB1*, and *TUBB3*, particularly relating to the TUBB3 E410K syndrome.[183,205] *HOXB1* mutations cause facial weakness, esotropia, and deafness, and thus do not meet criteria for Moebius syndrome.[253] Two genetic loci mapped in families with *autosomal dominant congenital facial weakness* who lack eye movement abnormalities were also initially reported as loci for Moebius syndrome,[254,255] but have subsequently been reclassified as hereditary congenital facial paresis 1 and 2 (HCFP1 and HCFP2). In addition, Moebius syndrome can also result from teratogens, in particular following *in utero* exposure to thalidomide[166,256] or misoprostol.[257,258] Finally, some have proposed that Moebius syndrome can result from an embryonic vascular disruption.[259]

SYNKINESIS SYNDROMES

Synkinesis refers to an involuntary action of one or more of the extraocular muscles associated with a voluntary eye or other movement, and typically results from aberrant innervation. As mentioned above, oculomotor synkinesis is common in CFEOM, and particularly in CFEOM1 resulting from mutations in *KIF21A*. In addition, 45% to 50% of congenital oculomotor palsies are associated with synkinesis, although some of these patients likely had CFEOM.[40,260] Acquired synkinesis commonly develops at least 6 weeks after trauma (including neurosurgery) or recurrent ophthalmoplegic migraine (see above), and is believed to result from aberrant regeneration of damaged nerves.[261] Ephaptic transmission may occasionally lead to transient synkinesis.[262,263] Recognized patterns of synkinesis with oculomotor palsies include synergistic convergence or divergence; lid elevation with adduction or abduction; and Y- or V-pattern exotropia, or globe retraction with attempted vertical gaze.[260,264,265] Surgical correction may be required.[266]

The most common misdirection resulting from abnormal development of the sixth nerve is Duane syndrome (see previous discussion); in addition, activation of the lateral rectus may be associated with superior rectus, levator, or contralateral lateral rectus activation ("synergistic divergence").

In individuals with ptosis, fibers from the trigeminal nucleus may become redirected to the levator palpebrae superioris, such that jaw movements lead to elevation of the ptotic lid (Marcus Gunn jaw-winking), or to exotropia (CN VI). More unusual congenital misdirection syndromes include pupil constriction with drinking, and ipsilateral tearing with abduction or lateral gaze deficits ("crocodile tears").[64,260] Both Marcus Gunn jaw-winking and crocodile tears may be seen in Duane syndrome and in CFEOM.

Aberrant innervation of the trochlear nerve does not seem to occur, but superior oblique myokymia has been speculated to follow trochlear palsy in some instances. This is intermittent, low amplitude, irregular but high frequency (up to 50 Hz) oscillation of the superior oblique muscle, which is usually monocular.[267–270] Bursts of such movement lasting seconds to minutes tend to occur intermittently throughout the day, for days to weeks at a time. Episodes may be precipitated by head tilt or by returning the eyes from downgaze. They may occur once or repeatedly, with symptom-free intervals lasting months, and generally resolve spontaneously.

OPHTHALMOPLEGIA IN PEDIATRIC NEUROMUSCULAR DISEASE

Neuropathic Ocular Weakness

Congenital

The EOMs are generally spared by the congenital neuropathies and hereditary motor and sensory neuropathies, but are affected in some specific genetic syndromes (see Chapter 15). Pupillary abnormalities are seen in several subtypes of congenital hypomyelinating neuropathy and Dejerine-Sottas syndrome. Ophthalmoplegia may be seen in several infantile-onset forms of demyelinating Charcot-Marie-Tooth disease (CMT4B, CMT4C, and CMT4E), and CFEOM and axonal Charcot-Marie-Tooth disease co-segregate with specific *TUBB3* mutations.[194,205] The combination of congenital axonal neuropathy and ophthalmoplegia should raise the question of mitochondrial diseases caused by mutations in *SCO2*, *C10orf2*, or *TK2*.[269,271,272] Progressive ptosis and ophthalmoplegia were also seen in a single case of atypical giant axonal neuropathy presenting as progressive gait abnormalities and bulbar dysfunction,[273] and strabismus and ophthalmoplegia can be seen in Andermann syndrome[274,275] and infantile neuroaxonal dystrophy.[276,277]

Spinal muscular atrophy is not usually associated with ptosis or external ophthalmoplegia. Most reports of these findings in spinal muscular atrophy predated molecular diagnosis, and likely described variants of this condition.[278–283] However, occasional atypical cases are linked to mutations in the survival motor neuron gene.[284] The Fazio-Londe/ Brown-Vialetto-Van Laere spectrum of conditions most commonly present between 12 and 36 months with stridor, facial weakness and ptosis. Optic atrophy is common but ophthalmoplegia rare.[285,286]

Acquired

Acquired ophthalmoplegia is quite common in pediatric Guillain-Barré syndrome.[287] A small proportion of affected children presents with the Miller Fisher syndrome of ataxia, external ophthalmoplegia, and areflexia.[288,289] Occasional cases presenting with isolated ophthalmoplegia may be diagnostically challenging.[290] Acquired external ophthalmoplegia is also seen in the neuropathic crises of Tangier disease and acute intermittent porphyria[291,292] and internal ophthalmoplegia may be identified during the bulbar phase of diphtheria and in acute sensory and autonomic neuropathy.[293,294] The extraocular muscles are generally spared in childhood chronic inflammatory demyelinating polyradiculoneuropathy, although diplopia is occasionally reported.[295,296]

Vitamin E deficiency may present with progressive external ophthalmoplegia (PEO), ptosis, and pigmentary retinopathy,[297] and occasionally causes third nerve palsies and aberrant regeneration.[298,299] Sarcoidosis may produce diplopia from involvement of any of the EOMs.[300]

Neuromuscular Junction Disorders

Congenital Myasthenic Syndromes

The congenital myasthenic syndromes present during the neonatal period or early infancy with ptosis, ophthalmoplegia, poor feeding, and hypoventilation. Numerous defects of presynaptic, synaptic, and postsynaptic function have been identified and are reviewed in detail in Chapter 26 (see Table 46.4).[301,302]

TABLE 46.4 Ophthalmologic Manifestations of Pediatric Muscle Diseases

	Ptosis	Strabismus	External Ophthalmoplegia	Pupillary Abnormalities	Retinopathy
Congenital myasthenic syndromes					
Presynaptic Choline acetyltransferase deficiency	+++	−	+	−	−
Synaptic Acetylcholinesterase deficiency	+	+	+	+++	−
Post-synaptic Receptor deficiencies, abnormalities of clustering or synaptogenesis AChR deficiency Rapsyn DOK7 GFPT1 DPAGT1 Kinetic abnormalities Slow channel Fast channel	+++	+	+++	−	−
Congenital myopathies					
Myotubular myopathy	+++	−	+++	−	−
Centronuclear myopathy	++	+	++	−	−
Minicore myopathy	+	+	+	−	−
Nemaline myopathy	+	+	−	−	−
Central core disease	+	−	−	−	−
Congenital fiber-type disproportion	++	−	++	−	−
Congenital muscular dystrophies					
Nesprin deficiency	−	−	+++	+	+
Congenital myotonic dystrophy	+	+	−	−	−
Mitochondrial myopathies					
Kearns-Sayre syndrome	+++	−	+++	−	+++
MELAS	+	−	+	−	++
MERRF	+	−	+	−	−
Muscular dystrophies					
Myotonic dystrophy	++	+	+	+	−
Epidermolysis bullosa and muscular dystrophy	++	−	++	−	−
Myasthenia gravis					
Transient neonatal myasthenia gravis	+	−	+	−	−
Autoimmune myasthenia gravis	++	+	++	−	−
Pediatric Lambert-Eaton myasthenic syndrome	++	+	−	−	−
Infant botulism	++	+	++	++	−
Organosphosphate toxicity	+	+	+	++	−
Thyroid myopathy	+	+	+	−	−

+, occasional; ++, frequent; +++, invariable.
Abbreviations: AChR, acetylcholine receptor; MELAS, syndrome of mitochondrial encephalopathy, lactic acidosis, and stroke-like episodes; MERRF, syndrome of myoclonus with epilepsy and ragged red fibers.

Acquired Disorders of Neuromuscular Transmission

Ptosis and external ophthalmoplegia are among the most common manifestations of myasthenia, and may develop in the absence of peripheral weakness. Although a variety of hypotheses attempt to explain the apparent targeting of the EOMs, selective involvement of the EOM in myasthenia is poorly understood.

Disease expression in acquired myasthenia gravis is governed by antigenic differences in the acetylcholine receptor (AChR) isoforms. The two muscle AChR isoforms are developmentally regulated and are comprised of alpha, beta, delta, and either gamma (embryonic) or epsilon (adult) subunits.[303,304] In peripheral muscle, maturational development of functional neuromuscular junctions results in the replacement of the embryonic by the adult AChR isoform. Mature EOMs, however, continue to express the fetal AChR subtype.[305] Myasthenics may develop antibodies preferentially binding fetal acetylcholine receptors, thereby selectively causing weakness of the EOM. A subset of multiply innervated fibers present only in the EOM contract tonically in proportional response to endplate depolarizations, with smaller endplate potentials resulting in decreased force of contraction.[22] Antibodies selective to the neuromuscular junctions of multiply innervated fibers specifically cause compromise of neuronal transmission in these fibers.

Other manifestations of ocular myasthenia are not explained by these hypotheses. Ptosis is one of the most common manifestations of myasthenia gravis, although the levator palpebrae superioris contains neither multiply innervated fibers nor fetal AChR subunits.[306] Preferential involvement of the EOMs may also relate to their physical qualities: the high firing rate, low acetylcholine receptor density, and lower mean quantal content of EOM fibers, the greater accessibility of their end plates to circulating antibodies, or the low tolerance of ocular alignment systems for errors that produce diplopia.[22,303,304]

The transient neonatal form of myasthenia gravis is seen in 10% to 15% of infants of mothers with acquired myasthenia gravis, and has been attributed to the transplacental passage of maternal anti-AChR antibodies.[307–309] Poor sucking, hypotonia, and facial diplegia are apparent within a few hours of birth in most cases. Ptosis (15%) and ophthalmoplegia (8%) are less common.

Symptoms are isolated to the ocular musculature at initial presentation in as many as 63% of children with acquired autoimmune myasthenia gravis.[310] Ptosis and ophthalmoplegia vary, but typically become more marked towards the end of the day. Ptosis may initially be unilateral but eventually becomes bilateral in most. Ptosis may be more apparent after repeated eyelid closures or saccades, prolonged upgaze, or passive elevation of the opposite eyelid. Cogan's eye twitch, with transient eyelid elevation on upgaze after sustained downgaze, is characteristic of myasthenic eye disease. Autoimmune myasthenia may involve one or all of the EOMs, the levator palpebrae superioris, and the orbicularis oculi. Ptosis and ophthalmoplegia are ameliorated by sleep or cooling of the affected muscles, or by administration of anticholinesterases. Eventual systemic involvement is seen in 85% of patients, usually within the first 2 years after disease onset.[311]

Rare pediatric cases of the Lambert-Eaton myasthenic syndrome are usually associated with an underlying neoplasm, and may present with ptosis or transient diplopia.[312]

Infant botulism causes ptosis, ophthalmoplegia, and sluggish pupillary reflexes, in addition to bulbar weakness and lethargy.[313] Abducens palsies and impaired accommodation have also been identified as early signs of foodborne botulism.[314]

Arthropod envenomation can result in acquired abnormalities of neuromuscular transmission. Involuntary conjugate roving eye movements, possibly associated with nystagmus, may accompany the systemic manifestations of scorpion envenomation. Presynaptic acetylcholine release is increased with both spider and scorpion bites. Tick paralysis initially causes an ascending paralysis. Tonic pupillary dilatation, external ophthalmoplegia, and facial weakness develop late in the disease course (see Chapter 19). Snake envenomation can cause acute neuromuscular blockade with ptosis, ophthalmoplegia, and descending bulbar dysfunction.

Iatrogenic causes of impaired neuromuscular transmission include medications acting presynaptically (such as corticosteroids, antiarrhythmics, aminoglycosides, anticonvulsants, beta-blockers, chloroquine, cisplatin, lithium, and magnesium), or postsynaptically (neuromuscular blockers, anticholinesterases, D-penicillamine, and phenothiazines). Ophthalmoplegia and ptosis may also accompany the systemic manifestations of organophosphate poisoning.[315]

Myopathies

Congenital Myopathies

Several congenital myopathies are associated with ophthalmoplegia (Table 46.4). Myotubular myopathy presents in the neonatal period with severe weakness, hypotonia, and ventilatory insufficiency. Ophthalmoplegia and ptosis are apparent in most boys with X-linked myotubular myopathy, and enable clinical differentiation from congenital myotonic dystrophy (see Case Example 46.3). The autosomal forms of this disorder, commonly known as centronuclear myopathy, present in infancy or childhood with facial diplegia, proximal muscle weakness, and more variable involvement of the external ocular musculature.[316]

CASE EXAMPLE 46.3

An 11-day-old boy was transferred from an outside hospital to Boston Children's Hospital for evaluation of hydrocephalus, profound hypotonia, and respiratory insufficiency. He was born at 31 weeks' gestational age after a pregnancy complicated by polyhydramnios and decreased fetal movements. There was no family history of neuromuscular disorders. At delivery, the infant was limp, apneic, and had no heart rate. He was intubated at 9 minutes of age and had Apgar scores of 0, 4, 6, at 1, 5, and 10 minutes, respectively. Continued ventilatory support was required. Neurologic evaluation revealed an expressionless facies with ptosis, limited abduction of both eyes, a high-arched palate, and tented upper lip. There was poor muscle bulk, generalized areflexia, equinovarus deformities of both ankles, and no antigravity movement. Sensation appeared intact. His examination was otherwise unremarkable but for undescended testes. Serial head ultrasounds showed stable ventriculomegaly with no parenchymal abnormalities. A tracheostomy was placed at 4 weeks of age, at which time muscle biopsy demonstrated increased fiber size variation, with a large number of both type 1 and type 2 fibers containing prominent central nuclei or clear central zones surrounded by a peripheral myofibrillar rim. DNA sequence analysis subsequently demonstrated a point mutation within intron 4 of the *MTM1* gene, which was predicted to cause aberrant splicing of MTM1 RNA. The same mutation was identified on screening of maternal DNA.

Comment

This child's presentation was consistent with a congenital myopathy, with the finding of ptosis and external ophthalmoplegia suggesting the diagnosis of myotubular myopathy. Ventriculomegaly has been reported in myotubular myopathy. DNA sequencing confirmed the diagnosis and will enable prenatal diagnosis in further pregnancies.

Minicore myopathy is characterized by hypotonia, slowly progressive proximal weakness, and paresis of the facial and neck musculature. Ptosis is uncommon, and external ophthalmoplegia is seen in only 10% of cases, generally those associated with ryanodine mutations.[317] Central core disease is not usually associated with abnormal EOM. Ptosis is most commonly seen in the subset of patients with the King-Denborough syndrome, and ophthalmoplegia in patients with recessive central core disease.[318,319] Progressive external ophthalmoplegia has also been reported in multicore myopathy in association with short-chain acyl-CoA dehydrogenase deficiency.[320]

Nemaline myopathy commonly causes facial weakness and ptosis. External ophthalmoplegia is not usually seen in primary nemaline myopathy,[321] and its presence should raise suspicion of mitochondrial disease.[322]

External ophthalmoparesis and ptosis are seen in a minority of children with congenital fiber-type disproportion, and may be predictors of severe weakness and early mortality in this condition.[323]

Childhood muscle disorders with external ophthalmoplegia and type 1 fiber predominance have been variably termed congenital myopathy with uniform type 1 fibers and minimal change myopathy. The genetic basis of these uncommon syndromes remains unclear (Cancilla et al., 1971; Bender et al., 2007; Sugie et al. 1982; Hansen et al., 1977; Lo et al., 1990).[324–330]

Muscular Dystrophies

The congenital muscular dystrophies tend to be associated with variable developmental and maturational abnormalities of the central nervous system and eyes, and may also affect the heart, retina, and cochlear system, but almost invariably spare the extraocular muscles. The necrosis, regeneration, and fibrosis seen in the peripheral muscles are absent from the EOMs, for reasons that remain unresolved but may relate to retention of developmental protein isoforms, variable innervation, fiber size, or typing, or the low absolute forces generated by the EOMs. Alternately, the EOMs may have the capacity for upregulation of utrophin, a sarcolemmal protein that may substitute for dystrophin in maintenance of sarcoplasmic integrity, and/or for adequate sequestration of increased calcium released by dystrophic sarcolemmal damage.[331]

Although "classical" congenital muscular dystrophy (CMD) spares the ocular movements, limitation of upgaze and lateral eye movement is occasionally seen in merosin-deficient CMD.[332] There are also several case reports of children with merosin-positive or undefined CMD with ptosis or ophthalmoplegia, or both.[333,334] Subsequent investigations have identified a nesprin mutation in one such kindred.[335] Some cases are likely to relate to mitochondrial pathology.[336] Ptosis and ophthalmoplegia are not seen in the Fukuyama, Walker-Warburg, or muscle-eye-brain forms of CMD, or in the dystrophinopathies, sarcoglycanopathies, or disorders of the dystrophin-associated glycoproteins.

The muscular dystrophy associated with autosomal recessive epidermolysis bullosa simplex (MD-EBS), and caused by mutations in plectin, is unusual in that it frequently affects the EOM.[337,338] This syndrome presents at birth with bullous skin eruptions and progressive nail dystrophy. Progressive limb-girdle muscle weakness and ptosis become apparent in infancy or later. In several cases, the presence of fatigability and fluctuating ptosis and ophthalmoplegia have raised the question of an associated disorder of neuromuscular transmission, and a positive response to treatment with 3,4-diaminopyridine was seen in one patient.[339]

Although peripheral and facial weakness are marked in congenital myotonic dystrophy, involvement of the EOMs is uncommon in this condition.[340] Ptosis is seen in up to 65% of older patients with myotonic dystrophy, but diplopia and external ophthalmoplegia are relatively uncommon.[341] Slowing of saccades and disturbances of smooth pursuit in this condition have been attributed to a combination of peripheral muscle weakness and lesions within the primary visual pathways. Other ocular findings may include cataracts, macular dystrophy, and pupillary abnormalities.[341]

Oculopharyngeal Muscular Dystrophy

First described by Taylor in 1915, oculopharyngeal muscular dystrophy (OPMD) is most commonly seen in French-Canadian families.[342] This uncommon muscular dystrophy is characterized by slowly progressive ptosis and dysphagia, which most commonly develop during the fifth decade of life. Limitation of ocular movements and development of moderate facial and proximal limb weakness are seen late in the disease course. OPMD results from stable expansions of a $(GCG)_6$ repeat sequence in the first exon of the poly(A) binding protein 2 (*PABP2*) gene on chromosome 14q11.2-13.[342]

The existence of a "childhood-onset" form of OPMD remains in doubt. The youngest genetically confirmed case reported to date, a $(GCG)_9$ homozygote, was symptomatic from 21 years of age.[343] A number of OPMD variants have been described, with early onset or atypical features including profound ophthalmoplegia, significant distal weakness, and early respiratory failure.[344–348] Marked elevation of the serum creatine kinase level is present in some cases. In most cases, muscle biopsies demonstrate dystrophic changes, rimmed vacuoles, and cytoplasmic or nuclear filaments, or both. These overlap syndromes have features of OPMD, oculopharyngodistal myopathy, and the inclusion body myopathies. Some have been shown not to link to *PABP2*.[344]

Other Myopathies

Oculopharyngodistal Myopathy

This rare myopathy can be inherited in both an autosomal recessive and dominant fashion. Affected patients have ptosis and progressive weakness of the extraocular, bulbar, facial, and distal muscles from the second or third decade (Figure 46.12). Creatine kinase levels are elevated and muscle biopsy specimens are markedly dystrophic with fiber degeneration, increased internal nuclei, and rimmed vacuoles containing cytoplasmic filaments. Oculopharyngodistal myopathy is distinguished from OPMD by the early development of ophthalmoplegia and the severity of distal weakness.[350,351]

Inclusion Body Myopathies

The inclusion body myopathies are a heterogeneous group of hereditary or sporadic disorders, the pathologic hallmarks of which are the presence of "rimmed vacuoles" (cytoplasmic vacuoles containing peripheral granular material) and cytoplasmic and nuclear inclusions, consisting of filaments measuring 15 to 21 nm. The autosomal recessive hereditary inclusion body myopathies may present in childhood with an unusual distribution of weakness and with foot drop and quadriceps sparing.[352] Autosomal dominant hereditary inclusion body myopathies can also present in childhood, usually with limb-girdle weakness.[353] Most hereditary inclusion body myopathies spare the EOMs, but inclusion body myopathy with joint contractures and ophthalmoplegia (IBM3) invariably causes progressive external ophthalmoplegia, which may present as early as the first decade.[354]

Endocrine Myopathies

Most cases of thyrotoxic myopathy do not result in ophthalmoparesis, and exophthalmic ophthalmoplegia, when present, is variably associated with thyrotoxicosis.[355] Graves' ophthalmopathy is more benign in children than adults.[356] Other endocrine myopathies spare the EOMs.

Channelopathies

Ptosis and ophthalmoplegia are not seen in myotonia or paramyotonia congenita, or in the periodic paralyses, although these conditions may cause lid lag, blepharospasm, and eyelid myotonia. Fluctuating ptosis and ophthalmoplegia were present in a single kindred with an X-linked syndrome of episodic muscle weakness precipitated by fever or anesthesia. A vacuolar myopathy similar to that seen in the periodic paralyses was identified on muscle biopsy.[357]

Mitochondrial Disorders

The elevated oxidative enzyme activity and high mitochondrial content of the EOMs may explain their high fatigue resistance and preferential involvement in mitochondrial disorders. Tissue-specific expression in these diseases is probably determined by the extent of mitochondrial heteroplasmy. Degeneration is more marked in the extraocular than peripheral muscles in Kearns-Sayre syndrome, but ragged red fibers, myofibrillary degeneration, and abnormal mitochondria have been identified in the EOMs of patients with other mitochondrial disorders (such as MELAS and MERRF), even in the absence of clinical involvement.[358,359] Ptosis and chronic progressive

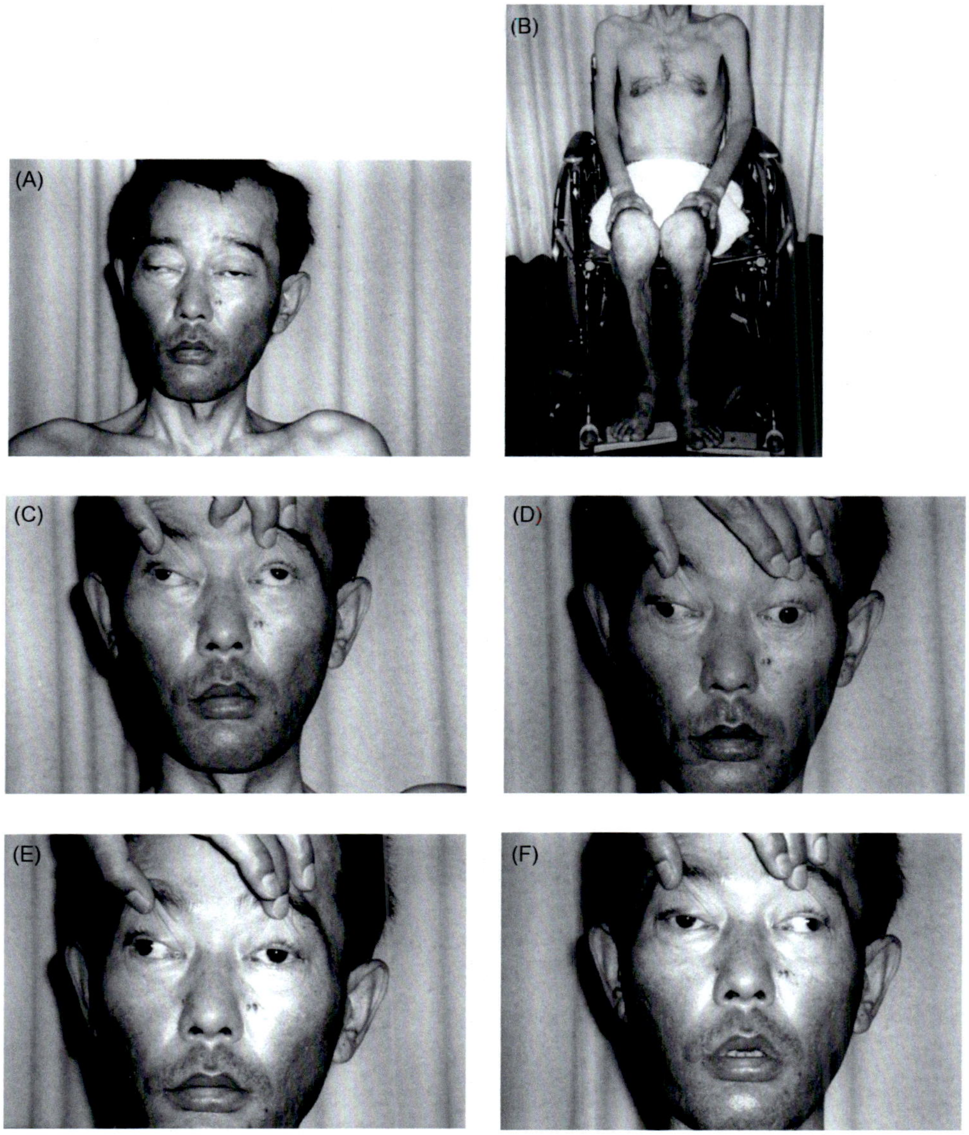

FIGURE 46.12 Oculopharyngodistal myopathy. Affected individual has (**A**) Bilateral ptosis on primary gaze; (**B**) Distal wasting of the upper and lower extremities; ophthalmoplegia on attempted (**C**) upgaze, (**D**) downgaze, (**E**) rightward gaze and (**F**) leftward gaze. *Photographs courtesy of Dr. Eiichiro Uyama. Reprinted from Uyama et al.,* Neuromuscular Disorders *vol. 8, 1998,*[349] *with permission from Elsevier.*

external ophthalmoplegia (PEO) are often the initial manifestations of Kearns-Sayre syndrome, which is also defined by the findings of pigmentary retinopathy, high cerebrospinal fluid protein content, heart block, and ataxia.[359]

Isolated chronic PEO manifests only with ptosis and limited extraocular movements, although involvement of the somatic musculature is commonly apparent on muscle biopsy. The autosomal dominant form of PEO is rarely symptomatic before 20 years of age, but recessive forms can present in childhood, sometimes in association with severe childhood-onset cardiomyopathy.[360] Autosomal recessive PEO is also seen with ptosis, sensorimotor peripheral neuropathy, and gastrointestinal pseudo-obstruction in the multisystem syndrome known as mitochondrial neurogastrointestinal encephalomyopathy (MNGIE).[361]

Arthrogryposis

Ptosis and PEO are uncommon in the arthrogryposis syndromes, but may be seen in forms associated with severe congenital myopathy,[362] and in some forms of distal arthrogryposis.[363,364]

Spinocerebellar Ataxias

Most spinocerebellar ataxias (SCAs) spare the EOMs. Marinesco-Sjögren syndrome may cause strabismus, ptosis, and cataracts[365] and ophthalmoplegia develops during childhood in "infantile onset" SCA and the autosomal recessive form of SCA 8. Of the autosomal dominant SCAs, myokymia is seen in SCAs 2 and 3, PEO, diplopia, and ptosis in SCA3, and ptosis in SCA7. Several of the autosomal dominant SCAs also result in supranuclear gaze disorders, nystagmus, and abnormal saccades.[366,367]

REFERENCES

1. Brodsky M, Baker R, Hamed L. *Pediatric Neuro-ophthalmology.* New York: Springer; 1996.
2. Engle E. The genetics of strabismus: Duane, Moebius, and fibrosis syndromes. In: Traboulsi EI, editor. *Genetic Diseases of the Eye: A Textbook and Atlas.* New York: Oxford University Press; 1998. pp. 477–512.
3. Porter JD, Baker RS. Muscles of a different "color": the unusual properties of the extraocular muscles may predispose or protect them in neurogenic and myogenic disease. *Neurology* 1996;**46**:30–7.
4. Drachman DA, Wetzel N, Wasserman M, Naito H. Experimental denervation of ocular muscles. A critique of the concept of "ocular myopathy". *Arch Neurol* 1969;**21**:170–83.
5. Martinez AJ, Biglan AW, Hiles DA. Structural features of extraocular muscles of children with strabismus. *Arch Ophthalmol* 1980;**98**:533–9.
6. Engle EC, Goumnerov BC, McKeown CA, Schatz M, Johns DR, Porter JD, et al. Oculomotor nerve and muscle abnormalities in congenital fibrosis of the extraocular muscles. *Ann Neurol* 1997;**41**:314–25.
7. Demer JL. Mechanics of the orbita. *Dev Ophthalmol* 2007;**40**:132–57.
8. Miller JM. Understanding and misunderstanding extraocular muscle pulleys. *J Vis* 2007;**7**(10):1–5.
9. Porter JD, Baker RS, Ragusa RJ, Brueckner JK. Extraocular muscles: basic and clinical aspects of structure and function. *Surv Ophthalmol* 1995;**39**:451–84.
10. Demer JL, Oh SY, Poukens V. Evidence for active control of rectus extraocular muscle pulleys. *Invest Ophthalmol Vis Sci* 2000;**41**:1280–90.
11. Spencer RF, Porter JD. Structural organization of the extraocular muscles. *Rev Oculomot Res* 1988;**2**:33–79.
12. Porter JD, Burns LA, May PJ. Morphological substrate for eyelid movements: innervation and structure of primate levator palpebrae superioris and orbicularis oculi muscles. *J Comp Neurol* 1989;**287**:64–81.
13. Porter JD, Israel S, Gong B, Merriam AP, Feuerman J, Khanna S, et al. Distinctive morphological and gene/protein expression signatures during myogenesis in novel cell lines from extraocular and hindlimb muscle. *Physiol Genomics* 2006;**24**:264–75.
14. Fischer MD, Budak MT, Bakay M, Gorospe JR, Kjellgren D, Pedrosa-Domellof F, et al. Definition of the unique human extraocular muscle allotype by expression profiling. *Physiol Genomics* 2005;**22**:283–91.
15. Khanna S, Cheng G, Gong B, Mustari MJ, Porter JD. Genome-wide transcriptional profiles are consistent with functional specialization of the extraocular muscle layers. *Invest Ophthalmol Vis Sci* 2004;**45**:3055–66.
16. Cheng G, Merriam AP, Gong B, Leahy P, Khanna S, Porter JD. Conserved and muscle-group-specific gene expression patterns shape postnatal development of the novel extraocular muscle phenotype. *Physiol Genomics* 2004;**18**:184–95.
17. Khanna S, Merriam AP, Gong B, Leahy P, Porter JD. Comprehensive expression profiling by muscle tissue class and identification of the molecular niche of extraocular muscle. *FASEB J* 2003;**17**:1370–2.
18. Porter JD, Khanna S, Kaminski HJ, Rao JS, Merriam AP, Richmonds CR, et al. Extraocular muscle is defined by a fundamentally distinct gene expression profile. *Proc Natl Acad Sci USA* 2001;**98**:12062–7.
19. Andrade FH, Porter JD, Kaminski HJ. Eye muscle sparing by the muscular dystrophies: lessons to be learned? *Microsc Res Tech* 2000;**48**:192–203.
20. Brueckner JK, Itkis O, Porter JD. Spatial and temporal patterns of myosin heavy chain expression in developing rat extraocular muscle. *J Muscle Res Cell Motil* 1996;**17**:297–312.
21. Oda K, Shibasaki H. Antigenic difference of acetylcholine receptor between single and multiple form endplates of human extraocular muscle. *Brain Res* 1988;**449**:337–40.
22. Kaminski HJ, Maas E, Spiegel P, Ruff RL. Why are eye muscles frequently involved in myasthenia gravis? *Neurology* 1990;**40**:1663–9.
23. Horton RM, Manfredi AA, Conti-Tronconi BM. The 'embryonic' gamma subunit of the nicotinic acetylcholine receptor is expressed in adult extraocular muscle. *Neurology* 1993;**43**:983–6.
24. Nakano M, Yamada K, Fain J, Sener EC, Selleck CJ, Awad AH, et al. Homozygous mutations in ARIX(PHOX2A) result in congenital fibrosis of the extraocular muscles type 2. *Nat Genet* 2001;**29**:315–20.
25. Hriso E, Masdeu JC, Miller A. Monocular elevation weakness and ptosis: an oculomotor fascicular syndrome? [see comments]. *J Clin Neuroophthalmol* 1991;**11**:111–13.
26. Guy J, Savino PJ, Schatz NJ, Cobbs WH, Day AL. Superior division paresis of the oculomotor nerve. *Ophthalmology* 1985;**92**:777–84.
27. Schumacher-Feero LA, Yoo KW, Solari FM, Biglan AW. Third cranial nerve palsy in children. *Am J Ophthalmol* 1999;**128**:216–21.
28. Holmes JM, Mutyala S, Maus TL, Grill R, Hodge DO, Gray DT. Pediatric third, fourth, and sixth nerve palsies: a population-based study. *Am J Ophthalmol* 1999;**127**:388–92.
29. Ng YS, Lyons CJ. Oculomotor nerve palsy in childhood. *Can J Ophthalmol* 2005;**40**:645–53.
30. Harley RD. Paralytic strabismus in children. Etiologic incidence and management of the third, fourth, and sixth nerve palsies. *Ophthalmology* 1980;**87**:24–43.
31. Ing EB, Sullivan TJ, Clarke MP, Buncic JR. Oculomotor nerve palsies in children. *J Pediatr Ophthalmol Strabismus* 1992;**29**:331–6.
32. Keith CG. Oculomotor nerve palsy in childhood. *Aust NZ J Ophthalmol* 1987;**15**:181–4.
33. Kodsi SR, Younge BR. Acquired oculomotor, trochlear, and abducent cranial nerve palsies in pediatric patients. *Am J Ophthalmol* 1992;**114**:568–74.
34. Hamed LM. Associated neurologic and ophthalmologic findings in congenital oculomotor nerve palsy. *Ophthalmology* 1991;**98**: 708–14.
35. Lindenberg R. Significance of the tentorium in head injuries from blunt forces. *Clin Neurosurg* 1964;**12**:129–42.
36. Balkan R, Hoyt CS. Associated neurologic abnormalities in congenital third nerve palsies. *Am J Ophthalmol* 1984;**97**:315–19.

37. Tsaloumas MD, Willshaw HE. Congenital oculomotor palsy: associated neurological and ophthalmological findings. *Eye (Lond)* 1997;**11**(Pt 4):500–3.
38. Flanders M, Watters G, Draper J, O'Gorman A. Bilateral congenital third cranial nerve palsy. *Can J Ophthalmol* 1989;**24**:28–30.
39. White WL, Mumma JV, Tomasovic JJ. Congenital oculomotor nerve palsy, cerebellar hypoplasia, and facial capillary hemangioma. *Am J Ophthalmol* 1992;**113**:497–500.
40. Pham RT, Levine MR, Marcotty A. Congenital synkinetic movements associated with elevation of the upper eyelid. *Ophthal Plast Reconstr Surg* 1994;**10**:30–3.
41. Yamada K, Andrews C, Chan WM, McKeown CA, Magli A, De Berardinis T, et al. Heterozygous mutations of the kinesin KIF21A in congenital fibrosis of the extraocular muscles type 1 (CFEOM1). *Nat Genet* 2003;**35**:318–21.
42. Cheng L, Desai J, Miranda C, Duncan JS, Qiu W, Nugent A, et al Human CFEOM1 mutations attenuate KIF21A autoinhibition and cause oculomotor axon stalling. *Neuron* 2014;**82**:334–49.
43. Smith JL. The "nuclear third" question. *J Clin Neuroophthalmol* 1982;**2**:61–3.
44. Pollard ZF. Diagnosis and treatment of inferior oblique palsy. *J Pediatr Ophthalmol Strabismus* 1993;**30**:15–18.
45. Roper-Hall G, Burde RM. Inferior rectus palsies as a manifestation of atypical IIIrd cranial nerve disease. *Am Orthopt J* 1975;**25**:122–30.
46. Van Dalen J, Van Mourik-Noodernbos A. Isolated inferior rectus paresis: A report of six cases. *Neuro-ophthalmology* 1984;**4**:89–94.
47. von Noorden GK, Hansell R. Clinical characteristics and treatment of isolated inferior rectus paralysis. *Ophthalmology* 1991;**98**:253–7.
48. Mizen TR, Burde RM, Klingele TG. Cryptogenic oculomotor nerve palsies in children. *Am J Ophthalmol* 1985;**100**:65–7.
49. Abdul-Rahim AS, Savino PJ, Zimmerman RA, Sergott RC, Bosley TM. Cryptogenic oculomotor nerve palsy. The need for repeated neuroimaging studies. *Arch Ophthalmol* 1989;**107**:387–90.
50. Dieterich M, Brandt T. Ocular torsion and perceived vertical in oculomotor, trochlear and abducens nerve palsies. *Brain* 1993;**116**:1095–104.
51. Bryan JS, Hamed LM. Levator-sparing nuclear oculomotor palsy. Clinical and magnetic resonance imaging findings. *J Clin Neuroophthalmol* 1992;**12**:26–30.
52. Kaminski HJ, Ruff RL. Ocular muscle involvement by myasthenia gravis. *Ann Neurol* 1997;**41**:419–20.
53. Wong V, Wong WC. Enhancement of oculomotor nerve: a diagnostic criterion for ophthalmoplegic migraine? *Pediatr Neurol* 1997;**17**:70–3.
54. Mark AS, Casselman J, Brown D, Sanchez J, Kolsky M, Larsen 3rd TC, et al. Ophthalmoplegic migraine: reversible enhancement and thickening of the cisternal segment of the oculomotor nerve on contrast-enhanced MR images. *AJNR Am J Neuroradiol* 1998;**19**:1887–91.
55. Gelfand AA, Gelfand JM, Prabakhar P, Goadsby PJ. Ophthalmoplegic "migraine" or recurrent ophthalmoplegic cranial neuropathy: new cases and a systematic review. *J Child Neurol* 2012;**27**:759–66.
56. Hupp SL, Kline LB, Corbett JJ. Visual disturbances of migraine. *Surv Ophthalmol* 1989;**33**:221–36.
57. Durkan GP, Troost BT, Slamovits TL, Spoor TC, Kennerdell JS. Recurrent painless oculomotor palsy in children. A variant of ophthalmoplegic migraine? *Headache* 1981;**21**:58–62.
58. Prats JM, Mateos B, Garaiza C. Resolution of MRI abnormalities of the oculomotor nerve in childhood ophthalmoplegic migraine. *Cephalgia* 1999;**19**(7):655–9.
59. Bharucha DX, Campbell TB, Valencia I, Hardison HH, Kothare SV. MRI findings in pediatric ophthalmoplegic migraine: a case report and literature review. *Pediatr Neurol* 2007;**37**:59–63.
60. Thompson HS. Adie's syndrome: some new observations. *Trans Am Ophthalmol Soc* 1977;**75**:587–626.
61. Jacobson DM, Vierkant RA. Comparison of cholinergic supersensitivity in third nerve palsy and Adie's syndrome. *J Neuroophthalmol* 1998;**18**:171–5.
62. Jeffery AR, Ellis FJ, Repka MX, Buncic JR. Pediatric Horner syndrome. *J AAPOS* 1998;**2**:159–67.
63. Mahoney NR, Liu GT, Menacker SJ, Wilson MC, Hogarty MD, Maris JM. Pediatric Horner syndrome: etiologies and roles of imaging and urine studies to detect neuroblastoma and other responsible mass lesions. *Am J Ophthalmol* 2006;**142**:651–9.
64. Miller N, Newman N. *Walsh and Hoyt's Clinical Neuro-ophthalmology*. 5th ed. Baltimore: Williams & Wilkins; 1998.
65. Brazis PW. Palsies of the trochlear nerve: diagnosis and localization—recent concepts. *Mayo Clin Proc* 1993;**68**:501–9.
66. Vanooteghem P, Dehaene I, Van Zandycke M, Casselman J. Combined trochlear nerve palsy and internuclear ophthalmoplegia. *Arch Neurol* 1992;**49**:108–9.
67. Burgerman RS, Wolf AL, Kelman SE, Elsner H, Mirvis S, Sestokas AK. Traumatic trochlear nerve palsy diagnosed by magnetic resonance imaging: case report and review of the literature. *Neurosurgery* 1989;**25**:978–81.
68. von Noorden GK, Murray E, Wong SY. Superior oblique paralysis. A review of 270 cases. *Arch Ophthalmol* 1986;**104**:1771–6.
69. Donahue SP, Lavin PJ, Hamed LM. Tonic ocular tilt reaction simulating a superior oblique palsy: diagnostic confusion with the 3-step test. *Arch Ophthalmol* 1999;**117**:347–52.
70. Ohtsuki H, Hasebe S, Kono R, Yamane T, Fujiwara H, Shiraga F. Large Bielschowsky head-tilt phenomenon and inconspicuous vertical deviation in the diagnostic positions in congenital superior oblique palsy. *Am J Ophthalmol* 2000;**130**:854–6.
71. Kushner BJ. Errors in the three-step test in the diagnosis of vertical strabismus [see comments]. *Ophthalmology* 1989;**96**:127–32.
72. Kushner BJ. Simulated superior oblique palsy. *Ann Ophthalmol* 1981;**13**:337–43.
73. Getman I, Goldstein JH. Diagnosis of an isolated vertical muscle palsy. *Acta Ophthalmol (Copenh)* 1983;**61**:85–93.
74. Cartwright MJ, Wyatt DB. The three-step test. *Ophthalmology* 1989;**96**:1447–8.
75. Demer JL, Clark RA, Engle EC. Magnetic resonance imaging evidence for widespread orbital dysinnervation in congenital fibrosis of extraocular muscles due to mutations in KIF21A. *Invest Opthalmol Vis Sci* 2005;**46**(2):530–9.
76. Wilson ME, Hoxie J. Facial asymmetry in superior oblique muscle palsy. *J Pediatr Ophthalmol Strabismus* 1993;**30**:315–18
77. Ozkan SB, Aribal ME, Sener EC, Sanac AS, Gurcan F. Magnetic resonance imaging in evaluation of congenital and acquired superior oblique palsy. *J Pediatr Ophthalmol Strabismus* 1997;**34**:29–34.
78. Younge BR, Sutula F. Analysis of trochlear nerve palsies. Diagnosis, etiology, and treatment. *Mayo Clin Proc* 1977;**52**:11–18.
79. Sydnor CF, Seaber JH, Buckley EG. Traumatic superior oblique palsies. *Ophthalmology* 1982;**89**:134–8.

80. Baker RS, Epstein AD. Ocular motor abnormalities from head trauma. *Surv Ophthalmol* 1991;**35**:245–67.
81. Burger LJ, Kalvin NH, Smith JL. Acquired lesions of the fourth cranial nerve. *Brain* 1970;**93**:567–74.
82. Guyton D, von Noorden G. Sensory adaptations to cyclodeviations. In: Reinecke R, editor. *Strabismus*. New York: Grune & Stratton; 1978. pp. 399–403.
83. Mansour AM, Reinecke RD. Central trochlear palsy. *Surv Ophthalmol* 1986;**30**:279–97.
84. Orwitz JI, Galetta SL, Teener JW. Bilateral trochlear nerve palsy and downbeat nystagmus in a patient with cephalic tetanus. *Neurology* 1997;**49**:894–5.
85. Archambault P, Wise JS, Rosen J, Polomeno RC, Auger N. Herpes zoster ophthalmoplegia. Report of six cases. *J Clin Neuroophthalmol* 1988;**8**:185–93.
86. Patel AS, Kelman SE, Duncan GW, Rismondo V. Painless diplopia caused by extraocular muscle sarcoid. *Arch Ophthalmol* 1994;**112**:879–80.
87. Acierno MD. Vertical diplopia. *Semin Neurol* 2000;**20**:21–30.
88. Hamed L. Cyclic, periodic ophthalmic disorders. In: Margo C, Hamed L, Mames R, editors. *Diagnostic Problems in Clinical Ophthalmology*. Philadelphia: Saunders; 1993. pp. 711–15.
89. Dickey CF, Scott WE, Cline RA. Oblique muscle palsies fixating with the paretic eye. *Surv Ophthalmol* 1988;**33**:97–107.
90. Brandt T, Dieterich M. Pathological eye-head coordination in roll: tonic ocular tilt reaction in mesencephalic and medullary lesions. *Brain* 1987;**110**(Pt 3):649–66.
91. Moster ML, Bosley TM, Slavin ML, Rubin SE. Thyroid ophthalmopathy presenting as superior oblique paresis. *J Clin Neuroophthalmol* 1992;**12**:94–7.
92. Rush JA, Shafrin F. Ocular myasthenia presenting as superior oblique weakness. *J Clin Neuroophthalmol* 1982;**2**:125–7.
93. Parkinson D, Johnston J, Chaudhuri A. Sympathetic connections to the fifth and sixth cranial nerves. *Anat Rec* 1978;**191**:221–6.
94. Gutman I, Levartovski S, Goldhammer Y, Tadmor R, Findler G. Sixth nerve palsy and unilateral Horner's syndrome. *Ophthalmology* 1986;**93**:913–16.
95. Benevento WJ, Tychsen L. Distinguishing compensatory head turn from gaze palsy in children with unilateral oculomotor or abducens nerve paresis. *Am J Ophthalmol* 1993;**115**:116–18.
96. Afifi AK, Bell WE, Menezes AH. Etiology of lateral rectus palsy in infancy and childhood. *J Child Neurol* 1992;**7**:295–9.
97. Robertson DM, Hines JD, Rucker CW. Acquired sixth-nerve paresis in children. *Arch Ophthalmol* 1970;**83**:574–9.
98. de Grauw AJ, Rotteveel JJ, Cruysberg JR. Transient sixth cranial nerve paralysis in the newborn infant. *Neuropediatrics* 1983;**14**:164–5.
99. Nixon RB, Helveston EM, Miller K, Archer SM, Ellis FD. Incidence of strabismus in neonates. *Am J Ophthalmol* 1985;**100**:798–801.
100. Archer SM, Sondhi N, Helveston EM. Strabismus in infancy. *Ophthalmology* 1989;**96**:133–7.
101. Afifi AK, Kaelber WW. Nociception: pathways and mechanisms—an update. *Middle East J Anesthesiol* 1989;**10**:107–31.
102. Chrousos GA, Dipaolo F, Kattah JC, Laws Jr. ER. Paresis of the abducens nerve after trivial head injury. *Am J Ophthalmol* 1993;**116**:387–8.
103. Heinze J. Cranial nerve avulsion and other neural injuries in road accidents. *Med J Aust* 1969;**2**:1246–9.
104. Rosa L, Carol M, Bellegarrigue R, Ducker TB. Multiple cranial nerve palsies due to a hyperextension injury to the cervical spine. Case report. *J Neurosurg* 1984;**61**:172–3.
105. Pomeroy SL, Holmes SJ, Dodge PR, Feigin RD. Seizures and other neurologic sequelae of bacterial meningitis in children. *N Engl J Med* 1990;**323**:1651–7.
106. Chu ML, Litman N, Kaufman DM, Shinnar S. Cranial nerve palsies in *Streptococcus pneumoniae* meningitis. *Pediatr Neurol* 1990;**6**:209–10.
107. Cohen HA, Nussinovitch M, Ashkenazi A, Straussberg R, Kaushansky A. Benign abducens nerve palsy of childhood. *Pediatr Neurol* 1993;**9**:394–5.
108. Bixenman WW, von Noorden GK. Benign recurrent VI nerve palsy in childhood. *J Pediatr Ophthalmol Strabismus* 1981;**18**:29–34.
109. Sternberg I, Ronen S, Arnon N. Recurrent, isolated, post-febrile abducens nerve palsy. *J Pediatr Ophthalmol Strabismus* 1980;**17**:323–4.
110. Reinecke RD, Thompson WE. Childhood recurrent idiopathic paralysis of the lateral rectus. *Ann Ophthalmol* 1981;**13**:1037–9.
111. Afifi AK, Bell WE, Bale JF, Thompson HS. Recurrent lateral rectus palsy in childhood. *Pediatr Neurol* 1990;**6**:315–18.
112. Kertes PJ, Baker JD, Noel LP. Neuro-ophthalmic complications of acute *varicella*. *Can J Ophthalmol* 1998;**33**:324–8.
113. Straussberg R, Cohen AH, Amir J, Varsano I. Benign abducens palsy associated with EBV infection. *J Pediatr Ophthalmol Strabismus* 1993;**30**:147–50.
114. Killer H, Matzkin D, Sternman D. Intracavernous carotid aneurysm as a rare cause of isolated sixth nerve palsy in an eight-year-old child. *Neuro-ophthalmology* 1993;**13**:147.
115. Raymond GL, Crompton JL. Spasm of the near reflex associated with cerebrovascular accident. *Aust NZ J Ophthalmol* 1990;**18**:407–10.
116. Berkovitz S, Belkin M, Tenenbaum A. Childhood myasthenia gravis. *J Pediatr Ophthalmol* 1977;**14**:269–73.
117. Kennerdell JS, Dresner SC. The nonspecific orbital inflammatory syndromes. *Surv Ophthalmol* 1984;**29**:93–103.
118. Hutcheson KA, Lambert SR. Cyclic esotropia after a traumatic sixth nerve palsy in a child. *J AAPOS* 1998;**2**:376–7.
119. Anderson RL, Baumgartner SA. Strabismus in ptosis. *Arch Ophthalmol* 1980;**98**:1062–7.
120. Anderson RL, Baumgartner SA. Amblyopia in ptosis. *Arch Ophthalmol* 1980;**98**:1068–9.
121. Kerr NC, Hoehn MB. Botulinum toxin for sixth nerve palsies in children with brain tumors. *J AAPOS* 2001;**5**:21–5.
122. Metz HS. 20th annual Frank Costenbader Lecture—muscle transposition surgery. *J Pediatr Ophthalmol Strabismus* 1993;**30**:346–53.
123. Maruo T, Iwashige H, Kubota N, Sakaue T, Ishida T, Honda M, et al. Results of surgery for paralytic exotropia due to oculomotor palsy. *Ophthalmologica* 1996;**210**:163–7.
124. Buckley EG, Townshend LM. A simple transposition procedure for complicated strabismus. *Am J Ophthalmol* 1991;**111**:302–6.
125. Mudgil AV, Repka MX. Ophthalmologic outcome after third cranial nerve palsy or paresis in childhood. *J AAPOS* 1999;**3**:2–8.

126. Schumacher-Feero LA, Yoo KW, Solari FM, Biglan AW. Results following treatment of third cranial nerve palsy in children. *Trans Am Ophthalmol Soc* 1998;**96**:455–72; discussion 72–4.
127. Robb RM. Idiopathic superior oblique palsies in children. *J Pediatr Ophthalmol Strabismus* 1990;**27**:66–9.
128. Lee JP. Superior oblique myokymia. A possible etiologic factor. *Arch Ophthalmol* 1984;**102**:1178–9.
129. Simons BD, Saunders TG, Siatkowski RM, Feuer WJ, Lavina AM, Capo H, et al. Outcome of surgical management of superior oblique palsy: a study of 123 cases. *Binocul Vis Strabismus Q* 1998;**13**:273–82.
130. King AJ, Stacey E, Stephenson G, Trimble RB. Spontaneous recovery rates for unilateral sixth nerve palsies. *Eye (Lond)* 1995;**9**(Pt 4):476–8.
131. Holmes JM, Droste PJ, Beck RW. The natural history of acute traumatic sixth nerve palsy or paresis. *J AAPOS* 1998;**2**:265–8.
132. Aroichane M, Repka MX. Outcome of sixth nerve palsy or paresis in young children. *J Pediatr Ophthalmol Strabismus* 1995;**32**:152–6.
133. Selezinka W, Sandall GS, Henderson JW. Rectus muscle union in sixth nerve paralysis. Jensen rectus muscle union. *Arch Ophthalmol* 1974;**92**:382–6.
134. Miller N, Newman N. *The Essentials: Walsh & Hoyt's Clinical Neuro-ophthalmology, Companion to 5th Edition*. Baltimore: Lippincott Williams & Wilkins; 1999.
135. Engle EC, Castro AE, Macy ME, Knoll JHM, Beggs AH. A gene for isolated congenital ptosis maps to a 3 cM region within 1p32-p34.1. *Am J Hum Genet* 1997;**60**:1150–7.
136. Johnson CC. Congenital ptosis. In: Albert DM, Jakobiec FA, editors. *Principles and Practice of Ophthalmology: Clinical Practice*. Philadelphia: W.B. Saunders Co; 1994. pp. 2839–46.
137. Lemagne JM, Colonval S, Moens B, Brucher JM. [Anatomical modification of the levator muscle of the eyelid in congenital ptosis]. *Bull Soc Belge Ophthalmol* 1992;**243**:23–7.
138. Engle EC, Marondel I, Houtman WA, et al. Congenital fibrosis of the extraocular muscles [autosomal dominant congenital external ophthalmoplegia]: Genetic homogeneity, linkage refinement, and physical mapping on chromosome 12. *Am J Hum Genet* 1995;**57**:1086–94.
139. McMullan TF, Collins AR, Tyers AG, Robinson DO. A novel X-linked dominant condition: X-linked congenital isolated ptosis. *Am J Hum Genet* 2000;**66**:1455–60.
140. McMullan TW, Crolla JA, Gregory SG, Carter NP, Cooper RA, Howell GR, et al. A candidate gene for congenital bilateral isolated ptosis identified by molecular analysis of a de novo balanced translocation. *Hum Genet* 2002;**110**:244–50.
141. Gilbert PW. The origin and development of the human extrinsic ocular muscles. *Contrib Embryol* 1957;**36**:59–78.
142. Sevel D. A reappraisal of the origin of human extraocular muscles. *Ophthalmology* 1981;**88**:1330–8.
143. Sevel D. Development of the nerves of the extraocular muscles. In: Reinecke RD, editor. *Strabismus II*. New York: Grune and Stratton; 1984. pp. 645–57.
144. Amati P, Chomel JC, Nivelon-Chevalier A, Gilgenkrantz S, Kitzis A, Kaplan J, et al. A gene for blepharophimosis-ptosis-epicanthus inversus syndrome maps to chromosome 3q23. *Hum Genet* 1995;**96**:213–15.
145. Harrar HS, Jeffery S, Patton MA. Linkage analysis in blepharophimosis-ptosis syndrome confirms localization to 3q21-24. *J Med Genet* 1995;**32**:774–7.
146. Small KW, Stalvey M, Fisher L, Mullen L, Dickel C, Beadles K, et al. Blepharophimosis syndrome is linked to chromosome 3q. *Hum Mol Genet* 1995;**4**:443–8.
147. Crisponi L, Deiana M, Loi A, Chiappe F, Uda M, Amati P, et al. The putative forkhead transcription factor FOXL2 is mutated in blepharophimosis/ptosis/epicanthus inversus syndrome. *Nat Genet* 2001;**27**:159–66.
148. Brown HW. Congenital structural muscle anomalies. In: Allen JH, editor. *Strabismus Ophthalmic Symposium*. St. Louis: C.V. Mosby Co; 1950. pp. 205–36.
149. Gutowski NJ, Bosley TM, Engle EC. 110th ENMC International Workshop: The congenital cranial dysinnervation disorders (CCDDs). Naarden, The Netherlands, 25–27 October, 2002. *Neuromuscul Disord* 2003;**13**:573–8.
150. Oystreck DT, Engle EC, Bosley TM. Recent progress in understanding congenital cranial dysinnervation disorders. *J Neuroophthalmol* 2011;**31**:69–77.
151. Bosley TM, Abu-Amero KK, Oystreck DT. Congenital cranial dysinnervation disorders: a concept in evolution. *Curr Opin Ophthalmol* 2013;**24**:398–406.
152. Graeber CP, Hunter DG, Engle EC. The genetic basis of incomitant strabismus: consolidation of the current knowledge of the genetic foundations of disease. *Semin Ophthalmol* 2013;**28**:427–37.
153. Danis P. Sur les anomalies congenitale de la motilite oculaire d'origine musculaire et en particular sur le sundrome de Stilling-Turk-Duane. *Ann d'Ocul* 1948;**1811**:148.
154. Kirkham TH. Inheritance of Duane's syndrome. *Br J Ophthalmol* 1970;**54**:323–9.
155. Duane A. Congenital deficiency of abduction, associated with impairment of adduction, retraction movements, contraction of the palpebral fissure and oblique movements of the eye. *Arch Ophthalmol* 1905;**34**:133–59.
156. Pfaffenbach DD, Cross HE, Kearns TP. Congenital anomalies in Duane's retraction syndrome. *Arch Ophthalmol* 1972;**88**:635–9.
157. Shauly Y, Weissman A, Meyer E. Ocular and systemic characteristics of Duane syndrome. *J Pediatr Ophthalmol Strabismus* 1993;**30**:178–83.
158. Huber A. Electrophysiology of the retraction syndrome. *Br J Ophthalmol* 1974;**58**:293–300.
159. Marshman WE, Schalit G, Jones RB, Lee JP, Matthews TD, McCabe S. Congenital anomalies in patients with Duane retraction syndrome and their relatives. *J AAPOS* 2000;**4**:106–9.
160. Apple C. Congenital abducens paralysis. *Am J Ophthalmol* 1939;**22**:169–73.
161. Gifford H. Congenital defects of abduction and other ocular movements and their relation to birth injuries. *Am J Ophthalmol* 1926;**9**:3.
162. Hotchkiss MG, Miller NR, Clark AW, Green WG. Bilateral Duane's retraction syndrome: a clinical-pathological case report. *Arch Ophthalmol* 1980;**98**:870–4.
163. Miller NR, Kiel SM, Green WR, Clark AW. Unilateral Duane's retraction syndrome (Type 1). *Arch Ophthalmol* 1982;**100**: 1468–72.

164. Gunderson T, Zeavin B. Observations on the retraction syndrome of Duane. *Arch Ophthalmol* 1956;**55**:576–80.
165. Huber A. Duane's retraction syndrome; consideration on pathophysiology and etiology. In: Reinecke R, editor. *Strabismus II*. Orlando: Grune & Stratton; 1984. pp. 345–61.
166. Miller M. Thalidomide embryopathy: a model for the study of congenital incomitant horizontal strabismus. *Trans Am Ophthalmol Soc* 1991;**89**:623–74.
167. Miller MT, Stromland KK. What can we learn from the thalidomide experience: an ophthalmologic perspective. *Curr Opin Ophthalmol* 2011;**22**:356–64.
168. Miyake N, Chilton J, Psatha M, Cheng L, Andrews C, Chan WM, et al. Human CHN1 mutations hyperactivate alpha2-chimaerin and cause Duane's retraction syndrome. *Science* 2008;**321**: 839–43.
169. Chan WM, Miyake N, Zhu-Tam L, Andrews C, Engle EC. Two novel CHN1 mutations in two families with Duane's retraction syndrome. *Arch Ophthalmol* 2011;**129**:649–52.
170. Demer JL, Clark RA, Lim KH, Engle EC. Magnetic resonance imaging evidence for widespread orbital dysinnervation in dominant Duane's retraction syndrome linked to the DURS2 locus. *Invest Ophthalmol Vis Sci* 2007;**48**:194–202.
171. Chung M, Stout JT, Borchert MS. Clinical diversity of hereditary Duane's retraction syndrome. *Ophthalmology* 2000;**107**:500–3.
172. Miyake N, Demer JL, Shaaban S, Andrews C, Chan WM, Christiansen SP, et al. Expansion of the CHN1 strabismus phenotype. *Invest Ophthalmol Vis Sci* 2011;**52**:6321–8.
173. Clark C, Austen O, Poparic I, Guthrie S. Alpha2-Chimaerin regulates a key axon guidance transition during development of the oculomotor projection. *J Neurosci* 2013;**33**:16540–51.
174. Miyake N, Andrews C, Fan W, He W, Chan WM, Engle EC. CHN1 mutations are not a common cause of sporadic Duane's retraction syndrome. *Am J Med Genet A* 2010;**152A**:215–17.
175. Al-Baradie R, Yamada K, St Hilaire C, Chan WM, Andrews C, McIntosh N, et al. Duane radial ray syndrome (Okihiro syndrome) maps to 20q13 and results from mutations in SALL4, a new member of the SAL family. *Am J Hum Genet* 2002;**71**:1195–9.
176. Kohlhase J, Heinrich M, Schubert L, Liebers M, Kispert A, Laccone F, et al. Okihiro syndrome is caused by SALL4 mutations. *Hum Mol Genet* 2002;**11**:2979–87.
177. Kohlhase J, Schubert L, Liebers M, Rauch A, Becker K. Mohammed SN, et al. Mutations at the SALL4 locus on chromosome 20 result in a range of clinically overlapping phenotypes, including Okihiro syndrome, Holt-Oram syndrome, acro-renal-ocular syndrome, and patients previously reported to represent thalidomide embryopathy. *J Med Genet* 2003;**40**:473–8.
178. Kohlhase J, Chitayat D, Kotzot D, Ceylaner S, Froster UG, Fuchs S, et al. SALL4 mutations in Okihiro syndrome (Duane-radial ray syndrome), acro-renal-ocular syndrome, and related disorders. *Hum Mutat* 2005;**26**:176–83.
179. Paradisi I, Arias S. IVIC syndrome is caused by a c.2607delA mutation in the SALL4 locus. *Am J Med Genet A* 2007;**143**:326–32.
180. Borozdin W, Graham Jr. JM, Bohm D, Bamshad MJ, Spranger S, Burke L, et al. Multigene deletions on chromosome 20q13.13–q13.2 including SALL4 result in an expanded phenotype of Okihiro syndrome plus developmental delay. *Hum Mutat* 2007;**28**:830.
181. Demer JL, Clark RA, Lim KH, Engle EC. Magnetic resonance imaging of innervational and extraocular muscle abnormalities in Duane-radial ray syndrome. *Invest Ophthalmol Vis Sci* 2007;**48**:5505–11.
182. Sakaki-Yumoto M, Kobayashi C, Sato A, Fujimura S, Matsumoto Y, Takasato M, et al. The murine homolog of SALL4, a causative gene in Okihiro syndrome, is essential for embryonic stem cell proliferation, and cooperates with Sall1 in anorectal, heart, brain and kidney development. *Development* 2006;**133**:3005–13.
183. Tischfield MA, Bosley TM, Salih MA, Alorainy IA, Sener EC, Nester MJ, et al. Homozygous HOXA1 mutations disrupt human brainstem, inner ear, cardiovascular and cognitive development. *Nat Genet* 2005;**37**:1035–7.
184. Holve S, Friedman B, Hoyme HE, Tarby TJ, Johnstone SJ, Erickson RP, et al. Athabascan brainstem dysgenesis syndrome. *Am J Med Genet* 2003;**120A**:169–73.
185. Bosley TM, Salih MA, Alorainy IA, Oystreck DT, Nester M, Abu-Amero KK, et al. Clinical characterization of the HOXA1 syndrome BSAS variant. *Neurology* 2007;**69**:1245–53.
186. Bosley TM, Alorainy IA, Salih MA, Aldhalaan HM, Abu-Amero KK, Oystreck DT, et al. The clinical spectrum of homozygous HOXA1 mutations. *Am J Med Genet A* 2008;**146**:1235–40.
187. Tischfield MA, Chan WM, Grunert JF, Andrews C, Engle EC. HOXA1 mutations are not a common cause of Duane anomaly. *Am J Med Genet A* 2006;**140**:900–2.
188. Baris HN, Chan WM, Andrews C, Behar DM, Donovan DJ, Morton CC, et al. Complex cytogenetic rearrangements at the DURS1 locus in syndromic Duane retraction syndrome. *Clin Case Rep* 2014;**1**:13.
189. Bosley TM, Salih MA, Jen JC, Lin DD, Oystreck D, Abu-Amero KK, et al. Neurologic features of horizontal gaze palsy and progressive scoliosis with mutations in ROBO3. *Neurology* 2005;**64**:1196–203.
190. Jen JC, Chan WM, Bosley TM, Wan J, Carr JR, Rub U, et al. Mutations in a human ROBO gene disrupt hindbrain axon pathway crossing and morphogenesis. *Science* 2004;**304**:1509–13.
191. Chan WM, Traboulsi EI, Arthur B, Friedman N, Andrews C, Engle EC. Horizontal gaze palsy with progressive scoliosis can result from compound heterozygous mutations in ROBO3. *J Med Genet* 2006;**43**:e11.
192. Sabatier C, Plump AS, Le M, Brose K, Tamada A, Murakami F, et al. The divergent Robo family protein rig-1/Robo3 is a negative regulator of slit responsiveness required for midline crossing by commissural axons. *Cell* 2004;**117**:157–69.
193. Sicotte NL, Salamon G, Shattuck DW, Hageman N, Rub U, Salamon N, et al. Diffusion tensor MRI shows abnormal brainstem crossing fibers associated with ROBO3 mutations. *Neurology* 2006;**67**:519–21.
194. Tischfield MA, Baris HN, Wu C, Rudolph G, Van Maldergem L, He W, et al. Human TUBB3 mutations perturb microtubule dynamics, kinesin interactions, and axon guidance. *Cell* 2010;**140**:74–87.
195. Cederquist GY, Luchniak A, Tischfield MA, Peeva M, Song Y, Menezes MP, et al. An inherited TUBB2B mutation alters a kinesin-binding site and causes polymicrogyria, CFEOM and axon dysinnervation. *Hum Mol Genet* 2012;**21**:5484–99.
196. Yamada K, Hunter DG, Andrews C, Engle EC. A novel KIF21A mutation in a patient with congenital fibrosis of the extraocular muscles and Marcus Gunn jaw-winking phenomenon. *Arch Ophthalmol* 2005;**123**:1254–9.

197. Demer JL, Clark R, Tischfield MA, Engle EC. Evidence of an asymmetrical endophenotype in congenital fibrosis of extraocular muscles type 3 resulting from TUBB3 mutations. *Invest Ophthalmol Vis Sci* 2010;**51**:4600–11.
198. Shen X, Meza-Carmen V, Puxeddu E, Wang G, Moss J, Vaughan M. Interaction of brefeldin A-inhibited guanine nucleotide-exchange protein (BIG) 1 and kinesin motor protein KIF21A. *Proc Natl Acad Sci USA* 2008;**105**:18788–93.
199. Kakinuma N, Kiyama R. A major mutation of KIF21A associated with congenital fibrosis of the extraocular muscles type 1 (CFEOM1) enhances translocation of Kank1 to the membrane. *Biochem Biophys Res Commun* 2009;**386**:639–44.
200. Lee KH, Lee JS, Lee D, Seog DH, Lytton J, Ho WK, et al. KIF21A-mediated axonal transport and selective endocytosis underlie the polarized targeting of NCKX2. *J Neurosci* 2012;**32**:4102–17.
201. van der Vaart B, van Riel WE, Doodhi H, Kevenaar JT, Katrukha EA, Gumy L, et al. CFEOM1-associated kinesin KIF21A is a cortical microtubule growth inhibitor. *Dev Cell* 2013;**27**:145–60.
202. Desai J, Velo MP, Yamada K, Overman LM, Engle EC. Spatiotemporal expression pattern of KIF21A during normal embryonic development and in congenital fibrosis of the extraocular muscles type 1 (CFEOM1). *Gene Expr Patterns* 2012;**12**:180–8.
203. Chan WM, Andrews C, Dragan L, Fredrick D, Armstrong L, Lyons C, et al. Three novel mutations in KIF21A highlight the importance of the third coiled-coil stalk domain in the etiology of CFEOM1. *BMC Genet* 2007;**8**:26.
204. Doherty E, Macy M, Wang S, et al. CFEOM3: A new extraocular congenital fibrosis syndrome that maps to 16q24.2–q24.3. *Invest Ophthalmol Vis Sci* 1999;**40**(8):1687–94.
205. Chew S, Balasubramanian R, Chan WM, Kang PB, Andrews C, Webb BD, et al. A novel syndrome caused by the E410K amino acid substitution in the neuronal beta-tubulin isotype 3. *Brain* 2013;**136**:522–35.
206. Wang SM, Zwaan J, Mullaney PB, Jabak MH, Al-Awad A, Beggs AH, et al. Congenital fibrosis of the extraocular muscles type 2 [CFEOM2], an inherited exotropic strabismus fixus, maps to distal 11q13. *Am J Hum Genet* 1998;**63**:517–25.
207. Yazdani A, Chung DC, Abbaszadegan MR, Al-Khayer K, Chan WM, Yazdani M, et al. A novel PHOX2A/ARIX mutation in an Iranian family with congenital fibrosis of extraocular muscles type 2 (CFEOM2). *Am J Ophthalmol* 2003;**136**:861–5.
208. Pattyn A, Morin X, Cremer H, Goridis C, Brunet JF. Expression and interactions of the two closely related homeobox genes Phox2a and Phox2b during neurogenesis. *Development* 1997;**124**:4065–75.
209. Guo S, Brush J, Teraoka H, Goddard A, Wilson SW, Mullins MC, et al. Development of noradrenergic neurons in the zebrafish hindbrain requires BMP, FGF8, and the homeodomain protein soulless/Phox2a. *Neuron* 1999;**24**:555–66.
210. Tukel T, Uzumcu A, Gezer A, Kayserili H, Yuksel-Apak M, Uyguner O, et al. A new syndrome, congenital extraocular muscle fibrosis with ulnar hand anomalies, maps to chromosome 21qter. *J Med Genet* 2005;**42**:408–15.
211. Aubourg P, Krahn M, Bernard R, Nguyen K, Forzano O, Boccaccio I, et al. Assignment of a new congenital fibrosis of extraocular muscles type 3 (CFEOM3) locus, FEOM4, based on a balanced translocation t(2;13) (q37.3;q12.11) and identification of candidate genes. *J Med Genet* 2005;**42**:253–9.
212. Berkman MD, Feingold M. Oculoauriculovertebral dysplasia (Goldenhar's syndrome). *Oral Surg Oral Med Oral Pathol* 1968;**25**:408–17.
213. Astle WF, Rosenbaum AL. Familial congenital fourth cranial nerve palsy. *Arch Ophthalmol* 1985;**103**:532–5.
214. Harris Jr. DJ, Memmen JE, Katz NN, Parks MM. Familial congenital superior oblique palsy. *Ophthalmology* 1986;**93**:88–90.
215. Botelho PJ, Giangiacomo JG. Autosomal-dominant inheritance of congenital superior oblique palsy. *Ophthalmology* 1996;**103**:1508–11.
216. Kim JH, Hwang JM. Absence of the trochlear nerve in patients with superior oblique hypoplasia. *Ophthalmology* 2010;**117** (2208–13):e1–2.
217. Yang HK, Kim JH, Hwang JM. Congenital superior oblique palsy and trochlear nerve absence: a clinical and radiological study. *Ophthalmology* 2012;**119**:170–7.
218. Jiang Y, Matsuo T, Fujiwara H, Hasebe S, Ohtsuki H, Yasuda T. ARIX gene polymorphisms in patients with congenital superior oblique muscle palsy. *Br J Ophthalmol* 2004;**88** 263–7.
219. Jiang Y, Matsuo T, Fujiwara H, Hasebe S, Ohtsuki H, Yasuda T. ARIX and PHOX2B polymorphisms in patients with congenital superior oblique muscle palsy. *Acta Med Okayama* 2005;**59**:55–62.
220. Parks MM. Doyne memorial lecture, 1977. The superior oblique tendon. *Trans Ophthalmol Soc UK* 1977;**97**:288–304.
221. Wilson ME, Eustis Jr HS, Parks MM. Brown's syndrome. *Surv Ophthalmol* 1989;**34**:153–72.
222. von Noorden G. *Binocular Vision and Ocular Motility: Theory and Management of Strabismus*. 5th ed. St. Louis: Mosby, Inc.; 1996.
223. Kaeser PF, Brodsky MC. Fourth cranial nerve palsy and Brown syndrome: two interrelated congenital cranial dysinnervation disorders? *Curr Neurol Neurosci Rep* 2013;**13**:352.
224. Ellis FJ, Jeffery AR, Seidman DJ, Sprague JB, Coussens T, Schuller J. Possible association of congenital Brown syndrome with congenital cranial dysinnervation disorders. *J AAPOS* 2012;**16**:558–64.
225. Möbius P. Uber infantilen Kernschwund. *Munch Med Wochenschr* 1892;**39**:17–58.
226. Hanson PA, Rowland LP. Mobius syndrome and facioscapulohumeral muscular dystrophy. *Arch Neurol* 1971;**24**:31–9.
227. Harper PS, Rudel R. Myotonic dystrophy. In: Engel AG, Franzini-Armstrong C, editors. *Myology*. New York: McGraw-Hill, Inc; 1995. pp. 1192–219.
228. Baraitser M. Genetics of Mobius syndrome. *J Med Genet* 1977;**14**:415–17.
229. Carta A, Mora P, Neri A, Favilla S, Sadun AA. Ophthalmologic and systemic features in mobius syndrome an italian case series. *Ophthalmology* 2011;**118**:1518–23.
230. Stromland K, Sjogreen L, Miller M, Gillberg C, Wentz E, Johansson M, et al. Mobius sequence—a Swedish multidiscipline study. *Eur J Paediatr Neurol* 2002;**6**:35–45.
231. Verzijl HT, van der Zwaag B, Cruysberg JR, Padberg GW. Mobius syndrome redefined: a syndrome of rhombencephalic maldevelopment. *Neurology* 2003;**61**:327–33.
232. Rogers GL, Hatch Jr. GF, Gray I. Mobius syndrome and limb abnormalities. *J Pediatr Ophthalmol* 1977;**14**:134–8.

233. Miller MT, Stromland K, Gillberg C, Johansson M, Nilsson EW. The puzzle of autism: an ophthalmologic contribution. *Trans Am Ophthalmol Soc* 1998;**96**:369–85; discussion 85–7.
234. Gadoth N, Biedner B, Torok G. Mobius syndrome and Poland anomaly: case report and review of the literature. *J Pediatr Ophthalmol Strabismus* 1979;**16**:374–6.
235. Briegel W. Neuropsychiatric findings of Mobius sequence—a review. *Clin Genet* 2006;**70**:91–7.
236. Miller MT, Stromland K. The Mobius sequence: a relook. *J AAPOS* 1999;**3**:199–208.
237. Amaya LG, Walker J, Taylor D. Mobius syndrome: a study and report of 18 cases. *Binocular Vis Q* 1990;**5**:119–32.
238. Cronemberger MF, de Castro MJB, Brunoni D, Mendonca TS, Alvarenga EH, Rizzo AM, et al. Ocular and clinical manifestations of Mobius' syndrome. *J Pediatr Ophthalmol Strabismus* 2001;**38**:156–62.
239. de Souza-Dias CR, Goldchmit M. Further considerations about the ophthalmic features of the Mobius sequence, with data of 28 cases. *Arq Bras Oftalmol* 2007;**70**:451–7.
240. Pitner SE, Edwards JE, McCormick WF. Observations on the pathology of Moebius syndrome. *J Neurol Neurosurg Psychiatry* 1965;**28**:362–74.
241. Towfighi J, Marks K, Palmer E, Vannucci R. Moebius syndrome. Neuropathologic observations. *Acta Neuropathol (Berl)* 1979;**48**:11–17.
242. Verzijl HT, van der Zwaag B, Lammens M, ten Donkelaar HJ, Padberg GW. The neuropathology of hereditary congenital facial palsy vs Mobius syndrome. *Neurology* 2005;**64**:649–53.
243. Sudarshan A, Goldie WD. The spectrum of congenital facial diplegia (Moebius syndrome). *Pediatr Neurol* 1985;**1**:180–4.
244. Meyer E, Ludatscher RM, Zonis S. Mobius syndrome. Electron microscopic study of the extraocular muscles. *Ophthalmic Paediatr Genet* 1984;**4**:91–5.
245. Traboulsi E, Maumenee I. Extraocular muscle aplasia in Moebius syndrome. *J Pediatr Ophthalmol Strabismus* 1986;**23**:120–2.
246. Saint-Martin C, Clapuyt P, Duprez T, Ghariani S, Verellen G. Mobius sequence and severe pons hypoplasia: a case report. *Pediatr Radiol* 1998;**28**:932.
247. Sano M, Kaga K, Takeuchi N, Kitazumi E, Kodama K. A case of Mobius syndrome—radiological and electrophysiological findings. *Int J Pediatr Otorhinolaryngol* 2005;**69**:1583–6.
248. Matsunaga Y, Amamoto N, Kondoh T, Ohtsuka Y, Miyazoe H, Kamimura N, et al. A severe case of Moebius syndrome with calcification on the fourth ventricular floor. *J Hum Genet* 1998;**43**:62–4.
249. Heidary G, Traboulsi EI, Engle EC. The genetics of strabismus and associated disorders. In: Traboulsi EI, editor. *Genetic Diseases of the Eye*. Second ed. Oxford: Oxford University Press; 2012. pp. 657–88.
250. MacKinnon S, Oystreck DT, Andrews C, Chan WM, Hunter DG, Engle EC. Diagnostic distinctions and genetic analysis of patients diagnosed with moebius syndrome. *Ophthalmology* 2014;**121**:1461–8.
251. Ziter FA, Wiser WC, Robinson A. Three-generation pedigree of a Mobius syndrome variant with chromosome translocation. *Arch Neurol* 1977;**34**:437–42.
252. Slee JJ, Smart RD, Viljoen DL. Deletion of chromosome 13 in Moebius syndrome. *J Med Genet* 1991;**28**:413–14.
253. Webb BD, Shaaban S, Gaspar H, Cunha LF, Schubert CR, Hao K, et al. HOXB1 founder mutation in humans recapitulates the phenotype of Hoxb1(-/-) mice. *Am J Hum Genet* 2012;**91**:171–9.
254. Kremer H, Kuyt LP, van den Helm B, van Reen M, Leunissen JA, Hamel BC, et al. Localization of a gene for Mobius syndrome to chromosome 3q by linkage analysis in a Dutch family. *Hum Mol Genet* 1996;**5**:1367–71.
255. Verzijl HT, van den Helm B, Veldman B, Hamel BC, Kuyt LP, Padberg GW, et al. A second gene for autosomal dominant Mobius syndrome is localized to chromosome 10q, in a Dutch family. *Am J Hum Genet* 1999;**65**:752–6.
256. Gonzalez CH, Vargas FR, Perez AB, Kim CA, Brunoni D, Marques-Dias MJ, et al. Limb deficiency with or without Mobius sequence in seven Brazilian children associated with misoprostol use in first trimester of pregnancy. *Am J Med Genet* 1993;**47**:59–64.
257. Gonzalez CH, Marques-Dias MJ, Kim CA, Sugayama SM, Da Paz JA, Huson SM, et al. Congenital abnormalities in Brazilian children associated with misoprostol misuse in first trimester of pregnancy. *Lancet* 1998;**351**:1624–7.
258. Goldberg AB, Greenberg MB, Darney PD. Misoprostol and pregnancy. *N Engl J Med* 2001;**344**:38–47.
259. Bouwes-Bavinck J, Weaver D. Subclavian artery supply disruption sequence: hypothesis of a vascular etiology for Poland, Klippel-Feil, and Mobius anomalies. *Am J Med Genet* 1986;**23**:903–18.
260. Freedman H, Kushner B. Congenital ocular aberrant innervation—new concepts. *J Pediatr Ophthalmol Strabismus* 1997;**34**:10–16.
261. Sibony P, Evinger C, Lessel S. Retrograde horseradish peroxidase transport after oculomotor nerve injury. *Invest Ophthalmol Vis Sci* 1986;**27**:975–80.
262. Lee SH, Yeow YK, Tan CB, Tjia H. Transient oculomotor nerve synkinesis in non-Hodgkin's lymphoma. *Clin Neuroophthalmol* 1992;**12**:203–6.
263. Lepore FE, Glaser JS. Misdirection revisited. A critical appraisal of acquired oculomotor nerve synkinesis. *Arch Ophthalmol* 1980;**98**:2206–9.
264. Guy J, Engel H, Lessner A. Acquired contralateral oculomotor synkinesis. *Arch Neurol* 1989;**46**:1021–3.
265. Jordan DR, Miller DG, Anderson RL. Acquired oculomotor-abducens synkinesis. *Can J Ophthalmol* 1990;**25**:149–51.
266. Manners RM, Rosser P, Collin JR. Levator transposition procedure: a review of 35 cases. *Eye* 1996;**10**:539–44.
267. Hayakawa Y, Takagi M, Hasebe H, Hasegawa S, Takada R, Usui T, et al. A case of superior oblique myokymia observed by an image-analysis system. *J Neuroophthalmol* 2000;**20**:163–5.
268. Leigh RJ, Tomsak RL, Seidman SH, Dell'Osso LF. Superior oblique myokymia. Quantitative characteristics of the eye movements in three patients. *Arch Ophthalmol* 1991;**109**:1710–13.
269. Rosenberg ML, Glaser JS. Superior oblique myokymia. *Ann Neurol* 1983;**13**:667–9.
270. Hoyt WF, Keane JR. Superior oblique myokymia. Report and discussion on five cases of benign intermittent uniocular microtremor. *Arch Ophthalmol* 1970;**84**:461–7.
271. Yiu EM, Ryan MM. Demyelinating prenatal and infantile developmental neuropathies. *J Peripher Nerv Syst* 2012;**17**:32–52.

272. Yiu EM, Ryan MM. Genetic axonal neuropathies and neuronopathies of pre-natal and infantile onset. *J Peripher Nerv Syst* 2012;**17**:285–300.
273. Larbrisseau A, Jasmin G, Hausser C, Brochu P, Geoffroy G. Generalized giant axonal neuropathy—a case with features of Fazio-Londe disease. *Neuropadiatrie* 1979;**10**:76–86.
274. Dupre N, Howard HC, Mathieu J, Karpati G, Vanasse M, Bouchard JP, et al. Hereditary motor and sensory neuropathy with agenesis of the corpus callosum. *Ann Neurol* 2003;**54**:9–18.
275. Howard HC, Mount DB, Rochefort D, Byun N, Dupre N, Lu J, et al. The K-Cl cotransporter KCC3 is mutant in a severe peripheral neuropathy associated with agenesis of the corpus callosum. *Nat Genet* 2002;**32**:384–92.
276. Aicardi J, Castelein P. Infantile neuroaxonal dystrophy. *Brain* 1979;**102**:727–48.
277. Kurian MA, Morgan NV, MacPherson L, Foster K, Peake D, Gupta R, et al. Phenotypic spectrum of neurodegeneration associated with mutations in the PLA2G6 gene (PLAN). *Neurology* 2008;**70**:1623–9.
278. Aberfeld DC, Namba T. Progressive ophthalmoplegia in Kugelberg-Welander disease. Report of a case. *Arch Neurol* 1969;**20**:253–6.
279. Pachter BR, Pearson J, Davidowitz J, Reuben R, Boal D, Carr R, et al. Congenital total external ophthalmoplegia associated with infantile spinal muscular atrophy. Fine structure of extraocular muscle. *Invest Ophthalmol* 1976;**15**:320–4.
280. Dubrovsky A, Taratuto AL, Martino R. Distal spinal muscular atrophy and ophthalmoparesis. A case with selective type 2 fiber hypotrophy. *Arch Neurol* 1981;**38**:594–6.
281. Hamano K, Tsukamoto H, Yazawa T, Yoshimura M, Takita H. Infantile progressive spinal muscular atrophy with ophthalmoplegia and pyramidal symptoms. *Pediatr Neurol* 1994;**10**:320–4.
282. Oka A, Matsushita Y, Sakakihara Y, Momose T, Yanaginasawa M. Spinal muscular atrophy with oculomotor palsy, epilepsy, and cerebellar hypoperfusion. *Pediatr Neurol* 1995;**12**:365–9.
283. Gordon PH, Hays AP, Rowland LP, Dickoff DJ, Schotland DL, Rosenberg RN, et al. Erroneous diagnosis corrected after 28 years. Not spinal muscular atrophy with ophthalmoplegia but minicore myopathy. *Arch Neurol* 1996;**53**:1194–6.
284. Korinthenberg R, Sauer M, Ketelsen UP, Hanemann CO, Stoll G, Graf M, et al. Congenital axonal neuropathy caused by deletions in the spinal muscular atrophy region. *Ann Neurol* 1997;**42**:364–8.
285. Bosch AM, Stroek K, Abeling NG, Waterham HR, Ijlst L, Wanders RJ. The Brown-Vialetto-Van Laere and Fazio Londe syndrome revisited: natural history, genetics, treatment and future perspectives. *Orphanet J Rare Dis* 2012;**7**:83.
286. Foley AR, Menezes MP, Pandraud A, Gonzalez MA, Al-Odaib A, Abrams AJ, et al. Treatable childhood neuronopathy caused by mutations in riboflavin transporter RFVT2. *Brain* 2014;**137**:44–56.
287. Bradshaw DY, Jones Jr. HR. Guillain-Barré syndrome in children: clinical course, electrodiagnosis, and prognosis. *Muscle Nerve* 1992;**15**:500–6.
288. Ropper AH. Miller Fisher syndrome and other acute variants of Guillain-Barré syndrome. *Baillieres Clin Neurol* 1994;**3**:95–106.
289. Wong V. A neurophysiological study in children with Miller Fisher syndrome and Guillain-Barré syndrome. *Brain Dev* 1997;**19**:197–204.
290. Lee SH, Lim GH, Kim JS, Oh SY, Kim JK, Cha JK, et al. Acute ophthalmoplegia (without ataxia) associated with anti-GQ1b antibody. *Neurology* 2008;**71**:426–9.
291. Albers J. Porphyric neuropathy. In: Mendell KJ, Cornblath DR, Kissel JT, editors. *Diagnosis and Management of Peripheral Nerve Disorders*. New York: Oxford University; 2001. pp. 344–66.
292. Mendell J. Lipoprotein disorders: Tangier disease and abetalipoproteinemia/familial hypobetalipoproteinemia. In: Mendell KJ, Cornblath DR, Kissel JT, editors. *Diagnosis and Management of Peripheral Nerve Disorders*. New York: Oxford University; 2001. pp. 518–31.
293. Hopkins IJ, Shield LK, Harris M. Subacute cholinergic dysautonomia in childhood. *Clin Exp Neurol* 1981;**17**:147–51.
294. McDonald W, Kocen R. Diphtheritic neuropathy. In: Dyck P, Thomas P, Griffin J, editors. *Peripheral Neuropathy*. Philadelphia: Saunders; 1993. pp. 1412–17.
295. Nevo Y, Pestronk A, Kornberg AJ, Connolly AM, Yee WC, Iqbal I, et al. Childhood chronic inflammatory demyelinating neuropathies: clinical course and long-term follow-up. *Neurology* 1996;**47**:98–102.
296. Ryan MM, Grattan-Smith PJ, Procopis PG, Morgan G, Ouvrier RA. Childhood chronic inflammatory demyelinating polyneuropathy: clinical course and long-term outcome. *Neuromuscul Disord* 2000;**10**:398–406.
297. Kohlschutter A. Vitamin E and neurological problems in childhood: a curable neurodegenerative process. *Dev Med Child Neurol* 1993;**35**:642–6.
298. Tomasi LG. Reversibility of human myopathy caused by vitamin E deficiency. *Neurology* 1979;**29**:1182–6.
299. Cohen DA, Bosley TM, Savino PJ, Sergott RC, Schatz NJ. Primary aberrant regeneration of the oculomotor nerve. Occurrence in a patient with abetalipoproteinemia. *Arch Neurol* 1985;**42**:821–3.
300. Le Luyer B, Devaux AM, Dailly R, Ensel P. [Polyradiculoneuritis as a manifestation of childhood sarcoidosis]. *Arch Fr Pediatr* 1983;**40**:175–8.
301. Engel AG. Current status of the congenital myasthenic syndromes. *Neuromuscul Disord* 2012;**22**:99–111.
302. Finlayson S, Beeson D, Palace J. Congenital myasthenic syndromes: an update. *Pract Neurol* 2013;**13**:80–91.
303. Serra A, Ruff RL, Leigh RJ. Neuromuscular transmission failure in myasthenia gravis: decrement of safety factor and susceptibility of extraocular muscles. *Ann NY Acad Sci* 2012;**1275**:129–35.
304. Liu R, Xu H, Wang G, Li J, Gou L, Zhang L, et al. Extraocular muscle characteristics related to myasthenia gravis susceptibility. *PLOS ONE* 2013;**8**:e55611.
305. Kaminski HJ, Ruff RL. Insights into possible skeletal muscle nicotinic acetylcholine receptor (AChR) changes in some congenital myasthenias from physiological studies, point mutations, and subunit substitutions of the AChR. *Ann NY Acad Sci* 1993;**681**:435–50.
306. Kaminski HJ, Kusner LL, Nash KV, Ruff RL. The gamma-subunit of the acetylcholine receptor is not expressed in the levator palpebrae superioris. *Neurology* 1995;**45**:516–18.
307. Lefvert AK, Osterman PO. Newborn infants to myasthenic mothers: a clinical study and an investigation of acetylcholine receptor antibodies in 17 children. *Neurology* 1983;**33**:133–8.

308. Morel E, Eymard B, Vernet-der Garabedian B, Pannier C, Dulac O, Bach JF. Neonatal myasthenia gravis: a new clinical and immunologic appraisal on 30 cases. *Neurology* 1988;**38**:138–42.
309. Niks EH, Verrips A, Semmekrot BA, Prick MJ, Vincent A, van Tol MJ, et al. A transient neonatal myasthenic syndrome with anti-musk antibodies. *Neurology* 2008;**70**:1215–16.
310. Snead 3rd OC, Benton JW, Dwyer D, Morley BJ, Kemp GE, Bradley RJ, et al. Juvenile myasthenia gravis. *Neurology* 1980;**30**:732–9.
311. Grob D, Arsura EL, Brunner NG, Namba T. The course of myasthenia gravis and therapies affecting outcome. *Ann NY Acad Sci* 1987;**505**:472–99.
312. Hajjar M, Markowitz J, Darras BT, Kissel JT, Srinivasan J, Jones HR. Lambert-Eaton syndrome, an unrecognized treatable pediatric neuromuscular disorder: three patients and literature review. *Pediatr Neurol* 2014;**50**:11–17.
313. Underwood K, Rubin S, Deakers T, Newth C. Infant botulism: a 30-year experience spanning the introduction of botulism immune globulin intravenous in the intensive care unit at Childrens Hospital Los Angeles. *Pediatrics* 2007;**120**:e1380–5.
314. Brook I. Botulism: the challenge of diagnosis and treatment. *Rev Neurol Dis* 2006;**3**:182–9.
315. Levy-Khademi F, Tenenbaum AN, Wexler ID, Amitai Y. Unintentional organophosphate intoxication in children. *Pediatr Emerg Care* 2007;**23**:716–18.
316. Romero NB, Bitoun M. Centronuclear myopathies. *Semin Pediatr Neurol* 2011;**18**:250–6.
317. Jungbluth H, Sewry CA, Muntoni F. Core myopathies. *Semin Pediatr Neurol* 2011;**18**:239–49.
318. D'Arcy CE, Bjorksten A, Yiu EM, Bankier A, Gillies R, McLean CA, et al. King-denborough syndrome caused by a novel mutation in the ryanodine receptor gene. *Neurology* 2008;**71**:776–7.
319. Klein A, Lillis S, Munteanu I, Scoto M, Zhou H, Quinlivan R, et al. Clinical and genetic findings in a large cohort of patients with ryanodine receptor 1 gene-associated myopathies. *Hum Mutat* 2012;**33**:981–8.
320. Tein I, Haslam RH, Rhead WJ, Bennett MJ, Becker LE, Vockley J. Short-chain acyl-CoA dehydrogenase deficiency: a cause of ophthalmoplegia and multicore myopathy. *Neurology* 1999;**52**:366–72.
321. Ryan MM, Schnell C, Strickland CD, Shield LK, Morgan G, Iannaccone ST, et al. Nemaline myopathy: a clinical study of 143 cases. *Ann Neurol* 2001;**50**:312–20.
322. Wright RA, Plant GT, Landon DN, Morgan-Hughes JA. Nemaline myopathy: an unusual cause of ophthalmoparesis. *J Neuroophthalmol* 1997;**17**:39–43.
323. Clarke NF. Congenital fiber-type disproportion. *Semin Pediatr Neurol* 2011;**18**:264–71.
324. Cancilla PA, Kalyanaraman K, Verity MA, Munsat T, Pearson CM. Familial myopathy with probable lysis of myofibrils in type I fibers. *Neurology* 1971;**21**:579–85.
325. Bender AN, Bender MB. Muscle fiber hypotrophy with intact neuromuscular junctions. A study of a patient with congenital neuromuscular disease and ophthalmoplegia. *Neurology* 1977;**27**:206–12.
326. Sugie H, Hanson R, Rasmussen G, Verity MA. Congenital neuromuscular disease with type I fibre hypotrophy, ophthalmoplegia and myofibril degeneration. *J Neurol Neurosurg Psychiatry* 1982;**45**:507–12.
327. Hanson PA, Mastrianni AF, Post L. Neonatal ophthalmoplegia with microfibers: a reversible myopathy? *Neurology* 1977;**27**:974–80.
328. Oh SJ, Danon MJ. Nonprogressive congenital neuromuscular disease with uniform type 1 fiber. *Arch Neurol* 1983;**40**:147–50.
329. Lo WD, Barohn RJ, Bobulski RJ, Kean J, Mendell JR. Centronuclear myopathy and type-1 hypotrophy without central nuclei. Distinct nosologic entities? *Arch Neurol* 1990;**47**:273–6.
330. Ohtaki E, Yamaguchi Y, Yamashita Y, Matsuishi T, Terasawa K, Katafuchi Y, et al. Complete external ophthalmoplegia in a patient with congenital myopathy without specific features (minimal change myopathy). *Brain Dev* 1990;**12**:427–30.
331. Porter JD. Commentary: extraocular muscle sparing in muscular dystrophy: a critical evaluation of potential protective mechanisms. *Neuromuscul Disord* 1998;**8**:198–203.
332. Philpot J, Muntoni F. Limitation of eye movement in merosin-deficient congenital muscular dystrophy. *Lancet* 1999;**353**:297–8.
333. Martinelli P, Gabellini AS, Ciucci G, Govoni E, Vitali S, Gulli MR. Congenital muscular dystrophy with central nervous system involvement: case report. *Eur Neurol* 1987;**26**:17–22.
334. Voit T, Parano E, Straub V, Schroder JM, Schaper J, Pavone P, et al. Congenital muscular dystrophy with adducted thumbs, ptosis, external ophthalmoplegia, mental retardation and cerebellar hypoplasia: a novel form of CMD. *Neuromuscul Disord* 2002;**12**:623–30.
335. Mercuri E, Muntoni F. The ever-expanding spectrum of congenital muscular dystrophies. *Ann Neurol* 2012;**72**:9–17.
336. Vondracek P, Hermanova M, Vodickova K, Fajkusova L, Blakely EL, He L, et al. An unusual case of congenital muscular dystrophy with normal serum CK level, external ophtalmoplegia, and white matter changes on brain MRI. *Eur J Paediatr Neurol* 2007;**11**:381–4.
337. Kletter G, Evans OB, Lee JA, Melvin B, Yates AB, Bock HG. Congenital muscular dystrophy and epidermolysis bullosa simplex. *J Pediatr* 1989;**114**:104–7.
338. Yiu EM, Klausegger A, Waddell LB, Grasern N, Lloyd L, Tran K, et al. Epidermolysis bullosa with late-onset muscular dystrophy and plectin deficiency. *Muscle Nerve* 2011;**44**:135–41.
339. Banwell BL, Russel J, Fukudome T, Shen XM, Stilling G, Engel AG. Myopathy myasthenic syndrome, and epidermolysis bullosa simplex due to plectin deficiency. *J Neuropathol Exp Neurol* 1999;**58**:832–46.
340. Aring E, Ekstrom AB, Tulinius M, Sjostrom A. Ocular motor function in relation to gross motor function in congenital and childhood myotonic dystrophy type 1. *Acta Ophthalmol* 2012;**90**:369–74.
341. Ter Bruggen JP, Bastiaensen LA, Tyssen CC, Gielen G. Disorders of eye movement in myotonic dystrophy. *Brain* 1990;**113** (Pt 2):463–73.
342. Brais B. Oculopharyngeal muscular dystrophy. *Handb Clin Neurol* 2011;**101**:181–92.
343. Blumen SC, Brais B, Korczyn AD, Medinsky S, Chapman J, Asherov A, et al. Homozygotes for oculopharyngeal muscular dystrophy have a severe form of the disease. *Ann Neurol* 1999;**46**:115–18.

344. Hill ME, Creed GA, McMullan TF, Tyers AG, Hilton-Jones D, Robinson DO, et al. Oculopharyngeal muscular dystrophy: phenotypic and genotypic studies in a UK population. *Brain* 2001;**124**:522–6.
345. Fukuhara N, Kumamoto T, Tsubaki T, Mayuzumi T, Nitta H. Oculopharyngeal muscular dystrophy and distal myopathy. Intrafamilial difference in the onset and distribution of muscular involvement. *Acta Neurol Scand* 1982;**65**:458–67.
346. Lacomis D, Kupsky WJ, Kuban KK, Specht LA. Childhood onset oculopharyngeal muscular dystrophy. *Pediatr Neurol* 1991;**7**:382–4.
347. Amato AA, Jackson CE, Ridings LW, Barohn RJ. Childhood-onset oculopharyngodistal myopathy with chronic intestinal pseudo-obstruction. *Muscle Nerve* 1995;**18**:842–7.
348. Rose MR, Landon DN, Papadimitriou A, Morgan-Hughes JA. A rapidly progressive adolescent-onset oculopharyngeal somatic syndrome with rimmed vacuoles in two siblings. *Ann Neurol* 1997;**41**:25–31.
349. Uyama E, Uchino M, Chateau D, Tome FMS. Autosomal recessive oculopharyngodistal myopathy in light of distal myopathy with rimmed vacuoles and oculopharyngeal muscular dystrophy. *Neuromuscul Disord* 1998;**8**:119–25.
350. van der Sluijs BM, Ter Laak HJ, Scheffer H, van der Maarel SM, van Engelen BG. Autosomal recessive oculopharyngodistal myopathy: a distinct phenotypical, histological, and genetic entity. *J Neurol Neurosurg Psychiatry* 2004;**75**:1499–501.
351. Durmus H, Laval SH, Deymeer F, Parman Y, Kiyan E, Gokyigiti M, et al. Oculopharyngodistal myopathy is a distinct entity: clinical and genetic features of 47 patients. *Neurology* 2011;**76**:227–35.
352. Argov Z, Yarom R. "Rimmed vacuole myopathy" sparing the quadriceps. A unique disorder in Iranian Jews. *J Neurol Sci* 1984;**64**:33–43.
353. Chutkow JG, Heffner Jr. RR, Kramer AA, Edwards JA. Adult-onset autosomal dominant limb-girdle muscular dystrophy. *Ann Neurol* 1986;**20**:240–8.
354. Martinsson T, Darin N, Kyllerman M, Oldfors A, Hallberg B, Wahlstrom J. Dominant hereditary inclusion-body myopathy gene (IBM3) maps to chromosome region 17p13.1. *Am J Hum Genet* 1999;**64**:1420–6.
355. Uretsky SH, Kennerdell JS, Gutai JP. Graves' ophthalmopathy in childhood and adolescence. *Arch Ophthalmol* 1980;**98**:1963–4.
356. Krassas GE, Gogakos A. Thyroid-associated ophthalmopathy in juvenile Graves' disease—clinical, endocrine and therapeutic aspects. *J Pediatr Endocrinol Metab* 2006;**19**:1193–206.
357. Ryan MM, Taylor P, Donald JA, Ouvrier RA, Morgan G, Danta G, et al. A novel syndrome of episodic muscle weakness maps to xp22.3. *Am J Hum Genet* 1999;**65**:1104–13.
358. Takeda S, Ohama E, Ikuta F. Involvement of extraocular muscle in mitochondrial encephalomyopathy. *Acta Neuropathol* 1990;**80**:118–22.
359. Gronlund MA, Honarvar AK, Andersson S, Moslemi AR, Oldfors A, Holme E, et al. Ophthalmological findings in children and young adults with genetically verified mitochondrial disease. *Br J Ophthalmol* 2010;**94**:121–7.
360. Bohlega S, Tanji K, Santorelli FM, Hirano M, al-Jishi A, DiMauro S. Multiple mitochondrial DNA deletions associated with autosomal recessive ophthalmoplegia and severe cardiomyopathy. *Neurology* 1996;**46**:1329–34.
361. Teitelbaum JE, Berde CB, Nurko S, Buonomo C, Perez-Atayde AR, Fox VL. Diagnosis and management of MNGIE syndrome in children: case report and review of the literature. *J Pediatr Gastroenterol Nutr* 2002;**35**:377–83.
362. Bharucha-Goebel DX, Santi M, Medne L, Zukosky K, Dastgir J, Shieh PB, et al. Severe congenital RYR1-associated myopathy: the expanding clinicopathologic and genetic spectrum. *Neurology* 2013;**80**:1584–9.
363. Friedman BD, Heidenreich RA. Distal arthrogryposis type IIB: further clinical delineation and 54-year follow-up of an index case. *Am J Med Genet* 1995;**58**:125–7.
364. Shaaban S, Duzcan F, Yildirim C, Chan WM, Andrews C, Akarsu NA, et al. Expanding the phenotypic spectrum of ECEL1-related congenital contracture syndromes. *Clin Genet* 2014;**85**:562–7.
365. Krieger M, Roos A, Stendel C, Claeys KG, Sonmez FM, Baudis M, et al. SIL1 mutations and clinical spectrum in patients with Marinesco-Sjogren syndrome. *Brain* 2013;**136**:3634–44.
366. Paulson HL. The spinocerebellar ataxias. *J Neuroophthalmol* 2009;**29**:227–37.
367. Parker JL, Santiago M. Oculomotor aspects of the hereditary cerebellar ataxias. *Handb Clin Neurol* 2012;**103**:63–83.

Chapter 47

Oromotor Dysfunction in Neuromuscular Disorders: Evaluation and Treatment

Manuel Roig-Quilis

INTRODUCTION

Sucking, chewing, and swallowing are vital functions that are dependent upon the coordinated interaction of a set of neural structures designated by several authors as the "oromotor system" (OMS).[1,2] Also dependent on this system are other not-so-vital but important functions such as language motor articulation and gestural communication by means of facial expression. Direct control of the OMS is located in the brainstem. Sensory inputs and motor actions are mediated through neuronal groups, the cranial nerve nuclei, located along the brainstem axis. These nuclei are, in turn, under the control of cortical and subcortical structures. This neuronal network allows the integration of the sensory information received and the voluntary or involuntary motor responses to these stimulations. The interaction of all these circuits allows the generation of complex reflexes and of voluntary actions that contribute to functions, such as sucking, chewing, swallowing, phonation, and facial nonverbal communication. The OMS can be disrupted in many ways and at different levels, giving rise to a multitude of disorders, most of them congenital, with clinical manifestations presenting and persisting throughout childhood.[3,4]

Many aspects of OMS dysfunction have been reviewed in the literature under different headings. The most common terms used in monographs or pediatric textbooks have been: congenital flaccid bulbar paresis,[5] congenital suprabulbar and pseudobulbar paresis,[6] feeding or swallowing disorders in children,[7] and dysphagia in childhood.[8,9] Surprisingly, despite the fact that in most instances these disorders are caused by nervous system diseases, no monographs or reviews have approached their study from a neurological perspective and much less from a neuromuscular disorder (NMD) viewpoint. Indeed, pediatricians and child neurologists are often faced with neonates or infants with facial diplegia, feeding difficulties, dysphagia, muscle hypotonia, respiratory distress, and joint contractures that sometimes are associated with other craniofacial malformations. The causes for this constellation of clinical signs and symptoms are manifold. Congenital tumors, vascular malformations, severe perinatal trauma or hypoxic-ischemic events and restricted prenatal disruptive lesions, or malformations of the cerebral hemispheres or the brainstem, as well as congenital NMDs, are known to produce the aforementioned clinical picture during the neonatal period or infancy.[10] Over the last 25 years, we have developed a concept and classification for disorders of the OMS (Table 47.1) that takes into consideration both the cause and the anatomical location of the underlying disease.[4,11] As can be deduced from Table 47.1, occasionally both feeding and/or speech problems are reversible because the defect causing these functional difficulties can be corrected or because the clinical manifestations subside after a more or less prolonged period of time, as is the case for patients with acute illnesses, central nervous system (CNS) maturation delay, or mild congenital neuromuscular disorders.[12,13] In this chapter, we will consider only those disorders giving rise to persistent or irreversible clinical manifestations.

Oromotor dysfunction (OMD) is among the possible consequences of perinatal hypoxic-ischemic injuries. These encephalopathies initially give rise to a suprabulbar syndrome with generalized hypotonia, lethargy, and facial diplegia with salivary incompetence that, subsequently, will result in feeding difficulties, respiratory complications, and clear signs of CNS involvement such as poor axial tone, spastic hypertonia or dystonia of the limbs, and dysarthria.[8,14] OMD arising from restricted bilateral hemispheric involvement produces a suprabulbar paresis syndrome, first reported by Worster-Drought and characterized by orofacial paresis, dysarthria, difficulties of

B.T. Darras, H. Royden Jones, Jr., M.M. Ryan & D.C. De Vivo (Eds): Neuromuscular Disorders of Infancy, Childhood and Adolescence, Second edition.
DOI: http://dx.doi.org/10.1016/B978-0-12-417044-5.00047-0

TABLE 47.1 Oromotor Disorders in Childhood: Classification and Causes

Reversible

Classification	Cause
Acute	Traumatic
	Infectious
	Inflammatory
Delayed development	

Irreversible

Classification	Subclassification / Anatomic Location	Cause / Syndrome
Craniofacial anatomic defects	Isolated or Nonsyndromic	Choanal atresia
		Oro-facial clefting
		Micro-retrognathia
		Velopalatine insufficiency
	Dysmorphological syndromes with craniofacial defects	Velocardiofacial syndrome (del 22q11)
		CHARGE association
		Branchial arch syndromes: Goldenhar Treacher-Collins Nager Miller
		Other syndromes: Opitz G Opitz-BBB Pallister-Hall Coffin-Siris Möbius Pierre Robin Carey-Zineman-Fiter Franceschetti Costello
Neurological disorders	Cerebral hemisphere involvement	Cortical Due to gene mutations Due to disruptive intrauterine lesions
		Subcortical white matter
		Mixed cortical-subcortical
	Brainstem involvement	Dysgenetic Due to gene mutations Due to disruptive intrauterine lesions
	Motor unit disorders	Congenital myotonic dystrophy
		Infantile spinal muscular atrophy
		Congenital myopathies
		Congenital myasthenic syndromes
		Congenital muscular dystrophies
		Other
	Other	Perinatal pathology
		Prolonged ventilation
		Tracheotomy
		Abnormal sensory input

voluntary tongue motor control, mild pyramidal signs, and, characteristically, a dissociation of cortical and subcortical motor control. Genetic factors, prenatal infections, and vascular lesions are considered the most likely causes of this disorder.[6,15,16] Newborns with congenital dysfunction of multiple cranial nerves are also a challenging diagnostic problem. Throughout the years, we have been able to identify a group of patients with a recognizable symptom complex characterized by a variable combination of congenital cranial nerve dysfunction, decreased muscle tone, and, in many cases, mild signs of pyramidal tract involvement due to prenatal disruptive brainstem lesions or to malformations of the brainstem, for whom we have proposed the term "brainstem dysgenesis" (BSD).[10,17,18] Recently, other authors have postulated that the orofacial malformations seen in Pierre Robin, Moebius, and CHARGE syndromes are also the product of embryonic defects in brainstem development.[19,20] In our experience, a clear-cut division can be drawn between the natural history of OMD arising from CNS diseases and that of OMD arising from NMDs. In the first case, when initial feeding and swallowing problems subside, learning difficulties, speech articulatory problems, and seizures (frequently) become apparent. In the natural course of most NMDs, as the initial muscle hypotonia, respiratory problems, and feeding difficulties improve, food sticking due to inadequate chewing, poor control of saliva, pulmonary aspiration of food or fluids, poor oral hygiene and dental decay, speech intelligibility, lack of facial expression, joint contractures, and skeletal deformities become the major clinical problems.[4,6,16]

Regardless of the disease causing the OMD, it is important to recognize that its impact on the three domains—body structures and functions, activity, and participation—established in the International Classification of Functioning, Disability, and Health[21] changes as children grow and are exposed to different environmental situations. In newborns and infants with NMDs, the greatest impact on health and well-being are respiratory complications and undernutrition due to poor sucking and swallowing. As a result, interventions must focus on preventing food aspiration and ensuring appropriate nourishment by providing alternative methods of ingestion. At the other end of childhood, the school setting may have its greatest impact on well-being and participation because teachers and peers may not be aware that the physical impairment and/or communication problems of children with OMD may prevent them from participating in most activities. Interventions may seek to alter the environment by training the peers and teachers to recognize and understand the child's communication, and adapt the physical surroundings and pattern of interaction, so the child can participate in all regular classroom activities. Because change in health issues occurs as children with OMD get older, it is crucial that a team of health professionals be involved in the care of these youngsters and their families. In practice, individual members of the health team become central at certain points in the patient's disease and then move into the background as other problems are encountered and addressed. Because health professionals revolve through the children's lives, parents may remain the only constant members of the team with recollection of the outcomes of all the interventions or treatments. As clinicians, we should bear in mind that each new health issue faced by a child demands readjustment. Information should be shared with parents to enable them to give their opinion and to make the decisions that will affect their child's immediate and long-term health and well-being.[4,20,22]

OROMOTOR DYSFUNCTION DUE TO NEUROMUSCULAR DISORDERS

The central symptom of NMDs is muscle weakness, which often has a progressive course. The degree and location of muscle group weakness can be unevenly distributed, depending on the underlying NMD, and, in a number of NMDs, both skeletal and smooth muscles are involved. Weakness in some NMDs and in most congenital NMDs involves facial, oral, and pharyngeal musculature and is responsible for the multiple OMS functional problems (Box 47.1). Other factors such as respiratory insufficiency, gastroesophageal reflux, esophageal dyskinesia, laryngomalacia or/and tracheomalacia, failure to thrive, and congenital heart disease are often associated with NMDs and, in addition to OMD, can have an impact on feeding, swallowing, and speech production, and must also be taken into consideration.[23,24]

Facial muscle weakness impairs eye closure, lip function, and facial expression. Characteristically, the lower lip is flaccid and rotated outwards, and the upper lip is retracted. An expressionless face affects the ability to express emotions as well as the use of facial expressions for nonverbal communication and social interaction. Weak masseter muscles often lead to an open mouth posture; the tongue has a low and forward position in the oral cavity and its range of movement and fine motor skills are often impaired. Chewing ability can be affected by weak masseter muscles, jaw contractures, and also by malocclusion. Altered pressures on the teeth and dental arches in children with weak and hypotonic perioral muscles have an impact on facial growth and increase the risk for developing malocclusion.[25] Decreased jaw mobility and restricted oral opening is sometimes present from birth, but in some NMDs, the progression of the disease will eventually lead to contractures in the

BOX 47.1 Functional Oromotor Difficulties Due to Neuromuscular Disorders

Sucking, Chewing, Swallowing

- Weak sucking leading to inefficient oral feeding
- Poor control of saliva by the tongue and lips resulting in oral and pharyngeal pooling of secretions and/or drooling
- Choking or food sticking in the throat as a result of inadequate chewing
- Difficulties biting and chewing; modification of dietary textures required to avoid hard or chewy consistencies that result in lengthy, unenjoyable mealtimes
- Poor oral hygiene; dental decay and build-up of plaque
- Weak swallowing; reduced laryngeal elevation and excursion and pharyngeal muscle weakness leading to a weak, delayed, piecemeal swallowing and post-swallow pharyngeal residue
- Insufficient oral intake leading to nutritional inadequacy resulting in faltering growth

Communication and Speech

- Weak or absent articulatory contacts resulting in compensatory articulatory patterns affecting speech clarity and intelligibility
- Hypernasality affecting resonance and clarity
- Quiet, breathy, and/or high-pitched vocal quality
- Reduced range of facial expression
- Disturbed nonverbal communication

Modified from Aloysius A, 2011,[24] with the permission of Viguera Editores.

temporomandibular joints. Buccinator muscle weakness contributes to food pooling into the cheeks during bolus preparation. These muscles, together with the tongue, help position the food bolus between the molars for efficient chewing. Weak oral and pharyngeal muscles interfere with all phases of swallowing. During the preparatory oral phase, lip closure is necessary to prevent food leakage and promote efficient swallowing. The lips are also active when taking food from the fork or emptying the spoon and when drinking from a cup, glass, or straw. Weakness of the muscles that elevate and retract the soft palate to accomplish velopharyngeal closure allows food to enter the nasal cavity and indirectly impairs swallowing efficiency. During the oral phase of swallowing, the tongue, the perioral, and the velopharyngeal muscles must be strong and coordinated to achieve effective swallowing and emptying of the mouth. The pharyngeal and esophageal phases are also dependent on strong, coordinated muscles for safe and effective swallowing. Weakness of the muscles of the floor of the mouth, including the mylohyoid, geniohyoid, and anterior belly of the digastric muscle, limits the upward and forward movement of the hyoid bone and larynx, thereby restricting the adequacy of the opening of the upper esophageal sphincter.[23,26]

A fine temporal coordination between respiration, chewing, swallow initiation, and airway closure is required to prevent aspiration. During swallowing, the larynx is closed to prevent aspiration as the bolus passes the airway. Between swallows, during chewing, and during bolus collection in the pharynx, the airway remains open and breathing continues. Newborns are one exception; in them, the oral phase is inseparable from the sucking reflex. On contact with the nipple, rhythmic contractions of the tongue occur and the lips are pressed together. The frequency of sucking varies from 0.6 to 2 per second. The pharyngeal phase occurs in a reflex manner after each suck so that the organization 1 suck/1 swallow and its coordination with respiration occur in an entirely automatic sequence. Deglutition apnea occurs simultaneously with the elevation of the soft palate, even before the closing of the glottis. Swallowing occurs during the natural pause between the end of inspiration and the onset of expiration. This apnea lasts 450 to 600 milliseconds. Thus, the newborn can feed itself without interrupting respiration, the rate of which is governed by the sucking rhythm and returns to its previous baseline rate when the meal is finished.[27,28]

Optimal phonation and speech production also require respiratory strength, breath support, and control to manage the airflow stream. Respiratory health and function will therefore relate to swallowing and speech performance. This interrelationship can be affected by weak and poorly coordinated muscle movements. Weak and hypotonic oral and velopharyngeal muscles and respiratory insufficiency are responsible for the dysarthria observed in individuals with NMDs. Common symptoms are a low voice, blurred speech, hypernasality, compensatory articulation of bilabial consonants due to impaired lip function, and weak pronunciation of consonants.[29,30]

Saliva is produced within the mouth and is swallowed automatically every few minutes throughout the day. Drooling can vary from minute to minute depending on factors such as hunger, thirst, fatigue, anxiety, excitement, and the circadian rhythm of saliva production over time. Drooling is not caused by hypersalivation but rather by OMD. Patients with NMDs and OMD are at risk of drooling and lung aspiration of oral secretions because its control is dependent upon effective lip closure and swallowing, as well as upon an effective sensory system able to react to the discrete inputs produced by the oral secretion volume.[31–33]

NEUROMUSCULAR DISORDERS WITH ASSOCIATED OROMOTOR DYSFUNCTION

Few studies have examined the relationships between OMD and NMDs. The lack of medical research on this subject probably reflects the fact that in the last 40 years, scientific interest has been centered on elucidating the definitive cause and pathological mechanisms of NMDs rather than on establishing the intimate relationship between specific groups of peripheral nervous system diseases and their regularly associated clinical dysfunctions in order to improve NMD management. OMD can be present in some acquired NMDs or be identified at some point of the natural course of many inherited, progressive NMDs. OMD is a serious clinical problem in the advanced phases of this last group of illnesses, when weakness hampers respiratory function and involves muscles of the OMS; also, the lack of coordination due to altered sensory input of inappropriate motor responses interferes with the correct functioning of the OMS. Yet, in the early phases of NMDs, OMD is also a particularly common and important clinical problem that interferes with fetal development and shows clinical symptoms at birth or during the first few months of life. The degree and number of clinical manifestations closely relate to the phenotypic severity of the underlying NMD. In NMDs with onset during fetal development, it is difficult to distinguish genuine OMD from other associated factors such as multiple joint contractures (including temporomandibular ankylosis), micrognathia, retrognathia, macroglossia or microglossia, and weakness of respiratory muscles that, unquestionably, have a major impact on proper oromotor function.[3,10,16]

The most common NMDs associated with OMD are depicted in Box 47.2. Two main groups of NMDs according to the age at onset of clinical manifestations have

BOX 47.2 Neuromuscular Disorders with Associated Oromotor Dysfunction

Newborn/Infancy

Myotonic disorders

- Myotonic dystrophy type 1 (Steinert disease)
- Schwartz-Jampel Syndrome
- Spinal muscular atrophies (SMAs)
  - SMA type I and type II (SMN1 gene mutations)
  - SMA with arthrogryposis (XL-SMA; SMAX2)
  - Congenital SMA (5q) type IA with arthrogryposis
  - Distal infantile spinal muscular atrophy with diaphragm paralysis (SMARD1)
  - Lethal congenital contracture syndrome
  - SMA and pontocerebellar hypoplasia type I

Congenital myopathies

- Nemaline rod myopathies
- Core myopathies
- Minicore myopathies
- Centronuclear myopathies
- Myofibrillar myopathies
- Early-onset myopathy, areflexia, respiratory distress & dysphagia (EMARDD)

Congenital muscular dystrophies

- O-Glycosylation disorders (genes involved)
  - *FKRP*
  - *DPM*
  - *FKTN*
  - *POMT 1, POMT 2*
  - *POMGNT1*
  - *GTDC2*
  - *LARGE*
  - *ISPD*
  - Merosin deficiency: *LAMA2*
- Extracellular matrix protein disorders
  - Ullrich (AR): COL6A1, COL6A2, COL6A3
  - Bethlem (AD): COL6A1, COL6A2, COL6A3
  - Integrin: ITGA7
  - Other: DNM2, TCAP, CHKB
- Facioscapulohumeral muscular dystrophy (4q35)
- Rigid Spine syndromes
  - SEPN1
  - FHL1
  - LMNA

Disorders of neuromuscular transmission

- Neonatal transient myasthenia gravis
- Congenital myasthenia syndromes
- Presynaptic type
  - CMS with episodic apnea
- Synaptic type
  - Endplate acetylcholinesterase deficiency
- Post-synaptic type
  - Deficiency of the acetylcholine receptor
  - Fast-channel syndrome
  - Rapsyn deficiency
  - CMS with centronuclear myopathy

Metabolic myopathies

- Glycogen storage diseases (Pompe disease)
- Mitochondrial encephalomyopathies

Childhood

- SMA
- Myotonic dystrophy
- Limb-girdle muscular dystrophies
- Facioscapulohumeral muscular dystrophy
- Metabolic myopathies
- Inflammatory myopathies
- Inflammatory polyneuropathies
- Myasthenia gravis

been considered. Although we are aware that a large number of NMDs, at some point in their natural course, give rise to OMD and therefore constitute a major management problem in the field of NMDs, in our experience the neonatal form of congenital myotonic dystrophy, spinal muscular atrophy (SMA) Type IA or type I, congenital myopathies, congenital muscular dystrophies, and congenital myasthenic syndromes are, in order of frequency, the most commonly found NMDs associated with OMD throughout childhood. All the aforementioned disorders have been extensively described in other chapters of this book, and therefore only a brief note pointing out the relationships between each group of NMDs and their most commonly associated OMD will be presented here. As suggested in the introduction of this chapter, all types of congenital NMDs should be differentiated from arthrogryposis, BSD, Worster-Drought or Prader Willi syndromes (see Chapters 6 and 7). All of the aforementioned conditions can present during the neonatal period with profound generalized hypotonia, respiratory problems, multiple cranial nerve involvement, and feeding difficulties requiring enteral tube nutrition.[10,16]

Slowly progressive muscle weakness and myotonia are cardinal symptoms of myotonic dystrophy type I. Severe hypotonia, respiratory insufficiency, and sucking difficulties are present in newborns who have the congenital form of myotonic dystrophy and particularly in those born prematurely. In this form of myotonic dystrophy, muscle tone and muscle strength improve during the first 2 years of life but, later on in the course of the disease, myotonia in the hands, jaws, and tongue appears and muscle weakness slowly returns. A moderate OMD is present in nearly all children with infantile myotonic dystrophy due to delayed oromotor development and weak hypotonic muscles. The face and jaw muscles are generally the most affected, causing impaired facial expression, an open mouth, and difficulties with chewing. Craniofacial development of many individuals with myotonic dystrophy type I is characteristic, with a more vertical cranial growth, narrower maxillary arches, and deeper palatal depth than healthy controls. Malocclusion is common, especially an anterior open bite and a cross bite. Impaired lip articulation and hypernasality due to velopalatine insufficiency contribute to dysarthria, although associated intellectual disabilities also influence speech and language development. Other commonly affected organs or systems in myotonic dystrophy type I are heart, smooth muscle, gastrointestinal tract, endocrine system, and CNS; their clinical involvement can aggravate the OMD problems.[34,35,36]

Infants with SMA type I (see Chapter 8) often have a weak cry and cough, paradoxical respirations due to the relative sparing of the diaphragm, and difficulties with breathing, feeding, and swallowing as well as tongue fasciculations, but this last sign is not known to cause any OMD. Speech is rarely affected in SMA types II–IV, but patients are at risk of developing a decrease in jaw-opening ability as the disease progresses. Increasing difficulties with chewing and swallowing in children with SMA type II have also been reported.[37,38] Unusual phenotypes of congenital motor neuron disorders include SMA with arthrogryposis (XL-SMA; SMAX2), congenital SMA (5q) with arthrogryposis (SMA type IA), distal infantile spinal muscular atrophy with diaphragmatic paralysis (SMARD1), early-onset myopathy with areflexia, respiratory distress, and dysphagia (EMARDD), congenital lethal contractures syndromes, and SMA with pontocerebellar hypoplasia type I; in these conditions, the onset of the disease occurs prenatally and the infants are born with multiple joint contractures (arthrogryposis) as well as with respiratory problems that contribute to their OMD and eventually lead to the patient's death. In congenital lethal contractures syndrome and acute forms of SMA type I, brainstem motor neurons are involved; in the latter condition, dysphagia may be present from birth but facial muscles are spared. Extraocular muscles are characteristically involved in congenital SMA with arthrogryposis. Again, all these rare forms of congenital motor neuron disorders are very difficult to differentiate clinically from BSD and arthrogryposis.[10] Progressive bulbar paralysis of childhood or Fazio-Londe disease is a rare inherited disorder. The age of onset ranges from 12 months to 12 years. The most frequent presenting symptoms are stridor, followed by ptosis, dysarthria, facial palsy, and dysphagia. Progressive involvement of cranial nerves and respiratory difficulties are responsible for the patient's death within a few years. Brown-Vialetto-van Laere syndrome 2 (BVVLS2), or progressive bulbar palsy with sensorineural deafness, is considered to be an allelic form of Fazio-Londe disease and is due to a riboflavin transporter defect. Deafness, gait disturbance, and exercise intolerance as well as the possibility of medical treatment are its distinctive traits.[39,40]

Congenital myopathies are defined by the early onset weakness together with characteristic pathological findings on muscle biopsy and normal to slightly elevated creatine kinase levels (see Chapter 28). Because of advances in molecular genetics, congenital myopathies are now understood to be a more diverse group of disorders than initially suspected. As a group, they show varied and overlapping clinical and histologic phenotypes, but they share more genotypes than anticipated. In fact, many of the histological features on muscle biopsies, which were used in the past to distinguish them, occur on a spectrum that crosses the boundaries of individual genetic entities. Children with congenital myopathies have muscle weakness and delayed motor development. Severe phenotypes often have significant respiratory problems that lead to

CASE EXAMPLE 47.1

This Caucasian male was the product of a second pregnancy of healthy nonconsanguineous parents. Gestation was complicated by decreased fetal movements, intrauterine growth failure, and polyhydramnios that required amniocentesis at 30 weeks of gestation. Delivery by Cesarean section was performed at 33 weeks due to increasing signs of fetal distress. Weight at birth was 1410 g. The patient was immediately transferred to the neonatal ICU owing to his prematurity and respiratory insufficiency. On admission, severe muscle hypotonia, facial diplegia, micrognathia with retrognathia, and bilateral distal joint contractures were detected. Respiratory support (nasopharyngeal CPAP) and NGT feeding were indicated since birth. Initial diagnosis of fetal akinesia sequence with arthrogryposis versus BSD were entertained and extensive laboratory investigations were performed, including CK, blood and urine testing for inborn errors of metabolism, EEG, EKG, EMG-NCV, and BAEPS, all of them with inconclusive results. Screening for SMN gene mutations was negative. Cranial MRI was normal with the exception of a reported mild loss of volume at dorsal level of the rhombencephalon.

On the second day of life the patient suffered a cardiorespiratory arrest, requiring resuscitation maneuvers, adrenaline, endotracheal intubation, and mechanical ventilation. Eighteen hours later, the patient was returned to CPAP, which was maintained until the 70th day of life. Successive physical examinations revealed good visual contact with his surroundings, severe OMD with constant drooling, severe muscle hypotonia, lack of head control, proximal limb weakness, and absent tendon reflexes. Percutaneous gastrostomy was performed at 1 month of life. Scopolamine patches as well as botulinum toxin A injections into the salivary glands were prescribed in order to decrease the salivary production and avoid repeated aspiration episodes and respiratory infections, without success. At 3 months of age, the patient was discharged with hospital support at home. The clinical course of this infant was characterized by repeated hospital admissions owing to respiratory complications, several of them requiring prolonged admission to the pediatric ICU. At 11 months of age, a second EMG study showed an unquestionable myogenic pattern, and the pattern of muscle involvement on MRI was suggestive of congenital myopathy. Repeat cranial MRI was normal and muscle biopsy showed central nuclei with loss of myofibrillar organization, suggestive of the recessive form of RYR1 mutation. Two pathogenic mutations were identified in the RYR1 gene (c.122T>C in exon 2 and c.4718C>T in exon 33); the parents were unaffected heterozygotes for each mutation.

Comment

Newborn presentations of congenital myopathies are very difficult to differentiate clinically from other congenital causes of OMD and particularly from BSD. Oral secretion management in these patients is also very difficult despite the use of all available therapeutic tools.

the patient's demise or that require prolonged ventilatory support. Feeding impairment, dysphagia, drooling, and dysarthria with hypernasal speech are frequent findings due to the profound weakness of the pharyngeal and facial muscles. Involvement of the extraocular, facial, and oromotor musculature is not infrequent in this group of myopathies.[41] See Case Example 47.1.

Congenital muscular dystrophies (CMD) are a group of genetic disorders in which weakness and abnormal muscle histologic features are present at birth (see Chapter 29). The mode of inheritance for most of them is autosomal recessive with significant genetic heterogeneity. Muscle weakness is slowly progressive or not progressive, depending on the individual disease, but complications of the dystrophy can become more prominent over time. From a practical clinical point of view, CMD can be divided into two subgroups, one with normal mental development and the other with associated intellectual disability. Magnetic resonance imaging of the brain, therefore, is imperative in the clinical approach to CMD as it reveals abnormalities of brain formation and neuronal migration or changes in the white matter. Most forms of CMD with abnormalities of brain formation that have been defined on a molecular level show alterations of dystroglycan O-linked glycosylation. The main phenotypes in this group have been delineated on clinical grounds and include Fukuyama CMD, muscle-eye-brain disease, and Walker-Warburg syndrome, but variations and overlapping phenotypes are frequent. Common clinical characteristics of this group of NMDs include severe muscular dystrophy, neuronal migration defects including lissencephaly type II (cobblestone lissencephaly), pachygyria, cerebellar and brainstem abnormalities, and variable ocular anomalies. OMD is present in most cases, congenital contractures are common, and some infants require ventilatory assistance and tube feeding.[42] Primary laminin-2 (merosin) deficiency causes one of the most common single forms of CMD. These patients usually present at birth or during the first month of life with muscular hypotonia, joint contractures, respiratory insufficiency, and feeding problems. Facial weakness is often prominent and motor development is markedly delayed, precluding independent ambulation. Most patients develop respiratory insufficiency requiring ventilatory support during the first decade of life. Gastroesophageal reflux, feeding difficulties, and poor nutrition are common. Even though abnormalities of the white matter are commonly found in laminin-2 mutations, they generally do not prevent normal mental development.[43] Collagen VI is a component of the extracellular matrix that does not belong to the group of dystrophin-associated proteins. Autosomal dominant mutations of collagen VI are known to cause Bethlem myopathy. Autosomal recessive mutations give rise to a more severe clinical phenotype known as Ullrich myopathy. Common clinical manifestations of patients with collagen VI deficiency are generalized muscle weakness, striking hypermobility of distal joints, and variable contractures of more proximal joints. Additional clinical findings include kyphoscoliosis, protruding calcanei, and follicular hyperkeratosis. Patients with the Ullrich phenotype may develop severe respiratory insufficiency but seldom have OMD.[44] CMD with rigid spine is characterized by early rigidity of the spine and a restrictive respiratory syndrome. These signs are often preceded by hypotonia and predominant axial muscle weakness in the first months of life. Mutations in at least three genes (*SEPN*, *FHL1*, and *LMNA*) have been linked to the rigid spine phenotype. *SEPN1* gene mutations are associated with multiminicores, which are typically smaller in size than those associated with recessive *RYR1* mutations. Clinically, patients with SEPN1-related disease manifest with a predominantly axial myopathy with weak neck flexors, spinal rigidity, and scoliosis as well as prominent respiratory compromise disproportionate to the extremity weakness. The respiratory compromise may be progressive even when limb weakness remains relatively mild and static. Bulbar weakness has been reported in some cases while extraocular muscle involvement does not seem to occur. The majority of SEPN1 patients are able to achieve and maintain independent ambulation. *FHL1* gene mutations give rise to a slowly progressive clinical picture that, due to the severe hyperextension of the neck, interferes with or impedes oral feeding in the late stages of disease.[42,45]

Facioscapulohumeral muscular dystrophy occurs in all age groups, but onset is most common during adolescence. About one third of patients have mild symptoms. The inheritance pattern is autosomal dominant, and the most common cause of the disease is a contraction of a repeat (D4Z4) on chromosome 4 (4q35) (see Chapter 32). A severe infantile form is often associated with hearing loss (Coats' syndrome). Asymmetric weakness is characteristic and is primarily located in the face, shoulder girdle, and upper arm muscles. The most commonly affected facial muscles are the sphincter muscles that close the eyes (orbicularis oculi) and the mouth (orbicularis oris). In addition to impaired facial expression and eye closure, facial weakness can affect saliva control and cause milder problems related to eating and drinking. The progression of muscle weakness and wasting is relatively slow. Speech is rarely affected during childhood.[46]

Disorders of neuromuscular transmission (see Chapters 25, 26, 27) have variable pathophysiology and expression. Symptoms usually appear during the first few months or years of life, and typical clinical findings include feeding difficulty, respiratory dysfunction, ophthalmoparesis, ptosis, and hypotonia. Symptoms may worsen with crying and physical activity. In neonatal transient myasthenia gravis there is no correlation between the severity of the maternal disease and that of the infant. Symptoms appear within the first few hours and the mean disease duration is 18 days. A prenatal onset results in a severe form with polyhydramnios, arthrogryposis, and severe hypotonia. Occasionally, maternal disease may be latent and acute choking episodes may be a presenting symptom in a floppy baby with associated bulbar weakness.[47,48] Congenital myasthenic syndromes (CMS) comprise a group of heterogeneous genetic diseases whose clinical manifestations vary depending on the syndrome subtype. Some patients present with signs at birth or shortly after, whereas others, especially those with mild presentations, go undiagnosed until adolescence or adulthood. CMS can be classified as presynaptic, synaptic, or postsynaptic, depending on the location of the primary defect within the neuromuscular junction. Postsynaptic defects comprise up to 75% to 80% of all cases. Abnormalities of presynaptic neuromuscular transmission constitute the rarest forms of CMS. The subtype of CMS with episodic apnea is attributed to deficiency of choline acetyl transferase, and its main clinical manifestations include sudden episodes of respiratory distress and bulbar weakness causing apnea. These episodes may be triggered by fever, infections, or stress. Endplate acetylcholinesterase deficiency begins in childhood and is characterized by weakness, muscle atrophy, and slow pupillary response to light stimulation after the administration of acetylcholinesterase inhibitors. Skeletal deformities (lordosis or scoliosis), ptosis, ophthalmoplegia, dysphagia, difficulty breathing, and limb weakness can occur with disease progression. Defects of the acetylcholine receptor genes can cause CMS that can begin in childhood, adolescence, or adulthood. In affected families, the pattern of inheritance is autosomal recessive. The main signs include hypotonia, delayed motor development, ptosis, ophthalmoplegia, weakness that may worsen on exertion, skeletal deformities (arthrogryposis, lordosis, or scoliosis), muscular atrophy, dysphagia, and respiratory difficulty. The fast-channel syndrome, due to the abnormal activity of the acetylcholine receptor ion channel, presents in neonates and infants, and its main clinical signs include ptosis, ophthalmoplegia, dysphagia, and weakness. The three main groups of postsynaptic subtypes of CMS with defects of the acetylcholine receptor complex are rapsyn deficiency, which is typically of neonatal onset and often associated with arthrogryposis and bulbar symptoms; Dok-7 myasthenia; and MuSK deficiency. Their clinical symptoms are similar to those of agrin defects and involve ptosis, ophthalmoplegia, and proximal weakness of the limbs. Other rare forms of postsynaptic CMS and OMD include Escobar syndrome, which represents the association of CMS with arthrogryposis multiplex congenita, pterygium, and difficulty breathing, and CMS with centronuclear myopathy.[49,50]

OMS DYSFUNCTION ASSESSMENT

The involvement of respiratory, feeding, and speech functions in children with NMDs is common and, quite often, persists throughout their lifespan. Therefore, the need for periodic OMD assessment and management revisions as the child grows and experiences changes in health status cannot be overemphasized. Moreover, oral motor skills develop within a system that changes rapidly during the first 3 years of life, in terms of both structural growth and neurological control. Thus, it is imperative to suspect and establish the diagnosis of the OMD as early as possible, not only to prevent further deterioration but to enhance appropriate development of the OMS.[4,7] The evaluation of children with OMD must be conducted in the setting of a multidisciplinary team. In essence, the OMS assessment of children with NMDs is the same as for all children with OMD, but emphasis is placed on muscle strength and coordination. The way to examine the OMS is largely dependent on the child's age, the level of collaboration, or both. When performing a qualitative assessment of oromotor function (Table 47.2), the child is asked to mimic oromotor activities or to perform a specific task on command. However, it is not unusual in the field of pediatric NMDs to care for infants younger than 3 years or children with associated developmental delay who have difficulty understanding verbal commands or who are unable to perform the instructions of the oromotor assessment on command. These children should be evaluated on the basis of observation of their spontaneous oromotor functions. The examination of OMD, much like the examination of movement disorders, should include video recording to allow for documentation, later evaluation, and comparison on follow-up. Currently, there are no recognized standardized pediatric assessment protocols of oromotor function. In any case, motility, muscle tone, and strength of facial muscles, tongue, jaw, and the velopharyngeal area must be evaluated. Estimation of muscle strength is performed using palpation of muscles while contracting, and experienced examiners make reliable evaluations by means of this technique. Another method to estimate the force of resistance is to use a spatula against the tongue, an oral screen inside the lips, or a hand on the mandible. Reduced endurance is suspected when speech or swallowing performance declines over a short time (for example, over the course of a conversation or meal). Although this inference may be valid, it does not provide detailed information about the specific muscle groups that are prone to weakness and fatigue. Dysphagia is easily suspected when

TABLE 47.2 Methods for Assessing Oromotor Function in NMD

Test Variable (Resting Position)	Qualitative Assessment (Observation)	Quantitative Assessment
Mobility	Speech examination	Computerized video-image analysis
Range of movement	Mealtime observation	Ultrasonography
Coordination	Oromotor tasks	Neurophysiological studies
Differentiation	Maximal muscle contraction	EMG (face, tongue, palate)
Praxis		Diadochokinetic testing
		Mandible scanner
Strength	Estimation of force against resistance	Force meter
Maximal force		MRI
Endurance		EMG
Sensibility	Stimulation: Tactile, Thermal, Vibratory	Two-point discrimination test
	Oral stereognosis	
Oral structures	Inspection	Cephalometry
	Palpation	Measuring jaw retractions
	Dental examination	Measuring jaw opening

Modified from Sjögreen L, 2011,[23] with the permission of Viguera Editores.

there are bouts of coughing or choking during swallowing, indicating leakage of food into the larynx, or nasal regurgitation due to velopharyngeal weakness.[23,26]

The face and oral cavity should be examined carefully to detect any minor facial malformations, facial hypotonia, oculomotor paresis, buccopharyngeal salivary stasis with drooling, asymmetry of the tongue or soft palate, weakness or flaccidity of the soft palate, asymmetry of palatal reflex, poor cough reflex, and tongue fasciculations and/or atrophy. Because impaired facial expression is common in children with NMDs, a thorough observation of facial gestures should be included in the assessment of this group of patients. Intraoral examination should also look for signs of malocclusion of the dentition with anterior open bite; a high, narrow arched palate; and/or a limited mouth opening range due to muscle weakness and contractures of the temporomandibular joint. The maximal jaw opening should be measured regularly to monitor for early detection or progression of temporomandibular joint contractures. Observation of muscle tone at rest should include body posture, head position, and the position of the lips, jaw, and tongue. Other aspects of oromotor function that should be assessed are symmetry of functions, range and direction of movement, coordination, differentiation, and praxis. Mealtime observations provide valuable information about the child's oromotor competence. Children with a not-so-severe NMD and moderate OMD, due to their difficulties with oral preparation and swallowing, may have to swallow three or four times to empty the mouth, may avoid drinking or, on the contrary, take a sip of water after each mouthful and thus need more time than other children to finish a meal. Flexion and extension of the neck to aid in projecting the food into the pharynx can also be seen.[23,26] Auscultation should search for respiratory sounds following ingestion of a small amount of water. It is important to note that swallowing disorders may be subclinical and may be revealed only by the onset of bronchopulmonary complications. In principle, any respiratory complaint in a child with an NMD and OMD should suggest food or saliva aspiration and therefore a swallowing disorder.[4,8]

Screening of orofacial sensation and perception, provided that the child is able to cooperate, must be added to the oromotor assessment if there is indication of sensory impairment. Oral stereognosis can be tested using test pieces of different forms, sizes, and surfaces.[51,52] Two-point discrimination gives information about the density of tactile sensory units on the face and in the mouth. It is important to identify oral sensory-perceptual deficits and impaired feedback from the oral structures because this may complicate the development of speech and feeding. The assessment of children who drool is complex due to each child's unique profile of contributing factors. The assessments focus on the oral motor features and functions that may contribute to drooling, its severity, and/or its frequency. Although there are both standardized and nonstandardized assessment tools for eating and drinking skills, few tools are specifically designed to assess drooling and none of them specifically addresses children with NMDs.[53,54]

Quantitative measures of oromotor functions have been developed (Table 47.2), but rarely have been used in clinical practice; further research and standardization are required before they can be included in the standard OMD assessment. One exception is counting syllable repetitions and other diadochokinetic tests.[55] Video-computer interactive systems for motion analysis of the facial muscles are now available, but it will be a long time before they are used in clinical practice because they are expensive and the analysis is complex and time-consuming.[56] Different instruments are available for measuring lip force,[57] tongue force,[58] and bite force,[59] but there is still a lack of normative data concerning the pediatric population. Swallowing in adults, children, and newborns has been extensively investigated.[50,52,60] In this chapter, we will only comment on the most commonly used test in clinical practice: videofluoroscopy and neurophysiological techniques.

Videofluoroscopy (VF) is an X-ray technique that, by simultaneous, real-time image viewing of the oral, pharyngeal, and esophageal stages of swallowing and the interaction between these stages, enables analysis of the swallowing process from the start of the oral anticipatory stage through to the onset of the esophageal stage. VF is widely regarded as the most reliable method of evaluating the pharyngeal and upper esophageal phases of swallowing and thus assessing airway protection in pediatric patients. The procedure is recorded into videotape or DVD and is available for review. The information that can be obtained from this technique is quite extensive and depends on the field of medicine for which it is used. As it refers to OMD and NMDs, the main types of information gained with VF are the existence of structural or motility disorders in the oral cavity or pharynx; the interaction between oral, pharyngeal, and esophageal stages of swallowing; a dynamic recording of how different food and liquid consistencies are swallowed, indicating the safest consistency; the adequacy of airway protection during swallowing; the deterioration of function associated with fatigue; and information confirming the presence of silent aspiration. A major benefit of VF is that, having identified the problem, it allows determination of whether any practical management is likely to optimize safety and facilitate the feeding process. Patients who are lethargic, orally defensive, or medically unstable are not able to participate adequately in a VF study. Likewise, children who do not have an established oral feeding regimen cannot be assessed by this technique.[60,61] For children who are mainly fed by nasogastric tube (NGT) or gastrostomy but have an oral component to their feeding, it is recommended that they become accustomed to taking small tastes of food for at least 1 to 2 weeks prior to assessment by VF. In most cases, it is not necessary to remove an NGT for the purposes of a VF study. Flexible endoscopic examination is increasingly used to assess dysphagia in adult practice, but the likelihood of using this technique other than with the very young or with competent and cooperative older children is doubtful. At present, it remains a restricted and very specialized procedure unlikely to replace the need for VF.[60,61]

Besides the invaluable contributions of conventional NCV and EMG studies to establish the diagnosis of the NMD causing OMD, several other methods of investigation using neurophysiological techniques that provide convergent information about brainstem nuclei and pathways involved in sucking and swallowing can be used in quantitative evaluation of OMD. Specific EMG studies of the face, tongue, and soft palate muscles provide information about the function of the motor units in the territory of the facial nerve, the pharyngeal plexus (IX-X nerves), and the hypoglossal nerve. The blink reflex study explores the integrity of fibers and nuclei of the V nerve, the V to VII internuclear pathways, and the motor nucleus of the VII nerve and its temporofacial fibers.[8] Simultaneous EMG recordings of the genioglossus muscle, specific for the oral phase, and of the thyrohyoid muscle, specific for the pharyngeal phase, while sucking the nipple of a feeding bottle allows the analysis of the activity of each muscle and the timing of their activation, enabling evaluation of the coordination of the first two phases of swallowing and providing information about the automatisms generated and regulated in the central pattern generator for swallowing. All of these EMG investigations target the disturbed motor organization, assess its severity, and show whether it is isolated or associated with lesions of other bulbar nuclei, thus helping to establish the differential diagnosis of other disorders causing OMD.[62,63]

Muscle biopsy may be needed to prove or exclude NMD in selected cases. A good example is Case Example 47.1 of this chapter, where widespread bulbar dysfunction, coupled with negative results of NCV and EMG studies performed during the first month of life, led to the tentative diagnosis of BSD. The differential diagnosis of all these congenital conditions that lead to OMD is difficult to establish in many instances. Routine MRI performed at birth or during infancy has limitations; however, its practice is essential in cases presenting with severe clinical manifestations in order to exclude brain abnormalities such as white matter disorders, lissencephaly, polymicrogyria, pachygyria, or brainstem malformations.[10,15,16,42]

TREATMENT OF OROMOTOR DYSFUNCTION DUE TO A NEUROMUSCULAR DISORDER

The management of OMD associated with an NMD is complex due to the protracted and varying courses of most NMDs, the diverse areas of health care involved, and the significant amount of resources that their treatment requires.[4,15,16] Patients with an NMD and OMD should be assessed and treated in light of available information about the underlying disease and the number and nature of their clinical manifestations. Two of the essential treatment aims of OMD due to NMD are: (i) to sustain life, and (ii) to maximize health and well-being. Interventions to sustain life are usually pharmacological or surgical and involve modifying functions or body structures to enable the patients to breathe and ingest nutrients. As a general rule, a severe NMD presenting at birth or in early infancy is most likely to put the child's survival at abrupt risk. The decision to sustain life is a particularly important and difficult issue in every patient. This management option is better taken under the shelter of a multidisciplinary professional team and with parental participation in the decision.[4,10,22] Certain life-threatening congenital NMDs have a better prognosis than initial clinical manifestations would indicate. If adequate support is given, respiratory problems improve, swallowing recovers its competence, and the strength of sucking increases, enabling the gradual and safe reintroduction of oral feeding. Case Example 47.2 illustrates this clinical setting.

OMD is the rule in congenital NMDs that present at birth with breathing problems requiring respiratory support. Airway stability and efficient respiratory function are mandatory before attempting any type of oral feeding; thus neonates with a severe NMD and respiratory distress require nasogastric tube feeding (NGT), at least until appropriate respiratory and nutritional management are determined.[18,65] Neonates and infants with NMDs and mild to moderate respiratory difficulties also have feeding problems that vary depending upon the extent and degree of the muscle weakness. Bilateral facial muscle weakness, velopalatine incoordination, and, at times, an associated temporomandibular ankylosis can lead to sucking and swallowing difficulties. Furthermore, concomitant difficulties in these infants in coughing and in handling oral secretions result in pooling of saliva in the pharynx, with the need for adequate head positioning and oropharyngeal suctioning to limit the risks of aspiration. All circumstances mentioned above interfere with the oral function of these infants, who are unable to accommodate the volume of food required to meet their caloric needs, and consequently, NGT feedings are also indicated to prevent undernutrition.[18,24,66–68]

CASE EXAMPLE 47.2

This patient was reported in 1987 as the first case in the literature who survived symptomatic nemaline myopathy presenting in the newborn period.[64] A Caucasian female was born severely hypotonic, with facial diplegia, respiratory insufficiency, accumulation of upper respiratory secretions, and absent tendon reflexes. Repeated micro-aspirations and pulmonary infections required endotracheal assisted ventilation and NGT feeding during the first week of life. EMG studies and muscle biopsy led to the diagnosis of nemaline myopathy. At 3 months of life, despite the prognosis at that time and at the express demand of her parents, the patient underwent tracheostomy, which allowed weaning from mechanical ventilation. At 17 months, feeding disturbances persisted; she was fed through NGT and was able to swallow only fluids and jelly. At 29 months she was able to lift herself up from the floor and walk alone. She expelled respiratory secretions by coughing and her language comprehension was considered normal for her age. At 36 months of age, surgical closure of the tracheostomy was undertaken. Bilateral myringotomy tubes were inserted because of severe hearing loss secondary to chronic middle ear serous fluid retention. By 4 years of age, she attended playschool where she adjusted well. No respiratory complications were reported, and she was able to eat some solid food. Weakness was confined to the proximal muscle groups. The patient has continued her clinical follow-up at the neuromuscular unit, and several years ago a mutation at the nebulin gene was identified. She has completed university education and presently, at age 29 years, is fully employed and drives her own car. Her major complaints in the last years have been nasal voice, hearing fluctuations in relationship with upper respiratory tract infections, and chronic serous otitis, fatigue with prolonged chewing and temporomandibular joint instability, proximal muscle weakness, mild scoliosis, and subcutaneous fat deposition in lower abdomen and thighs.

Comment

The publication of this patient's report in 1987 contributed to changes in the clinical treatment of newborns with congenital myopathies. Her ongoing follow-up reinforces the validity of those treatment options, and demonstrates the quality of life that patients with congenital myopathies may achieve as well as the fact that OMD may be the most bothersome sequel.

The consequences of dysphagia for the developing child are important and wide ranging. Dysphagia can result in chronic undernutrition, which is recognized to predispose to increased levels of irritability, decreased motivation, poor energy levels, and a greater incidence of longer-lasting infections through disturbances of the immune system.[24,69] Growth and nutritional status must, therefore, be monitored regularly in all infants with an NMD and OMD, as their food intake may fall half below calculated needs and their muscle mass and weight tend to remain 1 or 2 percentiles below their height on growth charts. According to most authors, NGT feeding is the first choice of enteral tube nutrition if the calculated period of time required for nutritional support does not exceed 6 weeks. NGT feeding should ensure energy and appropriate protein intake, optimize growth, and prevent osteoporosis. Continuous, slow, nocturnal delivery of nutrients and intermittent oral feeding during the day is the commonest recommended method of NGT food administration. While on enteral tube nutrition, maximizing positive oral and facial sensory stimulation is crucial in order to decrease the potential for developing orofacial hypersensitivity and aversion that may cause future food refusal. Nonnutritive sucking on a dummy or pacifier may give the opportunity to practice sucking and swallowing saliva and, if done during tube feeding, may reinforce the association between oral stimulation and satiation and may improve feeding outcomes.[24,70–72]

Medical treatment of OMD is often unsuccessful in patients with severe NMDs, and surgical procedures are necessary to avoid known complications such as aspiration pneumonia, respiratory infection, chronic lung injury, and undernutrition. Percutaneous endoscopic gastrostomy (PEG) is an effective intervention to supplement oral intake and should be offered proactively to maintain weight, rather than waiting for significant loss of weight and undernutrition, which then becomes impossible to regain.[73] There are opposing views in the medical literature about the subsequent increase in gastroesophageal reflux (GER) associated with the insertion of a gastrostomy device.[74,75] According to many authors, anti-reflux surgery concomitant with the placement of a PEG should be considered as part of management if GER is already present or strongly suspected before gastrostomy is recommended. Nissen's anti-reflux surgery is the most widely used technique; it repositions the intra-abdominal esophagus and leads to a plication of the gastric fundus surrounding the distal portion of the esophagus. If the condition of the patient permits, this procedure is currently performed by laparoscopy, which allows better recovery after the intervention. In the opinion of many pediatric gastroenterologists, Nissen's anti-reflux surgery is an extreme practice and increases the risk of complications. According to them, symptoms reappear in up to 40% of patients and require repeated surgery that delays recovery.[75–77] However, controlled studies trying to resolve this specific issue in children with NMDs are lacking.[78]

In severely undernourished infants or children, short-term NGT feeding while waiting for gastrostomy surgery can maintain health and reassure the patient and the parents that, once nutritional status improves, general well-being and mood will often improve. In children with NMDs, PEG typically is used for food supplementation during times of fatigue or illness as well as for delivery of medications. Small meals and snacks are maintained as part of the child's diet. In the case of a child who is at risk of aspirating on particular consistencies, these can be avoided orally and their nutrition and hydration can be supplied via gastrostomy.[24] PEG seldom has to be used to replace all oral intake. There are risks associated with having nothing to eat or drink by mouth, and wherever possible, some oral intake is maintained both for health and for pleasure. Feeding needs and strategies when using enteral nutrition intake by PEG must be evaluated to adapt to each stage of child's development. The use of PEG for enteral feeding of patients with OMD improved the quality of life of the patients and their caretakers, reduced feeding time, and facilitated the delivery of treatment.[69] PEG outcomes in NMDs have been evaluated as positive and effective in optimizing nutritional intake and consequent growth outcomes.[71] However, we must be aware that PEG and the subsequent decrease in length of the subdiaphragmatic esophagus, a liquid diet, and supine positioning are factors that increase GER; obesity has been described in patients with very low levels of movement when adequate caloric intake is provided by PEG.[18,69]

As mentioned earlier, poor handling of oral secretions in infants and children with NMDs results in pooling of saliva in the pharynx, increasing the risks of lung aspiration and respiratory tract infections. Infrequent swallowing, open-mouthed posture, and the inability to discriminate differences in volumes of secretions or between a wet or a dry chin are situations that contribute to drooling. The child's mental status and motivation level are important aspects to consider for effective treatment. Rehabilitation programs are more successful in children and adolescents who are conscious of their appearance. Although some programs focus on improving saliva control, guidelines are lacking to facilitate treatment of children with specific diagnoses, at specific ages, and with specific oromotor impairments.[33] In most instances, saliva control improves as children get older. However, if the disease is progressive, drooling can be a persistent and even a mounting problem, indicating OMD deterioration. Regulation of saliva production is mainly under the control of the parasympathetic nervous system. Anticholinergic drugs such as benztropine, glycopyrrolate, and trihexyphenidyl hydrochloride are used to treat drooling but, as a general rule, are not indicated in patients with NMDs. In addition to detrimental side effects related to the medications, the use of anticholinergic drugs to dry up saliva makes secretions thicker, and this, coupled with a weak cough, makes secretions harder to swallow or expectorate.[33] Transdermal scopolamine patches that provide a convenient method of delivering the drug at a constant rate over a 3-day period have been used in single case reports, especially in newborns with severe dysphagia, but with uncertain benefits.[18] Salivary gland injection of botulinum toxin type A diminishes drooling in the majority of young people.[79–81] However, with the exception of ALS, SMA type II, and nemaline myopathy, botulinum toxin type A has seldom been used in NMDs due to its reported adverse effects, particularly loss of motor control of the head, and because of chewing and swallowing problems that may arise from botulinum toxin diffusion into the surrounding muscle tissue.[82–83] Surgery is usually considered when other less invasive procedures to improve drooling have been exhausted, particularly in individuals 6 years of age or older. A meta-analysis of 50 articles showed that surgical management using a variety of procedures provided significant subjective relief in 81.6% of patients, and that bilateral submandibular gland excision and parotid duct rerouting appeared to have the highest subjective success rates.[84] Surgical options to ameliorate drooling must be considered in infants with a severe congenital NMD who survive the neonatal period. The first clinical example presented in this chapter illustrates that, in addition to the difficulties of reaching a diagnosis, despite feeding through gastrostomy, respiratory function compromise due to abnormal swallowing of the accumulated oral secretions was the cause of lung micro-aspirations and subsequent severe respiratory infections.

Interventions to promote health and well-being focus on reducing impairment of body structures and enhancing functions. Typically, these goals rely on improving strength and correcting abnormalities by means of rehabilitation programs, compensatory and maintenance interventions, and the use of alternative augmentative communication systems. All these interventions are designed to limit the effects of OMD due to an NMD on a patient's feeding, swallowing, functional speech abilities, and gestural communication. Rehabilitation programs should begin soon after diagnosis and should focus on improving oral motor skills through sensorimotor stimulation. It is crucial to evaluate each case individually and to establish a specific treatment plan. There is little benefit in treating the isolated symptoms of OMD, and little evidence exists that an approach focusing purely on improving strength and correcting abnormalities is effective. Instead, the whole child and family picture should be considered to optimize the best possible functional outcomes.[3,4,22,24] Understandably, in clinical practice many of the aforementioned interventions are carried on concurrently but, for simplicity of discussion, the intervention for each aspect of OMD will be described separately.

As mentioned previously, good respiratory management is essential throughout the natural course of all NMDs to optimize feeding and speech outcomes, maximize growth, and avoid respiratory infections. Again, this is best done by a multidisciplinary team approach that should include pediatricians, pediatric gastroenterologists and nutritionists, pediatric pulmonologists, and physical, respiratory, and speech and language therapists as well as social workers. Recommendations to optimize respiratory health in NMDs include regular monitoring of respiratory function, the use of prophylactic antibiotics and vaccinations, the use of cough-assist devices, and sleep-study monitoring to enable the early detection and treatment of nocturnal hypoventilation. In individuals without an NMD, a swallow is nearly always followed by an expiration that contributes to airway protection, and hypercapnia in healthy individuals has been shown to increase the incidence of inspiration after swallowing. If this pattern applies to NMD patients, swallowing difficulties will increase as respiratory function deteriorates. In fact, according to one study, only one-half of patients with an NMD have a pattern of post-swallow expiration and the other half have multiple swallows per bolus, often requiring several swallows over several breathing cycles. Glossopharyngeal breathing can occur when there is increasing respiratory weakness; this is particularly common in late-stage Duchenne muscular dystrophy. Air is gulped through the mouth and forced through the glottis into the lungs to expand the chest and aid breathing. For this group of patients and others with similar diseases, this breathing maneuver is very difficult to accomplish simultaneously with swallowing or speaking. Swallowing may deteriorate during an acute respiratory infection due to a general depression of function during the illness.[30,66,67]

Nocturnal hypoventilation is another respiratory complication leading to headaches, poor mood upon waking, increased fatigue, and reduced appetite frequently found in progressive long-lasting NMDs. Long-term, noninvasive ventilation (NIV) via a nasal or facial mask and tracheostomy are the more commonly used techniques for the delivery of ventilation in this situation. These two procedures, by reducing the breathing effort and the anxiety associated with breathlessness, allow the patients to concentrate on swallowing as well as to improve their performance.[85–86] A single study that examined breathing and swallowing interactions in NMD patients treated with tracheostomy found that swallowing improved when they were on mechanical ventilation rather than breathing spontaneously. According to the authors, mechanical ventilation via tracheostomy may reverse hypercapnia, restore subglottic pressure, and reduce the work of the respiratory muscles, thus allowing the upper airway muscles to serve only for swallowing.[85] When tracheostomy is performed, it is important that the communication needs of the child are anticipated and the use of alternative communication systems or speaking valves is foreseen. This anticipatory management is particularly important for the child who may have a short-term tracheostomy postsurgically, so that they are able to communicate effectively during their recovery. The use of NIV, alternatively, can improve appetite, mood, and energy levels; increase oral intake; and lead to a reversal of faltering growth. NIV may also improve swallowing by restoring subglottic air pressure and reducing glossopharyngeal breathing. Moreover, the use of NIV may support speech by reducing the effort of breathing and improving respiratory support and control, thereby increasing vocal volume and intelligibility. Complications from NIV include swallowing air into the stomach, producing bloating and discomfort, decrease in appetite, and, in small children with NMDs who require long-term NIV, the development of midface deformities. The constant pressure due to a face or nose mask that applies pressure to the middle face may inhibit facial growth and result in a retracted upper jaw. Careful mask fitting and monitoring is required.[23,25] In patients with NMDs, breathing problems in speech generally are not due to lack of available air but to a problem in maintaining airstream. Individuals with NMDs may have weak inspiratory muscle strength and reduced vital capacity that result in weak cough production and the inability to clear the lungs and airways effectively of residual food, oral secretions, or saliva. These difficulties may lead to reduced breath support for loud, sustained voice production. Children with NMDs may have limited breath support for connected speech and often require the use of compensatory strategies such as speaking in short phrases with pauses for breath between them. Weak and hypotonic oral and velopharyngeal muscles and respiratory insufficiency are responsible for the dysarthria observed in individuals with NMDs. Common symptoms are a low voice, blurred speech, hypernasality, compensatory articulation of bilabial consonants due to impaired lip function, and weak pronunciation of consonants.[22–24,29,30] Conductive hearing loss from the first months of life, due to chronic middle ear bilateral effusions, may cause abnormal phonological acquisition of words and contribute to the speech abnormalities of these groups of patients. Abnormal eustachian tube drainage secondary to a high-arched palate[64] or to weakness of selective OMS muscle groups[48] has been postulated as the cause of this type of chronic fluid retention in the middle ear. High index of suspicion and early treatment with myringotomy tubes or bone conduction hearing aids may prevent speech and language difficulties.

Feeding and swallowing issues are unusual in NMDs presenting in childhood but emerge later in adolescence, when the loss of ambulation and the onset of respiratory decline appear. Initially, short-term supplemental NGT feeding may be required during acute illness or surgery in order to maintain food intake, prevent weight loss, and provide the additional nutrition required for recovery. This support is particularly important post-scoliosis surgery when a substantial amount of weight loss may occur.[87] Major feeding problems usually become apparent after an acute respiratory infection or with prolonged choking episodes. In fact, respiratory tract complications are so important in the late stages of NMDs that the results of feeding and swallowing management interventions may be measured by reduction in the number and severity of chest infections. As a consequence of repeated respiratory infections or choking episodes, patients limit their oral intake due to fear of food becoming stuck in the throat, leading to a dramatic reduction of food intake and weight loss. Swallow assessments during recovery from acute respiratory infections should be interpreted cautiously, as VF may not reflect the child's usual swallowing performance when in good health. According to many experts, VF at this point may not be a helpful procedure, as it may not alter management and may heighten anxiety. The information obtained by VF is typically a weak pharyngeal swallow with associated post-swallow pharyngeal residue and no aspiration.[53,60,61,69] Anxiety can be exacerbated by coexisting learning difficulties that make counseling regarding fears more difficult. The fear of choking may be closely linked to fears regarding death, and a depressive state can also be present. All these respiratory complications have a tendency to occur at the time of transition out of school into adult services, and psychological support is invaluable at this stage.[24,88]

The prognosis for survival and quality of life in boys diagnosed with Duchenne muscular dystrophy and other childhood NMDs has improved dramatically in the past 20 years with advances in respiratory support techniques. The use of NIV during eating and drinking supports breathing and swallowing, reduces glossopharyngeal breathing, and may reduce anxiety and the fear of choking.[24,88] This has led to the need for long-term nutritional supplementation providing sufficient calories to sustain optimal growth. Initially, oral supplementation can be achieved by adding fat and oil to the diet or by using prescribed special formulas that provide a nutritionally balanced supplement. Commonly, a strategy to alter bolus volume and consistency is recommended. Such feeding changes have a good chance of being effective when applied to selected cases; however, formal evidence for this approach is limited and feeding modifications should be undertaken judiciously.[88,89] Trunk and neck position influence the oral and pharyngeal phases of swallowing in patients with dysphagia. In children with NMDs, motor deficits with reduced head and trunk control and poor stability of body posture have an impact on motor function, comfort, and performance. Correcting and maintaining optimal posture can have a beneficial effect on dysphagia as well as on limb function in this group of children.[66–68] All these interventions may be useful during a relatively short period of time, once there is clear indication that the patient will not be able to sustain his oral intake in the long term, and then a gastrostomy tube placement is the preferred intervention. When PEG is indicated in the advanced stages of a NMD, the issue of GER must also be taken into consideration. The function of the lower esophageal sphincter can be affected by alterations in pressure between the stomach and airways; it may therefore need to be considered in patients with respiratory problems, particularly those requiring respiratory support. Discomfort, pain, and ulcerative esophagitis can result from untreated GER, which, in turn, can cause an association of pain with eating and behavioral food refusal or aversion. Whereas NMDs typically affect skeletal muscles, the smooth muscle of the GI tract may also be involved in these disorders and contribute to constipation due to dysmotility of the lower GI tract. Constipation can be exacerbated by the fact that many children with NMDs are relatively immobile and spend most of their time sitting. Limited fluid intake can also contribute to dehydration and subsequent constipation. Constipation may lead to abdominal discomfort, poor appetite, and reduced oral intake. It is essential that constipation be managed proactively with abdominal massage, diet, and medication.[24]

Many patients with OMS dysfunction, particularly those with bilateral facial muscle weakness, have problems relating to their peers, especially when they reach adolescence. In fact, some of these patients are unjustifiably identified socially as intellectually impaired and find themselves offered fewer opportunities for social achievement and independence. Again, speech and language therapy as well as psychological support are invaluable at this stage.[4,22,24]

REFERENCES

1. Lazarov N. Neurobiology of orofacial proprioception. *Brain Res Rev* 2007;**56**:362–83.
2. Martinez S, Puelles E. Functional anatomy of the oromotor system. In: Roig-Quilis M, Pennington L, editors. *Oromotor Disorders in Childhood.* Barcelona: Viguera; 2011. pp. 5–21.
3. Roig M. Congenital oromotor disorders. *Dev Med Child Neurol* 2006;**48**:787.
4. Roig-Quilis M. Childhood oromotor disorders: definition, classification, and natural history. In: Roig-Quilis M, Pennington L, editors. *Oromotor Disorders in Childhood.* Barcelona: Viguera; 2011. pp. 117–24.
5. Graham PJ. Congenital flaccid bulbar palsy. *Br Med J* 1964;**2**:28–9.
6. Worster-Drought C. Congenital suprabulbar paresis. *J Lanryngol* 1956;**70**:453–63.

7. Illingworth RS. Sucking and swallowing difficulties in infancy diagnostic problem of dysphagia. *Arch Dis Child* 1969;**44**:655–65.
8. Renault F. Neurogenic dysphagia in newborns and infants In: Royden Jr JH, DeVivo D, Darras B, editors. *Neuromuscular Disorders of Infancy, Childhood and Adolescence. A Clinician's Approach.* Oxford: Butterworth/Heinemann; 2003. pp. 1127–40.
9. Lefton-Greif MA. Pediatric dysphagia. *Phys Med Rehabil Clin N Am* 2008;**19**:837–51.
10. Roig-Quilis M. Oromotor disorders due to brainstem lesions. In: Roig-Quilis M, Pennington L, editors. *Oromotor Disorders in Childhood.* Barcelona: Viguera; 2011. pp. 149–62.
11. Roig-Quilis M, Rodríguez-Palmero A. Oromotor disorders in a paediatric neurology unit. Their classification and clinical course. *Rev Neurol* 2008;**47**:509–16.
12. Leroy-Malherbe V, Husson B, Landrieu P, Tardieu M. Syndrome of delayed maturation of the sucking-swallowing reflex. *Arch Pediatr* 1994;**1**:20–5.
13. Frank MM, Gatewood OM. Transient pharyngeal incoordination in the newborn. *Am J Dis Child.* 1966;**111**:178–81.
14. Sarnat HB. Watershed infarcts in the fetal and neonatal brainstem. An aetiology of central hypoventilation, dysphagia, Möbius syndrome and micrognathia. *Eur J Paediatr Neurol* 2004;**8**:71–87.
15. Clark M, Carr L, Reilly S, Neville BG. Worster-Drought syndrome, a mild tetraplegic perisylvian cerebral palsy. Review of 47 cases. *Brain* 2000;**123**:2160–70.
16. Neville B, Clark M. Oromotor dysfunction due to cerebral hemispheric lesions. In: Roig-Quilis M, Pennington L, editors. *Oromotor Disorders in Childhood.* Barcelona: Viguera; 2011. pp. 139–45.
17. Roig M, Gratacos M, Vazquez E, Del Toro M, Foguet A, Ferrer I, et al. Brainstem dysgenesis: report of five patients with congenital hypotonia, multiple cranial nerve involvement, and ocular motor apraxia. *Dev Med Child Neurol* 2003;**45**:489–93.
18. Castilla-Fernández Y, Boix H, Macaya A, Vázquez E, Gratacòs M, Roig-Quilis M. Brainstem dysgenesis during the neonatal period: diagnosis and management. *J Perinat Med* 2013;**41**:437–45.
19. Abadie V, Morisseau-Durand M-P, Beyler C, Manach Y, Couly G. Brainstem dysfunction: a possible neuroembryological pathogenesis of isolated Pierre Robin sequence. *Eur J Pediatr* 2002;**161**: 275–80.
20. Verzijl HT, Van der Zwaag B, Cruysberg JR, Padberg GW. Möbius syndrome redefined: a syndrome of rhombencephalic maldevelopment. *Neurology* 2003;**61**:327–33.
21. WHO. International Classification of Functioning, Disability and Health: ICF. Geneva: WHO; 2001. URL: <http://www.who.Int/about/definition/en/print.html> [12.02.10].
22. Pennington L, Reilly S. Management of oromotor disorders in childhood. In: Roig-Quilis M, Pennington L, editors. *Oromotor Disorders in Childhood.* Barcelona: Viguera; 2011. pp. 215–24.
23. Sjögreen L. Oromotor disorders arising from neuromuscular diseases. In: Roig-Quilis M, Pennington L, editors. *Oromotor Disorders in Childhood.* Barcelona: Viguera; 2011. pp. 165–74.
24. Aloysius A. Treatment of oromotor difficulties in congenital neuromuscular disorders. In: Roig-Quilis M, Pennington L, editors. *Oromotor Disorders in Childhood.* Barcelona: Viguera; 2011. pp. 301–16.
25. Kiliaridis S, Katsaros C. The effects of myotonic dystrophy and Duchenne muscular dystrophy on the orofacial muscles and dentofacial morphology. *Acta Odontol Scand* 1998;**56**:369–74.
26. Guillén M. Clinical assessment of the oromotor system. In: Roig-Quilis M, Pennington L, editors. *Oromotor Disorders in Childhood.* Barcelona: Viguera; 2011. pp. 65–79.
27. Miller MJ, Kitchoosakum P. Relationship between respiratory control and feeding in the developing infant. *Semin Neonatol* 2004; **9**:221–7.
28. Rodriguez-Palmero A, Macaya A. Feeding disorders in premature infants. In: Roig-Quilis M, Pennington L, editors. *Oromotor Disorders in Childhood.* Barcelona: Viguera; 2011. pp. 177–90.
29. Love RJ. *Childhood Motor Speech Disability. Needham Heights.* Allyn & Bacon; 2000.
30. Clark HM. Neuromuscular treatments for speech and swallowing: a tutorial. *Am J Speech Lang Pathol* 2003;**12**:400–15.
31. Crysdale WS, McCann C, Roske L, Joseph M, Semenuk D, Chiat P. Saliva control issues in the neurologically damaged. A 30 year experience in team management. *Int J Pediatric Otorhinolaryngol* 2006;**70**:519–27.
32. Scully C, Limeres J, Gleeson M, Tomas I, Diz P. Drooling. *J Oral Pathol Med* 2009;**38**:321–7.
33. Johnson H, Desai M, Reddihough D. Saliva and drooling. In: Roig-Quilis M, Pennington L, editors. *Oromotor Disorders in Childhood.* Barcelona: Viguera; 2011. pp. 333–48.
34. Sjögreen L, Engvall M, Ekström AB, Lohmander A, Kiliaridis S, Tulinius M. Orofacial dysfunction in children and adolescents with myotonic dystrophy. *Dev Med Child Neurol* 2007;**49**:18–22.
35. Roig M, Balliu P, Navarro C, Brugera R, Losada M. Presentation clinical course and outcome of the congenital form of myotonic dystrophy. *Pediatr Neurol* 1994;**11**:208–13.
36. Hilbert JE, Johnson NE, Moxley III RT. New insights about the incidence, multisystem manifestations, and care of patients with congenital myotonic dystrophy. *J Pediatr* 2013;**163**:12–14.
37. Messina S, Pane M, De Rose P, Vasta I, Sorleti D, Aloysius A, et al. Feeding problems and malnutrition in spinal muscular atrophy type II. *Neuromuscul Disord* 2008;**18**:389–93.
38. Van Den Engel Hoek L, Erasmus CE, Van Bruggen HW, De Swart BJM, Sie LTL, Steenks MH, et al. Dysphagia in spinal muscular atrophy type II: more than a bulbar problem? *Neurology* 2009;**73**:1787–91.
39. Bosch AM, Abeling NG, IJlst L, Knoester H, van der Pol WL, Stroomer AEM, et al. Brown-Vialetto-Van Laere and Fazio Londe syndrome is associated with a riboflavin transporter defect mimicking mild MADD: a new inborn error of metabolism with potential treatment. *J Inherit Metab Dis* 2011;**34**:159–64.
40. Johnson JO, Gibbs JR, Megarbane A, Urtizberea JA, Hernandez DG, Foley AR, et al. Exome sequencing reveals riboflavin transporter mutations as a cause of motor neuron disease. *Brain* 2012;**135**:2875–82.
41. Nance JR, Dowling JJ, Gibbs EM, Bönnemann CG. Congenital Myopathies: An Update. *Curr Neurol Neurosci Rep* 2012;**12**:165–74.
42. Bertini E, D'Amico A, Gualandi F, Petrini S. Congenital muscular dystrophies: a brief review. *Semin Pediatr Neurol* 2011;**18**: 277–88.
43. Geranmayeh F, Clement E, Feng LH. Genotype-phenotype correlation in a large population of muscular dystrophy patients with LAMA2 mutations. *Neuromuscul Disord* 2010;**20**:241–50.
44. Nadeau A, Jimenez-Mallebrera C, Aloysius A. Natural history of Ullrich congenital muscular dystrophy. *Neurology* 2009;**73**:25–31.

45. Moghadaszadeh B, Petit N, Jaillard C, Brockington M, Quijano Roy S, Merlini L, et al. Mutations in SEPN1 cause congenital muscular dystrophy with spinal rigidity and restrictive respiratory syndrome. *Nat Genet* 2001;**29**:17–18.
46. Chen TH, Lai YH, Lee PL, Hsu JH, Goto K, Hayashi YK, et al. Infantile facioscapulohumeral muscular dystrophy revisited: Expansion of clinical phenotypes in patients with a very short EcoRI fragment. *Neuromuscul Disord* 2013;**23**:298–305.
47. Hoff JM, Daltveit AK, Gilhus NE. Asymptomatic myasthenia gravis influences pregnancy and birth. *Eur J Neurol* 2004;**11**:559–62.
48. Oskoui M, Jacobson L, Chung WK, Haddad J, Vincent A, Kaufmann P, et al. Fetal acetylcholine receptor inactivation syndrome and maternal myasthenia gravis. *Neurology* 2008;**71**:2010–12.
49. Lorenzoni PJ, Scola RH, Kamoi Kay CS, Werneck LC. Congenital myasthenic syndrome: a brief review. *Pediatr Neurol* 2012;**46**:141–8.
50. Engel AG. Current status of the congenital myasthenic syndromes. *Neuromuscul Disord* 2012;**22**:99–111.
51. Rogers B, Arvedson J. Assessment of infant oral sensorimotor and swallowing function. *Ment Retard Dev Disabil Res Rev* 2005;**11**:74–82.
52. Clark HM, Henson PA, Barber WD, Stierwalt JA, Sherrill M. Relationship among subjective and objective measures of tongue strength and oral phase swallowing impairments. *Am J Speech Lang Path* 2003;**112**:40–50.
53. Arvedson JC, Brodsky L. *Pediatric Swallowing and Feeding: Assessment and Management Albany*. NY: Singular; 2002.
54. Sonies BC, Cintas HL, Parks R, Miller J, Caggiano C, Gottshall SG, et al. Brief assessment of motor function: content validity and reliability of the oral motor scales. *Am J Phys Med Rehabil* 2009;**88**:464–72.
55. Ackermann H, Hertrich I, Hehr T. Oral diadochokinesis in neurological dysarthrias. *Folia Phoniatr Logop* 1995;**47**:15–23.
56. Wachtman GS, Cohn JF, VanSwearingen JM, Manders EK. Automated tracking of facial features in patients with facial neuromuscular dysfunction. *Plast Reconstr Surg* 2001;**107**:1124–33.
57. Hagg M, Olgarsson M, Anniko M. Reliable lip force measurement in healthy controls and in patients with stroke: a methodologic study. *Dysphagia* 2008;**23**:291–6.
58. Hewitt A, Hind J, Kays S, Nicosia M, Doyle J, Tompkins W, et al. Standardized instrument for lingual pressure measurement. *Dysphagia* 2008;**23**:16–25.
59. Boliek CA, Rieger JM, Li SY, Mohamed Z, Kickham J, Amundsen K. Establishing a reliable protocol to measure tongue sensation. *J Oral Rehabil* 2007;**34**:433–41.
60. Arvedson J, Lefton-Greif M. *Pediatric Videofluoroscopic Swallow Studies: A Professional Manual with Caregiver Guidelines*. San Antonio TX: Communication Skill Builders/Psychological Corporation; 1998.
61. Ryan M, Hiorns M. Videofluoroscopy swallow study. In: Roig-Quilis M, Pennington L, editors. *Oromotor Disorders in Childhood*. Barcelona: Viguera; 2011. pp. 97–112.
62. Renault F, Raimbault J. Electromyographie faciale, linguale et pharyngée chez l'enfant: une méthode d'étude des troubles de succion-déglutition et de leur physiopathologie. *Neurophysiol Clin* 1992;**22**:249–60.
63. Verzijl HTFM, Padberg GW, Zwarts MJ. The spectrum of Mobius syndrome: an electrophysiological study. *Brain* 2005;**128**:1728–36.
64. Roig M, Hernandez MA, Salcedo S. Survival from nemaline myopathy in the newborn period. *Ped Neurosci* 1987;**13**:95–7.
65. Wallgren-Petterssona C, Bushby K, Mellies U, Simonds A. 117th ENMC workshop: ventilatory support in congenital neuromuscular disorders—congenital myopathies, congenital muscular dystrophies, congenital myotonic dystrophy and SMA (II). *Neuromuscul Disord* 2004;**14**:56–69.
66. Tilton AH, Miller MD, Khoshoo V. Nutrition and swallowing in pediatric neuromuscular patients. *Semin Pediatr Neurol* 1998;**5**:106–15.
67. Hill M, Hughes T, Milford C. Treatment for swallowing difficulties (dysphagia) in chronic muscle disease. *Cochrane Database Syst Rev* 2004;2:CD004303.
68. Gisel E. Interventions and outcomes for children with dysphagia. *Dev Disabil Res Rev* 2008;**14**:165–73.
69. Sullivan PB, Lambert B, Rose M, Ford-Adams M, Johnson A, Griffiths P. Prevalence and severity of feeding and nutritional problems in children with neurological impairment: Oxford Feeding Study. *Dev Med Child Neurol* 2000;**42**:674–80.
70. Wang CH, Bonnemann CG, Rutkowski A. Consensus statement on standard of care for congenital muscular dystrophies. *J Child Neurol* 2010;**25**:1559–81.
71. Ramelli GP, Aloysius A, King C, Davis T, Muntoni F. Gastrostomy placement in paediatric patients with neuromuscular disorders: indications and outcome. *Dev Med Child Neurol* 2007;**49**:367–71.
72. Pinelli J, Symington A. Non-nutritive sucking for promoting physiologic stability and nutrition in preterm infants. *Cochrane Database Syst Rev* 2005;4:CD001071.
73. Gottrand F, Sullivan PB. Gastrostomy tube feeding: when to start, what to feed and how to stop. *Eur J Clin Nutr* 2010;**64**(Suppl. 1):S17–21.
74. Douzinas EE, Tsapalos A, Dimitrakopoulos A, Diamanti-Kandarakis E, Rapidis AD, Roussos C. Effect of percutaneous endoscopic gastrostomy on gastroesophageal reflux in mechanically-ventilated patients. *World J Gastroenterol* 2006;**12**:114–18.
75. Wilson GJ, Van der Zee DC, Bax NM. Endoscopic gastrostomy placement in the child with gastroesophageal reflux: is concomitant antirreflux surgery indicated? *J Pediatr Surg* 2006;**41**:1441–5.
76. Catto-Smith AG, Jiménez S. Morbidity and mortality after percutaneous endoscopic gastrostomy in children with neurological disability. *J Gastroenterol Hepatol* 2006;**21**:734–8.
77. McLean AA, Alvarez NR, Davies J, Lopez P, Pizano LR. Complications of percutaneous endoscopic and fluoroscopic gastrostomy tube insertion procedures in 378 patients. *Gastroenterol Nurs* 2007;**30**:337–41.
78. Morgan AT, Dodrill P, Ward EC. Interventions for oropharyngeal dysphagia in children with neurological impairment. *Cochrane Database Syst Rev* 2012;10:CD009456.
79. Reid SM, Johnstone BR, Westbury C, Rawicki B, Reddihough DS. Randomized trial of botulinum toxin injections into the salivary glands to reduce drooling in children with neurological disorders. *Dev Med Child Neurol* 2008;**50**:123–8.
80. Cotton RT, Reddihough DS, Erasmus CE, Johnson H, McKellar GMW, Jongerius P. Botulinum toxin assessment, intervention and aftercare for pediatric and adult drooling: international consensus statement. *Eur J Neurol* 2010;**17**(Suppl. 2):S109–21.
81. Chan KH, Liang C, Wilson P, Higgins D, Allen GC. Long-term safety and efficacy data on botulinum toxin type A: an injection for sialorrhea. *JAMA Otolaryngol Head Neck Surg* 2013;**139**:134–8.

82. Suzukia Y, Sano N, Shinonaga C, Fukuda M, Hyodo M, Morimoto T. Successful botulinum toxin treatment of dysphagia in a spinal muscular atrophy type 2 patient. *Brain Dev* 2007;**29**:662–5.
83. Restivo DA, Giuffrida S, Marchese Ragona R, Falsaperla R Successful botulinum toxin treatment of dysphagia in a young child with nemaline myopathy. *Dysphagia* 2001;**16**:228–9.
84. Reed J, Mans CK, Brietzke SE, Reed J, Mans CK, Brietzke SE. Surgical management of drooling: a meta-analysis. *Arch Otolaryngol Head Neck Surg* 2009;**135**:924–31.
85. Terzi N, Orlikowski D, Aegerter P, Lejaille M, Ruquet M, Zalcman G, et al. Breathing-swallowing interaction in neuromuscular patients: a physiological evaluation. *Am J Respir Crit Care Med* 2007;**175**:269–76.
86. Nishino T, Hasegawa R, Ide T, Isono S. Hypercapnia enhances the development of coughing during continuous infusion of water into the pharynx. *Am J Crit Care Med* 1998;**157**:815–21.
87. Mooney JF. Perioperative enteric nutritional supplementation in pediatric patients with neuromuscular scoliosis. *J South Orthop Assoc* 2000;**9**:202–6.
88. Aloysius A, Born P, Kinali M, Davis T, Pane M, Mercuri E. Swallowing difficulties in Duchenne muscular dystrophy: indications for feeding assessment and outcome of videofluoroscopic swallow studies. *Eur J Paediatr Neurol* 2008;**12**:239–45.
89. Pane M, Vasta I, Messina S, Sorleti D, Aloysius A, Sciarra F, et al. Feeding problems and weight gain in Duchenne muscular dystrophy. *Eur J Ped Neurol* 2006;**10**:231–6.

Chapter 48

Complex Regional Pain Syndromes I and II (Reflex Sympathetic Dystrophy, Causalgia)

Navil F. Sethna and Deirdre Logan

INTRODUCTION

The definitions of reflex sympathetic dystrophy (RSD) and causalgia syndromes have been revised recently to avoid implicating the sympathetic nervous system in treatment or mechanism; the syndromes have been renamed with the clinically descriptive terms of complex regional pain syndrome types I and II (CRPS type I and CRPS type II).[1] Given the incomplete understanding of these syndromes, restructuring the diagnostic criteria of RSD to CRPS nomenclature was based on a consensus of the contemporary knowledge about the nature of the disorder revised by international expert clinicians and scientists.[1] According to the International Association of Study of Pain (IASP) and the National Institute of Neurological Disorders and Stroke, CRPS is a chronic pain condition that is probably due to dysfunction in the central and/or peripheral nervous systems.[1] The initial consensus-based diagnostic criteria described a collection of clinical symptoms and signs affecting a region of the body, most commonly after trauma (Table 48.1). CRPS is characterized by abnormal sensory, autonomic, cutaneous and bone changes restricted to commonly distal limb regions. These new diagnostic criteria stimulated additional research to ensure their adequacy. The diagnostic accuracy of these criteria were further improved in a series of controlled trials by comparison to well-defined neuropathic conditions such as diabetic neuropathy (Table 48.1).[2] These revised criteria do not demonstrate high reliability and are not useful clinically in assessing the severity and prognosis or in treating subcategories of symptoms and signs.

Despite the standardization of the diagnostic criteria in adults, there are relatively few data on the validation of the clinical presentation in children. In children both the diagnosis and treatment remain challenging for any clinician. A multidisciplinary approach is often crucial for successful outcome, particularly in severe cases where the child is assessed and managed as an individual as well as a family and community member.

CRPS-I and CRPS-II, like their respective predecessors RSD and causalgia, raise a major question about possible mechanism(s). Cardinal symptoms of CRPS-I and CRPS-II are pain with allodynia (pain from an innocuous stimulus), hyperalgesia (more pain than expected from a painful stimulus), or both. CRPS-II develops following a distinct peripheral nerve injury, whereas CRPS-I usually follows a minor noxious event without obvious nerve injury and with the symptoms spread across multiple peripheral nerve territories (Figure 48.1). It is not clear whether CRPS-I and CRPS-II are different syndromes or represent different features in a continuum of nerve injury. Three stages (acute, dystrophic, and chronic atrophic) of CRPS are described in adults based on the severity or time course of the disorder.[3] In children, the duration of CRPS-I is usually shorter than in adults and so trophic changes are not a common presentation, although such changes are observed.[3–6] This chapter summarizes the current understanding of childhood CRPS type I (CRPS-I) assessment and management.

INCIDENCE AND PREVALENCE

There is evidence that CRPS in children is more common than was previously thought but is rare and the true incidence is unknown. In a rural population of 106,470 in the USA, the incidence of CRPS-I in children between the ages of 10 and 19 years was 1.6 per 10^5 and was nonexistent in younger age children.[7] In adults, depending on how data are gathered and

B.T. Darras, H. Royden Jones, Jr., M.M. Ryan & D.C. De Vivo (Eds): Neuromuscular Disorders of Infancy, Childhood and Adolescence, Second edition.
DOI: http://dx.doi.org/10.1016/B978-0-12-417044-5.00048-2

TABLE 48.1 Diagnostic Criteria for Complex Regional Pain Syndrome

Signs/Symptoms	Criteria
1. Continuing pain	Disproportionate to any inciting pain
2. Symptoms	Present in at least 1 of each of 4 of the following categories
(i) Sensory	Hyperalgesia and/or allodynia
(ii) Vasomotor	Temperature asymmetry and/or skin color change/asymmetry
(iii) Sudomotor/edema	Edema and/or sweating/sweating asymmetry
(iv) Motor/trophic	Decreased range of motion (weakness, tremor, dystonia) and/or trophic changes (hair, nail, skin)
3. Signs*	Presence of at least 2 or more of the following categories at the time of examination
(i) Sensory	Hyperalgesia (to pinprick) and/orallodynia (to light touch) and/or deep pressure and/orjoint movement
(ii) Vasomotor	Temperature asymmetry and/or skin color change/asymmetry
(iii) Sudomotor/edema	Edema and/or sweating/sweating asymmetry
(iv) Motor/trophic	Decreased range of motion (weakness, tremor, dystonia) and/or trophic changes (hair, nail, skin)
4. Other diagnoses	No other diagnosis that better explains the signs and symptoms

**A sign is counted only if observed at time of diagnosis.*
From Harden (2010).[2]

how diagnoses are made, as many as 5.5 per 10^5 person-years are at risk in a rural area of USA based on IASP diagnostic criteria, with a prevalence of 20.6 per 10^5.[7] The estimated overall incidence rate of adult CRPS is much higher in a large cross-section Dutch population, 26 per 10^5 person-years, using different diagnostic criteria.[8] In contrast to adult CRPS, childhood CRPS is marked by (1) predominance of preadolescent females (5–7:1 male); (2) predominance of lower extremity involvement (4–6:1 upper extremity) particularly the ankle and foot; (3) less frequently preceded by surgery or major trauma (in less than 10% spontaneous onset); (4) greater stress in family and school environment; and (5) favorable response to physical therapy.[5,6,9–12]

PATHOPHYSIOLOGY

The pathophysiological mechanisms of CRPS-I are not well understood. In adult trials, no unifying pathophysiological mechanisms have been identified to explain CRPS in all patients. In children, the data are very limited because pain testing trials that invoke pain with repeated application of noxious stimuli are objectionable ethically and are frequently opposed by patients and parents. Several mechanisms are postulated for the generation and/or maintenance of CRPS-I such as small fiber nociceptor axons dysfunction,[13] sympathetic dysregulation,[14,15] tissue hypoxia and reperfusion injury,[16] inflammation,[17,18,19] autoimmune antibodies,[20] and central and cortical reorganization.[21,22] Several studies have shown that immobilization impairs tactile perception and may shrink somatosensory cortical (S1) representation. Mobilization of the affected limb reverses these changes; however, whether pain and/or mobilization causes these cortical reorganizations or the cortical changes cause the pain remains an open question.[23] A preliminary study in children with CRPS showed that allodynic stimulation of the affected limb and comparison to the contralateral unaffected limb and to normal controls resulted in significant changes in CNS circuitry in children with CRPS and having features similar to those described in adults with CRPS.[24] These changes improved after treatment and symptoms resolution.[24]

The precise mechanisms of CRPS-I remain a research challenge. However, they may involve somatosensory and motor and sympathetic nervous systems in a subgroup of CRPS patients in whom the pain could be sympathetically maintained.[25,26] The latter mechanism may explain why in a few CRPS patients a series of temporary sympathetic ganglia blockades with local anesthetics leads to pain relief lasting beyond the pharmacological duration of the local anesthetic action.[25–27]

A recent controlled trial in children with CRPS, ages 7–17 years, demonstrated that thermal and mechanical quantitative sensory abnormalities are strongly suggestive of

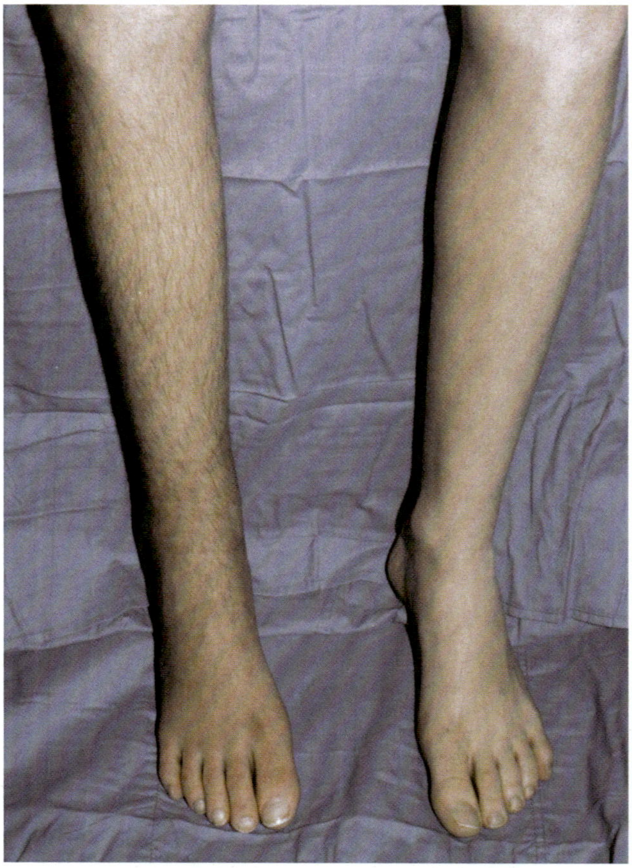

FIGURE 48.1 Unilateral chronic regional pain syndrome I showing swelling, skin discoloration, muscle atrophy, excessive hair growth, and brittle nails.

central sensitization.[28] A follow-up double blind placebo-controlled crossover trial by the same investigators compared intravenous versus lumbar sympathetic blockade with lidocaine in a small number of children (ages 10–18 years) and found that a single lumbar sympathetic blockade produced significant reduction in pain intensity in response to evoked dynamic and static mechanical stimulation, mechanical temporal summation, and spontaneous pain scores in approximately 40% of the patients. The findings of this study provide support for the concept that in some patients a component of pain involves a sympathetic efferent and somatic afferent cycle of activities and suggest that selective sympathetic ganglia blockade can alleviate pain.[27] The long-term benefit of sympathetic blockade in children and adults has not been investigated or demonstrated.[12,29,30]

Other contributory factors have been proposed to influence the development of CRPS, such as genetic association with different human leukocyte antigens (HLA),[31,32] and hereditary patterns among members of families. An observational study of 31 families identified two to five affected families and found that the phenotype of CRPS was more severe in the familial cases compared to sporadic cases.[33] Other studies suggest mitochondrial (female transmission) inheritance in families.[34,35] These preliminary reports suggest that the association of CRPS with genetic factors is multifactorial without a distinct inheritance pattern. The DNA profiling is yet to be determined.

Several psychological conditions have been advanced to explain the triggers and/or maintenance of the CRPS both in children and adults, but no consistent relationship has been demonstrated. As with adults, the psychological aspects are strongly associated with a wide range of chronic disorders in children.[36] Many coexisting factors such as emotional stress, anxiety, depression, fear of movement,[37] and other psychological disturbances have been reported in children with CRPS, as have other behavioral elements such as reinforcing effects of responses to pain,[38,39] pseudoneurologic presentation, and major life events.[6,36,38,40] However, most existing studies have been conducted in samples of children already experiencing chronic pain, thus preventing a thorough understanding of whether psychological distress precedes or follows the onset of the pain symptoms.

In the pediatric literature, research has not examined the role of psychological factors in CRPS as widely or as systematically as in the adult literature. Clinically, however, there exist similar viewpoints and assumptions that CRPS in children may represent primarily a manifestation of psychological distress.[41,42] This dichotomized view of CRPS as either a biological or nonbiological (i.e. psychological) entity is at odds with current conceptualizations of chronic pain as a biopsychosocial phenomenon—that is, an experience that is shaped by a confluence of integrated factors in biological, psychological and social realms.[43,44] Multiple case reports have focused on children with features of CRPS with comorbid psychiatric illness, including conversion disorder, post-traumatic stress disorder, somatization disorder, and anorexia.[40,45–49] A majority of these case reports implicate a psychiatric disorder as primary to the development of CRPS symptoms; however, these reports lack prospective data to support this claim, use single subject designs, rely solely on clinical interview, and fail to apply specific diagnostic criteria when labeling patients as having CRPS. These reports do underscore that psychological distress is a hallmark of pediatric CRPS.

A few studies have been published using larger samples of children with CRPS incorporating psychiatric clinical interviews[5,10,12,22,50,51] or standardized assessments of psychological distress.[6,36,41,52,53] All but one of these interview-based studies concluded that children and adolescents with CRPS had a high prevalence of elevated emotional distress compared to healthy peers and tended to be high-achievers living in high-stress families, citing these as factors that contribute to the syndrome.[5,10,22,50] Conversely, results from studies utilizing standardized assessment measures indicated that children with CRPS routinely displayed normative levels of emotional distress, depressive symptoms, intelligence, and family functioning.[6,36,41,52] A study by Logan and colleagues compared children with CRPS to children with other chronic pain disorders and found similar levels of psychological distress across groups, offering some evidence against the view that CRPS differs from other pediatric chronic pain problems in the severity of accompanying psychological distress.[54]

Clearly, psychological factors are important facets of the overall biopsychosocial experience of CRPS and should be thoroughly assessed and identified as targets of treatment when relevant. Importantly, however, many children presenting for tertiary evaluation of CRPS report having previously been told that the condition is due to stress, or is "all in their heads." While psychological factors in the child and family likely do influence the expression of CRPS in terms of its course, severity, and/or response to treatment, assuming that children fit a particular personality type or have significant psychiatric involvement in the presentation of their pain condition based on a CRPS diagnosis alone is unfounded and may be detrimental to children and their families seeking to understand an often difficult and confusing diagnosis. Certainly, the identification and treatment of psychological concomitants, such as symptoms of depression or anxiety, and of interpersonal factors that can influence the course of the disorder, such as family dynamics and parental responses to pain behaviors, are important goals in managing pediatric CRPS. Yet it is crucial to recognize that CRPS presents as a multifaceted biopsychosocial phenomenon; consequently, treatment should entail coordinated interdisciplinary efforts that address the biological, physical, and psychosocial aspects and sequelae of this complex condition.

A recent comprehensive review of literature of epidemiological, genetic, and experimental studies in adults with CRPS concluded that CRPS is a complex disorder and implicated multiple mechanisms in the pathophysiology that results in maladaptive neuroplasticity in most patients.[55]

DIAGNOSIS

CRPS is a clinical diagnosis based on the medical history and signs and symptoms listed by IASP criteria (Table 48.1). The symptoms and signs indicate a multifactorial disorder associated with aberrant inflammatory response to tissue injury, vasomotor dysfunction, and maladaptive neuroplasticity.[55] At present there are no specific pathologic, histologic, or biochemical markers for its diagnosis. The laboratory tests and radiological imaging studies are helpful to exclude underlying causes or conditions that might mimic the CRPS presentation, e.g. infection, stress fracture, ligaments or meniscal tear, arthritis, erythromelalgia, etc. Over time, the disorder may spread to adjacent noninjured areas, and ipsilateral and/or contralateral limb.[56,57]

Quantitative sensory testing has recently been introduced to assess the functioning of cutaneous somatic small fibers in adults with neuropathy and painful conditions, including CRPS. In a preliminary study in children with CRPS, the investigators found various abnormalities, which suggest considerable heterogeneity of evoked mechanical and thermal sensory abnormalities among children presenting with CRPS-related pain.[28] As with adult CRPS, the thermal and mechanical sensory abnormalities appear in different combinations in different patients with similar clinical presentations. In a majority of patients, the pathogenesis of pain is seemingly of central origin.[28] A large-scale adult study compared healthy adults with patients with various neuropathic pain syndromes and CRPS; it noted that although most patients with CRPS exhibited gain in sensory function (hyperesthesia, allodynia, hyperalgesia, hyperpathia) a number of patients presented with a mixed gain and loss of sensory function.[58]

Early diagnosis may improve prognosis and mitigate the unnecessary suffering from delayed diagnosis or misdiagnosis that may stigmatize the patient's condition as not "real" or "all in the head," and avoid disuse atrophy, poor physical functioning, and in some cases exposures to unwarranted and potentially harmful interventional procedures that have not clearly shown to provide additional benefit.[12,59]

TREATMENT

In view of lack of proven medical cure and involvement of multiple systems (skin, nerves, muscle, bone, emotional distress, somatosensory reorganization) this condition demands the expertise of multiple disciplines. Multidisciplinary treatment models have led to significant advances in clinical and scientific understanding as well as improvement of care guided by the principles of biopsychosocial treatment.

Patient Education

Provision of knowledge to patients and families regarding the nature of CRPS-I and factors exacerbating and sustaining the pain and disability is essential to alleviate concerns and fears, and improve acceptance and compliance with realistic expectations of the course of the condition and its responses to various therapies.

Collaboration between health-care providers and patients and their families is central in setting and achieving the child- and family-centered goals for successful management of the chronic pain and functional rehabilitation. Educating the patient and family about the complexity of the disorder can help them understand the symptoms, particularly the nonprotective nature of the chronic pain and concerns of harm with the mobilization of the affected painful limb. Conversely, immobility invariably results in muscle atrophy, bone osteopenia, and loss of function and general deconditioning.

Reassuring the patient and family of the absence of detectable pathology by clinical and laboratory studies may, in many instances, allay the child's and family's concerns of overlooked serious disease and decrease the overall healthcare utilization. Parents are educated to coach their children to assume appropriate independent responsibilities in their self-care and to reinforce the positive behaviors in resuming age-appropriate quality of life.

Pharmacological Treatment

In adults with CRPS, the level of evidence from a recent meta-analysis of the randomized controlled trials for efficacy of pharmacological therapies to alleviate pain is generally weak.[60] However, there are emerging novel treatments that show efficacy or lack of efficacy in small controlled trials with fewer than 50 patients per trial and which warrant further confirmatory evidence in better quality trials with a larger number of patients and looking at long-term outcomes.[60] The trials with positive outcomes included use of diverse agents: bisphosphonate therapy (four randomized controlled trials (RCTs) intravenous and oral), parecoxib, oral tadalafil, low-dose intravenous ketamine infusion (two RCTs of moderate evidence), lumbar sympathetic block (limited evidence), low-dose intravenous immunoglobulin, and memantine. There is moderate evidence in two RCTs for efficacy of vitamin C to prevent CRPS after dorsal radius facture in adults.[60]

Another more recent Cochrane review of a broad range of treatments of pharmacological and physical rehabilitation and interventional pain procedures including surgery, concluded that generally the quality of evidence was low for the various individual therapies and the results did not lead to clinically important and sustained benefit at one year follow-up.[61]

In children the data on the pharmacological interventions are limited to case reports and case series on the use of various analgesic agents and interventions that have been trialed in adults with CRPS with variable success.[5,9,12,45,50,52,62–64] There is only one RCT double blinded and crossover study on use of lumbar sympathetic blockade for management of CRPS with limited and short-term improvement.[27]

Multidisciplinary Management

In view of the multifactorial and multisystem nature of CRPS-I, it warrants an integrated multidisciplinary management individualized to the patient's needs. The overall goals are to minimize pain and restore functioning of the affected limb and return to normal activities appropriate for age.

There is strong evidence of effectiveness in six RCTs of traditional and graded motor imagery models that physical rehabilitation can reduce pain and improve function in adults with CRPS.[60] Similar rigorous trials have not been performed in children.

Remission of pediatric CRPS occurs in most patients with outpatient multidisciplinary care. In a few children the outpatient management may fail for a variety of reasons such as negative experience with healthcare providers, misconception despite education, anxiety and depression, poor coping skills, performance pressure in school and sports, and family and social positive reinforcement of the illness behavior (e.g. guarding, immobilization of the affected limb, over-use of assistive devices such as braces, crutches, wheelchairs, or rolling walkers, over-reliance on analgesics). In these circumstances, the interdisciplinary rehabilitation offers a holistic approach to optimize functioning in all domains of physical, psychological, cognitive, and social care, school participation, and normalization of the quality of life.[4,10]

Occupational and Physical Therapies

The goal of occupational therapy is to diminish the cutaneous and deep tissue hypersensitivity (allodynia) of the affected limb by a process of regulating the adaptability of neural processing through daily and graded exposure to various tactile stimuli and texture discrimination (e.g. touch, warm, cold, vibration, different textures, etc.) and gradual resumption of self-care and activities of daily living.

Several barriers to mobilization of the affected limb are movement allodynia, fear of movement, anxiety, poor or lack of coping skills, and learned disuse (neglect) leading to self-defeating behaviors and chronification of CRPS disability. For these reasons in some cases it is challenging to engage patients in outpatient multidisciplinary therapy but this strategy may fail even when the patient is offered frequent physical therapy, e.g. two to three times a week. The goal of physical therapy is a stepwise progression of mobilization of the affected limb by elimination of the assistive devices, application of the passive stretching initially and gradual exposure to isotonic followed by isometric exercises, strengthening, proprioception training, active use of the affected joints through normal range of mobility (i.e. desensitization of the movement allodynia), and general conditioning (aerobic fitness), posture correction, and gait normalization. Further advancement of goals is accomplished through graded restoration of independent activities of daily living and return to social and recreational activities and school attendance. The introduction of the novel physical therapy approach of graded motor imagery, including the mirror visual feedback therapy, has accelerated the self-initiation of the affected limb movement. The mirror creates a virtual reality feedback to the brain to initiate movement of the affected limb by visual feedback from the movement of the mirror symmetric location of the reverse image (an illusion of the contralateral unaffected limb). This therapy can hasten motor recovery by "resetting" the positive adaptive neural function.[65–67]

Psychological Support

CRPS is a complex condition best understood and treated through a biopsychosocial framework[68] in which a multitude of factors in the individual and environment influence the course of the condition, expression of symptoms, and the course of treatment. The psychological component of treatment typically follows a cognitive-behavioral (CBT) model, a highly effective, evidence based approach for pain rehabilitation.[69] Self-management of the pain is often emphasized, and the treatment focuses on skills such as relaxation techniques (e.g. deep breathing, relaxation, guided imagery, biofeedback training), problem-solving, and developing coping strategies for managing stressful life events. More recently, acceptance and commitment therapy based approaches have been incorporated into the psychological treatment of pediatric CRPS to help patients learn to return to valued activities in spite of pain.[70] There are varying philosophies on the role of parents and families in the treatment of pediatric CRPS, but there is some evidence to support the importance of incorporating parents as active participants in the psychological treatment.[71]

Long-Term Care of CRPS

Children and adolescents with CRPS often experience transient recurrent flares of pain spontaneously or more commonly following re-injuries with an incidence ranging from 20% to 36%.[5,6,52,72] These pain episodes could be alarming to the patients and families because of fear of recurrence or worsening of the CRPS. Therefore, follow-up care of these patients should be afforded by pain medicine clinicians, primary care clinics, or rehabilitation programs for providing reassurance, necessary care to manage the acute flares, and redirection of the patient to self-management strategies of desensitization, home exercises, and coping skills to maintain an active and independent life style.

PROGNOSIS

Childhood CRPS presents with a wide clinical spectrum that ranges from mild complaints and a self-limited course to severe pain and impairment of limb function. In contrast to adult CRPS, the time course of illness in childhood CRPS is shorter and prognosis is favorable in most children.[5,6,10,12,54] A chronic or transient relapsing course may also occur in a minority of patients. Early treatment offers better chances for shortening the course of the functional disability and associated pschological distress and may prevent osseous growth deformities and acquired limb deformities that may require surgical correction.[39,73] Initial outpatient conservative aggressive therapy of mobilization, psychological support, and a trial of pharmacologic and nonpharmacologic interventions appear effective in the majority of patients with childhood CRPS. A few children with longstanding CRPS and/or severe psychosocial distress will require a more intensive interdisciplinary rehabilitation approach for management.

REFERENCES

1. Stanton-Hicks M, Janig W, Hassenbusch S, Haddox JD, Boas R, Wilson P. Reflex sympathetic dystrophy: changing concepts and taxonomy. *Pain* 1995;**63**(1):127–33.
2. Harden RN. Objectification of the diagnostic criteria for CRPS. *Pain Med* 2010;**11**(8):1212–15.
3. Schwartzman RJ, Erwin KL, Alexander GM. The natural history of complex regional pain syndrome. *Clin J Pain* 2009;**25**(4):273–80.
4. Logan DE, Carpino EA, Chiang G, Condon M, Firn E, Gaughan VJ, et al. A day-hospital approach to treatment of pediatric complex regional pain syndrome: initial functional outcomes. *Clin J Pain* 2012;**28**(9):766–74.
5. Low AK, Ward K, Wines AP. Pediatric complex regional pain syndrome. *J Pediatr Orthop* 2007;**27**(5):567–72.
6. Sherry DD, Wallace CA, Kelley C, Kidder M, Sapp L. Short- and long-term outcomes of children with complex regional pain syndrome type I treated with exercise therapy. *Clin J Pain* 1999;**15**(3):218–23.
7. Sandroni P, Benrud-Larson LM, McClelland RL, Low PA. Complex regional pain syndrome type I: incidence and prevalence in Olmsted County, a population-based study. *Pain* 2003;**103**(1–2):199–207.
8. de Mos M, de Bruijn AG, Huygen FJ, Dieleman JP, Stricker BH, Sturkenboom MC. The incidence of complex regional pain syndrome: a population-based study. *Pain* 2007;**129**(1–2):12–20.
9. Dadure C, Motais F, Ricard C, Raux O, Troncin R, Capdevila X. Continuous peripheral nerve blocks at home for treatment of recurrent complex regional pain syndrome I in children. *Anesthesiology* 2005;**102**(2):387–91.
10. Stanton-Hicks M. Plasticity of complex regional pain syndrome (CRPS) in children. *Pain Med* 2010;**11**(8):1216–23.
11. Wesdock KA, Stanton RP, Singsen BH. Reflex sympathetic dystrophy in children. A physical therapy approach. *Arthritis Care Res* 1991;**4**(1):32–8.
12. Wilder RT, Berde CB, Wolohan M, Vieyra MA, Masek BJ, Micheli LJ. Reflex sympathetic dystrophy in children. Clinical characteristics and follow-up of seventy patients. *J Bone Joint Surg: Am Vol* 1992;**74**(6):910–19.
13. Oaklander AL, Rissmiller JG, Gelman LB, Zheng L, Chang Y, Gott R. Evidence of focal small-fiber axonal degeneration in complex regional pain syndrome-I (reflex sympathetic dystrophy). *Pain* 2006;**120**(3):235–43.
14. Wasner G, Heckmann K, Maier C, Baron R. Vascular abnormalities in acute reflex sympathetic dystrophy (CRPS I): complete inhibition of sympathetic nerve activity with recovery. *Arch Neurol* 1999;**56**(5):613–20.
15. Schattschneider J, Hartung K, Stengel M, Ludwig J, Binder A, Wasner G, et al. Endothelial dysfunction in cold type complex regional pain syndrome. *Neurology* 2006;**67**(4):673–5.
16. Coderre TJ, Bennett GJ. A hypothesis for the cause of complex regional pain syndrome-type I (reflex sympathetic dystrophy): pain due to deep-tissue microvascular pathology. *Pain Med* 2010;**11**(8):1224–38.

17. Huygen FJ, De Bruijn AG, De Bruin MT, Groeneweg JG, Klein J, Zijlstra FJ. Evidence for local inflammation in complex regional pain syndrome type 1. *Mediators Inflamm* 2002;**11**(1):47–51.
18. Kingery WS. Role of neuropeptide, cytokine, and growth factor signaling in complex regional pain syndrome. *Pain Med* 2010;**11**(8):1239–50.
19. Taha R, Blaise GA. Update on the pathogenesis of complex regional pain syndrome: role of oxidative stress. *Can J Anaesth* 2012;**59**(9):875–81.
20. Goebel A, Leite MI, Yang L, Deacon R, Cendan CM, Fox-Lewis A, et al. The passive transfer of immunoglobulin G serum antibodies from patients with longstanding Complex Regional Pain Syndrome. *Eur J Pain* 2011;**15**(5):504.e1–6.
21. Maihofner C, Handwerker HO, Neundorfer B, Birklein F. Patterns of cortical reorganization in complex regional pain syndrome. *Neurology* 2003;**61**(12):1707–15.
22. Maillard SM, Davies K, Khubchandani R, Woo PM, Murray KJ. Reflex sympathetic dystrophy: a multidisciplinary approach. *Arthritis Rheum* 2004;**51**(2):284–90.
23. Di Pietro F, McAuley JH, Parkitny L, Lotze M, Wand BM, Moseley GL, et al. Primary somatosensory cortex function in complex regional pain syndrome: a systematic review and meta-analysis. *J Pain* 2013;**14**(11):1270–88.
24. Lebel A, Becerra L, Wallin D, Moulton EA, Morris S, Pendse G, et al. fMRI reveals distinct CNS processing during symptomatic and recovered complex regional pain syndrome in children. *Brain: J Neurol* 2008;**131**(Pt 7):1854–79.
25. Krumova EK, Gussone C, Regeniter S, Westermann A, Zenz M, Maier C. Are sympathetic blocks useful for diagnostic purposes? *Reg Anesth Pain Med* 2011;**36**(6):560–7.
26. Price DD, Long S, Wilsey B, Rafii A. Analysis of peak magnitude and duration of analgesia produced by local anesthetics injected into sympathetic ganglia of complex regional pain syndrome patients. *Clin J Pain* 1998;**14**(3):216–26.
27. Meier PM, Zurakowski D, Berde CB, Sethna NF. Lumbar sympathetic blockade in children with complex regional pain syndromes: a double blind placebo-controlled crossover trial. *Anesthesiology* 2009;**111**(2):372–80.
28. Sethna NF, Meier PM, Zurakowski D, Berde CB. Cutaneous sensory abnormalities in children and adolescents with complex regional pain syndromes. *Pain* 2007;**131**(1–2):153–61.
29. Perez RS, Zollinger PE, Dijkstra PU, Thomassen-Hilgersom IL, Zuurmond WW, Rosenbrand KC, et al. Evidence based guidelines for complex regional pain syndrome type 1. *BMC Neurol* 2010;**10**:20.
30. Sethna NF, Berde CB. Sympathetic nerve blocks, pragmatic trials, and responder analysis. *Anesthesiology* 2012;**116**(1):12–14.
31. van Hilten JJ, van de Beek WJ, Roep BO. Multifocal or generalized tonic dystonia of complex regional pain syndrome: a distinct clinical entity associated with HLA-DR13. *Ann Neurol* 2000;**48**(1):113–16.
32. van de Beek WJ, van Hilten JJ, Roep BO. HLA-DQ1 associated with reflex sympathetic dystrophy. *Neurology* 2000;**55**(3):457–8.
33. de Rooij AM, de Mos M, Sturkenboom MC, Marinus J, van den Maagdenberg AM, van Hilten JJ. Familial occurrence of complex regional pain syndrome. *Eur J Pain* 2009;**13**(2):171–7.
34. Higashimoto T, Baldwin EE, Gold JI, Boles RG. Reflex sympathetic dystrophy: complex regional pain syndrome type I in children with mitochondrial disease and maternal inheritance. *Arch Dis Child* 2008;**93**(5):390–7.
35. Tan EC, Janssen AJ, Roestenberg P, van den Heuvel LP, Goris RJ, Rodenburg RJ. Mitochondrial dysfunction in muscle tissue of complex regional pain syndrome type I patients. *Eur J Pain* 2011;**15**(7):708–15.
36. Sherry DD, Weisman R. Psychologic aspects of childhood reflex neurovascular dystrophy. *Pediatrics* 1988;**81**(4):572–8.
37. Simons LE, Kaczynski KJ. The Fear Avoidance model of chronic pain: examination for pediatric application. *J Pain* 2012;**13**(9):827–35.
38. Agrawal SK, Rittey CD, Harrower NA, Goddard JM, Mordekar SR. Movement disorders associated with complex regional pain syndrome in children. *Dev Med Child Neurol* 2009;**51**(7):557–62.
39. Majumdar A, Lopez-Casas J, Poo P, Colomer J, Galvan M, Lingappa L, et al. Syndrome of fixed dystonia in adolescents—short term outcome in 4 cases. *Eur J Paediatr Neurol* 2009;**13**(5):466–72.
40. Pearson RD, Bailey J. Complex regional pain syndrome in an 8-year-old female with emotional stress during deployment of a family member. *Mil Med* 2011;**176**(8):876–8.
41. Cruz N, O'Reilly J, Slomine BS, Salorio CF. Emotional and neuropsychological profiles of children with complex regional pain syndrome type-I in an inpatient rehabilitation setting. *Clin J Pain* 2011;**27**(1):27–34.
42. Feliu MH, Edwards CL. Psychologic factors in the development of complex regional pain syndrome: history, myth, and evidence. *Clin J Pain* 2010;**26**(3):258–63.
43. Gatchel RJ, Peng YB, Peters ML, Fuchs PN, Turk DC. The biopsychosocial approach to chronic pain: scientific advances and future directions. *Psychol Bull* 2007;**133**(4):581–624.
44. Wilder RT. Management of pediatric patients with complex regional pain syndrome. *Clin J Pain* 2006;**22**(5):443–8.
45. Cimaz R, Matucci-Cerinic M, Zulian F, Falcini F. Reflex sympathetic dystrophy in children. *J Child Neurol* 1999;**14**(6):363–7.
46. Guler-Uysal F, Basaran S, Geertzen JH, Goncu K. A 2 1/2-year-old girl with reflex sympathetic dystrophy syndrome (CRPS type I): case report. *Clin Rehabil* 2003;**17**(2):224–7.
47. Jaworowski S, Allen RC, Finkelstein E. Reflex sympathetic dystrophy in a 12-year-old twin with comorbid conversion disorder in both twins. *J Paediatr Child Health* 1998;**34**(6):581–3.
48. Karakaya I, Coskun A, Agaoglu B, Iseri P, Inanir M, Canatay H. Psychiatric approach in the treatment of reflex sympathetic dystrophy in an adolescent girl: a case report. *Turk J Pediatr* 2006;**48**(4):369–72.
49. Silber TJ. Anorexia nervosa and reflex sympathetic dystrophy syndrome. *Psychosomatics* 1989;**30**(1):108–11.
50. Kachko L, Efrat R, Ben Ami S, Mukamel M, Katz J. Complex regional pain syndromes in children and adolescents. *Pediatrics Int* 2008;**50**(4):523–7.
51. Harden RN, Oaklander AL, Burton AW, Perez RS, Richardson K, Swan M, et al. Complex regional pain syndrome: practical diagnostic and treatment guidelines, 4th edition. *Pain Med* 2013;**14**(2):180–229.
52. Lee BH, Scharff L, Sethna NF, McCarthy CF, Scott-Sutherland J, Shea AM, et al. Physical therapy and cognitive-behavioral treatment for complex regional pain syndromes. *J Pediatr* 2002;**141**(1):135–40.

53. Sherry DD, McGuire T, Mellins E, Salmonson K, Wallace CA, Nepom B. Psychosomatic musculoskeletal pain in childhood: clinical and psychological analyses of 100 children. *Pediatrics* 1991;**88**(6): 1093–9.
54. Logan DE, Williams SE, Carullo VP, Claar RL, Bruehl S, Berde CB. Children and adolescents with complex regional pain syndrome: more psychologically distressed than other children in pain? *Pain Res Manag* 2013;**18**(2):87–93.
55. Marinus J, Perez RS, van Eijs F, van Gestel MA, Geurts JW, Huygen FJ, et al. The role of pain coping and kinesiophobia in patients with complex regional pain syndrome type 1 of the legs. *Clin J Pain* 2013;**29**(7):563–9.
56. Maleki J, LeBel AA, Bennett GJ, Schwartzman RJ. Patterns of spread in complex regional pain syndrome, type I (reflex sympathetic dystrophy). *Pain* 2000;**88**(3):259–66.
57. Rommel O, Gehling M, Dertwinkel R, Witscher K, Zenz M, Malin JP, et al. Hemisensory impairment in patients with complex regional pain syndrome. *Pain* 1999;**80**(1–2):95–101.
58. Maier C, Baron R, Tolle TR, Binder A, Birbaumer N, Birklein F, et al. Quantitative sensory testing in the German Research Network on Neuropathic Pain (DFNS): somatosensory abnormalities in 1236 patients with different neuropathic pain syndromes. *Pain* 2010;**150** (3):439–50.
59. Olsson GL, Meyerson BA, Linderoth B. Spinal cord stimulation in adolescents with complex regional pain syndrome type I (CRPS-I). *Eur J Pain* 2008;**12**(1):53–9.
60. Cossins L, Okell RW, Cameron H, Simpson B, Poole HM, Goebel A. Treatment of complex regional pain syndrome in adults: a systematic review of randomized controlled trials published from June 2000 to February 2012. *Eur J Pain* 2013;**17**(2):158–73.
61. O'Connell NE, Wand BM, McAuley J, Marston L, Moseley GL. Interventions for treating pain and disability in adults with complex regional pain syndrome. *Cochrane Database Syst Rev* 2013;4: CD009416.
62. Brooke V, Janselewitz S. Outcomes of children with complex regional pain syndrome after intensive inpatient rehabilitation. *PMR: J Inj Funct Rehabil* 2012;**4**(5):349–54.
63. Franklin A, Austin T. The use of a continuous brachial plexus catheter to facilitate inpatient rehabilitation in a pediatric patient with refractory upper extremity complex regional pain syndrome. *Pain Pract* 2013;**13**(2):109–13.
64. Kumar R, Aggarwal A, Faridi MM. Complex regional pain syndrome type 1 and scurvy. *Indian Pediatr* 2009;**46** (6):529–31.
65. McCabe CS, Haigh RC, Blake DR. Mirror visual feedback for the treatment of complex regional pain syndrome (type 1). *Curr Pain Headache Rep* 2008;**12**(2):103–7.
66. McCabe CS, Haigh RC, Ring EF, Halligan PW, Wall PD, Blake DR. A controlled pilot study of the utility of mirror visual feedback in the treatment of complex regional pain syndrome (type 1). *Rheumatology (Oxford)* 2003;**42**(1):97–101.
67. Moseley GL. Graded motor imagery for pathologic pain: a randomized controlled trial. *Neurology* 2006;**67**(12):2129–34.
68. Stanton-Hicks MD, Burton AW, Bruehl SP, Carr DB, Harden RN, Hassenbusch SJ, et al. An updated interdisciplinary clinical pathway for CRPS: report of an expert panel. *Pain Pract* 2002;**2** (1):1–16.
69. Eccleston C, Palermo TM, de C Williams AC, Lewandowski A, Morley S, Fisher E, et al. Psychological therapies for the management of chronic and recurrent pain in children and adolescents. *Cochrane Database Syst Rev* 2012;12:CD003968.
70. Wicksell RK, Melin L, Lekander M, Olsson GL. Evaluating the effectiveness of exposure and acceptance strategies to improve functioning and quality of life in longstanding pediatric pain—a randomized controlled trial. *Pain* 2009;**141** (3):248–57.
71. Gaughan V, Logan D, Sethna N, Mott S. Parents' perspective of their journey caring for a child with chronic neuropathic pain. *Pain Manag Nurs* 2014;**15**(1):246–57.
72. Murray CS, Cohen A, Perkins T, Davidson JE, Sills JA. Morbidity in reflex sympathetic dystrophy. *Arch Dis Child* 2000;**82**(3):231–3.
73. Rush PJ, Wilmot D, Saunders N, Gladman D, Shore A. Severe reflex neurovascular dystrophy in childhood. *Arthritis Rheum* 1985;**28**(8):952–6.

Chapter 49

Friedreich Ataxia

Massimo Pandolfo

INTRODUCTION

In 1863, Nicholaus Friedreich, professor of medicine in Heidelberg, Germany, wrote three articles "about a degenerative atrophy of the posterior columns of the spinal cord" causing progressive ataxia, sensory loss, and muscle weakness. The cause of what was called "Friedreich ataxia" (now abbreviated as FRDA) remained mysterious for more than 130 years thereafter. For many years, even the definition of this entity, as well as its inheritance, was somewhat blurred by confusing similarities with different inherited ataxias and neuropathies that were subsequently described. Landmark studies carried out in the late 1970s and early 1980s established the autosomal recessive pattern of inheritance and defined rigorous diagnostic criteria.[1,2] These studies were key for the collection of clinically homogeneous families to be used for biochemical studies, and, after 1985, for molecular genetic studies. Eventually, the FRDA gene (*FXN*), and its major mutation, an intronic GAA triplet repeat expansion, were discovered in 1996.[3] Thanks to this discovery, genetic testing, genotype-phenotype correlations, pathophysiological studies, and new approaches to treatment are becoming possible.

EPIDEMIOLOGY

In Caucasians, FRDA accounts for almost half of the cases of recessive ataxia.[4,5] It has a prevalence of around 2×10^{-5} in Western Europe. This figure declines as one moves towards the East and North, likely because of a distant founder effect.[6] Additionally, there are local clusters of FRDA due to a founder effect, such as those observed in Rimouski, in Québec, and in Kathikas-Arodhes, Cyprus. FRDA is almost nonexistent in Japan and possibly in all non-Caucasian populations (see below).

PATHOLOGY

Central Nervous System

Friedreich considered the degeneration of the posterior columns of the spinal cord as the hallmark of the disease.[7] The appearance of the posterior columns is shrunken, grayish, and translucent. Atrophy is most severe for the fibers originating more caudally, therefore affecting the Goll more than the Burdach tract. Clarke's column, which receives afferents from collaterals of axons in the dorsal columns, shows profound neuronal loss. Spinocerebellar tracts, which originate from Clarke's column, also are atrophic, dorsal elements more than ventral.[8] This finding may be the only known example of trans-synaptic degeneration in humans. The sensory system involvement continues rostrally into the brainstem, with atrophy of the gracilis and cuneate nuclei and of the medial lemnisci, particularly in the ventral portion deriving from the gracile nuclei. Atrophy extends to the cranial nerve sensory nuclei and the entering roots of cranial nerves V, IX, and X, the descending trigeminal tracts, and the solitary tracts.[8] The accessory cuneate nuclei, corresponding to Clarke's column in the spinal cord, are markedly atrophic. Therefore, the sensory systems providing information to the cerebrum and cerebellum regarding the position and speed of body segments, particularly the lower limbs, are severely compromised in FRDA.

In the motor system, the long crossed and uncrossed corticospinal motor tracts are atrophic, suggesting a "dying back" process because atrophy is more severe distally. Loss of pyramidal neurons in the motor cortex is variable, while motor neurons in the brainstem and in the ventral horns are preserved.[8]

In the cerebellum, atrophy of the dentate nuclei and the superior cerebellar pedunculi is most prominent,[8] whereas cortical atrophy tends to occur late and be mild.

B.T. Darras, H. Royden Jones, Jr., M.M. Ryan & D.C. De Vivo (Eds): Neuromuscular Disorders of Infancy, Childhood and Adolescence, Second edition.
DOI: http://dx.doi.org/10.1016/B978-0-12-417044-5.00049-4

Quantitative analysis of synaptic terminals suggests a loss of contacts over Purkinje cell bodies and proximal dendrites. Neuropathology studies in subjects affected with FRDA who died at different ages indicate that sensory system involvement occurs very early and may even be developmental, while the dentate nucleus and pyramidal tracts are affected later as disease progresses.[9]

Areas of variable involvement include the auditory system and the optic nerves and tracts. The basal ganglia may show moderate cell loss in the external pallidus and subthalamic nuclei in some cases. The thalamus, striatum, and substantia nigra do not appear to be directly involved by the disease. Finally, because many patients with FRDA die as a consequence of heart disease, widespread hypoxic changes and focal infarcts are often found in the central nervous system.

In the peripheral nervous system, the major finding is the loss of large primary sensory neurons in the dorsal root ganglia (DRG).[10,11] Peripheral nerves show loss of large myelinated sensory fibers,[12] while the fine, unmyelinated fibers are well preserved. Interstitial connective tissue is increased. This sensory axonal neuropathy is an early event in the course of the disease and is scarcely progressive.

In summary, FRDA causes a profound loss of large dorsal root ganglia neurons, leading to degeneration of the central and peripheral branches of their axons, and probably to trans-synaptic degeneration of Clarke's column with consequent atrophy of the spinocerebellar tracts. Primary sensory neurons in cranial nerves are sometimes atrophic as well, typically in the acoustic nerve where atrophy leads to sensorineural hearing loss. In addition, the dentate nucleus and the distal portion of the corticospinal motor tracts undergo progressive atrophy. Therefore, patients with FRDA have a deficit of deep sensory input to the cerebrum and the cerebellum as well as a deficient cerebellar output, producing a mixed cerebellar and sensory ataxia. The simultaneous presence of axonal sensory neuropathy and pyramidal tract degeneration leads to the typical combination of absent tendon reflexes and extensor plantar responses.

Heart

Heart pathology in FRDA typically consists of hypertrophic cardiomyopathy with thickening of ventricular walls and interventricular septum, evolving into a dilatative cardiomyopathy in the late stage of the disease.[13,14] Hypertrophic and normal-appearing cardiomyocytes intermingle with atrophic, degenerating, even necrotic fibers. Connective tissue is increased, and diffuse and focal inflammatory cell infiltration may be observed. A variable number of cardiomyocytes, from less than 1% to more than 10%, have intracellular iron deposits but do not necessarily show morphological changes indicative of degeneration. This finding is specific to FRDA and may reflect the basic biochemical defect.[15]

Other Organs

Kyphoscoliosis is observed in more than three-quarters of the patients, mostly as a double thoracolumbar curve resembling the idiopathic form of scoliosis (see Case Example 49.1). Pes cavus, pes equinovarus, and clawing of the toes are found in about half the cases and thought to be secondary to nerve degeneration. Late in the disease, patients show a loss of islet cells without the signs of autoimmune inflammatory reaction found in type 1 diabetes.[16,17]

CLINICAL ASPECTS

Onset

The onset of FRDA occurs most commonly around puberty, but may occur earlier as well as in adult life.[18,19] Molecular genetic studies have shown that late-onset patients (LOFA) carry mutations in the same gene as in typical FRDA cases.[20,21] Large variations in age of onset within a sibship are occasionally observed, and are now explained at least in part by the dynamic nature of the mutation.

The typical patient at onset is a child who recently started to sway when walking and who falls easily. As in many degenerative diseases, an acute illness or injury may precede the appearance of the symptoms. Sometimes the child was active in sports and there were no clues of the impending neurological illness. More commonly the child had been considered "clumsy" for some time before the overt appearance of symptoms. Presentations with gait instability or generalized clumsiness are most common, but non-neurological manifestations such as scoliosis (often considered to be idiopathic) or cardiomyopathy (5%) may precede the onset of ataxia. Rarely, patients are diagnosed with idiopathic hypertrophic cardiomyopathy and treated for up to 2–3 years before neurological symptoms appear.[14]

Neurological Signs and Symptoms

Ataxia with mixed cerebellar and sensory features is the hallmark of the disease,[18,19] usually beginning as truncal ataxia causing swaying, reduced balance, and occasional falls. The gait abnormality is initially subtle and may reveal itself only as difficulty in tandem gait. Subsequently, the ataxic gait becomes evident with irregular steps, veering, and difficulty in turning. Inability to stand on one foot progresses to instability when standing

CASE EXAMPLE 49.1

History

The patient is a 14-year-old right-handed boy who has two older sisters with a similar disease. He was born at term after an uneventful pregnancy. Development was normal and he was healthy until approximately age 5, when he started to sway when walking. His gait worsened gradually. He is able to run, but has suffered minor injuries from falls. He does not have numbness or tingling. He is able to write and has no difficulties using his arms. There is no bowel or bladder incontinence and he has no difficulty with swallowing or with his speech. He is bothered most by feeling "dizzy" every 2 to 3 weeks, which at times requires him to sit. He has a poor appetite.

The patient has no history of medical or surgical illnesses and does not take any medication.

Physical Examination

The patient is a thin boy in apparent good general health. The general physical examination is negative except for mild kyphoscoliosis.

Neurological Examination

Mental Status: The patient is well oriented. He can write a sentence and prints his name without difficulty. He recalls 3 out of 3 objects at 0 and 3 minutes.

Cranial Nerves II–XII: Speech is fluent without dysarthria. Extraocular movements are full; there is mild gaze-evoked nystagmus and fixation instability. The rest of the cranial nerve examination is normal.

Motor Examination: There is marked proximal and distal atrophy of all four extremities, especially the arms, without fasciculations. Strength is 5/5 proximally in the upper extremities with 5 − grips bilaterally and 5/5 strength distally in the lower extremities. Tone is normal in the axial muscles and in all four extremities.

Reflexes: Absent proximally and distally in the upper and lower extremities. The plantar responses are flexor bilaterally.

Sensation: Intact to light touch, pinprick, proprioception, temperature and vibration.

Coordination: There is mild ataxia on heel-to-shin testing more than finger-to-nose testing. Mild dysdiadochokinesia. Writing and fine hand movements are normal.

Gait: Romberg sign is present. He is able to walk on his toes and heels, and there is good stride and arm swing. He has a slightly ataxic gait and has difficulty with tandem gait.

Nerve Conduction Studies

Mild slowing of MCV of right common peroneal nerve (36.5 m/sec), left common peroneal nerve (38.5 m/sec), left tibial nerve (37.9 m/sec). Absent SNAPs of the left median, ulnar, superficial peroneal, and bilateral sural nerves with absent left H-reflex.

MRI of the Brain and Spine

The upper cervical cord is slightly reduced in size without definite intrinsic signal abnormality. The posterior fossa and supratentorial compartment structures are normal.

EKG

Diffuse anomalies of repolarization with inverted T-waves in most derivations.

Echocardiography

Thickness of the left ventricular wall and of the interventricular septum are at the upper limit of normal. Heart mass is increased (132 g/m^2, normal <120).

Laboratory Tests

Routine laboratories, thyroid studies, anti-GM1, phytanic acid, vitamin E levels were normal. A FRDA molecular test revealed two expanded alleles corresponding to 700 and 900 GAA triplets.

Comment

This 14-year-old boy has a 9-year history of progressive gait difficulty, producing falls. Examination is significant for mild ataxic gait, the presence of a Romberg sign, fixation instability, moderate distal wasting of all four extremities, absent tendon reflexes, and mild appendicular ataxia greater in the legs than the arms. The case is typical of early-stage FRDA, except for a slower than usual progression. It illustrates well the usefulness and limitations of clinical diagnostic criteria, which if strictly applied would not have allowed a definite diagnosis. Onset at 5 years is well within the spectrum, and earlier than in most cases. A positive family history with two affected sisters, but healthy parents, suggests recessive inheritance of the condition. There is progressive truncal and limb ataxia, and tendon reflexes are absent in all four limbs. There are no cardiac symptoms, but EKG and echocardiography reveal repolarization abnormalities and moderate heart hypertrophy, typical findings in FRDA. However, the clinical course has so far been very slow, and the patient is still able to run, has flexor plantar responses, and has no dysarthria 9 years after onset. Also, there is no evident deep sensory loss, an unusual finding this late in the clinical course. Conversely, amyotrophy is more prominent than usual. Neurophysiological findings suggest the diagnosis of FRDA, particularly because SNAPs are absent and the H-reflex cannot be obtained, but MCVs are a little less than 40 m/sec. MRI, excluding cerebellar atrophy and showing some thinning of the cervical spinal cord, is highly suggestive of the diagnosis. All things considered, there is a very clear indication for ordering molecular testing, which confirmed the diagnosis.

with feet close together worsened by eye closure (positive Romberg sign). With further progression, gait becomes more and more ataxic and broad-based, with frequent losses of balance that requires intermittent support (furniture, walls, an accompanying person's arm). Patients become completely unable to stand with feet close together, and then need support even when standing with their feet apart. A cane and then a walker become necessary. Finally, 10 to 15 years after onset, most individuals with FRDA lose the ability to walk, stand, and sit without support. Variability is remarkable, however; some patients are wheelchair dependent in their early teens while others are able to walk until their thirties and occasionally later.[18,19]

After trunk ataxia, limb incoordination with dysmetria and intention tremor appear. Fine motor skills become impaired, with increasing difficulty in activities such as writing, dressing, and handling utensils. Ataxia is progressive and there are no remissions, although at the beginning of the illness periods of stability are frequently observed. Within five years from clinical onset, a progressive dysarthria appears, consisting of slow, jerky speech with sudden utterances.[22] Speech deteriorates over years until it becomes almost unintelligible. Swallowing difficulties, particularly with liquids, commonly appear with advancing disease. Patients with very advanced disease frequently choke and may require modified foods and eventually a nasogastric tube or gastrostomy feedings.

Higher functions are well preserved; intellectual and learning disabilities do not seem to be more common in patients with FRDA than in the general population. However, subtle neuropsychological abnormalities do occur, mostly consisting of slowed information processing speed as well as altered motor planning.[23,24] If not properly managed FRDA may have a substantial impact on academic, professional, and personal development.

The neurological examination shows a characteristic pattern of abnormal findings. Axonal sensory neuropathy results in sensory loss and loss of deep tendon reflexes. Loss of position and vibration sense may not be evident at onset but is invariably found within a few years of onset. Perception of light touch, pain, and temperature is initially normal and tends to decrease with advancing disease. Loss of tendon reflexes at least in the lower limbs is a typical feature of the disease, but a minority of patients with FRDA retain tendon reflexes and sometimes have brisk reflexes with spasticity; this group of patients is referred to as FRDA with retained tendon reflexes (FARR). Some patients with retained reflexes eventually lose them with disease progression, but many retain reflexes late in the course of the disease. Pyramidal involvement causes progressive muscular weakness that becomes severe only late in the progression of the disease.[18,19] In fact, ataxia and not weakness is the primary cause for loss of ambulation; even when patients become wheelchair dependent, they maintain on average 70% of their normal strength in the lower limbs. Pyramidal involvement also results in extensor plantar responses, but exceptions to this rule are well known. Sensory neuropathy and pyramidal tract degeneration result most often in the typical mixed picture of areflexia associated with extensor plantar responses, but sometimes one component obscures the other. Such partial pictures are usually observed in milder cases of the disease.[19]

Muscle tone is often normal at onset. However, with advancing pyramidal involvement and particularly when ambulation becomes severely impaired, many patients complain of spasms in the lower limbs, mostly occurring at night. When patients are wheelchair dependent, disuse atrophy occurs.

The typical oculomotor abnormality of FRDA is fixation instability with square-wave jerks,[25] which may be seen early in the course of the disease. Various combinations of cerebellar, vestibular and brain-stem oculomotor signs may be observed, but nystagmus (gaze-evoked) is uncommon and ophthalmoparesis does not occur. Optic atrophy with or without visual impairment is detected in about 30% of patients, more as the disease advances.[26] Sensorineural hearing loss affects about 20% of patients. However, sophisticated studies of auditory function revealed abnormalities in patients with normal tonal audiometry, even early in the course of the disease, which correlate with overall clinical progression.[27,28] Auditory neuropathy and dysfunction of auditory pathways result in impaired speech perception.[29]

Heart Disease

In most patients with FRDA heart disease remains asymptomatic, but in a significant minority, particularly in patients with earlier age of onset, it contributes to disability and may cause premature death.[13,14] Cardiac mortality is excessive among FRDA patients, particularly before the age of 40, after which causes of death are more commonly related to advanced neurological disability.[30] The most common cardiac symptoms are shortness of breath (40%) and palpitations (11%).[2] The electrocardiogram (EKG) shows T-wave inversions, signs of ventricular hypertrophy, conduction disturbances in about 10% of patients, supraventricular ectopic beats, and atrial fibrillation. The latter is a negative prognostic sign. The frequency of EKG changes may be underestimated because studies vary over time and include occasional normal recordings. Repeated EKG recordings are the most sensitive test for FRDA-related cardiomyopathy. Echocardiography and Doppler echocardiography demonstrate concentric hypertrophy of the ventricles (62%) or asymmetric septal hypertrophy (29%), with diastolic function abnormalities.[13,14]

Diabetes Mellitus

About 10% of FRDA patients develop diabetes mellitus and 20% have carbohydrate intolerance. Beta-cell dysfunction as well as peripheral insulin resistance occurs.[16,17] Although some cases may be initially controlled by oral hypoglycemic drugs, insulin dependence is eventually the rule. FRDA patients should be checked regularly for the development of diabetes because it may increase the burden of disease, complicate the neurological picture, and even promote potentially fatal complications.

Other Clinical Manifestations

Kyphoscoliosis may cause pain and cardio-respiratory problems. Pes cavus and pes equinovarus may render ambulation even more difficult. Autonomic disturbances, most commonly cold and cyanotic legs and feet, increase in frequency as the disease advances. Parasympathetic abnormalities such as decreased heart rate variability have been reported. Urgency of micturition is relatively common and incontinence may occur.

Biochemical Investigations

Commonly tested biochemical parameters in blood, urine, and cerebrospinal fluid are normal in FRDA. There is an ongoing search for biomarkers of disease status, progression, and response to treatments. Measures of iron metabolism and oxidative stress have been investigated first. Blood iron, iron binding capacity, and ferritin are normal, but there is an increase in circulating transferrin receptors. Several markers of oxidative stress have been reported to be increased, including urinary 8-hydroxy-2′-deoxyguanosine (8OH2′dG), a marker of oxidative DNA damage and plasma malondialdehyde, a marker of lipid peroxidation.[31,32] However, measuring oxidative stress markers has proven difficult in clinical practice, and an attempt to use 8OH2′dG as an endpoint in a clinical trial has failed.[33] Proteomic studies of plasma and urine are under-way. Gene expression profiling in peripheral blood mononuclear cells (PBMCs) has detected a set of genes whose expression is significantly and consistently different in FRDA patients versus carriers and controls. In *ex vivo* experiments, some of these genes respond to treatments that normalize frataxin levels.[34]

Neuroimaging

Thinning of the cervical spinal cord with signal abnormalities in the posterior and lateral columns can be detected on sagittal and axial magnetic resonance (MRI) images in almost all FRDA patients. Brainstem, cerebellum, and cerebrum are less severely affected, but they tend to be smaller in FRDA patients compared with healthy controls. The extent of this diffuse atrophy correlates with clinical severity. Mild vermian and lobar cerebellar atrophy occurs in more severe and more advanced cases.

More recently, advanced imaging techniques have been used to detect structural changes in FRDA, which have been confirmed to extend beyond the cerebellum and spinal cord. Diffusion-based MRI studies showed significantly lower fractional anisotropy, higher mean diffusivity and increased radial diffusivity in dentato-rubral, dentato-thalamic, and thalamo-cortical tracts, correlating with clinical severity.[35,36] Altered cerebello-cerebral white matter connectivity, as shown by diffusion MRI and tractography,[37] is the likely basis of abnormalities detected by functional imaging studies, in particular decreased brain activation during specific cognitive tasks.[38]

Metabolic imaging studies have also revealed widespread abnormalities. Fluorodeoxyglucose (FDG) PET scan showed increased metabolism in the brain of FRDA patients who are still ambulatory. As disease progresses and the patients lose their ability to walk, glucose consumption decreases and eventually becomes subnormal.[39] Although not yet fully explained, this abnormality may relate to mitochondrial dysfunction.

Neurophysiological Investigations

Electromyography and nerve conduction studies reveal the underlying axonal sensory neuropathy, with sensory nerve action potentials (SNAPs) lost or severely reduced and motor and sensory nerve conduction velocities (NCVs) within or just below the normal range. These findings clearly distinguish an early case of FRDA from a case of hereditary demyelinating sensorimotor neuropathy. The severity of the neurophysiological abnormalities related to the axonal sensory neuropathy does not correlate with disease duration, indicating that the atrophy of sensory neurons is nonprogressive.[40] Degeneration of both peripheral and central sensory fibers also results in dispersion and delay of somatosensory evoked potentials (SEPs). Brainstem auditory evoked potentials (BAEPs), beginning from the most rostral component, wave V, progressively deteriorate in all patients.[41] Visual evoked potentials (VEPs) are commonly (50–90%) reduced in amplitude with normal P 100 latency. In all cases, central motor conduction is slower than normal; in contrast to the sensory involvement, this slowing becomes more severe with increasing disease duration.[41]

Variant Phenotypes: The Role of Molecular Testing

When direct molecular testing for FRDA became available, positive results were obtained in up to 10% of patients with recessive or sporadic ataxia who did not fulfill the diagnostic criteria for FRDA.[3,19] No one clinical finding or combination of findings exclusively characterizes positive cases or is necessarily present. Even neurophysiological evidence of axonal sensory neuropathy is occasionally absent in patients with a proven molecular diagnosis. However, presence of the "classical" features of FRDA, including cardiomyopathy, is highly predictive of a positive test, although a few negative cases have been reported. Absence of cardiomyopathy and moderate or severe cerebellar cortical atrophy, demonstrated by MRI, appear to be the best predictors of a negative test.

Specific variant phenotypes have been shown to be due to the same genetic mutation causing typical FRDA (GAA repeat expansion), including LOFA, FARR, and Acadian ataxia.[19] LOFA patients have onset after age 25 years and exhibit an overall milder, slowly evolving disease with fewer skeletal abnormalities. The frequency of cardiomyopathy may be lower in LOFA than in typical FRDA. FARR patients have retained deep tendon reflexes in the lower limbs after the onset of neurological symptoms. They also tend to be mild cases with later onset and slower course than typical FRDA patients. Acadian ataxia is found among the descendants of French colonizers who settled in Eastern Canada during the seventeenth century and were expelled by the British in the eighteenth century. Many moved to Louisiana, where they became known as "Cajuns" (a corruption of "Acadians"), and some later returned to the Canadian Atlantic provinces. The disease was introduced into this population by very few individuals, possibly a single couple, a founder effect. Acadian ataxia is milder and evolves more slowly than typical FRDA, with less severe cardiomyopathy.

Prognosis

On average, FRDA patients lose their ability to walk 15 years after onset, but variability is very wide. Early onset and left ventricular hypertrophy appear to be predictors of a faster rate of progression of the disease. The burden of neurological impairment, cardiomyopathy, and occasionally diabetes shortens life expectancy.[30] In a retrospective study of FRDA patients to determine cause of death, followed by a case-control analysis comparing characteristics of deceased patients with living, age- and sex-matched FRDA controls, cardiac dysfunction was the most frequent cause of death, most commonly congestive heart failure or arrhythmia.[30] In this study, median age of death due to cardiac cause was 17 years, while median age of death due to noncardiac cause was 29 years.[30] However, survival may be significantly prolonged with treatment of cardiac symptoms, treatment of diabetes, and prevention and control of complications resulting from prolonged disability. Carefully managed patients may live several more decades. Currently no treatments affect the degenerative process, but emerging knowledge from characterization of the FRDA gene product may change this picture in the future.

DIFFERENTIAL DIAGNOSIS

Nonhereditary diseases may also cause early-onset, chronic, progressive ataxia. Causes that may be revealed by MRI include posterior fossa tumors and malformations, including platybasia and basilar impression, as well as the lesions of multiple sclerosis. Toxic, autoimmune, and acquired metabolic disorders have to be considered in the appropriate clinical context.[42] When ataxia develops at any age due to an unidentified acquired condition, including in cases where both dominant transmission of FRDA and obvious cerebellar atrophy are absent, FRDA must be considered whether or not the diagnostic criteria for typical FRDA are fulfilled.[42,43] Rare genetic disorders that should be considered if FRDA testing is negative are discussed here.

Vitamin E Deficiency

Patients with ataxia and isolated vitamin E deficiency (AVED; OMIM 277460), a cause of ataxia with sensory neuronopathy similar to that found in FRDA, may show a phenotype that closely resembles typical FRDA, but more commonly have atypical features (see Case Example 49.2). Head titubation and dystonia, frequent in AVED, are unusual in FRDA. AVED patients may retain lower limb reflexes. Peripheral nerve conduction studies may be normal. Somatosensory evoked potentials show normal peripheral and delayed central conduction, indicating a disproportionate involvement of the central versus the peripheral axonal branch of primary sensory neurons. Serum vitamin E levels are markedly reduced in all patients and represent the diagnostic test. The disease is caused by mutations in the gene encoding alpha-tocopherol transfer protein, the liver enzyme that loads alpha-tocopherol onto lipoproteins. The disease is treatable with high oral doses of vitamin E.[44]

Vitamin E deficiency is also a feature of abetalipoproteinemia, a systemic metabolic disorder usually manifesting shortly after birth with vomiting, diarrhea, and steatorrhea, and later by ataxia and sensory neuronopathy. Laboratory tests reveal abetalipoproteinemia with decreased levels of cholesterol, triglycerides and lipid-soluble vitamins (A, D, E, K), and acanthocytosis. The disease is due to mutation in the *MTP* gene, encoding the microsomal triglyceride transfer protein.

Sensory Axonal Neuropathy with Dysarthria and Ophthalmoplegia

Sensory axonal neuropathy with dysarthia and ophthalmoplegia (SANDO) is a phenotype linked to mutations in the *POLG* gene, which encodes DNA polymerase gamma, the mitochondrial DNA polymerase. Age of onset overlaps

CASE EXAMPLE 49.2

History

The patient is a 17-year-old right-handed boy from Sicily who has an older sister with a similar disease and whose parents are second-degree cousins. He was born at term after an uneventful pregnancy. Development was normal. Since childhood, the patient has been considered clumsy and never participated in sports in school. At age 7, a mild kyphoscoliosis was diagnosed, which then showed very limited progression. At around the same age, the presence of bilateral genu recurvatum was noticed. No other major health problem was evident until approximately age 13, when the patient started to have balance problems, to sway when walking, and occasionally to fall. His gait has gradually worsened and he now requires support to prevent falling. In the last year, the mild dysarthria has been noticed. A nodding head tremor has also appeared during the last year.

The patient goes to a technical school; he is able to write without difficulty and to use his hands in skilled activities. There is no dysphasia and no bowel or bladder incontinence. The patient has a regular diet and does not drink or smoke. There is no history of major medical or surgical illnesses and he does not take any medications.

Physical Examination

The patient is an average-size adolescent in apparent good general health. The general physical examination reveals a mild kyphoscoliosis and a bilateral genu recurvatum. The rest of the examination is negative.

Neurological Examination

Mental status: The patient is well oriented. He can write a sentence and prints his name without difficulty. He recalls 3 out of 3 objects at 0 and 3 minutes.

Cranial Nerves II–XII: Speech is mildly dysarthric but easy to understand. There is a continuous nodding tremor of the head at a frequency of about 2–3 Hz. There is a mild bilateral eyelid retraction. Extraocular movements are full. There is mild gaze-evoked nystagmus. The rest of the cranial nerve examination is normal.

Motor examination: Muscle bulk and strength are normal in the four limbs. The muscles of all four extremities are markedly hypotonic.

Reflexes: Absent proximally and distantly in the upper and lower extremities. The plantar response is extensor bilaterally.

Sensation: Intact to light touch, pin prick, and temperature. Loss of vibration sense at the malleoli bilaterally and severe impairment at the knees bilaterally. Joint position sense is impaired to the ankles bilaterally.

Coordination: Finger to nose testing is within normal limits. Rapid alternating movements are very mildly impaired. Writing and fine movements of the hand appear normal. Mild ataxia at heel-to-shin testing.

Gait: Romberg is positive. Gait is moderately ataxic with polydirectional swaying, enlarged base, and irregular steps.

Nerve Conduction Studies

MCV is normal in the right common peroneal nerve, left common peroneal nerve, and left tibial nerve. SNAP is decreased in the left median nerve (3.1 μV); SCV is normal in the left median nerve.

MRI of the Brain and Spine

The upper cervical cord is slightly reduced in size. There is moderate diffuse atrophy of the cerebellum and of the cerebral hemispheres with mild increase in size of the ventricular system and dilatation of the sulci. There is no definite signal abnormality.

EKG

Sinus tachycardia (120 beats per second); no other EKG abnormality.

Echocardiography

Normal.

Laboratory Tests

Routine laboratory testing, including thyroid status and lipoproteins, was normal. An FRDA molecular test was negative. Plasma vitamin E level, determined by HPLC, was undetectable.

Comment

This young man is the offspring of consanguineous parents and has a similarly affected sister, indicating the recessive nature of his ailment. He presents with a 3-year history of gait instability, loss of balance, and occasional falls. More recently a mild dysarthria and head tremor were added to the clinical picture. There is deep sensory loss in the lower limbs, absent tendon reflexes in the four limbs, and bilateral extensor plantar responses. SNAPs are reduced but still measurable; MCV and SCV velocities are normal. There is no evidence of cardiomyopathy other than sinus tachycardia, in particular no repolarization abnormalities in the EKG and no heart hypertrophy. Overall, the clinical picture resembles an early case of FRDA. However, a few findings are unusual in this case: coordination is surprisingly well preserved in the upper limbs, and there is a continuous nodding head tremor and also eyelid retraction. Kyphoscoliosis is frequent in FRDA, but genu recurvatum is not typical and pes cavus foot deformities are absent. The molecular test for FRDA turned out to be negative. Plasma vitamin E measurement demonstrated profound deficiency of this vitamin, not accompanied by plasma lipid or lipoprotein abnormalities, as found in ataxia with isolated vitamin E deficiency (AVED). Mutation analysis of the tocopherol-transfer protein gene revealed that this patient and his sister are both homozygous for a common AVED mutation found in North Africa and Sicily.

The clinical features of this patient more commonly seen in AVED than in FRDA include the nodding tremor of the head, the minimal upper limb ataxia, the lack of hypertrophic cardiomyopathy, and the preservation of SNAPs. In this regard, vitamin E deficiency affects the central branch of the T-shaped axon of primary sensory neurons more than peripheral branch, while FRDA affects both branches equally. Genu recurvatum, eyelid retraction, and the diffuse mild cerebellar and cerebral atrophy on MRI are not typical of AVED and represent peculiar features of this patient.

The patient was treated with 500 mg of alpha-tocopherol daily. Plasma vitamin E measurement, repeated after two months, was normal. The patient has remained on treatment for several years and not only has there been no further deterioration of the neurological symptoms, but there have been some improvements of his dysarthria and of his gait.

with that for LOFA and may be as late as the seventh decade. SANDO is mainly an afferent ataxia due to sensory neuronopathy, but it has cerebellar features as well, in particular dysarthria, making it similar to FRDA. However, distinctive features of SANDO should suggest the diagnosis, such as ophthalmoparesis, ptosis, myoclonus, variable elevation of lactic acid, and MRI abnormalities that include some cerebellar atrophy along with white matter changes in the cerebellum and sometimes stroke-like lesions.

Autosomal Recessive Ataxias with Sensorimotor Axonal Neuropathy

This group includes ataxia telangiectasia (AT, OMIM 208900), AT-like disorder (ATLD, OMIM 208910), ataxia with oculomotor apraxia types 1 and 2 (AOA1, OMIM 208920; and AOA2, OMIM 606002), the autosomal recessive spastic ataxia of Charlevoix and Saguenay (ARSACS, OMIM 270550), and some of the metabolic disorders discussed in the following.[5] AT, ATLD, AOA1, and AOA2 are due to mutations in genes involved in DNA repair and DNA/RNA metabolism, respectively *ATM*, *MRE11A*, *APTX*, and *SETX*, while the mutated gene in ARSACS, *SACS*, is likely to be involved in protein quality control.[5,42] All these diseases have characteristic clinical features that help orient the diagnosis. Cerebellar atrophy is a common finding that distinguishes them from FRDA. In addition, specific biochemical abnormalities characterize AT (increased alpha-fetoprotein), AOA1 (increased LDL cholesterol and decreased albumin), and AOA2 (increased alpha-fetoprotein).[5,42] AOA1 is the most common recessive ataxia in Japan, where FRDA is virtually nonexistent.

Other Progressive Metabolic Ataxias

Other rare, usually autosomal recessive diseases are characterized by prominent cerebellar atrophy and sometimes extensive white matter abnormalities that distinguish them from FRDA. Sensorimotor neuropathy is present in some of these diseases.

Hexosaminidase A deficiency (OMIM 272800) may cause early onset cerebellar ataxia, followed by development of upper and lower motor neuron signs. Dementia, psychosis, and ophthalmoplegia may be present. Rectal biopsy specimens reveal lamellar cytoplasmic inclusions. Loss of serum and leukocyte enzyme activities varies from partial to complete. Elevated levels of serum lactate dehydrogenase have been present in several patients and may point to the diagnosis.

Cerebrotendinous xanthomatosis (cholestanolosis; OMIM 213700) has childhood onset. Xanthomata, especially of the Achilles tendon, and cataracts appear early, and neurological impairment develops later. The most prominent clinical feature is spastic ataxia, associated with pseudobulbar palsy, dementia, palatal myoclonus, and peripheral neuropathy. Serum cholestanol (a metabolite of cholesterol) is increased. Lack of the sterol 27-hydroxylase can be shown in fibroblast cultures.

Krabbe disease (OMIM 245200) usually has its onset in infancy, and in most patients by age 10 years. Intellectual disability, psychomotor deterioration, impaired vision, seizures, progressive spasticity, ataxia and a demyelinating peripheral neuropathy may be present. Most patients show rapid deterioration initially, followed by a more gradual progression lasting for years. Assays for galactosylceramidase (galactocerebroside β-galactosidase) in peripheral leukocytes or cultured fibroblasts offer the most reliable diagnostic tool.

Niemann-Pick disease type C (OMIM 257220), also called DAF (down gaze paralysis, ataxia, foam cell) syndrome, has onset between 6 months and 18 years, and is characterized by ataxia, mental impairment, supranuclear vertical gaze paralysis, dystonia, seizures, pyramidal signs, organomegaly, and pulmonary involvement. Lipid-laden macrophages (foam cells) are present in bone marrow aspiration and liver biopsy specimen. Sphingomyelinase is not deficient, but esterification of exogenous cholesterol is impaired in cultured fibroblasts. Filipin stain of fibroblasts reveals accumulation of non-esterified cholesterol and is diagnostic. The underlying genetic mutations are in the *NPC1* or *NPC2* genes.

The "pseudosclerotic" form of Wilson disease (OMIM 277900) presents with ataxia, dysarthria, and intention tremor. Onset usually occurs in the second decade. Signs of hepatic dysfunction, greatly reduced serum ceruloplasmin, and increased urinary copper excretion point to the diagnosis. Liver copper is greatly increased, and its measurement represents the most sensitive and accurate test for the disease.

Mitochondrial Disorders

Ataxia is a common feature in mitochondrial disorders such as Kearns-Sayre syndrome (KSS; OMIM 530000), myoclonic epilepsy with ragged-red fibers (MERRF; OMIM 545000), and neuropathy, ataxia, and retinitis pigmentosa (NARP; OMIM 551500). KSS is a form of sporadic chronic progressive ophthalmoplegia that begins before the age of 20 years and is characterized by pigmentary retinopathy, elevated levels of cerebrospinal fluid protein, ataxia, and heart block. Almost all patients have large, single deletions in mitochondrial DNA. MERRF is characterized by action myoclonus, myoclonic epilepsy, cerebellar ataxia, weakness, and short stature. Dementia and hearing loss may be present. Onset varies from the first to the fifth decade. Maternal inheritance may be evident. Plasma pyruvate, lactate, alanine, and creatine phosphokinase are increased, and muscle biopsy shows accumulation of abnormal subsarcolemmal and intermyofibrillar mitochondria (ragged-red fibers). Pathogenic point mutations have been shown in the *lysine tRNA* gene of the mitochondrial DNA. NARP is a maternally inherited multisystem disorder characterized by developmental delay, retinitis pigmentosa, dementia, seizures, ataxia, and sensory neuropathy. It is associated with a point mutation in the *ATPase 6* gene of mtDNA.

Hereditary Spastic Paraplegia

Cerebellar ataxia may occur in hereditary spastic paraplegias. Autosomal recessive spastic paraplegia type 7 (SPG7; OMIM 602783) in particular has to be considered in the differential diagnosis of recessive ataxia. It may be associated with optic and cerebellar atrophy, a characteristic that should differentiate it from FRDA. The mutated gene in SPG7 encodes paraplegin, a mitochondrial protein.

X-linked spastic paraplegia type 2 (SPG2; OMIM 312920), which is caused by mutations in the proteolipid protein gene, may present in pure form or be complicated with nystagmus, dysarthria, mild ataxia, intellectual disability, and optic atrophy. Mutations in this gene also cause Pelizaeus-Merzbacher disease.

X-linked and Autosomal Dominant Diseases

The possibility of X-linked recessive spinocerebellar degeneration may be raised in sibships where only males are affected. Adrenomyeloneuropathy (OMIM 300100) presents with spastic paraplegia and distal sensory loss in affected males, but cerebellar signs may be prominent. Hypoadrenalism and MRI findings of diffuse demyelination may lead to the diagnosis, which is confirmed by measurement of very long-chain fatty acids showing elevated C(26:0) levels in plasma and fibroblasts.

The major X-linked ataxia is fragile X-associated tremor and ataxia syndrome (FXTAS), a condition occurring in carriers of intermediate length (premutation) alleles (50–200 triplets) of the CGG repeat whose full expansion (>200 triplets) causes the fragile X syndrome of intellectual disability. FXTAS mostly affects men older than 50 years of age and is mainly characterized by ataxia and action tremor, the latter usually absent in FRDA. Typical MRI findings are fluid-attenuated inversion recovery (FLAIR) hyperintensities in middle cerebellar peduncle and cerebral periventricular white matter. However, MRI may look normal early in the course of the disease. Tremor may be essential-like, cerebellar, or parkinsonian; peripheral neuropathy is very common and may be length-dependent or not; and 60% of patients have parkinsonism. FXTAS also occurs in women, although the characteristic clinical manifestation of fragile X permutation in females is premature ovarian failure.

Ataxia with Hypogonadism

Ataxia with hypogonadism is a heterogeneous syndrome in which the age at onset and clinical picture vary widely. In addition to hypogonadism, the presence of cerebellar atrophy distinguishes this group of conditions from FRDA. Both hypergonadotrophic and hypogonadotrophic hypogonadism have been reported. Ataxia with hypogonadotrophic hypogonadism can be associated with brisk tendon reflexes (Gordon-Holmes syndrome) or chorioretinal dystrophy (Boucher-Neuhäuser syndrome; OMIM 215470). Neuroimaging studies show cerebellar atrophy. An axonal neuropathy is common. These two syndromes are allelic for recessive mutations in the gene *PNPLA6*, encoding patatin-like phospholipase domain-containing protein 6.[45] The same gene is mutated in a form of recessive spastic paraplegia, SPG39. Ataxia with hypogonadotrophic hypogonadism has also been linked to mutations in the *STUB1* gene, encoding the ubiquitin ligase carboxy terminus of Hsp70-interacting protein (CHIP), a co-chaperone and ubiquitin ligase involved in protein quality control.[46] We recently identified *STUB1* mutations in sibs with ataxia and brisk reflexes, but no hypogonadism.

The Richards-Rundle syndrome (OMIM 245100) is characterized by autosomal recessive ataxia, hearing loss, mental deterioration, ketoaciduria, and hypergonadotrophic hypogonadism.

Ataxia with Visual Loss

Visual loss in FRDA is usually late and consequent to optic nerve atrophy. Optic atrophy and retinitis pigmentosa are found in several hereditary cerebellar and afferent ataxias.

Marinesco-Sjögren syndrome (OMIM 248800) is an autosomal recessive disorder characterized by intellectual disability, cerebellar ataxia with cerebellar atrophy, congenital cataracts, short stature, and delayed sexual development due to hypergonadotrophic hypogonadism. Most patients have elevated serum creatine kinase levels and myopathic changes on muscle biopsy. Multiple skeletal abnormalities may be found. The causative mutations are found in the *SIL1* gene, encoding a nucleotide exchange factor for the endoplasmic reticulum resident chaperone BiP, involved in protein quality control in the endoplasmic reticulum.[47]

Posterior column ataxia and retinitis pigmentosa (PCARP) is a form of inherited afferent ataxia characterized by night blindness and areflexia in infancy, followed by onset of progressive ataxia in the second decade.

PCARP is due to mutations in the *FLVCR* gene, encoding a heme-transporter protein that maintains the intracellular heme concentration.[48]

Wolfram syndrome (OMIM 222300), also known by the acronym DIDMOAD (diabetes insipidus, diabetes mellitus, optic atrophy, and deafness), is an autosomal recessive disorder whose neurological manifestations include ataxia, myoclonus, peripheral neuropathy, intellectual disability, and psychiatric illnesses. The responsible gene, *WFS1*, encodes a predicted endoplasmic reticulum (ER) transmembrane protein, wolframin, thought to be involved in the regulation of ER stress.

Optic atrophy is also a feature of infantile-onset spinocerebellar ataxia (IOSCA; OMIM 271245), a severe autosomal recessive disorder with ataxia and intellectual disability.

Ataxia and retinitis pigmentosa are also found in metabolic diseases such as abetalipoproteinemia (OMIM 200100), AVED (OMIM 277460), Refsum's disease (OMIM 266500), NARP (OMIM 551500), and KSS (OMIM 530000). In neuronal ceroid lipofuscinoses with infantile (OMIM 256730) and juvenile onset (OMIM 204200), ophthalmological examination reveals optic atrophy and retinal and macular degeneration.[49]

Ataxia with Deafness

There are many reports of autosomal recessive ataxia associated with cochleovestibular degeneration as well as with other anomalies. The clinical presentation of these cases is sometimes similar to that of FRDA. The originally described patients with Lichtenstein-Knorr syndrome, characterized by progressive hearing loss, unsteady gait, and dysarthria, might in fact have been affected by FRDA, according to Harding. Additional features in Berman syndrome (OMIM 208850) are mental deficiency and signs of upper and lower motor neuron involvement. Intellectual disability and skin pigmentary changes are present in the Jeune-Tommasi disease.

Hereditary Motor and Sensory Neuropathies (HMSN)

Recessive hereditary neuropathies may cause afferent ataxia, in particular Dejerine-Sottas disease (CMT1E), due to recessive mutations in the *PMP22* gene; CMT4F, due to mutations in the periaxin gene; and the group of the hereditary sensory and autonomic neuropathies (HSAN I-V).[42]

MOLECULAR GENETICS

The FRDA gene (*FXN*) is localized in the proximal long arm of chromosome 9, just below the pericentromeric heterochromatic region, and is composed of seven exons spread over 95 kb of genomic DNA. The gene is transcribed in the centromere-to-telomere direction. The first exon harbors an unmethylated CpG island. The major and probably only functionally relevant mRNA has a size of 1.3 kb and corresponds to the first five exons, numbered 1 to 5a. The encoded protein, predicted to contain 210 amino acids, was named frataxin.[3]

The gene is expressed in all cells but at variable levels in different tissues and during development.[50] In adult humans, frataxin mRNA is most abundant in the heart and spinal cord, followed by liver, skeletal muscle, and pancreas. In mouse embryos, expression starts in the neuroepithelium at embryonic day 10.5 (E10.5), then reaches its highest level at E14.5 and into the postnatal period. In developing mice, the highest levels of frataxin mRNA are found in the spinal cord, particularly at the thoracolumbar level, and in the dorsal root ganglia. The developing brain is also very rich in frataxin mRNA, which is abundant in proliferating neural cells in the periventricular zone, the cortical plates, and the ganglionic eminence (precursor of the basal ganglia). Robust expression is also detected in heart, axial skeleton, and some epithelial (skin, teeth) and mesenchymal tissues (brown fat). Overall, frataxin expression is generally higher in mitochondria-rich cells, such as cardiomyocytes and neurons.

The GAA Expansion Mutation

Size and Genetic Instability

The most common mutation causing FRDA (98%) is the hyperexpansion of a GAA triplet repeat in the first intron of the frataxin gene (Figure 49.1). Disease-associated repeats contain from ~ 70 to more than 1000 triplets, most commonly 600–900. Because of the recessive nature of the disease, affected individuals have expansions in both homologs of chromosome 9, while heterozygous carriers are clinically normal. Like other repeat expansions that cause disease, this one is unstable during parent-child transmission. Paternal transmission is most often accompanied by a contraction of the repeat, while maternal transmission may result in further expansion or in contraction with about equal probability.[51] Instability also occurs during somatic cell division, leading to a degree of somatic mosaicism for expansion sizes that has been observed in cells and tissues in both FRDA patients[52] and animal models.[53] In particular, somatic expansions have been observed in specifically vulnerable cells such as primary sensory neurons in the dorsal root ganglia.[54]

Genetic Epidemiology

The FRDA-associated GAA repeat expansion is the most common disease-causing triplet repeat expansion

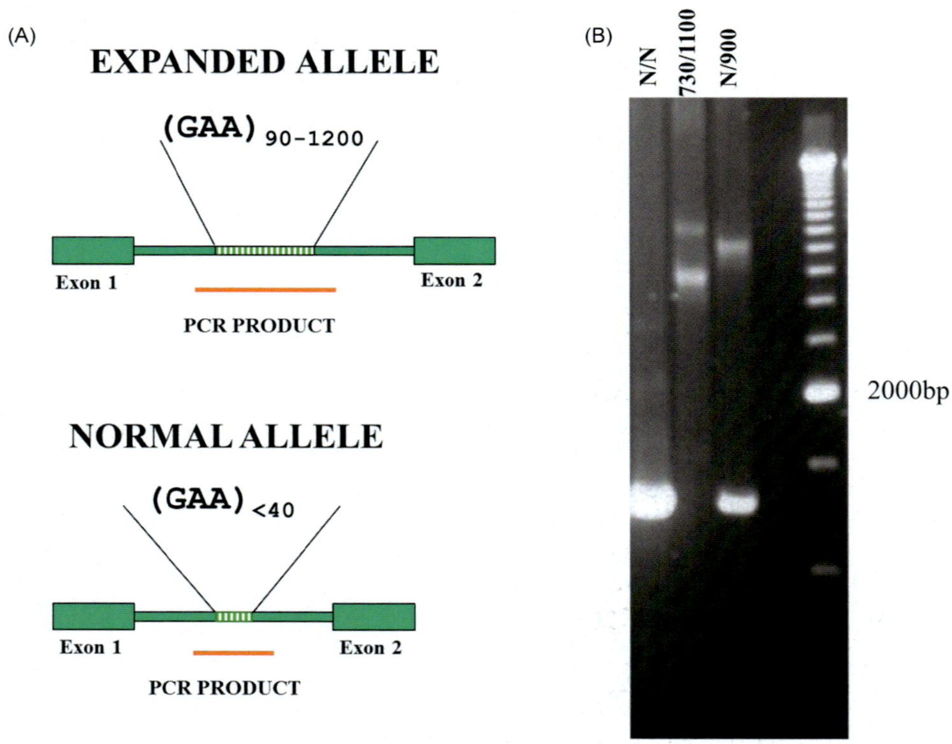

FIGURE 49.1 (**A**) Scheme showing the intronic GAA repeat in the frataxin gene. To determine its length, a PCR product containing the repeat is generated between two flanking primers and its size is estimated by gel electrophoresis. (**B**) Agarose gel electrophoresis of PCR products containing the GAA repeat in the frataxin gene. The sample in the first lane (N/N) is from a homozygous normal individual; the sample in the second lane is from an FRDA patient with two expanded GAA repeats (the numbers 730 and 1100 indicate the estimated number of GAA triplets in each allele); the third lane is from a heterozygous individual with one normal (N) and one expanded (estimated at 900 triplets) allele. Such an individual may be a healthy carrier or an FRDA patient, if the chromosome containing the normal size repeat carries a frataxin point mutation.

identified so far; 1 in 90 Europeans is a carrier.[55] In normal chromosomes, two classes of alleles of the GAA repeat can be distinguished. In Caucasians, short normal (SN) alleles contain 6–10 GAA triplets and account for 83% of chromosomes; long normal (LN) alleles contain more than 12 triplets and account for 17% of chromosomes. LN alleles constitute a reservoir of "at risk" alleles that may eventually expand into the disease range, as indicated by linkage dysequilibrium analysis and by the direct observation of catastrophic expansions of LN alleles to pathological repeats containing hundreds of triplets.[55,56] A few LN alleles are interrupted by a hexanucleotide repeat (GAGGAA), which probably has a stabilizing role. The GAA triplet repeat expansion that causes FRDA is found only in individuals of European, North African, Middle Eastern, or Indian origin. LN alleles are found in these populations as well as in Sub-Saharan Africans. Analysis of closely linked markers suggests that expansions arose through a unique two-step process, the first leading to the appearance of LN alleles in Africa, the second to larger LN alleles and pathological expansions in ancestors of Indo-Europeans and North Africans. A major implication of these findings is that FRDA may not exist among Sub-Saharan Africans, Amerindians, and peoples from China, Japan, and Southeast Asia.[57]

Pathogenic Mechanism of the GAA Expansion Mutation

The expanded GAA repeat inhibits the expression of the frataxin gene. Severely reduced levels of frataxin mRNA and protein have been demonstrated in tissue samples and cultured cells from FRDA patients. The reduction is proportional to the size of the expanded GAA repeats, particularly the smaller repeats.[58] A GAA triplet repeat is a long DNA segment containing only purines (R) on one strand and pyrimidines (Y) on the complementary strand. In a test tube, R•Y sequences can form triple helical structures capable of associating in unique bimolecular complexes (sticky DNA).[59] Remarkably, these structures form only for lengths of GAA triplet repeats associated with FRDA, and do not form for variant R•Y sequences occasionally found in the frataxin gene but not associated with disease. Triplexes are known to block transcription, providing a potential mechanism for inhibited gene

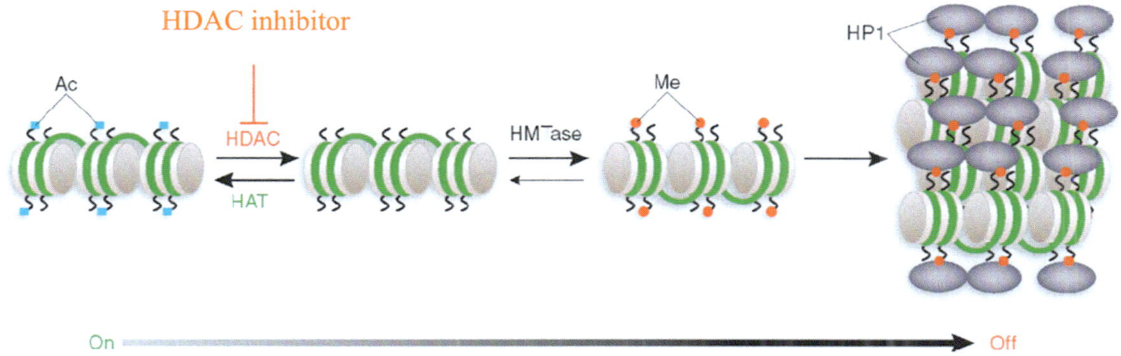

FIGURE 49.2 The expanded GAA repeats in the *FXN* gene cause chromatin remodeling by triggering histone deacetylation and HP1 recruitment. The resulting condensed chromatin conformation is not permissive for transcription. Specific HDAC inhibitors can reverse these changes and restore *FXN* expression.

expression in FRDA.[60] However, although a triple helical structure at long GAA repeat can form in a test tube in physiological conditions of salt concentration and pH, their occurrence at the *FXN* GAA repeat *in vivo* has not yet been demonstrated. The inhibitory effect of the GAT repeat on *FXN* gene expression is mediated by an epigenetic mechanism that eventually causes chromatin condensation (Figure 49.2).[61] The primary trigger of these changes, including any role of the specific structure possibly adopted by the GAA repeat, is not yet known. Studies in cells from FRDA patients and in animal models carrying long GAA repeats have provided evidence in the vicinity of the repeats of decreased acetylation of several lysine residues in core histones and of increased DNA methylation, which are epigenetic marks of transcriptionally silent chromatin. The key modification appears to be the deacetylation of lysine 9 of histone H3 (H3K9) and its trimethylation. Trimethylated H3K9 binds heterochromatin protein 1 (HP1), which is responsible for chromatin condensation and transcriptional silencing (Figure 49.2). Counteracting H3K9 deacetylation is being intensively explored as a therapeutic approach for FRDA.[62]

Genotype-phenotype Correlation for the GAA Expansion

Smaller expansions allow a higher residual gene expression,[58] so expansion sizes predictably have an influence on the severity of the phenotype.[63,64] A direct correlation has been firmly established between the size of GAA repeats and earlier age of onset, earlier age of wheelchair dependence, more rapid rate of disease progression, and presence of additional disease manifestations indicative of more widespread degeneration. However, differences in GAA expansions account for only about 50% of the variability in age of onset, indicating that other factors influence the phenotype. These factors may include somatic mosaicism for expansion sizes, variations in the frataxin gene itself, modifier genes, and environmental factors.

Point Mutations

About 2% of FRDA chromosomes carry normal-length GAA repeat but have missense, nonsense, or splice site mutations ultimately affecting the frataxin coding sequence.[65] (See Case Example 49.3.) All affected individuals with a point mutation so far identified are heterozygous for an expanded GAA repeat on the other homolog of chromosome 9. It is possible that homozygotes for point mutations have not yet been found because point mutations are rare, but it is more likely that homozygosity for frataxin point mutations would cause a lethal phenotype, as suggested by the observation that frataxin knockout mice die during embryonic development.

A few missense mutations are associated with milder atypical phenotypes with slow progression, suggesting that the mutated proteins preserve some residual function. Patients carrying the G130V mutation have early onset but slow progression, no dysarthria, mild limb ataxia, and retained reflexes. A similar phenotype occurs in individuals with the mutations D122Y and R165P. For unclear reasons, optic atrophy is more frequent in patients with point mutations of any kind (50%).

Frataxin Structure and Function

Subcellular Localization

Frataxin is highly conserved throughout evolution, with homologs in mammals, invertebrates, yeast, and plants. The protein is targeted to the mitochondrial matrix, where its N-terminal mitochondrial targeting sequence is proteolytically removed by the mitochondrial processing

CASE EXAMPLE 49.3

History

The patient is a 19-year-old woman with a background history of intellectual disability, a seizure disorder, and a progressive spastic paraparesis with deep sensory loss since the age of 15 years. She also has difficulty with bladder control. She has had developmental delay and attended special education schooling. The mixed seizure disorder included absence spells since childhood, a few grand mal seizures, and occasional myoclonic jerks, particularly upon awakening. Seizures have been well controlled of late. She has moderately severe asthma, with intermittent steroid use and one hospitalization.

The patient is a nonsmoker and nondrinker. Her family history reveals no neurological complaint; an older brother aged 30 is said to be well, her mother has had back problems but no gait disorder, and her father is well. Her current medications are carbamazepine 200 mg tid, clonazepam 1 mg bid, hormonal medications, hyoscyamine, NSAIDs for back pain, bronchodilators, and nasal sprays.

Recently, the patient has reported some tingling in her fingers and admits that she occasionally drops things, such as cups of water. The weakness in her legs appears progressive, so that she is now spending almost all of her time in a wheelchair; she is able to transfer independently to her bed and toilet. She complains of spasms currently bothering her several times a day. Baclofen has been prescribed in the past but was discontinued because of somnolence. Her bladder problems have been well controlled with hyoscyamine.

Neurological Examination

Mental status: The patient is well oriented. She can write a sentence and prints her name without difficulty. She recalls 3 out of 3 objects at 0 and 3 minutes.

Cranial Nerves II–XII: Speech is fluent without dysarthria. The cranial nerve examination, including ocular movements, is normal. Jaw jerk is absent.

Motor examination: There is some mild wasting of the calf and quadriceps muscles consistent with disuse atrophy. No fasciculations are noted. In the upper limbs, there is normal tone and power. Lower limb examination reveals marked spasticity but no clonus; muscle strength is grade 3 to 4+ in a pyramidal distribution symmetrically.

Reflexes: Brisk at the upper limbs, extremely brisk at the knees but only 1+ at the ankles bilaterally. Plantar responses are extensor bilaterally.

Sensation: Normal in the upper limbs, apart from some possible impairment of vibration sense at the distal interphalangeal joint. Loss of vibration sense at the medial malleolus bilaterally. Joint position sense is impaired to the ankle. There is normal sensation to soft touch and pin prick in both upper and lower limbs.

Coordination: Finger-to-nose testing is within normal limits. Normal rapid alternating movements. Heel-to-shin testing cannot be performed because of marked spasticity in the patient's lower limbs.

Gait: The patient is unable to stand by herself.

Nerve Conduction Studies

Slightly slowed MCV in the left common peroneal nerve (42 m/sec). The left peroneal F-wave is normal. Normal SNAP of the left superficial peroneal nerve (9.7 μV).

MRI of the Brain and Spine

In the right periventricular region, there is an abnormal area of increased signal intensity on the long TR images. This area demonstrates no enhancement on the post contrast images. The course, caliber, and signal intensity of the spinal cord are within normal limits. The scan is otherwise unremarkable.

Laboratory Tests

Routine laboratories, thyroid studies, anti-GM1, phytanic acid, vitamin E, cholestanol, beta galactosidase, long-chain fatty acids levels were normal. SCA1 and SCA3 molecular tests were negative. FRDA molecular test revealed one expanded allele corresponding to 950 GAA triplets. Further analysis of the frataxin gene revealed a nonsense mutation in the first exon causing premature truncation of the coding sequence.

Comments

This patient shows a clinical picture that is very atypical for FRDA, even considering that delayed psychomotor development and a seizure syndrome are common neurological problems which may be coincidental. The case appears sporadic. The patient has a progressive spastic paraparesis accompanied by sensory symptoms and a diminished ankle jerk that suggest a peripheral neuropathy. She has no dysarthria, her ocular movements are normal, and coordination seems normal in her upper limbs. Nerve conduction studies were almost normal. MRI did not provide a diagnosis, the findings being nonspecific or negative, but it did exclude cerebral, cerebellar, or brainstem atrophy, malformations, hydrocephalus, white matter degeneration, demyelination, and space-occupying lesions. Lacking other diagnostic clues, a test for FRDA was justified on the basis of reports of atypical cases resembling spastic paraparesis. The finding of heterozygosity for a GAA triplet repeat expansion and a frataxin point mutation confirms the diagnosis. Point mutations may be associated with atypical phenotypes; however, homozygous expansions remain the most frequent disease-causing genotype in this group.

---

peptidase (MPP) after the protein is imported into mitochondria. MPP first removes 40 aminoacids, and then 20 more amino acids are removed in a second cleavage to generate mature frataxin.[66]

The Function of Frataxin and FRDA Pathogenesis

A number of functions have been attributed to frataxin,[67] with studies suggesting a role in iron storage and

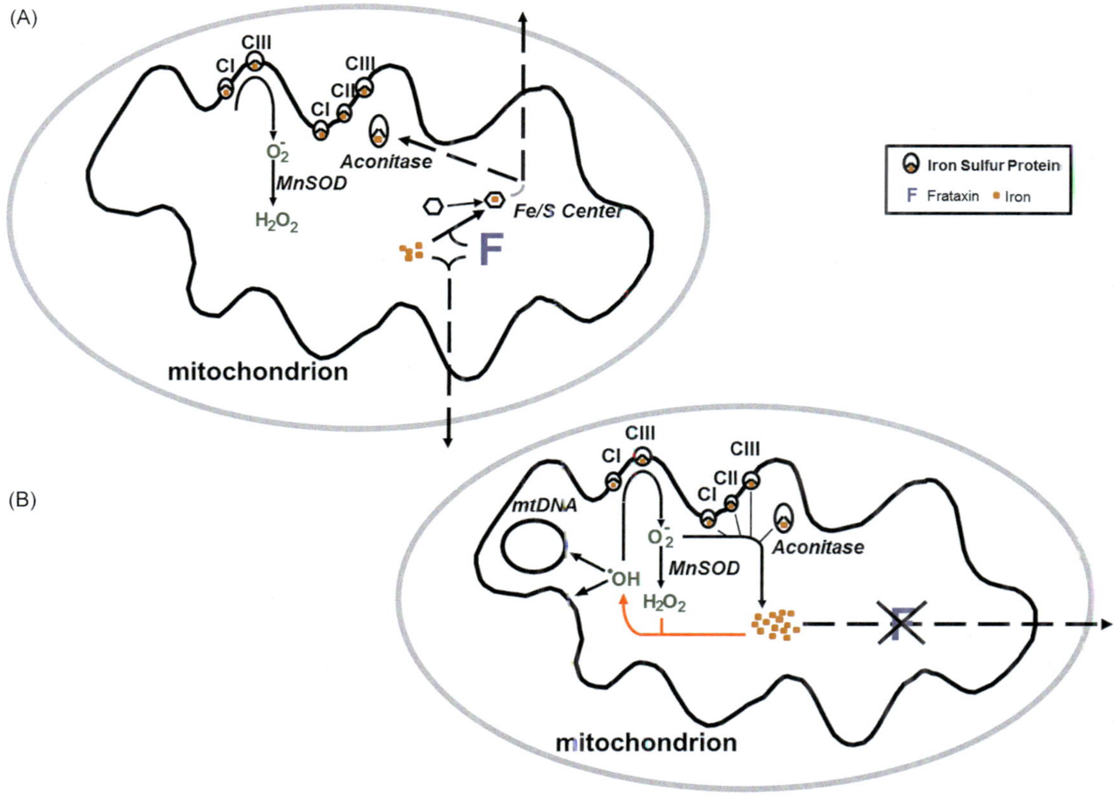

FIGURE 49.3 (**A**) Scheme of a normal mitochondrion containing a sufficient amount of frataxin. Frataxin activates the biogenesis of Fe-S clusters to be inserted into mitochondrial proteins or exported to the cytosol. (**B**) When frataxin is insufficient, impaired Fe-S cluster biogenesis leads to multiple deficiencies of Fe-S proteins, including those involved in energy metabolism, and causes free iron build-up in mitochondria. Excess free iron and increased ROS from an impaired respiratory chain trigger Fenton chemistry to generate the toxic hydroxy radical (•OH).

detoxification, stimulation of oxidative phosphorylation, activation of antioxidant defenses, regulation of iron metabolism, and protection against pro-apoptotic stimuli. Some of these postulated functions remain controversial; others may represent downstream effects rather than a primary activity of the protein. Genetic, biochemical, and structural data indicate that frataxin is a component of a multiprotein complex that assembles iron-sulfur clusters (Fe-S clusters) in the mitochondrial matrix (Figure 49.3). Within this complex, frataxin appears to act as an allosteric activator of a key enzyme in Fe-S cluster synthesis, the cysteine desulfurase Nsf1, so that Fe-S cluster synthesis proceeds at a very low rate in the absence of frataxin.[68] Fe-S clusters are cofactors for proteins with a variety of functions and subcellular localizations, including the Krebs cycle enzyme aconitase and several subunits of respiratory complexes I, II, and III in the mitochondria, enzymes of amino acid and purine metabolism in the cytosol, and DNA replication and repair factors in the nucleus.[69] In mammalian cells, Fe-S cluster assembly factors are also present in the cytosol, but to be active they need an essential substrate exported from mitochondria and assembled by the mitochondrial Fe-S cluster biogenesis machinery.[69]

By impairing Fe-S cluster biogenesis, frataxin deficiency results in energy deficit, altered iron metabolism, and oxidative damage. Fe-S cluster proteins in all cellular compartments show reduced activity, but aconitase and respiratory complexes I, II, and III in the mitochondria are particularly affected. Impairment of the respiratory chain results in reduced mitochondrial membrane potential, energy deficit, and leaking of electrons that may react with molecular oxygen and form reactive oxygen species (ROS), in particular superoxide (O_2^-).[70] A homeostatic response aiming to restore Fe-S cluster levels, a situation usually caused by iron deficiency, leads to increased cellular iron uptake and import into mitochondria. Iron responsive protein 1 (IRP1) is a cytosolic Fe-S cluster protein functioning as an aconitase when its Fe-S cluster is fully assembled and, when its Fe-S cluster is partially or fully disassembled, as an RNA binding protein that recognizes a sequence motif on mRNAs for proteins involved in iron metabolism. IRP1 stabilizes mRNAs for iron import proteins, such as the transferrin receptor (TfR), and blocks translation of mRNAs for proteins that need iron, such as aconitase, or that store iron, such as ferritin. Excessive cellular iron import and IRP1 RNA binding activity have indeed been demonstrated in frataxin-deficient cells. Other mechanisms must play a role, however, particularly in directing the flow of imported iron into mitochondria. When this response is induced by iron deficiency in a normal cell, increased mitochondrial iron import restores Fe-S cluster levels. In FRDA, because of impaired Fe-S cluster synthesis, iron cannot be efficiently utilized and progressively accumulates in mitochondria. Iron and ROS in mitochondria are the substrates for the Fenton reaction, which generates the toxic hydroxyl radical ($OH^{\bullet}$) that can damage nucleic acids, proteins, and lipids. Excess ROS also lead to the depletion of small antioxidant molecules, in particular glutathione, which is largely consumed by mitochondrial glutathione peroxidase.

Interestingly, several other genes whose mutations lead to defective Fe-S cluster biogenesis have been identified, including *ISCU* and *NFU1* but the corresponding phenotypes are entirely different from FRDA. The reasons for such heterogeneity are still unknown, as is the basis of specific cell vulnerability in FRDA.

Animal and Cellular Models

Although frataxin has homologs in all eukaryotes and many prokaryotes, the unique genetics of FRDA has made the development of animal models very difficult. The "perfect" model should carry large GAA repeats and suffer a chronic partial frataxin deficiency that results in FRDA-like pathology. *Fxn* knock-out mice die after 7 days of embryonic development (E7), near the time of gastrulation.[71] Similarly, complete frataxin knock-out results in early lethality in all multicellular organisms tested (*D. melanogaster*, *C. elegans*, *A. thaliana*).[72] Unicellular eukaryotes, such as *S. cerevisiae*, are viable without a frataxin homolog gene (*Yfh1* in *S. cerevisiae*) but show altered iron metabolism, mitochondrial iron accumulation, and oxidative stress, along with progressive loss of mitochondrial DNA and of oxidative phosphorylation.[73] Knock-in[74] and YAC transgenic[75] mice have been generated that carry expanded GAA repeats and show frataxin downregulation and the same epigenetic changes occurring in cells from FRDA patients, but their frataxin levels remain high enough to prevent the development of a full-blown FRDA-like phenotype. However, although none of the current animal models really recapitulates all aspects of the human disease, they have been valuable tools to explore specific aspects of FRDA pathogenesis and test potential therapeutics.[72] Conditional knock-out mice uniformly develop a degenerative process in the organs where frataxin has been deleted, which recapitulates many events in FRDA pathogenesis, including loss of Fe-S cluster enzyme activities, altered iron metabolism with mitochondrial iron accumulation, and mitochondrial dysfunction.[76] These animals have been and are used to test therapies that aim to correct these aspects of FRDA pathogenesis, as well as those that aim at protein and gene replacement. Conversely, animals carrying GAA repeats have allowed investigators to study the epigenetic transcriptional silencing of the frataxin gene and to test therapeutics aimed at restoring frataxin gene expression.[77]

Cellular models range from yeast cells with deleted or downregulated frataxin homolog, to cells derived from animal models and cells from FRDA patients, in particular fibroblasts and peripheral blood mononuclear cells (PBMCs). Other models involve transfection with antisense or inhibitory RNA (RNAi)-expressing constructs, or expression of mutated human frataxin on a mouse knock-out background. Again, all these models have been useful to investigate some aspects of pathogenesis and to test therapeutics, but none of them consists of specifically vulnerable cells carrying the GAA expansion mutation in human disease.[72] Recently, however, several groups have obtained induced pluripotent stem cells (iPSCs) from an individual with FRDA, which are being used to derive neurons and cardiomyocytes. These cells represent the most faithful *in vitro* model for the disease and are being increasingly used to investigate pathogenesis, specific vulnerability, and the effect of potential therapeutics. Initial studies have demonstrated that iPSCs and iPSC-derived differentiated cells from FRDA patients show GAA repeat instability, epigenetic changes at the *FXN* locus with reduced frataxin expression, and mitochondrial dysfunction.[78]

NEW APPROACHES TO TREATMENT

Attempts at symptomatic treatment have used drugs that affect neurotransmitters acting in the cerebellar circuitry. These include cholinergic agonists such as physostigmine and choline, neuropeptides such as TRH, serotoninergic agonists such as 5-OH tryptophan and buspirone, and the dopaminergic drug amantadine. These compounds have been tested only in small, uncontrolled studies, and overall results have not been encouraging.

More recently, based on the hypothesis that iron-mediated oxidative damage plays a major role in the pathogenesis of FRDA, removal of excess mitochondrial iron and/or antioxidant treatment has been attempted (Figure 49.4). The short chain coenzyme Q analog idebenone and the orally administered iron chelator deferiprone have been tested in randomized controlled trials (RCTs), but results have been negative or inconclusive.[79]

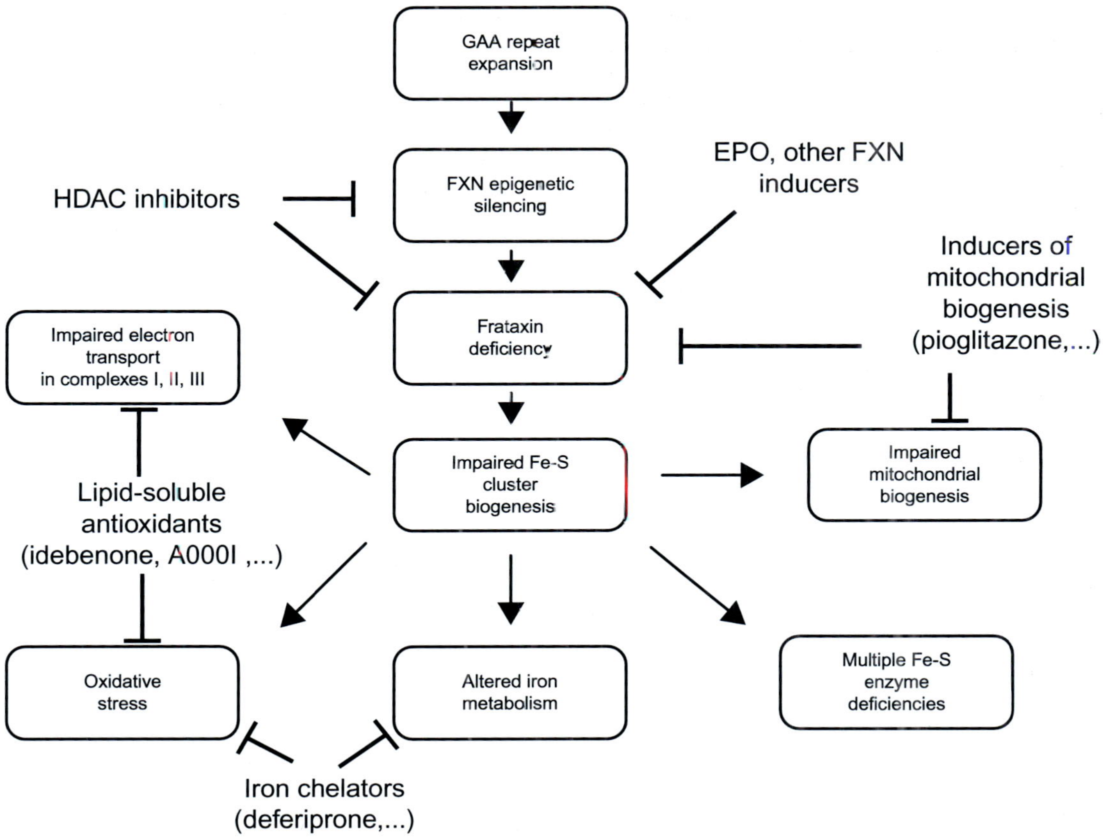

FIGURE 49.4 Potential therapeutic targets in the FRDA pathogenic cascade.

However, newer, more potent antioxidants and respiratory chain activators are now undergoing RCTs.[80] At the same time, approaches targeting the epigenetic silencing of the *FXN* gene are being explored. In particular, the efficacy of a specific class of histone deacetylase inhibitors has been proven in upregulating frataxin in cells from FRDA patients, including cultured neurons derived from iPSCs, as well as in GAA repeat-based animal models.[62] A phase I clinical trial in FRDA patients has shown that one of these drugs increases *FXN* mRNA histone acetylation in PBMCs, providing a proof of concept for this therapeutic approach.

As more potential therapeutics come to clinical development, clinical assessment tools and biomarkers must be identified to guide trial design.[81] Ongoing large natural history studies are critical for this purpose.

Rehabilitation remains a cornerstone of FRDA management. Programs should include exercises aimed at maximizing the residual capacity of motor control. Orthopedic interventions are sometimes necessary, including surgical correction of severe scoliosis and of foot deformity in patients who can still walk.

REFERENCES

1. Geoffroy G, Barbeau A, Breton G, Lemieux B, Aube M, Leger C, et al. Clinical description and roentgenologic evaluation of patients with Friedreich's ataxia. *Can J Neurol Sci* 1976;**3**(4):279–86.
2. Harding AE. Friedreich's ataxia: a clinical and genetic study of 90 families with an analysis of early diagnostic criteria and intrafamilial clustering of clinical features. *Brain* 1981;**104**(3):589–620.
3. Campuzano V, Montermini L, Moltò MD, Pianese L, Cossee M, Cavalcanti F, et al. Friedreich's ataxia: autosomal recessive disease caused by an intronic GAA triplet repeat expansion. *Science* 1996;**271**(5254):1423–7.
4. Anheim M, Fleury M, Monga B, Anheim M, Fleury M, Monga B, et al. Epidemiological, clinical, paraclinical and molecular study of a cohort of 102 patients affected with autosomal recessive progressive cerebellar ataxia from Alsace, Eastern France: implications for clinical management. *Neurogenetics* 2010;**11**(1):1–12.
5. Anheim M, Tranchant C, Koenig M. The autosomal recessive cerebellar ataxias. *N Engl J Med* 2012;**366**(7):636–46.
6. Vankan P. Prevalence gradients of Friedreich's ataxia and R1b haplotype in Europe co-localize, suggesting a common Palaeolithic origin in the Franco-Cantabrian ice age refuge. *J Neurochem* 2013;**126** (Suppl. 1):11–20.

7. Koeppen AH. Nikolaus Friedreich and degenerative atrophy of the dorsal columns of the spinal cord. *J Neurochem* 2013;**126** (Suppl. 1):4–10.
8. Koeppen AH, Davis AN, Morral JA. The cerebellar component of Friedreich's ataxia. *Acta Neuropathol* 2011;**122**(3):323–30.
9. Koeppen AH, Mazurkiewicz JE. Friedreich ataxia: neuropathology revised. *J Neuropathol Exp Neurol* 2013;**72**(2):78–90.
10. Koeppen AH, Morral JA, Davis AN, Qian J, Petrocine SV, Knutson MD, et al. The dorsal root ganglion in Friedreich's ataxia. *Acta Neuropathol* 2009;**118**(6):763–76.
11. Koeppen AH. Friedreich's ataxia: pathology, pathogenesis, and molecular genetics. *J Neurol Sci* 2011;**303**(1-2):1–12.
12. Morral JA, Davis AN, Qian J, Gelman BB, Koeppen AH. Pathology and pathogenesis of sensory neuropathy in Friedreich's ataxia. *Acta Neuropathol* 2010;**120**(1):97–108.
13. Payne RM, Wagner GR. Cardiomyopathy in Friedreich ataxia: clinical findings and research. *J Child Neurol* 2012;**27**(9):1179–86.
14. Weidemann F, Störk S, Liu D, Hu K, Herrmann S, Ertl G, et al. Cardiomyopathy of Friedreich ataxia. *J Neurochem* 2013;**126** (Suppl. 1):88–93.
15. Ramirez RL, Qian J, Santambrogio P, Levi S, Koeppen AH. Relation of cytosolic iron excess to cardiomyopathy of Friedreich's ataxia. *Am J Cardiol* 2012;**110**:1820–7.
16. Cnop M, Igoillo-Esteve M, Rai M, Begu A, Serroukh Y, Depondt C, et al. The central role and mechanisms of β-cell dysfunction and death in Friedreich's ataxia-associated diabetes. *Ann Neurol* 2012;**72**:971–82.
17. Cnop M, Mulder H, Igoillo-Esteve M. Diabetes in Friedreich ataxia. *J Neurochem* 2013;**126**(Suppl. 1):94–102.
18. Pandolfo M. Friedreich ataxia: the clinical picture. *J Neurol* 2009;**256**(Suppl. 1):3–8.
19. Parkinson MH, Boesch S, Nachbauer W, Mariotti C, Giunti P. Clinical features of Friedreich's ataxia: classical and atypical phenotypes. *J Neurochem* 2013;**126**(Suppl. 1):103–17.
20. De Michele G, Filla A, Cavalcanti F, Di Maio L, Pianese L, Castaldo I, et al. Late onset Friedreich's disease: clinical features and mapping of mutation to the FRDA locus. *J Neurol Neurosurg Psychiatr* 1994;**57**(8):977–9.
21. Klockgether T, Chamberlain S, Wüllner U, Fetter M, Dittmann H, Petersen D, et al. Late-onset Friedreich's ataxia. Molecular genetics, clinical neurophysiology, and magnetic resonance imaging. *Arch Neurol* 1993;**50**(8):803–6.
22. Folker JE, Murdoch BE, Cahill LM, Rosen KM, Delatycki MB, Corben LA, et al. Articulatory kinematics in the dysarthria associated with Friedreich's ataxia. *Motor Control* 2011;**15**(3):376–89.
23. Corben LA, Georgiou-Karistianis N, Fahey MC, Storey E, Churchyard A, Horne M, et al. Towards an understanding of cognitive function in Friedreich ataxia. *Brain Res Bull* 2006;**70**(3):197–202.
24. Corben LA, Georgiou-Karistianis N, Bradshaw JL, Hocking DR, Churchyard AJ, Delatycki MB. The Fitts task reveals impairments in planning and online control of movement in Friedreich ataxia: reduced cerebellar-cortico connectivity? *Neuroscience* 2011;**192**:382–90.
25. Spieker S, Schulz JB, Petersen D, Fetter M, Klockgether T, Dichgans J. Fixation instability and oculomotor abnormalities in Friedreich's ataxia. *J Neurol* 1995;**242**(8):517–21.
26. Seyer LA, Galetta K, Wilson J, Sakai R, Perlman S, Mathews K, et al. Analysis of the visual system in Friedreich ataxia. *J Neurol* 2013;**260**(9):2362–9.
27. Rance G, Corben L, Delatycki M. Auditory processing deficits in children with Friedreich ataxia. *J Child Neurol* 2012;**27** (9):1197–203.
28. Rance G, Corben LA, Delatycki MB. Auditory pathway changes mirror overall disease progress in individuals with Friedreich ataxia. *J Neurol* 2012;**259**(12):2746–8.
29. Rance G, Fava R, Baldock H, Chong A, Barker E, Corben L, et al. Speech perception ability in individuals with Friedreich ataxia. *Brain* 2008;**131**(Pt 8):2002–12.
30. Tsou AY, Paulsen EK, Lagedrost SJ, Perlman SL, Mathews KD, Wilmot GR, et al. Mortality in Friedreich ataxia. *J Neurol Sci* 2011;**307**(1–2):46–9.
31. Emond M, Lepage G, Vanasse M, Pandolfo M. Increased levels of plasma malondialdehyde in Friedreich ataxia. *Neurology [Internet]* 2000;**55**(11):1752–3.
32. Schulz JB, Dehmer T, Schöls L, Mende H, Hardt C, Vorgerd M, et al. Oxidative stress in patients with Friedreich ataxia. *Neurology* 2000;**55**(11):1719–21.
33. Di Prospero NA, Baker A, Jeffries N, Fischbeck KH. Neurological effects of high-dose idebenone in patients with Friedreich's ataxia: a randomised, placebo-controlled trial. *Lancet Neurol* 2007;**6** (10):878–86.
34. Coppola G, Burnett R, Perlman S, Versano R, Gao F, Palsterer H, et al. A gene expression phenotype in lymphocytes from Friedreich ataxia patients. *Ann Neurol* 2011;**70**(5):790–804.
35. Akhlaghi H, Yu J, Corben L, Georgiou-Karistianis N, Bradshaw JL, Storey E, et al. Cognitive deficits in Friedreich ataxia correlate with micro-structural changes in dentatorubral tract. *Cerebellum* 2013;**13**(2):187–98.
36. Lund C, Bremer A, Lossius MI, Selmer KK, Brodtkorb E, Nakken KO. Dravet syndrome as a cause of epilepsy and learning disability. *Tidsskr Nor Laegeforen* 2012;**132**(1):44–7.
37. Zalesky A, Akhlaghi H, Corben LA, Bradshaw JL, Delatycki MB, Storey E, et al. Cerebello-cerebral connectivity deficits in Friedreich ataxia. *Brain Struct Funct* 2014;**219**(3):969–81.
38. Georgiou-Karistianis N, Akhlaghi H, Corben LA, Delatycki MB, Storey E, Bradshaw JL, et al. Decreased functional brain activation in Friedreich ataxia using the Simon effect task. *Brain Cogn* 2012;**79**(3):200–8.
39. Gilman S, Junck L, Markel DS, Koeppe RA, Kluin KJ. Cerebral glucose hypermetabolism in Friedreich's ataxia detected with positron emission tomography. *Ann Neurol* 1990;**28**(6):750–7.
40. Santoro L, De Michele G, Perretti A, Crisci C, Cocozza S, Cavalcanti F, et al. Relation between trinucleotide GAA repeat length and sensory neuropathy in Friedreich's ataxia. *J Neurol Neurosurg Psychiatr* 1999;**66**(1):93–6.
41. Santoro L, Perretti A, Lanzillo B, Coppola G, De Joanna G, Manganelli F, et al. Influence of GAA expansion size and disease duration on central nervous system impairment in Friedreich's ataxia: contribution to the understanding of the pathophysiology of the disease. *Clin Neurophysiol* 2000;**111**(6):1023–30.
42. Pandolfo M, Manto M. Cerebellar and afferent ataxias. *Continuum (Minneap Minn)* 2013;**19**(5, Movement Disorders):1312–43.
43. Schulz JB, Boesch S, Bürk K, Durr A, Giunti P, Mariotti C, et al. Diagnosis and treatment of Friedreich ataxia: a European perspective. *Nat Rev Neurol* 2009;**5**(4):222–34.
44. Pandolfo M. Drug insight: antioxidant therapy in inherited ataxias. *Nat Clin Pract Neurol* 2008;**4**(2):86–96.

45. Synofzik M, Gonzalez MA, Lourenco CM, Coutelier M, Haack TB, Rebelo A, et al. PNPLA6 mutations cause Boucher-Neuhauser and Gordon Holmes syndromes as part of a broad neurodegenerative spectrum. *Brain* 2014;**137**:69–77.
46. Shi C-H, Schisler JC, Rubel CE, Tan S, Song B, McDonough H, et al. Ataxia and hypogonadism caused by the loss of ubiquitin ligase activity of the U box protein CHIP. *Hum Mol Genet* 2014;**23**:1013–24.
47. Anttonen A-K, Mahjneh I, Hämäläinen RH, Lagier-Tourenne C, Kopra O, Waris L, et al. The gene disrupted in Marinesco-Sjögren syndrome encodes SIL1, an HSPA5 cochaperone. *Nat Genet* 2005;**37**(12):1309–11.
48. Rajadhyaksha AM, Elemento O, Puffenberger EG, Schierberl KC, Xiang JZ, Putorti ML, et al. Mutations in FLVCR1 cause posterior column ataxia and retinitis pigmentosa. *Am J Hum Genet* 2010;**87**(5):643–54.
49. Schulz A, Kohlschütter A, Mink J, Simonati A, Williams R. NCL diseases—clinical perspectives. *Biochim Biophys Acta* 2013;**1832**(11):1801–6.
50. Jiralerspong S, Liu Y, Montermini L, Stifani S, Pandolfo M. Frataxin shows developmentally regulated tissue-specific expression in the mouse embryo. *Neurobiol Dis* 1997;**4**(2):103–13.
51. De Michele G, Cavalcanti F, Criscuolo C, Pianese L, Monticelli A, Filla A, et al. Parental gender, age at birth and expansion length influence GAA repeat intergenerational instability in the X25 gene: pedigree studies and analysis of sperm from patients with Friedreich's ataxia. *Hum Mol Genet* 1998;**7**(12):1901–6.
52. Montermini L, Kish SJ, Jiralerspong S, Lamarche JB, Pandolfo M. Somatic mosaicism for Friedreich's ataxia GAA triplet repeat expansions in the central nervous system. *Neurology* 1997;**49**(2):606–10.
53. Clark RM, De Biase I, Malykhina AP, Al-Mahdawi S, Pook M, Bidichandani SI. The GAA triplet-repeat is unstable in the context of the human FXN locus and displays age-dependent expansions in cerebellum and DRG in a transgenic mouse model. *Hum Genet* 2007;**120**(5):633–40.
54. De Biase I, Rasmussen A, Endres D, Al-Mahdawi S, Monticelli A, Cocozza S, et al. Progressive GAA expansions in dorsal root ganglia of Friedreich's ataxia patients. *Ann Neurol* 2007;**61**(1):55–60.
55. Cossée M, Schmitt M, Campuzano V, Reutenauer L, Moutou C, Mandel JL, et al. Evolution of the Friedreich's ataxia trinucleotide repeat expansion: founder effect and premutations. *Proc Natl Acad Sci USA* 1997;**94**(14):7452–7.
56. Montermini L, Andermann E, Labuda M, Richter A, Pandolfo M, Cavalcanti F, et al. The Friedreich ataxia GAA triplet repeat: premutation and normal alleles. *Hum Mol Genet* 1997;**6**(8):1261–6.
57. Labuda M, Labuda D, Miranda C, Poirier J, Soong BW, Barucha NE, et al. Unique origin and specific ethnic distribution of the Friedreich ataxia GAA expansion. *Neurology* 2000;**54**(12):2322–4.
58. Campuzano V, Montermini L, Lutz Y, Cova L, Hindelang C, Jiralerspong S, et al. Frataxin is reduced in Friedreich ataxia patients and is associated with mitochondrial membranes. *Hum Mol Genet* 1997;**6**(11):1771–80.
59. Sakamoto N, Chastain PD, Parniewski P, Ohshima K, Pandolfo M, Griffith JD, et al. Sticky DNA: self-association properties of long GAA.TTC repeats in R.R.Y triplex structures from Friedreich's ataxia. *Mol Cell* 1999;**3**(4):465–75.
60. Wells RD. DNA triplexes and Friedreich ataxia. *FASEB J* 2008;**22**(6):1625–34.
61. Yandim C, Natisvili T, Festenstein R. Gene regulation and epigenetics in Friedreich's ataxia. *J Neurochem* 2013;**126**(Suppl. 1):21–42.
62. Gottesfeld JM, Rusche JR, Pandolfo M. Increasing frataxin gene expression with histone deacetylase inhibitors as a therapeutic approach for Friedreich's ataxia. *J Neurochem* 2013;**126**(Suppl. 1):147–54.
63. Dürr A, Cossée M, Agid Y, Campuzano V, Mignard C, Penet C, et al. Clinical and genetic abnormalities in patients with Friedreich's ataxia. *N Engl J Med* 1996;**335**(16):1169–75.
64. Montermini L, Richter A, Morgan K, Justice CM, Julien D, Castellotti B, et al. Phenotypic variability in Friedreich ataxia: role of the associated GAA triplet repeat expansion. *Ann Neurol* 1997;**41**(5):675–82.
65. Cossée M, Dürr A, Schmitt M, Dahl N, Trouillas P, Allinson P, et al. Friedreich's ataxia: point mutations and clinical presentation of compound heterozygotes. *Ann Neurol* 1999;**45**(2):200–6.
66. Schmucker S, Argentini M, Carelle-Calmels N, Martelli A, Puccio H. The in vivo mitochondrial two-step maturation of human frataxin. *Hum Mol Genet* 2008;**17**(22):3521–31.
67. Pastore A, Puccio H. Frataxin: a protein in search for a function. *J Neurochem* 2013;**126**(Suppl. 1):43–52.
68. Tsai C-L, Barondeau DP. Human frataxin is an allosteric switch that activates the Fe-S cluster biosynthetic complex. *Biochemistry* 2010;**49**(43):9132–9.
69. Rouault TA. Biogenesis of iron-sulfur clusters in mammalian cells: new insights and relevance to human disease. *Dis Model Mech* 2012;**5**(2):155–64.
70. González-Cabo P, Palau F. Mitochondrial pathophysiology in Friedreich's ataxia. *J Neurochem* 2013;**126**(Suppl. 1):53–64.
71. Cossée M, Puccio H, Gansmuller A, Koutnikova H, Dierich A, LeMeur M, et al. Inactivation of the Friedreich ataxia mouse gene leads to early embryonic lethality without iron accumulation. *Hum Mol Genet* 2000;**9**(8):1219–26.
72. Perdomini M, Hick A, Puccio H, Pook MA. Animal and cellular models of Friedreich ataxia. *J Neurochem* 2013;**126**(Suppl. 1):65–79.
73. Babcock M, de Silva D, Oaks R, Davis-Kaplan S, Jiralerspong S, Montermini L, et al. Regulation of mitochondrial iron accumulation by Yfh1p, a putative homolog of frataxin. *Science* 1997;**276**(5319):1709–12.
74. Miranda CJ, Santos MM, Ohshima K, Smith J, Li L, Bunting M, et al. Frataxin knockin mouse. *FEBS Lett* 2002;**512**(1–3):291–7.
75. Al-Mahdawi S, Pinto RM, Varshney D, Lawrence L, Lowrie MB, Hughes S, et al. GAA repeat expansion mutation mouse models of Friedreich ataxia exhibit oxidative stress leading to progressive neuronal and cardiac pathology. *Genomics* 2006;**88**(5):580–90.
76. Puccio H. Conditional mouse models for Friedreich ataxia, a neurodegenerative disorder associating cardiomyopathy. *Handb Exp Pharmacol* 2007;**178**:365–75.
77. Rai M, Soragni E, Jenssen K, Burnett R, Herman D, Coppola G, et al. HDAC inhibitors correct frataxin deficiency in a Friedreich ataxia mouse model. *PLOS ONE* 2008;**3**(4):e1958.
78. Hick A, Wattenhofer-Donzé M, Chintawar S, Tropel P, Simard JP, Vaucamps N, et al. Neurons and cardiomyocytes derived from induced pluripotent stem cells as a model for mitochondrial defects in Friedreich's ataxia. *Dis Model Mech* 2013;**6**:608–21.

79. Kearney M, Orrell RW, Fahey M, Pandolfo M. Antioxidants and other pharmacological treatments for Friedreich ataxia. *Cochrane Database Syst Rev* 2012;4:CD007791.
80. Lynch DR, Willi SM, Wilson RB, Cotticelli MG, Brigatti KW, Deutsch EC, et al. A0001 in Friedreich ataxia: Biochemical characterization and effects in a clinical trial. *Mov Disord* 2012;**27**(8):1026–33.
81. Bürk K, Schulz SR, Schulz JB. Monitoring progression in Friedreich ataxia (FRDA): the use of clinical scales. *J Neurochem* 2013;**126**(Suppl. 1):118–24.

Chapter 50

Dominantly Inherited Spinocerebellar Syndromes

Stephen M. Maricich and Huda Y. Zoghbi

INTRODUCTION

The dominantly inherited ataxias are a heterogeneous group of neurologic disorders that currently include 30 spinocerebellar ataxias (SCAs), the related disorder dentatorubral-pallidoluysian atrophy (DRPLA), and 7 episodic ataxias (EAs). The number of these disorders will undoubtedly grow as only 60% to 70% of the dominantly inherited ataxias fall into one of the current categories. As a group they are rare: the estimated prevalence of SCA is $\sim$3 per 100,000 individuals.[1] Ataxia (axial, gait, limb) and dysarthria are typically the presenting symptoms and are common to all SCAs, and other cerebellar signs (diplopia, dysdiadochokinesia, dysmetria, hypotonia, nystagmus, rebound phenomenon) often accompany them. The presence of other neurological and nonneurological symptoms varies, and the clinical course can vary greatly among the different subtypes and even within the same family, ranging from a benign, relatively static ataxia to rapid deterioration of multiple neurological functions. This variability challenges the clinician's ability to distinguish between the ataxias on the basis of clinical features alone. Atrophy/degeneration of the cerebellum is a constant, with variable involvement of the spinal tracts and/or brain stem. In some cases, associated noncerebellar symptoms/signs can provide important clues to the underlying diagnosis; where these exist, we will highlight them in the sections to come.

Although the SCAs are typically considered as presenting in adulthood, virtually all of them have been documented in children, and even in supposed adult-onset cases subtle signs of cerebellar dysfunction may have been present for years before diagnosis. Infantile- and juvenile-onset SCAs typically show more rapid disease progression, greater severity, and a broader range of neurological features. A detailed genealogy will often reveal the mode of inheritance, but an apparently negative family history does not rule out a hereditary ataxia, because there may be *de novo* mutations, genetic anticipation causing onset of symptoms in a child prior to onset in a parent (see below), death of an adult relative before full symptom onset (or proper diagnosis), adoption, or potential alternate paternity.

The identification of responsible genes and causative alleles has contributed greatly to understanding the underlying pathogenesis of these disorders and provided a clearer nosology than clinical presentation. Therefore, in this chapter, we categorize the SCAs/DRPLA by genetic cause (polyglutamine expansions, noncoding expansions, missense mutations, and mutations not yet identified) and treat the EAs separately (Table 50.1). For each group, we describe common clinical features and the current understanding of molecular pathogenesis, and then highlight any aspects of the specific diseases that can help differentiate them clinically or pathogenetically. We then close with comments on diagnostic approach and treatment.

SCAs CAUSED BY POLYGLUTAMINE EXPANSIONS

Six SCAs (1, 2, 3, 6, 7, and 17) and one associated disease (DRPLA), representing about 40% to 60% of SCA diagnoses, are caused by the expansion of a series of CAG triplet repeats in the coding region of the affected gene. These expanded repeats then encode abnormally long polyglutamine (polyQ) tracts that disrupt protein function in various ways. Triplet repeat diseases share the following characteristics:

1. Penetrance is close to 100% (if you've got a disease allele, you're very likely to develop disease).
2. CAG tracts are typically interrupted by another trinucleotide sequence (such as CAA or CAT), but disease alleles tend to have an uninterrupted run of CAG repeats.

B.T. Darras, H. Royden Jones, Jr., M.M. Ryan & D.C. De Vivo (Eds): Neuromuscular Disorders of Infancy, Childhood and Adolescence, Second edition.
DOI: http://dx.doi.org/10.1016/B978-0-12-417044-5.00050-0

TABLE 50.1 Autosomal Dominant Cerebellar Ataxias

Ataxia	Chromosome	Gene	Gene Product	Mechanism	Age of Onset (Years)	Normal Repeat	Expanded Repeat
Polyglutamine expansion							
SCA1	6p23	*SCA1*	Ataxin-1	CAG repeat	6–60	6–44*	39–82*
SCA2	12q24	*SCA2*	Ataxin-2	CAG repeat	2–65	15–24	35–59
SCA3/MJD	14q24.3-q31	*MJD1*	Ataxin-3	CAG repeat	11–70	13–36	61–84
SCA6	19q13	*CACNA1A*	CACNA1A	CAG repeat	16–73	4–20	21–33
SCA7	3p21.1-p12	*SCA7*	Ataxin-7	CAG repeat	Birth–53	4–35	37–460
SCA17	6q27	*SCA17*	TBP	CAG repeat	3–48	25–42	45–66
DRPLA	12p13.31	*DRPLA*	Atrophin-1	CAG repeat	4 months–55	7–34	53–93
Atypical expansion							
SCA8	13q21	*SCA8*	SCA8 RNA	CTG repeat in 3′ UTR	18–72	2–91*	110–155*
SCA10	22q13	*SCA10*	Ataxin-10	ATTCT repeat in intron 9	14–45	10–22	750–4500
SCA12	5q31-q33	*SCA12*	P2R2B	CAG repeat in 5′ UTR	8–55	7–32	55–78
SCA31	16q22.1	*SCA31*	TK2, BEAN	TGGAA intronic repeat	30–80	0	>2.5kb
SCA36	20p13	*SCA36*	NOP56	GGCCTG intronic repeat	40–50	5–14	650–2500
Missense mutation							
EA6	5p13	*SLC1A3*	EAAT1	Missense mutation	Birth–Adult		
SCA5	11p11-q11	*SPTBN2*	b-III spectrin	Missense mutation	10–68		
SCA11	15q14-q21.3	*TTBK2*	TTBK2	Truncating mutation	15–43		
SCA14	19q13.4	*PKCγ*	PKCγ	Missense mutation	10–69		
SCA15	3p24.2-3pter	*ITPR1*	ITPR1	Missense mutation	Child–Adult		
SCA23	20p13-p12.2	*PDYN*	PDYN	Missense mutation	43–56		
SCA26	19p13.3	*eEF2*	eEF2	Missense mutation	26–60		
SCA27	13q34	*FGF14*	FGF14	Missense mutation	15–20		
SCA28	18p11.21	*AFG3L2*	AFG3L2	Missense mutation	3–36		
SCA35	20p13	*TGM6*	TGM6	Missense mutation	40–48		

(*Continued*)

TABLE 50.1 (Continued)

Ataxia	Chromosome	Gene	Gene Product	Mechanism	Age of Onset (Years)	Normal Repeat	Expanded Repeat
Channelopathy							
EA1	12p13	*EA1*	KCNA1	Channelopathy	Early Childhood		
EA2/FHM	19p13	*CACNA1A*	CACNA1A	Channelopathy: missense and nonsense mutations	4–30		
EA5	2q23.3	*CACNB4*	CACNB4	Missense mutation	20–30		
SCA13	19q13.3-q13.4	*KCNC3*	Kv3.3	Missense mutation	<1–45		
SCA19	1p21-q21	*KCND3*	Kv4.3	Missense mutation	10–45		
Mutation unknown							
EA3	Unknown	Unknown	Unknown	Unknown	23–42		
EA4	Unknown	Unknown	Unknown	Unknown	1–42		
SCA4	16q22	Unknown	Unknown	Unknown	19–59		
SCA18/SMNA	7q31-q32	Unknown	Unknown	Unknown	12–25		
SCA20	11	Unknown	Unknown	Unknown	19–64		
SCA21	7p21.3-p15.1	Unknown	Unknown	Unknown	6–30		
SCA25	2p	Unknown	Unknown	Unknown	1½–39		
SCA30	4q34.3-q35.1	Unknown	Unknown	Unknown	45–76		
SCA32	7q32-q33	Unknown	Unknown	Unknown	Adult		
SCA34	6p12.3-16.2	Unknown	Unknown	Unknown	Adult		

**there is overlap of pathogenic and nonpathogenic repeat length. See text for details.*

Long uninterrupted polyQ tracts are unstable and prone to slippage during DNA replication, which increases the likelihood of further expansion, though reduction can also occur.

3. There is an inverse correlation between polyQ tract length and age of onset/disease severity.
4. The dynamic nature of the mutation (the tendency of the repeat tract to expand, particularly if the disease allele is inherited from the father) results in *anticipation*, in which disease onset is earlier for each subsequent generation within a family.
5. CAG repeat length in the nonpathogenic allele and other heritable factors modulate age of onset and clinical severity.[2–6]

Clinically, polyglutamine expansion SCAs tend to present with ataxia, dysarthria, and nystagmus. Some SCAs tend to cause straightforward cerebellar ataxia; others produce a variety of noncerebellar features that can include bulbar dysfunction (dysphagia, facial muscle fasciculations/atrophy), oculomotor dysfunction, episodic vertigo, extrapyramidal signs (chorea, dystonia, tremor), pyramidal signs (hypertonia/spasticity, hyperreflexia), peripheral neuropathy, sleep disturbances (insomnia, restless leg syndrome), intellectual impairment, psychiatric symptoms, and seizures. Lifespan is typically shortened. Neuroimaging shows involvement of the cerebellum with progressive brainstem atrophy, findings that are corroborated on pathology specimens.

Molecular genetic studies of the polyglutamine expansion diseases have revealed some common themes in their pathogenesis. Nuclear inclusions containing the expanded proteins and components of the ubiquitin-proteasome system are found in all of these diseases except SCA6, supporting the notion that the mutant proteins misfold and resist degradation.[7–11] This degradation resistance may raise the effective levels of the protein, contributing to its toxic gain of function; interestingly, inclusion formation may serve a protective effect by sequestering the mutant protein.[12] The polyglutamine expansion appears to alter the interactions of the polyQ protein with its native interactors, which can result in both a gain and loss of normal function.[13,14] In addition to shared mechanisms, polyQ diseases also seem to share certain molecular pathways. For example, many of the same genes are dysregulated in mouse models of Huntington disease (another triplet repeat disease) and SCA7.[15] Similarly, "interactome" studies analyzing direct protein-protein interactions place many of the SCA proteins in the same pathways.[16,17]

Spinocerebellar Ataxia Type 1

Spinocerebellar ataxia type 1 (SCA1 – MIM 164400) typically causes a slowly progressive ataxia and dysarthria; over time, the disease evolves to include optic nerve atrophy, maculopathy,[18,19] retinal nerve fiber degeneration,[20] pyramidal and extrapyramidal symptoms, cognitive impairment, and peripheral nerve involvement. Bulbar dysfunction ensues, starting as dysphagia and progressing to choking spells and aspiration; eventual respiratory failure and pneumonia are common causes of death. The typical age of onset is in the third or fourth decade of life, but symptoms can start in childhood or as late as 60 years of age.[21] Childhood onset is associated with more rapid disease progression and broader symptomatology including mild mental retardation.[22] SCA1 has been described in multiple ethnic groups.[23]

SCA1 is caused by CAG repeat expansion in the *SCA1* gene.[24] Normal *SCA1* alleles have 6 to 44 CAG repeats, and those with 20 or more repeats are interrupted by 1 to 4 CAT trinucleotide units. In contrast, disease-causing alleles have 39 to 82 repeats and lack intervening CAT sequences.[25] More than 70 repeats tends to cause juvenile onset.[24,25] The *SCA1* gene encodes ataxin-1, a protein that is expressed in several tissues and that shuttles between the cytoplasm and nucleus in neurons.[7,26,27] Although its exact functions are unknown, native ataxin-1 interacts with multiple proteins involved in cell signaling, chromatin modulation, regulation of gene expression and RNA splicing.[14,28–34] CAG repeat expansions alter ataxin-1's interactions with its native protein partners, making some complexes more persistent and leading to wide-reaching changes in gene expression.[35–38] Both gain and loss of normal ataxin-1 function contribute to SCA1 pathogenesis.[14,38–40] Phosphorylation of ataxin-1 governs protein-protein interactions and is important for normal function and pathogenesis.[41–46]

Native ataxin-1 appears to play a role in controlling maturation of climbing fiber input to Purkinje cells.[47] Loss of ataxin-1 function impairs spatial and motor learning in mice,[48] and human patients with chromosomal deletions spanning SCA1 develop mental retardation and seizures[49]; neither develops ataxia or neuronal degeneration. An intriguing recent report suggests that the *SCA1* locus also encodes an out-of-frame transcript that interacts with ataxin-1[50]; what (if any) bearing this has on pathogenesis is still unclear. Intermediate expansions (>31 repeats) are associated with sporadic amyotrophic lateral sclerosis (ALS).[51]

SCA1 mouse models have revealed a great deal about the molecular pathogenesis of the disease and allowed the testing of several potential therapeutic avenues. For example, chaperone protein overexpression, which presumably aids in the folding and/or elimination of expanded ataxin-1, resulted in improved motor coordination and suppressed Purkinje cell degeneration in transgenic mice.[52] Genetic and pharmacological inhibition of ataxin-1 phosphorylation improved pathology and neurological function.[41,42,45] Lithium therapy, which is thought to induce beneficial changes in gene expression, improved motor function in transgenic mice.[53] Genetic overexpression and pharmacological administration of vascular endothelial growth factor (VEGF) mitigated pathogenesis.[54] It is particularly noteworthy that exercise extended the lifespan of transgenic mice.[55] Aminopyridine treatment also mitigates cerebellar dysfunction and pathology.[56]

Spinocerebellar Ataxia Type 2

Spinocerebellar ataxia type 2 (SCA2 – MIM 183090) is the most common SCA in Cuba and Mexico, though it also occurs in patients from other ethnic backgrounds.[57–64] The clinical findings are ataxia, dysarthria, tremor, nystagmus, and extremely slow saccades. There is also evidence for autonomic cardiovascular involvement in presymptomatic individuals.[65] Hyporeflexia and ophthalmoparesis occur in more than half of the patients. Pyramidal signs are present; deep tendon reflexes are brisk early on and are absent later in the course. Dystonia, chorea, and fasciculations have been reported; cognitive impairment and dementia occur in a significant minority of patients.[66] Periodic leg movements (PLMs) and REM sleep pathology are common and correlate with disease severity. Dysphagia and bulbar failure occur in the last stages of the disease. About 40% of patients develop symptoms before their 25th birthday. Childhood onset can present as a developmental regression syndrome[67]; infantile onset, which occurs with extreme CAG repeat expansion (>130 copies), presents with hypotonia, dysphagia, ocular signs, visual impairment, retinitis pigmentosa, autonomic dysfunction, global developmental delay, infantile spasms, and infant-onset epilepsy.[68–72]

SCA2 is caused by expansion of a CAG repeat tract in the coding region of the *SCA2* gene.[58,62,63] The normal and disease ranges of CAG repeats are 14 to 31 and 34 to 750 repeats, respectively.[60,71] All but the shortest repeat tracts in normal alleles are interrupted by 1–3 CAA. Mutant alleles are typically uninterrupted, although cases with a single CAA have been reported; these cases tend to be associated with sporadic or familial parkinsonism.[73–76] Mild SCA2 expansions can produce parkinsonism and multiple system atrophy (MSA) phenotypes.[77] Interestingly, intermediate-length (27–33) expansions are an independent risk factor for the development of amyotrophic lateral sclerosis.[78,79]

SCA2 encodes ataxin-2, a cytoplasmic protein whose mRNA is alternatively spliced and found in multiple tissues and all regions of the CNS, where it is present predominantly in neurons and is highly expressed in Purkinje cells.[58,62,63,80,81] Ataxin-2 is a cytoplasmic protein that is subcellularly localized to the Golgi apparatus and rough endoplasmic reticulum.[82–84] Nuclear localization of mutant ataxin-2 protein is not necessary for SCA2 pathogenesis, but expanded ataxin-2 does accumulate in the cytoplasm of affected neurons.[81,85,86]

At the molecular level, ataxin-2 associates with RNA-binding proteins, polyribosomes and stress granules, suggesting that it functions in RNA processing and translational regulation of protein production.[82,87–91] The *Drosophila* homolog of ataxin-2 has recently been shown to regulate translation of proteins involved in the circadian rhythm, an intriguing finding that could help explain sleep disturbances in SCA2 patients.[92,93] Ataxin-2 may also be involved in endocytosis of and trophic signaling mediated by receptor tyrosine kinases, and polyglutamine expansion changes the levels of associated proteins.[94–97] The morphological changes seen in diseased Purkinje cells may result from abnormal cytoskeletal architecture secondary to disruption of microtubule assembly[85,98]; whether these effects are mediated through RNA interactions is unknown. There is some evidence that ataxin-2 functions as a mediator of oxidative damage.[95] Parkin, a gene associated with Parkinson disease, interacts with ataxin-2 and can increase clearance of normal and expanded ataxin-2.[99] Mutant ataxin-2 has also been linked to Ca^{2+}-mediated cytotoxicity through association with the inositol 1,4,5-trisphosphate receptor; inhibition of signaling through this receptor decreased Ca^{2+} levels in a mouse model of SCA2, concomitantly mitigating incoordination and Purkinje cell death.[100,101] ALS-associated polyglutamine expansions in ataxin-2 enhance stress-induced caspase activation and could provide a mechanism for motor neuron death.[51,102] There is also evidence of interaction between ataxin-1 and ataxin-2.[17,103,104] Interestingly, the gene mutated in SCA6 (*CACNA1A*) is, when not mutated, a disease modifier for SCA2: longer repeat lengths in normal alleles are associated with earlier age of SCA2 onset.[105]

Spinocerebellar Ataxia Type 3

Spinocerebellar ataxia type 3 (SCA3 – MIM 109150), also called Machado-Joseph disease (MJD) after two affected families of Portuguese-Azorean origin,[106] accounts for 30–50% of dominantly inherited ataxias and is the most common hereditary spinocerebellar ataxia worldwide.[1,107–109] Originally regarded as separate entities, genetic mapping and subsequent molecular cloning revealed that SCA3 and MJD are caused by triplet repeat mutations in the same gene.[110,111] Typical clinical features of SCA3 include progressive ataxia, areflexia, peripheral amyotrophy, external ophthalmoplegia, bulging eyes, faciolingual myokymia, muscle atrophy, parkinsonian features, dystonia, spasticity, dementia, and dysautonomia. SCA3 expansions have also been associated with pure parkinsonism in the absence of ataxia.[112] Disease onset is usually in the second to fourth decades of life.

The *MJD1* gene encodes ataxin-3, the smallest polyglutamine protein. Predominantly a cytoplasmic protein, expanded ataxin-3 accumulates in ubiquitinated nuclear inclusions that colocalize with the proteasome.[113,114] Ataxin-3 functions in the ubiquitin proteasome pathway to regulate protein degradation.[115–118] Interestingly, one of the E3-ubiquitin ligases that interacts with ataxin-3 is parkin, providing a potential link to parkinsonian features of the disease.[119] Ataxin-3 also functions as a transcriptional repressor via chromatin binding activity and association with histone deacetylases, an interaction that is disrupted in expanded alleles.[13,120,121] Overexpression of expanded ataxin-3 in cultured cells induces apoptosis, suggesting that the mutant protein is either directly or indirectly involved with a cellular suicide pathway.[122] Other studies suggest that protein misfolding, presumably initiated by the expanded polyglutamine tract, leads to ubiquitination and subsequent formation of intranuclear inclusions.[114] Cleavage of ataxin-3 by cellular proteases releases the polyglutamine region and is important in pathogenesis[123,124]; phosphorylation of ataxin-3 plays a role in nuclear localization, nuclear inclusion formation, and protein stability.[125] Similar to ataxin-2, mutant ataxin-3 associates with the inositol 1,4,5-trisphosphate receptor; inhibition of Ca^{2+} release in a mouse model of SCA3 using dantrolene improved coordination.[126] Ataxin-3 interacts with a huntingtin-associated protein,[127] and ataxin-2 is a modifier of the ataxin-3 neurodegeneration phenotype in *Drosophila*.[128]

Recent work suggests that treatment of adult patients with the $\alpha4\beta2$ neuronal nicotinic acetylcholine receptor partial agonist varenicline improves some aspects of the ataxia.[129] The HDAC inhibitor sodium butyrate delayed the onset of neurological phenotypes and Purkinje cell degeneration in a mouse model of SCA3.[130] Evidence from a mouse model suggests that reducing levels of expanded ataxin-3 yields lasting improvements in neurological status.[131]

Spinocerebellar Ataxia Type 6

Spinocerebellar ataxia type 6 (SCA6 – MIM 183086) is among the most common SCAs, particularly in individuals of Asian descent.[132,133] Clinically, SCA6 typically presents as a slowly progressive ataxia and dysarthria with associated intention tremor and dysphagia. Patients can develop diplopia, hyperreflexia, extensor plantar responses, and nystagmus. As in SCA2 and SCA3, SCA6 can be associated with parkinsonian symptoms and dopaminergic dysfunction.[134] A significant minority of patients also develop dementia.[135]

SCA6 is caused by CAG repeat expansions in the C-terminal coding region of the *CACNA1A* gene.[135] These polyglutamine tracts are the shortest to be found among the triplet repeat diseases, with a mere 21 repeats being pathogenic. Homozygosity for expanded repeats does not alter phenotypic presentation or age of onset, suggesting that loss of function is not a primary pathogenic mechanism.[136] *CACNA1A* encodes the brain-specific, voltage-sensitive $\alpha\alpha_{1A}$ ($Ca_v2.1$) subunit of the P/Q-type calcium channel, which is highly expressed in Purkinje cells.[137] The expanded proteins aggregate in the cytoplasm of Purkinje cells, but these inclusions are not ubiquitinated and lack several other components found in the inclusions of other CAG repeat disorders.[137] Abnormalities of the endolysosomal protein degradation pathway, not changes in Ca^{2+}-channel function, appear to mediate disease pathogenesis.[138,139] Recent work has added a new level of complexity by showing that *CACNA1A* actually encodes a bicistronic message, where the C-terminus of the α1A subunit (α1ACT) can be translated independently of the channel and functions as a transcription factor.[140] Expression of α1ACT containing a triplet repeat expansion is cytotoxic and inhibits neurite outgrowth, while overexpression of α1ACT in mouse models of SCA6 actually rescues some pathologic features.[140]

While triplet repeat expansions cause SCA6, point mutations and deletions in the *CACNA1A* coding sequence cause episodic ataxia type 2 and familial hemiplegic migraine type 1 (see below, under *Episodic Ataxias*).[141] These mutations can present as early as infancy with isolated symptoms, such as nystagmus, and then progress to ataxia.[142] Some affected families display significant phenotypic overlap between the three entities.[143,144]

Spinocerebellar Ataxia Type 7

Spinocerebellar ataxia type 7 (SCA7 – MIM 164500) is the only SCA that commonly occurs with retinal degeneration. Decreased visual acuity secondary to progressive pigmentary macular degeneration begins as a cone dystrophy and progresses to involve the entire retina, often leading to complete blindness.[145–147] As in other SCAs, progressive cerebellar ataxia, dysarthria, dysphagia, dysmetria, dysdiadochokinesia, external ophthalmoplegia, hyperreflexia, ptosis, auditory hallucinations, and delusions are also common. Patients may present first with ataxia, visual loss, or both; those with cerebellar ataxia may have normal vision for decades, whereas those with primary visual symptoms usually develop ataxia within a few years. An aggressive infantile-onset form of SCA7 presents with hypotonia, dysphagia, myoclonic seizures, and visual disturbances that typically lead to rapid mental deterioration and severe physical disability, culminating in death by age 3 or earlier.[148,149] This occurs only with paternal transmission of the disease allele and is associated with cardiac abnormalities, particularly patent ductus arteriosus.[148,150,151] Childhood-onset SCA7, which may be associated with myoclonic seizures, is less aggressive than the infantile-onset form but more progressive than the adult-onset form.

The CAG expansions that cause SCA7 are the most unstable of the polyglutamine diseases, and can be upwards of 300 repeats.[152] As with other triplet repeat diseases, expansion length is inversely correlated with age of onset.[150] The SCA7 locus encodes ataxin-7, a protein required for normal neuronal and photoreceptor development that participates in epigenetic regulation of gene expression through its interactions with histone acetyltransferase complexes.[117,153–156] Expansion of the polyglutamine region in mutant ataxin-7 proteins disrupts these interactions and alters gene expression, findings that have been linked to retinal dystrophy and neuronal dysfunction.[157–161] In addition, protein interactome studies have identified genes associated with macular degeneration in the interactome of ataxin-7.[16] Pathology studies demonstrate morphological evidence for mitochondrial pathology,[162–164] and expanded alleles might activate a mitochondrial-mediated apoptotic cascade that results in neuronal death.[165] Ubiquitin-positive nuclear inclusions are found in patients' brains and SCA7 transgenic mouse models.[166–168] However, cellular dysfunction predates the appearance of neuronal inclusions and dysfunction of the ubiquitin proteasome pathway does not occur, suggesting that inclusions in and of themselves are not the primary cause of dysfunction and that they may in fact be neuroprotective.[159,169] Interestingly, mutant protein malfunction in multiple cell types, and excitotoxicity mediated by glial abnormalities, appear to play roles in pathogenesis.[170,171]

A very recent report suggests that administration of interferon-β can improve motor function in a transgenic mouse model of SCA7.[172] Another promising finding is that abolishing expression of mutant ataxin-7 protein halts and/or reverses disease progression in a transgenic mouse model, suggesting that there is hope for an eventual treatment that could slow or halt disease progression.[173]

Spinocerebellar Ataxia Type 17

Phenotypic variability in spinocerebellar ataxia type 17 (SCA17 – MIM 607136) is striking, even for a triplet repeat disease: the disease can present as cerebellar ataxia associated with dysarthria and extrapyramidal symptoms (e.g. parkinsonism and dystonia), or as a Huntington disease phenocopy with chorea as the major manifestation, or it can appear first in the form of cognitive decline with any of a variety of psychiatric symptoms from depression to hallucinations.[174–182] Individuals within the same family can present with dramatically different clinical features, and the same individual over time can manifest each of the various subtypes. Epilepsy can develop late in the disease course.[183,184] Neuroimaging shows diffuse cortical and cerebellar atrophy that is most pronounced in the vermis. Pathologic specimens reveal loss of Purkinje cells, anterior horn cells, and neurons of the inferior olivary nucleus, and overall brain atrophy.[176,177]

SCA17 is caused by CAG repeat expansions in the TATA-binding protein (*TBP*) gene.[176,185] Short CAA tracts interrupt the CAG repeats in both healthy and disease alleles, which distinguishes *TBP* from other SCA-associated genes. Repeat sizes of 42–49 repeats show variable penetrance.[186,187] Interestingly, small expansions of the SCA17 allele have also been associated with isolated parkinsonism in the absence of ataxia.[188]

TBP is the DNA binding subunit required for transcription initiation by all three eukaryotic RNA polymerases.[189] Mutant TBP interacts aberrantly with transcription factors, altering transcription of chaperone proteins important for protein folding and degradation.[190–192] Intranuclear inclusions containing expanded TBP and ubiquitin are found in several neuronal cell types including Purkinje cells.[176,177]

Dentatorubral-Pallidoluysian Atrophy

Dentatorubral-pallidoluysian atrophy (DRPLA; Haw River Syndrome – MIM 125370) is very rare except in Japan, where it accounts for a significant proportion of autosomal dominant SCAs.[193,194] It is characterized by progressive ataxia, myoclonus, epilepsy, choreoathetosis, dystonia, dementia, and psychiatric symptoms.[195,196] Haw River Syndrome, once thought to be a separate entity described in several generations of an African-American family without myoclonic epilepsy but with basal ganglia calcifications and neuroaxonal dystrophy, is caused by the same mutation. Clinical onset prior to 20 years of age tends to present as progressive epilepsy, myoclonus, cerebellar ataxia, and intellectual disability; onset after age 20 tends to present as cerebellar ataxia, choreoatheotosis, tremors, and dementia. Anticipation is prominent, and the phenotypic diversity within and among families is striking, as is the overlap with the clinical presentation of Huntington's chorea. Neuroimaging shows cerebellar, tegmental, and cerebral atrophy accompanied by white matter signal changes; late onset cases can also have signal changes in the pons, midbrain, thalamus, and globus pallidus.[197] Neuronal loss occurs in the dentate nucleus, red nucleus, globus pallidus, and subthalamic nucleus, and leukoencephalopathy occurs in the cerebral white matter.[197,198]

DRPLA is caused by CAG repeat expansions in the *atrophin-1* gene.[199–203] Anticipation occurs with both paternal and maternal transmission of the disease allele, and extreme expansions (>90 repeats) can lead to infantile disease.[204] Phenotypic severity is affected by gene dosage, with homozygosity for a pathogenic allele causing more severe clinical manifestations than would otherwise be predicted based on repeat length.[4] Atrophin-1 mRNA and protein are expressed ubiquitously in human tissue[203,205]; the number of CAG repeats varies in different tissues and tends to be larger in brain, suggesting somatic instability of the repeat.[206] However, the degree of expansion does not seem to parallel neuropathologic involvement. Intranuclear inclusions containing the polyglutamine-expanded protein are found in both neurons and glia; their presence in oligodendrocytes may cause the observed white matter lesions.[207]

The first protein found to interact with atrophin-1 was glyceraldehyde-3-phosphate dehydrogenase (GAPDH), which also interacts with the SCA1, Huntington, and SBMA proteins and thus seems likely to interact with the polyglutamine domain itself.[208] Atrophin-1 participates in transcriptional regulation, which may occur at the epigenetic level through alterations in histone acetylation[209–212]; administration of a histone deacetylase inhibitor improved symptoms and survival in a transgenic mouse model.[213,214] Atrophin-1 may also normally function in a pathway with insulin/IGF-1.[215] Proteolytic processing of the expanded protein appears to play a role in pathogenesis and the formation of nuclear inclusions.[216–218] Toxicity might also result from the sequestration of atrophin-1 interacting proteins from their normal sites of action.[219] Expanded atrophin-1 alleles do not appear to function in a dominant-negative way.[220]

SCAs CAUSED BY OTHER (NONPOLYGLUTAMINE) REPEAT EXPANSIONS

Spinocerebellar Ataxia Type 8

Spinocerebellar ataxia type 8 (SCA8 – MIM 608768) presents with ataxic dysarthria, nystagmus, limb and gait ataxia, limb spasticity, and diminished vibratory sense.[221] Cases can present in childhood.[222] MRI of the brain demonstrates cerebellar atrophy that may be accompanied by white matter hyperintensities in other brain areas.[223] The vast majority of SCA8 disease alleles in individuals of European descent share a common haplotype.[224] Neuropathology reveals intranuclear inclusions in both mouse models and human tissue.

SCA8 has several characteristics that make it unusual among dynamic repeat diseases. The SCA8 locus harbors a noncoding gene with a CTG repeat region that is transcribed into mRNA but not translated into protein and, in the opposite direction, a CAG repeat in a polyglutamine expansion protein designated ataxin-8.[221,225] Repeat length is highly variable in wild-type alleles and can be as large as 174, by far the largest number of wild-type repeats of any SCA; SCA8 is also unique in that there is also substantial overlap between pathogenic and nonpathogenic repeat lengths. Homozygosity for an expanded allele does not exacerbate the disease phenotype as it does in typical polyglutamine expansion SCAs and DRPLA. Penetrance is variable, and repeat lengths contract with paternal transmission but expand with maternal transmission, two more unusual aspects of this disease.[224,226] SCA8 expansions have been found in some SCA1 and SCA6 kindreds and in patients with Alzheimer's disease, Parkinson's disease, and vitamin E deficiency heterozygous for the TTPA mutation; the significance of this is unclear.[227–230]

Expanded CTG and CAG alleles may cause disease through both RNA- and protein-mediated mechanisms. It has therefore been suggested that the noncoding *SCA8* mRNA functions as an endogenous inhibitory RNA for *KLHL1*, a gene whose locus it partially overlaps, or as a regulator of RNA-binding proteins and splice factors.[231–234] KLHL1 appears to be necessary for proper Purkinje cell dendritic branching and has been implicated in modulating P/Q-type calcium channel function; what role, if any, this plays in SCA8 pathogenesis is unclear.[235,236]

Spinocerebellar Ataxia Type 10

The mutation that causes spinocerebellar ataxia type 10 (SCA10 – MIM 603516) appears to have arisen in New World Amerindians, with cases described in people of Brazilian, Mexican, and Portuguese descent.[237–239] The association of ataxia, dysarthria, and nystagmus with epilepsy distinguishes this SCA clinically. Mood disorders and evidence of polyneuropathy on nerve conduction studies can also be seen. Brain imaging shows cerebellar atrophy.

SCA10 is one of two human diseases (the other being SCA31) known to be caused by a pentanucleotide repeat. The ATTCT repeat sequence occurs in intron 9 of the *ataxin-10* gene and is highly unstable: normal individuals have 10 to 29 repeats, while affected individuals typically have 750 to 4500 repeats.[240] As with other repeat disorders, there is an inverse correlation between expansion size and age of onset. Genetic anticipation is present, with paternally inherited repeats being highly unstable while maternal transmission results in more stable repeats.

Ataxin-10 is widely expressed in the brain and other tissues, and is localized to the neuronal cytosol and perinuclear region.[241] Ataxin-10 transcripts may participate in nucleosome formation, which is altered in repeat expansions,[242] and the native protein may be involved in neuritogenesis through interactions with G-proteins and also in cytokinesis.[243,244] Data from cell culture and transgenic mouse models indicate that decreased *ataxin-10* RNA levels and expression of expanded RNAs both result in increased apoptosis, possibly through a ribonucleoprotein-related mechanism.[241,245,246] This suggests that SCA10 pathogenesis cannot be accounted for by a simple gain- or loss-of-function model.

Spinocerebellar Ataxia Type 12

The initial symptom of spinocerebellar ataxia type 12 (SCA12 – MIM 604326) is typically action tremor, a feature that distinguishes the disease from other SCAs.[247] SCA12 is slowly progressive, with symptoms including head tremor, gait ataxia, dysmetria, dysarthria, hyperreflexia, parkinsonian signs, abnormal eye movements, and occasionally dementia. Childhood-onset nystagmus and lower extremity dystonia have been reported.[248] Brain imaging reveals both cortical and cerebellar atrophy. Pathology is available only on a single brain, and revealed diffuse atrophy of cerebral and cerebellar cortices and loss of Purkinje cells. SCA12 is the third most common SCA in India but otherwise has been described in only a few families of German-American pedigree.[249,250]

SCA12 is caused by a noncoding triplet CAG expansion found 133 nucleotides upstream of the transcription start site for PPP2R2B, a brain-specific regulatory subunit of protein phosphatase 2A (PP2A). The expanded allele does not lead to a polyglutamine tract, and how it affects PPP2R2B function is unknown. The minimum repeat length necessary to cause disease has not yet been established, and no relationship between repeat size and age of onset has yet been discerned. PPP2RB subunits are thought to modulate PP2A function by regulating substrate specificity and intracellular targeting, and recent evidence suggests that mitochondrial impairment leading to oxidative stress might play a role in pathogenesis.[249,251]

Spinocerebellar Ataxia Type 31

Spinocerebellar ataxia type 31 (SCA31 – MIM 117210) is one of the most common SCAs in Japan but is rare elsewhere.[252,253] Symptoms include late-onset pure cerebellar ataxia consisting of gait and limb ataxia, nystagmus and hypotonia, typically accompanied by sensorineural hearing impairment.[254,255] Neuroimaging shows isolated cerebellar atrophy.

SCA31 is caused by TGGAA pentanucleotide repeat insertions in the introns of the *TK2* and *BEAN* genes, which are on opposite strands and transcribed in opposite directions.[256] The length of the insertion is inversely correlated with age of disease onset but not with rate of disease progression.[257] Mild anticipation is manifest, suggesting that the insertion might have a propensity towards expansion.[257] Notably, expression of *TK2* and *BEAN* is not altered by the expansions, but expanded RNAs form nuclear RNA foci and bind splicing factors; how this might lead to pathogenesis is unknown.[258]

Spinocerebellar Ataxia Type 36

Spinocerebellar ataxia type 36 (SCA36; Asidan – MIM 614153) is a cerebellar/motor neuron combination disease in which affected individuals present with ataxia, dysarthria, tongue fasciculations, and nystagmus in the fifth to sixth decade and subsequently develop muscle weakness, atrophy, and fasciculations.[259,260] Hyperreflexia is also common, and the disease can be associated with progressive sensorineural hearing loss and oromandibular dystonia.[261–263] Neuroimaging shows cerebellar atrophy. SCA36 maps to chromosome 20p13 and is caused by GGCCTG hexanucleotide repeats in intron 1 of the *NOP56* gene, which encodes an important component of the transcription and RNA splicing machinery.[259] Normal repeat number is 5–14, while pathogenic repeats can number 650–2500.[261] Levels of the NOP56 protein do not appear to be altered in affected individuals, but RNA foci are present, suggesting a toxic gain-of-function as the pathogenic mechanism.

SCAs CAUSED BY MISSENSE MUTATIONS

These SCAs tend to present as pure cerebellar ataxias that are minimally progressive; lifespan is typically not affected. The identified genes participate in various aspects of neuronal development and/or function, and recent work has shown that at least some of them function in pathways that include triplet repeat proteins.[17]

Spinocerebellar Ataxia Type 5

Spinocerebellar ataxia type 5 (SCA5 – MIM 600224) patients suffer from mild disturbance of gait and limb coordination that worsens with age. Some patients manifest increased reflexes, gazed-evoked nystagmus, facial myokymia, and decreased vibration sense.[264] Juvenile-onset SCA5 patients present with bulbar and pyramidal dysfunction as well as cerebellar dysfunction; this may shorten life. All four reported juvenile-onset cases resulted from maternal inheritance.[265]

SCA5 is caused by mutations in the *SPTBN2* gene that encodes β-III spectrin, a protein expressed throughout the brain and in high levels in Purkinje cells.[266] Pathogenesis is believed to occur secondary to decreased trafficking of proteins from the Golgi complex to the membrane, which affects stabilization of the excitatory amino acid transporter EAAT4 at the plasma membrane and disrupts other intracellular transport pathways.[267] The mutation may also exhibit dominant negative effects secondary to loss of native protein association with ARP1, a component of the dynactin/dynein complex.[268] β-III spectrin has also been implicated in normal neuronal development and in the recessive disorder spectrin-associated autosomal cerebellar ataxia type 1 (SPARCA1).[269,270] Unlike SCA5, which is typically an adult disease, SPARCA1 presents with childhood ataxia and cognitive impairment and is typically associated with truncating β-III spectrin mutations.

Spinocerebellar Ataxia Type 11

Spinocerebellar ataxia type 11 (SCA11 – MIM 604432) is a rare, relatively benign, late-onset, slowly progressive pure cerebellar syndrome occasionally associated with mild hyperreflexia and vertical nystagmus.[271] Truncating mutations in the tau tubulin kinase 2 (*TTBK2*) gene cause SCA11.[272,273] TTBK2 is involved in ciliogenesis[274]; how this relates to SCA11 pathology is unclear.

Spinocerebellar Ataxia Type 13

Spinocerebellar ataxia type 13 (SCA13 – MIM605259) is unusual in that it can start in early childhood and includes moderate mental retardation (IQ = 62–76) as a prominent feature.[275,276] Patients exhibit slowly progressive gait ataxia, dysarthria, moderate mental retardation, mild developmental motor delay, and, in some cases, epilepsy.[277] Nystagmus and pyramidal signs are sometimes observed. Patients older than 50 years of age show additional neurological signs such as dysphagia, urinary urgency, and bradykinesia. MRI shows moderate cerebellar and pontine atrophy.

SCA13 is caused by missense mutations in the voltage gated potassium channel Kv3.3 (KCNC3).[278] The mutations are dominant-negative in nature, with those affecting channel gating properties and neuronal excitability associated with early-onset phenotypes, whereas those affecting current amplitude are associated with late-onset phenotypes.[276,279,280] Mutations associated with infantile onset are also associated with axonal pathfinding errors in a zebrafish model, suggesting that this may play a role in early pathogenesis.[281]

Spinocerebellar Ataxia Type 14

Spinocerebellar ataxia type 14 (SCA14 – MIM605361) presents with slowly progressive gait ataxia, dysarthria, horizontal gaze nystagmus, and abnormal smooth pursuit movements with or without hyper- and hyporeflexia, myoclonus, dystonia, and/or peripheral neuropathy.[282–286] There is large variation in age of onset, with earlier onset tending to produce a broader range of symptoms. Imaging reveals atrophy of the cerebellar vermis and/or hemispheres. Limited pathologic specimens suggest a primary Purkinje cell defect.[282]

Molecular studies reveal that missense mutations in exon 4 of the protein kinase Cγ (PKCγ) gene are responsible.[283,287,288] Mutations alter kinase activity and membrane targeting of PKCγ in response to calcium.[289–291] Mutant PKCγ forms intracellular aggregates that appear to impair the ubiquitin proteasome system and lead to endoplasmic reticulum stress.[292–295] Overexpression of mutant protein in Purkinje cells causes abnormalities of dendritic branching and decreased synaptic pruning.[296–298] There also appears to be crosstalk between PKCγ and proteins involved in other human ataxias, as mutant PKCγ preferentially phosphorylates aprataxin, the mutated protein in ataxia with oculomotor apraxia type I (AOA1).[299]

Spinocerebellar Ataxia Type 15

Spinocerebellar ataxia type 15 (SCA15 – MIM 606658) is an autosomal dominant, slowly progressive ataxia associated with nystagmus and dysarthria.[300–302] Action and postural tremor, buccolingual dyskinesias, chorea, facial myokymia, pyramidal signs, and cognitive decline can also be part of the clinical spectrum.[303–305] Neuroimaging demonstrates cerebellar atrophy. Identification of the causative gene came not from positional cloning efforts but rather serendipitously through analysis of a spontaneous mutation that arose in a transgenic mouse colony.[306] SCA15 is caused by mutations in *ITPR1*, which encodes the inositol 1,4,5-triphosphate receptor type I, a calcium channel found predominantly in the endoplasmic reticulum. This locus had initially been excluded from the original family but was subsequently shown to be involved in that and other kindreds.[307,308] Further studies showed that SCA16 and SCA29, both originally thought to be distinct entities, were also caused by *ITPR1* mutations. The SCA16 and SCA29 designations are therefore left blank.[309–312]

Spinocerebellar Ataxia Type 19

Spinocerebellar ataxia type 19 (SCA19 – MIM 607346) can present with a pure cerebellar syndrome or ataxia accompanied by cognitive impairment, irregular postural tremor, and myoclonus.[313–316] Neuroimaging shows atrophy of the cerebellar hemispheres or vermis. SCA19 is caused by mutations in *KCND3*, which encodes the Kv4.3 voltage-gated potassium channel.[317,318] The mutations appear to prevent channel targeting to the plasma membrane and are accompanied by a complete lack of channel conductance. This molecular cloning also revealed that SCA19 and SCA22 are caused by the same mutations; the SCA22 nomenclature has subsequently been vacated.

Spinocerebellar Ataxia Type 23

Spinocerebellar ataxia type 23 (SCA23 – MIM 610245) is an infrequent (<0.5% of all cases) cause of ataxia.[319–321] Disease appears in midlife (>40), is slowly progressive, and includes gait and limb ataxia, disturbance of oculomotor control, dysarthria, hyperreflexia, and sometimes peripheral neuropathy.[322] Neuroimaging demonstrates severe cerebellar atrophy. Pathology in a single individual who died at 80 years of age showed loss of Purkinje cells, neurons in the dentate nuclei, and inferior olives; thinning of the cerebellopontine tracts; demyelination of the posterior and lateral columns in the spinal cord; and ubiquitin-positive, polyglutamine-negative intranuclear inclusions in nigral neurons that resembled Marinesco bodies. The pathogenic significance of this last finding is uncertain. SCA23 is caused by point mutations in *PDYN*, which encodes prodynorphin, the precursor protein for the opioid neuropeptides, α-neoendorphin, and dynorphins A and B.[319] Some of these mutations increase dynorphin A levels, which has been associated with increased glutamate-mediated neurotoxicity.[319,323]

Spinocerebellar Ataxia Type 26

Spinocerebellar ataxia type 26 (SCA26 – MIM 609306) was described in a single six-generation family of Norwegian descent with slowly progressive gait and upper limb ataxia accompanied by dysarthria.[324] Age of onset ranged from 26 to 60 years. Brain pathology showed significant loss of Purkinje cells with minimal effects in other brain regions.[325] Mutations in the gene encoding eukaryotic elongation factor 2 (eEF2) co-segregate with the disease phenotype.[325] These mutations interfere with translocation during protein translation, suggesting a novel pathogenic mechanism.

Spinocerebellar Ataxia Type 27

Spinocerebellar ataxia type 27 (SCA27 – MIM 609307), formerly fibroblast growth factor 14-SCA (MIM 601515), presents with trembling of the hands during childhood followed by slowly progressive limb ataxia, dysmetric saccades and smooth pursuit movements, nystagmus, psychiatric symptoms, and decreased cognitive performance.[326,327] Brain imaging is normal or shows cerebellar atrophy.

SCA27 is caused by mutations in the *FGF14* gene, and has also been associated with chromosomal translocation involving the *FGF14* locus.[326,328,329] Interestingly, a patient with a reciprocal chromosomal translocation involving the *FGF14* locus presented with a paroxysmal nonkinesigenic dystonia phenotype.[330] Mice lacking *FGF14* are ataxic and have paroxysmal dystonia, demonstrating the linkage between the two phenotypes.[331] FGF14 is transported into neuronal processes, and introduction of mutant protein into an *in vitro* cerebellar culture system impairs Ca^{2+} channels and synaptic transmission, suggesting a possible pathogenic mechanism.[332]

Spinocerebellar Ataxia Type 28

Spinocerebellar ataxia type 28 (SCA28 – MIM 610246) is a rare infantile-, childhood- or juvenile-onset, slowly progressive SCA that typically presents with mild gait incoordination and includes nystagmus, dysarthria, ophthalmoparesis, ptosis and hyperreflexia.[333–335] Neuroimaging shows cerebellar atrophy. SCA28 is caused by mutations in *AFG3L2*, which encodes a mitochondrial protein that is a component of the *m*-AAA metalloprotease complex involved in the maintenance of the mitochondrial proteome.[336] AFG3L2 forms this complex with paraplegin, a protein that is mutated in a type of recessive spastic paraplegia (SPG7)[337]; homozygous *AFG3L2* mutations cause a syndrome that combines features of SCA28 and SPG7.[338] In yeast, mutations in *AFG3L2* impair mitochondrial respiration, cytochrome oxidase-c activity, and the proteolytic activity of the *m*-AAA metalloprotease complex. These impairments likely lead to respiratory chain compromise and increased generation of reactive oxygen species, which could lead to Purkinje cell degeneration and cerebellar dysfunction.[339]

Spinocerebellar Ataxia Type 35

Spinocerebellar ataxia type 35 (SCA35 – MIM 613908) has been described in three Chinese families. The age of onset ranged from 40 to 48 years; presenting features were a slowly progressive gait ataxia and dysarthria (patients needed a walking aid or wheelchair after about 10 years); upper limb incoordination appeared later. With disease progression patients also developed pyramidal signs with hyperreflexia and extensor plantar responses, torticollis, tremor, dysmetria, position sense defects, and pseudobulbar palsy in the absence of nystagmus.[340,341] Neuroimaging shows cerebellar and brainstem atrophy, and pathology studies show loss of Purkinje cells, deep cerebellar nuclear neurons, and lower motor neurons. A combination of linkage analysis and exome sequencing demonstrated that SCA35 is caused by mutations in *TGM6*, which encodes a transglutaminase gene.[340] Pathogenesis could be linked to alterations in transglutaminase activity that lead to apoptosis and/or through interactions with polyglutamine proteins.[342,343]

SCAs OF UNKNOWN CAUSE

These SCAs have typically been described in single families where they usually present with pure cerebellar ataxia and dysarthria, although other neurological features can also be present. Disease progression is typically slow and does not affect lifespan. These SCAs are likely to be caused by missense mutations in genes that have not yet been identified. When known, genetic loci are listed.

Spinocerebellar Ataxia Type 4

The clinical phenotype of spinocerebellar ataxia type 4 (SCA4 – MIM 600223) consists of ataxia coupled with prominent axonal sensory neuropathy and extensor plantar reflexes.[344,345] Analysis of 34 candidate genes failed to reveal the causative locus.[346]

A phenotypically distinct disorder (16q-ADCA) consisting of late-onset, pure cerebellar ataxia (ataxia, dysarthria, and nystagmus) with or without sensorineural hearing impairment has been described in Japan.[254,255,347] 16q-ADCA maps to the same general chromosomal location as SCA4 and appears to be caused by single nucleotide substitutions in the 5′-untranslated region of puratrophin-1, a protein found in multiple tissues and predicted to be involved in intracellular signaling and Golgi apparatus targeting.[255] Puratrophin-1 functions as a guanine nucleotide exchange factor important for GTPase signaling.[348] Cytoplasmic aggregates containing puratrophin-1 are found exclusively in Purkinje cells of affected individuals.[255] Screening of 537 unrelated European ataxia patients excluded for SCA1, 2, 3, 6, and 17 failed to reveal the described substitution.[349] However, as the puratrophin-1 gene is large and complex, it remains undetermined whether mutations are causative in Europeans.

Spinocerebellar Ataxia Type 18

Spinocerebellar ataxia type 18 (SCA18 – MIM 607458), or sensory motor neuropathy with ataxia (SMNA), was described in a single five-generation American family of Irish descent.[282] The clinical presentation consists first of gait instability in the dark, followed by limb ataxia, nystagmus, decreased vibratory and position sense, decreased deep tendon reflexes, and proximal or distal muscle weakness and atrophy occasionally associated with upgoing plantar responses and/or pes cavus. Neuroimaging was normal or demonstrated mild cerebellar atrophy. EMG/NCV showed evidence of denervation and axonal sensory neuropathy. The disorder was mapped to chromosome 7q22-32; interferon-related developmental regulator gene 1 (*IFDR1*) has been suggested as a candidate gene.[350]

Spinocerebellar Ataxia Type 20

Spinocerebellar ataxia type 20 (SCA20 – MIM 608687) was described in a single kindred of Anglo-Celtic origin.[351] Dysarthria is often the initial symptom; palatal tremor and a characteristic dysphonia are other distinguishing symptoms. Imaging shows mild to moderate cerebellar atrophy and the unique finding of dentate calcification in the absence of basal ganglia calcification. SCA20 is associated with duplications of chromosome 11q12.2-q12.3.[352]

Spinocerebellar Ataxia Type 21

Spinocerebellar ataxia type 21 (SCA21 – MIM 607454) was described in a single French kindred with moderate gait and limb ataxia, extrapyramidal signs such as tremor, akinesia and cogwheeling, hyporeflexia, and mild mental impairment.[353–355] Eye movements are normal, and disease progression is slow. SCA21 maps to chromosome 7p21.3-p15.1. The phenotypes of several parent-child pairs suggest anticipation. Brain imaging shows marked atrophy of the cerebellum.

Spinocerebellar Ataxia Type 25

Spinocerebellar ataxia type 25 (spinocerebellar ataxia with sensory neuropathy; SCA25 – MIM 608703) was described in a single French kindred with variable clinical presentations ranging from pure sensory neuropathy with little cerebellar involvement to a Friedreich's ataxia-like phenotype.[356,357] Cerebellar ataxia, peripheral sensory neuropathy manifest as loss of vibratory, light touch and pain sensation, pes cavus, and scoliosis were present in different family members. The index case of the family started walking at age 17 months with frequent falls, and suffered repeated vomiting from the age of 2 years; gastrointestinal symptoms were initial features in two other patients. There is no evidence of anticipation for age of onset, but the two most incapacitated individuals were from the most recent generation. Brain imaging showed global cerebellar atrophy. SCA25 maps to chromosome 2p15-p21.

Spinocerebellar Ataxia Type 30

Spinocerebellar ataxia type 30 (SCA30 – MIM 613371), a relatively pure, midlife onset cerebellar ataxia, has been described in a single Australian family of Anglo-Celtic ethnicity.[358] SCA30 appears to have a particularly insidious onset, with slowly evolving gait ataxia and mild to moderate dysarthria. Four of the six affected individuals had lower-limb hyperreflexia; all had hypermetric saccades; one had slight gaze-evoked nystagmus. Neuroimaging of two patients showed cerebellar atrophy. The locus was mapped to chromosome 4q34.3-q35.1.

Spinocerebellar Ataxia Type 32

Spinocerebellar ataxia type 32 (SCA32 – MIM 613909) was described in a single Chinese kindred.[359] The presentation was of cerebellar ataxia, azoospermia with complete absence of testicular germ cells and progenitors in men, and intellectual impairment if the onset of symptoms was prior to age 40. Neuroimaging showed cerebellar atrophy. The locus maps to chromosome 7q32-q33.

Spinocerebellar Ataxia Type 34

Spinocerebellar ataxia type 34 (SCA34 – MIM 633190) is a neurocutaneous syndrome that was first reported in a five-generation French Canadian kindred.[360] Shortly after birth the affected individuals developed distinct papulosquamous erythematous ichthyosiform plaques on their skin, predominantly on the dorsal aspects of the hands, feet, and ears or at the joints (elbows, knees, ankles, wrists). These lesions tended to subside during the summer and were greatly attenuated after age 25; curiously, they would sometimes reappear after age 40. In their 20s, patients might display extremely subtle deficits such as slight gait hesitation after a long walk or diminished tendon reflexes. In their 40s, however, patients developed a slowly progressive neurological syndrome characterized by severely diminished tendon reflexes (without loss of position or vibration sense), limb paresthesias and muscle spasms, and, by advanced age, nystagmus, dysarthria, and a severe gait ataxia. A subsequent genetic study identified other affected families in Europe and mapped the locus to chromosome 6p12.3-p16.2.[361]

EPISODIC ATAXIAS

The episodic ataxias are rare dominant disorders characterized by intermittent ataxia, dysarthria, and nystagmus.

Attacks last from seconds to several days and can occur several times per day or can be spaced by years. Some patients also have interictal ataxia and/or eye movement abnormalities. Neuroimaging demonstrates cerebellar atrophy in some cases. Treatment with the carbonic anhydrase inhibitor acetazolamide decreases the frequency and duration of the attacks in all but episodic ataxia type 4 (EA4).[362]

Episodic ataxia type 1 (EA1; episodic ataxia with myokymia – MIM 160120) is characterized by intermittent episodes of ataxia that occur spontaneously or that are precipitated by exercise, fever, stress, or sudden movement. Attacks last seconds to minutes and can occur several times each day. *Myokymia*, a rippling movement of muscle, is typically observed in the periorbital and small hand muscles and persists between ataxic attacks. Identical twin studies demonstrate that as-yet-uncharacterized nongenetic factors play a role in age of onset and disease symptomatology.[363] Point mutations in *KCNA1*, which encodes the Kv1.1 potassium channel, cause EA1 by affecting channel gating properties and ultimately alter GABA release.[364–366]

Episodic ataxia type 2 (EA2; hereditary paroxysmal cerebellar ataxia – MIM 108500) is characterized by intermittent attacks of ataxia and dysarthria lasting minutes to days, followed by a period of fatigue. Nausea, migraine, weakness, vertigo, diplopia, oscillopsia, and/or dystonia may accompany attacks, which are precipitated by emotional stress or, in some cases, carbohydrate-rich meals. Nystagmus (gaze-evoked or downbeat), mild cerebellar ataxia and occasionally epilepsy can occur in interictal periods, and progressive cerebellar ataxia is seen in some kindreds. 4-aminopyridine may be a useful alternative if acetazolamide fails or loses effectiveness.[367,368] EA2, familial hemiplegic migraine type 1 (FHM1), and SCA6 (see above) are caused by mutations in the *CACNA1A* gene.[141,144] *CACNA1A* encodes the brain-specific, voltage-sensitive α_{1A} ($Ca_v2.1$) subunit of the P/Q-type calcium channel, which is highly expressed in Purkinje cells.[137] Several missense and nonsense mutations in the gene are associated with EA2[369]; the nature of the mutation is not necessarily predictive of whether an individual will clinically present with EA2 or FHM. Ataxia might result from imprecision of Purkinje cell pacemaking.[370]

EA3 (MIM 606554) was described in a large Canadian kindred whose members suffered attacks of vertigo, incoordination, imbalance, tinnitus, diplopia, and/or visual blurring lasting for minutes to hours.[371] Attacks occurred once or twice a day, followed by a refractory period, and were diminished by acetazolamide. Half the affected individuals experienced interictal myokymia. Generalized seizures occurred in some affected individuals. Genetic linkage was assigned to chromosome 1q42; the causative gene is unknown.[372]

Episodic ataxia type 4 (EA4; periodic vestibulocerebellar ataxia; PATX – MIM 606552) has been described in two Caucasian families from North Carolina.[373,374] Clinical findings include episodic attacks of vertigo, nausea, tinnitus, horizontal nystagmus, oscillopsia, and ataxia that begin in the third to sixth decade and may evolve into a constant condition. The visual sensation of objects moving past the patient, such as occurs while riding in a car, can exacerbate attacks, while lying quietly for 15 to 30 minutes can alleviate an attack. Acetazolamide was not therapeutic, but antihistamines decreased the frequency and severity of the attacks.[373] Known episodic ataxia loci were excluded, but the gene locus remains unknown.[375]

EA5 (MIM 613855) was described in a single French-Canadian kindred whose members experienced attacks of ataxia and vertigo that lasted for several hours. The interictal period was marked by mild dysarthria, ataxia, and nystagmus.[376] Acetazolamide prevented the attacks, which recurred when the drug was briefly withdrawn. Missense mutations in the calcium channel β4-subunit gene *CACNB4* are causative. Interestingly, the same mutations were associated with generalized epilepsy in the absence of ataxia in an unrelated German family, suggesting that other genetic loci/environmental factors play a role in determining disease phenotype.[376]

EA6 (MIM 612656) has been reported in one 10-year-old boy and three individuals in an unrelated Dutch family.[377,378] The boy had four episodes of ataxia and dysarthria triggered by febrile illness; at the age of six, he suffered a 5-day episode of severe right-sided headache, hemiparesis, and diminished consciousness. The Dutch family members did not experience significant headaches but, beginning in the first or second decade, had episodes of ataxia lasting 2 to 3 hours accompanied by nausea and vomiting, vertigo, photo- and phonophobia, diplopia, and dysarthria. Attacks can be provoked by infections, stress, fatigue, alcohol, or caffeine. EA6 is caused by mutations in *SLC1A3*, which encodes the excitatory amino acid transporter 1 (EAAT1) responsible for removing glutamate from the synaptic cleft.[377] Mutant proteins act both in a dominant negative fashion by preventing formation of normal EAAT1 channels in the cell membrane and in a gain-of-function fashion by increasing the anionic current of the channels.[377,379]

EA7 (MIM 611907) was described in a single four-generation cohort.[380] Clinically, EA7 is characterized by episodes of ataxia, weakness, and dysarthria that last hours to days and that can be triggered by excitement or exercise. Attacks occur with a frequency ranging from monthly to yearly and, fortunately, tend to decrease with age. Two affected individuals suffered migraine headaches that were not associated with the attacks; in fact, no interictal symptoms were detected in any of the family. The disorder maps to chromosome 19q2; the causative gene is unknown.

CLINICAL CASE DESCRIPTIONS

The reader is encouraged to read the following case descriptions (Case Examples 50.1–50.5) and attempt to make a diagnosis based on the clinical, laboratory, imaging, and autopsy findings.

CASE EXAMPLE 50.1

MM was one of six surviving children born uneventfully to unrelated parents (five other children died at birth or in infancy with no sign of neurologic illness). He developed gait instability and slurred speech at age 8 and dysphagia by age 11. By age 12, MM had prominent eyes and dystonia with facial twitching upon intentional movement but no muscle wasting. Upward gaze was slightly limited but there was no nystagmus; optic fundi and pupils were normal. Deep tendon reflexes were abnormally active, with sustained clonus at the knee and ankle and bilateral extensor plantar responses. Cogwheel rigidity was present in the upper limbs, and ataxia, dysmetria, and spasticity were prominent in the lower limbs. MM had normal intelligence and a resting heart rate of 120 bpm. Postmortem analysis after MM's death at the age of 15 revealed degeneration and gliosis of the dentate nuclei, pontine nuclei, cranial nerve nuclei, gracile and cuneate nuclei, substantia nigra, and medial lemniscus in the brain, and of anterior horn cells, intermediolateral columns, Clarke's columns, and posterior root ganglia in the spinal cord.

CASE EXAMPLE 50.2

This patient was the first child of 27-year-old nonconsanguineous parents. The birth was uneventful, although the mother had been in remission from Hodgkin's disease for 10 months prior to conception. As a 2-week-old infant, the patient suffered apneic episodes and frequent vomiting, which resolved spontaneously. At 3 months, her poor head control prompted head CT and MRI, but these and all other laboratory tests were normal. She showed bilateral epicanthal folds and a prominent nasal bridge, and both her height and weight were at the 10th percentile. Over the course of the next 7 months, the child remained hypotonic and developed dysphagia and more frequent apneas. Though she smiled and cooed, she showed severe developmental delay: she did not track visually or to sound, and her hands were almost constantly clenched. Ophthalmologic evaluation found bilateral optic atrophy and retinitis pigmentosa. She died at age 2 of respiratory failure.

CASE EXAMPLE 50.3

LW had a normal birth and neonatal period, and he achieved all developmental milestones. However, at 30 months of age, he began having episodes of tonic extension of the extremities with upward rolling eye movements in conjunction with a fever. LW was diagnosed with retinitis pigmentosa at age 9.5 and was legally blind by the age of 11. Ataxia, dysmetria, bilateral ptosis, and dysarthria were apparent at 10 years of age. He became wheelchair bound at age 13 and began to have myoclonic seizures several times a day (10 to 20 seconds of jerking in all extremities but most prominently in the legs). The seizures were treated with clonazepam. That same year, weight loss necessitated the placement of an enterogastric tube (by age 17, he weighed 48 kg and was 5 feet 4 inches tall). At 14, the seizures grew longer and occurred in the context of fever and vomiting. Progressive weakness led to respiratory failure and death 3 years later.

CASE EXAMPLE 50.4

A 6-year-old Japanese girl developed clonic convulsions of the right extremities and generalized tonic convulsions; electroencephalography (EEG) showed continuous spike-and-wave activity. By the age of 7, she displayed myoclonus, opsoclonus, atypical absence and complex partial seizures, incoordination, and dementia. At age 13, her IQ was 27; by this age, she showed an ataxic gait with hypotonia, athetosis, increased deep tendon reflexes, and ankle clonus. Intention tremor, dysdiadochokinesia, and dysmetria were noted. CT and MRI showed atrophy of the cerebrum, cerebellum, and mesencephalic tegmentum. Follow-up EEG showed continuous, irregular spike-and-wave complexes during atypical absence; a later EEG recorded frequent focal spikes in the right occipital region. Ten years after symptom onset, this patient developed myoclonic seizures upon photic stimulation.

Case Discussions

These cases were chosen because they emphasize the difficulties associated with clinical diagnosis of the dominantly inherited ataxias. Patient #1 came from an Azorean family in which both parents showed signs of SCA3 by

CASE EXAMPLE 50.5

This handsome and intelligent little boy was perfectly well until he became "nervous, emotional, irritable, and uninterested in playing with other children" at the age of 5. He began wetting his bed at night and became easily fatigued, falling into bed by midafternoon as though exhausted. He seemed uncertain on his feet when he walked, though he wasn't dizzy, and he complained of headaches. He entered school during his sixth year but had to withdraw after 2 months owing to the rapid progression of his illness; suddenly, he could no longer see objects in a store window or climb stairs erect. Initial examination noted pale retina with spotty degeneration; "bulbar" intonation; dysphagia; normal reflexes but explosive, jerky movements; and rhythmic tremors. The patient's height and weight were normal. As the illness progressed, the patient suffered constant choreiform movements and tachycardia (130 bpm), and he cried out frequently as though frightened. The patient died at 9 years of age from aspiration pneumonia.

the age of 27, though this case was described in 1982 when the disease was known only as Machado-Joseph disease and anticipation was not yet understood (and sometimes, therefore, dismissed).[381] This patient and his younger brother were the only two of the large number of children known to be affected. The infantile death of five other children, however, is suspicious, even though the report indicated that there had been no sign of neurologic disease. Because both parents had SCA3, it is virtually impossible to believe that they had only two affected children out of 11. It is possible that neurologic disease simply wasn't recognized in these children, either because neonatal SCA3 looks drastically different from early-onset disease, or because the children were never seen by a neurologist before their deaths. Without an understanding of anticipation, it would have been difficult to attribute the deaths to the same disease that was causing no or only mild symptoms in the young parents. Notice that the pathology is considerably broader than it would be in an adult-onset case—involvement of the sensory ganglia and the gracile and cuneate nuclei is not associated with adult SCA3—and that there may be anatomic lesions that produce little clinical correlate. Dysautonomia besides the tachycardia was absent, for example, despite degeneration of the dorsal nuclei of the vagus and the intermediolateral columns.

Patient #2, despite the retinal degeneration associated almost exclusively with SCA7, actually had infantile SCA2.[68] DNA testing revealed 220 CAG repeats. A correct diagnosis in this case was facilitated by an obvious family history: the father had an expanded SCA2 allele of 43 repeats. This was the first case reported of an SCA2 patient with pigmentary retinopathy, which had been considered pathognomonic for SCA7. As mentioned earlier, juvenile cases often show pathology in neural tissues quite distinct from those affected by the adult-onset disorders; the retina is neural tissue, but it may simply have a higher threshold for damage from CAG repeats than other brain tissues. As the authors of this report point out, visual impairment is the most common initial symptom in SCA2 patients with more than 59 repeats, whereas ataxia is the presenting symptom in those individuals with pathologic repeats of less than 59.

Patient #3[382] was diagnosed early on as having Kearns-Sayre syndrome (KSS). The presentation made this a reasonable diagnosis: ptosis, ophthalmoplegia, pigmentary degeneration of the retina, cerebellar ataxia, and short stature. However, the lab results were normal, and an MRI when LW was 11 years old revealed cerebellar atrophy and other findings that the neuroradiologist noted were consistent with a dominant ataxia. The apparent lack of family history of inherited ataxia probably caused the physicians to overlook an SCA diagnosis despite the absence of common KSS symptoms such as comcardiomyopathy, hearing loss, or hypoparathyroidism. It is not possible to tell from the case report whether the parents themselves were short, or whether the child's stature was unusual for the family (LW was cared for by his aunt because of "psychosocial problems"). It is common, however, for juvenile-onset SCA patients to be small in comparison with healthy children. When the autopsy showed dramatic loss of Purkinje cells and internal granule cells in the cerebellar cortex and anterior horn cells within the spinal cord, along with atrophy of the basis pontis, inferior olive, and cerebellum (Figure 50.1A through 1C), postmortem DNA analysis was performed. Molecular testing revealed that LW had juvenile-onset SCA7 with 70 to 87 CAG repeats. Family history can, of course, appear to be negative in cases of alternate paternity, adoption, or the death of an affected parent before symptom onset. It is important to remember, however, that individual members of even large families with multiple affected generations have often received numerous incorrect diagnoses. Thanks to the advent of DNA-based molecular genetic testing, this diagnostic confusion should become a rarer event.

Patient #4[383] was initially diagnosed with Lennox-Gastaut "syndrome" because of her progressive myoclonic epilepsy and mental retardation. It appears that another diagnosis was pursued when her younger brother also had a brief episode of generalized tonic clonic convulsions with fever at the age of 6; at some point it was

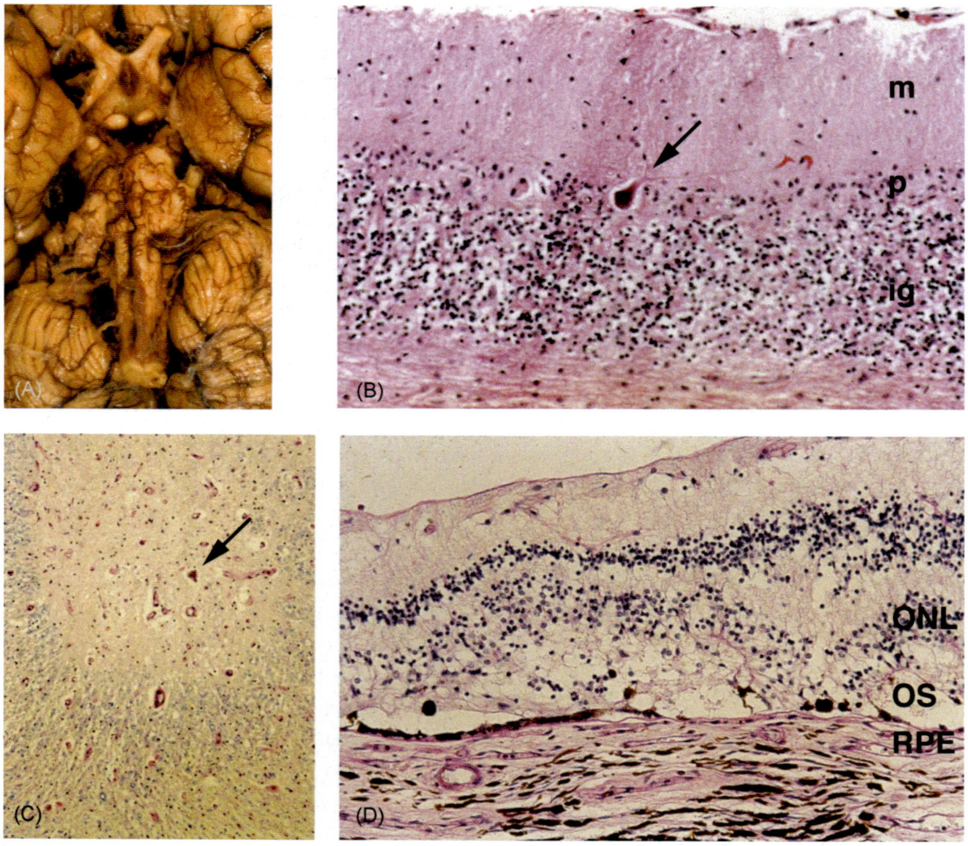

FIGURE 50.1 Pathology of Case Example 50.3. (**A**) Postmortem GROSS examination of the brain shows marked atrophy of the basis pontis and inferior olives and mild atrophy of the cerebellum. (**B**) Microscopically, there is a dramatic loss of Purkinje cells (p) and internal granule cells (ig) within the cerebellar cortex. One of the few remaining Purkinje cells is marked with an arrow. The molecular layer is designated (m). (**C**) Similarly, there is a severe loss of anterior horn cells within the spinal cord. An arrow points to a single anterior horn cell. (**D**) There is severe photoreceptor degeneration in the posterior pole of the retina with very rare intraretinal pigment deposits. Remaining photoreceptor nuclei are seen in the outer nuclear layer (ONL). The outer segments (OS) of the photoreceptors are almost completely destroyed. Patches of atrophic retinal pigment epithelium (RPE) are intermixed with hyperpigmented and hypertrophic retinal pigment epithelial cells.

realized that perhaps their father's "torticollis of psychosomatic nature" actually signaled an inherited disorder. The father, in fact, had two deceased older brothers: one developed an unsteady gait at 33 that progressed to ataxia, myoclonus, and dementia, and the other, who had been diagnosed in early childhood as having epilepsy and mental retardation, died at 19 with no other information about his medical background. DNA testing revealed that Patient #4 and her brother had juvenile-onset DRPLA with an expanded allele measuring about 300 bp. Apparently, their father refused to submit to EEG or gene analysis. Regarding the children's disease, it is interesting to note that the types of seizures experienced by juvenile DRPLA patients metamorphose over the course of the illness.

Patient #5,[384] who had SCA7, is noteworthy in that his initial symptoms were cognitive rather than physical. We do not mean to imply that this is an unusual occurrence (perhaps the psychosocial problems leading to patient #3 being transferred to his aunt's care were not attributable entirely to family conflicts). Two of Patient #5's brothers also had SCA7 and showed their first symptom, visual loss, at 14 and 15 years of age, respectively; the father noticed his first sign of illness, also visual loss, at 31 years and became demented 12 years into his illness. The case report on this family, published in 1959 and based on a clinical relationship spanning 13 years, is unusually detailed and nuanced—indeed, in our literature searches, we have invariably found the earliest clinical reports, especially those dating from the first three quarters of the twentieth century, to be the most thorough and sensitive to subtleties in presentation.

It is understandable that as biomedical science has advanced, clinicians' interests would turn to laboratory test results in hopes of gaining some clue as to the molecular effects of a disease, yet surely the cursory descriptions in much of the current literature are a loss to patients and physicians alike. As Refsum, quoted by Woodworth, wrote, "an investigation of heredity based purely on questioning is practically worthless, '. . .for by such procedure misleading information is so often obtained, partly because they may deliberately withhold information. Lastly, without a complete examination of the whole family we are bound to miss the rudimentary forms of the disease wherein diagnosis can only be made by an expert physician.' "[384] Such clinical rigor is, perhaps, most important in the face of unusual phenotypes for which there are a plethora of tests that could conceivably be ordered. Notice that the misdiagnoses of the patients described earlier all erred in the direction of more common disorders (more common, that is, than the SCAs), despite the absence of laboratory tests supporting that less-rare diagnosis or the presence of neurologic symptoms within the family. The trick is to recognize when one is dealing with a rare disorder, and to have at one's disposal the genetic tests most likely to establish the specific diagnosis.

DIAGNOSIS

When the constellation of symptoms and signs suggest a hereditary ataxia, the overlap of these disorders with each other and with other neurological diseases often precludes precise classification on clinical grounds alone. Still, important clues can be gleaned from the history and physical. The presence of noncerebellar neurological symptoms (bulbar dysfunction, extrapyramidal symptoms, peripheral nerve involvement, sleep disturbances, cognitive deficits, psychiatric symptoms, and seizures) associated with disease progression suggests a triplet repeat SCA, while "pure" cerebellar findings suggest one of the SCAs caused by point or unknown mutations. Within the triplet repeat category, bulging eyes suggest SCA3, migraine headaches suggest SCA6, visual disturbances suggest SCA7, and a Huntington disease phenocopy suggests SCA17 or DRPLA. Intermittent symptoms suggest SCA6 or an EA.

Many ataxia patients will have seen several physicians prior to evaluation by a child neurologist, and will have had many basic laboratory tests that are invariably normal. The single most important laboratory test to order when a patient presents with ataxia is measurement of serum vitamin E levels. Although not discussed in this chapter because it is a recessive disease, ataxia with vitamin E deficiency (AVED) is the only completely treatable inherited ataxia, and it responds to vitamin E supplementation. Ancillary tests such as EEG to evaluate seizures and nerve conduction studies/electromyography to characterize peripheral nerve involvement can confirm abnormalities but will offer little help toward a specific diagnosis. MRI of the brain and spinal cord can be useful to distinguish "pure" cerebellar ataxias from those that involve multiple CNS regions. None of these tests, however, can provide a specific diagnosis.

Accurate diagnosis of these disorders hinges upon genetic testing, which has the added benefit of providing information useful for genetic counseling and family planning. Mutation testing is available for all of the disorders presented in this chapter whose causative genes have been identified. Testing for all of the triplet repeat, most of the other repeats, and some of the point mutation disorders is available in the form of comprehensive panels offered by commercial vendors, while the remainder of the diseases are tested by research labs. The invaluable online resource www.genetests.org offers up-to-date information about the tests that are available and where they are performed. Commercial testing is expensive, particularly when done on a gene-by-gene basis. Our recommendation is to test for a specific disorder when the history, physical exam findings, and ancillary testing are strongly suggestive, and then to move to a commercially available ataxia panel if that test is negative. This approach can save a substantial amount of money if the first test provides the answer. In the absence of strong clinical suspicion, however, the most cost-effective strategy is to start with the panel. It is important to bear in mind that genetic testing is not foolproof. Large triplet repeat expansions of the types seen with infantile onset disease can be missed because of technical difficulties inherent in the testing,[69] and the significance of previously uncharacterized point mutations may be difficult to interpret. As the price of next-generation sequencing drops it will likely supplant testing for each of the nonrepeat diseases, and it has already been used successfully to identify SCA cases.[340,385–388] It is also important to remember the substantial overlap between the SCAs and other neurodegenerative diseases such as Huntington, Parkinson, and even Creutzfeldt-Jakob disease.[389] Regardless of how the results are obtained, all positive test results further dictate parental testing.

THERAPY

Unfortunately, current treatments for the SCAs and DRPLA are purely symptomatic and relatively limited. Parkinsonian features may be DOPA-responsive,[76] while piracetam may be useful for controlling myoclonus associated with end-stage disease.[390] Treatment with the dopamine agonist lisuride may relieve periodic leg movements and insomnia.[391] Unfortunately, tremor-controlling drugs do not work well for cerebellar tremors. Dietary factors do not curtail symptoms but vitamin supplements are recommended, particularly if caloric intake is reduced. The picture is more promising for the EAs, as acetazolamide and

4-aminopyridine are effective for decreasing the frequency and intensity of attacks. Potential side effects of acetazolamide therapy are paresthesias, anorexia, weight loss, marrow suppression, skin rash, and osteomalacia. Because of the danger of renal calculi, renal function and abdominal radiographs must be followed in patients on chronic acetazolamide treatment.[362]

Diligent exercise and physical therapy can improve motor coordination and perhaps slow disease progression: a study of the after-effects of intensive short-term physical therapy with patients with pure cerebellar ataxia (including SCA6 and SCA31 patients) demonstrated that one month of two-hour daily training produced improvements in gait, balance, and activities of daily living that persisted for six months.[392] Another recent study demonstrated that patients who exercise consistently have shown improvement in ataxia for one year.[393] Occupational therapy is invaluable in helping patients continue caring for themselves and helping family members understand their needs. Psychotherapy can benefit those patients whose symptoms include cognitive or mood disturbances as well as help individuals and their loved ones live fuller lives in the face of an inexorable disease.

Inasmuch as there is no known therapy to halt or cure these diseases, special attention and care are required from the family members of patients. Genetic counseling is available for family members and at-risk individuals, though consensus holds that children at risk for adult-onset disorders should not have testing in the absence of symptoms.

ACKNOWLEDGMENTS

Our clinical case description section would not have been conceived if not for Dr. Basil Darras alerting us to the very interesting case of Patient #3, who had been diagnosed with KSS; we are grateful to Dr. Margaret E. McLaughlin and Dr. Douglas Anthony for allowing us to include the results of their studies on Patient #3, including their figures, in this section; we also would like to thank Vicky Brandt, who on the basis of that case thought of adding the clinical case description section, analyzed the other clinical cases, and contributed an enlightening synthesis in the form of the case discussion.

REFERENCES

1. Schöls L, Bauer P, Schmidt T, Schulte T, Riess O. Autosomal dominant cerebellar ataxias: clinical features, genetics, and pathogenesis. *Lancet Neurol* 2004;**3**:291–304.
2. Lang AE, Rogaeva EA, Tsuda T, Hutterer J, St George-Hyslop P. Homozygous inheritance of the Machado-Joseph disease gene. *Ann Neurol* 1994;**36**:443–7.
3. Sobue G, Doyu M, Nakao N, Shimada N, Mitsuma T, Maruyama H, et al. Homozygosity for Machado-Joseph disease gene enhances phenotypic severity. *J Neurol Neurosurg Psychiatry* 1996; **60**:354–6.
4. Sato K, Kashihara K, Okada S, Ikeuchi T, Tsuji S, Shomori T, et al. Does homozygosity advance the onset of dentatorubral-pallidoluysian atrophy? *Neurology* 1995;**45**:1934–6.
5. van de Warrenburg BPC, Hendriks H, Durr A, van Zuijlen MCA, Stevanin G, Camuzat A, et al. Age at onset variance analysis in spinocerebellar ataxias: a study in a Dutch-French cohort. *Ann Neurol* 2005;**57**:505–12.
6. Globas C, Montcel du ST, Baliko L, Boesch S, Depondt C, DiDonato S, et al. Early symptoms in spinocerebellar ataxia type 1, 2, 3, and 6. *Mov Disord* 2008;**23**:2232–8.
7. Skinner PJ, Koshy BT, Cummings CJ, Klement IA, Helin K, Servadio A, et al. Ataxin-1 with an expanded glutamine tract alters nuclear matrix-associated structures. *Nature* 1997;**389**:971–4.
8. Cummings CJ, Mancini MA, Antalffy B, DeFranco DB, Orr HT, Zoghbi HY. Chaperone suppression of aggregation and altered subcellular proteasome localization imply protein misfolding in SCA1. *Nat Genet* 1998;**19**:148–54.
9. Cummings CJ, Reinstein E, Sun Y, Antalffy B, Jiang Y, Ciechanover A, et al. Mutation of the E6-AP ubiquitin ligase reduces nuclear inclusion frequency while accelerating polyglutamine-induced pathology in SCA1 mice. *Neuron* 1999;**24**:879–92.
10. Opal P, Zoghbi HY. The role of chaperones in polyglutamine disease. *Trends Mol Med* 2002;**8**:232–6.
11. Paulson HL. Protein fate in neurodegenerative proteinopathies: polyglutamine diseases join the (mis)fold. *Am J Hum Genet* 1999;**64**:339–45.
12. Klement IA, Skinner PJ, Kaytor MD, Yi H, Hersch SM, Clark HB, et al. Ataxin-1 nuclear localization and aggregation: role in polyglutamine-induced disease in SCA1 transgenic mice. *Cell* 1998;**95**:41–53.
13. Evert BO, Vogt IR, Vieira-Saecker AM, Ozimek L, de Vos RAI, Brunt ERP, et al. Gene expression profiling in ataxin-3 expressing cell lines reveals distinct effects of normal and mutant ataxin-3. *J Neuropathol Exp Neurol* 2003;**62**:1006–18.
14. Lim J, Crespo-Barreto J, Jafar-Nejad P, Bowman AB, Richman R, Hill DE, et al. Opposing effects of polyglutamine expansion on native protein complexes contribute to SCA1. *Nature* 2008;**452**:713–18.
15. Friedrich B, Euler P, Ziegler R, Kuhn A, Landwehrmeyer BG, Luthi-Carter R, et al. Comparative analyses of Purkinje cell gene expression profiles reveal shared molecular abnormalities in models of different polyglutamine diseases. *Brain Res* 2012;**1481**: 37–48.
16. Kahle JJ, Gulbahce N, Shaw CA, Lim J, Hill DE, Barabási A-L, et al. Comparison of an expanded ataxia interactome with patient medical records reveals a relationship between macular degeneration and ataxia. *Hum Mol Genet* 2011;**20**:510–27.
17. Lim J, Hao T, Shaw C, Patel AJ, Szabó G, Rual J-F, et al. A protein-protein interaction network for human inherited ataxias and disorders of Purkinje cell degeneration. *Cell* 2006;**125**:801–14.
18. Lebranchu P, Le Meur G, Magot A, David A, Verny C, Weber M, et al. Maculopathy and spinocerebellar ataxia type 1: A new association? *J Neuroophthalmol* 2013.
19. Vaclavik V, Borruat F-X, Ambresin A, Munier FL. Novel maculopathy in patients with spinocerebellar ataxia type 1 autofluorescence findings and functional characteristics. *JAMA Ophthalmol* 2013;**131**:536–8.

20. Stricker S, Oberwahrenbrock T, Zimmermann H, Schroeter J, Endres M, Brandt AU, et al. Temporal retinal nerve fiber loss in patients with spinocerebellar ataxia type 1. *PLOS ONE* 2011;**6**:e23024.
21. Zoghbi HY, Orr HT. Glutamine repeats and neurodegeneration. *Annu Rev Neurosci* 2000;**23**:217–47.
22. Zoghbi HY, Pollack MS, Lyons LA, Ferrell RE, Daiger SP, Beaudet AL. Spinocerebellar ataxia: variable age of onset and linkage to human leukocyte antigen in a large kindred. *Ann Neurol* 1988;**23**:580–4.
23. Ranum LP, Chung MY, Banfi S, Bryer A, Schut LJ, Ramesar R, et al. Molecular and clinical correlations in spinocerebellar ataxia type I: evidence for familial effects on the age at onset. *Am J Hum Genet* 1994;**55**:244–52.
24. Orr HT, Chung MY, Banfi S, Kwiatkowski TJ, Servadio A, Beaudet AL, et al. Expansion of an unstable trinucleotide CAG repeat in spinocerebellar ataxia type 1. *Nat Genet* 1993;**4**:221–6.
25. Chung MY, Ranum LP, Duvick LA, Servadio A, Zoghbi HY, Orr HT. Evidence for a mechanism predisposing to intergenerational CAG repeat instability in spinocerebellar ataxia type I. *Nat Genet* 1993;**5**:254–8.
26. Banfi S, Servadio A, Chung MY, Kwiatkowski TJ, McCall AE, Duvick LA, et al. Identification and characterization of the gene causing type 1 spinocerebellar ataxia. *Nat Genet* 1994;**7**:513–20.
27. Servadio A, Koshy B, Armstrong D, Antalffy B, Orr HT, Zoghbi HY. Expression analysis of the ataxin-1 protein in tissues from normal and spinocerebellar ataxia type 1 individuals. *Nat Genet* 1995;**10**:94–8.
28. Tong X, Gui H, Jin F, Heck BW, Lin P, Ma J, et al. Ataxin-1 and Brother of ataxin-1 are components of the Notch signalling pathway. *EMBO Rep* 2011;**12**:428–35.
29. de Chiara C, Menon RP, Strom M, Gibson TJ, Pastore A. Phosphorylation of S776 and 14-3-3 binding modulate ataxin-1 interaction with splicing factors. *PLOS ONE* 2009;**4**:e8372.
30. Lam YC, Bowman AB, Jafar-Nejad P, Lim J, Richman R, Fryer JD, et al. ATAXIN-1 interacts with the repressor Capicua in its native complex to cause SCA1 neuropathology. *Cell* 2006;**127**:1335–47.
31. Tsai C-C, Kao H-Y, Mitzutani A, Banayo E, Rajan H, McKeown M, et al. Ataxin 1, a SCA1 neurodegenerative disorder protein, is functionally linked to the silencing mediator of retinoid and thyroid hormone receptors. *Proc Natl Acad Sci USA* 2004;**101**:4047–52.
32. Tsuda H, Jafar-Nejad H, Patel AJ, Sun Y, Chen H-K, Rose MF, et al. The AXH domain of Ataxin-1 mediates neurodegeneration through its interaction with Gfi-1/Senseless proteins. *Cell* 2005;**122**:633–44.
33. Serra HG, Duvick L, Zu T, Carlson K, Stevens S, Jorgensen N, et al. RORalpha-mediated Purkinje cell development determines disease severity in adult SCA1 mice. *Cell* 2006;**127**:697–708.
34. Cvetanovic M, Rooney RJ, Garcia JJ, Toporovskaya N, Zoghbi HY, Opal P. The role of LANP and ataxin 1 in E4F-mediated transcriptional repression. *EMBO Rep* 2007;**8**:671–7.
35. Lin X, Antalffy B, Kang D, Orr HT, Zoghbi HY. Polyglutamine expansion down-regulates specific neuronal genes before pathologic changes in SCA1. *Nat Neurosci* 2000;**3**:157–63.
36. Serra HG, Byam CE, Lande JD, Tousey SK, Zoghbi HY, Orr HT. Gene profiling links SCA1 pathophysiology to glutamate signaling in Purkinje cells of transgenic mice. *Hum Mol Genet* 2004;**13**:2535–43.
37. Rodriguez-Lebron E, Liu G, Keiser M, Behlke MA, Davidson BL. Altered Purkinje cell miRNA expression and SCA1 pathogenesis. *Neurobiol Dis* 2013;**54**:456–63.
38. Crespo-Barreto J, Fryer JD, Shaw CA, Orr HT, Zoghbi HY. Partial loss of ataxin-1 function contributes to transcriptional dysregulation in spinocerebellar ataxia type 1 pathogenesis. *PLoS Genet* 2010;**6**:e1001021.
39. Burright EN, Clark HB, Servadio A, Matilla T, Feddersen RM, Yunis WS, et al. SCA1 transgenic mice: a model for neurodegeneration caused by an expanded CAG trinucleotide repeat. *Cell* 1995;**82**:937–48.
40. Watase K, Weeber EJ, Xu B, Antalffy B, Yuva-Paylor L, Hashimoto K, et al. A long CAG repeat in the mouse Sca1 locus replicates SCA1 features and reveals the impact of protein solubility on selective neurodegeneration. *Neuron* 2002;**34**:905–19.
41. Chen H-K, Fernandez-Funez P, Acevedo SF, Lam YC, Kaytor MD, Fernandez MH, et al. Interaction of Akt-phosphorylated ataxin-1 with 14-3-3 mediates neurodegeneration in spinocerebellar ataxia type 1. *Cell* 2003;**113**:457–68.
42. Emamian ES, Kaytor MD, Duvick LA, Zu T, Tousey SK, Zoghbi HY, et al. Serine 776 of ataxin-1 is critical for polyglutamine-induced disease in SCA1 transgenic mice. *Neuron* 2003;**38**:375–87.
43. Ju H, Kokubu H, Todd TW, Kahle JJ, Kim S, Richman R, et al. Polyglutamine disease toxicity is regulated by Nemo-like kinase in spinocerebellar ataxia type 1. *J Neurosci* 2013;**33**:9328–36.
44. Lagalwar S, Orr HT. Regulation of ataxin-1 phosphorylation and its impact on biology. *Methods Mol Biol* 2013;**1010**:201–9.
45. Park J, Al-Ramahi I, Tan Q, Mollema N, Diaz-Garcia JR, Gallego-Flores T, et al. RAS-MAPK-MSK1 pathway modulates ataxin 1 protein levels and toxicity in SCA1. *Nature* 2013;**498**:325–31.
46. Duvick L, Barnes J, Ebner B, Agrawal S, Andresen M, Lim J, et al. SCA1-like disease in mice expressing wild-type ataxin-1 with a serine to aspartic acid replacement at residue 776. *Neuron* 2010;**67** 929–35.
47. Ebner BA, Ingram MA, Barnes JA, Duvick LA, Frisch JL, Clark HB, et al. Purkinje cell ataxin-1 modulates climbing fiber synaptic input in developing and adult mouse cerebellum. *J Neurosci* 2013;**33**:5806–20.
48. Matilla A, Roberson ED, Banfi S, Morales J, Armstrong DL, Burright EN, et al. Mice lacking ataxin-1 display learning deficits and decreased hippocampal paired-pulse facilitation. *J Neurosci* 1998;**18**:5508–16.
49. Davies AF, Mirza G, Sekhon G, Turnpenny P, Leroy F, Speleman F, et al. Delineation of two distinct 6p deletion syndromes. *Hum Genet* 1999;**104**:64–72.
50. Bergeron D, Lapointe C, Bissonnette C, Tremblay G, Motard J, Roucou X. An out-of-frame overlapping reading frame in the ataxin-1 coding sequence encodes a novel ataxin-1 interacting protein. *J Biol Chem* 2013;**288**(30):21824–35.
51. Conforti FL, Spataro R, Sproviero W, Mazzei R, Cavalcanti F, Condino F, et al. Ataxin-1 and ataxin-2 intermediate-length PolyQ expansions in amyotrophic lateral sclerosis. *Neurology* 2012;**79**:2315–20.
52. Cummings CJ, Sun Y, Opal P, Antalffy B, Mestril R, Orr HT, et al. Over-expression of inducible HSP70 chaperone suppresses neuropathology and improves motor function in SCA1 mice. *Hum Mol Genet* 2001;**10**:1511–18.
53. Watase K, Gatchel JR, Sun Y, Emamian E, Atkinson R, Richman R, et al. Lithium therapy improves neurological function and

hippocampal dendritic arborization in a spinocerebellar ataxia type 1 mouse model. *PLoS Med* 2007;**4**:e182.

54. Cvetanovic M, Patel JM, Marti HH, Kini AR, Opal P. Vascular endothelial growth factor ameliorates the ataxic phenotype in a mouse model of spinocerebellar ataxia type 1. *Nat Med* 2011;**17**:1445–7.
55. Fryer JD, Yu P, Kang H, Mandel-Brehm C, Carter AN, Crespo-Barreto J, et al. Exercise and genetic rescue of SCA1 via the transcriptional repressor Capicua. *Science* 2011;**334**:690–3.
56. Hourez R, Servais L, Orduz D, Gall D, Millard I, de Kerchove d'Exaerde A, et al. Aminopyridines correct early dysfunction and delay neurodegeneration in a mouse model of spinocerebellar ataxia type 1. *J Neurosci* 2011;**31**:11795–807.
57. Wadia NH, Swami RK. A new form of heredo-familial spinocerebellar degeneration with slow eye movements (nine families). *Brain* 1971;**94**:359–74.
58. Pulst SM, Nechiporuk A, Nechiporuk T, Gispert S, Chen XN, Lopes-Cendes I, et al. Moderate expansion of a normally biallelic trinucleotide repeat in spinocerebellar ataxia type 2. *Nat Genet* 1996;**14**:269–76.
59. Orozco G, Estrada R, Perry TL, Araña J, Fernandez R, Gonzalez-Quevedo A, et al. Dominantly inherited olivopontocerebellar atrophy from eastern Cuba. Clinical, neuropathological, and biochemical findings. *J Neurol Sci* 1989;**93**:37–50.
60. Cancel G, Durr A, Didierjean O, Imbert G, Bürk K, Lezin A, et al. Molecular and clinical correlations in spinocerebellar ataxia 2: a study of 32 families. *Hum Mol Genet* 1997;**6**:709–15.
61. Belal S, Cancel G, Stevanin G, Hentati F, Khati C, Ben Hamida C, et al. Clinical and genetic analysis of a Tunisian family with autosomal dominant cerebellar ataxia type 1 linked to the SCA2 locus. *Neurology* 1994;**44**:1423–6.
62. Sanpei K, Takano H, Igarashi S, Sato T, Oyake M, Sasaki H, et al. Identification of the spinocerebellar ataxia type 2 gene using a direct identification of repeat expansion and cloning technique, DIRECT. *Nat Genet* 1996;**14**:277–84.
63. Imbert G, Saudou F, Yvert G, Devys D, Trottier Y, Garnier JM, et al. Cloning of the gene for spinocerebellar ataxia 2 reveals a locus with high sensitivity to expanded CAG/glutamine repeats. *Nat Genet* 1996;**14**:285–91.
64. Lopes-Cendes I, Andermann E, Attig E, Cendes F, Bösch S, Wagner M, et al. Confirmation of the SCA-2 locus as an alternative locus for dominantly inherited spinocerebellar ataxias and refinement of the candidate region. *Am J Hum Genet* 1994;**54**:774–81.
65. Montes-Brown J, Machado A, Estévez M, Carricarte C, Velázquez-Pérez L. Autonomic dysfunction in presymptomatic spinocerebellar ataxia type-2. *Acta Neurol Scand* 2012;**125**:24–9.
66. Bürk K, Globas C, Bösch S, Gräber S, Abele M, Brice A, et al. Cognitive deficits in spinocerebellar ataxia 2. *Brain* 1999; **122**(Pt 4):769–77.
67. Ramocki MB, Chapieski L, McDonald RO, Fernandez F, Malphrus AD. Spinocerebellar ataxia type 2 presenting with cognitive regression in childhood. *J Child Neurol* 2008;**23**:999–1001.
68. Babovic-Vuksanovic D, Snow K, Patterson MC, Michels VV. Spinocerebellar ataxia type 2 (SCA 2) in an infant with extreme CAG repeat expansion. *Am J Med Genet* 1998;**79**:383–7.
69. Mao R, Aylsworth AS, Potter N, Wilson WG, Breningstall G, Wick MJ, et al. Childhood-onset ataxia: testing for large CAG-repeats in SCA2 and SCA7. *Am J Med Genet* 2002;**110**:338–45.
70. Di Fabio R, Santorelli F, Bertini E, Balestri M, Cursi L, Tessa A, et al. Infantile childhood onset of spinocerebellar ataxia type 2. *Cerebellum* 2012;**11**:526–30.
71. Paciorkowski AR, Shafrir Y, Hrivnak J, Patterson MC, Tennison MB, Clark HB, et al. Massive expansion of SCA2 with autonomic dysfunction, retinitis pigmentosa, and infantile spasms. *Neurology* 2011;**77**:1055–60.
72. Vinther-Jensen T, Ek J, Duno M, Skovby F, Hjermind LE, Nielsen JE, et al. Germ-line CAG repeat instability causes extreme CAG repeat expansion with infantile-onset spinocerebellar ataxia type 2. *Eur J Hum Genet* 2013;**21**:626–9.
73. Costanzi-Porrini S, Tessarolo D, Abbruzzese C, Liguori M, Ashizawa T, Giacanelli M. An interrupted 34-CAG repeat SCA-2 allele in patients with sporadic spinocerebellar ataxia. *Neurology* 2000;**54**:491–3.
74. Furtado S, Farrer M, Tsuboi Y, Klimek ML, la Fuente-Fernández de R, Hussey J, et al. SCA-2 presenting as parkinsonism in an Alberta family: clinical, genetic, and PET findings. *Neurology* 2002;**59**:1625–7.
75. Charles P, Camuzat A, Benammar N, Sellal F, Destée A, Bonnet A-M, et al. Are interrupted SCA2 CAG repeat expansions responsible for parkinsonism? *Neurology* 2007;**69**:1970–5.
76. Socal MP, Emmel VE, Rieder CRM, Hilbig A, Saraiva-Pereira ML, Jardim LB. Intrafamilial variability of Parkinson phenotype in SCAs: novel cases due to SCA2 and SCA3 expansions. *Parkinsonism Relat Disord* 2009;**15**:374–8.
77. Kim J-M, Hong S, Kim GP, Choi YJ, Kim YK, Park SS, et al. Importance of low-range CAG expansion and CAA interruption in SCA2 Parkinsonism. *Arch Neurol* 2007;**64**:1510–18.
78. Elden AC, Kim H-J, Hart MP, Chen-Plotkin AS, Johnson BS, Fang X, et al. Ataxin-2 intermediate-length polyglutamine expansions are associated with increased risk for ALS. *Nature* 2010;**466**:1069–75.
79. Corrado L, Mazzini L, Oggioni GD, Luciano B, Godi M, Brusco A, et al. ATXN-2 CAG repeat expansions are interrupted in ALS patients. *Hum Genet* 2011;**130**:575–80.
80. Sahba S, Nechiporuk A, Figueroa KP, Nechiporuk T, Pulst SM. Genomic structure of the human gene for spinocerebellar ataxia type 2 (SCA2) on chromosome 12q24.1. *Genomics* 1998;**47**:359–64.
81. Huynh DP, Del Bigio MR, Ho DH, Pulst SM. Expression of ataxin-2 in brains from normal individuals and patients with Alzheimer's disease and spinocerebellar ataxia 2. *Ann Neurol* 1999;**45**:232–41.
82. Shibata H, Huynh DP, Pulst SM. A novel protein with RNA-binding motifs interacts with ataxin-2. *Hum Mol Genet* 2000;**9**:1303–13.
83. Huynh DP, Yang H-T, Vakharia H, Nguyen D, Pulst SM. Expansion of the polyQ repeat in ataxin-2 alters its Golgi localization, disrupts the Golgi complex and causes cell death. *Hum Mol Genet* 2003;**12**:1485–96.
84. van de Loo S, Eich F, Nonis D, Auburger G, Nowock J. Ataxin-2 associates with rough endoplasmic reticulum. *Exp Neurol* 2009;**215**:110–18.
85. Huynh DP, Figueroa K, Hoang N, Pulst SM. Nuclear localization or inclusion body formation of ataxin-2 are not necessary for SCA2 pathogenesis in mouse or human. *Nat Genet* 2000;**26**:44–50.
86. Koyano S, Uchihara T, Fujigasaki H, Nakamura A, Yagishita S, Iwabuchi K. Neuronal intranuclear inclusions in spinocerebellar ataxia type 2. *Ann Neurol* 2000;**47**:550.

87. Nihei Y, Ito D, Suzuki N. Roles of ataxin-2 in pathological cascades mediated by TAR DNA-binding protein 43 (TDP-43) and Fused in Sarcoma (FUS). *J Biol Chem* 2012;**287**:41310–23.
88. Kozlov G, Safaee N, Rosenauer A, Gehring K. Structural basis of binding of P-body-associated proteins GW182 and ataxin-2 by the Mlle domain of poly(A)-binding protein. *J Biol Chem* 2010;**285**:13599–606.
89. Ciosk R, DePalma M, Priess JR. ATX-2, the *C. elegans* ortholog of ataxin 2, functions in translational regulation in the germline. *Development* 2004;**131**:4831–41.
90. Satterfield TF, Pallanck LJ. Ataxin-2 and its *Drosophila* homolog, ATX2, physically assemble with polyribosomes. *Hum Mol Genet* 2006;**15**:2523–32.
91. Nonhoff U, Ralser M, Welzel F, Piccini I, Balzereit D, Yaspo M-L, et al. Ataxin-2 interacts with the DEAD/H-box RNA helicase DDX6 and interferes with P-bodies and stress granules. *Mol Biol Cell* 2007;**18**:1385–96.
92. Zhang Y, Ling J, Yuan C, Dubruille R, Emery P. A role for *Drosophila* ATX2 in activation of PER translation and circadian behavior. *Science* 2013;**340**:879–82.
93. Lim C, Allada R. ATAXIN-2 activates PERIOD translation to sustain circadian rhythms in *Drosophila*. *Science* 2013;**340**: 875–9.
94. Drost J, Nonis D, Eich F, Leske O, Damrath E, Brunt ER, et al. Ataxin-2 modulates the levels of Grb2 and SRC but not ras signaling. *J Mol Neurosci* 2013;**51**:68–81.
95. Bertoni A, Giuliano P, Galgani M, Rotoli D, Ulianich L, Adornetto A, et al. Early and late events induced by polyQ-expanded proteins: identification of a common pathogenic property of polYQ-expanded proteins. *J Biol Chem* 2011;**286**:4727–41.
96. Ralser M, Nonhoff U, Albrecht M, Lengauer T, Wanker EE, Lehrach H, et al. Ataxin-2 and huntingtin interact with endophilin-A complexes to function in plastin-associated pathways. *Hum Mol Genet* 2005;**14**:2893–909.
97. Nonis D, Schmidt MHH, van de Loo S, Eich F, Dikic I, Nowock J, et al. Ataxin-2 associates with the endocytosis complex and affects EGF receptor trafficking. *Cell Signal* 2008;**20**:1725–39.
98. Satterfield TF, Jackson SM, Pallanck LJ. A *Drosophila* homolog of the polyglutamine disease gene SCA2 is a dosage-sensitive regulator of actin filament formation. *Genetics* 2002;**162**:1687–702.
99. Huynh DP, Nguyen DT, Pulst-Korenberg JB, Brice A, Pulst SM. Parkin is an E3 ubiquitin-ligase for normal and mutant ataxin-2 and prevents ataxin-2-induced cell death. *Exp Neurol* 2007;**203**:531–41.
100. Kasumu AW, Liang X, Egorova P, Vorontsova D, Bezprozvanny I. Chronic suppression of inositol 1,4,5-triphosphate receptor-mediated calcium signaling in cerebellar Purkinje cells alleviates pathological phenotype in spinocerebellar ataxia 2 mice. *J Neurosci* 2012;**32**:12786–96.
101. Liu J, Tang T-S, Tu H, Nelson O, Herndon E, Huynh DP, et al. Deranged calcium signaling and neurodegeneration in spinocerebellar ataxia type 2. *J Neurosci* 2009;**29**:9148–62.
102. Hart MP, Gitler AD. ALS-associated ataxin 2 polyQ expansions enhance stress-induced caspase 3 activation and increase TDP-43 pathological modifications. *J Neurosci* 2012;**32**:9133–42.
103. Welzel F, Kaehler C, Isau M, Hallen L, Lehrach H, Krobitsch S. FOX-2 dependent splicing of ataxin-2 transcript is affected by ataxin-1 overexpression. *PLOS ONE* 2012;**7**:e37985.
104. Al-Ramahi I, Pérez AM, Lim J, Zhang M, Sorensen R, de Haro M, et al. dAtaxin-2 mediates expanded Ataxin-1-induced neurodegeneration in a *Drosophila* model of SCA1. *PLoS Genet* 2007;**3**: e234.
105. Pulst SM, Santos N, Wang D, Yang H, Huynh D, Velázquez L, et al. Spinocerebellar ataxia type 2: polyQ repeat variation in the CACNA1A calcium channel modifies age of onset. *Brain* 2005;**128**:2297–303.
106. Nakano KK, Dawson DM, Spence A. Machado disease. A hereditary ataxia in Portuguese emigrants to Massachusetts. *Neurology* 1972;**22**:49–55.
107. Bharucha NE, Bharucha EP, Bhabha SK. Machado-Joseph-Azorean disease in India. *Arch Neurol* 1986;**43**:142–4.
108. Sasaki H, Wakisaka A, Hamada K, Hamada T, Shima K, Tashiro K. Clinicopathological study of Joseph disease: report of 4 pedigrees and its nosological consideration. *Hokkaido Igaku Zasshi* 1992;**67**:174–90.
109. Sasaki H, Wakisaka A, Tashiro K, Hamada T, Katoh T. Linkage study of Machado-Joseph disease: genetic evidence for the locus different from SCA1. *Rinsho Shinkeigaku* 1992;**32**:13–16.
110. Schols L, Amoiridis G, Langkafel M, Büttner T, Przuntek H, Riess O, et al. Machado-Joseph disease mutations as the genetic basis of most spinocerebellar ataxias in Germany. *J Neurol Neurosurg Psychiatry* 1995;**59**:449–50.
111. Kawaguchi Y, Okamoto T, Taniwaki M, Aizawa M, Inoue M, Katayama S, et al. CAG expansions in a novel gene for Machado-Joseph disease at chromosome 14q32.1. *Nat Genet* 1994;**8**:221–8.
112. Gwinn-Hardy K, Singleton A, O'Suilleabhain P, Boss M, Nicholl D, Adam A, et al. Spinocerebellar ataxia type 3 phenotypically resembling Parkinson disease in a black family. *Arch Neurol* 2001;**58**:296–9.
113. Paulson HL, Perez MK, Trottier Y, Trojanowski JQ, Subramony SH, Das SS, et al. Intranuclear inclusions of expanded polyglutamine protein in spinocerebellar ataxia type 3. *Neuron* 1997;**19**:333–44.
114. Chai Y, Koppenhafer SL, Shoesmith SJ, Perez MK, Paulson HL. Evidence for proteasome involvement in polyglutamine disease: localization to nuclear inclusions in SCA3/MJD and suppression of polyglutamine aggregation in vitro. *Hum Mol Genet* 1999;**8**:673–82.
115. Burnett B, Li F, Pittman RN. The polyglutamine neurodegenerative protein ataxin-3 binds polyubiquitylated proteins and has ubiquitin protease activity. *Hum Mol Genet* 2003;**12**:3195–205.
116. Doss-Pepe EW, Stenroos ES, Johnson WG, Madura K. Ataxin-3 interactions with rad23 and valosin-containing protein and its associations with ubiquitin chains and the proteasome are consistent with a role in ubiquitin-mediated proteolysis. *Mol Cell Biol* 2003;**23**:6469–83.
117. Scheel H, Tomiuk S, Hofmann K. Elucidation of ataxin-3 and ataxin-7 function by integrative bioinformatics. *Hum Mol Genet* 2003;**12**:2845–52.
118. Scaglione KM, Zavodszky E, Todi SV, Patury S, Xu P, Rodriguez-Lebron E, et al. Ube2w and ataxin-3 coordinately regulate the ubiquitin ligase CHIP. *Mol Cell* 2011;**43**:599–612.
119. Tsai YC, Fishman PS, Thakor NV, Oyler GA. Parkin facilitates the elimination of expanded polyglutamine proteins and leads to preservation of proteasome function. *J Biol Chem* 2003;**278**:22044–55.

120. Evert BO, Araujo J, Vieira-Saecker AM, de Vos RAI, Harendza S, Klockgether T, et al. Ataxin-3 represses transcription via chromatin binding, interaction with histone deacetylase 3, and histone deacetylation. *J Neurosci* 2006;**26**:11474–86.
121. Chou A-H, Yeh T-H, Ouyang P, Chen Y-L, Chen S-Y, Wang H-L. Polyglutamine-expanded ataxin-3 causes cerebellar dysfunction of SCA3 transgenic mice by inducing transcriptional dysregulation. *Neurobiol Dis* 2008;**31**:89–101.
122. Ikeda H, Yamaguchi M, Sugai S, Aze Y, Narumiya S, Kakizuka A. Expanded polyglutamine in the Machado-Joseph disease protein induces cell death in vitro and in vivo. *Nat Genet* 1996;**13**:196–202.
123. Jung J, Xu K, Lessing D, Bonini NM. Preventing ataxin-3 protein cleavage mitigates degeneration in a *Drosophila* model of SCA3. *Hum Mol Genet* 2009;**18**:4843–52.
124. Goti D, Katzen SM, Mez J, Kurtis N, Kiluk J, Ben-Haïem L, et al. A mutant ataxin-3 putative-cleavage fragment in brains of Machado-Joseph disease patients and transgenic mice is cytotoxic above a critical concentration. *J Neurosci* 2004;**24**:10266–79.
125. Mueller T, Breuer P, Schmitt I, Walter J, Evert BO, Wüllner U. CK2-dependent phosphorylation determines cellular localization and stability of ataxin-3. *Hum Mol Genet* 2009;**18**:3334–43.
126. Chen X, Tang T-S, Tu H, Nelson O, Pook M, Hammer R, et al. Deranged calcium signaling and neurodegeneration in spinocerebellar ataxia type 3. *J Neurosci* 2008;**28**:12713–24.
127. Takeshita Y, Fujinaga R, Kokubu K, Islam MN, Jahan MR, Yanai A, et al. Interaction of ataxin-3 with huntingtin-associated protein 1 through Josephin domain. *Neuroreport* 2011;**22**:232–8.
128. Lessing D, Bonini NM. Polyglutamine genes interact to modulate the severity and progression of neurodegeneration in *Drosophila*. *PloS Biol* 2008;**6**:e29.
129. Zesiewicz TA, Greenstein PE, Sullivan KL, Wecker L, Miller A, Jahan I, et al. A randomized trial of varenicline (Chantix) for the treatment of spinocerebellar ataxia type 3. *Neurology* 2012;**78**:545–50.
130. Chou A-H, Chen S-Y, Yeh T-H, Weng Y-H, Wang H-L. HDAC inhibitor sodium butyrate reverses transcriptional downregulation and ameliorates ataxic symptoms in a transgenic mouse model of SCA3. *Neurobiol Dis* 2011;**41**:481–8.
131. Boy J, Schmidt T, Wolburg H, Mack A, Nuber S, Böttcher M, et al. Reversibility of symptoms in a conditional mouse model of spinocerebellar ataxia type 3. *Hum Mol Genet* 2009;**18**:4282–95.
132. Schols L, Krüger R, Amoiridis G, Przuntek H, Epplen JT, Riess O. Spinocerebellar ataxia type 6: genotype and phenotype in German kindreds. *J Neurol Neurosurg Psychiatry* 1998;**64**:67–73.
133. Mantuano E, Veneziano L, Jodice C, Frontali M. Spinocerebellar ataxia type 6 and episodic ataxia type 2: differences and similarities between two allelic disorders. *Cytogenet Genome Res* 2003;**100**:147–53.
134. Khan NL, Giunti P, Sweeney MG, Scherfler C, Brien MO, Piccini P, et al. Parkinsonism and nigrostriatal dysfunction are associated with spinocerebellar ataxia type 6 (SCA6). *Mov Disord* 2005;**20**:1115–19.
135. Zhuchenko O, Bailey J, Bonnen P, Ashizawa T, Stockton DW, Amos C, et al. Autosomal dominant cerebellar ataxia (SCA6) associated with small polyglutamine expansions in the α1A-voltage-dependent calcium channel. *Nat Genet* 1997;**15**:62–9.
136. Takiyama Y, Sakoe K, Namekawa M, Soutome M, Esumi E, Ogawa T, et al. A Japanese family with spinocerebellar ataxia type 6 which includes three individuals homozygous for an expanded CAG repeat in the SCA6/CACNL1A4 gene. *J Neurol Sci* 1998;**158**:141–7.
137. Ishikawa K, Fujigasaki H, Saegusa H, Ohwada K, Fujita T, Iwamoto H, et al. Abundant expression and cytoplasmic aggregations of [alpha]1A voltage-dependent calcium channel protein associated with neurodegeneration in spinocerebellar ataxia type 6. *Hum Mol Genet* 1999;**8**:1185–93.
138. Watase K, Barrett CF, Miyazaki T, Ishiguro T, Ishikawa K, Hu Y, et al. Spinocerebellar ataxia type 6 knockin mice develop a progressive neuronal dysfunction with age-dependent accumulation of mutant CaV2.1 channels. *Proc Natl Acad Sci USA* 2008;**105**:11987–92.
139. Unno T, Wakamori M, Koike M, Uchiyama Y, Ishikawa K, Kubota H, et al. Development of Purkinje cell degeneration in a knockin mouse model reveals lysosomal involvement in the pathogenesis of SCA6. *Proc Natl Acad Sci USA* 2012;**109**:17693–8.
140. Du X, Wang J, Zhu H, Rinaldo L, Lamar K-M, Palmenberg AC, et al. Second cistron in CACNA1A gene encodes a transcription factor mediating cerebellar development and SCA6. *Cell* 2013;**154**:118–33.
141. Ophoff RA, Terwindt GM, Vergouwe MN, van Eijk R, Oefner PJ, Hoffman SM, et al. Familial hemiplegic migraine and episodic ataxia type-2 are caused by mutations in the Ca2 + channel gene CACNL1A4. *Cell* 1996;**87**:543–52.
142. Self J, Mercer C, Boon EMJ, Murugavel M, Shawkat F, Hammans S, et al. Infantile nystagmus and late onset ataxia associated with a CACNA1A mutation in the intracellular loop between s4 and s5 of domain 3. *Eye (Lond)* 2009;**23**:2251–5.
143. Alonso I, Barros J, Tuna A, Coelho J, Sequeiros J, Silveira I, et al. Phenotypes of spinocerebellar ataxia type 6 and familial hemiplegic migraine caused by a unique CACNA1A missense mutation in patients from a large family. *Arch Neurol* 2003;**60**:610–14.
144. Jodice C, Mantuano E, Veneziano L, Trettel F, Sabbadini G, Calandriello L, et al. Episodic ataxia type 2 (EA2) and spinocerebellar ataxia type 6 (SCA6) due to CAG repeat expansion in the CACNA1A gene on chromosome 19p. *Hum Mol Genet* 1997;**6**:1973–8.
145. Neetens A, Martin J-J, Libert J, Van Den Ende P. Autosomal dominant cone dystrophy-cerebellar atrophy (ADCoCA) (modified ADCA Harding II). *Neuro-Ophthalmol* 1990;**10**(5):261–75.
146. Gouw LG, Digre KB, Harris CP, Haines JH, Ptácek LJ. Autosomal dominant cerebellar ataxia with retinal degeneration: clinical, neuropathologic, and genetic analysis of a large kindred. *Neurology* 1994;**44**:1441–7.
147. Gouw LG, Kaplan CD, Haines JH, Digre KB, Rutledge SL, Matilla A, et al. Retinal degeneration characterizes a spinocerebellar ataxia mapping to chromosome 3p. *Nat Genet* 1995;**10**:89–93.
148. Benton CS, de Silva R, Rutledge SL, Bohlega S, Ashizawa T, Zoghbi HY. Molecular and clinical studies in SCA-7 define a broad clinical spectrum and the infantile phenotype. *Neurology* 1998;**51**:1081–6.
149. Enevoldson TP, Sanders MD, Harding AE. Autosomal dominant cerebellar ataxia with pigmentary macular dystrophy. A clinical and genetic study of eight families. *Brain* 1994;**117**(Pt 3):445–60.

150. Johansson J, Forsgren L, Sandgren O, Brice A, Holmgren G, Holmberg M. Expanded CAG repeats in Swedish spinocerebellar ataxia type 7 (SCA7) patients: effect of CAG repeat length on the clinical manifestation. *Hum Mol Genet* 1998;**7**:171–6.
151. van de Warrenburg BP, Frenken CW, Ausems MG, Kleefstra T, Sinke RJ, Knoers NV, et al. Striking anticipation in spinocerebellar ataxia type 7: the infantile phenotype. *J Neurol* 2001;**248**:911–14.
152. David G, Abbas N, Stevanin G, Durr A, Yvert G, Cancel G, et al. Cloning of the SCA7 gene reveals a highly unstable CAG repeat expansion. *Nat Genet* 1997;**17**:65–70.
153. Helmlinger D, Hardy S, Sasorith S, Klein F, Robert F, Weber C, et al. Ataxin-7 is a subunit of GCN5 histone acetyltransferase-containing complexes. *Hum Mol Genet* 2004;**13**:1257–65.
154. Palhan VB, Chen S, Peng G-H, Tjernberg A, Gamper AM, Fan Y, et al. Polyglutamine-expanded ataxin-7 inhibits STAGA histone acetyltransferase activity to produce retinal degeneration. *Proc Natl Acad Sci USA* 2005;**102**:8472–7.
155. McMahon SJ, Pray-Grant MG, Schieltz D, Yates JR, Grant PA. Polyglutamine-expanded spinocerebellar ataxia-7 protein disrupts normal SAGA and SLIK histone acetyltransferase activity. *Proc Natl Acad Sci USA* 2005;**102**:8478–82.
156. Yanicostas C, Barbieri E, Hibi M, Brice A, Stevanin G, Soussi-Yanicostas N. Requirement for zebrafish ataxin-7 in differentiation of photoreceptors and cerebellar neurons. *PLOS ONE* 2012;**7**: e50705.
157. La Spada AR, Fu Y-H, Sopher BL, Libby RT, Wang X, Li LY, et al. Polyglutamine-expanded ataxin-7 antagonizes crx function and induces cone-rod dystrophy in a mouse model of SCA7. *Neuron* 2001;**31**:913–27.
158. Chen S, Peng G-H, Wang X, Smith AC, Grote SK, Sopher BL, et al. Interference of Crx-dependent transcription by ataxin-7 involves interaction between the glutamine regions and requires the ataxin-7 carboxy-terminal region for nuclear localization. *Hum Mol Genet* 2004;**13**:53–67.
159. Yoo S-Y, Pennesi ME, Weeber EJ, Xu B, Atkinson R, Chen S, et al. SCA7 knockin mice model human SCA7 and reveal gradual accumulation of mutant ataxin-7 in neurons and abnormalities in short-term plasticity. *Neuron* 2003;**37**:383–401.
160. Helmlinger D, Hardy S, Abou-Sleymane G, Eberlin A, Bowman AB, Gansmüller A, et al. Glutamine-expanded ataxin-7 alters TFTC/STAGA recruitment and chromatin structure leading to photoreceptor dysfunction. *PloS Biol* 2006;**4**:e67.
161. Chou A-H, Chen C-Y, Chen S-Y, Chen W-J, Chen Y-L, Weng Y-S, et al. Polyglutamine-expanded ataxin-7 causes cerebellar dysfunction by inducing transcriptional dysregulation. *Neurochem Int* 2010;**56**:329–39.
162. Forsgren L, Libelius R, Holmberg M, Döbeln von U, Wibom R, Heijbel J, et al. Muscle morphology and mitochondrial investigations of a family with autosomal dominant cerebellar ataxia and retinal degeneration mapped to chromosome 3p12-p21.1. *J Neurol Sci* 1996;**144**:91–8.
163. Cooles P, Michaud R, Best PV. A dominantly inherited progressive disease in a black family characterised by cerebellar and retinal degeneration, external ophthalmoplegia and abnormal mitochondria. *J Neurol Sci* 1988;**87**:275–88.
164. Modi G. Morphological abnormalities of hepatic mitochondria in two patients with spinocerebellar ataxia type 7. *J Neurol Neurosurg Psychiatry* 2000;**68**:393–4.
165. Wang H-L, Yeh T-H, Chou A-H, Kuo Y-L, Luo L-J, He C-Y, et al. Polyglutamine-expanded ataxin-7 activates mitochondrial apoptotic pathway of cerebellar neurons by upregulating Bax and downregulating Bcl-x(L). *Cell Signal* 2006;**18**:541–52.
166. Holmberg M, Duyckaerts C, Durr A, Cancel G, Gourfinkel-An I, Damier P, et al. Spinocerebellar ataxia type 7 (SCA7): a neurodegenerative disorder with neuronal intranuclear inclusions. *Hum Mol Genet* 1998;**7**:913–18.
167. Lindenberg KS, Yvert G, Müller K, Landwehrmeyer GB. Expression analysis of ataxin-7 mRNA and protein in human brain: evidence for a widespread distribution and focal protein accumulation. *Brain Pathol* 2000;**10**:385–94.
168. Yvert G, Lindenberg KS, Picaud S, Landwehrmeyer GB, Sahel JA, Mandel JL. Expanded polyglutamines induce neurodegeneration and trans-neuronal alterations in cerebellum and retina of SCA7 transgenic mice. *Hum Mol Genet* 2000;**9**:2491–506.
169. Bowman AB, Yoo S-Y, Dantuma NP, Zoghbi HY. Neuronal dysfunction in a polyglutamine disease model occurs in the absence of ubiquitin-proteasome system impairment and inversely correlates with the degree of nuclear inclusion formation. *Hum Mol Genet* 2005;**14**:679–91.
170. Custer SK, Garden GA, Gill N, Rueb U, Libby RT, Schultz C, et al. Bergmann glia expression of polyglutamine-expanded ataxin-7 produces neurodegeneration by impairing glutamate transport. *Nat Neurosci* 2006;**9**:1302–11.
171. Furrer SA, Mohanachandran MS, Waldherr SM, Chang C, Damian VA, Sopher BL, et al. Spinocerebellar ataxia type 7 cerebellar disease requires the coordinated action of mutant ataxin-7 in neurons and glia, and displays non-cell-autonomous Bergmann glia degeneration. *J Neurosci* 2011; **31**:16269–78.
172. Chort A, Alves S, Marinello M, Dufresnois B, Dornbierer J-G, Tesson C, et al. Interferon β induces clearance of mutant ataxin 7 and improves locomotion in SCA7 knock-in mice. *Brain* 2013;**136**:1732–45.
173. Furrer SA, Waldherr SM, Mohanachandran MS, Baughn TD, Nguyen K-T, Sopher BL, et al. Reduction of mutant ataxin-7 expression restores motor function and prevents cerebellar synaptic reorganization in a conditional mouse model of SCA7. *Hum Mol Genet* 2013;**22**:890–903.
174. Bauer P, Laccone F, Rolfs A, Wullner U, Bosch SB, Peters H, et al. Trinucleotide repeat expansion in SCA17/TBP in white patients with Huntington's disease-like phenotype. *J Med Genet* 2004;**41**:230–2.
175. Maltecca F, Filla A, Castaldo I, Coppola G, Fragassi NA, Carella M, et al. Intergenerational instability and marked anticipation in SCA-17. *Neurology* 2003;**61**:1441–3.
176. Nakamura K, Jeong SY, Uchihara T, Anno M, Nagashima K, Nagashima T, et al. SCA17, a novel autosomal dominant cerebellar ataxia caused by an expanded polyglutamine in TATA-binding protein. *Hum Mol Genet* 2001;**10**:1441–8.
177. Rolfs A, Koeppen AH, Bauer I, Bauer P, Buhlmann S, Topka H, et al. Clinical features and neuropathology of autosomal dominant spinocerebellar ataxia (SCA17). *Ann Neurol* 2003;**54**:367–75.
178. Stevanin G, Fujigasaki H, Lebre A-S, Camuzat A, Jeannequin C, Dode C, et al. Huntington's disease-like phenotype due to trinucleotide repeat expansions in the TBP and JPH3 genes. *Brain* 2003;**126**:1599–603.

179. Toyoshima Y, Yamada M, Onodera O, Shimohata M, Inenaga C, Fujita N, et al. SCA17 homozygote showing Huntington's disease-like phenotype. *Ann Neurol* 2004;**55**:281–6.
180. Koide R, Kobayashi S, Shimohata T, Ikeuchi T, Maruyama M, Saito M, et al. A neurological disease caused by an expanded CAG trinucleotide repeat in the TATA-binding protein gene: a new polyglutamine disease? *Hum Mol Genet* 1999;**8**:2047–53.
181. Bruni AC, Takahashi-Fujigasaki J, Maltecca F, Foncin JF, Servadio A, Casari G, et al. Behavioral disorder, dementia, ataxia, and rigidity in a large family with TATA box-binding protein mutation. *Arch Neurol* 2004;**61**:1314–20.
182. Lee W-W, Kim SY, Kim J-Y, Kim HJ, Park SS, Jeon BS. Extrapyramidal signs are a common feature of spinocerebellar ataxia type 17. *Neurology* 2009;**73**:1708–9.
183. Filla A, De Michele G, Cocozza S, Patrignani A, Volpe G, Castaldo I, et al. Early onset autosomal dominant dementia with ataxia, extrapyramidal features, and epilepsy. *Neurology* 2002;**58**:922–8.
184. Belluzzo M, Musho-Ilbeh S, Monti F, Pizzolato G. A case of nocturnal frontal lobe epilepsy in a patient with spinocerebellar ataxia type 17. *Seizure* 2012;**21**:805–6.
185. Fujigasaki H, Martin J-J, De Deyn PP, Camuzat A, Deffond D, Stevanin G, et al. CAG repeat expansion in the TATA box-binding protein gene causes autosomal dominant cerebellar ataxia. *Brain* 2001;**124**:1939–47.
186. Oda M, Maruyama H, Komure O, Morino H, Terasawa H, Izumi Y, et al. Possible reduced penetrance of expansion of 44 to 47 CAG/CAA repeats in the TATA-binding protein gene in spinocerebellar ataxia type 17. *Arch Neurol* 2004;**61**:209–12.
187. Zühlke C, Dalski A, Schwinger E, Finckh U. Spinocerebellar ataxia type 17: report of a family with reduced penetrance of an unstable Gln49 TBP allele, haplotype analysis supporting a founder effect for unstable alleles and comparative analysis of SCA17 genotypes. *BMC Med Genet* 2005;**6**:27.
188. Kim J-Y, Kim SY, Kim J-M, Kim YK, Yoon K-Y, Kim J-Y, et al. Spinocerebellar ataxia type 17 mutation as a causative and susceptibility gene in parkinsonism. *Neurology* 2009;**72**: 1385–9.
189. Kao CC, Lieberman PM, Schmidt MC, Zhou Q, Pei R, Berk AJ. Cloning of a transcriptionally active human TATA binding factor. *Science* 1990;**248**:1646–50.
190. Huang S, Ling JJ, Yang S, Li X-J, Li S. Neuronal expression of TATA box-binding protein containing expanded polyglutamine in knock-in mice reduces chaperone protein response by impairing the function of nuclear factor-Y transcription factor. *Brain* 2011;**134**:1943–58.
191. Friedman MJ, Shah AG, Fang Z-H, Ward EG, Warren ST, Li S, et al. Polyglutamine domain modulates the TBP-TFIIB interaction: implications for its normal function and neurodegeneration. *Nat Neurosci* 2007;**10**:1519–28.
192. Friedman MJ, Wang C-E, Li X-J, Li S. Polyglutamine expansion reduces the association of TATA-binding protein with DNA and induces DNA binding-independent neurotoxicity. *J Biol Chem* 2008;**283**:8283–90.
193. Watanabe H, Tanaka F, Matsumoto M, Doyu M, Ando T, Mitsuma T, et al. Frequency analysis of autosomal dominant cerebellar ataxias in Japanese patients and clinical characterization of spinocerebellar ataxia type 6. *Clin Genet* 1998;**53**:13–19.
194. Le Ber I, Camuzat A, Castelnovo G, Azulay J-P, Genton P, Gastaut J-L, et al. Prevalence of dentatorubral-pallidoluysian atrophy in a large series of white patients with cerebellar ataxia. *Arch Neurol* 2003;**60**:1097–9.
195. Smith JK, Gonda VE, Malamud N. Unusual form of cerebellar ataxia; combined dentato-rubral and pallido-Luysian degeneration. *Neurology* 1958;**8**:205–9.
196. Naito H, Oyanagi S. Familial myoclonus epilepsy and choreoathetosis: hereditary dentatorubral-pallidoluysian atrophy. *Neurology* 1982;**32**:798–807.
197. Kanazawa I. Dentatorubral-pallidoluysian atrophy or Naito-Oyanagi disease. *Neurogenetics* 1998;**2**:1–17.
198. Muñoz E, Campdelacreu J, Ferrer I, Rey MJ, Cardozo A, Gómez B, et al. Severe cerebral white matter involvement in a case of dentatorubropallidoluysian atrophy studied at autopsy. *Arch Neurol* 2004;**61**:946–9.
199. Koide R, Ikeuchi T, Onodera O, Tanaka H, Igarashi S, Endo K, et al. Unstable expansion of CAG repeat in hereditary dentatorubral-pallidoluysian atrophy (DRPLA). *Nat Genet* 1994;**6**:9–13.
200. Nagafuchi S, Yanagisawa H, Sato K, Shirayama T, Ohsaki E, Bundo M, et al. Dentatorubral and pallidoluysian atrophy expansion of an unstable CAG trinucleotide on chromosome 12p. *Nat Genet* 1994;**6**:14–18.
201. Burke JR, Wingfield MS, Lewis KE, Roses AD, Lee JE, Hulette C, et al. The Haw River syndrome: dentatorubropallidoluysian atrophy (DRPLA) in an African-American family. *Nat Genet* 1994;**7**:521–4.
202. Warner TT, Williams L, Harding AE. DRPLA in Europe. *Nat Genet* 1994;**6**:225.
203. Margolis RL, Li SH, Young WS, Wagster MV, Stine OC, Kidwai AS, et al. DRPLA gene (atrophin-1) sequence and mRNA expression in human brain. *Brain Res Mol Brain Res* 1996;**36**:219–26.
204. Shimojo Y, Osawa Y, Fukumizu M, Hanaoka S, Tanaka H, Ogata F, et al. Severe infantile dentatorubral pallidoluysian atrophy with extreme expansion of CAG repeats. *Neurology* 2001;**56**:277–8.
205. Nagafuchi S, Yanagisawa H, Ohsaki E, Shirayama T, Tadokoro K, Inoue T, et al. Structure and expression of the gene responsible for the triplet repeat disorder, dentatorubral and pallidoluysian atrophy (DRPLA). *Nat Genet* 1994;**8**:177–82.
206. Ueno S, Kondoh K, Kotani Y, Komure O, Kuno S, Kawai J, et al. Somatic mosaicism of CAG repeat in dentatorubral-pallidoluysian atrophy (DRPLA). *Hum Mol Genet* 1995;**4**:663–6.
207. Hayashi Y, Kakita A, Yamada M, Koide R, Igarashi S, Takano H, et al. Hereditary dentatorubral-pallidoluysian atrophy: detection of widespread ubiquitinated neuronal and glial intranuclear inclusions in the brain. *Acta Neuropathol* 1998;**96**:547–52.
208. Wood JD, Yuan J, Margolis RL, Colomer V, Duan K, Kushi J, et al. Atrophin-1, the DRPLA gene product, interacts with two families of WW domain-containing proteins. *Mol Cell Neurosci* 1998;**11**:149–60.
209. Shimohata T, Nakajima T, Yamada M, Uchida C, Onodera O, Naruse S, et al. Expanded polyglutamine stretches interact with TAFII130, interfering with CREB-dependent transcription. *Nat Genet* 2000;**26**:29–36.
210. Shimohata T, Onodera O, Tsuji S. Interaction of expanded polyglutamine stretches with nuclear transcription factors leads to aberrant transcriptional regulation in polyglutamine diseases. *Neuropathology* 2000;**20**:326–33.

211. Wood JD, Nucifora FC, Duan K, Zhang C, Wang J, Kim Y, et al. Atrophin-1, the dentato-rubral and pallido-luysian atrophy gene product, interacts with ETO/MTG8 in the nuclear matrix and represses transcription. *J Cell Biol* 2000;**150**:939–48.
212. Zhang S, Xu L, Lee J, Xu T. *Drosophila* atrophin homolog functions as a transcriptional corepressor in multiple developmental processes. *Cell* 2002;**108**:45–56.
213. Zoltewicz JS, Stewart NJ, Leung R, Peterson AS. Atrophin 2 recruits histone deacetylase and is required for the function of multiple signaling centers during mouse embryogenesis. *Development* 2004;**131**:3–14.
214. Ying M, Xu R, Wu X, Zhu H, Zhuang Y, Han M, et al. Sodium butyrate ameliorates histone hypoacetylation and neurodegenerative phenotypes in a mouse model for DRPLA. *J Biol Chem* 2006;**281**:12580–6.
215. Okamura-Oho Y, Miyashita T, Ohmi K, Yamada M. Dentatorubral-pallidoluysian atrophy protein interacts through a proline-rich region near polyglutamine with the SH3 domain of an insulin receptor tyrosine kinase substrate. *Hum Mol Genet* 1999;**8**:947–57.
216. Ellerby LM, Andrusiak RL, Wellington CL, Hackam AS, Propp SS, Wood JD, et al. Cleavage of atrophin-1 at caspase site aspartic acid 109 modulates cytotoxicity. *J Biol Chem* 1999;**274**:8730–6.
217. Schilling G, Wood JD, Duan K, Slunt HH, Gonzales V, Yamada M, et al. Nuclear accumulation of truncated atrophin-1 fragments in a transgenic mouse model of DRPLA. *Neuron* 1999;**24**:275–86.
218. Nucifora FC, Ellerby LM, Wellington CL, Wood JD, Herring WJ, Sawa A, et al. Nuclear localization of a non-caspase truncation product of atrophin-1, with an expanded polyglutamine repeat, increases cellular toxicity. *J Biol Chem* 2003;**278**:13047–55.
219. Nucifora FC, Sasaki M, Peters MF, Huang H, Cooper JK, Yamada M, et al. Interference by huntingtin and atrophin-1 with cbp-mediated transcription leading to cellular toxicity. *Science* 2001;**291**:2423–8.
220. Yu J, Ying M, Zhuang Y, Xu T, Han M, Wu X, et al. C-terminal deletion of the atrophin-1 protein results in growth retardation but not neurodegeneration in mice. *Dev Dyn* 2009;**238**:2471–8.
221. Koob MD, Moseley ML, Schut LJ, Benzow KA, Bird TD, Day JW, et al. An untranslated CTG expansion causes a novel form of spinocerebellar ataxia (SCA8). *Nat Genet* 1999;**21**:379–84.
222. Felling RJ, Barron TF. Early onset of ataxia in a child with a pathogenic SCA8 allele. *Pediatr Neurol* 2005;**33**:136–8.
223. Kumar N, Miller GM. White matter hyperintense lesions in genetically proven spinocerebellar ataxia 8. *Clin Neurol Neurosurg* 2008;**110**:65–8.
224. Ikeda Y, Dalton JC, Moseley ML, Gardner KL, Bird TD, Ashizawa T, et al. Spinocerebellar ataxia type 8: molecular genetic comparisons and haplotype analysis of 37 families with ataxia. *Am J Hum Genet* 2004;**75**:3–16.
225. Moseley ML, Zu T, Ikeda Y, Gao W, Mosemiller AK, Daughters RS, et al. Bidirectional expression of CUG and CAG expansion transcripts and intranuclear polyglutamine inclusions in spinocerebellar ataxia type 8. *Nat Genet* 2006;**38**:758–69.
226. Moseley ML, Schut LJ, Bird TD, Koob MD, Day JW, Ranum LP. SCA8 CTG repeat: en masse contractions in sperm and intergenerational sequence changes may play a role in reduced penetrance. *Hum Mol Genet* 2000;**9**:2125–30.
227. Izumi Y, Maruyama H, Oda M, Morino H, Okada T, Ito H, et al. SCA8 repeat expansion: large CTA/CTG repeat alleles are more common in ataxic patients, including those with SCA6. *Am J Hum Genet* 2003;**72**:704–9.
228. Sulek A, Hoffman-Zacharska D, Zdzienicka E, Zaremba J. SCA8 repeat expansion coexists with SCA1—not only with SCA6. *Am J Hum Genet* 2003;**73**:972–4.
229. Cellini E, Piacentini S, Nacmias B, Forleo P, Tedde A, Bagnoli S, et al. A family with spinocerebellar ataxia type 8 expansion and vitamin E deficiency ataxia. *Arch Neurol* 2002;**59**:1952–3.
230. Sobrido MJ, Cholfin JA, Perlman S, Pulst SM, Geschwind DH. SCA8 repeat expansions in ataxia: a controversial association. *Neurology* 2001;**57**:1310–12.
231. Mutsuddi M, Marshall CM, Benzow KA, Koob MD, Rebay I. The spinocerebellar ataxia 8 noncoding RNA causes neurodegeneration and associates with staufen in *Drosophila*. *Curr Biol* 2004;**14**:302–8.
232. Chen W-L, Lin J-W, Huang H-J, Wang S-M, Su M-T, Lee-Chen G-J, et al. SCA8 mRNA expression suggests an antisense regulation of KLHL1 and correlates to SCA8 pathology. *Brain Res* 2008;**1233**:176–84.
233. Daughters RS, Tuttle DL, Gao W, Ikeda Y, Moseley ML, Ebner TJ, et al. RNA gain-of-function in spinocerebellar ataxia type 8. *PLoS Genet* 2009;**5**:e1000600.
234. Nemes JP, Benzow KA, Moseley ML, Ranum LP, Koob MD. The SCA8 transcript is an antisense RNA to a brain-specific transcript encoding a novel actin-binding protein (KLHL1). *Hum Mol Genet* 2000;**9**:1543–51.
235. He Y, Zu T, Benzow KA, Orr HT, Clark HB, Koob MD. Targeted deletion of a single Sca8 ataxia locus allele in mice causes abnormal gait, progressive loss of motor coordination, and Purkinje cell dendritic deficits. *J Neurosci* 2006;**26**:9975–82.
236. Aromolaran KA, Benzow KA, Koob MD, Piedras-Rentería ES. The Kelch-like protein 1 modulates P/Q-type calcium current density. *Nsc* 2007;**145**:841–50.
237. Grewal RP, Tayag E, Figueroa KP, Zu L, Durazo A, Nunez C, et al. Clinical and genetic analysis of a distinct autosomal dominant spinocerebellar ataxia. *Neurology* 1998;**51**:1423–6.
238. Matsuura T, Ranum LPW, Volpini V, Pandolfo M, Sasaki H, Tashiro K, et al. Spinocerebellar ataxia type 10 is rare in populations other than Mexicans. *Neurology* 2002;**58**:983–4.
239. Almeida T, Alonso I, Martins S, Ramos EM, Azevedo L, Ohno K, et al. Ancestral origin of the ATTCT repeat expansion in spinocerebellar ataxia type 10 (SCA10). *PLOS ONE* 2009;**4**:e4553.
240. Matsuura T, Yamagata T, Burgess DL, Rasmussen A, Grewal RP, Watase K, et al. Large expansion of the ATTCT pentanucleotide repeat in spinocerebellar ataxia type 10. *Nat Genet* 2000;**26**:191–4.
241. März P, Probst A, Lang S, Schwager M, Rose-John S, Otten U, et al. Ataxin-10, the spinocerebellar ataxia type 10 neurodegenerative disorder protein, is essential for survival of cerebellar neurons. *J Biol Chem* 2004;**279**:35542–50.
242. Hagerman KA, Ruan H, Edamura KN, Matsuura T, Pearson CE, Wang Y-H. The ATTCT repeats of spinocerebellar ataxia type 10 display strong nucleosome assembly which is enhanced by repeat interruptions. *Gene* 2009;**434**:29–34.
243. Waragai M, Nagamitsu S, Xu W, Li YJ, Lin X, Ashizawa T. Ataxin 10 induces neuritogenesis via interaction with G-protein beta2 subunit. *J Neurosci Res* 2006;**83**:1170–8.

244. Li J, Wang J, Hou W, Jing Z, Tian C, Han Y, et al. Phosphorylation of Ataxin-10 by polo-like kinase 1 is required for cytokinesis. *Cell Cycle* 2011;**10**:2946–58.
245. White MC, Gao R, Xu W, Mandal SM, Lim JG, Hazra TK, et al. Inactivation of hnRNP K by expanded intronic AUUCU repeat induces apoptosis via translocation of PKCdelta to mitochondria in spinocerebellar ataxia 10. *PLoS Genet* 2010;**6**:e1000984.
246. White M, Xia G, Gao R, Wakamiya M, Sarkar PS, McFarland K, et al. Transgenic mice with SCA10 pentanucleotide repeats show motor phenotype and susceptibility to seizure: a toxic RNA gain-of-function model. *J Neurosci Res* 2012;**90**:706–14.
247. Holmes SE, O'Hearn EE, McInnis MG, Gorelick-Feldman DA, Kleiderlein JJ, Callahan C, et al. Expansion of a novel CAG trinucleotide repeat in the 5′ region of PPP2R2B is associated with SCA12. *Nat Genet* 1999;**23**:391–2.
248. O'Hearn E, Holmes SE, Calvert PC, Ross CA, Margolis RL. SCA-12: Tremor with cerebellar and cortical atrophy is associated with a CAG repeat expansion. *Neurology* 2001;**56**:299–303.
249. Holmes SE, O'Hearn E, Margolis RL. Why is SCA12 different from other SCAs? *Cytogenet Genome Res* 2003;**100**:189–97.
250. Srivastava AK, Choudhry S, Gopinath MS, Roy S, Tripathi M, Brahmachari SK, et al. Molecular and clinical correlation in five Indian families with spinocerebellar ataxia 12. *Ann Neurol* 2001;**50**:796–800.
251. Wang Y-C, Lee C-M, Lee L-C, Tung L-C, Hsieh-Li HM, Lee-Chen G-J, et al. Mitochondrial dysfunction and oxidative stress contribute to the pathogenesis of spinocerebellar ataxia type 12 (SCA12). *J Biol Chem* 2011;**286**:21742–54.
252. Ouyang Y, Sakoe K, Shimazaki H, Namekawa M, Ogawa T, Ando Y, et al. 16q-linked autosomal dominant cerebellar ataxia: a clinical and genetic study. *J Neurol Sci* 2006;**247**:180–6.
253. Ishikawa K, Durr A, Klopstock T, Müller S, De Toffol B, Vidailhet M, et al. Pentanucleotide repeats at the spinocerebellar ataxia type 31 (SCA31) locus in Caucasians. *Neurology* 2011;**77**:1853–5.
254. Nagaoka U, Takashima M, Ishikawa K, Yoshizawa K, Yoshizawa T, Ishikawa M, et al. A gene on SCA4 locus causes dominantly inherited pure cerebellar ataxia. *Neurology* 2000;**54**:1971–5.
255. Ishikawa K, Toru S, Tsunemi T, Li M, Kobayashi K, Yokota T, et al. An autosomal dominant cerebellar ataxia linked to chromosome 16q22.1 is associated with a single-nucleotide substitution in the 5′ untranslated region of the gene encoding a protein with spectrin repeat and Rho guanine-nucleotide exchange-factor domains. *Am J Hum Genet* 2005;**77**:280–96.
256. Sato N, Amino T, Kobayashi K, Asakawa S, Ishiguro T, Tsunemi T, et al. Spinocerebellar ataxia type 31 is associated with "inserted" penta-nucleotide repeats containing (TGGAA)n. *Am J Hum Genet* 2009;**85**:544–57.
257. Sakai H, Yoshida K, Shimizu Y, Morita H, Ikeda S-I, Matsumoto N. Analysis of an insertion mutation in a cohort of 94 patients with spinocerebellar ataxia type 31 from Nagano, Japan. *Neurogenetics* 2010;**11**:409–15.
258. Niimi Y, Takahashi M, Sugawara E, Umeda S, Obayashi M, Sato N, et al. Abnormal RNA structures (RNA foci) containing a penta-nucleotide repeat (UGGAA)n in the Purkinje cell nucleus is associated with spinocerebellar ataxia type 31 pathogenesis. *Neuropathology* 2013;**33**:600–11.
259. Kobayashi H, Abe K, Matsuura T, Ikeda Y, Hitomi T, Akechi Y, et al. Expansion of intronic GGCCTG hexanucleotide repeat in NOP56 causes SCA36, a type of spinocerebellar ataxia accompanied by motor neuron involvement. *Am J Hum Genet* 2011;**89**:121–30.
260. Ikeda Y, Ohta Y, Kobayashi H, Okamoto M, Takamatsu K, Ota T, et al. Clinical features of SCA36: a novel spinocerebellar ataxia with motor neuron involvement (Asidan). *Neurology* 2012;**79**:333–41.
261. García-Murias M, Quintáns B, Arias M, Seixas AI, Cacheiro P, Tarrí R, et al. "Costa da Morte" ataxia is spinocerebellar ataxia 36: clinical and genetic characterization. *Brain* 2012;**135**(Pt 5):1423–35.
262. Ikeda Y, Ohta Y, Kurata T, Shiro Y, Takao Y, Abe K. Acoustic impairment is a distinguishable clinical feature of Asidan/SCA36. *J Neurol Sci* 2013;**324**:109–12.
263. Miyashiro A, Sugihara K, Kawarai T, Miyamoto R, Izumi Y, Morino H, et al. Oromandibular dystonia associated with SCA36. *Mov Disord* 2013;**28**:558–9.
264. Stevanin G, Herman A, Brice A, Durr A. Clinical and MRI findings in spinocerebellar ataxia type 5. *Neurology* 1999;**53**:1355–7.
265. Ranum LP, Schut LJ, Lundgren JK, Orr HT, Livingston DM. Spinocerebellar ataxia type 5 in a family descended from the grandparents of President Lincoln maps to chromosome 11. *Nat Genet* 1994;**8**:280–4.
266. Ikeda Y, Dick KA, Weatherspoon MR, Gincel D, Armbrust KR, Dalton JC, et al. Spectrin mutations cause spinocerebellar ataxia type 5. *Nat Genet* 2006;**38**:184–90.
267. Lorenzo DN, Li M-G, Mische SE, Armbrust KR, Ranum LPW, Hays TS. Spectrin mutations that cause spinocerebellar ataxia type 5 impair axonal transport and induce neurodegeneration in *Drosophila*. *J Cell Biol* 2010;**189**:143–58.
268. Clarkson YL, Gillespie T, Perkins EM, Lyndon AR, Jackson M. Beta-III spectrin mutation L253P associated with spinocerebellar ataxia type 5 interferes with binding to Arp1 and protein trafficking from the Golgi. *Hum Mol Genet* 2010;**19**:3634–41.
269. Lise S, Clarkson Y, Perkins E, Kwasniewska A, Sadighi Akha E, Schnekenberg RP, et al. Recessive mutations in SPTBN2 implicate β-III spectrin in both cognitive and motor development. *PLoS Genet* 2012;**8**:e1003074.
270. Elsayed SM, Heller R, Thoenes M, Zaki MS, Swan D, Elsobky E, et al. Autosomal dominant SCA5 and autosomal recessive infantile SCA are allelic conditions resulting from SPTBN2 mutations. *Eur J Hum Genet* 2014;**22**:286–8.
271. Worth PF, Giunti P, Gardner-Thorpe C, Dixon PH, Davis MB, Wood NW. Autosomal dominant cerebellar ataxia type III: linkage in a large British family to a 7.6-cM region on chromosome 15q14-21.3. *Am J Hum Genet* 1999;**65**:420–6.
272. Houlden H, Johnson J, Gardner-Thorpe C, Lashley T, Hernandez D, Worth P, et al. Mutations in TTBK2, encoding a kinase implicated in tau phosphorylation, segregate with spinocerebellar ataxia type 11. *Nat Genet* 2007;**39**:1434–6.
273. Bauer P, Stevanin G, Beetz C, Synofzik M, Schmitz-Hubsch T, Wüllner U, et al. Spinocerebellar ataxia type 11 (SCA11) is an uncommon cause of dominant ataxia among French and German kindreds. *J Neurol Neurosurg Psychiatry* 2010;**81**:1229–32.
274. Goetz SC, Liem KF, Anderson KV. The spinocerebellar ataxia-associated gene Tau tubulin kinase 2 controls the initiation of ciliogenesis. *Cell* 2012;**151**:847–58.
275. Herman-Bert A, Stevanin G, Netter JC, Rascol O, Brassat D, Calvas P, et al. Mapping of spinocerebellar ataxia 13 to chromosome 19q13.3-q13.4 in a family with autosomal dominant

cerebellar ataxia and mental retardation. *Am J Hum Genet* 2000;**67**:229–35.

276. Figueroa KP, Minassian NA, Stevanin G, Waters M, Garibyan V, Forlani S, et al. KCNC3: phenotype, mutations, channel biophysics—a study of 260 familial ataxia patients. *Hum Mutat* 2010;**31**:191–6.
277. Bürk K, Strzelczyk A, Reif PS, Figueroa KP, Pulst SM, Zühlke C, et al. Mesial temporal lobe epilepsy in a patient with spinocerebellar ataxia type 13 (SCA13). *Int J Neurosci* 2013;**123**:278–82.
278. Waters MF, Minassian NA, Stevanin G, Figueroa KP, Bannister JPA, Nolte D, et al. Mutations in voltage-gated potassium channel KCNC3 cause degenerative and developmental central nervous system phenotypes. *Nat Genet* 2006;**38**:447–51.
279. Issa FA, Mazzochi C, Mock AF, Papazian DM. Spinocerebellar ataxia type 13 mutant potassium channel alters neuronal excitability and causes locomotor deficits in zebrafish. *J Neurosci* 2011;**31**:6831–41.
280. Minassian NA, Lin M-CA, Papazian DM. Altered Kv3.3 channel gating in early-onset spinocerebellar ataxia type 13. *J Physiol* 2012;**590**:1599–614.
281. Issa FA, Mock AF, Sagasti A, Papazian DM. Spinocerebellar ataxia type 13 mutation that is associated with disease onset in infancy disrupts axonal pathfinding during neuronal development. *Dis Model Mech* 2012;**5**:921–9.
282. Brkanac Z, Bylenok L, Fernandez M, Matsushita M, Lipe H, Wolff J, et al. A new dominant spinocerebellar ataxia linked to chromosome 19q13.4-qter. *Arch Neurol* 2002;**59**:1291–5.
283. van de Warrenburg BPC, Verbeek DS, Piersma SJ, Hennekam FAM, Pearson PL, Knoers NVAM, et al. Identification of a novel SCA14 mutation in a Dutch autosomal dominant cerebellar ataxia family. *Neurology* 2003;**61**:1760–5.
284. Yamashita I, Sasaki H, Yabe I, Fukazawa T, Nogoshi S, Komeichi K, et al. A novel locus for dominant cerebellar ataxia (SCA14) maps to a 10.2-cM interval flanked by D19S206 and D19S605 on chromosome 19q13.4-qter. *Ann Neurol* 2000;**48**:156–63.
285. Chen D-H, Cimino PJ, Ranum LPW, Zoghbi HY, Yabe I, Schut L, et al. The clinical and genetic spectrum of spinocerebellar ataxia 14. *Neurology* 2005;**64**:1258–60.
286. Ganos C, Zittel S, Minnerop M, Schunke O, Heinbokel C, Gerloff C, et al. Clinical and neurophysiological profile of four German families with spinocerebellar ataxia type 14. *Cerebellum* 2014;**13**:89–96.
287. Chen D-H, Brkanac Z, Verlinde CLMJ, Tan X-J, Bylenok L, Nochlin D, et al. Missense mutations in the regulatory domain of PKC gamma: a new mechanism for dominant nonepisodic cerebellar ataxia. *Am J Hum Genet* 2003;**72**:839–49.
288. Yabe I, Sasaki H, Chen D-H, Raskind WH, Bird TD, Yamashita I, et al. Spinocerebellar ataxia type 14 caused by a mutation in protein kinase C gamma. *Arch Neurol* 2003;**60**:1749–51.
289. Verbeek DS, Knight MA, Harmison GG, Fischbeck KH, Howell BW. Protein kinase C gamma mutations in spinocerebellar ataxia 14 increase kinase activity and alter membrane targeting. *Brain* 2005;**128**:436–42.
290. Adachi N, Kobayashi T, Takahashi H, Kawasaki T, Shirai Y, Ueyama T, et al. Enzymological analysis of mutant protein kinase Cgamma causing spinocerebellar ataxia type 14 and dysfunction in Ca2 + homeostasis. *J Biol Chem* 2008;**283**:19854–63.
291. Verbeek DS, Goedhart J, Bruinsma L, Sinke RJ, Reits EA. PKC gamma mutations in spinocerebellar ataxia type 14 affect C1 domain accessibility and kinase activity leading to aberrant MAPK signaling. *J Cell Sci* 2008;**121**:2339–49.
292. Seki T, Adachi N, Ono Y, Mochizuki H, Hiramoto K, Amano T, et al. Mutant protein kinase Cgamma found in spinocerebellar ataxia type 14 is susceptible to aggregation and causes cell death. *J Biol Chem* 2005;**280**:29096–106.
293. Seki T, Takahashi H, Adachi N, Abe N, Shimahara T, Saito N, et al. Aggregate formation of mutant protein kinase C gamma found in spinocerebellar ataxia type 14 impairs ubiquitin-proteasome system and induces endoplasmic reticulum stress. *Eur J Neurosci* 2007;**26**:3126–40.
294. Seki T, Abe-Seki N, Kikawada T, Takahashi H, Yamamoto K, Adachi N, et al. Effect of trehalose on the properties of mutant {gamma}PKC, which causes spinocerebellar ataxia type 14, in neuronal cell lines and cultured Purkinje cells. *J Biol Chem* 2010;**285**:33252–64.
295. Seki T, Takahashi H, Yamamoto K, Ogawa K, Onji T, Adachi N, et al. Congo red, an amyloid-inhibiting compound, alleviates various types of cellular dysfunction triggered by mutant protein kinase cγ that causes spinocerebellar ataxia type 14 (SCA14) by inhibiting oligomerization and aggregation. *J Pharmacol Sci* 2010;**114**:206–16.
296. Seki T, Shimahara T, Yamamoto K, Abe N, Amano T, Adachi N, et al. Mutant gammaPKC found in spinocerebellar ataxia type 14 induces aggregate-independent maldevelopment of dendrites in primary cultured Purkinje cells. *Neurobiol Dis* 2009;**33**:260–73.
297. Zhang Y, Snider A, Willard L, Takemoto DJ, Lin D. Loss of Purkinje cells in the PKCgamma H101Y transgenic mouse. *Biochem Biophys Res Commun* 2009;**378**:524–8.
298. Shuvaev AN, Horiuchi H, Seki T, Goenawan H, Irie T, Iizuka A, et al. Mutant PKCγ in spinocerebellar ataxia type 14 disrupts synapse elimination and long-term depression in Purkinje cells in vivo. *J Neurosci* 2011;**31**:14324–34.
299. Asai H, Hirano M, Shimada K, Kiriyama T, Furiya Y, Ikeda M, et al. Protein kinase C gamma, a protein causative for dominant ataxia, negatively regulates nuclear import of recessive-ataxia-related aprataxin. *Hum Mol Genet* 2009;**18**:3533–43.
300. Storey E, Gardner RJ, Knight MA, Kennerson ML, Tuck RR, Forrest SM, et al. A new autosomal dominant pure cerebellar ataxia. *Neurology* 2001;**57**:1913–15.
301. Novak MJU, Sweeney MG, Li A, Treacy C, Chandrashekar HS, Giunti P, et al. An ITPR1 gene deletion causes spinocerebellar ataxia 15/16: a genetic, clinical and radiological description. *Mov Disord* 2010;**25**:2176–82.
302. Marelli C, van de Leemput J, Johnson JO, Tison F, Thauvin-Robinet C, Picard F, et al. SCA15 due to large ITPR1 deletions in a cohort of 333 white families with dominant ataxia. *Arch Neurol* 2011;**68**:637–43.
303. Di Gregorio E, Orsi L, Godani M, Vaula G, Jensen S, Salmon E, et al. Two Italian families with ITPR1 gene deletion presenting a broader phenotype of SCA15. *Cerebellum* 2010;**9**:115–23.
304. Synofzik M, Beetz C, Bauer C, Bonin M, Sanchez-Ferrero E, Schmitz-Hubsch T, et al. Spinocerebellar ataxia type 15: diagnostic assessment, frequency, and phenotypic features. *J Med Genet* 2011;**48**:407–12.
305. Castrioto A, Prontera P, Di Gregorio E, Rossi V, Parnetti L, Rossi A, et al. A novel spinocerebellar ataxia type 15 family with involuntary movements and cognitive decline. *Eur J Neurol* 2011;**18**:1263–5.

306. van de Leemput J, Chandran J, Knight MA, Holtzclaw LA, Scholz S, Cookson MR, et al. Deletion at ITPR1 underlies ataxia in mice and spinocerebellar ataxia 15 in humans. *PLoS Genet* 2007;**3**:e108.
307. Knight MA, Kennerson ML, Anney RJ, Matsuura T, Nicholson GA, Salimi-Tari P, et al. Spinocerebellar ataxia type 15 (sca15) maps to 3p24.2-3pter: exclusion of the ITPR1 gene, the human orthologue of an ataxic mouse mutant. *Neurobiol Dis* 2003;**13**:147–57.
308. Hara K, Shiga A, Nozaki H, Mitsui J, Takahashi Y, Ishiguro H, et al. Total deletion and a missense mutation of ITPR1 in Japanese SCA15 families. *Neurology* 2008;**71**:547–51.
309. Miyoshi Y, Yamada T, Tanimura M, Taniwaki T, Arakawa K, Ohyagi Y, et al. A novel autosomal dominant spinocerebellar ataxia (SCA16) linked to chromosome 8q22.1-24.1. *Neurology* 2001;**57**:96–100.
310. Miura S, Shibata H, Furuya H, Ohyagi Y, Osoegawa M, Miyoshi Y, et al. The contactin 4 gene locus at 3p26 is a candidate gene of SCA16. *Neurology* 2006;**67**:1236–41.
311. Iwaki A, Kawano Y, Miura S, Shibata H, Matsuse D, Li W, et al. Heterozygous deletion of ITPR1, but not SUMF1, in spinocerebellar ataxia type 16. *J Med Genet* 2007;**45**:32–5.
312. Huang L, Chardon JW, Carter MT, Friend KL, Dudding TE, Schwartzentruber J, et al. Missense mutations in ITPR1 cause autosomal dominant congenital nonprogressive spinocerebellar ataxia. *Orphanet J Rare Dis* 2012;**7**:67.
313. Schelhaas HJ, Ippel PF, Hageman G, Sinke RJ, van der Laan EN, Beemer FA. Clinical and genetic analysis of a four-generation family with a distinct autosomal dominant cerebellar ataxia. *J Neurol* 2001;**248**:113–20.
314. Verbeek DS, Schelhaas JH, Ippel EF, Beemer FA, Pearson PL, Sinke RJ. Identification of a novel SCA locus (SCA19) in a Dutch autosomal dominant cerebellar ataxia family on chromosome region 1p21-q21. *Hum Genet* 2002;**111**:388–93.
315. Chung MY, Lu YC, Cheng NC, Soong BW. A novel autosomal dominant spinocerebellar ataxia (SCA22) linked to chromosome 1p21-q23. *Brain* 2003;**126**:1293–9.
316. Schelhaas HJ, van de Warrenburg BPC. Clinical, psychological, and genetic characteristics of spinocerebellar ataxia type 19 (SCA19). *Cerebellum* 2005;**4**:51–4.
317. Duarri A, Jezierska J, Fokkens M, Meijer M, Schelhaas HJ, Dunnen den WFA, et al. Mutations in potassium channel kcnd3 cause spinocerebellar ataxia type 19. *Ann Neurol* 2012;**72**:870–80.
318. Lee Y-C, Durr A, Majczenko K, Huang Y-H, Liu Y-C, Lien C-C, et al. Mutations in KCND3 cause spinocerebellar ataxia type 22. *Ann Neurol* 2012;**72**:859–69.
319. Bakalkin G, Watanabe H, Jezierska J, Depoorter C, Verschuuren-Bemelmans C, Bazov I, et al. Prodynorphin mutations cause the neurodegenerative disorder spinocerebellar ataxia type 23. *Am J Hum Genet* 2010;**87**:593–603.
320. Fawcett K, Mehrabian M, Liu Y-T, Hamed S, Elahi E, Revesz T, et al. The frequency of spinocerebellar ataxia type 23 in a UK population. *J Neurol* 2013;**260**:856–9.
321. Schicks J, Synofzik M, Beetz C, Schiele F, Schols L. Mutations in the PDYN gene (SCA23) are not a frequent cause of dominant ataxia in Central Europe. *Clin Genet* 2011;**80**:503–4.
322. Verbeek DS, van de Warrenburg BP, Wesseling P, Pearson PL, Kremer HP, Sinke RJ. Mapping of the SCA23 locus involved in autosomal dominant cerebellar ataxia to chromosome region 20p13-12.3. *Brain* 2004;**127**:2551–7.
323. Jezierska J, Stevanin G, Watanabe H, Fokkens MR, Zagnoli F, Kok J, et al. Identification and characterization of novel PDYN mutations in dominant cerebellar ataxia cases. *J Neurol* 2013;**260**:1807–12.
324. Yu G-Y, Howell MJ, Roller MJ, Xie T-D, Gomez CM. Spinocerebellar ataxia type 26 maps to chromosome 19p13.3 adjacent to SCA6. *Ann Neurol* 2005;**57**:349–54.
325. Hekman KE, Yu G-Y, Brown CD, Zhu H, Du X, Gervin K, et al. A conserved eEF2 coding variant in SCA26 leads to loss of translational fidelity and increased susceptibility to proteostatic insult. *Hum Mol Genet* 2012;**21**:5472–83.
326. van Swieten JC, Brusse E, de Graaf BM, Krieger E, van de Graaf R, de Koning I, et al. A mutation in the fibroblast growth factor 14 gene is associated with autosomal dominant cerebellar ataxia [corrected]. *Am J Hum Genet* 2003;**72**:191–9.
327. Brusse E, de Koning I, Maat-Kievit A, Oostra BA, Heutink P, van Swieten JC. Spinocerebellar ataxia associated with a mutation in the fibroblast growth factor 14 gene (SCA27): A new phenotype. *Mov Disord* 2006;**21**:396–401.
328. Dalski A, Atici J, Kreuz FR, Hellenbroich Y, Schwinger E, Zühlke C. Mutation analysis in the fibroblast growth factor 14 gene: frameshift mutation and polymorphisms in patients with inherited ataxias. *Eur J Hum Genet* 2005;**13**:118–20.
329. Misceo D, Fannemel M, Barøy T, Roberto R, Tvedt B, Jaeger T, et al. SCA27 caused by a chromosome translocation: further delineation of the phenotype. *Neurogenetics* 2009;**10**:371–4.
330. Shimojima K, Okumura A, Natsume J, Aiba K, Kurahashi H, Kubota T, et al. Spinocerebellar ataxias type 27 derived from a disruption of the fibroblast growth factor 14 gene with mimicking phenotype of paroxysmal non-kinesigenic dyskinesia. *Brain Dev* 2012;**34**:230–3.
331. Wang Q, Bardgett ME, Wong M, Wozniak DF, Lou J, McNeil BD, et al. Ataxia and paroxysmal dyskinesia in mice lacking axonally transported FGF14. *Neuron* 2002;**35**:25–38.
332. Yan H, Pablo JL, Pitt GS. FGF14 regulates presynaptic Ca^{2+} channels and synaptic transmission. *Cell Rep* 2013;**4**:66–75.
333. Cagnoli C, Mariotti C, Taroni F, Seri M, Brussino A, Michielotto C, et al. SCA28, a novel form of autosomal dominant cerebellar ataxia on chromosome 18p11.22-q11.2. *Brain* 2006;**129**:235–42.
334. Edener U, Wöllner J, Hehr U, Kohl Z, Schilling S, Kreuz F, et al. Early onset and slow progression of SCA28, a rare dominant ataxia in a large four-generation family with a novel AFG3L2 mutation. *Eur J Hum Genet* 2010;**18**:965–8.
335. Cagnoli C, Stevanin G, Brussino A, Barberis M, Mancini C, Margolis RL, et al. Missense mutations in the AFG3L2 proteolytic domain account for ~1.5% of European autosomal dominant cerebellar ataxias. *Hum Mutat* 2010;**31**:1117–24.
336. Di Bella D, Lazzaro F, Brusco A, Plumari M, Battaglia G, Pastore A, et al. Mutations in the mitochondrial protease gene AFG3L2 cause dominant hereditary ataxia SCA28. *Nat Genet* 2010;**42**: 313–21.
337. Casari G, De Fusco M, Ciarmatori S, Zeviani M, Mora M, Fernandez P, et al. Spastic paraplegia and OXPHOS impairment caused by mutations in paraplegin, a nuclear-encoded mitochondrial metalloprotease. *Cell* 1998;**93**:973–83.
338. Pierson TM, Adams D, Bonn F, Martinelli P, Cherukuri PF, Teer JK, et al. Whole-exome sequencing identifies homozygous

AFG3L2 mutations in a spastic ataxia-neuropathy syndrome linked to mitochondrial m-AAA proteases. *PLoS Genet* 2011;**7**: e1002325.

339. Maltecca F, Magnoni R, Cerri F, Cox GA, Quattrini A, Casari G. Haploinsufficiency of AFG3L2, the gene responsible for spinocerebellar ataxia type 28, causes mitochondria-mediated Purkinje cell dark degeneration. *J Neurosci* 2009;**29**:9244–54.
340. Wang JL, Yang X, Xia K, Hu ZM, Weng L, Jin X, et al. TGM6 identified as a novel causative gene of spinocerebellar ataxias using exome sequencing. *Brain* 2010;**133**:3510–18.
341. Li M, Pang SYY, Song Y, Kung MHW, Ho SL, Sham P-C. Whole exome sequencing identifies a novel mutation in the transglutaminase 6 gene for spinocerebellar ataxia in a Chinese family. *Clin Genet* 2013;**83**:269–73.
342. Guan W-J, Wang JL, Liu Y-T, Ma Y-T, Zhou Y, Jiang H, et al. Spinocerebellar ataxia type 35 (SCA35)-associated transglutaminase 6 mutants sensitize cells to apoptosis. *Biochem Biophys Res Commun* 2013;**430**:780–6.
343. Guan W-J, Xia K-D, Ma Y-T, Liu Y-T, Shi Y-T, Jiang H, et al. Transglutaminase 6 interacts with polyQ proteins and promotes the formation of polyQ aggregates. *Biochem Biophys Res Commun* 2013;**437**:94–100.
344. Flanigan K, Gardner K, Alderson K, Galster B, Otterud B, Leppert MF, et al. Autosomal dominant spinocerebellar ataxia with sensory axonal neuropathy (SCA4): clinical description and genetic localization to chromosome 16q22.1. *Am J Hum Genet* 1996;**59**:392–9.
345. Hellenbroich Y, Bubel S, Pawlack H, Opitz S, Vieregge P, Schwinger E, et al. Refinement of the spinocerebellar ataxia type 4 locus in a large German family and exclusion of CAG repeat expansions in this region. *J Neurol* 2003;**250**:668–71.
346. Hellenbroich Y, Pawlack H, Rüb U, Schwinger E, Zühlke C. Spinocerebellar ataxia type 4. Investigation of 34 candidate genes. *J Neurol* 2005;**252**:1472–5.
347. Onodera Y, Aoki M, Mizuno H, Warita H, Shiga Y, Itoyama Y. Clinical features of chromosome 16q22.1 linked autosomal dominant cerebellar ataxia in Japanese. *Neurology* 2006;**67**:1300–2.
348. Gupta M, Kamynina E, Morley S, Chung S, Muakkassa N, Wang H, et al. Plekhg4 is a novel dbl family Guanine nucleotide exchange factor protein for rho family GTPases. *J Biol Chem* 2013;**288**:14522–30.
349. Wieczorek SS, Arning LL, Alheite II, Epplen JTJ. Mutations of the puratrophin-1 (PLEKHG4) gene on chromosome 16q22.1 are not a common genetic cause of cerebellar ataxia in a European population. *J Hum Genet* 2006;**51**:363–7.
350. Brkanac Z, Spencer D, Shendure J, Robertson PD, Matsushita M, Vu T, et al. IFRD1 is a candidate gene for SMNA on chromosome 7q22-q23. *Am J Hum Genet* 2009;**84**:692–7.
351. Knight MA, Gardner RJM, Bahlo M, Matsuura T, Dixon JA, Forrest SM, et al. Dominantly inherited ataxia and dysphonia with dentate calcification: spinocerebellar ataxia type 20. *Brain* 2004;**127**:1172–81.
352. Knight MA, Hernandez D, Diede SJ, Dauwerse HG, Rafferty I, van de Leemput J, et al. A duplication at chromosome 11q12.2-11q12.3 is associated with spinocerebellar ataxia type 20. *Hum Mol Genet* 2008;**17**:3847–53.
353. Devos D, Schraen-Maschke S, Vuillaume I, Dujardin K, Nazé P, Willoteaux C, et al. Clinical features and genetic analysis of a new form of spinocerebellar ataxia. *Neurology* 2001;**56**:234–8.
354. Vuillaume I, Devos D, Schraen-Maschke S, Dina C, Lemainque A, Vasseur F, et al. A new locus for spinocerebellar ataxia (SCA21) maps to chromosome 7p21.3-p15.1. *Ann Neurol* 2002;**52**:666–70.
355. Delplanque J, Devos D, Vuillaume I, De Becdelievre A, Vangelder E, Maurage CA, et al. Slowly progressive spinocerebellar ataxia with extrapyramidal signs and mild cognitive impairment (SCA21). *Cerebellum* 2008;**7**:179–83.
356. Stevanin G, Bouslam N, Thobois S, Azzedine H, Ravaux L, Boland A, et al. Spinocerebellar ataxia with sensory neuropathy (SCA25) maps to chromosome 2p. *Ann Neurol* 2004;**55**:97–104.
357. Stevanin G, Broussolle E, Streichenberger N, Kopp N, Brice A, Durr A. Spinocerebellar ataxia with sensory neuropathy (SCA25). *Cerebellum* 2005;**4**:58–61.
358. Storey E, Bahlo M, Fahey M, Sisson O, Lueck CJ, Gardner RJM. A new dominantly inherited pure cerebellar ataxia, SCA 30. *J Neurol Neurosurg Psychiatry* 2009;**80**:408–11.
359. Jiang H, Zhu HP, Gomez CM. SCA32: an autosomal dominant cerebellar ataxia with azoospermia maps to chromosome 7q32-q33. *Mov Disord* 2010;**25**:S192.
360. Giroux JM, Barbeau A. Erythrokeratodermia with ataxia. *Arch Dermatol* 1972;**106**:183–8.
361. Turcotte Gauthier M. Étude clinique et génétique d'une nouvelle forme d'ataxie spinocerebelleuse pure associee a l'erythrokeratodermie. 2010. Available from: <http://hdl.handle.net/1866/4281>.
362. Griggs RC, Moxley RT, Lafrance RA, McQuillen J. Hereditary paroxysmal ataxia: response to acetazolamide. *Neurology* 1978;**28**:1259–64.
363. Graves TD, Rajakulendran S, Zuberi SM, Morris HR, Schorge S, Hanna MG, et al. Nongenetic factors influence severity of episodic ataxia type 1 in monozygotic twins. *Neurology* 2010;**75**:367–72.
364. Browne DL, Gancher ST, Nutt JG, Brunt ER, Smith EA, Kramer P, et al. Episodic ataxia/myokymia syndrome is associated with point mutations in the human potassium channel gene, KCNA1. *Nat Genet* 1994;**8**:136–40.
365. Adelman JP, Bond CT, Pessia M, Maylie J. Episodic ataxia results from voltage-dependent potassium channels with altered functions. *Neuron* 1995;**15**:1449–54.
366. Herson PS, Virk M, Rustay NR, Bond CT, Crabbe JC, Adelman JP, et al. A mouse model of episodic ataxia type-1. *Nat Neurosci* 2003;**6**:378–83.
367. Strupp M, Kalla R, Dichgans M, Freilinger T, Glasauer S, Brandt T. Treatment of episodic ataxia type 2 with the potassium channel blocker 4-aminopyridine. *Neurology* 2004;**62**:1623–5.
368. Strupp M, Kalla R, Claassen J, Adrion C, Mansmann U, Klopstock T, et al. A randomized trial of 4-aminopyridine in EA2 and related familial episodic ataxias. *Neurology* 2011;**77**:269–75.
369. Jen J, Kim GW, Baloh RW. Clinical spectrum of episodic ataxia type 2. *Neurology* 2004;**62**:17–22.
370. Walter JT, Alviña K, Womack MD, Chevez C, Khodakhah K. Decreases in the precision of Purkinje cell pacemaking cause cerebellar dysfunction and ataxia. *Nat Neurosci* 2006;**9**:389–97.
371. Steckley JL, Ebers GC, Cader MZ, McLachlan RS. An autosomal dominant disorder with episodic ataxia, vertigo, and tinnitus. *Neurology* 2001;**57**:1499–502.
372. Cader MZ, Steckley JL, Dyment DA, McLachlan RS, Ebers GC. A genome-wide screen and linkage mapping for a large pedigree with episodic ataxia. *Neurology* 2005;**65**:156–8.

373. Farmer TW, Mustian VM. Vestibulocerebellar ataxia. A newly defined hereditary syndrome with periodic manifestations. *Arch Neurol* 1963;**8**:471–80.
374. Vance JM, Pericak-Vance MA, Payne CS, Coin JT. Linkage and genetic analysis in adult onset periodic vestibulo-cerebellar ataxia: report of a new family. *Am J Hum Genet* 1984;**36**:78S.
375. Damji KF, Allingham RR, Pollock SC, Small K, Lewis KE, Stajich JM, et al. Periodic vestibulocerebellar ataxia, an autosomal dominant ataxia with defective smooth pursuit, is genetically distinct from other autosomal dominant ataxias. *Arch Neurol* 1996;**53**:338–44.
376. Escayg A, De Waard M, Lee DD, Bichet D, Wolf P, Mayer T, et al. Coding and noncoding variation of the human calcium-channel beta4-subunit gene CACNB4 in patients with idiopathic generalized epilepsy and episodic ataxia. *Am J Hum Genet* 2000;**66**:1531–9.
377. Jen JC, Wan J, Palos TP, Howard BD, Baloh RW. Mutation in the glutamate transporter EAAT1 causes episodic ataxia, hemiplegia, and seizures. *Neurology* 2005;**65**:529–34.
378. de Vries B, Mamsa H, Stam AH, Wan J, Bakker SLM, Vanmolkot KRJ, et al. Episodic ataxia associated with EAAT1 mutation C186S affecting glutamate reuptake. *Arch Neurol* 2009;**66**:97–101.
379. Winter N, Kovermann P, Fahlke C. A point mutation associated with episodic ataxia 6 increases glutamate transporter anion currents. *Brain* 2012;**135**:3416–25.
380. Kerber KA, Jen JC, Lee H, Nelson SF, Baloh RW. A new episodic ataxia syndrome with linkage to chromosome 19q13. *Arch Neurol* 2007;**64**:749–52.
381. Coutinho P, Guimarães A, Scaravilli F. The pathology of Machado-Joseph disease. Report of a possible homozygous case. *Acta Neuropathol* 1982;**58**:48–54.
382. ME M, D A. Personal communication via Dr. Basil Darras, 2001. n.d.
383. Saitoh S, Momoi MY, Yamagata T, Miyao M, Suwa K. Clinical and electroencephalographic findings in juvenile type DRPLA. *Pediatr Neurol* 1998;**18**:265–8.
384. Woodworth JA, Beckett RLS, Netsky MG. A composite of hereditary ataxias: a familial disorder with features of olivopontocerebellar atrophy, Leber's optic atrophy, and Friedreich's ataxia. *AMA Arch Intern Med* 1959;**104**:594–606.
385. Sailer A, Scholz SW, Gibbs JR, Tucci A, Johnson JO, Wood NW, et al. Exome sequencing in an SCA14 family demonstrates its utility in diagnosing heterogeneous diseases. *Neurology* 2012;**79**:127–31.
386. Hersheson J, Haworth A, Houlden H. The inherited ataxias: genetic heterogeneity, mutation databases, and future directions in research and clinical diagnostics. *Hum Mutat* 2012;**33**: 1324–32.
387. Németh AH, Kwasniewska AC, Lise S, Parolin Schnekenberg R, Becker EBE, Bera KD, et al. Next generation sequencing for molecular diagnosis of neurological disorders using ataxias as a model. *Brain* 2013;**136**:3106–18.
388. Tan E-K. *Genetics: Utility of Next-Generation Sequencing in Ataxias*. Nature Publishing Group; 2013.
389. Shatunov A, Fridman EA, Pagan FI, Leib J, Singleton A, Hallett M, et al. Small de novo duplication in the repeat region of the TATA-box-binding protein gene manifest with a phenotype similar to variant Creutzfeldt-Jakob disease. *Clin Genet* 2004;**66**:496–501.
390. De Rosa A, Striano P, Barbieri F, de Falco A, Rinaldi C, Tucci T, et al. Suppression of myoclonus in SCA2 by piracetam. *Mov Disord* 2006;**21**:116–18.
391. Velázquez-Pérez L, Rodríguez-Labrada R, Álvarez-González L, Aguilera-Rodríguez R, Álvarez Sánchez M, Canales-Ochoa N, et al. Lisuride reduces involuntary periodic leg movements in spinocerebellar ataxia type 2 patients. *Cerebellum* 2012;**11**: 1051–6.
392. Miyai I, Ito M, Hattori N, Mihara M, Hatakenaka M, Yagura H, et al. Cerebellar ataxia rehabilitation trial in degenerative cerebellar diseases. *Neurorehabil Neural Repair* 2012;**26**:515–22.
393. Ilg W, Synofzik M, Brötz D, Burkard S, Giese MA, Schols L. Intensive coordinative training improves motor performance in degenerative cerebellar disease. *Neurology* 2009;**73**:1823–30.

Part VIII

General Therapeutic Principles

Chapter 51

Principles and Practice of Molecular Therapies

Louise R. Rodino-Klapac and Jerry R. Mendell

INTRODUCTION

Treatment for the muscular dystrophy group of disorders continues to gain momentum based on improvements in therapeutic strategies, facilitated by a greater availability of funding for translational research. This review emphasizes the progress made for Duchenne muscular dystrophy (DMD) and related dystrophinopathies as well as the limb girdle muscular dystrophies (LGMDs). Clinical gene therapy trials for DMD, Becker muscular dystrophy (BMD), and LGMDs are underway.[1–3] Initial clinical trials used direct intramuscular gene delivery and are now moving forward in hopes of producing clinical benefit through regional vascular delivery.[4] Proof of principle studies have also shown that packaging genes exceeding the usual limits of adeno-associated virus (AAV) look promising and expand the disease repertoire previously thought to limit clinical translation.[5] Exon skipping using a phosphorodiamidate morpholino oligomer (PMO) by all measures enhances dystrophin expression and prevents disease progression.[6] It is an exciting time in muscle disease therapeutics with continued expansion of skills to overcome treatment obstacles.

NEWBORN SCREENING FOR MUSCULAR DYSTROPHY: THE KEY TO MAKING A THERAPEUTIC DIFFERENCE

It has become apparent, perhaps now more than ever, that our best chance of helping patients with muscular dystrophy is to treat as early as possible.[7] We have direct evidence that DMD is present at or shortly after birth. Creatine kinase (CK) is elevated at the time of delivery (>2000 U/L)[8] and functional motor assessment demonstrates low scores measured by the Bayley-III motor functional scale for gross motor and fine motor function.[9] In the neonatal period, most DMD boys perform more than a standard deviation below the mean and show loss of function and less maturational improvement with age. The time to treat DMD is at this early age, before muscle degeneration presents a challenge that is clearly difficult to overcome.

The stage for early intervention was set through an Ohio newborn screening study with the development of a 2-tier system for identifying DMD boys.[8] In a cohort of nearly 40,000 newborns, using the dried blood spot card otherwise taken for mandated testing (regulated state by state), additional blood spots were available for both CK and DNA testing. If the CK was found to be 3 standard deviations above normal, DNA was extracted for whole genome amplification with assessment of single-/multi-exon deletions/duplications in the *DMD* gene using multiplex ligation-dependent probe amplification (MLPA) and for point mutations. When applied throughout the state of Ohio, 6 single-exon or multi-exon deletions were found (Table 51.1). The Ohio system is more applicable than the screening paradigm previously used in virtually all other countries in the world, where a single-tier system had been used. In these countries, the single-tier approach first determines CK levels at birth; if elevated, infants are retested at 6 to 8 weeks of age or even later for validation. If the CK remains high, *DMD* gene mutation analysis is performed at the follow-up appointment. This system is most feasible in countries with a nationalized health care system with the capacity to track newborns through follow-up care. However, it is poorly adapted to a system without nationalized health care and where hospital discharges are planned within 24 to 48 hours postdelivery. In addition, privatized medicine brings less certainty for

B.T. Darras, H. Royden Jones, Jr., M.M. Ryan & D.C. De Vivo (Eds): Neuromuscular Disorders of Infancy, Childhood and Adolescence, Second edition.
DOI: http://dx.doi.org/10.1016/B978-0-12-417044-5.00051-2

time and place of follow-up care. It is noteworthy that an International Workshop on Newborn Screening for DMD also advocated the 2-tier system of screening.[10] Also of interest, the latest estimates of incidence of DMD in newborn males based on worldwide studies is approximately 1:5000, replacing the older estimates of 1:3500 (Table 51.2).

TABLE 51.1 DMD Boys Identified in Newborn Screening Study in the State of Ohio

Gender	CK U/L	Gene	Mutation	Frame
Male	2462	*DMD*	Del Ex 50	Out
Male	2675	*DMD*	Del Ex 5–41	In
Male	2003	*DMD*	Del Ex 8–9	Out
Male	2466	*DMD*	Del Ex 45	Out
Male	2791	*DMD*	Del Ex 45–48	Out
Male	2688	*DMD*	Del Ex 4–7	Out

GENE REPAIR THROUGH EXON SKIPPING AND MUTATION SUPPRESSION

Exon Skipping Using Antisense Oligonucleotides

DMD has been the disease target for exon skipping. It is the most common, severe childhood form of muscular dystrophy. The very large size of the DMD gene[11] results in spontaneous mutations and an unending trail of new cases and new carriers, emphasizing the compelling need to find a treatment. The importance of the early natural history data cannot be underestimated, establishing outcomes for clinical trials when intervention may be most advantageous.[12] On average, however, most DMD boys are not diagnosed until 4 to 5 years of age, and we know that muscle biopsies at this time point show florid hallmarks of the dystrophic process with fiber loss, necrosis, and fibrosis.

The response to glucocorticoids is important to emphasize. DMD is the only dystrophy showing unequivocal evidence of functional benefit to steroids, convincingly

TABLE 51.2 Incidence of Newborn Males Detected by Screening Studies Worldwide

Year	Location	Number Screened	Incidence
1979	New Zealand	10,000 2 DMD	1:5000
1986	W. Germany	358,000 78 DMD	1:4589
1988	Manitoba, CA	54,000 10 DMD	1:5400
1989	Lyon, Fr	37,312 7 DMD	1:5330
1991	W. Pennsylvania, USA	49,000 10 DMD	1:4900
1998	Cyprus	30,219 5 DMD	1:6002
2006	Antwerp, Be	281,214 51 DMD	1:5500
2011	Wales, UK	335,045 63 DMD	1:5266
2013	Ohio, USA	37,749 6 DMD	1:6291

*The only active newborn screening program is Antwerp; the approximate incidence of DMD newborn males is 1:5000.[9,10]

established in a randomized, double-blind, controlled prednisone trial.[13] Improved extremity muscle strength and function were observed, including time to rise from supine to standing, walking 9 m, and climbing stairs. Similar results were later reported with deflazacort, an alternative, sodium-sparing glucocorticoid.[14] A high-dose, weekend regimen of prednisone (10 mg/kg/wk: half on Saturday and half on Sunday) showed equal efficacy to daily steroids.[15] Glucocorticoids are accepted as the standard of care and are acceptable and encouraged for most clinical trials testing novel therapies such as exon skipping agents. It is generally accepted that glucocorticoids will extend ambulation by 2 years in the DMD population, usually wheelchair dependent by 10 to 12 years of age.

Exon skipping is targeted at the pre-mRNA level, allowing one or more exons to be omitted to restore the dystrophin reading frame. This is accomplished with splice-switching oligomers, typically 20 to 30 nucleotides in length and complementary in sequence to regions of the pre-mRNA transcript. Preclinical efficacy has been demonstrated in the mdx, dystrophin/utrophin knockout mouse, and the dystrophin-deficient dog using both 2′O-methyl-ribo-oligonucleoside-phosphorothioate (2′OMePS) and phosphorodiamidate morpholino oligomers (PMOs).[16–18] The objective of exon skipping is to restore the disrupted reading frame of the DMD gene with the production of functional dystrophin, recognizing that it will not be full length and clinically may result in a Becker-like phenotype. Clinically, it is estimated that 70% to 80% of DMD mutations can be corrected by exon skipping. The early emphasis for clinical trials has been to target the hotspot region for deletions between exons 42 and 55, where mutations are generally amenable to exon skipping. Exon 51 skipping is applicable to the largest group of all DMD patients (13%) inclusive of deletions: 45–50, 47–50, 48–50, 49–50, 50; 52, or 52–63.[19]

Proof-of-principle clinical trials in DMD used a 2′OMePS oligomer (PRO051/GSK2402968, Prosensa Therapeutics) or a PMO (AVI-4658/eteplirsen®; Sarepta Therapeutics) delivered directly to muscle, targeting exon 51.[20,21] Using either oligomer, exon skipping was validated by RT-PCR and by demonstrating newly synthesized dystrophin protein correctly localized to the sarcolemma. Phase I/II extension studies were performed using systemically delivered oligomers. In the PRO051 trial, dose-related efficacy was achieved with evidence of new dystrophin expression in approximately 60% to 100% of muscle fibers in 10 of 12 patients with modest improvement in the 6-minute walk test (6MWT).[22] In the phase II open-label study with AVI-4658 (eteplirsen®), 7 of 19 patients saw a modest response with a mean increase of sarcolemmal dystrophin from 8.9% to 16.4%.[23] The dosing results were variable with one patient responding to eteplirsen treatment with 2 mg/kg, while all other responders received the higher dosing levels of 10 and 20 mg/kg.

In the only double-blind randomized controlled trial of exon skipping using eteplirsen, high- and low-dose treated cohorts were compared to placebo for the first 24 weeks, after which all boys were on treatment.[6] Dystrophin expression and distance walked on the 6MWT were outcome measures. Enrollment included DMD boys, ages 7 to 13, with confirmed DMD gene deletions correctable by skipping exon 51. All boys were on a stable dose of glucocorticoids (prednisone or deflazacort) for at least 6 months. The design of the trial is depicted in Figure 51.1. The results of dystrophin production at all time points are shown in Figure 51.2. At 12 weeks' posttreatment, no significant dystrophin was found. Dystrophin production could be seen at week 24, and this further quantitatively increased after 48 weeks of eteplirsen. These findings are a clear indication that dystrophin production started sometime after week 12 and was more influenced by duration rather than dose of treatment. We also found restoration of sarcoglycans and nNOS binding accompanying dystrophin production (Figure 51.3). The distance walked in the 6MWT was remarkably stable for boys taking eteplirsen from the start of the study without decline. In contrast the placebo cohort declined until week 36, consistent with natural history studies.[24,25] Their walking distance stabilized 12 weeks after starting eteplirsen, consistent with dystrophin production; however, there had been a total decline in distance walked of 67.3 m compared to the treated group ($p \leq 0.001$) (Figure 51.4). An important lesson from the trial was the need to modify enrollment criteria for future exon-skipping trials; two treated patients had to be excluded from the final account of walking distance because they lost ambulation prior to the initiation of dystrophin production by exon skipping. Their pretreatment profile (the tallest with the lowest baseline 6MWT values)[25,26] predicted a rapid decline different from that of all others enrolled in the study.

The safety profile of eteplirsen was impeccable. Not a single significant adverse event occurred following treatment. The study has now extended for more than 1.5 years, and stability and safety continue without deviation. Two patients had transient proteinuria, but with rechecking this was no longer present. This very safe and promising method of treatment is now extending to include other exons.[27] None is as inclusive as targeting exon 51, but collectively a large part of the DMD population can be reached using this strategy.

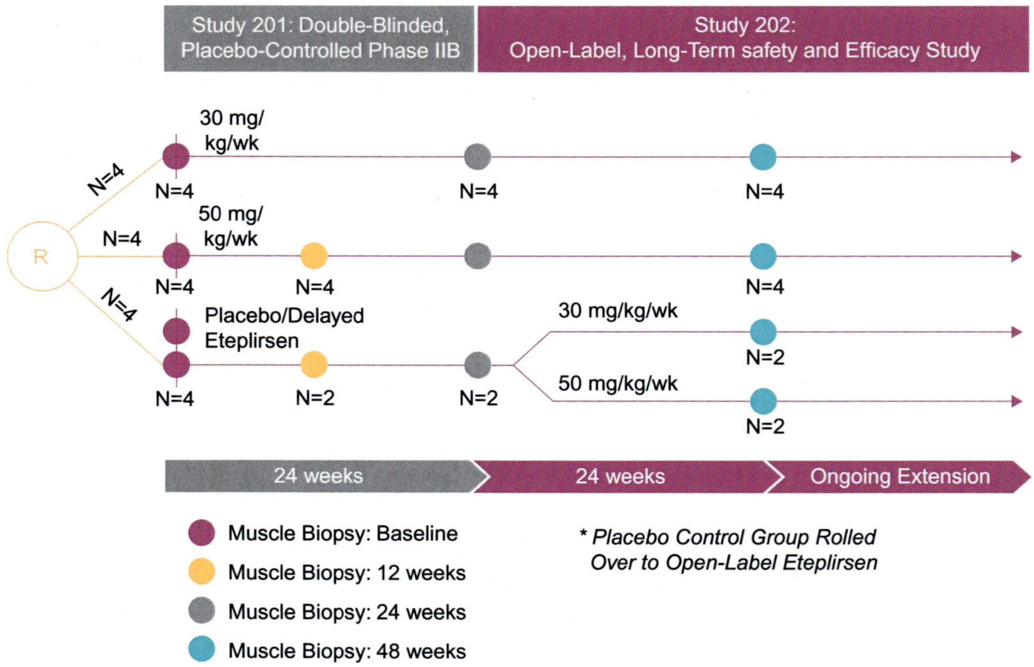

FIGURE 51.1 Eteplirsen exon skipping study design. Twelve subjects with DMD were randomized (R) to 1 of 3 eteplirsen-treated cohorts in study 201: cohort 1 received 30 mg/kg/wk; cohort 2, 50 mg/kg/wk; and cohort 3 was placebo-treated. At week 25, cohort 3 switched to open-label treatment, either 30 or 50 mg/kg/wk; thereafter referred to as "placebo-delayed." Patients were maintained on the same starting dose of eteplirsen under the open label extension Study 202. Biceps biopsies were obtained on all patients at baseline and deltoid biopsies at week 48 for analysis of dystrophin expression. At week 12 biceps biopsies were obtained from patients in cohort 2 (50 mg/kg/wk) and 2 placebo-treated patients in cohort 3. At week 24 biceps biopsies were obtained from patients in cohort 1 (30 mg/kg/wk) and 2 placebo-treated patients in cohort 3. *Reprinted from Mendell et al., 2013.*[6]

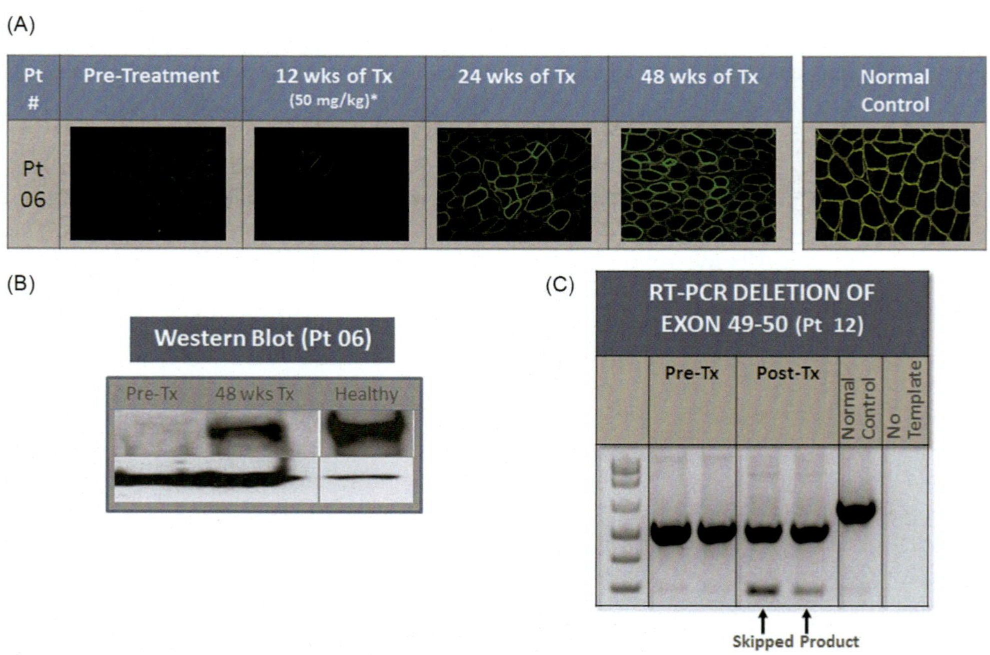

FIGURE 51.2 Effects of eteplirsen on dystrophin expression. **(A)** Representative example of time dependent increases in dystrophin-positive fibers throughout the study can be seen. Dystrophin expression begins between week 12 and week 24 of eteplirsen treatment, and continues to increase after that time point. **(B)** Western blot from a representative patient demonstrates measurable dystrophin at week 48 compared to baseline. **(C)** RT-PCR demonstrates presence of skipped mRNA product (289 bp) following eteplirsen treatment in the muscle of a representative patient with exons 49–50 deletion. *Modified from Mendell et al., 2013.*[6]

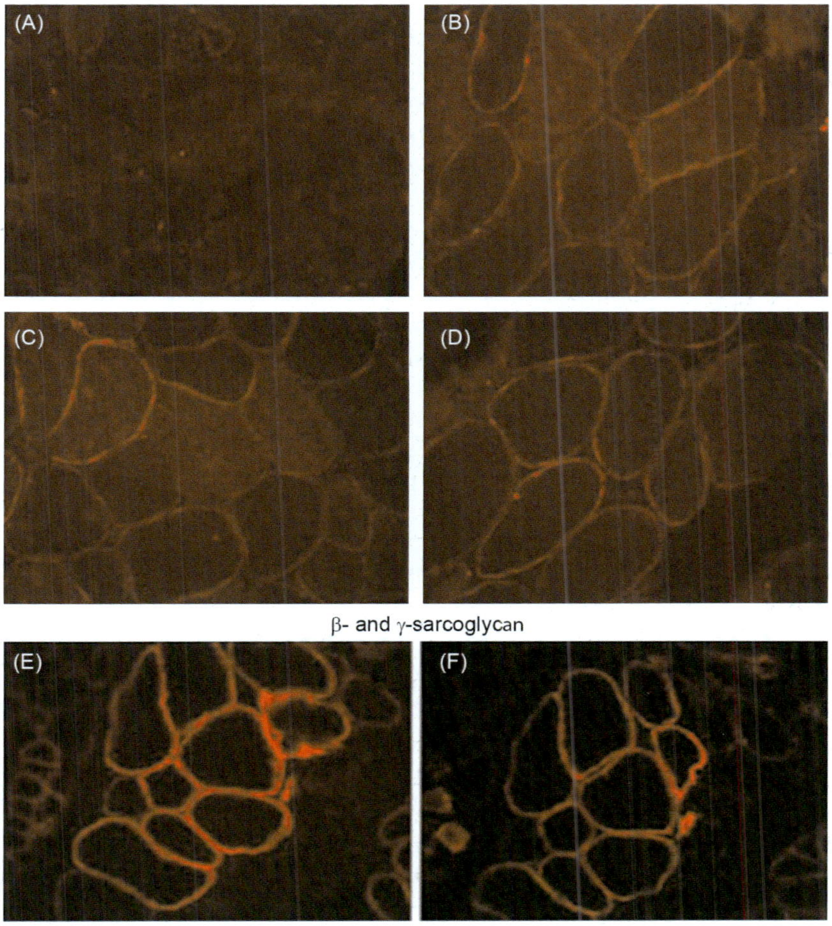

FIGURE 51.3 Neuronal nitric oxide synthase and sarcoglycans confirm dystrophin expression. **(A)** Neuronal nitric oxide synthase (nNOS) shows no staining in muscle from a DMD patient with deletion in exon 52. **(B, D)** Restoration of nNOS binding site following eteplirsen treatment from patient 6 at week 48. **(C)** Normal muscle showing nNOS expression (no treatment). **(E)** β-Sarcoglycan and **(F)** γ-sarcoglycan staining at week 48 (patient 6), demonstrating restoration of the sarcoglycan complex with eteplirsen. *Reprinted from Mendell et al., 2013.*[6]

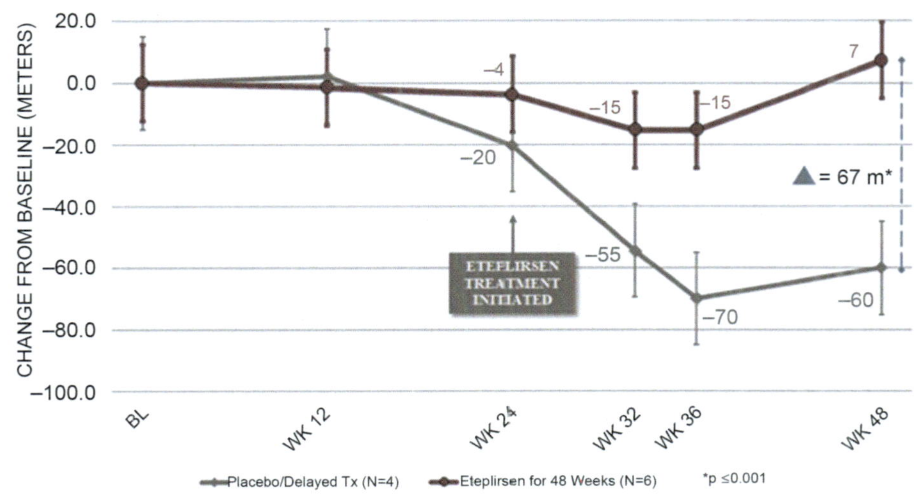

FIGURE 51.4 Functional efficacy of eteplirsen. The dark purple line shows the change from baseline (BL) in distance walked on the 6-minute walk test (6MWT) over time for the 6 evaluable patients who received eteplirsen from the start of Study 201. The dark gray line shows change from baseline in distance walked on the 6MWT for the 4 patients who received placebo for the first 24 weeks and eteplirsen for the last 24 weeks. It is estimated from dystrophin expression studies and multiple biopsies that significant dystrophin was produced after 12 weeks. This is clearly indicated in the placebo/delayed cohort starting on eteplirsen at week 25 and stabilizing in function at week 36. The eteplirsen-treated cohorts diverged from placebo/delayed patients at week 12, after which they no longer showed significant decline. Two of the treated patients starting at week 1 lost ambulation before they produced significant dystrophin and are not represented on this graph. *Reprinted from Mendell et al., 2013.*[6]

IGNORING STOP CODONS USING SMALL MOLECULES

Aminoglycoside Therapy

Stop codons account for approximately 13% to 15% of DMD mutations.[28,29] Effective treatment of this group has potential impact equivalent to exon skipping. Encouraging results were originally obtained in *mdx* mice where the aminoglycoside, gentamicin, successfully increased dystrophin expression to about 20%.[30] Gentamicin-treated mice showed improved histology and protection against contraction-induced injury. Plasma CK levels were also reduced by 70%. Molecular studies of mutation suppression in the *mdx* mouse served as a stimulus to attempt a similar approach using gentamicin in DMD patients. The initial clinical trial reported in four DMD patients carrying premature termination codons in the *DMD* gene reported no improvement in clinical measures or dystrophin expression. Of interest, serum CK dropped significantly.[31] A subsequent gentamicin trial reported more favorable results in a small cohort of four subjects.[32] When gentamicin was given over two cycles, each for 6 days with a 7-week hiatus between dosing, muscle biopsy revealed increased dystrophin expression in 3 out of 4 patients. Serum CK decreased from day 2 to day 6 of the treatment but subsequently came back to the initial levels. No clinically meaningful outcomes were obtained in any of these patients.

A third clinical trial conducted by our group at the Nationwide Children's Hospital consisted of a two-phase protocol.[33] An initial short-term gentamicin, 14-day dosing study was carried out in 10 DMD boys with proven stop codon mutations. A biological effect was established with a 50% reduction of CK by day 14. The results were stop codon-dependent since serum CK did not change in a control group with frameshift mutations. Having demonstrated a biological effect in the short-term trial, a 6-month study was designed with weekly or twice-weekly gentamicin administration. Enrollment was dependent on normal hearing and kidney function, and negative testing for the known A1555G mitochondrial DNA mutation in the 12S rRNA gene predisposing to aminoglycoside-induced hearing loss.[34] Renal function was monitored using cystatin C to avoid false negative findings related to decreased serum creatinine and creatinine clearance levels in the DMD population.[35] Sixteen subjects with documented stop codon mutations received weekly ($N = 12$) or twice weekly ($N = 4$) gentamicin. No patient demonstrated a decline in renal or hearing function except one patient given a miscalculated dose (125% of recommended) for the first 4 administrations, resulting in a transient high-frequency hearing loss that returned to normal within 3 months. Owing to the small sample size, weekly and biweekly treated patients were analyzed as one group. Twelve subjects agreed to pretreatment and posttreatment muscle biopsies. In 3 of 9 subjects, dystrophin protein increased to 13.0–15% of wild-type levels (Figure 51.5). In addition to dystrophin expression in muscle tissue, serum CK was mildly reduced after 6 months of gentamicin treatment (9851 ± 2109 U/L to 5316 ± 850 U/L, $p = 0.04$). Additional clinical outcome measures were variable. The average muscle score (AMS), which was expected to decline by 0.2 units over 6 months[13] showed little or no change (pretreatment AMS 5.37 ± 0.39 compared with posttreatment AMS 5.35 ± 0.38). However, no functional outcomes were improved, such as the time to walk 30 feet, climb four standard steps, or stand from a supine position on the floor.

Overall, the clinical trial validates that mutation suppression or readthrough can potentially be translated to patients but would require more frequent or higher dosing to produce clinically meaningful results.

PTC124 (Ataluren®)

Ataluren is a small molecule developed by PTC Therapeutics in an effort to advance an orally bioavailable product to bypass nonsense mutations and avoid potential renal- and ototoxicity of aminoglycosides. It was originally developed by means of an optimized high-throughput screening campaign. Dose dependent readthrough of all 3 nonsense codons was observed with the highest readthrough at UGA, followed by UAG and then UAA. PTC124 was a more potent nonsense-suppressing agent than gentamicin. Treatment of *mdx* mice restored dystrophin production in all skeletal muscles examined, including the diaphragm, and cardiac muscle. The dystrophin levels were found to be 20% to 25% those of control mouse muscles, and partially restored force generation and resistance against eccentric exercise, suggesting that PTC124 reduces muscle fragility.[36] PTC-treated *mdx* mice exhibited increased levels of sarcoglycans consistent with stabilization of the dystrophin-associated proteins. Levels of serum CK were reduced by treatment with PTC124. The rescue of skeletal muscle in *mdx* mice was seen within 2 to 8 weeks of exposure. Readthrough by PTC124 was selective and specific to disease-causing premature termination without evidence of changes in the ribosomal readthrough of normal stop codons.

Phase 1 studies of PTC124 in healthy adult volunteers assessed safety and tolerability, concluding that the drug was well tolerated with the exception of mild headaches, dizziness, and gastrointestinal discomfort at high dose. A phase IIb double-blind, placebo-controlled, dose-ranging, efficacy and safety study failed to demonstrate improvement in the 6MWT. Further detailed analysis of the Phase IIb data, however, suggested the possibility of a bell-shaped dose-response with clinical benefit to DMD boys receiving low dose, and in high-dose patients in whom serum levels of PTC124 were low.[37] In view of these tentative findings,

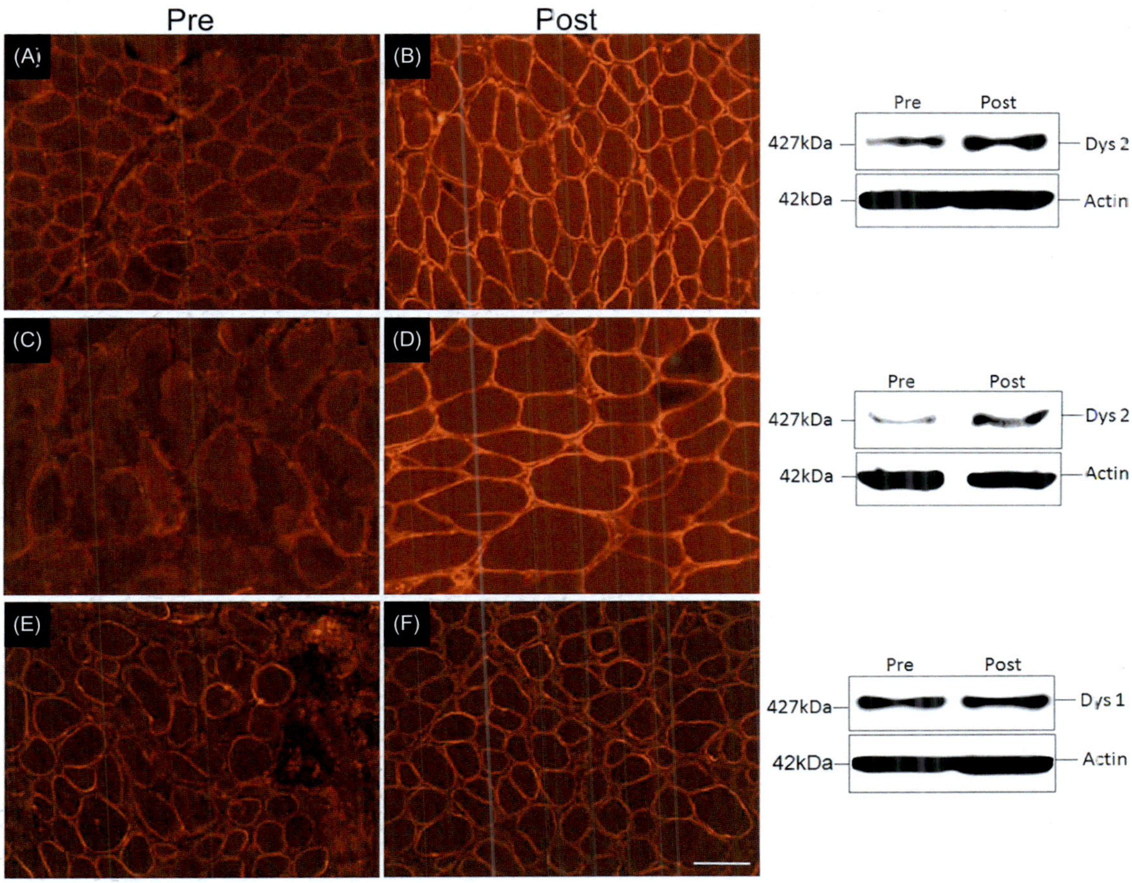

FIGURE 51.5 Dystrophin expression following gentamicin treatment. Pre- and post-gentamicin treatment muscle biopsies demonstrating dystrophin expression by immunofluorescence (IF) using DYS 2 Novacastra and Western blot (WB). (**A**, **B**) Pre- and post-treatment IF muscle biopsy sections (patient 6) stained with DYS2 antibody; adjacent WB shows full-length dystrophin at 427 kDa. Gentamicin–induced dystrophin expression to 15.44%. (**C**, **D**) Pre- and post-treatment muscle biopsy (patient 10) IF sections stained with DYS2 antibody; adjacent WB shows full-length dystrophin at 427 kDa. Gentamicin-induced dystrophin increased to 13.0%. (**E**, **F**) Pre- and post-treatment muscle biopsy sections from patient 5 without response to gentamicin showing negligible increase after treatment by IF and WB. Scale bar = 100 μm. *Reprinted from Malik et al., 2010.*[33]

PTC124 has moved ahead with a phase III, multicenter, randomized, double-blind, placebo-controlled study to determine efficacy and safety of ataluren in DMD boys 7 to 16 years of age with known stop codons. The study is currently underway and results are not yet available (ClinicalTrials.gov identifier: NCT01826487).

GENE REPLACEMENT STRATEGIES TARGETING SPECIFIC MUSCULAR DYSTROPHIES

Replacement of the DMD Gene for Duchenne Muscular Dystrophy

Dystrophin is a very large protein translated by the largest gene in the human genome (> 2.4 MBP) with an mRNA of 14 kb (11 kb coding region from start codon to stop codon of *DMD* gene).[11] Unfortunately, the size of the coding region creates problems for gene delivery using adeno-associated virus (AAV), a small vector with a cloning capacity of ≤ 5 kb.[38] Offsetting this limiting feature are clinical observations that large, in-frame deletions of the central rod domain often lead to a mild form of dystrophin deficiency, Becker muscular dystrophy (BMD).[39] These clinical observations set the stage for the construction of small dystrophins (mini- or micro-) to protect against the dystrophic process.[40] Most of the reduction in size is achieved by deletion of all but 4 or 5 spectrin repeats and 3 hinge regions in the rod domain, and preservation of the cysteine rich (CR) domain. Further size reduction can be achieved by removal of the C-terminal (CT) without forfeiting function. Preclinical studies in the *mdx* mouse using a mini-dystrophin gene (Δ3990 kb) under control of a cytomegalovirus (CMV) set the stage for clinical translation.[41,42] Robust and sustained gene expression was observed for 6 months, with protection of muscle fibers from uptake of Evans blue dye, and an increase in isometric force and resistance to contraction-induced injury.

In March 2006, six boys with frame-shifting deletions of the dystrophin gene were enrolled in the first clinical gene therapy trial for DMD at Nationwide Children's Hospital.[1] At enrollment, four patients were receiving standard glucocorticoid therapy that was continued throughout the course of the study. Asklepios Biopharmaceutical manufactured a modified serotype 2 capsid of AAV with insertion of 5 amino acids from AAV1 (rAAV2.5.CMV.Δ3990). The study design was randomized, double blind, and controlled, with gene transfer to one biceps while the opposite side received saline (cohort 1, N = 3) or empty capsids (cohort 2, N = 3). Patients 1, 2, and 3 received low dose vector (2.0×10^{10} vector genomes/kg body weight); patients 4, 5, and 6 received a dose 5-fold higher (1.0×10^{11} vg/kg body weight). Muscle biopsy specimens were assessed on day 42 (patients 1, 3, 4, and 6) or day 90 (patients 2 and 5). Transgene expression was virtually undetectable on the side of treatment with the exception of 4 dystrophin-positive fibers in patient 3 and one dystrophin-positive fiber in patient 6. Failure of expression was not due to vector delivery since all patients showed vector DNA in treated muscles.

The surprising finding in this clinical trial was a positive T-cell response in several patients in the clinical trial. Peripheral blood mononuclear cells (PBMCs) were tested in an interferon-γ enzyme-linked immunosorbent spot (ELISpot) assay for reactivity to mini-dystrophin. Three synthetic peptide pools spanning the entire mini-dystrophin sequence (designated MDP1, MDP2, and MDP3) were used to stimulate PBMCs in the ELISpot assay. Patient 2 showed a robust T-cell response against MDP2 beginning on the 15th day post gene transfer (Figure 51.6). The response peaked on day 30, slowly declined during the next 3 months, and was intermittently positive after day 120.

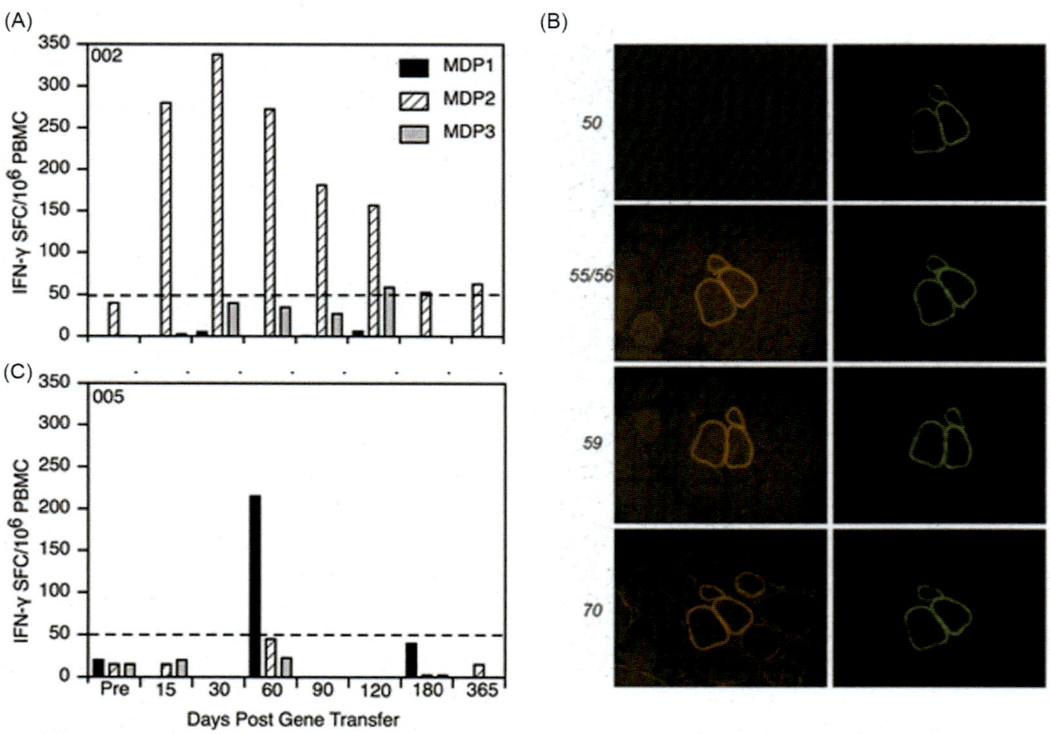

FIGURE 51.6 Dystrophin-specific T-cell response. Peripheral blood mononuclear cells (PBMCs) were tested in an interferon-γ enzyme-linked immunosorbent spot (ELISpot) assay for reactivity to mini-dystrophin in two patients in the clinical DMD gene therapy trial. Three synthetic peptide pools spanning the entire mini-dystrophin sequence (designated MDP1, MDP2, and MDP3) were used to stimulate PBMCs interferon-γ T cells in the ELISpot assay. **(A)** Patient with exon 50 deletion showed a showed a robust T-cell response against MDP2 beginning on the 15th day post gene transfer. The response peaked on day 30 and slowly declined during the next 3 months. **(B)** An unexpected mechanism of immunity was found for this patient with an exon 50 frame-shifting deletion. The immune response was triggered from amino acids 2809 to 2829 of exon 57, downstream from the patient's deletion. Exon specific antibodies against dystrophin showed that exon 57 was expressed in the revertant cluster (antibody stains showed expression of exons 55–56 and 59 implying that exon 57 was also expressed). **(C)** In a different patient with exons 49 through 54 deleted, T cells were found directed against exon 59, downstream of the frame-shifting mutation. These patients illustrate that misfolded protein expressed on revertant fibers can trigger a T-cell immune response. *From Mendell et al., 2010.*[43]

An unexpected mechanism of immunity was found in this patient. An MDP2-specific interferon-γ T-cell response was triggered by amino acids 2809 to 2829 of exon 57, downstream from the patient's exon-50 deletion. The patient's mutation predicted a frame-shift, thereby impeding dystrophin expression. However, in the muscle biopsy from patient 2, revertant fibers were visualized with the use of exon-specific antibodies against dystrophin expressed from exons 55–56 and 59 (Figure 51.6), implying that exon 57 was expressed in the correct reading frame. Thus, T cells were primed from exon 57; additionally, we were able to demonstrate a robust interferon-γ response prior to treatment in subsequent studies on PBMCs saved from a previous blood draw. A T-cell response was also seen in another patient on day 60.

In cohort 2, a different mechanism of immunity was elicited in patient 5 exhibiting a delayed, transient T-cell response detecting 3 discrete dystrophin epitopes, all in the region of the patient's deletion: an exon 7 epitope recognized CD8+ T cells; an epitope that localized to exon 8 was recognized by CD4+ T cells, and another in exon 6 recognized by CD4+ T cells. Because exons 3 through 17 were deleted from the dystrophin gene in this patient, it was clear that CD4+ and CD8+ T cells were primed by mini-dystrophin that expressed exons 3 through 12 in the region of the patient's deletion.

The findings from this study are important as we move forward with gene therapy for DMD. First we learned that we could not express mini-dystrophin in a deleted region of the patient's DMD gene. In retrospect, this seems intuitive but it was underappreciated as an inclusion/exclusion criterion for this study. The second important lesson learned is that revertant fibers, expressing dystrophin by restoring the reading frame in the DMD gene through alternative splicing or a second-site mutation, can express an immunogenic novel epitope never before encountered by the patient, most likely related to misfolding of the newly expressed epitope. This finding was very much unexpected given that most clinicians and scientists predicted that revertant fibers would induce immune tolerance. Validation of this principle was demonstrated in a follow-up study to determine the prevalence of such immunity within a DMD cohort from our Muscular Dystrophy Association clinic.[44] Dystrophin-specific T-cell immunity was evaluated in subjects with DMD who were either receiving the glucocorticoid prednisone (N = 24) or deflazacort (N = 29), or who were not receiving steroids (N = 17), as well as from normal age-matched control subjects (N = 21). Anti-dystrophin T-cell responses were detected in none of the normal control patients, and responses were found in 9 of 17 (52.9%) untreated patients with DMD. In contrast, they were found in only 11 of 53 (20.8%) steroid-treated patients with DMD. These findings confirm the presence of preexisting circulating T-cell immunity to dystrophin in a sizable proportion of patients with DMD, and emphasize the need to consider pretesting for dystrophin immunity as an enrollment criterion for DMD clinical gene therapy trials.

Alpha-Sarcoglycan Gene Replacement for Limb Girdle Muscular Dystrophy 2D

Our experience with gene transfer for limb girdle muscular dystrophy 2D (LGMD2D), alpha-sarcoglycan (α-SG) deficiency, was dramatically different from that with DMD.[2,3] This is the most common form of the sarcoglycanopathies. There were several factors that favored a satisfactory outcome. The predominant mutation in LGMD2D is a missense mutation, providing a basis for gene expression that reduces the risk of immunogenicity from transgene expression. The full length α-SG cDNA is <2 kb and is well within the packaging capacity of recombinant AAV. We also used a muscle-specific creatine kinase (tMCK) promoter for the first time in a clinical trial improving the safety profile of gene transfer targeting muscle and avoiding off target toxicity. Indications from preclinical studies supported the use of AAV1 to achieve sustained gene expression in dystrophic muscle.[45] The clinical trial was a double-blind, randomized, controlled study (vector on one side and saline injected in the other). In contrast to the DMD trial, the extensor digitorum brevis (EDB) muscle was the target for gene delivery enabling removal of the entire muscle post gene transfer without loss of function to the patient.

Six LGMD2D subjects with established SGCA mutations at both alleles were enrolled. A dose escalation study was initially planned but detection of robust gene expression following gene transfer to the first three subjects favored using the same dose of AAV1.tMCK.SGCA in all patients (3.25×10^{11} vg). Gene transfer using this vector, cassette, and mode of delivery proved to be safe, without any adverse effects and with robust gene expression for 5 months. Gene expression studies on muscle biopsies of EDB were done prior to breaking the blind. Relatively consistent findings were seen in 5 of the 6 subjects with restoration of α-SG on the side of gene transfer (Figure 51.7). Successful gene transfer was corroborated using PCR amplification of a unique 5′ untranslated leader sequence that was incorporated into the transgene cassette to distinguish endogenous from transgene α-SG expression.

The final participant in this study, subject 6, was the exception compared to all others. The muscle biopsy showed comparatively low level α-SG gene expression on the treated and untreated sides at the 6-month time point. Several features favored poor muscle fiber transduction. The muscle biopsy failed to show expression of major histocompatibility complex I (MHC I) antigen on any muscle fiber (Figure 51.8). We have learned that gene transfer leads to antigen presentation at the sarcolemma.

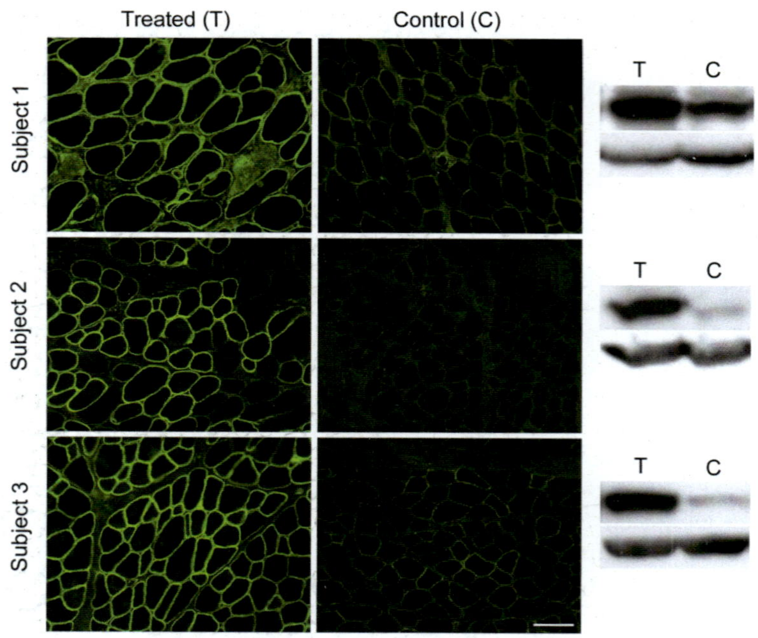

FIGURE 51.7 Alpha-Sarcoglycan (α-SG) expression following gene transfer in LGMD2D patients. Post-gene transfer tissue sections from extensor digitorum brevis muscles for subjects 1−3. α-SG gene expression on the side of vector injection (treated) is compared to muscle from the opposite side (control). **(Right)** Western blots from subjects 1−3 show increased α-SG gene expression on the side of gene transfer compared to the contralateral side (treated on the left, control on the right) showing residual gene expression from mutant protein. α-SG is normalized to actin (lower band). Scale bar = 100 μm. *Reprinted from Mendell et al. 2009.*[46]

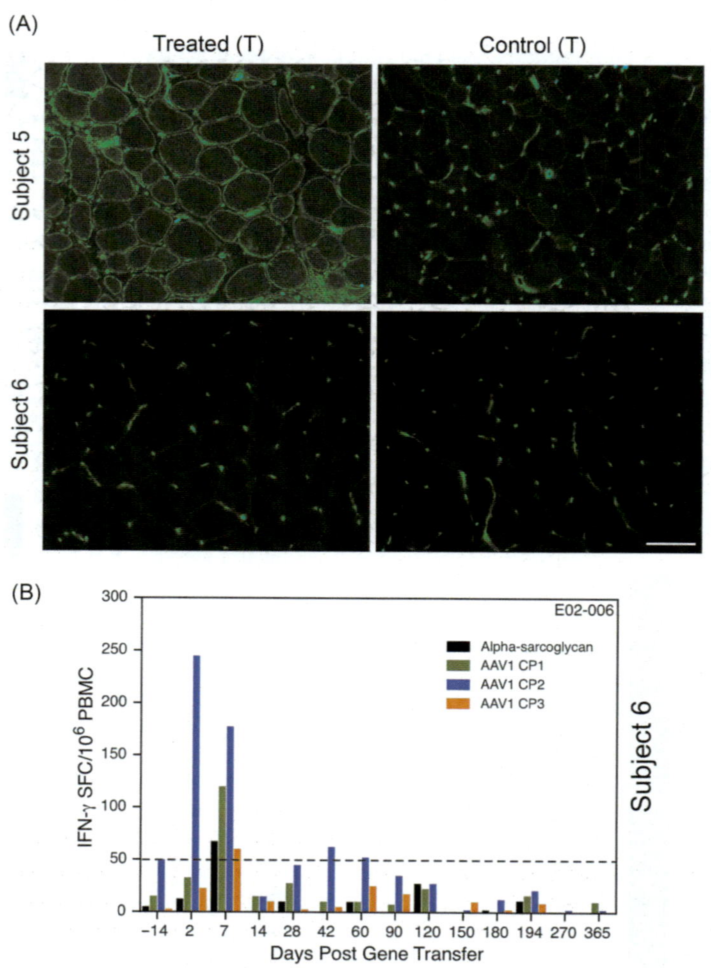

FIGURE 51.8 MHC class I staining of sarcolemmal membrane. **(A)** Gene transfer leads to antigen presentation at the sarcolemma. The absence of MHC I in subject 6 of the LGMD2D study was in direct contrast to the findings of all other participants in the clinical trial, illustrated by MHC I expression shown in subject 5 on the treated side. Blood vessels will show positive staining for MHC I that is background and nonspecific. **(B)** Subject 6 of the LGMD2D gene transfer study also had an immune response demonstrated in the IFN-γ ELISpot assay to AAV1 capsid pool 2 (AAV1 CP2). This was seen as early as day 2 (blue), and again on day 7. The positive IFN-γ ELISpot assay occurring shortly after gene transfer was indicative of an amnestic response suggesting preexisting immunity to AAV1. *Reprinted from Mendell et al., 2010.*[3]

The absence of MHC I was in direct contrast to the findings for all other participants in the clinical trial. The patient was also found to have both early humoral and T-cell responses to AAV1 capsid. The INF-γ ELISpot assay demonstrated T-cell activation to AAV1 capsid as early as day 2 following gene transfer (also present on day 7) (Figure 51.8). Accompanying the AAV capsid-induced T-cell response, we found a very rapid rise in AAV neutralizing antibody titers reaching levels more than 30 times greater than seen in other cases. Collectively these findings favor an amnestic response related to preexisting immunity to AAV.

The lessons learned from this trial, combined with our experience in the DMD clinical gene transfer trial, help refine approaches as we move forward with gene therapy for muscular dystrophies. AAV1 expresses well in muscle over long periods and we demonstrated robust expression using a muscle specific promoter. With proper vector cassette design we can differentiate endogenous versus transgene expressed product, an observation particularly relevant to other autosomal recessive disorders having a high percentage of missense mutations. The study also highlighted that we can never be overconfident regarding preexisting immunity to AAV capsid. The immune system can impede vector transduction, as we encountered in this study by the observation of a rapid rise in both T-cell and antibody responses.

Dysferlin Gene Replacement for LGMD2B

Similarly to gene replacement for the *DMD* gene, the 6.5 kb dysferlin cDNA also exceeds the packaging capacity of AAV. Different approaches have been explored for generating mini- or full-length dysferlin that could be considered for clinical trial.[5,47–49] A model that provides a small or miniDYSF simulates an approach used to treat the *mdx* mouse using mini- and micro-dystrophins.[48,49] In these studies, a dysferlin gene construct was based on a clinical observation whereby a late-onset, mild LGMD2B patient was found to have large genomic deletion in DYSF resulting in a truncated protein only one-third the size of full-length dysferlin. Using AAV to transfer this "natural miniDYSF" by intramuscular delivery to dysferlin deficient mice, researchers found improved membrane repair but without correction of the dystrophic histological pattern.

Alternatively to replacement with miniDYSF, novel ways are under investigation to replace the full-length protein. Lostal et al. initially used a dual vector approach, taking advantage of AAV's ability to concatamerize.[47] In principle, the 5′ AAV vector holds a cDNA inclusive of the DYSF N terminal (exons 1–28) and a splice donor site, while a second 3′ vector embraces a splice acceptor site and the DYSF C terminus (exons 29–55) followed by a polyadenylation signal (Figure 51.9). In dysferlin deficient mice, this strategy enabled efficient and improved membrane repair by following intramuscular injection. However, when given systemically, expression was limited to <5% of muscle fibers, adding to the challenges of bringing this to the clinic.

Another strategy, referred to as homologous recombination, was designed to take advantage of the unique plasticity demonstrated by the AAV5 serotype.[5] It was anticipated that this serotype could package an oversized cDNA inclusive of the full-length dysferlin cDNA (6.5 kb). What happened instead, however, was that AAV5 spontaneously packaged two halves of DYSF yielding virions with regions of homology that

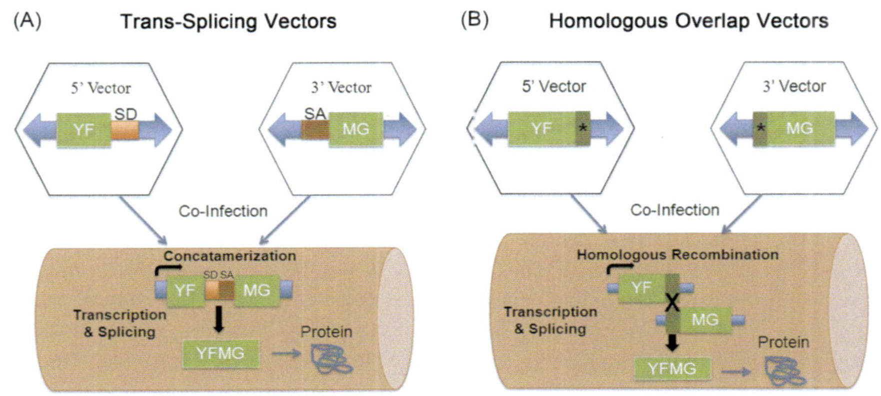

FIGURE 51.9 Strategies for delivery of large genes. (A) A trans-splicing approach is a dual vector strategy in which the transgene cassette is divided in two, whereby a splice donor (SD) site is added to the 5′ vector and a splice acceptor (SA) site is added to the 3′ vector. Upon co-delivery to muscle, the vectors concatamerize and undergo splicing to generate a full-length transcript. (B) A second dual vector strategy employs an overlap sequence that is shared by both vectors (asterisk). Following delivery to muscle, recombination between the two vectors occurs to generate full-length transcript. YFMG, Your Favorite Muscle Gene.

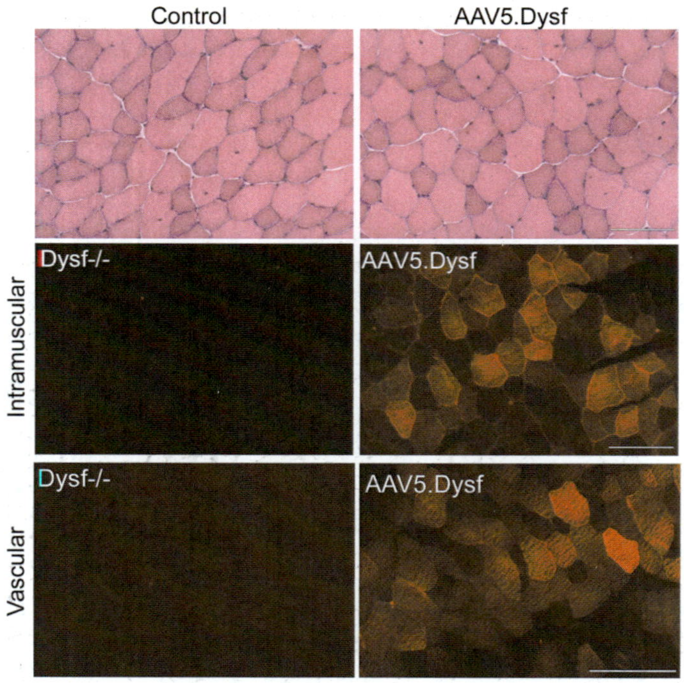

FIGURE 51.10 AAV5.Dysferlin delivery. Dysferlin deficient mice (Dysf −/−) mice were treated by intramuscular or vascular delivery via the femoral artery with AAV5.Dysf. One month after treatment, dysferlin expression was restored to the muscle. **Left panels:** Dysf −/− no treatment. **Right panels:** AAV5.Dysf treated muscle. The overexpression of dysferlin in the muscle fibers is typical of this method of gene transfer and is nontoxic to the cell. *Reprinted from Grose et al., 2012.*[5]

recombined resulting in a full-length transgene following delivery to muscle (Figure 51.9). Using this approach, our laboratory has shown that AAV5 delivery of DYSF leads to robust expression in muscle by both intramuscular and vascular routes (Figure 51.10). The resulting expression restores the capacity of dysferlin deficit muscle to repair the sarcolemma following injury and also restores force and resistance to fatigue in the diaphragm. Long-term studies demonstrate that dysferlin expression is sustained for at least 1 year in dysferlin-deficient mice without measurable decrease in protein levels. There is momentum pushing this approach toward clinical trial.

VASCULAR DELIVERY TO IMPROVE CLINICAL OUTCOMES

In order to achieve meaningful clinical improvement for muscular dystrophy using gene therapy, vascular delivery approaches to target isolated limbs or systemic distribution is imperative. Delivering AAV vectors through regional or systemic circulation allows for broad distribution across entire muscle groups versus intramuscular (IM) injection limited to a single muscle. Beyond the benefit of widespread distribution, vascular delivery holds many advantages. Several groups have shown that vascular delivery is less immunogenic than IM delivery.[50,51] This is due in part to the increased vector genome copy number per nucleus which has been shown to be 5 to 10 times higher in IM injected muscle versus muscles transduced via the vasculature. Since muscle is a multinucleated syncytium, equal transduction among nuclei results in more uniform protein expression across the entire length of the muscle fiber versus segmental overexpression.[4] Multiple strategies have been used to target isolated limbs in large animals through either the venous or the arterial circulation.[51–55]

Our group has preferred to use an arterial approach that avoids high volume, high pressure gene delivery that is required when approaching delivery through the venous circulation.[56] We have utilized fluoroscopy guided catheterization to target individual muscle groups such as the gastrocnemius and quadriceps or delivery of vector to the entire lower limb. To optimize arterial gene delivery and distribution, we have developed a method utilizing recirculation to increase and improve vector transduction. The method employs a closed circuit with a tourniquet placed at the level of the groin accompanied by catheterization of both the femoral artery and vein. Venous blood is first removed from the circuit and stored in a reservoir, followed by infusion of a saline-based solution to remove platelets and other cells from the extremity that could bind to the vector. The vector is then infused and recirculated using a mechanical pump, thereby offering transgene to many muscle groups in the extremity through multiple passes. We have established this method to be highly efficient in achieving widespread distribution throughout the limb. We are currently translating this recirculation approach to a clinical trial in LGMD2D patients with AAV.tMCK.SGCA. This will be the first vascular gene therapy trial for muscular dystrophy. The study has been approved by the FDA for clinical trial.

THERAPEUTIC ENHANCEMENT USING SURROGATE GENES

Follistatin Gene Therapy

Blocking the myostatin pathway shows dramatic muscle building properties that are highly conserved across species, including mice, sheep, cattle, canines, and humans.[57–60] Myostatin is a member of the transforming growth factor (TGF)-β superfamily, and numerous studies have demonstrated that it is a negative regulator of skeletal muscle growth. Follistatin is a powerful inhibitor of myostatin with the additional advantage of controlling muscle mass through pathways independent of the myostatin signaling cascade.[61] In studies of myostatin knockout mice crossed to mice carrying a follistatin transgene, there was a quadrupling of muscle mass compared with the doubling of muscle mass that is observed from lack of myostatin alone.

In preparation for taking this to clinical trial, we identified an alternatively spliced isoform of follistatin (FS344) that can be packaged in AAV.[62] In the *mdx* mouse, intramuscular injections of AAV1 carrying the FS344 gene increased muscle mass (histological measure) and quantitative strength, as assessed by hind- and forelimb grip strength. The benefits persisted for more than 2 years and there were no adverse effects. In nonhuman primates, direct muscle injections of AAV1.CMV.FS344 into the quadriceps muscle increased size and strength for as long as 15 months.[63] Necropsies showed no abnormalities in any organ, and clinical chemistries and hormone levels related to gonadal function including follicle stimulating hormone, luteinizing hormone, testosterone, and estradiol were normal.

We have now taken this to clinical trial in two patient groups who could benefit by increasing size and strength of the quadriceps muscle. In Becker muscular dystrophy (BMD) and in sporadic inclusion body myositis (sIBM) the quadriceps muscle is targeted by the disease, leading to frequent falls and loss of ambulation. A total of nine patients have been treated without encountering adverse events. The distance traveled in the 6MWT is being evaluated as the primary outcome measure. There appears to be improvement, but further testing will be required before any conclusions are established.

Combinational Therapy Using Follistatin and Micro-dystrophin

Although preclinical studies with micro-dystrophin in the *mdx* mouse have been very favorable, functional tests by multiple groups have demonstrated that tetanic force and resistance to contraction-induced injury is not fully restored following AAV delivery.[40] Micro-dystrophin may simply be unable to restore normal function in the absence of important structural and signaling domains that are not included in micro-dystrophin without the nNOS binding site and C-terminus.[64] In addition, treatment at later stages of disease faces the formidable task of attempting to improve function in muscle encumbered with extensive replacement by fat and fibrosis. We hypothesized that combinatorial treatment using follistatin and micro-dystrophin gene replacement would restore function of DMD muscle.[65] In an aged *mdx* mouse model with overt pathology, we co-delivered AAV.micro-dystrophin and AAV.follistatin with endpoints at either 13 or 20 months of age.[66] Remarkably, this combination therapy restored specific force and resistance to eccentric contraction, greater than either form of gene delivery could when used alone (Figure 51.11). This approach has clinical applicability and could extend to facilitating the function of molecular approaches considered for future studies.

Alpha7 Integrin Gene Therapy

We have described several challenges for AAV-mediated delivery of dystrophin for DMD including the size of the cDNA, potential for immune response elicited by the newly delivered transgene, the absence of functional domains in micro-dystrophin, and the connective tissue and fat replacement presenting obstacles in reversing dystrophic histopathology. In the face of these hurdles, investigating alternative mechanisms to compensate for loss of dystrophin becomes an attractive option. We tested one such approach using AAV-mediated delivery of α7 integrin (ITGA7).[67] Alpha7 is a laminin receptor that serves as a link between the extracellular matrix and intracellular actin cytoskeleton, much like dystrophin. While present throughout the sarcolemma, α7 is enriched at the neuromuscular and myotendinous junctions. In DMD, α7 is upregulated over the sarcolemma, giving the appearance of a cellular attempt to compensate for the absence of dystrophin.[68] Following early work in transgenic models, we hypothesized that AAV-mediated overexpression of ITGA7 would improve muscle stability and preserve force over time. We treated the hindlimb muscles of *mdx* mice by isolated limb perfusion with AAV.MCK.ITGA.[67] Following treatment, we found that α7 protected the muscle from eccentric contraction induced injury and improved muscle integrity as measured by reducing the number of fibers leaking Evan's blue dye. AAV.ITGA7 is an attractive treatment option for DMD for important reasons: the size of the full cDNA is accommodated by the packaging limitations of AAV and the protein is expressed in DMD muscle, eliminating the potential for an immune response to the transgene or its product. In addition, α7 integrin may also have potential therapeutic application for forms of muscular dystrophy other than DMD.

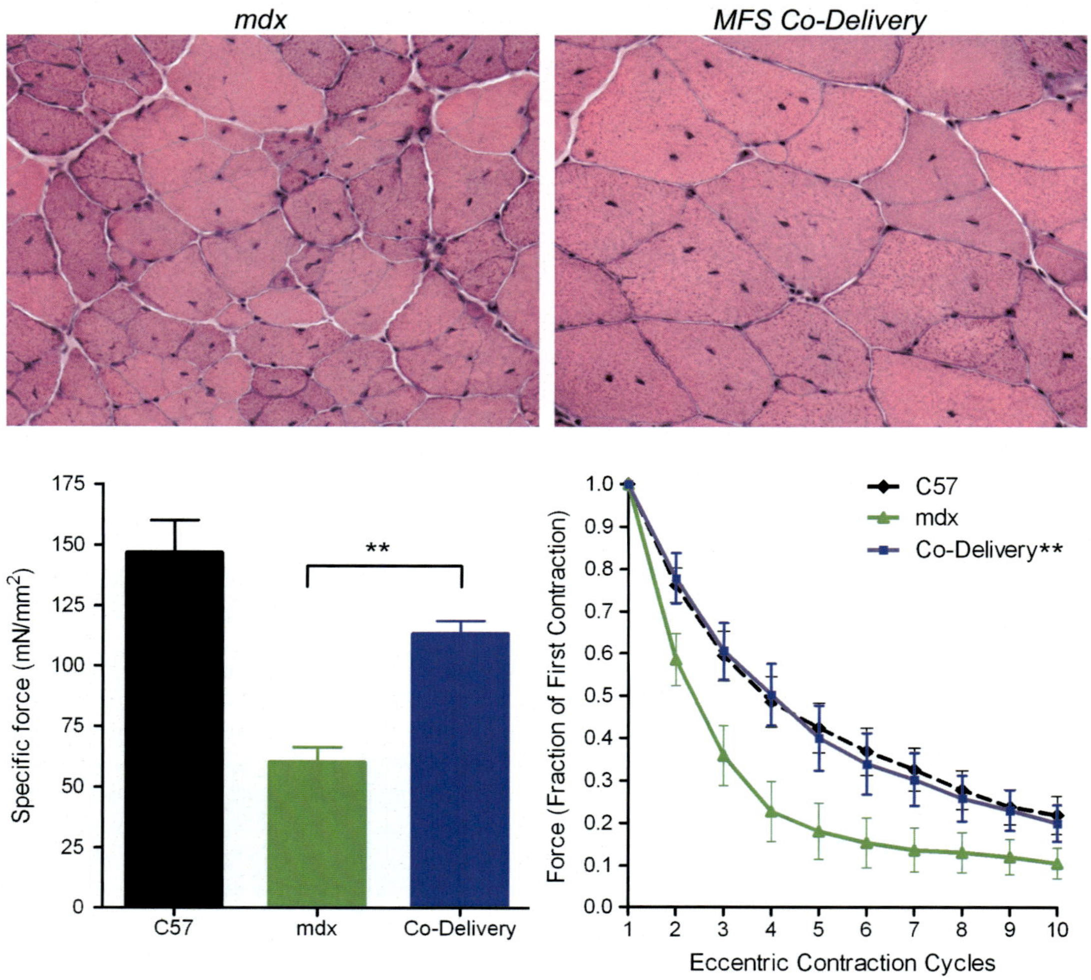

FIGURE 51.11 Co-delivery of micro-dystrophin and follistatin restores force deficits. *mdx* mice were treated at 6 months of age with AAV.micro-dystrophin and AAV.follistatin. There was a significant increase in muscle fiber size and reduction in inflammatory cell infiltrates. At 1 year, the mice were assessed for muscle physiology in the extensor digitorum longus muscle. Both specific force (**bottom left**) and resistance to eccentric contraction induced injury (**bottom right**) were restored in the mice. *Reprinted from Oxford University Press: Rodino-Klapac et al., 2013.*[66]

STRATEGIES RESPONSIVE TO IMMUNITY OF AAV AND/OR TRANSGENE

As the safety profile of AAV continues to expand through a growing number of clinical trials, there has been considerable emphasis on developing new generation vectors or methods to optimize delivery and efficacy. Although often touted for limited immunogenicity, AAV does in fact elicit an immune response; however, the intensity of the response and impact on transgene expression varies widely depending on the disease, vector, target organ, and route of delivery. T-cell responses directed against either capsid or transgene have been noted,[1,69] but the most common barrier identified in all trials is preexisting humoral immunity to AAV. Even very-low-titer antibodies have been shown to block or neutralize transduction,[2,70] and enrolling only antibody negative subjects may seriously limit recruitment for clinical trials.[71] Various strategies have been used or are in development to circumvent AAV antibodies including: (1) pharmacologic agents to target B cells[72]; (2) using alternative AAV serotypes[73]; (3) co-delivery of empty capsids or capsid decoys to neutralize or "soak up" antibodies[74]; and (4) utilizing plasmapheresis to reduce anti-AAV antibody titer.[75] Our group has shown in nonhuman primates with preexisting immunity that performing plasmapheresis immediately prior to gene transfer results in sustained expression levels equivalent to that achieved following gene transfer in an antibody naïve animal.[76] This commonly used procedure has immediate relevance to current gene therapy trials and could be easily integrated into the design of most clinical trials. Plasmapheresis is particularly well suited for re-delivery of AAV if the response is waning or if there is a need to deliver transgene using a different serotype.

Novel strategies are also in development to bypass innate immunity[77] or off-target transgene expression.[78] The single stranded AAV genome triggers an innate immune response through interaction with Toll-like receptors (TLRs) on the surface of innate immune cells such as macrophages and dendritic cells.[79] Specifically, TLR9 has been shown to sense major pathogenic DNA viruses by recognizing unmethylated bacterial and viral CpG motifs.[80] TLR signaling leads to the production of inflammatory cytokines, resulting in an enhanced innate and adaptive immune response following AAV delivery.[5,81] Faust and colleagues hypothesized that removal of CpG motifs from AAV vectors would help to evade this component of the immune response. They found CpG-depleted vectors enabled persistent gene expression and evaded immunity as compared to their non-depleted counterparts, thereby providing proof of principle that CpG-depleted vectors may improve clinical trial outcomes.[77] Utilizing a different approach to evade the immune system, Boisgerault and colleagues attempted to invoke immune tolerance by inserting a microRNA target sequence, miR142.3, into their vector cassette containing α-sarcoglycan.[78] miR142.3 is only present in immune cells and therefore expression would be blocked following off-target transduction. This strategy was highly effective in normal mice, although the response was reduced in α-sarcoglycan deficient mice. The method awaits optimization but the approach holds a promise for preventing immune reactions and off-target expression.

SUMMARY

The future looks bright for targeting clinical muscle diseases using molecular therapies. Multiple phase I and II trials are currently completed or are actively enrolling. Presently for muscular dystrophies the treatment is limited. We do have glucocorticoids as the only pharmacologic agent that changes the natural history of disease. The most notable benefits of glucocorticoids include the potential for prolonging ambulation, and possibly one that is relatively less appreciated, significantly reducing the incidence of scoliosis. On the other hand, the limitations of glucocorticoids, irrespective of regimen employed, are well known. The list of side effects is long and often painful to observe in the clinical setting (weight gain, cushingoid appearance, delayed puberty and growth retardation, cataracts, hypertension, glucose intolerance, etc.). The future for treatment lies in altering pathogenesis using molecular pathways. Many approaches have been discussed in this chapter, all of which have potential clinical application. It will, however, take time to implement them and to pick and choose the best methods to take forward.

ACKNOWLEDGMENTS

The studies discussed in this chapter were supported by the Muscular Dystrophy Association, Parent Project Muscular Dystrophy, Jesse's Journey Foundation, NIH/NINDS 7R01NS043186, NIH/NINDS 1U54NS055958, The Jain Foundation, and NIH NICHD 1U54HD066409-01.

REFERENCES

1. Mendell JR, Campbell K, Rodino-Klapac L, Sahenk Z, Shilling C, Lewis S, et al. Dystrophin immunity revealed by gene therapy in Duchenne muscular dystrophy. *N Engl J Med* 2010;**363**:1429–37.
2. Mendell JR, Rodino-Klapac LR, Rosales-Quintero X, Kota J, Coley BD, Galloway G, et al. LGMD2D gene therapy restores alpha-sarcoglycan and associated proteins. *Ann Neurol* 2009;**66**:290–7.
3. Mendell JR, Rodino-Klapac LR, Rosales XQ, Coley BD, Galloway G, Lewis S, et al. Sustained alpha-sarcoglycan gene expression following gene transfer in LGMD2D. *Ann Neurol* 2010;**68**:629–38.
4. Rodino-Klapac LR, Montgomery CL, Bremer WG, Malik V, Shontz KM, Davis N, et al. Persistent expression of FLAG tagged micro-dystrophin in non-human primates with intramuscular and vascular approaches. *Mol Ther* 2010;**18**:109–17.
5. Grose WE, Clark KR, Griffin D, Malik V, Shontz KM, Montgomery CL, et al. Homologous recombination mediates functional recovery of dysferlin deficiency following AAV5 gene transfer. *PLOS ONE* 2012;**6**:e39233.
6. Mendell JR, Rodino-Klapac LR, Sahenk Z, Roush K, Bird L, Lowes LP, et al. Eteplirsen for the treatment of Duchenne muscular dystrophy. *Ann Neurol* 2013;**74**:637–47.
7. Merlini L, Cicognani A, Malaspina E, Gennari M, Gnudi S, Talim B, et al. Early prednisone treatment in Duchenne Muscular Dystrophy. *Muscle Nerve* 2003;**27**:222–7.
8. Mendell JR, Shilling C, Leslie ND, Flanigan KM, al-Dahhak R, Gastier-Foster J, et al. Evidence based path to newborn screening for Duchenne muscular dystrophy. *Ann Neurol* 2012;**71** 304–13.
9. Connolly AM, Florence JM, Cradock MM, Malkus EC, Schierbecker JR, Siener CA, et al. Motor and cognitive assessment of infants and young boys with Duchenne muscular dystrophy; results from the Muscular Dystrophy Association DMD Clinical Research Network. *Neuromuscul Disord* 2013;**23**:529–39.
10. Ellis JA, Vroom E, Muntoni F. 195th ENMC international workshop: newborn screening for Duchenne muscular dystrophy 14–16th december, 2012, Naarden, the Netherlands. *Neuromuscul Disord* 2013;**23**:682–9.
11. Koenig M, Monaco AP, Kunkel LM. The complete sequence of dystrophin predicts a rod-shaped cytoskeletal protein. *Cell* 1988;**53**:219–28.
12. Brooke MH, Fenichel GM, Griggs RC, Mendell JR, Moxley R, Miller JP, et al. Clinical investigation in Duchenne Dystrophy: 2. Determination of the "Power" of therapeutic trials based on the natural history. *Muscle Nerve* 1983;**6**:91–103.
13. Mendell JR, Moxley RT, Griggs RC, Brooke MH, Fenichel GM, Miller JP, et al. Randomized, double-blind six-month trial of prednisone in Duchenne's muscular dystrophy. *N Engl J Med* 1989;**320**:1592–7.

14. Biggar WD, Gingras M, Fehlings DL, Harris VA, Steele CA. Deflazacort treatment of Duchenne muscular dystrophy. *J Pediatr* 2001;**138**:45–50.
15. Escolar DM, Hache LP, Clemens PR, Cnaan A, McDonald CM, Viswanathan V, et al. Randomized, blinded trial of weekend vs daily prednisone in Duchenne muscular dystrophy. *Neurology* 2011;**77**:444–52.
16. Goyenvalle A, Babbs A, Powell D, Kole R, Fletcher S, Wilton SD, et al. Prevention of dystrophic pathology in severely affected dystrophin/utrophin-deficient mice by morpholino-oligomer-mediated exon-skipping. *Mol Ther* 2010;**18**:198–205.
17. Honeyman K, Cheng AJ, Ly T, Lloyd F, Fletcher S, Morgan JE, et al. Antisense-induced exon skipping and synthesis of dystrophin in the mdx mouse. *Proc Natl Acad Sci USA* 2001;**98**:42–7.
18. Yokota T, Lu QL, Partridge T, Kobayashi M, Nakamura A, Takeda S, et al. Efficacy of systemic morpholino exon-skipping in Duchenne dystrophy dogs. *Ann Neurol* 2009;**65**:667–76.
19. Aartsma-Rus A, Fokkema I, Verschuuren J, Ginjaar I, van Deutekom J, van Ommen GJ, et al. Theoretic applicability of antisense-mediated exon skipping for Duchenne muscular dystrophy mutations. *Hum Mutat* 2009;**30**:293–9.
20. van Deutekom JC, Janson AA, Ginjaar IB, Frankhuizen WS, Aartsma-Rus A, Bremmer-Bout M, et al. Local dystrophin restoration with antisense oligonucleotide PRO051. *N Engl J Med* 2007;**357**:2677–86.
21. Kinali M, Kinali M, Arechavala-Gomeza V, Feng L, Cirak S, Hunt D, et al. Local restoration of dystrophin expression with the morpholino oligomer AVI-4658 in Duchenne muscular dystrophy: a single-blind, placebo-controlled, dose-escalation, proof-of-concept study. *Lancet Neurol* 2009;**8**:918–28.
22. Goemans NM, Tulinius M, van den Akker JT, Burm BE, Ekhart PF, Heuvelmans N, et al. Systemic administration of PRO051 in Duchenne's muscular dystrophy. *N Engl J Med* 2011;**364**:1513–22.
23. Cirak S, Arechavala-Gomeza V, Guglieri M, Feng L, Torelli S, Anthony K, et al. Exon skipping and dystrophin restoration in patients with Duchenne muscular dystrophy after systemic phosphorodiamidate morpholino oligomer treatment: an open-label, phase 2, dose-escalation study. *Lancet* 2011;**378**:595–605.
24. McDonald CM, Henricson EK, Abresch RT, Florence JM, Eagle M, Gappmaier E, et al. The 6-minute walk test and other clinical endpoints in severe dystrophinopathy: reliability, concurrent validity, and minimal clinically important differences from a multicenter study. *Muscle Nerve* 2013;**48**:342–56.
25. McDonald CM, Henricson EK, Han JJ, Abresch RT, Nicorici A, Atkinson L, et al. The 6-minute walk test in Duchenne/Becker muscular dystrophy: longitudinal observations. *Muscle Nerve* 2010;**42**:966–74.
26. Strober JB. Therapeutics in Duchenne muscular dystrophy. *NeuroRx* 2006;**3**:225–34.
27. Aartsma-Rus A, Muntoni F. 194th ENMC international workshop. 3rd ENMC workshop on exon skipping: Towards clinical application of antisense-mediated exon skipping for Duchenne muscular dystrophy, 8–10 December 2012, Naarden, The Netherlands. *Neuromuscul Disord* 2013;**23**:934–44.
28. Mendell JR, Buzin CH, Feng J, Yan J, Serrano C, Sangani DS, et al. Diagnosis of duchenne dystrophy by enhanced detection of small mutations. *Neurology* 2001;**57**:645–50.
29. Flanigan KM, von Niederausern A, Dunn DM, Alder J, Mendell JR, Weiss RB. Rapid direct sequence analysis of the dystrophin gene. *Am J Hum Genet* 2003;**72**:931–9.
30. Barton-Davis ER, Cordier L, Shoturma DI, Leland SE, Sweeney HL. Aminoglycoside antibiotics restore dystrophin function to skeletal muscles of mdx mice. *J Clin Invest* 1999;**104**:375–81.
31. Wagner KR, Hamed S, Hadley DW, Gropman AL, Burstein AH, Escolar DM, et al. Gentamicin treatment of Duchenne and Becker muscular dystrophy due to nonsense mutations. *Ann Neurol* 2001;**49**:706–11.
32. Politano L, Nigro G, Nigro V, Piluso G, Papparella S, Paciello O, et al. Gentamicin administration in Duchenne patients with premature stop codon. Preliminary results. *Acta Myol* 2003;**22**: 15–21.
33. Malik V, Rodino-Klapac LR, Viollet L, Wall C, King W, Al-Dahhak R, et al. Gentamicin-induced readthrough of stop codons in Duchenne muscular dystrophy. *Ann Neurol* 2010;**67**:771–80.
34. Tang HY, Hutcheson E, Neill S, Drummond-Borg M, Speer M, Alford RL. Genetic susceptibility to aminoglycoside ototoxicity: how many are at risk? *Genet Med* 2002;**4**:336–45.
35. Viollet L, Gailey S, Thornton DJ, Friedman N, Flanigan KM, Mahan JD, et al. Utility of Cystatin C to monitor renal function in Duchenne muscular dystrophy. *Muscle Nerve* 2009;**40**:438–42.
36. Welch EM, Barton ER, Zhuo J, Tomizawa Y, Friesen WJ, Trifillis P, et al. PTC124 targets genetic disorders caused by nonsense mutations. *Nature* 2007;**447**:87–91.
37. Peltz SW, Morsy M, Welch EM, Jacobson A. Ataluren as an agent for therapeutic nonsense suppression. *Annu Rev Med* 2013;**64**:407–25.
38. Wu Z, Yang H, Colosi P. Effect of genome size on AAV vector packaging. *Mol Ther* 2010;**18**:80–6.
39. Yazaki M, Yoshida K, Nakamura A, Koyama J, Nanba T, Ohori N, et al. Clinical characteristics of aged Becker muscular dystrophy patients with onset after 30 years. *Eur Neurol* 1999;**42**:145–9.
40. Harper SQ, Hauser MA, DelloRusso C, Duan D, Crawford RW, Phelps SF, et al. Modular flexibility of dystrophin: implications for gene therapy of Duchenne muscular dystrophy. *Nat Med* 2002;**8**:253–61.
41. Wang B, Li J, Xiao X. Adeno-associated virus vector carrying human minidystrophin genes effectively ameliorates muscular dystrophy in mdx mouse model. *Proc Natl Acad Sci USA* 2000;**97**:13714–19.
42. Watchko J, O'Day T, Wang B, Zhou L, Tang Y, Li J, et al. Adeno-associated virus vector-mediated minidystrophin gene therapy improves dystrophic muscle contractile function in mdx mice. *Hum Gen Ther* 2002;**13**:1451–60.
43. Mendell JR, Campbell K, Rodino-Klapac L, Sahenk Z, Shilling C, Lewis S, et al. Dystrophin immunity in Duchenne's muscular dystrophy. *N Engl J Med* 2010;**363**:1429–37.
44. Flanigan KM, Campbell K, Viollet L, Wang W, Gomez AM, Walker CM, et al. Anti-dystrophin T cell responses in Duchenne muscular dystrophy: prevalence and a glucocorticoid treatment effect. *Hum Gene Ther* 2013;**24**:797–806.
45. Rabinowitz JE, Rolling F, Li C, Conrath H, Xiao W, Xiao X, et al. Cross-packaging of a single adeno-associated virus (AAV) type 2 vector genome into multiple AAV serotypes enables transduction with broad specificity. *J Virol* 2002;**76**:791–801.
46. Mendell JR, Rodino-Klapac L, Rosales-Quintero X, Kota J, Coley BD, Galloway G, et al. Limb-girdle muscular dystrophy type 2D

gene therapy restores alpha-sarcoglycan and associated proteins. *Ann Neurol* 2009;**66**(3):290–7.

47. Lostal W, Bartoli M, Bourg N, Roudaut C, Bentaïb A, Miyake K, et al. Efficient recovery of dysferlin deficiency by dual adeno-associated vector-mediated gene transfer. *Hum Mol Genet* 2010;**19**:1897–907.
48. Krahn M, Wein N, Bartoli M, Lostal W, Courrier S, Bourg-Alibert N, et al. A naturally occurring human minidysferlin protein repairs sarcolemmal lesions in a mouse model of dysferlinopathy. *Sci Transl Med* 2010;**2**:50ra69.
49. Lostal W, Bartoli M, Roudaut C, Bourg N, Krahn M, Pryadkina M, et al. Lack of correlation between outcomes of membrane repair assay and correction of dystrophic changes in experimental therapeutic strategy in dysferlinopathy. *PLOS ONE* 2012;**7**:e38036.
50. Toromanoff A, Chérel Y, Guilbaud M, Penaud-Budloo M, Snyder RO, Haskins ME, et al. Safety and efficacy of regional intravenous (r.i.) versus intramuscular (i.m.) delivery of rAAV1 and rAAV8 to nonhuman primate skeletal muscle. *Mol Ther* 2008;**16**:1291–9.
51. Toromanoff A, Adjali O, Larcher T, Hill M, Guigand L, Chenuaud P, et al. Lack of immunotoxicity after regional intravenous (RI) delivery of rAAV to nonhuman primate skeletal muscle. *Mol Ther* 2010;**18**:151–60.
52. Rodino-Klapac LR, Montgomery CL, Mendell JR, Chicoine LG. AAV-mediated gene therapy to the isolated limb in rhesus macaques. *Methods Mol Biol* 2011;**709**:287–98.
53. Hegge JO, Wooddell CI, Zhang G, Hagstrom JE, Braun S, Huss T, et al. Evaluation of hydrodynamic limb vein injections in nonhuman primates. *Hum Gene Ther* 2010;**21**:829–42.
54. Arruda VR, Stedman HH, Nichols TC, Haskins ME, Nicholson M, Herzog RW, et al. Regional intravascular delivery of AAV-2-F.IX to skeletal muscle achieves long-term correction of hemophilia B in a large animal model. *Blood* 2005;**105**:3458–64.
55. Katwal AB, Konkalmatt PR, Piras BA, Hazarika S, Li SS, John Lye R, et al. Adeno-associated virus serotype 9 efficiently targets ischemic skeletal muscle following systemic delivery. *Gene Ther* 2013;**20**:930–8.
56. Rodino-Klapac L, Janssen PML, Montgomery CL, Coley BD, Chicoine LG, Clark KR, et al. A translational approach for limb vascular delivery of the micro-dystrophin gene without high volume or high pressure for treatment of Duchenne Muscular Dystrophy. *J Transl Med* 2007;**5**:45.
57. McPherron AC, Lawler AM, Lee SJ. Regulation of skeletal muscle mass in mice by a new TGF-beta superfamily member. *Nature* 1997;**387**:83–90.
58. McPherron AC, Lee SJ. Double muscling in cattle due to mutations in the myostatin gene. *Proc Natl Acad Sci USA* 1997;**94**:12457–61.
59. Mosher DS, Quignon P, Bustamante CD, Sutter NB, Mellersh CS, Parker HG, et al. A mutation in the myostatin gene increases muscle mass and enhances racing performance in heterozygote dogs. *PLoS Genet* 2007;**3**:e79.
60. Schuelke M, Wagner KR, Stolz LE, Hübner C, Riebel T, Kömen W, et al. Myostatin mutation associated with gross muscle hypertrophy in a child. *N Engl J Med* 2004;**350**:2682–8.
61. Lee SJ. Quadrupling muscle mass in mice by targeting TGF-beta signaling pathways. *PLOS ONE* 2007;**2**:e789.
62. Haidet AM, Rizo L, Handy C, Umapathi P, Eagle A, Shilling C, et al. Long-term enhancement of skeletal muscle mass and strength by single gene administration of myostatin inhibitors. *Proc Natl Acad Sci USA* 2008;**105**:4318–22.
63. Kota J, Handy CR, Haidet A, Montogomery CL, Eagle A, Rodino-Klapac LR, et al. Follistatin gene delivery enhances muscle growth and strength in nonhuman primates. *Sci Transl Med* 2009;**1**:6ra15.
64. Lai Y, Thomas GD, Yue Y, Yang HT, Li D, Long C, et al. Dystrophins carrying spectrin-like repeats 16 and 17 anchor nNOS to the sarcolemma and enhance exercise performance in a mouse model of muscular dystrophy. *J Clin Invest* 2009;**119**:624–65.
65. Dumonceaux J, Marie S, Beley C, Trollet C, Vignaud A, Ferry A, et al. Combination of myostatin pathway interference and dystrophin rescue enhances tetanic and specific force in dystrophic mdx mice. *Mol Ther* 2010;**18**:881–7.
66. Rodino-Klapac LR, Janssen PML, Shontz KM, Canan B, Montgomery CL, Griffin D, et al. Micro-dystrophin and follistatin co-delivery restores muscle function in aged DMD model. *Hum Mol Genet* 2013;**22**:4929–37.
67. Heller KN, Montgomery CL, Janssen PM, Clark KR, Mendell JR, Rodino-Klapac LR. AAV-mediated overexpression of human alpha7 integrin leads to histological and functional improvement in dystrophic mice. *Mol Ther 2013;21* 2013;520–5.
68. Rooney JE, Welser JV, Dechert MA, Flintoff-Dye NL, Kaufman SJ, et al. Severe muscular dystrophy in mice that lack dystrophin and alpha7 integrin. *J Cell Sci* 2006;**119**:2185–95.
69. Mingozzi F, High KA. Immune responses to AAV vectors: overcoming barriers to successful gene therapy. *Blood* 2013;**122**:23–36.
70. Manno CS, Pierce GF, Arruda VR, Glader B, Ragni M, Rasko JJ, et al. Successful transduction of liver in hemophilia by AAV-Factor IX and limitations imposed by the host immune response. *Nat Med* 2006;**12**:342–7.
71. Calcedo R, Morizono H, Wang L, McCarter R, He J, Jones D, et al. Adeno-associated virus antibody profiles in newborns, children, and adolescents. *Clin Vaccine Immunol* 2011;**18**:1586–8.
72. Mingozzi F, Chen Y, Edmonson SC, Zhou S, Thurlings RM, Tak PP, et al. Prevalence and pharmacological modulation of humoral immunity to AAV vectors in gene transfer to synovial tissue. *Gene Ther* 2013;**20**:417–24.
73. Bartel MA, Weinstein JR, Schaffer DV. Directed evolution of novel adeno-associated viruses for therapeutic gene delivery. *Gene Ther* 2012;**19**:694–700.
74. Mingozzi F, Anguela XM, Pavani G, Chen Y, Davidson RJ, Hui DJ, et al. Overcoming preexisting humoral immunity to AAV using capsid decoys. *Sci Transl Med* 2013;**5**:194ra192.
75. Monteilhet V, Saheb S, Boutin S, Leborgne C, Veron P, Montus MF, et al. A 10 patient case report on the impact of plasmapheresis upon neutralizing factors against adeno-associated virus (AAV) types 1, 2, 6, and 8. *Mol Ther* 2011;**19**:2084–91.
76. Chicoine LG, Montgomery CL, Bremer WG, Shontz KM, Griffin DA, Heller KN, et al. Plasmapheresis eliminates the negative impact of AAV antibodies on micro-dystrophin gene expression following vascular delivery. *Mol Ther* 2014;**22**:338–47.

77. Faust SM, Bell P, Cutler BJ, Ashley SN, Zhu Y, Rabinowitz JE, et al. CpG-depleted adeno-associated virus vectors evade immune detection. *J Clin Invest* 2013;**123**:2994–3001.
78. Boisgerault F, Gross DA, Ferrand M, Poupiot J, Darocha S, Richard I, et al. Prolonged gene expression in muscle is achieved without active immune tolerance using MicrorRNA 142.3p-regulated rAAV gene transfer. *Hum Gene Ther* 2012;**24**: 393–405.
79. Rathinam VA, Fitzgerald KA. Innate immune sensing of DNA viruses. *Virology* 2011;**411**:153–62.
80. Zhu J, Huang X. Yang. The TLR9-MyD88 pathway is critical for adaptive immune responses to adeno-associated virus gene therapy vectors in mice. *J Clin Invest* 2009;**119**:2388–98.
81. Hemmi H, Takeuchi O, Kawai T, Kaisho T, Sato S, Sanjo H, et al. A Toll-like receptor recognizes bacterial DNA. *Nature* 2000;**408**: 740–5.

Chapter 52

Orthopedic Management

David P. Roye, Evan D. Sheha, and Nanfang Xu

INTRODUCTION

Childhood-onset neuromuscular diseases are typically inherited genetic diseases characterized by general neurodevelopmental dysfunction, aberrant muscular function and development, and resulting pediatric deformity. This population often requires early orthopedic intervention to address mobility and motor deficits, and in some cases improve the patient's comfort, hygiene, and general quality of life. This chapter systematically addresses the various childhood-onset neuromuscular diseases that are most frequently seen and treated by orthopedic surgeons through an overview of the disease, a brief review of the orthopedic literature, and a summary of common practices.

SPINAL MUSCULAR ATROPHY

Clinical Overview

Spinal muscular atrophy (SMA) is an autosomal recessive disease affecting the α motor neurons of the anterior horn of the spinal cord.[1] The typical patterns of atrophy are the result of a deletion in one of the survival motor neuron genes (*SMN1* and *SMN2*),[2] which produce a protein necessary for the survival of α motor neurons. A homozygous insult to *SMN1* leads to SMA more than 95% of the time because the *SMN2* gene produces only 10% to 20% of the functional protein. SMA is one of the leading causes of childhood disability, with an estimated incidence of 1 in 6000 to 10,000 newborns and a carrier frequency of 1 in 40 to 60.[3]

First described by Werdnig and Hoffman, the common presentation of SMA is generalized muscle weakness in early life. Muscle involvement is prominent in muscles supporting the trunk and the lower extremities (proximal muscles are involved to a greater degree than distal muscles). This disease should be considered when children fail to reach developmental milestones, exhibit hypotonia and fasciculations, or demonstrate the typical patterns of muscle weakness with loss of deep tendon reflexes (DTRs). Creatine kinase levels, nerve conduction studies, and EMG may help elucidate the etiology, and if it remains in question, testing for the *SMN1* gene deletion can confirm the diagnosis. These patients typically have normal nerve conduction studies, but exhibit abnormal findings on EMG and muscle biopsy. Children with the more severe forms of SMA (I and II) will present early for orthopedic management of early onset scoliosis and hip deformities resulting from muscle imbalance.

The commonly used classification[4] of SMA is based on the age of initial clinical onset.[5] SMA I (also known as Werdnig-Hoffman disease) is the most severe type and has an age of onset between 0 and 6 months. These patients will never sit independently, typically have poor head control, and may also exhibit bulbar dysfunction, tongue atrophy, and an absence of DTRs. Individuals with SMA I have an increased risk of mortality secondary to respiratory failure in early childhood, typically before 2 years of age.[6] Weakness of the intercostal muscles often results in a bell-shaped trunk as the chest wall collapses and the abdomen expands to support breathing. However, recent development of various targeted therapeutic approaches to increase the amount of full-length SMN protein and regenerate or repair damaged motor neurons is reaching preclinical and clinical trial stages.[7]

SMA II has an intermediate age of onset of 7 to 18 months. This group is defined by their ability to maintain an unsupported upright seated position, though they may or may not be able to maintain independent sitting throughout later life. Some may be able to achieve standing with long-leg braces or support frames, though generally these patients are unable to walk. With improved pulmonary toilet and GI care, many patients are now expected to live into adulthood.[8] Joint contractures remain the chief orthopedic issue for this population, and similarly to SMA I patients, the development of kyphoscoliosis secondary to intercostal muscle weakness may lead to life-threatening respiratory complications.

B.T. Darras, H. Royden Jones, Jr., M.M. Ryan & D.C. De Vivo (Eds): Neuromuscular Disorders of Infancy, Childhood and Adolescence, Second edition.
DOI: http://dx.doi.org/10.1016/B978-0-12-417044-5.00052-4

SMA III, also known as Kugelberg-Welander disease or juvenile SMA, is a mild form of the disease with an average age of onset after 18 months old. All patients are ambulatory (with considerable difficulty), although the age at which patients achieve independent walking is variable. The ability to walk may be lost late in the first decade or during early adolescence, and while patients may be able to walk, proximal muscle and hip abductor weakness effectively prevents them from running. Patients are often of above-average intelligence and function well in social, school, and work environments. These patients may develop scoliosis, and musculoskeletal and joint overuse symptoms are common. Patients who maintain ambulation typically have long-term mortality rates similar to that of the general population.

SMA IV has an onset in the second or third decade of life and is characterized by mild motor impairment. These patients typically do not require special orthopedic intervention.

Scoliosis and hip subluxation/dislocation are two major orthopedic concerns in the SMA population (Figure 52.1). SMA I may present with torticollis associated with compensation for scoliosis, generalized muscle weakness, and flexion contracture of the hip, knee, and ankle,[9] necessitating orthotics, bracing, or surgery. The increased risk of lower extremity fracture in SMA II and upper extremity fracture in SMA III has been attributed to osteopenia and osteoporosis, and may be managed nonsurgically with immobilization and fall prevention.

While the current SMA classification system is widely accepted, it is used primarily to predict the spectrum and age of onset of clinical symptoms to be expected. The International SMA Consortium agrees that care should be tailored to address patients' current functional status rather than their particular classification. Dividing this group into nonsitters, sitters, and walkers may prove more useful with regard to guiding treatment.[4]

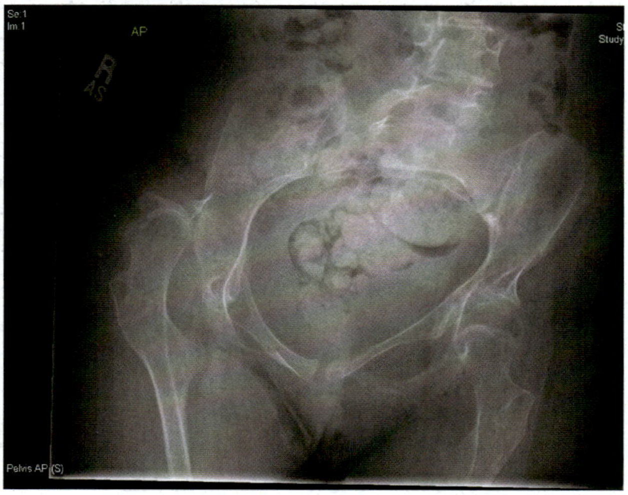

FIGURE 52.1 A patient with SMA demonstrating dislocation of the right hip and scoliosis of the lumbar spine.

Scoliosis

In one series of 43 patients, the prevalence of scoliosis was as high as 95%, and at 9 years of age the average curve measured 74, 54, and 23 degrees for SMA I, II, and III, respectively.[5] Granata et al. concluded that nonambulatory patients have larger and more rapidly progressing C-shaped curves than do type III patients,[10] and ambulatory status has been directly correlated to the severity of the curve.[11] Up to 80% of the curves seen in SMA patients are single thoracolumbar curves[11,12] and are frequently associated with other spinal deformities such as pelvic obliquity and kyphosis.[13]

Timing of intervention in scoliosis should take into account the balance between postponement of surgical intervention to preserve growth potential and the deterioration of respiratory function with curve progression. Pulmonary alveolar development happens primarily in early life, and by 8 years of age, children have approximately the same number of alveoli as they will in adulthood.[14] Robinson et al. reported a 4.7% decrease in predicted vital capacity and a 3.3% decrease in peak flow for every 10-degree increase in Cobb angle.[13] As a result of the high likelihood of progression, SMA patients require careful monitoring as there is a limited time window for optimal surgical outcomes.[15] Bracing of scoliosis has been attempted with variable success. In one series of 15 patients, the average curve was 88 degrees, and only three had curves less than 60 degrees; compliance was poor, as only 10 were able to follow the brace-wear prescription, and all were noted to have progression of scoliosis.[16] However, good results from bracing have been described for patients with SMA III.[1,17] For select patients, the author may opt to use wheelchair traction prior to definitive surgical management (Figure 52.2). For these patients, sitting supports may be cautiously considered as the benefits must be weighed against the potential detrimental effects on breathing and chest wall excursion.

The primary goal of spinal arthrodesis is to improve sitting position and balance during ambulation. Additional indications for surgery include progressive deformity and curve greater than 50 degrees, though in most series the curve has been significantly greater. With the use of Harrington and Luque rods now out of favor, posterior fixation using pedicle screws has shown good results. One study reported an average correction of 62% in a series of 9 patients, all with an improvement in function, although 4 of these had postoperative complications including pulmonary edema and painful instrumentation necessitating removal of hardware.[18] Patients and families have reported satisfaction with surgical outcomes regardless of the change in functional performance, which has been attributed to improved seated balance and cosmesis as well as fair improvement in pulmonary status and self-image.[19] This subjective improvement in pulmonary status has been quantified by Robinson

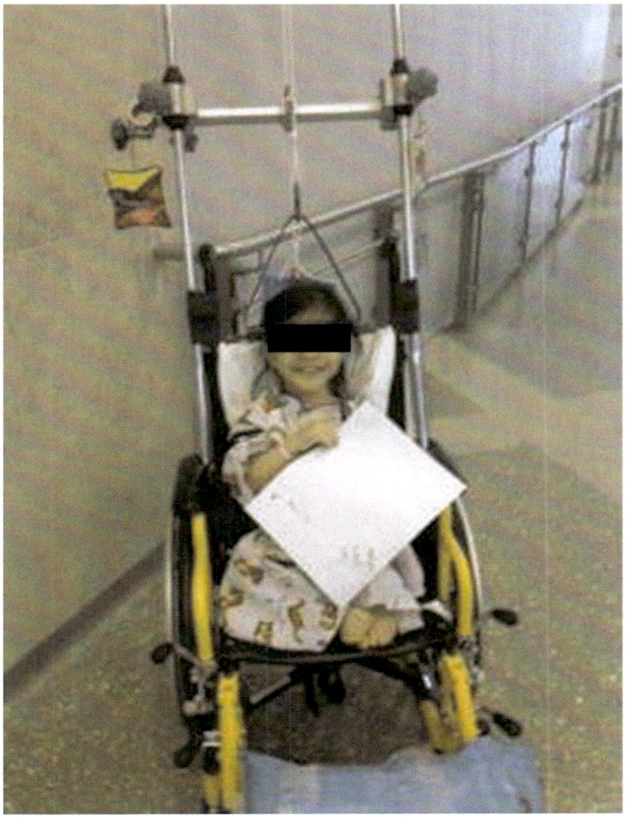

FIGURE 52.2 Wheelchair traction in a young patient with SMA. This modality can be useful to address severe deformities prior to definitive surgical management.

et al. as an increase in vital capacity in 8 of 9 patients undergoing posterior spinal fusion; however, data quantifying long-term changes are not available.[13] Nonetheless, early fusion in SMA and congenital scoliosis in general is not without pitfalls. In a study correlating quality of life, pulmonary function, and radiographic outcomes following early spinal fusion, the authors found that patients with early fusion had shorter spines, worse pulmonary function, and more pain than healthy peers.[20] As a result, growing constructs including chest-wall-based systems (the Vertical Expandable Prosthetic Titanium Rib), rib-to-pelvis, rib-to-spine, and spine-to-spine constructs have been gaining popularity in the SMA population as they allow for curve correction without arthrodesis. However, this modality may increase the risk of potential psychosocial effects given patients' young age and need for repetitive surgeries.[21] In a series of 15 patients, growing rods achieved a mean correction of 42%, though complications were frequent and were related to implant failure and prominence as well as the need for frequent reoperation to lengthen the implant.[22] Anterior surgery is contraindicated in this population due to the high frequency of progression.[16]

Special Considerations for SMA I

Marked hypotonia and significant pulmonary and nutritional difficulties are the primary contributors to morbidity and mortality. Generalized weakness may limit patients' abilities to participate in physical, occupational, and speech therapies. Devices to support and improve posture should be considered, and upper arm supports or slings to promote arm and hand function and range of motion may be helpful when used cautiously.[4] In general, however, orthopedic intervention is rarely indicated, and any discussion of surgical treatment should involve the family and multidisciplinary team. Of note, hypotonia may be associated with fractures at birth and suggestive of osteogenesis imperfecta, which may contribute to some diagnostic confusion though these heal quickly with appropriate nonoperative management.[23]

Perioperative Considerations

In general, management of SMA is multidisciplinary, involving pediatric neurology, genetics, orthopedics, physical therapy, and pulmonary and GI care. Pulmonary complications are common and result from the combination of weak intercostal musculature and strong diaphragmatic pull forming the bell-shaped torso, which leads to poor airway clearance and nighttime hypoventilation. Poor GI function, including bulbar dysfunction, poor cough, and swallowing difficulty, contributes to the high risk of pulmonary aspiration. Given the high risk of perioperative and postanesthesia pulmonary complications, patients' respiratory function should be optimized, and consultation with a pulmonologist may be indicated. While the ethical considerations surrounding noninvasive ventilation (NIV) in this population abound, evidence exists to suggest that NIV may actually alter time to onset of chest wall deformity.[24] Postoperative considerations include use of continuous ventilation, direct ICU transfer, and adequate analgesia to minimize hypoventilation and poor airway clearance secondary to pain.[4] Specific recommendations will vary based on a patient's functional status, and as a general rule, nonsitters are at highest risk for postoperative pulmonary complications while walkers typically require fewer and less invasive interventions.

HEREDITARY SENSORY AND MOTOR NEUROPATHIES

Background

Hereditary sensory and motor neuropathies (HSMNs) are characterized by the degradation of Schwann cells and the myelin sheath, resulting in progressive peripheral neuropathy. While this group of neuropathies includes a number of disease entities, e.g. congenital hypomyelinating neuropathy and Dejerine-Sottas disease, the most common of these is

Charcot-Marie-Tooth disease (CMT).[25] CMT is the most common hereditary neuromuscular disorder with a prevalence of approximately 10 to 82.3 per 100,000 individuals.[26,27] CMT encompasses a family of diseases with a variable clinical spectrum and patterns of inheritance, and has been linked to at least 40 different genes.[26] Characteristics of presentation and time of onset are dependent on the subtype, though CMT I (HSMN I) is most prevalent with patients typically developing symptoms within the first two decades of life.

Diagnosis and Presentation

Patients commonly present with nonspecific complaints of weakness, clumsiness, and unstable ankles resulting in frequent sprains.[28] The initial signs and symptoms on physical exam include distal weakness and muscle wasting, sensory loss, pes cavus or pes planus, foot drop with steppage gait, and reduced or absent DTRs.[27] Patients may also complain of pain in the lower limbs and lumbar spine. The diagnosis should be suspected based on the typical clinical presentation and a positive family history, and is supported by confirmatory molecular genetic testing.[29] For individuals with a negative family history, it should be noted that spontaneous mutations represent up to half of all cases.[30] The orthopedic concerns associated with this family of diseases most commonly involve the feet, manifesting as varus deformity, progressive pes cavus, and a host of foot deformities, though spinal deformity and hip dysplasia are also frequently seen. Careful examination is important in this population as the differential diagnosis of CMT based on physical exam includes poliomyelitis, spinal cord tumor, syringomyelia, and diastematomyelia.

Foot and Lower Extremity

Historically, CMT had been referred to by the general term "peroneal muscular atrophy," as the pattern of wasting typically involves the anterior compartment muscles and peroneus brevis while sparing the peroneus longus.[30–32] The foot deformities in CMT cover a spectrum of patterns including claw toes, forefoot and hindfoot cavus, hindfoot varus and foot drop. Loss of intrinsic foot muscles leads to claw toes early on, and continued unopposed pull from the peroneus longus and extensor hallucis longus (EHL) has been implicated as a possible cause of cavus foot.[33] Similarly, tibialis anterior muscle wasting may result in foot drop, and when the EHL is spared, this may result in a clawed hallux.[30] The presence of bilateral cavus feet is highly suggestive of CMT, and has been reported to be as high as 78%.[34]

A careful physical exam noting the presence or absence of specific muscle weakness and the rigidity or laxity of the deformities will inform surgical management. Clawtoes are an early-manifesting deformity and may be rigid at the time of initial presentation, though fore- and hindfoot deformities are oftentimes flexible and may remain so for years. Once plantar flexion of the first ray becomes rigid, the heel necessarily moves into a varus position during weight bearing. The Coleman block test, in which patients stand on a block and place their weight on the lateral border of the foot, can be used to evaluate the flexibility of this cavovarus deformity. Noting that the talus tilts out of varus and into valgus will indicate that the deformity is still flexible.[35] Conversely, ligamentous laxity is often present in this population and may lead to chronic inversion ankle sprain secondary to talofibular and calcaneofibular insufficiency.[30] Coleman block test results, in addition to ankle films to evaluate joint alignment and bony deformity as well as hindfoot alignment, are crucial to planning conservative or surgical management.[36]

Conservative Management

There is no evidence that bracing these children will ultimately prevent progression of foot deformities. However, functional bracing, such as ankle-foot orthoses (AFOs), may be helpful in addressing foot drop and chronic ankle sprain. Higher functioning and athletic patients may compensate on their own by adjusting stride height or increasing circumduction. For these patients, a simple ankle instability brace may be sufficient support to prevent frequent ankle sprain. Stretching the posterior compartment muscles may also help to decrease the rate of progression and allow for longer brace wear versus early surgery, and exercise and physical therapy may also be helpful to address weakness and contractures.[37] Use of botulinum toxin (Botox) to address the imbalance of agonist-antagonist muscles that are the ultimate cause of deformity has been preliminarily investigated and, thus far, found to be ineffective.[38]

Surgical Management

When considering surgery in this population, the surgeon must weigh the risks and benefits of early surgery while deformity is relatively flexible versus pursuing conservative management and attempting to stave off surgery until onset of a more rigid deformity. Furthermore, there are no strong data to inform the appropriate timing of surgery. Surgery must be individualized to patients based on the character and location of the deformity while taking into careful consideration the presence of deforming muscular imbalance. The author's preferred approach is to perform early soft tissue releases and rebalancing to prevent development of bony rigid deformity. Some argue that in patients with sensory deficits, fusion should be avoided.

Often, younger patients will do well with simple soft tissue release procedures. Release of the plantar fascia near its origin can help to correct cavus deformity. Additional soft tissue procedures include transfer of the peroneus longus to the peroneus brevis, which augments foot eversion while decreasing plantar flexion force on the first ray, and the Jones transfer, in which the EHL tendon is transferred to the first metatarsal neck, allowing this muscle to assist in dorsiflexion of the foot while avoiding claw toe deformity.[39] Transfer of the tibialis posterior tendon to the second or third cuneiform or around the second metatarsal may be used to centralize muscular forces and enhance dorsiflexion.[28] When strength of the tibialis anterior is spared, complete or split tendon transfer to the lateral cuneiform or cuboid can help to lateralize forces and augment eversion. This may be used in combination with lengthening of the tibialis posterior to further decrease inversion forces with good long-term results versus triple arthrodesis.[40] There is limited evidence regarding the success of soft tissue procedures when used alone, likely the result of the high diversity in patients' deformity and deforming forces at play, though in one study a small group of patients undergoing early operative soft tissue intervention avoided triple arthrodesis at long-term follow-up.[41]

Older children and those with rigid deformity often require osteotomies in combination with tendon transfers to achieve adequate correction. Options include dorsiflexion osteotomies for plantar-flexed first ray, midtarsal osteotomy, medial cuneiform osteotomy, and calcaneal osteotomies when hindfoot varus is present.[36] Triple arthrodesis is typically only used as a salvage procedure for fixed deformity as there is a high incidence of developing degenerative joint disease in the foot and ankle[36,42] and poor patient satisfaction related to appearance, function, pain, and need for reoperation.[43,44] Good functional results following triple arthrodesis at long-term follow-up have been reported, though in these studies soft-tissue balancing was used when indicated.[44,45] Despite the high frequency of ankle pain and laxity, ankle arthrodesis is rarely indicated and total ankle replacement is contraindicated due to failure related to uneven medial-sided wear patterns.[30]

Hip Dysplasia

Hip dysplasia in CMT was first reported in a case series by Kumar et al.[46] in the early 1980s, and has subsequently been reported at a rate of ~8%.[46,47] Hip dysplasia also has a greater association with CMT-1 than CMT-2, though the true prevalence remains unclear.[47] It has been suggested that the alteration in gait with resultant change in biomechanics, including increased external rotation and decreased adduction of the hips, may contribute to hip subluxation and dysplasia later in life.[48] While case reports exist documenting the incidence of hip dysplasia as a sentinel sign suggestive of CMT, the association between hip dysplasia and CMT is likely underreported.[47–50] Occasionally, patients have hip dysplasia present at birth, and in some cases have required proximal femoral or pelvic osteotomy to correct coxa valga and painful subluxation.[48,51]

Scoliosis

Scoliosis is not uncommon in individuals with CMT, and several studies have placed the prevalence of spinal deformity at up to 38%.[52,53] In one study, 33% of individuals with CMT and scoliosis had left-sided curves and spinal canal abnormalities, and half were hyperkyphotic.[54] However, curve pattern is highly variable in CMT. Patients most often exhibit right-sided single thoracic curves, though multiple curves are also frequently noted.[55] Management is similar to that of other neuromuscular scoliosis, and depends on the curve pattern, degree of deformity, and progression. Increased risk of progression has been associated with nonambulatory status, hyperkyphosis, and curves greater than 30 degrees. Bracing has been found to be ineffective at halting curve progression.[54] Fusion should be considered once the curve reaches 45 degrees.

Upper Extremity

While the lower limbs are typically affected earlier than the upper limbs, progressive decline in hand function has been noted in the literature as the decrease in manual function has a significant impact on disability and quality of life in CMT.[56–58] CMT manifests in the upper extremity as an intrinsic minus deformity with decreased stability, dexterity, grip strength, and tactile sensation,[56] and typically presents later in life, though there are some case reports of presentation of symptoms in the first decade.[59] Physical and occupational therapy may be helpful for these patients, and infrequently, tendon transfer, nerve decompression and soft tissue release, and joint arthrodesis may also be helpful.[60]

FRIEDREICH ATAXIA

Background

Friedreich's ataxia (FA) is caused by an autosomal recessive trinucleotide expansion of *frataxin.* The mutation results in progressive degeneration of the posterior columns of the spinal cord and the common clinical picture of ataxia, sensory loss, weakness, and loss of DTRs with onset of symptoms prior to age 25 years. It is the most common of the early onset hereditary ataxias, affecting approximately 1/50,000 Caucasians. The orthopedic manifestations of this disease include kyphoscoliosis and pes cavus, which typically develop in the second decade of life.[61]

Oftentimes, patients will exhibit scoliosis prior to onset of systemic symptoms, and the curve is typically more suggestive of adolescent idiopathic scoliosis (AIS) rather than neuromuscular disease. Labelle et al. identified scoliosis in each of a series of 56 patients and found that an early diagnosis of ataxia before age 10 and scoliosis before age 15 were independent risk factors for curve progression to greater than 60 degrees, at which point they were surgically stabilized. For those in whom scoliosis was not diagnosed until they had reached age 15, the curves typically did not progress beyond 40 degrees by skeletal maturity.[62]

In a recent case series of 31 consecutive patients with FA, 24 males and 7 females, all presented with scoliosis and a mean curve of 51 degrees.[63] The distribution of curves was nearly equal between single thoracic, thoracolumbar, and double thoracic or lumbar, and only two of the patients in the series exhibited the sweeping C-shaped curves typical of neuromuscular disease. The authors note that the high proportion of left-sided thoracic curves (9/20, 45%) help to differentiate scoliosis in FA versus that typically seen in AIS. Additionally, 29% of these patients (i.e. those who were observed to skeletal maturity and those who presented initially at skeletal maturity) did not require surgical intervention. This finding is supported in the literature, as individuals with an identifiable later onset of scoliosis did not necessarily progress to surgery.[62,64] Another study of 77 patients identified as having FA reported a 63% (49/77) prevalence of scoliosis. Of these 49 patients, 27 were male and 22 were female. The greater proportion of males reported is an additional factor that may serve to differentiate scoliosis seen in FA from that seen in AIS.[64,65] The study also reported 8 left-sided thoracic curves and 12 individuals exhibiting hyperkyphosis, which in the opinion of the authors provides further evidence supporting the difference between FA and AIS curves. Sixty percent of the patients in the series had progression of scoliosis greater than 6 degrees, though there were no associations between age, gender, or initial curve magnitude and risk of progression. An additional observation was the prevalence of hyperkyphosis, which has been reported in the literature at frequencies between 24.5% and 66%.[62,65] Thoracic hyperkyphosis beyond 40 degrees is also associated with the progression of a coronal curve beyond 40 degrees.

Bracing in FA

In two studies, bracing was attempted in a total of 19 patients with progression to a surgical curve occurring in 15 individuals.[64,65] Average progression of the curve despite bracing has been reported between 11 and 15 degrees per year.[64,66,67] Bracing in this population is fraught, and inability to control coronal progression is well documented. For many patients, bracing may make walking difficult as the brace interferes with balancing truncal movements, which may limit adherence to bracing to several hours per day or may prevent it altogether.[23] Given the possibility, though slight, of slowing progression, a trial of bracing in patients with FA should be considered on a case-by-case basis. Ambulatory patients with curves between 25 and 30 degrees may benefit from a brace, though it remains unlikely that these patients will ultimately avoid surgical intervention.

Surgical Intervention

Surgical intervention is indicated for curves greater than 60 degrees and in patients with 40 to 60 degree curves with an early presentation and high risk for progression. Consideration of the cardiomyopathy that frequently accompanies FA may warrant perioperative consultation with a cardiologist, and severe hypertrophic cardiomyopathy may be a relative contraindication for surgery in this population.[64] Intraoperative neurophysiologic monitoring is limited in this population, and SSEPs and MEPs have been found to be inconsistently reproducible. Prior to surgery, alerting the anesthesia clinicians to the likelihood of an intraoperative wake-up test may be indicated. Overall curve correction has been reported to be between 33% and 66% depending on curve type.[64,67,68]

Instrumentation typically extends to the lower lumbar spine distally, and fusion to the pelvis may be considered if the patient is nonambulatory. Fusion to T3/4 may prevent the development of proximal junctional kyphosis (PJK) secondary to poor truncal control and weak musculature. The authors recommend use of modern posterior spinal instrumentation techniques with pedicle screws and autogenous as well as allogenic bone graft.

Pes Cavus

The prevalence of both pes cavus and scoliosis in FA are associated with increasing GAA expansion size, which is thought to reflect the early onset of peripheral neuropathy seen in individuals with larger trinucleotide repeat sequences.[69] The fixed equinovarus foot deformity seen in FA has an impact on independence and mobility, particularly standing transfers. If soft tissue procedures and osteotomies fail to control the deformity, triple arthrodesis is a useful salvage procedure to straighten and stabilize the foot.[36] Given the high incidence of postoperative complications (3/7 subjects), preventative strategies (physical therapy, splinting, Botox injections) should be considered to avoid definitive surgical management, if possible. Aggressive postoperative rehabilitation will expedite recovery and healing, especially in patients who have been nonambulatory for extended periods of time prior to surgery.[69]

Muscle Spasms

In addition to progressive ataxia and kinetic tremor, other movement disorders may present in FA, including axial and limb dystonia, chorea, and myoclonus.[70] A number of case reports describe hyperkinetic movement disorders in FA, and the presence of dystonia has been noted in other spinocerebellar ataxias.[70–72] Muscle spasm is generally of short duration and can be effectively treated with conservative measures, including massage, application of heating pads, and analgesics. Case reports have described the successful application of spinal cord stimulators and intrathecal baclofen pumps.[73] In more severe cases, use of muscle relaxants such as diazepam may be warranted. Special consideration should be taken for patients undergoing operative management of equinovarus deformity with a history of muscle spasm and dystonia, as they have increased risk of postoperative exacerbation secondary to tendon transfers. These patients may require diazepam or baclofen for an extended period following surgery.[23] In nonambulatory adult patients, thigh adductor and quadriceps spasm may limit their ability to perform perineal hygiene or prevent them from maintaining a comfortable seated position. When unresponsive to medical management or Botox, tenotomy has been used to improve symptoms.[23]

DUCHENNE MUSCULAR DYSTROPHY

Duchenne muscular dystrophy (DMD) is the most common of the inherited muscular dystrophies with an approximate incidence of 1 in 3500 live male births.[74] It is an X-linked recessive disease presenting with the typical dystrophic syndrome secondary to an absence of dystrophin protein.[75] While it is typically an inherited mutation, up to 30% of cases result from spontaneous mutations.[76] The disease is characterized by the gradual development of proximal to distal weakness and dystrophic changes in cardiac and skeletal muscle. Typically, these patients have delayed motor development and exhibit early signs of gait abnormality resulting from the replacement of muscle by fibrofatty tissue. Many of these children will lose ambulatory ability by the end of the first decade, and death secondary to progressive respiratory weakness or cardiomyopathy frequently occurs before age 20.[77]

DMD should be suspected when children have delayed motor development and a typical wide-based gait once they do begin to ambulate. Pelvic girdle weakness causes a compensatory lumbar hyperlordosis when standing, and contributes to reliance on the Gowers' maneuver to rise from the floor.[78] Signs of proximal weakness in the setting of a creatine kinase >5000 units/L are highly suspicious for Duchenne and Becker muscular dystrophies, and the diagnosis may be confirmed by finding a mutation of the dystrophin gene on DNA analysis.

Extremity Contractures

As a result of the typical progressive muscle weakness, individuals with DMD may tend towards leaving joints in static positions of relative flexion predisposing them to the development of joint contractures.[79,80] Therapy focused on active and passive stretching may be useful to prevent or delay the development of contractures.[81] One of the first deformities to develop in individuals with DMD is ankle equinus, resulting from the predilection to stand and walk on the toes in order to maintain knee extension (secondary to quadriceps weakness) and aid in standing by centering body mass.[82] For this reason, in the setting of rigid contracture, percutaneous or open Z-lengthening of the Achilles with release of fibrotic tissue may be attempted, and posterior tibialis transfer to augment dorsiflexion and eversion may also aid in balancing the foot in the setting of hindfoot varus.[78] Flexion contractures of the knee greater than 15 degrees may be addressed through hamstring lengthening. Patients should be braced following surgery, and early mobilization is encouraged to prevent further lower extremity atrophy. Achilles lengthening should be used selectively as overlengthening may decrease the time to loss of ambulation, and evidence supporting definitive indications or a specific time to perform surgery in these individuals is lacking given the high variability in patients' particular ambulatory status. Surgery performed in the late ambulatory phase and after patients have ceased walking has been found to be generally ineffective and is contraindicated.

Spinal Deformity

Patients with DMD are at increased risk of spinal deformity owing to the loss of ambulation and dependent sitting that typically occurs between 9 and 14 years of age. Paraspinal weakness and a lack of support for the spine during the phase of maximal growth velocity results in development of a typical, c-shaped curve with rapid progression that severely impairs sitting ability and may affect cardiac function and respiratory vital capacity (Figure 52.3).[83,84] While long-term glucocorticoids have been shown to slow curve progression, they have also been associated with increased risk of fracture and osteoporosis.[79,85] The prevalence of scoliosis in this population approaches 90% and patients generally require early operative intervention.[86] Bracing in DMD has not been shown to prevent the progression of the curve, and should be considered as a means to improve sitting balance along with wheelchair chest wall support. However, because of poor patient tolerance of bracing and

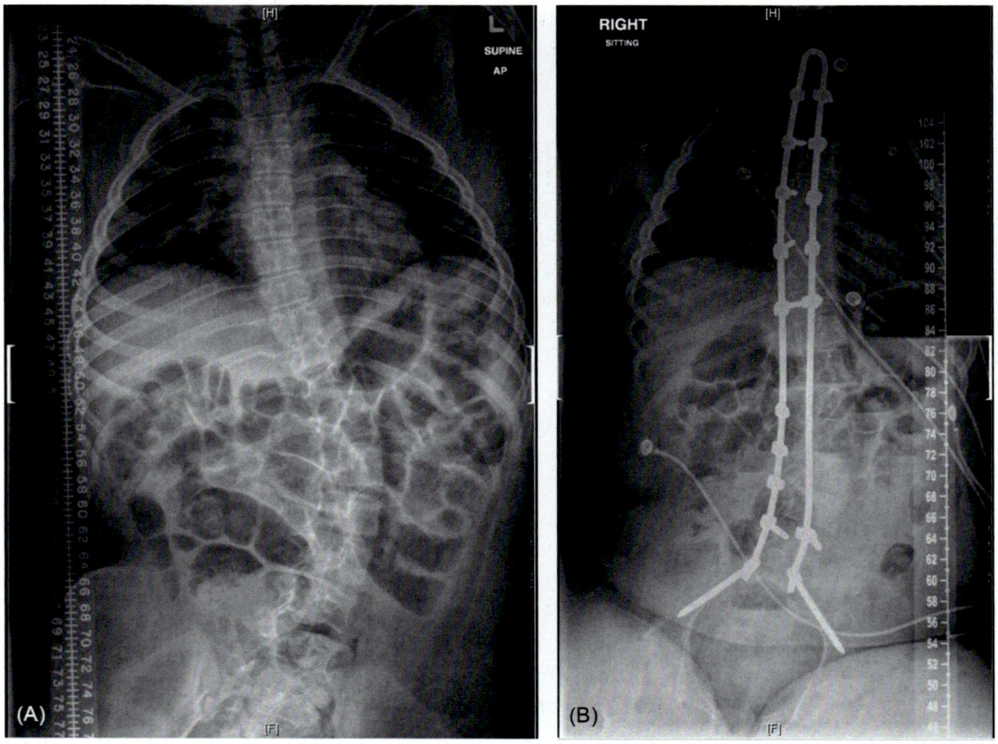

FIGURE 52.3 Preoperative (A) and postoperative (B) AP radiographs of a patient with DMD demonstrating the left-sided curve typical for this population. The patient was nonambulatory preoperatively, and spinal instrumentation was performed to improve sitting balance.

the risk of skin pressure sores, this modality should be used only in those individuals who are not surgical candidates.

At cessation of walking, patients should be regularly screened for scoliosis. Patients with a curve greater than 20 degrees should be closely monitored. Typical practice is to opt for early fusion, when the curve approaches 40 degrees. While several short-term and five-year-out studies failed to support this claim,[87,88] others have found that early spinal fusion can help maintain vital capacity for up to 8 years postoperatively.[89,90] A recent Cochrane review of scoliosis surgery in DMD failed to find evidence demonstrating positive effects of surgery on respiratory function or overall survival, and the attributable causes of mortality (respiratory infection or failure, progressive cardiomyopathy, sudden cardiac death) did not differ between operative and nonoperative groups.[87] However, correction of scoliosis did result in improved sitting and overall quality of life.[87] While the lack of randomized controlled trials curtails any definitive guidelines, patients with an FVC less than 35% have an increased risk of pulmonary insufficiency following surgery, and because loss of pulmonary function increases with magnitude of the curve and progressive muscle wasting, consideration of early correction is strongly recommended.[88,91] Additional considerations regarding surgical timing include the increased blood loss as fibrofatty tissue replaces paraspinal musculature in older individuals with DMD and the presence or likelihood of vertebral fractures secondary to chronic glucocorticoid use.[92]

BECKER MUSCULAR DYSTROPHY

Becker's muscular dystrophy (BMD) is an X-linked recessive, inherited mutation typically resulting in a qualitative mutation in dystrophin rather than the complete absence of the protein seen in DMD.[76] While the clinical manifestations of BMD are similar to those seen in DMD, onset of wheelchair dependence and cardiomyopathy typically occur later in life. One study found that, on average, patients with BMD became nonambulatory at age 27 years, and cardiac or respiratory-related death occurred early in the fourth decade, though wide age ranges were reported for onset of morbidity and mortality owing to the variability in dystrophin production.[93,94]

Orthopedic management of BMD and DMD is also similar. It has been suggested that patients with BMD may benefit from early bracing with AFOs or knee-ankle-foot orthoses (KAFOs) because they tend to retain muscle strength for a longer period of time and as a result are less prone to develop rigid contractures.[23] After loss of ambulatory status, patients should be monitored for the development of scoliosis and are candidates for early spinal arthrodesis to prevent progressive decline in respiratory status.

CONGENITAL MYOTONIC DYSTROPHY

Congenital myotonic dystrophy (CMD) is an inherited trinucleotide repeat disorder manifesting as profound muscle weakness during infancy and progressive distal weakness and myotonic contractions that appear during the second decade. Children with CMD present with early hypotonia and characteristic facies (long and narrow with bitemporal narrowing), and typically have multiple organ involvement including endocrine disturbances, conduction defects, and dysphagia.[95–97] Typical orthopedic manifestations involve the feet and spine, though hand weakness may also be seen in the second decade. In one retrospective series of 30 individuals, 73% required orthopedic surgery for lower extremity or spinal deformities, most commonly equinus, equinovarus foot, clubfoot, and scoliosis.[98]

Ambulation is typically delayed in these children, and distal weakness contributes to the development of foot deformity as well as the inability to walk. Clubfoot in CMD is less rigid than in other syndromes, such as arthrogryposis or amyoplasia. It can be effectively treated with early nonoperative intervention, including serial casting, heel stretching, and nighttime bracing, which may circumvent the need for surgery later in life.[23] When severe, these patients may require posteromedial surgical release or talectomy in order to achieve adequate correction.[99] Often, patients require AFOs to aid walking following stretching or surgical intervention, and concurrent use of Lofstrand crutches may be helpful. Truncal weakness is implicated in the development of spinal deformities, usually by age 8 to 10 years.[98]

Surgeons should be aware of the high risk for perioperative complications related to inhalational anesthetics, especially given the frequency of cardiac arrhythmia in this population. Careful dosing of anesthetics may prevent the need for prolonged postoperative assisted ventilation.[98]

CONGENITAL MYOPATHIES

The congenital myopathies are classified based on findings seen on muscle biopsy. Patients require multidisciplinary care for the various pulmonary, GI, and neurologic complications common to the family of congenital myopathies.[100] With regard to musculoskeletal involvement, patients suffer from generalized weakness or hypotonia, spinal and foot deformities, and poor skeletal muscle bulk.[101] Weakness typically manifests as proximal or limb-girdle weakness, though foot drop may also be seen.[102] Generalized weakness may manifest in other organ systems as evidenced by frequent and early onset ophthalmoparesis, and respiratory insufficiency requiring ventilation. In certain instances, and when appropriately managed, children with congenital myopathy do extremely well and go on to enjoy a childhood and adolescence approximate to that of the general population (Figure 52.4).

Though the clinical features of congenital myopathy often overlap with CMD, facial and bulbar weakness with distinct facial features including tented upper lip and high palate can differentiate this group of disorders from congenital muscular dystrophy.

Nemaline, or rod body, myopathy is characterized by variable time of onset and clinical severity. Initial presentation with neonatal hypotonia and weakness has been associated with high mortality rates. In one large multicenter study of 143 patients diagnosed with nemaline myopathy, several suffered from arthrogryposis and congenital fractures.[103] In early onset disease, patients may be slow to reach milestones such as independent sitting and walking, and may ultimately be wheelchair dependent. Poor motor function and trunk support may lead to development of contractures and thoracolumbar scoliosis in later life.[103]

Central core disease typically presents in infancy with hypotonia and motor delays. The common orthopedic manifestations of this disease include congenital hip dislocation, early-onset scoliosis, joint laxity, and foot deformity.[104] Scoliosis associated with central core disease tends to follow a pattern similar to AIS. From a perioperative management standpoint, the surgical team should be aware of the high susceptibility to malignant hyperthermia in these individuals.[104] The need for operative intervention can be circumvented or delayed through a program of physical therapy and exercise aimed at improving axial power, truncal stability, and muscle endurance.

ARTHROGRYPOSIS

Introduction

Arthrogryposis multiplex congenita represents a heterogeneous group of diseases with a common phenotype of at least two joint contractures present at birth. It is unclear in the literature as to how many specific disease entities arthrogryposis represents, though some surmise the number to be greater than 300.[105] This group of disorders encompasses neuromuscular disorders, genetic anomalies, and skeletal dysplasia, which share the common characteristic of periods of decreased fetal movement and resultant joint contractures at birth. These have typically been divided into three functional subgroups, including those disorders primarily affecting the limbs, disorders with involvement of both limb and other body parts, and arthrogryposis with CNS dysfunction.[106] Children from the first two subgroups represent the majority of living individuals, as neonatal mortality may approach 50% in the third subgroup.[107] While the exact and specific causes are largely unknown, this group of disorders may be thought of in terms of likely etiology that includes neurologic dysfunction, dysfunction at the level of muscular structure or function, connective tissue abnormalities, *in utero* space limitation, maternally transmitted disease, and failure of neurovascular structures to form *in utero*.[105] A separate entity in this spectrum, termed amyoplasia, is used to describe those individuals

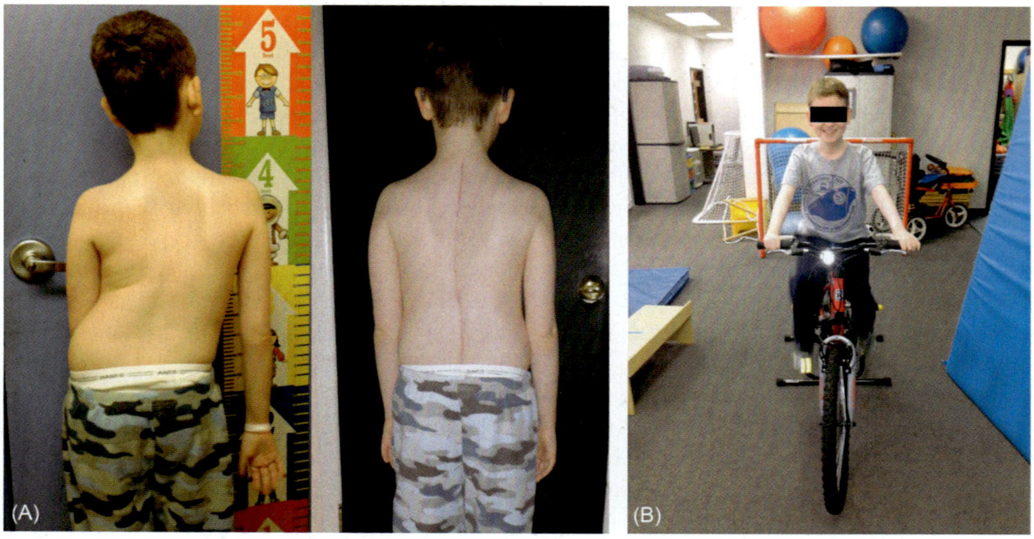

FIGURE 52.4 Preoperative (**A**) and postoperative (middle and **B**) pictures of a patient with centronuclear myopathy. The lumbar curve progressed to 60 degrees by the last preoperative visit, and the patient was having respiratory issues and pain. Spinal instrumentation and pelvic fixation was performed and the rightmost figure (**B**) shows the patient in rehab postoperatively.

with symmetric joint contractures related primarily to defective muscle development. It involves all four limbs greater than 80% of the time, and elbow and knee contractures, rigid flexion of the wrist, and flexion/abduction contractures of the hip are common.[108,109] Hip dysplasia and scoliosis develop in this population approximately 30% of the time, and abnormalities of the skin overlying affected joints as well as midline facial hemangioma may also be seen.[109]

While the diagnosis of amyoplasia is relatively straightforward and can be made with plain films documenting limb deformities, the presence of systemic or syndromic abnormalities may prompt further diagnostic and genetic testing including muscle biopsy, advanced imaging, or chromosomal studies. Treatment of deformities is performed to improve function and increase independence. Children with arthrogryposis require multidisciplinary care and early physical therapy involving range of motion and stretching exercises with posttherapy splinting. Many of these children will not have ambulatory potential, and clarifying the goals of therapy with families and patients is crucial in order to manage expectations and ensure a focus on augmenting functional independence.[110,111] With appropriate therapy, one study reported that 70% of children were able to perform activities of daily living and may preserve that function into adulthood, though more often than not some assistance will be required.[112] Surgery is performed to augment the gains made through physical therapy. Soft tissue release followed by casting is the first line of treatment; and tendon transfer and bony surgery may be indicated in cases of severe deformity.[23,109]

Perinatal Fractures

Clinicians should have a high suspicion for perinatal fractures, especially in the setting of soft tissue swelling and irritability shortly after birth. In one study of 36 newborns with arthrogryposis, there were 16 perinatal fractures in 9 individuals encompassing metaphyseal, diaphyseal, and Salter-Harris type III fracture patterns.[113]

Clubfoot

The incidence of clubfoot in arthrogryposis is 78% to 90%, and its management is similar to that utilized in idiopathic clubfeet.[114] Casting is the first line modality, though surgery may be required for definitive treatment as the deformity is often rigid and deformity frequently recurs postcasting.[111,114,115] Early posteromedial soft tissue release is indicated, and in the event that equinus deformity is not corrected, heel cord lengthening and tibiotalar and talocalcaneal capsulotomy may aid in correction of residual deformity.[23] External fixation and growth modulation has also been used to treat severe clubfeet (Figure 52.5). Several case series report success at long-term follow-up approximately half of the time with primary talectomy for arthrogrypotic clubfoot, and success rates of 82% have been reported when it is used as a revision procedure.[114,116,117] Care should be taken to ensure that the talus is completely excised as failure has been reported when part of the talus is left behind or becomes positioned beneath the tibia. Talectomy allows the foot to fall into dorsiflexion while maintaining the articular cartilage and growth potential.[99] Most authors prefer to use this operation as a salvage

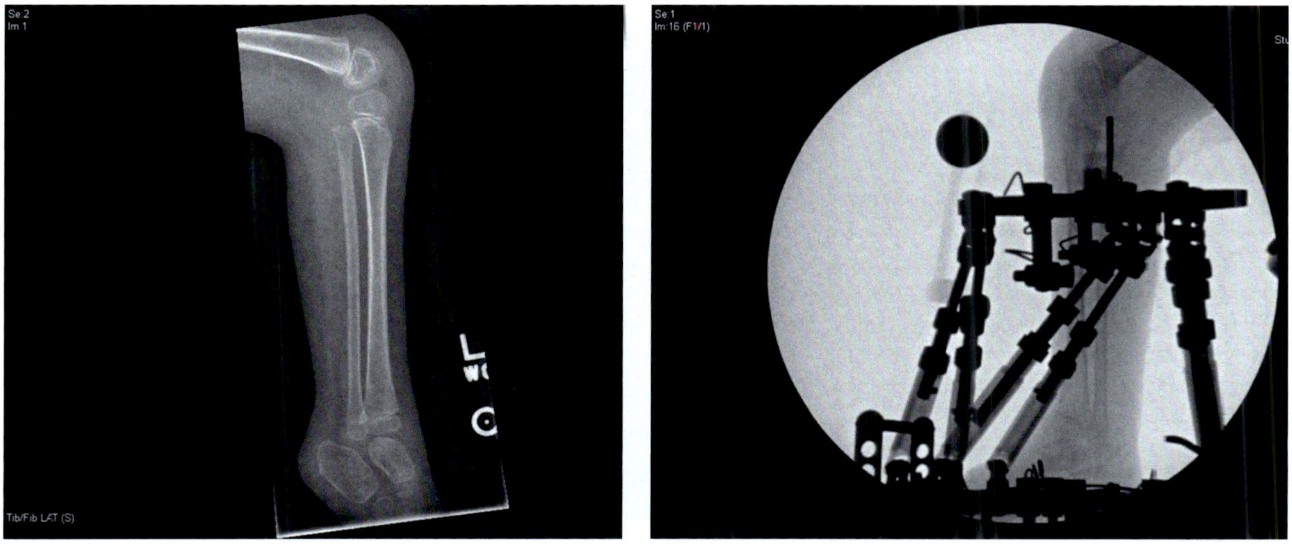

(A) (B)

FIGURE 52.5 Preoperative (A) and intraoperative (B) radiographs of a patient with arthrogryposis treated with an external fixator and growth modulation to address lower extremity contractures.

procedure as its revision options are limited.[109,118] Relapsed clubfoot has been treated with talectomy (if possible), decancellation of the talus and cuboid, or through use of an external fixator, which may help to achieve gradual correction.[119] Subtalar fusion may be indicated to correct hindfoot valgus, and in some a triple arthrodesis may be required to achieve a plantigrade foot.[109] Postoperative bracing with a molded AFO and nighttime brace wear is helpful in maintaining correction, though recurrence has been reported up to 73% of the time in individuals undergoing circumferential release.[116]

Knee Contractures

Up to 70% of patients have disabling flexion or extension contractures of the knees.[109] Flexion contractures occur more frequently (48%) and are more difficult to correct through nonoperative management than extension-type contractures, and patients with flexion contractures are half as likely to become community ambulators.[120] While most flexion deformities are mild, contractures greater than 90 degrees have been reported.[120]

Only 25% of flexion deformities will respond to flexion and manipulation, and the decision to pursue operative management should take into account the likelihood of ambulation in the future.[105] In mild contractures of less than 20 degrees, hamstring lengthening, posterior capsulotomy, and posterior cruciate ligament release may be adequate to allow for ambulation when used in conjunction with KAFO. If posterior release proves inadequate to achieve extension, there may be anterior adhesions requiring release. This procedure is typically done separately from the initial posterior operation as releasing both aspects of the knee risks significant stretch injury to the popliteal artery and nerve.[109] For flexion contractures between 20 to 40 degrees, serial casting and soft tissue procedures can typically address the deformity. When contractures exceed 40 degrees, a period of initial casting followed by tendon lengthening or extension osteotomy of the distal femur may be necessary to achieve adequate correction. A trapezoidal wedge osteotomy may also be performed, though this does not affect joint position and may be cosmetically unacceptable.[109] Extension osteotomy has been associated with recurrence and the need for reoperation as the child nears skeletal maturity, as well as the high risk of neurovascular insult.[23] Severe contractures greater than 80 degrees may be treated with soft tissue release or Ilizarov external fixation.[105,119]

Children with extension contractures are typically able to walk because the knee in extension is inherently stable; however, this deformity causes difficulty in sitting or rising from a seated position. Quadriceps lengthening or tenotomy in conjunction with passive range of motion exercises may help to address these issues, though the patient and family should be made aware of the likelihood of needing long-term orthotic support.[23] Care should be taken not to over-lengthen the knee in older children as this might convert the correction into a flexion contracture.

Hip Deformities

The characteristic hip deformity in arthrogryposis is a flexion-abduction-external rotation contracture with a resultant externally rotated gait. Unilateral or bilateral dislocation is common, occurring in 15% to 30% of individuals. When hips are not dislocated, early stretching exercises may help to improve positioning.[121] These dislocations are typically not amenable to closed reduction. Open reduction through an anterolateral or medial approach may be an option if the child has good hip range of motion, though there is considerable risk of redislocation, poor function, and AVN.[122,123] Regardless of risks, open reduction should be accompanied by soft tissue release and varus derotational osteotomies (VDROs).[23]

Surgery is indicated when hip contractures exceed 45 degrees, and intervention typically begins with soft tissue release and spica casting for several weeks. If deformity or subluxation persists, proximal femoral osteotomy may be used to address flexion and abduction and place the lower limbs in weight bearing position. Hip procedures are aimed at improving the child's ability to walk, and bracing in combination with the additional necessary lower extremity procedures to facilitate ambulation, extensive physical therapy, and crutches or walkers are an important aspect of postoperative care. In one study, the factors most closely associated with achieving ambulation included strong hip extension, knee contractures less than 20 degrees, plantigrade feet, and mild or corrected spinal deformity.[124]

Upper Extremity Deformities

Although these patients may have significant upper extremity deformity, often they are still able to perform general self-care and get around via wheelchair or crutches.[105] When performing surgery for upper extremity deformity, it is important not to lose sight of the ultimate goal—to preserve or improve function—and not sacrifice function for the sake of appearance. The classic upper extremity deformity is downsloping shoulders with internal rotation of the humerus, and elbow extension with wrist and finger flexion. The shoulders are frequently involved in amyoplasia. The typical upper extremity deformity makes feeding and grooming difficult as elbow extension with poor biceps function combined with rigid hand deformity make bringing the hands to the face and mouth particularly difficult. Despite these deformities, children prove to be highly adaptive, using the environment to their advantage in dressing, writing, and eating.[109]

Extensive use of passive range of motion in the very young is recommended, as this modality can improve wrist and finger contractures and may decrease the extent of or need for future surgery. Starting proximally, the humerus may require derotational osteotomy proximal to the insertion of the deltoid. Elbow extension contractures can be treated with posterior capsulotomy or triceps tenotomy, lengthening, or anterior transfer if early physiotherapy is not able to achieve significant gains in range of motion.[125–128] Transfer of the pectoralis major to improve elbow flexion may also be useful, though only if the muscle is adequately powerful to flex the elbow. Proximal row carpectomy may help address the flexion ulnar deviation, especially when combined with transfer of wrist flexors (the flexor carpi ulnaris in particular) to the dorsum of the hand as well as centralization of wrist extensors.[128] Improving passive wrist extension using serial casting prior to transfer can improve the chance of success with this procedure. Additionally, wrist fusion in older patients, and in some cases widening the first web space, may improve grasp or the ability to hold an object by weaving it through the fingers.

Scoliosis

Up to 66% of patients with arthrogryposis will develop scoliosis, though age of onset and pattern of deformity are highly variable and depend on the degree of muscle weakness.[129] A long C-shaped curve is seen in 30% of patients with amyoplasia and frequently presents in adolescence during periods of increased growth.[129] Early onset scoliosis in children with suspected amyoplasia should raise suspicion for another contributing etiology. Similarly to other neuromuscular-related scoliosis, bracing is typically ineffective, and instrumentation is often necessary.[130]

CEREBRAL PALSY

Introduction

The commonly accepted definition of cerebral palsy (CP) describes it as "a group of disorders of the development of movement and posture, causing activity limitation, that are attributed to non-progressive disturbances that occurred in the developing fetal or infant brain. The motor disorders of cerebral palsy are often accompanied by disturbances of sensation, cognition, communication, perception, and/or behavior, and/or by a seizure disorder."[131] CP often results from chromosomal aberrations, metabolic insult in the perinatal period, intrauterine infection, or fetal hypoxia. Individuals with CP have historically been under the care of orthopedists as the musculoskeletal insults may be severe and involve spinal deformity, hip subluxation and dislocation, and distal extremity deformity. The range of disability that exists within the spectrum of CP necessitates multidisciplinary care, and epidemiologic studies have shown that the types of specialists caring for this population have become more heterogeneous in the past several decades.

Scoliosis

The reported incidence of scoliosis in the population with CP is dependent on a number of variables, including ambulatory status, age, severity of motor disease, and gender, with males being more frequently affected than females.[132–134] Additionally, the curve type (i.e. resemblance to AIS versus the C-shaped curves typically seen in neuromuscular disease) is also associated with the degree of involvement, and poor truncal control is anecdotally considered to be a major contributor to the development of scoliosis.[135] There is evidence of an association with hip dislocation, though the causal relationship (whether the spinal deformity and fixed pelvic obliquity causes hip dislocation or vice versa) has not been fully elucidated.[136,137]

Patients typically present in the first decade with a flexible curve that may rapidly progress during adolescence. The risk of progression following peak growth has been reported to be between 0.8 and 4.4 degrees per year, with higher risk of progression in individuals with curves greater than 50 degrees and more severe musculoskeletal involvement.[138–140] Similarly to patterns seen in AIS, one study described that patients with a curve >40 degrees by age 15 progressed to a deformity >60 degrees 85% of the time, whereas only 13% of those with a curve <40 degrees progressed.[141]

Assessment of spinal deformity in the individual with CP should be geared toward determining optimal management strategies. Careful assessment of functional and ambulatory status, range of motion, tone, and motor strength are crucial components of the neuromuscular exam. With respect to the spine, evaluation should take note of presence of rib hump or Mongolian spots, curve flexibility, shoulder balance, and pelvic obliquity, as well as presence of hip pathology or contractures that may limit the extent of possible spinal correction as hip contracture may prevent postoperative sitting or lying.[132] In addition to evaluating curve magnitude, coronal and sagittal balance, and pelvic parameters, plain radiographs may reveal the presence of rotational deformity, rib fusion, and vertebral body wedging, and may help to elucidate the contribution of the deformity to pelvic obliquity. While many of these children will be unable to make use of standard standing scoliosis cassettes with side bending films, using traction, push-pull radiographs, or bracing can help to evaluate the flexibility of the curve.[135] MRI is not routinely necessary, though may be indicated in instances of rapid progression, severe deformity, or high suspicion of tethered cord.

Bracing in neuromuscular scoliosis is generally not indicated, though braces or wheelchair inserts may be used to improve sitting position and provide comfort.[142,143] While a baclofen pump may be applied for symptomatic treatment of spasticity, this modality does not increase the magnitude of surgical correction and has been associated with increased postoperative complications.[144,145]

Spine surgery should be considered in patients with curves greater than 40 degrees with risk of progression and interference with sitting, hygiene, or nursing care. Additional consideration should be given to an individual's Gross Motor Function Classification Scale (GMFCS) category. In a recent study, GMFCS Level V individuals progressed an average of 3.5 degrees/year in the coronal plane, versus 1 to 2 degrees/year observed in their GMFCS Levels I through IV counterparts.[146] Though there is mixed evidence to support the utility of spinal correction based on objective measures (pain relief, disease burden, risk for ulceration, functional gains),[147,148] parents and caregivers are generally pleased with results of surgery based on ease of care, comfort, and appearance.[149,150] The importance of patient-reported outcomes and the role of the family in surgical decision-making cannot be understated as caregivers are categorically excellent advocates for these children, and best understand their needs.[151]

Operative decision-making includes choice of construct, choice of fusion levels, and fusion to the pelvis. The author typically uses pedicle screws, which provide excellent fixation and correction in large, rigid curves. Furthermore, when used in patients who have not yet reached skeletal maturity, 3-column fixation avoids the crankshaft effect and its associated high rates of reoperation, and may also decrease the need for anterior release in hyperkyphosis.[152] In light of the high risk of PJK and end of construct failure in the setting of hyperkyphosis and poor trunk control, proximal fixation should extend to T1 or T2.[153–155] Extending the fusion to the pelvis using sacral and iliosacral screws or S-hooks is indicated in the setting of high pelvic obliquity in patients with high tone, and when the patient is nonambulatory. Failing to fuse to the pelvis in these individuals may lead to recurrent pelvic obliquity. In a study of 23 individuals with neuromuscular scoliosis undergoing instrumentation ending above the pelvis, patients with pelvic obliquity <15 degrees maintained adequate fusion at 5-year follow-up.[156] If the patient is able to ambulate or actively participate in transfers, fusion to the pelvis may result in loss of function.

These patients are at high risk for postoperative pulmonary and gastrointestinal complications, most frequently pulmonary insufficiency.[154,157,158] These patients may need to remain intubated following surgery and slowly weaned from intraoperative opioids in order to minimize the risk of hypoventilation and aspiration. Additionally, these patients have an increased risk for postoperative respiratory and urinary tract infections, and surgical site infection has been reported in up to 10% of cases.[159] Overall, postoperative complication rates have been reported between 44% and 80%, and the mortality risk is reported to be as high as 7%.[154]

Hip Disorders

In addition to scoliosis, hip displacement is one of the most common problems encountered in cerebral palsy.

The natural history of spastic hip disease in CP is the development of major muscular imbalance progressing to lateral hip displacement over the course of early childhood. The hip is normal at birth, but by 2 years of age increased spasticity and short musculature in the setting of bone growth results in progressive force on the hip joint and subluxation of the femur on the acetabulum. Action of the iliopsoas and adductors pull the femoral head into a posterior-superior orientation with respect to the acetabulum, which may also become permanently deformed as the femoral head is forced into the capsule.[160] Aberrant flexion and adduction and lack of compressive forces during walking and sitting causes persistent femoral anteversion and coxa valga, ultimately resulting in incomplete coverage of the femoral head. The incidence of hip displacement is correlated to more severe musculoskeletal involvement, and is present in up to 60% of nonambulatory patients.[137,161] Failure to address early dislocation can lead to chronically painful hips by early adulthood as well as problems with positioning, hygiene, mobility, and chronic pain.

Several factors related to severity of motor dysfunction predict the likelihood of hip dislocation. Seventy-five percent of children with spastic quadriplegia and GMFCS Level V will have hip migration, and children with Winters, Gage, and Hicks Group IV (WGH IV) hemiplegic gait classification (pelvic rotation, hip flexion, abduction, and internal rotation) are at higher risk for dislocation.[162,163] It is important to assess the likelihood that CP hips are intact, through a thorough physical exam assessing abduction with the hip and knee in extension, range of motion at knees and ankles, and the presence of scoliosis and pelvic obliquity. The 2008 Consensus Statement on Hip Surveillance provides a framework for identifying the critical indicators of progressive hip displacement based on GMFCS level.[164] Early identification and management with the goal of maintaining flexible, located, painless hips can help to decrease the need for surgery and disease burden for this population.[165] Well-located hips by age 5 predict increased probability of hip stability and appropriate acetabular development.

In addition to close follow-up and a careful physical exam, radiographic parameters can be useful in the observation of gradual hip subluxation. Migration percentage (MP) describes the amount of femoral head not covered by the acetabulum. In CP, it is common to see an average increase in the MP of 5.5% per year, while an increase greater than 10% over 12 months indicates hip joint instability and the likelihood of intervention.[164] Additional indicators of progression include deterioration in range of motion, change in muscle tone, altered gait or decreased tolerance of sitting or standing, or new onset hip pain or pain of unknown origin. Conservative measures for treating hip dislocation are few. Abduction bracing, which aims to maintain muscle length and prevent bony deformity of the femur, is not tolerated by this population secondary to the activation of a painful adductor stretch reflex. One study that looked at 46 children who received Botox and abduction bracing versus standard care to reduce progression to hip surgery found no difference in either group.[166]

Surgery is indicated when the MP progresses to greater than 30%, or when the patient develops hip pain or other pathologies (Legg-Calve-Perthes disease, AVN, slipped capital femoral epiphysis). The theory guiding surgery for hip displacement is to lengthen spastic muscles in order to prevent development of deformity that would necessitate bony reconstructive surgery or salvage. Tendon release and myotomy of the adductors is indicated in children less than 8 years old with hip abduction less than 30 degrees and an MP between 25% and 60% with a goal of hip abduction to 45 degrees.[167] Severely compromised hips often require reconstruction, which if possible should be delayed until after age 4, when remodeling of the femoral head is less likely and bone stock is more amenable to instrumentation. Soft tissue lengthening with shortening VDRO and pelvic osteotomy is indicated to reseat the femoral head within the acetabulum. This combined, one-stage approach results in excellent stability and pain relief.[168,169] In one study of 51 children receiving a total of 59 combined femoral and acetabular osteotomies, 95% of hips were stable after 2 years with resolution of pain in 82%.[170] Complete resection of the femoral head should be considered a salvage procedure only.

REFERENCES

1. Sucato DJ. Spine deformity in spinal muscular atrophy. *J Bone Joint Surg Am* 2007;**89**(Suppl. 1):148–54.
2. Lefebvre S, Burglen L, Reboullet S, Clermont O, Burlet P, Viollet L, et al. Identification and characterization of a spinal muscular atrophy-determining gene. *Cell* 1995;**80**:155–65.
3. Nurputra DK, Lai PS, Harahap NI, Morikawa S, Yamamoto T, Nishimura N, et al. Spinal muscular atrophy: from gene discovery to clinical trials. *Ann Hum Genet* 2013;**77**:435–63.
4. Wang CH, Finkel RS, Bertini ES, Schroth M, Simonds A, Wong B, et al. Consensus statement for standard of care in spinal muscular atrophy. *J Child Neurol* 2007;**22**:1027–49.
5. Evans GA, Drennan JC, Russman BS. Functional classification and orthopaedic management of spinal muscular atrophy. *J Bone Joint Surg Br* 1981;**63B**:516–22.
6. Gregoretti C, Ottonello G, Chiarini Testa MB, Mastella C, Rava L, et al. Survival of patients with spinal muscular atrophy type 1. *Pediatrics* 2013;**131**:e1509–14.
7. Cherry JJ, Androphy EJ. Therapeutic strategies for the treatment of spinal muscular atrophy. *Future Med Chem* 2012;**4**:1733–50.
8. Zerres K, Rudnik-Schoneborn S, Forrest E, Lusakowska A, Borkowska J, Hausmanowa-Petrusewicz I. A collaborative study on the natural history of childhood and juvenile onset proximal spinal muscular atrophy (type II and III SMA): 569 patients. *J Neurol Sci* 1997;**146**:67–72.

9. Wang HY, Ju YH, Chen SM, Lo SK, Jong YJ. Joint range of motion limitations in children and young adults with spinal muscular atrophy. *Arch Phys Med Rehabil* 2004;**85**:1689–93.
10. Granata C, Merlini L, Magni E, Marini ML, Stagni SB. Spinal muscular atrophy: natural history and orthopaedic treatment of scoliosis. *Spine (Phila Pa 1976)* 1989;**14**:760–2.
11. Schwentker EP, Gibson DA. The orthopaedic aspects of spinal muscular atrophy. *J Bone Joint Surg Am* 1976;**58**:32–8.
12. Shapiro F, Specht L. The diagnosis and orthopaedic treatment of childhood spinal muscular atrophy, peripheral neuropathy, Friedreich ataxia, and arthrogryposis. *J Bone Joint Surg Am* 1993;**75**:1699–714.
13. Robinson D, Galasko CS, Delaney C, Williamson JB, Barrie JL. Scoliosis and lung function in spinal muscular atrophy. *Eur Spine J* 1995;**4**:268–73.
14. Campbell Jr. RM, Smith MD. Thoracic insufficiency syndrome and exotic scoliosis. *J Bone Joint Surg Am* 2007;**89**(Suppl. 1):108–22.
15. Phillips DP, Roye Jr. DP, Farcy JP, Leet A, Shelton YA. Surgical treatment of scoliosis in a spinal muscular atrophy population. *Spine (Phila Pa 1976)* 1990;**15**:942–5.
16. Aprin H, Bowen JR, MacEwen GD, Hall JE. Spine fusion in patients with spinal muscular atrophy. *J Bone Joint Surg Am* 1982;**64**:1179–87.
17. Merlini L, Granata C, Bonfiglioli S, Marini ML, Cervellati S, Savini R. Scoliosis in spinal muscular atrophy: natural history and management. *Dev Med Child Neurol* 1989;**31**:501–8.
18. Modi HN, Suh SW, Hong JY, Cho JW, Park JH, Yang JH. Treatment and complications in flaccid neuromuscular scoliosis (Duchenne muscular dystrophy and spinal muscular atrophy) with posterior-only pedicle screw instrumentation. *Eur Spine J* 2010;**19**:384–93.
19. Bridwell KH, Baldus C, Iffrig TM, Lenke LG, Blanke K. Process measures and patient/parent evaluation of surgical management of spinal deformities in patients with progressive flaccid neuromuscular scoliosis (Duchenne's muscular dystrophy and spinal muscular atrophy). *Spine (Phila Pa 1976)* 1999;**24**:1300–9.
20. Vitale MG, Matsumoto H, Bye MR, Gomez JA, Booker WA, Hyman JE, et al. A retrospective cohort study of pulmonary function, radiographic measures, and quality of life in children with congenital scoliosis: an evaluation of patient outcomes after early spinal fusion. *Spine (Phila Pa 1976)* 2008;**33**:1242–9.
21. Matsumoto H, Williams BA, Corona J, Comer JS, Fisher PW, Neria Y, et al. Psychosocial effects of repetitive surgeries in children with early-onset scoliosis: are we putting them at risk? *J Pediatr Orthop* 2014;**34**:172–8.
22. McElroy MJ, Shaner AC, Crawford TO, Thompson GH, Kadakia RV, Akbarnia BA, et al. Growing rods for scoliosis in spinal muscular atrophy: structural effects, complications, and hospital stays. *Spine (Phila Pa 1976)* 2011;**36**:1305–11.
23. Shapiro F. Orthopedic treatment. In: Jones HR, De Vivo DC, Darras BT, editors. *Neuromuscular Disorders of Infancy, Childhood, and Adolescence: A Clinician's Approach*. Philadelphia: Butterworth Heinemann; 2003. pp. 1251–73.
24. Bach JR, Bianchi C. Prevention of pectus excavatum for children with spinal muscular atrophy type 1. *Am J Phys Med Rehabil* 2003;**82**:815–19.
25. Warner LE, Garcia CA, Lupski JR. Hereditary peripheral neuropathies: clinical forms, genetics, and molecular mechanisms. *Annu Rev Med* 1999;**50**:263–75.
26. Braathen GJ, Sand JC, Lobato A, Hoyer H, Russell MB. Genetic epidemiology of Charcot-Marie-Tooth in the general population. *Eur J Neurol* 2011;**18**:39–48.
27. Newman CJ, Walsh M, O'Sullivan R, Jenkinson A, Bennett D, Lynch B, et al. The characteristics of gait in Charcot-Marie-Tooth disease types I and II. *Gait Posture* 2007;**26**:120–7.
28. Yagerman SE, Cross MB, Green DW, Scher DM. Pediatric orthopedic conditions in Charcot-Marie-Tooth disease: a literature review. *Curr Opin Pediatr* 2012;**24**:50–6.
29. Szigeti K, Nelis E, Lupski JR. Molecular diagnostics of Charcot-Marie-Tooth disease and related peripheral neuropathies. *Neuromolecular Med* 2006;**8**:243–54.
30. Guyton GP. Current concepts review: orthopaedic aspects of Charcot-Marie-Tooth disease. *Foot Ankle Int* 2006;**27**:1003–10.
31. Guyton GP, Mann RA. The pathogenesis and surgical management of foot deformity in Charcot-Marie-Tooth disease. *Foot Ankle Clin* 2000;**5**:317–26.
32. Mann RA, Missirian J. Pathophysiology of Charcot-Marie-Tooth disease. *Clin Orthop Relat Res* 1988;221–8.
33. Gallardo E, Garcia A, Combarros O, Berciano J. Charcot-Marie-Tooth disease type 1A duplication: spectrum of clinical and magnetic resonance imaging features in leg and foot muscles. *Brain* 2006;**129**:426–37.
34. Nagai MK, Chan G, Guille JT, Kumar SJ, Scavina M, Mackenzie WG. Prevalence of Charcot-Marie-Tooth disease in patients who have bilateral cavovarus feet. *J Pediatr Orthop* 2006;**26**:438–43.
35. Paulos L, Coleman SS, Samuelson KM. Pes cavovarus. Review of a surgical approach using selective soft-tissue procedures. *J Bone Joint Surg Am* 1980;**62**:942–53.
36. Roye DP, Gomez JA. Cavus foot. In: Tolo VT, Skaggs DL, editors. *Master Techniques in Orthopaedic Surgery: Pediatrics*. Philadelphia: Lippincott Williams & Wilkins; 2008. pp. 235–51.
37. Burns J, Raymond J, Ouvrier R. Feasibility of foot and ankle strength training in childhood Charcot-Marie-Tooth disease. *Neuromuscul Disord* 2009;**19**:818–21.
38. Burns J, Scheinberg A, Ryan MM, Rose KJ, Ouvrier RA. Randomized trial of botulinum toxin to prevent pes cavus progression in pediatric Charcot-Marie-Tooth disease type 1A. *Muscle Nerve* 2010;**42**:262–7.
39. Tynan MC, Klenerman L. The modified Robert Jones tendon transfer in cases of pes cavus and clawed hallux. *Foot Ankle Int* 1994;**15**:68–71.
40. Ward CM, Dolan LA, Bennett DL, Morcuende JA, Cooper RR. Long-term results of reconstruction for treatment of a flexible cavovarus foot in Charcot-Marie-Tooth disease. *J Bone Joint Surg Am* 2008;**90**:2631–42.
41. Roper BA, Tibrewal SB. Soft tissue surgery in Charcot-Marie-Tooth disease. *J Bone Joint Surg Br* 1989;**71**:17–20.
42. Saltzman CL, Fehrle MJ, Cooper RR, Spencer EC, Ponseti IV. Triple arthrodesis: twenty-five and forty-four-year average follow-up of the same patients. *J Bone Joint Surg Am* 1999;**81**:1391–402.
43. Angus PD, Cowell HR. Triple arthrodesis. A critical long-term review. *J Bone Joint Surg Br* 1986;**68**:260–5.
44. Wukich DK, Bowen JR. A long-term study of triple arthrodesis for correction of pes cavovarus in Charcot-Marie-Tooth disease. *J Pediatr Orthop* 1989;**9**:433–7.

45. Santavirta S, Turunen V, Ylinen P, Konttinen YT, Tallroth K. Foot and ankle fusions in Charcot-Marie-Tooth disease. *Arch Orthop Trauma Surg* 1993;**112**:175–9.
46. Kumar SJ, Marks HG, Bowen JR, MacEwen GD. Hip dysplasia associated with Charcot-Marie-Tooth disease in the older child and adolescent. *J Pediatr Orthop* 1985;**5**:511–14.
47. Walker JL, Nelson KR, Heavilon JA, Stevens DB, Lubicky JP, Ogden JA, et al. Hip abnormalities in children with Charcot-Marie-Tooth disease. *J Pediatr Orthop* 1994;**14**:54–9.
48. Bamford NS, White KK, Robinett SA, Otto RK, Gospe Jr. SM. Neuromuscular hip dysplasia in Charcot-Marie-Tooth disease type 1A. *Dev Med Child Neurol* 2009;**51**:408–11.
49. Chan G, Bowen JR, Kumar SJ. Evaluation and treatment of hip dysplasia in Charcot-Marie-Tooth disease. *Orthop Clin North Am* 2006;**37**:203–9, vii.
50. McGann R, Gurd A. The association between Charcot-Marie-Tooth disease and developmental dysplasia of the hip. *Orthopedics* 2002; **25**:337–9.
51. Pailthorpe CA, Benson MK. Hip dysplasia in hereditary motor and sensory neuropathies. *J Bone Joint Surg Br* 1992;**74**:538–40.
52. Holmberg BH. Charcot-Marie-Tooth disease in northern Sweden: an epidemiological and clinical study. *Acta Neurol Scand* 1993; **87**:416–22.
53. Walker JL, Nelson KR, Stevens DB, Lubicky JP, Ogden JA, VandenBrink KD. Spinal deformity in Charcot-Marie-Tooth disease. *Spine (Phila Pa 1976)* 1994;**19**:1044–7.
54. Karol LA, Elerson E. Scoliosis in patients with Charcot-Marie-Tooth disease. *J Bone Joint Surg Am* 2007;**89**:1504–10.
55. Horacek O, Mazanec R, Morris CE, Kobesova A. Spinal deformities in hereditary motor and sensory neuropathy: a retrospective qualitative, quantitative, genotypical, and familial analysis of 175 patients. *Spine (Phila Pa 1976)* 2007;**32**:2502–8.
56. Eklund E, Svensson E, Hager-Ross C. Hand function and disability of the arm, shoulder and hand in Charcot-Marie-Tooth disease. *Disabil Rehabil* 2009;**31**:1955–62.
57. Vinci P, Villa LM, Castagnoli L, Marconi C, Lattanzi A, Manini MP, et al. Handgrip impairment in Charcot-Marie-Tooth disease. *Eura Medicophys* 2005;**41**:131–4.
58. Videler AJ, Beelen A, Nollet F. Manual dexterity and related functional limitations in hereditary motor and sensory neuropathy. An explorative study. *Disabil Rehabil* 2008;**30**:634–8.
59. Burns J, Bray P, Cross LA, North KN, Ryan MM, Ouvrier RA. Hand involvement in children with Charcot-Marie-Tooth disease type 1A. *Neuromuscul Disord* 2008;**18**:970–3.
60. Brown RE, Zamboni WA, Zook EG, Russell RC. Evaluation and management of upper extremity neuropathies in Charcot-Marie-Tooth disease. *J Hand Surg Am* 1992;**17**:523–30.
61. Pandolfo M. Friedreich ataxia: the clinical picture. *J Neurol* 2009;**256**(Suppl. 1):3–8.
62. Labelle H, Tohme S, Duhaime M, Allard P. Natural history of scoliosis in Friedreich's ataxia. *J Bone Joint Surg Am* 1986;**68**:564–72.
63. Tsirikos AI, Smith G. Scoliosis in patients with Friedreich's ataxia. *J Bone Joint Surg Br* 2012;**94**:684–9.
64. Tsirikos AI, Smith G. Scoliosis in patients with Friedreich's ataxia. *J Bone Joint Surg Br* 2012;**94**:684–9.
65. Milbrandt TA, Kunes JR, Karol LA. Friedreich's ataxia and scoliosis: the experience at two institutions. *J Pediatr Orthop* 2008; **28**:234–8.
66. Daher YH, Lonstein JE, Winter RB, Bradford DS. Spinal deformities in patients with Friedreich ataxia: a review of 19 patients. *J Pediatr Orthop* 1985;**5**:553–7.
67. Cady RB, Bobechko WP. Incidence, natural history, and treatment of scoliosis in Friedreich's ataxia. *J Pediatr Orthop* 1984;**4**:673–6.
68. Hensinger RN, MacEwen GD. Spinal deformity associated with heritable neurological conditions: spinal muscular atrophy, Friedreich's ataxia, familial dysautonomia, and Charcot-Marie-Tooth disease. *J Bone Joint Surg Am* 1976;**58**:13–24.
69. Durr A, Cossee M, Agid Y, Campuzano V, Mignard C, Penet C, et al. Clinical and genetic abnormalities in patients with Friedreich's ataxia. *N Engl J Med* 1996;**335**:1169–75.
70. Hou JG, Jankovic J. Movement disorders in Friedreich's ataxia. *J Neurol Sci* 2003;**206**:59–64.
71. Schols L, Szymanski S, Peters S, Przuntek H, Epplen JT, Hardt C, et al. Genetic background of apparently idiopathic sporadic cerebellar ataxia. *Hum Genet* 2000;**107**:132–7.
72. Sethi KD, Jankovic J. Dystonia in spinocerebellar ataxia type 6. *Mov Disord* 2002;**17**:150–3.
73. Ben Smail D, Jacq C, Denys P, Bussel B. Intrathecal baclofen in the treatment of painful, disabling spasms in Friedreich's ataxia. *Mov Disord* 2005;**20**:758–9.
74. Do T. Orthopedic management of the muscular dystrophies. *Curr Opin Pediatr* 2002;**14**:50–3.
75. Matsumura K, Lee CC, Caskey CT, Campbell KP. Restoration of dystrophin-associated proteins in skeletal muscle of mdx mice transgenic for dystrophin gene. *FEBS Lett* 1993;**320**:276–80.
76. Hoffman EP, Fischbeck KH, Brown RH, Johnson M, Medori R, Loike JD, et al. Characterization of dystrophin in muscle-biopsy specimens from patients with Duchenne's or Becker's muscular dystrophy. *N Engl J Med* 1988;**318**:1363–8.
77. van Essen AJ, Kneppers AL, van der Hout AH, Scheffer H, Ginjaar IB, ten Kate LP, et al. The clinical and molecular genetic approach to Duchenne and Becker muscular dystrophy: an updated protocol. *J Med Genet* 1997;**34**:805–12.
78. Sussman M. Duchenne muscular dystrophy. *J Am Acad Orthop Surg* 2002;**10**:138–51.
79. Bushby KM, Hill A, Steele JG. Failure of early diagnosis in symptomatic Duchenne muscular dystrophy. *Lancet* 1999;**353**: 557–8.
80. Emery AE. Population frequencies of inherited neuromuscular diseases—a world survey. *Neuromuscul Disord* 1991;**1**:19–29.
81. Bushby K, Finkel R, Birnkrant DJ, Case LE, Clemens PR, Cripe L, et al. Diagnosis and management of Duchenne muscular dystrophy, part 2: implementation of multidisciplinary care. *Lancet Neurol* 2010;**9**:177–89.
82. Khodadadeh S, McClelland MR, Patrick JH, Edwards RH, Evans GA. Knee moments in Duchenne muscular dystrophy. *Lancet* 1986; **2**:544–5.
83. Hsu JD. The natural history of spine curvature progression in the nonambulatory Duchenne muscular dystrophy patient. *Spine (Phila Pa 1976)* 1983;**8**:771–5.
84. Kurz LT, Mubarak SJ, Schultz P, Park SM, Leach J. Correlation of scoliosis and pulmonary function in Duchenne muscular dystrophy. *J Pediatr Orthop* 1983;**3**:347–53.
85. Alman BA, Raza SN, Biggar WD. Steroid treatment and the development of scoliosis in males with Duchenne muscular dystrophy. *J Bone Joint Surg Am* 2004;**86–A**:519–24.

86. Pandya S, Florence JM, King WM, Robison JD, Oxman M, Province MA. Reliability of goniometric measurements in patients with Duchenne muscular dystrophy. *Phys Ther* 1985;**65**: 1339–42.
87. Cheuk DK, Wong V, Wraige E, Baxter P, Cole A. Surgery for scoliosis in Duchenne muscular dystrophy. *Cochrane Database Syst Rev* 2013;**2**:CD005375.
88. Miller F, Moseley CF, Koreska J. Spinal fusion in Duchenne muscular dystrophy. *Dev Med Child Neurol* 1992;**34**:775–86.
89. Galasko CS, Williamson JB, Delaney CM. Lung function in Duchenne muscular dystrophy. *Eur Spine J* 1995;**4**:263–7.
90. Velasco MV, Colin AA, Zurakowski D, Darras BT, Shapiro F. Posterior spinal fusion for scoliosis in Duchenne muscular dystrophy diminishes the rate of respiratory decline. *Spine (Phila Pa 1976)* 2007;**32**:459–65.
91. Miller F, Moseley CF, Koreska J, Levison H. Pulmonary function and scoliosis in Duchenne dystrophy. *J Pediatr Orthop* 1988; **8**:133–7.
92. Bushby K, Finkel R, Birnkrant DJ, Case LE, Clemens PR, Cripe L, et al. Diagnosis and management of Duchenne muscular dystrophy, part 1: diagnosis, and pharmacological and psychosocial management. *Lancet Neurol* 2010;**9**:77–93.
93. Emery AE, Skinner R. Clinical studies in benign (Becker type) X-linked muscular dystrophy. *Clin Genet* 1976;**10**:189–201.
94. Holloway SM, Wilcox DE, Wilcox A, Dean JC, Berg JN, Goudie DR, et al. Life expectancy and death from cardiomyopathy amongst carriers of Duchenne and Becker muscular dystrophy in Scotland. *Heart* 2008;**94**:633–6.
95. Dietz FR, Mathews KD. Update on the genetic bases of disorders with orthopaedic manifestations. *J Bone Joint Surg Am* 1996; **78**:1583–98.
96. Roland EH. Muscular dystrophy. *Pediatr Rev* 2000;**21**:233–7, quiz 8.
97. Groh WJ, Groh MR, Saha C, Kincaid JC, Simmons Z, Ciafaloni E, et al. Electrocardiographic abnormalities and sudden death in myotonic dystrophy type 1. *N Engl J Med* 2008;**358**:2688–97.
98. Canavese F, Sussman MD. Orthopaedic manifestations of congenital myotonic dystrophy during childhood and adolescence. *J Pediatr Orthop* 2009;**29**:208–13.
99. Gross RH. The role of the Verebelyi-Ogston procedure in the management of the arthrogrypotic foot. *Clin Orthop Relat Res* 1985;99–103.
100. Wang CH, Dowling JJ, North K, Schroth MK, Sejersen T, Shapiro F, et al. Consensus statement on standard of care for congenital myopathies. *J Child Neurol* 2012;**27**:363–82.
101. North KN. Clinical approach to the diagnosis of congenital myopathies. *Semin Pediatr Neurol* 2011;**18**:216–20.
102. Clarke NF, Kolski H, Dye DE, Lim E, Smith RL, Patel R, et al. Mutations in TPM3 are a common cause of congenital fiber type disproportion. *Ann Neurol* 2008;**63**:329–37.
103. Ryan MM, Schnell C, Strickland CD, Shield LK, Morgan G, Iannaccone ST, et al. Nemaline myopathy: a clinical study of 143 cases. *Ann Neurol* 2001;**50**:312–20.
104. Jungbluth H. Central core disease. *Orphanet J Rare Dis* 2007;**2**:25.
105. Bevan WP, Hall JG, Bamshad M, Staheli LT, Jaffe KM, Song K. Arthrogryposis multiplex congenita (amyoplasia): an orthopaedic perspective. *J Pediatr Orthop* 2007;**27**:594–600.
106. Hall JG. Arthrogryposis multiplex congenita: etiology, genetics, classification, diagnostic approach, and general aspects. *J Pediatr Orthop B* 1997;**6**:159–66.
107. Hall JG. Genetic aspects of arthrogryposis. *Clin Orthop Relat Res* 1985;(194):44–53.
108. Robertson WL, Glinski LP, Kirkpatrick SJ, Pauli RM. Further evidence that arthrogryposis multiplex congenita in the human sometimes is caused by an intrauterine vascular accident. *Teratology* 1992;**45**:345–51.
109. Bernstein RM. Arthrogryposis and amyoplasia. *J Am Acad Orthop Surg* 2002;**10**:417–24.
110. Carlson WO, Speck GJ, Vicari V, Wenger DR. Arthrogryposis multiplex congenita. A long-term follow-up study. *Clin Orthop Relat Res* 1985;115–23.
111. Hahn G. Arthrogryposis. Pediatric review and habilitative aspects. *Clin Orthop Relat Res* 1985;104–14.
112. Sells JM, Jaffe KM, Hall JG. Amyoplasia the most common type of arthrogryposis: the potential for good outcome. *Pediatrics* 1996;**97**:225–31.
113. Diamond LS, Alegado R. Perinatal fractures in arthrogryposis multiplex congenita. *J Pediatr Orthop* 1981;**1**:189–92.
114. Zimbler S, Craig CL. The arthrogrypotic foot plan of management and results of treatment. *Foot Ankle* 1983;**3**:211–19.
115. Guidera KJ, Drennan JC. Foot and ankle deformities in arthrogryposis multiplex congenita. *Clin Orthop Relat Res* 1985; 93–8.
116. Niki H, Staheli LT, Mosca VS. Management of clubfoot deformity in amyoplasia. *J Pediatr Orthop* 1997;**17**:803–7.
117. Sølund K, Sonne-Holm S, Kjølbye JE. Talectomy for equinovarus deformity in arthrogryposis. A 13 (2–20) year review of 17 feet. *Acta Orthop Scand* 1991;**62**:372–4.
118. D'Souza H, Aroojis A, Chawara GS. Talectomy in arthrogryposis: analysis of results. *J Pediatr Orthop* 1998;**18**:760–4.
119. Brunner R, Hefti F, Tgetgel JD. Arthrogrypotic joint contracture at the knee and the foot: correction with a circular frame. *J Pediatr Orthop B* 1997;**6**:192–7.
120. Murray C, Fixsen JA. Management of knee deformity in classical arthrogryposis multiplex congenita (amyoplasia congenita). *J Pediatr Orthop B* 1997;**6**:186–91.
121. Palmer PM, MacEwen GD, Bowen JR, Mathews PA. Passive motion therapy for infants with arthrogryposis. *Clin Orthop Relat Res* 1985;54–9.
122. Szöke G, Staheli LT, Jaffe K, Hall JG. Medial-approach open reduction of hip dislocation in amyoplasia-type arthrogryposis. *J Pediatr Orthop* 1996;**16**:127–30.
123. Gruel CR, Birch JG, Roach JW, Herring JA. Teratologic dislocation of the hip. *J Pediatr Orthop* 1986;**6**:693–702.
124. Hoffer MM, Swank S, Eastman F, Clark D, Teitge R. Ambulation in severe arthrogryposis. *J Pediatr Orthop* 1983;**3**:293–6.
125. Van Heest A, Waters PM, Simmons BP. Surgical treatment of arthrogryposis of the elbow. *J Hand Surg Am* 1998;**23**: 1063–70.
126. Carroll RE, Hill NA. Triceps transfer to restore elbow flexion. A study of fifteen patients with paralytic lesions and arthrogryposis. *J Bone Joint Surg Am* 1970;**52**:239–44.
127. Carroll RE, Kleinman WB. Pectoralis major transplantation to restore elbow flexion to the paralytic limb. *J Hand Surg Am* 1979;**4**:501–7.

128. Axt MW, Niethard FU, Doderlein L, Weber M. Principles of treatment of the upper extremity in arthrogryposis multiplex congenita type I. *J Pediatr Orthop B* 1997;**6**:179–85.
129. Yingsakmongkol W, Kumar SJ. Scoliosis in arthrogryposis multiplex congenita: results after nonsurgical and surgical treatment. *J Pediatr Orthop* 2000;**20**:656–61.
130. Daher YH, Lonstein JE, Winter RB, Moe JH. Spinal deformities in patients with arthrogryposis. A review of 16 patients. *Spine (Phila Pa 1976)* 1985;**10**:609–13.
131. Bax M, Goldstein M, Rosenbaum P, Leviton A, Paneth N, Dan B, et al. Proposed definition and classification of cerebral palsy, April 2005. *Dev Med Child Neurol* 2005;**47**:571–6.
132. Jones-Quaidoo SM, Yang S, Arlet V. Surgical management of spinal deformities in cerebral palsy. A review. *J Neurosurg Spine* 2010;**13**:672–85.
133. Morrell DS, Pearson JM, Sauser DD. Progressive bone and joint abnormalities of the spine and lower extremities in cerebral palsy. *Radiographics* 2002;**22**:257–68.
134. Madigan RR, Wallace SL. Scoliosis in the institutionalized cerebral palsy population. *Spine (Phila Pa 1976)* 1981;**6**:583–90.
135. McCarthy JJ, D'Andrea LP, Betz RR, Clements DH. Scoliosis in the child with cerebral palsy. *J Am Acad Orthop Surg* 2006;**14**:367–75.
136. Cooke PH, Cole WG, Carey RP. Dislocation of the hip in cerebral palsy. Natural history and predictability. *J Bone Joint Surg Br* 1989;**71**:441–6.
137. Lonstein JE, Beck K. Hip dislocation and subluxation in cerebral palsy. *J Pediatr Orthop* 1986;**6**:521–6.
138. Majd ME, Muldowny DS, Holt RT. Natural history of scoliosis in the institutionalized adult cerebral palsy population. *Spine (Phila Pa 1976)* 1997;**22**:1461–6.
139. Thometz JG, Simon SR. Progression of scoliosis after skeletal maturity in institutionalized adults who have cerebral palsy. *J Bone Joint Surg Am* 1988;**70**:1290–6.
140. Ferguson RL, Allen Jr. BL. Considerations in the treatment of cerebral palsy patients with spinal deformities. *Orthop Clin North Am* 1988;**19**:419–25.
141. Saito N, Ebara S, Ohotsuka K, Kumeta H, Takaoka K. Natural history of scoliosis in spastic cerebral palsy. *Lancet* 1998; **351**:1687–92.
142. Olafsson Y, Saraste H, Al-Dabbagh Z. Brace treatment in neuromuscular spine deformity. *J Pediatr Orthop* 1999;**19**:376–9.
143. Renshaw TS, Green NE, Griffin PP, Root L. Cerebral palsy: orthopaedic management. *Instr Course Lect* 1996;**45**:475–90.
144. Hopf CG, Eysel P. One-stage versus two-stage spinal fusion in neuromuscular scolioses. *J Pediatr Orthop B* 2000;**9**:234–43.
145. Shilt JS, Lai LP, Cabrera MN, Frino J, Smith BP. The impact of intrathecal baclofen on the natural history of scoliosis in cerebral palsy. *J Pediatr Orthop* 2008;**28**:684–7.
146. Garg S, Engelman G, Yoshihara H, McNair B, Chang F. The relationship of Gross Motor Functional Classification Scale Level and hip dysplasia on the pattern and progression of scoliosis in children with cerebral palsy. *Spine Deformity* 2013; **1**:266–71.
147. Cassidy C, Craig CL, Perry A, Karlin LI, Goldberg MJ. A reassessment of spinal stabilization in severe cerebral palsy. *J Pediatr Orthop* 1994;**14**:731–9.
148. Larsson EL, Aaro S, Oberg B. Activities and functional assessment 1 year after spinal fusion for paralytic scoliosis. *Eur Spine J* 1999;**8**:100–9.
149. Comstock CP, Leach J, Wenger DR. Scoliosis in total-body-involvement cerebral palsy. Analysis of surgical treatment and patient and caregiver satisfaction. *Spine (Phila Pa 1976)* 1998;**23**:1412–24, discussion 24–5.
150. Tsirikos AI, Chang WN, Dabney KW, Miller F. Comparison of parents' and caregivers' satisfaction after spinal fusion in children with cerebral palsy. *J Pediatr Orthop* 2004;**24**:54–8.
151. Matsumoto H, Vitale MG, Hyman JE, Roye Jr. DP. Can parents rate their children's quality of life? Perspectives on pediatric orthopedic outcomes. *J Pediatr Orthop B* 2011;**20**:184–90.
152. Modi HN, Hong JY, Mehta SS, Srinivasalu S, Suh SW, Yi JW, et al. Surgical correction and fusion using posterior-only pedicle screw construct for neuropathic scoliosis in patients with cerebral palsy: a three-year follow-up study. *Spine (Phila Pa 1976)* 2009;**34**: 1167–75.
153. Benson ER, Thomson JD, Smith BG, Banta JV. Results and morbidity in a consecutive series of patients undergoing spinal fusion for neuromuscular scoliosis. *Spine (Phila Pa 1976)* 1998; **23**:2308–17, discussion 18.
154. Broom MJ, Banta JV, Renshaw TS. Spinal fusion augmented by luque-rod segmental instrumentation for neuromuscular scoliosis. *J Bone Joint Surg Am* 1989;**71**:32–44.
155. Gau YL, Lonstein JE, Winter RB, Koop S, Denis F. Luque-Galveston procedure for correction and stabilization of neuromuscular scoliosis and pelvic obliquity: a review of 68 patients. *J Spinal Disord* 1991;**4**:399–410.
156. Whitaker C, Burton DC, Asher M. Treatment of selected neuromuscular patients with posterior instrumentation and arthrodesis ending with lumbar pedicle screw anchorage. *Spine (Phila Pa 1976)* 2000;**25**:2312–18.
157. Sarwark J, Sarwahi V. New strategies and decision making in the management of neuromuscular scoliosis. *Orthop Clin North Am* 2007;**38**:485–96.
158. Rawlins BA, Winter RB, Lonstein JE, Denis F, Kubic PT, Wheeler WB, et al. Reconstructive spine surgery in pediatric patients with major loss in vital capacity. *J Pediatr Orthop* 1996;**16**:284–92.
159. Sponseller PD, Shah SA, Abel MF, Newton PO, Letko L, Marks M. Infection rate after spine surgery in cerebral palsy is high and impairs results: multicenter analysis of risk factors and treatment. *Clin Orthop Relat Res* 2010;**468**:711–16.
160. Chung CY, Park MS, Choi IH, Cho TJ, Yoo WJ, Lee KM. Morphometric analysis of acetabular dysplasia in cerebral palsy. *J Bone Joint Surg Br* 2006;**88**:243–7.
161. McCarthy RE, Simon S, Douglas B, Zawacki R, Reese N. Proximal femoral resection to allow adults who have severe cerebral palsy to sit. *J Bone Joint Surg Am* 1988;**70**:1011–16.
162. Soo B, Howard JJ, Boyd RN, Reid SM, Lanigan A, Wolfe R, et al. Hip displacement in cerebral palsy. *J Bone Joint Surg Am* 2006;**88**:121–9.
163. Vidal J, Deguillaume P, Vidal M. The anatomy of the dysplastic hip in cerebral palsy related to prognosis and treatment. *Int Orthop* 1985;**9**:105–10.
164. Wynter M, Gibson N, Kentish M, Love S, Thomason P, Kerr Graham H. The Consensus Statement on Hip Surveillance for

Children with Cerebral Palsy: Australian Standards of Care. *J Pediatr Rehabil Med* 2011;**4**:183–95.

165. Gordon GS, Simkiss DE. A systematic review of the evidence for hip surveillance in children with cerebral palsy. *J Bone Joint Surg Br* 2006;**88**:1492–6.

166. Willoughby K, Ang SG, Thomason P, Graham HK. The impact of botulinum toxin A and abduction bracing on long-term hip development in children with cerebral palsy. *Dev Med Child Neurol* 2012;**54**:743–7.

167. Flynn JM, Miller F. Management of hip disorders in patients with cerebral palsy. *J Am Acad Orthop Surg* 2002;**10**:198–209.

168. McNerney NP, Mubarak SJ, Wenger DR. One-stage correction of the dysplastic hip in cerebral palsy with the San Diego acetabuloplasty: results and complications in 104 hips. *J Pediatr Orthop* 2000;**20**:93–103.

169. Brunner R, Baumann JU. Clinical benefit of reconstruction of dislocated or subluxated hip joints in patients with spastic cerebral palsy. *J Pediatr Orthop* 1994;**14**:290–4.

170. Miller F, Girardi H, Lipton G, Ponzio R, Klaumann M, Dabney KW. Reconstruction of the dysplastic spastic hip with peri-ilial pelvic and femoral osteotomy followed by immediate mobilization. *J Pediatr Orthop* 1997;**17**:592–602.

Chapter 53

Rehabilitation of the Child with a Neuromuscular Disorder

Nancy E. Strauss, Jacqueline Montes, and Barry S. Russman

INTRODUCTION

Rehabilitation is that branch of medicine concerned with helping patients gain optimal function within the scope of their physical and cognitive limitations.[1] The development and implementation of a realistic and cost-effective rehabilitation program for the patient necessitates a "correct diagnosis" and an understanding of the natural course of disease. Furthermore, a team of professionals skilled in all areas of rehabilitation best develops a rehabilitation program.

In our experience, an interdisciplinary rather than a multidisciplinary approach, in which team members individually provide input to what a patient needs, is most effective. The advantage of the former approach is that each team member is not only skilled in a specialty but also has a working knowledge of and a respect for the skills of the other team members. Pediatric rehabilitation differs significantly from adult rehabilitation because growth and development must be considered when evaluating physical or cognitive disability.[2] Furthermore, a child's disability cannot be evaluated without taking into account the family and its commitment to participate in the rehabilitation program. As such, the rehabilitation team for a child with neuromuscular disease is made up of many different clinical specialists, family and community members, and dedicated organizations that provide services for children with neuromuscular disease. Clinical specialties with expertise applicable to the needs of patients with neuromuscular disease include, but are not limited to, the following: neurology, physiatry, orthopedics, pulmonology, cardiology, nutrition, genetic counseling, social work, physical therapy, occupational therapy, vocational therapy, speech therapy, respiratory therapy, psychology, recreational therapy, and special education.

Most childhood neuromuscular disorders have a known, relatively predictable course with respect to distribution of muscle weakness, progression of weakness, development of skeletal deformities, and impairment of mobility. With this understanding, the rehabilitation team must, to the best of their ability, anticipate the next stage of a disease and minimize or prevent complications.[3] In addition to the continued rehabilitation therapies, access and funding become important issues, especially in lifelong disabling conditions. The rehabilitation team must be aware of what resources are available to each patient and make the proper referrals. Third-party payers including commercial insurance companies and state-funded programs must be educated to ensure that needed services are covered and are being provided, and the physician must be a patient advocate in this situation.

This chapter will introduce the World Health Organization (WHO) model of describing a rehabilitation program for neuromuscular diseases, but we shall not fully integrate this model into our discussion of rehabilitation programs for neuromuscular diseases. The International Classification of Functioning, Disability and Health (ICF) is WHO's framework for measuring health and disability at both individual and population levels. The ICF was officially endorsed by all 191 WHO Member States in the Fifty-fourth World Health Assembly on 22 May 2001 (resolution WHA 54.21). The ICF domains are classified from body, individual, and societal perspectives by means of two lists: a list of body functions and structure, and a list of domains of activity and participation. Since an individual's functioning and disability occurs in a context, the ICF also includes a list of environmental factors (Figure 53.1). An extensive review of this topic is available on the ICF website.[4]

REHABILITATION GOALS AND PROGRAMS

The overriding principle is to maintain function within the scope of the individual's limitations, which include

B.T. Darras, H. Royden Jones, Jr., M.M. Ryan & D.C. De Vivo (Eds): Neuromuscular Disorders of Infancy, Childhood and Adolescence, Second edition.
DOI: http://dx.doi.org/10.1016/B978-0-12-417044-5.00053-6

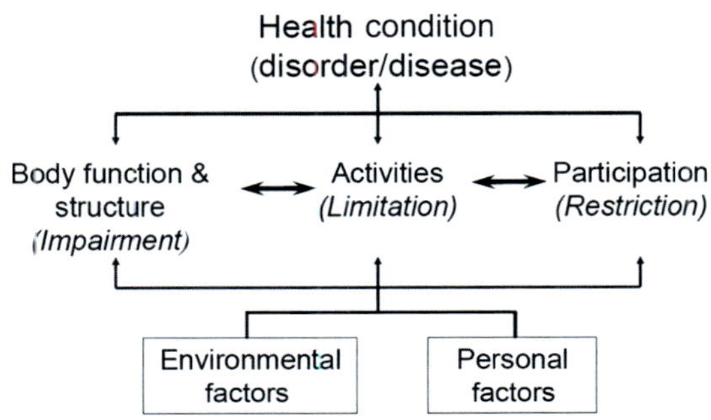

FIGURE 53.1 International Classification of Functioning, Disability and Health (ICF) framework for interaction of concepts. WHO 2001.18.

weakness, orthopedic deformities, cardiopulmonary problems, nutritional difficulties, and cognitive limitations. Assessing function is a complex, yet essential, component in the evaluation of any patient. As noted in the ICF model, in order to assess function adequately, the clinician must understand the functional demands of the individual patient relative to their particular lifestyle. The principles of rehabilitation for most childhood neuromuscular diseases are similar. While evidence-based medicine for rehabilitation programs for most of the neuromuscular diseases is limited, published consensus statements outlining standard of care are available for some disorders.[5,6]

The mainstays of treatment for individuals with neuromuscular diseases are (1) *rehabilitation therapies*, (2) *assistive devices*, and (3) *technology*. Regardless of the child's functional stage, (1) *ambulatory stage*, (2) *nonambulatory stage*, and (3) *stage of prolonged survival*,[7] the rehabilitation approach can be categorized into four domains, all of which must be considered when developing a rehabilitation plan for a child with neuromuscular disease: (1) strength, (2) orthopedic deformity, (3) cardiopulmonary dysfunction, and (4) cognitive dysfunction.[8] Early intervention is important to prevent the sequelae of weakness to the greatest extent possible and to promote achievement of developmental milestones.

REHABILITATION THERAPIES

Strength and Endurance

Muscle strength is decreased in the muscle groups significantly affected by a neuromuscular disorder. Usually, in myopathic disorders, muscle weakness is most pronounced in the proximal musculature. The reverse is true for neuropathies where weakness is most pronounced in distal musculature.[9,10] However, there are distal myopathies and proximal neuropathies. Whether muscle weakness has an effect on function will be dependent on the magnitude of the weakness, distribution of the weakness, and amount of strength required for a functional task. Endurance limitations resulting from muscle disease depend upon the functional reserve that every individual has for a given activity or task. Determining the amount of endurance required to complete a given task will aid in determining if the endurance limitation is such that task completion is (i) still possible independently; (ii) possible but only with more time given to complete it; (iii) possible, but only with assistance; or (iv) not possible at all. Even if independent task completion is still possible, patients will often choose dependence if it means not using up their functional reserve. Energy conservation becomes essential.

Muscle strength is maintained through frequent contractions, which must produce tension sufficient to exercise the muscle. Lack of physical activity will result in deconditioning, which will have marked deleterious effects in a child with neuromuscular disease, whose functional reserve is reduced even at baseline. Disuse atrophy may lead to joint malposition, nerve compression, and pain. The best way of preserving strength (or at least slowing the progression) in a patient with muscle disease is through performance of daily physical activity.

Strengthening exercise, or exercises that are used to maintain or improve a muscle's ability to generate force, helps to maintain function, avoid injury, and prevent disability. Skeletal muscle weakness is the cardinal sign and symptom of nearly all neuromuscular conditions. Strengthening exercises can be tailored for weak and strong muscle groups and can be performed with or without resistance.

In other neuromuscular diseases, largely with adults, there are several examples of the beneficial effects of exercise programs on function and strength.[11] Methodological limitations have contributed to the failure of consistent incorporation of exercise into clinical care of patients with neuromuscular diseases. A Cochrane Review of exercise studies in muscles disease deemed only 2 of 36 trials to be of acceptable methodological quality for inclusion into evidence-based recommendations.[12] For future studies, the Cochrane reviewers recommended a regularly supervised, well-defined exercise program that is consistent with the Physical Activity Guidelines for Healthy Americans.[13] Until more specific evidence is available, a moderate intensity, supervised program that includes strengthening exercises in combination with aerobic exercise is recommended.

Permit Weight Bearing

Functional weight bearing is important to maintaining the child's general health and requires a stable base of support and effective joint alignment. In patients with significant muscle weakness, this goal is often achieved by the use of external supports, such as orthoses, proper footwear,[14] assistive devices for standing, and gait aids.

Prolong Standing/Walking

The positive effects of standing and upright mobility involve multiple organ systems, such as the genitourinary, digestive, respiratory, cardiac, skeletal, and neuromuscular systems, as well as psychological wellbeing. A study questioned families attending neuromuscular clinics regarding quality of life and found that 89% named activities and health issues related to prolonging ambulation as most important to them.[15] However, the high energy costs of ambulation compared to the energy costs of using the wheelchair is often taken into consideration, particularly when navigating community distances. Attempts to prolong standing and ambulation, as long as it does not impair the child's safety or require too much use of their energy for their function, is recommended.

Orthopedic Deformities

All muscles do not weaken at the same rate or to the same degree. Therefore, muscle imbalance results from this variability in weakness. Muscle imbalance surrounding a joint causes joint contracture with resulting limitation in passive range of motion that leads to orthopedic deformity. When the lower extremity weight-bearing joints of the body are affected by contracture, abnormal body alignment occurs, changing the center of gravity, which can limit or arrest standing and or ambulation.[16] Upper extremity functional limitations are primarily due to proximal muscle weakness. Orthopedic deformity of the spine, such as scoliosis, can result in restriction of pulmonary function, impaired sitting balance, and cosmetic deformity.

Minimize and Prevent Joint Contractures

Muscle weakness and imbalance along with growth in height result in bony malalignment and contracture. Weakness and limited mobility can result in poor posture, decreased extensibility of muscle, and joint contractures. Axial muscle weakness and prolonged sitting further predispose to scoliosis. Patients with spinal muscular atrophy type 2, or "sitters," are particularly vulnerable to developing scoliosis. Maintaining flexibility and normal joint range of motion is essential to maximizing function and positioning and to minimizing pain and discomfort.

Children who retain the ability to walk are at lowest risk for joint contractures, but often show increased compensatory lordosis of the lumbar spine as a result of hip flexor weakness and altered gait mechanics. Hip flexor and knee extensor tightness as well as joint contractures are common in sitters or those unable to walk. Prone positioning is beneficial to counteract chronic sitting posture and the resulting hip flexor tightness. Weakness of the ankle dorsiflexors increases the risk of Achilles tendon tightness and equinovarus deformities. Shoulder girdle weakness can result in limitations at end-range shoulder motion and wrist and hand weakness resulting in contractures of the elbow, wrist, and fingers. Contractures can be the cause of cessation of standing and ambulation, pain, skin breakdown, and cosmetic deformity.

A physical therapy program is designed to minimize the development of contractures, which can produce difficulties with standing, sitting, lying in bed, and other activities. Bracing can be helpful as a program of night splinting for the feet and ankles to prevent contractures; as a program for scoliosis, at least to maintain the patient in an upright position, but not necessarily to prevent progression of the curve; or as a series of lower extremity orthoses to aid with ambulation. Finally, surgical interventions for a variety of different problems may be indicated.

Cardiopulmonary Dysfunction

With weakness of muscles of respiration, a restrictive pulmonary pattern on pulmonary function testing develops. This can be further affected by scoliosis and spinal deformities of the chest wall, ribs, and thoracic spine, thus further restricting vital capacity and resulting in thoracic insufficiency syndrome.[17] Limited endurance, fatigue, and poor concentration are among the negative effects of restrictive pulmonary disease. Limited functional reserve can result, leading to further disability.[18]

Cardiac involvement is a feature of a number of the myopathies and dystrophies including Duchenne muscular dystrophy (DMD) and some of the limb-girdle dystrophies, and can ultimately result in cardiomyopathy with conduction abnormalities, heart failure with decreased functional reserve, and ultimately death.[19,20]

Cognitive Impairment

Although not all patients with neuromuscular disease have cognitive impairment, it is a feature of a number of muscle diseases, for example DMD and myotonic dystrophy (DM).[21,22] Cognitive dysfunction has marked deleterious effects on every aspect of the individual's life. Rehabilitation interventions can be very successful in minimizing disability with respect to physical impairment; however, the compensatory techniques for impaired cognition are limited. Therefore, patients with cognitive dysfunction often have limited participation as described by the ICF model.

ASSISTIVE DEVICES

Maximizing Functional Mobility

To maximize functional mobility, assistive devices and gait aids are used as needed. An understanding of biomechanics is essential to determine which external devices, if any, will assist a patient. For example, a child with DMD may be ambulating independently with difficulty. Toe walking may enable the patient to walk by increasing the knee extension moment and preventing knee collapse. Hyperlordosis, which is compensating for the weak hip extensors, is assisting in keeping the patient upright. Bracing the child and restoring what appears to be proper alignment may actually deprive the patient of compensatory mechanisms and arrest their ability to walk. Devices that require upper extremity strength, such as crutches, canes, and walkers, may not be useful because of weakness of the proximal shoulder girdle region. Mobility aids, such as wheelchairs and scooters, can be liberating and a relief for the patient who is exhausted from walking.[23]

Maximizing Independence for Activities of Daily Living

When weakness is such that independence in age-appropriate activities of daily living becomes difficult or impossible, creative restructuring of the task with appropriate assistive devices is required.

Bracing

In addition to skilled rehabilitation, interventions including strengthening, stretching, range of motion, and functional activities, the use of bracing can be both preventative and corrective in childhood neuromuscular disorders. Incorporating experienced orthotists into the rehabilitation team is helpful for disease management. Static and dynamic ankle foot orthoses for nighttime use to prevent contractures are often used in both ambulant and nonambulant patients. In ambulant children, prolonged stretching during rest can counteract the development of contractures from toe-walking and altered gait mechanics. In nonambulant children, maintaining normal range of motion is important for positioning in both sitting and standing. Orthoses for walking may be too heavy and impair gait in a predominantly proximal condition. Orthoses that include both the ankle and the knee are infrequently used but may be used for supported standing more than for range of motion. Wrist orthoses, such as cock-up splints, can help maintain flexibility and also promote function.

Scoliosis is common in childhood neuromuscular disorders[24] and is most prevalent in nonambulant patients and worsens with increasing age and loss of ambulation.[25] Supportive bracing is often prescribed although there is limited evidence of its effectiveness; thoraco-lumbar-sacral orthoses may provide some benefit in slowing the progression and promote greater functional use of the upper extremities.

Mobility Devices

In neuromuscular disorders where independent walking never develops, early prescription of a mobility device is highly recommended. Early development relies in part on the interaction with and manipulation of one's environment. Unlike typically developing healthy children, infants and toddlers with neuromuscular disorders are not able to explore their environment and develop independent motor skills. Powered mobility promotes functional independence and has beneficial effects on cognitive and psychosocial development. Independent exploration and play as well as mobility helps offset the occurrence of learned helplessness in these children.

Powered mobility is feasible in children as young as 2 years old and results in developmental gains, increased independence, and decreased caregiver burden.[26–29] The timing of initiation with power mobility in children with neuromuscular disease should be similar to and correspond with the development of walking in the typically developing child. Practitioners should consider early power mobility for their patients and make determinations of readiness based on other factors not necessarily related to chronological age. Considerations for readiness include problem solving and spatial relation skills, which can be assessed using validated scales.[30] Moreover, clinical judgment of powered mobility readiness and factors such as the home environment should be considered. A detailed assessment and any data or videotapes addressing readiness for power mobility may be included in a letter to support medical necessity so that third-party payers have sufficient information to justify approval.

For patients who lose the ability to ambulate because of disease progression, mobility devices can restore independence. In this instance, unfortunately, there are no well-defined parameters for the timing of wheelchair prescription. Decisions regarding the type of wheelchair (e.g. powered versus manual) must be made. Additionally, structural barrier assessment, sometimes resulting in home modifications, must take place prior to wheelchair specification and delivery.

The appropriate timing of wheelchair referral and use is not clear. Variables that influence the need for a wheelchair include muscles affected, degree of weakness, and respiratory involvement. In adult motor neuron disease, larger strength losses distally are better tolerated with the least amount of changes in functional ambulation.[31] Moreover, knee flexor strength has been found to be an important factor in community ambulation and hip extensor strength for home ambulation. Clinically, symptoms or changes that may trigger a referral for a wheelchair evaluation include frequent falls, difficulty rising from a seated position, distal lower extremity weakness, inability to ambulate in the community, impaired mobility requiring assistance to perform activities of daily living, excessive fatigue, or declining pulmonary function.

Adaptive Equipment

Home adaptive equipment and assistive devices for activities of daily living are frequently used. The patient may use the problem-focused coping strategy of devices as well as tricks for better function to maximize their independence in daily activities such as eating, bathing, toileting, and dressing, particularly in older children and adolescents.[32] This equipment can include large, durable medical equipment such as positioning chairs for bathing and toileting as well as small devices such as adapted plates, cups, and utensils.

Mobile arm supports are available to aid the patient with marked upper extremity weakness in tasks such as feeding, driving a motorized wheelchair, and performing leisure activities.[33,34] One example of a commercially available device is the Wilmington Robotic Exoskeleton (WREX).[35] Passive devices, such as the WREX, normally use the elastic energy of rubber bands or springs to achieve antigravity motions. However, these devices are typically most useful in those with only moderate upper extremity weakness, relatively preserved distal muscle strength, minimal range of motion limitations, and sufficient axial strength to permit proper upright positioning. Often, use of a mobile arm support is less efficient in the early stages than the patient's well-adapted compensatory mechanisms and as such is not readily accepted.

ASSISTIVE TECHNOLOGY

The severe muscle weakness with often spared cognition in children with neuromuscular disorders often results in physical disability and functional dependence that limits patients from reaching their full potential. This in turn contributes to decreased quality of life and high personal and societal cost. Technology can equalize opportunities when productivity and functional independence are rendered less dependent on muscle strength.

Adaptive equipment can also include devices that allow a child to navigate a computer or control their environment. These can range from onscreen keyboards to alleviate the need to type on a traditional keyboard, to simple switches with adjustable or increased sensitivity, to more sophisticated eye-gaze or infrared technology. Touch screen technology such as that utilized in the iPad, when properly positioned, is usable even for very weak children. Voice-activated systems can also be incorporated both for navigating a computer or phone and for environmental controls such as lights and televisions. Referral to an assistive technology specialist is important so that the child can explore and incorporate the latest technology most appropriate to meet their needs.

ALTERNATIVE THERAPIES

We shall only mention alternative therapies without discussing them in detail as they are not available to many of the individuals with neuromuscular diseases. Aquatic therapy is highly recommended throughout the lifespan for most neuromuscular disorders because the buoyancy of water counteracts the effects of gravity on weak muscles and often makes movement possible. Swimming and nontraditional therapies such as yoga and hippotherapy are encouraged to augment standard therapeutic interventions because they promote flexibility and strength while being enjoyable to the patient.

CONCLUSION

Rehabilitation of children with neuromuscular diseases is based on consensus; controlled studies are lacking. A rehabilitation program can be divided into *therapies, assistive devices*, and *technology*. The program will change as the individual passes from the ambulatory stage to the nonambulatory and then to the maintenance stage of the condition. Rehabilitation therapies address issues such as strength and endurance, which most often are responsible for functional disabilities. A rehabilitation program helps patients compensate for the impairment so that function can be improved or maintained by using alternative mechanisms. It is almost inevitable that patients with neuromuscular diseases will develop orthopedic deformities, and this has to be addressed on an ongoing basis. Most children with neuromuscular disabilities will develop cardiac and/or pulmonary dysfunction, and thus cardiologists and pulmonologists are critical members of the neuromuscular program. Finally, some of the muscular diseases are associated with cognitive impairment, and how this relates to the disability must be addressed.

REFERENCES

1. Delisa JA, Martin GM, Currie DM. Rehabilitation medicine, past, present and future. In: Delisa JA, editor. *Rehabilitation Medicine*. Philadelphia: JB Lippincott; 1988.
2. Molnar GE. *Pediatric Rehabilitation*. Baltimore: Williams & Wilkins; 1985.
3. Dahlbom K, Ahlstrom G, Barany M, Kihlgren A, Gunnarsson LG. Muscular dystrophy in adults: a five-year follow-up. *Scand J Rehabil Med* 1999;**31**:178–84.
4. World Health Organization (WHO), International Classification of Functioning, Disability and Health (ICF). Available from: <www.who.int/classification/icf>.
5. Wang CH, Finkel RS, Bertini ES, Schroth M, Simonds A, Wong B, et al. Consensus statement for standard of care in spinal muscular atrophy. *J Child Neurol* 2007;**22**:1027–49.

6. Bushby K, Finkel R, Birnkrant DJ, Case LE, Clemens PR, Cripe L, et al. Diagnosis and management of Duchenne muscular dystrophy, part 2: implementation of multidisciplinary care. *Lancet Neurol* 2010;**9**:177–89.
7. Bach J, Lieberman J. Rehabilitation of the patient with a disease affecting the motor neuron. In: Delisa J, Gans B, editors. *Rehabiliation Medicine: Principles and Practice*. 2nd ed. Lippincott; 1993.
8. Lieberman JS, Strauss NE. Muscular dystrophy and other myopathies. In: Seltzer ME, editor. *Textbook of Neural Repair and Rehabilitation*. New York: Cambridge University Press; 2006. pp. 677–98.
9. Hilton-Jones D, Kissel JT. The examination and investigation of the patient with muscle disease. In: Karpati G, Hilton-Jones D, Griggs RC, editors. *Disorders of Voluntary Muscle*. 7th ed. New York: Cambridge; 2001.
10. Katirji B. Clinical assessment in neuromuscular disorders. In: Katirji B, Kaminski HJ, Preston DC, Ruff RI, Shapiro BF, editors. *Neuromuscular Disorders in Clinical Practice*. Boston: Butterworth Heinemann; 2002.
11. Cup EH, Pieterse AJ, Ten Broek-Pastoor JM, Munneke M, van Engelen BG, Hendricks HT, et al. Exercise therapy and other types of physical therapy for patients with neuromuscular diseases: a systematic review. *Arch Phys Med Rehabil* 2007;**88**:1452–64.
12. van der Kooi EL, Lindeman E, Riphagen I. Strength training and aerobic exercise training for muscle disease. *Cochrane Database Syst Rev* 2005;**1**:CD003907.
13. Garber CE, Blissmer B, Deschenes MR, Franklin BA, Lamonte MJ, Lee IM, et al. American College of Sports Medicine position stand. Quantity and quality of exercise for developing and maintaining cardiorespiratory, musculoskeletal, and neuromotor fitness in apparently healthy adults: guidance for prescribing exercise. *Med Sci Sports Exerc* 2011;**43**:1334–59.
14. Strauss NE, Angell DK. Foot conditions related to neuromuscular disorders in adults. In: Kim DJ, Wainapel SF, editors. *Foot and Ankle Rehabilitation*. Phildelphia: Hanley and Belfus; 2001.
15. Bothwell JE, Dooley JM, Gordon KE, MacAuley A, Camfield PR, MacSween J. Duchenne muscular dystrophy—parental perceptions. *Clin Pediatr (Phila)* 2002;**41**:105–9.
16. Johnson ER, Fowler Jr. WM, Lieberman JS. Contractures in neuromuscular disease. *Arch Phys Med Rehabil* 1992;**73**:807–10.
17. Campbell Jr. RM, Smith MD, Mayes TC, Mangos JA, Willey-Courand DB, Kose N, et al. The characteristics of thoracic insufficiency syndrome associated with fused ribs and congenital scoliosis. *J Bone Joint Surg Am* 2003;**85-A**:399–408.
18. Vitale MG, Matsumoto H, Roye Jr. DP, Gomez JA, Betz RR, Emans JB, et al. Health-related quality of life in children with thoracic insufficiency syndrome. *J Pediatr Orthop* 2008;**28**:239–43.
19. Melacini P, Vianello A, Villanova C, Fanin M, Miorin M, Angelini C, et al. Cardiac and respiratory involvement in advanced stage Duchenne muscular dystrophy. *Neuromuscul Disord* 1996;**6**:367–76.
20. Finsterer J, Stollberger C, Wahbi K. Cardiomyopathy in neurological disorders. *Cardiovasc Pathol* 2013;**22**:389–400.
21. Polakoff RJ, Morton AA, Koch KD, Rios CM. The psychosocial and cognitive impact of Duchenne's muscular dystrophy. *Semin Pediatr Neurol* 1998;**5**:116–23.
22. Axford MM, Pearson CE. Illuminating CNS and cognitive issues in myotonic dystrophy: Workshop report. *Neuromuscul Disord* 2013;**23**:370–4.
23. Wagner MB, Katirji B. Rehabilitation management and care of patients with neuromuscular disease. In: Katirji B, Kaminski HJ, Preston DC, Ruff RI, Shapiro BF, editors. *Neuromuscular Disorders in Clinical Practice*. Boston: Butterworth-Heinemann; 2002.
24. Vialle R, Thevenin-Lemoine C, Mary P. Neuromuscular scoliosis. *Orthop Traumatol Surg Res* 2013;**99**:S124–39.
25. Kinali M, Main M, Eliahoo J, Messina S, Knight RK, Lehovsky J, et al. Predictive factors for the development of scoliosis in Duchenne muscular dystrophy. *Eur J Paediatr Neurol* 2007;**11**:160–6.
26. Butler C. Effects of powered mobility on self-initiated behaviors of very young children with locomotor disability. *Dev Med Child Neurol* 1986;**28**:325–32.
27. Guerette P, Furumasu J, Tefft D. The positive effects of early powered mobility on children's psychosocial and play skills. *Assist Technol* 2013;**25**:39–48, quiz 9–50.
28. Tefft D, Guerette P, Furumasu J. The impact of early powered mobility on parental stress, negative emotions, and family social interactions. *Phys Occup Ther Pediatr* 2011;**31**:4–15.
29. Dunaway S, Montes J, O'Hagen J, Sproule DM, De Vivo DC, Kaufmann P. Independent mobility after early introduction of a power wheelchair in spinal muscular atrophy. *J Child Neurol* 2013;**28**:576–82.
30. Tefft D, Guerette P, Furumasu J. Cognitive predictors of young children's readiness for powered mobility. *Dev Med Child Neurol* 1999;**41**:665–70.
31. Slavin MD, Jette DU, Andres PL, Munsat TL. Lower extremity muscle force measures and functional ambulation in patients with amyotrophic lateral sclerosis. *Arch Phys Med Rehabil* 1998;**79**:950–4.
32. Natterlund B, Ahlstrom G. Problem-focused coping and satisfaction with activities of daily living in individuals with muscular dystrophy and postpolio syndrome. *Scand J Caring Sci* 1999;**13**:26–32.
33. Yasuda YL, Bowman K, Hsu JD. Mobile arm supports: criteria for successful use in muscle disease patients. *Arch Phys Med Rehabil* 1986;**67**:253–6.
34. Ambrosini E, Ferrante S, Schauer T, Klauer C, Gaffuri M, Ferrigno G, et al. A myocontrolled neuroprosthesis integrated with a passive exoskeleton to support upper limb activities. *J Electromyogr Kinesiol* 2014;**24**:307–17.
35. Jaeco Orthopedic. Product Description: WREX. Wilmington Robotic Exoskeleton Arm. [Website] <http://jaecoorthopedic.com/products/products/WREX%3A-Wilmington-Robotic-EXoskeleton-Arm.html>.

Chapter 54

Outcome Measures in Neuromuscular Diseases

Jacqueline Montes and Petra Kaufmann

INTRODUCTION: WHAT ARE OUTCOME MEASURES?

Good outcome measures are a critical component of any successful therapeutics evaluation program. Although some pediatric neuromuscular diseases are now seeing discoveries translated into early-phase clinical trials, there remains a great need for better therapeutics. Bringing these therapeutics to patients requires clinical trials using outcome measures that ideally are easy to implement, are sensitive to change, and reflect an important aspect of the disease.

Outcome measures can be biomarkers, meaning that they measure a biological characteristic of the disease in question. Biomarkers can be metabolites measured in blood or cerebrospinal fluid, electrophysiological measures, or imaging markers. Outcome measures can also be clinical measures, meaning that they measure something that can be reported by observers such as clinicians or caregivers, or by the patient directly.

The goal of using these biomarkers or clinical outcome measures is to assess changes that occur over the course of a clinical trial. Biomarkers or clinical measures can also be used in different contexts, such as when diagnostic or prognostic markers are sought. However, the overview in this chapter will focus only on measures of disease activity or progression, and will discuss them in the context of their potential usefulness in clinical research.

To set the stage for successful clinical trials, researchers need not only patients who are ready to participate, but also high-quality datasets on the natural history of the disease using outcome measures suitable for clinical trials. Collecting natural history data requires significant resources, and an uncommon degree of altruism on the part of patients and their families, because there is no potential for direct benefit to the participant. To lessen the burden added by the study visits, it can be helpful if outcome measures can be collected at the same time that clinical care is provided. Conversely, collecting a small set of outcome measures during clinical visits can also help patients and their care team with charting disease progression. The outcome measures information can help guide care, and can help better understand how the patient's disability affects function.

Developing and validating outcome measures requires time and resources, and is associated with burden to patients and research teams. Therefore, progress in a given field can be accelerated when research teams can agree on the use of a common set of outcome measures. Using a common set of measures across sites leads to larger datasets being available in a shorter timeframe, so that outcome measures can be developed and tested more efficiently. Avoiding the development and use of multiple outcome measures focused on the same concept and indication can save time and money. In situations when no good outcome measure is available to help address a question, it is preferable if researchers turn to existing outcome measures rather than starting anew in developing one. For example, using an existing scale, or part thereof, can lead to more rapid validation of a potential outcome measure, and can also create opportunities to address important questions across diseases, for example by comparing datasets for two different neuromuscular diseases.

Outcome measures are better delineated for some neuromuscular diseases than others; to a degree, this reflects how far along the respective fields are in terms of therapeutics development. This overview will focus on spinal muscular atrophy (SMA) and Duchenne muscular dystrophy (DMD), but several of the principles and in some cases even the outcome measures discussed may be applicable to a wider range of pediatric neuromuscular disorders.

B.T. Darras, H. Royden Jones, Jr., M.M. Ryan & D.C. De Vivo (Eds): Neuromuscular Disorders of Infancy, Childhood and Adolescence, Second edition.
DOI: http://dx.doi.org/10.1016/B978-0-12-417044-5.00054-8

OUTCOME MEASURES IN DIFFERENT STAGES OF THERAPEUTICS DEVELOPMENT

The hallmark of pediatric neuromuscular disorders is muscle weakness. Muscle weakness can be directly measured using manual or quantitative muscle testing. It can also be measured more indirectly in terms of its effect on motor function. For pivotal clinical trials, an outcome measure that is inherently related to how a patient functions or feels in their everyday life is more meaningful than a measure that quantitates strength but cannot be interpreted in terms of its functional consequences. Weakness can affect muscles other than skeletal muscles; respiratory muscle weakness, for example, is of interest as an outcome in most neuromuscular diseases. Similarly, involvement of bulbar muscles may prompt use of outcome measures assessing bulbar function. In a subset of pediatric neuromuscular disorders, cardiac muscle can be affected as well.

When muscle function or strength cannot be easily measured, or when it is not clear if clinical measures of muscle function would change within the observation period of a trial, then biomarkers may be useful. This is particularly relevant in early phase (Ib or II) trials. In this scenario, it is hoped that biomarkers may be more sensitive to early treatment effect than are clinical measures.

In rare instances in medicine, biomarkers can be surrogates of downstream clinical outcomes. Hypertension, for example, is thought to predict the downstream risk of cardiovascular events. In most instances, this kind of strong association cannot be established. The expectation for pediatric neuromuscular disorders at this time is not to establish a surrogate marker in the foreseeable future, but rather to use biomarkers in a therapeutics development program as an early signal of activity, to be followed by later stage trials with observations using clinical markers.

Another type of biomarker is not specific to the disease and its hallmark features, but is specific to the presumed mechanism of action of the intervention. These kinds of markers of biological activity can be very useful in early proof of mechanism trials when it is important to establish that the intervention has the desired biological effect in patients.

Clinical outcome measures ideally reflect an aspect of the disease that matters to patients. In other words, clinical outcome measures should reflect how a patient feels or functions. This information is best obtained from the patient directly. When assessments of functional status are obtained from an observer, such as a relative, caregiver, or clinician, then the outcome measure captures not only the actual function of the patient, but also the subjective perception and opinion of the observer. Therefore, any measure that captures information from the patient as a source is preferable to one that captures information indirectly

Clinical outcome measures are, ideally, easily interpretable and clinically meaningful. When an outcome measure focuses on a function relevant to the patient's daily living activities, the clinical meaning and relevance of any change can be easily understood and communicated. To ensure the meaningfulness of an outcome, experts in outcomes development often start with focus groups to determine which domains are most important to patients, rather than starting with what clinicians think is most important. When outcome measures contain a large number of disparate items, or when they mix functional information with clinical observation and physical examination, then the interpretation of any change is often unclear.

Different outcome measures have variable places in the spectrum of drug development. In early human trials (phase 0 and Ib), biomarkers of mechanism, to demonstrate that the intervention hits its target, are useful. In early exploratory efficacy trials (phase IIa), biomarkers that can signal an early effect of the intervention can be useful. Clinical efficacy trials (phase IIb or phase III) mandate clinical outcomes that ideally indicate how a patient feels, functions, or survives. Figure 54.1 summarizes the drug development process.

CLINICAL OUTCOME MEASURES

The ideal clinical outcome measure is age appropriate, easily administered, functionally meaningful, and imposes minimal patient burden. In rare neuromuscular conditions

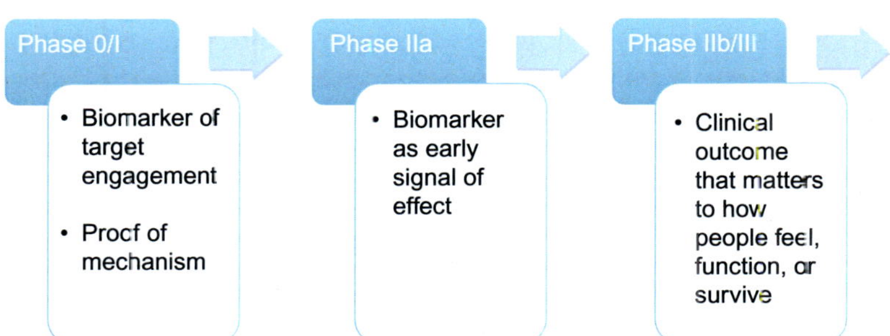

FIGURE 54.1 Outcome measures in the therapeutics evaluation process.

where enrollment in clinical trials requires multiple international sites, ideal outcome measures are those which require minimal training and equipment to allow participation of investigators with limited clinical resources.

It is best to build on existing tools when choosing an outcome measure because one can take advantage of any data available for the existing tools, compare data to published cohorts in the same or another condition, and take advantage of any regulatory pathway already established for the outcome measure. Outcome measures designed specifically for a particular condition are ideal but not mandatory since children with weakness have similar functional limitations.

There is a broad clinical phenotype both across different pediatric neuromuscular conditions as well as within individual disorders. For example, in SMA, there is a spectrum of disease severity and functional ability, ranging from infants and children never achieving the ability to sit unsupported, to those only able to sit but unable to stand and walk, and to those who can walk but have weakness limiting their ability to rise from the floor, run, and jump.[1] It is difficult to envision a single motor performance-based measure assessing functionally meaningful abilities for all SMA types. Similarly, boys with DMD transition from an ambulatory stage to a nonambulatory stage in which gross motor function is significantly reduced, and upper limb function is a more relevant potential outcome measure.[2] One must consider if the population of patients enrolled in a clinical trial will lose function during their participation that would not be captured on the outcome measure they choose because of a floor effect. For example, if a boy with DMD loses ambulation during a clinical trial where a walking test is the primary outcome measure, without a nonambulatory assessment any further changes in the participant's clinical function would not be identified.

We have categorized the different motor function outcome measures below into those for ambulatory and nonambulatory children. Careful consideration with regards to the best clinical outcome measure based on age, disease severity, functional level, and rate of progression must be made when designing a clinical trial.

Ambulatory: Walking, Endurance, Mobility

Gait assessments are desirable clinical outcome measures because they reflect weakness, the primary feature in neuromuscular diseases, and are functionally relevant. How a person or child is able to navigate their environment is meaningful and noticeable to both the patient and clinician. As a result of weakness, balance and endurance can be affected and should be quantified in the outcome measure (Table 54.1).

The Six Minute Walk Test (6MWT) is a commonly used assessment in childhood neuromuscular disease that has been used as the primary outcome measure in clinical trials involving conditions with ambulatory participants.[12–15] While the 6MWT was originally designed to measure exercise capacity in cardiopulmonary disease, it has been widely adopted as an assessment of functional walking ability in both adult and pediatric neurological and neuromuscular conditions.[16] The test is simply the distance walked in 6 minutes on a fixed linear course. It is easily administered and requires no special equipment and little training. Standardization is important and includes prespecifying the level of encouragement given to the patient so that any potentially confounding evaluator influences on performance are minimized.[17] In disorders like DMD, protocols have been established to make testing feasible in children where attentional difficulties may complicate the assessment.[18] Because there is a large body of normative data including children as young as 3 years old on the 6MWT,[19,20] comparisons across age groups, gender, and body size can be made. This is important because the performance in normal volunteers changes over the course of life due to development and aging, so that comparison to normative data may be needed in some circumstances, rather than just the measurement of absolute change over time. The 6MWT is currently the preferred measure of functional walking ability, endurance, and fatigue.

The 10-meter walk/run test is typically included with the 6MWT in clinical trials.[13,21] Unlike the 6MWT, children are permitted to walk or run, if possible, as fast as they can. The 10-meter walk/run is easy to administer and is a good complement to the 6MWT because it is more reflective of strength rather than endurance.

Other timed functional assessments for ambulatory participants include time to rise from the floor and time to ascend and descend stairs.[22] These tests typically include a qualitative score in addition to the timed evaluation, to best capture changes in strength and function. For example, time to rise from the floor will elicit a Gowers' maneuver in conditions like DMD and SMA, in which proximal weakness is the primary problem. If strength improves with a therapeutic intervention, understanding *how* subjects negotiate the transfer from floor to stand is as meaningful as how quickly they do this. Similarly, the pattern by which a child navigates stairs, such as the use of the railing and their step pattern, is important to evaluate, as it impacts how they function in their environment.

The timed up and go test (TUG) measures the time it takes a patient to stand up from an armchair, walk a distance of 3 meters, turn, walk back to the chair, and sit down.[23] It is a quick, meaningful, and applied objective measure of balance, gait speed, and functional mobility. While TUG was originally designed to identify the community-dwelling elderly at risk for falls, it has been widely adopted as a clinical outcome measure in both pediatric[24,25] and adult[26,27] neurological and neuromuscular

TABLE 54.1 Clinical Outcome Measures for Ambulant Patients

Outcome Measure	Age	Disease Specific	Assessment Characteristics	Published Clinical Trials
Time to rise from floor	> 5 years	No	Measures time to rise from supine; evaluate for Gowers' maneuver	gabapentin (SMA),[3] hydroxyurea (SMA),[4] pentoxifylline (DMD),[5] oxatomide (DMD),[6] creatine and glutamine (DMD),[7] prednisone (DMD)[8–10]
Time to walk/run 10 or 30 meters	> 5 years	No	From standing, time to walk/run on a straight path	gabapentin (SMA),[3] hydroxyurea (SMA),[4] pentoxifylline (DMD),[5] oxatomide (DMD),[6] creatine and glutamine (DMD),[7] prednisone (DMD)[8–10]
Time to climb steps	> 5 years	No	Measures time to ascend and descend stairs	gabapentin (SMA),[3] hydroxyurea (SMA),[4] pentoxifylline (DMD),[5] oxatomide (DMD),[6] creatine and glutamine (DMD),[7] prednisone (DMD)[8–10]
Six Minute Walk Test	> 4 years	No	Measures distance walked on a 25-meter course. Evaluates functional exercise capacity and fatigue	alglucosidase alfa (Pompe),[11] eteplirsen (DMD)[12]
Timed up and go test		No	Measures time to rise from a chair, walk 3 meters, turn around, and sit down	
North Star Ambulatory Assessment	> 5 years	DMD	Includes 17 items such as standing, running, jumping, hopping, and stair climbing scored on a 3-point qualitative scale. Some items are timed	

Abbreviation: DMD, Duchenne muscular dystrophy.

conditions. Unlike the 6MWT or 10-meter walk/run, TUG incorporates rising from and sitting in a chair, adding another functionally relevant aspect of mobility to the assessment.

So far, the tests described here are not specific to any one neuromuscular condition and are widely applied across clinical trials. Disease-specific measures, when available, may have some advantage as they focus on the unique characteristics and functional deficits of the disorder. The North Star Ambulatory Assessment (NSAA) is designed specifically for boys with DMD.[22] This 17-item assessment includes evaluations of standing, running, jumping, hopping, and stair climbing, scored on a 3-point qualitative scale. Some items on the NSAA are timed and include time to rise from the floor, time to climb stairs, and 10-meter walk/run.

Nonambulatory: Gross Motor Function Scales

Standardized gross motor function assessments are typically used as primary outcome measures in clinical trials where nonambulatory participants are included (Table 54.2). While ambulatory function is assessed in most of these exams, it is not required. This allows for inclusion and comparisons across a broad phenotypic spectrum of a disorder. Typically these assessments arrange items hierarchically so the child moves through activities from lower to higher functioning items. Gait assessments are typically included but seldom include an evaluation of endurance, making the 6MWT a good complement in ambulatory children. There are many motor

TABLE 54.2 Clinical Outcome Measures for Gross Motor Function

Outcome Measure	Age	Disease Specific	Assessment Characteristics	Published Clinical Trials
Gross Motor Functional Measure (GMFM)	2–18 years	No	66- or 88-item motor function assessment	hydroxyurea (SMA)[4]
Hammersmith Functional Motor Scale—Expanded (HFMSE)	2–45 years	SMA	33-item gross motor function assessment	phenylbutyrate* (SMA),[28,29] salbutamol* (SMA),[30] L-carnitine and valproic acid* (SMA)[31–33]
Motor Function Measure	1 – adults > 25 years	No	32 items in 3 domains of motor function	prednisone (DMD)[34]
Egen Klassifikation Scale (EK Scale)	teens – adults	No	evaluator administered, self-rated questionnaire of function	
Vignos Lower Extremity Scale		DMD	10-point scale of leg function and ambulation	prednisone (DMD)[8,9]
Wee Functional Independence Measure (Wee FIM)	6 months – 6 years	No	functional, evaluator administered questionnaire	
Test of Infant Motor Performance (TIMP) and Screening Items (TIMPSI)	infants to 4 months	No	observed and elicited behaviors and motor function	
Children's Hospital of Philadelphia Infant Test of Neuromuscular Disorders (CHOP INTEND)	infants and weak children	No	observed and elicited behaviors and motor function	

Abbreviation: DMD, Duchenne muscular dystrophy.

function assessments available but those tailored specifically to a single disorder are sometimes considered preferable as they focus on disease-specific problems. In infants, toddlers and young children, developmentally appropriate scales are necessary.

The Gross Motor Function Measure™ (GMFM) is a well-established assessment that was originally designed to assess children with cerebral palsy.[35] It has been adopted and used in many different neurological conditions, including SMA.[36,37] The GMFM takes approximately 45 to 60 minutes to complete in stronger, higher functioning children and contains 88 items in 5 dimensions: (A) lying and rolling, (B) sitting, (C) crawling and kneeling, (D) standing, and (E) walking, running, and jumping. This hierarchical organization permits patients to progress through each dimension, according to their ability, without a ceiling effect. A detailed manual for the GMFM is available including video instruction.

The Motor Function Measure (MFM) was developed specifically for people with neuromuscular disease to assess motor function.[38] Similarly to the GMFM, it is organized in 3 domains: (A) standing position and transfers, (B) axial and proximal motor function, and (C) distal motor function. Test items include a continuum of functional ability, organized in a hierarchical fashion, ranging from simple motor skills in supine to a 10-meter run. Also included in the MFM is a fine motor assessment of hand function. The MFM was designed specifically for neuromuscular conditions, and has been widely applied in SMA, DMD, and other dystrophinopathies, and other neuromuscular conditions.

The Hammersmith Functional Motor Scale (HFMS) was originally designed to assess motor function in patients with SMA type 2,[39] and was later revised to include higher functioning and ambulatory SMA patients in an expanded version (HFMSE).[40] This 33-item scale covers a continuum of functional assessments designed to delineate the patterns of weakness present in SMA patients. The test requires only standard equipment and takes less than 15 minutes on average to complete. The HFMSE is highly correlated with other motor function assessments, including the GMFM, and can discriminate between walkers and nonwalkers as well as respiratory function.[41] The scale is well-suited as a clinical outcome measure in SMA clinical trials because of its specificity and minimal patient burden.

While most functional assessments require the patient to perform each motor task, some scales are evaluator-administered questionnaires for which the examiner queries the child or parent on their abilities. The Wee Functional Independence Measure was designed to assess disability in children with developmental delay, based on the framework proposed by the World Health Organization in children between 6 months and 6 years.[42] It is organized in three main domains: (A) self-care, (B) mobility, and (C) cognition, and is scored on a 7-point scale ranging from total assistance to complete independence. While this assessment has been applied to some neuromuscular conditions,[43] it is best suited for those in which cognition may also be affected, which is not the case in SMA.

The Egen Klassifikation Scale is also an evaluator-administered questionnaire that was developed to assess motor function in older, nonambulatory children and teens, as well as adults, in disorders such as DMD and SMA.[44,45] It includes assessments of activities of daily living such as wheelchair use, ability to transfer, arm function and feeding, turning in bed, coughing, and speaking.

Scales incorporating developmentally appropriate motor function items should be included as outcome measures in clinical trials for infants and young children. The Test of Infant Motor Performance (TIMP) is a functional scale validated in preterm infants under 4 months of age.[46] The TIMP includes both observed and elicited motor assessments, is sensitive to age-related development, discriminates between those at low and high risk for motor problems,[47] and predicts delayed motor development in preterm infants.[48] In neuromuscular disease, specifically SMA, a shorter version of the TIMP that includes only the screening items has been applied, since these patients suffer such profound weakness and motor delay.[49]

Similarly, the Children's Hospital of Philadelphia Infant Test of Neuromuscular Disorders (CHOP INTEND) was developed specifically for weak infants.[50] It includes both active and elicited reflexive movement of the head, neck, and trunk, as well as proximal and distal limb strength. The CHOP INTEND is structured to move from easiest to most difficult, and includes gravity eliminated (lower scores) to antigravity movements (higher scores). Typically, motor function assessments are included as secondary outcome measures in clinical trials in SMA type 1 to measure change over time, but survival or time to respiratory failure are the ideal primary endpoints in this life-threatening form of the disorder.

Upper Limb Function

Assessments of upper limb function focus on fine motor function and activities of daily living. Moreover, since patients are nonambulatory in many neuromuscular conditions, assessments are needed that are tailored to their level of function and are meaningful to everyday life. The Jebsen Hand Function Test was designed to assess function in children with cerebral palsy in a way mimicking everyday activities.[51,52] This scale includes seven items that are performed bilaterally and timed. While not designed specifically for children with neuromuscular disorders, the test has been widely adopted in many pediatric as well as adult neurological and neuromuscular conditions, including DMD (Table 54.3).

TABLE 54.3 Clinical Outcome Measures for Upper Limb Function

Outcome Measure	Age	Disease Specific	Assessment Characteristics	Published Clinical Trials
Upper Limb Module for SMA (ULM)	>30 months	SMA	9-item assessment of arm and hand function	
PUL for DMD		DMD		
Brooke Upper Extremity Scale		DMD	6-point scale of arm function	prednisone (DMD)[8,9]
Jensen Hand Function Test		No	Time to complete 7 individual hand function assessments	
ABILHAND		No	22-item questionnaire on hand function and manual ability	
Upper-Limb Functional Ability Test		No	14-item assessment of proximal and distal arm function	
ACTIVLIMB		No		

Abbreviation: DMD, Duchenne muscular dystrophy.

The Upper Limb Module (ULM) is an assessment of arm function designed specifically for nonambulant SMA patients.[53] This nine-item scale can be performed reliably in children as young as 30 months old and takes no more than 10 minutes to complete. Items are scored on a 3-point scale according to the quality of performance: normal without assistance, modified method but without assistance, or unable to achieve the task without assistance. The ULM is intended to complement standard SMA-specific gross motor function measures such as the Hammersmith Functional Motor Scale (HFMS).

The Brooke Upper Extremity scale has been used to describe arm function and disease severity in patients with DMD.[54,55] This six-item scale ranges from raising arms above the head in a circle, for a maximum score, to no useful hand function. More recently, a more detailed assessment of upper limb function has been developed for individuals with dystrophinopathies such that the change of arm function from the ambulatory to nonambulatory stages of the disease can be measured. The Performance of the Upper Limb (PUL) Scale is designed to incorporate performance of shoulder, elbow, wrist, and hand function.[56]

Similarly to gross motor scales, there are self-reported questionnaires designed to describe upper limb function specific to neuromuscular conditions. The ACTIVLIMB is a 22-item questionnaire in which half of the items included evaluate upper extremity function in children and adults.[57] The ABIL HAND is a questionnaire tailored specifically to upper limb function, including items such as fastening a zipper, spreading butter, and counting money (bills).[58] Patient or self-reported outcome measures such as ACTIVLIMB and ABIL HAND, particularly when they target activities of daily living and function, are important to include as they may capture change that is meaningful to the individual.

Respiratory Function

Spirometry measures pulmonary function in patients and is used in clinical settings as well as in clinical research. Spirometry can provide information on the volume and speed of air flow during breathing. Spirometry devices are widely available in clinics and hospitals, and also come as portable devices so that spirometry can be performed not only in a pulmonary function laboratory, but also at the bedside and in the clinic. Spirometry results in a flow-volume curve, with the positive values representing expiration and the negative values representing inspiration. An experienced evaluator can assess the quality of the patient's effort and the technical quality of the spirometry by looking at the flow-volume curve. Spirometry is highly dependent on patient cooperation, so it can typically only be done in cooperative children when they are about 5 years old. Exceptionally, it can be successfully implemented at a younger age. In other cases, cognitive or behavioral factors delay interpretable results until the children are older.

The forced vital capacity is a widely used indirect measure of respiratory muscle strength in neuromuscular clinical research. Forced vital capacity, or FVC, is the maximum amount of air a person can exhale after a maximum inhalation. It is more than the volume of air exhaled after a regular breath (tidal volume) because it also includes the inspiratory reserve volume and the expiratory reserve volume. The forced vital capacity is associated with height, gender, and to a lesser degree with weight, age, and ethnicity. Therefore, the results are typically expressed as percent predicted by age, gender, and height. A spirometry reference value calculator is available on the Centers for Disease Control (CDC) website.[59,60]

FVC has been successfully implemented in several pediatric neuromuscular research studies and trials. It requires spirometers and repeated evaluator training to establish inter-rater reliability throughout the observation period. To enhance reliability, some groups have used a central evaluator to assess flow-volume curves and exclude data that do not meet a certain quality threshold.[37] Typically, three good attempts are required, with only the best one being used for data collection. The evaluator must be able to assess and document the participant's effort. In children, the instructions given and the encouragement received during the spirometry maneuvers are very important. Some spirometry systems also offer an incentive program; for example, in one system that shows a birthday cake with burning candles on a video screen, the candles progressively extinguish as the child exhales to provide an incentive. While small children often cannot initially cooperate with spirometry, it is often possible to gain the skill over time with repeated training sessions, or by providing a spirometer for home practice. In SMA, several cohort studies have included FVC as an outcome measure.[61] In a prospective study, FVC declined over time in patients with SMA types 2 and 3.[62] In DMD, FVC as a percent of predicted for healthy children declined over time in a prospective cohort study, as did several other respiratory measures including gastric pressure during cough and sniff nasal inspiratory pressure.[63]

Muscle Strength

Weakness is present in nearly all neuromuscular conditions and, as such, is an important assessment to include in clinical trials. Formal strength assessments are effort dependent and require cooperation, and are therefore typically not incorporated until the child is 5 years old. Because function is the primary concern in most neuromuscular disorders, strength assessments are typically included as secondary outcome measures or early activity markers in clinical trials. Validation of functional measures usually mandates determination of their relationship to strength.

Determining the force produced during a maximal voluntary isometric contraction (MVIC) is a commonly used method to assess quantitative strength. Both fixed and handheld assessments have been shown to be reliable in DMD and SMA.[61,64–66] Fixed systems incorporate a strain gauge, are typically costly, and require prolonged time for set-up and position changes. Handheld dynamometry is another method of quantifying strength. The examiner fixes a handheld device against a limb or body part while the patient performs a maximal isometric contraction. The limitation of handheld devices is that the examiner needs to be able to exert more force than the patient, making it difficult in stronger and older populations. The benefit of quantitative strength assessments is that they are thought to be more sensitive in stronger muscle groups or those with near normal strength.

Manual muscle testing (MMT) is typically included in a neurological exam, and in cooperative children can be included as a secondary outcome measure in clinical trials. Standardization and adherence to established scales such as the Medical Research Council (MRC) scale for grading is essential. MMT has been shown to be reliable in children with DMD, particularly in proximal muscle groups and muscles of the lower extremities.[67] The complement of both quantitative and qualitative strength assessments allows for a comprehensive evaluation of strength.

POTENTIAL BIOMARKERS

Ultrasound

Ultrasound is emerging as a potential biomarker in clinical research of pediatric neuromuscular disorders because it provides information on the affected end organ or muscle, and because it can be done noninvasively without requiring much cooperation from the patient. Ultrasound equipment is available in most settings where neuromuscular trials would be performed, and the equipment is typically easily portable. Several studies have looked at the changes on muscle ultrasound as a result of the neuromuscular disease, such as the dystrophic process in DMD. The echo intensity can be quantified by the different shades of gray on the image, and the muscle thickness can also be estimated.[68] In DMD, progressive abnormalities on muscle ultrasound were found in small, longitudinal observations that reported an increase of echo intensity with age, associated with disease progression.[69] In SMA, a cross-sectional ultrasound study found a correlation between strength and muscle luminosity, normalized to subcutaneous fat luminosity.[70]

Electrical Impedance Myography

Electrical impedance myography (EIM) is another noninvasive method to assess the severity of neuromuscular disease. This relatively novel technology quantitates the electrical properties of muscles. A handheld device is used to measure impedance of muscle, which is thought to be associated with membrane integrity and other muscle tissue properties. Therefore, as muscle fibers become atrophic or undergo necrosis in neuromuscular disorders, these tissue changes are reflected in changes in impedance. EIM has been used in a small cohort of healthy controls and boys with DMD, and has shown correlation with functional measures.[71] In SMA, children differed from healthy controls; whereas EIM parameters changed over time in healthy controls, they remained stable in SMA patients.[72]

Motor Unit Number Estimation

Motor unit number estimation is a noninvasive electrophysiological technique that has been used as a biomarker in neurogenic neuromuscular disorders such as SMA. It is based on determining the compound motor action potential (CMAP) amplitude or area under the curve of a distal muscle in response to supramaximal stimulation, and then dividing the result by the amplitude or area under the curve of a single motor unit action potential. If the amplitude is used, then two measurements (with μV or mV as units) are divided by each other, resulting in a number that represents the estimated count of the number of discrete units contributing to the compound motor unit action potential. There are several methods by which the single motor unit action potential (SMUAP) can be assessed, with the multipoint stimulation technique being the most commonly used in SMA. MUNE was found to be associated with SMN2 copy number, age of onset, and function in SMA.[73] In a longitudinal cohort study, SMA patients had evidence of motor neuron loss and reinnervation at baseline. Longitudinally, the CMAPs remained stable in SMA, but MUNE increased associated with a decrease in SMUAP, perhaps suggesting an increase in nascent units in SMA over time.[74]

MR Imaging

Noninvasive imaging of skeletal muscles using magnetic resonance imaging (MRI) has been used to obtain qualitative and quantitative information on the affected muscles in neuromuscular disorders. In DMD, MRI studies showed a characteristic pattern of pelvic girdle muscle involvement with fatty infiltration most commonly involving the gluteus and adductor magnus muscles and sparing of the gracilis muscle.[75] In a small cohort of DMD patients, the fatty infiltration seen on MRI correlated with the functional status found on clinical assessments.[76] In a study including patients with different forms of muscular dystrophy, MRI showed abnormalities that were not detectable in manual muscle strength testing, suggesting subclinical involvement of some muscles.[77] In SMA, thigh muscle MR images remained stable in volume over a 6-month follow-up period.[78]

PATIENT REPORTED OUTCOME MEASURES

Quality of Life and Caregiver Burden

Evaluating quality of life is important if a change in this parameter reflects a change detected by a clinical or biological measure. The World Health Organization defines quality of life as "an individual's perception of their position in life in the context of the culture and value systems in which they live and in relation to their goals, expectations, standards and concerns."[79] A person's quality of life is affected by multiple dimensions, including physical function and well-being as well as psychological and social factors. The suitability of an intervention should not be determined by its impact on physical function or disease status alone; the effect on quality of life also needs to be considered. It is possible to have an intervention with unwanted secondary effects that outweigh any positive benefits in physical function or disease state. Therefore, assessments of quality of life should be included as a secondary outcome measure in clinical trials.[80] The PedsQL™ Pediatric Quality of Life Inventory instrument, a proprietary test to measure quality of life, is a validated measure for use with healthy school and community populations, as well as with pediatric populations with acute and chronic health conditions.[66,81] The validity of the PedsQL Pediatric Quality of Life Inventory was demonstrated through known group comparisons and correlations with other measures of disease burden. Age-specific forms for children 5 to 18 years old, parent/caregiver forms for children 2 to 18 years old, and a neuromuscular disease specific module are available. The PedsQL assessments have been shown to be reliable in SMA[66,82] and DMD.[83] The NeuroQoL is a self-report instrument that can be used to evaluate a range of domains of health-related quality of life. The questions can be answered via a computer program or in paper format. The instrument was developed using modern psychometric methods for use in adult and pediatric patients, both in English and Spanish, and is available online.[84,85]

In addition to measuring quality of life, describing and quantifying the effect a condition has on caregivers and family members is an important outcome measure. An ideal intervention has a positive effect on the disease, and alleviates some of the burden perceived by the caregivers. In families with chronically ill children, the Pediatric Evaluation of Disability Inventory (PEDI) is a parent-reported measure of functional disability.[86,87] Part I of the PEDI allows the parent to evaluate the child's ability in three domains: self-care, mobility, and social function. Parts II and III assess the amount of caregiver assistance and modifications necessary to perform functional activities and activities of daily living, and have been shown to be an objective measure of caregiver burden.

CONCLUSIONS

The last three decades have seen remarkable progress for pediatric neuromuscular disorders. For two of the most prevalent diseases, DMD and SMA, the discovery of the underlying genetic etiology has led to an exponential increase in our understanding of disease biology and to several targets for therapeutic approaches. As a result, there are now several promising therapeutics in clinical development, some with a specific and well-understood mechanism and robust preclinical efficacy data. With novel therapeutics on the horizon and in development, it becomes imperative that the community of neuromuscular researchers strives to find the most useful outcome measures to evaluate new treatments. Given that most pediatric neuromuscular diseases are relatively rare, and that there is great urgency in translating the recent discoveries into better therapeutics for patients, it is critical that the research community collaborate across privately and publicly sponsored projects and harmonize the outcome measures used so that redundancies can be avoided and data from different studies can be compared and pooled. This will accelerate progress towards our common goal: effective treatments for pediatric neuromuscular disorders.

REFERENCES

1. Munsat TL, Davies KE. International SMA consortium meeting. (26–28 June 1992, Bonn, Germany). *Neuromuscul Disord* 1992;**2**:423–8.
2. Mercuri E, McDonald C, Mayhew A, Florence J, Mazzone E, Bianco F, et al. International workshop on assessment of upper limb function in Duchenne Muscular Dystrophy: Rome, 15–16 February 2012. *Neuromuscul Disord* 2012;**22**:1025–8.
3. Merlini L, Solari A, Vita G, Bertini E, Minetti C, Mongini T, et al. Role of gabapentin in spinal muscular atrophy: results of a multicenter, randomized Italian study. *J Child Neurol* 2003;**18**:537–41.
4. Chen TH, Chang JG, Yang YH, Mai HH, Liang WC, Wu YC, et al. Randomized, double-blind, placebo-controlled trial of hydroxyurea in spinal muscular atrophy. *Neurology* 2010;**75**:2190–7.
5. Escolar DM, Zimmerman A, Bertorini T, Clemens PR, Connolly AM, Mesa L, et al. Pentoxifylline as a rescue treatment for DMD: a randomized double-blind clinical trial. *Neurology* 2012;**78**:904–13.

6. Buyse GM, Goemans N, Henricson E, Jara A, van den Hauwe M, Leshner R, et al. CINRG pilot trial of oxatomide in steroid-naive Duchenne muscular dystrophy. *Eur J Paediatr Neurol* 2007; **11**:337–40.
7. Escolar DM, Buyse G, Henricson E, Leshner R, Florence J, Mayhew J, et al. CINRG randomized controlled trial of creatine and glutamine in Duchenne muscular dystrophy. *Ann Neurol* 2005; **58**:151–5.
8. Fenichel GM, Florence JM, Pestronk A, Mendell JR, Moxley 3rd RT, Griggs RC, et al. Long-term benefit from prednisone therapy in Duchenne muscular dystrophy. *Neurology* 1991;**41**:1874–7.
9. Mendell JR, Moxley RT, Griggs RC, Brooke MH, Fenichel GM, Miller JP, et al. Randomized, double-blind six-month trial of prednisone in Duchenne's muscular dystrophy. *N Engl J Med* 1989; **320**:1592–7.
10. Griggs RC, Moxley 3rd RT, Mendell JR, Fenichel GM, Brooke MH, Pestronk A, et al. Prednisone in Duchenne dystrophy. A randomized, controlled trial defining the time course and dose response. Clinical Investigation of Duchenne Dystrophy Group. *Arch Neurol* 1991;**48**:383–8.
11. van der Ploeg AT, Clemens PR, Corzo D, Escolar DM, Florence J, Groeneveld GJ, et al. A randomized study of alglucosidase alfa in late-onset Pompe's disease. *N Engl J Med* 2010;**362**:1396–406.
12. Mendell J, Rodino-Klapac LR, Sahenk Z, Roush K, Bird L, Lowes LP, et al. Eteplirsen for the treatment of duchenne muscular dystrophy. *Ann Neurol* 2013;**7**:637–47.
13. Montes J, McDermott MP, Martens WB, Dunaway S, Glanzman AM, Riley S, et al. Six-minute walk test demonstrates motor fatigue in spinal muscular atrophy. *Neurology* 2010;**74**:833–8.
14. Angelini C, Semplicini C, Ravaglia S, Bembi B, Servidei S, Pegoraro E, et al. Observational clinical study in juvenile-adult glycogenosis type 2 patients undergoing enzyme replacement therapy for up to 4 years. *J Neurol* 2012;**259**:952–8.
15. McDonald CM, Henricson EK, Abresch RT, Florence JM, Eagle M, Gappmaier E, et al. THE 6-minute walk test and other endpoints in Duchenne muscular dystrophy: Longitudinal natural history observations over 48 weeks from a multicenter study. *Muscle Nerve* 2013;**48**:343–6.
16. Solway S, Brooks D, Lacasse Y, Thomas S. A qualitative systematic overview of the measurement properties of functional walk tests used in the cardiorespiratory domain. *Chest* 2001;**119**:256–70.
17. ATS statement: guidelines for the six-minute walk test. *Am J Respir Crit Care Med* 2002;166:111–17.
18. McDonald CM, Henricson EK, Han JJ, Abresch RT, Nicorici A, Elfring GL, et al. The 6-minute walk test as a new outcome measure in Duchenne muscular dystrophy. *Muscle Nerve* 2010;**41**: 500–10.
19. Geiger R, Strasak A, Treml B, Gasser K, Kleinsasser A, Fischer V, et al. Six-minute walk test in children and adolescents. *J Pediatr* 2007;**150**:395–399.e1–2.
20. Lammers AE, Hislop AA, Flynn Y, Haworth SG. The 6-minute walk test: normal values for children of 4–11 years of age. *Arch Dis Child* 2008;**93**:464–8.
21. McDonald C, Henricson E, Abresch RT, Han JJ, Leshner R, Escolar D, et al. Functional motor performance characteristics of boys with duchenne muscular dystrophy by age groups and steroid use: one-year data from the CINRG longitudinal study project. *Neurology* 2010;**74**:A219.
22. Mazzone E, Martinelli D, Berardinelli A, Messina S, D'Amico A, Vasco G, et al. North Star Ambulatory Assessment, 6-minute walk test and timed items in ambulant boys with Duchenne muscular dystrophy. *Neuromuscul Disord* 2010;**20**:712–16.
23. Podsiadlo D, Richardson S. The timed "Up & Go": a test of basic functional mobility for frail elderly persons. *J Am Geriatr Soc* 1991;**39**:142–8.
24. Williams EN, Carroll SG, Reddihough DS, Phillips BA, Galea MP. Investigation of the timed 'up & go' test in children. *Dev Med Child Neurol* 2005;**47**:518–24.
25. Katz-Leurer M, Rotem H, Lewitus H, Keren O, Meyer S. Relationship between balance abilities and gait characteristics in children with post-traumatic brain injury. *Brain Inj* 2008;**22**:153–9.
26. Montes J, Cheng B, Diamond B, Doorish C, Mitsumoto H, Gordon PH. The timed up and go test: predicting falls in ALS. *Amyotroph Lateral Scler* 2007;**8**:292–5.
27. Dunaway S, Montes J, Garber CE, Carr B, Kramer SS, Kamil-Rosenberg S, et al. Performance of the timed "up & go" test in spinal muscular atrophy. *Muscle Nerve* 2014;**50**:273–7.
28. Mercuri E, Bertini E, Messina S, Pelliccioni M, D'Amico A, Colitto F, et al. Pilot trial of phenylbutyrate in spinal muscular atrophy. *Neuromuscul Disord* 2004;**14**:130–5.
29. Mercuri E, Bertini E, Messina S, Solari A, D'Amico A, Angelozzi C, et al. Randomized, double-blind, placebo-controlled trial of phenylbutyrate in spinal muscular atrophy. *Neurology* 2007;**68** 51–5.
30. Pane M, Staccioli S, Messina S, D'Amico A, Pelliccioni M, Mazzone ES, et al. Daily salbutamol in young patients with SMA type II. *Neuromuscul Disord* 2008;**18**:536–40.
31. Swoboda KJ, Scott CB, Reyna SP, Prior TW, LaSalle B, Sorenson SL, et al. Phase II open label study of valproic acid in spinal muscular atrophy. *PLOS ONE* 2009;**4**:e5268.
32. Lewelt A, Krosschell KJ, Scott C, Sakonju A, Kissel JT, Crawford TO, et al. Compound muscle action potential and motor function in children with spinal muscular atrophy. *Muscle Nerve* 2010;**42**: 703–8.
33. Kissel JT, Scott CB, Reyna SP, Crawford TO, Simard LR, Krosschell KJ, et al. SMA CARNIVAL TRIAL PART II: a prospective, single-armed trial of L-carnitine and valproic acid in ambulatory children with spinal muscular atrophy. *PLOS ONE* 2011;**6**:e21296.
34. Vuillerot C, Girardot F, Payan C, Fermanian J, Iwaz J, De Lattre C, et al. Monitoring changes and predicting loss of ambulation in Duchenne muscular dystrophy with the Motor Function Measure. *Dev Med Child Neurol* 2010;**52**:60–5.
35. Russell DJ, Rosenbaum PL, Cadman DT, Gowland C, Hardy S, Jarvis S. The gross motor function measure: a means to evaluate the effects of physical therapy. *Dev Med Child Neurol* 1989;**31**:341–52.
36. Nelson L, Owens H, Hynan LS, Iannaccone ST. The gross motor function measure is a valid and sensitive outcome measure for spinal muscular atrophy. *Neuromuscul Disord* 2006;**16**:374–80.
37. Kaufmann P, McDermott MP, Darras BT, Finkel R, Kang P, Oskoui M, et al. Observational study of spinal muscular atrophy type 2 and 3: functional outcomes over 1 year. *Arch Neurol* 2011;**68**:779–86.
38. Berard C, Payan C, Hodgkinson I, Fermanian J. A motor function measure for neuromuscular diseases. Construction and validation study. *Neuromuscul Disord* 2005;**15**:463–70.

39. Main M, Kairon H, Mercuri E, Muntoni F. The Hammersmith functional motor scale for children with spinal muscular atrophy: a scale to test ability and monitor progress in children with limited ambulation. *Eur J Paediatr Neurol* 2003;**7**:155–9.
40. O'Hagen JM, Glanzman AM, McDermott MP, Ryan PA, Flickinger J, Quigley J, et al. An expanded version of the Hammersmith Functional Motor Scale for SMA II and III patients. *Neuromuscul Disord* 2007;**17**:693–7.
41. Glanzman AM, O'Hagen JM, McDermott MP, Martens WB, Flickinger J, Riley S, et al. Validation of the Expanded Hammersmith Functional Motor Scale in spinal muscular atrophy type II and III. *J Child Neurol* 2011;**26**:1499–507.
42. Msall ME, DiGaudio K, Duffy LC, LaForest S, Braun S, Granger CV. WeeFIM. Normative sample of an instrument for tracking functional independence in children. *Clin Pediatr (Phila)* 1994;**33**:431–8.
43. Chung BH, Wong VC, Ip P. Spinal muscular atrophy: survival pattern and functional status. *Pediatrics* 2004;**114**:e548–53.
44. Steffensen B, Hyde S, Lyager S, Mattsson E. Validity of the EK scale: a functional assessment of non-ambulatory individuals with Duchenne muscular dystrophy or spinal muscular atrophy. *Physiother Res Int* 2001;**6**:119–34.
45. Steffensen BF, Lyager S, Werge B, Rahbek J, Mattsson E. Physical capacity in non-ambulatory people with Duchenne muscular dystrophy or spinal muscular atrophy: a longitudinal study. *Dev Med Child Neurol* 2002;**44**:623–32.
46. Campbell SK, Kolobe TH, Osten ET, Lenke M, Girolami GL. Construct validity of the test of infant motor performance. *Phys Ther* 1995;**75**:585–96.
47. Campbell SK, Hedeker D. Validity of the Test of Infant Motor Performance for discriminating among infants with varying risk for poor motor outcome. *J Pediatr* 2001;**139**:546–51.
48. Campbell SK, Kolobe TH, Wright BD, Linacre JM. Validity of the Test of Infant Motor Performance for prediction of 6-, 9- and 12-month scores on the Alberta Infant Motor Scale. *Dev Med Child Neurol* 2002;**44**:263–72.
49. Krosschell KJ, Maczulski JA, Scott C, King W, Hartman JT, Case LE, et al. Reliability and validity of the TIMPSI for infants with spinal muscular atrophy type I. *Pediatr Phys Ther* 2013;**25**:140–8, discussion 9.
50. Glanzman AM, Mazzone E, Main M, Pelliccioni M, Wood J, Swoboda KJ, et al. The Children's Hospital of Philadelphia Infant Test of Neuromuscular Disorders (CHOP INTEND): test development and reliability. *Neuromuscul Disord* 2010;**20**:155–61.
51. Jebsen RH, Taylor N, Trieschmann RB, Trotter MJ, Howard LA. An objective and standardized test of hand function. *Arch Phys Med Rehabil* 1969;**50**:311–19.
52. Taylor N, Sand PL, Jebsen RH. Evaluation of hand function in children. *Arch Phys Med Rehabil* 1973;**54**:129–35.
53. Mazzone E, Bianco F, Martinelli D, Glanzman AM, Messina S, De Sanctis R, et al. Assessing upper limb function in nonambulant SMA patients: development of a new module. *Neuromuscul Disord* 2011;**21**:406–12.
54. Brooke MH, Griggs RC, Mendell JR, Fenichel GM, Shumate JB, Pellegrino RJ. Clinical trial in Duchenne dystrophy. I. The design of the protocol. *Muscle Nerve* 1981;**4**:186–97.
55. Brooke MH, Fenichel GM, Griggs RC, Mendell JR, Moxley R, Florence J, et al. Duchenne muscular dystrophy: patterns of clinical progression and effects of supportive therapy. *Neurology* 1989;**39**:475–81.
56. Mayhew A, Mazzone ES, Eagle M, Duong T, Ash M, Decostre V, et al. Development of the Performance of the Upper Limb module for Duchenne muscular dystrophy. *Dev Med Child Neurol* 2013;**55**:1038–45.
57. Vandervelde L, Van den Bergh PY, Goemans N, Thonnard JL. ACTIVLIM: a Rasch-built measure of activity limitations in children and adults with neuromuscular disorders. *Neuromuscul Disord* 2007;**17**:459–69.
58. Vandervelde L, Van den Bergh PY, Penta M, Thonnard JL. Validation of the ABILHAND questionnaire to measure manual ability in children and adults with neuromuscular disorders. *J Neurol Neurosurg Psychiatry* 2010;**81**:506–12.
59. Wang X, Dockery DW, Wypij D, Fay ME, Ferris Jr. BG. Pulmonary function between 6 and 18 years of age. *Pediatr Pulmonol* 1993;**15**:75–88.
60. Centers for Disease Control (CDC). Spirometry Training Program. Workplace Safety and Health Topics. Available at: http://www.cdc.gov/niosh/topics/spirometry/refcalculator.html.
61. Iannaccone ST. Outcome measures for pediatric spinal muscular atrophy. *Arch Neurol* 2002;**59**:1445–50.
62. Kaufmann P, McDermott MP, Darras BT, Finkel RS, Sproule DM, Kang PB, et al. Prospective cohort study of spinal muscular atrophy types 2 and 3. *Neurology* 2012;**79**:1889–97.
63. Khirani S, Ramirez A, Aubertin G, Boule M, Chemouny C, Forin V, et al. Respiratory muscle decline in Duchenne muscular dystrophy. *Pediatr Pulmonol* 2014;**49**:473–81.
64. Escolar DM, Henricson EK, Mayhew J, Florence J, Leshner R, Patel KM, et al. Clinical evaluator reliability for quantitative and manual muscle testing measures of strength in children. *Muscle Nerve* 2001;**24**:787–93.
65. Mayhew JE, Florence JM, Mayhew TP, Henricson EK, Leshner RT, McCarter RJ, et al. Reliable surrogate outcome measures in multicenter clinical trials of Duchenne muscular dystrophy. *Muscle Nerve* 2007;**35**:36–42.
66. Iannaccone ST, Hynan LS. Reliability of 4 outcome measures in pediatric spinal muscular atrophy. *Arch Neurol* 2003;**60**:1130–6.
67. Florence JM, Pandya S, King WM, Robison JD, Baty J, Miller JP, et al. Intrarater reliability of manual muscle test (Medical Research Council scale) grades in Duchenne's muscular dystrophy. *Phys Ther* 1992;**72**:115–22 [discussion 22–6].
68. Brockmann K, Becker P, Schreiber G, Neubert K, Brunner E, Bonnemann C. Sensitivity and specificity of qualitative muscle ultrasound in assessment of suspected neuromuscular disease in childhood. *Neuromuscul Disord* 2007;**17**:517–23.
69. Jansen M, van Alfen N, Nijhuis van der Sanden MW, van Dijk JP, Pillen S, de Groot IJ. Quantitative muscle ultrasound is a promising longitudinal follow-up tool in Duchenne muscular dystrophy. *Neuromuscul Disord* 2012;**22**:306–17.
70. Wu JS, Darras BT, Rutkove SB. Assessing spinal muscular atrophy with quantitative ultrasound. *Neurology* 2010;**75**:526–31.
71. Rutkove SB, Darras BT. Electrical impedance myography for the assessment of children with muscular dystrophy: a preliminary study. *J Phys Conf Ser* 2013;**434**:012069.
72. Rutkove SB, Gregas MC, Darras BT. Electrical impedance myography in spinal muscular atrophy: a longitudinal study. *Muscle Nerve* 2012;**45**:642–7.

73. Swoboda KJ, Prior T, Scott C, McNaught T, Wride M, Reyna S, et al. Natural history of denervation in SMA: relation to age, SMN2 copy number, and function. *Ann Neurol* 2005;**57**:704–12.
74. Kang PB, Gooch CL, McDermott MP, Darras BT, Finkel RS, Yang ML, et al. The motor neuron response to SMN1 deficiency in spinal muscular atrophy. *Muscle Nerve* 2014;**49**:636–44.
75. Marden FA, Connolly AM, Siegel MJ, Rubin DA. Compositional analysis of muscle in boys with Duchenne muscular dystrophy using MR imaging. *Skeletal Radiol* 2005;**34**:140–8.
76. Kim HK, Merrow AC, Shiraj S, Wong BL, Horn PS, Laor T. Analysis of fatty infiltration and inflammation of the pelvic and thigh muscles in boys with Duchenne muscular dystrophy (DMD): grading of disease involvement on MR imaging and correlation with clinical assessments. *Pediatr Radiol* 2013;**43**(10):1327–35.
77. Sookhoo S, Mackinnon I, Bushby K, Chinnery PF, Birchall D. MRI for the demonstration of subclinical muscle involvement in muscular dystrophy. *Clin Radiol* 2007;**62**:160–5.
78. Sproule DM, Montgomery MJ, Punyanitya M, Shen W, Dashnaw S, Montes J, et al. Thigh muscle volume measured by magnetic resonance imaging is stable over a 6-month interval in spinal muscular atrophy. *J Child Neurol* 2011;**26**:1252–9.
79. WHOQOL Measuring quality of life. World Health Organization (WHO) 1997 Available from: http://www.who.int/mental_health/media/68.pdf.
80. Kaufmann P, Muntoni F. Issues in SMA clinical trial design. The International Coordinating Committee (ICC) for SMA subcommittee on SMA clinical trial design. *Neuromuscul Disord* 2007;**17**:499–505.
81. Varni JW, Seid M, Rode CA. The PedsQL: measurement model for the pediatric quality of life inventory. *Med Care* 1999;**37**:126–39.
82. Iannaccone ST, Hynan LS, Morton A, Buchanan R, Limbers CA, Varni JW. The PedsQL in pediatric patients with Spinal Muscular Atrophy: feasibility, reliability, and validity of the Pediatric Quality of Life Inventory Generic Core Scales and Neuromuscular Module. *Neuromuscul Disord* 2009;**19**:805–12.
83. McDonald CM, McDonald DA, Bagley A, Sienko Thomas S, Buckon CE, Henricson E, et al. Relationship between clinical outcome measures and parent proxy reports of health-related quality of life in ambulatory children with Duchenne muscular dystrophy. *J Child Neurol* 2010;**25**:1130–44.
84. Cella D, Nowinski C, Peterman A, Victorson D, Miller D, Lai JS, et al. The neurology quality-of-life measurement initiative. *Arch Phys Med Rehabil* 2011;**92**:S28–36.
85. NeuroQOL: Quality of Life in Neurological Disorders. 2013. Northwestern University. [Website] Available from: <http://www.neuroqol.org/>.
86. Feldman AB, Haley SM, Coryell J. Concurrent and construct validity of the pediatric evaluation of disability inventory. *Phys Ther* 1990;**70**:602–10.
87. Berg M, Jahnsen R, Froslie KF, Hussain A. Reliability of the pediatric evaluation of disability inventory (PEDI). *Phys Occup Ther Pediatr* 2004;**24**:61–77.

Index

Note: Page numbers followed by "*b*", "*f*" and "*t*" refer to boxes, figures and tables, respectively.

B

C

D

E

F

G

H

M

S

Printed in the United States
By Bookmasters

T

U

Printed in the United States
By Bookmasters